Nanostructures for Drug Delivery

Nanostructures in Therapeutic Medicine Series

Nanostructures for Drug Delivery

Edited by

Ecaterina Andronescu
University Politehnica of Bucharest, Bucharest, Romania

Alexandru Mihai Grumezescu
University Politehnica of Bucharest, Bucharest, Romania

ELSEVIER

Elsevier
Radarweg 29, PO Box 211, 1000 AE Amsterdam, Netherlands
The Boulevard, Langford Lane, Kidlington, Oxford OX5 1GB, United Kingdom
50 Hampshire Street, 5th Floor, Cambridge, MA 02139, United States

Library of Congress Cataloging-in-Publication Data
A catalog record for this book is available from the Library of Congress

British Library Cataloguing-in-Publication Data
A catalogue record for this book is available from the British Library

ISBN: 978-0-323-46143-6

For information on all Elsevier publications visit our website at
https://www.elsevier.com/books-and-journals

Publisher: Matthew Deans
Acquisition Editor: Simon Holt
Editorial Project Manager: Sabrina Webber
Production Project Manager: Julie-Ann Stansfield
Designer: Victoria Pearson Esser

Typeset by Thomson Digital

Contents

List of Contributors

Naveed Ahmed
Quaid-i-Azam University, Islamabad, Pakistan

Houman Alimoradi
University of Otago, Dunedin, New Zealand

Daniel A. Allemandi
National University of Salta-CONICET, Salta, Argentina

Mohammad Amani-Tehran
Amirkabir University of Technology, Tehran, Iran

Sepideh Amjad-Iranagh
Amirkabir University of Technology, Tehran, Iran

Xueqin An
East China University of Science and Technology, Shanghai, China

Ashleigh Anderson
University of Ulster, Jordanstown, Northern Ireland

Ecaterina Andronescu
University Politehnica of Bucharest, Bucharest, Romania

Roohollah Bagherzadeh
Amirkabir University of Technology, Tehran, Iran

Xavier Banquy
University of Montreal, Montreal, QC, Canada

José M. Bermúdez
National University of Córdoba, Córdoba, Argentina

Sourav Bhandari
ISF Educational Society, Moga, India

Najma Bibi
Quaid-i-Azam University, Islamabad, Pakistan

Alexandra Bolocan
Emergency University Hospital; "Carol Davila" University of Medicine and Pharmacy, Bucharest, Romania

Burhan Bora
Ege University, Bornova-Izmir, Turkey

Chong Cheng
University at Buffalo, State University of New York, Buffalo, NY, United States

Alicia Cid
National University of Córdoba, Córdoba, Argentina

Adrian G. Ciucă
University Politehnica of Bucharest, Bucharest, Romania

Sarha Cupri
University of Catania, Catania, Italy

Nily Dan
Drexel University, Philadelphia, PA, United States

James Davis
University of Ulster, Jordanstown, Northern Ireland

Gokcen B. Demirel
Gazi University, Ankara, Turkey

Marek K. Dobke
University of California San Diego, San Diego, CA, United States

Steven Dominguez
University of California San Diego, San Diego, CA, United States

Surabhi Dubey
Dr. H.S. Gour Vishwavidyalaya, Sagar, India

Safaa S. El-Gamal
Alexandria University, Alexandria, Egypt

Amal H. El-Kamel
Alexandria University, Alexandria, Egypt

Noha S. El-Salamouni
Pharos University in Alexandria, Alexandria, Egypt

Serap Evran
Ege University, Bornova-Izmir, Turkey

Ragwa M. Farid
Pharos University in Alexandria, Alexandria, Egypt

Virginia Fuochi
University of Catania, Catania, Italy

Pio Maria Furneri
University of Catania, Catania, Italy

Dinar Gabdrakhmanov
A.E. Arbuzov Institute of Organic and Physical Chemistry, Kazan Scientific Center, Russian Academy of Sciences, Kazan, Russian Federation

Allan B. Gamble
University of Otago, Dunedin, New Zealand

Chityal Ganesh Kumar
CSIR-Indian Institute of Chemical Technology, Hyderabad, India

Tarun Garg
ISF Educational Society, Moga, India

Farzaneh Ghasemkhah
Amirkabir University of Technology, Tehran, Iran

Irina Gheorghe
Research Institute of the University of Bucharest (ICUB), Bucharest, Romania

Gregory I. Giles
University of Otago, Dunedin, New Zealand

Amit K. Goyal
ISF Educational Society, Moga, India

Cristina I. Grecu
University Politehnica of Bucharest, Bucharest, Romania

Khaled Griesh
University of Otago, Dunedin, New Zealand; Arabian Gulf University, Manama, Kingdom of Bahrain

Alexandru Mihai Grumezescu
University Politehnica of Bucharest, Bucharest, Romania

Rijun Gui
East China University of Science and Technology, Shanghai; Qingdao University, Qingdao, China

Ufuk Gunduz
Middle East Technical University, Ankara, Turkey

Mehrdad Hamidi
Zanjan University of Medical Sciences, Zanjan, Iran

Patrice Hildgen
University of Montreal, Montreal, QC, Canada

Alina M. Holban
University Politehnica of Bucharest; Research Institute of the University of Bucharest (ICUB), Bucharest, Romania

Fatemeh Jahanmard
Amirkabir University of Technology, Tehran, Iran

Magdalena Jarosz
Jagiellonian University in Krakow, Krakow, Poland

Yousef Javadzadeh
Tabriz University of Medical Science, Tabriz, Iran

Gunjan Jeswani
Shri Shankaracharya Group of Institutions, SSTC, Bhilai, India

Arvind K. Jha
Shri Shankaracharya Group of Institutions, SSTC, Bhilai, India

Anjali Joshi
Panjab University, Chandigarh, India

Joanna Kapusta-Kołodziej
Jagiellonian University in Krakow, Krakow, Poland

Ruslan Kashapov
A.E. Arbuzov Institute of Organic and Physical Chemistry, Kazan Scientific Center, Russian Academy of Sciences, Kazan, Russian Federation

Gul Majid Khan
Quaid-i-Azam University, Islamabad, Pakistan

Sepideh Khoee
University of Tehran, Tehran, Iran

Filiz Kuralay
Ordu University, Ordu, Turkey

Gulbin Kurtay
Ankara University, Ankara, Turkey

Augustine Lalloz
University of Montreal, Montreal, QC, Canada

Masoud Latifi
Amirkabir University of Technology, Tehran, Iran

Pierre L. Latreille
University of Montreal, Montreal, QC, Canada

Hui Li
Shanghai Jiao Tong University, Shanghai, China

Gina A. Mackert
University of California San Diego, San Diego, CA, United States

Siddharth S. Matikonda
University of Otago, Dunedin, New Zealand

Nishi Mody
Dr. H.S. Gour Vishwavidyalaya, Sagar, India

Somayeh Mohamadi
University of Environment, Karaj, Iran

Canan Ozyurt
Ege University, Bornova-Izmir, Turkey

Santiago D. Palma
National University of Salta-CONICET, Salta, Argentina

Tatiana Pashirova
A.E. Arbuzov Institute of Organic and Physical Chemistry, Kazan Scientific Center, Russian Academy of Sciences, Kazan, Russian Federation

Swarnali D. Paul
Shri Shankaracharya Group of Institutions, SSTC, Bhilai, India

Anna Pawlik
Jagiellonian University in Krakow, Krakow, Poland

Giulio P. Petronio
IRCCS San Raffaele Pisana, BioBIM—Multidisciplinary Interinstitutional BioBank, Rome, Italy

Rosario Pignatello
University of Catania, Catania, Italy

Sujitha Pombala
CSIR-Indian Institute of Chemical Technology, Hyderabad, India

Yedla Poornachandra
CSIR-Indian Institute of Chemical Technology, Hyderabad, India

Daniela A. Quinteros
National University of Salta-CONICET, Salta, Argentina

Jean M. Rabanel
University of Montreal, Montreal, QC, Canada

Goutam Rath
ISF Educational Society, Moga, India

Soledad Ravetti
National University of Villa Maria, Córdoba, Argentina

Petre Rotărescu
University Politehnica of Bucharest, Bucharest, Romania

Harish Sharma
Shri Shankaracharya Group of Institutions, SSTC, Bhilai, India

Rajeev Sharma
Dr. H.S. Gour Vishwavidyalaya, Sagar, India

Ruchi Sharma
Lovely Professional University, Jalandhar, India

Tamilvanan Shunmugaperumal
Lovely Professional University, Jalandhar, India

Narinder Singh
Panjab University, Chandigarh, India

Oleg Sinyashin
A.E. Arbuzov Institute of Organic and Physical Chemistry, Kazan Scientific Center, Russian Academy of Sciences, Kazan, Russian Federation

Magdalena Stevanović
Institute of Technical Sciences of the Serbian Academy of Sciences and Arts, Belgrade, Serbia

Grzegorz D. Sulka
Jagiellonian University in Krakow, Krakow, Poland

Haotian Sun
University at Buffalo, State University of New York, Buffalo, NY, United States

Karolina Syrek
Jagiellonian University in Krakow, Krakow, Poland

Raja Sekharan Thenrajan
Sankaralingam Bhuvaneswari College of Pharmacy, Sivakasi, India

Amrit Pal Toor
Panjab University, Chandigarh, India

Ozge Ugurlu
Ege University, Bornova-Izmir, Turkey

Gozde Unsoy
Middle East Technical University, Ankara, Turkey

Gaurav Verma
Panjab University, Chandigarh, India; Massachusetts Institute of Technology, Cambridge, MA, United States

Suresh P. Vyas
Dr. H.S. Gour Vishwavidyalaya, Sagar, India

Zhiqiang Xie
Louisiana State University, Baton Rouge, LA, United States

Mehmet Yılmaz
Sinop University, Sinop, Turkey

Morteza Yaghoobian
Shahid Beheshti Medical University, Tehran, Iran

Shadi Yaqoubi
Tabriz University of Medical Science, Tabriz, Iran

Yun Yu
University at Buffalo, State University of New York, Buffalo, NY, United States

Weien Yuan
Shanghai Jiao Tong University, Shanghai, China

Lucia Zakharova
A.E. Arbuzov Institute of Organic and Physical Chemistry, Kazan Scientific Center, Russian Academy of Sciences, Kazan, Russian Federation

Fatemeh Zamani
Amirkabir University of Technology, Tehran; Hazrat-e Masoumeh University, Qom, Iran

Jiuhong Zhang
Kent State University, Kent, OH, United States

Nan Zhang
Shanghai Ocean University, Shanghai, People's Republic of China

Jian Zhong
Shanghai Ocean University, Shanghai, People's Republic of China

Foreword of the Series

Material science and engineering at the nanoscale has brought revolutionary advances to the biomedical sciences, overturning many of the traditionally known approaches. Nanotechnology has driven many of the most successful innovative technologies, and the impressive record of accomplishments in the field make nanostructures promising candidates for medical therapy applications. The advantages that nanomaterials have already provided to therapeutics, such as targeted and controlled delivery, wide accessibility, high specificity, low side effects, improved efficiency, and impressive versatility are currently considered key elements in designing personalized medicine approaches for prophylaxis, diagnosis, and therapy.

Therapeutic nanostructures can be greatly diverse, and their unique properties have led to the development of highly specialized biosensors, more efficient drug delivery vehicles, and controlled release targeting systems to fight severe or incurable diseases, such as cancer, infections, and cardiovascular disease.

In view of the astounding progress made in the field of therapeutic nanotechnology and its rapidly progressing expansion, this book aims to collect in one place all the recent and most innovative aspects of nanomaterials in both current and future therapy. The series is organized into five volumes, covering the main areas that are relevant for the design and implementation of nanostructures in medical therapies.

The first volume, *Nanostructures for Novel Therapy: Synthesis, Characterization, and Applications*, describes methods to obtain and characterize nanosystems, emphasizing their biomedical applications. Special attention is paid in this volume to modern synthesis methods to reduce side effects and limit the toxicity of nanomaterials in biomedical applications. Numerous examples of nanostructures designed for therapy, as well as the most efficient synthesis and characterization routes for these materials, are clearly described and critically analyzed.

The second volume, entitled *Nanostructures for Drug Delivery*, covers one of the most widely utilized and investigated applications of nanomaterials in the biomedical field, namely, drug delivery. Designing nanostructures to specifically and safely carry therapeutic agents to their final destination is an intriguing approach to future targeted therapies. This approach could provide a treatment for previously incurable diseases, as well as reducing the side effects of current drugs. Many highly active drugs are severely limited by side effects related to their unspecific sites of action. This volume introduces the readers to the amazing field of nanomedicine by discussing the versatility and variety of nanovehicles for drug delivery and targeting. Moreover, readers will find numerous examples and will learn about the currently used or investigational drug delivery agents for therapy, prophylaxis, and diagnosis.

Volume 3, *Nanostructures for Antimicrobial Therapy*, highlights the impressive progress made by nanotechnology in the design of novel antimicrobial approaches. Since microbial resistance to antibiotics is a real and increasingly worrying issue across all countries, the development of more efficient antimicrobial agents to provide control of future infections is at a high priority. Antimicrobial nanosystems have proved to be remarkably efficient against drug-resistant microorganisms, plus they are able to fight biofilm-associated infections and can control the social behavior of microbial communities. Nanostructures can also reduce microbial virulence factors and reduce pathogenesis mechanisms, offering a promising alternative for future therapy.

Volume 4, entitled *Nanostructures for Cancer Therapy*, covers the applications of nanomedicine in cancer diagnosis and treatment. The use of nanoparticles for cancer therapy is not in itself a new approach, but numerous recent advances have been made in this area. The aim of this volume is to cover the most interesting new approaches in the management of this deadly disease. Nanosized drugs are currently believed to represent the most efficient approach in cancer chemotherapy, and this volume provides coverage of the latest and most novel findings, while also discussing possible improvements in more established types of nanosystems that can increase the efficiency of cancer therapy.

The final volume of this series, entitled *Nanostructures for Oral Medicine*, covers the progress made in applications of nanotechnology in treating various diseases of the oral cavity as well as progress in nanotechnology applications in dentistry. Readers can learn about the most efficient modern materials used to treat or to prevent widely encountered oral diseases such as gingivitis, periodontitis, caries, and dental plaque. Moreover, restorative dentistry also now makes wide use of nanomaterials.

Overall, this book series provides a state-of-the-art compendium of knowledge, and a crystal ball for seeing into the future of biomedical nanotechology and nanomedicine. It has appeal for researchers, clinicians, engineers, pharmacologists, pharmacists, oncologists, infectious disease experts, and dentists. In addition, interested general readers will discover the impact, current progress, and future applications of nanotechnology in therapeutics and diagnosis. Taken together, nanoscale approaches will improve the efficiency of personalized medicine for better management of diseases in the 21st century.

Michael R. Hamblin

Wellman Center for Photomedicine, Massachusetts General Hospital, Boston, MA, United States

Department of Dermatology, Harvard Medical School, Boston, MA, United States

Harvard-MIT Division of Health Sciences and Technology, Cambridge, MA, United States

Preface

ABOUT THE SERIES (VOLUMES I–V)

In our permanently changing world, novel therapeutics are constantly required to manage health and well-being of population. Although numerous diseases are currently considered incurable, massive progress made in biomedicine but also in associated fields, such as chemistry, physics, engineering, pharmacology and materials science, offers a new light to the therapeutics domain. In this context, most physicists and researchers believe that a personalized and adequate treatment may significantly improve the outcome of severe diseases and ensure a faster healing. Nanotechnology offers great perspectives for personalized medicine because nanostructured therapeutics proved their efficiency and amazing impact in improving therapy, prophylaxis, and diagnosis. The emerging field of nanosized materials has numerous applications in the biomedical field, especially in therapy. This series of five volumes came out by the need of learning about recent progress of the science of nanostructured materials to improve current therapy and lead to the next level. The books offer an interesting and updated perspective regarding applications of nanomaterials in therapy of most investigated and difficult-to-treat diseases, such as cancer and severe infections. The presentation approach of each chapter contained in those five volumes is clear and easy to understand by most readers and of a great interest for biomedical specialists, researchers, and engineers. The series is organized in an attractive manner for students and academics on the field, starting with a volume dealing with synthesis, characterization, and main applications of nanostructures, emphasizing on their impact in therapy. Next volume reveals the most recent progress made on a very investigated field, considered a key element in personalized medicine and future therapy, namely nanostructured drug delivery systems. Their impact in antimicrobial therapy is also widely discussed, and suggestive examples are given and explained. Moreover, a whole volume is dedicated to the management of the disease of the century—cancer—revealing the huge value added by the utilization of nanosystems in the therapy of this deadly disease. Important aspects related to improved diagnosis and prophylaxis are highlighted. In the last volume, the progress and novel applications of nanotechnology in oral medicine is dissected. The field of oral diseases represents an interesting and a priority field because both physicists and researchers believe that they can be prevented and treated more easily with targeting systems and nanofunctional prosthetics. All chapters are clearly illustrated to highlight most important or difficult-to-understand aspects, and suggestive examples are often enumerated in organized tables, which are explained and discussed. Overall, the series contains very recent but accessible and very interesting information regarding the progress of nanostructures in therapeutics and gives a novel perspective about future therapy of severe diseases.

Alexandru Mihai Grumezescu

University Politehnica of Bucharest, Bucharest, Romania

ABOUT VOLUME II

The second volume of the series *Therapeutic Nanostructures* is entitled *Nanostructures for Drug Delivery*. This volume reveals the main types of nanosized drug delivery systems, dissecting their main advantages and drawbacks in the therapy of severe and difficult-to-treat diseases. Nanosized carriers are commented in all stages of their preparation and use, starting with suitable synthesis methods, appropriate characterization, and functionalization, and ending with their application in drug delivery and therapy. A special attention is given to green nanotechnology which enables the production of natural, biocompatible and less toxic nanosystems which may be efficiently used in delivering pharmacological agents. Numerous properties of such drug delivery systems are highly appreciated for future therapy, namely longtime circulation in the bloodstream, target specificity, stimuli responsively, possibility of intracellular delivery, contrast functionality, and increased stability. Different ways for the control of the load capacity and release profile are also analyzed in many of the chapters.

Volume II contains 30 chapters, prepared by outstanding researchers affiliated in the USA, Canada, Argentina, Ireland, Italy, New Zealand, Poland, Serbia, Romania, Russia, Turkey, Egypt, Iran, Pakistan, India, and China.

In Chapter 1, entitled *Therapeutic Nanomaterials: From a Drug Delivery Perspective*, C. Ganesh Kumar et al. focus on the synthetic routes and applications of nanomaterials as potential drug delivery agents. Different nanomaterial types including polymeric nanomaterials, dendrimers, nanoparticles, carbon nanotubes, quantum dots, and their applications have been discussed and a number of case studies have been presented in this chapter.

Nily Dan, in Chapter 2, entitled *Core–Shell Drug Carriers: Liposomes, Polymerosomes, and Niosomes*, reveals the three types of main formulations used as drug nanocarriers and summarizes studies of their performance in biomedical applications such as cancer therapy (tumor targeting) or transport through the blood–brain barrier.

Chapter 3, entitled *The New Nanocarriers Based on Graphene and Graphene Oxide for Drug Delivery Applications*, prepared by Somayeh Mohamadi and Mehrdad Hamidi, gives an overview about a new carbon-based carrier for drug delivery applications. The unique properties of graphene nanosheets such as two-dimensional planar structure with sp^2 hybridation, large surface area, chemical and mechanical stability, super conductivity, and good biocompatibility provide the opportunity to design drug carriers with dual-targeting function.

In Chapter 4, entitled *Nanostructured Nanoparticles for Improved Drug Delivery*, Jean Michel Rabanel et al. provide a comprehensive view of the different types of structures found in nanoparticles intended for medical use reported so far in the literature, together with some insight regarding their fabrication processes and their physicochemical properties and how this nanoparticles deliver active compounds in a controlled fashion to the human body or developed imaging strategies.

Chapter 5, entitled *Design of Functionalized Materials for Use in Micro and Nanoscale Drug Delivery Devices and Smart Patches*, prepared by Ashleigh Anderson and James Davis, provides a critical overview of the latest developments in functional materials and nanoscale device architectures that can facilitate controlled drug delivery.

Chapter 6, entitled *Niosomes: A Novel Approach in Modern Drug Delivery Systems*, prepared by Sepideh Khoee and Morteza Yaghoobian, discusses about new drug delivery vehicles and gives a brief description on the preparation, applications, advantages and disadvantages of particular delivery system—niosomes.

Fatemeh Zamani et al., in Chapter 7, entitled *Nanofibrous and Nanoparticle Materials as Drug-Delivery Systems*, emphasize on the emerging area of nanoparticles synthesis through electrospinning technique to generate biomimetic nanofibers and nanoparticles as drug delivery devices that are responsive to different stimuli, such as temperature, pH, light, and the electric/magnetic field for controlled release of therapeutic substances.

Yun Yu et al., in Chapter 8, entitled *Brush Polymer-Based Nanostructures for Drug Delivery*, give an overview on the advantages of brush polymer (BP)-based drug delivery systems. Also, future research directions of BP-based drug delivery systems are highlighted.

Magdalena Jarosz et al., in Chapter 9, entitled *Drug Delivery Systems Based on Titania Nanostructures*, focus on different titanium (TiO_2) nanostructures used as drug delivery systems, such as nanotubular and nanoporous TiO_2, TiO_2 nanoparticles, and nanowhiskers.

Houman Alimoradi et al., in Chapter 10, entitled *Redox Activated Polymeric Nanoparticles in Tumor Therapy*, focus on the biochemical basis for oxidative stress in tumors, its role in cell-signaling, the pathophysiology of tumor vasculature, and the differences in the redox-metabolism between cancer cells and nonmalignant tissues. The recent developments toward designing redox responsive drug delivery systems which have been classified as polysulfide, polyselenide, quinones, metal complexes, arylboronic esters, aryl oxalate, and other miscellaneous examples are also discussed.

Chapter 11, entitled *Polymeric Micro- and Nanoparticles for Controlled and Targeted Drug Delivery*, prepared by Magdalena Stevanović, reports the production and applications of polymeric micro- and nanoparticles with a special emphasis on obtaining polyester particles, the incorporation of different active substances within polymer matrix, the degradation and release process of active substances from the polymeric particles, the physiochemical and biological properties of such obtained systems, as well as their application as drug delivery systems.

Chapter 12, entitled *Novel Gels: Implications for Drug Delivery*, prepared by Swarnali D. Paul et al., describes all the emerging prospects of novel gels along with their formulation aspects, manufacturing technologies, and current applications, focusing on therapeutic potential.

Chapter 13, entitled *Nanosuspension Drug Delivery System: Preparation, Characterization, Postproduction Processing, Dosage Form, and Application*, prepared by Jiuhong Zhang et al., gives an overview about nanosuspension drug-delivery system (DDS) discussing preparation methods, characterization methods, postproduction process including the solidification techniques and surface modification process, common dosage forms, and clinical applications of nanosuspension DDS.

Chapter 14, entitled *Polymer-Based Nanocarriers for Therapeutic Nucleic Acids Delivery*, prepared by Weien W. Yuan and Hui H. Li, describes the challenges, advantages, and recent progress of polymer-based nanocarriers for therapeutic nucleic acids delivery. Considerations for manufacturing, safety issues, and regulatory requirements for these novel nanocarriers are also discussed.

Chapter 15, entitled *Multifunctional Therapeutic Hybrid Nanocarriers for Targeted and Triggered Drug Delivery: Recent Trends and Future Prospects*, prepared by Gulbin Kurtay et al., presents new generation and promising hybrid nanocarriers based on noble metal, porous silica, polymer, and core/shell therapeutics which exhibit active targeting, triggered release of cargo, and imaging capability for in vivo studies.

Ragwa M. Farid et al., in Chapter 16, entitled *Lipid-Based Nanocarriers for Ocular Drug Delivery*, give an up-to-date overview about lipid-based nanocarriers that can enhance the corneal absorption of both hydrophilic and lipophilic drugs and improve their ocular bioavailability. Nanostructured lipid carriers (NLCs) and lipid drug conjugates (LDCs) have emerged as a new generation of solid lipid nanoparticles (SLNs) to overcome problems of low entrapment efficiency and drug expulsion during storage.

Rajeev Sharma et al., in Chapter 17, entitled *Nanoparticulate Carrier(s): An Emerging Paradigm in New Generation Vaccine Development*, summarize the cutting edge technologies of nanoparticulate-carrier-based new generation and vaccine development, including design, trials, and clinical outcomes.

Chapter 18, entitled *Pathogen-Specific Nucleic Acid Aptamers as Targeting Components of Antibiotic and Gene Delivery Systems*, prepared by Canan Ozyurt et al., presents an overview of current cell-specific aptamer-conjugated nanoparticles. Also, the authors address two issues: nanoparticle-based antimicrobial gene delivery and modification of nanocarriers with aptamers.

Tamilvanan Shunmugaperumal et al., in Chapter 19, entitled *Multifunctional Nanosized Emulsions for Theragnosis of Life Threatening Diseases*, envision the use of the multifunctional oil-in-water (o/w) nanosized emulsions (NE) that carries imaging agents, anticancer or lipid lowering or antiatherosclerotic drug molecules and homing devices together for simultaneous imaging/diagnosing and treatment of cancer and atherosclerosis. A complete outline is given on the formulation of drug-loaded NE together with active and passive targeting moieties for accessing the unreachable organs present inside the human body.

Chapter 20, entitled *Therapeutic Nanostructures for Pulmonary Drug Delivery*, prepared by Yousef Javadzadeh and Shadi Yaqoubi, discusses about pulmonary drug delivery, inhalers, nanoparticles, and the advantages and fate of inhaled nanoparticles.

Najma Bibi et al., in Chapter 21, entitled *Nanostructures in Transdermal Drug Delivery Systems*, dissect the benefits of transdermal drug delivery over other delivery systems, arguing about the barriers in the skin to be faced by nanomaterials and their permeation pathways. Physicochemical characteristics required for the good penetration of nanostructures are also discussed.

Chapter 22, entitled *Advancement in Pulmonary Drug Delivery Systems for Treatment of Tuberculosis*, prepared by Tarun Garg et al., presents an overview about the importance of pulmonary drug delivery systems such as liposomes, niosomes, nanoparticles, microparticles, dendrimers, solid lipid nanoparticles, micelles, nanosuspensions, nanoemulsions, and microemulsions for an effective treatment of tuberculosis. The advances in delivery devices from conventional metered dose inhalers to dry powder inhalers are also discussed along with their applications.

Chapter 23, entitled *Nanosized Devices as Antibiotics and Antifungals Delivery: Past, News and Outlook*, prepared by Pio M. Furneri et al., discusses about developments in the field of (nanosized) delivery systems encapsulating antibacterial and antifungal drugs. A specific attention has been given to the pharmaceutical, microbiological, and clinical outcomes of systems for which authors have provided in vitro microbiological data and, more hopefully, in vivo results.

Chapter 24, entitled *Drug Delivery Mediated by Confined Nanosystems: Structure-Activity Relations and Factors Responsible for the Efficacy of Formulations*, prepared by Lucia Zakharova et al., presents different ways for the control of the load capacity and release profile. The chapter also discusses about the ability of formulated drug to integrate with cell membrane and penetrate through the blood-brain barrier.

Daniela Alejandra Quinteros et al., in Chapter 25, entitled *Therapeutic Use of Monoclonal Antibodies: General Aspects and Challenges for Drug Delivery*, identified and described the major issues associated with therapeutics approaches, formulating drawbacks, and delivering antibody drugs, particularly focused on the challenges and opportunities that these present for the future.

Chapter 26, entitled *Targeted Drug Delivery Via Chitosan-Coated Magnetic Nanoparticles*, prepared by Gozde Unsoy and Ufuk Gundu, presents an up-to-date review about polymer-coated magnetic nanoparticles, characterized by high surface-to-volume ratios, which are excellent scaffolds for loading targeting moieties, permeation enhancers, imaging tags, and drugs, simultaneously providing diagnostic and therapeutic capabilities.

Narinder Singh et al., in Chapter 27, entitled *Drug Delivery: Advancements and Challenges*, elaborate an updated idea regarding the effect of various drug delivery systems in contrast to their toxicity with a focus on nanocarriers and nanosystems.

Chapter 28, entitled *Stimuli-Responsive Liposome and Control Release Drug*, prepared by Xueqin An and Rijun Gui, presents an up-to-date review about the relative thermal, light, magnetism, and pH stimuli responsive properties of liposomes. Also, the mechanism of controlled drug release is discussed.

Chapter 29, entitled *Nanotechnology to Enhance Transdermal Delivery of Hydrophilic Humectants for Improved Skin Care: A Model for Therapeutic Applications*, prepared by Steven Dominguez et al., presents current status on research and development of some of the most promising agents, including nanotropocollagen and hyaluronic acid, which can serve as models for nanotechnology employed for transcutaneous active pharmaceutical ingredients (APIs) delivery.

Chapter 30, entitled *Nanostructures for Drug Delivery: Pharmacokinetic and Toxicological Aspects*, prepared by Adrian G. Ciucă et al., offers an extensive and updated report of the recent progresses and degree of toxicity that could be generated by a number of new medical therapies, with a great impact on medicine, namely "therapeutic devices" based on nanoparticles.

Ecaterina Andronescu
University Politehnica of Bucharest, Bucharest, Romania

Alexandru Mihai Grumezescu
University Politehnica of Bucharest, Bucharest, Romania

CHAPTER

1

THERAPEUTIC NANOMATERIALS: FROM A DRUG DELIVERY PERSPECTIVE

Chityal Ganesh Kumar, Yedla Poornachandra, Sujitha Pombala

CSIR-Indian Institute of Chemical Technology, Hyderabad, India

CHAPTER OUTLINE

1 INTRODUCTION

Nanotechnology is an emerging interdisciplinary field having confluence of various disciplines, such as physics, chemistry, material science and engineering devoted to the construction of structures in the nanometer regime (often 100 nm or smaller) with diverse physical and chemical properties. Although nanoscience and nanotechnology are new research areas, the use of nanomaterials by mankind was known since antiquity. Red colloidal gold tinctures as drugs for longevity, the so-called "Makaradhwaja" and "Jin Tu" were used in India and China, respectively, since ancient times (Hayat, 1989; Mahdihasan, 1985). The use of noble nanomaterials by glassmakers dates back to the Roman Empire, which is evidenced by the exotic Lycurgus Cup of 4th century AD that exhibited dichroic effect due to the presence of colloidal silver and gold particles (Barber and Freestone, 1990). Similar examples exist on the use of noble metallic nanomaterials in stained-glass windows of old European cathedrals and for coloring ceramic vases and ornaments by the Chinese (Burda et al., 2005). In the early 19th century, antibacterial preparations based on colloidal silver, such as Collargol (Argentum Crede) and Protargol (Argenti proteinatum) were marketed in Europe (Kolthoff, 1925). Paul Ehrlich first presented the concept of drug targeting or site-specific drug delivery in 1909. The "magic bullet" concept was put forth by him and his colleagues when they synthesized the first man-made antibiotic, arsphenamine, for the treatment of syphilis. This molecule after some chemical modifications inhibited the pathogens without affecting the host cells and acted as a "magic bullet." Furthermore, he also introduced the concepts, such as chemoreceptor and chemotherapy, and linked the chemical structure of the compound to their pharmacological activity (Bosch and Rosich, 2008).

In 2004, the concept of nanomedicine was put forth by the European Science Foundation, which represents a new area in the field of drug delivery research (Reis et al., 2007). A significant attention has been focused on the potentials of nanoscience and nanotechnology in drug delivery because they offer site-specific, time-controlled delivery of different molecular weight drugs and other bioactive molecules. The concept of drug delivery as a noninvasive system has recently emerged as a highly competitive and fast developing technology for treating medical infections and diseases. In this context, the nanomediated drug delivery systems, such as nanoparticles, dendrimers, carbon nanotubes, quantum dots, and so on are believed to have great potential to develop various drug carrier systems to address the growing burden of new diseases. One of the main challenges in drug delivery is to deliver the drug at the target site in the body to avoid potential side effects to the normal tissue. These nanomaterials can be modified for better efficiency to facilitate a large amount of drugs through nanocarriers, which can eventually reach the inaccessible areas, such as brain tissues, cancer cells, and other infected tissues. According to BCC Research Report (www.bccresearch.com/market-research/pharmaceuticals/advanced-drug-delivery-markets-phm006j.html), the global market for advanced drug delivery systems was valued at \$151.3 billion in 2013. It was forecasted that this market value will rise to reach nearly \$173.8 billion in 2018 at a 5-year compound annual growth rate of 2.8%. To meet this remarkable global demand, the synthesis of nanomaterials for diverse drug delivery applications has been a burgeoning area of research in the field of nanotechnology. This chapter focuses on the synthetic routes and characterization of various nanobiomaterials, and their drug delivery applications have been reviewed.

2 GENERAL METHODS OF SYNTHESIS OF NANOMATERIALS

Two types of approaches are generally applied for the fabrication of different types of nanomaterials (Fig. 1.1), namely the top–down approach, a process of stable division of bulk material into nanoparticles and the bottom–up approach, in which the nanoparticles are built up from the atomic level to nanolevel through clusters (Fig. 1.2). In top–down approach, the dissolution of metal from its respective bulk material is the basic principle. Several methods adopted in the top–down approach include laser ablation, solvated metal atom dispersion (Lin et al., 1986), high vacuum evaporation (Aiyer et al., 1994), electric arc reduction (Bradley, 1994), and electrochemical reduction (Reetz and Helbig, 1994). The combination of atoms or clusters to form nanoparticles is called bottom–up approach, which has gained popularity in recent years because there is a precise control over size and monodispersity of the nanoparticles (Toshima and Yonezawab, 1998). In this method, different reducing agents, such as

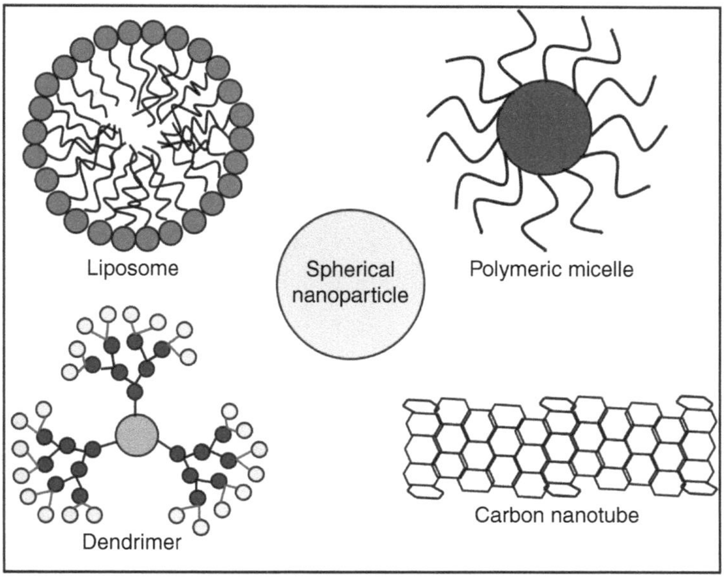

FIGURE 1.1 Different Types of Nanomaterials

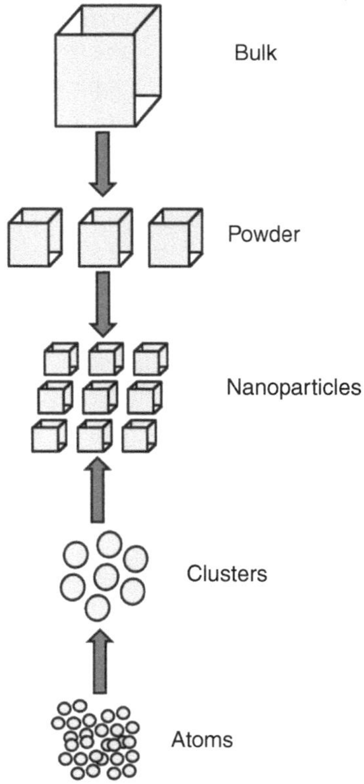

FIGURE 1.2 Top–Down and Bottom–Up Approach for Nanomaterial Synthesis

sodium borohydride, trisodium citrate, tannic acid, hydrazine, ascorbic acid, and tartaric acid were used to reduce the metal ions to form nanoparticles (Cushing et al., 2004). Metal nanoparticles, such as Au, Ag, Pt, Pd, and Rh can be prepared using mild reducing agents under ordinary conditions. The synthetic strategies of nanomaterials integrate both synthesis and assembly into a single method with better control over size, shape, and structure. The size, morphology, stability, and other properties of the nanomaterials are strongly influenced by the experimental conditions, interaction of reducing agents, and stabilizing agents. Therefore, the selection and design of a synthetic route plays a key role for determining the properties, stability, and applications of nanomaterials (Sharma et al., 2009). Currently, the techniques available for the production of nanomaterials essentially fall into three categories, such as physical, chemical, and biological methods.

2.1 PHYSICAL METHODS

A wide range of nanoparticles can be produced using different metals by employing physical methods with minor modifications. Several physical methods, such as sputter deposition, laser ablation or cluster beam deposition, microwave, evaporation–condensation, mechanical milling and pulsed wire discharge (PWD) are employed for the nanomaterial fabrication. Laser ablation and ball milling are the most commonly used methods for the preparation of nanoparticles using different kinds of solvents (Marine et al., 2000). However, the major limitation is that it is very difficult to produce ultrafine particles using these methods. Further, the physical synthetic methods involve costly equipments or vacuum systems for the preparation of nanomaterials. In case of mechanical milling, the main advantages are the simple operation, low cost of production and the possibility to scale up the process to produce large quantities (McCormick and Froes, 1998). Further, the important factors that affect the quality of the nanoparticles include the mill type, milling speed, temperature, time, atmosphere, shape, and size distribution of the grinding medium, weight ratio of ball to powder and the amount required for filling the vial (Suryanarayana, 2001). Sonolysis is also an efficient method for the synthesis of nanomaterials. For example, the iron oxide nanoparticles can be synthesized by decomposition (thermolysis or sonolysis) of organometallic precursors. Very high temperature hot spots generated by the rapid disintegration of sonically generated cavities allows the conversion of ferrous salts into magnetic nanoparticles. In this case, the different polymers, organic stabilizing agents, or structural hosts are used to limit growth of the nanomaterial (Dinega and Bawendi, 1999; Osuna et al., 1996; Puntes et al., 1999, 2002; Verelst et al., 1999). Some of the major advantages of physical methods include the use of no toxic chemicals, it is comparatively fast, lack of solvent contamination and the uniformity of nanomaterial distribution; however, the major disadvantage of these methods is the lesser quantity and poor quality of the nanomaterials produced as compared to chemical methods. These methods are also time-consuming and still under developmental stage.

2.2 CHEMICAL METHODS

Various chemical approaches have been employed for the fabrication of large amounts of nanomaterials within a shorter time period with fairly good control on the size distribution. Using chemical synthetic routes, different shapes of nanomaterials could be achieved by adjusting the concentration of the reacting chemicals and controlling the reaction environment (Murray et al., 2000). For metal nanoparticle synthesis, the most common approach employed is the chemical reduction method, which involves the reduction of an ionic metal salt in an appropriate medium in the presence of reducing agents. Different

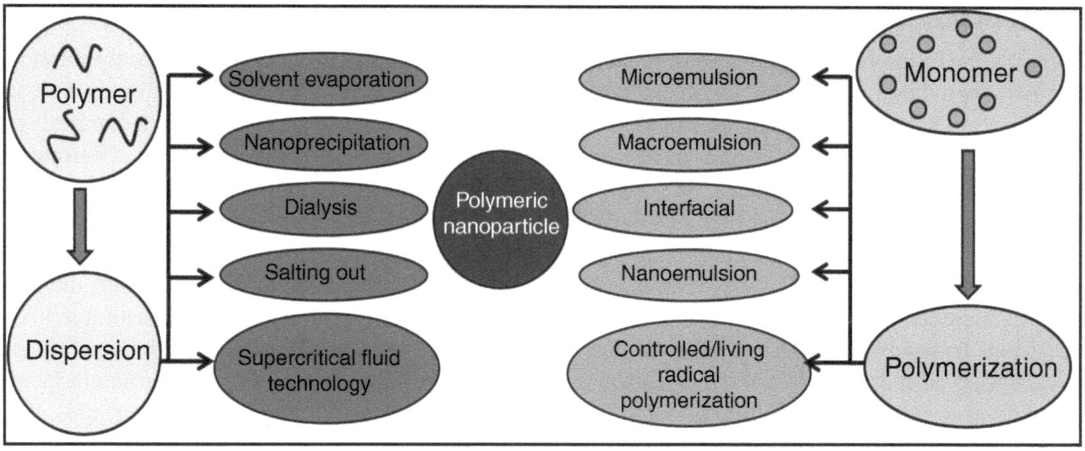

FIGURE 1.3 Different Methods Employed for the Synthesis of Polymeric Nanoparticles

organic and inorganic reducing agents, such as sodium citrate, ascorbic acid, sodium borohydride, dimethylformamide, have been used for the reduction of metal ions in aqueous or nonaqueous solutions. For example, in citrate reduction method, the reducing agent (citrate) reduces Au^+ ions resulting in the formation of colloidal gold (Au^0), followed by aggregation into oligomeric clusters, which eventually leads to size distribution and the formation of Au^0 nanoparticles (Turkevich et al., 1951). During synthesis, the capping agents play a key role in the stabilization of nanomaterials to prevent their aggregation (Oliveira et al., 2005).

Different polymers, such as polyethylene glycol (PEG) and poly(lactic-*co*-glycolic acid) copolymer (PLGA) and its derivatives are employed for the chemical synthesis of polymeric nanomaterials. Nanoprecipitation (solvent deposition) and oil/water emulsion are some of the methods used for the fabrication of functional (polyCOOH, polyCONHOH) polymeric nanoparticles using PLGA–COOH/PLGA–PEG–COOH polymers of different molecular weights (10–45 kDa). Some of the important methods employed for the fabrication of polymeric nanomaterials are depicted in Fig. 1.3. Dendrimers are highly branched three-dimensional nanostructures finding a wide range of therapeutic applications. Their synthesis interfaces between molecular and polymer chemistry methods. The fabrication of dendrimers involves the step-by-step controlled synthesis, and they relate to the polymer chemistry due to their repetitive structure comprising of monomers (Bosman et al., 1999; Frechet and Tomalia, 2001; Majoral and Caminade, 1999; Newkome et al., 2001). During the synthesis process, the dendrimer grows outward from a multifunctional core molecule. The core molecule reacts with monomers containing one reactive and two dormant molecules, resulting in the first-generation dendrimer. Later, the new fringe of the molecule is activated for reactions with further monomers. However, for the fabrication of dendrimers, two major synthetic strategies are employed, including divergent and convergent approaches. Both these strategies have relative advantages and disadvantages and the appropriate synthetic route depends mainly on the kind of monomer used and the target polymer structure (Kawaguchi et al., 1995; Labbe et al., 1996). Some of the major disadvantages observed with the use of chemical methods are the cost, energy intensive process, and toxic chemicals that are used as reducing agents, organic solvents used to generate hazardous waste that are potentially harmful to the environment and to biological systems.

2.3 BIOLOGICAL METHODS

Currently, with the renewed interest in implementation of green chemistry principles for the fabrication of nanomaterials for obtaining sustainable processes, minimization of waste, and environment-friendly for which simple strategies would be desirable. The synthesis of nanomaterials using biological methods has played a vital importance from a medicinal and technological perspective (Kattumuri et al., 2007; Zhang et al., 2007a). The secrets were originally inspired from nature, which has led to the development of biomimetic methods and advanced nanomaterials. Some of the notable examples are the formation of magnetic nanoparticles by magnetotactic bacteria (Moon et al., 2007) and biosilicification of silica nanospheres by diatoms (Kröger et al., 1999). Since the past few decades, the prokaryotic members were the only candidates being exploited for the capability to reduce insoluble toxic metal ions to soluble nontoxic metal salts due to their valence changes. However, in the recent past, it was observed that highly evolved organisms, such as plants, algae, diatoms, and other components of eukaryotes exhibited the reducing potential to convert the metal ions to metal nanoparticles. The major advantages of nanomaterial synthesis through biological approaches include its rapid synthesis at ambient temperatures, neutral pH, controlled size and morphology, less toxicity, and improved biocompatibility. Biodirected synthesis of metal nanoparticles has gained renewed attention, which uses different biomaterials such as enzymes, phytochemicals or other metabolites with reducing properties derived from various natural sources, such as plants, fungi, yeasts, actinomycetes, and bacteria as both reducing and stabilizing agents (Mohanpuria et al., 2008).

Plants are one of richest sources of natural products, including therapeutically useful compounds. Plants have evolved in the presence of natural nanomaterials. Nevertheless, the probability of plant exposure to nanomaterials has increased to a greater extent with the ongoing increase in the production and use of fabricated nanomaterials. Natural products of plant origin represent one of the most enduring approaches to achieve this target. The presence of diverse secondary metabolites, such as enzymes, proteins, and/or other reducing agents with electron-shuttling compounds, is usually employed in the fabrication of nanoparticles by plant metabolites (Akhtar et al., 2013). The role of plant metabolites for nanomaterial synthesis is directly related to the mechanisms of nanotechnology and green chemistry (Song and Kim, 2008). Several researchers reported that the use of dried leaves, buds, or other plant parts cut into pieces and extracted with different organic solvents resulted in different plant extract fractions. These extracts were incubated with the metal salts under the determined experimental conditions. The plant metabolites present in these extracts served as both reducing and stabilizing agents. The synthesized nanomaterials were reported to have different morphological shapes such as spherical, hexagonal, and triangular, which are mainly dependent on the chemical composition of the extract, concentration, and pH of the medium.

The biological synthesis of nanomaterials from living organisms, such as bacteria, fungi, and plants has huge potential and is encouraged due to the ease in rapid synthesis, controlled toxicity, control over size characteristics, reasonable, and an environment-friendly approach. Microbes can biosynthesize the nanomaterials by grasping target ions from their surrounding environment and then convert the metal ions into the elemental metal through cellular enzymes or metabolites. According to the location of nanoparticle formation they are classified into intracellular and extracellular synthesis (Faivre and Schüler, 2008; Roh et al., 2006). These microbial systems functioning as nanofactories have been investigated for different nanomaterial fabrication using metals, such as gold, silver, platinum, zirconium, zinc, cadmium, and iron (magnetite), which are available as oxides or metallic forms. The nanomaterial synthesis is based on the change of the oxidation state of the element either oxidation or reduction of the metal atoms. The synthesis procedure involves the mixing of metal precursors with the reactants present in crude or partially purified extracts derived from the cell-free supernatants under distinct experimental conditions. In addition, the synthetic conditions, such as pH, temperature, concentration, and reaction time need to be optimized to control the nanomaterial synthesis to obtain the desired size and morphology (Kumar et al., 2010). Biomolecules can also be exploited for the synthesis of metal nanoparticles. The synthesis of hybrid nanomaterials can be achieved using molecular precursors in the presence of biological templates (e.g., DNA, proteins) (Cauerhff and Castro, 2013). Most commonly involved biomolecules are viruses, proteins, peptides, and enzymes. Very few reports are available on the nanoparticle synthesis using this approach. Biomolecules with three-dimensional structures and vast diversity have several advantages and by using these precursors, the nanomaterial synthesis can possess several different characteristics and properties.

3 CHARACTERIZATION OF NANOMATERIALS

The characterization of nanomaterials is important to understand the physicochemical properties that determine their application in diverse research fields. After fabrication of nanomaterials, a series of analytical and imaging techniques are employed for the characterization of the prepared nanomaterials. The solubility, size, morphology, surface charge, surface chemistry, and surface plasmon resonance are the important physicochemical properties of the nanomaterials (Elzey and Grassian, 2010). Usually the nanomaterial properties differ in its size and shape, so the accurate measurement of size and shape is critical for their biomedical applications. The optical properties of nanomaterials are determined based on the measurement of surface plasmon resonance (Kreibig and Vollmer, 1995). In case of metal nanoparticles, the maximum absorbance, which determines the surface plasmon resonance, lies in the ultraviolet region of the electromagnetic spectrum. In the case of gold and silver nanoparticles, the resonance spans within the visible region because of interband transitions (Kumar et al., 2010). This absorption wavelength can be confirmed using UV-visible spectroscopy to identify the type of metal in the sample or to study any changes in the surrounding medium of the nanomaterials (Kreibig and Vollmer, 1995). Scanning electron microscopy (SEM), transmission electron microscopy (TEM), and high resolution transmission electron microscopy (HRTEM) are the different imaging techniques employed to determine the morphological characteristics of the nanomaterials. SEM generates high-resolution images of surface morphology of the different nanomaterials (Joshi and Viswanathan, 2006). TEM has much higher resolution as compared to traditional light microscopy since short wavelength of highly energetic electron beams are used to obtain images of the object under study (Williams and Carter, 1996). HRTEM is a high-magnification imaging mode of the transmission electron microscope that allows the high resolution imaging of the crystallographic structure of the nanomaterials at atomic scale level (O'Keefe et al., 1978). Atomic force microscopy (AFM) is one of the most popular tools to illustrate the material surfaces at nanoscale level. The principles, instrumentation aspects, and the applications of AFM to study the structural and physical properties of nanostructures, nanofibers, and for nondestructive characterization of textile materials are very well described (Dufrêne, 2002; Jalili and Laxminarayana, 2004).

The infrared spectrum of the nanomaterials is recorded using a Fourier transform spectrometer, employed for the identification of functional groups of the capping/stabilizing agents used in the synthesis of the nanomaterials (Silverstein et al., 2005). Similar to infrared spectroscopy, the Raman scattering can be used for the identification and quantification of the molecule (Eliasson et al., 2001). Surface-enhanced Raman scattering (SERS) is a powerful tool to obtain the signal from fluorescent chromophores (Nabiev et al., 1994). In SERS, an increased intensity at the near-field excitation was observed at the location of the analyte, which is adsorbed to the noble metal. It was recorded that the net surface enhancement increases by 10-order of magnitude (Bjerneld et al., 2000; Xu et al., 1999). Inductively coupled plasma-mass spectrometry (ICP-MS) is a sensitive analytical tool used for the quantification at a lower concentration (in parts per trillion) of both metals (Moldovan et al., 2004) and nonmetals (Wuilloud and Altamirano, 2006) used in the nanomaterial synthesis. ICP-MS is superior over other elemental analytical techniques, such as atomic absorption spectrometry (AAS) and inductively coupled plasma–optical emission spectrometry (ICP–OES), in view of its shorter measurement time, rapid sample processing and superior detection capabilities (Nelms, 2005). X-ray photoelectron spectroscopy (XPS) is used to analyze the atomic composition, chemical state, and electronic state of the elements present on the nanomaterial surfaces. It is also termed as electron spectroscopy for chemical analysis (ESCA) (Azoulay, 1983). Energy dispersive X-ray spectroscopy (EDX) analysis is used to determine the surface elemental composition and provides an overall metal profiling of the sample (Russ, 1984). Dynamic light scattering (DLS), also termed as photon correlation spectroscopy (PCS), is based on light scattering, which is used to determine the average particle size and charge of the nanomaterials (Berne and Pecora, 2000). Zeta potential measurement helps to understand the stability of the nanoparticles in a suspended medium that is dependent on the repulsion forces between the particles. The zeta potential is an indirect measurement of the overall surface charge that the particle acquires in a particular medium (Hunter, 1988). X-ray diffraction (XRD) analysis provides structural information of the crystalline nanomaterials. This technique is used for the identification and characterization of nanomaterials based on their diffraction pattern (Cullity and Stock, 2001).

4 APPLICATIONS OF NANOMATERIALS FOR DRUG DELIVERY

Drug delivery is a method or process of administering a pharmaceutical compound or a drug to achieve a therapeutic effect in humans or animals. The ultimate aim of drug delivery systems is to tailor-make the drug formulation to meet the individual requirements under the control of pathophysiological or in vivo conditions rather than the in vitro characteristics (Brannon-Peppas, 1995). Such delivery systems offer several advantages including improved efficacy, reduced toxicity, and improved patient compliance and convenience as compared to conventional dosage forms. Several drug delivery systems have been formulated and investigated for drug delivery through different routes of administration. In contrast to the traditional dosage forms, novel drug delivery systems aim to deliver the drug at a rate directed by the needs of the body during the treatment period, and channel the bioactive entity to the site of action thereby reducing the side effects in other locations (Misra et al., 1997). The application of nanomaterials in drug delivery has gained renewed interest in the recent years due to their unique properties, such as large surface area to volume ratio, which helps to achieve controlled and targeted drug delivery by encapsulation, entrapment, or conjugation of the drug. Using nanomaterials as drug carriers have several benefits, such as their size, which can allow penetrating into cell membranes and lysosomal escape after endocytosis. Apart from the size, surface chemical properties (capping agents) of the nanomaterials are crucial for particle uptake, binding, and stabilization of proteins and distribution.

Surface modification of nanomaterials offers several possibilities to improve the drug delivery efficiency in terms of cellular uptake, intracellular transport, and cellular binding. Currently, several types of nanoparticulate systems, including polymeric based drug carriers, dendrimers, nanoparticles, carbon nanotubes, and quantum dots have been used in drug delivery systems.

4.1 POLYMERIC NANOMATERIALS

Fabrication of polymeric nanomaterials, for example dendrimers, polymeric micelles, and hyperbranched polymers is a growing area of existing biomaterial science, due to their unique properties and huge potential in drug delivery (Gong et al., 2012; Kim et al., 2012b; Lim and Simanek, 2012; Xu et al., 2012). In polymeric nanomaterials, the polymer used should be biocompatible and often biodegradable. Polymeric nanoparticles are submicron-sized (<1000 nm) polymeric colloidal particles, composed of biodegradable or biostable polymers, in which a therapeutic agent/drug of interest can be entrapped or embedded within their polymeric matrix or conjugated on the surface of nanoparticles. In polymeric nanoparticles, the encapsulated drug molecules are protected from the enzymatic and hydrolytic degradation. In general, polymeric nanoparticles contain a core–shell structure and the core consists of a dense polymer matrix in which a hydrophobic drug can be encapsulated. The core–shell structure can be changed by varying the composition of hydrophilic and hydrophobic groups on the polymers. The halo structure is made up of hydrophilic polymers like different polysaccharides, PEG, and PVP. It serves to give steric stability to the polymeric nanoparticles upon intravenous administration. Poly(ε-caprolactone) (PCL), poly(lactic acid) (PLA), PLGA, alginic acid, gelatin, and chitosan are some of the polymers used from a therapeutic perspective. Due to biocompatibility or biodegradability, these polymers are the good candidates for drug delivery applications (Cheng et al., 2007a; Gan and Wang, 2007; Gu et al., 2008; Szlek et al., 2013). When compared to free drugs, the polymeric nanoparticles have several other advantages including improved drug bioavailability, high carrier capacity, ability to release the payload in a controlled manner and adapt to different routes of administration and concentrate in different infectious locations by virtue of their enhanced permeability and protection. Specific molecules conjugated to polymeric nanoparticles have shown enhancement in the targeting of specific cells and tissues (Bharatwaj et al., 2010). Polymeric nanoparticles are structurally stable and can be synthesized with a sharper size distribution. During the polymeric nanoparticle synthesis, different properties, such as size, zeta potential, and drug release profiles can be precisely tuned by selecting different polymers, polymer lengths, surfactants, and solvents (Umamaheshwari and Jain, 2003).

Over the past few decades, drug carriers containing stimulus responsive building blocks as a polymeric structure have drawn tremendous attention in drug delivery applications (Wel et al., 2006). By introducing these stimuli-responsive polymers in magnetite has provided several benefits for the fabrication of novel drug delivery system. Recently, much attention has been directed to the biomedical applications of magnetite nanoparticles, especially in targeted drug delivery devices (Li et al., 2012; Sahu et al., 2012). Kim et al. (2008) reported the synthesis of temperature-responsive magnetomicelles that contains functionalized magnetic core, Fe_3O_4–undecylenic acid (Fe_3O_4–UA) and an amphiphilic layer of temperature-responsive polymer. Amphiphilic poly(undecylenic acid-*co*-*N*-isopropyl acrylamide) was grafted to the Fe_3O_4–UA core as a temperature-responsive micellar surface layer to prepare well dispersed Fe_3O_4–UA-*g*-P (UA-*co*-NIPAAm) magnetomicelles. These nanosized magnetomicelles exhibited good potential for temperature-triggered controlled drug release.

4.2 POLYMERIC MICELLES

In general, the polymeric micelles are nanoscopic (>100 nm) and contain amphiphilic block copolymers with a core–shell structure. These polymeric micelles exhibit larger cores than surfactant micelles an important feature for higher solubilization capacity as compared to the regular micelles (He et al., 2011). Polymeric micelles have been investigated extensively for drug delivery applications. They are unique systems where aggregated amphiphilic copolymers are in dynamic equilibrium with free unimers. The use of micelles prepared from amphiphilic copolymers has attracted much attention for poorly soluble drugs due to their solubilization (Li et al., 2011; Xiong et al., 2011). The unique features of polymeric micelles including size, high stability, and low critical micelle concentration (CMC) along with core–shell arrangement qualify them as attractive candidates for use in drug delivery systems for clinical applications, especially for hydrophobic drugs having poor solubility in water (Butun et al., 2011; Chen and Liu, 2012). The polymeric micelles are divided into two groups based on the drug-loading pattern: physical drug entrapment or covalent drug conjugation. In physical drug entrapment micelles, the drug payloads are incorporated in the micelle core through hydrophobic interactions (Gong et al., 2012). The drug physicochemical stability and its release patterns are determined by the equilibrium rates, which are controlled time-dependent. In the case of covalent drug conjugation micelles, the drug binding linkers stably tether the drug molecules in the micelle core until the micelles gather at the site of the action and elicit the drug release because of various stimuli generated by ions, signal peptides, enzymes, and pH (Nishiyama et al., 2005). Several factors that favor the drug loading in polymeric micelles include the physicochemical properties of the drug and core-forming polymer, loading method, and the nature of the solvent used for the same and hydrophobic block length. Other factors that can alter drug loading and release kinetics include chemical composition of the micelle core, polymer–drug compatibility, and physical state of the micelle core (He et al., 2007; Jie et al., 2005).

Stimuli-responsive micelles are potential drug delivery systems, in which the polymeric matrices are synthesized bearing a stimulus-responsive moiety. These micelles are sensitive to small environmental changes that enable the release of drug molecules at the target sites (Rijcken et al., 2007; Schmaljohann, 2006). To date, several nanoscaled polymeric micelles have been developed for drug delivery based on various stimuli, such as temperature (Soppimath et al., 2005), pH (Zhao et al., 2012), and magnetic field (Park et al., 2008). Thermoresponsive polymers are smart nanomaterials that possess the ability to respond to changes in temperature. Thermoresponsive polymers are mainly of two types: lower critical solution temperature (LCST) and upper critical solution temperature (UCST). The LCST and UCST are the respective critical temperature points below or above at which the polymer and solvent are completely miscible. For example, a polymer solution below the LCST appears as a clear, homogenous solution, while a polymer, solution above the LCST looks cloudy and when the polymer solution leads to LCST, it is also referred to as cloud point (Ward and Georgiou, 2011). Different polymers with thermoresponsive properties, such as poly(*N*-isopropylacrylamide) (PNIPAAm) (Hoffman, 1987), poly(dimethyl acrylamide) (PDMAAm) (Martellini et al., 2003), PEG (Kiremitci et al., 2007; Yang and Kao, 2006), poly(ethylene oxide)–poly(propylene oxide)–poly(ethylene oxide) triblocks (PEO–PPO–PEO) (Niu et al., 2011) and poly(ethylene glycol)–poly(lactic acid)–poly(ethylene glycol) triblocks (PEG–PLA–PEG) (Ruan and Feng, 2003) were used in drug delivery. PNIPAAm was the most widely studied polymer with reference to biomedical applications, because it has an LCST of around 32°C, which is very close to the body temperature (37°C) and exhibited fast on-off switching. The LCST adjustment was achieved by the copolymerization of PNIPAAm with hydrophilic and hydrophobic monomers rendering the overall higher or lower hydrophilicity of the polymer, respectively (Feil et al., 1993; Yin et al., 2006). This polymer when crosslinked into hydrogels exhibited a coil to globule state resulting in a rapid decrease in the gel volume, which causes a diffusion controlled rapid release of the entrapped drug (Jones et al., 2007, 2008; Okuyama et al., 1993). Conetworks of PNIPAAm with hydroxyethyl methacrylate (HEMA) were prepared, which exhibited a LCST of 34°C that favored drug release and mechanical strength for in vivo application as light activated antimicrobial implants (Jones et al., 2007). In continuation to the earlier study, zinc tetraphenylporphyrin (Zn-TPP) was incorporated into PNIPAAm-*co*-PHEMA polymer and it was found that above the LCST, the Zn-TPP release rate was dramatically reduced. Further, the careful tailoring of the polymer composition and subsequently the LCST enabled the on–off drug release for various applications (Jones et al., 2008). A conetwork of PDMAAm-*co*-poly(methoxyethyl acrylate) was synthesized, which showed that this hydrogel released the drug at body temperature (Martellini et al., 2003). PNIPAAm crosslinked with polyamino acid chains were synthesized to produce thermoresponsive degradable hydrogels (Yoshida et al., 2003).

Elastin is a thermoresponsive polypeptide, which has a unique thermoresponsive property. Below its transition temperature, it is in a solvated state and gets extended in solution; however, above its LCST of 30°C, the chains fold to form nanoparticles, which entrap drug molecules inside their structure, thus they can be used as a drug delivery vector (Bessa et al., 2010; Rincon et al., 2006). A thermoresponsive PNIPAAm polymer with cleavable lactic acid and dextran groups were prepared to produce a biodegradable hydrogel (Xiao et al. 2004). PNIPAAm and PVCL–HEMA polymers were grafted onto a dextran chain to produce injectable and biodegradable hydrogels, which were capable of drug delivery over several days with negligible cytotoxicity (Wu et al., 2009). Hydrogels were prepared using modified xyloglucan polymers and by the removal of galactose units the LCST could be altered, which renders the polymer more hydrophobic (Brun-Graeppi et al., 2010). Hydrogels of PEG containing magnetic nanoparticles were synthesized for possible application in drug delivery and the gels exhibited a magnetic ability by internal heating on application of an alternating magnetic field, which results in the deswelling of the gels (Meenach et al., 2010). PEGMA hydrogels were prepared by inclusion of magnetite nanoparticles during the polymerization stage. These hydrogels showed a thermoresponsive ability with LCST of 60°C and deswell with an increase in temperature. Moreover, the tunable superparamagnetic property was dependant on the amount of magnetite added (Papaphilippou et al., 2009, 2011).

A series of amphiphilic terpolymers based on sodium 2-acrylamido-2-methyl-1-propane sulfonate (AMPS), *N*-isopropylacrylamide (NIPAM), and cinnamoyloxyethyl methacrylate (CEMA) were fabricated. These terpolymers were water soluble and stimulus responsive. They were sensitive to various factors, such as temperature, UV irradiation, and ionic strength (Szczubialka and Nowakowska, 2003). In another study, hyaluronic acid (HA), a natural anionic polysaccharide, was modified by introducing reduction-sensitive disulfide bonds between the carboxyl groups and the backbone of HA (HA–SS–COOH). Both reducible shielding (HA–SS–COOH) and stable hyaluronic acid shielding were introduced in the construction of DNA/polyethyleneimine (PEI) complexes through electrostatic interaction. The disulfide bridges of polymer and DNA complexes were proved to be susceptible to intracellular redox conditions. This complex exhibited lower cytotoxicity, higher gene transfection efficiency and also increased the cellular uptake by HA receptor, which is overexpressed in carcinoma cells. Additionally, the HA–SS–COOH shielding was found to be superior to HA because of extra reduction of the responsive deshielding function (He et al., 2012).

4.3 DENDRIMERS

The cascade synthesis of low molecular weight acrylates using acrylonitrile having iterative growth with branching was first conceptulized by the Vögtle group from the University of Bonn, Germany, which formed the basis of dendrimer chemistry (Buhleier et al., 1978). However, this method faced problems in the product purification and isolation resulting in low yields and the obtained molecules were not large enough to exhibit the unique dendrimer properties. The word *dendrimer* is of Greek origin, the word *dendron* means tree or branch and *meros* means part. Further, in 1993, two independent groups from Freiburg University (Wörner and Mühlhaupt, 1993) and from DSM (De Brabander-van den Berg and Meijer, 1993) performed vast and enhanced modification of the Vögtle approach to prepare poly(propylene imine) dendrimers. The DSM route was quite notable because it formed the viable commercial route for synthesis of aliphatic amine dendrimers. Later in 1981, Denkewalter and coworkers patented the synthesis of L-lysine-based dendrimers (Denkewalter et al., 1981). In 1985, Donald A. Tomalia and his group working with Dow Chemical Co. followed a parallel and independent approach for the divergent and macromolecular preparation of poly(amidoamine) (PAMAM) dendrimers, the so-called "true dendrimers," which became quite successful and even today it is the preferred commercial synthetic route for the Starburst (trademark of Dow) family of dendrimers (Tomalia et al., 1985). Later several groups made significant contributions toward the synthetic methodologies and the specific applications. Still continuous efforts are in progress to improve the efficiency and decrease the cost of synthesis of these macromolecules. Added to the synthetic developments for the acceptance of dendrimers as discrete entities, parallel efforts were in progress to characterize the molecules with structural precision using various techniques, such as mass spectrometric (MS) techniques for protein characterization. The precise protein molecular weights (up to 1 M Da) and polydispersity values were determined using electrospray ionization-mass spectrometry (ESI-MS) (Kallos et al., 1991; Tomalia and Durst, 1993) and matrix-assisted laser desorption/ionization time-of-flight mass spectrometry (MALDI–TOF–MS) (Müller et al., 2007) techniques. Based on these techniques, it was demonstrated that all dendrimer constructions obeyed mathematically defined mass-growth principles. Other techniques such as critical gel filtration chromatography (Dubin et al., 1992; Smith et al., 1987; Tomalia et al., 1990), light scattering/viscosity (Brothers et al., 1998), photophysical (Caminati et al., 1990; Frechet and Tomalia, 2001; Turro et al., 1991), electron microscopy (Jackson et al., 1998; Tomalia et al., 1985), gel capillary electrophoresis (Brothers et al., 1998), AFM (Frechet and Tomalia, 2001; Li et al., 2000) have been used to verify the principles of "dendritic growth and amplification" and unusual properties that result from the dendritic state.

Dendrimers are defined as globular, repeatedly branched, macromolecular structures having three major components, such as focal core, peripheral layer comprising of different functional groups, and interior layer composed of several building blocks. The multiple arms extend from a central core that is approximately 2–10 nm in diameter (Svenson and Tomalia, 2005). The highly branched nature of dendrimers also provides enormous surface area to size ratio. They are synthesized through an algorithmic step-by-step pattern with every iterative sequence of reactions producing a higher generation molecule. The size of the dendrimer is described as a function of generation (G_n), where n is 0, 0.5, 1.0, and 1.5 and G is the number of repetition cycles. The molecular weight of the dendrimer nearly doubles with each additional generation and a doubled (discrete) number of functional groups are added (Tomalia, 2005). Two seminal synthetic routes are available for the synthesis of dendrimers, including (1) *divergent approach*: the synthesis begins from the core and continues radially outward by successive stepwise activation and condensation steps with addition of successive layers of building blocks; (2) *convergent approach*: the synthesis begins at the periphery of the final molecule and proceeds inwards. A comparison of both these methods suggests that the convergent approach provides better control of the final dendrimer structure as compared to the divergent approach. On the other hand, the divergent approach is suitable for large-scale preparation of dendrimers. However, both these strategies follow step-wise processes that are tedious and time consuming. These methods also have relative advantages and disadvantages and the synthetic route depends mainly on the type of monomer employed and the structure of polymer (Kawaguchi et al., 1995; Labbe et al., 1996). A wide range of biocompatible polymers, including polyamidoamine (PAMAM), poly(amidoamine-organosilicon) (PAMAMOS), PEI, polypropylene imine (PPI), polyethylene glycol (PEG), have been used for the synthesis of dendrimers as a host for both hydrophilic and hydrophobic drugs (Gillies and Frechet, 2005). The PAMAM dendrimers have either amine or carboxylic moieties (Tomalia et al., 1990), while the PPI have multiple cationic amine moieties (Baars et al., 2000) present on the periphery.

In recent years, there has been a rapid increase in the use of dendrimers in medicinal and pharmaceutical chemistry for diverse applications. The well-defined molecular architecture, controllable size, predictable molecular weight, low polydispersity, inherent stability, biocompatibility, controllable "surface" functionalization to optimize biodistribution, and high penetration capacity of dendritic structures through the cell membrane are some of the unique properties, which make them suitable candidates as carriers for drug delivery (Bronstein and Shifrina, 2012; Gurdag et al., 2006). In addition, the presence of positive and negative charged functional groups on the surfaces of dendrimers makes it easy for attaching hydrophilic drug molecules through covalent conjugation or electrostatic interaction (Cheng and Xu, 2005; Florence, 2005; Gillies and Frechet, 2005). These noncovalent insertions or complexes offers diverse and promising advantages like better water solubility, stability of the drug, programmed drug release from the matrices and enhanced pharmacodynamic (PD) and pharmacokinetic (PK) behaviors (Gupta et al., 2006). Due to non- or low-immunogenicity, dendrimers are suggested to be the best alternative as compared to the synthesized peptide carriers and natural protein carriers (Yang et al., 2009). In addition, enhanced penetration and retention effect offers preferential cellular uptake of the materials by cancerous tissues (Imae and Hamaguchi, 2012). The use of dendrimers as potential drug delivery agents can be prepared in two different ways: (1) physical entrapment of drug molecules inside the dendritic structure, and (2) covalent attachment of drug molecules or other functionalities onto the surface to provide dendrimer-drug conjugates.

Physical entrapment: The internal cavity of the designed dendrimer is used for the entrapment of drugs and their later controlled release. Earlier studies showed that DNA complexed with PAMAM dendrimers were used for gene delivery applications (Haensler and Szoka, 1993), while the "dendritic boxes" and unimolecular micelles were used for encapsulating guest molecules, such as hydrophobic drugs and dye molecules (Hawker et al., 1993; Jansen et al., 1994, 1995; Naylor et al., 1989; Newkome et al., 1991). The advantage of using dendritic unimolecular micelles as compared to the conventional polymeric micelles is that the micellar structure is maintained at all concentrations as there are covalent connections between the hydrophobic segments. However, the major drawback is the difficulty to control the release of the molecules from the dendrimer core. Harsh conditions are required in some cases to release the encapsulated drug (Jansen et al., 1995), while in others the drug is not well retained and they get released rapidly (Kojima et al., 2000; Liu et al., 1999). Tailoring of these dendrimers can result in relatively hydrophobic interior for encapsulation (Watkins et al., 1997). Both hydrophobic and hydrophilic drugs can be loaded into the dendrimers. The interior side of dendritic macromolecule contains nonpolar cavities, which ensure them the ability to encapsulate hydrophobic drug molecules

(Gupta et al., 2006). The selective liberation of the encapsulated guest molecules could be achieved by either partial or complete removal of the shell under finely tuned hydrolytic conditions (Jansen et al., 1995). Further, the introduction of PEG grafting on the surface of the PAMAM dendrimer has enabled the dendritic micelles to encapsulate anticancer drugs like adriamycin and methotrexate (Kojima et al., 2000) and 5-fluorouracil (Bhadra et al., 2003) has been attempted. The drug encapsulation in the dendritic core cavity increases with increased generation and chain length of the attached PEG grafts. The drug release rates in these systems were slow in low ionic strength aqueous solution, while it was released rapidly in isotonic solutions. Another approach for controlled release of drugs from the micellar cavity of dendrimers, which are prepared using hybrids of PEO and dendrimers with pH-sensitive hydrophobic acetal groups on the periphery. Under mild acidic pH conditions, the hydrolysis of the hydrophobic acetal groups takes place, which causes the disruption of the micelle and release of the drug (Gillies et al., 2004). Drug-entrapped dendrimers are thought to be useful because the drug molecules stay intact within the dendrimers and the drug gets released rapidly under physiological conditions (Oliveira et al., 2010). Hence, the encapsulation ability of dendrimers needs further improvement for drug-delivery systems.

Covalent attachment: To develop dendrimer-drug conjugates it is necessary to couple drug molecules directly to the "surface" of the dendrimer. Due to the presence of multiple surface functionalities, one dendrimer molecule can accommodate multiple drug molecules, and the number of drug molecules per conjugate can be changed depending on the different generations of dendrimers or by changing the coupling conditions. Novel polyaryl ether dendrimers with two different surface functional moieties were designed and they were able to covalently bind model drugs and solubilizing groups (Liu et al., 1999; Liu and Frechet, 1999). The attachment of PEG units on the surface provides water solubility to the assembly serving as the first functionality, while the second functionality is to use the attached hydrophobic drug molecules. The conjugation of drugs to dendrimers through hydrophobic encapsulation or electrostatic interactions generally preserves the chemical integrity and pharmacological properties of the drugs. Whereas, the covalent linkage of drugs to the surface functional groups of dendrimers via chemical bonds is more suitable for a better control over drug release than can be achieved by simple encapsulation of drugs into the dendrimers or electrostatic conjugation of drugs with the dendrimers (Cheng et al., 2008; Li et al., 2012a,b). A single dendrimer under stable conditions can carry many drug molecules using various functional groups on the outer shell and reach the target site through enhanced permeation and retention (EPR) effects, which is a passive targeting approach. So, the conjugation of drug molecules to the dendrimer is an attractive strategy for drug delivery. The linkage stability and cleavability are the important features for the release of drug molecules from the drug conjugates at the target site. Several types of functional groups on the outer shell of dendrimer are responsible for its reactivity and they conjugate with different types of therapeutic agents, targeting moieties, and biocompatible molecules (Cheng et al., 2007b). Namazi and Adeli (2005) reported that the citric acid–polyethylene glycol–citric acid (CPEGC) triblock dendrimers as biocompatible compounds containing different drug molecules finds application in drug delivery. The amount of trapped molecules and drugs molecules, such as 5-aminosalicylic acid (5-ASA), pyridine, mefenamic acid, and diclofenac was measured and the release of these molecules and drugs were studied under in vitro conditions. An amphiphilic hyperbranched block copolymer conjugated with folate (H40-PLA-*b*-MPEG/PEG–FA) was prepared comprising of a dendritic Boltorn H40 core, a hydrophobic PLA inner shell and a hydrophilic methoxy PEG and folate-conjugated PEG (PEG–FA) outer shell, which acted as a tumor-targeted drug delivery carrier. The initial burst release of the drug from the doxorubicin

Table 1.1 Dendrimer–Drug Conjugates Used for Drug Delivery

Dendrimer Types	Drug/Application	Outcome/Highlights	References
PAMAM (polyamidoamine) dendrimer	5-fluorouracil/cancer therapy	Controlled release of 5-fluorouracil	Zhuo et al. (1999)
PAMAM dendrimer	Propranolol/hypertension	These prodrugs increase drug solubility and bypass drug efflux transporters and also increase the drug bioavailability	D'Emanuele et al. (2004)
G3 PAMAM lauroyl chain dendrimers	Colchicine/colon cancer therapy	Effectively cross cell monolayers and enhance the bioavailability of the drug	Jevprasesphant et al. (2004)
PAMAM dendrimers	Niclosmide/tapeworm	Increases the solubility and also offers the possible role for the controlled release of the drug from solid dosage form	Devarakonda et al. (2005)
Polylysine dendrimer	VivaGel (SPL7013 Gel)/ sexually transmitted infections	Effective against HIV virus and HSV in in vitro and animal models	Rupp et al. (2007)
PEGylated poly(L-lysine) dendrimer-camptothecin conjugates	Camptothecin/cancer therapy	Exhibited superior activity in both murine and human colon carcinoma cells	Fox et al. (2009)
Magnetic nanobeads embedded in PNIPAM shell	Doxorubicin/cancer therapy	Temperature responsive behavior for controlled loading and release of drug	Deka et al. (2011)
PAMAM-PLA-*b*-PEG-conjugated with anti-CD105 monoclonal antibody (TRC105)	Doxorubicin/cancer therapy	pH-controlled drug release and PET imaging capabilities	Guo et al. (2013)
Poly-(*N*-ε-carbobenzyloxy-L-lysine) (PZLL) and dendritic polyamidoamine (PAMAM) block copolymer	Doxorubicin/cancer therapy	Sustained in vitro release of drug at pH 7.4 and 5.8 in HeLa cells	Lin et al. (2014)

(DOX)-loaded micelles was up to 4 h, followed by a sustained release of entrapped DOX over a period of about 40 h. The cellular uptake of the DOX-loaded micelles was found to be higher due to the folate-receptor-mediated endocytosis, which contributed to higher cytotoxicity in 4T1 mouse mammary carcinoma cell line (Prabaharan et al., 2009a). Table 1.1 shows some examples of dendrimer-drug conjugates used for drug delivery.

4.4 POLYSACCHARIDE-BASED NANOMATERIALS

Natural polysaccharides can be derived from various sources, such as algae (alginate), plants (guar gum, pectin), microbes (dextran, pullulan, gellan gum, xanthan gum), animal (chitosan, chondroitin, hyaluronan) (Liu et al., 2008b; Nair and Laurencin, 2007a). The advantage of using polysaccharides as drug carriers is due to their availability as natural resources, low cost, biodegradable, nontoxic, stable, and hydrophilic (Liu et al., 2008b). These polysaccharides offer a wide structural diversity and properties due to their chemical composition and wide range of molecular weights. The presence of different reactive groups in their structure make them amenable for easy chemical and biochemical modification. The hydrophilic groups, such as hydroxyl, carboxyl, and amino groups present in the structure enhance the bioadhesion with biological tissues such as epithelial and mucous membranes, a useful approach to improve bioavailability of drugs in drug delivery systems (Lee et al., 2000). Some examples of different polysaccharide-based nanomaterials used for target-based drug delivery applications are shown in Table 1.2.

Alginate is a nontoxic, biodegradable polysaccharide obtained from marine brown algae and some bacteria. It is an anionic copolymer comprising of (1→4) linked β-D-mannuronic acid and α-L-guluronic acid units of different composition and sequence depending on the alginate source. Alginate shows a variable molecular weight, which is dependent on the enzymatic control during the production and its extraction decides the degree of polymerization. Commercial alginates exhibit a typical average molecular weight of approximately 200,000 Da, however, alginates of molecular weight as high as 400,000–500,000 Da are also available (Rehm, 2009). Alginates are used as excipient in drug products, which is dependent on its thickening, gel forming, and stabilizing properties. The different favorable properties like biocompatibility, biodegradability, pH sensitiveness, and mucoadhesiveness of these polymers have been explored by pharmacologists as oral delivery matrices for proteins (George and Abraham, 2006). Alginate can be designed as a controlled-release product based on its favorable pH sensitive properties for drug delivery. Alginate forms two types of gels dependent on pH, that is, an acid gel and an ionotropic gel, gives the polymer unique properties as compared to neutral macromolecules. At low pH, alginic acid hydrates resulting in the formation of a high-viscosity "acid gel," while ionotropic gelation is due to the divalent metal interaction, namely the calcium ion. This molecule can be tailor-designed for a number of applications. Alginate microspheres can be easily prepared based on the gelling properties; however, they exhibit several problems with regard to mechanical stability and the drug release observed is too fast. To circumvent these problems, polymer-coated alginate microspheres and nanoparticles have proved promising for modified drug delivery (Matricardi et al., 2008). Different drugs, such as indomethacin (Pillay et al., 1998), prednisolone (Wittayaareekul et al., 2006), metoclopramide hydrochloride, antiemetic drug (Gavini et al., 2005), protein drugs (George and Abraham, 2007; Lin et al., 2005), diclofenac (Nayak and Pal, 2011), antidiabetic drug, probucol (Mooranian et al., 2015), resveratrol (Kassem et al., 2015), and risperidone (Bera et al., 2015) have been investigated as drug delivery agents for prolonged and better drug release at the target site.

Guar gum is a nonionic, biodegradable, water-soluble, renewable, and low-cost natural polysaccharide obtained from the endosperm of guar beans, *Cyamopsis tetragonoloba*, belonging to the Leguminosae family. It is composed of linear chains of (1→4) linked β-D-mannopyranosyl units and short-side branched α-D-galactopyranosyl units linked at (1→6) positions (Sarmah et al., 2011). Guar gum hydrates and forms a highly viscous solution in cold water and remains stable in the pH range of 5–7. Therefore, it finds use as a thickener in cosmetics and food industry (Barbucci et al., 2008). Guar gum can be modified by derivatization, grafting, and network formation to improve its properties for diverse biomedical applications (Prabaharan, 2011). Guar gum nanoparticles were prepared by nanoprecipitation and crosslinking methods (Sarmah et al., 2009; Soumya et al., 2010). Different factors responsible for the nanoparticle formation are the molecular mass of the galactomannan, solvent,

Table 1.2 Polysaccharide-Based Nanomaterials for Target-Based Drug Delivery Applications

Nanomaterials	Synthesis Methods	Drug	Application	References
Alginate and chitosan				
Alginate gel beads-entrapped liposome	Encapsulation	Bee venom peptide	n.s.	Xing et al. (2003)
Chitosan-coated alginate beads	Ionotropic gelation	Ampicillin	Intestinal delivery	Anal and Stevens (2005)
Alginate/chitosan microparticles	Ionotropic gelation	Metoclopramide hydrochloride	Antiemetic drug nasal delivery	Gavini et al. (2005)
Chitosan microspheres	Spray-drying	Carbamazepine	Nasal delivery	Gavini et al. (2006)
Alginate/chitosan microparticles	Ionotropic gelation	Prednisolone	Colon delivery	Wittayaareekul et al. (2006)
Chitosan-coated alginate beads	Ionotropic gelation	Metronidazole	Anti-*H. pylori* therapy	Ishak et al. (2007)
Chitosan-coated alginate beads containing PNIPAAM	Ionotropic gelation and chemical cross-linking	Indomethacin	Sustained pH/temperature-sensitive drug delivery	Shi et al. (2008)
Polyelectrolyte-coated alginate microspheres	Ionotropic gelation	Dexamethasone	Drug carrier for implantable glucose biosensor	Jayant et al. (2009)
Hyaluronic acid–coupled chitosan nanoparticles	Ionotropic gelation	Oxaliplatin	Anticancer therapy	Jain et al. (2009a)
Chitosan arginine beads	Crosslinking	n.s.	Antibacterial therapy	Tang et al. (2010)
Alginate/chitosan microbeads	Ionotropic gelation	Amoxicillin	Oral delivery	Angadi et al. (2012)
Tamarind seed polysaccharide-alginate composite beads	Ionotropic gelation	Diclofenac	Oral delivery	Nayak and Pal (2011)
N-trimethyl chitosan-hyaluronan-cisplatin conjugate nanoparticles	Ionotropic gelation	Cisplatin	Anticancer therapy	Cafaggi et al. (2011)
Chitosan nanoconstructs	Ionotropic gelation	Heparin	Oral delivery	Paliwal et al. (2012)
Alginate coated chitosan core shell nanoparticles	Ionotropic gelation	Enoxaparin (low molecular weight heparin)	Oral enoxaparin delivery for antithrombosis	Bagre et al. (2013)
Chitosan (CS) and poly(gamma-glutamic acid) nanoparticles	Ionotropic gelation	Insulin	Diabetes therapy	Lin et al. (2007)

(*Continued*)

Table 1.2 Polysaccharide-Based Nanomaterials for Target-Based Drug Delivery Applications (*cont.*)

Nanomaterials	Synthesis Methods	Drug	Application	References
Alginate-dextran sulfate core with chitosan-PEG-albumin shell	Ionotropic gelation	Insulin	Oral insulin delivery	Woitiski et al. (2012)
Alginate/chitosan beads	Ionotropic gelation	Insulin	Oral insulin delivery	Tahtat et al. (2013)
Chitosan-alginate core–shell nanoparticles	Ionotropic pregelation	Insulin	Oral insulin delivery	Mukhopadhyay et al. (2015)
PLA-grafted-chitosan nanoparticles	Protection-graft-deprotection	Amphotericin B	Ocular drug delivery	Zhou et al. (2013)
Tragacanth gum-alginate microspheres	Orifice-ionic gelation	Diclofenac	Oral delivery	Amin et al. (2013)
Chitosan/alginate complexes	Ionotropic pregelation	Chlorhexidine digluconate	Vaginal delivery	Abruzzo et al. (2013)
Alginate/chitosan microparticles	Ionotropic pregelation	Rifampicin	Antituberculosis therapy	Lacerda et al. (2014)
Chenodeoxycholic acid–sodium alginate microcapsules	Microencapsulation	Probucol	Oral delivery for type 2 diabetes	Mooranian et al. (2015)
Thiolated alginate microbeads	Ionotropic gelation	Resveratrol	Oral delivery for periodontal disease	Kassem et al. (2015)
Chitosan-coated alginate nanoparticles	Ionotropic gelation	Daptomycin	Ocular delivery	Costa et al. (2015)
Alginate-locust bean gum IPM microspheres	Ionotropic gelation	Aceclofenac	Oral delivery	Jana et al. (2015)
Phytosterol-alginate nanoparticles	Ionotropic gelation and encapsulation	Doxorubicin	Anticancer therapy	Wang et al. (2015)
Lecithin/chitosan nanoparticles	Encapsulation	Melatonin	Transmucosal delivery	Hafner et al. (2009)
Lecithin/chitosan nanoparticles	Encapsulation	Melatonin	Transdermal delivery	Hafner et al. (2011)
Lecithin/chitosan nanoparticles	Encapsulation	Clobetasol-17-propionate	Epidermal delivery	Senyiğit et al. (2010)
Lecithin/chitosan nanoparticles	Entrapment	Quercetin	Topical delivery	Tan et al. (2011)
Lecithin/chitosan nanoparticles	Encapsulation	Natamycin	Ocular drug delivery	Bhatta et al. (2012)
Lecithin/chitosan nanoparticles	Ionic interaction	Diflucortolone valerate	Epidermal delivery	Özcan et al. (2013a)
Lecithin/chitosan nanoparticles	Entrapment	Betamethasone valerate	Topical delivery	Özcan et al. (2013b)

Table 1.2 Polysaccharide-Based Nanomaterials for Target-Based Drug Delivery Applications (*cont.*)

Nanomaterials	Synthesis Methods	Drug	Application	References
Lecithin/chitosan nanoparticles	Encapsulation	Amphotericin-B	Ocular drug delivery	Chhonker et al. (2015)
Dextran				
Dextran magnetite and conjugated cisplatin	Thermal ablation	Cisplatin	Anticancer therapy	Sonoda et al. (2010)
Folic acid-dextran-camptothecin nanoparticles	Supercritical antisolvent	Camptothecin	Anticancer therapy	Zu et al. (2011)
Magnetite-dextran composite nanoparticle	Crosslinking	Mitoxantrone	Anticancer therapy	Krukemeyer et al. (2012)
Gellan gum and xanthan gum				
Gellan gum-PVA hydrogel microspheres	Encapsulation	Carvedilol	Antihypertensive, Oral drug delivery	Agnihotri and Aminabhavi (2005)
Gellan gum beads	Ionotropic gelation	Cephalexin	Oral delivery	Agnihotri et al. (2006)
Gellan gum beads	Ionotropic gelation	Amoxicillin	Oral delivery	Babu et al. (2010)
Xanthan gum-PVA microspheres	Emulsion crosslinking	Diclofenac	Oral delivery	Ray et al. (2010)
Other polymers				
Guar gum microspheres	Emulsification	Metronidazole	Colon-targeted delivery	Chourasia and Jain (2004a)
BSA nanoparticles	Desolvation crosslinking; folate functionalized	Gemcitabine	Anticancer therapy	Dubey et al. (2015)

n.s., not specified.

surfactant, crosslinker, and agitation. Guar gum acts as stabilizer and capping agent during the green synthesis process to prepare iron nanoparticles (INPs) (Tiraferri et al., 2008) and silica nanoparticles (Singh et al., 2011). Guar gum and its derivatives can be used in controlled release formulations in various forms such as coatings, matrix tablets, hydrogels, microspheres, and nanoparticles for targeted drug delivery (Aminabhavi et al., 2014; Prabaharan, 2011). Guar gum and its derivatives has been used in various controlled release formulations as a carrier for different drugs such as trimetazidine dihydrochloride (Krishnaiah et al., 2002a), metoprolol tartrate (Krishnaiah et al., 2002b), celecoxib (Krishnaiah et al., 2002c), metronidazole (Krishnaiah et al., 2002d), tinidazole (Krishnaiah et al., 2002e), 5-fluorouracil (Krishnaiah et al., 2002f), ornidazole (Krishnaiah et al., 2003), methotrexate (Chourasia and Jain, 2004b), sennosides (Momin and Pundarikakshudu, 2004), curcumin (Elias et al., 2010), tamoxifen (Sarmah et al., 2014), and budesonide (Liu and Zhou, 2015).

Pectin is a plant cell wall structural polysaccharide comprising of poly(D-galacturonic acid) residues linked via α(1→4) glycosidic bonds. It has an average molecular weight ranging 50,000–180,000 Da (Sinha and Kumria, 2001). The carboxyl groups are present partially in the methyl ester form having different degrees of esterification (DE) and amidation (DA), which determines the content of carboxylic acid in pectin chains. Pectin from different sources exhibits different gelling abilities, due to variations in molecular size and chemical composition. Pectin finds ample uses in the food industry as a gelling and thickening agents and of late has been used as an excipient in pharmaceutical formulations (Liu et al., 2003). The salient property of pectin is its gelling activity and the gel process is dictated by different parameters such as type and concentration of pectin (DE, DA), modification of hydroxyl groups, pH, temperature, and divalent cations. Pectins with high DE (>50%) are known as high methoxy (HM) pectins, which enable gel formation with an increase in the amounts of hydrophobic groups and reducing the solubility of pectin. On the other hand, low methoxy (HM) pectins have DE value less than 50%, which makes them highly water soluble and the gel formation occurs only at an extremely low pH solution or in the presence of divalent cations causes the crosslinking of the galacturonic acids of the pectin chains. The hydrophilic property can be reduced with an increasing tendency to form gels by introducing amide groups in low DE pectin (Liu et al., 2003). Due to the high water-solubility property, pectin cannot protect the drug load effectively during its passage through the stomach and small intestine, so many research groups are attempting to develop water-resistant pectin derivatives, which can be degraded enzymatically (Semdé et al., 2000; Sinha and Kumria, 2001). Low methoxyl pectins are important from a drug delivery viewpoint as they can form gels with calcium ion (Ca^{2+}) by noncovalent associations with the carbohydrate chains of pectin and they find potential applications especially in nasal formulations (Morris et al., 2010).

Pectin has been used as an excipient or carrier for oral and colon-specific delivery in different formulations, such as microspheres, films, gels, or droplets (Liu et al., 2003; Sriamornsak, 2011). Pectin is suitable for application as colon-specific drug delivery vehicle because it is selectively digested by the colon microflora to release the drug with minimal degradation in upper gastrointestinal tract. The oral administration of drugs for colon-specific delivery system is expected to increase the drug bioavailability at the target site, reduce drug dose and systemic adverse effects. Thus, the pectin matrix has dual advantages as drug carrier and therapeutic agent for use in treatment of colon cancer (Wong et al., 2011). The combination of pectin with other polymers of either natural or synthetic origin can be used to prepare novel formulations (Chourasia and Jain, 2004b). Composites prepared using pectin in combination with another polymer may alter the degree of swelling, change mechanical properties, improve the drug stability, and enable controlled drug release (Liu et al., 2003; Wong et al., 2011). Pectin in combination with chitosan (Fernandez-Hervas and Fell, 1998), poly(lactide-*co*-glycolide) (Liu et al., 2004), zein, a corn protein (Liu et al., 2006), Eudragit S100 (Jain et al., 2007a), 4-aminothiophenol (Perera et al., 2010), and hyaluronic acid (Pliszczak et al., 2011; Tsai et al., 2013) acted as good controlled drug release devices. Different pectin-derived matrices used for colon-targeted drug delivery are tabulated in Table 1.3.

Chitosan is a hydrophilic biopolymer composed of glucosamine and *N*-acetylglucosamine units linked by (1→4) β-glucosidic bonds. Commercially, it is derived by the alkaline deacetylation of chitin, the main components of the shells of crabs, shrimps and lobsters, as well as from cell wall of fungi (Muzzarelli and Muzzarelli, 2005). After deacetylation, the chitosan is composed primarily of D-glucosamine repeating units. The degree of deacetylation (%DD) in commercial chitosan is generally in the range of 60%–100%. The average molecular weight of commercial chitosan is between 3,800 and 20,000 Da. Chitosan is nontoxic, biodegradable, and has biocompatible properties; however, its applications are limited due to its insolubility above pH 6 (Guerrero et al., 2010). It is a weak base and insoluble in water and organic solvents. However, its solubility can be increased in diluted aqueous acidic solution ($pH < 6.5$), which

Table 1.3 Pectin-Derived Matrices for Colon-Targeted Drug Delivery

Nanomaterials	Synthesis Method	Drug	References
Zn-pectinate gel microparticles	Crosslinking	Ketoprofen	El-Gibaly (2002)
Pectin/ethylcellulose film coating	Extrusion and spheronization	5-Aminosalicylic acid	Ahmed (2005)
Zinc pectinate beads	Ionotropic gelation	Ketoprofen	Dupuis et al. (2006); Chambin et al. (2006)
Eudragit-coated calcium pectinate gel	Ionotropic gelation	5-Fluorouracil	Jain et al. (2007a)
Eudragit-coated pectin microspheres	Emulsion dehydration	5-Fluorouracil	Paharia et al. (2007)
Pectin/ethylcellulose film coatings	Extrusion and spheronization	5-fluorouracil	Wei et al. (2007)
Pectin/ethylcellulose film coatings	Extrusion and spheronization	5-aminosalicylic acid	He et al. (2008)
Pectin-ethylcellulose-Eudragit L30D-55	Extrusion and spheronization	5-aminosalicylic acid	Fude et al. (2007)
Pectin/chitosan/Eudragit RS films	Blending and solvent casting	Theophylline	Ghaffari et al. (2007, 2008)
Calcium pectinate microspheres	Emulsification	Methotrexate	Chaurasia et al. (2008)
Enteric-coated calcium pectinate microspheres	Entrapment	Theophylline	Maestrelli et al. (2008, 2012)
Chitosan/pectin polyelectrolyte complexes	Spray-drying	Vancomycin	Bigucci et al. (2008)
Pectin microspheres	Emulsion dehydration	Metronidazole	Vaidya et al. (2009)
Pectin/chitosan hydrogels	Spray-drying	Vancomycin	Bigucci et al. (2009)
Pectin-4-aminothiophenol conjugate	Spray-drying	Metronidazole	Perera et al. (2010)
Zinc-pectinate beads	Entrapment	Resveratrol	Das et al. (2010)
Calcium pectinate beads	Entrapment	Resveratrol	Das and Ng (2010a,b)
Zinc-pectin-chitosan composite	Entrapment	Resveratrol	Das et al. (2011a)
Pectin microparticles	glutaraldehyde-crosslinking	Resveratrol	Das et al. (2011b)
Zinc-pectinate beads	Ionotropic gelation	Theophylline	Dhalleine et al. (2011)
Eudragit-coated pectin microspheres	Emulsion dehydration	Mesalamine	Jain et al. (2011a)
Eudragit S-100 based pectin-hydroxypropylmethylcellulose microsponges	Quasi-emulsion solvent diffusion	Dicyclomine	Jain et al. (2011b)
Eudragit S100 coated calcium pectinate microspheres	Emulsification-linkage	Curcumin	Zhang et al. (2011)
SPION encapsulated pectin nanocarriers	Probe sonication	5-Fluorouracil	Dutta and Sahu (2012a)
Hyaluronan-cisplatin conjugate nanoparticles entrapped in Eudragit S100-coated pectinate/alginate microbeads	Electrospraying	Cisplatin	Tsai et al. (2013)
Calcium pectinate/alginate microspheres	Electrospray	Cisplatin	Hsu et al. (2013)
Silica-coated calcium pectinate beads	Crosslinking	Theophylline	Assifaoui et al. (2013)

converts the glucosamine units into a soluble form with protonated amine groups (Sinha et al., 2004). Chitosan has received increased attention in nanomedicine as a controlled release matrix having polycationic character with reactive functionalities, easy for enzyme degradation, and generation of nontoxic degradation products. It has been evaluated as a carrier for cytarabine, an antineoplastic agent (Blanco et al., 2000). It finds use as a biomaterial for wound healing because it is biodegraded by enzyme action (Bernardo et al., 2003). Genipin, a crosslinking agent, was used to improve the structural reinforcement and antibacterial properties of Ag-chitosan nanoparticle bioconjugates, which find use in wound dressing applications (Liu and Huang, 2008). Chitosan and PVA as natural and synthetic polymers, respectively, are able to form crosslinked hydrogels by chemical crosslinking (Nuttelman et al., 2001) or freeze–thaw (Hassan and Peppas, 2000) methods and in the past few years they have been used in the preparation of drug-loaded microparticles (Davidenko et al., 2009; Muzzarelli and Muzzarelli, 2005). The bioadhesive properties of chitosan have been explored in novel bioadhesive drug delivery systems, which prolong the drugs residence time at the site of absorption and increase its bioavailability (Varum et al., 2008). Some of the drugs when included into chitosan-based drug delivery systems and when administered via nasal (Learoyd et al., 2008) or gastrointestinal (Guerrero et al., 2010) routes have shown improved therapeutic efficacy. Chitosan functionalized AuNPs were demonstrated to improve the transport of insulin across the intestinal track of diabetic rats (Bhumkar et al., 2007).

Hyaluronic acid (HA, also known as hyaluronan) is present in the soft connective tissues of animals, which is a linear, high-molecular-weight glycosaminoglycan comprised of 500 to several thousand repeating disaccharide units of D-glucuronic acid and *N*-acetylglucosamine residues linked by β (1→3) and β (1→4) glycosidic bonds and contains no sulfo groups. Solutions of high-molecular-weight HA find use in biomedical applications owing to its biological compatibility and useful rheological properties (Burdick and Prestwich, 2011). HA-based nanoparticle composites take advantage of the biological compatibility of HA (Choi et al., 2010). PEGylation of hyaluronic acid nanoparticles improved the target-based delivery of drugs to the tumor site (Choi et al., 2011a). PEG-conjugated HA nanoparticles showed promising potential as a drug carrier of doxorubicin and camptothecin for cancer therapy (Choi et al., 2011b). Hyaluronan-cisplatin conjugate nanoparticles were prepared for colon targeted drug delivery (Tsai et al., 2013). HA-capped silica nanoconjugates exhibited high colloidal stability, good biocompatibility, and precise cell-targeting ability for drug release, which makes them promising agents for biomedical applications (Chen et al., 2013). HA-coated docetaxel-loaded cholesteryl hemisuccinate vesicles were prepared, which showed pH-induced release of docetaxel in MCF-7 tumor-bearing mouse model (Song et al., 2014). HA functionalized gold nanoparticles showed targeted delivery of metformin drug in HepG2 cells (Kumar et al., 2015).

Dextran is a polysaccharide composed of many glucose units linked by α(1→6)-glucosidic bonds in the main chain. The average molecular weight of dextran is as high as 10^7–10^8 Da. Dextran is a water soluble, neutral, biodegradable polymer synthesized by different bacterial strains, such as *Leuconostoc mesentroides*, *Lactobacillus plantarum*, *Lactobacillus sanfrancisco*, and *Gluconobacter oxydans* (Heinze et al., 2006). They are available commercially in different molecular weights with a relatively narrow size distribution. Dextrans possess large number of hydroxyl groups, which can be easily conjugated to drugs and proteins by either direct attachment or through a linker (Mehvar, 2000). Dextran conjugation has resulted in new tailor-made polymers with different molecular weights, shapes, structures, and the functional groups required for coupling at the desired positions in the chain (Varshosaz, 2012). Dextrans find extensive uses in the pharmaceutical, food, and chemical industries (Heinze et al., 2006). Dextran has a long history for clinical use as a blood plasma volume expander and as heparin substitute in anticoagulant therapy. Dextran and its derivatives have potential application for use in modified drug delivery (Aumelas et al., 2007; Coviello et al., 2007). It is not only used as a carrier to prepare drug-nanoparticulate systems (Tiyaboonchai and Limpeanchob, 2007; Tiyaboonchai et al., 2003), but also employed to cover these systems (Gavory et al., 2011). Dextran prodrugs are considered as very useful systems for achieving controlled drug release and drug targeting. Several dextran-based prodrugs and conjugates have been prepared for various drugs, such as ketorolac (Vyas et al., 2007), etodolac (Vyas et al., 2009), budesonide (Varshosaz et al., 2009), and flufenamic acid (Lee et al., 2011). In particular, the dextran-drug conjugates and dextran nanoparticles efficiently deliver the chemotherapeutic agents and enhance the effectiveness and therapeutic property of the chemotherapeutic agents within the target site (Li et al., 2009).

Gellan gum is a high molecular weight anionic linear polysaccharide produced by fermentation from *Sphingomonas paucimobilis*. This polymer is a tetrasaccharide repeating unit of D-glucose, L-rhamnose, and D-glucuronate in a molar ratio of 2:1:1 (Fialho et al., 2008). Gellan gum is conventionally used in the food industry due to its distinct gelling, thickening, stabilizing, and texturizing properties (Sanderson and Clark, 1983). Chemical hydrogels of gellan are prepared using chemical crosslinking of networks to increase their mechanical properties and enable slower drug release profiles (Coviello et al., 1999). In view of its unique structure and beneficial properties, gellan is currently being used for the development of hydrogels for controlled release forms in different various pharmaceutical formulations including oral, ophthalmic, nasal, and other (Osmałek et al., 2014).

Xanthan gum is fermentation-derived natural, high molecular weight polysaccharide from *Xanthomonas campestris*. It consists of (1→4) linked β-D-glucose units, having a trisaccharide side chain attached to alternate D-glucosyl residues. Although the gum has many suitable properties desirable for drug delivery, it has limited practical use with respect to the unmodified forms due to slow dissolution and substantial swelling in biological fluids. Chemical modification of xanthan gum is achieved by different conventional chemical methods like carboxymethylation, and grafting, such as free radical, microwave-assisted, chemoenzymatic, and plasma-assisted chemical grafting has resulted in the alteration of different physicochemical properties for diverse biological applications (Badwaik et al., 2013).

4.5 NANOPARTICLES

In the past few decades, there has been an upsurge in research on the area of metal nanoparticles exhibiting significantly distinct physical, chemical, and biological properties as compared to their bulk counterparts. Further, the use of nanobioconjugates has become predominant in metal nanoparticles for biomedical applications (Bhattacharya and Mukherjee, 2008), because the first use of immunogold labeling by Faulk and Taylor (1971). These metal nanoparticles are used in drug delivery carrier for delivery of drugs, proteins, peptides, plasmids, DNAs, and so on, and also find use in detection, diagnosis, and therapy (targeted and nontargeted) (Bhattacharya and Mukherjee, 2008). In general, metal-based nanoparticles of different shapes and sizes have been examined as diagnostic and drug delivery systems. Gold, nickel, silver, iron oxide, zinc oxide, gadolinium, and titanium dioxide are the most common metals used for nanoparticle synthesis (Diaz and Vivas-Mejia, 2013). The metal nanoparticles are highly advantageous due to their small size, large specific surface area and high fraction of surface atoms, in vivo drug delivery characteristics and distinct physicochemical properties such as catalytic activity, optical, and magnetic properties, which make them attractive materials for therapeutic applications (Duran et al., 2005; Kowshik et al., 2003; Shahverdi et al., 2007). Some examples of various nanoparticles with different functionalization agents and their target-based applications are shown in Table 1.4.

Table 1.4 Nanoparticles with Different Functionalization Agents and Their Target-Based Applications

Functionalization Agent	Drug	Application	References
Gold nanoparticles			
Phthalocyanine	n.s.	Photodynamic therapy	Doubrovsky et al. (2010)
Porphyran	Doxorubicin	Anticancer therapy	Venkatpurwar et al. (2011)
Sophorolipid-gellan gum	Doxorubicin	Glioma therapy	Dhar et al. (2011a)
Gellan gum	n.s.	Anticancer therapy	Dhar et al. (2011b, 2012)
Eclipta alba leaf extract	Doxorubicin	Anticancer therapy	Mukherjee et al. (2012)
Pullulan	5-fluorouracil	Anticancer therapy	Ganeshkumar et al. (2014)
Chondroitin sulfate	Insulin	Diabetes therapy	Cho et al. (2014)
Oleic acid in toluene and polysiloxane polymer	n.s.	Antimicrobial coatings	Noimark et al. (2014)
Polyaspartic acid	Doxorubicin	Anticancer therapy	Khandekar et al. (2014)
Xanthan gum	Doxorubicin	Anticancer therapy	Pooja et al. (2014)
Gum karaya	Gemcitabine hydrochloride	Anticancer therapy	Pooja et al. (2015)
Hyaluronic acid	Metformin	Anticancer therapy	Kumar et al. (2015)
Silver nanoparticles			
Rhamnolipid	n.s.	Antimicrobial therapy	Kumar et al. (2010)
carboxymethylated–curdlan and fucoidan polymers	n.s.	Antimicrobial therapy	Leung et al. (2010)
sulphated polysaccharides from marine red algae, *Porphyra vietnamensis*	n.s.	Antimicrobial therapy	Venkatpurwar and Pokharkar (2011)
Bacterial cellulose	n.s.	Antimicrobial therapy	Yang et al. (2012a)
EPS from *Lactobacillus rhamnosus*	n.s.	Antimicrobial therapy	Kanmani and Lim (2013)
EPS from *B. subtilis*	n.s.	Antimicrobial therapy	Selvakumar et al. (2013)
Magnetic nanoparticles			
Dextran	Doxorubicin and paclitaxel	Anticancer therapy	Jain et al. (2008)
Polylactide (PLA)	Paclitaxel	Anticancer therapy	Chorny et al. (2010)
Polyethylene glycol (PEG)	Doxorubicin	Anticancer therapy	Yallapu et al. (2010)
Pectin	Oxaliplatin	Anticancer therapy	Dutta and Sahu (2012b)
Pectin and chitosan	Diclofenac	Anti-inflammatory	Dutta and Sahu (2012c)
Polyethylene glycol (PEG)	Curcumin	Breast cancer therapy	Yallapu et al. (2012)
Catechin-dextran conjugate	Catechin	Anticancer therapy	Vittorio et al. (2012, 2014)
dextran and cisplatin-bearing hyaluronic acid coating	Cisplatin	Anticancer therapy	Unterweger et al. (2014)
Aminosilane	Cathelicidin LL-37 peptide or cragenin CSA-13	Theranostic colon cancer therapy	Niemirowicz et al. (2015a)

n.s., not specified.

4.5.1 Gold nanoparticles

Gold nanoparticles (GNPs) are colloidal suspensions of nanosized gold particles. They are generally prepared on the basis of reduction of chloroauric acid in the presence of various capping and stabilizing agents. The different reducing agents used are trisodium citrate (Turkevich et al., 1951), toluene along with tetra-octanyl ammonium bromide as phase transfer reagent (Brust et al., 1994), different amino acids, such as aspartic acid (Mandal et al., 2002), L-tryptophan (Akbarzadeh et al., 2009), different biopolymers, such as natural gum (Dhar et al., 2008b), dextran (Jang et al., 2010), gellan gum (Dhar et al., 2011a,b), methanolic phytoconstituents of medicinal plants (Ramezani et al., 2008), and so on. The synthesis to obtain GNPs in the size range of 10–25 nm is dependent on various factors, such as temperature, salt concentration, and rate of addition of reactants. The different unique features of GNPs like small size, large surface area to volume ratio, high reactivity to the living cells, stability at high temperatures and translocation into the cells qualifies them as suitable candidates for drug delivery to the target site (Tiwari et al., 2011). GNPs find application for cancer therapy and its diagnosis, delivery of drug, gene, DNA, peptides, and so on. GNPs functionalized with specific biomolecules can effectively target and destroy the bacteria or cancer cells (Tiwari et al., 2011). The antimicrobial efficacy of GNPs was greatly enhanced by conjugating GNPs with different antibiotics to kill various antibiotic-resistant bacteria (Burygin et al., 2009). GNPs conjugated with vancomycin were able to specifically target and kill vancomycin-resistant enterococci (Gu et al., 2003). GNPs functionalized with aminoglycoside antibiotics like streptomycin, gentamicin, and neomycin exhibited remarkable antimicrobial activity against both Gram-negative and Gram-positive bacteria (Grace and Pandian, 2007). Cefaclor a second-generation β-lactam antibiotic acted as a reducing agent in gold NP synthesis and showed promising antimicrobial activity against both *Staphylococcus aureus* and *Eschericha coli* (Das et al., 2009). This antibiotic alone inhibited the peptidoglycan synthesis and caused the cell porous, whereas GNPs interacted with the cell wall and generated holes in the cell wall, resulting in intracellular cell material leakage with bacterial cell death (Saha, 2007). Gentamicin-gold nanospheres were prepared and the resulting carrier system was effective against *S. aureus* as compared to free gentamicin (Ahangari et al., 2013). Ampicillin-coated GNPs (15 ± 5 nm) were prepared, which showed promising antimicrobial activity against different Gram-positive and Gram-negative bacterial strains (Pender et al., 2013).

Pegylated SERS gold nanoparticles were prepared and conjugated to tumor-targeting ligands such as single-chain variable fragment (ScFv) antibodies. These conjugated nanoparticles targeted the tumor biomarkers such as epidermal growth factor receptors present on human cancer cells and in xenograft tumor models (Qian et al., 2008). Gold NPs were functionalized with a synthesized monolayer of folate-conjugated poly(L-aspartate-doxorubicin)-*b*-PEG copolymer (Au-P[LA-DOX]-*b*-PEG-OH/FA), which acted as a carrier for tumor-targeted drug delivery. An acid-cleavable hydrazone linkage enabled the covalent conjugation of the doxorubicin drug onto the hydrophobic inner shell. The drug loading level was estimated to be 17 wt.%. The conjugated drug was released much more rapidly at pH 5.3 and 6.6 as compared to pH 7.4 for exhibiting higher cytotoxicity against 4T1 mouse mammary carcinoma cell line (Prabaharan et al., 2009b). In one study, doxorubicin was functionalized on GNPs via an amino group. The reaction occurred under mild acidic conditions. The doxorubicin could not be adsorbed on the GNPs under alkaline conditions since the amino group was not protonated. However, under acidic conditions, protonation caused a positively charged amino group favouring the drug adsorption (Mirza and Shamshad, 2011). GNPs were PEGylated and then grafted with DOX using a cleavable disulfide linkage to obtain Au-PEG-SS-DOX nanoconjugates. These nanoconjugates released the drug at doses >15 μM very efficiently and showed cytotoxicity against multidrug resistant (MDR) cancer cells (Gu et al., 2012). GNPs (average particle size of 13 nm) were synthesized using porphyran as a reducing agent and the porphyran-capped GNPs were functionalized with doxorubicin. These NPs exhibited a pH-responsive drug release profile in human glioma cell line (LN-229) as compared with native porphyran, with significant drug release (sixfold higher) observed at pH 4.5 as compared to pH 7.4 (Venkatpurwar et al., 2011). GNPs (average particle size of 11.3 nm) were synthesized using cell-free culture supernatant of *Delftia* sp. strain KCM-006 and functionalized with resveratrol, an anticancer drug. The NPs exhibited a pH-responsive drug release profile in human lung cancer (A549) with significant drug release observed at pH 5.2 as compared to pH 7.4 (Kumar et al., 2014). Hybrid GNPs were produced by the interaction of highly reactive nascent GNP with chemical moieties present on specific molecules of biological interest such as peptides and proteins. Seven amino acid truncated bombesin molecule (BBN8–14) is a gastrin releasing peptide (GRP) conjugated to GNP and targeted the GRP receptors that are overexpressed in a diversity of neoplasma, such as small cell lung, prostate, breast, gastric, pancreatic, gastrointestinal carcinoid, and colon cancers (Ghasemali et al., 2013; Nejati-Koshki et al., 2013).

4.5.2 Silver nanoparticles

In the present context of increasing resistance to antibiotics and the dramatic fall in the development of new antibiotics has compelled the search for new antimicrobial agents. Silver nanoparticles has gained a renewed attention in the field of nanotechnology due to its unique physicochemical properties such as good conductivity, chemical stability, catalytic, antimicrobial, antifungal, antiviral, anticancer, and antiinflammatory, which have led to the recent upsurge of research in the synthesis and application of silver nanoparticles (Sharma et al., 2009). Several methods have been developed for the fabrication of AgNPs including microemulsion, organic-water two-phase synthesis, radiolysis, and reduction in aqueous solution is the most common protocol. The antimicrobial properties of silver nanoparticles and its use since ancient times have been well documented and find immense potential as a disinfectant for wound care (Lansdown, 2004). Among all metal nanoparticles, silver nanoparticles have demonstrated to be the most effective against bacteria, fungi, viruses, and other eukaryotic microorganisms (Sharma et al., 2009). Different factors known to affect the antimicrobial activity are the particle size, shape, crystallinity, surface chemistry, pH, ionic strength, divalent cations, ligands, and other macromolecules (Marambio-Jones and Hoek, 2010). From an antimicrobial mechanism perspective, silver nanoparticles (size of <20 nm diameter) attach to the sulfur-containing proteins of bacterial cell membrane, changing the membrane permeability resulting in pore formation in the bacterial cell walls. The silver ions infiltrate into the bacterial cells through these pores, where they undergo reduction as cell tries to pump-out these ions from the cytoplasm, eventually causing destruction of the cell membrane and bacterial cell death (Jung et al., 2008; Morones et al., 2005). At microquantity levels, silver ions affect the cell membrane physiology by inactivating the proteins, blocking the respiration and inhibits electron transfer, and subsequently kills the bacteria (Araujo et al., 2012).

One of the major applications of silver nanoparticles in therapeutics is in topical ointments containing nanosilver and other silver derivatives, which find use to prevent microbial infection of burns and open wounds. An alternative approach is the use of dressings impregnated with antimicrobial compounds, such as silver ions for suppression of microbial growth. The first nanosilver based wound dressing termed as "Acticoat" was developed in 1995 and this medicament was used in the management of burns enabling the healing process and removal of the formed scars, if any (Burrell et al., 1995; Dunn and Edwards-Jones, 2004). Antibacterial silver/sodium carboxymethyl cotton dressings were prepared by the particle cation exchange of sodium with silver and used for wounds caused by burns (Parikh et al., 2005). Nanosilver functionalized on cotton fabrics that finds use in antiseptic dressings showed excellent bactericidal activity against *S. aureus* found in wounds (Lee et al., 2007). The antimicrobial function exhibited is through the release of Ag^{+} ions from elemental silver or silver compounds that are incorporated into the dressing material that would enhance the healing process and prevent the infections. Silver nanoparticles impregnated into bacterial cellulose produced by *Acetobacter xylinum* (strain TISTR 975) was used as an antimicrobial wound dressing material for wound treatment (Maneerung et al. 2008). Novel wound dressings coated with silver–silica nanocomposite material (silica hydrogel and silver nanoparticles) showed enhanced bactericidal activity and were relatively safe to human health (Egger et al., 2009). A silver nanoparticle-based topical antimicrobial gel formulation was prepared for the treatment of burns and wounds, which showed antimicrobial activity against both Gram-positive and Gram-negative as well as *Candida albicans* (Jain et al., 2009b). Cotton fibers loaded with silver nanoparticles functionalized with natural extracts of *Eucalyptus citriodora* (neelagiri) and *Ficus bengalensis* (marri) showed excellent antimicrobial activity against *E. coli*, which find use in burn/wound dressings and in the production of antibacterial textiles and finishings (Ravindra et al., 2010). Researchers at the University of Bath and the burns team at the Southwest UK Paediatric Burns Centre at Frenchay Hospital in Bristol developed an advanced wound dressing, which contains nanocapsules comprising of antibiotics and dye. In the presence of pathogenic bacteria that produce toxins, the nanocapsules get triggered and the dressing changes its color when the antibiotic is released giving a signal that there is an infection in the wound. Experts suggest that this dressing has great potential to victims affected with burns and susceptible to bacterial toxic shock syndrome (Anonymous, 2010). Curcumin-encapsulated silver nanocomposite-bovine serum albumin hydrogels were prepared using silver nanoparticles and hydrophobic curcumin using acrylamide as crosslinking agent, which finds use as wound dressing agents (Vimala et al., 2014). Silver nanoparticles were also explored for its potential use as a preservative in cosmetic preparations (Kokura et al., 2010).

Bone cements are used in orthopaedic surgeries to join fractured bones and artificial joint replacements that find suitable place in dentistry and arthoplasty. Antibiotic-releasing bone cement (Kaplan et al., 2012) have shown success in the recent past; however, this has resulted in the generation of antibiotic-resistant bacterial strains. Silver nanoparticles added to bone cements showed antimicrobial action without impinging on the cytotoxicity of the biomaterial (Buckley et al., 2010; Ewald et al., 2011). Further, the tiopronin-capped silver nanoparticles impregnated into bone cement showed promising antimicrobial activity against methicillin-resistant *S. aureus* (MRSA) (Prokopovich et al., 2013). In the recent years, silver nanoparticles find wide usage in dental practice for their antimicrobial properties. Some reports indicated that silver nanoparticle preparations used as antibacterial coatings on titanium implants exhibited reduced bacterial adhesion and inhibited biofilm formation on the implant surfaces and prevented failure of dental implants (Liao et al., 2010; Secinti et al., 2011; Zhao et al., 2011). Biocompatible PMMA-silver nanoparticle discs were prepared, which inhibited *C. albicans* on denture base acrylic resins that can prevent the formation of denture stomatitis (Acosta-Torres et al., 2012; Li et al., 2014; Wady et al., 2012).

Antimicrobial polymers and coatings containing silver have gained renewed interest due to their long-term biocidal properties, high thermal stability, and low volatility (Kumar and Munstedt, 2005) for application in health care, biomedical devices, personal hygiene, and food industries. Traditional biomaterial surfaces functionalized with silver nanoparticles can serve as antimicrobial coatings to produce surfaces resistant to biofouling. Plastic catheter surfaces modified with silver nanoparticle coating prevented biofilm formation by both Gram-positive and Gram-negative bacteria as well as yeasts (Roe et al., 2008). Silver nanoparticle coatings on polyurethane and plastic catheters were demonstrated to exhibit antibacterial and antiadhesive properties against pathogenic strains of *E. coli, Enterococci, Pseudomonas aeruginosa*, and *S. aureus* (Monteiro et al., 2009). Surface charge and hydrophobicity played a key functional role in determining the surface chemistry of the coating material that influences the bacterial attachment and biofilm formation (Rzhepishevska et al., 2013). Kocuran-functionalized silver nanoparticles with potential antimicrobial and antiadhesive properties were demonstrated as antimicrobial coatings on silicone urethral catheters to prevent bacterial adhesion and biofilm formation in our laboratory (Kumar and Sujitha, 2014). Silver nanoparticles have been used as effective drug carrier agents and have attracted special attention due to their physiochemical properties.

Different reports have shown that silver nanoparticles conjugated with various drugs showed a significant increase in the efficacy of the tested antimicrobial agents. Silver nanoparticles conjugated with ampicillin successfully inhibited the growth of antibiotic-resistant isolates of *P. aeruginosa, S. aureus*, and *Enterobacter aerogenes* (Brown et al., 2012). Miconazole-conjugated silver nanoparticles demonstrated promising antifungal activity against various *Candida* strains and played a dual mechanism of action by inhibiting ergosterol biosynthesis and enabled ROS accumulation in mature *Candida* biofilms (Kumar and Poornachandra, 2015). Apart from its antimicrobial applications, the surface functionalization of silver nanoparticles with polysaccharides (PS-10 and PS-25) successfully inhibited the infectivity of monkey pox virus (MPV) (Rogers et al., 2008). In addition, the poly(*N*-vinyl-2-pyrrolidone) (PVP)-coated silver nanoparticles exhibited inhibitory activity against HIV type 1 virus (Lara et al., 2010a,b). Recent developments in the application of silver nanoparticles in cancer drug delivery were greatly enhanced due to its intrinsic anticancer activity. The nanosilver acts as a good drug delivery agent for targeting the cancerous cells (Nair and Laurencin, 2007b). The protein-conjugated silver sulfide nanoparticles showed good inhibition of glioma cell viability (Yang et al., 2008). Nanocarriers with a particle diameter of around 50 nm showed enhanced permeability into the cancerous cells without affecting the normal cell. Silver nanoparticles were functionalized with thiol-terminated photolabile DNA oligonucleotides and the resulting nanocarrier exhibited controlled gene silencing in cervical cancer cells (Brown et al., 2013). The surface of silver nanoparticles functionalized with starch was found to be nontoxic and greatly reduced the viability of cervical cancer cells (Kahrilas et al., 2014). A multifunctional nanocomposite comprising of silver nanoparticles and alisertib (drug) conjugated with polymeric nanoparticles was developed. The resulting targeted drug delivery system exhibited significant antiproliferative activity against glioblastoma (Locatelli et al., 2014). A recent study demonstrated that dual functionalized silver nanoparticles with anticancer drugs such as DOX and alendronate (Ald) showed successful intracellular delivery of the drugs into the cervical cancer cells and the synergistic combination greatly improved the efficacy of the drug-delivery system (Benyettou et al., 2015).

4.5.3 Iron nanoparticles

Magnetic nanoparticles (MNPs), such as iron (III) oxide (Fe_2O_3), is superparamagnetic in nature and also one of the three main oxides of iron, whereas other two are FeO and Fe_3O_4. Fe_3O_4 occurs naturally as the mineral magnetite and is also superparamagnetic in nature. INPs possess unique features, such as their ultrafine size, magnetic properties, and biocompatibility. Surface functionalization of the MNPs provides the opportunity for conjugation of different biomolecules, such as drugs or homing ligands, as well as combine the theranostic properties within the same structure (Dionigi et al., 2014). Some of the biomaterials, such as poly(L-glutamate) (Jeong et al., 2005), poly (D,L-lactide) (Akbarzadeh et al., 2012), hyaluronic acid (Kumar et al., 2007), and carboxyl-functionalized poly(amidoamine) dendrimers (Shi et al., 2007) have proved to be ideal shell material of choice due to their low toxicity, biocompatibility, and stability in reticuloendothelial system. The MNPs have been used as contrast agents in magnetic resonance imaging (MRI) for more than 20 years. In recent years, MNPs have emerged as potential candidates for various biomedical applications including targeted drug delivery and imaging, hyperthermia (treating tumors with heat), gene therapy, stem cell tracking, rapid DNA sequencing, early detection of inflammatory, cancer, diabetes, and atherosclerosis (Chorny et al., 2013; Ho et al., 2011; Hua et al., 2010; Jaffer et al., 2006; Kim et al., 2012a; Mou et al., 2015; Yang et al., 2012b). MNPs synthesized as drug carrier agents retain the ability to track their movement throughout the body. It is significant because it allows clinicians to examine the effect of injected therapeutics to arrive at their target sites (Veiseh et al., 2010). The INPs conjugated to several antibodies and other ligand molecules were tested for MRI imaging of tumors (Laurent et al., 2008). De and coworkers reported the targeted drug delivery of INPs in breast cancer cells by using magneto-fluorescent nanoparticles. The anticancer drug doxorubicin and herceptin1 antibody were entrapped into the INPs functionalized with poly(D,L-lactide-*co*-glycolide) (De et al., 2008). In another study, magnetic INPs conjugated to anticancer drug methotrexate were cytotoxic to breast and cervical cancer cells that were overexpressing the folate receptor. The INPs were accumulated at the folate receptor, released the drug within the cancer cells and finally killed the cancer cells (Kohler et al., 2005). Chitosan coated MNPs were prepared and this nanocarrier system was loaded with different anticancer drugs like doxorubicin (Unsoy et al., 2014a) and Bortezomib (Velcade) (Unsoy et al., 2014b). Results indicated that the drug released in cancer cells was pH-dependent and was more efficacious as compared to the free drug treatment.

Silica-coated magnetic nanoparticles (S_iO_2-MNPs) prepared by sol–gel method were further conjugated with tissue plasminogen activator (tPA, 0.5 mg/mL). The S_iO_2-MNP-tPA exhibited enhanced storage stability in buffer and operation stability in whole blood up to 9.5 and 2.8-fold, respectively. Further, the effective thrombolysis with S_iO_2-MNP-tPA under magnetic guidance was demonstrated in an ex vivo thrombolysis model, which showed 34% and 40% reduction in blood clot lysis time in the runs without magnetic targeting and with free tPA, respectively, using the same drug dosage. These biocompatible S_iO_2-MNP find use as a magnetic targeting drug carrier system to improve clinical thrombolytic therapy (Chen et al., 2012). Silica-coated magnetic nanoparticles (S_iO_2-MNPs) were prepared using a combination of controlled precipitation and hydrothermal methods. Streptokinase (SK) and tissue plasminogen activator (tPA) were conjugated to the synthesized S_iO_2-MNPs acted as nanocarriers for targeted therapeutic delivery, which find application in clinical thrombolytic therapy for acute myocardial infarction (Tadayon et al., 2015). Similar to gold and other metal nanoparticles, INP also show hyperthermia. INPs targeted to tumors can be used for simultaneous MRI imaging and damage the tumor cells by heating. The ability of INPs to target tumors, MRI contrast agent, delivery of anticancer drugs to tumor cells and hyperthermia makes them ideal candidates for anticancer therapy. A magnetic drug delivery system was developed for loading antibiotics, such as chloramphenicol (Thach et al., 2008), nystatin (Hussein-Al-Ali et al., 2014a), and streptomycin (Hussein-Al-Ali et al., 2014b) on chitosan-coated MNPs through chemical adsorption, which is based on nonspecific electrostatic or hydrophobic interactions. Ceragenins (CSA-13) represents the family of synthetic mimics of cationic antibacterial peptides exhibiting antibacterial properties (Chin et al., 2008). Magnetic nanoparticles functionalized with ceragenin (MNP-CSA-13) were prepared and it was observed that the CSA-13 was released from the nanoparticles at pH 5 that favors the imine bond hydrolysis, indicating its potential to eradicate *Helicobacter pylori* (Leszczynska et al., 2009). MNP-CSA-13 also exhibited bactericidal activity and prevented the formation of *P. aeruginosa* biofilms (Niemirowicz et al., 2015b).

4.5.4 Silica nanoparticles

Among all inorganic nanomaterials, silica nanomaterials have gained a significant position due to easy synthesis and surface chemistry. Rice husk is a renewable agricultural waste and a major source of amorphous silica (Chandrasekhar et al., 2003). Nanostructured silica powder was prepared from rice husk by acid or alkaline leaching, followed by calcination at higher temperature (Amutha et al., 2010; Liou, 2004; Liou and Yang, 2011; Real et al., 1996; Thudaji and Nuntiya, 2008), electrodeposition and reduction (Zaky et al., 2008), and acid precipitation (Yuvvakkumar et al., 2014). Silica is an inert chemical material and optically transparent, which makes it a good matrix for entrapment of drugs/biomolecules that are sensitive to changes in the surrounding environment. Monodispersed and spherical silica nanoparticles were first prepared by Stober et al. (1968). Silica nanoparticles of different sizes ranging between 50 nm and 1 μm were synthesized with aqueous alcohol solutions of silicon alkoxides using ammonia as a catalyst. The particle size was dependent on the type of silicon alkoxide and alcohol. Methanol gave the smallest particles, and the particle size increased with increasing alcohol chain length. Currently, the synthesis and applications of mesoporous silica nanomaterials have received much attention due to their highly ordered structures, larger pore size, and high surface area (Scott et al., 2001). These stable mesoporous structure and surface properties makes them ideal candidates for hosting and delivery of various pharmaceutical drugs, proteins, and other biomolecules (Hartmann, 2005; Li et al., 2012). MCM-41 is one of the important fabricated mesoporous silica nanomaterial reported for the first time (Kresge et al., 1992), and its application as a drug-delivery system was reported for the first time by Vallet-Regi et al. (2001). Ibuprofen, an anti-inflammatory drug was loaded into the mesopores of MCM-41 that showed high drug loading and sustained drug release. Apart from MCM-41, other groups of mesoporous nanomaterials with larger pore size including SBA-15, SBA-16, SBA-1, SBA-3, HMS, and MSU were used for drug delivery. In addition, the mesoporous silica has been found to be relatively nontoxic and biocompatible in nature (Vivero-Escoto et al., 2010). Zeng et al. (2005) demonstrated the use of mesoporous MCM-41 modified with aminopropyl groups to function as drug-controlled delivery system for aspirin. It was observed that the amount of aminopropyl groups present on the mesoporous MCM-41 dictated the releasing properties of this delivery system. Mesoporous SBA-15 functionalized with amine groups was demonstrated as a drug carrier for controlled drug release of ibuprofen (IBU) and bovine serum albumin (BSA) (Song et al., 2005), itraconazole and IBU (Mellaerts et al., 2008). Further, the use of mesoporous silica nanomaterials to deliver camptothecin (Lu et al., 2007a) and paclitaxel (Lu et al., 2007b) and DOX (Fan et al., 2011; Sahoo et al., 2013) to human cancer cells has been well demonstrated. In another study, the engineered mesoporous silica nanomaterials coated with PEI–PEG copolymer was used to deliver DOX to a KB-31 (a DOX-sensitive HeLa squamous carcinoma cell line) xenografted tumor model (Meng et al., 2011).

Silica nanomaterials can be used as promising candidates for improved drug delivery wherein the drug molecules are either loaded into the cavities of the hollow silica nanoparticles or the surface modifications of the nanoparticles with biorecognition entities to enable the location of specific cells or receptors in the body. Once the specific target is recognized, the nanoparticles release their drug payload at a precise controlled rate, which can be achieved by tailoring the internal structure of the nanoparticles for a desired diffusion or release profile (Barbé et al., 2004). They can also be used for other biological applications such as imaging and oxygen transfer (Liu et al., 2012a). These materials have proved to be an efficient antimicrobial drug carrier for the local release of antibiotics to address biofilm related infections, which is a great challenge from a biomedical perspective (Yokoyama et al., 2006). Some novel controlled release carriers have been developed recently using silica-based nanomaterials. Porous silica xerogel was prepared to entrap dexmedetomidine and the release rate of the drug from the matrix has been evaluated (Kortesuo et al., 2001). Porous hollow silica nanoparticles (PHSNs) were synthesized by direct coating of nanotemplates with silica film using a sol-gel approach for use as a novel controlled release carrier for controlled drug delivery (Li et al., 2004). Similarly, PHSNs were prepared by a sol–gel route and avermectin was loaded and the sustained release of the drug was studied (Li et al., 2006b). Silica-based nanomaterials contain large pore sizes and surface areas, which allows the absorption of huge amounts of drugs, consequently providing sufficient concentrations for local treatment. The surface of silica materials contains silanol groups, which allows silanization reactions and therefore opens possibilities for increasing the drug-loading amount and also for controlled drug release (He and Shi, 2011). Silica nanoparticles functionalized with amphotericin-B exhibited antifungal activity against a wide spectrum of *Candida* strains and thus these nanoparticle-conjugates can be used as antifungal coatings on medical devices (Paulo et al., 2010). Silica nanoparticles coated with a quaternary ammonium cationic surfactant, didodecyldimethylammonium bromide (DDAB) exhibited promising antibacterial and antifungal activities and can find use as antimicrobial surface coating (Botequim et al., 2012). Nitric oxide (NO)-releasing silica nanoparticles for sustained NO release were prepared by functionalizing an O^2-protected diazeniumdiolate NO donor within the silica network and these modified silica nanoparticles were doped into dental nanocomposites, which exhibited effective decrease in the number of viable adherent *Streptococcus mutans* from the dental restorative materials (Carpenter et al., 2013). A novel ε-caprolactam-silica nanoparticles conjugated with kanamycin improved the antimicrobial activity against *S. aureus* ATCC 25923 and *E. coli* ATCC 25922 strains and acted as a new antibiotic drug delivery systems to control bacterial infections (Voicu et al., 2013). Silica nanoparticles conjugated with bacitracin and kanamycin exhibited promising antistaphylococcal activity at lower doses as compared to the individual antibiotics. This improved delivery system of topical antibiotics was used for the treatment of staphylococcal skin infections (Grumezescu et al., 2014).

4.6 CARBON NANOTUBES

Carbon nanotubes (CNTs) are unique, one-dimensional molecules consisting of graphene sheets arranged in tubular structure, first discovered by Iijima (1991). This pioneering work has sparked a renewed interest toward the research and development of CNTs. Based on the arrangement of graphene sheets, CNTs have been categorized into single-walled CNT (SWCNT, diameter 0.4–2 nm) and multiwalled CNT (MWCNT) consisting of multiple concentric cylinders of increasing diameter 2–100 nm. The diameter of the CNT is typically less than 100 nm but the length can vary from few nanometers to several micrometers. Their distinct features such as excellent optical, magnetic, chemical, electrical, and mechanical surface properties make them ideal candidates for a broad range of applications such as structural materials (Goldberger et al., 2006; Guldi et al., 2006). Recently, CNTs have raised considerable attention due to its potential application in biotechnology like in sensors (Qureshi et al., 2012; Zhang et al., 2007b), biomedical devices (Li et al., 2011), and drug delivery systems (Karchemski et al., 2012; Zhang and Olin, 2012). The CNT has limited biological applications because pristine CNTs are completely insoluble in all solvents and thus limits their use in biological systems and some CNTs without any functionalization exhibited toxicity (Tasis et al., 2003; Warheit et al., 2004). Covalent modification by organic functionalization of end groups and side walls of functionalized CNTs showed dramatic increase in solubility in the range of solvents, including water (Georgakilas et al., 2002). Chemical modification by introduction of carboxylic or ammonium groups in carbon nanotubes can increase their solubility and functionalized so they can be used to transport different active biomolecules such proteins, peptides, nucleic acids, and other therapeutic molecules (Bianco et al., 2005a; Tasis et al., 2006).

Several studies have been reported that by functionalization of CNTs, the drugs are first attached to the carrier by either covalent or with noncovalent bonding, which enable the drug to be transported across the cell membranes and enter into cytoplasm and sometimes even in the nucleus (Kam et al., 2005; Pantarotto et al., 2004). Surface properties of CNTs can greatly influence their interaction with cells and consequently their internalization into the cells. The folate-conjugated CNTs have been explored to kill cancerous cells that are overexpressing the folate receptor (Kam et al. 2005). The ability of nanotubes to transport DNA across cell membrane is used in gene therapy studies. DNA can also be attached to the tips of nanotubes or they can be incorporated within the tubes (McDevitt et al., 2007). In addition, the bioactive agents that can serve as a carrier for drugs, vaccines, antigens, and gene delivery are conjugated to CNTs by functionalization (Bianco et al., 2005b; Tran et al., 2009). Antifungal agents (amphotericin B) or anticancer drugs (methotrexate) were covalently linked to carbon nanotubes with a fluorescent agent (FITC). It was observed that drugs bound to carbon nanotubes improved the solubility and decreased its toxic effects and also exhibited potent antifungal activity (Benincasa et al., 2011; Prato et al., 2008; Wu et al., 2005). Similar to metal nanoparticles, CNTs induce an increase in local temperature and this has been used to selectively destroy the malignant cancer cells without damaging the normal cells (Bianco et al., 2008). PEG-modified CNTs were used to delivery various anticancer molecules, such as platinum (IV) prodrug (Dhar et al., 2008a; Feazell et al., 2007), doxorubicin (Liu et al., 2007), and paclitaxel (Liu et al., 2008a). In addition, Sobhani et al. (2011) reported the anticancer drug delivery system based on carbon nanotube-dendrimer hybrid nanomaterials. In this, c-Fe_2O_3 nanoparticles were placed onto the surface of MWNTs and finally they acquired CNT/*g*-Fe_2O_3 hybrid nanomaterials. Subsequently, the block copolymers like poly(citric acid)–polyethylene glycol–poly(citric acid) (PCA–PEG–PCA), were synthesized and cisplatin was conjugated and finally anticancer prodrugs were prepared, which successfully targeted the tumors on application of the magnetic field. Functionalized carbon nanotubes (f-CNTs) conjugated with different bioactive compounds including amphotericin B (Prajapati et al., 2011, 2012) and betulin (Saudagar and Dubey, 2014) exhibited antileishmanial activity. Amphotericin B loaded mannosylated MWCNTs also functioned as an efficient nanocarrier for antileishmanial therapy (Pruthi et al., 2012).

4.7 QUANTUM DOTS

Quantum dots (QDs) are metallic particles of size less than 10 nanometers and they can emit different fluorescent colors when stimulated by light. They are formed from tiny semiconductor luminescent nanocrystals interfaced between different semiconductor atoms from groups II-VI or III-V, for example, CdS, CdSe, ZnS, ZnO, GaAs, and so on. The QD core is normally composed of CdSe and then overlayered by another semiconductor material having a larger spectral band-gap, for example ZnS, which enables the structure to have increased photostability and quantum yield for emission. The colloidal core or shell QDs, such as CdSe/CdS/ZnS, CdSe/ZnS, InP/ZnS, and CdTe/CdSe are the most commonly fabricated ones for therapeutic applications (Cai et al., 2006; Yong et al., 2009). QDs can be stabilized with different capping or functionalization agents to prevent aggregation (Smith et al., 2008). They lack solubility in aqueous solutions and with functionalization the solubility properties can be altered or targeting agents can be added (Akerman et al., 2002; Bharali et al., 2005). Functionalization can be achieved by direct chemical coupling or by noncovalent self-assembly. The solubility of QDs can be increased by the use of thiol compounds such as mercaptoacetic acid, thioalkylated oligoethylene glycols, thiolated peptides (cysteine residues). Moreover, the QDs can be coated with amphiphilic diblock or triblock copolymers, polysaccharides, silica, peptides/proteins, or they can be encapsulated in phospholipid micelles depending on different type of applications (Chan and Nie, 1998; Dubertret et al., 2002;

Karakoti et al., 2015; Mattoussi et al., 2000; Pinaud et al., 2004; Shan et al., 2008). QDs possess unique and attractive optical properties such as high quantum yields, high molar extinction coefficients, good photostability, as well as narrow and symmetric photoluminescence spectra, which makes them an indispensable tool in biomedical research, especially for drug delivery, fluorescence imaging, and detection (Alivisatos, 2004; Fahmi and Chang, 2013; Geszke et al., 2011; Liu et al., 2012b; Smith et al., 2006). So far, QDs with high superior core–shell were prepared using hot colloidal synthesis, in which the surface stabilized hydrophobic moieties, such as oleic acid and/or trioctyl phosphine prevented their dispersion in biological fluids. However, water dispersible QDs with multifunctional groups has made feasible the possible conjugation for biological applications (Dong et al., 2009). QDs act as multifunctional platforms for the synthesis of distinguishable drug carrier systems with high potential for improving pharmacological treatment of several diseases. Usually, the drug delivery mechanism of QD drug formulation depends on the properties and architecture of the nanostructures. In general, two approaches were used to conjugate drug molecules to the QDs in QD-drug formulation; the first one is linking or conjugating drug molecules to the QD surface and the second one is loading/entrapping the bioactive molecule in a polymer nanoparticle system that contains either hydrophobic or hydrophilic QDs, depending on the type of encapsulated polymer. QD can serve as the structural scaffold for conjugating different bioactive molecules, such as proteins, polymers and drug molecules, which can be adapted to various target biomarkers (Gangrade, 2011). Several drugs and biomolecules, such as peptides, sugars and DNA, can be tagged with QDs for use as highly sensitive probes (Zrazhevskiy et al., 2010).

QDs can also be used in cancer detection for imaging the tumor stage. This technology outlined some early success in the detection and treatment of breast cancer (Alivisatos et al., 2005; Gao and Dave, 2007). The new class of bioconjugated quantum dot contained an amphiphilic triblock copolymer for protection of in vivo systems and several PEG molecules for better biocompatibility and circulation, making them highly stable and producing bright signals. Hydrophobic drugs can be entrapped in between inorganic core and amphiphilic polymer coating layer of the QDs; whereas hydrophilic biomolecules, such as siRNA, antibodies, peptides, and aptamers can be conjugated onto the hydrophilic side of the amphiphilic polymer. This fully designed nanomaterial acts as a biomarker for identification of diseased cells and also acts as a therapeutic agent to treat the disease (Qi and Gao, 2008). Luminescent silicon QDs functionalized with 2-vinylpyridine acted as self-tracking vehicles for the delivery of small interfering RNA (siRNA) into tumor cells (Klein et al., 2009). In addition, these QDs can be used in viral diagnosis as antiviral drugs (Okamatsu et al., 2013; Wang et al., 2008). The rapid detection of respiratory syncytial virus (RSV) with high sensitivity can be achieved by using antibody-conjugated QDs and also used to estimate the expression levels of surface proteins (Agrawal et al., 2005). QD probes emit different wavelengths that can be used for imaging, as well as tracking for various cancer markers, potentially increasing the specificity and sensitivity of cancer detection (Jain et al., 2007b). Silicon QDs prepared using phospholipid micelles or antimesothelin were used as luminescent labels for pancreatic cancer cells (Erogbogbo et al., 2008, 2011). The biological molecules such as antibodies conjugated to QDs have shown promising application for detecting and quantifying a wide variety of cancer-associated molecules (Amiot et al., 2008). Lipid-encapsulated QDs were electroporated into *Bifidobacterium bifidum* and this QDs-carrying bacterium surface was further conjugated with folic acid that can specifically bind to the folic acid receptor overexpressed tumor cells. This method enabled the delivery of QDs specifically into solid tumors of living animals for imaging, diagnosis, and therapy (Liu et al., 2012b). Recent advances in the surface chemistry of QDs has expanded their use in receptor trafficking (Fichter et al., 2010; Sreenivasan et al., 2012; Vermehren-Schmaedick et al., 2014). Some of the examples of QDs used in various drug delivery applications are tabulated in Table 1.5.

Table 1.5 Quantum Dots Used in Various Drug Delivery Applications

Quantum Dot	Drug/Application	Highlights	References
Tri-*n*-octylphosphine oxide-coated ZnS-capped CdSe QDs	Peptide/cancer therapy	Disease sensing and site-specific delivery	Akerman et al. (2002)
CdS QDs	Daunorubicin/Cancer therapy	Enhances the drug efficiency and also inhibit the multidrug resistant leukemia cell proliferation	Li et al. (2006a)
CdSe/ZnS	Alpha-fetoprotein (AFP) monoclonal antibody/ Cancer therapy	Good stability, biocompatibility and specificity for ultrasensitive fluorescence imaging of molecular targets in liver cancer	Chen et al. (2008)
ZnO QDs	Chitosan and anticancer drugs/Cancer therapy	Controlled drug release and high colloidal stability	Yuan et al. (2010)
CdSe QD- hydroxypropyl-cellulose-poly(acrylic acid) (HPC–PAA) nanogel	Temozolomide/melanoma therapy	Highly stable and pH triggered sustained release of the drug	Wu et al. (2010)
Folic acid-conjugated carboxymethyl chitosan-Mn-doped ZnS QDs	5-Fluorouracil/Cancer therapy	Imaging of cancer cell and specific targeting to MCF-7 cell line	Mathew et al. (2010)
Carboxy terminated QDs	Saquinavir and transferrin/ antiretroviral therapy	Delivery system efficiently cross the blood brain barrier, and decreased the viral replication in HIV-1 infected peripheral blood mononuclear cells (PBMCs) within the brain	Mahajan et al. (2010)
QD-mucin 1 aptamer–doxorubicin conjugate	Doxorubicin/ovarian cancer therapy	Targeted DRU delivery and pH responsive drug releasing profile	Savla et al. (2011)
Liposome-QD hybrid incorporate with CdSe/ ZnS QDs	Doxorubicin/cancer therapy	Act as a multifunctional drug delivery vector for transporting different combinations of therapeutic molecules	Tian et al. (2011)
Alminoprofen-conjugated Si–QDs	Alminoprofen/Therapy for Rheumatism	Inhibition of COX-2 enzyme	Hanada et al. (2013)
DOX–QD–PEG-folic acid conjugates	Doxorubicin/Cancer therapy	Multifunctional nanovehicle for drug delivery	Chen et al. (2014)
CdTe quantum dots loaded nanogels	Methotrexate/Cancer therapy	Sustained release and enhanced the availability of drug	Li et al. (2015)
Palmitoylated peptide conjugated CdSe/ZnS QDs	WGDap(Palmitoyl) VKIKKP9GGH6 peptide/ Stem cell therapy	Peptide delivery to developing chick embryo brain	Agarwal et al. (2015)
Tacrolimus QDs	Tacrolimus/Stem cell therapy	Sustained release of drug	Yano et al. (2015)

5 CONCLUSIONS

The application of nanoscience and nanotechnology in medicine and more specifically in the realm of drug delivery is set to increase rapidly. There is an extensive range of nanomaterials and nanostructures being developed for the delivery of therapeutic compounds. Nanomaterials as drug carrier systems are fabricated to improve the therapeutic and pharmacological properties of the conventional drugs. The conjugation/integration of bioactive molecules to nanocarrier can protect the drug molecule from degradation, as well as it offers possibilities of targeting and controlled release of the drug. When compared to conventional drugs, the nanocarrier-drug conjugates are more effective and selective. In addition, the nanocarriers can reduce the toxicity and other adverse side effects in healthy cells by accumulating the drugs in target diseased tissues and also reduce the required dose of drug. There are several nanomaterials such as metal nanoparticles, polymeric nanomaterials, dendrimers, quantum dots, which are currently being developed as therapeutic agents and are under preclinical evaluation. A few nanomaterial drugs that are currently available in the pharmaceutical market are Doxil (doxorubicin) or DaunoXome (daunorubicin). Each nanomaterial has its own advantages, but as these nanomaterials become optimized for their specific application, the outcome will be a better therapy for the delivery of drug molecules to the target sites of the body. The use of biocompatible material, improved control of particle size, and better understanding of interaction between the nanomaterial and living cells, and their effects in in vivo conditions are the significant parameters for the development of protected drug delivery system. Real therapeutic advances can be achieved exclusively by carrying out meticulous studies in the field of nanomediated drug therapy. Application of nanomaterials in disease therapy may contribute to achieve an effective treatment for several diseases. Moreover, the conjugation of drugs like doxorubicin and miconazole, immobilization of different bioligands, such as folic acid and epidermal growth factor/antibodies to the surface of nanomaterials, improves the selectivity of the drug carriers. These nanocarriers have a bright future in the delivery of bioactive molecules and nanomediated drug delivery, which would result in a concomitant improvement in the efficacy and safety profiles of the conventional drugs. Limited information is currently available on the possible toxicity of various nanomaterials and their possible impact and risks on the environment and human health. Further, sufficient information needs to be generated on the safety and fate of these namomaterials in the human body, to enable the rational biological assessments for their timely translation to the clinics. Although many advancements have taken place in the area of nanomedicine and some of these new developments are now undergoing clinical trials for regulatory clearances; nevertheless it is the time factor that would decide the role of these new exciting paradigms being put to regular practice in the health care.

REFERENCES

Abruzzo, A., Bigucci, F., Cerchiara, T., Saladini, B., Gallucci, M.C., Cruciani, F., Vitali, B., Luppi, B., 2013. Chitosan/alginate complexes for vaginal delivery of chlorhexidine digluconate. Carbohydr. Polym. 91, 651–658.

Acosta-Torres, L.S., Mendieta, I., Nuñez-Anita, R.E., Cajero-Juárez, M., Castaño, V.M., 2012. Cytocompatible antifungal acrylic resin containing silver nanoparticles for dentures. Int. J. Nanomed. 7, 4777–4786.

Agarwal, R., Domowicz, M.S., Schwartz, N.B., Henry, J., Medintz, I., Delehanty, J.B., Stewart, M.H., Susumu, K., Huston, A.L., Deschamps, J.R., Dawson, P.E., Palomo, V., Dawson, G., 2015. Delivery and tracking of quantum dot peptide bioconjugates in an intact developing avian brain. ACS Chem. Neurosci. 6, 494–504.

Agnihotri, S.A., Aminabhavi, T.M., 2005. Development of novel interpenetrating network gellan gum-poly(vinyl alcohol) hydrogel microspheres for the controlled release of carvedilol. Drug Dev. Ind. Pharm. 31, 491–503.

Agnihotri, S.A., Jawalkar, S.S., Aminabhavi, T.M., 2006. Controlled release of cephalexin through gellan gum beads: effect of formulation parameters on entrapment efficiency, size, and drug release. Eur. J. Pharm. Biopharm. 63, 249–261.

Agrawal, A., Tripp, R.A., Anderson, L.J., Nie, S., 2005. Real-time detection of virus particles and viral protein expression with two-color nanoparticle probes. J. Virol. 79, 8625–8628.

Ahangari, A., Salouti, M., Heidari, Z., Kazemizadeh, A., Safari, A., 2013. Development of gentamicin-gold nanospheres for antimicrobial drug delivery to staphylococcal infected foci. Drug Deliv. 20, 34–39.

Ahmed, I.S., 2005. Effect of simulated gastrointestinal conditions on drug release from pectin/ethylcellulose as film coating for drug delivery to the colon. Drug Dev. Ind. Pharm. 31, 465–470.

Aiyer, H.N., Vijayakrishnan, V., Subbanna, G.N., Rao, C.N.R., 1994. Investigations of Pd clusters by the combined use of HREM, STM, high-energy spectroscopies and tunneling conductance measurements. Surface Sci. 313, 392–398.

Akbarzadeh, A., Zare, D., Farhangi, A., Mehrabi, M.R., Norouzian, D., Tangestaninejad, S., Moghadam, M., Bararpour, N., 2009. Synthesis and characterization of gold nanoparticles by tryptophane. Am. J. Appl. Sci. 6, 691–695.

Akbarzadeh, A., Mikaeili, H., Zarghami, N., Mohammad, R., Barkhordari, A., Davaran, S., 2012. Preparation and in vitro evaluation of doxorubicin-loaded Fe_3O_4 magnetic nanoparticles modified with biocompatible copolymers. Int. J. Nanomed. 7, 511–526.

Akerman, M.E., Chan, W.C., Laakkonen, P., Bhatia, S.N., Ruoslahti, E., 2002. Nanocrystal targeting in vivo. Proc. Natl. Acad. Sci. USA 99, 12617–12621.

Akhtar, M.S., Panwar, J., Yun, Y.-S., 2013. Biogenic synthesis of metallic nanoparticles by plant extracts. ACS Sust. Chem. Eng. 1, 591–602.

Alivisatos, P., 2004. The use of nanocrystals in biological detection. Nat. Biotechnol. 22, 47–52.

Alivisatos, A.P., Gu, W., Larabell, C., 2005. Quantum dots as cellular probes. Annu. Rev. Biomed. Eng. 7, 55–76.

Amin, M.L., Jesmeen, T., Sutradhar, K.B., Mannan, M.A., 2013. Development and in vitro evaluation of diclofenac sodium loaded mucoadhesive microsphere with natural gum for sustained delivery. Curr. Drug Deliv. 10, 765–770.

Aminabhavi, T.M., Nadagouda, M.N., Joshi, S.D., More, U.A., 2014. Guar gum as platform for the oral controlled release of therapeutics. Expert Opin. Drug Deliv. 11, 753–766.

Amiot, C.L., Xu, S., Liang, S., Pan, L., Zhao, J.X., 2008. Near-infrared fluorescent materials for sensing of biological targets. Sensors 8, 3082–3105.

Amutha, K., Ravibaskar, R., Sivakumar, G., 2010. Extraction, synthesis and characterization of nanosilica from rice husk ash. Int. J. Nano. Technol. Applicat. 4, 61–66.

Anal, A.K., Stevens, W.F., 2005. Chitosan-alginate multilayer beads for controlled release of ampicillin. Int. J. Pharm. 290, 45–54.

Angadi, S.C., Manjeshwar, L.S., Aminabhavi, T.M., 2012. Novel composite blend microbeads of sodium alginate coated with chitosan for controlled release of amoxicillin. Int. J. Biol. Macromol. 51, 45–55.

Anonymous, 2010. Revolutionary medical dressing uses nano-technology to fight infection. Available from: http://www.bath.ac.uk/news/2010/07/08/bacteriosafe-2/

Araujo, E., Andrade, N.J., Da Silva, L.H., Bernardes, P.C., Teixeira, A.V., de Sa, J.P., Fialho, Jr., J.F., Fernandes, P.E., 2012. Antimicrobial effects of silver nanoparticles against bacterial cells adhered to stainless steel surfaces. J. Food Protect. 74, 701–705.

Assifaoui, A., Bouyer, F., Chambin, O., Cayot, P., 2013. Silica-coated calcium pectinate beads for colonic drug delivery. Acta Biomater. 9, 6218–6225.

Aumelas, A., Serrero, A., Durand, A., Dellacherie, E., Leonard, M., 2007. Nanoparticles of hydrophobically modified dextrans as potential drug carrier systems. Colloids Surfaces B 59, 74–80.

Azoulay, J., 1983. Photoelectron spectroscopy: principles and applications. Vacuum 3, 211–213.

Baars, M.W.P.L., Kleppinger, R., Koch, M.H.J., Yeu, S.-L., Meijer, E.W., 2000. The localization of guests in water-soluble oligoethyleneoxy-modified poly(propylene imine) dendrimers. Angew. Chem. Int. Ed. 39, 1285–1288.

Babu, R.J., Sathigari, S., Kumar, M.T., Pandit, J.K., 2010. Formulation of controlled release gellan gum macro beads of amoxicillin. Curr. Drug Deliv. 7, 36–43.

Badwaik, H.R., Giri, T.K., Nakhate, K.T., Kashyap, P., Tripathi, D.K., 2013. Xanthan gum and its derivatives as a potential bio-polymeric carrier for drug delivery system. Curr. Drug Deliv. 10, 587–600.

Bagre, A.P., Jain, K., Jain, N.K., 2013. Alginate coated chitosan core shell nanoparticles for oral delivery of enoxaparin: in vitro and in vivo assessment. Int. J. Pharm. 456, 31–40.

Barbé, C., Bartlett, J., Kong, L., Finnie, K., Lin, H.Q., Larkin, M., Calleja, S., Bush, A., Calleja, G., 2004. Silica particles: a novel drug-delivery system. Adv. Mater. 16, 1959–1966.

Barber, D.J., Freestone, I., 1990. An investigation of the origin of the colour of the Lycurgus Cup by analytical transmission electron microscopy. Archaeometry 32, 33–45.

Barbucci, R., Pasqui, D., Favaloro, R., Panariello, G., 2008. A thixotropic hydrogel from chemically cross-linked guar gum: synthesis, characterization and rheological behaviour. Carbohydr. Res. 343, 3058–3065.

Benincasa, M., Pacor, S., Wu, W., Prato, M., Bianco, A., Gennaro, R., 2011. Antifungal activity of amphotericin B conjugated to carbon nanotubes. ACS Nano 5, 199–208.

Benyettou, F., Rezgui, R., Ravaux, F., Jaber, T., Blumer, K., Jouiad, M., Motte, L., Olsen, J.-C., Platas-Iglesias, C., Magzoub, M., Trabolsi, A., 2015. Synthesis of silver nanoparticles for the dual delivery of doxorubicin and alendronate to cancer cells. J. Mater. Chem. 3B, 7237–7245.

Bera, H., Boddupalli, S., Nandikonda, S., Kumar, S., Nayak, A.K., 2015. Alginate gel-coated oil-entrapped alginate-tamarind gum-magnesium stearate buoyant beads of risperidone. Int. J. Biol. Macromol. 78, 102–111.

Bernardo, M.V., Blanco, M.D., Sastre, R.L., Teijon, C., Teijon, J.M., 2003. Sustained release of bupivacaine from devices based on chitosan. Farmaco 58, 1187–1191.

Berne, B.J., Pecora, R., 2000. Dynamic Light Scattering. Dover Publications, New York.

Bessa, P.C., Machado, R., Nurnberger, S., Dopler, D., Banerjee, A., Cunha, A.M., Rodriguez-Cabello, J.C., Redl, H., van Griensven, M., Reis, R.L., Casal, M., 2010. Thermoresponsive self-assembled elastin-based nanoparticles for delivery of BMPs. J. Control. Release 142, 312–318.

Bhadra, D., Bhadra, S., Jain, S., Jain, N.K., 2003. A PEGylated dendritic nanoparticle carrier of fluorouracil. Int. J. Pharm. 257, 111–124.

Bharali, D.J., Lucey, D.W., Jayakumar, H., Pudavar, H.E., Prasad, P.N., 2005. Folate-receptor-mediated delivery of InP quantum dots for bioimaging using confocal and two-photon microscopy. J. Am. Chem. Soc. 127, 11364–11371.

Bharatwaj, B., Wu, L., Whittum-Hudson, J., Rocha, S., 2010. The potential for the noninvasive delivery of polymeric nanocarriers using propellant-based inhalers in the treatment of chlamydial respiratory infections. Biomaterials 31, 7376–7385.

Bhatta, R.S., Chandasana, H., Chhonker, Y.S., Rathi, C., Kumar, D., Mitra, K., Shukla, P.K., 2012. Mucoadhesive nanoparticles for prolonged ocular delivery of natamycin: in vitro and pharmacokinetics studies. Int. J. Pharm. 432, 105–112.

Bhattacharya, R., Mukherjee, P., 2008. Biological properties of "naked" metal nanoparticles. Adv. Drug Deliv. Rev. 60, 1289–1306.

Bhumkar, D.R., Joshi, H.M., Sastry, M., Pokharkar, V.B., 2007. Chitosan reduced gold nanoparticles as novel carriers for transmucosal delivery of insulin. Pharm. Res. 24, 1415–1426.

Bianco, A., Kostarelos, K., Prato, M., 2005a. Applications of carbon nanotubes in drug delivery. Curr. Opin. Chem. Biol. 9, 674–679.

Bianco, A., Kostarelos, K., Partidos, C.D., Prato, M., 2005b. Biomedical applications of functionalised carbon nanotubes. Chem. Commun. 5, 571–577.

Bianco, A., Kostarelos, K., Prato, M., 2008. Opportunities and challenges of carbon-based nanomaterials for cancer therapy. Expert Opin. Drug Deliv. 5, 331–342.
Bigucci, F., Luppi, B., Cerchiara, T., Sorrenti, M., Bettinetti, G., Rodriguez, L., Zecchi, V., 2008. Chitosan/pectin polyelectrolyte complexes: selection of suitable preparative conditions for colon-specific delivery of vancomycin. Eur. J. Pharm. Sci. 35, 435–441.
Bigucci, F., Luppi, B., Monaco, L., Cerchiara, T., Zecchi, V., 2009. Pectin-based microspheres for colon-specific delivery of vancomycin. J. Pharm. Pharmacol. 61, 41–46.
Bjerneld, E.J., Johansson, P., Kall, M., 2000. Single molecule vibrational fine-structure of tyrosine adsorbed on Ag nano-crystals. Single Mol. 1, 239–248.
Blanco, M.D., Gomez, C., Olmo, R., Muniz, E., Teijon, J.M., 2000. Chitosan microspheres in PLG films as devices for cytarabine release. Int. J. Pharm. 202, 29–39.
Bosch, F., Rosich, L., 2008. The contributions of Paul Ehrlich to pharmacology: a tribute on the occasion of the centenary of his Nobel Prize. Pharmacology 82, 171–179.
Bosman, A.W., Janssen, H.M., Meijer, E.W., 1999. About dendrimers: structure, physical properties, and applications. Chem. Rev. 99, 1665–1688.
Botequim, D., Maia, J., Lino, M.M.F., Lopes, L.M.F., Simoes, P.N., Ilharco, L.M., Ferreira, L., 2012. Nanoparticles and surfaces presenting antifungal, antibacterial and antiviral properties. Langmuir 28, 7646–7656.
Bradley, J.S., 1994. The chemistry of transition metal colloids. In: Schmid, G. (Ed.), Clusters and Colloids: From Theory to Applications. VCH Verlagsgesellschaft mbH, Weinheim, pp. 459–544.
Brannon-Peppas, L., 1995. Recent advances on the use of biodegradable microparticles and nanoparticles in controlled drug delivery. Int. J. Pharm. 116, 1–9.
Bronstein, L.M., Shifrina, Z.B., 2012. Dendrimers as encapsulating, stabilizing, or directing agents for inorganic nanoparticles. Chem. Rev. 111, 5301–5344.
Brothers, II, H.M., Piehler, L.T., Tomalia, D.A., 1998. Slab-gel and capillary electrophoretic characterization of polyamidoamine dendrimers. J. Chromatogr. A 814, 233–246.
Brown, A.N., Smith, K., Samuels, T.A., Lu, J., Obare, S.O., Scott, M.E., 2012. Nanoparticles functionalized with ampicillin destroy multiple-antibiotic-resistant isolates of *Pseudomonas aeruginosa* and *Enterobacter aerogenes* and methicillin-resistant *Staphylococcus aureus*. Appl. Environ. Microbiol. 78, 2768–2774.
Brown, P.K., Qureshi, A.T., Moll, A.N., Hayes, D.J., Monroe, W.T., 2013. Silver nanoscale antisense drug delivery system for photoactivated gene silencing. ACS Nano 7, 2948–2959.
Brun-Graeppi, A., Richard, C., Bessodes, M., Scherman, D., Narita, T., Ducouret, G., Merten, O.W., 2010. Study on the sol–gel transition of xyloglucan hydrogels. Carbohydr. Polym. 80, 555–562.
Brust, M., Walker, M., Bethell, D., Schiffrin, D.J., Whyman, R.J., 1994. Synthesis of thiol derivatized gold nanoparticles in a two phase liquid–liquid system. J. Chem. Soc. Chem. Commun. 7, 801–802.
Buckley, J.J., Lee, A.F., Olivi, L., Wilson, K., 2010. Hydroxyapatite supported antibacterial Ag_3PO_4 nanoparticles. J. Mater. Chem. 20, 8056–8063.
Buhleier, E., Wehner, W., Vögtle, F., 1978. Cascade and nonskid-chain-like synthesis of molecular cavity topologies. Synthesis 2, 155–158.
Burda, C., Chen, X., Narayanan, R., El-Sayed, M.A., 2005. Chemistry and properties of nanocrystals of different shapes. Chem. Rev. 105, 1025–1102.
Burdick, J.A., Prestwich, G.D., 2011. Hyaluronic acid hydrogels for biomedical applications. Adv. Mater. 23, H41–H56.
Burrell, R.E., McIntosh, C.L., Morris, L.R., 1995. Process of activating antimicrobial materials. US Patent 5,454,886.
Burygin, G.L., Khlebtsov, B.N., Shantrokha, A.N., Dykman, L.A., Bogatyrev, V.A., Khlebtsov, N.G., 2009. On the enhanced antibacterial activity of antibiotics mixed with gold nanoparticles. Nanoscale Res. Lett. 4, 794–801.
Butun, S., Ince, F.G., Erdugan, H., Sahiner, N., 2011. One-step fabrication of biocompatible carboxymethyl cellulose polymeric particles for drug delivery systems. Carbohydr. Polym. 86, 636–643.

Cafaggi, S., Russo, E., Stefani, R., Parodi, B., Caviglioli, G., Sillo, G., Bisio, A., Aiello, C., Viale, M., 2011. Preparation, characterisation and preliminary antitumour activity evaluation of a novel nanoparticulate system based on a cisplatin-hyaluronate complex and *N*-trimethyl chitosan. Invest. New Drugs 29, 443–455.

Cai, W., Shin, D.W., Chen, K., Gheysens, O., Cao, Q., Wang, S.X., Ghambhir, S.S., Chen, X., 2006. Peptide-labeled near-infrared quantum dots for imaging tumor vasculature in living subjects. Nano Lett. 6, 669–676.

Caminati, G., Turro, N.J., Tomalia, D.A., 1990. Photophysical investigation of starburst dendrimers and their interactions with anionic and cationic surfactants. J. Am. Chem. Soc. 112, 8515–8522.

Carpenter, A.W., Reighard, K.P., Saavedra, J.E., Schoenfisch, M.H., 2013. O^2-protected diazeniumdiolate-modified silica nanoparticles for extended nitric oxide release from dental composites. Biomater. Sci. 1, 456–459.

Cauerhff, A., Castro, G.R., 2013. Bionanoparticles, a green nanochemistry approach. Electronic J. Biotechnol. 16, 1–10.

Chambin, O., Dupuis, G., Champion, D., Voilley, A., Pourcelot, Y., 2006. Colon-specific drug delivery: influence of solution reticulation properties upon pectin beads performance. Int. J. Pharm. 321, 86–93.

Chan, W.C., Nie, S., 1998. Quantum dot bioconjugates for ultrasensitive nonisotropic detection. Science 281, 2016–2018.

Chandrasekhar, S., Sathyanarayana, K.G., Pramada, P.N., Raghavan, P., 2003. Review processing properties and applications of reactive silica from rice husk an overview. Mater. Sci. 38, 3159–3168.

Chaurasia, M., Chourasia, M.K., Jain, N.K., Jain, A., Soni, V., Gupta, Y., Jain, S.K., 2008. Methotrexate bearing calcium pectinate microspheres: a platform to achieve colon-specific drug release. Curr. Drug Deliv. 5, 215–219.

Chen, J.P., Yang, P.C., Ma, Y.H., Tu, S.J., Lu, Y.J., 2012. Targeted delivery of tissue plasminogen activator by binding to silica-coated magnetic nanoparticle. Int. J. Nanomed. 7, 5137–5149.

Chen, L.D., Liu, J., Yu, X.F., He, M., Pei, X.F., Tang, Z.Y., Wang, Q.Q., Pang, D.W., Li, Y., 2008. The biocompatibility of quantum dot probes used for the targeted imaging of hepatocellular carcinoma metastasis. Biomaterials 29, 4170–4176.

Chen, X., Tang, Y., Cai, B., Fan, H., 2014. "One-pot" synthesis of multifunctional GSH-CdTe quantum dots for targeted drug delivery. Nanotechnology 25, 235101.

Chen, Y., Liu, L., 2012. Modern methods for delivery of drugs across the blood–brain barrier. Adv. Drug Deliv. Rev. 64, 640–665.

Chen, Z., Li, Z., Lin, Y., Yin, M., Ren, J., Qu, X., 2013. Bioresponsive hyaluronic acid-capped mesoporous silica nanoparticles for targeted drug delivery. Chemistry 19, 1778–1783.

Cheng, J., Teply, B.A., Sherifi, I., Sung, J., Luther, G., Gu, F.X., Levy-Nissenbaum, E., Radovic-Moreno, A.F., Langer, R., Farokhzad, O.C., 2007a. Formulation of functionalized PLGA-PEG nanoparticles for in vivo targeted drug delivery. Biomaterials 28, 869–876.

Cheng, Y.Y., Xu, T.W., 2005. Dendrimers as potential drug carriers. Part I. Solubilization of non-steroidal anti-inflammatory drugs in the presence of polyamidoamine dendrimers. Eur. J. Med. Chem. 40, 1188–1192.

Cheng, Y.Y., Gao, Y., Rao, T.L., Li, Y.W., Xu, T.W., 2007b. Dendrimer-based prodrugs: design, synthesis, screening and biological evaluation. Comb. Chem. High Throughput Screen 10, 336–349.

Cheng, Y.Y., Wang, J.R., Rao, T.L., He, X.X., Xu, T.W., 2008. Pharmaceutical applications of dendrimers: promising nanocarriers for drug delivery. Front. Biosci. 13, 1447–1471.

Chhonker, Y.S., Prasad, Y.D., Chandasana, H., Vishvkarma, A., Mitra, K., Shukla, P.K., Bhatta, R.S., 2015. Amphotericin-B entrapped lecithin/chitosan nanoparticles for prolonged ocular application. Int. J. Biol. Macromol. 72, 1451–1458.

Chin, J.N., Jones, R.N., Sader, H.S., Savage, P.B., Rybak, M.J., 2008. Potential synergy activity of the novel ceragenin, CSA-13, against clinical isolates of *Pseudomonas aeruginosa*, including multidrug-resistant *P. aeruginosa*. J. Antimicrob. Chemother. 61, 365–370.

Cho, H.J., Oh, J., Choo, M.K., Ha, J.I., Park, Y., Maeng, H.J., 2014. Chondroitin sulfate-capped gold nanoparticles for the oral delivery of insulin. Int. J. Biol. Macromol. 63, 15–20.

Choi, K.Y., Chung, H., Min, K.H., Yoon, H.Y., Kim, K., Park, J.H., Kwon, I.C., Jeong, S.Y., 2010. Self-assembled hyaluronic acid nanoparticles for active tumor targeting. Biomaterials 31, 106–114.

Choi, K.Y., Min, K.H., Yoon, H.Y., Kim, K., Park, J.H., Kwon, I.C., Choi, K., Jeong, S.Y., 2011a. PEGylation of hyaluronic acid nanoparticles improves tumor targetability in vivo. Biomaterials 32, 1880–1889.

Choi, K.Y., Yoon, H.Y., Kim, J.H., Bae, S.M., Park, R.W., Kang, Y.M., Kim, I.S., Kwon, I.C., Choi, K., Jeong, S.Y., Kim, K., Park, J.H., 2011b. Smart nanocarrier based on PEGylated hyaluronic acid for cancer therapy. ACS Nano 5, 8591–8599.

Chorny, M., Fishbein, I., Yellen, B.B., Alferiev, I.S., Bakay, M., Ganta, S., Adamo, R., Amiji, M., Friedman, G., Levy, R.J., 2010. Targeting stents with local delivery of paclitaxel-loaded magnetic nanoparticles using uniform fields. Proc. Natl. Acad. Sci. USA 107, 8346–8351.

Chorny, M., Fishbein, I., Tengood, J.E., Adamo, R.F., Alferiev, I.S., Levy, R.J., 2013. Site-specific gene delivery to stented arteries using magnetically guided zinc oleate-based nanoparticles loaded with adenoviral vectors. FASEB J. 27, 2198–2206.

Chourasia, M.K., Jain, S.K., 2004a. Potential of guar gum microspheres for target specific drug release to colon. J. Drug Target. 12, 435–442.

Chourasia, M.K., Jain, S.K., 2004b. Polysaccharides for colon targeted drug delivery. Drug Deliv. 11, 129–148.

Costa, J.R., Silva, N.C., Sarmento, B., Pintado, M., 2015. Potential chitosan-coated alginate nanoparticles for ocular delivery of daptomycin. Eur. J. Clin. Microbiol. Infect. Dis. 34, 1255–1262.

Coviello, T., Dentini, M., Rambone, G., Desideri, P., Carafa, M., Murtas, E., Riccieri, F.M., Alhaique, F., 1999. A novel co-crosslinked polysaccharide: studies for a controlled delivery matrix. J. Control. Release 60, 287–295.

Coviello, T., Matricardi, P., Marianecci, C., Alhaique, F., 2007. Polysaccharide hydrogels for modified release formulations. J. Control. Release 119, 5–24.

Cullity, B.D., Stock, S.R., 2001. Elements of X-Ray Diffraction, third ed. Prentice Hall, Englewood Cliffs, NJ.

Cushing, B.L., Kolesnichenko, V.L., Connor, C.J.O., 2004. Recent advances in the liquid-phase syntheses of inorganic nanoparticles. Chem. Rev. 104, 3893–3946.

D'Emanuele, A., Jevprasesphant, R., Penny, J., Atwood, D., 2004. The use of a dendrimer-propranolol prodrug to bypass efflux transporters and oral bioavailability. J. Control. Release 95, 447–453.

Das, S., Chaudhury, A., Ng, K.Y., 2011a. Preparation and evaluation of zinc-pectin-chitosan composite particles for drug delivery to the colon: role of chitosan in modifying in vitro and in vivo drug release. Int. J. Pharm. 406, 11–20.

Das, S., Ng, K.Y., 2010a. Colon-specific delivery of resveratrol: optimization of multi-particulate calcium-pectinate carrier. Int. J. Pharm. 385, 20–28.

Das, S., Ng, K.Y., 2010b. Resveratrol-loaded calcium-pectinate beads: effects of formulation parameters on drug release and bead characteristics. Int. J. Pharm. 99, 840–860.

Das, S., Ng, K.Y., Ho, P.C., 2010. Formulation and optimization of zinc-pectinate beads for the controlled delivery of resveratrol. AAPS PharmSciTech. 11, 729–742.

Das, S., Ng, K.Y., Ho, P.C., 2011b. Design of a pectin-based microparticle formulation using zinc ions as the cross-linking agent and glutaraldehyde as the hardening agent for colonic-specific delivery of resveratrol: in vitro and in vivo evaluations. J. Drug Target. 19, 446–457.

Das, S.K., Das, A.R., Guha, A.K., 2009. Gold nanoparticles: microbial synthesis and application in water hygiene management. Langmuir 25, 8192–8199.

Davidenko, N., Blanco, M.D., Peniche, C., Becherán, L., Guerrero, S., Teijón, J.M., 2009. Effects of different parameters on characteristics of chitosan-poly(acrylic acid) nanoparticles obtained by the method of coacervation. J. Appl. Polym. Sci. 111, 2362–2371.

De Brabander-van den Berg, E.M.M., Meijer, E.W., 1993. Poly(propylene imine) dendrimers: large scale synthesis by heterogenous actalyzed hydrogenations. Angew. Chem. 105, 1370–1373.

De, M., Ghosh, P.S., Rotillo, V.M., 2008. Applications of nanoparticles in biology. Adv. Mater. 20, 4225–4241.

Deka, S.R., Quarta, A., Di Corato, R., Riedinger, A., Cingolani, R., Pellegrino, T., 2011. Magnetic nanobeads decorated by thermo-responsive PNIPAM shell as medical platforms for the efficient delivery of doxorubicin to tumour cells. Nanoscale 3, 619–629.

Denkewalter, R.G., Kolc, J., Lukasavage, W.J., 1981. Macromolecular highly branched homogenous compound based on lysine units. US Patent 4,289,872.

Devarakonda, B., Hill, R.A., Liebenberg, W., Brits, M., de Villiers, M.M., 2005. Comparison of the aqueous solubilization of practically insoluble niclosamide by polyamidoamine (PAMAM) dendrimers and cyclodextrins. Int. J. Pharm. 304, 193–209.

Dhalleine, C., Assifaoui, A., Moulari, B., Pellequer, Y., Cayot, P., Lamprecht, A., Chambin, O., 2011. Zinc-pectinate beads as an in vivo self-assembling system for pulsatile drug delivery. Int. J. Pharm. 414, 28–34.

Dhar, S., Liu, Z., Thomale, J., Dai, H., Lippard, S.J., 2008a. Targeted single-wall carbon nanotube-mediated Pt(IV) prodrug delivery using folate as a homing device. J. Am. Chem. Soc. 130, 11467–11476.

Dhar, S., Reddy, E.M., Shiras, A., Pokharkar, V., Prasad, B.L.V., 2008b. Natural gum reduced/stabilized gold nanoparticles for drug delivery formulations. Chem. Eur. J. 14, 10244–10250.

Dhar, S., Mali, V., Bodhankar, S., Shiras, A., Prasad, B.L., Pokharkar, V., 2011a. Biocompatible gellan gum-reduced gold nanoparticles: cellular uptake and subacute oral toxicity studies. J. Appl. Toxicol. 31, 411–420.

Dhar, S., Reddy, E.M., Prabhune, A., Pokharkar, V., Shiras, A., Prasad, B.L., 2011b. Cytotoxicity of sophorolipid-gellan gum-gold nanoparticle conjugates and their doxorubicin loaded derivatives towards human glioma and human glioma stem cell lines. Nanoscale 3, 575–580.

Dhar, S., Murawala, P., Shiras, A., Pokharkar, V., Prasad, B.L., 2012. Gellan gum capped silver nanoparticle dispersions and hydrogels: cytotoxicity and in vitro diffusion studies. Nanoscale 4, 563–567.

Diaz, M.R., Vivas-Mejia, P., 2013. Nanoparticles as drug delivery systems in cancer medicine: emphasis on RNAi-containing nanoliposomes. Pharmaceuticals 6, 1361–1380.

Dinega, D.P., Bawendi, M.G., 1999. Characterization of ligand binding by saturation transfer difference NMR spectroscopy. Angew. Chem. Int. Ed. 38, 1784–1788.

Dionigi, C., Lungaro, L., Goranov, V., Riminucci, A., Piñeiro-Redondo, Y., Bañobre-López, M., Rivas, J., Dediu, V., 2014. Smart magnetic poly(*N*-isopropylacrylamide) to control the release of bio-active molecules. J. Mater. Sci. Mater. Med. 25, 2365–2371.

Dong, W., Guo, L., Wang, M., Xu, S., 2009. CdTe QDs-based prostate-specific antigen probe for human prostate cancer cell imaging. J. Luminescence 129, 926–930.

Doubrovsky, V.A., Yanina, I.Y., Tuchin, V.V., 2010. In homogeneity of photo-induced fat cell lipolysis. In: Tuchin, V.V., Genina, E.A. (Eds.), Sartov Fall Meeting Sept. 1, 2010. International Society for Optics and Photonics, United Kingdom.

Dubertret, B., Skourides, P., Norris, D.J., Noireaux, V., Brivanlou, A.H., Libchaber, A., 2002. In vivo imaging of quantum dot encapsulated in phospholipid micelles. Science 298, 1759–1762.

Dubey, R.D., Alam, N., Saneja, A., Khare, V., Kumar, A., Vaidh, S., Mahajan, G., Sharma, P.R., Singh, S.K., Mondhe, D.M., Gupta, P.N., 2015. Development and evaluation of folate functionalized albumin nanoparticles for targeted delivery of gemcitabine. Int. J. Pharm. 492, 80–91.

Dubin, P.L., Edwards, S.L., Kaplan, J.I., Mehta, M.S., Tomalia, D.A., Xia, J., 1992. Carboxylated starburst dendrimers as calibration standards for aqueous size exclusion chromatography. Anal. Chem. 64, 2344–2347.

Dufrêne, Y.F., 2002. Atomic force microscopy, a powerful tool in microbiology. J. Bacteriol. 184, 5205–5213.

Dunn, K., Edwards-Jones, V., 2004. The role of Acticoat with nanocrystalline silver in the management of burns. Burns 30, S1–S9.

Dupuis, G., Chambin, O., Génelot, C., Champion, D., Pourcelot, Y., 2006. Colonic drug delivery: influence of cross-linking agent on pectin beads properties and role of the shell capsule type. Drug Dev. Ind. Pharm. 32, 847–855.

Duran, N., Marcato, P.D., Alves, O.L., Souza, G., 2005. Mechanistic aspects of biosynthesis of silver nanoparticles by several *Fusarium oxysporum* strains. J. Nanotechnol. 3, 8–14.

Dutta, R.K., Sahu, S., 2012a. Development of a novel probe sonication assisted enhanced loading of 5-FU in SPION encapsulated pectin nanocarriers for magnetic targeted drug delivery system. Eur. J. Pharm. Biopharm. 82, 58–65.

Dutta, R.K., Sahu, S., 2012b. Development of oxaliplatin encapsulated in magnetic nanocarriers of pectin as a potential targeted drug delivery for cancer therapy. Results Pharma Sci. 2, 38–45.

Dutta, R.K., Sahu, S., 2012c. Development of diclofenac sodium loaded magnetic nanocarriers of pectin interacted with chitosan for targeted and sustained drug delivery. Colloids Surfaces B 97, 19–26.

Egger, S., Lehmann, R.P., Height, M.J., Loessner, M.J., Schuppler, M., 2009. Antimicrobial properties of a noval silver–silica nanocomposite material. Appl. Environ. Microbiol. 75, 2973–2976.

El-Gibaly, I., 2002. Oral delayed-release system based on Zn-pectinate gel (ZPG) microparticles as an alternative carrier to calcium pectinate beads for colonic drug delivery. Int. J. Pharm. 232, 199–211.

Elias, E.J., Anil, S., Ahmad, S., Daud, A., 2010. Colon targeted curcumin delivery using guar gum. Nat. Prod. Commun. 5, 915–918.

Eliasson, C., Loren, A., Murty, K.V.G.K., Josefson, M., Kall, M., Abrahamsson, J., Abrahamsson, K., 2001. Multivariate evaluation of doxorubicin surface-enhanced Raman spectra. Spectrochim. Acta Part A 57, 1907–1915.

Elzey, S., Grassian, V.H., 2010. Agglomeration, isolation and dissolution of commercially manufactured silver nanoparticles in aqueous environments. J. Nanopart. Res. 12, 1945–1958.

Erogbogbo, F., Yong, K.T., Roy, I., Xu, G., Prasad, P.N., Swihart, M.T., 2008. Biocompatible luminescent silicon quantum dots for imaging of cancer cells. ACS Nano 2, 873–878.

Erogbogbo, F., Tien, C.A., Chang, C.W., Yong, K.T., Law, W.C., Ding, H., Roy, I., Swihart, M.T., Prasad, P.N., 2011. Bioconjugation of luminescent silicon quantum dots for selective uptake by cancer cells. Bioconjug. Chem. 22, 1081–1088.

Ewald, A., Hosel, D., Patel, S., Grover, L.M., Barralet, J.E., Gbureck, U., 2011. Silver-doped calcium phosphate cements with antimicrobial activity. Acta Biomater. 7, 8056–8063.

Fahmi, M.Z., Chang, J.Y., 2013. Forming double layer-encapsulated quantum dots for bio-imaging and cell targeting. Nanoscale 5, 1517–1528.

Faivre, D., Schüler, D., 2008. Magnetotactic bacteria and magnetosomes. Chem. Rev. 108, 4875–4898.

Fan, J., Fang, G., Wang, X., Zeng, F., Xiang, Y., Wu, S., 2011. Targeted anticancer prodrug with mesoporous silica nanoparticles as vehicles. Nanotechnology 22, 455102.

Faulk, W.P., Taylor, G.M., 1971. An immunogold method for the electron microscope. Immunochemistry 8, 1081–1083.

Feazell, R.P., Nakayama-Ratchford, N., Dai, H., Lippard, S.J., 2007. Soluble single-walled carbon nanotubes as longboat delivery systems for platinum(IV) anticancer drug design. J. Am. Chem. Soc. 129, 8438–8439.

Feil, H., Bae, Y.H., Feijen, J., Kim, S.W., 1993. Effect of comonomer hydrophilicity and ionization on the lower critical solution temperature of *N*-isopropylacrylamide copolymers. Macromolecules 26, 2496–2500.

Fernandez-Hervas, M., Fell, J., 1998. Pectin/chitosan mixtures as coatings for colon-specific drug delivery: an in vitro evaluation. Int. J. Pharm. 169, 115–119.

Fialho, A.M., Moreira, L.M., Granja, A.T., Popescu, A.O., Hoffmann, K., Sá-Correia, I., 2008. Occurrence, production, and applications of gellan: current state and perspectives. Appl. Microbiol. Biotechnol. 79, 889–900.

Fichter, K.M., Flajolet, M., Greengard, P., Vu, T.Q., 2010. Kinetics of G-protein-coupled receptor endosomal trafficking pathways revealed by single quantum dots. Proc. Natl. Acad. Sci. USA 107, 18658–18663.

Florence, A.T., 2005. Dendrimers: a versatile targeting platform. Adv. Drug Deliv. Rev. 57, 2104–2105.

Fox, M.E., Guillaudeu, S., Fréchet, J.M., Jerger, K., Macaraeg, N., Szoka, F.C., 2009. Synthesis and in vivo antitumor efficacy of PEGylated poly(L-lysine) dendrimer-camptothecin conjugates. Mol. Pharm. 6, 1562–1672.

Frechet, J.M.J., Tomalia, D.A., 2001. Dendrimers and Other Dendritic Polymers. John Wiley & Sons Ltd, Chichester, UK.

Fude, C., Lei, Y., Jie, J., Hongze, P., Wenhui, L., Dongmei, C., 2007. Preparation and in vitro evaluation of pH, time-based and enzyme-degradable pellets for colonic drug delivery. Drug Dev. Ind. Pharm. 33, 999–1007.

Gan, Q., Wang, T., 2007. Chitosan nanoparticle as protein delivery carrier—systematic examination of fabrication conditions for efficient loading and release. Colloids Surf. B 59, 24–34.

Ganeshkumar, M., Ponrasu, T., Raja, M.D., Subamekala, M.K., Suguna, L., 2014. Green synthesis of pullulan stabilized gold nanoparticles for cancer targeted drug delivery. Spectrochim. Acta A Mol. Biomol. Spectrosc. 130, 64–71.

Gangrade, S.M., 2011. Nanocrystals-a way for carrier free drug delivery. Pharma Buzz 6, 26–31.

Gao, X., Dave, S.R., 2007. Quantum dots for cancer molecular imaging. Adv. Exp. Med. Biol. 620, 57–73.

Gavini, E., Hegge, A.B., Rassu, G., Sanna, V., Testa, C., Pirisino, G., Karlsen, J., Giunchedi, P., 2006. Nasal administration of carbamazepine using chitosan microspheres: in vitro/in vivo studies. Int. J. Pharm. 307, 9–15.

Gavini, E., Rassu, G., Sanna, V., Cossu, M., Giunchedi, P., 2005. Mucoadhesive microspheres for nasal administration of an antiemetic drug, metoclopramide: *in-vitro/ex-vivo* studies. J. Pharm. Pharmacol. 57, 287–294.

Gavory, C., Durand, A., Six, J.L., Nouvel, C., Marie, E., Leonard, M., 2011. Polysaccharide-covered nanoparticles prepared by nanoprecipitation. Carbohydr. Polym. 84, 133–140.

Georgakilas, V., Kordatos, K., Prato, M., Guldi, D.M., Holzinger, M., Hirsch, A., 2002. Organic functionalization of carbon nanotubes. J. Am. Chem. Soc. 124, 760–761.

George, M., Abraham, T.E., 2006. Polyionic hydrocolloids for the intestinal delivery of protein drugs: alginate and chitosan: a review. J. Control. Release 114, 1–14.

George, M., Abraham, T.E., 2007. pH sensitive alginate-guar gum hydrogel for the controlled delivery of protein drugs. Int. J. Pharm. 335, 123–129.

Geszke, M., Murias, M., Balan, L., Medjahdi, G., Korczynski, J., Moritz, M., Lulek, J., Schneider, R., 2011. Folic acid-conjugated core/shell ZnS: Mn/ZnS quantum dots as targeted probes for two photon fluorescence imaging of cancer cells. Acta Biomater. 7, 1327–1338.

Ghaffari, A., Navaee, K., Oskoui, M., Bayati, K., Rafiee-Tehrani, M., 2007. Preparation and characterization of free mixed-film of pectin/chitosan/Eudragit RS intended for sigmoidal drug delivery. Eur. J. Pharm. Biopharm. 67, 175–186.

Ghaffari, A., Avadi, M.R., Moghimi, H.R., Oskoui, M., Bayati, K., Rafiee-Tehrani, M., 2008. Mechanistic analysis of drug release from theophylline pellets coated by films containing pectin, chitosan and Eudragit RS. Drug Dev. Ind. Pharm. 34, 390–402.

Ghasemali, S., Nejati-Koshki, K., Tafsiri, E., Rahmati-Yamchi, M., Akbarzadeh, A., Alizadeh, E., Abbasi, M., Barkhordari, A., Tozihi, M., Kordi, S., Zarghami, N., 2013. Inhibitory effects of β-cyclodextrin-helenalin complexes on H-TERT gene expression in T47D breast cancer cell line: results of real time quantitative PCR. Asian Pacific J. Cancer Prev. 14, 6949–6953.

Gillies, E., Frechet, J., 2005. Dendrimers and dendritic polymers in drug delivery. Drug Discov. Today 10, 35–43.

Gillies, E.R., Jonsson, T.B., Frechet, J.M., 2004. Stimuli-responsive supramolecular assemblies of linear-dendritic copolymers. J. Am. Chem. Soc. 126, 11936–11943.

Goldberger, J., Fan, R., Yang, P., 2006. Inorganic nanotubes: a novel platform for nanofluidics. Acc. Chem. Res. 39, 239–248.

Gong, J., Chen, M., Zheng, Y., Wang, S., Wang, Y., 2012. Polymeric micelles drug delivery system in oncology. J. Control. Release 159, 312–323.

Grace, A.N., Pandian, K., 2007. Antibacterial efficacy of aminoglycosidic antibiotics protected gold nanoparticles: a brief study. Colloids Surf. A 297, 63–70.

Grumezescu, A.M., Ghitulica, C.D., Voicu, G., Huang, K.S., Yang, C.H., Ficai, A., Vasile, B.S., Grumezescu, V., Bleotu, C., Chifiriuc, M.C., 2014. New silica nanostructure for the improved delivery of topical antibiotics used in the treatment of staphylococcal cutaneous infections. Int. J. Pharm. 463, 170–176.

Gu, F.X., Zhang, L., Teply, B.A., Mann, N., Wang, A., Radovic-Moreno, A.F., Langer, R., Farokhzad, O.C., 2008. Precise engineering of targeted nanoparticles by using self-assembled biointegrated block copolymers. Proc. Natl. Acad. Sci. USA 105, 2586–2591.
Gu, H., Ho, P.L., Tong, E., Wang, L., Xu, B., 2003. Presenting vancomycin on nanoparticles to enhance antimicrobial activities. Nano Lett. 3, 1261–1263.
Gu, Y.-J., Cheng, J., Man, C.W.-Y., Wong, W.-T., Cheng, S.H., 2012. Gold-doxorubicin nanoconjugates for overcoming multidrug resistance. Nanomed. Nanotechnol. Biol. Med. 8, 204–211.
Guerrero, S., Teijón, C., Muñiz, E., Teijón, J.M., Blanco, M.D., 2010. Characterization and in vivo evaluation of ketotifen-loaded chitosan microspheres. Carbohydr. Polym. 79, 1006–1013.
Guldi, D.M., Rahman, G.M.A., Sgobba, V., Ehli, C., 2006. Multifunctional molecular carbon materials: from fullerenes to carbon nanotubes. Chem. Soc. Rev. 35, 471–487.
Guo, J., Hong, H., Chen, G., Shi, S., Zheng, Q., Zhang, Y., Theuer, C.P., Barnhart, T.E., Cai, W., Gong, S., 2013. Image-guided and tumor-targeted drug delivery with radiolabeled unimolecular micelles. Biomaterials 34, 8323–8332.
Gupta, U., Agashe, H.B., Asthana, A., Jain, N.K., 2006. Dendrimers: novel polymeric nanoarchitectures for solubility enhancement. Biomacromolecules 7, 649–658.
Gurdag, S., Khandare, J., Stapels, S., Matherly, L.H., Kannan, R.M., 2006. Activity of dendrimer-methotrexate conjugates on methotrexate-sensitive and -resistant cell lines. Bioconjugate Chem. 17, 275–283.
Haensler, J., Szoka, F.C., 1993. Polyamidoamine cascade polymers mediate efficient transfection of cells in culture. Bioconjugate Chem. 4, 372–379.
Hafner, A., Lovrić, J., Pepić, I., Filipović-Grcić, J., 2011. Lecithin/chitosan nanoparticles for transdermal delivery of melatonin. J. Microencapsul. 28, 807–815.
Hafner, A., Lovrić, J., Voinovich, D., Filipović-Grcić, J., 2009. Melatonin-loaded lecithin/chitosan nanoparticles: physicochemical characterisation and permeability through Caco-2 cell monolayers. Int. J. Pharm. 381, 205–213.
Hanada, S., Fujioka, K., Futamura, Y., Manabe, N., Hoshino, A., Yamamoto, K., 2013. Evaluation of anti-inflammatory drug-conjugated silicon quantum dots: their cytotoxicity and biological effect. Int. J. Mol. Sci. 14, 1323–1334.
Hartmann, M., 2005. Ordered mesoporous materials for bioadsorption and biocatalysis. Chem. Mater. 17, 4577–4593.
Hassan, C.M., Peppas, N.A., 2000. Structure and application of poly(vinyl alcohol) hydrogels produced by conventional cross-linking or by freezing-thawing methods. Adv. Polym. Sci. 153, 37–66.
Hawker, C.J., Wooley, K.L., Fréchet, J.M.J., 1993. Unimolecular micelles and globular amphiphiles: dendritic macromolecules as novel recyclable solubilization agents. J. Chem. Soc. Perkin Trans. 1, 1287–1297.
Hayat, M.A., 1989. Colloidal Gold: Principles, Methods and Applications. Elsevier, New York.
He, G., Ma, L.L., Pan, J., Venkatraman, S., 2007. ABA and BAB type triblock copolymers of PEG and PLA: a comparative study of drug release properties and "stealth" particle characteristics. Int. J. Pharm. 334, 48–55.
He, Q., Shi, J., 2011. Mesoporous silica nanoparticle based nano drug delivery systems: synthesis, controlled drug release and delivery, pharmacokinetics and biocompatibility. J. Mater. Chem. 21, 5845–5855.
He, W., Qing, D., Cao, D.Y., Fan, L.F., Xiang, B., 2008. Selective drug delivery to the colon using pectin-coated pellets. PDA J. Pharm. Sci. Technol. 62, 264–272.
He, X., Wua, X., Gao, C., Wanga, K., Lin, S., Huang, W., Xie, M., Yan, D., 2011. Synthesis and self-assembly of a hydrophilic, thermo-responsive poly(ethylene oxide) monomethyl ether-block-poly(acrylic acid)-block-poly(*N*-isopropylacrylamide) copolymer to form micelles for drug delivery. React. Funct. Polym. 71, 544–552.
He, Y., Cheng, G., Xie, L., Nie, Y., He, B., Gu, Z., 2012. Polyethyleneimine/DNA polyplexes with reduction-sensitive hyaluronic acid derivatives shielding for targeted gene delivery. Biomaterials 34, 1235–1245.
Heinze, T., Liebert, T., Heublein, B., Hornig, S., 2006. Functional polymers based on dextran. Adv. Polym. Sci. 205–206, 199–291.

Ho, D., Sun, X., Sun, S., 2011. Monodisperse magnetic nanoparticles for theranostic applications. Acc. Chem. Res. 44, 875–882.
Hoffman, A.S., 1987. Applications of thermally reversible polymers and hydrogels in therapeutics and diagnostics. J. Control. Release 6, 297–305.
Hsu, F.Y., Yu, D.S., Huang, C.C., 2013. Development of pH-sensitive pectinate/alginate microspheres for colon drug delivery. J. Mater. Sci. Mater. Med. 24, 317–323.
Hua, M.Y., Yang, H.W., Chuang, C.K., Tsai, R.Y., Chen, W.J., Chuang, K.L., Chang, Y.H., Chuang, H.C., Pang, S.T., 2010. Magnetic-nanoparticle-modified paclitaxel for targeted therapy for prostate cancer. Biomaterials 31, 7355–7363.
Hunter, R.J., 1988. Zeta Potential in Colloid Science: Principles and Applications. Academic Press, London.
Hussein-Al-Ali, S.H., El Zowalaty, M.E., Kura, A.U., Geilich, B., Fakurazi, S., Webster, T.J., Hussein, M.Z., 2014a. Antimicrobial and controlled release studies of a novel nystatin conjugated iron oxide nanocomposite. Biomed. Res. Int. 2014, 651831.
Hussein-Al-Ali, S.H., El Zowalaty, M.E., Hussein, M.Z., Ismail, M., Webster, T.J., 2014b. Synthesis, characterization, controlled release, and antibacterial studies of a novel streptomycin chitosan magnetic nanoantibiotic. Int. J. Nanomed. 9, 549–557.
Iijima, S., 1991. Helical microtubules of graphitic carbon. Nature 354, 56–58.
Imae, T., Hamaguchi, S.I., 2012. Network of sodium hyaluronate with nano-knots junction of poly(amido amine) dendrimer. Carbohydr. Polym. 88, 352–360.
Ishak, R.A., Awad, G.A., Mortada, N.D., Nour, S.A., 2007. Preparation, in vitro and in vivo evaluation of stomach-specific metronidazole-loaded alginate beads as local anti-*Helicobacter pylori* therapy. J. Control. Release 119, 207–214.
Jackson, C.L., Chanzy, H.D., Booy, F.B., Drake, B.J., Tomalia, D.A., Bauer, B.J., Amis, E.J., 1998. Visualization of dendrimer molecules by Transmission Electron Microscopy (TEM): staining methods and Cryo-TEM of vitrified solutions. Macromolecules 31, 6259–6265.
Jaffer, F.A., Nahrendorf, M., Sosnovik, D., Kelly, K.A., Aikawa, E., Weissleder, R., 2006. Cellular imaging of inflammation in atherosclerosis using magnetofluorescent nanomaterials. Mol. Imaging 5, 85–92.
Jain, A., Gupta, Y., Jain, S.K., 2007a. Potential of calcium pectinate beads for target specific drug release to colon. J. Drug Target. 15, 285–294.
Jain, A., Jain, S.K., Ganesh, N., Barve, J., Beg, A.M., 2009a. Design and development of ligand-appended polysaccharidic nanoparticles for the delivery of oxaliplatin in colorectal cancer. Nanomedicine 6, 179–190.
Jain, J., Arora, S., Rajwade, J.M., Omray, P., Khandelwal, S., Paknikar, K.M., 2009b. Silver nanoparticles in therapeutics: development of an antimicrobial gel formulation for topical use. Mol. Pharm. 6, 1388–1401.
Jain, S., Shukla, K., Jain, V., Saraf, S., 2007b. Nanoparticles: emerging carriers for delivery of bioactive agents. Pharma Times 39, 30–35.
Jain, T.K., Richey, J., Strand, M., Leslie-Pelecky, D.L., Flask, C.A., Labhasetwar, V., 2008. Magnetic nanoparticles with dual functional properties: drug delivery and magnetic resonance imaging. Biomaterials 29, 4012–4021.
Jain, V., Prasad, D., Jain, D., Mishra, S.K., Singh, R., 2011a. Factorial design-based development of measlamine microspheres for colonic delivery. Biomatter 1, 182–188.
Jain, V., Jain, D., Singh, R., 2011b. Factors effecting the morphology of Eudragit S-100 based microsponges bearing dicyclomine for colonic delivery. J. Pharm. Sci. 100, 1545–1552.
Jalili, N., Laxminarayana, K., 2004. A review of atomic force microscopy imaging systems: application to molecular metrology and biological sciences. Mechatronics 14, 907–945.
Jana, S., Gandhi, A., Sheet, S., Sen, K.K., 2015. Metal ion-induced alginate-locust bean gum IPN microspheres for sustained oral delivery of aceclofenac. Int. J. Biol. Macromol. 72, 47–53.
Jang, H., Kim, Y.-K., Ryoo, S.-R., Kim, M.-H., Min, D.-H., 2010. Facile synthesis of robust and biocompatible gold nanoparticles. Chem. Commun. 46, 583–585.
Jansen, J.F.G.A., de Brabander-van den Berg, E.M.M., Meijer, E.W., 1994. Encapsulation of guest molecules into a dendritic box. Science 266, 1226–1229.

Jansen, J.F.G.A., Meijer, E.W., Brabander-van den Berg, E.M.M., 1995. The dendritic box: shape-selective liberation of encapsulated guests. J. Am. Chem. Soc. 117, 4417–4418.

Jayant, R.D., McShane, M.J., Srivastava, R., 2009. Polyelectrolyte-coated alginate microspheres as drug delivery carriers for dexamethasone release. Drug Deliv. 16, 331–340.

Jeong, Y.-I., Seo, S.-J., Park, I.-K., Lee, H.-C., Kang, I.-C., Akaike, T., Cho, C.-S., 2005. Cellular recognition of paclitaxel-loaded polymeric nanoparticles composed of poly(γ-benzyl L-glutamate) and poly(ethylene glycol) diblock copolymer endcapped with galactose moiety. Int. J. Pharm. 296, 151–161.

Jevprasesphant, R., Penny, J., Attwood, D., D'Emanuele, A., 2004. Transport of dendrimer nanocarriers through epithelial cells via the transcellular route. J. Control. Release 97, 259–267.

Jie, P., Venkatraman, S.S., Min, F., Freddy, B.Y., Huat, G.L., 2005. Micelle-like nanoparticles of star-branched PEO-PLA copolymers as chemotherapeutic carrier. J. Control. Release 110, 20–33.

Jones, D.S., Lorimer, C.J., Andrews, G.P., McCoy, C.P., Gorman, S.P., 2007. An examination of the thermorheological and drug release properties of zinc tetraphenylporphyrin-containing thermoresponsive hydrogels, designed as light activated antimicrobial implants. Chem. Eng. Sci. 62, 990–999.

Jones, D.S., Lorimer, C.P., McCoy, C.P., Gorman, S.P., 2008. Characterization of the physicochemical, antimicrobial, and drug release properties of thermoresponsive hydrogel copolymers designed for medical device applications. J. Biomed. Mater. Res. 85B, 417–426.

Joshi, M., Viswanathan, V., 2006. High-performance filaments from compatibilized polypropylene/clay nanocomposites. J. Appl. Polym. Sci. 102, 2164–2174.

Jung, W.K., Koo, H.C., Kim, K.W., Shin, S., Kim, S.H., Park, Y.H., 2008. Antibacterial activity and mechanism of action of the silver ion in *Staphylococcus aureus* and *Escherichia coli*. Appl. Environ. Microbiol. 74, 2171–2178.

Kahrilas, G.A., Haggren, W., Read, R.L., Wally, L.M., Fredrick, S.J., Hiskey, M., Prieto, A.L., Owens, J.E., 2014. Investigation of antibacterial activity by silver nanoparticles prepared by microwave-assisted green syntheses with soluble starch, dextrose, and arabinose. ACS Sust. Chem. Eng. 2, 590–598.

Kallos, G.J., Tomalia, D.A., Hedstrand, D.M., Lewis, S., Zhou, J., 1991. Molecular weight determination of a polyamidoamine Starburst polymer by electrospray ionization mass spectrometry. Rapid Commun. Mass Spectrom. 5, 383–386.

Kam, N.W.S., O'Connell, M., Wisdom, J.A., Dai, H.J., 2005. Carbon nanotubes as multifunctional biological transporters and near-infrared agents for selective cancer cell destruction. Proc. Nat. Acad. Sci. USA 102, 11600–11605.

Kanmani, P., Lim, S.T., 2013. Synthesis and structural characterization of silver nanoparticles using bacterial exopolysaccharide and its antimicrobial activity against food and multidrug resistant pathogens. Process Biochem. 48, 1099–1106.

Kaplan, L., Kurdziel, M., Baker, K.C., Verner, J., 2012. Charactrerization of daptomycin-loaded antibiotic cement. Orthopedics 35, E503–E509.

Karakoti, A.S., Shukla, R., Shanker, R., Singh, S., 2015. Surface functionalization of quantum dots for biological applications. Adv. Colloid Interface Sci. 215, 28–45.

Karchemski, F., Zucker, D., Barenholz, Y., Regev, O., 2012. Carbon nanotubes-liposomes conjugate as a platform for drug delivery into cells. J. Control. Release 160, 339–345.

Kassem, A.A., Farid, R.M., Issa, D.A., Khalil, D.S., Abd-El-Razzak, M.Y., Saudi, H.I., Eltokhey, H.M., El-Zamarany, E.A., 2015. Development of mucoadhesive microbeads using thiolated sodium alginate for intrapocket delivery of resveratrol. Int. J. Pharm. 487, 305–313.

Kattumuri, V., Katti, K., Bhaskaran, S., Boote, E.J., Casteel, S.W., Fent, G.M., Robertson, D.J., Chandrasekhar, M., Kannan, R., Katti, K.V., 2007. Gum arabic as a phytochemical construct for the stabilization of gold nanoparticles: in vivo pharmacokinetics and X-ray-contrast-imaging studies. Small 3, 333–341.

Kawaguchi, T., Walker, K.L., Wilkins, C.L., Moore, J.S., 1995. Double exponential dendrimer growth. J. Am. Chem. Soc. 117, 2159–2165.

Khandekar, S.V., Kulkarni, M.G., Devarajan, P.V., 2014. Polyaspartic acid functionalized gold nanoparticles for tumor targeted doxorubicin delivery. J. Biomed. Nanotechnol. 10, 143–153.

Kim, G.C., Li, Y.Y., Chu, Y.F., Cheng, S.X., Zhuo, R.X., Zhang, X.Z., 2008. Nanosized temperature-responsive Fe_3O_4-UA-*g*-P(UA-*co*-NIPAAm) magnetomicelles for controlled drug release. Eur. Polym. J. 44, 2761–2767.

Kim, J.E., Shin, J.Y., Cho, M.H., 2012a. Magnetic nanoparticles: an update of application for drug delivery and possible toxic effects. Arch. Toxicol. 86, 685–700.

Kim, J.H., Li, Y., Kim, M.S., Kang, S.W., Jeong, J.H., Lee, D.S., 2012b. Synthesis and evaluation of biotin-conjugated pH-responsive polymeric micelles as drug carriers. Int. J. Pharm. 427, 435–442.

Kiremitci, A.S., Ciftci, A., Ozalp, M., Gumuesderelioglu, M., 2007. Novel chlorhexidine releasing system developed from thermosensitive vinyl ether-based hydrogels. J. Biomed. Mater. Res. 83B, 609–614.

Klein, S., Zolk, O., Fromm, M.F., Schrodl, F., Neuhuber, W., Kryschi, C., 2009. Functionalized silicon quantum dots tailored for targeted siRNA delivery. Biochem. Biophys. Res. Commun. 387, 164–168.

Kohler, N., Sun, C., Wang, J., Zhang, M.Q., 2005. Methotrexate-modified superparamagnetic nanoparticles and their intracellular uptake into human cancer cells. Langmuir 21, 8858–8864.

Kojima, C., Kono, K., Maruyama, K., Takagishi, T., 2000. Synthesis of polyamideamine dendrimers having poly(ethylene glycol) grafts and their ability to encapsulate anticancer drugs. Bioconjugate Chem. 11, 910–917.

Kokura, S., Handa, O., Takagi, T., Ishikawa, T., Naito, Y., Yoshikawa, T., 2010. Silver nanoparticles as a safe preservative for use in cosmetics. Nanomedicine 6, 570–574.

Kolthoff, I.M., 1925. The state of silver in Protargol (Argenti proteinatum) and Collargol (colloid silver; argentum crédé). J. Pharm. Sci. 14, 183–189.

Kortesuo, P., Ahola, M., Kangas, M., Leino, T., Laakso, L., Vuorilehto, L., 2001. Alkyl-substituted silica gel as a carrier in the controlled release of dexmedetomidine. J. Control. Release 76, 227–238.

Kowshik, M., Ashtaputre, S., Kharrazi, S., 2003. Extracellular synthesis of silver nanoparticles by a silver-tolerant yeast strain MKY3. Nanotechnology 14, 95–100.

Kreibig, U., Vollmer, M., 1995. Optical Properties of Metal Clusters. Springer-Verlag, Berlin-Heidelberg, Germany.

Kresge, C.T., Leonowicz, M.E., Roth, W.J., Vartuli, J.C., Beck, J.S., 1992. Ordered mesoporous molecular sieves synthesized by a liquid-crystal template mechanism. Nature 359, 710–712.

Krishnaiah, Y.S., Karthikeyan, R.S., Gouri Sankar, V., Satyanarayana, V., 2002a. Three-layer guar gum matrix tablet formulations for oral controlled delivery of highly soluble trimetazidine dihydrochloride. J. Control. Release 81, 45–56.

Krishnaiah, Y.S., Karthikeyan, R.S., Satyanarayana, V., 2002b. A three-layer guar gum matrix tablet for oral controlled delivery of highly soluble metoprolol tartrate. Int. J. Pharm. 241, 353–366.

Krishnaiah, Y.S., Satyanarayana, V., Kumar, B.D., Karthikeyan, R.S., 2002c. Studies on the development of colon-targeted delivery systems for celecoxib in the prevention of colorectal cancer. J. Drug Target. 10, 247–254.

Krishnaiah, Y.S., Bhaskar Reddy, P.R., Satyanarayana, V., Karthikeyan, R.S., 2002d. Studies on the development of oral colon targeted drug delivery systems for metronidazole in the treatment of amoebiasis. Int. J. Pharm. 236, 43–55.

Krishnaiah, Y.S., Muzib, Y.I., Rao, G.S., Bhaskar, P., Satyanarayana, V., 2002e. Design and in vitro evaluation of oral colon targeted drug delivery systems for tinidazole. J. Drug Target. 10, 579–584.

Krishnaiah, Y.S., Satyanarayana, V., Kumar, B.D., Karthikeyan, R.S., 2002f. In vitro drug release studies on guar gum-based colon targeted oral drug delivery systems of 5-fluorouracil. Eur. J. Pharm. Sci. 16, 185–192.

Krishnaiah, Y.S., Muzib, Y.I., Rao, G.S., Bhaskar, P., Satyanarayana, V., 2003. Studies on the development of colon targeted oral drug delivery systems for ornidazole in the treatment of amoebiasis. Drug Deliv. 10, 111–117.

Kröger, N., Deutzmann, R., Sumper, M., 1999. Polycationic peptides from diatom biosilica that direct silica nanosphere formation. Science 286, 1129–1132.

Krukemeyer, M.G., Krenn, V., Jakobs, M., Wagner, W., 2012. Magnetic drug targeting in a rhabdomyosarcoma rat model using magnetite-dextran composite nanoparticle-bound mitoxantrone and 0.6 tesla extracorporeal magnets: sarcoma treatment in progress. J. Drug Target. 20, 185–193.
Kumar, A., Sahoo, B., Montpetit, A., Behera, S., Lockey, R.F., Mohapatra, S.S., 2007. Development of hyaluronic acid-Fe_2O_3 hybrid magnetic nanoparticles for targeted delivery of peptides. Nanomedicine 3, 132–137.
Kumar, C.G., Mamidyala, S.K., Das, B., Sridhar, B., Devi, G.S., Karuna, M.S.L., 2010. Synthesis of biosurfactant-based silver nanoparticles with purified rhamnolipids isolated from *Pseudomonas aeruginosa* BS-161R. J. Microbiol. Biotechnol. 20, 1061–1068.
Kumar, C.G., Poornachandra, Y., 2015. Biodirected synthesis of miconazole-conjugated bacterial silver nanoparticles and their application as antifungal agents and drug delivery vehicles. Colloids Surf. B 125, 110–119.
Kumar, C.G., Poornachandra, Y., Mamidyala, S.K., 2014. Green synthesis of bacterial gold nanoparticles conjugated to resveratrol as delivery vehicles. Colloids Surf. B 123, 311–317.
Kumar, C.G., Sujitha, P., 2014. Green synthesis of Kocuran-functionalized silver glyconanoparticles for use as anti-biofilm coatings on silicone urethral catheters. Nanotechnology 25, 325101.
Kumar, C.S., Raja, M.D., Sundar, D.S., Antoniraj, M.G., Ruckmani, K., 2015. Hyaluronic acid co-functionalized gold nanoparticle complex for the targeted delivery of metformin in the treatment of liver cancer (HepG2 cells). Carbohydr. Polym. 128, 63–74.
Kumar, R., Munstedt, H., 2005. Silver ion release from antimicrobial polyamide/silver composites. Biomaterials 26, 2081–2088.
Labbe, G., Forier, B., Dehaen, W., 1996. A fast double-stage convergent synthesis of dendritic polyethers. Chem. Commun. 18, 2143–2144.
Lacerda, L., Parize, A.L., Fávere, V., Laranjeira, M.C., Stulzer, H.K., 2014. Development and evaluation of pH-sensitive sodium alginate/chitosan microparticles containing the antituberculosis drug rifampicin. Mater. Sci. Eng. C 39, 161–167.
Lansdown, A.B.G., 2004. A review of the use of silver in would care: facts and fallacies. Brit. J. Nurs. 13, S6–S19.
Lara, H.H., Ayala-Nuñez, N.V., Ixtepan-Turrent, L., Rodriguez-Padilla, C., 2010a. Mode of antiviral action of silver nanoparticles against HIV-1. J. Nanobiotechnol. 8, 1.
Lara, H.H., Ixtepan-Turrent, L., Garza-Treviño, E.N., Rodriguez-Padilla, C., 2010b. PVP-coated silver nanoparticles block the transmission of cell-free and cell-associated HIV-1 in human cervical culture. J. Nanobiotechnol. 8, 15.
Laurent, S., Forge, D., Port, M., Roch, A., Robic, C., Elst, L.V., Muller, R.N., 2008. Magnetic iron oxide nanoparticles: synthesis, stabilization, vectorization, physicochemical characterizations, and biological applications. Chem. Rev. 108, 2064–2110.
Learoyd, T.P., Burrows, J.L., French, E., Seville, P.C., 2008. Chitosan-based spray-dried respirable powders for sustained delivery of terbutaline sulfate. Eur. J. Pharm. Biopharm. 68, 224–234.
Lee, H.Y., Park, H.K., Lee, Y.M., Kim, K., Park, S.B., 2007. A practical procedure for producing silver nanocoated fabric and its antibacterial evaluation for biomedical applications. Chem. Commun. 2007, 2959–2961.
Lee, J.W., Park, J.H., Robinson, J.R., 2000. Bioadhesive-based dosage forms: the next generation. J. Pharm. Sci. 89, 850–866.
Lee, Y., Kim, I.H., Kim, J., Yoon, J.H., Shin, Y.H., Jung, Y., Kim, Y.M., 2011. Evaluation of dextran-flufenamic acid ester as a polymeric colon-specific prodrug of flufenamic acid, an anti-inflammatory drug, for chronotherapy. J. Drug Target. 19, 336–343.
Leszczynska, K., Naniot, A., Fein, D.E., Wen, Q., Naniot, Z., Savage, P.B., Diamond, S., Janmey, P.A., Bucki, R., 2009. Bactericidal activities of the cationic steroid CSA-13 and the cathelicidin peptide LL-37 against *Helicobacter pylori* in simulated gastric juice. BMC Microbiol. 9, 187.
Leung, T.C.Y., Wong, C.K., Xie, Y., 2010. Green synthesis of silver nanoparticles using biopolymers, carboxymethylated: curdlan and fucoidan. Mater. Chem. Phys. 121, 402–405.

Li, J., Piehler, L.T., Qin, D., Baker, Jr., J.R., Tomalia, D.A., Meier, D.J., 2000. Visualization and characterization of poly(amidoamine) dendrimers by atomic force microscopy. Langmuir 16, 5613–5616.

Li, J., Wu, C., Gao, F., Zhang, R., Lv, G., Fu, D., Chen, B., Wang, X., 2006a. In vitro study of drug accumulation in cancer cells via specific association with CdS nanoparticles. Bioorg. Med. Chem. Lett. 16, 4808–4812.

Li, M.H., Choi, S.K., Thomas, T.P., Desai, A., Lee, K.H., Kotlyar, A., Holl, M.M.B., Baker, Jr., J.R., 2012a. Dendrimer-based multivalent methotrexates as dual acting nanoconjugates for cancer cell targeting. Eur. J. Med. Chem. 47, 560–572.

Li, X., Li, H., Liu, G., Deng, Z., Wu, S., Li, P., Xu, Z., Xu, H., Chu, P.K., 2012b. Magnetite-loaded fluorine-containing polymeric micelles for magnetic resonance imaging and drug delivery. Biomaterials 33, 3013–3024.

Li, X., Liu, X., Huang, J., Fan, Y., Cui, F.Z., 2011. Biomedical investigation of CNT based coatings. Surf. Coat. Technol. 206, 759–766.

Li, Y.L., Zhu, L., Liu, Z., Cheng, R., Meng, F., Cui, J.H., Ji, S.J., Zhong, Z., 2009. Reversibly stabilized multifunctional dextran nanoparticles efficiently deliver doxorubicin into the nuclei of cancer cells. Angew. Chem. Int. Ed. Engl. 48, 9914–9918.

Li, Z., Barnes, J.C., Bosoy, A., Stoddart, J., Zink, J.I., 2012. Mesoporous silica nanoparticles in biomedical applications. Chem. Soc. Rev. 41, 2590–2605.

Li, Z., Sun, J., Lan, J., Qi, Q., 2014. Effect of a denture base acrylic resin containing silver nanoparticles on *Candida albicans* adhesion and biofilm formation. Gerodontology 33 (2), 209–216.

Li, Z., Xu, W., Wang, Y., Shah, B.R., Zhang, C., Chen, Y., Li, Y., Li, B., 2015. Quantum dots loaded nanogels for low cytotoxicity, pH-sensitive fluorescence, cell imaging and drug delivery. Carbohydr. Polym. 121, 477–485.

Li, Z.Z., Wen, L.X., Shao, L., Chen, J.F., 2004. Fabrication of porous hollow silica nanoparticles and their applications in drug release control. J. Control. Release 98, 245–254.

Li, Z.-Z., Xu, S.-A., Wen, L.-X., Liu, F., Liu, A.-Q., Wang, Q., Sun, H.-Y., Yu, W., Chen, J.-F., 2006b. Controlled release of avermectin from porous hollow silica nanoparticles: influence of shell thickness on loading efficiency, UV-shielding property and release. J. Control. Release 111, 81–88.

Liao, J., Anchun, M., Zhu, Z., Quan, Y., 2010. Antibacterial titanium plate deposited by silver nanoparticles exhibits cell compatibility. Int. J. Nanomed. 5, 337–342.

Lim, J., Simanek, E.E., 2012. Triazine dendrimers as drug delivery systems: from synthesis to therapy. Adv. Drug Deliv. Rev. 64, 826–835.

Lin, J.T., Zou, Y., Wang, C., Zhong, Y.C., Zhao, Y., Zhu, H.E., Wang, G.H., Zhang, L.M., Zheng, X.B., 2014. Cationic micellar nanoparticles for DNA and doxorubicin co-delivery. Mater. Sci. Eng. C 44, 430–439.

Lin, S.T., Franklin, M.T., Klabunde, K.J., 1986. Nonaqueous colloidal gold: clustering of metal atoms in organic media. Langmuir 2, 259–260.

Lin, Y.H., Liang, H.F., Chung, C.K., Chen, M.C., Sung, H.W., 2005. Physically crosslinked alginate/*N*,O-carboxymethyl chitosan hydrogels with calcium for oral delivery of protein drugs. Biomaterials 26, 2105–2113.

Lin, Y.H., Mi, F.L., Chen, C.T., Chang, W.C., Peng, S.F., Liang, H.F., Sung, H.W., 2007. Preparation and characterization of nanoparticles shelled with chitosan for oral insulin delivery. Biomacromolecules 8, 146–152.

Liou, T.H., Yang, C.C., 2011. Synthesis and surface characteristics of nanosilica produced from alkali-extracted rice husk ash. Mater. Sci. Eng. B 176, 521–529.

Liou, T.-H., 2004. Preparation and characterization of nano structured silica from rice husk. Mat. Sci. Eng. 364A, 313–323.

Liu, B.S., Huang, T.B., 2008. Nanocomposites of genipin-crosslinked chitosan/silver nanoparticles: structural reinforcement and antimicrobial properties. Macromol. Biosci. 8, 932–941.

Liu, L., Fishman, M.L., Kost, J., Hicks, K.B., 2003. Pectin-based systems for colon-specific drug delivery via oral route. Biomaterials 24, 3333–3343.

Liu, L., Fishman, M.L., Hicks, K.B., Kende, M., Ruthel, G., 2006. Pectin/zein beads for potential colon-specific drug delivery: synthesis and in vitro evaluation. Drug Deliv. 13, 417–423.

Liu, L., Won, Y.J., Cooke, P.H., Coffin, D.R., Fishman, M.L., Hicks, K.B., Ma, P.X., 2004. Pectin/poly(lactide-*co*-glycolide) composite matrices for biomedical applications. Biomaterials 25, 3201–3210.

Liu, M., Frechet, J.M.J., 1999. Designing dendrimers for drug delivery. Pharm. Sci. Technol. Today 2, 393–401.

Liu, M., Kono, K., Frechet, J.M.J., 1999. Water-soluble dendrimer-poly(ethylene glycol) starlike conjugates as potential drug carriers. J. Polym. Sci. Part A 37, 3492–3503.

Liu, M., Gan, L., Chen, L., Zhu, D., Xu, Z., Hao, Z., Chen, L., 2012a. A novel liposome-encapsulated hemoglobin/silica nanoparticle as an oxygen carrier. Int. J. Pharm. 427, 354–357.

Liu, Y., Zhou, H., 2015. Budesonide-loaded guar gum microspheres for colon delivery: preparation, characterization and in vitro/in vivo evaluation. Int. J. Mol. Sci. 16, 2693–2704.

Liu, Y., Zhou, M., Luo, D., Wang, L., Hong, Y., Yang, Y., Sha, Y., 2012b. Bacteria-mediated in vivo delivery of quantum dots into solid tumor. Biochem. Biophys. Res. Commun. 425, 769–774.

Liu, Z., Chen, K., Davis, C., Sherlock, S., Cao, Q., Chen, X., Dai, H., 2008a. Drug delivery with carbon nanotubes for in vivo cancer treatment. Cancer Res. 68, 6652–6660.

Liu, Z., Jiao, Y., Wang, Y., Zhou, C., Zhang, Z., 2008b. Polysaccharides-based nanoparticles as drug delivery systems. Adv. Drug Deliv. Rev. 60, 1650–1662.

Liu, Z., Sun, X., Nakayama, N., Dai, H., 2007. Supramolecular chemistry on water-soluble carbon nanotubes for drug loading and delivery. ACS Nano 1, 50–56.

Locatelli, E., Naddaka, M., Uboldi, C., Loudos, G., Fragogeorgi, E., Molinari, V., Pucci, A., Tsotakos, T., Psimadas, D., Ponti, J., Franchini, M.C., 2014. Targeted delivery of silver nanoparticles and alisertib: in vitro and in vivo synergistic effect against glioblastoma. Nanomedicine 9, 839–849.

Lu, J., Liong, M., Zink, J.I., Tamanoi, F., 2007a. Mesoporous silica nanoparticles as a delivery system for hydrophobic anticancer drugs. Small 3, 1341–1346.

Lu, J., Liong, M., Sherman, S., Xia, T., Kovochich, M., Nel, A.E., Zink, J.I., Tamanoi, F., 2007b. Mesoporous silica nanoparticles for cancer therapy: energy-dependent cellular uptake and delivery of paclitaxel to cancer cells. Nanobiotechnology 3, 89–95.

Maestrelli, F., Cirri, M., Corti, G., Mennini, N., Mura, P., 2008. Development of enteric-coated calcium pectinate microspheres intended for colonic drug delivery. Eur. J. Pharm. Biopharm. 69, 508–518.

Maestrelli, F., Cirri, M., Mennini, N., Bragagni, M., Zerrouk, N., Mura, P., 2012. Influence of cross-linking agent type and chitosan content on the performance of pectinate-chitosan beads aimed for colon-specific drug delivery. Drug Dev. Ind. Pharm. 38, 1142–1151.

Mahajan, S.D., Roy, I., Xu, G., Yong, K.-T., Ding, H., Aalinkeel, R., Reynolds, J., Sykes, D., Nair, B.B., Lin, E.Y., Prasad, P.N., Schwartz, S.A., 2010. Enhancing the delivery of anti retroviral drug "Saquinavir" across the blood brain barrier using nanoparticles. Curr. HIV Res. 9, 396–404.

Mahdihasan, S., 1985. Cinnabar-gold as the best alchemical drug of longevity, called Makaradhwaja in India. Am. J. Chin. Med. 13, 93–108.

Majoral, J.P., Caminade, A.M., 1999. Dendrimers containing heteroatoms (Si, P, B, Ge, or Bi). Chem. Rev. 99, 845–880.

Mandal, S., Selvakannan, P.R., Phadtare, S., Pasricha, R., Sastry, M., 2002. Synthesis of a stable gold hydrosol by the reduction of chloroaurate ions by the amino acid, aspartic acid. Proc. Indian Acad. Sci. Chem. Sci. 114, 513–520.

Maneerung, T., Tokura, S., Rujiravanit, R., 2008. Impregnation of silver nanoparticles into bacterial cellulose for antimicrobial wound dressing. Carbohydr. Polym. 72, 43–51.

Marambio-Jones, C., Hoek, E.M.V., 2010. A review of the antibacterial effects of silver nanomaterials and potential implications for human health and the environment. J. Nanopart. Res. 12, 1531–1551.

Marine, W., Patrone, L., Luk'yanchuk, B., Sentis, M., 2000. Strategy of nanocluster and nanostructure synthesis by conventional pulsed laser ablation. Appl. Surf. Sci. 154, 345–352.

Martellini, F., Mei, L.H.I., Balino, J.L., Carenza, M., 2003. Water and drug transport in radiation-crosslinked poly(2-methoxyethylacrylate-*co*-dimethyi acrylamide) and poly(2-methoxyethylacrylate-*co*-acrylamide) hydrogels. Radiat. Phys. Chem. 66, 155–159.

Mathew, M.E., Mohan, J.C., Manzoor, K., Nair, S.V., Tamura, H., Jayakumar, R., 2010. Folate conjugated carboxymethyl chitosan-manganese doped zinc sulphide nanoparticles for targeted drug delivery and imaging of cancer cells. Carbohydr. Polym. 80, 442–448.

Matricardi, P., Meo, C.D., Coviello, T., Alhaique, F., 2008. Recent advances and perspectives on coated alginate microspheres for modified drug delivery. Expert Opin. Drug Deliv. 5, 417–425.

Mattoussi, H., Mauro, J.M., Goldman, E.R., Anderson, G.P., Sundar, V.C., Mikulec, F.C., Bawendi, M.G., 2000. Self-assembly of CdSe-ZnS quantum dot bioconjugates using an engineered recombinant protein. J. Am. Chem. Soc. 122, 12142–12150.

McCormick, P.G., Froes, F.H., 1998. The fundamentals of mechanochemical processing. J. Miner. Met. Mater. Soc. 50, 61–65.

McDevitt, M.R., Chattopadhyay, D., Kappel, B.J., Jaggi, J.S., Schiffman, S.R., Antczak, C., Njardarson, J.T., Brentjens, R., Scheinberg, D.A., 2007. Tumor targeting with antibody functionalized, radiolabelled carbon nanotubes. J. Nucl. Med. 48, 1180–1189.

Meenach, S.A., Anderson, K.W., Hilt, J.Z., 2010. Synthesis and characterization of thermoresponsive poly(ethylene glycol)-based hydrogels and their magnetic nanocomposites. J. Polym. Sci. 48A, 3229–3235.

Mehvar, R., 2000. Dextrans for targeted and sustained delivery of therapeutic and imaging agents. J. Control. Release 69, 1–25.

Mellaerts, R.J., Van Speybroeck, J., Chen, M., Van Humbeeck, H., Augustijns, J., Van den Mooter, P.G., Martens, J., 2008. Physical state of poorly water soluble therapeutic molecules loaded into SBA-15 ordered mesoporous silica carriers: a case study with itraconazole and ibuprofen. Langmuir 24, 8651–8659.

Meng, H., Xue, M., Xia, T., Ji, Z., Tarn, D., Zink, J.I., Nel, A.E., 2011. Use of size and a copolymer design feature to improve the biodistribution and the enhanced permeability and retention effect of doxorubicin-loaded mesoporous silica nanoparticles in a murine xenograft tumor model. ACS Nano 5, 4131–4144.

Mirza, A.Z., Shamshad, H., 2011. Preparation and characterization of doxorubicin functionalized gold nanoparticles. Eur. J. Med. Chem. 46, 1857–1860.

Misra, A., Pal, R., Majumdar, S.S., Talwar, G.P., Singh, O., 1997. Biphasic testosterone delivery profile observed with two different transdermal formulations. Pharm. Res. 14, 1264–1268.

Mohanpuria, P., Rana, K.N., Yadav, S.K., 2008. Biosynthesis of nanoparticles: technological concepts and future applications. J. Nanopart. Res. 10, 507–517.

Moldovan, M., Krupp, E.M., Holliday, A.E., Donard, O.F., 2004. High resolution sector field ICP-MS and multicollector ICP-MS as tools for trace metal speciation in environmental studies: a review. J. Anal. Atomic Spectrom. 19, 815–822.

Momin, M., Pundarikakshudu, K., 2004. In vitro studies on guar gum based formulation for the colon targeted delivery of sennosides. J. Pharm. Pharm. Sci. 7, 325–331.

Monteiro, D.R., Gorup, L.F., Takamiya, A.S., Ruvollo, A.C., de Camargo, E.R., Barbosa, D.B., 2009. The growing importance of materials that prevent microbial adhesion: antimicrobial effect of medical devices containing silver. Int. J. Antimicrob. Agents 34, 103–110.

Moon, J.-W., Roh, Y., Lauf, R.J., Vali, H., Yeary, L.W., Phelps, T.J., 2007. Microbial preparation of metal-substituted magnetite nanoparticles. J. Microbiol. Methods 70, 150–158.

Mooranian, A., Negrulj, R., Mikov, M., Golocorbin-Kon, S., Arfuso, F., Al-Salami, H., 2015. Novel chenodeoxycholic acid-sodium alginate matrix in the microencapsulation of the potential antidiabetic drug, probucol: an in vitro study. J. Microencapsul. 32, 589–597.

Morones, J.R., Elechiguerra, J.L., Camacho, A., Holt, K., Kouri, J.B., Ramirez, J.T., Yacaman, M.J., 2005. The bactericidal effect of silver nanoparticles. Nanotechnology 16, 2346–2353.

Morris, G., Kök, S., Harding, S., Adams, G., 2010. Polysaccharide drug delivery systems based on pectin and chitosan. Biotechnol. Genet. Eng. Rev. 27, 257–284.

Mou, X., Ali, Z., Li, S., He, N., 2015. Applications of magnetic nanoparticles in targeted drug delivery system. J. Nanosci. Nanotechnol. 15, 54–62.

Mukherjee, S., Sushma, V., Patra, S., Barui, A.K., Bhadra, M.P., Sreedhar, B., Patra, C.R., 2012. Green chemistry approach for the synthesis and stabilization of biocompatible gold nanoparticles and their potential applications in cancer therapy. Nanotechnology 23, 455103.

Mukhopadhyay, P., Chakraborty, S., Bhattacharya, S., Mishra, R., Kundu, P.P., 2015. pH-sensitive chitosan/alginate core–shell nanoparticles for efficient and safe oral insulin delivery. Int. J. Biol. Macromol. 72, 640–648.

Müller, R., Laschober, C., Szymanski, W.W., Allmaier, G., 2007. Determination of molecular weight, particle size, and density of high number generation PAMAM dendrimers using MALDI–TOF–MS and nES–GEMMA. Macromolecules 40, 5599–5605.

Murray, C.B., Kagan, C.R., Bawendi, M.G., 2000. Synthesis and characterization of monodisperse nanocrystals and close-packed nanocrystals assemblies. Ann. Rev. Mater. Sci. 30, 545–610.

Muzzarelli, R.A.A., Muzzarelli, C., 2005. Chitosan chemistry: relevance to the biomedical sciences Polysaccharides 1: structure, characterization and use. Adv. Polym. Sci. 186, 151–209.

Nabiev, I., Chourpa, I., Manfait, M., 1994. Comparative studies of antitumor DNA intercalating agents, aclacinomycin and saintopin, by means of surface-enhanced Raman scattering spectroscopy. J. Phys. Chem. 98, 1344–1350.

Nair, L.S., Laurencin, C.T., 2007a. Biodegradable polymers as biomaterial. Progr. Polym. Sci. 6, 762–798.

Nair, L.S., Laurencin, C.T., 2007b. Silver nanoparticles: synthesis and therapeutic applications. J. Biomed. Nanotechnol. 3, 301–316.

Namazi, H., Adeli, M., 2005. Dendrimers of citric acid and poly (ethylene glycol) as the new drug-delivery agents. Biomaterials 26, 1175–1183.

Nayak, A.K., Pal, D., 2011. Development of pH-sensitive tamarind seed polysaccharide-alginate composite beads for controlled diclofenac sodium delivery using response surface methodology. Int. J. Biol. Macromol. 49, 784–793.

Naylor, A.M., Goddard, III, W.A., Kiefer, G.E., Tomalia, D.A., 1989. Starburst dendrimers. 5. Molecular shape control. J. Am. Chem. Soc. 111, 2339–2341.

Nejati-Koshki, K., Akbarzadeh, A., Pourhasan-Moghadam, M., Abhari, A., Dariushnejad, H., 2013. Inhibition of leptin and leptin receptor gene expression by silibinin–curcumin combination. Asian Pacific J. Cancer Prev. 14, 6595–6599.

Nelms, S.M., 2005. Inductively Coupled Plasma Mass Spectrometry Handbook. Blackwell Publishing Ltd., Oxford, UK.

Newkome, G.R., Moorefield, C.N., Baker, G.R., Saunders, M.J., Grossman, S.H., 1991. Unimolecular micelles. Angew. Chem. Int. Ed. Engl. 30, 1178–1180.

Newkome, G.R., Moorefield, C.N., Vögtle, F., 2001. Dendrimers and Dendrons: Concepts, Syntheses, Applications. WILEY-VCH Verlag GmbH, Weinheim, FRG.

Niemirowicz, K., Prokop, I., Wilczewska, A.Z., Wnorowska, U., Piktel, E., Wątek, M., Savage, P.B., Bucki, R., 2015a. Magnetic nanoparticles enhance the anticancer activity of cathelicidin LL-37 peptide against colon cancer cells. Int. J. Nanomed. 10, 3843–3853.

Niemirowicz, K., Surel, U., Wilczewska, A.Z., Mystkowska, J., Piktel, E., Gu, X., Namiot, Z., Kułakowska, A., Savage, P.B., Bucki, R., 2015b. Bactericidal activity and biocompatibility of ceragenin-coated magnetic nanoparticles. J. Nanobiotechnol. 13, 32.

Nishiyama, N., Bae, Y., Miyata, K., Fukushima, S., Kataoka, K., 2005. Smart polymeric micelles for gene and drug delivery. Drug Discov. Today 2, 21–26.

Niu, G., Djaoui, A.B., Cohn, D., 2011. Crosslinkable PEO–PPO–PEO triblocks as building blocks of thermo-responsive nanoshells. Polymer 52, 2524–2530.

Noimark, S., Page, K., Bear, J.C., Sotelo-Vazquez, C., Quesada-Cabrera, R., Lu, Y., Allan, E., Darr, J.A., Parkin, I.P., 2014. Functionalized gold and titania nanoparticles and surfaces for use as antimicrobial coatings. Faraday Discuss. 175, 273–287.

Nuttelman, C.R., Mortisen, D.T., Henery, S.M., Anseth, K.S., 2001. Attachment of fibronection to poly (vinyl alcohol) hydrogels promotes NIH3T3 cell adhesion, proliferation, and migration. J. Biomed. Mater. Res. 52, 217–223.

Okamatsu, M., Feng, F., Ohyanagi, T., Nagahori, N., Someya, K., Sakoda, Y., Miura, N., Nishimura, S., Kida, H., 2013. Fluorescence polarization-based assay using *N*-glycan-conjugated quantum dots for screening in hemagglutinin blockers for influenza A viruses. J. Virol. Methods 187, 390–394.

O'Keefe, M.A., Buseck, P.R., Iijima, S., 1978. Computed crystal structure images for high resolution electron microscopy. Nature 274, 322–324.

Okuyama, Y., Yoshida, R., Sakai, K., Okano, T., Sakurai, Y., 1993. Swelling controlled zero-order and sigmoidal drug release from thermoresponsive poly(*N*-isopropylacrylamide-*co*-butyl methacrylate) hydrogel. J. Biomater. Sci. Polym. Ed. 4, 545–556.

Oliveira, J.M., Salgado, A.J., Sousa, N., Mano, J.F., Reis, R.L., 2010. Dendrimers and derivatives as a potential therapeutic tool in regenerative medicine strategies: a review. Progr. Polym. Sci. 35, 1163–1194.

Oliveira, M., Ugarte, D., Zanchet, D., Zarbin, A., 2005. Influence of synthetic parameters on the size, structure and stability of dodecanethiol-stabilized silver nanoparticles. J. Colloid Interface Sci. 292, 429–435.

Osmałek, T., Froelich, A., Tasarek, S., 2014. Application of gellan gum in pharmacy and medicine. Int. J. Pharm. 466, 328–340.

Osuna, J., de Caro, D., Amiens, C., Chaudret, B., Snoeck, E., Respaud, M., Broto, J.M., Fert, A., 1996. Synthesis, characterization, and magnetic properties of cobalt nanoparticles from an organometallic precursor. J. Phys. Chem. 10, 14571–14574.

Özcan, I., Azizoğlu, E., Senyiğit, T., Özyazıcı, M., Özer, Ö., 2013a. Enhanced dermal delivery of diflucortolone valerate using lecithin/chitosan nanoparticles: in-vitro and in-vivo evaluations. Int. J. Nanomed. 8, 461–475.

Özcan, I., Azizoğlu, E., Senyiğit, T., Özyazici, M., Özer, Ö., 2013b. Comparison of PLGA and lecithin/chitosan nanoparticles for dermal targeting of betamethasone valerate. J. Drug Target. 21, 542–550.

Paharia, A., Yadav, A.K., Rai, G., Jain, S.K., Pancholi, S.S., Agrawal, G.P., 2007. Eudragit-coated pectin microspheres of 5-fluorouracil for colon targeting. AAPS PharmSciTech. 8, E87–E93.

Paliwal, R., Paliwal, S.R., Agrawal, G.P., Vyas, S.P., 2012. Chitosan nanoconstructs for improved oral delivery of low molecular weight heparin: in vitro and in vivo evaluation. Int. J. Pharm 422, 179–184.

Pantarotto, D., Briand, J.P., Prato, M., Bianco, A., 2004. Translocation of bioactive peptides across cell membranes by carbon nanotubes. Chem. Commun. 2004, 16–17.

Papaphilippou, P., Loizou, L., Popa, N.C., Han, A., Vekas, L., Odysseos, A., Krasia-Christoforou, T., 2009. Superparamagnetic hybrid micelles, based on iron oxide nanoparticles and well-defined diblock copolymers possessing β-ketoester functionalities. Biomacromolecules 10, 2662–2671.

Papaphilippou, P.C., Pourgouris, A., Marinica, O., Taculescu, A., Athanasopoulos, G.I., Vekas, L., Krasia-Christoforou, T., 2011. Fabrication and characterization of superparamagnetic and thermoresponsive hydrogels based on oleic-acid-coated Fe_3O_4 nanoparticles, hexa(ethylene glycol) methyl ether methacrylate and 2-(acetoacetoxy)ethyl methacrylate. J. Magn. Magn. Mater. 323, 557–563.

Parikh, D.V., Fink, T., Rajasekharan, K., Sachinvala, N.D., Sawhney, A.P.S., Calamari, T.A., Parikh, A.D., 2005. Antimicrobial silver/sodium carboxymethyl cotton dressings for burn wounds. Text. Res. J. 75, 134–138.

Park, J.H., von Maltzahn, G., Ruoslahti, E., Bhatia, S.N., Sailor, M.J., 2008. Micellar hybrid nanoparticles for simultaneous magnetofluorescent imaging and drug delivery. Angew. Chem. Int. Ed. 47, 7284–7288.

Paulo, C.S.O., Vidal, M., Ferreira, L.S., 2010. Antifungal nanoparticles and surfaces. Biomacromolecules 11, 2810–2817.

Pender, D.S., Vagala, L.M., Badwaik, V.D., Thompson, H., Paripelly, R., Dakshinamurthy, R., 2013. A new class of gold nanoantibiotics: direct coating of ampicillin on gold nanoparticles. Pharm. Nanotechnol. 1, 126–135.

Perera, G., Barthelmes, J., Bernkop-Schnürch, A., 2010. Novel pectin-4-aminothiophenole conjugate microparticles for colon-specific drug delivery. J. Control. Release 145, 240–246.

Pillay, V., Dangor, C.M., Govender, T., Moopanar, K.R., Hurbans, N., 1998. Ionotropic gelation: encapsulation of indomethacin in calcium alginate gel discs. J. Microencapsul. 15, 215–226.

Pinaud, F., King, D., Moore, H.-P., Weiss, S., 2004. Bioactivation and cell targeting of semiconductor CdSe/ZnS nanocrystals with phytochelatin-related peptides. J. Am. Chem. Soc. 126, 6115–6123.

Pliszczak, D., Bourgeois, S., Bordes, C., Valour, J.P., Mazoyer, M.A., Orecchioni, A.M., Nakache, E., Lanteri, P., 2011. Improvement of an encapsulation process for the preparation of pro- and prebiotics-loaded bioadhesive microparticles by using experimental design. Eur. J. Pharm. Sci. 44, 83–92.

Pooja, D., Panyaram, S., Kulhari, H., Rachamalla, S.S., Sistla, R., 2014. Xanthan gum stabilized gold nanoparticles: characterization, biocompatibility, stability and cytotoxicity. Carbohydr. Polym. 110, 1–9.

Pooja, D., Panyaram, S., Kulhari, H., Reddy, B., Rachamalla, S.S., Sistla, R., 2015. Natural polysaccharide functionalized gold nanoparticles as biocompatible drug delivery carrier. Int. J. Biol. Macromol. 80, 48–56.

Prabaharan, M., 2011. Prospective of guar gum and its derivatives as controlled drug delivery systems. Int. J. Biol. Macromol. 49, 117–124.

Prabaharan, M., Grailer, J.J., Pilla, S., Steeber, D.A., Gong, S., 2009a. Folate-conjugated amphiphilic hyperbranched block copolymers based on Boltorn H40, poly(L-lactide) and poly(ethylene glycol) for tumor-targeted drug delivery. Biomaterials 30, 3009–3019.

Prabaharan, M., Grailer, J.J., Pilla, S., Steeber, D.A., Gong, S., 2009b. Gold nanoparticles with a monolayer of doxorubicin-conjugated amphiphilic block copolymer for tumor-targeted drug delivery. Biomaterials 30, 6065–6075.

Prajapati, V.K., Awasthi, K., Gautam, S., Yadav, T.P., Rai, M., Srivastava, O.N., Sundar, S., 2011. Targeted killing of *Leishmania donovani* in vivo and in vitro with amphotericin B attached to functionalized carbon nanotubes. J. Antimicrob. Chemother. 66, 874–879.

Prajapati, V.K., Awasthi, K., Yadav, T.P., Rai, M., Srivastava, O.N., Sundar, S., 2012. An oral formulation of amphotericin B attached to functionalized carbon nanotubes is an effective treatment for experimental visceral leishmaniasis. J. Infect. Dis. 205, 333–336.

Prato, M., Kostarelos, K., Bianco, A., 2008. Functionalized carbon nanotubes in drug design and discovery. Acc. Chem. Res. 41, 60–68.

Prokopovich, P., Leech, R., Carmalt, C.J., Parkin, I.P., Perni, S., 2013. A novel bone cement impregnated with silver: tiopronin nanoparticles: its antimicrobial, cytotoxic, and mechanical properties. Int. J. Nanomed. 8, 2227–2237.

Pruthi, J., Mehra, N.K., Jain, N.K., 2012. Macrophages targeting of amphotericin B through mannosylated multiwalled carbon nanotubes. J. Drug Target. 20, 593–604.

Puntes, V.F., Krishnan, K.M., Alivisatos, A.P., 1999. Colloidal nanocrystal shape and size control: the case of cobalt. Science 291, 2115–2117.

Puntes, V.F., Krishnan, K.M., Alivisatos, A.P., 2002. Synthesis of colloidal cobalt nanoparticles with controlled size and shapes. Top. Catal. 19, 145–148.

Qi, L., Gao, X., 2008. Emerging application of quantum dots for drug delivery and therapy. Expert Opin. Drug Deliv. 5, 63–67.

Qian, X., Peng, X.H., Ansari, D.O., Yin-Goen, Q., Chen, G.Z., Shin, D.M., Yang, L., Young, A.N., Wang, M.D., Nie, S., 2008. In vivo tumor targeting and spectroscopic detection with surface-enhanced Raman nanoparticle tags. Nat. Biotechnol. 26, 83–90.

Qureshi, A., Roci, I., Gurbuz, Y., Niazi, J.H., 2012. An aptamer based competition assay for protein detection using CNT activated gold-interdigitated capacitor arrays. Biosens. Bioelectron. 34, 165–170.

Ramezani, N., Ehsanfar, N., Shamsa, F., Amin, G., Shahverdi, H.R., Esfahani, H.M., Shamsaie, A., Bazaz, R.D., Shahverdi, A.R., 2008. Screening of medicinal plant methanol extracts for the synthesis of gold nanoparticles by their reducing potential. Z. Naturforsch. 63B, 903–908.

Ravindra, S., Murali Mohan, Y., Reddy, N.N., Raju, K.M., 2010. Fabrication of antibacterial cotton fibres loaded with silver nanoparticles via "green approach". Colloids Surf. A 367, 31–40.

Ray, S., Banerjee, S., Maiti, S., Laha, B., Barik, S., Sa, B., Bhattacharyya, U.K., 2010. Novel interpenetrating network microspheres of xanthan gum-poly(vinyl alcohol) for the delivery of diclofenac sodium to the intestine: in vitro and in vivo evaluation. Drug Deliv. 17, 508–519.

Real, C., Alteala, M.D., Criado, J.M., 1996. Preparation of silica from rice husk. J. Am. Ceramic Soc. 79, 2012–2016.

Reetz, M.T., Helbig, W., 1994. Size-selective synthesis of nanostructured transition metal clusters. J. Am. Chem. Soc. 116, 7401–7402.

Rehm, B.H.A., 2009. Alginates: Biology and Applications. Springer-Verlag, Berlin, Heidelberg.

Reis, C.P., Ribeiro, A.J., Houng, S., Veiga, F., Neufeld, R.J., 2007. Nanoparticulate delivery system for insulin: design, characterization and in vitro/in vivo bioactivity. Eur. J. Pharm. Sci. 30, 392–397.

Rijcken, C.J.F., Soga, O., Hennink, W.E., van Nostrum, C.F., 2007. Triggered destabilisation of polymeric micelles and vesicles by changing polymers polarity: an attractive tool for drug delivery. J. Control. Release 120, 131–148.

Rincon, A.C., Molina-Martinez, I.T., de Las Heras, B., Alonso, M., Bailez, C., Rodriguez-Cabello, J.C., Herrero-Vanrell, R., 2006. Biocompatibility of elastin-like polymer poly(vpavg) microparticles: in vitro and in vivo studies. J. Biomed. Mater. Res. 78A, 343–351.

Roe, D., Karandikar, B., Bonn-Savage, N., Gibbins, B., Roullet, J.B., 2008. Antimicrobial surface functionalization of plastic catheters by silver nanoparticles. J. Antimicrob. Chemother. 61, 869–876.

Rogers, J.V., Parkinson, C.V., Choi, Y.W., Speshock, J.L., Hussain, S.M., 2008. A preliminary assessment of silver nanoparticles inhibition of monkey pox virus plaque formation. Nanoscale Res. Lett. 3, 129–133.

Roh, Y., Vali, H., Phelps, T.J., Moon, J.-W., 2006. Extracellular synthesis of magnetite and metal-substituted magnetite nanoparticles. J. Nanosci. Nanotechnol. 6, 3517–3520.

Ruan, G., Feng, S.S., 2003. Preparation and characterization of poly(lactic acid)-poly(ethylene glycol)-poly(lactic acid) (PLA-PEG-PLA) microspheres for controlled release of paclitaxel. Biomaterials 24, 5037–5044.

Rupp, R., Rosenthal, S.L., Stanberry, L.R., 2007. VivaGel (SPL7013 Gel): a candidate dendrimer–microbicide for the prevention of HIV and HSV infection. Int. J. Nanomed. 2, 561–566.

Russ, J.C., 1984. Fundamentals of Energy Dispersive X-ray Analysis. Butterworth-Heinemann, London.

Rzhepishevska, O., Hakobyan, S., Ruhal, R., Gautrot, J., Barbero, D., Ramstedt, M., 2013. The surface charge of anti-bacterial coatings alters motility and biofilm architecture. Biomater. Sci. 1, 589–602.

Saha, B., 2007. In vitro structural and functional evaluation of gold nanoparticles conjugated antibiotics. Nanoscale Res. Lett. 2, 614–622.

Sahoo, B., Devi, K.S.P., Sahu, S.K., Nayak, S., Maiti, T.K., Dhara, D., Pramanik, P., 2013. Facile preparation of multifunctional hollow silica nanoparticles and their cancer specific targeting effect. Biomater. Sci. 1, 647–657.

Sahu, S.K., Maiti, S., Pramanik, A., Ghosh, S.K., Pramanik, P., 2012. Controlling the thickness of polymeric shell on magnetic nanoparticles loaded with doxorubicin for targeted delivery and MRI contrast agent. Carbohydr. Polym. 87, 2593–2604.

Sanderson, G.R., Clark, R.C., 1983. Laboratory-produced microbial polysaccharide has many potential food applications as a gelling, stabilizing and texturizing agent. Food Technol. 37, 63–70.

Sarmah, J.K., Bhattacharjee, S.K., Mahanta, R., Mahanta, R., 2009. Preparation of cross-linked guar gum nanospheres containing tamoxifen citrate by single step emulsion in situ polymer cross-linking method. J. Incl. Phenomena Macrocyclic Chem. 65, 329–334.

Sarmah, J.K., Mahanta, R., Bhattacharjee, S.K., Mahanta, R., Biswas, A., 2011. Controlled release of tamoxifen citrate encapsulated in cross-linked guar gum nanoparticles. Int. J. Biol. Macromol. 49, 390–396.

Sarmah, J.K., Bhattacharjee, S.K., Roy, S., Mahanta, R., Mahanta, R., 2014. Biodegradable guar gum nanoparticles as carrier for tamoxifen citrate in treatment of breast cancer. J. Biomater. Nanobiotechnol. 5, 220–228.

Saudagar, P., Dubey, V.K., 2014. Carbon nanotube based betulin formulation shows better efficacy against *Leishmania* parasite. Parasitol. Int. 63, 772–776.

Savla, R., Taratula, O., Garbuzenko, O., Minko, T., 2011. Tumor targeted quantum dot-mucin 1 aptamer-doxorubicin conjugate for imaging and treatment of cancer. J. Control. Release 153, 16–22.
Schmaljohann, D., 2006. Thermo- and pH-responsive polymers in drug delivery. Adv. Drug Deliv. Rev. 58, 1655–1670.
Scott, B.J., Wirnsberger, G., Stucky, G.D., 2001. Mesoporous and mesostructured materials for optical applications. Chem. Mater. 13, 3140–3150.
Secinti, K.D., Özalp, H., Attar, A., Sargon, M.F., 2011. Nanoparticle silver ion coatings inhibit biofilm formation on titanium implants. J. Clin. Neurosci. 18, 391–395.
Selvakumar, R., Aravind, S., Anuradha, M.A., Balachandran, Y.L., 2013. A facile synthesis of silver nanoparticle with SERS and antimicrobial activity using *Bacillus subtilis* exopolysaccharides. J. Exp. Nanosci. 9, 1075–1087.
Semdé, R., Amighi, K., Devleeschouwer, M.J., Moës, A.J., 2000. Effect of pectinolytic enzymes on the theophylline release from pellets coated with water insoluble polymers containing pectin HM or calcium pectinate. Int. J. Pharm. 197, 169–179.
Senyiğit, T., Sonvico, F., Barbieri, S., Ozer, O., Santi, P., Colombo, P., 2010. Lecithin/chitosan nanoparticles of clobetasol-17-propionate capable of accumulation in pig skin. J. Control. Release 142, 368–373.
Shahverdi, A.R., Fakhimi, A., Shahverdi, H.R., Minaian, S., 2007. Synthesis and effect of silver nanoparticles on the antibacterial activity of different antibiotics against *Staphylococcus aureus* and *Escherichia coli*. Nanomed. Nanotechnol. Biol. Med. 3, 168–171.
Shan, Y., Wang, L., Shi, Y., Zhang, H., Li, H., Liu, H.I., Yang, B., Li, T., Fang, X., Li, W., 2008. NHS-mediated QDs-peptide/protein conjugation and its application for cell labeling. Talanta 75, 1008–1014.
Sharma, V.K., Yngard, R.A., Liu, Y., 2009. Silver nanoparticles: green synthesis and their antimicrobial activities. Adv. Colloid Interface Sci. 145, 83–96.
Shi, J., Liu, L., Sun, X., Cao, S., Mano, J.F., 2008. Biomineralized polysaccharide beads for dual-stimuli-responsive drug delivery. Macromol. Biosci. 8, 260–267.
Shi, X., Thomas, T.P., Myc, L.A., Kotlyar, A., Baker, Jr., J.R., 2007. Synthesis, characterization, and intracellular uptake of carboxyl-terminated poly(amidoamine) dendrimer-stabilized iron oxide nanoparticles. Phys. Chem. Chem. Phys. 9, 5712–5720.
Silverstein, R., Webster, F., Kiemle, D.J., 2005. Spectrometric Identification of Organic Compounds, seventh ed. John Wiley & Sons, New York.
Singh, V., Singh, S.K., Pandey, S., Sanghi, R., 2011. Synthesis and characterization of guar gum templated hybrid nano silica. Int. J. Biol. Macromol. 49, 233–240.
Sinha, V.R., Kumria, R., 2001. Polysaccharides in colon-specific drug delivery. Int. J. Pharm. 224, 19–38.
Sinha, V.R., Singla, A.K., Wadhawan, S., Kaushik, R., Kumria, R., Bansal, K., Dhawan, S., 2004. Chitosan microspheres as a potential carrier for drugs. Int. J. Pharm. 274, 1–33.
Smith, A.M., Dave, S., Nie, S., True, L., Gao, X., 2006. Multicolor quantum dots for molecular diagnostics of cancer. Expert Rev. Mol. Diag. 6, 231–244.
Smith, A.M., Duan, H., Mohs, A.M., Nie, S., 2008. Bioconjugated quantum dots for in vivo molecular and cellular imaging. Adv. Drug Deliv. Rev. 60, 1226–1240.
Smith, P.B., Martin, S.J., Hall, M.J., Tomalia, D.A.A., 1987. Characterization of the structure and synthetic reactions of polyamidoamine "STARBURST" polymers. In: Mitchell, Jr., J. (Ed.), Applied Polymer Analysis Characterization. Hanser, Munich, pp. 357–385.
Sobhani, Z., Dinarvand, R., Atyabi, F., Ghahremani, M., Adeli, M., 2011. Increased paclitaxel cytotoxicity against cancer cell lines using a novel functionalized carbon nanotube. Int. J. Nanomed. 6, 705–719.
Song, J.Y., Kim, B.S., 2008. Rapid biological synthesis of silver nanoparticles using plant leaf extracts. Bioproc. Biosyst. Eng. 32, 79–84.
Song, S., Chen, F., Qi, H., Li, F., Xin, T., Xu, J., Ye, T., Sheng, N., Yang, X., Pan, W., 2014. Multifunctional tumor-targeting nanocarriers based on hyaluronic acid-mediated and pH-sensitive properties for efficient delivery of docetaxel. Pharm. Res. 31, 1032–1045.

Song, S.W., Hidajat, K., Kawi, S., 2005. Functionalized SBA-15 materials as carriers for controlled drug delivery: influence of surface properties on matrix-drug interactions. Langmuir 21, 9568–9575.

Sonoda, A., Nitta, N., Nitta-Seko, A., Ohta, S., Takamatsu, S., Ikehata, Y., Nagano, I., Jo, J., Tabata, Y., Takahashi, M., Matsui, O., Murata, K., 2010. Complex comprised of dextran magnetite and conjugated cisplatin exhibiting selective hyperthermic and controlled-release potential. Int. J. Nanomed. 5, 499–504.

Soppimath, K.S., Tan, D.C.-W., Yang, Y.-Y., 2005. pH-triggered thermally responsive polymer core–shell nanoparticles for drug delivery. Adv. Mater. 17, 318–323.

Soumya, R.S., Ghosh, S., Abraham, E.T., 2010. Preparation and characterization of guar gum nanoparticles. Int. J. Biol. Macromol. 46, 267–269.

Sreenivasan, V.K., Kim, E.J., Goodchild, A.K., Connor, M., Zvyagin, A.V., 2012. Targeting somatostatin receptors using in situ-bioconjugated fluorescent nanoparticles. Nanomedicine 7, 1551–1560.

Sriamornsak, P., 2011. Application of pectin in oral drug delivery. Expert Opin. Drug Deliv. 8, 1009–1023.

Stober, W., Fink, A., Bohn, E., 1968. Controlled growth of monodisperse silica spheres in the micron size range. J. Colloid Interface Sci. 26, 62–69.

Suryanarayana, C., 2001. Mechanical alloying and milling. Progr. Mater. Sci. 46, 1–184.

Svenson, S., Tomalia, D.A., 2005. Dendrimers in biomedical applications: reflections on the field. Adv. Drug Deliv. Rev. 57, 2106–2129.

Szczubialka, K., Nowakowska, M., 2003. Response of micelles formed by smart terpolymers to stimuli studied by dynamic light scattering. Polymer 44, 5269–5274.

Szlek, J., Paclawski, A., Lau, R., Jachowicz, R., Mendyk, A., 2013. Heuristic modeling of macromolecule release from PLGA microspheres. Int. J. Nanomed. 8, 4601–4611.

Tadayon, A., Jamshidi, R., Esmaeili, A., 2015. Delivery of tissue plasminogen activator and streptokinase magnetic nanoparticles to target vascular diseases. Int. J. Pharm. 495 (1), 428–438.

Tahtat, D., Mahlous, M., Benamer, S., Khodja, A.N., Oussedik-Oumehdi, H., Laraba-Djebari, F., 2013. Oral delivery of insulin from alginate/chitosan crosslinked by glutaraldehyde. Int. J. Biol. Macromol. 58, 160–168.

Tan, Q., Liu, W., Guo, C., Zhai, G., 2011. Preparation and evaluation of quercetin-loaded lecithin-chitosan nanoparticles for topical delivery. Int. J. Nanomed. 6, 1621–1630.

Tang, H., Zhang, P., Kieft, T.L., Ryan, S.J., Baker, S.M., Wiesmann, W.P., Rogelj, S., 2010. Antibacterial action of a novel functionalized chitosan arginine against Gram negative bacteria. Acta Biomater. 6, 2562–2571.

Tasis, D., Tagmatarchis, N., Georgakilas, V., Prato, M., 2003. Soluble carbon nanotubes. Chem. Eur. J. 9, 4000–4008.

Tasis, D., Tagmatarchis, N., Bianco, A., Prato, M., 2006. Chemistry of carbon nanotubes. Chem. Rev. 106, 1105–1136.

Thach, C.V., Hai, N.H., Chau, N., 2008. Size controlled magnetite nanoparticles and their drug loading ability. J. Korean Phys. Soc. 52, 1332–1335.

Thudaji, N., Nuntiya, A., 2008. Preparation of nanosilica powder from rice husk ash by precipitation method. Chiang Mai J. Sci. 35, 206–211.

Tian, B., Al-Jamal, W.T., Al-Jamal, K.T., Kostarelos, K., 2011. Doxorubicin-loaded lipid-quantum dot hybrids: surface topography and release properties. Int. J. Pharm. 416, 443–447.

Tiraferri, A., Chen, K.L., Sethi, R., Elimelech, M., 2008. Reduced aggregation and sedimentation of zero-valent iron nanoparticles in the presence of guar gum. J. Colloid Interface Sci. 324, 71–79.

Tiwari, P.M., Vig, K., Dennis, V.A., Singh, S.R., 2011. Functionalized gold nanoparticles and their biomedical applications. Nanomaterials 1, 31–63.

Tiyaboonchai, W., Limpeanchob, N., 2007. Formulation and characterization of amphotericin B-chitosan-dextran sulfate nanoparticles. Int. J. Pharm. 329, 142–149.

Tiyaboonchai, W., Woiszwillo, J., Sims, R.C., Middaugh, C.R., 2003. Insulin containing polyethylenimine-dextran sulfate nanoparticles. Int. J. Pharm. 255, 139–151.

Tomalia, D.A., 2005. Birth of a new macromolecular architecture: dendrimers as quantized building blocks for nanoscale synthetic polymer chemistry. Progr. Polym. Sci. 30, 294–324.

Tomalia, D.A., Baker, H., Dewald, J., Hall, M., Kallos, G., Martin, S., Roeck, J., Ryder, J., Smith, P., 1985. A new class of polymers Starburst-dendritic macromolecules. Polym. J. 7, 117–132.

Tomalia, D.A., Durst, H.D., 1993. Supramolecular chemistry I: directed synthesis and molecular recognition. Top. Curr. Chem. 165, 193–313.

Tomalia, D.A., Naylor, M., Goddard, III, W.A., 1990. Starburst dendrimers: molecular-level control of size, shape, surface chemistry, topology, and flexibility from atoms to macroscopic matter. Angew. Chem. Int. Ed. Engl. 29, 138–175.

Toshima, N., Yonezawab, T., 1998. Bimetallic nanoparticles novel materials for chemical and physical applications. New J. Chem. 22, 1179–1201.

Tran, P.A., Zhang, L., Webster, T.J., 2009. Carbon nanofibers and carbon nanotubes in regenerative medicine. Adv. Drug Deliv. Rev. 61, 1097–1114.

Tsai, S.W., Yu, D.S., Tsao, S.W., Hsu, F.Y., 2013. Hyaluronan-cisplatin conjugate nanoparticles embedded in Eudragit S100-coated pectin/alginate microbeads for colon drug delivery. Int. J. Nanomed. 8, 2399–2407.

Turkevich, J., Stevenson, P.C., Hillier, J., 1951. A study of the nucleation and growth process in the synthesis of colloidal gold. Discuss. Faraday Soc. 11, 55–75.

Turro, N.J., Barton, J.K., Tomalia, D.A., 1991. Molecular recognition and chemistry in restricted reaction spaces: photophysics and photoinduced electron transfer on the surfaces of micelles, dendrimers, and DNA. Acc. Chem. Res. 24, 332–340.

Umamaheshwari, R.B., Jain, N.K., 2003. Receptor mediated targeting of lectin conjugated gliadin nanoparticles in the treatment of *Helicobacter pylori*. J. Drug Target. 11, 415–423.

Unsoy, G., Khodadust, R., Yalcin, S., Mutlu, P., Gunduz, U., 2014a. Synthesis of doxorubicin loaded magnetic chitosan nanoparticles for pH responsive targeted drug delivery. Eur. J. Pharm. Sci. 62, 243–250.

Unsoy, G., Yalcin, S., Khodadust, R., Mutlu, P., Onguru, O., Gunduz, U., 2014b. Chitosan magnetic nanoparticles for pH responsive Bortezomib release in cancer therapy. Biomed. Pharmacother. 68, 641–648.

Unterweger, H., Tietze, R., Janko, C., Zaloga, J., Lyer, S., Dürr, S., Taccardi, N., Goudouri, O.M., Hoppe, A., Eberbeck, D., Schubert, D.W., Boccaccini, A.R., Alexiou, C., 2014. Development and characterization of magnetic iron oxide nanoparticles with a cisplatin-bearing polymer coating for targeted drug delivery. Int. J. Nanomed. 9, 3659–3676.

Vaidya, A., Jain, A., Khare, P., Agrawal, R.K., Jain, S.K., 2009. Metronidazole loaded pectin microspheres for colon targeting. J. Pharm. Sci. 98, 4229–4236.

Vallet-Regi, M., Ramila, A., del Real, R.P., Perez-Pariente, J., 2001. A new property of MCM-41: drug delivery system. Chem. Mater. 13, 308–311.

Varshosaz, J., 2012. Dextran conjugates in drug delivery. Expert Opin. Drug Deliv. 9, 509–523.

Varshosaz, J., Emami, J., Tavakoli, N., Fassihi, A., Minaiyan, M., Ahmadi, F., Dorkoosh, F., 2009. Synthesis and evaluation of dextran-budesonide conjugates as colon specific prodrugs for treatment of ulcerative colitis. Int. J. Pharm. 365, 69–76.

Varum, F.J., McConnell, E.L., Sousa, J.J., Veiga, F., Basit, A.W., 2008. Mucoadhesion and the gastrointestinal tract. Crit. Rev. Ther. Drug Carrier Syst. 25, 207–258.

Veiseh, O., Gunn, J.W., Zhang, M., 2010. Design and fabrication of magnetic nanoparticles for targeted drug delivery and imaging. Adv. Drug Deliv. Rev. 62, 284–304.

Venkatpurwar, V., Pokharkar, V., 2011. Green synthesis of silver nanoparticles using marine polysaccharide: study of *in-vitro* antibacterial activity. Mater. Lett. 65, 999–1002.

Venkatpurwar, V., Shiras, A., Pokharkar, V., 2011. Porphyran capped gold nanoparticles as a novel carrier for delivery of anticancer drug: in vitro cytotoxicity study. Int. J. Pharm. 409, 314–320.

Verelst, M., Ely, T.O., Amiens, C., Snoeck, E., Lecante, P., Mosset, A., Respaud, M., Broto, J.M., Chaudret, B., 1999. Synthesis and characterization of CoO, Co_3O_4, and mixed Co/CoO nanoparticles. Chem. Mater. 11, 2702–2708.

Vermehren-Schmaedick, A., Krueger, W., Jacob, T., Ramunno-Johnson, D., Balkowiec, A., Lidke, K.A., Vu, T.Q., 2014. Heterogeneous intracellular trafficking dynamics of brain-derived neurotrophic factor complexes in the neuronal soma revealed by single quantum dot tracking. PLoS One 9, e95113.

Vimala, K., Varaprasad, K., Sadiku, R., Ramam, K., Kanny, K., 2014. Development of novel protein-Ag nanocomposites for drug delivery and inactivation of bacterial applications. Int. J. Biol. Macromol. 63, 75–82.

Vittorio, O., Cirillo, G., Iemma, F., Di Turi, G., Jacchetti, E., Curcio, M., Barbuti, S., Funel, N., Parisi, O.I., Puoci, F., et al., 2012. Dextran-catechin conjugate: a potential treatment against the pancreatic ductal adenocarcinoma. Pharm. Res. 29, 2601–2614.

Vittorio, O., Voliani, V., Faraci, P., Karmakar, B., Iemma, F., Hampel, S., Kavallaris, M., Cirillo, G., 2014. Magnetic catechin-dextran conjugate as targeted therapeutic for pancreatic tumour cells. J. Drug Target. 22, 408–415.

Vivero-Escoto, J.L., Slowing, I.I., Trewyn, B.G., Lin, V.S-.Y., 2010. Mesoporous silica nanoparticles for intracellular controlled drug delivery. Small 6, 1952–1967.

Voicu, G., Grumezescu, V., Andronescu, E., Grumezescu, A.M., Ficai, A., Ficai, D., Ghitulica, C.D., Gheorghe, I., Chifiriuc, M.C., 2013. Caprolactam-silica network, a strong potentiator of the antimicrobial activity of kanamycin against gram-positive and Gram-negative bacterial strains. Int. J. Pharm. 446, 63–69.

Vyas, S., Trivedi, P., Chaturvedi, S.C., 2007. Ketorolac-dextran conjugates: synthesis, in vitro and in vivo evaluation. Acta Pharm. 57, 441–450.

Vyas, S., Trivedi, P., Chaturvedi, S.C., 2009. Dextran-etodolac conjugates: synthesis, in vitro and in vivo evaluation. Acta Pol. Pharm. 66, 201–206.

Wady, A.F., Machado, A.L., Zucolotto, V., Zamperini, C.A., Berni, E., Vergani, C.E., 2012. Evaluation of *Candida albicans* adhesion and biofilm formation on a denture base acrylic resin containing silver nanoparticles. J. Appl. Microbiol. 112, 1163–1172.

Wang, C.H., Hsu, Y.S., Peng, C.A., 2008. Quantum dots encapsulated with amphiphilic alginate as bioprobe for fast screening anti-dengue virus agents. Biosens. Bioelectron. 24, 1018–1025.

Wang, J., Wang, M., Zheng, M., Guo, Q., Wang, Y., Wang, H., Xie, X., Huang, F., Gong, R., 2015. Folate mediated self-assembled phytosterol-alginate nanoparticles for targeted intracellular anticancer drug delivery. Colloids Surf. B 129, 63–70.

Ward, M.A., Georgiou, T.K., 2011. Thermoresponsive polymers for biomedical applications. Polymers 3, 1215–1242.

Warheit, D.B., Laurence, B.R., Reed, K.L., Roach, D.H., Reynolds, G.A.M., Webb, T.R., 2004. Comparative pulmonary toxicity assessment of single-wall carbon nanotubes in rats. Toxicol. Sci. 77, 117–125.

Watkins, D.M., Sayed-Sweet, Y., Klimash, J.W., Turro, N.J., Tomalia, D., 1997. Dendrimers with hydrophobic cores and the formation of supramolecular dendrimer-surfactant assemblies. Langmuir 13, 3136–3141.

Wei, H., Qing, D., De-Ying, C., Bai, X., Fang, F., 2007. Pectin/ethylcellulose as film coatings for colon-specific drug delivery: preparation and in vitro evaluation using 5-fluorouracil pellets. PDA J. Pharm. Sci. Technol. 61, 121–130.

Wel, H., Zhang, X.Z., Zhou, Y., Chang, S.X., Zhuo, R.X., 2006. Self-assembled thermoresponsive micelles of poly(*N*-isopropylacrylamide-*b*-methyl methacrylate). Biomaterials 27, 2028–2034.

Williams, D.B., Carter, C.B., 1996. Transmission Electron Microscopy: A Text Book for Material Science. Plenum Press, New York.

Wittayaareekul, S., Kruenate, J., Prahsarn, C., 2006. Preparation and in vitro evaluation of mucoadhesive properties of alginate/chitosan microparticles containing prednisolone. Int. J. Pharm. 312, 113–118.

Woitiski, C.B., Neufeld, R.J., Soares, A.F., Figueiredo, I.V., Veiga, F.J., Carvalho, R.A., 2012. Evaluation of hepatic glucose metabolism via gluconeogenesis and glycogenolysis after oral administration of insulin nanoparticles. Drug Dev. Ind. Pharm. 38, 1441–1450.

Wong, T.W., Colombo, G., Sonvico, F., 2011. Pectin matrix as oral drug delivery vehicle for colon cancer treatment. AAPS PharmSciTech. 12, 201–214.

Wörner, C., Mühlhaupt, R., 1993. Polynitril- und polyaminfunktionalisierte poly(trimethylenimin)-dendrimere. Angew. Chem. 105, 1367–1372.
Wu, D.Q., Qiu, F., Wang, T., Jiang, X.J., Zhang, X.Z., Zhuo, R.X., 2009. Toward the development of partially biodegradable and injectable thermoresponsive hydrogels for potential biomedical applications. ACS Appl. Mater. Interfaces 1, 319–327.
Wu, W., Aiello, M., Zhou, T., Berliner, A., Banerjee, P., Zhou, S., 2010. In-situ immobilization of quantum dots in polysaccharide-based nanogels for integration of optical pH-sensing, tumor cell imaging, and drug delivery. Biomaterials 31, 3023–3031.
Wu, W., Wieckowski, S., Pastorin, G., Benincasa, M., Klumpp, C., Brland, J.P., Gennaro, R., Prato, M., Bianco, A., 2005. Targeted delivery of amphotericin B to cells by using functionalized carbon nanotubes. Angew. Chem. Int. Ed. Engl. 44, 6358–6362.
Wuilloud, R.G., Altamirano, J.C., 2006. Speciation analysis of non-metallic elements using plasma-based atomic spectrometry for detection. Curr. Anal. Chem. 2, 353–377.
Xiao, H., Nayak, B.R., Lowe, T.L., 2004. Synthesis and characterization of novel thermoresponsive-*co*-biodegradable hydrogels composed of *N*-isopropylacrylamide, poly(L-lactic acid), and dextran. J. Polym. Sci. 42A, 5054–5066.
Xing, L., Dawei, C., Liping, X., Rongqing, Z., 2003. Oral colon-specific drug delivery for bee venom peptide: development of a coated calcium alginate gel beads-entrapped liposome. J. Control. Release 93, 293–300.
Xiong, X.B., Falamarzian, A., Garg, S.M., Lavasanifar, A., 2011. Engineering of amphiphilic block copolymers for polymeric micellar drug and gene delivery. J. Control. Release 155, 248–261.
Xu, H., Bjerneld, E.J., Kall, M., Borjesson, L., 1999. Spectroscopy of single hemoglobin molecules by surface enhanced Raman scattering. Phys. Rev. Lett. 83, 4357–4360.
Xu, Q., Liu, Y., Su, S., Li, W., Chen, C., Wu, Y., 2012. Anti-tumor activity of paclitaxel through dual-targeting carrier of cyclic RGD and transferrin conjugated hyperbranched copolymer nanoparticles. Biomaterials 33, 1627–1639.
Yallapu, M.M., Foy, S.P., Jain, T.K., Labhasetwar, V., 2010. PEG-functionalized magnetic nanoparticles for drug delivery and magnetic resonance imaging applications. Pharmaceut. Res. 27, 2283–2295.
Yallapu, M.M., Othman, S.F., Curtis, E.T., Bauer, N.A., Chauhan, N., Kumar, D., Jaggi, M., Chauhan, S.C., 2012. Curcumin-loaded magnetic nanoparticles for breast cancer therapeutics and imaging applications. Int. J. Nanomed. 7, 1761–1779.
Yang, G., Xie, J., Deng, Y., Bian, Y., Hong, F., 2012a. Hydrothermal synthesis of bacterial cellulose/AgNPs composite: a "green" route for antibacterial application. Carbohydr. Polym. 87, 2482–2487.
Yang, H., Kao, W.Y.J., 2006. Thermoresponsive gelatin/monomethoxy poly(ethylene glycol)-poly(D,Llactide) hydrogels: formulation, characterization, and antibacterial drug delivery. Pharm. Res. 23, 205–214.
Yang, H.W., Hua, M.Y., Liu, H.L., Huang, C.Y., Wei, K.C., 2012b. Potential of magnetic nanoparticles for targeted drug delivery. Nanotechnol. Sci. Appl. 5, 73–86.
Yang, L., Wang, H.J., Yang, H.Y., Liu, S.H., Zhang, B.F., Wang, K., Ma, X.M., Zheng, Z., 2008. Shape-controlled synthesis of protein-conjugated silver sulfide nanocrystals and study on the inhibition of tumor cell viability. Chem. Commun. 2008, 2995–2997.
Yang, W., Cheng, Y., Xu, T., Wang, X., Wen, L.P., 2009. Targeting cancer cells with biotin-dendrimer conjugates. Eur. J. Med. Chem. 44, 862–868.
Yano, S., Mori, S., Saito, T., Yokoyama, H., Machishima, T., Shimada, T., Yahagi, Y., Sugiyama, K., Ogasawara, Y., Takahara, S., Kasama, K., Katsube, A., Kamiyama, Y., Suzuki, K., Inui, Y., Usui, N., Aiba, K., Yamashita, T., 2015. Pharmacokinetics for once-daily modified release formulation of tacrolimus hydrate in unrelated hematopoietic stem cell transplantation. Ann. Hematol. 94, 491–496.
Yin, X., Hoffman, A.S., Stayton, P.S., 2006. Poly(*N*-isopropylacrylamide-*co*-propylacrylic acid) copolymers that respond sharply to temperature and pH. Biomacromolecules 7, 1381–1385.
Yokoyama, R., Suzuki, S., Shirai, K., Yamauchi, T., Tsubokawa, N., Tsuchimochi, M., 2006. Preparation and properties of biocompatible polymer-grafted silicananoparticle. Eur. Pol. J. 42, 3221–3229.

Yong, K.-T., Hu, R., Roy, I., Ding, H., Vathy, L.A., Bergey, E.J., Mizuma, M., Maitra, A., Prasad, P.N., 2009. Tumor targeting and imaging in live animals with functionalized semiconductor quantum rods. ACS Appl. Mater. Interfaces 1, 710–719.

Yoshida, T., Aoyagi, T., Kokufuta, E., Okano, T., 2003. Newly designed hydrogel with both sensitive thermoresponse and biodegradability. J. Polym. Sci. 41A, 779–787.

Yuan, Q., Hein, S., Misra, R.D.K., 2010. New generation of chitosan-encapsulated ZnO quantum dots loaded with drug: synthesis, characterization and in vitro drug delivery response. Acta Biomat. 6, 2732–2739.

Yuvvakkumar, R., Elango, V., Rajendran, V., Kannan, N., 2014. High-purity nano silica powder from rice husk using a simple chemical method. J. Exp. Nanosci. 9, 272–281.

Zaky, R.R., Hessien, M.M., El-Midany, A.A., 2008. Preparation of silica nanoparticles from semi-burned rice straw ash. Powder Technol. 185, 31–35.

Zeng, W., Qian, X.F., Zhang, Y.B., Yin, J., Zhu, Z.K., 2005. Organic modified mesoporous MCM-41 through solvothermal process as drug delivery system. Mater. Res. Bull. 40, 766–772.

Zhang, G., Keita, B., Dolbecq, A., Mialane, P., Secheresse, F., Miserque, F., Nadjo, L., 2007a. Green chemistry-type one-step synthesis of silver nanostructures based on Mo V-Mo VI mixed valence polyoxometalates. Chem. Mater. 19, 5821–5823.

Zhang, L., Cao, F., Ding, B., Li, Q., Xi, Y., Zhai, G., 2011. Eudragit® S100 coated calcium pectinate microspheres of curcumin for colon targeting. J. Microencapsul. 28, 659–667.

Zhang, R., Olin, H., 2012. Carbon nanomaterials as drug carriers: real time drug release investigation. Mater. Sci. Eng. C Biomimetic Mater. Sens. Sys. 32, 1247–1252.

Zhang, Y.B., Kanungo, M., Ho, A.J., Freimuth, P., van der Lelie, D., Chen, M., Khamis, S.M., Datta, S.S., Johnson, A.T., Misewich, J.A., Wong, S.S., 2007b. Functionalized carbon nanotubes for detecting viral proteins. Nano Lett. 7, 3086–3091.

Zhao, B.X., Zhao, Y., Huang, Y., Luo, L.M., Song, P., Wang, X., Chen, S., Yu, K.F., Zhang, X., Zhang, Q., 2012. The efficiency of tumor-specific pH-responsive peptide-modified polymeric micelles containing paclitaxel. Biomaterials 33, 2508–2520.

Zhao, L., Wang, H., Huo, K., Cui, L., Zhang, W., Ni, H., Zhang, Y., Wu, Z., Chu, P.K., 2011. Antibacterial nano-structured titania coating incorporated with silver nanoparticles. Biomaterials 32, 5706–5716.

Zhou, W., Wang, Y., Jian, J., Song, S., 2013. Self-aggregated nanoparticles based on amphiphilic poly(lactic acid)-grafted-chitosan copolymer for ocular delivery of amphotericin B. Int. J. Nanomed. 8, 3715–3728.

Zhuo, R.X., Du, B., Lu, Z.R., 1999. In vitro release of 5-fluorouracil with cyclic core dendritic polymer. J. Control. Release 57, 249–257.

Zrazhevskiy, P., Sena, M., Gao, X., 2010. Designing multifunctional quantum dots for bioimaging, detection, and drug delivery. Chem. Soc. Rev. 39, 4326–4354.

Zu, Y., Wang, D., Zhao, X., Jiang, R., Zhang, Q., Zhao, D., Li, Y., Zu, B., Sun, Z., 2011. A novel preparation method for camptothecin (CPT) loaded folic acid conjugated dextran tumor-targeted nanoparticles. Int. J. Mol. Sci. 12, 4237–4249.

CHAPTER

CORE–SHELL DRUG CARRIERS: LIPOSOMES, POLYMERSOMES, AND NIOSOMES

2

Nily Dan
Drexel University, Philadelphia, PA, United States

CHAPTER OUTLINE

1 INTRODUCTION

Medicinal drugs have been used to cure diseases or alleviate symptoms since prehistoric ages. However, despite tremendous successes, the efficacy of many therapies is limited by the available administration routes, which (until the 1900s when intravenous methodologies were developed) were restricted to oral—through the digestive system—or topical—through the skin.

Oral delivery requires passage through the harsh environment of the digestive system, which can degrade or denature biochemical agents. A classic example is the degradation, by proteolytic enzymes, of insulin in the gastrointestinal (GI) tract. The need to administer insulin using nonoral, invasive methods results in reduced quality of life of diabetics and low compliance rates (Hamman et al., 2005; Khafagy et al., 2007; Morishita and Peppas, 2006). Topical, transdermal applications are limited since only few drugs can cross the dermal barrier, and their effects are largely localized to tissue near the skin (Anissimov et al., 2013; Schaefer, 1993; Singh and Roberts, 1994).

Other challenges to successful drug therapies abound even in intravenous methods: Drugs in the circulatory system are deposited indiscriminately in all types of tissue, which may cause unpleasant or even harmful side effects, such as those associated with chemotherapy drugs (Dodd, 1987). Also, molecules or objects recognized as "foreign" in the blood are rapidly degraded and/or cleared by the immune system. This not only limits the amount that reaches its target tissue, but also leads to concentration of drugs in organs like the liver and spleen, with potentially toxic results (Sherwood, 2015).

Recent developments have led, in addition to the development of novel drugs, to a reexamination of their delivery mechanisms. The goal of drug delivery systems, or carriers, is to enhance therapeutic efficiency by:

1. targeting specific tissue (Andresen et al., 2005; Beduneau et al., 2007; Chilkoti et al., 2002; Immordino et al., 2006; Jain and Jain, 2015; Kedar et al., 2010; Maeda et al., 2009; Myerson et al., 2015; Narang and Mahato, 2010; Olivier, 2005; Perez-Herrero and Fernandez-Medarde, 2015; Samad et al., 2007; Sudimack and Lee, 2000; Thanh-Huyen and Amiji, 2015; Toporkiewicz et al., 2015; Torchilin, 2004; Torchilin, 2005), and
2. providing control over the rate of drug release (Freiberg and Zhu, 2004; Gabizon, 1995; McCoy et al., 2010; Mohammadi-Samani and Taghipour, 2015; Robinson, 1978; Timko and Kohane, 2012; Toan et al., 2015; Varde and Pack, 2004; Vasir et al., 2003; Zhu and Chen, 2015).

This combination yields higher therapeutic efficacy and reduced toxicity. In addition, the carrier encapsulates and protects the drugs from environmental degradation agents in vitro and immunoagents in vivo (e.g., free radicals, enzymes), thereby increasing the effective drug lifetime. They may also be used as "theranostics" that combine drug delivery with imaging capabilities (Bardhan et al., 2011; Janib et al., 2010; Li et al., 2012a; McCarthy and Weissleder, 2008; Xie et al., 2010).

Drug delivery systems may be adapted to a variety of delivery methods: In addition to oral and topical/transdermal administration, these systems may be used for intravenous (parenteral) administration or inhalation/pulmonary ones (Antosova et al., 2009; Breitkreutz and Boos, 2011; Chetty and Chien, 1998; Cleland et al., 2001; Fogueri and Singh, 2009; Kumar et al., 2006; Leucuta, 2012; Mathias and Hussain, 2010; Swaminathan and Ehrhardt, 2012; Ulrich, 2002).

Cancer therapy illustrates the need for drug delivery: Classic chemotherapy agents affect the entire body, leading to numerous adverse side effects (Dodd, 1987). Some are temporary (hair loss, fatigue), but others pose significant and long-term risks such as heart or kidney failure. As a result, the maximal dosage used is lower than optimal for tumor treatment. Also, the effects of a dose last for only a few hours or days (Barpe et al., 2010), so patients must undergo several cycles of treatment. These issues can be addressed by an effective drug delivery system that can circulate for a long period without being cleared by the immune system, includes a mechanism for specific targeting of cancer tissue, and provides a rate of release tailored for maximal efficacy (Amoozgar and Goldberg, 2015; Bahrami et al., 2015; Brannon-Peppas and Blanchette, 2004; Brigger et al., 2002; Byrne et al., 2008; Cho et al., 2008; Danhier et al., 2010; Davis et al., 2008; Fernandes et al., 2015; Maeda et al., 2009; Nie et al., 2007; Peer et al., 2007; Tong and Langer, 2015; Topete et al., 2015).

A number of drug delivery systems have been developed and tested (Allen and Cullis, 2004; Anderson and Shive, 1997; Cheng et al., 2014; Cho et al., 2008; Kawaguchi, 2000; Langer, 1998; LaVan et al., 2003; Lawrence and Rees, 2000; Moghimi et al., 2001; Muller et al., 2000; Peppas et al., 2006; Torchilin, 2005). Extensive in vitro studies linked delivery efficiency and the drug release profile to parameters, such as the carrier size, shape, and surface chemistry and charge (Elnaggar et al., 2014; Eloy et al., 2014; Kumar, 2000; Ma et al., 2013; Mallick and Choi, 2014; Manjunath et al., 2005; Mundargi et al., 2008; Nkabinde et al., 2012; Qing et al., 2013; Rana et al., 2013; Sinha et al., 2004). These parameters have also been found to affect in vivo delivery (Froehlich, 2012; Gill et al., 2007; Kettler et al., 2014; Mailaender and Landfester, 2009; Salatin et al., 2015; Truong et al., 2015; Win and Feng, 2005).

Drug carriers can be roughly divided into two categories: solid nanoparticles and core–shell ones. Solid nanoparticles include inorganic (e.g., gold) particles that are typically used for imaging but may also bind and carry some therapeutics (Boisselier and Astruc, 2009; Gao et al., 2004; Ghosh et al., 2008; Giljohann et al., 2010; Gupta and Gupta, 2005; Laurent et al., 2008; Mornet et al., 2004; Pankhurst et al., 2003; Sperling et al., 2008; Sun et al., 2008), and organic ones (micelles, solid lipid nanoparticles) that solubilize and deliver hydrophobic drugs (Adams et al., 2003; Allen et al., 1999; Almeida and Souto, 2007; Joshi and Mueller, 2009; Lavasanifar et al., 2002; Mehnert and Mader, 2001; Muller et al., 2000; Nishiyama and Kataoka, 2006; Torchilin, 2007; Wissing et al., 2004).

Core–shell carriers consist of vesicles: aqueous cores surrounded by a bilayer shell. They are designed to primarily deliver hydrophilic drugs, although a small amount of hydrophobic ones can be incorporated into the bilayer. This category, which is the focus of this chapter, includes liposomes, polymersomes, and niosomes (colloidosomes, where the shell is composed of nanosized colloidal shells, are not used for drug delivery applications, although they are of interest for foods and cosmetics). Inspired by cellular membranes, they are composed of an aqueous core surrounded by a self-assembled bilayer (Fig. 2.1). In liposomes, the shell is based on phospholipids; polymersomes utilize synthetic, biocompatible copolymers; and niosomes consist of nonionic and ionic surfactants.

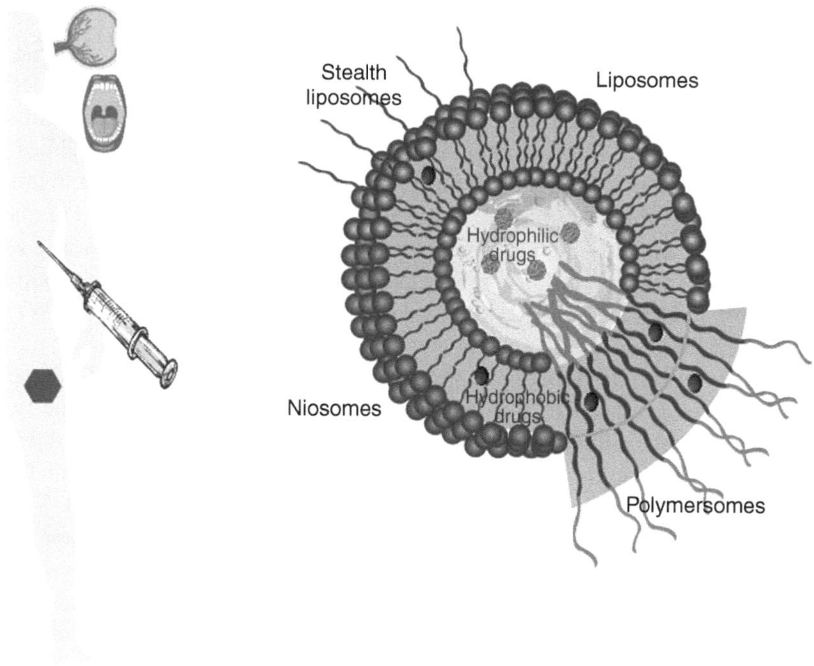

FIGURE 2.1

Core–shell, vesicle-based carriers (*right*) and their potential delivery routes (*left*): Liposomes are composed of phospholipid bilayers; stealth liposomes incorporate PEG-carrying lipids so that the polymer chain extends and provides steric stabilization and enhanced circulation time. Niosomes are composed of bilayers of nonionic surfactants, whereas polymersomes consist of diblock copolymer shells, so that their bilayer thickness is much greater than liposomes or niosomes. The carriers carry hydrophilic drugs in their aqueous cores, but can also dissolve some amount of hydrophilic drugs in the bilayer. Typical administration routes include ocular, oral, transdermal, or intravenous.

Vesicles of all types have been extensively studied and shown to be effective drug carriers. Several liposomal formulations have had US Food and Drug Administration (FDA) approval for two decades; for example, the liposomal preparation Doxil was approved in 1995 for the delivery of doxorubicin, a broadly used chemotherapy agent, while the liposomal form of the antifungal agent amphotericin B, Ambisome, received FDA approval in 1997. More are currently in phase II or phase III clinical trials (Lytton-Jean et al., 2015; Marchal et al., 2015). In addition to their encapsulation capacity, the self-assembled bilayer shell can incorporate stabilizing agents that prolong circulation time in vivo, targeting moieties for selective delivery, and environmental triggers that induce release under tightly defined conditions. As noted, they can incorporate any type of hydrophilic drug in the core, and—to some degree—also hydrophobic ones dissolved into the bilayer, thereby delivering a dual cargo for more effective impact (Ahmed et al., 2006). They can also be used to deliver nucleic acids, as discussed in detail in another chapter and elsewhere (Al-Dosari and Gao, 2009; Dan and Danino, 2014; Li and Szoka, 2007; Merdan et al., 2002; Neu et al., 2005; Opalinska and Gewirtz, 2002; Patil et al., 2005; Wasungu and Hoekstra, 2006; Zhang et al., 2004).

2 SELF-ASSEMBLED BILAYERS

Amphiphiles are molecules composed of a hydrophilic "head" group, and a hydrophobic "tail" region. The head groups may contain ionic, polar, or hydrogen-bond-forming groups that render them soluble in water. The tail regions are saturated or unsaturated hydrocarbon chains. The term *surfactant*—a constriction of surface active agent—is commonly used for small-molecule amphiphiles that have one hydrocarbon tail. Polymeric amphiphiles are copolymers composed of hydrophobic tail block(s) and hydrophilic head block(s). While most copolymeric amphiphiles are diblock-based (a macromolecular equivalent of surfactants), triblocks and other geometries show amphiphilic behavior as well. Phospholipids are an amphiphilic category of lipid molecules where the hydrophilic head group is attached to two hydrocarbon chain tails, which may be identical or differ in molecular weight (MW) and/or degree of saturation (Israelachvili, 2011).

Despite significant differences between the types of amphiphiles, they have some common features. When their concentration in a single solvent reaches a critical value, the critical micelle concentration (CMC), amphiphiles assemble into structures where the soluble group shields the insoluble group from the surrounding environment (Israelachvili, 2011). In water, the CMC increases with the effective size of the head groups, and decreases with the MW of the tail. The geometry of the aggregate is set by an effective volume ratio between the heads and tails, where molecules with relatively small tails and large head groups (or head groups with a large effective size such as ionically charged ones) will form spherical micelles in water. Molecules with similar-sized groups would form flat, sheet-like aggregates bilayers (Israelachvili, 2011), which close to form vesicles with an aqueous core.

Bilayers are composed of a "core" or inner layer of the hydrophobic tails, surrounded by two layers of the hydrophilic heads (Fig. 2.1). They are also referred to as membranes or lamellae (Israelachvili, 2011). Bilayers are characterized by the density of molecular packing at the water interface (as given, for example, by the interfacial area per molecule, Σ) and the thickness h (Fig. 2.2), which is usually taken to be the thickness of only the hydrophobic regions (see later). Bilayer thickness increases with tail MW. In lipid bilayers, h increases linearly with the tail MW, while Σ either is

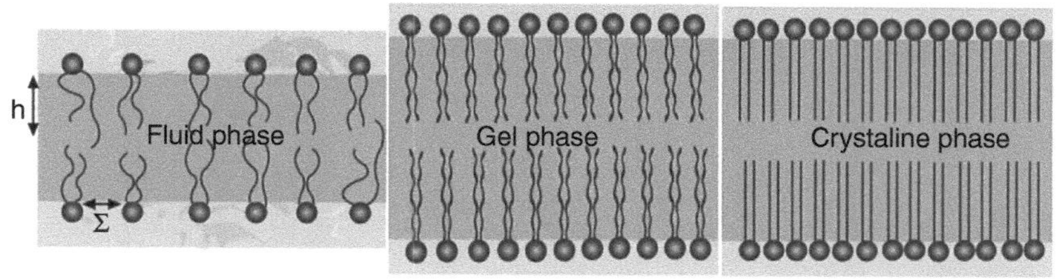

FIGURE 2.2 Phospholipid Bilayer Phases

At relatively high temperatures, the bilayer consists of fluids, and molecules can move freely in the bilayer plane, and occasionally "flip-flop" between the two monolayers. The packing density is low, so that Σ, the area per chain, is high, and the thickness is less. At lower temperatures, the bilayer undergoes a transition to the gel phase, where the lipids are more closely packed and motion is inhibited. In the crystalline phase, the tails adopt a liquid-crystal configuration. The transition temperature depends on the bilayer composition, a function, for example, of cholesterol content.

insensitive or weakly decreases with the tail MW (Kučerka et al., 2011). This is because in these layers, the area per molecule is dominated by the head groups, so that h is proportional to the tail volume—a function of its length. However, in diblock copolymer assemblies, the packing density is set by a balance between the head–head and tail–tail interactions. As a result, in the polymersome bilayers h scales as the square root of the hydrophobic block MW (Bermudez et al., 2002).

Bilayers, and in particular phospholipid bilayers, display several states. In the liquid phase, amphiphilic molecules move freely within the bilayer plane, and the tails are disordered (Fig. 2.2). Decreasing the temperature leads to a transition to a gel state where lateral mobility is suppressed and the tails are highly packed. Further cooling induces a crystalline state, as sketched in Fig. 2.2. The gel transition temperature depends on the bilayer composition and the components' properties, such as tail MW and degree of saturation (Blume, 1979; Koynova and Caffrey, 1998; Mantsch and McElhaney, 1991; Suurkuusk et al., 1976).

The optimal configuration of bilayers is locally flat, with a preferred surface density and thickness set by the amphiphilic structure. However, the self-assembled bilayers are both bendable and stretchable: Their overall surface area can be increased (for example, using micropipette aspiration), and they can bend either as a result of spontaneous fluctuations or in response to external forces. The area expansion modulus K_A accounts for the energy required to perturb the area per molecule from its optimal value. Several definitions have been used for the stretching/compression modulus, but a generally useful form is (Iglic, 2012):

$$E_{exp} = \frac{1}{2} K_A \left(\frac{\Sigma - \Sigma_0}{\Sigma_0} \right)^2 \tag{2.1}$$

where E_{exp} is the energy required to perturb the membrane area per molecule from the equilibrium, optimal value Σ_0 to a new value Σ. K_A is typically in the range of 0.1–0.3 J/m^2 for phospholipid bilayers (Bouvrais, 2012). In comparison, red blood cells, whose membranes include a multitude of proteins and other components, have a K_A value of order 0.45 J/m^2 (Bouvrais, 2012). In polymeric bilayers, K_A increases with the MW, as discussed subsequently (Eq. 2.3).

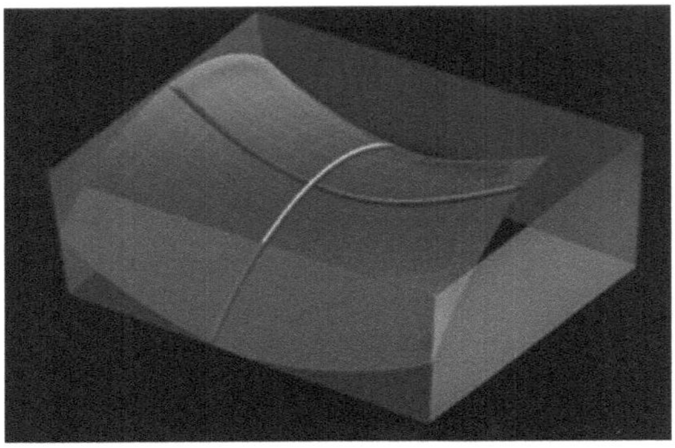

FIGURE 2.3 Principal Curvatures of a Surface

The curvature, C, is defined by h/R, where h is the thickness of the surface and R the radius of curvature. For a flat surface, both radii are infinite, so both curvatures are zero. In a sphere, both radii are the same and so are the two principal curvatures. In a cylinder, one curvature is zero (along the cylinder axis), and one is finite.

As noted, bilayers can also bend—either because of spontaneous fluctuations or as response to an external force. Two moduli describe the energetic penalty due to bending: the mean bending modulus κ and the Gaussian modulus κ_G (Bouvrais, 2012; Helfrich, 1973).

$$H = \frac{\Sigma_0}{2} K_A \left(\frac{\Sigma - \Sigma_0}{\Sigma_0} \right)^2 + \frac{1}{2}\kappa(C_1 + C_2 - C_0)^2 + \kappa_G C_1 C_2 \tag{2.2}$$

where H is the elastic free energy per molecule, and C_1 and C_2 are the principal curvatures (inversely proportional to the radii, as sketched in Fig. 2.3). C_0 defines the preferred curvature of the layer, which is taken to be zero for bilayer-forming systems. Experimental measurements yield, as a rule, the value of the mean bending modulus since the imposed deformation by techniques such as micropipette aspiration are spherical. They find that κ for phospholipid bilayers are of order 10^{-19} J (i.e., in the range of 10–100 k_BT) (Iglic, 2012).

The specific values of bilayer moduli depend on the amphiphile properties (tail MW and degree of saturation, head group interactions), solution conditions (ionic strength, temperature) and—in mixed bilayers—the composition (Bouvrais, 2012; Bouvrais et al., 2014; Claessens et al., 2004). A self-consistent model for polymers attached to a surface suggests correlations between bilayer moduli and the polymer MW. The model predicts that the optimal thickness of the tail region, that is, bilayer core, h_0, scales as the tail MW N_t divided by $\Sigma_0^{1/3}$ (Milner et al., 1988) and the moduli are (Ball et al., 1991)

$$\begin{aligned} K_A &= \frac{a_1}{\Sigma^2} N_t \\ \kappa &= \frac{a_2}{\Sigma^4} N_t^3 \\ \bar{\kappa} &= -\frac{a_3}{\Sigma^4} N_t^3 \end{aligned} \tag{2.3}$$

where a_1, a_2, and a_3 are constants of order unity. Although this model was developed for high-MW, strongly stretched chains, it has been shown to apply also to small-molecule systems and lipid bilayers: For example, Rawicz et al. (2000) found that in saturated phospholipid bilayers the compressibility modulus is proportional to h_0/Σ_0, which, using the correlation that $h_0 \sim N_t/\Sigma_0$, means $K_A \sim N_t/\Sigma_0^{4/3}$. They also find that ($\kappa/K_A$) is linearly proportional to h_0, in agreement with Eq. 2.3. As a result, these moduli were used to calculate features of phospholipid bilayers, such as the interactions between trans-membrane proteins (ArandaEspinoza et al., 1996; Dan et al., 1993; Dan and Safran, 1998; Wrenn et al., 2013), where the theoretical model predictions were verified by molecular simulations (Brannigan and Brown, 2006).

In mixed bilayers, the composition affects the moduli. For example, κ was shown to increase nearly linearly with cholesterol content in 1,2-dimyristoyl-glycero-3-phosphatidylcholine (DMPC) bilayers (above the fraction of cholesterol of 10%) (Bruning et al., 2014). Similar trends were found in other systems (Henriksen et al., 2006; Méléard et al., 1997; Song and Waugh, 1993). However, this effect is not universal: For example, in dioleoylphosphatidylcholine (DOPC), the bending modulus does not change significantly with cholesterol content (Gracia et al., 2010; Pan et al., 2009). In phospholipids, tail chain asymmetry also affects the moduli (Dan, 2007; Dimova, 2014; Illya et al., 2005).

3 LIPOSOMES AS DRUG CARRIERS

Liposomes are the earliest and most widely studied drug delivery vesicle system. They are based on phospholipid bilayers, whose inherent biocompatibility makes them a preferred option for in vivo applications. Liposomes are biologically inert, nontoxic, biodegradable, and relatively immunogenic. In addition to encapsulation of hydrophilic drugs in their aqueous center, they can incorporate lipophilic ones in small quantities (Allen and Cullis, 2013; Andresen et al., 2005; Barenholz, 2001; Drummond et al., 1999, 2000; Gregoriadis, 1995; Langer, 1998; Lasic, 1998; Lian and Ho, 2001; Malam et al., 2009; Senior, 1987; Sharma and Sharma, 1997; Torchilin, 2005).

A variety of phospholipids have been studied in liposomal drug delivery systems. The range includes molecules with neutral, anionic, or zwitterionic (i.e., contain both an anionic and cationic charges) head groups. Commonly used phospholipids in FDA-approved systems include phosphatidylcholine (PC), phosphatidylethanol-amine (PE), and sphingomyelin. Anionic phospholipids (e.g., phosphatidylglycerol or phosphatidylserine) are rarely used, since the negative charge reduces cellular uptake (Hillaireau and Couvreur, 2009; Opanasopit et al., 2002; Semple et al., 1998). Just as in cell membranes, however, optimal performance requires mixing charged phospholipids with other components, such as cholesterol, whose content affects the bilayer phase, fluidity, and moduli (Giddam et al., 2012; Henri et al., 2015; Patel et al., 2015; van Hoogevest and Wendel, 2014), or fusogenic peptides that promote endosomal escape (Shete et al., 2014) (Fig. 2.4).

Several categories of liposomes have been investigated for drug delivery applications: Small unilamellar vesicles (SUVs) are coated by a single bilayer, with a size range of order 100 nm. Giant unilamellar vesicles (GUVs) are similarly coated by a single bilayer, but are of order 10 μm, a size range that is ineffective for clinical applications (Bae and Park, 2011) but useful for in vitro investigations. Multilamellar vesicles (MLVs) are coated by several bilayers, and are typically larger than SUVs because of the rigidity of multilayered shell, of order 1 μm. This size is somewhat larger than the cut-off limits for cellular uptake, but can still be taken up by some tissue and is below the limit for rigid

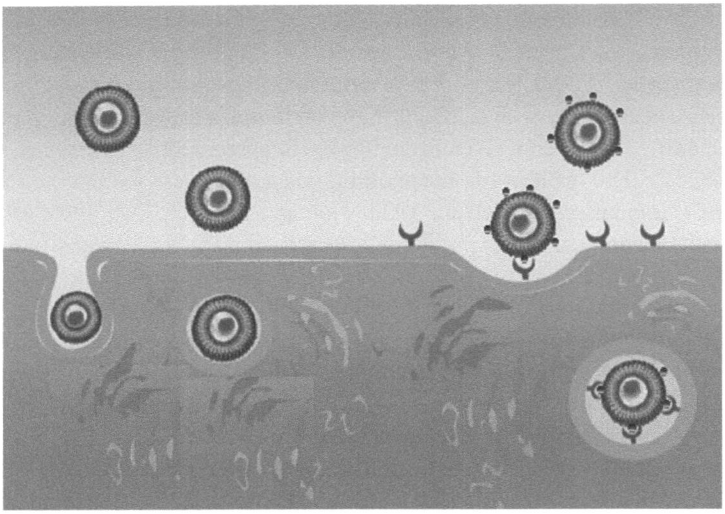

FIGURE 2.4 Cellular Internalization of Core–Shell Vesicle Carriers

The carrier is incorporated into an endosome, whose formation may be triggered by carrier approach or by binding to surface ligands.

particles' circulation in capillaries (Bae and Park, 2011). The advantage of the multilayered shell is that it enables incorporation and delivery of some hydrophobic drugs.

Despite their biological components, liposomes are captured by phagocytes and cleared from blood circulation within hours of administration (Blume and Cevc, 1990; Ishida et al., 2002; Senior et al., 1985). Liposome circulation time was found to depend on the vesicle size (Juliano and Stamp, 1975), where smaller vesicles circulate longer than larger ones (Senior et al., 1983, 1985). Surface charge and bilayer fluidity also affect circulation, although no simple correlations were found (Gabizon and Papahadjopoulos, 1992; Levchenko et al., 2002; Nishikawa et al., 1990).

The immunoclearance process is triggered by the adsorption of serum opsonin proteins such as beta 2-macroglobulin, immunoglobulins, or fibronectin onto the liposome's outer surface (Immordino et al., 2006). Thus, suppression or retardation of protein adsorption can enhance circulation time (Chonn et al., 1992). However, the variety of protein types means that it is difficult, if not impossible, to suppress adsorption through characteristics such as the bilayer's surface charge or specific chemistry. "Stealth liposomes" or "PEGylated liposomes" were developed to address this issue (Allen, 1994; Barenholz, 2001; Gref et al., 2000; Immordino et al., 2006; Lasic and Needham, 1995; Lian and Ho, 2001; Moghimi and Szebeni, 2003; Nag and Awasthi, 2013; Sharma and Sharma, 1997). In these, a fraction of the phospholipids is replaced by PEG-carrying lipids, namely, polyethylene glycol chains attached to a lipid molecule, most often through the reactive primary amine of PE (Woodle, 1998). The lipidic tail inserts into the bilayer, while the water-soluble polymer chain extends out (Fig. 2.1). The PEG layer provides steric stability and, more important, slows down opsonin adsorption and thus the rate of phagocytic uptake (Allen, 1994; Gref et al., 2000; Immordino et al., 2006; Woodle, 1998).

The effect of PEG-carrying lipids on circulation is significant: For example, Woodle (1998) has shown that after 4 h, about 80% of liposomes injected into rats were cleared, in comparison to only ~30% in the case of "stealth" liposomes. The enhanced circulation times also affect the distribution of the stealth liposomes in tissue (Allen and Hansen, 1991; Allen et al., 1995; Cattel et al., 2004; Gabizon and Martin, 1997; Oku and Namba, 2005; Woodle and Lasic, 1992). However, it should be emphasized that the PEG layer does not eliminate protein adsorption (and thus clearance); the effect is only to slow down the adsorption kinetics (Halperin, 1999).

Circulation time was found to increase with the PEG-lipid content in the bilayer, and with the PEG chain MW (Gref et al., 2000; Moghimi and Szebeni, 2003; Mosqueira et al., 2001); however, due to the self-assembled nature of the bilayer, the two are related: Increasing the MW of the PEG chain increases the effective head-group size, which can cause a transition to micellar aggregates (Hristova et al., 1995) so that the maximal fraction of PEG in the bilayer decreases with increasing chain MW. Therefore, circulation time increases with either PEG density of MW only up to a point, at which it reaches a plateau (Gref et al., 2000; Moghimi and Szebeni, 2003; Mosqueira et al., 2001). (It should also be noted that excess PEG-carrying lipids form micelles. Therefore, the fraction of PEG-carrying lipids in the bilayer may differ from that of the original mixture.)

More recently, other types of water-soluble polymers have been used to enhance liposomal circulation and stability. Although clearance rate (as given, for example, by the circulation half-time) may vary as a function of the polymer chain chemistry and MW, all are found to reduce clearance significantly when compared to standard liposomes, since they reduce the rate of opsonin adsorption and binding (Amoozgar and Yeo, 2012; Cao and Jiang, 2012; Hu et al., 2014; Kierstead et al., 2015; Kohli et al., 2014; Mohr et al., 2014; Nag et al., 2015; Reibel et al., 2015; Sun et al., 2015).

Other components may be included in lipid bilayers to allow targeting (Bazak et al., 2015; Caracciolo, 2015; Fernandes et al., 2015; Khare et al., 2014; Lin et al., 2015; Sakurai et al., 2015; Werengowska-Ciecwierz et al., 2015; Zhang et al., 2015). For example, it has been shown that liposomal uptake by rat alveolar macrophages was enhanced by inclusion of O-steroyl amylopectin and maleylated bovine serum albumin (Vyas et al., 2004) or mannose (Wijagkanalan et al., 2008) in the bilayer. Acid-catalyzed hydrolysis of vinyl ethers has been used to induce the degradation of liposomes (Boomer et al., 2003).

The relative stability of liposomes is an advantage during circulation, but is also a detriment: As noted, while effective drug carriers protect their cargo and must therefore remain stable in circulation, they need eventually to release the encapsulated drug in a predetermined way. Various methods have been developed to trigger degradation of the membrane (e.g., via temperature, external fields, or pH) (Blenke et al., 2013; Felber et al., 2012; Kneidl et al., 2014; Moussa et al., 2015; Movahedi et al., 2015; Paliwal et al., 2015), allowing release based on altered conditions in diseased tissue (e.g., tumor hypoxia) in stimuli-responsive liposomes. Alternately, external fields may be applied to induce liposomal breakup (e.g., ultrasound, heat) to induce local release (Bibi et al., 2012; Moussa et al., 2015; Movahedi et al., 2015; Ta and Porter, 2013; Tila et al., 2015; Torchilin, 2014; Zhao and Rodriguez, 2013).

Most liposomes designed to degrade in specific tissue environments are pH-sensitive, since tumors or inflamed tissue are more acidic than healthy tissue (Drulis-Kawa and Dorotkiewicz-Jach, 2010; Shi et al., 2002). These liposomes are stable under physiological conditions, but destabilize when the environment is acidic—thereby releasing the encapsulated drug. Unsaturated phosphatidylethanolamine such as dioleoyl phosphatidyl ethanolamine (DOPE) or diacetylenic-phosphatidyl-ethanolamine (DAPE) can impart pH sensitivity to liposomes, mixed with molecules such as oleic acid for stability (Karanth and Murthy, 2007). Alternately, pH-sensitive peptides or proteins may be incorporated into the bilayer (Simões et al., 2004).

Another potential release trigger is temperature: Thermally sensitive liposomes are ones where either the naturally occurring hyperthermia of a diseased tissue or an external application of heat causes release (Dabbagh et al., 2015; Kneidl et al., 2014; Kono et al., 2015). These types of liposomes are often based on dipalmitoylphosphatidylcholine (DPPC), since its gel-to-liquid crystalline phase transition—which induces leakage—occurs at 41°C (Kim et al., 2014; Lee et al., 2014; May et al., 2013).

Ultrasound (US) is a promising method for triggered release (Ahmed et al., 2015; Husseini et al., 2014; Liu et al., 2015; Moussa et al., 2015; Novell et al., 2015; Sirsi and Borden, 2014; Udroiu, 2015; Wallace and Wrenn, 2015). It is benign, has been used clinically for imaging for decade, and can penetrate into deep tissue that cannot be accessed easily via thermal fields (Husseini et al., 2014). US-induced release from liposomes occurs through two mechanisms: (1) the absorption of acoustic energy causes local heating, which can trigger release from thermo-sensitive liposomes, and (2) cavitation, where gas-filled bubbles oscillate in response to the pressure caused by the US field, leading to rupture of bilayers (Dijkmans et al., 2004; Schroeder et al., 2009; Wrenn et al., 2012). Unfortunately, these mechanisms also act on cellular membranes (Krasovitski et al., 2011), which limits the strength and duration of safe US application. To address this issue, liposomes were made US-sensitive by inclusion of gas bubbles (Geers et al., 2012; Schroeder et al., 2009; Stride and Coussios, 2010; Wrenn et al., 2012, 2013). The field causes the bubbles to undergo volume changes, resulting in shear forces that porate or rupture the bilayer and allow drug release while not harming cellular membranes. Although no commercial formulations are available at this time, some US-sensitive liposomal drug carriers are currently undergoing clinical trials (Husseini et al., 2014).

US-sensitive liposomal drug delivery systems are developed to perform in the frequency range of the clinically approved diagnostic fields, which are in the range of 1–18 MHz. However, studies have shown that low-frequency US (LFUS), with frequencies >1 MHz effectively and transiently increases the permeability of lipid membranes without the need for thermal sensitivity or added gas bubbles (Enden and Schroeder, 2009; Pong et al., 2005, 2006; Schroeder et al., 2007; Small et al., 2011, 2012). It has been suggested that the poration of bilayers is due to partitioning of gases from water into the hydrophobic bilayer, leading to the formation of microbubbles that respond to the LFUS (Dan, 2015).

Like US, magnetic fields are benign and can penetrate deeply into tissue. This has led to investigation of magnetically doped liposomes as drug carriers, which are triggered to release the drug when under a magnetic field (Liu et al., 2015). The sensitivity to magnetic fields is achieved by the inclusion of magnetic nanoparticles in the bilayer, and the release may occur either because of heating or as the result of field-induced stresses.

The ability to synthesize drug carriers in a controlled and sterile manner is essential for their commercial production. Liposomes form spontaneously when dissolving phospholipids in water. However, the result is polydisperse, multilamellar vesicles (MLVs) that need to be homogenized by sonication or high-pressure value homogenization. Unilamellar liposomes can be obtained through rehydration (Taylor et al., 2005). Drug loading can take place during the synthesis step by dissolving the compound(s) in the aqueous solution, but the resulting loading efficiency is low (McClements, 2015). The loading efficiency of some compounds can be increased by using external fields, but the method is not broadly applicable. Recently, microfluidics have been explored for the production of drug-carrying liposomes, since the method yields monodisperse, unilamellar liposomes of any size, with 100% loading efficiency (Hassan et al., 2014; Jahn et al., 2004, 2010; Matosevic, 2012; Vladisavljevic et al., 2012; Yu et al., 2009).

4 POLYMERSOMES AS DRUG CARRIERS

Amphiphilic diblock copolymers are the macromolecular equivalent of small-molecule surfactants, with a hydrophobic section (block) and a hydrophilic one (Letchford and Burt, 2007). Amphiphilic multiblock polymers combine several units of each type, in geometries that may include alternating, random, or branched. Like other amphiphilic molecules, amphiphilic block copolymers form aggregates in aqueous solutions, with a bilayer architecture favored when the effective volumes of the hydrophobic and hydrophilic sections are similar. Polymersomes are vesicles whose shells are composed of such block copolymers.

As a rule, polymersomes were synthesized using diblock copolymers, although some other architectures such as triblocks have also been used (Discher and Ahmed, 2006; Discher et al., 1999; Guan et al., 2015; Krishnamoorthy et al., 2014; Lee and Feijen, 2012; Letchford and Burt, 2007; Meng et al., 2009b; Messager et al., 2014). Various copolymer chemistries have been used to explore drug delivery applications: poly(ethylene glycol)-*b*-poly(ethyl ethylene) (PEG-PEE) (Bermudez et al., 2002), poly(ethylene glycol)-*b*-poly(butadiene) (PEG-PBD) (Bermudez et al., 2002; Li et al., 2007), poly(ethylene glycol)-*b*-poly(styrene) (PEG-PS) (Kabanov et al., 1998), poly(acrylic acid)-*b*-poly(styrene) (PAA-PS) (Shen and Eisenberg, 2000), polyethylene oxide-block-polycaprolactone (PEO-PCL) (Levine et al., 2008), poly(butadiene)-*b*-PEO (Demirgoez et al., 2009; Li et al., 2007), PEO-block-poly(γ-methyl-ε-caprolactone) (Petersen et al., 2013), poly(γ-benzyl-L-glutamate)-block-hyaluronan (Upadhyay et al., 2012), and similar compositions of triblock copolymers (Cao et al., 2015; Gallon et al., 2015; Nardin et al., 2000; Qin et al., 2015).

In aggregates of small molecules such as lipid bilayers, the incorporation of active moieties (e.g., targeting ligands, proteins) can disrupt the aggregate structure by changing the preferred packing (Israelachvili, 2011). This limits the degree of bilayer functionalization. In polymeric species, the high MW of the blocks masks the effects of functionalization on self-assembly, so that these can be effectively modified for either targeting or triggered degradation without affecting the bilayer structure (Felber et al., 2012; Guan et al., 2015; He et al., 2012; Krishnamoorthy et al., 2014; Liu et al., 2013; Meng et al., 2009b; Onaca et al., 2009; Pawar et al., 2013; Petersen et al., 2013; Rijcken et al., 2007; Thevenot et al., 2013). A number of targeting moieties have been incorporated into polymersomes, including antibodies, peptides, and small molecules such as folate or biotin (Egli et al., 2011; Tanner et al., 2011). Indeed, studies find that polymersomes can target specific tissue, in particular tumors (Licciardi et al., 2010; Pang et al., 2010; Prabhu et al., 2015; Upadhyay et al., 2010b, 2012; Zhu et al., 2015a).

Another advantage of polymersomes, when compared to liposomes, is that the polymeric water-soluble head group is an inherently effective barrier for immunoprotein adsorption. As a result, polymersomes are "stealth-like" with circulation times that usually exceed those of PEGylated liposomes (Lee et al., 2011a; Photos et al., 2003). Circulation time was shown to depend on the MW of the hydrophilic block (Photos et al., 2003) and the size of the polyersomes (Brinkhuis et al., 2012). The latter is especially interesting, since it seems like there is a sharp transition from short (~hours) to long (~day) circulation time around polymersome sizes of order 100 nm (Brinkhuis et al., 2012).

The longer circulation time of polymersomes, when compared to stealth liposomes, is likely due to the efficiency of the polymeric layer as a barrier to opsonin adsorption: In stealth liposomes, the small size of the lipid anchors limits the MW and density of the PEG chains. In polymersomes, the polymeric hydrophobic anchor allows crowding of the hydrophilic blocks, creating a dense impenetrable layer that is highly effective at prohibiting adsorption. However, as in the case of stealth liposomes, the effect is kinetic, so that phagocytosis does occur eventually. Indeed, polymersomes were found to accumulate in the liver—a sign of phagocyte clearance (Lee et al., 2011a; Photos et al., 2003).

The high MW of the polymers leads to thicker bilayers (Section 2). As a result, their area and bending moduli are higher than those of liposomes, with better mechanical stability and resistance to shear (Discher and Ahmed, 2006; Discher et al., 1999; Letchford and Burt, 2007; Meng et al., 2009b). Also, the thick hydrophobic core can solubilize appreciable amounts of hydrophobic drugs, thereby enabling polymersomes to deliver a combination of hydrophilic and hydrophobic therapeutics (Ahmed et al., 2006; Pramod et al., 2014; Surnar and Jayakannan, 2013; Thambi et al., 2012; Wu et al., 2014). In addition, the hydrophobic layer can be cross-linked to form a stable capsule (Cheng et al., 2011; Gaitzsch et al., 2011; Li et al., 2013a; Thibault et al., 2006; Xu et al., 2009).

As noted, the stability of drug carriers, which is an advantage during storage and circulation, is a disadvantage when considering release. This is especially important in polymersomes. Agents such as pH, temperature, or the presence of a specific biochemical are required to disrupt the polymeric bilayer, with the goal of increasing the solubility of the hydrophobic block in water. This removes the driving force for self-assembly and dissociates the bilayer (Gillies and Fréchet, 2005; Jeong et al., 2015; Kim et al., 2012; Qing et al., 2013; Qiu et al., 2013; Rodríguez-Hernández and Lecommandoux, 2005). A similar effect is achieved by breaking the bond between the hydrophobic and hydrophilic blocks using, for example, disulfide cleavage (Cerritelli et al., 2007). It should be noted, however, that although light-induced cleavage has also been used for this purpose (Cabane et al., 2010; Li et al., 2013b), it would not be useful for in vivo drug delivery applications. External fields have also been used to trigger release: Few studies examined US-induced release, but magnetic ones were shown to be effective (Meng et al., 2009a; Thevenot et al., 2013).

Most triggered release methods disrupt the bilayer completely, so that the drug release follows a burst profile. If a slower release profile is needed, it can be obtained by using biodegradable polymers where the rate of release depends on the rate of polymer degradation (Kim et al., 2006; Meng et al., 2005, 2009a).

As in liposomes, polymersomes form spontaneously through self-assembly in water, but achieving efficient encapsulation of drugs requires more directed methodologies (Messager et al., 2014). An interested method for polymersome synthesis is based on a double emulsion process where the inner, aqueous solution contains the hydrophilic drug, and the intermediate oil phase the hydrophobic drug. The emulsion is stabilized by the amphiphilic copolymer. Subsequent removal of the oil phase yields the polymersomes (Lorenceau et al., 2005). This method is especially effective for use in microfluidic devices, where (1) the encapsulation efficiency is ~100%, and (2) the polymersomes formed are highly monodisperse, with sizes that can be exactly controlled over a large range (~10 nm–100 μm) (Brown et al., 2010; Duncanson et al., 2012; Kim et al., 2011, 2013; Liao et al., 2012; Shum et al., 2008, 2011; Thiele et al., 2010). However, some issues can arise, such as inhomogeneities in the bilayer (Hayward et al., 2006).

Unlike liposomes whose bilayer constituents are mostly biological in origin, polymersome bilayers are purely synthetic. Studies have demonstrated their safety in vivo, mostly in murine models (Chen et al., 2013; Jain et al., 2011; Petersen et al., 2013; Petkar et al., 2011; Upadhyay et al., 2010a, 2012; Xu et al., 2014). However, their potential accumulation in different organs (e.g., liver, spleen) must be taken into account.

5 NIOSOMES AS DRUG CARRIERS

Nonionic surfactant-based vesicles, or niosomes, are the small-molecule synthetic analogue of liposomes and polymersomes (Schreier and Bouwstra, 1994; Uchegbu and Florence, 1995; Uchegbu and Vyas, 1998). They are lower in cost than liposomes (Uchegbu and Florence, 1995; Uchegbu and Vyas, 1998), which renders them an attractive option for applications requiring large quantities, such as cosmetics (Aljuffali et al., 2015; Mazda et al., 1997; Tavano et al., 2014; Wu and Guy, 2009). In addition, extensive studies focus on the development of niosomes for drug delivery applications (Drummond and Fong, 1999; Fang et al., 2001; Gaudana et al., 2009; Kaur et al., 2004; Uchegbu and Vyas, 1998).

Various types of nonionic amphiphiles have been used for niosome formation. However, unlike phospholipids, which were designed by nature to favor bilayers, nonionic surfactants display a rich range of aggregate geometries (Israelachvili, 2011). Moreover, mixing may significantly affect the structure. As a result, selection of amphiphilic constituents must be done carefully to ensure vesicle formation. Niosomes based on natural or synthetic sugar-based surfactants with head groups that range from mono- to polysaccharide (e.g., glucose, lactose, maltose, or sucrose) have been formed into drug-carrying niosomes (Ei-Laithy et al., 2011; Faivre and Rosilio, 2010; Muzzalupo et al., 2013). Another class of surfactants used for this purpose is sorbitan esters (Spans), fatty esters of the cyclized derivatives of sugar alcohol sorbitol-sorbitan (Balakrishnana et al., 2009; Guinedi et al., 2005; Hao et al., 2002; Uchegbu et al., 1995; Varshosaz et al., 2003; Yoshioka et al., 1994). Polyethoxylated sorbitan esters (Tweens), sometimes mixed with cholesterol, have also been studied (Alsarra et al., 2005; Bayindir and Yuksel, 2010; Di Marzio et al., 2011; Manosroi et al., 2003), as has a range of other nonionic surfactants (Salim et al., 2014).

Encapsulation efficiency in niosomes depends on the type of surfactant, the mixture composition (e.g., percent cholesterol), and the synthesis methodology. It is commonly in the range of 10%–40% (Abd-Elbary et al., 2008; Guinedi et al., 2005; Yoshioka et al., 1994), although some formulations can achieve ~75%–90% (Abdelbary and El-gendy, 2008; Waddad et al., 2013).

Although niosomes are thought to be stable, studies show that—unlike liposomes or polymersomes—the drug "leaks" from the vesicles, with an exponential release profile in vitro characteristic of diffusive processes (Abdelbary and El-gendy, 2008; Abd-Elbary et al., 2008; Akbari et al., 2015; Guinedi et al., 2005; Gopinath et al., 2004; Waddad et al., 2013; Yoshioka et al., 1994). As in the degree of encapsulation efficiency, the rate of release varies with noisome composition and method of preparation.

One issue that is not always addressed in niosome studies is potential interactions between the encapsulated drug and the surfactant shell (Uchegbu and Vyas, 1998): Comparison between niosomes composed of the same surfactants encapsulating different drugs suggests that in some cases these may either incorporate into the bilayer or affect the head–head interactions (Uchegbu and Vyas, 1998).

6 BIOMEDICAL APPLICATIONS OF CORE–SHELL VESICULAR DRUG CARRIERS

Drug carriers need to effectively target specific tissue in vivo, and release their cargo in that location in a controlled manner. Achieving these goals requires the ability to circulate in a stable manner in the vasculature until reaching the target organ or tissue, and then destabilize to release the drug. As will be discussed later, there are pros and cons to the different types of vesicular carriers that depend on the different needs of the specific applications.

One issue that is common to practically all drug carriers regardless of the medical application is cellular uptake: Although some therapies are based on drug release in the vicinity of the target cells, carrier uptake is generally preferable. It is frequently suggested that transport through cellular membranes decreases with increasing particle size, and is suppressed for those larger than ~500 nm (Bae and Park, 2011; Win and Feng, 2005). However, studies of gold nanoparticles show that maximal uptake peaks at a specific particle size that depends on the cell type (Chithrani et al., 2006; Levy et al., 2010; Trono et al., 2011). Similar trends were found for polymeric particle uptake by macrophages (Tabata and Ikada, 1988); quite surprisingly, the optimal uptake can be of order 1 μm (Hillaireau and Couvreur, 2009). It should also be noted that "optimal uptake;; may be defined as the maximal number of particles per cell, or as the total mass (or volume) of particles taken up by the cell. Most studies report the former, but the latter may be more relevant for drug delivery applications. Since the volume increases as the radius cubed, the differences can be significant. For example, Tabata and Ikada (1988) found uptake rates of 6 particles/cell for ~2 μm particles, and 3 particles/cell for 5 μm ones. Although the number of the smaller particles was twice as much, the *volume* of the 5 μm ones was about 7.5 times larger, thereby delivering nearly 10 times more drug. Besides the dependence on particle size and cell type, cellular uptake was shown to vary for different particle surface chemistries, net charge, and shape (Kettler et al., 2014; Ma et al., 2013; Morachis et al., 2012; Zhao et al., 2011).

The most widely studied application of drug delivery systems in general, and vesicle-based ones in particular, is that of cancer therapy. Traditional drug administration methods are low in specificity, a significant issue in anticancer drugs that tend to be highly toxic to healthy tissue as well: Whereas some side effects are transient (e.g., hair loss), others may be life-threatening (cardiotoxicity). To manage these side effects, dosages are typically kept at values lower than the optimal level (Bocci and Francia, 2014; Ewer and Yeh, 2006). Encapsulation of anticancer drugs in a drug delivery carrier can address most, if not all, of these issues: The carrier shields the toxic therapeutic, preventing it from interacting with any tissue until the time of release. Prolonged circulation allows effective biodistribution and may reduce the frequency of administration. Most important, targeting strategies could allow concentration of the drug in the tumor tissue at concentrations that are much higher than those that could be otherwise achieved safely. Indeed, the first liposome-based drug delivery systems to receive FDA approval and commercial distribution were Doxil/Caelyx (Janssen Pharmaceutica NV, Beerse, Belgium), a stealth liposomes–based system, and Myocet (GP Pharm SA, Barcelona, Spain/Teva Pharmaceutical Industries Ltd, Krakow, Poland). These were introduced in the 1990s for treatment of Kaposi's sarcoma, ovarian cancer, multiple myeloma, and metastatic breast cancer.

Several targeting methods have been developed for the delivery of chemotherapy agents to tumors. Passive targeting utilizes the enhanced permeability and retention effect (EPR) of tumors (Brigger et al., 2002; Cho et al., 2008; Danhier et al., 2010; Rapoport, 2007): The poor histological organization of cancer cells causes hyperpermeability, enhanced by limited lymphatic drainage (the result of uncontrolled growth). As a result, particles in a specific size range of 380–780 nm selectively accumulate in tumors (Hobbs et al., 1998). It should be noted, however, that while small particles (>10 nm) are cleared quickly through the renal system, particles of this size tend to accumulate in the liver, spleen, or bone marrow (Almeida et al., 2011; De Jong et al., 2008; Simon and Sabliov, 2014; Sonavane et al., 2008). The ensuing concentration of drugs may induce toxic side effects (Bae and Park, 2011). Thus, utilizing the "leaky vasculature" effect to target tumors requires a delicate balance of particle size and dosage.

Table 2.1 Some Commercially Available Liposome-Based Drug Delivery Products (Zou et al., 2015).

Name	Type	Drug	Indication	Year Approved
Doxil/Caelyx	PEGylated liposome	Doxorubicin	Kaposi's sarcoma Ovarian cancer Breast cancer	1995 1999 2003
DaunoXome	Liposome	Daunorubicin	Kaposi's sarcoma	1996
Ambisome	Liposome	Amphotericin B	Fungal infection	1997
Depocyt	Liposome	Cytarabine	Neoplastic meningitis, lymphomatous meningitis	1999
Myocet	Liposome	Doxorubicin	Breast cancer	2000
Visudyne	Liposome	Verteporfin	Age-related macular degeneration	2000
Definity	Lipid microsphere	Octafluoropropane	Suboptimal echocardiograms	2001
DepoDur	Liposome	Morphine	Postoperative pain	2004
Octocog alfa	Liposome	Factor VIII	Hemophilia A	2008
Marqibo	Liposome	Vincristine	Acute lymphoblatic leukemia	2012
Lipodox	PEGylated liposome	Doxorubicin	Kaposi's sarcoma, ovarian cancer, breast cancer	2013

Active targeting utilizes ligands that bind to overexpressed surface molecules or receptors that are unique to cancer cells. The binding can lead to triggered release, or more often to receptor-mediated endocytosis (Katsogiannou et al., 2011; Trapani et al., 2012). For example, CD44 has been found to be an effective target for liposomal binding (Dalla Pozza et al., 2013; Dufay Wojcicki et al., 2012; Platt and Szoka, 2008; Qhattal et al., 2014; Wojcicki et al., 2012).

A large number of clinical trials investigated liposomal drug delivery systems as treatment for cancer, and several formulations have been approved and are commercially used (Table 2.1 for examples). Patients receiving the liposomal-encapsulated drugs showed comparable or higher overall response rate when compared to unmodified drugs or traditional therapy methods, in a range of studies on breast cancer, advanced or recurrent ovarian cancer, multiple myeloma, prostate cancer, and others (Allen and Cullis, 2013; Batist et al., 2001; Beedassy and Cardi, 1999; Gabizon et al., 2003; Gokhale et al., 2001; Harris et al., 2002; Hubert et al., 2000; Lyass et al., 2000; Marchal et al., 2015; Muggia et al., 1997; Ranson et al., 1997; Singla et al., 2002; Tejada-Berges et al., 2002; Tila et al., 2015; Zuccari et al., 2015). Most important, the studies find that liposomal-based administration reduces toxicity when compared with free drugs, although some specific side effects, such as skin and hypersensitivity reactions were reported, in addition nausea, neutropenia, or vomiting (Iwamoto, 2013).

The most extensive area of polymersome drug delivery research is also that of cancer therapies. Biodegradable polymersomes containing a mixture of anticancer drugs were shown to increase the maximum tolerated dosage and shrink tumors when compared to the free drugs in vivo (Ahmed et al., 2006; Bakalova et al., 2015) or polymeric micelle carriers in vitro (Chen et al., 2010), even without active targeting. Most studies of polymersomes, though, utilize their capacity for functionalization to achieve cancer-targeted delivery (Guan et al., 2015; Krishnamoorthy et al., 2014; Oltra et al., 2014; Prabhu et al., 2015), for example, with anti-EGFR antibodies (Lee et al., 2011b; Li et al., 2012b) or a PR_b integrin-targeting peptide (Demirgoez et al., 2009; Petersen et al., 2013). Polymersomes have also been studied for targeting of leukocytes (Robbins et al., 2010) and macrophages (Broz et al., 2005), the delivery of antibiotics (Lane et al., 2015), or cochlear targeting for inner ear diseases (Buckiova et al., 2012; Roy et al., 2010; Surovtseva et al., 2012; Zhang et al., 2012). However, to date there are no commercial polymersome drug carriers.

Despite the intense focus on cancer, liposomes and polymersomes have been investigated for other therapies, and several liposomal formulations are now available for therapies ranging from fungal infections to pain (Table 2.1). Several developments capitalize on a perceived weakness of these carriers, namely their relatively rapid uptake by macrophages, as a means to target these cells. For example, in the case of antiparasitic and antimicrobial drugs used to treat mononuclear phagocytic infections, rapid clearing is an advantage (Karlowsky and Zhanel, 1992; Pinto-Alphandary et al., 2000). The process also allows targeting organs such as the liver, spleen, and bone marrow (Maesaki, 2002; Nobili et al., 2006; Vyas and Gupta, 2006). In another example, liposomal uptake by macrophages, coupled to the fact that the HIV primarily infects the immune and central nervous systems cells, was used to directly introduce antiretroviral drugs to infected cells while reducing their content in noninfected tissue (Desormeaux and Bergeron, 1998; Gupta and Jain, 2010; Mallipeddi and Rohan, 2010; Ojewole et al., 2008; Ramana et al., 2014; Sagar et al., 2014). Unfortunately, the limited hydrophobic drug-loading capacity, short shelf life, and cost of liposomes are likely to hinder widespread use for this purpose (Gupta and Jain, 2010; Mallipeddi and Rohan, 2010; Ojewole et al., 2008; Ramana et al., 2014; Sagar et al., 2014).

Liposomal uptake by activated macrophages can also be used to access inflamed tissue, such as rheumatoid arthritis (vanRooijen and Sanders, 1997). In addition to passive targeting, active targeting through, for example, folate receptors, minimizes nonspecific binding to receptor-negative tissues (Nogueira et al., 2015; Varghese et al., 2014). Similarly, liposomes have been used to target unresolved inflammation following myocardial infarction to delay the onset of heart failure in mice (Kain et al., 2015), or for attenuating atherosclerosis and reducing atherosclerosis-dependent myocardial infarction and stroke (Hosseini et al., 2015), or to treat neuropathic pain (Kobayashi et al., 2015).

Another exciting area in drug delivery research focuses on carriers that can cross the blood–brain barrier (BBB). The barrier limits or inhibits transport of many types of drugs (including antibiotics) into the brain. Therefore, developing carriers that can cross it effectively would enable treatment of cerebral diseases such as brain tumors, Alzheimer's disease, Parkinson's disease, and epilepsy. Although liposomes and other nano-carriers have been investigated for this purpose (Chen and Liu, 2012; Jain and Jain, 2015; Larsen et al., 2014; Tajes et al., 2014; Tiwari and Amiji, 2006; van Tellingen et al., 2015), functionalized polymersomes may be especially effective in improving the delivery of therapeutics to the brain (Chen et al., 2014; Larsen et al., 2014; Pang et al., 2008; Stojanov et al., 2012; Xin et al., 2011; Yu et al., 2012). Niosomes have also been investigated as carriers for transport through the BBB (Bragagni et al., 2014; Dufes et al., 2004; Jain and Jain, 2015).

Most of the delivery methods studied for cancer, crossing the BBB, or infections rely on IV-based administration. However, there is significant interest in noninvasive methodologies such as transdermal or ocular administration.

In transdermal delivery, drugs enter the bloodstream after crossing the dermal (skin) barrier. Transdermal delivery is of interest as a noninvasive method that can replace oral administration when the drugs may degrade in the digestive system or cause unpleasant side effects when digested. However, few drugs can cross the stratum corneum (SC) barrier, and the flux of those who do is low. The current use of technologies such as ultrasound to enhance transdermal transport is effective, but requires specialized instrumentation that is not available for home use (Barry, 2001; Brown et al., 2006; Prausnitz and Langer, 2008; Prausnitz et al., 2004). Niosome-based formulations have been tested for transdermal delivery of several drugs, showing enhancement in the dermal permeation rate when compared to either unmodified drugs or liposomal carriers (Abdelbary and AbouGhaly, 2015; Akhtar et al., 2015; Alomrani et al., 2015; Azeem et al., 2009; Barry, 2001; Bragagni et al., 2015; Bseiso et al., 2015; Cevc, 1996; Choi and Maibach, 2005; Fang et al., 2001; Ioele et al., 2015; Pando et al., 2015; Schreier and Bouwstra, 1994; Singh et al., 2015; Sinico and Fadda, 2009; Thong et al., 2007). The technique is not limited to drugs; antigens and adjuvants encapsulated in niosomes have been investigated as a noninvasive immunization method (Maheshwari et al., 2011; Mahor et al., 2007; Vyas et al., 2005). Niosome efficacy was shown to vary (as may be expected) with the size, shape, and chemistry of the surfactants (Marianecci et al., 2014).

The localized nature of ocular diseases means that drugs administered systemically reach the afflicted tissue in very low doses. However, the eye structure poses a number of physiological barriers to direct ocular administration. Niosomal formulations have been tested for treatment of corneal disorders such as those associated with diabetes mellitus (Abdelkader et al., 2012), showing that elastic niosomes are more effective than either liposomes or conventional treatments (Abdelkader et al., 2014; Gaafar et al., 2014; Kaur et al., 2012). In particular, the use of mucoadhesives, which are known to improve ocular delivery (Bernkop-Schnurch, 2005; Kaur and Smitha, 2002; Khutoryanskiy, 2011; Ludwig, 2005), can enhance niosomal ocular delivery (Aggarwal and Kaur, 2005; Kaur et al., 2010; Sah and Suresh, 2015).

Oral delivery is clearly the most preferable noninvasive drug administration route, whose use can increase patient compliance and outcomes. However, enzymatic degradation of compounds and especially peptide and protein drugs such as insulin in the gastrointestinal tract leads to extremely low absorption rates. While "some" drug carriers can protect such drugs, their transport through the GI system requires them to be stable at low pH and resist hydrolysis by intestinal enzymes or degradation by bile salts (Zhou and Neutra, 2002). On the other hand, uptake by Peyer's patches in the gut can occur for particles of order 1 μm, much larger than those used for IV administration (Ulrich, 2002). A number of liposomal formulations have been studied for oral delivery (Chen and Langer, 1998; Chen et al., 1996; Deng et al., 2015; Jensen et al., 2015; Kisel et al., 2001; Mansourpour et al., 2015; Sakuma et al., 2001; Takahashi et al., 2009; Takeuchi et al., 2005a; Yao et al., 2015). Coating liposomes with mucoadhesive polymers can enhance efficiency (Abd El Azim et al., 2015; Park et al., 2015; Takeuchi and Sugihara, 2010; Takeuchi et al., 1994, 1996, 2003, 2005b; Zhu et al., 2015b). Fewer studies utilize polymersomes for oral delivery (Chen and Wang, 2014; Surnar and Jayakannan, 2013).

Niosomes are often made from food-grade compounds, which led to interest in their use for oral delivery. The stability of orally administered drugs encapsulated in niosomes was found to increase when compared to traditional forms such as tablets (Ahmed et al., 2014; Akhter et al., 2012; Jin et al., 2013), as did their bioavailability (Akhter et al., 2012; Aqil et al., 2013; Arzani et al., 2015; Attia et al., 2007; Hasan et al., 2013; Jadon et al., 2009; Jin et al., 2013; Sezgin-Bayindir et al., 2013, 2015; Zhong et al., 2014). Niosomes were shown to stabilize proteins and peptides in the GI system (Di Marzio et al., 2013; Jain et al., 2005, 2006, 2013, 2014a,b; Pardakhty et al., 2007, 2011; Sinha et al., 2007; Wilkhu et al., 2013), and deliver vaccines orally (Jain and Vyas, 2006; Katare et al., 2006; Mann et al., 2006; Rentel et al., 1999; Wilkhu et al., 2013, 2014), an area with great potential for better compliance and broad reach.

7 DISCUSSION

Medical research has led to numerous drugs that can mitigate or cure diseases that, just a few decades ago, would be considered fatal. For example, survival rate in pediatric leukemia have increased from nearly zero in the 1950s to 50%–90% in the 2010s (Napper and Watson, 2013). Retroviral drugs have reduced the number of deaths from AIDS-related causes such as non-Hodgkin's lymphoma significantly (Bower et al., 2005). However, these novel drugs require delivery methods that are similarly sophisticated.

The two overarching issues that drug delivery systems need to address are tissue targeting and controlled release. These arise since the potency of many drugs can cause adverse side reactions if they act on healthy tissue. Other benefits of drug delivery systems include long-term release, which reduces the need for frequent dosing, and protection of sensitive drugs from environmental degradation agents ranging from free radical oxidizing agents to enzymatic hydrolysis.

Bilayer-based core–shell drug carriers utilize the tendency of amphiphilic molecules to form vesicles with an aqueous core to deliver hydrophilic drugs. Despite their differences, liposomes (based on phospholipids), polymersomes (based on block copolymers), and niosomes (composed of nonionic surfactants) provide similar benefits as drug carriers: the ability to incorporate high loads of hydrophilic drugs, circulate for (relatively) long times, and incorporate functional moieties that would both target specific tissue and induce degradation and drug release in the target tissue.

Liposomes are the most practical drug delivery system, as can be seen from the fact that to date only liposomal formulations have been approved by regulatory agencies and made commercially available (Table 2.1). Their largest advantages are that their main constituents—phospholipids and cholesterol—are inherently biocompatible, nontoxic even at high concentrations, and moderately priced. The rich variety of phospholipids, and their phase behavior, allows tailoring of a number of properties such as bilayer state (Fig. 2.2). They can be functionalized using targeting ligands to bind to specific cell receptors, and incorporate pH- or temperature-sensitive components to trigger release in the target locations. Methodologies for liposome formation are well established (Maherani et al., 2011). The two main issues of manufacturing—low encapsulation efficiency (which increases the cost due to loss of expensive drugs) and polydispersity (since effective performance is highly sensitive to size)—can be addressed using microfluidic methods (Maherani et al., 2011). However, the biocompatibility of the constituent phospholipids also means that liposomes are highly susceptible to many serum agents, from the adsorption of opsonins to enzymatic hydrolysis. These are reduced in stealth liposomes, but the restricted capacity of the phospholipid bilayer to carry PEG chains limits their effect.

The synthetic variability of block copolymers yields polymersome bilayers that are more robust and can resist mechanical deformations effectively, and can solubilize significant amounts of hydrophobic drugs in addition to the hydrophilic ones in the core. The dense hydrophilic block layer resists immunoprotein adsorption, enhancing circulation time significantly. They can be functionalized, and even cross-linked. Although their enhanced stability could be considered a negative when drug release is required, they can be engineered with pH or temperature sensitivity to trigger release. Arguably, the main issues that limit the use of polymersomes are the synthetic nature of the polymers, which can potentially be toxic, and the high cost of their manufacturing.

Niosomes are composed of the cheapest and most abundant and versatile components, nonionic (and ionic) surfactants. Many of the surfactants used in niosomes are natural in origin and highly biocompatible. Indeed, a number of the surfactants used in niosomes are utilized in foods and cosmetics.

However (contrary to some reports), they are not very stable: While liposomes can hold their encapsulated cargo for at least days to months (especially if stored at temperatures below the gel transition point) (Lasic and Papahadjopoulos, 1998), and polymersomes can do so even longer, niosomes tend to lose encapsulated drugs over a period of hours. In addition, the mechanical stability of niosomes is low.

One enormous obstacle to the optimization of drug carriers is the large numbers of parameters. As discussed in the Sections 3–5, carrier performance depends on size, composition (including functional groups and targeting ligands), surface charge, and method of administration. Environmental features play a role as well: Temperature controls the bilayer phase, while ionic strength and pH affect head–head interactions, and therefore the packing density. In liposomes, the number of bilayer shells is also important. Studies need to account for size polydispersity of the vesicles, and, since they are formed via self-assembly, the inability to fully control composition: The composition of the formed vesicles is not necessarily the same as that of the original mixture. For example, lipid bilayers are limited in the amount of PEG-carrying lipids they can carry before undergoing a transition to micelles. In addition, delivery is highly sensitive to cell or tissue type, so that optimization for one application is unlikely to be informative regarding another.

Another issue that should be accounted for, but is sometimes neglected, is interactions between the encapsulated drugs and the bilayer components. Amphiphilic drugs can incorporate in the bilayer, thereby affecting the carrier stability and interactions. Charged compounds can screen and distort head–head interactions. Therefore, results obtained for one drug may not translate to another. In addition, interactions with some bilayer components can affect and degrade certain drugs—for example, inducing denaturation of pharmaceutical proteins.

Many drug delivery studies examine the functionality of the carrier: biodistribution, targeting, and/or release rate. However, increasing functionality by conjugation or incorporation of active components may reduce biocompatibility. Furthermore, comparing studies is difficult; for example, uptake as a function of size can be reported by the number of particles per cell or by the volume taken up per cell. The latter is clearly more relevant when considering the amount of drug that could be delivered (e.g., 125 particles with a radius of 20 nm are comparable in volume to one with a radius of 100 nm), but not always presented. When examining pharmacodynamics, drug carriers are sometimes compared to other carriers, and sometimes to the unmodified free drug. This lack of a standard approach complicates analysis.

In summary, while vesicle-based drug delivery systems have the potential to advance therapies significantly, their development requires the ability to optimize multiple factors of physiochemical features (size, charge, chemistry, degradation/release), physiological features, and cellular features. In addition, one type of system could not address the multiple needs of administration routes and therapeutic targets.

8 CONCLUSIONS

Careful design of core–shell, vesicle-based drug carriers (liposomes, polymersomes, and niosomes) can address the challenges to successful therapies based on hydrophilic drugs: targeting of specific tissue, and release in predetermined manner, time, and location. They offer therapeutic efficiencies that are much higher than the unmodified drugs. In addition, carriers protect the encapsulated drug from chemical and biological degradation agents, and can be adapted to different administration methods.

Extensive studies link aspects of delivery efficiency and the drug release profile to parameters, such as the carrier size, shape, and surface chemistry and charge, as well as environmental conditions (temperature, pH) and cell type. Some liposome-based formulations are currently in clinical use, but, despite extensive studies, progress requires more research combining physical, chemical, biological, and physiological approaches.

CONFLICT OF INTERESTS

The author declares that there is no conflict of interests regarding the publication of this paper.

REFERENCES

Abd El Azim, H., Nafee, N., Ramadan, A., Khalafallah, N., 2015. Liposomal buccal mucoadhesive film for improved delivery and permeation of water-soluble vitamins. Int. J. Pharm. 488, 78–85.

Abdelbary, A.A., AbouGhaly, M.H.H., 2015. Design and optimization of topical methotrexate loaded niosomes for enhanced management of psoriasis: application of Box-Behnken design, in-vitro evaluation and in-vivo skin deposition study. Int. J. Pharm. 485, 235–243.

Abdelbary, G., El-gendy, N., 2008. Niosome-encapsulated gentamicin for ophthalmic controlled delivery. AAPS PharmSciTech 9, 740–747.

Abd-Elbary, A., El-Laithy, H.M., Tadros, M.I., 2008. Sucrose stearate-based proniosome-derived niosomes for the nebulisable delivery of cromolyn sodium. Int. J. Pharm. 357, 189–198.

Abdelkader, H., Wu, Z., Al-Kassas, R., Alany, R.G., 2012. Niosomes and discomes for ocular delivery of naltrexone hydrochloride: morphological, rheological, spreading properties and photo-protective effects. Int. J. Pharm. 433, 142–148.

Abdelkader, H., Alani, A.W.G., Alany, R.G., 2014. Recent advances in non-ionic surfactant vesicles (niosomes): self-assembly, fabrication, characterization, drug delivery applications and limitations. Drug Deliv. 21, 87–100.

Adams, M.L., Lavasanifar, A., Kwon, G.S., 2003. Amphiphilic block copolymers for drug delivery. J. Pharm. Sci. 92, 1343–1355.

Aggarwal, D., Kaur, I.P., 2005. Improved pharmacodynamics of timolol maleate from a mucoadhesive niosomal ophthalmic drug delivery system. Int. J. Pharm. 290, 155–159.

Ahmed, F., Pakunlu, R.I., Brannan, A., Bates, F., Minko, T., Discher, D.E., 2006. Biodegradable polymersomes loaded with both paclitaxel and doxorubicin permeate and shrink tumors, inducing apoptosis in proportion to accumulated drug. J. Control. Rel. 116, 150–158.

Ahmed, S., El-Setouhy, D.A., Badawi, A.A.E.-L., El-Nabarawi, M.A., 2014. Provesicular granisetron hydrochloride buccal formulations: in vitro evaluation and preliminary investigation of in vivo performance. Eur. J. Pharm. Sci. 60, 10–23.

Ahmed, S.E., Martins, A.M., Husseini, G.A., 2015. The use of ultrasound to release chemotherapeutic drugs from micelles and liposomes. J. Drug Target. 23, 16–42.

Akbari, V., Abedi, D., Pardakhty, A., Sadeghi-Aliabadi, H., 2015. Release studies on ciprofloxacin loaded non-ionic surfactant vesicles. Avicenna J. Med. Biotechnol. 7, 69–75.

Akhtar, N., Arkvanshi, S., Bhattacharya, S.S., Verma, A., Pathak, K., 2015. Preparation and evaluation of a buflomedil hydrochloride niosomal patch for transdermal delivery. J. Liposome Res. 25, 191–201.

Akhter, S., Kushwaha, S., Warsi, M.H., Anwar, M., Ahmad, M.Z., Ahmad, I., Talegaonkar, S., Khan, Z.I., Khar, R.K., Ahmad, F.J., 2012. Development and evaluation of nanosized niosomal dispersion for oral delivery of Ganciclovir. Drug Dev. Ind. Pharm. 38, 84–92.

Al-Dosari, M.S., Gao, X., 2009. Nonviral gene delivery: principle, limitations, and recent progress. AAPS J. 11, 671–681.

Aljuffali, I.A., Hsu, C.-Y., Lin, Y.-K., Fang, J.-Y., 2015. Cutaneous delivery of natural antioxidants: the enhancement approaches. Curr. Pharm. Design 21, 2745–2757.

Allen, T.M., 1994. Long-circulating (sterically stabilized) liposomes for targeted drug-delivery. Trends Pharmacol. Sci. 15, 215–220.

Allen, T.M., Cullis, P.R., 2004. Drug delivery systems: entering the mainstream. Science 303, 1818–1822.

Allen, T.M., Cullis, P.R., 2013. Liposomal drug delivery systems: from concept to clinical applications. Adv. Drug Deliv. Rev. 65, 36–48.

Allen, T.M., Hansen, C., 1991. Pharmacokinetics of stealth versus conventional liposomes—effect of dose. Biochim. Biophys. Acta 1068, 133–141.

Allen, T.M., Hansen, C.B., Demenezes, D.E.L., 1995. Pharmacokinetics of long-circulating liposomes. Adv. Drug Deliv. Rev. 16, 267–284.

Allen, C., Maysinger, D., Eisenberg, A., 1999. Nano-engineering block copolymer aggregates for drug delivery. Colloids Surf. B 16, 3–27.

Almeida, A.J., Souto, E., 2007. Solid lipid nanoparticles as a drug delivery system for peptides and proteins. Adv. Drug Deliv. Rev. 59, 478–490.

Almeida, J.P.M., Chen, A.L., Foster, A., Drezek, R., 2011. In vivo biodistribution of nanoparticles. Nanomedicine 6, 815–835.

Alomrani, A.H., Al-Agamy, M.H., Badran, M.M., 2015. In vitro skin penetration and antimycotic activity of itraconazole loaded niosomes: various non-ionic surfactants. J. Drug Deliv. Sci. Tec. 28, 37–45.

Alsarra, I.A., Bosela, A.A., Ahmed, S.M., Mahrous, G.M., 2005. Proniosomes as a drug carrier for transdermal delivery of ketorolac. Eur. J. Pharm. Biopharm. 59, 485–490.

Amoozgar, Z., Goldberg, M.S., 2015. Targeting myeloid cells using nanoparticles to improve cancer immunotherapy. Adv. Drug Del. Rev. 91, 38–51.

Amoozgar, Z., Yeo, Y., 2012. Recent advances in stealth coating of nanoparticle drug delivery systems. Wiley Interdisciplin. Rev. Nanomed. Nanobiotechnol. 4, 219–233.

Anderson, J.M., Shive, M.S., 1997. Biodegradation and biocompatibility of PLA and PLGA microspheres. Adv. Drug Deliv. Rev. 28, 5–24.

Andresen, T.L., Jensen, S.S., Jorgensen, K., 2005. Advanced strategies in liposomal cancer therapy: problems and prospects of active and tumor specific drug release. Prog. Lipid Res. 44, 68–97.

Anissimov, Y.G., Jepps, O.G., Dancik, Y., Roberts, M.S., 2013. Mathematical and pharmacokinetic modelling of epidermal and dermal transport processes. Adv. Drug Deliv. Rev. 65, 169–190.

Antosova, Z., Mackova, M., Kral, V., Macek, T., 2009. Therapeutic application of peptides and proteins: parenteral forever? Trends Biotechnol. 27, 628–635.

Aqil, F., Munagala, R., Jeyabalan, J., Vadhanam, M.V., 2013. Bioavailability of phytochemicals and its enhancement by drug delivery systems. Cancer Lett. 334, 133–141.

ArandaEspinoza, H., Berman, A., Dan, N., Pincus, P., Safran, S., 1996. Interaction between inclusions embedded in membranes. Biophys. J. 71, 648–656.

Arzani, G., Haeri, A., Daeihamed, M., Bakhtiari-Kaboutaraki, H., Dadashzadeh, S., 2015. Niosomal carriers enhance oral bioavailability of carvedilol: effects of bile salt-enriched vesicles and carrier surface charge. Int. J. Nanomed. 10, 4797–4813.

Attia, I.A., El-Gizawy, S.A., Fouda, M.A., Donia, A.M., 2007. Influence of a niosomal formulation on the oral bioavailability of acyclovir in rabbits. AAPS PharmSciTech 8, E106.

Azeem, A., Anwer, M.K., Talegaonkar, S., 2009. Niosomes in sustained and targeted drug delivery: some recent advances. J. Drug Target. 17, 671–689.

Bae, Y.H., Park, K., 2011. Targeted drug delivery to tumors: myths, reality and possibility. J. Control. Rel. 153, 198–205.

Bahrami, B., Mohammadnia-Afrouzi, M., Bakhshaei, P., Yazdani, Y., Ghalamfarsa, G., Yousefi, M., Sadreddini, S., Jadidi-Niaragh, F., Hojjat-Farsangi, M., 2015. Folate-conjugated nanoparticles as a potent therapeutic approach in targeted cancer therapy. Tumor Biol. 36, 5727–5742.

Bakalova, R., Lazarova, D., Nikolova, B., Atanasova, S., Zlateva, G., Zhelev, Z., Aoki, I., 2015. Delivery of size-controlled long-circulating polymersomes in solid tumours, visualized by quantum dots and optical imaging in vivo. Biotechnol. Biotec. Equip. 29, 175–180.

Balakrishnana, P., Shanmugam, S., Lee, W.S., Lee, W.M., Kim, J.O., Oh, D.H., Kim, D.-D., Kim, J.S., Yoo, B.K., Choi, H.-G., Woo, J.S., Yong, C.S., 2009. Formulation and in vitro assessment of minoxidil niosomes for enhanced skin delivery. Int. J. Pharm. 377, 1–8.

Ball, R.C., Marko, J.F., Milner, S.T., Witten, T.A., 1991. Polymers grafted to a convex surface. Macromolecules 24, 693–703.

Bardhan, R., Lal, S., Joshi, A., Halas, N.J., 2011. Theranostic nanoshells: from probe design to imaging and treatment of cancer. Acc. Chem. Res. 44, 936–946.

Barenholz, Y., 2001. Liposome application: problems and prospects. Curr. Opin. Colloid Interf. Sci. 6, 66–77.

Barpe, D.R., Rosa, D.D., Froehlich, P.E., 2010. Pharmacokinetic evaluation of doxorubicin plasma levels in normal and overweight patients with breast cancer and simulation of dose adjustment by different indexes of body mass. Eur. J. Pharm. Sci. 41, 458–463.

Barry, B.W., 2001. Novel mechanisms and devices to enable successful transdermal drug delivery. Eur. J. Pharm. Sci. 14, 101–114.

Batist, G., Ramakrishnan, G., Rao, C.S., Chandrasekharan, A., Gutheil, J., Guthrie, T., Shah, P., Khojasteh, A., Nair, M.K., Hoelzer, K., Tkaczuk, K., Park, Y.C., Lee, L.W., Myocet Study, G., 2001. Reduced cardiotoxicity and preserved antitumor efficacy of liposome-encapsulated doxorubicin and cyclophosphamide compared with conventional doxorubicin and cyclophosphamide in a randomized, multicenter trial of metastatic breast cancer. J. Clinical Oncol. 19, 1444–1454.

Bayindir, Z.S., Yuksel, N., 2010. Characterization of niosomes prepared with various nonionic surfactants for paclitaxel oral delivery. J. Pharm. Sci. 99, 2049–2060.

Bazak, R., Houri, M., El Achy, S., Kamel, S., Refaat, T., 2015. Cancer active targeting by nanoparticles: a comprehensive review of literature. J. Cancer Res. Clin. Oncol. 141, 769–784.

Beduneau, A., Saulnier, P., Benoit, J.-P., 2007. Active targeting of brain tumors using nanocarriers. Biomaterials 28, 4947–4967.

Beedassy, A., Cardi, G., 1999. Chemotherapy in advanced prostate cancer. Semin. Oncol. 26, 428–438.

Bermudez, H., Brannan, A.K., Hammer, D.A., Bates, F.S., Discher, D.E., 2002. Molecular weight dependence of polymersome membrane structure, elasticity, and stability. Macromolecules 35, 8203–8208.

Bernkop-Schnurch, A., 2005. Thiomers: a new generation of mucoadhesive polymers. Adv. Drug Deliv. Rev. 57, 1569–1582.

Bibi, S., Lattmann, E., Mohammed, A.R., Perrie, Y., 2012. Trigger release liposome systems: local and remote controlled delivery? J. Microencapsul. 29, 262–276.

Blenke, E.O., Mastrobattista, E., Schiffelers, R.M., 2013. Strategies for triggered drug release from tumor targeted liposomes. Expert Opin. Drug Del. 10, 1399–1410.

Blume, A., 1979. A comparative study of the phase transitions of phospholipid bilayers and monolayers. Biochim. Biophys. Acta 557, 32–44.

Blume, G., Cevc, G., 1990. Liposomes for the sustained drug release invivo. Biochim. Biophys. Acta 1029, 91–97.

Bocci, G., Francia, G., 2014. Metronomic Chemotherapy: Pharmacology and Clinical Applications. Springer, Berlin, Heidelberg.

Boisselier, E., Astruc, D., 2009. Gold nanoparticles in nanomedicine: preparations, imaging, diagnostics, therapies and toxicity. Chem. Soc. Rev. 38, 1759–1782.

Boomer, J.A., Inerowicz, H.D., Zhang, Z.Y., Bergstrand, N., Edwards, K., Kim, J.M., Thompson, D.H., 2003. Acid-triggered release from sterically stabilized fusogenic liposomes via a hydrolytic DePEGylation strategy. Langmuir 19, 6408–6415.

Bouvrais, H., 2012. Bending ridities of lipid bilayers: their determination and main inputs in biophysical studies. In: Iglic, A. (Ed.), Advances in Planar Lipid Bilayers and Liposomes, vol. 15, Elsevier, Totowa, NJ, pp. 1–76.
Bouvrais, H., Duelund, L., Ipsen, J.H., 2014. Buffers affect the bending rigidity of model lipid membranes. Langmuir 30, 13–16.
Bower, M., Gazzard, B., Mandalia, S., Newsom-Davis, T., Thirlwell, C., Dhillon, T., Young, A.M., Powles, T., Gaya, A., Nelson, M., Stebbing, J., 2005. A prognostic index for systemic AIDS-related non-Hodgkin lymphoma treated in the era of highly active antiretroviral therapy. Ann. Intern. Med. 143, 265–273.
Bragagni, M., Mennini, N., Furlanetto, S., Orlandini, S., Ghelardini, C., Mura, P., 2014. Development and characterization of functionalized niosomes for brain targeting of dynorphin-B. Eur. J. Pharm. Biopharm. 87, 73–79.
Bragagni, M., Scozzafava, A., Mastrolorenzo, A., Supura, C.T., Mura, P., 2015. Development and ex vivo evaluation of 5-aminolevulinic acid-loaded niosomal formulations for topical photodynamic therapy. Int. J. Pharm. 494, 258–263.
Brannigan, G., Brown, F.L.H., 2006. A consistent model for thermal fluctuations and protein-induced deformations in lipid bilayers. Biophys. J. 90, 1501–1520.
Brannon-Peppas, L., Blanchette, J.O., 2004. Nanoparticle and targeted systems for cancer therapy. Adv. Drug Deliv. Rev. 56, 1649–1659.
Breitkreutz, J., Boos, J., 2011. Drug delivery and formulations. Handb. Exp. Pharmacol. 205, 91–107.
Brigger, I., Dubernet, C., Couvreur, P., 2002. Nanoparticles in cancer therapy and diagnosis. Adv. Drug Deliv. Rev. 54, 631–651.
Brinkhuis, R.P., Stojanov, K., Laverman, P., Eilander, J., Zuhorn, I.S., Rutjes, F.P.J.T., Van Hest, J.C.M., 2012. Size dependent biodistribution and SPECT imaging of In-111-labeled polymersomes. Bioconjug. Chem. 23, 958–965.
Brown, M.B., Martin, G.P., Jones, S.A., Akomeah, F.K., 2006. Dermal and transdermal drug delivery systems: current and future prospects. Drug Deliv. 13, 175–187.
Brown, L., McArthur, S.L., Wright, P.C., Lewis, A., Battaglia, G., 2010. Polymersome production on a microfluidic platform using pH sensitive block copolymers. Lab Chip 10, 1922–1928.
Broz, P., Benito, S.M., Saw, C., Burger, P., Heider, H., Pfisterer, M., Marsch, S., Meier, W., Hunziker, P., 2005. Cell targeting by a generic receptor-targeted polymer nanocontainer platform. J. Control. Rel. 102, 475–488.
Bruning, B.-A., Prevost, S., Stehle, R., Steitz, R., Falus, P., Farago, B., Hellweg, T., 2014. Bilayer undulation dynamics in unilamellar phospholipid vesicles: effect of temperature, cholesterol and trehalose. Biochim. Biophys. Acta 1838, 2412–2419.
Bseiso, E.A., Nasr, M., Sammour, O., Abd El Gawad, N.A., 2015. Recent advances in topical formulation carriers of antifungal agents. Indian J. Dermatol. Venereol. Leprol. 81, 457–463.
Buckiova, D., Ranjan, S., Newman, T.A., Johnston, A.H., Sood, R., Kinnunen, P.K.J., Popelar, J., Chumak, T., Syka, J., 2012. Minimally invasive drug delivery to the cochlea through application of nanoparticles to the round window membrane. Nanomedicine 7, 1339–1354.
Byrne, J.D., Betancourt, T., Brannon-Peppas, L., 2008. Active targeting schemes for nanoparticle systems in cancer therapeutics. Adv. Drug Deliv. Rev. 60, 1615–1626.
Cabane, E., Malinova, V., Meier, W., 2010. Synthesis of photocleavable amphiphilic block copolymers: toward the design of photosensitive nanocarriers. Macromol. Chem. Phys. 211, 1847–1856.
Cao, Z., Jiang, S., 2012. Super-hydrophilic zwitterionic poly(carboxybetaine) and amphiphilic non-ionic poly(ethylene glycol) for stealth nanoparticles. Nano Today 7, 404–413.
Cao, X., Chen, Y., Chai, W., Zhang, W., Wang, Y., Fu, P.-F., 2015. Thermoresponsive self-assembled nanovesicles based on amphiphilic triblock copolymers and their potential applications as smart drug release carriers. J. Appl. Polym. Sci. 132, 41361.
Caracciolo, G., 2015. Liposome-protein corona in a physiological environment: challenges and opportunities for targeted delivery of nanomedicines. Nanomed.-Nanotechnol. 11, 543–557.
Cattel, L., Ceruti, M., Dosio, F., 2004. From conventional to stealth liposomes: a new frontier in cancer chemotherapy. J. Chemother. 16, 94–97.

Cerritelli, S., Velluto, D., Hubbell, J.A., 2007. PEG-SS-PPS: reduction-sensitive disulfide block copolymer vesicles for intracellular drug delivery. Biomacromolecules 8, 1966–1972.
Cevc, G., 1996. Transfersomes, liposomes and other lipid suspensions on the skin: permeation enhancement, vesicle penetration, and transdermal drug delivery. Crit. Rev. Ther. Drug Carrier Syst. 13, 257–388.
Chen, H.M., Langer, R., 1998. Oral particulate delivery: status and future trends. Adv. Drug Deliv. Rev. 34, 339–350.
Chen, Y., Liu, L., 2012. Modern methods for delivery of drugs across the blood–brain barrier. Adv. Drug Deliv. Rev. 64, 640–665.
Chen, C.-Y., Wang, H.-L., 2014. Dual thermo- and pH-responsive zwitterionic sulfobataine copolymers for oral delivery system. Macromol. Rapid Comm. 35, 1534–1540.
Chen, H.M., Torchilin, V., Langer, R., 1996. Lectin-bearing polymerized liposomes as potential oral vaccine carriers. Pharm. Res. 13, 1378–1383.
Chen, W., Meng, F., Cheng, R., Zhong, Z., 2010. pH-sensitive degradable polymersomes for triggered release of anticancer drugs: a comparative study with micelles. J. Control. Rel. 142, 40–46.
Chen, J., Shao, R., Zhang, X.D., Chen, C., 2013. Applications of nanotechnology for melanoma treatment, diagnosis, and theranostics. Int. J. Nanomed. 8, 2677–2688.
Chen, Y.-C., Chiang, C.-F., Chen, L.-F., Liang, P.-C., Hsieh, W.-Y., Lin, W.-L., 2014. Polymersomes conjugated with des-octanoyl ghrelin and folate as a BBB-penetrating cancer cell-targeting delivery system. Biomaterials 35, 4066–4081.
Cheng, R., Meng, F., Ma, S., Xu, H., Liu, H., Jing, X., Zhong, Z., 2011. Reduction and temperature dual-responsive crosslinked polymersomes for targeted intracellular protein delivery. J. Mater. Chem. 21, 19013–19020.
Cheng, W., Gu, L., Ren, W., Liu, Y., 2014. Stimuli-responsive polymers for anti-cancer drug delivery. Mater. Sci. Eng. C 45, 600–608.
Chetty, D.J., Chien, Y.W., 1998. Novel methods of insulin delivery: an update. Crit. Rev. Ther. Drug Carrier Syst. 15, 629–670.
Chilkoti, A., Dreher, M.R., Meyer, D.E., Raucher, D., 2002. Targeted drug delivery by thermally responsive polymers. Adv. Drug Deliv. Rev. 54, 613–630.
Chithrani, B.D., Ghazani, A.A., Chan, W.C.W., 2006. Determining the size and shape dependence of gold nanoparticle uptake into mammalian cells. Nano Lett. 6, 662–668.
Cho, K., Wang, X., Nie, S., Chen, Z., Shin, D.M., 2008. Therapeutic nanoparticles for drug delivery in cancer. Clin. Cancer Res. 14, 1310–1316.
Choi, M.J., Maibach, H.I., 2005. Liposomes and niosomes as topical drug delivery systems. Skin Pharmacol. Physi. 18, 209–219.
Chonn, A., Semple, S.C., Cullis, P.R., 1992. Association of blood proteins with large unilamellar liposomes invivo—relation to circulation lifetimes. J. Biol. Chem. 267, 18759–18765.
Claessens, M., Van Oort, B.F., Leermakers, F.A.M., Hoekstra, F.A., Stuart, M.A.C., 2004. Charged lipid vesicles: effects of salts on bending rigidity, stability, and size. Biophys. J. 87, 3882–3893.
Cleland, J.L., Daugherty, A., Mrsny, R., 2001. Emerging protein delivery methods. Curr. Opin. Biotechnol. 12, 212–219.
Dabbagh, A., Abdullah, B.J.J., Abdullah, H., Hamdi, M., Abu Kasim, N.H., 2015. Triggering mechanisms of thermosensitive nanoparticles under hyperthermia condition. J. Pharm. Sci. 104, 2414–2428.
Dalla Pozza, E., Lerda, C., Costanzo, C., Donadelli, M., Dando, I., Zoratti, E., Scupoli, M.T., Beghelli, S., Scarpa, A., Fattal, E., Arpicco, S., Palmieri, M., 2013. Targeting gemcitabine containing liposomes to CD44 expressing pancreatic adenocarcinoma cells causes an increase in the antitumoral activity. Biochim. Biophys. Acta 1828, 1396–1404.
Dan, N., 2007. Lipid tail chain asymmetry and the strength of membrane-induced interactions between membrane proteins. Biochim. Biophys. Acta 1768, 2393–2399.
Dan, N., 2015. Drug release through liposome pores. Colloids Surf. B 126, 80–86.

Dan, N., Danino, D., 2014. Structure and kinetics of lipid-nucleic acid complexes. Adv. Colloid Interf. Sci. 205, 230–239.

Dan, N., Safran, S.A., 1998. Effect of lipid characteristics on the structure of transmembrane proteins. Biophys. J. 75, 1410–1414.

Dan, N., Pincus, P., Safran, S.A., 1993. Membrane-induced interactions between inclusions. Langmuir 9, 2768–2771.

Danhier, F., Feron, O., Preat, V., 2010. To exploit the tumor microenvironment: passive and active tumor targeting of nanocarriers for anti-cancer drug delivery. J. Control. Rel. 148, 135–146.

Davis, M.E., Chen, Z., Shin, D.M., 2008. Nanoparticle therapeutics: an emerging treatment modality for cancer. Nat. Rev. Drug Discov. 7, 771–782.

De Jong, W.H., Hagens, W.I., Krystek, P., Burger, M.C., Sips, A.J.A.M., Geertsma, R.E., 2008. Particle size-dependent organ distribution of gold nanoparticles after intravenous administration. Biomaterials 29, 1912–1919.

Demirgoez, D., Pangburn, T.O., Davis, K.P., Lee, S., Bates, F.S., Kokkoli, E., 2009. PR_b-targeted delivery of tumor necrosis factor-alpha by polymersomes for the treatment of prostate cancer. Soft Matter 5, 2011–2019.

Deng, J., Zhang, Z., Liu, C., Yin, L., Zhou, J., Lv, H., 2015. The studies of *N*-octyl-*N*-arginine-chitosan coated liposome as an oral delivery system of cyclosporine A. J. Pharm. Pharmacol. 67, 1363–1370.

Desormeaux, A., Bergeron, M.G., 1998. Liposomes as drug delivery system: a strategic approach for the treatment of HIV infection. J. Drug Target. 6, 1–15.

Di Marzio, L., Marianecci, C., Petrone, M., Rinaldi, F., Carafa, M., 2011. Novel pH-sensitive non-ionic surfactant vesicles: comparison between Tween 21 and Tween 20. Colloids Surf. B 82, 18–24.

Di Marzio, L., Esposito, S., Rinaldi, F., Marianecci, C., Carafa, M., 2013. Polysorbate 20 vesicles as oral delivery system: in vitro characterization. Colloids Surf. B 104, 200–206.

Dijkmans, P.A., Juffermans, L.J.M., Musters, R.J.P., Van Wamel, A., Ten Cate, F.J., Van Gilst, W., Visser, C.A., De Jong, N., Kamp, O., 2004. Microbubbles and ultrasound: from diagnosis to therapy. Eur. J. Echocardiogr. 5, 245–256.

Dimova, R., 2014. Recent developments in the field of bending rigidity measurements on membranes. Adv. Colloid Interf. Sci. 208, 225–234.

Discher, D.E., Ahmed, F., 2006. Polymersomes. Annu. Rev. Biomed. Eng. 8, 323–341.

Discher, B.M., Won, Y.Y., Ege, D.S., Lee, J.C.M., Bates, F.S., Discher, D.E., Hammer, D.A., 1999. Polymersomes: tough vesicles made from diblock copolymers. Science 284, 1143–1146.

Dodd, M.J., 1987. Managing the Side Effects of Chemotherapy and Radiation. Prentice Hall Press, New York, NY, USA.

Drulis-Kawa, Z., Dorotkiewicz-Jach, A., 2010. Liposomes as delivery systems for antibiotics. Int. J. Pharm. 387, 187–198.

Drummond, C.J., Fong, C., 1999. Surfactant self-assembly objects as novel drug delivery vehicles. Curr. Opin. Colloid Interf. Sci. 4, 449–456.

Drummond, D.C., Meyer, O., Hong, K.L., Kirpotin, D.B., Papahadjopoulos, D., 1999. Optimizing liposomes for delivery of chemotherapeutic agents to solid tumors. Pharmacol. Rev. 51, 691–743.

Drummond, D.C., Zignani, M., Leroux, J.C., 2000. Current status of pH-sensitive liposomes in drug delivery. Prog. Lipid Res. 39, 409–460.

Dufay Wojcicki, A., Hillaireau, H., Nascimento, T.L., Arpicco, S., Taverna, M., Ribes, S., Bourge, M., Nicolas, V., Bochot, A., Vauthier, C., Tsapis, N., Fattal, E., 2012. Hyaluronic acid-bearing lipoplexes: physico-chemical characterization and in vitro targeting of the CD44 receptor. J. Control. Rel. 162, 545–552.

Dufes, C., Gaillard, F., Uchegbu, I.F., Schatzlein, A.G., Olivier, J.C., Muller, J.M., 2004. Glucose-targeted niosomes deliver vasoactive intestinal peptide (VIP) to the brain. Int. J. Pharm. 285, 77–85.

Duncanson, W.J., Lin, T., Abate, A.R., Seiffert, S., Shah, R.K., Weitz, D.A., 2012. Microfluidic synthesis of advanced microparticles for encapsulation and controlled release. Lab Chip 12, 2135–2145.

Egli, S., Schlaad, H., Bruns, N., Meier, W., 2011. Functionalization of block copolymer vesicle surfaces. Polymers 3, 252–280.
Ei-Laithy, H.M., Shoukry, O., Mahran, L.G., 2011. Novel sugar esters proniosomes for transdermal delivery of vinpocetine: preclinical and clinical studies. Eur. J. Pharm. Biopharm. 77, 43–55.
Elnaggar, Y.S.R., El-Refaie, W.M., El-Massik, M.A., Abdallah, O.Y., 2014. Lecithin-based nanostructured gels for skin delivery: an update on state of art and recent applications. J. Control. Rel. 180, 10–24.
Eloy, J.O., De Souza, M.C., Petrilli, R., Abriata Barcellos, J.P., Lee, R.J., Marchetti, J.M., 2014. Liposomes as carriers of hydrophilic small molecule drugs: strategies to enhance encapsulation and delivery. Colloids Surf. B 123, 345–363.
Enden, G., Schroeder, A., 2009. A mathematical model of drug release from liposomes by low frequency ultrasound. Ann. Biomed. Eng. 37, 2640–2645.
Ewer, M.S., Yeh, E., 2006. Cancer and the Heart. BC Decker, ON, Canada.
Faivre, V., Rosilio, V., 2010. Interest of glycolipids in drug delivery: from physicochemical properties to drug targeting. Expert Opin. Drug Del. 7, 1031–1048.
Fang, J.Y., Hong, C.T., Chiu, W.T., Wang, Y.Y., 2001. Effect of liposomes and niosomes on skin permeation of enoxacin. Int. J. Pharm. 219, 61–72.
Felber, A.E., Dufresne, M.-H., Leroux, J.-C., 2012. pH-sensitive vesicles, polymeric micelles, and nanospheres prepared with polycarboxylates. Adv. Drug Deliv. Rev. 64, 979–992.
Fernandes, E., Ferreira, J.A., Peixoto, A., Lima, L., Barroso, S., Sarmento, B., Santos, L.L., 2015. New trends in guided nanotherapies for digestive cancers: a systematic review. J. Control. Rel. 209, 288–307.
Fogueri, L.R., Singh, S., 2009. Smart polymers for controlled delivery of proteins and peptides: a review of patents. Recent Pat. Drug Deliv. Formulat. 3, 40–48.
Freiberg, S., Zhu, X., 2004. Polymer microspheres for controlled drug release. Int. J. Pharm. 282, 1–18.
Froehlich, E., 2012. The role of surface charge in cellular uptake and cytotoxicity of medical nanoparticles. Int. J. Nanomed. 7, 5577–5591.
Gaafar, P.M.E., Abdallah, O.Y., Farid, R.M., Abdelkader, H., 2014. Preparation, characterization and evaluation of novel elastic nano-sized niosomes (ethoniosomes) for ocular delivery of prednisolone. J. Liposome Res. 24, 204–215.
Gabizon, A.A., 1995. Liposome circulation time and tumor targeting—implications for cancer-chemotherapy. Adv. Drug Deliv. Rev. 16, 285–294.
Gabizon, A., Martin, F., 1997. Polyethylene glycol coated (pegylated) liposomal doxorubicin—rationale for use in solid tumours. Drugs 54, 15–21.
Gabizon, A., Papahadjopoulos, D., 1992. The role of surface-charge and hydrophilic groups on liposome clearance invivo. Biochim. Biophys. Acta 1103, 94–100.
Gabizon, A., Shmeeda, H., Barenholz, Y., 2003. Pharmacokinetics of pegylated liposomal doxorubicin—review of animal and human studies. Clin. Pharmacokinet. 42, 419–436.
Gaitzsch, J., Appelhans, D., Graefe, D., Schwille, P., Voit, B., 2011. Photo-crosslinked and pH sensitive polymersomes for triggering the loading and release of cargo. Chem. Comm. 47, 3466–3468.
Gallon, E., Matini, T., Sasso, L., Mantovani, G., Arminan De Benito, A., Sanchis, J., Caliceti, P., Alexander, C., Vicent, M.J., Salmaso, S., 2015. Triblock copolymer nanovesicles for pH-responsive targeted delivery and controlled release of siRNA to cancer cells. Biomacromolecules 16, 1924–1937.
Gao, X.H., Cui, Y.Y., Levenson, R.M., Chung, L.W.K., Nie, S.M., 2004. In vivo cancer targeting and imaging with semiconductor quantum dots. Nat. Biotechnol. 22, 969–976.
Gaudana, R., Jwala, J., Boddu, S.H.S., Mitra, A.K., 2009. Recent perspectives in ocular drug delivery. Pharm. Res. 26, 1197–1216.
Geers, B., Dewitte, H., De Smedt, S.C., Lentacker, I., 2012. Crucial factors and emerging concepts in ultrasound-triggered drug delivery. J. Control. Rel. 164, 248–255.
Ghosh, P., Han, G., De, M., Kim, C.K., Rotello, V.M., 2008. Gold nanoparticles in delivery applications. Adv. Drug Deliv. Rev. 60, 1307–1315.

Giddam, A.K., Zaman, M., Skwarczynski, M., Toth, I., 2012. Liposome-based delivery system for vaccine candidates: constructing an effective formulation. Nanomedicine 7, 1877–1893.

Giljohann, D.A., Seferos, D.S., Daniel, W.L., Massich, M.D., Patel, P.C., Mirkin, C.A., 2010. Gold nanoparticles for biology and medicine. Angew. Chem. Int. Ed. 49, 3280–3294.

Gill, S., Lobenberg, R., Ku, T., Azarmi, S., Roa, W., Prenner, E.J., 2007. Nanoparticles: characteristics, mechanisms of action, and toxicity in pulmonary drug delivery—a review. J. Biomed. Nanotechnol. 3, 107–119.

Gillies, E.R., Fréchet, J.M.J., 2005. pH-responsive copolymer assemblies for controlled release of doxorubicin. Bioconjug. Chem. 16, 361–368.

Gokhale, P.C., Pei, J., Zhang, C.B., Ahmad, I., Rahman, A., Kasid, U., 2001. Improved safety, pharmacokinetics and therapeutic efficacy profiles of a novel liposomal formulation of mitoxantrone. Anticancer Res. 21, 3313–3321.

Gopinath, D., Ravi, D., Rao, B.R., Apte, S.S., Renuka, D., Rambhau, D., 2004. Ascorbyl palmitate vesicles (aspasomes): formation, characterization and applications. Int. J. Pharm. 271, 95–113.

Gracia, R.S., Bezlyepkina, N., Knorr, R.L., Lipowsky, R., Dimova, R., 2010. Effect of cholesterol on the rigidity of saturated and unsaturated membranes: fluctuation and electrodeformation analysis of giant vesicles. Soft Matter 6, 1472–1482.

Gref, R., Luck, M., Quellec, P., Marchand, M., Dellacherie, E., Harnisch, S., Blunk, T., Muller, R.H., 2000. 'Stealth' corona-core nanoparticles surface modified by polyethylene glycol (PEG): influences of the corona (PEG chain length and surface density) and of the core composition on phagocytic uptake and plasma protein adsorption. Colloids Surf. B 18, 301–313.

Gregoriadis, G., 1995. Engineering liposomes for drug delivery: progress and problems. Trends Biotechnol. 13, 527–537.

Guan, L., Rizzello, L., Battaglia, G., 2015. Polymersomes and their applications in cancer delivery and therapy. Nanomedicine 10, 2757–2780.

Guinedi, A.S., Mortada, N.D., Mansour, S., Hathout, R.M., 2005. Preparation and evaluation of reverse-phase evaporation and multilamellar niosomes as ophthalmic carriers of acetazolamide. Int. J. Pharm. 306, 71–82.

Gupta, A.K., Gupta, M., 2005. Synthesis and surface engineering of iron oxide nanoparticles for biomedical applications. Biomaterials 26, 3995–4021.

Gupta, U., Jain, N.K., 2010. Non-polymeric nano-carriers in HIV/AIDS drug delivery and targeting. Adv. Drug Deliv. Rev. 62, 478–490.

Halperin, A., 1999. Polymer brushes that resist adsorption of model proteins: design parameters. Langmuir 15, 2525–2533.

Hamman, J.H., Enslin, G.M., Kotze, A.F., 2005. Oral delivery of peptide drugs—barriers and developments. Biodrugs 19, 165–177.

Hao, Y.M., Zhao, F.L., Li, N., Yang, Y.H., Li, K., 2002. Studies on a high encapsulation of colchicine by a niosome system. Int. J. Pharm. 244, 73–80.

Harris, L., Batist, G., Belt, R., Rovira, D., Navari, R., Azarnia, N., Welles, K., Winer, E., Grp, T.D.S., 2002. Liposome-encapsulated doxorubicin compared with conventional doxorubicin in a randomized multicenter trial as first-line therapy of metastatic breast carcinoma. Cancer 94, 25–36.

Hasan, A.A., Madkor, H., Wageh, S., 2013. Formulation and evaluation of metformin hydrochloride-loaded niosomes as controlled release drug delivery system. Drug Deliv. 20, 120–126.

Hassan, N., Oyarzun-Ampuero, F., Lara, P., Guerrero, S., Cabuil, V., Abou-Hassan, A., Kogan, M.J., 2014. Flow chemistry to control the synthesis of nano and microparticles for biomedical applications. Curr. Top. Med. Chem. 14, 676–689.

Hayward, R.C., Utada, A.S., Dan, N., Weitz, D.A., 2006. Dewetting instability during the formation of polymersomes from block-copolymer-stabilized double emulsions. Langmuir 22, 4457–4461.

He, C., Zhuang, X., Tang, Z., Tian, H., Chen, X., 2012. Stimuli-sensitive synthetic polypeptide-based materials for drug and gene delivery. Adv. Healthcare Mater. 1, 48–78.

Helfrich, W., 1973. Elastic properties of lipid bilayers: theory and possible experiments. Z. Naturforsch. C 28, 693–703.

Henri, P., Ospital, R., Teissie, J., 2015. Content delivery of lipidic nanovesicles in electropermeabilized cells. J. Memb. Biol. 248, 849–855.

Henriksen, J., Rowat, A.C., Brief, E., Hsueh, Y.W., Thewalt, J.L., Zuckermann, M.J., Ipsen, J.H., 2006. Universal behavior of membranes with sterols. Biophys. J. 90, 1639–1649.

Hillaireau, H., Couvreur, P., 2009. Nanocarriers' entry into the cell: relevance to drug delivery. Cell. Mol. Life Sci. 66, 2873–2896.

Hobbs, S.K., Monsky, W.L., Yuan, F., Roberts, W.G., Griffith, L., Torchilin, V.P., Jain, R.K., 1998. Regulation of transport pathways in tumor vessels: role of tumor type and microenvironment. Proc. Natl. Acad. Sci. USA 95, 4607–4612.

Hosseini, H., Li, Y., Kanellakis, P., Tay, C., Cao, A., Tipping, P., Bobik, A., Toh, B.-H., Kyaw, T., 2015. Phosphatidylserine liposomes mimic apoptotic cells to attenuate atherosclerosis by expanding polyreactive IgM producing B1a lymphocytes. Cardiovasc. Res. 106, 443–452.

Hristova, K., Kenworthy, A., McIntosh, T.J., 1995. Effect of bilayer composition on the phase-behavior of liposomal suspensions containing poly(ethylene glycol)-lipids. Macromolecules 28, 7693–7699.

Hu, C.-M.J., Fang, R.H., Luk, B.T., Zhang, L., 2014. Polymeric nanotherapeutics: clinical development and advances in stealth functionalization strategies. Nanoscale 6, 65–75.

Hubert, A., Lyass, O., Pode, D., Gabizon, A., 2000. Doxil (Caelyx): an exploratory study with pharmacokinetics in patients with hormone-refractory prostate cancer. Anti-Cancer Drug 11, 123–127.

Husseini, G.A., Pitt, W.G., Martins, A.M., 2014. Ultrasonically triggered drug delivery: breaking the barrier. Colloids Surf. B 123, 364–386.

Iglic, A., 2012. Advances in Planar Lipid Bilayers and Liposomes. Elsevier Science.

Illya, G., Lipowsky, R., Shillcock, J.C., 2005. Effect of chain length and asymmetry on material properties of bilayer membranes. J. Chem. Phys., 122.

Immordino, M.L., Dosio, F., Cattel, L., 2006. Stealth liposomes: review of the basic science, rationale, and clinical applications, existing and potential. Int. J. Nanomed. 1, 297–315.

Ioele, G., Tavano, L., De Luca, M., Ragno, G., Picci, N., Muzzalupo, R., 2015. Photostability and ex-vivo permeation studies on diclofenac in topical niosomal formulations. Int. J. Pharm. 494, 490–497.

Ishida, T., Harashima, H., Kiwada, H., 2002. Liposome clearance. Biosci. Rep. 22, 197–224.

Israelachvili, J.N., 2011. Intermolecular and Surface Forces, Revised third ed. Elsevier Science, Amsterdam, The Netherlands.

Iwamoto, T., 2013. Clinical application of drug delivery systems in cancer chemotherapy: review of the efficacy and side effects of approved drugs. Biol. Pharm. Bull. 36, 715–718.

Jadon, P.S., Gajbhiye, V., Jadon, R.S., Gajbhiye, K.R., Ganesh, N., 2009. Enhanced oral bioavailability of griseofulvin via niosomes. AAPS PharmSciTech 10, 1186–1192.

Jahn, A., Vreeland, W.N., Gaitan, M., Locascio, L.E., 2004. Controlled vesicle self-assembly in microfluidic channels with hydrodynamic focusing. J. Am. Chem. Soc. 126, 2674–2675.

Jahn, A., Stavis, S.M., Hong, J.S., Vreeland, W.N., Devoe, D.L., Gaitan, M., 2010. Microfluidic mixing and the formation of nanoscale lipid vesicles. ACS Nano 4, 2077–2087.

Jain, A., Jain, S.K., 2015. Ligand-appended BBB-targeted nanocarriers (LABTNs). Crit. Rev. Ther. Drug Carrier Syst. 32, 149–180.

Jain, S., Vyas, S.P., 2006. Mannosylated niosomes as adjuvant-carrier system for oral mucosal immunization. J. Liposome Res. 16, 331–345.

Jain, S., Singh, P., Mishra, V., Vyas, S.P., 2005. Mannosylated niosomes as adjuvant-carrier system for oral genetic immunization against hepatitis B. Immunol. Lett. 101, 41–49.

Jain, S., Sharma, R.K., Vyas, S.P., 2006. Chitosan nanoparticles encapsulated vesicular systems for oral immunization: preparation, in-vitro and in-vivo characterization. J. Pharm. Pharmacol. 58, 303–310.

Jain, J.P., Jatana, M., Chakrabarti, A., Kumar, N., 2011. Amphotericin-B-loaded polymersomes formulation (PAMBO) based on (PEG)(3)-PLA copolymers: an in vivo evaluation in a murine model. Mol. Pharm. 8, 204–212.

Jain, A., Jain, A., Gulbake, A., Shilpi, S., Hurkat, P., Jain, S.K., 2013. Peptide and protein delivery using new drug delivery systems. Crit. Rev. Ther. Drug Carrier Syst. 30, 293–329.

Jain, S., Harde, H., Indulkar, A., Agrawal, A.K., 2014a. Improved stability and immunological potential of tetanus toxoid containing surface engineered bilosomes following oral administration. Nanomed.-Nanotechnol. Biol. Med. 10, 431–440.

Jain, S., Indulkar, A., Harde, H., Agrawal, A.K., 2014b. Oral mucosal immunization using glucomannosylated bilosomes. J. Biomed. Nanotechnol. 10, 932–947.

Janib, S.M., Moses, A.S., Mackay, J.A., 2010. Imaging and drug delivery using theranostic nanoparticles. Adv. Drug Deliv. Rev. 62, 1052–1063.

Jensen, S.M., Christensen, C.J., Petersen, J.M., Treusch, A.H., Brandl, M., 2015. Liposomes containing lipids from Sulfolobus islandicus withstand intestinal bile salts: an approach for oral drug delivery? Int. J. Pharm. 493, 63–69.

Jeong, E.S., Park, C., Kim, K.T., 2015. Doubly responsive polymersomes towards monosaccharides and temperature under physiologically relevant conditions. Polym. Chem. 6, 4080–4088.

Jin, Y., Wen, J., Garg, S., Liu, D., Zhou, Y., Teng, L., Zhang, W., 2013. Development of a novel niosomal system for oral delivery of Ginkgo biloba extract. Int. J. Nanomed. 8, 421–430.

Joshi, M.D., Mueller, R.H., 2009. Lipid nanoparticles for parenteral delivery of actives. Eur. J. Pharm. Biopharm. 71, 161–172.

Juliano, R.L., Stamp, D., 1975. The effect of particle size and charge on the clearance rates of liposomes and liposome encapsulated drugs. Biochem. Biophys. Res. Comm. 63, 651–658.

Kabanov, A.V., Bronich, T.K., Kabanov, V.A., Yu, K., Eisenberg, A., 1998. Spontaneous formation of vesicles from complexes of block ionomers and surfactants. J. Am. Chem. Soc. 120, 9941–9942.

Kain, V., Ingle, K.A., Colas, R.A., Dalli, J., Prabhu, S.D., Serhan, C.N., Joshi, M., Halade, G.V., 2015. Resolvin D1 activates the inflammation resolving response at splenic and ventricular site following myocardial infarction leading to improved ventricular function. J. Mol. Cell. Cardiol. 84, 24–35.

Karanth, H., Murthy, R.S.R., 2007. pH-sensitive liposomes—principle and application in cancer therapy. J. Pharm. Pharmacol. 59, 469–483.

Karlowsky, J.A., Zhanel, G.G., 1992. Concepts on the use of liposomal antimicrobial agents—applications for aminoglycosides. Clin. Infect. Dis. 15, 654–667.

Katare, R., Gupta, P.N., Mahor, S., Rawat, A., Khatri, K., Katare, Y., Panda, A.K., Vyas, S.P., 2006. Development of polysaccharide-capped niosomes for oral immunization of tetanus toxoid. J. Drug Deliv. Sci. Tec. 16, 167–172.

Katsogiannou, M., Peng, L., Catapano, C.V., Rocchi, P., 2011. Active-targeted nanotherapy strategies for prostate cancer. Curr. Cancer Drug Tar. 11, 954–965.

Kaur, I.P., Smitha, Rs., 2002. Penetration enhancers and ocular bioadhesives: two new avenues for ophthalmic drug delivery. Drug Dev. Ind. Pharm. 28, 353–369.

Kaur, I.P., Garg, A., Singla, A.K., Aggarwal, D., 2004. Vesicular systems in ocular drug delivery: an overview. Int. J. Pharm. 269, 1–14.

Kaur, I.P., Aggarwal, D., Singh, H., Kakkar, S., 2010. Improved ocular absorption kinetics of timolol maleate loaded into a bioadhesive niosomal delivery system. Graef. Arch. Clin. Exp. 248, 1467–1472.

Kaur, I.P., Rana, C., Singh, M., Bhushan, S., Singh, H., Kakkar, S., 2012. Development and evaluation of novel surfactant-based elastic vesicular system for ocular delivery of fluconazole. J. Ocul. Pharmacol. Ther. 28, 484–496.

Kawaguchi, H., 2000. Functional polymer microspheres. Prog. Polym. Sci. 25, 1171–1210.

Kedar, U., Phutane, P., Shidhaye, S., Kadam, V., 2010. Advances in polymeric micelles for drug delivery and tumor targeting. Nanomed.-Nanotechnol. Biol. Med. 6, 714–729.

Kettler, K., Veltman, K., Van De Meent, D., Van Wezel, A., Hendriks, A.J., 2014. Cellular uptake of nanoparticles as determined by particle properties, experimental conditions, and cell type. Environ. Toxicol. Chem. 33, 481–492.
Khafagy, E.-S., Morishita, M., Onuki, Y., Takayama, K., 2007. Current challenges in non-invasive insulin delivery systems: a comparative review. Adv. Drug Deliv. Rev. 59, 1521–1546.
Khare, V., Alam, N., Saneja, A., Dubey, R.D., Gupta, P.N., 2014. Targeted drug delivery systems for pancreatic cancer. J. Biomed. Nanotechnol. 10, 3462–3482.
Khutoryanskiy, V.V., 2011. Advances in mucoadhesion and mucoadhesive polymers. Macromol. Biosci. 11, 748–764.
Kierstead, P.H., Okochi, H., Venditto, V.J., Chuong, T.C., Kivimae, S., Frechet, J.M.J., Szoka, F.C., 2015. The effect of polymer backbone chemistry on the induction of the accelerated blood clearance in polymer modified liposomes. J. Control. Rel. 213, 1–9.
Kim, Y., Tewari, M., Pajeroski, D.J., Sen, S., Jason, W., Sirsi, S., Lutz, G., Discher, D.E., 2006. Efficient nuclear delivery and nuclear body localization of antisense oligo-nucleotides using degradable polymersomes. Conference Proceedings: Annual International Conference of the IEEE Engineering in Medicine and Biology Society, 1, 4350–4353.
Kim, S.-H., Kim, J.W., Cho, J.-C., Weitz, D.A., 2011. Double-emulsion drops with ultra-thin shells for capsule templates. Lab Chip 11, 3162–3166.
Kim, H., Kong, Y.J., Jeong, E.S., Kong, S., Kim, K.T., 2012. Glucose-responsive disassembly of polymersomes of sequence-specific boroxole-containing block copolymers under physiologically relevant conditions. ACS Macro Lett. 1, 1194–1198.
Kim, S.-H., Nam, J., Kim, J.W., Kim, D.-H., Han, S.-H., Weitz, D.A., 2013. Formation of polymersomes with double bilayers templated by quadruple emulsions. Lab Chip 13, 1351–1356.
Kim, M.S., Lee, D.-W., Park, K., Park, S.-J., Choi, E.-J., Park, E.S., Kim, H.R., 2014. Temperature-triggered tumor-specific delivery of anticancer agents by cRGD-conjugated thermosensitive liposomes. Colloids Surf. B 116, 17–25.
Kisel, M.A., Kulik, L.N., Tsybovsky, I.S., Vlasov, A.P., Vorob'Yov, M.S., Kholodova, E.A., Zabarovskaya, Z.V., 2001. Liposomes with phosphatidylethanol as a carrier for oral delivery of insulin: studies in the rat. Int. J. Pharm. 216, 105–114.
Kneidl, B., Peller, M., Winter, G., Lindner, L.H., Hossann, M., 2014. Thermosensitive liposomal drug delivery systems: state of the art review. Int. J. Nanomed. 9, 4387–4398.
Kobayashi, Y., Kiguchi, N., Fukazawa, Y., Saika, F., Maeda, T., Kishioka, S., 2015. Macrophage-T cell interactions mediate neuropathic pain through the glucocorticoid-induced tumor necrosis factor ligand system. J. Biol. Chem. 290, 12603–12613.
Kohli, A.G., Kierstead, P.H., Venditto, V.J., Walsh, C.L., Szoka, F.C., 2014. Designer lipids for drug delivery: from heads to tails. J. Control. Rel. 190, 274–287.
Kono, K., Takashima, M., Yuba, E., Harada, A., Hiramatsu, Y., Kitagawa, H., Otani, T., Maruyama, K., Aoshima, S., 2015. Multifunctional liposomes having target specificity, temperature-triggered release, and near-infrared fluorescence imaging for tumor-specific chemotherapy. J. Control. Rel. 216, 69–77.
Koynova, R., Caffrey, M., 1998. Phases and phase transitions of the phosphatidylcholines. Biochim. Biophys. Acta 1376, 91–145.
Krasovitski, B., Frenkel, V., Shoham, S., Kimmel, E., 2011. Intramembrane cavitation as a unifying mechanism for ultrasound-induced bioeffects. Proc. Natl. Acad. Sci. USA 108, 3258–3263.
Krishnamoorthy, B., Karanam, V., Chellan, V.R., Siram, K., Natarajan, T.S., Gregory, M., 2014. Polymersomes as an effective drug delivery system for glioma—a review. J. Drug Target. 22, 469–477.
Kučerka, N., Nieh, M.-P., Katsaras, J., 2011. Fluid phase lipid areas and bilayer thicknesses of commonly used phosphatidylcholines as a function of temperature. Biochim. Biophys. Acta 1808, 2761–2771.
Kumar, M., 2000. Nano and microparticles as controlled drug delivery devices. J. Pharm. Pharm. Sci. 3, 234–258.

Kumar, T.R.S., Soppimath, K., Nachaegari, S.K., 2006. Novel delivery technologies for protein and peptide therapeutics. Curr. Pharm. Biotechnol. 7, 261–276.
Lane, D.D., Su, F.Y., Chiu, D.Y., Srinivasan, S., Wilson, J.T., Ratner, D.M., Stayton, P.S., Convertine, A.J., 2015. Dynamic intracellular delivery of antibiotics via pH-responsive polymersomes. Polym. Chem. 6, 1255–1266.
Langer, R., 1998. Drug delivery and targeting. Nature 392, 5–10.
Larsen, J.M., Martin, D.R., Byrne, M.E., 2014. Recent advances in delivery through the blood–brain barrier. Curr. Top. Med. Chem. 14, 1148–1160.
Lasic, D.D., 1998. Novel applications of liposomes. Trends Biotechnol. 16, 307–321.
Lasic, D.D., Needham, D., 1995. The "stealth" liposome: a prototypical biomaterial. Chem. Rev. 95, 2601–2628.
Lasic, D.D., Papahadjopoulos, D., 1998. Medical Applications of Liposomes. Elsevier Science, Amsterdam, The Netherlands.
Laurent, S., Forge, D., Port, M., Roch, A., Robic, C., Elst, L.V., Muller, R.N., 2008. Magnetic iron oxide nanoparticles: synthesis, stabilization, vectorization, physicochemical characterizations, and biological applications. Chem. Rev. 108, 2064–2110.
LaVan, D.A., Mcguire, T., Langer, R., 2003. Small-scale systems for in vivo drug delivery. Nat. Biotechnol. 21, 1184–1191.
Lavasanifar, A., Samuel, J., Kwon, G.S., 2002. Poly(ethylene oxide)-block-poly(L-amino acid) micelles for drug delivery. Adv. Drug Deliv. Rev. 54, 169–190.
Lawrence, M.J., Rees, G.D., 2000. Microemulsion-based media as novel drug delivery systems. Adv. Drug Deliv. Rev. 45, 89–121.
Lee, J.S., Feijen, J., 2012. Polymersomes for drug delivery: design, formation and characterization. J. Control. Rel. 161, 473–483.
Lee, J.S., Ankone, M., Pieters, E., Schiffelers, R.M., Hennink, W.E., Feijen, J., 2011a. Circulation kinetics and biodistribution of dual-labeled polymersomes with modulated surface charge in tumor-bearing mice: comparison with stealth liposomes. J. Control. Rel. 155, 282–288.
Lee, J.S., Groothuis, T., Cusan, C., Mink, D., Feijen, J., 2011b. Lysosomally cleavable peptide-containing polymersomes modified with anti-EGFR antibody for systemic cancer chemotherapy. Biomaterials 32, 9144–9153.
Lee, H., Kim, H.R., Park, J.C., 2014. Dynamics and stability of lipid bilayers modulated by thermosensitive polypeptides, cholesterols, and PEGylated lipids. Phys. Chem. Chem. Phys. 16, 3763–3770.
Letchford, K., Burt, H., 2007. A review of the formation and classification of amphiphilic block copolymer nanoparticulate structures: micelles, nanospheres, nanocapsules and polymersomes. Eur. J. Pharm. Biopharm. 65, 259–269.
Leucuta, S.E., 2012. Drug delivery systems with modified release for systemic and biophase bioavailability. Curr. Clin. Pharm. 7, 282–317.
Levchenko, T.S., Rammohan, R., Lukyanov, A.N., Whiteman, K.R., Torchilin, V.P., 2002. Liposome clearance in mice: the effect of a separate and combined presence of surface charge and polymer coating. Int. J. Pharm. 240, 95–102.
Levine, D.H., Ghoroghchian, P.P., Freudenberg, J., Zhang, G., Therien, M.J., Greene, M.I., Hammer, D.A., Murali, R., 2008. Polymersomes: a new multi-functional tool for cancer diagnosis and therapy. Methods 46, 25–32.
Levy, R., Shaheen, U., Cesbron, Y., See, V., 2010. Gold nanoparticles delivery in mammalian live cells: a critical review. Nano Rev., 1.
Li, W., Szoka, Jr., F.C., 2007. Lipid-based nanoparticles for nucleic acid delivery. Pharm. Res. 24, 438–449.
Li, S., Byrne, B., Welsh, J., Palmer, A.F., 2007. Self-assembled poly(butadiene)-*b*-poly(ethylene oxide) polymersomes as paclitaxel carriers. Biotechnol. Prog. 23, 278–285.
Li, Z., Barnes, J.C., Bosoy, A., Stoddart, J.F., Zink, J.I., 2012a. Mesoporous silica nanoparticles in biomedical applications. Chem. Soc. Rev. 41, 2590–2605.
Li, Z., Wu, L., Hu, P., Han, S., Zhang, T., Fan, H., Jin, W., Jin, Q., Mu, Y., 2012b. Soft nanomaterial-based targeting polymersomes for near-infrared fluorescence multispectral in vivo imaging. Nanoscale 4, 7097–7105.

Li, G., Guo, L., Wen, Q., Zhang, T., 2013a. Thermo- and pH-sensitive ionic-crosslinked hollow spheres from chitosan-based graft copolymer for 5-fluorouracil release. Int. J. Biol. Macromol. 55, 69–74.

Li, J., Jiang, H., Hu, W., Xia, H., Zou, G., Zhang, Q., 2013b. Photo-controlled hierarchical assembly and fusion of coumarin-containing polydiacetylene vesicles. Macromol. Rapid Commun. 34, 274–279.

Lian, T., Ho, R.J.Y., 2001. Trends and developments in liposome drug delivery systems. J. Pharm. Sci. 90, 667–680.

Liao, J., Wang, C., Wang, Y., Luo, F., Qian, Z., 2012. Recent advances in formation, properties, and applications of polymersomes. Curr. Pharm. Design 18, 3432–3441.

Licciardi, M., Paolino, D., Celia, C., Giammona, G., Cavallaro, G., Fresto, M., 2010. Folate-targeted supramolecular vesicular aggregates based on polyaspartyl-hydrazide copolymers for the selective delivery of antitumoral drugs. Biomaterials 31, 7340–7354.

Lin, Y.-S., Lee, M.-Y., Yang, C.-H., Huang, K.-S., 2015. Active targeted drug delivery for microbes using nano-carriers. Curr. Top. Med. Chem. 15, 1525–1531.

Liu, G., Liu, W., Dong, C.-M., 2013. UV- and NIR-responsive polymeric nanomedicines for on-demand drug delivery. Polym. Chem. 4, 3431–3443.

Liu, J., Detrembleur, C., Mornet, S., Jerome, C., Duguet, E., 2015. Design of hybrid nanovehicles for remotely triggered drug release: an overview. J. Mater. Chem. B 3, 6117–6147.

Lorenceau, E., Utada, A.S., Link, D.R., Cristobal, G., Joanicot, M., Weitz, D.A., 2005. Generation of polymerosomes from double-emulsions. Langmuir 21, 9183–9186.

Ludwig, A., 2005. The use of mucoadhesive polymers in ocular drug delivery. Adv. Drug Deliv. Rev. 57, 1595–1639.

Lyass, O., Uziely, B., Ben-Yosef, R., Tzemach, D., Heshing, N.I., Lotem, M., Brufman, G., Gabizon, A., 2000. Correlation of toxicity with pharmacokinetics of pegylated liposomal doxorubicin (Doxil) in metastatic breast carcinoma. Cancer 89, 1037–1047.

Lytton-Jean, A.K.R., Kauffman, K.J., Kaczmarek, J.C., Langer, R., 2015. Cancer nanotherapeutics in clinical trials. Cancer Treat. Res. 166, 293–322.

Ma, N., Ma, C., Li, C., Wang, T., Tang, Y., Wang, H., Mou, X., Chen, Z., He, N., 2013. Influence of nanoparticle shape, size, and surface functionalization on cellular uptake. J. Nanosci. Nanotechnol. 13, 6485–6498.

Maeda, H., Bharate, G.Y., Daruwalla, J., 2009. Polymeric drugs for efficient tumor-targeted drug delivery based on EPR-effect. Eur. J. Pharm. Biopharm. 71, 409–419.

Maesaki, S., 2002. Drug delivery system of anti-fungal and parasitic agents. Curr. Pharm. Design 8, 433–440.

Maherani, B., Arab-Tehrany, E., Mozafari, M.R., Gaiani, C., Linder, M., 2011. Liposomes: a review of manufacturing techniques and targeting strategies. Curr. Nanosci. 7, 436–452.

Maheshwari, C., Pandey, R.S., Chaurasiya, A., Kumar, A., Selvam, D.T., Prasad, G.B.K.S., Dixit, V.K., 2011. Non-ionic surfactant vesicles mediated transcutaneous immunization against hepatitis B. Int. Immunopharmacol. 11, 1516–1522.

Mahor, S., Gupta, P.N., Rawat, A., Vyas, S.P., 2007. A needle-free approach for topical immunization: antigen delivery via vesicular carrier system(s). Curr. Med. Chem. 14, 2898–2910.

Mailaender, V., Landfester, K., 2009. Interaction of nanoparticles with cells. Biomacromolecules 10, 2379–2400.

Malam, Y., Loizidou, M., Seifalian, A.M., 2009. Liposomes and nanoparticles: nanosized vehicles for drug delivery in cancer. Trends Pharmacol. Sci. 30, 592–599.

Mallick, S., Choi, J.S., 2014. Liposomes: versatile and biocompatible nanovesicles for efficient biomolecules delivery. J. Nanosci. Nanotechnol. 14, 755–765.

Mallipeddi, R., Rohan, L.C., 2010. Progress in antiretroviral drug delivery using nanotechnology. Int. J. Nanomed. 5, 533–547.

Manjunath, K., Reddy, J.S., Venkateswarlu, V., 2005. Solid lipid nanoparticles as drug delivery systems. Method. Find. Exp. Clin. 27, 127–144.

Mann, J.F.S., Scales, H.E., Shakir, E., Alexander, J., Carter, K.C., Mullen, A.B., Ferro, V.A., 2006. Oral delivery of tetanus toxoid using vesicles containing bile salts (bilosomes) induces significant systemic and mucosal immunity. Methods 38, 90–95.

Manosroi, A., Wongtrakul, P., Manosroi, J., Sakai, H., Sugawara, F., Yuasa, M., Abe, M., 2003. Characterization of vesicles prepared with various non-ionic surfactants mixed with cholesterol. Colloids Surf. B 30, 129–138.
Mansourpour, M., Mahjub, R., Amini, M., Ostad, S.N., Shamsa, E.S., Rafiee-Tehrani, M., Dorkoosh, F.A., 2015. Development of acid-resistant alginate/trimethyl chitosan nanoparticles containing cationic beta-cyclodextrin polymers for insulin oral delivery. AAPS PharmSciTech 16, 952–962.
Mantsch, H.H., McElhaney, R.N., 1991. Phospholipid phase-transitions in model and biological-membranes as studied by infrared-spectroscopy. Chem. Phys. Lipids 57, 213–226.
Marchal, S., El Hor, A., Millard, M., Gillon, V., Bezdetnaya, L., 2015. Anticancer drug delivery: an update on clinically applied nanotherapeutics. Drugs 75, 1601–1611.
Marianecci, C., Di Marzio, L., Rinaldi, F., Celia, C., Paolino, D., Alhaique, F., Esposito, S., Carafa, M., 2014. Niosomes from 80s to present: the state of the art. Adv. Colloid Interf. Sci. 205, 187–206.
Mathias, N.R., Hussain, M.A., 2010. Non-invasive systemic drug delivery: developability considerations for alternate routes of administration. J. Pharm. Sci. 99, 1–20.
Matosevic, S., 2012. Synthesizing artificial cells from giant unilamellar vesicles: state-of-the art in the development of microfluidic technology. Bioessays 34, 992–1001.
May, J.P., Ernsting, M.J., Undzys, E., Li, S.-D., 2013. Thermosensitive liposomes for the delivery of gemcitabine and oxaliplatin to tumors. Mol. Pharm. 10, 4499–4508.
Mazda, F., Ozer, A.Y., Ercan, M.T., Hincal, A.A., 1997. Preparation and characterization of urea niosomes—in vitro and in vivo studies. STP Pharma Sci. 7, 205–214.
McCarthy, J.R., Weissleder, R., 2008. Multifunctional magnetic nanoparticles for targeted imaging and therapy. Adv. Drug Deliv. Rev. 60, 1241–1251.
McClements, D.J., 2015. Encapsulation, protection, and release of hydrophilic active components: potential and limitations of colloidal delivery systems. Adv. Colloid Interf. Sci. 219, 27–53.
McCoy, C.P., Brady, C., Cowley, J.F., Mcglinchey, S.M., Mcgoldrick, N., Kinnear, D.J., Andrews, G.P., Jones, D.S., 2010. Triggered drug delivery from biomaterials. Expert Opin. Drug Deliv. 7, 605–616.
Mehnert, W., Mader, K., 2001. Solid lipid nanoparticles—production, characterization and applications. Adv. Drug Deliv. Rev. 47, 165–196.
Méléard, P., Gerbeaud, C., Pott, T., Fernandez-Puente, L., Bivas, I., Mitov, M.D., Dufourcq, J., Bothorel, P., 1997. Bending elasticities of model membranes: influences of temperature and sterol content. Biophys. J. 72, 2616–2629.
Meng, F.H., Engbers, G.H.M., Feijen, J., 2005. Biodegradable polymersomes as a basis for artificial cells: encapsulation, release and targeting. J. Control. Rel. 101, 187–198.
Meng, F., Hennink, W.E., Zhong, Z., 2009a. Reduction-sensitive polymers and bioconjugates for biomedical applications. Biomaterials 30, 2180–2198.
Meng, F., Zhong, Z., Feijen, J., 2009b. Stimuli-responsive polymersomes for programmed drug delivery. Biomacromolecules 10, 197–209.
Merdan, T., Kopecek, J., Kissel, T., 2002. Prospects for cationic polymers in gene and oligonucleotide therapy against cancer. Adv. Drug Deliv. Rev. 54, 715–758.
Messager, L., Gaitzsch, J., Chierico, L., Battaglia, G., 2014. Novel aspects of encapsulation and delivery using polymersomes. Curr. Opin. Pharmacol. 18, 104–111.
Milner, S.T., Witten, T.A., Cates, M.E., 1988. Theory of the grafted polymer brush. Macromolecules 21, 2610–2619.
Moghimi, S.M., Szebeni, J., 2003. Stealth liposomes and long circulating nanoparticles: critical issues in pharmacokinetics, opsonization and protein-binding properties. Prog. Lipid Res. 42, 463–478.
Moghimi, S.M., Hunter, A.C., Murray, J.C., 2001. Long-circulating and target-specific nanoparticles: theory to practice. Pharmacol. Rev. 53, 283–318.
Mohammadi-Samani, S., Taghipour, B., 2015. PLGA micro and nanoparticles in delivery of peptides and proteins: problems and approaches. Pharm. Dev. Technol. 20, 385–393.

Mohr, K., Mueller, S.S., Mueller, L.K., Rusitzka, K., Gietzen, S., Frey, H., Schmidt, M., 2014. Evaluation of multifunctional liposomes in human blood serum by light scattering. Langmuir 30, 14954–14962.
Morachis, J.M., Mahmoud, E.A., Almutairi, A., 2012. Physical and chemical strategies for therapeutic delivery by using polymeric nanoparticles. Pharmacol. Rev. 64, 505–519.
Morishita, M., Peppas, N.A., 2006. Is the oral route possible for peptide and protein drug delivery? Drug Discov. Today 11, 905–910.
Mornet, S., Vasseur, S., Grasset, F., Duguet, E., 2004. Magnetic nanoparticle design for medical diagnosis and therapy. J. Mater. Chem. 14, 2161–2175.
Mosqueira, V.C.F., Legrand, P., Morgat, J.L., Vert, M., Mysiakine, E., Gref, R., Devissaguet, J.P., Barratt, G., 2001. Biodistribution of long-circulating PEG-grafted nanocapsules in mice: effects of PEG chain length and density. Pharm. Res. 18, 1411–1419.
Moussa, H.G., Martins, A.M., Husseini, G.A., 2015. Review on triggered liposomal drug delivery with a focus on ultrasound. Curr. Cancer Drug Tar. 15, 282–313.
Movahedi, F., Hu, R.G., Becker, D.L., Xu, C., 2015. Stimuli-responsive liposomes for the delivery of nucleic acid therapeutics. Nanomed.-Nanotechnol. 11, 1575–1584.
Muggia, F.M., Hainsworth, J.D., Jeffers, S., Miller, P., Groshen, S., Tan, M., Roman, L., Uziely, B., Muderspach, L., Garcia, A., Burnett, A., Greco, F.A., Morrow, C.P., Paradiso, L.J., Liang, L.J., 1997. Phase II study of liposomal doxorubicin in refractory ovarian cancer: antitumor activity and toxicity modification by liposomal encapsulation. J. Clin. Oncol. 15, 987–993.
Muller, R.H., Mader, K., Gohla, S., 2000. Solid lipid nanoparticles (SLN) for controlled drug delivery—a review of the state of the art. Eur. J. Pharm. Biopharm. 50, 161–177.
Mundargi, R.C., Babu, V.R., Rangaswamy, V., Patel, P., Aminabhavi, T.M., 2008. Nano/micro technologies for delivering macromolecular therapeutics using poly(D,L-lactide-*co*-glycolide) and its derivatives. J. Control. Rel. 125, 193–209.
Muzzalupo, R., Tavano, L., La Mesa, C., 2013. Alkyl glucopyranoside-based niosomes containing methotrexate for pharmaceutical applications: evaluation of physico-chemical and biological properties. Int. J. Pharm. 458, 224–229.
Myerson, J.W., Brenner, J.S., Greineder, C.F., Muzykantov, V.R., 2015. Systems approaches to design of targeted therapeutic delivery. Wiley Interdiscip. Rev. Syst. Biol. Med. 7, 253–265.
Nag, O.K., Awasthi, V., 2013. Surface engineering of liposomes for stealth behavior. Pharmaceutics 5, 542–569.
Nag, O.K., Yadav, V.R., Croft, B., Hedrick, A., Awasthi, V., 2015. Liposomes modified with superhydrophilic polymer linked to a nonphospholipid anchor exhibit reduced complement activation and enhanced circulation. J. Pharm. Sci. 104, 114–123.
Napper, A.D., Watson, V.G., 2013. Targeted drug discovery for pediatric leukemia. Frontiers Oncol. 3, 170.
Narang, A.S., Mahato, R.I., 2010. Targeted Delivery of Small and Macromolecular Drugs. CRC Press, Boca Raton, Fl, USA.
Nardin, C., Hirt, T., Leukel, J., Meier, W., 2000. Polymerized ABA triblock copolymer vesicles. Langmuir 16, 1035–1041.
Neu, M., Fischer, D., Kissel, T., 2005. Recent advances in rational gene transfer vector design based on poly(ethylene imine) and its derivatives. J. Gene Med. 7, 992–1009.
Nie, S., Xing, Y., Kim, G.J., Simons, J.W., 2007. Nanotechnology applications in cancer. Annu. Rev. Biomed. Eng. 9, 257–288.
Nishikawa, K., Arai, H., Inoue, K., 1990. Scavenger receptor-mediated uptake and metabolism of lipid vesicles containing acidic phospholipids by mouse peritoneal-macrophages. J. Biol. Chem. 265, 5226–5231.
Nishiyama, N., Kataoka, K., 2006. Current state, achievements, and future prospects of polymeric micelles as nanocarriers for drug and gene delivery. Pharmacol. Ther. 112, 630–648.
Nkabinde, L.A., Shoba-Zikhali, L.N.N., Semete-Makokotlela, B., Kalombo, L., Swai, H.S., Hayeshi, R., Naicker, B., Hillie, T.K., Hamman, J.H., 2012. Permeation of PLGA nanoparticles across different in vitro models. Curr. Drug Deliv. 9, 617–627.

Nobili, S., Landini, I., Giglioni, B., Mini, E., 2006. Pharmacological strategies for overcoming multidrug resistance. Curr. Drug Targets 7, 861–879.

Nogueira, E., Mangialavori, I.C., Loureiro, A., Azoia, N.G., Sarria, M.P., Nogueira, P., Freitas, J., Harmark, J., Shimanovich, U., Rollett, A., Lacroix, G., Bernardes, G.J.L., Guebitz, G., Hebert, H., Moreira, A., Carmo, A.M., Rossi, J.P.F.C., Gomes, A.C., Preto, A., Cavaco-Paulo, A., 2015. Peptide anchor for folate-targeted liposomal delivery. Biomacromolecules 16, 2904–2910.

Novell, A., Al Sabbagh, C., Escoffre, J.-M., Gaillard, C., Tsapis, N., Fattal, E., Bouakaz, A., 2015. Focused ultrasound influence on calcein-loaded thermosensitive stealth liposomes. Int. J. Hyperther. 31, 349–358.

Ojewole, E., Mackraj, I., Naidoo, P., Govender, T., 2008. Exploring the use of novel drug delivery systems for antiretroviral drugs. Eur. J. Pharm. Biopharm. 70, 697–710.

Oku, N., Namba, Y., 2005. Glucuronate-modified, long-circulating liposomes for the delivery of anticancer agents. Methods Enzymol. 391, 145–162.

Olivier, J.-C., 2005. Drug transport to brain with targeted nanoparticles. NeuroRx 2, 108–119.

Oltra, N.S., Nair, P., Discher, D.E., 2014. From stealthy polymersomes and filomicelles to "self" peptide-nanoparticles for cancer therapy. In: Prausnitz, J.M., Doherty, M.F., Segalman, R.A. (Eds.), Ann. Rev. Chem. Biomol. Eng., vol. 5, pp. 281–299.

Onaca, O., Enea, R., Hughes, D.W., Meier, W., 2009. Stimuli-responsive polymersomes as nanocarriers for drug and gene delivery. Macromol. Biosci. 9, 129–139.

Opalinska, J.B., Gewirtz, A.M., 2002. Nucleic-acid therapeutics: basic principles and recent applications. Nat. Rev. Drug Discov. 1, 503–514.

Opanasopit, P., Nishikawa, M., Hashida, M., 2002. Factors affecting drug and gene delivery: effects of interaction with blood components. Crit. Rev. Ther. Drug Carrier Syst. 19, 191–233.

Paliwal, S.R., Paliwal, R., Vyas, S.P., 2015. A review of mechanistic insight and application of pH-sensitive liposomes in drug delivery. Drug Deliv. 22, 231–242.

Pan, J., Tristram-Nagle, S., Nagle, J.F., 2009. Effect of cholesterol on structural and mechanical properties of membranes depends on lipid chain saturation. Phys. Rev. E, 80.

Pando, D., Matos, M., Gutierrez, G., Pazos, C., 2015. Formulation of resveratrol entrapped niosomes for topical use. Colloids Surf. B 128, 398–404.

Pang, Z., Lu, W., Gao, H., Hu, K., Chen, J., Zhang, C., Gao, X., Jiang, X., Zhu, C., 2008. Preparation and brain delivery property of biodegradable polymersomes conjugated with OX26. J. Control. Rel. 128, 120–127.

Pang, Z., Feng, L., Hua, R., Chen, J., Gao, H., Pan, S., Jiang, X., Zhang, P., 2010. Lactoferrin-conjugated biodegradable polymersome holding doxorubicin and tetrandrine for chemotherapy of glioma rats. Mol. Pharm. 7, 1995–2005.

Pankhurst, Q.A., Connolly, J., Jones, S.K., Dobson, J., 2003. Applications of magnetic nanoparticles in biomedicine. J. Phys. D Appl. Phys. 36, R167–R181.

Pardakhty, A., Varshosaz, J., Rouholamini, A., 2007. In vitro study of polyoxyethylene alkyl ether niosornes for delivery of insulin. Int. J. Pharm. 328, 130–141.

Pardakhty, A., Moazeni, E., Varshosaz, J., Hajhashemi, V., Najafabadi, R.A., 2011. Pharmacokinetic study of niosome-loaded insulin in diabetic rats. Daru-J. Pharm. Sci. 19, 404–411.

Park, J.W., Kim, S.J., Kwag, D.S., Kim, S., Park, J., Youn, Y.S., Bae, Y.H., Lee, E.S., 2015. Multifunctional delivery systems for advanced oral uptake of peptide/protein drugs. Curr. Pharm. Design 21, 3097–3110.

Patel, B., Gupta, N., Ahsan, F., 2015. Particle engineering to enhance or lessen particle uptake by alveolar macrophages and to influence the therapeutic outcome. Eur. J. Pharm. Biopharm. 89, 163–174.

Patil, S.D., Rhodes, D.G., Burgess, D.J., 2005. DNA-based therapeutics and DNA delivery systems: a comprehensive review. AAPS J. 7, E61–E77.

Pawar, P.V., Gohil, S.V., Jain, J.P., Kumar, N., 2013. Functionalized polymersomes for biomedical applications. Polym. Chem. 4, 3160–3176.

Peer, D., Karp, J.M., Hong, S., Farokhzad, O.C., Margalit, R., Langer, R., 2007. Nanocarriers as an emerging platform for cancer therapy. Nat. Nanotechnol. 2, 751–760.

Peppas, N.A., Hilt, J.Z., Khademhosseini, A., Langer, R., 2006. Hydrogels in biology and medicine: from molecular principles to bionanotechnology. Adv. Mater. 18, 1345–1360.

Perez-Herrero, E., Fernandez-Medarde, A., 2015. Advanced targeted therapies in cancer: drug nanocarriers, the future of chemotherapy. Eur. J. Pharm. Biopharm. 93, 52–79.

Petersen, M.A., Hillmyer, M.A., Kokkoli, E., 2013. Bioresorbable polymersomes for targeted delivery of cisplatin. Bioconjug. Chem. 24, 533–543.

Petkar, K.C., Chavhan, S.S., Agatonovik-Kustrin, S., Sawant, K.K., 2011. Nanostructured materials in drug and gene delivery: a review of the state of the art. Crit. Rev. Ther. Drug 28, 101–164.

Photos, P.J., Bacakova, L., Discher, B., Bates, F.S., Discher, D.E., 2003. Polymer vesicles in vivo: correlations with PEG molecular weight. J. Control. Rel. 90, 323–334.

Pinto-Alphandary, H., Andremont, A., Couvreur, P., 2000. Targeted delivery of antibiotics using liposomes and nanoparticles: research and applications. Int. J. Antimicrob. Agents 13, 155–168.

Platt, V.M., Szoka, Jr., F.C., 2008. Anticancer therapeutics: targeting macromolecules and nanocarriers to hyaluronan or CD44, a hyaluronan receptor. Mol. Pharm. 5, 474–486.

Pong, M., Umchid, S., Lewin, P.A., Wheatley, M.A., Wrenn, S.P., 2005. Ultrasound mediated leakage from PEG incorporated liposomes. Abstr. Pap. Am. Chem. S. 230, U1250.

Pong, M., Umchid, S., Guarino, A.J., Lewin, P.A., Litniewski, J., Nowicki, A., Wrenn, S.P., 2006. In vitro ultrasound-mediated leakage from phospholipid vesicles. Ultrasonics 45, 133–145.

Prabhu, R.H., Patravale, V.B., Joshi, M.D., 2015. Polymeric nanoparticles for targeted treatment in oncology: current insights. Int. J. Nanomed. 10, 1001–1018.

Pramod, P.S., Shah, R., Chaphekar, S., Balasubramanian, N., Jayakannan, M., 2014. Polysaccharide nano-vesicular multidrug carriers for synergistic killing of cancer cells. Nanoscale 6, 11841–11855.

Prausnitz, M.R., Langer, R., 2008. Transdermal drug delivery. Nat. Biotechnol. 26, 1261–1268.

Prausnitz, M.R., Mitragotri, S., Langer, R., 2004. Current status and future potential of transdermal drug delivery. Nat. Rev. Drug Discov. 3, 115–124.

Qhattal, H.S.S., Hye, T., Alali, A., Liu, X., 2014. Hyaluronan polymer length, grafting density, and surface poly(ethylene glycol) coating influence in vivo circulation and tumor targeting of hyaluronan-grafted liposomes. ACS Nano 8, 5423–5440.

Qin, J., Liu, Q., Zhang, J., Chen, J., Chen, S., Zhao, Y., Du, J., 2015. Rationally separating the corona and membrane functions of polymer vesicles for enhanced T-2 MRI and drug delivery. ACS Appl. Mater. Interf. 7, 14043–14052.

Qing, G., Li, M., Deng, L., Lv, Z., Ding, P., Sun, T., 2013. Smart drug release systems based on stimuli-responsive polymers. Mini-Rev. Med. Chem. 13, 1369–1380.

Qiu, F., Wang, D., Wang, R., Huan, X., Tong, G., Zhu, Q., Yan, D., Zhu, X., 2013. Temperature-induced emission enhancement of star conjugated copolymers with poly(2-(dimethylamino)ethyl methacrylate) coronas for detection of bacteria. Biomacromolecules 14, 1678–1686.

Ramana, L.N., Anand, A.R., Sethuraman, S., Krishnan, U.M., 2014. Targeting strategies for delivery of anti-HIV drugs. J. Control. Rel. 192, 271–283.

Rana, S., Chaudhary, H., Arora, V., Kholi, K., Kumar, V., 2013. Effect of physicochemical properties of biodegradable polymers on nano drug delivery. Polym. Rev. 53, 546–567.

Ranson, M.R., Carmichael, J., Obyrne, K., Stewart, S., Smith, D., Howell, A., 1997. Treatment of advanced breast cancer with sterically stabilized liposomal doxorubicin: results of a multicenter phase II trial. J. Clin. Oncol. 15, 3185–3191.

Rapoport, N., 2007. Physical stimuli-responsive polymeric micelles for anti-cancer drug delivery. Prog. Polym. Sci. 32, 962–990.

Rawicz, W., Olbrich, K.C., McIntosh, T., Needham, D., Evans, E., 2000. Effect of chain length and unsaturation on elasticity of lipid bilayers. Biophys. J. 79 (1), 328–339.

Reibel, A.T., Mueller, S.S., Pektor, S., Bausbacher, N., Miederer, M., Frey, H., Roesch, F., 2015. Fate of linear and branched polyether-lipids in vivo in comparison to their liposomal formulations by F-18-radiolabeling and positron emission tomography. Biomacromolecules 16, 842–851.

Rentel, C.O., Bouwstra, J.A., Naisbett, B., Junginger, H.E., 1999. Niosomes as a novel peroral vaccine delivery system. Int. J. Pharm. 186, 161–167.

Rijcken, C.J.F., Soga, O., Hennink, W.E., Van Nostrum, C.F., 2007. Triggered destabilisation of polymeric micelles and vesicles by changing polymers polarity: an attractive tool for drug delivery. J. Control. Rel. 120, 131–148.

Robbins, G.P., Saunders, R.L., Haun, J.B., Rawson, J., Therien, M.J., Hammer, D.A., 2010. Tunable leuko-polymersomes that adhere specifically to inflammatory markers. Langmuir 26, 14089–14096.

Robinson, J.R., 1978. Sustained and Controlled Release Drug Delivery Systems. Marcel Dekker, New York, NY.

Rodríguez-Hernández, J., Lecommandoux, S., 2005. Reversible inside-out micellization of pH-responsive and water-soluble vesicles based on polypeptide Diblock copolymers. J. Am. Chem. Soc. 127, 2026–2027.

Roy, S., Johnston, A.H., Newman, T.A., Glueckert, R., Dudas, J., Bitsche, M., Corbacella, E., Rieger, G., Martini, A., Schrott-Fischer, A., 2010. Cell-specific targeting in the mouse inner ear using nanoparticles conjugated with a neurotrophin-derived peptide ligand: potential tool for drug delivery. Int. J. Pharm. 390, 214–224.

Sagar, V., Pilakka-Kanthikeel, S., Pottathil, R., Saxena, S.K., Nair, M., 2014. Towards nanomedicines for neuroAIDS. Rev. Med. Virol. 24, 103–124.

Sah, A.K., Suresh, P.K., 2015. Recent advances in ocular drug delivery, with special emphasis on lipid based nanocarriers. Recent Pat. Nanotechnol. 9, 94–105.

Sakuma, S., Hayashi, M., Akashi, M., 2001. Design of nanoparticles composed of graft copolymers for oral peptide delivery. Adv. Drug Deliv. Rev. 47, 21–37.

Sakurai, Y., Kajimoto, K., Hatakeyama, H., Harashima, H., 2015. Advances in an active and passive targeting to tumor and adipose tissues. Expert Opin. Drug Del. 12, 41–52.

Salatin, S., Dizaj, S.M., Khosroushahi, A.Y., 2015. Effect of the surface modification, size, and shape on cellular uptake of nanoparticles. Cell Biol. Int. 39, 881–890.

Salim, M., Minamikawa, H., Sugimura, A., Hashim, R., 2014. Amphiphilic designer nano-carriers for controlled release: from drug delivery to diagnostics. MedChemComm 5, 1602–1618.

Samad, A., Sultana, Y., Aqil, M., 2007. Liposomal drug delivery systems: an update review. Curr. Drug Deliv. 4, 297–305.

Schaefer, H., 1993. Penetration and percutaneous-absorption of topical retinoids—a review. Skin Pharmacol. 6, 17–23.

Schreier, H., Bouwstra, J., 1994. Liposomes and niosomes as topical drug carriers—dermal and transdermal drug-delivery. J. Control. Rel. 30, 1–15.

Schroeder, A., Avnir, Y., Weisman, S., Najajreh, Y., Gabizon, A., Talmon, Y., Kost, J., Barenholz, Y., 2007. Controlling liposomal drug release with low frequency ultrasound: mechanism and feasibility. Langmuir 23, 4019–4025.

Schroeder, A., Kost, J., Barenholz, Y., 2009. Ultrasound, liposomes, and drug delivery: principles for using ultrasound to control the release of drugs from liposomes. Chem. Phys. Lipids 162, 1–16.

Semple, S.C., Chonn, A., Cullis, P.R., 1998. Interactions of liposomes and lipid-based carrier systems with blood proteins: relation to clearance behaviour in vivo. Adv. Drug Deliv. Rev. 32, 3–17.

Senior, J.H., 1987. Fate and behavior of liposomes invivo—a review of controlling factors. CRC Crit. Rev. Ther. Drug Carrier Syst. 3, 123–193.

Senior, J., Gregoriadis, G., Mitropoulos, K.A., 1983. Stability and clearance of small unilamellar liposomes—studies with normal and lipoprotein-deficient mice. Biochim. Biophys. Acta 760, 111–118.

Senior, J., Crawley, J.C.W., Gregoriadis, G., 1985. Tissue distribution of liposomes exhibiting long half-lives in the circulation after intravenous-injection. Biochim. Biophys. Acta 839, 1–8.

Sezgin-Bayindir, Z., Onay-Besikci, A., Vural, N., Yuksel, N., 2013. Niosomes encapsulating paclitaxel for oral bioavailability enhancement: preparation, characterization, pharmacokinetics and biodistribution. J. Microencapsul. 30, 796–804.

Sezgin-Bayindir, Z., Antep, M.N., Yuksel, N., 2015. Development and characterization of mixed niosomes for oral delivery using candesartan cilexetil as a model poorly water-soluble drug. AAPS PharmSciTech 16, 108–117.
Sharma, A., Sharma, U.S., 1997. Liposomes in drug delivery: progress and limitations. Int. J. Pharm. 154, 123–140.
Shen, H.W., Eisenberg, A., 2000. Block length dependence of morphological phase diagrams of the ternary system of PS-*b*-PAA/dioxane/H_2O. Macromolecules 33, 2561–2572.
Sherwood, L., 2015. Human Physiology: From Cells to Systems. Cengage Learning, Boston, MA, USA.
Shete, H.K., Prabhu, R.H., Patravale, V.B., 2014. Endosomal escape: a bottleneck in intracellular delivery. J. Nanosci. Nanotechnol. 14, 460–474.
Shi, G., Guo, W., Stephenson, S.M., Lee, R.J., 2002. Efficient intracellular drug and gene delivery using folate receptor-targeted pH-sensitive liposomes composed of cationic/anionic lipid combinations. J. Control. Rel. 80, 309–319.
Shum, H.C., Kim, J.-W., Weitz, D.A., 2008. Microfluidic fabrication of monodisperse biocompatible and biodegradable polymersomes with controlled permeability. J. Am. Chem. Soc. 130, 9543–9549.
Shum, H.C., Zhao, Y.-J., Kim, S.-H., Weitz, D.A., 2011. Multicompartment polymersomes from double emulsions. Angew. Chem. Int. Ed. 50, 1648–1651.
Simões, S., Moreira, J.N., Fonseca, C., Düzgüneş, N., De Lima, M.C.P., 2004. On the formulation of pH-sensitive liposomes with long circulation times. Adv. Drug Deliv. Rev. 56, 947–965.
Simon, L.C., Sabliov, C.M., 2014. The effect of nanoparticle properties, detection method, delivery route and animal model on poly(lactic-*co*-glycolic) acid nanoparticles biodistribution in mice and rats. Drug Metab. Rev. 46, 128–141.
Singh, P., Roberts, M.S., 1994. Skin permeability and local tissue concentrations of nonsteroidal antiinflammatory drugs after topical application. J. Pharmacol. Exp. Ther. 268, 144–151.
Singh, D., Pradhan, M., Nag, M., Singh, M.R., 2015. Vesicular system: versatile carrier for transdermal delivery of bioactives. Artif. Cell. Nanomed. Biotechnol. 43, 282–290.
Singla, A.K., Garg, A., Aggarwal, D., 2002. Paclitaxel and its formulations. Int. J. Pharm. 235, 179–192.
Sinha, V.R., Singla, A.K., Wadhawan, S., Kaushik, R., Kumria, R., Bansal, K., Dhawan, S., 2004. Chitosan microspheres as a potential carrier for drugs. Int. J. Pharm. 274, 1–33.
Sinha, V.R., Singh, A., Kumar, R.V., Singh, S., Kumria, R., Bhinge, J.R., 2007. Oral colon-specific drug delivery of protein and peptide drugs. Crit. Rev. Ther. Drug 24, 63–92.
Sinico, C., Fadda, A.M., 2009. Vesicular carriers for dermal drug delivery. Expert Opin. Drug Del. 6, 813–825.
Sirsi, S.R., Borden, M.A., 2014. State-of-the-art materials for ultrasound-triggered drug delivery. Adv. Drug Deliv. Rev. 72, 3–14.
Small, E.F., Willy, M.C., Lewin, P.A., Wrenn, S.P., 2011. Ultrasound-induced transport across lipid bilayers: influence of phase behavior. Colloids Surf. A 390, 40–47.
Small, E.F., Dan, N.R., Wrenn, S.P., 2012. Low-frequency ultrasound-induced transport across non-raft-forming ternary lipid bilayers. Langmuir 28, 14364–14372.
Sonavane, G., Tomoda, K., Makino, K., 2008. Biodistribution of colloidal gold nanoparticles after intravenous administration: effect of particle size. Colloids Surf. B 66, 274–280.
Song, J., Waugh, R.E., 1993. Bending rigidity of SOPC membranes containing cholesterol. Biophys. J. 64, 1967–1970.
Sperling, R.A., Rivera Gil, P., Zhang, F., Zanella, M., Parak, W.J., 2008. Biological applications of gold nanoparticles. Chem. Soc. Rev. 37, 1896–1908.
Stojanov, K., Georgieva, J.V., Brinkhuis, R.P., Van Hest, J.C., Rutjes, F.P., Dierckx, R.A.J.O., De Vries, E.F.J., Zuhorn, I.S., 2012. In vivo biodistribution of prion- and GM1-targeted polymersomes following intravenous administration in mice. Mol. Pharm. 9, 1620–1627.
Stride, E.P., Coussios, C.C., 2010. Cavitation and contrast: the use of bubbles in ultrasound imaging and therapy: proceedings of the Institution of Mechanical Engineers Part H. J. Eng. Med. 224, 171–191.
Sudimack, J., Lee, R.J., 2000. Targeted drug delivery via the folate receptor. Adv. Drug Deliv. Rev. 41, 147–162.

Sun, C., Lee, J.S.H., Zhang, M., 2008. Magnetic nanoparticles in MR imaging and drug delivery. Adv. Drug Deliv. Rev. 60, 1252–1265.

Sun, R., Du, X.-J., Sun, C.-Y., Shen, S., Liu, Y., Yang, X.-Z., Bao, Y., Zhu, Y.-H., Wang, J., 2015. A block copolymer of zwitterionic polyphosphoester and polylactic acid for drug delivery. Biomater. Sci. 3, 1105–1113.

Surnar, B., Jayakannan, M., 2013. Stimuli-responsive poly(caprolactone) vesicles for dual drug delivery under the gastrointestinal tract. Biomacromolecules 14, 4377–4387.

Surovtseva, E.V., Johnston, A.H., Zhang, W., Zhang, Y., Kim, A., Murakoshi, M., Wada, H., Newman, T.A., Zou, J., Pyykko, I., 2012. Prestin binding peptides as ligands for targeted polymersome mediated drug delivery to outer hair cells in the inner ear. Int. J. Pharm. 424, 121–127.

Suurkuusk, J., Lentz, B.R., Barenholz, Y., Biltonen, R.L., Thompson, T.E., 1976. A calorimetric and fluorescent probe study of the gel-liquid crystalline phase transition in small, single-lamellar dipalmitoylphosphatidylcholine vesicles. Biochemistry 15, 1393–1401.

Swaminathan, J., Ehrhardt, C., 2012. Liposomal delivery of proteins and peptides. Expert Opin. Drug Del. 9, 1489–1503.

Ta, T., Porter, T.M., 2013. Thermosensitive liposomes for localized delivery and triggered release of chemotherapy. J. Control. Rel. 169, 112–125.

Tabata, Y., Ikada, Y., 1988. Effect of the size and surface-charge of polymer microspheres on their phagocytosis by macrophage. Biomaterials 9, 356–362.

Tajes, M., Ramos-Fernandez, E., Weng-Jiang, X., Bosch-Morato, M., Guivernau, B., Eraso-Pichot, A., Salvador, B., Fernandez-Busquets, X., Roquer, J., Munoz, F.J., 2014. The blood–brain barrier: structure, function and therapeutic approaches to cross it. Mol. Membr. Biol. 31, 152–167.

Takahashi, M., Uechi, S., Takara, K., Asikin, Y., Wada, K., 2009. Evaluation of an oral carrier system in rats: bioavailability and antioxidant properties of liposome-encapsulated curcumin. J. Agr. Food Chem. 57, 9141–9146.

Takeuchi, H., Sugihara, H., 2010. Absorption of calcitonin in oral and pulmonary administration with polymer-coated liposomes. Yakugaku Zasshi 130, 1135–1142.

Takeuchi, H., Yamamoto, H., Niwa, T., Hino, T., Kawashima, Y., 1994. Mucoadhesion of polymer-coated liposomes to rat intestine in-vitro. Chem. Pharmaceut. Bull. 42, 1954–1956.

Takeuchi, H., Yamamoto, H., Niwa, T., Hino, T., Kawashima, Y., 1996. Enteral absorption of insulin in rats from mucoadhesive chitosan-coated liposomes. Pharm. Res. 13, 896–901.

Takeuchi, H., Matsui, Y., Yamamoto, H., Kawashima, Y., 2003. Mucoadhesive properties of carbopol or chitosan-coated liposomes and their effectiveness in the oral administration of calcitonin to rats. J. Control. Rel. 86, 235–242.

Takeuchi, H., Matsui, Y., Sugihara, H., Yamamoto, H., Kawashima, Y., 2005a. Effectiveness of submicron-sized, chitosan-coated liposomes in oral administration of peptide drugs. Int. J. Pharm. 303, 160–170.

Takeuchi, H., Thongborisute, J., Matsui, Y., Sugihara, H., Yamamoto, H., Kawashima, Y., 2005b. Novel mucoadhesion tests for polymers and polymer-coated particles to design optimal mucoadhesive drug delivery systems. Adv. Drug Deliv. Rev. 57, 1583–1594.

Tanner, P., Baumann, P., Enea, R., Onaca, O., Palivan, C., Meier, W., 2011. Polymeric vesicles: from drug carriers to nanoreactors and artificial organelles. Accounts Chem. Res. 44, 1039–1049.

Tavano, L., Muzzalupo, R., Picci, N., De Cindio, B., 2014. Co-encapsulation of lipophilic antioxidants into niosomal carriers: percutaneous permeation studies for cosmeceutical applications. Colloids Surf. B 114, 144–149.

Taylor, T.M., Davidson, P.M., Bruce, B.D., Weiss, J., 2005. Liposomal nanocapsules in food science and agriculture. Crit. Rev. Food Sci. 45, 587–605.

Tejada-Berges, T., Granai, C.O., Gordinier, M., Gajewski, W., 2002. Caelyx/Doxil for the treatment of metastatic ovarian and breast cancer. Expert Rev. Anticanc. 2, 143–150.

Thambi, T., Deepagan, V.G., Ko, H., Lee, D.S., Park, J.H., 2012. Bioreducible polymersomes for intracellular dual-drug delivery. J. Mater. Chem. 22, 22028–22036.

Thanh-Huyen, T., Amiji, M.M., 2015. Targeted delivery systems for biological therapies of inflammatory diseases. Expert Opin. Drug Del. 12, 393–414.
Thevenot, J., Oliveira, H., Lecommandoux, S., 2013. Polymersomes for theranostics. J. Drug Deliv. Sci. Tec. 23, 38–46.
Thibault, R.J., Uzun, O., Hong, R., Rotello, V.M., 2006. Recognition-controlled assembly of nanoparticles using photochemically crosslinked recognition-induced polymersomes. Adv. Mater. 18, 2179–2183.
Thiele, J., Abate, A.R., Shum, H.C., Bachtler, S., Foerster, S., Weitz, D.A., 2010. Fabrication of polymersomes using double-emulsion templates in glass-coated stamped microfluidic devices. Small 6, 1723–1727.
Thong, H.Y., Zhai, H., Maibach, H.I., 2007. Percutaneous penetration enhancers: an overview. Skin Pharmacol. 20, 272–282.
Tila, D., Ghasemi, S., Yazdani-Arazi, S.N., Ghanbarzadeh, S., 2015. Functional liposomes in the cancer-targeted drug delivery. J. Biomater. Appl. 30, 3–16.
Timko, B.P., Kohane, D.S., 2012. Materials to clinical devices: technologies for remotely triggered drug delivery. Clin. Ther. 34, S25–S35.
Tiwari, S.B., Amiji, M.M., 2006. A review of nanocarrier-based CNS delivery systems. Curr. Drug Deliv. 3, 219–232.
Toan, T., Hernandez, M., Patel, D., Wu, J., 2015. Temperature and pH responsive microfibers for controllable and variable ibuprofen delivery. Adv. Mater. Sci. Eng. 2015, 6.
Tong, R., Langer, R., 2015. Nanomedicines targeting the tumor microenvironment. Cancer J. 21, 314–321.
Topete, A., Barbosa, S., Taboada, P., 2015. Intelligent micellar polymeric nanocarriers for therapeutics and diagnosis. J. Appl. Polym. Sci. 132, 42650.
Toporkiewicz, M., Meissner, J., Matusewicz, L., Czogalla, A., Sikorski, A.F., 2015. Toward a magic or imaginary bullet? Ligands for drug targeting to cancer cells: principles, hopes, and challenges. Int. J. Nanomed. 10, 1399–1414.
Torchilin, V.P., 2004. Targeted polymeric micelles for delivery of poorly soluble drugs. Cell. Mol. Life Sci. 61, 2549–2559.
Torchilin, V.P., 2005. Recent advances with liposomes as pharmaceutical carriers. Nat. Rev. Drug Discov. 4, 145–160.
Torchilin, V.P., 2007. Micellar nanocarriers: pharmaceutical perspectives. Pharmaceut. Res. 24, 1–16.
Torchilin, V.P., 2014. Multifunctional, stimuli-sensitive nanoparticulate systems for drug delivery. Nat. Rev. Drug Discov. 13, 813–827.
Trapani, G., Denora, N., Trapani, A., Laquintana, V., 2012. Recent advances in ligand targeted therapy. J. Drug Target. 20, 1–22.
Trono, J.D., Mizuno, K., Yusa, N., Matsukawa, T., Yokoyama, K., Uesaka, M., 2011. Size, concentration and incubation time dependence of gold nanoparticle uptake into pancreas cancer cells and its future application to X-ray drug delivery system. J. Radiat. Res. 52, 103–109.
Truong, N.P., Whittaker, M.R., Mak, C.W., Davis, T.P., 2015. The importance of nanoparticle shape in cancer drug delivery. Expert Opin. Drug Del. 12, 129–142.
Uchegbu, I.F., Florence, A.T., 1995. Nonionic surfactant vesicles (niosomes)—physical and pharmaceutical chemistry. Adv. Colloid Interf. Sci. 58, 1–55.
Uchegbu, I.F., Vyas, S.P., 1998. Non-ionic surfactant based vesicles (niosomes) in drug delivery. Int. J. Pharm. 172, 33–70.
Uchegbu, I.F., Double, J.A., Turton, J.A., Florence, A.T., 1995. Distribution, metabolism and tumoricidal activity of doxorubicin administered in sorbitan monostearate (Span-60) niosomes in the mouse. Pharmaceut. Res. 12, 1019–1024.
Udroiu, I., 2015. Ultrasonic drug delivery in oncology. J. Buon 20, 381–390.
Ulrich, A.S., 2002. Biophysical aspects of using liposomes as delivery vehicles. Biosci. Rep. 22, 129–150.

Upadhyay, K.K., Bhatt, A.N., Castro, E., Mishra, A.K., Chuttani, K., Dwarakanath, B.S., Schatz, C., Le Meins, J.-F., Misra, A., Lecommandoux, S., 2010a. In vitro and in vivo evaluation of docetaxel loaded biodegradable polymersomes. Macromol. Biosci. 10, 503–512.

Upadhyay, K.K., Bhatt, A.N., Mishra, A.K., Dwarakanath, B.S., Jain, S., Schatz, C., Le Meins, J.-F., Farooque, A., Chandraiah, G., Jain, A.K., Misra, A., Lecommandoux, S., 2010b. The intracellular drug delivery and anti tumor activity of doxorubicin loaded poly(gamma-benzyl L-glutamate)-*b*-hyaluronan polymersomes. Biomaterials 31, 2882–2892.

Upadhyay, K.K., Mishra, A.K., Chuttani, K., Kaul, A., Schatz, C., Le Meins, J.-F., Misra, A., Lecommandoux, S., 2012. The in vivo behavior and antitumor activity of doxorubicin-loaded poly(gamma-benzyl L-glutamate)-block-hyaluronan polymersomes in Ehrlich ascites tumor-bearing BALB/c mice. Nanomed.-Nanotechnol. 8, 71–80.

van Hoogevest, P., Wendel, A., 2014. The use of natural and synthetic phospholipids as pharmaceutical excipients. Eur. J. Lipid Sci. Tech. 116, 1088–1107.

van Tellingen, O., Yetkin-Arik, B., De Gooijer, M.C., Wesseling, P., Wurdinger, T., De Vries, H.E., 2015. Overcoming the blood–brain tumor barrier for effective glioblastoma treatment. Drug Resist. Update. 19, 1–12.

vanRooijen, N., Sanders, A., 1997. Elimination, blocking, and activation of macrophages: three of a kind? J. Leukocyte Biol. 62, 702–709.

Varde, N.K., Pack, D.W., 2004. Microspheres for controlled release drug delivery. Expert Opin. Biol. Th. 4, 35–51.

Varghese, B., Vlashi, E., Xia, W., Lopez, W.A., Paulos, C.M., Reddy, J., Xu, L.-C., Low, P.S., 2014. Folate receptor-beta in activated macrophages: ligand binding and receptor recycling kinetics. Mol. Pharm. 11, 3609–3616.

Varshosaz, J., Pardakhty, A., Hajhashemi, V.I., 2003. Development and physical characterization of sorbitan monoester niosomes for insulin oral delivery. Drug Deliv. 10, 251–262.

Vasir, J.K., Tambwekar, K., Garg, S., 2003. Bioadhesive microspheres as a controlled drug delivery system. Int. J. Pharm. 255, 13–32.

Vladisavljevic, G.T., Kobayashi, I., Nakajima, M., 2012. Production of uniform droplets using membrane, microchannel and microfluidic emulsification devices. Microfluid. Nanofluid. 13, 151–178.

Vyas, S.P., Gupta, S., 2006. Optimizing efficacy of amphotericin B through nanomodification. Int. J. Nanomed. 1, 417–432.

Vyas, S.P., Kannan, M.E., Jain, S., Mishra, V., Singh, P., 2004. Design of liposomal aerosols for improved delivery of rifampicin to alveolar macrophages. Int. J. Pharm. 269, 37–49.

Vyas, S.P., Singh, R.P., Jain, S., Mishra, V., Mahor, S., Singh, P., Gupta, P.N., Rawat, A., Dubey, P., 2005. Non-ionic surfactant based vesicles (niosomes) for non-invasive topical genetic immunization against hepatitis B. Int. J. Pharm. 296, 80–86.

Waddad, A.Y., Abbad, S., Yu, F., Munyendo, W.L.L., Wang, J., Lv, H., Zhou, J., 2013. Formulation, characterization and pharmacokinetics of Morin hydrate niosomes prepared from various non-ionic surfactants. Int. J. Pharm. 456, 446–458.

Wallace, N., Wrenn, S.P., 2015. Ultrasound triggered drug delivery with liposomal nested microbubbles. Ultrasonics 63, 31–38.

Wasungu, L., Hoekstra, D., 2006. Cationic lipids, lipoplexes and intracellular delivery of genes. J. Control. Rel. 116, 255–264.

Werengowska-Ciecwierz, K., Wisniewski, M., Terzyk, A.P., Furmaniak, S., 2015. The chemistry of bioconjugation in nanoparticles-based drug delivery system. Adv. Condensed Matter Phys. 2015, 27.

Wijagkanalan, W., Kawakami, S., Takenaga, M., Igarashi, R., Yamashita, F., Hashida, M., 2008. Efficient targeting to alveolar macrophages by intratracheal administration of mannosylated liposomes in rats. J. Control. Rel. 125, 121–130.

Wilkhu, J.S., Mcneil, S.E., Anderson, D.E., Perrie, Y., 2013. Characterization and optimization of bilosomes for oral vaccine delivery. J. Drug Target. 21, 291–299.
Wilkhu, J.S., Mcneil, S.E., Anderson, D.E., Perrie, Y., 2014. Consideration of the efficacy of non-ionic vesicles in the targeted delivery of oral vaccines. Drug Deliv. Translat. Res. 4, 233–245.
Win, K.Y., Feng, S.S., 2005. Effects of particle size and surface coating on cellular uptake of polymeric nanoparticles for oral delivery of anticancer drugs. Biomaterials 26, 2713–2722.
Wissing, S.A., Kayser, O., Muller, R.H., 2004. Solid lipid nanoparticles for parenteral drug delivery. Adv. Drug Deliv. Rev. 56, 1257–1272.
Wojcicki, A.D., Hillaireau, H., Nascimento, T.L., Arpicco, S., Taverna, M., Ribes, S., Bourge, M., Nicolas, V., Bochot, A., Vauthier, C., Tsapis, N., Fattal, E., 2012. Hyaluronic acid-bearing lipoplexes: physico-chemical characterization and in vitro targeting of the CD44 receptor. J. Control. Rel. 162, 545–552.
Woodle, M.C., 1998. Controlling liposome blood clearance by surface-grafted polymers. Adv. Drug Deliv. Rev. 32, 139–152.
Woodle, M.C., Lasic, D.D., 1992. Sterically stabilized liposomes. Biochim. Biophys. Acta 1113, 171–199.
Wrenn, S.P., Dicker, S.M., Small, E.F., Dan, N.R., Mleczko, M., Schmitz, G., Lewin, P.A., 2012. Bursting bubbles and bilayers. Theranostics 2, 1140–1159.
Wrenn, S.P., Small, E., Dan, N., 2013. Bubble nucleation in lipid bilayers: a mechanism for low frequency ultrasound disruption. Biochim. Biophys. Acta 1828, 1192–1197.
Wu, X., Guy, R.H., 2009. Applications of nanoparticles in topical drug delivery and in cosmetics. J. Drug Deliv. Sci. Tec. 19, 371–384.
Wu, J.-L., Wang, C.-Q., Zhuo, R.-X., Cheng, S.-X., 2014. Multi-drug delivery system based on alginate/calcium carbonate hybrid nanoparticles for combination chemotherapy. Colloids Surf. B 123, 498–505.
Xie, J., Lee, S., Chen, X., 2010. Nanoparticle-based theranostic agents. Adv. Drug Deliv. Rev. 62, 1064–1079.
Xin, H., Jiang, X., Gu, J., Sha, X., Chen, L., Law, K., Chen, Y., Wang, X., Jiang, Y., Fang, X., 2011. Angiopep-conjugated poly(ethylene glycol)-*co*-poly(epsilon-caprolactone) nanoparticles as dual-targeting drug delivery system for brain glioma. Biomaterials 32, 4293–4305.
Xu, H., Meng, F., Zhong, Z., 2009. Reversibly crosslinked temperature-responsive nano-sized polymersomes: synthesis and triggered drug release. J. Mater. Chem. 19, 4183–4190.
Xu, J., Zhao, Q., Jin, Y., Qiu, L., 2014. High loading of hydrophilic/hydrophobic doxorubicin into polyphosphazene polymersome for breast cancer therapy. Nanomed.-Nanotechnol. 10, 349–358.
Yao, X., Bunt, C., Cornish, J., Quek, S.-Y., Wen, J., 2015. Oral delivery of bovine lactoferrin using pectin- and chitosan-modified liposomes and solid lipid particles: improvement of stability of lactoferrin. Chemical Biol. Drug Des. 86, 466–475.
Yoshioka, T., Sternberg, B., Florence, A.T., 1994. Preparation and properties of vesicles (niosomes) of sorbitan monoesters (Span-20, Span-40, Span-60 and Span-80) and a sorbitan triester (Span-85). Int. J. Pharm. 105, 1–6.
Yu, B., Lee, R.J., Lee, L.J., 2009. Microfluidic methods for production of liposomes. Methods Enzymol. 465, 129–141.
Yu, Y., Pang, Z., Lu, W., Yin, Q., Gao, H., Jiang, X., 2012. Self-assembled polymersomes conjugated with lactoferrin as novel drug carrier for brain delivery. Pharm. Res. 29, 83–96.
Zhang, S.B., Xu, Y.M., Wang, B., Qiao, W.H., Liu, D.L., Li, Z.S., 2004. Cationic compounds used in lipoplexes and polyplexes for gene delivery. J. Control. Rel. 100, 165–180.
Zhang, Y., Zhang, W., Johnston, A.H., Newman, T.A., Pyykko, I., Zou, J., 2012. Targeted delivery of Tet1 peptide functionalized polymersomes to the rat cochlear nerve. Int. J. Nanomed. 7, 1015–1022.
Zhang, Y., Xing, S.-G., Wang, Z., Kang, Q.-H., Ling, Y., Yao, M.-Y., He, Y.-P., Jin, Y., Chu, X.-G., 2015. Recent research of aptamer in target drug delivery. Prog. Biochem. Biophys. 42, 236–243.
Zhao, G., Rodriguez, B.L., 2013. Molecular targeting of liposomal nanoparticles to tumor microenvironment. Int. J. Nanomed. 8, 61–71.

Zhao, F., Zhao, Y., Liu, Y., Chang, X., Chen, C., Zhao, Y., 2011. Cellular uptake, intracellular trafficking, and cytotoxicity of nanomaterials. Small 7, 1322–1337.

Zhong, M., Feng, Y., Liao, H., Hu, X., Wan, S., Zhu, B., Zhang, M., Xiong, H., Zhou, Y., Zhang, J., 2014. Azithromycin cationic non-lecithoid nano/microparticles improve bioavailability and targeting efficiency. Pharm. Res. 31, 2857–2867.

Zhou, F., Neutra, M.R., 2002. Antigen delivery to mucosa-associated lymphoid tissues using liposomes as a carrier. Biosci. Rep. 22, 355–369.

Zhu, Y.-J., Chen, F., 2015. pH-responsive drug-delivery systems. Chem.-Asian J. 10, 284–305.

Zhu, J., Xu, X., Hu, M., Qiu, L., 2015a. Co-encapsulation of combretastatin-A4 phosphate and doxorubicin in polymersomes for synergistic therapy of nasopharyngeal epidermal carcinoma. J. Biomed. Nanotechnol. 11, 997–1006.

Zhu, Q., Li, X., Xia, D., Yu, H., Chen, D., Fan, W., Gan, Y., 2015b. Lipid-based formulations for oral drug delivery: effects on drug absorption and metabolism. Curr. Drug Metab. 16, 200–210.

Zou, H., Wang, Z., Feng, M., 2015. Nanocarriers with tunable surface properties to unblock bottlenecks in systemic drug and gene delivery. J. Control. Rel. 214, 121–133.

Zuccari, G., Milelli, A., Pastorino, F., Loi, M., Petretto, A., Parise, A., Marchetti, C., Minarini, A., Cilli, M., Emionite, L., Di Paolo, D., Brignole, C., Piaggio, F., Perri, P., Tumiatti, V., Pistoia, V., Pagnan, G., Ponzoni, M., 2015. Tumor vascular targeted liposomal-bortezomib minimizes side effects and increases therapeutic activity in human neuroblastoma. J. Control. Rel. 211, 44–52.

CHAPTER

THE NEW NANOCARRIERS BASED ON GRAPHENE AND GRAPHENE OXIDE FOR DRUG DELIVERY APPLICATIONS

3

Somayeh Mohamadi*, Mehrdad Hamidi**

**University of Environment, Karaj, Iran*

***Zanjan University of Medical Sciences, Zanjan, Iran*

CHAPTER OUTLINE

1 INTRODUCTION

Carbon is an attractive element due to its ability to form all sp, sp^2, and sp^3 hybridization of electronic orbitals, which causes the formation of a large variety of compounds and structures with a wide range of properties (Delhaes, 2011; Gogotsi, 2011). Carbon is known as a basis of life, but different forms of carbonic structures, such as graphite, diamond, fullerenes, carbon nanotubes, and graphene have also received a lot of attention recently. The term *graphene* first appeared in 1987 to describe single sheets of graphite, and graphene was presumed not to exist in the free state, being described as an academic material; however, the experimental discovery of graphene goes back to the Geim and Novoselov report in 2004 (Fradkin, 1986; Geim and Novoselov, 2007; Mouras et al., 1987). With the finding of graphene, the interest of researchers shifted from other carbon materials toward graphene, and many scientists tried to manipulate the surface of graphene through the wide array of modification methods.

Graphene is a two-dimensional (2D) single layer of carbon atoms with a hexagonal packed structure. The carbon bonds are sp^2 hybridized, where the in-plane σ_{C-C} bond is one of the strongest bonds in nature, and the out-of-plane π bond, which contributes to a delocalized network of electrons, is responsible for the electron conduction of graphene and provides the weak interaction among graphene layers or between graphene and other molecules. Also, as discussed above, it was believed that 2D crystals were thermodynamically unstable and could not exist. In fact, the melting temperature of thin films rapidly diminishes with decreasing thickness, and they become unstable (segregate into islands or decompose) at a thickness of, typically, atomic layers (Evans et al., 2006; Venables et al., 1984). However, the perfecting viewpoint is the extracted 2D crystals become intrinsically stable by mild crumpling and surface ripples. In other words, graphene exists in the 3D space and hence the atoms can oscillate out-of-plane and cause flexibility in this 2D crystal.

Graphene and its derivative graphene oxide (GO) have also shown important potential in biotechnology applications (Yang et al., 2013), particularly in DNA detection (Lu et al., 2009), cellular probing (Sun et al., 2008), gene therapy (Feng et al., 2011), and photodynamic therapy (Tian et al., 2011).

Based on the success of carbon nanotubule (CNT)–based drug delivery systems, the idea of using graphene sheets as drug carriers was born and has grown rapidly in the past several years. Graphene with a high surface area, unique structure and geometry, and flexibility possesses remarkable physicochemical properties, including high Young's modulus, high fracture strength, excellent electrical and thermal conductivity, fast mobility of charge carriers, and biocompatibility. Graphene is also known for its excellent chemical resistance, but one major exception is poor resistance to the elements of the VI group (oxygen, sulfur, selenium, and tellurium), so it is possible to achieve graphene oxide particles by the oxidation process. Introducing oxygen functional groups onto the graphene surface provides this opportunity to manipulate the surface chemistry and open a large area for drug carrier designing. The surface chemistry of nanoparticles is a fundamental parameter in the biological interaction with tissues, both normal and tumoral cells. For this propose, the precise understanding of chemical properties and the functional groups that exist on the graphene surface and edges, as well as modification of graphene surface is essential. In this chapter, first, the chemistry of graphene will be described and then followed by discussion about different surface modifications and functionalization methods for drug carriers. Second, their toxicity and biocompatibility will be presented.

2 CHEMISTRY OF THE GRAPHENE SURFACE

Different preparation methods lead to graphene sheets with different chemical properties and thicknesses. The methods can be broadly classified as colloidal suspension, arc discharge, detonation technique, and thermal, chemical, or mechanical exfoliation. In mechanical exfoliation methods, large-scale production is not possible but the methods are the most compatible with electronic device fabrication. Since no chemical reaction is used in these methods, it is expected the functional groups that exist on the graphene surface would be the same as in pristine graphite. The term *graphite* by itself describes an ideal material with a perfect graphite structure, including no defects and functional groups. In the chemical oxidation procedure, in general, GO is synthesized by different methods. The oxidation of graphite to GO breaks up the sp^2-hybridized structure of the stacked graphene sheets, generating defects that manifest as clear wrinkles in the stack and increase the distance between adjacent sheets from 3.35 Å in graphite powder to ≈6.8 Å for GO powder (Bourlinos et al., 2003). During an oxidation reaction depends on the oxidation intensity basal plane of the sheet, which is decorated with different oxygen functional groups. An NMR study of ^{13}C-labeled graphene oxide indicated hydroxyl and epoxy (1,2-ether) groups are introduced on the basal plane of the sheet. Carbonyl groups can present as carboxylic acids along the sheet edge and organic carbonyl defects within the sheet (He et al., 1998; Lerf et al., 1998). These functional groups provide reactive handles for a variety of surface-modification reactions, which can be used to develop functionalized graphene oxide- and graphene-based materials.

From the chemical point of view, production of graphene oxide that possesses many reactive oxygen-containing groups is suitable for further functionalization and tuning properties of GO sheets; in other words, the reactivity of GO is significant. But during an oxidation, spatially at the highest oxidation levels graphene's π-system is interrupted, and the aqueous dispersion changes to brown color (Compton and Nguyen, 2010). In this state, its electrical conductivity drops off dramatically. It is also often essential that the GO can be transformed back into a conductive graphitic material by rapid heating (thermal), microwave irradiation (MWI), photo irradiation, photocatalyst, electrochemical, and chemical reagent reduction (Pei and Cheng, 2012). Though numerous strategies have been proposed to reduce GO, there is still this question: can the functional groups of a GO sheet be fully eliminated? For the elimination of functional groups, there are also two effects that should be considered: whether the oxygen-containing groups can be removed and whether the areas after removal can be restored to a long-range conjugated structure, so that there are pathways for carrier transport within the rGO sheet.

For conductivity restoring, it can be said that the reduction of GO must be mainly aimed at eliminating epoxy and hydroxyl groups on the plane, while other groups, for example carboxyl, carbonyl, and ester groups, present at the edges or defective areas have only a limited influence on the conductivity of the rGO sheet (Pei and Cheng, 2012). Because of the low binding energy, a single hydroxyl group attached to the interior aromatic domain is not stable; it is expected that the mentioned groups dissociate or migrate to the edges of aromatic domains at low temperatures (Pei and Cheng, 2012). According to Gao's calculations, the critical dissociation temperature (T_c) of hydroxyl groups attached to the edges of GO is 650°C, and only above this temperature can hydroxyl groups be fully removed (Gao et al., 2009; Xu et al., 2006). For dehydroxylation, a hydroxyl group is believed to more favorably leave the graphene sheet directly, producing an OH radical and a graphene radical, which does not result in the formation of a lattice defect within the plane. On the basis of this calculation, after thermal annealing at temperatures of 700–1200°C in a vacuum, the hydroxyl groups can be fully eliminated, while the epoxy groups are retained. As the experimental work has discovered, a large number of functional groups can be removed by moderate heating above 200°C with enough time, but the full deoxygenation of GO solely by thermal annealing is rather difficult even at temperatures as high as 1200°C (Pei and Cheng, 2012).

Like thermal reduction, chemical reduction can also not fully remove the functional groups in GO (Fig. 3.1). The reduction mechanism by hydrazine was first proposed by Stankovich et al. (2007). This reduction process starts from the ring opening of epoxy groups and forms hydroxyl groups on the original sites. Their results show that the hydrazine reduction can only result in reducing epoxy groups, while no reaction path was found for the reduction of the hydroxyl, carbonyl, and carboxyl groups of GO. According to DFT simulation, the formed hydroxyl groups attached within an aromatic domain are not stable even at moderate temperatures, and can be removed or can migrate to the edges of aromatic

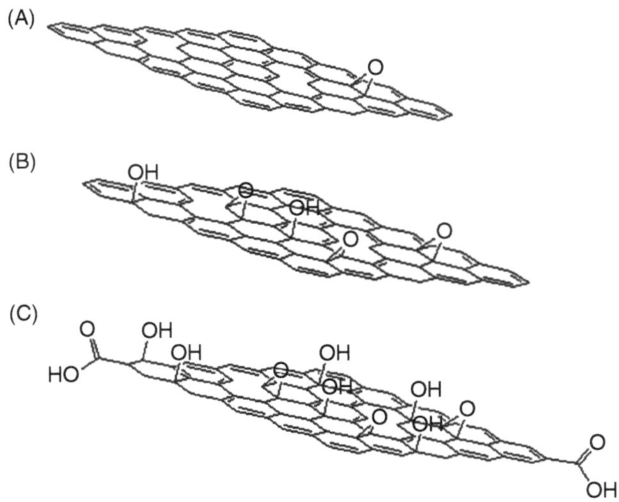

FIGURE 3.1 Schematic Model of a Graphene

(A) Reduced graphene oxide and (B) graphene oxide sheet.

domains and restore the conjugated structure after dehydroxylation (Stankovich et al., 2007). Depending on the field in which graphene is considered to be used, the additional process should be included.

3 MODIFICATION OF GRAPHENE SURFACE FOR DRUG DELIVERY APPLICATIONS

The exploration of graphene in drug delivery and biomedicine was first initiated by the Hongjie Dai group at Stanford University (CA, USA) in 2008 (Feng et al., 2011). The question was whether graphene can be considered as an interesting material in drug delivery. Since graphene comprises fused six-member aromatic cycles, the aromatic anticancer drugs can be effectively loaded on the graphene surface via π–π stacking for intracellular drug delivery. The published data in the literature showed ultrahigh drug-loading efficiency achieved due to the extremely large surface area of graphene, which has every atom exposed on its surface (Feng et al., 2011). Besides high loading capacity, the ability of dispersion in the aqueous environment to have a stable suspension is also important.

In view of the fact that pristine graphene is hydrophobic in nature, with the water contact angle in the range of 95–100 degree, achieving stable suspension in water requires surfactants or other stabilizing agents. On the other hand, graphene oxide with water contact angle of 30.7 degree is capable of forming hydrogen bonds due to the presence of hydroxide groups on the basal plane and carboxylate groups on the edge site. In reduced graphene oxide (rGO), some oxygen functional groups are still retained, making it less hydrophobic than graphene, and it exhibits less basal reactivity than graphene oxide (Goenka et al., 2014).

In designing a nanocarrier for drug delivery, besides physicochemical stability in the physiological environment, biocompatibility and toxic issues, the interactions with biological systems at the molecular level, diffusivity, blood circulation half-life, drug loading and release characteristics, drug transport ability from the bloodstream to target cells within tissues, and immunogenicity should be considered. Graphene nanosheets tend to aggregate in physiological solutions due to electrostatic charges and nonspecific binding to protein. Therefore, development of functionalized graphene nanosheets has improved their solubility and biocompatibility, and has reduced cytotoxicity and genotoxicity. Two main strategies, covalent conjugation and noncovalent physisorption, are commonly used for surface modification to build desired functionalized graphene nanosheets. In this section, various designed drug nanocarriers based on graphene nanosheets have been discussed.

3.1 GRAPHENE MODIFICATION BY PEG

Surface functionalization of GO with polyethylene glycol (PEG) for drug carrier purposes has been explored by several research groups since PEG functionalization has prolonged the circulation time of the pharmaceuticals in blood and avoid rapid uptake by macrophages of the mononuclear phagocyte system (MPS). As a result, high drug accumulation at tumor sites is attained through the enhanced permeability and retention (EPR) effect (Pelaz et al., 2015).

The Hongjie Dai group for the first time used an amine-terminated, branched PEG to functionalize GO to be highly stable in physiological solutions. They started with rGO suspension, converted the OH groups to COOH via chloroacetic acid, and followed by addition of 6-arm polyethylene glycol–amine solution (Fig. 3.2) (Sun et al., 2008).

Then, doxorubicin (DOX) was loaded onto this synthesis carrier. Biocompatibility, photoluminescence, drug loading, and delivery studies suggested promising applications of graphene materials in biological medical areas. This group also showed that PEGylated nanographene oxide (NGO–PEG) readily complexes with a water-insoluble aromatic molecule SN38, a camptothecin (CPT) analogue, via noncovalent Van der Waals interaction (Fig. 3.3). They found that the loading ratio (the weight ratio of loaded drug to carriers) of GO could reach 200%, which is much higher than that of other nanocarriers that usually have a loading ratio lower than 100%. The NGOPEG–SN38 complex exhibited excellent aqueous solubility in all biological solutions. They found that SN38 on NGO–PEG exhibited negligible release from NGO in phosphate buffered saline (PBS) solution and ~30% release in serum in 3 days. This suggested strong noncovalent binding of SN38 on graphene oxide sheets. The slow but finite release of SN38 in serum was likely caused by the binding of SN38 by serum proteins (Liu et al., 2008).

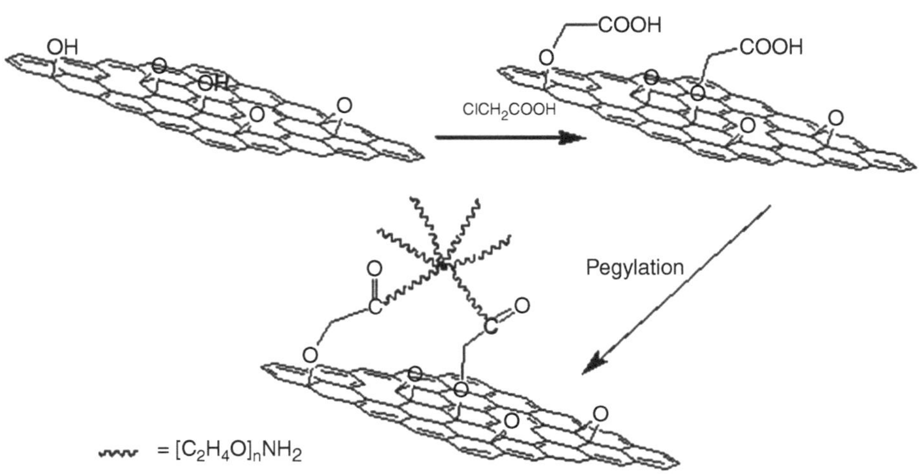

FIGURE 3.2 Functionalization of rGO by Six-Arm Polyethylene Glycol-Amine

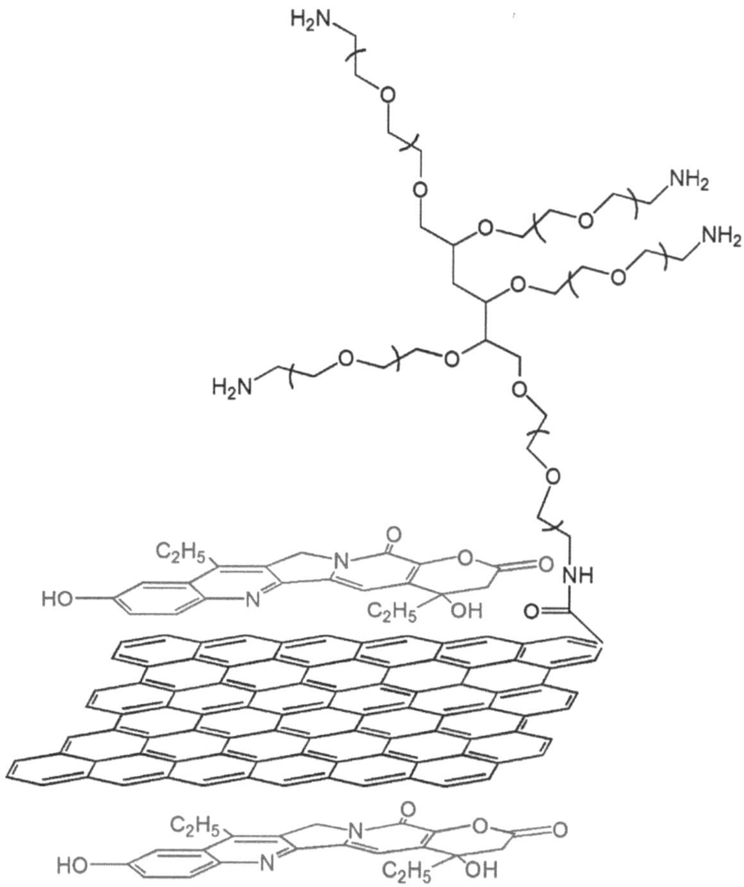

FIGURE 3.3 Schematic Illustration of Pegylation of Graphene Oxide by PEG Stars

As mentioned before, surface functionalization of GO with PEG for drug carrier application has been investigated by several research groups; however, the benefit of the PEG coating is limited by its diffusion barrier, which adversely affects the drug release (Romberg et al., 2008). For effective therapy, rapid intracellular drug release is highly preferred upon immediate arrival of the delivery system. Therefore, responding to a specific stimulus, such as a reduction environment on the tumor site, to release the loaded drug can be useful.

In a report by Wen et al. (2012), the carrier was composed of GO with a sheddable PEG shell attached via a disulfide linkage (NGO–SS–mPEG; mPEG = methoxy polyethylene glycol) that can respond to GSH changes for intracellular drug delivery. The cytosolic GSH (glutathione) level in some tumor cells has been found to be at least fourfold higher compared with normal cells (Meng et al., 2009). This sharp contrast in GSH concentration may serve as an ideal stimulus for a PEG release mechanism. PEGylation of the NGO nanocarrier was done by conjugating amino-terminated PEG to carboxylated GO nanosheets via a disulfide bond (Fig. 3.4). GO–SS–mPEG exhibits high solubility and stability in both PBS and cell medium. In contrast, GO–COOH experiences fast aggregation in

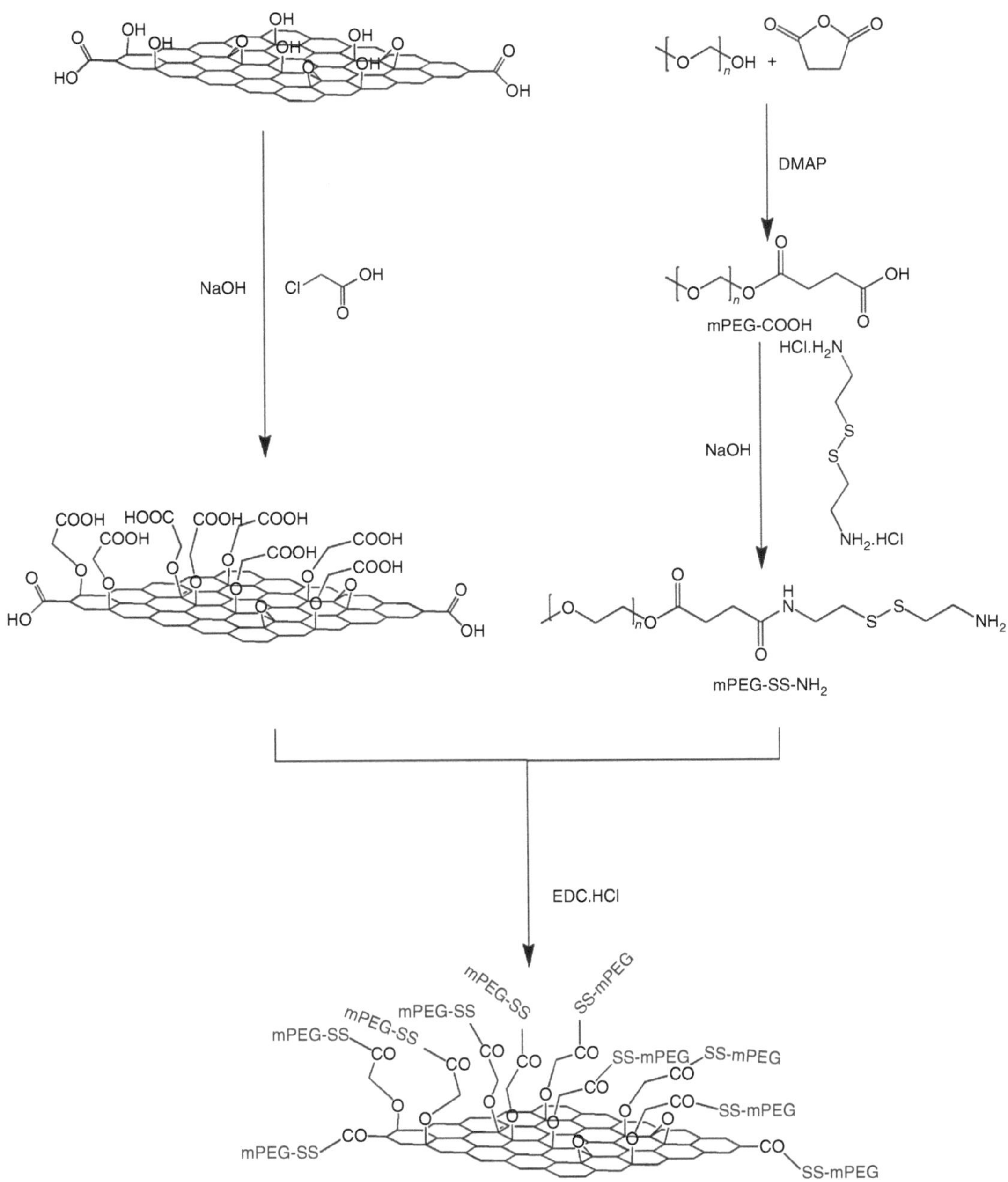

FIGURE 3.4 Synthesis Pathway of Disulfide-Linked NGO-SS-mPEG

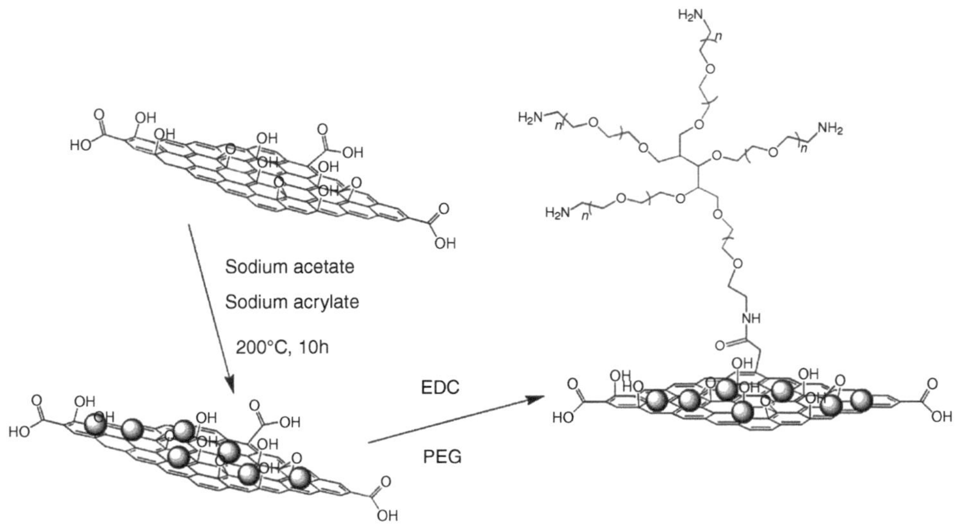

FIGURE 3.5 Synthesizing Strategy of Multifunctional Superparamagnetic Graphene Oxide–Iron Oxide Hybrid Nanocomposite

PBS or cell medium. However, cleaving the disulfide bond (S—S) by GSH as a reducing agent on the tumor site causes instability of GO–SS–mPEG solution in 0.1 mm PBS due to S–S bone cleavage and the detachment of PEG coating from the planes of the GO sheet. For drug loading the result indicated successful loading of DOX onto NGO–SS–mPEG through noncovalent interaction, including π–π stacking and hydrophobic interactions. The release behavior studies showed less than 35% DOX is released in 48 h for pH of 5.5 without GSH. However, upon exposure to 10 mm GSH, 55% DOX is released in 48 h at the same pH. While at pH 7.4 without GSH, only less than 5% of the anticancer agent is released in a period of 72 h. The pH-controlled release relies on the protonation effect of DOX at low pH, which could make DOX more hydrophilic and water soluble, thus leading to the increased release of DOX in aqueous solution (Wen et al., 2012).

Ma et al. (2012) synthesized a multifunctional superparamagnetic graphene oxide–iron oxide hybrid nanocomposite (GO–IONP), which enables magnetically targeted drug delivery and then is functionalized by a biocompatible six-arm polyethylene glycol–amine to acquire high stability in physiological solutions (Fig. 3.5). Many studies have shown that magnetic nanoparticles can be retained at tumor sites by a locally applied external magnetic field, thereby enabling magnetic tumor targeting (Ma et al., 2012).

The synthesized GO–IONP–PEG exhibited excellent stability in various physiological solutions, including saline, cell medium, and serum. DOX was loaded onto GO–IONP–PEG, and it was found it was found that the DOX loading efficiency at pH 8.0 increased from 44% to ≈220% (the weight ratio of DOX to GO) with increasing amounts of added DOX. Related analysis showed that within 360 min about 20% of DOX was released from the nanocomposite at pH 7.4, while nearly 50% of DOX was released in the acidic buffer at pH 5.0. The obtained data from in vitro experiment showed that the magnetically targeted drug delivery by GO–IONP–PEG–DOX was able to selectively kill cells that were localized close to the magnet, without affecting the viability of cells outside the magnetic field.

The photothermal effect of GO–IONP–PEG for damaging tumor cells was also investigated, and it was found that, after being exposed to the near-infrared (NIR) laser for 20 min, cells near the magnet were effectively destroyed, while those far from the magnet were essentially unaffected (Ma et al., 2012).

3.2 MODIFICATION WITH FOLIC ACID

Zhang et al. (2010) synthesized functional GO with good biocompatibility and physiological stability by introducing sulfonic acid groups into the GO surface and attaching, folic acid (FA) molecules (Fig. 3.6).

Camptothecin and DOX, both of which are water-insoluble aromatic molecules, were loaded together onto the GO in a controlled way. The zeta potentials of the NGO, FA–NGO and NGO–SO_3H, were determined to be –86.8, –119.7, and –116 mV, respectively, indicating a significant difference in the surface charge, and thus the colloidal stability, of the synthesized graphene derivatives. It was found that the drug loading ratio for DOX was 400%, which was remarkably higher than that of CPT with only ≈4.5% loading ratio. The difference in the loading ability of the two drugs was ascribed to the difference in their chemical structures, and thus the difference in the π–π stacking and hydrophobic interactions with graphene oxide nanosheets. In addition, the release of the drugs from the FA–NGO with change of pH was investigated. It was found that ≈35% of DOX and ≈17% of loaded CPT were released after 48 h at pH 5.0, while much less DOX or CPT was released after the same time at pH 7.0 (Fig. 3.7) (Zhang et al., 2010).

Depan et al. (2011) also reported a folate-decorated and graphene-mediated drug delivery system. This nanocarrier system was synthesized by attaching DOX to graphene oxide via strong π–π stacking interaction and followed by encapsulation of graphene oxide with FA-conjugated chitosan (Fig. 3.8).

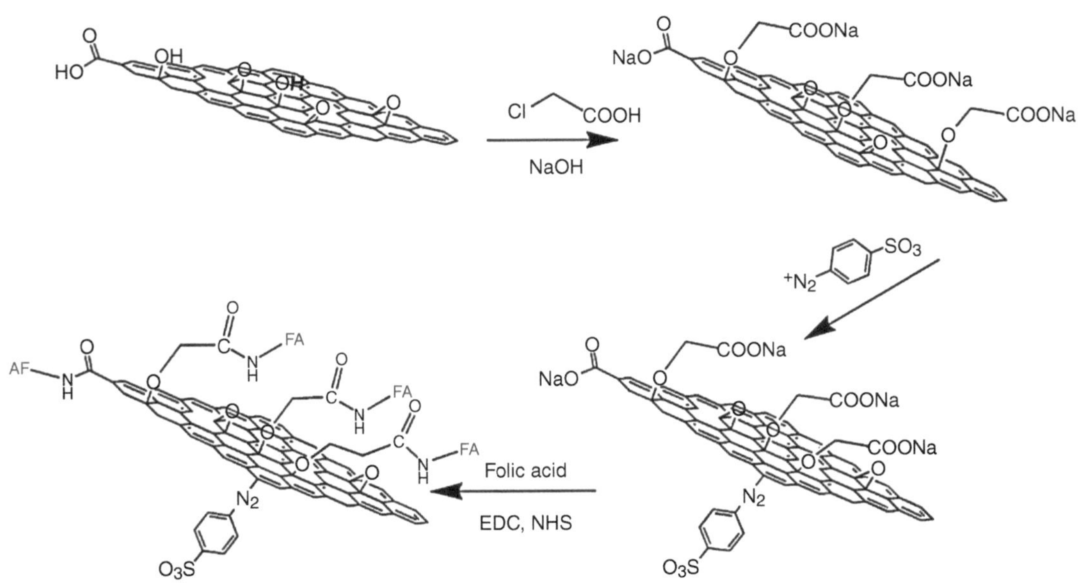

FIGURE 3.6 Schematic Illustration of the Preparation of FA–NGO Conjugates

EDC, 1-Ethyl-3-(3 dimethylaminopropyl)carbodiimide; NHS,*N*-hydroxysuccinimide.

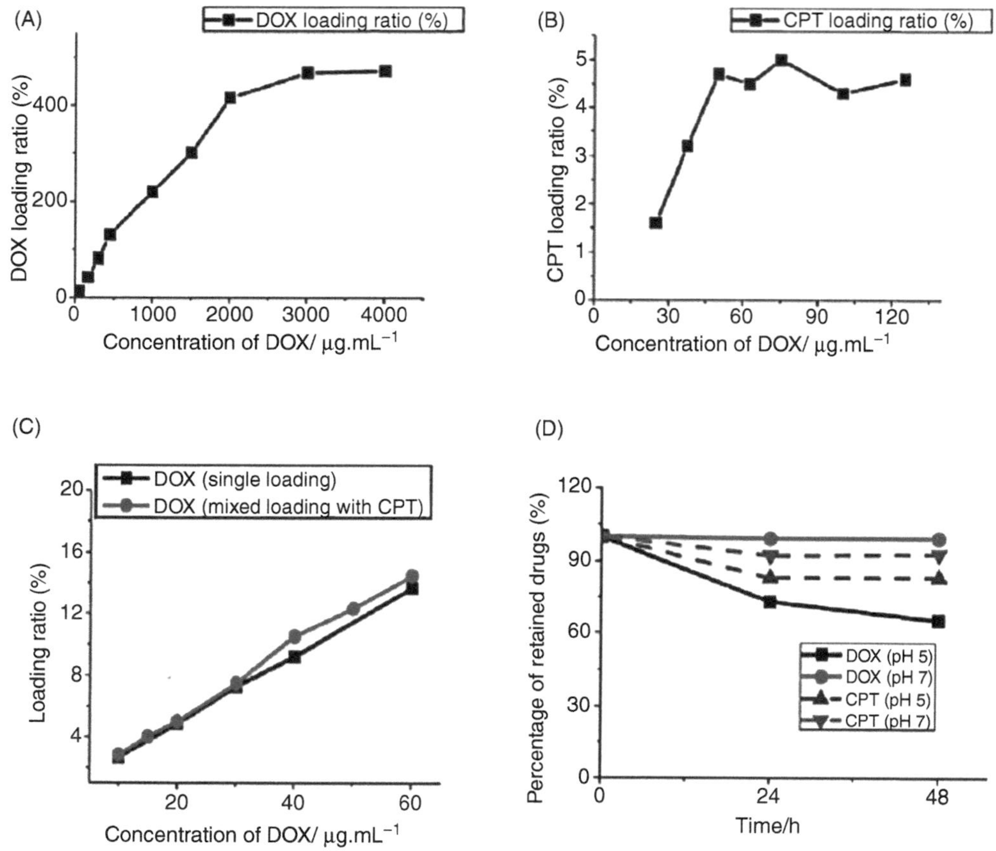

FIGURE 3.7

(A) Plot of loading ratio of DOX to FA–NGO versus the concentration of DOX; (B) plot of loading ratio of CPT to FA–NGO versus the concentration of CPT; (C) plot of loading ratio of DOX to FA–NGO versus concentration of DOX in a mixture of DOX and CPT (concentration of CPT 100 mg/mL); and (D) plot of the release of DOX and CPT from FA–NGO at pH5 and 7, respectively.

Reprinted with permission from Zhang, L., Xia, J., Zhao, Q., Liu, L., Zhang, Z., 2010. Functional graphene oxide as a nanocarrier for controlled loading and targeted delivery of mixed anticancer drugs. Small 6, 537–544. Copyright 2010, John Wiley & Sons.

Analysis showed that in this procedure the DOX-loading efficiency was ≈96%. Encapsulation of DOX–GO with FA–chitosan conjugate was preceded for enhancing the stability of the nanocarrier system in aqueous medium and targeted delivery of anticancer drug because of folate receptors.

Here also it was observed that at pH 7.4, DOX released in a slower and controlled manner from the DOX–GO and DOX–GO–CHI–FA systems (≈7 and 11%, respectively) after 72 h. As well, the amount of drug released from the chitosan–FA–encapsulated carrier was lower by ≈4% than from the nonencapsulated DOX–GO carrier. On the other hand, the release pattern under acidic media indicated that the amount of DOX released in the 72 h was much higher than at neutral conditions (≈18% and 24% for encapsulated and nonencapsulated carriers, respectively). Actually, under acidic conditions, two kinds

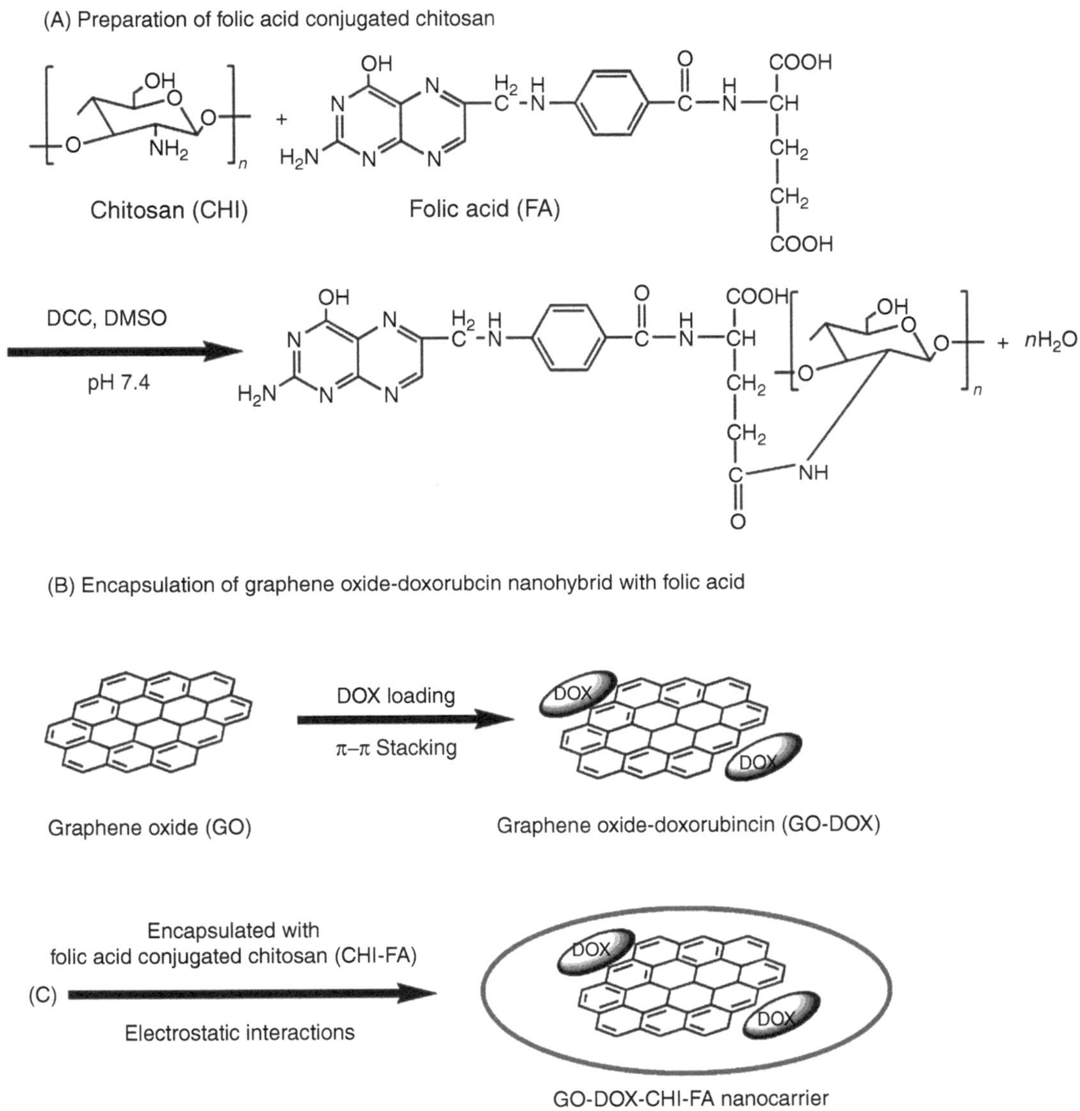

FIGURE 3.8 Schematic Illustration of DOX Loading Pathway Onto Graphene Oxide With Folic Acid Conjugated Chitosan

of hydrogen bonding can be formed between (1) —COOH of GO and —OH of DOX, and (2) between —OH of GO and —OH of DOX. It may be noted that —NH_2 of DOX forms NH_3^+ with H^+ and cannot participate in hydrogen bonding. Furthermore, at low pH, the H^+ in solution would compete with the hydrogen bond forming group and weaken the previously outlined hydrogen bonding interaction, which may lead to a greater release of DOX (Depan et al., 2011).

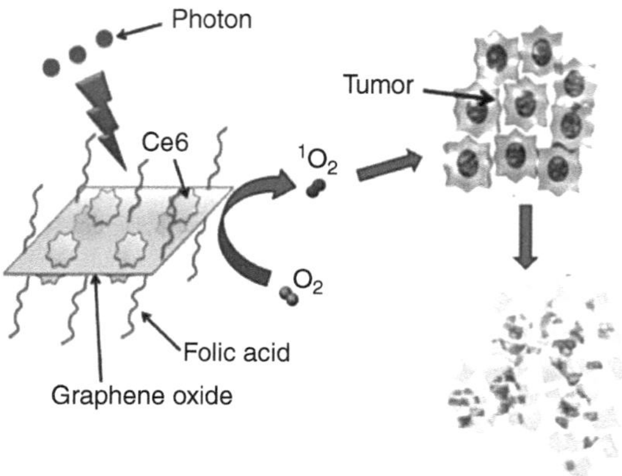

FIGURE 3.9 Photosensitizers Ce6 Loaded by Folic Acid–Conjugated Graphene Oxide

Reprinted with permission from Huang, P., Xu, C., Lin J., Wang, C., Wang, C., Zhang, C., Zhou, X., Guo, S., Cui, D., 2011. Folic acid (FA)-conjugated graphene oxide loaded with photosensitizers for targeting photodynamic therapy. Theranostics 1, 240–250. Copyright 2011, John Wiley & Sons.

The lower amount of drug releasing in the chitosan–FA–encapsulated GO–DOX than in the non-encapsulated carrier can be explained by a strong intermolecular force between DOX and the carrier and the presence of an outer shell of biopolymeric CHI that offers an additional path that the drug has to travel across the macromolecular chains. The low release rate of DOX from encapsulated GO–DOX revealed the strong interaction between DOX and FA through hydrogen bonding and subsequent diffusion of DOX through the CHI shell during the CHI chains' degradation. So, since under acidic medium chitosan chains degrade faster, this faster degradation will result in faster diffusion of drug to target place in a promising manner (Depan et al., 2011).

Huang et al. (2011) designed FA-conjugated GO as drug delivery system for targeting photodynamic therapy (PDT). In Huang's strategy, sulfonic acid groups were introduced to GO, which caused the stabilization of GO under physiological conditions, and then FA molecules were attached to the GO for targeting specific cells with folate receptors (Fig. 3.9). Chlorine6 (Ce6), a promising photosensitizer (PS) with a high sensitizing efficacy, was loaded onto the GO via hydrophobic interactions and π–π stacking.

Zeta potential analysis showed that an aqueous suspension of GO exhibits a zeta potential of ca. –40.28, indicative of negatively charged surfaces caused by the presence of hydrophilic carboxyl groups. With the increase of Ce6 concentration, the zeta potential of the GO–Ce6 composites becomes more negative (–50.69) due to the aggregation of negatively charged Ce6 on the surface of GO. Huang et al. (2011) suggested that there are two different hybrid structures in the solution, including GO–Ce6–GO and Ce6–GO–Ce6 sandwiches, depending on the concentration of Ce6. The measurements showed that loading efficiency was ≈80%. They also observed nanocarriers without FA conjugation didn't have the targeting function, meanwhile, FA–GO–Ce6 and free FA did not exhibit the targeting function, which suggests that free FA can block the binding of FA–GO–Ce6 with folate receptors expressed in the tumor cells. Selective accumulation for FA–GO–Ce6 could be observed in tumor cells (Huang et al., 2011).

Yang et al. (2012) used adamantanyl porphyrin as a linker between graphene oxide and FA-modified β-cyclodextrin (CD) and designed a new drug carrier based on graphene. In their work, porphyrin moiety acts as a binding group, and an adamantane moiety is encapsulated in the cavity of cyclodextrin. By taking advantage of the mutual cooperation of FA-modified b-cyclodextrin (FACD), adamantane-grafted porphyrin, and GO, they preserved the structural integrity of GO, and the resultant supramolecular assemblies could specifically carry DOX into folate-receptor-positive malignant cells without imparting an enhanced toxicity (Fig. 3.10). The loading ratio of DOX on the surface of GO was calculated to be 168% according to the photometric standard curve of DOX at $l = 490$ nm and GO at $l = 680$ nm. By taking advantage of the strong hydrophobic interaction between the CD cavity and adamantane moiety, FA-modified-β-CD was easily introduced to the DOX/adamantanyl porphyrin/GO system to improve the biocompatibility and target effect. Consequently, the resulting nanosupramolecular assembly could specifically recognize folate receptor-positive tumor cells owing to the very high affinity between FA and folate receptor. Moreover, DOX can be released from GO under acidic conditions, which is attributed to the increased hydrophilicity and solubility of DOX and reduced hydrogen-bonding interactions between DOX and GO (Yang et al., 2012).

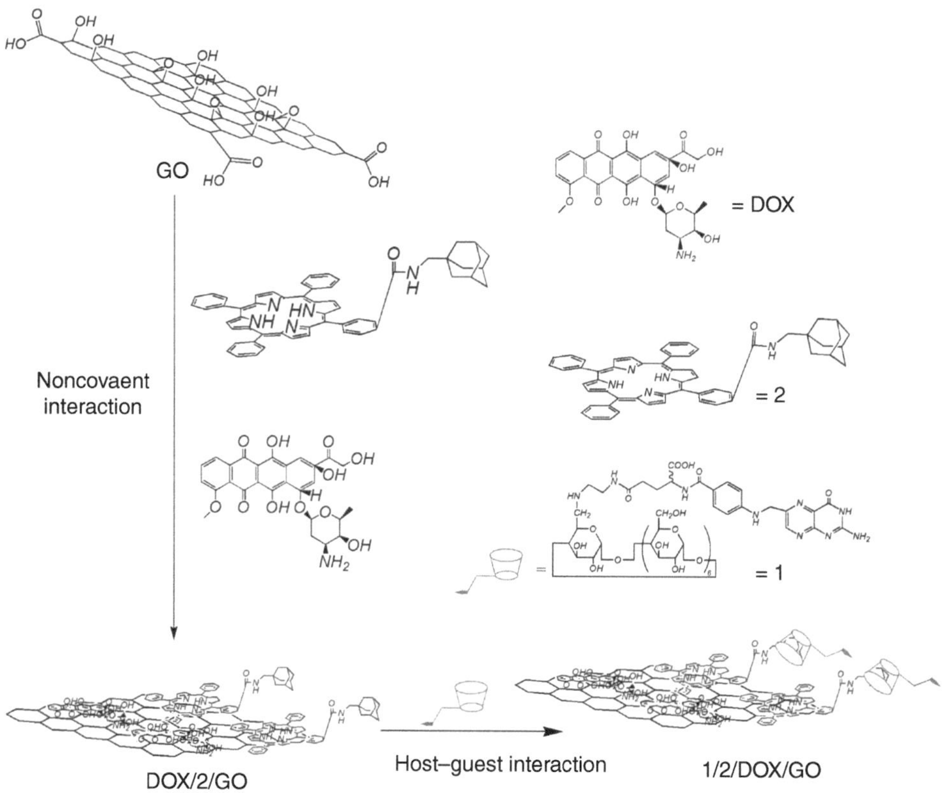

FIGURE 3.10 Synthesis of 1/2/DOX/GO From Graphene Oxide

As mentioned before, because the microenvironments of tumor tissues are acidic, this pH-controlled release process is significant in drug delivery applications. The anticancer activity tests showed that DOX and FA-modified β-cyclodextrin/adamantanyl porphyrin/DOX/GO displayed similar anticancer activity toward HeLa cell lines, a type of human cervical carcinoma cell widely used as a model for folate receptor-positive cells. However, FA-modified b-CD plays a very important role in the anticancer effect of the nanosupramolecular assembly (Yang et al., 2012).

3.3 MODIFICATION BY CHITOSAN

Chitosan, a naturally occurring linear cationic polysaccharide, has found widespread applications in drug and gene delivery. Regarding the intrinsic properties of chitosan, it offers the following advantages in drug delivery applications: (1) Since many surfaces are negatively charged and chitosan is positively charged, promoting electrostatic interaction and attachment of chitosan to the nanoparticles does not require chemical binding; (2) cross-linking and ligand attachment for targeted drug delivery in chitosan are facilitated due to the presence of amino ($—NH_2$) groups; (3) vesicles based on cationic chitosan can effectively attach to the negatively charged phospholipid bilayer of the cellular membrane; (4) the presence of lysosomes in cellular endocytosis can facilitate the degradation of chitosan to release the encapsulated drug at higher efficiency upon intracellular uptake of nanoparticles; (5) solubility of chitosan in pH of 5.3 (endosomal pH) and insolubility in pH of 7.4 (physiological pH) avoids early release of encapsulated drug before the target site is reached; and (6) its mucoadhesive and bacteriostatic properties extend the retention to targeted substrates, and discourage bacterial uptake in the gastrointestinal tract (Fang et al., 2010).

Fang et al. (2010) prepared stable graphene suspension by using chitosan (CS) as a mediating agent. In Fang et al.'s method, the GO suspension simply was added to the solution of chitosan under stirring and chitosan molecules physically adsorbed into graphene oxide surface. They showed that the graphene in the as-prepared suspension can be switched reversibly between a well-dispersed and a more aggregated state with pH as a stimulus (Fig. 3.11) (Fang et al., 2010).

Bao et al. (2011) functionalized graphene oxide (GO) with chitosan (CS) via a facile amidation process. In their method 64% of chitosan molecules were attached to the graphene oxide surface. They also observed stimuli-responsive behavior for chitosan-grafted graphene. Fig. 3.12 shows variation in particle size and zeta potential as a function of pH value for GO—CS. Actually, at pH < 4, the amino groups on CS chains are fully protonated and the GO—CS composites individually disperse in water due to the strong electrostatic repulsion. The increased average size is ascribed to the extended and semirigid CS branches on the GO sheets. With increasing the pH to 5–7, the majority of CS segments are deprotonated and the intra- and interpolymer repulsions decrease, resulting in the rolling up of the polymer chains. Also, loading the water-insoluble anticancer drug camptothecin showed the loading capacity of the synthesized nanocarrier is about 20%. The in vitro release profile of CPT from GO—CS in PBS buffer at 37°C showed that 17.5% CPT release was reached upon 72 h (Bao et al., 2011).

3.4 MODIFICATION BY GELATIN

Gelatin is a linear polypeptide that consists of different amounts of 18 amino acids, which is the thermally and hydrolytically denatured product of collagen and is commonly used for pharmaceutical and medical applications because of its biodegradability and biocompatibility in physiological environments. Gelatin both has the mild reductive effect to graphene oxide sheets and plays an important role

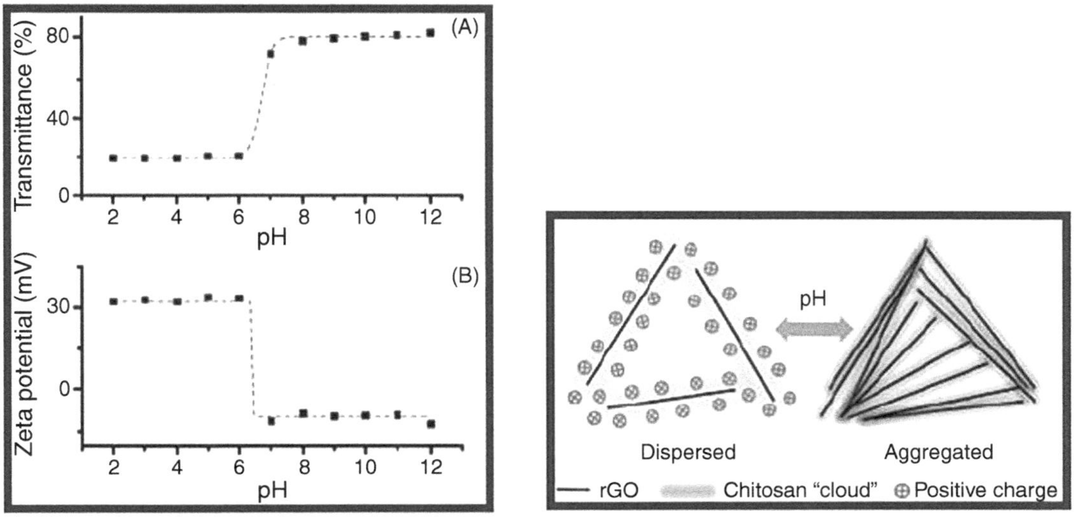

FIGURE 3.11 pH Responsiveness of the rGO/Chitosan Suspension

(A) Transmittance of the supernatants of rGO/chitosan suspensions with different pH values after sedimentations. (B) ζ potential of the rGO/chitosan suspension at different pH (*left*). Schematic illustration of the reversible changes of the chitosan/rGO in aqueous media with pH as a stimulus (*right*).

Reprinted with permission from Fang, M., Long, J., Zhao, W., Wang, L., Chen, G., 2010. pH-Responsive chitosan-mediated graphene dispersions. Langmuir 26, 16771–16774. Copyright 2010, American Chemical Society.

as a functionalized reagent to prevent aggregation. Liu et al. (2011) functionalized graphene nanosheets by gelatin. The obtained biocompatible gelatin–GNS exhibited excellent stability in water and various physiological fluids, including cellular growth media, as well as serum, which were critical prerequisites for the biomedicine application of graphene.

The DOX, a commonly used anticancer drug, was loaded on gelatin–graphene nano sheets to form the DOX@gelatin–GNS complex. The saturated adsorption of gelatin–GNS toward DOX was 16 μg of DOX to 120 μg of gelatin–GNS, which was larger than iron oxide nanoparticles and hollow chitosan/poly(acrylic acid) nanospheres. Also, the release behavior of DOX from DOX@gelatin–GNS experienced a sustained release process due to a time-consuming DOX release. Liu et al. (2011) supposed this can be due to encapsulation of DOX@gelatin–GNS in endosome after the endocytosis of cancer cells, and then merged with lysosome, which contained many kinds of hydrolase, such as phosphatase, cathepsin, and collagenase. With the degradation effect of collagenase to gelatin on the GNS, the DOX was released gradually from the GNS and then exhibited a sustained release effect (Liu et al., 2011).

In another study, which was reported by An et al. (2013), gelatin functionalized graphene nanosheets were used as a carrier of methotrexate (MTX). The loading of MTX on gelatin-GNS was calculated to be 21.1 μg MTX to 75 μg gelatin-GNS, which exhibited a high drug-loading capacity of MTX via strong π–π stacking interaction between the aromatics portion of MTX and the conjugated structure of graphene nanosheets. The MTX release data from MTX@gelatin–GNS at temperature of 37°C in the PBS exhibited an excellent ability of synthesized carrier for controlled drug release. The pH-dependent release behavior of MTX from MTX@gelatin–GNS showed that the release amount under

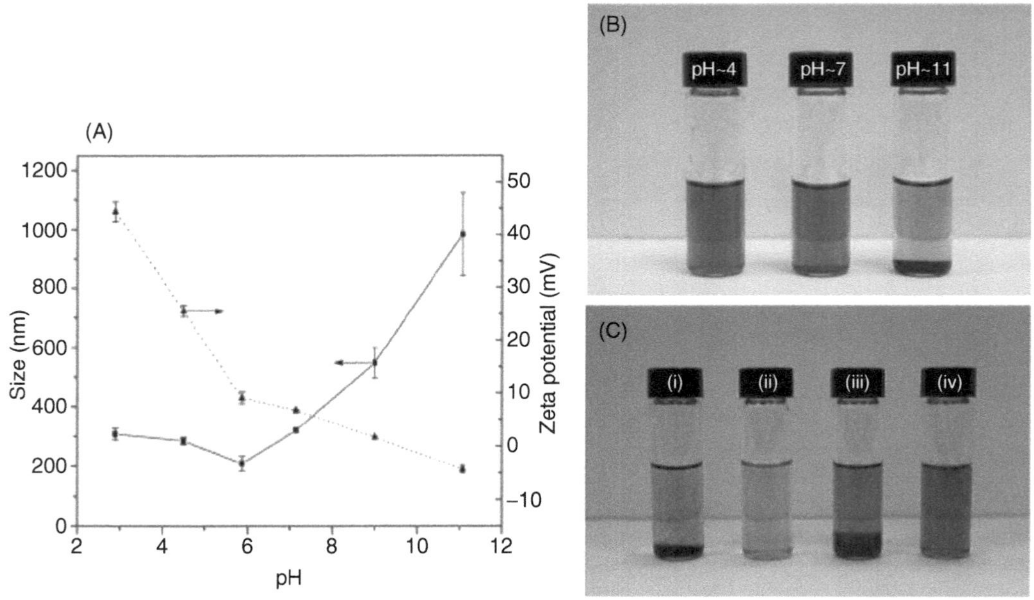

FIGURE 3.12

(A) Variation in particle size and zeta potential as a function of pH value for GO–CS. (B) Digital image for the long-term dispersion status of 0.1 wt.% GO–CS in solutions with different pHs. (C) Digital image for the short-term dispersion status of 0.03 wt.% GO (i–ii) and GO–CS (iii–iv) in PBS buffer and cell culture medium, respectively.

Reprinted with permission from Bao, H., Pan, Y., Ping, Y., Sahoo, N.G., Wu, T., Li, L., Li, J., Gan, L.H., 2011. Chitosan-functionalized graphene oxide as a nanocarrier for drug and gene delivery. Small 7, 1569–1578. Copyright 2011, John Wiley & Sons.

acidic conditions is much higher than that under neutral conditions, which is important for clinical medicine, since the microenvironments in extracellular tissues of tumors and intracellular lysosomes and endosomes are acidic (Fig. 3.13). The release behavior at acidic conditions has been attributed to:

- the much higher solubility of MTX under acidic conditions than under neutral conditions,
- protonation of the amine ($-NH_2$) groups of MTX and the partial dissociation of hydrogen-bonding interactions, and
- the decomposition of the gelatin under acidic conditions (An et al., 2013).

3.5 MODIFICATION BY PEI

Zhang et al. (2011) grafted polyethyleneimine (PEI) chains into the surface of GO. Covalent linking of PEI to GO was done through formation of an amide bond using EDC chemistry (EDC = *N*-ethyl-*N*'-(3-(dimethylamino)propyl)carbodiimide hydrochloride), as illustrated in Fig. 3.14. The resultant PEI–GO had a stable dispersion in saline solution (10% NaCl) and did not precipitate even after centrifugation at 13,000 rpm for 2 h. The zeta potential of PEI–GO was determined to be + 55.5 mV, which facilitates cellular uptake. Zhang et al. (2011) studied loading and delivery of DOX by PEI–GO into HeLa cells.

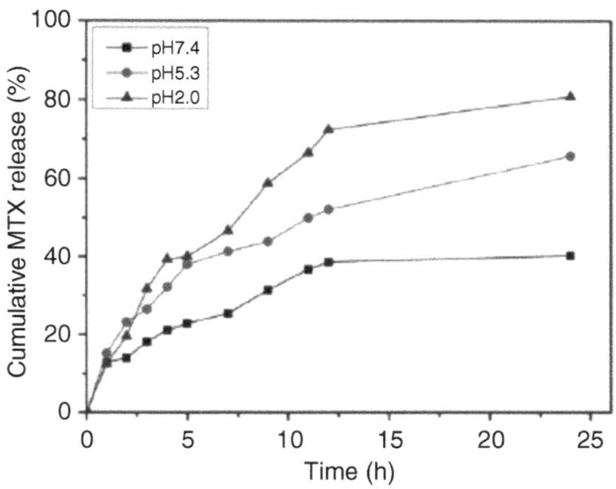

FIGURE 3.13 The Release of MTX From MTX@gelatin–GNS at Different pH Values

Reprinted with permission from An, J., Gou, Y., Yang, C., Hu, F., Wang, C., 2013. Synthesis of a biocompatible gelatin functionalized graphene nanosheets and its application for drug delivery. Mater. Sci. Eng. C 33 2827–2837. Copyright 2013, Elsevier.

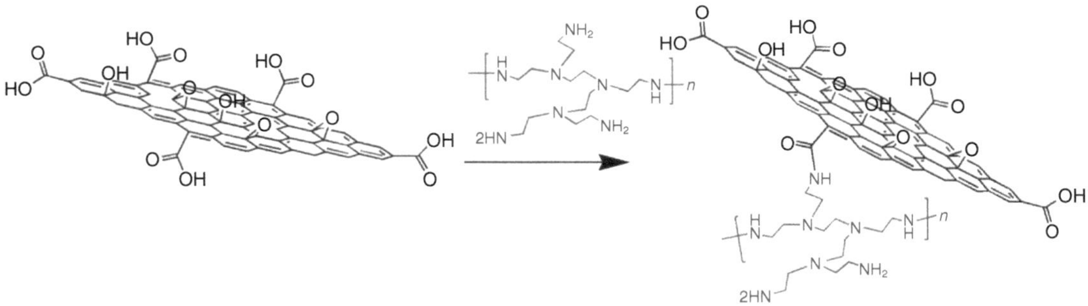

FIGURE 3.14 Schematic Illustration of Graphene Oxide Modification by PEI

Loading of DOX onto PEI-GO was examined by using DOX fluorescence. Adsorption of fluorescent molecules to PEI–GO usually results in quenching of the fluorescence due to fluorescence energy transfer. It was shown that loading of DOX onto PEI–GO leads to nearly complete quenching of DOX fluorescence. The release tests indicated that the PEI–GO/DOX complexes were stable in the cell medium and little DOX was released before cellular uptake. Intracellular release of DOX from the PEI–GO was necessary to enter the nuclei and bind with DNA to show its anticancer (Zhang et al., 2011).

Preparation of PEI–GO by noncovalent adsorption of PEI to GO by electrostatic and hydrogen-bonding interactions was also conducted by Feng et al. (2011). The conjugation of GO with positively charged PEI causes condensation of plasmid DNA onto the surface of GO sheet through electrostatic interaction and can be used as a gene delivery system (Feng et al., 2011).

Covalent functionalization of rGO with *p*-aminobenzoic acid (rGO–C_6H_4–COOH) and conjugating by PEI and biotin was reported by Wei et al. (2014). In this procedure, β-cyclodextrin (b-CD) molecules were introduced to reduce the cytotoxicity of the drug delivery system and to improve the biocompatibility, and were used as host molecules for accommodating guest molecules (DOX) (Fig. 3.15).

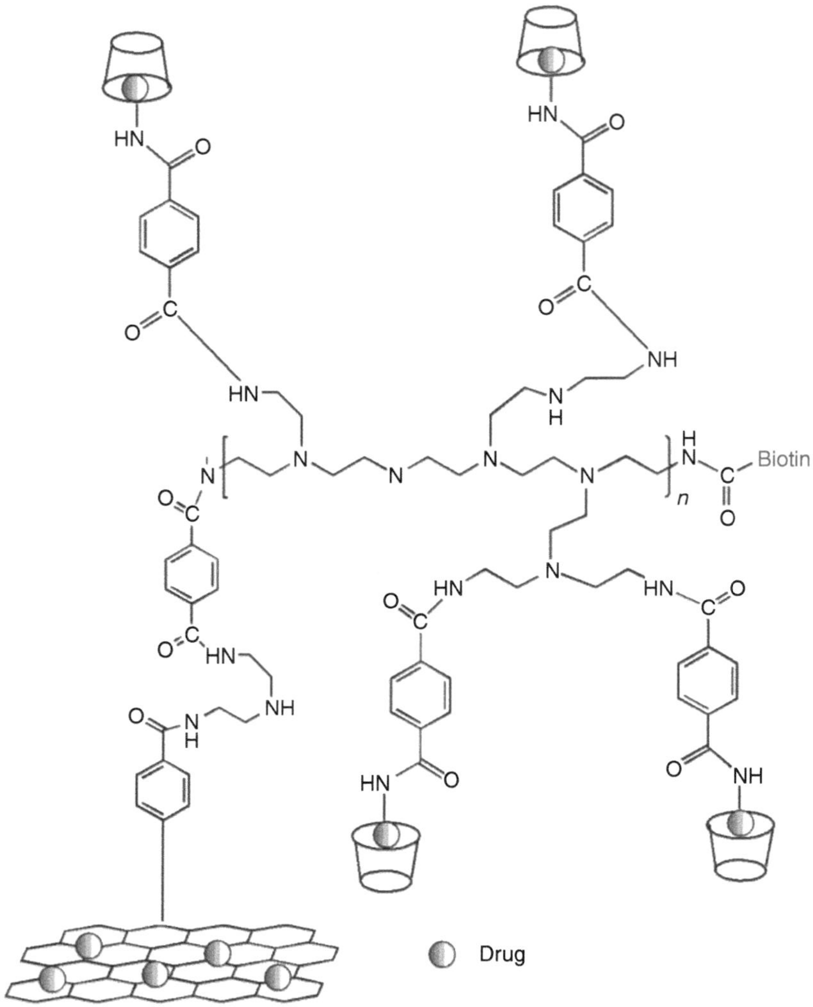

FIGURE 3.15 Schematic Illustration of the Drug Delivery Materials Based on rGO–C_6H_4–COOH and b-CD

The DOX loading capacity was determined to be ≈24.64% (the DOX loading ratio on rGO–PEI–CD-Biotine). They related the drug loading to the synergetic effect of the hydrophobic interior of b-CD and the plane of graphene sheets together. The driving force was attributed to mainly the balance of the host–guest and hydrophobic interactions, the π–π stacking, and hydrophobic interactions between DOX and the drug delivery materials. The release experiments were done at pH = 7.4 PBS and at pH = 5.5 acetate buffered saline (ABS) solution, and the researchers concluded that the drug release behavior is pH dependent at high DOX concentrations and salt dependent at low DOX concentrations. The rapid uptake by HepG2 cells may be the reason for the specific conjugation of biotin of rGOPEI–CD–biotin and biotin-specific receptors, which are often overexpressed on the HepG2 cell surface. The

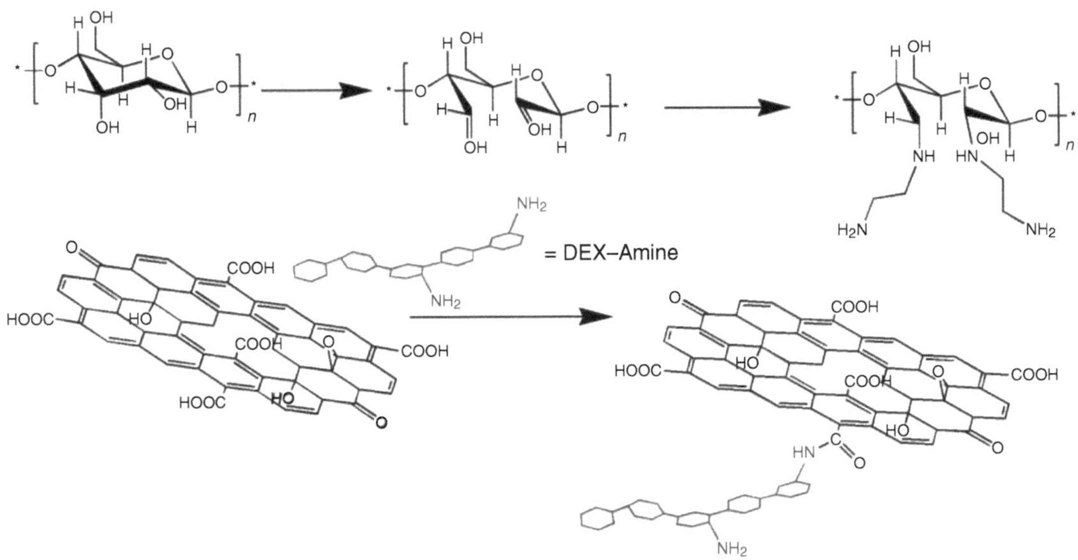

FIGURE 3.16 Functionalization of Graphene Oxide Surface with DOX

biotin content in cancerous tumors is significantly higher than in normal tissue, and rapid proliferations of cancer cells require extra biotin (Wei et al., 2014).

3.6 MODIFICATION BY DEXTRAN

Dextran (DEX) is a hydrophilic natural polymer that can be fully degraded in living biological systems. Zhang et al. (2011) functionalized a graphene oxide surface with DOX and examined its physicochemical stability in the physiological solutions. For chemical attachment of DEX into the graphene surface, the amino groups were introduced into DEX by oxidizing the glycol bond in dextran monomer and giving two aldehyde groups, which were available for conjugation to ethanediamine. Then the synthesized DEX–NH_2 was attached to the graphene oxide surface through the amide linkage (Fig. 3.16).

Fig. 3.17 shows the stability of GO and GO–DOX with concentration of 0.15 mg/mL in different solvents. Compared to GO, which precipitated in the presence of salts and was unstable in various biological solutions, including saline, serum, and cell medium, GO–DEX was rather stable in the above solutions without obvious agglomeration (Zhang et al., 2011).

3.7 MODIFICATION BY POLY(*N*-ISOPROPYLACRYLAMIDE)

Poly(*N*-isopropylacrylamide) (PNIPAAm) is a thermosensitive polymer that shows that low critical solution temperature (LCST) at 33°C can be a good candidate for a targeted drug delivery system. Liu et al. (2010) reported the modification of graphene basal planes with well-defined thermosensitive pyrene-terminal-PNIPAAm via π–π stacking interactions (Fig. 3.18). The modification of graphene using π–π stacking should have advantages over previous attachment strategies, as the electrical

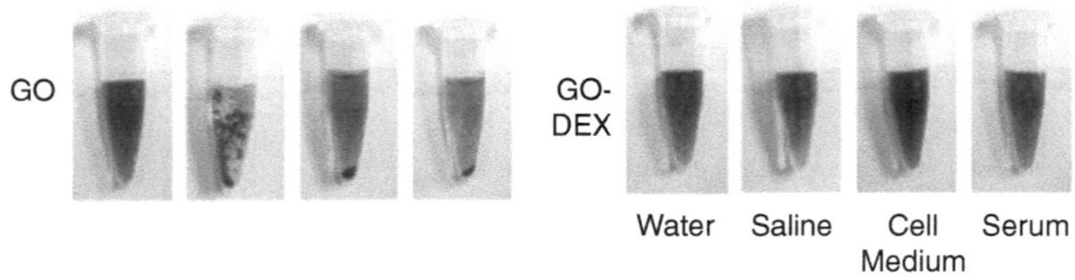

FIGURE 3.17 Stability of GO and GO–DOX in Different Solvents

Reprinted with permission from Zhang, S., Yang, K., Feng, L., Liu, Z., 2011. In vitro and in vivo behaviors of dextran functionalized graphene. Carbon 9, 4040–4049. Copyright 2011, American Chemical Society.

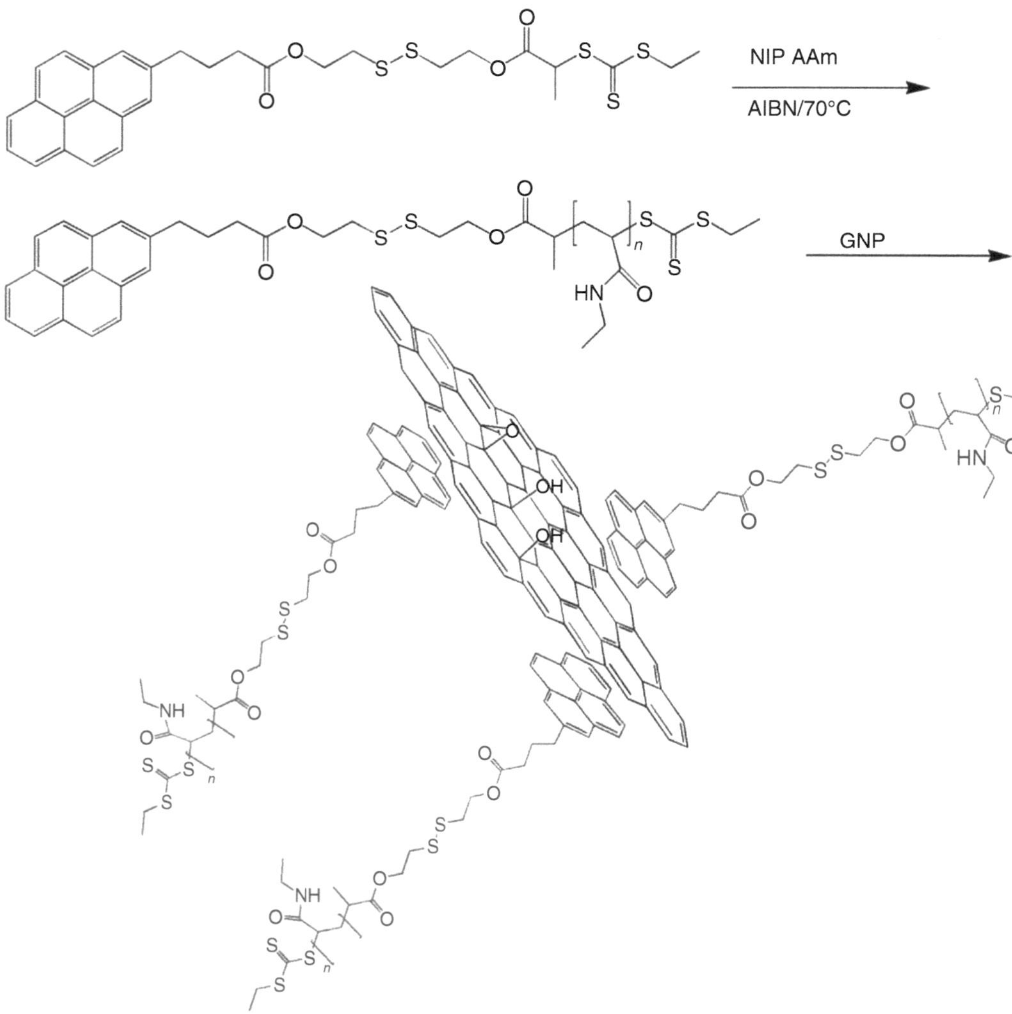

FIGURE 3.18 A Schematic Depicting the Synthesis of Pyrene-Terminated PNIPAAm Using a Pyrene-Functional RAFT Agent and the Subsequent Attachment of the Polymer to Graphene

conductivity will be retained after modification. A thermal sensitivity test of an aqueous solution of graphene-modified PNIPAAm showed that a decreased LCST for immobilized PNIPAAm polymer chains had occurred and at temperatures above 24°C the graphene–PNIPAAm composite suspension aggregated and precipitated.

However, the synthesized system is not suitable for in vivo applications as its LCST is below body temperature, and graphene with pNIPAAm as a surface modifier is expected to rapidly aggregate upon injection. Therefore, NIPAAm was copolymerized with hydrophilic monomers to yield polymers with an LCST above 37°C.

Pan et al. (2011) modified the graphene surface by PNIPAAm through click chemistry. The PNIPAM–N_3 homopolymer was first prepared by the atomic transfer radical polymerization (ATRP) method, and GO sheets were functionalized by propargyl amine and followed by the click reaction between alkynyl-GO and PNIPAM–N_3 (Fig. 3.19). The researchers found the phase transition temperature (coil-to-globule transition) of $PNIPAM_{32}$-N_3 (M_n,NMR = 3800 g mol^{-1}) is 37.8°C, but modified graphene exhibited stable dispersion in water even at 40°C. They related this observation to the fact that the GO sheets were only partially reduced during the PNIPAM–GS preparation and their hydrophilicity remained. This group used the synthesized PNIPAM–GS as a carrier of camptothecin (CPT), a potent topoisomerase I inhibitor, and found that CPT could be complexed with PNIPAMGS by directly mixing a CPT/dimethyl sulfoxide (DMSO) solution with a PNIPAM–GS aqueous solution. The analyzed

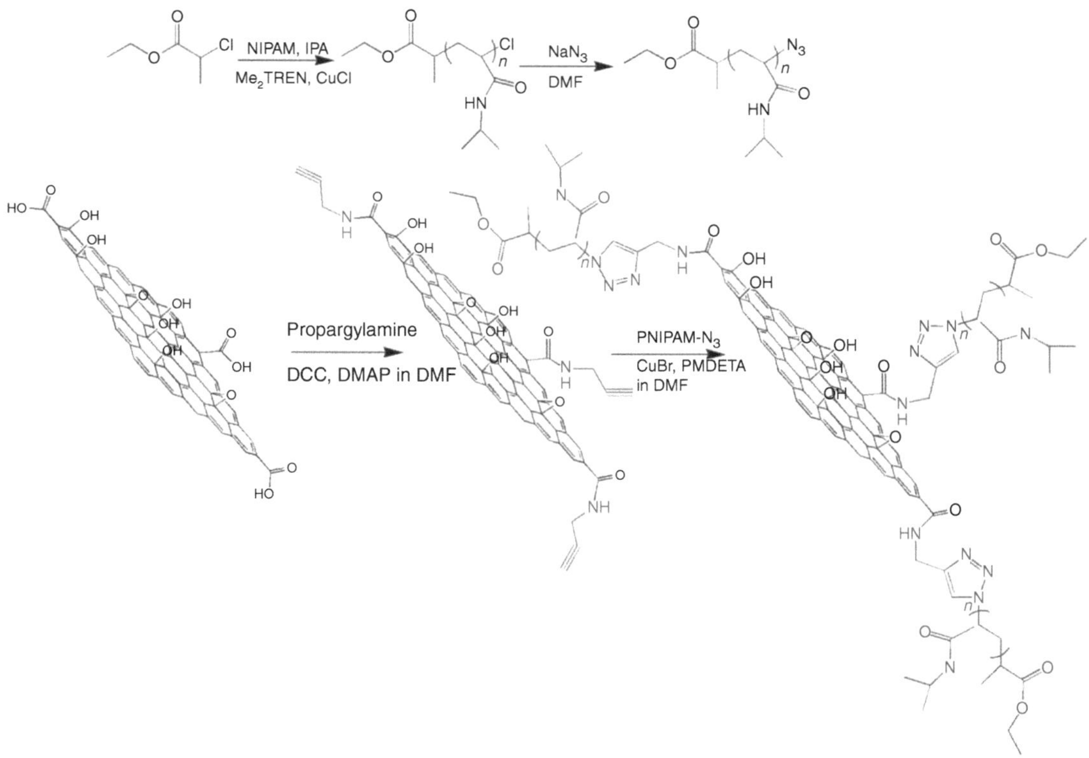

FIGURE 3.19 Synthesis Route of PNIPAM-GS

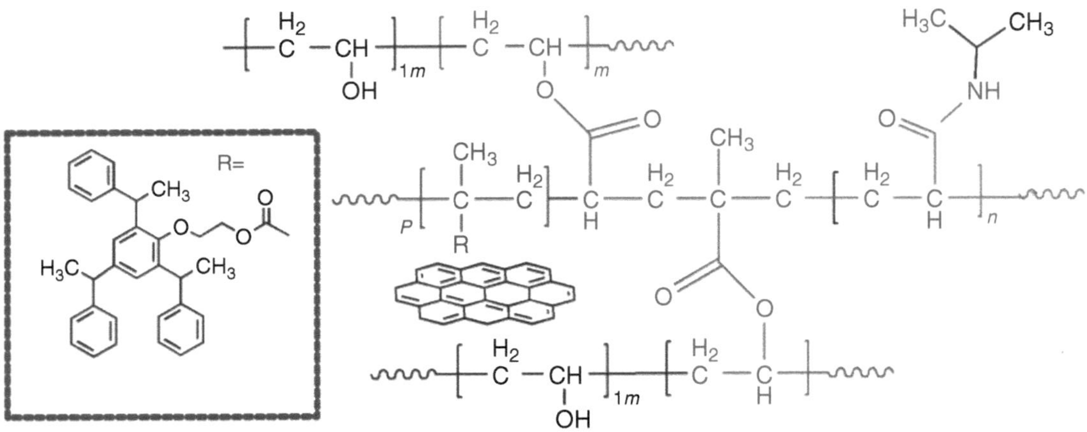

FIGURE 3.20 Average Structure of Graphene –PVAMA Network, G/PVAMA

data showed that 1 g of PNIPAM–GS bound about 0.185 g of CPT. The obtained data from a release test of PNIPAM–GS–CPT in water at 37°C showed the release of 16.9% within the first 72 h (Pan et al., 2011).

Toumia et al. (2015) modified the graphene surface by pNiPAAm, which displays an LCST close to physiological temperature. Modification was done by the photo-cross-linking of methacrylated PVA chains and followed by copolymerizing of NiPAAm with the methacrylic grafts of PVA (Fig. 3.20). The release properties of the GPVA/*p*(MA-*co*-NiPAAm) hydrogel were influenced by the temperature using as model molecule the anticancer drug DOX. Permeability of the matrix is affected by the volume phase transition, whereas stacking of DOX on the graphene is stabilized by $\pi-\pi$ interactions. The data analysis showed that the amount of DOX released at physiological temperature was higher than at room temperature (Toumia et al., 2015).

In the work done by Gollavelli and Ling (2012), graphene sheets with magnetic, water dispersibility, and fluorescent properties synthesized as a multifunctional graphene drug carrier. Magnetic graphene was obtained by using GO and ferrocene, which irradiated inside of the focus microwave oven. Further covalent modification was accomplished by polyacrylic acid (PAA) and fluorescein *o*-methacrylate (in order to increase water dispensability and induce fluorescent properties) via conventional radical polymerization using the sonication and microwave irradiation method (Fig. 3.21) (Gollavelli and Ling, 2012).

3.8 MODIFICATION BY POLY(ALLYLAMINE)

Zhou et al. (2014) synthesized a pH-responsive nanocarrier based on modification of GO surface by charge-reversal polyelectrolyte and formation of a complex with integrin $\alpha_V\beta_3$ monoantibody for targeted delivery and controlled release of DOX into cancer cells. Citraconic anhydride-functionalized poly(allylamine) (PAH–Cit) is another common charge-reversal polyelectrolyte that has been used recently. This type of polyelectrolyte can be readily converted back to cationic poly(allylamine) by amide hydrolysis upon exposure to mild acidic environments. In Zhou's method, PAH–Cit was conjugated with a cationic polyelectrolyte (PEI)–coated GO by electrostatics (Fig. 3.22). The DOX loading

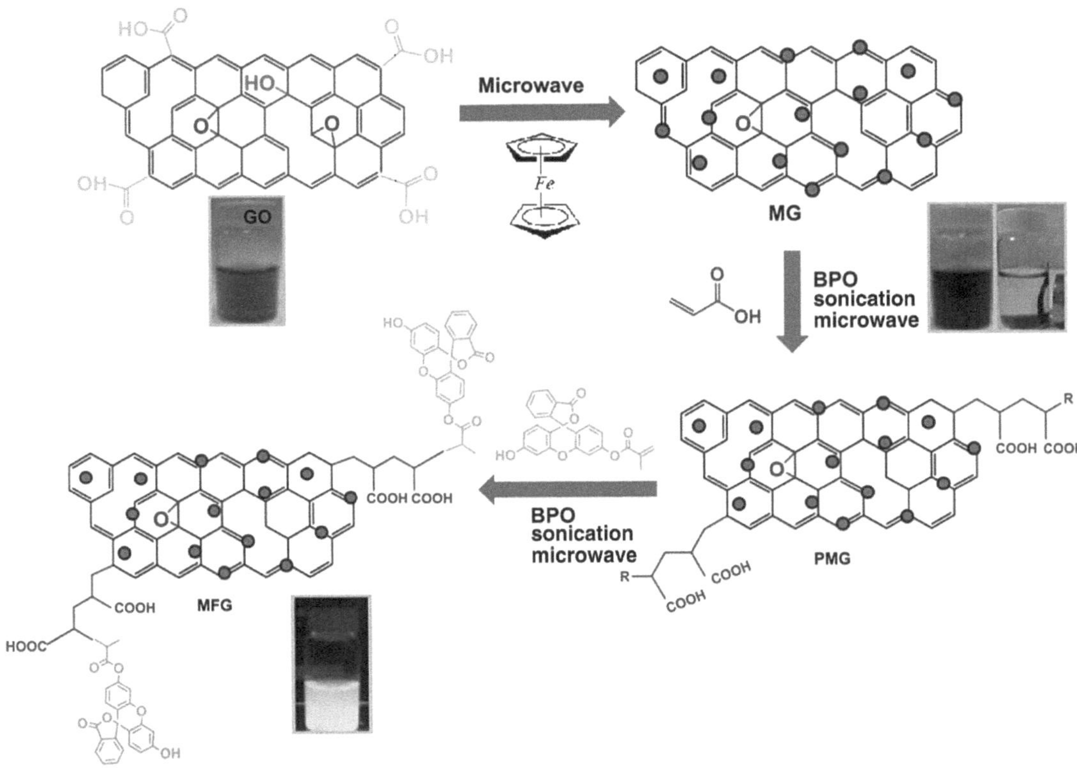

FIGURE 3.21 Schematic Representation for the Synthesis of MFG Using Microwave-Heated Sonication-Assisted Process and Surface Functionalization With AA and FMA

capacity was determined to be over 90%. At neutral conditions DOX was released very slowly (about 28.75% of the total bound released for 24 h). With decreasing the pH to 6.8 (the pH of tumor tissues), the release efficiency of DOX was enhanced to 50.25% of the total DOX after 24 h. The release efficiency was even up to 81.25% under the endosomal pH of cancer cells (pH = 5.0).

The related analysis showed that the synthesized nanocarrier can specifically transport DOX to the targeted cancer cells with high efficiency. Also, it was confirmed that GO–Abs/PEI/PAH–Cit/DOX could be efficiently taken up by U87 MG cells with subsequent intracellular release of DOX and followed by transporting of DOX into the nucleus with the nanocarrier left in the cytoplasm (Zhou et al., 2014).

3.9 PHOTOTHERMAL ANTICANCER ACTIVITY OF GRAPHENE

The use of NIR light in the 700–1100 nm range for the induction of hyperthermia is principally attractive, because biological systems mostly lack chromophores that absorb in the NIR region. Hence, NIR light can effectively and safely penetrate to the normal tissue and selectively destroy cancer cells

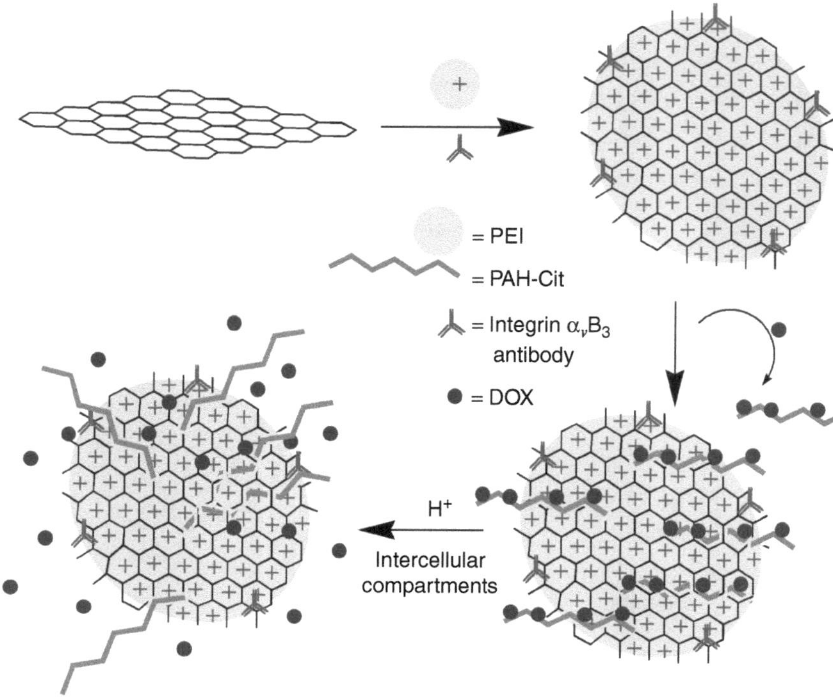

FIGURE 3.22 Construction of the Targeted Charge-Reversal Nanocarrier (GO–Abs/PEI/PAH–Cit/DOX) and Controlled Release of DOX at Endosome or Lysosome, Stimulated by the pH-Dependent Charge Reversal of the Charge-Reversal Polyelectrolytes on GO

to which the nanoparticles are attached so lesions can be treated without the need for direct access to the tumor site. Yang et al. (2012) utilized nanographene sheets (NGS) with polyethylene glycol (PEG) coating as a strong optical absorber in the NIR region for in vivo photothermal therapy, achieving ultra-efficient tumor ablation after intravenous administration of NGS and low-power NIR laser irradiation on the tumor. PEGylated NGS (NGS–PEG) was prepared by using GO and conjugation with amine terminated six-arm branched PEG (Fig. 3.23). The performance of PEGylated NGS compared with gold nanomaterials appears to be comparable in terms of administration routes (intravenous). To verify the potential of using NGS in photothermal therapy, a NGS–PEG solution was exposed to an 808 nm NIR laser at a power density of 2 W/cm^2 with water as the control. The NGS–PEG solution showed a rapid increase of temperature when exposed to the laser within a short time (Yang et al., 2012).

Markovic et al. (2011) also studied on the possibility of using graphene as photothermal anticancer in the range of the NIR spectrum. A stable water suspension of graphene nanoparticles was prepared by exfoliation of graphite in the aqueous solution of nonionic/nontoxic macromolecule polyvinylpyrrolidone (PVP). Comparing graphene with single-wall carbon nanotube (SWCNT), the experimental results showed that the mass-specific heat capacity of SWCNT (0.65 J/gK) (Hone et al., 2000) is three-fold lower than that of graphene (2 J/gK) (Mounet and Marzari, 2005), indicating that nanotubes should generate heat faster than graphene. However, the fact that graphene nanoparticles were presumably

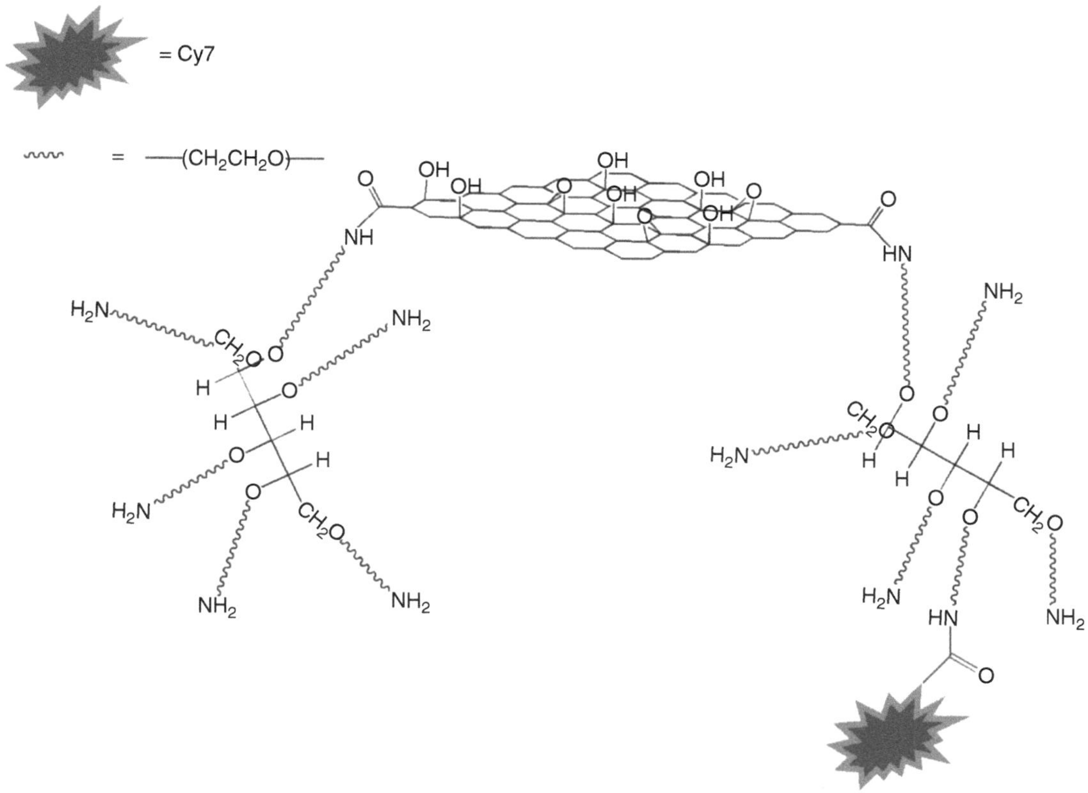

FIGURE 3.23 A Scheme of an NGS With PEG Functionalization and Labeled by Cy7

better dispersed than SWCNT apparently could influence the photothermal sensitivity of the two carbon family nanomaterials (Markovic et al., 2011).

Besides photothermal anticancer activity of graphene, loading of an anticancer drug onto graphene particles can be very efficient in cancer treatment. Kim et al. (2013) reported functionalization of GO by branched polyethylenimine (BPEI) and PEG. In their procedure BPEI was covalently conjugated with the carboxylic acid group of GO, and the resulting BPEI–GO has been reduced by hydrazine monohydrate (rGO), followed by PEGylation through the coupling reaction (Fig. 3.24).

The synthesized carrier was used for DOX loading. The PEG–BPEI–rGO and DOX-loaded PEG–BPEI–rGO suspension was stable in water. It was found that the drug-loading capacity for PEGBPEI-rGO was remarkably higher than that of the unreduced form, PEG–BPEI–GO (about 100% for PEG–BPEI–rGO and only about 10% for PEG–BPEI–GO). This was attributed to the difference in the π–π stacking and hydrophobic interactions with DOX due to restoration of fused aromatic clusters upon reduction. The release analysis exhibited that the DOX loading onto PEG–BPEI–rGO has excellent stability in physiological solutions, such as saline and cell medium over 48 h and no drug has been released. But the presence of ethanol is destructive to the noncovalent interaction between DOX and PEG–BPEI–rGO, meaning desorption of DOX from PEG–BPEI–rGO. However, here also, loaded DOX could be efficiently

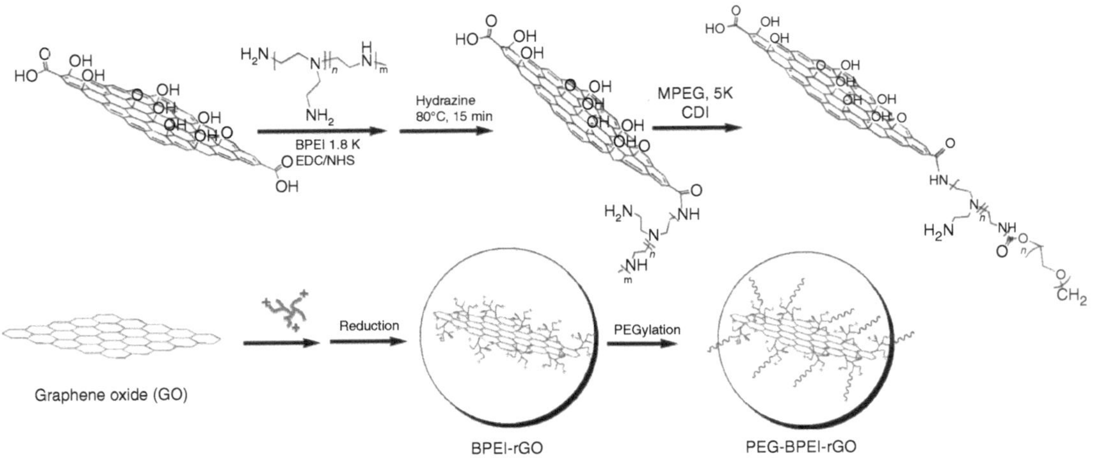

FIGURE 3.24 Schematic Illustration of the Preparation of PEG--BPEI–rGO

released by glutathione (GSH) and the photothermal effect of irradiated NIR in test tubes. A comparison of drug-releasing ability of stimuli showed a drug release upon NIR irradiation, GSH, and acidic conditions of about 15, 37, and 10% for 180 min, respectively, while DOX release was increased by ~45% upon application of multiple stimuli to the PEG–BPEI–rGO/DOX solution (Kim et al., 2013).

Experiments data indicated that PEG–BPEI–rGO successfully escaped from the endosome by NIR irradiation before lysosomal degradation, which subsequently could induce high drug efficacy by GSH-mediated drug release in the cytosol. Actually, upon NIR irradiation, the photothermal heating caused a change in binding energy between PEG–BPEI–rGO and DOX, leading to desorption of DOX from the carrier and diffusion of DOX into the cytosol of the cells. It has been observed that 10 min NIR irradiation led to the significant release of drug into the cells, whereas longer NIR irradiation did not cause the further increase of DOX release into cells (Kim et al., 2013).

3.10 MODIFICATION BY POLY *N*-VINYL CAPROLACTAM

Kavitha et al. (2014) functionalized graphene oxide covalently with poly *N*-vinyl caprolactam (GO–PVCL) via in situ ATRP. In their procedure, amine-functionalized GO (GO-NH_2) was produced by reacting of GO with 1,3-diaminopropane and followed by introducing 2-bromo-2-methylpropionly bromide as a surface initiator. Surface-initiated polymerization of *N*-vinyl caprolactam was accomplished through ATRP (Fig. 3.25).

The dispersion stability of GO–PVCL in water and PBS buffer at room temperature and 43°C was examined using a Turbiscan. The obtained data showed that GO is stable in water due to electrostatic repulsion but unstable in PBS, probably due to the screening effect of the added salts (Kam and Dai, 2005). But for GO–PVCL no discernible change in transmission with time was observed in both water and PBS at room temperature. In contrast, above LCST, GO–PVCL aggregated and settled due to hydrophobic interactions. These results suggest that grafting PVCL enhanced the dispersion and stability under physiological conditions. The aggregation above LCST paves the way for the targeted delivery at the tumor site if an active target element for site-specific heating can be introduced. The

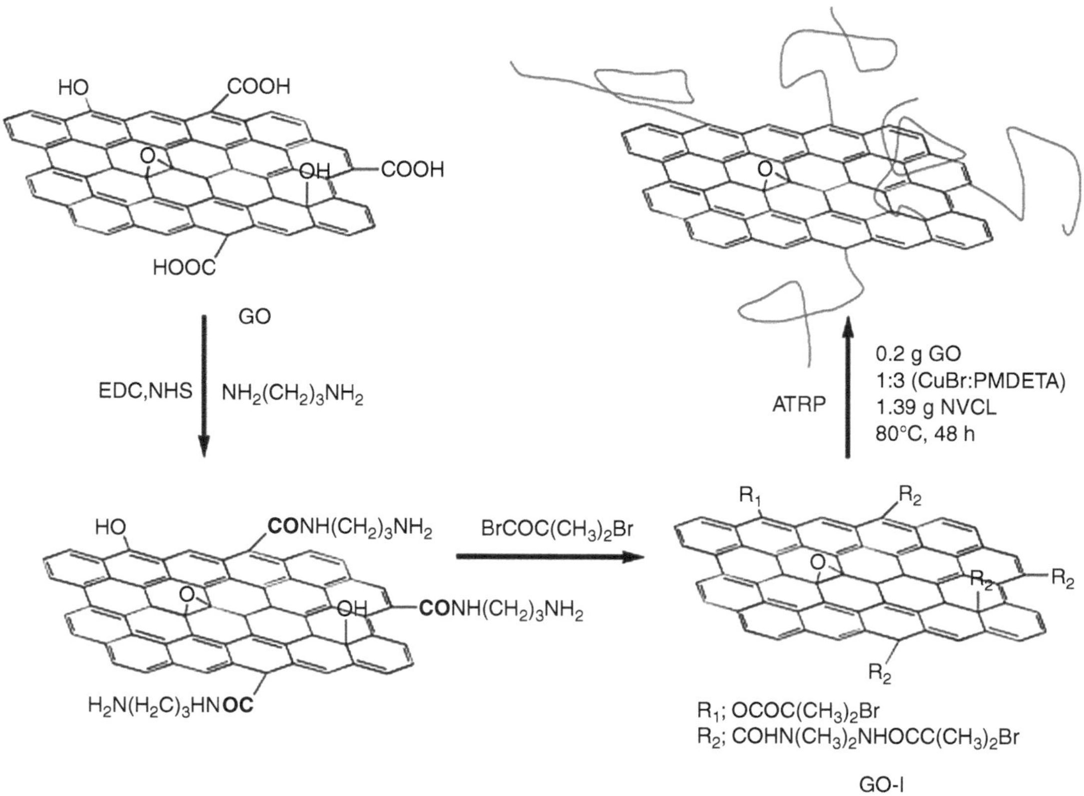

FIGURE 3.25 Schematic Diagram Showing the Synthesis of GO–PVCL

sizes of GO–PVCL as a function of temperature and their measured reversibility with changing from 25 to 45°C and vice versa over five cycles has been shown in Fig. 3.26 (Kavitha et al., 2014).

Actually, below the LCST of PVCL, H-bonding occurred between the amide group of the caprolactam ring and water, causing GO dispersion with the mean size of approximately 438.5 nm. While above the LCST, hydrophobic interactions predominated, resulting in PVCL meso-globules ranging from a hydrophobic core to a hydrophilic surface, leading to GO–PVCL with a mean size of 5 μm. PVCL is a typical example of type I classical Flory–Huggins miscibility behavior, leading to a continuous change in the LCST from 30 to 50°C as its molecular weight increases (Kavitha et al., 2014).

The drug loading onto GO–PVCL was achieved by the simple mixing of a GO–PVCL aqueous solution with camptothecin (CPT) dissolved in DMSO, followed by centrifuging, washing, and filtration. The UV–Vis spectrometry analysis indicated 1 g of GO–PVCL could load ~0.2 g of CPT. The release behavior of the synthesized carrier could be investigated by fluorescence spectroscopy due to the high degree of CPT fluorescence quenching after adsorption onto GO–PVCL. The obtained data revealed that drug adsorbed physically onto GO–PVCL retained its biological activity and the system has a stable solution in an acetate buffer (pH 5.5), water, and PBS at 37°C. Small amounts of CPT were released in water and PBS over a 72 h period, which is lower than that from polymer micellar systems

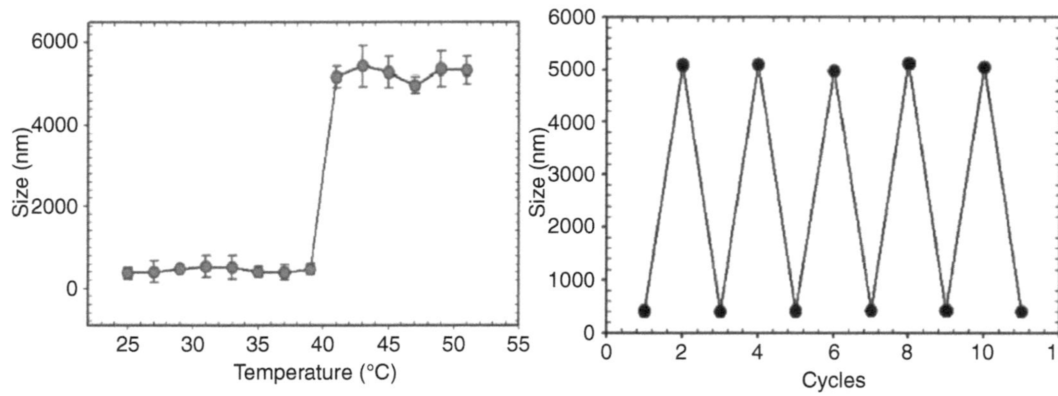

FIGURE 3.26

Z-average size of GO–PVCL in water as a function of the temperature measured by DLS; the concentration of GO–PVCL was 1 mg mL^{-1}. The error bars are based on triplicate samples (*left*) and reversible change in the size of GO–PVCL, as determined by DLS, during the heating and cooling cycles (*right*).

Reprinted with permission from Kavitha, T., Kang, I., Park, S., 2014. Poly(N-vinyl caprolactam) grown on nanographene oxideas an effective nanocargo for drug delivery. Colloids Surf. B Biointerfaces 115, 37–45. Copyright 2014, Elsevier.

because CPT was attached close to the surface of the graphene sheets by π–π stacking instead of being encapsulated in polymer micelles or nanoparticles. This system has several advantages over micelles and liposome nanoparticles, such as retention of biological activity of the drug, slow and steady drug release, outstanding stability, and a high loading capacity compared to micelles. The GO–PVCL–CPT complex exhibited outstanding stability and a high loading capacity at neutral pH, while in an acidic solution (pH 5.5), the appreciable release of CPT (~60%) over 3 days was observed. This was attributed to the increased hydrophilicity and solubility of CPT at this pH, which indicates the pH-triggered release of the drug. Again, stability of the GO–PVCL–CPT complex under normal physiological conditions at 45°C was observed whereas the drug was released at pH 5.5, but the amount released decreased to ~30% compared to 60%. The drug released above the LCST was hindered because the predominant hydrophobic interactions led to the coiled conformation of the polymer chains above the LCST, which hindered easier drug release. The LCST of PVCL 40°C, which is greater than the physiological body temperature (37°C) but lower than the temperature of heated tumors (42°C), can be more promising for active targeting of drug delivery since, when injected in vivo, they become insoluble and accumulate in a locally heated tumor (Kavitha et al., 2014).

3.11 MODIFICATION BY POLY *N*-ISOPROPYL ACRYLAMIDE–POLYETHYLENE GLYCOL COPOLYMER

Wang et al. (2014a) synthesized the PNIPAM–PEO by free radical polymerization reaction as smart polymeric nanoparticles (PNP) with an average diameter of about 50 ± 12 nm and then assembled these PNPs onto the 1-pyrenebutyric acid *N*-hydroxysuccinimide ester (PNHS)–functionalized GO nanosheets (Fig. 3.27).

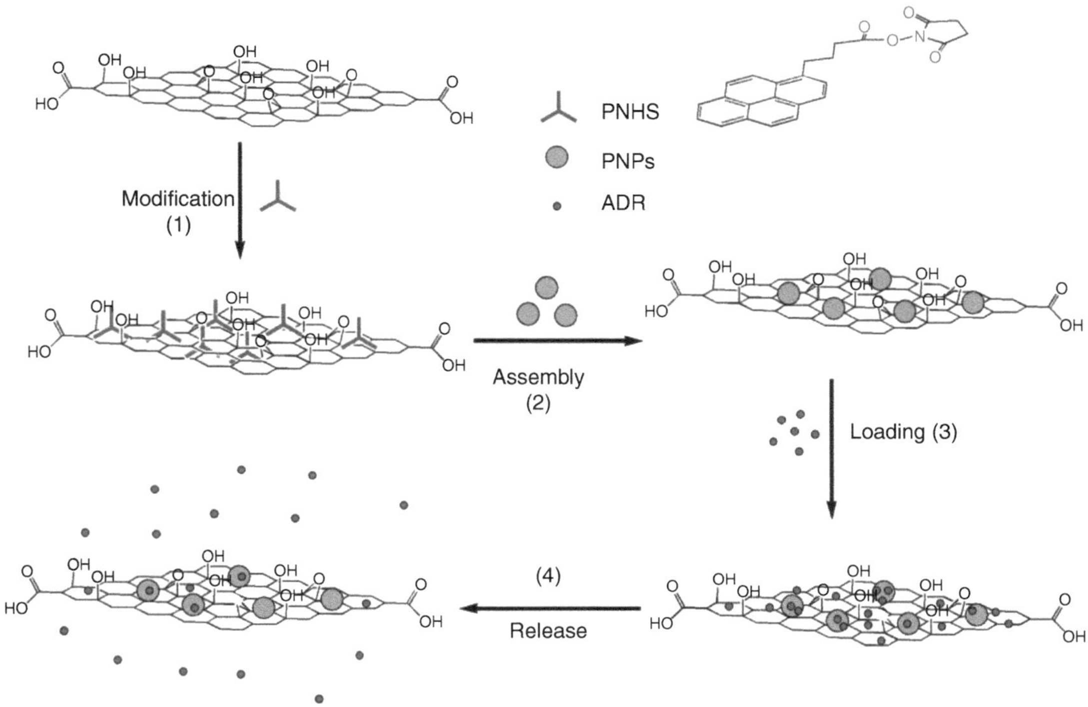

FIGURE 3.27 Schematic Representation of the Synthesis of GO–PNP Hybrids and the Subsequent Loading and Release of Drug Molecules

In this procedure GO as an ultrathin nanosheet offers a large surface area for the assembly of PNPs, and a classic aromatic anticancer drug, Adriamycin (ADR), was used to explore the performance of the synthesized GO–PNP hybrids in drug loading and release. For the synthesis of PNIPAM–PEO PNPs, NIPAM and PEO were used as monomer and macromonomer, respectively, and *N,N'*-methylenebisacrylamide (BIS) as the cross-linker. The reaction was done through the congenital radical polymerization by adding $K_2S_2O_8$ as the initiator. Assembly of the produced PNPs onto GO–PNHS nanosheets is based on the covalent interaction between —NH— groups on the PNPs and the NHS groups on GO–PNHS nanosheets (Wang et al., 2014b).

The LCST of synthesized PNPs was determined about 34.8°C, so the temperature of the loading process was set at 37°C. When the temperature of the GO–PNP aqueous solution was set at room temperature (<LCST of PNPs), the PNPs in the aqueous solution kept stretching and exhibited hydrophilic characteristics. However, addition of the solution into the ADR solution (DMF, 37°C, >LCST) led to the PNPs in the hybrid starting to shrink due to phase transition, and the PNIPAM chains in the PNPs became hydrophobic cores. With the continued dropping, the aqueous phase became the major solution and the hydrophobic ADR was wrapped into the hydrophobic cores of PNPs. At the same time, some ADR molecules were adsorbed onto the GO surface by π–π stacking and hydrophobic interaction, 17,26,48; thus ADR molecules were loaded into the PNPs and onto the GO surface. The loading efficiency of ADR (depending on PNP covering the GO surface) typically for the GO–PNP hybrid with a mass GO/PNPs ratio of 1:5 (≈22 PNPs per μm^2) was calculated to be about 87%. Actually, occupation of the GO surface by both PNHS and PNPs decreases the surface area and the loading efficiency of GO. The synthesized carrier exhibited stable and continuous drug release during the in vitro releasing, and the released amount of drugs remained stable after 100 h incubation.

The release efficiency of the synthesized hybrids depended on the PNP covering on the GO surface. When the PNP covering on the GO surface is about 85 PNPs mμ^2 (GO/PNPs) 1:20, m/m), the drug release efficiency can even reach 22% (Wang et al., 2014b).

3.12 USING GRAPHENE FOR GLIOMA TREATMENT

Gliomas are the most common primary malignant brain tumor and account for about 40% of the incidence. It is impossible to thoroughly eradicate glioma by surgery for infiltration growth. Furthermore, the blood–brain barrier (BBB) restricts the penetration of drugs systemically administrated to glioma. G. Liu designed a GO-based nanocarrier loading DOX for glioma targeting. Transferrin (Tf) is a monomeric glycoprotein that can transport one (monoferric Tf) or two (diferric Tf) iron atoms. In Liu's method, Tf was chemically conjugated to the surfaces of GO (Fig. 3.28), onto which DOX was then adsorbed.

In their procedure, six-arm polyethylene glycol–amine was used for PEGylation of graphene oxide and followed via coupling of Tf as a classic ligand targeting to PEG–GO by using 1-ethyl-3-(3-dimethyl-aminopropyl) carbodiimide (EDC) chemistry. The synthesized system was highly stable in physiological solutions. DOX was loaded onto Tf–PEG–GO via physisorption, and the loading ratio (the weight ratio of loaded drug to carriers) was estimated to be 115.4%, indicating efficient loading of DOX by Tf–PEG–GO. The related analysis showed that Tf–PEG–GO–DOX can enter the brain by a TfR-mediated endocytosis pathway and further accumulate at the tumor site. Moreover, Tf–PEG–GO–DOX's ability for transportation of more DOX to the tumor site and improvement of its antitumor effect was also demonstrated.

Chlorotoxin, a 36-amino acid peptide isolated from the venom of the giant Israeli scorpion (*Leiurus quinquestriatus*), has attracted increasing attention due to its high selectivity for glioma and other tumors of neuroectodermal origin by binding to matrix metalloproteinase-2. Wang et al. (2014b) explored the use of chlorotoxin-conjugated graphene oxide (CTX–GO) as a high specific and high drug loading efficiency system for glioma-specific drug delivery (Fig. 3.29).

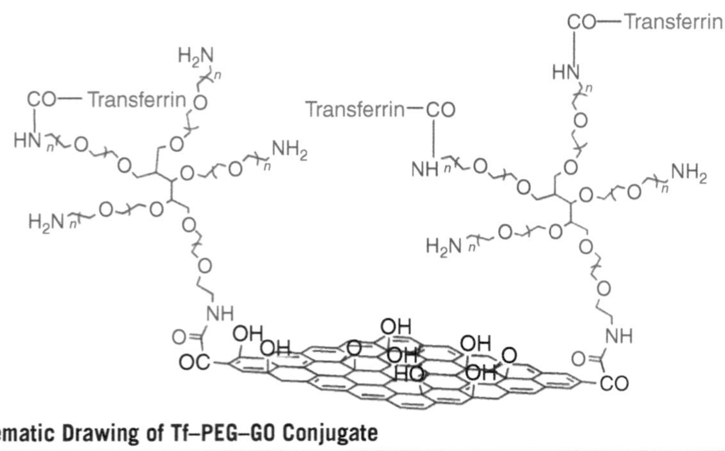

FIGURE 3.28 Schematic Drawing of Tf–PEG–GO Conjugate

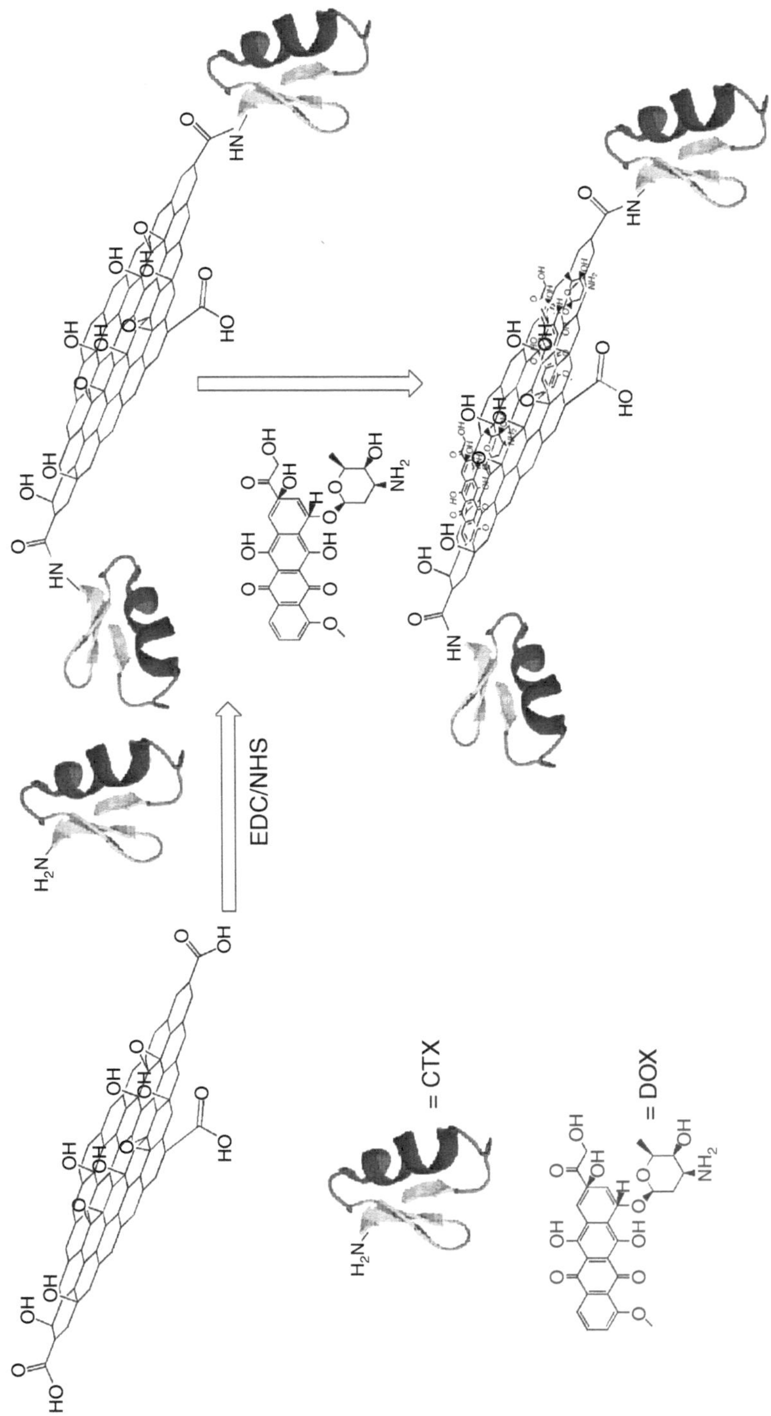

FIGURE 3.29 Schematic Illustration of Chlorotoxin-Conjugated Graphene Oxide Preparation

In their procedure, chlorotoxin was attached –COOH of graphene oxide edges by amid linkage through the carbodiimide-mediated amidation. Next, the CTX–GO sheets were loaded with DOX. In this system, besides π–π stacking interactions between GO and DOX, the amine (—NH_2) and hydroxyl (—OH) groups on DOX can also induce a strong hydrogen-bonding interaction with the —COOH and —OH groups on GO. The loading efficiency of DOX was found to be around 570 mg DOX per gram CTX–GO. This value is higher than the loading efficiency reported for DOX onto folate-conjugated GO. Here also the release of DOX from CTX–GO was clearly pH-dependent. However, the percentage of unreleased DOX on CTX–GO sheets after 60 h was still more than 70%, even under acidic conditions, confirming the presence of strong π–π stacking interactions between CTX–GO and DOX in addition to hydrogen bonding (Wang et al., 2014b).

A tumor-targeting delivery system based on hyaluronyl rGO nanosheets was reported by Miao et al. (2013). CD44 receptors, which have been observed in many types of tumor cells, can bind to hyaluronic acid (HA) (ligand molecule) and act as a target for improved delivery of anticancer therapeutics. In Miao's study, cholesteryl hyaluronic acid (CHA) was synthesized by linking the amino group of cholesteryl-2-aminoethylcarbamate (CAEC) with the carboxyl group of HA and used to coat rGO nanosheets. CHA–rGO nanosheets showed increased colloidal stability under physiological conditions and improved in vivo safety, with a survival rate of 100% after intravenous administration of 40 mg/kg compared to rGO. The DOX loading capacity of CHA–rGO was ≈90%, fourfold greater than that of rGO. Also, uptake of DOX by CD44-overexpressing KB cells was higher for CHA–rGO than for rGO. After intravenous administration in tumor-bearing mice, CHA–rGO/DOX showed higher tumor accumulation than rGO/DOX. The in vivo antitumor efficacy of DOX delivered by CHA–rGO was significantly increased compared with free DOX or rGO/DOX, and in CHA–rGO/DOX-treated mice, tumor weights were reduced to 14.1% ± 0.1% of those in untreated mice (Mia et al., 2013).

Shim et al. (2014) reported rGO nanosheets coated with an antiangiogenic anticancer taurocholate derivative of low-molecular-weight heparin (LMWH) (LHT7) as a tumor-targeting nano delivery platform for anticancer drugs. Indeed, derivatives of LMWH, a highly sulfated polysaccharide, have been considered as potential anticancer agents. Subsequently, surface coating of rGO with LHT7 would improve the dispersion stability of these nanosheets and enhance the anticancer efficacy. In this method, LHT7 physically adsorbed onto the graphene surface and modified it. DOX was loaded onto LHT–rGO nanosheets by physical adsorption (Fig. 3.30). As shown in Fig. 3.31, LHT–rGO showed greater dispersion stability in buffers and serum than did rGO nanosheets (Shim et al., 2014).

Immediately after dispersion of LHTrGO in PBS, rGO and LHT–rGO nanosheets with LHT7:rGO weight ratios less than 0.5:1 exhibited aggregation. Twenty-four hours after dispersion, LHT–rGO with a LHT7:rGO weight ratio of 1:1 formed aggregates; in contrast, LHT–rGO with LHT7:rGO weight ratios greater than 5:1 exhibited a stable dispersion at 24 h that was maintained up to 3 d (Shim et al., 2014).

LHT–rGO nanosheets showed a higher capacity for DOX loading (~90%) compared with rGO nanosheets. In vivo tumor tissue distribution analysis revealed at 1 h after dosing, the distribution at tumor sites was greater in LHT–rGO/DOX-treated mice than in rGO/DOX-treated mice, a difference that was maintained at 24 and 48 h post dose. Indeed, at 48 h post dose, the tumor retention of LHT–rGO/DOX was sevenfold higher than that of rGO/DOX. Also, it was concluded that LHT–rGO/DOX delivered to tumor tissues diffuses from the tumor vasculature into the surrounding tumor cells (Shim et al., 2014).

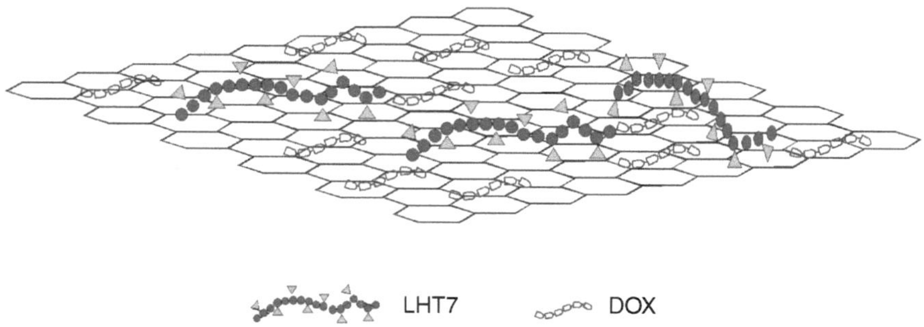

FIGURE 3.30 A Schematic Illustration of the Loading of DOX on the LHT–rGO Complex

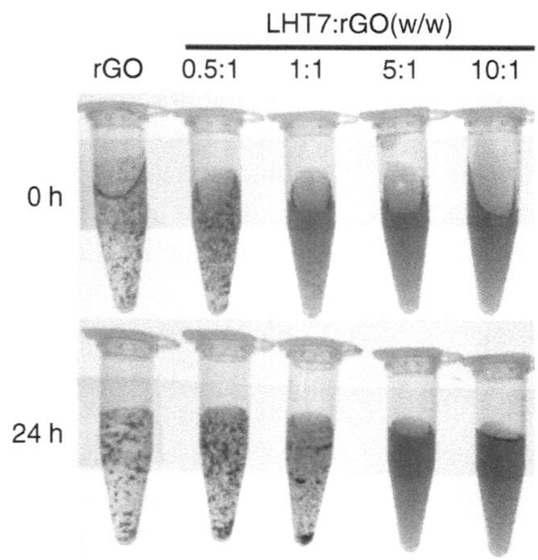

FIGURE 3.31 Dispersion Stability of LHT–rGO Nanosheets in PBS

Reprinted with permission from Shim, G., Kima, J.Y., Han, J., Chung, S.W., Lee, S., Byun, Y., Oh, Y.K., 2014. Reduced graphene oxide nanosheets coated with an anti-angiogenic anticancer low-molecular-weight heparin derivative for delivery of anticancer drugs. J. Control. Rel. 189, 80–89. Copyright 2014, Elsevier.

4 IN VITRO TOXICOLOGY AND BIOSAFETY OF GRAPHENE DERIVATIVE

It is very important to investigate the physicochemical interaction of the nanoparticles with the in vitro and in vivo organelles before applying or considering for biological application. Due to the increasing importance of GO, there is a need for more detailed and accurate in vitro and in vivo studies regarding the toxicity of the GO. Due to the variety of toxicity studies in this section, in vitro studies will be discussed. Graphene nanosheets (GNs) vary in shape, size, surface area, number of layers, lateral dimensions, surface chemistry, stiffness, defect density or quality of the individual graphene sheets, and purity; and all these properties significantly influence the interaction of graphene nanosheets with biological systems. Generally, GNs with small size, sharp edges, and rough surfaces easily internalize into the cell as compared to larger and smoother sheets. Graphene microsheets enter cells through spontaneous membrane penetration at edge asperities and corner sites (Guo and Mei, 2014).

Reports indicate that graphene nanosheets exert measurable cytotoxicity in both in vitro and in vivo studies in various types of bacteria, mammalian cells, and animal models. Graphene platelets had a strong tendency to localize close to the cells, but not enter into the cells. At a concentration of 100 μg/mL, a 24 h treatment caused about 50% cell death and loss of membrane integrity and apoptosis. Layered graphene platelets (1–10 layers) at concentrations $\geq 5\ \mu g/cm^2$ also significantly increased the release of lactase dehydrogenase (LDH) in immortalized human acute monocytic leukemia cells (THP-1), indicating loss of membrane integrity. The depletion of reduced glutathione and higher expression of a panel of cytokines (e.g., monocyte chemotactic protein-1, interleukin (IL)-1, macrophage inflammatory protein-1R, and IL–1β) were also observed after graphene platelet exposure (Schinwald et al., 2012).

Sasidharan et al. (2011) investigated the possible mechanisms of differential toxicities exerted by pristine graphene compared to carboxyl functionalized graphene (COOH–G). They observed that, although high concentrations of graphene accumulated on the cell membrane, cell death was not caused by any physical damage of the plasma membrane but exerted internal stress and forced the cells to undergo apoptosis. Actually, the accumulation of graphene on the plasma membrane caused significant intracellular oxidative stress associated with reactive oxygen species (ROS) stress. Also, it is possible that accumulation of p–G on the cell membrane might have hindered the uptake of essential nutrients and proteins and blockaded ion channels, such as that of Ca^{2+}, leading to intracellular stress (Sasidharan et al., 2011).

Most published studies have evaluated GO and rGO due to their better solubility, dispersibility, and/or stability in water and under physiological conditions compared to other graphenes.

Wang et al. (2013) reported that GO could induce dose- and time-dependent cytotoxicity in and also can enter into the cytoplasm and nucleus, decreasing cell adhesion and inducing cell floating and apoptosis. But it should be mentioned that the dose of less than 20 μg/mL of graphene oxide did not exhibit toxicity to human fibroblast cells, whereas the dose of more than 50 μg/mL exhibited obvious cytotoxicity (Wang et al., 2013).

This group also examined the effects of GO on organs of mice. It was observed that the GO accumulations were primarily in the lungs, liver, and spleen, but no GO was found in the brain tissues, which definitely suggests that GO cannot get through the blood–brain barrier. The chronic toxicity responses were observed occurring in the lungs and liver after tail vein injection. Also, histopathological analysis revealed that pulmonary exposures to GO produced a dose-dependent lung inflammatory response, which can be ascribed by neutrophils and foamy alveolar macrophage accumulation. With the increase in GO dose, the toxicity reaction of the lungs of mice becomes more and more severe (Wang et al., 2013).

Chang et al. (2011) investigated the toxicity of GO at cell level. They observed that the viability loss is dose-related. Size is another factor on viability. The influence of large GO (l-GO) and medium GO (m-GO) on the viability of A549 cells is tiny. Even at the highest concentration of 200 μg/mL, more than 80% of the cell viability remains. However, small GO (s-GO) induces more viability loss than l-GO and m–GO. At 200 μg/mL of GO, the cell viability is 67% at 24 h post exposure.

Yuan et al. (2012) explored cytotoxicity of oxidized single-walled carbon nanotubes (SWCNT) and graphene oxide on human hepatoma hepG2 cells. They found that GO was less cytotoxic in comparison to oxidized SWCNTs and suggested that GO was more biocompatible and could be a promising candidate for bio-related applications (Yuan et al., 2012).

Since GO could enter into human bodies through different routes, such as injection, dermal penetration, ingestion, and especially inhalation, normal human lung fibroblast (HLF) cells were considered as an optimal model in vitro in Wang's study to assess the toxicity of GO and some GO derivatives. Their analysis showed that no cell viability reduction was observed when the HLF cells were treated with GO at the concentration of 1 and 10 mg/mL. However, the cytotoxicity of GO became more obvious (<0.01) than the untreated controls at the concentration beyond 50 mg/mL along with increasing the treating time. Furthermore, the results indicated that GO could damage the membrane at a negligible extent (Wang et al., 2013).

They also investigated on the GO effect on DNA damage in HLF cells, including single-strand breaks, double-strand breaks, alkaline labile sites, or a combination of these in a cell. They observed that there were statistically significant increases in tail length and the percentage of DNA in tail after 24-h treatment with the tested concentrations of GO (1, 50 and 100 mg/mL[1]). As well, the results reflected that oxidative stress occurred in GO-treated HLF cells and also Antioxidant (*N*-acetylcysteine) ($P < 0.05$ at a concentration of 50 mg/mL GO and $P < 0.01$ at concentration of 100 mg/mL GO) could reverse the cytotoxic effect of GO on HLF cells (Wang et al., 2013).

Studies revealed that functionalization of GO can alter cytotoxicity of this carbon-based material. Modification of GO by PEI, PEG, and LA–PEG was considered by Wang et al. (2013). It was observed the cytotoxic effect on HLF cells induced by PEG–GO was decreased, whereas PEI-modified GO could induce a more severe cytotoxic effect on HLF cells compared with GO, and the effect of LA–PEG-modified GO on the HLF cells viability was in between. It should be noted that the effect of GO derivatives on cell viability was concentration dependent (Wang et al., 2013).

The results showed that compared with the two other GO derivatives, DNA damage induced by LA–PEG-modified GO was mild in a significant way, whereas PEI-modified GO induced DNA damage of HLF cells in a comparable way with unmodified GO and the genotoxic effect caused by PEG-modified GO was in between that induced by PEI- and LA–PEG-modified GO. A novel finding from this study was that the surface charge of GO and GO derivatives changed their aggregation status, as well as their ability to be internalized. The researchers concluded that the ideal GO derivatives should have a lower positive electronic charge to reduce their toxic effect on cells (Wang et al., 2013).

DCs play a key role in the immune system since they orchestrate initiation, amplification, and suppression of immune response. In particular, maturation of DCs is crucial for the initiation of immunity. In a study by Zhi et al. (2013), analysis revealed that GO can induce markedly the dose-dependent DCs' maturation. TNF-a, IL-1b, and IL-6 are three kinds of primary proinflammatory cytokines secreted by DCs whose levels were increased with the increase of the dose of GO. Apoptosis is a strictly programmed cell death process responsible for the ordered removal of superfluous, aged, or damaged cells induced by GO. Also, GO sheets were more susceptibly ingested by macrophages and formed black, dense aggregates within the cells after 48 h exposure at the concentration 25 mg/mL, while GO coated with polyvinylpyrrolidone exhibited improved immunological biocompatibility in vitro (Zhi et al., 2013).

Studies by Liao et al. (2011) showed that the blood compatibility and cytotoxicity of graphene-based materials depend on sizes and oxygen content. Sonicated (smaller) GO exhibited higher hemolytic activity than untreated (larger) GO. Compared to individually dispersed GO sheets having higher surface oxygen content, the aggregated graphene sheets showed lower hemolytic activity (Fig. 3.32). Covering the GO sheets with chitosan eliminated their hemolytic activity. Moreover, compared to reversibly aggregated GO, the aggregated graphene sheets generated more ROSin human skin fibroblast cells and strongly associated to the cell surface (Liao et al., 2011).

Investigation by Singh et al. (2011) showed that the addition of GO to a suspension of freshly isolated human platelets ($0.5–0.8 \times 10^9$ cells/mL) induced cell aggregation upon stirring at 37°C in a concentration-

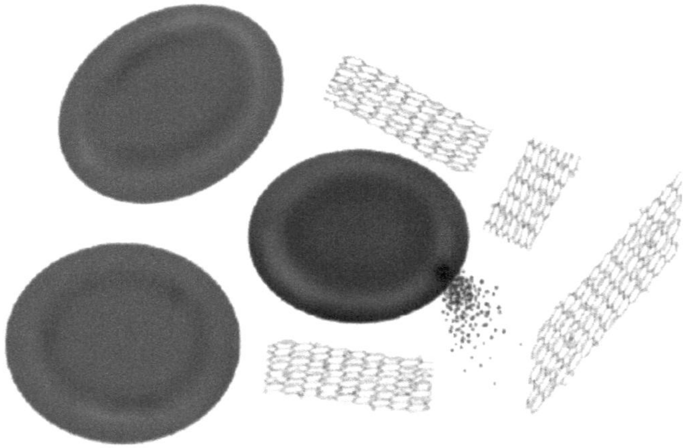

FIGURE 3.32 The Hemolytic Activity of Graphene Oxide PBS

Reprinted with permission from Liao, K.H., Lin, Y.S., Macosko, C.W., Haynes, C.L., 2011. Cytotoxicity of graphene oxide and graphene in human erythrocytes and skin fibroblasts. ACS Appl. Mater. Interfaces 3, 2607–2615. Copyright 2011, Elsevier.

dependent manner. At 2 μg/mL, aggregation induced by GO was even stronger than that elicited by thrombin (1 U/mL), one of the most potent platelet agonists. But rGO (2 μg/mL) elicited a minor aggregatory response, which was about 10% of that induced by GO at the same concentration, which may be correlated with the reduced charge density on the graphene surface. They also observed that GO at high concentration (20 μg/mL) elevated platelet cytosolic ROS to a level comparable to that achieved with H_2O_2 (10 μM), which was completely scavenged by the reductant *N*-acetylcysteine (NAC) (Singh et al., 2011).

This group also investigated the effect of amine functionalized graphene on platelet reactivity. Their results revealed that modification by amine (G–NH_2) did not have totally the stimulatory effect on human platelets, nor did it induce pulmonary thromboembolism in mice following intravenous administration. With exposure of platelets to either 2 or 5 μg/mL G–NH_2 no significant change in ROS level was observed while at higher concentration (10 μg/mL), a twofold rise in cytosolic ROS was recorded. Further, G–NH_2 did not exhibit any hemolytic activity even at high concentration (50 μg/mL); it did not evoke lysis of erythrocytes, another major cellular component in blood. The researchers believed that negatively charged graphene sheets physically interacted with platelets and activated them, whereas positively charged sheets did not (Fig. 3.33) (Singh et al., 2012).

In the case of rGO, Akhavan et al. (2012) also observed the size- and concentration-dependent cytotoxicity and genotoxicity. Analysis showed significant cytotoxic effects for rGO with an average lateral dimension of 11 nm (the smallest rGO in this study), even at the lowest concentration of 1 μg/mL and after a 1 h exposure. rGOs with an average lateral dimension of 3.8 μm, the largest rGO used in this study, exhibited lower cytotoxicity as compared to rGOs with average lateral dimensions of 91 and 418 nm.

Assays for RNA efflux from cells, an indirect indicator of membrane damage, determined there was a consistent size- and concentration-dependent response in rGO-treated hMSCs. The smaller rGO sheets induced higher RNA effluxes than did the larger rGO sheets. Moreover, rGOs generated 13–26-fold higher levels of ROS when compared to the control. This suggests that oxidative stress is one of the mechanisms involved in rGO cytotoxicity. In the genotoxicity study, 1 h exposure to rGOs with average lateral dimensions of 11 nm and 91 nm initiated significant increases in DNA damage and

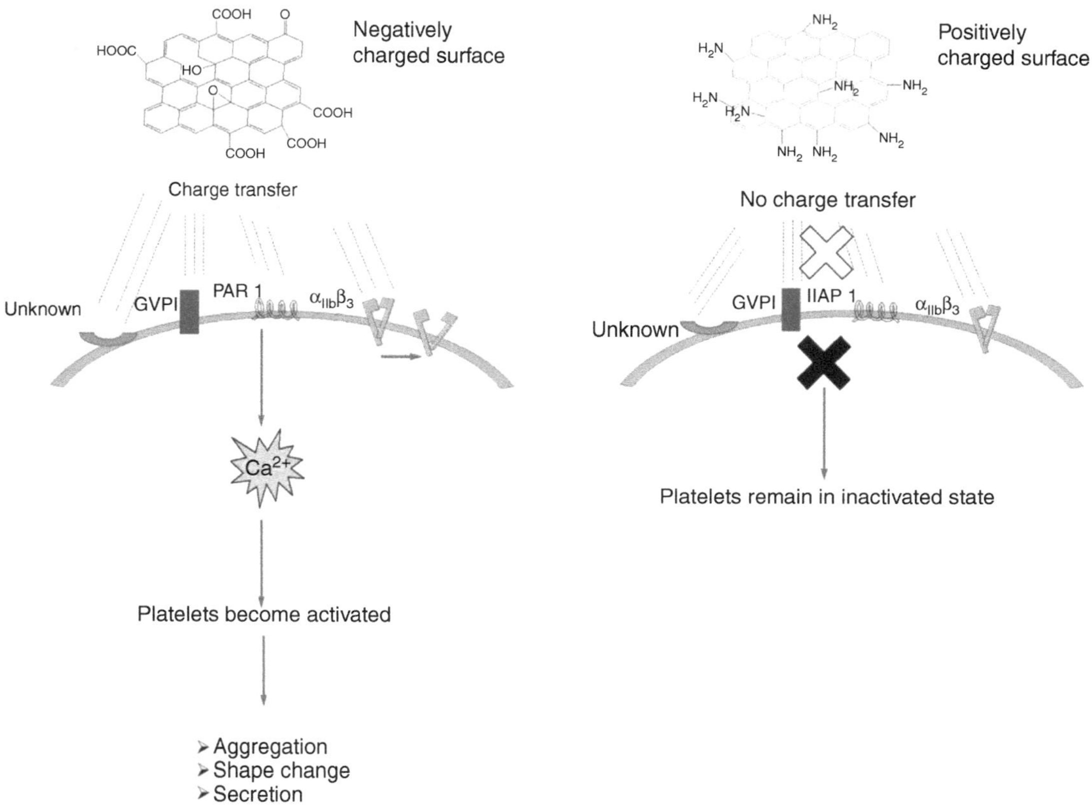

FIGURE 3.33 Proposed Model for Differential Effect of GO and G–NH_2 on Platelet Functions

Surface charge distribution on graphene determines its interaction with platelets.

chromosomal aberration frequency at concentrations as low as 0.1 μg/mL and 1.0 μg/mL, respectively. These concentrations were 10 times lower than the threshold concentration observed in the cell viability test. The two larger rGO sheets induced only slight DNA fragmentation at the highest concentration of 100 μg/mL and after a longer exposure time of 24 h (Akhavan et al., 2012).

Tan et al. (2013) investigated the effects of PEGylation of GO nanosheets on serum behaviors. nGO–PEG showed dramatic reductions in both protein binding in general and the ability to induce complement C3 cleavage compared with bare GO. Also, data revealed certain selectivity and increased binding capacities of nGO–PEG toward a few serum proteins, six of which were identified, including C3a/C3a(des–Arg), a proinflammatory/obesity-related adipogenic cytokine produced from C3 cleavage during complement activation (Tan et al., 2013). In another study, which was done by Zhang et al. (2011) cellular experiments performed on human cervical cancer HeLa cells showed that covalently conjugated GO with dextran remarkably reduced cellular toxicity . Functionalization, however, does not always eliminate the toxicity of GFNs. The reported IC_{50} of noncovalent PEGylated nanosized rGO was about 80–85 μg/mL in both the human breast cancer cell line MCF-7 and human glioblastoma U87MG cells (Robinson et al., 2011).

5 CONCLUSIONS

The applications of graphene nanosheets derivatives as drug nanocarriers have developed rapidly in the past few years. The unique physical properties of graphene/GO, such as π-stacking with aromatic drugs, unique NIR absorption, biodistribution, and pharmacokinetic properties that can be tuned by controlling the size, the surface chemistry, and the targeting ligand have made them good candidates for drug delivery applications. Thus, a comprehensive understanding of the surface chemistry and the interaction of GFNs with living systems and their adverse effects in vitro and in vivo are essential for further development and safe use of graphene-based nanomaterials. Based on the previous toxicity, investigations of graphene, graphene oxide, and their derivatives all demonstrate in vitro toxicity. However, it was concluded that cells are very sensitive to size, shape, solubility, and concentration of graphene nanomaterials. Since GO is more soluble and dispersible than graphene in the physiological environment, it can be considered more biocompatible, which results in less damage and toxicity in human cell types, such as skin fibroblasts and red blood cells. Despite of the large improvement in the designing of drug carriers on the basis of graphene, systematic studies on how this nanomaterial affects the immune systems, reproductive systems, and nerve systems are needed. Many more preclinical toxicity studies should be done before graphene-based cancer therapies can be finally translated into the clinic.

REFERENCES

Akhavan, O., Ghaderi, E., Akhavan, A., 2012. Size-dependent genotoxicity of graphene nanoplatelets in human stem cells. Biomaterials 33, 8017–8025.

An, J., Gou, Y., Yang, C., Hu, F., Wang, C., 2013. Synthesis of a biocompatible gelatin functionalized graphene nanosheets and its application for drug delivery. Mater. Sci. Eng. C 33, 2827–2837.

Bao, H., Pan, Y., Ping, Y., Sahoo, N.G., Wu, T., Li, L., Li, J., Gan, L.H., 2011. Chitosan-functionalized graphene oxide as a nanocarrier for drug and gene delivery. Small 7, 1569–1578.

Bourlinos, A.B., Gournis, D., Petridis, D., Szabo, T., Szeri, A., Dekany, I., 2003. Graphite oxide: chemical reduction to graphite and surface modification with primary aliphatic amines and amino acids. Langmuir 19, 6050–6055.

Chang, Y., Yang, S.T., Liu, J.H., Dong, E., Wang, Y., Cao, A., Liu, Y., Wang, H., 2011. In vitro toxicity evaluation of graphene oxide on A549 cells. Toxicol. Lett. 200, 201–210.

Compton, O.C., Nguyen, S.T., 2010. Graphene oxide, highly reduced graphene oxide, and graphene: versatile building blocks for carbon-based materials. Small 6 (6), 711–723.

Delhaes, P., 2011. Carbon-Based Solids and Materials. John Wiley & Sons, Hoboken, NJ.

Depan, D., Shah, J., Misra, R.D.K., 2011. Controlled release of drug from folate-decorated and graphene mediated drug delivery system: synthesis, loading efficiency, and drug release response. Mater. Sci. Eng. C 31, 1305–1312.

Evans, J.W., Thiel, P.A., Bartelt, M.C., 2006. Morphological evolution during epitaxial thin film growth: formation of 2D islands and 3D mounds. Sur. Sci. Rep. 61, 1–128.

Fang, M., Long, J., Zhao, W., Wang, L., Chen, G., 2010. pH-responsive chitosan-mediated graphene dispersions. Langmuir 26, 16771–16774.

Feng, L., Zhang, S., Liu, Z., 2011. Graphene based gene transfection. Nanoscale 3, 1252–1257.

Fradkin, E., 1986. Critical behavior of disordered degenerate semiconductors. Phys. Rev. B 33, 3263–3268.

Gao, X., Jang, J., Nagase, S., 2009. Hydrazine and thermal reduction of graphene oxide: reaction mechanisms, product structures, and reaction design. J. Phys. Chem. C 114 (2), 832–842.

Geim, A.K., Novoselov, K.S., 2007. The rise of graphene. Nat. Mater. 6, 183–191.

Goenka, S., Sant, V., Sant, S., 2014. Graphene-based nanomaterials for drug delivery and tissue engineering. J. Control Rel. 173, 75–88, 10.

Gogotsi, Y., 2011. Controlling graphene properties through chemistry. J. Phys. Chem. Lett. 2, 2509–2510.

Gollavelli, G., Ling, Y.C., 2012. Multi-functional graphene as an in vitro and in vivo imaging probe. Biomaterials 33, 2532–2545.

Guo, X., Mei, N., 2014. Assessment of the toxic potential of graphene family nanomaterials. J. Food Drug Anal. 22, 105–115.

He, H.Y., Klinowski, J., Forster, M., Lerf, A., 1998. A new structural model for graphite oxide. Chem. Phys. Lett. 287, 53–56.

Hone, J., Batlogg, B., Benes, Z., Johnson, A.T., Fischer, J.E., 2000. Quantized phonon spectrum of single-wall carbon nanotubes. Science 289, 1730–1733.

Huang, P., Xu, C., Lin, J., Wang, C., Wang, C., Zhang, C., Zhou, X., Guo, S., Cui, D., 2011. Folic acid-conjugated graphene oxide loaded with photosensitizers for targeting photodynamic therapy. Theranostics 1, 240–250.

Kam, N.W.S., Dai, H.J., 2005. Carbon nanotubes as intracellular protein transporters: generality and biological functionality. J. Am. Chem. Soc. 127, 6021–6026.

Kavitha, T., Kang, I., Park, S., 2014. Poly(*N*-vinyl caprolactam) grown on nanographene oxide as an effective nanocargo for drug delivery. Colloids Surf. B 115, 37–45.

Kim, H., Lee, D., Kim, J., Kim, T., Kim, W.J., 2013. Photothermally triggered cytosolic drug delivery via endosome disruption using a functionalized reduced graphene oxide. Nano. Am. Chem. Soc. 7, 6735–6746.

Lerf, A., He, H., Forster, M., Klinowski, J., 1998. Structure of graphite oxide revisited. J. Phys. Chem. B 102, 4477–4482.

Liao, K.H., Lin, Y.S., Macosko, C.W., Haynes, C.L., 2011. Cytotoxicity of graphene oxide and graphene in human erythrocytes and skin fibroblasts. ACS Appl. Mater. Interf. 3, 2607–2615.

Liu, Z., Robinson, J.T., Sun, X., Dai, H., 2008. PEGylated nanographene oxide for delivery of water-insoluble cancer drugs. J. Am. Chem. Soc. 130, 10876–10877.

Liu, J., Yang, W., Tao, L., Li, D., Boyer, C., Davis, T.P., 2010. Thermosensitive graphene nanocomposites formed using pyrene-terminal polymers made by RAFT polymerization. J. Polym. Sci. A 48, 425–433.

Liu, K., Zhang, J.J., Cheng, F.F., Zheng, T.T., Wang, C., Zhu, J.J., 2011. Green and facile synthesis of highly biocompatible graphene nanosheets and its application for cellular imaging and drug delivery. J. Mater. Chem. 21, 12034–12040.

Lu, C.H., Yang, H.H., Zhu, C.L., Chen, X., Chen, G.N., 2009. A graphene platform for sensing biomolecules. Angew. Chem. Int. Ed. 48, 4785–4787.

Ma, X., Tao, H., Yang, K., Feng, L., Cheng, L., Shi, X., Li, Y., Guo, L., Liu, Z., 2012. A functionalized graphene oxide–iron oxide nanocomposite for magnetically targeted drug delivery, photothermal therapy, and magnetic resonance imaging. Nano Res. 5 (3), 199–212.

Markovic, Z.M., Harhaji-Trajkovic, L.M., Todorovic-Markovic, B.M., Kepi, D.P., Arsikin, K.M., Jovanovic, S.P., Pantovic, A.C., Dramicanin, M.D., Trajkovic, V.S., 2011. In vitro comparison of the photothermal anticancer activity of graphene nanoparticles and carbon nanotubes. Biomaterials 32, 1121–1129.

Meng, F.H., Hennink, W.E., Zhong, Z.Y., 2009. Reduction-sensitive polymers and bioconjugates for biomedical applications. Biomaterials 30, 2180–2198.

Miao, W., Shim, G., Kang, C.M., Lee, S., Choe, Y.S., Choi, H.G., Oh, Y.K., 2013. Cholesteryl hyaluronic acid-coated, reduced graphene oxide nanosheets for anti-cancer drug delivery. Biomaterials 34 (37), 9638–9647.

Mounet, N., Marzari, N., 2005. First-principles determination of the structural, vibrational and thermodynamic properties of diamond, graphite, and derivatives. Phys. Rev. B 71, 205–214.

Mouras, S., Hamm, A., Djdoura, D., Cousseins, J.C., 1987. Synthesis of first stage graphite intercalation compounds with fluorides. Rev. Chim. Miner. 24, 572–582.

Pan, Y., Bao, H., Sahoo, N.G., Wu, T., Li, L., 2011. Water-soluble poly(*N*-isopropylacrylamide)–graphene sheets synthesized via click chemistry for drug delivery. Adv. Funct. Mater. 21, 2754–2763.

Pei, S., Cheng, H.M., 2012. The reduction of graphene oxide. Carbon 50 (9), 3210–3228.

Pelaz, B., Pino, P.D., Maffre, P., Hartmann, R., Gallego, M., Rivera-Fernandez, S., De La Fuente, J.M., Nienhaus, G.U., Parak, W.J., 2015. Surface functionalization of nanoparticles with polyethylene glycol (PEG): effects on protein adsorption and cellular uptake. ACS Nano 9 (7), 6996–7008.

Robinson, J.T., Tabakman, S.M., Liang, Y., Casalongue, H.S., Vinh, D., Dai, H., 2011. Ultrasmall reduced graphene oxide with high near-infrared absorbance for photothermal therapy. J. Am. Chem. Soc. 133, 6825–6831.

Romberg, B., Hennink, W.E., Storm, G., 2008. Sheddable coatings for long-circulating nanoparticles. Pharm. Res. 25 (1), 55–71.

Sasidharan, A., Panchakarla, L.S., Chandran, P., Menon, D., Nasir, S., Rao, C.N., Koyakotty, M., 2011. Differential nano-bio interactions and toxicity effects of pristine versus functionalized graphene. Nanoscale 3, 2461–2464.

Schinwald, A., Murphy, F.A., Jones, A., MacNee, W., Donalson, K., 2012. Graphene-based nanoplatelets: a new risk to the respiratory system as a consequence of their unusual aerodynamic properties. ACS Nano 6, 736–746.

Shim, G., Kima, J.Y., Han, J., Chung, S.W., Lee, S., Byun, Y., Oh, Y.K., 2014. Reduced graphene oxide nanosheets coated with an anti-angiogenic anticancer low-molecular-weight heparin derivative for delivery of anticancer drugs. J. Control. Rel. 189, 80–89.

Singh, S.K., Singh, M.K., Nayak, M.K., Kumari, S., Shrivastava, S., Grácio, J.J.A., Dash, D., 2011. Thrombus inducing property of atomically thin graphene oxide sheets. ACS Nano 5, 4987–4996.

Singh, S.K., Singh, M.K., Nayak, M.K., Kumari, S., Shrivastava, S., Grácio, J.J.A., Dash, D., 2012. Amine-modified graphene: thrombo-protective safer alternative to graphene oxide for biomedical applications. ACS Nano 6, 2731–2740.

Stankovich, S., Dikin, D.A., Piner, R.D., Kohlhaas, K.A., Kleinhammes, A., Jia, Y., Wu, Y., Nguyen, S.T., Ruoff, R.S., 2007. Synthesis of graphene-based nanosheets via chemical reduction of exfoliated graphite oxide. Carbon 45 (7), 1558–1565.

Sun, X.M., Liu, Z., Welsher, K., Robinson, J.T., Goodwin, A., Zaric, S., Dai, H., 2008. Nano-graphene oxide for cellular imaging and drug delivery. Nano Res. 1 (3), 203–212.

Tan, X., Feng, L., Zhang, J., Yang, K., Zhang, S., Liu, Z., Peng, R., 2013. Functionalization of graphene oxide generates a unique interface for selective serum protein interactions. ACS Appl. Mater. Interfaces 5 (4), 1370–1377.

Tian, B., Wang, C., Zhang, S., Feng, L., Liu, Z., 2011. Photothermally enhanced photodynamic therapy delivered by nano-graphene oxide. ACS Nano 5 (9), 7000–7009.

Toumia, Y., Orlanducci, S., Basoli, F., Licoccia, S., Paradossi, G., 2015. Soft confinement of graphene in hydrogel matrixes. J. Phys. Chem. B 119 (5), 2051–2061.

Venables, J.A., Spiller, G.D.T., Hanbucken, M., 1984. Nucleation and growth of thin films. Rep. Prog. Phys. 47, 399–459.

Wang, A., Pu, K., Dong, B., Liu, Y., Zhang, L., Zhang, Z., Duand, W., Zhua, Y., 2013. Role of surface charge and oxidative stress in cytotoxicity and genotoxicity of graphene oxide towards human lung fibroblast cells. J. Appl. Toxicol. 33, 1156–1164.

Wang, H., Gu, W., Xiao, N., Ye, L., Xu, Q., 2014a. Chlorotoxin-conjugated graphene oxide for targeted delivery of an anticancer drug. Int. J. Nanomed. 18 (9), 1433–1442.

Wang, H., Sun, D., Zhao, N., Yang, X., Shi, Y., Li, J., Su, Z., Wei, G., 2014b. Thermo-sensitive graphene oxide–polymer nanoparticle hybrids: synthesis, characterization, biocompatibility and drug delivery. J. Mater. Chem. B 2, 1362–1370.

Wei, G., Dong, R., Wang, D., Feng, L., Dong, S., Songa, A., Hao, J., 2014. Functional materials from the covalent modification of reduced graphene oxide and β-cyclodextrin as a drug delivery carrier. New J. Chem. 38, 140–145.

Wen, H., Dong, C., Dong, H., Shen, A., Xia, W., Cai, X., Song, Y., Li, X., Li, Y., Shi, D., 2012. Engineered redox-responsive PEG detachment mechanism in PEGylated nano-graphene oxide for intracellular drug delivery. Small 8, 760–769.

Xu, S.C., Irle, S., Musaev, D.G., Lin, M.C., 2006. Quantum chemical study of the dissociative adsorption of OH and H_2O on pristine and defective graphite (0001) surfaces: reaction mechanisms and kinetics. J. Phys. Chem. C 111 (3), 1355–1365.

Yang, Y., Zhang, Y.M., Chen, Y., Zhao, D., Chen, J.T., Liu, Y., 2012. Construction of a graphene oxide based noncovalent multiple nanosupramolecular assembly as a scaffold for drug delivery. Chem. Eur. J. 18, 4208–4215.

Yang, M., Yao, J., Duan, Y., 2013. Graphene and its derivatives for cell biotechnology. Analyst 138, 72–86.

Yuan, J., Gao, H., Sui, J., Duan, H., Chen, W.N., Ching, C.B., 2012. Cytotoxicity evaluation of oxidized single-walled carbon nanotubes and graphene oxide on human hepatoma HepG2 cells: an iTRAQ-coupled 2D LC-MS/MS proteome analysis. Toxicol. Sci. 126, 149–161.

Zhang, L., Xia, J., Zhao, Q., Liu, L., Zhang, Z., 2010. Functional graphene oxide as a nanocarrier for controlled loading and targeted delivery of mixed anticancer drugs. Small 6, 537–544.

Zhang, S., Yang, K., Feng, L., Liu, Z., 2011. In vitro and in vivo behaviors of dextran functionalized graphene. Carbon 9, 4040–4049.

Zhi, X., Fang, H., Bao, C., Shen, G., Zhang, J., Wang, K., Guo, S., Wan, T., Cui, D., 2013. The immunotoxicity of graphene oxides and the effect of PVP-coating. Biomaterials 34, 5254–5261.

Zhou, T., Zhou, X., Xing, D., 2014. Controlled release of doxorubicin from graphene oxide based charge-reversal nanocarrier. Biomaterials 35, 4185–4194.

CHAPTER

NANOSTRUCTURED NANOPARTICLES FOR IMPROVED DRUG DELIVERY

4

Jean M. Rabanel, Pierre L. Latreille, Augustine Lalloz, Patrice Hildgen, Xavier Banquy
University of Montreal, Montreal, QC, Canada

CHAPTER OUTLINE

ABBREVIATIONS

bFGF	Bovine fibroblast growth factor
BSA	Bovine serum albumin
CaALg	Calcium alginate
CS	Chitosan sulfate
Cy7	Cyanine 7
DNA	Desoxyribonucleic acid
DOX	Doxorubicin
DSPC	1,2-dioctadecanoyl-sn-glycero-3-phosphocholine
DSPE	1,2-distearoyl-*sn*-glycero-3-phosphoethanolamine
GMO	Glycerol mono-oleate
GOD	Glucose oxidase enzyme

Nanostructures for Drug Delivery

HA	Hyaluronic acid
HA20C10	Decylamino hydrazide derivative of hyaluronic acid
LbL	Layer-by-layer assembly
LCST	Lower critical solubility temperature
MEND	Multifunctional envelop nanodevice
MP(s)	Microparticle(s)
MRI	Magnetic resonance imaging
MRTN	Mesoporous ratttle type nanoparticle
MSN	Mesoporous silica nanoparticle
NC(s)	Nanocapsule(s)
NFH	1-naphtyl perfluoroheptanyl ketone
NIR	Near infrared
NIRF	Near-infrared fluorescence
NP(s)	Nanoparticle(s)
NSAID	Nonsteroidal antiinflammatory drug
PAA	Poly(acrylic acid)
PAA-*co*-PB	Poly(acrylic acid)-*co*-poly(butadiene)
PAA-*co*-PS	Poly(acrylic acid)-*co*-poly(styrene)
***p*ATP-capped AuNC**	*P*-aminothiophenol-capped gold nanocages
PDLLA	Poly(D,L-lactic acid)
PEE	Poly(ethylethylene)
PEG:	Poly(ethylene glycol)
PEI	Poly(ethyleneimine)
PFPO	Polyperfluopropylene oxide
PGOHMA	Poly(glycerol methacrylate)s
PGSD	Poly(styrene-*co*-glycidyl methacrylate-*co*-divinylbenzene)
PIBCA	Poly(isobutyl cyanoacrylate)
PLA	Poly(lactic acid)
PLGA	Poly(lactic-*co*-glycolic acid)
PLLA	Poly(L-lactic acid)
POD	Peroxidase enzyme
PS-*b*-P4VP	Polystyrene-*b*-poly(4-vinylpyridine)
PTX	Paclitaxel
Py	Pyrene
RNA	Ribonucleic acid
SEM	Scanning electron microscopy
siRNA	Small interfering ribonucleic acid
SC	Sodium cholate
SPC	Soybean phosphatidylcholine
TEM	Transmission electron microscopy

1 INTRODUCTION

Nowadays, nanomaterials intended for nanomedicine applications tend to be increasingly sophisticated and complex in order to respond to precise specifications and applications. A lot of efforts have been recently dedicated to the development of combined therapies, involving encapsulation of two or more active molecules in the same nanocarrier or the development of dual imaging and drug delivery systems (theranostic). Additionally, increased knowledge regarding interactions between foreign materials and host have led to a demand for controlled surface properties and inclusion of stimuli-sensitive materials in order to optimize delivery and/or imaging.

As a consequence, nanomedicine tend to integrate different materials responsible for these different functionalities. One striking feature of the most recent nanomedicine concepts, in contrast with the first generation of drug nanocarriers (plain liposomes, micelles and nanospheres), is the increasing complexity of particle organization resulting from nanostructurations and arrangements of different components.

This increased complexity is also the result of the impetus to obtain more biomimetic and biocompatible carriers, based on bioinspired materials: "virus-like" particles (Lee et al., 2008; Ren et al., 2011), mimetics of lipoprotein particles (De Miguel et al., 1995; Fischer et al., 2014), and surfaces mimicking cell membrane properties (Hu et al., 2011a).

The development of several technologies allowing an increased precision in the manufacture of nanosized objects also contributes to this situation, such as more controlled self-assembly of one (Zhang and Eisenberg, 1996), two or more amphiphilic molecules (Liu et al., 2015), molding processes at the micro and nanoscale (Rolland et al., 2005), combination of multiple nano-objects into hierarchical assemblies (Yuan et al., 2014), and assembly by host–guest interactions (Mishra et al., 2004; Rodell et al., 2015). This is also the result of the development of analytic instruments able to probe the matter with an unprecedented precision.

This chapter is divided into three sections covering the most commonly encountered particle structures in nanomedicines intended for drug delivery and/or imaging, such as (1) particles exhibiting a core–shell structure, (2) capsules and vesicles, and (3) particles with nanocompartments (Fig. 4.1). Our objective is not to present an exhaustive bibliographic review, but rather to highlight a selection of works that are presenting the most significant contributions and future directions in the field of nanostructured nanoparticles (NPs) aimed for drug delivery and imaging.

2 THE CORE–SHELL NANOSTRUCTURE

2.1 CORE–SHELL: A COMMON NANOSTRUCTURE IN DRUG DELIVERY NANOCARRIERS

Core–shell particles are composed of two domains, respectively identified as the inner part of the particle, the core, and an outer part, the shell (Fig. 4.1). The term "core–shell nanoparticles" was initially used in early studies on surfactant micelles and block copolymer micelle structures (Allen et al., 1999; Riley et al., 2003). With time, it has been used to describe the structure of many other particle types as well, such as solid polymer particles, dendrimers, metallic particles, and so forth.

The core is designed with the view to optimize the encapsulation of drug or/and imaging molecules. The shell can have different functions: (1) to protect the core content and regulate its release (He et al., 2015a); (2) to provide suitable surface properties for NP stability in biological media (Rabanel et al., 2014); (3) to provide optimal interactions with proteins, cells and tissues, either by nonspecific adhesion (Bravo-Osuna et al., 2007) or ligand-receptor recognition (Peer et al., 2007); (4) to encapsulate a second drug with different chemical properties than the one encapsulated in the core, particularly in applications involving dual or synergistic therapies using a single drug nanocarrier (Parhi et al., 2012).

The frontier between the shell and the core can be continuous, that is, a gradient of materials from the center to the surface, discontinuous that is, a layer of material grafted or physically adsorbed on the NP, or a layer resulting from a phase separation of different polymer or copolymers (Fig. 4.2). The thickness of the shell can vary with the nature of the material or its function: from a few nanometers (monolayer of a surfactant molecule), with the sole function to render the surface more biocompatible, to a thick hydrogel wall, which could have multiple functions, such as surface modification, drug entrapment, or control release.

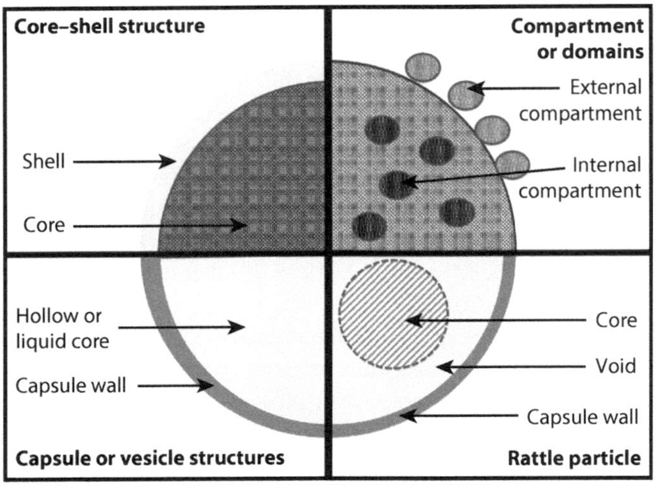

FIGURE 4.1

Summary of Reviewed Nanostructures Found in Drug Nanocarriers

For the purpose of this chapter, core–shell particles definition will be limited to particles exhibiting a solid core, that is, polymers, lipids, or inorganic cores associated with a continuous layer of different materials interacting simultaneously with the core's boundary and the external media. Particles with a liquid core, a drug core, or a void (hollow particle or gas core) will be discussed and examined in the Section 3.

2.2 INVENTORY OF STRUCTURES AND SELECTED EXAMPLES

In Table 4.1 a list of drug nanocarriers with a core–shell structure is presented. The list, although not exhaustive, presents the main classes of core–shell particles recently described in the literature for which data regarding encapsulation (and/or release) of drugs or imaging molecules have been published.

Part A, Adapted from Ran, Z., Sun, Y., Chang, B., Ren, Q., Yang, W. 2013. Silica composite nanoparticles containing fluorescent solid core and mesoporous shell with different thickness as drug carrier. J. Colloid Interface Sci., 410, 94–101; Part B, Adapted from Thevenot, J., Troutier, A.-L., David, L., Delair, T., Ladavière, C., 2007. Steric stabilization of lipid/polymer particle assemblies by poly(ethylene glycol)-lipids. Biomacromolecules, 8, 3651–3660; Part C, Adapted from Hu, C.-M. J., Zhang, L., Aryal, S., Cheung, C., Fang, R. H., Zhang, L., 2011a. Erythrocyte membrane-camouflaged polymeric nanoparticles as a biomimetic delivery platform. Proc. Natl. Acad. Sci., 108, 10980–10985; Part D, Adapted from Yue, Z., You, Z., Yang, Q., Lv, P., Yue, H., Wang, B., Ni, D., Su, Z., Wei, W., Ma, G., 2013. Molecular structure matters: PEG-b-PLA nanoparticles with hydrophilicity and deformability demonstrate their advantages for high-performance delivery of anti-cancer drugs. J. Mater. Chem. B, 1, 3239–3247; Part E, Adapted from Sahiner, N., Alb, A.M., Graves, R., Mandal, T., McPherson, G.L., Reed, W.F., John, V.T., 2007. Core-shell nanohydrogel structures as tunable delivery systems. Polymer, 48, 704–711; Part F, Adapted from Wu, J., Kong, T., Yeung, K. W.K., Shum, H.C., Cheung, K.M.C., Wang, L., To, M.K.T., 2013. Fabrication and characterization of monodisperse PLGA–alginate core–shell microspheres with monodisperse size and homogeneous shells for controlled drug release. Acta Biomaterialia, 9, 7410–7419; Part G, Adapted from Go, D.P., Gras, S.L., Mitra, D., Nguyen, T.H., Stevens, G.W., Cooper-White, J.J., O'Connor, A.J., 2011. Multilayered microspheres for the controlled release of growth factors in tissue engineering. Biomacromolecules, 12, 1494–1503; Part H, Adapted from Lim, M.P.A., Lee, W.L., Widjaja, E., Loo, S.C.J., 2013. One-step fabrication of core-shell structured alginate-PLGA/PLLA microparticles as a novel drug delivery system for water soluble drugs. Biomaterials Science, 1, 486–493.

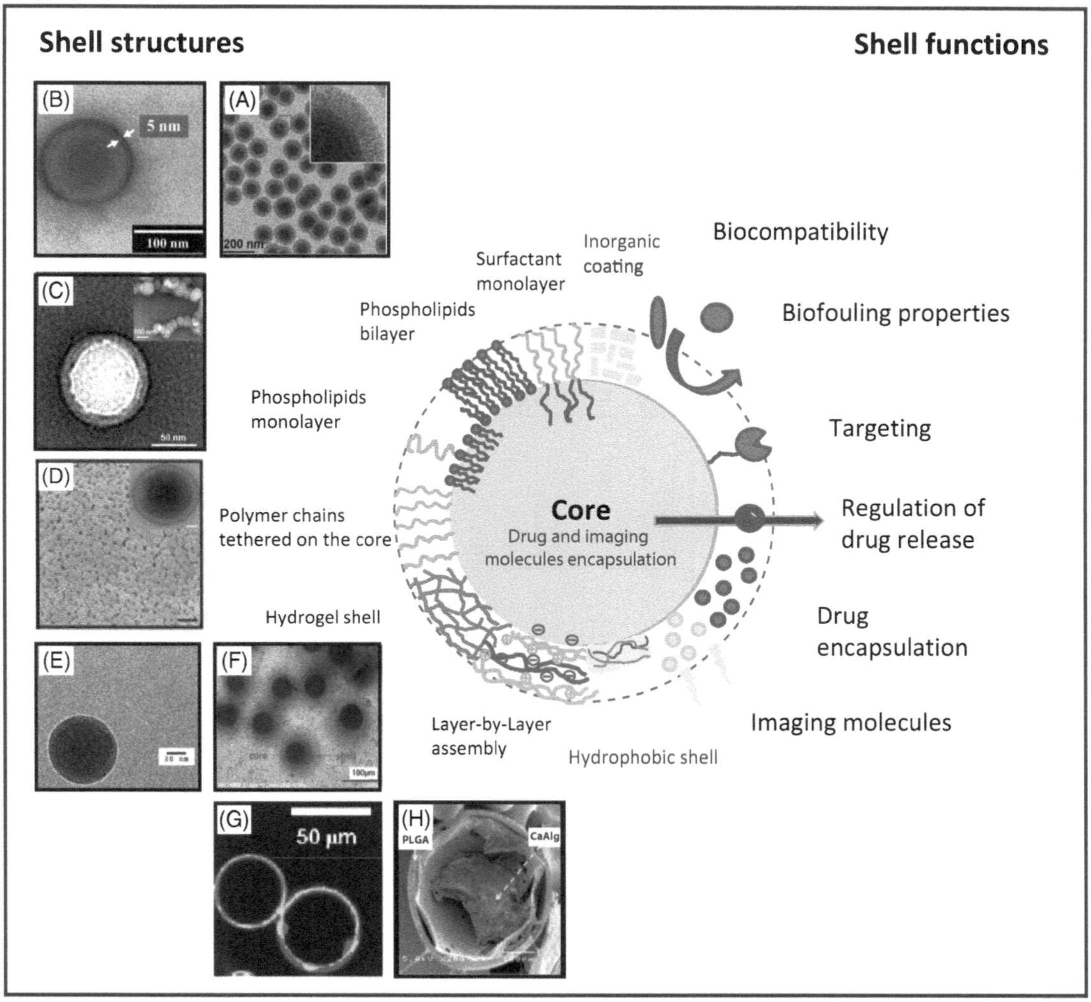

FIGURE 4.2 Schematic Representation of Core–Shell Structures (on the Left); Major Functions of the Shell (on the Right)

(A) High-resolution TEM image of fluorescent solid SiO_2 core with a mesoporous silica shell of different thickness for drug release control. The insert is an enlarged image of the marked area. (B) TEM images of LipoParticles (PEG45-lipid/DPPC/DPTAP 10/40/50 mol%) synthesized in pure water and observed by TEM after staining with sodium silico-tungstate. *White arrows* indicate the lipid bilayer thickness. (C) TEM image of red blood cell membrane-coated PLGA NP. The particles were negatively stained with uranyl acetate. (D) Scanning electron microscopy (SEM) image of PEG-*b*-PLA NP, insert: TEM Image. (E) TEM images of 0.5% ethylene glycol dimethacrylate cross-linked poly(AN-*co*-NIPAM) nanohydrogel at mole ratios of 1:1. The scale bar is 20 nm. (F) SEM images of PLGA–alginate core–shell microspheres. The scale is 100 μm. (G) Confocal scanning light microscopy image of PLGA microspheres coated with a layer-by-layer assembly of hyaluronic acid (HA) and chitosan sulfate (CS). Rhodamine-B-labeled CS appears [dark gray (red in the web version)] and Alexa-Fluor-labeled CS appears [light gray (green in the web version)] in the overlay image. The scale bars are 50 μm in length; (H) SEM image of a cross-sectioned alginate–PLGA particle. Alginate hydrogel appears collapsed in SEM condition. Scale bar is 100 μm. Original version of the figure is colored.

Table 4.1 Core–Shell Particles in Drug Delivery

Shell Material	Core Material	Shell Functions	Core Functions	Type of Shell	Applications	References
Hydrophilic polymer tethered on core surface						
PEG chains	PLGA	Stability, stealthiness	Drug encapsulation	Polymer brush	Cytotoxic drug delivery	Gref et al. (1995); Hrkach et al. (1997)
Chitosan	PIBCA	Mucoadhesion	Drug encapsulation	Covalent coating	Oral delivery	Bravo-Osuna et al. (2007)
Polyetheneimine	Poly(methacrylic acid)	Cell uptake, nucleic acid complexation	Structural	Polymer brush	Gene delivery	Zhu et al. (2005)
Phospholipids and lipid conjugates shell						
Phospholipids (monolayer)	Polysaccharide	Stability, biomimetic	Hydrophilic drug encapsulation	Monolayer	Peptide/Vaccine delivery	Peyrot et al. (1994)
Lipid and lipids-PEG (monolayer)	PLGA	Stability (decrease of burst release	Drug encapsulation	Monolayer	Hydrophobic drug delivery	Zhang et al. (2008)
Lecithin-PEG (monolayer)	PLGA	Stability, biomimetic	Drug encapsulation	Monolayer	Hydrophobic drug delivery	Chan et al. (2009)
Phospholipids	PLGA-doxorubicin	Drug encapsulation	Drug encapsulation	Bilayer	Dual drug delivery	Sengupta et al. (2005)
Phospholipids	PLGA	Targeted delivery	Targeted delivery	Bilayer	Drug and gene codelivery	Wang et al. (2010)
Phospholipid (bilayer)	MSN	Stability, biomimetic	Drug encapsulation Scaffold	Bilayer	Drug delivery	Ashley et al. (2011)
Cell membrane	PLGA	Biomimetic	Biocompatibility, long circulation, targeting	Bilayer	Drug delivery	Hu et al. (2011a)
Phospholipids, PEG	Nucleic acids/ Polycations	Stability, targeting	Nucleic acids encapsulation	Bilayer	Nucleic acids delivery	Hatakeyama et al. (2011)
Hydrogel shell						
Alginate	PLGA	Drug encapsulation	Release by diffusion	Hydrogel	Drug delivery	Wu et al. (2013)
Alginate	Pectin	Drug encapsulation	Release by diffusion	Hydrogel	Drug delivery	Del Gaudio et al. (2014)

PEG/BSA	Polypeptide	Drug encapsulation, pH responsive release	Biocompatibility	Protein	Drug delivery	Lee et al. (2008)
BSA	Cross-linked Chitosan/ PNIPPAm	Drug encapsulation, pH responsive release	Biocompatibility	Protein	Drug delivery	Ju et al. (2014)
Poly(*N*-isopropylacrylamide) hydrogel	Gel	pH responsive release	pH- and temperature-controlled release		Drug delivery	Soppimath et al. (2005)
PEG hydrogel	Ag core	Drug encapsulation controlled release	Imaging	Hydrogel	Drug delivery	Wu et al. (2010)
PAA shell	MSN	pH responsive drug release	Drug encapsulation	Hydrogel	Drug delivery	Hong et al. (2009)
Layer-by-layer						
Alginate/chitosan	MSN	Doxrubicin (DOX) encapsulation	pH-dependant release	LbL assembly	Cancer	Feng et al. (2014)
Lys, HA, EDA-modified Poly(glycerol methacrylic acid)	MSN	Amoxicillin encapsulation	Enzyme dependant release	LbL assembly	Antibacterial	Wu et al. (2015)
HA, chitosan	PLGA	Structural	bFGF encapsulation and release	LbL assembly	Tissue engineering	Go et al. (2011)
Inorganic shell						
CaPhosphate	MSN	Imaging	Drug encapsulation and pH-regulated drug delivery	Calcium layer	Theranostic	Rim et al. (2011)
Silica	MSN	Controlled release	Drug encapsulation	Porous scaffold	Drug delivery	Ran et al. (2013)
Hydrophobic shell						
PLGA	BSA	Controlled release	Hydrophilic drug encapsulation	Polymeric wall	Hydrophilic drug delivery	Chitkara and Kumar (2013)
PLGA or PLA	Alginates	Controlled release	Hydrophilic drug encapsulation	Polymeric wall	Hydrophilic drug delivery	Lim et al. (2013)

2.2.1 Polymer brushes as nanoparticle shell: the PEG paradigm

Most of the particulate systems designed for drug delivery or blood circulation have a poly(ethylene glycol) (PEG) layer at the surface, that includes anchored chains on the core surface. The PEG shell greatly reduces the adsorption of opsonins and other serum proteins on particle surface, preventing macrophage uptake, modifying the particle pharmacokinetic and biodistribution (Owens and Peppas, 2006). The PEG shell can be added either as a physiosorbed layer on the NP surface (Moghimi and Szebeni, 2003) or as a chemisorbed layer on a core material. It is generally accepted that the PEG shell should reach a certain chain surface density (number of chain/nm^2), in order for PEG chains to adopt a "brush" conformation. In this conformation, the exposure of the core surface (often hydrophobic or charged) to opsonins is limited (Vonarbourg et al., 2006). Linear PEG chains with a molecular weight varying from 2 to 10 kilodaltons are the most commonly reported molecular weights, but other conformations have been tested, such as the "loop" conformation or mixed brush conformation (Rabanel et al., 2014). Usually the thin PEG shell is not easily observed by transmission electronic microscopy due to the low electronic density of PEG brushes. Heavy metal salts are used to negatively stain the particles to visualize the shell.

2.2.2 Other alternatives to PEG

While PEG remains the gold standard to achieve stealthiness and biocompatibility, other neutral hydrophilic polymers have been tested to prepare hydrophilic shell, such as polysaccharides (Lira et al., 2011), poly(vinyl alcohol), PEG-PVA combinations, polyvinylpyrolidone (PVP), and lately poly(carboxybetaine), a zwitterionic and antifouling polymer (Elsabahy et al., 2013).

Hydrophilic shell can also be added for different purposes, such as for conferring mucoadhesive properties to the NP. The potential bioadhesive behavior of chitosan and thiolated chitosan grafted on poly(isobutyl cyanoacrylates) (PIBCA) NPs was evaluated by Bravo-Osuna et al. Both the molecular weight of chitosan and its concentration in the NP shell influenced the NP diameter and hence the transport through the mucus layer (Bravo-Osuna et al., 2007) (Sosnik et al., 2014).

Polyetheneimine (PEI) shells have also been used for gene delivery (Zhu et al., 2005). The positive charges on the shell brought by PEI allowed the complexation of nucleic acids (polyanions) for gene or RNA delivery or/and provided positive charges to carrier surfaces that are favorable for interactions with cell membranes but may have negative impact on residence time and toxicity (Nel et al., 2009).

2.2.3 Phospholipids and lipids conjugates

In order to obtain more biocompatible drug nanocarriers, a phospholipids layer (mono-, bi-, or multilayer) made of one or several lipids associated with lipid-PEG conjugates has been used as NPs shell. Interestingly, the lipid shell can provide an additional compartment useful to encapsulate a second hydrophobic active molecule.

2.2.3.1 The phospholipid monolayer shell

Zhang et al. (2008) described a one-step preparation method to produce core–shell particles with a polymeric hydrophobic core and a shell composed of a phospholipid monolayer able to stabilize the particles during their formation. The authors showed an increased drug loading and a decreased burst release of Docetaxel for surface-modified NP compared to poly(lactic-*co*-glycolic acid) (PLGA) or PEG-PLGA NPs (Zhang et al., 2008).

The success in the preparation of those systems relies on the optimization of the ratio lipid/core material and lipid/lipid-PEG conjugates (Chan et al., 2009). Similar studies using PEG-1,2-distearoyl-*sn*-glycero-3-phosphoethanolamine (PEG-DSPE) mixed with lecithin show that it is possible to prepare the hybrid core–shell particles using a single sonication step using a mixture of lipids and PLGA in water (Fang et al., 2010).

2.2.3.2 The phospholipid bilayers shell

The preparation of NP with phospholipid bilayer shell has been reported using a two-step method involving the separate preparation of the polymer core and the liposomes (or membrane vesicles) followed by the mixing of the two particle populations. Excess of lipid was shown to decrease drug loading due to drug–lipid interactions (Troutier and Ladavière, 2007).

More than two decades ago, Peyrot et al. (1994) reported the preparation of "supra-molecular biovectors" composed of a cross-linked polysaccharide hydrophilic core modified on the surface by acylation of fatty acids and stabilized by a phospholipid monolayer shell. The work of Sengupta et al. (2005) is one of the first studies reporting dual drug delivery from polymer-phospholipid core–shell NP. In this study, core–shell NPs were prepared with doxorubicin (DOX)-conjugated PLGA, coated with a shell composed of a mixture of PEG-DSPE, and phospholipid encapsulating a lipophilic drug (Sengupta et al., 2005). The so-called "nanocell" showed typical morphological features of a core–shell particle and controlled release of their dual cargo over several days.

The phospholipid bilayer associated to a PEG layer is also found in multifunctional envelope nanodevice (MEND). MEND is a vehicle developed by Pr H. Harashima research group, consisting of a nucleic acid complex with a polycationic polymer and a lipid envelope structure (bilayer) functionalized with PEG chains and specific target ligands (Hatakeyama et al., 2011).

Mesoporous silica NPs (MSN) have been extensively studied for drug delivery and imaging applications. Thanks to their high porosity, MSN can potentially accommodate high drug payload, exhibit excellent stability, and allow simple surface modification. However, MSN have several limitations such as limited control over drug release, and biocompatibility issues. This results in the necessity to modify their surface in order to prevent interaction of surface silanols with cell membrane and to control the release of the encapsulated drug. Nanoporous silica particles modified with a phospholipid bilayer (Protocell) showed a better encapsulation and an enhanced targeting efficacy over liposomes (Ashley et al., 2011). Combined with the high loading capacity of MSN, the cytotoxic effect of DOX-loaded Protocell was improved 10^6 times over liposomes. The synergistic delivery of PTX (from the shell) and gemcitabine (from the core) by hybrid silica NP was made possible. As a result, the shell is acting not only for the stabilization of the particles, but also as a drug reservoir (Meng et al., 2015).

2.2.3.3 Cell membrane mimetic shells

The Zhang research group has also proposed to use not only phospholipids (Fang et al., 2010, 2015), but also erythrocytes membranes (Hu et al., 2011a) or platelets membrane vesicles, as well (Hu et al., 2015a), with the view to create cell-membrane mimetic shells.

2.2.4 Hydrogel shells

Hydrogel shells can accommodate different types of hydrophilic cargo, such as large molecules, charged active molecules, and biologics. Hydrogels usually exhibit a very high release rate (Hoare and Kohane, 2008). They can be designed to respond (by swelling or shrinking) to environmental cues such as pH, ionic force changes, or temperature.

2.2.4.1 Diffusion-controlled release from hydrogel shells

PLGA–alginate core–shell microparticles (MPs) were fabricated using a microfluidic technique (Wu et al., 2013). Rifampicin was loaded into both PLGA MPs and PLGA–alginate core–shell MPs. The shell was shown to modulate the drug release kinetics and a near-zero-order release pattern was observed. In another study, a model nonsteroidal antiinflammatory drug (NSAID) was encapsulated in a hydrogel core–shell particle prepared with pectin (core) and alginate (shell). The core/shell structure allowed a controlled release of the model drug in simulated gastric and intestinal fluids (Del Gaudio et al., 2014).

2.2.4.2 Stimuli-responsive hydrogel shells

In order to achieve deeper tumor penetration of drug carriers, Ju et al. (2014) have developed a core–shell nanogel NP with a cross-linked polyelectrolyte core (composed of *N*-lysinal-*N′*-succinyl chitosan and poly(*N*-isopropylacrylamide), and a cross-linked bovine serum albumin (BSA) shell. The core of the particle was able to swell under acidic conditions, typical of a tumor environment. Swelling of the NP resulted in a rapid release of the encapsulated cargo. The BSA shell provided a biocompatible negatively charged shell, maintaining the stability of the nanogel and the pH-responsive properties. A similar approach has been proposed by Lee et al. (2008). The particle core was made of poly(l-histidine-*co*-phenylalanine) loaded with DOX. The shell of the NPs was formed by BSA proteins tethered to the core via PEG chains. Upon pH change, the polypeptide core swelled, allowing the release of the drug.

Hydrogel shells have been conjugated to inorganic cores. Silver NPs have been covered with a glucose-sensitive copolymer shell to create a hybrid nanogel (Wu et al., 2010). The Ag core (10 ± 3 nm) provided fluorescence imaging capacity while the responsive polymer shell regulated the release of loaded insulin.

pH-dependent release of DOX encapsulated in MSN was allowed by grafting polyacrylic acid (PAA) on the core surface (Hong et al., 2009). Tang et al. (2011) developed a chitosan/polymethacrylic acid shell on MSN particles. Release of DOX from the functionalized MSN was found to occur at a faster rate at acidic pH compared to pH 7.4 .

Polymer-coated (polymerized PEG-diacrylate and NIPAm) MSN have also been proposed for stimuli responsive delivery of DOX in response to proteases present in tumor environment or heat (Singh et al., 2011).

2.2.5 Layer-by-layer shell

The "layer-by-layer" technique (LbL) consists of the alternate deposition of polymers of different charges (polyanions/polycations). LbL deposition provides significant flexibility in shell thickness, which can be modulated according to the number of deposition/purification cycles. The technique has been used to stabilize Au core NPs (Schneider and Decher, 2004) and to be a powerful method to produce core–shell particles. LbL can also be useful to encapsulate polyanionic drugs such as polynucleotides.

MSN, modified on their surface with amine groups, were transformed into pH-responsive drug delivery systems by LbL deposition of an alginate/chitosan polyelectrolyte multilayer coating on their surface. The multilayer shell conferred biocompatibility to the NPs and pH-dependent release of DOX (Feng et al., 2014).

Particles with a gelatin core and an LbL shell of polyelectrolytes (polystyrene sulfonate/polyamine hydrochloride; polyglutamic acid/poly-L-Lysine; dextran sulfate/protamine sulfate) exhibited controlled release of different active polyphenols, including curcumim (Shutava et al., 2009).

LbL technique was also used to coat surface-modified PLGA MP with alternate layers of hyaluronic acid and chitosan. This coating "sandwiched" a layer of basic fibroblast growth factor (bFGF) noncovalently associated to heparin (Fig. 4.2G). The system provided sustained delivery of biologically relevant amount of bFGF over a period of 40 days, preventing its rapid degradation and preserving its activity (Go et al., 2011). Similarly, small interfering RNA (siRNA) has been encapsulated and delivered from LbL assembly of oppositely charged polyelectrolytes siRNA and PEI alternatively deposited as a shell on gold NPs cores (Elbakry et al., 2009).

2.2.6 Hydrophobic shells

Although shells tend to be prepared from hydrophilic biocompatible materials, in some recent studies hydrophobic shells have been proposed to increase the encapsulation of hydrophilic actives and to provide controlled release. Alginate hydrogel NPs with PLGA or poly(L-lactic acid) (PLLA) shell were compared to pure calcium alginate NP (CaAlg) for encapsulation and delivery of a model hydrophilic drug (Fig. 4.2H). The core–shell particle showed a sustained release over 4 days (PLGA) and 7 days (PLLA) in contrast to "naked" CaAlg NPs that exhibited complete release within 1 day. The polymer shell was found to limit the rate of water influx and acted as a rate-limiting membrane for drug diffusion (Lim et al., 2013).

2.2.7 Inorganic shells

Inorganic materials can also be used as shell, owing for their stability and biocompatibility. MSN covered with a calcium phosphate shell have been shown to release their content intracellularly within the acidic environment of endosome/lysosome (Rim et al., 2011). Drug-loaded MSN were coated with a shell of silica, with thickness that was tuned to obtain an optimal drug release (Fig. 4.2A for a TEM image of the carrier) (Ran et al., 2013). Silica shell was also used to control the release of DOX from amorphous calcium carbonate/DOX core particles at acidic pH (Zhao et al., 2015a).

3 THE CAPSULE AND VESICLE NANOSTRUCTURE

Capsules and vesicles are colloidal systems in which drug is entrapped in a cavity surrounded by a membrane. This section aims to overview three different kinds of capsule/vesicle structures: nanocapsules (NCs), yolk-shell NPs, and colloidosomes (Fig. 4.3) with a special emphasis on their biomedical properties.

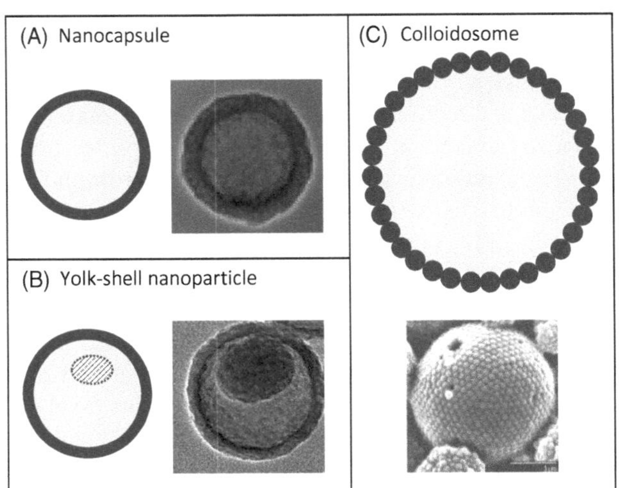

FIGURE 4.3 Schematic Representations and Microscopic Images of Different Capsule Types

(A) NC (Chen et al., 2011); (B) yolk-shell NP (Liu et al., 2013); (C) colloidosome (Yow and Routh, 2009).

3.1 NANOCAPSULES

Unlike nanospheres in which drug is dispersed in a continuous matrix, NCs are vesicular systems in which drug is confined in a reservoir surrounded by a membrane (Brigger et al., 2002; Mora-Huertas et al., 2010; Zhang et al., 2015). The composition of NC wall is diverse and can be divided in three major categories: soft materials (polymers, polypeptides, lipids), hard materials (silica, carbon, gold), and hybrid materials, a combination of both soft and hard materials (Ariga et al., 2012). Composition of the core can also be finely tuned to fit the cargo properties with the properties of the capsule. Depending on the materials chosen and the desired structure of the capsule core and wall, various methods including nanoprecipitation, double emulsion, LbL coating, and sacrificial core template approach can be used for the preparation of NCs (del Mercato et al., 2014; Mora-Huertas et al., 2010; Zhang et al., 2015). Table 4.2 provides a nonexhaustive list of the different classes of NC walls and cores recently described in the literature as well as their potential biomedical applications.

3.1.1 Encapsulation of lipophilic or hydrophilic compounds as a function of core type

NC structure confers a significant advantage over the conventional matrix-type nanospheres for encapsulation of lipophilic or hydrophilic compounds due to the optimized solubility of active in oily or aqueous reservoirs constituting the core of the capsule (Anton et al., 2009; Mora-Huertas et al., 2010).

The oil-core capsule is the most reported NC and is designed for high encapsulation of lipophilic compounds (Almouazen et al., 2013; Anton et al., 2009; Gonzalo et al., 2013). However, the encapsulation of hydrophilic compounds remains a challenge, due to a poor entrapment efficiency in oil-filled NCs. The improvement of the stability and the absorption rate of fragile hydrophilic macromolecules, such as peptides/proteins or nucleotides, is nevertheless crucial. It is also of importance to modulate biodistribution and reduce toxicity of small hydrophilic drugs such as anticancer agents (Vrignaud et al., 2011). The poor entrapment efficiency of hydrophilic compounds led therefore to the preparation of aqueous core NCs. Couvreur's laboratory was one of the first to develop aqueous core polymeric capsules dispersed in an aqueous medium. Lambert and coworkers prepared aqueous-core NCs containing antisense oligonucleotides by interfacial polymerization in water in oil emulsion followed by resuspension in water. As a result, oligonucleotides encapsulated in NCs were more efficiently protected against degradation by serum nucleases than oligonucleotides absorbed onto the nanosphere surface (Lambert et al., 2000). Hollow capsules prepared by the sacrificial template approach also have proved their ability to encapsulate hydrophilic compounds (De Koker et al., 2011) as recently demonstrated with the encapsulation of insulin in glucose-sensitive polyelectrolyte capsule (Guo et al., 2014) or with the encapsulation of BSA in pH-sensitive polyelectrolyte capsule (Gu et al., 2013).

Coencapsulation and delivery of lipophilic and hydrophilic drugs can also be achieved with NCs. Qian et al. (2015) recently reported the preparation of hybrid protein lipid polymer NCs composed of three layers: an outer aqueous phase, an intermediate oil phase and an inner aqueous phase. Protein transcription factor p53 and lipophilic drug paclitaxel (PTX) were respectively encapsulated in the inner aqueous core and in the oil phase. Cell viability assays on HeLa cells showed low cell apoptosis, with PTX (40%) or p53 (40%) single drug-loaded NCs. The combination of both drugs in the same NC, synergistically increased the apoptosis rate to 100% (Qian et al., 2015).

Table 4.2 Recent Examples of Nanocapsules for Biomedical Applications

Wall Material	Core Material	Wall Functions	Core Functions	Applications	References
Polymers					
PLA	Oil (Mygliol 829)	Biodegradable	Encapsulation of calcitriol	Drug delivery in chemotherapy	Almouazen et al. (2013)
PLA with PEG coating	Aqueous	Biodegradable + long circulating behavior	Encapsulation of gemcitabine hydrochloride[a]	Drug delivery in chemotherapy	Cosco et al. (2015)
HA	Aqueous	Biocompatible + hyaluronidase sensitive	Encapsulation of polyhexanide[a] and fluorescent dye	Enzyme-triggered release for prevention of infections	Baier et al. (2013)
Poly-L-lysine/Poly-glutamic acid (polyelectrolyte multilayers)	Oil (linseed oil/chloroform)	Biocompatible, decreasing cytotoxic effect	Encapsulation of quantum dots	Imaging	Adamczak et al. (2013)
Tertiary amine and hydrazide grafted polyaspartamide/ carboxyl and aldehyde grafted polyaspartamide (polyelectrolyte multilayers)	Hollow	Biodegradable + pH sensitive	Encapsulation of BSA[a]	pH-triggered protein delivery	Gu et al. (2013)
Modified chitosan/poly(D-gluconamidoethyl methacrylate-r-3-acrylamidophenylboronic acid) (polyelectrolyte multilayers)	Hollow	Biodegradable + glucose sensitive	Encapsulation of insulin[a]	Glucose-triggered protein delivery	Guo et al. (2014)

(continued)

Table 4.2 Recent Examples of Nanocapsules for Biomedical Applications (*Cont.*)

Wall Material	Core Material	Wall Functions	Core Functions	Applications	References
Proteins/peptides					
BSA + folic acid+ Fe_3O_4 particles embedded	Aqueous ?	Magnetic responsive + selectivity to tumor cells	Encapsulation of rhodamine B[a]	Targeted delivery in chemotherapy	Li et al. (2015b)
Arginylglycylaspartic acid-Ferritin	Hollow	Non immunogenic cage-like nanostructure of 24 subunits + selectivity to tumor cells	Encapsulation of photosensitizers	Targeted photodynamic therapy	Zhen et al. (2013)
Dihydrolipolyl acyltransferase	Hollow + cysteines	Biodegradable virus like structure of 60 subunits	Encapsulation of DOX through conjugation with cysteines	Drug delivery in chemotherapy	Ren et al. (2011)
Lipids					
Mineral oil + polyurea film	Aqueous	Encapsulation of red sudan III	Encapsulation of BSA-fluorescein isothiocyanate[a]/methylene blue[a]	Coencapsulation/drug delivery	Anton et al. (2009)
Inorganic material					
Silica	Hollow	Large pores for release of macromolecular compounds	Encapsulation of siRNA[a] and Fe_3O_4 particles	Drug delivery	Chen et al. (2011)
Carbon + fluorescent carbon dots embedded	Hollow	Upconversion fluorescence, absorption near infrared light, interactions with drug	Encapsulation of DOX	Imaging/photothermal therapy/pH and light responsive drug delivery for cancer treatment	Wang et al. (2015)

[a]*hydrophilic compound.*

3.1.2 Permeability control as a function of wall structure

An important attribute of NC is the thickness of the wall, which provides fine-tuning of drug permeability. Several studies have explored the relationship between the number of polyelectrolyte layers composing the wall and the release of a compound encapsulated in the core. Results showed that capsules with only a few layers release their content in a less sustained manner than those with a larger number of layers (Antipov et al., 2001; Cuomo et al., 2015). In another study, Jovanovic and coworkers investigated the impact of wall thickness for intravenous drug detoxification. Silica NCs with an oily core were synthesized and partitioning studies were performed to evaluate the capacity to efficiently remove quinoline, a model lipophilic compound, from aqueous media. Results showed that the amount of quinoline sequestrated inside capsules was inversely proportional to wall thickness, signifying that for detoxification applications, a thin wall was more appropriate (Jovanovic et al., 2005).

Wall porosity also affects the permeability properties of NCs. Porous NCs are combining the advantages of a reservoir structure able to efficiently encapsulate molecules with the advantages of porous materials able to provide controlled release of active compounds (Liu et al., 2009). Pore size is an important parameter influencing the diffusion across the capsule wall, which is limiting the size of molecules able to diffuse through the porous wall (Dergunov and Pinkhassik, 2011; Gustafson et al., 2013; Shmakov et al., 2011). The pore size can be controlled to tune the drug release. For instance, the release rate of DOX from porous silica NCs was significantly slowed down by decreasing the pore size using PEG molecules on the surface (Yang et al., 2008). Large pore size may also find applications, notably for the delivery of biomacromolecules, as described by Chen and coworkers. The authors used silica NCs with large pores size to encapsulate siRNA for transfection and to collect iron oxide NPs for magnetic resonance imaging (Chen et al., 2011). Charges localized inside pores affect drug release as well. Porous polymeric capsules containing carboxylic acids at the pore entrance were recently proposed for pH-mediated encapsulation and release of anionic compounds. In acidic conditions, "opened" neutral pores allowed free diffusion out of the capsule, while in neutral or basic conditions, negatively charged pores hindered anions release due to electrostatic repulsions (Dergunov et al., 2014).

The use of stimuli responsive material for the wall, in particular polyelectrolyte multilayers, is another way to control the drug release from NCs. A variety of stimuli may initiate changes in capsule wall properties, such as chemical stimuli (pH, redox, enzyme, glucose) or physical stimuli (thermal, magnetic, electric, mechanic, light) (De Koker et al., 2011; Delcea et al., 2011; del Mercato et al., 2014; Esser-Kahn et al., 2011; Zhao et al., 2015b).

3.2 YOLK-SHELL NPs

The last decade has seen the development of an original structure called "yolk-shell" or "nanorattle" structure. This structure is characterized by one or more mobile cores encapsulated in a hollow capsule. It differs from core–shell nanostructure, reviewed in Section 2 of this chapter, due to the presence of free space between the core and the shell. The most common way to produce yolk-shell structures is based on a selective etching process, which involves core coating by two layers of different materials followed by a selective removal of the middle layer. The structure and chemical composition of the core, the shell and the hollow interior can be modulated depending on the desired application (Priebe and Fromm, 2015). Due to their unique structure, yolk-shell NPs appear to be a promising class of multifunctional nanocarriers.

3.2.1 Yolk-shell structure for targeted drug delivery

Drug delivery is an obvious application of yolk-shell NPs due to the hollow space, which has been proven to efficiently carry higher drug concentration compared to solid NPs (Hu et al., 2015b; Jiao et al., 2015; Liu et al., 2013). Compared to simple NC, a magnetic core may provide added value for targeted transportation. For instance, Zhang et al. designed a yolk-shell NP with a Fe_3O_4 core and a polyethyleimine functionalized fluorescent mesoporous silica shell to carry siRNA (siRNA-Fe_3O_4@fmSiO_2). Enhanced transfection efficiency was found in cancer cells treated with siRNA-Fe_3O_4@fmSiO_2 under external magnetic field (Zhang et al., 2012).

3.2.2 Yolk-shell structure for multifunctional treatments

One of the first examples of multifunctional theranostic yolk-shell NPs is the work of Gao et al. in which yolk-shell NPs were synthesized with a FePt core as an anticancer agent and a Fe_2O_3 shell as a potent MRI contrast agent (Gao et al., 2008b). Compared to FePt@Fe_3O_4 core–shell NP, yolk-shell structure showed a better contrast enhancement because of the larger interface between the iron oxide hollow NP and the water phase. In addition, cytotoxic assays on HeLa cells demonstrated that yolk-shell NPs had the highest cytotoxicity compared to core–shell NP. Indeed, contrary to yolk-shell NPs, compact Fe_3O_4 shells of core–shell NPs prevent the oxidative species from reaching the FePt core and so prevented disruption of the core and release of platinum ions.

Recently Jiao et al. (2015) developed a multifunctional theranostic silica nanorattle combining chemotherapy, targeting, and diagnosis (Fig. 4.4). PTX loading into the nanorattle hollow space was improved over PTX loading into mesoporous silica NP. Tumors could be imaged by MRI using the Fe_3O_4 core, a MRI contrast agent and by fluorescence using Cy7 (NIRF dye) loaded in the shell. Moreover, the shell was functionalized with transferrin for specific accumulation of the theranostic particle in tumor tissues overexpressing the transferrin receptor.

Nanorattle structures containing a metallic core have proved to be effective for photothermal therapy as well. In a recent study, *p*-aminothiophenol-capped gold nanocages (*p*ATP-capped AuNC) were encapsulated into a porous silica capsule. *p*ATP-capped AuNC core was used as an efficient near-infrared absorbent to produce heat for photothermal therapy as well as a surface-enhanced Raman scattering (SERS) substrate for imaging. Chemotherapy was combined to photothermal therapy and imaging by encapsulating DOX into the hollow space between core and shell (Hu et al., 2015b).

Finally, similarly to NC, yolk-shell structure was found appropriate for pH-triggered release (Chen et al., 2014a; Du and Liu, 2014), temperature-triggered release (Du and Liu, 2014; Kang et al., 2012), light-induced release (Gandra et al., 2014; Zhao et al., 2014), or magnetic-induced release applications (Hu et al., 2011b).

3.3 COLLOIDOSOMES: AN OUTLOOK OF THEIR BIOMEDICAL APPLICATIONS

Colloidosomes are microcapsules whose external shell is composed of a close-packed layer of colloidal particles locked together (Dinsmore et al., 2002; Rossier-Miranda et al., 2009; Velev et al., 1996). Colloidosomes have demonstrated efficient delivery of molecular as well as macromolecular compounds, thanks to their shell permeability arising from the interstices between the shell particles. Control over passive diffusion can be achieved by modulating absorbed NP size or shell thickness (Kim et al., 2007;

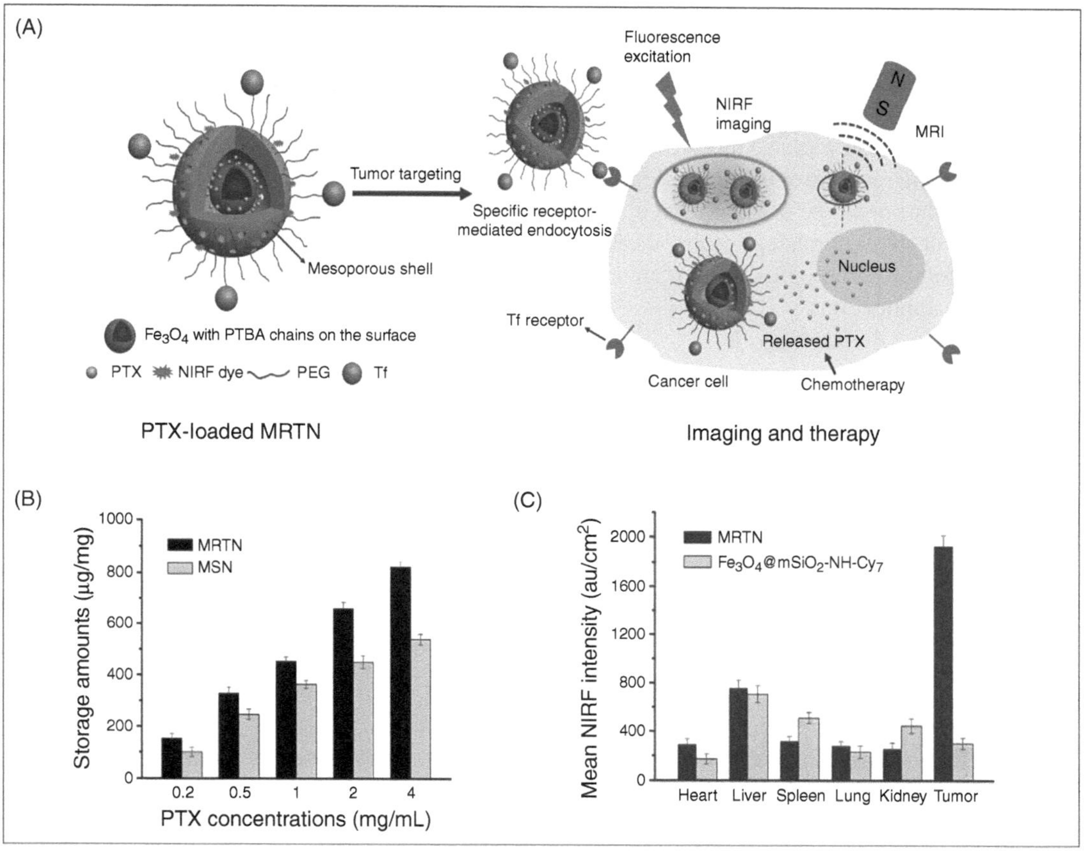

FIGURE 4.4

(A) Schematic representation of PTX-loaded multifunctional rattle type NP (MRTN) for tumor targeting, MRI, fluorescence imaging, and chemotherapy; (B) storage amounts of PTX in MRTN and MSN at different PTX concentrations; (C) mean NIRF intensity of the organs and tumor tissues from the mice treated with MRTN and with rattle type NP without transferin (Fe_3O_4@$mSiO_2$ -NH-Cy7) at 1 h postinjection.

Adapted from Jiao, Y., Sun, Y., Tang, X., Ren, Q., Yang, W., 2015. Tumor-targeting multifunctional rattle-type theranostic nanoparticles for MRI/NIRF bimodal imaging and delivery of hydrophobic drugs. Small, 11, 1962–1974.

Rosenberg and Dan, 2013). Commonly used to ensure structural stability, sintering or cross-linking of adsorbed particles modify the shell morphology as well as the permeability properties (Li et al., 2015a; Yow and Routh, 2009). For instance, Yow and Routh (2009) described the impact of the sintering process on the shell structure and permeability. It has been found that longer sintering time produced smoother shells, decreased their permeability and therefore allowed for longer release rate (Yow and Routh, 2009).

In addition, different strategies were explored to create a smart generation of colloidosomes equipped with stimuli responsive properties for active release. Sander and Studart (2013) developed complex colloidosomes by incorporating functional NPs within the hollow core for on-demand delivery. On-demand release was successfully achieved by different strategies: (1) pH-triggered release upon desorption of cargo from entrapped NPs, (2) pH-triggered release induced by capsule shell disruption, or (3) on-off release mediated by magnetic NPs inside the shell (Sander and Studart, 2013). Light-triggered release was also reported with the use of a stable colloidosome shell made of negatively and positively charged nitrophenylene-doped silica NPs that disassembled upon light exposure due to charge reversal of one type of NPs (Li et al., 2015a).

However, despite promising results, investigations have remained limited to physicochemical studies. Further complete biological studies are expected soon to capture the potential of such structure for drug delivery applications.

4 MULTICOMPARTMENT NANOPARTICLES

Multicompartment NPs designed for pharmaceutical purposes are receiving increasing interests, fuelled by the need for the simultaneous delivery of active compounds of different hydrophobicity and the development of theranostic applications. However, not every class of nanocarrier can simultaneously possess all these properties. In the following section we reviewed the most promising structures and their most interesting pharmaceutical applications. A brief summary of the reviewed structures including their compositions, structures, and applications can be found in Table 4.3.

4.1 MULTICOMPARTMENT LIPID VESICLES

Lipid vesicles such as liposomes have already demonstrated their importance in clinical applications. One of its most famous achievements is the delivery of doxorubicin (DOX) with the liposomal preparation Doxil marketed in 1995 (Barenholz, 2012). Recent efforts to tune the design of lipid NPs in order to introduce compartmentalization have been made for applications that require a highly biocompatible, multifunctional, and universal lipid nanocarrier (Paleos et al., 2013).

The development of "vesosome," a multicompartment liposome, was an example of an attempt to increase liposome versatility. A vesosome is a liposome enclosing smaller liposomes as secondary compartments (Fig. 4.5A). Basically, small liposomes encapsulating the drug (about 100 nm in diameter) are produced, followed by their aggregation under controlled conditions. Then, the small liposome aggregates are encapsulated into large vesicles by the mean of interactions of cochleate cylinders of multilamellar lipid tubules. Vesosomes were evaluated only in a few studies but researchers (Paleos et al., 2013). The multiple bilayers protecting the encapsulated drug were found to avoid efficiently early leakage of the carrier and to achieve delayed release (Boyer and Zasadzinski, 2007). Vesosomes encapsulating a model drug (carboxyfluorescein) were found to release only between 20% and 30% of their content during a 20 h incubation time, even in presence of lipases. In comparison simple bilayer vesicles showed a release of about 60% in 20 h without the presence of lipase and up to 100% release within 8 h in presence of lipase. In calf serum, the release was delayed up to 12 h before leakage the outer layer initiating the release of caboxyfluorescein. Tested as a topical vaccine (without needle), the vesosomel system encapsulating the tetanus toxin, produced a sustained immune response over 8 weeks in rats and proved to significantly penetrate the skin layers (Mishra et al., 2006).

Table 4.3 Structure and Applications of Multicompartment NPs as a Biomedical and Pharmaceutical Tool

Nanostructure	Applications	Nanodomains	Possible Bulk Materials	Preparation Method	References
Lipid vesicles					
Autoasembled lipid NPs	Macromolecule entrapment	Hydrophilic	DMAP-BLP; PEG-C_{14}/C_{18}	Microfluidic; autoassembled	Leung et al. (2012)
Vesosomes	Liposome shielding; topical vaccine	Hydrophilic	Liposome materials	Central liposomes aggregation and bilayer coverage	Paleos et al. (2013)
Liquid crystalline lipid nanocarrier					
Lamellar (L_2), cubosome (pn3m, la3d), hexosome (H_2), janus	Macromolecule delivery; simultaneous drug delivery; theranostic; gene therapy	Hydrophobic; Hydrophilic	MO, GMO, DOTAP; phytantriol; pluronic; capric acid; vitamin E	Autoassembled via sonication	Angelova et al. (2011); Tran et al. (2015); Zhai et al. (2015a,b)
Phase-separated and growth-induced multicompartment NPs					
Janus, multicompartment, patchy particles, hamburger-like	Controlled NP shape; control on compartment organization; elective encapsulation	Hydrophobic; fluorophilic; hydrophilic	PAA-PS/PAA-PB; PS-P4VP/Au-P4VP; PEG-PEE-PFPO	Phase-separated micelles	Caltagirone et al. (2015); Du and O'Reilly (2011); Lodge et al. (2005)
Growth-induced multicompartment NPs					
Surface compartment, dumbbell-like, cubic crystalline	Nonspherical NP; selective encapsulation; simultaneous drug delivery	Hydrophobic; fluorophilic; hydrophilic	PDMA-b-PS-b-P4VP; PGSD; silica; organo-silica	Sequential synthesis, controlled synthesis (RAFT, ATRP)	Doermbach and Pich (2015); Li et al. (2014); Suteewong et al. (2013)
Multicompartment nanoreservoirs					
Multiple hydrophobic nanodomains	Enzyme triggered release; high drug loading and solubility; controlled release; selective targeting	Hydrophobic; hydrophilic	$CaCO_3$; acrylates; PLL + chitosan; HA20C10; nanodiamonds	Encapsulation of hydrophobic core	Boudou et al. (2012); Kim et al. (2014); Moore et al. (2013)
Multiple hydrophilic nanodomains	High hydrophilic drug loading; controlled release; macromolecule entrapment	Hydrophobic; hydrophilic	Palmitic/stearic acid; SC/SPC; PDLLA/PGOHMA	Double emulsion oil-water-oil	Gao et al. (2008a); Liu et al. (2007)
Shell-in-shell micro/nanocapsules	Biomolecule reactor; enzyme/substrate delivery; triggered release	Hydrophilic	$CaCO_3$/PSS/PAH; gold NP; magnetite NP	Core dissolution of ball-in-ball particles	Kreft et al. (2007a,b)

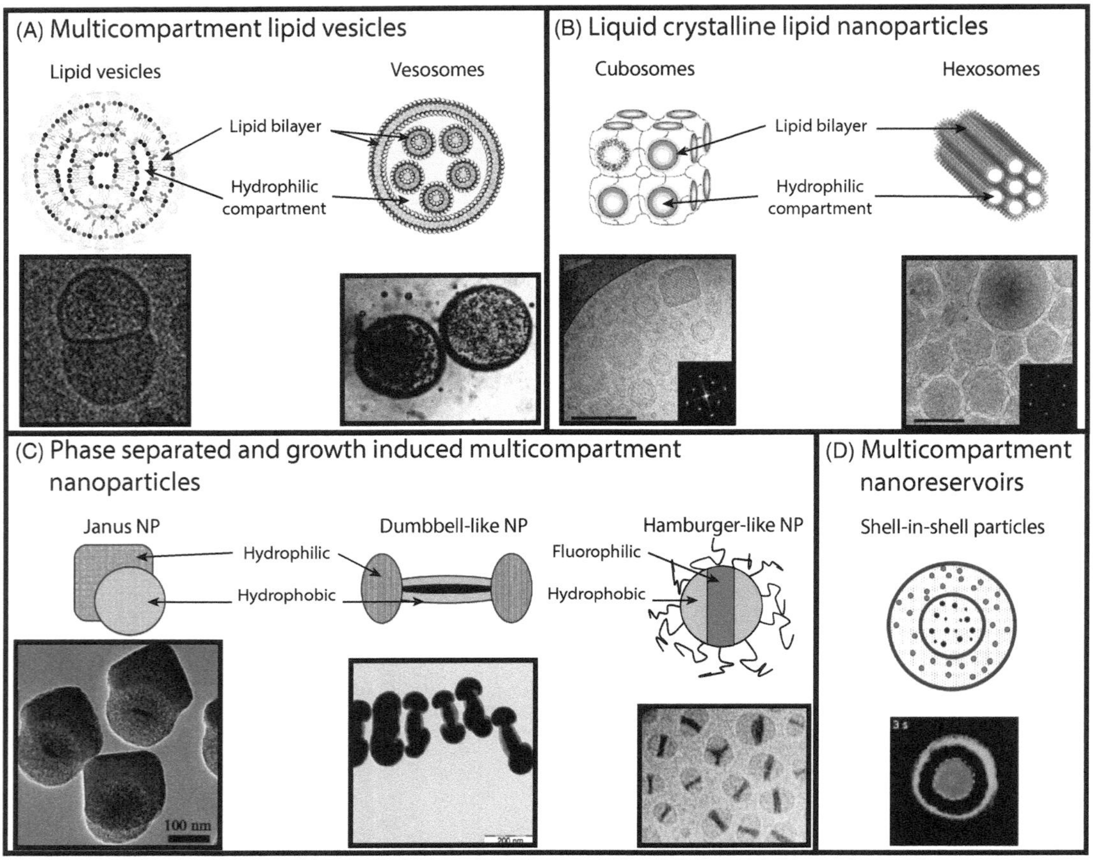

FIGURE 4.5

Schematic representation of (A) multicompartment lipid vesicles, on left lipid vesicles (Leung et al., 2015) and on right vesosomes (Paleos et al., 2013). (B) Liquid crystalline lipid NPs schemes (Zabara and Mezzenga, 2014) and TEM pictures (Zhai et al., 2015b), on left represent cubosomes (Q229) and on right hexosomes. (C) Phase-separated and growth-induced multicompartment NPs, on left janus MSN (Suteewong et al., 2013), in middle dumbbell-like NPs (Doermbach and Pich, 2015), and on right hamburger-like NPs (Marsat et al., 2011). (D) Multicompartment nanoreservoir expressed as shell-in-shell particle containing enzymes (Kreft et al., 2007a).

Micelles are another class of lipid vesicles NPs, where the drug is entrapped within the lipid core. Typically, 3-(dimethylamino)propyl(12Z,15Z)-3-[(9Z,12Z)-octadeca-9,12-dien-1-yl]henicosa-12,15-dienoate (DMAP-BLP), DSPC, cholesterol, and PEG-C_{14} (PEG-DMG) or PEG-C_{18} (PEG-DSG) are forming such vesicles (Table 4.3). This system was found quite efficient for gene silencing following siRNA administration at doses of 1 and 5 mg/kg (Chen et al., 2014b). However, large formulation volumes are required to reach these dose levels to compensate for poor siRNA encapsulation. It was found that the use of a cationic lipid, 1,2-dilinoleyl-4-(2-dimethylaminoethyl)-[1,3]-dioxolane (DLinkC2-DMA) was promoting the formation of multicompartment lipid vesicles (Fig. 4.5A). The hydrophilic domains were containing siRNA/cationic lipids complexes, increasing the loading of the hydrophilic macromolecules (Leung et al., 2012). The observed high drug loading suggested that the interaction between the polar heads of DLinkC2-DMA and the siRNA was very strong. Moreover, DLinkC2-DMA have high membrane permeation and low immunogenicity, which makes it an interesting candidate for lipid NPs preparation (Basha et al., 2011). A microfluidic technique was adapted to prepare this type of vesicles. Efficient encapsulation of siRNA, mRNA, and DNA were reported, indicating an interesting potential for a scale-up production (Leung et al., 2015).

4.2 LIQUID CRYSTALLINE LIPID CARRIERS

4.2.1 Lyotropic drug carriers

The use of liquid crystal has found its way as an emerging concept in the field of drug delivery systems. Prepared from lipids, lyotropic carriers form in water a liquid crystalline structure auto-assemble into different crystals or "phases" (Fig. 4.5B). They adopt phases according to the lipids chemistry and composition, the water content, and the temperature. Known phases or lipid organizations is including lamellar structures, cubic structures (cubosomes) and hexagonal array of cylinders structures (hexosomes) (Zabara and Mezzenga, 2014). The system is based on reverse micelles, where the water is "entrapped" within the lipid structure, resulting in a liquid crystal or mesophases. The lamellar structure is made of successive bilayers of amphiphilic lipid where hydrophobic domains remain in contact with each other and polar head groups are facing each other with a thin film of water in-between. The lamellar system is basically a 2-D structure, the hexosomes are more like a 3D structure. Its unit cell is made of "hollowed" amphiphilic lipid cylinders containing water. This structure is stacked multiple times to form the liquid crystal lipid NP (Fig. 4.5B). Finally, the cubosome is an even more complex structure than the previous ones with a 3D crystal. The cubosome is mostly formed of amphiphilic lipid bilayers creating water channels within the structure in different patterns such as the bicontinuous cubic phases (Pn3m) or the reversed micellar cubic phase (Fd3m), for example, (Fig. 4.5B). By its unique structure, it allows to create lipid nanocarriers with multiple compartments, which all possess a very high lipid-water interface (500 m^2/g surfactant) (Zabara and Mezzenga, 2014). This unique characteristic favors interfacial interactions such as adsorption or the segregation of pharmaceutical compounds into water nanopockets or even their incorporation within the lipidic compartmentof the mesophase (Angelova et al., 2011).

4.2.2 Liquid crystalline lipid carriers, composition and synthesis

As shown in Table 4.3, liquid crystalline lipid carriers are mainly made from amphiphilic lipids, similar to glyceryl monooleate (or monoolein-based lipids). Control of the lipid-water ratio is important to obtain specific crystalline architectures. The lamellar structure is formed using low water content, while cubosomes and hexosomes structures are prepared using higher water content. Temperature can also affect the crystal phase. For example, hexosomes are thermodynamically stable at higher temperature than cubosomes, generally at temperatures exceeding 50°C. However, a third component can be added to increase the stability of the lipid crystal, such as a cosurfactant or an oil (Zabara and Mezzenga, 2014).

4.2.3 Cytotoxicity of liquid crystalline lipid carriers

Since this type of carrier is quite in its infancy, studies using this type of carrier have been mostly focusing on the proof of concept as a pharmaceutical carrier. Regarding small body of studies have touched the safety and biodegradability of these carriers, showing encouraging results. In studies evaluating the safety of the lipid crystals, phytantriol was found more toxic (toxic dose twice lower) than glycerol mono-oleate (GMO) in terms of cellular viability, hemolytic activity, inflammatory stress response, and complement activation in human plasma (Hinton et al., 2014; Mat Azmi et al., 2015). These findings suggest a clear role of the bulk material on the safety of the crystalline lipid carrier. Although, incorporation of PEGylated lipids in phytantriol cubosomes was able to reduce their cytotoxicity (Zhai et al., 2015a). The use of 1,2-distearoyl-*sn*-glycero-3-phosphoethanolamine (DSPE) conjugated with high molecular weight PEG (DSPE-PEG) was found to significantly reduce the cytotoxicity effect compared to the use of Pluronic polymer F-127. Studies regarding the safety of different structures showed that hexosome and lamellar mesophases were the safest structures (Zhai et al., 2015b). However, it remains unclear if the improved safety was a direct impact of the mesophase structure or an indirect impact of the lipids composition.

4.2.4 Drug-loaded liquid crystalline lipid carriers

An interesting property of lyotropic carriers is their ability to entrap a vast spectrum of pharmaceutical compounds, ranging from small molecules to macromolecules such as nucleic acids or proteins. Such advantage over other drug nanocarriers is made possible thanks to their unique structure of lipids interpenetrated with water nanochannels, providing multiple compartments for the efficient loading of hydrophilic, lipidic, and amphiphilic molecules (Caltagirone et al., 2015; Linkevičiūtė et al., 2015).

A study comparing the release profiles of two nonsteroidal antiinflammatory drugs (NSAID) demonstrated insightful results on the role of liquid crystals compartments and its structure on drug release kinetics (Cohen-Avrahami et al., 2014). The hydrophilic NSAID (sodium diclofenac) was clearly affected by the water content in the carrier with an increased release rate with the increase of water nanochannel size. In terms of structure, the lamellar mesophase released more slowly (~60% in 50 h) than the cubic mesophase structure (~85% in 50 h). On the other hand, the hydrophobic drug (celecoxib) was found unaffected by the mesophase structure and the water content, releasing about 20% in 50 h in both mesophases. These results indicate that the carrier structure is controlling the release of hydrophilic drugs, but not the release of hydrophobic drugs. This result could be of interest to synchronize the simultaneous release rate of hydrophilic and hydrophobic compounds.

Similar effects were observed with macromolecules (BSA and Cytochrome C) (Chemelli et al., 2012). A very high loading of 5 mg of protein per gram of cubosomes was achieved. The most hydrophobic protein, Cytochrome C, seemed to be less hindered by the NP structure than BSA, suggesting that Cytochrome C is associated with the lipid phase, while BSA remained mostly in the water compartment. These results confirm that the multicompartment nanostructure is playing a key role in the release kinetics of molecule having various physicochemical properties.

A similar strategy was also used to deliver siRNA. PEGylated cubosomes showed the highest gene silencing effect due to a higher cell membrane penetration (75%) compared to liposomes (38%) and commercial lipofectamine (53%) (Kim and Leal, 2015).

Lipid crystals are also drawing attentions for theranostic application with their ability to simultaneously host an imaging agent and an active compound. A targeted cubosome containing folate-conjugated PF108 as a targeting ligand was produced for the coencapsulation of a fluorescent probe (rhodamine-PF108) and docetaxel, an anticancer drug (Meli et al., 2015). The drug-loaded cubosomes efficacy was slightly superior to efficacy of free docetaxel (dose of free docetaxel 12.5-fold higher). Moreover HeLa cells having internalized cubosomes were identified by fluorescence.

Multiple studies have been performed on lyotropic lipid nanocarriers for different purposes such as cancer cells targeting with antigen fragment moieties (Zhai et al., 2015a). Interestingly, a study showed the simultaneous integration of different structure (such as Im3m, pn3m, and lamellar) within one nanocarrier (Tran et al., 2015). This new possibility could eventually lead to more versatility and fine-tuning of these nanocarriers.

4.3 PHASE SEPARATED AND GROWTH-INDUCED MULTICOMPARTMENT NPs

Recent studies describing multicompartmentalized solid NPs have shown great advances in their feasibility and their potential use as pharmaceutical nanocarriers. Multiple preparation routes have arisen and now allow for the preparation of a wide variety of structures as patchy, janus, anisotropic, hamburger-like, and dumbbell-like NPs (Fig. 4.5C) (Moughton et al., 2012). These NPs structures could be advantageous for a selective or simultaneous loading of hydrophilic or hydrophobic active compounds into specific compartment. The addition of another component such as an outer layer sensitive to magnetic fields, light, or pH may create a more personalized nanocarrier able to deliver drug on demand or at specific time and location. There are two major ways to produce multicompartment solid NPs. The phase separation technique of immiscible polymer is the method of choice along with other methods based on the growth-induced or sequential synthesis. (Du and O'Reilly, 2011).

The principle of the phase-separated NPs resides in the insolubility of each block-polymers used for the preparation of the NP core. As a result, the polymers separate and form distinct compartments or structures within the particles. As mentioned, multiple structures can be obtained by the use of this technique, one important example being the core–shell structure described in Section 2 of this chapter. A study reported the selective synthesis of janus and hamburger-like multicompartment NPs by the use of two copolymer blocks of polyacrylic acid-*co*-polystyrene (PAA-*co*-PS) and polyacrylic acid-*co*-polybutadiene (PAA-*co*-PB) (Pochan et al., 2011). The key parameters to control the structure of the multicompartment NPs were the speed of water addition in the organic solution of polymers and the block polymer ratios (PS/PB). Polystyrene-*b*-poly(4-vinylpyridine) (PS-*b*-P4VP) can produce patchy NPs by emulsion solvent-evaporation. Those NPs can dissociate into janus nanoparticles if the P4VP is selectively cross-linked and dissolve in THF (Deng et al., 2015). Cautious choices should be made regarding the choice of the polymer architectures (dibloc, triblock, terpolymers, star-shaped, blend of diblock), their hydrophobicity and the ratio of polymers, since all of these parameters can affect the structure and the intrinsic properties of the multicompartment NPs.

Similar processes are at work to create multicompartment micelles using triblock-copolymer. Lodge et al. (2005) proposed to prepare micelles from a triblock polymer, where each block are having a different affinity: an hydrophilic block made of polyethylene oxide (PEO), a hydrophobic block made of polyethylethylene (PEE) and a fluorophilic block made of polyperfluopropylene oxide (PFPO). The PEO block was on the surface of the NP stabilizing the particle while PFPO and PEE were found separated within the core with PEE on the extremities. The ability of the formed micelle to selectively load hydrophobic and fluorophilic molecules was demonstrated. Pyrene (Py), a lipophilic molecule, and 1-naphtyl perfluoroheptanyl ketone (NFH), a fluorinated molecule, were simultaneously loaded and were found in the PEE and PFPO compartments, respectively. Removal of the lipophilic (PEE) compartment produced selective loading of NFH with very low encapsulation for Py. Significant efforts have been made to improve the biocompatibility of nanostructured particles by mimicking biological structures. This effort was carried out with dual lipophilic and fluoriphilic compartments, using biological polymers and modifying them in order to preserve the encapsulation capability of the carriers. For example, poly(3-hydroxyalkanoate)-based multicompartment micelles were found highly biocompatible and exhibited a fluorinated compartment and a lipophilic compartment for dual drug loading (Babinot et al., 2013).

NPs can also be structured during the polymerization process, without going through a step of precipitation or phase separation. Polymerization can be used to create a triblock polymer of different block affinities (Table 4.3), (He et al., 2015b). Sequential polymerization of poly(styrene-*co*-glycidyl methacrylate-*co*-divinylbenzene) (PGSD) on a ellipsoidal hematite (Fe_2O_3) central core followed with a silicate deposition can be used to produce dumbbell-like nanoparticles as represented in Fig. 4.5C (Doermbach and Pich, 2015).

Mesoporous silica NPs (MSN), are also a promising avenue to develop NPs containing multiple compartment. Recent studies have developed complex MSN structures by branching hexagonal mesoporous silica on the sides of the MSN (Suteewong et al., 2013). Using a similar strategy, a hydrophobic compartment made of mesoporous organosilica was deposited on hydrophilic NIR sensitive NPs coated with a double layer of SiO_2, functionalized with azo-isocyanatopropylethoxysilane (azo-ICPES). The targeted application was the simultaneous delivery of DOX (hydrophilic) and paclitaxel (hydrophobic). In combination with NIR treatment, the MNS released simultaneously paclitaxel and doxorubicin, increasing the overall cytoxicity efficacy up to 50% (Li et al., 2014).

4.4 MULTICOMPARTMENT NANORESERVOIRS

An alternate technique to create multicompartment NPs with multiple domains is by the assembly of smaller NP(s) and incorporating them within a larger particle with different properties. Many structures can be designed such as shell-in-shell capsule, nanomaterials within liposomes or even capsules for a broad range of pharmaceutical applications.

4.4.1 Structures with multiple hydrophobic nanodomains

Nanocarriers with multiple hydrophobic domains offer more control over hydrophobic drug loading and release. Boudou et al. (2012) described a hydrophilic capsule of $CaCO_3$ with nanodomains of poly-L-lysine or quaternized chitosan, containing PTX, alternated with decylamino hydrazide derivative of HA (HA20C10). The core containing the hydrophobic drug was sensitive to hyaluronidase found in tumor environments, which is able to trigger the dissolution of the core and the release of the drug. Cytotoxicity studies showed that the multicompartment capsule was highly efficient in killing tumor cells, but slightly less efficient compared to 0.168 μM of free paclitaxel. The reduced efficacy can be explained by the size of the particles (4.2 μm) exceeding the cell uptake capability.

Nanodiamonds have also be used as a drug nanocarrier due to their hydrophobic properties (Mochalin et al., 2012). Even if nanodiamonds were shown to be nontoxic and well tolerated in vivo, nanodiamonds seem not fully stable in suspension. However, they have demonstrated good carrying capabilities and have demonstrated to be an effective tool in cancer therapies (Chow et al., 2011). To prevent possible colloidal aggregation of nonmodified or chemically surface modified nanodiamonds and to compensate the lack of targeting moieties, nanodiamonds were encapsulated into liposomes containing EFGR targeting moieties (Moore et al., 2013). Using nanodiamonds-adsorbed epirubicin in liposomes, improved tumor volume regression was observed. The liposomal shell was found to offer a selective targeting and a possible increased stability.

4.4.2 Structures with Multiple Hydrophilic Nanodomains

Gao et al. (2008a) evaluated reverse micelles from a star-shaped poly(glycerol methacrylate)s (PGOHMA) incorporated into polymeric PDLLA NPs of 300–350 nm for the encapsulation of vasopressin and a model hydrophilic dye. With increasing concentration of reverse micelles, the loaded hydrophilic dyes content were found to preferentially increase within the hydrophilic compartments. The vasopressin load was released in vitro in almost 10 days with a burst release of nearly 60% within the first day. The study reported a slower burst effect when micelles incorporation was reduced, indicating that reverse micelles increased the PDLLA permeability for vasopressin. Reverse micelles of sodium cholate (SC) and soybean phosphatidylcholine (SPC) within stearic acid and palmitic acid solid lipid NPs were designed for insulin encapsulation (Liu et al., 2007). By increasing components forming the reverse micelles as SC, the authors observed an increase in entrapment efficiency, which is in accordance with Gao et al. (2008a) observations. The optimal affinity for insulin was determined at the SC/SPC ratio of 0.5, where an impressive entrapment efficiency of 97.8% and drug loading of 18.9% was achieved, making also an effective tool for large hydrophilic molecule encapsulation.

4.4.3 Shell-in-shell micro/nanocapsules

A shell-in-shell MP and NP is another innovative concept where a smaller capsule is entrapped within a bigger capsule (Fig. 4.5D). The first capsule serves to prevent its content from mixing within the content of the bigger particles (Fig. 4.5D). The particle design allows for bioenzymatic reactions in which two proteins or enzymes sequentially react to a stimulus in order to produce biologically active molecules. This approach is useful to deliver biomolecules or inducing enzymatic reactions at a specific location by delivering a substrate and its enzyme or multiple enzymes (Chandrawati et al., 2011).

Using the same approach, the concept proposed by the Prs. Sukhorukov and Möhwald is to induce a bioreaction using an external trigger. Glucose oxidase enzyme (GOD) was encapsulated in the outer compartment of the NP, while the peroxidase enzyme (POD) was encapsulated in the inner compartment. After addition of glucose (the trigger) and amplex red (enzymatic substrate), GOD produced peroxide in the presence of glucose, which slowly diffused through the inner layer and activated POD. In the presence of amplex red, POD produced resorufin, a fluorescent molecule (Kreft et al., 2007a). Since not every application would benefit from a molecule-triggered activation of the capsule, a remotely triggered system was designed by making use of gold NPs. The gold NPs, embedded in the polyelectrolyte inner shell, were stimulated by NIR laser (500–530 nm), causing the disruption of the inner shell and the mixing of inner and outer content by mechanical force (Kreft et al., 2007b). Reaction occurring in 3-compartment capsules was detected within 60 s, while it took 300–400 s for it to occur in 5-compartment capsules (Bäumler and Georgieva, 2010).

5 CONCLUSIONS AND PERSPECTIVES

In this chapter, we have reviewed the nanostructures in NPs used or intended to be used in drug delivery and imaging or as multitask systems. A lot of new nanostructured materials with increasing levels of complexity have been put forward in the past few years. Intensive progresses were made in preparation, structure control, and structure analysis. However, few systems were actually developed for drug delivery and tested extensively.

There are still several issues to address before these promising carriers can be translated from the bench to the clinical stage. The main challenge ahead with complex NP architectures remains in controlling the preparation process for scaling-up in order to obtain highly reproducible and monodisperse structures with a low level of defects. For example, self-assembly processes are simultaneously controlled by thermodynamic and kinetic processes, which may be highly difficult to control during a scale-up process. Other complex challenges await with the change of morphologies of block polymers micelles with their preparation methods (Mai and Eisenberg, 2012), as well as the evaluation of biocompatibility and toxicity of their components, structures, and the active-loaded NP.

Nanomedicines are confronted to several biological barriers, which are to be overcome before a drug cargo can be efficaciously delivered. Several other challenges lie in the understanding of the interactions between nanostructured carriers and biological systems, long-term stability under physiological conditions, tissue penetration, targeting, and the use of responsive materials. One approach to overcome these limitations is to design these particles with the view to obtain structure closer or similar to natural systems, biomimetic, or bioinspired materials: virus, lipoproteins, cell membranes, and so forth.

ACKNOWLEDGMENTS

XB acknowledges the financial support of CIHR (CRC tier 2). PH thanks FRQNT (Team projects) and NSERC (Discovery program) for their financial support.

REFERENCES

Adamczak, M., Krok, M., Pamuła, E., Posadowska, U., Szczepanowicz, K., Barbasz, J., Warszyński, P., 2013. Linseed oil based nanocapsules as delivery system for hydrophobic quantum dots. Colloids Surf. B 110, 1–7.

Allen, C., Maysinger, D., Eisenberg, A., 1999. Nano-engineering block copolymer aggregates for drug delivery. Colloids Surf. B 16, 3–27.

Almouazen, E., Bourgeois, S., Jordheim, L., Fessi, H., Briançon, S., 2013. Nanoencapsulation of vitamin D3 active metabolites for application in chemotherapy: formulation study and in vitro evaluation. Pharma. Res. 30, 1137–1146.

Angelova, A., Angelov, B., Mutafchieva, R., Lesieur, S., Couvreur, P., 2011. Self-assembled multicompartment liquid crystalline lipid carriers for protein, peptide, and nucleic acid drug delivery. Acc. Chem. Res. 44, 147–156.

Antipov, A.A., Sukhorukov, G.B., Donath, E., Möhwald, H., 2001. Sustained release properties of polyelectrolyte multilayer capsules. J. Phys. Chem. B 105, 2281–2284.

Anton, N., Saulnier, P., Gaillard, C., Porcher, E., Vrignaud, S., Benoit, J.-P., 2009. Aqueous-core lipid nanocapsules for encapsulating fragile hydrophilic and/or lipophilic molecules. Langmuir 25, 11413–11419.

Ariga, K., Ji, Q., Richards, G.J., Hill, J.P., 2012. Soft capsules, hard capsules, and hybrid capsules. Soft Mater. 10, 387–412.

Ashley, C.E., Carnes, E.C., Phillips, G.K., Padilla, D., Durfee, P.N., Brown, P.A., Hanna, T.N., Liu, J., Phillips, B., Carter, M.B., Carroll, N.J., Jiang, X., Dunphy, D.R., Willman, C.L., Petsev, D.N., Evans, D.G., Parikh, A.N., Chackerian, B., Wharton, W., Peabody, D.S., Brinker, C.J., 2011. The targeted delivery of multicomponent cargos to cancer cells by nanoporous particle-supported lipid bilayers. Nat. Mater. 10, 389–397.

Babinot, J., Renard, E., Le Droumaguet, B., Guigner, J.-M., Mura, S., Nicolas, J., Couvreur, P., Langlois, V., 2013. Facile synthesis of multicompartment micelles based on biocompatible poly(3-hydroxyalkanoate). Macromol. Rapid Comm. 34, 362–368.

Baier, G., Cavallaro, A., Vasilev, K., Mailänder, V., Musyanovych, A., Landfester, K., 2013. Enzyme responsive hyaluronic acid nanocapsules containing polyhexanide and their exposure to bacteria to prevent infection. Biomacromolecules 14, 1103–1112.

Barenholz, Y., 2012. Doxil®—the first FDA-approved nondrug: lessons learned. J. Control. Rel. 160, 117–134.

Basha, G., Novobrantseva, T.I., Rosin, N., Tam, Y.Y.C., Hafez, I.M., Wong, M.K., Sugo, T., Ruda, V.M., Qin, J., Klebanov, B., Ciufolini, M., Akinc, A., Tam, Y.K., Hope, M.J., Cullis, P.R., 2011. Influence of cationic lipid composition on gene silencing properties of lipid nanoparticle formulations of siRNA in antigen-presenting cells. Mol. Ther. 19 (12), 2186–2200.

Bäumler, H., Georgieva, R., 2010. Coupled enzyme reactions in multicompartment microparticles. Biomacromolecules 11, 1480–1487.

Boudou, T., Kharkar, P., Jing, J., Guillot, R., Pignot-Paintrand, I., Auzely-Velty, R., Picart, C., 2012. Polyelectrolyte multilayer nanoshells with hydrophobic nanodomains for delivery of Paclitaxel. J. Control. Rel. 159, 403–412.

Boyer, C., Zasadzinski, J.A., 2007. Multiple lipid compartments slow vesicle contents release in lipases and serum. ACS Nano 1, 176–182.

Bravo-Osuna, I., Vauthier, C., Farabollini, A., Palmieri, G.F., Ponchel, G., 2007. Mucoadhesion mechanism of chitosan and thiolated chitosan-poly(isobutyl cyanoacrylate) core-shell nanoparticles. Biomaterials 28, 2233–2243.

Brigger, I., Dubernet, C., Couvreur, P., 2002. Nanoparticles in cancer therapy and diagnosis. Adv. Drug Deliv. Rev. 54, 631–651.

Caltagirone, C., Arca, M., Falchi, A.M., Lippolis, V., Meli, V., Monduzzi, M., Nylander, T., Rosa, A., Schmidt, J., Talmon, Y., Murgia, S., 2015. Solvatochromic fluorescent BODIPY derivative as imaging agent in camptothecin-loaded hexosomes for possible theranostic applications. RSC Adv. 5, 23443–23449.

Chan, J.M., Zhang, L., Yuet, K.P., Liao, G., Rhee, J.-W., Langer, R., Farokhzad, O.C., 2009. PLGA–lecithin–PEG core–shell nanoparticles for controlled drug delivery. Biomaterials 30, 1627–1634.

Chandrawati, R., van Koeverden, M.P., Lomas, H., Caruso, F., 2011. Multicompartment particle assemblies for bioinspired encapsulated reactions. J. Phys. Chem. Lett. 2, 2639–2649.

Chemelli, A., Maurer, M., Geier, R., Glatter, O., 2012. Optimized loading and sustained release of hydrophilic proteins from internally nanostructured particles. Langmuir 28, 16788–16797.

Chen, Y., Chu, C., Zhou, Y., Ru, Y., Chen, H., Chen, F., He, Q., Zhang, Y., Zhang, L., Shi, J., 2011. Reversible pore-structure evolution in hollow silica nanocapsules: large pores for siRNA delivery and nanoparticle collecting. Small 7, 2935–2944.

Chen, H., Qi, B., Moore, T., Wang, F., Colvin, D.C., Sanjeewa, L.D., Gore, J.C., Hwu, S.J., Mefford, O.T., Alexis, F., Anker, J.N., 2014a. Multifunctional yolk-in-shell nanoparticles for pH-triggered drug release and imaging. Small 10, 3364–3370.

Chen, S., Tam, Y.Y.C., Lin, P.J.C., Leung, A.K.K., Tam, Y.K., Cullis, P.R., 2014b. Development of lipid nanoparticle formulations of siRNA for hepatocyte gene silencing following subcutaneous administration. J. Control. Rel. 196, 106–112.

Chitkara, D., Kumar, N., 2013. BSA-PLGA-based core-shell nanoparticles as carrier system for water-soluble drugs. Pharm. Res. 30, 2396–2409.

Chow, E.K., Zhang, X.-Q., Chen, M., Lam, R., Robinson, E., Huang, H., Schaffer, D., Osawa, E., Goga, A., Ho, D., 2011. Nanodiamond therapeutic delivery agents mediate enhanced chemoresistant tumor treatment. Sci. Transl. Med. 3, 73ra21.

Cohen-Avrahami, M., Shames, A.I., Ottaviani, M.F., Aserin, A., Garti, N., 2014. On the correlation between the structure of lyotropic carriers and the delivery profiles of two common NSAIDs. Colloids Surf. B 122, 231–240.

Cosco, D., Paolino, D., De Angelis, F., Cilurzo, F., Celia, C., Di Marzio, L., Russo, D., Tsapis, N., Fattal, E., Fresta, M., 2015. Aqueous-core PEG-coated PLA nanocapsules for an efficient entrapment of water soluble anticancer drugs and a smart therapeutic response. Euro. J. Pharm. Biopharm. 89, 30–39.

Cuomo, F., Lopez, F., Piludu, M., Miguel, M.G., Lindman, B., Ceglie, A., 2015. Release of small hydrophilic molecules from polyelectrolyte capsules: effect of the wall thickness. J. Colloid Interf. Sci. 447, 211–216.

De Koker, S., De Cock, L.J., Rivera-Gil, P., Parak, W.J., Auzély Velty, R., Vervaet, C., Remon, J.P., Grooten, J., De Geest, B.G., 2011. Polymeric multilayer capsules delivering biotherapeutics. Adv. Drug Deliv. Rev. 63, 748–761.

De Miguel, I., Ioualalena, K., Bonnefous, M., Peyrot, M., Nguyen, F., Cervilla, M., Soulet, N., Dirson, R., Rieumajou, V., Imbertie, L., Solers, C., Cazes, S., Favre, G., Samain, D., 1995. Synthesis and characterization of supramolecular biovector (SMBV) specifically designed for the entrapment of ionic molecules. Biochimica et Biophysica Acta (BBA) 1237, 49–58.

Del Gaudio, P., Auriemma, G., Russo, P., Mencherini, T., Campiglia, P., Stigliani, M., Aquino, R.P., 2014. Novel co-axial prilling technique for the development of core–shell particles as delayed drug delivery systems. Euro. J. Pharm. Biopharm. 87, 541–547.

del Mercato, L.L., Ferraro, M.M., Baldassarre, F., Mancarella, S., Greco, V., Rinaldi, R., Leporatti, S., 2014. Biological applications of LbL multilayer capsules: from drug delivery to sensing. Adv. Colloid Interf. Sci. 207, 139–154.

Delcea, M., Möhwald, H., Skirtach, A.G., 2011. Stimuli-responsive LbL capsules and nanoshells for drug delivery. Adv. Drug Deliv. Rev. 63, 730–747.

Deng, R., Liang, F., Qu, X., Wang, Q., Zhu, J., Yang, Z., 2015. Diblock copolymer-based janus nanoparticles. Macromolecules 48, 750–755.

Dergunov, S.A., Pinkhassik, E., 2011. Synergistic coentrapment and triggered release in hollow nanocapsules with uniform nanopores. J. Amer. Chem. Soc. 133, 19656–19659.

Dergunov, S.A., Durbin, J., Pattanaik, S., Pinkhassik, E., 2014. pH-mediated catch and release of charged molecules with porous hollow nanocapsules. J. Amer. Chem. Soc. 136, 2212–2215.

Dinsmore, A.D., Hsu, M.F., Nikolaides, M.G., Marquez, M., Bausch, A.R., Weitz, D.A., 2002. Colloidosomes: selectively permeable capsules composed of colloidal particles. Science 298, 1006–1009.

Doermbach, K., Pich, A., 2015. Facile synthesis of dumbbell-shaped multicompartment nanoparticles. Nanoscale 7, 9169–9173.

Du, P., Liu, P., 2014. Novel smart yolk/shell polymer microspheres as a multiply responsive cargo delivery system. Langmuir 30, 3060–3068.

Du, J., O'Reilly, R.K., 2011. Anisotropic particles with patchy, multicompartment, and janus architectures: preparation and application. Chem. Soc. Rev. 40, 2402–2416.

Elbakry, A., Zaky, A., Liebl, R., Rachel, R., Goepferich, A., Breunig, M., 2009. Layer-by-layer assembled gold nanoparticles for siRNA delivery. Nano Lett. 9, 2059–2064.

Elsabahy, M., Li, A., Zhang, F., Sultan, D., Liu, Y., Wooley, K.L., 2013. Differential immunotoxicities of poly(ethylene glycol)- vs. poly(carboxybetaine)-coated nanoparticles. J. Control. Rel. 172, 641–652.

Esser-Kahn, A.P., Odom, S.A., Sottos, N.R., White, S.R., Moore, J.S., 2011. Triggered release from polymer capsules. Macromolecules 44, 5539–5553.

Fang, R.H., Aryal, S., Hu, C.-M.J., Zhang, L., 2010. Quick synthesis of lipid–polymer hybrid nanoparticles with low polydispersity using a single-step sonication method. Langmuir 26, 16958–16962.

Fang, R.H., Luk, B.T., Hu, C.-M.J., Zhang, L., 2015. Engineered nanoparticles mimicking cell membranes for toxin neutralization. Adv. Drug Deliv. Rev. 90, 69–80.

Feng, W., Nie, W., He, C., Zhou, X., Chen, L., Qiu, K., Wang, W., Yin, Z., 2014. Effect of pH-responsive alginate/chitosan multilayers coating on delivery efficiency, cellular uptake, and biodistribution of mesoporous silica nanoparticles-based nanocarriers. ACS Appl. Mater. Interf. 6, 8447–8460.

Fischer, N.O., Weilhammer, D.R., Dunkle, A., Thomas, C., Hwang, M., Corzett, M., Lychak, C., Mayer, W., Urbin, S., Collette, N., Chiun Chang, J., Loots, G.G., Rasley, A., Blanchette, C.D., 2014. Evaluation of nanolipoprotein particles (NLPs) as an in vivo delivery platform. PLoS One 9, e93342.

Gandra, N., Portz, C., Singamaneni, S., 2014. Multifunctional plasmonic nanorattles for spectrum-guided locoregional therapy. Adv. Mater. 26, 424–429.

Gao, H., Jones, M.-C., Chen, J., Liang, Y., Prud'homme, R.E., Leroux, J.-C., 2008a. Hydrophilic nanoreservoirs embedded into polymeric micro/nanoparticles: an approach to compatibilize polar molecules with hydrophobic matrixes. Chem. Mater. 20, 4191–4193.

Gao, J., Liang, G., Cheung, J.S., Pan, Y., Kuang, Y., Zhao, F., Zhang, B., Zhang, X., Wu, E.X., Xu, B., 2008b. Multifunctional yolk− shell nanoparticles: a potential MRI contrast and anticancer agent. J. Amer. Chem. Soc. 130, 11828–11833.

Go, D.P., Gras, S.L., Mitra, D., Nguyen, T.H., Stevens, G.W., Cooper-White, J.J., O'Connor, A.J., 2011. Multilayered microspheres for the controlled release of growth factors in tissue engineering. Biomacromolecules 12, 1494–1503.

Gonzalo, T., Lollo, G., Garcia-Fuentes, M., Torres, D., Correa, J., Riguera, R., Fernandez-Megia, E., Calvo, P., Avilés, P., Guillén, M.J., 2013. A new potential nanoncological therapy based on polyamino acid nanocapsules. J. Control. Rel. 169, 10–16.

Gref, R., Domb, A., Quellec, P., Blunk, T., Müller, R.H., Verbavatz, J.M., Langer, R., 1995. The controlled intravenous delivery of drugs using PEG-coated sterically stabilized nanospheres. Adv. Drug Deliv. Rev. 16, 215–233.

Gu, X., Wang, J., Wang, Y., Wang, Y., Gao, H., Wu, G., 2013. Layer-by-layer assembled polyaspartamide nanocapsules for pH-responsive protein delivery. Colloids Surf. B 108, 205–211.

Guo, H., Guo, Q., Chu, T., Zhang, X., Wu, Z., Yu, D., 2014. Glucose-sensitive polyelectrolyte nanocapsules based on layer-by-layer technique for protein drug delivery. J. Mater. Sci, 25, 121–129.

Gustafson, T.P., Dergunov, S.A., Akers, W.J., Cao, Q., Magalotti, S., Achilefu, S., Pinkhassik, E., Berezin, M.Y., 2013. Blood-triggered rapid-release porous nanocapsules. RSC Adv. 3, 5547–5555.

Hatakeyama, H., Akita, H., Harashima, H., 2011. A multifunctional envelope type nano device (MEND) for gene delivery to tumors based on the EPR effect: a strategy for overcoming the PEG dilemma. Adv. Drug Deliv. Rev. 63, 152–160.

He, D., Wang, S., Lei, L., Hou, Z., Shang, P., He, X., Nie, H., 2015a. Core–shell particles for controllable release of drug. Chem. Eng. Sci. 125, 108–120.

He, X., Qu, Y., Gao, C., Zhang, W., 2015b. Synthesis of multicompartment nanoparticles of a triblock terpolymer by seeded RAFT polymerization. Polym. Chem. 6, 6386–6393.

Hinton, T.M., Grusche, F., Acharya, D., Shukla, R., Bansal, V., Waddington, L.J., Monaghan, P., Muir, B.W., 2014. Bicontinuous cubic phase nanoparticle lipid chemistry affects toxicity in cultured cells. Toxicol. Res. 3, 11–22.

Hoare, T.R., Kohane, D.S., 2008. Hydrogels in drug delivery: progress and challenges. Polymer 49, 1993–2007.

Hong, C.-Y., Li, X., Pan, C.-Y., 2009. Fabrication of smart nanocontainers with a mesoporous core and a pH-responsive shell for controlled uptake and release. J. Mater. Chem. 19, 5155–5160.

Hrkach, J.S., Peracchia, M.T., Bomb, A., Lotan, n., Langer, R., 1997. Nanotechnology for biomaterials engineering: structural characterization of amphiphilic polymeric nanoparticles by 1H NMR spectroscopy. Biomaterials 18, 27–30.

Hu, C.-M.J., Zhang, L., Aryal, S., Cheung, C., Fang, R.H., Zhang, L., 2011a. Erythrocyte membrane-camouflaged polymeric nanoparticles as a biomimetic delivery platform. Proc. Natl. Acad. Sci. 108, 10980–10985.

Hu, S.-H., Chen, Y.-Y., Liu, T.-C., Tung, T.-H., Liu, D.-M., Chen, S.-Y., 2011b. Remotely nonrupturable yolk/shell capsules for magnetically triggered drug release. Chem. Comm. 47, 1776–1778.

Hu, C.-M.J., Fang, R.H., Wang, K.-C., Luk, B.T., Thamphiwatana, S., Dehaini, D., Nguyen, P., Angsantikul, P., Wen, C.H., Kroll, A.V., Carpenter, C., Ramesh, M., Qu, V., Patel, S.H., Zhu, J., Shi, W., Hofman, F.M., Chen, T.C., Gao, W., Zhang, K., Chien, S., Zhang, L., 2015a. Nanoparticle biointerfacing by platelet membrane cloaking. Nature 526, 118–121.

Hu, F., Zhang, Y., Chen, G., Li, C., Wang, Q., 2015b. Double-walled Au nanocage/SiO2 nanorattles: integrating SERS imaging, drug delivery, and photothermal therapy. Small 11, 985–993.

Jiao, Y., Sun, Y., Tang, X., Ren, Q., Yang, W., 2015. Tumor-targeting multifunctional rattle-type theranostic nanoparticles for MRI/NIRF bimodal imaging and delivery of hydrophobic drugs. Small 11, 1962–1974.

Jovanovic, A.V., Underhill, R.S., Bucholz, T.L., Duran, R.S., 2005. Oil core and silica shell nanocapsules: toward controlling the size and the ability to sequester hydrophobic compounds. Chem. Mater. 17, 3375–3383.

Ju, C., Mo, R., Xue, J., Zhang, L., Zhao, Z., Xue, L., Ping, Q., Zhang, C., 2014. Sequential intra-intercellular nanoparticle delivery system for deep tumor penetration. Angewandte Chemie Int. Ed. 53, 6253–6258.

Kang, X., Cheng, Z., Yang, D., Ma, P.a., Shang, M., Peng, C., Dai, Y., Lin, J., 2012. Design and synthesis of multifunctional drug carriers based on luminescent rattle-type mesoporous silica microspheres with a thermosensitive hydrogel as a controlled switch. Adv. Funct. Mater. 22, 1470–1481.

Kim, H., Leal, C., 2015. Cuboplexes: topologically active siRNA delivery. ACS Nano 9, 10214–10226.

Kim, J.W., Fernandez-Nieves, A., Dan, N., Utada, A.S., Marquez, M., Weitz, D.A., 2007. Colloidal assembly route for responsive colloidosomes with tunable permeability. Nano Lett. 7, 2876–2880.

Kim, H.-J., Zhang, K., Moore, L., Ho, D., 2014. Diamond nanogel-embedded contact lenses mediate lysozyme-dependent therapeutic release. ACS Nano 8, 2998–3005.

Kreft, O., Prevot, M., Möhwald, H., Sukhorukov, G.B., 2007a. Shell-in-shell microcapsules: a novel tool for integrated, spatially confined enzymatic reactions. Angewandte Chemie Int. Ed. 46, 5605–5608.

Kreft, O., Skirtach, A.G., Sukhorukov, G.B., Möhwald, H., 2007b. Remote control of bioreactions in multicompartment capsules. Adv. Mater. 19, 3142–3145.

Lambert, G., Fattal, E., Pinto-Alphandary, H., Gulik, A., Couvreur, P., 2000. Polyisobutylcyanoacrylate nanocapsules containing an aqueous core as a novel colloidal carrier for the delivery of oligonucleotides. Pharm. Res. 17, 707–714.

Lee, E.S., Kim, D., Youn, Y.S., Oh, K.T., Bae, Y.H., 2008. A virus-mimetic nanogel vehicle. Angewandte Chemie Int. Ed. 47, 2418–2421.

Leung, A.K.K., Hafez, I.M., Baoukina, S., Belliveau, N.M., Zhigaltsev, I.V., Afshinmanesh, E., Tieleman, D.P., Hansen, C.L., Hope, M.J., Cullis, P.R., 2012. Lipid nanoparticles containing siRNA synthesized by microfluidic mixing exhibit an electron-dense nanostructured core. J. Phys. Chem. CJ. Phys. Chem. C 116, 18440–18450.

Leung, A.K.K., Tam, Y.Y.C., Chen, S., Hafez, I.M., Cullis, P.R., 2015. Microfluidic mixing: a general method for encapsulating macromolecules in lipid nanoparticle systems. J. Phys. Chem. B 119, 8698–8706.

Li, X., Zhou, L., Wei, Y., El-Toni, A.M., Zhang, F., Zhao, D., 2014. Anisotropic growth-induced synthesis of dual-compartment janus mesoporous silica nanoparticles for bimodal triggered drugs delivery. J. Amer. Chem. Soc. 136, 15086–15092.

Li, S., Moosa, B.A., Croissant, J.G., Khashab, N.M., 2015a. Electrostatic assembly/disassembly of nanoscaled colloidosomes for light-triggered cargo release. Angewandte Chemie Int. Ed. 54, 6804–6808.

Li, Z., Yang, T., Lin, C., Li, Q., Liu, S., Xu, F., Wang, H., Cui, X., 2015b. Sonochemical synthesis of hydrophilic drug-loaded multifunctional bovine serum albumin nanocapsules. ACS Appl. Mater. Interf. 7, 19390–19397.

Lim, M.P.A., Lee, W.L., Widjaja, E., Loo, S.C.J., 2013. One-step fabrication of core-shell structured alginate-PLGA/PLLA microparticles as a novel drug delivery system for water soluble drugs. Biomater. Sci. 1, 486–493.

Linkevičiūtė, A., Misiūnas, A., Naujalis, E., Barauskas, J., 2015. Preparation and characterization of quercetin-loaded lipid liquid crystalline systems. Colloids Surf. B 128, 296–303.

Lira, M.C.B., Santos-Magalhães, N.S., Nicolas, V., Marsaud, V., Silva, M.P.C., Ponchel, G., Vauthier, C., 2011. Cytotoxicity and cellular uptake of newly synthesized fucoidan-coated nanoparticles. Euro. J. Pharm. Biopharm. 79, 162–170.

Liu, J., Gong, T., Wang, C., Zhong, Z., Zhang, Z., 2007. Solid lipid nanoparticles loaded with insulin by sodium cholate-phosphatidylcholine-based mixed micelles: preparation and characterization. Int. J. Pharm. 340, 153–162.

Liu, J., Liu, F., Gao, K., Wu, J., Xue, D., 2009. Recent developments in the chemical synthesis of inorganic porous capsules. J. Mater. Chem. 19, 6073–6084.

Liu, T., Liu, H., Fu, C., Li, L., Chen, D., Zhang, Y., Tang, F., 2013. Silica nanorattle with enhanced protein loading: a potential vaccine adjuvant. J. Colloid Interf. Sci. 400, 168–174.

Liu, Y., Liu, B., Nie, Z., 2015. Concurrent self-assembly of amphiphiles into nanoarchitectures with increasing complexity. Nano Today 10, 278–300.

Lodge, T.P., Rasdal, A., Li, Z., Hillmyer, M.A., 2005. Simultaneous, segregated storage of two agents in a multicompartment micelle. J. Amer. Chem. Soc. 127, 17608–17609.

Mai, Y., Eisenberg, A., 2012. Self-assembly of block copolymers. Chem. Soc. Rev. 41, 5969–5985.

Marsat, J.-N., Heydenreich, M., Kleinpeter, E., Berlepsch, H.v., Böttcher, C., Laschewsky, A., 2011. Self-assembly into multicompartment micelles and selective solubilization by hydrophilic–lipophilic – fluorophilic block copolymers. Macromolecules 44, 2092–2105.

Mat Azmi, I.D., Wu, L., Wibroe, P.P., Nilsson, C., Østergaard, J., Stürup, S., Gammelgaard, B., Urtti, A., Moghimi, S.M., Yaghmur, A., 2015. Modulatory effect of human plasma on the internal nanostructure and size characteristics of liquid-crystalline nanocarriers. Langmuir 31, 5042–5049.

Meli, V., Caltagirone, C., Falchi, A.M., Hyde, S.T., Lippolis, V., Monduzzi, M., Obiols-Rabasa, M., Rosa, A., Schmidt, J., Talmon, Y., Murgia, S., 2015. Docetaxel-loaded fluorescent liquid-crystalline nanoparticles for cancer heranostics. Langmuir 31, 9566–9575.

Meng, H., Wang, M., Liu, H., Liu, X., Situ, A., Wu, B., Ji, Z., Chang, C.H., Nel, A.E., 2015. Use of a lipid-coated mesoporous silica nanoparticle platform for synergistic gemcitabine and paclitaxel delivery to human pancreatic cancer in mice. ACS Nano 9, 3540–3557.

Mishra, S., Webster, P., Davis, M.E., 2004. PEGylation significantly affects cellular uptake and intracellular trafficking of non-viral gene delivery particles. Euro. J. Cell Biol. 83, 97–111.

Mishra, V., Mahor, S., Rawat, A., Dubey, P., Gupta, P.N., Singh, P., Vyas, S.P., 2006. Development of novel fusogenic vesosomes for transcutaneous immunization. Vaccine 24, 5559–5570.

Mochalin, V.N., Shenderova, O., Ho, D., Gogotsi, Y., 2012. The properties and applications of nanodiamonds. Nat. Nano. 7, 11–23.

Moghimi, S.M., Szebeni, J., 2003. Stealth liposomes and long circulating nanoparticles: critical issues in pharmacokinetics, opsonization, and protein-binding properties. Prog. Lipid Res. 42, 463–478.

Moore, L., Chow, E.K.-H., Osawa, E., Bishop, J.M., Ho, D., 2013. Diamond-lipid hybrids enhance chemotherapeutic tolerance and mediate tumor regression. Adv. Mater. 25, 3532–3541.

Mora-Huertas, C.E., Fessi, H., Elaissari, A., 2010. Polymer-based nanocapsules for drug delivery. Int. J. Pharm. 385, 113–142.

Moughton, A.O., Hillmyer, M.A., Lodge, T.P., 2012. Multicompartment block polymer micelles. Macromolecules 45, 2–19.

Nel, A.E., Madler, L., Velegol, D., Xia, T., Hoek, E.M., Somasundaran, P., Klaessig, F., Castranova, V., Thompson, M., 2009. Understanding biophysicochemical interactions at the nonbio interface. Nat. Mater. 8, 543–557.

Owens, 3rd, D.E., Peppas, N.A., 2006. Opsonization, biodistribution, and pharmacokinetics of polymeric nanoparticles. Int. J. Pharm. 307, 93–102.

Paleos, C.M., Tsiourvas, D., Sideratou, Z., Pantos, A., 2013. Formation of artificial multicompartment vesosome and dendrosome as prospected drug and gene delivery carriers. J. Control. Rel. 170, 141–152.

Parhi, P., Mohanty, C., Sahoo, S.K., 2012. Nanotechnology-based combinational drug delivery: an emerging approach for cancer therapy. Drug Discov. Today 17, 1044–1052.

Peer, D., Karp, J.M., Hong, S., Farokhzad, O.C., Margalit, R., Langer, R., 2007. Nanocarriers as an emerging platform for cancer therapy. Nat. Nano. 2, 751–760.

Peyrot, M., Sautereau, A.M., Rabanel, J.M., Nguyen, F., Tocanne, J.F., Samain, D., 1994. Supramolecular biovectors (Smbv): a new family of nanoparticulate drug-delivery systems—synthesis and structural characterization. Int. J. Pharm. 102, 25–33.

Pochan, D.J., Zhu, J., Zhang, K., Wooley, K.L., Miesch, C., Emrick, T., 2011. Multicompartment and multigeometry nanoparticle assembly. Soft Matter 7, 2500–2506.

Priebe, M., Fromm, K.M., 2015. Nanorattles or yolk-shell nanoparticles--what are they, how are they made, and what are they good for? Chemistry 21, 3854–3874.

Qian, K., Wu, J., Zhang, E., Zhang, Y., Fu, A., 2015. Biodegradable double nanocapsule as a novel multifunctional carrier for drug delivery and cell imaging. Int. J. Nanomed. 10, 4149–4157.

Rabanel, J.-M., Hildgen, P., Banquy, X., 2014. Assessment of PEG on polymeric particles surface, a key step in drug carrier translation. J. Control. Rel. 185, 71–87.

Ran, Z., Sun, Y., Chang, B., Ren, Q., Yang, W., 2013. Silica composite nanoparticles containing fluorescent solid core and mesoporous shell with different thickness as drug carrier. J. Colloid Interf. Sci. 410, 94–101.
Ren, D., Kratz, F., Wang, S.-W., 2011. Protein nanocapsules containing doxorubicin as a pH-responsive delivery system. Small 7, 1051–1060.
Riley, T., Heald, C.R., Stolnik, S., Garnett, M.C., Illum, L., Davis, S.S., King, S.M., Heenan, R.K., Purkiss, S.C., Barlow, R.J., Gellert, P.R., Washington, C., 2003. Core–shell structure of PLA–PEG nanoparticles used for drug delivery. Langmuir 19, 8428–8435.
Rim, H.P., Min, K.H., Lee, H.J., Jeong, S.Y., Lee, S.C., 2011. pH-tunable calcium phosphate-covered mesoporous silica nanocontainers for intracellular controlled release of guest drugs. Angewandte Chemie Int. Ed. 50, 8853–8857.
Rodell, C.B., Mealy, J.E., Burdick, J.A., 2015. Supramolecular guest–host interactions for the preparation of biomedical materials. Bioconjug. Chem. 26 (12), 2279–2289.
Rolland, J.P., Maynor, B.W., Euliss, L.E., Exner, A.E., Denison, G.M., DeSimone, J.M., 2005. Direct fabrication and harvesting of monodisperse, shape-specific nanobiomaterials. J. Amer. Chem. Soc. 127, 10096–10100.
Rosenberg, R.T., Dan, N., 2013. Designing nanoparticle colloidal shells for selective transport. Soft Mater. 11, 143–148.
Rossier-Miranda, F.J., Schroën, C.G.P.H., Boom, R.M., 2009. Colloidosomes: versatile microcapsules in perspective. Colloids Surf. A 343, 43–49.
Sander, J.S., Studart, A.R., 2013. Nanoparticle-filled complex colloidosomes for tunable cargo release. Langmuir 29, 15168–15173.
Schneider, G., Decher, G., 2004. From functional core/shell nanoparticles prepared via layer-by-layer deposition to empty nanospheres. Nano Lett. 4, 1833–1839.
Sengupta, S., Eavarone, D., Capila, I., Zhao, G., Watson, N., Kiziltepe, T., Sasisekharan, R., 2005. Temporal targeting of tumor cells and neovasculature with a nanoscale delivery system. Nature 436, 568–572.
Shmakov, S.N., Dergunov, S.A., Pinkhassik, E., 2011. Ship-in-a-bottle entrapment of molecules in porous nanocapsules. Chem. Comm. 47, 8223–8225.
Shutava, T.G., Balkundi, S.S., Vangala, P., Steffan, J.J., Bigelow, R.L., Cardelli, J.A., O'Neal, D.P., Lvov, Y.M., 2009. Layer-by-layer-coated gelatin nanoparticles as a vehicle for delivery of natural polyphenols. ACS Nano 3, 1877–1885.
Singh, N., Karambelkar, A., Gu, L., Lin, K., Miller, J.S., Chen, C.S., Sailor, M.J., Bhatia, S.N., 2011. Bioresponsive mesoporous silica nanoparticles for triggered drug release. J. Amer. Chem. Soc. 133, 19582–19585.
Soppimath, K.S., Tan, D.C.W., Yang, Y.Y., 2005. pH-triggered thermally responsive polymer core–shell nanoparticles for drug delivery. Adv. Mater. 17, 318–323.
Sosnik, A., das Neves, J., Sarmento, B., 2014. Mucoadhesive polymers in the design of nondrug delivery systems for administration by nonparenteral routes: a review. Prog. Polym. Sci. 39, 2030–2075.
Suteewong, T., Sai, H., Hovden, R., Muller, D., Bradbury, M.S., Gruner, S.M., Wiesner, U., 2013. Multicompartment mesoporous silica nanoparticles with branched shapes: an epitaxial growth mechanism. Science 340, 337–341.
Tang, H., Guo, J., Sun, Y., Chang, B., Ren, Q., Yang, W., 2011. Facile synthesis of pH sensitive polymer-coated mesoporous silica nanoparticles and their application in drug delivery. Int. J. Pharm. 421, 388–396.
Tran, N., Mulet, X., Hawley, A.M., Conn, C.E., Zhai, J., Waddington, L.J., Drummond, C.J., 2015. First direct observation of stable internally ordered janus nanoparticles created by lipid self-assembly. Nano Lett. 15, 4229–4233.
Troutier, A.-L., Ladavière, C., 2007. An overview of lipid membrane supported by colloidal particles. Adv. Colloid Interf. Sci. 133, 1–21.
Velev, O., Furusawa, K., Nagayama, K., 1996. Assembly of latex particles by using emulsion droplets as templates. 1. Microstructured hollow spheres. Langmuir 12, 2374–2384.

Vonarbourg, A., Passirani, C., Saulnier, P., Benoit, J.P., 2006. Parameters influencing the stealthiness of colloidal drug delivery systems. Biomaterials 27, 4356–4373.

Vrignaud, S., Benoit, J.-P., Saulnier, P., 2011. Strategies for the nanoencapsulation of hydrophilic molecules in polymer-based nanoparticles. Biomaterials 32, 8593–8604.

Wang, H., Zhao, P., Su, W., Wang, S., Liao, Z., Niu, R., Chang, J., 2010. PLGA/polymeric liposome for targeted drug and gene codelivery. Biomaterials 31, 8741–8748.

Wang, H., Sun, Y., Yi, J., Fu, J., Di, J., del Carmen Alonso, A., Zhou, S., 2015. Fluorescent porous carbon nanocapsules for two-photon imaging, NIR/pH dual-responsive drug carrier, and photothermal therapy. Biomaterials 53, 117–126.

Wu, W., Mitra, N., Yan, E.C.Y., Zhou, S., 2010. Multifunctional hybrid nanogel for integration of optical glucose sensing and self-regulated insulin release at physiological pH. ACS Nano 4, 4831–4839.

Wu, J., Kong, T., Yeung, K.W.K., Shum, H.C., Cheung, K.M.C., Wang, L., To, M.K.T., 2013. Fabrication and characterization of monodisperse PLGA–alginate core–shell microspheres with monodisperse size and homogeneous shells for controlled drug release. Acta Biomaterialia 9, 7410–7419.

Wu, Y., Long, Y., Li, Q.-L., Han, S., Ma, J., Yang, Y.-W., Gao, H., 2015. Layer-by-layer (LBL) self-assembled biohybrid nanomaterials for efficient antibacterial applications. ACS Appl. Mater. Interf. 7, 17255–17263.

Yang, J., Lee, J., Kang, J., Lee, K., Suh, J.-S., Yoon, H.-G., Huh, Y.-M., Haam, S., 2008. Hollow silica nanocontainers as drug delivery vehicles. Langmuir 24, 3417–3421.

Yow, H.N., Routh, A.F., 2009. Release profiles of encapsulated actives from colloidosomes sintered for various durations. Langmuir 25, 159–166.

Yuan, C., Raghupathi, K., Popere, B.C., Ventura, J., Dai, L., Thayumanavan, S., 2014. Composite supramolecular nanoassemblies with independent stimulus sensitivities. Chem. Sci. 5, 229–234.

Zabara, A., Mezzenga, R., 2014. Controlling molecular transport and sustained drug release in lipid-based liquid crystalline mesophases. J. Control. Rel. 188, 31–43.

Zhai, J., Hinton, T.M., Waddington, L.J., Fong, C., Tran, N., Mulet, X., Drummond, C.J., Muir, B.W., 2015a. Lipid–PEG conjugates sterically stabilize and reduce the toxicity of phytantriol-based lyotropic liquid crystalline nanoparticles. Langmuir 31, 10871–10880.

Zhai, J., Scoble, J.A., Li, N., Lovrecz, G., Waddington, L.J., Tran, N., Muir, B.W., Coia, G., Kirby, N., Drummond, C.J., Mulet, X., 2015b. Epidermal growth factor receptor-targeted lipid nanoparticles retain self-assembled nanostructures and provide high specificity. Nanoscale 7, 2905–2913.

Zhang, L., Eisenberg, A., 1996. Multiple morphologies and characteristics of "crew-cut" micelle-like aggregates of polystyrene-*b*-poly(acrylic acid) diblock copolymers in aqueous solutions. J. Amer. Chem. Soc. 118, 3168–3181.

Zhang, L., Chan, J.M., Gu, F.X., Rhee, J.-W., Wang, A.Z., Radovic-Moreno, A.F., Alexis, F., Langer, R., Farokhzad, O.C., 2008. Self-assembled lipid–polymer hybrid nanoparticles: a robust drug delivery platform. ACS Nano 2, 1696–1702.

Zhang, L., Wang, T., Li, L., Wang, C., Su, Z., Li, J., 2012. Multifunctional fluorescent-magnetic polyethyleneimine functionalized Fe3O4-mesoporous silica yolk-shell nanocapsules for siRNA delivery. Chem. Comm. 48, 8706–8708.

Zhang, Y., Hsu, B.Y.W., Ren, C., Li, X., Wang, J., 2015. Silica-based nanocapsules: synthesis, structure control, and biomedical applications. Chem. Soc. Rev. 44, 315–335.

Zhao, L., Peng, J., Huang, Q., Li, C., Chen, M., Sun, Y., Lin, Q., Zhu, L., Li, F., 2014. Near-infrared photoregulated drug release in living tumor tissue via yolk-shell upconversion nanocages. Adv. Funct. Mater. 24, 363–371.

Zhao, Y., Luo, Z., Li, M., Qu, Q., Ma, X., Yu, S.-H., Zhao, Y., 2015a. A preloaded amorphous calcium carbonate/doxorubicin@silica nanoreactor for pH-responsive delivery of an anticancer drug. Angewandte Chemie 127, 933–936.

Zhao, Y., Lv, L.-P., Jiang, S., Landfester, K., Crespy, D., 2015b. Advanced stimuli-responsive polymer nanocapsules with enhanced capabilities for payloads delivery. Polym. Chem. 6, 4197–4205.

Zhen, Z., Tang, W., Guo, C., Chen, H., Lin, X., Liu, G., Fei, B., Chen, X., Xu, B., Xie, J., 2013. Ferritin nanocages to encapsulate and deliver photosensitizers for efficient photodynamic therapy against cancer. ACS Nano 7, 6988–6996.

Zhu, J., Tang, A., Law, L.P., Feng, M., Ho, K.M., Lee, D.K.L., Harris, F.W., Li, P., 2005. Amphiphilic core–shell nanoparticles with poly(ethylenimine) shells as potential gene delivery carriers. Bioconjug. Chem. 16, 139–146.

CHAPTER

5 DESIGN OF FUNCTIONALIZED MATERIALS FOR USE IN MICRONANOSCALE DRUG DELIVERY DEVICES AND SMART PATCHES

Ashleigh Anderson, James Davis

University of Ulster, Jordanstown, Northern Ireland

CHAPTER OUTLINE

1 INTRODUCTION

Smart devices that can routinely monitor a patient's health with little or no intervention from clinical staff have slowly begun to emerge over recent years and have arisen from significant advances in sensing and communication technologies. The integration of vital signs monitoring systems within phone/watch systems, in addition to a range of discrete wearable medical appliances, are but the first of a new generation of connected health systems that will have considerable impact on future preventative screening measures within healthcare practice. An extensive effort has been expended in research targeted at the chemical and electronic integration of sensors for the detection of a vast array of disease or injury biomarkers and, in principle, it can be expected that a proportion will be capable of successful translation to the emerging remote/point of care formats. At present, the core rationale is a "detect-alert" sequence informing either the patient or a central clinic (Chan et al., 2012). In cases where a rapid response may be required, a much more effective alternative would be the development of a truly smart system that could autonomously act on the information received from the sensing component and, where appropriate, deliver a therapeutic agent in a controlled manner. While the sensing strategies may be entering a mature phase of development, the designs required for the development of a "detect–act" system enabling precise control of a therapeutic agent are still in their relative infancy and represent a considerable scientific challenge.

In recent years, there has been tremendous progress in drug delivery and therapeutic treatment strategies and it could be postulated that these positive results have been achieved as a result of the development of new materials with specific functionalities that can facilitate the release process. The emergence of so-called smart drug delivery systems relates to their ability to selectively release drugs in a regulated manner. This may involve keeping the drug concentration at desirable levels in the body and avoiding the need for frequent dosing or it may be to provide a rapid response upon the detection of a catastrophic event, such as the release of an antidote following the onset of anaphylaxis. Irrespective of the end application, there is a desire for autonomous devices that can monitor and, where appropriate, act to alter the treatment profile for patients, but acquiring such devices requires solutions to significant technological challenges. The UK Engineering and Physical Sciences Research Council (EPSRC) recently prioritized a number of "Healthcare Grand Challenges," of which the following were identified as key targets:

- Design of robust bioelectronics sensors and devices for long-term sensing and control within biomedical applications.
- Production of novel, low-cost diagnostic devices, for timely and accurate diagnosis that will ultimately inform appropriate and timely interventions, thereby reducing costs-improving outcomes.
- Optimizing treatment plans and care through more effective diagnosis, encompassing more robust evidence-based and timely interventions. This would be supported by the acquisition of real-time telemetry enabling patient self-management of health and well-being.
- Procedures for detecting abnormal patterns in the physiological time course data, which could provide an early warning of the onset of complications that may be deleterious to patient health.
- Clear decision pathways and dashboards to enable rapid assessment by healthcare professionals. The removal of ambiguities in sensor data will support safe and effective community management of patients suffering with chronic conditions or following early discharge.
- Innovative approaches to data collection that can be adapted to the needs of individual patients and allow more efficient interactions with clinicians. These should also embrace direct autonomous reporting capabilities to provide greater care through minimizing intrusion to the patient's normal life.

There is a pronounced emphasis on the design of new sensing platforms for the identification and monitoring of disease biomarkers, but the critical question is: what to do with the information when you have it? Buried within the proposed future healthcare requirements is the key challenge, which effectively embodies the sum of the others:

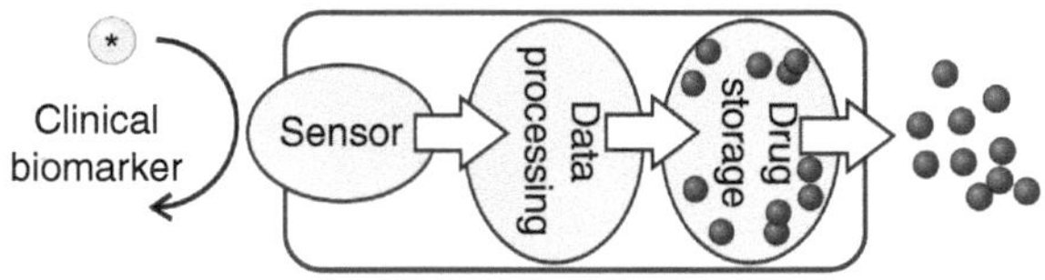

FIGURE 5.1 Combined Sensor–Actuator Device

- The design of new sensing technologies and platforms that can enable the acquisition of health telemetry data that can facilitate autonomous device intervention. A typical example may be the controlled release of a drug that could provide greater control over a particular disease (i.e., through increased compliance).

In order to facilitate the development of such devices, there has to be a holistic approach to their design encompassing improvements in the technological properties, such as sensor electronics, signal processing, communications, and biocompatibility, as well as investigations of clinical efficacy. The design of functional materials that can respond, on demand, to the dictates of the sensor-processing units within a miniaturized device has effectively opened up a wealth of new research directions. A schematic of the ideal smart release mechanism is detailed in Fig. 5.1, where drug release is achieved upon the detection of an appropriate biomarker.

All too often, there can be significant delays between the onset of a disease and the application of the therapeutic. One particular scenario is the treatment of extreme allergic reactions, in which the prompt delivery of epinephrine can often be the difference between a swift recovery or admission to a hospital. Turner et al. (2015) recently discovered that there has been a 615% increase in hospital admissions caused by anaphylaxis in the UK over the period 1992–2012. Could the availability of a smart transdermal patch capable of detecting the onset of anaphylaxis and responding accordingly with a controlled release of epinephrine improve patient outcome? There are obvious cost-benefits to this scenario, as indeed there are to other conditions, where delivering appropriate treatments will minimize central healthcare costs associated with the hospitalization of patients.

Such devices are equally important to the effective management of chronic diseases, such as diabetes, heart disease, and colitis, where there is often a pressing need for well-defined concentration profiles. In these cases, the ability of the smart system to continuously adapt to the prevailing blood level would allow the swift delivery of a more appropriate therapeutic dose and thereby maintain better effective control. This is particularly important in the management of diabetes, where rigorous glycemic control is an important factor in minimizing diabetic complications. Examples of systems that fall within such a design description presently are insulin pumps (Renard et al., 2013). Although these are widely available, they remain an expensive option for diabetic patients and their operation is far from ideal and it is recognized that there remains considerable opportunities for improvements to be made to the size and wearability of such devices. Micro devices that can routinely monitor a particular aspect of a patient's health with minimal intervention from clinical staff are slowly becoming more prevalent in a healthcare setting (Bandodkar and Wang, 2014; Chan et al., 2012; Clifton et al., 2013; Florea and Diamond, 2015; Weinstein et al., 2014). The challenge will be to complement such devices with an actuation facility to complete the "connected health" vision of future healthcare.

A variety of micro devices that have been designed for use within a number of different drug delivery applications is indicated in Fig. 5.2.

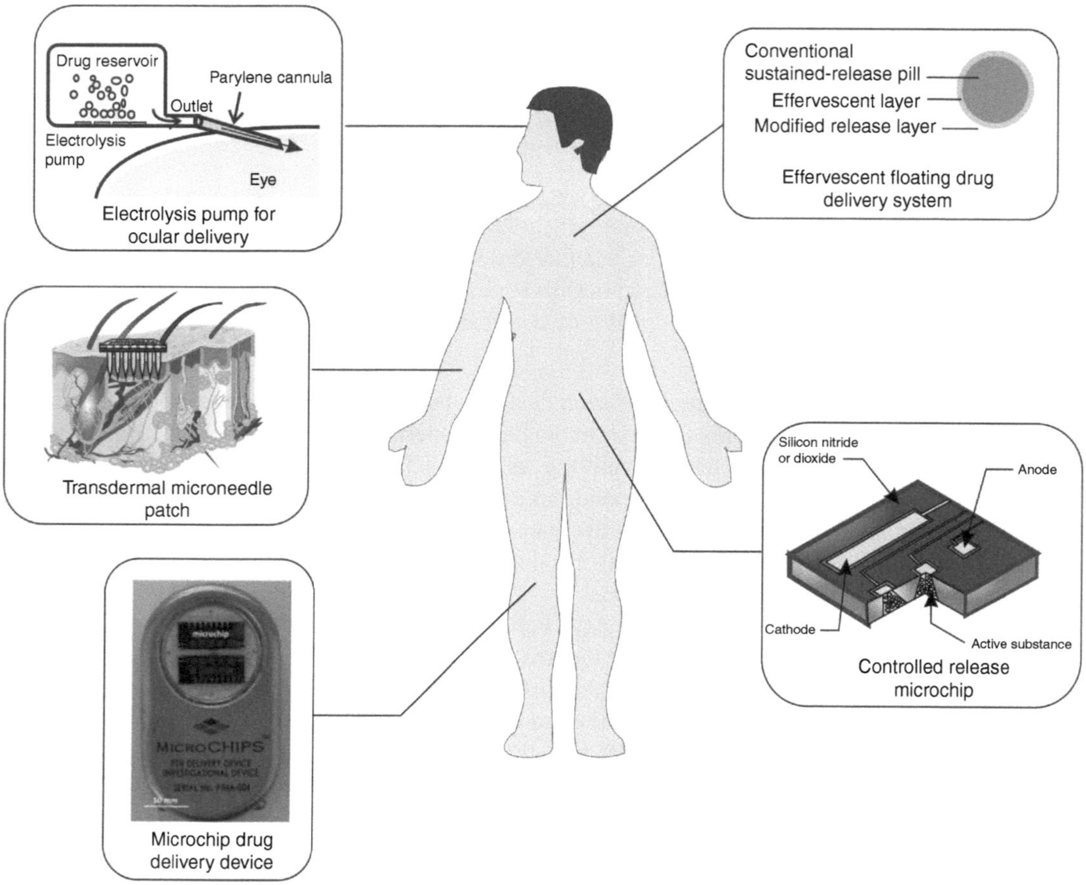

FIGURE 5.2 Application of Micro Devices at Various Sites in the Body

The development of novel approaches to smart drug delivery has provided significant improvements in treatment therapies for numerous biomedical applications. This review, therefore, seeks to provide a summary and a critical analysis of the current status and future developments of functional materials that can be used to control the release of drug therapies via the imposition of an electronic trigger and that could eventually be suitable for incorporation with a miniaturized "smart" device.

2 TRADITIONAL APPROACHES TO DRUG DELIVERY

At present, drug delivery is most commonly achieved through a number of clinical means, which include: oral delivery (Choonara et al., 2014), inhaler (Casettari and Illum, 2014; Illum, 2012; Kozlovskaya et al., 2014), injection (Prausnitz and Langer, 2008), or transdermal absorption (Escobar-Chávez et al., 2011), and have allowed the treatment of a truly extensive range of conditions. Each of these

Table 5.1 Conventional Drug Delivery Approaches and Associated Barriers

Route of Delivery	Advancements to Delivery	Barriers to Delivery	References
Oral	Simplicity of use, ease of treatment termination, dose accuracy.	GI tract, acidity in stomach can lead to drug degradation, adherence regime.	Choonara et al. (2014); Sattar et al. (2014); Thanki et al. (2013)
Ocular	Eyes drops provide convenient delivery, high patient acceptance.	Ineffective delivery to the posterior segment, tear washout.	Cima et al. (2014); Kompella et al. (2013); Yasin et al. (2014)
Nasal	Ease of use, direct route to the blood stream, bypasses gastrointestinal degeneration.	Dosage control, nasal mucosal irritation.	Casettari and Illum (2014); Illum (2012); Kozlovskaya et al. (2014)
Transdermal	Ease of use, can be worn for long periods of time, less invasive.	Poor permeability, location irritation.	Arora et al. (2008); Indermun et al. (2014); Paudel et al. (2010)

delivery strategies presents its own inherent limitations of which response time, dosage degradation, and invasiveness are among the more apparent. While there have been many advances, their effectiveness at treating chronic illnesses that are dose dependent and require frequent, precise, and targeted delivery regimes have been regularly questioned (Tiwari et al., 2012). Associated complications with these routes to delivery can consist of incomplete dosage, either as a consequence of compliance or through drug degradation and therein generate ambiguities over the dosage concentration. A summary of the most common routes to delivery and the associated progressions and obstacles for each approach to delivery is detailed in Table 5.1.

Conventional release strategies, particularly oral delivery systems, predominantly revolve around encapsulants and have been comprehensively reviewed (Choonara et al., 2014; Maroni et al., 2013; Sattar et al., 2014; Thanki et al., 2013). The most common approaches to drug delivery are timed/slow release coatings that gradually dissolve and thereby release the therapeutic component. These can, with little modification, offer temporal control over the release profile but can also be adapted to employ enteric coatings (i.e., cellulose acetate phthalate), which are designed to pass through the acid conditions of the stomach and then slowly release the therapeutic payload in the more neural/alkaline intestines.

Oral delivery possesses many positive attributes, such as the ease of use and the convenience of dose control and will clearly be attractive in terms of cost. This approach, however, is not always the most appropriate, especially where there is a need for rapid absorption or where the therapeutic agent could be substantially degraded as a consequence of the action of gastrointestinal enzymes and where it can act as a barrier to protein and peptide absorption (Choonara et al., 2014). One of the key disadvantages of oral delivery is adenosuppression (Philip and Philip, 2010). Therefore, it could be envisaged that selective delivery would allow drug dose to be reduced, which could reduce these negative side effects. Due to these adverse side effects, there is an evident need to develop site-specific delivery devices to the GI tract, specifically the colon (Van den Mooter and Kinget, 1995). Drug administration via needle (typically subcutaneous injection) can alleviate some of the issues that limit oral efficacy but can also possess a number of inherent drawbacks—particularly the degree of invasiveness, which can engender an inevitable reluctance on the part of the patient to self-administer. There are also numerous concerns over health and safety in terms of use and disposal.

2.1 TRANSDERMAL DRUG DELIVERY

The development of transdermal systems have allowed modern treatments to bypass the limitations traditionally associated with conventional routes to delivery mentioned in previous sections and, while the majority are essentially passive systems, there is an increasing interest in adapting the foundations for smart systems (Arora et al., 2008; Indermun et al., 2014; Paudel et al., 2010). The most famous among the transdermal patches is undoubtedly the nicotine patch, which was approved for widespread use by the FDA in 1992. The transdermal effects of nicotine and the associated success rates in smoking cessation have been widely documented since their inception (Rose et al., 1985; Stead et al., 2012). The success of the nicotine replacement therapy patches could be due to a number of factors including: ease of use, strength of dosage, wearability, and cost factors—all of which would be key factors when considering the translation of the approach to an electronic system. The success of a number of other transdermal delivery systems has also been widely acknowledged and some are detailed in Table 5.2. The advantages and disadvantages of transdermal drug delivery systems are detailed in Table 5.3.

The development and use of fentanyl transdermal patches for chronic pain management has been widely reported (Lane, 2013; Margetts and Sawyer, 2007) and a large number of instances of abuse have been documented (Moon and Chun, 2011; Mounteney et al., 2015). In particular, case studies involving the extraction and intravenous injection of lethal doses of fentanyl where the outcomes have been fatal have been reported (Tharp et al., 2004). It is key to note that the potency of fentanyl is around 100–200 times the potency of that compared to morphine, depending on the variety evaluated (Stanley, 2014). Efforts have recently been made to develop a nanostructured patch to reduce the ability to tamper with the fentanyl reservoir (Cai et al., 2015). The approach utilized geopolymers, a type of ceramic (typically composed of nanoparticulate silica or kaolin networks), within which the fentanyl is

Table 5.2 Transdermal Patch Systems Currently on the Market

Product Name	Active Ingredient	Indication
BuTrans transdermal patch	Buprenorphine	Analgesia
Catapres-TTS	Clonidine	High blood pressure (hypertension)
Evorel patch	Estradiol	Hormone replacement therapy (HRT)
EVRA transdermal patch	Norelgestromin (NGMN) and ethinyl estradiol (EE)	Female contraception
Exelon transdermal patch	Rivastigmine	Alzheimer's disease
FENCINO transdermal patch	Fentanyl	Analgesia
Neupro transdermal patch	Rotigotine	Parkinson's disease
NICORETTE INVISIPATCH	Nicotine	Nicotine replacement therapy
Nitro-Dur transdermal patch	Glyceryl trinitrate	Prophylaxis of angina
Transderm Scop patch	Scopolamine	Motion sickness

Table 5.3 Advantages and Disadvantages of Transdermal Delivery

Advantages
Less invasive then hypodermic needles Ease of use Reduce frequency of dosing No negative gastrointestinal side effects Reduction of dosage peaks and troughs Removal of patch will terminate treatment
Disadvantages
Barrier presented by the skin Opioid transdermal drug delivery systems can be subject to abuse Skin sensitivity Limited availability of pharmacological agents that can be delivered this way Drug permeability Issue of power

placed. The physical properties of the nano particles can be manipulated to attenuate the release profiles as required. The latter, as with most transdermal systems, is by passive diffusion, which, as the limiting step of release, minimizes the effectiveness of the extraction processes necessary for acquiring sufficient quantities for abuse (Jämstorp et al., 2010). One question that arises is whether the emergence of smart devices could further reduce the potential for abuse. The answer will depend on the technologies involved and how the drug is sequestered within the system and, as will be discussed in subsequent sections, may involve a degree of tethering that can be broken only by the action of an electronic signal.

3 COMPLIANCE

There are a number of factors that influence the success of a drug therapy: of key importance is the issue of patient compliance (Dailey et al., 2001; Tarrants et al., 2010). Inadequate adherence by patients to a treatment regime as a result of apathy in the administration of oral medicines, a fear or reluctance to inject subcutaneously or simple errors in routine can all adversely affect the dosage and concentration of a drug therapy. It is possible to envisage that through insufficient adherence, health practitioners could easily assume that the treatment has been ineffective. The course of action that follows could lead to a change in the patient's treatment—side effects associated with an increase in dosage can lead to further complications. This can significantly increase healthcare costs as a result of incorrect treatment regime adherence (NICE, 2007). There is overwhelming evidence corroborating the fact that patient compliance is a key issue that can interfere with treatment outcomes. It is noted that this problem presents an urgent need to develop smart drug delivery systems that have the ability to autonomously deliver a drug therapy without the need for intervention from the patient or clinician.

A qualitative review conducted by the National University of Singapore (Jin et al., 2008) evaluated the issues surrounding patient compliance to specific treatments and the factors that can cause deviations from the prescribed regime. While compliance with short-term therapies was estimated at between 70% and 80%, it was found that, generally, the rate of compliance for long-term therapies typically fell to between 40% and 50%. In other compliance studies it has been estimated that the rate of compliance for the management of diabetes through the administration of insulin via hypodermic needle is around 20% (Bloom Cerkoney and Hart, 1980).

It is important to take into account patient compliance when considering outpatient care and the administration of drug treatments, therefore, for this kind of system, the most desirable outcome would be to completely remove patient intervention from the equation. The administration process is key when we consider the overall success of a drug therapy and failure to maintain routine adherence can often compromise the effectiveness of the treatment. The removal of responsibility from the patient, as shown in the smart system concept in Fig. 5.1, would avoid delays in clinical assessment and treatment outcome. It could also be predicted that through passive sensing, greater control over drug release and faster response time could ultimately be achieved.

4 EVOLUTION OF SMART DRUG DELIVERY DEVICES

The core rationale behind the development of functional materials for use in drug delivery applications relates to the fact that the eventual delivery device will be placed within an environment that is subject to change and that the system itself will respond in a predesigned way to specific changes. The release of the encapsulated drug is achieved when the coating, particle, or film is subjected to the appropriate physiochemical signal. Early approaches relied on factors, such as an increase in pH or a change in redox status (either a more reducing or oxidizing environment) and typically employed a responsive gel or polymeric binder as an encapsulant (Abbaspourrad et al., 2013; Bertz et al., 2013; Cheng et al., 2013; Siegel, 2014).

More recently, and as indicated by the idealized smart device, there has been a considerable push to design systems that respond specifically to a particular disease biomarker. In order to achieve optimum drug performance and outcome, it may be a requirement for the patient to administer multiple doses throughout the day, which would be considered a hindrance. To combat this, recent advancements have been made to the approaches and methods for drug delivery. This has enabled dosages to achieve a number of critical advancements, including controlled delivery rate of an appropriate concentration and control of multiple doses. These techniques are capable of adjusting or controlling optimal drug delivery rate, sustaining the time interval of therapeutic action, and/or directing the delivery of drug to an exact location or tissue will be explored in the following sections.

4.1 MICROELECTROMECHANICAL DEVICES (MEMs)

Electrochemically modified materials and electrochemically controlled drug delivery systems have increasingly been seen as the prime methodology for the development of microdevice release systems and prevail over stimuli-responsive delivery approaches as they can be easily miniaturized and can allow greater control over the drug delivery process. One significant area of interest within drug delivery is the development of microelectromechanical systems (MEMs). These devices are particularly effective at the long-term delivery of a drug conveniently from within a micro-chip system as shown in Fig. 5.3. For many therapies that require daily injections or doses of a drug treatment, implantable devices can improve patient compliance and efficacy of treatment (Kleiner et al., 2014). Currently, implantable drug delivery systems are being developed that use highly sophisticated methods, such as

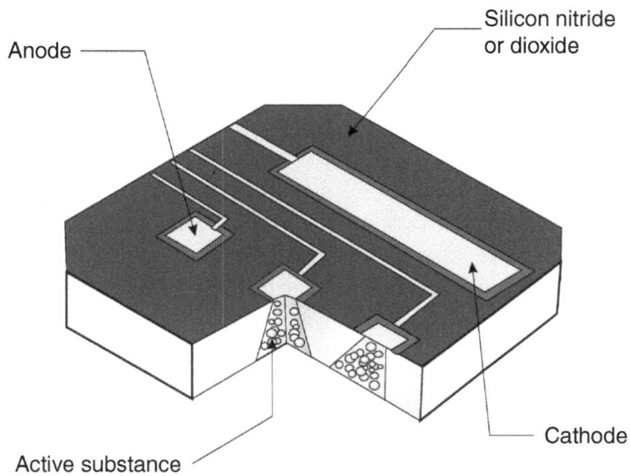

FIGURE 5.3 A Solid State Microchip Drug Delivery System

microfabrication technology. These microfabricated drug delivery devices have been designed to be fortified with precisely measured geometry in order to distribute drugs in a highly favorable and regulated manner. The drug delivery device shown in Fig. 5.3 is a solid-state microchip system that has the capability to delivery individual or multiple reagent release from the microscale reservoir location in the microchip (Santini et al., 1999). The gold anode membrane acts as a seal, until the electrochemical trigger is activated, resulting in the controlled dissolution of the gold film. Upon receiving the electrochemical signal (potential at +1.04V for several seconds), the gold thin film anode begins to degrade, allowing the entrapped molecules from the reservoir to freely pass from the device in a controlled manner. Further development of the device in question could see this prototype become part of the next generation of "smart" drug delivery devices.

MEMs drug delivery systems provide a large profile of benefits; however, the contrasting limitations of implantable drug delivery systems have been discussed, where the invasive issues associated with implantable drug delivery systems were highlighted (Kleiner et al., 2014). Of course, with any implantable device a minor surgical procedure is required to introduce the new drug therapy. This may cause concern to some patients and the result may be distressing. The size of an implantable device, as small as possible to reduce patient discomfort, also means that there is limited capacity for storing the drug treatment; therefore, there is the question of how much drug can be delivered before the treatment is exhausted. Battery size and battery life are another cause for concern within MEMs systems. The battery itself occupies quite a large volume within the system, therefore restricting the capacity of the device to hold the drug therapy.

In addition to this, MEMs have been developed for a large array of applications including: controlled release applications (Pirmoradi et al., 2011), lab-on-a-chip systems (Jung et al., 2015), and microfluidic devices (Yoon et al., 2014). MEMs devices offer minimally invasive treatment and can allow control of the delivery site and amount. For example, some drugs, such as hormones, may be more effective when released in a manner similar to the way they would be produced naturally. Through the use of MEMS technology and micro-fabrication techniques, drug delivery research has been able to make a significant departure from traditional methods. These miniaturized systems have been successfully demonstrated as suitable drug carriers for implantation (Gensler et al., 2012; Kleiner et al., 2014). It must be noted, however, that the aim of this overview is not to consider the fabrication of the MEMs devices themselves, but rather to investigate the release methodologies that are being employed within them and those techniques that could, in principle, be adapted to the ever-decreasing device formats.

The materials properties at the device-body interface are key when it comes to encouraging interactions between a biological system and the material interface itself. Important considerations take into account the analysis of physiochemical parameters, which include hydrophilicity, protein adhesion, electrical and electrochemical parameters of the material and overall suitability. Appropriate material considerations for sensing and monitoring are furthering the development of novel delivery applications for smaller, less intrusive drug delivery systems.

4.2 BARRIERS TO DELIVERY

Two critical problems known to affect drug delivery are response time and physical barriers to delivery. As discussed previously, transdermal systems can provide solutions to many of the drawbacks associated with traditional approaches to drug delivery. It is commonplace for treatments to be administered by a healthcare professional or the patient themselves and typically require repeated doses over a period of days or weeks. While such procedures can usually be accommodated with relative ease, it can become much more problematic when dealing with a chronic condition in which timescales can extend to months or years. The transdermal approach can provide sustained release for longer periods of time without assistance (Prausnitz and Langer, 2008) and can help to alleviate patient compliance issues. The success of a transdermal drug delivery system, however, can be limited due to associated obstacles to delivery (primarily transport across the skin barrier) and, as a consequence, the number of drugs that have been translated to this system is comparatively small in relation to more conventional methods. The skin is made up of three major layers: the epidermis, dermis, and the hypodermis as shown in Fig. 5.4.

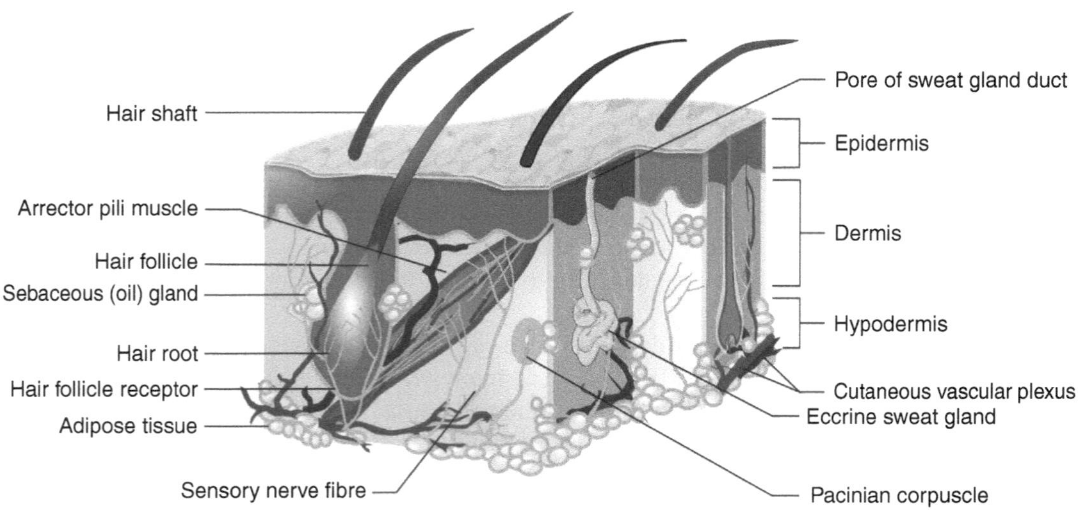

FIGURE 5.4 Schematic Demonstrating the Layers of the Skin

The difficulty of trying to transport drugs across the skin barrier can be attributed to the low permeability of the skin tissue in relation to drugs. When examining the structure of the skin, the outer most protective layer is the stratum corneum (SC), which is hydrophobic in nature. The function of the SC is to prevent water loss and to protect the underlying layers from mechanical and chemical stress. The SC measures around 10–20 μm in depth but varies at different sites and consists of around 25 layers of dead squamous cells, which further limit drug penetration (Escobar-Chávez et al., 2011; Russell et al., 2008).

The epidermis is the most superficial layer of the skin, it provides the outermost barrier to transdermal drug delivery and consists mostly of epithelial tissue. It has been suggested that the inclusion of microneedles to the transdermal delivery system can offer the potential to overcome this obstruction. The dimensions of microneedles can range from 400 to 1500 μm and therefore they are, in principle, capable of penetrating the epidermis, whereby the drug could be released to the dermis, a highly vascularized site, which is the primary location for the initial of the therapeutic agent. It should also be noted that they are also deemed minimally invasive, therefore reducing nervous stimulation, skin trauma, or patient discomfort that is known to occur with the use of hypodermic needles (Indermun et al., 2014). Taking into consideration the numerous advantages that transdermal delivery patches provide, there is also significant room for further improvements in terms dosage. Moreover, when considering the addition of electronics to produce a "smart" patch, power requirements come to the fore that would need to be fully considered—especially in relation to the size of the device and site at which it would be placed.

5 ELECTROCHEMICAL APPROACHES TO DRUG RELEASE

Electrochemical drug release strategies are arguably in their infancy, but a surprising number have emerged in recent year and can be attributed to the emerging interest in smart devices and the recognition that electrochemical methodologies are particularly suited to MEMs type architectures. A large variety of procedurally different approaches have been explored and include conducting polymers (Alizadeh and Shamaeli, 2014; Svirskis et al., 2010a;), carbon nanotube composite structures (Luo et al., 2011; Prakash et al., 2011; Riggio et al., 2009), redox gels (Dong et al., 2006; Jin et al., 2012; Liakos et al., 2013; Narayanan et al., 2012), conductive electrode substrates for pH manipulation (Anderson et al., 2014), and pump systems (Ai et al., 2010; Li et al., 2010; Sheybani et al., 2013; Shin et al., 2011). The electrochemical transitions and chemical structures that allow their adaption to drug release are examined in the following subsections.

There are a number of physiological barriers present that must be taken into consideration when designing novel materials for use in drug delivery devices in order to transport and protect the drug. Therefore, it is key to develop systems that have the ability to take advantage of the physiological tissue encountered during delivery by responding to stimuli in this environment, promoting retention time with improved permeation and absorption at the target delivery site. The success of electrically controllable devices and their electrochemical properties and transitions that enable application to drug delivery similar to those mentioned earlier, are measured in the following sections.

5.1 CONDUCTING POLYMERS

Historically, conducting polymers have a rich background in sensing applications and have shown to be promising candidates for use in drug release applications, such as electrochemical actuators. Polymers, such as polypyrrole (George et al., 2006; Svirskis et al., 2010b) and poly(phenylene-vinylene)

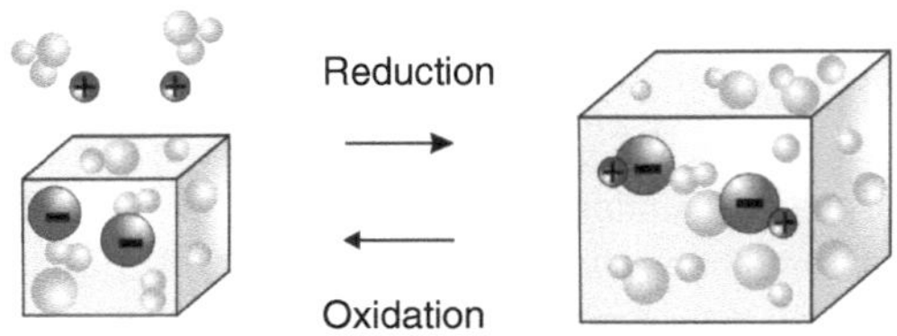

FIGURE 5.5 A Polypyrrole System Prepared With Immobilized Anions That Swells and Incorporates Cations Upon Reduction; Cations Can Then Be Expelled on Oxidation

(Tang et al., 2009), have been previously demonstrated as successful contenders in controlled release applications (Wadhwa et al., 2006). The ability to have control over the redox state of polymers to allow for physical swelling or dissolution, via the capture or expulsion of anions (detailed in Fig. 5.5) provides the opportunity for drug loading or entrapment. Classic examples can include the use of the following: neurotrophin (Richardson et al., 2009; Wise and Gillespie, 2012), dexamethasone (Bohl et al., 2012; Leprince et al., 2010), risperidone (Svirskis et al., 2010b), and several NMDA receptor inhibitors (Stauffer et al., 2011).

Early applications of polymer systems had the drug loaded as counter ions within the film, which can be released upon reduction of the polymeric film. While the methodology behind the use of polymer films as drug releasing actuators is relatively simple, the approach has become one of significant importance in the management of Sensorineural Hearing Loss (SNHL). SNHL is related to damage or deterioration of spiral ganglion neurons (SGNs) or the loss of hair cells that change sounds into electrical signals, which are then sent to the brain (Wise and Gillespie, 2012). The damage to the sensory hair cells (HCs) is detailed in Fig. 5.6, where a normal cochlea and a deaf cochlea are presented, complete with a cochlear implant and the notable absence of hair cells in the figure to the right. The advantages of polypyrrole over conventional polymer systems are that it can be formed through electrochemical oxidation, allowing controlled deposition at the electrode surface. This is particularly significant where there is a need to develop microfabricated systems at the nano-micro scale, whereby the site of deposition can be spatially controlled with a high degree of precision.

Current research appears to validate that the SGNs can be protected from damage via implantation of a cochlear implant, which can deliver neurothrophin to the cochlea and provide electrical stimulation to the cochlea. Drug eluting electrodes for the delivery of neurotrophins to cochlear neurons for the preservation SGNs have been designed (Richardson et al., 2009). The electrodes developed are coated with a conductive polymer: polypyrrole, which has the ability to aid in the protection and preservation of SGNs without altering or imposing a detrimental threat on the function of the cochlear implant. The advantage of using a device, that is, drug eluting is particularly effective in this case because of the inherent inaccessibility of the anatomy of interest, therefore a localized form of drug delivery to the cochlear is ideal. The polymeric coating loaded with the drug inside its network allows a degree of control of the drug release profile. This method of protection from degeneration performs in direct contrast to direct injection, which has the possibility of fluid leakage and can often lead to inconsistent treatment dosage.

In the most common cases, the film itself has been used as the primary drug carrier. However, this can bring about evident issues over the quantity of the drug, which can be entrapped within the film. As a consequence, this approach will impact on the scope of applications, which can often require high doses or constant delivery. In recent years, there have been noticeable advancements in the use of composite approaches, which have increased drug-carrying abilities within controlled release applications.

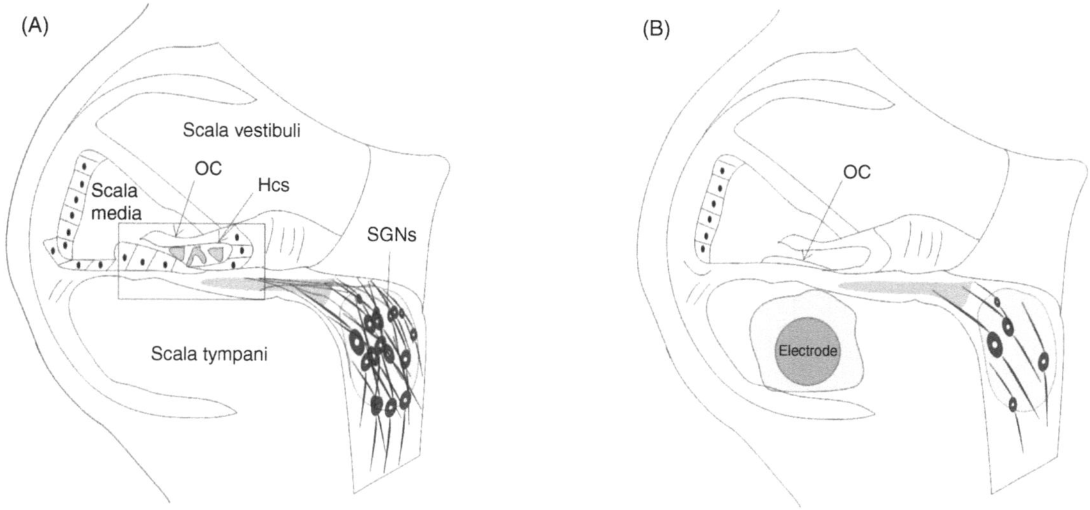

FIGURE 5.6 (A) Normal cochlea. (B) Deaf cochlea electrode array can be implanted to electrically activate SGN

This has led to a change in the way that polymers can be used. Instead of being a controllable sponge that can be electrically manipulated to release the drug, increasingly, conducting polymers are being utilized as a molecular valve that serves to encase a drug within a reservoir. In particular, the use of polypyrrole to stopper the ends of carbon nanotubes (CNT) containing the drug dexamethasone has been demonstrated. The release of the drug molecules from inside the loaded CNTs was accomplished through the electrochemical reduction of the polymer plugs at –0.5V to open them and at an opposite potential of +0.5V to close them, as indicated in Fig. 5.7 (Luo et al., 2011). Similar approaches

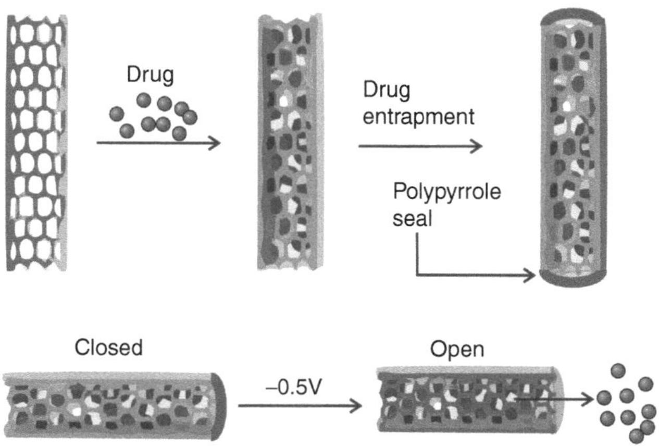

FIGURE 5.7 Electrochemically Controlled Release of Drugs from Loaded Nanotubes

From Luo, X. et al., 2011. Carbon nanotube nanoreservior for controlled release of anti-inflammatory dexamethasone. Biomaterials, 32(26), pp. 6316–6323.

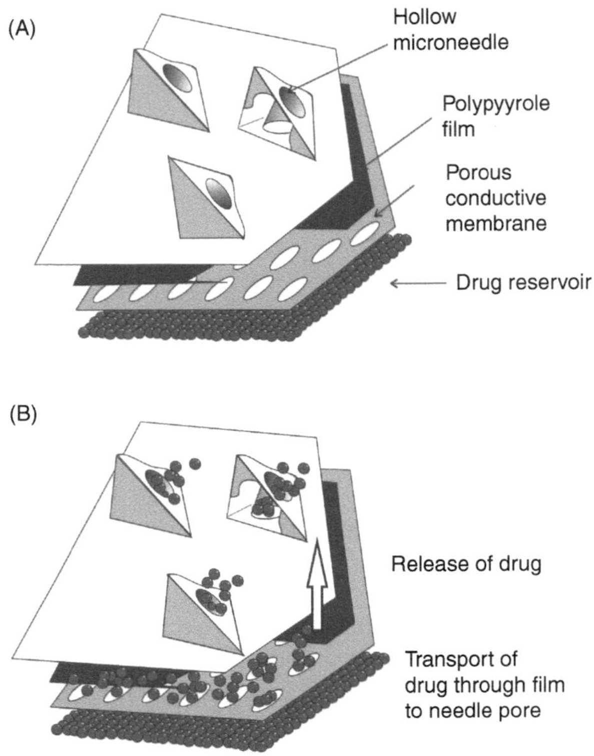

FIGURE 5.8 Microneedle Patch With Polypyyrole Release Film

(A) Before release (–1.1V) and (B) after release (+0.5V).

involving the use of CNTs in the application of chemotherapy treatments have also been investigated (Prakash et al., 2011; Vashist et al., 2011).

In an approach that is methodologically similar to CNT release, the tunable qualities of the conducting films have been exploited in the development of transdermal microneedle systems. As mentioned in previous sections, the majority of transdermal drug systems operate through passive transfer from the reservoir to and across the skin barrier but there is an increasing number of smart stimuli responsive approaches beginning to appear. One such system has been designed by Valdés-Ramírez et al. (2012) involving a microneedle patch complete with a drug reservoir. The basic system is highlighted in Fig. 5.8. In this case, polypyrrole is utilized as the electronic gate. A laser-patterned membrane possessing small micron-sized holes is combined with polypyrrole thin film to seal the drug reservoir. In this particular case, the polypyrrole film acts as a barrier restricting the drug molecules within the reservoir; the film is then deposited onto porous polycarbonate film, which has been sputtered with gold. Micromachining of the patch is used to produce the hollow microneedle component and is followed by adhesion onto the polypyrrole/polycarbonate composite layer. Drug release can be achieved when a reducing potential is applied to the electrode substrate, which induces a change in the redox state of the film and a consequent change in morphology that enables the drug molecules to pass through the film.

While the open-close transitions at the needles are at the microscale, the drug reservoir however is not limited to similar dimensions and can be decidedly macroscale, allowing a more substantial reservoir and sustained release. Critically, the reservoir can be designed to contain multiple reagents, allowing the device to possess individually addressable reservoirs, allowing the device to achieve multireagent release from a single patch.

5.2 MANIPULATION OF REDOX HYDROGELS

Sodium alginate, a naturally occurring water-soluble polymer, has been extensively used in drug delivery (Almeida and Almeida, 2004; Dong et al., 2006). Alginate gels possess many desirable characteristics in terms of mechanical properties including: tunability, and biodegradability in physiological conditions, allowing them to perform favorably in many biomedical applications. In particular, they have been previously used in controlled release applications (Liakos et al., 2013), wound dressings (Liakos et al., 2014), and tissue engineering (Venkatesan et al., 2014). Cross-linking through interaction with metal ions, such as Mg^{2+} and Ca^{2+} have long been exploited (Jin et al., 2012) but, from an electrochemical perspective, neither system offers an opportunity for electrochemical control. Substituting in Fe^{2+} however, has been shown to open up a number of options through which electrochemical manipulation can involve both the entrapment of the therapeutic and its subsequent release. The presence of ferrous ion (Fe^{2+}) has little effect on dissolved alginate, but its electrochemical oxidation to the ferric state (Fe^{3+}) induces cross-linking (Fig. 5.9) and offers the possibility of controlling the nucleation of the hydrogel, its spatial location, and the material entrapped within the 3D polysaccharide network.

The ferric ions (Fe^{3+}) are typically regarded as hard metal cations, which bind preferentially to the negatively charged carboxyl groups within the alginate effectively binding much more strongly compared to Fe^{2+}. The ferrous ion ($Fe^{2+)}$ is considered a "soft" metal cation and more often binds to neutral ligands. It can be determined that Fe^{3+} binding to the carboxyl groups will be substantially stronger in comparison to the Fe^{2+} cations, which have a lower ability for cross-linking to the alginate and which will directly affect the alginate gel stability. Electrochemical reduction of the ferric coordinating ion leads to the disruption and dissolution of the alginate layers, resulting in the release of the entrapped drug molecules. It could be envisaged that this system could be used as an alternative to the polypyrrole systems described for the hollow microneedles. After receiving an appropriate electrochemical signal, the dissolution of the hydrogel thin film will occur; this will result in the controlled release and transport of the drug molecules from the reservoir, through the microneedles to the desired site of delivery.

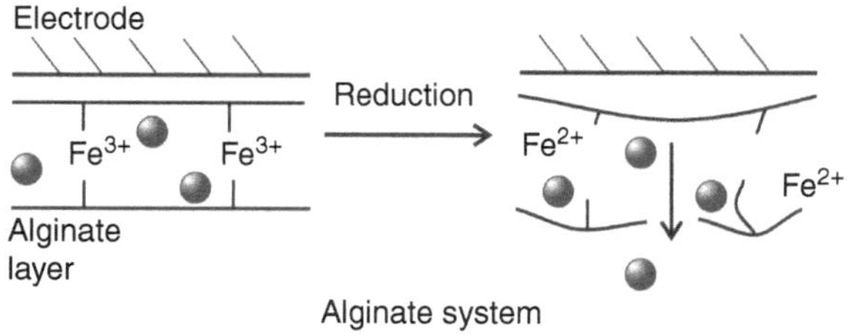

FIGURE 5.9 Electrochemical Reduction of the Ferric Coordinating Ions Leads to the Disruption and Dissolution of the Alginate Layers Resulting in the Release of the Entrapped Drug Molecules

The concept of "artificial muscle" (Low et al., 2000) originates from early work on the electrochemical properties of conducting polymers—particularly polypyrrole. In more recent designs related to drug delivery, the "artificial muscle" involves the use of a chemo mechanical actuator comprising the combination of a hydrogel and a redox polymer, which has electronically conducting capabilities. Under certain conditions the polymer blend demonstrates a swelling feature that can be controlled via an appropriate electronic trigger (potential control), resulting in swelling or a decrease in size in order to open and close, much like a controllable valve (Low et al., 2000).

5.3 CONTROL OF LOCAL pH

A critical challenge encountered when first designing formulations for delivering drugs to the colon was the ability to preserve the drug formation in its original state. This was particularly difficult given variation in pH and enzyme/microbe composition found within the colon, and it was all too easy for drug integrity to be compromised and early release and premature absorption in the upper section of the GI tract to occur. Historically, pH-sensitive films have shown to be effective at enabling drug delivery to a specific part of the gut and to release the drug in a manner that is consistent with the local pH. The pH in the area inside the stomach typically ranges between pH 1 and pH 2 (during fasting); however, after eating, the pH increases markedly and upon transfer of the chyme from stomach to small intestine, it is common for the pH to increase to around 6.6 and higher (Philip and Philip, 2010; Evans et al., 1988). As the pH of the target site can be relatively accurately predetermined, there is scope for some degree of control over the nature of release mechanism. The use of such "enteric" coatings in oral tablet formulations has long been established (Maroni et al., 2013), whereby the coating dissolves upon reaching an area of higher pH. A similar approach utilizes microbe triggering to achieve drug release, where biodegradable polymers protect the drug from the local environment until it reaches the colon, whereby microbial enzymes degrade the films, leading to the disruption of the polymer backbone and the release of the drug.

An alternative approach that may be more suitable for microelectronic applications has recently been described by Anderson and coworkers (Anderson and Davis, 2015; Anderson et al., 2015). The rationale has been to retain the barrier properties of the enteric coating (in this case cellulose acetate phthalate) but exploit the electrochemical manipulation of pH within micropores, to induce the breakdown of the protective film and therein enable a more refined release profile. A reservoir-based drug-delivery system based on this approach is highlighted in Fig. 5.10. The system is based on a conductive carbon composite (polycarbonate) layer that has been laser-patterned to create pores.

It is possible to control the release of the drug by holding the potential at the negative end of the potential window (–2V, 60 s). As a result, this electronic trigger initiates a change in pH within the pore, which causes the pH-sensitive polymer layer to dissolve and release the encased drug in a controlled manner. The change in pH is only in the immediate vicinity of the holes in the membrane and thus won't affect the surrounding tissue. Through holding the electrode at –1V, the electrode reaction proton reduction/hydrogen evolution ($H^+_{(aq)} + e > \frac{1}{2}H_{2(g)}$) occurs with the removal of H^+, resulting in an increase in the pH. Once the protons are consumed, the reaction is $H_2O + e > \frac{1}{2}H_2 + OH^-$, and at this point the local pH increases above 7 and the protective film begins to degrade. As polyesters are unstable in alkaline solutions, we can then control the release of the drug at a predetermined time, which can be initiated by the conductive layer.

The disadvantage of trying to raise the pH is the fact that it is necessary to impose large negative potentials. In order to combat this issue, it is possible to use other chemical modifiers to then control the

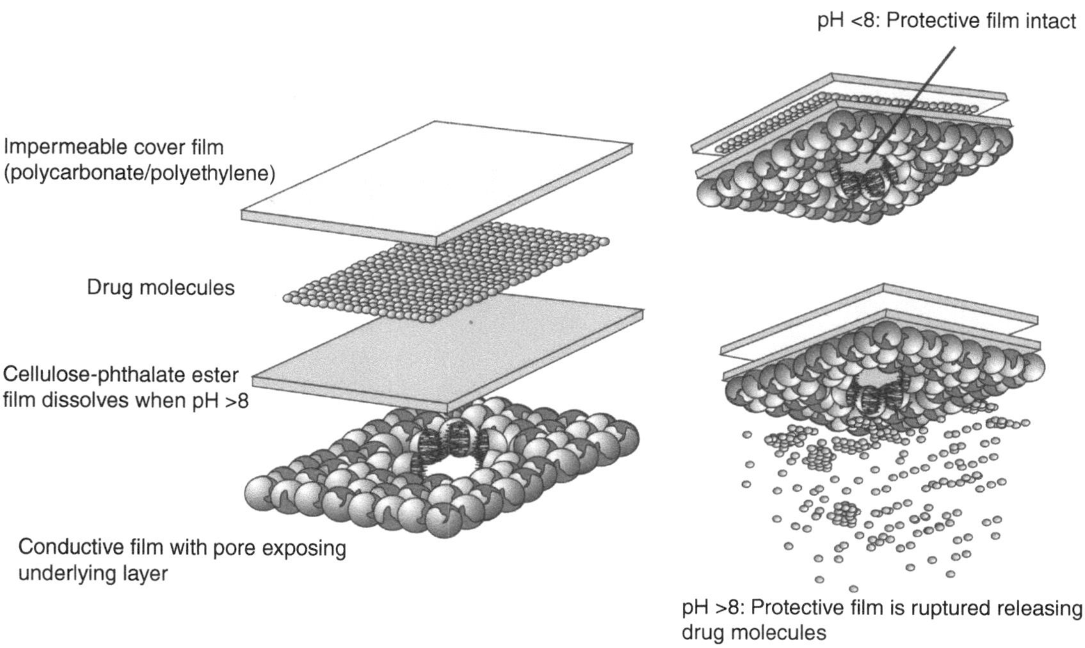

FIGURE 5.10 Configuration of the Electrochemical Release System

pH as shown by the quinone system in Fig. 5.11 (Frasconi et al., 2010; Lawrence and Robinson, 2007). Upon oxidizing the hydroquinoid form, two protons are released, thereby lowering the pH. Similarly, reducing the quinoid form could result in an increase in the pH. The key point is that the redox position (and hence protonation) can be controlled at potentials that are much less extreme than those necessary for hydrogen evolution

5.4 ELECTROCHEMICAL PUMP SYSTEMS

The approaches to drug release that have been described thus far have generally focused on the diffusive movement of the drug from a reservoir to the desired site of delivery—often this does not suffice when it comes to direct delivery to more complicated or restricted anatomy (Ai et al., 2010). In order to achieve a more significant and sustained drug movement, a number of approaches have been developed to directly induce the movement of the drug during the release process. Electrolysis has been investigated as a means to manipulate the local pH (Anderson and Davis, 2015; Cheng and Chang, 2011), but had an associated drawback of gaseous evolution; however, this result can be used in a different context to achieve release within a pump system (Sheybani et al., 2013).

This practicalities of this methodology have been investigated and the general strategy is outlined in Fig. 5.12 where the generation of gas within the drug reservoir causes the drug molecules to become displaced and travel to the delivery site (A) and the initiation of electrolysis (B) and after (C) to displace the drug from the reservoir (Sheybani et al., 2013). This simple approach has been applied to treat various conditions of the eye that require discrete continuous delivery as a result of the intricate anatomy of the delivery site.

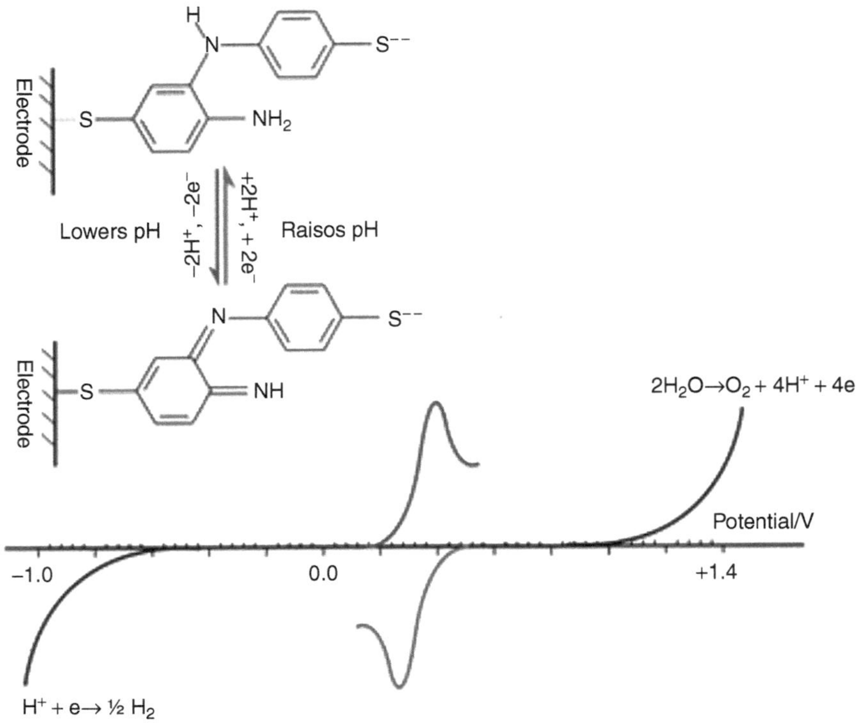

FIGURE 5.11 Manipulating pH Through the Addition of Other Chemical Modifiers at Lower Potentials

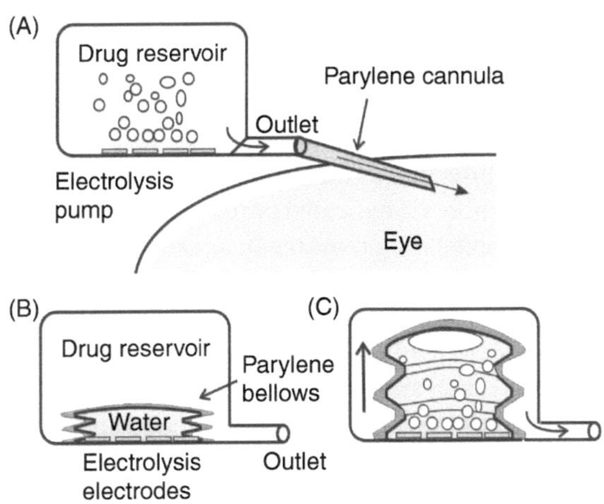

FIGURE 5.12 Electrolysis Pumps Exploiting Gaseous Evolution to Displace the Drug From the Device

Simple one pot system (A) and the use of bellows to isolate the electrodes from the drug before (B) and after (C) initiation of the electrolysis process.

Conventional drug delivery to the eye includes eye drops, oral delivery, and intraocular hypodermic injection (Cima et al., 2014; Kompella et al., 2013; Li et al., 2008; Yasin et al., 2014). The electrochemical intraocular micro device proffers many advantages over traditional drug delivery, such as compactness at the anatomical site, along with continuous sustained delivery. One further system that can provide more benefits over the bellows system is an approach using electroosmotic pumping (Ai et al., 2010). These systems have no moving parts, offer simpler construction rationales and rely on the flow from the drug reservoir through two electrodes, which are separated by a porous layer, such as alumina or silica (Chen et al., 2010). Luft et al. (1978) were among the first to implement this approach for the novel development and pursuit of one of the first insulin pumps. In more recent times, this approach has been used in a number of bioanalytical systems and microfluidic platforms; however, there are substantial concerns with the functioning voltages where electrolysis of water can occur, leading to gas production. The resulting generation of O_2 and H_2 can become confined within the membrane and at the surface, serving to insulate part of the electrode and resulting in irregular flow patterns.

5.5 COVALENTLY IMMOBILIZED DRUGS

The approaches that have been outlined before generally rely on the drug particles being free—albeit sealed within the drug compartment. The approach utilizing covalently immobilized drugs aligns with the electrochemical cleavage of drugs covalently bound to the electrode surface. The ability to covalently fix a drug to an electrode substrate eliminates the possibility of gradual leakage. Leakage of any sort could become apparent where the long-term storage of an active compound is required. Hydrogels containing drugs covalently immobilized to the gel framework through azoaromatic linkages had initially been proposed as an innovative approached to the treatment of Inflammatory Bowel Disease (IBD). The hydrogel structure contains a pH-sensitive monomer and also the addition of an azo cross-linking agent, to protect the drug from digestion in the stomach and degradation in the small intestine. The azo linkage is selectively broken through the reductive action of azoreductases found in the colon and the drug component is released. Unfortunately, the production of toxic by-products as a result of the degradation of azo-polymeric matrices has been discovered, therefore making such systems unsuitable for long-term treatment (Shantha et al., 1995; Stubbe et al., 2001; Yamaoka et al., 2000). An alternative approach to the hydrogel system, however, has recently been developed and is much more amendable to devices at the micro-nano scale. Drug components possessing diazonium salt functionalities can be directly adhered to carbon surfaces through the formation of a covalent azo linkage. Rather than rely on enzymes to release the drug component, the release can be affected through direct electrochemical reduction of the azo functional group at the carbon interface (Wildgoose et al., 2005). The toxicological consequences of this strategy have yet to be investigated but it could clearly provide a facile mechanism through which to functionalize carbon nanostructures, such as carbon nanotubes and graphene nanoflakes.

6 INNOVATION AND APPLICATION

One of the earliest MEMs pill devices for controlled release began with the development GI microcapsule. Contrary to its given name, the GI microcapsule was a tubelike device that could facilitate real-time drug release while sampling fluid from the GI tract (Cui et al., 2008). The polycarbonate-based smart capsule had a location tracking unit, a receiver, a reservoir for fluid sampling and drug loading, and a battery. In the capsule, the drug delivery reservoir was sealed with an actuating polymer membrane. When the sensor, located inside the capsule, detects changes in the tissue environment, the chip processes this signal and causes a drug-releasing signal to be generated, this results in the disruption of the polymer membrane and the deployment of the necessary amount of drug to the surrounding tissue.

Staying within the theme of smart capsule systems, there have been a number of recent advancements in the development of highly controllable "intelligent" pill systems, which can provide real-time feedback coupled with a delivery mechanism. The "IntelliCap" electronic pill system has been developed by Medimetrics (Söderlind et al., 2015). The pill system comprises monitoring and delivery elements including a reservoir, microcontroller, pump, and wireless communication capabilities. The pill system can collect measurements of local pH when it is traveling through the gastrointestinal tract and can provide site-specific delivery, which is controlled through the inbuilt microelectronics. This kind of system allows easy accessibility to fluid sampling from a previously difficult-to-examine area.

There have also been notable advancements in the development of transdermal patch technologies. A strain-triggered drug-delivery patch has been engineered (Di et al., 2015), whereby the encased drug molecules can be released via the application of stretching forces or compression. The patch consists of a stretchable elastomer that has microcapsules located on the surface, which can be loaded with the drug. Sustained drug release can be achieved by bodily movements or instantaneous release can be achieved by applying an intentional force upon the device. Microneedles have also been engineered onto the patch system to facilitate a discrete route to delivery.

Furthermore, research is continually progressing to provide better forms of insulin delivery for diabetic patients, such as a transdermal microneedle system capable of automatically releasing insulin when high levels of glucose are detected (Yu et al., 2015). Under normal physiological conditions, the device is stable, where the microneedle vesicles have glucose oxidase and a dose of insulin encapsulated within their structures. However, when there is an elevation of the glucose levels, the glucose oxidase converts the excess glucose to gluconic acid, while depleting the oxygen during this process. Due to the low levels of oxygen, a hypoxic environment is created and this causes the vesicles within the microneedles to change from hydrophobic to hydrophilic, ultimately causing the vesicle to dissociate and release the insulin.

7 CONCLUSIONS

While many of the materials and approaches to controlled release are promising candidates for successful drug delivery, relatively few have been used in conjunction with micro-nanofabrication technologies. By combining the research from these "smart" materials into suitable microdevices, it may be possible to tune the specificity of drug loading and control over release during the delivery phase. It is clear that patients suffering from chronic illnesses would be among the prime beneficiaries from the use controllable release systems, by delivering medication through a more discrete and more convenient route. At present, there is a need to further develop current transdermal patch technologies and explore the potential for the addition of feedback systems, instigating the development of smart patch drug-delivery systems. The availability of an electrochemically controlled smart device that could be worn by an outpatient could address many of the issues associated with delays in drug delivery and in compliance. These proffer much better outcomes, particularly when it comes to patient compliance where all too often, human error can result in the treatment regime becoming ineffective and which can compromise the overall condition of the patient.

Advances in the manufacture of electronic devices for drug delivery have now achieved a size that can be similar to those present within the physiological environment of the body. This possibility creates exciting potential for merging microelectronics and biosensor technology to acquire novel mechanisms with unparalleled ability and potential. The improvement of device size and design of these drug delivery systems means they have the ability to release a multifaceted drug therapy profile in a controlled manner in order to optimize the therapeutic outcome of the drug treatments. These would have the ability to provide a more effective method to drug release, taking into account the amount of drug to be delivered and at the correct time in response to an appropriate electrochemical trigger. It could also be invaluable in weaning patients off opioid medications in a much more controlled manner. The outcome would be to develop a platform technology that is at the forefront of electronic delivery.

REFERENCES

Abbaspourrad, A., et al., 2013. Polymer microcapsules with programmable active release. J. Amer. Chem. Soc. 135 (20), 7744–7750.

Ai, Y., et al., 2010. A low-voltage nano-porous electroosmotic pump. J. Colloid Interf. Sci. 350 (2), 465–470.

Alizadeh, N., Shamaeli, E., 2014. Electrochemically controlled release of anticancer drug methotrexate using nanostructured polypyrrole modified with cetylpyridinium: release kinetics investigation. Electrochim. Acta 130, 488–496.

Almeida, P.F., Almeida, A.J., 2004. Crosslinked alginate gelatine beads a new matrix for controlled release of pindolol. J. Control. Release. 97 (3), 431–439.

Anderson, A., Davis, J., 2015. Next generation transdermal drug delivery: an electrochemical approach to pH manipulation for controlled release within smart patch technologies. IFMBE Proceedings: World Congress on Medical Physics and Biomedical Engineering 51, 919–922.

Anderson, A., et al., 2014. Investigating the use of endogenous quinoid moieties on carbon fibre as means of developing micro pH sensors. Mater. Sci. Eng. C 43, 533–537.

Anderson, A., McConville, A., Davis, J., 2015. Electrochemical bubble rip: a new approach to controlled drug release. Electrochem. Comm. 60, 88–91.

Arora, A., Prausnitz, M.R., Mitragotri, S., 2008. Micro-scale devices for transdermal drug delivery. Int. J. Pharm. 364 (2), 227–236.

Bandodkar, A.J., Wang, J., 2014. Noninvasive wearable electrochemical sensors: a review. Trends Biotech. 32 (7), 363–371.

Bertz, A., et al., 2013. Encapsulation of proteins in hydrogel carrier systems for controlled drug delivery: influence of network structure and drug size on release rate. J. Biotechnol. 163 (2), 243–249.

Bloom Cerkoney, K.A., Hart, L.K., 1980. The relationship between the health belief model and compliance of persons with diabetes mellitus. Diabetes Care 3 (5), 594–598.

Bohl, A., et al., 2012. Development of a specially tailored local drug delivery system for the prevention of fibrosis after insertion of cochlear implants into the inner ear. J. Mater. Sci. 23 (9), 2151–2162.

Cai, B., Engqvist, H., Bredenberg, S., 2015. Development and evaluation of a tampering resistant transdermal fentanyl patch. Int. J. Pharm. 488 (1–2), 102–107.

Casettari, L., Illum, L., 2014. Chitosan in nasal delivery systems for therapeutic drugs. J. Control. Rel. 190, 189–200.

Chan, M., et al., 2012. Smart wearable systems: current status and future challenges. Artif. Intell. Med. 56 (3), 137–156.

Chen, Y.-F., et al., 2010. Surface modification of nanoporous anodic alumina membranes and its use in electroosmotic flow. Sens. Act. B 145 (1), 575–582.

Cheng, L.-J., Chang, H.-C., 2011. Microscale pH regulation by splitting water. Biomicrofluidics 5 (4), 046502.
Cheng, R., et al., 2013. Dual and multistimuli responsive polymeric nanoparticles for programmed site-specific drug delivery. Biomaterials 34 (14), 3647–3657.
Choonara, B.F., et al., 2014. A review of advanced oral drug delivery technologies facilitating the protection and absorption of protein and peptide molecules. Biotechnol. Adv. 32 (7), 1269–1282.
Cima, M.J., et al., 2014. Single compartment drug delivery. J. Control. Rel. 190, 157–171.
Clifton, L., et al., 2013. Gaussian processes for personalized e-health monitoring with wearable sensors. IEEE Trans. Biomed. Eng. 60 (1), 193–197.
Cui, J., et al., 2008. The study of a remote-controlled gastrointestinal drug delivery and sampling system. Telemed. J. 14 (7), 715–719.
Dailey, G., Kim, M.S., Lian, J.F., 2001. Patient compliance and persistence with antihyperglycemic drug regimens: evaluation of a medicaid patient population with type 2 diabetes mellitus. Clin. Therap. 23 (8), 1311–1320.
Di, J., et al., 2015. Stretch-triggered drug delivery from wearable elastomer films containing therapeutic depots. ACS Nano 9 (9), 9407–9415.
Dong, Z., Wang, Q., Du, Y., 2006. Alginate/gelatin blend films and their properties for drug controlled release. J. Memb. Sci. 280 (1–2), 37–44.
Escobar-Chávez, J.J., et al., 2011. Microneedles: a valuable physical enhancer to increase transdermal drug delivery. J. Clin. Pharm. 51 (7), 964–977.
Evans, D.F., et al., 1988. Measurement of gastrointestinal pH profiles in normal ambulant human subjects. Gut 29 (8), 1035–1041.
Florea, L., Diamond, D., 2015. Advances in wearable chemical sensor design for monitoring biological fluids. Sens. Actuat. B 211, 403–418.
Frasconi, M., et al., 2010. Electrified selective “sponges” made of Au nanoparticles. J. Amer. Chem. Soc. 132 (27), 9373–9382.
Gensler, H., et al., 2012. An implantable MEMS micropump system for drug delivery in small animals. Biomed. Microdev. 14 (3), 483–496.
George, P.M., et al., 2006. Electrically controlled drug delivery from biotin-doped conductive polypyrrole. Adv. Mater. 18 (5), 577–581.
Illum, L., 2012. Nasal drug delivery: recent developments and future prospects. J. Control. Rel. 161 (2), 254–263.
Indermun, S., et al., 2014. Current advances in the fabrication of microneedles for transdermal delivery. J. Control. Rel. 185 (1), 130–138.
Jämstorp, E., et al., 2010. Mechanically strong geopolymers offer new possibilities in treatment of chronic pain. J. Control. Rel. 146 (3), 370–377.
Jin, J., et al., 2008. Factors affecting therapeutic compliance: a review from the patient’s perspective. Ther. Clin. Risk Manage. (1), 269–286.
Jin, Z., et al., 2012. Electrochemically controlled drug-mimicking protein release from iron-alginate thin-films associated with an electrode. ACS Appl. Mater. Interf. 4 (1), 466–475.
Jung, W., et al., 2015. Point-of-care testing (POCT) diagnostic systems using microfluidic lab-on-a-chip technologies. Microelectron. Eng. 132, 46–57.
Kleiner, L.W., Wright, J.C., Wang, Y., 2014. Evolution of implantable and insertable drug delivery systems. J. Control. Rel. 181 (1), 1–10.
Kompella, U.B., et al., 2013. Nanomedicines for back of the eye drug delivery, gene delivery, and imaging. Prog. Retinal Eye Res. 36, 172–198.
Kozlovskaya, L., Abou-Kaoud, M., Stepensky, D., 2014. Quantitative analysis of drug delivery to the brain via nasal route. J. Control. Rel. 189, 133–140.
Lane, M.E., 2013. The transdermal delivery of fentanyl. Eur. J. Pharm. Biopharm. 84 (3), 449–455.
Lawrence, N.S., Robinson, K.L., 2007. Molecular anchoring of anthracene-based copolymers onto carbon nanotubes: enhanced pH sensing. Talanta 74 (3), 365–369.
Leprince, L., et al., 2010. Dexamethasone electrically controlled release from polypyrrole-coated nanostructured electrodes. J. Mater. Sci. 21 (3), 925–930.

Li, P.-Y., et al., 2008. An electrochemical intraocular drug delivery device. Sensor. Actuat. A-Phys. 143 (1), 41–48.
Li, P.Y., et al., 2010. A parylene bellows electrochemical actuator. J. Microelectromech. S. 19 (1), 215–228.
Liakos, I., et al., 2013. Controlled antiseptic release by alginate polymer films and beads. Carbohyd. Polym. 92 (1), 176–183.
Liakos, I., et al., 2014. All-natural composite wound dressing films of essential oils encapsulated in sodium alginate with antimicrobial properties. Int. J. Pharm. 463 (2), 137–145.
Low, L.-M., et al., 2000. Microactuators toward microvalves for responsive controlled drug delivery. Sens. Actuat. B-Chem. 67 (1–2), 149–160.
Luft, G., Kuehl, D., Richter, G.J., 1978. Electro-osmotic valve for the controlled administration of drugs. Med. Biol. Eng. Comput. 16 (1), 45–50.
Luo, X., et al., 2011. Carbon nanotube nanoreservior for controlled release of anti-inflammatory dexamethasone. Biomaterials 32 (26), 6316–6323.
Margetts, L., Sawyer, R., 2007. Transdermal drug delivery: principles and opioid therapy. Cont. Ed. Anaesth. Crit. Care 7 (5), 171–176.
Maroni, A., et al., 2013. Film coatings for oral colon delivery. Int. J. Pharm. 457 (2), 372–394.
Moon, J.M., Chun, B.J., 2011. Fentanyl intoxication caused by abuse of transdermal fentanyl. J. Emerg. Med. 40 (1), 37–40.
Mounteney, J., et al., 2015. Fentanyls: are we missing the signs? Highly potent and on the rise in Europe. Int. J. Drug Pol. 26 (7), 626–631.
Narayanan, R.P., et al., 2012. Photodegradable iron(III) cross-linked alginate gels. Biomacromolecules 13 (8), 2465–2471.
NICE, 2007. Costing statement: medicines adherence: involving patients in decisions about prescribed medicines and supporting adherence.
Paudel, K.S., et al., 2010. Challenges and opportunities in dermal/transdermal delivery. Ther. Delivery 1 (1), 109–131.
Philip, A.K., Philip, B., 2010. Colon-targeted drug delivery systems: a review on primary and novel approaches. Oman Med. J. 25 (2), 79–87.
Pirmoradi, F.N., et al., 2011. On-demand controlled release of docetaxel from a batteryless MEMS drug delivery device. Lab Chip 11 (16), 2744.
Prakash, S., et al., 2011. Polymeric nanohybrids and functionalized carbon nanotubes as drug delivery carriers for cancer therapy. Adv. Drug Deliv. Rev. 63 (14–15), 1340–1351.
Prausnitz, M.R., Langer, R., 2008. Transdermal drug delivery. Nat. Biotechnol. 26 (11), 1261–1268.
Renard, E., Cobelli, C., Kovatchev, B.P., 2013. Closed loop developments to improve glucose control at home. Diabetes Res. Clin. Pr. 102 (2), 79–85.
Richardson, R.T., et al., 2009. Polypyrrole-coated electrodes for the delivery of charge and neurotrophins to cochlear neurons. Biomaterials 30 (13), 2614–2624.
Riggio, C., et al., 2009. Combination of polymer technology and carbon nanotube array for the development of an effective drug delivery system at cellular level. Nanoscale Res. Lett. 4 (7), 668–673.
Rose, J.E., et al., 1985. Transdermal nicotine reduces cigarette craving and nicotine preference. Clin. Pharm. Therapeut. 38 (4), 450–456.
Russell, L.M., Wiedersberg, S., Delgado-Charro, M.B., 2008. The determination of stratum corneum thickness: an alternative approach. Euro. J. Pharma. Biopharma. 69 (3), 861–870.
Santini, J.T., Cima, M.J., Langer, R., 1999. A controlled-release microchip. Nature 397 (6717), 335–338.
Sattar, M., Sayed, O.M., Lane, M.E., 2014. Oral transmucosal drug delivery: current status and future prospects. Int. J. Pharm. 471 (1–2), 498–506.
Shantha, K.L., Ravichandran, P., Rao, K.P., 1995. Azo polymeric hydrogels for colon-targeted drug delivery. Biomaterials 16 (17), 1313–1318.
Sheybani, R., Gensler, H., Meng, E., 2013. A MEMS electrochemical bellows actuator for fluid metering applications. Biomed. Microdev. 15 (1), 37–48.
Shin, W., et al., 2011. A Miniature, nongassing electroosmotic pump operating at 0.5 v. J. Amer. Chem. Soc. 133 (8), 2374–2377.

Siegel, R.A., 2014. Stimuli sensitive polymers and self-regulated drug delivery systems: a very partial review. J. Control. Rel. 190, 337–351.
Söderlind, E., et al., 2015. Validation of the IntelliCap® system as a tool to evaluate extended release profiles in human GI tract using metoprolol as model drug. J. Control. Rel. 217, 300–307.
Stanley, T.H., 2014. The Fentanyl story. J. Pain Symptom Manag. 15 (12), 1215–1226.
Stauffer, W.R., et al., 2011. Rapid modulation of local neural activity by controlled drug release from polymer-coated recording microelectrodes. J. Neural Eng. 8 (4), 044001.
Stead, L.F., et al., 2012. Nicotine replacement therapy for smoking cessation. Cochrane Data. Sys. Rev. 11, CD000146.
Stubbe, B., et al., 2001. The in vitro evaluation of "azo containing polysaccharide gels" for colon delivery. J. Control. Rel. 75 (1–2), 103–114.
Svirskis, D., Travas-Sejdic, J., et al., 2010a. Electrochemically controlled drug delivery based on intrinsically conducting polymers. J. Control. Rel. 146 (1), 6–15.
Svirskis, D., Wright, B.E., et al., 2010b. Evaluation of physical properties and performance over time of an actuating polypyrrole based drug delivery system. Sens. Actuat. B-Chem. 151 (1), 97–102.
Tang, H., et al., 2009. Fluorescent DNA-poly(phenylenevinylene) hybrid hydrogels for monitoring drug release. Chem. Comm. 6, 641–643.
Tarrants, M.L., et al., 2010. Drug therapies for Parkinson's disease: A database analysis of patient compliance and persistence. Am. J. Geriatric Pharmacol. 8 (4), 374–383.
Thanki, K., et al., 2013. Oral delivery of anticancer drugs: challenges and opportunities. J. Control. Rel. 170 (1), 15–40.
Tharp, A.M., Winecker, R.E., Winston, D.C., 2004. Fatal intravenous fentanyl abuse: four cases involving extraction of fentanyl from transdermal patches. Am. J. Forensic Med. 25 (2), 178–181.
Tiwari, G., et al., 2012. Drug delivery systems: an updated review. Int. J. Pharma. Invest. 2 (1), 2–11.
Turner, P.J., Gowland, M.H., Sharma, V., Ierodiakonou, D., Harper, N., Garcez, T., Boyle, R.J., 2015. Increase in anaphylaxis-related hospitalizations but no increase in fatalities: an analysis of United Kingdom national anaphylaxis data, 1992-2012. J. Allergy Clin. Immunol. 135 (4), 956e1–963e1.
Valdés-Ramírez, G., Windmiller, J.R., Claussen, J.C., Martinez, A.G., Kuralay, F., Zhou, M., Wang, J., 2012. Multiplexed and switchable release of distinct fluids from microneedle platforms via conducting polymer nanoactuators for potential drug delivery. Sens. Actuators B. Chem. 161 (1), 1018–1024.
Van den Mooter, G., Kinget, R., 1995. Oral colon-specific drug delivery: a review. Drug Deliv. 2 (2), 81–93.
Vashist, S.K., et al., 2011. Delivery of drugs and biomolecules using carbon nanotubes. Carbon 49 (13), 4077–4097.
Venkatesan, J., et al., 2014. Alginate composites for bone tissue engineering: a review. Int. J. Biol. Macromol. 72C, 269–281.
Wadhwa, R., Lagenaur, C.F., Cui, X.T., 2006. Electrochemically controlled release of dexamethasone from conducting polymer polypyrrole coated electrode. J. Control. Rel. 110 (3), 531–541.
Weinstein, R.S., et al., 2014. Telemedicine, telehealth, and mobile health applications that work: opportunities and barriers. Am. J. Med. 127 (3), 183–187.
Wildgoose, G.G., et al., 2005. X-Ray photoelectron spectroscopy studies of graphite powder and multiwalled carbon nanotubes covalently modified with Fast Black K: evidence for a chemical release mechanism via electrochemical reduction. J. Mater. Chem. 15 (9), 953.
Wise, A.K., Gillespie, L.N., 2012. Drug delivery to the inner ear. J. Neural Eng. 9 (6), 065002.
Yamaoka, T., et al., 2000. Linear type azo-containing polyurethane as drug-coating material for colon-specific delivery: its properties, degradation behavior, and utilization for drug formulation. J. Control. Rel. 66 (2–3), 187–197.
Yasin, M.N., et al., 2014. Implants for drug delivery to the posterior segment of the eye: a focus on stimuli-responsive and tunable release systems. J. Control. Rel. 196, 208–221.
Yoon, Y.-J., et al., 2014. Microfluidics biosensor chip with integrated screen-printed electrodes for amperometric detection of nerve agent. Sens. Actuat. B-Chem. 198, 233–238.
Yu, J., et al., 2015. Microneedle-array patches loaded with hypoxia-sensitive vesicles provide fast glucose-responsive insulin delivery. Proc. Natl. Acad. Sci. USA 112 (27), 8260–8265.

CHAPTER

6

NIOSOMES: A NOVEL APPROACH IN MODERN DRUG DELIVERY SYSTEMS

Sepideh Khoee*, Morteza Yaghoobian**

**University of Tehran, Tehran, Iran*

***Shahid Beheshti Medical University, Tehran, Iran*

CHAPTER OUTLINE

1 INTRODUCTION

Rapid and significant progress in the use of nanotechnology in treatment and diagnosis of diseases has made a new field called nanomedicine and related subfields, such as pharmaceutical nanocarriers known as new branches of medical science. Nanostructures can be made from various materials including polymers, metals, metal oxides, nanogel, lipid-based carriers (liposomes) and surfactant based carriers (niosomes). Reducing the size of drug carriers to a nanoscale has many benefits including: (1) improving the pharmacokinetics and biodistribution of therapeutic agents, (2) reducing toxicity by accumulation of the drug in the target site, (3) facilitating drug passage between the cells and (4) increasing their retention time in biological systems that increase the efficacy of the drug. Novel drug-delivery systems play major roles in drugs development. This technology is an approach to improve the drug release profile, absorption, distribution, and elimination of drugs. To create a new drug-delivery system, two requirements must be considered: First, the drug should be released in accordance to a programmed rate. Second, it should release effective amount of the active ingredient at the target site; doses of traditional forms are not able to meet these goals. Vesicles systems with a bilayer membrane and hollow space in the interior have attracted more attention as drug-delivery systems. The main purpose of the development of vesicular structures is modification of distribution profiles, to control the release of drug in a sustained way and targeted delivery of the drug to the particular body site. Vesicular systems that are able to accommodate larger amounts of drug and create an acceptable surface for targeting enable the drugs to carrying both hydrophilic and lipophilic components. The unique structure of these vesicles is niosome that was produced from nonionic surfactants in the aqueous phase, a system with the advantages of liposomes and the permeability of membranes. An important prerequisite for the use of niosomes as a drug carrier is their integrity in biological fluids. As a carrier, niosomes must be able to circulate in the body and also protect the drug for specified periods of time and must be able to interact with the target site and delivery of their contents to the interior of cells. Niosomes are preferred over other bilayer structures due to chemical stability, biodegradability, biocompatibility, low production cost, low toxicity, and easy storage and handling. Niosomes have been studied with different methods of administration including oral, intramuscular, intravenous, transdermal, and so forth. These vesicles also increase the absorption rate of some drugs from the cell membranes. Production of niosomes in the last decade is an efficient tool in drug delivery due to the development of drug-delivery systems.

2 FACTORS AFFECTING THE FORMATION OF NIOSOMES

The preparation of niosomes and achievement of their desired structure are dependent on sufficient knowledge and a deep understanding of the physicochemical properties of their components. Several factors are involved in the formulation of niosomes. These factors can depend on method of preparation, route of administration, and material consumption. In this chapter, the most important parameters affecting noisome structure will be discussed.

2.1 TYPE OF SURFACTANTS

The main components of niosomes are surfactants. These compounds have a unique chemical structure and this structure gives them dual properties. Surfactants have a lipophilic tail and a hydrophilic head; accordingly, they have two regions with different solubility. The head group involves cationic, anionic, and amphoteric molecules easily dissolved in a water medium. The lipophilic region may consist of alkanes, fluorocarbons, aromatic, or other nonpolar groups. Based on the characteristics of the head group region, surfactants are classified as cationic, anionic, amphoteric, and nonionic. Nonionic surfactants are the most common ones in the preparation of niosomes because of their stability, compatibility, and low toxicity compared to other surface active agents (Hall, 1987; Jiao, 2008; Zografi, 1995). A variety of nonionic surfactants, such as derivatives of alkyl ethers, alkyl esters, and sorbitan fatty acid esters are used in niosomes formation (Giddi et al., 2007; Sankhyan and Pawar, 2012). Based on the head groups, alkyl ethers can be divided into two subgroups: alkyl ethers whose hydrophilic head groups consist of glycerol derivative and those that contain ethylene oxide subunits. Alkyl ethers and alkylglyceryl ethers make relatively stable niosomes, so these surfactants can be used to entrap proteins and peptides, but when they are used in combination with cholesterol, their encapsulation efficiency is reduced (Pardakhty et al., 2007). An example of this group of surfactants is the Brij family. Among the members of the Brij family is Brij30 (Polyoxyethelene 4 lauryl ether) with HLB 9.7 and a phase transition temperature less than 10°C, which provide large unilamellar vesicles with high entrapment efficiency. Unlike other alkyl ether derivatives, it was found that the presence of cholesterol cannot modify formation of vesicles with Brij30. Another limitation of this surfactant is its incompatibility with compounds, such as polyoxyethylene alkylethers, benzocaine, retinoid and oxidizable drugs since, with such substances, it causes oxidation leading to discoloration of the product (Manconi et al., 2003; Raymond et al., 2006). Brij58 (Polyoxyethylene cetyl ether) with HLB 15.7 has an important feature, since reverse vesicles can be formed easily from this type of surfactant, and those vesicles are useful to study active transport of ions across the cell membrane. Brij72 and Brij 76 are stearyl derivatives of polyoxyethylene ether (Polyoxyethylenestearyl ethers) that show favorable properties for the formation of niosomes. Brij72 (HLB= 4.9) forms multilamellar vesicles with higher entrapment efficiency than Brij 76 (HLB= 12.4) (Raymond et al., 2006).

Another group of nonionic surfactants that has been used widely is alkyl esters, such as sorbitan fatty acid esters (Spans) and polyoxyethylene sorbitan fatty acid esters (Tweens). They are regarded as nonirritant and nontoxic substances. Gel transition temperature of spans increases as the length of the acyl chain increases. For example, Span 20 (sorbitan monolaurate) with 9 carbon atoms in the carbon chain is liquid at room temperature. Gel transition temperature of Span 40 (sorbitan monopalmitate) with 13 carbon atoms in the carbon chain is about 47°C and Span 60 with 15 carbon atoms has a gel transition temperature of about 58°C (Bouwstra et al., 1997; Guinedi et al., 2005). Interestingly, with increasing carbon-chain length, entrapment efficiency increases (Hao et al., 2002). Tweens, which is derived from polyoxyethylene reacted with sorbitan fatty acid esters (Spans), have high hydrophilic-lipophilic balance (HLB). In addition to the previously mentioned surfactants, new groups of surfactants are used in the preparation of niosomes. New surfactants, such as Bola-surfactants and Gemini nonionic surfactants are newer classes of surface-active agents that have been synthesized and used to prepare innovative noisome systems. Characteristics of Gemini surfactants are lower CMC value, in addition to being nontoxic, more stable, nonirritating, and nonhemolytic, whereas Bola-Surfactants have higher solubility, a lower aggregation number, and higher CMC (Hait and Moulik, 2002; Ikeda 2003; Yan et al., 2009).

Tyloxapol is a liquid polymer of 4-(1,1,3,3-tetramethylbutyl) phenol that is composed of ethylene oxide and formaldehyde. It is a nonionic biological surfactant of the alkyl aryl polyether alcohol type. It is used in marketed ophthalmic products and as a mucolytic agent for treating pulmonary diseases. Tyloxapol assemble in aqueous solution to form niosomes (Regev and Zana, 1999). Mehta and Jindal reported a formulation of Tyloxapol niosomes for encapsulation of antitubercular drugs (ATD). The high stability and compatibility of the ATD loaded niosomes suggested long shelf life (Mehta and Jindal, 2013).

2.2 THERMODYNAMIC FEATURE

Certainly, formation of vesicles by surfactants is not a coincidence. Thermodynamically stable niosomes are created only when a certain amount of surfactants and additives are mixed together. The energy required for the formation of niosomes occurs through interaction energies (mechanical energy, chemical potential excess energy, and surface energy). The aggregation of nonionic surfactant monomers and the formation of niosomes is the result of two opposing forces: first, a high interfacial tension between lipophilic groups and water that causes these groups to associate, and second, repulsive forces between head groups (the steric and hydrophilic repulsion) (Uchegbu and Florence, 1995). Niosomes in the range of 1–10 mm have been found to be more stable than those in the submicron size. Thermodynamically, bigger niosomes have lower surface free energy and tend to segregate more than smaller ones in order to increase the excess free energy (Moazeni et al., 2010).

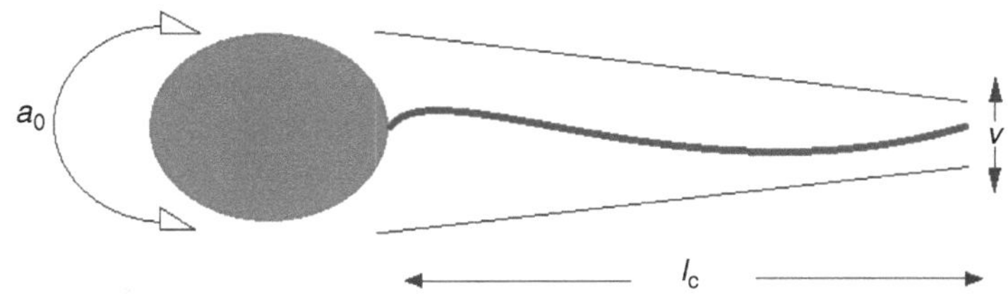

FIGURE 6.1 Critical Packing Parameter (CPP) of Surfactant Where *v* is Lipophilic Tail Volume, *l*c the Critical Lipophilic Tail Length, and a_0 the Polar Head Surface Area

2.3 HYDROPHILIC–LIPOPHILIC BALANCE (HLB)

HLB is a dimensionless critical parameter in the selection of surfactant for the formation of niosomes. This parameter affects the size and amount of drug loading. Surfactants with a HLB number between 4 and 8 are compatible to vesicle formation. Surfactants with HLB values higher than 6 are barely able to form vesicles; thus, for the formation of niosomes, cholesterol must be added (Yoshida et al., 1992).

2.4 GEOMETRIC FEATURES OF AMPHIPHILIC MOLECULE

An important factor in determining the type of association formed in aqueous environments is geometric features. The morphologies of the aggregate formed may be predicted by nominal geometric parameters of the surfactant molecule. The critical packing parameter (CPP) is defined as $v/a_0 l_c$, where (a_0) is the polar head surface area per molecule at the aggregate surface, (V) is the lipophilic tail volume, and (l_c) is the critical lipophilic tail length (Fig. 6.1). According to the CPP value, the shape and size of the equilibrium aggregate would evolve from spherical micelles (CPP $\leq$ 0.33) to cylindrical micelles (1/3 $\leq$ CPP $\leq$ 0.5), bilayers (0.5 $\leq$ CPP $\leq$ 1), or inverse micelles (CPP $>$ 1) (Israelachvili, 2011).

2.5 GEL–LIQUID TRANSITION TEMPERATURE (*Tc*)

One of the important parameters influencing the preparation of niosomes is phase transition temperature (*Tc*) that affects entrapment efficiency, membrane permeability, bilayer rigidity, and stability. Phase transition temperature and the length of alkyl chain of nonionic surfactants are interconvertible factors that affecting the niosomal formulation since both are mutually dependent. It was found that niosome formed from surfactants with the lower phase transition temperature yields vesicles with more flexibility (Alves and Loh, 2012). Usually, decreasing the number of carbons in the alkyl chain length leads to lessening in the gel-to- liquid phase transition temperature, and the lower the gel-to-liquid phase transition temperature results in the production of more leaky niosomes. In other words, the membrane permeability of short chain length surfactants is more than that of surfactants with longer alkyl chain length. Encapsulation efficiency is another property of nonionic niosomes, which is strongly affected by the phase transition temperature. A higher gel-to-liquid phase transition temperature results in a higher encapsulation efficiency. Phase transition temperature depends on the degree of unsaturation in the alkyl chain, which in turn changes the chain flexibility, membrane permeability, and drug release rate. Lack of saturation in the chain lowers the phase transition temperatures and consequently increases chain fluidity and niosomes permeability.

Temperature of the hydration medium should always be above *Tc* of the system, because it will have a direct effect on the formation of niosomes and also induces changes in bilayer shape. As an example, span 60 is the desirable surfactant because it has high phase transition temperature and low HLB value, so it form niosomes of acceptable size (Mahale et al., 2012).

PBS (phosphate buffer saline) often used for hydration in formulation of niosomes with different pH (the pH of the hydration medium depends on the physicochemical properties of the drug). However, other factors, such as volume of hydration medium and time of hydration, as well as temperature are also critical factors that should be considered (Uchegbu and Florence, 1995).

2.6 ADDITIVE AGENTS

Niosome additives have been used for several reasons, including altering the biodistribution, increasing the physical stability, and improving permeability of niosomal formulations. Various additive agents have been used to improve the formulation of niosomes. Cholesterol is the most important compound among these additives; it is a steroid derivative (Fig. 6.2). Cholesterol provides greater stability by increasing the T_C of the vesicles. The presence of cholesterol in the structure of niosomes increases hydrodynamic diameter of the niosomes and increases entrapment efficiency of drugs thus the amount of drug can be changed by various value of cholesterol. Adding cholesterol to the structure of vesicles affect other properties, such as rigidity, membrane permeability, ease of rehydration, and toxicity of niosomes. Cholesterol interacts with surfactants through hydrogen bonding between its OH group and the surfactant alkyl chain, which increase the mechanical inflexibility of the membranes and membrane cohesion (Nasseri, 2005). The amount of cholesterol that is consumed depends on the HLB value of the surfactants. By increasing the HLB value (above 10), the hydrophilic head group is large. Thus, to compensate for this change, it is necessary to increase the minimum amount of cholesterol to be added (Yoshioka et al., 1994).

One series of the other common additives are ionic compounds. These materials prevent vesicles sticking together and increase their stability by electrostatic repulsion. Negatively charged ionic compounds, such as dicetyl phosphate (DCP) and positively charged molecules stearyl amine (STR) or stearyl pyridinium chloride are used in the preparation of niosomes. Use of ionic surfactants was restricted because the formation of niosomes will be prevented if more than a certain amount of these compounds is used. The absence or presence of DCP in the preparation of gentamicin sulphate niosomes was evaluated and optimized by Abdelbary and El-gendy. It was found that the incorporation of DCP in niosomes significantly increased gentamicin sulfate encapsulation efficiency and significantly improved its ocular bioavailability compared with the neutral niosomal systems (Abdelbary and El-gendy, 2008).

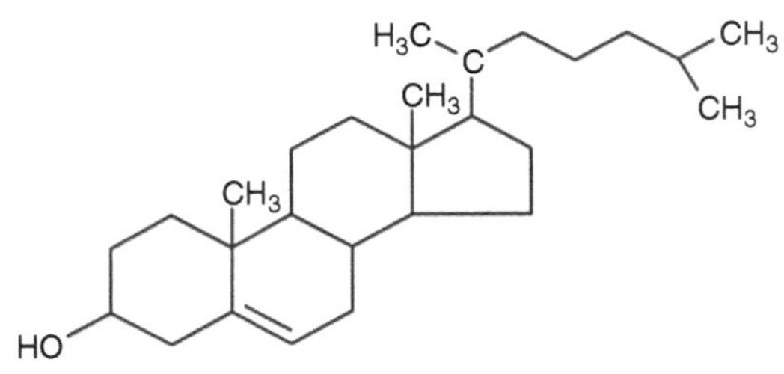

FIGURE 6.2 Chemical Structure of Cholesterol

The other additive is nonionic compounds in which cholesteryl poly-24-oxyethylene ether (SC24) is added in niosomal formulations to provide the steric barrier to vesicle membranes, which prohibits vesicle aggregation and leads to improvement of physical stability of niosomes (Dimitrijevic et al., 1997).

2.7 THE NATURE OF THE DRUG

The nature of the drug has an important role in drug entrapment in niosomes. The drug entrapment is affected by multiple factors, such as interaction between drug and membrane of the noisome, chemical structure, molecular weight, and hydrophilicity and lipophilicity of the drug. For example, a water-soluble drug, such as diclofenac sodium show the maximum loading in niosomes with Tween 60, and it is interesting to see that methotrexate as a hydrophobic drug has maximum entrapment in noisome with Span 60 (Chandraprakash et al., 1993; Raja et al., 1994; Stafford et al., 1988).

3 TYPES OF NIOSOMES

Many types of niosomes are mentioned in the literature, They can be classified according to the size of niosomes (SUV, LUV, etc.) or on the basis of the number of bilayers (SUV, MLV, etc.). Based on these factors, niosomes can be divided into the categories described in the following subsections.

3.1 MULTILAMELLAR VESICLES (MLV)

MLV consists of a number of bilayer surrounding the aqueous lipid compartment separately. The approximate sizes of these vesicles are 0.5–10 μm in diameter (Fig. 6.3). These niosomes are the most widely used because of their simple preparation and mechanical stability when stored for long periods of time. These vesicles are highly suited as drug carriers for lipophilic compounds. Pardakhty et al. prepared multilamellar vesicles (MLVs) by the film-hydration method for encapsulation of human insulin. Three different surfactants with different physical states—that is, Span 60 (solid), Brij 52 (semisolid) and Brij 92 (liquid state)—were employed to form stable niosomal suspensions in the presence of

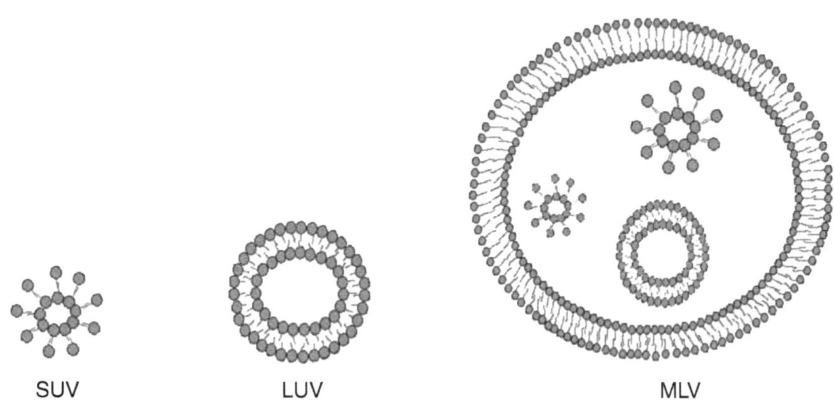

FIGURE 6.3 Structure of SUV, LUV, and MLV

Cholestrol. The mean volume diameters (dv) of prepared vesicles and insulin entrapment efficiencies were measured and it was found that by increasing the HLB value from 4.7 (Span 60) to 5.3 (Brij 52), the mean volume diameter of vesicles increased significantly. Moreover, it was observed that the extent and rate of insulin release in simulated gastric fluid (SGF) and simulated intestinal fluid (SIF) were low for both Brij 92 and Span 60 formulations, whereas, Brij 52 niosome showed almost complete depletion of entrapped insulin in these media (Pardakhty et al., 2011).

3.2 LARGE UNILAMELLAR VESICLES (LUV)

This type of noisome (100–3000 nm) has a high aqueous phase-to-surfactants compartment ratio; accordingly the bioactive material can be entrapped with a very economical use of membrane surfactants (Fig. 6.3). Span 20-based niosome was prepared by the film hydration method and loaded with Newcastle disease vaccine. Newcastle disease is a contagious bird disease affecting many domestic and wild avian species. Adjuvanticity, the ability to modify the immune response, was evaluated using hemagglutination inhibition test (HI). HI was assessed by immunizing orally sixty 3-week-old specific-pathogen-free chickens with the vaccine (0.2 mL/dose) with a 2-week interval between each immunization. The hemagglutination inhibition test showed a 71% increment in immune response over that of the marketed La Sota vaccine which had a 60% increment in immune response (Okore et al., 2011)

3.3 SMALL UNILAMELLAR VESICLES (SUV)

Generally, small unilamellar vesicles (SUV) are produced from multilamellar vesicles (MLV) by various methods, such as sonication, extrusion under high pressure, and high-pressure homogenization. The approximate sizes of small unilamellar vesicles are 10–100 nm. There are three disadvantages of SUVs: (1) they are highly unstable thermodynamically, (2) they tend to aggregate and (3) generally, the amount of loaded hydrophilic drugs in this group is low. (Fig. 6.3)

3.4 BOLA-NIOSOMES

Bola-surfactant compounds have two hydrophilic heads that are connected by one or two long lipophilic spacers. The early 1980s researchers found them in the membrane of archaebacteria (Benvegnu et al., 2004; Meister and Blume, 2007). The result of studies show that bola-surfactants have much lower surface activity but create strong assembling ability, which is determined by much higher surface tension and lower CMC (critical micelle concentration) than their conventional counterparts with the same composition ratio (Ikeda et al., 1989; Meguro et al., 1987). Bola-niosomes showed a suitable tolerability both in in-vitro and in-vivo, and this group of vesicles are made of alpha omega-hexadecyl-bis-(1-aza-18-crown-6) (Bola-surfactant)-Span 80-cholesterol (2:3:1 molar ratio) (Paolino et al., 2008).

3.5 PRONIOSOMES

Proniosomes obtained from a water-soluble carrier is coated by a thin film of dry nonionic surfactant. The carriers should be safe and nontoxic, free flowing, poorly soluble in the loaded mixture solution, and have good water solubility for ease of hydration. Examples include maltodextrin, sorbitol, mannitol, glucose monohydrate, lactose monohydrate and sucrose stearate. The proniosomes are dry niosome formulations that should be hydrated immediately before use. An important advantage of proniosome is the reduction of niosome physical instability, such as aggregation, fusion, and leakage of drug in overtime. On the other hand, it is easier to work with proniosomes as powder (Hu and Rhodes, 2000). Because of the importance of proniosomes, consider them deeply and thoroughly in this section.

3.5.1 Method for Preparation of Proniosomes

Proniosomes can be prepared in a number ways, such as slurry, slow spray coating, and coacervation phase separation methods.

1. *Slurry Method*
   The slurry method is used to prepare *proniosomes* from a stock solution of surfactants and cholesterol in a suitable solvent. In this method, the entire volume of surfactant solution is added to the carrier powder (like sorbitol, maltodextrin, etc.) in a rotary evaporator flask. Additional organic solvent was added to the solution to form a slurry, and it was attached to a rotary flash evaporator (50–70 rpm, 45 ± 2°C). A vacuum should be continued to ensure that the solvent has been completely evacuated and the powder becomes completely dry (Radha et al., 2013). The time required to produce proniosomes by this simple method is independent of the ratio of surfactant solution to carrier material, and it appears to be scalable. Time consumption and the need for specific equipment for the vacuum and nitrogen gas are the limitations of the slurry method (Blazek-Welsh and Rhodes, 2001).
2. *Slow Spray Coating Method*
   In this method, surfactants in organic solvents are sprayed onto carrier powder, and then the solvent is evaporated (under vacuum at 65–70°C for 15–30 min). The surfactant coating on the carrier comes out to be very thin, and hydration of this coating produces multilamellar niosomes. The main advantage of this method is to provide suitable technique for lipophilic drug without instability or tendency of drug to hydrolysis. Drawback of this method is the degradation of carrier particles that occurred when the coating of surfactant solution is applied too quickly (Mishra et al., 2011). The niosomes can be derived from drugs containing proniosome by adding the drug to the surfactant mixture prior to spraying the solution on to the carriers or by adding the drug to the aqueous solution used to hydrate the proniosomes. The advantage of this method is the uniform coating on the carrier, so the active substances and surfactants are kept from degradation; on the other hand, rehydration process is more efficient because the surfactant coating is thinner.
3. *Coacervation Phase Separation Method*
   Proniosomal gels can be prepared by this method. The surfactant, carrier, lipid, and drug can be put in a wide mouthed glass vial and a small amount of solvent added to it. The mixture is heated over a water bath at 60–70°C for 5 min until the surfactant is dissolved completely. Then the aqueous phase is added to the vial and warmed until a clear solution is formed. Then the mixture is cooled down to room temperature until the dispersion gets converted to a proniosome (Singh et al., 2010).

3.5.2 Types of Proniosomes

Depending on the method of preparation, the proniosomes exist in two forms: dry granular proniosomes and liquid crystalline proniosomes.

3.5.2.1 Dry Granular Proniosome

The dry granular proniosome is divided as sorbitol-based proniosome (SBP) and maltodextrin-based proniosomes (MBP), depending on the type of carrier used. SBP is a dry formulation that involves sorbitol as a carrier and is prepared by the slow spraying method. These are made by spraying the surfactant mixture, prepared in an organic solvent, onto the sorbitol powder and then the solvent is evaporated. Although the sorbitol carrier is soluble in the organic solvent, the process needs to be repeated until the desired surfactant coating has been obtained. It should be noted that the residual sorbitol decreases entrapment efficiency. MBP has been used to produce proniosome by the slurry method. In this formulation, maltodextrin is used as the carrier substance. Although maltodextrin is a polysaccharide, it has minimal solubility in organic solvents. Thus, it was possible to coat the maltodextrin particles by simply adding surfactant to the organic solvent to dry the maltodextrin and evaporating the solvent. Hollow blown maltodextrin particles can be used for a significant gain in surface area. Accordingly, the proniosome rehydration process is more efficient because the higher surface area results in a thinner surface coating (Blazek-Welsh and Rhodes, 2001).

3.5.2.2 Liquid Crystalline Proniosomes

Ibrahim et al. (2008) developed a formulation of proniosomes for flurbiprofen delivery by transdermal carrier systems using "Span 20, Span 40, Span 60, and Span 80" as the common, nonirritant, safe, and available nonionic surfactants with and without cholesterol. It was found that, in the case of Span 20 and Span 80, the added amount of cholesterol affected the preparation type to be either a proniosomal alcoholic solution or a liquid crystalline gel system. The lowest transition temperatures of Span 20 and Span 80 (16°C and −12°C, respectively) caused these materials to be liquid at room temperature and cannot form gels at lower concentrations of cholesterol. On the other hand, Span 40 (Tc = 42°C) and Span 60 (Tc = 53°C) produced white creamy gels in the presence or absence of cholesterol because they have high transition temperatures and are solids at room temperature, so they act as gelators by themselves. Gel formation will start at 20% and 30% (molar percent) of cholesterol for Span 20 and Span 80, respectively. Since Span 80 is more hydrophobic (HLB = 4.3) than Span 20 (HLB = 8.6), it produced two phases of proniosomal liquids at cholesterol concentrations below 30%. The formed proniosomal gels were found to be thermo-reversible. The prepared systems were hypothesized to have controlled release for flurbiprofen over an extended period of time (Ibrahim et al., 2008).

3.6 APSOSOMES

Apsosome is a combination of cholesterol, acorbylpalmitate, and highly charged lipid diacetyl phosphate. This component was first hydrated in an aqueous solvent and then sonicated to production niosomes. Apsosomes can be used to increase the transdermal permeation of drugs and reduce the disorder caused by oxygen (Bhaskaran and Lakshmi, 2009).

3.7 DISCOMES

Under certain conditions of the phase diagram of nonionic surfactant, some large discoid structures have been created, which are discomes (large disclike niosomes). Uchegbu et al., provided niosomes from hexadecyl diglycerol ether (C16), cholesterol, and dicetyl phosphate by mechanical shaking and sonication. They reported that niosomes in the discome phase were large and gradually increased in size after sonication. Discomes are thermoresponsive niosomes, which become leaky when temperature is raised above 37°C. This type of niosomes has been proposed as a potential as an ocular drug delivery system (Uchegbu et al., 1992; Uchegbu and Vyas, 1998). Recently, Abdelkader et al. prepared niosome discomes for encapsulation of naltrexone hydrochloride (NTX), a promising agent in treatment of the diabetic complications affecting the cornea (diabetic Keratopathy). In this study, niosomes were prepared by reverse phase evaporation method. The formed niosomes were capable of entrapping NTX up to 61.5% (w/w) (Abdelkader et al., 2011; McLaughlin et al., 2010).

3.8 ELASTIC NIOSOMES

Elastic niosomes have the ability to be flexible without destroying their structure, and they pass through pores smaller than their size. These vesicles are composed of nonionic surfactants, water, and ethanol. Their structural flexibility efficiently improves their penetration capability through intact skin layers. Manosroi et al. (2008a) prepared elastic niosomes from Tween 61, Span 60, and ethanol for entrapment of diclofenac diethyl ammonium so that their deformability index values were 13.76 times higher than the conventional niosomes.

4 NIOSOME PREPARATION

Theoretically, niosome formation requires the presence of particular surfactants, additives, and aqueous solvent. The aggregation of surfactant monomers into vesicles on hydration is the result of a high interfacial tension between water and lipophilic portion of the surfactant. The surfactant molecules tend to be oriented in such a way that the hydrophilic head groups of surfactants point outward, whereas the lipophilic tails face each other to form the bilayer (Fig. 6.4). The opposite forces lead to a

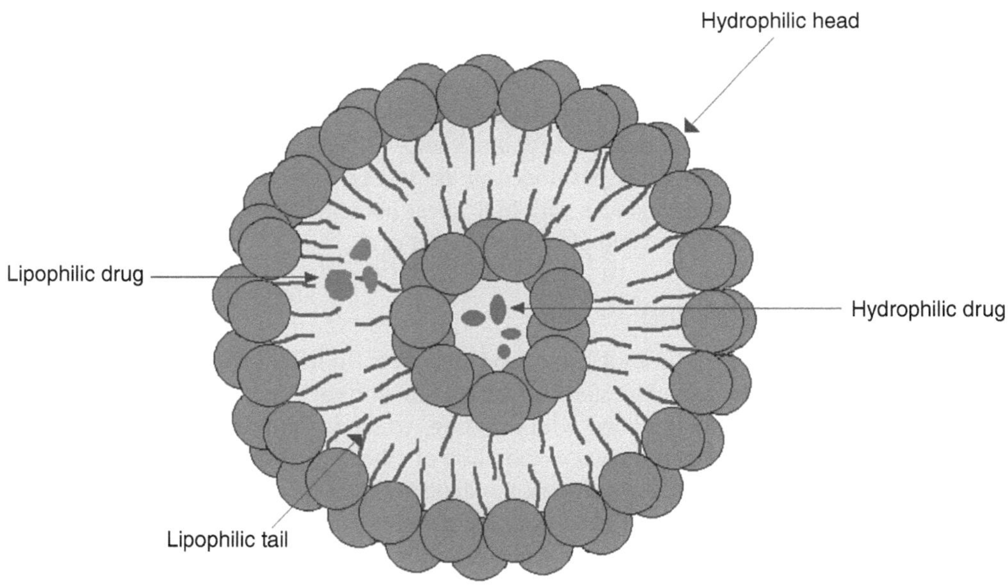

FIGURE 6.4 Niosome Structure

supramolecular assembly. The other essential components in the preparation of niosomes are additives, such as cholesterol. Several methods have been designed to prepare niosomes. In this section we refer only to the number of important procedures to prepare niosomes.

4.1 THIN FILM HYDRATION METHOD

The thin film hydration method is a highly efficient simple method that is commonly used in the preparation of niosomes. The surfactants as well as additives, such as cholesterol are dissolved in organic solvents (if the drug is lipophilic, it should be added simultaneously). The solution is mixed in a round-bottom flask under vacuum and at a certain temperature. A thin film is formed on the wall of the balloon by removing the organic solvent under this condition. For hydration, water or PBS (Phosphate-buffered saline) is added to mixture at a temperature above the phase transition temperature of the surfactant (as a hydrophilic drug is added to the mixture at this point). Tavano et al. incorporated antioxidants into niosomal carriers by the thin film hydration method. In this study accurately weighed amounts of Tween 60 were dissolved in chloroform in a round-bottom flask; then solvent was evaporated under reduced pressure and constant rotation to form a thin lipid film (Tavano et al., 2014a). In another study, Muzzalupo et al. prepared alkyl glucopyranoside-based niosomes containing methotrexate for pharmaceutical applications, MLV were obtained according to the thin film hydration method. Niosomes size ranges were between 300 and 500 nm, with low polydispersity indexes (Muzzalupo et al., 2013). This method forms MLVs that have been used for a variety of pharmaceutical carriers, such as preparing niosomes for encapsulation of nimesulide (Shahiwala and Misra, 2002), insulin (Pardakhty et al., 2007), hydroxyl camptothecin (Hong et al., 2009), minoxidil (Balakrishnan et al., 2009; Mura et al., 2007), glucocorticoid (Marianecci et al., 2010), salicylic acid and p-hydroxyl benzoic acid (Hao and Li, 2011), 5-fluorouracil (Wang et al., 2012), transferrin (Tavano et al., 2013a), methotrexate (Muzzalupo et al., 2013), antioxidants (Tavano et al., 2014a), and so on.

4.2 THE "BUBBLE" METHOD

In this method, organic solvents are not used. Surfactants, additives, and PBS transferred to a three-necked glass reactor. To control the temperature, the reactor is placed in a water bath. Nitrogen gas enters the first neck; the thermometer is positioned in the second neck, and water reflux in the third neck. These components are mixed for 15 s with high shear homogenizer and immediately afterward bubbled with nitrogen gas at 70°C to yield niosomes (Pawar et al., 2012).

4.3 ETHER INJECTION METHOD

This method was introduced in 1976 by Deamer and Bangham. In this method a solution containing a particular ratio of surfactant and cholesterol in an organic solvent, such as diethyl ether is slowly injected into an aqueous solution supplied with the drug, which is kept at a temperature above the boiling point of the organic solvent. Gradual evaporation of ether causes the formation of single-layer niosomes. Depending on the reaction conditions, the diameter of niosomes will be achieved between 50 and 1000 nm. This method also has been used for the entrapment of drugs, such as Rifampicin. It was found that the niosomes prepared by ether injection method demonstrated entrapment efficiencies slightly higher than the niosomes prepared by the hand-shaking method. On the other hand, increasing the cholesterol content in the formulation resulted in decreased entrapment efficiency of microparticles. This may be explained by the fact that every fractional addition of cholesterol produced gradual increases in the rigidity of the vesicles. The uniformity of shape depends on hydration time and temperature. By increasing the hydration time, the shape and size of the niosomes become more uniform. According to this phenomenon, the optimum hydration time was found to be 45 min. In-vitro release rate studies revealed that the time taken for 90% leaching of the drug was 20 h for niosomes of plain surfactant and 23 h for niosomes prepared using surfactant and cholesterol in a ratio of 75:25 mol fraction. When surfactant and cholesterol were in a ratio of 50:50 the time for 90% leaching of drug was 36 h, whereas when surfactant and cholesterol were taken in the ratio of 25:75, the time taken for 90% leaching of the drug was more than 48 h (Jain and Vyas, 1995).

4.4 HAND-SHAKING METHOD

This method is very similar to the thin film hydration method and produces multilamellar vesicles. In this method, the components (surfactants and additives) are dissolved in an organic solvent that was evaporated by a rotary evaporator to form a thin film on the inside wall of the flask. Dry film was hydrated with aqueous solution (PBS, pH = 7.4) containing the drug for about 1 h with mechanical shaking to form niosomes. Waddad et al., prepared niosomes by the hand-shaking method for entrapment of Morin hydrate (a bioflavonoid with antioxidant and anticancer activity). In this study, three different nonionic surfactants (Span 60, Span 80, and Tween 60) were used for different niosomal systems. Three optimized formulations of Morin hydrate (MH) niosomes were successfully prepared as a result of the Taguchi orthogonal array analysis. The optimized MH-Tween 60 niosomes proved to have the highest entrapment efficiency, greater stability, and better release compared to the others MH niosomes (Waddad et al., 2013). In another investigation, niosomal formulation studied as carriers for delivery of diclofenac sodium, niosomes were prepared by the hand-shaking method with sorbitan monoesters (Span 20, 40, 60, and 80) and sorbitan tri-oleate (Span 85) and co-surfactants of polyoxyethelene fatty acid esters (Tween 20, 40, 60, and 80), with or without cholesterol and charged lipids like stearylamine or dicetylphosphate (Marwa et al., 2013).

4.5 HEATING METHOD

Mixtures of nonionic surfactant and additives, such as cholesterol or charge-inducing molecules are added to a hydration solution (PBS, distilled water, etc.) in the presence of a polyol, such as glycerol under nitrogen atmosphere at room temperature. The mixture is heated (about 120°C) on a hot-plate stirrer to dissolve cholesterol while stirring; then the temperature is scaled down and niosomes are left at room temperature (Mozafari et al., 2005).

4.6 SONICATION METHOD

Sonication is an effective method for the production of small vesicles. In this method, providing high energy by sonication can reduce the average size of niosomes. There are two ways to sonication: the probe sonication is employed for preparation of suspensions in small volume, whereas bath sonication is more suitable for the formation of a large volume of dispersions. The solution of drug in the buffer is added to the surfactant/cholesterol mixture in a 10ml glass vial. This dispersion is then probe sonicated at 60°C for 3 min (Verma et al., 2010). Novel niosomes-based formulations of diallyl disulfide (DADS) were evaluated by Alam et al. for their potential to treat disseminated candidi in a mouse model. In this study, they used the sonication method for preparation of Span-based niosomal formulations. The data of their study offered an efficient, negligibly toxic, and cost-effective alternative to the presently used allopathic drug formulations that have undesirable side effects when administered at pharmacologically effective dosage regimens. The promising findings of this study corroborate the approach, whereby therapeutic efficacy of DADS can be enhanced by incorporation into niosomes (Alam et al., 2013). In another study, cefdinir niosomes were provided by the sonication method using Span 60, cholesterol with and without soya lecithin (Bansal et al., 2013).

4.7 MICROFLUIDIZATION METHOD

Microfluidization method is based on the submerged jet principle. In this method, two fluidized streams (one containing surfactant and the other drug) interact at ultra-high velocities. The fluid collected can be recycled through the pump until vesicles of spherical dimensions are obtained. This method resulted in unilamellar niosomes with more uniformity, better reproducibility, and smaller size (Khandare et al., 1994).

4.8 REVERSE PHASE EVAPORATION METHOD

The reverse phase evaporation technique was first described by Szoka and Papahadjopoulos. Controlling the size of niosomes is the advantage of this method, whereas solubility of drug in ether and difficulty of complete removal of ether from the final formulation are drawbacks of this method. The fundamental trick of this method is the elimination of the solvent from the emulsion by evaporation. In order to form an emulsion, a solution of cholesterol and surfactant is prepared in a mixture of ether and chloroform (1:1), an aqueous phase containing the drug is added to this mixture, and the resulting two phases are sonicated at 4–5°C. After sonication, buffer was added and, consequently, a clear gel was formed. The organic phase was evaporated at 40–60°C and at low pressure. The viscose suspension of niosomes was diluted with PBS and heated in water bath at 60°C for 10 min. The hydration process continues until complete evaporation. LUVs are formed during the evaporation of organic solvent (Reddy and Udupa, 1993). Zarei et al. prepared niosomes by reverse phase evaporation for entrapped paclitaxel pegylated and nonpegylated with Span 60 and span 20. (Zarei et al., 2013).

4.9 TRANS-MEMBRANE pH GRADIENT METHOD

Equal proportions of surfactant and cholesterol are dissolved in chloroform. Then the solvent is evaporated under reduced pressure to produce a thin film on the wall of the round bottom flask similar to the thin film hydration method. Hydration of film is done by a solution of an acidic compound (generally citric acid, pH = 4.0) with the help of vortex mixing. Multilamellar vesicles are formed and then are subjected to freeze–thaw cycles and sonication, and then an aqueous solution containing the drug is added to this niosomal suspension and vortexed. In order to create a pH gradient across the niosomal bilayer, the pH of the sample is then raised to 7–7.2 using disodium hydrogen phosphate and the mixture is heated at 60°C for 10 min to give niosomes. A neutral exterior pH concludes in a mixture of both unprotonated and protonated forms of the drug, which are membrane permeable and membrane impermeable, respectively. The unprotonated neutral form of the drug tends to pass the niosome bilayer, which becomes protonated after entering into the acidic medium and trapped inside the vesicles. This diffuses across the membrane and continues until the interior and exterior concentrations of drug are in equilibrium (Bhaskaran and Lakshmi, 2009; Pawar et al., 2012).

4.10 THE ENZYMATIC METHOD

In this method, esterases cleave ester links, and the surfactants (polyoxyethylenestearyl derivatives and polyoxyethylene cholesteryl sebacetate diacetate) and cholesterol in combination with dicetyl phosphate and other lipids produce multilamellar niosomes (Uchegbu and Vyas, 1998).

4.11 SUPERCRITICAL CARBON DIOXIDE FLUID ($scCO_2$)

Supercritical carbon dioxide ($scCO_2$) has been used as an acceptable method to prepare niosomes because it is a one-step method and does not use organic solvent. The $scCO_2$ has good solvating properties (Tc = 31.1°C, Pc = 73.8 bar), because it is inexpensive and does not adversely affect the environment (Manosroi et al., 2008b).

5 CHARACTERIZATION

Physicochemical characterization of niosomes is needed to control the quality of the products. Parameters, such as morphology, size, bilayer specification, charge of vesicle, zeta potential, stability, entrapment efficiency, and in vitro release must be evaluated.

5.1 MORPHOLOGY AND SIZE

Determination of niosome size is a necessary quality control assay because this factor is very effective on the biodistribution and plasma pharmacokinetics of niosomes. Niosomes are usually spherical, and their size can be determined using several techniques, such as microscopy or dynamic light scattering (DLS).The microscopic studies give us information about the size and shape (morphology and geometry) of the niosomes that formed from various vesicle compositions. Several electron microscopy techniques used in the study of niosomes, such as transmission electron microscopy (TEM) (Manconi et al., 2006), negative-stained transmission electron microscopy (NS-TEM) (Arunothayanun et al., 2000), freeze fracture transmission electron microscopy (FF-TEM), atomic force microscopy (AFM) (Cortesi et al., 2007), scanning electron microscopy (SEM), and scanning tunneling microscopy (STM) (Zareýe et al., 1997). Niosomes size and polydispersity index (PdI) can be determined by using a multifunctional zeta potential analyzer. The repulsive forces between the bilayers and entrapped drug play a decisive role in the final size of the niosomes. Zetasizer also provides useful information about the product uniformity of the noisome. PdI can be indicative of the polydispersity of the niosomes suspension. PdI value less than 0.4 demonstrates a monodisperse sample population. A problem of determining size by DLS is that this will not offer information about the shape of the niosome (Manosroi et al., 2010; Muzzalupo et al., 2008).

5.2 CHARGE OF VESICLE AND ZETA POTENTIAL

Zeta potential is very important in the behavior and properties of vesicles; thus, it should be considered carefully in the preparation of niosomes (Agrati et al., 2011). The HLB values of surfactants are very effective on zeta potential. As a result, a trend toward more negative values is noticed with increasing the HLB values (Fathi Azarbayjani et al., 2009). Neutral niosomes are more unstable against aggregation and fusion than charged niosomes.

5.3 NIOSOME STABILITY

Stability of the niosomes is important to maintain the structure of niosomes and drug leakage during storage (before administration) and in biological fluids (after administration). Factors affecting the stability of the niosomes are the type of surfactant, additive content, concentration and nature of encapsulated drug, and condition of storage (Verma et al., 2010). Stability is determined by niosome size, size distribution, zeta potential variations, and entrapment efficiency or by the evaluation of the release rate in absence or in the presence of plasma fluids. Niosomes are more stable than liposomes against chemical degradation. Physical changes that will cause instability in vesicles are drug leakage, aggregation, and bilayer fusion. Types of surfactants and additives determine the fluidity and flexibility of the bilayer membrane. Accordingly, at a higher gel–liquid transition temperature, higher drug maintenance will be observed. Researchers have shown that smaller niosomes are more unstable than larger niosomes, because smaller niosomes have larger surface and higher free energy, so they tend to accumulate and fuse more than larger ones. Temperature is also a very important factor; any change in temperature may cause changes in bilayer membranes. Light can be effective on the unstability of the niosomes. It is well known that exposure to UV irradiation and fluorescent light could promote photo degradation of niosomes (Manconi et al., 2003). The biological sustainability of niosomes is a complex issue because it depends on a variety of factors that interact with the noisome. On the other hand, administration routes of niosomes are different; therefore, the stability is largely a function of the prescribed route.

5.4 ENTRAPMENT EFFICIENCY

To use niosomes for drug delivery, the most important factor is the entrapment efficiency. Entrapment efficiency of niosomes depends on the method of preparation, the nature of the drug, the type of surfactant, and additives. Niosomes prepared by ether injection method have better entrapment efficiency than those prepared by the thin film hydration method. The Tween surfactants with a long alkyl chain and a large HLB value have the highest entrapment efficiency for water-soluble drugs. Cholesterol as a common additive can stabilize bilayers, prevents leakiness, and retards permeation of solutes enclosed in the aqueous core of the vesicles. On the other hand, dicetyl phosphate (DCP) as a charged molecule, is often used to prevent niosome aggregation and increase the stability of niosome dispersion. Consequently, the addition of cholesterol or dicetyl phosphate (DCP) to nonionic surfactants alter the entrapment efficiency. The entrapment efficiency may be defined as the total amount of active ingredient that is released by niosomes. The following equation is used to determine the entrapment efficiency (EE):

$$(\%\text{EE}) = (\text{Amount of drug entrapped} / \text{total amount of drug}) \times 100$$

One of the common methods to determine the amount of entrapped drug in niosomes is, first, separation of unentrapped drug (dialysis, centrifugation, or gel filtration); second, digestion of noisome by using suitable medium (isopropanol, methanol, etc.) and finally, analyzing the resultant solution by appropriate assay method to calculate the entrapped drug. (Balakrishnan et al., 2009; Kumar and Rajeshwarrao, 2011; Shahiwala and Misra, 2002).

The pH of the hydration medium greatly affects the entrapment efficiency of niosomes. In one investigation Mokhtar et al. (2008) provided flurbiprofen niosomes by formation of proniosomes.

High entrapment of flurbiprofen was observed at acidic pH with a maximum encapsulation efficiency of 94.6% at pH 5.5. At pH 7.4 and 8, the lowest encapsulation of flurbiprofen occurred with no significant difference between them. The high encapsulation efficiency of flurbiprofen at acidic pH may be due to ionization of their carboxylic acid groups at this pH. Thus, at lower pH, the proportion of unionized flurbiprofen increases and partitions more readily into the bilayer than the ionized species (Mokhtar et al., 2008).

5.5 IN VITRO RELEASE

The release of drugs from niosomal suspensions can be affected by several factors, such as drug concentration and hydration volume. Generally, the release of the drug from vesicles is studied by the use of dialysis membrane (cut off dialysis bag depends on MW of drug). After removal of the free drug, the niosomal suspension containing the drug equivalent to drug content is transferred into the dialysis bag, which is already washed several times with distilled water and tied at both the ends. Subsequently, it was placed in PBS solution (100–300 mL, pH =7.4) and kept with constant agitation on a magnetic stirrer maintaining a temperature of 37°C. Sample solutions are collected from the beaker at certain time intervals and replaced with the same volume of fresh dissolution media, then analyzed using an appropriate assay method (Erdogan et al., 2005; Manosroi et al., 2010)

6 ADVANTAGES OF NIOSOMES

1. Niosome act as a reservoir to release drug in a gradual and controlled manner.
2. Niosomes are osmotically active and stable. Also, the stability of the drug over long storages time is much higher compared to liposomes.
3. They are improving the medication-taking behavior in one of the following ways: the clearance of the drug is delayed, the drug is protected from the biological environment, and the drug effect is restricted to the target site.
4. Niosomes have low toxicity and high compatibility with biological systems because the commonly used nonionic surfactants are biocompatible, biodegradable, and non-immunogenic.
5. The administration routes of niosomes are diverse (intravenous, oral, etc.).
6. The oral bioavailability of drugs with low absorption has been improved by niosomes and increases permeability of the drug into skin.
7. A wide variety of drugs (hydrophilic and lipophilic ones) with different solubility can be entrapped in the niosomes.
8. Niosomes are effectively use to deliver various drugs to targeted organs.
9. Niosomes are flexible and easily modified because of the functional groups on their hydrophilic heads.
10. Storage and handling of surfactant does not require special conditions.

7 LIMITATION OF NIOSOMAL DRUG DELIVERY SYSTEM

However, there is not enough research on the toxicity of niosomes, but the toxic effect of some surfactants (not niosomes) was evaluated. Based on these studies, ester surfactants are less toxic than the ether surfactants (this is due to the effect of enzymes on the ester bond) and nonionic surfactants have more compatibility and low toxicity compared to other types of surfactants. The results show that increase in alkyl chain length of surfactants cause a decrease in toxicity, whereas increase in the polyoxyethylene chain length enhances cytotoxicity (Hofland et al., 1992). There are a few factors that limit niosome usage in drug delivery systems, such as sterilization of niosomes, instability of aqueous suspension of niosomes (may inadvertently lead to a decrease in the shelflife), and, in some cases, the preparations of some types of niosomes require special instruments. Sterilization heat is harmful for niosomal delivery systems. Dry heat and steam sterilization are unsuitable for niosomal formulations because *Tc*, which is relatively lower than temperature required for heat sterilization, may cause extensive drug leakage from the niosome membrane.

8 COMPARISON OF NIOSOMES VERSUS LIPOSOMES

Niosomes and liposomes have the same structure (bilayer vesicles) but they are different in unit components and chemical properties. Nonionic surfactants are the basic structural components of niosomes, whereas the basic structural components of liposomes are phospholipids. Both structures have the same performance as carriers and can be used in sustained and targeted drug delivery system. Like any other system, their functionality and performance depends on composition of bilayer and methods of preparation. Studies have shown that both systems display almost similar behavior and interaction in the living cells (prolonging the circulation time of entrapped drug and altering distribution and metabolic stability). Due to all these similarities, niosomes have advantages over liposomes that make them an alternative to liposomes. The niosome shows better chemical stability because liposome ingredients (phospholipids) are chemically unstable. Liposomes require special method for storage, handling, and purification of phospholipids, and niosomes are less expensive than liposomes (Khan et al., 2011; Verma and Bindal, 2012).

9 IN-VIVO BEHAVIOR OF NIOSOMES

The size of niosomes is very impressive on tissue distribution or accumulation. The large size niosomes stick to alveoli and will stop in the lungs, whereas the smaller size can easily pass through fenestrations in liver sinusoidal epithelium and thus have better access to spleen. Due to the structural similarity liposomes and niosomes can be predicted that both obey the same mechanisms to face with cells, niosomes behave in-vivo like liposomes. Interaction of cells and niosomes may be described by five mechanisms (Carter et al., 1989; Gregoriadis, 1985):

1. Intermembrane transfer:
   Intermembrane transfer to components that are between the two layers can be done without destroying the structure of niosomes. Interactions between niosomes and lipoproteins may be continued to demolish whole vesicle.
2. Contact release:
   Contact release of niosomes occurs when contact vesicles with the cell increases permeability of the niosome membrane, this leads to release of high concentration of water soluble components in the area adjacent to the cell. In fact, it is an effective way to release the active ingredient without destroying the structure of all niosomes.
3. Endocytosis:
   The niosome enveloped by the cell and lysozyme digests and destroys the bilayer structure of niosomes. In this way, the entrapped material is released into the medium.

4. Fusion:
   In some cases, the property of the niosomal membrane allows the noisome to stick firmly to the cell; these fusion can cause niosomes to easily transfer their contents into the cytoplasm of the cell.
5. Adsorption:
   Adsorption can result from the interaction of physical forces between the cell surface and niosomes or the result of specific binding ligands on the surface of the niosomes and receptors of cell.

10 PHARMACEUTICAL APPLICATIONS AND ADMINISTRATION OF NIOSOMES

Although niosomes are used in a variety of fields (medicine, diagnostics, and cosmetics), it seems that most of the applications have been in drug delivery. Niosome in drug delivery may be examined from two perspectives: the route of administration and the type of active ingredient that should be carried by the niosome. Various routes of administration used in niosomes have been investigated, and it was found that the route of administration strongly influences the design of their formulation. Here, we try to introduce their most important routes and applications.

10.1 ORAL DELIVERY

In this route of administration, drugs deal with problems, such as acids and digestive enzymes in the stomach and small intestine, poor absorption, and variable bioavailability of drug. Therefore, to improve the bioavailability, drugs were administered via a new drug delivery system, such as niosomes. The first oral administration of niosome was reported by Azmin et al. (1985) in a study involving the noisome containing methotrexate. This formulation of noisome significantly increased bioavailability of methotrexate. Niosome prepared from Span 40, Span 60, and cholesterol was proposed as an oral delivery system for Ganciclovir. In vivo studies in rats demonstrate a significant increase in the bioavailability of the drug after oral administration of an optimized formulation when compared to a tablet dosage form (Akhter et al., 2012). Also, other study shows that noisome improved poor and variable oral bioavailability of Cefdinir; it is classified in the Biopharmaceutics Classification Scheme (BCS) as a class IV drug (Bansal et al., 2013). Recently, in one study, mixed niosomes were prepared for oral delivery of Candesartan Cilexetil as a model of a poorly water-soluble drug. The in vitro drug release from niosomes improved after niosomal entrapment compared to free Candesartan Cilexetil (Sezgin-Bayindir et al., 2015).

10.2 TRANSDERMAL DELIVERY

Drug delivery through the skin protect is the drug from the hepatic first-pass metabolism. The main advantage of a transdermal drug delivery route over other types of medication delivery is that the patch provides a controlled release of the medication into the patient, usually through either a porous membrane covering a reservoir of medication or through body heat melting thin layers of medication embedded in the adhesive. The main disadvantage of these systems stems from the fact that the skin is a very effective barrier; as a result, only medications whose molecules are small enough to penetrate the stratum corneum (SC) of skin can be delivered by this method. The intercellular lipid barrier in the stratum corneum is so loose to be permeable against vesicles, such as niosomes. Absorption of a drug into and across skin occurs by passive diffusion when the driving force is the concentration difference between drug carriers and blood. During the last decade, several studies have been performed with niosomes to improve drug delivery through the skin. Some of these efforts include the transdermal or dermal delivery of: Gallidermin (Manosroi et al., 2010), Curcumin (Gupta and Dixit, 2011), Elagic acid (Junyaprasert et al., 2012), Clomipramine (Mohawed et al., 2014), Salidroside (Zhang et al., 2015).

Several mechanisms have been proposed to explain the ability of niosomes in transdermal and dermal drug delivery, such as: (1) niosomes diffuse from the stratum corneum layer of the skin; (2) smaller niosomes are formed in skin (reformation of noisome vesicles). The water content of skin is a critical determinative issue for interpreting and establishing this mechanism, because the stratum corneum has a smaller diameter of lipid lamellar spaces than the noisome bilayer; (3) niosomes interact with stratum corneum with fusion, aggregation, and adhesion to the cell surface, which causes a high thermodynamic activity gradient of the drug at the niosome-stratum corneum surface, which is the driving force for the permeation of lipophilic drugs across the stratum corneum, (4) niosomes may change the properties of the stratum corneum, which makes the intercellular lipid barrier of stratum corneum weaker so permeability increases; (5) nonionic surfactant acts as a permeation enhancer thus improving drug permeation from niosomes (Fang et al., 2001; Javadzadeh et al., 2010; Ogiso et al., 1996; Sahin, 2007).

10.3 OCULAR DELIVERY

From the conventional form of ocular dosage (ophthalmic solution, suspension, and ointment), it is difficult to achieve high bioavailability, because there are many barriers, such as impermeability of corneal epithelium and precorneal tear film that prevent absorption of the drug. Therefore, the bioavailability of drugs administrated by ocular route is typically less than 5% (Abdelkader et al., 2011). To minimize the problems associated with conventional eye drops, different ocular drug delivery systems have been investigated, such as niosomes. Ocular drug delivery by niosomes is preferable because, first, the size of the noisome is enough to resist drainage by reflex tearing and eye blinking, and, second the shape of the niosomes is such that it remains on the eye surface (Aggarwal et al., 2004). Niosomes have been evaluated as ocular drug delivery, such as Tacrolimus (Li et al., 2014) and Gatifloxacin (Zubairu et al., 2015)

10.4 PULMONARY DELIVERY

In recent years, drug delivery via the respiratory system is considered by researchers because it enables the drug to be targeted directly to the lung, both for local and systemic treatment. The pulmonary route is very interesting because the high permeability and large surface area of the alveolar region. Despite the benefits of pulmonary administration, there are also some limitations, such as low efficiency of inhalation system, less drug mass per puff, and poor formulation stability for drug. Niosomes could be used to decrease these problems for protection and more effective delivery of different drugs to the respiratory tract. So far, researchers have recommended different formulations of niosomes for the pulmonary route of drug delivery (Jatav et al., 2011; Moazeni et al., 2010). Proniosomes of the antiasthma steroid beclometasone dipropionate were developed to generate niosomes that were suitable for aerosolization by either air-jet or vibrating-mesh nebulization methods (Elhissi et al., 2013).

10.5 NASAL ROUTE

Nasal mucosa has been considered as a potential administration route to achieve faster and higher level of drug absorption because it is permeable to more drugs than the gastrointestinal tract due to the lack of pancreatic and gastric enzymatic activity, neutral pH of the nasal mucus, and less dilution by gastrointestinal contents (Kisan et al., 2007). The advantages of nasal a drug delivery system include: fast onset of action, avoidance of first pass metabolism thereby improving bioavailability over oral and rectal doses, delivery to the systemic circulation through nasal drug delivery for drugs that cannot be absorbed orally, and exhibition of poor oral absorption for polar compounds may be particularly suited for this route of delivery. The drawbacks of this route include the followings: many medications are not adequately concentrated to achieve ideal dosing volumes, absorption surface area is less when compared to gastrointestinal, and there is a risk of local side effects and irreversible damage of the cilia on the nasal mucosa (Chajed et al., 2011). Priprem et al. (2012) prepared melatonin encapsulated niosomes composed of Span 60, cholesterol, sodium deoxycholate. Size reduction of melatonin niosomes was performed by extrusion through 100 nm polycarbonate membrane, and intranasally administered nanoniosomes could distribute melatonin to the liver, hypothalamus, and testis of male rats (Priprem et al., 2012).

10.6 CANCER THERAPY

New drug delivery systems are designed to increase drug bioavailability and to reduce drug degradation and harmful side effects of anticancer drugs. The niosomes have been shown to be a promising carriers. Most antineoplastic agents cause boisterous side effects. Niosomes, through changes in metabolism (prolong circulation and half -life of drug), can reduce the side effects of drug. It is expected that niosomes containing anti-cancer drugs may have a good accumulation within tumors if designed to fit well. Paclitaxel (PCT) is an antineoplastic agent that can be successfully entrapped in a variety of formulations of niosomes. Niosomes containing Span 40, cholesterol, and dicetyl phosphate (0.0475:0.0475:0.005) showed increase in the stability of PCT against the gastrointestinal enzymes and high entrapment efficiency. The slow release observed from these formulations could be advantageous for reducing the toxic side effects of PCT (Bayindir and Yuksel, 2010). Gude et al., showed that as compared to free drugs, niosomes containing Cisplatin effectively prevented weight loss and reduced bone marrow toxicity (Gude et al., 2002). Niosomes containing methotrexate is designed by Azmin et al. (1985) to have a much higher AUC (Area under the concentration –time curve) than methotrexate solution. Paolino et al. (2008) designed bola-niosomes, which contain 5-fluorouracil to treat skin cancer, this carrier significantly enhanced penetration of the drug compared to an aqueous solution of 5-fluorouracil. Doxorubicin is often used in combination chemotherapy and classified as an anthracycline antibiotic and the niosomal delivery of this drug to mice bearing S-180 tumor was evaluated by Rogerson and coworkers. Doxorubicin was administered as a bolus injection, via the tail vein in mice, at a dose of 5 mg/kg, and the concentration of the drug in S180 sarcoma has been increased, even after a single injection. Doxorubicin profiles following administration of free and niosomal drug were investigated, and it was found that after the initial peaks (30 min), levels of niosomal-DOX prepared from surfactant alone were slightly higher than that of free drug administration, though only to a significant degree following administration in cholesterol-containing niosomes. The increase in the contention of drug in the plasma is due to the fact that the circulation process and slow release from niosomes is more possible than entrapment of vesicles in the liver and accumulation in this organ (Rogerson et al., 1988). However, its previous work in mice has suggested that niosomes may suffer uptake by the reticuloendothelial system, and methotrexate was accumulated in the liver following administration in niosomes (Azmin et al., 1985). In another work, adriamycin has been entrapped within vesicles prepared from a monoalkyl triglycerol ether, and its activity was compared with adriamycin solution in human lung tumor cells grown in monolayer and spheroid culture and in tumor xenografted nude mice. The activity of the encapsulated adriamycin in vitro is maintained with similar clonogenic survival curves following treatment of monolayers and identical growth delays following spheroid exposure. The pharmacokinetics of adriamycin are altered in vivo in human lung tumor-bearing nude mice, when it is administered in niosomal form. There is a prolonged release of drug from the plasma compartment with significantly lower peak levels; lower peak cardiac adriamycin concentrations with a shorter tissue half-life and decreased cardiac AUC and a greater degree of hepatic metabolism to inactive 7-deoxyaglycones. The tumor peak drug level and AUC was similar regardless of the mode of administration of adriamycin. The growth delay (i.e., the time it takes for the tumor volume to double) was significantly longer for adriamycin (15 days) and niosomal adriamycin (11 days) than for the control (5.8 days). It is possible that the therapeutic ratio of adriamycin could be enhanced by administration in niosomal form. It can be concluded that noisome-encapsulated adriamycin has a different tissue distribution from that of free adriamycin, with the predominant effect being lowered cardiac drug concentrations. It is possible, as total dose limitation in antineoplastic therapy is related to development of cardiomyopathy, that even a modest reduction in cardiotoxicity could enable an increase in the total dose of adriamycin administered for a reduction in the likelihood of toxicity (Kerr et al., 1988). Tavano et al., vehiculated doxorubicin by novel magneto-niosomes in order to evaluate the physico–chemical properties of a drug. Tween 60 and Pluronic L64 were used as surfactants in this study, which showed that magneto-niosome formulations were stable for long periods of time and performed a controlled drug release (Tavano et al., 2013b). Tamoxifen Citrate (TMC) niosomes were applied to localized cancer therapy through in vitro breast cancer cytotoxicity as well as in vivo solid antitumor efficacy. The in vivo experiment showed greater reduction in tumor size of niosomal TMC compared to the free drug (Shaker et al., 2015). In one study, niosomal formulations based on dodecylglucuronamide surfactant used to entrap Doxorubicin and 5-fluorouracil as model drugs showed that dodecylglucuronamide surfactant was able to produce niosomes with presence or absence of cholesterol (Tavano et al., 2014b).

10.7 GENE DELIVERY

The success of gene therapy depends a lot on the characteristics of vector. Although it has taken niosomes nearly three decades to enter the pharmaceutical science, only a few studies have focused on their role as vectors for gene delivery. Niosomes have been used as cutaneous gene delivery system especially for the treatment of a variety of skin diseases (Geusens et al., 2011). Niosomes must be charged in their bilayer structure to be able to act as a vector (Huang et al., 2011; Manosroi et al., 2008c). Negatively charged niosomes were designed for gene delivery by Basiri et al. In this study, niosomes were used for DNA (PUC18 supercoiled plasmid) complexation. Different amounts of Span, Tween, and cholesterol with or without dicetylphosphate were used for niosome preparation by the film hydration method (Basiri et al., 2012). Puras et al. used a safe and easy way to prepare a novel niosome formulation based on the 2,3-di(tetradecyloxy)propan-1-amine cationic lipid and combined with squalene and polysorbate 80 to evaluate the transfection efficiency in rat retinas to treat inherited retinal diseases. The chemical structure of the 2,3-di(tetradecyloxy) propan-1-amine cationic lipid offers hope for gene delivery applications, since it contains four domains (polar head, nonpolar hydrophobic chains, linker, and backbone) that rule the transfection processes. However, it was found that the ether bond of the linker domain is more toxic to the cells than the counterpart ester bond. To overcome this problem, the nonionic surfactant polysorbate 80 has been combined with the cationic lipid to form niosomes, since it has been reported that the presence of polysorbate 80 decreases the proportion of the cationic lipid in the formulation, which enhances the tolerance of the cells. In addition, polysorbate 80 acts as an emulsifier creating a steric barrier that avoids aggregation, and the presence of polyethylene glycol (PEG) chains in its structure improves transfection efficiency of liposome formulations. In order to enhance the stability and rigidity of the vesicles, squalene has been incorporated into the noisome formulation, instead of the classical cholesterol. Squalene, a natural lipid belonging to the terpenoid family, is a biochemical precursor of cholesterol and other steroids. Niosomes provided by the solvent emulsification evaporation technique were mixed with the pCMS-EGFP plasmid to form lipoplexes. This novel niosomal system displays a promising approach to deliver genetic substance into the retina to treat inherited retinal diseases (Puras et al., 2014).

10.8 DRUG TARGETING

Targeted drug delivery by niosomes can be examined in two parts. First, the cells of Reticulo Endothelial System (RES) preferentially take up niosomes− is played by one type of circulating serum factor called opsonin−. These factors mark the niosomes for clearance by RES macrophages, so niosomes are beneficial carriers for targeted drug delivery in cases in which infecting organism reside in the organs of RES, in one study showing that Tween 20 niosomes have interinsic selectivity to macrophages (Agrati et al., 2011). Second, the drug can be lead to the target and direct organs, such as brain and liver. Bragagni et al. studied functionalized niosomes performed for brain targeting of dynorphin-B. In this work, niosomal formulation was functionalized with the glucose derivative N-palmitoyl glucosamine. The optimized niosomal formulation of dynorphin-B was administered intravenously (IV) to mice and showed a pronounced antinociceptive effect, clearly more than that given by IV administration of the simple solution of the peptide at the same concentration (Bragagni et al., 2014). Ribavirin niosomes were prepared by thin film hydration method using Span 60, cholesterol and DCP with different molar ratio for liver targeting. The results showed that the optimized niosomal formulation can significantly increase entrapment percentage of ribavirin in comparison to the other molar ratios. Niosomal ribavirin dispersion with optimized formulation and free ribavirin solution were administered as a single dose of 30 mg/kg by intraperitoneal injection into two groups of rats to compare the liver ribavirin concentration. The obtained results show more than sixfold accumulation of ribavirin in liver compared with ribavirin-free solution (Hashim et al., 2010). In another work, Bragagni et al. designed a niosomal formulation functionalized with the glucose-derivative *N*-palmitoylglucosamine (NPG) to prepare a brain targeted delivery system for encapsulation of doxorubicin. Their selected formulation consisted of Span, cholesterol, Solulan, and NPG, obtained via thin-layer evaporation. Intravenous administration of this formulation to rats resulted in the reducing drug accumulation in the heart in comparison to the commercial formulation (Bragagni et al., 2012).

10.9 IMMUNOLOGICAL APPLICATIONS

Immune niosomes are vesicles with antibodies attached to their surface, as potent adjuvant in terms of immunological selectivity, low toxicity, and stability. In-vivo studies showed that the niosomes containing Leishmania major (ALM) have a moderate effect in the prevention of cutaneous Leishmanias in BALB/c mice (Brewer and Alexander, 1992). Formulation of niosomes prepared from Span 85/cholesterol for encapsulating of DNA encoding and HBsAg have been developed and its immune stimulating activity was studied by topical application of niosomes in Balb/c mice. Results were compared with naked DNA and liposomes encapsulated DNA (Vyas et al., 2005). Pardakhty et al. (2012) studied a series of positively charged micron-sized niosomal formulations containing sorbitan esters, cholesterol, and CTAB (Cetyl trimethyl ammonium bromide) for the entrapment of autoclaved ALM.

10.10 DELIVERY OF PEPTIDE DRUGS

Oral administration of peptide drugs is faced with many challenges, such as enzyme digestion in gastrointestinal tract. Different studies have shown that niosomes are capable of protecting this group of drugs against breakdown. In one study, niosomal carrier for hemoglobin (Hb) was prepared from PEG 6000, Tween 80, and Span 80. Their results show that Hb can be adsorbed and outspread on the surface of the niosome membrane and stabilized through this system (Liu et al., 2007). Researchers prepared a type of insulin entrapped in niosomes, which were stable in solubilizing bile salt solutions. The results of this study indicated that niosomes may be utilized as oral carriers of insulin, because they could effectively prolong the release of insulin in both SGF (simulated gastric fluid) and SIF (simulated intestinal fluid) (Pardakhty et al., 2011).

10.11 DIAGNOSTIC IMAGING

Nanometric size of the carriers is an important factor in the diagnosis of tumors at an early stage. The niosomes are used as carriers of Iobitridol as a diagnostic agent used for X-ray imaging. These niosomes were prepared from polyoxyethylene glycol 4000 stearate, D-alpha tocopheryl polyethylene glycol 1000 succinate, sorbitan monostearate, cholesterol, and dicetylphosphate (Muller et al., 2000). Luciani et al. evaluated a magnetic resonance (MR) imaging contrast agent for tumor detection based on combination of polyethyleneglycol (PEG) and glucose conjugates to the surface of niosomes for the targeting of overexpressed glucose receptors, this niosomal system significantly improved tumor targeting of an encapsulated paramagnetic agent assessed with MR imaging in a human carcinoma xenograft model (Luciani et al., 2004).

11 FINAL CONSIDERATION AND OUTLOOK

Niosomes are some of the best new drug delivery carriers that have great potential for targeted drug delivery. The advantage of niosomes as drug delivery carriers is their ability to create various structures, such as proniosome, discome, and aspasome. In recent years, in the field of nanomedicine and drug delivery, a lot of attentions has been paid to vesicular systems, especially niosomes and liposomes. Niosomes change the plasma clearance, tissue distribution, metabolism, bioavailability, and cellular interaction of the drug. Niosomes have advantages over other drug delivery carriers and can be used in various fields of pharmaceutical sciences. They are a very useful tool for targeted drug delivery and diagnostic approaches. Despite numerous studies and with respect to the fact that niosomes have a long way to go to pass the stage of clinical reality, there are still multiple and serious challenges regarding the niosomes. For several reasons (stability, cost, etc.), niosomes are better than liposomes for drug delivery. Surfactants as building components of niosomes have the most important role in the formation and properties of these carriers, thus any developments in the synthesis of new surfactants that are nontoxic, low-cost, biocompatible, and biodegradable will increase efficiency of niosomes. Therefore, it can be hoped that the special structures of the niosomes arise in the coming years.

REFERENCES

Abdelbary, G., El-gendy, N., 2008. Niosome-encapsulated gentamicin for ophthalmic controlled delivery. AAPS Pharm. Sci. Tech. 9 (3), 740–747.

Abdelkader, H., Ismail, S., Kamal, A., Alany, R .G., 2011. Design and evaluation of controlled release niosomes and discomes for naltrexone hydrochloride ocular delivery. J. Pharm. Sci. 100 (5), 1833–1846.

Aggarwal, D., Garg, A., Kaur, I.P., 2004. Development of a topical niosomal preparation of acetazolamide: preparation and evaluation. J. Pharm. Pharmacol. 56, 1509–1517.

Agrati, C., Marianecci, C., Sennato, S., Carafa, M., Bordoni, V., Cimini, E., Tempestilli, M., Pucillo, L.P., Turchi, F., Martini, F., Borioni, G., Bordi, F., 2011. Multicompartment vectors as novel drug delivery systems: selective activation of Tγδ lymphocytes after zoledronic acid delivery. Nanomedicine 7, 153–161.

Akhter, S., Kushwaha, S., Warsi, M.H., Anwar, M., Ahmad, M.Z., Ahmad, I., Talegaonkar, S., Khan, Z.I., Khar, R.K., Ahmad, F.J., 2012. Development and evaluation of nanosized niosomal dispersion for oral delivery of Ganciclovir. Drug Dev. Ind. Pharm. 38, 84–92.

Alam, M., Zubair, S., Farazuddin, Ahmad, E., Khan, A., Zia, Q., Malik, A., Mohammad, O., 2013. Development, characterization and efficacy of niosomal diallyl disulfide in treatment of disseminated murine candidiasis. Nanomedicine 9, 247–256.

Alves, F.R., Loh, W., 2012. Vesicles prepared with the complex salts dioctadecyldimethyl ammonium polyacrylates. J. Colloid Interface Sci. 368, 292–300.

Arunothayanun, P., Bernard, M.S., Craig, D.Q., Uchegbu, I.F., Florence, A.T., 2000. The effect of processing variables on the physical characteristics of non-ionic surfactant vesicles (niosomes) formed from a hexadecyl diglycerol ether. Int J. Pharm. 201, 7–14.

Azmin, M.N., Florence, A.T., Handjani-Vila, R.M., Stuart, J.F., Vanlerberghe, G., Whittaker, J.S., 1985. The effect of non-ionic surfactant vesicle (niosome) entrapment on the absorption and distribution of methotrexate in mice. J. Pharm. Pharmacol. 37, 237–242.

Balakrishnan, P., Shanmugam, S., Lee, W.S., Lee, W.M., Kim, J.O., Oh, D.H., Kim, D.D., Kim, J.S., Yoo, B.K., Choi, H.G., Woo, J.S., Yong, C.S., 2009. Formulation and in vitro assessment of minoxidil niosomes for enhanced skin delivery. Int. J. Pharm. 244, 73–80.

Bansal, S., Aggarwal, G., Chandel, P., Harikumar, S.L., 2013. Design and development of cefdinir niosomes for oral delivery. J. Pharm. Bioallied. Sci. 5 (4), 318–325.

Basiri, M., Pardakhty, A., Mozafari, R., Mottaghi, A., 2012. Preparation and characterization of negatively-charged niosomes as gene-delivery vectors. Res. Pharm. Sci. 7 (5), S365.

Bayindir, Z.S., Yuksel, N., 2010. Characterization of niosomes prepared with various nonionicsurfactants for paclitaxel oral delivery. J. Pharm. Sci. 99, 2049–2060.

Benvegnu, T., Brard, M., Plusquellec, D., 2004. Archaeabacteria bipolar lipid analogues: structure, synthesis and lyotropic properties. Curr. Opin. Colloid Interface Sci. 8, 469–479.

Bhaskaran, S., Lakshmi, P.K., 2009. Comparative evaluation of niosomes formulations prepared by different techniques. Acta Pharm. Sci. 51, 27–32.

Blazek-Welsh, A.I., Rhodes, D.G., 2001. Maltodextrin-based proniosomes. AAPS Pharm. Sci. 3 (1), E1.

Bouwstra, J.A., Van Hal, D.A., Hofland, H.E.J., 1997. Preparation and characterization of nonionic surfactant vesicles. Colloid Surf. A. 123-124, 71–80.

Bragagni, M., Mennini, N., Ghelardini, C., Mura, P., 2012. Development and characterization of niosomal formulations of doxorubicin aimed at brain targeting. J. Pharm. Pharm. Sci. 15, 184–196.

Bragagni, M., Mennini, N., Furlanetto, S., Orlandini, S., Ghelardini, C., Mura, P., 2014. Development and characterization of functionalized niosomes for brain targeting of dynorphin-B. Eur. J. Pharm. Biopharm. 87, 73–79.

Brewer, J.M., Alexander, J., 1992. The adjuvant activity of non-ionic surfactant vesicles (niosomes) on the BALB/c humoral responseto bovine serum albumin. Immunology 75 (4), 570–575.

Carter, K.C., Dolan, T.F., Baillie, A.J., MacColgan, C., 1989. Visceral leishmaniasis: drug carrier system characteristics and the ability to clear parasites from the liver, spleen and bone marrow in *Leishmania donovani* infected BALB/c mice. J. Pharm. Pharmcol. 41 (2), 87–91.

Chajed, S., Sangle, S., Barhate, S., 2011. Advantageous nasal drug delivery system: a review. Int. J. Pharm. Sci. Res. 2 (6), 1322–1336.

Chandraprakash, K.S., Udupa, N., Devi, P.U., Pillai, G.K., 1993. Effect of niosome encapsulation of methotrexate. Macrophage activation on tissue distribution of methotrexate and tumour size. Int. J. Pharm. 1, 133–137.

Cortesi, R., Esposito, E., Corradini, F., Sivieri, E., Drechsler, M., Rossi, A., Scatturin, A., Menegatti, E., 2007. Non-phospholipid vesicles as carriers for peptides and proteins: production, characterization and stability studies. Int. J. Pharm. 339, 52–60.

Dimitrijevic, D., Lamandin, C., Uchegbu, I.F., Shaw, A.J., Florence, A.T., 1997. The effect of monomers and of micellar and vesicular forms of non-ionic surfactants (Solulan C24 and Solulan 16) on Caco-2 cell monolayers. J. Pharm. Pharmacol. 49, 611–616.

Elhissi, A., Hidayat, K., Phoenix, D.A., Mwesigwa, E., Crean, S., Ahmed, W., Faheem, A., Taylor, K.M., 2013. Air-jet and vibrating-mesh nebulization of niosomes generated using a particulate-based proniosome technology. Int. J. Pharm. 444, 193–199.

Erdogan, S., Ozer, A.Y., Bilgili, H., 2005. In vivo behavior of vesicular urokinase. Int. J. Pharm. 295, 1–6.

Fang, J.Y., Hong, C.T., Chiu, W.T., Wang, Y.Y., 2001. Effect of liposomes and niosomes on skin permeation of enoxacin. Int. J. Pharm. 219, 61–72.

Fathi Azarbayjani, A., Tan, E.H., Chan, Y.W., Chan, S.Y., 2009. Transdermal delivery of haloperidol by proniosomal formulations with non-ionic surfactants. Biol. Pharm. Bull. 32 (8), 1453–1458.

Geusens, B., Strobbe, T., Bracke, S., Dynoodt, P., Sanders, N., Van Gele, M., Lambert, J., 2011. Lipid-mediated gene delivery to the skin. Eur. J. Pharm. Sci. 43 (4), 199–211.

Giddi, H.S., Arunagirinathan, M.A., Bellare, J.R., 2007. Self-assembled surfactant nano-structures important in drug delivery. Indian J. Exp. Biol. 45 (2), 133–159.

Gregoriadis, G., 1985. Liposomes as carriers for drugs and vaccines. Trends Biotechnol. 3, 235.

Gude, R.P., Jadhav, M.G., Rao, S.G., Jagtap, A.G., 2002. Effects of niosomal cisplatin and combination of the same with theophylline and with activated macrophages in murine B16F10 melanoma model. Cancer Biother. Radiopharm. 17 (2), 183–192.

Guinedi, A.S., Mortada, N.D., Mansour, S., Hathout, R.M., 2005. Preparation and evaluation of reverse-phase evaporation and multi-lamellar niosomes as ophthalmic carriers of acetazolamide. Int. J. Pharm. 306, 71–82.

Gupta, N.K., Dixit, V.K., 2011. Development and evaluation of vesicular system for curcumin delivery. Arch. Dermatol. Res. 303 (2), 89–101.

Hait, S.K., Moulik, S.P., 2002. Gemini surfactants: a distinct class of self-assembling molecules. Curr. Sci. 82, 1101–1111.

Hall, D.G., 1987. Thermodynamics of micelle formation. In: Schick, J. (Ed.), Nonionic Surfactants: Physical Chemistry. In: Surfactant Science Series. Marcel Dekker, New York, pp. 233–296.

Hao, Y.M., Li, K.A., 2011. Entrapment and release difference resulting from hydrogen bonding interactions in noisome. Int. J. Pharm. 403, 245–253.
Hao, Y., Zhao, F., Li, N., Yang, Y., Li, K., 2002. Studies on a high encapsulation of colchicine by a niosome system. Int. J. Pharm. 244, 73–80.
Hashim, F., El-Ridy, M., Nasr, M., Abdallah, Y., 2010. Preparation and characterization of niosomes containing ribavirin for liver targeting. Drug Deliv. 17 (5), 282–287.
Hofland, H.E., Bouwstra, J.A., Verhoef, J.C., Buckton, G., Chowdry, B.Z., Ponec, M., Junginger, H.E., 1992. Safety aspects of non-ionic surfactant vesicles—a toxicity study related to the physicochemical characteristics of non-ionic surfactants. J. Pharm. Pharmacol. 44, 287–294.
Hong, M., Zhu, S., Jiang, Y., Tang, G., Pei, Y., 2009. Efficient tumor targeting of hydroxyl camptothecin loaded PEGylated niosomes modified with transferrin. J. Control. Release 133, 96–102.
Hu, C., Rhodes, D.G., 2000. Proniosomes: a novel drug carrier preparation. Int. J. Pharm. 206, 110–122.
Huang, Y., Rao, Y., Chen, J., Yang, V.C., Liang, W., 2011. Polysorbate cationic synthetic vesicle for gene delivery. J. Biomed. Mater. Res. A. 96, 513–519.
Ibrahim, M.M.A., Sammour, O.A., Hammad, M.A., Megrab, N.A., 2008. In-vitro evaluation of proniosomes as a drug carrier for flurbiprofen. AAPS Pharm. Sci. Tech. 9 (3), 782–790.
Ikeda, I., 2003. Synthesis of Gemini (Dimeric) and related surfactants. In: Zana R., Xia J. (Eds.), Gemini Surfactants: Synthesis, Interfacial and Solution-Phase Behavior, and Applications. Surfactant Science Series, Marcel Dekker, New York, pp. 9–35.
Ikeda, K., Yasuda, M., Ishikawa, M., Esumi, K., Meguro, K., Binana-Limbele, W., Zana, R., 1989. Physicochemical properties of α, ω-type bolaform surfactant in aqueous solution. Eicosane-1,20- bis(triethylammonium bromide). Colloid Polym. Sci. 267, 825–830.
Israelachvili, J.N., 2011. Intermolecular and Surface Forces, third ed. Academic Press, London, pp. 536–542.
Jain, C.P., Vyas, S.P., 1995. Preparation and characterization of niosomes containing rifampicin for lung targeting. J. Microencapsul. 12 (4), 401–407.
Jatav, V.S., Singh, S.K., Khatri, P., Sharma, A.K., Singh, R., 2011. Formulation and in-vitro evaluation of rifampicin-loaded niosomes. J. Chem. Pharm. Res. 3, 199–203.
Javadzadeh, Y., Shokri, J., Hallaj-Nezhadi, S., Hamishehkar, H., Nokhodchi, A., 2010. Enhancement of percutaneous absorption of Finasteride by cosolvents, cosurfactant and surfactants. Pharm. Dev. Technol. 15 (6), 619–625.
Jiao, J., 2008. Polyoxyethylated nonionic surfactants and their applications in topical ocular drug delivery. Adv. Drug Deliv. Rev. 60, 1663–1673.
Junyaprasert, V.B., Singhsa, P., Suksiriworapong, J., Chantasart, D., 2012. Physicochemical properties and skin permeation of Span 60/Tween 60 niosomes of ellagic acid. Int. J. Pharm. 423, 303–311.
Kerr, D.J., Rogerson, A., Morrison, G.J., Florence, A.T., Kaye, S.B., 1988. Antitumour activity and pharmacokinetics of niosome encapsulated adriamycin in monolayer, spheroid and xenograft. Br. J. Cancer 58, 432–436.
Khan, A., Sharma, P.K., Visht, S., Malviya, R., 2011. Niosomes as colloidal drug delivery system: a review. J. Chronotherapy Drug Deliv. 2 (1), 15–21.
Khandare, J.N., Madhavi, G., Tamhankar, B.M., 1994. Niosomes novel drug delivery system. East Pharmacist. 37, 61–64.
Kisan, R.J., Manoj, N.G., Ishaque, M.S., Vilarsrao, J.K., Sambjahi, S.P., 2007. Nasal drug delivery system-factors affecting and applications. Curr. Drug Therapy 2, 27–38.
Kumar, G.P., Rajeshwarrao, P., 2011. Nonionic surfactant vesicular systems for effective drug delivery—an overview. Acta Pharm. Sin. B. 1 (4), 208–219.
Li, Q., Li, Z., Zeng, W., Ge, S., Lu, H., Wu, C., Ge, L., Liang, D., Xu, Y., 2014. Proniosome-derived niosomes for tacrolimus topical ocular delivery: In vitro cornea permeation, ocular irritation, and in vivo anti-allograft rejection. Eur. J. Pharm. Sci. 62, 115–123.

Liu, T., Guo, R., Hua, W., Qui, J., 2007. Structure behaviors of hemoglobin in PEG 6000/Tween 80/Span 80/H2O noisome system. Colloid Surf. A. 293, 255–261.
Luciani, A., Olivier, J.C., Siauve, N., Brillet, P.Y., Bessoud, B., Gazeau, F., Uchegbu, I.F., Kahn, E., Frija, G., Cuenod, C.A., 2004. Glucose receptor MR imaging of tumors: study in mice with PEGylated paramagnetic niosomes. Radiology (2), 135.
Mahale, N.B., Thakkar, P.D., Mali, R.G., Walunj, D.R., Chaudhari, S.R., 2012. Niosomes: novel sustained release nonionic stable vesicular systems—n overview. Adv. Colloid Interface Sci. 183–184, 46–54.
Manconi, M., Valenti, D., Sinico, C., Lai, F., Loy, G., Fadda, A.M., 2003. Niosomes as carriers for tretinoin II. Influence of vesicular incorporation on tretinoin photost ability. Int. J. Pharm. 260, 261–272.
Manconi, M., Sinico, C., Valenti, D., Lai, F., Fadda, A.M., 2006. Niosomes as carriers for tretinoin III. A study into the in vitro cutaneous delivery of vesicle-incorporated tretinoin. Int. J. Pharm. 311 (1-2), 11–19.
Manosroi, A., Jantrawut, P., Manosroi, J., 2008a. Anti-inflammatory activity of gel containing novel elastic niosomes entrapped with diclofenac diethylammonium. Int. J. Pharm. 360, 156–163.
Manosroi, A., Chutoprapat, R., Abe, M., Manosroi, J., 2008b. Characteristics of niosomes prepared by supercritical carbon dioxide ($scCO_2$) fluid. Int. J. Pharm. 352, 248–255.
Manosroi, A., Thathang, K., Werner, R.G., Schubert, R., Manosroi, J., 2008c. Stability of luciferase plasmid entrapped in cationic bilayer vesicles. Int. J. Pharm. 356, 291–299.
Manosroi, A., Khanrin, P., Lohcharoenkal, W., Werner, R.G., Manosroi, W., Götz, F., Manosroi, W., Manosroi, J., 2010. Transdermal absorption enhancement through rat skin of gallidermin loaded in niosomes. Int. J. Pharm. 392, 304–310.
Marianecci, C., Celia, C., Fresta, M., Carafa, M., Alhaique, F., 2010. Non-ionic surfactant vesicles in pulmonary glucocorticoid delivery: characterization and interaction with human lung fibroblast. J. Control. Release 147 (1), 127–135.
Marwa, A., Omaima, S., Hanaa, E.G., Mohammed, A.S., 2013. Preparation and in-vitro evaluation of diclofenac sodium niosomal formulations. Int. J. Pharm. Sci. Res. 4 (5), 1757–1765.
McLaughlin, P.J., Sassani, J.W., Klocek, M.S., Zagon, I.S., 2010. Diabetic keratopathy and treatment by modulation of the opioid growth factor (OGF)–OGF receptor (OGFr) axis with naltrexone. Brain Res. Bull. 81, 236–247.
Meguro, K., Ikeda, K., Otsuji, A., Taya, M., Yasuda, M., Esumi, K., 1987. Physicochemical properties of α, ω-type surfactant in aqueous solution. J. Colloid Interface Sci. 118, 372.
Mehta, S.K., Jindal, N., 2013. Formulation of Tyloxapol niosomes for encapsulation, stabilization and dissolution of anti-tubercular drugs. Colloids Surf. B Biointerfaces 101, 434–441.
Meister, A., Blume, A., 2007. Self-assembly of bipolar amphiphiles. Curr. Opin. Colloid Interface Sci. 12, 138–147.
Mishra, A., Kapoor, A., Bhargava, S., 2011. Proniosoml gel as a carrier for improved transdermal drug delivery. Asian J. Pharm. Life Sci. 1 (4), 370–379.
Moazeni, E., Gilani, K., Sotoudegan, F., Pardakhty, A., Najafabadi, A.R., Ghalandari, R., Fazeli, M.R., Jamalifar, H., 2010. Formulation and in vitro evaluation of ciprofloxacin containing niosomes for pulmonary delivery. J. Microencapsul. 27, 618–627.
Mohawed, O.A.M., El-Ashmoony, M.M., Elgazayerly, O.N., 2014. Niosome encapsulated clomipramine for transdermal controlled delivery. Int. J. Pharm. Pharm. Sci. 6, 567–575.
Mokhtar, M., Sammour, O.A., Hammad, M.A., Megrab, N.A., 2008. Effect of some formulation parameters on flurbiprofen encapsulation and release rates of niosomes prepared from proniosomes. Int. J. Pharm. 361, 104–111.
Mozafari, M.R., Reed, C.J., Rostron, C., Hasirci, V., 2005. Construction of a gene delivery vector using anionic liposomes. In: Nanoliposomes: from fundamentals to recent developments, Trafford Publishing Ltd., Oxford, UK, pp. 101–111.
Muller, D., Foulon, M., Bonnemain, B., Vandamme, T.F., 2000. Niosomes as carriers of radiopaque contrast agents for X-ray imaging. J. Microencapsul. 17, 227–243.

Mura, S., Pirot, F., Manconi, M., Falson, F., Fadda, A.M., 2007. Liposomes and niosomes as potential carriers for dermal delivery of minoxidil. J. Drug Target. 15 (2), 101–108.

Muzzalupo, R., Tavano, L., Trombino, S., Cassano, R., Piccia, N., La Mesab, C., 2008. Niosomes from α,ω-trioxyethylene-bis(sodium 2-dodecyloxy-propylenesulfonate): preparation and characterization. Colloids Surf. B: Biointerfaces 64, 200–207.

Muzzalupo, R., Tavano, L., La Mesa, C., 2013. Alkyl glucopyranoside-based niosomes containing methotrexate for pharmaceutical applications: evaluation of physico-chemical and biological properties. Int. J. Pharm. 458 (1), 224–229.

Nasseri, B., 2005. Effect of cholesterol and temperature on the elastic properties of niosomal membranes. Int. J. Pharm. 300 (1–2), 95–101.

Ogiso, T., Niinaka, N., Iwaki, M., 1996. Mechanism for enhancement effect of lipid disperse system on percutaneous absorption. J. Pharm. Sci. 85, 57–64.

Okore, V.C., Attama, A.A., Ofokansi, K.C., Esimone, C.O., Onuigbo, E.B., 2011. Formulation and Evaluation of Niosomes. Indian J. Pharm Sci. 73 (3), 323–328.

Paolino, D., Cosco, D., Muzzalupo, R., Trapasso, E., Picci, N., Fresta, M., 2008. Innovative bola-surfactant niosomes as topical delivery system of 5-fluorouracil for the treatment of skin cancer. Int. J. Pharm. 353, 233–242.

Pardakhty, A., Varshosaz, J., Rouholamini, A., 2007. In vitro study of polyoxyethylene alkyl ether niosomes for delivery of insulin. Int. J. Pharm. 328, 130–141.

Pardakhty, A., Moazeni, E., Varshosaz, J., Hajhashemi, V., Rouholamini Najafabadi, A., 2011. Pharmacokinetic study of niosome-loaded insulin in diabetic rats. Daru 19 (6), 404–411.

Pardakhty, A., Shakibaie, M., Daneshvar, H., Khamesipour, A., Mohammadi-Khorsand, T., Forootanfar, H., 2012. Preparation and evaluation of niosomes containing autoclaved Leishmania major: a preliminary study. J. Microencapsul. 29 (3), 219–224.

Pawar, S.D., Pawar, R.G., Kodag, P.P., Waghmare, A.S., 2012. Niosome: an unique drug delivery system. Int. J. Bio. Pharm. Alli Sci. 3, 409–412.

Priprem, A., Limphirat, W., Lim sitthi chaikoon, S., Johns, J., Mahakunakorn, P., 2012. Intranasal delivery of nanosized melatonin encapsulated niosomes in rats. Open Access Sci. Rep. 1 (4), 232–237.

Puras, G., Mashal, M., Zárate, J., Agirre, M., Ojeda, E., Grijalvo, S., Eritja, R., Diaz-Tahoces, A., Martínez Navarrete, G., Avilés-Trigueros, M., Fernández, E., Pedraz, J.L., 2014. A novel cationic niosome formulation for gene delivery to the retina. J. Control. Release 174, 27–36.

Radha, G.V., Sudha Rani, T., Sarvani, B., 2013. A review on proniosomal drug delivery system for targeted drug action. J. Basic Clin. Pharm. 4 (2), 42–48.

Raja, N.R.A., Pillai, G.K., Udupa, N., Chandrashekhar, G., 1994. Antiinflammatory activity of niosome encapsulated diclofenac sodium in arthitic rats. Ind. J. Pharmacol. 26, 46–48.

Raymond C.R., Paul J.S., Sian C.O., 2006. Polyoxyethylene alkyl ethers. In: Handbook of Pharmaceutical Excipients, fifth ed. Pharmaceutical Press, London, pp. 564–571.

Reddy, D.N., Udupa, N., 1993. Formulation and evaluation of oral and transdermal preparations of flurbiprofen and piroxicam incorporated with different carriers. Drug Dev. Ind. Pharm. 19, 843–852.

Regev, O., Zana, R., 1999. Aggregation behavior of tyloxapol, a nonionic surfactant oligomer, in aqueous solution. J. Colloid Interface Sci. 210 (1), 8–17.

Rogerson, A., Cummings, J., Willmott, N., Florence, A.T., 1988. The distribution of doxorubicin in mice following administration in niosomes. J. Pharm. Pharmacol. 40 (5), 337–342.

Sahin, N.O., 2007. Niosomes as nanocarrier systems. In: Nanomaterials and Nanosystems for Biomedical Applications. Springer Netherlands, pp. 67–81.

Sankhyan, A., Pawar, P., 2012. Recent trends in niosome as vesicular drug delivery system. J. Appl. Pharm. Sci. 02 (06), 20–32.

Sezgin-Bayindir, Z., Antep, M.N., Yuksel, N., 2015. Development and characterization of mixed niosomes for oral delivery using candesartan cilexetil as a model poorly water-soluble drug. AAPS Pharm. Sci. Tech. 16 (1), 108–117.

Shahiwala, A., Misra, A., 2002. Studies in topical application of niosomally entrapped nimesulide. J. Pharm. Pharm. Sci. 5 (3), 220–225.
Shaker, D.S., Shaker, M.A., Hanafy, M.S., 2015. Cellular uptake, cytotoxicity and in-vivo evaluation of Tamoxifen citrate loaded niosomes. Int. J. Pharm. 493, 285–294.
Singh, M.N., Hemant, K.S.Y., Ram, M., Shivakumar, H.G., 2010. Microencapsulation: a promising technique for controlled drug delivery. Res. Pharm. Sci. 5 (2), 65–77.
Stafford, S., Baillie, A.J., Florence, A.T., 1988. Drug effects on the size of chemically defined nonionic surfactant vesicles. J. Pharm. Pharmacol. 40, 26.
Tavano, L., Muzzalupo, R., Mauro, L., Pellegrino, M., Andò, S., Picci, N., 2013a. Transferrin-conjugated pluronic niosomes as a new drug delivery system for anticancer therapy. Langmuir 29, 12638–12646.
Tavano, L., Vivacqua, M., Carito, V., Muzzalupo, R., Caroleo, M.C., Nicoletta, F., 2013b. Doxorubicin loaded magneto-niosomes for targeted drug delivery. Colloids Surf. B Biointerfaces 102, 803–807.
Tavano, L., Muzzalupo, R., Picci, N., Cindio, B.D., 2014a. Co-encapsulation of antioxidants into niosomal carriers: gastrointestinal release studies for nutraceutical applications. Colloids Surf. B: Biointerfaces 114, 82–88.
Tavano, L., Aiello, R., Ioele, G., Picci, N., Muzzalupo, R., 2014b. Niosomes from glucuronic acid-based surfactant as new carriers for cancer therapy: preparation, characterization and biological properties. Colloids Surf. B Biointerfaces 118, 7–13.
Uchegbu, I.F., Florence, A.T., 1995. Nonionic surfactant vesicles (niosomes) physical and pharmaceutical chemistry. Adv. Colloid Interface Sci. 58, 1–55.
Uchegbu, I.F., Vyas, S.P., 1998. Non-ionic surfactant based vesicles (niosomes) in drug delivery. Int. J. Pharm. 172, 33–70.
Uchegbu, I.F., Bouwstra, J.A., Florence, A.T., 1992. Large disk-shaped structures (Discomes) in non-ionic surfactant vesicles to micelle transitions. J. Phys. Chem. 96, 10548–10553.
Verma, A.K., Bindal, M., 2012. A review on niosomes: an ultimate controlled and novel drug delivery carrier. Int. J. Nanoparticles 5 (1), 73–87.
Verma, S., Singh, S.K., Syan, N., Mathur, P., Valecha, V., 2010. Nanoparticle vesicular systems: a versatile tool for drug delivery. J. Chem. Pharm. Res. 2 (2), 496–509.
Vyas, S., Singh, R., Jain, S., Mishra, V., Mahor, S., Singh, P., Gupta, P.N., Rawat, A., Dubey, P., 2005. Non-ionic surfactant based vesicles (niosomes) for non-invasive topical genetic immunization against hepatitis B. Int. J. Pharm. 296, 80–86.
Waddad, A.Y., Abbad, S., Yu, F., Munyendo, W.L.L., Wang, J., Lv, H., Zhou, J., 2013. Formulation, characterization and pharmacokinetics of Morin hydrate niosomes prepared from various non-ionic surfactants. Int. J. Pharm. 456, 446–458.
Wang, M., Yuan, Y., Gao, Y., Ma, H.M., Xu, H.T., Zhang, X.N., Pan, W.S., 2012. Preparation and characterization of 5-fluorouracil pH-sensitive niosome and its tumor-targeted evaluation: in vitro and in vivo. Drug Dev. Ind. Pharm. 38 (9), 1134–1141.
Yan, Y., Lu, T., Huang, J., 2009. Recent advances the mixed systems of bolaamphiphiles and oppositely charged conventional surfactants. J. Colloid Interface Sci. 337, 1–10.
Yoshida, H., Lehr, C.M., Kok, W., Junginger, H.E., Verhoef, J.C., Bouwistra, J.A., 1992. Niosomes for oral delivery of peptide drugs. J. Control. Release 21, 145–153.
Yoshioka, T., Stermberg, B., Florence, A.T., 1994. Preparation and properties of vesicles (niosomes) of sobitan monoesters (Span 20, 40, 60, and 80) and a sorbitan triester (Span 85). Int. J. Pharm. 105, 1–6.
Zarei, M., Norouzian, D., Honarvar, B., Mohammadi, M., Shamabadi, H.E., Akbarzadeh, A., 2013. Paclitaxel loaded niosome nanoparticle formulation prepared via reverse phase evaporation method: an in vitro evaluation. Pak. J. Biol. Sci. 16 (6), 295–298.
Zareÿe, M.H., Mozafari, M.R., Hasirci, V., Pýkýn, E., 1997. Scanning tunnelling microscopy investigation of liposome-DNA-Ca2+ complexes. J. liposome Res. 7, 491–502.

Zhang, Y., Zhang, K., Wu, Z., Guo, T., Ye, B., Lu, M., Zhao, J., Zhu, C., Feng, N., 2015. Evaluation of transdermal salidroside delivery using niosomes via in vitro cellular uptake. Int. J. Pharm. 478, 138–146.

Zografi, G., 1995. Interfacial phenomena. In: Gennaro, A.R. (Ed.), Remington: The Science and Practice of Pharmacy. nineteenth ed., Mark Publishing, Easton, Pennsylvania, pp. 241–251.

Zubairu, Y., Negi, L.M., Iqbal, Z., Talegaonkar, S., 2015. Design and development of novel bioadhesive niosomal formulation for the transcorneal delivery of anti-infective agent: in-vitro and ex-vivo investigations. Asian J. Pharma. Sci. 10, 322–330.

CHAPTER

7

NANOFIBROUS AND NANOPARTICLE MATERIALS AS DRUG-DELIVERY SYSTEMS

Fatemeh Zamani*,, Fatemeh Jahanmard*, Farzaneh Ghasemkhah*, Sepideh Amjad-Iranagh*, Roohollah Bagherzadeh*, Mohammad Amani-Tehran*, Masoud Latifi***

**Amirkabir University of Technology, Tehran, Iran*

***Hazrat-e Masoumeh University, Qom, Iran*

CHAPTER OUTLINE

1 INTRODUCTION

The interdisciplinary field of engineering drug-delivery systems, which combines biology, materials engineering, and medicine, is a growing research field dealing with the manipulation of materials at different scales for targeting treatments. These materials and systems are engineered to improve the efficiency and to decrease the side effects of traditional drugs. Therefore, a drug-delivery system is aimed to present a therapeutic agent in the body while having control of the rate, time, and site of release within a predictable scheme. Nanoscale drug-delivery systems have the specialty of enhancing the therapeutic index of a given drug with a controllable amount of release and drug exposure. It is believed that current drug-delivery systems are designed to release a specific amount of drugs in a controlled regime, although the scope is limited to targeting only tissues rather than individual cells (Goldberg et al., 2007). Micronanoscale drug systems have greater surface area-to-volume ratios, enabling them to overcome solubility-limited bioavailability (Goldberg et al., 2007; Gonsalves et al., 2007; Jian et al., 2015).

It is also well documented (Goldberg et al., 2007) that the preferred size range for nanoparticles in cell-sized drug delivery is around 10–100 nm. Nanoscale drug-delivery systems have also very interesting properties from their micro and macro scale, which makes them suitable for many different applications and treatments. More interestingly, these properties can be tailored for specific responsibilities, such as controlling, solubility, biodegradability, and so on. Various fabrication methods and approaches have been introduced for micro/nanostructures drug-delivery systems. Meanwhile, in recent decades, electrospinning and electrospraying have been two techniques utilized for the fabrication of continuous ultrafine fibers and particles for diverse applications, including drug-delivery systems. This chapter presents nanobiomaterial structures developed for drug delivery, and focuses on the advanced applications of electrospinning and electrospraying techniques for incorporation of therapeutic agents into the polymeric carriers in order to develop novel drug-delivery systems and devices. Furthermore, the chapter intends to characterize morphology, describing structural features, fabrication methods, materials, and recent applications, considering the novel nanoscale particle-based drug-delivery systems that have been invented recently.

2 NANOPARTICLE MATERIALS IN DRUG-DELIVERY SYSTEMS

Advances in multifunctional nanoscale drug-delivery systems have recently received great attention. The application of nanoparticle systems for medicine and biology are more important, such as dendrimers, liposomes, micelles, carbon nanotubes, magnetic nanoparticles, and others.

2.1 DENDRIMERS

Dendrimers are a class of highly branched, monodispersed, and synthetic macromolecules with a well-defined structure and composition (Parveen et al., 2012; Tomalia et al., 2007). Dendrimers can be conveniently fabricated by convergent or divergent synthesis (Ihre et al., 2001; Tomalia et al., 1985).

The size of dendrimers is sufficiently small, like PAMAM G5, which is 5.4 nm. Numerous classes of dendrimers, including polyamidoamine (PAMAM), polypropyleneimine (PPI), poly(glycerol-*co*-succinic acid), poly-L-lysine (PLL), melamine, triazine, poly(glycerol), poly[2,2-bis(hydroxymethyl)-propionic acid], and poly(ethylene glycol) (PEG), as well as carbohydrate-based and citric-acid-based ones, have been developed for drug delivery (Crampton and Simanek, 2007; Tomalia et al., 2007).

The guest molecules, such as drugs and genes can encapsulate in the cavities of dendrimers. The biocompatibility of dendrimers can be improved by modifying the final groups of dendrimers, such as acetylation, carboxylation, and PEGylation.

G5 PAMAM dendrimers (with acetyl termini) covalently linked with fluorescein isothiocyanate (FI) and folic acid (FA) can encapsulate doxorubicin (DOX) for targeted delivery to cancer cells (Wang et al., 2011). In another study, amine-terminated G5 dendrimers were used as a platform to be sequentially modified with fluorescein isothiocyanate (FI) via a thiourea linkage and RGD peptide via a polyethylene glycol (PEG) spacer, followed by acetylation of the remaining dendrimer terminal amines for anticancer drug encapsulation (He et al., 2015). DOX was delivered using folate-conjugated fifth-generation poly(propyleimine) (PPI) dendrimers (Gupta et al., 2010). Superpara-magnetic-iron-oxide-nanoparticle–dendrimer complexes have been reported for siRNA delivery (Taratula et al., 2011). In another study, folate-conjugated poly L-lysine (PLL) dendrimers have been used for the delivery of DOX (Jain et al., 2014).

2.2 LIPOSOMES

Lipids have two parts in their structure: hydrophilic and hydrophobic. When a lipid is in contact with water, the hydrophobic parts of the molecule do not have interaction with the solvent, and they gather together in self-assembly and create liposomes. Liposomes consist of an aqueous core surrounded by a lipid bilayer. The classification of liposomes is generally based on their size and number of lipidic bilayers. We have different methods for liposome preparation, and these methods can be classified as mechanical agitation, solvent evaporation, solvent injection, and detergent solubilization (Çağdaş et al., 2014). Liposomes show high uptake into cells via endocytosis. Liposomes have a potential as drug carriers in the treatment of cancer (Allen and Cullis, 2013) and in antimicrobial-delivery systems. Hydrophobic drugs have been placed inside the bilayer, and hydrophilic drugs are entrapped in the aqueous core.

Nanoliposomes have immunosuppressive and angiogenetic properties (Sabnani et al., 2015). Encapsulating the albumin-paclitaxel nanoparticles into the PEGylated liposomal bilayer is one of the other applications of liposome in drug delivery. One limitation of albumin nanoparticles is their poor colloidal stability in vivo. Albumin-paclitaxel nanoparticles encapsulate into the liposomes by a thin-film hydration-extrusion technique (Ruttala and Ko, 2015). One type of pH-sensitive liposome that has been developed is targeted at the CD19 epitope on B-lymphoma cells that showed enhanced DXR delivery into the target cells (Ishida et al., 2001). Encapsulation of DXR in the stealth liposomes showed significant accumulation and therapeutic efficacy and reduced the toxic effect against human pancreatic carcinoma AsPC-1 (Vaage et al., 1997). In a study it was reported that a peptide-functionalized liposome for enhanced targeted ovarian carcinoma (the peptide is T7) that specifically binds to the transferrin receptor TfR overexpressed on ovarian carcinoma cells (Wu et al., 2015). The freeze-dried empty liposomes (FDEL) method for mass production of liposomal products was developed. This method was used for preparation of liposomes containing DNA with a particular attention to the stability of DNA (Kikuchi et al., 1999). In another study, liposomes were modified with avidin (avidin liposomes) for recognizing cancer cells and then modified with fusogenic protein (HVJ liposomes) for fusing the cancer cells. Then an emulsion of perfluoropentane (ePFC5) was loaded in the modified liposome. This liposome was used for targeting and ultrasound-triggered injury of cancer cells (Ninomiya et al., 2016). In another project, folic acid was modified with dodecylamine and embedded in a bilayer by alkane chain and then folate-PDA-liposome composite with targeted delivery and fluorescence tracing was prepared. This system was performed in cancer cells and normal cells in vitro (Lee et al., 2015).

2.3 MICELLES

The self-assembly of amphiphilic block copolymers in water is based on nonpolar and hydrophobic interactions among the lipophilic core-forming polymer chains. The process is concomitantly driven by a gain in entropy of the solvent molecules as the hydrophobic components withdraw from the aqueous media (Gaucher et al., 2005). Most amphiphilic copolymers contain either a polyester or a poly(amino acid) derivative as the hydrophobic segment employed for drug delivery purposes. Many anticancer agents are poorly soluble in water, which can cause serious problems for their clinical application. Micelles are a very good choice for delivering these drugs.

One application of micelles in drug-delivery systems is explained here. A novel receptor-targeted micelle was synthesized and an anticancer drug, paclitaxel (PTX), was loaded. The micelle composition was based on tocopherol succinate-chitosan-polyethylene glycol-folic acid (TS-CS-PEG-FA) (Emami et al., 2015). In another study, an α-tocopheryl succinate (α-TOS) containing a diblock copolymer micellar system was synthesized by conjugation of α-TOS molecules, and an mPEG-*b*-PHEMA hydrophilic diblock copolymer by ester bonds was loaded with doxorubicin (DOX) for HCT116 colon cancer therapy. DOX-loaded polymeric micelles enhanced the cytotoxicity of DOX and α-TOS to cancer cells in vitro and an exceptional tumor-inhibiting effect in vivo (Lee et al., 2015). One of the other applications of micelles in drug-delivery systems is poly(ethylene glycol)-block-poly(2-methyl-2-carboxyl-propylene carbonate-graft-dodecanol) (PEG-PCD) lipopolymers. In the hydrophobic core of PEG-PCD micelles loaded a model drug embelin and significantly improved its solubility (Li et al., 2010a).

2.4 CARBON NANOTUBES

Carbon nanotubes (CNTs) are very frequently used in medical research and are efficient in drug delivery. The use of CNTs in drug delivery and biosensing technology has the potential to revolutionize medicine. Functionalization of single-walled nanotubes (SWNTs) has proven to enhance solubility and allow efficient tumor targeting/drug delivery.

Using the SWNTs for PTX delivery into xenograft tumors in mice is one application of CNTs in medicine. PTX is insoluble in water but when it is conjugated to PEGylated SWNTs it shows high water solubility. SWNT-PTX affords much longer blood circulation time of PTX. PEGylated PTX leads to high tumor uptake of the drug through the EPR effect (Liu et al., 2008). Emulsion polymerization in the presence of sodium methacrylate and *N,N′*-ethylenebisacrylamide was used to produce hydrogels composed of gelatin and multiwalled carbon nanotubes, and was proposed as drug delivery microspheres for the electroresponsive release of diclofenac sodium salt (Spizzirri et al., 2013). CNTs have the potential for transdermal drug delivery. Buckypapers based on different types of CNTs were prepared, and clonidine hydrochloride was loaded. This system was used for the design of novel transdermal drug-delivery systems (Schwengber et al., 2015).

2.5 Fe_3O_4 NANOPARTICLES

Magnetic composite nanospheres have attracted much attention due to their potential in the fields of catalysis, sensor, water purification, enzyme immobilization, DNA extraction, and drug delivery. These nanoparticles usually synthesize via the coprecipitation method. Fe (II) and (III) salts and NH_4OH are needed for Fe_3O_4 preparation.

Silane-coated magnetic nanoparticles/glycidylmethacrylate-grafted-maleatedcyclodextrin composite hydrogel was synthesized, and it was used as a drug carrier for the controlled delivery of 5-fluorouracil (5-FU). This compound is biocompatible, and it indicates sustained and controlled release of the drug to the targeted site. The drug release mechanism was found to obey the non-Fickian diffusion (n = 0.709) method. Swelling of the polymer and drug diffusion play important roles in the drug release mechanism. In this system, advantages of magnetic core and host–guest interactions of β-CD were mixed for the controlled delivery of 5-FU to sustain the therapeutic index of the drug

(Anirudhan et al., 2015). Fluorescent Fe_3O_4/PPy nanospheres were prepared by in situ polymerization, and this compound has high magnetization. Methotrexate (MTX) was conjugated onto Fe_3O_4/PPy nanospheres by the EDC/NHS coupling chemistry. It has been shown that the nanoparticles can effectively inhibit the growth of HeLa cells with low particle concentration (Jiang et al., 2015). In another study, Fe_3O_4-Au nanoparticles were synthesized, and then folic acid was used as the targeting agent. L-cysteine was used as linker for the attachment of folic acid to the gold surface. Cysteine was attached to the surface of nanoparticles via the SH groups (Karamipour et al., 2015). A new hydrogel was synthesized by using chitosan (CS) cross-linked carboxymethyl-β-cyclodextrin (CM-β-CD) polymer-modified Fe_3O_4 magnetic nanoparticles. This system was used for delivering hydrophobic anticancer drug 5-FU (CS-CDpoly-MNPs). The hydrophobic cavity of carboxymethyl-β-cyclodextrin is the best place for insoluble anticancer drugs through host–guest interactions (Ding et al., 2015).

3 NANOFIBROUS MATERIALS AS DRUG-DELIVERY SYSTEMS

Drug delivery is the process of administering a drug to achieve a therapeutic effect in humans or animals. In this system drugs will release their active ingredients only in the target area of the body and in a controlled manner without affecting healthy tissues and organs. There are various routes of drug administration, such as the oral, parenteral, transdermal, and inhalation routes (Ignatious et al., 2010; Kenawy et al., 2009) and different mechanisms of drug release, including dissolution, diffusion, partitioning, osmosis, swelling, erosion, and targeting (Siegel and Rathbone, 2012).

Properties of the carrier material are one of the important issues to control the rate of release from a drug carrier (Gonsalves et al., 2007). Nanofibers are considered as a potential drug carrier for the drug-delivery system due to their unique characteristics (Maleki et al., 2013). Electrospinning, because of its appealing features, including the high loading capacity, ease of operation, high encapsulation efficiency, simultaneous delivery of diverse therapies, and cost-effectiveness, is used in drug delivery (Hu et al., 2014; Son et al., 2014). It has great flexibility in material selection of either polymers (biodegradable or nondegradable materials) or drugs (antibiotics, anticancer drugs, proteins, aptamer, DNA, and RNA) for drug delivery applications. Additionally, a number of different drug loading methods can also be utilized, such as coatings and encapsulated drugs (coaxial and emulsion electrospinning) by using the various electrospinning techniques. Electrospun nanofibers have been successfully used to achieve different controlled drug release profiles, such as immediate, smooth, and delayed releases (Son et al., 2014). The drug-loaded nanofibrous membrane can be applied for wound healing or as a postoperation implant for antibiotic, antifungal, antimicrobial, and anticancer drug delivery (Toshkova et al., 2010; Vargas et al., 2010; Zong et al., 2004).

3.1 ELECTROSPINNING

This section deals with the process of producing ultrafine fibers, with cross-sectional diameters approximately in the range of 100 nm to 1 μm, using the electrospinning technique (Bagherzadeh et al., 2012; Ko and Wan, 2014; Yarin et al., 2013). Electrospinning, with three major components consisting of a high-voltage source, a syringe pump, and a collector, has been widely investigated for fabrication of nanoscale fibers, attributable to its advantages containing the simple generation of nanofibers and the ability to spin many kinds of polymers. The electrospinning process involves a strong electrical potential

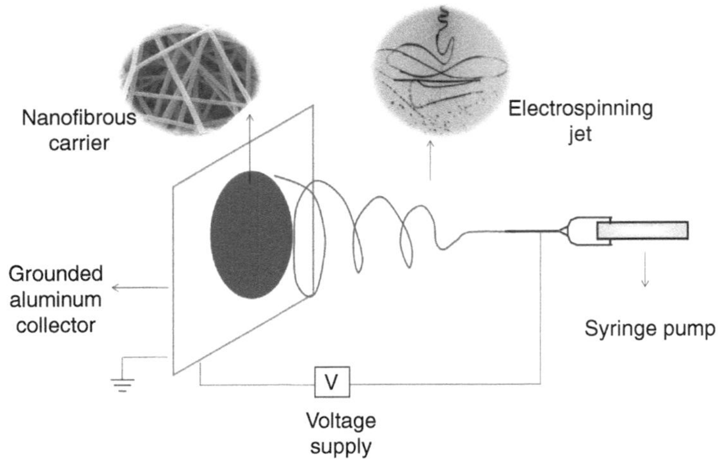

FIGURE 7.1 Schematic Setup of the Electrospinning Process

applied to the polymer solution in a syringe. The resulted electric field between a capillary tip and a grounded collector causes charge induction on the solution. When charges within the fluid reached a critical amount and could overcome the surface tension, a charged jet of the solution in the spinning region and a Taylor cone at the tip of the needle were formed. As this jet travels in a spiral path in the air, and the solvent evaporates from the jet, continuous and nanoscale fibers are formed and deposited on the collection plate (Ramakrishna et al., 2005) (Fig. 7.1).

There are many parameters, including (1) polymer solution parameters, such as viscosity, conductivity, surface tension, and elasticity; (2) processing conditions, such as applied voltage, electrospinning distance, feed rate of solution, and type of collector; and (3) ambient conditions, such as temperature and humidity, affecting electrospinning of the polymer solution. Therefore, it is possible to create a nanofiber with a different morphology by varying the parameters (Bagherzadeh et al., 2013a; Ko and Wan, 2014; Ramakrishna et al., 2005; Sill and von Recum, 2008; Yarin et al., 2013). For example, the stronger electric field or longer flight time of the electrospinning jet can effectively reduce the fiber diameter. Increasing voltage also influences bead formation and the shape of the beads (spindle-like or spherical) (Ramakrishna et al., 2005). Furthermore, conductivity of the collector affects the porosity of the nanofibrous web, and nonconductive collectors usually have a lower porosity. When the collector is a cylindrical drum, the rate of fiber collection changes the structural properties of the web, such as fiber orientation, web porosity and even fiber diameter (Ramakrishna et al., 2005; Zamani, 2013; Zamani et al., 2013b). Zamani et al. (2013a, 2014) demonstrated that the surface roughness of PLGA nanofibrous substrate increased at a nano- and micrometer scale using various solvents in different humidity, to exploit the breath figure method and utilize the vapor-induced phase separation (VIPS) phenomenon. Jahanmard et al. (2013) also fabricated nanoporous fibers by electrospinning PCL solution in various ambient humidities (30%–70%) and temperatures (20–30°C) by using these two mechanisms (phase separation and breath figure).

Two- and three-dimensional electrospun nanofibrous substrates by exploiting the advantage of combining high specific surface area, interconnected porosity, and absorption capacity are also possibilities for efficient surface functionalization that have had great potential in many biomedical applications, such as tissue engineering, wound dressing, implant, and drug delivery (Bagherzadeh et al., 2013b, 2014a; Jahanmard et al., 2013; Maleki et al., 2013; Zamani, 2013; Zamani et al., 2013a, 2014, 2013b).

3.1.1 Recent modifications in conventional electrospinning setups

There are different ways to produce 2D and 3D nanofibrous structures using electrospinning. The basic setup of conventional electrospinning was widely utilized to fabricate 2D/sheet membranes with randomly oriented fibers from a single polymer solution on the surface of a flat collector. In order to expand the universality of electrospinning, various modifications have been made on electrospinning, which were usually based on the collector, the syringe, and polymeric solution.

A number of collector modifications, such as using a rotating cylinder collector, a frame collector, a wheel-like disk, or an auxiliary electrical field have been carried out to produce the electrospun membrane with ordered architecture for special applications (Huang et al., 2003; Li et al., 2003; Wang et al., 2013). Zamani and coworkers electrospun full-aligned nanofibers with a cylindrical drum collector by controlling the rotation speed of the collector (Zamani et al., 2013b).

To electrospin a variety of polymers into nanofibers, several electrospinning systems, such as co-electrospinning, as well as coaxial, side-by-side, and layer-by-layer electrospinning were developed.

Coelectrospinning is a method to promote some properties of polymeric nanofibers, such as physical, mechanical, and biological functionality, which electrospin blends or a mixture of polymers in the same solvent. For example, researchers could coelectrospin conductive nanofibers by blending polyaniline with PCL and PLGA in chloroform. To fabricate composite nanofibers containing different polymers that have properties from each of the polymeric components and that cannot be dissolved in the same solvent or kept in the same container, techniques of side-by-side and layer-by-layer electrospinning are practical. In layer-by-layer electrospinning, each polymer is electrospun to form a layer and sequentially collected on the same collector. In side-by-side electrospinning, two different polymer solutions are simultaneously electrospun from different syringes to form a mixed layer on the same collector (Bagherzadeh et al., 2014b; Liang et al., 2007). Kenawy et al. (2009) reported the preparation of blend nanofibrous mats comprising PCL and PU via electrospinning.

Recently, coaxial electrospinning has received great attention because of its potential in electrospinning some difficult-to-process polymer solutions, also encapsulating various drugs and bioactive agents. The polymers can be coelectrospun without direct mixing, in the form of a core within the shell of the other polymer. In this technique, two different polymer solutions are added to a concentric spinneret with two different diameters. Electric forces act on both polymer solutions and result in significant stretching of the polymer jets. By adjusting the solution feed rates and applied voltage, core–shell nanofibers are electrospun. Furthermore, by electrospinning two immiscible liquids through coaxial electrospinning and removing the cores, hollow nanofibers are prepared (Cui et al., 2010; Yang et al., 2008a, 2008b; Zhang et al., 2008).

3.1.2 Biomaterial systems for drug delivery

Various polymers from natural polymers to synthetic polymers and copolymers have been electrospun to fabricate membranes with desired properties for different applications (Bagherzadeh et al., 2014b; Cui et al., 2010; Gheibi et al., 2014; Hu et al., 2014). To determine proper biomedical applications for biomaterials, several important properties, such as structure, biocompatibility, biodegradability, and mechanical strength must be considered. One main interest of electrospun nanofibers is for biomedical applications. Biomaterials have been widely used in many areas, including tissue engineering, wound dressing, drug delivery, biosensors, medical prostheses, dental applications, sutures, stent coatings, implants, molecular filtration, biological agent preservation, health supplements, and cosmetic facial masks. The performance of a biomaterial greatly depends on its properties and biocompatibility, the health condition of the receiver, and the experience of the surgeon. Biomaterials can be categorized according to their source. There are several main classes of biomaterials: natural and synthetic polymers, and inorganic and composite materials (Ikada, 2011; Park and Lakes, 2007; Ratner et al., 2004).

Researchers have successfully electrospun these biocompatible materials, such as collagen, silk fibroin, hyaluronic acid, PCL, PLLA, PLGA, PU, and a blend of them to produce nanofibrous carriers (Sill and von Recum, 2008; Zamani et al., 2013c).

3.2 DRUG LOADING MECHANISMS IN ELECTROSPUN SYSTEMS

According to the type of drug, such as proteins, genes, and small molecule drugs, there are different approaches to obtain a successful release profile for drug loading. Drugs can be loaded on or into the nanofiber. By using various electrospinning techniques, a number of different drug-loading methods, such as coelectrospinning, layer-by-layer, coaxial, surface immobilization, and hybridization of nanofibers have been developed (Son et al., 2014).

In drug-delivery systems, drugs can be loaded within/on the nanofibers. For incorporation of drugs within nanofibers, coelectrospinning of drugs and the polymer is the simplest method in which the drugs and polymer dissolve in the same solvent and coelectrospin. Drug solubility in the polymer solution is one of the main factors in this method. Hydrophilic drugs, proteins, and peptides have been dissolved with PVA or PEO in the same solvent and electrospun (Gatti et al., 2013; Tang et al., 2012). Hydrophobic drugs, such as DOX and PTX have electrospun well with organic solvents (Xu et al., 2009). Kenawy et al. (2009), with the coelectrospinning of ketoprofen and polycaprolactone (PCL), polyurethanes, and their blends, prepared fibrous membranes and investigated their drug release at body temperature (37°C) and at room temperature (20°C). Researchers have fabricated polysulphone/nanohydroxyapatite (PSu/nHA) composite nanofibers as a carrier for the drug, which could be used in postsurgical treatment of cancer (Saravanabhavan and Dharmalingam, 2013).

Surface migration of the insoluble drug molecules that dispersed inside the solution causes burst release. Sustained release is a way to improve drug loading and reduce the burst release of drugs. Sustained release is accessible by blending of hydrophilic and hydrophobic polymers. For example, the adding of hydrophilic polymers, such as gelatin (Meng et al., 2011), polyethylene glycol (PEG) (Cao et al., 2010), and polyvinyl alcohols (PVAs) has an important effect on improving the sustained release of drugs (Jannesari et al., 2011).

The layer-by-layer electrospinning technique was developed for the purpose of time-controlled drug release, with subsequent electrospinning of two different solutions. Two kinds of polymeric solutions containing two different drugs were prepared and sequentially electrospun to fabricate nanofiber meshes. Nogueira et al. (2010) used the layer-by-layer method by preparing chitosan/silk fibroin films with oriented nanofibers. Okuda et al. (2010) presented multilayered drug-loaded PLCL electrospun nanofibrous meshes with the time-programmed dual-drug sustained-release system that were fabricated using sequential electrospinning. They also revealed that the drug release speed is controllable using the fiber diameter and mesh thickness as morphological features of electrospun products (Okuda et al., 2010).

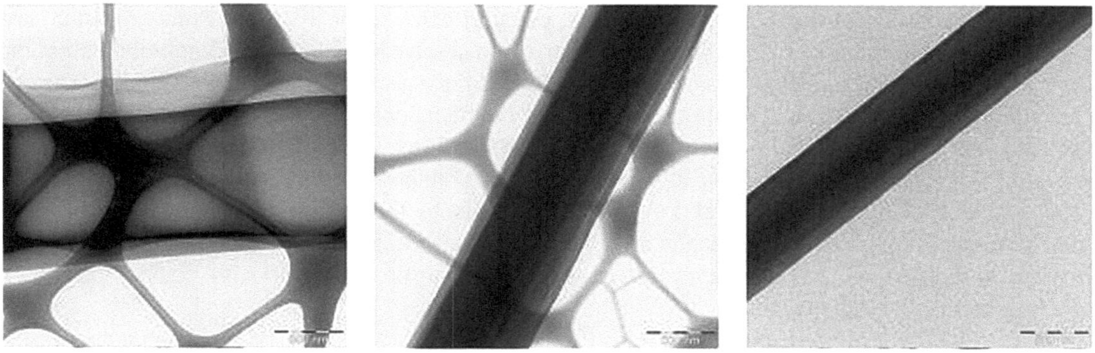

(A) (B) (C)

FIGURE 7.2 SEM Micrographs of Nanofibrous Core–Sheath Carrier

(A) Shell feed rate: 1 mL/min, core feed rate: 0.5 mL/min, shell diameter: 1422 ± 78 nm, and core diameter: 972 ± 71 nm; (B) shell feed rate: 3 mL/min, core feed rate: 0.3 mL/min, shell diameter: 750 ± 77 nm, and core diameter: 512 ± 50 nm; (C) shell feed rate: 0.7 mL/min, core feed rate: 0.05 mL/min, shell diameter: 437 ± 160 nm, and core diameter: 299 ± 79 nm.

Also, the coaxial electrospinning system has been also developed for core–sheath nanofiber formation, which can electrospin two immiscible polymer solutions containing drugs in the core of the nanofiber (Wang et al., 2012b; Yu et al., 2011). Zhang et al. (2006) investigated the possibility of encapsulating a nonelectrospinnable protein in conjunction with PEG, as a water-soluble polymer, into biodegradable PCL nanofibers by the coaxial electrospinning system.

It has been well documented that both hydrophobic and hydrophilic drugs, such as PTX and tetracycline hydrochloride, respectively, and biomacromolecules, such as proteins and DNA, can be encapsulated to the electrospun fibers (Zamani et al., 2013c). Fig. 7.2 shows SEM images of nanofibrous core–sheath carrier.

Researchers indicated that coaxial electrospun PCL^{Col}/PVA^{HA} core–sheath nanofibers encapsulating drugs have a great potential in enhancing implant osseointegration and preventing implant infection. PCL^{Col} and PVA^{HA} solution were prepared by mixing type I collagen and hydroxyapatite nanorods with polycaprolactone and polyvinyl alcohol for core and sheath of the nanofiber, respectively. Then doxycycline (Doxy) and dexamethasone (Dex) were successfully incorporated into the PCL^{Col}/PVA^{HA} nanofibers for controlled release (Song et al., 2013).

Jiang et al. (2012) investigated the preparation of core–sheath nanofibers for providing biphasic drug release profiles. Ketoprofen (KET) (Volpato et al., 2012), as the model drug, was loaded into both the sheath and core parts of the nanofibers. Polyvinylpyrrolidone (PVP), a hydrophilic polymer, was selected as the sheath part for immediate drug release. Zein was exploited as the core matrix for the sustained release of KET (Jiang et al., 2012). Yu et al. (2013b) also prepared core–sheath nanofibers with biphasic drug (Volpato et al., 2012) release profiles, using polyvinylpyrrolidone (PVP) as the sheath polymer and ethyl cellulose (EC) as the core matrix. Drug release consisting of an immediate and a sustained release, in the first and second phases, was tailored by adjusting the sheath flow rate and a typical diffusion mechanism, respectively (Yu et al., 2013b).

For incorporation of drugs on the nanofibers, physical and chemical surface immobilization can avoid drug denaturation caused by high voltage or organic solvents. For physical immobilization on the surface of nanofibers, heparin as a negatively charged polysaccharide has a strong binding affinity for growth factors (Lee et al., 2012b), and chemical modification of the nanofiber surface is carried out with amine, carboxyl, and hydroxyl (Cho et al., 2010). Electrospinning technique can be modified for multidrug-delivery systems and can individually control the release kinetics of each drug. Various drugs can be loaded by using several drug loading methods, by the coaxial system and coelectrospinning or surface immobilization. Wang et al. (2012b) investigated the enzyme immobilization on electrospun poly[bis(*p*-ethylphenoxy)]phospha-zene/polyacrylonitrile (PMPPh/PAN) coaxial nanofiber membranes by physical adsorption.

In addition to the aforementioned methods for the drug-delivery system, the electrospun nanofibers have been hybridized with other types of devices, nano- and microsized carriers and hydrogel (Lee et al., 2011, 2013; Son and Yoo, 2012). The multidrug-delivery system has been achieved by loading both micelles containing the hydrophobic drug (curcumin) and the hydrophilic drug (DOX) by the use of a coelectrospinning process. The multiple drugs embedded within the fiber matrix and showed a time-programmed release (Yang et al., 2014). Lee et al. (2013) generated a composite combination of PCL electrospun nanofibers within the PEG hydrogel micropattern and demonstrated the potential of the microarchitectured nanofibers to drug-delivery systems.

Researchers demonstrated that drug-loaded electrospun nanofibers were able to provide sustained release profiles for different types of active pharmaceutical ingredients. For most nonbiodegradable polymeric nanofibers, diffusion plays an important role in the release behavior, and it becomes complicated if the carrier is biodegradable. The diffusion of drug molecules from the fiber and the degradation of the carrier should be considered together (Hu et al., 2014).

When the drug is loaded within the nanofibers, drug release is a result of diffusion, and in the later stages it will be a combination of the diffusion and polymer degradation. When drugs are located around the nanofiber surface, chemical methods are better than physical ones in controlling the amount of incorporated drug, and exhibit slow drug-release kinetics with reduced initial burst release (Cho et al., 2010). Many studies showed reduced initial burst release kinetics using layer-by-layer devices (Poon et al., 2011).

By using the core–sheath nanofibers solutions in drug-delivery systems, the initial burst release is decreased using sheaths of nanofibers and drug release occurs after swelling or dissolving of the core polymer (Liao et al., 2006; Song et al., 2013). Maleki and coworkers showed the potential of core–shell nanostructures for sustained drug release, and suppressing the burst release effect (Maleki et al., 2013).

3.3 CONTROLLED RELEASE OF DRUG FROM NANOFIBROUS CARRIER

Controlled release drug delivery can be introduced as the delivery of a predefined amount of drug, over an identified period of time, in an expectable manner. The purpose of all controlled release systems is to develop the efficiency of drug therapy (Brouwers, 1996; Langer, 1998). This development decreases the side effects, reduces the amount of drug administrations required for the duration of treatment, and eliminates the need for particular drug administration (e.g., repeated injections) (Panyam and Labhasetwar, 2003; Uhrich et al., 1999).

Drugs loaded in electruspun nanofiber systems have been formulated in two basic designs: matrices or reservoirs (Romero-Cano and Vincent, 2002; Tarcha, 1990).

In the matrix-type produced with a single nozzle electrospinning system, a therapeutic agent is homogeneously dispersed throughout a polymer matrix in nanofibrous mats or on the surface of the nanofibers. In the first method, the rate of drug diffusion through the polymer matrix ordinarily decreases with time, because the drug with longer distance to travel requires a longer diffusion time to release, and for the second one the functional molecules on the fiber surface are directly exposed to the ambient environment, and the immobilized drug experiences an initial burst release (Xu et al., 2008).

Therefore, in order to develop electrospun nanofibers as suitable nanobiomaterials, the surfaces of electrospun nanofibers have been modified by physical adsorption of diverse bioactive molecules on electrospun nanofiber surfaces or chemically functionalized for producing reactive functional groups to the immobilized bioactive agent onto the surfaces.

In physical adsorption of bioactive molecules on the surface of the nanofiber, the instant release of drugs from the nanofiber surface and the controlled release of some therapeutic agents are suitable for some applications, such as inhibition of bacterial contamination during the few hours after surgery (Bölgen et al., 2007). In addition, the adsorption of drug-loaded nanoparticles on the nanofibers can cause unique drug releasing profiles, which the nanofiber itself cannot obtain, in addition to the high drug loading capacity (Chen et al., 2007; Park et al., 2008). In addition, therapeutic proteins and nucleic acids must be physically immobilized onto the surface for controlled delivery, because their activities after the release decrease if they were encapsulated within the bulk phase of nanofibers by coelectrospinning.

There are three different modes of the physical drug loading method on the surface of the electrospun nanofibers for drug delivery application. The first is the simplest approach for loading drugs on the nanofibrous mesh and driving force for surface adsorption, which is electrostatic interaction, hydrogen bonding, hydrophobic interaction, and van der Waals interaction (Yoo et al., 2009).

The second mode is the assembly of nanoparticles on the surface of nanofibers (Kalra et al., 2008; Zhang, 2003). The nanoparticles on the nanofibers structure were organized using attractive forces between opposite charges produced by electrospinning and electrospray (Park et al., 2008). The third mode is layer-by-layer multilayer assembly (Tang et al., 2006; Thierry et al., 2003; Zhang et al., 2005a). This method was driven by an electrostatic force on charged substrates between polyanions and polycations layers, and then self-assembled multilayer coating was produced (Thierry et al., 2005).

In order to immobilize bioactive agents on the nanofiber surface, chemical modification must be done to create reactive functional groups. Chemical immobilization of the bioactive agent onto the surface of electrospun nanofibers is preferred over physical immobilization, since the covalently attached molecules are not easily leached out from the surface-modified nanofibers when incubated over a long period. However, it should be noted that inactivation of the immobilized molecules can occur when active sites are chemically modified by the covalent modification (Yoo et al., 2009).

Further, coaxial electrospinning provides a versatile one-step method to produce core–sheath structured fibers (reservoir type) with excellent control of the nanofiber carrier properties, which is favorable for controlling drug release characteristics (Jiang et al., 2005). Fig. 7.3 illustrates drug release of two different types of carriers. Therefore, in the reservoir type, the drug is enclosed in the polymer matrix and two possible structures are classified by the core composition: In the first case the pure drug in the solid or in the solution state is used as the core surrounded by a polymer sheath. In the other case the drug molecules are dispersed in a polymer matrix and the whole dispersion system is used as the core. In the first case, if the polymer sheath in comparison with the core is uniform and thick enough, the diffusion rate of the drug will be completely stable during the life of the delivery system. In the second

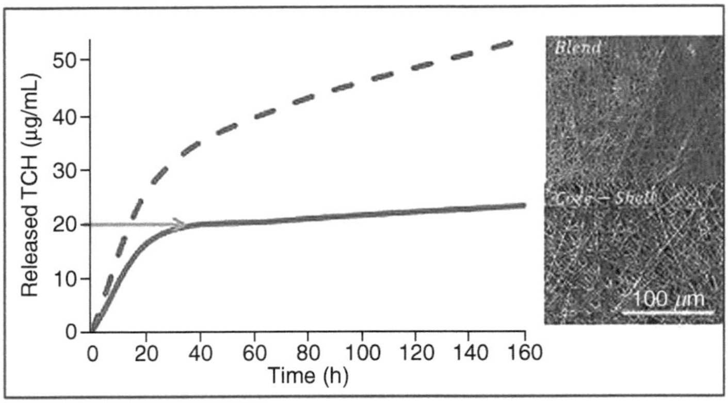

FIGURE 7.3 Drug Release Illustration of Two Different Types of Carriers

Dashed line, core–sheath blend: PLGA 90% + TCH 10%; *continuous line*, PLGA 100%, with the same core materials.

case, the release behavior of the drug is more complicated, and it does not behave in a constant manner (Xu et al., 2008).

3.3.1 Effective parameters on drug release from electrospun fibers

There are several factors that can affect the drug release from electrospun nanofibers: nanofibrous carrier structure (e.g., fiber geometry, thickness, diameter, and porosity) (Cui et al., 2006; Okuda et al., 2010); drug loading (Buschle-Diller et al., 2007); fiber crystallinity (Kenawy et al., 2002); drug state (Zamani et al., 2010); drug solubility in the release medium (Natu et al., 2010; Xie and Buschle-Diller, 2010); drug molecular weight (Taepaiboon et al., 2006); and drug–polymer–solvent interactions (Chew et al., 2005).

The release kinetics of the nanofibrous carrier are highly influenced by drug loading. For example, higher loadings will produce faster release. On one hand, at high loadings, there is a more surface-segregated drug that dissolves fast and on the other hand, there is an increase in porosity during drug elution proportional to the initial amount of the drug (Buschle-Diller et al., 2007).

The state of the drug and the structure of the polymer that forms the fiber are other factors that can affect the drug release (Zamani et al., 2010). The crystallinity of the polymer controls the rate of drug release as semicrystalline polymers revealed in common a higher extent of burst due to the immediate release of the drug deposited at the fiber surface, and then the hindered release of the drug from the fiber bulk because of limited water uptake in the semicrystalline regions. The drug state in the fibers is also an essential factor because a drug in a crystalline state will mainly be deposited outside the fibers and cause burst release, whereas the drug in an amorphous form will be loaded inside the fibers and have a sustained release (Kenawy et al., 2002; Thakur et al., 2008).

Drug compatibility with a polymer solution was also an important factor in controlling release; when lipophilic drugs were incorporated in lipophilic polymers and hydrophilic drugs were incorporated in hydrophilic polymers, drug deposition outside the fibers decreased and subsequently eliminated burst release (Zeng et al., 2005). Fig. 7.4 demonstrates different release kinetics of two biodegradable polymers.

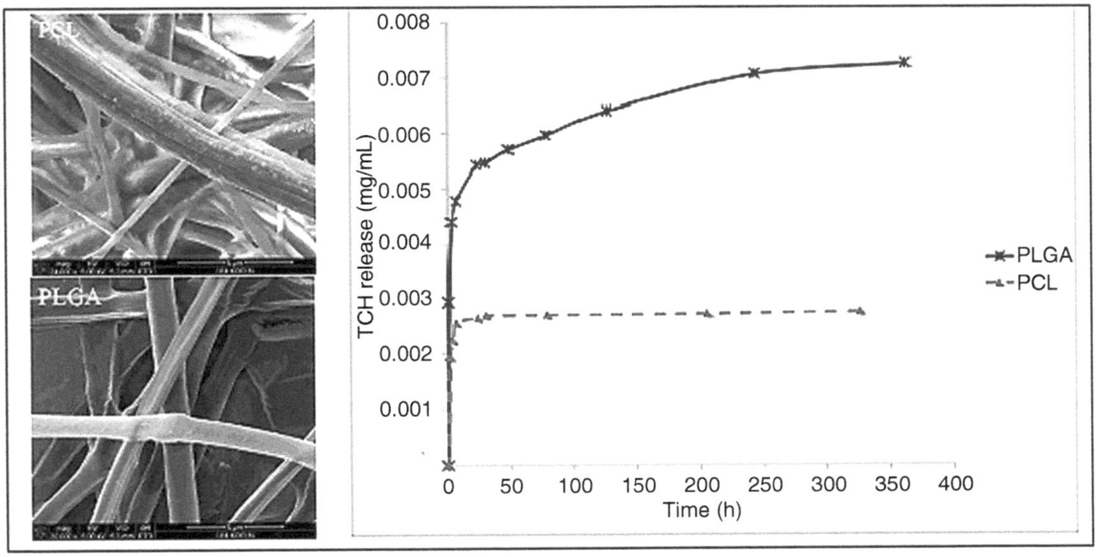

FIGURE 7.4 Different Release Kinetics of Two Biodegradable Polymers: PLGA and PCL

Moreover, the interaction between the drug and the polymer avoid the crystallization of the drug in the fibers and can even control sustained release of drugs in the crystalline state due to chemical interaction with the polymer (Chew et al., 2005).

3.3.2 Various drug release kinetics

In drug-delivery systems, electrospun nanofiber carriers showed different controlled drug release profiles, such as sustained drug release (Yu et al., 2013a), time-controlled release (Okuda et al., 2010), and biphasic releases (Yu et al., 2013b).

Among them, sustained drug release is an important method of administering and maintaining preferred drug concentrations in the blood within an identified therapeutic window, or in target tissues within a specified period of drug delivery.

Drug-loaded nanofibers can be prepared by codissolving solutions of the biomedical active agent and the polymer by traditional single-fluid electrospinning. For these nanofibers, the initial burst effect is unavoidable due to the large nanofiber surfaces or drug distribution on the nanofiber surfaces. Since the initial burst effect is unfavorable for sustained release, it should be mitigated or even eliminated (Chunder et al., 2007; Liu et al., 2012b). Therefore, posttreatments of the electrospun nanofiber carriers (Taepaiboon et al., 2007a) or encapsulations of the drug in the core part of core–sheath nanofibers can be accomplished (Wu et al., 2011). However, any additional posttreatment on electrospun nanofibers increases the complexity of the process and may damage the bioactive agent. In addition, the core–sheath nanofiber can be produced when the sheath fluids have good electrospinnability (Moghe and Gupta, 2008).

For example, Yu and coworkers investigated a new type of electrospun nanofiber carrier prepared by a modified coaxial electrospinning process. They used unspinnable Zein solution as sheath fluid, and the coaxial electrospinning process could be smoothly conducted to generate ketoprofen-loaded Zein nanofibers coated with a thin layer of blank Zein. In vitro dissolution tests showed that the nanofibers coated with blank Zein did not exert any initial burst release effect, and could enable linear drug release over a period of 16 h via a diffusion mechanism (Yu et al., 2009).

Biphasic release is another important profile of drug release in which an initial release of a fraction of drug in the shortest time after administration is accomplished for relieving the disease, and then sustained release of the remaining drug during a specified time can improve the therapy and avoid repeated administration (Jha et al., 2011; Ofori-Kwakye and Fell, 2003).

In summary, during biphasic release, a drug is released at two stages: an immediate drug release followed by a constant release. For example, Yu et al. (2013b) could provide biphasic drug release profiles by coaxial electrospinning to prepare core–sheath nanofibers, while polyvinylpyrrolidone is the sheath polymer, ethyl cellulose is the core matrix, and ketoprofen is the model drug (Volpato et al., 2012).

The KET was compatible with the sheath polymer and core matrix due to hydrogen bonding. In vitro dissolution tests showed that these nanofibers could provide typical biphasic release profiles, including a two-stage release. The amount of drug released in the first stage could be adjusted by the sheath flow rate, and in the second stage the remaining drug released was controlled by a diffusion mechanism (Yu et al., 2013b).

While biphasic releases can be very useful in the improvement of advanced pharmaceutics, they are unsuitable if a drug has a high first-pass effect or if therapy duration needs very time-specific delivery. Therefore, time-controlled pulsatile release is a profile of release that is appropriate for drug release after a predefined time period. Since a number of diseases follow biological rhythms (e.g., asthma attacks occur mainly late at night, or pain from rheumatoid arthritis increases in the morning), precise time-controlled release profiles are required. Pulsatile-delivery systems consequently enhance patient compliance and improve clinical outcomes (Bussemer et al., 2001).

Kaassis et al. (2014) have prepared novel and highly tunable pulsatile drug-delivery systems through the electrospinning of a blend of poly(ethylene oxide) (PEO), sodium alginate (SA), and sodium ibuprofen (SI). These fibers contain crystallites of SI embedded in a PEO–SA matrix. The release profile of this system includes two phases. This comprises a rapid burst release, and then after 120–150 min a further drug is released. The amount of release in the initial phase, and the length of time between the first and final drug release phases, can be adjusted by the SI and SA contents of the fibers, respectively (Kaassis et al., 2014).

In another study, Okuda et al. (2010) fabricated multilayered drug-loaded electrospun nanofiber meshes characterized with time-programmed dual-release formulation. Time-programmed controlled release of dual drugs was achieved using tetra-layered nanofiber meshes. These tetra layers were fabricated using sequential electrospinning of different (PLCL) poly(L-lactide-*co*-ε-caprolactone) solutions, in which the first and third layers of this mesh contain drugs of ChroB and TPPS, respectively; the second layer is used as a barrier mesh, and the last layer as a basement mesh. Furthermore, using multilayered electrospun nanofiber meshes developed the combination therapies in time-programmed dual-release systems. In vitro release tests showed that the drug release rate and sustained release period are controllable using morphological features of electrospun fibers, such as the fiber diameter and mesh thickness (Okuda et al., 2010).

3.4 TARGETED DRUG DELIVERY

Drug-delivery systems are formulations, approaches, and engineered technologies for the controlled release and/or the targeted delivery of therapeutic agents. The target of a delivery method may be body organs (lung, brain, and heart), tissue types (nerve, muscle), structures inside of cells, or disease-specific structures (tumor cells). A high concentration of drugs in the cells is necessary for the desired therapeutic effect, but this exacerbates the problem of toxicity. Thus, attempts are being made to target the drugs selectively to the affected sites in order to increase their efficiency and decrease their side effects. The expansion of novel nanoparticles and nanofibrous structures for targeted drug delivery is an exciting and challenging research field, in particular for the delivery of emerging cancer therapies. Nanoparticles are suitable for targeted delivery of therapeutic or diagnostic compounds due to their size and narrow size distribution, and nanofibers, because of their unique characteristics, such as higher loading efficiency, superior mechanical performance, controlled release behavior, and excellent stability, could also be used in the delivery of various drugs to the target site in the future.

Electrospun nanofibrous structures have been successfully used as targeted and controlled carriers for drug/cell/gene delivery into the various body organs through different routes. Fabrication of electrospun nanofibers from stimuli-responsive polymers is one of the methods for producing a targeted drug release system, especially in oral drug delivery. These electrospun nanofibers (also called smart nanofibers) show physiochemical responses to small changes in their surrounding environment, such as pH, temperature, light, and electrical and magnetic fields (Alvarez-Lorenzo and Concheiro, 2014; Mano, 2008; Mura et al., 2013; Weng and Xie, 2015). The human body is regulated by acid–base homeostasis, which keeps the pH of the arterial blood between 7.38 and 7.42. Also, many tissues or cell compartments have their own distinctive pH environments for normal functioning. The use of the pH-responsive electrospun nanofibers results in a release at the characteristic pH of the disease, and when the pH shifts to the normal value in improved conditions, such nanofibers could reduce the release rate or completely cease the release. For example, Eudragits are a category of copolymers of methacrylic esters, and their solubility is water dependent on a pH of the mediate. These pH-sensitive Eudragit polymers are used for targeted delivery of the drugs to human organs, in particular for various parts of the human gastrointestinal (GI) tract (as a result of increasing the pH of the GI from 1–3 in the stomach to 5–7 in the lower part of the GI tract). Ignatious et al. (2010) obtained Eudragit L100-55 electrospun nanofibers that were loaded with a poorly water-soluble drug to produce drug-encapsulated nanofibers. This research indicated that the drug did not release at a lower pH. In fact, Eudragit L100-55 is a copolymer that is water insoluble in a pH lower than 5.5 and soluble only in water above a pH of 5.5. When drug-encapsulated Eudragit L100-55 nanofibers are used for oral administration, they are prevented from drug release in the stomach. However, as this nanofibrous system reaches the lower GI (where pH is higher than 5.5) it releases the drug. Thus, such polymers can be used for the targeted delivery of drugs to a lower part of the GI tract. Shen et al. (2011) prepared Eudragit L100-55 nanofibers loaded with diclofenac sodium (a nonsteroidal antiinflammatory drug) using the electrospinning process, and demonstrated that the drug-loaded nanofibers had pH-dependent release profiles. They claimed that the drug-loaded Eudragit L100-55 nanofibers have the potential for oral colon–targeted delivery systems. Also, Yu et al. (2013a, 2014) reported preparation of colon-targeted delivery systems using drug-encapsulated nanofibers by the coaxial electrospinning process. In their study, pH-sensitive Eudragit L100 copolymer was used as a shell fluid, and a solution consisting of PVP and diclofenac sodium was used as a core solution. In vitro dissolution tests demonstrated that drug-loaded Eudragit L100 nanofibers have very limited release at a mediate with a pH of 2 (simulated gastric fluid) but there is a quick and burst release at a pH of 7 (simulated intestinal fluid). These results and other similar reports (Heshmati and Makhmalzadeh, 2012; Yu et al., 2014) verify the colon-targeted drug release property of pH-sensitive nanofibers. Similarly, various reports were published in which therapeutic agents have been incorporated with electrospun nanofibers responsive to temperature, light, electric field, and magnetic field; these are discussed in detail in Weng and Xie's report (Weng and Xie, 2015). Wang et al. (2012a) also obtained targeted drug-delivery systems by adding Fe_3O_4 nanoparticles and drugs (e.g., indomethacin and aspirin) to electrospun dehydroxypropyl methyl cellulose phthalate and CA nanofibers. In order to move nanofibers to the target site under the guidance of an external magnetic field, the presence of Fe_3O_4 nanoparticles had no influence on the release profiles.

There are various studies (Boas et al., 2015; Chunder et al., 2007; Ge et al., 2007) that prepared the polyelectrolyte fibers combined with stimuli-responsive polymers using the layer-by-layer multilayer assembly technique by the electrospinning process. These polyelectrolyte multilayer assemblies could be used as a promising drug-delivery system for modulating drug release. The layer-by-layer method involves sequential adsorption and deposition of oppositely charged materials that are driven by intermolecular forces, including electrostatic force, hydrogen bonding, covalent bonding, and so on. Thus, the incorporation of therapeutic agents into polyelectrolyte electrospun layers provides versatile controlled release systems. Multilayered polyelectrolyte nanofibrous mats were prepared by the electrospinning process from a polyelectrolyte mixture (in pH 2.1) consisting of poly(acrylic acid) (PAA) and poly(allylamine hydrochloride) (PAH) (Chunder et al., 2007). The result of fibrous mats was insoluble in water at a higher pH because PAA and PAH that are weakly charged polyelectrolytes form a complex under neutral pH and low salt concentrations as a result of chemical changes in their functional groups (amine and carboxylic acid). In other words, the carboxylic acid is deprotonated to carboxylate and the amine is protonated to ammonium under neutral pH. Therefore, ionic cross-links are generated and then polyelectrolyte complexes are formed. Subsequently, first, in order to prevent fiber dissolution in high salt concentrations, the fiber was cross-linked at 140°C, and after that, methylene blue (MB, cationic molecule) was electrostatically adsorbed onto the polyelectrolyte fibers. Release profiles showed a slow release or zero release of MB at higher pH nonbuffered solutions, whereas MB release at lower pH solutions was very quick. Therefore, the release of MB from PAA/PAH fibers could be controlled by changing the pH of the nonbuffered solutions. On the other hand, MB release from PAA/PAH fibers was fast in a phosphate buffered saline (PBS) solution and could not be sustained. The sustained release profiles in the PBS solution were obtained by depositing hydrophobic perfluorosilane coatings on the fiber surfaces. This also demonstrated that temperature-controlled release of MB could be achieved by coating the PAA/PAH fibers with temperature-sensitive PAA/poly(*N*-isopropylacrylamide) (PNIPAAM) multilayers at a pH of 6.5 (Chunder et al., 2007).

Loading of target molecules and drugs on the surface of electrospun nanofibers (called postprocessing modification) is a facile method for preparation of nanofibrous drug-delivery systems with targeted applications. A typical approach is functionalization of electrospun mats with heparin, perlecan, or cationized gelatin at the fiber surface for loading growth factors or other proteins by specific interactions (Casper et al., 2005, 2007; Lu et al., 2009). For example, heparin, a negatively charged polysaccharide, has strong binding affinity for various growth factors, such as the vascular endothelial growth factor (VEGF), fibroblast growth factor (FGF), transforming growth factor-β (TGF-β), and heparin-binding epidermal growth factor (HBEGF). By considering the importance of the strategy of growth factors delivery to targeted specific sites with controlled release for therapeutic applications, heparin immobilization of the nanofiber surface and subsequent attachment of the growth factor can be an effective way for growth factors delivery (Joung et al., 2008). There have been many studies (Casper et al., 2005; Joung et al., 2008; Kurpinski et al., 2010; Lee et al., 2007, 2012a, 2012b; Patel et al., 2007; Rajangam et al., 2006) that have prepared the heparin-functionalized electrospun mats to permit adsorption of growth factors and subsequently long-term delivery of GFs. Casper and coworkers (Casper et al., 2005) incorporated poly(ethylene glycol) functionalized with low-molecular-weight heparin (PEG-LMWH) into poly(ethylene oxide) (PEO) or poly(lactide-*co*-glycolide) (PLGA) fibers and compared it with free LMWH. As the incorporation of PEG-LMWH resulted in improvements in the binding growth factors to the electrospun nanofibers, the PEG-LMWH retention in the fibers was improved compared with LMWH alone. In fact, incorporation of PEG-LMWH into the fibers permits retention of the heparin for at least 14 days, in contrast to free LMWH, which is completely released from the fibers after 24 h. The slower release of heparin-binding growth factors would allow for drug delivery over a period of time that was needed for tissue repair or regeneration (Casper et al., 2005). Moreover, another method for immobilizing target biomolecules or drugs on the nanofiber surface is chemical modification of the nanofiber surface with amine, carboxyl, hydroxyl, or thiol groups. Chemical immobilization of target biomolecules on electrospun fibers hinders the release of drugs, so this method can be a suitable and applicable delivery of growth factors and genes where a slow and prolonged release of the therapeutic agent is required (Kim and Yoo, 2010; Volpato et al., 2012; Zou et al., 2011).

3.5 TYPE OF DRUGS LOADED IN ELECTROSPUN NANOFIBERS

As electrospinning affords great flexibility in selecting materials for drug delivery applications, a broad range of drugs, from antibiotics and anticancer agents to proteins, DNA, and RNA, can be delivered. A number of different loading methods can be used for incorporation of drugs into electrospun nanofibers using the typical and modified electrospinning technique. In this part, some of the representative drugs loaded into electrospun fibers for drug delivery are presented.

3.5.1 Antibiotics, antifungal, antimicrobial, and antibacterial agents

Antibiotics and various antibacterial agents can be loaded in nanofibers to afford antibacterial properties for different applications. Ignatova et al. (2013) summarized various antibiotics, such as tetracycline hydrochloride (Kenawy et al., 2002; Zahedi et al., 2012), ciprofloxacin, levofloxacin, and moxifloxacin, and antibacterial agents (e.g., 8-hydroxyquinoline derivatives, itraconazolebenzalkonium chloride (BC), fusidic acid (Gilchrist et al., 2013), or silver nanoparticles) encapsulated in nanofibers. Most polymers that were used as the carrier polymers for antibiotic and antibacterial agents were PCL, PLA, and PLGA, and other synthetic or natural polymers were added to regulate the fibers' properties, such as their biodegradability and hydrophobicity. Kenawy et al. (2002) demonstrated that drug release behavior was influenced by the nature of the polymeric carrier and drug content.

3.5.2 Anticancer drugs

One of the applications of drug-loaded electrospun fibers is postsurgical chemotherapy, of which their implantation into cavities prevented tumor recurrence after tumor resection. Various anticancer drugs, such as DOX or doxorubicin hydrochloride (Liu et al., 2013; Lu et al., 2012; Xu et al., 2008, 2009), PTX (Ma et al., 2011; Ranganath and Wang, 2008; Xie and Wang, 2006; Xu et al., 2009), platinum complexes (Xie et al., 2008), dichloroacetate (Liu et al., 2012a), and BCNU (Xu et al., 2006), titanocene dichloride (Chen et al., 2010), and others, have been electrospun into fibers and used for postoperative local chemotherapy.

3.5.3 Bioactive molecules: protein, growth factor, and gene (DNA, siRNA)

Various biomolecules can be incorporated into electrospun fibers to enhance their functional properties for biomedical applications. The most frequently used biomolecules are proteins, growth factors, and growth factor coding genes. The bioactivity preservation of the biomolecules incorporated with electrospun nanofibers during the electrospinning process is a requisite for the successful growth factor delivery that needs to be considered. Various studies have been carried out to explore strategies for the incorporation of biomolecules into electrospinning fibers to deliver therapeutic proteins or genes. Various methods have been used for loading bioactive molecules into electrospun fibers, such as physical adsorption, blend electrospinning, coaxial electrospinning, emulsion electropinning, and covalent immobilization. A comprehensive review (Ji et al., 2011) has provided detailed information for delivering growth factors and genes for tissue engineering applications.

The physical adsorption approach gives little interference with the activity of loaded biomolecules, but this approach had uncontrolled release profiles. Therefore, it is seldom used to load protein or genes in electrospun scaffolds. Some studies reported that bioactive molecules, such as bone morphogenic protein-2 (BMP2) (Nie et al., 2008), fibroblast growth factor (bFGF) (Montero et al., 2012, 2014), or VEGF (Del Gaudio et al., 2013) was loaded into the fibers using the physical adsorption approach. Also, similar evidence is available for gene delivery using this approach, but the released gene exhausted within a short time, and over 95% of incorporated DNA was released within 10 days (Nie and Wang, 2007; Nie et al., 2009).

Blend electrospinning allows more sustained release profiles compared to physical adsorption due to localization biomolecules within the fibers. In general, a sustained release profile can be obtained over several weeks using this technique. Researchers have used blend electrospinning to incorporate various types of proteins and genes in scaffolds, including bovine serum albumin (BSA) (Kowalczyk et al., 2008; Yang et al., 2008a; Zeng et al., 2005), lysozyme (Li et al., 2008; Yang et al., 2008b), and growth factors, such as BMP2 (Li et al., 2006; Nie et al., 2008) or epidermal growth factor (EGF) (Schneider et al., 2009). In this part, emulation electrospinning was assumed to belong to the blend electrospinning approach. Biomolecules may lose their activity in the blend elctrospinning approach due to conformational changes in the organic solution environment, especially emulation electrospinning. In some studies, several strategies have been applied to improve protein stability (Casper et al., 2005; Li et al., 2008; Michel et al., 2005; Wang and Annunziata, 2007). The first study from using blend electrospinning for gene delivery was reported by Luu and coworkers (Luu et al., 2003). In this study, pCMVβ plasmid was incorporated into PLA-PEG/PLGA fibers. Then, other groups (Li et al., 2006; Nie and Wang, 2007; Nie et al., 2009) have used this approach and have shown that DNA kept its structure.

Coaxial electrospinning has gained much attention in the protein delivery field recently due to their great potential in preserving proteins during the electrospinning process and controlled release of proteins through the shell barrier. Many studies have achieved encapsulation of different types of proteins into core–shell fibers using this approach, such as BSA (Jiang et al., 2005, 2006; Xiaoqiang et al., 2009; Zhang et al., 2006); lysozyme (Jiang et al., 2005); VEGF (Jia et al., 2011; Li et al., 2015; Seyednejad et al., 2012); platelet-derived growth factor-bb (PDGF-bb) (Li et al., 2010b; Liao et al., 2006); nerve growth factor (NGF) (Wang et al., 2008); and fibroblast growth factor (bFGF) (Joung, 2011; Sahoo et al., 2010). Some studies reported gene delivery using coaxial electrospinning (Liao et al., 2009; Saraf et al., 2010). Other researchers published that biomolecules were immobilized onto the fiber surface via chemical bond, including BSA (Choi and Yoo, 2007), EGF (Choi et al., 2008), and gene (Kim and Yoo, 2010).

3.6 ROUTES OF DRUG DELIVERY

To date, the drug-loaded nanostructure systems can be applied via different routes, such as implantation, injection, and topical or oral administration, for treatment of a wide range of human diseases. Many drugs are the first-pass metabolism after oral administration due to susceptibility to the GI tract, in particular proteins and peptides. On the other hand, some drugs have low penetration across the skin through the transdermal delivery route. In fact, when drugs with low bioavailability, such as anticancer drugs (5%–20% oral bioavailability) (Thanki et al., 2013), proteins, and peptides (less than 1%–2% oral bioavailability) (Renukuntla et al., 2013) are taken orally, some patients have immediate or delayed allergy and site irritation to intravenous injection (Eichenbaum et al., 2013).

The drug-loaded incorporated electrospun nanofibers were mainly applied topically for skin and wound healing and transdermally or as implanted drug-delivery systems, and reports on the oral administration of electrospun nanofibers are also available.

3.6.1 Implantation administration

Drug-loaded nanofibers can be applied as a postoperation implant for delivery of antibiotic, antifungal, antimicrobial, and anticancer agents. Local delivery of drug-loaded electrospun nanofibers has advantages, such as decreasing the minimum required dosage of drug (leading to reducing side effects due to less systemic absorption), significantly improving drug-encapsulation efficiency, and reducing the burst release in comparison to other drug carriers. Also, electrospinning mats can facilitate the diffusion of drug agents into the surrounding medium (Taepaiboon et al., 2006).

One of the applications of drug-loaded electrospun fibers is postsurgical chemotherapy for which their implantation into cavities has prevented tumor recurrence after tumor resection. The typical procedure for reduction of recurrence after surgical tumor removal is the subsequent chemotherapy. Local recurrence around the original tumor resection area possibly causes the poor median survival time postdiagnosis. The electrospun mats loaded with anticancer drugs can provide high local dosage by implantation administration through covering of the solid tumor site and result in reducing the restriction of frequent administrations. Therefore, sustained local delivery of chemotherapy agents from an intracranial implant may prolong survival after the primary tumor resection. Liu et al. (2013) prepared DOX-encapsulated PLA nanofibers using the electrospinning process and observed that the majority of the loaded DOX in the fibers was released and diffused into the tumor site underneath the fiber mats. The result showed that DOX-encapsulated PLA nanofibers had a great inhibitory effect on tumor growth and little damage to other organs. In a similar report, Liu et al. (2012a) prepared the polylactide (PLA) mats incorporated with dichloroacetate (DCA) by electrospinning, and these electrospun mats were directly implanted to cover the solid tumor. Results indicated that solid subcutaneous tumors completely disappeared from 50% of the tumor-bearing mice, and a tumor suppression degree of 96% was achieved in less than 19 days. Also, Ranganath and coworkers (Ranganath and Wang, 2008) implanted paclitaxel-loaded electrospun matrices into the tumor and compared it to a commercial paclitaxel formulation directly injected into the tumor. Results indicated a stronger reduction in brain tumor growth after 24 days. Other studies (Ignatova et al., 2013; Luo et al., 2012; Pradilla et al., 2006; Xu et al., 2009) on the delivery of chemotherapeutics from electrospun matrices are reported and show that using drug-loaded electrospun nanofibers in local chemotherapy provides an encouraging prospect, especially for those patients receiving complete tumor resection or cytoreductive surgery.

3.6.2 Transdermal administration

Transdermal delivery is to apply the drugs or medicines to the skin. The human skin is an impermeable barrier that provides a strong protection against the external substances, such as fungi, bacteria, viruses, dust, allergens, and large molecules. Targeted drug delivery to skin sites alleviates the physical pain and also promotes the patient's convenience and compliance for drug treatments (Khafagy et al., 2007; Prausnitz et al., 2004; Thomas and Finnin, 2004). Transdermal administration can be avoided due to the drawbacks of oral administration, such as enzymatic degradation and rapid clearance in the gastrointestinal tract or the first-pass metabolism. Therefore, overcoming the strong skin barrier is the major challenge of transdermal drug delivery, because this barrier leads to slow drug penetration rates and subsequently limited drug uptake rates and lack of dosage flexibility.

Electrospun nanofibrous mats have recently shown huge capabilities for wound dressing, such as regulating of wound moisture and efficiently absorbing exudates due to the high surface area of nanofibers (Jannesari et al., 2011; Kenawy et al., 2007). Wound dressing helps in preserving the wounds against infections and microorganisms, absorbing exudates, and also accelerating the wound healing process (Khil et al., 2003; Zhang et al., 2005b). The small pore size of nanofibrous mats that have high porosity can preserve the wound from bacterial infections. Numerous studies have reported that drug-loaded nanofibers were applied for wound dressing, especially bioactive dressing. Bioactive dressings containing antibiotics are usually required in the initial period of wound healing, and drug-loaded nanofiber mats can offer desired drug release for bioactive dressing (Jannesari et al., 2011). Jannesari et al. (2011) used the electrospun nanofiber from a blend of poly(vinyl acetate) and poly(vinyl alcohol) loaded with ciprofloxacin hydrochloride for wound dressing application, and demonstrated that the blend of poly(vinyl acetate) and PVA nanofibers results in decreasing the burst release rate and extending the total drug release. In a study by Said et al. (2011), electrospun nanofiber loaded with fusidic acid was applied, and the interactions with wound bacteria were explored. It was observed that bacterial colonization of fibers formed a thick layer of biofilm. Schneider et al. (2009) demonstrated that silk nanofibers loaded with epidermal growth factors decreased the time of wound closure and accelerated the wound healing process.

Electrospun nanofibers can be also used as drug-delivery systems in transdermal administrations. These transdermal systems, which assist the delivery of target agents into the body through the skin, can be a good alternative for oral administration of sensitive drugs (Prausnitz and Langer, 2008; Schneider et al., 2009). Furthermore, controlled drug release of such transdermal systems results in great improvement of drug efficiency due to reduction of variations in dosage of the therapeutic agents in the body. One of the limitations of transdermal drug–delivery systems is the limitation in the number of drugs that can pass through the skin due to its low permeability. To date, electrospun nanofibers were successfully used as drug carriers for transdermal delivery of various types of vitamins, antioxidants, and antiinflammatory drugs (Ignatova et al., 2013; Im et al., 2010; Ngawhirunpat et al., 2009; Suwantong et al., 2007, 2008; Taepaiboon et al., 2007b; Yun et al., 2011). Herbal pharmaceutical compounds were also successfully incorporated in electrospun nanofibers as topical/transdermal or wound-dressing devices (Suwantong et al., 2008).

3.6.3 Oral administration

Oral drug delivery is considered as an attractive route and perhaps the most preferred route for drug administration because oral administration provides some advantages for patients, such as the ease of drug administration by patients and a high degree of flexibility on dosages (Jain, 2008). In oral administration, drugs should be protected under unstable biological environments including drug degradation induced by the GI tract and first-pass liver effects after oral administration before reaching the targeted sites (Jain, 2008).

Whereas oral administration is a preferable route for drug administration due to the aspects of convenience, patient compliance, and cost-effectiveness, a few studies (Im et al., 2012; Shen et al., 2011) have published the application of drug-loaded nanofibers as oral drug–delivery systems. To date, several studies have reported the fabrication of drug-loaded Eudragit L100-55 nanofibers for oral delivery purposes. Shen and coworkers (Shen et al., 2011) successfully prepared Eudragit L100-55 nanofibers loaded with diclofenac sodium (a nonsteroidal antiinflammatory drug) using the electrospinning process for oral drug delivery purposes. The pH-dependent release profiles of drug-loaded nanofibers indicated the ability of drug-loaded Eudragit L100-55 nanofibers to be used as an oral colon–targeted drug-delivery system. Yu et al. (2013a, 2014) reported the preparation of colon-targeted delivery systems using drug-encapsulated nanofibers by the coaxial electrospinning process. In another study, Heshmati and Makhmalzadeh (2012) reported the preparation of electrospun indomethacin-loaded Eudragit S100 and Eudragit RS100 nanofibers for colon-targeted drug delivery. This study used a combination of ES (pH-dependent polymer) and ERS (time-dependent polymer) as a single coating for the design of colon-delivery systems of indomethacin electrospinning. The release characteristic indicated that drug-loaded ES and ERS nanofibers have the potential for oral-targeted drug-delivery systems (Heshmati and Makhmalzadeh, 2012). In Im and coworkers' report (Im et al., 2012), PVA electrospun fibers were coated with alginate and CS to preserve *Lactobacillus* to the large intestine. When PVA nanofibers reached the stomach and small intestine, CS and alginate were removed due to their solubility in acid and neutral conditions, respectively, and thus *Lactobacillus* was released into the large intestine.

A great majority of drugs are highly hydrophobic and poorly water-soluble. The oral absorption of these drugs is often limited due to their poor solubility or slow dissolution in the human GI contents. In order to enhance the oral absorption of such drugs, several approaches are being developed to improve the dissolution rates and solubility of drug molecules. Electrospinning is being considered as one of the technologies that can produce nanosized drugs incorporated in polymeric nanofibers. In other words, electrospinning of drug solutions in polymers produces nanofibers with a high surface area. This extremely high surface area has a significant influence on the bioavailability of a poorly soluble drug, because the increased surface area can lead to an increased dissolution rate. Verreck et al. (2003a, 2003b) have applied the electrospinning technique for transdermal and oral delivery of poorly soluble drugs (Ignatious and Baldoni, 2001; Ignatious and Sun, 2003). They reported the electrospinning of a combination of hydroxypropyl methylcellulose (HPMC) and itraconazole (a poorly soluble drug) and showed the drug release from nanofibers depended on the drug/polymer ratio and fiber diameter (Verreck et al., 2003a). In their latter report, Verreck et al. (2003b) reported the electrospinning of itraconazole and ketanserin in a nonbiodegradable matrix (segmented polyurethane) in order to improve the transdermal delivery of these poorly soluble drugs. In this report, it was shown that the rate of drug release from the nanofibers depended on the drug-to-polymer ratio. Since these initial publications, there have been several reports on electrospun fibers containing drugs. Ignatious et al. (2010) published an expert review in this field and discussed in detail the potential of electrospinning for enhancing the oral delivery of poorly soluble drugs.

4 FUTURE DIRECTIONS

In spite of diverse benefits of different methods of fabrication of nanoscale materials as drug-delivery systems, electrospinning, and electrospraying techniques show great potential for these applications, especially in the cases of nanofibrous systems and structures. These techniques have the powerful engineering capability of manipulating the processing conditions along with selection of desire materials, to encapsulate suitable therapeutic agents within the specific structure. Further investigation on employing more suitable carriers for better preservation of therapeutic agents is required. It is also urgently required to investigate using different modifications in the conventional electrospinning setups for new concepts of drug-delivery systems, more interestingly for multidrug-delivery systems. Coaxial techniques have reasonable potentials for multidrug-delivery systems, since they can provide the possibility of controlling each individual drug-release kinetic, designing the different surface and core properties, and better bioactivity protection of unstable drugs. Besides all benefits of electrospinning and electrospraying techniques in drug-delivery systems, there still are a few weak points that limit the complete success of these techniques, more importantly on an industrial scale, even though new generations of these techniques have the forecast view of scaling-up procedures, such as the method used in nozzleless setups.

REFERENCES

Allen, T.M., Cullis, P.R., 2013. Liposomal drug delivery systems: from concept to clinical applications. Adv. Drug Deliv. Rev. 65, 36–48.

Alvarez-Lorenzo, C., Concheiro, A., 2014. Smart drug delivery systems: from fundamentals to the clinic. Chem. Commun. 50, 7743–7765.

Anirudhan, T.S., Divya, P.L., Nima, J., 2015. Synthesis and characterization of silane coated magnetic nanoparticles/glycidylmethacrylate-grafted-maleated cyclodextrin composite hydrogel as a drug carrier for the controlled delivery of 5-fluorouracil. Mater. Sci. Eng. C 55, 471–481.

Bagherzadeh, R., Latifi, M., Najar, S.S., Tehran, M.A., Gorji, M., Kong, L., 2012. Transport properties of multi-layer fabric based on electrospun nanofiber mats as a breathable barrier textile material. Tex. Res. J. 82, 70–76.

Bagherzadeh, R., Latifi, M., Najar, S.S., Tehran, M.A., Kong, L., 2013a. Three-dimensional pore structure analysis of nano/microfibrous scaffolds using confocal laser scanning microscopy. J. Biomed. Mater. Res. A 101A, 765–774.

Bagherzadeh, R., Najar, S.S., Latifi, M., Tehran, M.A., Kong, L., 2013b. A theoretical analysis and prediction of pore size and pore size distribution in electrospun multilayer nanofibrous materials. J. Biomed. Mater. Res. A 101A, 2107–2117.

Bagherzadeh, R., Latifi, M., Kong, L., 2014a. Three-dimensional pore structure analysis of polycaprolactone nano-microfibrous scaffolds using theoretical and experimental approaches. J. Biomed. Mater. Res. A 102, 903–910.

Bagherzadeh, R., Latifi, M., Najar, S.S., Tehran, M.A., Kong, L., 2014b. The application of Cd Se/ZnS quantum dots and confocal laser scanning microscopy for three-dimensional imaging of nanofibrous structures. J. Ind. Text. 43, 496–510.

Boas, M., Gradys, A., Vasilyev, G., Burman, M., Zussman, E., 2015. Electrospinning polyelectrolyte complexes: pH-responsive fibers. Soft Matter 11, 1739–1747.

Bölgen, N., Vargel, I., Korkusuz, P., Menceloğlu, Y.Z., Pişkin, E., 2007. In vivo performance of antibiotic embedded electrospun PCL membranes for prevention of abdominal adhesions. J. Biomed. Mater. Res. B 81, 530–543.

Brouwers, J., 1996. Advanced and controlled drug delivery systems in clinical disease management. Pharm. World Sci. 18, 153–162.

Buschle-Diller, G., Cooper, J., Xie, Z., Wu, Y., Waldrup, J., Ren, X., 2007. Release of antibiotics from electrospun bicomponent fibers. Cellulose 14, 553–562.

Bussemer, T., Otto, I., Bodmeier, R., 2001. Pulsatile drug-delivery systems. Crit. Rev. Ther. Drug, 18.

Çağdaş, M., Demir Sezer, A., Bucak, S., 2014. Application of Nanotechnology in Drug Delivery. InTech, Available from: http://www.intechopen.com/books/application-of-nanotechnology-in-drug-delivery/application-of-nanotechnology-in-drug-delivery.

Cao, H., Jiang, X., Chai, C., Chew, S.Y., 2010. RNA interference by nanofiber-based siRNA delivery system. J. Control. Release 144, 203–212.

Casper, C.L., Yamaguchi, N., Kiick, K.L., Rabolt, J.F., 2005. Functionalizing electrospun fibers with biologically relevant macromolecules. Biomacromolecules 6, 1998–2007.

Casper, C.L., Yang, W., Farach-Carson, M.C., Rabolt, J.F., 2007. Coating electrospun collagen and gelatin fibers with perlecan domain I for increased growth factor binding. Biomacromolecules 8, 1116–1123.

Chen, C., Lv, G., Pan, C., Song, M., Wu, C., Guo, D., Wang, X., Chen, B., Gu, Z., 2007. Poly(lactic acid) (PLA) based nanocomposites—a novel way of drug-releasing. Biomed. Mater. 2, L1.

Chen, P., Wu, Q.-S., Ding, Y.-P., Chu, M., Huang, Z.-M., Hu, W., 2010. A controlled release system of titanocene dichloride by electrospun fiber and its antitumor activity in vitro. Eur. J. Pharm. Biopharm. 76, 413–420.

Chew, S.Y., Wen, J., Yim, E.K., Leong, K.W., 2005. Sustained release of proteins from electrospun biodegradable fibers. Biomacromolecules 6, 2017–2024.

Cho, Y.I., Choi, J.S., Jeong, S.Y., Yoo, H.S., 2010. Nerve growth factor (NGF)-conjugated electrospun nanostructures with topographical cues for neuronal differentiation of mesenchymal stem cells. Acta Biomater. 6, 4725–4733.

Choi, J.S., Yoo, H.S., 2007. Electrospun nanofibers surface-modified with fluorescent proteins. J. Bioact. Compat. Poly. 22, 508–524.

Choi, J.S., Leong, K.W., Yoo, H.S., 2008. In vivo wound healing of diabetic ulcers using electrospun nanofibers immobilized with human epidermal growth factor (EGF). Biomaterials 29, 587–596.

Chunder, A., Sarkar, S., Yu, Y., Zhai, L., 2007. Fabrication of ultrathin polyelectrolyte fibers and their controlled release properties. Colloids Surf. B 58, 172–179.

Crampton, H.L., Simanek, E.E., 2007. Dendrimers as drug delivery vehicles: non-covalent interactions of bioactive compounds with dendrimers. Polym. Int. 56, 489–496.

Cui, W., Li, X., Zhu, X., Yu, G., Zhou, S., Weng, J., 2006. Investigation of drug release and matrix degradation of electrospun poly(DL-lactide) fibers with paracetanol inoculation. Biomacromolecules 7, 1623–1629.

Cui, W., Zhou, Y., Chang, J., 2010. Electrospun nanofibrous materials for tissue engineering and drug delivery. Sci. Technol. Adv. Mater. 11, 014108.

Del Gaudio, C., Baiguera, S., Boieri, M., Mazzanti, B., Ribatti, D., Bianco, A., Macchiarini, P., 2013. Induction of angiogenesis using VEGF releasing genipin-crosslinked electrospun gelatin mats. Biomaterials 34, 7754–7765.

Ding, Y., Shen, S.Z., Sun, H., Sun, K., Liu, F., Qi, Y., Yan, J., 2015. Design and construction of polymerized-chitosan coated Fe_3O_4 magnetic nanoparticles and its application for hydrophobic drug delivery. Mater. Sci. Eng. C 48, 487–498.

Eichenbaum, G., Zhou, J., De Smedt, A., De Jonghe, S., Looszova, A., Arien, T., Van Goethem, F., Vervoort, I., Shukla, U., Lammens, L., 2013. Methods to evaluate and improve the injection site tolerability of intravenous formulations prior to first-in-human testing. J. Pharmacol. Toxicol. 68, 394–406.

Emami, J., Rezazadeh, M., Hasanzadeh, F., Sadeghi, H., Mostafavi, A., Minaiyan, M., Rostami, M., Davies, N., 2015. Development and in vitro/in vivo evaluation of a novel targeted polymeric micelle for delivery of paclitaxel. Int. J. Biol. Macromol. 80, 29–40.

Gatti, J., Smithgall, M., Paranjape, S., Rolfes, R., Paranjape, M., 2013. Using electrospun poly(ethylene-oxide) nanofibers for improved retention and efficacy of bacteriolytic antibiotics. Biomed. Microdevices 15, 887–893.

Gaucher, G., Dufresne, M.-H., Sant, V.P., Kang, N., Maysinger, D., Leroux, J.-C., 2005. Block copolymer micelles: preparation, characterization and application in drug delivery. J. Control. Release 109, 169–188.

Ge, L., Pan, C., Chen, H., Wang, X., Wang, C., Gu, Z., 2007. The fabrication of hollow multilayered polyelectrolyte fibrous mats and its morphology study. Colloids Surf. A 293, 272–277.
Gheibi, A., Latifi, M., Merati, A., Bagherzadeh, R., 2014. Piezoelectric electrospun nanofibrous materials for self-powering wearable electronic textiles applications. J. Polym. Res. 21, 1–7.
Gilchrist, S.E., Lange, D., Letchford, K., Bach, H., Fazli, L., Burt, H.M., 2013. Fusidic acid and rifampicin co-loaded PLGA nanofibers for the prevention of orthopedic implant associated infections. J. Control. Release 170, 64–73.
Goldberg, M., Langer, R., Jia, X., 2007. Nanostructured materials for applications in drug delivery and tissue engineering. J. Biomat. Sci.-Polym. E. 18, 241–268.
Gonsalves, K., Halberstadt, C., Laurencin, C.T., Nair, L., 2007. Biomedical Nanostructures. John Wiley & Sons.
Gupta, U., Dwivedi, S.K.D., Bid, H.K., Konwar, R., Jain, N., 2010. Ligand anchored dendrimers based nanoconstructs for effective targeting to cancer cells. Int. J. Pharm. 393, 186–197.
He, X., Alves, C.S., Oliveira, N., Rodrigues, J., Zhu, J., Bányai, I., Tomás, H., Shi, X., 2015. RGD peptide-modified multifunctional dendrimer platform for drug encapsulation and targeted inhibition of cancer cells. Colloids Surf. B 125, 82–89.
Heshmati, A.A., Makhmalzadeh, B.S., 2012. Preparation and evaluation of electrospun indomethacin loaded Eudragit S100 and Eudragit RS100 nanofibers for colon-targeted drug delivery. Res. Pharm. Sci. 7, S245.
Hu, X., Liu, S., Zhou, G., Huang, Y., Xie, Z., Jing, X., 2014. Electrospinning of polymeric nanofibers for drug delivery applications. J. Control. Release 185, 12–21.
Huang, Z.-M., Zhang, Y.Z., Kotaki, M., Ramakrishna, S., 2003. A review on polymer nanofibers by electrospinning and their applications in nanocomposites. Compos. Sci. Technol. 63, 2223–2253.
Ignatious, F., Baldoni, J., 2001. Electrospun pharmaceutical compositions. Google Patents.
Ignatious, F., Sun, L., 2003. Electrospun amorphous pharmaceutical compositions. Google Patents.
Ignatious, F., Sun, L., Lee, C.-P., Baldoni, J., 2010. Electrospun nanofibers in oral drug delivery. Pharm. Res. 27, 576–588.
Ignatova, ., Rashkov, I., Manolova, N., 2013. Drug-loaded electrospun materials in wound-dressing applications and in local cancer treatment. Expert Opin. Drug Deliv. 10, 469–483.
Ihre, H., Padilla De Jesús, O.L., Fréchet, J.M., 2001. Fast and convenient divergent synthesis of aliphatic ester dendrimers by anhydride coupling. J. Am. Chem. Soc. 123, 5908–5917.
Ikada, Y., 2011. Tissue Engineering: Fundamentals and Applications. Academic Press.
Im, J.S., Bai, B.C., Lee, Y.-S., 2010. The effect of carbon nanotubes on drug delivery in an electro-sensitive transdermal drug delivery system. Biomaterials 31, 1414–1419.
Im, J.S., Lee, S.K., Bai, B.C., Lee, Y.-S., 2012. Prediction and characterization of drug release in a multi-drug release system. J. Indust. Eng. Chem. 18, 325–330.
Ishida, T., Kirchmeier, M., Moase, E., Zalipsky, S., Allen, T., 2001. Targeted delivery and triggered release of liposomal doxorubicin enhances cytotoxicity against human B lymphoma cells. BBA-Biomembranes 1515, 144–158.
Jahanmard, F., Tehran, M.A., Zamani, F., Nematollahi, M., Mobarakeh, L.G., Nasr-Esfahani, M.H., 2013. Effect of nanoporous fibers on growth and proliferation of cells on electrospun poly(ε-caprolactone) scaffolds. Int. J. Polym. Mater. Polym. Biomater. 63, 57–64.
Jain, K.K., 2008. Drug delivery systems—an overview. Springer, Totowa, NJ.
Jain, K., Gupta, U., Jain, N.K., 2014. Dendronized nanoconjugates of lysine and folate for treatment of cancer. Eur. J. Pharm. Biopharm. 87, 500–509.
Jannesari, M., Varshosaz, J., Morshed, M., Zamani, M., 2011. Composite poly(vinyl alcohol)/poly(vinyl acetate) electrospun nanofibrous mats as a novel wound dressing matrix for controlled release of drugs. Int. J. Nanomed. 6, 993–1003.
Jha, M.K., Rahman, M.H., Rahman, M.M., 2011. Biphasic oral solid drug delivery system: a review. Int. J. Pharm. Sci. Res 2, 1108–1115.
Ji, W., Sun, Y., Yang, F., Van Den Beucken, J.J., Fan, M., Chen, Z., Jansen, J.A., 2011. Bioactive electrospun scaffolds delivering growth factors and genes for tissue engineering applications. Pharm. Res. 28, 1259–1272.

Jia, X., Zhao, C., Li, P., Zhang, H., Huang, Y., Li, H., Fan, J., Feng, W., Yuan, X., Fan, Y., 2011. Sustained release of VEGF by coaxial electrospun dextran/PLGA fibrous membranes in vascular tissue engineering. J. Biomat. Sci.-Polym. E. 22, 1811–1827.

Jian, Y.-T., Yang, Y., Tian, T., Stanford, C., Zhang, X.-P., Zhao, K., 2015. Effect of pore size and porosity on the biomechanical properties and cytocompatibility of porous NiTi alloys. PLoS One 10, e0128138.

Jiang, H., Hu, Y., Li, Y., Zhao, P., Zhu, K., Chen, W., 2005. A facile technique to prepare biodegradable coaxial electrospun nanofibers for controlled release of bioactive agents. J. Control. Release 108, 237–243.

Jiang, H., Hu, Y., Zhao, P., Li, Y., Zhu, K., 2006. Modulation of protein release from biodegradable core–shell structured fibers prepared by coaxial electrospinning. J. Biomed. Mater. Res. B 79, 50–57.

Jiang, Y.-N., Mo, H.-Y., Yu, D.-G., 2012. Electrospun drug-loaded core–sheath PVP/Zein nanofibers for biphasic drug release. Int. J. Pharm. 438, 232–239.

Jiang, H., Zhao, L., Gai, L., Wang, Y., Hou, Y., Liu, H., 2015. Conjugation of methotrexate onto dedoped Fe_3O_4/PPy nanospheres to produce magnetic targeting drug with controlled drug release and targeting specificity for HeLa cells. Synthetic Met. 207, 18–25.

Joung, Y.K., 2011. Controlled release of growth factors from core–shell structured PLGA microfibers for tissue engineering. Biomater. Res. 15, 78.

Joung, Y.K., Bae, J.W., Park, K.D., 2008. Controlled release of heparin-binding growth factors using heparin-containing particulate systems for tissue regeneration. Expert Opin. Drug Deliv. 5 (11), 1173–1184.

Kaassis, A.Y., Young, N., Sano, N., Merchant, H.A., Yu, D.-G., Chatterton, N.P., Williams, G.R., 2014. Pulsatile drug release from electrospun poly(ethylene oxide)–sodium alginate blend nanofibres. J. Mater. Chem. B 2, 1400–1407.

Kalra, V., Lee, J., Lee, J.H., Lee, S.G., Marquez, M., Wiesner, U., Joo, Y.L., 2008. Controlling nanoparticle location via confined assembly in electrospun block copolymer nanofibers. Small 4, 2067–2073.

Karamipour, S., Sadjadi, M., Farhadyar, N., 2015. Fabrication and spectroscopic studies of folic acid-conjugated Fe_3O_4@ Au core–shell for targeted drug delivery application. Spectrochim. Acta A 148, 146–155.

Kenawy, E.-R., Bowlin, G.L., Mansfield, K., Layman, J., Simpson, D.G., Sanders, E.H., Wnek, G.E., 2002. Release of tetracycline hydrochloride from electrospun poly(ethylene-*co*-vinylacetate), poly(lactic acid), and a blend. J. Control. Release 81, 57–64.

Kenawy, E.-R., Abdel-Hay, F.I., El-Newehy, M.H., Wnek, G.E., 2007. Controlled release of ketoprofen from electrospun poly(vinyl alcohol) nanofibers. Mater. Sci. Eng. A 459, 390–396.

Kenawy, E.-R., Abdel-Hay, F.I., El-Newehy, M.H., Wnek, G.E., 2009. Processing of polymer nanofibers through electrospinning as drug delivery systems. Mater. Chem. Phys. 113, 296–302.

Khafagy, E.-S., Morishita, M., Onuki, Y., Takayama, K., 2007. Current challenges in non-invasive insulin delivery systems: a comparative review. Adv. Drug Deliv. Rev. 59, 1521–1546.

Khil, M.S., Cha, D.I., Kim, H.Y., Kim, I.S., Bhattarai, N., 2003. Electrospun nanofibrous polyurethane membrane as wound dressing. J. Biomed. Mater. Res. B 67, 675–679.

Kikuchi, H., Suzuki, N., Ebihara, K., Morita, H., Ishii, Y., Kikuchi, A., Sugaya, S., Serikawa, T., Tanaka, K., 1999. Gene delivery using liposome technology. J. Control. Release 62, 269–277.

Kim, H.S., Yoo, H.S., 2010. MMPs-responsive release of DNA from electrospun nanofibrous matrix for local gene therapy: in vitro and in vivo evaluation. J. Control. Release 145, 264–271.

Ko, F.K., Wan, Y., 2014. Introduction to Nanofiber Materials. Cambridge University Press, Cambridge.

Kowalczyk, T., Nowicka, A., Elbaum, D., Kowalewski, T.A., 2008. Electrospinning of bovine serum albumin: optimization and the use for production of biosensors. Biomacromolecules 9, 2087–2090.

Kurpinski, K.T., Stephenson, J.T., Janairo, R.R.R., Lee, H., Li, S., 2010. The effect of fiber alignment and heparin coating on cell infiltration into nanofibrous PLLA scaffolds. Biomaterials 31, 3536–3542.

Langer, R., 1998. Drug delivery and targeting. Nature 392, 5–10.

Lee, J.S., Go, D.H., Bae, J.W., Lee, S.J., Park, K.D., 2007. Heparin conjugated polymeric micelle for long-term delivery of basic fibroblast growth factor. J. Control. Release 117, 204–209.

Lee, H.J., Kim, H.-S., Kim, H.O., Koh, W.-G., 2011. Micropatterns of double-layered nanofiber scaffolds with dual functions of cell patterning and metabolite detection. Lab Chip 11, 2849–2857.
Lee, J., Yoo, J., Atala, A., Lee, S., 2012a. Controlled heparin conjugation on electrospun poly(ε-caprolactone)/gelatin fibers for morphology-dependent protein delivery and enhanced cellular affinity. Acta Biomater. 8, 2549–2558.
Lee, J., Yoo, J.J., Atala, A., Lee, S.J., 2012b. The effect of controlled release of PDGF-BB from heparin-conjugated electrospun PCL/gelatin scaffolds on cellular bioactivity and infiltration. Biomaterials 33, 6709–6720.
Lee, H.J., Park, Y.H., Koh, W.-G., 2013. Fabrication of nanofiber microarchitectures localized within hydrogel microparticles and their application to protein delivery and cell encapsulation. Adv. Funct. Mater. 23, 591–597.
Lee, K.-Y., Chiang, Y.-T., Hsu, N.-Y., Yang, C.-Y., Lo, C.-L., Ku, C.-A., 2015. Vitamin E containing polymer micelles for reducing normal cell cytotoxicity and enhancing chemotherapy efficacy. Acta Biomater. 24, 286–296.
Li, D., Wang, Y., Xia, Y., 2003. Electrospinning of polymeric and ceramic nanofibers as uniaxially aligned arrays. Nano Lett. 3, 1167–1171.
Li, C., Vepari, C., Jin, H.-J., Kim, H.J., Kaplan, D.L., 2006. Electrospun silk-BMP-2 scaffolds for bone tissue engineering. Biomaterials 27, 3115–3124.
Li, Y., Jiang, H., Zhu, K., 2008. Encapsulation and controlled release of lysozyme from electrospun poly(ε-caprolactone)/poly(ethylene glycol) non-woven membranes by formation of lysozyme–oleate complexes. J. Mater. Sci. 19, 827–832.
Li, F., Danquah, M., Mahato, R.I., 2010a. Synthesis and characterization of amphiphilic lipopolymers for micellar drug delivery. Biomacromolecules 11, 2610–2620.
Li, H., Zhao, C., Wang, Z., Zhang, H., Yuan, X., Kong, D., 2010b. Controlled release of PDGF-bb by coaxial electrospun dextran/poly(L-lactide-*co*-ε-caprolactone) fibers with an ultrafine core/shell structure. J. Biomat. Sci.-Polym. E. 21, 803–819.
Li, Z., Song, L., Huang, X., Wang, H., Shao, H., Xie, M., Xu, Y., Zhang, Y., 2015. Tough and VEGF-releasing scaffolds composed of artificial silk fibroin mats and a natural acellular matrix. RSC Adv. 5, 16748–16758.
Liang, D., Hsiao, B.S., Chu, B., 2007. Functional electrospun nanofibrous scaffolds for biomedical applications. Adv. Drug Deliv. Rev. 59, 1392–1412.
Liao, I., Chew, S., Leong, K., 2006. Aligned core–shell nanofibers delivering bioactive proteins. Nanomedicine-UK 1, 465–471.
Liao, I.-C., Chen, S., Liu, J.B., Leong, K.W., 2009. Sustained viral gene delivery through core–shell fibers. J. Control. Release 139, 48–55.
Liu, Z., Chen, K., Davis, C., Sherlock, S., Cao, Q., Chen, X., Dai, H., 2008. Drug delivery with carbon nanotubes for in vivo cancer treatment. Cancer Res. 68, 6652–6660.
Liu, D., Liu, S., Jing, X., Li, X., Li, W., Huang, Y., 2012a. Necrosis of cervical carcinoma by dichloroacetate released from electrospun polylactide mats. Biomaterials 33, 4362–4369.
Liu, W., Thomopoulos, S., Xia, Y., 2012b. Electrospun nanofibers for regenerative medicine. Adv. Healthcare Mater. 1, 10–25.
Liu, S., Zhou, G., Liu, D., Xie, Z., Huang, Y., Wang, X., Wu, W., Jing, X., 2013. Inhibition of orthotopic secondary hepatic carcinoma in mice by doxorubicin-loaded electrospun polylactide nanofibers. J. Mater. Chem. B 1, 101–109.
Lu, Y., Jiang, H., Tu, K., Wang, L., 2009. Mild immobilization of diverse macromolecular bioactive agents onto multifunctional fibrous membranes prepared by coaxial electrospinning. Acta Biomater. 5, 1562–1574.
Lu, T., Jing, X., Song, X., Wang, X., 2012. Doxorubicin-loaded ultrafine PEG-PLA fiber mats against hepatocarcinoma. J. Appl. Polym. Sci. 123, 209–217.
Luo, X., Xie, C., Wang, H., Liu, C., Yan, S., Li, X., 2012. Antitumor activities of emulsion electrospun fibers with core loading of hydroxycamptothecin via intratumoral implantation. Int. J. Pharm. 425, 19–28.

Luu, Y., Kim, K., Hsiao, B., Chu, B., Hadjiargyrou, M., 2003. Development of a nanostructured DNA delivery scaffold via electrospinning of PLGA and PLA–PEG block copolymers. J. Control. Release 89, 341–353.

Ma, G., Liu, Y., Peng, C., Fang, D., He, B., Nie, J., 2011. Paclitaxel loaded electrospun porous nanofibers as mat potential application for chemotherapy against prostate cancer. Carbohydr. Polym. 86, 505–512.

Maleki, M., Latifi, M., Amani-Tehran, M., Mathur, S., 2013. Electrospun core–shell nanofibers for drug encapsulation and sustained release. Polym. Eng. Sci. 53, 1770–1779.

Mano, J.F., 2008. Stimuli-responsive polymeric systems for biomedical applications. Adv. Eng. Mater. 10, 515–527.

Meng, Z.X., Xu, X.X., Zheng, W., Zhou, H.M., Li, L., Zheng, Y.F., Lou, X., 2011. Preparation and characterization of electrospun PLGA/gelatin nanofibers as a potential drug delivery system. Colloids Surf. B 84, 97–102.

Michel, R., Pasche, S., Textor, M., Castner, D.G., 2005. Influence of PEG architecture on protein adsorption and conformation. Langmuir 21, 12327–12332.

Moghe, A., Gupta, B., 2008. Co-axial electrospinning for nanofiber structures: preparation and applications. Polym. Rev. 48, 353–377.

Montero, R.B., Vial, X., Nguyen, D.T., Farhand, S., Reardon, M., Pham, S.M., Tsechpenakis, G., Andreopoulos, F.M., 2012. bFGF-containing electrospun gelatin scaffolds with controlled nano-architectural features for directed angiogenesis. Acta Biomater. 8, 1778–1791.

Montero, R.B., Vazquez-Padron, R.I., Pham, S.M., D'Ippolito, G., Andreopoulos, F.M., 2014. Electrospun gelatin constructs with tunable fiber orientation promote directed angiogenesis. Open J. Regenerative Med. 3, 1–12.

Mura, S., Nicolas, J., Couvreur, P., 2013. Stimuli-responsive nanocarriers for drug delivery. Nat. Mater. 12, 991–1003.

Natu, M.V., De Sousa, H.C., Gil, M., 2010. Effects of drug solubility, state and loading on controlled release in bicomponent electrospun fibers. Int. J. Pharm. 397, 50–58.

Ngawhirunpat, T., Opanasopit, P., Rojanarata, T., Akkaramongkolporn, P., Ruktanonchai, U., Supaphol, P., 2009. Development of meloxicam-loaded electrospun polyvinyl alcohol mats as a transdermal therapeutic agent. Pharm. Dev. Technol. 14, 73–82.

Nie, H., Wang, C.-H., 2007. Fabrication and characterization of PLGA/HAp composite scaffolds for delivery of BMP-2 plasmid DNA. J. Control. Release 120, 111–121.

Nie, H., Soh, B.W., Fu, Y.C., Wang, C.H., 2008. Three-dimensional fibrous PLGA/HAp composite scaffold for BMP-2 delivery. Biotechnol. Bioeng. 99, 223–234.

Nie, H., Ho, M.-L., Wang, C.-K., Wang, C.-H., Fu, Y.-C., 2009. BMP-2 plasmid loaded PLGA/HAp composite scaffolds for treatment of bone defects in nude mice. Biomaterials 30, 892–901.

Ninomiya, K., Yamashita, T., Tanabe, Y., Imai, M., Takahashi, K., Shimizu, N., 2016. Targeted and ultrasound-triggered cancer cell injury using perfluorocarbon emulsion-loaded liposomes endowed with cancer cell-targeting and fusogenic capabilities. Ultrason. Sonochem. 28, 54–61.

Nogueira, G.M., Swiston, A.J., Beppu, M.M., Rubner, M.F., 2010. Layer-by-layer deposited chitosan/silk fibroin thin films with anisotropic nanofiber alignment. Langmuir 26, 8953–8958.

Ofori-Kwakye, K., Fell, J.T., 2003. Leaching of pectin from mixed films containing pectin, chitosan and HPMC intended for biphasic drug delivery. Int. J. Pharm. 250, 251–257.

Okuda, T., Tominaga, K., Kidoaki, S., 2010. Time-programmed dual release formulation by multilayered drug-loaded nanofiber meshes. J. Control. Release 143, 258–264.

Panyam, J., Labhasetwar, V., 2003. Biodegradable nanoparticles for drug and gene delivery to cells and tissue. Adv. Drug Deliv. Rev. 55, 329–347.

Park, J., Lakes, R.S., 2007. Biomaterials: An Introduction. Springer Science & Business Media, New York, NY.

Park, C.H., Kim, K.-H., Lee, J.-C., Lee, J., 2008. In-situ nanofabrication via electrohydrodynamic jetting of countercharged nozzles. Polym. Bull. 61, 521–528.

Parveen, S., Misra, R., Sahoo, S.K., 2012. Nanoparticles: a boon to drug delivery, therapeutics, diagnostics and imaging. Nanomed. Nanotechnol. 8, 147–166.

Patel, S., Kurpinski, K., Quigley, R., Gao, H., Hsiao, B.S., Poo, M.-M., Li, S., 2007. Bioactive nanofibers: synergistic effects of nanotopography and chemical signaling on cell guidance. Nano Lett. 7, 2122–2128.

Poon, Z., Chang, D., Zhao, X., Hammond, P.T., 2011. Layer-by-layer nanoparticles with a pH-sheddable layer for in vivo targeting of tumor hypoxia. ACS Nano 5, 4284–4292.

Pradilla, G., Wang, P.P., Gabikian, P., Li, K., Magee, C.A., Walter, K.A., Brem, H., 2006. Local intracerebral administration of paclitaxel with the paclimer delivery system: toxicity study in a canine model. J. Neurooncol. 76, 131–138.

Prausnitz, M.R., Langer, R., 2008. Transdermal drug delivery. Nat. Biotechnol. 26, 1261–1268.

Prausnitz, M.R., Mitragotri, S., Langer, R., 2004. Current status and future potential of transdermal drug delivery. Nat. Rev. Drug Discov. 3, 115–124.

Rajangam, K., Behanna, H.A., Hui, M.J., Han, X., Hulvat, J.F., Lomasney, J.W., Stupp, S.I., 2006. Heparin binding nanostructures to promote growth of blood vessels. Nano Lett. 6, 2086–2090.

Ramakrishna, S., Fujihara, K., Teo, W.-E., Lim, T.-C., Ma, Z., 2005. An Introduction to Electrospinning and Nanofibers. World Scientific, Singapore.

Ranganath, S.H., Wang, C.-H., 2008. Biodegradable microfiber implants delivering paclitaxel for post-surgical chemotherapy against malignant glioma. Biomaterials 29, 2996–3003.

Ratner, B.D., Hoffman, A.S., Schoen, F.J., Lemons, J.E., 2004. Biomaterials Science: An Introduction to Materials in Medicine. Academic Press, Oxford.

Renukuntla, J., Vadlapudi, A.D., Patel, A., Boddu, S.H., Mitra, A.K., 2013. Approaches for enhancing oral bioavailability of peptides and proteins. Int. J. Pharm. 447, 75–93.

Romero-Cano, M., Vincent, B., 2002. Controlled release of 4-nitroanisole from poly(lactic acid) nanoparticles. J. Control. Release 82, 127–135.

Ruttala, H.B., Ko, Y.T., 2015. Liposome encapsulated albumin-paclitaxel nanoparticle for enhanced antitumor efficacy. Pharm. Res. 32, 1002–1016.

Sabnani, M.K., Rajan, R., Rowland, B., Mavinkurve, V., Wood, L.M., Gabizon, A.A., La-Beck, N.M., 2015. Liposome promotion of tumor growth is associated with angiogenesis and inhibition of antitumor immune responses. Nanomed. Nanotechnol. 11, 259–262.

Sahoo, S., Ang, L.T., Goh, J.C.H., Toh, S.L., 2010. Growth factor delivery through electrospun nanofibers in scaffolds for tissue engineering applications. J. Biomed. Mater. Res. A 93, 1539–1550.

Said, S.S., Aloufy, A.K., El-Halfawy, O.M., Boraei, N.A., El-Khordagui, L.K., 2011. Antimicrobial PLGA ultrafine fibers: interaction with wound bacteria. Eur. J. Pharm. Biopharm. 79, 108–118.

Saraf, A., Baggett, L.S., Raphael, R.M., Kasper, F.K., Mikos, A.G., 2010. Regulated non-viral gene delivery from coaxial electrospun fiber mesh scaffolds. J. Control. Release 143, 95–103.

Saravanabhavan, S.S., Dharmalingam, S., 2013. Fabrication of polysulphone/hydroxyapatite nanofiber composite implant and evaluation of their in vitro bioactivity and biocompatibility towards the post-surgical therapy of gastric cancer. Chem. Eng. J. 234, 380–388.

Schneider, A., Wang, X., Kaplan, D., Garlick, J., Egles, C., 2009. Biofunctionalized electrospun silk mats as a topical bioactive dressing for accelerated wound healing. Acta Biomater. 5, 2570–2578.

Schwengber, A., Prado, H.J., Zilli, D.A., Bonelli, P.R., Cukierman, A.L., 2015. Carbon nanotubes buckypapers for potential transdermal drug delivery. Mater. Sci. Eng. C 57, 7–13.

Seyednejad, H., Ji, W., Yang, F., Van Nostrum, C.F., Vermonden, T., Van Den Beucken, J.J., Dhert, W.J., Hennink, W.E., Jansen, J.A., 2012. Coaxially electrospun scaffolds based on hydroxyl-functionalized poly(ε-caprolactone) and loaded with VEGF for tissue engineering applications. Biomacromolecules 13, 3650–3660.

Shen, X., Yu, D., Zhu, L., Branford-White, C., White, K., Chatterton, N.P., 2011. Electrospun diclofenac sodium loaded Eudragit L 100-55 nanofibers for colon-targeted drug delivery. Int. J. Pharm. 408, 200–207.

Siegel, R.A., Rathbone, M.J., 2012. Overview of controlled release mechanisms. Fundamentals and Applications of Controlled Release Drug Delivery. Springer, New York, NY.

Sill, T.J., von Recum, H.A., 2008. Electrospinning: applications in drug delivery and tissue engineering. Biomaterials 29, 1989–2006.

Son, Y.J., Yoo, H.S., 2012. Dexamethasone-incorporated nanofibrous meshes for antiproliferation of smooth muscle cells: thermally induced drug-loading strategy. J. Biomed. Mater. Res. A 100A, 2678–2685.

Son, Y., Kim, W., Yoo, H., 2014. Therapeutic applications of electrospun nanofibers for drug delivery systems. Arch. Pharm. Res. 37, 69–78.

Song, W., Yu, X., Markel, D.C., Shi, T., Ren, W., 2013. Coaxial PCL/PVA electrospun nanofibers: osseointegration enhancer and controlled drug release device. Biofabrication 5, 035006.

Spizzirri, U.G., Hampel, S., Cirillo, G., Nicoletta, F.P., Hassan, A., Vittorio, O., Picci, N., Iemma, F., 2013. Spherical gelatin/CNTs hybrid microgels as electro-responsive drug delivery systems. Int. J. Pharm. 448, 115–122.

Suwantong, O., Opanasopit, P., Ruktanonchai, U., Supaphol, P., 2007. Electrospun cellulose acetate fiber mats containing curcumin and release characteristic of the herbal substance. Polymer 48, 7546–7557.

Suwantong, O., Ruktanonchai, U., Supaphol, P., 2008. Electrospun cellulose acetate fiber mats containing asiaticoside or *Centella asiatica* crude extract and the release characteristics of asiaticoside. Polymer 49, 4239–4247.

Taepaiboon, P., Rungsardthong, U., Supaphol, P., 2006. Drug-loaded electrospun mats of poly(vinyl alcohol) fibres and their release characteristics of four model drugs. Nanotechnology 17, 2317.

Taepaiboon, P., Rungsardthong, U., Supaphol, P., 2007a. Effect of cross-linking on properties and release characteristics of sodium salicylate-loaded electrospun poly(vinyl alcohol) fibre mats. Nanotechnology 18, 175102.

Taepaiboon, P., Rungsardthong, U., Supaphol, P., 2007b. Vitamin-loaded electrospun cellulose acetate nanofiber mats as transdermal and dermal therapeutic agents of vitamin A acid and vitamin E. Eur. J. Pharm. Biopharm. 67, 387–397.

Tang, Z., Wang, Y., Podsiadlo, P., Kotov, N.A., 2006. Biomedical applications of layer-by-layer assembly: from biomimetics to tissue engineering. Adv. Mater. 18, 3203.

Tang, C., Ozcam, A.E., Stout, B., Khan, S.A., 2012. Effect of pH on protein distribution in electrospun PVA/BSA composite nanofibers. Biomacromolecules 13, 1269–1278.

Taratula, O., Garbuzenko, O., Savla, R., Andrew Wang, Y., He, H., Minko, T., 2011. Multifunctional nanomedicine platform for cancer specific delivery of siRNA by superparamagnetic iron oxide nanoparticles–dendrimer complexes. Curr. Drug Deliv. 8, 59–69.

Tarcha, P.J., 1990. Polymers for Controlled Drug Delivery. CRC Press, Boston.

Thakur, R., Florek, C., Kohn, J., Michniak, B., 2008. Electrospun nanofibrous polymeric scaffold with targeted drug release profiles for potential application as wound dressing. Int. J. Pharm. 364, 87–93.

Thanki, K., Gangwal, R.P., Sangamwar, A.T., Jain, S., 2013. Oral delivery of anticancer drugs: challenges and opportunities. J. Control. Release 170, 15–40.

Thierry, B., Winnik, F.M., Merhi, Y., Silver, J., Tabrizian, M., 2003. Bioactive coatings of endovascular stents based on polyelectrolyte multilayers. Biomacromolecules 4, 1564–1571.

Thierry, B., Kujawa, P., Tkaczyk, C., Winnik, F.M., Bilodeau, L., Tabrizian, M., 2005. Delivery platform for hydrophobic drugs: prodrug approach combined with self-assembled multilayers. J. Am. Chem. Soc. 127, 1626–1627.

Thomas, B.J., Finnin, B.C., 2004. The transdermal revolution. Drug Discov. Today 9, 697–703.

Tomalia, D.A., Baker, H., Dewald, J., Hall, M., Kallos, G., Martin, S., Roeck, J., Ryder, J., Smith, P., 1985. A new class of polymers: starburst-dendritic macromolecules. Polym. J. 17, 117–132.

Tomalia, D., Reyna, L., Svenson, S., 2007. Dendrimers as multi-purpose nanodevices for oncology drug delivery and diagnostic imaging. Biochem. Soc. T. 35, 61.

Toshkova, R., Manolova, N., Gardeva, E., Ignatova, M., Yossifova, L., Rashkov, I., Alexandrov, M., 2010. Antitumor activity of quaternized chitosan-based electrospun implants against Graffi myeloid tumor. Int. J. Pharm. 400, 221–233.

Uhrich, K.E., Cannizzaro, S.M., Langer, R.S., Shakesheff, K.M., 1999. Polymeric systems for controlled drug release. Chem. Rev. 99, 3181–3198.

Vaage, J., Donovan, D., Uster, P., Working, P., 1997. Tumour uptake of doxorubicin in polyethylene glycol-coated liposomes and therapeutic effect against a xenografted human pancreatic carcinoma. Br. J. Cancer 75, 482.

Vargas, E.T., Do Vale Baracho, N., De Brito, J., De Queiroz, A., 2010. Hyperbranched polyglycerol electrospun nanofibers for wound dressing applications. Acta Biomater. 6, 1069–1078.

Verreck, G., Chun, I., Peeters, J., Rosenblatt, J., Brewster, M.E., 2003a. Preparation and characterization of nanofibers containing amorphous drug dispersions generated by electrostatic spinning. Pharm. Res. 20, 810–817.

Verreck, G., Chun, I., Rosenblatt, J., Peeters, J., Van Dijck, A., Mensch, J., Noppe, M., Brewster, M.E., 2003b. Incorporation of drugs in an amorphous state into electrospun nanofibers composed of a water-insoluble, nonbiodegradable polymer. J. Control. Release 92, 349–360.

Volpato, F.Z., Almodóvar, J., Erickson, K., Popat, K.C., Migliaresi, C., Kipper, M.J., 2012. Preservation of FGF-2 bioactivity using heparin-based nanoparticles, and their delivery from electrospun chitosan fibers. Acta Biomater. 8, 1551–1559.

Wang, Y., Annunziata, O., 2007. Comparison between protein–polyethylene glycol (PEG) interactions and the effect of PEG on protein–protein interactions using the liquid–liquid phase transition. J. Phys. Chem. B 111, 1222–1230.

Wang, J.G., Liu, J.J., Fan, C.Y., Mo, X.M., He, C.L., Chen, F., 2008. Preparation of coaxial electrospinning nanofibers for nerve growth factor delivery. J. Clin. Rehab. Tiss. Eng. Res. 12 (23), 4440–4444.

Wang, Y., Cao, X., Guo, R., Shen, M., Zhang, M., Zhu, M., Shi, X., 2011. Targeted delivery of doxorubicin into cancer cells using a folic acid–dendrimer conjugate. Polym. Chem. 2, 1754–1760.

Wang, L., Wang, M., Topham, P.D., Huang, Y., 2012a. Fabrication of magnetic drug-loaded polymeric composite nanofibres and their drug release characteristics. RSC Adv. 2, 2433–2438.

Wang, S.-G., Jiang, X., Chen, P.-C., Yu, A.-G., Huang, X.-J., 2012b. Preparation of coaxial-electrospun poly[bis(p-methylphenoxy)]phosphazene nanofiber membrane for enzyme immobilization. Int. J. Mol. Sci. 13, 14136–14148.

Wang, X., Ding, B., Li, B., 2013. Biomimetic electrospun nanofibrous structures for tissue engineering. Mater. Today 16, 229–241.

Weng, L., Xie, J., 2015. Smart electrospun nanofibers for controlled drug release: recent advances and new perspectives. Curr. Pharm. Design 21, 1944–1959.

Wu, X.-M., Branford-White, C.J., Yu, D.-G., Chatterton, N.P., Zhu, L.-M., 2011. Preparation of core–shell PAN nanofibers encapsulated α-tocopherol acetate and ascorbic acid 2-phosphate for photoprotection. Colloids Surf. B 82, 247–252.

Wu, H., Yao, L., Mei, J., Li, F., 2015. Development of synthetic of peptide-functionalized liposome for enhanced targeted ovarian carcinoma therapy. Int. J. Clin. Exp. Pathol. 8, 207.

Xiaoqiang, L., Yan, S., Rui, C., Chuanglong, H., Hongsheng, W., Xiumei, M., 2009. Fabrication and properties of core–shell structure P (LLA-CL) nanofibers by coaxial electrospinning. J. Appl. Polym. Sci. 111, 1564–1570.

Xie, Z., Buschle-Diller, G., 2010. Electrospun poly(D, L-lactide) fibers for drug delivery: the influence of cosolvent and the mechanism of drug release. J. Appl. Polym. Sci. 115, 1–8.

Xie, J., Wang, C.-H., 2006. Electrospun micro- and nanofibers for sustained delivery of paclitaxel to treat C6 glioma in vitro. Pharm. Res. 23, 1817–1826.

Xie, J., Tan, R.S., Wang, C.H., 2008. Biodegradable microparticles and fiber fabrics for sustained delivery of cisplatin to treat C6 glioma in vitro. J. Biomed. Mater. Res. A 85, 897–908.

Xu, X., Chen, X., Xu, X., Lu, T., Wang, X., Yang, L., Jing, X., 2006. BCNU-loaded PEG–PLLA ultrafine fibers and their in vitro antitumor activity against glioma C6 cells. J. Control. Release 114, 307–316.

Xu, X., Chen, X., Ma, P.A., Wang, X., Jing, X., 2008. The release behavior of doxorubicin hydrochloride from medicated fibers prepared by emulsion-electrospinning. Eur. J. Pharm. Biopharm. 70, 165–170.

Xu, X., Chen, X., Wang, Z., Jing, X., 2009. Ultrafine PEG–PLA fibers loaded with both paclitaxel and doxorubicin hydrochloride and their in vitro cytotoxicity. Eur. J. Pharm. Biopharm. 72, 18–25.

Yang, Y., Li, X., Cui, W., Zhou, S., Tan, R., Wang, C., 2008a. Structural stability and release profiles of proteins from core–shell poly(DL-lactide) ultrafine fibers prepared by emulsion electrospinning. J. Biomed. Mater. Res. A 86, 374–385.

Yang, Y., Li, X., Qi, M., Zhou, S., Weng, J., 2008b. Release pattern and structural integrity of lysozyme encapsulated in core–sheath structured poly(DL-lactide) ultrafine fibers prepared by emulsion electrospinning. Eur. J. Pharm. Biopharm. 69, 106–116.

Yang, G., Wang, J., Li, L., Ding, S., Zhou, S., 2014. Electrospun micelles/drug-loaded nanofibers for time-programmed multi-agent release. Macromol. Biosci. 14, 965–976.

Yarin, A.L., Pourdeyhimi, B., Ramakrishna, S., 2013. Fundamentals and Applications of Micro-and Nanofibers. MRS Bulletin, vol. 40 . Cambridge University Press.

Yoo, H.S., Kim, T.G., Park, T.G., 2009. Surface-functionalized electrospun nanofibers for tissue engineering and drug delivery. Adv. Drug Deliv. Rev. 61, 1033–1042.

Yu, D.-G., Shen, X.-X., Branford-White, C., White, K., Zhu, L.-M., Bligh, S.A., 2009. Oral fast-dissolving drug delivery membranes prepared from electrospun polyvinylpyrrolidone ultrafine fibers. Nanotechnology 20, 055104.

Yu, D.-G., Branford-White, C., Bligh, S.W.A., White, K., Chatterton, N.P., Zhu, L.-M., 2011. Improving polymer nanofiber quality using a modified co-axial electrospinning process. Macromol. Rapid Comm. 32, 744–750.

Yu, D.-G., Chian, W., Wang, X., Li, X.-Y., Li, Y., Liao, Y.-Z., 2013a. Linear drug release membrane prepared by a modified coaxial electrospinning process. J. Membr. Sci. 428, 150–156.

Yu, D., Wang, X., Li, X., Chian, W., Li, Y., Liao, Y., 2013b. Electrospun biphasic drug release polyvinylpyrrolidone/ethyl cellulose core/sheath nanofibers. Acta Biomater. 9, 5665–5672.

Yu, D.-G., Xu, Y., Li, Z., Du, L.-P., Zhao, B.-G., Wang, X., 2014. Coaxial electrospinning with mixed solvents: from flat to round Eudragit L100 nanofibers for better colon-targeted sustained drug release profiles. J. Nanomater. 2014, 8.

Yun, J., Im, J.S., Lee, Y.-S., Kim, H.-I., 2011. Electro-responsive transdermal drug delivery behavior of PVA/PAA/MWCNT nanofibers. Eur. Polym. J. 47, 1893–1902.

Zahedi, P., Karami, Z., Rezaeian, I., Jafari, S.H., Mahdaviani, P., Abdolghaffari, A.H., Abdollahi, M., 2012. Preparation and performance evaluation of tetracycline hydrochloride loaded wound dressing mats based on electrospun nanofibrous poly(lactic acid)/poly(ε-caprolactone) blends. J. Appl. Polym. Sci. 124, 4174–4183.

Zamani, F., 2013. Engineering of structural properties of PLGA nanofibrous scaffold for neural cell culture. PhD thesis, Amirkabir University of Technology.

Zamani, M., Morshed, M., Varshosaz, J., Jannesari, M., 2010. Controlled release of metronidazole benzoate from poly ε-caprolactone electrospun nanofibers for periodontal diseases. Eur. J. Pharm. Biopharm. 75, 179–185.

Zamani, F., Amani-Tehran, M., Latifi, M., Shokrgozar, M., 2013a. The influence of surface nanoroughness of electrospun PLGA nanofibrous scaffold on nerve cell adhesion and proliferation. J. Mater. Sci. 24, 1551–1560.

Zamani, F., Latifi, M., Amani-Tehran, M., Shokrgozar, M., 2013b. Effects of PLGA nanofibrous scaffolds structure on nerve cell directional proliferation and morphology. Fibers Polym. 14, 698–702.

Zamani, M., Prabhakaran, M.P., Ramakrishna, S., 2013c. Advances in drug delivery via electrospun and electrosprayed nanomaterials. Int. J. Nanomed. 8, 2997–3017.

Zamani, F., Amani-Tehran, M., Latifi, M., Shokrgozar, M.A., Zaminy, A., 2014. Promotion of spinal cord axon regeneration by 3D nanofibrous core–sheath scaffolds. J. Biomed. Mater. Res. A 102, 506–513.

Zeng, J., Aigner, A., Czubayko, F., Kissel, T., Wendorff, J.H., Greiner, A., 2005. Poly(vinyl alcohol) nanofibers by electrospinning as a protein delivery system and the retardation of enzyme release by additional polymer coatings. Biomacromolecules 6, 1484–1488.

Zhang, S., 2003. Fabrication of novel biomaterials through molecular self-assembly. Nat. Biotechnol. 21, 1171–1178.

Zhang, J., Senger, B., Vautier, D., Picart, C., Schaaf, P., Voegel, J.-C., Lavalle, P., 2005a. Natural polyelectrolyte films based on layer-by-layer deposition of collagen and hyaluronic acid. Biomaterials 26, 3353–3361.
Zhang, Y., Lim, C.T., Ramakrishna, S., Huang, Z.-M., 2005b. Recent development of polymer nanofibers for biomedical and biotechnological applications. J. Mater. Sci. 16, 933–946.
Zhang, Y., Wang, X., Feng, Y., Li, J., Lim, C., Ramakrishna, S., 2006. Coaxial electrospinning of (fluorescein isothiocyanate-conjugated bovine serum albumin)-encapsulated poly(ε-caprolactone) nanofibers for sustained release. Biomacromolecules 7, 1049–1057.
Zhang, D., Chang, J., Zeng, Y., 2008. Fabrication of fibrous poly(butylene succinate)/wollastonite/apatite composite scaffolds by electrospinning and biomimetic process. J. Mater. Sci. 19, 443–449.
Zong, X., Li, S., Chen, E., Garlick, B., Kim, K.-S., Fang, D., Chiu, J., Zimmerman, T., Brathwaite, C., Hsiao, B.S., 2004. Prevention of postsurgery-induced abdominal adhesions by electrospun bioabsorbable nanofibrous poly(lactide-*co*-glycolide)-based membranes. Ann. Surg. 240, 910.
Zou, J., Yang, Y., Liu, Y., Chen, F., Li, X., 2011. Release kinetics and cellular profiles for bFGF-loaded electrospun fibers: effect of the conjugation density and molecular weight of heparin. Polymer 52, 3357–3367.

CHAPTER

8 BRUSH POLYMER-BASED NANOSTRUCTURES FOR DRUG DELIVERY

Yun Yu, Haotian Sun, Chong Cheng

University at Buffalo, State University of New York, Buffalo, NY, United States

CHAPTER OUTLINE

1 INTRODUCTION

Although direct administration of drug has been broadly utilized in clinic practice, drug delivery using nanoscopic scaffolds may increase residence time of drugs in systemic circulation, protect drugs from physiological degradation, and help drugs to overcome various biological barriers and multidrug resistance, thereby leading to enhanced drug bioavailability for disease treatment with reduced dose and side effects (Allen and Cullis, 2004). Among varieties of nanomaterials with diverse composition and structural features, polymer-based nanoscopic delivery scaffolds have been broadly studied for drug delivery application because of their tailorable structures and versatile properties (Duncan, 2003). As an important category of nanostructured polymers, brush polymers (BPs; also termed molecular brushes or densely grafted polymers) can possess a broad variety of well-defined unimolecular nanostructures, as well as intermolecularly assembled nanostructures (Sheiko et al., 2008). The nanoscopic

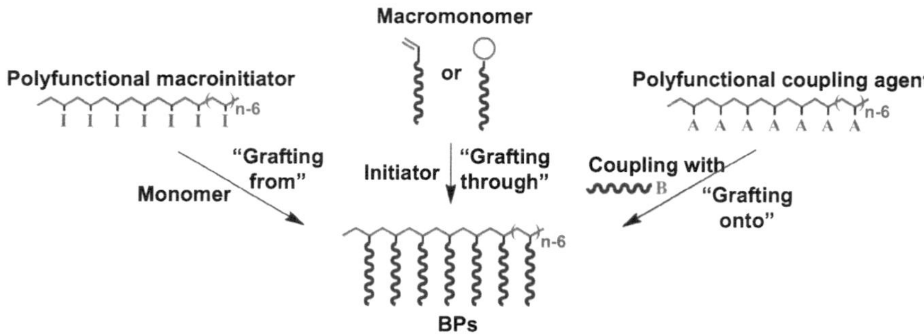

FIGURE 8.1 General Approaches for the Synthesis of BPs

dimensions of BP-based nanostructures (~ten to hundreds of nm) can fit very well with the preferred size range of 10–200 nm for elongated blood circulation of nano-objects (Petros and DeSimone, 2010). Moreover, BPs can be prepared without using the repetitive and tedious procedure that is typically required for the synthesis of dendrimers, another important category of polymers with unimolecular nanostructures. Therefore, drug delivery using BP-based scaffolds has attracted significant interest in recent years.

With backbones densely grafted with polymeric grafts (or long side chains), BPs can be prepared by three basic approaches, "grafting from," "grafting through," and "grafting onto" (Fig. 8.1) (Sheiko et al., 2008). The "grafting from" method is based on polymerization using a polyfunctional macroinitiator or macro-chain transfer agents (macro-CTAs) with numerous chain-forming sites. In the "grafting through" approach (also termed the macromonomer approach), BPs are prepared by the polymerization of macromonomers with one polymerizable monomer group. In the "grafting onto" approach, BPs are obtained by the coupling reaction between a polymeric reagent having numerous side reactive functionalities (1) as the backbone precursor with another polymeric reagent having one terminal reactive group (2) as the graft precursor. Moreover, combinations or repetitive utilization of these approaches can be employed for the preparation of BPs with complicated grafting structures. Furthermore, BPs with reactive groups can be converted into superstructures with bigger dimensions through chemical reaction (Liu et al., 2012).

BPs can possess diverse structures of backbones and grafts; the concept of BPs, if broadly defined, can also cover macromolecules with BP-based building blocks. Based on their macromolecular architectural features, BPs can be divided into several major categories, including (1) mono-BPs with only one type of homopolymeric grafts, (2) brush–coil block copolymers with both brush-like block(s) and linear coil-like main-chain block(s), (3) core–shell BPs with multiblock grafts, (4) statistically heterografted BPs having multiple types of grafts statistically distributed along the backbone, and (5) block BPs with multiple backbone blocks carrying different types of grafts (Fig. 8.2). Based on backbone and graft structures, BPs can possess well-defined unimolecular nanostructures with various shapes: globular (for BPs with short backbone relative to graft), cylindrical (for BPs with long backbone relative to graft), star-like (for BPs with long, star-like backbone relative to graft), or cyclic (for BPs with long, cyclic backbone relative to graft) (Schappacher and Deffieux, 2008; Sheiko et al., 2008). With multiple immiscible polymeric components, they can also form self-assembled unimolecular or multimolecular core–shell or Janus nanostructures (Li et al., 2012a; Sheiko et al., 2008).

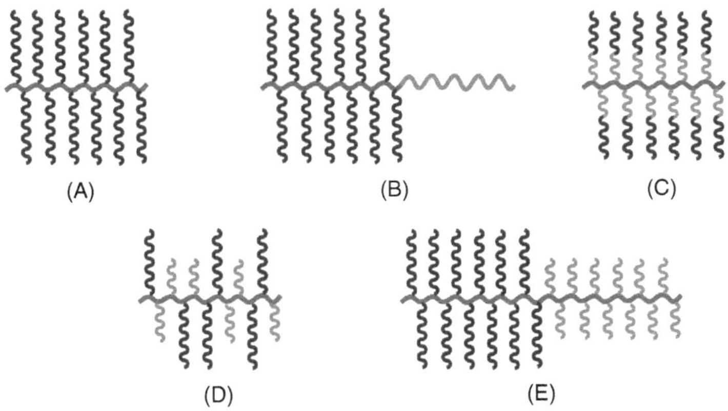

FIGURE 8.2 Schematic Illustrations of BPs

(A) Mono-BP, (B) brush–coil block copolymer, (C) core–shell BP, (D) statistically heterografted BP, and (E) block BP.

Over the past decade, a broad variety of BPs with diverse structural features and properties have been integrated with drugs through different approaches to construct BP-based drug-delivery systems. In this chapter, an overview on these BP-based drug-delivery systems is presented. The merits and limitations of using BP-based nanostructures for drug delivery applications are discussed. The future directions in the research area of BP-based drug-delivery systems are commented on. It should be noted that BP-based nanostructures, particularly those with positively charged cationic components, have been utilized for gene delivery application based on their electrostatic interactions with negatively charged genetic materials, although the corresponding studies will not be reviewed here (Carlson et al., 2013; Hörtz et al., 2015; Jiang et al., 2007; Li et al., 2011; Liu et al., 2010; Shi et al., 2013a, 2013b; Yu et al., 2009; Zhai et al., 2012; Zhang et al., 2014; Zou et al., 2011).

2 STRUCTURAL DESIGN OF BP-BASED DRUG-DELIVERY SYSTEMS

To act as effective scaffolds for drug delivery through systemic circulation, polymeric nanostructures should possess: (1) significant water solubility or dispersibility, (2) appropriate nanoscopic sizes, (3) remarkable structural stability during systemic circulation, (4) stimuli-responsive structures to exert control over drug release process, (5) degradability to minimize long-term side effects, and (6) other functional groups as chemical handles (Tong and Cheng, 2007). In addition, other structural features allowing for the integration of drug with delivery scaffolds and subsequent drug release are also needed. According to these structural requirements, the structural design of BP-based drug-delivery systems will be discussed in the subsequent paragraphs.

Water-soluble or dispersible BP-based polymeric nanostructures are needed for the corresponding drug-delivery systems that can be circulated in the systemic system. Except for BPs with relatively short grafts (Mao and Gan, 2009), backbones have little mass fraction and even less steric accessibility as compared to grafts, and therefore water solubility or dispersibility of BPs largely depends on their grafts. Mono-BPs with hydrophilic grafts often can be unimolecularly dissolved in aqueous solutions (Johnson et al., 2011). Core–shell BPs with amphiphilic diblock grafts, each consisting of a hydrophobic first block (close to backbone) and a hydrophilic second block, can form water-dispersible assemblies in aqueous systems. These assemblies can be either unimolecular or multimolecular in nature, while long hydrophilic blocks in the diblock grafts promote the formation of unimolecular micelles (Zhao et al., 2012). In order to gain water solubility or dispersibility, statistically hetero-grafted BPs and block BPs generally should have at least one type of grafts that are hydrophilic (Li et al., 2010, 2012b); brush–coil block copolymers typically have either brush domain or coil domain to be hydrophilic. In addition, water-dispersible cross-linked nanoparticles (NPs) with BP-based units should have hydrophilic peripheral domains, typically derived from hydrophilic BPs (Liu et al., 2012).

Among various BP-based nanostructures, poly(ethylene glycol) (PEG) is the most commonly used hydrophilic polymer component because of its nontoxicity, biocompatibility, and low nonspecific binding (Greenwald et al., 2003). With the ability to escape the surveillance of the reticuloendothelial system better than common controls, "stealth" nanoparticles can be created through PEG protection, and the effects are highly dependent on the conformation and density of the surface PEG chains (Li et al., 2010). Recently, Wang, Ye, and coworkers reported a detailed study on the effects of surface PEG densities on the biological behaviors of nanoparticles under both in vivo and in vitro environments. A library of nanoparticles with similar sizes around 100 nm and different surface PEG densities was established using PEG-*block*-poly(ε-caprolactone) (PEG-*b*-PCL). It was demonstrated that higher surface PEG density can lead to greater accumulation of nanoparticles in tumor tissue, resulting in advanced antitumor activities (Du et al., 2015). Other hydrophilic polymers, such as poly(methacrylic acid) (PMAA) and poly(hydroxyethyl methacrylate) (PHEMA), have also been employed to serve as the hydrophilic components of BP-based scaffolds. With temperature-sensitive properties, poly(*N*-isopropylacrylamide) (PNIPAAM) can also serve as the hydrophilic component of BP-based scaffolds at temperatures below ~30°C (Li et al., 2004). Recently, as a special category of hydrophilic polymers that carry both cations and anions within the same macromolecules, zwitterionic polymers have attracted significant interest for biomedical applications because of their remarkable antibiofouling and "stealth" properties (Jiang and Cao, 2010). However, BPs with zwitterionic polymeric components have not been reported to date.

Because NPs smaller than ~10 nm (corresponding to molecular weight (MW) of less than ~40 kDa) can be quickly cleared from the bloodstream through the kidneys, and NPs larger than 200 nm can also be readily cleared through the reticuloendothelial system (Petros and DeSimone, 2010), BP-based nanostructures should have hydrodynamic sizes of 10–200 nm for elongated blood circulation. Specifically, NPs with dimensions of 10–100 nm are considered the most effective for passive targeting of tumor tissues as a result of the enhanced permeability and retention (EPR) effect (Duncan, 2006), and therefore BP-based nanostructures with hydrodynamic sizes of 10–100 nm can be further preferred for the delivery of anticancer drugs. Unimolecular nanostructures of BPs typically have (radial) dimensions within the range of 10–100 nm, but BPs with long backbones may have length dimensions larger than 200 nm. In order to control the hydrodynamic sizes of assembled multimolecular BP-based nanostructures within a specific range, one needs to ensure the dimensions of individual BP macromolecules well below the upper limit of the dimension range. For instance, to target core–shell micelles of amphiphilic diblock BPs with sizes less than 100 nm, the length dimensions of individual diblock BP macromolecules should be less than 50 nm. It should also be noted that cellular uptake behavior highly depends on the dimension of scaffolds. Within the size range of 10–200 nm, BP-based nanostructures typically can be internalized to the cytoplasm through endocytolysis. However, without specific modification, nanostructures in this size range could not be transported into nuclei through the nuclear pore complex (NPC), which allows only objects smaller than ~5 nm to cross unassisted (Kapon et al., 2010). Therefore, drug delivery to nuclei via BP-based nanostructures typically would rely on drug release outside the nuclei, followed by passive diffusion of the drug to internalize nuclei.

BP-based nanostructures should maintain remarkable structural stability during systemic circulation. Relative to unimolecular nanostructures of BPs, which are maintained essentially through covalent bonds, assembled multimolecular nanostructures of BPs are more sensitive to environmental conditions and may need to be applied in concentrations higher than the critical micelle concentration (CMC) in order to preserve the preferred dimensions and properties. BP-based nanostructures consisting of a number of BP units with sizes less than 10 nm may be stabilized through cross-linking (Liu et al., 2012). It should be noted that cross-linking can further enhance structural stability of unimolecular structures of BPs (Cheng et al., 2007), and cross-linking of BPs in their unimolecular form can also serve as the initial step for the preparation of BP-derived nanocages or nanotubes (Cheng et al., 2006; Huang and Rzayev, 2009).

On the other hand, it is highly preferred that triggered drug release from drug-delivery systems can occur at disease sites. Stimuli-responsive structures can be employed in the BP-based drug carriers to exert control over the drug release process (Tong et al., 2014). Since the concept of stimuli-responsive drug delivery was first suggested with the design of thermosensitive liposomes decades ago (Yatvin et al., 1978), numerous stimuli-responsive drug carriers have been designed and studied, including exogenous stimuli-responsive drug carriers (responsive to changes of temperature, magnetic fields, ultrasounds, light, and electric fields), and endogenous stimuli-responsive drug carriers (responsive to variations in pH, redox potential, the concentrations of enzymes) (Mura et al., 2013). Recently, BP-based drug carriers with stimuli-responsive structures, especially these pH-sensitive BP-based drug carriers, have been widely studied (Chang et al., 2013; Yang et al., 2011, 2012a, 2012b; Yu et al., 2015; Zou et al., 2011). Specifically, the BP-based drug carriers with pH-responsive conjugation linkages effectively took advantage of the slight difference of pH existing between the extracellular microenvironment of tumors (pH = 6.5–7.2) and healthy tissues (pH ~ 7.4) (Mura et al., 2013). The low pH around the tumors is mainly caused by irregular angiogenesis in pathological fast-growing tumors, and then the rapid consumption of both oxygen and nutrients triggers a shift toward glycolytic metabolism, eventually leading to the accumulation of acidic metabolites. Other types of stimuli-responsive drug-delivery systems were also reported, including thermo-responsive drug carriers (Hu et al., 2015), redox-responsive drug carriers (Sowers et al., 2014; Su et al., 2015), light-responsive drug carriers (Burts et al., 2014), and multistimuli-responsive drug carriers (Wang et al., 2012).

Moreover, after the major occurrence of a drug release event, BP-based nanostructures should eventually be able to form residues with sizes smaller than 10 nm, in order to facilitate kidney clearance of the scaffold materials and to minimize their long-term side effects. This can be a dissembling process for assembled BP-based nanostructures. For BP-based nanostructures based on covalent stabilization, cleavable bonds should be present in backbones, grafts, or cross-linkages (if the nanostructures are cross-linked), allowing for slow chain-breaking events. For instance, degradability can be incorporated with BPs by using degradable polymers, such as polylactide (PLA) and poly(ε-caprolactone) (PCL), as their backbones or grafts (Du et al., 2009; Yu et al., 2013); degradability can also be introduced to BP-based cross-linked NPs by using stimuli-responsive degradable cross-linkages (Liao et al., 2014).

As chemical handles, functional groups with specific reactivity should be possessed in BP-based scaffolds for drug delivery application. These functional groups can be utilized not only to modify physicochemical properties of BP-based nanostructures, but also to enhance their biomedical functions through multifunctionalization (Bühler et al., 2014; Sowers et al., 2014). In particular, functional groups at peripheral hydrophilic domains of BP-based nanostructures can be employed to tether the biologically related moieties for biosurface interactions, such as targeting ligands and cell-penetrating peptides. Imaging functions can also be integrated into BP-based scaffolds. Optical imaging and magnetic resonance imaging (MRI) are commonly employed in both research and clinical laboratories. Moreover, near-infrared (NIR) optical imaging has also been emerging as a promising tool for image-guided therapy (Sowers et al., 2014).

Besides the aforementioned general design considerations of BP-based nanostructures for drug delivery applications, there are also some published results that may provide guidance for the design of specific types of BP-based nanostructures. According to the recent study of Caruso and coworkers on water-soluble cylindrical BPs with adjustable length and tunable rigidity, the increase of the aspect ratio or rigidity of these BPs decreases their plasma residence time, and the increase of the brush aspect ratio also leads to increased deposition in mononuclear phagocytic system organs (Müllner et al., 2015). The recent work of Gianneschi and coworkers demonstrated that formulating peptides into well-defined BPs not only can preserve bioactivity but also can protect the peptides from proteolysis, and might suggest the benefits of formulating a broad variety of bioactive moieties into BP-based nanostructures (Blum et al., 2014).

It should be noted that BP-based scaffolds can also be utilized for immobilized drug delivery in which systemic circulation of the drug-delivery systems is not needed. Accordingly, significant water solubility or dispersibility, appropriate nanoscopic sizes, and remarkable structural stability during systemic circulation are not required for these systems. On the other hand, it is still preferred that the BP-based scaffolds have biodegradable structures to facilitate drug release and minimize side effects, as well as functional groups to enable biofunctionalization.

Drugs can be integrated with BP-based scaffolds via either physical loading or conjugation. Encapsulation is the most commonly used physical loading approach. Because most drugs are hydrophobic, typically hydrophobic domains are required within BP-based scaffolds to allow for drug encapsulation. Although drug encapsulation is relatively facile, in principle unfavorable burst release of drug cannot be fully avoided. On the other hand, drug conjugation requires reaction between drug and functional group on BP-based scaffolds to construct BP–drug conjugates (BPDCs). For these drug-conjugated systems, drug release is realized through cleavage of the conjugation linkage followed by diffusion, thereby leading to sustained drug release behavior (Duncan, 2006; Zou et al., 2015). In the following two sections, BP-based drug-delivery systems based on encapsulation and conjugation of drugs, respectively, will be discussed.

3 BP-BASED ENCAPSULATED DRUG-DELIVERY SYSTEMS

In typical BP-based encapsulated drug-delivery systems, BP-based scaffolds allow drugs to be physically loaded through hydrophobic–hydrophobic interaction, electrostatic interaction or molecular entrapment. For drug delivery through systemic circulation, BP-based polymers generally need to be amphiphilic and form unimolecular or assembled multimolecular micelles for drug encapsulation. Amphiphilicity is not required for BP-based scaffolds for immobilized drug-delivery systems. In the following subsections, representative examples of BP-based polymers for encapsulated drug delivery applications will be reviewed based on their macromolecular architectural types.

3.1 MONO-BPs

Mono-BPs (i.e., BPs with only one kind of grafts) are the most common type of BPs (Fig. 8.2A). An early study reported by Kissel et al. (1991) gave seminal examples of mono-BP drug-delivery systems. In this study, polyol macromolecules, such as glucose, mannitol, poly(vinyl alcohol) (PVA), dextran acetate, and hydroxypropyl cellulose (HPC), were used to initiate ring-opening polymerization (ROP) to produce BPs with poly(lactic-*co*-glycolic acid) (PLGA) grafts. After the "grafting from" process, the hydrophilic component of the backbone (i.e., hydroxyl group) was converted to hydrophobic ester bonds and resulted in entirely hydrophobic mono-BP macromolecules. Microparticles were subsequently prepared using these hydrophobic polymers to encapsulate a hydrophobic drug, bromocriptine. Specifically, the microparticles made of glucose-initiated PLGA-based BPs were developed as a parenteral depot formulation of Parlodel entering clinical trials (Kissel et al., 1991).

Kissel and coworkers also reported several studies on amphiphilic BPs with PLA or PLGA grafts and charge-modified hydrophilic dextran backbones (such as dextran sulfate sodium and diethylaminoethyl dextran chloride) or water-soluble amine-modified PVA backbones as platforms for parenteral-delivery systems of hydrophilic substances, such as proteins and peptides based on electrostatic interaction (Li et al., 1997, 1998). With degradation mechanism strongly influenced by the used polyol, the degradation of PLA and PLGA grafts within BP structures was very different from that of linear PLA and PLGA. Hydrolysis behavior of these polyester grafts could be adjusted from random to nonrandom, and accordingly improved control over the drug release profile could be achieved for such mono-BP encapsulation systems.

Different with the aforementioned examples, most of BPs for encapsulation were designed to construct self-assembled micellar structures for drug delivery. As presented in the work of Gan and coworkers, amphiphilic mono-BPs with hydrophilic poly(ethylene glycol-*co*-glycidol) (PEG-*co*-GL) backbones and short hydrophobic PCL grafts were synthesized by living anionic polymerization and ROP (Mao and Gan, 2009). By changing the backbone composition (through the EG-to-GL ratio) and the PCL graft length, a series of BPs were studied in terms of micelle formation and stability. The results indicated that the graft length can affect micellization, and short PCL grafts can result in compact cores and relatively small micelles, which might be favorable for drug loading and controlled release.

Very recently, mono-BPs with hydrophobic backbones and hydrophilic grafts have been developed to form micelles and carry hydrophobic small molecule drug by Liu and coworkers (Fig. 8.3A) (Yuan et al., 2015). In their work, a chemotherapeutic drug doxorubicin (DOX) and photothermal conjugated polymer were coloaded into micelles of a NIR laser-responsive amphiphilic BP. Hydrophobic 2-diazo-1,2-naphthoquinones (DNQ) moieties were conjugated to a BP with poly(L-lysine) (PLL) backbone and PEG grafts to introduce hydrophobicity to the BP core for facilitating drug encapsulation, and DNQ could also be transformed to the hydrophilic form upon NIR two-photon laser irradiation, thereby leading to triggered drug release. It was confirmed that, with concentration higher than its CMC of 0.024 mg/mL, PLL-*g*-PEG/DNQ can self-assemble into multimolecular micelles in aqueous solution, with an average size of 78.5 ± 6.4 nm as measured by dynamic light scattering (DLS). Cumulative release profile of DOX indicated only 10% burst release within 40 min in buffer 7.4 at 25°C, and about 80% release triggered by laser irradiation at 800 nm in pH 5.5 at 50°C.

3.2 BRUSH–COIL BLOCK COPOLYMERS

The brush–coil block copolymers discussed in this section consist of two parts, one BP part and one linear coil-like polymer part (Fig. 8.2B). For amphiphilic brush–coil block copolymers, the two parts can

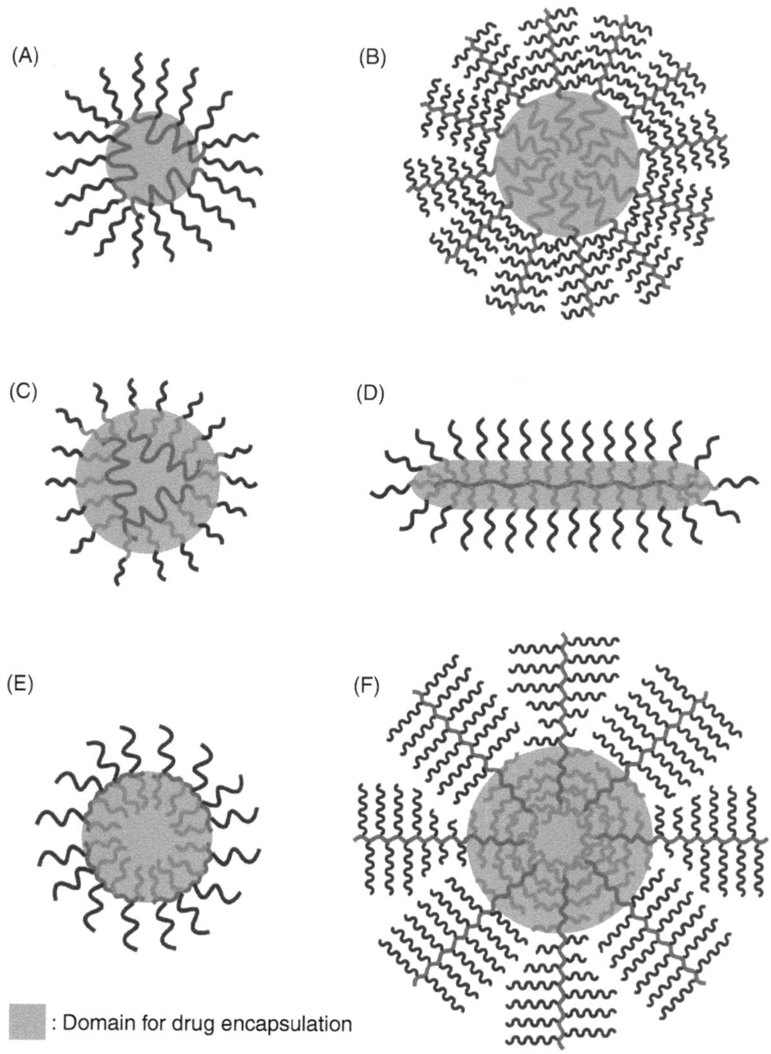

FIGURE 8.3 Schematic Illustrations of BP-Based Nanostructures for Drug Encapsulation

(A) Multimolecular assemblies of mono-BP, (B) multimolecular assemblies of brush–coil diblock copolymer, (C) multimolecular assemblies of core–shell BPs, (D) unimolecular nanostructures of core–shell BPs, (E) multimolecular assemblies of statistically heterografted BPs, and (F) multimolecular assemblies of diblock BPs.

be hydrophobic and hydrophilic, respectively. Specifically, amphiphilic brush–coil block copolymers with hydrophilic BP part and hydrophobic coil part can form assembled multimolecular micelles with hydrophobic cores which can be utilized for drug encapsulation (Fig. 8.3B). For example, a biodegradable amphiphilic brush–coil diblock copolymer consisting of polyphosphoester-*g*-PEG (PPEG) brush and PCL coil was synthesized by Zhang and coworkers through ROP. With CMCs around 10^{-3} mg/mL in water, PPEG-*b*-PCL brush–coil diblock copolymers with PPEG/PCL ratios of 17.1:82.3 and

38.7:61.3 formed core–shell micellar structures with average diameters of 74 and 85 nm, respectively, showing remarkable potential to encapsulate hydrophobic drugs (Du et al., 2006). In addition, another advantageous feature of such BP-based encapsulated systems is that all the main-chain components of PPEG-*b*-PCL (i.e., polyphosphate and PCL) are fully degradable, and therefore, the side effects of scaffolds can be minimized for in vivo applications. PEG can also be employed as "coil" in the brush–coil block copolymers. Deng, Dong, and coworkers reported a novel pH-sensitive brush–coil block copolymer, methoxy poly(ethylene glycol)-*block*-(polycaprolactone-*graft*-poly(methacrylic acid)) (mPEG-*b*-(PCL-*g*-PMAA)), which could form PEGylated nanoparticles (Chang et al., 2013). The copolymer was synthesized by combining atom transfer radical polymerization (ATRP) and ROP. With CMCs of 6.8×10^{-4} and 9.6×10^{-4} mg/mL, both of the two copolymer samples with different ratios of PCL and PMMA could self-assemble into nanoparticles in aqueous solution with average sizes of 104–129 nm. TEM confirmed the spherical morphology of the nanoparticles. A hydrophobic drug ibuprofen (IBU) was encapsulated into the nanoparticles with a high drug loading content of 13%, and a high drug loading efficiency of 75%. The pH-sensitive property was demonstrated by in vitro drug release studies. The results showed that more than 95% of the drug was released within 12 h at a high pH of 7.4, whereas the cumulative release percentage was only about 40% within the same time period at a low pH of 3.0.

Moreover, Yang et al. (2011) reported a novel pH-sensitive amphiphilic brush–coil diblock copolymer, poly(methyl methacrylate-*co*-methacrylic acid)-*b*-poly(PEG methyl ether monomethacrylate) (P(MMA-*co*-MAA)-*b*-PPEGMA), which was prepared by "grafting from" using the ATRP technique. With CMC of 1–4 mg/L, this copolymer was able to self-assemble into micelles in aqueous solution with an average size of 120–250 nm as determined by DLS. Ibuprofen (IBU), a poorly water-soluble drug, was selected as the model drug and encapsulated into the P(MMA-*co*-MAA)-based core domains of micelles with 90% loading efficiency via a dialysis method. In the drug release study, about 10% of burst release of drug was observed, and it was also found that pH-sensitive drug release could be enhanced by the increase of the MAA fraction in the coil block. For an optimized formulation, cumulative release of IBU was less than 20% in a simulated gastric fluid (pH 1.2) over 12 h but 90% in a simulated intestinal fluid (pH 7.4) within 6 h, indicating its potential as an oral drug-delivery system (Yang et al., 2011).

Incorporating partial degradability with macromolecular structures, they also developed a pH-sensitive amphiphilic brush–coil block copolymer PLA-*b*-PMAA-*b*-PPEGMA by the combination of activators regenerated by electron transfer (ARGET) ATRP and ROP techniques (Yang et al., 2012b). The resulting copolymer self-assembled to micelles with sizes around 140–250 nm as determined by DLS. By selecting nifedipine (NFD) as the model hydrophobic drug, in vitro release behavior of the drug-encapsulated micelles was found to be pH-dependent; however, burst releases were as high as 40%, presumably due to the low hydrophobic-to-hydrophilic ratio of the copolymer.

It should be noted that, as analogues of brush–coil block copolymers with reduced grafting density on grafted blocks, block–graft copolymers have also been utilized for drug delivery applications, and the results may provide helpful guidance for the design of drug-delivery systems based on brush–coil block copolymers. To improve paclitaxel (PTXL) delivery, Wooley, Gustafson, and coworkers designed and synthesized a block–graft copolymer, poly(2-ethylbutoxy phospholane)-*block*-poly(2-butynyl phospholane)-*graft*-poly(ethylene glycol) (PEBP-*b*-PBYP-*g*-PEG), and further converted it to PEGylated polyphosphoester-based micelles and shell cross-linked knedel-like nanoparticles (SCKs) (Zhang et al., 2015). The poor solubility of PTXL limited the ability to deliver maximally effective doses; their previous studies had successfully loaded PTXL into the hydrophobic cores of SCKs, and increased the water solubility of PTXL while maintaining its cytotoxicity against cancer cells (Samarajeewa et al., 2013). However, the relatively low drug loading percentages still made them less promising for clinical applications. Recently, by employing biodegradable polyphosphoester backbone, they designed a PEBP-*b*-PBYP-*g*-PEG copolymer, and the micelles and SCKs prepared from this block–graft copolymer could effectively incorporate high loading of PTXL. Either micelles or SCKs could reach encapsulate 10 wt.% PTXL, resulting in a high concentration of PTXL in aqueous suspension up to 4.8 mg/mL, while the free drug had aqueous solubility of only <2.0 μg/mL. Drug release studies showed that the half-life of sustained PTXL release from the SCKs was nearly double that from the micelles. By forming an additional physicochemical layer through the shell part, the cross-linking in SCKs could mediate PTXL release. In vivo studies revealed the relationship between structure and function of these nanoscopic drug-delivery systems, and the $t_{1/2}$ and biodistribution results demonstrated the ability of both micelles and SCKs to undergo extravasation from the lungs within a reasonable time line, thereby enabling repeated dosing.

In addition to diblock copolymers having one brush or grafted polymer block and one linear polymer block, triblock copolymers with two side BP blocks and one central linear polymer block have also been studied for drug encapsulation. Li and coworkers reported a thermo-responsive triblock copolymer, poly[2-(2-methoxyethoxy) ethyl methacrylate-*co*-oligo(ethylene glycol) methacrylate]-*b*-poly(L-lactide)-*b*-poly[2-(2-methoxyethoxy) ethyl methacrylate-*co*-oligo(ethylene glycol) methacrylate] [P(MEO_2MA-*co*-OEGMA)-*b*-PLLA-*b*-P(MEO_2MA-*co*-OEGMA)] (Hu et al., 2015). The brush-like amphiphilic copolymers were synthesized by ATRP using a Br-PLLA-Br macroinitiator. These copolymers formed micelles by self-assembly in aqueous medium. The sizes of the micelles increased from 20.7 to 102.5 nm, and the CMC increased from 0.0113 to 0.0130 mg/mL, with the increase of content of OEGMA. Curcumin, a hydrophobic anticancer drug, was encapsulated in micelles with loading efficiency higher than 90%. The LCST of the drug-loaded micelles was lower than that of blank micelles, resulting in a thermo-responsive activity. However, in vitro drug release studies showed initial burst release behavior.

3.3 CORE–SHELL BPs

Recently, core–shell BPs with amphiphilic diblock grafts have attracted remarkable attention (Fig. 8.2C). In a typical core–shell BP for drug encapsulation, the first block of grafts (close to the backbone) is hydrophobic and forms the packed hydrophobic inner core of micelles for encapsulation of the hydrophobic drug, while the second block is hydrophilic and forms the outer shell or corona of micelles in an aqueous environment after self-assembling. The studies of Wang's group and Chen's group on core–shell BPs with polymethacrylate-based backbones and PCL-*b*-PEG amphiphilic diblock grafts presented interesting comparisons, which showed that block ratios of diblock grafts can critically affect self-assembly, drug encapsulation, and release behavior of core–shell BPs.

Core–shell BPs having a polymethacrylate backbone with degree of polymerization (DP) of 70 or 200 and amphiphilic PCL_{20}-*b*-PEG_{45} or PCL_{22}-*b*-PEG_{45} diblock grafts were synthesized by a combined "grafting from" and coupling approach by Wang and coworkers (Du et al., 2009). With CMC values twofold lower than linear PCL_{19}-*b*-PEG_{45} diblock copolymer, these core–shell BPs assembled into multimolecular micelles with sizes below 100 nm in aqueous solutions (Fig. 8.3C). The micelles exhibited higher DOX loading capacity as compared to the micelles from linear PCL_{19}-*b*-PEG_{45} diblock copolymer, and the burst release of free DOX from the BP micelles was significantly suppressed. The results suggested that DOX might be mainly loaded into the intermolecular space in the multimolecular micelles of BPs, and drug release was retarded due to the higher diffusion resistance required for the encapsulated drug moieties to cross the BP domains.

On the other hand, another core–shell BP having a polymethacrylate backbone with a DP of 940 and dense PCL_{17}-*b*-PEO_{113} grafts was synthesized through the "grafting onto" approach by Chen and coworkers (Zhao et al., 2012). With much longer PEG block than PCL block in grafts, the BP self-assembled to unimolecular cylindrical micellar worms of ~230 nm in length (Fig. 8.3D). DOX was loaded into its PCL core by hydrophobic interaction; however, the BPs demonstrated a lower loading efficiency but a faster release rate of DOX compared with the spherical micelles from linear PCL_{17}-*b*-PEO_{113}. The results suggested that the PCL-based core domains with PCL segments fixed by backbone of BP could not reorganize to host more drug molecules, and also be less favorable to retaining cargoes.

Salt/pH dual-responsive micelles based on amphiphilic core–shell BPs, PHEMA-*g*-(PCL-adenine:uracil-PEG), was developed by Zhu and coworkers. In particular, PCL and PEG were paired through two hydrogen bonds between two nucleobases (adenine and uracil), which are moderately strong, highly directional, and sensitive to not only acidic pH, but also salt. DOX was loaded into the micelles of around 50 nm in diameter with an efficiency of 70%. The cumulative release profile of the DOX-loaded micelles exhibited a controlled drug release behavior responsive to both salt and pH changes. As revealed by in vitro study, the DOX-loaded micelles exhibited high anticancer efficacy to kill HeLa cancer cells, while the blank micelles showed a very limited cytotoxicity (Zhai et al., 2012).

3.4 HETEROGRAFTED BPs

With high complexity in structure, statistically heterografted BPs and block BPs with different types of grafts have been developed (Fig. 8.2D–E), and showed remarkable potential to overcome challenges in drug delivery applications. Statistically heterografted BPs with poly(*t*-butyl methacrylate) (PBMA) and PNIPAAm grafts were developed by Xu and coworkers via a simple free radical polymerization route. The polymer brushes can be spontaneously assembled into spherical core–shell micelles (Fig. 8.3E) with well-dispersed individual NPs in aqueous solutions in the range of 36–93 nm by TEM and less than 200 nm by DLS. The BPs were temperature-sensitive with low critical solution temperature (LCST) about 28.8–29.4°C, and the self-assembled micelles exhibited temperature-triggered drug release behavior by using caffeine as a model drug (Luo et al., 2012). Bagheri and Bigdeli (2013) reported the synthesis of a novel biodegradable brush copolymers methacrylate backbone, as well as cholesteryl-modified PLA and PEG side chains, via both ATRP and free radical polymerization (FRP). The amphiphilic copolymers were used to form micelles with hydrophilic PEG shells and hydrophobic PLA-based cores (Fig. 8.3E). The CMCs of the micelles of copolymers prepared via ATRP and FRP were 0.7 and 0.9 mg/L, respectively. TEM showed that the spherical micelles had average diameters of 32 nm and 111 nm, respectively. Naproxen was used as a hydrophobic model drug, and these micelles exhibited loading efficiency of 85%–89%, with loading capacity of 17%–18%.

Kasi, Lu, and coworkers synthesized amphiphilic diblock BPs comprised of polymethacrylate-*g*-PEG brush in one block and polymethacrylate-*g*-cholesterol mesogen brush with hemitelechelic thiol graft ends in the other block (Nguyen et al., 2014). To achieve dual encapsulation, gold NPs (AuNPs) were templated in situ with cholesterol mesogen chains through covalent binding with their hemitelechelic thiol graft ends, while DOX was encapsulated by self-assembly triggered by dropwise injection of DMF solution of AuNPs template BPs into distilled water. The resulting dually encapsulated drug-delivery system exhibited size around 60–80 nm by TEM and ~157 nm by DLS, indicating the formation of multimolecular micelles (Fig. 8.3F). High DOX encapsulation efficiency (85.6%) and high DOX loading (21.4%) were achieved due to the hydrophobic interaction between DOX and cholesterol moieties. The in vitro release study showed that dually encapsulated micelles released DOX in a sustained release pattern without considerable initial burst release (about 4% release after the first day and 45.3% release after 10 days), which may be attributed to strong hydrophobic interaction between DOX and cholesterol moieties. It was also expected that drug release may be accelerated inside cells because of the enzyme-assisted degradation of ester bonds between the polymethacrylate backbone and the cholesterol mesogen side chains.

The research team also reported a series of diblock BPs (P(NBCh9-*b*-NBPEG)) with a polynorbornene (PNB) backbone carrying PEG grafts in one block and cholesterol grafts in the other block (Tran et al., 2014). These BPs self-assembled to uniform NPs with an average particle diameter below 200 nm in aqueous media (Fig. 8.3F), due to the narrowly distributed MW of the copolymers well controlled by ring-opening metathesis polymerization (ROMP). With high DOX encapsulation efficiency (88.4%) and loading content (22.1 wt.%), in vitro release of free DOX from the NPs was nearly linear (less than 2% per day) in phosphate buffered saline (PBS) solution without noticeable burst effect. In vivo studies on A549 lung tumor-bearing SCID mice showed that DOX-encapsulated NPs had significantly increased the duration of the drug in circulation and resulted in greater tumor accumulation with significantly reduced toxicity compared to that with free DOX.

Zhang and coworkers developed a diblock BP, P(PLAMA-*co*-MAA)-*b*-PPEGMA consisting of a block with a PMAA-based backbone carrying PLA grafts and a block with a polymethacrylate backbone carrying PEG grafts. These BPs can form multimolecular assemblies with sizes around 200 nm in aqueous solutions as measured by DLS (Fig. 8.3F). By using the micelles of BPs as carriers, as high as 20% loading of IBU was achieved with nearly 90% loading efficiency, indicating that the PLA-based hydrophobic domains of the assemblies may readily encapsulate IBU. In vitro drug release study revealed well-controlled pH-dependent release of IBU from the micelles, which did not show obvious burst release and could be expedited as the content of MAA in the BPs increased (Yang et al., 2012a).

Convertine and coworkers prepared a series of copolymers consisting of poly(ethylene glycol) methyl ether methacrylate (O950) and 2-hydroxyethyl methacrylate (HEMA) via reversible addition-fragmentation chain transfer (RAFT) polymerization (Lane et al., 2015b). These copolymers were functionalized with trithiocarbonate-based RAFT chain transfer agents (CTAs) to form graft chain transfer agents (gCTAs). The gCTAs were employed in the RAFT polymerization of polymerize *N*-(2-hydroxypropyl)methacrylamide (HPMA) and dimethylaminoethyl methacrylate (DMAEMA). Carrying both PEG grafts and P(HPMA-*co*-DMAEMA) copolymer grafts, the resulting pH-responsive endosomalytic heterografted BPs could self-assemble to polymersomes, which were proved to be effective carriers for encapsulation of potent antibiotic drugs.

4 BP-BASED CONJUGATED DRUG-DELIVERY SYSTEMS

Within BP-based conjugated drug-delivery systems, drug molecules are covalently bonded to BP scaffolds through linkages that can be cleaved for drug release under desirable circumstances. As compared with drug encapsulated systems using assembled scaffolds, drug-conjugated systems have better stability in terms of storage and resistance to the biological environment. With well-defined structures,

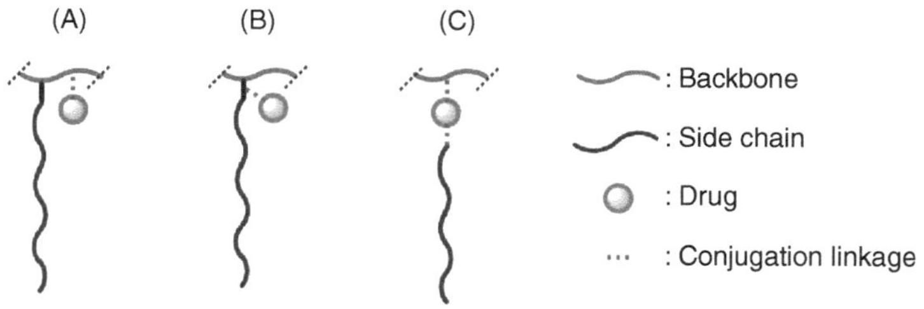

FIGURE 8.4 Schematic Illustrations of Grafting Structures of BPDCs With Drug Conjugated With BP Backbone

(A) Monovalent grafting sites to link with graft and drug respectively, (B) divalent grafting sites to link with both graft and drug, and (C) drug serving as a divalent bridge to link with both backbone and side chain.

polymer–drug conjugates (PDCs) generally have precisely controlled drug loading and can provide sustained drug release without burst effect. The structural design of BP-based PDCs may vary, but each design requires thoughtful consideration. To provide good shielding of drug moieties, three types of grafting structures may allow the linking of drug moieties with backbone of BP scaffolds, including (1) monovalent grafting sites to link with graft and drug, respectively (Fig. 8.4A), (2) divalent grafting sites to link with both drug and side chain (Fig. 8.4B), and (3) drug serving as a divalent bridge to link with both backbone and side chain (Fig. 8.4C). In terms of a synthetic route, similar to BP synthesis, "grafting onto" and "grafting through" can also be applied to conjugate drug moieties for the preparation of BPDCs. From a synthetic point of view, the preparation of conjugated drug-delivery systems typically is more challenging than the preparation of encapsulated drug-delivery systems; therefore, mono-BPs are commonly usedas base polymers for BPDCs, and the research results that have been published to date are largely focused on structural design and synthesis. Therefore, the literature on BPDCs will be reviewed based on their backbone structures, which are determined through synthesis.

4.1 BPDCs WITH PNB BACKBONES

Based on the merits of ROMP, several types of BPDCs with PNB backbones have been developed. By utilizing the "grafting through" strategy, ROMP using Grubbs' catalyst can enable facile and well-controlled polymerization of norbornene (NB)-functionalized macromonomers and prodrugs. The high tolerance of functional groups of the ROMP systems can also facilitate the incorporation of functionalities with the brush-like macromolecular structures, allowing the conjugation of drug moieties via the "grafting onto" strategy. To date, both Johnson et al. and our group have studied BPDCs with PNB backbones.

Johnson et al. (2010, 2011) published two articles to report the facile synthesis of clickable PNB-*g*-PEG BPs, facilitated by ROMP "graft-through" of either an alkyne-bearing or an azide-bearing clickable NB-PEG macromonomer, and their transformation to BPDCs bearing DOX and/or camptothecin (CT). As shown in Fig. 8.5, alkyne-bearing NB monomer along with azide-bearing drug moiety, and "azide-convertible" chloride-bearing NB monomer along with alkyne-bearing drug moiety, respectively, were prepared through multistep organic synthesis. In the first article, two azide-bearing prodrugs with photocleavable nitrobenzyloxylcarbonyl (NBOC) linkages, DOX-NBOC-N_3 and CT-NBOC-N_3, were successfully conjugated onto bivalent NB-PEG/alkyne macromonomer through quantitative copper-catalyzed azide-alkyne cycloaddition (CuAAC) "click" reaction followed by ROMP to yield BPDCs, including PNB-*g*-PEG/DOX (A), PNB-*g*-PEG/CT, and PNB-*g*-PEG/DOX/CT (Fig. 8.5A) (Johnson

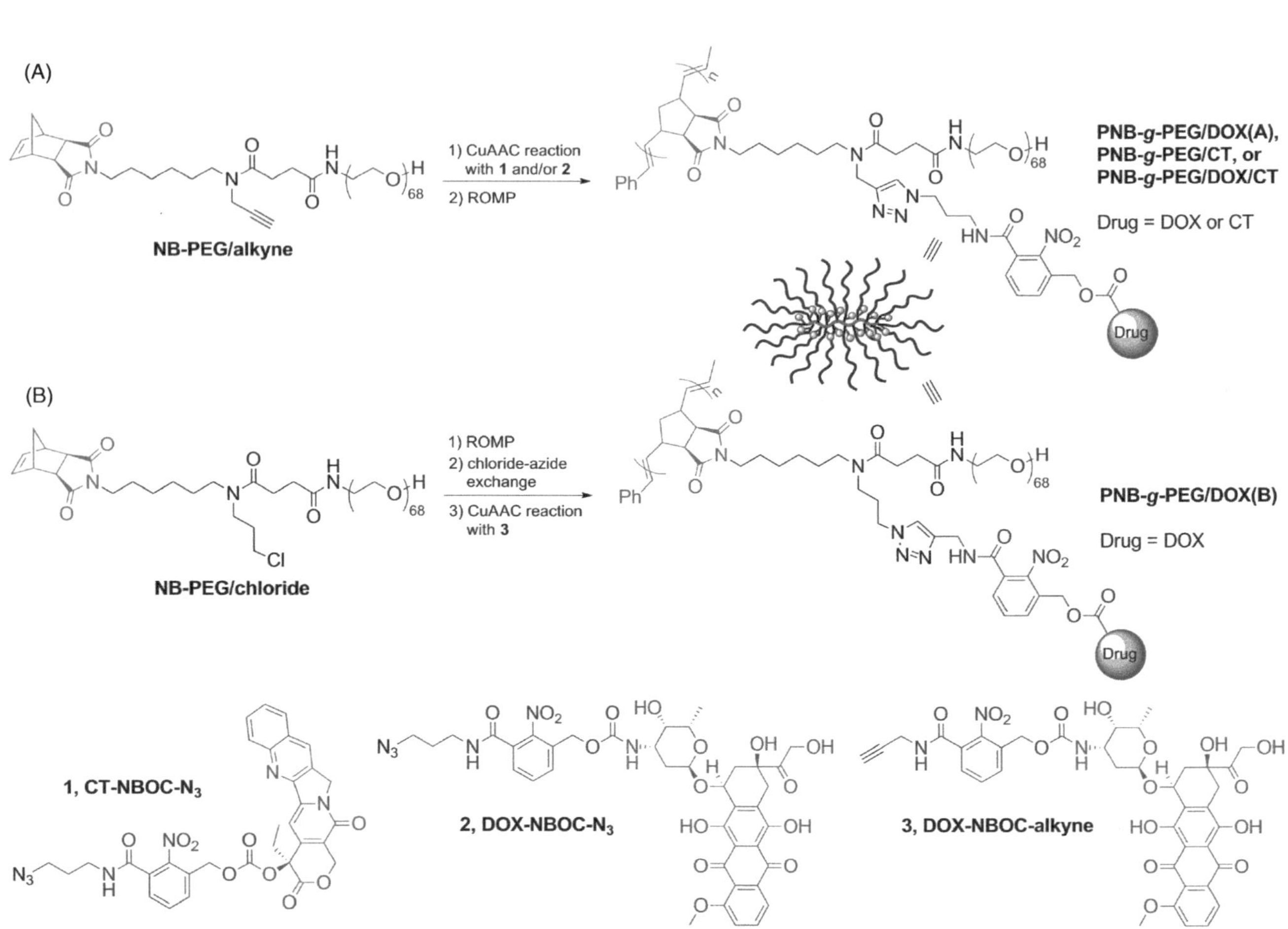

FIGURE 8.5 Synthesis of PNB-*g*-PEG/Drug BPDCs With Divalent Grafting Sites

(A) PNB-*g*-PEG/DOX, PNB-*g*-PEG/CT, and PNB-*g*-PEG/DOX/CT by "grafting from", and (B) PNB-*g*-PEG/DOX by "grafting onto".

et al., 2010). In the second article, clickable PNB-*b*-PEG/azide was prepared through ROMP of NB-PEG/chloride, followed by halide-azide exchange. Then a click conjugation reaction between DOX-NBOC-alkyne and PNB-*b*-PEG/azide was performed, yielding PNB-*g*-PEG/DOX (B) (Fig. 8.5B). All of the resulting BPDCs possessed drug moieties at the core near the PNB backbone with PEG grafts extended into solution. Hydrodynamic radii of nanostructures formed by the PDCs increased with DP of PNB backbone, suggesting that they reassembled to form unimolecular micelles. Despite their structural similarity, two DOX-containing BPDCs, PNB-*g*-PEG/DOX (A and B), displayed different nanoscopic size with the same DP values. After UV irradiation at 365 nm for 10 min, 50 and 70% of free DOX were released from BPDCs A and B, respectively. Meanwhile, the release rates of free drug, such as DOX and CT through NBOC linkage of BP conjugates were also varied by synthetic approaches and drug species. Statistical copolymerization of NB-DOX and NB-CT monomers yielded a potential codelivery system for combination therapy. In vitro cell viability experiments indicated that all these conjugates did not exhibit significant cytotoxicity without UV irradiation, but became therapeutically effective against MCF-7 human cancer cells based upon UV-triggered drug release (Johnson et al., 2010, 2011).

In early 2011, our group reported novel pH-sensitive BPDCs with PNB backbone carrying PEG grafts and PTXL moieties (Zou et al., 2011). As statistical copolymers, these BPDCs were synthesized by "graft-through" of mixtures of PTXL-bearing NB monomer and PEG-based NB macromonomer via ROMP (Fig. 8.6, route *a*). Acid-sensitive cycloacetal-based linkage was created to attach PTXL to NB for smart and controlled drug delivery. Relative to first-generation Grubbs' catalyst,

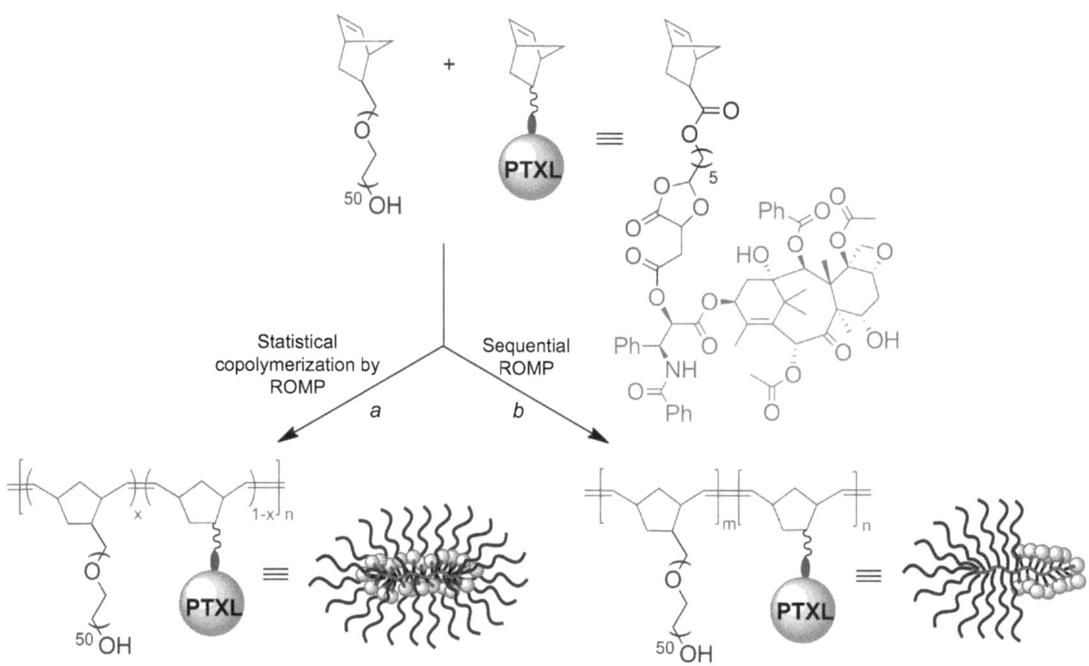

FIGURE 8.6 Synthesis of Statistical BPDC and Diblock BDPCs With Monovalent Grafting Sites Carrying PEG Grafts and PTXL Moieties

the third-generation catalyst as the ROMP initiator led to well-controlled MWs and low molar-mass dispersity ($Đ$) values of resulting BPDCs. As determined by DLS analysis of their aqueous solutions, these BPDCs formed nanostructures (11.8 nm for backbone DP of 104, and 28.6 nm for backbone DP of 390) with narrow monomodal size distribution, suggesting their unimolecular structures in water. With greatly enhanced water solubility relative to PTXL, the BPDCs exhibited selective and fast drug release under acidic conditions due to their acid-sensitive cycloacetal-based linkages between PTXL moieties and BP backbones.

Based on our previous studies, using the same NB-PTXL monomer and NB-PEG macromonomer, diblock BPDCs were synthesized by sequential ROMP (Fig. 8.6, route *b*) (Zou et al., 2015). The diblock BPDCs self-assembled into spherical multimolecular nanostructures in aqueous solutions. As compared with statistical BPDCs, the release of PTXL moiety from the assemblies of diblock BPDCs was relatively slow and exhibited quite limited acid sensitivity, indicating a significant influence of the macromolecular structure and assembly of BPDCs on their drug release behavior. Specifically, resistance of drug diffusion after cleavage of conjugation linkage might be much higher in assembled multimolecular nanostructures of BPs than in unimolecular nanostructures of BPs. The cytotoxicity study showed that the diblock BPDCs are therapeutically effective against MCF-7 human cancer cells, but less cytotoxic than the statistical BPDCs, presumably due to the retarded drug release from the diblock BPDCs.

Although well-defined BPDCs with PNB backbones can be synthesized, a shortcoming of these BPDCs is that their macromolecular structures are not biodegradable and may result in side effects for in vivo applications when their hydrodynamic sizes are larger than the size limit (~10 nm) for kidney clearance. To address such a concern, very recently Johnson and coworkers have successfully introduced degradability into BP-based PDCs with PNB backbones through a series of stimuli cleavable multifunctional NB cross-linkers. They developed a "brush-first" ROMP method, which yields brush–arm star polymer (BASP) nanostructures via the sequential polymerization of NB macromonomers and multifunctional NB cross-linkers (Liu et al., 2012, 2013). A photocleavable cross-linker NB-NBOC-NB was first introduced to prepare azide-bearing N_3-BASP nanostructure (Burts et al., 2014). Because PEG chains would extend into aqueous solutions to afford water solubility of BASPs and provide shielding for their hydrophobic cores, the azide groups from N_3-BASPs must be very sterically hindered. As a result, an optimized CuAAC protocol by using large excesses of reagents relative to azide was used to achieve high conversions of DOX-BASPs. Based on the structural design, the BASP corona was connected entirely through photocleavable NBOC groups while the DP of the nondegradable PNB-*g*-PEG BP unit was around 15. UV irradiation should, in principle, lead to the degradation of NBOC and regeneration of PNB-*g*-PEG with DP around 15. GPC traces before and after 4 h of UV irradiation of N_3-BASPs in water showed that the degradation products overlap with the parent PNB-*g*-PEG BPs, indicating nearly quantitative degradation in the case that 10 equivalent of cross-linker relative to PNB-*g*-PEG was used for the preparation of BASPs. But degradation efficiency of such BASPs under UV irradiation decreased with the increase of cross-linker feeding amount (Burts et al., 2014).

Another cleavable cross-linker selected for BASP formation was a Pt(IV) diester derivatives, Pt(IV) bis-NB (PT-XL), in which labile Pt–O bonds could be cleaved upon glutathione-induced intracellular reduction and thus lead to particle degradation to yield both ~5 nm BPs and free cisplatin (Liao et al., 2014). With using DOX-conjugated NB-PEG macromonomers, CT-conjugated NB-PEG macromonomers, and NB-PEG macromonomers, as well as PT-XL cross-linker, the resulting multidrug-conjugated BASP NPs capable of controlled loading and synchronized release of multiple therapeutic agents (DOX, CT, and cisplatin) for combination therapy were successfully synthesized through the "brush-first" ROMP method with an optimized macromonomer/cross-linker ratio of 7:3. The nanoscopic size of the multidrug-conjugated BASP NPs was determined as 103 nm by TEM and 180 nm by DLS. Although the release kinetics of each free drug was not explored in this work, effective drug release from the BASP NPs under experimental conditions was inferred through cell internalization and cytotoxicity study against OVCAR3 ovarian carcinoma cells, and the potential application of the BASP for combination therapy was suggested.

Furthermore, acid-labile bis-NB cross-linkers were also synthesized and evaluated for the formation of BASPs via the brush-first ROMP method to meet the needs for cancer drug delivery. Thus, PEGylated BASPs would degrade in the mildly acidic tumor microenvironment. An acetal-based cross-linker was identified, and it could lead to highly controlled BASP formation reactions. A combination of this cross-linker with a DOX-conjugated NB-PEG macromonomer resulted in DOX-conjugated BASP nanostructures (56 nm by TEM and 48 nm by DLS) with drug loading of 11.4% that would simultaneously degrade into smaller PNB-*g*-PEG (~40 kDa) through acetal cleavage in the core, and release cytotoxic DOX through hydrolysis of ester bonds under an in vitro condition. However, incubation of the BASPs in pH 6.0 and 7.4 PBS buffers for 24 h showed that small amounts of DOX were release, with a greater release at neutral pH as compared with pH 6.0, indicating that the systems need to be further modified to introduce acid-triggered drug release behavior (Gao et al., 2014).

It should be mentioned that all the aforementioned studies of BPDCs focus on the synthetic chemistry and proof-of-concept in vitro study, and the safety profile of polymeric scaffolds with PNB backbones for in vivo applications has not been fully understood and needs to be further investigated.

4.2 BPDCs WITH ALIPHATIC POLYESTER BACKBONES

Although many synthetic polymers are found biocompatible and nonimmunogenic, they are not degradable and therefore can be used safely only when their MWs are higher than the renal clearance threshold (~40 kDa) (Duncan, 2006, 2009; Knop et al., 2010). BP scaffolds can easily have MWs over 40 kDa due to their grafting structure, and therefore degradability of BPs is critically needed when in vivo applications are considered. Because the hydrodynamic sizes and MWs of BPs can dramatically decrease through backbone cleavage, having a biodegradable backbone would be ideal for the development of BP-based drug conjugation systems.

Motivated by the tremendous advantages of integrating biocompatibility and biodegradability into drug-delivery systems, functional aliphatic polyesters have attracted extensive attention in the past decades (Jiang et al., 2008; Mecerreyes et al., 2000; Parrish et al., 2005; Riva et al., 2007; Trollsås et al., 2000). BPDCs with aliphatic polyester backbones have also been explored in recent years. A very first example of a conjugate with polyester backbone and grafting structure was reported by Duncan and coworkers in 2005. Chain extension reactions of PEG4000 with bis-(4-hydroxytetramethylene) maleate (DBM) yielded copolymers DBM2-PEG4000 with biodegradable carbonate linkages and number-average MW (M_n) values of 37–53 kDa and $Đ$ values of 3.0–3.7. Via thioether linkages, DBM2-PEG4000 was subsequently conjugated with HS-PEG3000-Gly-Phe-Lue-Gly DOX (HS-PEG3000-GFLG-DOX) grafts, and the resulting DBM2-PEG4000-S-PEG3000-GFLG-DOX has 3–4 wt.% of DOX loading. This GFLG linker between DOX and BP scaffold allows specific cleavage by the lysosomal thiol-dependent proteases. During incubation with isolated lysosomal enzymes, DOX release was relatively slow with <5% release over 5 h, presumably because the steric accessibility of the GFLG linker was restricted by the limited solubility of the conjugate, which may lead to the formation of multimolecular micelles in aqueous solution. However, in vitro and in vivo assessments showed that the conjugate was less appealing than PEG-peptide-DOX or HPMA copolymer-GFLG-DOX as a candidate for further development as an antitumor conjugate (Andersson et al., 2005).

In 2007, Emrick and coworker reported a BPDC with a polylactone backbone that was prepared through CuAAC click reaction of an alkyne-functionalized polylactone with monoazide-functionalized CT followed by further reacting with azide-terminated PEG (Parrish and Emrick, 2007). The conjugate had grafting densities of ~18% CT and ~33% PEG, in terms of percentage of backbone repeat units containing grafts and conjugated drug moieties. They found that PEG grafting resulted in a tremendous improvement on the solubility properties in water and in organic solvents (e.g., acetone, DMSO, and CH_2Cl_2), although biological assessment of the toxicity and efficacy of this conjugate was not reported in the article.

More recently, our group successfully developed several well-defined BPDCs with PLA-based backbones, as shown in Fig. 8.7. As a pioneer work in our group, a functional PLA-*g*-PTXL-PEG BPDC, with grafting density of 30%, 23 wt.% of PTXL, and 54 wt.% of PEG, was synthesized and studied. PTXL, as the divalent agent, was chemically modified to bridge between the degradable PLA-based backbone and hydrophilic PEG side chains (Fig. 8.7A) (Yu et al., 2011). As the backbone precursor, the acetylene-functionalized PLA with a M_n of 9.3 kDa and a low $Đ$ of 1.23 was prepared by organocatalyzed ROP of an acetylene-functionalized lactide (LA) monomer with L-LA. The graft precursor, azide-PTXL-PEG, was prepared by multistep organic synthesis and then grafted onto the functional PLA through a CuAAC click reaction with 95% grafting efficiency. According to DLS analysis, the BPDC exhibited average hydrodynamic sizes of 18.4 nm in water that is a selective solvent for PEG, but 10.6 nm in DMF that is a good solvent for all types of structural components, indicating that the BPDC assembled into multimolecular structures in aqueous solution. The release of PTXL requires the successive cleavages of two types of conjugation bonds through hydrolysis, and could not be readily controlled because of different sensitivities of the conjugation linkages. Relative to the ester linkage between PTXL and PEG, the cycloacetal group between PTXL and PLA-based backbone showed a faster rate in hydrolysis under acidic conditions.

To further improve the drug release profile of PTXL, we subsequently developed a BPDC through one-pot click reaction of acetylene-functionalized PLA with azide-functionalized PTXL and azide-functionalized PEG, in which PTXL as a monovalent agent was chemically bonded to the backbone through single cycloacetal linkage (Fig. 8.7B) (Yu et al., 2013). Well-defined acetylene-functionalized PLA, with a M_n of 9.1 kDa and a $Đ$ of 1.15, was prepared through organocatalyzed ROP of the acetylene-functionalized LA. The resulting BPDC had PTXL grafting density of 47%, PEG grafting density of 52%, M_n of 92.2 kDa, and PTXL loading of 23.2 wt.%. Consistent with our previous finding, the BPDC assembled to form multimolecular nanostructures with sizes in the range of 10–30 nm in water. In vitro drug release study showed that PTXL moiety was released through the cleavage of the cycloacetal linkage, with a half-life of about 22 h in pH 7.4 at 37°C. The cytotoxicity assessment revealed that the BPDC was more therapeutically effective toward MCF-7 breast cancer cells than free PTXL at drug concentrations higher than 0.1 μg/mL, while the BP scaffolds were not cytotoxic. Cell internalization study showed that the BPDC scaffold was readily taken up into the cytoplasm of MCF-7 cells via endocytosis.

In a combination of biodegradable PLA-based backbones and hydrophilic PEG grafts, our BP-based platform not only exhibited nanoscopic size and significant water solubility, but also demonstrated

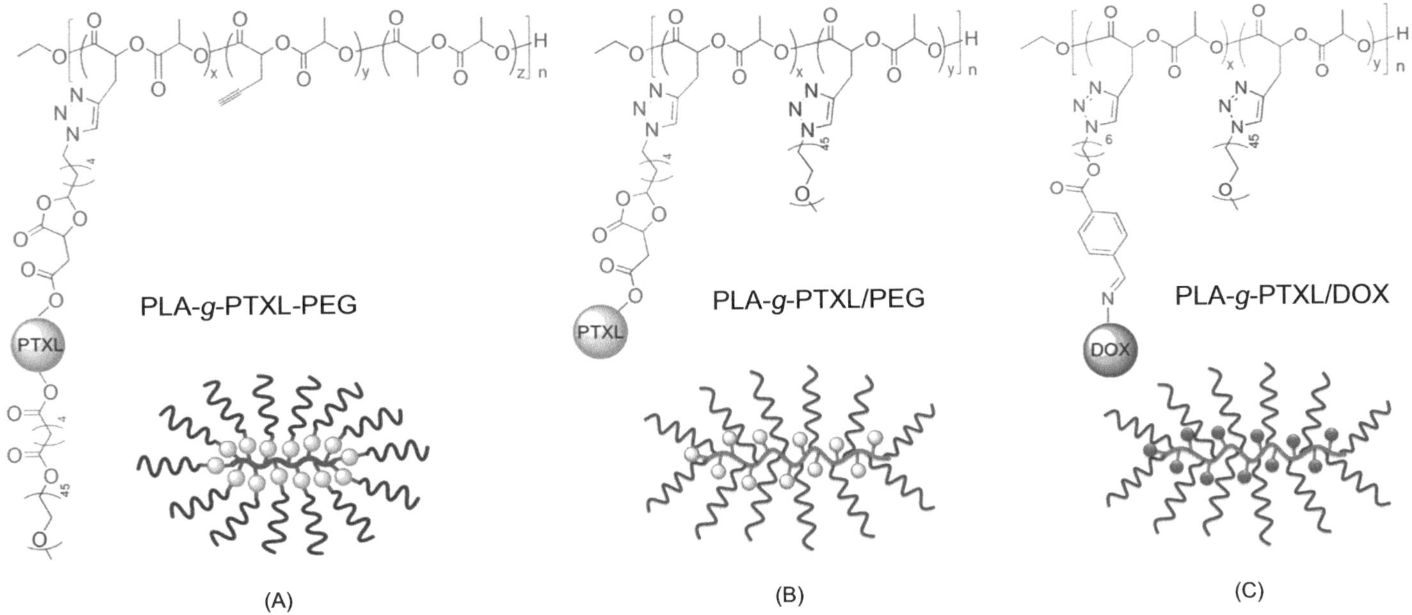

FIGURE 8.7 Schematic Illustrations of BPDCs With PLA-Based Backbones

(A) PLA-*g*-PTXL-PEG, (B) PLA-*g*-PTXL/PEG with monovalent grafting sites, and (C) PLA-*g*-DOX/PEG with monovalent grafting sites.

remarkable degradability and noncytotoxicity, and therefore could be appealing for the delivery of other drugs as well. Because a tumor tissue is more acidic than a normal tissue (Helmlinger et al., 2002), pH-responsive drug release is highly preferred in targeted drug release within tumors. Thus, we further applied our platform to conjugate another model cancer drug, DOX, through an acid-labile linkage. The biodegradable scaffold, PLA-*g*-PEG/aldehyde with PEG grafting density of 54% and aldehyde grafting density of 46%, was first prepared via a quantitative one-pot CuAAC click reaction of acetylene-functionalized PLA with 6-azidohexyl 4-formylbenzoate and azide-functionalized PEG (Fig. 8.7C). Subsequently, the PLA-*g*-PEG/DOX BPDC, with M_n of 76.7 kDa, DOX loading of 9.2 wt.% and hydrodynamic size around 13 nm, was obtained by conjugating DOX with the scaffold through an acid-sensitive Schiff base linkage. With Schiff base linkages, the BPDC showed remarkable pH-responsive drug release behavior with the half-life of ~7 h at pH 5.5 and only ~20% release after 48 h at pH 7.4. It was found that the BPDC internalized the cytoplasm of MCF-7 cells within 4 h, and delivered DOX moieties into the nuclei within 24 h. Moreover, the cytotoxicity assessment revealed that the BPDC was therapeutically effective toward MCF-7 cells (Yu et al., 2015).

4.3 BPDCs WITH OTHER POLYMER BACKBONES

Besides BPDCs with PNB or aliphatic polyester backbones, several BP-based conjugated drug-delivery systems with other types of polymer backbones have been explored recently. For instance, Yang and coworkers synthesized a BPDC with polyoxetane backbone (DP ~30) by cationic ROP of an acetylene-functionalized 3-ethyl-3-(hydroxymethyl)oxetane (EAMO), following by click conjugation reactions of azide-functionalized PEG and CT (Zolotarskaya et al., 2012). With improved water solubility, the PEAMO-*g*-PEG/CT BPDC demonstrated dose-dependent cytotoxicity to human glioma cells while the carrier itself was not cytotoxic, indicating its potential application in anticancer therapy. Moreover, the researchers performed detailed studies about these clickable cytocompatible PEG-grafted polyoxetane BPs (Zolotarskaya et al., 2013). Although PEAMO was insoluble, all PEG-grafted PEAMO BPs were water-soluble. They also synthesized a fluorescein-carrying PEAMO-*g*-PEG by click coupling 6-azidohexanoic acid (AHA) to the backbone, followed by coupling 5-(aminoacetamido)fluorescein (AAF) as a model imaging agent with the acid terminal of AHA throughamidation chemistry.

BPDCs with polyacrylamide backbone were reported by Convertine and coworkers very recently (Lane et al., 2015a). They prepared linear polyacrylamides through reversible addition-fragmentation chain transfer (RAFT) polymerization of *N,N*-dimethylacrylamide (DMA) and 2-hydroxyethylacrylamide (HEAm) at first, and then introduced RAFT side functionalities to these linear polymers via the hydroxyl groups of HEAm units, followed by RAFT "grafting from" process to polymerize vinyl monomers. Subsequently, they further employed the resulting BPs with hydroxyl-containing grafts for functional group transformation to introduce RAFT functionalities, followed by RAFT "grafting from" preparation of the second generation of grafts. The finally obtained second-generation BPs, with high MW (over 1000 kDa) and capable of forming signal polymer NPs, were used as scaffolds for further "grafting onto" incorporation of biomedical groups, such as DOX moieties, through conjugation reactions. With hydrolyzable conjugate linkages, the DOX-conjugated BPDC NPs exhibited evident chemotherapeutic activity toward SKOV3 ovarian cancer cells.

It should be noted that with macromolecular structures similar to BPDCs, PDCs with relatively long side chains have also been studied for drug delivery. For instance, by using chitosan as the backbone biodegradable polymer in the structural design of PDCs, Hu and coworkers reported a biodegradable PDC, in which DOX was conjugated onto a stearic acid-grafted chitosan oligosaccharide (CSO-SA) via disulfide linkers (Su et al., 2015). The resulting conjugate DOX-SS-CSO-SA could form micelles by self-assembly in aqueous medium with a low CMC of 49.1 μg/mL. DLS showed that the micelles had narrowly distributed sizes of 62.8 ± 5.4 nm, which was in accordance with the TEM results. The drug loading content was 5.5 wt.% indicated by UV-vis spectroscopy. In a reductive environment, the disulfide bonds were cleaved rapidly within 20 min, resulting in rapid drug release as compared to the situation in a low reductive environment. In vitro antitumor activity study showed that the conjugate DOX-SS-CSO-SA had higher cytotoxicity against DOX-resistant cells than free DOX. In vivo distribution study showed a selective accumulation of DOX-SS-CSO-SA in tumors. In addition, in vivo antitumor efficacy study confirmed that the conjugate could efficiently suppress tumor growth.

5 MULTIFUNCTIONAL BP-BASED DRUG-DELIVERY SYSTEMS

With integrated imaging, targeting, and/or diagnosis functions besides drug administration, multifunctional drug-delivery systems have attracted significant interest and may possess tremendous advantages for in vivo applications. Although most of the reported studies on BP-based drug-delivery systems focus on synthesis, nanostructure formation, and drug release profile, it is an important research direction to multifunctionalize BP-based drug-delivery systems to meet the comprehensive requirements for disease treatment. Among various biomedical functionalities, imaging agents may be incorporated with BPs by bonding with either backbone or grafts; on the other hand, targeting and diagnosis agents typically need to be tethered with the peripheral domains of the BP-based nanostructures by conjugating them at graft ends (Fig. 8.8).

Using well-established BP with PNB backbone as the scaffold, Johnson and coworkers recently reported interesting work on redox-responsive cylindrical BPs for in vivo MRI and fluorescence imaging (Sowers et al., 2014). Azide-functionalized spirocyclohexyl nitroxide as a metal-free MRI contrast agent and azide-functionalized Cy5.5 as a NIR dye were conjugated to alkyne-functionalized NB-PEG via click reaction, respectively, to form two NB-based macromonomers. The final redox-responsive BPs having both MRI contrast agent and NIR dye linked with PNB backbones (Fig. 8.8A), also termed ORCAFluors, were synthesized by ROMP of the two macromonomers. In vivo study revealed that ORCAFluor displayed MRI contrast enhancement in the regions associated with the renal pelvis, renal artery, renal vein, and aorta. In addition, it was confirmed through both in vitro and in vivo studies that exposure of ORCAFluor to ascorbate leads to nitroxide reduction and enhanced fluorescence emission.

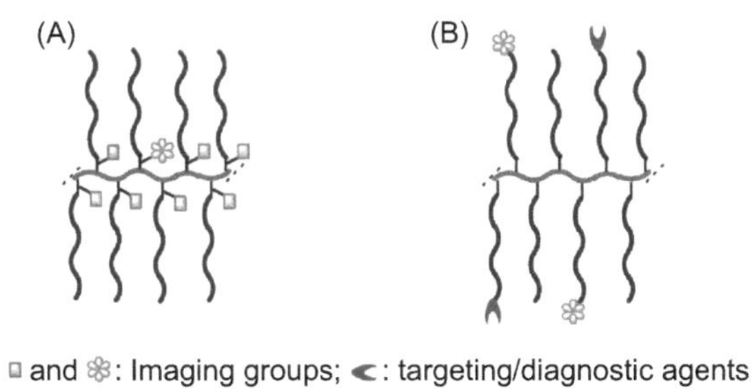

FIGURE 8.8 Schematic Illustrations of Multifunctionalization of BP Scaffolds (Drug Can Be Loaded in the Core Domains of BP-Based Nanostructures)

(A) BPs with imaging and targeting/diagnostic agents linked with backbones, and (B) BPs with imaging and targeting/diagnostic agents linked at graft ends.

Schmidt and coworkers recently reported a BP-based immune cancer therapeutic system based on multifunctionalization of BPs (Bühler et al., 2014). Cylindrical BPs with polymethacrylamide backbone and poly(2-oxazoline) grafts having azide terminals were prepared by the "graft through" approach via radical polymerization of poly(2-oxazoline)-based azide-functionalized methacrylamide macromonomer. The multifunctional system with click conjugated dye and an anti-DEC205 antibody at the graft ends of the cylindrical BPs (Fig. 8.8B) demonstrated antibody-mediated specific binding and uptake through in vitro studies. The multifunctional BP-based system obtained through further conjugation of an antigen peptide exhibited a significantly stronger antigen-specific T-cell stimulatory capacity than the systems conjugated with just antigen or polymer–antigen.

Weck and coworkers reported membranotropic peptide-functionalized polylactide-*graft*-oligo(ethylene glycol) (PLA-*g*-OEG) BPs, and investigated them for intracellular drug delivery (Borchmann et al., 2015). ROP of an azido OEG-functionalized lactide monomer was performed using an initiator carrying a protected aldehyde functionality. The resulting PLA-*g*-OEG was functionalized with a fluorescent dye (7-nitrobenzoxadiazole, NBD) by copper-free strain-promoted azide-alkyne click chemistry, and a cell internalization peptide (gH625) by reductive amination. Both of the reactions were highly efficient, resulting in high yields. In vitro studies with HeLa cells demonstrated that the conjugate showed enhanced cellular uptake. Because of the function of gH625 peptide, localization of the conjugate in close proximity of the nuclei was observed.

Recently Gong, Cai, and coworkers also reported a multifunctional BP-based drug-delivery system (Guo et al., 2014). Multifunctional theranostic BP-based micelles with sizes of ~30 nm were formed by the conjugation reaction of an amphiphilic core–shell BP, PHEMA-*g*-(PLLA-*b*-PEG), with an anti-CD105 monoclonal antibody (TRC105) and a macrocyclic chelator to enable ^{64}Cu labeling. The DOX-loaded BP-based micelles showed a very sensitive pH-dependent drug release profile. In vitro study exhibited a significantly higher cellular uptake of the TRC105-conjugated micelles by CD105-positive human umbilical vein endothelial cells as a result of CD105-mediated endocytosis. Furthermore, as illustrated through positron emission tomography (PET) imaging and ex vivo biodistribution studies, a higher level of tumor accumulation was confirmed in 4T1 murine breast tumor–bearing mice that were treated with ^{64}Cu-labeled targeted micelles relative to those treated with nontargeted controls. With the merit to offer pH-triggered drug release, noninvasive PET diagnosis, and tumor-targeting ability simultaneously, this BP-based multifunctional system may have promising potential to serve as a nanoplatform for cancer theranostics.

The PLL-*g*-PEG/DNQ dual encapsulation systems reported by Liu and coworkers were reviewed in Section 3.1. The research team also further functionalized the system by incorporating cyclic arginine-glycine-aspartic acid (cRGD) tripeptide by thiol-ene reaction of thiolated cRGD with maleimide terminals of PEG grafts (Yuan et al., 2015). The cRGD-functionalized NP micelles showed significantly enhanced accumulation within integrin $\alpha v\beta 3$ overexpressed cancer cells.

6 SUMMARY

BP-based drug-delivery systems have gained increasing attention over the past decade. With densely grafted macromolecular architectures, BP-based polymers can form unimolecular and multimolecular nanostructures within the appropriate size range to avoid fast systemic clearance. General design considerations of BP-based scaffolds for drug delivery applications have been established. Based on comprehensive structural considerations, many types of new monomers or compounds can be further studied for the construction of BP-based drug-delivery systems. For example, as substitutes for PEG, zwitterionic materials have been widely investigated to obtain stable and long-circulating nanocarriers (Jiang and Cao, 2010). Integration of zwitterionic groups with BP-based drug carriers potentially may lead to a novel class of polymer scaffolds with optimal properties for therapeutic delivery.

Drug-encapsulated systems using a broad variety of BP-based scaffolds have been prepared and studied. With specific structural design of the core domains of BP-based nanostructures, burst release can be effectively suppressed in some of these systems. BPDCs, especially those with mono-BPs as base polymers, have also been synthesized and investigated. By incorporating stimuli-responsive conjugation linkages to link drug moieties with the scaffolds, some of these BPDC systems have exhibited triggered drug release behavior. For both encapsulated and conjugated drug-delivery systems based on BPs, their drug release behavior is also influenced significantly by the assembly behavior of the BP-based scaffolds. Multifunctional BP-based drug-delivery systems have also been reported very recently. By incorporating targeting ligands, imaging, and/or diagnostic agents with BP-based scaffolds, the resulting multifunctional systems can possess improved biomedical functions.

It can be expected that BP-based drug-delivery systems will remain as a vivid research area in the next decades. The studies on these systems that have been reported so far are concentrated on material design and preparation, as well as preliminary biomedical assessment. The development of well-defined biocompatible, water-dispersible BP-based scaffolds that carry reactive functional groups, have controlled nanostructures, and can be degraded gradually under biological aqueous conditions will attract continued interest. BP-based drug-conjugated systems, as well as BP-based drug-encapsulated systems with suppressed burst release, may be intensively studied due to their relevance to in vivo applications, and multifunctionalization of these systems will be highly preferred in order to promote accuracy and efficacy of drug delivery. While promising in vitro results should be obtained at first, the biomedical merits of these BP-based delivery systems need to be further investigated and verified through extensive in vivo studies before their translation to the clinic can be planned.

ACKNOWLEDGMENTS

The authors thank the U.S. National Science Foundation for the financial support (DMR-1206715 and DMR-1609914). The authors would also like to acknowledge the collaborations and discussions with past and present coworkers, whose contributions can be found in the list of references.

REFERENCES

Allen, T.M., Cullis, P.R., 2004. Drug delivery systems: entering the mainstream. Science 303, 1818–1822.

Andersson, L., Davies, J., Duncan, R., Ferruti, P., Ford, J., Kneller, S., Mendichi, R., Pasut, G., Schiavon, O., Summerford, C., Tirk, A., Veronese, F.M., Vincenzi, V., Wu, G., 2005. Poly(ethylene glycol)-poly(ester-carbonate) block copolymers carrying PEG-peptidyl-doxorubicin pendant side chains: synthesis and evaluation as anticancer conjugates. Biomacromolecules 6, 914–926.

Bagheri, M., Bigdeli, E., 2013. Preparation of stealth micellar nanoparticles of novel biodegradable and biocompatible brush copolymers with cholesteryl-modified PLA and PEG side chains. J. Polym. Res. 20, 1–11.

Blum, A.P., Kammeyer, J.K., Yin, J., Crystal, D.T., Rush, A.M., Gilson, M.K., Gianneschi, N.C., 2014. Peptides displayed as high density brush polymers resist proteolysis and retain bioactivity. J. Am. Chem. Soc. 136, 15422–15437.

Borchmann, D.E., Tarallo, R., Avendano, S., Falanga, A., Caraerry, T.P., Galdiero, S., Weck, M., 2015. Membranotropic peptide-functionalized poly(lactide)-*graft*-poly(ethylene glycol) brush copolymers for intracellular delivery. Macromolecules 48, 942–949.

Bühler, J., Gietzen, S., Reuter, A., Kappel, C., Fischer, K., Decker, S., Schäffel, D., Koynov, K., Bros, M., Tubbe, I., Grabbe, S., Schmidt, M., 2014. Selective uptake of cylindrical poly(2-oxazoline) brush-antiDEC205 antibody–OVA antigen conjugates into DEC-positive dendritic cells and subsequent T-cell activation. Chem. A Eur. J. 20, 12405–12410.

Burts, A.O., Liao, L., Lu, Y.Y., Tirrell, D.A., Johnson, J.A., 2014. Brush-first and click: efficient synthesis of nanoparticles that degrade and release doxorubicin in response to light. Photochem. Photobiol. 90, 380–385.

Carlson, P.M., Schellinger, J.G., Pahang, J.A., Johnson, R.N., Pun, S.H., 2013. Comparative study of guanidine-based and lysine-based brush copolymers for plasmid delivery. Biomater. Sci. 1, 736–744.

Chang, L., Liu, J., Zhang, J., Deng, L., Dong, A., 2013. pH-sensitive nanoparticles prepared from amphiphilic and biodegradable methoxy poly(ethylene glycol)-*block*-(polycaprolactone-*graft*-poly(methacrylic acid)) for oral drug delivery. Polym. Chem. 4, 1430–1438.

Cheng, C., Qi, K., Khoshdel, E., Wooley, K.L., 2006. Tandem synthesis of core–shell brush copolymers and their transformation to peripherally cross-linked and hollowed nanostructures. J. Am. Chem. Soc. 128, 6808–6809.

Cheng, C., Qi, K., Germack, D.S., Khoshdel, E., Wooley, K.L., 2007. Synthesis of core-crosslinked nanoparticles with controlled cylindrical shape and narrowly-dispersed size via core–shell brush block copolymer templates. Adv. Mater. 19, 2830–2835.

Du, J.-Z., Chen, D.-P., Wang, Y.-C., Xiao, C.-S., Lu, Y.-J., Wang, J., Zhang, G.-Z., 2006. Synthesis and micellization of amphiphilic brush–coil block copolymer based on poly(ε-caprolactone) and PEGylated polyphosphoester. Biomacromolecules 7, 1898–1903.

Du, J.-Z., Tang, L.-Y., Song, W.-J., Shi, Y., Wang, J., 2009. Evaluation of polymeric micelles from brush polymer with poly(ε-caprolactone)-*b*-poly(ethylene glycol) side chains as drug carrier. Biomacromolecules 10, 2169–2174.

Du, X.-J., Wang, J.-L., Liu, W.-W., Yang, J.-X., Sun, C.-Y., Sun, R., Li, H.-J., Shen, S., Luo, Y.-L., Ye, X.-D., Zhu, Y.-H., Yang, X.-Z., Wang, J., 2015. Regulating the surface poly(ethylene glycol) density of polymeric nanoparticles and evaluating its role in drug delivery in vivo. Biomaterials 69, 1–11.

Duncan, R., 2003. The dawning era of polymer therapeutics. Nat. Rev. Drug Discov. 2, 347–360.

Duncan, R., 2006. Polymer conjugates as anticancer nanomedicines. Nat. Rev. Cancer 6, 688–701.

Duncan, R., 2009. Development of HPMA copolymer–anticancer conjugates: clinical experience and lessons learnt. Adv. Drug Deliv. Rev. 61, 1131–1148.

Gao, A.X., Liao, L., Johnson, J.A., 2014. Synthesis of acid-labile PEG and PEG-doxorubicin-conjugate nanoparticles via brush-first ROMP. ACS Macro Lett. 3, 854–857.

Greenwald, R.B., Choe, Y.H., McGuire, J., Conover, C.D., 2003. Effective drug delivery by PEGylated drug conjugates. Adv. Drug Deliv. Rev. 55, 217–250.

Guo, J., Hong, H., Chen, G., Shi, S., Nayak, T.R., Theuer, C.P., Barnhart, T.E., Cai, W., Gong, S., 2014. Theranostic unimolecular micelles based on brush-shaped amphiphilic block copolymers for tumor-targeted drug delivery and positron emission tomography imaging. ACS Appl. Mater. Interf. 6, 21769–21779.

Helmlinger, G., Sckell, A., Dellian, M., Forbes, N.S., Jain, R.K., 2002. Acid production in glycolysis-impaired tumors provides new insights into tumor metabolism. Clin. Cancer Res. 8, 1284–1291.

Hörtz, C., Birke, A., Kaps, L., Decker, S., Wächtersbach, E., Fischer, K., Schuppan, D., Barz, M., Schmidt, M., 2015. Cylindrical brush polymers with polysarcosine side chains: a novel biocompatible carrier for biomedical applications. Macromolecules 48, 2074–2086.

Hu, Y., Darcos, V., Monge, S., Li, S., 2015. Thermo-responsive drug release from self-assembled micelles of brush-like PLA/PEG analogues block copolymers. Int. J. Pharm. 491, 152–161.

Huang, K., Rzayev, J., 2009. Well-defined organic nanotubes from multicomponent bottlebrush copolymers. J. Am. Chem. Soc. 131, 6880–6885.

Jiang, S., Cao, Z., 2010. Ultralow-fouling, functionalizable, and hydrolyzable zwitterionic materials and their derivatives for biological applications. Adv. Mater. 22, 920–932.

Jiang, X., Lok, M.C., Hennink, W.E., 2007. Degradable-brushed pHEMA-pDMAEMA synthesized via ATRP and click chemistry for gene delivery. Bioconjugate Chem. 18, 2077–2084.

Jiang, X., Vogel, E.B., Smith III, M.R., Baker, G.L., 2008. Clickable polyglycolides: tunable synthons for thermoresponsive, degradable polymers. Macromolecules 41, 1937–1944.

Johnson, J.A., Lu, Y.-Y., Burts, A.O., Lim, Y.-H., Finn, M.G., Koberstein, J.T., Turro, N.J., Tirrell, D.A., Grubbs, R.H., 2011. Core-clickable PEG-branch-azide bivalent-bottle-brush polymers by ROMP: grafting-through and clicking-to. J. Am. Chem. Soc. 133, 559–566.

Johnson, J.A., Lu, Y.Y., Burts, A.O., Xia, Y., Durrell, A.C., Tirrell, D.A., Grubbs, R.H., 2010. Drug-loaded, bivalent-bottle-brush polymers by graft-through ROMP. Macromolecules 43, 10326–10335.

Kapon, R., Naim, B., Zbaida, D., Nevo, R., Tsabari, O., Reich, Z., 2010. Permeating the nuclear pore complex. Nucleus 1, 475–480.

Kissel, T., Brich, Z., Bantle, S., Lancranjan, I., Nimmerfall, F., Vit, P., 1991. Parenteral depot-systems on the basis of biodegradable polyesters. J. Control. Release 16, 27–41.

Knop, K., Hoogenboom, R., Fischer, D., Schubert, U.S., 2010. Poly(ethylene glycol) in drug delivery: pros and cons as well as potential alternatives. Angew. Chem. Int. Ed. 49, 6288–6308.

Lane, D.D., Chiu, D.Y., Su, F.Y., Srinivasan, S., Kern, H.B., Press, O.W., Stayton, P.S., Convertine, A.J., 2015a. Well-defined single polymer nanoparticles for the antibody-targeted delivery of chemotherapeutic agents. Polym. Chem. 6, 1286–1299.

Lane, D.D., Su, F.Y., Chiu, D.Y., Srinivasan, S., Wilson, J.T., Ratner, D.M., Stayton, P.S., Convertine, A.J., 2015b. Dynamic intracellular delivery of antibiotics via pH-responsive polymersomes. Polym. Chem. 6, 1255–1266.

Li, C., Gunari, N., Fischer, K., Janshoff, A., Schmidt, M., 2004. New perspectives for the design of molecular actuators: thermally induced collapse of single macromolecules from cylindrical brushes to spheres. Angew. Chem. Int. Ed. 43, 1101–1104.

Li, C., Guo, T., Zhou, D., Hu, Y., Zhou, H., Wang, S., Chen, J., Zhang, Z., 2011. A novel glutathione modified chitosan conjugate for efficient gene delivery. J. Control. Release 154, 177–188.

Li, Z., Ma, J., Cheng, C., Zhang, K., Wooley, K.L., 2010. Synthesis of hetero-grafted amphiphilic diblock molecular brushes and their self-assembly in aqueous medium. Macromolecules 43, 1182–1184.

Li, Y., Nothnagel, J., Kissel, T., 1997. Biodegradable brush-like graft polymers from poly(D,L-lactide) or poly(D,L-lactide-*co*-glycolide) and charge-modified, hydrophilic dextrans as backbone—synthesis, characterization and in vitro degradation properties. Polymer 38, 6197–6206.

Li, Y., Themistou, E., Zou, J., Das, B.P., Tsianou, M., Cheng, C., 2012a. Facile synthesis and visualization of Janus double-brush copolymers. ACS Macro Lett. 1, 52–56.

Li, Y., Volland, C., Kissel, T., 1998. Biodegradable brush-like graft polymers from poly(D,L-lactide) or poly(D,L-lactide-coglycolide) and charge-modified, hydrophilic dextrans as backbone—in-vitro degradation and controlled releases of hydrophilic macromolecules. Polymer 39, 3087–3097.

Li, Y., Zou, J., Das, B.P., Tsianou, M., Cheng, C., 2012b. Well-defined amphiphilic double-brush copolymers and their performance as emulsion surfactants. Macromolecules 45, 4623–4629.

Liao, L., Liu, J., Dreaden, E.C., Morton, S.W., Shopsowitz, K.E., Hammond, P.T., Johnson, J.A., 2014. A convergent synthetic platform for single-nanoparticle combination cancer therapy: ratiometric loading and controlled release of cisplatin, doxorubicin, and camptothecin. J. Am. Chem. Soc. 136, 5896–5899.

Liu, J., Burts, A.O., Li, Y., Zhukhovitskiy, A.V., Ottaviani, M.F., Turro, N.J., Johnson, J.A., 2012. "Brush-first" method for the parallel synthesis of photocleavable, nitroxide-labeled poly(ethylene glycol) star polymers. J. Am. Chem. Soc. 134, 16337–16344.

Liu, X.-Q., Du, J.-Z., Zhang, C.-P., Zhao, F., Yang, X.-Z., Wang, J., 2010. Brush-shaped polycation with poly(ethylenimine)-*b*-poly(ethylene glycol) side chains as highly efficient gene delivery vector. Int. J. Pharm. 392, 118–126.

Liu, J., Gao, A.X., Johnson, J.A., 2013. Particles without a box: brush-first synthesis of photodegradable PEG star polymers under ambient conditions. J. Vis. Exp., e50874.

Luo, Y.-L., Zhang, L.-L., Xu, F., 2012. Synthesis, micellization and caffeine drug release kinetics of novel PBMA-*b*-PNIPAAm block polymer brushes. Chem. Eng. J. 189-190, 431–442.

Mao, J., Gan, Z., 2009. Amphiphilic PEG-*co*-PGL-*g*-PCL copolymer brushes: synthesis, micellization and controlled drug delivery. Macromol. Chem. Phys. 210, 2078–2086.

Müllner, M., Dodds, S.J., Nguyen, T.-H., Senyschyn, D., Porter, C.J.H., Boyd, B.J., Caruso, F., 2015. Size and rigidity of cylindrical polymer brushes dictate long circulating properties in vivo. ACS Nano 9, 1294–1304.

Mecerreyes, D., Miller, R.D., Hedrick, J.L., Detrembleur, C., Jérôme, R., 2000. Ring-opening polymerization of 6-hydroxynon-8-enoic acid lactone: novel biodegradable copolymers containing allyl pendent groups. J. Polym. Sci. A 38, 870–875.

Mura, S., Nicolas, J., Couvreur, P., 2013. Stimuli-responsive nanocarriers for drug delivery. Nat. Mater. 12, 991–1003.

Nguyen, C.T., Tran, T.H., Lu, X., Kasi, R.M., 2014. Self-assembled nanoparticles from thiol functionalized liquid crystalline brush block copolymers for dual encapsulation of doxorubicin and gold nanoparticles. Polym. Chem. 5, 2774–2783.

Parrish, B., Breitenkamp, R.B., Emrick, T., 2005. PEG- and peptide-grafted aliphatic polyesters by click chemistry. J. Am. Chem. Soc. 127, 7404–7410.

Parrish, B., Emrick, T., 2007. Soluble camptothecin derivatives prepared by click cycloaddition chemistry on functional aliphatic polyesters. Bioconjugate Chem. 18, 263–267.

Petros, R.A., DeSimone, J.M., 2010. Strategies in the design of nanoparticles for therapeutic applications. Nat. Rev. Drug Discov. 9, 615–627.

Riva, R., Schmeits, S., Jérôme, C., Jérôme, R., Lecomte, P., 2007. Combination of ring-opening polymerization and "click chemistry": toward functionalization and grafting of poly(ε-caprolactone). Macromolecules 40, 796–803.

Samarajeewa, S., Shrestha, R., Elsabrahy, M., Karawa, A., Li, A., Zentay, R.P., Kostel, J.G., Dorshow, R.B., Wooley, K.L., 2013. In vitro efficacy of paclitaxel-loaded dual-responsive shell cross-linked polymer nanoparticles having orthogonally degradable disulfide cross-linked corona and polyester core domains. Mol. Pharmaceutics 10, 1092–1099.

Schappacher, M., Deffieux, A., 2008. Synthesis of macrocyclic copolymer brushes and their self-assembly into supramolecular tubes. Science 319, 1512–1515.

Sheiko, S.S., Sumerlin, B.S., Matyjaszewski, K., 2008. Cylindrical molecular brushes: synthesis, characterization, and properties. Prog. Polym. Sci. 33, 759–785.

Shi, J., Choi, J.L., Chou, B., Johnson, R.N., Schellinger, J.G., Pun, S.H., 2013a. Effect of polyplex morphology on cellular uptake, intracellular trafficking, and transgene expression. ACS Nano 7, 10612–10620.

Shi, J., Schellinger, J.G., Johnson, R.N., Choi, J.L., Chou, B., Anghel, E.L., Pun, S.H., 2013b. Influence of histidine incorporation on buffer capacity and gene transfection efficiency of HPMA-*co*-oligolysine brush polymers. Biomacromolecules 14, 1961–1970.

Sowers, M.A., McCombs, J.R., Wang, Y., Paletta, J.T., Morton, S.W., Dreaden, E.C., Boska, M.D., Ottaviani, M.F., Hammond, P.T., Rajca, A., Johnson, J.A., 2014. Redox-responsive branched-bottlebrush polymers for in vivo MRI and fluorescence imaging. Nat. Commun. 5, 5460.

Su, Y., Hu, Y., Du, Y., Huang, X., He, J., You, J., Yuan, H., Hu, F., 2015. Redox-responsive polymer–drug conjugates based on doxorubicin and chitosan oligosaccharide-*g*-stearic acid for cancer therapy. Mol. Pharmaceutics 12, 1193–1202.

Tong, R., Cheng, J., 2007. Anticancer polymeric nanomedicines. Polym. Rev. 47, 345–381.

Tong, R., Tang, L., Ma, L., Tu, C., Baumgartner, R., Cheng, J., 2014. Smart chemistry in polymeric nanomedicine. Chem. Soc. Rev. 43, 6982–7012.

Tran, T.-H., Nguyen, C.T., Gonzalez-Fajardo, L., Hargrove, D., Song, D., Deshmukh, P., Mahajan, L., Ndaya, D., Lai, L., Kasi, R.M., Lu, X., 2014. Long circulating self-assembled nanoparticles from cholesterol-containing brush-like block copolymers for improved drug delivery to tumors. Biomacromolecules 15, 4363–4375.

Trollsås, M., Lee, V.Y., Mecerreyes, D., Löwenhielm, P., Möller, M., Miller, R.D., Hedrick, J.L., 2000. Hydrophilic aliphatic polyesters: design, synthesis, and ring-opening polymerization of functional cyclic esters. Macromolecules 33, 4619–4627.

Wang, D., Huan, X., Zhu, L., Liu, J., Qiu, F., Yan, D., Zhu, X., 2012. Salt/pH dual-responsive supramolecular brush copolymer micelles with molecular recognition of nucleobases for drug delivery. RSC Adv. 2, 11953–11962.

Yang, Y.Q., Guo, X.D., Lin, W.J., Zhang, L.J., Zhang, C.Y., Qian, Y., 2012a. Amphiphilic copolymer brush with random pH-sensitive/hydrophobic structure: synthesis and self-assembled micelles for sustained drug delivery. Soft Matter 8, 454–464.

Yang, Y.Q., Lin, W.J., Zhao, B., Wen, X.F., Guo, X.D., Zhang, L.J., 2012b. Synthesis and physicochemical characterization of amphiphilic triblock copolymer brush containing pH-sensitive linkage for oral drug delivery. Langmuir 28, 8251–8259.

Yang, Y.Q., Zheng, L.S., Guo, X.D., Qian, Y., Zhang, L.J., 2011. pH-sensitive micelles self-assembled from amphiphilic copolymer brush for delivery of poorly water-soluble drugs. Biomacromolecules 12, 116–122.

Yatvin, M., Weinstein, J., Dennis, W., Blumenthal, R., 1978. Design of liposomes for enhanced local release of drugs by hyperthermia. Science 202, 1290–1293.

Yu, J.-H., Quan, J.-S., Kwon, J.-T., Xu, C.-X., Sun, B., Jiang, H.-L., Nah, J.-W., Kim, E.-M., Jeong, H.-J., Cho, M.-H., Cho, C.-S., 2009. Fabrication of a novel core–shell gene delivery system based on a brush-like polycation of α, β-poly(L-aspartate-graft-PEI). Pharm. Res. 26, 2152–2163.

Yu, Y., Chen, C.-K., Law, W.-C., Mok, J., Zou, J., Prasad, P.N., Cheng, C., 2013. Well-defined degradable brush polymer–drug conjugates for sustained delivery of paclitaxel. Mol. Pharmaceutics 10, 867–874.

Yu, Y., Chen, C.-K., Law, W.-C., Sun, H., Prasad, P.N., Cheng, C., 2015. A degradable brush polymer–drug conjugate for pH-responsive release of doxorubicin. Polym. Chem. 6, 953–961.

Yu, Y., Zou, J., Yu, L., Ji, W., Li, Y., Law, W.-C., Cheng, C., 2011. Functional polylactide-*g*-paclitaxel–poly(ethylene glycol) by azide–alkyne click chemistry. Macromolecules 44, 4793–4800.

Yuan, Y., Wang, Z., Cai, P., Liu, J., Liao, L.-D., Hong, M., Chen, X., Thakor, N., Liu, B., 2015. Conjugated polymer and drug co-encapsulated nanoparticles for chemo- and photo-thermal combination therapy with two-photon regulated fast drug release. Nanoscale 7, 3067–3076.

Zhai, X., Wang, W., Wang, C., Wang, Q., Liu, W., 2012. PDMAEMA-*b*-polysulfobetaine brushes-modified 3-polylysine as a serum-resistant vector for highly efficient gene delivery. J. Mater. Chem. 22, 23576–23586.

Zhang, F., Zhang, S., Pollack, S.F., Li, R., Gonzalez, A.M., Fan, J., Zou, J., Leiniger, S.E., Pavia-Sanders, A., Johnson, R., Nelson, L.D., Raymond, J.E., Elsabrahy, M., Hughes, D.M.P., Lenox, M.W., Gustafson, T.P., Wooley, K.L., 2015. Improving paclitaxel delivery: in vitro and in vivo characterization of PEGylated polyphosphoester-based nanocarriers. J. Am. Chem. Soc. 137, 2056–2066.

Zhang, M., Xiong, Q., Wang, Y., Zhang, Z., Shen, W., Liu, L., Wang, Q., Zhang, Q., 2014. A well-defined coil-comb polycationic brush with "star polymers" as side chains for gene delivery. Polym. Chem. 5, 4670–4678.

Zhao, P., Liu, L., Feng, X., Wang, C., Shuai, X., Chen, Y., 2012. Molecular nanoworm with PCL core and PEO shell as a non-spherical carrier for drug delivery. Macromol. Rapid Commun. 33, 1351–1355.

Zolotarskaya, O.Y., Wagner, A.F., Beckta, J.M., Valerie, K., Wynne, K.J., Yang, H., 2012. Synthesis of water-soluble camptothecin–polyoxetane conjugates via click chemistry. Mol. Pharmaceutics 9, 3403–3408.

Zolotarskaya, O.Y., Yuan, Q., Wynne, K.J., Yang, H., 2013. Synthesis and characterization of clickable cytocompatible poly(ethylene glycol)-grafted polyoxetane brush polymers. Macromolecules 46, 63–71.
Zou, J., Jafr, G., Themistou, E., Yap, Y., Wintrob, Z.A.P., Alexandridis, P., Ceacareanu, A.C., Cheng, C., 2011. pH-sensitive brush polymer–drug conjugates by ring-opening metathesis copolymerization. Chem. Comm. 47, 4493–4495.
Zou, J., Yu, Y., Li, Y., Ji, W., Chen, C.-K., Law, W.-C., Prasad, P.N., Cheng, C., 2015. Well-defined diblock brush polymer–drug conjugates for sustained delivery of paclitaxel. Biomater. Sci. 3, 1078–1084.

CHAPTER

9 DRUG DELIVERY SYSTEMS BASED ON TITANIA NANOSTRUCTURES

Magdalena Jarosz, Joanna Kapusta-Kołodziej, Anna Pawlik, Karolina Syrek, Grzegorz D. Sulka
Jagiellonian University in Krakow, Krakow, Poland

CHAPTER OUTLINE

1 INTRODUCTION

Nowadays, the most commonly used drug administration paths are oral and intravenous. They provide a very high initial level of drug in the bloodstream; however, the effective dosage drastically decreases below the therapeutic window after a very short time. This effect is called a peak-and-valley effect, and may cause some serious side effects. First of all, drug concentration after the administration may reach a toxic level. Afterward, it falls below the therapeutic window causing the whole treatment to be ineffective. Another drawback of the aforementioned ways of drug delivery is the fact that most of the currently used antibiotics and other therapeutic molecules are inactivated in the gastrointestinal system (e.g., liver, kidney, or stomach). That is why the effectiveness of the delivery of drug molecules to the targeted organ is very poor (Langer, 2001). Therefore, drugs must be frequently delivered via injections or intravenous infusions, both very uncomfortable for patients. In order to overcome those obstacles, new, more efficient, localized, and selective delivery systems are needed. It is especially important when it comes to the different kinds of implants.

An implant is a medically inert device that is used to replace or support a fraction of the body in order to help recovery. Implantable devices may be divided into two groups: temporal or permanent. The first group consists of patches, capsules, or microchips that possess a time-limited functionality because of either therapeutic or technical reasons. Permanent implants, such as bone implants, stents, or dental implants are supposed to function in the body much longer without the loss of their properties.

New achievements in biomedical science along with the development in creating new materials and technologies have influenced the expansion of controlled drug delivery systems (Langer and Peppas, 2003). As mentioned before, the main purpose of controlled drug delivery is to provide the exact amount of the administered drug into the specific sites inside the human body, and to maintain prolonged release for the maximum therapeutic effect.

The newest developments in drug delivery systems are based on inorganic nanomaterial platforms. Among different compounds, titania nanostructures have recently gained a huge interest in the area of novel drug delivery systems, especially in case of implantable devices. It is mainly because of very good biocompatibility of nanostructured TIO_2, as well as its antibacterial and mechanical properties (Peng et al., 2009; Zhang et al., 2014).

In this chapter, we will focus on different titania nanostructures used as drug delivery systems, such as nanotubular and nanoporous TiO_2, titania nanoparticles, and nanowhiskers.

2 NANOSTRUCTURED TIO_2 USED FOR DRUG DELIVERY SYSTEMS

It is widely known that many properties of nanosized titania may significantly improve the performance of promising drug delivery systems. The essential factor for tailoring nanostructured TiO_2 properties is its fabrication technique. To date, significant progress has been achieved in the fabrication of TiO_2 nanomaterials. In most cases, nanostructured TiO_2 materials can be synthesized by either dry or wet processes. Nowadays a variety of methods, such as sol–gel (Chao et al., 2012; Ganesh et al., 2012; Heredia-Cervera et al., 2009; Jiang et al., 2012; López et al., 2006), hydrothermal and solvothermal processes (Huang et al., 2014; Mund et al., 2014), or electrochemical anodization (Ghicov and Schmuki, 2009; Sulka et al., 2010), have been developed to control the size, morphology, and uniformity of TiO_2 nanostructures. The following sections further elaborate on some of the aforementioned preparation procedure to obtain TiO_2 nanostructures.

2.1 SOL–GEL PROCESS

One of the most frequently used methods for fabrication of nanostructured TiO_2 with diverse morphologies is the sol–gel process. Typically, in the sol–gel route, a colloidal suspension or sol is synthesized by the hydrolysis and polymerization reactions of the precursors, which are usually inorganic metal salts or metal organic compounds, such as metal alkoxides. In the next step, the liquid sol is converted into gel phase by the complete polymerization and evaporation of the solvent. Moreover, this procedure provides the ability to obtain thin films on diverse substrates by deposition techniques, such as spin-coating, dip-coating, spray pyrolysis, electrophoretic, and a template assisted deposition onto a conducting glass substrate [e.g., fluorine-doped tin oxide (FTO) or indium tin oxide (ITO) glasses] or other preferred surfaces.

Fabrication of nanostructured TiO_2 via sol–gel technique can be accomplished by hydrolysis of a titanium precursor, which is usually achieved by acid-catalyzed hydrolysis of titanium(IV) alkoxide [e.g., Ti(IV) isopropoxide] (Heredia-Cervera et al., 2009; López et al., 2006), titanium tetraisopropoxide TTIP (Chen, 2012) followed by a condensation process. As a result, TiO_2 microtubes (López et al., 2006), mesoporous ultrathin TiO_2 films (Chen, 2012) or reservoirs containing aggregated titania nanoparticles with macro- and mesopores (Heredia-Cervera et al., 2009) may be formed. An example of such structures is shown in Fig. 9.1.

Liu et al. (2011) synthesized TiO_2 microtubes with a very large specific surface area using the sol–gel method followed by a hydrothermal treatment of samples. They used glass fibers as solid templates for the fabrication of hierarchical nano-/microstructured TiO_2. Tetrabutyltitanate was used as a

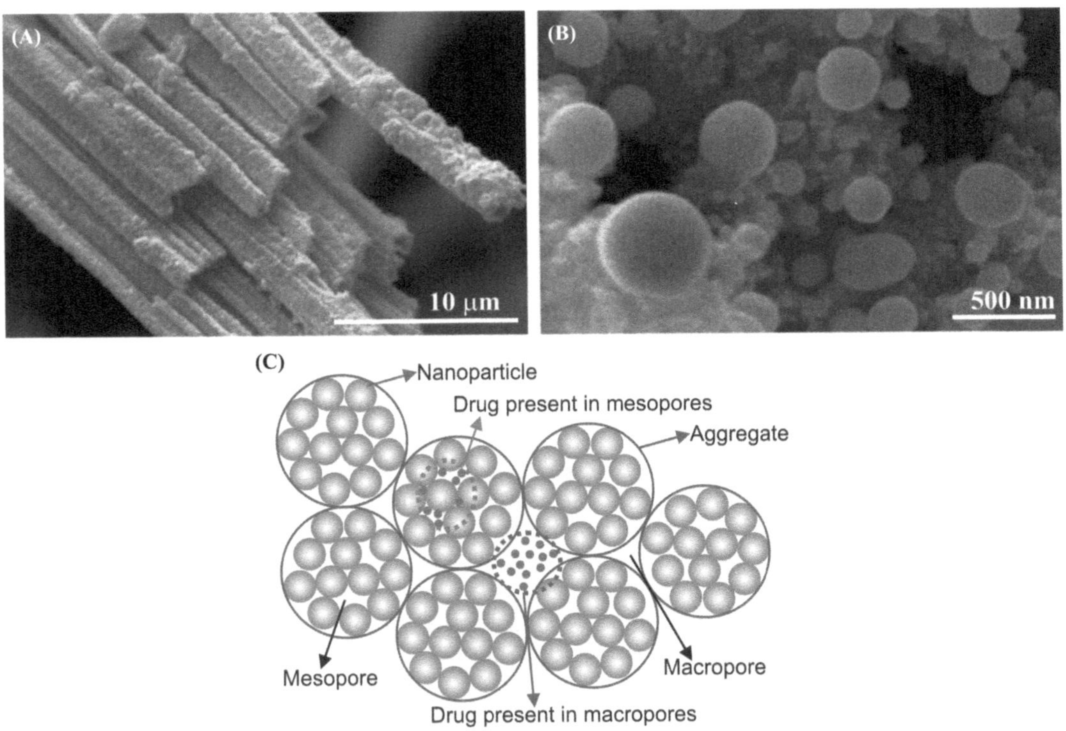

FIGURE 9.1 SEM Micrographs of TiO_2-Microtubes Synthesized via Sol–Gel Method Using Tetrahydrofuran (THF) Solution of Ti(IV) Isopropoxide as a Precursor and Calcined at 500°C

(A) Titania microtubes formed by the agglomeration of titania particles. Magnification 1000×. (B) Nonspherical shape of the titania particles forming microtubes. Magnification 50,000× . (C) Scheme of the morphology of the TiO_2 reservoirs prepared using sol–gel method in an aqueous solution using titanium(IV) isopropoxide and acetic acid. Nanoparticles, aggregates and macro- and mesopores along with the entrapped drug were marked.

Part A and B reprinted with permission from López, T., Ortiz-Islas, E., Manjarrez, J., Reinoso, F.R., Sepulveda, A., Gonzalez, R.D., 2006. Biocompatible titania microtubes formed by nanoparticles and its application in the drug delivery of valproic acid. Opt. Mater. 29, 70–74.

precursor for the sol–gel process. First, a TiO_2 semitransparent sol was formed, in which glass fibers were soaked, and then annealed at 100°C for 1 h. The procedure was repeated four times in order to obtain the TiO_2 gel coating on glass fibers. Afterward, template-supported structures were put in the autoclave containing a $Ti(OBu)_4$–HCl–H_2O solution. For synthesizing hierarchical TiO_2 rod arrays, they were treated hydrothermally at 150°C for different periods of time. Finally, the glass fiber support was removed by dissolution in hydrofluoric acid. As a result, a double-layer structured TiO_2 microtubes were obtained. Microtubes were 8 μm in diameter and walls were 2–3 μm thick. The outer layer contained a thick nanorod-like TiO_2 layer, whereas the inner layer was formed from the TiO_2 gel. It was found that the nanorod crystal TiO_2 array was formed from the flower-like clusters. Moreover, each rod composed of parallel arranged nanowires with the diameter of 10 nm. Such hierarchical structures possess a very high specific surface area, which enables them to be applied as a more effective catalyst, sensor, sorbent, or drug delivery system.

Another example of the template-supported synthesis of TiO_2 porous network by using a sol–gel method was described by McMaster et al. (2012). In this work, type I collagen gel and titanium(IV) isopropoxide (TIP) were used as a template and precursor, respectively. TiO_2-collagen hybrids were obtained by soaking a collagen gel in a TIP solution in isopropanol, followed by hydrolysis and condensation reactions. In order to remove the template and crystallize the TiO_2 network, a heat treatment in flowing air at 450°C was applied. Prior to the calcination process, a part of samples underwent a solvothermal treatment. The samples were placed in an autoclave with ethanol and water and heated for 16 h at 160°C. As-prepared samples were assessed for their biocompatibility and drug delivery potential. TiO_2 networks were immersed in a simulated body fluid (SBF) solution in order to deposit apatite particles. It was found out that apatite clusters grew on all of the samples, proving its biocompatibility and bioactivity in in vitro conditions. This result indicated that TiO_2 networks may bond to the living tissue also in an in vivo environment. Owing to the high surface area and porous structure, the synthesized TiO_2 network was used as a drug delivery system. Ibuprofen was used as a model drug for release tests. It was shown that the release of the drug from the porous network consists of two steps: an initial burst release, followed by a sustained release of the drug. A Higuchi model was ascribed to the release profiles, indicating that the drug release is consistent with a Fickian diffusion model, however, the initial stage of the release was non-Fickian because of the unbounded drug molecules to the TiO_2 network. The drug release tests showed also that the adsorption and release of the ibuprofen depend on both the surface area, and size and volume of mesopores. A higher amount of the loaded and released drug was obtained for the samples treated solvothermally, where the specific surface area increase was observed. Therefore, biocompatible TiO_2 porous networks may also be used for storage and release of drug molecules in various biomedical applications.

More recently a new strategy was adopted by Wang et al. (2013) for a synthesis of mesoporous titanium zirconium oxide nanospheres for application to the drug delivery systems. They used a sol–gel process to synthesize TiZr oxide nanospheres. As precursors, zirconium propoxide and titanium isopropoxide with different Ti to Zr atomic ratios were used. As-prepared nanospheres were then mixed with ethanol and water, and underwent a solvothermal treatment for 16 h at 160°C. Afterward, TiZr nanospheres were used as carriers of three model drugs: ibuprofen, dexamethasone, and erythromycin. Drug release profiles of all three drug molecules indicated sustained release with the initial burst release during first 24 h. The kinetics of the release was described using the Higuchi model, but with the initial rapid release. Careful inspection revealed that such structures exhibited a very high stability and biocompatibility in in vitro conditions, and therefore may be used as potential drug delivery systems in the future.

2.2 HYDROTHERMAL/SOLVOTHERMAL METHOD

Another common technique used for preparation of variety of TiO_2 nanomaterials is a hydrothermal (Mund et al., 2014) and solvothermal (Huang et al., 2014) route. In this procedure, the synthesis is generally conducted in aqueous/organic solutions that contain required material precursors inside temperature-controlled steel pressure vessels called autoclaves. The reaction temperature can be elevated above the boiling point of the water/organic solvent, reaching the pressure of vapor saturation. The pressure produced in the autoclave is affected by the temperature and amount of solution. The solvothermal procedure is almost the same as a hydrothermal method, but in the first technique solvents used for synthesis are nonaqueous, while water is used in hydrothermal processes. Therefore, the temperature of solvothermal synthesis can be increased much higher than that in the hydrothermal method. However, the solvothermal route allows better control over the size, morphology, and crystallinity of TiO_2 nanomaterials, compared to hydrothermal methods. As a result, TiO_2 nanoparticles and nanotubes may be synthesized and further used as drug delivery systems. An example of such a structure is presented in Fig. 9.2.

In the work of Mund et al. (2014), a comparison between the synthesis of TiO_2 nanoparticles by using a conventional drying and propanol drying method was reported. The presented results revealed that nanoparticles prepared by an air-drying method have a broad range of size; however, the use of propanol ensured that over 50% of the nanoparticles had the hydrodynamic diameter less than 100 nm. Moreover, the nanostructures were very smooth and highly spherical. In further steps, both kinds of nanoparticles were used to evaluate their drug delivery properties. TiO_2 nanoparticles were loaded with paclitaxel (PTX), the drug that was proved to have therapeutic efficacy in the treatment of a number of types of cancer including: breast, ovarian, lung, and pancreatic cancers. The drug was loaded to the nanoparticles by pouring the PTX solution into the aqueous suspension of TiO_2 nanoparticles. As a result, PTX–TiO_2 nanocomposites were formed. It was found out that the loading and encapsulation efficiencies are slightly increased for the samples dried in propanol conditions. The release of the PTX was conducted under different pH values. The release profiles showed the two-stage release kinetics. At the beginning, the release of the drug was rapid (the burst effect), and this fact was observed especially for the low pH values (6.0 and 5.2). Then, the release proceeded much slower, especially in a

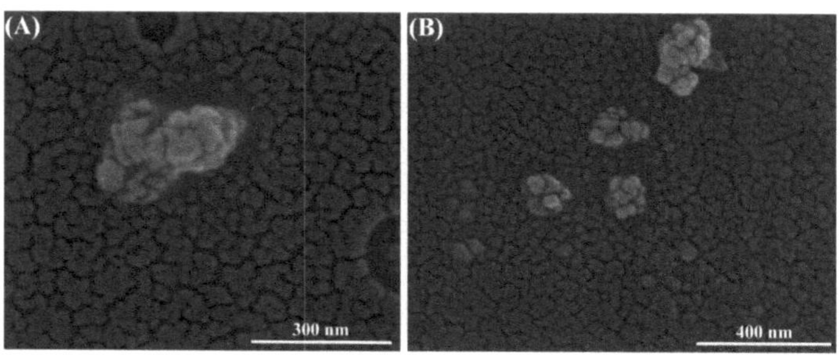

FIGURE 9.2 SEM Images of TiO_2 Nanoparticles and PTX-TiO_2 Nanocomposites Synthesized via Hydrothermal Method Using TiO_2 Powder as a Precursor

(A) TiO_2 nanoparticles; (B) PTX-TiO_2 nanocomposites. After a freeze-drying step, samples were annealed at 700°C for 2 h in order to obtain nanoparticles.

Reprinted with permission from Mund, R., Panda, N., Nimesh, S., Biswas, A., 2014. Novel titanium oxide nanoparticles for effective delivery of paclitaxel to human breast cancer cells. J. Nanopart. Res. 16, 2739(1)–2739(12).

buffer solution with a pH of 5.2. At the physiological pH conditions, the release of the drug from TiO_2 nanoparticles was significantly slower than that observed for acidic pH values. It was ascribed to the fact that in an acidic environment the PTX–TiO_2 nanocomposite dissociates, releasing drug molecules into the medium. In vivo studies of anticancer activity of prepared nanocomposites showed that cancer cells viability drastically decreases when PTX–TiO_2 nanocomposites were used, proving their better anticancer efficacy. The results presented in this work showed also that by changing the preparation conditions of the hydrothermal method, the properties of the synthesized nanostructures can be significantly enhanced.

To sum up, it is well known that properties of the obtained TiO_2 nanomaterials, such as: morphology, surface area, porosity, and crystal phase strongly depend on the reaction conditions, including the temperature, reaction time, type of solvent and precursor, solution pH, drying conditions, and posttreatment process. The sol–gel process is a very attractive method for the design of nanostructured titania because it allows to control the textural properties of the final product, simply by choosing the proper synthetic conditions. Hydrothermal and solvothermal methods also offer a wide range of possibilities in order to achieve different types of TiO_2 nanostructures. Moreover, by combining the sol–gel method with a hydrothermal or solvothermal treatment, samples with a significantly increased specific surface area may be obtained. Those facts become particularly important for the possible use of nanostructured carriers in the drug delivery systems.

2.3 ANODIZATION PROCESS

One of the most promising and cost-effective methods to synthesize TiO_2 nanostructures is anodization. The process was first applied by Zwilling et al. (1999), who obtained nanotubular titanium oxide using an electrolyte based on a mixture of chromic and hydrofluoric acids. Anodization is an electrochemical oxidation of Ti foil that leads to nanotubular or nanoporous TiO_2 layers (Ghicov and Schmuki, 2009; Sulka et al., 2010). The obtained oxide layers are formed as a result of the two competitive processes: the growth of compact TiO_2 layer and the partial dissolution of the formed oxide structure by fluoride ions. The method allows to control the pore/tube size, geometry, and thickness of the obtained structure (Moseke et al., 2012). The tube diameters depend on the applied voltage, while the length is determined by the charge passed (consequently duration of the process), and the type of the electrolyte (Peng et al., 2009; Moseke et al., 2012). Anodization of Ti can be carried out in both aqueous and nonaqueous electrolytes. Geometrical parameters of anodic TiO_2 nanostructures vary in length from 300 nm even up to 50 μm, while the nanotube diameter can be in the range of 30–150 nm (Aninwene et al., 2008; Aw et al., 2011a; Chennell et al., 2013; Popat et al., 2007a; Shokuhfar et al., 2013; Xiao et al., 2009; Yao and Webster, 2009).

The nanostructured TiO_2 obtained via anodization in the aqueous and organic electrolytes are presented in Figs. 9.3 and 9.4, respectively.

If the voltage is periodically changed during the anodization step, the obtained TiO_2 nanotubes possess a periodically modulated tube diameter (Gulati et al., 2015). Gulati et al. (2015) performed the anodization, which included three steps: (1) constant potential step—100 V, 10 min; (2) step with periodically changed potential—drop from 100 to 60 V and ramp from 60 to 100 V, repeated 15 times; and (3) constant potential step—100 V, 10 min.

In addition, the water content in the anodizing electrolyte influences the obtained TiO_2 structure. By using the electrolyte based on ethylene glycol with fluoride ions (NH_4F) and small amount of water (1.79 wt.%), Sulka et al. (2010, 2013) obtained nanoporous anodic titanium oxide (ATO) layers on Ti

FIGURE 9.3 SEM Images of TiO_2 Nanotubes Prepared by Anodization in the Aqueous Electrolyte Containing 0.5 vol.% of HF

Anodization was conducted under the constant voltage of 20 V for 45 min. (A) Cross-sectional view; (B–C) top views.

Reprinted with permission from Popat, K.C., Eltgroth, M., LaTempa, T.J., Grimes, C.A., Desai, T.A., 2007a. Titania nanotubes: a novel platform for drug-eluting coatings for medical implants? Small 3, 1878–1881.

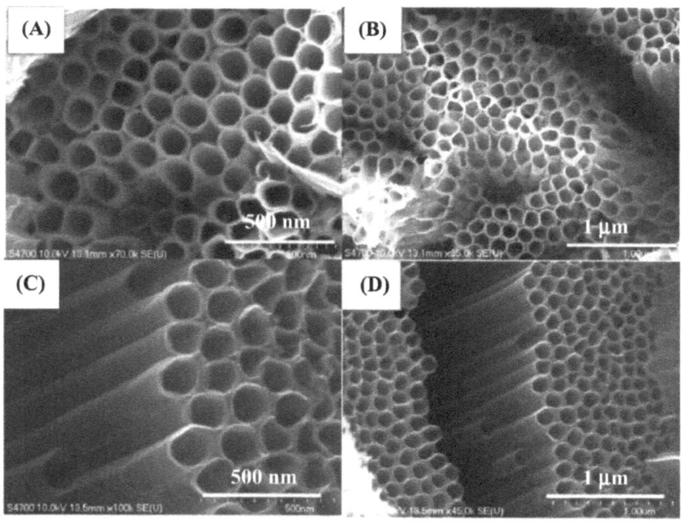

FIGURE 9.4 SEM Images of TiO_2 Nanotubes Prepared by Anodization in an Electrolyte Based on Ethylene Glycol Solution Containing 0.2 wt.% NH_4F and 1 mL of Water Under a Constant Voltage 60 V for 1 h

(A–B) Top views of the as-prepared TiO_2 nanotubes; (C–D) top views of the annealed in air at 450°C for 3 h TiO_2 nanotubes.

Reprinted with permission from Shokuhfar, T., Sinha-Ray, S. Sukotjo, C., Yarin, A.L., 2013. Intercalation of anti-inflammatory drug molecules within TiO_2 nanotubes. RSC Adv. 3, 17380–17386.

foil. Applying different voltages they obtained ATO structures with various pore diameters (60–80 nm) and oxide thicknesses (1.3–7.3 μm). Fig. 9.5 shows the top, bottom, and side view of the synthesized ATO layer.

The as-prepared nanotubes and nanoporous TiO_2 layers are usually amorphous. A number of studies have indicated that the amorphous structure of TiO_2 can be changed to anatase or rutile by annealing samples at different temperatures, namely 450–500 or above 600°C, respectively. The main problem

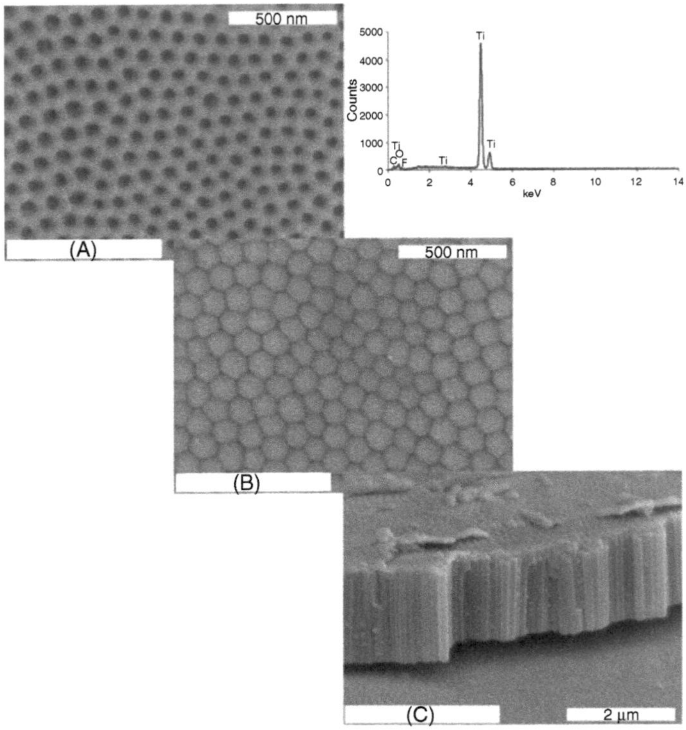

FIGURE 9.5

(A) Top, (B) bottom, and (C) side views of anodic TiO_2 with an EDX spectrum of the ATO surface obtained in an ethylene glycol solution containing NH_4F (0.38 wt.%) and H_2O (1.79 wt.%) at 40 V and 20°C. The three-step anodizing procedure was used. After the first and second anodizations the oxide layers were removed, and finally the third anodization was performed. The duration of the third step was 10 min.

Reprinted with permission from Sulka, G.D., Kapusta-Kołodziej, J., Brzózka, A., Jaskuła, M., 2010. Fabrication of nanoporous TiO_2 by electrochemical anodization. Electrochim. Acta 55, 4359–4367.

encountered during annealing of nanotubular titanium oxide layers is the fact that at the heating temperature above 700°C, nanotubes and nanopores start to collapse, and therefore can no longer be used as drug reservoirs (Jarosz et al., 2015; Yao and Webster, 2009). Changing crystalline structure of TiO_2 along with the anodization conditions may significantly alter the drug release profiles.

Titanium dioxide nanotubes can be also obtained on titanium wires. Gulati et al. (2011) reported the formation of the nanostructured TiO_2 on Ti wires via a two-step anodization process. The anodizing conditions for both steps were as follows: potential—100 V; temperature—20°C; electrolyte composition—3 wt.% H_2O; and 0.3 wt.% NH_4F in ethylene glycol. The first anodizing step lasted 1 h, whereas the duration of the second step was 1–2 h. Although the anodization conditions were similar to ones applied for the titanium foil, the created layers were thicker (70 ± 2 μm) and the pore diameters were larger (170 ± 10 nm) (Aw et al., 2011b; Gulati et al., 2011, 2012a).

Recent studies indicated that such nanostructured Ti wires may be used in orthopedics as drug-releasing implants (Aw et al., 2012c) and as an alternative delivery system of chemotherapeutic agents to brain tumors (Gulati et al., 2012b). Aw et al. tested the distribution of drugs from the titania nanotube (TNT) layer on Ti wire in an ex vivo environment. They used a Zetos bone bioreactor developed by David Jones from the Department of Experimental Orthopaedics and Biomechanics, Philipps University of Marburg in Germany, and Everett Smith from the Medical Sciences Center, Madison, in the United States. This three-dimensional (3D) bone reactor allows use of discs of cancellous bone for at least 3 weeks by providing a continuous perfusion with a medium and a contact with the strains that are physiologically relevant (David et al., 2008; Endres et al., 2009). The TNT-Ti wires with rhodamine B, a fluorescent dye used as a model drug molecule, were placed inside the bone bioreactor. A continuous perfusion of the culture medium was provided. Titania nanotubes on the Ti wire were obtained via anodization process previously described elsewhere (Gulati et al., 2011). Rhodamine B was loaded inside the nanotubes by immersing wires in a water solution of the drug for 1–3 days. It is worth mentioning that the wires were rotated every 2–4 h in order to provide an even distribution of the drug. Both in vitro and ex vivo drug releases were tested in detail. The in vitro studies showed a two-phase release of rhodamine B. First, the burst release during the first 6 h was observed, followed by a slow and constant release of the remaining drug. First-order kinetics was ascribed to the burst release, and then a diffusion process took place. Authors showed also that the best-fitting kinetic model was the Higuchi model, followed by a zero-order release, which confirmed the proper characteristics of the drug-releasing titania nanotube arrays on Ti surfaces. Due to the fact that the bone structure is significantly more complex, and in vitro models cannot imitate the conditions that are present in the living organism, the release of the drug from TNT-Ti wires was tested ex vivo. The Ti wires with titania nanotubes enriched with model drug molecules were put inside bovine trabecular bone cores. Using bioluminescence, the drug release from the nanotubes to the bone core was monitored. It was shown that the drug concentration changed considerably in all of the directions across the bone (x, y, and z). The results showed also that at the beginning of the process we can observe the diffusion of the drug from the TNT-Ti wire, followed by the diffusion of rhodamine into the bone. The release of the model drug was described by the burst effect (confirming the in vitro studies). On the other hand, the distribution of the drug throughout the bone is a very slow process, depending on the bone structure and the flow rate of the culture medium.

One of the most remarkable accomplishments of application of the nanostructured titanium oxide as a drug delivery system was proposed recently by Gulati et al. (2012b). They used the Ti wire with titania nanotubes as a drug delivery system for chemotherapeutics in brain tumors therapy (Gulati et al., 2012b). The main problem concerning delivery of drugs to the brain tumor is the blood–brain barrier that blocks most of the active molecules to pass to the brain. The approach that is commonly used nowadays, which employs chemically modified drugs for overcoming the blood–brain barrier, is not fully successful. Therefore, local drug delivery systems are a promising alternative. Using TNT-Ti wires in brain surgeries may be beneficial mainly because of the simplicity of its implementation. In this work authors used two model drugs typically administrated during central nervous system diseases: a water-soluble neurotransmitter agent dopamine (DOPA) used to treat Parkinson's and Alzheimer diseases, for example, and an insoluble-in-water anticancer drug, doxorubicin (DOXO). The drugs were loaded inside the nanotubes using an immersing method. As a solvent for DOPA and DOXO, water and toluene were used, respectively. For both drugs, two-phase release profiles were observed. The burst release during the first 6 h of the experiment was followed by a slow and constant release of the remaining drug. During the burst stage 60% of DOPA and 25% of DOXO was released. The observed differences were ascribed to the different solubility of drugs in a buffer solution used as a medium in release tests and the disparity in the size of molecules. Slightly different release profiles were also observed for the second stage of the release process. The second release phase was described as a diffusion process and it is strictly related with a high surface area of nanotubes along with the long capillary structures. Therefore, the size and solubility of the drugs are also important factors to be considered during drug release studies. The kinetic model that best fitted to the release profiles was the Higuchi model followed by the zero-order release. Such behavior is desired when a pharmaceutical dosage of the drug must be delivered. This chapter proved that titanium implants with nanotubular or nanoporous TiO_2 layers may be used as drug delivery systems not only in bone implants but also as carriers of chemotherapeutics in brain tumor treatments.

Recently, the anodization of titanium alloys have attracted attention of scientists (Zhang et al., 2014). The nanoporous titanium dioxide layers on Ti alloys can be fabricated employing electrochemical techniques. The anodizing conditions are similar to those applied to pure Ti (Hamlekhan et al., 2015; Zhang et al., 2014). The anodization conditions and the geometrical parameters of the resulting TiO_2 nanotubes on titanium alloy are summarized in Table 9.1.

Recently, Doadrio et al. (2015) used the anodized Ti6Al4V alloy as a drug carrier for ibuprofen. They obtained three different morphologies of nanotubes by using different anodizing electrolytes. A mixture of sulfuric and hydrofluoric acids were used to obtain TiO_2 nanotubes (TNT). On the other hand, high-aspect-ratio titania (TNT+) nanotubes were synthesized using an electrolyte containing $NH_4H_2PO_4$ and NH_4F. In both cases the anodization process was carried out at 20 V for 1 h. The third type of TiO_2 structure, bottle-shaped titania nanotubes (TNTb) were obtained in the same electrolyte as TNT+ but two different voltage steps (10 V for 30 min and 20 V for 1 h) were applied. Ibuprofen as a model drug molecule was loaded inside nanotubes by soaking the samples in the *n*-pentane drug solution with the continuous stirring for 2 days, followed by drying process at 37°C for 24 h. The release studies showed that regardless of the shape and length of the obtained TiO_2 nanostructures, the release kinetics of ibuprofen was very similar to each other. Only for TNTb structures, a slower release was observed, probably because of the differences in the pore diameters at the bottom and top of the titania layers. The release kinetics for all tested samples showed a zero-order kinetics, so the most desired release profile when it comes to therapeutic application.

Table 9.1 Anodization Conditions and Nanotube Geometrical Parameters for TiO_2 Layers Formed on Different Alloys

Alloy	Electrolyte Composition	Anodizing Conditions	TiO_2 Nanotubes Geometrical Parameters
Ti90Al6V4 Hamlekhan et al. (2015)	4 wt.% H_2O, 0.2 wt.% NH_4F, ethylene glycol	Voltage: 60, 70, 90, and 120 V; time: 15, 30, 120 min	Length: 400 nm–70 μm; diameter: 60–170 nm
β-type Ti25Nb3Mo2Sn3Zr (TLM) Zhang et al. (2014)	1 wt.% H_2O, 0.5 wt.% NH_4F, ethylene glycol	Voltage: 30 V; time: one-step process: 1 h; two-step process: 4 h/10 min	Length: 470 nm; diameter: 50 nm
Ti4Zr22Nb2Sn Liang et al. (2011)	0.31 M NH_4F, water/glycerol (50:50 vol.%)	Voltage: 10–56 V; time: 30, 60, 120, 180 min	Length: 2.1–4.1 μm

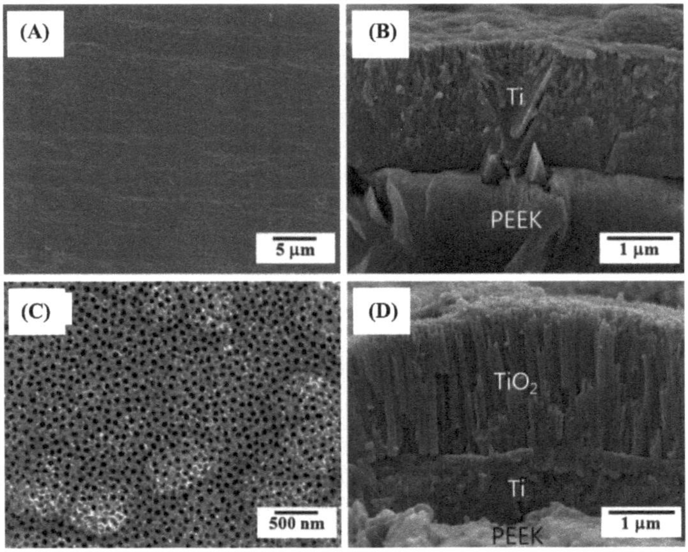

FIGURE 9.6 SEM Images of Ti-Coated Polyetheretherketone, PEEK

(A, top view; B, cross-sectional view) and nanoporous anodic TiO_2 layers (C, top view; D, cross-sectional view).

Reprinted with permission from Han, C.-M., Jang, T.-S., Kim, H.-E., Koh, Y.-H., 2013. Creation of nanoporous TiO_2 surface onto polyetheretherketone for effective immobilization and delivery of bone morphogenetic protein. J. Biomed. Mater. Res. A 102, 793–800.

It is possible to fabricate nanoporous TiO_2 layers on Ti-coated substrates. Han et al. (2013) reported the anodization of Ti-coated polyetheretherketone (PEEK). Before the anodization, PEEK substrate was coated with the Ti layer using e-beam evaporation. The anodizing process was carried out under the similar conditions as described previously (ethylene glycol-based electrolyte with 2 wt.% H_2O and 0.25 wt.% NH_4F, potential—60 V; duration—30 min). The SEM images of as-prepared nanoporous TiO_2 layers are presented in Fig. 9.6.

3 DRUG LOADING

Molecules of various size, solubility, and charge can be loaded inside TiO_2 nanotubes (Hu et al., 2012; Hamlekhan et al., 2015; Popat et al., 2007a). The most popular method used for drugs loading in porous matrix, like nanoporous/nanotubular titanium oxide, is physical adsorption (Aninwene et al., 2008). In general, there are two ways used for loading pharmaceuticals inside TiO_2 nanostructures. The first approach involves the soaking of titania samples in the drug solution (Chennell et al., 2013). Alternatively, a proper volume of drug/protein solution is pipetted onto the TiO_2 surface once (Hu et al., 2012) or repeatedly (Gulati et al., 2011; Popat et al., 2007a) to load an appropriate amount of the compound. The process can be carried out under vacuum conditions in order to allow the drug solution to penetrate the nanotubes more effectively (Moseke et al., 2012; Zhang et al., 2014).

Aninwene et al. (2008) proposed the alternative approach for drug loading, which involves the use of the SBF. The method provides the formation of calcium phosphate crystals on the TiO_2 surface after soaking the porous template in SBF. Such coating is supposed to provide titania nanotubes with an optimal surface tension that in turn enables a prolonged release of the drug from the pores (Aninwene et al., 2008).

The important parameter that describes a drug loading process is loading efficiency (Popat et al., 2007a; Xiao et al., 2009). This parameter can be calculated according to Eq. (9.1) (Xiao et al., 2009):

$$\eta = \frac{\rho_0 - \rho_r}{\rho_0} \cdot 100\% \tag{9.1}$$

where η is the loading efficiency, ρ_0—the amount of the drug originally loaded in the nanopores/nanotubes, and ρ_r—the amount of the drug in the solution which was rinsed after the loading process (Xiao et al., 2009). Alternatively, the loading efficiency can be calculated using the similar formula if the drug concentration in the original and rinse solutions are known (Popat et al., 2007a).

The amount of drug loaded in nanopores or nanotubes can be determined using physicochemical methods, such as thermo-gravimetric analysis (TGA) (Gulati et al., 2011, 2012a) or Fourier-transform infrared (FTIR) spectroscopy (López et al., 2007; Zhang et al., 2014).

It was found that TiO_2 nanotubes on Ti have improved the loading efficiency of titanium-based implants (Çalışkan et al., 2014). The loading process conditions and geometrical parameters of nanotubes have an impact on the drug amount loaded inside the nanotubes (Gulati et al., 2015). Therefore, the loading capacity can be controlled by anodizing conditions, such as electrolyte concentration, potential, and time. It is reported that the pore/tube diameter has less impact on the amount of loaded drug compared to the nanotubes length (Çalışkan et al., 2014). If the nature of interaction between drug molecule and titania surface is physical, the loading process can be optimized by changing loading conditions, like drug concentration, duration of the process, and postloading treatment (Çalışkan et al., 2014; Gulati et al., 2012a). It was reported that bigger molecules with a negative charge are loaded in greater amount than smaller ones with a positive charge (Popat et al., 2007a). If the concentration of drug solution was increased, the loading efficiency increased (Xiao et al., 2009).

3.1 MECHANISM OF DRUG LOADING INSIDE TIO_2 NANOTUBES

The mechanism of drug loading into nanoporous/nanotubular TiO_2 was proposed by Zhang et al. (2014). The scheme of this mechanism is presented in Fig. 9.7. When the nanoporous TiO_2 layers are soaked in the drug solution, hydrogen bonds are formed between drug molecules and the surface of the oxide film. The solvent evaporation allows the drug to form crystalline grains inside the nanopores. With increasing the drug-loading amount, the grains become bigger and are in contact with each other through weak intermolecular bonding forces. When the drug delivery system with loaded drug is soaked in PBS, the drug particles will dissolve during the release study. This step is associated with a burst release and is called the burst effect. The next step, which is typically observed in drug release profiles, is called a sustained release effect. At this stage, the release rate decreases significantly due to the presence of hydrogen bonds and possible steric hindrance. It is noticed that a larger amount of the drug can be loaded in pores with greater length than width. The greater the diameters of pores/tubes, the greater the reduction in the amount of the released drug during the burst effect (Zhang et al., 2014).

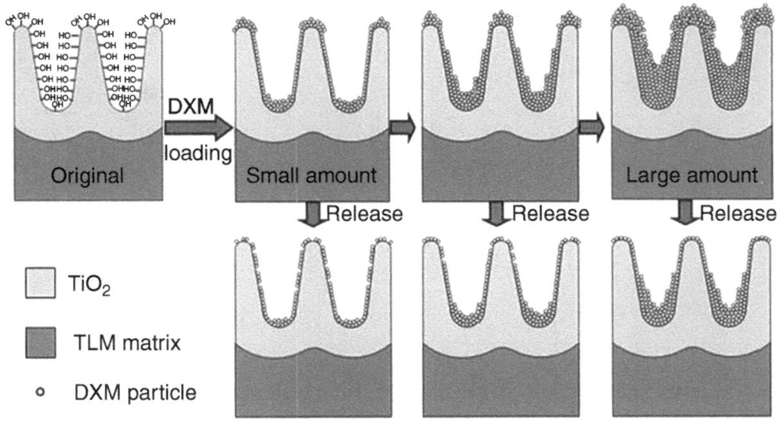

FIGURE 9.7 The Mechanism of the Drug Loading and Releasing from TiO_2 Nanotubes on β Type Ti25Nb3Mo2Sn3Zr Alloy (TLM) Matrix

Dexamethasone (DXM) was used as a model drug.

Reprinted with permission from Zhang, Y., He, F., Sun, Z., Li, L., Huang, Y., 2014. Controlled delivery of dexamethasone from TiO_2 film with nanoporous structure on Ti–25Nb–3Mo–2Sn–3Zr biomedical alloy without polymeric carrier. Mater. Lett. 128, 384–387.

4 DRUG RELEASE

The drug release studies are typically performed by soaking drug-loaded samples in either deionized water (Chennell et al., 2013; Han et al., 2011), PBS (Çalıs¸kan et al., 2014; Gulati et al., 2015; Popat et al., 2007a; Yao and Webster, 2009) or physiological serum (López et al., 2007) at 37°C in order to simulate the physiological conditions of the human body. At appropriate time periods, a portion (Chennell et al., 2013) or the whole volume of medium (Yao and Webster, 2009) is collected and replaced with the same volume of the same solution. Alternatively, the process can be carried out in a shaken 6- or 24-well plate and the medium is collected from different wells after given time periods (Popat et al., 2007a,b; Xiao et al., 2009). The drug concentration in collected samples is determined by using appropriate analytical methods, such as UV-Vis spectrophotometry (Gulati et al., 2011, 2012a; Mandal et al., 2014; Moseke et al., 2012; Popat et al., 2007b; Yao and Webster, 2009; Zhang et al., 2014) or flame atomic absorption spectrometer (FAAS) (Xiao et al., 2009).

The drug release profile is presented as a plot, being a relationship between release percentage and time. Release percentage is expressed as a quotient of amount of the drug released after the selected time period and the amount of the drug totally released (Gulati et al., 2011). The typical drug release profiles are presented in Fig. 9.8.

As can be seen, the release process consists of two steps: (1) an initial burst release and (2) a steady release (Gulati et al., 2011, 2015; López et al., 2007). The burst effect can be ascribed to the release of drug molecules that are adsorbed on the nanotube surface. This release step is very fast because of the high concentration gradient between the nanotubes and surrounding medium (Gulati et al., 2015). The amount of drug released during that stage should be high enough to prevent, for example, bacterial infections occurring after the surgery (Gulati et al., 2011, 2012a). After that initial step, the drug is delivered slowly and linearly (Gulati et al., 2011; López et al., 2007). During the second step, drug

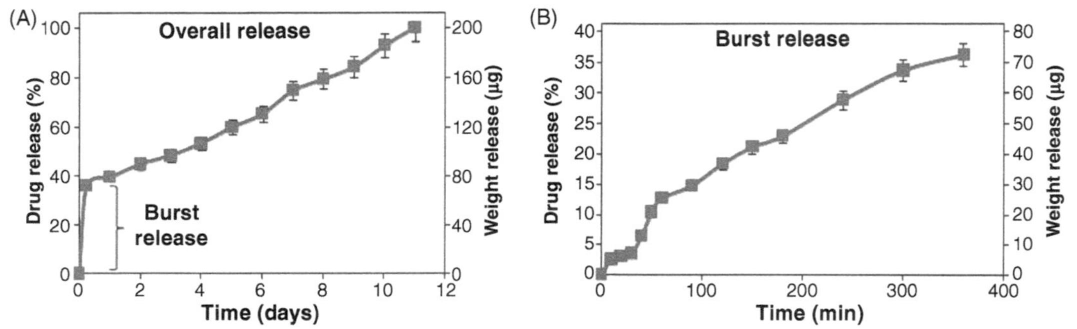

FIGURE 9.8

Total release (A) and burst release (B) profiles of gentamicin from titania nanotubes on Ti wires. Drug release was investigated in phosphate-buffered saline (PBS).

Reprinted with permission from Gulati, K., Aw, M.S., Losic, D., 2011. Drug-eluting Ti wires with titania nanotube arrays for bone fixation and reduced bone infection. Nanoscale Res. Lett. 6, 571–576.

molecules diffuse outward from the interior of nanotubes. The drug is released slowly because the molecules have to move from the nanotube bottoms to the medium (López et al., 2015). The drug delivery process can last even up to 3 weeks (Peng et al., 2009; Yao and Webster, 2009).

It is reported that geometrical parameters of nanopores/nanotubes not only have an impact on the amount of drug loaded inside the pores/tubes, but also on the release process. If the nanotubes are longer and have smaller diameter, a larger amount of the drug can be loaded and, consequently, released from the nanotubes (Çalıs¸kan et al., 2014; Moseke et al., 2012; Xiao et al., 2009). Obviously, longer nanotubes are characterized by prolonged release (Çalıs¸kan et al., 2014). A nonlinear relationship between the release time or amount of the released drug and the length of nanotubes can be attributed to a heterogeneous cylindrical shape of nanotubes (Çalıs¸kan et al., 2014). It is worth pointing out that the type of electrolyte used for anodization has an impact on the morphology the of obtained TiO_2 nanotubes. The oxide layers formed in organic electrolytes provide a network of free-standing nanotubes with large diameters. Drug molecules can be adsorbed also on the outer surface of nanotubes (Moseke et al., 2012). If the aqueous electrolyte is used, the anodic structure consists of an interconnected array of small pores. Therefore, the total surface area is increased, and more drugs can be loaded.

The amount of the drug loaded in nanotubes affects drug-release kinetics. The mechanism of drug release is directly related to the concentration gradient between the nanotube surface and surrounding medium (López et al., 2009). The increase in the amount of loaded substance causes the decrease in the rate of release (Hamlekhan et al., 2015; Popat et al., 2007a,b). In addition, the charge of molecules has an effect on the kinetics of release process. It is assumed that hydroxyl groups on the surface of TiO_2 layers cause the surface to be negatively charged. Therefore, the positively charged molecules have a stronger ionic interaction with the negatively charged surface, and the release occurs slower (Hamlekhan et al., 2015; Popat et al., 2007a). The opposite effect is observed for negatively charged molecules (Xin and Chu, 2011).

Knowledge of the mathematical formulation that describes the drug release is helpful in order to predict the amount of drug released after an appropriate period of time. Therefore, different functions can be fitted to the release curves (Moseke et al., 2012). In the literature, there are theoretical and empirical models that are used to describe the drug release process from different types of DDSs. Among them, the first-order kinetics (López et al., 2009; Yao and Webster, 2009), Higuchi (Gulati et al., 2011, 2012a), Korsmeyer–Peppas (Moseke et al., 2012), and Hill (Mandal et al., 2014) models are considered to be important, as they fit well to profiles of the drug release from TiO_2 nanotubes. Researchers indicated that the diffusion of drugs from TiO_2 nanostructures, which is of course a nondegradable drug-carrier system, can be described mathematically using Fick's law (Gulati et al., 2011, 2012a; Gulati et al., 2015; Hamlekhan et al., 2015). The release process can be attributed to mass transfer from nanotubes, where the drug concentration is higher, to the medium in which the drug concentration is lower. A high total surface area and capillary-like structure of nanotubes indicate that phenomena occurring on the surface affect the release process. The electrostatic interactions between the oxide surface and drug molecules are supposed to cause the prolonged drug release from TiO_2 nanotubes (Gulati et al., 2011).

5 FUNCTIONALIZATION OF TiO_2 NANOSTRUCTURES

5.1 SURFACE MODIFICATIONS

Surface modification is a feasible approach to improve the surface properties of the nanostructures and offers a high drug loading and sustained release with remarkable in vivo biocompatibility and bioavailability. Currently, research interest is focused mainly on modifying TiO_2 nanomaterials with semiconductors, and decorating with noble metal nanoparticles, for example, Au (Moon et al., 2014), Ag (Wang et al., 2012). Moreover, TiO_2 nanostructures can be modified by covalent attachment of organic monolayers, such as tris-(nitrilotrisacetic acid) (Tris-NTA) and biotin (Jana et al., 2013).

For instance, in the study by Wang et al. (2015b), titania nanotube arrays were decorated with Degussa P25 nanoparticles (TiO_2 NPs) using a hydrothermal method, a delivery of ibuprofen (IBU)—a nonsteroidal anti-inflammatory drug—was investigated in detail. The results of this research indicated that P25 NPs were successfully deposited on the surface of titania nanotubes with an optimal P25 decoration solution concentration of 7.5×10^{-4} M (Fig. 9.9).

Uniformly distributed P25 NPs on the nanotube surface led to a significant increase in the specific surface area of nanotubes, which resulted in an improved loading efficiency of IBU. Since nanotube diameters were evidently smaller (100 ± 10 nm) and the roughness and effective surface area of oxide were considerably increased, prolonged drug release was observed. These results showed that as-prepared decorated nanotubes have potential biomedical application in the development of a drug delivery system.

Huang et al. (2014) investigated two kinds of novel drug delivery nanosystems, such as Enro-NH_2-TNTs and Enro-SH-TNTs, which were obtained by combining the characteristic pH-adjusted recrystallized loading process of enrofloxacin (Enro) and surface silane coupling agent–modified titania nanotubes (NH_2-TNTs and SH-TNTs). The mechanism of synthesis of TNTs and further synthesis of NH_2-TNTs and SH-TNTs is shown in Fig. 9.10.

Enrofloxacin (Enro), belonging to fluoroquinolone broad-spectrum antibiotics, was chosen as a model drug. Through X-ray diffraction pattern (XRD), FTIR, BET surface area analysis, and transmission electron microscopy (TEM) imaging, Enro molecules were found to interact with modified titania nanotubes through ionic or hydrogen bonds. In vitro drug release and cell experiments showed that the facile surface modification could improve the biocompatibility and lower the cytotoxicity of nanotubular material (Huang et al., 2014). Furthermore, these modified drug delivery nanosystems provided

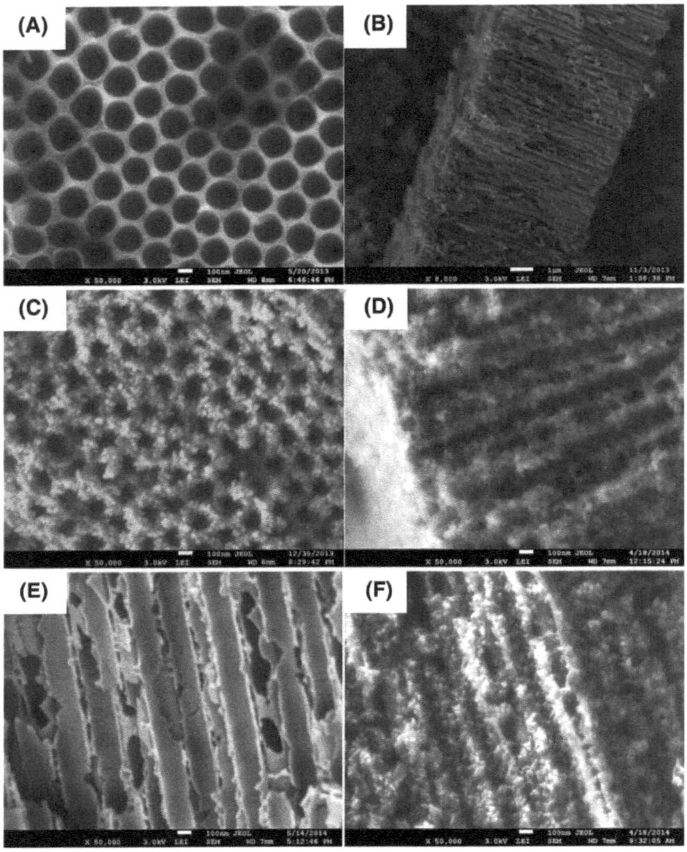

FIGURE 9.9 SEM Images of TiO_2 Nanotube Arrays Synthesized Using Anodization Process Under a Constant Voltage of 60 V for 24 h in an Ammonium Fluoride/Glycerol Electrolyte (10 vol.% H_2O and 0.5 wt.% NH_4F) at About 30°C.

(A–B) Before and (C–D) after decoration with P25 nanoparticles; (E–F) cross-section view of high-magnification before and after decoration with P25 nanoparticles.

Reprinted with permission from Wang, Z., Xie, C., Luo, F., Li, P., Xiao, X., 2015b. P25 nanoparticles decorated on titania nanotubes arrays as effective drug delivery system for ibuprofen. App. Surf. Sci. 324, 621–626.

a higher drug bioavailability and a longer effect of in vivo administration of the drug to chickens. It was concluded that the surface modification of titania nanotubes combined with a pH-adjusted loading process has an immense potential for the preparation of long-acting drug delivery systems, especially for hydrochloride drugs.

Shrestha et al. (2009) showed a simple way of fabricating magnetic TiO_2 nanotubes by using a permanent magnet, which allows an efficient and quick filling of the nanotubes with Fe_3O_4 nanoparticles. Furthermore, these magnetic nanotubes (with loaded Fe_3O_4 nanoparticles) were used as magnetically guided photocatalysts to a photoinduced site-selective release of drugs or killing cancer cells.

The alternative work (Ge et al., 2015) describes hydrothermal growth of TiO_2 nanorods on Ta modified with a mesoporous bioactive glass (MBG). TiO_2 nanorod films with incorporated titania-substituted

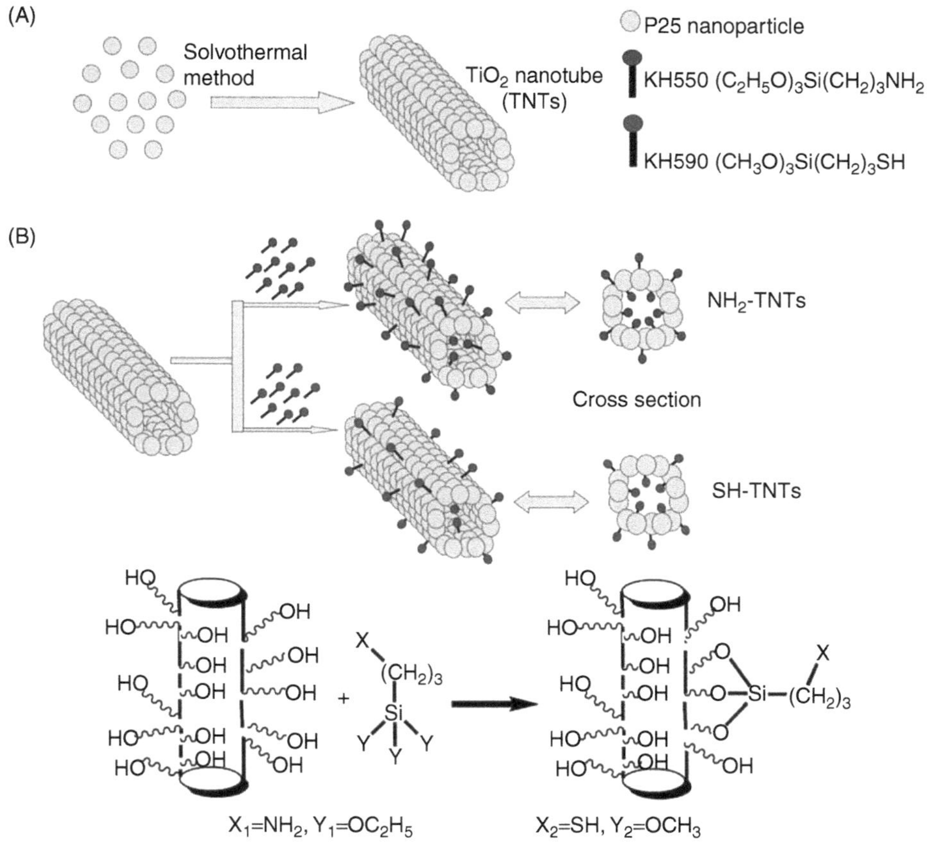

FIGURE 9.10

Mechanism of (A) synthesis of TNTs through a solvothermal method and (B) synthesis of NH_2-TNTs and SH-TNTs through a condensation method.

Reprinted with permission from Huang, P., Wang, J., Lai, S., Liu, F., Ni, N., Cao, Q., Liu, W., Deng, D.Y.B., Zhou, W., 2014. Surface modified titania nanotubes containing anti-bacterial drugs for controlled delivery nanosystems with high bioactivity. J. Mater. Chem. B 2, 8616–8625.

MBG (T-MBG) were prepared on the tantalum substrate by the sol–gel technique, and the degree of T-MBG incorporation was adjusted to achieve the height of exposed TiO_2 nanorods in the range of 100–300 nm. The incorporated T-MBG cohered tightly to TiO_2 nanorods and was stable during the soaking in PBS for 7 days. The incorporation of T-MBG played appositive role in the cytocompatibility of the resulting films. Vancomycin hydrochloride (VH) release profiles exhibited different behavior for TiO_2 nanorods and T-MGB films. For the nonmodified films, a rapid burst release followed by a subsequent 2 h long release of VH was observed. The total amount of the released drug was 71%. On the other hand, when the T-MGB film was used, the burst release process was inhibited, while the sustained release was significantly extended with an increasing T-MGB incorporation (even up to 96 h). Kinetic analysis proved that enhanced drug loading and its prolonged release are correlated with the degree of an incorporation of T-MGB. Those results present a new, promising alternative for surface modification techniques used for implantable devices.

5.2 SURFACE COATINGS FOR IMPROVED DRUG RELEASE

In order to increase the efficiency of drug delivery systems based on anodic titanium oxide (ATO) layers, new approaches have been provided to allow the use of different drugs, regardless of their solubility. For this purpose, ATO layers were modified with different coatings, for example, polymers or chitosan (Gulati et al., 2012a; Kumeria et al., 2015). Also polymer micelles were used to entrap drugs and load them inside TiO_2 nanopores. On the other hand, for an enhanced and more controllable delivery of drugs, additional molecules, such as Fe_3O_4 nanoparticles were used in order to trigger the release of drugs using for example, ultrasonic waves (Aw and Losic, 2013), radiofrequency (Bariana et al., 2014), or magnetic field (Aw et al., 2012a).

5.2.1 Micelles

It is known that an important factor that affects the amount of released drug is the structure of the implant. In the studies carried out by Aw et al. (2011b, 2012a,b), titanium oxide nanostructures were synthesized using a two-step anodization process in an electrolyte based on ethylene glycol containing water and ammonium fluoride. The electrolysis was performed using the anodizing potential of 100 V for 2 h (for both anodizing steps). The resulting diameters and length of nanotubes were found to be in the range of 110–140 nm and 40–50 μm, respectively. The initial studies involved also comparison between implants made of anodic aluminum oxide (AAO) nanopores and anodic titanium oxide (ATO) nanotubes.

Polymeric micelles are considered to be ideal drug carriers. This is due to their ability to load a substance inside the micelle (protecting it from external factors) and release the content in the appropriate biological environment. The micelles, which are obtained by a self-assembling formation, have an outer hydrophilic and inner hydrophobic layer (Aw et al., 2012b). Micelles of d-α-tocopheryl polyethylene glycol 1000 (TPGS) and poly(ethylene oxide)–poly(propylene oxide)–poly(ethylene oxide) (PEO–PPO–PEO) were used for the delivery of imdomicine (Aw et al., 2012a). The prepared micelle solution was pipetted on ATO samples and left for 20–30 min to dry. The excess of the solution was wiped, and the whole process was repeated 25 times. The study showed that the loading and releasing process of polymeric micelles, which differ in size, charge, chemical composition of hydrophobic tails, are mainly affected by the size of micelles and the pore/tube diameter of the oxide structure. It was shown that for smaller pore diameters the drug release is longer. The release of polymeric micelles from the substrate consists of a two-step release kinetics with the burst release over the first 6–8 h (31%–55% of released drug), followed by the slow release step from 8 h up to 22 days (Aw et al., 2011b).

5.2.2 Magnetic field–induced drug release

In order to use a magnetic field to release drug encapsulated polymer micelles from nanoporous/nanotubular ATO, prior to release, the material should be appropriately modified with a drug-releasing agent, for example, dopamine-modified magnetic iron oxide nanoparticles (Fe_3O_4-DOPA) (Shen et al., 2014). The drug encapsulated polymer micelles at the top of ATO nanotubes act as drug carriers, while magnetic nanoparticles at the bottom of ATO nanotubes move after applying an external magnetic field and force the release polymer micelles out from the ATO. The role of dopamine modification is to improve both: biocompatibility of the drug releasing agent and loading capacity of ATO. It was shown that complete release of the drug can be achieved within a short period of time (1–1.5 h). The release of polymer micelles from ATO layers was studied for three types of amphiphilic micelles and also for indomethacin, a hydrophobic drug entrapped inside the lipophilic core of micelles (Aw et al., 2012b).

Another example of the magnetic field–induced drug release was proposed by Da Silva et al. (2014). They synthesized TiO_2-crosslinked pectin microspheres enriched with Fe_3O_4 nanoparticles using an ultrasound-induced crosslinking/polymerization reaction. Then, amoxicillin was used as a model drug molecule for release tests. A detailed comparison between pure pectin microspheres and the TiO_2-crosslinked pectin microspheres with Fe_3O_4 nanoparticles was performed. It was found that the release of amoxicillin from the nanostructured pectin microspheres was slower than for pure micropsheres. It was attributed to the fact that the release from the nanostructured surfaces is abnormal and disturbed by the reduction of the polymer motion and tortuosity effect. Moreover, when the magnetic field was applied, the release of the drug molecules drastically decreased because of the enhanced influence of the anomalous transport of the drug in the medium. Additionally, it was shown that in acidic conditions, the release of amoxicillin from microspheres is sustained, which is extremely beneficial when the acidic environment of the stomach (a narrow therapeutic window for most antibiotics) is considered. In general, the release of drug molecules may be strictly controlled by a magnetic field preventing a premature release of the drug.

5.2.3 Ultrasound and radiofrequency-induced drug release

The use of ultrasound results in an abrupt change in concentration of the drug under the influence of a pulse. Important parameters that should be monitored during this process are: the pulse length, amplitude, and its power. All these parameters affect the release profile (Aw and Losic, 2013). The disadvantage of this system is a fact that the patient could be accidentally exposed to ultrasounds. To avoid this, a radiofrequency-triggered release may be a better solution. Typically, Au nanoparticles are used as energy transducers, and allow the release system to be externally controlled (Bariana et al., 2014). In this case, before loading of the drug or drug micelles, nanotubes should be coated with Au nanoparticles. It is believed that the heating of Au nanoparticles, induced by radiofrequency stimulation, accelerates transport of drug carriers from ATO. It was shown that by using noninvasive radiofrequency, the amount of the released drug was increased several times (Bariana et al., 2014).

5.2.4 Polymer coatings

As mentioned before, the release of drugs from anodic TiO_2 at the early stage of the process is associated with the fast diffusion of drugs from the ATO surfaces. To inhibit this burst release, covering the ATO surface with polymers is often considered. Jia and Kerr (2015) studied the effect of high and low molecular weight of poly(lactic-*co*-glycolic acid) (MW-PLGA) on the release of ibuprofen (IBU) in SBF at 37°C. Both drug and polymer were dissolved in dichloromethane (6 mg/mL IBU, PLGA 12 mg/mL) and left for 3 days at 40°C. In order to check how deep the penetration of nanotubes with polymer occurs, an SEM analysis was performed. It was found that a thick polymer layer was deposited on the ATO surface, while a thin polymer layer was observed at the bottom of nanotubes. Comparing the amount of released drug from the ATO layers covered with low and high MW-PLGA, it was revealed that a 20% decrease in the released drug was found for the ATO covered with high MW-PLGA. In addition, the surface modification with low and high MW-PLGA resulted in prolonged ibuprofen release until 5 and 9 days, respectively. Moreover, the mechanical properties of the implant were also improved (Sirivisoot et al., 2011).

Using other drugs, kaprofen and lidocaine loaded in the same way as mentioned previously, the effect of the medium pH on the release process was studied (Jia and Kerr, 2015). The rate of drug release from PLGA-modified TiO_2 nanotubes is determined by the drug solubility, degree of polymer swelling/degradation, and by electrostatic interactions between polymer and drug. It was suggested that the controlled release of drug carriers from polymer-coated ATO layers can be effectively done by changing pK_a values of polymer and drug solubility (Jia and Kerr, 2015).

Another idea for the controlled release of drugs is based on the use of electrically controlled drug eluting coatings, particularly conductive polymer coatings, for example, polypyrrole. Sirivisoot et al. (2011) tested TiO_2 nanotubes with multiwalled carbon nanotubes covered with a drug-modified polypyrrole thin layer. During the electropolymerization of polypyrrole, penicillin/streptomycin and dexamethasone were codeposited. Drugs can be released from the polypyrrole film on demand by applying an appropriate voltage. The electrochemical release of drugs was carried out in phosphate buffer with pH 7.4 using cyclic voltammetry. Data showed that after five cycles (potential range between −1.0 and 1.0 V vs. Ag/AgCl, scan rate of 100 mV/s), up to 80% of the drugs was released.

Kumeria et al. (2015) conducted research on anodic TiO_2 having a diameter of 100 ± 20 nm and coated with 2.5 μm layer of chitosan and poly(lactic-*co*-glycolic acid) PLGA. Gentamicin as a model antibiotic was loaded inside nanotubes with and without the coating polymer layer. In the case when the uncoated nanotubular ATO layer was used, gentamicin was loaded directly into the nanotubes, and a 100% release of the antibiotic was observed after 2 weeks. The modification of the ATO surface with the polymer layer resulted in an extended release of the drug up to 22 and 26 days for chitosan and PLGA, respectively. The best adhesive properties for the osteoblast growth showed samples coated with the chitosan layer. In that case, the number of cells adhered to the substrate after 1.5 and 24 h was the same.

On the other hand, Mandal et al. (2014) found, that the amount of loaded drug is reduced while a hydrophobic modification is applied, whereas it is increased when the hydrothermally synthesized titania nanotube surface is more hydrophilic. Consequently, the modification with various organosilanes affects the rate of drug release. A fast release is observed for hydrophobically functionalized surfaces and a slow release for a hydrophilic functionalization. It was also suggested that the nature of the functional groups on titania nanotubes (–OH, $–NH_2$, $–CH_3$) plays a crucial role in the release process, and these groups may act as inbuilt regulators for achieving different drug releases.

6 OTHER TYPES OF NANOSTRUCTURED TIO_2 USED AS DRUG DELIVERY SYSTEMS

In order to fulfill the requirements of titania nanostructures in area of drug delivery systems, various architectures of nanostructured TiO_2 have been fabricated and investigated, including TiO_2 nanopowders (Signoretto et al., 2011), nanoparticles (Kim et al., 2012; Qin et al., 2011; Ren et al., 2013; Wang et al., 2015b), mesoporous titania (Chao et al., 2012; Ganesh et al., 2012; Heredia-Cervera et al., 2009; Wu et al., 2011), titania microtubes (López et al., 2006), nanorods (Ge et al., 2015), TiO_2 nanospheres (Jiang et al., 2012), core–shell nanostructures (Shen et al., 2014; Wang et al., 2012), TiO_2 whiskers (Xu et al., 2015) and modified titania nanoparticles (Jana et al., 2013; Manaia et al., 2015) and nanotubes (Huang et al., 2014; Moon et al., 2014; Oh et al., 2013; Shrestha et al., 2009; Wang et al., 2015a).

For example, in order to overcome the multidrug resistance of breast cancer cells, Ren et al. (2013) modified TiO_2 nanoparticles with doxorubicin (DOX). The electrostatic interactions hold both the drug and nanoparticles together. The research confirmed the enhanced DOX transporting to human breast cancer cells (MCF-7/ADM). The results suggested also that TiO_2/DOX nanocomposite carriers transport the drug into the intracellular cytoplasm via internalization into endosomes, bypassing the P-glycoprotein-mediated efflux of doxorubicin. Most probably, the drug is released in acidic endosomes, and then accumulated in the intracellular nucleus. The TiO_2/DOX carriers reverse the multidrug resistance of MCF-7/ADM cells leading to the enhanced anticancer activity of DOX up to 2.4 times when compared with that of DOX alone.

Another solution that enhanced the performance of doxorubicin was proposed by Qin et al. (2011). His group synthesized highly water-dispersible TiO_2 nanoparticles with abundant carboxyl groups using a ligand exchange method. Briefly, the oleic acid coating on the surface of TiO_2 was exchanged with carboxylic silane and modified nanoparticles were used for anticancer drug delivery (Fig. 9.11).

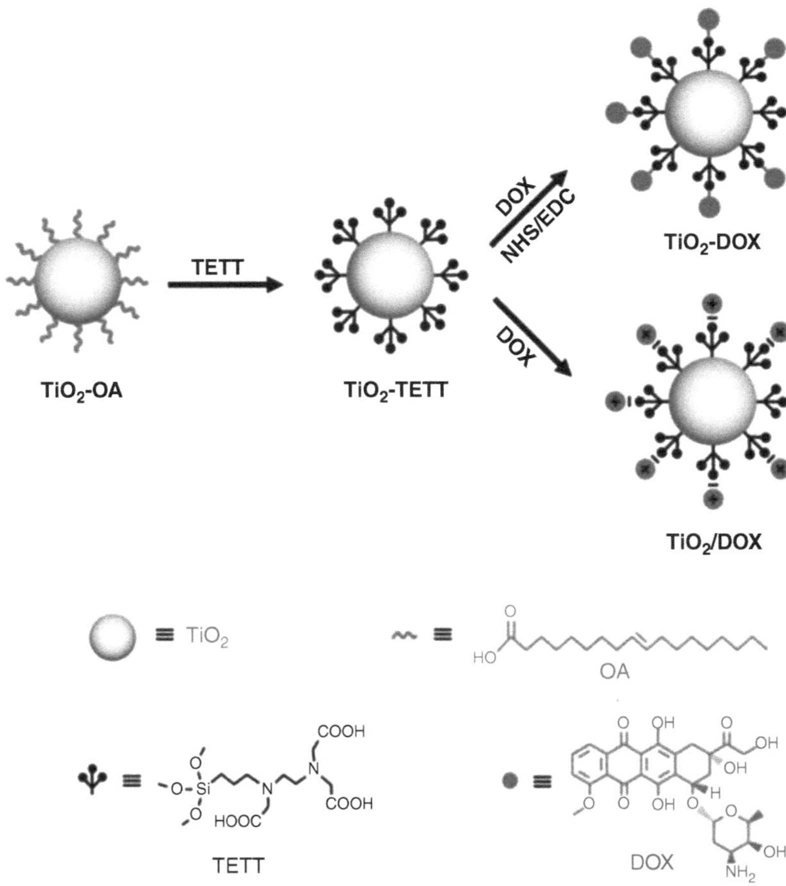

FIGURE 9.11 Scheme of the Synthesis of Highly Water-Dispersible TiO_2 Nanoparticles and Loading of DOX via a Noncovalent Complexation (TiO_2/DOX) or Covalent Conjugation (TiO_2/DOX)

Reprinted with permission from Qin, Y., Sun, L., Li, X., Cao, Q., Wang, H., Tang, X., Ye, L., 2011. Highly water-dispersible TiO_2 nanoparticles for doxorubicin delivery: effect of loading mode on therapeutic efficacy. J. Mater. Chem. 21, 18003–18010.

They systematically explored the effect of drug loading mode in DDSs, on chemotherapeutic efficiency of doxorubicin (DOX) in terms of cytotoxicity, cellular uptake, and intracellular distribution using C6 glioma cells. Their work showed that TiO_2 conjugates (TiO_2–DOX) exhibited a significantly higher cytotoxicity toward C6 cells than free DOX due to enhanced cellular uptake, while TiO_2 complexes (TiO_2/DOX) led to a decreased antitumor activity of DOX due to the limited release of the drug in its active form.

The study indicates that the therapeutic efficacy is strongly dependent on the nature of interaction between TiO_2 nanoparticle surface and loaded drug. That provides very important information for the potential applications of TiO_2 nanoparticles as drug carriers.

Another use of titania nanoparticles in drug delivery systems was proposed by Galkina et al. (2015). They prepared a cellulose nanofiber–titania nanocomposite for the delivery of two drugs: tetracycline (TC) and phosphomycin (Phos). They obtained cellulose nanofibers from cotton, using a copper ammonium complex to convert cotton into the solution, followed by acid hydrolysis to obtain cellulose. Then, titania nanoparticles were cross-linked with previously synthesized cellulose nanofibers using sodium hypophosphite (SHP) as a catalyst and 1,2,3,4-butanetetracarboxylic acid (BTCA) as a spacer. Drug loading was performed using three different techniques, where the order of reagent mixing was changed. The drug release was tested in the in vitro conditions in a citrate buffer at the constant temperature of 37°C and with constant stirring rate (100 rpm). The results showed that different modifications of the nanocomposite with loaded drugs lead to differences in the kinetics of drug release. For example, Phos molecule binds to titania via a complexation process which results in the formation of the complex that is a very highly dissociation-stable. The release of this drug can be described by the first-order kinetics with the burst release, and the total release time was extended up to several weeks. On the other hand, TC is bonded to the surface of titania by forming quite stable phenoxide complexes. The release profile of this molecule differs with different loading methods. When a drug powder was initially dissolved in water, the release profile showed the burst release stage. This is related with a physical adsorption of the drug to cellulose nanofibers. However, when the drug powder was dissolved first in a titania nanosol, the samples showed a long-term release profile without the burst release effect. Therefore, it was shown that the nature of bonds between titania nanoparticles and drug molecules plays an important role in the drug release studies.

In other works (Chao et al., 2012), ultrathin mesoporous TiO_2 coatings with a wormhole-like architecture were synthesized via an evaporation-induced self-assembly (EISA) method. The film thickness was about 120 nm. The mesoporous TiO_2 thin films were employed as carriers for ibuprofen and vancomycin, and a controlled release of those drugs was studied. It appeared that the mesoporous worm-like nanostructured TiO_2 coatings have a good chemical affinity to ibuprofen. Both the chemical affinity and mesoporosity dominate the resulting release profiles for ibuprofen and vancomycin molecules. Moreover, using those structures, an excellent compatibility and bioactivity properties toward osteoblast cells adhesion, and in vitro apatite formation were demonstrated. These results imply that the mesoporous TiO_2 ultrathin film with its worm-like architecture may be used as drug delivery systems for bone implant coatings.

What is more, Ganesh et al. (2012) obtained mesoporous TiO_2 using a solvothermal method and loaded titania with a serotonin reuptake inhibitor (duloxetine, DX) by a wet method. UV-Vis spectroscopy showed about 11% loading efficiency. The drug release exhibited typical two-stage performance: the burst release between 0 and 12 h, followed by a slow extended release of DX. The amount of the drug released during the burst stage equaled to 20%, whereas the total amount of the released drug was about 90%. Thereby, it was proved that mesoporous TiO_2 could perform as a DX reservoir. Consequently, it can be used as a carrier of the potentially toxic drugs that require small dosage with an extended release.

In the following work (Jiang et al., 2012), cotton-like amorphous and anatase mesoporous titania (APMT and ATMT, respectively) nanoparticles with 3D net-like pores were fabricated by a facile and reproducible technique using pluronic F127 as a template and titanium(IV) isopropoxide as a titania source. The influence of different crystal forms of titania on drug loading and release behaviors was investigated. The APMT nanoparticles were explored as carriers for a poorly water-soluble drug—carvedilol (CAR). The monodispersed amorphous mesoporous titania spheres with the diameter of 200–300 nm are shown in Fig. 9.12A and B. The TEM images of cotton-like APMT structures, presented in Fig. 9.12C and D show interconnected pores with the size of 2–3 nm.

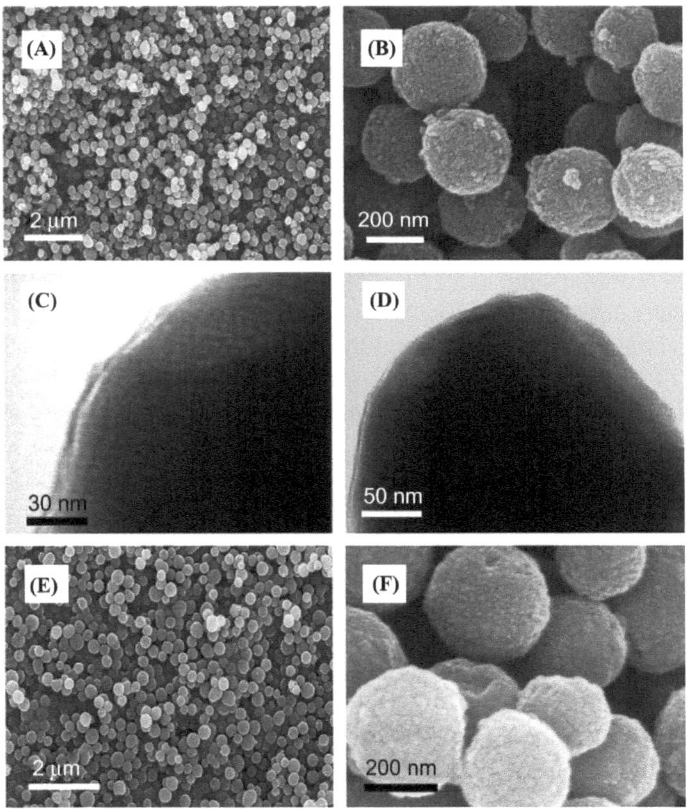

FIGURE 9.12

SEM microphotographs of APMT synthesized by dropping titanium(IV) isopropoxide precursor to the 0.1 M KCl with ethanol solution (A, B) and CAR–APMT prepared by loading carvedilol into the TiO_2 pores by a solvent deposition method (E, F). TEM microphotograph of APMT (C, D).

Reprinted with permission from Jiang, H., Wang, T., Wang, L., Sun, C., Jiang, T., Cheng, G., Wang, S., 2012. Development of an amorphous mesoporous TiO_2 nanosphere as a novel carrier for poorly water-soluble drugs: effect of different crystal forms of TiO_2 carriers on drug loading and release behaviors. Micropor. Mesopor. Mater. 153, 124–130.

The model drug, carvedilol, was effectively loaded into the pores of APMT and ATMT structures through a solvent deposition method. According to the results of differential scanning calorimetry and TGA both carriers can improve the dissolution rate of CAR. The results of X-ray diffraction, N_2 adsorption/desorption, SEM/TEM, and in vitro dissolution tests definitely exhibited, that the drug loading and release behaviors were strongly dependent on the crystal forms of the carriers, presumably attributed to the succeeding benefits of amorphous material:

1. The spatial confinement of small, net-like mesopores (2–3 nm) makes drugs exist in the carrier (APMT) in a stable amorphous form.
2. The mobility of the drug molecules were elevated due to the adsorbed water in the amorphous mesoporous titania. This could be a valid factor for the acceleration of the drug release rate.
3. The diffusion resistance of drug release was efficiently reduced by a large number of interconnected pores.
4. The hydrophilic surface of APMT could increase its dispersibility in the solution.

The results of this study emphatically confirmed that the drug loading and release behaviors were heavily dependent on the crystal form of the carriers, which will ensure a new thread for the research regarding delivery of poorly water-soluble drugs.

Another titania nanostructures employed for DDS systems are whiskers. Xu et al. (2015) investigated a new strategy of coupling the titanium oxide whiskers with the anticancer drug gambogic acid (GA) used in photodynamic therapy. The TiO_2 nanomaterial demonstrated a needle-like morphology, with an average size of 81.7 nm in width and 200–1000 nm in length, which indicated that TiO_2 whiskers have a uniform diameter distribution. It was shown that nanocomposites GA–TiO_2 induced an excellent enhancement in an antitumor activity estimated by a colorimetric assay used to assess the metabolic activity of cells: MTT (3-(4,5-dimethylthiazol-2-yl)-2,5-diphenyltetrazolium bromide) assay, nuclear DAPI (4',6-diamidino-2-phenylindole) staining, and flow cytometry. These findings identified TiO_2 whiskers as a material with good biocompatibility and photocatalytic activity. It was observed that in human leukemia cells (K562 cells), TiO_2 whiskers could increase the intracellular concentration of GA and its virtual antitumor efficiency. Therefore, TiO_2 whiskers may act as real drug delivery carriers targeting GA to carcinoma cells. Furthermore, photodynamic GA–TiO_2 nanocomposites could induce reinforcement in an antitumor activity with UV illumination. The previously mentioned results demonstrated a new and hopeful proposal for targeted cancer therapy.

7 CONCLUSIONS

Titania nanostructures, such as nanotubes, nanoporous layers, nanoparticles, nanowhiskers, and so on offer a wide range of solutions when it comes to drug delivery systems. Using simple methods described in this chapter, scientists may obtain structures that vary considerably in shape and surface morphology. These in turn influence the loading capacity of nanostructures that may be controlled by synthesis parameters. Drug molecules that vary in size, charge, and solubility may be effectively loaded and then released in a controllable way from the titania nanoreservoirs. TiO_2 nanostructures may be functionalized using different coatings that maintain the controlled release on the one hand, and increases biocompatibility of the implantable device on the other hand. Drug release may be also controlled using external sources, such as magnetic field, ultrasound, or radiofrequency.

To sum up, titanium oxide nanostructures have been proved to be an excellent modification of titanium and titanium alloys. Such surfaces enable a variety of possibilities to load different kinds of drug molecules and then release them in a very controllable way. Moreover, the drug release from titania nanostructures may be adjusted to a specific kind of implantable devices and to diverse applications, such as infection treatment, fracture healing, or even enhanced attachment and proliferation of bone cells. Certainly, more research must be conducted in order to overcome some drawbacks connected with such structures, like toxicity of coatings or stability of titania nanolayers in in vivo conditions. However, advantages of the application of titania nanostructures as a drug delivery system in implantable devices are beyond dispute.

REFERENCES

Aninwene, G.E., Yao, C., Webster, T.J., 2008. Enhanced osteoblast adhesion to drug-coated anodized nanotubular titanium surfaces. Int. J. Nanomed. 3, 257–264.

Aw, M.S., Losic, D., 2013. Ultrasound enhanced release of therapeutics from drug-releasing implants based on titania nanotube arrays. Int. J. Pharm. 443, 154–162.

Aw, M.S., Simovic, S., Addai-Mensah, J., Losic, D., 2011a. Polymeric micelles in porous and nanotubular implants as a new system for extended delivery of poorly soluble drugs. J. Mater. Chem. 21, 7082–7089.

Aw, M.S., Gulati, K., Losic, D., 2011b. Controlling drug release from titania nanotube arrays using polymer nanocarriers and biopolymer coating. J. Biomater. Nanobiotechnol. 02, 477–484.

Aw, M.S., Addai-Mensah, J., Losic, D., 2012a. Magnetic-responsive delivery of drug-carriers using titania nanotube arrays. J. Mater. Chem. 22, 6561–6563.

Aw, M.S., Addai-Mensah, J., Losic, D., 2012b. Polymer micelles for delayed release of therapeutics from drug-releasing surfaces with nanotubular structures. Macromol. Biosci. 12, 1048–1052.

Aw, M.S., Khalid, K.A., Gulati, K., Atkins, G.J., Pivonka, P., Findlay, D.M., Losic, D., 2012c. Characterization of drug-release kinetics in trabecular bone from titania nanotube implants. Int. J. Nanomed. 7, 4883–4892.

Bariana, M., Aw, M.S., Moore, E., Voelcker, N.H., Losic, D., 2014. Radiofrequency-triggered release for on-demand delivery of therapeutics from titania nanotube drug-eluting implants. Nanomedicine 9, 1263–1275.

Çalıs¸kan, N., Bayram, C., Erdal, E., Karahalilog˘lu, Z., Denkbas¸, E.B., 2014. Titania nanotubes with adjustable dimensions for drug reservoir sites and enhanced cell adhesion. Mater. Sci. Eng. C 35, 100–105.

Chao, C.-S., Liu, K.-H., Tung, W.-L., Chen, S.-Y., Liu, D.-M., Chang, Y.-P., 2012. Bioactive TiO_2 ultrathin film with wormlike mesoporosity for controlled drug delivery. Micropor. Mesopor. Mater. 152, 58–63.

Chen, G., 2012. Nanotube-based controlled drug delivery. Pharma. Anal. Acta. 3, e136(1)–e136(2).

Chennell, P., Feschet-Chassot, E., Devers, T., Awitor, K.O., Descamps, S., Sautou, V., 2013. In vitro evaluation of TiO_2 nanotubes as cefuroxime carriers on orthopaedic implants for the prevention of periprosthetic joint infections. Int. J. Pharm. 455, 298–305.

Da Silva, E.P., Sitta, D.L.A., Fragal, V.H., Cellet, T.S.P., Mauricio, M.R., Garcia, F.P., Nakamura, C.V., Guilherme, M.R., Rubira, A.F., Kunita, M.H., 2014. Covalent TiO_2/pectin microspheres with Fe_3O_4 nanoparticles for magnetic field-modulated drug delivery. Int. J. Biol. Macromol. 67, 43–52.

David, V., Guignandon, A., Martin, A., Malaval, L., Lafage-Proust, M.H., Rattner, A., Mann, V., Noble, B., Jones, D.B., Vico, L., 2008. Ex vivo bone formation in bovine trabecular bone cultured in a dynamic 3D bioreactor is enhanced by compressive mechanical strain. Tissue Eng. Part A. 14, 117–126.

Doadrio, A.L., Conde, A., Arenas, M.A., Hernández-López, J.M., de Damborenea, J.J., Pérez-Jorge, C., Esteban, J., Vallet-Regí, M., 2015. Use of anodized titanium alloy as drug carrier: ibuprofen as model of drug releasing. Int. J. Pharm. 492, 207–212.

Endres, S., Kratz, M., Wunsch, S., Jones, D.B., 2009. Zetos: a culture loading system for trabecular bone. Investigation of different loading signal intensities on bovine bone cylinders. J. Musculoskelet. Neuronal Interact. 9, 173–183.

Galkina, O.L., Önneby, K., Huang, P., Ivanov, V.K., Agafanov, A.V., Seisenbaeva, G.A., Kessler, V.G., 2015. Antibacterial and photochemical properties of cellulose nanofiber-titania nanocomposites loaded with two different types of antibiotic medicines. J. Mater. Chem. 3, 7125–7134.

Ganesh, M., Hemalatha, P., Peng, M.M., Cha, W.S., 2012. Drug release evaluation of mesoporous TiO_2: a nano carrier for duloxetine. Computer Applications for Modeling, Simulation and AutomobileSpringer-Verlag, Berlin Heidelberg, pp. 237–243.

Ge, F., Lin, J., Huang, X., Cheng, K., Wang, H., Weng, W., 2015. Preparation and drug release behavior of TiO_2 nanorod films with incorporating mesoporous bioactive glass. Thin Solid Films 584, 2–8.

Ghicov, A., Schmuki, P., 2009. Self-ordering electrochemistry: a review on growth and functionality of TiO_2 nanotubes and other self-aligned MO(x) structures. Chem. Commun. 20, 2791–2808.

Gulati, K., Aw, M.S., Losic, D., 2011. Drug-eluting Ti wires with titania nanotube arrays for bone fixation and reduced bone infection. Nanoscale Res. Lett. 6, 571–576.

Gulati, K., Ramakrishnan, S., Aw, M.S., Atkins, G.J., Findlay, D.M., Losic, D., 2012a. Biocompatible polymer coating of titania nanotube arrays for improved drug elution and osteoblast adhesion. Acta Biomater. 8, 449–456.

Gulati, K., Aw, M.S., Losic, D., 2012b. Nanoengineered drug-releasing Ti wires as an alternative for local delivery of chemotherapeutics in the brain. Int. J. Nanomed. 7, 2069–2076.

Gulati, K., Kant, K., Findlay, D., Losic, D., 2015. Periodically tailored titania nanotubes for enhanced drug loading and releasing performances. J. Mater. Chem. B 3, 2553–2559.

Hamlekhan, A., Sinha-Ray, S., Takoudis, C., Mathew, M.T., Sukotjo, C., Yarin, A.L., Shokuhfar, T., 2015. Fabrication of drug eluting implants: study of drug release mechanism from titanium dioxide nanotubes. J. Phys. D 48, 275401(1)–275401(11).

Han, C.-M., Lee, E.-J., Kim, H.-E., Koh, Y.-H., Jang, J.-H., 2011. Porous TiO_2 films on Ti implants for controlled release of tetracycline-hydrochloride (TCH). Thin Solid Films 519, 8074–8076.

Han, C.-M., Jang, T.-S., Kim, H.-E., Koh, Y.-H., 2013. Creation of nanoporous TiO_2 surface onto polyetheretherketone for effective immobilization and delivery of bone morphogenetic protein. J. Biomed. Mater. Res. A 102, 793–800.

Heredia-Cervera, B.E., González-Azcorra, S.A., Rodríguez-Gattorno, G., López, T., Ortiz-Islas, E., Oskam, G., 2009. Controlled release of phenytoin from nanostructured TiO_2 reservoirs. Sci. Adv. Mater. 1, 63–68.

Hu, Y., Cai, K., Luo, Z., Xu, D., Xie, D., Huang, Y., Yang, W., Liu, P., 2012. TiO_2 nanotubes as drug nanoreservoirs for the regulation of mobility and differentiation of mesenchymal stem cells. Acta Biomater. 8, 439–448.

Huang, P., Wang, J., Lai, S., Liu, F., Ni, N., Cao, Q., Liu, W., Deng, D.Y.B., Zhou, W., 2014. Surface modified titania nanotubes containing anti-bacterial drugs for controlled delivery nanosystems with high bioactivity. J. Mater. Chem. B 2, 8616–8625.

Jana, B., Mondal, G., Biswas, A., Chakraborty, I., Ghosh, S., 2013. Functionalized TiO_2 nanoparticles deliver oligo-histidine and avidin tagged biomolecules simultaneously into the cell. RSC Adv. 3, 8215–8219.

Jarosz, M., Syrek, K., Kapusta-Kołodziej, J., Mech, J., Małek, K., Hnida, K., Łojewski, T., Jaskuła, M., Sulka, G.D., 2015. Heat treatment effect on crystalline structure and photoelectrochemical properties of anodic TiO_2 nanotube arrays formed in ethylene glycol and glycerol based electrolytes. J. Phys. Chem. C 119, 24182–24191.

Jia, H., Kerr, L.L., 2015. Kinetics of drug release from drug carrier of polymer/TiO_2 nanotubes composite-pH dependent study. J. Appl. Polym. Sci. 132, 41570(1)–41570(11).

Jiang, H., Wang, T., Wang, L., Sun, C., Jiang, T., Cheng, G., Wang, S., 2012. Development of an amorphous mesoporous TiO_2 nanosphere as a novel carrier for poorly water-soluble drugs: effect of different crystal forms of TiO_2 carriers on drug loading and release behaviors. Micropor. Mesopor. Mater. 153, 124–130.

Kim, M., Seo, J.H., Jeon, W.I., Kim, M.-Y., Cho, K., Lee, S.Y., Joo, S.-W., 2012. Real-time monitoring of anticancer drug release in vitro and in vivo on titania nanoparticles triggered by external glutathione. Talanta 88, 631–637.

Kumeria, T., Seo, J.H., Jeon, W.I., Kim, M.-Y., Cho, K., Lee, S.Y., Joo, S.-W., 2015. Advanced biopolymer-coated drug-releasing titania nanotubes (TNTs) implants with simultaneously enhanced osteoblast adhesion and antibacterial properties. Colloids Surf. B 130, 255–263.

Langer, R., 2001. Drugs on target. Science 293, 58–59.

Langer, R., Peppas, N.A., 2003. Advances in biomaterials, drug delivery, and bionanotechnology. AIChE J. 49, 2990–3006.

Liang, Y.Q., Cui, Z.D., Zhu, S.L., Yang, X.J., 2011. Characterization of self-organized TiO_2 nanotubes on Ti–4Zr–22Nb–2Sn alloys and the application in drug delivery system. J. Mater. Sci. Mater. Med. 22, 461–467.

Liu, W., Zhang, L., Cao, L.-X., Su, G., Wang, Y.-G., 2011. Glass fibers templated preparation of TiO_2 microtubes assembled from nano/micro hierarchical TiO_2 crystals. J. Alloy Comp. 509, 3419–3424.

López, T., Ortiz-Islas, E., Manjarrez, J., Reinoso, F.R., Sepulveda, A., Gonzalez, R.D., 2006. Biocompatible titania microtubes formed by nanoparticles and its application in the drug delivery of valproic acid. Opt. Mater. 29, 70–74.

López, T., Ortiz, E., Quintana, P., González, R.D., 2007. A nanostructured titania bioceramic implantable device capable of drug delivery to the temporal lobe of the brain. Colloids Surf. A 300, 3–10.

López, T., Alexander-Katz, R., Castillo, P., González, M., Manjarrez, J., Gonzalez, R.D., Ilharco, L., Fidalgo, A., Rieumont, J., 2009. Kinetic study of controlled release of VPA and DPH antiepileptic drugs using biocompatible nanostructured sol–gel TiO_2. J. Mater. Sci. 44, 5459–5468.

López, T., Ortiz-Islas, E., Guevara, P., Rodríguez-Reinoso, F., Gómez, E., Cuevas, J.L., Novaro, O., 2015. Release of copper complexes from a nanostructured sol–gel titania for cancer treatment. J. Mater. Sci. 50, 2410–2421.

Manaia, E.B., Kaminski Kiatkoski, R.C., de Oliveira, A.G., Corrêa, M.A., Chiavacci, L.A., 2015. Multifunction hexagonal liquid-crystal containing modified surface TiO_2 nanoparticles and terpinen-4-ol for controlled release. Int. J. Nanomed. 10, 811–819.

Mandal, S.S., Jose, D., Bhattacharyya, A.J., 2014. Role of surface chemistry in modulating drug release kinetics in titania nanotubes. Mater. Chem. Phys. 147, 247–253.

McMaster, W.A., Wang, X., Caruso, R.A., 2012. Collagen-templated bioactive titanium dioxide porous networks for drug delivery. ACS Appl. Mater. Interf. 4, 4717–4725.

Moon, K., Bae, J.-M., Jin, S., Oh, S., 2014. Infrared-mediated drug elution activity of gold nanorod-grafted TiO_2 nanotubes. J. Nanomater. 2014, 1–8.

Moseke, C., Bae, J.-M., Jin, S., Oh, S., 2012. TiO_2 nanotube arrays deposited on Ti substrate by anodic oxidation and their potential as a long-term drug delivery system for antimicrobial agents. Appl. Surf. Sci. 258, 5399–5404.

Mund, R., Panda, N., Nimesh, S., Biswas, A., 2014. Novel titanium oxide nanoparticles for effective delivery of paclitaxel to human breast cancer cells. J. Nanopart. Res. 16, 2739(1)–2739(12).

Oh, S., Moon, K.-S., Moon, J.-H., Bae, J.-M., Jin, S., 2013. Visible light irradiation-mediated drug elution activity of nitrogen-doped TiO_2 nanotubes. J. Nanomater. 2013, 1–7.

Peng, L., Mendelsohn, A.D., LaTempa, T.J., Yoriya, S., Grimes, C.A., Desai, T.A., 2009. Long-term small molecule and protein elution from TiO_2 nanotubes. Nano Lett. 9, 1932–1936.

Popat, K.C., Eltgroth, M., LaTempa, T.J., Grimes, C.A., Desai, T.A., 2007a. Titania nanotubes: a novel platform for drug-eluting coatings for medical implants? Small 3, 1878–1881.

Popat, K.C., Eltgroth, M., Latempa, T.J., Grimes, C.A., Desai, T.A., 2007b. Decreased *Staphylococcus epidermis* adhesion and increased osteoblast functionality on antibiotic-loaded titania nanotubes. Biomaterials 28, 4880–4888.

Qin, Y., Sun, L., Li, X., Cao, Q., Wang, H., Tang, X., Ye, L., 2011. Highly water-dispersible TiO_2 nanoparticles for doxorubicin delivery: effect of loading mode on therapeutic efficacy. J. Mater. Chem. 21, 18003–18010.

Ren, W., Zeng, L., Shen, Z., Xiang, L., Gong, A., Zhang, J., Mao, C., Li, A., Paunesku, T., Woloschak, G.E., Hosmane, N.S., Wu, A., 2013. Enhanced doxorubicin transport to multidrug resistant breast cancer cells via TiO_2 nanocarriers. RSC Adv. 3, 20855–20861.

Shen, S., Guo, X., Wu, L., Wang, M., Wang, X., Kong, F., Shen, H., Xie, M., Ge, Y., 2014. Dual-core@shell-structured Fe_3O_4–$NaYF_4$@TiO_2 nanocomposites as a magnetic targeting drug carrier for bioimaging and combined chemo-sonodynamic therapy. J. Mater. Chem. B 2, 5775–5784.

Shokuhfar, T., Sinha-Ray, S., Sukotjo, C., Yarin, A.L., 2013. Intercalation of anti-inflammatory drug molecules within TiO_2 nanotubes. RSC Adv. 3, 17380–17386.

Shrestha, N.K., Macak, J.M., Schmidt-Stein, F., Hahn, R., Mierke, C.T., Fabry, B., Schmuki, P., 2009. Magnetically guided titania nanotubes for site-selective photocatalysis and drug release. Angew. Chem. Int. Ed. 121, 987–990.

Signoretto, M., Ghedini, E., Nichele, V., Pinna, F., Crocellà, V., Cerrato, G., 2011. Effect of textural properties on the drug delivery behaviour of nanoporous TiO_2 matrices. Micropor. Mesopor. Mater. 139, 189–196.

Sirivisoot, S., Pareta, R., Webster, T.J., 2011. Electrically controlled drug release from nanostructured polypyrrole coated on titanium. Nanotechnology 22, 085101(1)–085101(15).

Sulka, G.D., Kapusta-Kołodziej, J., Brzózka, A., Jaskuła, M., 2010. Fabrication of nanoporous TiO_2 by electrochemical anodization. Electrochim. Acta 55, 4359–4367.

Sulka, G.D., Kapusta-Kołodziej, J., Brzózka, A., Jaskuła, M., 2013. Anodic growth of TiO_2 nanopore arrays at various temperatures. Electrochim. Acta 104, 526–535.

Wang, Y., Chen, L., Liu, P., 2012. Biocompatible triplex Ag@SiO_2@mTiO_2 core-shell nanoparticles for simultaneous fluorescence-SERS bimodal imaging and drug delivery. Chem. Eur. J. 18, 5935–5943.

Wang, X., Chen, D., Caou, L., Li, Y., Boyd, B.J., Caruso, R.A., 2013. Mesoporous titanium zirconium oxide nanospheres with potential for drug delivery applications. ACS Appl. Mater. Interf. 5, 10926–10932.

Wang, T., Jiang, H., Wan, L., Zhao, Q., Jiang, T., Wang, B., Wang, S., 2015a. Potential application of functional porous TiO_2 nanoparticles in light-controlled drug release and targeted drug delivery. Acta Biomater. 13, 354–363.

Wang, Z., Xie, C., Luo, F., Li, P., Xiao, X., 2015b. P25 nanoparticles decorated on titania nanotubes arrays as effective drug delivery system for ibuprofen. App. Surf. Sci. 324, 621–626.

Wu, K.C.-W., Yamauchi, Y., Hong, C.-Y., Yang, Y.-H., Liang, Y.-H., Funatsu, T., Tsunoda, M., 2011. Biocompatible, surface functionalized mesoporous titania nanoparticles for intracellular imaging and anticancer drug delivery. Chem. Commun. 47, 5232–5234.

Xiao, X., Yang, L., Guo, M., Pan, C., Cai, Q., Yao, S., 2009. Biocompatibility and in vitro antineoplastic drug-loaded trial of titania nanotubes prepared by anodic oxidation of a pure titanium. Sci. China Ser. B 52, 2161–2165.

Xin, Y., Chu, P.K., 2011. Nanostructured titania coatings for biological applications: fabrication and characterization. In: Zhang, S. (Ed.), Biological and Biomedical Coatings Handbook: Processing and Characterization. CRC Press, USA, pp. 203–237.

Xu, P., Wang, R., Ouyang, J., Chen, B., 2015. A new strategy for TiO_2 whiskers mediated multimode cancer treatment. Nanoscale Res. Lett. 10, 94(1)–94(11).

Yao, C., Webster, T.J., 2009. Prolonged antibiotic delivery from anodized nanotubular titanium using a coprecipitation drug loading method. J. Biomed. Mater. Res. B. 91, 587–595.

Zhang, Y., He, F., Sun, Z., Li, L., Huang, Y., 2014. Controlled delivery of dexamethasone from TiO_2 film with nanoporous structure on Ti–25Nb–3Mo–2Sn–3Zr biomedical alloy without polymeric carrier. Mater. Lett. 128, 384–387.

Zwilling, V., Aucouturier, M., Darque-Ceretti, E., 1999. Anodic oxidation of titanium and TA6V alloy in chromic media. An electrochemical approach. Electrochim. Acta 45, 921–929.

CHAPTER

REDOX ACTIVATED POLYMERIC NANOPARTICLES IN TUMOR THERAPY

10

Houman Alimoradi*, Siddharth S. Matikonda*, Allan B. Gamble*, Gregory I. Giles*, Khaled Griesh*,**

**University of Otago, Dunedin, New Zealand*

***Arabian Gulf University, Manama, Kingdom of Bahrain*

CHAPTER OUTLINE

1 INTRODUCTION

The effectiveness of chemotherapeutic agents is particularly limited by their intrinsic toxicity to most, if not all, tissues. Conventional chemotherapy delivers the toxic drug indiscriminately to tumors and normal tissues (Xie et al., 2004). One of the various approaches to avoid undesirable systemic side effects and other off-target cytotoxicity is to change the biodistribution of anticancer drugs using cancer-selective drug-delivery systems (DDS). DDS are developed to improve the safety profile of toxic drugs and could protect them from degradation and metabolism, thus safeguarding their delivery to the targeted tissue. This strategy not only minimizes their penetration to normal tissues but also aids in overcoming other pharmaceutical limitations, such as hydrophobicity (e.g., paclitaxel) and membrane impermeability (e.g., gemcitabine) (Ernsting et al., 2012). Anatomical and pathological abnormalities of tumor tissue (such as higher permeability of tumor vascular) are generally exploited to selectively target solid tumors.

The enhanced permeability and retention (EPR) effect of macromolecules has been recognized as "a royal gate for targeted anticancer nanomedicines" (Iyer et al., 2006). The polymeric macromolecules that are traditionally used for drug delivery include liposomes, polymer nanoparticles, micelles, dendrimers, and inorganic nanoparticles made of iron oxide, gold, quantum dots, or metal oxide frameworks. Despite the significant progress of EPR-based DDS in solid tumor therapy, only a few nanocarriers have reached the clinic so far (Taurin et al., 2012). Hence, more advances in the drug delivery are required to increase the selectivity and efficacy of tumor therapy. It would be highly beneficial to develop drug carriers such that they can precisely and selectively sense the signal caused by tumor cells, which differentiate them from normal cells (e.g., low pH, higher temperature, and higher production of reactive oxygen species [ROS])). This strategy would eventually deliver their cargo according to the pathophysiology of tumor tissues, with such systems termed as stimuli responsive DDS (Mura et al., 2013). Not only can these DDS gain the benefit of EPR effect for targeting tumors but they can also sense the unique characteristics of the tumor microenvironment (such as low pH), actively delivering their cargo at the tumor site.

Recently, stimuli-responsive DDS sensitive to specific endogenous stimuli (low interstitial pH, high glutathione concentration in some types of tumors, such as hepatocellular carcinoma and high level of certain enzymes, such as matrix metalloproteinases) has been widely researched (Cheng et al., 2011; Huang et al., 2001; Li et al., 2013). At the cellular level, pH sensitivity can lead to the release of the drug cargo from the DDS system into late endosomes or lysosomes, hence promoting the escape of the drug carriers into the cytoplasm. At the tissue level, specific microenvironmental changes associated with pathological conditions, such as ischemia, inflammatory diseases, or infections are observed. Physical stimuli, such as thermo-, light- or ultrasound-sensitive triggers can be employed to achieve targeted delivery of pharmacologically active molecules to a diseased area in the body.

Cancer is a condition that evidences the inadequate control over many physiological mechanisms, leading to an improper functioning of the tumor cells, among which redox imbalance plays a very important role in both cancer aggression and resistance to cancer treatment (Giles, 2006). Due to a high metabolism rate, immature tumor vascular, and poor blood supply, many tumor cells are under cyclic hypoxia, increased intrinsic oxidative stress, and/or elevated antioxidant molecules in comparison to their nonmalignant counterparts (Acharya et al., 2010; Giles, 2006; Trachootham et al., 2009). In addition, different parts of tumor tissues show different level of oxygenation. The peripheral zones of tumors are usually better oxygenated than the innermost zones. As tumor volume expands, the central region of tumors usually suffer inadequate blood supply and tumors beyond certain volume frequently develop central necrosis. Unfortunately, not all the cells in the central necrotic region are dead. A small number of central cells can resist this sever insult and usually are the source for development of more resistant tumor cells. Such heterogeneity in the redox state of cancer cells can be targeted via redox-responsive DDS (RDDS) (Trachootham et al., 2009). This chapter highlights: (1) the biochemical foundation of oxidative stress and its role in cell signaling; (2) pathophysiology of tumor vascular and the possible reasons for heterogeneity in redox state of tumor cells; (3) the redox heterogeneity of cancer cells, and differences in redox metabolism between cancer cells and their nonmalignant cells of origin; (4) a brief comment on the concept of active and passive targeting of polymeric nanoparticles; and (5) finally, the text focuses on RDDS, which include six main classes of redox responsive nanocarriers; polysulfides, polyselenides, quinones, metal complexes, arylboronic esters, aryl oxalates, and other miscellaneous examples.

2 REACTIVE CHEMICAL SPECIES IN HUMAN PHYSIOLOGY

The reactive chemical species, derived from oxidative metabolism, are divided into two major types: radicals that have one or more unpaired electron(s) in their outer molecular orbitals (e.g., superoxide and nitric oxide), and nonradical compounds that do not contain unpaired electron(s) (e.g., hydrogen peroxide, ozone, peroxynitrite, and hydroxyl radical) but are chemically active and can be converted to free radicals. Based on the central atom, which carries an unpaired electron, there are four main types of free radicals: reactive oxygenated species (ROS), reactive nitrogen species (RNS), reactive sulfur species (RSS), and reactive chloride species (RCS) (Sosa et al., 2013). Of all reactive compounds, ROS and RNS play the most important roles in human physiology (Giles and Jacob, 2002). ROS include hydroxyl radical, singlet oxygen, superoxide anion, hydrogen peroxide (H_2O_2), and ozone (O_3). The major RNS is nitric oxide (NO), which is generated from arginine by nitric oxide synthase (NOS), and is capable of reacting with some ROS producing a range of subsequent RNS including peroxynitrite and nitrous anhydride (Sosa et al., 2013).

Generation of such reactive chemical compounds occurs under normal physiological conditions as an integral part of cell signaling. However, under pathological conditions, a severe increase in ROS levels can lead to irreversible oxidative damage. Therefore, perturbation to cellular redox homeostasis is implicated in the pathology of many diseases including cardiovascular disorders, diabetes, chronic inflammation, and cancer. The major mechanism by which oxidative stress regulates cell signaling proteins is via the oxidation of cysteine, tyrosine, and methionine amino acids, or other redox sensitive prosthetic groups, such as iron-sulfur clusters. These oxidative reactions can affect protein signaling at three levels. First, by regulating protein biosynthesis via alterations at the transcription, translation, and folding stages. Second, by posttranslational modifications to existing proteins, and third, via alteration of protein stability (Jacob et al., 2003). On the other hand, cells have a variety of defense mechanisms to inhibit the damage induced by ROS. For example, superoxide dismutase (SOD) converts superoxide to hydrogen peroxide and oxygen; catalase converts H_2O_2 to water and oxygen; whereas glutathione peroxidase converts hydrogen peroxide to water or lipid peroxides to alcohols. These defense mechanisms are paramount to the survival of all aerobic life forms. There are also several nonenzymatic detoxifying molecules, such as ascorbic acid, uric acid, and glutathione that effectively defend cells against the deleterious impact of ROS.

3 ORIGINS OF ENHANCED FREE RADICAL PRODUCTION IN TUMORS

The tumor vasculature is characterized by defective endothelial cells, lack of a smooth muscle layer, and wide fenestrations ranging from 300 to 4700 nm (Graves et al., 2010). Furthermore, vessels are often spread far apart and organized nonhierarchically (Dewhirst et al., 1999, 2008; Secomb et al., 2004; Vaupel and Harrison, 2004). The tumor's chaotic microvascular architecture and insufficient blood flow leads to tumor hypoxia, which is one of the main causes of enhanced free radical production in tumor cells. Hypoxia is a state in which partial pressure of oxygen (PaO_2) goes below 14 mmHg, whereas PaO_2 in normal tissues is on average 40 mmHg (Yeh and Kim, 2015). Clinically, around 50%–60% of all solid tumors irrespective of size and histopathological features are reported to be hypoxic (Vaupel, 2004). Vaupel and coworkers measured the median PaO_2 values of normal breast tissue and breast cancer tissue in patients, and showed that average PaO_2 in normal breast tissue was 65 mmHg, with no readings below 12.5 mmHg. In contrast, breast cancers had a median PaO_2 value at 30 mmHg, and in 18% of investigated regions within tumors it was between 0 and 5 mmHg (Vaupel et al., 1991).

Under hypoxia, cells rely on anaerobic respiration, which occurs only in the cytoplasm. Two molecules of ATPs are synthesized in the electron transport chain, using inorganic molecules other than O_2, and in such reactions CO_2 is released; however, as mentioned previously, O_2 is not required to provide energy (Sosa et al., 2013). Although anaerobic respiration is energetically less efficient, this process is faster than that the aerobic respiration. It is of interest that cancer cells display a high anaerobic respiration rate even in the presence of oxygen; this phenomena is termed as the Warburg effect (Kim and Dang, 2006). Consequently, high levels of free radicals are generated through this pathway in cancer cells, even in the presence of O_2. Moreover, under hypoxic conditions some pathways [like upregulation of NOS (Burke et al., 2013), NADPH oxidase (NOX) (Ushio-Fukai and Nakamura, 2008) and xanthine oxidase (XO) (Griguer et al., 2006)] are activated more than those under normal physiological situations, which results in higher levels of free radicals.

On the other hand, tumor hypoxia in many cells is cyclic in nature with two frequencies: (1) a few cycles per hour, due to the macrovesicles randomly getting blocked by aggregation of blood cells and (2) variations on a timescale of hours to days due to rapid growth of tumor and compression of vessels by surrounding tissue. Cyclic hypoxia increases the levels of free radicals more than stable hypoxic condition as a result of hypoxia and reperfusion injury. Within the tumor, NOX and XO are thought to play an important role in this phenomenon (Dewhirst et al., 2008). Enhanced free radicals force tumors to adapt by enriching the population of cells that can evade apoptosis and survive at higher ROS levels (Brown and Wilson, 2004; Subarsky and Hill, 2003).

The first detoxification strategy is to upregulate glutathione (GSH) and thioredoxin, which function as highly potent redox-buffering systems in cells (Jacob et al., 2003; Trachootham et al., 2009). In vivo, some tumor tissues, such as hepatocellular carcinoma have fourfold higher GSH concentrations than those of normal tissues (Kuppusamy et al., 2002). Cells with higher concentration of GSH, not only exist in different types of tumor, but also coexist in different regions within the same tumor. It must be emphasized that every single tumor cell may have a different amount and ratio of GSH/ROS levels at different stages (Fang et al., 2009; Lee et al., 1989; Ray et al., 2012).

Upregulating ROS-scavenging enzymes, such as SOD, catalase, and the peroxidases is the second adaptation mechanism by which cancer cells maintain redox homeostasis (Jacob et al., 2003; Murawaki et al., 2008). However, such adaptation processes appear to be of limited capacity in cancer cells, which, under sustained ROS stress conditions, tend to heavily utilize adaptation mechanisms and can also exhaust ROS-buffering capacity.

The third adaptive mechanism involves the regulation of cell signaling pathways, which are frequently modified as a result of oxidative stress. The hypoxia-inducible family of transcription factors (HIFs) bind to the hypoxia-response element (HRE) in the enhancer region of downstream genes and turn on the transcription of several genes which modulate energy production metabolism, glucose transporters, the expression of signaling molecules including growth factors (GF) and angiogenesis (Sosa et al., 2013; Wilson and Hay, 2011). HIF-1, which is known as the "master regulator of oxygen homeostasis" in cells contains two subunits: HIF-1α and HIF-1β (Semenza, 2003). HIF-1β is normally active; however, HIF-1α is degraded in normal PO_2 in the cells via two main inhibitory enzymes: prolyl hydroxylases (PHDs) and factor inhibiting HIF-1 (FIH-1). In almost all cell types under hypoxia (the average PaO_2 less than 10 mmHg), these enzymes get deactivated and consequently HIF-1α can evade degradation and binds to the HRE in the enhancer region of downstream genes and turns on the transcription of genes in order to protect the cell against the hypoxia (Semenza, 2014). Beyond hypoxia, other factors, such as enhanced level of ROS, cytokines, and other signaling molecules stimulate HIF-1α synthesis via activation of the phosphatidylinositol 3-kinase (PI3K) or mitogen-activated protein kinase (MAPK) pathways (Semenza, 2003). One of the main compensatory mechanisms against the hypoxia is increasing angiogenesis and permeability of the vessels, in which vascular endothelial growth factors (VEGFs) play the most important role. VEGFs enhance matrix metalloproteinases (MMPs) expression of the endothelial cells to break down the extracellular matrix and facilitate the migration of endothelial cells driving them toward the hypoxic tissues to improve the oxygenation of tissues under hypoxic insult. VEGF family, consist of five subtypes of; VEGF-A, VEGF-B, VEGF-C VEGF-D, and VEGF-E. Except VEGF D, which mainly controls lymphangiogenesis, they are instrumental in the formation of new blood vessels. Furthermore, VEGF increases the endothelial vascular permeability and hence it was originally known as the vascular permeability factor. In vascular permeability, VEGF A VEGF-A is reported as a key factor in vascular permeability (Leung et al., 1989; Nishida et al., 2006; Rafii and Skobe, 2003).

Immunohistochemical analysis demonstrates that the VEGF family and their receptors are overexpressed in most of the human solid tumors (Salven et al., 1998). Though angiogenesis in tumor tissues occur more than in normal tissues, it does not occur in an efficient manner, which is responsible for the formation of immature tumor vasculature (Folkman, 1992, 2002; Goldmann, 1907). Several clinical and preclinical studies have shown that, in most malignant solid tumors the expression of HIF-1α and VEGF are correlated with tumor metastasis and are negative predictors of survival rate of the patients (Adams et al., 2000; Lissoni et al., 2000). In addition to VEGFs, the permeability of tumor vessels is further amplified by other vascular mediators including bradykinin (BK), nitric oxide (NO), prostaglandins (PGs), and interleukin-2 (Fang et al., 2011).

The fourth adaptive mechanism of tumors in response to high level of ROS is the ROS induced mutations in DNA. As a result, cells with successful mutations can adapt to increased ROS, such as cells that lose sensitivity to p53-mediated apoptosis and can survive at significantly higher ROS concentration. Cells with ROS induced mutations show less sensitivity to chemotherapy and radiotherapy (Lowe et al., 1994; Teicher et al., 1981).

In brief, most of cancer cells in solid tumors exist under a state of hypoxia, higher oxidative stress, and upregulated cellular antioxidants, all of which are distinctly different from normal tissues. The adaptions required for tumor cells to survive and proliferate in this environment, including increases in antioxidants and altered metabolism and gene transcription, endows the tumor with a specific redox state. An emerging area of drug discovery is to exploit this imbalance in the redox state as stimuli for active targeting of cancer cells.

4 ACTIVE AND PASSIVE DRUG-DELIVERY SYSTEMS

Macromolecules and nanoparticles, larger than 40 kDa, can selectively leave the leaky tumor vascular bed and accumulate inside the interstitial space of tumor tissues (Fang et al., 2011). Furthermore, lymphatic drainage of tumor tissues is usually deficient (Iyer et al., 2006). These factors alter the molecular and fluid transport dynamics in the tumor and result in increase of the delivery and retention of macromolecules, a phenomena termed as enhanced permeability and retention "EPR" effect of macromolecules (Maeda et al., 2000). This concept was introduced as a basis of passive delivery of polymeric

DDS in cancer therapy and it is possible to achieve three to ten times higher concentration of polymeric drugs at the tumor site (Iyer et al., 2006). EPR effect has significantly influenced the design, development, and improvement of delivery platforms. One of the main limitations of EPR based DDS is less EPR effect of macromolecules in the hypoxic area of tumors, particularly the central area of metastatic cancers, which could be responsible for their limited applications in the clinic (Greish, 2007; Iyer et al., 2006).

Therefore, the factors [including vascular permeability (VPF), bradykinin, prostaglandins, and nitric oxide] that affect the EPR effect of macromolecules have been modified to control the tumor growth and or increase the passive DDS. As described earlier, the corner stone of EPR effect is VEGFA, which has a significant role in vascular permeability (Leung et al., 1989; Nishida et al., 2006; Rafii and Skobe, 2003). However, various permeability enhancers have been found to contribute to the capability of nanoparticles to concentrate in tumor tissues. Bradykinin (BK) and prostaglandins (PGs) are important mediators of vascular permeability during tissue injury and inflammation (Matsukado et al., 1998; Unterberg et al., 1984). Bradykinin enhances tumor growth (Ikeda et al., 2004) and overexpression of BK receptors in various solid tumors has been reported (Iyer et al., 2006). Maeda et al. have discovered that bradykinin antagonists (such as HOE-140), and cyclo-oxygenase (COX) inhibitors (such as indomethacin) significantly repressed vascular permeability of tumors (Wu et al., 1998). The effect of NO as a mediator on vasodilatation, vascular permeability, angiogenesis, cell proliferation are well known (Doi et al., 1996; Fang et al., 2011). Hence, scavenging of NO, such as an imidazolineoxyl A-oxide derivative, or inhibition of its generation, using a NOS inhibitor Nv-monomethyl-L-arginine, was found to suppress vascular permeability and, thereby, the EPR effect and, hence, tumor growth (Maeda et al., 1994). NO donors have been used to enhance EPR effect and improve the tumor targeting of anticancer drugs (Fang et al., 2011; Yasuda, 2008). For instance, Seki et al. (2009) have shown that, by applying nitroglycerin (NG) ointment on the skin of breast cancer-bearing rats, the EPR effect of macromolecules was significantly enhanced. To determine EPR effect, Evans blue dye (EBD) was used. EBD binds to serum albumin and make a macromolecule. Hence, the concentration of the dye in any specific tissue (such as tumor) shows the vascular permeability of the tissue. EBD concentration in tumors of the animals that were treated by NG ointment was found to be 2–3 times greater than nontreated tumor bearing animals. Moreover, similar studies have shown that NO donors, such as isosorbide dinitrate can increase the efficacy of radiotherapy by increasing blood flow of the tumor (Jordan et al., 2000). Matrix metalloproteinases (collagenases) have also been implicated in the targeting of nanoparticles to tumor tissues. Matrix metalloproteinases MMPs are a series of enzymes that degrades the extracellular matrix, allowing tumor cells to expand and grow and have a pivotal role in cell migration and specially tumor metastasis (Coussens et al., 2002). Moreover, MMPs increase the EPR effect of macromolecules (Iyer et al., 2006). All these mediators implicated in the EPR phenomenon can potentially be used to further improve the targeting of nanoparticles to tumor tissues.

Though the EPR effect can safeguard the delivery of nanosystems to tumor tissues, it does not guarantee the uptake of the system by target cells.

Eventually the activity of DDS in tumor therapy relies entirely on the entry of drug carriers/cargo into target cells. In this step, tumor-specific properties (e.g., low pH and tumor specific ligands) can be utilized to promote the cytotoxic effect of these systems. However, the active targeting process cannot be entirely viewed as a separate process to diffusion, because it occurs only after passive accumulation of DDS in tumors (Bae, 2009). Over the last few decades, the use of nanocarriers for controlled and targeted delivery of anticancer drugs has shifted toward stimuli-responsive delivery systems (Peer et al., 2007). Various stimuli including pH (Gao et al., 2010), light (Kang et al., 2011), temperature (Chung et al., 1999), and ultrasound have been utilized for targeted activation/ release of cytotoxic compounds at tumor site (Mura et al., 2013). In this chapter, we mainly focus on active DDS, which rely on the disrupted redox state of cancer cells for selective targeting of solid tumors.

5 REDOX RESPONSIVE DRUG-DELIVERY SYSTEMS

Ideally, a redox responsive nanocarrier includes a functional group(s), which, upon interaction with oxidizing and/or reducing agents in cancer cells (such as H_2O_2 and GSH), undergoes electronic changes. This can lead to a resonance shift of electrons and results in the chemical bond cleavage. It may also undergo a change in the hydrophobicity of the polymer, which subsequently changes the integrity of the NP and releases the cargo. For instance, in polysulfide based RDDS, disulfide bonds act as a redox sensitive motif and, following its reaction with H_2O_2 (oxidation into sulfone and later to disulfone; Song et al., 2014), the hydrophobicity of the polymers change leading to the collapse of NPs and drug release.

The design of a successful redox responsive NP must meet the following requirements. First, the NPs must be adequately stable to metabolism in normal cells, and they must have the minimal toxicity. Second, the NPs pharmacokinetics must allow the agent to diffuse through tumor tissue before activation. Third, the NPs must have suitable pharmacological and solubility properties, as well as high encapsulation or loading capacity. Fourth, because tumor contains heterogeneous redox environment and all cells do not have significantly different redox stat from normal cells, the loaded drug (s) should also have the ability to back-diffuse a limited distance from the host cell; this phenomenon is called bystander effect (Brown and Wilson, 2004; Freeman et al., 1993). Furthermore, an ideal redox activated prodrug responds to the stimuli in a nonlinear behavior. In this way, activation of the prodrug should allow a prompt reaction to high concentration of stimuli bursts, but not to low concentrations of a strong stimuli, nor to the presence of weak stimulation (Vo et al., 2009).

With respect to these properties, there are six major classes of redox responsive functional groups (polysulfides, polyselenides, quinones, metal complexes, boronic esters, and aryloxalates), which are generally recruited to develop RDDS. In the following subsections, redox-responsive polymers containing these functional groups will be discussed.

5.1 SULFUR (II)-CONTAINING POLYMERS AS REDOX RESPONSIVE DRUG-DELIVERY SYSTEMS

Sulfur-containing molecules have long attracted attention as potential redox activated moieties due to their range of biologically accessible oxidation states. They can also act as an indicator for cellular redox status (Jacob et al., 2003). The first polysulfide-containing redox-responsive nanocarriers were designed in 2004 (Napoli et al., 2004), and currently they are the most widely used and advanced RDDS (Vo et al., 2009). As mentioned earlier, thioether or disulfide chemical bonds act as the sensor for the NPs because they can be oxidized by H_2O_2, and/or reduced by GSH. As a good example of polysulfide-based RDDS, Wang has designed a nanocarrier that utilizes the heterogeneous redox state of tumors for selective degradation. Briefly, a potent topoisomerase-I inhibitor, 7-ethyl-10-hydroxyl-camptothecin (SN38), was conjugated to an oligo ethylene glycol (OEG) by a thioether molecule to make OEG-2S-SN38 nanocapsules. The NPs in PBS at pH 7.4 released less than 40% SN38 within 2 days. Upon addition of 5 mM H_2O_2 (these levels will not be found in vivo but demonstrate a switching effects), the nanocapsules quickly degraded and release the SN38 (more than 80% within 15 min). The antitumor activity of the nanocapsule was then evaluated in colon cancer bearing mice, and 22 weeks after the first treatment the survival rate of animals treated by the nanocapsules was 77%, whereas it was 14% for the animals treated with the free drug, providing evidence for effective in vivo drug release (Wang et al., 2013).

Sulfur (II)-based RDDS, can be divided into two major subgroups poly(alkylene sulfide) and disulfide incorporated polymers. There are three main methods to prepare poly(alkylene sulfides); first, step-growth polymerization; second, anionic ring-opening polymerization (ROP); and third, emulsion anionic ROP combined with subsequent cross-linking or functionalization (Song et al., 2014). Step-growth polymerization can be used to synthesize both poly(alkylene sulfide) and poly(arylene sulfide) via nucleophilic substitution and nucleophilic or radical addition reactions. However, it is difficult to control the molecular weight and (poly dispersity index) PDI of the polymers. In the ROP approach, five-membered rings or larger cyclic alkylene sulfides undergo either cationic or anionic ROP to obtain poly(alkylene sulfide) with higher molecular weights, whereas, in the third strategy, oxidation-responsive NPs consisting of a polypropylene sulfide core covered by a poly(ethylene glycol) (PEG) corona are prepared by emulsion based anionic ROP of propylene sulfide using pluronic F-127 as an emulsifier, 1,3-propanedithiol as a bifunctional initiator, and 1,8-diazabicyclo[5.4.0]undec-7-ene as a basic catalyst. The success of this facile one-pot approach is due to the hydrophobic character of the monomer (propylene sulfide) and PPS, and the tolerance of the propagating species (thiolates) to water at basic pH values ($pH < 11$).

For synthesis of polymers that incorporate disulfide(s), an exchange reaction is an efficient method to incorporate the disulfide linkage with other molecules and polymers (Huo et al., 2014). In the process of cross-linking structures, such as shell cross-linked micelles, interlayer cross-linked micelles and gels, disulfide-containing cross-linkers play an important role (O'Reilly et al., 2006).

Sulfur-based RDDS have been extensively used for delivery of different anticancer agents [including, conventional chemotherapeutic drugs, such as doxorubicin (Wang et al., 2011), anticancer proteins, such as caspases (Zhao et al., 2011) and genes, such as pGL-3 (Cai et al., 2011)]. In the following section, a few of these examples are discussed.

Gao et al. designed an poly[oligo(ethyleneglycol)thiomalicate] (POEGTMA) based amphiphilic tri-block copolymer (PCL-b-POEGTMA-b-PCL) using a combination of condensation and ROP in one pot. They used an oligomer (ethylene glycol) as a dialcohol monomer to copolymerize with D/L-thiomalic acid via polycondensation at 100°C to make POEGTMA. Sequentially, the polymerization system was cooled to 65°C, and 3-caprolactone (CL) was added to the reaction mixture to produce the triblock copolymer. In aqueous solution, a micellar system was prepared from the polymer and paclitaxel, followed by oxidation of mercapto groups of the polymer to make a micelle with disulfide cross-links (Gao et al., 2013). The paclitaxel release from the micelle in PBS solution was less than 10%, within 100 h, whereas in the presence of 10 mM dithiothreitol a 100% release was observed during the same time period.

A positively charged polymeric shell interconnected by disulfide-containing cross-linkers was recently synthesized for intracellular anticancer protein delivery (Zhao et al., 2011). Caspase-3 was noncovalently loaded in a redox-responsive nanocapsule, which, under reducing conditions, could release the proapoptotic caspase and induce cell death. By condensation of acrylamide (AAm) and

N-(3-aminopropyl) methacrylamide (APMAAm) and different disulfide cross-linkers (such as *N,N*-methylene bisacrylamide and *N,N*-bis(acryloyl)cystamine), a poly-aminoester was produced. Then a Rhodamine-tagged caspase-3 was encapsulated in the polymer system and redox responsive NPs were produced. Using an enhanced green fluorescence protein (eGFP), nanocapsules were showed to be efficiently internalized into the cells and also to release the protein in the reducing cytosolic environment. Caspase-3 loaded NPs induced apoptosis in three human cancer cell lines HeLa, MCF-7 and U-87 MG (Zhao et al., 2011).

Further, several NPs were developed for gene delivery using polypeptides with disulfide cross-links. For instance, McKenzie et al. (2000) designed a polypeptide by inserting multiple cysteine residues into short synthetic peptides, to make a 20-mer peptide; Cys-Trp-Lys18 (CWK18). Following the binding of the peptide into DNA (through ionic binding of the peptide to the DNA backbone) the thiol group of cysteine residues was oxidized and formed disulfide bonds, which stabilized the plasmid DNA condensates. Disulfide bond formation led to a decrease in particle size of the DNA complex and prevented dissociation. The cross-linked polypeptide/DNA complex elevated the expression of the plasmid DNA from 2.5-to 10-fold in COS-7 cell line and from 4- to 60-fold in HepG2 (human liver carcinoma) cells compared to the uncross-linked alkylated CWK18 (AlkCWK18)/DNA complex (Fig. 10.1).

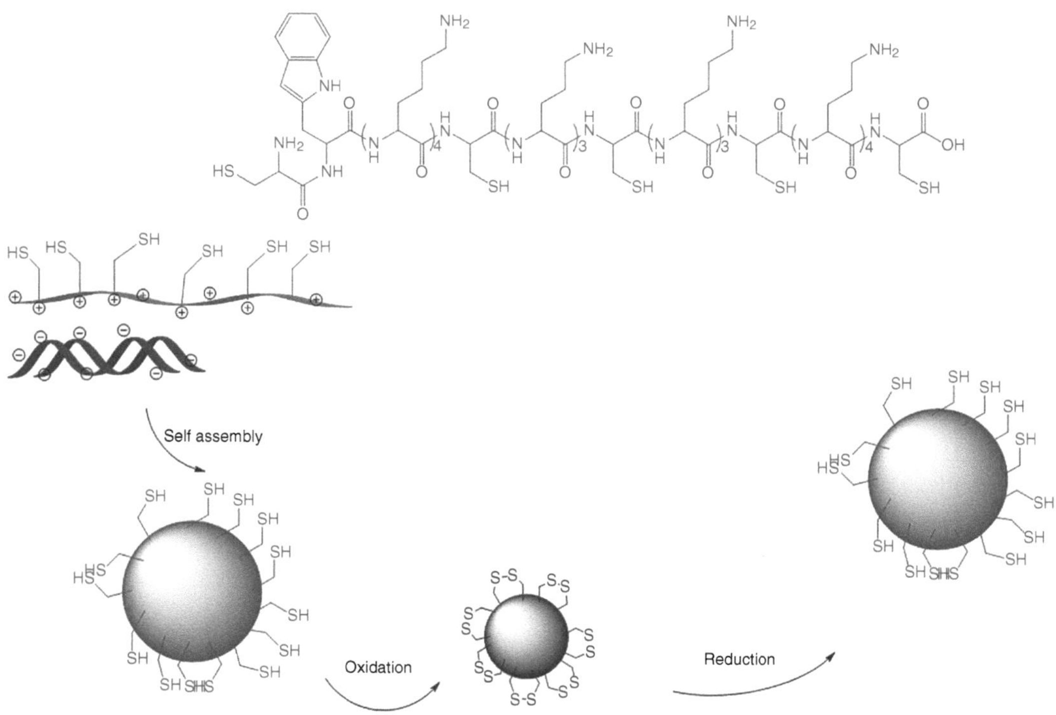

FIGURE 10.1 Ionic Interaction of the Peptide (Positive Charge) With DNA Backbone (Negative Charge) Make Peptide DNA Condensates.

Following the oxidation of thiol groups in the peptide and formation of disulfide cross-links, the DNA condensates became stable, which could shield DNA during the delivery process

Sufficient cell internalization and successful endosomal escape of the catiomer/DNA condensates are two critical stages in developing polymeric gene–delivery systems (Park et al., 2010). To overcome these problems, Shi and coworkers designed a dual redox and acid responsive NP in which two PEG strains were connected to an acid-labile poly L-lysine (PLL) segment cross-linked by imine linkers (PLL-glutaraldehyde-star) using a disulfide bond to make mPEG-SS-PLL-star structure (Cai et al., 2012). PEG-SS as redox responsive and external PEG shell was used to increase the internalization of the NPs to the cells in the presence of tumor relevant GSH concentration, and the acid sensitive part is to accelerate the release of genetic cargo following successful internalization into targeted cells. The mPEG-SS-PLL-star showed high DNA binding affinity, and formed stable DNA condensates, which underwent rapid destabilization in the presence of 10 mM GSH (its size increased from 200 to 650 nm within 2 h). In addition, at pH 5 the size of the condensates increased from 200 to 1000 nm indicating the hydrolysis of imine groups of the polymer, which could not effectively condense to the DNA. Moreover, the plasmid DNA (pDNA) release rate from the condensates was investigated using agarose gel-electrophoresis. Migration of pDNA encapsulated by the polymer at pH 7.4 was completely inhibited, however incubation for 30 min in the presence of 10 mM GSH or pH 5.0 effectively allowed migration.

Park and coworkers developed disulfide-cross-linked hyaluronic acid (HA) nanogels physically encapsulating green fluorescence protein (GFP) siRNA for tumor-targeted intracellular delivery of siRNA. These HA nanogels had a size distribution ranging from 200 to 500 nm with a siRNA loading efficiency of approximately 50%. They were prepared from thiol-derivatized HA via an inverse water-in-oil emulsion method followed by self-cross-linking through the formation of disulfide linkages. HA nanogels effectively protected the encapsulated siRNA from enzymatic degradation. The entrapped siRNA was released completely within 1h in a system with 10 mM GSH. The HA/siRNA nanogels were readily taken up by HCT-116 cells, which possess HA-specific CD44 receptors on their surface, but were not taken up by CD44 deficient NIH-3T3 cells. Gene silencing studies showed that HA nanogels encapsulating GFP siRNA exerted a significant level of GFP gene inhibition in HCT-116 cells, in both 10% serum (62.1, 6.1% inhibition efficiency) and nonserum (68.5, 10.6% inhibition efficiency) conditions. Stable PEG/DNA nanogels were prepared by using a thiol-functionalized six-arm branched PEG for DNA solubilization in DMSO followed by oxidative cross-linking.

5.2 SELENIUM-CONTAINING REDOX-RESPONSIVE DRUG-DELIVERY SYSTEMS

Selenium (II)-containing organic compounds actively participate in redox reactions with a higher sensitivity to oxidants compared with sulfur-containing analogues (Song et al., 2014). Selenium -containing polymers have the same mechanism of action as sulphide-containing polymers (Table 10.1). The application of selenium-containing polymers in drug delivery is still in its infancy due to the poor aqueous solubility and the difficulty in the incorporation of the diselenide bond (Song et al., 2014). Similar to sulfur-containing micelles, the oxidation of selenide to selenone group, a more hydrophilic functionality, results in the oxidation-responsive disassembly of the formed aggregates. A family of selenide- or diselenide-containing polymers have been utilized as redox sensitive biomaterials for controlled release and the predominant synthetic approach toward these polymers is through step-growth polymerization either in the main backbone or as a side chain (Xu et al., 2013).

Ma et al. have constructed a selenium or sulfur-containing amphiphilic block copolymer (PEG-PUSe-PEG) or (PEG-PUS-PEG) (PUSe stands for polyurethane selenide, whereas PUS refers to polyurethane sulfide) using PEG as hydrophilic motifs and a polyselenide hydrophobic block. This

Table 10.1 The Major Redox Responsive Functional Groups used in Drug Delivery Systems

	Functional Group		References
1	Thioether or disulfide	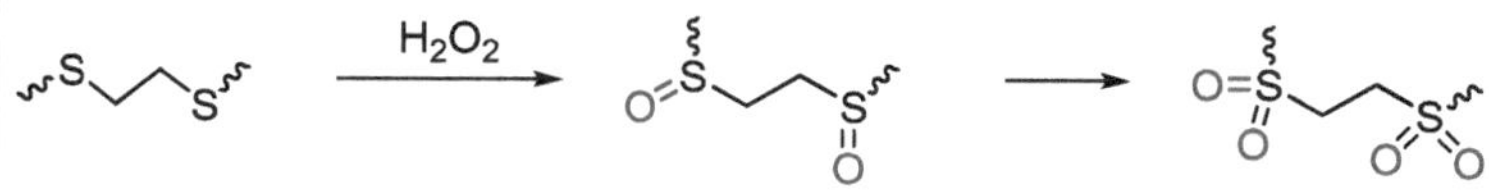	Song et al. (2014)
		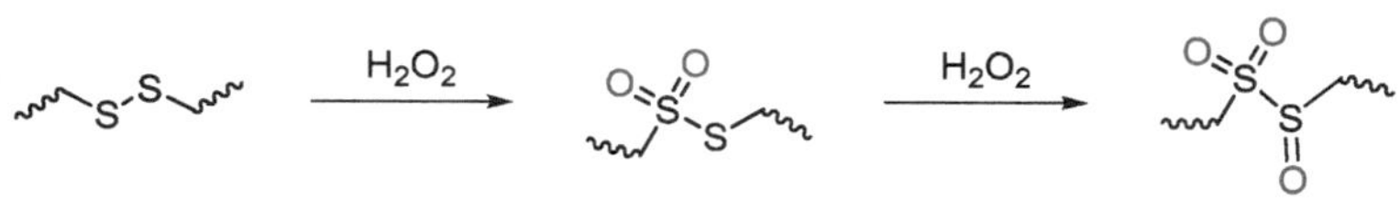	Huo et al. (2014)
			Sun et al. (2013)
2	Diselenide	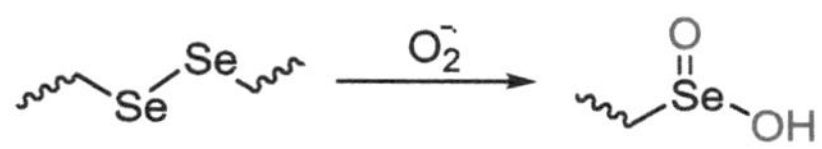	Song et al. (2014)
		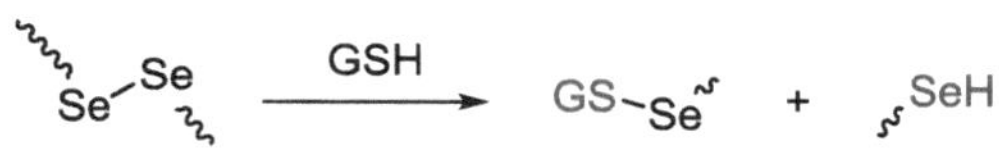	Huo et al. (2014)
3	Quinone	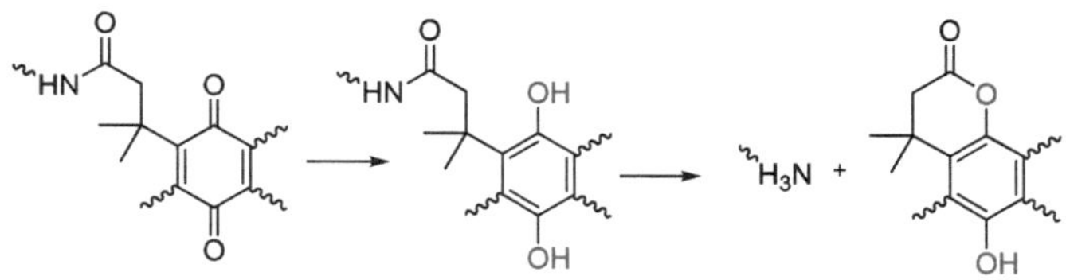	McCarley (2012)
4	Metal complex	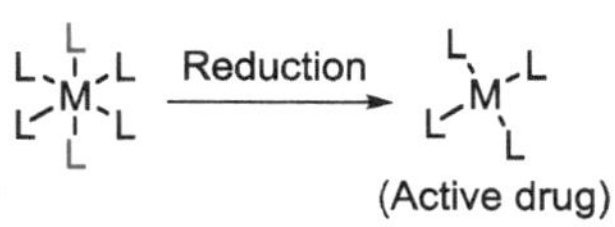	Graf and Lippard (2012)
		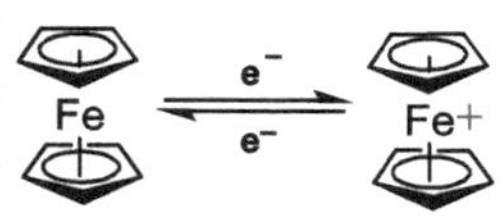	McCarley (2012)
5	Boric acid	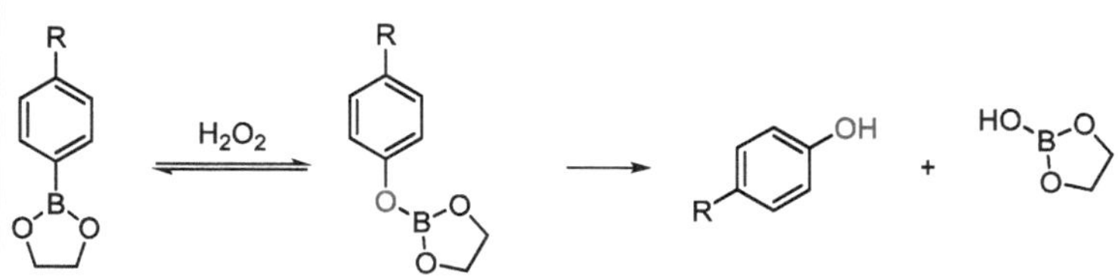	de Gracia Lux et al. (2012)
6	Poly-aryloxalate	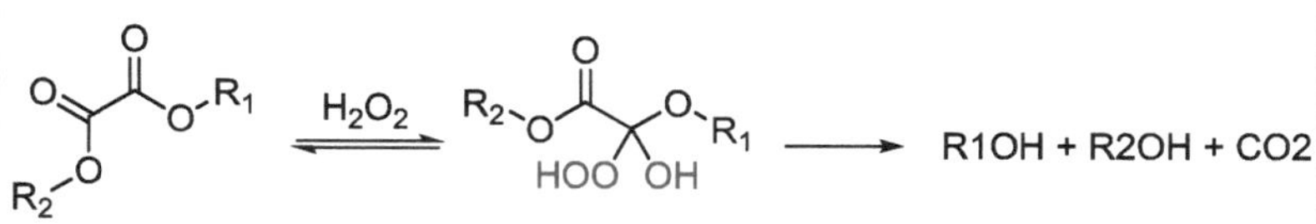	Kim et al. (2010)

hydrophobic block was synthesized via polymerization of toluene diisocyanate with monoselenide or sulfur-containing diols. Finally, doxorubicin was encapsulated into the polymeric system to make micelles. PEG-PUSe-PEG showed more sensitivity to a mild oxidative environment (such as 0.1% H_2O_2 v/v) compared with the sulfide analogue PEG-PUS-PEG micelles (Ma et al., 2010). An exposure time of 10 h to 0.1% H_2O_2 and oxidation of thioether or selenoether groups, PEG-PUSe-PEG released almost 72% of its doxorubicin content compared to 41% in PEG-PUS-PEG.

Selenium-containing redox-responsive NPs were made from self-assembly of a series of side-chain selenium-containing amphiphilic poly (ethylene oxide-b-acrylic acid) block copolymers PEO-b-PAA-Se in aqueous solution (PAA refers to polyacrylic acid; Ren et al., 2012). The NPs were shown to be redox responsive in the presence of 0.1% H_2O_2. The selenide group oxidized into hydrophilic selenoxide causing the disassembly of the spherical micellar aggregates. Interestingly, oxidation of selenide into selenoxide moiety can be reversed by Vitamin C (as a reducing agent), and this reaction can be repeated at least 7 times (Ren et al., 2012).

From the polycondensation of toluene diisocyanate with diselenide-containing diols, and termination with PEG monomethyl ether at both ends, Xu and coworkers made tri-block copolymers PEG-PUSeSe-PEG (Han et al., 2013) by self-assembly of the polymer and DOX. Selenium-containing NPs, which showed responsiveness to singlet oxygen, as well as H_2O_2 were developed. Singlet oxygen can be produced via a solution containing a photosensitizer when irradiated with red light. The reaction of diselenide with singlet oxygen results in formation of selenic acid (Fig. 10.2). In another report from the same group, they synthetized a novel coordination-responsive system for the controlled release of

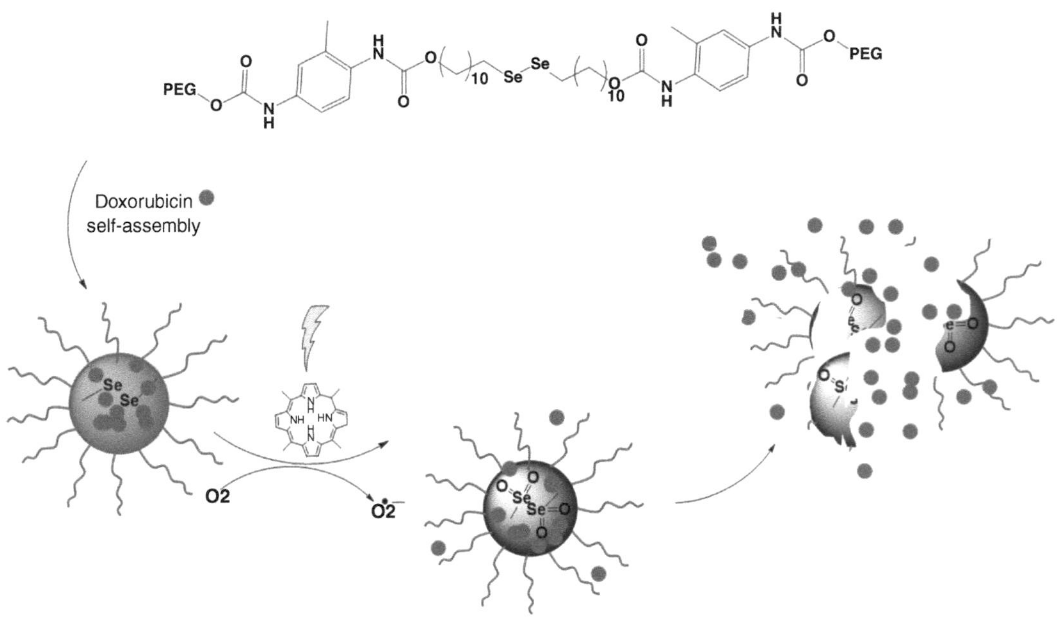

FIGURE 10.2 Schematic Illustration of a Diselenide-Containing Redox-Responsive Drug-Delivery System

A porphyrin-based photosensitizer was used to generate superoxide radicals that can subsequently react with diselenide bond. This results in cleavage of the diselenide bond leading to the enhanced release of loaded doxorubicin.

doxorubicin using a complex of platinum cations and PEG–PUSe–PEG polymers (Cao et al., 2012). Pt-coordinating micelles of PEG–PUSe–PEG released the encapsulated doxorubicin content in a controlled manner. The release of doxorubicin in PBS solution at 37°C were doubled in the presence 10 mM GSH in the solution.

5.3 QUINONE CONTAINING REDOX-RESPONSIVE DRUG-DELIVERY SYSTEM

Special physical, electronic, and chemical properties of quinones have generated intense interest to make intelligent quinone based RNPPs for cancer therapy, and this has been the subject of a recent review (Ma and Long, 2014). Two main types of quinone-based NPs have been used for delivery systems. First, micellar systems employed to increase solubility and other pharmacokinetic properties of redox quinone-based drugs, such as β-lapachone (Ma et al., 2015). Second, a quinone-based structure is used as a redox responsive moiety of the NP (Black et al., 2011). β-lapachone is an anticancer prodrug that can be activated by quinone oxidoreductase-1 (NQO1) using NADP(H) (Pink et al., 2000). NQO1 is an flavo-enzyme that has shown to be overexpressed in many human cancers by up to 20-fold excess compared to normal tissues including lung (Belinsky and Jaiswal, 1993), pancreatic (Logsdon et al., 2003), and breast tissues (Siegel and Ross, 2000). A β-Lapachone NP has been designed using a core of gold NPs, PEG shell, and per-6-thio-β-cyclodextrin as a drug pocket. The xenograft A549 tumor has shown that intravenous injection of the NPs followed by irradiation of tumor sites caused long-lasting increase of NQO1 activity in cancer cells. This is responsible for more synergistic effect and a significant delay (at least 8 days posttreatment) in tumor growth of NP. The NPs showed radio-sensitization properties, as well as redox sensitivity. In mice bearing xenograft human tumors, the intravenous injection of gold NPs exhibited highly enhanced radiotherapeutic efficacy. (Jeong et al., 2009).

Ubiquinone or Coenzyme Q10 is a redox component of the respiratory chain of cells, and its potent antioxidant activity and poor aqueous solubility makes it a popular candidate for drug delivery in several diseases including cancer (Jolliet et al., 1998). Coenzyme Q10 NPs showed higher antioxidant and antiinflammatory or hepatoprotective activities compared to the free coenzyme Q10 (Sharma et al., 2011; Swarnakar et al., 2011). In 2008, McCarley and coworkers developed stable 100 nm diameter redox-sensitive liposomes from quinone-dioleoyl phosphatidylethanolamine (Q-DOPE) lipids that encapsulated fluorescent dye (Ong et al., 2008). Stable liposomes with only DOPE could not be successfully made under physiological conditions; however, on adding a trimethyl-locked quinone switch (as a quinone head group) to the *N*-terminus of DOPE lipids, stable liposomes could be made. Upon two-electron reduction, the quinone head group, which is the redox responsive moiety in this system, is converted to the respective hydroquinone. The generated phenolic substituent undergoes an intramolecular cyclization reaction via nucleophilic attack on the carbonyl substituent of the acid side chain, cleaving the quinone head group from DOPE, which destabilizes the liposomes and increases the release of fluorescent dye (Fig. 10.3).

It has already been reported that the rate of the intramolecular cyclization reaction (lactonization) in free ortho-hydroxyhydrocinnamic acids is rapid (in seconds) if the quinone propionic acid has a trimethyl-lock motif, in which two methyls are at the geminal position and a third methyl ortho to the site of attachment (Amsberry and Borchardt, 1990; Milstien and Cohen, 1972). This strategy has been employed to design redox-responsive biodegradable polymer for the development of paclitaxel-incorporated nanoparticles in cancer therapy (Cho et al., 2012).

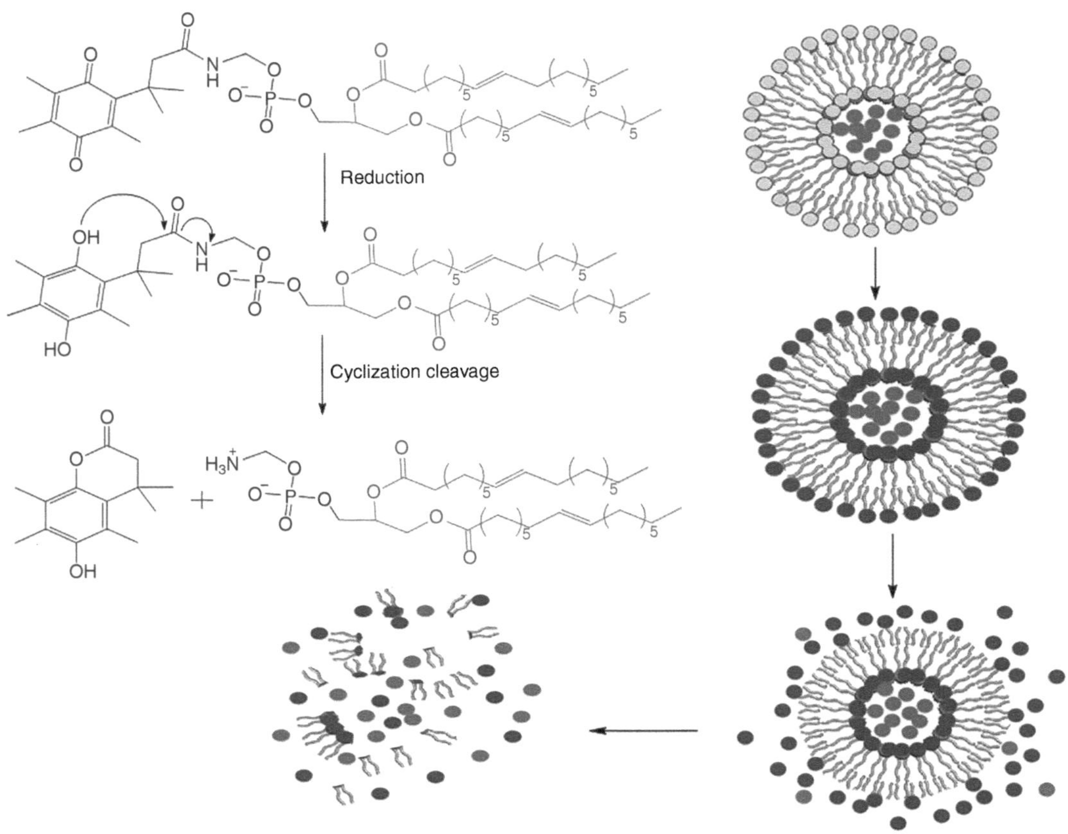

FIGURE 10.3 Schematic Illustration of a Quinone-Containing Redox Responsive Drug Delivery System

Two-electron reduction and removal of quinone propionic acid units destabilizes the liposomes and results in increased release of fluorescent dye. The light gray (light blue in the web version) color indicates the inactive quinone-containing polymer, whereas the dark gray (dark blue in the web version) shows the activated hydroquinone containing polymer (which can undergo the cyclization cleavage)

5.4 REDOX ACTIVATED METAL COMPLEXES IN DRUG-DELIVERY SYSTEMS

Metal-based redox activated prodrugs have a relatively small contribution in cancer therapy in comparison to other DDS. However the discovery and successful clinical applications of some modified prodrugs based on cisplatin has shown their potential application in cancer treatment. To design a redox-responsive metal complex-containing delivery system there are two main strategies (Table 10.1) as described next. These involve the metal complex acting as the redox active functional group and nanostructure protecting it from premature reduction, eventually increasing their selectivity and efficacy either by encapsulation or surface tethering (for review see Graf and Lippard, 2012) (Fig. 10.4A).

1. First, the metal ion of the complex must possess two accessible oxidation states with a large difference in the stability of the complexes between the two oxidation states (Hall et al., 2003; Mellor et al., 2005). Some complexes of platinum, ruthenium, iron, cobalt, copper, and chromium,

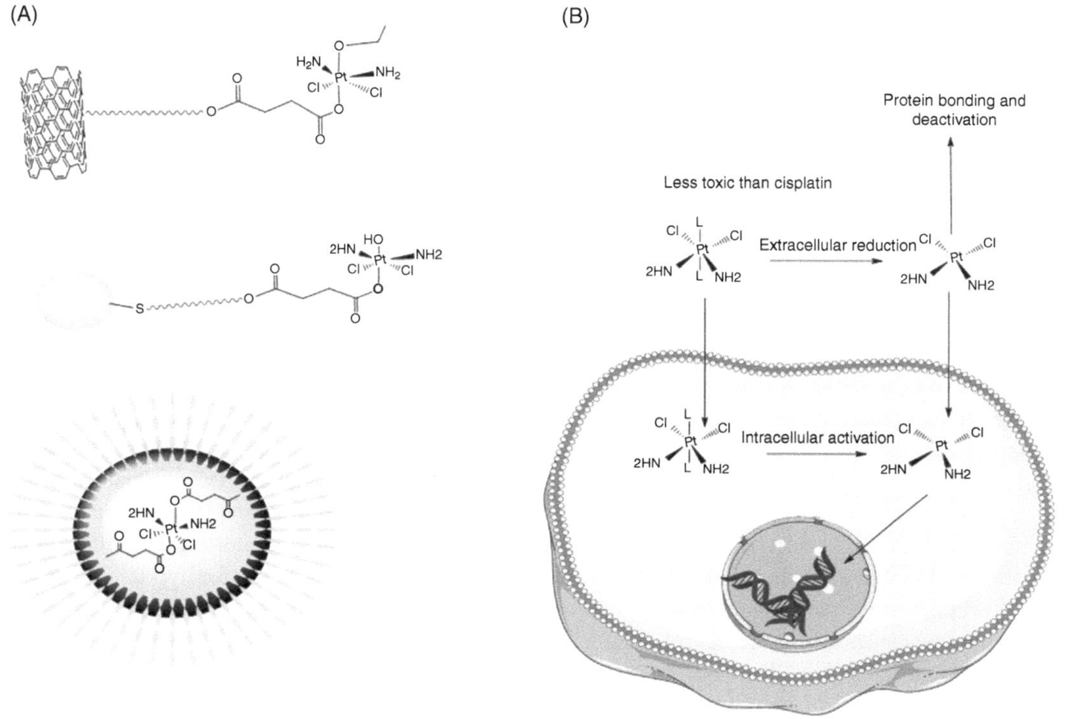

FIGURE 10.4

(A) Three main types of metal complex-containing redox-responsive drug-delivery systems (attachment of the complex into carbon nanotubes, gold NPs, and encapsulation of the complexes). (B) Illustration showing how metal complex-containing redox-responsive drug-delivery systems can be used for intracellular delivery of a toxic drug, such as cisplatin. "L" refers to nanostructure.

have such a property, and several RDDS have been synthesized based on these metal ions. Using cisplatin (IV) derivatives functionalized with a dicarboxylic anhydride, which is subsequently attached to a polymer, is widely used in metal complex based RDDS (Graf and Lippard, 2012). This strategy is advantageous for the metal-based prodrug because it can provide a higher stability and protection from premature reduction and a more efficient targeting ability through attachment to the polymers/nanoparticles. Some of the examples are, cisplatin (IV) attached to single-walled carbon nanotubes (Feazell et al., 2007), single-walled carbon nanotubes and folate (Dhar et al., 2008b) and gold nanoparticles (Kumar et al., 2014), as well as some nanoparticles containing Pt (IV) complex covalently conjugated to amphiphilic block copolymers (Cheng et al., 2014). For example, a hydrophobic prodrug of cisplatin was prepared using a Pt (IV) precursor, in which cisplatin was attached to two alkylesters resulting in a diester substituted cisplatin. Then the prodrug was encapsulated into prostate-specific membrane antigen (PSMA) targeted NPs made of poly(D,L-lactic-*co*-glycolic acid) (PLGA)-poly(ethylene glycol) (PEG). The encapsulation could overcome the problems associated with cisplatin encapsulation, such as reactivity of the cisplatin with polymers, and allows controlled delivery of a cisplatin upon intracellular reduction (Fig. 10.4B). The effectiveness of the NPs against the prostatic cancer cell line (LNCaP) is approximately an order of magnitude greater than that of free cisplatin (Dhar et al., 2008a).

2. Second, Ferrocene-containing polymers can also be used in RDDS. Following oxidation, the neutral charged ferrocene converts to a ferricenium cation, and this lead to structure instability of the NP and increased the drug release (Golan et al., 2011; Hardy et al., 2014; Upton et al., 2014).

Another major phase of treating cancer involves being able to diagnose it in early stages of its incidence. Positron emission tomography and magnetic resonance imaging are some of the widely used diagnostic tools that are limited by the use of radioactives and less than optimum detection limits. Optical imaging is one of the widely used strategies to overcome these drawbacks. Using hypoxia-sensitive groups that rely on the reducing conditions of the tumor microenvironment is prone to show false negative results due to their reduction by other reducing agents upregulated in tumors. Also, it is desirable to use that portion of electromagnetic spectra that has relatively higher penetration to target deep-solid tumors, such as near infrared (NIR). However, NIR active fluorescent probes are reported to have low quantum yields and low photostability. Addressing all these limitations, alternate strategies have been proposed that use antenna-like species that take up the NIR energy and transfer the energy to produce light that can in turn be used for the excitation of other hypoxia indicators. One of the recently published works that use this strategy to develop a highly sensitive nanosensors for hypoxia imaging in vivo was reported by Jianlin and coworkers.(Liu et al., 2014). In their study, the authors reported novel LRET (luminescence resonance energy transfer process) nanosensor containing two parts. The upconversion nanoparticles (UNCP) core (hollow mesoporous silica nanoparticles) and an oxygen indicator tris-(4,7-diphenyl-1,10-phenanthroline) ruthenium (II) dichloride ($[Ru(dpp)_3]^{2+}Cl_2$). Upon irradiation with NIR, the upconverted visible light emitted by UCNPs can be used as a donor for $[Ru(dpp)_3]^{2+}Cl_2$, and the normally quenched indicator illuminates under oxygen-deficient hypoxic conditions. The researchers also mention that, in contrast to free $[Ru(dpp)_3]^{2+}Cl_2$, the designed nanosensors exhibit a significantly high sensitivity for the detection of oxygen in hypoxic regions. Moreover, as mentioned earlier the developed nanosensor was more sensitive to the changes in the amount of oxygen and was not indifferent toward other redox conditions. Another example of a similar strategy being used was published in 2011 in Joanna Napp et al., where the researchers developed an NIR-emissive polystyrene nanoparticles (PS-NPs). The novelty in their work was the sensor ability of their nanoparticle that was suitable for ratiometric luminescence intensity imaging and lifetime imaging of oxygen distribution. The nanoparticle developed included an oxygen sensitive dye and a reference dye, both of which can be excited with a 635 nm laser diode. The polystyrene matrix was reported to provide high colloidal stability. The authors used a Pd(II) mesotetraphenyl-tetrabenzoporphyrin (PdTPTBP) as oxygen-sensitive probe incorporated into 100-nm sized PS-NPs and referenced it against an oxygen-insensitive cyanine dye (DY-635). The targeting feature of the nanoparticles was achieved by coating PS-NPs with maleimide-modified nontoxic polyethylene glycol (PEG) and subsequent functionalization of the surface with Herceptin (humanized IgG1 monoclonal antibody), which targets human epidermal growth factor receptor 2 (Her2/neu). The oxygen sensitive PdTPTBP dye undergoes quenching of phosphorescence by the presence of oxygen. This happens via the transfer of energy from the phosphor in the course when oxygen molecules collide with molecules of the dye in the excited triplet state. Alternatively, in hypoxic conditions (low levels of oxygen), this quenching is absent, consequently leading to phosphorescence. The proof of concept experiments showed that the NIR emissive metalloporphyrin PdTPTBP showed an approximately linear oxygen response of its phosphorescence intensity. Its lifetime is in the bioanalytical-relevant range of tissue and arterial blood, which is in the range of 0 and 130 hPa. The referenced signal was found to increase up to 4.5-fold between normoxia and hypoxia proving its effectiveness.

One of the major limitations of relying on oxygen dependent probes in hypoxia is that the binding of these probes to cancerous tissue is largely impaired by the low pH condition in the tumor microenvironment. Another major limitation is the water insoluble nature and poor tissue penetration, preventing their application to hypoxia imaging in vivo. Addressing these limitations, Xianchuang Zeng et al. recently published their work on developing a near-infrared optical imaging probe highly specific to the hypoxic tumor microenvironment to detect tumor and cancer cells with the sensitivity to a few thousands cancer cells. The nanoparticle designed included a poly(*N*-vinylpyrrolidone) (PVP)-conjugated iridium (III) complex probe that can extend its phosphorescence emission spectra to the NIR-region. The ability to conduct a ratiometric analysis of hypoxia in the tumor was achieved by using an O_2 sensitive, NIR emitting macromolecule as a second component on the PVP. This oxygen-sensitive, NIR emitting and water-soluble phosphorescent macromolecular probe was designed to detect the hypoxic tumor environment of various cancer models, including metastatic tumors in vivo. It was also shown to detect a few cancer cells before the formation of the tumor based on the increased oxygen consumption during cancer cell proliferation. With the application of their strategy, the authors demonstrated that solid tumors and cancer cells metastasized to the lymph nodes and could be detected with high sensitivity and low nonspecificity. Further, the proliferation of cancer cells could also be monitored with the reported imaging probe and NIR imaging before the formation of the solid tumor. This pioneering work introduced a hypoxia-specific imaging tool for early detection of cancer cells and also the ability to assess tumor microenvironment in vivo using the NIR optical imaging method.

5.5 BORONIC ESTER-CONTAINING PRODRUGS AND DRUG-DELIVERY SYSTEMS

Kuivila et al. (1954) reported that molecules with phenylboronic ester groups can react with H_2O_2 and generate phenol and boronic acid as the oxidation products (Table 10.1). Some unique properties of boronic acids and esters, such as mild acidity, better stability, sensitivity to H_2O_2, coupled with ease of handling and biocompatibility make them an attractive class of redox-responsive triggers (Das et al., 2013; Kaupp et al., 2003). Furthermore, boronic acids and esters do not appear to have any serious intrinsic toxicity issues, and the end by-products, borate ester, are regarded as "green" compounds (Baker et al., 2011).

Chemically, all polymers containing arylboronic ester motifs are redox responsive (Song et al., 2014), and there are various reports of polymers with phenylboronic acid or phenylboronic ester groups used to make ROS activated DDS (Li et al., 2011; Song et al., 2013a). Almutairi's group have synthetized new phenylboronic ester containing polymeric NPs that were stable in aqueous solutions. The exposure to peroxide transferred boronic ester protecting groups into boronic acid, which is prone to hydrolysis at physiological pH and temperature. The boronic ester hydrolysis leads to the formation of phenols on the polymer backbone. These phenol groups undergo quinone methide rearrangement, which finally leads to polymer degradation. Hence, upon exposure to H_2O_2, the degradation of the NPS was significantly faster (de Gracia Lux et al., 2012), and the release of an encapsulated dye was twofold higher upon incubation with activated neutrophils, which generate higher levels of ROS, than control neutrophils.

Fu et al. (2015) have prepared novel biocompatible redox responsive micellar systems from amphiphilic triblock polycarbonates bearing either aromatic or aliphatic boronic ester-containing pendent groups, which could sense biological H_2O_2. They encapsulated a florescent dye into the micelles and evaluated the redox disassembly of the NPs. They showed that the disassembly behavior of NPs depends on boronic ester blocks of the NPs, and for complete disassembly of the NPs following H_2O_2 treatment a substantial reversal in polarity of the boronic ester groups is required. Upon exposure to H_2O_2, aliphatic boronic ester-containing micelles swell to loosely aggregated microsized nanostructures, whereas the aromatic boronic ester-containing micelles undergo a complete disassembly giving small fragments.

Recently, pH/oxidation dual-responsive micelles were formed using pendent orthoester and phenylboronic ester groups, which showed a high sensitivity to H_2O_2 (50 μM), and can potentially be used in the fields of polymer-based vaccine, H_2O_2-detection, in vivo imaging, and polymer–drug conjugates with on-demand release of antiinflammatory and chemotherapeutic drugs. To prepare this polymer, phenylboronic pinacol ester-containing diacrylate and N-aminoethylpiperazine were polymerized via a Michael-type addition followed by postmodification with methoxy-PEG-succinimide ester. Upon exposure of to H_2O_2, the side chain phenylboronic esters will be oxidized into phenol groups, which leads to sequential degradation of the polymer. In addition, the amino groups on the polymer can react with quinone methides generated in situ during the oxidative degradation of the polymers (Song et al., 2013b).

5.6 ARYL OXALATE-CONTAINING PRODRUGS AND DRUG-DELIVERY SYSTEMS

Chandross (1963) discovered a new chemiluminescence system using oxalyl chloride and hydrogen peroxide. A unique feature of this system is its high specificity to H_2O_2, and this makes it very promising for in vitro or in vivo H_2O_2 detection (Lee et al., 2007; Lim et al., 2010). Murthy's group have designed new nanoparticles formulated from peroxalate esters and fluorescent dye, which could image H_2O_2 in vivo with high specificity and sensitivity (Lee et al., 2007, 2008). The facile controllability of physicochemical properties, such as biodegradability, crystallinity and mechanical strength are a few of the main advantages of poly aryl oxalate NPs (Hong et al., 2011). Lee's group have used polyoxalate as a RDDS for various ischemia induced injuries, and demonstrated that such NPs show antioxidant and antiinflammatory effects (Kim et al., 2011, 2012; Lee et al., 2013) (Table 10.1). The nano- or micropolyoxalate-containing particles showed excellent cytotoxicity profiles and fast drug-release kinetics, which potentially can be utilized for drug delivery of antiinflammatory agents, as well as anticancer agents (Hong et al., 2011). Hong et al. (2011) have recently reported H_2O_2-responsive nanoparticles formulated from copolyoxalate containing vanillyl alcohol (VA) termed poly(vanillin oxalate) PVO as a novel redox-responsive NPs (Kwon et al., 2013). The nanoparticle was acid and redox responsive and finally after the degradation, VA was released, which itself is potent and natural antiinflammatory and antioxidant. To prepare such a nanostructures, they reacted VA with 2-(Hydroxymethyl)-2-methylpropane-1,3-diol and made 4-(5-(hydroxymethyl)-5-methyl-1,3-dioxan-2-yl)-2-methoxyphenol. Then, PVO was synthesized from the monomer and oxalyl chloride. The NPs made of PVO polymer showed potent antioxidant activities by scavenging H_2O_2 and inhibiting the generation of ROS. It also could reduce the expression of proinflammatory cytokines in activated macrophages in vitro and in vivo. H_2O_2 reacted with peroxalate ester bonds in the polymer and released bioactive vanillin from the NPs. Therefore, the NPs, not only scavenge H_2O_2, but also release vanillin which has potent antioxidant and antiinflammatory effects.

6 MISCELLANEOUS SYSTEMS

In addition to the six main strategies of redox-responsive delivery systems described earlier, other strategies, such as using the thioketal, nitroimidazole, azide, nitroxide, and tellurium-containing polymers have been employed.

For example, telerium-containing polymers have been used as new redox-responsive carries in drug delivery. In comparison to selenium and sulfur analogues, tellurium-containing compounds have higher sensitivity to oxidation properties (Schneider et al., 2006). Similar to phenylselenocysteine, phenyltellurocysteine compounds also inhibit of several human cytochrome P450 isoenzymes, and the potency of inhibition is Te>Se (Nogueira et al., 2004). Despite the design of several tellurium-containing anticancer drugs, such as tellurium-containing redox catalysts, which showed to be selectively cytotoxic at low micromolar concentrations in primary tumor cell cultures and animal models (Huang et al., 2006), occurrence of tellurium in redox-responsive anticancer prodrugs (either RCPs or RNPPs) is poorly explored. Nevertheless, recently some novel tellurium-containing polymers have been synthetized, which showed responsiveness to the stimuli of biologically relevant concentration of GSH, which potentially can be used to design redox-responsive prodrugs, as well as a drug delivery vehicle. Briefly, a dialkyl telluride, was polymerized with 2,4-toluenediisocyanate (TDI) to obtain a modified tellurium-containing polyurethane (PUTe), next PUTe was attached to PEG chains from both ends to make a PEG–PUTe–PEG polymer. Cisplatin was loaded to the polymer mainly based on the coordination of cisplatin and Te of PUTe. A competitive coordination of biomolecules (such as GSH) with Te enabled the controlled release from the nanoparticles. In a later report, such nanoparticle (Cao et al., 2015) have been reported to be responsive to 100 μM H_2O_2 (Cao et al., 2014; Fang et al., 2015).

Murthy's group has synthesized a delivery vehicle, using thioketal as a redox-sensitive functional group, to localize the delivery of siRNA to diseased intestinal tissue (Wilson et al., 2010). They designed a new polymer composed of thioketal linkages from 1,4-benzenedimethanethiol and 2,2-dimethoxypropane using a step-growth polymerization. Then, to formulate siRNA-loaded particles, TNF-α–siRNA was mixed with a cationic lipid 1,2-dioleoyl-3-trimethylammonium-propane (DOTAP), and the complex was added to an organic solution containing thioketal polymer to make siRNA containing redox-responsive nanoparticles. The specificity of the nanoparticles for ROS was evaluated by incubation with a superoxide solution. Exposure to superoxide decreases its molecular weight from almost 9000 Da to approximately 810 Da within 8 h indicating the degradation of nanoparticles.

2-Nitroimidazole has been used as a hypoxia-targeting moiety. It plays an important role when conjugated to polymeric carriers containing imaging agents, therapeutic agents, or therapeutic adjuvants for treatment and detection of hypoxic areas of cancers (Giri et al., 2013). Recently, doxorubicin was loaded into a nanoparticle made of a modified 2-nitroimidazole attached to carboxymethyl-dextran backbone (Thambi et al., 2014). The NPs released doxorubicin in a sustained manner under the normoxic condition (49% release of doxorubicin over 12 h), whereas the drug release rate remarkably increased under hypoxia condition (to afford complete release within 12 h). They showed that under hypoxia, NADPH acts as a reducing agent and converts 2-nitroimidazole (poorly soluble in water) to 2-aminoimidazole (a highly water-soluble group), changing the system's stability and leading to the release of the cargo. Live animal near-infrared fluorescence imaging demonstrated that the nanoparticles can be effectively accumulated in tumor sites. The nanoparticles were administered to the SCC7 tumor-bearing mice and its antitumor effect was compared with free doxorubicin. Sixteen days posttreatment, tumor weights of animals treated by these nanoparticles were 3 times less than that of free doxorubicin treated animals.

Nagasaki's group designed a new redox-sensitive nanoparticle system using nitroxide-containing polymers (Yoshitomi and Nagasaki, 2014). 4-Hydroxy TEMPO or 4-amino TEMPO was used as a redox-responsive element confined to the core of the polymeric micelles, because TEMPO can undergo either oxidation or reduction to generate hydrophilic groups (Fu et al., 2010). The oxidation of TEMPO changes the assembly of polymers in the nanoparticle structure releasing the cargo (Yoshitomi and Nagasaki, 2011). These nanoparticles showed a remarkable therapeutic effect for several oxidative stress injuries ranging from gastrointestinal inflammation and nephron-protection to tumor suppression (Yoshitomi et al., 2011). Preadministration of TEMPO-containing nanoparticle prior to treatments with doxorubicin in tumor-bearing mice significantly suppressed the progression of tumor size in comparison to free doxorubicin (Yoshitomi et al., 2013; Yoshitomi and Nagasaki, 2014).

Proline, an amino acid, can be oxidized by hydroxyl free radicals and get converted to pyrrolidone intermediate and finally after hydrolysis to an amino-butyric acid (Stadtman and Levine, 2003). Yu S. et al. have constructed a bio-reductive porous scaffold composed of PEG-β-polycaprolactone copolymers cross-linked by proline-containing linkers. These polymeric scaffolds can be loaded with drugs, and respond to oxidative environments through gradual degradation for the controlled release of therapeutics targeted to diseases (Yu et al., 2011).

Azobenzene derivatives can be reduced into aniline derivatives both enzymatically (Zbaida and Levine, 1991) and nonenzymatically (Kiyose et al., 2010; Nam and Renganathan, 2000). Like nitroaromatics, the azo-based derivatives are reduced stepwise, so that first an electron is accepted by the azo group, and under low oxygen condition the product of one-electron reduction that is prone to further reduction to the corresponding hydrazine compound. The resulting hydrazine analogue can then undergo further reduction to form an amine. Liu et al. initially conjugated combretastatin A-4, a potent anticancer and angiogenesis inhibitor, to a poly ethylene glycol (PEG) hexanethiol through an azobenzene linked amphiphilic molecule (Kiyose et al., 2010; Liu et al., 2015; Piao et al., 2013). These amphiphilic molecules were shown to self-assemble into azo-based micelles in an aqueous solution. With the presence of the azo group in the micellar structure, it was expected to disassemble the micelles under the reductive environment of hypoxia.

It is worth mentioning that several polymeric gene–delivery systems have been used for targeting cells with imbalance redox state. These polymeric themselves may not be redox responsive, but they carry genes that can regulate imbalance redox state of the cells. As previously noted, the most important adaptive responses to hypoxia and enhanced ROS in the cells are controlled through a specific gene expression programs, in which HIF-1α is the main regulatory protein. Because most human solid tumors have significantly higher level of HIF-1α expression, genes containing HRE promoter regions can be overexpressed in tumor cells, and as a consequence the concentration of the respective protein must be higher than other cells (Talks et al., 2000; Zhong et al., 1999), which either directly induce toxicity or reduce the activity of vital proteins in the cells (Humbert et al., 2012). Hence, toxic genes containing HRE promoter (such as BAX expression genes) were loaded to a carrier and delivered to tumor cells (Kim et al., 2009). Though several major methods have been used to deliver genes to tumor cells, probably the main limitations of gene therapy are the lack of safe and efficient delivery system. Generally speaking, two types of vehicles have been used to target hypoxic cells in tumors: synthetic vectors (polypeptides, lipids, and polymers) and recombinant biological carrier (viruses, bacteria, and macrophages). Polymeric vehicles mainly are cationic-based nanoparticles that electrostatically can interact with anionic nucleic acids and trap the gene, and this can protect the gene against potential degradation. Most of synthetic nanoparticles accumulate passively in tumors as a result of leaky vascular of tumors

(Kirtane and Panyam, 2013). However, bioreducible polymers possess mainly disulfide linkages in the polymer structures and can carry genes and release them at a reducing environment, such as hypoxic tumor cells (Kim and Kim, 2011).

7 CONCLUSIONS

Free radicals are important mediators in normal physiology and contribute in various vital cell-signaling pathways. Cell functions, including gene expression, energy metabolism, and cell death processes, are controlled by multiple signaling pathways that can be regulated by the redox environment. Cancer cells exhibit an abnormal redox state coupled to compensatory redox adaptation that promotes cell survival. These phenomena have a crucial role in tumor development, metastasis, and drug resistance. However, at the same time, such abnormalities make tumor cells different from their nonmalignant counterparts, and a current focus of research is to utilize this significant difference for controlled and active delivery of anticancer drugs. So far, six main redox-responsive delivery systems have been used: polysulfides, polyselenides, quinones, metal complexes, arylboronic esters, and aryl oxalates. Taking an inspiration from nature, sulfur-containing polymers could be the most promising option to make a redox-responsive delivery system because they are biocompatible, biodegradable, hydrophobic in nature, and upon oxidation or reduction their hydrophilicity increases, aiding in delivering drugs at the site of action. Though, there is a wide range of redox-responsive delivery systems, so far, none of them have advanced to clinical trials. Hence, more research is required in this direction.

REFERENCES

Acharya, A., Das, I., Chandhok, D., Saha, T., 2010. Redox regulation in cancer: a double-edged sword with therapeutic potential. Oxid. Med. Cel. Longev. 3, 23–34.

Adams, J., Carder, P.J., Downey, S., Forbes, M.A., Maclennan, K., Allgar, V., Kaufman, S., Hallam, S., Bicknell, R., Walker, J.J., 2000. Vascular endothelial growth factor (VEGF) in breast cancer: comparison of plasma, serum, and tissue VEGF and microvessel density and effects of tamoxifen. Cancer Res. 60, 2898–2905.

Amsberry, K.L., Borchardt, R.T., 1990. The lactonization of 2'-hydroxyhydrocinnamic acid amides: a potential prodrug for amines. J. Org. Chem. 55, 5867–5877.

Bae, Y.H., 2009. Drug targeting and tumor heterogeneity. J. Control. Release 133, 2.

Baker, S.J., Tomsho, J.W., Benkovic, S.J., 2011. Boron-containing inhibitors of synthetases. Chem. Soc. Rev. 40, 4279–4285.

Belinsky, M., Jaiswal, A.K., 1993. NAD (P) H: quinone oxidoreductase1 (DT-diaphorase) expression in normal and tumor tissues. Cancer Metastasis Rev. 12, 103–117.

Black, K.C., Liu, Z., Messersmith, P.B., 2011. Catechol redox induced formation of metal core- polymer shell nanoparticles. Chem. Mater. 23, 1130–1135.

Brown, J.M., Wilson, W.R., 2004. Exploiting tumour hypoxia in cancer treatment. Nat. Rev. Cancer 4, 437–447.

Burke, A.J., Sullivan, F.J., Giles, F.J., Glynn, S.A., 2013. The yin and yang of nitric oxide in cancer progression. Carcinogenesis 34, 503–512.

Cai, X.-J., Dong, H.-Q., Xia, W.-J., Wen, H.-Y., Li, X.-Q., Yu, J.-H., Li, Y.-Y., Shi, D.-L., 2011. Glutathione-mediated shedding of PEG layers based on disulfide-linked catiomers for DNA delivery. J. Mater. Chem. 21, 14639–14645.

Cai, X., Dong, C., Dong, H., Wang, G., Pauletti, G.M., Pan, X., Wen, H., Mehl, I., Li, Y., Shi, D., 2012. Effective gene delivery using stimulus-responsive catiomer designed with redox-sensitive disulfide and acid-labile imine linkers. Biomacromolecules 13, 1024–1034.

Cao, W., Li, Y., Yi, Y., Ji, S., Zeng, L., Sun, Z., Xu, H., 2012. Coordination-responsive selenium-containing polymer micelles for controlled drug release. Chem. Science 3, 3403–3408.

Cao, W., Gu, Y., Meineck, M., Li, T., Xu, H., 2014. Tellurium-containing polymer micelles: competitive-ligand-regulated coordination responsive systems. J. Am. Chem. Soc. 136, 5132–5137.

Cao, W., Gu, Y., Li, T., Xu, H., 2015. Ultra-sensitive ROS-responsive tellurium-containing polymers. Chem. Commun. 51, 7069–7071.

Chandross, E.A., 1963. A new chemiluminescent system. Tetrahedron Lett. 4, 761–765.

Cheng, R., Feng, F., Meng, F., Deng, C., Feijen, J., Zhong, Z., 2011. Glutathione-responsive nano-vehicles as a promising platform for targeted intracellular drug and gene delivery. J. Control. Release 152, 2–12.

Cheng, Z., Dai, Y., Kang, X., Li, C., Huang, S., Lian, H., Hou, Z., Ma, P., Lin, J., 2014. Gelatin-encapsulated iron oxide nanoparticles for platinum (IV) prodrug delivery, enzyme-stimulated release and MRI. Biomaterials 35, 6359–6368.

Cho, H., Bae, J., Garripelli, V.K., Anderson, J.M., Jun, H.-W., Jo, S., 2012. Redox-sensitive polymeric nanoparticles for drug delivery. Chem. Commun. 48, 6043–6045.

Chung, J., Yokoyama, M., Yamato, M., Aoyagi, T., Sakurai, Y., Okano, T., 1999. Thermo-responsive drug delivery from polymeric micelles constructed using block copolymers of poly (N-isopropylacrylamide) and poly (butylmethacrylate). J. Control. Release 62, 115–127.

Coussens, L.M., Fingleton, B., Matrisian, L.M., 2002. Matrix metalloproteinase inhibitors and cancer–trials and tribulations. Science 295, 2387–2392.

Das, B.C., Thapa, P., Karki, R., Schinke, C., Das, S., Kambhampati, S., Banerjee, S.K., Van Veldhuizen, P., Verma, A., Weiss, L.M., 2013. Boron chemicals in diagnosis and therapeutics. Future Med. Chem. 5, 653–676.

de Gracia Lux, C., Joshi-Barr, S., Nguyen, T., Mahmoud, E., Schopf, E., Fomina, N., Almutairi, A., 2012. Biocompatible polymeric nanoparticles degrade and release cargo in response to biologically relevant levels of hydrogen peroxide. J. Am. Chem. Soc. 134, 15758–15764.

Dewhirst, M., Ong, E., Braun, R., Smith, B., Klitzman, B., Evans, S., Wilson, D., 1999. Quantification of longitudinal tissue pO_2 gradients in window chamber tumours: impact on tumour hypoxia. Brit. J. Cancer 79, 1717.

Dewhirst, M.W., Cao, Y., Moeller, B., 2008. Cycling hypoxia and free radicals regulate angiogenesis and radiotherapy response. Nat. Rev. Cancer 8, 425–437.

Dhar, S., Gu, F.X., Langer, R., Farokhzad, O.C., Lippard, S.J., 2008a. Targeted delivery of cisplatin to prostate cancer cells by aptamer functionalized Pt (IV) prodrug-PLGA–PEG nanoparticles. Proc. Natl. Acad. Sci. 105, 17356–17361.

Dhar, S., Liu, Z., Thomale, J., Dai, H., Lippard, S.J., 2008b. Targeted single-wall carbon nanotube-mediated Pt (IV) prodrug delivery using folate as a homing device. J. Am. Chem. Soc. 130, 11467–11476.

Doi, K., Akaike, T., Horie, H., Noguchi, Y., Fujii, S., Beppu, T., Ogawa, M., Maeda, H., 1996. Excessive production of nitric oxide in rat solid tumor and its implication in rapid tumor growth. Cancer 77, 1598–1604.

Ernsting, M.J., Tang, W.L., Maccallum, N.W., Li, S.D., 2012. Preclinical pharmacokinetic, biodistribution, and anti-cancer efficacy studies of a docetaxel-carboxymethylcellulose nanoparticle in mouse models. Biomaterials 33, 1445–1454.

Fang, J., Seki, T., Maeda, H., 2009. Therapeutic strategies by modulating oxygen stress in cancer and inflammation. Adv. Drug Deliv. Rev. 61, 290–302.

Fang, J., Nakamura, H., Maeda, H., 2011. The EPR effect: unique features of tumor blood vessels for drug delivery, factors involved, and limitations and augmentation of the effect. Adv. Drug Deliv. Rev. 63, 136–151.

Fang, R., Xu, H., Cao, W., Yang, L., Zhang, X., 2015. Reactive oxygen species (ROS)-responsive tellurium-containing hyperbranched polymer. Polym. Chem. 6, 2817–2821.

Feazell, R.P., Nakayama-Ratchford, N., Dai, H., Lippard, S.J., 2007. Soluble single-walled carbon nanotubes as longboat delivery systems for platinum (IV) anticancer drug design. J. Am. Chem. Soc. 129, 8438–8439.

Folkman, J., 1992. The role of angiogenesis in tumor growth. Semin. Cancer Biol. 3 (2), 65–71.

Folkman, J., 2002. Role of Angiogenesis in Tumor Growth and Metastasis. Semi. Onco. 29, 15–18.

Freeman, S.M., Abboud, C.N., Whartenby, K.A., Packman, C.H., Koeplin, D.S., Moolten, F.L., Abraham, G.N., 1993. The "bystander effect": tumor regression when a fraction of the tumor mass is genetically modified. Cancer Res. 53, 5274–5283.

Fu, H., Policarpio, D.M., Batteas, J.D., Bergbreiter, D.E., 2010. Redox-controlled 'smart' polyacrylamide solubility. Polym. Chem. 1, 631–633.

Fu, Y.-H., Chen, C.-Y., Chen, C.-T., 2015. Tuning of hydrogen peroxide-responsive polymeric micelles of biodegradable triblock polycarbonates as a potential drug delivery platform with ratiometric fluorescence signaling. Polym. Chem. 6, 8132–8143.

Gao, W., Chan, J.M., Farokhzad, O.C., 2010. pH-responsive nanoparticles for drug delivery. Mol. Pharm. 7, 1913–1920.

Gao, L., Luo, Q., Wang, Y., Du, H., Li, X., Shen, Z., Zhu, W., 2013. Facile preparation of shell crosslinked micelles for redox-responsive anticancer drug release. RSC Adv. 4, 4177–4180.

Giles, G.I., 2006. The redox regulation of thiol dependent signaling pathways in cancer. Curr. Pharm. Des. 12, 4427–4443.

Giles, G.I., Jacob, C., 2002. Reactive sulfur species: an emerging concept in oxidative stress. Biol. Chem. 383, 375–388.

Giri, B.P., Gregg, K., Singh, P., Dagli, D.J., Giri, A., 2013. Hypoxia-Targeted Polymeric Micelles for Cancer Therapy and Imaging. Google Patents.

Golan, S., Aytar, B.S., Muller, J.P., Kondo, Y., Lynn, D.M., Abbott, N.L., Talmon, Y., 2011. Influence of biological media on the structure and behavior of ferrocene-containing cationic lipid/DNA complexes used for DNA delivery. Langmuir 27, 6615–6621.

Goldmann, E., 1907. The growth of malignant disease in man and the lower animals: with special reference to the vascular system. Lancet 170, 1236–1240.

Graf, N., Lippard, S.J., 2012. Redox activation of metal-based prodrugs as a strategy for drug delivery. Adv. Drug Deliv. Rev. 64, 993–1004.

Graves, E.E., Maity, A., Le, Q.-T., 2010. The Tumor Microenvironment in Non–Small-Cell Lung Cancer. Semin. Radiat. Oncol. 20, 156–163.

Greish, K., 2007. Enhanced permeability and retention of macromolecular drugs in solid tumors: a royal gate for targeted anticancer nanomedicines. J. Drug Targeting 15, 457–464.

Griguer, C.E., Oliva, C.R., Kelley, E.E., Giles, G.I., Lancaster, J.R., Gillespie, G.Y., 2006. Xanthine oxidase–dependent regulation of hypoxia-inducible factor in cancer cells. Cancer Res. 66, 2257–2263.

Hall, M.D., Dillon, C.T., Zhang, M., Beale, P., Cai, Z., Lai, B., Stampfl, A.P., Hambley, T.W., 2003. The cellular distribution and oxidation state of platinum (II) and platinum (IV) antitumour complexes in cancer cells. JBIC, J. Biol. Inorg. Chem. 8, 726–732.

Han, P., Li, S., Cao, W., Li, Y., Sun, Z., Wang, Z., Xu, H., 2013. Red light responsive diselenide-containing block copolymer micelles. J. Mater. Chem. B 1, 740–743.

Hardy, C.G., Zhang, J., Yan, Y., Ren, L., Tang, C., 2014. Metallopolymers with transition metals in the side-chain by living and controlled polymerization techniques. Prog. Polym. Sci. 39 (10), 1742–1796.

Hong, D., Song, B., Kim, H., Kwon, J., Khang, G., Lee, D., 2011. Biodegradable polyoxalate and copolyoxalate particles for drug-delivery applications. Ther. Deliv. 2, 1407–1417.

Huang, Z.-Z., Chen, C., Zeng, Z., Yang, H., Oh, J., Chen, L., Lu, S.C., 2001. Mechanism and significance of increased glutathione level in human hepatocellular carcinoma and liver regeneration. FASEB J. 15, 19–21.

Huang, X., Dong, Z., Liu, J., Mao, S., Luo, G., Shen, J., 2006. Tellurium-based polymeric surfactants as a novel seleno-enzyme model with high activity. Macromol. Rapid Commun. 27, 2101–2106.

Humbert, O., Davis, L., Maizels, N., 2012. Targeted gene therapies: tools, applications, optimization. Crit. Rev. Biochem. Mol. Biol. 47, 264–281.
Huo, M., Yuan, J., Tao, L., Wei, Y., 2014. Redox-responsive polymers for drug delivery: from molecular design to applications. Polymer Chem. 5, 1519–1528.
Ikeda, Y., Hayashi, I., Kamoshita, E., Yamazaki, A., Endo, H., Ishihara, K., Yamashina, S., Tsutsumi, Y., Matsubara, H., Majima, M., 2004. Host stromal bradykinin B2 receptor signaling facilitates tumor-associated angiogenesis and tumor growth. Cancer Res. 64, 5178–5185.
Iyer, A.K., Khaled, G., Fang, J., Maeda, H., 2006. Exploiting the enhanced permeability and retention effect for tumor targeting. Drug Discov. Today 11, 812–818.
Jacob, C., Giles, G.I., Giles, N.M., Sies, H., 2003. Sulfur and selenium: the role of oxidation state in protein structure and function. Angew. Chem. Int. Edit. 42, 4742–4758.
Jeong, S.-Y., Park, S.-J., Yoon, S.M., Jung, J., Woo, H.N., Yi, S.L., Song, S.Y., Park, H.J., Kim, C., Lee, J.S., 2009. Systemic delivery and preclinical evaluation of Au nanoparticle containing β-lapachone for radiosensitization. J. Control. Release 139, 239–245.
Jolliet, P., Simon, N., Barre, J., Pons, J., Boukef, M., Paniel, B., Tillement, J., 1998. Plasma coenzyme Q10 concentrations in breast cancer: prognosis and therapeutic consequences. Int. J. Clin. Pharm.Ther. 36, 506–509.
Jordan, B.F., Misson, P.-D., Demeure, R., Baudelet, C., Beghein, N., Gallez, B., 2000. Changes in tumor oxygenation/perfusion induced by the no donor, isosorbide dinitrate, in comparison with carbogen: monitoring by EPR and MRI. Int. J. Radiat. Oncol. Biol. Phys. 48, 565–570.
Kang, H., Trondoli, A.C., Zhu, G., Chen, Y., Chang, Y.-J., Liu, H., Huang, Y.-F., Zhang, X., Tan, W., 2011. Near-infrared light-responsive core–shell nanogels for targeted drug delivery. ACS Nano 5, 5094–5099.
Kaupp, G., Naimi-Jamal, M.R., Stepanenko, V., 2003. Waste-free and facile solid-state protection of diamines, anthranilic acid, diols, and polyols with phenylboronic acid. Chem. Eur. J. 9, 4156–4161.
Kim, J.-W., Dang, C.V., 2006. Cancer's molecular sweet tooth and the Warburg effect. Cancer Res. 66, 8927–8930.
Kim, T.-I., Kim, S.W., 2011. Bioreducible polymers for gene delivery. React. Funct. Polym. 71, 344–349.
Kim, H.A., Mahato, R.I., Lee, M., 2009. Hypoxia-specific gene expression for ischemic disease gene therapy. Adv. Drug Deliv. Rev. 61, 614–622.
Kim, S., Park, H., Song, Y., Hong, D., Kim, O., Jo, E., Khang, G., Lee, D., 2011. Reduction of oxidative stress by< i> p-hydroxybenzyl alcohol-containing biodegradable polyoxalate nanoparticulate antioxidant. Biomaterials 32, 3021–3029.
Kim, H., Kim, Y., Guk, K., Yoo, D., Lim, H., Kang, G., Lee, D., 2012. Fully biodegradable and cationic poly (amino oxalate) particles for the treatment of acetaminophen-induced acute liver failure. Int. J. Pharm. 434, 243–250.
Kim, S., Seong, K., Kim, O., Kim, S., Seo, H., Lee, M., Khang, G., Lee, D., 2010. Polyoxalate nanoparticles as a biodegradable and biocompatible drug delivery vehicle. Biomacromolecules 11, 555–560.
Kirtane, A.R., Panyam, J., 2013. Polymer nanoparticles: weighing up gene delivery. Nat. Nanotechnol. 8, 805–806.
Kiyose, K., Hanaoka, K., Oushiki, D., Nakamura, T., Kajimura, M., Suematsu, M., Nishimatsu, H., Yamane, T., Terai, T., Hirata, Y., Nagano, T., 2010. Hypoxia-sensitive fluorescent probes for in vivo real-time fluorescence imaging of acute ischemia. J. Am. Chem. Soc. 132, 15846–15848.
Kuivila, H.G., Keough, A.H., Soboczenski, E.J., 1954. Areneboronates from diols and polyols1. J. Org. Chem. 19, 780–783.
Kumar, A., Huo, S., Zhang, X., Liu, J., Tan, A., Li, S., Jin, S., Xue, X., Zhao, Y., Ji, T., 2014. Neuropilin-1-targeted gold nanoparticles enhance therapeutic efficacy of platinum (IV) drug for prostate cancer treatment. ACS Nano.
Kuppusamy, P., Li, H., Ilangovan, G., Cardounel, A.J., Zweier, J.L., Yamada, K., Krishna, M.C., Mitchell, J.B., 2002. Noninvasive imaging of tumor redox status and its modification by tissue glutathione levels. Cancer Res. 62, 307–312.
Kwon, J., Kim, J., Park, S., Khang, G., Kang, P.M., Lee, D., 2013. Inflammation-responsive antioxidant nanoparticles based on a polymeric prodrug of vanillin. Biomacromolecules 14, 1618–1626.

Lee, F., Vessey, A., Rofstad, E., Siemann, D., Sutherland, R., 1989. Heterogeneity of glutathione content in human ovarian cancer. Cancer Res. 49, 5244–5248.

Lee, D., Khaja, S., Velasquez-Castano, J.C., Dasari, M., Sun, C., Petros, J., Taylor, W.R., Murthy, N., 2007. In vivo imaging of hydrogen peroxide with chemiluminescent nanoparticles. Nat. Mater. 6, 765–769.

Lee, D., Erigala, V.R., Dasari, M., Yu, J., Dickson, R.M., Murthy, N., 2008. Detection of hydrogen peroxide with chemiluminescent micelles. Int. J. Nanomed. 3, 471.

Lee, D., Bae, S., Hong, D., Lim, H., Yoon, J.H., Hwang, O., Park, S., Ke, Q., Khang, G., Kang, P.M., 2013. H_2O_2-responsive molecularly engineered polymer nanoparticles as ischemia/reperfusion-targeted nanotherapeutic agents. Sci. Rep. 3, 1–8.

Leung, D.W., Cachianes, G., Kuang, W.J., Goeddel, D.V., Ferrara, N., 1989. Vascular endothelial growth factor is a secreted angiogenic mitogen. Science 246, 1306–1309.

Li, C., Hu, J., Liu, T., Liu, S., 2011. Stimuli-triggered off/on switchable complexation between a novel type of charge-generation polymer (CGP) and gold nanoparticles for the sensitive colorimetric detection of hydrogen peroxide and glucose. Macromolecules 44, 429–431.

Li, H., Yu, S.S., Miteva, M., Nelson, C.E., Werfel, T., Giorgio, T.D., Duvall, C.L., 2013. Matrix metalloproteinase responsive, proximity-activated polymeric nanoparticles for siRNA delivery. Adv. Funct. Mater. 23, 3040–3052.

Lim, C.K., Lee, Y.D., Na, J., Oh, J.M., Her, S., Kim, K., Choi, K., Kim, S., Kwon, I.C., 2010. Chemiluminescence-generating nanoreactor formulation for near-infrared imaging of hydrogen peroxide and glucose level in vivo. Adv. Funct. Mater. 20, 2644–2648.

Lissoni, P., Malugani, F., Bonfanti, A., Bucovec, R., Secondino, S., Brivio, F., Ferrari-Bravo, A., Ferrante, R., Vigore, L., Rovelli, F., 2000. Abnormally enhanced blood concentrations of vascular endothelial growth factor (VEGF) in metastatic cancer patients and their relation to circulating dendritic cells, IL-12 and endothelin-1. J. Biol. Regul. Homeost. Agents 15, 140–144.

Liu, J., Liu, Y., Bu, W., Bu, J., Sun, Y., Du, J., Shi, J., 2014. Ultrasensitive nanosensors based on upconversion nanoparticles for selective hypoxia imaging in vivo upon near-infrared excitation. J. Am. Chem. Soc. 136, 9701–9709.

Liu, H., Zhang, R., Niu, Y., Li, Y., Qiao, C., Weng, J., Li, J., Zhang, X., Xiao, Z., Zhang, X., 2015. Development of hypoxia-triggered prodrug micelles as doxorubicin carriers for tumor therapy. RSC Adv. 5, 20848–20857.

Logsdon, C.D., Simeone, D.M., Binkley, C., Arumugam, T., Greenson, J.K., Giordano, T.J., Misek, D.E., Hanash, S., 2003. Molecular profiling of pancreatic adenocarcinoma and chronic pancreatitis identifies multiple genes differentially regulated in pancreatic cancer. Cancer Res. 63, 2649–2657.

Lowe, S.W., Bodis, S., Mcclatchey, A., Remington, L., Ruley, H.E., Fisher, D.E., Housman, D.E., Jacks, T., 1994. p53 status and the efficacy of cancer therapy in vivo. Science, 807–1807.

Ma, W., Long, Y.-T., 2014. Quinone/hydroquinone-functionalized biointerfaces for biological applications from the macro-to nano-scale. Chemical Soc. Rev. 43, 30–41.

Ma, N., Li, Y., Ren, H., Xu, H., Li, Z., Zhang, X., 2010. Selenium-containing block copolymers and their oxidation-responsive aggregates. Polym. Chem. 1, 1609–1614.

Ma, X., Huang, X., Moore, Z., Huang, G., Kilgore, J.A., Wang, Y., Hammer, S., Williams, N.S., Boothman, D.A., Gao, J., 2015. Esterase-activatable β-lapachone prodrug micelles for NQO1-targeted lung cancer therapy. J. Control. Release 200, 201–211.

Maeda, H., Noguchi, Y., Sato, K., Akaike, T., 1994. Enhanced vascular permeability in solid tumor is mediated by nitric oxide and inhibited by both new nitric oxide scavenger and nitric oxide synthase inhibitor. Cancer Sci. 85, 331–334.

Maeda, H., Wu, J., Sawa, T., Matsumura, Y., Hori, K., 2000. Tumor vascular permeability and the EPR effect in macromolecular therapeutics: a review. J. Control. Release 65, 271–284.

Matsukado, K., Sugita, M., Black, K.L., 1998. Intracarotid low dose bradykinin infusion selectively increases tumor permeability through activation of bradykinin B2 receptors in malignant gliomas. Brain Res. 792, 10–15.

Mccarley, R.L., 2012. Redox-responsive delivery systems. Ann. Rev. Anal. Chem. 5, 391–411.

McKenzie, D.L., Kwok, K.Y., Rice, K.G., 2000. A potent new class of reductively activated peptide gene delivery agents. J. Biol. Chem. 275, 9970–9977.

Mellor, H., Snelling, S., Hall, M., Modok, S., Jaffar, M., Hambley, T., Callaghan, R., 2005. The influence of tumour microenvironmental factors on the efficacy of cisplatin and novel platinum (IV) complexes. Biochem. Pharmacol. 70, 1137–1146.

Milstien, S., Cohen, L.A., 1972. Stereopopulation control. I. Rate enhancement in the lactonizations of o-hydroxyhydrocinnamic acids. J. Am. Chem. Soc. 94, 9158–9165.

Mura, S., Nicolas, J., Couvreur, P., 2013. Stimuli-responsive nanocarriers for drug delivery. Nat. Mater. 12, 991–1003.

Murawaki, Y., Tsuchiya, H., Kanbe, T., Harada, K., Yashima, K., Nozaka, K., Tanida, O., Kohno, M., Mukoyama, T., Nishimuki, E., 2008. Aberrant expression of selenoproteins in the progression of colorectal cancer. Cancer Lett. 259, 218–230.

Nam, S., Renganathan, V., 2000. Non-enzymatic reduction of azo dyes by NADH. Chemosphere 40, 351–357.

Napoli, A., Valentini, M., Tirelli, N., Müller, M., Hubbell, J.A., 2004. Oxidation-responsive polymeric vesicles. Nat. Mater. 3, 183–189.

Nishida, N., Yano, H., Nishida, T., Kamura, T., Kojiro, M., 2006. Angiogenesis in cancer. Vasc. Health Risk Manag. 2, 213.

Nogueira, C.W., Zeni, G., Rocha, J.B., 2004. Organoselenium and organotellurium compounds: toxicology and pharmacology. Chem. Rev. 104, 6255–6286.

Ong, W., Yang, Y., Cruciano, A.C., Mccarley, R.L., 2008. Redox-triggered contents release from liposomes. J. Am. Chem. Soc. 130, 14739–14744.

O'Reilly, R.K., Hawker, C.J., Wooley, K.L., 2006. Cross-linked block copolymer micelles: functional nanostructures of great potential and versatility. Chemical Soc. Rev. 35, 1068–1083.

Park, I.K., Singha, K., Arote, R.B., Choi, Y.J., Kim, W.J., Cho, C.S., 2010. pH-responsive polymers as gene carriers. Macromol. Rapid Commun. 31, 1122–1133.

Peer, D., Karp, J.M., Hong, S., Farokhzad, O.C., Margalit, R., Langer, R., 2007. Nanocarriers as an emerging platform for cancer therapy. Nat. Nano 2, 751–760.

Piao, W., Tsuda, S., Tanaka, Y., Maeda, S., Liu, F., Takahashi, S., Kushida, Y., Komatsu, T., Ueno, T., Terai, T., Nakazawa, T., Uchiyama, M., Morokuma, K., Nagano, T., Hanaoka, K., 2013. Development of azo-based fluorescent probes to detect different levels of hypoxia. Angew. Chem. 125, 13266–13270.

Pink, J.J., Planchon, S.M., Tagliarino, C., Varnes, M.E., Siegel, D., Boothman, D.A., 2000. NAD (P) H: quinone oxidoreductase activity is the principal determinant of β-lapachone cytotoxicity. J. Biol. Chem. 275, 5416–5424.

Rafii, S., Skobe, M., 2003. Splitting vessels: keeping lymph apart from blood. Nat. Med. 9, 166–168.

Ray, P.D., Huang, B.-W., Tsuji, Y., 2012. Reactive oxygen species (ROS) homeostasis and redox regulation in cellular signaling. Cell. Signal. 24, 981–990.

Ren, H., Wu, Y., Ma, N., Xu, H., Zhang, X., 2012. Side-chain selenium-containing amphiphilic block copolymers: redox-controlled self-assembly and disassembly. Soft Matter. 8, 1460–1466.

Salven, P., Lymboussaki, A., Heikkila, P., Jaaskela-Saari, H., Enholm, B., Aase, K., Von Euler, G., Eriksson, U., Alitalo, K., Joensuu, H., 1998. Vascular endothelial growth factors VEGF-B and VEGF-C are expressed in human tumors. Am. J. Pathol. 153, 103–108.

Schneider, A., Rodrigues, O.E., Paixão, M.W., Appelt, H.R., Braga, A.L., Wessjohann, L.A., 2006. Stereoselective synthesis of Boc-protected L-seleno-and tellurolanthionine, L-seleno-and tellurocystine and derivatives. Tetrahedron Lett. 47, 1019–1021.

Secomb, T.W., Hsu, R., Park, E.Y., Dewhirst, M.W., 2004. Green's function methods for analysis of oxygen delivery to tissue by microvascular networks. Ann. Biomed. Eng. 32, 1519–1529.

Seki, T., Fang, J., Maeda, H., 2009. Enhanced delivery of macromolecular antitumor drugs to tumors by nitroglycerin application. Cancer Sci. 100, 2426–2430.

Semenza, G.L., 2003. Targeting HIF-1 for cancer therapy. Nat. Rev. Cancer 3, 721–732.
Semenza, G.L., 2014. Oxygen sensing, hypoxia-inducible factors, and disease pathophysiology. Annu. Rev. Pathol. 9, 47–71.
Sharma, A., Soliman, G.M., Al-Hajaj, N., Sharma, R., Maysinger, D., Kakkar, A., 2011. Design and evaluation of multifunctional nanocarriers for selective delivery of coenzyme Q10 to mitochondria. Biomacromolecules 13, 239–252.
Siegel, D., Ross, D., 2000. Immunodetection of NAD (P) H: quinone oxidoreductase 1 (NQO1) in human tissues. Free Radic. Biol. Med. 29, 246–253.
Song, C.-C., Ji, R., Du, F.-S., Li, Z.-C., 2013a. Oxidation-responsive poly(amino ester) s containing arylboronic ester and self-immolative motif: synthesis and degradation study. Macromolecules 46, 8416–8425.
Song, C.-C., Ji, R., Du, F.-S., Liang, D.-H., Li, Z.-C., 2013b. Oxidation-accelerated hydrolysis of the ortho ester-containing acid-labile polymers. ACS Macro. Lett. 2, 273–277.
Song, C.-C., Du, F.-S., Li, Z.-C., 2014. Oxidation-responsive polymers for biomedical applications. J. Mater. Chem. B 2, 3413–3426.
Sosa, V., Moliné, T., Somoza, R., Paciucci, R., Kondoh, H., Lleonart, M.E., 2013. Oxidative stress and cancer: an overview. Ageing Res. Rev. 12, 376–390.
Stadtman, E., Levine, R., 2003. Free radical-mediated oxidation of free amino acids and amino acid residues in proteins. Amino Acids 25, 207–218.
Subarsky, P., Hill, R.P., 2003. The hypoxic tumour microenvironment and metastatic progression. Clin. Exp. Metastasis 20, 237–250.
Sun, H., Meng, F., Cheng, R., Deng, C., Zhong, Z., 2013. Reduction-sensitive degradable micellar nanoparticles as smart and intuitive delivery systems for cancer chemotherapy. Expert Opin. Drug Deliv. 10, 1109–1122.
Swarnakar, N.K., Jain, A.K., Singh, R.P., Godugu, C., Das, M., Jain, S., 2011. Oral bioavailability, therapeutic efficacy and reactive oxygen species scavenging properties of coenzyme Q10-loaded polymeric nanoparticles. Biomaterials 32, 6860–6874.
Talks, K.L., Turley, H., Gatter, K.C., Maxwell, P.H., Pugh, C.W., Ratcliffe, P.J., Harris, A.L., 2000. The expression and distribution of the hypoxia-inducible factors HIF-1α and HIF-2α in normal human tissues, cancers, and tumor-associated macrophages. Am. J. Pathol. 157, 411–421.
Taurin, S., Nehoff, H., Greish, K., 2012. Anticancer nanomedicine and tumor vascular permeability; where is the missing link? J. Control. Release 164, 265–275.
Teicher, B.A., Lazo, J.S., Sartorelli, A.C., 1981. Classification of antineoplastic agents by their selective toxicities toward oxygenated and hypoxic tumor cells. Cancer Res. 41, 73–81.
Thambi, T., Deepagan, V., Yoon, H.Y., Han, H.S., Kim, S.-H., Son, S., Jo, D.-G., Ahn, C.-H., Suh, Y.D., Kim, K., 2014. Hypoxia-responsive polymeric nanoparticles for tumor-targeted drug delivery. Biomaterials 35, 1735–1743.
Trachootham, D., Alexandre, J., Huang, P., 2009. Targeting cancer cells by ROS-mediated mechanisms: a radical therapeutic approach? Nat. Rev. Drug Discov. 8, 579–591.
Unterberg, A., Wahl, M., Baethmann, A., 1984. Effects of bradykinin on permeability and diameter of pial vessels in vivo. J. Cerebr. Blood Flow Metab. 4, 574–585.
Upton, B.M., Gipson, R.M., Duhovic´, S., Lydon, B.R., Matsumoto, N.M., Maynard, H.D., Diaconescu, P.L., 2014. Synthesis of ferrocene-functionalized monomers for biodegradable polymer formation. Inorg. Chem. Front. 1, 271–277.
Ushio-Fukai, M., Nakamura, Y., 2008. Reactive oxygen species and angiogenesis: NADPH oxidase as target for cancer therapy. Cancer Lett. 266, 37–52.
Vaupel, P., 2004. Tumor Microenvironmental Physiology and Its Implications for Radiation Oncology. Semin. Radiat. Oncol. 14, 198–206.
Vaupel, P., Harrison, L., 2004. Tumor hypoxia: causative factors, compensatory mechanisms, and cellular response. Oncologist 9, 4–9.

Vaupel, P., Schlenger, K., Knoop, C., Höckel, M., 1991. Oxygenation of human tumors: evaluation of tissue oxygen distribution in breast cancers by computerized O_2 tension measurements. Cancer Res. 51, 3316–3322.

Vo, C.D., Kilcher, G., Tirelli, N., 2009. Polymers and sulfur: what are organic polysulfides good for? Preparative strategies and biological applications. Macromol. Rapid Commun. 30, 299–315.

Wang, Y.-C., Wang, F., Sun, T.-M., Wang, J., 2011. Redox-responsive nanoparticles from the single disulfide bond-bridged block copolymer as drug carriers for overcoming multidrug resistance in cancer cells. Bioconjugate Chem. 22, 1939–1945.

Wang, J., Sun, X., Mao, W., Sun, W., Tang, J., Sui, M., Shen, Y., Gu, Z., 2013. Tumor redox heterogeneity-responsive prodrug nanocapsules for cancer chemotherapy. Adv. Mater. 25, 3670–3676.

Wilson, W.R., Hay, M.P., 2011. Targeting hypoxia in cancer therapy. Nat. Rev. Cancer 11, 393–410.

Wilson, D.S., Dalmasso, G., Wang, L., Sitaraman, S.V., Merlin, D., Murthy, N., 2010. Orally delivered thioketal nanoparticles loaded with TNF-α–siRNA target inflammation and inhibit gene expression in the intestines. Nat. Mater. 9, 923–928.

Wu, J., Akaike, T., Maeda, H., 1998. Modulation of enhanced vascular permeability in tumors by a bradykinin antagonist, a cyclooxygenase inhibitor, and a nitric oxide scavenger. Cancer Res. 58, 159–165.

Xie, H., Audette, C., Hoffee, M., Lambert, J.M., Blattler, W.A., 2004. Pharmacokinetics and biodistribution of the antitumor immunoconjugate, cantuzumab mertansine (huC242-DM1), and its two components in mice. J. Pharmacol. Exp. Ther. 308, 1073–1082.

Xu, H., Cao, W., Zhang, X., 2013. Selenium-containing polymers: promising biomaterials for controlled release and enzyme mimics. Account. Chem. Res. 46, 1647–1658.

Yasuda, H., 2008. Solid tumor physiology and hypoxia-induced chemo/radio-resistance: novel strategy for cancer therapy: nitric oxide donor as a therapeutic enhancer. Nitric Oxide 19, 205–216.

Yeh, J.J., Kim, W.Y., 2015. Targeting tumor hypoxia with hypoxia-activated prodrugs. J. Clin. Oncol. 33, 1505–1508.

Yoshitomi, T., Nagasaki, Y., 2011. Nitroxyl radical-containing nanoparticles for novel nanomedicine against oxidative stress injury. Nanomedicine 6, 509–518.

Yoshitomi, T., Nagasaki, Y., 2014. Reactive oxygen species-scavenging nanomedicines for the treatment of oxidative stress injuries. Adv. Healthc. Mater. 3 (8), 1149–1161.

Yoshitomi, T., Hirayama, A., Nagasaki, Y., 2011. The ROS scavenging and renal protective effects of pH-responsive nitroxide radical-containing nanoparticles. Biomaterials 32, 8021–8028.

Yoshitomi, T., Ozaki, Y., Thangavel, S., Nagasaki, Y., 2013. Redox nanoparticle therapeutics to cancer—increase in therapeutic effect of doxorubicin, suppressing its adverse effect. J. Control. Release 172, 137–143.

Yu, S.S., Koblin, R.L., Zachman, A.L., Perrien, D.S., Hofmeister, L.H., Giorgio, T.D., Sung, H.-J., 2011. Physiologically relevant oxidative degradation of oligo (proline) cross-linked polymeric scaffolds. Biomacromolecules 12, 4357–4366.

Zbaida, S., Levine, W.G., 1991. A novel application of cyclic voltammetry for direct investigation of metabolic intermediates in microsomal azo reduction. Chem. Res.Toxicol. 4, 82–88.

Zhao, M., Biswas, A., Hu, B., Joo, K.-I., Wang, P., Gu, Z., Tang, Y., 2011. Redox-responsive nanocapsules for intracellular protein delivery. Biomaterials 32, 5223–5230.

Zhong, H., De Marzo, A.M., Laughner, E., Lim, M., Hilton, D.A., Zagzag, D., Buechler, P., Isaacs, W.B., Semenza, G.L., Simons, J.W., 1999. Overexpression of hypoxia-inducible factor 1α in common human cancers and their metastases. Cancer Res. 59, 5830–5835.

CHAPTER

11 POLYMERIC MICRO- AND NANOPARTICLES FOR CONTROLLED AND TARGETED DRUG DELIVERY

Magdalena Stevanović

Institute of Technical Sciences of the Serbian Academy of Sciences and Arts, Belgrade, Serbia

CHAPTER OUTLINE

1 INTRODUCTION

Controlled drug delivery is a multidisciplinary field compared to conventional delivery systems. It also offers numerous advantages, such as improved efficacy, better stability of the incorporated (encapsulated or immobilized) substances against (e.g., enzymatic) degradation, reduced toxicity, easy administration, and so on (Feng et al., 2004; Kumar et al., 2004; Langer, 2001; Mu and Feng, 2003; Ratnam et al., 2006). Polymers are often used as carriers for the active substances. Polymers in which active substances can be successfully encapsulated are usually polyesters, polysaccharides, etc. The controlled delivery of active substances requires that the encapsulated or immobilized material retain its performance as a biological activity or the activity of acceptable degradation products (Besenbache et al., 2007; Moghimi et al., 2005; Sahoo et al., 2007; Stevanović and Uskoković, 2009a,b; Sumer and Gao, 2008; Wong et al., 2007). Methods for the encapsulation/immobilization of the active substances is commonly performed to retain and/or to slowly release them, and to provide a more stable environment for the encapsulated/immobilized species (Ishida et al., 2003). In this way, treatments and procedures that would not otherwise be possible are now in use.

Particles, whether nano or micro, occupy a unique position in the controlled release of active substances because of their different and specific properties from the bulk material (Mainardes and Evangelista, 2005; Stevanović et al., 2011b, 2012a). The possibility to control particle size, surface properties, and the release of active substances in order to achieve the site-specific action of the substances at the optimal rate and dose regimen is one of the major challenges in designing micro- and nanoparticles as a delivery system (Ito et al., 2007). Micro- and nanoparticles can be used to deliver a wide variety of active substances, such as hydrophilic or hydrophobic drugs, vitamins, minerals, different nutrients, and so on. The ways to administrate them in the body are different (D'Emanuele et al., 2004; Kozubek et al., 2001; Lian and Ho, 2001; Liu and Frechet, 1999; Medina et al., 2004; Nafee et al., 2007; Stevanović et al., 2007a,b; Yoo, 2006).

One often hears of vitamins, minerals, or other nutrients deficit in human body, which are crucial for normal physiologic functioning of the human body. For example, a system for the controlled delivery of vitamins can bring to the more balanced and efficient concentration of vitamins over an extended period of time (Stevanović et al., 2009, 2014). These systems can be prepared from a variety of materials, such as polysaccharides, liposomes, synthetic polymers, and so on. (D'Emanuele et al., 2004; Kozubek et al., 2001; Lian and Ho, 2001; Liu and Frechet, 1999; Medina et al., 2004; Nafee et al., 2007; Stevanović et al., 2007a,b; Yoo, 2006). Active substances can be delivered in systems that would otherwise be unacceptable or aggressive to them (Greene, 2007; Vasir et al., 2005). For instance, water-soluble substances can be delivered in nonaqueous systems, such as ointments. The nature and matrix of the carriers can be chosen from a wide variety of materials (usually selected from a list of materials "generally recognized as safe" (GRAS) for applications in pharmaceutical or food industry (Rados, 2004) to meet the needs of the application, as well as regulatory demands, for example, biodegradability. Such particles can be multifunctional and serve to deliver different substances at the same time. Actually, mutually chemically incompatible actives can be formulated together. An additional way to exploit this technology is to encapsulate one substance that can facilitate, or stabilize, the second (Stevanović et al., 2007a,b; Vandervoort and Ludvig, 2002). By isolating some substances, formulation and processing issues are minimized, undesirable properties can be masked, and so on.

The purpose of this chapter is to point out some important issues that are related to various methods employed to encapsulate the active substances within polymeric carriers, such as thermal phase separation, melt dispersion, solvent evaporation, spray drying, homogenization of water and organic phases, and so on This review also gives a comparison of the characteristics of polymeric micro- and nanoparticles prepared by different methods. Encapsulation efficiency, the release rate, the size distribution of particles with encapsulated active substances are some of the parameters that are used for the evaluation of the encapsulation system characteristics.

2 POLYMERIC MICRO- AND NANOPARTICLES FOR A CONTROLLED DELIVERY OF ACTIVE SUBSTANCES

Polymers have been extensively studied as carriers of active substances in the pharmacy, food industry, and so on (Legrand et al., 1999). For controlled drug delivery, polymers that are biodegradable and bioerodible represent the most important type of materials. Polymers in which vitamins and other active substances can be successfully encapsulated are polyesters, polysaccharides, and so on. Polyesters are synthetic polymers that have ester linkages. They are used as fibers, films, in composites and elastomers, and so on. Polyesters, such as polylactides, polyglycolides, poly(lactide-*co*-glycolides) (PLGAs), poly(ε-caprolactone) are approved by the World Health Organization (WHO) and Food and Drug Administration (FDA) as substances that can be used in medicine and pharmacy. They are biodegradable polymers with numerous advantages, such as excellent processing characteristics, biocompatibility, and biodegradation at rates that can be tailored for the intended application. In degradation, covalent bond cleavage by chemical reactions while erosion occurs by the dissolution of chain fragments in non-crosslinked systems without chemical alterations to the molecular structure (Liechty et al., 2010). For dissolution to occur, the polymer must absorb the surrounding solution and must interact with solution via charge interactions or hydrogen bonding mechanisms. Degradation, as well as erosion can occur from the surface or from the bulk of the material (Stevanović et al., 2011a). When degradation starts from the surface, the polymeric material is rapidly eliminated from the surface of the particle, but the polymer volume fraction stays quite unchanged. Inversely, when the degradation starts from the bulk, not enough change occurs in the size of the polymeric particle until it is nearly completely degraded, but the fraction of polymer residual in the carrier decreases during time. The most dominant process is defined by the rates of solution penetration within the polymeric particle, diffusion of the degradation products, and the degradation or dissolution of the polymer structure. Polyester micro- and nanoparticles are used for the controlled delivery of several classes of medicaments, such as anticancer agents, growth factors, antibiotics, antimicrobial agents, and so on. Different reviews in the literature can be found on this topic, such as the nanoparticle formation mechanisms (Moinard-Checot et al., 2006), the classification of particulated systems (Letchford and Burt, 2007), methods for the preparation of carriers (Vauthier and Bouchemal, 2008), and so on. It is commonly assumed that the system described as a microsphere is comprised of a fairly homogeneous mixture of polymer and active agents, whereas microcapsules have at least one discrete domain of active agent and sometimes more (Singh et al., 2010). Some of the variants on microparticle structures are given in Fig. 11.1.

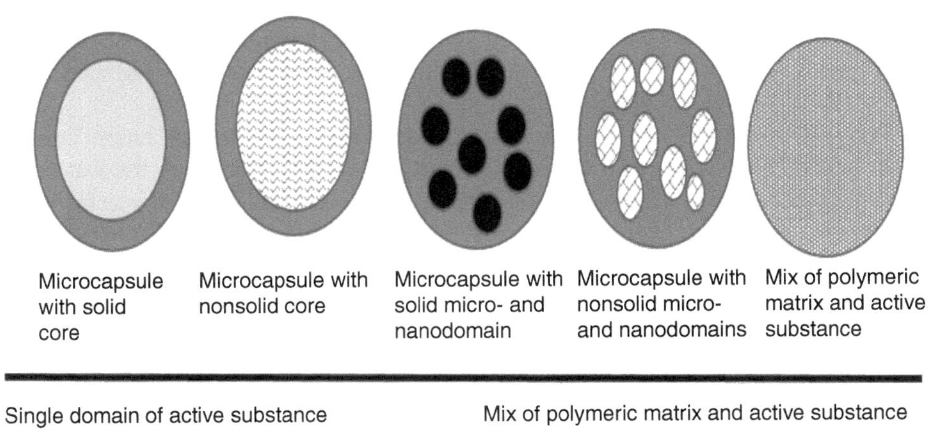

FIGURE 11.1 Variants on Microparticle Structures

3 FORMULATION OF MICRO- AND NANOPOLYMER PARTICLES

Based on the selected material, the techniques for obtaining polymeric particles for the delivery of active substances can be commonly divided into several groups: the dispersion of preformed polymers, the polymerization of monomers, and the ionic gelation or coacervation of hydrophilic polymers. In the literature, other methods, such as supercritical fluid technology or particle replication in non-wetting templates, have also been described (Gratton et al., 2007; Kang et al., 2008; Mishima, 2008). For example, for the preparation of particles from the polyester polymer PLGA many approaches are proposed. The emulsification–evaporation method (Astete et al., 2007; Prabha and Labhasetwar, 2004; Sahoo and Labhasetwar, 2005; Song et al., 2006), emulsification–solvent diffusion method (Lee et al., 2005; Zhang et al., 2006), nanoprecipitation (Bilati et al., 2005; Govender et al., 1999), and spray-drying (Atuah et al., 2003; Rivera et al., 2004; Takashima et al., 2007) are all widely used to make PLGA micro- and nanoparticles of various sizes.

These methods have comparable primary steps, where an aqueous medicament solution is emulsified in an organic solution in order to form a water/oil dispersion. The medicament can be also dispersed as a powder (solid particles) in an organic polymeric solution, or codissolved in a solvent together with the polymer. Afterward the solution or dispersion is treated according to the one of the aforementioned techniques.

In the solvent extraction or evaporation procedure, the polymer is dissolved in an organic solvent, such as dichloromethane, acetone, acetonitrile, chloroform, or ethyl acetate. These solvents are also used as a solvent for the hydrophobic drugs. The mixture that contains polymer and drug is then emulsified in a water solution containing a surfactant, stabilizing, or emulsifying agent to produce an oil-in-water emulsion. When a stable emulsion is formed, the organic solvent is evaporated either by decreasing the pressure or by continuous stirring. Different parameters have influence on the particle's size and morphology, such as the type and concentration of stabilizer, speed of the homogenization, and the polymer concentration (Kwon et al., 2001). For the purposes of obtaining a small particle size, high-speed homogenization or centrifugation can be employed (Stevanović et al., 2007a,b,c,d; Zambaux et al., 1998). Stevanović et al. (2007d) have shown that by varying parameters like aging time (after the nonsolvent is added) and the time of centrifugation, it is possible to influence the morphology, size, agglomeration, and uniformity of PLGA particles. For example, PLGA particles produced by a physicochemical solvent/nonsolvent method with the shortest aging time with the nonsolvent, and the longest time and highest velocity of the centrifugal processing, has the smallest particles and the highest uniformity (Stevanović et al., 2007c).

This solvent–diffusion technique is an adapted version of solvent–evaporation process (Niwa et al., 1993). In this procedure, a water-miscible solvent is used together with the water-immiscible organic solvent as an oil phase. Thanks to a spontaneous diffusion of solvents, an interfacial turbulence arises between the two phases, leading to the creation of small particles. Solvent evaporation, as well as solvent diffusion, methods can be employed for hydrophobic or hydrophilic drugs. In the case of hydrophilic drugs, a multiple water/oil/water emulsion needs to be formed with the drug dissolved in the internal aqueous phase (Stevanović and Uskoković, 2009a).

The coacervation method is a technique that consists of three parts that are under continuous agitation: the first part, in which a solution must be made with three immiscible phases: the core material (i.e., active substance), the coating material (i.e., polymeric material), and a solvent; the second part, in which the liquid coating is placed around the core material, which is achieved by mixing the coating phase with the solvent phase (in which the active substances reside); and the third part, when the coating is rigidized thermally or by desolvation (Thomasin et al., 1998a,b).

Recently, Guan et al. (2015) have described coaxial electrospray method to produce multilayer PLGA micro- and nanoparticles with well controlled shape and size. ALA-PDT is a technology used in the treatment for skin and other cancers. In their work, ALA-loaded PLGA nanoparticles with high drug loading efficiency were produced via coaxial electrospray and used for the PDT treatment in HSC-3 cells.

Spray drying gives an attractive and relatively simple alternative to the aforementioned processes. In this case, the solution containing an antigen or water/oil emulsion is atomized in a flow of drying air at an elevated temperature. The organic solvent is quickly evaporated and leaves behind solid micro- or nanoparticles that are separated from the drying air and collected in a deposition chamber (Gander et al., 1996; Tamber et al., 2005).

Park et al. (2007) described methods based on filling micromolds with polymeric particles, in contrast to polymer melts, to prepare microstructures composed of multiple materials, with complex geometries, and produced using mild processing conditions. Polymer particles with sizes of about 1–30 μm have been made from poly(lactide), poly(glycolide), and PLGA by using spray drying and emulsion methods either with or without the incorporated drug. These polymeric particles have been filled into micromolds at room temperature and melted or attached together to form microstructures according to different protocols (Park et al., 2007).

The emulsion method produces PLGA spheres of 100–250 μm (Choi et al., 2005), 45 μm (Martinez-Sancho et al., 2004), and 30 μm (Daugherty et al., 1997) in diameter. Changing the parameters of the emulsion procedure led to obtaining spherical particles with smaller diameters up to 10 μm (Jeong et al., 2003).

Usually, the emulsion process is most appropriate for water-insoluble medicaments, such as steroids, while the double-emulsion process is most often used to incorporate water-soluble medicaments, such as peptides (Makadia and Siegel, 2011).

PLGA spherical microparticles with the encapsulated drug paclitaxel have been prepared by the spray-drying method. These particles have a size range of 1–8 μm (Liu et al., 2007). Xu et al. (2011) described a microemulsion method to encapsulate Nile Red dye (and LY294002 in carboxylic acid, group-terminated PLGA). The average size of the microspheres was of about 1 μm.

Further modifications of the process with additional evaporation produced spherical particles with diameters in the submicron scale. The first submicron spherical particles obtained were 570–970 nm (Feng et al., 2004) and 244–260 nm (Murakami et al., 2000) in diameter. PLGA particles obtained by the physicochemical solvent/nonsolvent method were in the size range of 110–170 nm (Stevanović et al., 2007b, 2009).

PLGA nanoparticles can also be synthesized by the nanoprecipitation method. In the study of Bilati et al. (2005), the mean particle size was closely dependent on the type of nonsolvent selected. When alcohols were used, the final mean size increased in the sequence: methanol $<$ ethanol $<$ propanol. The nanoparticles obtained ranged from about 85 to 560 nm in size (Bilati et al., 2005).

Recently, Alshamsan (2014) has shown that nanoprecipitation is more efficient than the emulsion solvent evaporation method to encapsulate cucurbitacin I in PLGA nanoparticles. It is likely that cucurbitacin I escapes with the organic solvent after the emulsification step to the aqueous phase, leading to its ineffective entrapment in the polymeric matrix. Avoiding emulsification seems efficient in increasing cucurbitacin I disposition in the instantly precipitating NPs. Therefore, the nanoprecipitation method increases cucurbitacin I entrapment in PLGA NPs, and possibly in other water-insoluble polar drugs.

When spray drying is applied for the preparation of cationic PLGA nanospheres as gene-delivery vectors, in order to minimize aggregation and loss of gene transfection efficiency, the mean particle diameter was 100–250 nm (Takashima et al., 2007).

Jin et al. (2007) examined PLGA nanoparticles with encapsulated paclitaxel, etanidazole, or paclitaxel + etanidazole prepared by o/w and w/o/w emulsification-solvent evaporation method. The prepared nanoparticles were spherical, reaching between 80 and 150 nm in size. The drug encapsulation efficiency was higher for paclitaxel and lower for etanidazole. With the emulsion evaporation method using sodium dodecyl sulfate as a surfactant, the size of the obtained particles ranged from 40 to 70 nm (Astete et al., 2007).

4 THE MOST WIDELY USED INCORPORATION TECHNIQUES

In the pharmaceutical, medical, and food industries encapsulation, as well as immobilization techniques are widely used (Kumar, 2000; Stupar et al., 2014; Wieland-Berghausen et al., 2002). Encapsulation protects components that are sensitive to the environment or to provide time-released delivery of active substances (Fig. 11.2). Certain ingredients are easy to encapsulate using conventional techniques. Others, such as sensitive water-soluble components like vitamin C, are very difficult to encapsulate so controlled protection and release is necessary. Fig. 11.2 represents a microparticle with an encapsulated active substance.

The choice of a particular method of encapsulation is mainly determined by drug solubility and molecular stability considerations. Many studies have been done on vitamin C (ascorbic acid) with variables optimized to determine the most stable way of encapsulating this vitamin. The commonly utilized techniques for encapsulation of the active substances, such as vitamins, minerals, or other nutrients within micro- or nanoparticles are thermal phase separation (coacervation), spray chilling and cooling, melt dispersion, solvent evaporation, fluidized bed coating, spray drying, homogenization of water and organic phases, rotational suspension separation, extrusion and inclusion complexation, and so on. Some of them will be described here.

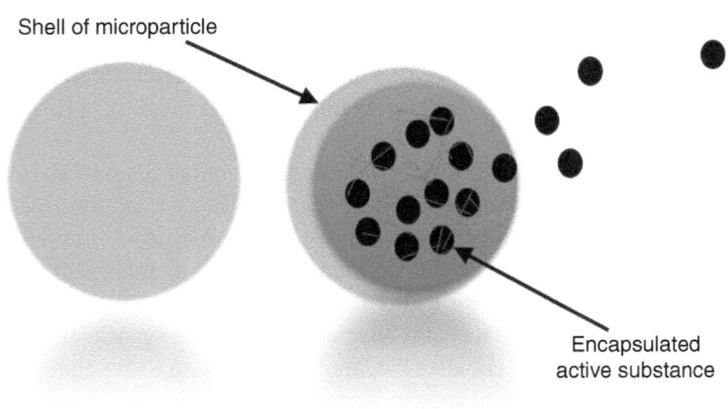

FIGURE 11.2 A Microparticle With an Encapsulated Active Substance

4.1 THERMAL PHASE SEPARATION

By use of coacervation (thermal phase separation), the active substance is coated with a polymer shell leading to the formation of microcapsules (Wieland-Berghausen et al., 2002). Ethyl cellulose is the most widely used polymer in the coacervation process (Benita and Donbrow, 1982; Miller and Anderson, 1964; Uddin et al., 2001; Yalabik-Kas, 1983; Yoshiyuki et al., 1983). Several factors—such as homogenization speed, cooling rate, concentration of phase-separation inducing agent and viscosity/molecular weight of ethyl cellulose—show an effect on the encapsulation efficiency and the release behavior of the encapsulated drug. Uddin et al. (2001) have been encapsulating vitamins using this technique. Ethyl cellulose has been used as the wall-forming material. The molecular weight of the ethyl cellulose was varied and it was determined that the molecular weight of ethyl cellulose and the addition of polyisobutylene significantly influenced the aggregation and release rate of microcapsules that contain ascorbic acid. It was determined that microencapsulation product size decreased as the molecular weight of ethyl cellulose increased.

Bachtsi and Kiparissides (1996) described how they made poly(vinyl alcohol) (PVA) microcapsules that were prepared by the coacervation method. This is followed by the chemical crosslinking of the coacervated PVA membrane with glutaraldehyde. They examined the influence of several parameters—such as the size of the particles, the degree of crosslinking of the coacervated PVA membrane, the influence of the amount of the desolvating agent, and the concentration of the surfactant used in the synthesis. The ionic strength of the release medium on the release rate of the oil from the microcapsules into a model surfactant solution has also been studied. It was found that the oil release rate exhibited a first-order kinetic model. The permeability coefficient of the santosol oil for the PVA coacervated/crosslinked membrane was shown to vary from 10^{-4} to 10^{-6} cm/s and was strongly dependent both on the environmental conditions of the release medium and the physical and morphological characteristics of the microcapsules.

In the work of Nihant et al. (1995), the phase separation of PLGA solutions in dichloromethane (CH_2Cl_2) is described. The phase separation of PLGA is induced by the addition of a silicone oil to promote protein microencapsulation. The process is very fast and that is the reason why the system is out of equilibrium. In this paper the influence of the main processing parameters on the encapsulation process has been examined and has underlined that the kinetics of the main encapsulation steps has a great effect on the appearances of the final microspheres.

A phase separation is a method that is also used for the encapsulation of the medicament chlorpromazine hydrochloride. Chlorpromazine hydrochloride is a highly water-soluble drug. In the procedure described in the work of Lin and Yang (1986), ethylene-vinyl acetate was used as agent to induce coacervation. The influence of ethylene-vinyl acetate concentration on the drug release properties of chlorpromazine hydrochloride microcapsules and tabletted microcapsules was examined. Microcapsules were obtained by the deposition of ethyl cellulose round solid chlorpromazine hydrochloride particles and by using 0%–6% ethylene-vinyl acetate as a coacervation-inducing agent. It was shown that the size of the particles depends on the concentration of the coacervation agent and decrease when the concentration increases. The concentration of coacervation agent also has an influence on the surface topography of the microcapsules, atmospheric stability, powder flowability, and the release rate from untabletted or tabletted microcapsules.

Simovic et al. (2010) have prepared a porous silica–lipid hybrid (SLH) microcapsule for delivery of the drug indomethacin. SLH is an oral-delivery system for indomethacin fabricated from Pickering emulsion templates. Indomethacin forms an electrostatic complex with cationic lipid present in the oil phase. The SLH microcapsules have been prepared either by spray drying (range of about 1–5 μm) or phase coacervation (range of about 20–50 μm) showed a specific internal porous structure with pores in the range of 20–100 nm. Experiments of dissolution under sink conditions, as well as in the presence of electrolytes show a decreased extent of dissolution; this is confirmation of the lipophilic nature of the drug–lipid complex, as well as its location in the oil phase. In vivo experiments exhibited complete medicament absorption and statistically higher fasted-state bioavailability than the water suspensions and oil/water submicron emulsions of indomethacin. It is hypothesized that the SLH microcapsule improves oral absorption via complete solubilization of drug–lipid electrostatic complexes during enzymatic lipolysis in the gastrointestinal tract.

Poulain et al. (2003) have used coacervation method for preparing biocompatible polymeric microspheres with sizes of about 0.5–5 μm as carriers for drugs for the treatment of thrombosis. In the case of the drugs used for the treatment of thrombosis, the pharmacological activity of serine protease inhibitors is often linked to the presence of amidine functions. In this paper, inulin and inulin acetate—with or without 1,12-dodecanedicarboxylic acid—were selected for the preparation of the microspheres. (E,E)-bis(amidinobenzylidene) cycloheptanone [(E,E)-BABCH] is used as model drug. The influence of different parameters, such as drug mass, velocity, and time of stirring during synthesis have been examined. It was shown that encapsulation efficiency reached 65% irrespective to the nature of the polymer when a stirring time of 30 min, a high stirring speed, and a centrifugation during 15 min is used. The release of the model drug was studied in vitro and exhibited three characteristic phases: the initial phase, for example, a rapid "drug burst" (in the first 5 min, 58%–62% of the drug was delivered), then a slow diffusion phase that lasted 33 h, and finally an increasing rate until complete drug release was observed for 2.5 days.

Mao et al. (2001) examined obtaining of chitosan–DNA nanoparticles by complex coacervation. They studied the influence of different parameters during synthesis—such as the concentrations of ingredients, the temperature of the solutions, and the pH and molecular weights of chitosan and DNA. The particle size was optimized to be from 100 to 250 nm when the ratio between the amino and phosphate group was adjusted to be from 3 to 8 while the concentration of the chitosan was 100 μg/mL. The particles were within the narrow size distribution. These particles had a positive zeta potential of about 12–18 mV. By the electrophoretic mobility method, it has been shown that chitosan–DNA nanoparticles protect the encapsulated plasmid DNA from degradation. It has been also shown that the transfection efficiency of prepared nanoparticles depends on the cell type. The presence of fetal bovine serum in the concentration of 10% did not interfere with the transfection ability. The authors developed three different schemes to conjugate transferrin or KNOB protein to the nanoparticle surface. When the transferrin was conjugated to the particles, a maximum of a four-fold increase in their transfection efficiency in HEK293 cells and HeLa cells was obtained. In the case of KNOB conjugation to the nanoparticles, the gene expression level in HeLa cells improved by 130-fold. The conjugation of polyethylene glycol on the nanoparticles provides lyophilization without aggregation. Also, the bioactivity of the particles was preserved for at least 1 month in storage. The clearance of the PEGylated nanoparticles in mice after intravenous administration was slower than in the case of unmodified nanoparticles at 15 min, and with higher depositions in kidney and liver. However, no difference was observed at the 1 h time point.

Stott et al. (1996) have used complex coacervation technique for the coacervation of cationic antidepressants and counter-ions of anionic bile salts. Complex coacervation represents the technique of the separation of an aqueous mixture of oppositely charged ions into a dense coacervate oil phase and a dilute equilibrium phase (Stott et al., 1996). Stott et al. (1996) also investigated the coacervation of cationic tricyclic antidepressants and counter-ions of anionic bile salts sodium cholate or sodium deoxycholate, and the surfactant sodium lauryl sulfate. As cationic tricyclic antidepressants they used amitriptyline, imipramine, and doxepin. The samples were characterized by different techniques, such as microscopy, particle size analysis chromatography, titration, and thermogravimetric analysis. For the purpose of enhancing the transdermal flux of charged species, the authors selected two systems: amitriptyline with sodium deoxycholate, which separates into two distinct phases, and amitriptyline with sodium lauryl sulfate, which remains as a sol. Amitriptyline with sodium deoxycholate produced an 18-fold increase and amitriptyline with sodium lauryl sulfate 22-fold in comparison with aqueous solution. The experiments were conducted using human epidermal membrane with an aqueous receptor and the flux from a 0.025 M aqueous solution that is above the critical micelle concentration of 3.0 ± 0.54 μg/cm^2/h. The flux from an amitriptyline–sodium deoxycholate coacervate donor was 6.6 ± 0.71 μg/cm^2/h, which is a 2.2-fold increase. Amitriptyline with sodium lauryl sulfate, on the other hand, reduced the flux in comparison with the aqueous solution. Based on the results from the study it was concluded that the increased lipophilicity of the coacervate oil phase can increase the transdermal flux of charged species.

The complex coacervate method is also used in the study of Jin and Kim (2008) to obtain injectable and thermo-reversible gels. These gels are formed in an aqueous solution with two oppositely charged biomacromolecules. High molecular weight gelatin type A is used as a positively charged biomacromolecule while chondroitin 6-sulfate is used as negatively charged. A negative, thermo-sensitive polysaccharide, methylcellulose containing a salting-out salt, ammonium sulfate, has been coformulated with previously mentioned compounds. By using complex coacervation and a thermo-reversible gel they confirmed its synergistic effects on the complex formation of the release rates of model proteins and in situ gel depot formation. The gels showed sustained release patterns of the model protein over 25 days with slight initial bursts. Innovative in situ gel carrier systems exhibit the advantages of complex coacervation and temperature responsiveness for efficient protein drug delivery since it achieves high protein loading, sustained protein release, simplicity of administration, an aqueous environment without toxic organic solvents, and a simple preparation method.

Matrigel hydrogels and PLGA scaffolds have been examined as carriers of angiogenic factors (des Rieux et al., 2011) To achieve tissue regeneration, one of the most important requirements is a good vascularization. The delivery of the angiogenic factors, such as VEGF, has been widely examined for creating a vascular network within the tissue. des Rieux et al. (2011) examined if the incorporation of VEGF within nanoparticles could enhance angiogenesis in vivo as compared to free VEGF. By a complex formation with dextran sulfate and coacervation by chitosan, nanoparticles with encapsulated VEGF were synthesized. After 2 weeks, the release of the VEGF that was incorporated within hydrogels was reduced from 30% to 1%, while the release of the VEGF that was incorporated within the PLGA scaffolds increased from 20% to 30% compared to the free VEGF. In vivo experiments showed that VEGF encapsulation improved angiogenesis in both type of 3D matrices. In the case of Matrigel hydrogels more endothelial and red blood cells were observed, as well as in the case of PLGA scaffolds. Encapsulation of VEGF in Matrigel hydrogels, as well as in PLGA scaffolds as a carriers enhanced VEGF efficiency. The encapsulation of VEGF within hydrogels, as well as scaffolds is a promising approach for tissue regeneration engineering.

There are studies that deal with the association of bioactive molecules to casein, the major milk protein, and that investigate the evidence of their efficacy in adapting the release, and/or enhancing the bioavailability of the associated compounds (Elzoghby et al., 2011). Casein has a lot of interesting properties that make it a good candidate for traditional and innovative drug-delivery systems. In the literature, the capacity of casein to change drug dissolution from compacts has been reported. The high tensile strength of casein films allows its use as a film coating for tablets. For the preparation of innovative casein-based systems, such as hydrogels for the controlled release of bioactives, crosslinkers used a naturally occurring genipin and a natural tissue enzyme, transglutaminase. Since casein has emulsifying and bubble-forming properties it is used to develop casein-floating beads. These beads are used to increase the residence time of drugs in the stomach. Also, microparticles of casein are prepared by the emulsification method and by crosslinking with glutaraldehyde, and enzymatic crosslinking by transglutaminase, coacervation, and electrostatic complexation. The methods used for the preparation of different casein nanosystems for the delivery of active substances are also enzymatic crosslinking, graft copolymerization, heat gelation, and polyelectrolyte ionic complexation. It can be concluded that casein-based formulations are promising materials for controlled drug delivery.

The coacervation method has been used for the preparation of chitosan nanoparticles with encapsulated small interfering RNA (siRNA) (Lee et al., 2009). siRNA has been widely examined as a potential therapeutic for the treatment of different diseases. Its fast degradation and low intracellular association, however, limit the usage of siRNA, in vitro and in vivo. Chitosan nanoparticles with encapsulated small interfering RNA have been synthesized in the presence of polyguluronate. Polyguluronate is isolated from alginate. The samples are characterized by different methods to evaluate their properties, such as size, surface charge, morphology, and interaction with siRNA. It was found that the mean diameter of the particles depends on the ratio of chitosan and siRNA. The size of the particles was 110–430 nm. The nanoparticles exhibited low cytotoxicity and were suitable for delivery of siRNA to HEK 293FT and HeLa cells. The chitosan nanoparticles were considered good candidates for siRNA delivery.

In the work of Carrasquillo et al. (2001), the coacervation method was used for the preparation of poly(D,L-lactide-*co*-glycolide) microspheres with encapsulated bovine serum albumin as a model protein. Bovine serum albumin in powder form was obtained by spray-freeze drying and after that suspended in methylene chloride containing poly(D,L-lactide-*co*-glycolide). Thereafter, coacervation is performed by adding silicon oil and microsphere hardening in heptane. The structure of the bovine serum albumin is characterized by Fourier-transform infrared spectroscopy (FTIR). Structural perturbations in bovine serum albumin were reduced during the spray-freeze drying step by using the excipient trehalose. The encapsulation of bovine serum albumin within poly(D,L-lactide-*co*-glycolide) microspheres was performed without significant structural perturbations of the protein. The encapsulated bovine serum albumin had a similar monomer content and the same secondary structure as non-encapsulated. There was a small initial burst release and a prolonged and continuous sustained release phase of the protein from the polymeric matrix.

A key factor in determining the bioavailability of intranasally administered drugs is often rapid mucociliary clearance. The usage of mucoadhesive micro- and nanoparticles provides a potential strategy for improving the retention of drugs within the nasal cavity, and thus improve the resulting pharmacokinetic profile. A study by Lim et al. (2000) describes the comparison of a number of novel, potentially mucoadhesive microspheres, prepared by solvent evaporation, composed of hyaluronic acid, chitosan glutamate, and a combination of the two with microcapsules of hyaluronic acid and gelatin prepared by complex coacervation. The particles had a mean diameter of about 20 μm (hyaluronic acid), 28 μm (hyaluronic acid/chitosan glutamate), and 30 μm (chitosan glutamate). The encapsulation of a model drug, gentamicin sulfate was 46.90 ± 0.53% (hyaluronic acid), 28.04 ± 1.21 (hyaluronic acid/chitosan glutamate), and 13.32 ± 1.04 (chitosan glutamate). The release of the drug was characterized and it was shown that the release of gentamicin from hyaluronic acid and hyaluronic acid/chitosan glutamate was 50% longer than in the case of chitosan glutamate and was best modeled as a release from a carrier. The mucociliary transport rate of the particles across an isolated frog palate has been used to determine the degree of mucoadhesion of each system. The rank order of mucoadhesion for the microspheres and the microparticles was hyaluronic acid = hyaluronic acid/chitosan glutamate > chitosan glutamate > hyaluronic acid/gelatin > chitosan glutamate. The encapsulation of gentamicin did not affect the mucoadhesive properties. The combination of hyaluronic acid with chitosan may afford additional advantages in combining the mucoadhesive potential of hyaluronic acid with the penetration-enhancing effect of chitosan.

4.2 MELT DISPERSION

In the melt-dispersion technique, the drug-containing molten wax phase is emulsified into a heated, emulsifier-containing external phase. Depending on the solubility of the drug, the external phase can be either aqueous (for water-insoluble drugs) or nonaqueous (for water-soluble drugs). On cooling the emulsion, the liquid droplets congeal and a suspension of the wax microparticles is formed. The microparticles are then separated, mostly by filtration or centrifugation, sometimes washed to remove free drug crystals and surfactants, dried, and sized. Carnauba wax was used in the melt dispersion method for the encapsulation of ascorbic acid (Uddin et al., 2001).

In the study of Mathiowitz and Langer (1987) a "hot melt" microencapsulation is described as a method for obtaining polyanhydride microspheres with encapsulated dyes (acid-orange), *p*-nitroaniline, or insulin. Poly[bis(*p*-itcarboxy phenoxy)propane anhydride] copolymerized with sebacic acid has been used as a model polymer. The degradation of the microspheres depends on the amount of the hydrophilic drugs that are used in the synthesis. Insulin-incorporated microspheres implanted into diabetic rats resulted in normoglycemia for a period of 3–4 days.

4.3 SOLVENT EVAPORATION

The emulsification-evaporation method (Astete et al., 2007; Prabha and Labhasetwar, 2004; Sahoo and Labhasetwar, 2005; Song et al., 2006), spontaneous emulsification-solvent diffusion method (Lee et al., 2005; Zhang et al., 2006), nanoprecipitation method (Bilati et al., 2005; Govender et al., 1999), solvent evaporation, and spray drying (Atuah et al., 2003; Rivera et al., 2004; Takashima et al., 2007), are all widely used to prepare micro- and nanoparticles of various sizes. Each of these methods employs a similar first step, where an aqueous drug solution is emulsified in an organic polymer solution to form a water-in-oil dispersion (w/o) (O'Donnell and McGinity, 1997). If appropriate, the ingredient may also be dispersed as a solid substance in an organic polymer solution, or dissolved in a common solvent together with the polymer. In the solvent evaporation technique, the drug substance is dispersed homogeneously in the polymer. The particle size is influenced by the type and concentration of stabilizer, homogenizer speed, and the polymer concentration (Kwon et al., 2001). However, this technique leads to low encapsulation efficiencies for water-soluble drugs (Bodmeier et al., 1994; Wieland-Berghausen et al., 2002).

The solvent-evaporation method is used to explore the influence of changing the temperature during the experiments, the core-to-wall ratio, and the influence of the presence of the plasticizers (triethyl citrate) on the release rate of the microencapsulated ascorbic acid (Uddin et al., 2001). Results exhibit that the presence of plasticizer reduced the release rate. The ratios of the core to wall used were 1:1 and 3:1. This parameter had no significant effect on the release rate. Also, the two temperatures used in the experiments of 28 and 55°C, exhibited insignificant influence on the release rate. In the experiments of spray drying, four different polymer-coating formulations, whether singly or as a mixture, were used. These were gel, starch, ethyl-cellulose, and β-cyclodextrin. The loss of ascorbic acid during spray drying was 20%. Different coating materials resulted in particle capsules with sizes in the range from 90 to 280 μm. The ratio of encapsulated ascorbic acid was less than 50%. This indicates and means that less than 50% of ascorbic acid was actually microencapsulated.

Moderately uniform-sized, biodegradable, poly(lactide) microcapsules have been successfully produced by combining a shirasu porous glass membrane emulsification method and multiple emulsion-solvent evaporation method (Liu et al., 2005). An aqueous phase containing lysozyme was used as the internal water phase. Poly(lactide) and arlacel 83 were dissolved in a mixture solvent of dichloromethane and toluene that was used as the oil phase. These two solutions were emulsified by a homogenizer to form a water/oil primary emulsion. This primary emulsion was filtrated through the uniform pores of a shirasu porous glass membrane into the external water phase by the pressure of nitrogen gas to form the uniform water/oil/water droplets. Afterward the solid polymer microparticles were obtained by evaporating the solvent. It is essential to avoid the phase separation of primary emulsion during the membrane emulsification. It was shown that when the density alteration of the internal water phase and oil phase was reduced to nearly zero and arlacel 83 was employed as the oil emulsifier, the phase separation was not observed within 24 h. The drug encapsulation efficiency was found to be dependent on poly(lactide) molecular weight, additive type and its concentration in the internal water phase, the emulsifier type and concentration in the oil phase, the NaCl concentration and the pH value in the external water phase. In comparison with the stirring method, it was found that the size was more uniform and the drug encapsulation efficiency was much higher when the microcapsules were prepared by the shirasu porous glass membrane emulsification method and the highest drug encapsulation efficiency of 92.20% was obtained.

The emulsification is the major and first step of the emulsification solvent evaporation method (Rosca et al., 2004). However, the second step, which is the solvent transport out from the emulsion droplets, has great influence on the particles' morphology, as well as on the microparticles' encapsulation and release behavior. Rosca et al. (2004) studied the mechanism of solvent elimination from the emulsion droplets and its influence on particle morphology, encapsulation, and release behavior. Very often, the solvent is highly volatile and makes the solvent elimination process very fast—thus difficult to observe. The initial emulsion was monitored by an optical microscope under controlled solvent evaporation conditions. In this study, the results from the optical microscopic observations—validated with laser diffractometry analysis—showed that in single emulsion formulations, spherical microparticles are formed by accelerated solvent elimination due to the combined effects of high solvent volatility and polymer precipitation. The solvent expulsion is accompanied by significant shrinkage that generates a thin layer of nanoparticles on the microparticle surface; this is confirmed by electron microscopy and laser diffractometry. During the intense solvent removal, the encapsulated substance is drained, affecting the loading efficiency. Therefore, it will concentrate near the microparticle surface contributing to the initial burst release. In the procedure such as in the case of double emulsion formulations, particles with different morphologies are generated due to the presence of the aqueous-phase droplets inside the emulsion droplet. During the solvent elimination, these microdroplets normally coalesce under the pressure of the precipitating polymer. Depending mostly on the polymer concentration and emulsification energies, the final microparticles will be a mixture of honeycomb, capsule, or plain structure. During the shrinkage due to the incompressibility of the inner droplets, the precipitating polymer wall around them may break and form holes through which the encapsulated substance is partly expulsed. Through these holes, the encapsulated substance is further partitioning with the external aqueous phase during solvent evaporation and contributes to the initial burst release during the application.

4.4 SPRAY DRYING

Spray drying is a well-known technique that generates solid formulations from a solution or suspension and has special applications in the pharmaceutical and food industry. Spray drying is a single-step process that involves atomization of the liquid feed, which is then injected into a drying chamber containing hot air or nitrogen. The droplets instantly dry into solid particles that are subsequently collected in the drying chamber. The two factors that can influence the properties of the output material are the spray-dried particle formulation factors and the spray-drying parameters (Seville, 2007). Spray drying was used as a preparation method for vitamin C/Eudragit microspheres (Esposito et al., 2002). Vitamin C/Eudragit microspheres obtained by this method showed potential for delivery of vitamin C by oral rout. Spray drying was used by Finotelli and Rocha-Leao (2005) to obtain microcapsules with different contents of ascorbic acid for application in the food industry as fortification ().

Jensen et al. (2010) examined the influence of different parameters on the poly(D,L-lactide-*co*-glycolide) nanoparticles with encapsulated small interfering RNA that have been synthesized by the spray-drying technique. Poly(D,L-lactide-*co*-glycolide) nanoparticles with encapsulated small interfering RNA have been converted into nanocomposite microparticles intended for inhalation. The spray-drying procedure has been optimized by the statistical design of the experiment and by characterizing powder appearances after systematic variation of the formulation parameters. The parameters that were varied were concentration, carbohydrate excipient, and the ratio of nanoparticles to excipient. Trehalose, lactose, and mannitol were used as carbohydrate excipients. The parameters were varied to observe their influence on moisture content, size of the particles and their morphology, and powder yield. The recognized optimum conditions were applied for spray drying the small interfering encapsulated nanocomposite microparticles. The result of this condition was a system with a low water content (0.78% w/w) and with an optimal size for inhalation. When mannitol was used as the carbohydrate excipient, a significantly lower moisture content was obtained than when trehalose or lactose were used. Also, larger amounts of nanoparticles lead to changes in the surface morphology of the spray-dried particles. The integrity and biological activity of the small interfering RNA were preserved during the spray-drying process. Jensen et al. (2010) have concluded that spray drying is a very good method for the synthesis of nanocomposite microparticles comprising small interfering RNA within poly(D,L-lactide-*co*-glycolide) nanoparticles for potential use in inhalation therapy.

4.5 SPRAY CHILLING AND SPRAY COOLING

Spray chilling and spray cooling are similar methods to spray drying in that the core material is dispersed in a liquefied coating or wall material and atomized, but there is generally no water to be evaporated. The core and wall mixture are atomized into either cool of chilled air that causes the wall to solidify around the core (Risch, 2009a). In spray chilling, the coating is typically a fractionated or hydrogenated oil. In spray cooling, the wall is typically oil, although other materials can be used. These methods actually differ only in the melting point of the wall material used and are most often used to encapsulate solid materials, such as vitamins, minerals, or acidulants.

4.6 HOMOGENIZATION OF WATER AND ORGANIC PHASES

Spherical nanoparticles of the poly(D,L-lactide-*co*-glycolide) were produced using a physicochemical method with solvent/nonsolvent systems (Stevanović et al., 2007a,b,c). The encapsulation of the ascorbic acid, folic acid, silver nanoparticles (AgNp), or AgNp together with ascorbic acid in the PLGA polymer matrix have been successfully performed by homogenization of water and organic phases (Stevanović and Uskoković, 2009b; Stevanović et al., 2007a,b, 2008a,b, 2012b, 2013, 2014). Typically, the PLGA commercial granules have been dissolved in acetone and after that the aqueous solution of the ascorbic acid has been added in PLGA solution in acetone while continuously being homogenized. The concentration of ascorbic acid in water was varied in order to obtain particles with different ratios of PLGA and ascorbic acid (PLGA/ascorbic acid 85/15 wt.%, PLGA/ascorbic acid 70/30 wt.%, PLGA/ascorbic acid 50/50 wt.%, and PLGA/ascorbic acid 30/70 wt.%) (Stevanović et al., 2007a,b). The introduction of 15% of ascorbic acid does not influence the size, while a further increase of ascorbic acid concentration increases the size of PLGA particles. After the precipitation with alcohol methanol or ethanol, the particles are stabilized due to the zeta potential created by the stabilizer. This was followed by centrifugation, decantation, and drying of the particles. Polyvinyl pyrrolidone (povidone, PVP) or PVA were used as a stabilizer of the particles.

Stevanović et al. (2014) described how they obtained PLGA particles with encapsulated AgNp stabilized by polyglutamic acid (PLGA/AgNpPGA) together ascorbic acid (PLGA/AgNpPGA/ascorbic acid particles). PLGA/AgNpPGA/ascorbic acid particles have been synthesized by a physicochemical method with solvent/nonsolvent systems that are spherical, have a mean diameter of 775 nm, and a narrow size distribution with a polydispersity index of 0.158. The encapsulation efficiency of AgNpPGA/ascorbic acid within PLGA was determined to be greater than 90%. The entire amount of encapsulated ascorbic acid was released in 68 days, and the entire amount of AgNpPGAs was released in 87 days of degradation (Stevanović et al., 2014). A schematic representation of degradation and release of the PLGA/AgNpPGA/ascorbic acid particles is given in Fig. 11.3.

Desai and Park (2005b) reports the encapsulation of vitamin C within sodium alginate beads by an alternative approach. The alternative encapsulation procedure mostly includes immobilization of vitamin C in hydrated zinc oxide layers and subsequently encapsulation of as prepared particles in sodium

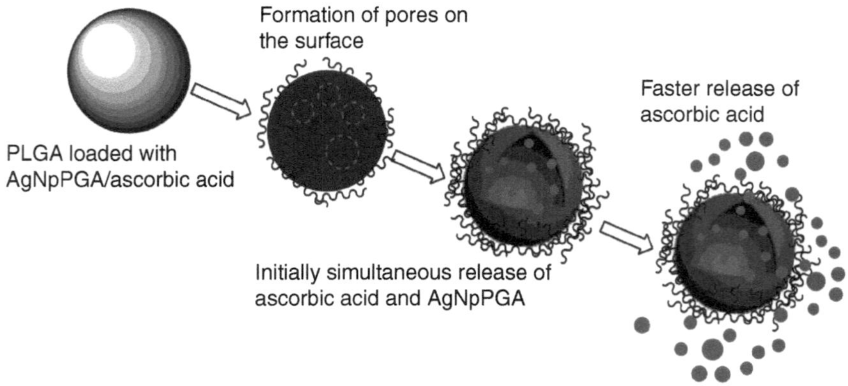

FIGURE 11.3 A Schematic Representation of the Degradation and Release of PLGA/AgNpPGA/Ascorbic Acid Particles

alginate beads. Vitamin C has been immobilized in hydrated zinc oxide layers by a coprecipitation method. FTIR exhibited that the vitamin C has been found to be stable during and after its immobilization.

Hydrated zinc oxide as an inorganic matrix for the immobilization of vitamin C was used by Yang et al. (2003). The encapsulation of vitamin C within a biocompatible layered inorganic material was achieved by a coprecipitation reaction, in which the layered inorganic lattice and its intercalate of vitamin C are simultaneously formed. The nanometer-sized powders of vitamin C intercalate; thus prepared, it was again encapsulated with silica nanosol to form a nanoporous shell structure. This ternary nanohybrid of vitamin C layered inorganic core-SiO_2 shell exhibited an enhanced storage stability and a sustained releasing of vitamin C. Furthermore, the encapsulation of vitamin C with inorganic mineral was very helpful in delivering vitamin C molecules into the skin through stratum corneum, facilitating transdermal penetration of vitamin C in topical applications.

4.7 EXTRUSION

Extrusion is a relatively new process compared to spray drying and it is often used in the food industry (Risch, 2009b). This process begins by forming a low moisture (5%–10%) carbohydrate melt. An emulsifier is added to the melt and then the material for the encapsulation is added with vigorous agitation. This molten emulsion is forced through a die into a cold isopropanol bath. The melt solidifies into an amorphous structure that is broken into small, rod-shaped pieces by mechanical agitation. The pieces are recovered by centrifugation, dried under a vacuum, mixed with a free-flowing agent, and stored.

Hot-melt extrusion produces thermodynamically stable dispersions of amorphous quinine, a model drug, within an amorphous polymeric platform (Jones et al., 2014). Characterization of the pre-extrudates and extrudates was performed using differential scanning calorimetry, X-ray diffraction, and Raman spectroscopy. Water uptake by the raw materials was determined using the dynamic vapor sorption technique. Additionally, the presence or absence of crystalline drug following storage at 25°C/60% relative humidity and 40°C/75% relative humidity in a sealed glass jar, and at 40°C/75% relative humidity in an open glass jar for 3 months was determined using powder X-ray diffraction. Amorphous quinine was generated in situ during extrusion from both quinine base (5, 10, and 20% w/w drug loading) and from quinine hydrochloride (5 and 10% w/w drug loading) and remained thermodynamically stable as a solid-in-solid dispersion within the polymeric extrudates. When processed with polymer, quinine hydrochloride (20% w/w) was converted to amorphous quinine hydrochloride. Though stable for up to 3 months when stored under sealed conditions, this amorphous form was unstable, resulting in recrystallization of the hydrochloride salt following storage for 1 month at 40°C/75% relative humidity in an open glass jar. The behavior of the amorphous quinine hydrochloride (20% w/w) extrudate was related, at least in part, to the lower stability and the hygroscopic properties of this amorphous form.

Carriers for the sustained release of ibuprofen have been produced by the hot-melt extrusion method. Ibuprofen was used as the model drug and ethyl cellulose as the sustained release agent. Since release of ibuprofen has been slow from ibuprofen/ethyl cellulose matrices at a ratio of 60:40 w/w, respectively, Brabander et al. (2003) modified the procedure in such a way that excipient hydroxypropyl methylcellulose and xanthan gum has been added to the formulation. This affected the ibuprofen release. Initially, a burst release was observed for the preparations containing hydroxypropyl methylcellulose and a nearly time-independent release was seen for the xanthan gum based mini-matrices. The ibuprofen release from the mini-matrices was generally diffusion controlled. Also, swelling of the matrices played an imperative role to achieve complete drug release. The preparations were stable during 1 year of storage at 60% relative humidity and room temperature. At an elevated humidity and temperature the ibuprofen release increased.

Hot-melt extrusion was also used in the work of Follonier et al. (1995) for the preparation of the sustained-release pellets. Diltiazem hydrochloride was used as the model drug in this study. The obtained preparations, pellets, can be packed into hard, gelatin capsules. With this method it achieved in vitro release at rates that are slow enough to enrich the therapeutic plasma levels for the drug with a once or twice daily administration. The release profile was dependent on the polymer-to-drug ratio, and this was examined by a double exponential decay equation to distinguish between the surface release and diffusion release phases. In order to achieve optimal release of the drug, different parameters—such as type of the polymer, addition of additives for the purpose of forming pores, addition of hydrophilic polymers, and so on—were studied. The probability of reaching a nearly zero-order release without any coating has been investigated with poly(ethylene-*co*-vinyl acetate). Poly(ethylene-*co*-vinyl acetate) is an inert polymer. Assuming that this could be reached by including polymers with pH-dependent behavior, the influence of cationic and enteric additives has been evaluated. At the end, swelling agents were included into the pellets that reduced the burst release.

Fule and Amin (2014) examined the influence of different surfactants on the formulations of solid dispersions and a model antiulcer drug lafutidine was prepared by hot-melt extrusion. PEG 400, Lutrol F127, Lutrol F68 were used as surfactants. Synthesized amorphous glassy solid dispersion was found to be thermodynamically and physicochemically stable. Traces of the crystalline antiulcer drug lafutidine were not observed in the extrudates according to differential scanning calorimetry, scanning electron microscopy, X-ray diffraction, and Raman spectroscopy. This method revealed drug–polymer molecular miscibility and surface interaction at the microlevel. By the ^{1}H–COSY NMR spectroscopy method, it confirmed miscibility and interaction among antiulcer drug lafutidine solid dispersions and a solubilizing agent (amphiphilic Soluplus), with chemical shift drifting and line broadening. Also, computational modeling (MD simulation) showed intermolecular interaction between molecules. The dissolution rate and solubility of the antiulcer drug lafutidine was enhanced remarkably in developed solid dispersions systems. Also, it was shown that the ratio between polymer and surfactants is very important and has a significant role in the dissolution-rate enhancement of antiulcer drug lafutidine solid dispersions. This system has promising potential for oral delivery and might be an effective approach for improving the therapeutic potential of the antiulcer drug lafutidine.

5 CHARACTERISTICS OF POLYMERIC MICRO- AND NANOPARTICLES WITH AN ENCAPSULATED VITAMIN AS AN ACTIVE SUBSTANCE AND THEIR COMPARISON

Encapsulation efficiency, release rate, and the size distribution of particles are some of the parameters that are used for evaluating encapsulation system characteristics.

Uddin et al. (2001) evaluated the influence of process perimeters on ascorbic acid properties. Different methods for the encapsulation of ascorbic acid were chosen. Uddin et al. selected the spray-drying technique, thermal phase separation, the melt-dispersion method, as well as solvent evaporation. It is well known that ascorbic acid is a highly oxidative substance and can cause difficulties in food systems. During the processing steps, ascorbic acid can change color from white to yellow, which can have an effect on food colors. Also, ascorbic acid can react with other constituents and come to undesirable variations in the color and perception of the food. The results confirmed that the encapsulation of ascorbic acid will prevent color change, delay its core release rate, and also cover its acid taste. Regarding the thermal phase separation method, it was shown that the molecular weight of ethyl cellulose and the addition of polyisobutylene significantly affected the agglomeration and release rate of microparticles. There is a noteworthy change in release ratios with various molecular weights of ethyl cellulose. The initial release rate of ascorbic acid was lower when the ethyl cellulose with higher molecular weight was used. The molecular rate was not a parameter for the complete release of ascorbic acid from the microcapsules, since high or low molecular weight gave a release of ~1.0 in 20 min.

In the melt-dispersion method, spherical capsules were prepared with carnauba wax (Uddin et al., 2001). The ascorbic acid release rate was slower in this case than in the case when the capsules were prepared with ethyl cellulose. This shows that carnauba wax keeps the ascorbic acid more stable. In the solvent-evaporation method, a higher molecular weight of ethyl cellulose and the addition of plasticizer have been found to be significant for encapsulation. During the spray-drying process, the loss of ascorbic acid during microencapsulation has been found to be least. Uddin et al. (2001) showed that the loss of ascorbic acid encapsulated by spray drying was 20%. It was also shown that starch and β-cyclodextrin have influence on the degradation of the capsules with encapsulated ascorbic acid.

Trindade and Grosso (2000) examined the stability of ascorbic acid encapsulated in granules made of rice starch and also in granules made of gum arabic. The capsules prepared from starch were larger than those prepared from gum arabic. Microcapsules prepared from starch had sizes of about and less than 57 μm and with a mean size of 20.5 μm while 90% of gum arabic capsules had sizes less than 27 μm with mean size of 8.0 μm.

Using this simple method, spherical and uniform particle powder were obtained from the commercial granules of poly(D,L-lactide-*co*-glycolide), where the mean particle size ranges from 110 to 170 nm. The various concentrations of the vitamin C were encapsulated within PLGA particles. The results of the determination of the particle yield for various PLGA/ascorbic acid ratios were similar for each of the samples and in all cases greater than 50% (Stevanović et al., 2007a,b). The loading efficiency was determined to be greater than 90% in all ratios of PLGA/ascorbic acid particles (Stevanović et al., 2007a,b). The degradation of the PLGA without and with ascorbic acid has been followed, as well as morphological changes that occurred during the degradation (Stevanović et al., 2007a, 2008a,b). The degradation were tracked for 8 weeks and it was determined that PLGA completely degrades within this period, fully releasing all encapsulated ascorbic acid. In the first 24 days, the samples degrade slower, while later the pace of the degradation increases. In the first 24 days of the degradation, for all samples, less than the 10% of the encapsulated ascorbic acid was released (Stevanović et al., 2007a,b). At the beginning the particles maintain their initial shape, but after 24 days the particles start to agglomerate and create the porous film, where the porosity increases until the complete degradation of the samples. By the end of the experiment the nanoparticles were fully degraded and there were no more traces of them in the solution. PLGA degrades via backbone hydrolysis (bulk erosion) and the degradation products are the monomers, lactic acid, and glycolic acid. It could be expected that the faster degradation of the lower molar mass fraction, present in copolymer, increases the local acidity, thereby, accelerating the hydrolysis of higher molar mass species. In other words, when acid accumulation creates a local pH drop, catalytic degradation of the polymer itself occurs. The different ionized forms of ascorbic acid have different redox properties, so that the redox-chemistry of the ascorbic acid is highly pH dependent (Bors and Buettner, 1997; Buettner, 1988; Rozario, 2005). Ascorbic acid decomposes into biologically inactive compounds by autooxidation only at alkaline pH (Kim et al., 1996). In the solution with low pH, decomposition of the ascorbic acid can happen, for example, under the influence of the enzymes (enzymatic oxidation) (Kim et al., 1996).

The biological behavior of PLGA nanospheres without and with encapsulated ascorbic acid is discussed in terms of in vitro toxicity in human hepatoma cells and in vivo biodistribution in rats after intravenous injection (Stevanović et al., 2009). Neither PLGA nanospheres nor PLGA/ascorbic acid 85%/15% nanoparticles significantly affected the viability of the HepG2 cells (Stevanović et al., 2009). PLGA nanospheres with encapsulated ascorbic acid exhibit prolonged blood circulation, accompanied by a time-dependent reduction in lung, liver and spleen, and addition in kidney, stomach, and intestine (Stevanović et al., 2009).

Vitamin C/Eudragit microspheres obtained by the spray-drying method are not able to provide slower release of the drug with respect to the free form of ascorbic acid. However, these microparticles have a morphology and size distribution that make these particles promising for the delivery of ascorbic acid in the treatment of colorectal cancer (Esposito et al., 2002).

The procedure of obtaining microparticles with encapsulated ascorbic acid by spray drying was described by Finotelli and Rocha-Leao (2005). The morphology of the Capsul/vitamin C particles was observed by a scanning electron microscopy, whose analysis showed a tendency of agglomeration. Particle size analysis showed a multimodal particle size distribution, but with a main mode in intermediate diameters range (4–8 μm). The particle yield was 52%. Ascorbic acid stability was studied for particles stored at room temperature and at 45°C, showing 100% of retention at the beginning. Microcapsules containing 20% of ascorbic acid recovered by a mixture presented 7% of ascorbic acid reduction in samples for up to 60 days stored at 28°C temperature.

The spray-drying technique was also used by Desai and Park (2005a) for encapsulation of ascorbic acid in tripolyphosphate crosslinked chitosan microparticles (Desai et al., 2005; Desai and Park, 2006a,b) The particles had spherical shapes and a mean size of about 6 μm–9.0. The ascorbic acid encapsulation efficiency was 58%. Encapsulation efficiency was decreased as the amount of tripolyphosphate solution was increased. Also, the quantity of crosslinking influenced the release rate and the size of the particles.

6 CONCLUSIONS

This chapter reports on polymers for a controlled delivery of active substances. It focuses on the most widely used encapsulation techniques for the encapsulation of active substances within polymeric carriers, such as thermal phase separation, melt dispersion, solvent evaporation, spray drying, homogenization of water and organic phases, and so on. This chapter also gives a comparison of the characteristics of polymeric micro- and nanoparticles that are prepared by different methods. Presently, there is great interest in polymer-carrier systems in the fields of medicine, pharmacy, and the food industry based on the need for a system with a controlled and balanced release of active substances. It is foreseen that applications of different methods in micro- and nanoparticle research for this purposes will expand in the near future.

ACKNOWLEDGMENTS

MS acknowledges the support from the Ministry of Education, Science and Technological Development of the Republic of Serbia, under grant project III45004.

REFERENCES

Alshamsan, A., 2014. Nanoprecipitation is more efficient than emulsion solvent evaporation method to encapsulate cucurbitacin I in PLGA nanoparticles. Saudi Pharm. J. 22 (3), 219–222.
Astete, C.E., Kumar, C.S.S.R., Sabliov, C.M., 2007. Size control of poly(D,L-lactide-*co*-glycolide) and poly(D,L-lactide-*co*-glycolide)-magnetite nanoparticles synthesized by emulsion evaporation technique. Colloids Surf. A 299, 209–216.
Atuah, K.N., Walter, E., Merkle, H.P., Alpar, H.O., 2003. Encapsulation of plasmid DNA in PLGA-stearylamine microspheres: a comparison of solvent evaporation and spray-drying methods. J. Microencapsul. 20 (3), 387–399.
Bachtsi, A.R., Kiparissides, C., 1996. Synthesis and release studies of oil-containing poly(vinyl alcohol) microcapsules prepared by coacervation. J. Control. Rel. 38 (1), 49–58.
Benita, S., Donbrow, M., 1982. Effect of polyisobutylene on ethylcellulose-walled microcapsules: wall structure and thickness of salicylamide and theophylline microcapsules. J. Pharmaceut. Sci. 71 (2), 205–210.
Besenbache, F., Sutherland, D.S., Hovgaard, M.B., 2007. From nanoscience to nanotechnology: utilizing the nanoscale. Toxicol. Lett. 172 (1), S34.
Bilati, U., Allémann, E., Doelker, E., 2005. Development of a nanoprecipitation method intended for the entrapment of hydrophilic drugs into nanoparticles. Eur. J. Pharmacol. 24, 67–75.
Bodmeier, R., Wang, H., Hermann, J., 1994. Microencapsulation of chlorpheniramine maleate, a drug with intermediate solubility properties, by a non-aqueous solvent evaporation technique. STP Pharma. Sci. 4 (4), 275–281.
Bors, W., Buettner, G.R., 1997. The vitamin C radical and its reactions. In: Packer, L., Fuchs, J. (Eds.), Vitamin C in Health and Disease. Marcel Dekker, New York, p. 75.
Brabander, C.De., Vervaet, C., Remon, J.P., 2003. Development and evaluation of sustained release mini-matrices prepared via hot melt extrusion. J. Control. Rel. 89 (2), 235–247.
Buettner, G.R., 1988. In the absence of catalytic metals ascorbate does not autooxidaze at pH 7: ascorbate as test for catalytic metals. J. Biochem. Biophys. Methods 16 (1), 27–40.
Carrasquillo, K.G., Stanley, A.M., Aponte-Carro, J.C., De Jésus, P., Costantino, H.R., Bosques, C.J., Griebenow, K., 2001. Non-aqueous encapsulation of excipient-stabilized spray-freeze dried BSA into poly(lactide-*co*-glycolide) microspheres results in release of native protein. J. Control Rel. 76 (3), 199–208.
Choi, Y., Park, S., Suh, H., 2005. Adipose tissue engineering using mesenchymal stem cells attached to injectable PLGA spheres. Biomaterials 26, 5855–5863.
D'Emanuele, A., Attwood, D., Abu-Rmaileh, R., 2004. The use of a dendrimer-propranolol prodrug to bypass efflux transporters and enhance oral bioavailability. J. Control. Rel. 95, 447–453.
Daugherty, A., Cleland, J., Duenas, E., Mrsny, R., 1997. Pharmacological modulation of the tissue response to implanted polylactic-*co*-glycolic acid microspheres. Eur. J. Pharm. Biopharm. 44 (1), 89–102.
des Rieux, A., Ucakar, B., Mupendwa, B.P.K., Colau, D., Feron, O., Carmeliet, P., Préat, V., 2011. 3D systems delivering VEGF to promote angiogenesis for tissue engineering. J. Control. Rel. 150 (3), 272–278.
Desai, K.G.H., Park, H.J., 2005a. Encapsulation of vitamin C in tripolyphosphate cross-linked chitosan microspheres by spray drying. J. Microencapsul. 22 (2), 179–192.
Desai, K.G.H., Park, H.J., 2005b. Recent developments in microencapsulation of food ingredients. Drying Technol. 23 (7), 1361–1394.
Desai, K.G.H., Park, H.J., 2006a. Effect of manufacturing parameters on the characteristics of vitamin C encapsulated tripolyphosphate-chitosan microspheres prepared by spray-drying. J. Microencapsul. 23 (1), 91–103.
Desai, K.G.H., Park, H.J., 2006b. Characteristics of vitamin C encapsulated tripolyphosphate-chitosan microspheres as affected by chitosan molecular weight. J. Microencapsul. 23 (1), 79–90.
Desai, K.G., Liu, C., Park, H.J., 2005. Characteristics of vitamin C immobilized particles and sodium alginate beads containing immobilized particles. J. Microencapsul. 22 (4), 363–376.

Elzoghby, A.O., El-Fotoh Wael, S.A., Elgindy Nazik, A., 2011. Casein-based formulations as promising controlled release drug delivery systems. J. Control. Rel. 153 (3), 206–216.

Esposito, E., Cervellati, F., Menegatti, E., Nastruzzi, C., Cortesi, R., 2002. Spray dried Eudragit microparticles as encapsulation devices for vitamin C. Int. J. Pharm. 242, 329–334.

Feng, S.S., Mu, L., Yin Win, K., Huang, G., 2004. Nanoparticles of biodegradable polymers for clinical administration of paclitaxel. Curr. Med. Chem. 11, 413–424.

Finotelli, P.V., Rocha-Leao, M.H.M., 2005. Microencapsulation of ascorbic acid in maltodextrin and capsul using spray-drying. 2nd Mercosur Congress on Chemical Engineering and 4th Mercosur Congress on Process Systems Engineering, Rio de Janeiro, Brazil, pp. 1–11.

Follonier, N., Doelker, E., Cole, E.T., 1995. Various ways of modulating the release of diltiazem hydrochloride from hot-melt extruded sustained release pellets prepared using polymeric materials. J. Control. Rel. 36 (3), 243–250.

Fule, R., Amin, P., 2014. Development and evaluation of lafutidine solid dispersion via hot melt extrusion: investigating drug-polymer miscibility with advanced characterization. Asian J. Pharmaceut. 9 (2), 92–106.

Gander, B., Johansen, P., Nam-Tran, H., Merkle, H.P., 1996. Thermodynamic approach to protein microencapsulation into poly(D,L-lactide) by spray drying. Int. J. Pharm. 129, 51–61.

Govender, T., Stolnik, S., Garnett, M.C., Illum, L., Davis, S.S., 1999. PLGA nanoparticles prepared by nanoprecipitation: drug loading and release studies of a water-soluble drug. J. Control. Rel. 57, 171–185.

Gratton, S.E.A., Pohlhaus, P.D., Lee, J., Guo, J., Cho, M.J., De Simone, J.M., 2007. Nanofabricated particles for engineered drug therapies: a preliminary biodistribution study of PRINT nanoparticles. J. Control. Rel. 121 (1–2), 10–18.

Greene, M.E., 2007. Understanding cell interactions is a snap: nanobiotechnology. Nano Today 2 (3), 11.

Guan, Y., Zhou, Y., Sheng, J., Ying, J.Z., Jumahan, H., Hu, Y., 2015. Preparation of ALA-loaded PLGA nanoparticles and its application in PDT treatment. J. Chem. Technol. Biotechnol. 91 (4), 1128–1135.

Ishida, M., Sakai, H., Sugihara, S., Aoshima, S., Yokoyama, S., Abe, M., 2003. Controlled release of vitamin E from thermo-responsive polymeric physico-gel. Chem. Pharm. Bull. 51 (11), 1348–1349.

Ito, F., Fujimori, H., Makino, K., 2007. Incorporation of water-soluble drugs in PLGA microspheres. Colloids Surf. B 54, 173–178.

Jensen, D.M.K., Cun, D., Maltesen, M., Frokjaer, J.S., Nielsen, H.M., Foged, C., 2010. Spray drying of siRNA-containing PLGA nanoparticles intended for inhalation. J. Control. Rel. 142 (1), 138–145.

Jeong, Y., Song, J., Kang, S., Ryu, H., Lee, Y., Choi, C., Shin, B.-A., Kim, K.-K., Ahn, K.-Y., Jung, S., 2003. Preparation of poly(D,L-lactide-*co*-glycolide) microspheres encapsulating all-trans retinoic acid. Int. J. Pharm. 259, 79–91.

Jin, K.-M., Kim, Y.-H., 2008. Injectable, thermo-reversible and complex coacervate combination gels for protein drug delivery. J. Control. Rel. 127 (3), 249–256.

Jin, C., Bai, L., Wu, H., Tian, F., Guo, G., 2007. Radiosensitization of paclitaxel, etanidazole and paclitaxel + etanidazole nanoparticles on hypoxic human tumor cells in vitro. Biomaterials 28 (25), 3724–3730.

Jones, D.S., Margetson, D.N., McAllister, M.S., Yu, T., Shu, L., McCoy, C.P., Andrews, G.P., 2014. Thermodynamically stable amorphous drug dispersions in amorphous hydrophilic polymers engineered by hot melt extrusion. Chem. Eng. Res. Des. 92 (12), 3046–3054.

Kang, Y., Yin, G., Ouyang, P., Huang, Z., Yao, Y., Liao, X., Chen, A., Pu, X., 2008. Preparation of PLLA/PLGA microparticles using solution enhanced dispersion by supercritical fluids (SEDS). J. Colloid Interf. Sci. 322 (1), 87–94.

Kim, Y.-R., Yu, S.-W., Lee, S.-R., Hwang, Y.-Y., Kang, S.-O., 1996. A heme-containing ascorbate oxidase from *Pleurotus ostereatus*. J. Biol. Chem. 271 (6), 3105–3111.

Kozubek, A., Gubernator, J., Przeworska, E., Stasiuk, M., 2001. Liposomal drug delivery, a novel approach: PLARosomes. Acta Biochim. Pol. 47 (3), 639–649.

Kumar, R.M.N.V., 2000. Nano and microparticles as controlled drug delivery devices. J. Pharm. Pharm. Sci. 3 (2), 234–258.

Kumar, R.M.N.V., Bakowsky, U., Lehr, C.M., 2004. Preparation and characterization of cationic PLGA nanospheres as DNA carriers. Biomaterials 25, 1771–1777.

Kwon, H.Y., Lee, J.Y., Choi, S.W., Jang, Y., Kim, J.H., 2001. Preparation of PLGA nanoparticles containing estrogen by emulsification-diffusion method. Colloids Surf. A 182, 123–130.

Langer, R., 2001. Perspectives: drug delivery. Drugs on target. Science 293, 58–59.

Lee, S.J., Jeong, J.R., Shin, S.C., Kim, J.C., Chang, Y.H., Lee, K.H., Kim, J.D., 2005. Magnetic enhancement of iron oxide nanoparticles encapsulated with poly(D,L-lactide-*co*-glycolide). Colloids Surf. A 255, 19–25.

Lee, D.W., Yun, K.-S., Ban, H.-S., Choe, W., Lee, S.K., Lee, K.Y., 2009. Preparation and characterization of chitosan/polyguluronate nanoparticles for siRNA delivery. J. Control. Rel. 139 (2), 146–152.

Legrand, P., Barratt, G., Mosqueira, V., Fessi, H., Devissaguet, J.D., 1999. Polymeric nanocapsules as drug delivery systems. A review. S. T. P. Pharma. Sci. 9, 411–418.

Letchford, K., Burt, H., 2007. A review of the formation and classification of amphiphilic block copolymer nanoparticulate structures: micelles, nanospheres, nanocapsules, and polymersomes. Eur. J. Pharm. Biopharm. 65, 259–269.

Lian, T., Ho, R.J.Y., 2001. Trends and developments in liposome drug delivery systems. J. Pharmceut. Sci. 90 (6), 667–680.

Liechty, W.B., Kryscio, D.R., Slaughter, B.V., Peppas, N.A., 2010. Polymers for drug delivery systems. Annu. Rev. Chem. Biomol. Eng. 1, 149–173.

Lim, S.T., Martin, G.P., Berry, D.J., Brown, M.B., 2000. Preparation and evaluation of the in vitro drug release properties and mucoadhesion of novel microspheres of hyaluronic acid and chitosan. J. Control. Rel. 66 (2–3), 281–292.

Lin, S.-Y., Yang, J.-C., 1986. Studies on microencapsulation. Part IV: effect of ethylene-vinyl acetate as a coacervation-inducing agent on the production and release behavior of chlorpromazine hydrochloride microcapsules and tabletted microcapsules. J. Control. Rel. 3 (1-4), 221–228.

Liu, M., Frechet, J.M., 1999. Designing dendrimers for drug delivery. Pharm. Sci. Technol. Today 10, 393–401.

Liu, R., Ma, G.H., Meng, F.-T., Su, Z.-G., 2005. Preparation of uniform-sized PLA microcapsules by combining Shirasu Porous Glass membrane emulsification technique and multiple emulsion-solvent evaporation method. J. Control. Rel. 103 (1), 31–43.

Liu, J., Meisner, D., Kwong, E., Wu, X.Y., Johnston, M.R., 2007. A novel trans-lymphatic drug delivery system: implantable gelatin sponge impregnated with PLGA–paclitaxel microspheres. Biomaterials 28 (21), 3236–3244.

Mainardes, R., Evangelista, R., 2005. PLGA nanoparticles containing praziquantel: effect of formulation variables on size distribution. Int. J. Pharma. 290, 137–144.

Makadia, H.K., Siegel, S.J., 2011. Poly lactic-*co*-glycolic acid (PLGA) as biodegradable controlled drug delivery carrier. Polymers 3, 1377–1397.

Mao, H.-Q., Roy, K., Troung-Le, V.L., Janes, K.A., Lin, K.Y., Wang, Y., August, J.T., Leong, K.W., 2001. Chitosan-DNA nanoparticles as gene carriers: synthesis, characterization and transfection efficiency. J. Control. Rel. 70 (3), 399–421.

Martinez-Sancho, C., Herrero-Vanrell, R., Negro, S., 2004. Study of gamma-irradiation effects on aciclovir poly(D,L-lactic-*co*-glycolic) acid microspheres for intravitreal administration. J. Control. Rel. 99, 41–52.

Mathiowitz, E., Langer, R., 1987. Polyanhydride microspheres as drug carriers I. Hot-melt microencapsulation. J. Control. Rel. 5 (1), 13–22.

Medina, O.P., Zhu, Y., Kairemo, K., 2004. Targeted liposomal drug delivery in cancer. Curr. Pharm. Des. 10 (24), 2981–2989.

Miller, R.E., Anderson, J.L., 1964. Encapsulation process and its products. US Patent 3,155,590 (November).

Mishima, K., 2008. Biodegradable particle formation for drug and gene delivery using supercritical fluid and dense gas. Adv. Drug Deliv. Rev 60 (3), 411–432.

Moghimi, S.M., Hunter, A.C., Murray, J.C., 2005. Nanomedicine: current status and future prospects. FASEB J. 19, 311–330.

Moinard-Checot, D., Chevalier, Y., Briançon, S., Fessi, H., Guinebretière, S., 2006. Nanoparticles for drug delivery: review of the formulation and process difficulties illustrated by the emulsion–diffusion process. J. Nanosci. Nanotechnol. 6, 2664–2681.

Mu, L., Feng, S.S., 2003. A novel controlled release formulation for the anticancer drug paclitaxel (Taxol(R)): PLGA nanoparticles containing vitamin E TPGS. J. Control. Rel. 86, 33–48.

Murakami, H., Kobayashi, M., Takeuchi, H., Kawashima, Y., 2000. Further application of a modified spontaneous emulsification solvent diffusion method to various types of PLGA and PLA polymers for preparation of nanoparticles. Powder Technol. 107, 137–143.

Nafee, N., Taetz, S., Schneider, M., Schaefer, U.F., Lehr, C.-M., 2007. Chitosan-coated PLGA nanoparticles for DNA/RNA delivery: effect of the formulation parameters on complexation and transfection of antisense oligonucleotides. Nanomedicine 3 (3), 173–183.

Nihant, N., Grandfils, C., Jérôme, R., Teyssié, P., 1995. Microencapsulation by coacervation of poly(lactide-*co*-glycolide) IV: effect of the processing parameters on coacervation and encapsulation. J. Control. Rel. 35 (2–3), 117–125.

Niwa, T., Takeuchi, H., Hino, T., Kunou, N., Kawashima, Y., 1993. Preparation of biodegradable nanoparticles of water-soluble and insoluble drugs with D,L-lactide/glycolide copolymer by a novel spontaneous emulsification solvent diffusion method, and the drug release behavior. J. Control. Rel. 25, 89–98.

O'Donnell, P.B., McGinity, J.W., 1997. Preparation of microspheres by the solvent evaporation technique. Adv. Drug Deliv. Rev. 28 (1), 25–42.

Park, J.H., Choi, S.O., Kamath, R., Yoon, Y.K., Allen, M., Prausnitz, M., 2007. Polymer particle-based micromolding to fabricate novel microstructures. Biomed. Microdev. 9 (2), 223–234.

Poulain, N., Dez, I., Perrio, C., Lasne, M.-C., Prud'homme, M.-P., Nakache, E., 2003. Microspheres based on inulin for the controlled release of serine protease inhibitors: preparation, characterization and in vitro release. J. Control. Rel. 92 (1–2), 27–38.

Prabha, S., Labhasetwar, V., 2004. Nanoparticle-mediated wild-type p53 gene delivery results in sustained antiproliferative activity in breast cancer cells. Mol. Pharmacol. 1, 211–219.

Rados, C., 2004. GRAS: time-tested and trusted food ingredients. FDA Consumer 38, 2.

Ratnam, V.D., Ankola, D., Bhardwaj, V., Sahana, D.K., Ravi Kumar, M.N.V., 2006. Role of antioxidants in prophylaxis and therapy: a pharmaceutical perspective. J. Control. Rel. 113 (3), 189–207.

Risch, S.J., 2009a. Encapsulation: overview of uses and techniques, encapsulation and controlled release of food ingredients. ACS Symposium Series, Vol. 590, American Chemical Society. Chapter 1, pp. 2–7.

Risch, S.J., 2009b. Encapsulation of flavors by extrusion, flavor encapsulation. ACS Symposium Series, Vol. 370, American Chemical Society. Chapter 11, pp. 103–109.

Rivera, P.A., Martinez-Oharriz, M.C., Rubio, M., Irache, J.M., Espuelas, S., 2004. Fluconazole encapsulation in PLGA microspheres by spray-drying. J. Microencapsul. 21 (2), 203–211.

Rosca, I.D., Watari, F., Uo, M., 2004. Microparticle formation and its mechanism in single and double emulsion solvent evaporation. J. Control. Rel. 99 (2), 271–280.

Rozario, F., 2005. In vivo and in vitro studies on heavy doses and fluorescent-labelled vitamin C using ESR and fluorescence spectroscopy. Doctoral Dissertation, Vom Fachbereich Chemie der Technischen Universität Kaiserslautern.

Sahoo, S.K., Labhasetwar, V., 2005. Enhanced antiproliferative activity of transferrin-conjugated paclitaxel-loaded nanoparticles is mediated via sustained intracellular drug retention. Mol. Pharmacol. 2, 373–383.

Sahoo, S.K., Parveen, S., Panda, J.J., 2007. The present and future of nanotechnology in human health care. Nanomedicine 3 (1), 20–31.

Seville, P.C., 2007. Spray-dried powders for pulmonary drug delivery. Crit. Rev. Therapeut. Drug Carrier Syst. 24 (4), 307–360.

Simovic, S., Hui, H., Song, Y., Davey, A.K., Rades, T., Prestidge, C.A., 2010. An oral delivery system for indomethicin engineered from cationic lipid emulsions and silica nanoparticles. J. Control. Rel. 143 (3), 367–373.

Singh, M.N., Hemant, K.S.Y., Ram, M., Shivakumar, H.G., 2010. Microencapsulation: a promising technique for controlled drug delivery. Res. Pharm. Sci. 5 (2), 65–77.

Song, K.C., Lee, H.S., Choung, I.Y., Cho, K.I., Ahn, Y., Choi, E.J., 2006. The effect type of organic phase solvents on the particle size of poly(D,L-lactide-*co*-glycolide) nanoparticles. Colloids Surf. A 276, 162–167.

Stevanović, M., Uskoković, D., 2009a. Poly(D,L-lactide-*co*-glycolide) for controlled drug delivery of vitamins. Curr. Nanosci. 5 (1), 1–15.

Stevanović, M., Uskoković, D., 2009b. Influence of different degradation medium on release of ascorbic acid from poly(D,L-lactide-*co*-glycolide) nano- and microspheres. Russ. J. Phys. Chem. A 83 (9), 1457–1460.

Stevanovic, M., Pavlovic, V., Petkovic, J., Filipic, M., Uskokovic, D., 2011a. ROS-inducing potential, influence of different porogens and in vitro degradation of poly(D,L-lactide-*co*-glycolide)-based material. Express Polym. Lett. 5, 996–1008.

Stevanović, M.M., Jordović, B., Uskoković, D.P., 2007a. Preparation and characterization of poly(D,L-lactide-*co*-glycolide) nanoparticles containing ascorbic acid. J. Biomed. Biotechnol., 8.

Stevanović, M., Savić, J., Jordović, B., Uskoković, D., 2007b. Fabrication, in vitro degradation and the release behaviours of poly(D,L-lactide-*co*-glycolide) nanospheres containing ascorbic acid. Colloids Surf. B 59 (2), 215–223.

Stevanović, M., Ignjatović, N., Jordović, B., Uskoković, D., 2007c. Stereological analysis of the poly(D,L-lactide-*co*-glycolide) submicron sphere prepared by solvent/non-solvent chemical methods and centrifugal processing. J. Mater. Sci. Mater. Med. 18 (7), 1339–1344.

Stevanović, M., Jordović, B., Nedić, Z., Miličević, D., 2007d. The stabilizer influence on morphological characteristics of poly(D,L-lactide-*co*-glycolide) nanospheres. Mater. Sci. 555, 447–452.

Stevanović, M.M., Jordović, B., Uskoković, D.P., 2008a. Morphological changes of poly(D,L-lactide-*co*-glycolide) nanoparticles containing ascorbic acid during in vitro degradation process. J. Microsc. 532 (3), 511–516.

Stevanović, M., Radulović, A., Jordović, B., Uskoković, D., 2008b. Poly(D,L-lactide-*co*-glycolide) nanospheres for the sustained release of folic acid. J. Biomed. Nanotechnol. 4 (3), 349–358.

Stevanović, M., Maksin, T., Petković, J., Filipič, M., Uskoković, D., 2009. An innovative, quick and convenient labeling method for the investigation of pharmacological behavior and the metabolism of poly(D,L-lactide-*co*-glycolide) nanospheres. Nanotechnology 20 (33), 335102.

Stevanović, M., Kovačević, B., Petković, J., Filipič, M., Uskoković, D., 2011b. Effect of poly-alpha, gamma, L-glutamic acid as a capping agent on morphology and oxidative stress-dependent toxicity of silver nanoparticles. Int. J. Nanomed. 6, 2837–2847.

Stevanović, M., Savanović, I., Uskoković, V., Škapin, S., Bračko, I., Jovanović, U., Uskoković, D., 2012a. A new, simple, green, and one-pot four-component synthesis of bare and poly(alpha,gamma, L-glutamic acid)-capped silver nanoparticles. Colloid Polym. Sci. 290 (3), 221–231.

Stevanović, M.M., Škapin, S.D., Bračko, I., Milenković, M., Petković, J., Filipič, M., Uskoković, D.P., 2012b. Poly(lactide-*co*-glycolide)/silver nanoparticles: synthesis, characterization, antimicrobial activity, cytotoxicity assessment, and ROS-inducing potential. Polymer 53 (14), 2818–2828.

Stevanović, M., Uskoković, V., Filipović, M., Škapin, S.D., Uskoković, D., 2013. Composite PLGA/AgNpPGA/AscH nanospheres with combined osteoinductive, antioxidative and antimicrobial activities. ACS Appl. Mater. Interfaces 5 (18), 9034–9042.

Stevanović, M., Bračko, I., Milenković, M., Filipović, N., Nunić, J., Filipič, M., Uskoković, D.P., 2014. Multifunctional PLGA particles containing poly(L-glutamic acid)-capped silver nanoparticles and ascorbic acid with simultaneous antioxidative and prolonged antimicrobial activity. Acta Biomater. 10 (1), 151–162.

Stott, P.W., William, A.C.S., Barry, B.W., 1996. Characterization of complex coacervates of some tricyclic antidepressants and evaluation of their potential for enhancing transdermal flux. J. Control. Rel. 41 (3), 215–227.

Stupar, P., Pavlović, V., Nunić, J., Cundrič, S., Filipič, M., Stevanović, M., 2014. Development of lyophilized spherical particles of poly(epsilon-caprolactone) and examination of their morphology, cytocompatibility and influence on the formation of reactive oxygen species. J. Drug Deliv. Sci. Technol. 24 (2), 191–197.

Sumer, B., Gao, J., 2008. Theranostic nanomedicine for cancer. Nanomedicine 3 (2), 137–140.
Takashima, Y., Saito, R., Nakajima, A., Oda, M., Kimura, A., Kanazawa, T., Okada, H., 2007. Spray-drying preparation of microparticles containing cationic PLGA nanospheres as gene carriers for avoiding aggregation of nanospheres. Int. J. Pharm. 343 (1–2), 262–269.
Tamber, H., Johansen, P., Merkle, H.P., Gander, B., 2005. Formulation aspects of biodegradable polymeric microspheres for antigen delivery. Adv. Drug Deliv. Rev. 57 (3), 357–376.
Thomasin, C., Nam-Tran, H., Merkle, H.P., Gander, B., 1998a. Drug microencapsulation by PLA/PLGA coacervation in the light of thermodynamics. 1. Overview and theoretical considerations. J. Pharm. Sci. 87 (3), 259–268.
Thomasin, C., Merkle, H.P., Gander, B., 1998b. Drug microencapsulation by PLA/PLGA coacervation in the light of thermodynamics. 2. Parameters determining microsphere formation. J. Pharmceut. Sci. 87 (3), 269–275.
Trindade, M.A., Grosso, C.R.F., 2000. The stability of ascorbic acid microencapsulated in granules of rice starch and in gum arabic. J. Microencapsul. 17 (2), 169–176.
Uddin, M.S., Hawlader, M.N.A., Zhu, H.J., 2001. Microencapsulation of ascorbic acid: effect of process variables on product characteristics. J. Microencapsul. 18 (2), 199–209.
Vandervoort, J., Ludvig, A., 2002. Biocompatibile stabilizers in the preparation of PLGA nanoparticles:a factorial design study. Int. J. Pharma. 238 (1–2), 77–92.
Vasir, J.K., Reddy, M.K., Labhasetwar, V.D., 2005. Nanosystems in drug targeting: opportunities and challenges. Curr. Nanosci. 1, 47–64.
Vauthier, C., Bouchemal, K., 2008. Methods for the preparation and manufacture of polymeric nanoparticles. Pharm. Res. 26, 1025–1058.
Wieland-Berghausen, S., Schote, U., Frey, M., Schmidt, F., 2002. Comparison of microencapsulation techniques for the water-soluble drugs nitenpyram and clomipramine HCl. J. Control. Rel. 85 (1–3), 35–43.
Wong, S.Y., Pelet, J.M., Putnam, D., 2007. Polymer systems for gene delivery-past, present, and future. Prog. Polym. Sci. 32 (8–9), 799–837.
Xu, R.X., Xu, J.S., Zuo, T., Shen, R., Huang, T.H., Tweedle, M.F., 2011. Drug-loaded biodegradable microspheres for image-guided combinatory epigenetic therapy in cells. J. Biomed. Opt. 16 (2), 020507.
Yalabik-Kas, H.S., 1983. Microencapsulation and in vitro dissolution of oxazepam from ethylcellulose microcapsules. Drug Dev. Ind. Pharm. 9 (6), 1047–1060.
Yang, J.H., Lee, S.Y., Han, Y.S., Park, K.C., Choy, J.H., 2003. Efficient transdermal penetration and improved stability of L-ascorbic acid encapsulated in an inorganic nanocapsule. Bull. Korean Chem. Soc. 24 (4), 499–503.
Yoo, H.S., 2006. Preparation of biodegradable polymeric hollow microspheres using O/O/W emulsion stabilized by Labrafil. Colloids Surf. B 52, 47–51.
Yoshiyuki, K., Hirata, G., Samejima, M., 1983. Studies on microcapsules II: influence of molecular weight of ethylcellulose in the microencapsulation of ascorbic acid. Chem. Pharmaceut. Bull. 31, 4476–4482.
Zambaux, M., Bonneaux, F., Gref, R., Maincent, P., Dellacherie, E., Alonso, M., Labrude, B., Vigneron, C., 1998. Influence of experimental parameters on the characteristics of poly(lactic acid) nanoparticles prepared by double emulsion method. J. Control. Rel. 50, 31–40.
Zhang, H., Cui, W., Bei, J., Wang, S., 2006. Preparation of poly(lactide-*co*-glycolide-*co*-caprolactone) nanoparticles and their degradation behaviour in aqueous solution. Polym. Degrad. Stabil. 91 (9), 1929–1936.

FURTHER READING

Cheng, F.Y., Wang, S.P.H., Su, C.H., Tsai, T.L., Wu, P.C., Shieh, D.B., Chen, J.-H., Hsieh, P.C.-H., Yeh, C.-S., 2008. Stabilizer-free poly(lactide-*co*-glycolide) nanoparticles for multimodal biomedical probes. Biomaterials 29, 2104–2112.
Lee, J.-B., Ahn, J., Lee, J., Kwak, H.-S., 2004. L-ascorbic acid microencapsulated with polyacyglycerol monostearate for milk fortifaction. Biosci. Biotechnol. Biochem. 68 (3), 495–500.

CHAPTER

12 NOVEL GELS: IMPLICATIONS FOR DRUG DELIVERY

Swarnali D. Paul, Harish Sharma, Gunjan Jeswani, Arvind K. Jha
Shri Shankaracharya Group of Institutions, SSTC, Bhilai, India

CHAPTER OUTLINE

1 INTRODUCTION

Gels are semisolid, homogeneous formulations comprising an external solution or dispersion of medications in a suitable hydrophobic or hydrophilic three-dimensional network base. The uses of gels are quite broad range, including applications in food, cosmetics, biotechnology, pharmaceuticals, and s o forth. Gels are popular delivery systems because they are simple to produce and appropriate for controlling drugs through different routes since it provides an intimate contact between the therapeutic agent and the site of action or absorption. Typically, gels can be distinguished in two types: hydrogels and organogels, according to the physical state of the gelling agent in the dispersion. Hydrogels are prepared with water, water-soluble ingredients, or dispersible colloids. Organogels are prepared using water insoluble oleaginous materials. According to the nature of the solvent, it is also further classified in two types, aqueous gels and xerogels. Aqueous gels are, of course, water based. Solid gels with low solvent concentration are known as Xerogels. Xerogels are often produced by evaporation of a solvent, leaving the gel framework behind. They can be returned to the gel state by the introduction of an agent that, on imbibition, swells the gel matrix. Examples of xerogels include dry gelatin, tragacanth ribbons, and acacia tears and dry cellulose, and polystyrene.

With the advances in drug-delivery systems, development of novel gel systems also has been triggered for delivery of drugs locally or systemically in sustained, controlled, and targetable manner. Hence, gel based systems that respond to specific biological or external stimuli like pH, temperature, ionic strength, enzymes, antigens, light, magnetic field, ultrasound, and electric current are being designed and appraised as "smart gel systems" for various applications.

1.1 ADVANCES IN GEL DELIVERY SYSTEMS

In this section, we discuss the incorporation of different novel drug delivery systems into gel. These gels are also known as "novel gels" due to their unique properties, formulation aspects, manufacturing technologies, and mechanism of drug release. Fig. 12.1 illustrates different novel gels with their image.

Novel gels are a superior drug-delivery system than conventional gel because:

- Drug release can be sustained or modified at the target site.
- They respond to biological or external stimuli in case of smart gel systems.
- Therapeutic efficacy can be improved with reduced adverse effects.
- High drug loading is achieved without any chemical interactions.
- Nanogels, which have a size range in nanometers, have the ability to reach the small capillaries or tissues (Gonçalves et al., 2010).
- Nanogels possess the advantage to circumvent quick clearance by phagocytic cells due to their small size and unique surface properties, which permits both active and passive drug targeting.
- They are highly biocompatible and biodegradable.

1.2 CLASSIFICATION OF NOVEL GELS

Novel gels are more commonly classified into two major ways.

1. Based on their responsive behavior, they are of two types: stimuli-responsive and nonresponsive. Nonresponsive gels simply swell after absorbing water. But stimuli-responsive gels swell upon

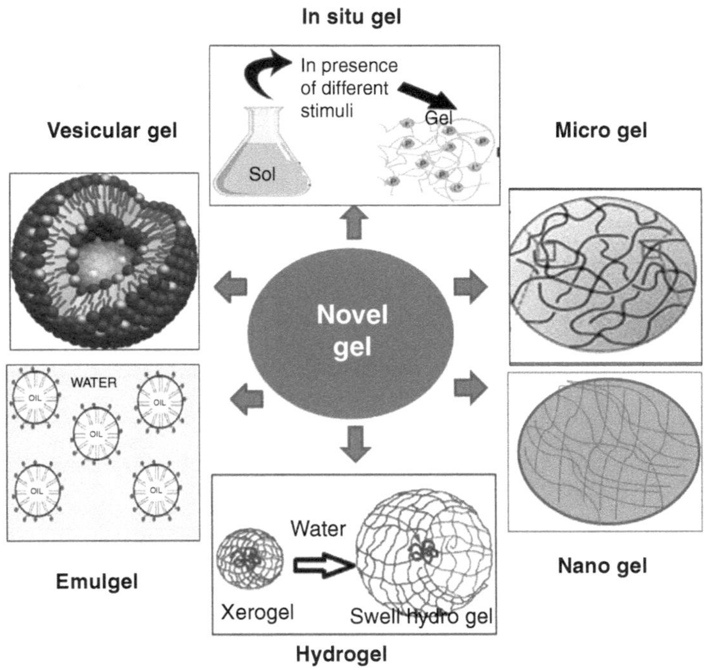

FIGURE 12.1 Different Novel Gels With Their Pictorial Image

exposure to external stimuli, such as temperature, pH, ionic strength, and magnetic field. Gels that respond to more than one environmental stimulus are known as multiresponsive gels (Gupta et al., 2002; Sasaki and Akiyoshi, 2010).

2. Based on the type of linkages present in the three-dimensional structure of the gel network, they are of two types: chemically cross-linked gels and physically cross-linked gels.

1.3 CHARACTERIZATION OF NOVEL GEL

Due to the versatile nature and wide applications of novel gels, their characterization parameters are also extensive. Characterization or evaluation of different novel gels depends on their physical and chemical properties. Commonly those properties include size and shape, water swelling capacity, surface charge, rheological properties, and so forth. Physical examination is mandatory for any gel to check its color, physical state, consistency, homogeneity, and phase separation (for emulgel). Spreading coefficiency, extrudability, drug-release property, drug-permeation study, and drug content should be checked for any gel. Many techniques and principles are involved behind the evaluation of any novel gel. Dynamic light scattering (DLS) technique is used for measuring the size range of nanogels, whereas the differential scanning calorimetry (DSC) technique is used for determining the interaction between the excipients in the gel preparation and phase behavior. Scanning electron microscopes (SEM) and transmission electron microscopes (TEM) are used to check the surface morphology and

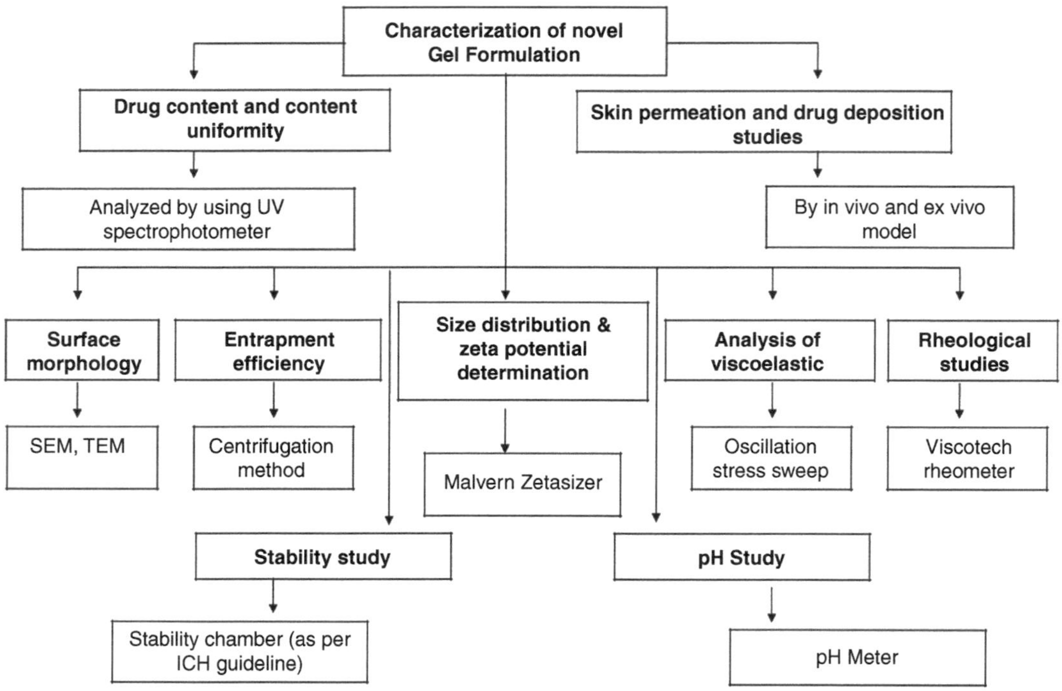

FIGURE 12.2 Characterization Parameters of Novel Gels With Their Method of Evaluation

gel network. Dynamic oscillation stress sweep is performed to determine the linear viscoelastic region (LVR), especially for vesicular gels. Centrifugation method (< 7000X g), ultracentrifugation method (15000X g), exhaustive dialysis method, and gel permeation are carried out to determine drug entrapment efficiency of gels. Lamellarity of vesicular gel can be studied by small angle X-ray scattering and 31P-NMR. Ex-vivo bio-adhesive strength is an important parameter for any muco adhesive gel. This study is generally performed on the isolated skin of animals like mice, cadavers, rats, and so forth. Stability study of different novel gels are carried out according to ICH guidelines to understand their shelf life and degradation pattern. For this, gels are subjected to different temperature and humidity for a specified time period. Then they are examined for drug content, drug entrapment, and drug release property in order to check their constancy. Fig. 12.2 represents a graphical illustration of different characterization methods of novel gels.

Different Novel Gels are described in this chapter with their manufacturing, applications, and other aspects.

2 HYDROGELS

Hydrogels are three-dimensional networks of hydrophilic polymers that swell in the presence of water while maintaining the structural integrity. Hydrogels have had an enormous contribution in the development of the different drug delivery systems for decades. But their contribution in smart drug-delivery systems demanded that researchers categorize them as novel gels. "Intelligent" or "smart" hydrogels are those that respond to different physical stimuli including temperature, electric fields, solvent composition, light, pressure, sound, and magnetic fields, as well as different chemical or biochemical stimuli like pH, ions and specific molecular recognition events (Bae, 1997). These smart hydrogels are exploited widely in self-regulated drug-delivery systems. Many researchers have claimed patents for these smart hydrogels for their different applications. Some important patents related with different novel gels are compiled in Table 12.1.

2.1 TYPES OF HYDROGEL

Conventional hydrogels can be classified in many ways based on their sources, composition, configurations, and so forth. Here we have described the classification of different hydrogels.

1. Based on sources
   a. Natural origin: Examples are gelatine, collagen, and different polysaccharides like starch, alginate, and agarose.
   b. Synthetic origin: They are prepared using chemical polymerization methods like poly(*N*-isopropylacrylamide). The synthetic polymers are chemically stronger and lipophilic in nature compared to natural polymers.
2. Based on polymeric composition
   a. Homopolymeric hydrogels: They are polymer networks derived from a single species of monomer.
   b. Copolymeric hydrogels: They consist of two or more different species of monomers with at least one water-soluble component, arranged randomly.
   c. Multipolymer interpenetrating polymeric hydrogel (IPN): They consist of two self-governing cross-linked artificial and/or natural polymer modules.
3. Based on configuration
   a. Amorphous: Noncrystalline structure.
   b. Semicrystalline: They are compound mixtures of crystalline and amorphous phases.
   c. Crystalline: Crystal structure.

Novel hydrogels or smart hydrogels can be classified as follows:

1. Temperature-sensitive hydrogels: These hydrogels are stimulated under specific temperature conditions. Polymers exhibit a temperature-responsive phase-transition property that is used for developing these types of hydrogel. Examples of these polymers include poly(*N*-isopropylacrylamide) (PNIPAAm), copolymers of NIPAAm, and so forth. Poly(*N*,*N*-diethylacrylamide) (PDEAAm) is widely used because of its lower critical solution temperature (LCST) in the range of 25–32°C, close to body temperature. Commonly, water-solubility of polymers is increased with increasing temperature, but polymers with LCST are exceptional. They decrease their water-solubility as the temperature increases. Hydrogels made of LCST polymers shrink as the temperature increases above the LCST (Qiu and Park., 2012). This type of swelling behavior is known as inverse (or negative) temperature-dependence. Temperature-sensitive hydrogels are subclassified into three categories:
   a. Negatively thermosensitive: Examples are cross-linked P(NIPAAm–co-BMA), interpenetrating polymer networks of P(NIPAAm) and poly(tetramethyleneether glycol) (PTMEG).

Table 12.1 Patent Information on Different Novel Gels

Type of Novel Gel	Patent Description	Patent No.	Inventors	Publication Year
Hydrogel	Transparent wound dressing as thin-film, with a nonadhesive central portion containing hydrogel for wound healing process	US Patent 5,106,629	Cartmell and Sturtevant	1992
	Wound dressing composed of cotton gauze, or other fibrous substrate, impregnated with a thermoplastic hydrogel forming polymer	US Patent 8,409,606 B2	Shah et al.	1996
	Delivery device capable of releasing a drug enclosed within a hydrogel, which contracts responding to a chemical or physical stimulation	US patent 5,226,902	Bae et al.	1993
	Biodegradable and nontoxic multiblock hydrogel copolymers as drug delivery matrices	US Patent 5,514,380	S.S. Song, H.H. Kim, Y.W. Yi	1996
	Hydrogel-based composition for manufacturing porous scaffolds	EP 1 664 168 B1	E. Jabbari, M.J. Yaszemski, B.L. Currier	2006
	Use of pure cross-linked keratin-based hydrogels for tissue engineering cell scaffolds	US Patent 6,379,690 B2	Blanchard et al.	2002
	Method of manufacturing a hydrogel of antimicrobial nonwoven fibrous dressing with controlled release of silver ions	US Patent 8,431,151 B2	P. Mather, J. Wu, D. Ren, S. Hou	2013
In situ gel	Medical uses of in situ formed gels	US 5,587,175 A	Tacey X. Viegas, Lorraine E. Reeve, Raymond L. Henry	1996
	A novel in situ gel forming solution for ocular drug delivery	WO 2011018800 A2	Nandan Mohan Chandavarkar, Kour Chand Jindal, Rajkumar Malayandi	2011
	Recipe for in situ gel, and implant, drug delivery system formed thereby.	US 20,150,111,834 A1	Felice Cheng, Maggie J.M. Lu, Yi-JuKo, Min-Ying Lin, Shuen-Hsiang Chou	2015
	In situ gel formation of pectin	US 6,777,000 B2	Yawei Ni, Kenneth M. Yates	2004
	Delivery of physiologixal agents with in situ gels comprising anionic polysaccharides	EP 1663111 A2	Yawei Ni, Kenneth M. Yates	2006
	Methods for localized in situ detection of mRNA	US 8,551,710 B2	Mats Nilsson Bernitz, Chatarina Larsson, Ida Grundberg	2013

Microgel	Process for microgel preparation	US6,545,095	David Henry Solomon	2003
	Free radical polymerization process for microgel preparation	US20,050,272,868	Greg G. Qiao	2005
	Polymerization from a diallylamine and a compound containing a macromolecular chain consisting of units derived from such an amine	WO2008000766A2	Mathias Destarac	2008
	Microgels and process for their preparation	US6,355,718	Charles T. Berge	2002
	Acrylic resin-acrylic microgel compositions	US 4,290,932 A	Howard J. Wright, David P. Leonard, Roger A. Etzell	1981
	Microgel drug delivery system	US 5,078,994 A	Mridula Nair, Julia S. Tan	1992
	Microgel compositions for delivery of substances to the brain	WO2015027342A1	Justin E. Piazza	2015
	Microgel compositions and methods	WO2010147632A1	Todd R. Hoare	2010
	Microgel particles for the delivery of bioactive materials	WO2003082316A1	Jean M.J. Frechet	2003
Nanogel	Lipid nano carrier based nanogel	WO 2012176212 A1	Narayana Kala, Aurora Sundeep	2012
	Nanogel networks including polyion polymer fragments and biological agent	US20,020,136,769, WO2003082348A1	Alexander V. Kabanov, Sergey V. Vinogradov	2004
	Mucosal vaccine using cationic nanogel	US 20,110,206,729 A1	Kazunari Akiyoshi, Hiroshi Kiyono, Yoshikaz U. Yuki, Tomonori Nochi	2011
	Nanogel-based contrast agents for optical molecular imaging	US 20,070,237,821 A1	Jeffrey Leon, James Bennett, Tiecheng Qiao, John Harder, Thomas Mourey, Gary Slater, Lijun Dai	2007
	Degradable nanogel for drug delivery	US 20,070,224,164 A1	Maciej Radosz, Youqing Shen, Edward VanKirk, William Murdoch	2007
	Glucose-sensitive nanogel	CN 103694479 A	Chenxue Si, Li Zhao, Ding Jianxun, Xiao Chunsheng	2014
	Nanogel mediated tumor drug delivery	US 20,140,100,182 A1	Vinod Labhasetwar, Sivakumar Vijayaraghavalu	2014
	Hydrophobic polysaccharide nanogel with an exosome	WO 2014054588 A1	Akiyoshi et al.	2014
Emulgel	Transdermal formulations containing thiocolchicoside and ibuprofene in emulgel	EP 2019666 A2		
	Thiocolchicoside or salts thereof is formulated into nano size droplets in emulgel system	WO 2012053012 A2		
	Nano size droplets of etodolac in the form of emulgel exhibit greater permeability, and improved bioavailability	US 20,130,295,145 A1		
	A novel emulsion gel ointment for topical application	EP 1588697 A1	Kurt H. Bauer	2005

b. Positively thermosensitive: They swell at high temperatures and shrink at low temperatures. Examples are IPNs of poly(acrylic acid) and polyacrylamide (PAAm) or P(AAm–co-BMA).
c. Examples of thermally reversible gels include Pluronics and Tetronics. Pluronics have been approved by FDA and EPA for applications in food additives, pharmaceutical ingredients, and agricultural products.

2. pH-sensitive hydrogels
These hydrogels either accept or release protons in response to changes in environment pH. Poly(acrylic acid) (PAA) and become ionized at high pH, whereas poly(*N,N′*-diethylaminoethyl methacrylate) (PDEAEM) become ionized at low pH. Polycationic hydrogels show minimum swelling at neutral pH. These pH-sensitive hydrogels were studied after being incorporated in capsules (Gutowska et al., 1997) or silicone matrices (Bilia et al., 1996) to modulate the drug release. Results were promising.
3. Glucose-sensitive hydrogels
Hydrogels can be stimulated depending on the glucose concentration in the environment. This phenomenon requires glucose-responsive cross-linking. These hydrogels find application in the development of self-modulated insulin delivery systems.
4. Electric signal-sensitive hydrogels
These hydrogels undergo contraction or swelling in the presence of an applied electric field. Sometimes, they show swelling on one side and contraction on the other side, resulting in deformation of the hydrogels. They are usually made of polyelectrolytes. Hydrogels made of poly (2-acrylamido-2-methylpropane sulfonic acid-*co*-*n*-butylmethacrylate) were able to release edrophonium chloride and hydrocortisone in a pulsatile manner in the presence of an electric field (Kwon et al., 1991).
5. Light-sensitive hydrogels
These hydrogels are activated or deactivated in the presence of light. These hydrogels can be further classified in two categories: UV-sensitive and visible light-sensitive hydrogels. They have wide applications in electrical fields like developing of display units, optical switches, as well as in drug-delivery systems like ophthalmic drug delivery devices.
6. Other stimuli sensitive hydrogels

Hydrogels also respond to other stimuli, such as pressure, specific ions, thrombin, and antigen.

2.2 PREPARATION OF HYDROGELS

Usually, a hydrogel is a hydrophilic polymeric system cross-linked in a manner to produce an elastic structure. Thus, any method that is used to generate a cross-linked polymer can be used to produce a hydrogel. Copolymerization or cross-linking is based on reaction of hydrophilic monomers with multifunctional cross-linkers to produce hydrogel. Hydrophilic linear polymers obtained from both natural and synthetic origin are used to cross-link to form hydrogels in a number of ways:

1. Linking polymer chains via chemical reaction.
2. Using ionizing radiation to generate "main-chain free radicals," which recombine as cross-link junctions.
3. Physical interactions, such as entanglements, electrostatics, and crystallite formation.

The most common and most versatile technique for the production of synthetic hydrogels is the free-radical multifunctional vinyl monomers.

2.3 APPLICATIONS OF HYDROGEL

The use of novel hydrogels is not only limited to drug and nutraceutical delivery but also extended to reformative medicine. Different successful marketed products of hydrogel along with other novel gel are described in Table 12.2.

2.3.1 Contact Lenses

Hydrogels have been successfully used in the preparation of contact lenses for years. Bausch & Lomb, the pioneer company in this field, got approval from the Food and Drug Administration for their poly-2-hydroxyethylmethacrylate (PHEMA) lenses in 1971. A hydrogel should have the minimum luminous transmittance value for contact lenses of 95% (Port, 1999). One of the US Patent (No. 4,472,327 A) proposed cosmetic hydrogel contact lenses which serves dual purpose: to correct the visual defects and to change the apparent color of the eye (Neefe, 1984).

2.3.2 Wound Healing

Hydrogels act as a moist wound dressing material due to the ability to absorb and hold water in its network assembly. They have the ability to absorb and retain the wound exudates along with the foreign materials within its network structure. In addition to this, they also reduce the fluid loss from the wound surface and protect the wound for rapid healing.

2.3.3 Tissue Engineering

Tissue engineering helps to develop biological substitutes (tissue/organ). Recently, the use of resorbable hydrogels in this field has gained much importance. Resorbable polymers like polylactic acid (PLA), polyglycolic acid (PGA), and their copolymers (PLA-*co*-PGA; PLGA) have been successfully used in different biomedical application for a long time.

2.3.4 Gene Delivery

Gene delivery is associated with the incorporation of foreign genetic materials into the host cells. Polymers are employed for this purpose by different researchers. These polymers include poly-L-lysine (PLL), polyamidoamine dendrimer (PAMAM), polyethylenimmine (PEI), PGA, PLA, and PLGA. PAMAM and PEI polymers provide high transfection efficiency, but they have poor degradability. Thus, biodegradable polymers like PLA, PLGA, and PGA have gained importance for this purpose.

2.3.5 Applications of Hydrogels in Drug Delivery

Among all other application of hydrogel, controlled drug-delivery systems is the most popular. As discussed earlier, hydrogels may change their equilibrium water uptake because of the change in the environmental stimuli like temperature, pH, and ionic strength of the release media. Thus, drug release from this hydrogel can be altered. The drugs susceptible to high first-pass metabolism can also be delivered systemically by sublingual/buccal route using this type of delivery system. Enteric polymers like Eudragits have been used to deliver drugs at various sites of the intestine or to protect acid-labile drugs, to avoid the contact of the gastric mucosa with the gastric-irritant drugs.

Table 12.2 Information on Marketed Product of Different Novel Gel

Type of Novel Gel	Marketed Product	Drug	Manufacturer	Activity
In situ gel	Timoptic-XE	Timolol maleate	Merck & Co.	Nonselective beta-adrenergic receptor blocking agent
	AzaSite	Azithromycin	InSite Vision	Macrolide antibiotic for the treatment of bacterial conjunctivitis
	Pilopine HS	Pilocarpine hydrochloride	Alcon Laboratories Inc.	Treatment of glaucoma
	Akten	Lidocaine hydrochloride	Akorn Inc	Ocular surface anesthesia
	Virgan	Ganciclovir	Spectrum Thea Pharmaceutical	Ophthalmic antiviral for the treatment of acute herpetic keratitis
Emulgel	Miconaz-H- emugel	Miconazole nitrate, hydrocortisone	Medical Union Pharmaceuticals	Antibacterial effect on gram-positive bacteria
	Lupigyl gel	Metronidazole	Lupin Pharma	Antiinfectives and antiseptics for local oral treatment
	Clinagel	Clindamycin phosphate and Allantoin	Stiefel Pharma	Lincosamides
	Avindo gel	Azithromycin	Cosme Pharma Laboratories	Anti-infective
	Voveran Emulgel	Diclofenac	Novartis	Pain relief
	Nucoxia Emulgel	Etoricoxib	Zydus Cadila	Rheumatoid arthritis, psoriatic arthritis, osteoarthritis, ankylosing spondylitis
Liposomal gel	Celadrin		Imagenetix, Inc.	
	Optisome	Tetracaine		Local anesthetic
	Lipo CTM	Vitamin C, vitamin E, and zinc		Skin cream
	Lipo-Gest			A natural balancing cream
Niosomal gel	Lancome		Lancome, Paris and L'OrealCompany	Antiaging, face Whitening activity

Nanogel	Nioret Nano Gel	Tretinoin	Zydica Healthcare	Used in acne vulgaris
	NanoGel	Concentrated nano particulate silver	Organix Labs	Advanced acne and rash therapy
	Oxalgin Nanogel	Diclofenac	Zydus Cadila	Reduce inflammation and as an analgesic reducing pain in conditions, such as arthritis or acute injury
	Zyflex Nano Gel	Thiocolchicoside	Zydus (Synovia)	Used in joint and muscular pain
	CariFree HA nanogel	Nano hydroxyapatite, xylitol	Essology Pty Ltd, Australia	Offers an improved level of preventative care while aiding in remineralization
Hydrogel	Gentle 59		Product of Vista Optics Ltd., marketed at Budapest	Soft contact lens
	Hydrofiber, hydrogel		ConvaTec Company	Used in wound dressing
	Woundtab		First Water	Superabsorbent hydrogel used as wound dressing
	Granugel, a clear, viscous hydrogel		ConvaTec	For the management of partial and full-thickness wounds
	Pilobuc, a modified release hydrogel buccal patch	Pilocarpine		Used in the symptoms of Sjögren's syndrome
	SUPPRELIN LA		Endo Pharmaceuticals Solutions Inc	Subcutaneous insert used for the release of histrelin acetate
	Dynarex	Glycerine	Dynarex Corporation	Protectant for dressing and management of stasis ulcers, pressure ulcers
	Ocular therapeutics	Poly(ethylene glycol), dexamethasone		In postoperative inflammation and pain

3 IN SITU GEL

In situ gel technology is one of the smart drug-delivery systems, and it has stimulated development of many medical and biomedical controlled release systems. These systems are liquid at room temperature and spontaneously transformed into gel in the presence of suitable environment like any body fluids, change in pH, or change in temperature (Miyazaki et al., 1999). In situ is a Latin word that indicates “in its original place or in position.” In situ gelation is an automatic process of gel formation at the desired site after its application (Bajpai, 2014). In situ gel is one of the versatile gel technologies among other novel gels.

There are several advantages of this system, including good thixotropic property, increased residence time of drug, decrease frequency of drug administration, rapid absorption and onset of effect, low dose required, minimized local and systemic side effects, direct transport into systemic circulation and CNS, and possible ease for administered by several routes, such as oral, ocular, rectal, vaginal, injectable, and intraperitoneal routes (Cho et al., 2008; Dondeti and Zia, 1996; Talasahaz and Ali, 2008).

3.1 PROCESS OF IN SITU GEL FORMATION

Several researchers have described the principle of formation of these gels. Simply stated, in situ gel occurs due to different stimuli like ion activation (e.g., sodium alginate polymer responds in ion activation), by temperature change (e.g., poloxamer, chitosan), pH change (e.g., carbopol), environmental change, and solvent exchange. In situ formation is based on two mechanisms: physical mechanism and chemical mechanism. Physical mechanism involves swelling and diffusion. In swelling, in situ formation occurs when material absorbs water from surrounding environment, and expands in the desired space. Diffusion involves penetration of solvent from polymer solution into nearby tissue and the consequences are precipitation or solidification of polymer. Table 12.3 represents different polymers used for in situ gelling. Different chemical reactions that result in situ gelation may involve precipitation of inorganic solids from different mechanism like ionic cross-linking, enzymatic cross-linking, and photo-polymerization.

3.1.1 Ionic Cross-Linking

This is based on the principle of phase transition of polymers in occurrence with different ions. Carrageenan, gellan gum, alginic acid, and so forth are known as ion-sensitive polymers (Bhardwaj et al., 2000). Carrageenan polymers are found in different forms in nature, and they convert into gel in the presence of different ions. K-carrageenan turns into brittle, rigid gels in the presence of potassium ions, but i-carrageenan convert into elastic gels in the presence of calcium ions. Another such example is Gellan gum. This is commercially available as Gelrite. It forms spontaneous gelling in the presence of different cations including Na^+, Mg^{2+}, Ca^{2+}, and K^+ (Guo et al., 1998).

3.1.2 Enzymatic Cross-Linking

In this process, in situ gel formation is catalyzed by natural enzymes. One interesting example of such a system is intelligent stimuli-responsive delivery of insulin. In this system, release of insulin is modulated in response to blood glucose level from cationic pH-sensitive polymers containing immobilized insulin and glucose oxidase in a pulsatile fashion. Rate of gel formation can be adjusted by the amount of enzyme (Podual et al., 2000).

Table 12.3 Different Polymers Used in Formulation of In Situ Gelling System

Polymer	Drug	Diseases/Uses	Result	References
N-stearoyl L-alanine methyl ester (SAM)	Rivastigmine	Alzheimer's disease	Drug with SAM showed a prolonged therapeutic efficacy for 11 days	Anda et al. (2008)
Pluronic F127, carbopol 934P, and HPMC	Midazolam hydrochloride	Epileptic seizure	Increased bioavailability of the drug	Shyamoshree and Amalkumar (2011)
Pluronic F127, carbopol 974	Sumatriptan succinate	Headache and migraine	Controlled drug release with higher permeability rate of 250 min	Shyam et al. (2007)
Carbopol, sodium alginate, HPMC	Diclofenac sodium	Ophthalmic anti-inflammatory	Drug release prolonged for 8 h	Padma et al. (2010)
Sodium alginate, polaxamer, polyethylene glycol, HPMC	Nimesulide	Rectal anti-inflammatory	Addition of PEG resulted in satisfactory drug release rate and rectal retention	Yuan et al. (2012)
Sodium alginate, calcium chloride, sodium bicarbonate, sodium citrate, HPMC-K100	Amoxillin	Antimicrobial activity	Drug release was found to be 6–10 h	Dasharath et al. (2011)
Pluronic, hydroxyl propyl β-cyclodextrin, HPMC	Artemether	Cerebral malaria	Better stability for 90 days	Hitendra et al. (2011)

3.1.3 Photo-Polymerization

Photo-polymerization is commonly used for in situ formation of organic materials. In this technique, reactive macromers or solution of monomers and initiators used to inject into a tissue site. Then electromagnetic radiation is applied to form gel (Jones et al., 2002). Acrylate or similar polymerizable functional groups are typically used with individual monomers and macromers due to their rapid photo-polymerization ability in the presence of appropriate photo initiator (Burkoth and Anseth, 2000).

3.1.4 Temperature Dependent Systems

In this process, gel formation occurs in the presence of thermo reversible hydrogels. Examples are chitosan, pluronics, tetronics, xyloglucans, hydroxypropylmethyl cellulose, or hypromellose (HPMC).

3.1.5 pH-triggered Systems

There are few polymers that can transform into gel with change in pH. Examples are cellulose acetate phthalate (CAP) latex, carbopol, polymethacrilicacid (PMMA), polyethylene glycol (PEG), pseudolatexes.

3.2 APPLICATIONS OF IN SITU GEL SYSTEM

Enormous research works are being carried out with different category of drug, to prove the significance of in situ gelling systems. As in situ gels have been applied through various routes like oral, rectal, ocular, injectable, vaginal, nasal, parental, and intraperitoneal so their applications are categorized as per route of administration.

3.2.1 Oral Delivery

Different antifungal, antiinflammatory, antihistaminic drugs have been formulated in in situ gel. Clotrimazole, an antifungal drug formulated in in situ gel with Carbopol 934P, gellan gum, HPMC in order to achieve zero order release kinetics with the sustained release for 8 h (Harish et al., 2012). An in situ oral formulation of the drug dexomethorphan, an antitussive agent was formulated along with sodium alginate, sodium citrate, calcium chloride, and chitosan. Sustained drug release pattern was found in both the simulated intestinal and gastric condition (El Maghraby et al., 2012). Commonly used natural polymers that are used for in situ forming oral drug delivery are pectin, gellan gum, and xyloglucan. The potential of an orally administered in situ gel formed by pectin has been evaluated for the sustained delivery of drugs (Wataru et al., 2004). Another formulation of theophylline prepared by gellan gum with calcium chloride and sodium citrate complex administered orally has also been evaluated. In the acidic environment of the stomach, calcium ions were released leading to gelation of gellan, and in situ gel was formed (Miyazaki et al., 1999). Moink et al. (2010) reported a formulation of famotidine with calcium chloride and sodium alginate, which resulted in extraordinary formation of gel in the stomach region and sustained the release of the drug. Similarly, results were obtained with cimetidine with gellan gum, sodium alginate, and xyloglucan for treating ulcers (Miyazaki et al., 2011). Theophylline in situ gel to treat asthma was formulated with sodium alginate and 2%–1% w/v concentration of alginate. When administered in rats, it showed increased bioavailability of the drug (Miyazaki et al., 2000). Salbutamol sulfate in situ gel was formulated with HPMC and carbopol 934Pto treat chronic diseases showed a sustained drug release for 8 h (Tanagi et al., 2008).

3.2.2 Ocular Delivery

Various researchers have described efficiency of in situ gel vehicles for ophthalmic drug delivery. They prolonged the precorneal resident time and improve ocular bioavailability of the drug. Local drug delivery to the eye is very popular for many compounds like anti-inflammatory agents, antimicrobial agents, antiglaucoma medication, and so forth. Precorneal residence time of these in situ gels is extended, and drug release rate is also prolonged because of their viscosity compared with conventional eye drops (Miyazaki et al., 2001). These authors also tried to prepare in situ gels for ocular delivery using 1.5% w/w xyloglucan as the natural polymer to determine mitotic response. Sultana, et al. in 2006 reported a better ophthalmic delivery system for perfloxacinmesylate based on in situ gel of gelrite. For in situ gel-based ocular delivery, natural polymers, such as gellan gum, alginic acid, and xyloglucan are the most commonly used polymers (Mikkelson, 1984,). Among synthetic polymers, Pluronic F127-poly (acrylic acid) copolymers, solutol HS15, and Poloxamer 188 are widely used.

3.2.3 Parenteral Drug Delivery

In situ drug-delivery systems have gained significant attention over the years. Ha et al. (2006) prepared thermo responsive and injectable hydrogels based on hyaluronic acid and poly (n-isopropylacrylamide). A novel, thermosensitive parenteral system was established with β-glycerophosphate for tumor treatment. Bupivacaine hydrochloride was delivered parenterally by poly(D,L-lactide) and poly(D,L-lactideco-glycolide)(PLGA) in situ gel for prolonged analgesic effect of the drug (Heiko et al., 2008).

3.2.4 Rectal and Vaginal Delivery

In situ gels also possess a potential application for drug delivery by vaginal and rectal route. Miyazaki et al. (2001) developed a thermoreversible gel of indomethacin with xyloglucan for rectal delivery. They found better results with this system as compared to commercial suppositories. Athermosensitive, prolonged release, muco-adhesive vaginal gel was formulated for the treatment of vaginitis with Clotrimazole-β-cyclodextrin complex.

Table 12.4 Different Applications of In Situ Gel

Drug	Specification	Formulation/Polymer	Result	References
Diltiazem hydrochloride	Floating in situ gel	HPMC, sodium alginate, calcium chloride, sodium methyl paraben.	Alternate for conventional dosage form	Prathiba et al. (2011)
Prilocaine hydrochloride	Dental administration	Chitosan and HPMC	Responsible for controlled release of drug for longer period than sodium alginate due to its viscous nature	Gupta et al. (2008); Vinay et al. (2011)
Dexomethorphan	Antitussive agent	Sodium citrate, calcium chloride, sodium alginate, and chitosan	The release pattern of this formulation in both the intestinal and gastric condition showed sustained release	Miyazaki et al. (2001)
Peurarin	Antiangiogenicagent	Polaxamer 407,188 and carbopol	Diffusion-controlled in vitro drug release for about 8 h	Qi et al. (2007)
Fluoroquinolone and moxifloxacin	Antibiotic	—	To overcome the drawbacks associated with its eye drops	Shashank et al. (2012)

Exploitation of polymeric in situ gels for controlled release of various drugs has been mentioned for years. The result was enthusiastic. Other applications of in situ gel are described in Table 12.4. Re-gel is a product of Macromed. It is a novel, peritumoral, injectable depot formulation of interleukin-2 (IL-2) for cancer immunotherapy. Cytoryn, another product of in situ gel enhances the immunological response in IL-2 therapy. The Atrigel system developed by Dunn and co-workers in 1987 is used for both parenteral and site-specific drug delivery. When this liquid preparation is inserted into the body the water miscible organic solvent dissipates and water breaches into the organic phase. In this way, the phases get separated and the polymer gets precipitated as a depot at the site of injection. Marketed products available with this technology are Atridox, which contains 8.5 percent doxycycline and is used for periodontal treatment product with subgingival delivery; Lupron depot and Eligard are used for the treatment of advanced prostate cancer. Both of them contain leuprolide acetate. This technology is protected by 33 patents in the United States and 35 patents in the rest of the world. Other marketed products are listed in Table 12.2. There are several patents that are claimed with the in situ gel system and they are described with other novel gels in Table 12.1.

4 EMULSION GELS

Emulsion gels are popular as emulgel. As the name implies, Emulgel is the system in which emulsion technology and gel technology are combined together. They are widely used as a controlled topical delivery system for their dual drug release technique of gel and emulsion. Although classic gels have several advantages, the major limitation is entrapment of a hydrophobic drug. Similarly, conventional emulsions have wide applications, but their limitations are stability and controlled drug-release capacity. Thus by incorporating a gelling agent in the water phase of emulsion, a conventional emulsion can be converted into emulgel. An emulgel has several advantages over other gels. Some of them are listed here:

- Good thixotropic property.
- Better drug entrapment capacity than conventional gel.

- Controlled release of drug at site of action.
- Less greasy.
- Easily spreadable and washable.
- Longer shelf life than emulsions.
- Ease of production and lower production cost than other vesicular gels.

4.1 PREPARATION OF EMULSION GEL

Emulgels are prepared by two basic methods. The first step is formulation of an emulsion by using water phase and oil phase by emulsification method. Drugs are incorporated either in water or oil phase according to their solubility. The second step involves preparation of gel using hydrocolloids by soaking in warm water. After cooling, the prepared emulsion is incorporated into preformed gel and stir to disperse emulsion uniformly into gel base (Chaudhari et al., 2009).

Emulgel formulation contains aqueous phase, oils, gelling agents, penetration enhancers, and emulsifiers for preparation of emulsion (Zhang et al., 1995). Emulgel contains an emulsion that acts as a vehicle for dissolving drugs. Emulgel contains both aqueous and oily vehicles, so both hydrophilic, as well as hydrophobic drugs can be incorporated. For externally applied emulsions, mineral oils—either alone or combined with paraffin—are extensively used as the carrier for the drug (Mortazavi and Aboofazeli, 2003). Fixed oils like fish liver oils or other oils of natural origin (e.g., castor oils, arachis oil, cottonseed oil, and maize oils) are also used in the oily phase in preparation of emulgel (Bonacucina et al., 2009).

Emulgel mostly contains polyethylene glycol, sorbitan mono-oleate (Span 80), polyoxyethylenesorbitanmonooleate (Tween 80), stearic acid and sodium stearate as emulsifying agents (Rutrer, 1987). Commonly used gelling agents in preparation of emulgel are Carbopol-934 (1%), Carbopol-940 (1%), HPMC-2910 (2.5%). Penetration enhancers that are used in emulgels are oleic acid (1%), clove oil (8%), menthol (5%), and so forth.

4.2 APPLICATIONS OF EMULGEL

Among all the marketed products of emulgel, "Voltarenemulgel" of Diclofenac diethyl ammonium, marketed by Novartis Pharma, is most popular.

4.3 MICROEMULSION GELS

Emulsion gels have been further improved as microemulsion gels for the first time by Chen et al. (2006) for the delivery of ibuprofen. Suthar et al. (2009) developed a microemulsion gel of tretinoin for the treatment of acne. Microemulsion gels are very similar to emulsion gels with the standing property of improved bioavailability for their size range in micron. Generally, a microemulsion is clear, stable, and isotropic in nature. The small droplets size of microemulsions provide better observance to biomembranes and easy transport of drugs. This microemulsion can be converted to gel by incorporating gelling agents like gelatin, poloxamers, carbomer, gellan gum and so forth. These microemulsion gels have proved to be better than emulsion gels due to their better skin penetration, stability, washability, and bioavailability.

Popularity of this gel technology has boosted applications for patenting. US patent no. 5,336,432 A revealed "a composition for microemulsion gel having bleaching and antiseptic properties." The advantages of this microemulsion gel are its colloidal nature, which allows users to select areas of the body where the gel can be used as an antiseptic or in the hair for applications as streaking and frosting. Another patent (WO 2011157428 A2) disclosed in situ lecithin microemulsion gel formulation proposed for use as a diagnostic or therapeutic agent in ophthalmology and in the ENT field in rhinology. A gel-microemulsion formulation is patented for its spermicidal and antimicrobial action by Osmond et al., 2000. Although a number of research studies have been reported for this technology, marketed products are limited. Lubrasilmicroemulsions gels are recommended for hair treatment, manufactured for Ashland Inc. by Guardian Laboratories, a division of United-Guardian, Inc. This is prepared by clathrates of glyceryl acrylate and polyacrylic acid.

5 MICROGEL

The first time Baker coined the term microgel to describe cross-linked polybutadiene latex particles (Baker, 1949). Frequently, the term microgel is used interchangeably to define different other gel systems, which are different in size, physicochemical property, and structure, such as nanogel and macrogel. Typically microgels are commonly referred to as those gels that have a size range in microns. Macrogels have large cross-linked structures with a size range greater than a micron. A nanogel, by comparison, possesses a size range in nanometers. Further, microgels are colloidal dispersions of gel particles in a solvent. Microgels possess a number of similar characteristics to macrogels but differ in physical molecular arrangement (Joanna et al., 2011). Most importantly, both of them swell in the presence of water or other solvents. Swelling of these gels is determined by a few factors like cross-link density, the polymer/water compatibility, and the presence of electrical charges. Microgels have steady constructions due to its internal covalent or strong physical forces (Pollack, 2001). Like other novel gels, they are also classified in two categories: stimuli sensitive and stimuli nonsensitive.

Microgels offer several advantages over conventional macrogels, including free-flowing capacity, rapid response under environmental changes, very high surface-to-volume ratios that enable mass transport, fine-tuning properties, and dynamic light scattering and small-angle light scattering ability, the ability to assemble into useful larger objects, such as 2D assemblies at the air–water and oil–water interfaces (Fujii et al., 2006; Zhang and Pelton, 1999), controlled and sustained drug release at the desired site, improved therapeutic efficacy, and reduced side effects.

5.1 PREPARATION OF MICROGEL

There are three possible types for microgel preparation: (1) from monomers like vinyl monomers, bifunctional cross- linking monomer *N,N*-methylenebisacrylamide (MBA); (2) from polymers like *N*-isopropyl acrylamide (POLYNIPAM); and (3) from macrogels by mechanically grinding to form microgels (Griffin et al., 2007).

Microgel preparation techniques can be simply classified into three types based on the particle formation mechanism: homogenous nucleation, emulsification, and complexation. Homogeneous nucleation refers to those preparations in which microgel particles are generated from initially homogeneous (or nearly so) solutions. Emulsification refers to those methods in which aqueous droplets of a pregel solution are produced in an oil or brine phase. In the next step, these droplets are cross-linked or polymerized into a microgel. Finally, microgels can be prepared by mixing two dilute, water-soluble polymers that form complexes in water (Miao et al., 2007). A completely different approach to microgel synthesis involves mixing dilute solutions of oppositely charged polyelectrolytes to form colloidally dispersed, polyelectrolyte complexes.

5.2 APPLICATIONS OF MICROGEL

Microgels have a wide application in different fields. They have been studied for their mucoadhesive property, biodegradable property, hydrophilic property, stimuli-responsive property, and many more. Microgels are applied for their targeted and nontargeted delivery systems by different researchers. Retin-A Micro is a microgel product used to treat acne and other skin problems and it contains Tretinoin that is manufactured by Ortho Pharmaceuticals, USA. Applications of microgels in drug delivery are compiled in Table 12.5.

Table 12.5 Applications of Different Microgels

Type of Microgel	Drug	Application/Activity	References
Mucoadhesive microgels	—	Oral and nonoral drug delivery	Chowdary and Rao (2004); Alpar et al. (2005); Peppas and Huang (2004)
Metal-containing microgels	—	Polymer microgels as templates for the synthesis of metal nanoparticles impregnated into the polymer network	Zhang et al. (2004); Sonvico et al. (2005); Arya et al. (2005)
Hydrophilic and biocompatible microgels	Doxorubicin, 5 flurouracil	Targeted delivery of encapsulated drugs via blood circulation	Vinogradov et al. (2002); De Jaeghere et al. (2000); Oupicky et al. (2002)
Microgel decorating the backbone of Carboxymethylcellulose (CMC) and NIPA n	Lysozyme	Topical application	Lu et al. (2011)
PNIPA-*co*-acrylic acid microgel	5-Aminolevulinic acid	Microgel was synthesized via microemulsion polymerization in oil-in-water system for topical application	Sahiner et al. (2011)
NIPA microgel	Diclofenac and naproxen	Topical application	Cirillo et al. (2011) Gomez et al. (2012)
Glucose responsive microgels	Aminophenylboronic acid to a pre-existing carboxylatedmicrogel via EDC coupling	In water treatment, biological separation, glucose sensors, self-regulating insulin delivery devices, and fluid switches	Todd and Robert (2006)
Biodegradable Alginate and gelatin-based microgel		For cell encapsulation	Li et al. (2014)
Microgels based on poly(ethylene glycol) (PEG)		The colors of these hydrogel films change with changes of environment, such as temperature, pH, salt concentration, and so on	Zhibing et al. (2010)

6 NANOGELS

Development of nanogels is the result of successful application of nanotechnology in the field of gel technology to improve applicability. Nanogels are hydrogel polymeric network dispersions with particle size in the nanometer range. The term *nanogel* (NanoGel) was first time announced by Alexander and Serguei (2008) to describe cross-linked bifunctional systems of a nonionic polymer and a polyion for transport of polynucleotides. Nanogels are basically a carrier system for delivering drugs or they can be chemically modified to incorporate various ligands for targeted drug delivery and triggered drug release (Vinogradov et al., 2002). Due to their exclusive construction, nanogel polymer networks possess dual properties of cross-linked polymer gels and dispersed colloidal particles.

Nanogel-based systems offers different advantages including highly biocompatablity and biodegradability, rapid swelling/contracting characteristics, higher drug-loading capacity, effectiveness in avoiding the rapid renal exclusion but small enough to avoid the uptake by the reticuloendothelial system (Labhasetwar and Leslie-Pelecky, 2007), ability to solubilize hydrophobic molecules in their networks of gel, better stability over the surfactant micelles, exhibition of lower critical micelle concentrations, and slower rates of dissociation. They are good carriers for intracellular delivery (Toita et al., 2011). They also enhance cellular accumulation significantly, without causing observable cytotoxic processes (Rigogliuso et al., 2012).

They are also categorized as hydrogel nanoparticles. However, with respect to the classic nanoparticles, they have an elevated degree of encapsulation and offers an ideal tridimensional microenvironment for many macromolecules (Rigogliuso et al., 2012). Nanogels also display abundant quicker response as compared to the conventional hydrogels.

6.1 APPROACHES FOR SYNTHESIS AND PREPARATION OF NANOGELS

Nanogels can be synthesized by three basic methods: Photo polymerization, Chemical cross linking by emulsion polymerization and Pulse radiolysis method. Photo polymerization involves synthesis of nanogels directly in the presence of light. Here Polymerization of nanogels from monomers molecules initiated directly with UV. In this method no initiator, cross-linkers and surfactant are used. Thus pure nanogels can be obtained. N-isopropyl acrylamide nanogels can be prepared by this method. Chemical cross-linking by emulsion polymerization method involves addition of cross linking agent with monomers, organic solvent, surfactant, and water to prepare organic phase. This organic phase is then irradiated and purified to obtained cross-linked nanogel. Polyacrylamide, poly acrylic acid, polyethyleneimine, polyvinyl alcohol nanogels are synthesized by this method. In pulse radiolysis method the polymer is subjected to ionizing radiation in dilute aqueous solution to promote intramolecular recombination of polymer radicals to form nanogels.

Other approaches that are also used for preparation of nanogels are presented in the following subsections:

6.1.1 Physical self-assembly of interactive polymers

This method usually employs the accumulation of hydrophilic polymers by forming hydrophobic or electrostatic interactions and/or hydrogen bonding in a mild aqueous media. Self-associating hydrophilic polymers allow encapsulation of large molecules thereby being a useful preparation of protein-loaded nanogels (Alexander and Vinogradov, 2009). Akiyoshi et al., 1998 reported "hydrogels by hydrophobic association of cholesterol-modified pullulan in presence of insulin."

6.1.2 Chemical cross-linking of preformed polymers

The cross-linking method was widely applied to synthesize a variety of functional nanogels with large pore size (Hennink and van Nostrum, 2002). In particular, it was used to synthesize the first cross-linked cationic nanogel for polynucleotide delivery. The chemical cross-linking method is further subdivided into carbodiimide coupling, the Michael addition reaction, and free radical polymerization. Free radical polymerization of monomers involves hydrophilic or water-soluble monomers to prepare well-defined synthetic nanogels in the presence of different cross-linkers. This method is further subdivided into four type's of polymerization procedure. They are precipitation, inverse (mini) emulsion, inverse micro-emulsion, and dispersion polymerization.

6.1.3 Heterogeneous controlled/living radical polymerization

Recently, controlled/living radical polymerization has been discovered as a device for the production of different bio-conjugates like polymer–protein/peptide. Different methods have been developed for this purpose; however, atom-transfer radical polymerization (ATRP), stable free radical polymerization (SFRP), and reversible addition fragmentation chain transfer (RAFT) polymerization (Jung et al., 2008) have been the most fruitful techniques.

6.1.4 Template-assisted nanofabrication of nanogel particles

Rolland et al. (2005) developed a novel method for fabrication of nanogels known as imprint photolithographic technique or PRINT (particle replication in nonwetting templates). This method uses nonwetting elastomeric moulds of a low surface energy prepared on patterned silicon master templates by photochemical cross-linking of dimethacrylate-functionalized perfluoropolyether oligomers.

Another method based on this principle is the micromolding method. There are many advantages to this method, including circumvention of clean-room facilities and costly lithographic equipment (Farhana et al., 2013).

6.2 APPLICATIONS OF NANOGELS

Nanogels are very promising carriers among other novel drug delivery vehicles for utilization in different therapeutics, diagnostics, macromolecules, and others. Here we have compiled some of the recent developments in the use of nanogels.

Nanogels found a versatile application in cancer treatment. Multistimuli responsive nanogels are very effective in targeted therapy of cancer as compared to single responsive nanogels (Morinloto et al., 2008; Rijcken et al., 2007). A wide number of research works have been carried out with nanogels for the treatment of cancer. Dickerson et al. recommend targeted delivery of siRNAs, a gene-regulating tool by nanogels to increase the efficacy of chemotherapy drugs (Dickerson et al., 2010). Another study revealed that paclitaxel (PTX) loaded nanogel offers greater cytotoxicity in HEPG-2 cells than free drugs (Li et al., 2011). However, an injectable dual-responsive micellar nanogel system for controlled delivery of paclitaxel in cancer therapy also showed improved solubility and stability of the drug (Chen et al., 2013). Oh et al., 2010 utilized glycol chitosan to prepare a novel pH-responsive nanogel system that recognized tumor pH. This nanogel loaded with doxorubicin showed higher concentration of drug at cancer sites than free drug.

Nanogels are established with great efficacy in the treatment of autoimmune disease, neurodegenerative disorders, diabetes, inflammatory disorders, and so many other disorders. They are effective for intracellular delivery also. Some of their applications are listed in Table 12.6. Patents are primarily filed with nanogel for its synthesis method. For example, polyacrylic acid nanogel was prepared for

Table 12.6 Applications of Nanogel

Type of Nanogel	Drug	Disease	Activity	References
PAMA-DMMA nanogels	Doxorubicin	Cancer	Increase in the release rate as the pH value decreased. Higher cytotoxicity at pH 6.8 in cell-viability studies	Du et al. (2010)
Chitosan-based nanogels decorated with hyaluronate	Photosensitizers like tetra-phenyl-porphyrin-tetra-sulfonate (TPPS4), tetra-phenyl-chlorin-tetra-carboxylate (TPCC4), and chlorin e6 (Ce6)	Rheumatic disorders	Rapidly taken up (<4 h) by the macrophages and accumulated in their cytoplasm and organelles	Schmitt et al. (2010)
PCEC nanoparticles in Pluronic hydrogels	Lidocaine	Local anesthesia	Produced long-lasting infiltration anaesthesia of about 360 min	Yin et al. (2009)
Poly(lactide-*co*-glycolic acid) and chitosan nanoparticle dispersed in HPMC and Carbopol gel	Spantide II	Allergic contact dermatitis and other skin inflammatory disorders	Nanogelinncreases potential for the percutaneous delivery of spantide II	Punit et al. (2012)
pH-sensitive polyvinyl pyrrolidone-poly (acrylic acid) (PVP/PAAc) nanogels	Pilocarpine		Maintain an adequate concentration of the pilocarpine at the site of action for prolonged period of time	Abd El-Rehim et al. (2013)
Cross-linked poly (ethylene glycol) and polyethylenimine	Oligonucleotides	Neurodegenerative diseases	Effectively transported across the BBB. The transport efficacy is further increased when the surface of the nanogel is modified with transferrin or insulin	Vinogradov et al. (2004)
Cholesterol bearing pullulan nanogels	Recombinant murine interleukine-12	Tumor immunotherapy	Sustained release nanogel	Farhana et al. (2013)
Poly(*N*-isopropylacrylamide) and chitosan		Hyperthermia cancer treatment and targeted drug delivery	Thermosensitive magnetically modalized	Farhana et al., 2013
Cross-linked branched network of polyethyleneimine and PEG Polyplexnanogel	Fludarabine	Cancer	Elevated activity and reduced cytotoxicity	Farhana et al. (2013)
Biocompatible nanogel of cholesterol-bearing pululan	As artificial chaperone	Treatment of Alzheimer's disease	Inhibit aggregation of amyloid β-protien	Ikeda et al. (2006)
DNA nanogel with photo cross-linking	Genetic material	Gene therapy	Controlled delivery of plasmid DNA	Lee et al. (2009)

dispensing biologically active molecules by the free radical polymerization method by using gamma radiations for cross-linking of polymers (Somasundaran and Chakraborty, 2006). In Table 12.1, we have compiled some patent information on nanogels for its application as delivery system.

7 VESICULAR GEL

Vesicles are small-scale carriers designed with a variety of amphiphilic molecules, including lipids, surfactants, and block copolymers (Pawar and Edgar, 2012). They consist of a hydrophilic core surrounded by a bilayer of the amphiphiles. These vesicles have gained attention for their unique delivery system of drugs in pharmaceuticals, nutrients in foods, fragrances and dyes in cosmetics and textiles, and so forth. These vesicular gels facilitate topical delivery of drugs carried by liposomes, niosomes, novasomes, and other vesicles exhibit interesting applications for dermal delivery of locally acting drugs, as well as for enhancing transdermal delivery of drugs intended for systemic use. Thus non-invasive routes are gaining importance as alternative route to oral administration. Vesicular gels can be prepared simply by mixing preformed vesicular systems with the hydrogel matrix. A flow diagram is presented in Fig. 12.3 for depicting the common procedure of vesicular gel formation. In this section we have discussed different vesicular gel system for their drug delivery aspects.

7.1 LIPOSOMAL GEL

Liposomes are a very popular carrier system in the field of drug delivery. Because liposomes possess many exclusive properties, such as structural similarity with biological membranes, ability to encapsulate both water-soluble and lipophilic substances in their different phases, biodegradability and non-toxicity, and the ability to act as drug localizers as well as drug carriers. This natural similarity allows them to infiltrate the epidermal barrier to a greater extent as compared to other dosage forms. Although liposomes show promising result in transdermal drug delivery, topical applications of these formulations are fewer due to rheological problem. Thus, they can be incorporated into the gels that are applied to the skin.

7.1.1 Classification of Liposomal Gel

Liposomal gel can be classified into four types according to their method of preparation.

1. Classic liposomal hydrogels
   Liposomal hydrogels can be prepared by two methods: either by integration of the prepared liposomes after separation of the un-entrapped drug into the suitable gel matrix or by addition of gelling agents in liposome dispersions without separation of un-entrapped drug.
2. Modified liposomal gels
   Conventional liposomal gels are modified to improve their transdermal delivery known as modified liposomal gels. They are including transferosomal gel, ethosomal gel, phytosomal gel, and so forth.
3. Vesicular phospholipid gels (VPGs)
   Unlike liposomal hydrogels VPGs obtain their gel like structure by the strongly packed arrangement of the vesicles. Brandl et al., describe VPGs as semisolid, aqueous phospholipid-dispersions, in which vesicles or liposomes represent the lipid phase. VPGs offer

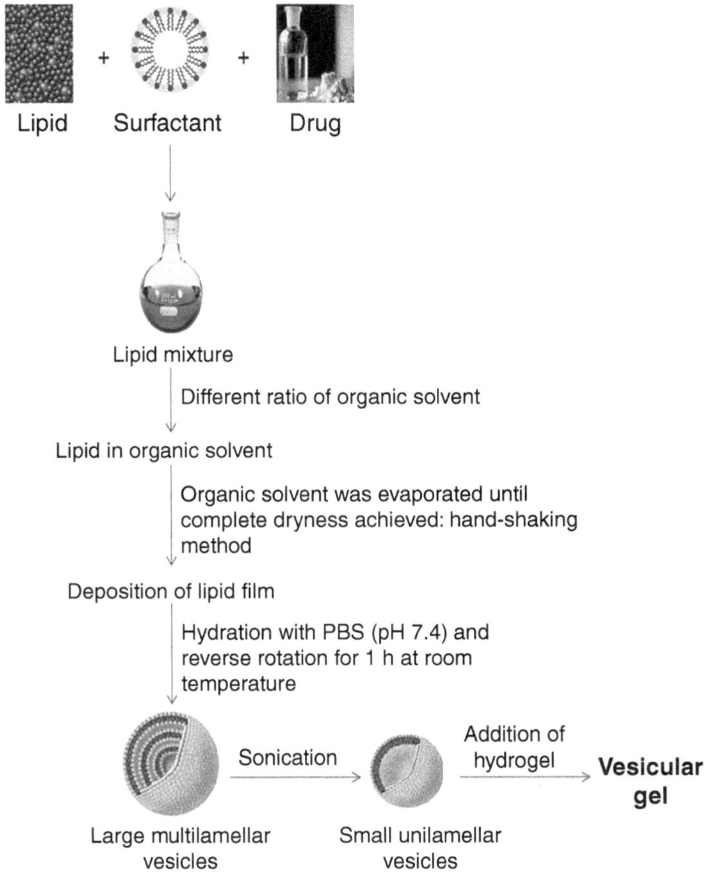

FIGURE 12.3 Schematic Diagram of Preparation of Vesicular Gels

potential as local depot systems because they appear to retain entrapped drugs when in contact with excess aqueous medium.

4. Microgel in liposomes (M-i-L)

One of the latest liposomal gel technologies is M-i-L. It is also known as liposome-nanogel assembly, gel core liposomes, lipobeads, and hydrogel supported lipid bilayer. M-i-L is a phospholipid double layer vesicle that includes a hydrogel (microgel) polymer inside its core, which helps to prevent their rapid degradation inside the human body and early drug release. Thus, it combines the advantages of both compartments structure and capacity of mimicking the natural cell structures. Different methods are approved for the preparation of M-i-L. The commonly used method is to prepare the microgel first and then coat it with the phospholipid bi-layer. This can be achieved either by mechanical adsorption of lipids onto gel particles or electrostatic adsorption simply by mixing oppositely charged liposomes and hydrogels. M-i-L has shown promising results in different studies when administered intramuscularly and intranasally (Tiwari et al., 2009b), and more specifically for vaccine delivery.

Liposomes can be prepared by different traditional method including lipid film hydration, the solvent spherule method, the French pressure cell method, and solvent injection methods. These liposomes can be entrapped into gel by the addition of a gelling agent into aqueous phase. These gelling agents include carbopol 934 and 940, carbopol 934 NF, gellan gum, xanthan gum, hydroxyl propyl methyl cellulose, poly vinyl alcohol, polymethacrylic acid, poloxamer, and so forth.

Different researchers have mentioned liposomal gel as a drug delivery vehicle. Patel et al., 2009 described topical application of tretinoin and ketoconazole by carbopol liposomal gel. Budai et al. (2007) reported poly vinyl alcohol and polymethacrylic acid based liposomal gel for ocular delivery of ciprofloxacin. Kannaiyan et al., 2009 described topical delivery of carbopol-based liposomal gel for the delivery of paracetamol. Mitkari et al. (2010) developed topically applied liposomal gel of fluconazole. Recently, use of liposomal gels in skin cosmetics products has been generated. Vegetable phospholipids are extensively used for this purpose since they have a high content of esterified essential fatty acids, which is used to upsurge the barrier function of the skin and eliminate water loss. Liposomal gel has been used for the ocular application of many drugs including gatifloxacin, indomethacin, ofloxacin, linezolid, levofloxacin, ciprofloxacin. Oral route is also exploited for the delivery of these gels. Drugs that are applied through the oral route include clotrimazole, clarithromycin, and amoxicillin. Fluconazole and nonoxynol have been applied through liposomal gel for vaginal delivery.

A US patent (US 4,708,861 A) for liposome-gel compositions was published in 1987 and declared "gel matrix inhibited the dispersion and clearance of the sequestered liposomes without interfering with the ability of the liposomes to release the entrapped bioactive agent." Another US patent (US 4,761,288 A) described a "novel multiphase drug delivery system which comprises lipid vesicles with a biologically active compound, dispersed in a hydrocolloid gel."

Marketed formulations of topical liposomal gel include the following: Celadrin is a topical liposomal lotion that is registered under the trademark of Imagenetix, Inc.; Optisome™ liposomal gel is used to deliver local anesthetic, tetracaine; Lipo CTM Liposome contains vitamin C, vitamin E and zinc; Lipo-Gest™ is a natural balancing cream and liposome containing progesterone is also available as a cream.

"Novasome" is a patented technology of IGI laboratories Novavax. Novasomes are modifications of the liposomal drug delivery system. Novasomes are multilamellar vesicles with a size range of 200–700 nanometers, consisting of 2–7 bilayered membrane each composed of an amorphous core and amphiphilic molecule. Like liposomes they are also prepared by molecules that include long chain acids, long chain amino and glycerolipids, and long chain fatty alcohol derivatives. The delivery of mahoniaaquifolium extract by Novasome technology is patented by Mills et al., for the treatment of psoriasis, eczema, and other dry skin conditions. Weiner et al. (1998) patented topical penetration of minoxidil by encapsulating it in a Novasome lipid vesicle having glycerol dilaurate as a primary wall component. Wright (1997) formulated oral vaccine utilizing Novasome-WFI diluents for prevention against gram-negative bacterial infections.

7.2 NIOSOMAL GEL

Niosomes are vesicles having a bilayer structure like liposome. Unlike liposome, niosomes are prepared by nonionic surfactants like span. Thus they can overcome the stability problem that liposomes cause due to the presence of phospholipids in conventional liposome. The bilayer may be multilamellar or unilamellar, and it encircles the hydrophilic component, whereas lipophilic components remain in the bilayer itself. The capability of a nonionic surfactant to produce bilayer vesicles is reliant on the HLB value of the surfactant, the critical packing parameter, and the chemical structure of the components.

Nonionic surfactants that are used for the formation of niosomes are of following types:

1. Alkyl Ethers: Surfactants containing alkyl ethers are of three types. Surfactant-1, having molecular weight 473, is "C16 monoalkyl glycerol ether with average of three glycerol units." Surfactant-2 is a diglycerol ether having molecular weight 972 with an average of seven glycerol units. Surfactant 3 is an ester-linked surfactant with molecular weight 393.
2. Alkyl Esters: Most popular surfactant used for the preparation of niosomes among alkyl esters are sorbitan esters or span.
3. Alkyl Amides: Galactosides and glucosides.
4. Long chain fatty acids and amino acid moieties have also been used in some niosomes preparation.

Niosomes are of different types: proniosome, aspasome, gel noisome, emulsion noisome, and bolaniosome. Proniosomes are the initial stage of noisomes that contain a carrier and surfactants only. They need to be hydrated for preparing niosomes. Unlike niosomes they can eliminate the problem of aggregation, leaking, and fusion. Bola surfactant, an extraordinary type of surfactants, is made of [omegahexadecylbis-(1-aza-18 crown-6)] along with span-80 and cholesterol in 2:3:1 molar ratio are investigated widely. This chemical moiety is commercially known as bola surfactant (Yoshida et al., 1992). Aspasomes are modified noisome that are prepared by acorbyl palmitate and highly charged lipid along with cholesterol. They are used to improve the transdermal permeation of drugs (Bhaskaran and Lakshmi, 2009). Gel niosomes are also known as niosomal gel. They are simply niosomes dispersed in a gelling media like carbopol, HPMC, and others. Emulsion niosomes are niosomal vesicles dispersed in water in an oil emulsion (v/w/o). Hu and Rhodes (1999) reported that the emulsification of a water-soluble niosomes into an oily vehicle produce "vesicle in water in oil emulsion."

Niosomal gels are prepared primarily by using drug, nonioninc surfactant, and cholesterol to prepare noisome vesicles. Then, obtained niosomes are incorporated in a hydrogel base for the formation of gel. In-vitro diffusion studies of such niosomal gels with conventional gels has revealed the mean flux value and diffusion coefficient to be 5–7 times lower for niosomal gel (Yoshida et al., 1992).

The applications of niosomes are diversified in many fields; however, three main categories are: targeting and retention of drug in blood circulation; improving the physical properties and stability of the drugs; and controling the release of drugs. Alcantar et al. in 2007 patented a delivery system for application within a brain cavity involving niosome embedded in a biodegradable hydrogel (chitosan polymer). Further, in 2010 again they patented a delivery system of noisome gel that was temperature and pH sensitive (Alcantar et al., 2010). A cross-linked chitosan was used as the hydrogel, which remains liquid at room temperature and transforms into gels inside the body temperature and pH. Shirsand et al. (2012) reported niosomes gal loaded with antifungal drug ketoconazole, showed prolonged drug action as compare to non-niosomal form. These researchers again reported in a recent publication that niosomal gel encapsulated with clotrimazole provided better patient compliance by achieving prolonged drug release to the dermis (Shirsand et al., 2015). Ujwala and Shivkumar, 2014 described a niosomal gel of serratiopeptidase. They found a fourfold increase in steady state flux in an ex vivo skin permeation study and comparable topical antiinflammatory activity like dicolfenac gel. Gupta et al. (2014) reported a topical niosomal gel containing a combination of benzoyl peroxide and tretinoin for antiacne activity. In-vivo studies of this niosomal gel showed that the niosomal gel was more efficacious than other anti-acne creams with a 4.16-fold lower dose of benzoyl peroxide provided the same therapeutic index at targeted sites. A variety of cosmetics like anti-ageing products and face whitening products are based on niosomes and are manufactured by Lancome, Paris and L'Oreal Company.

7.3 TRANSFEROSOMAL GEL

Although liposomes and niosomes have shown promising applications in different fields, their poor skin permeability, leakage of drug, aggregation, breaking, and fusion of vesicles makes them unsuitable for transdermal delivery (Cevc et al., 1997). To overcome these, a modified liposome carrier system called "transferosome," has been introduced. The name *transferosome* was derived from two words. The Latin word *transfer,* meaning "to carry across," and the Greek word *soma,* meaning "body" combines to generate the word *transferosome* (Prajapati et al., 2011).

Like liposomes, each transferosome consists of at least one inner aqueous compartment, which is surrounded by a lipid bilayer. But this lipid bilayer has specially tailored properties, due to the incorporation of "edge activators," which make them deformable. Surfactants, such as sodium cholate, sodium deoxycholate, span 80, and tween 80 have been used as edge activators. Materials commonly used for the preparation of transferosomes are phospholipids (egg phosphatidyl choline, soya phosphatidyl choline), alcohol (methanol, chloroform, ethanol) as a solvent, edge activators (tween 80, sodium cholate) for providing flexibility, and buffering agent (saline phosphate buffer pH 7.4), as a hydrating medium. For preparation of transferosomal gel, hydrocolloids are incorporated into the aqueous dispersion of prepared transferosomes.

Transferosomes gels are used as carriers for protein and peptides, such as insulin, bovine serum albumin, vaccines, and so forth, which are completely degraded in the GI tract when administered orally (Nanda et al., 2005). Transferosomal gel of ketoprofen (Diractin) is a good pain reliever formulation for local application in knee osteoarthritis (Rother et al., 2007). A tranferosomal gel containing curcumin showed higher entrapment efficiency and higher permeation of drug (Patel et al., 2009). Maurya et al. (2010) formulated transferosomal gel of indinavir sulfate by adding transferosomes into a gel base, which decreased lag time and high transdermal flux. Malakar et al. (2012) reported transferosomal gel containing insulin has demonstrated prolonged hypoglycemic effect when administered transdermally. Transferosomal formulation of paromomycin sulfate showed good result in topical therapy of cutaneous leishmaniasis (Bavarsad et al., 2012).

8 FUTURE PROSPECTS

Although these novel gels are proving their efficacy in various applications, more work is to be carried out to establish their actual potency in the field of drug delivery. Environment-sensitive or stimuli-sensitive gels are employing for smart drug delivery because some of these variables are found in the body itself, such as low pH and elevated temperatures. Thus, they can be used for site-specific controlled drug delivery. Pressure-responsive, light-sensitive, and electro-sensitive gels also can be used in diagnosis, drug delivery, and bioseparation. Vesicular novel gels are in their early stage. Effectiveness of these vesicles in gel form should be explored in details by modifying their components, formulation, and delivery techniques. Moreover, new manufacturing techniques should be established for the successful design and application of these novel gels.

REFERENCES

Abd El-Rehim, H.A., Swilem, A.E., Klingner, A., Hegazy, el-S.A., Hamed, A.A., 2013. Developing the potential ophthalmic applications of pilocarpine entrapped into polyvinylpyrrolidone-poly (acrylic acid) nanogel dispersions prepared by γ radiation. Biomacromolecules 11, 688–698.

Akiyoshi, K., Kobayashi, S., Shichibe, S., Mix, D., Baudys, M., Kim, S.W., Sunamoto, J.J., 1998. Self-assembled hydrogel nanoparticle of cholesterol-bearing pullulan as a carrier of protein drugs: complexation and stabilization of insulin. Cont. Rel. 54, 313–320.

Alcantar, N., Williams, E.C., Ryan, T., 2010. Niosome-hydrogel drug delivery system. US 20,100,068,264 A1.

Alexander, K.V., Serguei, V.V., 2008. Nanogels as Pharmaceutical Carriers, Multifunctional Pharmaceutical Nanocarriers. Springer Science, New York, pp. 67–80.

Alexander, V.K., Vinogradov, S.V., 2009. Nanogels as pharmaceutical carriers: finite networks of infinite capabilities. Angew. Chem. Int. Ed. Engl. 48 (30), 5418–5429.

Alpar, H.O., Somavarapu, S., Atuah, K.N., Bramwell, V.W., 2005. Biodegradable mucoadhesive particulates for nasal and pulmonary antigen and DNA delivery. Adv. Drug. Del. Rev. 57, 411–430.

Anda, V., Michel, L., Guillaume, B., Jean-Christophe, L., 2008. In-situ-forming oleogel implant for Rivastigmine delivery. Pharm. Res. 25, 845–852.

Arya, H., Kaul, Z., Wadhwa, R., Taira, K., Hirano, T., Kaul, S.C., 2005. Quantum dots in bio-imaging: revolution by the small. Biochem. Biophys. Res. Commun. 329, 1173–1177.

Bae, Y.H., 1997. Stimuli-sensitive drug delivery. In : Park, K. (Ed.), Controlled Drug Delivery: Challenge and Strategies. American Chemical Society, Washington, DC, pp. 147–160.

Bajpai, V., 2014. In-situ gel nasal drug delivery system—a review. Int. J. Pharma. Sci. 4, 577–580.

Baker, W.O., 1949. Microgel, a new macromolecule. Ind. Eng. Chem. 41, 511–520.

Bavarsad, N., Bazzaz, B.S.F., Khamesipour, A., Jaafari, M.R., 2012. In-vitro and in-vivo anti-leishmanial properties of transfersomes containing paromomycin sulfate in susceptible balb/C mice. Acta Tropica 124, 33–41.

Bhardwaj, T.R., Kanwar, M., Lal, R., Gupta, A., 2000. Natural gums and modified natural gums as sustained release carriers. Drug Devel. Ind. Pharm. 26, 1025–1038.

Bhaskaran, S., Lakshmi, P.K., 2009. Niosomes: a controlled and novel drug delivery system. Acta Pharm. Sci. 51, 27–32.

Bilia, A., Carelli, V., Di Colo, G., Nannipieri, E., 1996. In vitro evaluation of a pH-sensitive hydrogel for control of GI drug delivery from silicone-based matrices. Int. J. Pharm. 130, 83–92.

Bonacucina, G., Cespi, M., Palmieri, G.F., 2009. Characterization and stability of emulsion gels based on acrylamide/sodium acryloyldimethyltaurate copolymer. AAPS Pharm. Sci. Tech. 10, 34–45.

Budai, L., Hajdú, M., Budai, M., Gróf, P., Béni, S., Noszál, B., Klebovich, I., Antal, I., 2007. Gels and liposomes in optimized ocular drug delivery: studies on ciprofloxacin formulations. Int. J. Pharm. 343 (1–2), 34–40.

Burkoth, A.K., Anseth, K.S., 2000. A review of photo crosslinked poly anhydrides: in-situ forming degradable networks. Biomaterials 21, 2395–2404.

Cevc, G., Blume, G., Schatzlein, A., 1997. Transferosomes mediated transepidermal delivery improves the regiospecificity and biological activity of corticosteroids in-vivo. J. Control. Release 45, 211.

Chaudhari, P., Ajab, A., Malpure, P., Kolsure, P., Sanap, D., 2009. Development and in-vitro evaluation of thermo reversible nasal gel formulations of Rizatriptan benzoate. Indian J. Pharm. Edu. Res. 43, 55–62.

Chen, H., Du, D., Li, L., et al., 2006. Microemulsion-based hydrogel formulation of ibuprofen for topical delivery. Int. J. Pharm. 315, 52–58.

Chen, D., Yu, H., Sun, K., Liu, W., Wang. H., 2013. Dual thermos responsive and pH-responsive self-assembled micellar nanogel for anticancer drug delivery. Drug Del. Available from: http://dx.doi.org/10.3109/10717544.2013.838717

Cho, E., Gwak, H., Chun, I., 2008. Formulation and evaluation of ondansetron nasal delivery systems. Int. J. Pharm. 349, 101–107.

Cirillo, G., Iemma, F., Spizzirri, U.G., Puoci, F., Curcio, M., Parisi, O.I., Picci, N., 2011. Imprinted hydrophilic nanospheres as drug delivery systems for 5-fluorouracil sustained release. J. Drug Target 17, 72–77.

Dasharath, M.P., Divyesh, K.P., Chhagan, N.P., 2011. Formulation and evaluation of floating oral in-situ gelling. Int. Scholarly Res. Netw. 2011, 1–8.

De Jaeghere, F., Allemann, E., Feijen, J., Kissel, T., Doelker, E., Gurny, R., 2000. Cellular uptake of PEO surface-modified nanoparticles: evaluation of nanoparticles made of PLA:PEO diblock and triblock copolymers. J. Drug Target 8, 143–153.

Dickerson, E.B., Blackburn, W.H., Smith, M.H., Kapa, L.B., Lyon, L.A., McDonald, J.F., 2010. Chemosensitization of cancer cells by siRNA using targeted nanogel delivery. BMC Cancer 10, 10.

Dondeti, P., Zia, H., 1996. Bioadhesive and formulation parameters affecting nasal absorption. Int. J. Pharm. 127, 115–133.

Du, J.Z., Sun, T.M., Song, W.J., Wu, J., Wang, J., 2010. A tumor-acidity-activated charge-conversional nanogel as an intelligent vehicle for promoted tumoral-cell uptake and drug delivery. Angew. Chem. Int. Ed. 49, 3621–3626.

Dunn, R.L., English, J.P., Stoner, Jr., W.C., Potter, A.G., Perkins, B.H., 1987. Biodegradable fibers for the controlled release of tetracycline in treatment of peridontal disease. Proc. Int. Symp. Control Release Bioact. Mater. 14, 289.

El Maghraby, G.M., Elzayat, E.M., Alanazi, F.K., 2012. Development of modified in-situ gelling oral liquid sustained release formulation of Dextromethorphan. Drug Dev. Ind. Pharm. 38, 971–978.

Farhana, S., Manirujjaman, Md.I., Mohammad, A., Sanjida, S., 2013. An overview of nanogel drug delivery system. J. Appl. Pharma. Sci. 3 (8 Suppl. 1), S95–S105.

Fujii, S., Ryan, A.J., Armes, S.P., 2006. Long-range structural order, Moirepatterns, and iridescence in latexstabilizedfoams. J. Am. Chem. Soc. 128, 7882–7886.

Gomez, C., Benito, M., Katime, I., Teijon, J.M., Blanco, M.D., 2012. In vitro transdermal and biological evaluation of ALA-loaded poly(N-isopropylacrylamide) and poly(N-isopropylacrylamide-co-acrylic acid) microgels for photodynamic therapy. J. Microencapsul. 29, 626.

Gonçalves, C., Pereira, P., Gama, M., 2010. Self Assembled hydrogel nanoparticles for drug delivery applications. Materials 3, 1420–1460.

Griffin, J.M., Robb, I., Bismarck, A., 2007. Preparation and characterization of surfactant-free stimuli-sensitive microgel dispersions. J. Appl. Polym. Sci. 104, 1912–1919.

Guo, J.H., Skinner, G.W., Harcum, W.W., Barnum, P.E., 1998. Pharmaceutical applications of naturally occurring water-soluble polymers. Pharm. Sci. Technol. Today 1, 254–261.

Gupta, A., Singh, S., Kotla, N.G., 2014. Formulation and evaluation of a topical niosomal gel containing a combination of benzoyl peroxide and tretinoin for antiacneactivity. Int. J. Nanomedicine 10, 171–182.

Gupta, P., Vermani, K., Garg, S., 2002. Hydrogels: from controlled release to pH-responsive drug delivery. Drug Discov. Today. 7 (10), 569–579.

Gutowska, A., Bark, J.S., Kwon, I.C., Bae, Y.H., Kim, S.W., 1997. Squeezing hydrogels forcontrolled oral drug delivery. J. Cont. Rel. 48, 141–148.

Ha, D.I., Lee, S.B., Chong, M.S., et al., 2006. Preparation of thermo-responsive and injectable hydrogels based on hyaluronic acid and poly(*N*-isopropylacrylamide) and their drug release behaviors. Macromol. Res. 14, 87–93.

Harish, N.M., Prabhu, P., Charluyu, R.N., Subramanyam, E.V.S., 2012. Formulation and evaluation of in-situ gel containing Clotrimazole for oral candidiasis. J. Pharm. Sci. 4, 1885–1889.

Heiko, K., Erol, Y., Gayle, A.B., Roland, B., 2008. In-vitro and in-vivo drug release from a novel in-situ forming drug delivery system. Pharm. Res. 25, 1347–1354.

Hennink, W.E., van Nostrum, C.F., 2002. Novel crosslinking methods to design hydrogels. Adv Drug Deliv Rev. 54, 13–36.

Hitendra, S.M., Saurabh, K.S., Sanjay, J., Surana, J., 2011. Nasal in-situ gel containing hydroxy propyl β-cyclodextrin inclusion complex of Artemether: development and in-vitro evaluation. J. Incl. Phenom. Macrocycl Chem. 70, 49–58.

Hu, C., Rhodes, D.G., 1999. Proniosomes As Drug Carriers. Int. J. Pharm. 185, 23–35.

Ikeda, K., Okada, T., Sawada, S.C., Akiyoshi, K., Matsuzaki, K., 2006. Inhibition of the formation of amyloid β protein fibrils using biocompatible nanogel as artificial chaperones. FEBS 580, 6587–6595.

Joanna, B., ThorneGeorge, J., VineMartin, J., Snowden, 2011. Microgel applications and commercial considerations. Colloid Polym. Sci. 289, 625–646.

Jones, M.R., Philip, M.S., Massersmith, B., 2002. In-situ forming biomaterials. Oral Maxillofac. Surg. Clin. N. Am. 14, 29–38.

Jung, K.O., Ray, D., Daniel, J., Siegwart, K.M., 2008. The development of microgels/nanogels for drug delivery applications. Prog. Polym. Sci. 33, 448–477.

Kannaiyan, S., Muthuprasanna, P., Vardhareddy, V., 2009. Effect of carbopol gel in stable liposomes and their enhanced antipyretic effect. Asian J. Pharma., 257–260.

Kwon, I.C., Bae, Y.H., Okano, T., Kim, S.W., 1991. Drug release from electric current sensitive polymers. J. Cont. Rel. 17, 149–156.

Labhasetwar, V., Leslie-Pelecky, D.L., 2007. Nanogels: Chemistry to Drug Delivery. Wiley–Blackwell, Hoboken, NJ, USA, pp. 131–172.

Lee, J.I., Kim, J.S., Yoo, H.S., 2009. DNA nanogels composed of chitosan and pluronic with thermo-sensitive and photo-cross-linking properties. Int. J. Pharm. 373, 93–99.

Li, N., Wang, J., Yang, X., Li, L., 2011. Novel nanogels as drug delivery systems for poorly soluble anticancer drugs. Colloids Surf. B. 83, 237–244.

Li, J., Zhang, Z., Liu, C., 2014. Gelatin-based microgels. US 20,140,170,224 A1.

Lu, S., Liu, M., Ni, B., 2011. Degradable, injectable poly(*N*-isopropylacrylamide)-based hydrogels with low gelation concentrations for protein delivery application. Chem. Eng. J. 173, 241–250.

Malakar, J., Sen, S.O., Nayak, A.K., Sen, K.K., 2012. Formulation, optimization and evaluation of transferosomal gel for transdermal insulin delivery. Saudi Pharm. J. 20, 355–363.

Maurya, S.D., Prajapati, S.K., Gupta, A.K., Saxena, G.K., Dhakar, R.C., 2010. Formulation Development and evaluation of ethosome of stavudine. Ind. J. Pharm. Edu. Res. 44, 102–108.

Miao, C., Chen, X., Pelton, R., 2007. Adhesion of poly(vinyl amine) microgels to wet cellulose. Ind. Eng. Chem. Res. 46, 6486–6493.

Mikkelson, T.J., 1984. Ophthalmic drug delivery. Pharm. Tech. 8, 90–98.

Mitkari, B.V., Korde, S.A., Mahadik, K.R., Kokare, C.R., 2010. Formulation and evaluation of topical liposomal gel for fluconazole. Indian J. Pharma. Educ. 44 (4), 324–333.

Miyazaki, S., Aoyama, H., Kawasaki, N., Kubo, W., Attwood, D., 1999. In-situ gelling gellan formulations as vehicles for oral drug delivery. J Cont. Rel. 60, 287–295.

Miyazaki, S., Kubo, W., Attwood, D., 2000. Oral sustained delivery of Theophylline using in-situ gelation of sodium alginate. J. Cont. Rel. 67, 275–280.

Miyazaki, S., Suzuki, S., Kawasaki, N., Endo, K., Takahashi, A., Attwood, D., 2001. In-situ gelling xyloglucan formulations for sustained release ocular delivery of pilocarpine hydrochloride. Int. J. Pharm. 229, 29–36.

Miyazaki, S., Kawasaki, N., Kubo, W., Endo, K., Attwood, D., 2011. Comparison of in-situ gelling formulations for the oral delivery of Cimetidine. Int. J. Pharm. 220, 161–168.

Moink, Mahajan M., Bhupendrag, P., Vishnum, P., Jayvadan, K.P., 2010. Sodium alginate based in-situ gelling system of Famotidine: preparation and in-vivo characterizations. E-j of Sci. Tech. 5, 1.

Morinloto, N., Qiu, X.P., Winnik, F.M., Akiyoshi, K., 2008. Dual stimuli-responsive nanogels by self-assembly of polysaccharides lightly grafted with thiol-terminated poly(Nisopropylacrylamide) chains. Macromolecules 41, 5985–5987.

Mortazavi, S.A., Aboofazeli, R., 2003. An investigation into the effect of various penetration enhancers on percutaneous absorption of Piroxicam. Iranian J. Pharma. Res. 2, 135–140.

Nanda, A., Nanda, S., Dhall, M., Rao, R., 2005. Transferosomes a novel ultradeformable vesicular carrier for T.D.D.S. Pharma Info Net. 5, 395.

Neefe, C.W., 1984. Method of making hydrogel cosmetic contact lenses. US 4,472,327.

Oh, N.M., Oh, K.T., Baik, H.J., Lee, B.R., Lee, A.H., Youn, Y.S., Lee, E.S., 2010. A self-organized 3-diethylaminopropyl-bearing glycol chitosan nanogel for tumor acidic pH targeting: In vitro evaluation. Coll. Surf. B. Biointerfaces 78, 120–126.

Osmond, D.C.,Mingshu, Li., Uckun, F.M., Yiv, S., 2000. Gel-microemulsion formulations. WO 2000056366 A1.

Oupicky, D., Ogris, M., Howard, K.A., Dash, P.R., Ulbrich, K., Seymour, L.W., 2002. Importance of lateral and steric stabilization of polyelectrolyte gene delivery vectors for extended systemic circulation. Mol. Ther. 5, 463–472.

Padma, P.J., Karthika, K., Rekha, N.R., Khalid, E., 2010. Formulation and evaluation of *in-situ* ophthalmic gels of Diclofenac sodium. J. Chem. Pharm. Res. 2, 528–535.

Patel, R., Singh, S.K., Singh, S., Sheth, N.R., Gendle, R., 2009. Development and characterization of curcumin loaded transferosome for transdermal delivery. J. Pharm. Sci. Res. 1, 71–80.

Pawar, S.N., Edgar, K.J., 2012. Alginate derivatization: a review of chemistry, properties and applications. Biomaterials 33, 3279–3305.

Peppas, N.A., Huang, Y., 2004. Nanoscale technology of mucoadhesive interactions. Adv. Drug. Del. Rev. 56, 1675–1687.

Podual, K., Doyle, F.J., Peppas, N.A., 2000. Dynamic behaviour of glucose oxidase-containing microparticles of poly(ethylene)-grafted cationic hydrogels in an environment of changing pH. Biomaterials 21, 1439–1450.

Pollack, G.H., 2001. Cells, Gels, and the Engines of Life. Ebner and Sons, Seattle, WA, pp. 305.

Port, M.J.A., 1999. Contact lens surface properties and interactions. Optometry Today 30, 27–36.

Prajapati, S.T., Patel, C.G., Patel, C.N., 2011. Transfersomes: A vesicular carrier system for transdermal drug delivery. Asian J. Biochem. Pharm. Res. 1, 507–524.

Prathiba, P., Deepak, C., Gaurav, J., Shoerey, R.V., 2011. Formulation and evaluation of oral in-situ gel of Diltiazem HCl. Int. J. Novel Drug Deliv. 2, 1.

Punit, S., Pinaki, D., Apurva, P., Mandip, S., 2012. Skin permeating nanogel for the cutaneous co-delivery of two anti-inflammatory drugs. Biomaterials 33 (5), 1607–1617.

Qi, H., Chen, W., Huang, C., Li, L., Chen, C., Li, W., 2007. Development of a poloxameranalogs/carbopol-based in-situ gelling and mucoadhesive ophthalmic delivery system for puerarin. Int. J. Pharm. 337, 178–187.

Qiu, Y., Park, K., 2012. Environment-sensitive hydrogels for drug delivery. Adv. Drug Del. Rev 64, 49–60.

Rigogliuso, S., Sabatino, M., Adamo, G., Grimaldi, N., Dispenza, C., Ghersi, G., 2012. Polymeric nanogels: Nanocarriers for drug delivery application. Chem. Eng. Trans. 27, 247–252.

Rijcken, C.J., Soga, O., Hennink, W.E., Van Nostrum, C.F., 2007. Triggered destabilisation of polymeric micelles and vesicles by changing polymers polarity: an attractive tool for drug delivery. J. Control. Release 120, 131–148.

Rolland, J.P., Maynor, B.W., Euliss, L.E., Exner, A.E., Denison, G.M., De Simone, J.M., 2005. Direct fabrication and harvesting of monodisperse, shape-specific nanobiomaterials. J. Am. Chem. Soc. 127, 10096–10100.

Rother, M., Lavins, B.J., Kneer, W., Lehnhardt, K., Seidel, E.J., Mazgareanu, S., 2007. Efficacy and safety of epicutaneous ketoprofen in Transfersome (IDEA-033) versus oral celecoxib and placebo in osteoarthritis of the knee: multicentre randomised con-trolled trial. Ann. Rheum. Dis. 66, 1178–1183.

Rutrer, N., 1987. Drug absorption through the skin: a mixed blessing. Arch. Dis. Child. 62, 220–221.

Sahiner, N., Ozay, O., Aktas, N., 2011. 4-Vinylpyridine-based smart nanoparticles with N-isopropylacrylamide, 2-hydroxyethyl methacrylate, acrylic acid, and methacrylic acid for potential biomedical applications. Curr. Nanosci. 7 (3), 453–462.

Sasaki, Y., Akiyoshi, K., 2010. Nanogel engineering for new nanobiomaterials: from chaperoning engineering to biomedical applications. Chem. Rec. 10 (6), 366–376.

Schmitt, F., Lagopoulos, L., Käuper, P., Rossi, N., Busso, N., Barge, J., Wagnières, G., Laue, C., Wandrey, C., Juillerat-Jeanneret, L., 2010. Chitosan-based nanogels for selective delivery of photosensitizers to macrophages and improved retention in and therapy of articular joints. J. Cont. Rel. 144, 242–250.

Shashank, N.N., Bharani, S.S., Thakur, R.S., 2012. Formulation and evaluation of ph triggered in situ ophthalmic gel of moxifloxacin hydrochloride. Int. J. Pharm. Pharm. Sci. 4 (2), 452–459.

Shirsand, S.B., Para, M.S., Nagendrakumar, D., Kanani, K.M., Keerthy, D., 2012. Formulation and evaluation of Ketoconazole niosomal gel drug delivery system. Int. J. Pharm. Investig. 2, 201–207.

Shirsand, S.B., Kumar, G.R., Keshavshett, G.G., Bushetti, S.S., Swamy, P.V., 2015. Formulation and evaluation of clotrimazole niosomal gel for topical application. RGUHS. J. Pharm. Sci. 5, 32–38.

Shyam, D.B., Manish, A.S., Dhananjay, R.N., Ritesh, R.K., Tupkar, S.V., Barhate, S.D., 2007. Formulation and evaluation of Sumatriptan succinate nasal in-situ gel using fulvic acid as novel permeation enhancer. Int. J. Pharm. Res. Dev. 2 (8), 7.

Shyamoshree, B., Amalkumar, B., 2011. Development and characterization of mucoadhesive in-situ nasal gel of Midazolam prepared with Ficus Carica mucilage. AAPS Pharm. Sci. Tech. 11 (3), 1223–1231.

Somasundaran, P., Chakraborty, S., 2006. Polymeric: nanoparticles and nanogels for extraction and release of compounds. WO/2006/052285.

Sonvico, F., Dubernet, C., Colombo, P., Couvreur, P., 2005. Metallic colloid nanotechnology, applications in diagnosis and therapeutics. Curr. Pharm. Des. 11, 2095–2105.

Sultana, Y., Aqil, M., Ali, A., Zafar, S., 2006. Evaluation of carbopolmethyl cellulose based sustained release ocular delivery system for pefloxacin mesylate using rabbit eye model. Pharm. Dev. Technol. 11 (3), 313–319.

Suthar, A.M., Modi, J.D., Patel, M.P., Baria, A.H., 2009. Microemulsion-based gel formulation and evaluation of tretinoin for topical delivery. Int. J. Pharm. Res. 1, 28–34.

Talasahaz, A.H.H., Ali, A., 2008. In-situ gel forming systems of poloxamer 407 and hydroxypropyl cellulose or hydroxypropyl methyl cellulose mixtures for controlled delivery of vancomycin. J. Appl. Polym. Sci. 109, 2369–2374.

Tanagi, N., Rahul, T., Nitin, J., Pradip, D., Vivek, C., Nitin, H., 2008. Formulation and evaluation of pH induced nasal in-situ gel of Salbutamol sulphate. Int. J. Pharma. Sci. Nanotechnol. 1 (2), 177–182.

Tiwari, S., Goyal, A.K., Mishra, N., Vaidya, B., Mehta, A., Dube, D., Vyas, S.P., 2009b. Liposome in-situ gelling system: novel carrier based vaccine adjuvant for intranasal delivery of recombinant protein vaccine. Procedia Vaccinol. 1, 148–163.

Todd, H., Robert, P., 2006. Glucose responsive microgels. WO 2006102762 A1.

Toita, S., Sawada, S., Akiyoshi, K., 2011. Polysaccharide nanogel gene delivery system with endosome-escaping function: co-delivery of plasmid DNA and phospholipase A2. J. Con. Rel. 155, 54–59.

Ujwala, A.S., Shivkumar, S.K., 2014. Serratiopeptidase niosomal gel with potential in topical delivery. J. Pharma. 2014, 1–9.

Vinay, W., Mohan, V.M., Manjunath, S.Y., 2011. Formulation and evaluation of stomach specific in-situ gel of Metoclopramide using natural bio-degradable polymers. Int. J. Pharm. Biomed. Sci. 2, 2.

Vinogradov, S.V., Bronich, T.K., Kabanov, A.V., 2002. Nanosized cationic hydrogels for drug delivery: preparation, properties and interactions with cells. Adv. Drug Del. Rev. 54, 135–147.

Vinogradov, S.V., Batrakova, E.V., Kabanov, A.V., 2004. Nanogels foroligo nucleotide delivery to the brain. Bioconjug Chem. 15, 50–60.

Wataru, K., Yasuhiro, K., Miyazaki, S., Attwood, D., 2004. In-situ gelling pectin formulations for oral sustained delivery of paracetamol. Drug Develop. Ind. Pharm. 30, 593–599.

Weiner, N., Wallach, D.F.H., Egbaria, K., Chandrashekharan, R., 1998. Stimulation of hair follicles. US patent 5,834,014.

Wright, D.C., 1997. Lipid vesicles having a bilayer containing a surfactant with anti-viral and spermicidal delivery of peptide drugs. US 5,700,679.

Yin, Q.Q., Wu, L., Gou, M.L., Qian, Z.Y., Zhang, W.S., Liu, J., 2009. Long-lasting infiltration anaesthesia by lidocaine-loaded biodegradable nanoparticles in hydrogel in rats. Acta Anaesthesiol. Scand. 53, 1207–1213.

Yoshida, H., Lehr, C.M., Kok, W., Junginger, H.E., Verhoef, J.C., Bouwistra, J.A., 1992. Niosomes: a controlled and novel drug delivery system. J. Cont. Rel. 21, 145–153.

Yuan, Y., Cui, Y., Zhang, L., Zhu, H.P., Guo, Y.S., Zhong, B., et al., 2012. Thermosensitive and mucoadhesive in-situ gel based on poloxamer as new carrier for rectal administration of Nimesulide. Int. J. Pharm. 430, 114–119.

Zhang, J., Pelton, R., 1999. Poly(*N*-isopropylacrylamide) microgels at the air–water interface. Langmuir, 18032–18036.

Zhang, X.L., Zhao, R., Qian, W., 1995. Preparation of an emulgel for treatment of aphthous ulcer on the basis of carbomers. Chin. Pharm. J. 30, 417–418.

Zhang, J., Xu, S., Kumacheva, E., 2004. Polymer microgels: reactors for semiconductor, metal, and magnetic nanoparticles. J. Am. Chem. Soc. 126, 7908–7914.

Zhibing, H., Tong, C., Chenglin, C., Manuel, M., 2010. Monodisperse thermo-responsive microgels of poly(ethylene glycol) analogue-based biopolymers, their manufacture, and their applications. US 20,100,076,105 A1.

FURTHER READING

Badgujar, S.D., Sontakke, M.A., Narute, D.R., Karmarkar, R.R., Tupkari, S.V., Barhate, S.D., 2007. Formulation and evaluation of Sumatriptan succinate nasal in-situ gel using fulvic acid as novel permeation enhancer. Int. J. Pharm. Res. Dev. 2, 7.

Bae, Y.H., Kim, S.W., Valuev, L.I., 1993. Pulsatile drug delivery device using stimuli sensitive hydrogel. US patent 5,226,902.

Basu, S., Bandyopadhyay, A.K., 2011. Development and characterization of mucoadhesive in-situ nasal gel of Midazolam prepared with Ficus Carica mucilage. AAPS. Pharm. Sci. Tech. 11, 1223–1231.

Blanchard, C.R., Timmons, S.F., Smith, R.A., 2002. Keratin-based hydrogel for biomedical applications and method of production. US 6,379,690 B2.

Caló, E., Khutoryanskiy, V.V., 2015. Biomedical applications of hydrogels: a review of patents and commercial products. Eur. Pol. J. 65, 252–267.

Cartmell, J.V., Sturtevant, W.R., 1992. Transparent hydrogel wound dressing. US 5,106,629.

Cavallari, C., Pietra, B., Di, A.M., Rodriguez. L., 2009. Pharmaceutical preparations for transdermal use. EP 2019666 A2.

Chowdary, K.P., Rao, Y.S., 2004. Mucoadhesive microspheres for controlled drug delivery. Biol. Pharm. Bull. 27, 1717–1724.

Gupta, H., Singh, R.M., Singh, G.N., Kaushik, D., Sharma, A., 2008. pH-induced in-situ gel for Prilocarpine anaesthesia. Indian J. Pharm. Sci. 70, 776–778.

Hazot, P., Delair, T., Pichot, C., Chapel, J.P., Elaissari, A., 2003. Poly(*N*-ethyl methacrylamide)thermally sensitive microgel latexes: effect of the nature of the cross linker on the polymerization kinetics and physicochemical properties. C. R. Chim. 6, 1417–1424.

Holzer, D., 2011. In-situ lecithin microemulsion gel formulation. WO 2011157428 A2.

Hu, Z., Cai, T., Chi, C., Marquez, M., 2010. Monodisperse thermo-responsive microgels of poly(ethylene glycol) analogue-based biopolymers, their manufacture, and their applications. US 20,100,076,105 A1.

Jabbari, E., Yaszemski, M.J., Currier, B.L., 2006. Hydrogel porogens for fabricating biodegradable scaffolds. EP 1664168 B1.

Jadon, P.S., Gajbhiye, V., Rajesh, S.J., Kavita, R., Narayanan, G., 2009. A controlled and novel drug delivery system. AAPS. Pharm. Sci. Tech. 10, 1187–1192.

Kabanov, A.V., Vinogradov, S.V., Alakhov, Y.V., 2001. Nanogel networks and biological agent compositions. US 6,333,051.

Kabanov, A.V., Vinogradov, S.V., 2004. Nanogel networks including polyion polymer fragments and biological agent compositions. US 6,696,089.

Kothari, J.S., Pancholi, J.S., Patel, J.D., Sheik, S., 2012. Topical pharmaceutical compositions comprising thiocolchicoside. WO 2012053012 A2.

Lawrence, D.B., Cai, T., Hu, Z., Marquez, M., Dinsmore, A.D., 2007. Temperature-responsive semipermeable capsules composed of colloidal microgelspheres. Langmuir. 23, 395–398.

Mahajan, H.S., Shah, K.S., Surana, S.J., 2011. Nasal in-situ gel containing hydroxy propyl β-cyclodextrin inclusion complex of Artemether: Development and *in-vitro* evaluation. Incl. Phenom. Macrocycl. Chem. 70, 49–58.

Manirujjaman, F.S., Imran-Ul-Haque, M., Arafat, M., Sharmin, S., 2013. An overview of nanogel drug delivery system. J App. Pharm. Sci. 3, S95–S105.

Mather, P., Wu, J., Ren, D., Hou, S., 2013. Antimicrobial nanostructured hydrogel web containing silver. US Patent 8,431,151 B2.

Michael, M., 1998. Multiphase liposomal drug delivery system. US 4,761,288 A.

Mircea, C. Popescu, Alan L., Weiner, Sharon, S. Carpenter-Green, 1987. Liposome-gel compositions. US 4,708,861 A.

Nandgude, N., Thube, R., Jaiswal, N., Deshmukh, P., Chatap, V., Hire, N., 2008. Formulation and evaluation of pH induced nasal in-situ gel of Salbutamol sulphate. Int. J. Pharm. Sci. Nanotech. 1, 177–183.

Nayak, S.N., Sogali, B.S., Thakur, R.S., 2012. Formulation and evaluation of pH triggered in-situ ophthalmic gel of moxifloxacin hydrochloride. Int. J. Pharm. Sci. 4, 452–459.

Ngai, T., Auweter, H., Behrens, S.H., 2006. Environmental responsiveness ofmicrogel particles and particle-stabilizedemulsions. Macromolecules 39, 8171–8177.

Norma, A., Kristina, D., Elizabeth, H., Ryan, T., Vanauker, M., 2007. Niosome-hydrogel drug delivery. WO 2007123993 A2.

Norman, W., Donald, F.H., Wallach, K.E., Chandrasekharan, R., 1988. Stimulation of hair follicles. US 5,834,014 A.

Oh, J.K., Ray, D., Daniel, J.S., Krzysztof, M., 2008. The development of microgels/nanogels for drug delivery applications. Prog. Polym. Sci. 33, 448–477.

Osmond, D.C., Mingshu, L., Fatih, U.M., Seang, Y., 2000. Gel-microemulsion formulations. WO 2000056366 A1.

Patel, D.M., Patel, D.K., Patel, C.N., 2011. Formulation and evaluation of floating oral in-situ gelling. Int. Sch. Res. Net. 2011, 1–8.

Petchul, J., Gaudreault, R., 1992. Composition for microemulsion gel having bleaching and antiseptic properties. US 5,336,432 A.

Preetha, J.P., Karthika, K., Rekha, N.R., Khalid Elshafie, P., 2010. Formulation and evaluation of in-situ ophthalmic gels of Diclofenac sodium. J. Chem. Pharm. Res. 2, 528–535.

Rathi, R., Zentner, C., Gaylen, M., 1999. MacroMed, Inc. Biodegradable low molecular weight triblock poly (lactide-*co*-glycolide) polyethylene glycol copolymers having reverse thermal gelation properties. US 6,004,573.

Rathi, R., Zentner, C., Gaylen, M., Jeong, B., 2000. MacroMed, Inc. Biodegradable low molecular weight triblock poly (lactide-co-glycolide) polyethylene glycol copolymers having reverse thermal gelation properties. US 6,117,949.

Roy, S.B., Kothari, J.S., Sheikh, S., Patel, J.D., Pancholi, J.S., 2013. Topical pharmaceutical compositions comprising etodolac. US 20,130,295,145 A1.

Sawhney, A.S., Pathak, C.P., Hubbell, J.A., Hill, J.L., Desai, N.P., 1995. Photopolymerizable biodegradablehydrogels as tissue contacting materials and controlled release carriers. US 5,410,016.

Sawhney, A.S., Jarrett, P., Bassett, M., Blizzard, C., 2013. Drug delivery through hydrogel plugs. US 8,409,606 B2.

Schatzlein, A., Cevc, G., 1995. In: Cevc G., Paltauf E. (Eds.), Phospholipids: Characterization, Metabolism, and Novel Biological Applications. AOCS press, Champaign, p. 191.

Shah, K.R., Kydonieus, A., Jamshidi, K., Decker, S.C., Chang, T., 1996. Thermoplastic hydrogel impregnated composite material. US 5,527,271.

Shinde, U.A., Kanojiya, S.S., 2014. Serratiopeptidase niosomal gel with potential in topical delivery. J. Pharm. 2014, 1–9.

Song, S.S., Kim, H.H., Yi, Y.W., 1996. Biodegradable hydrogel copolymer as drug delivery matrix. US 5,514,380.
Suzuki, D., Tsuji, S., Kawaguchi, H., 2007. Janus microgels prepared bysurfactant-free pickering emulsion-basedmodification and their self-assembly. J. Am. Chem. Soc. 129, 8088–8089.
Thorne, J.B., Vine, G.J., Snowden, M.J., 2011. Microgel applications and commercial considerations. Colloid Polym. Sci. 289, 625–646.
Tiwari, S., Goyal, A.K., Mishra, N., Khatri, K., Vaidya, B., Mehta, A., Wu, Y., Vyas, S.P., 2009a. Development and characterization of novel carrier gel core liposomes based trans-mission blocking malaria vaccine. J. Control. Release 140, 157–165.

CHAPTER

13

NANOSUSPENSION DRUG DELIVERY SYSTEM: PREPARATION, CHARACTERIZATION, POSTPRODUCTION PROCESSING, DOSAGE FORM, AND APPLICATION

Jiuhong Zhang, Zhiqiang Xie[†], Nan Zhang*, Jian Zhong***

**Shanghai Ocean University, Shanghai, People's Republic of China*

***Kent State University, Kent, OH, United States*

[†]Louisiana State University, Baton Rouge, LA, United States

CHAPTER OUTLINE

1 INTRODUCTION

Currently more than 40% of new chemical entities are reported to be lipophilic compounds and one-third of drugs recognized by United States Pharmacopeia are poorly soluble (Chen et al., 2011). During the process of drug exploration and investigation, more and more new drug candidates have been found to be insoluble or poorly soluble in aqueous medium or organic solvent (Grau et al., 2000). Poor solubility may trigger issues, such as poor bioavailability due to uncontrollable precipitation and an erratic absorption profile (Merisko-Liversidge and Liversidge, 2008; Rabinow, 2004). Bioavailability refers to the percentage of drug that has been incorporated into circulation of human body (Mercadante and Vitrano, 2010). Therefore, developing new formulating strategies for drug molecules with poor solubility to obtain an adequate bioavailability has become a challenge for the scientific and industrial professionals. Several methods have been developed, such as complexation by surfactants or cyclodextrines (Hörter and Dressman, 2001; Loftsson and Brewster, 1996), solid dispersions (Leuner and Dressman, 2000), solubilization using co-solvents (Jeon et al., 2000), salt formation (Serajuddin, 2007), screening of soluble drug analogs or prodrugs (Müller, 2009), and application of permeation enhancers, oily solutions, and surfactant dispersions (Ginoya and Atara, 2013). Even though these methods have achieved some reasonable success, their limitation, such as a large amount of additives that may induce stability and toxicity issues. Therefore, a unique strategy is urgently needed that can resolve the formulation-related challenges so as to improve their clinical efficacy and maximally optimize the pharmaco-economics (Peters et al., 2000).

The nanosuspension drug delivery system (DDS) was firstly reported in 1994 (Müller et al., 1999) and a lot of research efforts were devoted to it so as to generate a general formulation for the poorly soluble drugs. Nanosuspension is a colloidal dispersion of submicron drug particles. A pharmaceutical nanosuspension is usually defined as very finely dispersed and biphasic colloid containing solid drug particles of a size less than 1 μm. The suspension contains no matrix material and is stabilized by surfactants and polymers. It is prepared via suitable routes as DDS for the application of oral, topical, parenteral, ocular, and pulmonary administration (Krishna and Prabhakar, 2011). Nanosuspensions with the drug particle reduced to nano-size exhibit excellent advantages (Kocbek et al., 2006b; Liu et al., 2012a; Müller et al., 2011; Patravale et al., 2004; Sastry et al., 2000). Reducing the particle size can greatly increase the solubility, thus improving the bioavailability. Additionally, there are other distinguished advantages, such as easy formulation, ease of scale-up, narrow size distribution of drug particle, controllable drug quantity, and little batch-to-batch variation. Moreover, this strategy can be generally applied to drugs with poor solubility in both aqueous and nonaqueous media. The nanosuspension DDS also makes it possible for the drug to be applied as a liquid dosage form or transformed into solid dosage form, such as tablet or capsule (Junyaprasert and Morakul, 2015). Therefore, the drug can be administrated via oral, pulmonary, ocular, dermal, and intravenous routes (Lakshmi and Kumar, 2010; Liu et al., 2012a; Patravale et al., 2004; Pawar et al., 2014; Shegokar and Müller, 2010). A decent number of nanosuspension DDS products have been marketed (Chen et al., 2015). This chapter will give an overview on the various reported preparation methods and characterization of nanosuspension DDS. Besides, various solidification techniques of nanosuspension and the application of nanosuspension DDS in different administration routes will be discussed.

2 PREPARATION OF LIQUID NANOSUSPENSION DDS

Liquid nanosuspension DDS is a liquid colloidal dispersion of crystalline or amorphous drug nanoparticles with average size below 1000 nm, stabilizers and liquid dissolution medium (Rabinow, 2004). Stabilizers are polymer surfactants utilized to maintain the stability of a nanosuspension. The liquid dissolution medium is water, aqueous solution, nonaqueous solution, or organic solvent. A variety of factors, such as the crystalline structure and particle size of the drug nanoparticle can affect its saturation solubility. Other factors, such as dissolution medium and temperature also play a role (Junyaprasert and Morakul, 2015; Kesisoglou and Mitra, 2012). According to the Kelvin and the Ostwald-Freundlich equations (Junyaprasert and Morakul, 2015), the saturation solubility increases with the decrease of particle size when the size of the drug particles falls below the size of 1 μm. Drug particle size reduction also results in an increase in surface area, thus accelerating the rate of dissolution and improving the bioavailability. This can be explained by the Nernst-Brunner and Levich modification of the Noyes–Whitney equation (Sutradhar et al., 2013). Many fabrication techniques have been developed to prepare the liquid nanosuspension DDS by research laboratories and pharmaceutical professionals. Those methods are generally classified into three categories as (1) top-down technology, (2) bottom-up technology, and (3) combination technology (Müller et al., 2011; Verma et al., 2009; Zhang et al., 2014). Other preparation techniques, such as supercritical fluid technology, emulsification-solvent evaporation method, and melt emulsification method exhibit some excellent advantages and will be discussed as well in the following section.

2.1 TOP-DOWN TECHNOLOGY

In top-down technology, large drug crystals decrease into micrometer range and then go down to nanodimension in a stabilizer solution. The principle methods utilized in the top-down technology include the high-energy process called high-pressure homogenization and the low-energy process called media milling.

2.1.1 High-pressure homogenization

High-pressure homogenization is commonly used for preparation of nanosuspensions of poorly soluble drugs (Yadollahi et al., 2015). Fig. 13.1 shows the illustration of a typical high-pressure homogenization process. There are two types of high-pressure homogenization that are categorized as dissocubes

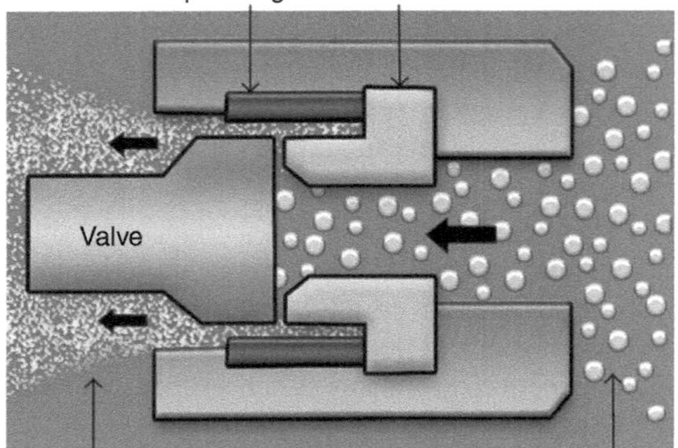

FIGURE 13.1 Illustration of the High-Pressure Homogenization Process

Reprinted from Sutradhar, K.B., Khatun, S., Luna, I.P. 2013. Increasing possibilities of nanosuspension. J. Nanotechnol. 2013, 12.

and nanopure. Homogenization process of dissocubes occurs in aqueous media, whereas homogenization process of nanopure occurs in water mixtures or nonaqueous media. In the dissocubes process, a suspension is forced through a small gap when a pressure as high as 1500 bar is applied. The resulting decrease in the static pressure and increase in the dynamic pressure can cause the water to boil at room temperature. Water boiling at room temperature will generate a lot of gas bubbles. The gas bubbles will lead to cavitation phenomenon in which the gas bubbles implode after the suspension departs the gap and the pressure goes back to atmospheric level. The implosion combined with collisions and high shear causes the drug particles to fragment into nano-size (Gao et al., 2008). Factors including drug particle hardness, number of homogenization cycles, homogenization pressure applied, and temperature may influence the physical characteristics of the resulting nanosuspensions, such as the particle size distribution.

Dissocubes process exhibits several advantages. No erosion of processed materials was incurred. Nanosuspensions with extremely low and high concentration can be prepared by handling 1 mg/mL to 400mg/mL drug quantity (Krause and Müller, 2001). Besides, it allows aseptic production of nanosuspensions for parenteral administration. When a high pressure of 1500 bar was applied, even less than 1 ppm metal contamination can be detected under 20 homogenization cycles (Gao et al., 2008). However, the major limitation of this method is that many cycles of homogenization and pretreatment are needed to obtain microparticles before the homogenization process (Kumar et al., 2011). Another drawback is that the high cost of instruments that will increase the cost of dosage formation.

Another high-pressure homogenization process is the nanopure, which is conducted at low temperature and often mentioned as the deep-freeze method. For drug nanocrystals dispersed in water mixtures or nonaqueous media, the decrease in pressure is not sufficient for cavitation to occur due to the low vapor pressure of oils or oily fatty acids and their high boiling point. Owing to insufficient decrease in pressure, the drug nanosuspensions in the nonaqueous media were homogenized at 0°C or below the freezing point (Keck and Müller, 2006). Therefore, the nanopure process can be used for thermo-labile compounds. Also, results achieved from the nanopure process are comparable to those from the dissocubes process.

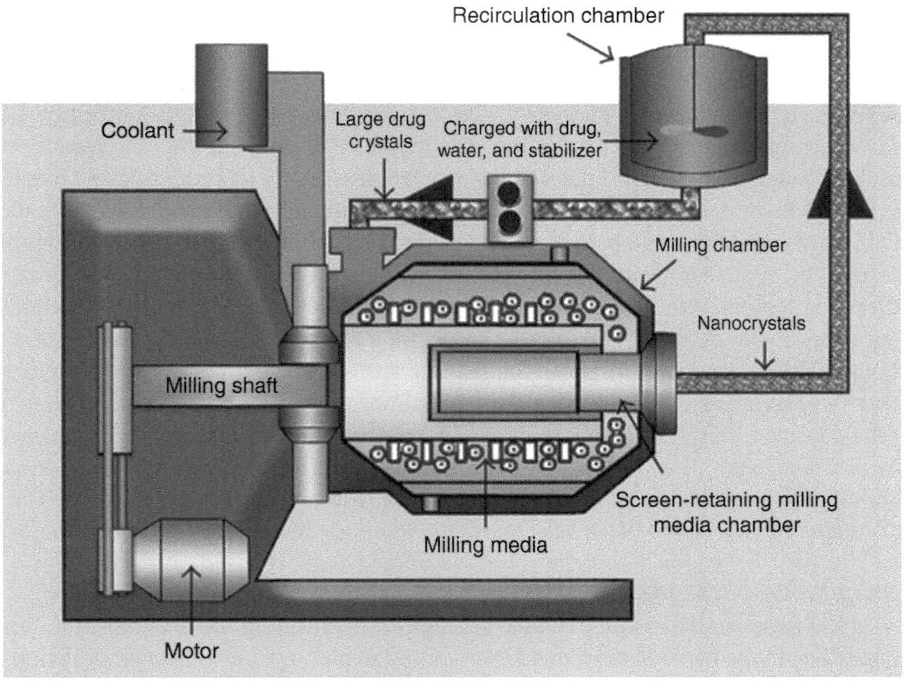

FIGURE 13.2 Illustration of the Media Milling Process

Reprinted from Sutradhar, K.B., Khatun, S., Luna, I.P. 2013. Increasing possibilities of nanosuspension. J. Nanotechnol. 2013, 12.

2.1.2 Media milling

The media milling method was discovered by Liversidge et al. (1992). In this method, nanosuspension is generated by pearl mills or high-shear media mills. The typical media mill is illustrated in Fig. 13.2. During the shearing process, the drug particles are broken down into nanoparticles. And because of its connection to the recirculating chamber, continuous production is maintained. Both batch operation and continuous operation can be performed through media milling, and the particle size can be reduced to smaller than 200 nm in 30–60 min. The milling media or balls are usually made of polystyrene resin with hardness or ceramic-sintered aluminum oxide. The aqueous suspension of the drug and stabilizer is filled in the milling chamber and the milling media or pearls rotate at a very high shear rate, thus forming drug nanoparticles from friction and collisions. Shearing can generate a lot of heat, and this procedure can be undertaken with controlled temperature. Additionally, the main advantages of media milling are subtle batch-to-batch variation and facile scale-up. However, this method may cause the erosion of pearls that might contaminate drug nanoparticle product (Chingunpituk, 2011).

2.2 BOTTOM-UP TECHNOLOGY

In the bottom-up process, the drug is usually dissolved in an organic solvent and an antisolvent is added to form precipitation in the present of a stabilizer. The commonly utilized bottom-up techniques are antisolvent precipitation and precipitation-ultrasonication.

2.2.1 Antisolvent precipitation method

Antisolvent precipitation has been utilized for years to prepare drug particles in micro- and nano-size (Yang et al., 2011). Generally, the drug is first dissolved in a solvent, to which the antisolvent is then added quickly in the presence of a surfactant. The rapid formation of the drug precipitate results in the sudden supersaturation of drug in the mixed solution, thus forming the ultrafine crystalline or amorphous drug nanoparticles. This process involves two phases: nuclei formation and crystal growth. The nanocrystals grown under supersaturation condition usually have a broad particle size distribution (Zabihi et al., 2015). In order to prepare nanosuspension with suitable particle size distribution factors, such as the ratio of organic solvent and antisolvent, the anti-solvent addition rate, the drug precursor concentration, the temperature of the nanoparticle formation, the stabilizer utilized should be taken into account (Phromviyo et al., 2014). In order to obtain a more stable nanosuspension, the stabilizer utilized should exhibit excellent affinity to the particle surface and a high diffusivity so as to rapidly cover the newly generated surface. Besides, there should be enough stabilizer to completely cover the surface of the particles. Temperature has an effect on the formation of nanoparticle, thus changing the size distribution. Generally, the solubility of drugs increases and the level of supersaturation decreases with the increase of temperature, and the available crystallization particles are reduced. The resulting increase of solute molecules will benefit the rapid growth of the nanocrystal (Sipoli et al., 2015).

2.2.2 Precipitation-ultrasonication method

In recent years, ultrasonication as an effective strategy for controlling the crystallization process has been developed due to the fact that ultrasound irradiation can promote the molecular diffusion and mass transfer (Zeng and Weber, 2014). It was found that the crystal size can be reduced by simply increasing the applied ultrasonic power. Nevertheless, when the ultrasonic power input is between 400 and 580 W, no significant change in the particle size was observed. Therefore, 400 W power input is enough to achieve the optimal particle size reduction. In addition, the time that the drug particle is exposed to ultrasonication has an effect on the nanoparticle size. It was found that the particle size is significantly decreased if the ultrasonication time was extended to 15 min. However, no obvious change in the particles size was observed even with a longer sonication time. Therefore, 15 min is the optimal time that can be utilized to generate nanosuspension. Besides, polydispersity index and the nanoparticle quality can be significantly improved with the application of ultrasonication to a milled suspension (Patel et al., 2014a).

2.2.3 Flash nanoprecipitation

Recently, flash nanoprecipitation (FNP) as a bottom-up process has been developed to prepare the organic nanoparticles (Chow et al., 2014). Confined impinging jet mixers (CIJMs) and multi-inlet vortex mixers (MIVMs) are designed to effectively mix all components in the solvent mixture. The rapid precipitation of solutes and a higher supersaturation level in flash nanoprecipitation generally accelerate the nanoparticle formation. Usually CIJM, utilized as a tool to achieve FNP, can efficiently promote the product conversion. However, in this method identical momenta and equivalent volumetric flow rates of the antisolvent streams are required. This requirement has limited the build-up of the supersaturation level inside the mixer.

In order to resolve the limitation of CIJM, MIVM is designed for FNP to achieve rapid micromixing and ease of utilization and scaling up (Thorat and Dalvi, 2012). The self-sustained micromixing of MIVMs is driven by mixing each stream with the momentum. The mixing of streams with unequal volumetric flows is allowed in the four-stream MIVM. And the supersaturation and solvent composition can be tuned by changing the content and flow velocity of each individual stream. Therefore it is still possible to achieve good micromixing of streams, even with one or more at high flow volume and the rest at low flow volume. In addition, MIVM can also be utilized to prepare nanoparticles with multiple active species and, by changing the solvent/antisolvent input ratios, the nanoparticle stability can be controlled (Shen et al., 2011). Hence, the condition of rapid micromixing and high-supersaturation levels can be utilized to prepare nanoparticles with controlled size distribution and high DLR.

2.3 COMBINATION TECHNOLOGY

The combination technology generally undergoes a bottom-up process, such as precipitation, which is followed by top-down process, such as high-pressure homogenization. NANOEDGE technology is a commonly utilized integrated technology and is a registered trademark by Baxter International, Inc. with serial number 76322804. NANOEDGE is designed to enable water-insoluble drugs to be useful medications. A combination of precipitation and homogenization techniques usually give a better particle size distribution and better stability (Ginoya and Atara, 2013). Water-miscible solvents including methanol, ethanol, and isopropanol are used to perform the precipitation. It is necessary to remove the solvents leftover in the precipitate even though trace amount of solvent is allowed in the formulation process. However, in the NANOEDGE technology, an evaporation step is included to yield solvent-free starting material, which is further processed by high-pressure homogenization. The application of NANOEDGE technology can overcome the drawback of both the precipitation technique and the homogenization technology. In the NANOEDGE method, the precipitated suspension is further homogenized, thus causing the particle size to decrease and further crystal growth to be impeded. Particles in the nano range with improved thermodynamics are obtained after the homogenization step because this step exercises a check on crystal growth of nanoparticles.

2.4 OTHER TECHNOLOGIES

2.4.1 Supercritical fluid technology

Supercritical fluid technology is commonly utilized to prepared nanoparticles from drug solutions. The various methods including the supercritical solution process (RESS), the supercritical antisolvent process and precipitation with compressed antisolvent process (PCA) can be utilized. In the RESS method, the drug solution is expanded in supercritical fluid with a nozzle, which results in the loss of solvent power of the supercritical fluid and precipitates the drug as fine particles. It was reported that cyclosporine nanoparticles in the size range of 400-700 nm were prepared by using this method (Young et al., 2000). In the PCA method, the drug solution is atomized into a chamber with compressed carbon dioxide. Fine crystals precipitate out with the solution getting supersaturated after the solvent is removed. In the supercritical antisolvent process, a supercritical fluid in which a drug is poorly soluble is used. The drug is dissolved into a solvent that is miscible with the supercritical fluid. After injecting the drug solution into the supercritical fluid, the supercritical fluid extracts all the solvent and the drug solution turns supersaturated. Then the drug will precipitate as fine crystals. Compared with other techniques, the major drawback of the discussed methods is that most of the solvents utilized are hazardous. Also, large amount of surfactants and stabilizers are used during the preparation process.

2.4.2 Emulsification-solvent evaporation technique

In the emulsification-solvent evaporation technique, the solution of a drug is first prepared. Then, the emulsification was performed in another liquid in which the drug was poorly soluble. After evaporating all the solvent, the drug nanocrystals precipitate out. The high-speed stirrer can be utilized to regulate the crystal growth and particle aggregation by creating high-shear forces (Ginoya and Atara, 2013).

2.4.3 Emulsion diffusion method

The emulsion cannot only be utilized as a drug-delivering vehicle but also can be used as templates to prepare nanosuspensions. For the drugs that are soluble in volatile organic solvent and partially water-miscible organic, the emulsion diffusion method can be used. The drug particles are dissolved in the organic solvent or the mixture of solvents. Then, the organic solution is dispersed in the aqueous phase containing suitable surfactants with stirring to form an emulsion. High-pressure homogenization is commonly applied to further homogenize the emulsion after the emulsion was obtained. After homogenization, the emulsion was diluted with water and homogenized by homogenizer to diffuse the organic solvent, which was followed by conversion of the droplets into solid particles. The particle size of the nanosuspension can be tuned by controlling the size of the emulsion droplet since one particle is formed in each emulsion droplet. In the emulsion diffusion method, methanol, ethanol, ethyl acetate, and chloroform are commonly used as the organic solvents (Paun and Tank, 2012).

The advantage of the emulsion diffusion method is that no specialized equipment is required. Particle size can be easily be tuned by controlling the emulsion droplet size. Also, the formulation is easy to scale up. However, the major drawback for the emulsion-diffusion method is that nanosuspensions of drugs with poor solubility in both aqueous and organic media cannot be prepared via this method. Besides, hazardous solvents are used in the preparation process, which will cause safety concerns. Di-ultrafiltration was required for the purification of the drug nanosuspension, which will elevate the cost of the whole process. A large amount of surfactants and stabilizers are required to be utilized in the fabrication process (Verma and Bindal, 2012). All these disadvantages may limit its widespread application.

2.4.4 Melt emulsification method

In the melt emulsification method, the drug material is firstly dispersed in the aqueous solution with stabilizer. Then an emulsion was obtained after the solution was heated to a temperature above the melting point of the drug and homogenized. During this process, the emulsion was maintained at a temperature above the melting point of the drug. Then the drug emulsion was cooled to room temperature slowly or with an ice bath. Compared with the previously discussed solvent-diffusion method, the advantage of the melt-emulsification method is that organic solvents are not necessary in the preparation process. It was reported that ibuprofen nanosuspension prepared from the melt-emulsification method exhibits better dissolution rate than the nanosuspension formulated with the solvent-diffusion method (Kocbek et al., 2006a).

2.4.5 Lipid emulsion/micro-emulsion template

Another way to prepare nanosuspensions is by dispersing the drug particles in a partially water miscible solvent to form an emulsion and diluting the emulsion (Patravale and Kulkarni, 2004). Microemulsions can also be used as templates to prepare nanosuspensions. Thermodynamically stable and isotropically clear microemulsions can be prepared by mixing two immiscible liquids, such as oil and water in the presence of an interfacial film of surfactant and cosurfactant. The drug can be dispersed in the preformed micro-emulsion or loaded into the internal phase. After appropriate dilution of the micro-emulsion, the drug nanosuspension will be obtained. The griseofulvin nanosuspension is one of the reported examples that utilized the microemulsion method. In the preparation process, the sodium salt of taurodeoxycholate, water, butyl lactate, and lecithin were used (Shah et al., 2007). The advantage for the lipid emulsions as templates for nanosuspension is that the particle size of the nanosuspension is controllable. Also, the preparation is easy to scale up. However, the major drawback is the utilization of organic solvents that may cause toxicity. Additionally, a large amount of surfactants and stabilizers are required to be used.

2.4.6 Nanojet technology

Nanojet technology is an opposite stream technique. In this technology, a chamber is utilized in which the stream of suspension is separated into two or more parts and the particles colloid with each other at high pressure. The high shear force generated from the high-pressure process can help reduce the particle size. Commonly utilized microfluidizers M110L and M110S (microfluidics) are based on this principle. The major disadvantage of this technique is that a large portion of microparticles may be formed with a large mass passing through the microfluidizer.

2.5 FORMULATION CONSIDERATIONS

The formulation of nanosuspension is a complicated process and many factors affecting the formulation should be considered.

2.5.1 Temperature

It is important to maintain the optimal temperature while formulating the nanosuspension. In the emulsion technique, homogenization process occurs at a lower temperature after the drug-containing organic solution is added to the aqueous surfactant solution. Maintaining a low temperature is an important precaution to be taken because of the fact that organic solvents are involved in the formulation and a higher temperature will result in irregular particles formation because of rapid removal of organic solvent. On the other hand, if a lower temperature is maintained during formulation, the solvent diffuses out of the system at a slow rate and the formation of spherical and uniform nanoparticles is favored (Ginoya and Atara, 2013).

2.5.2 Stabilizer and other additives

During the formulation process of nanosuspension, the stabilizer plays a critical role. The stabilizer is able to wet the nanoparticles completely and supply steric or ionic barriers to impede the Ostwald's ripening and agglomeration of nanosuspensions, thus producing a stable formulation. In some formulation processes, a mixture of stabilizers is required and the drug-to-stabilizer ratio in the formulation usually varies from 1:20 to 20:1. Specific nanosuspension requires specific investigation. Povidones, cellulosics, polysorbates, poloxamers, and lecithin are the most widely investigated stabilizers (Nakarani et al., 2010). To better formulate nanosuspensions, other additives, such as cosurfactants, buffers, salts, and polyols are added. However, the administration route of the formulated drug or the properties of the drug particles should be considered when choosing the additives. Similarly, different cosurfactants will be used when different formulation methods were chosen to prepare nanosuspensions from emulsions.

It was observed that cosurfactants have an effect on the phase behavior, the emulsion composition, and the drug loading into the nanosuspension. However, transcutol, glycofurol, ethanol, and isopropanol can still be used as safe cosurfactants when the nanosuspension was formulated via emulsion method (Ginoya and Atara, 2013).

2.5.3 Organic solvents

In the formulation process of nanosuspensions, sometimes organic solvents must be utilized. Therefore, potential toxicity and the ease of their removal from the formulation should be thoroughly considered when picking the organic solvent. Generally the water-miscible solvents, such as ethanol and isopropanol, and partially water-miscible solvents, such as ethyl formate, propylene carbonate, triacetin, butyl lactate, and benzyl alcohol are considered to be pharmaceutically acceptable and less hazardous. Those solvents are preferred over the conventional hazardous halogenated solvents, such as chloroform and dichloromethane in the formulation process (Ginoya and Atara, 2013).

2.5.4 Stirring speed

It is easy to neglect the stirring speed as a key variable in the nanosuspension formulation. High-pressure homogenization (HPH) or high shear homogenization (HSH) of nanosuspensions can result in suspensions with smaller nanoparticles. Increasing the speed of stirring in the HSH process or increasing the number of cycles in the HPH process can easily obtain the nanosized range of drug particles. However, a huge amount of foam in the suspension will be generated with the instruments operating at a high speed. This can lower the efficacy in size reduction and the formation of uniform nanoparticle. Therefore, it is important that the instruments run at a recommended speed and with an optimal cycle number (Gao et al., 2007).

3 CHARACTERIZATION OF NANOSUSPENSION DDS

The characterization of nanosuspension DDS generally include chemical, physical, and biological tests. However, mean particle size and particle size distribution, crystalline state and particle morphology, nanosuspension charge, saturation solubility and dissolution velocity, stability and in vivo biological performance are the essential characterizations for the nanosuspension.

3.1 MEAN PARTICLE SIZE AND PARTICLE-SIZE DISTRIBUTION

The mean particle size and particle-size distribution are of great significance for nanosuspensions since these two key parameters usually determine some critical properties of nanosuspensions, such as the physical stability, dissolution velocity, saturation solubility, as well as their biological performance. Moneghini et al., (2001); Serajuddin, (1999) reported that different particle sizes of the drug display significant variation of saturation solubility and dissolution velocity. Photon correlation spectroscopy (PCS) is a powerful technique that can be applied to quickly and accurately determine the mean particle size of nanosuspensions. Furthermore, the width of the particle-size distribution (polydispersity index (*PI*)) can also be determined by the PCS technique. It should be noted that the *PI* is another important parameter that controls the physical stability of nanosuspensions. To make nanosuspensions of long-term stability, the *PI* should be as low as possible.

Unfortunately, due to the low measuring range of PCS technique, it is still challenging to detect the possible contamination of nanosuspensions with particle size greater than 3 μm. Therefore, laser diffractometry (LD) analysis is necessary because it can quantify the possible drug microparticles of nanosuspensions. Laser diffractometry can be carried out to measure particles with the size ranging from 0.05 to 80 μm. In addition, LD has become critical for characterizing nanosuspensions since even a minor fraction of particles greater than 5–6 μm (the size of the smallest blood capillary) would probably lead to emboli formation or capillary blockage. For the PCS or LD analysis, deionized water can be used to dilute the nanosuspensions. Regarding some nanosuspensions intended for intravenous administration, the Coulter counter technique is also used for particle size analysis. This technique usually yields the absolute number of particles per volume unit, therefore, it has been considered more appropriate for determining the contamination of nanosuspensions (Arias et al., 1997).

3.2 PARTICLE MORPHOLOGY AND CRYSTALLINE STATE

The characterizations of the particle morphology and crystalline state can help us better understand the possible morphological or structural changes of drugs when subjected to nano-sizing. It is important to study the amount of amorphous drug nanoparticles within the nanosuspensions since they are likely to be produced during the preparation of nanosuspensions. The amorphous form is thermodynamically unstable and tends to transform into a crystalline form during storage. Such transformation over storage should be taken into consideration when preparing or formulating nanosuspensions. Therefore, X-ray powder diffraction (XRD) and scanning electron microscopy (SEM) are usually used to determine the crystalline structure and particle morphology of drugs, respectively (Müller et al., 1999). In addition, differential scanning calorimetry (DSC) is another useful technique for characterizing the crystalline structure and determining the amorphous fractions of drugs.

3.3 PARTICLE SURFACE CHARGE

Particle surface charge is important for ensuring the stability of nanosuspensions. The electro-static repulsion provided by charges around particle surfaces can prevent drug nanoparticles from precipitation and likely severe aggregation. Because it is impossible to measure the surface charge directly, the zeta potential of a nanosuspension can be utilized as an indicator of the stability of the nanosuspension and can be controlled by the drug and stabilizer. Zeta potential is the electrical potential at the surface of the electric double layers between the Stern layer and the diffusion layer of oppositely charged ions (Nel et al., 2009). A minimum zeta potential of ±30 mV is usually required to ensure an electrostatically stabilized nanosuspension. However, ±20 mV is sufficient if a steric stabilizer was added to the nanosuspension (Jacobs and Müller, 2002). It is important to carefully maintain the zeta potential while the nanosuspension was in production and storage. A Zetasizer can be used to measure the zeta potential. The drug is dissolved in Milli-Q water to give a suitable concentration in which the electrophoretic cell can be dipped (Sameut Bouhaik et al., 2013). The measurement of zeta potential is usually performed in the original dispersion medium, physiological salt solutions, or buffers with different molarities (Nagelreiter and Valenta, 2013). It can also be performed by diluting the drug in water. Such dilution may slightly alter the surface charge and give a zeta potential different from the true value (Eldridge et al., 2014). The extent of adsorption and binding strength of stabilizer molecules to the particle surface are also measured by the zeta potential. The zeta potential can be measured by laser light scattering (Deshiikan and Papadopoulos, 1998) or by electroacoustic means (O'brien et al., 1995). In these technologies, the measurement of the electrophoretic mobility of the particles in an electric field was performed. The electroacoustic method has an advantage over laser the light diffraction by not being restricted to measurements in diluted dispersions.

3.4 SATURATION SOLUBILITY AND DISSOLUTION VELOCITY

The saturation solubility and dissolution velocity are two important parameters that can be used to analyze changes in the in vivo performances of drugs, such as bioavailability, plasma peak, and blood profile (Sutradhar et al., 2013). It is well known that a nanosuspension is commonly used to improve the saturation solubility of the drug; therefore, it is necessary to investigate the saturation solubility of a nanosuspension. In addition, the dissolution velocity is another important parameter for a nanosuspensions, which shows the advantages over conventional formulations. Therefore it is necessary to assess the saturation solubility and dissolution velocity of the drug nanosuspensions under different conditions, such as in different physiological buffers and at different temperatures.

3.5 STABILITY

During the preparation process of nanosuspension, the particle size is reduced, which leads to increased surface energy. Higher surface energy will destabilize the colloidal suspension. Hence, in order to minimize the particle agglomeration and the possibility of Ostwald ripening, the stabilizer needs to be used (Chaurasia et al., 2012). Polysorbates, povidones, poloxamer, lecithin, polyoleate, and cellulose polymers are commonly used to stabilize the nanosuspension. It was observed that the mixture of surfactants and polymers are useful for long-term stabilization of nanosuspensions (Rabinow, 2004). The stabilizers play the role as an ionic barrier and inhibit the close interaction between particles. The surfactants can change the zeta potential and increase the electrostatic repulsion between the nanoparticles, thus giving better stability. When discussing the stability of the nanosuspension, the precipitation of particles should be considered. According to Stoke's law, the decrease in precipitation velocity can be achieved by decreasing particle size, reducing the density difference of solid phase, and increasing the viscosity of the medium (Liu et al., 2012b). Increasing the uniformity of particle sizes can also increase the stability of nanosuspension.

3.6 IN VIVO BIOLOGICAL PERFORMANCE

The study of in vivo biological performance of the drug is of particular importance without considering the delivery system or route employed. For intravenously injected nanosuspensions, it is very important to study the in vivo biological performance because it usually depends on the organ distribution. The protein absorption pattern can be observed after the intravenous injection of drug nanoparticles and its composition can yield the information of organ distribution (Blunk et al., 1993, 1996). Therefore, to better understand the in vivo behavior, appropriate techniques, such as hydrophobic interaction chromatography and 2D PAGE are commonly used to investigate the surface hydrophobicity (Wallis and Müller, 1993) and measure the qualitative and quantitative composition of protein absorption (Patravale and Kulkarni, 2004), respectively.

4 POSTPRODUCTION PROCESSING

4.1 SOLIDIFICATION OF LIQUID NANOSUSPENSION

Liquid dosage forms are considered to be suitable for children and elderly patients in clinical treatment (Junyaprasert and Morakul, 2015); however, they are not convenient for shipping and storage. Moreover, although stabilizers are commonly added to the nanosuspension to prevent the drug nanoparticles from precipitation, crystalline transformation, and severe aggregation (Wang et al., 2013b), the nanosuspension is still not stable over the long-term storage since it is thermodynamically unstable. However, solid dosage forms usually can address these issues and offer more advantages over liquid dosage forms. For example, solid dosage forms can improve the stability of nanosuspensions over storage. In addition, it is also convenient for patients to take and carry the drugs from a marketing perspective. In fact, almost two-thirds of prescriptions are dispensed as solid dosage forms (Mahida and Gupta, 2013). Hence, the transformation from liquid dosage forms into solid dosage forms has been studied (Van Eerdenbrugh et al., 2008b). In terms of the transformation, the nanosuspension is firstly transformed into a solid and that process is defined as solidification of nanosuspension. Then, the resultant solid products can be formulated as nebulized dosage for pulmonary delivery, sterile powder for injection, and other dosage forms, such as capsules, pellets, and tablets for oral delivery (Wang et al., 2013b).

As we discussed earlier, the nanosuspensions usually have some disadvantages including the stability problems for long-term storage process. However, solid dosage forms can effectively address these issues because of their good stability and patient convenience. Therefore, it is highly desirable to transform the nanosuspensions into solid dosage forms. Currently, many solidification methods of liquid nanosuspensions have been studied, such as spray drying, freeze drying, lyophilization, granulation, and palletization (Abdelwahed et al., 2006). It is well known that the nanoparticulate system has rapid dissolution. Therefore, it requires rapid disintegration of the solid and good redispersion of the nanoparticles. However, the drying process of the liquid nanosuspension usually creates stress on the particles, resulting in severe aggregation. In addition, the drying process can also introduce some thermal stresses that might destabilize the nanoparticles. When taking into consideration the aforementioned issues, it is necessary to add matrix-formers to the liquid nanosuspensions prior to solidification process. For example, Van Eerdenbrugh et al. (2008a) reported that the microcrystalline cellulose can work as a matrix former during freeze-drying of itraconazole nanosuspensions.

4.1.1 Freeze drying

In fact, freeze drying has been an established method for solidification of liquid nanosuspensions. Briefly, the solvent water can be removed by sublimation and adsorption from a frozen liquid nanosuspension under high vacuum conditions (Peters et al., 2014; Varshosaz et al., 2012). It is advantageous that the redispersibility of drug nanoparticles can be retained by freeze drying. Cryoprotectant and freezing rate are two important parameters controlling the process of freeze drying. To reduce or avoid the freeze damage and prevent aggregation of drug nanoparticles, a cryoprotectant is usually added to the as-prepared liquid nanosuspension (Yue et al., 2014). In terms of nanoparticles in the nanosuspension, cryoprotectants can be used to maintain their functional properties after the freeze drying process, as well as to overcome the instability of the freezing-drying process (Ali and Lamprecht, 2014). For example, trehalose, as one of the cryoprotectants, can help disintegrate nanoparticle agglomerates to some degree. Generally speaking, the nanoparticles tend to agglomerate during the freeze-drying process without a cryoprotectant. The freezing rate is the other important parameter influencing the stability of the nanosuspension and it is mainly determined by the temperature gradient. It is found that rapid freezing usually leads to smaller ice crystals with a larger specific surface area, thereby effectively reducing the mechanical stress on the particles (Wang et al., 2013a). In addition, we can observe the irreversible aggregation during drying process rather than freezing. The nanoparticles are more redispersible when subjected to a more homogeneous distribution of the drug nanosuspension and cryoprotectant. (Grohganz et al., 2013).

4.1.2 Spray drying

Since the 1980s, spray drying has been widely used to produce nanoparticles. There was increasing research interest in the production of pharmaceuticals by spray drying in the early 1990s (Wanning et al., 2015). Generally speaking, spray drying is usually advantageous over freeze drying (Nuchuchua et al., 2014). Briefly, an aqueous or organic solution is first atomized into ultrafine droplets and then solidified into solid particles via evaporation under a warm airflow (Kumar et al., 2014b). Unfortunately, spray drying is considered inappropriate for the production of nanoparticles since it is too difficult to collect the ultrafine nanoparticles by the cyclone. It should be noted that the parameters in spray drying are closely correlated. In other words, any change in one of the spray-drying process parameters will lead to the change of the others.

4.1.3 Gel method

The surface coating of nanoparticles can significantly alter their physical properties and are widely applied in both science and technology. These coatings can protect, solubilize, and functionalize the nanoparticles. High-resolution of TEM can be used to estimate the thickness of the surface coating layers and the diameters of particles. Factors, such as the particle size, dispersants, solvent, pH, and temperature can affect the stability of coated nanoparticle colloidal systems. The stability of a colloidal system can be described via Turbiscan backscattering and visual inspection of colloidal precipitation. Besides, wet ball milling can be used to prepare metallic nano-colloidal solutions (Kim et al., 2013). In the drug-delivery field, nanoparticles loaded with anticancer drug can be utilized as the treatment of hyperthermia. Grafting polymers onto the surface of nanoparticles enables the preparation of stable colloidal suspensions, which can be further stabilized by either synthetic polymers including polyvinyl alcohol and polyethylene glycol or natural polymers, such as chitosan and dextran. For example, iron oxide particles can be stable in an aqueous medium for a long time by coating their surfaces with charged surfactants (Serra et al., 2009). Also, hydrogels can be incorporated into the nanosuspension to improve its stability with slightly increasing the particle size.

4.2 SURFACE MODIFICATION

Nanosuspension is a commonly utilized strategy to increase the dissolution rate and saturation solubility for the poorly soluble drugs. However, toxicity and other severe side effects may be incurred due to rapid release of nanosuspensions. Therefore, the surface must be modified to control drug release and prolong the drug's action at the destination. For example, in the treatment of lymphatic mediated diseases (Charman and Stella, 1992), a nanosuspension is utilized to target the monocyte phagocytic system (MPS). The drug can accumulate at the action cite and cause toxicity. Conclusions can be made that surface modification is necessary if drug-release results are compared between the coated and uncoated surfaces. It was reported that layer-by-layer nanogel coating of procaine hydrochloride can be used to control the release of a drug (Tan et al., 2008). It was also reported that by adding mucoadhesive polymers to buparvaquone nanosuspensions, the infective score of *Cryptosporidium parvum* after oral administration of buparvaquone nanosuspensions was significantly decreased. This was due to the adhesion and prolonged action of drug particles at the absorption sites in the gastrointestinal tract (Kayser, 2001). In the targeted DDS the surface engineering by surface coating plays an important role. PEG (polyethylene glycol) is utilized to modify nanoparticle surface so as to decrease the protein adsorption and opsonization of nanoparticles, thus prolonging the circulation time of the drug nanoparticle (Alexis et al., 2008). Longer circulation time will allow the drug nanoparticles to permeate out of vasculature in infective and inflammatory areas and cancer tissues (Wu et al., 1993).

5 DOSAGE FORM OF NANOSUSPENSION DDS

Both liquid dosage form and solid dosage form of poorly soluble drugs can be applied for administration. Up to now, several liquid or solid (powder, tablet, pellet, capsule, and film) dosage forms of nanosuspension DDS have been developed.

5.1 LIQUID DOSAGE FORM

In the liquid nanosuspension, the drug nanoparticles are dispersed in a proper solvent medium. When comparing with other dosage forms, liquid dosage form exhibit advantages, such as less irritation, faster absorption, easier divided dose, and higher drug bioavailability (Chen et al., 2015). The poorly soluble drugs can be prepared into liquid nanosuspension for oral administration. Recent examples included aripiprazole (Xu et al., 2012), febuxostat (Ahuja et al., 2015), paclitaxel, (Patel et al., 2014c) herpetrione (Guo et al., 2013), myricetin (Hong et al., 2014), pranlukast (Baek et al., 2014), albendazole (Li et al., 2014), fenofibrate (Li et al., 2014), itraconazole (Li et al., 2014), probucol (Li et al., 2014), and revaprazan hydrochloride (Li et al., 2014).

5.2 POWDER DOSAGE FORM

The powder dosage is another dosage form of nanosuspension DDS that can be prepared from liquid nanosuspension of potential drug candidates. The dry powder form has a long-term stability and can be easily redispersed in water or another liquid medium prior to use. Besides, the dry drug powders have high dissolution rate and are easy to be redispersed to nanosuspension. The in vivo pharmacokinetic investigation suggests that the redispersed nanosuspensions have good oral bioavailability (Figueroa and Bose, 2013).

Recently, the powder dosage strategy was applied for many poorly soluble drugs, such as scutellarin (Yang et al., 2014) (Fig. 13.3), naproxen (Figueroa and Bose, 2013), Novartis Compound A (Figueroa and Bose, 2013), itraconazole (Mou et al., 2011), and bexarotene (Chen et al., 2014). For example, Yang et al. (2014) transformed a liquid nanosuspension of aglycone scutellarein into dry powder by the freeze-drying method. Compared with coarse drug, the dissolution rate of the dry powder dosage increased. Moreover, the drug powder dosage was easily wetted and redispersed into a liquid medium. The dry powder had a high dissolution rate and was easy to be wetted and redispersed to nanosuspension. The in vivo pharmacokinetic study of the redispersed dosage in rats showed that it is an active and rapid orally absorbed precursor nanosuspension DDS of its BCS class IV glycoside scutellarin.

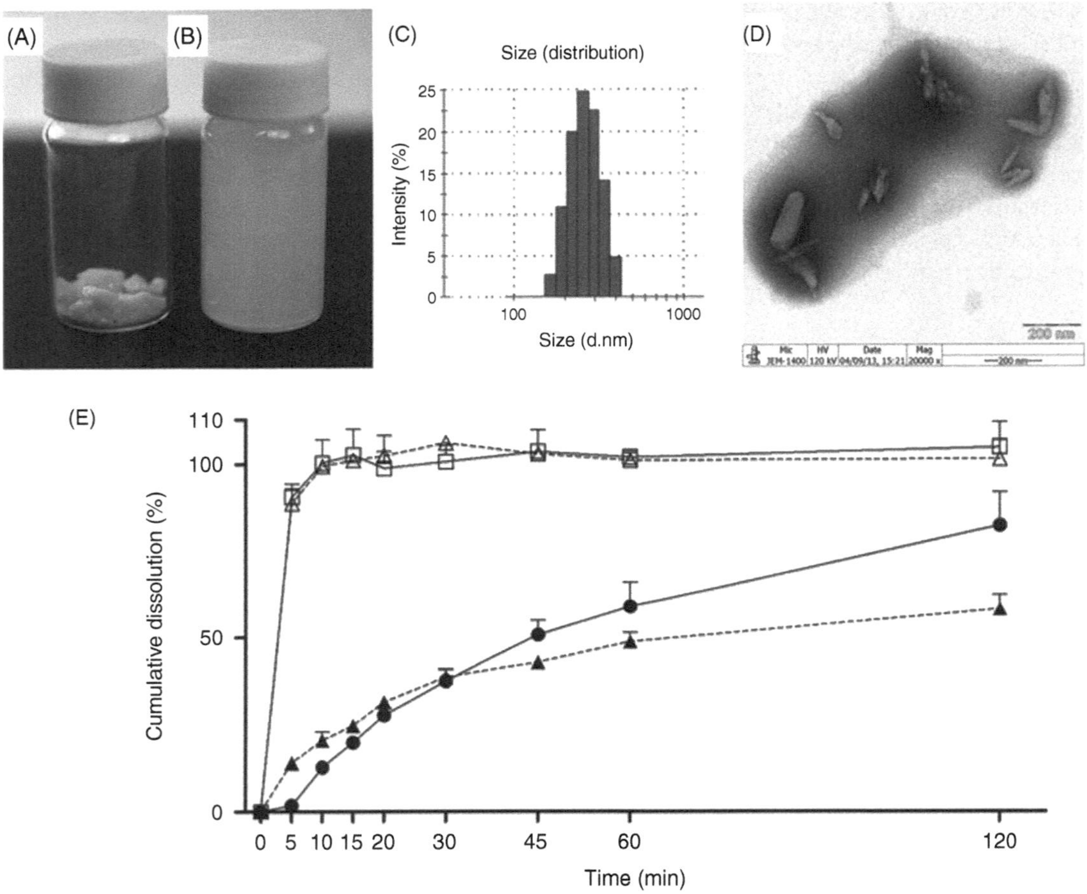

FIGURE 13.3 Morphology, Particle-Size Distribution, and Dissolutions of Scutellarein Nanosuspension DDS

(A) Lyophilized nanosuspension powder with HP-β-CD as protectant; (B) Liquid nanosuspension after redispersion; (C) Size distribution of redispersed liquid nanosuspension; (D) TEM micrograph of redispersed liquid nanosuspension (magnification ×20,000); (E) Dissolution profiles of redispersed liquid nanosuspension with HP-β-CD (SP-HP-β-CD) and Scutellarein coarse suspension with 1% SDS. -□-: SP-HP-β-CD in mimic gastric (pH 1.2) media; -△-: SP-HP-β-CD in mimic intestinal (pH 6.8) media; -●-: S coarse suspension in mimic gastric (pH 1.2); -▲-: S coarse suspension in mimic intestinal (pH 6.8) media.

Reprinted with permission from Yang, X., Miao, X., Cao, F., Li, S., Ai, N., Chang, Q., Lee, S.M.Y., Zheng, Y., 2014. Nanosuspension development of scutellarein as an active and rapid orally absorbed precursor of its BCS Class IV glycoside scutellarin. J. Pharm. Sci. 103, 3576–3584, Copyright 2014 John Wiley & Sons, Ltd.

5.3 PELLET DOSAGE FORM

Another solid dosage form, Pellet, exhibits many advantages including minimal gastrointestinal tract irritation and less sensitivity to the effect of stomach emptying (Kibria et al., 2008; Li et al., 2012). Liquid nanosuspensions can be directly converted into pellet dosage using fluid-bed coating technology. For example, He et al. (2013) prepared liquid indomethacin nanosuspension by the precipitation-ultrasonication method, and then the liquid nanosuspensions were coated onto nonpareil pellet core using fluid-bed coating technology . The drug-loaded pellets could be redispersed in water and they maintained the original size and size distribution of drug nanocrystals. In addition, the drug-loaded pellets also showed faster dissolution rate compared with the coarse drugs. The drug-loaded pellets were dependent on many parameters, such as binder types (Möschwitzer and Müller, 2013). Möschwitzer et al. prepared liquid hydrocortisone acetate nanosuspension and then converted it into pellets with sugar beads as pellet cores using fluid-bed coating technology. Different binder solutions, such as chitosan chloride, polyvinyl alcohol, hydroxypropyl methylcellulose, or polyvinylpyrrolidone were used during the preparation. All binders tested except for polyvinylpyrrolidone were compatible with the nanosuspension. Moreover, the zeta potential, the stability, and the dissolution profiles of the final pellet dosages were obviously affected by the binder types.

Pellet dosage can also be transformed from dry powder dosage by an extrusion-spheronization method. In the multiple-step preparation of extrusion-spheronization process, drug nanoparticle powder are first mixed with excipients followed by wet granulation, extrusion, spheronization, and drying. Typical example was shown by Mauludin et al.'s work (Mauludin et al., 2012). In their work, liquid ibuprofen nanosuspension was prepared and then dried into powder. After that, the dry ibuprofen nanosuspension powder was processed into pellets using the extrusion-spheronization method. The Ibuprofen nanocrystal in the pellets could be redispersed completely in water and showed good behaviors similar to the powder dosage.

5.4 TABLET DOSAGE FORM

Nowadays, the compressed tablet dosage form is one of the most popular and accounts for almost one-thirds of all prescriptions dispensed (Mahida and Gupta, 2013). The tablet dosage form is in great popularity not only because of the fact that an accurate drug dosage and uniform content to a specific site can easily delivered, but also because it can be used for sublingual, buccal, rectal, or intravaginal administration other than oral administration. The preparation of the tablet dosage form follows a few steps. A liquid nanosuspension containing drug particles was first solidified into drug nanoparticle powder, which was then mixed with excipients. The resulting mixture can be pressed into tablet dosage form with a machine (Sarnes et al., 2014). An alternative strategy to prepare tablet dosage form is to mix the drug and water-soluble excipients in a PVC blister and freeze it (Lai et al., 2014).

Tablets for poorly soluble drugs have attracted a lot of attention. Sun et al. (2015) prepared itraconazole tablets and then the formulation was optimized. The tablets were stable in terms of drug content and in vitro dissolution profiles. In vivo results demonstrated the oral absorption of the itraconazole tablets was comparable to that of the lead market product. Patel et al. (2014b) prepared nanosized telmisartan-loaded tablets. The in vitro drug-release study revealed the tablets had a remarkable improvement in the dissolution rate compared with the marketed telmisartan tablets. Moreover, the nanosized telmisartan-loaded tablets had 1.5-fold enhancement in oral bioavailability against the marked telmisartan tablets according to an in vivo pharmacokinetic study. In addition, tablets loaded with nanosized fenofibrate (Zuo et al., 2013), tadalafil (Obeidat and Sallam, 2014), and silybin nanoparticles (Xu et al., 2015) were reported to exhibit the improved bioavailability. The results of these studies suggests that the tablet is an excellent dosage form for the application of poorly water soluble drugs in oral administration (Fig. 13.4).

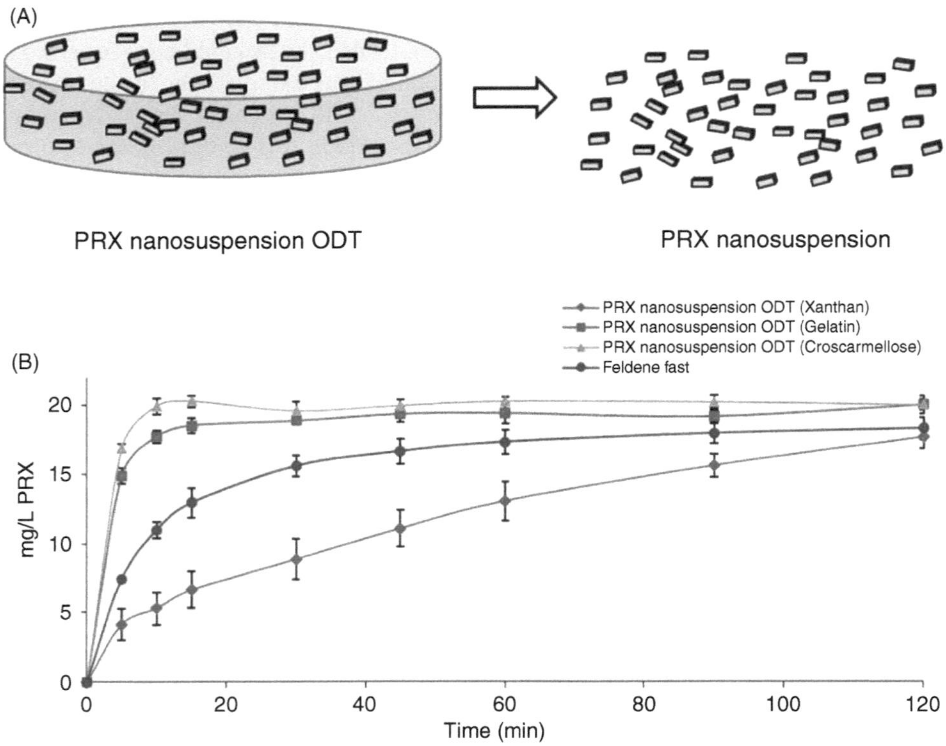

FIGURE 13.4

Schematic representation (A) and dissolutions (B) of piroxicam nanosuspension orally disintegrating tablets (ODT).

Adapted with permission from Lai, F., Pini, E., Corrias, F., Perricci, J., Manconi, M., Fadda, A.M., Sinico, C., 2014. Formulation strategy and evaluation of nanocrystal piroxicam orally disintegrating tablets manufacturing by freeze-drying. Int. J. Pharmaceut. 467, 27–33, Copyright 2014 Elsevier.

5.5 CAPSULE DOSAGE FORM

In the capsule dosage form, the drug nanoparticles are placed in the hollow hard capsule or sealed in the elastic soft capsule. The capsule solid dosage form has an advantage of being odor-free and having good stability.

Drug nanoparticle-loaded capsule dosage can be prepared by filling dry drug nanoparticle powder into a capsule. Bose et al., 2012 reported the preparation of capsules for oral administration use by first using a wet milling method to prepare the nanosuspension and then converting it into powder using a wet milling method and then converted it into powder via fluid-bed spray granulation method. The obtained powder was then placed into a hard gelatin capsule. Kumar et al. (2014a) prepared dried crystallized itraconazole powder and then enclosed the powder into capsules using a capsule filling kit. Du et al. (2013) prepared dried glimepiride powder dosage and then filled it into hard gelatin capsules. Compared with micronized drug loaded and commercial capsules, the drug nanocrystal-loaded capsules had a higher dissolution rate and a significant enhancement of in vivo bioavailability.

An alternative route to prepare capsule dosage with drug nanoparticle can be realized by filling the pellets containing the drug nanoparticle into a capsule. A liquid lutein nanosuspension was first prepared by Mitri et al. (2011b) via high-pressure homogenization, and the nanosuspension was subsequently converted into pellets by utilizing the extrusion-spheronization technique. Then the pellets were placed into the gelatin capsules for oral administration use Fig. 13.5. The prepared capsules loaded with lutein drug nanoparticles exhibit a superior in vitro release (usually 3–4 factor).

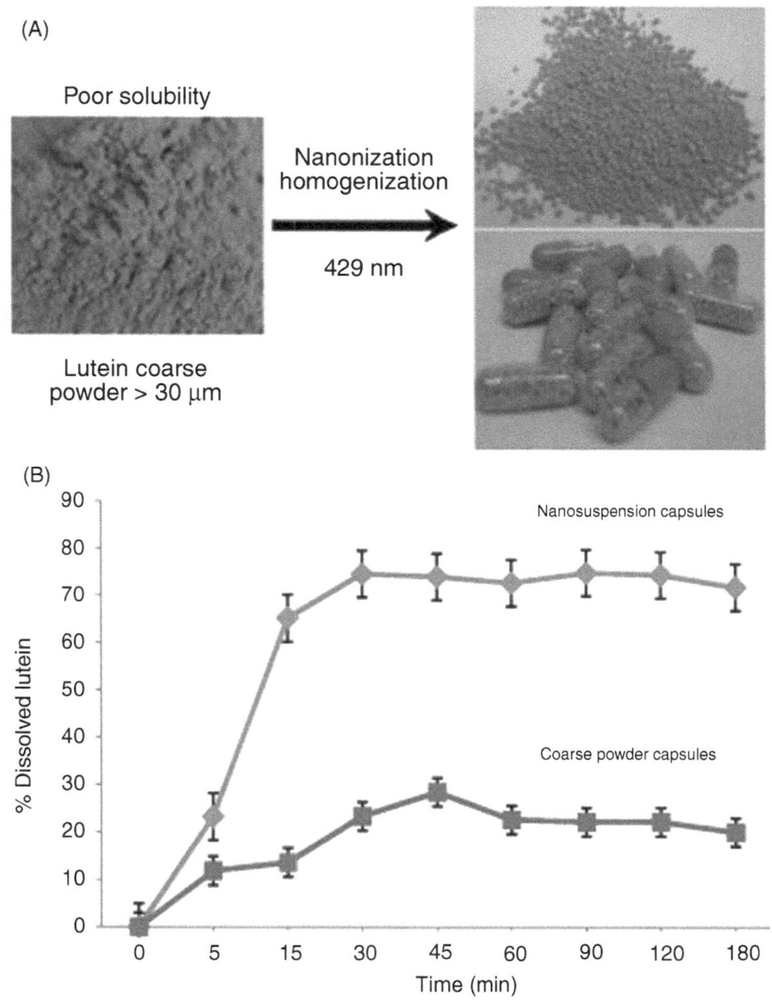

FIGURE 13.5

(A) Transformation of orange lutein coarse powder to golden colored lutein nanosuspension loaded spherical lactose pellets and pellets filled in capsules. (B) Dissolution profiles of lutein from nanosuspension loaded pellets and from coarse lutein powder pellets, both inserted into hard gelatin capsule.

Adapted with permission from Mitri, K., Shegokar, R., Gohla, S., Anselmi, C., Muller, R.H., 2011b. Lutein nanocrystals as antioxidant formulation for oral and dermal delivery. Int. J. Pharmaceut. 420, 141–146, Copyright 2011 Elsevier.

Osmotic capsule DDS has also been explored for controlled drug delivery in oral and implant administration. The osmotic capsule DDS uses osmotic pressure as a driving force to deliver drugs. It is a promising DDS for the delivery of poorly soluble drugs. Liu et al. (2014) prepared liquid carvedilol nanosuspension and then freeze-dried it into powder. The resultant dry drug powder was mixed with some osmotic agents and suspending agents, and then the mixes were filled into semipermeable cellulose acetate capsule shell with a 0.4 mm diameter orifice. The osmotic pump capsule after oral administration had a higher in vivo bioavailability in beagle dogs than the commercial formulation (Fig. 13.6). Moreover, it successfully controlled the drug release and minimized the fluctuation of plasma concentration.

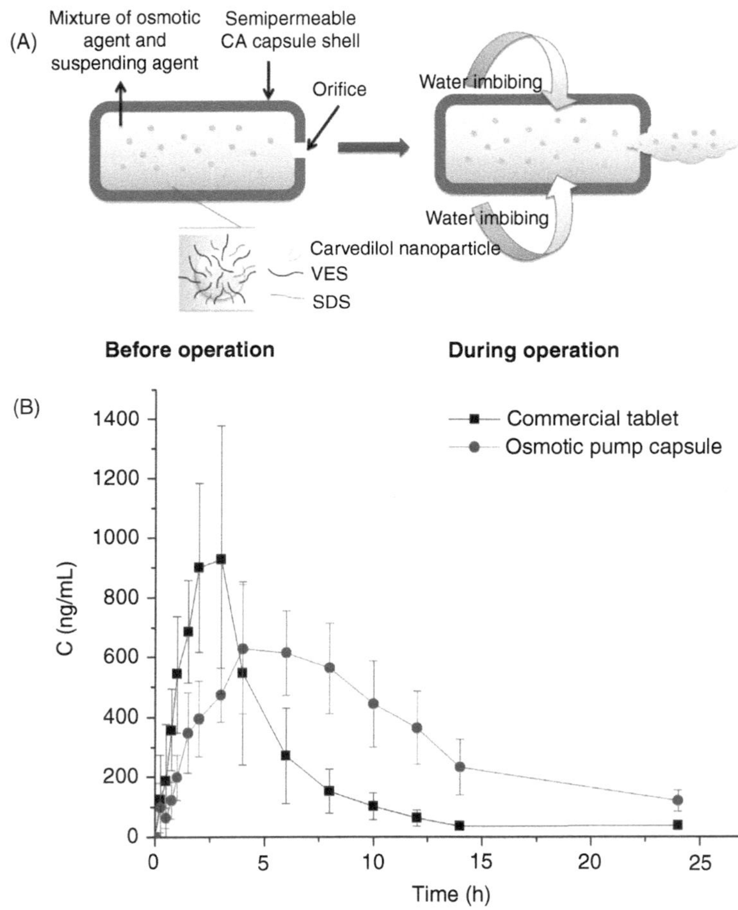

FIGURE 13.6

(A) Schematic of the process of drug release from the osmotic capsule system containing drug nanosuspension. (B) In vivo pharmacokinetics profiles of carvedilol in beagle dogs from the novel osmotic pump capsule and the commercial tablet ($n = 8$).

Adapted with permission from Li, W., Quan, P., Zhang, Y., Cheng, J., Liu, J., Cun, D., Xiang, R., Fang, L., 2014. Influence of drug physicochemical properties on absorption of water insoluble drug nanosuspensions. Int. J. Pharmaceut. 460, 13–23, Copyright 2014 Elsevier.

5.6 FILM DOSAGE FORM

Another interesting solid dosage form is the film dosage in which a drug nanoparticle is homogeneously dispersed in the film materials. It exhibit many advantages, such as the lack of powder generated during the preparation process and a simple preparation process.

The commonly used method to prepare film dosage is casting and drying the liquid nanosuspension. A novel orodispersible film loaded with drug nanoparticles was reported by Shen et al. (2013). The liquid herpetrione nanosuspensions was transformed into a solid film dosage and the oral bioavailability was also improved. The drug nanoparticles could be quickly redispersed in water from the orodispersible film. The orodispersible film showed a significant increase in the oral bioavailability compared with the liquid herpetrione coarse suspension. Further, the same group optimized the formulation of the herpetrione nanoparticle-loaded orodispersible film by Box-Behnken design-response surface methodology (Shen et al., 2014) (Fig. 13.7).

Film dosage can be prepared via the solidification of a hydrogel loaded with drug nanoparticle (Rana and Murthy, 2013). They prepared liquid carvedilol nanosuspension, and then mixed it with hydroxy propyl methyl cellulose and carbopol 934P to form a gel. PEG400 was added to the resultant gel as a plasticizer and then dried to form the mucoadhesive buccal film. The film had a fast in vitro drug release and a significant increase of the in vivo bioavailability compared with the commercial oral tablet.

6 APPLICATION OF NANOSUSPENSION DDS

Due to the particle-size reduction and surface-area increase of nanosuspension, it has revealed the potential to solve the problems associated with the delivery of poorly water-soluble and poorly lipid-soluble drugs. The nanosuspension DDS has found wide application in the area of oral pulmonary, parenteral and ocular drug delivery and they have exhibited excellent advantage over traditional drug administration route (Velmula et al., 2015). The application of nanosuspension DDS in different areas will be discussed in the following section.

6.1 ORAL DRUG ADMINISTRATION

The oral administration route is preferred over the various other administration routes of drug delivery due to the many advantages it exhibits. These advantages include safety, good patient compliance, ease of ingestion, pain avoidance, and versatility to accommodate various types of drugs (Sastry et al., 2000). Oral administration of nanosuspension DDS is propitious to improve the bioavailability of poorly soluble drugs. The main reasons are the many advantages of nanosuspension DDS for oral administration, such as enhanced dissolution rate and solubility of poorly soluble drugs, high adhesiveness of drug nanocrystals on the epithelial gut wall, prolonged absorption time of drug nanocrystals due to long gastrointestinal tract, and reduced variability caused by food (Liu et al., 2012a). Both liquid dosage form and solid dosage form, such as powder, tablet, pellet, capsule, and film can be utilized in oral administration. The prepared liquid nanosuspensions can be directly utilized in liquid dosage form for oral administration.

6.2 PARENTERAL ADMINISTRATION

Parenteral administration is preferred at times over other drug-administrations routes, such as in emergency situations of cardiac arrest and anaphylactic shock (Shi et al., 2009). This type of administration

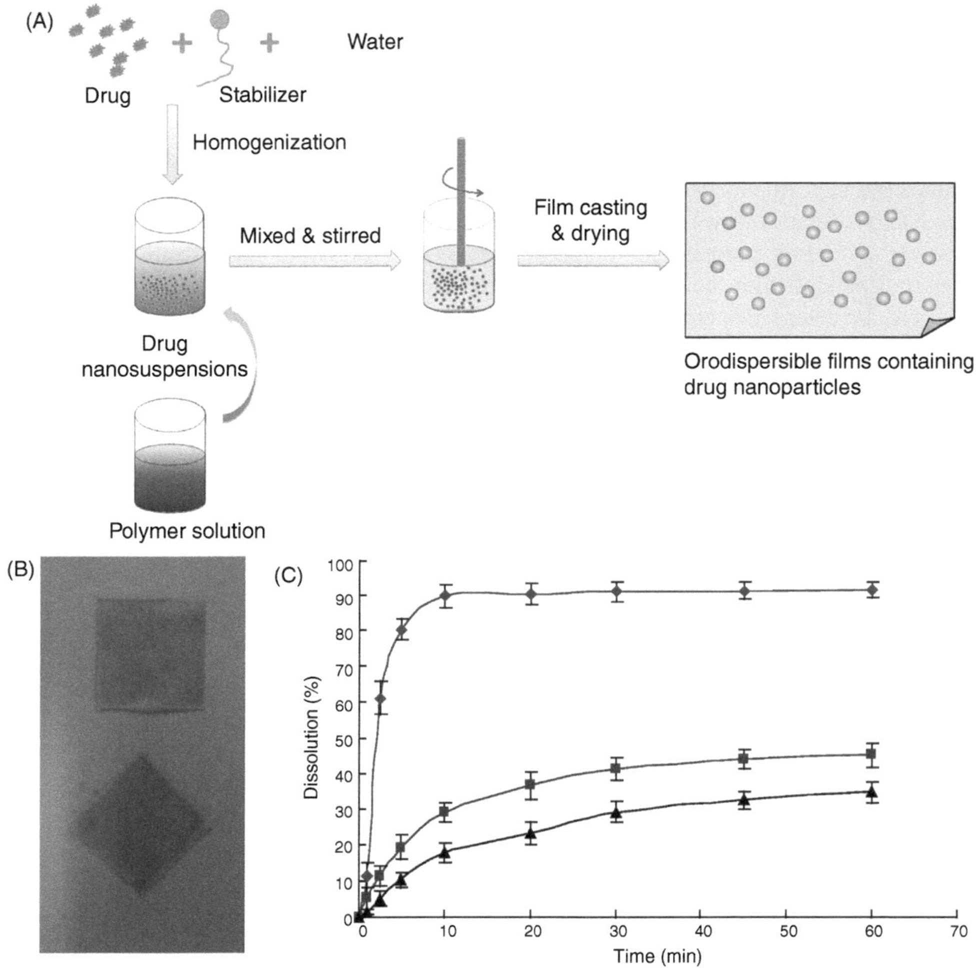

FIGURE 13.7

(A) Schematic for preparation of orodispersible film containing drug nanoparticles. (B) Appearance of orodispersible films containing herpetrione nanoparticles. (C) Dissolution profiles of the orodispersible films containing HPE nanoparticles (—◆—), the orodispersible films containing HPE particles (—■—) and HPE raw material (—▲—) (mean ± SD, $n = 3$).

Adapted with permission from Shen, B.D., Shen, C.Y., Yuan, X.D., Bai, J.X., Lv, Q.Y., Xu, H., Dai, L., Yu, C., Han, J., Yuan, H.L., 2013. Development and characterization of an orodispersible film containing drug nanoparticles. Eur. J. Phar. Biopharm. 85, 1348–1356, Copyright 2013 Elsevier.

route exhibits several advantages, such as first-pass metabolism avoidance, better bioavailability, and reliable dosage. Compared with oral administration, parenteral administration has control over the dose and rate, thus generating more predictable pharmaco-dynamic and pharmacokinetic profiles (Gibaldi et al., 2007). Generally, to avoid capillary blockage, the size of drug particles administered should be below 5 μm (Gibaldi et al., 2007). In a study on the investigation of oridonin nanosuspension's ability to inhibit tumor growth, it was observed that Oridonin in the form of nanosuspension can greatly elevate the rate of tumor inhibition by almost 20% when compared with the conventional form of Oridonin (Lou et al., 2009). The efficiency of therapy is improved while the cost is greatly reduced with the aid of nanosuspension.

6.3 PULMONARY DRUG DELIVERY

Pulmonary drug administration is very effective in solving several respiratory problems, such as chronic obstructive pulmonary diseases and asthma (Courrier et al., 2002). Compared to aforementioned oral and parenteral drug administrations, pulmonary drug administration possesses its own advantages. In pulmonary drug administration, the drug is directly delivered to the site of action. A decreased dosage is required and side effects are improved. Conventional pulmonary drug administration has a lot of problems, such as poor drug residence time, rapid drug release, and a lack of selectivity (Beck-Broichsitter et al., 2011). By using the nanosuspension strategy, the problems of poor drug solubility and a lack of selectivity can be resolved by delivering the drug directly to affected pulmonary cells. Nanosuspensions have a stronger adhesiveness to mucosal surfaces, which will results in prolonged residence time at target site, thus improving the selectivity and maximally reducing the drug loss (Jacobs and Müller, 2002). Pulmonary administration routes usually lead to an increased bioavailability because the nanosuspensions can enhance drug dissolution and diffusion rate and thus impede undesirable drug deposition in the pharynx and mouth.

6.4 OCULAR ADMINISTRATION

Traditional methods in ocular therapy have some disadvantages, such as bad drug solubility in lachrymal fluids, many side effects generated from systematic drug absorption, as well as repeated instillation (Gaudana et al., 2009). These limitations posed by the conventional administration route can be overcome with the application of nanosuspensions. This nanosuspension strategy exhibits many advantages, such as prolonging the ocular drug residence time and excellent improvement of bioavailability. In addition, positively charged nanoparticles in nanosuspension have strong adhesion to negatively charged mucin and the drug release can be extended. For example, in ocular drug delivery, chitosan as a mucoadhesive cationic polymer is utilized to adhere to negatively charged mucin and thus the drug residence time is greatly prolonged (Nagarwal et al., 2009). This natural adhesiveness of drug nanoparticles can decrease the amount of drug loss.

6.5 DERMAL DELIVERY

Due to an increased penetration of nanocrystals into a membrane, the nanocrystals exhibit improved permeation and bio-adhesiveness. In order to develop intravenous formulations, the injectability and fast dissolution needs to be investigated. However, for years researchers failed to take advantage of the use of adhesion, fast dissolution, or increased penetration for dermal and mucosal applications (Mitri et al., 2011a). The concentration of the poorly soluble drug was increased and this may result in an increased concentration gradient between the formulation and the skin, thus increasing the penetration of drug nanocrystal (Padois et al., 2011). In investigation of a penetration barrier, it was observed that the penetration ability of lutein nanocrystals through cellulose nitrate membranes are fourteen times better than that of raw powder. However, these lutein nanocrystals stayed in the skin of pig ear after its entry because of its lipophilicity (Schwarz et al., 2012).

6.6 TARGETED DRUG DELIVERY

The uptake in vivo behavior of drug nanoparticles can be tuned by changing the particle size. The change in particle size results in the change of surface properties. Thus, the nanosuspension system can be used as targeted delivery system. Smart crystal such drug particles below 100 nm can be used in the DDS to avoid the phagocytotic uptake of nanocrystals. Development of nanosuspension for targeted drug delivery is feasible due to its simplicity. It was reported that mucoadhesive nanosuspension was used for targeting of cryptosporidium parvum (Kayser, 2001). The organ distribution of the drug particles depends on the surface properties of the drug particles, such as surface hydrophobicity, charge, and the presence and concentration of certain functional groups. Therefore it is surprising to see that Tween 80-coated nanocrystals can be used for brain targeting. The application of atovaquone nanocrystals coated with Tween 80 to treat toxoplasmosis is a good example.

6.7 BIOAVAILABILITY ENHANCEMENT

Factors including the poor solubility, the poor permeability, and poor stability of the drugs in the gastrointestinal tract (GIT) may lead to poor bioavailability. By transforming the drug particles into nanosuspensions, the problem of poor solubility and poor permeability across the membrane can be resolved. For example, Oral administration of the Danazol nanosuspension as gonadotrophin inhibitor can give an absolute bioavailability of 82.3% compared with the only 5.2% of conventional dispersion (Danocrine). Kayser et al. also reported that nanosuspension of Amphotericin B showed a significant improvement in its oral absorption compared with the conventional formulation (Kayser et al., 2009). Oleanolic acid as a hepatoprotective agent is poorly soluble and bioavailability was improved with the use of a nanosuspension. The therapeutic effect was greatly elevate, which indicated elevated bioavailability. The nanosuspension exhibits faster dissolution (90% in 20 min) compared with the dissolution from a coarse powder (15% in 20 min) (Bhowmik et al., 2013).

7 SUMMARY AND OUTLOOK

This chapter presents a general overview on the different preparation techniques, characterization parameters, postprocessing techniques and the application of nanosuspensions in DDS. Nanosuspension DDS has exhibit enormous advantages and potential for resolving the administration issue of poorly soluble drugs. The dosage form developments of nanosuspension DDS and its application have attracted more and more research attention from the pharmaceutical and industrial community. The nanosuspension DDS exhibits attractive characteristics, such as uniform nanosized particles, improved solubility in biological media and improved bio-adhesiveness, increased drug concentrations, and prolonged residence time at the action sites. Those excellent characteristics can propel the innovative design and development of a new class of drug delivery systems. These nano-sized drug have a number of benefits, such as high surface area, controllable nano-size dimensions, and tailorable surface chemistry. In spite of all the advantages that nanosuspension DSS exhibits, there are still limits and disadvantages existing, such as absence of guidelines for the selection of stabilizer, cosurfactant, and other additives; unclear mechanism of nanosuspension stabilization and solidification; and incomparable particle-size distributions. Also, a controlled drug release and an appropriate bio-adhesion is expected to be achieved. Because the uptake of drug nanoparticles depends on the particle size, working on the size optimization of drug nanosuspension might enable us to prepare a nanosuspension formulation with better diffusion through the mucus gel layer. In addition, incorporation of polymers on the particle surface and size reduction can be regarded as the future step in nanosuspension research. All in all, the aim of future studies remains to resolve the challenges associated with poorly soluble drugs so as to achieve better bioavailability, dissolution velocity, and bio-adhesion of the drug.

ACKNOWLEDGMENTS

This work has been supported by research grants from Special National Key Research and Development Plan (2016YFD0400202-8) and the National Science and Technology Support Program of China (2015BAD17B02-2).

REFERENCES

Abdelwahed, W., Degobert, G., Fessi, H., 2006. A pilot study of freeze drying of poly (epsilon-caprolactone) nanocapsules stabilized by poly (vinyl alcohol): formulation and process optimization. Int. J. Pharm. 309, 178–188.

Ahuja, B.K., Jena, S.K., Paidi, S.K., Bagri, S., Suresh, S., 2015. Formulation, optimization and in vitro–in vivo evaluation of febuxostat nanosuspension. Int. J. Pharm. 478, 540–552.

Alexis, F., Pridgen, E., Molnar, L.K., Farokhzad, O.C., 2008. Factors affecting the clearance and biodistribution of polymeric nanoparticles. Mol. Pharm. 5, 505–515.

Ali, M.E., Lamprecht, A., 2014. Spray freeze drying for dry powder inhalation of nanoparticles. Eur. J. Pharm. Biopharm. 87, 510–517.

Arias, M., Moyano, J., Gines, J., 1997. Investigation of the triamterene–β-cyclodextrin system prepared by co-grinding. Int. J Pharm. 153, 181–189.

Baek, I.-H., Kim, J.-S., Ha, E.-S., Choo, G.-H., Cho, W., Hwang, S.-J., Kim, M.-S., 2014. Dissolution and oral absorption of pranlukast nanosuspensions stabilized by hydroxypropylmethyl cellulose. Int. J. Biol. Macromol. 67, 53–57.

Beck-Broichsitter, M., Schmehl, T., Seeger, W., Gessler, T., 2011. Pulmonary drug delivery with nanoparticles. Nanomedicine in Health and Disease. CRC Press, Hoboken, NJ, United States, 229–248.

Bhowmik, D., Harish, G., Duraivel, S., Kumar, B.P., Raghuvanshi, V., Kumar, K.P.S., 2013. Nanosuspension-A novel approach in drug delivery system. Pharm. Innov. J. 1, 50–63.

Blunk, T., Hochstrasser, D.F., Sanchez, J.C., Müller, B.W., Müller, R.H., 1993. Colloidal carriers for intravenous drug targeting: plasma protein adsorption patterns on surface-modified latex particles evaluated by two-dimensional polyacrylamide gel electrophoresis. Electrophoresis 14, 1382–1387.

Blunk, T., L Ck, M., Calv R, A., Hochstrasser, D., Sanchez, J.-C., Müller, B., Müller, R., 1996. Kinetics of plasma protein adsorption on model particles for controlled drug delivery and drug targeting. Eur. J. Pharm. Biopharm. 42, 262–268.

Bose, S., Schenck, D., Ghosh, I., Hollywood, A., Maulit, E., Ruegger, C., 2012. Application of spray granulation for conversion of a nanosuspension into a dry powder form. Eur. J. Pharm. Sci. 47, 35–43.

Charman, W.N., Stella, V.J., 1992. Lymphatic transport of drugs. CRC Press, Boca Raton, FL, United States.

Chaurasia, T., Singh, D., Nimisha, D.S., 2012. A review on nanosuspensions promising drug delivery strategy. Curr. Pharma. Res. 3, 764–776.

Chen, H., Khemtong, C., Yang, X., Chang, X., Gao, J., 2011. Nanonization strategies for poorly water-soluble drugs. Drug Discov. Today 16, 354–360.

Chen, L., Wang, Y., Zhang, J., Hao, L., Guo, H., Lou, H., Zhang, D., 2014. Bexarotene nanocrystal—Oral and parenteral formulation development, characterization and pharmacokinetic evaluation. Eur. J. Phar. Biopharm. 87, 160–169.

Chen, A., Shi, Y., Yan, Z., Hao, H., Zhang, Y., Zhong, J., Hou, H., 2015. Dosage form developments of nanosuspension drug delivery system for oral administration route. Curr. Pharm. Des. 21, 4355–4365.

Chingunpituk, J., 2011. Nanosuspension technology for drug delivery. Walailak J. Sci. Tech 4, 139–153.
Chow, S.F., Sun, C.C., Chow, A.H.L., 2014. Assessment of the relative performance of a confined impinging jets mixer and a multi-inlet vortex mixer for curcumin nanoparticle production. Eur. J. Pharm. Biopharm. 88, 462–471.
Courrier, H., Butz, N., Vandamme, T.F., 2002. Pulmonary drug delivery systems: recent developments and prospects. Crit. Rev. in Therapeut. Drug Carrier Sys., 19, 425–498.
Deshiikan, S., Papadopoulos, K., 1998. Modified Booth equation for the calculation of zeta potential. Colloid Polym. Sci. 276, 117–124.
Du, B., Shen, G., Wang, D., Pang, L., Chen, Z., Liu, Z., 2013. Development and characterization of glimepiride nanocrystal formulation and evaluation of its pharmacokinetic in rats. Drug Deliv. 20, 25–33.
Eldridge, J.A., Willmott, G.R., Anderson, W., Vogel, R., 2014. Nanoparticle ζ-potential measurements using tunable resistive pulse sensing with variable pressure. J. Colloid Interface Sci. 429, 45–52.
Figueroa, C.E., Bose, S., 2013. Spray granulation: Importance of process parameters on in vitro and in vivo behavior of dried nanosuspensions. Eur. J. Pharm. Sci. 85, 1046–1055.
Gao, L., Zhang, D., Chen, M., Zheng, T., Wang, S., 2007. Preparation and characterization of an oridonin nanosuspension for solubility and dissolution velocity enhancement. Drug Dev.Indust. Pharm. 33, 1332–1339.
Gao, L., Zhang, D., Chen, M., 2008. Drug nanocrystals for the formulation of poorly soluble drugs and its application as a potential drug delivery system. J. Nanopart. Res. 10, 845–862.
Gaudana, R., Jwala, J., Boddu, S.H., Mitra, A.K., 2009. Recent perspectives in ocular drug delivery. Pharm. Res. 26, 1197–1216.
Gibaldi, M., Lee, M., Desai, A., 2007. Gibaldi's drug delivery systems in pharmaceutical care. American Society of Health-System Pharmacists, Bethesda, MD, United States.
Ginoya, S., Atara, S., 2013. Nanosuspension: a novel approach towards the drug delivery system. Pharma Sci. Monitor 4, 100–122.
Grau, M.J., Kayser, O., Müller, R.H., 2000. Nanosuspensions of poorly soluble drugs—reproducibility of small scale production. Int. J. Pharmaceut. 196, 155–159.
Grohganz, H., Lee, Y.-Y., Rantanen, J., Yang, M., 2013. The influence of lysozyme on mannitol polymorphism in freeze-dried and spray-dried formulations depends on the selection of the drying process. Int. J. Pharm. 447, 224–230.
Guo, J.-J., Yue, P.-F., Lv, J.-L., Han, J., Fu, S.-S., Jin, S.-X., Jin, S.-Y., Yuan, H.-L., 2013. Development and in vivo/in vitro evaluation of novel herpetrione nanosuspension. Int. J. Pharm. 441, 227–233.
He, W., Lu, Y., Qi, J., Chen, L., Yin, L., Wu, W., 2013. Formulating food protein-stabilized indomethacin nanosuspensions into pellets by fluid-bed coating technology: physical characterization, redispersibility, and dissolution. Int. J. Nanomed. 8, 3119–3128.
Hong, C., Dang, Y., Lin, G., Yao, Y., Li, G., Ji, G., Shen, H., Xie, Y., 2014. Effects of stabilizing agents on the development of myricetin nanosuspension and its characterization: An in vitro and in vivo evaluation. Int. J. Pharm. 477, 251–260.
Hörter, D., Dressman, J.B., 2001. Influence of physicochemical properties on dissolution of drugs in the gastrointestinal tract1. Adv. Drug Deliv. Rev. 46, 75–87.
Jacobs, C., Müller, R.H., 2002. Production and characterization of a budesonide nanosuspension for pulmonary administration. Pharm. Res. 19, 189–194.
Jeon, H.-J., Jeong, Y.-I., Jang, M.-K., Park, Y.-H., Nah, J.-W., 2000. Effect of solvent on the preparation of surfactant-free poly(dl-lactide-co-glycolide) nanoparticles and norfloxacin release characteristics. Int. J. Pharm. 207, 99–108.
Junyaprasert, V.B., Morakul, B., 2015. Nanocrystals for enhancement of oral bioavailability of poorly water-soluble drugs. Asian J. Pharm. Sci. 10, 13–23.
Kayser, O., 2001. A new approach for targeting to Cryptosporidium parvum using mucoadhesive nanosuspensions: research and applications. Int. J. Pharm. 214, 83–85.
Kayser, O., Olbrich, C., Yardley, V., Kiderten, A.F., Croft, S.L., 2009. Formulation of amphotericin-B as nanosuspension for oral administration. Int. J. Pharm. 254, 73–75.

Keck, C.M., Müller, R.H., 2006. Drug nanocrystals of poorly soluble drugs produced by high pressure homogenisation. Eur. J. Pharm. Biopharm. 62, 3–16.
Kesisoglou, F., Mitra, A., 2012. Crystalline nanosuspensions as potential toxicology and clinical oral formulations for BCS II/IV compounds. AAPS J. 14, 677–687.
Kibria, G., Roni, M., Absar, M., Jalil, R.-U., 2008. Effect of plasticizer on release kinetics of diclofenac sodium pellets coated with Eudragit RS 30 D. AAPS Pharm. Sci.Tech. 9, 1240–1246.
Kim, H.G., Park, J.-I., Lee, G.H., 2013. Surface coating of Al nanoparticles by using a wet ball milling method: A facile synthesis and characterization of colloidal stability. Curr. Appl. Phys. 13, 1218–1224.
Kocbek, P., Baumgartner, S., Kristl, J., 2006a. Preparation and evaluation of nanosuspensions for enhancing the dissolution of poorly soluble drugs. Int. J. Pharm. 312, 179–186.
Kocbek, P., Baumgartner, S., Kristl, J., 2006b. Preparation and evaluation of nanosuspensions for enhancing the dissolution of poorly soluble drugs. Int. J. Pharmaceut. 312, 179–186.
Krause, K., Müller, R., 2001. Production and characterisation of highly concentrated nanosuspensions by high pressure homogenization. Int. J Pharm. 214, 21–24.
Krishna, K.B., Prabhakar, C., 2011. A review on nanosuspensions in drug delivery. Int. J. Pharm. Bio. Sci. 2, 549–558.
Kumar, A., Sahoo, S.K., Padhee, K., Kochar, P., Satapathy, A., Pathak, N., 2011. Review on solubility enhancement techniques for hydrophobic drugs. Pharm. Glob. 3, 001–007.
Kumar, S., Shen, J., Burgess, D.J., 2014a. Nano-amorphous spray dried powder to improve oral bioavailability of itraconazole. J. Control Rel. 192, 95–102.
Kumar, S., Xu, X., Gokhale, R., Burgess, D.J., 2014b. Formulation parameters of crystalline nanosuspensions on spray drying processing: A DoE approach. Int. J. Pharm. 464, 34–45.
Lai, F., Pini, E., Corrias, F., Perricci, J., Manconi, M., Fadda, A.M., Sinico, C., 2014. Formulation strategy and evaluation of nanocrystal piroxicam orally disintegrating tablets manufacturing by freeze-drying. Int. J. Pharm. 467, 27–33.
Lakshmi, P., Kumar, G.A., 2010. Nano-suspension technology: a review. Int. J. Pharm. Pharm. Sci. 2, 35–40.
Leuner, C., Dressman, J., 2000. Improving drug solubility for oral delivery using solid dispersions. Eur. J. Pharm. Biopharm. 50, 47–60.
Li, J., Liu, P., Liu, J.-P., Zhang, W.-L., Yang, J.-K., Fan, Y.-Q., 2012. Novel tanshinone II A ternary solid dispersion pellets prepared by a single-step technique: in vitro and in vivo evaluation. Eur. J. Pharm. Biopharm. 80, 426–432.
Li, W., Quan, P., Zhang, Y., Cheng, J., Liu, J., Cun, D., Xiang, R., Fang, L., 2014. Influence of drug physicochemical properties on absorption of water insoluble drug nanosuspensions. Int. J. Pharm. 460, 13–23.
Liu, Y., Xie, P., Zhang, D., Zhang, Q., 2012a. A mini review of nanosuspensions development. J. Drug Target 20, 209–223.
Liu, Y., Xie, P., Zhang, D., Zhang, Q., 2012b. A mini review of nanosuspensions development. J. Drug Target. 20, 209–223.
Liu, D., Yu, S., Zhu, Z., Lyu, C., Bai, C., Ge, H., Yang, X., Pan, W., 2014. Controlled delivery of carvedilol nanosuspension from osmotic pump capsule: In vitro and in vivo evaluation. Int. J. Pharm. 475, 496–503.
Liversidge, G.G., Cundy, K.C., Bishop, J.F., Czekai, D.A., 1992. Surface modified drug nanoparticles. Google Patents.
Loftsson, T., Brewster, M.E., 1996. Pharmaceutical applications of cyclodextrins. 1. Drug solubilization and stabilization. J. Pharm. Sci. 85, 1017–1025.
Lou, H., Zhang, X., Gao, L., Feng, F., Wang, J., Wei, X., Yu, Z., Zhang, D., Zhang, Q., 2009. In vitro and in vivo antitumor activity of oridonin nanosuspension. Int. J. Pharm. 379, 181–186.
Mahida, M.V., Gupta, M.M., 2013. Immediate release tablet of antihypertensive drug olmesartan medoxomile. J. Drug Deliv. Ther. 3, 186–195.
Mauludin, R., Möschwitzer, J., Müller, R.H., 2012. Fast dissolving ibuprofen nanocrystal loaded solid dosage forms. Int. J. Pharm. Pharm. Sci. 4, 543–549.

Mercadante, S., Vitrano, V., 2010. Pain in patients with lung cancer: pathophysiology and treatment. Lung Cancer 68, 10–15.

Merisko-Liversidge, E.M., Liversidge, G.G., 2008. Drug nanoparticles: formulating poorly water-soluble compounds. Toxicol. Pathol. 36, 43–48.

Mitri, K., Shegokar, R., Gohla, S., Anselmi, C., Müller, R.H., 2011a. Lipid nanocarriers for dermal delivery of lutein: preparation, characterization, stability and performance. Int. J. Pharm. 414, 267–275.

Mitri, K., Shegokar, R., Gohla, S., Anselmi, C., Müller, R.H., 2011b. Lutein nanocrystals as antioxidant formulation for oral and dermal delivery. Int. J. Pharm. 420, 141–146.

Müller, C.E., 2009. Prodrug approaches for enhancing the bioavailability of drugs with low solubility. Chem. Biodiv. 6, 2071–2083.

Müller, R.H., Becker, R., Kruss, B., Peters, K., 1999. Pharmaceutical nanosuspensions for medicament administration as system of increased saturation solubility and rate of solution, US 08/836,305.

Müller, R.H., Gohla, S., Keck, C.M., 2011. State of the art of nanocrystals – special features, production, nanotoxicology aspects and intracellular delivery. Eur. J. Pharm. Biopharm. 78, 1–9.

Moneghini, M., Kikic, I., Voinovich, D., Perissutti, B., Filipović-Grčić, J., 2001. Processing of carbamazepine–PEG 4000 solid dispersions with supercritical carbon dioxide: preparation, characterisation, and in vitro dissolution. Int. J. Pharm. 222, 129–138.

Mou, D., Chen, H., Wan, J., Xu, H., Yang, X., 2011. Potent dried drug nanosuspensions for oral bioavailability enhancement of poorly soluble drugs with pH-dependent solubility. Int. J. Pharm. 413, 237–244.

Möschwitzer, J.P., Müller, R.H., 2013. Factors influencing the release kinetics of drug nanocrystal-loaded pellet formulations. Drug Dev. Indust. Pharm. 39, 762–769.

Nagarwal, R.C., Kant, S., Singh, P., Maiti, P., Pandit, J., 2009. Polymeric nanoparticulate system: a potential approach for ocular drug delivery. J. Control. Rel. 136, 2–13.

Nagelreiter, C., Valenta, C., 2013. Size analysis of nanoparticles in commercial O/W sunscreens. Int. J. Pharm. 456, 517–519.

Nakarani, M., Patel, P., Patel, J., Patel, P., Murthy, R.S., Vaghani, S.S., 2010. Cyclosporine a-nanosuspension: formulation, characterization and in vivo comparison with a marketed formulation. Scientia Pharm. 78, 345.

Nel, A.E., Mädler, L., Velegol, D., Xia, T., Hoek, E.M., Somasundaran, P., Klaessig, F., Castranova, V., Thompson, M., 2009. Understanding biophysicochemical interactions at the nano–bio interface. Nat. Mater. 8, 543–557.

Nuchuchua, O., Every, H., Hofland, G., Jiskoot, W., 2014. Scalable organic solvent free supercritical fluid spray drying process for producing dry protein formulations. Eur. J. Pharm. Biopharm. 88, 919–930.

Obeidat, W., Sallam, A.-S., 2014. Evaluation of tadalafil nanosuspensions and their PEG solid dispersion matrices for enhancing its dissolution properties. AAPS Pharm. Sci. Tech. 15, 364–374.

O'brien, R., Cannon, D., Rowlands, W., 1995. Electroacoustic determination of particle size and zeta potential. J. Colloid Interf. Sci. 173, 406–418.

Padois, K., Canti Ni, C., Bertholle, V., Bardel, C., Pirot, F., Falson, F., 2011. Solid lipid nanoparticles suspension versus commercial solutions for dermal delivery of minoxidil. Int. J. Pharm. 416, 300–304.

Patel, C.M., Chakraborty, M., Murthy, Z., 2014a. Preparation of fenofibrate nanoparticles by combined stirred media milling and ultrasonication method. Ultrason. Sonochem. 21, 1100–1107.

Patel, J., Dhingani, A., Garala, K., Raval, M., Sheth, N., 2014b. Design and development of solid nanoparticulate dosage forms of telmisartan for bioavailability enhancement by integration of experimental design and principal component analysis. Powder Tech. 258, 331–343.

Patel, K., Patil, A., Mehta, M., Gota, V., Vavia, P., 2014c. Oral delivery of paclitaxel nanocrystal (PNC) with a dual Pgp-CYP3A4 inhibitor: Preparation, characterization and antitumor activity. Int. J. Pharm. 472, 214–223.

Patravale, V., Kulkarni, R., 2004. Nanosuspensions: a promising drug delivery strategy. J. Pharm. Pharm. 56, 827–840.

Patravale, V.B., Date, A.A., Kulkarni, R.M., 2004. Nanosuspensions: a promising drug delivery strategy. J. Pharm. Pharm. 56, 827–840.

Paun, J., Tank, H., 2012. Nanosuspension: an emerging trend for bioavailability enhancement of poorly soluble drugs. Asian J. Pharm. Tech. 2, 157–168.

Pawar, V.K., Singh, Y., Meher, J.G., Gupta, S., Chourasia, M.K., 2014. Engineered nanocrystal technology: In-vivo fate, targeting and applications in drug delivery. J. Control. Rel. 183, 51–66.

Peters, K., Leitzke, S., Diederichs, J.E., Borner, K., Hahn, H., Müller, R.H., Ehlers, S., 2000. Preparation of a clofazimine nanosuspension for intravenous use and evaluation of its therapeutic efficacy in murine Mycobacterium avium infection. J. Antimicrob. Chemother. 45, 77–83.

Peters, B.-H., Moln R, F., Ketolainen, J., 2014. Structural attributes of model protein formulations prepared by rapid freeze-drying cycles in a microscale heating stage. Eur. J. Pharm. Biopharm. 87, 347–356.

Phromviyo, N., Swatsitang, E., Chompoosor, A., 2014. Effect of a surface stabilizer on the formation of polyoxalate nanoparticles and their release profiles. Vacuum 107, 208–212.

Rabinow, B.E., 2004. Nanosuspensions in drug delivery. Nat. Rev. Drug Discov. 3, 785–796.

Rana, P., Murthy, R.S.R., 2013. Formulation and evaluation of mucoadhesive buccal films impregnated with carvedilol nanosuspension: a potential approach for delivery of drugs having high first-pass metabolism. Drug Deliv. 20, 224–235.

Sameut Bouhaik, I., Leroy, P., Ollivier, P., Azaroual, M., Mercury, L., 2013. Influence of surface conductivity on the apparent zeta potential of TiO2 nanoparticles: Application to the modeling of their aggregation kinetics. J. Colloid Interf. Sci. 406, 75–85.

Sarnes, A., Kovalainen, M., Häkkinen, M.R., Laaksonen, T., Laru, J., Kiesvaara, J., Ilkka, J., Oksala, O., Rönkkö, S., Järvinen, K., Hirvonen, J., Peltonen, L., 2014. Nanocrystal-based per-oral itraconazole delivery: Superior in vitro dissolution enhancement versus Sporanox is not realized in in vivo drug absorption. J. Control. Rel. 180, 109–116.

Sastry, S.V., Nyshadham, J.R., Fix, J.A., 2000. Recent technological advances in oral drug delivery—a review. Pharm. Sci. Technol. Today 3, 138–145.

Schwarz, J.C., Weixelbaum, A., Pagitsch, E., L w, M., Resch, G.P., Valenta, C., 2012. Nanocarriers for dermal drug delivery: Influence of preparation method, carrier type and rheological properties. Int. J. Pharm. 437, 83–88.

Serajuddin, A., 1999. Solid dispersion of poorly water-soluble drugs: early promises, subsequent problems, and recent breakthroughs. J. Pharm. Sci. 88, 1058–1066.

Serajuddin, A.T.M., 2007. Salt formation to improve drug solubility. Adv. Drug Deliv. Rev. 59, 603–616.

Serra, L., Dom Nech, J., Peppas, N.A., 2009. Engineering design and molecular dynamics of mucoadhesive drug delivery systems as targeting agents. Eur. J. Pharm. Biopharm. 71, 519–528.

Shah, T., Patel, D., Hirani, J., Amin, A., 2007. Nanosuspensions as a drug delivery system: A comprehensive review. Drug. Deliv. Tech. 7, 42–53.

Shegokar, R., Müller, R.H., 2010. Nanocrystals: industrially feasible multifunctional formulation technology for poorly soluble actives. Int. J. Pharm. 399, 129–139.

Shen, H., Hong, S., Prud'homme, R.K., Liu, Y., 2011. Self-assembling process of flash nanoprecipitation in a multi-inlet vortex mixer to produce drug-loaded polymeric nanoparticles. J. Nanopart. Res. 13, 4109–4120.

Shen, B.-D., Shen, C.-Y., Yuan, X.-D., Bai, J.-X., Lv, Q.-Y., Xu, H., Dai, L., Yu, C., Han, J., Yuan, H.-L., 2013. Development and characterization of an orodispersible film containing drug nanoparticles. Eur. J. Pharm. Biopharm. 85, 1348–1356.

Shen, C., Shen, B., Xu, H., Bai, J., Dai, L., Lv, Q., Han, J., Yuan, H., 2014. Formulation and optimization of a novel oral fast dissolving film containing drug nanoparticles by Box–Behnken design–response surface methodology. Drug Dev. Indust. Pharm. 40, 649–656.

Shi, Y., Porter, W., Merdan, T., Li, L.C., 2009. Recent advances in intravenous delivery of poorly water-soluble compounds. Expert Opin. Drug Del. 6, 1261–1282.

Sipoli, C.C., Radaic, A., Santana, N., De Jesus, M.B., De La Torre, L.G., 2015. Chitosan nanoparticles produced with the gradual temperature decrease technique for sustained gene delivery. Biochem. Eng. J. 103, 114–121.

Sun, W., Ni, R., Zhang, X., Li, L.C., Mao, S., 2015. Spray drying of a poorly water-soluble drug nanosuspension for tablet preparation: formulation and process optimization with bioavailability evaluation. Drug Dev. Indust. Pharm, 41, 927–933.

Sutradhar, K.B., Khatun, S., Luna, I.P., 2013. Increasing possibilities of nanosuspension. J. Nanotechnol. 2013, 12.

Tan, J.P., Wang, Q., Tam, K.C., 2008. Control of burst release from nanogels via layer by layer assembly. J.Control. Rel. 128, 248–254.

Thorat, A.A., Dalvi, S.V., 2012. Liquid antisolvent precipitation and stabilization of nanoparticles of poorly water soluble drugs in aqueous suspensions: recent developments and future perspective. Chem. Eng. J. 181, 1–34.

Van Eerdenbrugh, B., Froyen, L., Van Humbeeck, J., Martens, J.A., Augustijns, P., Van Den Mooter, G., 2008a. Drying of crystalline drug nanosuspensions—the importance of surface hydrophobicity on dissolution behavior upon redispersion. Eur. J. Pharm. Sci. 35, 127–135.

Van Eerdenbrugh, B., Van Den Mooter, G., Augustijns, P., 2008b. Top-down production of drug nanocrystals: Nanosuspension stabilization, miniaturization and transformation into solid products. Int. J. Pharm. 364, 64–75.

Varshosaz, J., Eskandari, S., Tabbakhian, M., 2012. Freeze-drying of nanostructure lipid carriers by different carbohydrate polymers used as cryoprotectants. Carbohydrate Polym. 88, 1157–1163.

Velmula, M., Pavuluri, P., Rajashekar, S., Rao, V.U.M., 2015. Nanosuspension technology for poorly soluble drugs—a review. World J. Pharm. Pharm. Sci. 4, 1612–1625.

Verma, A., Bindal, M., 2012. Nanosuspensions: Advantages and disadvantages. Indian J. Novel Drug Del. 4, 179–188.

Verma, S., Gokhale, R., Burgess, D.J., 2009. A comparative study of top-down and bottom-up approaches for the preparation of micro/nanosuspensions. Int. J. Pharm. 380, 216–222.

Wallis, K., Müller, R., 1993. Determination of the surface hydrophobicity of colloidal dispersions by mini-hydrophobic interaction chromatography. Pharmazeutische Indust. 55, 1124–1128.

Wang, Y., Zheng, Y., Zhang, L., Wang, Q., Zhang, D., 2013a. Stability of nanosuspensions in drug delivery. J. Control. Rel. 172, 1126–1141.

Wang, Y., Zheng, Y., Zhang, L., Wang, Q., Zhang, D., 2013b. Stability of nanosuspensions in drug delivery. J Control Rel 172, 1126–1141.

Wanning, S., Süverkrüp, R., Lamprecht, A., 2015. Pharmaceutical spray freeze drying. Int. J Pharm. 488, 136–153.

Wu, N.Z., Da, D., Rudoll, T.L., Needham, D., Whorton, A.R., Dewhirst, M.W., 1993. Increased microvascular permeability contributes to preferential accumulation of Stealth liposomes in tumor tissue. Cancer Res. 53, 3765–3770.

Xu, Y., Liu, X., Lian, R., Zheng, S., Yin, Z., Lu, Y., Wu, W., 2012. Enhanced dissolution and oral bioavailability of aripiprazole nanosuspensions prepared by nanoprecipitation/homogenization based on acid–base neutralization. Int. J. Pharm. 438, 287–295.

Xu, D., Ni, R., Sun, W., Li, L.C., Mao, S., 2015. In vivo absorption comparison of nanotechnology-based silybin tablets with its water-soluble derivative. Drug Dev. Ind. Pharm. 41, 552–559.

Yadollahi, R., Vasilev, K., Simovic, S., 2015. Nanosuspension technologies for delivery of poorly soluble drugs. J. Nanomat. 2015, 216375.

Yang, L., Huang, J.-M., Zu, Y.-G., Ma, C.-H., Wang, H., Sun, X.-W., Sun, Z., 2011. Preparation and radical scavenging activities of polymeric procyanidins nanoparticles by a supercritical antisolvent (SAS) process. Food Chem. 128, 1152–1159.

Yang, X., Miao, X., Cao, F., Li, S., Ai, N., Chang, Q., Lee, S.M.Y., Zheng, Y., 2014. Nanosuspension development of scutellarein as an active and rapid orally absorbed precursor of its BCS Class IV glycoside scutellarin. J. Pharm. Sci. 103, 3576–3584.

Young, T.J., Mawson, S., Johnston, K.P., Henriksen, I.B., Pace, G.W., Mishra, A.K., 2000. Rapid expansion from supercritical to aqueous solution to produce submicron suspensions of water-insoluble drugs. Biotech. Prog. 16, 402–407.

Yue, P.-F., Li, G., Dan, J.-X., Wu, Z.-F., Wang, C.-H., Zhu, W.-F., Yang, M., 2014. Study on formability of solid nanosuspensions during solidification: II novel roles of freezing stress and cryoprotectant property. Int. J Pharm. 475, 35–48.

Zabihi, F., Yang, M., Leng, Y., Zhao, Y., 2015. PLGA–HPMC nanoparticles prepared by a modified supercritical anti-solvent technique for the controlled release of insulin. J. Super. Fluids 99, 15–22.

Zeng, L., Weber, A.P., 2014. Aerosol synthesis of nanoporous silica particles with controlled pore size distribution. J. Aerosol. Sci. 76, 1–12.

Zhang, X., Chiu Li, L., Mao, S., 2014. Nanosuspensions of poorly water soluble drugs prepared by top-down technologies. Curr. Pharm. Des. 20, 388–407.

Zuo, B., Sun, Y., Li, H., Liu, X., Zhai, Y., Sun, J., He, Z., 2013. Preparation and in vitro/in vivo evaluation of fenofibrate nanocrystals. Int. J. Pharm. 455, 267–275.

CHAPTER

POLYMER-BASED NANOCARRIERS FOR THERAPEUTIC NUCLEIC ACIDS DELIVERY

14

Weien Yuan, Hui Li
Shanghai Jiao Tong University, Shanghai, China

CHAPTER OUTLINE

1 INTRODUCTION

Nucleic acids, including DNA and RNA, are the basic genetic material of all life forms on Earth. The studies of nucleic acids have also paved the way for the development of biochemistry, molecular biology, biotechnology and modern medicine. DNA molecules were firstly isolated by Swiss physician Friedrich Miescher in 1869 (Dahm, 2008). The double helix structure of DNA was firstly proposed and described by James D. Watson and Francis Crick in 1953 (Watson and Crick, 1953). Moreover, the genetic codons were decoded in 1961 by Marshall Nirenberg and Heinrich J. Matthaei (Leder and Nirenberg, 1964). Reflecting upon the history of biomedical research, it is obvious that discoveries in nucleic acids have played indispensable roles in biomedical advances.

In recent years, with the rapid progress in genetics and genomic sciences, scientists have identified that many diseases including certain devastating cancers are actually caused by genetic mutations. One of the most important applications of therapeutic nucleic acids is gene therapy (Kay, 2011; Parekh-Olmedo et al., 2005). Theoretically, if the genetic mutation in diseased cells can be corrected or replaced by directly delivering nucleic acids to the target cells, the disease can be cured at the genetic level. Moreover, many previously undruggable targets such as microRNA and long non-coding RNA can be modulated by nucleic acids with high specificity and low toxicity. Since nucleic acid drugs hold enormous therapeutic potential to treat a variety of diseases in a different manner compared with small molecule and protein drugs, both basic and clinical studies of therapeutic nucleic acids have been carried out extensively (Aiuti et al., 2007; Alexander et al., 2007; Kanasty et al., 2013).

In general, the common mechanisms of action of therapeutic nucleic acids include (1) increasing the amount of naturally-occurring proteins through the delivery of protein-coding nucleic acids; (2) producing a cell-killing effect through the delivery of nucleic acids coding a cytotoxic protein; (3) producing a protein with the capacity to elicit immune response through the delivery of nucleic acids (Liu, 2003); (4) producing an enzyme through the delivery of enzyme-coding nucleic acids; (5) manipulating the splicing of mRNA by splice-switching oligonucleotides (Bauman et al., 2009); (6) modulating protein expression through the delivery of siRNA (Elbashir et al., 2001), miRNA (He and Hannon, 2004), anti-miRNA or antisense oligonucleotides (Weiss et al., 1999); and (7) stimulating the innate immune system through the delivery of immunostimulatory nucleic acids such as CpG DNA (Klinman, 2004).

However, despite their spectacular therapeutic potential, many nucleic-acid-based drugs also have poor drug-like properties such as low bioavailability, sensitivity to enzyme-mediated degradation, short half-life in vivo, and low permeability to cross the cell membrane. In addition, for the successful management of various chronic diseases, long-term stable and controllable gene expression or interference will be needed. Moreover, the safety of the nucleic-acid drugs is also critically important. Similar to small molecule and protein drugs, the toxicities and adverse side effects of nucleic acid drugs must be controlled within an acceptable range when they are used in the clinic.

To overcome these challenges, various delivery approaches and formulations were used to optimize the efficacy and safety profiles of nucleic-acid-based drugs such as virus-mediated delivery, lipid-mediated delivery, polymer-mediated delivery, and physical delivery methods. Among these, polymer-based nanocarriers have emerged as one of the most promising approaches for delivering therapeutic nucleic acids both in vitro and in vivo. Many successful examples have shown that synthetic polymers can be rationally designed to produce desired structure and function for optimizing the performance of therapeutic nucleic acids (Pack et al., 2005).

In this chapter, we will first discuss the challenges and barriers for therapeutic nucleic acids delivery, and then we will elaborate on the advantages, uniqueness, and recent progress of polymer-based nanocarriers for therapeutic nucleic acids delivery. Moreover, we will also discuss the considerations for manufacturing, safety issues, and regulatory requirements for these novel nanocarriers.

2 CURRENT CHALLENGES AND BARRIERS FOR THERAPEUTIC NUCLEIC ACIDS DELIVERY

2.1 STABILITY OF NUCLEIC ACIDS WITHIN A BIOLOGICAL ENVIRONMENT

Due to the abundance of various deoxyribonucleases (DNases) and ribonucleases (RNases) within the biological environment, naturally-occurring nucleic acids including DNA and RNA are notorious for their relatively low stability in vivo. DNases or RNases are enzymes capable of degrading DNA or RNA by catalyzing the hydrolytic cleavage of phosphodiester bonds in the DNA or RNA backbone. These enzymes are distributed everywhere in the body and play vital roles in maintaining the normal function of the body. In a short time, these enzymes can easily degrade foreign nucleic acids. For example, naked plasmid DNA without protection only has an apparent half-life of 50–90 min in cells (Lechardeur et al., 1999). However, studies have found the enzymatic stability of nucleic acids can be dramatically improved by chemical modification of the nucleic acids structure. Common modifications include phosphorothioate (PS) backbone modification (Iannitti et al., 2014), 2′-O-Methyl modification (Majlessi et al., 1998), 2′-Fluoro modification (Ono et al., 1997), locked nucleic acids (Jepsen et al., 2004), etc. In addition, nucleic acids encapsulated in liposomes or complexed with cationic polymers also have improved stability in the body since the liposomes or polymers can physically protect the nucleic acids from being degraded by the nucleases.

2.2 IN VIVO PHARMACOKINETICS AND BIODISTRIBUTION

The second challenge for therapeutic nucleic acids is their poor in vivo pharmacokinetics and biodistribution profile. From the injection site to the cell membrane of target cells, nucleic acid drugs will encounter a series of barriers. These barriers include: (1) physical and chemical barriers, such as whether the nucleic acid drug and its delivery vehicle can maintain its normal morphology and structural integrity and not disassociate within a certain period of time in the biological environment; (2) in vivo barriers, such as the ability to avoid the uptake by the reticuloendothelial system (RES) (Saba, 1970) or macrophage system, the ability to cross the vessel wall, the ability to pass through the surrounding tissue, and the ability to enter target cells specifically.

2.3 INTRACELLULAR BARRIERS FOR NUCLEIC-ACID DELIVERY

Even after therapeutic nucleic acids successfully enter the target cell, they will still face a number of intracellular barriers. For example, studies have shown that although more than 95% of the cells can perform endocytosis on DNA vectors (Tseng et al., 1997), only a portion (typically less than 50%) of the cells could successfully express the target gene. For polymer-based nucleic-acid-delivery vehicles, they usually enter cells by endocytosis. Once inside the endocytic pathway, they may be transported to the endosomes and lysosomes (Mellman, 1996). Endosomes and lysosomes have a strong acidic

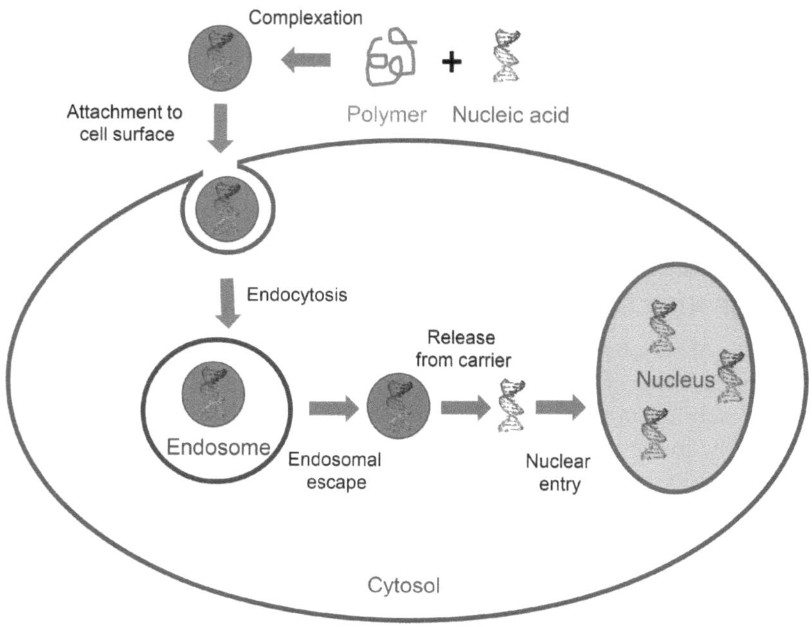

FIGURE 14.1 Barriers to Successful Nucleic Acid Therapeutics Delivery using Polymer-Based Nanocarriers

Adapted with some modification from Chen, W., Li, H., Liu, Z., Yuan, W., 2016. Lipopolyplex for Therapeutic Gene Delivery and Its Application for the Treatment of Parkinson's Disease. Front. Aging. Neurosci. 5, 68, with permission.

environment (pH about 4.5), and are filled with various enzymes. Nucleic acids and their carriers may be degraded in the endosomes and lysosomes before reaching their therapeutic targets. As a result, how to achieve the escape from the entrapment of endosomes and lysosomes is a key to leverage the therapeutic potential of nucleic acids therapeutics. Moreover, nucleic acid drugs also need to be released from their carriers to fulfill their therapeutic effect. In addition, for plasmids-based therapeutics, they also need to be transported across the nuclear membrane to enter into nucleus for successful gene expression. A summary of the intracellular barriers for nucleic acids delivery is illustrated in Fig. 14.1.

3 ADVANTAGES AND UNIQUENESS OF POLYMER-BASED NANOCARRIERS FOR ADDRESSING EXISTING CHALLENGES AND BARRIERS

3.1 CONDENSATION AND PROTECTION OF THERAPEUTIC NUCLEIC ACIDS BY POLYMERS

It is well known that cationic polymers can bind to nucleic acids via electrostatic interaction to form polyplex spontaneously, which is also called condensation. This is a thermodynamically favorable process because nucleic acids have phosphate backbones that are negatively charged, and binding with cationic polymer can neutralize the charge and stabilize and polyplex. Cationic polymers can be rationally designed to control specific properties such as molecular structure, molecular weight, length, surface charge, charge density, hydrophilicity, and biodegradability of the polymer. In general, the positive charge comes from amines including primary, secondary, tertiary, and/or quaternary amines. The pK_a of the amines need to be carefully studied to make sure that the amine group can have positive charge to form polyplex with nucleic acids under physiological conditions. After condensation, nucleic acids usually form nanoparticles or macroparticles with positive charge, which will facilitate their cellular uptake as well as protect the nucleic acids from nuclease-mediated degradation. Moreover, cationic polymers can also function as proton sponges to promote the endosomal escape of nucleic acids.

3.2 PASSIVE AND ACTIVE TARGETING OF POLYMER-BASED NANOCARRIERS

For nanoparticle-based drug-delivery systems, there are two kinds of targeting approaches: passive and active targeting. Passive targeting means the nanoparticles have a tendency to target certain organs or tissues without the effect of any targeting ligand. For example, nanoparticles with certain sizes tend to accumulate in the tumors more than they do in normal organs or tissues, which is named as the enhanced permeability and retention (EPR) effect (Maeda, 2001).On the other hand, active targeting refers to the approaches that involve the utilization of targeting ligands such as small molecules, proteins, antibodies, and nucleic acid aptamers. For example, small molecules such as folate can be conjugated to polymers for targeting folate-positive cancer cells (Zwicke et al., 2012). Transferrin can also be linked to nanocarriers for targeting tumor cells that express high levels of transferrin receptors (Daniels et al., 2012).

3.3 CONTROLLED RELEASE OF THERAPEUTIC NUCLEIC ACIDS

Another remarkable advantage of polymer-based nanocarriers is that the biodegradability of the polymer can be rationally adjusted to achieve the controlled release of its cargo in the physiological environment with therapeutic dosages. A wide range of biodegradable and biocompatible polymers such as poly(glycolic acid) (PGA) and poly(lactic-*co*-glycolic acid) (PLGA) have been extensively investigated for biomedical applications. These biodegradable polymers also can be used for controlled release of nucleic acids. For example, PLGA nanoparticles can be used to encapsulate siRNA for sustained release of siRNA (Pantazis et al., 2012). Moreover, nucleic acids can also be adsorbed on the cationic surface of chemically modified PLGA nanoparticles via the electrostatic interaction (Cun et al., 2010). The sustained release of therapeutic nucleic acids via formulation with biodegradable polymers has the potential to maintain the therapeutic effect of these nucleic acids in an extended time frame and reduce the number of required administrations, which has the potential to improve the patient compliance in the future.

4 RECENT EXAMPLES OF POLYMER-BASED NANOCARRIERS FOR THERAPEUTIC NUCLEIC-ACIDS DELIVERY

4.1 POLYETHYLENIMINE

Polyethylenimine (PEI) is a cationic polymer that contains repeating units composed of an amine group and a CH_2CH_2 spacer. Notably, PEI is an organic polymer that has the highest positive charge density potential as every third atom of this polymer is a protonable amino nitrogen. Two forms of PEI, including linear PEI and branched PEI, have been reported. It was first tested by Boussif et al. (1995) as a vector for gene and oligonucleotide transfection both in cell culture and in vivo. Due to its high charge density, it usually forms positive-charged complex with nucleic acids with high efficiency and provides efficient transfection and protection against nuclease-mediated degradation in cells. In addition, because of the large number of protonable amino nitrogen in PEI, it can also serve as a proton sponge to buffer the pH in the endosome and cause subsequent osmotic swelling, endosomal membrane rupture, and the eventual escape of the polymer-nucleic complex (polyplex) into the cytoplasm.

Both linear and branched forms of PEI have been tested for nucleic acid delivery. Wightman et al. (2001) reported the transfection efficiency of linear PEI22/DNA complex in vitro was generally greater than that of branched PEI/DNA complexes under salt conditions. They also proposed that the greater transfection efficiency of linear PEI might be a result of an inherent kinetic instability.

However, the safety of PEI is always a concern for its clinical translation. It has been reported that the positive charges of PEI could induce necrotic cell death and apoptosis and cause toxicity both in vitro and in vivo, which significantly limits its clinical usage. To reduce the toxicity as well as further improve the transfection efficiency of PEI, different chemical modifications of PEI have been explored. For example, Zintchenko et al. (2008) synthesized a series of derivates of branched PEI via modification of amines by ethyl acrylate, acetylation of primary amines, or conjugation of negatively charged chemical groups such as propionic acid or succinic acid groups. The resulting polymers showed improved gene silencing efficiency in siRNA-mediated gene knockdown. Moreover, the authors also found that succinylation of branched PEI could reduce the polymer toxicity up to tenfold compared with unmodified PEI. PEI with biodegradable link to facilitate the degradation of polymer also has the potential to provide reduced cytotoxicity. Xu et al. (2008) reported a PEI biscarbamate conjugate (PEIC) with a low molecular weight (MW = 2800, Mn = 910), which was synthesized by reacting low molecular weight PEI (MW = 800) with 1,4-butanediol bis(chloroformate) to generate biodegradable carbamate linkages. This modified PEI was shown to have lower cytotoxicity compared with 2 kDa or 25 kDa PEIs. This polymer synthesis strategy can also be extended to other biodegradable links. Recently, Wang et al. (2012), Che et al. (2016), Ge et al. (2014), Song et al. (2016), Xiang et al. (2012) also reported a small-molecular-weight PEI derivative with ethylene biscarbamate linkages, which also exhibited significantly lower cytotoxicity compared to unmodified 25 kDa PEI.

4.2 SPERMINE-BASED CATIONIC POLYMERS

Spermine is an endogenous polyamine bearing multiple amino groups. It has been found to play important roles in cellular metabolism in all eukaryotic cells. It also can condense DNA in sperm due to its positive charge in the physiological condition. Because spermine is an endogenous molecule widely present in the biological environment, it is expected that spermine will have better biocompatibility then other synthetic cationic molecules.

Recently, Du et al. (2012) reported the use of spermine as the basic building block to synthesize polycationic polymers via condensation with linker molecules including bischloroformate, succinyl chloride, and glyoxal (Fig. 14.2A). Each synthesized polymer could complex with the luciferase plasmid and form a polyplex with sizes of 150–200 nm and with a zeta potential of +15–30 mV when the polymer-to-DNA mass ratio was larger than 8/1. The authors also found that the choice of linker played a significant role in determining the biodegradability, toxicity, and transfection efficiency of the synthesized polymers. Another interesting example is polyspermine imidazole-4, 5-imine, which was

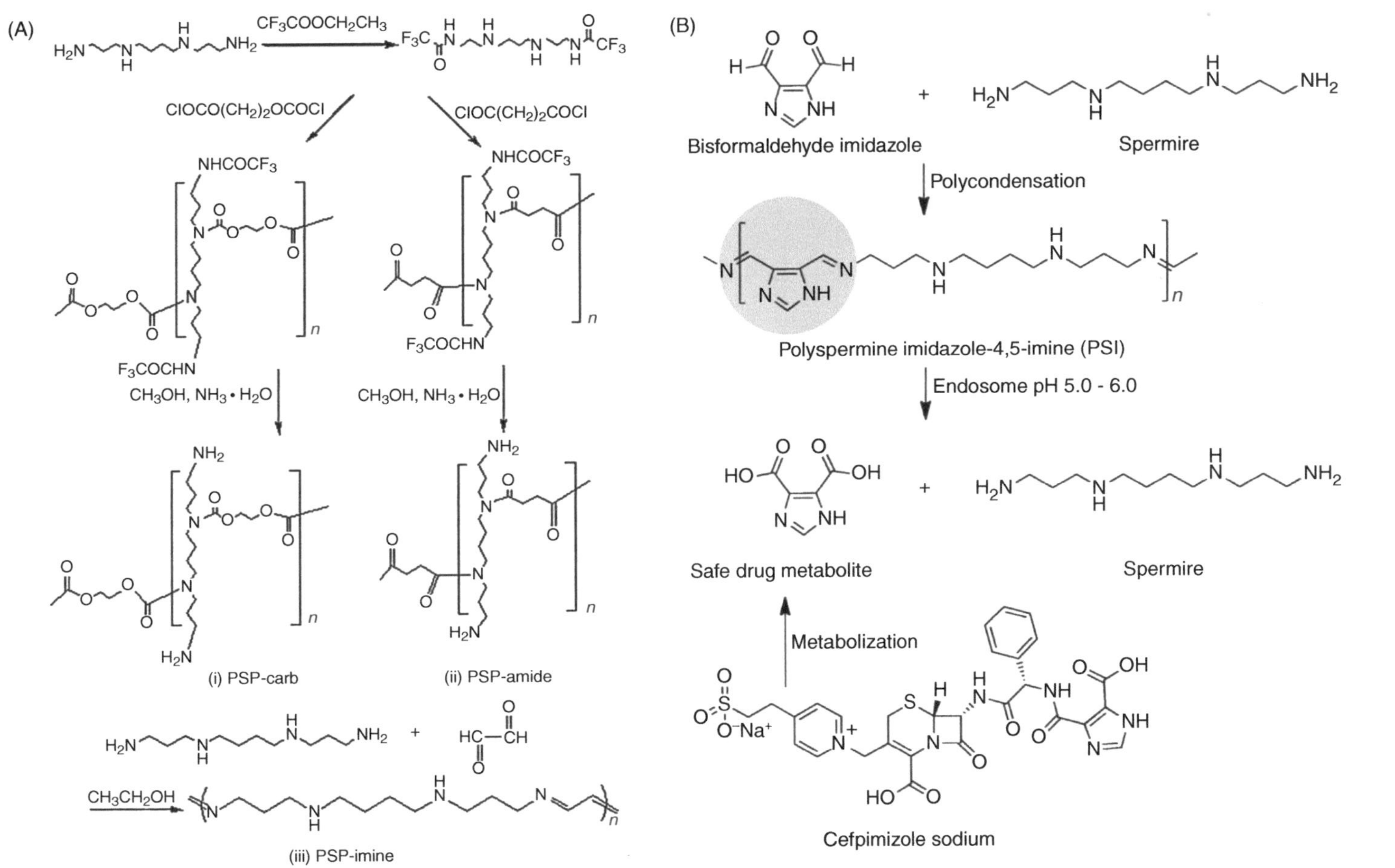

FIGURE 14.2

(A) Schemes for the synthesis of three cationic polymers that are based on spermine: (i) PSP-carb, (ii) PSP-amide, and (iii) PSP-imine. (B) Schemes for the synthesis and metabolism of polyspermine imidazole-4,5-imine.

Part A: Adapted from Du, Z.X., Chen, M.Y., He, Q.Q., Zhou, Y., Jin, T., 2012. Polymerized spermine as a novel polycationic nucleic acid carrier system. Int. J. Pharm. 434, 437–443, with permission, Copyright 2012 Elsevier B.V.; Part B: Adapted from Duan, S.Y., Yuan, W.E., Wu, F., Jin, T., 2012. Polyspermine imidazole-4,5-imine, a chemically dynamic and biologically responsive carrier system for intracellular delivery of siRNA. Angew. Chemie. Int. Ed. 51, 7938–7941, with permission, Copyright 2012 WILEY-VCH Verlag GmbH & Co.

chemically synthesized by condensing spermine with bisformaldehyde imidazole to form a pH-responsive linkage. This polymer could also efficiently condense siRNAs into nanoparticles when the ratio of amino groups of the polymer to phosphates of the nucleic acid (N/P ratio) was over 20. Moreover, both in vitro and in vivo assays showed that this spermine-based siRNA carrier can efficiently silence target genes with low toxicity (Fig. 14.2B) (Duan et al., 2012b). In 2012 and 2014, Duan et al. (2012a) and Du et al. (2014) reported another cationic polymer based on spermine as the basic building block: polyspermine imine, which is formed by condensing spermine and glyoxal and forming $\pi-\pi$ conjugated carbon–nitrogen double bonds in the polymer backbone. Biological assays demonstrated that this polymer could escort siRNA and induce gene silencing in target cells without visible cytotoxicity (Du et al., 2014).

4.3 CHITOSAN

Chitosan is a naturally-occurring linear cationic biopolymer composed of β-(1-4)-linked D-glucosamine and *N*-acetyl-D-glucosamine. It can be produced from shrimp and other crustacean shells. In general, it is considered to have good biocompatibility, biodegradability, and low toxicity.

Because chitosan has positive charge, similar to other cationic polymers, it is also able to form complexes with nucleic acids and protect them from nuclease-mediated degradation. However, because of its relative low positive charge density, unmodified chitosan also suffered from poor water solubility and relatively low transfection efficiency compared with PEI. As a result, numerous approaches for modifying chitosan to improve its capability for nucleic acids delivery have been explored. For example, Jiang et al. reported the use of polyethylene glycol (PEG)-grafted chitosan to form complexes with DNA and deliver the DNA cargo to the liver via bile duct and portal vein infusions. The authors found that the higher grafting degree of PEG to chitosan could induce higher transgene expression in other organs after portal vein infusion. In addition, toxicity evaluation also demonstrated that PEG grafting could reduce the acute liver toxicity caused by the polymer-DNA complexes (Jiang et al., 2006). Another interesting example for modified chitosan is polyethylenimine-graft-chitosan (PEI-g-chitosan) reported by Wong et al. (2006) (Fig. 14.3). The authors synthesized this copolymer by performing cationic polymerization of aziridine with oligo-chitosan. The authors found that the synthesized PEI-g-chitosan had a lower cytotoxicity but higher transfection efficiency than that of PEI 25 KDa.

4.4 CYCLODEXTRIN-BASED CATIONIC POLYMERS

Cyclodextrins are a family of molecules composed of cyclic oligomers of glucose. Cyclodextrins have been widely used in food, pharmaceutical, and chemical industries because of their low toxicity, low immunogenicity, and the ability to form noncovalent complexes with guest molecules. For example, cyclodextrins have been widely used as carriers to improve the water solubility of lipophilic vitamins and hormones.

Davis's group first demonstrated the usage of cyclodextrin-based cationic polymers for nucleic acids delivery in 1999 (Fig. 14.4A) (Gonzalez et al., 1999). The authors showed that positively charged β-cyclodextrin-containing polymers could be synthesized by copolymerizing difunctionalized β-cyclodextrin monomers with other difunctionalized comonomers. The synthesized β-cyclodextrin polymers could condense the pDNA into nanoparticles with 100–150 nm in size, which could transfect cells in vitro with comparable efficiency to PEI and Lipofectamine. Furthermore, Davis's group

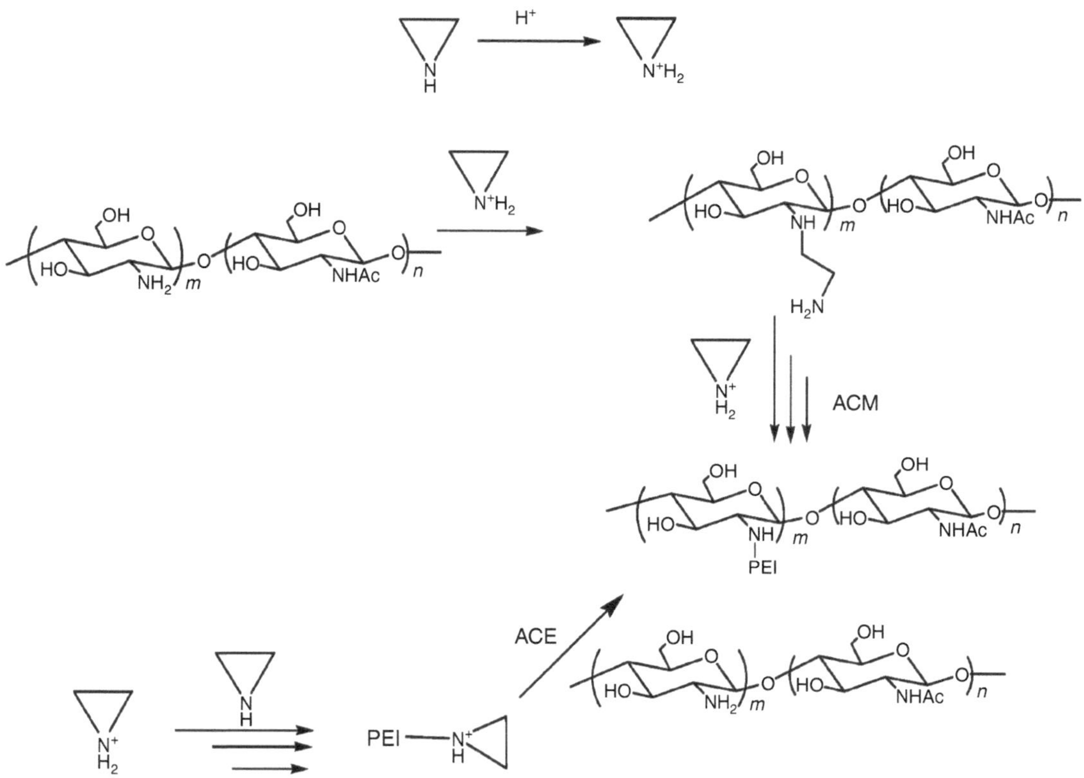

FIGURE 14.3 Scheme for the Synthesis of PEI-g-Chitosan

Adapted from Wong, K., Sun, G.B., Zhang, X.Q., Dai, H., Liu, Y., He, C.B., Leong, K.W., 2006. PEI-g-chitosan, a novel gene delivery system with transfection efficiency comparable to polyethylenimine in vitro and after liver administration in vivo. Bioconjugate Chem. 17, 152–158, with permission. Copyright 2006, American Chemical Society.

modified the cationic β-cyclodextrin polymers by complexing with adamantane−PEG conjugates or adamantane−PEG-ligand conjugates, which improved the stability of the polyplexes in physiological media and provided targeting ability to the polyplexes (Fig. 14.4B) (Pun and Davis, 2002). In 2010, Davis et al. (2010) also demonstrated that systemically administered nanoparticles composed of a cyclodextrin-based polymer, a transferrin targeting ligand, a PEG polymer and RRM2 siRNA could induce a specific gene knockdown in both mRNA and protein level in human, which clearly showed the potential for translating the cyclodextrin-based nucleic acids carriers from the bench to the bedside.

4.5 DENDRIMERS

Dendrimers are a kind of chemically synthesized spherical macromolecule with repetitively branched structures. Dendrimers are usually synthesized by either divergent or convergent strategies. The number of repeated cycles involved in the synthesis of the dendrimer is defined as the generation of the dendrimer. The properties of dendrimer greatly depend on their molecular weight and the functional group

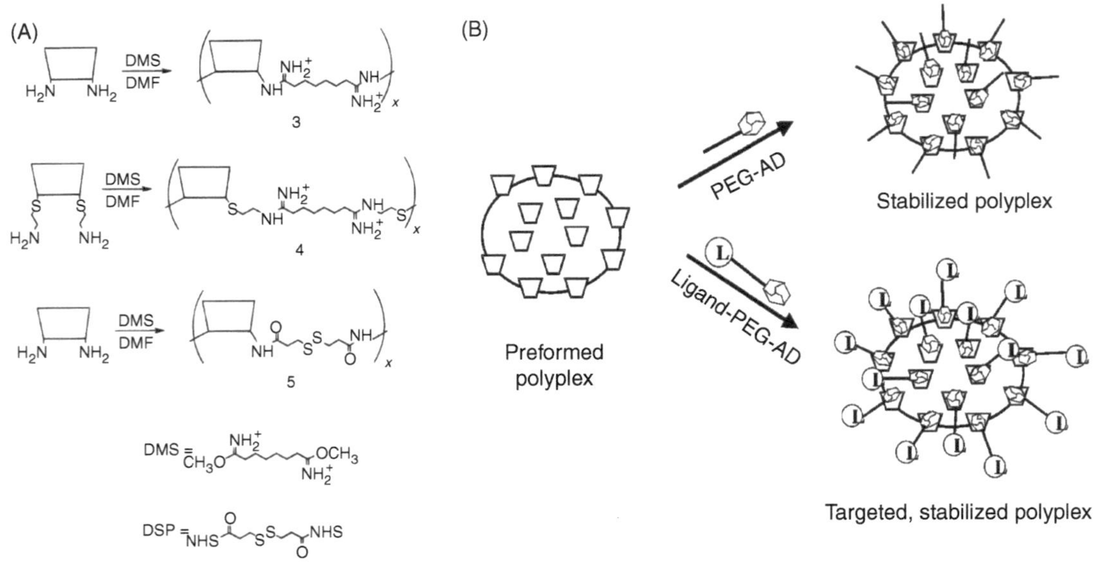

FIGURE 14.4

(A) Scheme for the synthesis of the cationic β-cyclodextrin polymers. (B) Scheme for post-DNA-complexation PEGylation of the cationic β-cyclodextrin polymers by complexing with adamantane–PEG conjugates or adamantane–PEG-ligand conjugates.

Part A: Adapted from Gonzalez, H., Hwang, S.J., Davis, M.E., 1999. New class of polymers for the delivery of macromolecular therapeutics. Bioconjugate Chem. 10, 1068–1074, with permission. Copyright 1999, American Chemical Society; Part B: Adapted from Pun, S.H., Davis, M.E., 2002. Development of a nonviral gene delivery vehicle for systemic application. Bioconjugate Chem. 13, 630–639, with permission, Copyright 2002, American Chemical Society.

embedded in their chemical structures. Due to their precise chemical composition and high chemical versatility, dendrimers have also been widely explored for drug delivery, as well as for the delivery of nucleic acids. Similar to other cationic polymers, positively charged dendrimers can form a complex with nucleic acids and increase their cellular uptake. In 1993, Szoka's group reported the polyamidoamine dendrimer-mediated plasmids transfection in both suspension and adherent cultured mammalian cells (Haensler and Szoka, 1993). The authors showed that maximal transfection efficiency is achieved by using a 68 Å dendrimer and a terminal amine to phosphate ratio of 6:1. It was proposed that the low pK_a's of the amine groups in the polymer played important roles to buffer the pH change in the endosomes. However, cytotoxicity was also observed after the transfection.

Poly-L-lysine dendrimers are also attractive carriers for nucleic acids delivery due to the well-known positive charge of lysine. Kaneshiro et al. (2007) synthesized a series of poly-L-lysine dendrimers with a cubic octa (3-aminopropyl)silsesquioxane (OAS) core (Fig. 14.5). Matrix-assisted laser desorption time of flight (MALDI-TOF) mass spectrometry confirmed the precisely defined molecular structures of the dendrimer. In vitro studies showed that the synthesized dendrimers could efficiently deliver plasmids to MDA-MB-231 breast cancer cells while possessing lower cytotoxicity than that of linear poly-L-lysine.

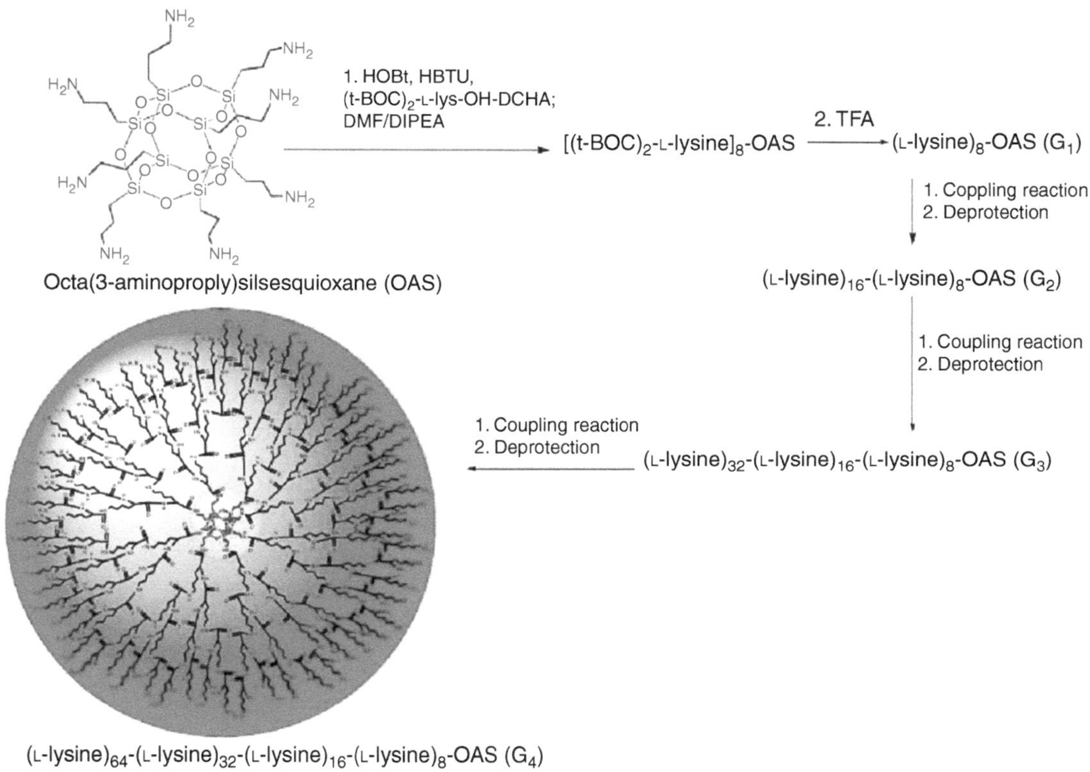

FIGURE 14.5 Scheme for the Synthesis of Poly-L-Lysine Dendrimers with a Cubic Octa(3-Aminopropyl) Silsesquioxane (OAS) Core

Adapted from Kaneshiro, T.L., Wang, X., Lu, Z.R., 2007. Synthesis, characterization, and gene delivery of poly-L-lysine octa(3-aminopropyl)silsesquioxane dendrimers: nanoglobular drug carriers with precisely defined molecular Architectures. Mol. Pharm. 4, 759–768, with permission, Copyright 2007, American Chemical Society.

4.6 LIPOPOLYPLEX

In general, lipopolyplexes are referred to nanoparticles formed by the combination of cationic polymer, cationic lipid, and nucleic acids (Chen et al., 2016). Lipopolyplexes have drawn attention in the nucleic acid therapeutics community due to their versatility and relatively high transfection efficiencies. Lipopolyplexes have the potential to combine both the advantages of lipid- and polymer-based delivery systems. For example, Pelisek et al. (2006) studies the formulation of lipopolyplexes composed of different cationic polymers (linear PEI, branched PEI, and poly-L-lysine) and cationic lipids (DOCSPER, DOSPER, and DOTAP). They showed that optimized lipopolyplex formulations with medium sizes of 140–220 nm were not only stable against aggregation at physiological ionic strength but also exhibited an up to 400-fold enhanced gene transfer efficiency compared with the corresponding lipoplexes or 25 kDa branched PEI polyplexes. Recently, Munye et al. (2015) studied the synergistic effect of liposome and peptide in enhancing the transfection efficiency with a lipopolyplex optimized for airway gene delivery. The authors showed that the peptide components could promote efficient cellular uptake, and the liposome component could facilitate endosomal escape. So lipopolyplexes might represent a new generation of nanocarriers for therapeutic nucleic acids as lipopolyplexes have both desired features from lipoplex and polyplex for gene transfer.

5 OTHER CONSIDERATIONS FOR POLYMER-BASED NANOCARRIERS FOR THERAPEUTIC NUCLEIC ACIDS DELIVERY

5.1 LARGE SCALE MANUFACTURING

It is important to note that large scale manufacturing of polymer-based nanocarriers is definitely required for their successful clinical translation, since a large amount of the material will be needed for clinical trial. However, how to identify a most efficient approach to produce these nanocarriers on large scales in a reproducible manner can be a very challenging task. The manufacturing of polymer-based nanocarriers is significantly different from that of traditional small-molecule or protein-based drugs. More pilot research should be performed to gain critical insights regarding the manufacturing of these nanomedicines. Quality control, impurity analysis, automation, and standardization of the manufacturing process must be carefully designed in a case-by-case manner to guarantee the final products will have the desired physiochemical and biological properties. Moreover, in order to minimize the use and generation of hazardous chemicals, the concept of green chemistry should also be taken into consideration when designing the process of large-scale manufacturing (Mulvihill et al., 2011).

5.2 SAFETY ISSUES

Safety is another major bottleneck for the successful clinical translation of polymer-based nanocarriers for nucleic acids. Since polymer-based nanocarriers are a new category of therapeutics, there might be some unexpected adverse side effects. To ensure the safety of using these nanocarriers in vivo, the short-term toxicity, long-term toxicity, immunogenicity, and the maximum tolerated dose must be carefully and systemically evaluated. The potential interaction of these nanocarriers with other drugs in vivo should also be considered and studied, since nanocarriers may alter both pharmacokinetics and pharmacodynamics of other drugs. Moreover, the development of novel polymers with better safety profiles and biocompatibility should also be continually pursued. In addition, biodegradable polymers that can be degraded into nontoxic metabolites inside the body might also be good candidates for nucleic acids delivery.

5.3 REGULATORY REQUIREMENTS AND APPROVAL

Before every new therapeutic agent goes to the market, it must fulfill various regulatory requirements and be approved by the administration agency such as the Food and Drug Administration (FDA) and the European Medicines Agency (EMA). To reach the regulatory requirements, good laboratory practice (GLP), good manufacturing practice (GMP), and good clinical practice (GCP) regulations must be strictly followed in various stages of the drug discovery and development process. For polymer-based nanocarriers, the regulatory requirements and approval decision-making process might be different from traditional therapeutics. A survey published in 2013 showed that there are more than 247 nanomedicine products have been approved by FDA and are currently in clinical trials (Etheridge et al., 2013). Studying these successful examples might provide valuable insights for future investigational new drug (IND) filing and clinical development of new polymer-based nanocarriers for nucleic acids.

6 CONCLUSIONS

Because of the key role nucleic acids played in both normal physiological and pathophysiological processes, nucleic acid therapeutics has recently become one of the hottest research areas in drug discovery and development. Nucleic acid drugs have the potential to emerge as the third generation of new drugs following traditional small molecule and protein drugs. Numerous efforts have been made to overcome the barriers for successful nucleic acids delivery both in vitro and in vivo. The utilization of polymer-based nanocarriers is one of the most promising approaches for optimizing the "drug-like" properties of nucleic acids such as pharmacokinetics, bioavailability, and tissue targeting. The ultimate goal will be the successful translation of nucleic-acid therapeutics from basic research to the clinic. Future research into large-scale manufacturing, safety profiles, and regulatory requirements of these nanomedicines should expedite their approval and clinical translation. It is also expected that there is still plenty of room for further optimizing existing polymers and/or inventing novel polymers for systemic therapeutic nucleic acids delivery.

ACKNOWLEDGMENTS

This study is financially supported by the grants of National Natural Science Foundation of China (No. 81570992), Projects of the Shanghai Committee of Science and Technology, China (No.15ZR1432500) and Funds for Interdisciplinary Projects of Medicine and Engineering by Shanghai JiaoTong University (No.YG2015MS06, YG2013MS52, YG2013MS62, and YG2014QN06).

REFERENCES

Aiuti, A., Bachoud-Levi, A.C., Blesch, A., Brenner, M.K., Cattaneo, F., Chiocca, E.A., Gao, G., High, K.A., Leen, A.M., Lemoine, N.R., Mcneish, I.A., Meneguzzi, G., Peschanski, M., Roncarolo, M.G., Strayer, D.S., Tuszynski, M.H., Waxman, D.J., Wilson, J.M., 2007. Progress and prospects: gene therapy clinical trials (part 2). Gene Ther. 14, 1555–1563.

Alexander, B.L., Ali, R.R., Alton, E.W., Bainbridge, J.W., Braun, S., Cheng, S.H., Flotte, T.R., Gaspar, H.B., Grez, M., Griesenbach, U., Kaplitt, M.G., Ott, M.G., Seger, R., Simons, M., Thrasher, A.J., Thrasher, A.Z., Yla-Herttuala, S., 2007. Progress and prospects: gene therapy clinical trials (part 1). Gene Ther. 14, 1439–1447.

Bauman, J., Jearawiriyapaisarn, N., Kole, R., 2009. Therapeutic potential of splice-switching oligonucleotides. Oligonucleotides 19, 1–13.

Boussif, O., Lezoualch, F., Zanta, M.A., Mergny, M.D., Scherman, D., Demeneix, B., Behr, J.P., 1995. A versatile vector for gene and oligonucleotide transfer into cells in culture and in-Vivo—polyethylenimine. Proc.Natl. Acad. Sci. USA 92, 7297–7301.

Che, J., Tao, A., Chen, S., Li, X., Zhao, Y., Yuan, W., 2016. Biological response carrier-mediated anti-angiogenesis shRNA for tumor treatment. Sci. Rep. 6, 35661.

Chen, W., Li, H., Liu, Z., Yuan, W., 2016. Lipopolyplex for therapeutic gene delivery and its application for the treatment of parkinson's disease. Front. Aging. Neurosci. 5, 68.

Cun, D., Foged, C., Yang, M., Frokjaer, S., Nielsen, H.M., 2010. Preparation and characterization of poly(D,L-lactide-*co*-glycolide) nanoparticles for siRNA delivery. Int. J. Pharm. 390, 70–75.

Dahm, R., 2008. Discovering DNA: Friedrich Miescher and the early years of nucleic acid research. Hum. Genet. 122, 565–581.

Daniels, T.R., Bernabeu, E., Rodriguez, J.A., Patel, S., Kozman, M., Chiappetta, D.A., Holler, E., Ljubimova, J.Y., Helguera, G., Penichet, M.L., 2012. The transferrin receptor and the targeted delivery of therapeutic agents against cancer. Biochim. Biophys. Acta 1820, 291–317.

Davis, M.E., Zuckerman, J.E., Choi, C.H.J., Seligson, D., Tolcher, A., Alabi, C.A., Yen, Y., Heidel, J.D., Ribas, A., 2010. Evidence of RNAi in humans from systemically administered siRNA via targeted nanoparticles. Nature 464, 1067–1070.

Du, Z.X., Chen, M.Y., He, Q.Q., Zhou, Y., Jin, T., 2012. Polymerized spermine as a novel polycationic nucleic acid carrier system. Int. J. Pharm. 434, 437–443.

Du, Z.X., Xiang, S.N., Zang, Y., Zhou, Y., Wang, C.D., Tang, H.L., Jin, T., Zhang, X.L., 2014. Polyspermine imine, a pH responsive polycationic siRNA carrier degradable to endogenous metabolites. Mol. Pharm. 11, 3300–3306.

Duan, S., Ge, X., Lu, N., Wu, F., Yuan, W., Jin, T., 2012a. Synthetic, polyspermine imidazole-4, 5-amide as an efficient and cytotoxicity-free gene delivery system. Int. J. Nanomedicine 7, 3813–3822.

Duan, S.Y., Yuan, W.E., Wu, F., Jin, T., 2012b. Polyspermine imidazole-4,5-imine, a chemically dynamic and biologically responsive carrier system for intracellular delivery of siRNA. Angewandte Chemie-Int. Ed. 51, 7938–7941.

Elbashir, S.M., Harborth, J., Lendeckel, W., Yalcin, A., Weber, K., Tuschl, T., 2001. Duplexes of 21-nucleotide RNAs mediate RNA interference in cultured mammalian cells. Nature 411, 494–498.

Etheridge, M.L., Campbell, S.A., Erdman, A.G., Haynes, C.L., Wolf, S.M., Mccullough, J., 2013. The big picture on nanomedicine: the state of investigational and approved nanomedicine products. Nanomedicine 9, 1–14.

Ge, X., Feng, J., Chen, S., Zhang, C., Ouyang, Y., Liu, Z., Yuan, W., 2014. Biscarbamate cross-linked low molecular weight Polyethylenimine polycation as an efficient intra-cellular delivery cargo for cancer therapy. J. Nanobiotech. 12, 13.

Gonzalez, H., Hwang, S.J., Davis, M.E., 1999. New class of polymers for the delivery of macromolecular therapeutics. Bioconjugate Chem. 10, 1068–1074.

Haensler, J., Szoka, F.C., 1993. Polyamidoamine cascade polymers mediate efficient transfection of cells in culture. Bioconjugate Chem. 4, 372–379.

He, L., Hannon, G.J., 2004. MicroRNAs: small RNAs with a big role in gene regulation. Nat. Rev. Genet. 5, 522–531.

Iannitti, T., Morales-Medina, J.C., Palmieri, B., 2014. Phosphorothioate oligonucleotides: effectiveness and toxicity. Curr. Drug Targ. 15, 663–673.

Jepsen, J.S., Sorensen, M.D., Wengel, J., 2004. Locked nucleic acid: a potent nucleic acid analog in therapeutics and biotechnology. Oligonucleotides 14, 130–146.

Jiang, X., Dai, H., Leong, K.W., Goh, S.H., Mao, H.Q., Yang, Y.Y., 2006. Chitosan-g-PEG/DNA complexes deliver gene to the rat liver via intrabiliary and intraportal infusions. J. Gene Med. 8, 477–487.

Kanasty, R., Dorkin, J.R., Vegas, A., Anderson, D., 2013. Delivery materials for siRNA therapeutics. Nat Mater 12, 967–977.

Kaneshiro, T.L., Wang, X., Lu, Z.R., 2007. Synthesis, characterization, and gene delivery of Poly-L-lySine octa(3-aminopropyl)silsesquioxane dendrimers: nanoglobular drug carriers with precisely defined molecular Architectures. Mol.Pharm. 4, 759–768.

Kay, M.A., 2011. State-of-the-art gene-based therapies: the road ahead. Nat. Rev. Genet. 12, 316–328.

Klinman, D.M., 2004. Immunotherapeutic uses of CpG oligodeoxynucleotides. Nat. Rev. Immunol. 4, 249–258.

Lechardeur, D., Sohn, K.J., Haardt, M., Joshi, P.B., Monck, M., Graham, R.W., Beatty, B., Squire, J., O'brodovich, H., Lukacs, G.L., 1999. Metabolic instability of plasmid DNA in the cytosol: a potential barrier to gene transfer. Gene Ther. 6, 482–497.

Leder, P., Nirenberg, M.W., 1964. RNA codewords and protein synthesis, 3. On the nucleotide sequence of a cysteine and a leucine RNA codeword. Proc. Natl. Acad. Sci.USA 52, 1521–1529.

Liu, M.A., 2003. DNA vaccines: a review. J Intern. Med. 253, 402–410.

Maeda, H., 2001. The enhanced permeability and retention (EPR) effect in tumor vasculature: the key role of tumor-selective macromolecular drug targeting. Adv. Enzyme Regul. 41, 189–207.

Majlessi, M., Nelson, N.C., Becker, M.M., 1998. Advantages of 2'-O-methyl oligoribonucleotide probes for detecting RNA targets. Nucleic Acids Res. 26, 2224–2229.

Mellman, I., 1996. Endocytosis and molecular sorting. Annu. Rev. Cell Dev. Biol. 12, 575–625.

Mulvihill, M.J., Beach, E.S., Zimmerman, J.B., Anastas, P.T., 2011. Green chemistry and green engineering: a framework for sustainable technology development. Ann. Rev. Environ. Resources 36 (36), 271–293.

Munye, M.M., Ravi, J., Tagalakis, A.D., Mccarthy, D., Ryadnov, M.G., Hart, S.L., 2015. Role of liposome and peptide in the synergistic enhancement of transfection with a lipopolyplex vector. Sci. Rep. 5, 9292.

Ono, T., Scalf, M., Smith, L.M., 1997. 2'-Fluoro modified nucleic acids: polymerase-directed synthesis, properties and stability to analysis by matrix-assisted laser desorption/ionization mass spectrometry. Nucleic Acids Res. 25, 4581–4588.

Pack, D.W., Hoffman, A.S., Pun, S., Stayton, P.S., 2005. Design and development of polymers for gene delivery. Nat. Rev. Drug Discov. 4, 581–593.

Pantazis, P., Dimas, K., Wyche, J.H., Anant, S., Houchen, C.W., Panyam, J., Ramanujam, R.P., 2012. Preparation of siRNA-encapsulated PLGA nanoparticles for sustained release of siRNA and evaluation of encapsulation efficiency. Methods Mol. Biol. 906, 311–319.

Parekh-Olmedo, H., Ferrara, L., Brachman, E., Kmiec, E.B., 2005. Gene therapy progress and prospects: targeted gene repair. Gene Ther. 12, 639–646.

Pelisek, J., Gaedtkel, L., Derouchey, J., Walkerl, G.F., Niko, S., Wagner, E., 2006. Optimized lipopolyplex formulations for gene transfer to human colon carcinoma cells under in vitro conditions. J. Gene Med. 8, 186–197.

Pun, S.H., Davis, M.E., 2002. Development of a nonviral gene delivery vehicle for systemic application. Bioconjugate Chem. 13, 630–639.

Saba, T.M., 1970. Physiology and physiopathology of the reticuloendothelial system. Arch. Intern. Med. 126, 1031–1052.

Song, J., Chen, Y., Jiang, S., Yang, K., Li, X., Zhao, X., Ouyang, Y., Fan, C., Yuan, W., 2016. Efficient and nontoxic biological response carrier delivering TNF-α shRNA for gene silencing in a murine model of rheumatoid arthritis. Front Immunol. 7, 305.

Tseng, W.C., Haselton, F.R., Giorgio, T.D., 1997. Transfection by cationic liposomes using simultaneous single cell measurements of plasmid delivery and transgene expression. J. Biol. Chem. 272, 25641–25647.

Wang, Y.Q., Su, J., Wu, F., Lu, P., Yuan, L.F., Yuan, W.E., Sheng, J., Jin, T., 2012. Biscarbamate cross-linked polyethylenimine derivative with low molecular weight, low cytotoxicity, and high efficiency for gene delivery. Intl. J. Nanomed. 7, 693–704.

Watson, J.D., Crick, F.H., 1953. Molecular structure of nucleic acids; a structure for deoxyribose nucleic acid. Nature 171, 737–738.

Weiss, B., Davidkova, G., Zhou, L.W., 1999. Antisense RNA gene therapy for studying and modulating biological processes. Cell Mol. Life Sci. 55, 334–358.

Wightman, L., Kircheis, R., Rossler, V., Carotta, S., Ruzicka, R., Kursa, M., Wagner, E., 2001. Different behavior of branched and linear polyethylenimine for gene delivery in vitro and in vivo. J. Gene Med. 3, 362–372.

Wong, K., Sun, G.B., Zhang, X.Q., Dai, H., Liu, Y., He, C.B., Leong, K.W., 2006. PEI-g-chitosan, a novel gene delivery system with transfection efficiency comparable to polyethylenimine in vitro and after liver administration in vivo. Bioconjugate Chem. 17, 152–158.

Xiang, S., Su, J., Tong, H., Yang, F., Tong, W., Yuan, W., Wu, F., Wang, C., Jin, T., Dai, K., Zhang, X., 2012. Small molecular cross-linked poly(ethylene imine) through biscarbamate linkage, a promising tool to deliver IL-1Ra gene to synoviral cells for arthritis therapy. Biomaterials 33, 6520–6532.

Xu, S.L., Chen, M.Y., Yao, Y., Zhang, Z.Q., Jin, T., Huang, Y., Zhu, H., 2008. Novel poly(ethylene imine) biscarbamate conjugate as an efficient and nontoxic gene delivery system. J. Control. Rel. 130, 64–68.

Zintchenko, A., Philipp, A., Dehshahri, A., Wagner, E., 2008. Simple modifications of branched PEI lead to highly efficient siRNA carriers with low toxicity. Bioconjugate Chem. 19, 1448–1455.

Zwicke, G.L., Mansoori, G.A., Jeffery, C.J., 2012. Utilizing the folate receptor for active targeting of cancer nanotherapeutics. Nano. Rev. 3, 18496.

CHAPTER

15 MULTIFUNCTIONAL THERAPEUTIC HYBRID NANOCARRIERS FOR TARGETED AND TRIGGERED DRUG DELIVERY: RECENT TRENDS AND FUTURE PROSPECTS

Gulbin Kurtay*, Mehmet Yılmaz[†], Filiz Kuralay[‡], Gokcen B. Demirel**

**Ankara University, Ankara, Turkey*

***Gazi University, Ankara, Turkey*

[†]Sinop University, Sinop, Turkey

[‡]Ordu University, Ordu, Turkey

CHAPTER OUTLINE

ABBREVIATIONS

ADPC Acryloyloxy dodecyl phosphorylcholine
αvβ3 Alpha-v beta-3
RGD Arginine-glycine-aspartic acid
BGE Benzyl-functional glycidyl ether
BPDC Brush polymer scaffold drug carrier
CNP Capillary nonperfusion

CNT	Carbon nanotube
CTAB	Cetyltrimethylammonium bromide
Ch	Chitlac
CCDP	2-Chloro-3′,4′-dihydroxyacetophenone
Cit	Citraconic amide
CT	Computed tomography
α-CD	α-cyclodextrin
CytC	Cytochrome C
DNA	Deoxyribonucleic acid
DBCO	Dibenzocyclooctynes
DMAEMA	*N*,*N*-Dimethylaminoethyl methacrylate
MTT	3-(4,5-Dimethylthiazol-2-yl)-2,5-diphenyltetrazolium bromide
DSmPEG	Disulfide methylene oxide poly(ethylene glycol)
DDA	Dodecylamine
DOX	Doxorubicin
EDA	Ethylenediamine
BMSN	Bi_2S_3-Embedded mesaporous silica nanoparticle
EPR	Enhanced permeability and retention
GFLG	Glycylphenylalanylleucylglycine
GSH	Glutathione
AS1411	G-rich oligonucleotide DNA aptamer
AuNC	Gold nanocage
AuNP	Gold nanoparticle
AuNR	Gold nanorod
AuNS	Gold nanoshell
Gy	Gray
HE	Hematoxylin and eosin
HGE	Histamine glycidyl ether
HSMSN	Hollow mesoporous silica nanoparticle
HUVEC	Human umbilical vein endothelial cells
MiaPaCa-2	Human pancreas carcinoma
HYD	Hydrazine
HPMA	*N*-(2-Hydroxypropyl) methacrylamide
ICG	Indocyanine green
LSPR	Localized surface plasmon resonance
Mal	6-Maleimidocaproyl
MR	Magnetic resonance
MRI	Magnetic resonance imaging
MCN-41	Mobil Composition of Matter No. 41
MSN	Mesoporous silica nanoparticle
mPEG	Methoxy poly(ethylene glycol)
mUCNP	Mesoporous core–shell structured up-conversion nanoparticles
PEG-silane	2-Methoxy(polyethyleneoxy)propyl-trimethoxysilane
NIR	Near infrared region
NP	Nanoparticle
NP-shell	Nanoparticle-in shell
NR-shell	Nanorod-in shell
NV-MSN	Nanovalve-mesoporous silica nanoparticle

pK_a	Negative Log of K_a
OGE	Octyl glycidyl ether
PAE	Poly(β-aminoester)
PASP	Poly(aspartate)
PDT	Photodynamic therapy
PEG	Polyethylene glycol
PLA	Poly(lactic acid)
PLGA	Poly(lactic-*co*-glycolic acid)
PSN	Porous silica nanoparticle
PET	Positron emission tomography
pH	Power of hydrogen
Plu	Pluronic
PTT	Photothermal therapy
QD	Quantum dot
ROS	Reactive oxygen species
rGO	Reduced graphene oxide
RES	Reticuloendothelial system
RNA	Ribonucleic acid
SCID	Severe combined immunodeficiency
AgNP	Silver nanoparticle
SDT	Sonodynamic therapy
siRNA	Small interfering RNA
SP	Spiropyran
SPMA	Spiropyran (SP) and phenylboronic acid conjugated poly(dimethylamino ethyl methacrylate-*co*-methyl acrylic acid)
SPAAC	Strain-promoted alkyne azide cycloaddition
SUN	Sunitinib
TUNEL	Terminal deoxynucleotidyl transferase dUTP nick end labeling
Tf	Transferrin
TRIS	Tris(hydroxymethyl)aminomethane
UCNP	Upconversion nanoparticle
US	Ultrasound
UV	Ultraviolet

1 INTRODUCTION

Cancer is an umbrella term given to an assemblage of related diseases and defined as uncontrolled growth of the abnormal cells in the body (American Cancer Society, 2015). This sneaky disease can develop various strategies such as self-sufficiency in growth signals, metastasis, and unlimited replication to guarantee its survival (Hanahan and Weinberg, 2000).

In cancer therapy, the world-accepted standard approach includes two crucial steps: diagnosis and treatment. Early stage cancer diagnosis is drastically important to enhance the treatment efficiency and to prolong the possibility of patient survival. Therefore, bioimaging is considered as a very powerful diagnosis tool for the early stage detection of cancer, and this technique is based on tracking the exact location of cancer cells or tumor tissues during the therapy regimens. Different bioimaging methods are used in clinical trials such as magnetic resonance imaging (MRI), positron emission tomography

(PET), medical optical imaging, and so forth. Accordingly, researchers have focused on developing assorted type of nanoparticle-based contrast agents, which are usable in different bioimaging processes. Subsequent to detection of the disease, an efficient therapy must be implemented to the patients. Unluckily, the most used traditional cancer treatment methods such as surgery, radiotherapy, and chemotherapy (Sudhakar, 2009) can cause several irreversible side effects to human body (Cho et al., 2008).

Nowadays, many newly developed drugs are being commercialized as anticancer agents. However, most of the drugs fail in clinical tests due to the therapeutic and toxicological limitations, such as drug resistance of cell and the cell's barrier effect (Lamprecht, 2007). Basically, the transportation of anticancer drugs within a cancer cell is regulated by the physicochemical nature of the interstitium region and characteristics of the drug molecule (i.e., hydrophobicity, size, and charge) (Jain, 2001). In the body or in a specific tissue, the distribution of drugs mainly is governed by their physiochemical properties including hydrophilicity, pK_a, and electrostatic charges. Unfortunately, in most cases, anticancer drugs cannot fulfill some critical requirements for the efficient therapy of tumor cites. This insufficiency leads to a great deal of distribution of drug in healthy tissue rather than in cancerous tissue, and that is the basic limitation of conventional chemotherapy procedures (Brigger et al., 2002). To overcome the nonspecific distribution, molecularly targeted chemotherapeutic agents has been proposed as a second generation of anticancer drugs (Sagnella and Drummond, 2012). Despite numerous attempts, similar to their first generation counterparts, second generation drug systems still have main limitations including nonspecific biodistribution, rapid clearance, and rapid degradation. Furthermore, most of the chemotherapeutic agents are hydrophobic and have quite low solubility in water. The hydrophobic nature of drug molecules results in formulation problems in drug preparation stages and in systemic administration. The employment of pharmaceutical solvents has been proposed to circumvent this problem. The life-threatening effects of these solvents restrict their usage in clinical applications (Feng and Chien, 2003; Torchilin, 2004).

Conventional anticancer agents, with their relatively low molecular mass, are rapidly distributed throughout the body within the systemic circulation. In most cases, the drug system is opsonized and rapidly cleared from the bloodstream by the reticuloendothelial system's (RES) defense mechanism and rapidly metabolized by the liver and/or rapidly excreted by the kidney (Moghimi and Patel, 1998; Torchilin, 2004; Woodle, 1998). Therefore, it is clear that novel drug delivery systems should be developed to provide effective targeting before being excreted from the body. An ideal drug delivery system should have a proper a pharmacokinetic profile to allow the interaction between the drug and its target (Perrie and Rades, 2012; Zahr and Pishko, 2009).

For this purpose, researchers focus on developing a new generation of nanoparticle-based biomaterials to reduce these side effects. Many nanoparticle-based delivery systems have been reported in the literature, and they provide several advantages compared to the traditional methods such as (1) optimal use of drug (2) improvement in the efficacy of the drug (3) reduction of several side effects (4) maintenance of the drug levels and finally, (5) improved patient compliance due to the fewer administration (Winey and Discher, 2006).

Unfortunately, among the many nanoparticles, only a few are suitable for clinical applications. The human body is unquestionably among the most complex systems that has perfect barriers protecting itself from external materials. In this manner, after the nanoparticles have been injected into blood circulation, several obstacles need to be overcome such as biological barriers; unwanted protein adsorption; accumulation; and/or removal from body without delivering the drug ingredients to the designated cell, tissue, or organ. In order to overcome all these handicaps and bring the pharmaceutical directly to the target side, the nanoparticles must be diverted, and they should possess multifunctional properties. In other words, the main expectation from the nanoparticles, which could be used in clinical essays, is that they should be capable of transporting drug molecules to solely infected tissues or organs without causing any side effects to healthy ones. Another important issue is the releasing efficiency of the particles. According to the scientists, these systems signify the directed transportation of the drug molecules at a predetermined release rate and in a specified period of time to the infected tissues or organs (Hoffman, 2008).

In consideration of the advantages and disadvantages of the reported nanoparticles, a new generation of nanoparticle-based therapeutics have been designed for treatment, diagnosis, monitoring, and control of biological systems. The main aim of these systems is to combine all these unique properties of the nanomaterials into a single system that is called *theranostics*. *Theranostics* can be defined as the combination of both therapeutic and diagnostic capability in one step. In this manner, theranostics have promising potential to develop personalized treatment approaches, which purpose to monitor the response to the treatment, to increase drug efficacy and safety, and to eliminate the unnecessary treatment of patients, resulting in significant cost savings for the overall healthcare system (Choi et al., 2012). Theranostics can be localized in a specific manner to the site of disease and eliminate the side effects. There are many reported studies are about to develop a new generation of theranostics to minimize these possible side effects.

This chapter mainly described the theranostic research conducted with nanomaterials based on noble metal, porous silica, polymeric, and core–shell nanocarrier systems, which exhibited active targeting, triggered release of pharmaceutic ingredients, and imaging capability including in vivo studies. Moreover, we conclude by giving a simple comparison of the ideal and desired future theranostic system with the current state of theranostic nanoparticles.

2 METAL NANOPARTICLES

In the past few decades, advances in nanomedicine and nanotechnology have enabled the development nanoparticles with multifunctional properties and their employment to overcome the shortcomings of traditional disease diagnostic and therapeutic agents (Farokhzad and Langer, 2006; Ferrari, 2005a; Sanvicens and Marco, 2008). Delivery of drugs from nano and microparticles is attracting remarkable attention in the treatment of ailments especially in cancer therapy.

Controlled delivery of drugs from nanostructured functional materials, especially metallic nanoparticles, attracts increasing attention because of the opportunities in cancer therapy and the treatment of other diseases (Arruebo et al., 2007). The potential of nanoparticle-based drug-delivery systems can be attributed to their significant advantages over traditional methods, such as (1) the ability to target specific locations in the body, (2) the decrease in the quantity of drug required to provide a particular concentration near the target, and (3) the decrease in the concentration of the drug at nontarget regions to lower side effects.

Metal nanoparticle drug-delivery systems with their small size enhance permeability and retention effects (EPR) allowing them to accumulate in targeted tumor tissues to higher extents than in normal tissue due to the leaky tumor blood vasculature (Ferrari, 2005a; Iyer et al., 2006). In addition, the high surface area of nanoparticles provides unique opportunities to load drugs. In this way, remarkable progress can be attained in drug's solubility, stability, and pharmacokinetic parameters (Ghosh et al., 2008). Both the small size and high surface area of nanoparticles would result in low dosing of the patient and high local concentration of drug in the targeted tumor tissues with fewer side effects to healthy tissue

(Alkilany et al., 2009). Moreover, metallic nanoparticles can be synthesized through well-defined procedures and easily functionalized, as well as conjugated various molecules with different functions. In metallic nanocarrier systems, the metallic core provides stability in the physiological medium while conjugated molecules determine surface properties, such as hydrophobicity and charge. Last, metallic nanoparticles as a result of their inherent characteristics, may represent unique optical and magnetic properties that are exploited in numerous therapeutic and diagnostic applications. To employ these major advantages of metallic nanoparticles in drug delivery system, different metallic nanoparticles including plasmonics (gold and silver); oxides (zinc oxide, titanium oxide, alumina, and silicon oxide); magnetics (iron oxides); and semiconductors were investigated in detail.

In the context of this chapter, we will mainly concentrate on drug delivery of magnetic and plasmonic nanoparticles.

2.1 MAGNETIC NANOPARTICLES

In biomedical applications, magnetic nanoparticles have obtained remarkable attention with their unique magnetic properties in numerous research areas especially in magnetic resonance imaging (MRI) (Sun et al., 2008). Magnetic nanoparticles can be in the form of different chemical substances including iron oxide, metallic, and bimetallic (McCarthy and Weissleder, 2008; Veiseh et al., 2010). However, due to their special features such as being low toxicity and having properly reactive surface to be modified with different materials, iron oxide nanoparticles have gained wide biomedical applications (Laurent et al., 2010; Lewinski et al., 2008). These attractive advantages have paved the way for different research interests covering medical imaging, hyperthermia, biosensor, magnetic separation, and, drug delivery (Durán et al., 2008; Frullano and Meade, 2007).

In most cases, therapeutic agents are attached to magnetic nanoparticles directly or encapsulated within a mixture containing a polymer and nanoparticles. The coating procedure provides several advantages including enhancing nanoparticle stability in different medium conditions, reducing leaching of magnetic cores, providing functional groups to allow the binding of biomolecules, and retarding clearance by RES (Arruebo et al., 2007). For example, silica coatings both prevent aggregate formation due to electrostatic repulsion and serve numerous opportunities for binding of biomolecules onto nanoparticle surface. Additionally, polyethylene glycol (PEG) molecules on magnetic nanoparticles maintain delaying the action of the RES and increase the EPR effect (Ferrari, 2005b). Here, it should be noted that the nonmagnetic coating material resulted in lower magnetic moments of final nanoparticle system. Moreover, Chen et al. (2010a) reported a novel drug delivery system containing magnetite nanoparticles chemically modified with doxorubicin molecules. Afterward, drug loaded magnetite nanoparticles functionalized with PEG and the porous silica layer. In this strategy, the silica layer served as a protective layer for both nanoparticles and chemotherapeutic agent and a thin barrier to control doxorubicin release over time. They also reported that this composite drug-delivery system represents slower doxorubicin release than bare magnetite nanoparticles, and the addition of hydrophilic PEG layer could reduce the effect of RES system and prolong the administration period of drug. A similar approach was employed to target nanoparticles across blood brain barrier (Liu et al., 2010). For this, magnetite nanoparticles were first coated with the chemotherapeutic agent, epirubicin, and then encapsulated with a thin layer of a polymer, poly[aniline-*co*–*N*-(1-one-butyric acid)] aniline. This nanoparticle system was targeted to the brain by employing both magnetic and ultrasound effect. This synergistic system led to remarkable increase in accumulation of nanoparticles. In the case of

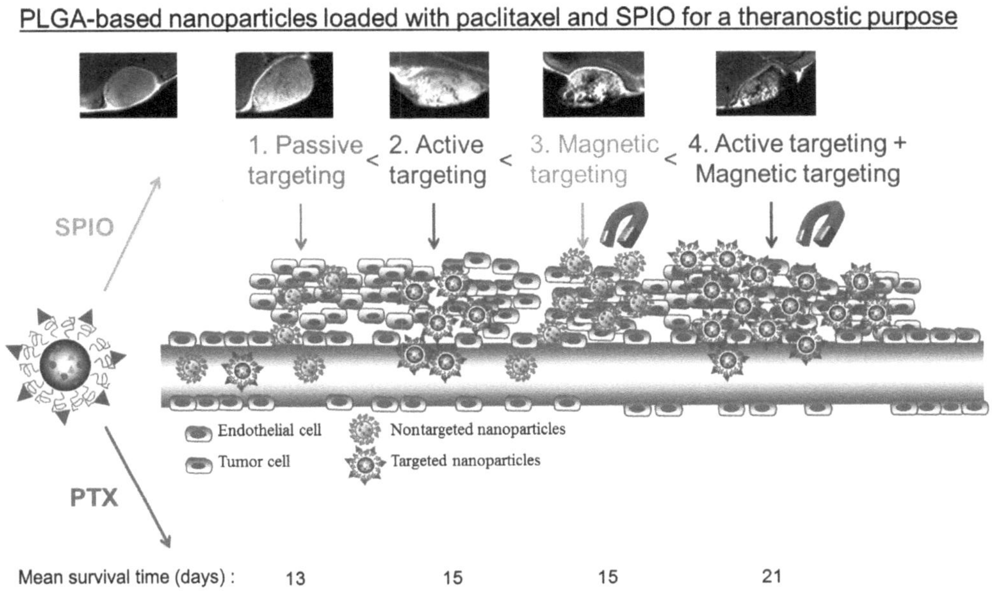

FIGURE 15.1 Schematic Representation of Different Tumor Targeting Strategies of Multifunctional Paclitaxel/SPIO-Loaded Nanoparticles for Tumor Imaging and Therapy

Reprinted from Schleich, N. et al., 2014. Comparison of active, passive and magnetic targeting to tumors of multifunctional paclitaxel/SPIO-loaded nanoparticles for tumor imaging and therapy. J. Control. Rel. 194, 82–91, with permission from Elsevier 2014.

external magnetic field applications, 15-fold higher epirubicin molecules were detected compared to the control groups. Also, a similar strategy was employed to create polyethylenimine loaded magnetic nanoparticles for brain tumors (Chertok et al., 2010). In vitro study depicted low cellular toxicity and high cell association. In addition, to increase the therapeutic efficacy and reduce toxic side effects of hydrophilic drug, 5-fluorouracil, Arias et al. proposed biodegradable shell of polyalkylcyanoacrylates onto magnetite core (2008). Nowadays, multifunctional nanoparticle systems loaded with therapeutic drugs and diagnostic agents into one single platform is a hot-topic research field due to their unique advantages (Janib et al., 2010). To employ multifunctional properties of magnetic nanoparticles, several theranostic nanocarrier strategies have been developed. For instance, Tabatabaei et al. (2015) reported that the thermal energy generated by magnetic heating of available magnetic nanoparticles in the brain capillaries of animal models can increase barrier permeability. They also used fluorescent dye to verify the blood brain barrier integrity. The data presented in this work is highly encouraging for a novel theranostic and targeted drug delivery system. In another work, Schleich et al. (2014) employed dual paclitaxel/superparamagnetic iron oxide-loaded PEGylated PLGA nanoparticles-based in vivo studies as an antitumor agent (Fig. 15.1). For this, different strategies including passive targeting via the EPR effect, active targeting of $\alpha v\beta 3$ integrin via RGD grafting, magnetic targeting via a magnet placed on the tumor, and the combination of the magnetic targeting and the active targeting of integrins were used and compared. The data were analyzed in terms of anticancer efficacy, biodistribution, and MRI contrast agent efficiency, and the combination of active strategy using RGD-grafted nanoparticles and magnetic targeting was detected as the most proper targeting approach.

Singh et al. (2014) investigated carbon nanotube@magnetic nanoparticle@mSiO_2 system as a multifunctional hybrid carrier for drug delivery and imaging. The material represented a high mesoporosity and excellent magnetic properties. Mesoporous and magnetic nanocarriers showed high loading capacity for biological molecules, including gentamicin, cytochrome C (CytC) and small interfering RNA (siRNA), and attained high cellular uptake rate through magnetism without any remarkable toxicity to cells. This report showed that this novel multifunctional nanocarrier system may be potentially applicable as theranostic systems.

2.2 PLASMONIC NANOPARTICLES

Although many nanoparticles have been developed for drug delivery systems, plasmonic gold nanoparticles have attracted enormous attention compared to the other nanoparticles. Similar to many traditional nanocarrier systems, gold nanoparticles enhance the EPR effect with their high surface area as well. However, upon interaction with the light in the visible or near infrared (NIR) region of the electromagnetic spectrum, gold nanoparticles represent distinctive optical properties (Murphy et al., 2005, 2008a,b). These properties mainly depend on the size and shape of nanoparticles. In addition, as a result of their stability, inert nature, noncytotoxicity, and high biocompatibility, these nanoparticles have been employed in several biomedical applications including drug delivery strategies. In addition to their general advantages, as a drug nanocarrier system gold nanoparticles can be functionalized with numerous materials including small biomolecules, proteins, enzymes, DNA, and can be easily synthesized by various methods with controlled dispersivity. In addition to spherical gold nanoparticles, gold nanorods provide numerous opportunities due to their unique optical properties, which can be manipulated with their aspect ratio. For gold nanorods, apart from the absorption peak associated to diameter of nanoparticles, an additional peak is observed in NIR region of electromagnetic spectrum. The characteristics of second peak can be manipulated by changing the aspect ratio of nanorods during the synthesis procedure. For these reasons, gold nanorods have attracted great attention in several biomedical research fields including the drug delivery systems. The main concern in the gold nanorod mediated drug delivery system is the loading of the therapeutic agent with proper affinity to system for efficient release. To achieve this goal, several innovative methods have been developed for different agents. In general the most common method to create a gold nanorod carrier system is to displace CTAB bilayer molecules with proper materials. In most cases, DNA, plasmid DNA, and RNA are employed as drug molecules (Chen et al., 2006; Somin et al., 2009; Wijaya et al., 2009). Alternatively, biocompatible polyelectrolytes were used to coat the CTAB layer, thereby reducing the toxicity of CTAB molecules (Alkilany et al., 2012). In addition, drug molecules thermo-sensitive shell of poly (*N*-isopropylacrylamide) hydrogel was coated on gold nanorods and upon irradiation with NIR light heat was generated and hydrogel was expanded and released the drug (Kawano et al., 2009).

As a theranostic approach, gold nanoparticles can be used to diagnose and treat cancer simultaneously. For this, Chen et al. (2013) proposed doxorubicin loaded gold nanoparticles through Au-S bond by employing a peptide molecule that can be cleaved by protease. In vivo study depicted that the rapid release of doxorubicin from the functionalized gold nanoparticles was observed as a result of overexpressed protease in tumor tissue and intracellular glutathione. With this approach, developed efficacy on tumor growth inhibition and fluorescent imaging was performed simultaneously. In another work, Deng et al. (2014) fabricated controllable self-assembly of gold nanoparticles through amphiphilic mixed polymers to increase interparticle plasmonic coupling in NIR region for enhanced photothermal therapy. The absorption spectrometry of gold nanoparticles self-assembly was drastically red-shifted

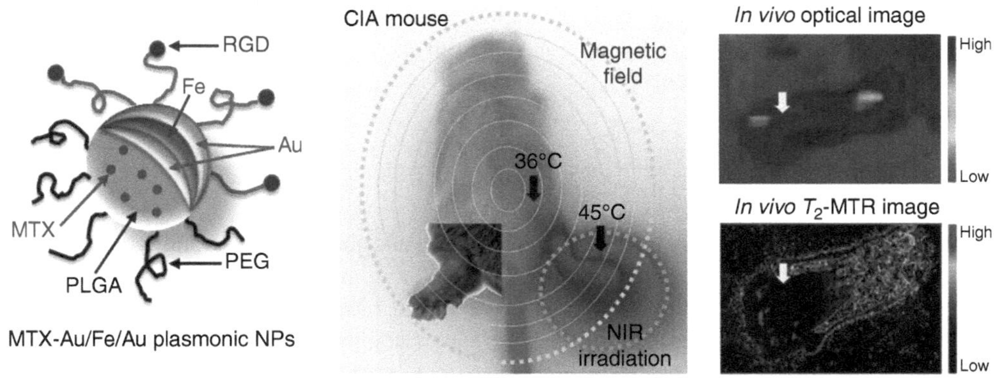

FIGURE 15.2 Schematic Depiction of the Fabrication and Bioimaging Mechanism of the Hybrid Au/Fe/Au Plasmonic NPs

Reprinted from Kim, H.J. et al., 2015. Drug-loaded gold/iron/gold plasmonic nanoparticles for magnetic targeted chemo-photothermal treatment of rheumatoid arthritis. Biomaterials 61, 95–102, with permission from Elsevier 2015.

from 520 to 830 nm, which maintained the NIR photo-thermal conversion ability. Additionally, the presence of gold with high X-ray absorption coefficient qualifies its usage as a potential CT contrast agent. This work proved that with the evidence of in vitro and in vivo studies, the proposed gold assembly of nanoparticles can be utilized as a potential photothermal therapeutic and CT contrast agent. Ke et al. (2014) prepared multifunctional nanocapsules that were fabricated through loading perfluorooctyl bromide and magnetic nanoparticles into poly(lactic acid), followed by the formation of PEGylated gold nanoshell on the surface. These nanocapsule systems proved to act as nanotheranostic agents to achieve bimodal ultrasound/MRI guided photothermal ablation in human tumor xenograft models. Kim et al. (2015) proposed methotrexate, the most common antirheumatic drug, loaded poly(lactic-*co*-glycolic acid) gold/iron/gold half-shell nanoparticles conjugated with arginine-glycine-aspartic acid for magnetic targeted chemo-photothermal treatment, and also in vivo multimodal imaging of rheumatoid arthritis (Fig. 15.2). As a result of NIR irradiation and its resonance of gold, local heat is generated at the inflammation region and drug release from composite nanoparticle system is accelerated. Additionally, the presence of iron in nanoparticle system enables in vivo T_2-magnetic MRI. Animal models depicted that combined NIR irradiation and external magnetic field application enhanced the therapeutic effects of the drug molecule.

In another work, to overcome major problems of brain cancer associated with the blood-brain barrier, Dixit et al. (2015) investigated the delivery of a photodynamic prodrug, Pc 4, from transferrin peptide targeted gold nanoparticles to brain cancer cell lines. Compared to untargeted nanoparticles, remarkable increments were detected in cellular uptake of targeted nanoparticles due to overexpressing the transferrin receptor in in vitro studies of cancer lines. This work provides clear evidence for both delivering chemotherapy to brain tumors and their noninvasive imaging simultaneously. Recently, Muthu et al. (2015) developed micelle system conjugated with transferrin for drug delivery of docetaxel and gold nanoclusters as theranostic system for cancer. The authors reported successful imaging in xenograft model and inhibition of tumor growth.

In conclusion, the emergence of metallic nanoparticles in biomedical application especially in drug delivery and theranostic systems remarkably impacted the developments in the relevant researches to cure many ailments. The unique characteristics of metallic nanoparticles paved the way for their use in all objectives of nanomedicine. Metallic nanotechnology with their rapid and multidirectional development will have an important role in therapy of numerous ailments especially in cancer therapy with novel engineered nanocarrier systems.

3 SILICA NANOPARTICLES

Silica is a major component of glass and sand, and it has a very good biocompatibility and nontoxicity. Having these desired properties has made silica a very important material in terms of different applications (Abdelaal et al., 2014; Liu et al., 2007). Silica structures have been well known for over 40 years with the advantage of well-defined and uniform porosity (Roggers et al., 2014). Silica nanoparticles have been very vital tools of biosensors, drug carriers, catalysis, and adsorbents because of their accessibility for surface modification including both inner and outer walls, hydrophilic character, large surface area, good chemical and thermal stabilities, permeability, and low density (Baeza et al., 2015; Feng et al., 2008; Kim et al., 2002). Various fabrication methodologies have been developed for porous silica nanoparticles. In general, these include emulsion/interfacial polymerization, heterophase polymerization, surface polymerization, spray-drying methods, colloidal templating, and leaching of clay minerals with acid (Graf and Van Blaaderen, 2002; Kosuge et al., 1995; Liu et al., 2009; Lu et al., 1999).

Advances in nanotechnology have led to the nanoscale materials utilized for various important areas such as drug delivery, diagnostics, medical imaging, and cell targeting (Lee et al., 2014a; Wang et al., 2015). Porous silica nanoparticles (PSNs) are a unique class of nanomaterials that have generated much interest for drug delivery due to their good chemical and thermal properties (Chakravarty et al., 2015; Chen et al., 2004; He and Shi, 2011; Huang et al., 2011; Kim et al., 2008; Yidirim et al., 2013). Specifically, the use of porous silica nanoparticles as the drug carriers has led to new leaps in targeted and triggered delivery for cancer treatment. It does not only improve the loading capacity but also provides the combination of high specific area and porosity. These nanoparticles provide a viable environment for drug molecule immobilization by retaining their biological activity with the desired amount and for facilitating targeted and triggered drug release. Herein, we discuss various studies for fabrication of porous silica nanoparticles and their applications in drug delivery developed for in vivo cancer therapy. The performances of nanohybrids are compared and examples of different approaches are discussed. In all cases, effective surface area, functionality, and loading capability directly impact the resultant carrier capabilities. Ultimately, these attractive features of nanoporous silica particles are expected to find applications in cancer therapy. A well-desired system should emphasize the biocompatibility and possible toxicity of the nanoparticles after its function in vivo. In addition, surface modifications can help to control the physicochemical, toxicological, and pharmacological properties of PSNs. Addition of functional groups on the surface allows the particles to interact with the anticancer drugs in a more efficient way for the targeted imaging and cancer treatment (Biju, 2014). Thus silica can be defined as a well-known host material for the combination of different modifications (Burns et al., 2006). Also, some of the pharmaceutical ingredients are poorly soluble in water. Thus oral drug delivery remains inefficient for this type of therapeutic agents. PSNs-based drug-delivery systems can highly improve the dissolution of these agents. Inorganic porous silica nanoparticles can controllably load hydrophilic or hydrophobic drugs (Du et al., 2013; Giménez et al., 2015; Zhang et al., 2013). For example, in the study of Cao et al. poorly soluble drug-loaded PSNs were prepared to release the drug (Cao et al., 2013). They developed a highly effective, long-acting and slow release (72 h) formulation for the delivery of silybin.

The drug delivery system consisted of three major components: Silybin, hydrophilic gel matrix and PSNs. They also focused on the in vitro dissolution kinetics of drug-loaded PSNs and performed correlations between in vitro and in vivo absorption of the formulations in beagle dogs. Zhang et al. (2014b) synthesized a β-cyclodextrin modified mesoporous silica nanoparticles with hydroxyl, amino, and thiol. This study is a good example for presenting the accessibility of silica nanoparticles for different surface modifications. In their study, a comparison of the mucoadhesive properties and the potential of nanocarrier as a drug delivery system for superficial bladder cancer therapy were performed. They found that the thiol-functionalized nanoparticles exhibited significantly higher mucoadhesivity on the urothelium compared to the hydroxyl- and amino-functionalized nanoparticles. It was shown that doxorubicin (DOX) was loaded into the pores and its release was triggered by acidic pH.

Another issue for the cancer therapy is the fact that most of the anticancer drugs trigger apoptosis, which leads to cell death. Therefore, it is important to inhibit this fact by targeting all intracellular molecular targets (Chen et al., 2009). Porous silica nanoparticles also offer advantages on this issue. There are many promising applications of mesoporous silica nanoparticles used for targeted and triggered anticancer drug delivery in the literature. Most of the mesoporous silica nanoparticles are in the form of a well-ordered hexagonal array of tubular pores called MCN-41type (Mobil Composition of Matter No. 41). In an example, Chen et al. (2015) fabricated multifunctional envelope-type mesoporous silica nanoparticles (MEMSN) for pH-responsive drug delivery and magnetic resonance imaging. The system combined the pH responsiveness, nontoxicity and biological specificity. The drug nanocarriers consisted of acetals immobilization on the surface of porous silica, and also lanthanide doped upconverting nanoparticle (Fig. 15.3).

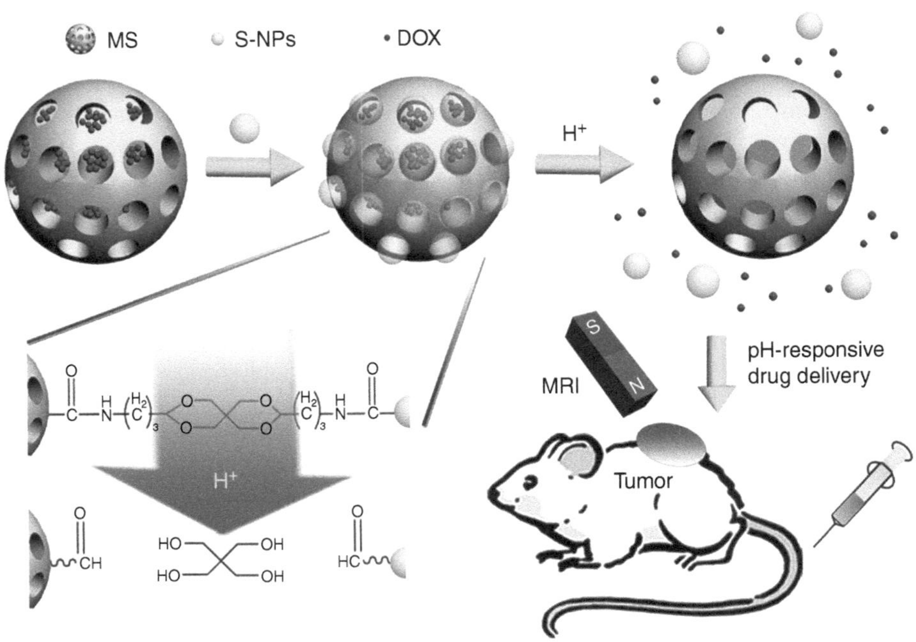

FIGURE 15.3 Schematic Illustration of the MEMSN System for pH-Responsive Drug Delivery and Magnetic Resonance Imaging

Reproduced from Chen, Y. et al., 2015. Multifunctional envelope-type mesoporous silica nanoparticles for pH-responsive drug delivery and magnetic resonance imaging. Biomaterials 60, 111, with permission from Elsevier 2015.

Anticancer drug, doxorubicin (DOX) was loaded in the pores of the nanocarrier, and it released the DOX under the acidic environment due to the hydrolysis of acetals in the structure. In the study, the functionalized ultra-small lanthanide doped nanoparticles with ultrathin TaO_x layer was grafted onto the orifices of silica nanoparticles with the acid-labile acetals. DOX loading was performed by soaking the nanoparticles into the DOX solution (1 mg/mL) in tris(hydroxymethyl)aminomethane (TRIS) buffer (pH 7.4) and stirring for 24 h. Then, the drug release system was used efficiently both in vitro and in vivo for cancer treatment. For in vivo application, the outer surface of the nanocomposite was modified with 2-methoxy(polyethyleneoxy) propyl-trimethoxysilane (PEG-silane). The system was allowed for passive tumor targeting. In addition, it increased tumor accumulation of the probes by the EPR effect, which was visualized by MRI in vivo. The evaluation of the cytotoxicity was carried out using 3-(4,5-dimethylthiazol-2-yl)-2,5-diphenyltetrazolium bromide (MTT) assay on the HeLa cells. The results showed that the nanocarrier could be harmlessly metabolized and degraded into nontoxic products within a few days.

The nanovalve-mesoporous silica nanoparticle (NV-MSN)-based drug delivery system that contained fluorescent molecules, targeting proteins, and pH-sensitive nanovalves was fabricated by Hwang et al. (2015). The nanovalves activated and delivered DOX when exposed to a pH lower than 6 since they consisted of an aniline-based stalk that was grafted on the surface of the nanoparticle. The stalk formed an inclusion complex with α-cyclodextrin (α-CD) via hydrophilic and hydrophobic interactions to seal DOX inside the pores at neutral conditions. When exposed to acidic conditions, the phenyl amine has been protonated and the binding constant between the aniline-based stalk and α-CD decreased by allowing the DOX to release. Nanocarrier performance was enhanced with biologically active targeting molecule, transferrin (Tf), which increased the endocytosis of mesoporous silica nanoparticles into the cancer cells. Iron was carried out into the cell by Tf, which could be exploited to increase the delivery payload and maintain the localized region of treatment. In vitro experiments performed with human pancreatic cancer cells (MiaPaCa-2) and showed the improved delivery of DOX and cell killing. In vivo studies were performed using severe combined immunodeficiency (SCID) mice and also depicted the enhanced DOX delivery. In vivo experiments demonstrated that the NV-MSNs developed in the study autonomously released DOX in xenograft tumors.

CytC end-capped mesoporous silica nanoparticles were fabricated by Zhang et al. Redox-responsive nanocarriers were used for the liver tumor-targeted therapy both in vitro and in vivo (2014a). The pores of silica nanoparticles containing DOX were capped by CytC via a linker of redox-cleavable disulfide bonds to mesoporous silica nanoparticles. Then, a G-rich oligonucleotide DNA aptamer (ApT), AS1411, was employed as specific targeting to the tumor cells. This aptamer was chosen since it had a high affinity to nucleoin, which was a protein implicated in tumorgenesis and angiogenesis. The DOX loaded nanocarriers could be triggered by reductant, such as glutathione within cellular environment. The nanocarrier exhibited important potential for tumor targeted and achieved triplex therapy effects on the tumor inhibition in vivo. These experiments were evaluated by the change of tumor size, TUNEL (terminal deoxynucleotidyl transferase dUTP nick end labeling) staining, and hematoxylin and eosin (HE) staining assays. For the in vivo part of the study, 18 six-week-old nude mice (Athymic BALB/c nu/nu, female) have been used. Each mouse was raised in a single sterile plastic cage with hardwood chips as beddings in a clean animal laboratory (oriental CRF-1) irradiated with 30 Gy rays.

PEG-incorporated mesoporous silica nanoparticles functionalized with amino-β-cyclodextrin were synthesized for targeting MDA-MB-231 xenograft model breast cancer in vivo (Zhang et al., 2014b). The functionalization of the nanoparticle was achieved by cleavable disulfide bonds where amino-β-cyclodextrin blocked drugs inside the pores. Afterward, incorporation of active folate targeting ligand, which improved the selective uptake of nanoparticles into tumor, was carried out. Targeted DOX delivery capability of nanocarriers were demonstrated by significant inhibition of the tumor growth in mice treated with DOX-loaded nanoparticles. DOX release was triggered by intracellular acidic pH and glutathione (GSH). The effect of nanoparticle size on in vivo biodistribution of the nanocarriers was investigated by injecting them with MDA-MB-231 tumor-bearing BALB/c mice for 24, 48, and 72 h, respectively. The fastest DOX release rate (86% of DOX within 48 h) was achieved when the dual stimuli of pH and GSH were utilized simultaneously.

Hollow mesoporous silica nanoparticles (HSMSNs) attracted another attention for drug delivery studies (Gao et al., 2011) HSMSNs (~150 nm in size) were functionalized for tumor targeted drug delivery by Chen et al. (2014). They synthesized the nanoparticles by hard-template method and conjugated to an antibody. Both hydrophobic and hydrophilic anticancer drugs sunitinib (SUN) and DOX, respectively, were loaded inside the hollow mesoporous nanoparticles via PEG linkers. In vivo positron emission tomography (PET) imaging, biodistribution, and blocking studies in 4T1 murine breast tumor-bearing mice were carried out to evaluate the tumor-targeting capability and drug delivery efficiency, which were validated by in vitro, in vivo, and ex vivo studies. The loading capacity for SUN and DOX were found to be 482 and 1129 mg g^{-1}, respectively, which was higher than that of regular mesoporous silica nanoparticle. Studies in human umbilical-vein endothelial cells HUVEC (CD105-positive) and MCF-7 human breast cancer cells (CD105-negative) showed the strong and specific binding of nanoconjuage to CD105-positive HUVEC cells with negligible non-specific binding. In vivo tumor targeting and PET imaging demonstrated CD105-specific accumulation of nanoconjugate in 4T1 tumor-bearing mice with peak tumor uptake of 5.5 ± 0.4%ID/g ($n = 3$) at 3 h postinjection. This value was found significantly higher than that of nontargeted group. Furthermore, improved drug delivery was achieved. Amine-functionalized hollow mesoporous silica nanoparticles (HMSNs) were also used by Goel et al. (2015), for tumor targeted imaging and drug delivery. Tumor targeted drug delivery in glioma xenografts was imaged by PET. Amine-functionalized HMSNs (150-250 nm in diameter) was conjugated with macrocyclic chelator, PEG and cyclo(Arg-Gly-Asp-D-Tyr-Lys). Hydrophobic, anti-angiogenic drug, SUN was loaded into the nanoconjugate and then in vitro/in vivo/ex vivo experiments were carried out to assess the drug-loading capacity, release, and tumor-targeted drug delivery. The results demonstrated the advantage of tumor integrin $\alpha_v\beta_3$ targeting for drug nanocarriers. Silica nanoparticles labeled with ^{89}Zr ($t_{1/2} = 78.4$ h) were developed for testing their in vivo stability to use them in drug delivery and molecular imaging (Chen et al., 2015).

Bi_2S_3-embedded mesoporous silica nanoparticles (BMSN) were presented as drug carriers for cancer treatment (Ma et al., 2015). Bi_2S_3 nanoparticles were coated with mesoporous silica and Ma and his co-workers donated the nanocarrier as BMSN in their study. It is known that bismuth-based nanoparticles have received great attention as sensitizers in radiotherapy (Graf and Van Blaaderen, 2002). It was shown that BMSNs exhibited a high DOX loading efficiency (45 wt.%). pH-responsive controlled DOX release was occurred due to the electrostatic interaction between the silanol surface and DOX. The drug loading was achieved by adding BMSNs into phosphate buffer solution including DOX under vigorous agitation. DOX content was calculated using UV-vis absorption peak intensity at 485 nm. The cell viability experiments showed that the encapsulation of the drug into BMSNs led to the improved therapeutic effect against multidrug-resistance cancer cells. In vitro and in vivo experiments were performed to demonstrate the radiosensitization effect of the nanoparticles in both X-ray external beam and P-32 radionuclide interstitial radiotherapy. The in vivo P-32 radiotherapy in conjunction with BMSNs presented a significant eradication of the tumor volumes by an average of 21% reduction to its initial values, in comparison to 2.01-fold increase in the case of P-32 treatment alone.

As a conclusion, porous silica nanoparticles have taken a long way for the design of newly generated drug delivery systems, in particular, for cancer therapy. Currently, porous silica nanoparticles are under investigation to achieve long-term in vivo targeted drug delivery on cancer cell lines. They are promising carriers for transporting and delivering chemotherapeutic agents in an efficient and specific way. The main goal of porous silica nanoparticles-based tumor-targeted drug-delivery systems with stimuli-responsive character is to improve the efficiency and to minimize the side effects of anticancer drugs used for therapy. There are still certain issues to overcome for increasing the accessibility of mesoporous silica nanoparticles for clinical applications, such as maintaining colloidal stability in physiological fluids and improving the penetration capability of the particles within tumor tissues. However, the integrated capability of silica nanoparticles makes them very important candidates for cancer diagnosis and therapy even in their present forms.

4 POLYMERIC NANOPARTICLES

Polymer-based theranostics are some of the most skillful supporters to help scientists in their fight against cancer. Similar to other groups of theranostics, the polymeric ones are also comprised of image-guided drug-carrier units. By the way, these materials are not to be limited just as drug delivery agents. They could also be used in gene delivery, photodynamic and radiation therapies. If we look at the composition of the polymeric theranostics, we can depict that these advanced materials consist of four fundamental components (three of them are essential and one is not necessary but favorable); the first part includes the therapeutic component that could be a drug or a gene carrier. Moreover, this component could be a photosensitizer in order to provoke the photodynamic therapy. The latter part involves polymer component, which brings in to the polymeric conjugate: stability, solubility, and biocompatibility. The third component possesses the imaging or diagnosis component, which could be a magnetic compound or a contrast agent for MRI, a photoluminescent agent—especially a fluorescent group—and even a microbubble, and finally, the last component, which is not obligatory in whole polymeric theranostic systems, consists of functional groups responsible for specific cell targeting. As mentioned in the previous subsections, an ideal drug-carrier system should find a solution to the inappropriate and uncontrollable distribution of the drug ingredients into the body, but, of course, in a clever way.

The classification of the polymeric nanocarrier could be made by the particle size and structure diverging nano to microcapsules or spheres. The fine particles are 100–2500 nm and ultrafine particles have been defined as particles that are 100 nm or less in diameter. These natural or synthetic polymer therapeutics including polymeric drugs, polymeric micelles, polymer-drug or polymer-protein conjugates and multicomponent polyplexes constitute an emerging topic in bio-nanocarrier systems at the macroscale to nanoscale range, and they are proliferating as a prominent tool for cancer therapy due to their ease of functionalization (Duncan, 2003) and modification of the polymer backbone by attaching numerous targeting ligands in order to achieve the intended site and tumor-specific drug delivery (Qiu and Bae, 2006). Moreover, polymeric carriers settled their efficacy in protecting and stabilizing the drug molecules (Bennet and Kim, 2014). The preparation technique of the polymeric nanoparticle is also a direct determinant in order to design a nanocarrier with desirable properties to provide proper drug delivery and targeting. Consequently, selection of the preparation technique depends on (1) required size of nanoparticles; (2) inherent properties of drug (e.g., stability and aqueous solubility); (3) surface properties (e.g., permeability and charge); (4) desired drug release profile; (5) attempted degree of biodegradability, biocompatibility, and toxicity; and, finally, (6) antigenicity of the final product (Khanbabaie and Jahanshahi, 2012).

Several studies have documented the usage of different kind of polymers for the preparation of the polymeric nanoparticles including linear (diblock, triblock, multiblock), graft and radial (simple star, miktoarm star) copolymers (Demirel et al., 2011; Demirel et al., 2013; Konkolewicz et al., 2011; Riess, 2003). These multifunctionalized nanoparticles could be in the form of micelle, fiber, capsule, platelet, spheroid, colloid, dendrimer, core–shell, polymer-coated nanocrystals, and nanoparticle-incorporated polymer matrixes. The majority of the techniques based on the top-down and bottom-up processes include polymerization, usage of preformed polymers, or ionic gelation. They profit from different approaches such as using organic solvents and the syntheses are conducted at elevated temperature under mechanical agitation or ultrasonication, which have the potential to degrade the drug ingredients. For these reasons, it a technique should be chosen that does not damage the active pharmaceuticals (Bennet and Kim, 2014). Thus far, previous studies have demonstrated that polymeric nanocarriers need to accomplish some requirements in order to be suitable for in vivo studies. A preferable polymeric nanocarrier must be biodegradable or at least it should be totally eliminated from the corpse in a specific and acceptable period of time to overcome the risk of unwanted accumulation and toxicity. Recent trends in cancer treatment include the development of endogenous (pH, salt concentration, redox, enzyme, etc.) or exogenous (photo, electric field, magnetic field, thermo, ultrasound, etc.) stimuli-responsive polymers. These macromolecules mimic biological systems by responding to external stimulus. Stimuli-responsive biocompatible polymeric materials are important tools for generating anticancer drug-delivery systems due to their prolonged circulation time and increased accumulation of drug concentration at the target site by EPR. As mentioned earlier, tumor tissues (pH < 7.2 for most solid tumors) differ from normal tissues (pH 7.4) by their prominently lower pH values. In connection with the proton sponge effect, positively charged polymer units can cause endosomal or lysosomal rupture. Therefore, it is important to design and synthesize different type of multifunctional and stimuli-sensitive polymeric systems. In this chapter, we aimed to reconcile this goal with recent and inspiring examples of these new-generation materials. For instance, Zhang et al. (2015) reported a detailed study about the organelle-specific targeting and investigation of the properties of pH-stimuli polymeric micelles, it was shown that histamine-functionalized block copolymers were able to partly escape from endo-lysosomes and relocate the active drug ingredients directly into the mitochondria. Target histamine-functionalized amphiphilic copolymers were prepared via ring opening polymerization reactions followed with postfunctionalization through UV-initiated thiolene click chemistry by using histamine glycidyl ether (HGE), octyl glycidyl ether (OGE), and benzyl-functional glycidyl ether (BGE) thiols. All synthesized polymers were found to be nontoxic in three breast cancer cell lines and successfully encapsulated doxorubicin into the mitochondria section. Another good example for the subcellular targeting research was carried out by Zhong et al., and they designed a stimulus-responsive polymeric conjugate platform in order to release the drug molecule to the tumor-cell nucleus following a multistage pathway. The conjugates are formed from a *N*-(2-hydroxypropyl) methacrylamide (HPMA) polymer backbone and charge-convertible nucleus transport subunits, which function in a pH-responsive way, and glycylphenylalanylleucylglycine (GFLG) was used as the linker. This unit could be detached in an enzyme-responsive way (Fig. 15.4). A 14-amino acid therapeutic H1-S6A, F8A peptide (H1) was selected as the model drug (Zhong et al., 2015).

Wang et al. (2014) prepared an interesting model of biodegradable and endosomal pH-sensitive polycarbonate-based prodrug: poly(5-methyl-5-allyloxycarbonyl-1,3- dioxan-2-one)-graft-

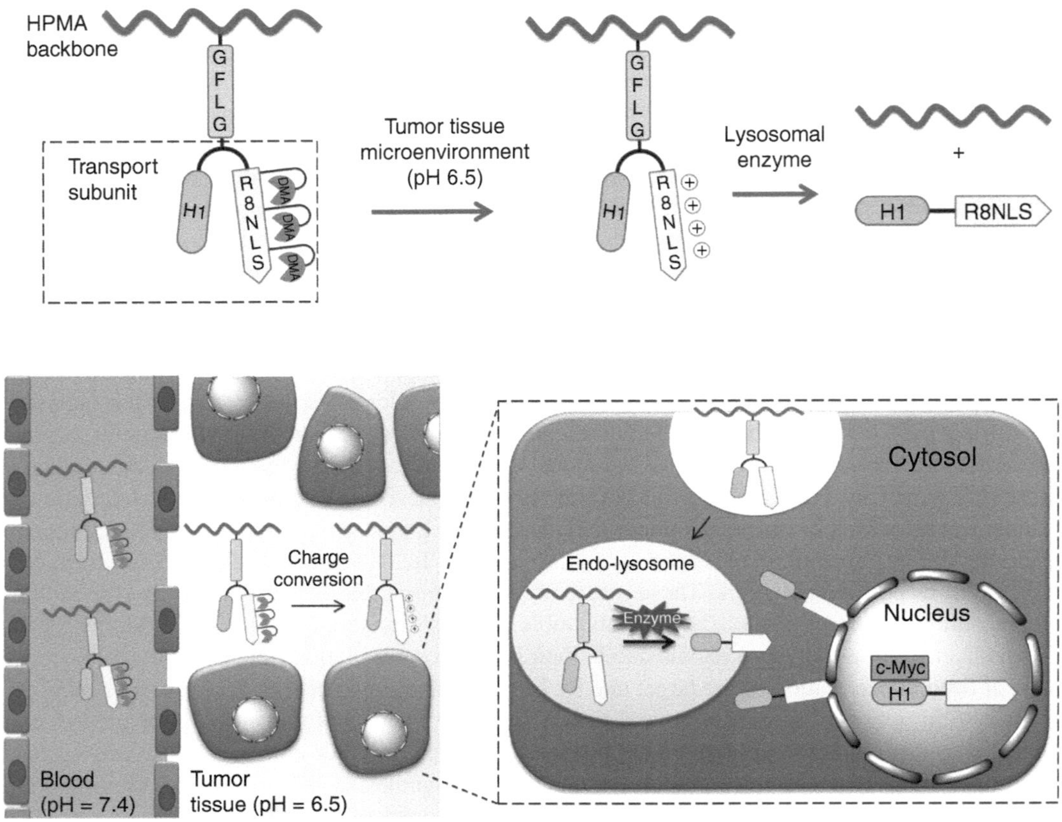

FIGURE 15.4 Schematic Illustration of the Multistage Nuclear Targeting P-GFLG-R8NLS-DMA-H1 System for Dual-Responsive HPMA Copolymer

Reproduced from Zhong, J. et al., 2015. A smart polymeric platform for multistage nucleus-targeted anticancer drug delivery. Biomaterials 65, 43–55, with permission from Elsevier 2015.

12-acryloyloxy dodecyl phosphorylcholine-*co*-6-maleimidocaproyl-doxorubicin [PMAC-graft-(ADPC-*co*-Mal-DOX)] by concomitant ring-opening polymerization and click reaction. The drug molecule (doxorubicin) was linked to the polymer backbone via a pH-sensitive hydrazone bond. Researchers found that the prodrug could form a self-assembled micellar structure, and cytotoxicity studies revealed that the system exhibited significant selectivity to HepG2 and HeLa cells (Wang et al., 2014). Huang et al. (2014) investigated citraconic amide, which was prepared from the reaction between a α-methyl derivative of maleic anhydride (citraconic anhydride) and a primary amine (ethylene diamine). The target amide had negative charge due to the carboxylate (—COO) groups as its end group, and it was stable at both basic and neutral pH environments. On the other hand, it became unstable and degraded to its former cationic amine at acidic conditions. In other words, the polymer could achieve charge reversal from negative to positive by the cleavage of unstable amide linkage. Moreover, this positive- charged system could interact with the anionic tumor cell membrane. The synthesized conjugate poly(aspartate)-graft-poly(ethyleneglycol)-dodecylamine-(hydrazine-doxoru bicin)-(ethylenediamine-citraconic amide) (PASP-g-PEG-DDA-(Hyd-DOX)-(EDA-Cit) was examined in order to show the charge regeneration of primary amine functional groups at different pH conditions. The results showed that when the conjugate was incubated at acidic (pH 6.6) and blood environment (pH 7.4), after a 5 h of incubation, the relative amounts of exposed amine groups were 72 and 42 percent, respectively. This research indicated that only in acidic conditions did the net charge in the studied conjugate change from negative to positive (Huang et al., 2014). Yu et al. (2015) designed a novel brush polymer scaffold drug carrier (BPDC) that reacted through a Schiff-base linkage to doxorubicin. Their synthetic procedure included azide-alkyne click reaction of alkyne-functionalized poly(lactic acid) (PLA) with 6-azidohexyl 4-formylbenzoate and azide-PEG_{2000}. The target PLA-g-ALD/PEG polymer scaffold was then conjugated with doxorubicin, and this water-soluble brush polymer exhibited a 10–30 nm when dispersed in water. As a result of its Schiff-base linkage, it showed a remarkable pH-stimuli drug release behavior with the half-life of 7 h at pH 5.5 (Yu et al., 2015). Photothermal therapy (PTT) is another important tool in the area of localized cancer treatment in which photon energy is converted into heat, and this energy is generally sufficient to destroy cancer cells. Composition of the organized cells is rich in terms of water ratio, and tumor cells most likely show disorganized proliferation when compared to normal cells. Thereby, healthy tissues are transparent to NIR light and as a consequence of this phenomenon, they are protected from photothermal damages. In clinical trials, NIR light and a light-absorbing agent are used together. Sharker et al. (2015) have studied poly(*N*-isopropylacrylamide) and poly(2-dimethyl amino ethyl metacrylate) as drug carriers because of their conformational changes related to environmental stimuli. Indocyanine Green (ICG) was selected as the NIR sensitive photothermal agent. Because of this, ICG was ionically complexed with *N,N*-dimethylaminoethyl methacrylate (DMAEMA) in the presence of CPPDN/rGO hybrid nanocomposites. [2-Chloro-3′,4′-dihydroxyacetophenone (CCDP) quaternized poly(ethylene glycol) grafted poly(DMAEMA-*co*-NIPAAm) (CPPDN) with reduced graphene oxide (rGO) [CPPDN/rGO]] (Sharker et al., 2015). In subsequent periods, overexpressed levels of the glutathione in the intracellular matrix of the tumor cells (~100–1000 times higher concentration than the extracellular environment) prompted researchers to develop disulfide-containing polymeric micelles in which the disulfide bonds could be broken to achieve fast intracellular release in high GSH concentration environments. Wei et al. (2006) prepared multistimuli-responsive copolymers based on poly(*N*-isopropylacrylamide), poly(*N,N*-dimethylacrylamide) and poly(10-undecenoic acid) as pH- and thermal-responsive drug carriers. These copolymers generated stable core–shell nanostructures, which disassembled and released the drug under acidic conditions (Wei et al., 2006). In another example, Bui et al. (2015) prepared a polymeric micelle based on poly(β-aminoester)-grafted disulfide methylene oxide poly(ethylene glycol) (PAE-g-DSmPEG) as a redox- and pH-sensitive drug-carrier platform. Their system consisted of a pH-sensitive core PAE and hydrophilic methoxy poly(ethylene glycol) (mPEG) shell units that could assemble into biocompatible micelles and be simply destroyed in an acidic environment or under high concentrations of GSH by allowing the release of the doxorubicin (Bui et al., 2015). The common survey method was used to obtain a redox-sensitive polymeric system root in the usage of a linker with a disulfide bond. By contrast, some drug ingredients that do not have reactive sites (carboxylic acid, amine, hydroxyl, etc.) for the conjugation are not convenient for such a system. Furthermore, in the case of some hydrophilic drugs, the presence of these active sides could also deteriorate the stability of the conjugate micelles. Following this approach, a redox-sensitive polymer-drug conjugate based on methoxy poly(ethylene glycol)-poly(lactic acid) copolymer (mPEG-PLA) was prepared and conjugated with curcumin via a disulfide bond and for the control purpose with an ester bond, respectively (Cao et al., 2015). It was noticed that under a mimicked redox environment (10 mM dithiothreitol), the extent of curcumin release at 48 h from disulfide bond-linked micelles was nearly 3 times higher than the ester-bonded control micelles. Pluronic (Plu) is known as a biocompatible block copolymer, but it is not highly stable because of its structural integrity. For this reason, it is crucial to generate stable Pluronic micelles by chemical modification of its former backbone. Pioneer research was done by Park et al. and comprised the synthesis of thiol terminal Pluronic micelles, which were stabilized by shells cross-linked with heparin, gold, or magnetic nanoparticles. Lee et al. (2014b) synthesized a pH and redox multistimuli-responsive cross-linked Pluronic-Chitlac (Plu-Ch) polymer between Plu-SS-BA

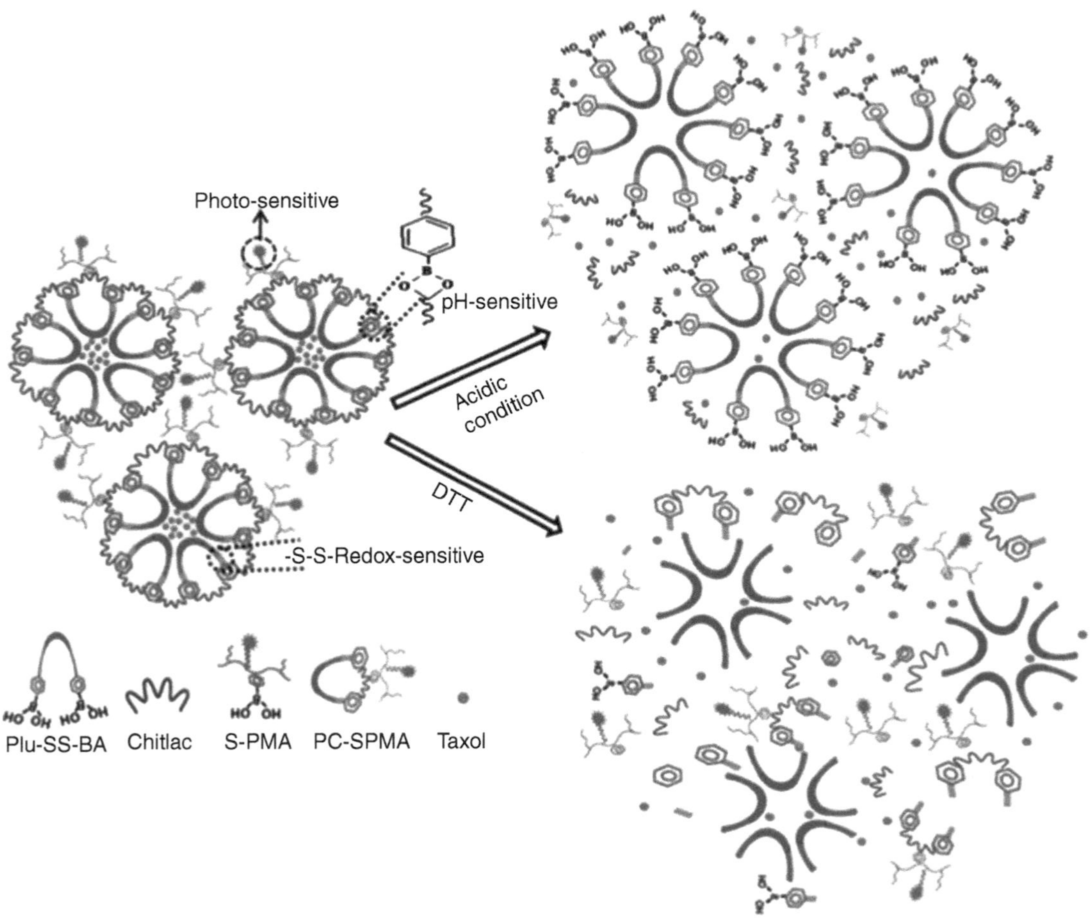

FIGURE 15.5 Schematic Illustration of the Taxol-Loaded PC-SPMA Micelles and Release of Taxol under Various Conditions

Reproduced from Lee, S.Y. et al., 2014. PH/redox/photo responsive polymeric micelle via boronate ester and disulfide bonds with spiropyran-based photochromic polymer for cell imaging and anticancer drug delivery. Eur. Polym. J. 57, 1–10, with permission from Elsevier 2014.

and chitlac in order to form stable micelles in aqueous solution (Fig. 15.5). The following step included the addition of the spiropyran (SP) and phenylboronic acid conjugated poly(dimethylamino ethyl methacrylate-*co*-methyl acrylic acid) (SPMA) onto the Plu-Ch micelles for photo-responsive behavior through boronate ester bonding to obtain PC-SPMA. The resulting micelles exhibited a high capacity to entrap Taxol and control the release of the drug under different physiological conditions. Recent studies showed that boronic-acid-containing polymers could be accepted as promising candidates for smart drug-delivery systems. These boronic-acid derivative polymers (e.g., phenylboronic acid) are able to bind reversibly to glucose, lactose, and catechol-like cis-diol compounds, which generate boronate esters that exhibit fast responsiveness to external pH.

These up-to-date findings have significant implications for understanding how stimuli-responsive polymeric therapeutics exhibit their curative effects in tissues or organs affected by cancer. As a conclusion, these promising studies show that continued efforts are needed to make these advanced materials more accessible and more affordable to patients.

5 CORE/SHELL NANOPARTICLES

Core/shell nanoparticles are provided to fabricate multifunctional systems in a nanoscopic- length scale. Core/shell nanoparticles can be especially engineered with multifunctional properties shell coating a core material. The core generally entraps the cargo molecules such as drugs, genes, contrast agents, and so forth and releases it in a desired way while preventing its direct contact with the biological environment. The shell provides an interface especially designed to tune particle interaction with biological environment. Theranostic core/shell NPs can be designed to obtain desired nanocarrier system. Theranostic core–shell NPs have very important advantages compared with simple NPs. First, core/shell NPs have very good biocompatibility and they are mainly used for bioimaging (Chen et al., 2010b). The core material of the core/shell NPs is responsible for contrasting ability of the system. Various types of core materials can be used as the contrast agent such as magnetic nanoparticles, upconversion nanoparticles, gold (AuNPs), silver (AgNPs) nanoparticles, and quantum dots (QDs) (Chatterjee et al., 2014). By contrast, the shell layer can adjust the surface properties of NPs that are important for biocompatibility, targeted delivery, specific bindings, and biosensing, and so forth (Ghosh Chaudhuri and Paria, 2012). Next, we will discuss the pioneering research of the core/shell NPs in two parts: diagnosis and theranostic properties of the core/shell systems.

5.1 CORE/SHELL NANOPARTICLES IN BIOIMAGING

Recently research has focused on the early detection of diseases to improve the health problem immediately. For instance, early stage detection of diseases such as cancer may prolong patient life. Therefore, several molecular imaging techniques such as magnetic resonance imaging (MRI), positron emission tomography (PET), and optical imaging have been extensively used in clinics. Molecular imaging allows the characterization and detection of biological processes at the cellular and even subcellular stage in organisms. Multifunctional core/shell nanoparticles are well suited for this purpose by using contrasting agents having adjustable properties. Various core/shell NPs have been reported as contrast agents for different bio-imaging methods. Next we will discuss recently developed different core/shell contrast agents with related bio-imaging techniques for in vivo applications.

5.1.1 Magnetic resonance imaging (MRI)

Magnetic Resonance Imaging (MRI) is one of the most powerful and common bioimaging tools to produce high-quality images of the internal organs of the human body (Chatterjee et al., 2014). The MRI technique is based on the basic principles of nuclear magnetic resonance and radio frequency pulses. These pulses provide detailed digital pictures of organs, soft tissues, and bones (Chatterjee et al., 2014). In the MRI process, contrast agents are injected into the patients before imaging to facilitate viewing the difference between the normal and abnormal tissues (Chatterjee et al., 2014). Various core/shell NP based MRI contrast agents have been proposed to track the labeled cancer cells in vivo.

Gadolinium (Gd) based compounds are commonly used as MRI contrast agents in clinics. A milestone study was reported by Bridot et al. (2007) who reported the preparation of hybrid Gd_2O_3 NPs for using multimodal contrast agent for in vivo imaging. They prepared luminescent polysiloxane shell coated Gd_2O_3 NPs for both in vivo fluorescent and MR imaging. Bridot et al. (2007) showed that the core/shell hybrid Gd_2O_3 NPs were well suited for in vivo dual modality magnetic resonance and fluorescence imaging. As seen in Fig. 15.6, the core/shell hybrid Gd_2O_3 NPs was injected into mice, and the NPs were monitored efficiently by fluorescence imaging. After 180 min of the injection, the accumulated core/shell hybrid Gd_2O_3 NPs in the kidneys and in the bladder were detected by fluorescence imaging (Fig. 15.6A). Moreover, after 1 h, injection of the core/shell hybrid Gd_2O_3 NPs became brighter and these accumulated NPs were detected by MR imaging (Fig. 15.6B).

Choi et al. (2010) designed lactobionic acid coated Gd_2O_3@MnO core/shell NPs to enhance the biocompatibility and water solubility besides contrasting properties of gadolinium. Biocompatible and water soluble Gd_2O_3@MnO core/shell NPs showed a high contrast enhancement in a mouse kidney after 90 min of injection of the NPs. Although Gd-based contrast agents provide excellent MR images, the main problem of Gd-based core/shell nanoparticles is their toxicity in vivo. Gd is very toxic in its ionic form and not biocompatible and also easily accumulates in the body. Therefore, Gd-based contrast agents may lead to harmful side effects for patients. Recent studies showed that Fe (III) based nanoparticles have promising potentials due to their intrinsic physical and chemical properties. Iron oxide NPs can be easily modified with different type of shell materials such as polymers and silica. Therefore, Fe (III)-based core/shell nanoparticles can be an alternative contrast agent against Gd. In a recent study, Zhou et al. (2014b) developed multifunctional PEGylated Fe@Fe_3O_4 NPs for targeting, photothermal therapy, and imaging. These nanoparticles exhibited photothermal conversion efficiency and much higher photothermal stability compared to Au nanorods. PEGylated Fe@Fe_3O_4 NPs also provided selective magnetic targeting for cancer cells in vitro and in vivo. Yin et al. (2015) reported new multifunctional fluorescent magnetic nanoparticles for multimodal imaging in vivo. They showed that these nanoparticles accumulated in tumor tissues to allow dual-modal detection of tumors in a living body. As a different approach, Zoppellaro et al. (2014) prepared biocompatible magnetic nanocrystal clusters for imaging and sensing applications. They showed that these condensed-type magnetic nanostructures exhibited higher magnetophoretic responses compared to other types of magnetic colloids and clustered systems. Recently, Xiao et al. (2015) reported a facile and green synthesis method to develop magnetic nanocrystals. They coated water-soluble magnetic Fe_3O_4 nanoparticles with green tea extracts. These nanocrystals showed high-uptake efficiencies by cells without any additional transfection agent and efficient contrast agent properties.

5.1.2 Optical imaging

Optical imaging is commonly used for in vivo and in vitro applications due to its simplest modality. The process works based on photon emission from fluorescent and bioluminescent probes. This is

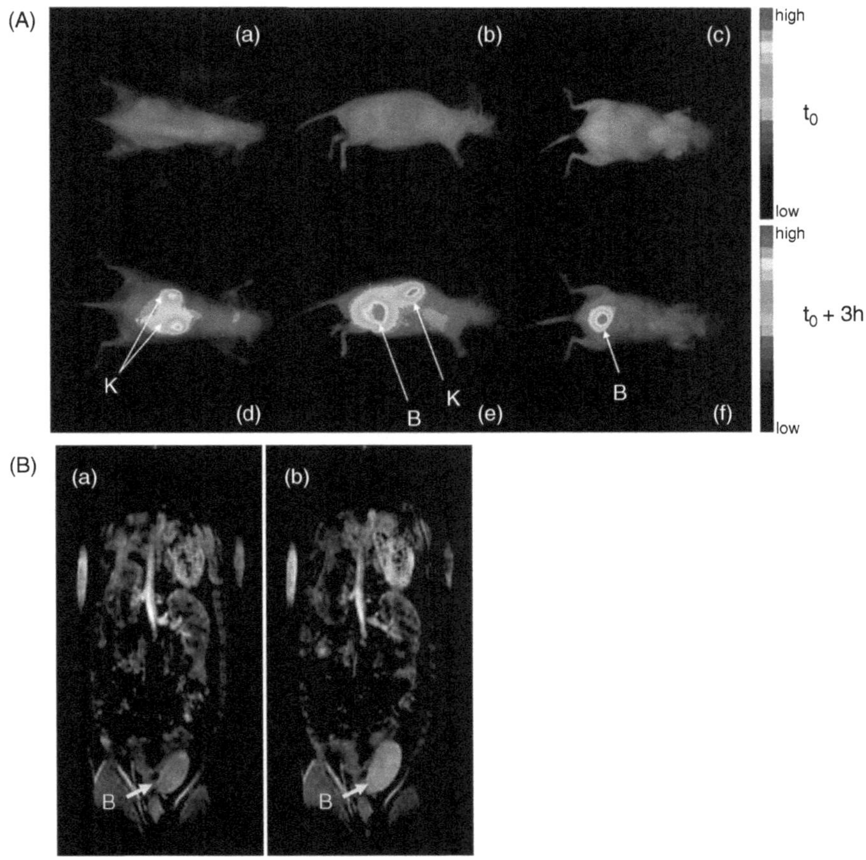

FIGURE 15.6 (A) Fluorescence reflectance imaging of (a, b, c) before and (d, e, f) 3 h after the injection of hybrid Gd_2O_3 NPs (*K*, kidneys; *B*, bladder). (B) T1-weighted images of rat (a) before and (b) 1 h after injection of hybrid Gd_2O_3 NPs (*B* for bladder).

Copyright with permission from Bridot, J.-L. et al., 2007. Hybrid Gd oxide nanoparticle multimodal contrast agents for in vivo imaging. J. Am. Chem. Soc. 129 (16), 5076–5084.

a significant advantage compared to other imaging modalities because the detection of low energy photons is relatively inexpensive. Different contrast agents such as gold nanoparticles, quantum dots, and upconversion particles are commonly used for targeted optical imaging of cancer cells in vivo. This technique can be used from visible to near-infrared (NIR) spectrum with a good spatial resolution. Noble metal nanoparticles have been used for biosensing and imaging in vivo due to their unique optical properties. Gold nanoparticles (rods, cages, stars etc.) and nanoshells provide localized surface plasmon resonance (LSPR) peaks in the NIR region (700–900 nm) (Anker et al., 2008; Li et al., 2012). Zhu et al. (2013) synthesized PEGylated gold nanorods (AuNRs) for NIR in vivo imaging. The PEGylated AuNRs exhibited much higher optical contrast at NIR than in the visible region in vivo. Gold nanoshells (AuNSs) have a great promising potential in cancer diagnosis and treatment in vivo, because AuNSs have tunable LSPR peaks in the NIR region. Vankayala et al. (2014) reported a very good study about the preparation of lipid-coated different core/shell gold nanomaterials including nanocages

(AuNCs), nanorod-in shell (NR shell) and nanoparticle-in shell (NP shell) for in vivo imaging. They demonstrated that all gold nanoshells can exert bimodal photothermal effects and photodynamic therapeutic effects upon NIR light excitation. They also reported that these Au nanoshells can be used as fluorescence cellular markers. Combining AuNPs with multiple discrete components such as magnetic NPs and quantum dots (QDs) into one system provides multimodality imaging ability. Li et al. (2015) developed a multifunctional system composed of hyaluronic acid (HA)-targeted Fe_3O_4@Au core/shell NPs (Fe_3O_4@Au-HA) for trimode imaging magnetic resonance/computed tomography/thermal imaging (MR/CT/thermal) and photo-thermal therapy in vivo. They reported that this multifunctional core/shell Fe_3O_4@Au-HA imaging and treatment agent showed the desired water dispersibility, colloidal stability, and biocompatibility in a given concentration range. Fe_3O_4@Au-HA core/shell NPs exhibited successful imaging performance for trimode MR/CT/thermal imaging in vitro and in vivo.

Nanocrystal QDs have been used for in vivo optical imaging because QDs exhibit bright and stable a wide wavelength range fluorescence. Panthani et al. (2013) synthesized zinc sulfide-coated copper indium sulfur selenide ($CuInSe_xS_{2-x}$/ZnS core/shell) nanocrystals and tested these nanocrystals as an imaging agent to track a microparticle-based oral vaccine administered to mice. They showed that these core/shell nanocrystals displayed strong fluorescence imaging to track the location of nanoparticle in vivo (Fig. 15.7).

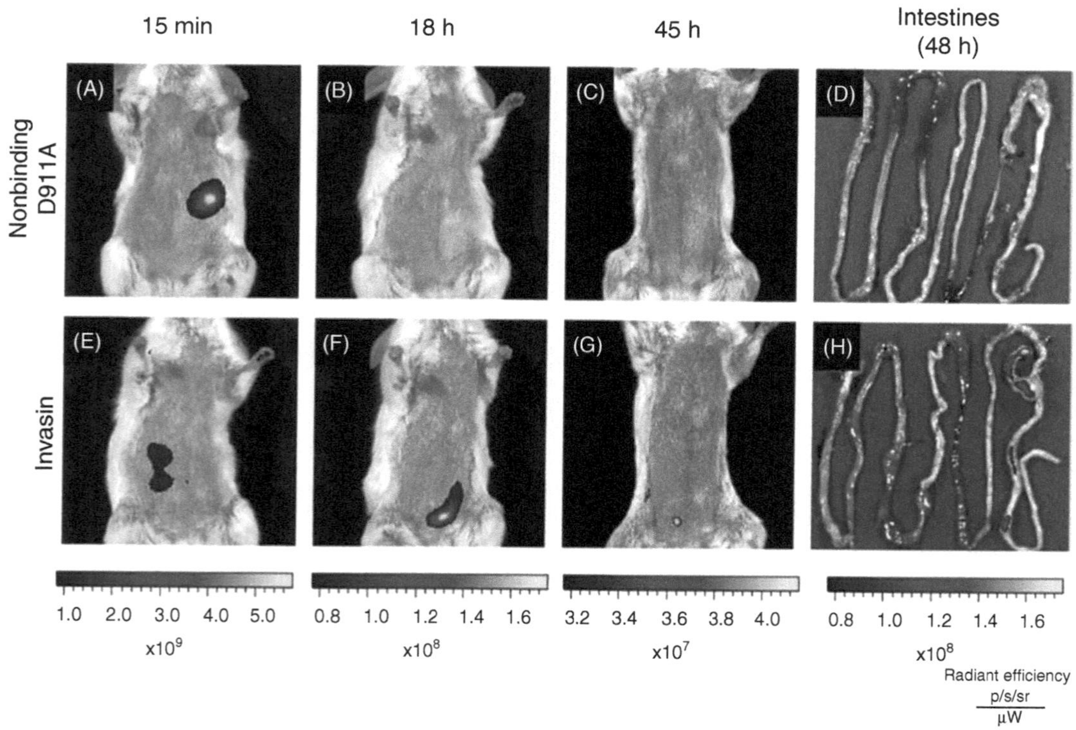

FIGURE 15.7 Representative in vivo fluorescence images tracking the $CuInSe_xS_{2-x}$/ZnS core–shell nanocrystals at 15 min (A, E), 18 h (B, F), and 45 h (C, G) after oral gavage and the intestines after dissection at 48 h (D, H).

QDs can quickly be quenched in water so that they can be used limited in biological applications. Core/shell-based systems provide long-term stability because the shell protects the core materials from oxidative environments. Aubert et al. (2014) fabricated silica coated luminescent CdSe/CdS (CdSe/CdS@SiO_2) nanoparticles to obtain long-term stable fluorescence probes for in vivo imaging. They observed that the silica coating enhanced the colloidal stability of core/shell NPs and protected optical properties for in vivo fluorescence imaging.

Up-conversion nanoparticles (UCNPs) have been also used for biomedical applications because they have unique advantages for optical imaging such as deep penetration into biological tissue, background-free fluorescence, and minimal photo-damage to living organisms (Deng and Wang, 2014; Wang et al., 2010a, 2011). The up-conversion process allows production of higher energy emission in the visible region from adsorption of two or more low-energy photons from NIR light (Deng and Wang, 2014; Wang et al., 2010a, 2011). Because of these unique properties, UCNPs are very good candidates for optical bioimaging and photothermal therapy in vivo. Zhou et al. (2014a) developed a mesoporous core/shell structured up-conversion nanoparticles (mUCNPs), composed of a NaYbF4:2%Er core and a mesoporous NaGdF4 shell for multimodal in vivo imaging and targeted drug delivery. They reported that these core/shell UCNPs exhibited good biocompability both in vivo and in vitro. Moreover NaYbF4:2%Er core and a mesoporous NaGdF4 shell UCNPs exhibited a good potential for multimodality bio-imaging.

In a recent work, Lv et al. (2015) integrated photodynamic therapy (PDT) with photothermal therapy (PTT) and chemotherapy to enhance antitumor efficacy. For this purpose, they prepared yolklike multifunctional UCNPs composed of up-conversion luminescent UCL GdOF:Ln core and mesoporous silica shell. Mesoporous silica shell provided large hallow cavities to load both multiple imaging agents and drug. The UCNPs were modified with zinc(II)-phthalocyanine (ZnPc) photosensitizers to product singlet oxygen when excited by red emission derived from NIR irradiation. To enhance the photothermal therapy ability, carbon dots were adapted to UCNs. The prepared GdOF:Ln@SiO_2-ZnPc-CDs core/shell nanoparticles were provided efficient DOX release. Moreover GdOF:Ln@SiO_2-ZnPc-CDs yolk-like core/shell nanoparticles allowed multimode imaging (MRI, X-ray CT, UCL, photothermal) and multiple therapies (PDT, PTT, and chemotherapy) in vitro and in vivo.

5.1.3 Positron emission tomography (PET)

Positron emission tomography (PET) is one of the important in vivo bioimaging method. PET provides the highest sensitivity and tissue penetration and give biological information at the molecular level in a living system with excellent sensitivity (Mackewn et al., 2005; Marsden et al., 2002). The radiolabeled probe molecules are used for PET imaging process (Ametamey et al., 2008). Among the radiolabeled molecules, fluorine-18 (^{18}F) is the most popular and most used positron emitter due to its unique nuclear and physical characteristics (Kim et al., 2006; Miller et al., 2008). Unfortunately, the half-life of ^{18}F is short to require essential blood circulation time for the labeled molecules. To enhance the half-life of the ^{18}F-labeled molecules and to provide specific uptake into the target cell or organs, different core–shell nanocarriers can be designed. Lee et al. (2013) developed aza-dibenzocyclooctynes (DBCO)-functionalized PEGylated mesoporous silica nanoparticles (MSNs) for PET imaging. The prepared system has worked based on the highly bio-orthogonal and rapid strain-promoted alkyne azide cycloaddition (SPAAC) conjugation reaction in a living system using ^{18}F. They observed a successful accumulation in the tumor site via the EPR effect. The nanoparticles reacted with an ^{18}F-labeled azide synthon and exhibited an efficient PET imaging proper properties for in vivo. The other popular isotope atom is ^{64}Cu due to its 12.7 h half-life, which is sufficient to study the biodistribution of long-circulating nanoparticles (Jensen et al., 2014). Jensen et al. designed ^{64}Cu-labeled core/shell triblock (PEG-PHEMA-PCMA) micelles using two different ^{64}Cu chelators in order to compare the biodistribution of the micelles with different chelating agents. The synthesized micelles exhibited long blood-circulation time and good accumulation in tumors without any significant difference for different chelating agents. They concluded that different chelating agents showed similar in vivo stability. Cui et al. (2015) described multimodal core/shell $Co_xFe_{3-x}O_4$@$NaYF_4$ nanoparticles with different cation dopants in the shell or core including $Co_{0.16}Fe_{2.84}O_4$@$NaYF_4$(Yb, Er) and Fe_3O_4@ $NaYF_4$(Yb, Tm). They observed that the lanthanide doped shell provided optical imaging and iron-oxide based core allowed MR imaging. They proposed that the prepared core/shell nanoparticles have promising potential to combine different imaging agents for PET, MRI and optical imaging in vivo.

5.2 CORE/SHELL NPS IN THERAPEUTIC

Core/shell nanoparticles are not just useful for bioimaging but also allow the fabricating multifunctional and targeted drug delivery systems. The core/shell systems have several advantages compared to other systems. First, the core material may be responsible for both imaging and triggered drug delivery. Theranostics core/shell nanoparticles can be designed for desired therapy methods such as photothermal therapy, sonodynamic therapy, and siRNA-therapy, and so forth. For this purpose, the various combination of imaging techniques and therapy methods can be fabricated using different core and shell materials.

Among these therapy strategies, photothermal therapy (PTT) has been extensively investigated for cancer treatment in clinical researches during the past several years (Cheng et al., 2014). In this strategy, NIR (650–900 nm) irradiation is converted into heat energy by the photothermal conversion (PTC) agents to heat and kill cancer cells locally without the risk of destroying healthy tissues (Oh et al., 2013). In the recent literature, there are several crucial reports about photothermal therapy in vivo. For example, Wu et al. (2015) fabricated poly(dopamine)-coated superparamagnetic iron oxide core/shell nanoparticles (SPION@PDA) to obtain an efficient therapeutic nanoparticles. PDA has been recently reported as an effective PTC agent due to its strong NIR absorption and high PTC efficiency. In this study, the SPION cluster core exhibited an efficient sensitivity for MRI and the PDA shell layer induced cancer cell death under NIR irradiation. They reported that SPION@PDA core/shell nanoparticles were easily uptaken by cancer cells under an external magnetic field. They reported that SPION@PDA core/shell nanoparticles have a promising theranostic agent potential for MRI-guided cancer photothermal therapy. Zheng et al. (2015) prepared core/shell PDA coated MNPs to enhance photothermal effect of nanoparticles compare to MNPs. According to their results, the synthesized core/shell NPs exhibited an excellent enhanced photothermal effect and therapeutic efficiency for in vivo tumor ablation.

Noble metals such as Au and Ag nanoparticles can be used to develop photothermal therapy agents due to their excellent plasmonic responses (Millstone et al., 2009; Skrabalak et al., 2008; Sotiriou et al., 2014). Recent studies showed that hybrid plasmonic nanomaterials have a promising potential to combine different theranostic modalities into a single nanoplatform (Jin, 2014; Jin et al., 2010) for cancer treatments with real-time diagnosis (Zhang et al., 2012). Recently, a multifunctional and targeted therapeutic PEGylated Au/Ag core/shell nanoshells (AuHNSs) were fabricated for photothermally induced drug delivery under NIR irradiation (Noh et al., 2015). They conjugated the PEGylated AuHNSs with anti-EGFR and then loaded antitumor drug, DOX (AuHNS-EGFR-DOX), for targeted drug release in NIR light. Noh et al. reported that targeted accumulation of AuHNS-EGFR-DOX core/shell nanoparticles on the cell membrane was more effective than receptor-mediated endocytosis and free DOX. Moreover, these multifunctional core/shell hybrid nanoparticles exhibited an efficient photothermal activity for targeted and rapid DOX release under low pH and NIR-irradiation conditions.

Sonodynamic therapy (SDT) can be defined ultrasound-triggered delivery system. In this process ultrasound activated sonosensitizers are used to generate reactive oxygen species (ROS). ROS directly damage cancer cells and tumors. Moreover, SDT is able to enhance the chemotherapeutic drug uptake (Siu et al., 2007). Generally, porphyrin derivatives (Yumita et al., 2012; Wang et al., 2010b), 5-aminolevulinic acid (Song et al., 2011) and TiO_2 (Ninomiya et al., 2012; Shen et al., 2015) are used as sonosensitizers for SDT-induced cancer treatments. Recent studies showed that TiO_2 nanoparticles are nontoxic and highly stable in biological environments (Yamaguchi et al., 2011) and very suitable to use in cancer SDT (Shen et al., 2015). In a recent study, Shen et al. (2015) developed TiO_2-encapsulated magnetic nanoparticles (Fe_3O_4@TiO_2 NPs) as a sonosensitizer for SD therapy of cancer. They showed that this fabricated core/shell sonosensitizer nanoparticles exhibited pH-dependent loading DOX loading and release in vitro. These Fe_3O_4@TiO_2 NPs generated ROS efficiently under ultrasound (US) irradiation and showed efficient cytotoxicity against tumors due to their enhancing membrane permeability as seen in Fig. 15.8. They also reported that the intracellular DOX uptake was increased by US and the cytotoxic effects of ROS. The developed Fe_3O_4@TiO_2-DOX NPs were reported as efficient agents for chemo-sonodynamic cancer therapy.

Therapeutic core/shell NPs can also be applied to small interfering RNA (siRNA) therapy because these NPs can effectively enhance the stability of siRNA in blood circulation and minimized the

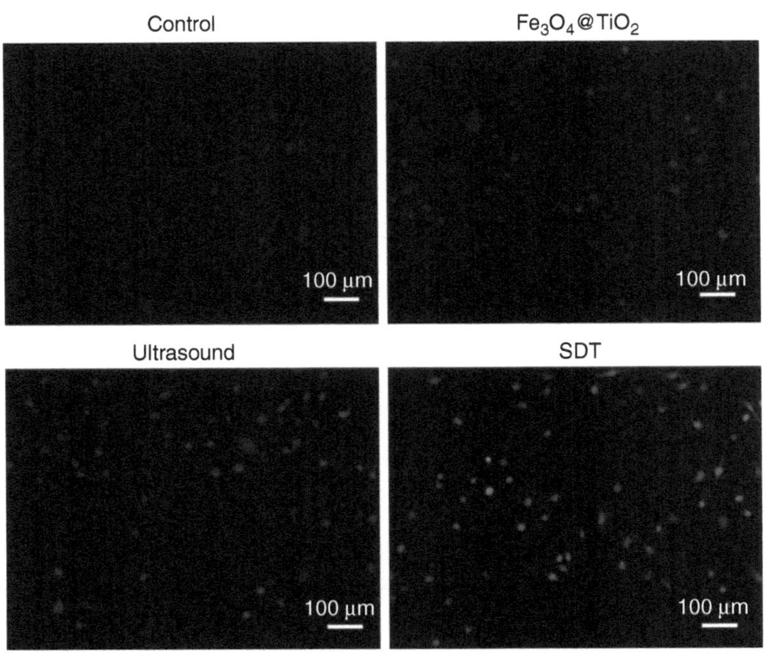

FIGURE 15.8 Fluorescent Images of MCF-7 Cells Stained by Fe_3O_4@TiO_2-DOX NPs Showing Intracellular ROS

Reprinted from Shen, S. et al., 2015. Core–shell structured Fe_3O_4@TiO_2-doxorubicin nanoparticles for targeted chemo-sonodynamic therapy of cancer. Int. J. Pharm. 486(1–2), 380–388, with permission from Elsevier 2015.

degradation of siRNA by enzymes (Ryu et al., 2014). siRNA is one of the most promising agents for cancer treatment due to its specific silencing of the targeted gene (Gao et al., 2014). However, siRNA cannot enter the target cells and silence specific genes due to its negative charge, large molecular weight, and rapid elimination from the systemic circulation (Gao et al., 2014). For this purpose, siRNA are encapsulated into nanoparticles to overcome these obstacles. Gao et al. reported core/shell lipid/rPAA-Chol hybrid NPs (PEG-LP/siRNA NPs and T7-LP/siRNA NPs) for improving in vivo siRNA delivery. As seen in Fig. 15.9, they first covered rPAA-Chol/siRNA nanoplex core with a lipid shell

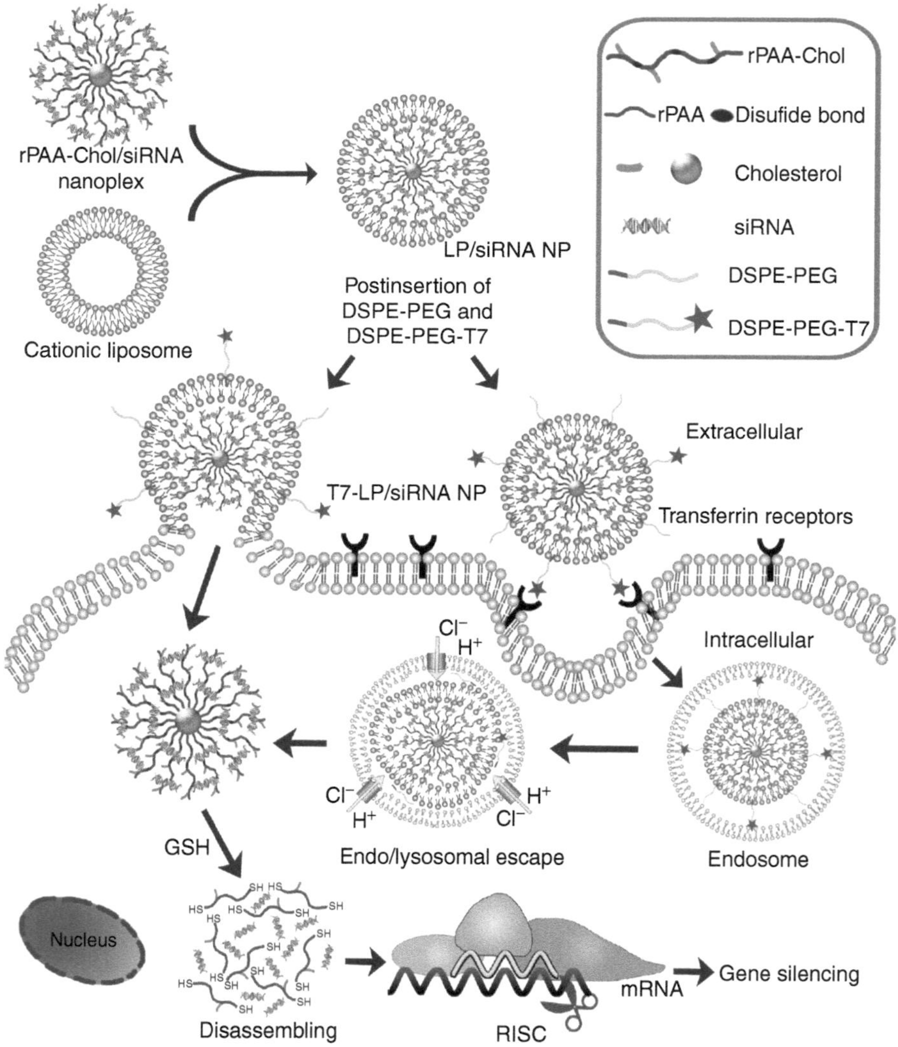

FIGURE 15.9 Schematic Illustration of the Assembling Process of Targeted Core–Shell Type Lipid/rPAA-Chol Hybrid Nanoparticles (T7-LP/siRNA NPs) and Their Intracellular Trafficking Pathway

Reprinted from Gao, L.Y. et al., 2014. Core-Shell type lipid/rPAA-Chol polymer hybrid nanoparticles for invivo siRNA delivery. Biomaterials 35(6), 2066–2078, with permission from Elsevier 2014.

and then modified this lipid shell with polyethylene glycol phospholipid (DSPE-PEG) and/or peptide (T7) to obtain desired core/shell hybrid NPs. These hybrid nanoparticles exhibited very significant advantages such as low cytotoxicity, stability in serum, higher cell uptake, and higher gene silencing to be used for theranostic platform.

As a conclusion, we have discussed the recently published various types of core/shell NPs for drug delivery and molecular imaging in vivo. These core/shell systems have promising potential to develop new theranostic platforms.

6 CONCLUSION AND FUTURE PROSPECTS

Over the past ten years, researchers have focused on the development of multifunctional nanoparticle-based therapeutic platforms that could be used in both medical imaging and therapy applications. These all-in-one and multifunctional therapeutic platforms are fabricated through the combination of different diagnosis and therapeutic techniques. These types of dual-purpose biomaterials are defined as *theranostics* and theranostic biomaterials are useful platforms allowing image-guided therapy of cancer.

In this chapter, we discussed recently published and promising noble metal, magnetic, silica, polymer and core/shell nanoparticle-based theranostics used in both diagnosis and treatment in vivo protocols. Latterly, researchers are working on these nanoparticle-based platforms to fabricate a new generation of personalized medicine that possesses desired properties such as biocompatibility, lower cytotoxicity, bio-imaging capability, targeted delivery, and ease of availability. Unfortunately, for clinical applications, these theranostic platforms are accepted as a future expectation because they still have numerous challenges such as solubility, biodistribution, toxicity, and stability. Therefore, herein we tried to mention the recent studies that were focused on solving these obstacles for the translation of these multifunctional nanoparticles-based platforms to clinical applications as personalized medicine.

As a conclusion, we are expecting that the future of personalized medicine will provide the selective detection of the cancer cells with real time in vivo imaging and destroy them without high-level cytotoxicity. In our belief, these new-generation theranostic nanoparticles have valuable and promising potential and will define as a mirror picture the future of personalized medicine.

REFERENCES

Abdelaal, H.M., Zawrah, M.F., Harbrecht, B., 2014. Facile one-pot fabrication of hollow porous silica nanoparticles. Chem - Euro. J. 20 (3), 673–677.

Alkilany, A.M., et al., 2009. Cellular uptake and cytotoxicity of gold nanorods: molecular origin of cytotoxicity and surface effects. Small 5 (6), 701–708.

Alkilany, A.M., et al., 2012. Gold nanorods: Their potential for photothermal therapeutics and drug delivery, tempered by the complexity of their biological interactions. Adv. Drug Deliv. Rev. 64 (2), 190–199.

American Cancer Society, 2015. Cancer Facts & Figures 2015. American Cancer Society, Atlanta.

Ametamey, S.M., Honer, M., Schubiger, P.A., 2008. Molecular imaging with PET. Chem. Rev. 108 (5), 1501–1516.

Anker, J.N., et al., 2008. Biosensing with plasmonic nanosensors. Nat Mater. 7 (6), 442–453.

Arias, J.L., et al., 2008. Magnetite/poly(alkylcyanoacrylate) (core–shell) nanoparticles as 5-Fluorouracil delivery systems for active targeting. Euro. J. Pharm. Biopharm. 69 (1), 54–63.

Arruebo, M., et al., 2007. Magnetic nanoparticles for drug delivery. Nano Today 2 (3), 22–32.

Aubert, T., et al., 2014. Bright and stable CdSe/CdS @ SiO 2 nanoparticles suitable for long-term cell labeling. ACS Appl. Mater. Interf. 6 (14), 11714–11723.

Baeza, A., Colilla, M., Vallet-Regí, M., 2015. Advances in mesoporous silica nanoparticles for targeted stimuli-responsive drug delivery. Exp. Opin. Drug Deliv. 12 (2), 319–337.

Bennet, D., Kim, S., 2014. Polymer nanoparticles for smart drug delivery. InTech, pp. 257–310.

Biju, V., 2014. Chemical modifications and bioconjugate reactions of nanomaterials for sensing, imaging, drug delivery and therapy. Chem. Soc. Rev. 43 (3), 744–764.

Bridot, J.-L., et al., 2007. Hybrid Gd oxide nanoparticle multimodal contrast agents for in vivo imaging. J. the Am. Chem.l Soc. 129 (16), 5076–5084.

Brigger, I., Dubernet, C., Couvreur, P., 2002. Nanoparticles in cancer therapy and diagnosis. Adv. Drug Del. Rev. 54 (5), 631–651.

Bui, Q.N., et al., 2015. Redox- and pH-sensitive polymeric micelles based on poly(β-amino ester)-grafted disulfide methylene oxide poly(ethylene glycol) for anticancer drug delivery. Macromolecules 48 (12), 4046–4054.

Burns, A., Ow, H., Wiesner, U., 2006. Fluorescent core–shell silica nanoparticles: towards "Lab on a Particle" architectures for nanobiotechnology. Chem. Soc. Rev. 35 (11), 1028–1042.

Cao, X., et al., 2013. Seventy-two-hour release formulation of the poorly soluble drug silybin based on porous silica nanoparticles: in vitro release kinetics and in vitro/in vivo correlations in beagle dogs. Euro. J. Pharm. Sci. 48 (1–2), 64–71.

Cao, Y., et al., 2015. Triggered-release polymeric conjugate micelles for on-demand intracellular drug delivery. Nanotech. 26 (11), 115101.

Chakravarty, R., et al., 2015. Hollow mesoporous silica nanoparticles for tumor vasculature targeting and PET image-guided drug delivery. Nanomed. 10 (8), 1233–1246.

Chatterjee, K., et al., 2014. Core/shell nanoparticles in biomedical applications. Adv. Colloid Interf. Sci. 209, 8–39.

Chen, J.-F., et al., 2004. Preparation and characterization of porous hollow silica nanoparticles for drug delivery application. Biomaterials 25 (4), 723–727.

Chen, C.C., et al., 2006. DNA-gold nanorod conjugates for remote control of localized gene expression by near infrared irradiation. J. Am. Chem. Soc. 128 (11), 3709–3715.

Chen, A.M., et al., 2009. Co-delivery of Doxorubicin and Bcl-2 siRNA by mesoporous silica nanoparticles enhances the efficacy of chemotherapy in multidrug-resistant cancer cells. Small 5 (23), 2673–2677.

Chen, F.-H., et al., 2010a. Synthesis of a novel magnetic drug delivery system composed of doxorubicin-conjugated Fe3O4 nanoparticle cores and a PEG-functionalized porous silica shell. Chem. Commun. 46 (45), 8633–8635.

Chen, S., et al., 2010b. Engineered biocompatible nanoparticles for in vivo imaging applications. J. Am. Chem. Soc. 132 (42), 15022–15029.

Chen, W.-H., et al., 2013. Therapeutic nanomedicine based on dual-intelligent functionalized gold nanoparticles for cancer imaging and therapy in vivo. Biomaterials 34 (34), 8798–8807.

Chen, F., et al., 2014. Tumor targeted drug delivery with hollow mesoporous silica nanoparticles. J.Nuc. Med. 55 (1), 547.

Chen, Y., et al., 2015. Multifunctional envelope-type mesoporous silica nanoparticles for pH-responsive drug delivery and magnetic resonance imaging. Biomaterials 60, 111–120.

Cheng, L., et al., 2014. Functional nanomaterials for phototherapies of cancer. Chem. Rev. 114 (21), 10869–10939.

Chertok, B., David, A.E., Yang, V.C., 2010. Polyethyleneimine-modified iron oxide nanoparticles for brain tumor drug delivery using magnetic targeting and intra-carotid administration. Biomat. 31 (24), 6317–6324.

Cho, K., et al., 2008. Therapeutic nanoparticles for drug delivery in cancer. Clin. Cancer Res. 14 (5), 1310–1316.

Choi, E.S., et al., 2010. Water-soluble ultra-small manganese oxide surface doped gadolinium oxide (Gd2O3@ MnO) nanoparticles for MRI contrast agent. Euro.J. Inorgan. Chem. 2010 (28), 4555–4560.

Choi, K.Y., et al., 2012. Theranostic nanoplatforms for simultaneous cancer imaging and therapy: current approaches and future perspectives. Nanoscale 4 (2), 330.

Cui, X., et al., 2015. Synthesis, characterization, and application of core–shell $Co_{0.16}Fe_{2.84}O_4$@$NaYF_4$(Yb, Er) and Fe_3O_4@$NaYF_4$(Yb, Tm) nanoparticle as trimodal (MRI, PET/SPECT, and optical) imaging agents. Bioconjug. Chem. 27 (2), 319–328.
Demirel, et al., 2011. Nanoporous polymeric nanofibers based on selectively etched PS-b-PDMS block copolymers. Appl. Mater. Interfaces 4 (1), 280–285.
Demirel, et al., 2013. A new multiresponsive drug delivery system using smart nanogels. Chem. Phys. Chem. 14 (12), 2833–2840.
Deng, M., Wang, L., 2014. Unexpected luminescence enhancement of upconverting nanocrystals by cation exchange with well retained small particle size. Nano Res. 7 (5), 782–793.
Deng, H., et al., 2014. Theranostic self-assembly structure of gold nanoparticles for NIR photothermal therapy and X-Ray computed tomography imaging. Theranostics 4 (9), 904–918.
Dixit, S., et al., 2015. Transferrin receptor-targeted theranostic gold nanoparticles for photosensitizer delivery in brain tumors. Nanoscale 7 (5), 1782–1790.
Du, X., et al., 2013. Developing functionalized dendrimer-like silica nanoparticles with hierarchical pores as advanced delivery nanocarriers. Adv. Mat. 25 (41), 5981–5985.
Duncan, R., 2003. The dawning era of polymer therapeutics. Nat. Rev. Drug Discov. 2 (5), 347–360.
Durán, J.D.G., et al., 2008. Magnetic colloids as drug vehicles. J. Pharm. Sci. 97 (8), 2948–2983.
Farokhzad, O.C., Langer, R., 2006. Nanomedicine: Developing smarter therapeutic and diagnostic modalities. Adv. Drug Del. Rev. 58 (14), 1456–1459.
Feng, S.-S., Chien, S., 2003. Chemotherapeutic engineering: Application and further development of chemical engineering principles for chemotherapy of cancer and other diseases. Chem. Eng. Sci. 58 (18), 4087–4114.
Feng, Z., et al., 2008. A facile route to hollow nanospheres of mesoporous silica with tunable size. Chem. Commun. 23, 2629–2631.
Ferrari, M., 2005a. Cancer nanotechnology: opportunities and challenges. Nat. Rev. Cancer 5 (3), 161–171.
Ferrari, M., 2005b. Nanovector therapeutics. Curr. Opin. Chem. Biol. 9 (4), 343–346.
Frullano, L., Meade, T.J., 2007. Multimodal MRI contrast agents. J. Biol. Inorgan. Chem. 12 (7), 939–949.
Gao, Y., et al., 2011. Controlled intracellular release of doxorubicin in multidrug-resistant cancer cells by tuning the shell-pore sizes of mesoporous silica nanoparticles. ACS Nano 5 (12), 9788–9798.
Gao, L.Y., et al., 2014. Core–shell type lipid/rPAA-Chol polymer hybrid nanoparticles for invivo siRNA delivery. Biomat. 35 (6), 2066–2078.
Ghosh Chaudhuri, R., Paria, S., 2012. Core/shell nanoparticles: classes, properties, synthesis mechanisms, characterization, and applications. Chem. Rev. 112 (4), 2373–2433.
Ghosh, P., et al., 2008. Gold nanoparticles in delivery applications. Adv. Drug Del. Rev. 60 (11), 1307–1315.
Giménez, C., et al., 2015. Gated mesoporous silica nanoparticles for the controlled delivery of drugs in cancer cells. Langmuir 31 (12), 3753–3762.
Goel, S., et al., 2015. Multifunctional hollow mesoporous silica nanoparticles for tumor targeted, PET image-guided drug delivery. J. Nuc. Med. 56 (3), 1108.
Graf, C., Van Blaaderen, A., 2002. Metallodielectric colloidal core–shell particles for photonic applications. Langmuir 18 (2), 524–534.
Hanahan, D., Weinberg, R.a., 2000. The hallmarks of cancer. Cell 100, 57–70.
He, Q., Shi, J., 2011. Mesoporous silica nanoparticle based nano drug delivery systems: synthesis, controlled drug release and delivery, pharmacokinetics and biocompatibility. J. Mat. Chem. 21 (16), 5845.
Hoffman, A.S., 2008. The origins and evolution of "controlled" drug delivery systems. J. Control. Rel. 132 (3), 153–163.
Huang, X., et al., 2011. The shape effect of mesoporous silica nanoparticles on biodistribution, clearance, and biocompatibility in vivo. ACS Nano. 5 (7), 5390–5399.
Huang, H., et al., 2014. A smart drug delivery system from charge-conversion polymer-drug conjugate for enhancing tumor therapy and tunable drug release. Macromol. Biosci. 14 (4), 485–490.

Hwang, A.a., et al., 2015. Functional nanovalves on protein-coated nanoparticles for in vitro and in vivo controlled drug delivery. Small. 11 (3), 319–328.

Iyer, A.K., et al., 2006. Exploiting the enhanced permeability and retention effect for tumor targeting. Drug Discov. Today 11 (17–18), 812–818.

Jain, R.K., 2001. Delivery of molecular medicine to solid tumors: lessons from in vivo imaging of gene expression and function. J. Control. Rel. 74, 7–25.

Janib, S.M., Moses, A.S., MacKay, J.A., 2010. Imaging and drug delivery using theranostic nanoparticles. Adv. Drug Del. Rev. 62 (11), 1052–1063.

Jensen, A.I., et al., 2014. Positron emission tomography based analysis of long-circulating cross-linked triblock polymeric micelles in a U87MG mouse xenograft model and comparison of DOTA and CB-TE2A as chelators of copper-64. Biomacromol. 15 (5), 1625–1633.

Jin, Y., 2014. Multifunctional compact hybrid Au nanoshells: a new generation of nanoplasmonic probes for biosensing, imaging, and controlled release. Acc. Chem. Res. 47 (1), 138–148.

Jin, Y., et al., 2010. Multifunctional nanoparticles as coupled contrast agents. Nat. Comm. 1, 41.

Kawano, T., et al., 2009. PNIPAM gel-coated gold nanorods for targeted delivery responding to a near-infrared laser. Bioconjug. Chem. 20 (2), 209–212.

Ke, H., et al., 2014. Gold nanoshelled liquid perfluorocarbon magnetic nanocapsules: a nanotheranostic platform for bimodal ultrasound/magnetic resonance imaging guided photothermal tumor ablation. Theranostics 4 (1), 12–23.

Khanbabaie, R., Jahanshahi, M., 2012. Revolutionary impact of nanodrug delivery on neuroscience. Curr. Neuropharm. 10 (4), 370–392.

Kim, S.W., et al., 2002. Fabrication of hollow palladium spheres and their successful application to the recyclable heterogeneous catalyst for suzuki coupling reactions. J. Am. Chem. Soc. 124 (26), 7642–7643.

Kim, D.W., et al., 2006. A new class of SN2 reactions catalyzed by protic solvents: facile fluorination for isotopic labeling of diagnostic molecules. J. Am. Chem. Soc. 128 (50), 16394–16397.

Kim, J., et al., 2008. Multifunctional uniform nanoparticles composed of a magnetite nanocrystal core and a mesoporous silica shell for magnetic resonance and fluorescence imaging and for drug delivery. Angew. Chem. Int. Ed. 47 (44), 8438–8441.

Kim, H.J., et al., 2015. Drug-loaded gold/iron/gold plasmonic nanoparticles for magnetic targeted chemo-photothermal treatment of rheumatoid arthritis. Biomaterials 61, 95–102.

Konkolewicz, D., Monteiro, M.J., Perrier, S., 2011. Dendritic and hyperbranched polymers from macromolecular units: elegant approaches to the synthesis of functional polymers. Macromolecules 44 (18), 7067–7087.

Kosuge, K., Shimada, K., Tsunashima, A., 1995. Micropore formation by acid treatment of antigorite. Chem. Mater. 7, 2241–2246.

Lamprecht, A., 2007. Nanotherapeutics: Drug Delivery Concepts in Nanoscience. Pan Stanford Publishing Pte. Ltd., Singapore.

Laurent, S., et al., 2008. Magnetic iron oxide nanoparticles: synthesis, stabilization, vectorization, physicochemical characterizations, and biological applications. Chem. Rev. 108 (6), 2064–2110.

Lee, S.B., et al., 2013. Mesoporous silica nanoparticle pretargeting for PET imaging based on a rapid bioorthogonal reaction in a living body. Angew. Chem. Int. Ed. 52 (40), 10549–10552.

Lee, D., et al., 2014a. Ca^{2+}-dependent intracellular drug delivery system developed with "raspberry-type" particles-on-a-particle comprising mesoporous silica core and R synuclein-coated gold nanoparticles. ACS Nano. 8 (9), 8887–8895.

Lee, S.Y., et al., 2014b. PH/redox/photo responsive polymeric micelle via boronate ester and disulfide bonds with spiropyran-based photochromic polymer for cell imaging and anticancer drug delivery. Euro. Polym. J. 57, 1–10.

Lewinski, N., Colvin, V., Drezek, R., 2008. Cytotoxicity of nanoparticles. Small 4 (1), 26–49.

Li, Y., et al., 2012. Resonance scattering particles as biological nanosensors in vitro and in vivo. Chem. Soc. Rev. 41 (2), 632.

Li, J., et al., 2015. Hyaluronic acid-modified Fe3O4@Au core/shell nanostars for multimodal imaging and photothermal therapy of tumors. Biomaterials 38, 10–21.
Liu, Y., Miyoshi, H., Nakamura, M., 2007. Novel drug delivery system of hollow mesoporous silica nanocapsules with thin shells: preparation and fluorescein isothiocyanate (FITC) release kinetics. Colloid Surf. B: Biointerfaces 58 (2), 180–187.
Liu, J., et al., 2009. Recent developments in the chemical synthesis of inorganic porous capsules. J. Mater. Chem. 19 (34), 6073.
Liu, H.-L., et al., 2010. Magnetic resonance monitoring of focused ultrasound/magnetic nanoparticle targeting delivery of therapeutic agents to the brain. Proc.Natl. Acad. Sci. USA 107 (34), 15205–15210.
Lu, Y., et al., 1999. Aerosol-assisted self-assembly of mesostructured spherical nanoparticles. Nature 398, 223–227.
Lv, R., et al., 2015. Yolk-like multifunctional platform for multimodal imaging and synergistic therapy triggered by a single near-infrared light. ACS Nano. 9 (2), 1630–1647.
Ma, M., et al., 2015. Bi2S3-embedded mesoporous silica nanoparticles for efficient drug delivery and interstitial radiotherapy sensitization. Biomaterials 37, 447–455.
Mackewn, J.E., et al., 2005. Design and development of an MR-compatible PET scanner for imaging small animals. IEEE Trans. Nuc. Sci. 52 (5), 3271–3274.
Marsden, P.K., et al., 2002. Simultaneous PET and NMR. Br. J. Radiol. 75 (9), S53–S59.
McCarthy, J.R., Weissleder, R., 2008. Multifunctional magnetic nanoparticles for targeted imaging and therapy. Adv. Drug Del. Rev. 60 (11), 1241–1251.
Miller, P.W., et al., 2008. Synthesis of11C, 18F, 15O, and 13N radiolabels for positron emission tomography. Angew. Chem. Int. Ed. 47 (47), 8998–9033.
Millstone, J.E., et al., 2009. Colloidal gold and silver triangular nanoprisms. Small 5 (6), 646–664.
Moghimi, S.M., Patel, H.M., 1998. Serum-mediated recognition of liposomes by phagocytic cells of the reticuloendothelial system—The concept of tissue specificity. Adv. Drug Del. Rev. 32 (1–2), 45–60.
Murphy, C.J., et al., 2005. Anisotropic metal nanoparticles: Synthesis, assembly, and optical applications. J. Phys. Chem. B 109 (29), 13857–13870.
Murphy, C.J., Gole, A.M., Hunyadi, S.E., et al., 2008a. Chemical sensing and imaging with metallic nanorods. Chem. Commun. 5, 544–557.
Murphy, C.J., Gole, A.M., Stone, J.W., et al., 2008b. Gold nanoparticles in biology: Beyond toxicity to cellular imaging. Acc. Chem. Res. 41 (12), 1721–1730.
Muthu, M.S., et al., 2015. Theranostic vitamin E TPGS micelles of transferrin conjugation for targeted co-delivery of docetaxel and ultra bright gold nanoclusters. Biomater. 39, 234–248.
Ninomiya, K., et al., 2012. Targeted sonodynamic therapy using protein-modified TiO 2 nanoparticles. Ultrason. Sonochem. 19 (3), 607–614.
Noh, M.S., et al., 2015. Target-specific near-IR induced drug release and photothermal therapy with accumulated Au/Ag hollow nanoshells on pulmonary cancer cell membranes. Biomaterials 45, 81–92.
Oh, J., Yoon, H., Park, J.H., 2013. Nanoparticle platforms for combined photothermal and photodynamic therapy. Biomed. Eng. Lett. 3 (2), 67–73.
Panthani, M.G., et al., 2013. In vivo whole animal fluorescence imaging of a microparticle-based oral vaccine containing (CuInSe(x)S(2-x))/ZnS core–shell quantum dots. Nano. Lett. 13 (9), 4294–4298.
Perrie, Y., Rades, T., 2012. FASTtrack: Pharmaceutics—Drug Delivery and Targeting. Pharmaceutical Press, London and Chicago.
Qiu, L.Y., Bae, Y.H., 2006. Polymer architecture and drug delivery. Pharm. Res. 23 (1), 1–30.
Riess, G., 2003. Micellization of block copolymers. Prog. Polym. Sci. 28 (7), 1107–1170.
Roggers, R., et al., 2014. The practicality of mesoporous silica nanoparticles as drug delivery devices and progress toward this goal. AAPS Pharm. Sci. Tech. 15 (5), 1163–1171.
Ryu, J.H., et al., 2014. Theranostic nanoparticles for future personalized medicine. J. Control. Rel. 190, 477–484.
Sagnella, S., Drummond, C., 2012. Drug Del.: Nanomedi. Appr. Australian Biochemist 43 (3), 5–9.

Sanvicens, N., Marco, M.P., 2008. Multifunctional nanoparticles—properties and prospects for their use in human medicine. Trends Biotech. 26 (8), 425–433.

Schleich, N., et al., 2014. Comparison of active, passive and magnetic targeting to tumors of multifunctional paclitaxel/SPIO-loaded nanoparticles for tumor imaging and therapy. J. Control. Rel. 194, 82–91.

Sharker, S.M., et al., 2015. pH triggered in vivo photothermal therapy and fluorescence nanoplatform of cancer based on responsive polymer-indocyanine green integrated reduced graphene oxide. Biomaterials 61, 229–238.

Shen, S., et al., 2015. Core–shell structured Fe3O4@TiO2-doxorubicin nanoparticles for targeted chemo-sonodynamic therapy of cancer. Int. J. Pharm. 486 (1-2), 380–388.

Singh, R.K., et al., 2014. Multifunctional hybrid nanocarrier: Magnetic CNTs ensheathed with mesoporous silica for drug delivery and imaging system. ACS Appl. Mater. Interf. 6 (4), 2201–2208.

Siu, T., et al., 2007. Drug uptake enhancement using sonodynamic effects at 4 MHz—a potential application for micro-ultrasonic-transducers. IEEE Trans. Biomed. Eng. 54 (6), 1153–1156.

Skrabalak, S.E., et al., 2008. Gold nanocages: Synthesis, properties, and applications. Acc. Chem. Res. 41 (12), 1587–1595.

Somin, E.L., et al., 2009. Biologically functional cationic phospholipid-gold nanoplasmonic carriers of RNA. J. Am. Chem. Soc. 131 (39), 14066–14074.

Song, W., et al., 2011. Apoptosis of SAS cells induced by sonodynamic therapy using 5-aminolevulinic acid sonosensitizer. Anticancer Res. 31 (1), 39–43.

Sotiriou, G.a., et al., 2014. Engineering safer-by-design silica-coated ZnO nanorods with reduced DNA damage potential. Environ. Sci.: Nano. 1 (2), 144.

Sudhakar, A., 2009. History of cancer, ancient and modern treatment methods. J.Cancer Sci. Ther. 1 (2), 1–4.

Sun, C., Lee, J.S.H., Zhang, M., 2008. Magnetic nanoparticles in MR imaging and drug delivery. Adv. Drug Del. Rev. 60 (11), 1252–1265.

Tabatabaei, S.N., et al., 2015. Remote control of the permeability of the blood-brain barrier by magnetic heating of nanoparticles: a proof of concept for brain drug delivery. J. Control. Rel. 206, 49–57.

Torchilin, V.P., 2004. Targeted polymeric micelles for delivery of poorly soluble drugs. Cell. Mol. Life Sci. 61 (19-20), 2549–2559.

Vankayala, R., et al., 2014. Gold nanoshells-mediated bimodal photodynamic and photothermal cancer treatment using ultra-low doses of near infra-red light. Biomaterials 35 (21), 5527–5538.

Veiseh, O., Gunn, J., Zhang, M., 2010. Design and fabrication of magnetic nanoparticles for targeted drug delivery and imaging. Adv. Drug Deliv. Rev. 62 (3), 284–304.

Wang, F., et al., 2010a. Upconversion nanoparticles in biological labeling, imaging, and therapy. The Analyst 135 (8), 1839–1854.

Wang, P., et al., 2010b. Sonodynamic effects of protoporphyrin IX disodium salt on Ehrlich ascetic tumor cells. Ultrason. 50 (6), 634–638.

Wang, G., Peng, Q., Li, Y., 2011. Lanthanide-doped nanocrystals: synthesis, optical-magnetic properties, and applications. Acc. Chem. Res. 44 (5), 322–332.

Wang, Z., Niu, G., Chen, X., 2014. Polymeric materials for theranostic applications. Pharm. Res. 31 (6), 1358–1376.

Wang, Y., et al., 2015. Mesoporous silica nanoparticles in drug delivery and biomedical applications. Nanomed.: Nanotech., Biol. Med. 11 (2), 313–327.

Wei, H., et al., 2006. Self-assembled thermo- and pH responsive micelles of poly(10-undecenoic acid-b-*N*-isopropylacrylamide) for drug delivery. J. Control. Rel. 116 (3), 266–274.

Wijaya, A., et al., 2009. Selective release of multiple DNA oligonucleotides from gold nanorods. ACS Nano. 3 (1), 80–86.

Winey, K., Discher, D., 2006. Achieving long-term controlled drug delivery by using biodegradable polymeric nanoparticles. Dissertations available from ProQuest. Paper AAI3225435.

Woodle, M.C., 1998. Controlling liposome blood clearance by surface-grafted polymers. Adv. Drug Del. Rev. 32 (1–2), 139–152.

Wu, C., et al., 2015. Negatively charged magnetite nanoparticle clusters as efficient MRI probes for dendritic cell labeling and in vivo tracking. Adv. Func. Mater. 25 (23), 3581–3591.

Xiao, L., et al., 2015. Enhanced in vitro and in vivo cellular imaging with green tea coated water-soluble iron oxide nanocrystals. ACS Appl. Mater. Interf.

Yamaguchi, S., et al., 2011. Sonodynamic therapy using water-dispersed TiO2-polyethylene glycol compound on glioma cells: Comparison of cytotoxic mechanism with photodynamic therapy. Ultrason. Sonochem. 18 (5), 1197–1204.

Yidirim, A., et al., 2013. Pluronic polymer capped biocompatible mesoporous silica nanocarriers. Chem. Comm. 49, 9782–9784.

Yin, C., et al., 2015. Fluorescent oligo(p-phenyleneethynylene) contained amphiphiles-encapsulated magnetic nanoparticles for targeted magnetic resonance and two-photon optical imaging in vitro and in vivo. Nanoscale 7 (19), 8907–8919.

Yu, Y., et al., 2015. A degradable brush polymer–drug conjugate for pH-responsive release of doxorubicin. Polym. Chem. 6 (6), 953–961.

Yumita, N., et al., 2012. Involvement of reactive oxygen species in sonodynamically induced apoptosis using a novel porphyrin derivative. Theranostics 2 (9), 880–888.

Zahr, A.S., Pishko, M.V., 2009. Nanotechnology for Cancer Chemotherapy. In Nanotechnology in Drug Delivery. Springer, New York, pp. 491–518.

Zhang, Z., et al., 2012. Mesoporous silica-coated gold nanorods as a light-mediated multifunctional theranostic platform for cancer treatment. Adv. Mater. 24 (11), 1418–1423.

Zhang, J., et al., 2013. Multifunctional envelope-type mesoporous silica nanoparticles for tumor-triggered targeting drug delivery. J. Am. Chem. Soc. 135 (13), 5068–5073.

Zhang, B., et al., 2014a. Cytochrome c end-capped mesoporous silica nanoparticles as redox-responsive drug delivery vehicles for liver tumor-targeted triplex therapy in vitro and in vivo. J. Control. Rel. 192, 192–201.

Zhang, Q., et al., 2014b. Biocompatible, uniform, and redispersible mesoporous silica nanoparticles for cancer-targeted drug delivery in vivo. Adv. Func. Mater. 24 (17), 2450–2461.

Zhang, Y., et al., 2015. Histamine-functionalized copolymer micelles as a drug delivery system in 2D and 3D models of breast cancer. J. Mater. Chem. B 3 (12), 2472–2486.

Zheng, J., et al., 2015. Polydopamine-mediated immobilization of phenylboronic acid on magnetic microspheres for selective enrichment of glycoproteins and glycopeptides. Sci. China Chem. 58 (6), 1056–1064.

Zhong, J., et al., 2015. A smart polymeric platform for multistage nucleus-targeted anticancer drug delivery. Biomater. 65, 43–55.

Zhou, L., et al., 2014a. Mesoporous NaYbF4@NaGdF4 core–shell up-conversion nanoparticles for targeted drug delivery and multimodal imaging. Biomater. 35 (26), 7666–7678.

Zhou, Z., et al., 2014b. Iron/iron oxide core–shell nanoparticles for magnetic targeting MRI and near-infrared photothermal therapy. Biomater. 35 (26), 7470–7478.

Zhu, J., et al., 2013. Synthesis of PEGylated gold nanorods (Au NRs) as absorption nanoprobes for near-infrared optical imaging. RSC Adv. 3 (30), 12280.

Zoppellaro, G., et al., 2014. Theranostics of epitaxially condensed colloidal nanocrystal clusters, through a soft biomineralization route. Chem. Mater. 26 (6), 2062–2074.

CHAPTER

LIPID-BASED NANOCARRIERS FOR OCULAR DRUG DELIVERY

16

Ragwa M. Farid*, Noha S. El-Salamouni*, Amal H. El-Kamel, Safaa S. El-Gamal****

**Pharos University in Alexandria, Alexandria, Egypt*

***Alexandria University, Alexandria, Egypt*

CHAPTER OUTLINE

1 INTRODUCTION

The ocular bioavailability of topically applied drugs is highly limited due to the efficient protective mechanisms ensuring the proper functioning of the eye. These barriers are difficult to overcome by the instilled drugs. Another problem that must be taken into account is the attainment of an optimal drug concentration at the site of action (Souto et al., 2010).

Ocular diseases are usually treated by topical application of eye drops, which represent nearly 90% of the marketed formulations, due to their safety, simplicity, and acceptance by patients. Poor bioavailability of drugs from eye drops is mainly due to the precorneal loss factors including: tear dynamics, nonproductive absorption, transient residence time, and relative impermeability of the cornea

(Gan et al., 2013). As a consequence, only a small fraction (<5%) of the instilled dose is absorbed through the cornea and reaches the intraocular tissues, whereas the major portion of the administered dose drains into the nasolacrimal duct. This may induce unwanted systemic side effects (Pal Kaur and Kanwar, 2002).

Thus, strategies to improve the efficacy of topical treatments and overcome the ocular barriers remain a major goal for ocular drug delivery. In general, the majority of the successful delivery systems are present on the ocular surface over an extended period of time, and these systems typically improve drug bioavailability in the anterior chamber (Rawas-Qalaji and Williams, 2012).

During the last decades, several drug delivery systems, such as liposomes, nanoemulsions, microemulsions, nanoparticles, and dendrimers have emerged as novel strategies to improve ocular drug bioavailability (Souto et al., 2010).

Nanoparticle technologies in general gather several benefits,—for example, solubilization of hydrophobic and poorly soluble drugs; improvement of bioavailability and pharmacokinetic properties; in addition to drug protection from physical, chemical, and biological degradation. Moreover, the submicrometer size of these systems allows efficient transport and crossing of biological barriers protecting the eye, thus enabling appropriate drug delivery to the target site.

Recently, solid lipid nanoparticles (SLNs) and nanostructured lipid carriers (NLCs) were introduced as drug delivery systems for the eye. Several drugs—for example, antibiotics, plasmids, antiglaucoma, anti-inflammatory, and immuno-suppressive agents—have been loaded into SLNs for the treatment of ophthalmic diseases. However, an important parameter to take into account in ocular formulations is the particle size, since human eyes can tolerate particles up to 10 μm.

1.1 ANATOMY AND PHYSIOLOGY OF THE EYE

The eye consists of three chambers; anterior, posterior, and vitreous (Fig. 16.1). It also has three main layers: fibrous, vascular, and neural. The aqueous humor fills the anterior and posterior chambers and is partially responsible for maintaining the shape of the eyeball, in addition to moisturizing and providing nutrients to structures such as the cornea, which is avascular (Rawas-Qalaji and Williams, 2012).

Aqueous humor is produced at a constant rate by the ciliary body and to maintain the intraocular pressure (IOP) within the normal physiological levels, a constant amount is removed through the canal of Schlemm. The vitreous is the largest of the three chambers and it is filled with a transparent jelly-like fluid called vitreous humor. It helps maintaining the shape of the eyeball, holding the lens and retina in place and maintaining the IOP (Rawas-Qalaji and Williams, 2012).

1.2 LAYERS OF THE EYE (RAWAS-QALAJI AND WILLIAMS, 2012)

The eye's three layers are fibrous, vascular, and neural.

1.2.1 The fibrous layers

The outer fibrous layer of the eye consists of the sclera, cornea, and conjunctiva.

1.2.1.1 The sclera

It is the posterior five-sixths of the outer eyeball. It is made of dense collagenous connective tissue with elastic fibers, which make it firm and appear opaque white. Hydrophilic compounds with low molecular weight can permeate through the sclera's pores and intracellular spaces.

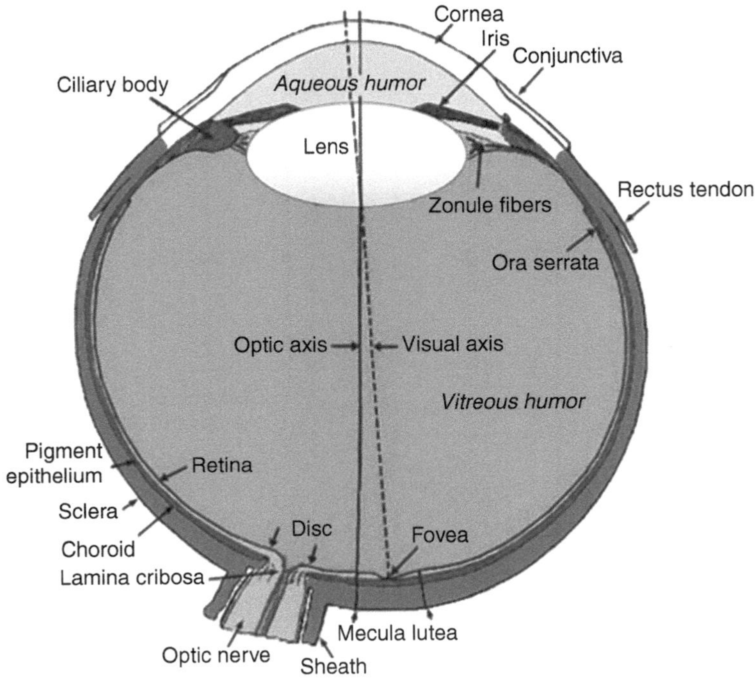

FIGURE 16.1 Anatomy of the Eye

From Kolb H., Fernandez E., Nelson R., 2007. Gross anatomy of the eye. The Organization of the Retina and Visual System, Salt Lake City.

1.2.1.2 The cornea

It is the transparent remaining one-sixth of the outer fibrous layer of the eye. The cornea consists of five layers (Fig. 16.2):

- *The outer epithelium:* it has 5 sublayers of 50–100 μm thickness. Its lipophilic character confers 90% of the eye's barrier to the entry of hydrophilic drugs and only 10% of its barrier to hydrophobic drugs.
- *Bowman's membrane:* is much thinner than the epithelium and does not present a barrier for drug entry across the cornea.
- *The stroma:* is the hydrophilic layer of the cornea and constitutes around 90% of the cornea's volume. Approximately 80% of the stroma is water, so it is considered the main barrier to absorption of lipophilic drugs across the cornea.
- *Descemet's membrane:* is a thin regenerative membrane, which does not impact ocular drug absorption.
- *The endothelium:* is a single layer of cells that covers the entire posterior surface of the cornea. Gap junctions between its cells make this layer 200-fold more permeable than the epithelium layer.

1.2.1.3 The conjunctiva

This is a double or triple epithelial layer of cells. Although these cells are bound by tight junctions like the cornea, the conjunctiva is more permeable to hydrophilic compounds due to larger paracellular epithelial pores and greater number of epithelial pores.

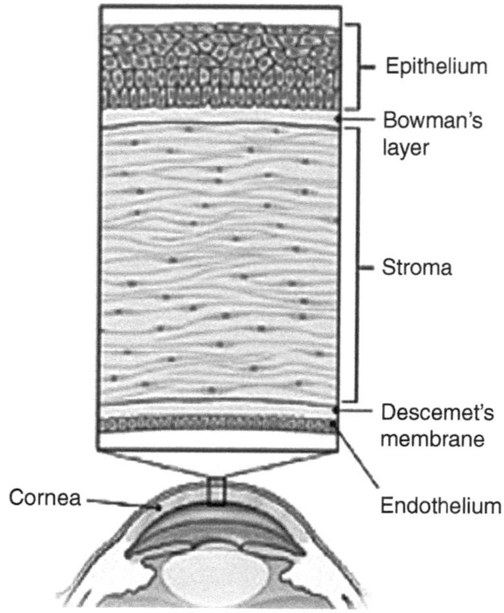

FIGURE 16.2 Layers of the Cornea

From Longstaff, A., 2009. The nervous system. In: Naish, J., Court, D.S., Revest, P., (Eds.), Medical Sciences. Elsevier Health Sciences, UK.

1.2.2 The vascular layer

The vascular layer is black due to the presence of melanin pigments. The posterior section of the vascular layer is called the choroid and it lines the inner surface of the sclera. Most of the eye's blood vessels are located in the vascular layer. This layer forms the blood-ocular barrier, which tightly regulates the entry of substances into the eye from the blood.

1.2.3 The neural layer

The innermost layer of the posterior section of the eyeball is the retina. The neural layer is the most challenging area in ocular drug delivery for treatment of disorders affecting the retina due to the possibility of drug metabolism and excretion.

2 ROUTES OF DELIVERY (THAKUR AND KASHIV, 2011)

There are three main routes for drug delivery to the eye; topical, intraocular, and systemic routes.

2.1 TOPICAL ROUTE

It is the most common route for ocular drug delivery. Introducing the drug directly to the conjunctival sac localizes the drug effect, facilitates drug entry that is hard to achieve with systemic delivery, and avoids first-pass metabolism.

2.2 INTRAOCULAR ROUTE

It is more difficult to achieve practically. Now research is concentrating on the development of intravitreal injections and the use of intraocular implants to improve delivery to the eye.

2.3 SYSTEMIC ROUTE

Several studies have shown that some drugs can distribute into ocular tissues following systemic administration. Oral administration of carbonic anhydrase inhibitors including Acetazolamide and Methazolamide demonstrate the capacity of a systemic drug to distribute into the eye.

3 MAJOR PHYSIOLOGICAL BARRIERS OF THE EYE TO TOPICAL OCULAR DRUG DELIVERY

Many intrinsic barriers and clearance mechanisms restrict ocular drug delivery (Fig. 16.3).

3.1 TEAR FILM BARRIER

Each eye has a lachrymal gland, which produces tears to lubricate and clean the cornea. Tears are a mixture of water, antimicrobial enzymes, lysozymes, salts, and mucus (Rawas-Qalaji and Williams, 2012).

There are several possible causes of the barrier behavior of the tear film:

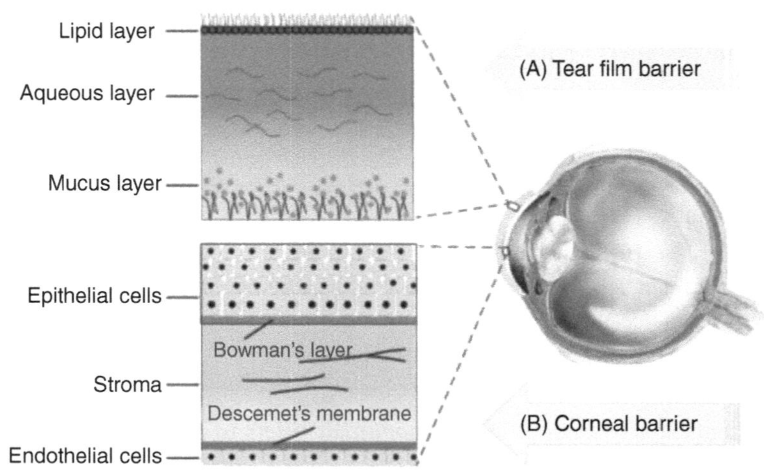

FIGURE 16.3 Two Major Barriers to Topical Ocular Drug Delivery

From Gan L., Wang J., Jiang M., Bartlett H., Ouyang D., Eperjesi F., et al., 2013. Recent advances in topical ophthalmic drug delivery with lipid-based nanocarriers. Drug Discov. Today18(5), 290–297.

3.1.1 High turnover rate

The basal tear flow is 1.2 mL/min. Reflex stimulation may increase lachrymation 100-fold. Topical administration in the form of eye drops is quickly washed away by the tear film after application (Gan et al., 2013).

3.1.2 Gel-like mucus layer

Approximately 2–3 mL mucus is secreted daily by the eye. Mucin in the tear film has a protective role by forming a hydrophilic gel layer over the ocular surface and clears foreign bodies and pathogens. In addition, it acts as a barrier to drug delivery (Gan et al., 2013).

3.1.3 Tears pH

Tears have an average pH of 7.4 and lack a strong buffering system. To maintain homeostasis, a reflex increase in tear production occurs to remove the irritating acidic or alkaline eye drops through the nasolacrimal duct, which further reduces drug ocular bioavailability. Therefore, the ideal ocular drug should have a pH between 7 and 7.7 (Rawas-Qalaji and Williams, 2012).

3.2 CORNEAL BARRIER (GAN ET AL., 2013)

The epithelium of the cornea consists of 5–6 layers of cells closely packed and connected by tight junctions. These layers form an effective barrier to most microorganisms and also to therapeutic drugs. The cornea is composed of five layers, each of alternating polarity. This sandwich-like structure makes the cornea a crucial barrier to most lipophilic and hydrophilic drugs. Although intercellular spaces in the conjunctival epithelium are wider than in the corneal epithelium, both the cornea and the conjunctiva are rate-limiting factors for water-soluble drugs.

4 REQUISITES OF CONTROLLED OCULAR DELIVERY (WILSON, 2004)

- To circumvent the protective barriers like drainage, lacrimation, and diversion of exogenous chemicals into the systemic circulation by the conjunctiva.
- To increase the ocular bioavailability of drugs by increasing the corneal contact time. This can be achieved by effective adherence to the corneal surface so that the released drug effectively reaches the anterior chamber.
- To overcome the side effects of pulsed dosing produced by conventional systems.
- To provide comfort and compliance to the patient and improve the therapeutic performance of the drug over conventional systems.
- To provide sustained and controlled drug delivery.
- To provide targeting within the ocular globe.

In the pharmaceutical sciences, the term *nanoparticle* refers to a particulate drug delivery system where particle size is in the nanometer range (1–1000 nm). Nanoparticles are being investigated extensively to develop drug delivery systems capable of allowing penetration through physiological barriers. Nanoparticles are either in the form of matrix-dispersion or a membrane-reservoir type, where drugs can be dissolved, entrapped, encapsulated, and dispersed in the particles or adsorbed on the surface of these particles (Seyfoddin et al., 2010).

Many nanostructured systems such as SLNs (Vyas et al., 2008), niosomes (Prabhu et al., 2010), nanocapsules (Calvo et al., 1996), nanosuspensions (Pignatello et al., 2006), liposomes (Chetoni et al., 2007) and others have been employed in ocular drug delivery (Seyfoddin et al., 2010).

5 LIPID-BASED NANOCARRIERS AS TOPICAL OCULAR DELIVERY SYSTEMS (GAN ET AL., 2013)

Lipids have been used as a drug delivery vehicle for accommodating lipophilic drugs and elevating their poor water solubility (Chen, 2008). For hydrophilic drugs, water-in-oil (w/o) emulsions and microemulsions were studied to solubilize them for ocular drug delivery. Despite their simple production, easy sterilization, and modest stability, their use in ocular drug delivery is limited by the choice of ingredients possessing good ocular tolerability (Vandamme, 2002). They require a high emulsifier concentration to ensure formulation stability. This limits their ocular applicability as emulsifiers are usually not well tolerated (Souto et al., 2010).

The specific physiological structure of the eyeball results in two major issues that determine ocular drug bioavailability; these are duration of residence on the ocular surface and corneal permeability. Lipid based nanocarriers can enhance ocular drug bioavailability by addressing both of these issues. Since the residence volume of the eye is generally 7–10 μL, most topically administered solutions are washed away within just 15–30 s of application (Fuente et al., 2010). Lipid components in the lipid-based nanocarriers might interact with the lipid layer of the tear film, enabling the carriers to remain in the conjunctival sac for a long time, where they act as a drug depot (Alany et al., 2006). Also, lipid-based nanocarriers can be modified with cationic material to form positively charged vehicles. After application, electrostatic interactions between cationic nanocarriers and the anionic surface of the cornea can reduce the contact angle and improve the drug residence.

In addition, the corneal epithelium protects the eye from external stimuli and renders it impermeable to topically administrated drugs. Lipid-based nanocarriers, can inhibit the activity of P-glycoprotein, a prominent efflux transporter on ocular tissues that limit the bioavailability of ocular drugs. In addition, lipid-based nanocarriers can open the tight junctions through the effects of nonionic emulsifiers to enhance drug penetration (Katragadda et al., 2006).

Lipid nanoparticles have gained interest since the last decade as a novel particulate drug delivery systems. Nowadays, they are considered as a strategy to enhance the bioavailability of topical drugs. Lipid nanoparticles of diameter 50–500 nm are suitable for ocular administration (Souto et al., 2010).

5.1 CLASSIFICATION OF LIPID NANOPARTICLES

5.1.1 SLNs

SLNs are structured as a solid lipid core in nanometer range accommodating the drug and stabilized by a layer of emulsifier (Sawant and Dodiya, 2008) (Fig. 16.4). In comparison to aqueous eye drops, ocular SLNs formulations can have extended residence time on the ocular surface and conjunctival sac (Gaudana et al., 2009) leading to a sustained drug release. Drug particles entrapped in SLNs can effectively cross the epithelium due to the SLNs lipophilic properties. Moreover, the epithelium is slightly negatively charged, thus, cationic SLNs can increase the corneal residence time of the drug and increase its absorption level (Seyfoddin et al., 2010).

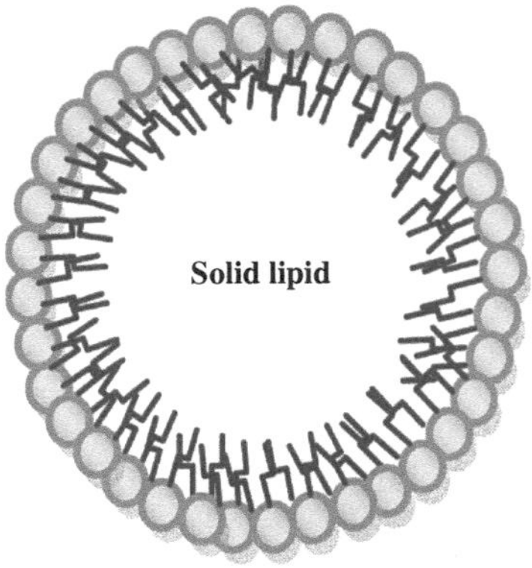

FIGURE 16.4 Structure of Solid Lipid Nanoparticles

SLNs offer several advantages as a drug delivery system (Seyfoddin et al., 2010) including; controlled drug release, drug targeting, long-term stability, longer shelf life, protection of labile compounds against chemical degradation, low cost, easy large-scale production, good loading of both hydrophilic and lipophilic drugs according to the production process, and absence of biotoxicity due to the use of physiological lipids. Moreover, SLNs can be sterilized by autoclaving and their biodegradability and free dispersibility in aqueous media, enable them to be formulated as modified eye drops.

5.1.1.1 Materials used in the preparation of SLNs

The typical composition of SLNs is comprised of physiological and biodegradable lipids, which are suitable for the incorporation of lipophilic and hydrophilic drugs within the lipid matrix. The matrix is stabilized by emulsifier(s) in aqueous dispersion. Also, a co-emulsifier may be used if required—for example, butanol and butyric acid (Souto et al., 2010).

5.1.1.1.1 Lipids

In general, lipids have already been approved by the European and United States regulatory authorities for topical applications and they are generally regarded as safe (GRAS). Their similarity to physiological lipids is the reason for anticipating excellent tolerability and reduced irritancy. The crystalline modifications, structure, concentration, and hydrophobicity of lipids are important factors in SLNs formation (Vyas et al., 2008). Super cooling of SLNs is an important must be ensured after melt homogenization, otherwise the final product will appear as a nanoemulsion. The solid matrix is expected to remain in the solid state at body temperature so that controlled and sustained drug release properties can be achieved (Bunjes et al., 1996).

Lipids, which form a highly ordered crystalline lattice, are unfavorable as drug expulsion occurs. Therefore, it is desirable to employ lipids with complex structures containing fatty acid chains of different lengths, which are able to accommodate the drug efficiently. However, lipids with long fatty acid side chains usually form large particles. A combination of long-chain and short-chain fatty acids provide better quality, stability, and high drug loading (DL) capacities.

Solid lipids used in the preparation of lipid nanoparticles for ocular drug delivery include; Compritol 888 ATO (C 888), Precirol ATO 5 (P ATO 5), glyceryl monostearate, Gelucire 44/14, Phospholipon 90G, stearylamine, tripalmitin (Dynasan 116) and stearic acid (Seyfoddin et al., 2010). Compritol ATO 888, is generally regarded as safe, it is a biocompatible, nonirritant and nontoxic partial triglyceride of behenic acid. It is mainly composed of glyceryl dibehenate with variable quantities of glyceryl monobehenate and tribehenate; (15%–23% monoglycerides, 40%–60% diglycerides, 21%–35% triglycerides, m.p. 69–74°C). The solid lipid glycerylbehenate is commonly used to prepare SLNs. It has a very small amount of thermodynamically unstable α-form, which disappears under thermal stress. C 888 usually crystallizes in its β'-modification, which is very sensible at elevated temperatures (Vyas et al., 2008). Studies showed that, C 888 has high loading capacity and can be used to avoid recrystallinity problems associated with other solid lipid used (Zur Mühlen et al., 1998). In addition, C 888 is associated with an amphiphilic characteristic with a high chemical stability that resulted in a good ability to encapsulate a lipophilic and/or hydrophilic drug (Kipp, 2004; Sangsen et al., 2013).

Precirol ATO 5 (glycerylpalmitostearate), is composed of a mixture of monoglycerides, diglycerides and triglycerides of palmitic acid (C_{16}) and stearic acid (C_{18}); (8%–22% monoglycerides, 40%–60% diglyceride, 25%–35% triglycerides; m.p. 50–60°C). It has an HLB value of 2 (Pharmacopoeia, 2007). Glycerylmonostearate (Imwitor 900, Geleol) is composed of glycerides of saturated fatty acids (Pandya et al., 2013), and it is a glycerol of stearic acid: 40%–55% monoglycerides; 30%–45% diglyceride; melting point 54.5–58.5°C. GMS has an HLB value in the range of 2.8–3.5 (Chavkin and Mackles, 1998). The drug entrapment in GMS was reported to be high, since esterification of glycerol by long chain fatty acids is responsible for high hydrophobicity of glycerides (Pandya et al., 2013). Gelucires are semisolid lipids often used in the preparation of NLCs for ocular drug delivery. They consist of a mixture of monoglycerides, diglycerides, and triglycerides and di-fatty acid esters of polyethylene glycol. Gelucires showed promising results for the corneal permeation of drugs (Li et al., 2008).

Cationic lipids—for example, a mixture of octadecylamine and *N*-[1-(2,3-Dioleoyloxy) propyl]-*N, N, N* trimethylammonium chloride (DOTAP), stearic acid and octadecylamine —are often used to induce a positive charge on the SLNs. Cationic SLNs have shown better corneal penetration (Basaran et al., 2010), which is probably due to ionic interaction of cationic particles with negatively charged epithelial cells (Cortesi et al., 2006). These ionic interactions prolong drug residence time inside the cornea and possibly induce corneal morphological changes resulting in enhanced drug penetration and absorption (Seyfoddin et al., 2010).

Triglycerides produce much more stable SLNs formulations than monoglycerides and diglycerides. Long chain triglycerides have higher melting points (m.p.), more stable, and produce better SLNs (Radomska-Soukharev, 2007). Many triglycerides have been used in the preparation of SLNs (Seyfoddin et al., 2010), such as tricarpin, trilaurin, trimyristin, tripalmitin, and tristearin. Triglycerides are a suitable core material because their melting point is above body temperature to allow controlled drug release but sufficiently low to assure solidification after melt homogenization (Bunjes et al., 1996).

During production and storage, glycerides may exist under various polymorphic forms. Such changes may lead to a transition to a more stable crystalline form, which potentiates drug expulsion. So, the use of a mixture of triglycerides produces crystals with a poorly ordered matrix (Bunjes et al., 1996), which can accommodate more drug quantity while avoiding drug expulsion during storage.

5.1.1.1.2 Emulsifiers

Lipid nanoparticles need low emulsifier concentration, and do not require the use of co-emulsifiers, compared to other lipid carriers such as microemulsion and nanoemulsions. There are many factors to be considered in choosing an emulsifier for SLN production for ocular use (Radomska-Soukharev, 2007):

- Hydrophilic-lipophilic balance (HLB) value, where emulsifiers with HLB values between 8 and 18 are suitable.

- Emulsifier should cause minimal or no irritation to the eye. Nonionic emulsifiers are preferred over ionic ones for ocular preparations because of their lower toxicity and irritancy.
- Emulsifier composition affects the in-vivo biodegradation of the lipid matrix. Nonionic emulsifiers are more effective in inhibiting the in-vivo degradation of lipid matrix.
- Both the type and concentration of the emulsifier can affect the chemical stability of physiological lipids used to prepare SLNs. The emulsifier type affects the particle size of SLNs, where SLNs made of the same lipid may have different sizes because of using different emulsifiers. Also, studies show that, high concentrations of emulsifiers produce small sized particles and low concentrations produce larger particles. Usually, a mixture of emulsifiers yields smaller particles and the system shows a lower polydispersity index (PI) and a higher lipid stability.
- Charge of the emulsifier (Li et al., 2008) is very important in ocular drug delivery. Using a cationic emulsifier is favored to induce a positive charge to the formulated SLNs, such as stearylamine. This allows a longer residence time and a better penetration to the posterior chamber of the eye.

Several emulsifiers are suitable for the preparation of SLNs for ocular drug delivery including (Seyfoddin et al., 2010) the following: polyoxyls (Cremophor EL), Transcutol P, stearylamine, polysorbate 80 (Tween 80), soya phosphatidylcholine (Epikuron 200), sodium taurocholate, Poloxomer 188, and bile salts (colic acid). Many of these agents are also used as corneal penetration enhancers (Pal Kaur and Kanwar, 2002).

Poloxamer 188 (Kolliphor P 188, Lutrol F 68, Pluronic F68), a hydrophilic emulsifier; is a tri-block copolymer of polyoxyethylene–polyoxypropylene. It has an HLB value of 29 (Sangsen et al., 2013). It was previously reported by Leonardi A. et al., in a study done to assess the properties and ocular tolerability of some emulsifiers commonly used to prepare ocular SLNs, that P 188 was the most valuable, showing no irritant effect on the ocular surface up to the highest tested emulsifier concentration of 0.4% w/v (Leonardi et al., 2014).

5.1.1.1.3 The drug

Both hydrophilic and lipophilic drugs can be incorporated in SLNs (Seyfoddin et al., 2010). Drugs incorporated in SLNs are usually released by diffusion through the lipid matrix and/or by biodegradation and surface erosion of the lipid matrix (Sawant and Dodiya, 2008). This allows a sustained and controlled release of the drug from the colloidal system thereby reducing the required dose and reducing the risk of exposure to toxic doses. SLNs also increase bioavailability of poorly penetrating drugs, and they can extend the drug's shelf-life and half-life by protecting it from the external environment. The drug incorporation into the SLNs can be described by three models (Muéller et al., 2000): homogenous matrix model, drug-enriched shell model, or drug-enriched core model (Fig. 16.5).

The morphological differences between these models depend on (Souto et al., 2010) the following: the chemical nature of the drug and the lipid, the solubility of the drug in the melted lipid, the type of production method (hot or cold) high pressure homogenization, the production temperature, and both the nature and concentration of emulsifiers.

5.1.1.2 Problems of SLNs

5.1.1.2.1 Drug expulsion during storage

Crystalline behavior of lipids in the nanoscale is very complex. In SLNs, the solidification temperature is usually lower (Bunjes and Koch, 2005) and less-ordered lipid modifications are often formed compared to bulk lipid (Müller et al., 2002a). Drug expulsion during storage is one of the most common

FIGURE 16.5 Models of Drug Incorporation in SLNs

issues associated with the lipid crystal transformation to a more stable β-modification with a more perfect crystalline lattice, leaving behind less space for drug accommodation, thus leading to drug expulsion (Müller et al., 2002a). Perfect lipid crystals can be formed when the lipid molecules are chemically identical. Prerequisite for good drug accommodation are larger distances between fatty acid chains of the glycerides and general imperfections in the crystal to accommodate the drug. Distances between fatty acid chains can be increased by using glycerides composed of different fatty acids (e.g., mixture of saturated and unsaturated fatty acids).

5.1.1.2.2 Drug-enriched shell and burst release

The burst effect is a common problem associated with SLNs release profiles. It is explained by the room temperature phase separation effect. It is proposed that when the hot oil-in-water (o/w) emulsion containing the drug is cooled to solidify, lipids crystallize to form SLNs with a drug-free core and a drug-enriched shell (Fig. 16.6). This is because the aqueous solubility of most drugs is enhanced at high temperatures. This changes the partition coefficient of the drug and its position in the SLNs. Therefore,

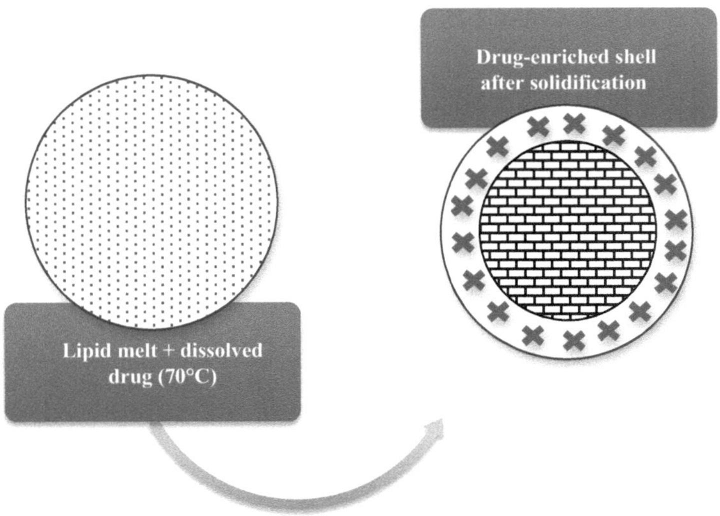

FIGURE 16.6 Burst Release Associated With the Release Profile of SLNs

high production temperatures promote drug localization on the surface of SLNs, causing a burst effect (Sawant and Dodiya, 2008). To achieve a prolonged release profile of SLNs, lower temperatures and emulsifier concentrations during production are advised. However, despite applying these conditions for some drug and lipid mixtures, burst release could not be avoided. Imperfect and amorphous type NLCs provide more flexibility to achieve the desired prolonged drug release (Müller et al., 2002a). It was previously reported that the burst effect is commonly associated with hydrophilic drugs (Kaur et al., 2008). It occurs due to drug adsorption on the nanoparticle surface rather than encapsulation, which results in releasing a major fraction of the drug in a short period of time. The presence of drugs on the surface of the particles promote the chemical degradation of sensitive drugs, whereas, drug orientation in the center inhibits the degradation process (Helgason et al., 2009).

5.1.1.2.3 Biotoxicity

SLNs include a wide range of compositions of complicated characters and functionality. They are made up of long or medium chain triglycerides usually mixed with monoglycerides and diglycerides and individual or mixed ionic and/or nonionic emulsifiers.

SLNs are usually made up of physiological lipids that are biodegradable with simple natural processes as enzyme digestion. This is one of the most important advantages of SLNs as a drug-delivery vehicle. These lipids and their degradation products are considered nontoxic to human cells. However, due to their nanosize range, their biotoxicity is an important issue as the human body reacts very differently to nanoparticles as compared to larger particles of the same material. Certain tests (e.g., EPISKIN test and HET-CAM test) can be performed to evaluate topical and ocular toxicity and irritation caused by SLNs. Considering the results obtained by many studies, SLNs are very well tolerated at the cellular level (Pardeike et al., 2009). However, all emulsifiers are irritants and poorly tolerated, hence their biotoxicity remains an important issue. The choice of emulsifier used is an important consideration, since water insoluble emulsifiers penetrate and fluidize biological membranes, and water-soluble emulsifiers can solubilize the membrane components. In addition, cationic emulsifiers are more toxic than anionic and nonionic ones, due to their ionic interaction with the negatively charged ocular cellular membranes (Pouton and Porter, 2008).

5.1.2 NLCs (Müller et al., 2002a)

SLNs are associated with a few drawbacks. Limited DL capacity, the possibility of drug expulsion during storage, and high water content of SLNs aqueous dispersions (70%–90%) have led to the introduction of a new, improved generation of lipid nanoparticles.

NLCs are novel lipid nanoparticles, produced by controlled addition of a liquid lipid (oil) to the solid lipid component to accommodate a larger quantity of the drug and achieve a better release profile. This can be achieved by increasing the space between the fatty acid chains of the glycerides, allowing more drugs to be accommodated. Also, admixtures of liquid lipid with solid lipids lead to the formation of imperfect lipid crystals that can allow enough space to accommodate drug molecules, avoiding drug expulsion during storage. Although NLCs contain up to 30% liquid lipids, the final product is in the solid state.

Since the solubility of drugs in oils is generally higher than in solid lipids, NLCs show higher DL capacity. Oils used for formulation of ocular NLCs include (Souto et al., 2010): sesame oil, castor oil, soybean oil, and corn oil.

Depending on the production method and lipid blend composition, there are different types of NLCs: imperfect, multiple type, and structureless NLCs (Fig. 16.7).

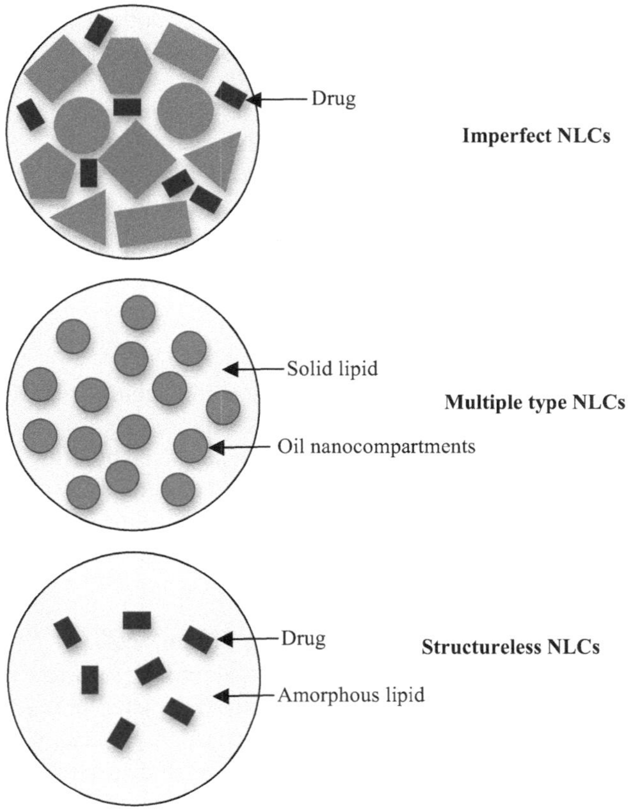

FIGURE 16.7 Types of NLCs

5.1.2.1 Imperfect NLCs (Sawant and Dodiya, 2008)

Systems in which a liquid lipid is added to the solid lipid to induce a larger distance between the fatty acid chains of the main lipid core are called imperfect NLCs.This causes imperfections in the crystalline structure of the lipid in which a very large amount of the drug can be incorporated.

5.1.2.2 Multiple type NLCs (oil-in-solid fat-in-water) (Müller et al., 2002a)

In this type, liquid lipids are mixed in excess with solid lipids, resulting in nanocompartments of liquid lipid formed within the solid core, which is already dispersed in the aqueous medium (Sawant and Dodiya, 2008). Such small nanocompartments within a nanoparticle are generated by a phase separation process during particle production. In the hot state, the two lipids form one phase, and during cooling, separation occurs leading to formation of small oil droplets. In these oil compartments, the drug solubility is higher, thus, increasing the DL capacity (Müller et al., 2002a). The liquid lipid nanocompartments are dispersed uniformly in the matrix of solid lipids that are protected from degradation and this allows prolonged release behavior of the system. This system is beneficial for drugs that have higher solubility in liquid oils (Sawant and Dodiya, 2008).

5.1.2.3 Structureless NLCs (amorphous) (Sawant and Dodiya, 2008)

Another approach is to carefully mix and melt together the selected liquid and solid lipids, which upon cooling solidify but do not crystallize. This type can avoid drug expulsion caused by crystallization.

5.1.3 Lipid drug conjugates (Seyfoddin et al., 2010)

Lipid drug conjugates (LDCs) are developed by converting a hydrophilic drug to a lipophilic drug conjugate or a prodrug by the addition of an ester or amide group. These are either used alone or combined with other solid lipids and are expected to enhance biological transport and allow targeting of hydrophilic drugs.

5.2 PREPARATION OF SLNs

The preparation of SLNs is based on solidified emulsion technologies. It is easy to incorporate lipophilic drugs into SLNs; however, hydrophilic drugs, especially proteins and peptides tend to partition in the aqueous phase during preparation. This problem can be solved by using emulsifiers as emulsion stabilizers (Almeida and Souto, 2007). Several methods can be used for the preparation of SLNs, such as high pressure homogenization, modified high shear homogenization, water-in-oil-in-water (w/o/w) double emulsion, solvent emulsification-evaporation, solvent emulsification-diffusion, o/w microemulsion, and preparation in coflowing microchannels.

5.2.1 High pressure homogenization (Mehnert and Mäder, 2001)

During the homogenization process, the dispersion is passed through a narrow gap by applying a high pressure. Passage through a short distance combined with the applied pressure produces high shear stress, breaking down the accelerated particles to submicron size. A preparatory step is involved in both hot and cold homogenization where the drug has to be dissolved or dispersed in the lipid melt. This step eliminates the use of organic solvent to dissolve the lipids, which might cause toxicity if residues remain.

5.2.1.1 Hot homogenization (Mehnert and Mäder, 2001)

The drug inside the lipid melt is first emulsified with a solution of emulsifier(s) using a high-speed stirrer. The primary emulsion affects the quality of the final particles and should ideally produce particles in micrometer range or lower. The primary emulsion is then subjected to high pressure homogenization in temperatures above the melting point of the lipid. The homogenization can be repeated a few times to achieve the desired particle size, taking into account that extra cycles can actually yield bigger particle sizes due to increased kinetic energy causing agglomeration. Finally, the hot o/w emulsion is cooled to room temperature to produce a supercooled melt to yield SLNs.

The elevated temperatures required to melt the lipid, which might further increase while applying high shear stress, can affect the chemical stability of thermo-labile drugs, antibiotics, peptides, and proteins used in ophthalmic delivery systems. However, if the drug is thermo-stable, the high shear and temperature used in this technique can help producing small-sized particles.

5.2.1.2 Cold homogenization (Mehnert and Mäder, 2001)

Excessive heating is avoided by solidifying the lipid–drug mixture in liquid nitrogen and grinding it to a powder with particle sizes below 100 nm before dispersing it in a cold aqueous solution containing the emulsifier(s). However, the preparatory step, which involves melting of the lipid and dispersing or dissolving the drug in it, is still necessary.

Although cold homogenization can control the crystallization behavior of lipids during super-cooling, it usually yields particles with higher average particle sizes.

5.2.2 Modified high-shear homogenization method (Elnaggar et al., 2011)

High-pressure homogenization is considered the optimum method for SLN preparation because it requires low emulsifier concentration and avoidance of organic solvent incorporation. Nevertheless, high-pressure homogenization results in high particle-surface energy and low stability. Modified high shear homogenization method combines the privileges of high-pressure homogenization while circumventing high-pressure implications.

In this method, the drug is dispersed in the emulsifier/coemulsifier mixture by sonication. Dispersion is aided by heating in the shaking water bath at 70°C. Then, the lipid is added to the blend and melted at 70°C. Water is heated separately to the same temperature and then added to the component blend under high shear homogenization, at 70°C. The hot emulsion formed was poured into cold, distilled water under high shear homogenization. The nanoparticle dispersion formed was then filtered through membrane filter.

5.2.3 Water-in-oil-in-water double emulsion

A recent method involves dissolving a drug in an aqueous solvent (Lv et al., 2009). The aqueous solution is then emulsified in an oil phase containing the lipid(s) dissolved in an organic solvent to form a primary w/o emulsion. The primary emulsion is then dispersed into an aqueous solution containing the emulsifier. The w/o/w double emulsion is mechanically agitated to evaporate the organic solvent completely until SLNs are formed. Heating is not involved during the process; so it is suitable for thermo-labile substances. SLNs of good quality can be prepared by adjusting the sonication intensity. The only disadvantage is the use of organic solvent and the possible presence of metal impurities from the sonicator.

5.2.4 Solvent emulsification-evaporation (Sjöström and Bergenståhl, 1992)

In this method, the lipid component is dissolved in a water immiscible organic solvent and the drug is either dissolved or dispersed in the lipid solution. The organic phase is then emulsified in an aqueous solution containing a biocompatible coemulsifier as bile salts. Emulsification is achieved using a high-speed homogenization followed by high-pressure homogenization. The system is mechanically stirred at room temperature until the solvent completely evaporates.

For preparation of cationic SLNs, the main lipid core is dissolved in an organic solvent and the cationic lipid is dispersed in the aqueous phase containing the emulsifier (Del Pozo-Rodríguez et al., 2008).

The quality of the primary emulsion directly affects the final SLNs size. The smaller the droplets in the primary emulsions, the smaller will be the final SLNs. Hence, it is important to optimize all parameters that influence the emulsification process such as the lipid and emulsifier's concentration and type. If optimized, this method can produce particles with an average size below 100 nm by precipitation into an o/w emulsion.

5.2.5 Solvent emulsification-diffusion (Trotta et al., 2005)

This method is based on the water miscibility of certain organic solvents such as butyl lactate or benzyl alcohol as the oil phase. A primary o/w emulsion is prepared containing the drug and the lipid-phase solution in a water-miscible solvent. This emulsion is then transferred into water. The hypothesis is that, upon introduction to an aqueous phase, the water miscible solvent will diffuse. So the lipophilic material dissolved in the solvent will solidify due to the diffusion of solvent from droplets to the continuous phase. This method is reliable for the entrapment of lipophilic and hydrophilic drugs if optimized.

5.2.6 Oil-in-water microemulsion (Marengo et al., 2000)

This method is easy and does not require organic solvents. However, a preheating step is required, which may be disadvantageous to thermo-labile drugs. The lipid components are melted at approximately 10°C above their melting point and the drug is mixed with the lipid melt. The lipid melt is then dispersed and homogenized in a hot aqueous phase containing the emulsifier heated at the same temperature of the lipid melt.

The hot o/w microemulsion is then cooled rapidly using an ice bath or it can be placed into cold water directly while maintaining mechanical stirring until SLNs are formed. Upon sudden decrease of temperature, the oil droplets in the hot o/w microemulsion crystallize rapidly, forming SLNs.

5.2.7 Preparation of SLNs in coflowing microchannels (Yun et al., 2009)

The basic principle of this method is solvent displacement. The apparatus consists of a co-flowing microchannel system with inner and outer capillaries. The lipid is dissolved in a watermiscible solvent, which is injected into the inner capillary. The aqueous phase containing the emulsifier is injected into the outer capillary simultaneously.

Displacement of the solvent from the lipid phase to the aqueous phase causes local super-saturation and solidification of lipids resulting in the formation of SLNs. Particles with different size and morphology can be prepared by altering parameters such as the velocity of liquid flow. The small diameter of microchannels is often blocked by SLNs hindering continuous production, which can be overcome by gas-liquid slug flow. Using this method, the critical procedures of the other production methods, such as high speed, high pressure, high temperature, and the use of toxic organic solvents, might be avoided.

5.3 POSTPREPARATION PROCEDURES

5.3.1 Purification and separation of SLNs (Seyfoddin et al., 2010)

It is an important step to avoid toxicity associated with extra emulsifiers. Purification methods include ultrafiltration, ultracentrifugation, and dialysis.

5.3.2 Sterilization of SLNs (Seyfoddin et al., 2010)

All ocular preparations must be sterilized; they have a short shelf life and should be used within 1 month of opening due to contamination risks. Sterilization techniques must not affect the physical formulation and chemical stability nor the drug release kinetics (Mehnert and Mäder, 2001). It should ensure formulation sterility without degradation or aggregation of the solid lipids to avoid toxicity and instability (Mehnert and Mäder, 2001). Common sterilization techniques include filtration, autoclaving, γ-radiation, and aseptic production.

An advantage of SLNs, compared to other colloidal systems, is that they can be sterilized by autoclaving, during which SLNs melt and recrystallize in a controlled manner. However, some structural features of the SLNs will be lost during the autoclaving process (Muéller et al., 2000). Sterilization by heating can induce physical instability, particle aggregation, and particle-size increase. Moreover, an important issue to be considered is the degradation of lipids, which can cause potential toxicity. It is important to know that heat-induced or radiation-induced degradation is not always associated with an increase in the particle size. In some cases; a decreased particle size results from the breakdown of lipids (Mehnert and Mäder, 2001).

Critical parameters include sterilization temperature and SLNs composition. The correct choice of the emulsifier is of significant importance for the physical stability of the sample at high temperatures. Increased temperatures, affect the mobility and the hydrophilicity of all emulsifiers to different extents. Steam sterilization causes the formation of an o/w emulsion due to the melting of the lipid particles; then solid particles are formed after re-crystallization.

A decrease in the sterilization temperature from 121 to 110°C can reduce sterilization-induced particle aggregation to a large extent (Schwarz and Mehnert, 1995). It was reported that, a large increase in the particle size (>500 nm) was observed for high amount of loaded drug. These results indicate that drug-related phenomena contribute to destabilization processes, in addition to changes of the emulsifier film. This destabilizing effect has also been observed in steam sterilization of fat emulsions (Lucks, 1993). This destabilization was attributed to a distortion of the mechanical properties of the emulsifier film. This explanation is also valid for SLNs due to the molten state of the lipid at 121°C. Similar findings were obtained by Freitas (1998), who indicated that lowering the lipid content prevents the particle increase to a large extent and avoids gelation following autoclaving.

Additionally, purging the sample with nitrogen showed a protective effect during sterilization. This observation suggests that chemical reactions could contribute to particle destabilization. For example, the removal of solubilized carbon dioxide by nitrogen purging might increase the pH value and decrease the rate of lipid hydrolysis, which, in turn, affects the re-crystallization of the lipid (Mori, 1988). However, further studies are necessary to investigate the mechanism of the protective effect of nitrogen purging.

Degradation does not always cause an increase in particle sizes. Sometimes, slysophosphatides or free fatty acids are formed, which might have small particle sizes but can cause toxicological problems.

Further studies that involve the aspects of chemical stability are necessary to determine the possibilities of SLNs sterilization (Mehnert and Mäder, 2001)

5.3.3 Freeze-drying of SLNs (Vighi et al., 2007)

Lyophilization involves removal of water by sublimation to convert the lipid dispersion to a solid state to extend the stability and avoid particle aggregation. Freeze-drying of SLNs improves their long-term chemical and physical stability, so degradation reactions, such as hydrolysis, are prevented and particle aggregation is avoided. Cooling may be done either rapidly or slowly. Rapid cooling involves dipping the vial containing the preparation into liquid nitrogen or adding the SLN dispersion dropwise into liquid nitrogen. This results in small, heterogeneous crystals. By comparison, slow cooling involves placing the vials in a freeze drier having a temperature of –25°C for 24 h; this results in large crystals (Mehnert and Mäder, 2001).

Freeze-drying may have some unfavorable effects on the SLNs, so certain protective materials known as cryoprotectors, such as sorbitol, mannose, trehalose, glucose, and polyvinylpyrrolidone, are used. These additives decrease particle aggregation and improve redispersion of the dry product. Trehalose was found to be the most effective cryoprotector (Vighi et al., 2007).

5.3.4 Spray-drying of SLNs (Freitas and Müller, 1998)

It is an alternative method to increase SLN stability by converting a solution or suspension into a dry product. Although spray drying is cheaper than lyophilization, it is rarely used for SLN formulations.

Several factors, such as spray-drying parameters, the chemical nature of the lipid, and type of redispersion medium, need to be considered to obtain a product of optimum particle size and redispersion.

5.4 RECENT STUDIES USING SLNS AND NLCS FOR OCULAR DRUG DELIVERY

Nanoparticulate delivery systems have recently been under consideration for topical ophthalmic drug delivery because they offer the possibility of modulating drug release, by facilitating its transport to the different compartments of the eye (Badawi et al., 2008). Nanoparticulate systems can increase corneal penetration and prolong residence time at the ocular surface, thus becoming attractive vehicles for the treatment of some ocular diseases (Hao et al., 2011). In 1991, SLNs appeared as a promising, alternative delivery system to traditional colloidal systems, such as emulsions, liposomes, polymeric microparticles, and nanoparticles (Freitas and Müller, 1999; Müller et al., 2008; Üner and Yener, 2007). It was proved that SLNs can prolong the ocular retention of drugs due to their lipophilic character and thus enhance their corneal penetration (Cavalli et al., 2002) and improve the ocular bioavailability (Attama et al., 2008).

Studies showed that SLNs are bioadhesive (Müller-Goymann, 2004), and their small size and particulate nature prevent tear wash out. Thus, they could prolong the eye-residence time and increase ocular bioavailability. Moreover, due to the stability as aqueous dispersion, SLNs can be formulated as eye drops of colloidal suspension. Due to the small size of the SLNs, this formulation is comfortable and avoids blurred vision (Gokce et al., 2008). All these properties make SLNs a promising delivery system to achieve therapeutic action with a smaller dose and fewer systemic and ocular side effects.

During the last decade, several ocular drugs have been incorporated into SLNs, and showed promising results and enhanced ocular bioavailability.

Cavalli et al. (2002) prepared SLNs using stearic acid for topical delivery of Tobramycin. The prepared SLNs were retained for longer time on the corneal surface and in the conjunctival sac of the rabbits compared to the aqueous solution. Also, SLNs produced a significantly higher Tobramycin bioavailability in the aqueous humor. Thus, it appeared as a promising vehicle for topical Tobramycin administration to replace subconjunctival injections necessary for resistant pseudomonal keratitis.

Lallemand et al. (2003) reported that SLNs can increase corneal uptake and intraocular half-lives of Cyclosporine A for topical ophthalmic delivery.

Attama et al. (2008) evaluated the ocular delivery of Diclofenac sodium SLNs using bio-engineered human cornea construct. The high entrapment efficiency (EE) and high permeation achieved in this study indicated that the performance of this analgesic drug could be improved by formulation as SLNs. In another study, Gokce et al. (2008) used high shear homogenization and ultrasound method to prepare Cyclosporine A-loaded SLNs. Ex-vivo studies showed an improved penetration of Cyclosporine A across/into excised corneal tissues. Also, SLNs were found to be able to penetrate into both the biological rabbit corneal epithelial cell substrate and excised pig cornea by capitation/internalization mechanisms into the cells. Thus, it was a promising formulation to target the cornea.

Basaran et al. (2010) successfully incorporated Cyclosporine A into cationic SLNs for ocular application using Dynasan 116. The prepared SLN formulation proved to remain stable at different temperatures during the 6 month storage period. Cyclosporine A was detected in both aqueous and vitreous humor for 48 h showing good ocular penetration of SLNs. Also, a controlled and prolonged release from the positively charged SLN formulation was determined due to increased residence time in the eyes. Also, Kalam et al. (2010) prepared a lipid/mixed lipid matrix consisting of stearic acid and a mixture (4:1) of stearic acid and C 888 to incorporate hydrophilic Gatifloxacin sesquihydrate for bacterial eye infections. This study proved that SLNs composed of stearic acid and C 888 is a good ocular delivery system considering the small particle size, particle-size stability, and physiologically tolerable components.

Liu et al. (2011) prepared and evaluated anti-inflammatory, antibacterial Baicalin SLNs for ocular delivery using triglycerides. SLNs were found to prolong Baicalin release and enhance its permeation coefficient. Also, pharmacokinetic studies in rabbits showed an enhanced ocular bioavailability. Also, Hao et al. (2011) optimized an SLN formulation using Box-Behnken design and successfully prepared the optimized formulation, loaded with Chloramphenicol, using a modified method of melt-emulsion ultrasonication and low-temperature solidification technique using glyceryl monostearate (GMS) as the solid lipid. The results indicated that SLNs could be potentially exploited as an ocular delivery system with improved entrapment efficiency (EE) and controlled drug release. In another study, Li et al. (2011) suggested that methazolamide-loaded SLNs prepared using P ATO 5, have great potential as a new therapeutic system for topical treatment of glaucoma with better patient compliance. The prepared SLNs showed a high therapeutic efficacy, a late occurrence of maximum action, and a more prolonged effect compared to the commercial product.

Hippalgaonkar et al. (2013) developed and characterized Indomethacin-loaded SLNs for ocular inflammation, using C 888 as a solid lipid. Results from this study suggested that topical SLNs could significantly improve the ocular bioavailability of Indomethacin to the posterior ocular tissues. Li et al. (2013) prepared Tetrandrine cationic SLNs by emulsion evaporation-solidification using the low-temperature method to treat opacification of the posterior lens capsule, glaucoma, retinopathy, and ocular inflammations. SLNs retained the drug better and enhanced the pharmacokinetic parameters compared to the ophthalmic solution. Also, cytotoxicity study showed that no significant toxicity was caused by SLNs at low concentrations.

The release properties of the SLNs are affected by the crystallinity of solid lipids. After preparation, the lipids crystallize partially in high-energy modifications with several imperfections in the crystal lattice (Müller et al., 2007). Then, during storage, these high-energy modifications may transform into low energy. The transformation into a highly ordered crystal structure causes a reduction in the number of imperfections in the crystal lattice, and thus drug expulsion occurs (Gonzalez-Mira et al., 2011; Müller et al., 2002b).

NLCs were produced as a second generation derived from SLNs to overcome the limitation caused by the high crystallinity of the SLNs matrices. NLCs exhibit many advantages for application as a drug carrier including the following: increased loading capacity, controlled release of incorporated drugs, and improved physical and chemical stability (Müller et al., 2002a,b; Tian et al., 2012).

In contrast to highly ordered SLN produced using solid lipids or blends of solid lipids, NLCs are produced by controlled mixing of solid lipids with spatially incompatible liquid lipids. These systems are characterized by a less-crystallized matrix, leading to improved drug payload and reduced drug expulsion during storage (Gonzalez-Mira et al., 2011; Jenning et al., 2000). In addition, by controlling the amount of liquid lipid added to the formulation, NLCs remain in the solid state at body temperature, thereby controlling the drug release (Hu et al., 2005).

Lin et al. (2007) successfully modified the inner structure of monocaprate SLNs by including a medium chain triglyceride, as liquid oil into the system to form NLCs using o/w microemulsion technique. The obtained results suggested a better drug encapsulation capacity of NLCs compared to SLNs.

Li et al. (2008) developed Ibuprofen-loaded NLCs using Gelucire 44/14 as solid lipid and Miglyol 812N as liquid lipid. The effects of Transcutol P as permeation enhancer and stearylamine as charge inducing agent were also studied. The results showed that the prepared NLCs formulation enhanced the bioavailability of Ibuprofen, compared to eye drops. Both Gelucire 44/14 and Transcutol P could enhance the drug corneal permeability to some extent. In addition, stearylmine enhanced the precorneal retention in-vivo.

Shen et al. (2010a) prepared thiolated NLCs for topical ocular delivery of Cyclosporine A using cysteine polyethylene glycol monostearate. The pre-corneal residence time was prolonged, and it was found that, Cyclosporine A was delivered to both the ocular surface and the anterior chamber. In another study (Shen et al., 2010b), Cyclosporine A-loaded NLCs was prepared using Precifac ATO 5 as solid lipid and Miglyol 840 as liquid lipid. It was found that the incorporation of liquid lipid could improve the ocular retention and penetration of therapeutics.

Gonzalez-Mira et al. (2011) prepared Flurbiprofen-loaded NLCs for miosis induced during cataract surgery, using stearic acid as solid lipid and a mixture of Miglyol 812 N and castor oil as liquid lipids. The results showed a high encapsulation efficiency and a sustained drug release. In another study, Tian et al. (2012) prepared novel surface modified NLCs with partially deacetylated water soluble chitosan, loaded by Flurbiprofen. This novel application of NLCs demonstrated a high potential for ocular drug delivery, with improved transcorneal penetration, being 1.4 and 1.75-folds of the nonmodified NLCs and the drug solution, respectively. In addition, the precorneal retention was found to be 1.3-folds of the NLCs and 2.4-folds of the drug solution.

Youshia et al. (2012) prepared cationic methazolamide-loaded nanostructured lipid matrices using heterolipids of novel mixtures of C 888 and cetostearyl alcohol for glaucoma. NLMs provided a higher encapsulation efficiency and a slower release rate compared to SLNs, as well as a marked decrease in the IOP compared to methazolamide solution. In another study, Liu et al. (2012) prepared Magniferin-loaded NLCs using GMS, Gelucire 44/14 and Miglyol 812 N, for the treatment of a cataract. The prepared NLCs showed a prolonged drug retention on the cornea, and a 5.69-fold increase in the ocular bioavailability. Also, Pardeike et al. (2012) succeeded in preparing a physically stable NLC formulation for poorly soluble drugs by applying hot high-pressure homogenization and autoclaving.Tian et al. (2013) investigated the transport mechanism of NLCs labeled with fluorescent marker molecules through the corneal epithelium and evaluated the in-vitro release from this colloidal delivery system. The results demonstrated that the collision-induced process leads to the deceptive particle uptake, in which the transfer of hydrophobic fluorescent marker was promoted by forming a complex between the nanoparticles and the biomembranes of the cornea.

Seyfoddin and Al-Kassas (2013), improved the ocular bioavailability of acyclovir by incorporating it into SLNs and NLCs. The type of lipid, drug-to-lipid ratios, type and concentration of surfactants, and type and amount of liquid lipids used in the formulations were optimized. SLNs and NLCs were prepared by the modified hot oil in using the water-microemuslion method. The prepared formulations were evaluated for their particle size, zeta potential, entrapment efficiency, solid-state characteristics, surface morphology, in vitro drug release, in addition to permeation through the excised cornea. It was found that the prepared nanoparticles were spherical and within the size range suitable for ocular drug delivery (400–777.56 nm). Incorporating liquid oil in the SLNs matrix resulted in NLCs with high entrapment efficiency (25%–91.64%) compared to SLNs (11.14%). Moreover, NLCs showed faster permeation through the excised corneas compared to SLNs. Thus, this study suggested the successful conversion of SLNs to physically superior NLCs, to enhance its ocular drug bioavailability.

Hippalgaonkar et al. (2013) developed and characterized indomethacin-loaded SLNs for ocular delivery. Various lipids, homogenization pressures/cycles, Tween 80 fraction in the mixture of surfactants (Poloxamer 188 and Tween 80, total surfactant concentration at 1% w/v), and pH were investigated in the preparation of the indomethacin-loaded SLNs. Compritol 888 ATO was selected as the lipid phase and homogenization at 15,000 psi for six cycles resulted in the smallest particle size, moreover, the increase in the Poloxamer 188 fraction resulted in a decrease in the indomethacin entrapment efficiency. The optimized formulation showed a mean particle size, polydispersity index, zeta potential, and entrapment efficiency of 140 nm, 0.16, –21 mV, and 72.0%, respectively. In addition, the optimized SLNs formulation were physically stable following sterilization process for a period of 1 month and showed an increased in vitro corneal permeability of indomethacin compared to solution, suggesting a significant improvement of the ocular bioavailability of indomethacin.

Zhang et al. (2014) evaluated the ocular performance of a cationic Eudragit RS 100 coated NLCs, loaded by Genistein for treatment of posterior capsular opacification following extracapsular cataract surgery. This study revealed that, surface modified NLCs significantly improved the NLC properties and exhibited a promising ophthalmic delivery system.

Leonardi et al. (2014) assessed the properties and the ocular safety profile of some surfactants commonly used to produce lipid nanoparticles. Using Dynasan 114 as the lipid matrix and a number of surfactants produced stable and small size SLNs. SLN formulations produced using Kolliphor P188 were superior to other formulations. No ocular irritation was observed with any of the tested SLNs up to the highest surfactant concentration of 0.4% w/v. However, the SLNs produced using Lipoid S100 or Cremophor A25 were only tolerated up to a surfactant concentration of 0.2% w/v, and for SLNs containing Kolliphor HS 15 and Tween 80 a maximum concentration of 0.05% were considered to be totally nonirritant to the ocular tissues.

Fangueiro et al. (2014) developed lipid nanoparticle dispersions based on a multiple-emulsion technique for encapsulation of either hydrophilic drugs or proteins by a full factorial design. A cationic lipid—namely, cetyltrimethylammonium bromide—was added in the lipid matrix of the optimized formulation to increase the ocular retention time and mucoadhesion by electrostatic attraction. Since the choice of the cationic agent concentration is critical to formulate a safe and stable lipid nanoparticle, the cationic lipid was included in the lipid matrix, in four different concentrations (0.25, 0.5, 0.75, and 1.0%). The toxicity of the cationic dispersion was tested by exposure to human retinoblastoma cell line Y-79 and the best cetyltrimethylammonium bromide concentration was found to be 0.5%.

Apaolaz et al. (2014) developed nanoparticles consisting on SLNs, protamine, and hyaluronic acid as carriers for gene therapy. Stable complexes positively charged and with a particle size ranging from 240 nm to 340 nm were obtained. Transfection studies in ARPE-19 and HEK-293 cells showed the versatility of vectors to efficiently transfect cells with different division rate, widening the potential applications of SLN-based vectors. In ARPE 19 cells, the incorporation of P and HA induced almost a 7-fold increase in the transfection capacity of SLNs.

Hao et al. (2014) prepared a thermosensitive in situ gelling formulation containing SLNs among poloxamer-based hydrogels as a potential carrier for novel ocular drug delivery. Resina Draconis-loaded SLNs were prepared using a melt-emulsion ultrasonication method. The physicochemical properties of the SLNs were characterized before and after their incorporation into the hydrogels. The optimized formulation consisted of 27.8% (w/v) poloxamer 407 and 3.55% (w/v) poloxamer 188 and showed particle size within the colloidal range. The hen's egg test-chorio allantoic membrane indicated that all the tested systems were nonirritant. In addition, coumarin-6 labeled SLNs in hydrogel system displayed a deep penetration intensity into the cornea at various times.

Wang et al. (2014) optimized methazolamide-loaded SLNs using three level Box-Behnken design. In this study, the amount of glycerylmonostearate, the amount of phospholipid, and the concentration of surfactant were chosen as the independent variables. On the other hand, the dependent variables were entrapment efficiency, drug loading, and particle size. The optimized SLNs formulation was prepared with 100 mg glycerylmonostearate, 150 mg phospholipid, and 1% Tween 80 and PEG 400 (1:1, w/v). Results showed a small particle size of 197.8 ± 4.9 nm and 68.39% entrapment efficiency. Moreover, in vivo studies of the prepared formulation showed a significantly prolonged intraocular-pressure lowering effect without any signs of irritation compared to methazolamide solution.

Leonardi et al. (2015) evaluated whether the melatonin ocular hypotensive effect was enhanced by its encapsulation in cationic SLNs, and determined the ocular tolerability of these formulations. Melatonin was loaded in cationic SLNs suitable for ophthalmic use. SLNs were prepared using Softisan 100 as the lipid matrix, in addition to either palmitic acid or stearic acid as lipid modifiers.

A cationic lipid (dodecyl dimethyl ammonium bromide) was added to provide a fixed positive charge to the SLNs and the intraocular pressure was measured in albino rabbits during 24 h to evaluate the ocular hypotensive effect of the prepared SLNs. It was found that melatonin exhibited a significant ($P < 0.01$) intraocular pressure reduction in rabbit eye and all the tested formulations showed good in vivo ocular tolerability. The prepared cationic SLNs containing stearic acid showed the maximum intraocular pressure reduction, reaching -7 mmHg, which lasted for approximately 24 h. Thus, cationic SLNs loaded with melatonin represented a potent and safe antiglaucoma treatment.

El-Salamouni et al. (2015) prepared brimonidine base-loaded SLN and NLC formulations using glycerylmonostearate as solid lipid matrix and evaluated the physical stability of the formulations following autoclave sterilization at 121°C for 15 min. The physical appearance, particle size, morphology, polydispersity index, zeta potential, and entrapment efficiency were evaluated. Physically stable formulations, of particle size below 500 nm suitable for ocular use were obtained following autoclaving at 121°C for 15 min. In addition, the autoclaved samples showed increased zeta potential values, indicating a higher physical stability compared to nonautoclaved ones. Moreover, an increased amount of brimonidine base was entrapped in the autoclaved formulations.

Üstündağ-Okur et al. (2015) developed novel ocular ofloxacin-loaded NLC-based inserts for treatment of bacterial keratitis. Ofloxacin-loaded NLCs were prepared by high shear homogenization and 0.75% chitosan oligosaccharide lactate was added. Glycerin or polyethylene glycol 400 at concentration range of 1%–15% was added to it as a plasticizer, and solvent casting evaporation developed inserts. The prepared inserts were characterized, also in vitro release, microbiological, ex vivo and in vivo studies were performed.

The optimal inserts developed were found to be those containing glycerin. This formulation showed release of ofloxacin by a combination of diffusion and swelling. In addition, this formulation was more bioadhesive with enhanced preocular retention time up to 24 h and an increased C_{max} to almost 6 times when compared to commercial product. Moreover, rabbits suffering from keratitis after infection with *Staphylococcus aureus* were treated with the optimized formulation and recovered in 7 days with no significant sign of conjunctival redness and corneal opacity. Thus, the prepared NLC-based inserts offered an appropriate vehicle for ocular drug delivery.

Zorzi et al. (2015) reviewed the most important biomaterials used for preparing ocular nanostructured drug delivery systems, especially those designated to delivery to the ocular surface and their impact on the final activity of such systems.

Kakkar et al. (2015) prepared ketoconazole loaded SLNs using Compritol 888 ATO and polyethylene glycol 600 matrix, using the hot high-pressure homogenization technique. The prepared formulation aimed at local treatment of ocular mycoses including corneal keratitis or endophthalmitis. Characterization of the prepared SLNs showed that polyethylene glycol endows the lipid matrix with amorphousness and imperfections, rigidity, and stability to aggregation, on storage and autoclaving. Without its inclusion, ketoconazole could not be loaded into SLNs. It ensured high ketoconazole entrapment efficiency of about 70%; small particle size (126 nm); and better permeation into the eye. Pharmacokinetic studies indicated 2.5- and 1.6-fold higher bioavailability (area under the curve) in the aqueous humor and the vitreous humor, respectively. Moreover, the biocompatibility, in vitro (both in corneal and retinal cell lines) and in vivo (in rabbits) ocular safety of the developed formulation were assessed and showed excellent results.

Wang et al. (2015) reviewed the recent lipid nanoparticle prepared from both cationic liposomes and anionic protamine-DNA complex. The liposome protamine/DNA lipoplex nanoparticles consists of a condensed DNA core surrounded by lipid bilayers. These nanoparticles are used for targeting the retina to treat retinitis pigmentosa, age-related macular degeneration, and diabetic retinopathy. Moreover, liposome protamine/DNA lipoplex nanoparticles have shown recently promising results to deliver functional genes and micro RNA for the treatment of retinal diseases.

Sah and Suresh (2015) reviewed the potential feasibility of lipid component in formulation of nanomedicine for ocular drug delivery and reported the clinical findings in order to improve the ocular pharmacotherapy.

Liu et al. (2015) developed and characterized a thermosensitive ocular in situ nanogel of cationic nanostructured lipid carriers loaded with curcumin to enhance preocular retention and ocular permeation. The prepared formulation was evaluated both in vitro and in vivo, and pharmacokinetic study in the aqueous humor was investigated by microdialysis technique.

The solution–gel transition temperature of the optimized formulation diluted by simulated tear fluid was $34 \pm 1.0°C$. Both the apparent permeability coefficient and the area under the curve of the optimized formulation were 1.56-fold and 9.24-fold, respectively, those of the curcumin solution. Moreover, the maximal concentration was significantly improved ($p < 0.01$) and the prolonged mean residence time ($p < 0.01$) indicated a controlled release formulation. The results of this study showed that the developed formulation is of great potential to increase the bioavailability of curcumin in the aqueous humor by enhancing corneal permeation and retention capacity.

Thus, based on previous studies, NLCs proved to be a promising alternative to SLNs to overcome the limitation of drug expulsion during storage.

6 CONCLUSIONS

It could be concluded that the emergence of lipid-based nanocarriers has provided a viable means of enhancing the bioavailability of ophthalmic formulations. This chapter highlighted that lipid-based nanocarriers can enhance the corneal absorption of both hydrophilic and lipophilic drugs and improved their ocular bioavailability. NLCs and LDCs have emerged as a new generation of SLNs to overcome problems of low entrapment efficiency and drug expulsion during storage.

REFERENCES

Alany, R.G., Rades, T., Nicoll, J., Tucker, I.G., Davies, N.M., 2006. W/O microemulsions for ocular delivery: evaluation of ocular irritation and precorneal retention. J. Control Rel. 111 (1), 145–152.

Almeida, A.J., Souto, E., 2007. Solid lipid nanoparticles as a drug delivery system for peptides and proteins. Adv. Drug Deliv. Rev. 59 (6), 478–490.

Apaolaza, P.S., Delgado, D., del Pozo-Rodríguez, A., Gascón, A.R., Solinís, M.Á., 2014. A novel gene therapy vector based on hyaluronic acid and solid lipid nanoparticles for ocular diseases. Int J. Pharm. 465 (1), 413–426.

Attama, A.A., Reichl, S., Müller-Goymann, C.C., 2008. Diclofenac sodium delivery to the eye: In vitro evaluation of novel solid lipid nanoparticle formulation using human cornea construct. Int. J. Pharm. 355 (1), 307–313.

Badawi, A.A., El-Laithy, H.M., Qidra, R.K.E., Mofty, H.E., 2008. Chitosan based nanocarriers for indomethacin ocular delivery. Arch. Pharm. Res. 31 (8), 1040–1049.

Basaran, E., Demirel, M., Sirmagül, B., Yazan, Y., 2010. Cyclosporine-A incorporated cationic solid lipid nanoparticles for ocular delivery. J. Microencapsul. 27 (1), 37–47.

Bunjes, H., Koch, M.H.J., 2005. Saturated phospholipids promote crystallization but slow down polymorphic transitions in triglyceride nanoparticles. J. Control Rel. 107 (2), 229–243.

Bunjes, H., Westesen, K., Koch, M.H.J., 1996. Crystallization tendency and polymorphic transitions in triglyceride nanoparticles. Int. J. Pharm. 129 (1), 159–173.

Calvo, P., Sánchez, A., Martínez, J., López, M.I., Calonge, M., Pastor, J.C., et al., 1996. Polyester nanocapsules as new topical ocular delivery systems for cyclosporin A. Pharm. Res. 13 (2), 311–315.

Cavalli, R., Gasco, M.R., Chetoni, P., Burgalassi, S., Saettone, M.F., 2002. Solid lipid nanoparticles (SLN) as ocular delivery system for tobramycin. Int. J. Pharm. 238 (1), 241–245.

Chavkin, L., Mackles, L., 1998. Chewable Molded Tablet Containing Medicinally Active Substances. Google Patents.

Chen, M.L., 2008. Lipid excipients and delivery systems for pharmaceutical development: a regulatory perspective. Adv. Drug Deliv. Rev. 60 (6), 768–777.

Chetoni, P., Burgalassi, S., Monti, D., Najarro, M., Boldrini, E., 2007. Liposome-encapsulated mitomycin C for the reduction of corneal healing rate and ocular toxicity. J. Drug Deliv. Sci. Tec. 17 (1), 43–48.

Cortesi, R., Argnani, R., Esposito, E., Dalpiaz, A., Scatturin, A., Bortolotti, F., et al., 2006. Cationic liposomes as potential carriers for ocular administration of peptides with anti-herpetic activity. Int. J. Pharm. 317 (1), 90–100.

Del Pozo-Rodríguez, A., Delgado, D., Solinís, M.A., Gascon, A.R., Pedraz, J.L., 2008. Solid lipid nanoparticles for retinal gene therapy: transfection and intracellular trafficking in RPE cells. Int. J. Pharm. 360 (1), 177–183.

Elnaggar, Y.S.R., El-Massik, M.A., Abdallah, O.Y., 2011. Fabrication, appraisal, and transdermal permeation of sildenafil citrate-loaded nanostructured lipid carriers versus solid lipid nanoparticles. Int. J. Nanomed. 6, 3195–3205.

El-Salamouni, N.S., Farid, R.M., El-Kamel, A.H., El-Gamal, S.S., 2015. Effect of sterilization on the physical stability of brimonidine-loaded solid lipid nanoparticles and nanostructured lipid carriers. Int. J. Pharm. 496, 976–983.

Fangueiro, J.F., Andreani, T., Egea, M.A., Garcia, M.L., Souto, S.B., Silva, A.M., et al., 2014. Design of cationic lipid nanoparticles for ocular delivery: development, characterization and cytotoxicity. Int. J. Pharm. 461 (1), 64–73.

Freitas, C., 1998. Feste Lipid-Nanopartikel (SLN): Mechanismen der physikalischen Destabilisierung und Stabilisierung PhD Thesis.

Freitas, C., Müller, R.H., 1998. Spray-drying of solid lipid nanoparticles (SLNTM). Eur. J. Pharm. Biopharm. 46 (2), 145–151.

Freitas, C., Müller, R.H., 1999. Correlation between long-term stability of solid lipid nanoparticles (SLN™) and crystallinity of the lipid phase. Eur. J. Pharm. Biopharm. 47 (2), 125–132.

Fuente, M.D.1., Raviña, M., Paolicelli, P., Sanchez, A., Seijo, B., Alonso, M.J., 2010. Chitosan-based nanostructures: a delivery platform for ocular therapeutics. Adv. Drug Deliv. Rev. 62 (1), 100–117.

Gan, L., Wang, J., Jiang, M., Bartlett, H., Ouyang, D., Eperjesi, F., et al., 2013. Recent advances in topical ophthalmic drug delivery with lipid-based nanocarriers. Drug Discov. Today 18 (5), 290–297.

Gaudana, R., Jwala, J., Boddu, S.H.S., Mitra, A.K., 2009. Recent perspectives in ocular drug delivery. Pharm. Res. 26 (5), 1197–1216.

Gokce, E.H., Sandri, G., Bonferoni, M.C., Rossi, S., Ferrari, F., Güneri, T., et al., 2008. Cyclosporine A loaded SLNs: evaluation of cellular uptake and corneal cytotoxicity. Int. J. Pharm 364 (1), 76–86.

Gonzalez-Mira, E., Egea, M.A., Souto, E.B., Calpena, A.C., Garcia, M.L., 2011. Optimizing flurbiprofen-loaded NLC by central composite factorial design for ocular delivery. Nanotech. 22 (4), 1–15.

Hao, J., Fang, X., Zhou, Y., Wang, J., Guo, F., Li, F., et al., 2011. Development and optimization of solid lipid nanoparticle formulation for ophthalmic delivery of chloramphenicol using a Box-Behnken design. Int. J. Nanomed. 6, 683–692.

Hao, J., Wang, X., Bi, Y., Teng, Y., Wang, J., Li, F., et al., 2014. Fabrication of a composite system combining solid lipid nanoparticles and thermosensitive hydrogel for challenging ophthalmic drug delivery. Colloids Surf. B: Biointerf. 114, 111–120.

Helgason, T., Awad, T.S., Kristbergsson, K., McClements, D.J., Weiss, J., 2009. Effect of surfactant surface coverage on formation of solid lipid nanoparticles (SLN). J. Colloid Interf. Sci. 334 (1), 75–81.

Hippalgaonkar, K., Adelli, G.R., Hippalgaonkar, K., Repka, M.A., Majumdar, S., 2013. Indomethacin-loaded solid lipid nanoparticles for ocular delivery: development, characterization, and in vitro evaluation. J. Ocul. Pharmacol. Ther. 29 (2), 216–228.

Hu F.Q., Jiang S.P., Du Y.Z., Yuan H., Ye Y.Q., Zeng S., 2005. Preparation and characterization of stearic acid nanostructured lipid carriers by solvent diffusion method in an aqueous system. Colloids Surf. B Biointerfaces 45(3),167–173.

Jenning, V., Gysler, A., Schäfer-Korting, M., Gohla, S.H., 2000. Vitamin A loaded solid lipid nanoparticles for topical use: occlusive properties and drug targeting to the upper skin. Eur. J. Pharm. Biopharm. 49 (3), 211–218.

Kakkar, S., Karuppayil, S.M., Raut, J.S., Giansanti, F., Papucci, L., Schiavone, N., et al., 2015. Lipid-polyethylene glycol based nano-ocular formulation of ketoconazole. Int. J. Pharm. 495 (1), 276–289.

Kalam, M.A., Sultana, Y., Ali, A., Aqil, M., Mishra, A.K., Chuttani, K., 2010. Preparation, characterization, and evaluation of gatifloxacin loaded solid lipid nanoparticles as colloidal ocular drug delivery system. J. Drug Target 18 (3), 191–204.

Katragadda, S., Talluri, R.S., Mitra, A.K., 2006. Modulation of P-glycoprotein-mediated efflux by prodrug derivatization: an approach involving peptide transporter-mediated influx across rabbit cornea. J. Ocul. Pharmacol. Ther. 22 (2), 110–120.

Kaur, I.P., Rana, C., Singh, H., 2008. Development of effective ocular preparations of antifungal agents. J. Ocul. Pharmacol. Ther. 24 (5), 481–494.

Kipp, J.E., 2004. The role of solid nanoparticle technology in the parenteral delivery of poorly water-soluble drugs. Int. J. Pharm. 284 (1), 109–122.

Lallemand, F., Felt-Baeyens, O., Besseghir, K., Behar-Cohen, F., Gurny, R., 2003. Cyclosporine A delivery to the eye: A pharmaceutical challenge. Eur. J. Pharm. Biopharm. 56 (3), 307–318.

Leonardi, A., Bucolo, C., Romano, G.L., Platania, C.B.M., Drago, F., Puglisi, G., et al., 2014. Influence of different surfactants on the technological properties and in vivo ocular tolerability of lipid nanoparticles. Int. J. Pharm. 470 (1), 133–140.

Leonardi, A., Bucolo, C., Drago, F., Salomone, S., Pignatello, R., 2015. Cationic solid lipid nanoparticles enhance ocular hypotensive effect of melatonin in rabbit. Int. J. Pharm. 478 (1), 180–186.

Li, X., Nie, S.F., Kong, J., Li, N., Ju, C.Y., Pan, W.S., 2008. A controlled-release ocular delivery system for ibuprofen based on nanostructured lipid carriers. Int. J. Pharm. 363 (1), 177–182.

Li, R., Jiang, S., Liu, D., Bi, X., Wang, F., Zhang, Q., et al., 2011. A potential new therapeutic system for glaucoma: solid lipid nanoparticles containing methazolamide. J. Microencapsul. 28 (2), 134–141.

Li, J., Guo, X., Liu, Z., Okeke, C.I., Li, N., Zhao, H., et al., 2013. Preparation and evaluation of charged solid lipid nanoparticles of tetrandrine for ocular drug delivery system: pharmacokinetics, cytotoxicity and cellular uptake studies. Drug Dev. Indus. Pharm. (0), 1–8.

Lin, X., Li, X., Zheng, L.Q., Yu, L., Zhang, Q., Liu, W., 2007. Preparation and characterization of monocaprate nanostructured lipid carriers. Colloids Surf A Physicochem. Eng. Asp. 311 (1), 106–111.

Liu, Z., Zhang, X., Wu, H., Li, J., Shu, L., Liu, R., et al., 2011. Preparation and evaluation of solid lipid nanoparticles of baicalin for ocular drug delivery system in vitro and in vivo. Drug Dev. Indus. Pharm. 37 (4), 475–481.

Liu, R., Liu, Z., Zhang, C., Zhang, B., 2012. Nanostructured lipid carriers as novel ophthalmic delivery system for mangiferin: Improving in vivo ocular bioavailability. J. Pharm. Sci. 101 (10), 3833–3844.

Liu, R., Sun, L., Fang, S., Wang, S., Chen, J., Xiao, X., et al., 2015. Thermosensitive in situ nanogel as ophthalmic delivery system of curcumin: development, characterization, in vitro permeation and in vivo pharmacokinetic studies. Pharm. Dev. Tech. 0, 1–7.

Lucks, J.S., 1993. Parenterale Fettemulsionen als Arzneistoffträger: Herstellung, Charakterisierung und Stabilität. Christian-Albrechts-Universität.

Lv, Q., Yu, A., Xi, Y., Li, H., Song, Z., Cui, J., et al., 2009. Development and evaluation of penciclovir-loaded solid lipid nanoparticles for topical delivery. Int. J. Pharm. 372 (1), 191–198.

Marengo, E., Cavalli, R., Caputo, O., Rodriguez, L., Gasco, M.R., 2000. Scale-up of the preparation process of solid lipid nanospheres. Part I. Int. J. Pharm. 205 (1), 3–13.

Mehnert, W., Mäder, K., 2001. Solid lipid nanoparticles: production, characterization and applications. Adv. Drug Deliv. Rev. 47 (2), 165–196.

Mori, H. (Ed.), 1988. Solidification Problems in Preparation of Fats. Dekker, New York.

Muéller, R.H., Maéder, K., Gohla, S., 2000. Solid lipid nanoparticles (SLN) for controlled drug delivery—a review of the state of the art. Eur. J. Pharm. Biopharm. 50 (1), 161–177.

Müller, R.H., Radtke, M., Wissing, S.A., 2002a. Nanostructured lipid matrices for improved microencapsulation of drugs. Int. J. Pharm. 242 (1), 121–128.

Müller, R.H., Radtke, M., Wissing, S.A., 2002b. Solid lipid nanoparticles (SLN) and nanostructured lipid carriers (NLC) in cosmetic and dermatological preparations. Adv. Drug Deliv. Rev. 54, S131–S155.

Müller, R.H., Petersen, R.D., Hommoss, A., Pardeike, J., 2007. Nanostructured lipid carriers (NLC) in cosmetic dermal products. Adv. Drug Deliv. Rev. 59 (6), 522–530.

Müller, R.H., Runge, S.A., Ravelli, V., Thünemann, A.F., Mehnert, W., Souto, E.B., 2008. Cyclosporine-loaded solid lipid nanoparticles (SLN): drug-lipid physicochemical interactions and characterization of drug incorporation. Eur. J. Pharm. Biopharm. 68 (3), 535–544.

Müller-Goymann, C.C., 2004. Physicochemical characterization of colloidal drug delivery systems such as reverse micelles, vesicles, liquid crystals and nanoparticles for topical administration. Eur. J. Pharm. Biopharm. 58 (2), 343–356.

Pal Kaur, I., Kanwar, M., 2002. Ocular preparations: the formulation approach. Drug. Dev. Ind. Pharm. 28 (5), 473–493.

Pandya, J.B., Parmar, R.D., Soniwala, M.M., Chavda, J.R., 2013. Solid lipid nanoparticles: overview on excipients. AJPTI 1 (3), 1–9.

Pardeike, J., Hommoss, A., Müller, R.H., 2009. Lipid nanoparticles (SLN, NLC) in cosmetic and pharmaceutical dermal products. Int. J. Pharm. 366 (1), 170–184.

Pardeike, J., Weber, S., Matsko, N., Zimmer, A., 2012. Formation of a physical stable delivery system by simply autoclaving nanostructured lipid carriers (NLC). Int. J. Pharm. 439 (1), 22–27.

Pharmacopoeia B., 2007. British Pharmacopoeia Commission Office, London, UK.

Pignatello, R., Ricupero, N., Bucolo, C., Maugeri, F., Maltese, A., Puglisi, G., 2006. Preparation and characterization of eudragit retard nanosuspensions for the ocular delivery of cloricromene. AAPS PharmSciTech. 7 (1), E192–E198.

Pouton, C.W., Porter, C.J.H., 2008. Formulation of lipid-based delivery systems for oral administration: materials, methods and strategies. Adv. Drug Deliv. Rev. 60 (6), 625–637.

Prabhu, P., Koland, M., Vijaynarayan, K., 2010. Preparation and evaluation of Niosomes of Brimonidine tartrate as ocular drug delivery system. Asian J. Pharmaceut. Res. Health Care 2 (4), 293–301.

Radomska-Soukharev, A., 2007. Stability of lipid excipients in solid lipid nanoparticles. Adv. Drug Deliv. Rev. 59 (6), 411–418.

Rawas-Qalaji, M., Williams, C.A., 2012. Advances in ocular drug delivery. Curr. Eye Res. 37 (5), 345–356.

Sah A.K., Suresh P.K., 2015. Recent advances in ocular drug delivery, with special emphasis on lipid based nanocarriers. Recent Pat Nanotechnol. 9(2), 94–105.

Sangsen, Y., Likhitwitayawuid, K., Sritularak, B., Wiwattanawongsa, K., Wiwattanapatapee, R., 2013. Novel solid lipid nanoparticles for oral delivery of oxyresveratrol: effect of the formulation parameters on the physicochemical properties and in vitro release. Int. J. Med. Pharm. Sci. Eng. 7 (12), 506–513.

Sawant, K.K., Dodiya, S.S., 2008. Recent advances and patents on solid lipid nanoparticles. Recent Pat. Drug Deliv. Formul. 2 (2), 120–135.

Schwarz, C., Mehnert, W. (Eds.), 1995. Sterilization of drug-free and tetracaine-loaded solid lipid nanoparticles (SLN). Proc. First World Meeting APGI/APV, Budapest.

Seyfoddin, A., Al-Kassas, R., 2013. Development of solid lipid nanoparticles and nanostructured lipid carriers for improving ocular delivery of acyclovir. Drug Dev. Indus. Pharm. 39 (4), 508–519.

Seyfoddin, A., Shaw, J., Al-Kassas, R., 2010. Solid lipid nanoparticles for ocular drug delivery. Drug Deliv. 17 (7), 467–489.

Shen, J., Deng, Y., Jin, X., Ping, Q., Su, Z., Li, L., 2010a. Thiolated nanostructured lipid carriers as a potential ocular drug delivery system for cyclosporine A: improving in vivo ocular distribution. Int. J. Pharm. 402 (1), 248–253.

Shen, J., Sun, M., Ping, Q., Ying, Z., Liu, W., 2010b. Incorporation of liquid lipid in lipid nanoparticles for ocular drug delivery enhancement. Nanotec. 21 (2), 1–10.

Sjöström, B., Bergenståhl, B., 1992. Preparation of submicron drug particles in lecithin-stabilized o/w emulsions I. Model studies of the precipitation of cholesteryl acetate. Int. J. Pharm. 88 (1), 53–62.

Souto, E.B., Doktorovova, S., Gonzalez-Mira, E., Egea, M.A., Garcia, M.L., 2010. Feasibility of lipid nanoparticles for ocular delivery of anti-inflammatory drugs. Curr. Eye Res. 35 (7), 537–552.

Thakur, R.R., Kashiv, M., 2011. Modern delivery systems for ocular drug formulations: a comparative overview WRT conventional dosage form. Int. J. Res. Pharm. Biomed. Sci. 1, 8–18.

Tian, B., Luo, Q., Song, S., Liu, D., Pan, H., Zhang, W., et al., 2012. Novel surface modified nanostructured lipid carriers with partially deacetylated water soluble chitosan for efficient ocular delivery. J. Pharm. Sci. 101 (3), 1040–1049.

Tian, B.C., Zhang, W.J., Xu, H.M., Hao, M.X., Liu, Y.B., Yang, X.G., et al., 2013. Further investigation of nanostructured lipid carriers as an ocular delivery system: in vivo transcorneal mechanism and in vitro release study. Colloids Surf. B Biointerf. 102, 251–256.

Trotta, M., Cavalli, R., Carlotti, M.E., Battaglia, L., Debernardi, F., 2005. Solid lipid micro-particles carrying insulin formed by solvent-in-water emulsion–diffusion technique. Int. J. Pharm. 288 (2), 281–288.

Üner, M., Yener, G., 2007. Importance of solid lipid nanoparticles (SLN) in various administration routes and future perspectives. Int. J. Nanomed. 2 (3), 289–300.

Üstündağ-Okur, N., Gökçe, E.H., Bozbıyık, D., Eğrilmez, S., Ertan, G., Özer, Ö., 2015. Novel nanostructured lipid carrier-based inserts for controlled ocular drug delivery: evaluation of corneal bioavailability and treatment efficacy in bacterial keratitis. Expert Op. Drug Del. 12 (11), 1791–1807.

Vandamme, T.F., 2002. Microemulsions as ocular drug delivery systems: recent developments and future challenges. Prog. Retin. Eye Res. 21 (1), 15–34.

Vighi, E., Ruozi, B., Montanari, M., Battini, R., Leo, E., 2007. Re-dispersible cationic solid lipid nanoparticles (SLNs) freeze-dried without cryoprotectors: characterization and ability to bind the pEGFP-plasmid. Eur. J. Pharm. Biopharm. 67 (2), 320–328.
Vyas, S.P., Rai, S., Paliwal, R., Gupta, P.N., Khatri, K., Goyal, A.K., et al., 2008. Solid lipid nanoparticles (SLNs) as a rising tool in drug delivery science: One step up in nanotechnology. Curr. Nanosci. 4 (1), 30–44.
Wang, F., Chen, L., Jiang, S., He, J., Zhang, X., Peng, J., et al., 2014. Optimization of methazolamide-loaded solid lipid nanoparticles for ophthalmic delivery using Box–Behnken design. J. Lipo. Res. 24 (3), 171–181.
Wang, Y., Rajala, A., Rajala, R.V., 2015. Lipid nanoparticles for ocular gene delivery. J.Func. Biomat. 6 (2), 379–394.
Wilson, C.G., 2004. Topical drug delivery in the eye. Exp. Eye Res. 78 (3), 737–743.
Youshia, J., Kamel, A.O., Shamy, A.E., Mansour, S., 2012. Design of cationic nanostructured heterolipid matrices for ocular delivery of methazolamide. Int. J. Nanomed. 7, 2483–2496.
Yun, J., Zhang, S., Shen, S., Chen, Z., Yao, K., Chen, J., 2009. Continuous production of solid lipid nanoparticles by liquid flow-focusing and gas displacing method in microchannels. Chem. Eng. Sci. 64 (19), 4115–4122.
Zhang, W., Li, X., Ye, T., Chen, F., Yu, S., Chen, J., et al., 2014. Nanostructured lipid carrier surface modified with Eudragit RS 100 and its potential ophthalmic functions. Int. J. Nanomed. 9, 4305–4315.
Zorzi G.K., Parraga J.E., Seijo B., Sánchez A. On the biomaterials for nanostructured ocular therapeutics. Curr. Org. Chem. 19(15), 1443–1459.
Zur Mühlen, A., Schwarz, C., Mehnert, W., 1998. Solid lipid nanoparticles (SLN) for controlled drug delivery–drug release and release mechanism. Eur. J. Pharm. Biopharm. 45 (2), 149–155.

FURTHER READING

Ekambaram, P., Sathali, A.A.H., Priyanka, K., 2012. Solid lipid nanoparticles: a review. Sci. Rev. Chem. Commun. 2 (1), 80–102.
Üner, M., 2006. Preparation, characterization and physico-chemical properties of solid lipid nanoparticles (SLN) and nanostructured lipid carriers (NLC): their benefits as colloidal drug carrier systems. Pharmazie 61 (5), 375–386.

CHAPTER

17

NANOPARTICULATE CARRIER(S): AN EMERGING PARADIGM IN NEW GENERATION VACCINE DEVELOPMENT

Rajeev Sharma, Nishi Mody, Surabhi Dubey, Suresh P. Vyas

Dr. H. S. Gour Vishwavidyalaya, Sagar, India

CHAPTER OUTLINE

1 INTRODUCTION

Traditional vaccines are mainly composed of live attenuated viruses, whole inactivated pathogens, or inactivated bacterial toxins. Progress in the field of biotechnology (recombinant DNA technology) have allowed the development of DNA and subunit vaccines, highly purified recombinant proteins, synthetic oligopeptides, synthetic oligosaccharides, or viral subunits, as well as conjugate vaccines in which a weak antigen is linked to a stronger immunogen such as a protein or membrane complex. Unlike live attenuated vaccines, these are safe since these do not revert to a virulent form and lack contaminant remains from the original pathogenic organism. However, they suffer from a number of limitations such as weak immunogenicity requiring the addition of an adjuvant, intrinsic instability, in vivo toxicity, and the need for multiple administrations. Immunological adjuvants were originally described by Ramon as "substances used in combination with a specific antigen that produced a more robust immune response than the antigen alone" (Ramon, 1924). A crucial aspect in addressing the challenges in vaccine development is vaccine delivery. Vaccine delivery encompasses (1) administration of the vaccine formulation to specific sites of the body and (2) delivery of the antigens to predefined site and activation of relevant cells of the immune system. Administration of vaccine formulations to specific sites of the body can be achieved by various routes, including intramuscular (i.m.) subcutaneous (s.c.), transcutaneous, oral, sublingual/ buccal, nasal, or pulmonary delivery. Each of these immunization routes requires specially designed formulations and specially designed delivery devices. Induction of desired immune response furthermore requires antigen delivery to professional antigen-presenting cells (APCs) and activation of these cells (Amorij et al., 2012). Therefore, there exists a critical need for fabricating vaccine delivery vehicles that could convey antigens as well as adjuvant(s) and elicit humoral, cellular, and mucosal immunity as desired. Recently, various nanoparticulate vaccine delivery systems have been incorporated for improving the aspects of antigen and adjuvant delivery and an opportunity to enhance the humoral and cellular immune responses (Fig. 17.1). Antibody-specific response in response to an antigen is

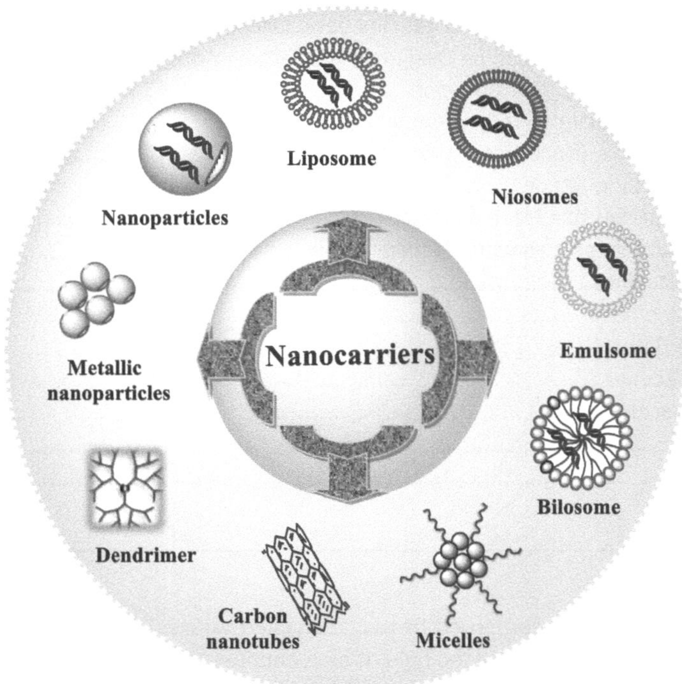

FIGURE 17.1 Various Types of Nanocarriers Explored in New Generation Vaccine Development

evoked by the generation of antibodies by the plasma B cells. The nanotechnology-based approaches developed for use in the field of vaccines encompass nanocarriers having a variety of composition, size, and surface properties. Nanoparticulate systems can be used both for delivery of antigen and/or as adjuvant for enhancing immune responses for therapeutic and prophylactic vaccines. Nanovaccines function by stimulating a potential immune response by generating antibodies for both the nanoparticle carrier itself and the integrated antigen.

2 ADVANTAGES OF NANOPARTICULATE-BASED VACCINE DELIVERY

In the search for better vaccines, particulate vaccines represent a promising concept to surmount the aforementioned obstacles (Storni et al., 2005). Formulation of antigens in particles in the viral or bacterial size range offers some attractive features, including protection of the antigen against degradation, facilitated uptake by APCs through passive or active targeting, depot formation, and codelivery of antigens and adjuvants to the same APCs that could assist in directing the type of immune response desired. Particulate antigen delivery systems are a strategy that satisfy many criteria of vaccine development. Particle characteristics influence the underlying mechanisms involved in the internalization, processing, and antigen presentation by APCs, which all impact the outcome of the immune response.

- Particulate carriers can serve as a depot and facilitate sustained release of antigens or adjuvants to antigen-presenting cells (Rice-Ficht et al., 2009). The frequent exposure of immunological cells leads to exaggerated immune response.
- Small-size particulates (smaller than 10 μm) are taken up more actively by phagocytic cells, such as dendritic cells and macrophages. This characteristic has been utilized to enhance the cellular uptake of antigens, thus increasing the efficiency of antigen recognition and presentation (Oyewumi et al., 2010).
- Solid particulates can protect the structural integrity and biological activity of antigen vaccines from degradation and facilitate internalization into the mucosa-associated lymphoid tissues (MALT), gut-associated lymphoid tissue (GALT), nasal-associated lymphoid tissues (NALT), and so forth, rendering them suitable for oral or mucosal vaccination (Borges et al., 2005).
- Particulate vaccines can potentially cross-present antigen, and antigen cross-presentation plays a significant role in the generation of CD8+ T-cell-mediated immune responses against viral infections (Jain et al., 2005a).
- Surface engineering of particulates can be done in order to improve targeted delivery of antigens and biocompatibility. Ligands specific for antigen-presenting cells surface receptors can be appended onto the surface of particle-based vaccines to increase specific delivery. Immune cells express a variety of surface receptors, including the toll-like receptors (TLRs), mannose receptors, and scavenger receptors (Apostolopoulos et al., 2013). Nanoparticulate carriers surface engineered with immune cell-targeting molecules, such as peptides (Misumi et al., 2009) antibodies (Raghuwanshi et al., 2012), and carbohydrates (Lepenies et al., 2013) may target these surface receptors to improve the antigen processing and/or act as an immunostimulant adjuvant to trigger or enhance immune response.

3 GENERATION OF IMMUNE RESPONSE

Antigen-presenting cells are the major cells of the immune system involved in the protection against invading antigens (pathogen or particulate carrier; Trombetta and Mellman, 2005). These are groups of cells that are widely distributed in tissues includes B cells, macrophage, and dendritic cells (DCs). In the early immune response, the macrophages and DCs capture the pathogen of both exogenous and endogenous origin. DCs are the most potent immune-initiator cells and exhibit antigen presentation because these cells always expresses high levels of both class I and II MHC molecule and co-stimulatory molecule (B7) causing activation of naive CD8 and CD4 T cells (Demento et al., 2009). The mechanisms for antigen presentations are studied at molecular levels, however. Fig. 17.2 gives a brief introduction of these pathways. Endogenous antigens like viral and tumor antigens/proteins or normal cell proteins are processed in cytosol and degraded by proteasomes into peptides that are transported to the endoplasmic reticulum (ER). In the ER, peptides associate stabilizes the newly formed complex of class I MHC and beta-2-microglobulin (Haana den et al., 2014). This complex containing peptide-loaded MHC I then moves to the cell surface for being processed by CD8 cytotoxic T cells T. Class I MHC is presented by all nucleated cells, whereas class II MHC is presented by only the antigen-presenting cells. Exogenous antigenic peptides that are inhaled, ingested, or injected like particulate antigen carriers are processed by class II or endocytic pathway and is presented by MHC II molecule on APCs. Antigens are taken up by the cells by either of the processes, whereas MHC class I molecules are constitutively expressed in all nucleated cells, MHC class II molecules are only expressed in APCs. Antigen is encapsulated into endosome and is subjected to a series of intracellular compartments of increasing acidity—early endosome (pH 6.5–6.0), late endosome (pH 6.0–5.0), and phagolysosome (pH <5.0). The endosome fuses with lysosomes to form phagolysosomes, which contain large amounts of lysoszymes. Degradation of material taken up by the endosome occurs through acidification and action of degradable enzymes of phagolysosomes. Class II MHC molecule is synthesized by ER but unlike MHC I it does not bind to the antigenic peptide in ER because its binding site is blocked by an invariant chain (li). MHC II is transferred to the endosomal compartment where the invariant chain is cleaved to the CLIP (Class II-associated invariant chain peptide) fragment that is removed by HLA-DM (human leukocyte antigen DM; is an intracellular protein involved in peptide presentation by MHC class II). This process is accompanied by the exchange of antigenic peptide with CLIP fragment and presented on APCs surface to be processed by CD4 helper T cell. MHC class I pathway expresses in all nucleated cells, whereas only a limited group of cells like antigen presenting cells express in MHC class II pathway (Chen and Jensen, 2008). A number of APCs have the capability to divert the endocytosed exogenous antigen to class I MHC pathway for its recognition and processing by the cytotoxic T cells, which generates a stronger immune response. Cross-presentation is highly functional and significant for antiviral and antitumor vaccines. A number of APCs like DCs have the capacity to deliver and present the endocytosed exogenous antigens into the MHC class I pathway, for their presentation leading to the activation of antigen specific CD8+ T cells. Cytotoxic T lymphocyte (CTL) releases toxic particles upon detection of infected cell and causes the death of the such cells hence preventing the further expansion of infection or abnormalities (Heath and Carbone, 2001; Kurts, 2000).

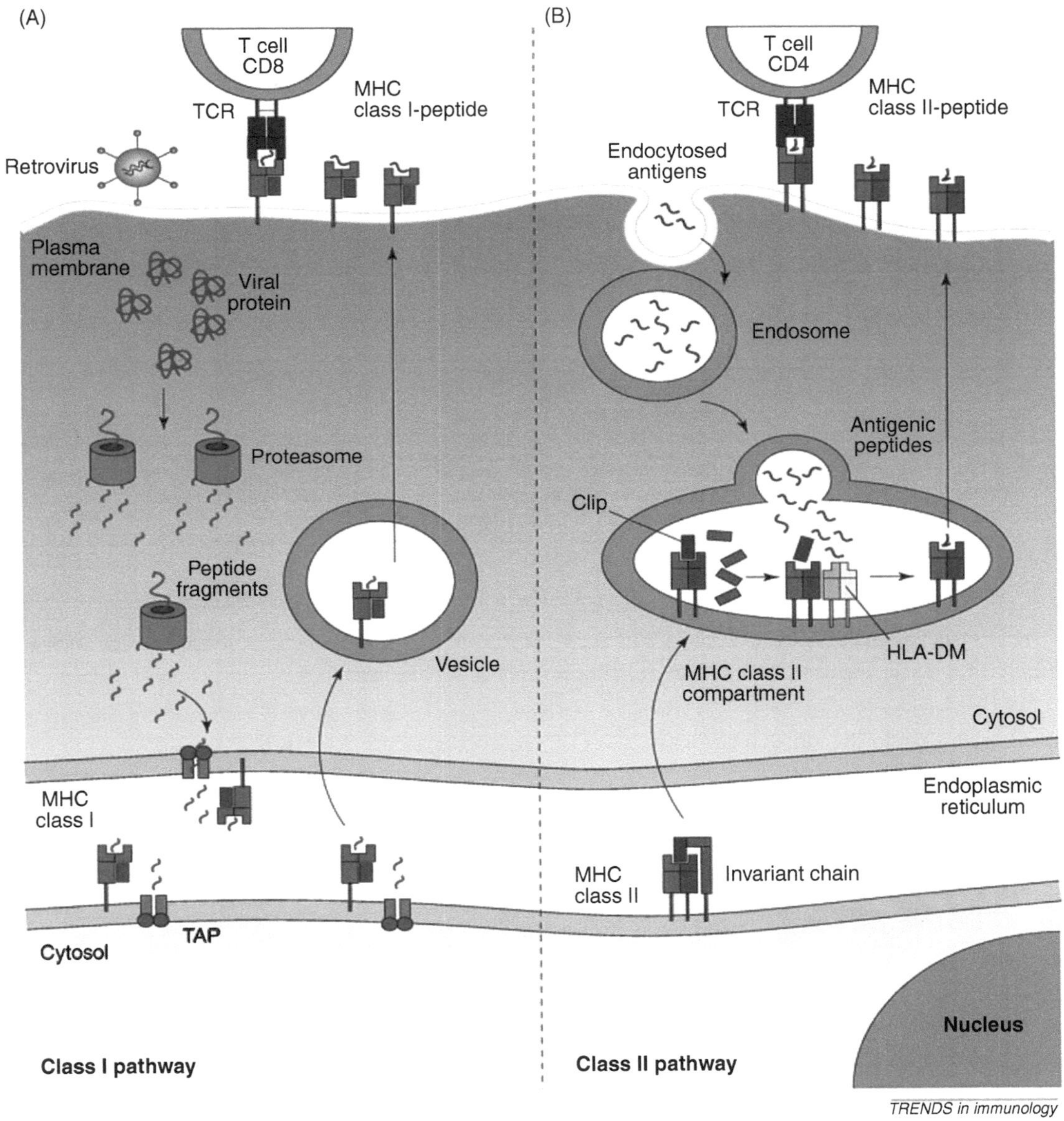

FIGURE 17.2 Schematic Diagram Showing the Pathways Through Which Peptides Are Presented on the Surface of Antigen-Presenting Cells for T-Cell Binding

Reproduced from Halling-Brown, M., Sansom, C.E., Davies, M., Titball, R.W., Moss, D.S., 2008. Are bacterial vaccine antigens T-cell epitope depleted? Trends Immunol. 29(8), 374–379, with permission of Elsevier.

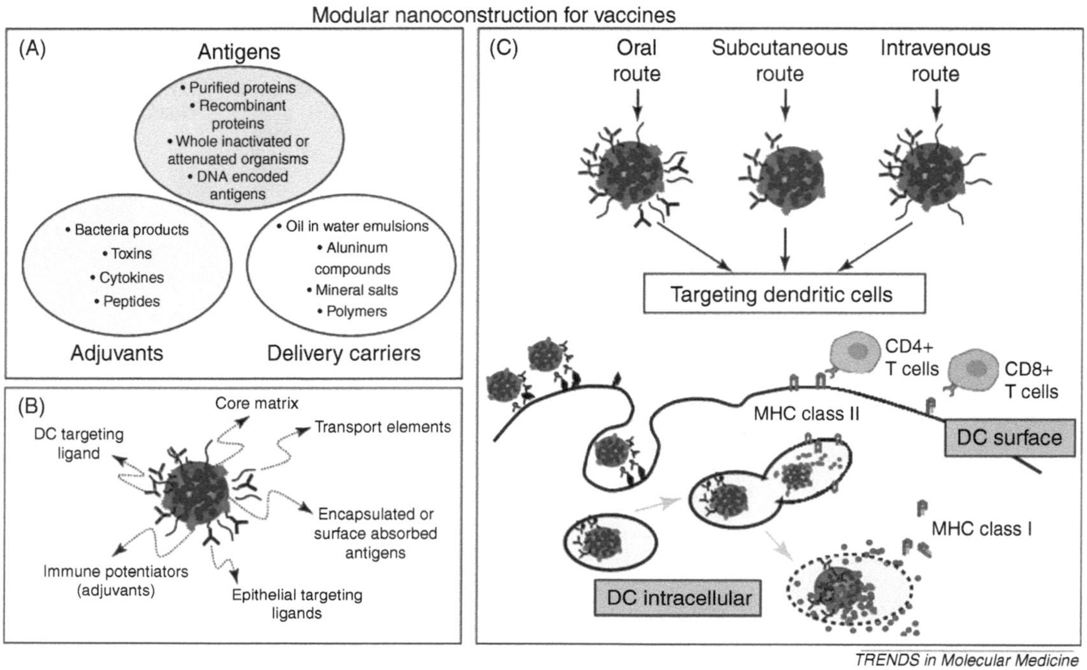

FIGURE 17.3 Basic Components and Modular Nanoconstruction for Vaccines

(A) Major components of vaccines, (B) a schematic of a modular vaccine nanoparticle, (C) schematic of nanoparticle vaccine administration by various routes and mode of DC intracellular antigen delivery.

Reproduced from Metcalfe, S.M., Fahmy, T.M., 2012. Targeted nanotherapy for induction of therapeutic immune responses. Trends Mol. Med. 18(2), 72–80, with permission of Elsevier.

4 DESIGN ASPECTS IN THE NEW GENERATION OF VACCINE DEVELOPMENT

Researchers working on the development of vaccines face an inherent dilemma: maximizing immunogenicity without compromising safety and tolerability. Early vaccines often induced long-lived protective immune responses, but tolerability was a major problem. Newer vaccines have very few side effects but can be of limited immunogenicity. One way to tackle this problem is to design vaccines that have all the properties of pathogens with the exception of causing disease. Key features of pathogens that can be mimicked by vaccine delivery systems are their size, shape, and surface molecule organization (Fig. 17.3). In addition, pathogen-associated molecular patterns can be used to induce innate immune responses that promote adaptive immunity. The factors currently being used to optimize the delivery of antigens and enhance vaccine efficacy are discussed next.

4.1 IMPORTANCE OF SIZE IN VACCINE DESIGN: THE EFFECT ON ANTIGEN UPTAKE AND PROCESSING

The size of antigen used for vaccine formulation differs significantly. Protein and viral subunit antigens are the smallest (<10 nm). These are formulates with adjuvants like alum and Freund's adjuvants to form larger particulates. Virosomes, which resemble small liposomes, are about 100–2000 nm and VIP (virus-imprinted particle) nanoparticles comprising supramolecular particulates are larger in size, ranging from 20–200 nm and largest size in range of 100 nm–20 μm is rendered by antigens presented in the context of microparticles, liposomes, water in oil emulsions (Freund's adjuvants), oil in water emulsions (MF59 adjuvant), mineral salts (alum adjuvants), and whole-cell vaccines. The major effector APCs responsible for induction of T cell responses are DCs and macrophages, both of which are highly specific for antigen uptake and processing. Key steps in the generation of potent immune responses includes uptake of antigens by DCs, trafficking of the DCs to the lymph nodes, and triggering of DC maturation, which are required to be assessed for the diverse vaccine delivery systems. Several antigen-associated properties, such as size, shape, surface charge, hydrophobicity and hydrophilicity, as well as receptor interactions influence the uptake of antigens by APCs (Ahsan et al., 2002). Ideal for targeted delivery of antigens, smaller particles are considered more promising and efficient to larger ones because smaller particles are more efficient in permeating biological barriers, passing through capillaries after injection, and achieving stability in blood circulation (Monfardini and Veronese, 1998). Thus, for vaccine delivery, nanoparticles with a diameter of 100 nm or less are preferred (Kim and Nie, 2005; Mumper et al., 2003). Subsequent to the uptake by APCs, antigens will arrive at the endolysosomal compartments. Here, they are degraded into peptides, which are laden onto MHC class II molecules and then transferred to the cell surface for activation of CD4+ T helper cells. Cross-presentation can further allow the exposure of antigen to MHC class I molecules leading to stimulation of CD8+ T cell responses, an important feature for therapeutic vaccines against cancer and chronic viral infections (Yewdell, 2010). Although soluble antigens are incapable of cross-presentation, particulate carriers/antigens can be prepared in the size range similar to or larger than viruses that get loaded on MHC I without intracellular replication. Processing of particulate antigens that are 20 nm–3 μm in diameter has been documented, and there is no clear indication that there are preferred sizes within this range.

It is, therefore, possible that APCs have evolved to effectively process any antigen with dimensions that are similar to pathogens, ranging from viruses (20–100 nm) to bacteria and even cells (in the micrometer range). So generating particulate vaccines may enhance the uptake, processing, and presentation of the antigens by professional APCs (Bachmann et al., 2010).

4.2 EFFECT OF SIZE ON ANTIGEN TRANSPORT

Vaccines are delivered through oral, subcutaneous, or intramuscular route. Secondary lymphoid organs are principle effectors of adaptive immune response, so it is essential that the injected antigen should reach the lymphoid organs from the site of delivery. The tissue fluid balance in the lymphatic system is maintained by permitting the movement of fluids and serum components that leaked from blood capillaries into the interstitial space back to the blood. Peripheral immune cells and pathogens also enter into the secondary lymphoid organ by this pathway. Soluble antigens are capable of entering the lymph capillaries impulsively and then move to lymph nodes where DCs reside. Spontaneous drainage of particulate vaccine is size limited, and only nanoparticles (<100 nm) can depart the interstitial matrix via the interstitial flow and be transported through the lymphatic vessels into the lymph nodes (Nishioka and Yoshino, 2001; Reddy et al., 2006). In view of the preceding criteria, researchers explored the optimum size range for lymphatic delivery using 25 nm and 100 nm polypropylene sulfide nanoparticles loaded with ovalbumin (OVA). Such particulate vaccines were injected intradermally, and they found that 25 nm particles reached the lymph nodes with a 10-fold higher efficiency than the 100 nm particles. Intradermal injection of the 25 nm particles in mice elicited humoral and cell-mediated immunity with the anti-OVA IgG levels being similar to the titers achieved with an intradermally administered OVA alum formulation, whereas the antibody titers elicited with the 100 nm particles were significantly lower (Reddy et al., 2007). Particulate in the size range of 20–200 nm can enter the lymphatic vessels efficiently with an optimal size being ~40 nm (Swartz, 2001). The distal initial lymphatic vessels have gaps and clefts between endothelial cells and hence, are partly open structures; by contrast the more complex proximal vessels are essentially impermeable to proteins that are more than 70 kDa (corresponding to a dynamic radius of 5−6 nm). Consequently, both cells and macromolecules that are smaller than 100−200 nm enter the lymphatic system at the distal sites of the lymph vessels. Nevertheless, the aforementioned approach for passive targeting is attractive and feasible because most of the nanoparticulates are too big for spontaneous passage to the lymph node. Particles that are larger than 200−500 nm do not efficiently enter lymph capillaries in a free form (Reddy et al., 2007). Instead, large particles need to be carried into the lymphatic system by specialized cells, such as DCs, which can squeeze through openings between overlapping endothelial cells. The influence of particle size on the type, level, and quality of the immune response may be ascribed to differences in pathways and mechanisms for cell uptake and for antigen presentation and processing. The publication by Xiang et al. (2006) is a good starting point for initial discussion, where it was reported that virus-sized particles in the size range of 20 –200 nm are usually taken up by endocytosis, resulting in a cellular-based immune response. It was also explained that particles with sizes between 500 nm and 5 μmare mainly taken up by phagocytosis and/or macro-pinocytosis and are more likely to promote a humoral immune response.

4.3 ABILITY OF PARTICULATE TO TARGET ANTIGENS TO DCs

The objective of vaccination with any formulation is to emulate the innate and adaptive responses of the immune system to infection (Bachmann and Jennings, 2010). The predominant interface between the innate and adaptive immune responses is antigen-presenting cells, and particularly dendritic cells (O'Hagan, 1997; Storni et al., 2005). Immature DCs present in the periphery capture the foreign antigen by endocytic internalization mechanism, which is a size-dependent phenomenon. Smaller particles of viral dimensions (20–200 nm) are taken up by receptor-mediated processes, whereas larger ones (>0.5 mm) are phagocytosed (receptor-ligand interaction) or taken up by macropinocytosis (a receptor-independent uptake mechanism) (Brewer et al., 2004). DCs expresses a variety of receptors on its their surfaces like mannose receptors, NOD-like receptors (NLRs), retinoic acid-inducible gene (RIG)-I-like receptors (RLRs) and C-type lectin receptors (CLRs), and number of toll toll-like receptors (TLR), which are utilized as one of the efficient strategy for facilitating specific uptake of nanoparticulate vaccine by DCs. The ligand specific to each of these receptors can be couples coupled to antigen or nanoparticulate carriers. Delivery of particulate vaccine directly to DCs enhances the immune response; for example, subcutaneously injected targeted liposomes induce stronger humoral and CTL responses than nontargeted liposomes (Tacken et al., 2006). However, most particulate vaccines are probably too large to diffuse freely through the peripheral tissue matrix, which results in inefficient targeting. Despite the many interesting reports on targeted antigen delivery, the potential of targeted particulate vaccine delivery has to be further explored and validated. Targeted antigen delivery might enhance not just efficacy but also the specificity of interaction between antigen-loaded particles and the surface receptors of DCs.

4.4 EFFECT OF SURFACE PROPERTIES

Surface properties of the carriers also serve an important factor that influences the interaction with immune cells and subsequently the propensity of immune response generated. It has been found that cationic charged nanoparticles are phagocytosed more efficiently than the negative nanoparticles (Roser et al., 1998; Tabata and Ikada, 1990). Similarly positive liposomes were phagocyotosed more effectively than negatively charge ones and induced higher CTL response in mice than negative and neutral liposomes (Nakanishi et al., 1999). Cationic entities showed higher uptake by APCs because of electrostatic interaction with negatively charged surface of APCs. However, charge itself is not the governing factor for phagocytosis since studies have suggested that microspheres having similar size but different charge were taken up to the comparable extent by the macrophages (O'Hagan et al., 1997). It reveals that, in addition to particle size and surface charge, chemical and physical properties of material used for nanoparticle formation also affect the phagocytosis mechanism of uptake and intracellular fate of antigen. Hydrophilic microparticles are engulfed by APCs by sink-into-the cell-like mechanism, which do not involve the formation of pseudopod and result in loose contact with the phagosomal membrane, whereas the uptake of positive and hydrophobic nanoparticulate results from formation of pseudopod because of the high affinity of the particle surface for the cell membrane, such as latex bead and lipid vesicles (Swanson and Baer, 1995). Kinetics of APCs maturation depends on the interaction of particles surface and phagosome membrane, which decides the intracellular fate of the antigen. Phagosomes vacuole matures by kiss-and-run hypothesis after the particles uptake. Subsequently acidification of phagosome and activation of proteolytic enzymes take place. This process is highly significant because the generation of epitope for further antigen processing depends on phagosomes maturation. Hydrophilic and negatively charged nanoparticles allow faster acidification due to rapid interaction of phagosomes with lysosomes (dechastellier and Thilo, 1997) leading to lysosomal degradation of the contents. The engulfment of hydrophobic and positive particles within closed membrane vesicles obstrucst maturation and interaction with lysosome, resulting in minor acidification resulting in lysosomal escape. This difference in processing with respect to surface properties suggests the view that the generation of TH_1 or TH_2 response depends on the nature of the nanoparticle vaccine. Lysosmal degradation pathway will lead to TH_2 mediated humoral immune response, and lysosomal escape will render cross-presentation and generation of TH_2 mediated cellular response (cytotoxic cell mediated), which is more desirable for therapeutic or cancer vaccines.

4.5 EFFECT OF PARTICLE SHAPE

Shape of particles is another essential parameter and has an important role in mitigating cellular responses. Infection efficacy and the subsequent replication of natural pathogens, like versatile virus or bacterial, depend on their shape. Numerous investigations have reported the particle shape plays a significant role in antigen uptake. Shape of particles plays a significant role in the interaction between antigen-presenting cells (APCs) and nanoparticulate-carrier systems. In case of larger particles (>1 μm), particle shape plays a dominant role in phagocytosis by macrophages because the internalization of particles is strongly dependent on the local geometry at the point of cell attachment, the interface between APCs and particles. Wormlike particles exhibited negligible phagocytosis compared to spherical particles (Champion and Mitragotri, 2006) but remained in the blood for a longer period of time than spherical particles. In another study, researchers investigated that spherical gold nanoparticles (AuNPs) were more efficient in eliciting antibody-mediated immune response and secretion of pro-inflammatory cytokines than other shapes (cube and rod) or the 20 nm-sized AuNPs, even though the rods (40 nm × 10 nm) were more efficiently taken up by APCs than the cubic and spherical AuNPs. Rod-shaped particles are quickly internalized by immune cells suggesting their resemblance to cell-infecting capability (Niikura et al., 2013).

5 NANOCARRIER(S) PROPERTIES THAT MAKE SUITABLE CARGO FOR VACCINE DELIVERY

Over the past few decades, development of nanovaccines using nanotechnology is a rapidly growing research area of nanomedicine and has gained much interest. *Nanovaccines* refers to nano-scale-based delivery systems in which desired antigen(s)/DNA/proteins/peptides have been attached or encapsulated. Nanovaccines have revealed the ability to enhance the body's immune system to kill infective organisms and thus prevent infections from spreading. However, to date, no effective vaccine has come out for the prevention of numerous lethal diseases, like tuberculosis, malaria, HIV, anthrax, hepatitis C, and so forth. It is clear that some rationale approaches as well as improved methodologies to boost the normal functions of the immune system against these infectious pathogens are required to combat consequent diseases. Nanovaccines can have a major role to play in alleviating the aforementioned diseases and offer several potential advantages including the enhanced bioavailability of antigens, site-specific delivery of antigens, and a reduced side-effect profile over the traditional/conventional vaccines (Zolnik et al., 2010). Today, the development of nanoparticulate carriers mediated vaccines has been highly sought worldwide owing to their reported ability to induce effective immunization through enhanced targeting and by triggering desirable immune responses (Fig. 17.4) (Nandedkar, 2009). They have been intensively investigated as delivery systems and adjuvants for development of nanovaccine and for effective therapeutic treatment of a number of infectious diseases and cancer (Dobrovolskaia et al., 2008; Hubbell et al., 2009; Kasturi et al., 2011; Moon et al., 2011). The nanoparticulate carriers such as nanoparticles, liposomes, niosomes, emulsomes, virosomes, SLNs, micelles, and so forth have been studied as vaccine delivery systems. Many parameters directly or indirectly govern the ability of the nanocarrier system to deliver the antigen to the APCS by modulating the rate of antigen release profile and nanocarriers uptake by APCs or M cells. Particle size, potential, hydrophobicity/ hydrophilicity, crystallinity of polymers, stability, bioadhesiveness, and the nature of additives are the factors that govern the performance of nanocarrier systems. The salient features that make the nanocarriers ideal for the delivery of vaccines are antigen protection, transport across the cell, and delivery of antigen(s) to APCs. The molecular aspects of designing nanocarriers together with the knowledge of the mechanism of immunity generation, receptors involved in internalization of specific antigen, and molecules involved in the process are taken into account to ensure effective generation of immune response against the antigen (Amorij et al., 2012; Leo et al., 2011). Although the nanocarriers are closer in size to the pathogenic size of particles, they are selectively taken up by the APCs/ M cells and are transcytosed to underlaying lymphoid tissue. Their interaction and processing in the lymphoid tissue ultimately depends on their nature and composition.

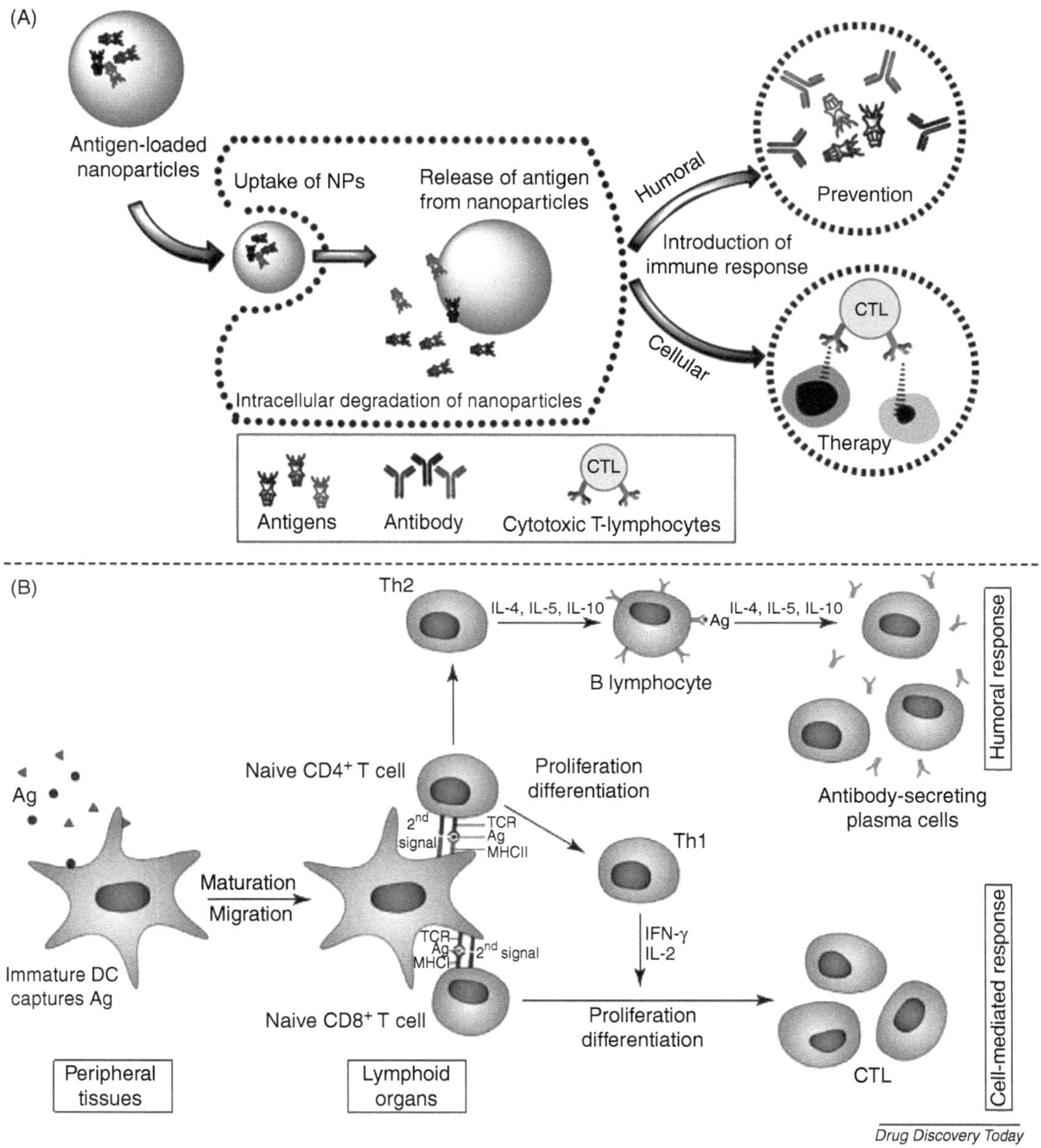

FIGURE 17.4

Schematic overview of (A) induction of immune responses by nanoparticle-based vaccine and (B) antigen presentation and Th priming. Immature DCs screen the peripheral tissues; upon encounter of DCs with antigen and PAMPs, migration to the lymph nodes and maturation is induced. In the lymphoid organs, Ag presentation via MHC class II and class I molecules accompanied by a second signal leads to activation, proliferation, and differentiation of respectively Ag-specific CD4+ and CD8+ T-cells. Activation of CD4+ helper T-cells results in secretion of cytokines and subsequent activation of CD8+ T-cells or B lymphocytes. The humoral branch of the immune response comprises activation of B lymphocytes followed by differentiation into antibody secreting plasma cells. Whereas the cellular response involves activation of CD8+ T-cells followed by proliferation and differentiation into cytotoxic T-cells.

Reproduced from De Temmerman M.L., Rejman, J., Demeester, J., Irvine, D.J., Gander, B., De Smedt, S.C., 2011. Particulate vaccines: on the quest for optimal delivery and immune response. Drug Discov Today. 16(13–14), 569–582; Sharma R., Agrawal U., Mody N., Vyas S.P., 2015. Polymer nanotechnology based approaches in mucosal vaccine delivery: challenges and opportunities. Biotechnol. Adv. 33(1), 64–79, with permission of Elsevier.

6 NANOPARTICULATE CARRIERS AS A VERSATILE PLATFORM FOR VACCINE DELIVERY

Over the past few decades, significant efforts have been made for the development of a new generation of vaccines using nanotechnology-based approaches. Nanovaccines consist of nanoscale-based particles formulated or attached with components to which an immune response is desired. Nanovaccines offer numerous potential advantages including enhanced bioavailability and site-specific delivery of immune components/antigens. The development of nanocarrier-based vaccines have been highly sought after owing to their reported ability to induce effective and desirable immune responses through selective antigen delivery of APCs (Nandedkar, 2009; Zolnik et al., 2010).

Some other advantages of a nanocarrier-based delivery system over a conventional system include the following:

- Owing to their nanosize, these carriers are better taken up by the APCs, thereby facilitating the antigen presentation and processing by APCs.
- Their versatile nature permits all types of immune components including antigens, proteins, peptides, nucleic acids, carbohydrates, and small molecules, such as haptens, to be encapsulated or loaded in nanocarriers with desirable modifications.
- They have the ability to generate higher titers of functional antibodies as well as to extend the duration of these antibodies.
- Apart from working as carriers, nanoparticle vaccines also work as an adjuvant to provoke the immune response and thus produce a larger response as compared to conventional vaccines.
- They are easily coformulated and combined with immunomodulators, such as pattern recognition receptor (PRR) agonists and Toll-like receptor (TLR) agonists or others.
- They have the ability to modulate the immune system.
- They have the flexibility to couple with site-specific ligands to achieve active targeting.
- They provide a unique opportunity to deliver pharmaceuticals into cells or even inside individual cellular compartments.

With these enormous benefits, in the past few years various nanoparticulate delivery systems such as nanoparticles, liposomes, noisomes, virosomes, carbon nanotubes, dendrimers, and so forth have emerged as potential and attractive candidates for delivery of immune components, and they have been intensively investigated as an immunopotentiator/adjuvant for vaccine development. Particularly, attention is given to the progress for rational design of nanocarriers, which have been receiving great interest and hold promise to generate a novel platform for single-dose vaccination and needle-free administration (O'Hagan et al., 2006; Zhao et al., 2014). Table 17.1 summarizes the various types of nanoparticulate carriers currently being studied in vaccine delivery.

7 NEW GENERATION VACCINES

Immunization is the process of boosting up the body's immune system to fight the foreign invaders, mainly disease-causing infectious agents. Conventional vaccines generally contain risk-free microbes either killed or attenuated microbes or parts of microbes that are nonpathogenic but is capable of stimulating an immune response against the causative organism. This vaccine-induced memory enables the immune system

Table 17.1 Summary of the Various Types of Nanocarriers Being Investigated for Their Use as Vaccine Delivery Systems

Nanocarriers	Type	Comments	Composition	Antigen	References
Polymer based	Polymeric nanoparticles	Nanosized particulate carriers which either encapsulate or entrap antigen(s), proteins, and DNA, and so on; biodegradable, biocompatible, and less toxic; strong adjuvant property, robust structural stability, and sustained and controlled release behavior; induce both cellular and humoral immune responses	Natural and synthetic. *Natural:* polysaccharide, inulin, pullulan, chitosan, dextran, gelatin, alginate, and so on *Synthetic:* poly(D,L-lactide-*co*-glycolide) (PLGA), poly(D,L-lactide) (PLA), poly(ε-caprolactone) (PCL), polystyrene, poly(g-glutamic acid) (g-PGA), poly(ethylene glycol) (PEG), poly(methyl methacrylate), polyethylenimine (PEI), poly (cyanoacrylates) (PCA), and so on	Hepatitis B surface antigen; synthetic bovine *parainfluenza* virus type-3 (BPI3V), ovalbumin (OVA), BSA tumor associated Ags, and so on	Kammona and Kiparissides (2012); Mansoor et al. (2014); Thomas et al. (2011); Sharma et al. (2015); Cevher et al. (2015); Gamazo et al. (2015)
	Dendrimer	Synthetic, well-defined hyper-branched structure; increase the specificity of vaccines; not induce adverse host responses, including inflammatory reactions; exhibit standalone adjuvanticity	Polyamidoamine (PAMAM), polypropylene imine (PPI), polyacrylate	Streptococcus pyogenes, GAS, tumor antigens, hepatitis B surface antigen, influenza, and so on	Heegaard et al. (2010); Skwarczynski et al. (2010); Dutta et al. (2008); Wright (1998)
Lipid based	Liposomes	Spherical vesicles consisting of one or more phospholipid bilayers surrounding aqueous core; elicitation of both humoral (antibody mediated) and cellular (cytotoxic T lymphocytes) immune response; biocompatible, biodegradable, safe and well tolerated; highly versatile nature; able to modulate the immune system	Phospholipids and cholesterol cationic lipids; DOTAP: 1,2-Dioleoyloxy-3-trimethylammonium propane chloride; DOTMA: 1,2-Dioleyloxy-3-trimethylammonium propane chloride. Others: poly(ethylene glycol)–distearoylphosphatidylethanolamine (DSPE-PEG), α-galactosylceramide, lysophosphatidylcholine, and so on	Hepatitis B surface antigen, influenza H1N1, diphtheria, tetanus, *P. falciparum* circumsporozoite protein, vibrio cholera, tumor-associated antigens (TAAs), and so on	Tiwari et al. (2009a); Paliwal et al. (2015); Tiwari et al. (2011); Gupta and Vyas (2011); Tiwari et al. (2010); Horiuchi et al. (2015); Liu et al. (2014); Tiwari et al. (2009b)

(*Continued*)

Table 17.1 Summary of the Various Types of Nanocarriers Being Investigated for Their Use as Vaccine Delivery Systems (*cont.*)

Nanocarriers	Type	Comments	Composition	Antigen	References
	Niosomes	Nonionic surfactant-based closed bilayered vesicles composed of nonionic amphiphiles; greater physical and chemical stability, more advantageous than liposome; higher flexibility to the vesicular structure; significant augmentation in immune response	Hydrated nonionic surfactant (i.e., Tween and Span) with cholesterol or its derivatives	Merozoite surface protein 1, MSP1, hepatitis B surface antigen, influenza H1N1, diphtheria, tetanus, and so on	Jain et al. (2006); Vyas et al. (2005); Jain et al. (2005b); Uchegbua and Vyas (1998)
	Bilosomes	Bile salt stabilized vesicles stabilized against the detrimental effect of bile acids; stability and potential as an oral delivery system; enormous potential as an effective carrier system for oral immunization	Phospholipids, cholesterol, nonionic surfactant with bile salts	*Clostridium tetani*, influenza hemaglutinin, tetanus toxoid, and so on	Singh et al. (2004); Shukla et al. (2008); Shukla et al. (2010); Conacher and Alexander (2001)
	Virosome	Virus-like particles consisting of unilamellar phospholipids bilayered vesicle with reconstructed virus envelopes (viral genetic material absent); size 150–200 (nm) with greater affinity for mucosal surface receptors; as adjuvant stimulate a more balanced T helper 1 versus T helper 2 response and eliciting more potent T helper 1 immune response	Phospholipids, viral surface glycoproteins	Influenza hemaglutinin; tetanus toxoid (TTx), ovalbumin, and so on	Mishra et al. (2006); Huckriede et al. (2003); Bungener et al. (2005); Bungener, et al. (2000); Arkema et al. (2000)
	Emulsome	Lipid-cored carriers stabilized with high percentage of soya lecithin and cholesterol; structural resemblance with the chylomicrons (natural lipoprotein of the body) and so expected to mimic these lipoproteins in behavior	Triglycerides, fatty acids, soya lecithin, cholesterol, α-tocopherol, butylated hydroxytouline (BHT), oleic acid or negatively charged phospholipids, such as phosphatidic acid, phosphatidylinositol, and phosphatidylsterine, spans	HBsAg, cholera toxin B subunit (CTB), HIV-1451(o-gp160451), and so on	Gill et al. (2012); Paliwal et al. (2009); Lowell et al. (1997); VanCott et al. (1998)

	Transferosome	A special type of lipid bilayered deformable/elastic vesicles consisting of phosphatidyl choline and an edge activator; highly stress responsive, adaptable, complex aggregate system; shape and composition of the bilayer makes the vesicle both self-optimizing and self- regulating; specially designed carriers system for noninvasive immunization	Phospholipids, surfactant, alcohol, buffering agent, and so on	Hepatitis B surface antigen (HBsAg), tetanus toxoid (TTx), and so on	Rajan et al. (2011); Xu et al. (2008); Gupta and Vyas (2004); Gupta et al. (2005)
	Solid lipid nanoparticles (SLNs)	Can be exploited as an effective nonviral transfection agent; act as a strong adjuvant; high monodispersity, long temporal stability, and relatively good stability	Solid lipids (e.g., triglycerides, fatty acids, etc.)	Foot and mouth disease virus (FMDV), visceral leishmaniasis, Hepatitis B, Plasmid DNA, and so on	Chen et al. (2014); Saljoughian et al. (2013); Doroud et al. (2010); Jin and Kim (2014)
Lipid-polymer hybrid nanoparticles		Nanosized particulate carriers with unique advantages of both polymeric nanoparticles and liposomes; novel, strong, and versatile nature; act as strong adjuvant/ immunopotentiator	Polymers (PLGA, PCL, PLA, and so on) and lipids (lecithin, DSPE, DOTMA, and so on)	*Chlamydia trachomatis* (CT), Mycobacterium tuberculosis (MTB), and Group B *Streptococcus* (GBS), ovalbumin, keyhole limpet hemocyanin (KLH), tumor-associated antigens, and so on	Rose et al. (2015); Hu et al. (2014); Cho et al. (2011); Silva et al. (2014)

to act quickly to protect the body if attacked by the same microbes in the future. Nowadays, number of disease associated antigens has emerged out as potential targets for immune therapy. These antigens are used to generate immunization in the manner similar to that of attenuated or killed microbes, and vaccines carrying such type of antigens are termed as a new generation or next generation vaccines. Next generation vaccine therapy aims to combine the best antigen with optimal delivery strategy in order to develop most effective immunotherapy for the management of disease. The challenge faced in developing a new generation vaccines is to achieve strong immunogenicity without increasing reactogenicity. In developing countries, where the infectious disease mortality and morbidity burden remains high, a different risk-benefit ratio prevails. In such venues, generally mild untoward effects and serious but rare adverse events attributable to vaccines (such as vaccine-associated paralytic poliomyelitis) are considered an acceptable price for the prevention of death and debilitating disease for the masses (Koff et al., 2013; Riedmann, 2011).

Here are some promising and proven strategies for the development of new generation vaccine:

- Conjugate vaccines.
- Rational attenuation of known pathogens by inactivation of specific genes.
- Bacterial live-vector vaccines.
- Viral live-vector vaccines.
- Subunit vaccines.
- Reverse vaccinology (genomics-based vaccines).
- Nonliving antigen delivery system based on such things as proteosomes, virosomes, virus-like particles, liposomes, and microspheres.
- DNA vaccines and replicons.
- Heterologous prime-boost vaccination strategies.
- Well tolerated adjuvants-based vaccines.
- Needle-free vaccines.

These advances in vaccine delivery systems are providing a new horizon for the development of efficient vaccine, which can be scale up from bench side to market (Vyas and Dixit, 1998; Vyas and Kumar, 2010). Some of the potential new generation vaccines are listed as follows:

1. Cancer vaccines
2. Malaria vaccines
3. Hepatitis B
4. Anthrax vaccines

7.1 CANCER VACCINES

Cancer vaccines are intended to boost up the body's natural ability to protect itself, through the immune system, from risk posed by abnormal or damaged cells, as in case of cancer. Cancer vaccine belongs to the class of biological response modifiers in the medical field. Biological response modifiers are those medicines/therapeutic agents that stimulate or restore the ability of the immune system to fight against diseases and infections. These can be broadly classified as:

- *Prophylactic or preventive vaccines*, which are aimed to prevent the development of cancer in healthy people. Cancer preventive vaccines target infectious agents that cause or contribute to the development of cancer (Frazer et al., 2007). They are similar to traditional vaccines, which help prevent infectious diseases, such as measles or polio, by protecting the body against infection.

Both cancer-preventive vaccines and traditional vaccines are based on antigens that are carried by infectious agents and that are relatively easy for the immune system to recognize as foreign.

- *Therapeutic or Treatment vaccines*, which are aimed to strengthen the natural defense of body against the cancer and thus help in treating an existing cancer (Lollini et al., 2006). The success of therapeutic vaccines largely depends on the ability of the immune system to circumvent immune suppression produced as a result of age or tumor or therapy. Also, the outgrowth of tumor cells can affect the efficiency of these vaccines. An additional factor that influences the effectiveness of therapeutic vaccines will be the outgrowth of tumor cells that, for one reason or another, can evade the immune response.

7.1.1 Challenges for cancer vaccines

Although several approaches are under continuous exploration for development of successful cancer vaccines, there are some major issues that should be addressed to increase their effectiveness and acceptability. These challenges include:

- An aging immune system
- Host-derived immunosuppressive effects
- Tumor escape from immune effector function

7.1.2 Cancer vaccine strategies

Numerous strategies have been developed to formulate cancer vaccines. These are illustrated in Fig. 17.5 and are briefly discussed in the following section. Cancer vaccines basically activate the

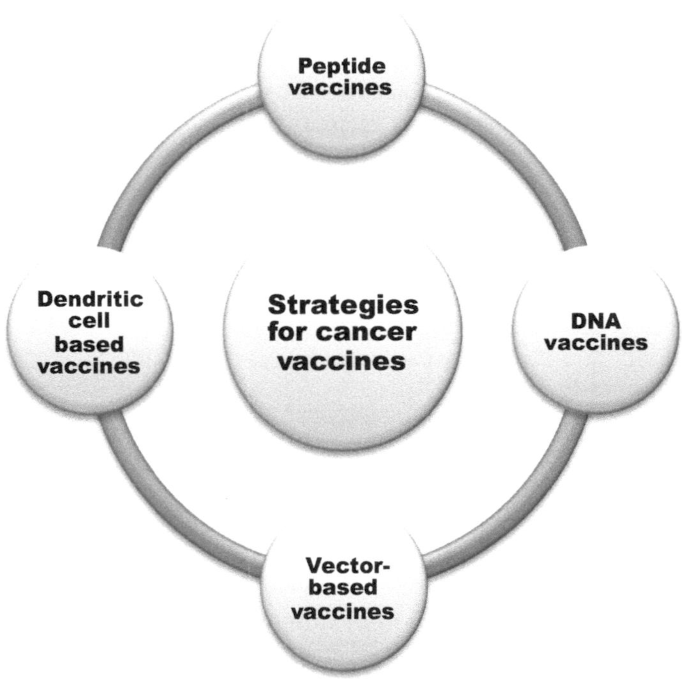

FIGURE 17.5 Various Strategies of Cancer Vaccine Development

soldiers of our immune system – that is, macrophages, natural killer cells, cytotoxic T cells, and DCs to operate against the cancer cells. Macrophages and DCs perform the specialized function of antigen presentation and thus are professional antigen presenting cells (APCs) of the immune system. Macrophages and DCs process the foreign substance and present the antigen derived from them onto their surface, and that is why they are called APCs. Such APCs then interact with the helper and cytotoxic T cells followed by their activation and multiplication. These activated T cells further activate other phagocytes (inactive macrophages and NK cells) and T cells, which can now directly target and destroy the tumor mass phagocytosis or by cellular toxins. Apart from T cells, B lymphocytes also play a role in fighting the tumor cells. They produce antibodies specific to the tumor cells, which mark the cancer cells for demolition and support the NK cells and macrophages. Additionally an immune memory is also generated in order to evade reformation of the same tumor (Dhake, 2013; Kumar et al., 2010).

7.1.3 Nano approaches for cancer immunotherapy

It is well established that nanoparticulate carriers improve the efficacy of bioactive(s), and many such nanocarriers-based therapeutic approaches have been successfully investigated to date. In cancer immunotherapy, the primary payload of nanocarriers will usually be tumor-associated antigens (TAAs), tumor-specific antigens (TSAs), peptides, or DNA encoding, and so forth. From an immunotherapeutic perspective, it would be desirable to develop novel carriers that can specifically target professional APCs via either active or passive targeting and deliver an immunomodulatory payload in a bioactive manner, resulting in the generation of a strong tumor-specific cytotoxic CD8+ T lymphocyte (CTL) response (Krishnamachari et al., 2011). The targeting and subsequent activation of professional APCs (dendritic cells, macrophages, etc.) is of particular importance because professional APCs are potent initiators of immune responses. In this manner, to attain a potent antitumor CTL immune response, it is essential to activate DCs to generate a pro-inflammatory response (Th1-type), interleukin-12 (IL-12), and gamma interferon (IFN-γ). In addition, nanoparticulate carriers can be designed to target the tumor itself, resulting in a site-specific accumulation of the TAAs and/or adjuvants and provide a controlled release for development of long-term antigenic memory (Fig. 17.6) (Zhang et al., 2007). Nanoparticulate carriers have the potential to improve tumor immunotherapy by addressing all the aforementioned goals due to their targetability, specificity, and physicochemical properties. In the last five years, various engineered nanomaterial based approaches are being developed as a versatile and efficient mediator of cancer immunotherapy/vaccine (Almeida et al., 2014; Cho et al., 2011; Matsuo et al., 2010; Xiang et al., 2015).

7.2 HEPATITIS B

Hepatitis is a chronic necro-inflammatory liver disease caused by hepatitis B virus. The inflammation can further exaggerate liver cancer or liver cirrhosis. Existing vaccines of hepatitis B are composed of hepatitis B surface antigen, one of the proteins present in the viral envelope, which provide protection against the infection by developing a humoral immune response. The drawback of such a vaccine was failure to develop the cellular immune response and serosal immune response. Cellular response is desirable for virus elimination from the infected hepatocytes, and serosal response is required for the protection of mucosal surfaces. The limitations of existing vaccine against hepatitis B are addressed using a number of approaches including mucosal immunization, the use of adjuvants, and development of surface functionalized novel antigen carriers (Paliwall et al., 2010). Nowadays, oral route of

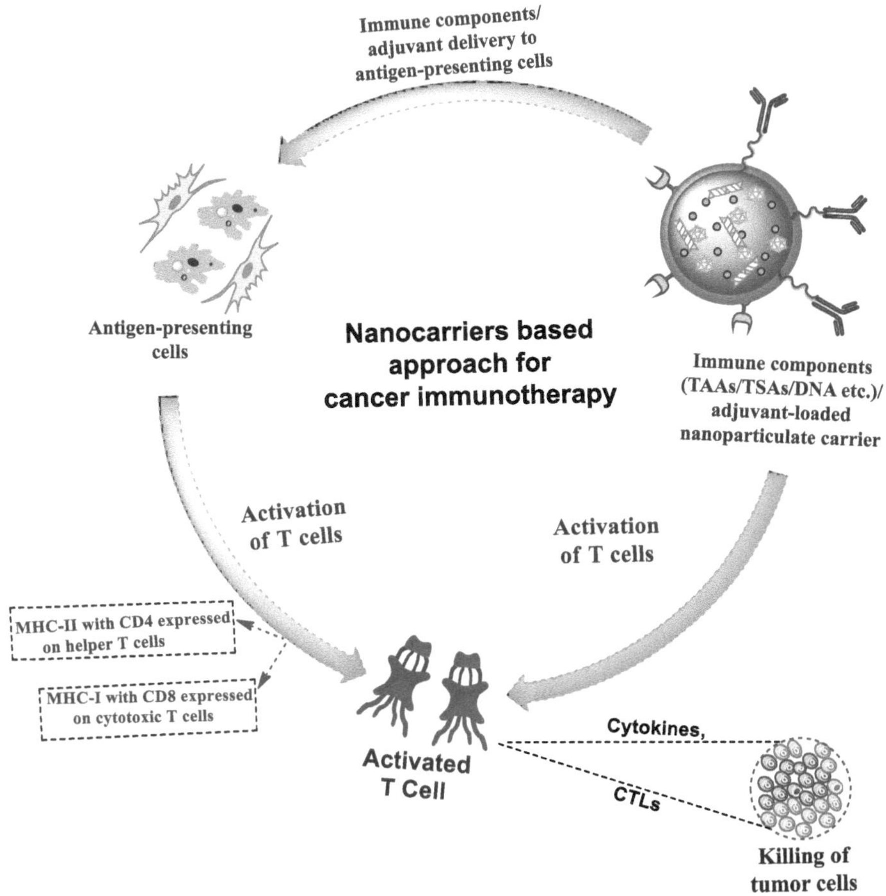

FIGURE 17.6 Schematic Overview of Nanocarriers-Based Approach for Cancer Therapy

immunization is seeking much attraction of researchers owing to its capability of inducing humoral as well as systemic immunity, which is not possible with parenteral vaccines. Nevertheless, in case of oral route, there are possibilities of antigen degradation because of the hostile gastrointentinal tract environment. As a result, higher payload and dosing frequency are required, and this could result in antigen tolerance. To overcome such problems, carrier mediated antigen delivery systems were formulated with the aim of antigen protection and subsequently improved immunization. Shivani et al. have synthesized bilosomes, bile salt stabilized vesicular carriers that are capable of transmucosal uptake followed by the generation of immune response against the entrapped antigen. In this study, HBsAg was used as model antigen and bilosomes were formed by a lipid cast film method. Fluorescence study revealed that the bilosomes were taken up by gut-associated lymphoid tissue (GALT). Anti-HBsAg IgG levels were measured and it was found that the bilosomes were capable of producing a comparable response as compared to that of intramuscular administration of alum adsorbed HBsAg. Additionally, bilosomes have elicited measurable sIgA in mucosal secretions, which is absent in the case of intramuscular administration of alum adsorbed HBsAg. The authors concluded that both systemic as well as mucosal immunity was generated upon oral administration of HbsAg loaded bilosomes (Shukla et al., 2008a,b). Poly lactic acid (PLA) is one of the extensively used biodegradable polymers used in vaccine delivery but it possesses the drawback of being hydrophilic and acidic in nature, which restricts its use as an antigen-delivering carrier system. Author research group has synthesized block copolymer of PLA and PEG in three different compositions: AB, ABA, and BAB, where A is PLA and B is PEG, to overcome these drawbacks. HBsAg-loaded nanoparticles were formulated and characterized against hepatitis B. The immunogenicity elicited by plain PLA nanoparticles, nanoparticles of block polymer, and conventional alum HBsAg-based vaccine were compared, and results suggested that BAB nanoparticles produced superior humoral response as compared to other polymeric nanoparticles. Block copolymeric nanoparticles also produced better sI_gA levels at all local and distal mucosal sites as compared with PLA nanoparticles, where alum-based formulations failed to give any considerable response. These data demonstrate the potential of BAB nanoparticles as mucosal vaccine delivery system capable of eliciting high and prolonged immune response (Jain et al., 2009).

7.3 MALARIA VACCINE

Malaria is one of the most widespread diseases with high death rates throughout the world. The causative organism of malaria is the parasite *Plasmodium* and four species of it—*falciparum, vivax, ovale,* and *malariae*—of which *falciparum* is the most common. Although chemotherapeutics is trying to control the disease, the cases of drug resistance are increasing day by day thereby raising the need for the development of successful vaccine against malaria. Efforts have been made for identification and testing a number of antigens present on the surface of parasites. Still, formulating a successful vaccine is a big challenge for immunologists. The low immunogenicity of antigenic candidates, toxicity of immunological adjuvant, and the life cycle of the protozoan parasite further add to the complexity of the situation, since it involves two hosts—human and mosquito (Tiwari et al., 2009b). Although the human host appears only to acquire natural immunity to blood stages, development of malaria vaccine is based on three life cycle stages: first, the sporozoite and liver stage (*preerythrocytic vaccines*), second, the asexual blood stage (*blood stage vaccines*), and finally the gametocyte/gamete stage (*transmission blocking vaccines*). Indeed, since the number of sexual stages and preerythrocytic stages are so low, they constitute bottlenecks or population choke points in the life cycle and are thus, conceptually, excellent targets for vaccine design (Fig. 17.7). Exploration in the field of carrier-mediated vaccine delivery suggests that as compared to the soluble form, antigens in particulate form can elicit higher immune responses and thus nanocarriers of optimum size can serve as good platforms for the delivery of antigens and adjuvants. They can be used to deliver protein antigens, plasmid DNA- and RNA-based vaccines, and subunit vaccines (Khatri et al., 2008). It is reported that the APCs assists the migration of antigen encapsulated nanocarrier to the lymph node where the antigen is exposed to the cells of immune system to trigger the immune response. The efficiency of the microparticles used for antigen delivery can be further enhanced by designing the molecular topology in a manner similar to so-called pathogen-associated molecular patterns (PAMPs). PAMPs correspond to small molecular sequences that are consistently found on pathogens. PAMPS are recognized by APCs by toll-like receptors and other pattern-recognition receptors (Ketloy et al., 2008). Peptidoglycans lipospolysaccharide, nucleic acid variants normally associated with bacteria or viruses, such as DNA rich in cytosine guanosine (CpG) and double-stranded RNA, respectively, and lipoteichoic acid from gram-positive bacteria are a

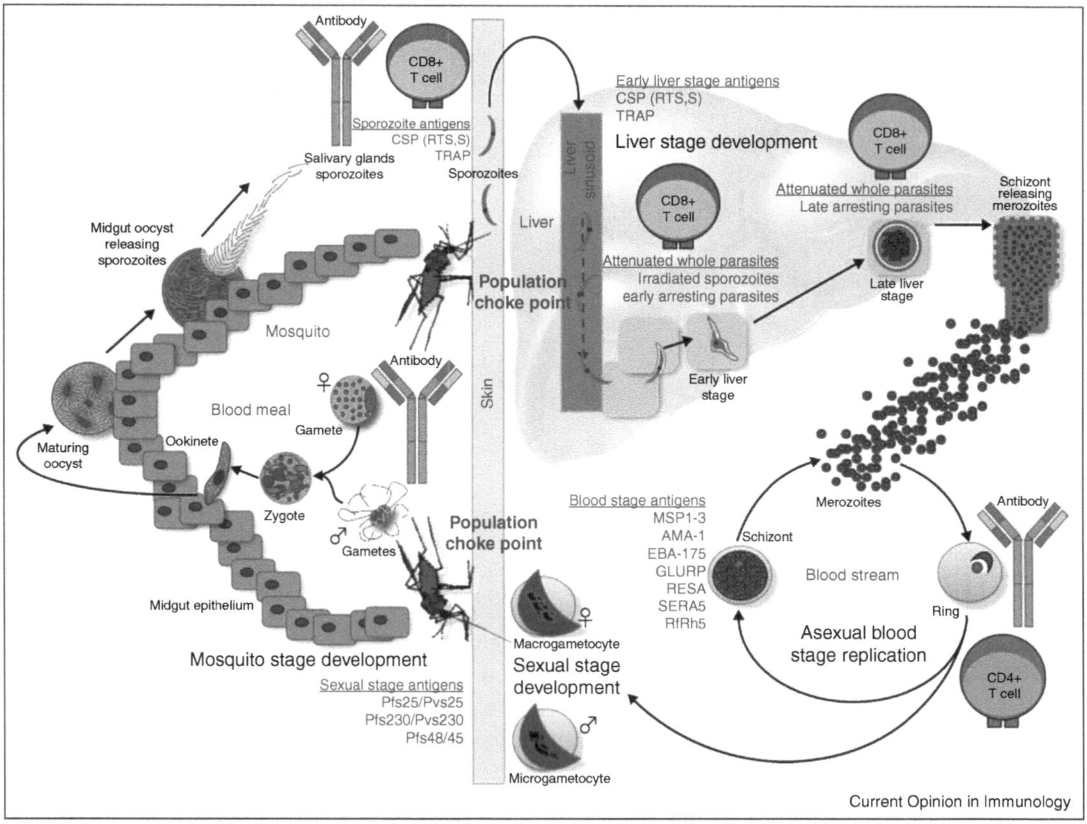

FIGURE 17.7 The Malaria Life Cycle, Parasites, and Parasite Targets Used for Vaccine Development

Reproduced from Vaughan, A.M., Kappe, S.H., 2012. Malaria vaccine development: persistent challenges 24(3), 324–331, with permission of Elsevier.

few example of PAMPs [11,12]. CpG causes stimulation of B cells followed by their proliferation and propagation; synthesis of nonspecific immunoglobulin; and activation of monocytes, macrophages, and dendritic cells to generate high levels of a number of cytokines (Gupta and Vyas, 2011). Once the cytokine level is increased, NK cells are activated, which secrets gamma interferon (IFN-γ) and shows enhanced activity of cell lysis (Chen et al., 2014; Gupta and Vyas, 2004; Haana den et al., 2014). With these immunostimulatory characteristics, CpG has emerged as a new class of adjuvant for immune enhancement of synthetic peptide and recombinant protein-based vaccines. Tiwari et al. (2009a,b) have developed an adjuvant-based transmission-blocking malaria vaccine loaded with Pfs25 protein antigen in gel core liposomes for controlled delivery of antigen to cells of the immune system (M1).

7.4 ANTHRAX VACCINE

Anthrax is a highly virulent zoonotic disease caused by the organism *Bacillus anthracis,* which produces two binary toxins: edema toxin and lethal toxin (LT). It invades the host through mucosal (respiratory,

Table 17.2 Potential Anthrax Vaccine Strategies

PA-Based Vaccination Strategies	Alternative Vaccination Strategies	Passive Antibody Therapy
First generation (AVA)	Bacilli capsular antigens	Anti-PA
2nd generation (PA102)	Attenuated whole spores	Anti-LF
3rd generation (alternative rPA-based)	Spore antigens	Anticapsule
Modified rPA	BclA	Other
Alternative adjuvants	BxpB	
Needle-free delivery	Other	
Intradermal injection		
Transcutaneous administration		
Mucosal administration		
Intranasal delivery		
Oral delivery (attenuated bacterial vector)		
Live, attenuated bacilli injection		
DNA vaccination		
Viral vector		
Plasmid injection		

Reproduced from Cybulski, J.Jr., Sanz, P., O'Brien, A.D., 2009. Anthrax vaccination strategies. Mol. Aspects Med. 30(6), 490–502, with permission from Elsevier.

gastrointestinal, and cutaneous) routes. Anthrax is a significant public health concern because of its potential as a bioterrorist and biowarfare agent. Biothrax (AVA) is the only available anthrax vaccine licensed for human use in the United States. Currently available anthrax vaccine has some drawbacks such as generating a protective immune response in a slow manner, requiring a lengthy vaccination schedule and multiple immunizations, and it has considerable side effects (Pittman et al., 2002; Turnbull, 2002). Therefore, there is a critical need for fabricating an effective and safe new-generation anthrax vaccine for protecting humans from disease. Recently, several nanotechnology-based vaccine delivery vehicles that could improve delivery of antigen and adjuvant(s) to APCs and elicit significant cellular, humoral, and mucosal immune response are under investigation. Additionally, there is a need for a robust adjuvant, to decrease the required antigen dose for vaccination and to improve the immunogenicity of a vaccine. (Bielinska et al., 2007; Mangal et al., 2014). Table 17.2 summarizes the various novel approaches for the development of new generation vaccine against *B. anthracis*.

8 CONCLUSIONS

In the area of immunology, nanoparticulate-carrier mediated approaches have shown great potential for the development of new-generation vaccines. Recently, various engineered nanocarriers have emerged as a revolutionary approach for selective/ targeted delivery of immunomodulatory payloads to the predefined site of action. These carriers are also capable of activating APCs, especially dendritic cells because of their active and passive uptake by DCs. The evidence quoted in the chapter revels that nanocarriers have promise for the development of effective vaccine against HIV, malaria, influenza HBV, anthrax, and cancer due to their potential for activating immune cells. Although enormous positive beneficial properties of nanocarriers, these are extensively the most admired choice for nanovaccines. Although extensive research have been carried out on nanoparticulate systems for vaccine development, only a few nanoparticulate vaccine formulations have been approved by the US Federal Drug Administration and European Medicines Agency for vaccination. Large numbers of natural and synthetic polymers are used for the fabrication of nanocarriers, each of which can be tailored as per the purpose of vaccine delivery. Such systems can also be fabricated to release vaccine based on the environment surrounding or external stimulus (stimuli responsive) like pH, temperature, presence of lytic enzymes, and so forth. Hence, nanoparticulate carriers offer a versatile and reliable platform for new generation vaccine development.

REFERENCES

Ahsan, F., Rivas, I.P., Khan, M.A., Torres Suarez, A.I., 2002. Targeting to macrophages: role of physicochemical properties of particulate carrier's liposomes and microspheres on the phagocytosis by macrophages. J. Control. Rel. 79, 29–40.

Almeida, J.P., Figueroa, E.R., Drezek, R.A., 2014. Gold nanoparticle mediated cancer immunotherapy. Nanomed. 10 (3), 503–514.

Amorij, J.P., Kersten, G.F.A., Saluja, V., et al., 2012. Towards tailored vaccine delivery: needs, challenges and perspectives. J.Control. Rel. 161, 363–376.

Apostolopoulos, V., Thalhammer, T., Tzakos, A.G., et al., 2013. Targeting antigens to dendritic cell receptors for vaccine development. J. Drug Deliv., 419–425.

Arkema, A., Huckriede, A., Schoen, P., Wilschut, J., Daemen, T., 2000. Induction of cytotoxic T lymphocyte activity by fusion-active peptide-containing virosomes. Vaccine 18 (14), 1327–1333.

Bachmann, M.F., Jennings, G.T., 2010. Vaccine delivery: a matter of size, geometry, kinetics and molecular patterns. Natures Rev. 10, 787–793.

Bielinska, A.U., Janczak, K.W., Landers, J.J., Makidon, P., Sower, L.E., Peterson, J.W., Baker, J.R., 2007. Mucosal immunization with a novel nanoemulsion-based recombinant anthrax protective antigen vaccine protects against Bacillus anthracis spore challenge. Infect Immun. 75 (8), 4020–4029.

Borges, O., Borchard, G., Verhoef, J.C., et al., 2005. Preparation of coated nanoparticles for a new mucosal vaccine delivery system. Int. J. Pharm. 299, 155–166.

Brewer, J.M., et al., 2004. Vesicle size influences the trafficking, processing, and presentation of antigens in lipid vesicles. J. Immun. 173, 6143–6150.

Bungener, L., Serre, K., Bijl, L., Leserman, L., Wilschut, J., Daemen, T., Machy, P., 2000. Virosome-mediated delivery of protein antigens to dendritic cells. Vaccine 20 (17–18), 2287–2295.

Bungener, L., Huckriede, A., de Mare, A., de Vries-Idema, J., Wilschut, J., Daemen, T., 2005. Virosome-mediated delivery of protein antigens in vivo efficient induction of class I MHC-restricted cytotoxic T lymphocyte activity. Vaccine 23 (10), 1232–1241.

Cevher, E., Salomon, S.K., Somavarapu, S., Brocchini, S., Alpar, H.O., 2015. Development of chitosan-pullulan composite nanoparticles for nasal delivery of vaccines: in vivo studies. J Microencapsul. 32 (8), 769–783.

Champion, J.A., Mitragotri, S., 2006. Role of target geometry in phagocytosis. Proc. Nat. Acad. Sci. USA 103, 4930–4934.

Chen, X., Jensen, P.E., 2008. MHC class II antigen presentation and immunological abnormalities due to deficiency of MHC class II and its associated genes. Exp. and Mol. Pathol. 85, 40–44.

Chen, G., Zeng, S., Jia, H., He, X., Fang, Y., Jing, Z., Cai, X., 2014. Adjuvant effect enhancement of porcine interleukin-2 packaged into solid lipid nanoparticles. Res. Vet. Sci. 96 (1), 62–68.

Cho, N.H., Cheong, T.C., Min, J.H., Wu, J.H., Lee, S.J., Kim, D., Yang, J.S., Kim, S., Kim, Y.K., Seong, S.Y., 2011. A multifunctional core-shell nanoparticle for dendritic cell-based cancer immunotherapy. Nat. Nanotech. 6 (10), 675–682.

Conacher, M.J., Alexander, J.M., 2001. Brewer Oral immunisation with peptide and protein antigens by formulation in lipid vesicles incorporating bile salts (bilosomes). Vaccine 19 (20–22), 2965–2974.

dechastellier, C., Thilo, L., 1997. Phagosome maturation and fusion with lysosomes in relation to surface property and size of the phagocytic particle. Euro. J. Cell. Biol. 74, 49–62.

Demento, S.L., Eisenbarth, S.C., Foellmer, H.G., Platt, C., Caplan, M.J., Mark Saltzman, W., Mellman, I., Ledizet, M., Fikrig, E., Flavell, R.A., Fahmy, T.M., 2009. Inflammasome-activating nanoparticles as modular systems for optimizing vaccine efficacy. Vaccine 27, 3013–3027.

Dhake K. 2013. Strategies for cancer vaccine development. buzzle.com, 1–8.

Dobrovolskaia, M.A., Aggarwal, P., Hall, J.B., McNeil, S.E., 2008. Preclinical studies to understand nanoparticle interaction with the immune system and its potential effects on nanoparticle biodistribution. Mol. Pharm. 5 (4), 487–495.

Doroud, D., Vatanara, A., Zahedifard, F., Gholami, E., Vahabpour, R., Rouholamini Najafabadi, A., Rafati, S., 2010. Cationic solid lipid nanoparticles loaded by cysteine proteinase genes as a novel anti-leishmaniasis DNA vaccine delivery system characterization and in vitro evaluations. J. Pharm. Pharm. Sci. 13 (3), 320–335.

Dutta, T., Garg, M., Jain, N.K., 2008. Poly(propyleneimine) dendrimer and dendrosome mediated genetic immunization against hepatitis B. Vaccine 26 (27–28), 3389–3394.

Frazer, I.H., Lowy, D.R., Schiller, J.T., 2007. Prevention of cancer through immunization: prospects and challenges for the 21st century. Eur. J. Immunol. 37 (Suppl. 1), S148–S155, 2007.

Gamazo, C., Bussmann, H., Giemsa, S., Camacho, A.I., Unsihuay, D., Martín-Arbella, N., Irache, J.M., 2015. Interactions of poly (anhydride) nanoparticles with macrophages in light of their vaccine adjuvant properties. Int. J. Pharm. 496 (2), 922–930.

Gill, B., Singh, J., Sharma, V., Hari Kumar, S.L., 2012. Emulsomes: an emerging vesicular drug delivery system. Asian J. Pharm. 6, 87–94.

Gupta, P.N., Vyas, S.P., 2004. Transfersomes for vaccine delivery; a potential approach for topical immunization. Med.Chem. Res. 13, 414–426, 6–7.

Gupta, P.N., Vyas, S.P., 2011. Investigation of lectinized liposomes as M-cell targeted carrier-adjuvant for mucosal immunization. Colloids Surf. B Biointerfaces 82 (1), 118–125.

Gupta, P.N., Mishra, V., Rawat, A., Dubey, P., Mahor, S., Jain, S., Chatterji, D.P., Vyas, S.P., 2005. Non-invasive vaccine delivery in transfersomes, niosomes and liposomes: a comparative study. Int. J. Pharm. 293 (1–2), 73–82.

Haana den, M.M., Arens, R., Menno, C., van Zelmc, 2014. The activation of the adaptive immune system: cross-talk between antigen-presenting cells, T cells and B cells. Immun. Lett. 162, 103–112.

Heath, W.R., Carbone, F.R., 2001. Cross-presentation, dendritic cells, tolerance and immunity. Annu. Rev. Immunol. 19, 47–56.

Heegaard, P.M., Boas, U., Sorensen, N.S., 2010. Dendrimers for vaccine and immunostimulatory uses. A review. Bioconjug. Chem. 17 (3), 405–418, 21.

Horiuchi, Y., Takagi, A., Uchida, T., Akatsuka, T., 2015. Targeting cryptic epitope with modified antigen coupled to the surface of liposomes induces strong antitumor CD8 T-cell immune responses in vivo. Oncol. Rep. 34 (6), 2827–2836.

Hu, Y., Ehrich, M., Fuhrman, K., Zhang, C., 2014. In vitro performance of lipid-PLGA hybrid nanoparticles as an antigen delivery system lipid composition matters. Nanoscale Res. Lett. 9 (1), 434.

Hubbell, J.A., Thomas, S.N., Swartz, M.A., 2009. Materials engineering for immunomodulation. Nature 462 (7272), 449–460.

Huckriede, A., Bungener, L., Daemen, T., Wilschut, J., 2003. Influenza virosomes in vaccine development. Methods Enzymol. 373, 74–91.

Jain, S., Singh, P., Mishra, V., Vyas, S.P., 2005a. Mannosylated niosomes as adjuvant-carrier system for oral genetic immunization against hepatitis B. Immun. Lett. 101 (1), 41–49.

Jain, S., Yap, W.T., Irvine, D.J., 2005b. Synthesis of protein-loaded hydrogel particles in an aqueous two-phase system for coincident antigen and CpG oligonucleotide delivery to antigen-presenting cells. Biomacromolecules 6 (5), 2590–2600.

Jain, S., Sharma, R.K., Vyas, S.P., 2006. Chitosan nanoparticles encapsulated vesicular systems for oral immunization preparation in-vitro and in-vivo characterization. J. Pharm. Pharmacol. 58 (3), 303–310.

Jain, A.K., Goyal, A.K., Gupta, P.N., Khatri, K., Mishra, N., Mehta, A., Mangal, S., Vyas, S.P., 2009. Synthesis, characterization and evaluation of novel triblock copolymer based nanoparticles forvaccine delivery against hepatitis B. J. Control Rel. 136 (2), 161–169.

Jin, S.E., Kim, C.K., 2014. Charge-mediated topical delivery of plasmid DNA with cationic lipid nanoparticles to the skin. Colloids Surf B Biointerf. 116, 582–590.

Kammona, O., Kiparissides, C., 2012. Recent advances in nanocarrier-based mucosal delivery of biomolecules. J. Control Rel. 161 (3), 781–794.

Kasturi, S.P., Skountzou, I., Albrecht, R.A., Koutsonanos, D., Hua, T., Nakaya, H.I., Ravindran, R., Stewart, S., Alam, M., Kwissa, M., Villinger, F., Murthy, N., Steel, J., Jacob, J., Hogan, R.J., García-Sastre, A., Compans, R., Pulendran, B., 2011. Programming the magnitude and persistence of antibody responses with innate immunity. Nature. 470 (7335), 543–547.

Ketloy, C., Engering, A., Srichairatanakul, U., Limsalakpetch, A., Yongvanitchit, K., Pichyangkul, S., Ruxrungtham, K., 2008. Expression and function of Toll-like receptors on dendritic cells and other antigen presenting cells from non-human primates. Vet. Immunol. Immunopathol. 125, 18–30.

Khatri, K., Goyal, A.K., Gupta, P.N., Mishra, N., Mehta, A., Vyas, S.P., 2008. Surface modified liposomes for nasal delivery of DNA vaccine. Vaccine 26, 2225–2233.

Kim, G.J., Nie, S., 2005. Targeted cancer nanotherapy. Mater. Today 8, 28–33.

Koff, W.C., Burton, D.R., Johnson, P.R., Walker, B.D., King, C.R., Nabel, G.J., Ahmed, R., Bhan, M.K., Plotkin, S.A., 2013. Accelerating next-generation vaccine development for global disease prevention. Sci. 340 (6136), 1232910, 2013.

Krishnamachari, Y., Geary, S.M., Lemke, C.D., Salem, A.K., 2011. Nanoparticle delivery systems in cancer vaccines. Pharm. Res. 28 (2), 215–236.

Kumar, V.P., Prasanthi, S., Lakshmi, V.R.S., Sai Santosh, M.V., 2010. Cancer vaccines: a promising role in cancer therapy. Acad. J. Cancer Res. 3 (2), 16–21.

Kurts, C., 2000. Cross-presentation: inducing CD8 T cell immunity and tolerance. J. Mol. Med. 78, 326–332.

Leo, O., Cunningham, A., Stern, P.L., 2011. Vaccine immunology, understanding modern vaccines, Persp. Vaccinol. 1(1), 25–59.

Lepenies, B., Lee, J., Sonkaria, S., 2013. Targeting C-type lectin receptors with multivalent carbohydrate ligands. Adv. Drug Deliv. Rev. 65, 1271–1281.

Liu, J., Wu, J., Wang, B., Zeng, S., Qi, F., Lu, C., Kimura, Y., Liu, B., 2014. Oral vaccination with a liposome-encapsulated influenza DNA vaccine protects mice against respiratory challenge infection. J. Med. Virol. 86 (5), 886–894.

Lollini, P.L., Cavallo, F., Nanni, P., Forni, G., 2006. Vaccines for tumour prevention. Nature Rev. Cancer 6 (3), 204–216.

Lowell, G.H., Kaminski, R.W., VanCott, T.C., Slike, B., Kersey, K., Zawoznik, E., Loomis-Price, L., Smith, G., Redfield, R.R., Amselem, S., Birx, D.L., 1997. Proteosomes; emulsomes., and cholera toxin B improve nasal immunogenicity of human immunodeficiency virus gp160 in mice induction of serum intestinal vaginal and lung IgA and IgG. J. Infect. Dis. 175 (2), 292–301.

Mangal, S., Pawar, D., Agrawal, U., Jain, A.K., Vyas, S.P., 2014. Evaluation of mucoadhesive carrier adjuvant, toward an oral anthrax vaccine. Artif. Cells Nanomed. Biotechnol. 42 (1), 47–57.

Mansoor, F., Earley, B., Joseph, P., Cassidy, Markey, B., et al., 2014. Intranasal delivery of nanoparticles encapsulating BPI3V proteins induces an early humoral immune response in mice. Res. Vet. Sci. 96 (3), 551–557.

Matsuo, K., Ishii, Y., Matsuo, K., Yoshinaga, T., Akashi, M., Mukai, Y., Yoshioka, Y., Okada, N., Nakagawa, S., 2010. The utility of poly(γ-glutamic acid) nanoparticles as antigen delivery carriers in dendritic cell-based cancer immunotherapy. Biol. Pharm. Bull. 33 (12), 2003–2007.

Mishra, V., Mahor, S., Rawat, A., Dubey, P., Gupta, P.N., Singh, P., Vyas, S.P., 2006. Development of novel fusogenic vesosomes for transcutaneous immunization. Vaccine 24 (27–28), 5559–5570.

Misumi, S., Masuyama, M., Takamune, N., et al., 2009. Targeted delivery of immunogen to primate M cells with tetragalloyl lysine dendrimer. J. Immunol. 182, 6061–6070.

Monfardini, C., Veronese, M., 1998. Stabilization of substances in circulation. Bioconjug. Chem. 9, 418–450.

Moon, J.J., Suh, H., Bershteyn, A., Stephan, M.T., Liu, H., Huang, B., Sohail, M., Luo, S., Um, S.H., Khant, H., Goodwin, J.T., Ramos, J., Chiu, W., Irvine, D.J., 2011. Inter bilayer-cross linked multilamellar vesicles as synthetic vaccines for potent humoral and cellular immune responses. Nat. Mater. 10, 243–251.

Mumper, R.J., Cui, Z., Oyewumi, M.O., 2003. Nanotemplate engineering of cell specific nanoparticles. J. Dispers. Sci. Technol. 24 (3), 569–588.

Nakanishi, T., et al., 1999. Positively charged liposome functions as an efficient immunoadjuvant in inducing cell-mediated immune response to soluble proteins. J. Control. Rel. 61, 233–240.

Nandedkar, T.D., 2009. Nanovaccines, recent developments in vaccination. J. Biosci. 34 (6), 995–1003.

Niikura, K., Matsunaga, T., Suzuki, T., Kobayashi, S., Yamaguchi, H., Orba, Y., et al., 2013. Gold nanoparticles as a vaccine platform: influence of size and shape on immunological responses as a vaccine in vitro and in vivo. ACS Nano 7, 3926–3938.

Nishioka, Y., Yoshino, H., 2001. Lymphatic targeting with nanoparticulate system. Adv. Drug Deliv. Rev. 47, 55–64.

O'Hagan, D.T., Ott, G.S., Van Nest, G., 1997. Recent advances in vaccine adjuvants: the development of MF59 emulsion and polymeric microparticles. Mol. Med. Today 3, 69–75.

O'Hagan, D.T., Singh, M., Ulmer, J.B., 2006. Microparticle-based technologies for vaccines. Methods 40 (1), 10–19.

Oyewumi, M.O., Kumar, A., Cui, Z., 2010. Nano-microparticles as immune adjuvants: correlating particle sizes and the resultant immune responses. Expert Rev. Vaccines 9, 1095–1107.

Paliwal, R., Paliwal, S.R., Mishra, N., Mehta, A., Vyas, S.P., 2009. Engineered chylomicron mimicking carrier emulsome for lymph targeted oral delivery of methotrexate. Int. J. Pharm. 380 (1–2), 181–188.

Paliwal, S.R., Paliwal, R., Vyas, S.P., 2015. A review of mechanistic insight and application of pH-sensitive liposomes in drug delivery. Drug Deliv. 22 (3), 231–242.

Paliwall, R., Paliwal, S.R., Vyas, S.P., 2010. Advances in effective vaccine development against hepatitis B: focus on mucosal vaccine delivery strategies. Therapeut. Deliv. 1 (3), 397–410.

Pittman, P.R., Kim-Ahn, G., Pifat, D.Y., Coonan, K., Gibbs, P., Little, S., Pace-Templeton, J.G., Myers, R., Parker, G.W., Friedlander, A.M., 2002. Anthrax vaccine, immunogenicity and safety of a dose-reduction, route-change comparison study in humans. Vaccine 20, 1412–1420.

Raghuwanshi, D., Mishra, V., Suresh, M.R., et al., 2012. A simple approach for enhanced immune response using engineered dendritic cell targeted nanoparticles. Vaccine 30, 7292–7299.

Rajan, R., Shoma, Jose, Mukund, V.P.B., Vasudevan, Deepa T., 2011. Transferosomes—A vesicular transdermal delivery system for enhanced drug permeation. J. Adv. Pharm. Technol. Res. 2 (3), 138–143.

Ramon, G., 1924. Sur la toxine et surranatoxine diphtheriques. Ann. Inst. Pasteur 38, 1–10.

Reddy, S.T., et al., 2006. In vivo targeting of dendritic cells in lymph nodes with poly(propylene sulfide) nanoparticles. J. Control Rel. 112, 26–34.

Reddy, S.T., et al., 2007. Exploiting lymphatic transport and complement activation in nanoparticle vaccines. Nat. Biotech. 25, 1159–1164.

Rice-Ficht, A.C., Arenas-Gamboa, A.M., Kahl-McDonagh, M.M., Ficht, T.A., 2009. Polymeric particles in vaccine delivery. Curr. Opin. Microbiol. 13 (1), 106–112.

Riedmann, E.M., 2011. Next generation vaccines. Hum. Vaccine 7 (7), 718–722.

Rose, F., Wern, J.E., Ingvarsson, P.T., van de Weert, M., Andersen, P., Follmann, F., Foged, C., 2015. Engineering of a novel adjuvant based on lipid-polymer hybrid nanoparticles; A quality-by-design approach. J. Control Rel. 210, 48–57.

Roser, M., et al., 1998. Surface modified biodegradable albumin nano and microspheres II: effect of surface charges on in vitro phagocytosis and biodistribution in rats. Eur. J. Pharm. Biopharm. 46, 255–263.

Saljoughian, N., Zahedifard, F., Doroud, D., Doustdari, F., Vasei, M., Papadopoulou, B., Rafati, S., 2013. Cationic solid-lipid nanoparticles are as efficient as electroporation in DNA vaccination against visceral leishmaniasis in mice. Parasite Immunol. 35 (12), 397–408.

Sharma, R., Agrawal, U., Mody, N., Vyas, S.P., 2015. Polymer nanotechnology based approaches in mucosal vaccine delivery: challenges and opportunities. Biotechnol. Adv. 33 (1), 64–79.

Shukla, A., Khatri, K., Gupta, P.N., Goyal, A.K., Mehta, A., Vyas, S.P., 2008a. Oral immunization against hepatitis B using bile salt stabilized vesicles (bilosomes). J. Pharm. Pharm. Sci. 11 (1), 59–66.

Shukla, A., Khatri, K., Gupta, P.N., Goyal, A.K., Mehta, A., Vyas, S.P., 2008b. Oral immunization against hepatitis B using bile salt stabilized vesicles (bilosomes). J. Pharm. Pharm. Sci. 11 (1), 58–65.

Shukla, A., Vyas, S.P., Katare, O.P., 2010. M-cell targeted delivery of recombinant hepatitis S surface antigen using cholera toxin B subunit conjugated bilosomes. Int. J. Pharm. 385 (1-2), 47–52.

Silva, J.M., Vandermeulen, G., Oliveira, V.G., Pinto, S.N., Rodrigues, C., Salgado, A., Afonso, C.A., Viana, A.S., Jérôme, C., Silva, L.C., Graca, L., Préat, V., Florindo, H.F., 2014. Development of functionalized nanoparticles for vaccine delivery to dendritic cells; a mechanistic approach. Nanomedicine 9 (17), 2639–2656.

Singh, P., Prabakaran, D., Jain, S., Mishra, V., Jaganathan, K.S., Vyas, S.P., 2004. Cholera toxin B sub-unit conjugated bile salt stabilized vesicles (bilosomes) for oral immunization. Int. J. Pharm. 278 (2), 379–390.

Skwarczynski, M., Zaman, M., Urbani, C.N., Lin, I.C., Jia, Z., Batzloff, M.R., Good, M.F., Monteiro, M.J., Toth, I., 2010. Polyacrylate dendrimer nanoparticles; a self-adjuvanting vaccine delivery system. Angew. Chem. Int. Ed. 49, 5742–5745.

Storni, T., Kündig, T.M., Senti, G., Johansen, P., 2005. Immunity in response to particulate antigen-delivery systems. Adv. Drug Deliv. Rev. 57, 333–355.

Swanson, J.A., Baer, S.C., 1995. Phagocytosis by zippers and triggers. Trends Cell Biol. 5, 89–93.

Swartz, M.A., 2001. The physiology of the lymphatic system. Adv. Drug Deliv. Rev. 50, 3–20.

Tabata, Y., Ikada, Y., 1990. Phagocytosis of polymer microspheres by macrophages. Adv Polym. Sci. 94, 107–141.

Tacken, P.J., et al., 2006. Targeting antigens to dendritic cells in vivo. Immunobiol. 211, 599–608.

Thomas, C., Rawat, A., Hope-Weeks, L., Ahsan, F., 2011. Aerosolized PLGA nanoparticles enhance humoral., mucosal and cytokine responses to hepatitis B vaccine. Mol. Pharm. 8, 405–415.

Tiwari, S., Goyal, A.K., Mishra, N., Vaidya, B., Mehta, A., Dube, D., Vyas, S.P., 2009a. Liposome in situ gelling system. Novel carrier based vaccine adjuvant for intranasal delivery of recombinant protein vaccine. Procedia in Vaccinology 1, 148–163.

Tiwari, S., Goyal, A.K., Mishra, N., Khatri, K., Vaidya, B., Mehta, A., Wu, Y., Vyas, S.P., 2009b. Development and characterization of novel carrier gel core liposomes based transmission blocking malaria vaccine. J. Control Rel. 140 (2), 157–165.

Tiwari, S., Agrawal, G.P., Vyas, S.P., 2010. Molecular basis of the mucosal immune system., from fundamental concepts to advances in liposome-based vaccines. Nanomedicine 5 (10), 1617–1640.

Tiwari, S., Verma, S.K., Agrawal, G.P., Vyas, S.P., 2011. Viral protein complexed liposomes for intranasal delivery of hepatitis B surface antigen. Int. J. Pharm. 413 (1–2), 211–219.

Trombetta, E.S., Mellman, I., 2005. Cell biology of antigen processing in vitro and in vivo. Annu. Rev. Immunol. 23, 975.

Turnbull, P.C., 2002. Current status of immunization against anthrax, old vaccines may be here to stay for a while. Curr. Opin. Infect Dis. 13, 113–120.

Uchegbua, I.F., Vyas, S.P., 1998. Non-ionic surfactant based vesicles (niosomes) in drug delivery. Int. J. Pharm. 172 (1–2), 33–70.

VanCott, T.C., Kaminski, R.W., Mascola, J.R., Kalyanaraman, V.S., Wassef, N.M., Alving, C.R., Ulrich, J.T., Lowell, G.H., Birx, D.L., 1998. HIV-1 neutralizing antibodies in the genital and respiratory tracts of mice intranasally immunized with oligomeric gp160. J. Immunol. 160 (4), 2000–2012.

Vyas, S.P., Dixit, V.K., 1998. Pharmaceutical Biotechnology. CBS Publishers, New Delhi, India.

Vyas, S.P., Kumar, H.D., 2010. Advances in Pharmaceutical Biotechnology. CBS Publishers, New Delhi, India.

Vyas, S.P., Singh, R.P., Jain, S., Mishra, V., Mahor, S., Singh, P., Gupta, P.N., Rawat, A., Dubey, P., 2005. Non-ionic surfactant based vesicles (niosomes) for non-invasive topical genetic immunization against hepatitis B. Int. J. Pharm. 296 (1–2), 80–86.

Wright, C.D. 1998. Adjuvant properties of poly (amidoamine) dendrimers. Google Patents. Available from: http://www.freepatentsonline.com/5795582.html

Xiang, S.D., Scholzen, A., Minigo, G., et al., 2006. Pathogen recognition and development of particulate vaccines: does size matter? Methods 40 (1), 1–9.

Xiang, J., Xu, L., Gong, H., Zhu, W., Wang, C., Xu, J., Feng, L., Cheng, L., Peng, R., Liu, Z., 2015. Antigen-loaded upconversion nanoparticles for dendritic cell stimulation, tracking, and vaccination in dendritic cell-based immunotherapy. ACS Nano. 9 (6), 6401–6411.

Xu, J., Ding, Y., Yang, Y., 2008. Enhancement of mucosal and cellular immune response in mice by vaccination with respiratory syncytial virus DNA encapsulated with transfersome. Viral Immunol. 21 (4), 483–489.

Yewdell, J.W., 2010. Designing CD8+ T cell vaccines: it's not rocket science (yet). Curr. Opin. Immunol. 22, 402–410.

Zhang, X.Q., Dahle, C.E., Weiner, G.J., Salem, A.K., 2007. A comparative study of the antigen-specific immune response induced by codelivery of CpG ODN and antigen using fusion molecules or biodegradable microparticles. J. Pharm. Sci. 96, 3283–3292.

Zhao, L., Seth, A., Wibowo, N., Zhao, C.X., Mitter, N., Yu, C., Middelberg, A.P., 2014. Nanoparticle vaccines. Vaccine 32 (3), 327–337.

Zolnik, B.S., González-Fernández, A., Sadrieh, N., Dobrovolskaia, M.A., 2010. Nanoparticles and the immune system. Endocrinol. 151 (2), 458–465.

FURTHER READING

Bonifaz, L.C., et al., 2004. In vivo targeting of antigens to maturing dendritic cell DEC 205 receptor improves T cell vaccination. J. Exp. Med. 199, 815–824.

Connacher, M.J., Alexander, J.M., 2000. Brewer Niosomes as immunological adjuvants. In: Uchegbu, I.F (Ed.), Synthetic Surfactant Vesicles. International Publishers Distributors Ltd, Singapore, pp. 185–205.

Jain, S., Vyas, S.P., 2006. Mannosylated niosomes as adjuvant-carrier system for oral mucosal immunization. J. Liposome Res. 16, 331–345.

Liau, J.J., Hook, S., Prestidge, C.A., Barnes, T.J., 2015. A lipid based multi-compartmental system liposomes-in-double emulsion for oral vaccine delivery. Eur. J. Pharm. Biopharm. 97 (Pt. A), 15–21.

Mahor, S., Rawat, A., Dubey, P.K., Gupta, P.N., Khatri, K., Goyal, A.K., Vyas, S.P., 2007. Cationic transfersomes based topical genetic vaccine against hepatitis B. Int. J. Pharm. 340 (1–2), 13–19.

Tada, R., Hidaka, A., Iwase, N., Takahashi, S., Yamakita, Y., Iwata, T., Muto, S., Sato, E., Takayama, N., Honjo, E., Kiyono, H., Kunisawa, J., Aramaki, Y., 2015. Intranasal immunization with DOTAP cationic liposomes combined with DC-cholesterol induces potent antigen-specific mucosal and systemic immune responses in mice. PLoS One 10 (10), e0139785.

CHAPTER

18 PATHOGEN-SPECIFIC NUCLEIC ACID APTAMERS AS TARGETING COMPONENTS OF ANTIBIOTIC AND GENE DELIVERY SYSTEMS

Canan Ozyurt, Burhan Bora, Ozge Ugurlu, Serap Evran

Ege University, Bornova-Izmir, Turkey

CHAPTER OUTLINE

1 SELEX FOR SELECTION OF SPECIFIC AND HIGH-AFFINITY APTAMERS

Aptamers are single-stranded DNA (ssDNA) or RNA structures that have the capability to selectively bind to different targets (Darmostuk et al., 2015; Groff et al., 2015). The use of aptamers as molecular recognition elements has become very common in the past two decades. Aptamers have several superiorities over other recognition molecules, such as antibodies that are often used for the purpose of biorecognition. Aptamers can be selected by using in vitro methods, which eliminate the need for

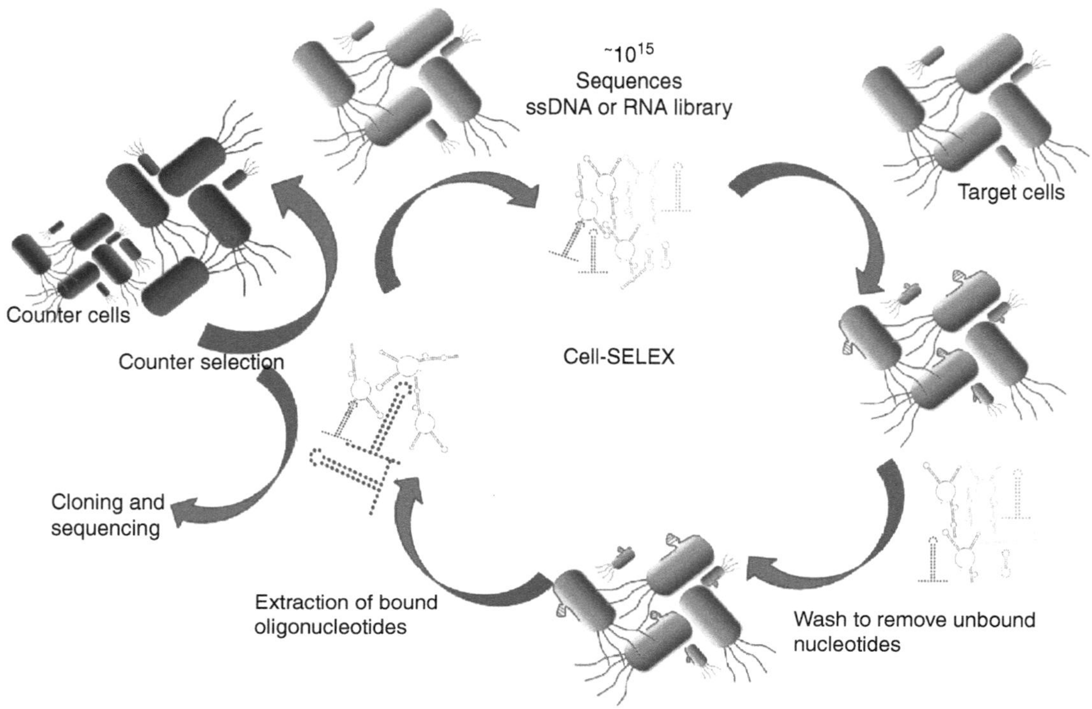

FIGURE 18.1 General Scheme for Cell-SELEX Steps

Counter-SELEX could be included to enhance the specificity of the selected aptamers against target cells.

animal cell culture, or by using in vivo methods. Moreover, aptamers can be chemically modified according to the intended purpose (Mayer, 2009). The ease of modification makes aptamers ideal tools for nanobiotechnological applications, such as drug or gene delivery systems.

Systematic evolution of ligands by exponential enrichment (SELEX) technology was first described by Tuerk and Gold (1990). SELEX technology enables researchers to develop DNA and RNA aptamers for several small molecules, proteins, cells, viruses, and tissues (Mayer, 2016).

In vitro selection of aptamers in order to recognize and capture the whole pathogenic cell basically relies on two possible strategies. According to the first strategy, the whole cell can be targeted during SELEX steps. According to the second strategy, the purified special components of the pathogenic cells, such as lipopolysaccharide layer, outer membrane proteins, glycoproteins, and receptors can be targeted. Protein-SELEX studies are focused on the purified outer membrane proteins (Dua et al., 2011).

A specialized SELEX strategy called whole cell–SELEX is the basic procedure of developing aptamers against viruses, whole cells, and unicellular organisms (Shangguan et al., 2015). According to this approach, the cells are grown in an appropriate growth medium, and then they are incubated with a random RNA or ssDNA oligonucleotide library (Fig. 18.1). The oligonucleotide library includes $\sim 10^{15}$ different sequences with a random region of 20–80 nucleotides. The variable sequences are flanked by the constant sequences, which function as primer binding sites during polymerase chain reaction

(PCR) amplification of the selected aptamer pool. In order to eliminate the nonspecific aptamers and increase the specificity for the target, counter-selection steps can be employed. Typically, cells having similar structural or surface features are used as the counter targets in SELEX rounds. For example, counter-SELEX could be a discriminative solution in order to select DNA or RNA aptamers against only the pathogenic strains of the same organism. Here, we present a brief summary of the cell-SELEX approach, which is frequently used for developing pathogen-specific aptamers.

2 APTAMERS FOR PATHOGENIC CELLS

Several research groups developed specific aptamers for various pathogenic organisms by using the cell-SELEX method. Table 18.1 summarizes some of those published studies, which are focused on developing pathogen-specific DNA or RNA aptamers.

2.1 IN VITRO SELECTION OF APTAMERS AGAINST *ESCHERICHIA COLI* SPECIES

One of the first studies that aimed to develop *Escherichia coli*–specific aptamer was published by So et al. (2008). In that study, nonpathogenic *E. coli* DH5α cells were targeted, and an RNA aptamer was used to develop a single-walled carbon nanotube field-effect transistor (FET) analysis system (So et al., 2008). In 2008, the lipopolysaccharide layer of pathogenic *E. coli* O:111 B4 was targeted by Bruno et al. (2008). In that study, binding properties of the selected DNA aptamers were confirmed by the colorimetric peroxidase assay.

The aptamer for *E. coli* K88 cells was selected by using whole-cell SELEX following 13 successive rounds (Peng et al., 2014). A counter-SELEX round was performed against ETEC K99, *Salmonella enteritidis*, and *Staphylococcus aureus* strains. Binding affinities of the candidate aptamers were determined by flow cytometry. The aptamer 5′-TGG GAG CTC AGA ATA AAC GCT CAA GGC ACA CAG GAC TAT ACA GTG TTG CAG TGT TGC TGT TCG ACA TGA GGC CCG GAT CA-3′ was found to show the highest binding affinity with a K_d value of 15 ± 4 nM. The study involved 11 SELEX rounds, and it was aimed to select the aptamers for fimbria proteins of the *E. coli* K88 strain (Li et al., 2011).

Another study was focused on Förster resonance energy transfer (FRET)-based screening of *E. coli* strain 8739 by employing outer membrane-specific aptamers (Bruno et al., 2010). In vitro selection of DNA aptamers against the purified outer membrane proteins was performed by using magnetic bead-SELEX (MB-SELEX).

The RNA aptamer specific for *E. coli* O157:H7 was selected by using the subtractive cell-SELEX method (Lee et al., 2012). In order to avoid the nonspecific interactions between aptamer and target cells, the RNA library was first incubated with *E. coli* K-12 cells. The RNA sequences bound to *E. coli* K-12 cells were then discarded by centrifugation. The RNA aptamers that did not bind to *E. coli* K-12 cells were subsequently incubated with the target *E. coli* O157:H7 cells. That approach was referred to as subtractive cell-SELEX. Binding affinity parameters were found to be consistent with the quantitative real-time PCR data. The selected RNA aptamers were also evaluated in terms of their binding affinity against the lipopolysaccharide (LPS) layers of *E. coli* O157:H7 and *E. coli* K-12. The aptamer folding into a two-arm structure was shown to bind specifically to the LPS of the *E. coli* O157:H7 strain.

The whole-cell SELEX strategy was used by Kim et al. (2013), and DNA aptamers were selected against the *E. coli* (KCTC 2571) strain that was isolated from fecal matter. In that study, *E. coli* KCTC

Table 18.1 The Published Studies Regarding the Selection of DNA and RNA Aptamers Against Pathogenic Bacteria

Cell type	Aptamer type	K_d	Counter cells used in counter-SELEX steps	References by year
Salmonella enterica serovar *typhi*	RNA	6.08 nM and 8.56 nM	None	Pan et al. (2005)
Mycobacterium tuberculosis	DNA	1–10 μM	*Calmette–Guérin bacillus*	Chen et al. (2007)
Campylobacter jejuni	DNA	292.8 ± 53.1 nM	*Pseudomonas fluorescens* (ATCC13525), *Pseudomonas aeruginosa* (ATCC23993), *Shigella flexneri* (ATCC12022), *E. coli* O157:H7, *Bacillus cereus* (ATCC49063), *Staphylococcus aureus* (ATCC23235), *B. cereus* (strain T), *Listeria monocytogenes* (ATCC19115), *L. monocytogenes* Scott A, *Salmonella enterica* subsp. *Enterica* (ATCC4931), *E. coli* (ATCC33625), *Shigella sonnei* (ATCC25931), *Streptococcus gallolyticus* (ATCC9809), *Enterococcus faecium* (ATCC19434), *Enterococcus faecalis* (ATCC51299), *S. enterica* serovar *enteritidis* (ME 46), and *Salmonella enterica* serovar *typhimurium*	Dwivedi et al. (2010)
Vibrio alginolyticus	DNA	Not determined	None	Jumin (2011)
Streptococcus pyogenes	DNA	9 nM, 10 nM	None	Hamula et al. (2011)
Pseudomonas aeruginosa	DNA	17.27 ± 5.00 nM 57.63 ± 11.64 nM	*S. maltophilia* and *A. baumannii*	Wang et al. (2011)
Salmonella O8	DNA	32.04 nM	*E. coli* and *S. choleraesuis*	Liu et al. (2012)
Vibrio parahaemolyticus	DNA	16.88 ± 1.92 nM	*S. aureus* (ATCC 29213), *L. monocytogenes*, *E. coli* (ATCC 25922), *Cronobacter sakazakii*, *Streptococcus pneumoniae*, and *S. typhimurium*	Duan et al. (2012a)
Escherichia coli O157:H7	RNA	110 nM	*E. coli K12*	Lee et al. (2012)
Listeria monocytogenes	DNA	48.74 ± 3.11 nM	*S. aureus*, *S. typhimurium*, *E. coli*, and *S. pneumoniae*	Duan et al. (2013b)
E. coli	DNA	12.4 to 25.2 nM	*K. pneumonia*, *C. freundii*, *E. aerogenes*, *S. epidermidis*, and *B. subtilis*	Kim et al. (2013)
Salmonella typhimurium	DNA	1.73 ± 0.54 μM	Mixture of *L. monocytogenes* Scott A, *E. coli* O157:H7 (ATCC43895), *B. cereus* (ATCC49063), and *E. faecalis* (ATCC29212)	Dwivedi et al. (2013)
Salmonella typhimurium	DNA	6.33 ± 0.58 nM	*L. monocytogenes*, *E. coli*, *S. aureus*, *S. pneumoniae*, *V. parahemolyticus*, and *C. sakazakii*	Duan et al. (2013c)

Table 18.1 The Published Studies Regarding the Selection of DNA and RNA Aptamers Against Pathogenic Bacteria (*cont.*)

Cell type	Aptamer type	K_d	Counter cells used in counter-SELEX steps	References by year
Salmonella paratyphi	DNA	47 ± 3 nM	*S. enteritidis*, *S. typhimurium*, *S. cholerae-suis*, and *S. arizonae*	Yang et al. (2013)
Vibrio alginolyticus	DNA	27.5 ± 9.2 nM	*V. alginolyticus*, *E. tarda*, *A. hydrophila*, *V. harveyi*	Tang et al. (2013)
Listeria monocytogenes	DNA	35.7 ± 8.02 μM	*E. coli* O157:H7 (ATCC43895), *Bacillus cereus* (ATCC49063), *Salmonella enterica* subsp. *Enterica* serovar *enteritidis* (ATCC13076), *Staphylococcus aureus* (ATCC23235), *Pseudomonas aeruginosa* (ATCC23993), *Shigella flexneri* (ATCC12022, *Brochothrix thermosphacta* (ATCC11509), *Lactococcus lactis* MG1363, and *Listeria strains*	Suh and Jaykus (2013)
Listeria monocytogenes	DNA	48.74 ± 3.11 nM	*S. aureus*, *S. typhimurium*, *E. coli*, and *Streptococcus pneumoniae*	Duan et al. (2013b)
Shigella dysenteriae	DNA	23.47 ± 2.48 nM	*Staphylococcus aureus*, *Salmonella typhimurium*, *Escherichia coli*, *Listeria monocytogenes*, and *Vibrio parahaemolyticus*	Duan et al. (2013a)
Staphylococcus aureus	DNA	35, 129 nM	*Staphylococcus epidermidis*	Chang et al. (2013)
Proteus mirabilis	DNA	7.7 nM, 4.1 nM	*ISM* improved the binding affinity instead of counter-SELEX	Savory et al. (2013)
E.coli K88	DNA	15 ± 4 nM	ETEC K99, *S. enteritidis*, and *S. aureus*	Peng et al. (2014)
E. coli NSM59	DNA	110 nM	*M. morganii*, *P. aeruginosa*, *B. subtilis*, uropathogenic strains of *E. coli*, and *E. coli* laboratory strains DH5α, SM10, and BW20338	Savory et al. (2014)
Salmonella typhimurium enteritidis	DNA	nM to μM	*S. enteritidis*, *S. typhimurium*, *P. aeruginosa*, and *E. coli*	Park et al. (2014)
Listeria monocytogenes	DNA	60.01 nM	*B. cereus* (ATCC49063) and *L. innocua* (ATCC19115)	Liu et al. (2014)
Mycobacterium tuberculosis	DNA	5.09 ± 1.43 nM	*M. bovis*	Mozioglu et al. (2016)
Shigella sonnei	DNA	5.980 ± 0.835 nM and 14.32 ± 2.19 nM	*S. dysenteriae*, *S. exneri*, *S. boydii*, *S. typhimurium*, and *E. coli*	Gong et al. (2015)

K_d *value indicates the dissociation constant. Counter-SELEX targets are also indicated, if the study involves counter-SELEX.*

1681, KCTC 2617, and KCTC 2618 cells, as well as *Klebsiella pneumonia* (KCTC 2208), *Citrobacter freundii* (KCTC 2006), *Enterobacter aerogenes* (KCTC 2190), *Staphylococcus epidermidis* (KCTC 1917), and *Bacillus subtilis* (KCTC 1022) cells were used for the counter-selection steps. Binding properties of the selected aptamers were evaluated by fluorescence measurements (Kim et al., 2013), and K_d values were found to range between 12.4 and 25.2 nM. As the counter-SELEX was applied in that study, the aptamers were able to bind selectively to *E. coli* cells. However, those aptamers were not able to discriminate among *E. coli* strains KCTC 1681, KCTC 2617, and KCTC 2618.

RNA aptamers were selected for three food pathogens (Maeng et al., 2012). Instead of using live cells, cell wall components were used as the targets of SELEX. Specific aptamers for *E. coli* O157:H7 lipopolysaccharide (LPS), teichoic acid from the cell wall of *S. aureus*, and outer membrane protein C (OmpC) of *Salmonella typhimurium* were developed. Then, the selected aptamers were used for monitoring the live cells on a slide system.

Uropathogenic *E. coli* NSM59—specific DNA aptamers were selected in 2014 by Savory et al. Among the six final candidate aptamers, the K_d value of the highest-affinity aptamer was found to be 110 nM. In that study, counter targets were gram-negative uropathogenic *Morganella morganii*, *Pseudomonas aeruginosa*, and Gram-positive nonpathogenic species of *B. subtilis* (Savory et al., 2014).

2.2 IN VITRO SELECTION OF APTAMERS AGAINST *LISTERIA* SPECIES

In 2013, a DNA aptamer was characterized for specific detection of *Listeria monocytogenes* (Duan et al., 2013b). In that whole cell–SELEX study, researchers used 5-carboxyfluorescein (FAM)-labeled aptamers to identify aptamer binding properties. Following eight selection steps, an aptamer with a K_d value of 48.74 ± 3.11 nM was defined.

Internalin A (InlA) is a surface protein that plays an important role in the invasion of *L. monocytogenes*. Ohk et al. (2010) developed the DNA aptamer against InlA protein of *L. monocytogenes* by using the filter plate–based SELEX. According to the fluorescence microscopy results, it was concluded that InlA-specific aptamer A8 was able to interact with the whole *L. monocytogenes* cells, whereas no interaction with *E. coli* or *Salmonella* cells was observed (Ohk et al., 2010).

Whole cell–SELEX approach was used to select aptamers against *Listeria* spp. Binding specificity was evaluated by flow cytometry measurements. The aptamers for *L. monocytogenes* were shown to exhibit nearly the same binding affinity for other members of the *Listeria* genus. However, only a weak binding was observed for the counter non-*Listeria* cells (Suh and Jaykus, 2013). Almost 1 year after that study, Suh et al. (2014a) performed another in vitro selection study for *Listeria* species. The second attempt for developing aptamers differed from the former one in terms of cell-SELEX. The target cells used in SELEX rounds were harvested from different growth stages, which was not a considered a parameter in many other standard cell-SELEX examples.

In vitro selection of aptamers for *L. monocytogenes* was also performed by Liu et al. (2014). *L. monocytogenes* cells were incubated with the random DNA library, and then used for SELEX rounds. Whole-cell SELEX was interrupted after nine rounds, and the ssDNA pool was cloned into *E. coli* cells.

Lee et al. (2015) used the whole cell-SELEX strategy in order to develop specific aptamers for *L. monocytogenes* cells. Binding affinities of the candidate aptamers were determined by surface plasmon resonance (SPR) measurements. It was concluded that nine SELEX rounds were enough for enrichment of the specific aptamers. Following the fourth and sixth rounds, two additional counter-SELEX steps against *B. subtilis* and *E. coli* were included. The aptamer candidates were characterized by aptamer-based sandwich assay (ABSA). Aptamers with the highest binding affinities were referred to as LMCA2 and LMCA26, which had K_d values of 2.01×10^{-12} and 1.56×10^{-10}, respectively.

2.3 IN VITRO SELECTION OF APTAMERS AGAINST *SALMONELLA* SPECIES

Joshi et al. (2009) developed DNA aptamers for the members of very dangerous food-borne pathogens *Salmonella enterica* serovars. In that study, outer membrane proteins (OMP) of *Salmonella* serovars and other *Enterobacteriacae* species were isolated, and then those proteins were used for the in vitro selection of candidate aptamers. For this aim, the purified OMPs of *Salmonella* serovars were incubated with the aptamer library. In order to provide a high specificity, the OMPs of *Enterobacteriacae* species were used as the counter targets during the counter-selection step. Following seven rounds of SELEX, the candidate aptamers were cloned and sequenced. The aptamers were used for magnetic capture assay of *S. typhimurium* from different specimens. In a later study by Duan et al. (2012b), that specific aptamer was also employed in a magnetic nanoparticle-based analysis system for *S. typhimurium*.

Moon et al. (2013) described specific DNA aptamers for *S. typhimurium*. The random DNA library was constructed to include 40 random and 60 constant nucleotides. After 10 rounds of SELEX, 6 rounds of counter-SELEX were performed to enhance the selectivity over *E. coli*, *S. enteritidis*, and *S. aureus*. A fluorescence-activated cell sorting (FACS)–based assay was used to analyze the binding characteristics (Moon et al., 2013).

A specific aptamer for *S. typhimurium* with a K_d value of 6.33 ± 0.58 nM was developed following 9 rounds of cell-SELEX (Duan et al., 2013c). The random DNA library included the sequences with a 40 nucleotide random region, as well as 22 nucleotide and 25 nucleotide primer binding sites. A counter-selection step against potential interfering bacterial cells, such as *L. monocytogenes*, *E. coli*, *S. aureus*, *S. pneumoniae*, *V. parahemolyticus*, and *C. sakazakii*, was also included. On the basis of flow cytometry measurements, binding affinities were determined. The highest binding affinity aptamer, named ST2P, had a K_d value of 6.33 ± 0.58 nM.

In another study, the aim was an RNA aptamer-based detection of *S. enteritidis* (Hyeon et al., 2012). *S. enteritidis* cells were immobilized onto anti-*Salmonella* magnetic beads. Then, the cell-immobilized magnetic beads were incubated with a random RNA library at a certain concentration ratio. According to the results of real-time PCR, the appropriate aptamer candidates were identified. After 10 rounds of SELEX, 22 of the aptamers were sequenced and the structural properties of these motifs were analyzed. The aptamer that exhibited the highest binding affinity was also used in a negative selection step involving other *Salmonella* serovars.

Yang et al. (2013) generated aptamers for a highly dangerous food-borne pathogen, *S. paratyphi* A. The selected aptamers were used to design a single-wall carbon nanotube system for detecting the pathogen of interest. In order to develop aptamers, the cells were grown in an appropriate culture medium, and 5×10^7 of the cells were incubated with the DNA aptamer library containing 40 nucleotide random sequences. Following 13 SELEX rounds and four counter-SELEX rounds, the final aptamer library was obtained. After cloning and sequencing of the candidate aptamers, binding affinity assay was performed by using Fluorescein isothiocyanate (FITC)-labeled DNA. The highest binding affinity aptamer had a K_d value of 47 ± 3 nM. For counter-selection steps, *S. enteritidis*, *S. typhimurium*, *S. cholera*, and *S. arizonae* serovars were used.

2.4 IN VITRO SELECTION OF APTAMERS AGAINST OTHER BACTERIAL CELLS

Campylobacter jejuni cells were targeted by Dwivedi et al. (2010), and an aptamer with a K_d value of 292.8 ± 53.1 nM was selected. In that study, *P. fluorescens* (ATCC13525), *P. aeruginosa* (ATCC23993), *S. flexneri* (ATCC12022), *E. coli* O157:H7, *B. cereus* (ATCC49063), *S. aureus* (ATCC23235), *B. cereus* (strain T), *L. monocytogenes* (ATCC19115), *L. monocytogenes* Scott A, *S. enterica* subsp. *Enterica* (ATCC4931), *E. coli* (ATCC33625), *S. sonnei* (ATCC25931), *S. gallolyticus* (ATCC9809), *E. faecium* (ATCC19434), *E. faecalis* (ATCC51299), *S. enterica* serovar *enteritidis* (ME 46), and *S. enterica* serovar *typhimurium* species were used as counter cells. Ten rounds of cell-SELEX and two rounds of counter-SELEX were performed to develop specific aptamers. Binding affinities of the aptamers were analyzed by flow cytometry. The DNA aptamer named ONS-23 was referred to as the first published aptamer for *C. jejuni*. The first capillary electrophoresis (CE)–based aptamer immunoassay for bacterial cells was carried out by using dead *C. jejuni* cells (Stratis-Cullum et al., 2009).

Cao et al. (2009) generated an aptamer, and a detection panel employing that aptamer was constructed for *S. aureus*. The random ssDNA library was constructed to include the sequences with a 45-nucleotide random region. A subtractive SELEX approach was used, and binding properties were evaluated by radiolabeling experiments. *Streptococcus* sp. and *S. epidermidis* cells were used as the counter targets.

As summarized in Table 18.1, there is a growing interest for developing specific aptamers for pathogenic cells. There are several cell-SELEX studies aiming to develop aptamers for different bacterial strains. Studies on in vitro selection of aptamers for specific pathogenic microorganisms are reviewed elsewhere (Davydova et al., 2015; Hong and Sooter, 2015).

3 NANOPARTICLE-BASED ANTIMICROBIAL GENE DELIVERY

Gene delivery has a great potential for introducing the genetic materials into the target cells, which is a promising tool for treatment of various diseases and disorders, such as autoimmune diseases, cancer, diabetes, and cardiovascular diseases (Dizaj et al., 2014a; Shu et al., 2015). In particular, the aim of gene therapy is replacing a defective or missing gene (Nayerossadat et al., 2012).

There are different types of gene delivery systems that could be employed in gene therapy (Nayerossadat et al., 2012). The vector is defined as the gene delivery vehicle, which is generally used to carry the genetic material into the cells (Johnson-Saliba and Jans, 2001). There are two different kinds of such systems: *viral vectors* and *nonviral vectors* (Robbins and Ghivizzani, 1998; Zhang et al., 2012b).

The success of a gene delivery approach depends on an efficient transfer of genes into the target cell. For that reason, carrier systems are generally used to increase the transfer efficiency and the release of genetic materials inside the cell. These carriers are important tools to transfer foreign genetic materials, such as plasmid DNA (or small nucleic acids), antisense oligonucleotides, and small interfering RNAs. The carrier systems are beneficial with regard to the controlled release of the foreign genetic material and the protection against nucleases. Upon conjugation of the genetic material, in conjunction with endosomal escape, the genetic material could enter into the nucleus (Yin et al., 2014).

Here, we present an overview of viral vectors that are used for the release of antimicrobial genes. We also discuss the potential use of nonviral gene delivery systems to transfer antimicrobial genes in order to enhance host defense mechanisms against infection.

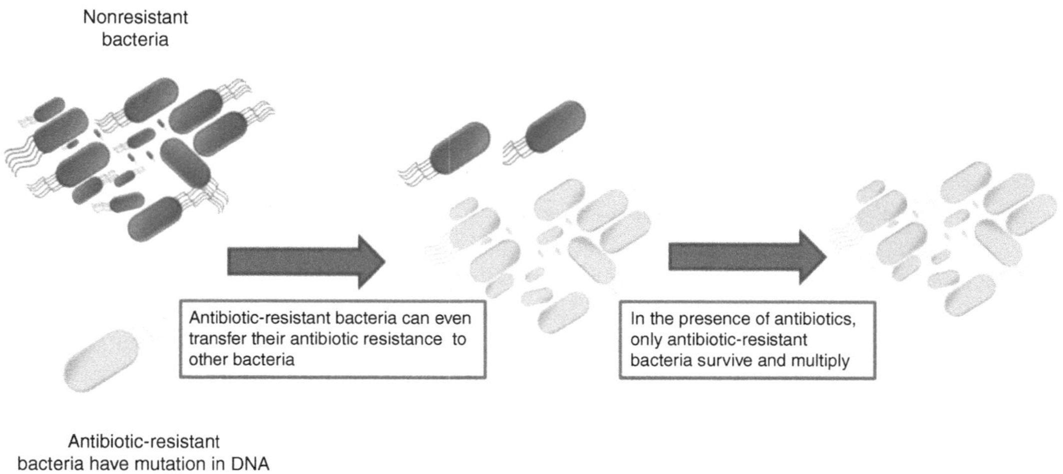

FIGURE 18.2 Antibiotic Overuse Causes Antibiotic Resistance in Pathogenic Bacteria

Genetic material responsible for the antibiotic resistance could be transferred from antibiotic-resistant bacteria to antibiotic-sensitive bacteria through conjugation, transformation, or transduction.

3.1 ANTIMICROBIAL PEPTIDES AND SMALL RNAs

An antibiotic is a kind of antimicrobial reagent that is used for the treatment and prevention of bacterial infection. However, misuse of antibiotics is a major and growing problem, which frequently results in development of antimicrobial drug resistance in pathogenic bacteria. Antibiotic-resistant bacteria could emerge because of de novo mutation or horizontal gene transfer of resistance factors among bacteria (Fig. 18.2). In addition, antibiotic use causes selective pressure for the emergence and amplification of antibiotic-resistant bacteria (Andersson and Hughes, 2010; Tenover, 2006).

From a historical viewpoint, the first antibiotic, penicillin, was discovered by Alexander Fleming in 1928 (Fleming, 1929). Since then, novel types of antibiotics were introduced to struggle with pathogens. Antibiotics are still the most commonly used drugs for treatment of various bacterial infections (Andersson and Hughes, 2010; Lu et al., 2004). However, compared to the rate of emergence of antibiotic-resistant bacteria, the rate of antibiotic discovery remains slower.

Recently, antimicrobial gene delivery systems have gained much importance. These systems have a significant potential to overcome several important clinical problems in terms of protecting against infection during the wound healing process, protecting immunocompromised patients from infection, and decreasing the need for antibiotics (Huang et al., 2002; Supp and Neely, 2008). Unfortunately, bacteria develop novel resistance mechanisms against antibiotics, and conventional antibiotics could induce drug resistance in pathogens. For that reason, attention has focused on the use of genetic materials, such as small interfering RNAs (siRNAs) and the genes coding for antimicrobial peptides.

Unlike conventional antibiotics, the gene delivery approach could be used for administration of antimicrobial peptides (AMPs). AMPs are encoded by genes, and therefore the genes coding for the AMP could be introduced into the cell (Izadpanah and Gallo, 2005). AMPs are widely found in host defense systems of many plants, insects, and animals (Nawrot et al., 2014; Izadpanah and Gallo, 2005). AMPs are small molecules that exhibit broad-spectrum antimicrobial activity against Gram-negative and

Gram-positive bacteria, viruses, and fungi. These peptides are amphipathic; that is, they have both a hydrophilic and a hydrophobic part. This feature enables peptides to be soluble in aqueous environments. AMPs are generally positively charged, and therefore they can penetrate the phospholipid bilayer of target cells by interacting with negatively charged microbial membrane (Izadpanah and Gallo, 2005). Cathelicidins and defensins are human antimicrobial peptides, and they are natural components of the defense system in skin. Cathelicidins and defensins have an extensive spectrum of antimicrobial activity against many microorganisms, such as bacteria, viruses, fungi, and some parasites. Defensins are expressed by keratinocytes and mucosal epithelial cells (Ahluwalia and Tarnawski, 2013; Supp and Neely, 2008). They are small, cationic, cysteine-rich β-sheet peptides. (Huang et al., 2002; Izadpanah and Gallo, 2005). Cathelicidins are synthesized by neutrophils and macrophages, as well as by epithelial cells. Cathelicidins and defensins are important antimicrobial peptides, and they regulate the innate immune system. Moreover, they modulate microbial growth, inflammation, and wound healing by inhibiting the growth of various microorganisms (Ahluwalia and Tarnawski, 2013). Owing to several beneficial effects, AMPs are significant candidates for antimicrobial gene delivery systems, and the number of studies in this area is rapidly increasing.

Small RNAs are efficiently used as the components of antimicrobial gene delivery systems for controlling bacterial infections. MicroRNAs (miRNAs) and small interfering RNAs (siRNAs) are small, noncoding regulatory RNA molecules, and they are generally 15–30 nucleotides long. RNA interference (RNAi) provides a therapeutic potential for specific suppression of the gene expression. RNAi blocks expression of the target gene by hybridizing to the target messenger RNA (mRNA). The mRNA is then degraded by ribonuclease H (RnaseH), and synthesis of the encoded protein is prevented. The mechanism of RNAi provides many advantages, such as target specificity, low toxicity, stability, and resistance to nucleases (Chiu et al., 2004; Wang et al., 2010).

3.2 VIRAL VECTORS FOR ANTIMICROBIAL GENE DELIVERY

Gene therapy has a great potential for delivery of defective genes, and it has the potential to treat human diseases like Parkinson's disease and different types of cancers. Efficient gene therapy depends on the efficient transfer of genes into the cell. Viruses have the necessary mechanisms for delivery of genes. Viral vectors, such as retrovirus, adenovirus (types 2 and 5), adeno-associated virus, herpes virus, pox virus, human foamy virus (HFV), and lentivirus, generally display a high degree of efficiency in gene delivery studies. Therefore, they are commonly used in gene therapy and clinical trials (Nayerossadat et al., 2012). However, there is much concern regarding the use of viral vectors in clinical trials. The most critical concern is that viral vectors could cause immunological problems. For this reason, the recent studies are directed toward using nonviral vectors in gene delivery. Despite the significant progress in nonviral systems, viral vector systems are more efficient than nonviral systems.

Carretero et al. (2004) designed a gene delivery approach for introducing AMPs. Human keratinocyte cell line (HaCaT) was transduced by adenoviral vectors containing the sequence of human antimicrobial peptides HBD-2, HBD-3 (human β-defensin 2/3), and LL-37 (human cathelicidin). The aim of that study was to enable the keratinocyte to release antimicrobial peptides into the cell culture medium, which was contaminated by *E. coli*, *S. aureus*, *P. aeruginosa*, and *E. faecalis*. The inhibitory effect of AMPs was evaluated. Moreover, antimicrobial activities of different AMPs were compared.

Huang et al. (2002) designed a model for an antimicrobial gene delivery system by using a retroviral vector carrying the cDNA sequence for HBD-2 (human β-defensin 2). The aim of that study was to test whether HBD-2 was functional in vivo. As a first step, the activity of the secreted HBD-2 from selected cell lines was tested, and it was found to display efficient antimicrobial activity. Then the researchers tested HBD-2 antimicrobial activities in vivo. For that aim, tumor formation was allowed in a model mouse that was nonobese diabetic/severely compromised immunodeficient (NOD/SCID). Then *E. coli* was injected into each tumor mass. The results were compared with control tumors (vector alone). The HBD-2 group was observed to contain only 7.8 ± 3.3% viable bacteria, whereas the control group contained 100% viable bacteria. Thus, it was concluded that the approach of HBD-2 gene therapy could be an effective candidate for enhancing antibacterial host defense (Huang et al., 2002; Tai et al., 2007).

Otitis media (OM) is an inflammatory disease of the middle ear, and antibiotics are still widely used for the treatment. Woo et al. (2015) used an adenoviral vector to transfer the gene coding for human antimicrobial peptide β-defensin 2 (DEFB4) into the middle-ear epithelial cells. Their experiments showed that adenovirus-mediated overexpression of β-defensin 2 was protective for OM by suppressing nontypeable *Haemophilus influenzae* (NTHi) adhesion to human middle-ear epithelial cells (Woo et al., 2015). That study also promises to be used as an alternative nonantibiotic approach, which would be critical for the treatment against OM pathogens *S. pneumoniae*, *H. influenzae* (NTHi), and *M. catarrhalis* (Moon and Lim, 2015).

The *rps*L wild-type gene is a dominant sensitive allele responsible for streptomycin resistance, and mutations in that gene result in antibiotic resistance. One interesting study was reported by Edgar et al. (2012). In that study, the researchers aimed to reverse the antibiotic resistance by using pathogens. For that purpose, they constructed phages including *rps*L genes to reintroduce the antibiotic sensitivity via the lysogenization.

The CRISPR-Cas9 system is one of the attractive gene delivery systems. This system is composed of a Cas9 nuclease to cleave the double-stranded DNA, and a sequence-specific RNA to guide Cas9 for the target sequence. The genes responsible for antibiotic resistance or virulence factors could be selected as the target sequences (Bikard et al., 2014; Citorik et al., 2014). CRISPR-Cas technology is an attractive delivery system that could be used for both Gram-positive and Gram-negative bacteria.

Although viral vector systems are very efficient tools, they present a significant risk to patients (Räty et al., 2008). Bacteriophages, which are used as gene delivery systems to treat bacterial infections, can secrete harmful endotoxins and cause side effects in patients. For that reason, Krom et al. (2015) constructed bacterial phagemids to express AMPs, as well as protein toxins that damage intracellular processes and that cause nonlytic bacterial death. The antimicrobial activity of the designed phagemid was studied in vivo by using a murine peritonitis infection model. The murine model was infected, and then it was treated with the designed phagemid. The result was a significant reduction in the viability of bacteria in vitro and over 80% survival in a virulent mouse model of peritonitis.

An ideal vector system to transfer the genetic materials must be stable, and it must be targeted into the tissue or cell of interest. Moreover, the system must be specific, and it should not cause any side effects. Although viral vector systems display high transfection efficiency, these systems have some limitations because of their possible mutagenic and carcinogenic properties. Moreover, they could also induce immune response problems. These systems may be a significant risk to patients (Nayerossadat et al., 2012; Räty et al., 2008). Unlike viral vectors, nonviral vectors have the potential to overcome the safety problems.

3.3 NONVIRAL VECTORS FOR ANTIMICROBIAL GENE DELIVERY

Conjugating biomolecules or genes to nanoparticles provides functional nanocarriers with unique physicochemical properties, which make the nanoparticles attractive for nonviral delivery systems.

In recent years, several functionalized nanoparticles have been constructed by conjugation of biomolecules and nucleic acids. In addition, discovery of cell-penetrating, nuclear-localized peptide, proteins, as well as targeting aptamers contributes to the success of nanoparticle-based gene delivery systems. For an effective gene therapy, the nanocarrier must pass through the cellular barriers and transport its DNA cargo into the nucleus. The properties of nonviral gene delivery systems still need to be improved with regard to the problems of a low level of transfection efficiency, targeting to the nucleus, and degradation of DNA. Plasmid DNA, antisense oligonucleotides, or small interfering RNAs could be used as the components of nonviral vector systems. The genetic material could be conjugated to different types of nanoparticles, such as lipid complexes, polymeric materials, magnetic nanoparticles, quantum dots, carbon nanotubes, and gold nanoparticles (Dizaj et al., 2014a).

Although the number of studies related to the nucleic acid–conjugated nanoparticles is rapidly increasing, there are only a few published studies that are particularly focused on antimicrobial gene delivery systems.

It was reported that the antimicrobial peptide cathelicidin could potentially be used for the treatment of ulcerative colitis (Tai et al., 2007). In that study, the therapeutic effect of the mouse cathelin-related antimicrobial peptide (mCRAMP) was investigated in mice. The mCRAMP was shown to prevent the initiation, and to regulate the healing, of formed lesions from inflammatory colitis. In a related study by Tai et al. (2013), a nonviral antimicrobial gene therapy was designed based on the mCRAMP-encoding plasmid (Ahluwalia and Tarnawski, 2013; Tai et al., 2013). Similar to the previous results, mCRAMP could efficiently prevent intestinal inflammation. Therapeutic efficiency of the nonviral plasmid vector system coding for mCRAMP (pcDNA3.1) was tested on dextran sulfate sodium (DSS)–induced cathelicidin-knockout (Cnlp$^{-/-}$) mice. In some of the disease models, the levels of interleukin-1b and tumor necrosis factor-a, as well as the number of apoptotic cells, were found to be increased. Following the intrarectal administration of mCRAMP or mCRAMP-encoding plasmid, all these abnormalities were significantly repaired. Moreover, the numbers of fecal bacteria and apoptotic cells and neutrophils were reduced. Mucus secretion was increased by impairing the expression of mucin.

Coagulase is the most important enzyme for pathogenesis of methicillin-resistant *S. aureus* (MRSA). A novel strategy related to the coagulase was introduced by Yanagihara et al. (2006), who showed how siRNA affected the expression of coagulase. They designed an siRNA molecule against the mRNA of coagulase. The inhibitory effect of siRNA on coagulase mRNA expression was evaluated by real-time PCR and radiolabeled siRNA experiments. A murine infection model was used to evaluate the therapeutic potential of siRNA. When the expression of *S. aureus* coagulase was analyzed, siRNA introduction was observed to result in a higher inhibitory effect than the control. This was the first study demonstrating the possible use of siRNA against bacteria.

In addition, siRNA was also used as an alternative for the antiviral drug resistance problem. Neff et al. (2011) used an RNA aptamer specific for gp120 protein, and constructed aptamer-siRNA complexes. Those complexes could be effectively used against HIV-1 infection (Neff et al., 2011).

It was shown that the aptamer was useful to enhance the transfection efficiency of the polyethylenimine (PEI)-based nonviral system, and without a significant cytotoxicity. The pDNA/PEI/sgc-8c aptamer polyplexes were designed to enable the targeted delivery of plasmid DNA (Shahidi-Hamedani et al., 2013). In that study, a DNA aptamer for the surface marker protein tyrosine kinase 7 (PTK7) was used to target human acute lymphoblastic leukemia T-cells.

McNamara et al. (2006) designed aptamer-siRNA chimeric RNAs for cell type–specific siRNA delivery into a xenograft model of prostate cancer. They showed that the chimeras were effective in reducing the expression of the target proteins and thereby inhibiting tumor growth. They demonstrated an effective approach displaying a therapeutic potential (McNamara et al., 2006). Thus, the combination of aptamers with siRNA, DNA, or some nanoparticles could provide an effective and nontoxic delivery approach to deal with bacterial infections.

Nonviral vectors based on nanoparticles could be easily conjugated with aptamers, which result in rapid, ultrasensitive, and effective applications (Chang et al., 2013). Nanoparticle carriers, namely nanocarriers, could cross biological barriers more efficiently than viral vectors. Moreover, for the purpose of gene delivery, nanoparticles have unique properties such as biodegradability, biocompatibility, controlled release, modular structure, larger surface area, and protection against serum endonucleases. Therefore, nanoparticles loaded with the genetic material of interest could be further modified with pathogen-specific aptamers to enhance their therapeutic efficiency. Although present gene delivery studies are more focused on targeting cancer cells, pathogen-specific aptamers could also function to deliver therapeutic genes into the infected tissues. In addition, several nanoparticles, such as silver (Ag), copper (Cu), gold (Au), titanium (Ti), and zinc (Zn) nanoparticles, as well as cationic lipids and chitosan nanoparticles, were found to show antimicrobial activity (Chávez de Paz et al., 2011; Dizaj et al., 2014b; Malarkodi et al., 2014).

Cell-specific aptamers attract interest as targeting components of nonviral gene delivery systems. As shown in Fig. 18.3, specific and selective binding of aptamers to cell surface markers could be used to design more efficient gene delivery systems.

4 FUNCTIONALIZATION OF NANOCARRIERS

Aptamers attract attention of researchers for development of highly sensitive and selective nanobiosensors. In addition to their selectivity and high affinity for their target analytes, ease of chemical modification, high stability over a wide range of pH and temperature changes, and target-triggered cascade amplification for generating sensitive signals make aptamers appropriate tools in several applications. Considering the individual advantages of aptamers and nanomaterials, combination of those two systems results in highly sophisticated and efficient platforms. There are recent and comprehensive reviews about this exciting and rapidly developing area (Kim et al., 2016; Ozalp et al., 2015; Sharma et al., 2015).

Gold nanoparticles (AuNPs) employ simple and efficient conjugation chemistry procedures for immobilization of nucleic acids onto the surface. In addition, AuNPs have some unique features that make them very attractive to be utilized in the nanobiosensor area. When aptamers are conjugated with AuNPs, protection against nucleases is provided. This feature enables researchers to chemically modify the free aptamers with substantial resistance against nucleases, especially for the applications involving serum (Chung et al., 2013). Aggregation-induced color change is another attractive feature of AuNPs. Salt-induced, cationic polymer-induced, and, finally, for significant improvement of biosensor sensitivity, target-triggered signal amplification cascade-induced AuNP aggregation were used by researchers for colorimetric detection of analytes (Ahirwar and Nahar, 2016; Chang et al., 2016; Chen et al., 2014; Yuan et al., 2016). Peroxidase-like activity of AuNPs is another promising feature of this nanomaterial for development of colorimetric biosensors. Sharma et al. (2014) used the peroxidase-like activity of AuNPs for development of signal-on nanozyme for colorimetric detection of kanamycin. Wang et al. (2016) improved the peroxidase-like activity using gold nanoparticle–doped iron oxide particles for sensitive detection of ochratoxin A (OTA).

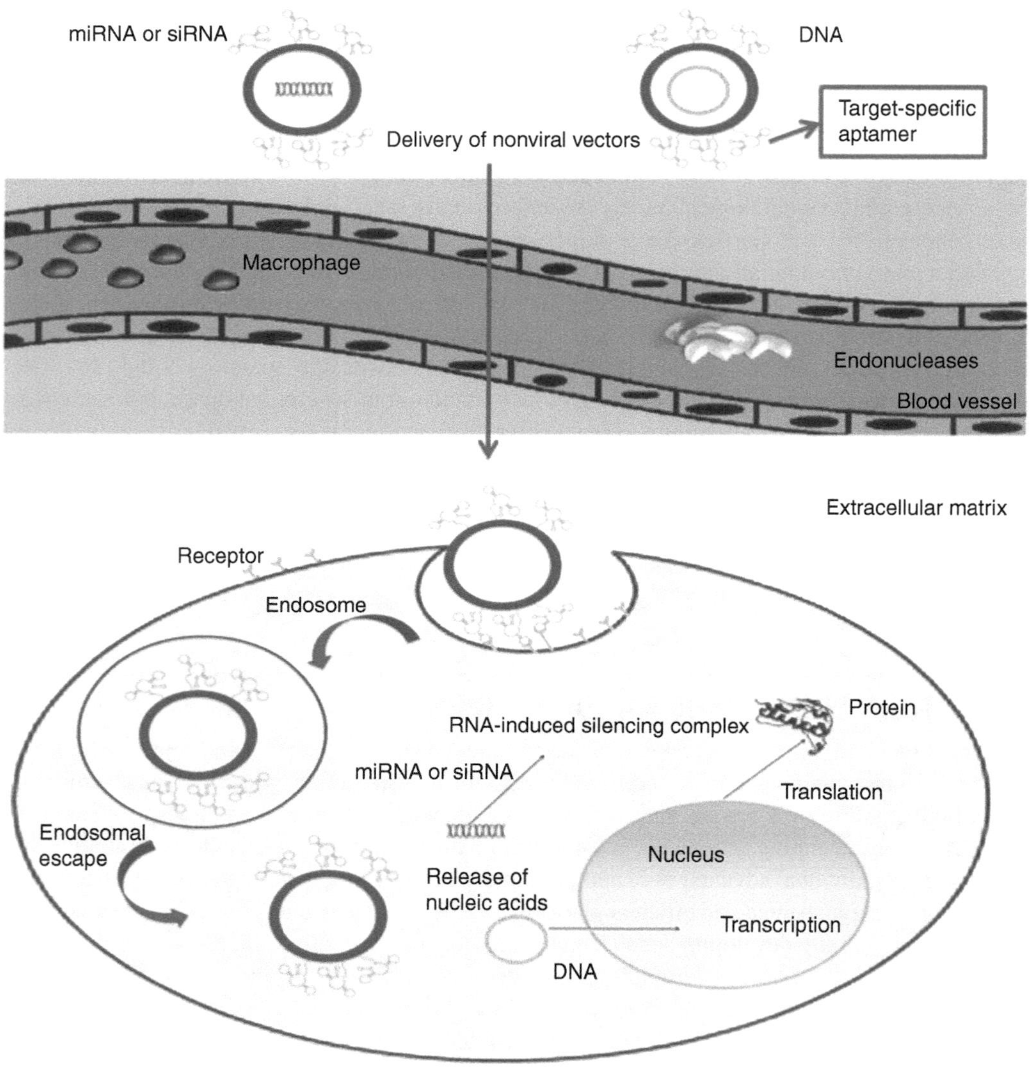

FIGURE 18.3 General Scheme Indicating the Possible Use of Aptamers as Targeting Components of Nonviral Vectors

It is also possible to combine fluorescent quenching activity of AuNPs with dye-labeled aptamers. Such studies resulted in signal-on biosensors for kanamycin and dual sensing Pb^{2+} and Hg^{2+} ions in serum, even in the presence of nucleases (Chang et al., 2013; Chen et al., 2015). Finally, as an inventive example of nanobiosensor design strategy, researchers combined advantages of aptamers, gold nanoparticles, and graphene-modified glassy carbon electrode for the development of a highly sensitive microfluidic nanobiosensor. They achieved a highly sensitive cortisol detection with wide linear range (Sanghavi et al., 2016).

Due to several unique properties, AuNPs are the most commonly used nanomaterials for biosensor development. However, magnetic nanoparticles, gold nanorods, single-walled carbon nanotubes (SWCN), graphenes, silica nanoparticles, and nanocrystal quantum dots are also frequently used because of their functional properties. Magnetic nanoparticles (MNPs) provide analyte separation with the aid of their magnetic property. Magnetic separation provides both signal amplification and sample wash in order to prevent any interference of contaminants from food or biological samples. Moreover, intrinsic peroxidase like activity of MNPs has enabled researchers to design colorimetric biosensors (Kim et al., 2016; Wang et al., 2016).

Aptamer-functionalized nanomaterials were developed for detection of *E. coli* in a microfluidic device (Chung et al., 2015). DNA aptamers were selected against *S. paratyphi* A, and the developed DNA aptamer was labeled with a DNAzyme probe. Then it was used in combination with SWCN for the selective chemiluminescent detection of bacteria (Yang et al., 2013). The bioninylated DNA aptamer was conjugated to streptavidin-labeled magnetic particle for selective capture and preenrichment of *L. monocytogenes*. A magnetic separation step was followed by real-time PCR for sensitive detection of bacteria (Suh and Jaykus, 2013). A similar strategy was also used by Suh et al. for RT-coupled sensitive detection of *C. jejuni* (Suh et al., 2014b). Ozalp et al. (2014) achieved one step forward by using MNPs grafted with glycidylmethylacrylate (GMA) brushes with high quantities of amino groups. The aim was to increase aptamer load, and subsequently improve the efficiency of capture of bacteria (Ozalp et al., 2014). A localized surface plasmon resonance (LSPR)–based sensor was developed for label-free and accurate detection of three different bacteria. The LSPR was designed by using silica nanoparticles with a gold layer deposition. Then the surface was modified by aptamers (Yoo et al., 2015). The aptamers conjugated to AuNPs were employed for colorimetric detection of *E. coli* or *S. typhimirium*. The principle of detection relied on the presence of target bacteria and a color change due to AuNP aggregation. That system enabled the visual detection of bacteria (Wu et al., 2012).

Aptamers and nanomaterials are commonly used together for the purpose of targeted drug delivery and therapy. This mutual relationship has resulted in many inspiring and inventive approaches, stimulating hope for future therapies. Gold nanoparticle-DNA aptamer conjugate was developed as a universal carrier for affinity tagged proteins. As a proof of concept, researchers examined the therapeutic efficiency of His-tagged and GST-tagged proteins bound to AuNP–aptamer conjugates. In vitro and in vivo experiments proved the efficiency of this universal nanocarrier system with almost no systemic toxic effect to organisms (Ryou et al., 2014). An antiepidermal growth factor receptor aptamer was conjugated to PEGylated gold nanorods. As a cancer biomarker, epidermal growth factor receptor (EGFR) was targeted by the specific aptamer, and photothermal therapy was employed by utilizing the heat generation property of gold nanorods upon near-infrared irradiation (NIR) (Choi et al., 2014). The nanoplatform was shown to be efficient for in vivo treatment of epidermal carcinoma. One of the most popular aptamers investigated as a promising probe for cancer therapy is the antinucleolin aptamer AS1411. Here 5 nm gold nanospheres were modified by using the AS1411 aptamer to improve the cell internalization. The selective cytotoxicity was proved by fluorescent images and silver staining (Malik et al., 2015). Fluorescence emission upon aptamer binding to a porphyrine derivative resulted in design of another efficient theranostic nanoplatform (Ai et al., 2014). Doxorubicin (DOX) intercalation to AS1411, as well as in combination with porphyrin, was a forward step to utilize the synergistic effect of photodynamic therapy and chemotherapeutic drug targeting (Shiao et al., 2014). Photothermal therapy, relying on the intrinsic property of gold nanorods to produce heat upon NIR, was selectively targeted to adenocarcinoma cells.

The DNA aptamer specific for adenocarcinoma cells was conjugated to silver–gold nanoparticles, and the resulting shell–core nanostructure enabled an effective photothermal therapy (PTT) even at a very low level of irradiation. Additionally, the Raman reporter was conjugated to the surface of the nanostructure to enable both surface-enhanced Raman scattering (SERS) imaging of cancer cells and monitoring of the PTT process (Wu et al., 2013). An Sgc8 aptamer was used in combination with AuNPs to improve drug loading capacity and to decrease the side effects (Danesh et al., 2015). Another strategy, PTT in combination with HSP70 inhibitor, is effective to prevent HSP70 response of target cells to PTT. A nanocomposite involving graphene oxide–attached gold nanoparticles was targeted to cancer cells by using the aptamer. The system was shown to have an efficient light-to-heat ratio as a result of massive AuNP targeting (Yang et al., 2015). Chandrasekaran et al. (2015) designed a modified cell-SELEX process to identify the DNA aptamer (KW16-13) capable of target cell recognition and internalization. The KW16-13-conjugated gold nanorods were used for PTT treatment of cancer cells. Due to the high specificity and cell internalization efficiency of KW16-13-conjugated gold nanorods, that system resulted in 96% cell death among cancer cells but less than 1% loss of nontarget healthy cells (Chandrasekaran et al., 2015).

Shiang et al. (2013) conjugated two different aptamers targeting polymerase and RNase H enzymes of HIV-1 RT. Employing aptamers enabled a targeting strategy at the molecular level, and a 60% decrease in HIV-1 viral infectivity was obtained (Shiang et al., 2013). AuNPs, which protected the aptamers from nucleases enhanced the success of aptamer-targeted delivery system.

There are several examples proving the success of using aptamers and nanomaterials together for enhanced therapeutic output: pH-dependent manner of intracellular drug release, computational tomography imaging of prostate cancer, DOX-loaded nanopolymerosomes, DOX-loaded unimolecular micelles targeted to PSMA, DOX-intercalated PSMA aptamer conjugated to dendrimer and PEI-grafted polyethylene glycol (PEG), and utilization of apoptosis inducing shRNA in combination with a chemotherapeutic drug and SWCNTs are some examples of that strategy (Alibolandi et al., 2015; Kim et al., 2010; Lee et al., 2011; Li et al., 2013; Mohammadi et al., 2015; Shiang et al., 2013; Taghdisi et al., 2011; Zhao et al., 2013).

Targeted delivery of antibiotics is of great importance with regard to developing efficient treatment strategies for bacterial infections. Since multidrug resistance of bacteria is a common health problem, an inventive and efficient solution is demanded urgently (Moreno-Gamez et al., 2015). Nanoparticle-based antibiotic delivery systems, which are defined as nanoantibiotics, are a promising approach to solve the antibiotic resistance problem (Huh and Kwon, 2011). Improving the solubility of drugs, prolonged drug half-life, and sustained and stimuli-responsive drug release are the advantages of nanoparticle-based antibiotic delivery. In addition, minimizing side effects of antibiotics because of poor targeting, and the opportunity of combining drug loading with alternative therapies such as photothermal and photodynamic therapies, could be proposed as other advantages of that kind of delivery system.

Although targeted drug delivery offers great potential to use several nanomaterials in combination with antibiotics, unfortunately this is still far away from being an exact antibiotic targeting strategy because of the lack of a specific targeting ligand in most cases. Liposomes, polymeric nanoparticles, dendrimers, and various inorganic nanoparticles like silica have been used for targeted drug delivery (Gao et al., 2014). In some of those studies, localization of nanomaterials to the infection site was achieved by vascular permeability as a result of inflammatory mediators. Other studies used polycationic nanomaterials to target the negative surface of bacteria (Azzopardi et al., 2013; Dillen et al., 2008; Sambhy et al., 2008).

Another approach to enhance the targeting of antibiotics to the pathogens is to attach antibiotics to several nanomaterials, such as dendrimers, iron oxide nanoparticles, gold nanoparticles, and porous silica nanoparticles (Gao et al., 2014). Lectins and antibodies have been utilized for selective targeting of antibiotics to pathogenic bacteria in several studies. However, pathogen-specific aptamers with the ability for specific targeting by antibiotics have been used in only a few approaches (Chen et al., 2012; Huang et al., 2009).

Here, we aim to discuss the advantages and benefits of specific aptamers to target pathogenic bacteria in combination with nanomaterials. It is obvious that beneficial properties of aptamer use are related to high selectivity and affinity of DNA/RNA aptamers developed against pathogenic cells (Darmostuk et al., 2015).

4.1 APTAMERS AS ANTIBIOTIC BINDING COMPONENTS OF HYDROGEL

To our knowledge, the strategy of nucleic acid functionalization of nanomaterials in order to achieve an enhanced drug loading capacity and controlled drug release was first introduced by Zhang et al. (2012a). In that study, functionalization with nucleic acid was aimed to introduce the chemical interactions between drug and nanomaterial. Unlike former studies focused on enhancing physical characteristics of hydrogels, nucleic acid functionalization of hydrogel resulted in several beneficial outcomes, such as enhanced drug loading capacity and sustained drug release up to 24 h. Nucleic acid functionalization was achieved by chemical conjugation of 5'-acrydite-modified DNA oligonucleotides to PEG hydrogel. The oligonucleotide sequences used in this study were developed by Niazi et al. (2008), and those were specific for tetracycline. Loading of tetracycline into the nucleic acid functionalized hydrogel was achieved by incubating the hydrogel in a solution containing known concentrations of tetracycline. Chemical interaction–driven tetracycline loading resulted in a threefold higher drug loading, compared to physical interaction–aided drug absorption to the nonfunctionalized native hydrogel. As a result of specific interaction between DNA aptamer and tetracycline, not only was drug loading enhanced, but also a sustained drug release was achieved. This method is the most likely to meet the urgent demand for in vivo applications. After 1 h, a 60% release of tetracycline from native gel was observed, whereas only 30% of the loaded drug was released from the functionalized hydrogel. Moreover, when drug release was tested related to the bacterial growth inhibition, the superior properties of the functionalized hydrogel could be identified clearly. The amount of tetracycline released from the native hydrogel was only sufficient to inhibit bacterial growth during the first 1 h, whereas sustained drug release from nucleic acid functionalized hydrogel achieved a prolonged bacterial growth inhibition with a 97% efficiency within 24 h. With their pioneering study, Zhang et al. (2012a) clearly demonstrated that nucleic acid functionalization would be beneficial to prevent loss of valuable drug with regard to increased partition coefficiency, enhanced drug loading capacity, and sustained drug release without affecting the physical properties of hydrogels.

4.2 NANOSCALE DRUG DELIVERY

A novel aptamer-mediated targeted antibiotic delivery system was developed by Sundaram et al. (2012). In that practical and efficient approach, the researchers used gold nanoparticles as drug delivery carriers, and they functionalized AuNPs by using neomycin-specific DNA aptamer, which was hybridized via single-stranded anchor DNA. Neomycin-binding DNA aptamer was used instead of an RNA aptamer for several reasons. In particular, the stability of DNA aptamers makes them superior to

RNA aptamers for in vivo applications. For this purpose, the researchers used a DNA molecule that had the same sequence as that of the original RNA aptamer. According to the SPR analysis, the binding affinity of the novel DNA aptamer did not show any change when compared to that of the RNA aptamer. In that study, The 5'-SH modified single-stranded anchor DNA was attached to AuNPs. Then, neomycin-binding DNA aptamer was bound to single-stranded anchor DNA via Watson–Crick base pairing. Attachment of the single-stranded anchor DNA to AuNPs makes this approach flexible so that one can easily adopt this nanocarrier by simply changing the sequence of the anchor DNA. Temperature triggered the drug release from this nanocarrier system, which was a beneficial feature of that approach. Drug release was achieved simply by disrupting the noncovalent interactions between the DNA aptamer and its target neomycin. The temperature-triggered release of neomycin from this drug carrier was achieved for an extended time period of 24 h. Due to the use of single-stranded anchor DNA, this system can be modified to achieve aptamer targeted/mediated delivery for any molecule, if a DNA/RNA aptamer is present for the target of interest.

4.3 ANTIBIOTIC-LOADED NANOPARTICLES

In order to meet the demand for target-specific delivery of encapsulated antibiotics to bacteria, Kavruk et al. (2015) developed a novel aptamer-mediated/targeted drug delivery nanocarrier. In that study, *S. aureus*–specific DNA aptamer was used to specifically target the bacteria for specific antibiotic delivery. Not only the specific delivery and target recognition triggered release of the antibiotic, but also the advantages of using mesoporous silica nanoparticles as drug carriers were combined in that study. *S. aureus*–specific DNA hairpin probe SA20hp was previously developed by Cao et al. (2009). That specific probe was used by Kavruk et al. (2015) to serve two purposes: First, SA20hp was used as a gatekeeper in order to ensure that the cargo molecule was entrapped in the pores until target recognition. Second, the SA20hp specificity to *S. aureus* contributed to the target recognition triggered release of the cargo molecule. The antibiotic vancomycin against *S. aureus* was loaded to mesoporous silica nanoparticles via a simple absorption mechanism. The gatekeeper, target-specific SA20hp molecule, was attached via its 5'NH_2 moiety to the sulfo-linker-modified, drug-loaded nanoparticles. Then 15.5 pmol of vancomycin was loaded and 11.8 ± 1.3 pmol of aptamers were covalently attached per mg of silica nanoparticles. Drug release was monitored by UV spectrum monitoring experiments in the absence and presence of target *S. aureus*. Those experiments clearly showed that the drug release was specifically triggered by the presence of the target, and leakage was only 5%. In order to determine the specificity resulting from DNA aptamer, drug release was also monitored in the presence of *S. epidermidis*. At the end of 24 h, drug release was only 15% compared to drug release by *S. aureus*. The minimal inhibitory concentrations (MICs) of vancomycin nanoparticles against *S. aureus* and *S. epidermidis* were found to be 0.420 μg/mL and 6.295 μg/mL, respectively. Considering that the MIC value of free vancomycin was in the range of 1.1–2.4 μg/mL, the results clearly indicated that MIC value is significantly lowered against the target bacteria *S. aureus*. *S. epidermidis* cells were affected only by higher amounts of vancomycin nanoparticles, and this also indicated that the antibiotic-loaded nanocapsule approach had almost no threat for nontarget bacteria. Time-kill studies were performed against *S. aureus* and *S. epidermidis*, which revealed the specificity and efficiency of vancomycin-loaded aptamer-gated nanocapsules. Vancomycin-loaded nanocapsules showed a strong prolonged bactericidal activity, while the growth of *S. epidermidis* was only slightly affected in the presence of nanocapsules.

The approach of the vancomycin-loaded aptamer-gated nanocapsule system clearly demonstrated the suitability and efficiency of pathogen-specific aptamers for the era of targeted delivery and release of antibiotics against pathogenic bacteria.

4.4 THERANOSTIC MULTIFUNCTIONAL NANOPLATFORMS

In a study, a theranostic multifunctional nanoplatform was composed of a spherical iron nanoparticle core, gold coating, PEG coating, methylene blue–bound DNA aptamer, and M3038 antibody specific for MDRB *Salmonella* DT104 (Dai et al., 2013). A spherical iron nanoparticle core provided the magnetic feature of that nanoplatform, which was useful for magnetic separation and collection of target bacteria prior to diagnosis and destruction. The simple and effective conjugation of biomolecules on the surface of the gold coating was achieved via gold–sulfur chemistry and high temperature generation in the presence of NIR light. Those features enabled using the gold coating as an optical nanoheater. Stabilization of high magnetic moment nanoparticles in blood and eliminating the toxicity of iron nanoparticles were proposed as reasons and advantages of gold coating. Star-shaped magnetic core–gold nanoparticles were coated with thiolated PEG in order to minimize the nonspecific interaction with blood cells. Methylene blue, which was bound to aptamer, was used for fluorescence imaging of the collected bacteria and for its photodynamic destruction property under NIR light. The capture efficiency and specificity of this nanoplatform were provided by multidrug-resistant *Salmonella* DT104–specific antibody, which was bound to the gold coating of the nanoplatform.

Magnetic core–plasmonic shell nanoparticle is a brilliant example of combining the advantages of smart nanomaterials and highly specific biomolecules. As a result of that successful design, a multifunctional nanoplatform was shown to be useful for magnetic separation of target bacteria from infected blood sample, fluorescence imaging, photothermal therapy (PTT), and photodynamic therapy (PDT). Magnetic capture and separation efficiency of the nanoplatform was reported as 97%, and the platform had the capability of separating antibiotic-resistant bacteria from the blood sample. The possibility for a combined synergistic therapy was verified to be highly effective for that nanoplatform. In the presence of NIR light (670 nm), the gold coating generated a high temperature, namely PTT, and methylene blue generated reactive oxygen species (ROS), namely PDT. As a result of the synergistic combination of PDT and PTT, 100% of antibiotic-resistant bacteria were dead after just 12 min of treatment. Although researchers did not test the specificity of that nanoplatform in the presence of any other bacteria, the specificity was reported to be acceptable to capture those bacteria in the blood sample.

The antibody was used to introduce target selectively and capture efficiency to the designed theranostic multifunctional nanoplatform. With their smaller size, higher stability against temperature and pH changes, and ease of chemical modification, DNA/RNA aptamers can also be used as targeting and affinity capturing in theranostic multifunctional nanoplatforms. As mentioned earlier, the antibiotic-loaded nanocapsule study involving the DNA aptamer against *S. aureus* was shown to be highly efficient and selective in the presence of *S. epidermidis*. Although the studies related to the use of aptamers in antibiotic delivery systems are very limited, the success of the published study by Kavruk et al. (2015) could be a promising starting point for employing more DNA/RNA aptamers in such nanoplatforms.

5 CONCLUSIONS AND FUTURE PERSPECTIVES

One of the most important health problems worldwide is antimicrobial drug resistance. Unfortunately, bacteria are continuously developing novel resistance mechanisms against antibiotics in use. For example, MRSA is responsible for MRSA infections in humans. Resistance to beta-lactam antibiotics, including penicillins, derivatives of penicillins, and cephalosporins, are reported. In particular, antibiotic resistance is a highly risky situation for people who have any open wounds and weakened immune systems (Bukharie, 2010; Yanagihara et al., 2006). The increase in the number of antibiotic-resistant bacteria and the undesired slow progress in discovering novel antimicrobial agents force the researchers to search and develop more efficient antimicrobial strategies. As an alternative, the efficiency of current treatments could be improved by employing different strategies.

In this chapter, we have discussed the possible use of novel antimicrobial strategies, which are mainly based on gene delivery or antibiotic delivery. Antimicrobial gene delivery is one of the recent attractive strategies. As mentioned previously, the genes coding for antimicrobial peptides could be delivered to contribute to the host's defense mechanisms. Another approach for antimicrobial gene delivery is the RNA interference mechanism, which relies on blocking the expression of a bacterial virulence factor. Although viral vectors are more efficient tools for gene delivery, they cause some undesired side effects. Therefore, nanoparticle-based nonviral vectors gain much interest. The modified nanoparticle carriers have been successfully applied for several in vitro and in vivo applications. Nanoparticles are small, and are easily modified and functionalized to develop several different delivery systems. Although the antimicrobial activities of some nanoparticles have been identified, nanoparticles have not been attempted as the carriers of antimicrobial gene delivery systems.

Controlled drug delivery is an attractive research area. However, the antibiotic delivery systems based on nanoparticles are limited to a few studies. Despite the need for more attempts, the published studies regarding controlled antibiotic release show promising results.

In this chapter, we also discussed the possible impact of aptamers on the efficiency of novel antimicrobial gene or drug delivery systems. Aptamers are short, single-stranded DNA or RNA oligonucleotides that can specifically bind to their targets. Aptamers are excellent recognition and targeting elements. Due to ease of production and their stability against environmental conditions that are not suitable for proteins, they are superior to monoclonal antibodies. Aptamer-based platforms, particularly biosensors, have gained interest in the past 25 years because of selectivity, sensitivity, and stability of aptamers. Ease of chemical synthesis and modification, high stability against harsh temperature and pH conditions, and low cost make DNA aptamers attractive. Compared with antibodies, DNA aptamers are more flexible to suit different applications, such as targeted drug delivery, quantification of target molecules, and in vivo imaging of several diseases. The global aptamers market is estimated to reach approximately $2.1 billion by 2018.

Surface modification of nanoparticles with antibodies, aptamers, and cell-penetrating peptides is particularly important for specific delivery of the genetic materials into the target cells. The combination of aptamers with other genetic materials or different nanoparticles could provide an effective and nontoxic delivery approach to contend with antibiotic resistance.

DNA aptamers are selected by a technique called SELEX, which includes selection of aptamers with desired binding properties from a random single-stranded DNA (ssDNA) library. Following the selection process in vitro, the identified aptamers showing significant binding specificity and affinity can be further synthesized in large amounts. Compared to conventional antibody production protocols, aptamer production involves oligonucleotide synthesis in vitro, and thus it is a more rapid and less expensive process. Identification and application of DNA or RNA aptamers against several pathogenic cells have been achieved (Aimaiti et al., 2015; Chang et al., 2013), but there is no published study regarding the use of aptamers in antimicrobial gene delivery systems.

Since aptamers can selectively bind to their target cells, they are promising tools to be used for targeting purposes. DNA or RNA aptamers, which are specific for their target pathogenic cells, could be used for selective, controlled delivery of antimicrobial genes or antibiotics. Aptamers could provide an alternative for replacement of antibodies in several applications, where recognition of and binding to the target cell are required.

In conclusion, aptamers display great potential to enhance the therapeutic efficiency of nanoparticle-based delivery systems that could be designed against pathogenic bacteria. In conjunction with the progress in aptamer technology, the frequency of aptamer use in nanoparticle-based antimicrobial gene or antibiotic delivery could be expected to increase in the near future.

REFERENCES

Ahirwar, R., Nahar, P., 2016. Development of a label-free gold nanoparticle-based colorimetric aptasensor for detection of human estrogen receptor alpha. Anal. Bioanal. Chem. 408, 327–332.

Ahluwalia, A., Tarnawski, A.S., 2013. Cathelicidin gene therapy: a new therapeutic option in ulcerative colitis and beyond? Gene Ther. 20 (2), 119–120.

Ai, J., Xu, Y., Lou, B., Li, D., Erkang, W., 2014. Multifunctional AS1411-functionalized fluorescent gold nanoparticles for targeted cancer cell imaging and efficient photodynamic therapy. Talanta 118, 54–60.

Aimaiti, R., Qin, L., Cao, T., Yang, H., Wang, J., Lu, J., Huang, X., Hu, Z., 2015. Identification and application of ssDNA aptamers against H37Rv in the detection of *Mycobacterium tuberculosis*. Appl. Microbiol. Biotechnol. 99 (21), 9073–9083.

Alibolandi, M., Ramezani, M., Sadeghi, F., Abnous, K., Hadizadeh, F., 2015. Epithelial cell adhesion molecule aptamer conjugated PEG-PLGA nanopolymersomes for targeted delivery of doxorubicin to human breast adenocarcinoma cell line in vitro. Int. J. Pharm. 479 (1), 241–251.

Andersson, D.I., Hughes, D., 2010. Antibiotic resistance and its cost: is it possible to reverse resistance? Nat. Rev. Microbiol. 8 (4), 260–271.

Azzopardi, E.A., Ferguson, E.L., Thomas, D.W., 2013. The enhanced permeability retention effect: a new paradigm for drug targeting in infection. J. Antimicrob. Chemoth. 68 (2), 257–274.

Bikard, D., Euler, C.W., Jiang, W., Nussenzweig, P.M., Goldberg, G.W., Duportet, X., Fischetti, V.A., Marraffini, L.A., 2014. Exploiting CRISPR-Cas nucleases to produce sequence-specific antimicrobials. Nat. Biotechnol. 32 (11), 1146–1150.

Bruno, J.G., Carrillo, M.P., Phillips, T., 2008. In vitro antibacterial effects of antilipopolysaccharide DNA aptamer-C1qrs complexes. Folia Microbiol. 53 (4), 295–302.

Bruno, J.G., Carrillo, M.P., Phillips, T., Andrews, C.J., 2010. A novel screening method for competitive FRET-aptamers applied to *E. coli* assay development. J. Fluoresc. 20 (6), 1211–1223.

Bukharie, H.A., 2010. A review of community-acquired methicillin-resistant *Staphylococcus aureus* for primary care physicians. J. Fam. Commun. Med. 17 (3), 117–120.

Cao, X., Li, S., Chen, L., Ding, H., Xu, H., Huang, Y., Li, J., Liu, N., Cao, W., Zhu, Y., Shen, B., Shao, N., 2009. Combining use of a panel of ssDNA aptamers in the detection of *Staphylococcus aureus*. Nucleic Acids Res. 37 (14), 4621–4628.

Carretero, M., Del Río, M., García, M., Escámez, M.J., Mirones, I., Rivas, L., Balague, C., Jorcano, J.L., Larcher, F., 2004. A cutaneous gene therapy approach to treat infection through keratinocyte-targeted overexpression of antimicrobial peptides. FASEB J. 18 (15), 1931–1933.

Chandrasekaran, R., Lee, A.S.W., Yap, L.W., Jans, D.A., Wagstaff, K.M., Cheng, W., 2015. Tumor cell-specific photothermal killing by SELEX-derived DNA aptamer-targeted gold nanorods. Nanoscale 8 (1), 187–196.

Chang, Y.C., Yang, C.Y., Sun, R.L., Cheng, Y.F., Kao, W.C., Yang, P.C., 2013. Rapid single cell detection of *Staphylococcus aureus* by aptamer-conjugated gold nanoparticles. Scientific Rep. 3, 1863.

Chang, C.C., Chen, C.Y., Chuang, T.L., Wu, T.H., Wei, S.C., Liao, H., Lin, C.W., 2016. Aptamer-based colorimetric detection of proteins using a branched DNA cascade amplification strategy and unmodified gold nanoparticles. Biosens. Bioelectr. 78, 200–205.

Chávez de Paz, L.E., Resin, A., Howard, K.A., Sutherland, D.S., Wejse, P.L., 2011. Antimicrobial effect of chitosan nanoparticles on *Streptococcus mutans* biofilms. Appl. Environ. Microbiol. 77 (11), 3892–3895.

Chen, F., Zhou, J., Luo, F., Mohammed, A.-B., Zhang, X.-L., 2007. Aptamer from whole-bacterium SELEX as new therapeutic reagent against virulent *Mycobacterium tuberculosis*. Biochem. Biophys. Res. Commun. 357, 743–748.

Chen, J., Zhang, C., Liu, Q., Shao, X., Feng, C., Shen, Y., Zhang, Q., Jiang, X., 2012. Solanum tuberosum lectin-conjugated PLGA nanoparticles for nose-to-brain delivery: in vivo and in vitro evaluations. J. Drug Target. 20 (2), 174–184.

Chen, Z., Tan, Y., Zhang, C., Yin, L., Ma, H., Ye, N., Qiang, H., Lin, Y., 2014. A colorimetric aptamer biosensor based on cationic polymer and gold nanoparticles for the ultrasensitive detection of thrombin. Biosens. Bioelectr. 56, 46–50.

Chen, J., Li, Z., Ge, J., Yang, R., Zhang, L., Qu, L.B., Wang, H.Q., Zhang, L., 2015. An aptamer-based signal-on bio-assay for sensitive and selective detection of kanamycin A by using gold nanoparticles. Talanta 139, 226–232.

Chiu, Y.L., Ali, A., Chu, C.Y., Cao, H., Rana, T.M., 2004. Visualizing a correlation between siRNA localization, cellular uptake, and RNAi in living cells. Chem. Biol. 11 (8), 1165–1175.

Choi, J., Park, Y., Choi, E.B., Kim, H.O., Kim, D.J., Hong, Y., Ryu, S.H., Lee, J.H., Suh, J.S., Yang, J., Huh, Y.M., Haam, S., 2014. Aptamer-conjugated gold nanorod for photothermal ablation of epidermal growth factor receptor-overexpressed epithelial cancer. J. Biomed. Opt. 19 (5), 51203.

Chung, C.H., Kim, J.H., Jung, J., Chung, B.H., 2013. Nuclease-resistant DNA aptamer on gold nanoparticles for the simultaneous detection of Pb^{2+} and Hg^{2+} in human serum. Biosens. Bioelectr. 41, 827–832.

Chung, J., Kang, J.S., Jurng, J.S., Jung, J.H., Kim, B.C., 2015. Fast and continuous microorganism detection using aptamer-conjugated fluorescent nanoparticles on an optofluidic platform. Biosens. Bioelectr. 67, 303–308.

Citorik, R.J., Mimee, M., Lu, T.K., 2014. Sequence-specific antimicrobials using efficiently delivered RNA-guided nucleases. Nat. Biotechnol. 32 (11), 1141–1145.

Dai, X., Fan, Z., Lu, Y., Ray, P.C., 2013. Multifunctional nanoplatforms for targeted multidrug-resistant-bacteria theranostic applications. ACS Appl. Mater. Interf. 5 (21), 11348–11354.

Danesh, N.M., Lavaee, P., Ramezani, M., Abnous, K., Taghdisi, S.M., 2015. Targeted and controlled release delivery of daunorubicin to T-cell acute lymphoblastic leukemia by aptamer-modified gold nanoparticles. Int. J. Pharm. 489 (1–2), 311–317.

Darmostuk, M., Rimpelova, S., Gbelcova, H., Ruml, T., 2015. Current approaches in SELEX: an update to aptamer selection technology. Biotechnol. Adv. 33 (6), 1141–1161.

Davydova, A., Vorobjeva, M., Pyshnyi, D., Altman, S., Vlassov, V., Alya, V., 2015. Aptamers against pathogenic microorganisms. Crit. Rev. Microbiol. 42 (6), 847–865.

Dillen, K., Bridts, C., Van der Veken, P., Cos, P., Vandervoort, J., Augustyns, K., Stevens, W., Ludwig, A., 2008. Adhesion of PLGA or Eudragit/PLGA nanoparticles to *Staphylococcus* and *Pseudomonas*. Int. J. Pharm. 349 (1–2), 234–240.

Dizaj, S.M., Jafari, S., Khosroushahi, A.Y., 2014a. A sight on the current nanoparticle-based gene delivery vectors. Nanoscale Res. Lett. 9 (1), 252.

Dizaj, S.M., Lotfipour, F., Barzegar-Jalalia, M., Zarrintana, M.H., Adibkia, K., 2014b. Antimicrobial activity of the metals and metal oxide nanoparticles. Mater. Sci. Eng. C 44, 278–284.

Dua, P., Kim, S., Lee, D.K., 2011. Nucleic acid aptamers targeting cell-surface proteins. Methods 54 (2), 215–225.

Duan, N., Wu, S., Chen, X., Huang, Y., Wang, Zhouping, 2012a. Selection and identification of a DNA aptamer targeted to *Vibrio parahemolyticus*. J. Agr. Food Chem. 60, 4034–4038.

Duan, N., Wu, S., Zhu, C., Ma, X., Wang, Z., Yu, Y., Jiang, Y., 2012b. Dual-color upconversion fluorescence and aptamer-functionalized magnetic nanoparticles-based bioassay for the simultaneous detection of *Salmonella typhimurium* and *Staphylococcus aureus*. Anal. Chim. Acta 723, 1–6.

Duan, N., Ding, X., Wu, S., Xia, Y., Ma, X., Wang, Z., Chen, J., 2013a. In vitro selection of a DNA aptamer targeted against *Shigella dysenteriae*. J. Microbiol. Meth. 94, 170–174.

Duan, N., Ding, X., He, L., Wu, S., Wei, Y., Wang, Z., 2013b. Selection, identification and application of a DNA aptamer against *Listeria monocytogenes*. Food Control 33 (1), 239–243.

Duan, N., Wu, S., Chen, X., Huang, Y., Xia, Y., Ma, X., Wang, Z., 2013c. Selection and characterization of aptamers against *Salmonella typhimurium* using whole-bacterium systemic evolution of ligands by exponential enrichment (SELEX). J. Agr. Food Chem. 61 (13), 3229–3234.

Dwivedi, H.P., Smiley, R.D., Jaykus, L.A., 2010. Selection and characterization of DNA aptamers with binding selectivity to *Campylobacter jejuni* using whole-cell SELEX. Appl. Microbiol. Biotechnol. 87 (6), 2323–2334.

Dwivedi, H.P., Smiley, R.D., Jaykus, L.-A., 2013. Selection of DNA aptamers for capture and detection of *Salmonella typhimurium* using a whole-cell SELEX approach in conjunction with cell sorting. Appl. Microbiol. Biotechnol. 97, 3677–3686.

Edgar, R., Friedman, N., Molshanski-Mor, S., Qimron, U., 2012. Reversing bacterial resistance to antibiotics by phage-mediated delivery of dominant sensitive genes. Appl. Environ. Microbiol. 78 (3), 744.

Fleming, A., 1929. On the antibacterial action of cultures of a penicillium, with special reference to their use in the isolation of *B. influenzae*. Br. J. Exp. Pathol. 10 (3), 226–236.

Gao, W., Thamphiwatana, S., Angsantikul, P., Zhang, L., 2014. Nanoparticle approaches against bacterial infections. Wiley Interdisciplin. Rev. Nanomed. Nanobiotechnol. 6 (6), 532–547.

Gong, W., Duan, N., Wu, S., Huang, Y., Chen, X., Wang, Z., 2015. Selection, identification, and application of dual DNA aptamers against *Shigella sonnei*. Anal. Methods 7, 3625–3631.

Groff, K., Brown, J., Clippinger, A.J., 2015. Modern affinity reagents: recombinant antibodies and aptamers. Biotechnol. Adv. 33, 1787–1798.

Hamula, C.L.A., Le, X.C., Li, X.-F., 2011. DNA aptamers binding to multiple prevalent M-types of *Streptococcus pyogenes*. Anal. Chem. 83, 3640–3647.

Hong, K.L., Sooter, L.J., 2015. Single-stranded DNA aptamers against pathogens and toxins: identification and biosensing applications. BioMed Res. Int. 2015, 419318.

Huang, G.T.J., Zhang, H.B., Kim, D., Liu, L., Ganz, T., 2002. A model for antimicrobial gene therapy: demonstration of human β-defensin 2 antimicrobial activities in vivo. Hum. Gene Ther. 13 (17), 2017–2025.

Huang, P.J., Tay, L.L., Tanha, J., Ryan, S., Chau, L.K., 2009. Single-domain antibody-conjugated nanoaggregate-embedded beads for targeted detection of pathogenic bacteria. Chemistry 15 (37), 9330–9334.

Huh, A.J., Kwon, Y.J., 2011. Nanoantibiotics: a new paradigm for treating infectious diseases using nanomaterials in the antibiotics resistant era. J. Control. Rel. 156 (2), 128–145.

Hyeon, J.Y., Chon, J.W., Choi, I.S., Park, C., Kim, D.E., Seo, K.H., 2012. Development of RNA aptamers for detection of *Salmonella enteritidis*. J. Microbiol. Meth. 89, 79–82.

Izadpanah, A., Gallo, R.L., 2005. Antimicrobial peptides. J. Am. Acad. Dermatol. 52 (3), 381–390.

Johnson-Saliba, M., Jans, D.A., 2001. Gene therapy: optimising DNA delivery to the nucleus. Curr. Drug Targets 2 (4), 371–399.

Joshi, R., Janagama, H., Dwivedi, H.P., Senthil Kumar, T.M., Jaykus, L.A., Schefers, J., Sreevatsan, S., 2009. Selection, characterization, and application of DNA aptamers for the capture and detection of *Salmonella enterica* serovars. Mol. Cell. Probe. 23 (1), 20–28.

Jumin, H., 2011. A simplified system without purification for selection of aptamers against *Vibrio alginolyticus*. Afr. J. Microbiol. Res. 5, 3564–3568.

Kavruk, M., Celikbicak, O., Ozalp, V.C., Borsa, B.A., Hernandez, F.J., Bayramoglu, G., Salih, B., Aricag, M.Y., 2015. Antibiotic loaded nanocapsules functionalized with aptamer gates for targeted destruction of pathogens. Chem. Comm. 51 (40), 8492–8495.

Kim, D., Jeong, Y.Y., Jon, S., 2010. A drug-loaded aptamer–gold nanoparticle bioconjugate for combined CT imaging and therapy of prostate cancer. ACS Nano 4 (7), 3689–3696.

Kim, Y.S., Song, M.Y., Jurng, J., Kim, B.C., 2013. Isolation and characterization of DNA aptamers against *Escherichia coli* using a bacterial cell-systematic evolution of ligands by exponential enrichment approach. Anal. Biochem. 436 (1), 22–28.

Kim, Y.S., Raston, N.H.A., Gu, M.B., 2016. Aptamer-based nanobiosensors. Biosens. Bioelectr. 76, 2–19.

Krom, R.J., Bhargava, P., Lobritz, M.A., Collins, J.J., 2015. Engineered phagemids for nonlytic, targeted antibacterial therapies. Nano Lett. 15 (7), 4808–4813.

Lee, I.H., An, S., Yu, M.K., Kwon, H.K., Im, S.H., Jon, S., 2011. Targeted chemoimmunotherapy using drug-loaded aptamer–dendrimer bioconjugates. J. Control. Rel. 155 (3), 435–441.

Lee, Y.J., Han, S.R., Maeng, J.S., Cho, Y.J., Lee, S.W., 2012. In vitro selection of *Escherichia coli* O157:H7-specific RNA aptamer. Biochem. Bioph. Res. Co. 417, 414–420.

Lee, S.H., Ahn, J.Y., Lee, K.A., Um, H.J., Sekhon, S.S., Sun Park, T., Min, J., Kim, Y.H., 2015. Analytical bioconjugates, aptamers, enable specific quantitative detection of *Listeria monocytogenes*. Biosens. Bioelectr. 68, 272–280.

Li, H., Ding, X., Peng, Z., Deng, L., Wang, D., Chen, H., He, Q., 2011. Aptamer selection for the detection of *Escherichia coli* K88. Can. J. Microbiol. 57 (6), 453–459.

Li, X., Zhao, Q., Qiu, L., 2013. Smart ligand: aptamer-mediated targeted delivery of chemotherapeutic drugs and siRNA for cancer therapy. J. Control. Rel. 171 (2), 152–162.

Liu, G., Yu, X., et al., 2012. Screening and preliminary application of a DNA aptamer for rapid detection of *Salmonella* O8. Microchim. Acta 178, 237–244.

Liu, G.Q., Lian, Y.-Q., Gao, C., Yu, X.F., Zhu, M., Zong, K., Chen, X.J., Yan, Y., 2014. In vitro selection of DNA aptamers and fluorescence-based recognition for rapid detection *Listeria monocytogenes*. J. Integr. Agr. 13 (5), 1121–1129.

Lu, K., Asano, R., Davies, J., 2004. Antimicrobial resistance gene delivery in animal feeds. Emerg. Infect. Dis. 10 (4), 679–683.

Maeng, J.S., Kim, N., Kim, C.T., Han, S.R., Lee, Y.J., Lee, S.W., Lee, M.H., Cho, Y.J., 2012. Rapid detection of food pathogens using RNA aptamers-immobilized slide. J. Nanosci. Nanotechnol. 12 (7), 5138–5142.

Malarkodi, C., Rajeshkumar, S., Paulkumar, K., Vanaja, M., Gnanajobitha, G., Annadurai, G., 2014. Biosynthesis and antimicrobial activity of semiconductor nanoparticles against oral pathogens. Bioinorg. Chem. Appl. 2014, 1–10.

Malik, M.T., O'Toole, M.G., Casson, L.K., Thomas, S.D., Bardi, G.T., Reyes-Reyes, E.M., Ng, C.K., Kang, K.A., Bates, P.J., 2015. AS1411-conjugated gold nanospheres and their potential for breast cancer therapy. Oncotarget 6 (26), 22270–22281.

Mayer, G., 2009. The chemical biology of aptamers. Angew. Chem. Int. Ed. 48 (15), 2672–2689.

Mayer, G., 2016. Nucleic Acid Aptamers Selection Characterization and Application. Springer, New York.

McNamara, J.O., Andrechek, E.R., Wang, Y., Viles, K.D., Rempel, R.E., Gilboa, E., Sullenger, B.A., Giangrande, P.H., 2006. Cell type–specific delivery of siRNAs with aptamer siRNA chimeras. Nat. Biotechnol. 24 (8), 1005–1015.

Mohammadi, M., Salmasi, Z., Hashemi, M., Mosaffa, F., Abnous, K., Ramezani, M., 2015. Single-walled carbon nanotubes functionalized with aptamer and piperazine-polyethylenimine derivative for targeted siRNA delivery into breast cancer cells. Int. J. Pharm. 485 (1–2), 50–60.

Moon, S.K., Lim, D.J., 2015. Intratympanic gene delivery of antimicrobial molecules in otitis media. Curr. Allergy Asthma Rep. 15 (4), 1–6.

Moon, J., Kim, G., Lee, S., Park, S., 2013. Identification of *Salmonella typhimurium*-specific DNA aptamers developed using whole-cell SELEX and FACS analysis. J. Microbiol. Meth. 95 (2), 162–166.

Moreno-Gamez, S., Hilla, A.L., Rosenblooma, D.I.S., Petrovd, D.A., Nowaka, M.A., Pennings, P.S., 2015. Imperfect drug penetration leads to spatial monotherapy and rapid evolution of multidrug resistance. Proc. Natl. Acad. Sci. USA 112 (22), 2874–2883.

Mozioglu, E., Gokmen, O., Tamerler, C., Kocagoz, Z.T., Akgoz, M., 2016. Selection of nucleic acid aptamers specific for *Mycobacterium tuberculosis*. Appl. Biochem. Biotechnol. 178 (4), 849–864.

Nawrot, R., Barylski, J., Nowicki, G., Broniarczyk, J., Buchwald, W., Goździcka-Józefiak, A., 2014. Plant antimicrobial peptides. Folia Microbiol. 59 (3), 181–196.

Nayerossadat, N., Maedeh, T., Ali, P.A., 2012. Viral and nonviral delivery systems for gene delivery. Adv. Biomed. Res. 1, 27.

Neff, C.P., Zhou, J., Remling, L., Kuruvilla, J., Zhang, J., Li, H., Smith, D.D., Swiderski, P., Rossi, J.J., Akkina, R., 2011. An Aptamer-siRNA chimera suppresses HIV-1 viral loads and protects from helper $CD4^+$ T cell decline in humanized mice. Sci. Translat. Med. 3 (66), 66ra6.

Niazi, J.H., Lee, S.J., Gu, M.B., 2008. Single-stranded DNA aptamers specific for antibiotics tetracyclines. Bioorg. Med. Chem. 16 (15), 7245–7253.

Ohk, S.H., Koo, O.K., Sen, T., Yamamoto, C.M., Bhunia, A.K., 2010. Antibody-aptamer functionalized fibre-optic biosensor for specific detection of *Listeria monocytogenes* from food. J. Appl. Microbiol. 109, 808–817.

Ozalp, V.C., Bayramoglu, G., Kavruk, M., Keskin, B.B., Oktem, H.A., Arica, M.Y., 2014. Pathogen detection by core–shell type aptamer-magnetic preconcentration coupled to real-time PCR. Anal. Biochem. 447, 119–125.

Ozalp, V.C., Kavruk, M., Dilek, O., Tahir Bayrac, A., 2015. Aptamers: molecular tools for medical diagnosis. Curr. Topics Med. Chem. 15 (12), 1125–1137.

Pan, Q., Zhang, X.-L., et al., 2005. Aptamers that preferentially bind type IVB pili and inhibit human monocytic-cell invasion by *Salmonella enterica* serovar *typhi*. Antimicrob. Agents Chemother. 49, 4052–4060.

Park, H.-C., Baig, I.A., Lee, S.-C., Moon, J.-Y., Yoon, M.-Y., 2014. Development of ssDNA aptamers for the sensitive detection of *Salmonella typhimurium* and *Salmonella enteritidis*. Appl. Biochem. Biotech. 174, 793–802.

Peng, Z., Ling, M., Ning, Y., Deng, L., 2014. Rapid fluorescent detection of *Escherichia coli* K88 based on DNA aptamer library as direct and specific reporter combined with immuno-magnetic separation. J. Fluoresc. 24, 1159–1168.

Räty, J.K., Pikkarainen, J.T., Wirth, T., Ylä-Herttuala, S., 2008. Gene therapy: the first approved gene-based medicines, molecular mechanisms and clinical indications. Curr. Mol. Pharmacol. 1 (1), 13–23.

Robbins, P.D., Ghivizzani, S.C., 1998. Viral vectors for gene therapy. Pharmacol. Therapeut. 80 (1), 35–47.

Ryou, S.M., Yeom, J.H., Kang, H.J., Won, M., Kim, J.S., Lee, B., 2014. Gold nanoparticle-DNA aptamer composites as a universal carrier for in vivo delivery of biologically functional proteins. J. Control. Rel. 196, 287–294.

Sambhy, V., Peterson, B.R., Sen, A., 2008. Antibacterial and hemolytic activities of pyridinium polymers as a function of the spatial relationship between the positive charge and the pendant alkyl tail. Angew. Chem. 120, 1270–1274.

Sanghavi, B.J., Moore, J.A., Chávez, J.L., Hagen, J.A., Kelley-Loughnane, N., Chou, C.F., Swami, N.S., 2016. Aptamer-functionalized nanoparticles for surface immobilization-free electrochemical detection of cortisol in a microfluidic device. Biosens. Bioelectr. 78, 244–252.

Savory, N., Lednor, D., Tsukakoshi, K., Abe, K., Yoshida, W., Ferri, S., Jones, B.V., Ikebukuro, K., 2013. In silico maturation of binding-specificity of DNA aptamers against *Proteus mirabilis*. Biotechnol. Bioeng. 110, 2573–2580.

Savory, N., Nzakizwanayo, J., Abe, K., Yoshida, W., Ferri, S., Dedi, C., Jones, B.V., Ikebukuro, K., 2014. Selection of DNA aptamers against uropathogenic *Escherichia coli* NSM59 by quantitative PCR controlled Cell-SELEX. J. Microbiol. Meth. 104, 94–100.

Shahidi-Hamedani, N., Shier, W.T., Moghadam Ariaee, F., Abnous, K., Ramezani, M., 2013. Targeted gene delivery with noncovalent electrostatic conjugates of sgc-8c aptamer and polyethylenimine. J. Gene Med. 15 (6–7), 261–269.

Shangguan, D., Bing, T., Zhang, N., 2015. Cell-SELEX: aptamer selection against whole cells. In: Tan, W., Fang, X. (Eds.), Aptamers Selected by Cell-SELEX for Theranostics. Springer, Berlin, Heidelberg, pp. 13–33.

Sharma, T.K., Ramanathan, R., Weerathunge, P., Mohammadtaheri, M., Daima, H.K., Shukla, R., Bansal, V., 2014. Aptamer-mediated turn-off/turn-on nanozyme activity of gold nanoparticles for kanamycin detection. Chem. Comm. 50, 15856–15859.

Sharma, R., Ragavan, K.V., Thakur, M.S., Raghavarao, K.S.M.S., 2015. Recent advances in nanoparticle based aptasensors for food contaminants. Biosens. Bioelectr. 74, 612–627.

Shiang, Y.C., Ou, C.M., Chen, S.J., Ou, T.Y., Lin, H.J., Huang, C.C., Chang, H.T., 2013. Highly efficient inhibition of human immunodeficiency virus type 1 reverse transcriptase by aptamers functionalized gold nanoparticles. Nanoscale 5, 2756–2764.

Shiao, Y.S., Chiu, H.H., Wu, P.H., Huang, Y.F., 2014. Aptamer-functionalized gold nanoparticles as photoresponsive nanoplatform for co-drug delivery. ACS Appl. Mater. Interf. 6, 21832–21841.

Shu, S.A., Wang, J., Tao, M.H., Leung, P.S.C., 2015. Gene therapy for autoimmune disease. Clin. Rev. Allerg. Immu. 49 (2), 163–176.

So, H.M., Park, D.W., Jeon, E.K., Kim, Y.H., Kim, B.S., Lee, C.K., Choi, S.Y., Kim, S.C., Chang, H., Lee, J.O., 2008. Detection and titer estimation of *Escherichia coli* using aptamer-functionalized single-walled carbon-nanotube field-effect transistors. Small 4 (2), 197–201.

Stratis-Cullum, D.N., McMasters, S., Pellegrino, P.M., 2009. Evaluation of relative aptamer binding to *Campylobacter jejuni* bacteria using affinity probe capillary electrophoresis. Anal. Lett. 42, 2389–2402.

Suh, S.H., Jaykus, L.A., 2013. Nucleic acid aptamers for capture and detection of *Listeria* spp. J. Biotechnol. 167, 454–461.

Suh, S.H., Dwivedi, H.P., Choi, S.J., Jaykus, L.A., 2014a. Selection and characterization of DNA aptamers specific for *Listeria* species. Anal. Biochem. 459, 39–45.

Suh, S.H., Dwivedi, H.P., Jaykus, L.A., 2014b. Development and evaluation of aptamer magnetic capture assay in conjunction with real-time PCR for detection of *Campylobacter jejuni*. LWT-Food Sci. Technol. 56 (2), 256–260.

Sundaram, P., Wower, J., Byrne, M.E., 2012. A nanoscale drug delivery carrier using nucleic acid aptamers for extended release of therapeutic. Nanomed.-Nanotechnol. 8, 1143–1151.

Supp, D.M., Neely, A.N., 2008. Cutaneous antimicrobial gene therapy: engineering human skin replacements to combat wound infection. Expert Rev. Dermatol. 3 (1), 73–84.

Taghdisi, S.M., Lavaee, P., Ramezani, M., Abnous, K., 2011. Reversible targeting and controlled release delivery of daunorubicin to cancer cells by aptamer-wrapped carbon nanotubes. Eur. J. Pharm. Biopharm. 77 (2), 200–206.

Tai, E.K., Wu, W.K., Wong, H.P., Lam, E.K., Yu, L., Cho, C.H., 2007. A new role for cathelicidin in ulcerative colitis in mice. Exp. Biol. Med. 232 (6), 799–808.

Tai, E.K., Wu, W.K., Wang, X.J., Wong, H.P.S., Yu, L., Li, Z.J., Lee, C.W., Wong, C.C., Yu, J., Sung, J.J., Gallo, R.L., Cho, C.H., 2013. Intrarectal administration of mCRAMP-encoding plasmid reverses exacerbated colitis in Cnlp$^{-/-}$ mice. Gene Ther. 20 (2), 187–193.

Tang, X., Zheng, J., Yan, Q., Li, Z., Li, Y., 2013. Selection of aptamers against inactive *Vibrio alginolyticus* and application in a qualitative detection assay. Biotechnol. Lett. 35, 909–914.

Tenover, F.C., 2006. Mechanisms of antimicrobial resistance in bacteria. Am. J. Med. 119 (6), 3–10.

Tuerk, C., Gold, L., 1990. Systematic evolution of ligands by exponential enrichment: RNA ligands to bacteriophage T4 DNA polymerase. Science 249, 505–510.

Wang, J., Lu, Z., Wientjes, M.G., Au, J.L.S., 2010. Delivery of siRNA therapeutics: barriers and carriers. AAPS J. 12 (4), 492–503.

Wang, K.-Y., Zeng, Y.-L., Yang, X.-Y., Li, W.-B., Lan, X.-P., 2011. Utility of aptamer-fluorescence in situ hybridization for rapid detection of *Pseudomonas aeruginosa*. Eur. J. Clin. Microbiol. 30, 273–278.

Wang, C., Qian, J., Wang, K., Yang, X., Liu, Q., Hao, N., Wang, C., Dong, X., Huang, X., 2016. Colorimetric aptasensing of ochratoxin A using Au@Fe_3O_4 nanoparticles as signal indicator and magnetic separator. Biosens. Bioelectr. 77, 1183–1191.

Woo, J.I., Kil, S.H., Brough, D.E., Lee, Y.J., Lim, D.J., Moon, S.K., 2015. Therapeutic potential of adenovirus-mediated delivery of betadefensin 2 for experimental otitis media. Innate Immun. 21 (2), 215–224.

Wu, W.H., Li, M., Wang, Y., Ouyang, H.X., Wang, L., Li, C.X., Cao, Y.C., Meng, Q.H., Lu, J.X., 2012. Aptasensors for rapid detection of *Escherichia coli* O157:H7 and *Salmonella typhimurium*. Nanoscale Res. Lett. 7 (1), 658.

Wu, P., Gao, Y., Lu, Y., Zhang, H., Cai, C., 2013. High specific detection and near-infrared photothermal therapy of lung cancer cells with high SERS active aptamer–silver–gold shell–core nanostructures. Analyst 138, 6501–6510.

Yanagihara, K., Tashiro, M., Fukuda, Y., Ohno, H., Higashiyama, Y., Miyazaki, Y., Hirakata, Y., Tomono, K., Mizuta, Y., Tsukamoto, K., Kohno, S., 2006. Effects of short interfering RNA against methicillin-resistant *Staphylococcus aureus* coagulase in vitro and in vivo. J. Antimicrob. Chemo. 57 (1), 122–126.

Yang, M., Peng, Z., Ning, Y., Chen, Y., Zhou, Q., Le, D., 2013. Highly specific and cost-efficient detection of *Salmonella paratyphi* A combining aptamers with single-walled carbon nanotubes. Sensors 13, 6865–6881.

Yang, L., Tseng, Y.T., Suo, G., Chen, L., Yu, J., Chiu, W.J., et al., 2015. Photothermal therapeutic response of cancer cells to aptamer–gold nanoparticle-hybridized graphene oxide under NIR illumination. ACS Appl. Mater. Interfaces 7 (9), 5097–5106.

Yin, H., Kanasty, R.L., Eltoukhy, A.A., Vegas, A.J., Dorkin, J.R., Anderson, D.G., 2014. Non-viral vectors for gene-based therapy. Nat. Rev. Genet. 15 (8), 541–555.

Yoo, S.M., Kim, D.K., Lee, S.Y., 2015. Aptamer-functionalized localized surface plasmon resonance sensor for the multiplexed detection of different bacterial species. Talanta 132, 112–117.

Yuan, L., Wang, X., Fan, Y., Liu, C., Jiang, D., Wo, X., et al., 2016. Digitizing gold nanoparticle-based colorimetric assay by imaging and counting single nanoparticles. Anal. Chem. 88 (4), 2321–2326.

Zhang, X., Soontornworajit, B., Zhang, Z., Chen, N., Wang, Y., 2012a. Enhanced loading and controlled release of antibiotics using nucleic acids as an antibiotic-binding effector in hydrogels. Biomacromolecules 13, 2202–2210.

Zhang, Y., Satterlee, A., Huang, L., 2012b. In vivo gene delivery by nonviral vectors: overcoming hurdles? Mol. Ther. 20 (7), 1298–1304.

Zhao, N., You, J., Zeng, Z., Li, C., Zu, Y., 2013. An ultra pH-sensitive and aptamer-equipped nanoscale drug-delivery system for selective killing of tumor cells. Small 9, 3477–3484.

CHAPTER

19

MULTIFUNCTIONAL NANOSIZED EMULSIONS FOR THERAGNOSIS OF LIFE THREATENING DISEASES

Tamilvanan Shunmugaperumal*, Raja Sekharan Thenrajan, Ruchi Sharma***

**Lovely Professional University, Jalandhar, India*

***Sankaralingam Bhuvaneswari College of Pharmacy, Sivakasi, India*

CHAPTER OUTLINE

ABBREVIATIONS

apos Apolipoproteins
ADMA Asymmetric dimethylarginine
AEBP1 Adipocyte enhancer-binding protein 1
AFP Alpha-fetoprotein
AS Atherosclerosis
BNP Brain natriuretic peptide
CA Cancer antigen
CEA Carcinoembryonic antigen
CK Creatinine kinase

CO	Cholestryloleate
CRP	C-Reactive protein
CT	Computed tomography
CVD	Cardiovascular disease
DDS	Drug delivery system
DHA	Decosahexaenoic acid
DOPE	1,2-dioleoyl-*sn*-glycero-3-phosphotidylethanolamine
DOTAP	1, 2-dioleoyl-*sn*-glycero-3-trimethylammonium-propane
DPPC	Dipalmitoyl phosphatidylcholine
DWI	Diffusion weighted imaging
ED	Endothelial dysfunction
EGFR	Epidermal growth factor receptor
EPA	Eicosapentaenoic acid
EPR	Enhanced permeation and retention
EYP	Egg yolk phosphatides
FDA	Food drug administration
FDG	Fluorodeoxyflucose
FMT	Fluorescence molecular tomography
GRAS	Generally regarded as safe
HE	Human epididymis
HCO	Hydrogenated castor oil
HIFU	High intensity focused ultrasound
ICI	Imperial chemical industries
LCT	Long-chain triglycerides
LPL	Lipoprotein lipase
Lp-PLA$_2$	Lipoprotein associated phospholipase A_2
mAb	monoclonal antibody
MCP-1	Monocyte chemoattractant protein-1
MCT	Medium-chain triglycerides
MPO	Myeloperoxidase
MPS	Mononuclear phagocyte system
MRI	Magnetic resonance imaging
MUFA	Mono unsaturated fatty acids
NE	Nanosized emulsions
NIR	Near infrared
Nm	Nanomolar
o/w	Oil-in-water
PEG	Polyethylene glycol
PEG-PE	Phosphatidylethanolamine derivative with polyethylene glycol
PET	Positron emission tomography
PFC	Perflorocarbons
POE	Polyoxyethylene
PSA	Prostate specific antigen
PUFA	Poly unsaturated fatty acids
QDs	Quantum dots
RANO	Response Assessment in Neuro-Oncology
RECIST	Response evaluation criteria in solid tumors
RES	Reticuloendothelial system

SFA	Saturated fatty acids
SM	Sphingomyelin
SO	Soybean
Tg	Thyroglobulin
TO	Triolein
VLDL	Very low density lipoproteins
w/o	Water-in-oil
WHO	World Health Organization

1 INTRODUCTION

To be healthy with the quality lifestyle is every human's desire. However, the modern day's lifestyle makes humans to live/lead their life in a somewhat clandestine manner. This means a human's health quality is knowingly or unknowingly affected by both acute and chronic diseases. Whereas, acute diseases show the symptoms visibly easile, chronic diseases render the quality of life inferior in a society because it impairs functional activities, working capacity, and therefore financial security. In this context, the role of a health-care professional has recently shifted from simply eradicating the disease to improve health. According to the World Health Organization (WHO), health is "A state of complete physical, mental and social well being and not merely the absence of disease or infirmity" (WHO, 1948). This definition, however, fails to explain how to preserve health. In this context, the modern medical treatment strategies are constantly revolutionized to provide satisfaction with the medical care process and improve the health-related quality of life. Having said that, the talent of health-care professionals in research and in practice are centered/concentrated on relieving the patient's sufferings, alleviating the severity of a disease by amelioration of symptoms, and rehabilitating the individual for work.

In the era of personalized medicine, it is highly imperative that diagnostic approaches are helping pharmaceutical and biotechnology sponsors to streamline their clinical trial process. Hence, both molecular assays and diagnostic imaging are routinely being used to stratify patients for treatment, monitor disease, and provide reliable early clinical phase assessments. Moreover, the importance of diagnostic approaches in drug development is highlighted by the rapidly expanding global cancer diagnostics market and the emergent attention of regulatory agencies worldwide, who are beginning to offer more structured platforms and guidance for this area. Aside from the earlier mentioned diagnosis along with amelioration, the medical strategies also utilize rhythm-based medical treatments (i.e., chronopharmaceutics) as well as newly introduced medical strategy, which is called theranosis, theragnosis, or pharmacodiagnostics.

1.1 THERANOSIS, THERAGNOSIS, OR PHARMACODIAGNOSTICS

In 2004, the U.S. Food and Drug Administration (FDA) released an important report entitled "Innovation/Stagnation: Challenges and Opportunities on the Critical Path to New Medical Products" (Kim et al., 2015). In this report, the FDA strongly expressed its interest in modernizing the manufacturing processes (or scientific tools) for medical products. Theranostics was indicated as one of the best candidates that can epochally change such scientific tools toward a critical path to new medical products. Since the concept was first introduced in 1998 by John Finkhouser, theranostics has been generally considered as combining therapy and diagnosis into one approach in order to develop an efficient new targeted therapy for personalized medicine (Kim et al., 2015; Xie et al., 2010). Therefore, theranostics is a term created by combining the two words, therapy and diagnostics, into one to better present the early detection of various diseases and their respective personalized treatments. According to Webster et al. (2015), the following phrase also seems to reflect the core of theranostics—have your cake and eat it, too—combining these two excellent ideas of nanotechnology and theranostics into inspiration for innovative creativities. One further step in this direction is, therefore, the introduction of the term "nano-theranostics," which extends further into the convergence with nanotechnology and personalized medicine, where many researchers and scientists have been developing cutting-edge nanotechnologies and applying them into tailored medicines for each individual. A concerted effort has been intensively invested to develop theranostic nanomaterials that can effectively combine therapeutic agents, targeting moieties, and imaging agents. For practical and clinical applications of nanotheranostics, all three components need to be cunningly integrated into nanoparticle-based carriers because of the three components feasible tailorability and functional diversity. The main concern behind the encroachment is to offer early detection of diseases and a personalized remedy for the patients suffering from diseases. Simultaneous diagnosis of disease, targeted drug delivery with minimum toxicity, and monitoring of treatment are, therefore, the novel strategies (in the near future) in the modern medical management of life-threatening atherosclerosical and oncological diseases (Pericleous et al., 2012). Although these strategies are complimentary to each other, the diagnosis is usually performed first and then the therapy follows to contain the progression of the cancer cells. Thus, theragnostics is a new concept or treatment strategy that represents a wide range of subjects, including personalized medicine, pharmacogenomics, and molecular imaging to develop efficient new targeted therapies with adequate benefit/risk to patients and a better molecular understanding of how to optimize drug selection (Ozdemir et al., 2006). Furthermore, it becomes clear that the nanomedicine system has contributed to the imaging as well as to the therapeutic responses by the utilization of nanoparticles as imaging agents. The role of various imaging agents like silica, gold, iron oxide nanoparticles, carbon nanotubes, and quantum dots (QDs) have been investigated. They have been found to promote the theranostic ability of the medicinal scheme for treatment (Xie et al., 2010). The following section explores the importance of imaging agents on the treatment/management of life-threatening diseases: atherosclerosis (AS) and cancer.

1.2 IMAGING

Basically, the imaging may be of two types: anatomical/physiological imaging and biomedical imaging. Whereas the anatomical/physiological imaging helps to visualize the internal structures of various organs of the human body, the biomedical imaging plays a vital role to explore the essential information regarding management of several diseases. Specifically for management of cancer, the imaging techniques help to derive estimation of morphological characters, structure, metabolic, and functional information about the site to be examined. The purpose of the detection techniques is to make the therapy effective and to reduce the mortality rate by grabbing the exact cause along with origin of a disease like cancer (Fass, 2008). The steps involved in cancer management include: prediction, screening, staging, prognosis, therapy planning, and response. Table 19.1 shows the various steps involved for cancer management along with the research reported in each step. The importance of diagnostic approaches in drug development is highlighted by the growing global cancer diagnostics market, which is expected to reach an estimated value of $168.6 billion by the year 2020 (Transparency Market Research, 2014).

Table 19.1 Steps Involved for Cancer Management Along With the Research Reported in Each Steps

S. No.	Stages of Cancer Management	References
1	Prediction	de Torres et al. (2007)
2	Screening	Lehman et al. (2007); Paajanen (2006); Sarkeala et al. (2008)
3	Biopsy guidance for detection	Nelson et al. (2007)
4	Staging	Kent et al. (2004); Brink et al. (2005); Shim et al. (2005)
5	Prognosis	Lee et al. (2004)
6	Therapy planning	Ferme et al. (2005); Ciernik et al. (2003)
7	Therapy guidance	Ashamalla et al. (2005)
8	Therapy response	Neves and Brindle (2006); Stroobants et al. (2003); Aboagye et al. (1998); Brindle (2008)
9	Recurrence	Keidar et al. (2004)
10	Palliation	Belfiore et al. (2004); Tam and Ahra (2007)

The currently available noninvasive biomedical technologies for macroscopically visualizing the tumors comprise of both electromagnetic and optical-based imaging techniques such as X-ray computed tomography (CT) scans, magnetic resonance imaging (MRI) scans, positron emission tomography (PET) scans, single photon emission CT, ultrasound scans, and optical imaging. It is unfortunate to specify that all the earlier mentioned imaging techniques have limitations, like sensing of the unusual growth of tissue in a nomothetic manner, that is, microscopic. In order to achieve information at a molecular level, many biomarkers have also been associated with the imaging systems. These biological agents can selectively target the tumors beside amplification of imaging signals; it also helps to identify the stage and aggressiveness of the tumor (Mishra and Verma, 2010). Understanding of the imaging and response requires spatial mapping at the level of gene expression, which makes the techniques not only complicated but it also leads to extraction of extraordinary information regarding stages of the disease (Smith et al., 2003). Many of the advanced imaging techniques are based on the interaction of the electromagnetic waves, ultrasonic waves, microwaves, X-rays, and other radiation either with the body tissues or fluid-surrounding cells (Table 19.2). Sometimes, the hybrid of the techniques is also employed to collect all the desired information about cell proliferation, gene expression, angiogenesis, and vascular infunctionalities. The selection of techniques for imaging depends on the sensitivity, specificity, targeting, and availability of the imaging agent (Fass, 2008).

As the use of diagnostic imaging techniques became widespread in clinical trials, a set of standardized imaging assessment criteria from the WHO were established (Miller et al., 1981). In the year 2000, a modified set of criteria called the response evaluation criteria in solid tumors (RECIST 1.0) was introduced as part of collaborative efforts between the European Organization for Research and Treatment of Cancer, the National Cancer Institute in the USA, and the National Cancer Institute of Canada Clinical Trials Group (Therasse et al., 2000).

RECIST refined an objective set of criteria that defined when tumor lesions in cancer patients improve (partial or complete response), remain unchanged (stable disease), or worsen (progressive disease)

Table 19.2 Imaging Techniques for Diagnosis of Cancer			
Techniques	**Principle**	**Applications**	**References**
Ultrasound elastography	Based on less elastic cancer tissues	Diagnosis of breast cancer, prostate cancer, and liver fibrosis	Hui et al. (2007); Lerner et al. (1990); Miyanaga et al. (2006); Pallwein et al. (2007); Tsutsumi et al. (2007)
Endoscopic ultrasound elastography	Electromagnetic radiation	Imaging of lymph nodes, pancreatic masses, adrenal, and submucosal tumors	Saftoiu and Vilman (2006)
Near infrared (NIR) spectroscopy	Nonionizing electromagnetic radiation	Breast imaging	Poplack et al. (2004, 2007)
Electrical impedance spectroscopy and tomography	Nonionizing electromagnetic radiation	Breast imaging	Tromberg et al. (2000)
Microwave imaging spectroscopy	Nonionizing electromagnetic radiation	Breast imaging	Pogue et al. (2001)
Photoacoustic and thermoacoustic imaging	Nonionizing electromagnetic radiation	Breast imaging	Franceschini et al. (1997); Grosenick et al. (1999)
Clinical optical imaging	Fluorescence and bioluminescence based imaging	Cancer imaging	He et al. (2007)
Magnetic resonance imaging (MRI)	Magnetic radiations (superior to mammography and ultrasound)	Cancer detection, breast cancer	Ross et al. (1982); Bhattacharyya et al. (2008)
Contrast-enhanced MRI	Gadolinium as contrasting agent (darkening of image in highly perfused tissues indicative of tumor)	Evaluation of microcalcifications before biopsy and evaluation of neo-angiogenesis	Takayoshi et al. (2007); Folkman (1992)
Diffusion weighted imaging (DWI)	Image contrast due to difference in the movement of water molecules within the tissues	Detection of cytotoxic edema in stroke, identification of tumor and metastases; tumor evaluation in abdomen and pelvis and characterization of benign and malignant breast lesions	Le Bihan et al. (1985)
DWI MRI	Image contrast due to difference in the movement of water molecules within the tissues and magnetic radiations	Detection and evaluation of pancreatic, bladder carcinomas, prostate cancer, lung carcinomas, and brain tumors	Liapi et al. (2008); Matsuki et al. (2007); Matoba et al. (2007); Provenzale et al. (2006)
Proton magnetic resonance spectroscopy	MR spectroscopic imaging	Localization of tumors and guiding biopsies in breast, brain, and prostate ductal carcinomas	Kurhanewicz et al. (2000); Yeung et al. (2002)
High-intensity focused ultrasound (HIFU)	Acoustic wave induced hyperthermia	Treatment of localized cancer	Medel et al. (2012)
Positron-emitting radioisotopes	Radio-isotope imaging	Used for imaging of bones, proteins, and tracing cancer	Langstrom et al. (2007)

during treatment. Since its introduction, RECIST has been updated (RECIST 1.1) to introduce standards for the assessment of lymph nodes, redefine "measurable" lesions, and assess disease progression, as well as establish recommendations for standardized image acquisition (Eisenhauer et al., 2009).

Today, a large number of oncology clinical trials employ RECIST to objectively assess cancer treatment response in solid tumors. Advances in imaging technologies and our understanding of disease have resulted in additional consortia guidelines for standardizing diagnostic imaging in oncology clinical trials. Most notably, the Cheson criteria (1999, 2007, 2014) have established guidelines for the use of diagnostic imaging using CT, MRI, and fluorodeoxyglucose (FDG)-PET as well as clinical findings for the assessment of lymphoma patients (Cheson et al., 2006).

In addition, the Response Assessment in Neuro-Oncology (RANO) criteria have been established for gliomas, and a number of other criteria have been introduced to specifically assess hepatocellular carcinoma, acute myeloid leukemia, prostate cancer, and the effects of immunotherapies on tumor responses (Cheson et al., 2003; Lencioni and Llovet, 2010; Scher et al., 2008; Van den Bent et al., 2011; Wolchok et al., 2009).

As these criteria have evolved, it has become clearer that conventional anatomical imaging techniques, although very useful, have not supplied the entire objective assessments needed to make accurate early phase go/no go decisions based on clinical trial results by US-FDA. Initiatives by the Radiological Society of North America, including the Quantitative Imaging Biomarker Alliance, to advance volumetric assessments of tumor lesions continue to gain momentum, and researchers are showing increased interest in developing tools for the evaluation of metrics derived from CT and MRI studies. As an example, techniques such as dual energy CT and spectral CT imaging are being used to better differentiate and characterize certain cancers. These types of image analysis in conjunction with efforts to assess the relationship of CT and MRI to the molecular biology of various tumors, is helping to shape the new fields of radiomics and radiogenomics. Although these approaches hold great potential for oncology clinical trials, it is likely to be several years before they can be implemented in a clinical environment.

On the other hand, nanometer-sized semiconductor particles, such as QDs with unique photochemical and photophysical properties, are of considerable interest in many research areas. Because QDs can be covalently linked to peptides, antibodies, nucleic acids, or small molecule ligands as fluorescent probes (Bailey et al., 2004; Bruchez et al., 1998; Chan and Nie, 1998; Iga et al., 2007; Medintz et al., 2005; Michalet et al., 2005). Moreover, fluorescent QDs have higher levels of brightness and photo-stability compared with conventional organic fluorophores, and are well suited for optical encoding and multiplexing applications due to their broad excitation profiles and narrow/symmetric emission spectra. In addition, the new generations of QDs have far-reaching potential for the study of intracellular processes at the single molecule level, high-resolution cellular imaging, long-term in vivo observation of cell trafficking, tumor targeting, and diagnostics (Alivisatos, 1996, 2004). Despite all of the aforementioned advantages, the direct use of QDs for drug delivery remains questionable due to their potential long-term toxicity (Iga et al., 2007). However, if the QD core could be encapsulated with other drug delivery vehicles like lipid- and polymer-based nanoparticle systems, then its inherent toxicity should significantly be reduced.

To make a complete structure for nanomedicine-based theragnostic approaches, the selection of appropriate drug delivery carrier is very important. In this regard, the potential of long-circulating and multifunctional oil-in-water (o/w) nanosized emulsions (NE) on the AS and cancer theragnostic procedures is being discussed one by one in this chapter in later sections.

1.3 CONCEPT OF MULTIFUNCTIONAL NANOSIZED EMULSIONS

Emulsions are biphasic and liquid-retentive formulation, which consists of two immiscible liquids: oil and water. If the amount of oil phase is significantly low when compared to the amount of water phase, then the final emulsion is termed as o/w emulsion. Conversely, when the amount of water phase is significantly lower than the oil phase, the resulting emulsion system appears to be somewhat more viscous and is called as water-in-oil (w/o) emulsion. Both o/w- and w/o-type of emulsion systems are stabilized against the aggregation, coalescence, and separation of dispersed oil or water phase by the addition of a third component called as emulsifying agent or emulgent or emulgator. In fact, therapeutic emulsions are being stabilized by the addition of more than one emulgent molecule in order to prevent random collision of—and then the coalescence of—dispersed oil or water phase of the o/w or w/o emulsion. Mixing of appropriate amounts of oil, water, and emulgent leads to the formation of an emulsion and this whole process is called emulsification. Apart from the amount of dispersed oil or water phase, which determines the type of final emulsion formed (whether o/w or w/o), the amount of single or multiple emulgent molecule added during the emulsification process will obviously control the type of emulsion produced. In addition, the size-reduction equipments such as high-energy or low-energy homogenizer used to mix the oil and water phases along with single or multiple emulgent molecules will also direct the final emulsion produced in terms of mean size of the dispersed phase in the final emulsion. Interestingly, both high- and low-energy homogenizers will generate emulsions with nanorange droplet particle sizes of the dispersed phase. However, an emulsion generated by the low-energy or spontaneous emulsification process is called microemulsion. On the other hand, an emulsion produced by means of the high-energy homogenizers is called nanoemulsions or nanosized emulsions (NE). Again, there are three basic differences between microemulsion and NE. First, the microemulsion possesses the dispersed phase droplet diameter value well below the range of 100 nm, typically even below 10 nm, as well. But the dispersed phase droplet diameter of the NE lies above 100 nm that, too, in the 300–500 nm level. Second, the notable difference between these two emulsions is their physical appearance. While the microemulsion creates a solution-like transparent appearance, the NE looks like a slightly bluish-colored white milk. Third, a moderately higher amount of single or multiple emulgent molecules is used for producing the microemulsion when compared to the emulsifier molecules' amount, which is being utilized to prepare the NE. In spite of these three basic differences between microemulsion and NE, these two terms (micro- and nanoemulsions) are, however, used interchangeably in many medical and pharmaceutical literatures. This chapter deals only with the o/w emulsion consisted of relatively low amounts of single or multiple emulgent molecules (in comparison to the high emulgent amount used to produce the microemulsion), and produced by the high-energy homogenizer equipments.

1.4 LIFE-THREATENING DISEASES: ATHEROSCLEROSIS AND CANCERS

1.4.1 Atherosclerosis (AS)

It is a chronic and degenerative disease of the large artery walls. It is a leading cause of mortality and morbidity (Libby et al., 2011) and is the single most important cause of cardiovascular disease (CVD), a predominant health problem worldwide (Murray and Lopez, 1997). AS is characterized by dysfunctioning of endothelial layer of cells, which activates the inflammatory mediators followed by cell proliferation and plaque development. As time passes, the plaque tends to get expanded, then ruptures, which in turn causes thrombosis and infarction of tissues and organs (Alexander and Dzau, 2000; Baralle and Baralle, 2000; Gimbrone, 1995; Lamberts et al., 1997; Libby et al., 1997; Markowitz, 1997). The molecular pathways for disease progression depict many causes behind its development. Oxidative stress, infections, and inflammation, apoptosis, effect of sex hormones, intake of toxins or mutagens, and DNA instability are likely to have a profound effect on development and aggravation in the disease. Clinical manifestations of AS include myocardial infarction, heart failure, stroke, and peripheral artery disease, all of which result in irreversible organ damage (Werner and Nickenig, 2006). The incidence rate of atherosclerotic disease has been on the rise due to dietary habits that originated in the West, insufficient exercise brought about by persistent motorization, increased mental stress from social downturn, and the rapid progression of aging in society (Fuster et al., 2011). As the disease AS has a long asymptomatic phase and the first manifestation of the disease may be sudden cardiac death, it is essential to find effective strategies to prevent it (Napoli et al., 2006). Current guidelines for the prevention of atherosclerotic diseases focus on treatment of established cardiovascular risk factors to attenuate the subsequent endothelial dysfunction (ED) and damage (Mensah et al., 2007). Because ED is an early event in AS, it plays a pivotal role in the atherogenesis process (Bonetti et al., 2003).

1.4.2 Biomarkers used for detecting artherosclerosis

The intensive research from past 2 decades has suggested the probability for early detection of many CVDs by understanding the pathology and cycle of events. Development of atherosclerotic plaque and other cardiovascular conditions like myocardial infarction and ischemia include activation of inflammatory cascade along with upregulation of many peptides and hormones. The estimation and measurement of such elevated responses dissimilar to the usual state could be alarming but constructive for early diagnosis of cardiac disorders (Koenig and Khuseyinova, 2007). The presence of mediators like cytokines, chemokines, growth-factors, and enzymes are clear indicators of cell proliferation, which if taken critically could be a life saver for many patients. From various studies, it has been observed that these mediators could serve as the potential indicators for the life-threatening CVDs. The measurement of these biomarkers is being expected to provide essential information about the presence and progression of the disease. The advancement in the techniques has offered many ways to analyze these responses with sensitivity. This could help lead the direction of diagnosis and treatment toward improved therapy with minimum of risk. Thus, the altered clinical responses are found to proceed as biomarkers. Many of these biomarkers along with their applications in the diagnosis of AS are presented in the Table 19.3. These biomarkers are able to allocate the changed conditions during AS or other CVDs either by upregulation or by initiating complementary reactions. Many of the markers are still under clinical testing and only few have been approved for routine practice. Regular testing of many more mediators and enzymes are still under consideration for development of best possible prognostic and diagnostic tools for AS (Brown and Bittner, 2008).

1.4.3 Cancers

Cancers are one of the most severe lethal diseases in today's world that kills millions of people every year. The uncontrolled spreading of this deadly disease throughout the world has become a major health concern for its treatment. From the statistical data it has appeared as the second leading cause of death in the United States and been found to be associated with AS (Wingo et al., 1998). These studies have depicted that AS and cancer develop correspondingly from a clonal proliferation of transformed cells at

Table 19.3 Biomarkers Along With Their Applications in the Diagnosis of Atherosclerosis

Biomarkers	Application	Current Status of the Biomarker	References
F2 isoprostanes	In vivo markers of oxidant injury in cardiovascular pathologies such as atherosclerosis and hypertension	Clinical biomarker	Milne et al. (2005); Nourooz (2008); Leleiko et al. (2009)
Brain natriuretic peptide (BNP)	Prognosis of patients with acute coronary syndrome	FDA approved	Kambayashi et al. (1990a,b)
NT-proBNP (amino terminal probrain natriuretic peptide)	Prognosis of patients with congestive heart failure and acute coronary syndrome	FDA approved	Shapiro et al. (2003)
Asymmetric dimethylarginine (ADMA)	Associated with cardiovascular events, atherosclerosis, hypertension, coronary artery disease, and chronic heart failure	Clinical biomarker	Boger (2003, 2005); Tousoulis et al. (2011); Wolf et al. (2010)
Myeloperoxidase (MPO)	Heart failure, acute coronary syndrome, and atherosclerosis	Clinical biomarker	Sinning et al. (2008); Schindhelm et al. (2009)
Creatinine kinase (CK)	Rise in level is indicative of acute myocardial infarction	Clinical biomarker	Jurlander et al. (2000); Newby et al. (2001); Hedges et al. (1994)
Lipoprotein associated phospholipase A_2 (Lp-PLA$_2$)	Increased amount is indicative of presence of atherogenicity	Clinical biomarker	Morrisett (2000); Stubbs et al. (1998)
C-reactive protein (CRP)	Prediction of peripheral arterial disease, artherosclerosis, myocardial infarction, and ischemic stroke	Clinical biomarker	Ridker (2007)
Cardiac troponin-I	Diagnosis of myocardial infarction, peripheral arterial disease, and artherosclerosis	Clinical biomarker	Thygesen et al. (2007); Antman et al. (2008)
Cystatin C	Predictor of coronary artery calcium progression	Clinical biomarker	Maahs et al. (2007)
Cytokines (interleukin 6 and 18)	Predictive of atherosclerotic plaque	Clinical biomarker	Maier et al. (2005); Gracie et al. (2003)
Monocyte chemoattractant protein-1 (MCP-1)	Expression of MCP-1 relates to the extent of atherosclerosis and macrophage infiltration into the atherosclerotic lesion in animal model	Clinical biomarker	Namiki et al. (2002)
Placental growth factor	Upregulation is representative of advanced atherosclerotic lesions	Clinical biomarker	Luttun et al. (2002)
Ghrelin (peptide hormone from stomach)	Upregulation is indicative of presence of atherosclerotic lesions	Clinical biomarker	Poykko et al. (2006)
Adipocyte enhancer-binding protein 1 (AEBP1)	Indicative of inflammation and development of atherosclerosis	Clinical biomarker	Majdalawieh et al. (2007)

the site of local tissue injury, inflammation, and genomic instability (Alexander and Dzau, 2000; Baralle and Baralle, 2000; Gimbrone, 1995; Lamberts et al., 1997; Libby et al., 1997; Markowitz, 1997). The etiologic and molecular pathways are found to be common for AS and cancer. The alteration starts at the molecular level either in the form of activation of oncogenes or inactivation of tumor suppressor genes. This results in the propagation of cancer cells, tissue infiltration, and organ dysfunctioning (Sarkar et al., 2007). The presence of a tumor is characterized by the presence of high vascular density and active angiogenesis for supplying blood to the site for growth. There is one more vital phenomenon called enhanced permeation and retention (EPR) effect, which is the outcome of the defective vascular architecture (Byrne et al., 2008; Iyer et al., 2006). The changes that take place at the molecular level thus can serve as the criteria for early diagnosis of tumor growth. For the purpose of detection and treatment, many advanced imaging agents and biomarkers are been developed and studied.

Current clinical cancer treatment strategies include surgery, radiotherapy, and chemotherapy. Surgery, which is the most important approach to remove solid tumors from patients, is highly invasive and can hardly eliminate all tumor cells especially when metastasis has already happened. Chemotherapy and radiotherapy, on the other hand, also have many limitations, such as severe toxic side effects, limited therapeutic efficacy, being able to induce drug resistance, and they can cause long-term damage of immune systems (Peer et al., 2007). The development of new generations of cancer treatment methods has thus been an urgent need. Among many different new cancer therapeutic strategies being explored in the past decade, therapies that are responsive to external physical stimuli usually by utilizing nanoscale agents that are capable of responding to light, magnetic field, ultrasound, radiofrequency, or X-ray, have received substantial interests from many research groups, particularly those working in the field of nanomedicine.

1.4.4 Biomarkers used for detecting cancers

A biomarker is an attribute that is a symbolic characteristic of any biologic, pathogenic, or pharmacologic response to a remedial intervention. A biomarker is any substance that indicates the presence of cancer in the body. It could be either a molecule secreted by a tumor or it could be the responsive interaction of body with cancer. There can be many types of biomarkers, such as genetic, epigenetic, proteomic, glycomic, and imaging biomarkers for cancer diagnosis, prognosis, and epidemiology. The assay of such biomarkers can be carried out by using the body fluids and thus, the process becomes entirely noninvasive. Although these biomarkers are an inevitable part of the diagnosis process but still some consequences related to stability and sensitivity has been observed. Considering the appropriate safety, stability, and theragnostic potential, there are nine of the biomarkers that have been approved from US-FDA (Table 19.4). In the near future many of the new advanced biomarkers are expected to be included on the list.

1.5 OBJECTIVES

This chapter envisions the use of the multifunctional o/w NE that carries imaging agents, anticancer, or lipid-lowering drug molecules and homing devices or targeting ligand molecules together for simultaneous imaging/diagnosing and treatment of AS and cancers. A complete outline is given on the formulation of drug-loaded NE together with active and passive targeting moieties for accessing the unreachable organs present inside the human body. In order to make a better nanosized drug targeting systems for reachable and unreachable organs/parts of the body and thus permitting the target

Table 19.4 US-FDA Approved Cancer Biomarkers

Biomarkers	Cancer Type	References
Carcinoembryonic antigen (CEA)	Malignant pleural effusion and peritoneal cancer dissemination	Li et al. (2000)
Bladder tumor antigen	Urothelial cell carcinoma	Mian et al. (2000)
Thyroglobulin (Tg)	Thyroid cancer metastasis	Lima et al. (2002)
Alpha-fetoprotein (AFP)	Hepatocellular carcinoma	De Masi et al. (2005)
Prostate specific antigen (PSA)	Prostate cancer	Gann et al. (1995)
Cancer antigen (CA 125)	Nonsmall cell lung cancer, hepatocellular carcinoma	Dabrowska et al. (2004)
Cancer antigen CA 19-9	Pancreatic cancer	Yamaguchi et al. (2004)
Cancer antigen CA 15-3 and HER2/neu	Breast cancer	Ciambellotti et al. (1993)
Leptin, prolactin, osteopontin, IF-II, and Human epididymis protein (HE4)	Ovarian cancer	Mor et al. (2005)
CD 98, Fascin, sPlgR[4], and 14-3-3 eta	Lung cancer	Xiao et al. (2005)

place-precise drug delivery and/or improved drug bioabsorption, the NE is classified into three generations based on the research conducted for over the past 20 years.

2 O/W-TYPE NANOSIZED EMULSIONS (NE) GENERATIONS

The o/w NE are heterogeneous dispersions of o/w, where the effective size of the oil droplets is reduced to less than 500 nm in diameter. The hydrophobic drugs can successfully be incorporated within the inner oil phase of an o/w NE or at the o/w interface of the system. Simple modifications can be made on surface/interface structures of NE. For instance, incorporating an emulsifier molecule alone or in a specific combination that is capable of producing either positive or negative charges over the emulsified oil droplets surface will lead to the formation of surface (charge)-modified emulsions. Based on these surface modifications, the o/w NE can be divided into cationic and anionic emulsions. Furthermore, based on the past and current performances decades, o/w NE can conveniently be classified into three generations (Fig. 19.1).

2.1 FIRST GENERATION NE

According to documented Indian scriptures, dating back to 5000 BC, nutritional status has always been associated with health (Chandra, 1985). Because any kind of nutritional depletion due to either changes in the quality and amount of dietary fat intake or abnormalities in lipid metabolism results in immunosuppression and therefore host defense impairment, favoring increased infection and mortality rates.

Traditionally, depletion in dietary fats in a malnourished or hypercatabolic patient is compensated through intravenous feeding using a solution containing amino acids, glucose, electrolytes, and

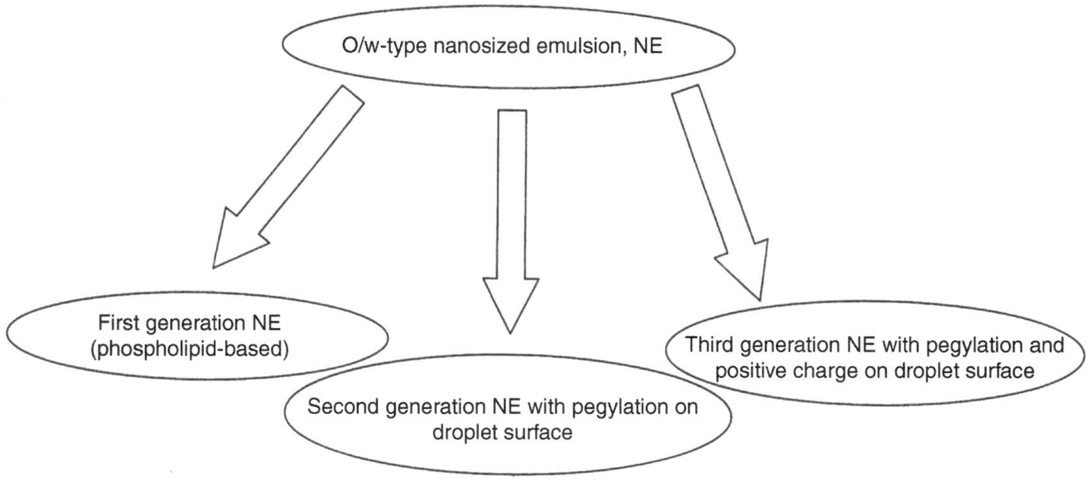

FIGURE 19.1 Flow Chart of Three Generations of Emulsion

vitamins as well as NE. Structurally, NE (o/w type emulsion) is triglyceride droplets enveloped with a stabilizing superficial layer of phospholipids. NEs for parenteral use are complex nutrient sources composed not only of fatty acids but also include substances other than triglycerides, such as phosphatidylcholine, glycerol, and α-tocopherol in variable amounts. The NE also has a complex inner structure and consists of particles with different structures, namely oil droplets covered by an emulsifier monolayer, oil droplets covered by emulsifier oligolayers, double-emulsion droplets, and possibly small unilamellar vesicles. Commercially available NE used as intravenous high-calorie nutrient fluids have particle sizes normally around 160–400 nm in diameter and typically, their surfaces are normally negatively charged. The triglycerides used to prepare NE may be presented structurally in long- or medium-chain forms, such as long-chain triglycerides (LCT) and medium-chain triglycerides (MCT). LCT contains fatty acid chains with 14, 16, 18, 20, and 22 carbon atoms and sometimes with double bonds. The number of double bonds present defines the fatty acids in LCT as saturated, mono- or polyunsaturated. If the first double bond is on carbon number 3,6, or 9 from the methyl end of the carbon chain then the fatty acid is n-3, n-6, or n-9, respectively. Purified soybean oil (SO) or safflower oil contains LCT with a high proportion of n-6 polyunsaturated fatty acids (PUFA), whereas olive oil has LCT with n-9 monounsaturated fatty acids (MUFA). Fish oil includes LCT with 20 or more carbon atoms where the first double bond is located between the third and the fourth carbons from the methyl terminal of the fatty acids chain (omega-3 or n-3). On the other hand, MCT is derived from coconut oil and has saturated fatty acids (SFA) with chains containing carbon atoms at 6, 8, 10, or 12 positions. Fatty acids that are important in parenteral nutrition and their sources are listed in Table 19.5. Fatty acids have common names (Table 19.5) and systematic names. They are also referred to by a shorthand nomenclature that denotes the number of C in the chain, the number of double bonds, and the position of the first double bond relative to the methyl-C (n; also termed ω; Table 19.5).

The simplest n-6 fatty acid is linoleic acid (18:2n-6) and the simplest n-3 fatty acid is α-linolenic acid (18:3n-3). Although mammalian cells cannot synthesize linoleic and α-linolenic acids, they can metabolize them by further desaturation and elongation. Linoleic acid can be converted to γ-linolenic (18:3n-6), then to dihomo-γ-linolenic acid (20:3n-6) and then to arachidonic acid (20:4n-6). Using

Table 19.5 Common Names, Short-Hand Nomenclature, and Sources of Fatty Acids Used in Parenteral Lipid Emulsions

Common Name	Shorthand Nomenclature	Typical Source
Caprylic acid	8:0	Coconut oil
Capric acid	10.0	Coconut oil
Myristic acid	14.0	Coconut oil
Palmitic acid	16.0	Olive oil, SO, fish oil
Oleic acid	18:1n-9	Olive oil, SO
Linoleic acid	18:2n-6	SO
α-Linolenic acid	18:3n-3	SO
Eicosapentaenoic acid (EPA)	20:5n-3	Fish oil
Docosahexaenoic acid (DHA)	22:6n-3	Fish oil

the same series of enzymes α-linolenic acid is converted to eicosapentaenoic acid (EPA) (20:5n-3). A complex pathway for further conversion of EPA to docosahexaenoic acid (DHA) (22:6n-3) exists (Sprecher, 2002).

The lipid typically used in parenteral nutrition is SO, in which linoleic acid comprises about 50% of the fatty acids present. Commercially available SO–based NEs include: Intralipid (Fresenius Kabi, Bad Homberg, Germany); Lipovenoes (Fresenius Kabi); Lipofundin (B Braun, Melsugen, Germany); Ivelip (Baxter Healthcare, Maurepas, France). A study of patients following major gastrointestinal surgery has identified that the amount of n-6 PUFA (i.e., linoleic acid) infused is one of two predictors of the length of hospital stay (increased by 1.6 d/100 g n-6 PUFA infused), the other being the delay in the onset of initiating nutritional support (Koch and Heller, 2005). Furthermore, the concern about potential harm, based mainly on the notion that n-6 PUFA might be "proinflammatory, immunosuppressive and procoagulatory," has led to the development of alternative first generation NEs for parenteral applications. Two alternative philosophies to reducing the amount of linoleic acid have been adopted. The first is to simply dilute SO with another oil that is fairly inert. Examples of this strategy include the use of so-called MCT and the use of olive oil. With MCT/LCT combinations in a specific ratio, NE appears to provide a more readily metabolizable source of energy (Rubin et al., 1991). Moreover, for drug solubilization purpose, MCT is reported to be 100 times more soluble in water than LCT and thus has an enhanced solubilizing capability. The second approach has been to partially replace SO with another oil that is believed to exert benefits in its own right. An example of this strategy is the use of fish oil. SO is often referred to as "LCT," but this nomenclature is an incorrect use of this term because the lipids found in olive oil, fish oil, and other oils not used in parenteral nutrition also contain LCT. The following FDA-approved NEs are available as alternatives to pure SO emulsions: Lipofundin MCT/LCT (B Braun), a 50:50 (v/v) mixture of MCT (in the form of coconut oil) and SO; Lipovenoes MCT (Fresenius Kabi), a 50:50 (v/v) mixture of MCT (in the form of coconut oil) and SO; Structolipid (Fresenius Kabi), produced by interesterification of a 50:50 (v/v) mixture of MCT (in the form of coconut oil) and SO; ClinOleic (Baxter Healthcare), an 80:20 (v/v) mixture of olive and SOs; Lipoplus (also known as Lipidem; B Braun), a 50:40:10 (by vol.) mixture of coconut, SO and fish oils; SMOFLipid (Fresenius Kabi), a 30:30:25:15 (by vol.) mixture of coconut, SO, olive, and fish oils. In addition, the product Omegaven (Fresenius Kabi), which is 100% fish oil, is available for use as a supplement to be diluted with another lipid emulsion of choice. In Europe NE containing LCT/MCT enriched with fish oil became available for research. Fish oil consists of two major omega-3 fatty acids, namely EPA and DHA. Both EPA and DHA have been shown to be active in a number of biological processes, including retinal and brain development, immune function, blood clotting, and prevention of cardiac arrhythmias (Horrocks and Yoe, 1999). Thus, omega-3 fatty acids laden NEs made from fish oil are likely to be increasingly used not only for nutrition support but also for modification of a number of biological and pathological processes.

The rationale for using newer first generation NEs, prepared from olive oil and fish oil, as parenteral nutrition and the clinical trials performed using these NEs in adult patients post surgery (mainly gastrointestinal) or critically ill adults was reviewed recently by Calder (2009).

2.2 SECOND GENERATION NE

An easy and substantial associability of lipophilic bioactive compound with the MCT or other vegetable oil–based NEs, however, makes the NE a vehicle/carrier for the formulation and delivery of drugs with a broad range of applications. These applications extend right from enhanced solubilization or stabilization of the entrapped drug to sustained release and site-specific delivery. Hence, the NE used for these applications are termed as second generation NE. Fittingly, the second generation NE can be administered by almost all available routes including topical (percutaneous) (Amselem and Friedman, 1998) and ocular (Tamilvanan and Benita, 2004), parenteral, oral, nasal (Tirucherai et al., 2002), and even aerosolization to the lungs (Mizushima et al., 1983). Typical o/w NE basically comprises about 0.5%–50% of a first component of an oil or oil mixture, about 0.1%–10% of a second component of an emulsifier, about 0.05%–5% of a nonionic surfactant and an aqueous component with the mean droplet size being in the submicron range, that is, below about 500 nm and preferably between about 100 and 300 nm.

Apart from serving as excellent vehicles for drug encapsulation, the o/w NEs also mitigate the toxicities associated with surfactant and ethoxylated castor oil (Cremophor EL)-based formulations. Cremophor EL is associated with nephrotoxicity, hypotension, and bronchospasms and can produce anaphylactic reaction. For example, propofol injectable solution was prepared in Cremophor EL by Imperial Chemical Industries (ICI) was gone into clinical usage but later withdrawn due to toxicity of Cremophor EL. Propofol was reformulated in o/w NE using generally regarded as safe (GRAS)-grade excipients, such as SO, glycerol, egg lecithin, and disodium edetate and launched with the trade name Diprivan by ICI (now Astra Zeneca) (Cockshott, 1985). Diprivan is used in intensive care medicine as a short-acting, intravenous sedative and known to have low toxicity, controlled sedation effect, rapid onset and quick recovery despite prolonged usage (Cockshott, 1985). Emulsion prepared using the edible lipidic substances (SO and safflower oils) have been extensively used in clinical practice to administer nutritive substances parenterally (e.g., Intralipid). These edible lipids serve as a resource of essential fatty acids, such as omega-3 and omega-6 PUFA, nonglucose-based calories, and vitamins E and K. Similarly, a Cremophor EL-free formulation of paclitaxel was prepared using TOCOSOL nanoemulsion to overcome anaphylactic reactions associated with Cremophor formulation. Paclitaxel-loaded TOCOSOL formulation was approved by FDA in 2003 for the treatment of nonsuperficial urothelial cancer. In another example, a Cremophor EL/ethanol-based cyclosporine (Sandimmune Injection) was reformulated into nanoemulsion using SO and egg lecithin composition. These studies indicated that a change in the vehicle may reduce the acute nephrotoxic side effects associated with cyclosporine in the Cremophor EL formulation (Tibell et al., 1993). Alprostadil palmitate, amphotericin B, dexamethasone, flurbiprofen axetil, and vitamins A, D, E, and K are some other examples of therapeutically relevant compounds that have been formulated in nanoemulsions for clinical applications.

2.2.1 Opsonization of second generation NE

The lipid-induced enhancement in oral bioavailability of many drugs having poor water solubility is a well-known documented fact when the drugs are incorporated into NE. However, direct intravascularly or locally administered conventional first- and second generation NEs could be taken up rapidly by the circulating monocytes for clearance by the reticuloendothelial cells (through organs such as the liver, spleen, and bone marrow). Regardless of the residence time of NEs within the vascular system, much of an injected dose is taken up via endocytosis by the cells of reticuloendothelial systems (RES) to end up in the lysosomal apparatus. NEs can be considered as being artificial chylomicrons and enter the fat metabolism pathway through the adsorption of apolipoproteins (apos) and the subsequent action of lipoprotein lipase (LPL). It appears that uptake of conventional NEs by the RES cells is lysosomotropic, resulting in the localization of emulsion droplets inside the lysosomes where they are degraded by local enzymes. Furthermore, the extent of clearance is enhanced by the adsorption of opsonic plasma proteins onto NE surfaces. However, hydrophobic particles like NEs are also taken up by macrophages without the necessity of opsonization.

An opsonization process is the adsorption of protein entities capable of interacting with specific plasma membrane receptors on monocytes and various subsets of tissue macrophages, thus promoting particle recognition by these cells. Classical examples of opsonic molecules include various subclasses of immunoglobulins, complement proteins like C1q and generated C3 fragments (C3b, iC3b), apos von Willebrand factor, thrombospondin, fibronectin, and mannose-binding protein. Reports that the coemulsification or incubation of NE with gelatin prior to intravenous injection enhanced its rate of clearance from the blood by fixed RES cells of the liver, lung, spleen, and bone marrow have led to the development of an artificial NE for RES functionality tests (Illum et al., 1989). The mechanisms employed are interactions with circulating plasma opsonin proteins, thereby exploiting a receptor-mediated process involving fibronectin. It should also be emphasized that the interaction of particles with blood protein may have effects beyond opsonization. These may include interference with the blood-clotting cascade, a process that may lead to fibrin formation, and anaphylaxis because of complement activation.

When given other parenteral routes, for example, intraperitoneally, subcutaneously, or intramuscularly, the majority of NE droplets enter the lymphatic system and eventually the blood circulation where particles behave as if given intravenously. Liver, spleen, and bone marrow uptake is significantly lower. Indeed, relative to the NEs size, lymph nodes take up a much greater (over 100-fold) proportion than any other RES tissue. Lymphatic transport was predominantly associated with chylomicron-based transport (Porter et al., 1996).

2.2.2 Long circulation concept on second generation NE

There is an increasing interest in developing injectable NE that is not cleared quickly from the circulation when NE is designed to reach non-RES tissues in the vascular system, extravascular sites of action, or to act as circulating drug reservoirs. It was thought that initial clearance rate by the RES can be affected by the presence of large numbers of NE particles occupying available RES receptors or exhausting opsonizing factors (Saba, 1970). Overloading or saturating the RES by single large doses or repeated administration may lead to subsequent remain of injected LE particles in the circulation and could be used to affect distribution patterns. Davis et al. (1992) confirmed that the infusion of NEs into the rabbit can cause slight temporary impairment of the RES as determined by subsequent administration of a radiolabeled colloid probe. An infused NE can cause RES blockade by one or both of two mechanisms. Immediately after infusion some of the NE particle may be recognized as foreign and are then cleared by the RES (largely the Kupffer cells of the liver). In addition, or alternatively, the Kupffer cells of the liver may become overloaded by NE remnants that will result from normal metabolism of the NE by tissue lipases (Davis et al., 1992). The study by Ueda et al. (2001) supports, in a way, the RES overloading or saturation concept. They prepared NEs consisting of SO and egg yolk phosphatides and studied the influence of the volume of intravenously injected NEs on the pharmacokinetic parameters of the dispersed oil particles and incorporated cargo molecules like menatetrenone (vitamin K_2). At 3.0 mL/kg injection volume, which equates to 180 mL for a person weighing 60 kg, these NEs showed a prolonged plasma half-life in rats following intravenous administration. In contrast, at 0.1 mL/kg injection volume, which equates to 6 mL for a 60 kg person, these NEs were shown to disappear from the circulation soon after intravenous administration to rats.

2.2.3 Approaches for making long circulating second generation NE

To augment substantially the NE half-life in blood circulation, the two different approaches that are being investigated most actively so far on second generation NE involve the use of either structured lipids as oil core in final NE or emulsion droplet surface modification using a coemulsifier with highly hydrophilic chains like polyoxyethylene (POE) and amphipathic polyethylene glycol (PEG) derivatives. Deckelbaum et al. (1990) explored how enzyme affinity and enzyme activity (LPL and hepatic lipase) regulate hydrolysis of phospholipid-stabilized emulsions of MCT versus LCT. It was shown that MCT NE is more readily hydrolyzed by LPL and hepatic lipase than LCT NE because of greater MCT solubility and mobility at the oil–water interface. The major factors involved are the affinity of the lipases for the interface and accessibility of individual substrate molecules to the lipases. In mixtures of LCT and MCT NEs, a higher affinity for the LCT-containing particle results in partitioning of the lipases away from the MCT NE with consequently diminished MCT hydrolysis. Hedeman et al. (1996) prepared NE using structured lipids and explored its potential to prolong the in vivo circulation time. Takino et al. (1994) developed NE for lipophilic drugs with the potential for prolonged circulation in the blood or hepatic targeting. A coating with sphingomyelin (SM) in the surface of the oil droplets resulted in avoidance of the RES (Takino et al., 1994). In an another study Takino et al. (1993) demonstrated further that ^{14}C cholesteryl oleate administered with oil droplets of a conventional NE was rapidly taken up from the circulation by the RES cells, while those containing SM survived in the circulation for a considerably longer period. Similarly, it has further been confirmed that the presence of SM in NE delays the removal of emulsion particles from rat plasma (Arimoto et al., 1998). The results of Kurihara et al. (1996) indicate that hydrogenated castor oil (HCO)-60 emulsions, when compared with conventional lecithin-stabilized emulsions, are more stable to LPL and show low uptake by RES organs, long circulations in the plasma and high distribution in tumors. Lin et al. (1992) confirmed that HCO-60 is a good emulsifier for the preparation of NE with better stability and prolonged and selective delivery properties. Thus, these sterically stabilized NEs could show potential as effective carriers for highly lipophilic antitumor agents to enhance the drug delivery in tumors. This was confirmed by Sakaeda et al. (1994) who found that the rate of selective delivery of Sudan II to liver, lungs, and spleen could be suppressed by using HCO-60-based NE. Conversely, the use of saturated MCT in NE was the most effective way to increase blood concentration of Sudan II, resulting in higher distribution to liver, lungs, spleen, and brain (Sakaeda and Hirano, 1995). Furthermore, an o/w-type NE containing HCO-60 was shown to be superior in the selective distribution of adriamycin-HCl to the liver and in decreasing concentration in heart and kidney (Yamaguchi et al., 1995). Again, Ueda et al. (2003) reported the effect of using a series of HCOs having different oxyethylene numbers such as HCO10, HCO20, HCO30, HCO60, and HCO100 on the pharmacokinetics of menatetrenone (vitamin K_2) incorporated in SO (SO)–based NE in rats. Plasma half-life of menatetrenone after administration as the NE prepared by HCO with 10 oxyethylene units (SO/HCO10) was similar to that after the administration as SO/egg yolk phosphatides (SO/EYP), but was shorter than that as the NEs prepared by HCOs with >20 oxyethylene units (SO/HCO20, SO/HCO30, SO/HCO60, and SO/HCO100). These findings clearly demonstrate that 20 oxyethylene units in HCOs are the minimum requirement for the prolongation of the plasma circulation time of the incorporated drug in SO/HCOs NEs. The earlier described studies suggest the involvement of oil or structured lipids in the enhancement of systemic circulation of the NE.

Using the established formulation approaches by which the emulsion droplet surfaces could be altered might, however, be more realistic and even further useful for a wide array of drug targeting purposes. Steric barrier or enhanced hydrophilicity effect exerted by POE chain having surfactants when added as coemulsifier into the phospholipid-stabilized NE allows, to some extent, the passive/inverse drug targeting to the lung, kidneys, and areas of inflammation (Lee et al., 1995; Liu and Liu, 1995). The addition of POE-based surfactants into the otherwise hydrophobic phospholipid-stabilized NE is particularly effective against plasma protein adsorption onto NE surfaces because of the hydrophilicity and unique solution properties of POE-based surfactants, including minimal interfacial free energy with water, high aqueous solubility, high mobility, and large exclusion volume (Lee et al., 1995). In addition, colloidal particles presenting hydrophilic surfaces with a low contact angle will be almost ignored by phagocytic cells (Davis and Hansrani, 1985), although NEs are not supposed to be recognized as foreign by the body to some extent. Examples of the POE chain containing surfactants employed so far in NE are Tween 80, Span 80, Brij, and poloxamer 188. The effectiveness of these polymeric surfactant molecules to intercalate at the oil–water interface with strong bounding to the phospholipid molecules could also be checked/judged through an in vitro monolayer experiment (Levy et al., 1991).

In general, the modification of particulate carriers using amphipathic PEG-containing molecules results in a prolongation of their blood circulation time (Bhadra et al., 2002; Harris et al., 2001). Therefore, similar to POE, a phosphatidylethanolamine derivative with polyethylene glycol (PEG-PE) is also incorporated as a coemulsifier into NE (termed as PEGylated NE) to augment its circulation half-life time (Wheeler et al., 1994). Liu and Liu (1995) studied the biodistribution of NEs coated with phosphatidylethanolamine derivatives with three different molecular weights PEG (m.w. 1000, 2000, and 5000). Among them, the PEG-2000 was able to prolong the circulation time of NE, probably due to the increased hydrophilicity of the droplet surface and/or the formation of a steric barrier. A dipalmitoyl phosphatidylcholine (DPPC) stabilized NE was prepared by Lundberg et al. (1996) and the effect of addition of PEG-PE, polysorbate 80, or Pluronic F-68 on the metabolism of DPPC-stabilized NE was studied. They employed two radioactive markers, [^{14}C] triolein (TO), which is susceptible to the action of LPL and [^{3}H] cholesteryl oleate (CO) ether, which is not, in order to study the fate of NE following injection into tail vein of female BALB/c inbred mice. Hence, the removal of ^{14}C-TO represents the triglyceride metabolism, whereas the other one is a core marker to represent whole particle removal by RES organ uptake. The NEs with DPPC as sole emulsifier were rapidly cleared from the blood with only 10%–11% of CO or TO left in circulation after 1 h. However, the addition of PEG-PE gave a prolonged clearance rate especially during first 3 h. A further addition of cosurfactant polysorbate 80 or Pluronics F-68 resulted in a marked extension of the circulation lifetime during the first 6 h. The notable effects of polysorbate 80 and Pluronic F-68 can apparently be attributed mainly to the decrease in droplet size, although an additional influence due to the increased hydrophilicity may not be ruled out.

2.2.4 Antibody conjugation onto long circulating second generation NE

In order to bring the colloidal carrier closer to otherwise inaccessible pathological target tissues, homing devices such as antibodies, cell recognizing proteins, etc are usually linked somehow onto the particle surfaces. Antibodies are proteins able to specifically recognize an antigen (e.g., a pathogen or a tumor agent). It has been observed over the past 2 decades that progression of cancer is often accompanied by the overexpression of one or several proteins, called tumor antigens. The use of monoclonal antibodies (mAb) for the treatment of cancer has been suggested as a means of targeting cancer cells while sparing normal cells (Farah et al., 1998). To avoid interference of PEG chains of PEG-PE-containing NE with antibody localization on the surface of the colloidal carrier, the coupling of antibodies to the terminal ends of the PEG chains has attracted much attention in recent years. The methods used to link the homing devices such as antibodies onto microspheres [another drug delivery systems (DDS)] consist of simple, direct adsorption to covalent bonds. However, the PEG-PE emulsifier-containing NE can usually be linked only through a conjugation process (through a covalent link) with antibodies. Such an anchorage of antibodies achieved by conjugation process at the distal ends of PEG chain emulsifiers orienting from oil-water interface of NE may provide the "active" targeting of biological drug compounds to life-threatening diseases including various forms of malignancy (Song et al., 1996).

Decker et al. (1995) defined two fundamental properties dominating the delivery of drugs from NEs: first, the concentration of the compound in the lipid phase of the emulsion is directly proportional to the concentration of the compound in the cell at equilibrium, and second, the rate of transfer is directly proportional to the concentration of particles in contact with cells. Moreover, the transfer is consistent with direct partitioning from the lipid phase of the emulsion to cells and occurs by the direct collision of emulsion particles with cells. Essential requirements of this "active" targeting approach include identification of recognition features (receptors) on the surface of the target, and the corresponding molecules (ligands) that can recognize the surface. Indeed, NE with appropriate ligands anchored on their surface must be able to access the target, bind to its receptors and, if needed, enter it. In the case of cells, intracellular entry will possibly be carried out by receptor mediated endocytosis or through an adsorptive pinocytosis to a minor extent. Access to the target is expected to occur easily in the intravascular space, as, for instance, is the case of populations of circulating cells, or where cells (e.g., certain malignant cells) are separated from the blood or other biological fluids by leaky capillaries. Obviously, because of the submicron size range (175–400 nm in diameter) of the NE, the more they circulate, therefore the greater their chance of reaching respective targets. Thus, antibody-associated NEs are more likely to offer a means to introduce drugs into the desired intracellular sites. Conjugation of an anti-B-cell lymphoma mAb, LL2, to long-circulating drug-carrier NEs has been reported by Lundberg et al. (1999).

Benita and coworkers have prepared antiferritin mAb (AMB8LK) conjugated cationic nanoemulsions (Goldstein et al., 2005). Neither the physicochemical properties of the prepared immunoemulsions (average droplet size of 120 nm with a zeta potential of +50 mV) nor the immunoreactivity of the mAb was altered during the emulsion preparation. In the subsequent study from the same research group, trastuzumab-conjugated cationic nanoemulsion was prepared and the emulsion was found to be attached significantly to the breast cancer cell line SKBR-3 overexpressing the HER2 receptor (Goldstein et al., 2007). Here, too, the emulsion stability at room temperature conditions and other physicochemical properties, such as size, zeta potential, pH, and drug content were followed up during the whole period of the animal experimentation and no marked changes were noticed.

For the treatment of highly advanced and refractory cancers occurring at breast and ovarian regions of the body, Cremophor EL–based paclitaxel i.v. formulation (Taxol) was initially developed. However, the presence of Cremophor EL (as a solubility enhancer) led to the problems of safety and tolerance. Therefore, reformulation strategy could be applied in order to improve the safety and efficacy of paclitaxel-based anticancer therapy. Because paclitaxel is a hydrophobic compound and has poor aqueous solubility, Lundberg et al. (2003) incorporated the drug into the dispersed oil phase of an emulsion. Unfortunately, the paclitaxel separated from the emulsion system after i.v. injection and mixed with the blood serum because the drug possesses log P value of 4.7. Due to this drug partitioning effect from emulsion to blood serum following the i.v. injection of drug-loaded emulsion, there was no difference in pharmacokinetic parameters of paclitaxel-loaded emulsion and the commercial paclitaxel formulation (Lee et al., 2005). If a drug needs to remain inside, the dispersed oil droplets of the emulsion following i.v. administration, then, the log P value of the selected drug should be >7 (Takino et al., 1994). If not, upon i.v. administration, the emulsion formulation will dilute infinitely and then release the drug quickly in the blood serum. This results in losing the advantage of long blood circulation and possibility of organ-passive targeting. Therefore, it becomes necessary to enhance the lipophilicity of the paclitaxel molecules by esterification reaction with a fatty acid as shown by Lundberg et al. (2003). These authors incorporated the esterified paclitaxel molecules (paclitaxel oleate) into a negatively charged (anionic) emulsion. By following the similar strategy, Goldstein et al. (2007) have produced paclitaxel palmitate (calculated log P value of 9) and entrapped the synthesized esterified paclitaxel into the positively charged (cationic) emulsion before antibody conjugation. Furthermore, the complement system activation following the i.v. administration of anti-HER2 immunoemulsions loaded with paclitaxel palmitate was also studied in order to show that the antiangiogenic mAb will diminish the resistance of the cancer cells and should enhance the cell uptake of paclitaxel palmitate.

2.2.5 Second generation NE for RES-related disease treatment

This section describes the use of second generation NEs for RES-related disease treatment. Upon parenteral administration of unmodified second generation NEs into the human body always results in uptake of the drug carriers by mononuclear phagocyte system (MPS) cells of RES in the liver. This particular and natural phenomenon can be exploited to improve the treatment of parasitic, fungal, viral, and bacterial diseases, such as leishmaniasis, AIDS, and hepatitis B. In general, antimicrobial agents-laden o/w NE contain superior therapeutic effect in comparison to the free antimicrobial agents both in terms of distribution to the relevant intracellular sites and in treating disseminated diseased states effectively. The o/w NE incorporating an antiviral prodrug (nucleoside analogue, iodo-deoxyuridine) and ligands like recombinant apoE was able to reach liver site of Wistar rat (Rensen et al., 1995). According to these authors, the parent drug did not show any affinity for emulsion due to hydrophilic property. However, derivatization with hydrophobic anchors allows not only incorporating at least 130 prodrug molecules per emulsion particle but also does not impart any effect on the NE structure and apo E association to NE. Although without being bound by any theory, the attachment of apo E components onto the NE particle helps the emulsified droplets not to be recognized as foreign particles. But it may allow the body to perceive the emulsified droplets as native chylomicrons or very low density lipoproteins (VLDL). Because of the nonrecognition as foreign particles, the o/w NE circulates more time in the blood and accumulates the emulsion laden cargo nearer to the target tissues, that is, up to 700 Nm drug concentration in liver parenchymal cells. In addition, the clearance rate of the drug decreases, thereby reducing the likelihood of toxic effects of the drug on clearance tissues since accumulation of the drug in other part of the clearance tissues is reduced. In this way, specific organs of the human body may be targeted by using the o/w NE.

Apart from using the apo E receptors, the saccharide moieties of glycolipids and glycoproteins on the cell surface also play an important role in various intercellular recognition processes. For example, mannan, amylopectin, and pullulan polysaccharides may be decorated onto the oil droplets of the NEs (Iwamoto et al., 1991). They have observed a higher accumulation of mannan-decorated NE in the lung in guinea pigs.

Today, the second generation NE can be exploited for both passive and active drug targeting purposes after attachment of targeting ligands and employment of surface modification/PEGylation.

2.3 THIRD GENERATION NE

To increase cellular uptake, cationization strategy is one of the changes that progressively occurred in last decades, particularly on the surfaces of nonviral, colloidal carrier systems, such as liposomes, nano- and microparticulates and nanocapsules. For making the surface of these lipidic and polymeric carrier systems a cationic property, some cationic lipids/polymers are usually added into these systems during/after preparation. But, adding alone the cationic substances in phospholipid stabilized first generation NE does not help to obtain a physically stabilized emulsion on a prolonged storage period. However, using different cationic lipids as emulsifier and additional helper lipids as coemulsifier, for example, 1,2-dioleoyl-*sn*-glycero-3-trimethylammonium-propane (DOTAP), 1,2-dioleoyl-*sn*-glycero-3-phosphoethanolamine (DOPE) and 1-palmitoyl-2oleoyl-*sn*-glycero-3-phosphoethanolamine-*N*-[poly(ethylene glycol)$_{2000}$] (PEG$_{2000}$PE), reports are available to prepare NEs with positive charges on their droplets surface (Kim et al., 2001a,b). Alternatively, after the inclusion of cation-forming substances like lipids (stearyl or oleyl chain having primary amines), polymers (chitosan), and surfactants (cetyltrimethylammonium bromide) during the preparation of second generation NE allows the formation of a stabilized system with positive charges on it. Further, the positive charge caused by stearylamine was also confirmed by a selective adsorption of thiocyanate. Its adsorption was correlated with increasing stearylamine concentration (Elbaz et al., 1993). So, NE consisting of complex emulsifiers, that is, phospholipid-polyoxyethylene surfactant-cationized primary amine or polymer combination can conveniently be called third generation NE.

2.3.1 Gene therapy using third generation NE

When the small molecular drugs are added extemporaneously into the preformed first- and second-generation NEs, the integrity of the emulsion is somewhat compromised in terms of oil and water phases separation or breakdown. Both DNA and single-stranded oligos are water soluble due to their polyanionic character. In case of high- and medium-sized DNA or single-stranded oligos, the only way, up to now, is the direct addition of these compounds, in aqueous solution form, into the preformed third generation NEs. However, the emulsion integrity could be maintained by the electrostatic interaction of the anionic DNA or single-stranded oligos with the cationic emulsion droplets and even these compounds can associate/link superficially at the oil–water interface of the emulsion (Choi et al., 2002; Teixeira et al., 1999). On the other hand, after administration of the DNA- or oligos-laden third generation NEs via parenteral and ocular routes, the release of the DNA and oligos from the associated emulsion droplet surfaces should therefore initially be dependent solely on the affinity between the physiological anions of the biological fluid and cationic surface of the emulsion droplets. The examples of biofluids containing mono- and divalent anions, a multitude of macromolecules and nucleases include both blood plasma and ocular fluids such as tear, aqueous humor, and vitreous. Therefore, there is a possibility that all these anionic endogenous biofluid's components could dissociate the DNA and oligos from cationic NEs following their parenteral or ocular administration. Indirectly, this type of displacement reaction is an indicator of the strength of the interaction (electrostatic) occurred between DNA or oligo and the NE.

Yet it would be of worthwhile to know the displacement reaction at an in vitro cell culture condition/models in the presence of serum. Interestingly, Yi et al. (2000) reported the serum stability of DNA-loaded NE, however, the origin of the serum stability of this emulsion needs further investigation. A comparative study involving both in vitro cell culture and i.v. administration into Balb/c mice of DNA-loaded NEs was conducted by Kim et al. (2003) using a DOTAP-based and DNA-loaded cationic NE. These authors found that the DNA-loaded NE yielded on high transfection levels even in the medium containing 80% serum. It is interesting to see the results of the emulsion developed by Kim et al. (2003), which suggested that in vitro cell culture system containing 80% serum is well mimicking the in vivo situation.

2.3.2 Unique property of the third generation NE

Extracting some beneficial property from the excipients added in the formulations is an attractive way to enhance drug targeting efficacy of colloidal carriers like nanospheres and liposomes. Prolongation in plasma residence time through an escaping tendency from RES uptake and penetration of therapeutic agents into cell surface possibly via endocytotic mechanism are the two main facts that is needed for parenterally administered colloidal carrier systems. Conferring positive-charge (cationization) onto the surface of colloidal carrier systems will enhance the endocytosis of therapeutic agents into cell. Similarly, making hydrophilic surface (pegylation) onto the surface of colloidal carriers will allow the particles to escape from RES uptake and thus prolong the longevity of the injected particles in the blood plasma. Therefore, the longer the time the drug-loaded particles circulate in the blood plasma, the better will be the chance of particles to reach ultimately the target site in the human body. When talking about nanospheres and liposomes, the cationization and pegylation were performed as a separate entity. That is, both nanospheres and liposomes did not possess these two facts together at a single time because of their structural conformation. However, the third generation NEs reported so far have differed significantly in a way that they do hold the combination of cationic and pegylatic surface properties onto them (Floyd, 1999).

The surface properties of the third generation NEs were experimentally tested by many researchers and a few of the studies are being discussed to better understand the impact of such properties on the efficiency of the NEs. The significance of cationization of NEs has been reported in a study based on ocular pharmacokinetics of cyclosporine A. The drug was assimilated with deoxycholic acid-based anionic- and stearylamine-based cationic NEs. When the NEs were applied topically into the rabbit eyes, it was observed that the cationic NE provided prolonged retention of more than 8 h in cornea and conjunctival tissues as compared to anionic NE (Abdulrazik et al., 2001). The reason behind such a response could be attributed to the anionic nature of cornea and conjunctiva at physiological pH (Rojanasakul and Robinson, 1989). The hypothesis behind the reservoir effect of cyclosporine A was the electrostatic interaction of tissues and drug-loaded cationic NE in the eye tissues. The observation was reinforced by another ex vivo study on a rabbit's cornea, where cationic NE carrier showed better wettability than either saline or anionic NE carrier (Klang et al., 2000). From the observations, it could be noticed that modifications on the surface of drug-embedded nanocarriers can influence various pharmacokinetic parameters of the drug. This fact is supported by studies conducted by various investigators as mentioned later. The impact of changed physical properties of NEs on the elimination rate from the blood was studied in order to understand the effect of cationic or anionic charge of NEs' surface on the circulation time and retention of drug in the body organs (Davis, 1982; Wretlind, 1981). It included administration of stearylamine-based cationic or deoxycholic acid-based anionic NE and Intralipid (a commercial anionic NE, which contains [^{14}C] cholesteryl oleate) into the tail vein of male Balb/c mice (20–26 g) at a volume of 5 mL/kg (Klang et al., 1998). The lipophilicity of cholesteryl oleate is very high (log P value 18.3) and it is resistant to degradation in the body, the in vivo behavior imitates the early phase of injected NEs (Takino et al., 1994, 1998). After intravenous administration of [^{14}C] labeled cholesteryl oleate NEs (cationic and anionic) into mice, it was found that the labeled cholesteryl oleate was accumulated in organs such as lung and liver. The accumulation pattern suggested the initial decrease followed by elevation in the radio-labeled compound in the liver over time. In the case of stearylamine-based cationic and deoxycholic acid-based anionic NEs, the concentration of [^{14}C] cholesteryl oleate in the lung was found to be low for both types of formulations as compared to Intralipid. However, the stearylamine-based cationic NEs depicted a longer retention time of [^{14}C] cholesteryl oleate in the plasma as compared to anionic NE. This clearly indicates that the cationic NE can offer long circulation and better reservoir effects than anionic NEs.

The aforementioned studies indicate that the third generation NEs hold exceptional features as compared to first and second generation NEs. Furthermore, the unique properties of the third generation NEs can help to enhance the drug bioavailability and better retention of the drug in the blood due to long circulation time. The reason behind such property of these NEs is recognized as the excess positive charge at the oil–water interface in conjunction with the projection of highly hydrophilic POE chain toward aqueous phase of o/w type emulsion. Thus, the surface modification makes these NEs better than the first and second generation NEs. Still, however detailed structural and stability studies must be undertaken for better understanding of the physical uniqueness in biological efficiency of such NEs when administered as parenteral and ocular dosage forms.

3 FUTURE PERSPECTIVES

The imaging sciences have developed enormously in conjunction with the principles and systems similar to those used in drug delivery and drug targeting. Targeted nanometric contrast agents have been developed based on polymers, lipids, and/or proteins that carry radionuclides, paramagnetic elements, and/or fluorescent probes suitable for imaging (Koning and Krijger, 2007).

The development of targeted nanocarriers in which therapeutic and imaging agents are merged into a single carrier will certainly be of importance in the context of clinically relevant AS and cancer researches. Indeed, scientists active in the field of imaging (e.g., nuclear and magnetic resonance imaging) have already started to exploit nanocarriers for molecular imaging. Image-guided drug delivery

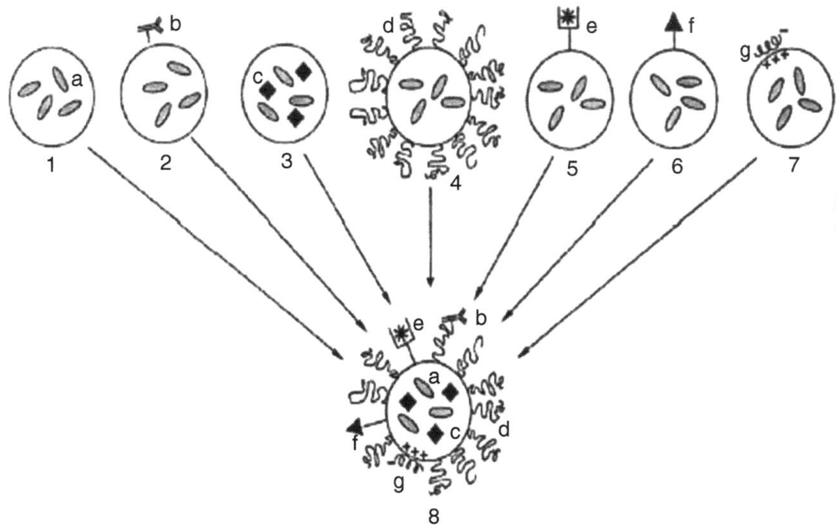

FIGURE 19.2 Systematic Designs of the Multifunctional Nanosized Emulsions

1, Conventional "basic" nanosystem (a: nanosystem-carrying drug cargo); *2*, Conjugated nanosystem or immunosyatem (b: mAb-conjugated nanosystems); *3*, Magnetic nanosystem (c: nanosystem containing magnetic particles and drug that allows detection of nanosystem sensitivity via using an external magnetic field and therefore, this nanosystem can be used as a contrasting agent during magnetic resonance imaging); *4*, Stealth nanosystem (d: hydrophillic PEG decorated nanosystem permitting the extended longevity of the nanosystem in the blood plasma; *5*, Contrast nanosystem for imaging functions (e: nanosystem containing heavy metal atoms such as ^{111}In, ^{99m}Tc, Gd, and Mn, as well as the nanosystem-entrapped-chelating moiety for gamma- or MR imaging application); *6*, Cell-penetrating nanosystem (f: nanosystem-carrying cell-penetrating peptide, CPP, that allows the uptake of carrier by the cells); *7*, Nanosystem-loaded DNA cargo like lipoplex (g: positively charged nanosystem complexed with DNA); *8*, Theoretically proposed and pioneering multifunctional pharmaceutical nanosystem linking all these designs (*1–7*).

using these multifunctional nanocarriers, containing therapeutic and imaging agents, will ultimately allow for online monitoring of tumor location, tumor targeting levels, intratumoral localization and drug release kinetics prior and during radio- and/or chemotherapeutic treatment. This system will contribute to a more personalized approach in cancer therapy, as processes in individual patients can be monitored, and physicians can decide on the best suited therapy based on the individual patient's status and response. However, an extremely interesting result of this parallel and potential development using the o/w-type NE within image-guided drug delivery is not yet explored fully.

Fig. 19.2 depicts the possible systematic designs of the multifunctional NE. From Fig. 19.2, it is clear that the potential of making magnetic nanocarrier (design 3) and contrast nanocarrier for imaging functions (design 5), cell-penetrating nanocarrier for intracellular drug delivery functions (design 6) and DNA-carrying nanocarrier such as lipoplex for correcting genetically determined diseases (design 7) are not scrutinized yet by using NE. All the selected four designs (3, 5, 6, and 7) possess the high potential to provide pioneering research work in the medical field because of their very specific environments/capabilities to visualize the diseased condition at cellular levels. One of the intentions of this chapter is to show the recent results obtained with long-circulating NE based on design 5, wherein the imaging/contrast agent was incorporated successfully to monitor noninvasively the progression of the diseases.

3.1 ATHEROSCLEROSIS THERAGNOSIS

As outlined in the introduction section of this chapter, the imaging part of the theragnostic nanomedicine/approach may conventionally be classified into two types. The first type of imaging includes ex vivo assessment of diseased tissues postmortem and noninvasive imaging primarily of the structural and anatomic features of atherosclerotic plaques. This preliminary and preclinical imaging helps the formulator to identify the best nanoscale formulation platform for investigating and applying the theragnostic principle. For instance, the utility of near infrared (NIR) protease-sensing, ProSense750 fluorescent preclinical imaging agent in combination with FMT (Fluorescence Molecular Tomography) 2500 quantitative preclinical imaging system for the noninvasive quantitative measurement of atherosclerotic disease biology and its response to therapy in apolipoprotein (apo) E-deficient mice in vivo was studied by Peterson (2012). Significant progression of disease was detected in apo E-deficient animals even after ingestion of high cholesterol diet over 20–30 weeks. However, treatment by using atorvastatin has shown an anticipated modest decrease in plasma cholesterol level. This study indicates the progression of AS in conjunction with high fatty food intake might be monitored by FMT and NIR in apo E-deficient mice model of AS.

The second type of imaging explores how to use dual-modal imaging in a single nanoscale delivery system for diagnosing the life-threatening diseases. For instance, macrophages have become widely recognized as a key target for AS imaging because they contribute significantly to the progression of AS. Dual-modal imaging contrast agents with unique X-ray CT and optical imaging capabilities have great potential in disease diagnosis because of complementary combination of the high spatial resolution of CT with the high sensitivity of optical imaging. In order to combine different imaging modalities like magnetic resonance imaging (MRI), X-ray CT and optical imaging into one single nanomaterial, Ding et al. (2013) have developed a multifunctional and dual-modal nanoemulsion platform to investigate atherosclerotic plaques. These authors have developed a kind of QDs(QDs)-iodinated oil nanoemulsion having 80 nm droplet size as a CT/fluorescence dual-modal contrast agents. They have selected hydrophobic QDs and embedded it in iodinated oil phase, which subsequently is dispersed in water phase to form the o/w nanoemulsion. The PEGylated lipids were used as one of the mixed emulsifier molecule in order to improve the circulation time of nanoemulsion. The longer the circulation time of the nanoscale delivery system allows more of the nanoemulsion droplets to accommodate in macrophages of the atherosclerotic plaques for providing a better contrast effect. In this way, the developed nanoemulsion-based QD encapsulated hybrid formulation could make an efficient optical/CT contrast enhancement for cell imaging. Further studies have also shown that the nanoemulsion has great potential of targeting and visualizing atherosclerotic plaques with uptake in vivo by macrophages. Therefore, this nanoemulsion holds a great potential for future applications.

When analyzing these two different research works, the work by Peterson (2012) clearly indicated that the developed delivery system did contain an antiatherosclerotic drug molecule in the formulation while, the work by Ding et al. (2013) did not incorporate any antiatherosclerotic drug molecule in their hybrid nanoscale formulation. However, both of these research works have contributed much on the future direction of managing the AS disease progression, monitoring and treating of the same by theragnostic principle. When compared to the polymer-based nanoparticle platform, which is considered to have a more rigid structural conformation, the more flexible structural components of lipid-based nanoscale formulation like o/w NEs should be used for the development of theragnostic nanoemulsion to manage atherosclerotic plaques. It should be mentioned that Ding et al. (2013) have already made a strong foundation and, in fact, this foundation should gave an impetus to work with nanoemulsion containing the good combination of contrasting agents. One more step forward is just including the antiatherosclerotic agent into the nanoemulsion system. Although currently many research laboratories are already working in this direction, the fruitful results are yet to come for the effective management of atherosclerotic plaques.

3.2 CANCER THERAGNOSIS

Many more research papers, patents, and investigational products at an advanced clinical study are currently available for o/w NE with cancer theragnosis potential. Only a few developmental success stories are narrated in brief in this chapter. At the very outset the multifunctional o/w NEs combine the following structural components:

1. Core material for drug encapsulation usually the oil phase
2. Surface modification for passive or active targeting together with PEGylated lipids for long blood circulation and
3. Imaging component for disease visualization
4. Conjugating the biomarker, homing device or targeting ligands onto the emulsion surfaces

Having mentioned the structural components of multifunctional o/w NEs, the conceptual use of this hybrid system as a multimodal platform for target-specific and image-guided drug delivery for cancer theragnosis is discussed in the next section. Before describing the conceptual use of NEs, the microenvironment of tumor cells needs to be understood. The tumor microenvironment is composed of extracellular matrix, activated fibroblast, immune cells, pericytes, adipocytes, epithelial cells, glial cells, vascular cells, endothelial cells, and proteins. To promote the tumor to progress into a malignant phenotype, all the earlier mentioned components of the tumor microenvironment support the physiology, structure, and function of the tumor-nurturing environment. There is an ample opportunity to exploit the tumor microenvironmental components that support cancer growth, such as angiogenesis, vascular abnormalities, oxygenation, perfusion, pH, and metabolic states as a platform for the delivery of therapeutic agents to tumor tissue.

Angiogenesis is a physiological mechanism that promotes formation of new blood vessels from existing ones. When the core size of the tumor reaches over 2 mm^3, the tumor tissue is exposed to hypoxic conditions. In order to survive in these hypoxic conditions, the cancer cells stimulate themselves angiogenesis to maintain the presence of oxygen and nutrients for their further growth. In these circumstances, cancer cells are simply dependent on the process of angiogenesis for the supply of oxygen and nutrients. If the angiogenesis process is inhibited, then, there is a potential for treating cancer. There are so many natural and synthetic angiogenesis inhibitors utilized for cancer therapy. However, most of these angiogenesis inhibitors have been reported to possess toxicity or delivery problems. To circumvent this problem, the angiogenesis inhibitors are simply incorporated into NE delivery systems. For example, the angiogenesis inhibitor, such as betulinic acid and fumagillin, were successfully incorporated into the NE delivery system to attain potent antiinflammatory activity and to suppress neovasculature with reduced exposure of these inhibitors to vital organs (Dehelean et al., 2013; Winter et al., 2008). The NE delivery system can thus be used to overcome some of the shortcomings of potential angiogenesis inhibitors and therefore exploit a phenomenon that is highly exhibited in the tumor microenvironment compared with normal tissues.

Since the tumor proliferation is depending on the angiogenesis, the surrounding tumor vasculature is poorly developed with pericytes deficiency. This leads to the presence of fenestrations in the vicinity of tumor tissues, which ultimately induce increased vascular permeability at the tumorized area. Similarly tumor tissues often have no or nonfunctional lymphatic vessels, which lead to poor lymphatic drainage from the tumorized area. By combining these two, a new term has been coined, which is called the enhanced permeability and retention (EPR) effect. This term has been considered as a gold standard to explain the delivery mechanism of several nanotherapeutic agents to tumor tissues (Maeda et al., 2000).

Depending on the type of tumor, growth rate, and microenvironment, most of the solid tumors possess a vascular pore size that has ranged between 380 and 780 nm. However, an optimal size range between 10 and 100 nm is the most critical for achieving higher therapeutic accumulation in the tumor microenvironment by the EPR effect (Danhier et al., 2010). Because any drug delivery system that consists of particle sizes below 10 nm might undergo rapid renal filtration and particles higher than 100 nm and gets easily recognized by the cells of the MPS, the developed nanoformulation having the size less than 400 nm could easily extravasate into the tumor tissue (Caron et al., 2013). From the unique property of third generation NE discussed elsewhere in this chapter, it becomes clearer that surface charge is highly critical to evade clearance. In addition, the third generation NEs also consist of a longer circulation time to facilitate ultimate reaching of the drug cargo into the tumor site. Furthermore, a variety of targeting ligands (homing devices) can be attached to the surface of nanoemulsions to ensure the attachment of the emulsion droplets to the receptors, which are overexpressed on tumor cells. In this context, epidermal growth factor receptor (EGFR), transferrin, and folate are the most studied targeting ligands on the nanosized delivery systems. The antiangiogenic, antiproliferative, and proapoptotic activity of EGFR-based targeting of nanoscale carriers have been reported to improve cellular cytotoxicity both at in vitro and in vivo (Acharya et al., 2009; Milane et al., 2011). Similarly a report by Talekar et al. (2012) showed that EGFR and folic acid receptor-mediated targeting of phosphotidyl-inositol-3-kinase inhibitor in ovarian adenocarcinoma cell line shows the possibility of using the NE carrier system. These authors have observed that the targeted NEs enhanced intracellular uptake by receptor-mediated mechanisms together with improved cellular cytotoxicity compared with nontargeted NEs.

It should be stressed that all of the aforementioned examples of targeted NEs could be monitored and visualized at an in vivo condition when an imaging agent is incorporated into the emulsion system. Aside from X-ray CT, MRI, and NIR, ultrasound is also commonly used for theragnostic purposes. Basically, ultrasound is commonly used as an external stimulus due to its accessibility, cost effectiveness, and ability to be used together with multimodal systems, which can incorporate ultrasound contrast agents such as microbubbles. For example, microbubbles containing perfluorocarbons (PFC) NEs can be developed. PFCs are synthetic organic compounds in which all or most of the hydrogen atoms have been replaced with fluorine atoms. The ^{19}F isotope of PFCs is biologically and chemically inert and thus provides excellent sensitivity in vivo. Following intravenous administration of microbubbles containing PFC NE, the emulsion droplets accumulate in the tumor cells by the EPR effect. Under the action of ultrasound externally, the tumor cell–accumulated emulsion droplets containing the ultrasound contrast agent can be converted into bubbles. Using this mechanism, cytotoxic agents-loaded nanoemulsions have been found to follow "post-droplet to bubble transition" and thus released the encapsulated drug with enhanced intracellular uptake and effective tumor regression (Gao et al., 2008; Kaneda et al., 2009; Rapoport et al., 2007, 2009). Similarly Bae et al. (2013) have recently reported the development of bimodal imaging contrast agent, PFC/rhodamine nanoemulsions with MRI and optical imaging capabilities. This emulsion was also infused with folate molecules, which allow localization into folate receptors expressing tumors and thus improving tumor detection and providing excellent signal sensitivity and high specificity. These authors have further concluded that the folate PFC/rhodamine NE could be utilized for early detection of disease, accurate diagnosis, and targeted tumor therapy for clinical applications.

Apart from PFC-based NEs for image-guided therapy, a multifunctional NE containing iron oxide nanocrystal for MRI, Cy7 for NIR fluorescence imaging and prednisolone acetate valerate for therapeutic purpose has recently been reported by Gianella et al. (2011). These authors concluded that the developed multimode nanoemulsion system could be applied for image-guided therapy for cancer.

4 CONCLUSIONS

Based on the unrelenting research input performed during last decades, the multifunctional activities from the o/w NEs are exploited both at laboratory and at commercial level. Commercial exploitations were achieved to improve the systemic bioavailability, including ocular bioavailability of the water nonsoluble drugs using the o/w NE approach. Some of the other functionalities like diagnostic agent–carrier function together with contrasting functionality, and even target-specific imaging functionality receive much attention in different multicentric studies in recent years. On the other hand, theranostic approach by the parenterally administered o/w NEs especially mainly for AS and cancer is an interesting avenue for the interdisciplinary research at laboratories across the world. In the similar fashion, personalized approach is also progressing in the direction for detecting the atherosclerotic plaque at coronary arteries by using multifunctional NEs. Considering the o/w NEs as basic nutrient, lipid drug carrier, stealth carrier, and cationic carriers were the research of previous decades. But, extending the utility of NEs as a multifunctional carrier property such as the image guided drug delivery is—and will be—the research for potential application to detect, manage, and monitor the life-threatening diseases like AS and cancer.

REFERENCES

Abdulrazik, M., Tamilvanan, S., Khoury, K., Benita, S., 2001. Ocular delivery of cyclosporin A. II. Effect of submicron emulsion's surface charge on ocular distribution of topical cyclosporine A. STP Pharm. Sci. 11 (6), 427–432.

Aboagye, E.O., Bhujwalla, Z.M., Shungu, D.C., Glickson, J.D., 1998. Detection of tumour response to chemotherapy by ^{1}H nuclear magnetic resonance spectroscopy: effect of 5-fluorouracil on lactate levels in radiation-induced fibrosarcoma I tumours. Cancer Res. 58 (5), 1063–1067.

Acharya, S., Dilnawaz, F., Sahoo, S.K., 2009. Targeted epidermal growth factor receptor nanoparticle bioconjugates for breast cancer therapy. Biomaterials 30 (29), 5737–5750.

Alexander, R.W., Dzau, V.J., 2000. Vascular biology: the past 50 years. Circulation 102 (20), 112–116.

Alivisatos, A.P., 1996. Perspectives on the physical chemistry of semiconductor nanocrystals. J. Phys. Chem. 100 (31), 13226–13239.

Alivisatos, A.P., 2004. The use of nanocrystals in biological detection. Nat. Biotechnol. 72 (1), 47–52.

Amselem, S., Friedman, D., 1998. Submicron emulsions as drug carriers for topical administration. Benita, S. (Ed.), Submicron Emulsions in Drug Targeting and Delivery, vol 9, Harwood Academic Publishers, Amsterdam, The Netherlands, pp. 153–173.

Antman, E.M., Anbe, D.T., Armstrong, P.W., et al., 2008. ACC/AHA guidelines for the management of patients with ST-elevation myocardial infarction: a report of the American College of Cardiology/American Heart Association Task Force on Practice Guidelines.(Committee to Revise the 1999 Guidelines for the Management of Patients With Acute Myocardial Infarction).

Arimoto, I., Matsumoto, C., Tanaka, M., Okuhira, K., Saito, H., Handa, T., 1998. Surface composition regulates clearance from plasma and triolein lipolysis of lipid emulsions. Lipids 33 (8), 773–779.

Ashamalla, H., Rafla, S., Parikh, K., Mokhtar, B., Goswami, G., Kambam, S., Abdel-Dayem, H., Guirguis, A., Ross, P., Evola, A., 2005. The contribution of integrated PET/CT to the evolving definition of treatment volumes in radiation treatment planning in lung cancer. Int J Radiat Oncol Biol Phys. 63 (4), 1016–1023.

Bae, P.K., Jung, J., Lim, S.J., Kim, D., Kim, S.K., Chung, B.H., 2013. Bimodal perfluorocarbon nanoemulsions for nasopharyngeal carcinoma targeting. Mol. Imaging Biol. 15 (4), 401–410.

Bailey, R.E., Smith, A.M., Nie, S., 2004. Quantum dots in biology and medicine. Physica E 25 (1), 1–12.

Baralle, M., Baralle, F.E., 2000. Genetics and molecular biology. Curr. Opin. Lipidol. 11 (6), 653–656.

Belfiore, G., Tedeschi, E., Moggio, G., Coffi, R., Cincotti, F.C., Rossi, R., 2004. CT-guided radiofrequency palliative ablation of unresectable lung cancer: clinical and imaging findings at one year follow-up. Radiology 233, 513.

Bhadra, D., Bhadra, S., Jain, P., Jain, N.K., 2002. Pegnology: a review of PEG-ylated systems. Pharmazie 57 (1), 5–29.

Bhattacharyya, M., Ryan, D., Carpenter, R., Vinnicombe, S., Gallagher, C.J., 2008. Using MRI to plan breast conserving surgery following neoadjuvant chemotherapy for early breast cancer. Br. J. Cancer 98 (2), 289–293.

Boger, R.H., 2003. Association of asymmetric dimethylarginine and endothelial dysfunction. Clin. Chem. Lab. Med. 41 (11), 1467–1472.

Bonetti, P.O., Lerman, L.O., Lerman, A., 2003. Endothelial dysfunction: a marker of atherosclerotic risk. Arterioscler. Thromb. Vasc. Biol. 23 (2), 168–175.

Boger, R.H., 2005. Asymmetric dimethylarginine (ADMA) and cardiovascular disease: insights from prospective clinical trials. Vasc. Med. 10 (Suppl. 1), S19–25.

Brindle, K., 2008. New approaches for imaging tumour responses to treatment. Nat. Rev. Cancer 8 (2), 94–107.

Brink, I., Schumacher, T., Mix, M., Ruhland, S., Stoelbent, E., Digel, W., 2005. Impact of [18F] FDG-PET on the primary staging of small-cell lung cancer. Eur. J. Nucl. Med. Mol. Imaging 31, 1614–1620.

Brown, T.M., Bittner, V., 2008. Biomarkers of atherosclerosis: clinical applications. Curr. Cardiol. Rep. 10 (6), 497–504.

Bruchez, Jr., M., Moronne, M., Gin, P., Weiss, S., Alivisatos, A.P., 1998. Semiconductor nanocrystals as fluorescent biological labels. Science 281 (5385), 2013–2016.

Byrne, J.D., Betancourt, T., Brannon-Peppas, L., 2008. Active targeting schemes for nanoparticle systems in cancer therapeutics. Adv. Drug Deliv. Rev. 60 (15), 1615–1626.

Calder, P.C., 2009. Hot topics in parenteral nutrition. Rationale for using new lipid emulsions in parenteral nutrition and a review of the trials performed in adults. Proc. Nutr. Soc. 68 (3), 252–260.

Caron, W.P., Lay, J.C., Fong, A.M., La-Beck, N.M., Kumar, P., Newman, S.E., Zhou, H., Monaco, J.H., Pearson, D.L.C., Brewster, W.R., Le, L.V., Jump, V.L.B., Gehrig, P.A., Zambon, W.C., 2013. Translational studies of phenotypic probes for the mononuclear phagocyte system and liposomal pharmacology. J. Pharmacol. Exp. Ther. 347 (3), 599–606.

Chan, W.C.W., Nie, S., 1998. Quantum dot bioconjugates for ultrasensitive nonisotopic detection. Science 281 (5385), 2016–2018.

Chandra, R.K., 1985. Grace A. Goldsmith Award lecture. Trace element regulation of immunity and infection. J. Am. Coll. Nutr. 4 (1), 5–16.

Cheson, B.D., Bennett, J.M., Kopecky, K.J., Buchner, T., Willman, C.L., Estey, E.H., Schiffer, C.A., Doehner, H., Tallman, M.S., Lister, T.A., 2003. Revised recommendations of the international working group for diagnosis, standardization of response criteria, treatment outcomes, and reporting standards for therapeutic trials in acute myeloid leukemia. J. Clin. Oncol. 21, 4642–4649.

Cheson, B.D., Greenberg, P.L., Bennett, J.M., Lowenberg, B., Wijermans, P.W., Nimer, S.D., Pinto, A., Beran, M., de Witte, T.M., Stone, R.M., Mittelman, M., Sanz, G.F., Gore, S.D., Schiffer, C.A., Kantarjian, H., 2006. Clinical application and proposal for modification of the International Working Group (IWG) response criteria in myelodysplasia. Blood 108 (2), 419–425.

Choi, B.Y., Chung, J.W., Park, J.H., Kim, K.H., Kim, Y.I., Koh, Y.H., Kwon, J.W., Lee, K.H., Choi, H.J., Kim, T.W., Kim, Y.J., Chung, H., Kwon, I.C., Jeong, S.Y., 2002. Gene delivery to the rat liver using cationic lipid emulsion/DNA complex: comparison between intra-arterial, intraportal and intravenous administration. Korean J. Radiol. 3 (3), 194–198.

Ciambellotti, E., Coda, C., Lanza, E., 1993. Determinations of CA 15-3 in the control of primary and metastatic breast carcinoma. Minerva Med. 84 (3), 107–112.

Ciernik, I.F., Dizendorf, E., Baumert, B.G., Reiner, B., Burger, C., Davis, J.B., Lutolf, U.M., Steinert, H.C., Von, S.G.K., 2003. Radiation treatment planning with an integrated positron emission and computer tomography (PET/CT): a feasibility study. Int. J. Radiat. Oncol. Biol Phys. 57 (3), 853–863.

Cockshott, I.D., 1985. Propofol ('Diprivan') pharmacokinetics and metabolism: an overview. Postgrad. Med. J. 61 (Suppl. 3), 45–50.

Dabrowska, M., Grubek-Saworska, H., Domagala-Kulawik, J., Bartoszewicz, Z., Kondracka, A., Krenke, R., Nejman, P., Chazan, R., 2004. Diagnostic usefulness of selected tumor markers (CA125, CEA, CoFRA 21-1) in bronchoalveolar lavage fuid in patients with nonsmall cell lung cancer. Pol. Arch. Med. Wewn. 111 (6), 659–665.

Danhier, F., Feron, O., Préat, V., 2010. To exploit the tumor microenvironment: passive and active tumor targeting of nanocarriers for anti-cancer drug delivery. J. Control. Release 148 (2), 135–146.

Davis, S.S., 1982. Emulsions system for the delivery of drugs by the parenteral route. In: Bundgaard, H., Hansen, B.A., Kofod, H. (Eds.), Optimization of Drug Delivery. Munksgaard, Copenhagen, Denmark, pp. 333–346.

Davis, S.S., Hansrani, P., 1985. The influence of emulsifying agents on the phagocytosis of lipid emulsions by macrophages. Int. J. Pharm. 23, 69–77.

Davis, S.S., Illum, L., Washington, C., Harper, G., 1992. Studies on the interaction of charge-reversed emulsions with the reticuloendothelial system. Int. J. Pharm. 82 (1–2), 99–105.

Dehelean, C.A., Feflea, S., Gheorgheosu, D., Ganta, S., Cimpean, A.M., Muntean, D., Amiji, M.M., 2013. Anti-angiogenic and anti-cancer evaluation of betulin nanoemulsion in chicken chorioallantoic membrane and skin carcinoma in Balb/c mice. J. Biomed. Nanotechnol. 9 (4), 577–589.

De Masi, S., Tosti, M.E., Mele, A., 2005. Screening for hepatocellular carcinoma. Dig. Liver Dis. 37 (4), 260–268.

de Torres, J.P., Bastarrika, G., Wisnivesky, J.P., Alcaide, A.B., Campo, A., Seijo, L.M., Pueyo, J.C., Villanueva, A., Lozana, M.D., Montes, U., Montuenga, L., Zulueta, J.J., 2007. Assessing the relationship between lung cancer risk and emphysema detected on low-dose CT of the chest. Chest 132 (6), 1932–1938.

Deckelbaum, R.J., Hamilton, J.A., Moser, A., Bengtsson-Olivecrona, G., Butbul, E., Carpentier, Y.A., Gutman, A., Olivecrona, T., 1990. Medium-chain versus long-chain triacylglycerol emulsion hydrolysis by lipoprotein lipase and hepatic lipase: implications for the mechanisms of lipase action. Biochemistry 29 (5), 1136–1142.

Decker, D.E., Vroegop, S.M., Goodman, T.G., Peterson, T., Buxser, S.E., 1995. Kinetics and thermodynamics of emulsion delivery of lipophilic antioxidants to cells in culture. Chem. Phys. Lipids 76 (1), 7–25.

Ding, J., Wang, Y., Ma, M., Zhang, Y., Lu, S., Jiang, Y., Qi, C., Luo, S., Dong, G., Wen, S., An, Y., Gu, M., 2013. CT/fluorescence dual-modal nanoemulsion platform for investigating atherosclerotic plaques. Biomaterials 34 (1), 209–216.

Eisenhauer, E.A., Therasse, P., Bogaerts, J., Schwatz, L.H., Sargent, D., Ford, R., Dancey, J., Arbuck, S., Gwyther, S., Mooney, M., Rubinstein, L., Shankar, L., Dodd, L., Kaplan, R., Lacombe, D., Verweij, J., 2009. New response evaluation criteria in solid tumours: revised RECIST guideline (version 1.1). Eur. J. Cancer 45 (2), 228–247.

Elbaz, E., Zeevi, A., Klang, S., Benita, S., 1993. Positively charged submicron emulsions: a new type of colloidal drug carrier. Int. J. Pharm. 96, R1–R6.

Farah, R.A., Clinchy, B., Herrera, L., Vitetta, E.S., 1998. The development of monoclonal antibodies for the therapy of cancer. Crit. Rev. Eukaryot. Gene Expr. 8 (3–4), 321–356.

Fass, L., 2008. Imaging and cancer: a review. Mol. Oncol. 2 (2), 115–152.

Ferme, C., Vanel, D., Ribrag, V., Girinsk, T., 2005. Role of imaging to choose treatment. Cancer Imag. 5 (Spec No A), S113–S119.

Floyd, A.G., 1999. Top ten considerations in the development of parenteral emulsions. Pharm. Sci. Technol. Today 4 (2), 134–143.

Folkman, J., 1992. The role of angiogenesis in tumor growth. Semin. Cancer Biol. 3 (2), 65–71.

Franceschini, M.A., Moesta, K.T., Fantini, S., Gaida, G., Gratton, E., Jess, H., Mantulin, W.W., Seeber, M., Schlag, P.M., Kaschke, M., 1997. Frequency-domain techniques enhance optical mammography: initial clinical results. Proc. Natl. Acad. Sci. USA 94 (12), 6468–6473.

Fuster, V., Kelly, B.B., Vedanthan, R., 2011. Promoting global cardiovascular health: moving forward. Circulation 123, 1671–1678.

Gann, P.H., Hennekens, C.H., Stampfer, M.J., 1995. A prospective evaluation of plasma prostate-specific antigen for detection of prostatic cancer. JAMA 273 (4), 289–294.

Gao, Z., Kennedy, A.M., Christensen, D.A., Rapoport, N.Y., 2008. Drug loaded nano/microbubbles for combining ultrasonography and targeted chemotherapy. Ultrasonics 48 (4), 260–270.

Gianella, A., Jarzyna, P.A., Mani, V., Ramachandran, S., Calcagno, C., Tang, J., Kann, B., Dijk, W.J., Thijssen, V.L., Griffioen, A.W., Storm, G., Fayad, Z.A., Mulder, W.J., 2011. Multifunctional nanoemulsion platform for imaging guided therapy evaluated in experimental cancer. ACS Nano 5 (6), 4422–4433.

Gimbrone, M.A., 1995. Vascular endothelium: an integrator of pathophysiologic stimuli in atherosclerosis. Am. J. Cardiol. 75 (6), 67B–70B.

Goldstein, D., Gofrit, O., Nyska, A., Benita, S., 2007. Anti-HER2 cationic immunoemulsion as a potential targeted drug delivery system for the treatment of prostate cancer. Cancer Res. 67 (1), 269–275.

Goldstein, D., Nassar, T., Lambert, G., Kadouche, J., Benita, S., 2005. The design and evaluation of a novel targeted drug delivery system using cationic emulsion-antibody conjugates. J. Control. Release 108 (2–3), 418–432.

Gracie, J.A., Robertson, S.E., McInnes, I.B., 2003. Interleukin-18. J. Leukoc. Biol. 73 (2), 213–224.

Grosenick, D., Wabnitz, H., Moesta, K.T., Schlag, P., 1999. Development of a time-domain optical mammograph and first invivo appliations. Appl. Opt. 38 (13), 2927–2943.

Harris, J.M., Martin, N.E., Modi, M., 2001. Pegylation: a novel process for modifying pharmacokinetics. Clin. Pharmacokinet. 40 (7), 539–551.

He, W., Wang, H.F., Hartmann, L.C., Cheng, J.X., Low, P.S., 2007. In vivo quantitation of rare circulating tumor cells by multiphoton intravital flow cytometry. Proc. Natl. Acad. Sci. USA 104 (28), 11760–11765.

Hedeman, H., Brondsted, H., Müllertz, A., Frokjaer, S., 1996. Fat emulsions based on structured lipids (1,3-specific triglycerides): an investigation of the in vivo fate. Pharm. Res. 13 (5), 725–728.

Hedges, J.R., Young, G.P., Henkel, G.F., Gibler, W.B., Green, T.R., Swanson, J.R., 1994. Early CK-MB elevations predict ischemic events in stable chest pain patients. Acad. Emerg. Med. 1 (1), 9–16.

Horrocks, L.A., Yoe, Y.K., 1999. Health benefits of docosahexaenoic acid (DHA). Pharmacol. Res. 40 (3), 211–225.

Hui, Z., Bing, O., Bao, M., Xia, F., Yan, L.W., Hai, Y.Y., et al., 2007. Comparison of ultrasound elastography, mammography, and sonography in the diagnosis of solid breast lesions. J. Ultrasound Med. 26 (6), 807–815.

Iga, A.M., Robertson, J.H.P., Winslet, M.C., Seifalian, A.M., 2007. Clinical potential of quantum dots. J. Biomed. Biotech. 1, 1–10.

Illum, L., West, P., Washington, C., Davis, S.S., 1989. The effect of stabilising agents on the organ distribution of lipid emulsions. Int. J. Pharm. 54 (1), 41–49.

Iwamoto, K., Kato, T., Kawahara, M., Koyama, N., Watanabe, S., Miyake, Y., Sunamoto, J., 1991. Polysaccharide-coated oil droplets in oil-in-water emulsions as targetable carriers for lipophilic drugs. J. Pharm. Sci. 80 (3), 219–224.

Iyer, A.K., Khaled, G., Fang, J., Maeda, H., 2006. Exploiting the enhanced permeability and retention effect for tumor targeting. Drug Discov. Today 11 (17–18), 812–818.

Medel, R., Monteith, S.J., Elias, W.J., Eames, M., Snell, J., Sheehan, J.P., Wintermark, M., Jolesz, F.A., Kassell, N.F., 2012. Magnetic resonance image-guided focused ultrasound surgery: Part 2: a review of current and future applications. Neurosurgery 71 (4), 755–763.

Jurlander, B., Clemmensen, P., Wagner, G.S., Grande, P., 2000. Very early diagnosis and risk stratification of patients admitted with suspected acute myocardial infarction by the combined evaluation of a single serum value of cardiac troponin-T, myoglobin, and creatine kinase MB(mass). Eur. Heart J. 21 (5), 382–389.

Kambayashi, Y., Nakao, K., Kimura, H., Kawabata, T., Nakamura, M., Inouye, K., Yoshida, N., Imura, H., 1990a. Biological characterization of human brain natriuretic peptide (BNP) and rat BNP: species-specific actions of BNP. Biochem. Biophys. Res. Commun. 173 (2), 599–605.

Kambayashi, Y., Nakao, K., Mukoyama, M., Saito, Y., Ogawa, Y., Shiono, S., Inouye, K., Yoshida, N., Imura, H., 1990b. Isolation and sequence determination of human brain natriuretic peptide in human atrium 37. FEBS Lett. 259 (2), 341–345.

Kaneda, M., Caruthers, S., Lanza, G.M., Wickline, S.A., 2009. Perfluorocarbon nanoemulsions for quantitative molecular imaging and targeted therapeutics. Ann. Biomed. Eng. 37 (10), 1922–1933.

Keidar, Z., Haim, N., Guralnik, L., Wollner, M., Bar-Shalom, R., Ben-Nun, A., Israel, O., 2004. PET/CT using FDG in suspected lung cancer recurrence: diagnostic value and impact on patient management. J. Nucl. Med. 45 (10), 1640–1646.

Kent, M.S., Port, J.L., Altorki, N.K., 2004. Current state of imaging for lung cancer staging. Thorac. Surg. Clin. 14 (1), 1–13.

Kim, T.W., Chung, H., Kwon, I.C., Sung, H.C., Jeong, S.Y., 2001a. Optimization of lipid composition in cationic emulsion as in vitro and in vivo transfection agents. Pharm. Res. 18 (1), 54–60.

Kim, Y.J., Kim, T.W., Chung, H., Kwon, I.C., Sung, H.C., Jeong, S.Y., 2001b. Counterion effects on transfection activity of cationic lipid emulsion. Biotechnol. Bioprocess Eng. 6 (4), 279–283.

Kim, Y.J., Kim, T.W., Chung, H., Kwon, I.C., Sung, H.C., Jeong, S.Y., 2003. The effects of serum on the stability and the transfection activity of the cationic lipid emulsion with various oils. Int. J. Pharm. 252 (1–2), 241–252.

Kim, E.K., Kim, T., Paik, S., Haam, S., Huh, Y.M., Lee, K., 2015. Nanomaterials for theranostics: recent advances and future challenges. Chem. Rev. 115 (1), 327–394.

Klang, S., Abdulrazik, M., Benita, S., 2000. Influence of emulsion droplet surface charge on indomethacin ocular tissue distribution. Pharm. Dev. Technol. 5 (4), 521–532.

Klang, S.H., Parnas, M., Benita, S., 1998. Emulsions as drug carriers-possibilities, limitations, and future perspectives. In: Muller, R.H., Benita, S., Bohm, H.L. (Eds.), Emulsions and Nanosuspensions for the Formulation of Poorly Soluble Drugs. Medpharm, Stuttgart, pp. 31–65.

Koch, T., Heller, A.R., 2005. Auswirkungen einer parenteralen ernahrung mit n-3-fettsauren auf das therapieergebnis-eine multizentrische analyse bei 661 patienten (Effects of parenteral nutrition with n-3-fatty acids on the result of therapy- a multicentre analysis with 661 patients). Akt. Ernahrungs 30, 15–22.

Koenig, W., Khuseyinova, N., 2007. Biomarkers of artherosclerosis plaque instability and rupture. Arterioscler. Thromb. Vasc. Biol. 27 (1), 15–26.

Koning, G.A., Krijger, G.C., 2007. Targeted multifunctional lipid-based nanocarriers for image-guided drug delivery. Anti-Cancer Agents Med. Chem. 7 (4), 425–440.

Kurhanewicz, J., Vigneron, D.B., Nelson, S.J., 2000. Three dimensional magnetic resonance spectroscopic imaging of brain and prostate cancer. Neoplasia 2 (1–2), 166–189.

Kurihara, A., Shibayama, Y., Mizota, A., Yasuna, A., Ikeda, M., Sasagawa, K., Kobayashi, T., Hisaoka, M., 1996. Lipid emulsions of palmitoylrhizoxin: effects of composition on lipolysis and biodistribution. Biopharm. Drug Dispos. 17 (4), 331–342.

Lamberts, S.W., Van den Beld, A.W., Van der Lely, A.J., 1997. The endocrinology of aging. Science 278 (5337), 419–424.

Langstrom, B., Itsenko, O., Rahman, O., 2007. [^{11}C] Carbon monoxide, a versatile and useful precursor in labelling chemistry for PET-ligand development. J. Label. Comp. Radiopharm. 50 (9–10), 794–810.

Le Bihan, D., Breton, E., 1985. Imagerie of diffusion in-lifepar Résonance magnétique nucléaire. C. R. Acad. Sci. (Paris) 301 (15), 1109–1112.

Lee, I.H., Park, Y.T., Ron, K., Chung, H., Kwon, I.C., Jeong, S.Y., 2005. Stable paclitaxel formulations in oily contrast medium. J. Control. Release 102 (2), 415–425.

Lee, J.H., Lee, H.B., Andrade, J.D., 1995. Blood compatibility of polyethylene oxide surfaces. Prog. Polym. Sci. 20 (6), 1043–1079.

Lee, K.S., Jeong, Y.J., Han, J., Kim, B.T., Kim, H., Kwon, O.J., 2004. T1 non-small cell lung cancer: imaging and histopathologic findings and their prognostic implications. Radiographics 24 (6), 1617–1636.

Lehman, C.D., Issacs, C., Schnall, M.D., Pisano, E.D., Ascher, S.M., Weatherall, P.T., Blumke, D.A., Bowen, D.J., Marcom, P.K., Armstrong, D.K., Domchek, S.M., Tomlinson, G., Skates, S.J., Gatsonis, C., 2007. Cancer yield of mammography, MR and US in high-risk women: prospective multi-institution breast cancer screening study. Radiology 244 (2), 381–388.

Leleiko, R.M., Vaccari, C.S., Sola, S., et al., 2009. Usefulness of elevations in serum choline and free F2-isoprostane to predict 30-day cardiovascular outcomes in patients with acute coronary syndrome. Am. J. Cardiol. 104 (5), 638–643.

Lencioni, R., Llovet, J.M., 2010. Modified RECIST (mRECIST) assessment for hepatocellular carcinoma. Semin. Liver Dis. 30 (1), 52–60.

Lerner, R.M., Huang, S.R., Parker, K.J., 1990. "Sonoelasticity" images derived from ultrasound signals in mechanically vibrated tissues. Ultrasound Med. Biol. 16 (3), 231–239.

Levy, M.Y., Benita, S., Baszkin, A., 1991. Interactions of a non-ionic surfactant with mixed phospholipid-oleic acid monolayers. Studies under dynamic conditions. Colloids Surf. 59, 225–241.

Li, C., Guo, B., wilson, P.B., Stewart, A., Byrne, G., Bundred, N., Kumar, S., 2000. Plasma levels of soluble CD105 correlate with metastasis in patients with breast cancer. Int. J. Cancer 89 (2), 122–126.

Liapi, E., Geschwind, J.F., Vossen, J.A., Buijs, M., Georgiades, C.S., Blumke, D.A., 2008. Functional MRI evaluation of tumor response in patients with neuroendocrine hepatic metastases treated with transcatheter arterial chemoembolization. Am. J. Roentgenol. 190 (1), 67–73.

Libby, P., Ridker, P.M., Hansson, G.K., 2011. Progress and challenges in translating the biology of atherosclerosis. Nature 473 (7347), 317–325.

Libby, P., Sukhova, G., Lee, R.T., Liao, J.K., 1997. Molecular biology of atherosclerosis. Int. J. Cardiol. 62 (Suppl. 2), S23–S29.

Lima, N., Cavaliere, H., Tomimori, E., Knobel, M., Medeiros, N.G., 2002. Prognostic value of serial serum thyroglobulin determinations after total thyroidectomy for differentiated thyroid cancer. J. Endocrinol. Invest. 25 (2), 110–115.

Lin, S.Y., Wu, W.H., Lui, W.Y., 1992. In vitro release, pharmacokinetic and tissue distribution studies of doxorubicin hydrochloride (Adriamycin HCl) encapsulated in lipiodolized w/o emulsions and w/o/w multiple emulsions. Pharmazie 47 (6), 439–443.

Liu, F., Liu, D., 1995. Long-circulating emulsions (oil-in-water) as carriers for lipophilic drugs. Pharm. Res. 12 (7), 1060–1064.

Lundberg, B.B., Griffiths, G., Hansen, H.J., 1999. Conjugation of an anti-B-cell lymphoma monoclonal antibody, LL2, to long-circulating drug-carrier lipid emulsions. J. Pharm. Pharmacol. 51 (10), 1099–1105.

Lundberg, B.B., Mortimer, B.C., Redgrave, T.G., 1996. Submicron lipid emulsions containing amphipathic polyethylene glycol for use as drug-carriers with prolonged circulation time. Int. J. Pharm. 134 (1–2), 119–127.
Lundberg, B.B., Risovic, V., Ramaswamy, M., Wasan, K.M., 2003. A lipophilic paclitaxel derivative incorporated in a lipid emulsion for parenteral administration. J. Control. Release 86 (1), 93–100.
Luttun, A., Tjwa, M., Moons, L., Wu, Y., Angelillo-Scherrer, A., Liao, F., Nagy, J.A., Hooper, A., Priller, J., De Klerck, B., Compernolle, V., Daci, E., Bohlen, P., Dewerchin, M., Herbert, J.M., Fava, R., Matthys, P., Carmeliet, G., Collen, D., Dvorak, H.F., Hicklin, D.J., Carmeliet, P., 2002. Revascularization of ischemic tissues by PlGF treatment, and inhibition of tumor angiogenesis, arthritis and atherosclerosis by anti-Flt1. Nat. Med. 8 (8), 831–840.
Maahs, D.M., Ogden, L.G., Kretowski, A., Bergeon, S.J.K., Kinney, G.L., Berl, T., Rewers, M., 2007. Serum cystatin C predicts progression of subclinical coronary atherosclerosis in individuals with type I diabetes. Diabetes 56 (11), 2774–2779.
Maeda, H., Wu, J., Sawa, T., Matsumura, Y., Hori, K., 2000. Tumor vascular permeability and the EPR effect in macromolecular therapeutics: a review. J. Control. Release 65 (1–2), 271–284.
Maier, W., Altwegg, L.A., Corti, R., Gay, S., Hersberger, M., Maly, F., Sutsch, E., Roffi, G., Neidhart, M., Eberli, M., Tanner, F.R., Gobbi, F.C., Von, S., Eckardstein, A., Luscher, T.F., 2005. Inflammatory markers at the site of ruptured plaque in acute myocardial infarction: locally increased interleukin-6 and serum amyloid A but decreased C-reactive protein. Circulation 111 (11), 1355–1361.
Majdalawieh, A., Zhang, l., Ro, H.S., 2007. Adipocyte enhancer-binding protein-1 promotes macrophage inflammatory responsiveness by up-regulating NF-B via IB negative regulation. Mol. Biol. Cell. 18 (3), 930–942.
Markowitz, S.D., 1997. Atherosclerosis: just another cancer? J. Clin. Invest. 100 (9), 2143–2145.
Matoba, M., Tonami, H., Kondou, T., Yokota, H., Higashi, K., Toga, H., Sakuma, T., 2007. Lung carcinoma: diffusion-weighted MR imaging—preliminary evaluation with apparent diffusion coefficient. Radiology 243 (2), 570–577.
Matsuki, M., Inada, Y., Nakai, G., 2007. Diffusion-weighed MR imaging of pancreatic carcinoma. Abdom. Imaging 32 (4), 481–483.
Medintz, I.L., Uyeda, H.T., Goldman, E.R., Mattoussi, H., 2005. Quantum dot bioconjugates for imaging, labeling and sensing. Nat. Mater. 4, 435–446.
Mensah, G.A., Ryan, U.S., Hooper, W.C., Engelgau, M.M., Callow, A.D., 2007. Vascular endothelium summary statement II: cardiovascular disease prevention and control. Vascul. Pharmacol. 46 (5), 318–320.
Mian, C., Lodde, M., Haitel, A., Egarter, V.E., Marberger, E., Pycha, A., 2000. Comparison of two qualitative assays, the UBC rapid test and the BTA stat test, in the diagnosis of urothelial cell carcinoma of the bladder. J. Urol. 56 (2), 228–231.
Michalet, X., Pinaud, F.F., Bentolila, L.A., Tsay, J.M., Doose, S., Li, J.J., Sundaresan, G., Wu, A.M., Gambhir, S.S., Weiss, S., 2005. Quantum dots for live cells, in vivo imaging, and diagnostics. Science 307 (5709), 538–544.
Milane, L., Duan, Z., Amiji, M., 2011. Development of EGFR-targeted polymer blendnanocarriers for combination paclitaxel/lonidamine delivery to treat multi-drug resistance in human breast and ovarian tumor cells. Mol. Pharm. 8 (1), 185–203.
Miller, A.B., Hoogstraten, B., Staquet, M., Winkler, A., 1981. Reporting results of cancer treatment. Cancer 47 (1), 207–214.
Milne, G.L., Musiek, E.S., Morrow, J.D., 2005. F2-isoprostanes as markers of oxidative stress in vivo: an overview. Biomarkers 10 (Suppl. 1), S10–S23.
Mishra, A., Verma, M., 2010. Cancer biomarkers: are we ready for the prime time. Cancers 2 (1), 190–208.
Miyanaga, N., Akaza, H., Yamakawa, M., Oikawa, T., Sekido, N., Hinotsu, S., Kawai, K., Shimazui, T., Shiina, T., 2006. Tissue elasticity imaging for diagnosis of prostate cancer: a preliminary report. Int. J. Urol. 13 (12), 1514–1518.

Mizushima, Y., Hoshi, K., Aihara, H., Kurachi, M., 1983. Inhibition of bronchoconstriction by aerosol of a lipid emulsion containing prostaglandin E1. J. Pharm. Pharmacol. 35 (6), 397.

Mor, G., Visintin, I., Lai, Y., Zhao, H., Schwatz, P., Rutherford, T., Yue, L., Bray-Ward, P., Ward, C., 2005. Serum protein markers for early detection of ovarian cancer. Proc. Natl. Acad. Sci. 102 (21), 7677–7682.

Morrisett, J.D., 2000. The role of lipoprotein[a] in atherosclerosis. Curr. Atheroscler. Rep. 2 (3), 243–250.

Murray, C.J., Lopez, A.D., 1997. Global mortality, disability, and the contribution of risk factors: global burden of disease study. Lancet 349 (9063), 1436–1442.

Namiki, M., Kawashima, S., Yamashita, T., Ozaki, M., Hirase, T., Ishida, T., Inoue, N., Hirata, K., Matsukawa, A., Morishita, R., Kaneda, Y., Yokoyama, M., 2002. Local overexpression of monocyte chemoattractant protein-1 at vessel wall induces infiltration of macrophages and formation of atherosclerotic lesion: synergism with hypercholesterolemia. Arterioscler. Thromb. Vasc. Biol. 22 (1), 115–120.

Napoli, C., Lerman, L.O., de Nigris, F., Gossl, M., Balestrieri, M.L., Lerman, A., 2006. Rethinking primary prevention of atherosclerosis-related diseases. Circulation 114 (23), 2517–2527.

Nelson, E.D., Slotoroff, C.B., Gomella, L.G., Halpern, E.J., 2007. Targeted biopsy of the prostate: the impact of color Doppler imaging and elastography on prostate cancer detection and Gleason score. J. Urol. 70 (6), 1136–1140.

Neves, A.A., Brindle, K.M., 2006. Assessing responses to cancer therapy using molecular imaging. Biochim. Biophysi. Acta 1766 (2), 242–261.

Newby, L.K., Storrow, A.B., Gibler, W.B., Garvey, J.L., Tucker, J.F., Kaplan, A.L., Schreiber, D.H., Tuttle, R.H., McNulty, S.E., Ohman, E.M., 2001. Bedside multimarker testing for risk stratification in chest pain units: the chest pain evaluation by creatine kinase-MB, myoglobin, and troponin I (CHECKMATE) study. Circulation 103 (14), 1832–1837.

Nourooz, Z.J., 2008. Key issues in F2-isoprostane analysis. Biochem. Soc. Trans. 36 (5), 1060–1065.

Ozdemir, V., Williams, J.B., Glatt, S.J., Tsuang, M.T., Lohr, J.B., Reist, C., 2006. Shifting emphasis from pharmacogenomics to theragnostics. Nat. Biotechnol. 24 (8), 942–946.

Paajanen, H., 2006. Increasing use of mammography improves the outcome of breast cancer in Finland. Breast J. 12 (1), 88–90.

Pallwein, L., Mitterberger, M., Struve, P., Pinggera, G., Horninger, W., Bartsch, G., Aigner, F., Lorenz, A., Pedross, F., Frauscher, F., 2007. Real-time elastography for detecting prostate cancer: preliminary experience. BJU Int. 100 (1), 42–46.

Peer, D., Karp, J.M., Hong, S., Farokhzad, O.C., Margalit, R., Langer, R., 2007. Nanocarriers as an emerging platform for cancer therapy. Nat. Nanotechnol. 2 (12), 751–760.

Pericleous, P., Gazouli, M., Lyberopoulou, A., Rizos, S., Nikiteas, N., Efstathopoulos, E.P., 2012. Quantum dots hold promise for early cancer imaging and detection. Int. J. Cancer 131 (3), 519–528.

Peterson, J.D., 2012. Non-invasive quantitative in vivo imaging of atherosclerosis disease progression and treatment response in apo E deficient mice using fluorescent molecular tomography and NIR fluorescent preclinical imaging agents. Perkin Elmer Leaflet, 1–7.

Pogue, B.W., Poplack, S.P., McBride, T.O., Wells, W.A., Osterman, K.S., Osterberg, U.L., Paulsen, K.D., 2001. Quantitative hemoglobin tomography with diffuse near infrared spectroscopy: pilot results in the breast. Radiology 218 (1), 261–266.

Poplack, S.P., Paulsen, K.D., Hartov, A., Meaney, P.M., Pogue, B.W., Tosteson, T.D., Grove, M.R., Soho, S.K., Wells, W.A., 2004. Electromagnetic breast imaging: average tissue property values in women with negative clinical findings. Radiology 231 (2), 571–580.

Poplack, S.P., Tosteson, T.D., Wells, W.A., Pogue, B.W., Meaney, P.M., Hartov, A., Kogel, C.A., Soho, S.K., Gibson, J.J., Paulsen, K.D., 2007. Electromagnetic breast imaging: results of a pilot study in women with abnormal mammograms. Radiology 243 (2), 350–359.

Porter, C.J., Charman, S.A., Charman, W.N., 1996. Lymphatic transport of halofantrine in the triple-cannulated anesthetized rat model: effect of lipid vehicle dispersion. J. Pharm. Sci. 85 (4), 351–356.

Poykko, S.M., Kellokoski, E., Ukkola, O., Kauma, H., Paivansalo, M., Kesaniemi, Y.A., Horkko, S., 2006. Plasma ghrelin concentrations are positively associated with carotid artery atherosclerosis in males. J. Intern. Med. 260 (1), 43–52.
Provenzale, J.M., Mukundan, S., Barboriak, D., 2006. Diffusion-weighted and perfusion MR imaging for brain tumor characterization and assessment of treatment response. Radiology 239 (3), 632–649.
Rapoport, N., Gao, Z., Kennedy, A., 2007. Multifunctional nanoparticles for combining ultrasonic tumor imaging and targeted chemotherapy. J. Natl. Cancer Inst. 99 (14), 1095–1106.
Rapoport, N.Y., Kennedy, A.M., Shea, J.E., Scaife, C.L., Nam, K.H., 2009. Controlled and targeted tumor chemotherapy by ultrasound-activated nanoemulsions/microbubbles. J. Control. Release 138 (3), 268–276.
Rensen, P.C.N., Van Dijk, M.C.M., Havenaar, E.C., Bijsterbosch, M.K., Kruijt, J.K., Van Berkel, T.J.C., 1995. Selective liver targeting of antivirals by recombinant chylomicrons: a new therapeutic approach to hepatitis. Nat. Med. 1 (3), 221–225.
Ridker, P.M., 2007. C-reactive protein and the prediction of cardiovascular events among those at intermediate risk. J. Am. Coll. Cardiol. 49 (21), 2129–2138.
Rojanasakul, Y., Robinson, J.R., 1989. Transport mechanisms of the cornea: characterization of barrier permselectivity. Int. J. Pharm. 55 (2–3), 237–246.
Ross, R.J., Thompson, J.S., Kim, K., Bailey, R.A., 1982. Nuclear magnetic resonance imaging and evaluation of human breast tissue: preliminary clinical trials. Radiology 143 (1), 195–205.
Rubin, M., Harell, D., Naor, N., Moser, A., Wielunsky, E., Merlob, P., Lichtenberg, D., 1991. Lipid infusion with different triglyceride cores (long-chain vs medium-chain/long-chain triglycerides): effect on plasma lipids and bilirubin binding in premature infants. J. Parenter. Enteral. Nutr. 15 (6), 642–646.
Saba, T.M., 1970. Physiology and physiopathology of the reticuloendothelial system. Arch. Intern. Med. 126 (6), 1031–1052.
Saftoiu, A., Vilman, P., 2006. Endoscopic ultrasound elastography–a new imaging technique for the visualization of tissue elasticity distribution. J. Gastrointestin. Liver Dis. 15 (2), 161–165.
Sakaeda, T., Hirano, K., 1995. O/W lipid emulsions for parenteral drug delivery. II. Effect of composition on pharmacokinetics of incorporated drug. J. Drug Target. 3 (3), 221–230.
Sakaeda, T., Takahashi, K., Nishihara, Y., Hirano, K., 1994. O/W lipid emulsions for parenteral drug delivery. I. Pharmacokinetics of the oil particles and incorporated sudan II. Biol. Pharm. Bull. 17 (11), 1490–1495.
Sarkar, F.H., Banerjee, S., Li, Y.W., 2007. Pancreatic cancer: pathogenesis, prevention and treatment. Toxicol. Appl. Pharm. 224 (3), 326–336.
Sarkeala, T., Heinavaara, S., Anttila, A., 2008. Breast cancer mortality with varying invitational policies in organised mammography. Br. J. Cancer 98 (3), 641–645.
Scher, H.I., Halabi, S., Tannock, I., Morris, M., Sternberg, C.N., Carducci, M.A., Eisenberger, M.A., Higano, C., Bubley, G.J., Dreicer, R., Petrylak, D., Kantoff, P., Basch, E., Kelly, W.K., Figg, W.D., Small, E.J., Beer, T.M., Wilding, G., Martin, A., Hussain, M., 2008. Design and end points of clinical trials for patients with progressive prostate cancer and castrate levels of testosterone: recommendations of the Prostate Cancer Clinical Trials Working Group. J. Clin. Oncol. 26 (11), 1148–1159.
Schindhelm, R.K., Van der Zwan, L.P., Teerlink, T., Scheffer, P.G., 2009. Myeloperoxidase: a useful biomarker for cardiovascular disease risk stratification? Clin. Chem. 55 (8), 1462–1470.
Shapiro, B.P., Chen, H.H., Burnett, J.C., Redfield, M.M., 2003. Use of plasma brain natriuretic peptide concentration to aid in the diagnosis of heart failure. Mayo Clin. Proc. 78 (4), 481–486.
Shim, S.S., Lee, K.S., Kim, B., Lee, E.J., Choi, J.Y., Chung, M.J., 2005. Non-small cell lung cancer: prospective comparison ofintegrated FDG PET/CT and CT alone for preoperativestaging. Radiology 236 (3), 1011–1019.
Sinning, C., Schnabel, R., Peacock, W.F., Blankenberg, S., 2008. Up-and-coming markers: myeloperoxidase, a novel biomarker test for heart failure and acute coronary syndrome application? Congest. Heart Fail. 14 (4 Suppl. 1), 46–48.
Smith, J.J., Sorensen, A.G., Thrall, J.H., 2003. Biomarkers in imaging: realizing radiology's future. Radiology 227 (3), 633–638.

Song, Y.K., Liu, D.X., Maruyama, K., Takizawa, T., 1996. Antibody mediated lung targeting of long-circulating emulsions. PDA J. Pharm. Sci. Tech. 50 (6), 372–377.

Sprecher, H., 2002. The roles of anabolic and catabolic reactions in the synthesis and recycling of polyunsaturated fatty acids. Prostaglandins Leukot. Essent. Fatty Acids 67 (2–3), 79–83.

Stroobants, S., Geominne, J., Seegers, M., Dimitrijevic, S., Dupont, P., Nuyts, J., Martens, M., van den Borne, B., Cole, P., Sciot, R., Dumez, H., Silberman, S., Mortelmans, L., van Oosterom, A., 2003. (18) FDG-Positron emission tomography for the early prediction of response in advanced soft tissue sarcoma treated with imatinib mesylate (Glivec). Eur. J. Cancer 39 (14), 2012–2020.

Stubbs, P., Seed, M., Lane, D., Collinson, P., Kendall, F., Noble, M., 1998. Lipoprotein(a) as a risk predictor for cardiac mortality in patients with acute coronary syndromes. Eur. Heart J. 19 (9), 1355–1364.

Takayoshi, U., Yuen, S., Kasami, M., Uchida, Y., 2007. Dynamic contrast enhanced screening of microcalcifications in the breast. Is there any value. Breast Cancer Res. Treat. 103 (3), 269–281.

Takino, T., Konishi, K., Takakura, Y., Hashida, M., 1994. Long circulating emulsion carrier systems for highly lipophilic drugs. Biol. Pharm. Bull. 17 (1), 121–125.

Takino, T., Nakajima, C., Takakura, Y., Sezaki, H., Hashida, M., 1993. Controlled biodistribution of highly lipophilic drugs with various parenteral formulations. J. Drug Target. 1 (2), 117–124.

Takino, T., Koreeda, N., Nomura, T., Sakaeda, T., Yamashita, F., Takakura, Y., Hashida, M., 1998. Control of plasma cholesterol-lowering action of probucol with various lipid carrier systems. Biol. Pharm. Bull. 21 (5), 492–497.

Talekar, M., Ganta, S., Singh, A., Amiji, M., Kendall, J., Denny, W.A., Garg, S., 2012. Phosphatidyl-inositol 3-kinase inhibitor (PIK75) containing surface functionalized nanoemulsion for enhanced drug delivery, cytotoxicity and pro-apoptotic activity in ovarian cancer cells. Pharm. Res. 29 (10), 2874–2886.

Tam, A., Ahra, K., 2007. Palliative interventions for pain in cancer patients. Sem. Intervent. Radiol. 24 (4), 419–429.

Tamilvanan, S., Benita, S., 2004. The potential of lipid emulsion for ocular delivery of lipophilic drugs. Europ. J. Pharm. Biopharm. 58 (2), 357–368.

Teixeira, H., Dubernet, C., Puisieux, F., Benita, S., Couvreur, P., 1999. Submicron cationic emulsions as a new delivery system for oligonucleotides. Pharm. Res. 16 (1), 30–36.

Therasse, P., Arbuck, S.G., Eisenhauer, E.A., Wanders, J., Kaplan, R.S., Rubinstein, L., Verweij, J., Van, G.M., Van, O.A.T., Christian, M.C., Gwyther, S.G., 2000. New guidelines to evaluate the response to treatment in solid tumors (RECIST guidelines). J. Natl. Cancer Inst. 92 (3), 205–216.

Thygesen, K., Alpert, J.S., White, H.D., 2007. Joint ESC/ACCF/AHA/WHF task force for the redefinition of myocardial infarction. Universal definition of myocardial infarction. Eur. Heart. J. 28 (20), 2525–2538.

Tibell, A., Larsson, M., Alvestrand, A., 1993. Dissolving intravenous cyclosporin A in a fat emulsion carrier prevents acute renal side effects in the rat. Transpl. Int. 6 (2), 69–72.

Tirucherai, G.S., Pezron, I., Mitra, A.K., 2002. Novel approaches to nasal delivery of peptides and proteins. STP Pharm. Sci. 12, 3–12.

Tousoulis, D., Siasos, G., Oikonomou, E., Oikonomou, E., Stougianos, P., Papageorgiou, N., Papavassiliou, A.G., Stefanadis, C., 2011. Asymmetric dimethylarginine (ADMA): is really a biomarker for cardiovascular prognosis? Int. J. Cardiol. 153 (2), 123–125.

Transparency Market Research, 2014. Cancer diagnostics market (tumor biomarker tests, imaging, endoscopy, and biopsy): global industry analysis, size, share, growth, trends and forecast, 2014–20 (Oct 31). Available from: http://www.transparencymarketresearch.com/cancer-diagnostics-market.html

Tromberg, B.J., Shah, N., Lanning, R., Cerussi, A., Espinoza, J., Pham, T., Svaasand, L., Butler, J., 2000. Noninvasive in vivo characterization of breast tumors using photon migration spectroscopy. Neoplasia 2 (1–2), 26–40.

Tsutsumi, M., Miyagawa, T., Matsumura, T., Kawazoe, N., Ishikawa, S., Shimokama, T., Shiina, T., Miyanaga, N., Akaza, H., 2007. The impact of real-time tissue elasticity imaging (elastography) on the detection of prostate cancer: clinicopathological analysis. Int. J. Clin. Oncol. 12 (4), 250–255.

Ueda, K., Ishida, M., Inoue, T., Fujimoto, M., Kawahara, Y., Sakaeda, T., Iwakawa, S., 2001. Effect of injection volume on the pharmacokinetics of oil particles and incorporated menatetrenone after intravenous injection as O/W lipid emulsions in rats. J. Drug Target. 9 (5), 353–360.

Ueda, K., Yamazaki, Y., Noto, H., Teshima, Y., Yamashita, C., Sakaeda, T., Iwakawa, S., 2003. Effect of oxyethylene moieties in hydrogenated castor oil on the pharmacokinetics of menatetrenone incorporated in O/W lipid emulsions prepared with hydrogenated castor oil and soybean oil in rats. J. Drug Target. 11 (1), 37–43.

Van den Bent, M.J., Wefel, J.S., Schiff, D., Taphoom, M.J., Jaeckle, K., Junck, L., Armstrong, T., Choucair, A., Waldman, A.D., Gorlia, T., Chamberlain, M., Baumert, B.G., Vogelbaum, M.A., Macdonald, D.R., Reardon, D.A., Wen, P.Y., Chang, S.M., Jacobs, A.H., 2011. Response assessment in neuro-oncology (a report of the RANO group): assessment of outcome in trials of diffuse low-grade gliomas. Lancet Oncol. 12 (6), 583–593.

Webster, T.J., Lee, S., An, S.S.A., 2015. Today's diverse nano-theranostic applications and tomorrow's promises. Int. J. Nanomed. 10 (Spec Iss), 1–2.

Werner, N., Nickenig, G., 2006. Influence of cardiovascular risk factors on endothelial progenitor cells: limitations for therapy? Arterioscler. Thromb. Vasc. Biol. 26 (2), 257–266.

Winter, P.M., Schmieder, A.H., Caruthers, S.D., Keene, J.L., Zhang, H., Wickline, S.A., Lanza, G.M., 2008. Minute dosages of $\alpha_V\beta_3$-targeted fumagillin nanoparticles impair Vx-2 tumor angiogenesis and development in rabbits. FASEB J. 22 (8), 2758–2767.

Wheeler, J.J., Wong, K.F., Ansell, S.M., Masin, D., Bally, M.B., 1994. Polyethylene glycol modified phospholipids stabilize emulsions prepared from triacylglycerol. J. Pharm. Sci. 83 (11), 1558–1564.

WHO, 1948. Preamble to the Constitution of the World Health Organization as adopted by the International Health Conference, New York, 19–22 June 1946, and entered into force on 7 April 1948.

Wingo, P.A., Ries, L.A., Rosenberg, H.M., Miller, D.S., Edward, B.K., 1998. Cancer incidence and mortality, 1973–1995: a report card for the U.S. Cancer 82 (6), 1197–1207.

Wolchok, J.D., Hoos, A., O'Day, S., Webber, J.S., Hamid, O., Lebbe, C., Maio, M., Binder, M., Bohnsack, O., Nichol, G., Humphrey, R., Hodi, F.S., 2009. Guidelines for the evaluation of immune therapy activity in solid tumors: immune-related response criteria. Clin. Cancer Res. 15 (23), 7412–7420.

Wolf, C., Lorenzen, J.M., Stein, S., Tsikas, D., Störk, S., Weidemann, F., Ertl, G., Anker, S.D., Bauersachs, J., Thum, T., 2010. Urinary asymmetric dimethylarginine (ADMA) is a predictor of mortality risk in patients with coronary artery disease. Int. J. Cardiol. 156 (30), 289–294.

Wretlind, A., 1981. Parenteral nutrition. Nutr. Rev. 39 (7), 257–265.

Xiao, T., Ying, W., Li, L., Hu, Z., May, Y., Jiao, L., Ma, J., Cai, Y., Lin, D., Guo, S., Han, N., Di, X., Li, M., Zhang, D., Su, K., Yuan, J., Zheng, H., Gao, M., He, J., Shi, S., Li, W., Xu, N., Zhang, H., Liu, Y., Zhang, K., Gao, Y., Qian, X., Cheng, S., 2005. An approach to studying lung cancer related proteins in human blood. Mol. Cell Proteomics 4 (10), 1480–1486.

Xie, J., Lee, S., Chen, X., 2010. Nanoparticle-based theranostic agents. Adv. Drug Deliv. Rev. 62 (11), 1064–1079.

Yamaguchi, K., Nagano, M., Torada, N., Hamasaki, N., Kawakita, M., Tanaka, M., 2004. Urine diacetylspermine as a novel tumor marker for pancreatobiliary carcinomasj. Rinsho Byori 52 (4), 336–339.

Yamaguchi, T., Nishizaki, K., Itai, S., Hayashi, H., Ohshima, H., 1995. Physicochemical characterization of parenteral lipid emulsion: influence of cosurfactants on flocculation and coalescence. Pharm. Res. 12 (9), 1273–1278.

Yeung, D.K.W., Yang, W.T., Tse, G.M.K., 2002. Breast cancer: in vivo proton MR spectroscopy in the characterization of histopathologic subtypes and preliminary observations in axillary node metastases. Radiology 225 (1), 190–197.

Yi, S.W., Yune, Y., Kim, T.W., Chung, H., Choi, Y.W., Kwon, I.C., Lee, E.B., Jeong, S.Y., 2000. A cationic lipid emulsion/DNA complex as a physically stable and serumresistant gene delivery system. Pharm. Res. 17 (3), 314–320.

FURTHER READING

Klang, S.H., Benita, S., 1998. Design and evaluation of submicron emulsions as colloidal drug carriers for intravenous administration. In: Benita, S. (Ed.), Submicron Emulsion in Drug Targeting & Delivery. Harwood Academic, The Netherlands, pp. 119–152.

Libby, P., 2000. Changing concepts of atherogenesis. J. Intern. Med. 247 (3), 349–358.

Wretlind, A., 1972. Complete intravenous nutrition. Theoretical and experimental background. Nutr. Metab. 14 (Suppl.), 1–57.

CHAPTER

20

THERAPEUTIC NANOSTRUCTURES FOR PULMONARY DRUG DELIVERY

Yousef Javadzadeh, Shadi Yaqoubi
Tabriz University of Medical Science, Tabriz, Iran

CHAPTER OUTLINE

1 INTRODUCTION

According to the hieroglyphs carved in Egypt in 1500 BC, inhalation is one of the most historic drug delivery routes (Anderson, 2005). For centuries the pulmonary approach has been known as the optimal route to deliver pharmacologically active agents to treat local diseases of the respiratory tract (such as asthma, chronic obstructive pulmonary disease, infections of respiratory tract, etc.). Over time, inhalation offered some superior qualities, which led researchers to study the lungs as a delivery route for systemic purposes; the increased surface of lung epithelium as a result of several divisions of airways together with the thin epithelial layer of alveolar sacs allows significant absorption of inhaled particles to the lung capillaries. Increased rate and extent of absorption means the minimizing of administered dose, which in turn will lead to decreased systemic side effects. Combination of the noninvasive characteristic of pulmonary delivery with the fact that there is no need for health care staff when using this delivery method leads to increasing patient compliance.

Nanoparticulate systems are widely studied for various drug delivery purposes. These fine systems have the ability to enhance the rate and extent of dissolution of poorly soluble drugs. Furthermore they are capable of creating systems with modified physicochemical properties.

Combining the advantages of pulmonary drug delivery with the outstanding attributes of nanosized particles offers a drug delivery approach with promising potential.

There are in vivo and in vitro methods available in order to determine the fate and deposition profile of inhaled formulations. The fate of inhaled particles is highly dependent on their size and morphological aspects, and the best aerodynamic size for pulmonary deposition is in the range of 1–5 μm; particles larger than 5 μm are mostly deposited in upper airways where the inspired air moves with a high velocity, and the throat, where the direction of air flow changes, is the most likely site among the upper airways to deposit large particles. Particles with an aerodynamic diameter below 0.5 μm are entrapped by Brownian motion and suspended for a long time in the inspired air; these particles are likely to be exhaled and leave the respiratory tract before deposition (Gonda, 2003; Williams et al., 2007). Thus nanoparticulate systems need to form a micron-size particle in the suitable range. In order to manufacture nanoparticle aerosol systems, there are several methods, such as spray drying, spray freeze drying, and so forth that improve inhalable drug particles in appropriate size range.

Pulmonary delivery devices are used in order to provide particles or droplets with inhalable characteristics; nebulizers, metered-dose inhalers, and dry powder inhalers are three main types of aerosols used for human inhalation.

This chapter will discuss the benefits of nanoparticles in the case of pulmonary drug delivery, various drugs, and nanocarrier systems proper for inhalation, pulmonary delivery devices, and the deposition and fate of inhaled particles in the respiratory tract.

2 NANOPARTICLES

The area of nanoscaled particulate systems that is related to medical science is referred to as nanomedicine by the National Institutes of Health (Bethesda, MD, USA) (Moghimi et al., 2005; Park et al., 2008; Webster, 2006). This term covers a vast applicability area in medical science including monitoring the changes with biological systems and treatment, diagnosis, and control of disease. This wide range of applicability is the main reason of increasing attention to this science. Diagnostic uses, such as imaging and high potential drug delivery efficiency (Steichen et al., 2013; Wang et al., 2012) of nanoparticles are the two most common applications of such systems.

There are many advantageous with nanomaterial making them very useful for medical purposes; reduced size of nanomedicine is the most important characteristic, which makes it a different system. The large surface to volume ratio associated with the small size is the foremost attention-grabbing specificity of nanoparticles; an increased surface area means a larger amount of molecules and atoms on the surface when compared with the inside of particles; the higher the proportion of the molecules or atoms on the surface of a particle, the higher the interaction with surrounding materials. In the case of application of nanoparticles as drug delivery carriers, this interaction will occur with pharmaceutically active agents and the larger surface area will lead to an increased drug loading efficiency (Hans and Lowman, 2002). Furthermore, when the surrounding area is the dissolving medium, an enhancement in the rate and extent of dissolution will be observed. The other advantage of nanoparticles is the enhanced uptake by tissue cells, which is also a sequel of surface enlargement (Lesniak et al., 2013; Walkey et al., 2012).

Compared with larger drug particles, nanosizing leads to improvement of physical and chemical characteristics. Modified drug release properties, a controllable release pattern, and many other advantages are examples of improved properties by nanoparticles (Adibkia et al., 2011, 2013; Javadzadeh et al., 2010). Other advantages of nanosized particulate systems are the ability to target the site of action, which is very valuable in the case of cancer therapy (Brannon-Peppas and Blanchette, 2012; Wang et al., 2012; Xiao et al., 2012) or diagnosis and monitoring of malignant or premalignant tissues (Rosen et al., 2012; Xie et al., 2011).

2.1 ADVANTAGES OF NANOPARTICLES IN PULMONARY DRUG DELIVERY

The effect of particle size on efficiency of pulmonary drug delivery is negligible. It is well known that to avoid particle exhalation it must be large enough, and on the other hand inhaled particles must be fine enough to avoid the precipitation in the upper respiratory tract, before reaching the alveolus. Table 20.1 shows the fate of various inhaled particles in the respiratory tract. Therefore the best size for inhalation therapy is in the range of 1–5 μm (Hamishehkar et al., 2012; Paranjpe and Müller-Goymann, 2014).

On the other hand, the mentioned outstanding attributes of nanoparticles make them the optimum size for pulmonary absorption. Thus, in order to obtain an optimized drug delivery system with maximum inhalable portion and maximum epithelial absorption, aggregated nanoparticle systems are

Table 20.1 Mechanism of Particles Deposition in the Respiratory Tract (Yang et al., 2008; Pilcer et al., 2012)

Size Range	Deposition Mechanism	Site of Deposition
Larger than 5 μm	Impaction	Upper airways (e.g., throat)
1–5 μm	Inertial impaction/gravitational sedimentation	Lower respiratory tract (e.g., bronchial tree and alveoli)
1–3 μm	Inertial impaction/gravitational sedimentation	Alveolus tissue
Below 0.5 μm	Diffusion	Mostly exhaled by the expiratory air follow

preferable. Such systems can be achieved via self-aggregation of nanoparticles during the preparation of inhalable powders, or they may be prepared by adsorption of nanoparticles on the surface of micron-sized carriers.

The advantages of inhalation of pharmaceutically active agents along with the outstanding attributes of nanoparticles make the pulmonary delivery of nanoparticles an interesting drug delivery approach for many purposes.

Nanosized pharmaceuticals offer several advantages compared with formulations containing larger particles; some of these advantages are:

1. Modified absorption to lung epithelium, as mentioned earlier the fine particle size and enlarged surface to mass ratio of nanoparticles is an opportunity to have a great absorption profile to alveolar epithelial cells.
2. Decreased velocity of dissolution associated with a decrease in particle size is confirmed by the Noyes–Whitney equation. It is more valuable in the case of nanostructured systems due to the information that says the size dependent enhancement of dissolution rate occurs with particles of a size below 1 μm (Adibkia et al., 2013; Dizaj et al., 2015).
3. Nanosized particulate systems are the most unlikely systems to remove via pulmonary macrophage (especially particles with size below 250 nm) (Shoyele and Cawthorne, 2006).
4. In the case of systemic drug delivery via lungs nanoparticles have the advantages of rapid absorption to blood vessels, which in turn leads to a short lag time to onset of action.

3 PULMONARY DRUG DELIVERY

Pulmonary administration of pharmaceutically active agents is now of burgeoning interest due to its several advantages. Some of the most important outstanding attributes of pulmonary drug delivery that makes it a unique and preferable drug delivery approach are (1) noninvasive characteristic of inhalation therapy; (2) avoidance of the first metabolic pathway; (3) the large surface area available by lung alveoli, which makes it suitable for systemic absorption; (4) the short lag time before the drug onset of action; and (5) the ability to reach blood supply to the lungs, and so forth.

Generally there are two main factors affecting the efficacy of inhaled medicine; the first and most important parameter is the physiological specificities of the respiratory tract, and the second very important factor is the characteristics of the inhaled formulation. Each of these parameters will be explained later. As mentioned earlier, pulmonary drug delivery is the preferable route of administration in order to produce both local and systemic effects: the site where drug particles will be deposited, drug absorption and the blood supplement of the deposition site, and different mechanisms that have a potential role in drug clearance.

There are a number of factors to consider during the formulation of a drug to deliver to the lungs; the physiology of human respiratory tract, and the size and morphology of particles are the three key factors determining the fate of inhaled particles.

3.1 PHYSIOLOGY OF PULMONARY SYSTEM

In order to design optimized drug delivery system, it is important to have complete information about the human respiratory tract and to follow the fate of the inhaled drug due to the physiological properties of pulmonary system (Beck-Broichsitter et al., 2012; Yu and Chien, 1997). Upper and lower airways connecting to each other via the larynx form the human respiratory tract. Upper airways include the nasal cavity, nasopharynx, oropharynx, and laryngopharynx. After the larynx there are lower airways consisting of trachea, bronchi, and terminal bronchioles. Lungs located on either side of the chest are the main part of the pulmonary system. Each of right and left lungs is divided into lobes by pleural membranes; three lobes on the left and two lobes on the right (Effros, 2006). Subsequent divisions of each lobe create lobules and hexagonal divisions of lungs, which are the smallest divisions of the lungs. Once inhaled, drug particles pass through the upper airways and larynx to reach the trachea, and after that there will be a series of bronchi and bronchioles. Twenty-three generations of bronchioles are available: in the 17th generation, alveolar ducts become apparent, and alveolar sacs are available at the end of the 23rd generation (Effros, 2006; Kleinstreuer et al., 2008a,b; Kunda et al., 2013; Rytting et al., 2008). With the movement from the trachea to deep within the lungs there are several changes in the epithelial layer: the type and thickness of epithelial cells and the mucosal layers vary in different stages of bronchial branches. Three available types of cells in the airways are ciliated cells, secretory cells, and basal cells. And two types of alveolar cells (type I and type II) form the alveoli (Steimer et al., 2005). As the airways become narrower in deep lungs the epithelial layer, which acts as a barrier against drug absorption, becomes thinner to about 500 nm in the alveolar sacs (compared with 40 μm in the bronchioles). Moreover, the thickness of the surfactant layer is thinner in lower airways and reaches to 50–80 nm in alveolar sacs (while the thickness of mucosal layer in upper airways is about 5–10 μm) (Rytting et al., 2008; Scheuch et al., 2006). The gas exchange in the alveolar sacs supplies the required oxygen to the bloodstream, which carries it to all cells in the body.

Pulmonary drug delivery is the inhalation of drug formulation through mouth and the further deposition of inhaled pharmacological agent in lower airways is the main purpose of this drug delivery route. The site of the respiratory tract in which drug particles are deposited is the determinative factor; in the case of systemic drug delivery via lungs it is necessary for particles to reach the terminal bronchi and be absorbed by the blood vessels present in the alveolar ducts. On the other hand, in the case of pulmonary drug delivery for treatment of local respiratory disease the absorption of drug particles into the blood vessels is not favorable.

3.2 ADVANTAGES OF PULMONARY DRUG DELIVERY

The pulmonary route has unique characteristics, making it an interesting approach for drug delivery. Some of the most important advantages of pulmonary drug delivery are listed here:

1. Direct delivery of pharmaceutically active agent to the site of action; in the case of local respiratory disease, pulmonary administration benefits several attributes, such as lowering or even elimination of systemic side effects, allowing for a lower administered dose, which in turn leads to fewer side effects and enhanced patient compliance (Azarmi et al., 2008). Furthermore, direct drug delivery has the superiority of gathering all of administered drug in the site of injury.
2. Rapid onset of action.
3. Bypass of the first hepatic metabolism.
4. Large available surface associated with several bifurcations and subdivisions of lung that end in the alveolar sacs, allowing for efficient absorption by the pulmonary route (Al-Qadi et al., 2012).
5. Low enzymatic activity of respiratory tract, which avoids degradation or disruption of drug particles before the absorption.

6. Very thin epithelial layer of the alveolar sacs, providing an appropriate area for absorption of drug particles.
7. Access to the blood supply of lungs, because of of the great number of arterioles available in the epithelium of alveolar sacs.
8. Less phagocytic uptake of nanoparticles in the respiratory tract; it has been reported that the particles with diameters less than 250 nm show minimal uptake by the phagocytic defense of the respiratory tract (Rytting et al., 2008; Shoyele and Cawthorne, 2006).

4 DRUGS SUITABLE FOR PULMONARY DRUG DELIVERY

Various investigations confirm the applicability of inhalation therapy for a wide range of drugs, including large molecular proteins and peptides, and either gene for both systemic and local drug delivery purposes. In order to treat local disease of the respiratory tract, pulmonary delivery is an ideal approach and many investigations have reported its potential efficiency, for example, application of antiinflammatory agents to cure chronic local disease of the respiratory tract, such as asthma or chronic obstructive pulmonary disease (COPD), which is a common example of the application of drugs with small molecules (Bassler, 2015; Cazzola et al., 2012; Chauhan and Ducharme, 2012; Loukides et al., 2013). Antibiotics are another common example of pulmonary drug delivery in case of local disease (Adi et al., 2010; De Jesús Valle et al., 2007; Le Brun et al., 2002; Pilcer et al., 2013a; Stass et al., 2013; Sung et al., 2009; Ungaro et al., 2012a; Westerman et al., 2007). Various investigations have illustrated the potential of pulmonary delivery to deliver proteins and peptides for either systemic or local targeting (Dombu and Betbeder, 2013; Ógáin et al., 2011). On the other hand, the main focus of gene delivery is on the local disease by lungs and respiratory tract (Beck-Broichsitter et al., 2012; Chow and Lam, 2015).

5 NANOSTRUCTURES FOR DRUG DELIVERY TO THE LUNGS

Various types of nonstructural systems are available that can be used in order to deliver different drugs to the lungs. One can choose the most suitable nanomedicine type according to the physical and chemical properties of the pharmacologic agent.

5.1 NANOSUSPENSIONS FOR PULMONARY DRUG DELIVERY

Nanosuspensions, also known as nanocrystals, are nanostructured systems that totally are composed of the active agent (Müller et al., 2011). Fig. 20.1 illustrates the schematic view of a nanocrystal. The 100% drug loading of these nanoparticles makes them specific systems to overcome the problem associated with delivery of poorly soluble drugs; in case of drugs that are poorly soluble in both water and organic solvents, nanocrystals are the proper solution (Muller and Keck, 2004; Patravale and Kulkarni, 2004). Nanosuspensions are the colloidal systems in which nanosized drug particles are dispersed and stabilized in order to avoid crystal growth (Junghanns and Müller, 2008). In the case of pulmonary drug delivery, nanosuspensions are preferable due to their high drug-loading efficiency, which is in conformity with limitations of pulmonary delivery in the use of excipients. Many investigations

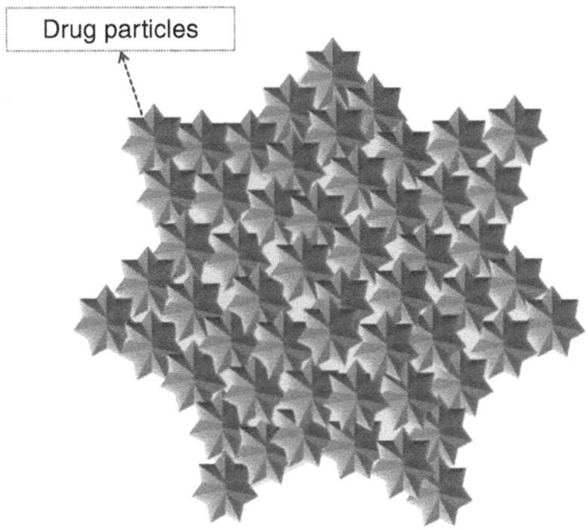

FIGURE 20.1 A Schematic View of Nanocrystals

have observed the supremacy of nanosuspensions for administrating antibacterial (Pilcer et al., 2013b), antifungal (Rundfeldt et al., 2009, 2013), antiinflammatory agents (Chiang et al., 2009), and so forth.

5.2 SOLID LIPID NANOPARTICLES FOR PULMONARY DRUG DELIVERY

Solid lipid nanoparticles (SLNs) composed of solid lipids (at room temperature) were developed in the 1990s as alternatives for polymeric-, liposomal-, and emulsion-based nanosized systems (Mansour et al., 2009; Müller et al., 2002). A schematic view of SLNs is shown in Fig. 20.2, while Fig. 20.3 illustrates a scheme of nanoemulsions. Their superior characteristics, compared to polymeric systems, are reduced irritability and less toxicity for the patient (Heydenreich et al., 2003). Furthermore natural solid lipids have more rapid biodegradability compared with even degradable polymers. Physical and chemical stability of SLNs is another advantageous factor that makes them a preferable alternative for

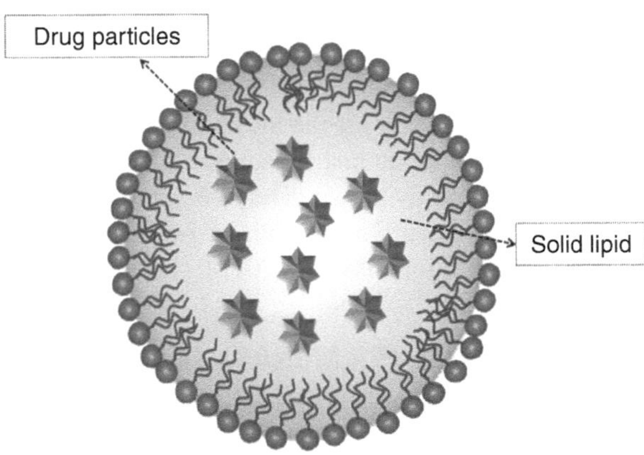

FIGURE 20.2 A Scheme of Solid Lipid Nanoparticle (SLN)

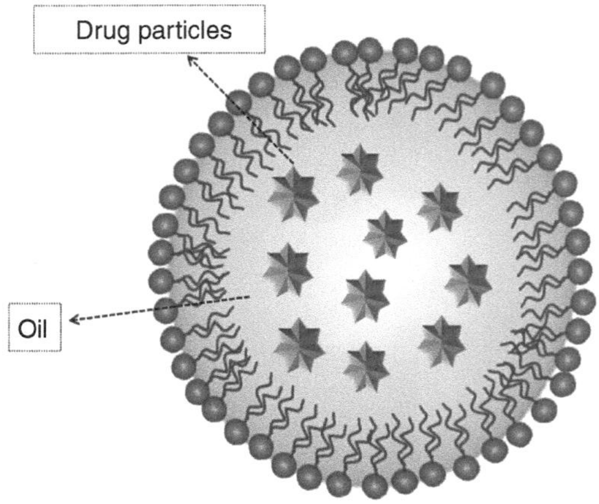

FIGURE 20.3 A Schematic View of an Emulsion-Based Nanoscaled System

emulsions and liposomes. Various researchers have reported the applicability of SLN systems for both systemic and local drug delivery via the respiratory tract. More common lipid-based nanoparticles, such as SLN and nanostructured lipid carriers (NLCs) are prepared in the form of nebulizers (Paranjpe et al., 2014; Pardeike et al., 2011; Pastor et al., 2014) or dry powder inhalers (Bi et al., 2009; Ezzati Nazhad Dolatabadi et al., 2014; Murthy et al., 2012; Rosiere et al., 2015; Weber et al., 2014).

5.3 LIPOSOMAL PULMONARY DRUG DELIVERY

In the case of pulmonary delivery of therapeutics, liposomes are one of the most favorable vehicles (Schreier et al., 1993) due to the nature of materials applied in the formulation of these nanosized systems; it has been reported that self-assembly of phospholipids is the basic principle of the formation of liposomes (Tikshdeep et al., 2012). On the other hand it is well known that phospholipids are the main part of lung surfaces (Wauthoz and Amighi, 2014), thus the endogenous nature of liposomes makes them a very attractive drug delivery system for lungs. Another advantage of liposomes in pulmonary delivery is the uniform deposition of locally active drugs (Manca et al., 2012; Shaji and Bhatia, 2013). Fig. 20.4 presents a schematic picture of liposomal systems. In spite of the extensive application of liposomal formulations in the form of nebulizers, they are not very appreciated now because of the questionable stability of the vesicles in liquid systems (Manca et al., 2012) and the leakage risk of loaded therapeutics from liposomes. In order to overcome the instability problem of nebulizers, nowadays dry powder forms of liposomes are of growing interest (Islam and Cleary, 2012; Willis et al., 2012). Due to the self-assembly of cationic liposomes with negatively charged genetic materials, simple production procedure, cost-effectiveness, and scalability there is growing interest in providing gene-loaded cationic liposomes (Boulikas, 2013; Zarogouldis et al., 2012).

5.4 POLYMERIC NANOPARTICULATE PULMONARY DELIVERY

Nanosized systems consist of biodegradable and biocompatible polymers are used as pulmonary delivery vehicles (Rytting et al., 2008) with the advantages of protecting the therapeutic against degradation and enzymatic metabolism. Additionally polymeric carriers have the ability to produce a controllable

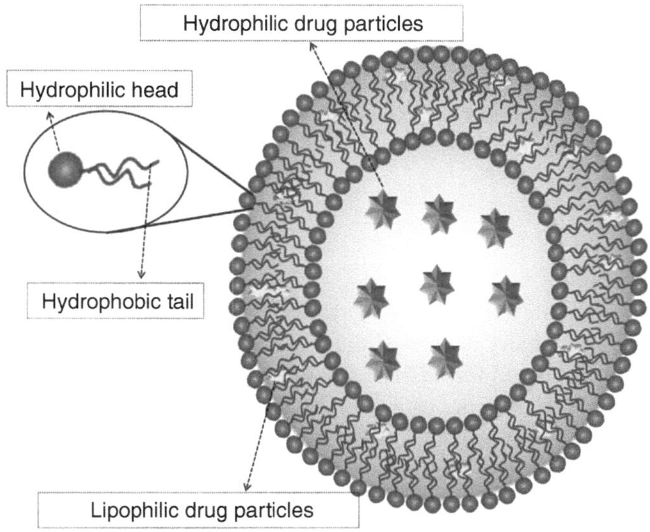

FIGURE 20.4 A Scheme of a Liposome Containing Both Hydrophilic and Lipophilic Drugs in Hydrophil Core and Hydrophobic Shell, Respectively

pattern of drug release. Some of the most common polymers used to prepare nanostructured carriers are poly(lactic-*co*-glycolic acid) (PLGA) (Emami et al., 2014; Ohashi et al., 2009; Rytting et al., 2008; Ungaro et al., 2012a,b), poly(lactic acid) (PLA), poly(alkyl-cyano-acrylate), chitosan (Makhlof et al., 2010), and alginate (Rytting et al., 2008). In spite of the mentioned advantages there are some limitations with polymeric carriers in the case of pulmonary delivery: the irritant and immunogenic specificity of polymers. In order to avoid any cytotoxicity or immunogenicity, any biodegradable polymer or copolymer should be completely studied before the application (Harush-Frenkel et al., 2010). And both in vivo and in vitro tests have to confirm the safety of the polymer itself and its degradation products.

Like the liposomes, polymeric systems with a positively charged surfaces are promising gene delivery vehicles (Bivas-Benita et al., 2004; Borchard, 2001; De Smedt et al., 2000; Gary et al., 2007), but first the limitations of polymer safety for the respiratory tract has to be overcome. The addition of modifier agents, such as poly(ethylene glycol) (PEG) or liposomes may help to solve this problem (Mansour et al., 2009).

6 PULMONARY DELIVERY DEVICES

Generally there are three main types of medical inhalers used to deliver therapeutic agents to the lungs; these devices include nebulizer, metered-dose inhaler (MDI), and dry powder inhaler (DPI). Each type of device has an exclusive drug delivery mechanism and requires a specific type of drug formulation. Furthermore, each inhaler device has a range of several advantages and disadvantages, providing an optional area for the clinician to choose from. Although in most cases all three types are applicable but knowledge about the inhaler devices will be very helpful in choosing the most suitable one. In order to make a choice about the combination of inhaler device and drug, there are some factors to keep in mind: type of the formulation that the pharmacological agent will be provided with is one of the most important factors, the other very important parameter is physiological and pathological circumstances in which they affect the inhalation and physical ability of the patient, for example, the age of patient. Two other very important parameters—cost effectiveness and patient compliance—are factors to take into account (Geller, 2005).

6.1 NEBULIZER

Nebulizers are classic pulmonary drug delivery devices. Glass nebulizers were an innovation of the late 19th century as the first inhaler devices (Labiris and Dolovich, 2003). These types of pulmonary drug delivery devices require liquid formulations to produce polydispersed particles in the range of an inhalable size. There are two main types of nebulizers available: the first and most common type is jet nebulizer that works based on changes in air pressure. The liquid formulation containing drug particles along with the compressed air within the inhaler chamber pass through a very narrow tube and enter a wide area; the increase in the volume of compressed air leads to reduction of its pressure and atomizes the liquid into micron size droplets. In order to catch large and noninhalable particles there is a baffle within the inhaler. Finally, small inhalable particles of drug will be sucked into the respiratory track to reach the site of action or the site of systemic absorption.

The other type of nebulizer utilizes ultrasonic power in order to produce micron-scaled inhalable particles; ultrasonic nebulizers break down the drug solution into small droplets via piezoelectric vibration. There is a baffle within the inhaler, as well as jet nebulizers, to remove large droplets and return them to the main chamber of device.

6.2 METERED DOSE INHALER (MDI)

Metered-dose inhaler is an inhaler device with specified doses that provides a certain dose of the active agent in each puff. MDIs consist of a suspension or dispersion of one or more active ingredient in propellant or a mixture of propellants or a mixture of solvent and propellants, are available since 1950 (Hindle et al., 1995). The major problem of MDIs that was a limitation to use this type of aerosols was the biohazard nature of their propellant; chlorofluorocarbon (CFC) propellants are known as damaging to the environment (Bohm, 1997; Manzer, 1990; Tripp, 1987). Therefore, hydrofluoroalkane (HFA) was chosen as an alternative compound for CFCs. The role of propellant in an MDI system is to provide the required pressure to atomize drug formulation into micron-scaled droplets. Pressurized metered-dose inhalers (pMDI) are a type of MDIs that works based on the pressurized propellant in the aerosol chamber. The advantage of MDIs is the accurate and repeatable dosing, which minimizes the errors.

6.3 DRY POWDER INHALER

Dry powder inhalers (DPI) are the third type of inhaler device that seems to be the most promising devices for future use (Hamishehkar et al., 2012). The breath actuating possibility of these devices is an advanced feature that overcomes the problem of coordination of patient breathing and the activation of the device (Malcolmson and Embleton, 1998). There are many other outstanding attributes of DPIs that make the most preferable pulmonary drug delivery device. Promising physical stability associated with

DPIs is due to the solid form of the formulation. In addition these types of pulmonary drug delivery devices are easy to use, portable and applicable for a wide range of pharmacologically active agents. Although pMDIs were the most commonly used devices for many years, after the Montreal Protocol in 1987 to limit the use of CFCs in products, DPIs became the most preferable devices. There are two types of classifications about DPI systems; the first one divides them based on the dose into two classes of single and multidose inhalers. According to this classification, single-dose DPIs are systems supplying each dose in an individual hard gelatin capsule, and the multiple-dose devices are those with a multiple-dosed reservoir within the inhaler chamber. The second type classification divides DPIs into passive and active systems; passive inhalers are the forms of DPIs that are completely breath actuated once the patient starts to inhale; a dry powder formulation comes out of the device and enters the respiratory tract. The second type, called an active inhaler, has an activating bottom that must be pressed when starting to inhale (Daniher and Zhu, 2008; Malcolmson and Embleton, 1998).

7 DEPOSITION OF INHALED PARTICLES IN RESPIRATORY TRACT

Morphological properties of particles (including size, shape, surface characteristics, and density), physiological specialties of human respiratory tract, and the parameters based on ventilation (breathing pattern) are three main factors affecting the deposition of inhaled particles in different regions of respiratory tract. As is obvious in Table 20.1, there are three main deposition patterns for inhaled particles: impaction, gravitational sedimentation, and diffusion (Paranjpe and Müller-Goymann, 2014; Sung et al., 2007; Yang et al., 2008). If the diameter of inhaled particle is above 5 μm, most probably it will impact in the upper airways and oropharynx (Paranjpe and Müller-Goymann, 2014; Yang et al., 2008). Impaction is mostly observed in the case of DPIs where the deposition profile of particles is significantly dependent on the breathing pattern of the patient (Paranjpe and Müller-Goymann, 2014). Gravitational sedimentation occurs for particles in the range of 1–5 μm as a result of gravity force; these particles deposit in the lower airways and bronchioles. In this case the role of ventilation parameters is negligible, where a slow breathing pattern gives the sufficient time for sedimentation to occur. The third phenomenon is diffusion, which is a result of Brownian motion of lung surfactant in deep lungs; particles possess a diameter equal to or below 1 μm up to 0.5 μm will be affected by this motion and after a lag time of floating in lungs they will be satisfactory deposited and dissolve in alveolar sacs (Hatch and Gross, 2013; Heyder, 2004).

In the case of nanosized inhaled particles, it has been reported that, depending on thermal motion of the air molecules within the respiratory area, inhaled particles will be deposited and the deposition is more significant when the diameter size of particles decreases to below 500 nm (Yang et al., 2008).

8 FATE OF INHALED PARTICLES

As mentioned earlier inhaled particles, if not exhaled, will be deposited in various regions of the pulmonary system. Factors affecting the type and site of deposition are discussed in the previous section. After the deposition particles may be absorbed into pulmonary epithelial cells or they will be removed upon pulmonary clearance. This will be determined according to the site of deposition, size, and morphological properties of the particle, and the physiological and pathological aspects of the deposition site; characteristic of the mucosal layer lining the entire respiratory tract is the foremost physiologic property of this system, which is significantly influenced by disease and pathologic conditions. After the deposition, inhaled particles contact the mucosal layer. This connection may lead to dissolving the particles. The dependency of dissolution rate on the size of particles is observed; the smaller the particle diameter, the higher the interaction with lining layer, which in turn will lead to a higher dissolution rate. High concentration associated with nanoparticulate systems (comparing with microparticles) (Kreyling et al., 2006) in combination with the rapid dissolution rate of them may cause an enhancement in the bioavailability of active pharmaceutical agent (Geiser et al., 2003; Zhang et al., 2011).

Natural characteristic of deposited particles is another aspect that plays an essential role in determining the fate of a particle (either absorption or nonabsorptive clearance). It is well known that small molecular drugs with lipid nature are the most likely particles to simply be absorbed into epithelial cells via cell membrane. Particles having rapid dissolution rate can undergo nonabsorptive clearance. The very thin mucosal and epithelial layer of alveolar ducts make these regions suitable for passive diffusion; therefore a large proportion of such lipophilic compounds will absorb through the epithelial membrane of the alveolar ducts. On the other hand, hydrophilic drugs with low molecular weight need to absorb through epithelial tight junctions or actively transport via cell membrane transporters. Large molecular active agents are the other group that is absorbed in upper airways via receptor-mediated transcytosis (Patton, 1996; Patton and Byron, 2007; Yang et al., 2008).

There are two main nonabsorptive clearance mechanisms to eliminate unabsorbed particles from respiratory tract: while the mucociliary clearance is the main cleaning mechanism of upper airways, macrophages play the most important role in alveolar defense (Groneberg et al., 2003; Paranjpe and Müller-Goymann, 2014; Patton, 1996; Patton et al., 2010; Yang et al., 2008). The epithelial layer of upper airways consists of ciliated cells covering with a layer of mucus lining on them, nonsoluble or slowly soluble drug particles that have incorporated with the mucus layer entrapped in the mucus and then eliminated from the respiratory tract as a result of movement of the cilia. The procedure of removing nonsoluble inhaled particles takes about two days, that is, a long time compared with the required time for dissolution of nanoparticles with large surface area and fast dissolution rate.

In the case of particles that could reach the alveolar area, the main defence system is the presence of macrophage cells; it has been reported that the uptake of particles possessing diameter about 1–3 μm occurs most often and the particles equal to or below 250 nm are the most unlikely particles to uptake via macrophage cells.

Another mechanism that causes the removal macromolecular particles from respiratory and most especially the alveolar area to capillary blood of lung is caveolae, which is presented in the cell membrane of type I alveolar epithelium and capillaries.

9 EVALUATION OF THE FATE OF INHALED PARTICLES

9.1 IN VIVO EVALUATION

Determining the deposition profile and fate of particles in animal respiratory tract has some difficulties; the trouble with administration of inhalable formulations to animals is one of these difficulties associated with in vivo evaluation of such formulations. This problem is seen especially in the case of DPIs, which are breath actuated. There is an invasive pathway used to deliver drug formulation to the respiratory tracts of animals; intratracheal intubation or instillation is the most widely used method in order to deliver drug particles to the lower airways of the animal (Al-Qadi et al., 2012; Beyerle et al., 2011; Kreyling et al., 2002; Morimoto et al., 2012; Naya et al., 2012; Sakagami, 2006; Sayes et al., 2007; Schreier et al., 1993). Another method is to develop an inhalation chamber that holds the animal and gives it the ability of breathing and inhaling the drug formulation (McConville et al., 2005; Sinha and Mukherjee, 2012).

After the formulation was administered it must be accurately followed up to determine the fate of inhaled particles within the body. Detectable inhaled particles can be followed after inhalation and the deposited amount of distributed drug in lungs, lung capillary, and secondary organs, such as liver, kidney, brain, heart, and so forth is available via different imaging or sampling approaches. Among the several methods to make drug particles detectable in the body, radiolabelling (Devadason et al., 2012; Kreyling et al., 2002; Videira et al., 2002) and fluorescence (Meers et al., 2008; Taratula et al., 2013; Zatta et al., 1993) are the two most common ones. These labeled systems are useful in order to visualize the pulmonary delivery of administered formulations (Mansour et al., 2009; Yang et al., 2008).

Pharmacokinetic studies are another means to determine lung deposition and drug absorption via the whole lung; in the case of drugs that are inhaled in order to reach systemic circulation, it seems that pharmacokinetic studies will be of a potential relevance to predict the therapeutic effect (Newman et al., 2000).

9.2 IN VITRO EVALUATIONS

Since the fate of inhaled particles is highly dependent to the morphological properties of particles and according to the importance of inertial impaction in pulmonary drug delivery, and regarding the difficulties of in vivo methods, in vitro approaches are preferable in most cases. There are various testing apparatuses that can be used to assess the behavior of the particles.

Impactor is the term used to define the apparatuses working in dry conditions, while if the stages of assessing use any liquid, the term *impinger* will be more suitable. Both impingers and impactors are commonly used to control the aerosolization efficacy of the inhaler and also to predict the in vivo deposition of particles. This testing equipment works based on the principles of inertial impaction: deposition in each stage of testing illustrates the aerodynamic diameter size of particle. Since the aerodynamic diameter is dependent on density, size, and shape it will be influenced mostly by inertial force and the role of gravitational force or diffusion will be very low (Martonen, 1993; Mitchell et al., 2007). Each testing device has several stages that simulate different parts of the human respiratory tract and lungs. Once the formulation leaves the inhaler and indrafts into the testing device, particles will be deposited in different stages, the impaction of particles in each stage is based on their aerodynamic diameter and inertia, which are illustrating the deposition profile of particles in human respiratory tract (Hamishehkar et al., 2012). Thus it can be concluded that the data derived from testing apparatus is highly relative to in vivo deposition behavior of the particles (Mitchell et al., 2007).

Among the several testing equipment Anderson cascade impactor (with and without preseparator), liquid impinger, and next generation impactor (with or without preseparator) are the mostly used devices, which are accepted by United States and European pharmacopeia. Pre separator is a single part that equips in the case of DPIs; it places before remaining stages in order to catch any large and noninhalable particles.

In order to obtain a quantitative and numerical factor to judge the combination of aerosol formulation and inhaler device European and United States pharmacopeias have introduced some expressions to discuss the aerosol efficiency. These factors explain the in vitro inhalation efficiency of particles based on inertial impaction (Hamishehkar et al., 2012; Pilcer et al., 2012). Fine particle dose (FPD) is the foremost factor, which dictates the amount of particles capable to deposit in deep lungs (it is evidence that particles with an aerodynamic diameter below 5 μm are of the suitable distribution to deposit in lower airways via inertial impaction method). There is another factor with the same meaning; it illustrates the percentage of emitted dose that is in the inhalable range. In fact, fine particle fraction (FPF) illustrates the fraction of particles that succeed in reaching the end stages of impactor or impinger methods compared to the whole drug collected in all stages of the apparatus. In other words FPF can be described as the percentage of FPD (Pilcer et al., 2012). The ability of the formulation to leave the inhaler device and indraft to the impactor or either impinger is defined via another expression: "emitted dose (ED)." This is more valuable in case of breathe actuating DPIs that the formulation is more likely to remain in the inhaler device. The quotient of dividing the FPD to ED is called "dispersibility" (Zeng et al., 2000). "Mass Median Aerodynamic Diameter (MMAD)" can be calculated from the cumulative curve of aerodynamic size distribution in 50% point and, as is obvious from the title, it has the concept of the particle diameter that has 50% of the aerosol mass residing above and 50% of its mass below it (Labiris and Dolovich, 2003). "Geometric standard deviation (GSD)" illustrates the degree of dispersion in a log normally distributed aerosol that the GSD value of 1 indicates an ideal aerosolized system with monodisperse particles (Marple et al., 2003).

In the case of nanoscaled material there are some other comprehensive testing methods that may be helpful in order to predict the particle behavior in patient body. Scanning electron Microscopy (SEM) is an in vitro microscopy method that can determine the size and morphological aspects of produced particles. Surface characteristics of an inhaled particle are another important parameter that can be seen via SEM, that is, fluffy and feathery particles tend to be suspended on the inspired air flow and reach the deep lungs. Atomic force microscopy (AFM) is another microscopic method used for surface nanotopographical imaging, measurement of surface energy, and measurement of interparticulate forces (Mansour et al., 2009). X-ray powder diffraction (XRPD) is another very important characterization tool to determine the crystallinity of the particles; crystallinity is a structural aspect illustrating information about the atomic pack and interatomic distance and angle (Mansour et al., 2009). There are many other methods, such as differential scanning calorimetry (DSC), inverse gas chromatography (IGC), and Karl Fischer titration, and so forth in order to evaluate thermal behavior, surface characteristics, and other aspects of nanoparticulate aerosols.

Finally it must be kept in mind that clinical evidence and therapeutic responses are not dependent only to the morphological and physical aspects of the particles and such in vitro evaluations are just a guideline for clinical practice to be considered along with other influencing factors (e.g. patient inhalation pattern, patient training, etc.).

10 CONCLUSIONS

Nanoparticulate drug delivery systems are of the optimum size and specifications for pulmonary absorption; on the other hand, it is well known that drug particles in the size range of 1–5 μm are the best for pulmonary deposition; thus aggregated nanoparticles that have the advantage of both proper deposition and superfine epithelial absorption will be very intensive for pulmonary drug delivery. Nanoparticles could be inhaled to the respiratory tract using three main pulmonary drug delivery devices: nebulizer, metered-dose inhaler, and dry powder inhaler. Each of mentioned devices has advantages and disadvantages and the choice of suitable delivery device depends on the pathologic specifications of the patient and the characteristics of the formulation. Regarding the difficulties of in vivo methods, and since the morphological characteristics of particles play a major role, the fate of inhaled particles can be best evaluated by in vitro methods. These methods properly judge the in vivo behavior of inhaled drug using quantitative factors defined by European and US pharmacopeias.

REFERENCES

Adi, H., Young, P.M., Chan, H.-K., Agus, H., Traini, D., 2010. Co-spray-dried mannitol–ciprofloxacin dry powder inhaler formulation for cystic fibrosis and chronic obstructive pulmonary disease. Eur. J. Pharm. Sci. 40, 239–247.

Adibkia, K., Barzegar-Jalali, M., Maheri-Esfanjani, H., Ghanbarzadeh, S., Shokri, J., Sabzevari, A., Javadzadeh, Y., 2013. Physicochemical characterization of naproxen solid dispersions prepared via spray drying technology. Powder Technol. 246, 448–455.

Adibkia, K., Javadzadeh, Y., Dastmalchi, S., Mohammadi, G., Niri, F.K., Alaei-Beirami, M., 2011. Naproxen–Eudragit® RS100 nanoparticles: preparation and physicochemical characterization. Colloids Surf. B 83, 155–159.

Al-Qadi, S., Grenha, A., Carrión-Recio, D., Seijo, B., Remuñán-López, C., 2012. Microencapsulated chitosan nanoparticles for pulmonary protein delivery: in vivo evaluation of insulin-loaded formulations. J. Control. Release 157, 383–390.

Anderson, P.J., 2005. History of aerosol therapy: liquid nebulization to MDIs to DPIs. Respir. Care 50, 1139–1150.

Azarmi, S., Roa, W.H., Löbenberg, R., 2008. Targeted delivery of nanoparticles for the treatment of lung diseases. Adv. Drug Deliv. Rev. 60, 863–875.

Bassler, D., 2015. Inhalation or instillation of steroids for the prevention of bronchopulmonary dysplasia. Neonatology 107, 358–359.

Beck-Broichsitter, M., Merkel, O.M., Kissel, T., 2012. Controlled pulmonary drug and gene delivery using polymeric nano-carriers. J. Control. Release 161, 214–224.

Beyerle, A., Braun, A., Merkel, O., Koch, F., Kissel, T., Stoeger, T., 2011. Comparative in vivo study of poly (ethylene imine)/siRNA complexes for pulmonary delivery in mice. J. Control. Release 151, 51–56.

Bi, R., Shao, W., Wang, Q., Zhang, N., 2009. Solid lipid nanoparticles as insulin inhalation carriers for enhanced pulmonary delivery. J. Biomed. Nanotechnol. 5, 84–92.

Bivas-Benita, M., Romeijn, S., Junginger, H.E., Borchard, G., 2004. PLGA–PEI nanoparticles for gene delivery to pulmonary epithelium. Eur. J. Pharm. Biopharm. 58, 1–6.

Bohm, P., 1997. Efficiency issues and the Montreal Protocol on CFCs. Environ. Emerg. Dev. Issues 2, 308–338.

Borchard, G., 2001. Chitosans for gene delivery. Adv. Drug Deliv. Rev. 52, 145–150.

Boulikas, T., 2013. Encapsulation of Plasmid DNA (Lipogenes™) and Therapeutic Agents with Nuclear Localization Signal/Fusogenic Peptide Conjugates into Targeted Liposome Complexes. Google Patents.

Brannon-Peppas, L., Blanchette, J.O., 2012. Nanoparticle and targeted systems for cancer therapy. Adv. Drug Deliv. Rev. 64, 206–212.

Cazzola, M., Page, C.P., Calzetta, L., Matera, M.G., 2012. Emerging anti-inflammatory strategies for COPD. Eur. Respir. J. 40, 724–741.

Chauhan, B.F., Ducharme, F.M., 2012. Anti-leukotriene agents compared to inhaled corticosteroids in the management of recurrent and/or chronic asthma in adults and children. Cochrane Database Syst. Rev. 5, CD002314.

Chiang, P.-C., Alsup, J.W., Lai, Y., Hu, Y., Heyde, B.R., Tung, D., 2009. Evaluation of aerosol delivery of nanosuspension for pre-clinical pulmonary drug delivery. Nanoscale Res. Lett. 4, 254–261.

Chow, M., Lam, J., 2015. Dry Powder Formulation of Plasmid DNA and siRNA for Inhalation. Curr. Pharm. Des. 21 (27), 3854–3866.

Daniher, D.I., Zhu, J., 2008. Dry powder platform for pulmonary drug delivery. Particuology 6, 225–238.

De Jesús Valle, M.J., López, F.G., Hurlé, A.D.-G., Navarro, A.S., 2007. Pulmonary versus systemic delivery of antibiotics: comparison of vancomycin dispositions in the isolated rat lung. Antimicrob. Agents Chemoth. 51, 3771–3774.

De Smedt, S.C., Demeester, J., Hennink, W.E., 2000. Cationic polymer-based gene delivery systems. Pharm. Res. 17, 113–126.

Devadason, S.G., Chan, H.-K., Haeussermann, S., Kietzig, C., Kuehl, P.J., Newman, S., Sommerer, K., Taylor, G., 2012. Validation of radio-labeling of drug formulations for aerosol deposition assessment of orally inhaled products. J. Aerosol Med. 25, S-6–S-9.

Dizaj, M., Vazifehasl, Z., Salatin, S., Adibkia, K., Javadzadeh, Y., 2015. Nanosizing of drugs: effect on dissolution rate. Res. Pharm. Sci. 10, 95–108.

Dombu, C.Y., Betbeder, D., 2013. Airway delivery of peptides and proteins using nanoparticles. Biomaterials 34, 516–525.

Effros, R.M., 2006. Anatomy, development, and physiology of the lungs. GI Motility Online, doi:10.1038/gimo73.

Emami, J., Pourmashhadi, A., Sadeghi, H., Varshosaz, J., Hamishehkar, H., 2014. Formulation and optimization of celecoxib-loaded PLGA nanoparticles by the Taguchi design and their in vitro cytotoxicity for lung cancer therapy. Pharm. Dev. Technol. 20 (7), 1–10.

Ezzati Nazhad Dolatabadi, J., Hamishehkar, H., Valizadeh, H., 2014. Development of dry powder inhaler formulation loaded with alendronate solid lipid nanoparticles: solid-state characterization and aerosol dispersion performance. Drug Dev. Ind. Pharm. 41 (9), 1–7.

Gary, D.J., Puri, N., Won, Y.-Y., 2007. Polymer-based siRNA delivery: perspectives on the fundamental and phenomenological distinctions from polymer-based DNA delivery. J. Control. Release 121, 64–73.

Geiser, M., Schürch, S., Gehr, P., 2003. Influence of surface chemistry and topography of particles on their immersion into the lung's surface-lining layer. J. Appl. Physiol. 94, 1793–1801.

Geller, D.E., 2005. Comparing clinical features of the nebulizer, metered-dose inhaler, and dry powder inhaler. Respir. Care 50, 1313–1322.

Gonda, I., 2003. Targeting by Deposition: Pharmaceutical Inhalation Aerosol Technology, second ed. Marcel Dekker, New York, 65-88.

Groneberg, D., Witt, C., Wagner, U., Chung, K., Fischer, A., 2003. Fundamentals of pulmonary drug delivery. Respir. Med. 97, 382–387.

Hamishehkar, H., Rahimpour, Y., Javadzadeh, Y., 2012. The Role of Carrier in Dry Powder Inhaler, Recent Advances in Novel Drug Carrier Systems. INTECH Open Access Publisher.

Hans, M., Lowman, A., 2002. Biodegradable nanoparticles for drug delivery and targeting. Curr. Opin. Solid State Mater. Sci. 6, 319–327.

Harush-Frenkel, O., Bivas-Benita, M., Nassar, T., Springer, C., Sherman, Y., Avital, A., Altschuler, Y., Borlak, J., Benita, S., 2010. A safety and tolerability study of differently-charged nanoparticles for local pulmonary drug delivery. Toxicol. Appl. Pharm. 246, 83–90.

Hatch, T.F., Gross, P., 2013. Pulmonary Deposition and Retention of Inhaled Aerosols. Elsevier Science.

Heydenreich, A., Westmeier, R., Pedersen, N., Poulsen, H., Kristensen, H., 2003. Preparation and purification of cationic solid lipid nanospheres—effects on particle size, physical stability and cell toxicity. Int. J. Pharm. 254, 83–87.

Heyder, J., 2004. Deposition of inhaled particles in the human respiratory tract and consequences for regional targeting in respiratory drug delivery. Proc. Am. Thoracic Soc. 1, 315–320.

Hindle, M., Newton, D.A., Chrystyn, H., 1995. Dry powder inhalers are bioequivalent to metered-dose inhalers: a study using a new urinary albuterol (salbutamol) assay technique. CHEST J. 107, 629–633.

Islam, N., Cleary, M.J., 2012. Developing an efficient and reliable dry powder inhaler for pulmonary drug delivery: a review for multidisciplinary researchers. Med. Eng. Physics 34, 409–427.

Javadzadeh, Y., Ahadi, F., Davaran, S., Mohammadi, G., Sabzevari, A., Adibkia, K., 2010. Preparation and physicochemical characterization of naproxen: PLGA nanoparticles. Colloids Surf. B 81, 498–502.

Junghanns, J.-U.A., Müller, R.H., 2008. Nanocrystal technology, drug delivery, and clinical applications. Int. J. Nanomed. 3, 295.

Kleinstreuer, C., Zhang, Z., Donohue, J., 2008a. Targeted drug-aerosol delivery in the human respiratory system. Annu. Rev. Biomed. Eng. 10, 195–220.

Kleinstreuer, C., Zhang, Z., Li, Z., 2008b. Modeling airflow and particle transport/deposition in pulmonary airways. Respir. Physiol. Neurobiol. 163, 128–138.

Kreyling, W., Semmler, M., Erbe, F., Mayer, P., Takenaka, S., Schulz, H., Oberdörster, G., Ziesenis, A., 2002. Translocation of ultrafine insoluble iridium particles from lung epithelium to extrapulmonary organs is size dependent but very low. J. Toxicol. Environ. Health A 65, 1513–1530.

Kreyling, W.G., Semmler-Behnke, M., Möller, W., 2006. Ultrafine particle-lung interactions: does size matter? J. Aerosol. Med. 19, 74–83.

Kunda, N.K., Somavarapu, S., Gordon, S.B., Hutcheon, G.A., Saleem, I.Y., 2013. Nanocarriers targeting dendritic cells for pulmonary vaccine delivery. Pharm. Res. 30, 325–341.

Labiris, N.R., Dolovich, M.B., 2003. Pulmonary drug delivery. Part II: the role of inhalant delivery devices and drug formulations in therapeutic effectiveness of aerosolized medications. Brit. J. Clin. Pharm. 56, 600–612.

Le Brun, P., De Boer, A., Mannes, G., De Fraîture, D., Brimicombe, R., Touw, D., Vinks, A., Frijlink, H., Heijerman, H., 2002. Dry powder inhalation of antibiotics in cystic fibrosis therapy: part 2: inhalation of a novel colistin dry powder formulation: a feasibility study in healthy volunteers and patients. Eur. J. Pharm. Biopharm. 54, 25–32.

Lesniak, A., Salvati, A., Santos-Martinez, M.J., Radomski, M.W., Dawson, K.A., Åberg, C., 2013. Nanoparticle adhesion to the cell membrane and its effect on nanoparticle uptake efficiency. J. Am. Chem. Soc. 135, 1438–1444.

Loukides, S., Bartziokas, K., Vestbo, J., Singh, D., 2013. Novel anti-inflammatory agents in COPD: targeting lung and systemic inflammation. Curr. Drug Targets 14, 235–245.

Makhlof, A., Werle, M., Tozuka, Y., Takeuchi, H., 2010. Nanoparticles of glycol chitosan and its thiolated derivative significantly improved the pulmonary delivery of calcitonin. Int. J. Pharm. 397, 92–95.

Malcolmson, R.J., Embleton, J.K., 1998. Dry powder formulations for pulmonary delivery. Pharm. Sci. Technol. Today 1, 394–398.

Manca, M.L., Manconi, M., Valenti, D., Lai, F., Loy, G., Matricardi, P., Fadda, A.M., 2012. Liposomes coated with chitosan–xanthan gum (chitosomes) as potential carriers for pulmonary delivery of rifampicin. J. Pharm. Sci. 101, 566–575.

Mansour, H.M., Rhee, Y.-S., Wu, X., 2009. Nanomedicine in pulmonary delivery. Int. J. Nanomed. 4, 299.

Manzer, L.E., 1990. The CFC-ozone issue: progress on the development of alternatives to CFCs. Science 249, 31–35.

Marple, V.A., Olson, B.A., Santhanakrishnan, K., Mitchell, J.P., Murray, S.C., Hudson-Curtis, B.L., 2003. Next generation pharmaceutical impactor (a new impactor for pharmaceutical inhaler testing). Part II: Archival calibration. J. Aerosol Med. 16, 301–324.

Martonen, T., 1993. Mathematical model for the selective deposition of inhaled pharmaceuticals. J. Pharm. Sci. 82, 1191–1199.

McConville, J.T., Carvalho, T.C., Iberg, A.N., Talbert, R.L., Burgess, D., Peters, J.I., Johnston, K.P., Williams Iii, R.O., 2005. Design and evaluation of a restraint-free small animal inhalation dosing chamber. Drug Dev. Ind. Pharm. 31, 35–42.

Meers, P., Neville, M., Malinin, V., Scotto, A., Sardaryan, G., Kurumunda, R., Mackinson, C., James, G., Fisher, S., Perkins, W., 2008. Biofilm penetration, triggered release and in vivo activity of inhaled liposomal amikacin in chronic Pseudomonas aeruginosa lung infections. J. Antimicrob. Chemother. 61, 859–868.

Mitchell, J., Newman, S., Chan, H.-K., 2007. In vitro and in vivo aspects of cascade impactor tests and inhaler performance: a review. AAPS Pharmsci. 8, 237–248.

Moghimi, S.M., Hunter, A.C., Murray, J.C., 2005. Nanomedicine: current status and future prospects. FASEB J. 19, 311–330.

Morimoto, Y., Hirohashi, M., Ogami, A., Oyabu, T., Myojo, T., Todoroki, M., Yamamoto, M., Hashiba, M., Mizuguchi, Y., Lee, B.W., 2012. Pulmonary toxicity of well-dispersed multi-wall carbon nanotubes following inhalation and intratracheal instillation. Nanotoxicology 6, 587–599.

Muller, R.H., Keck, C.M., 2004. Challenges and solutions for the delivery of biotech drugs: a review of drug nanocrystal technology and lipid nanoparticles. J. Biotechnol. 113, 151–170.

Müller, R.H., Radtke, M., Wissing, S.A., 2002. Solid lipid nanoparticles (SLN) and nanostructured lipid carriers (NLC) in cosmetic and dermatological preparations. Adv. Drug Deliv. Rev. 54, S131–S155.

Müller, R.H., Gohla, S., Keck, C.M., 2011. State of the art of nanocrystals–special features, production, nanotoxicology aspects and intracellular delivery. Eur. J. Pharm. Biopharm. 78, 1–9.

Murthy, R.R., Khattar, H., Singh, S., 2012. Formulation and characterization of nano lipid carrier dry powder inhaler containing ciprofloxacin hydrochloride and n-acetyl cysteine. Int. J. Drug Del. 4, 316–325.

Naya, M., Kobayashi, N., Ema, M., Kasamoto, S., Fukumuro, M., Takami, S., Nakajima, M., Hayashi, M., Nakanishi, J., 2012. In vivo genotoxicity study of titanium dioxide nanoparticles using comet assay following intratracheal instillation in rats. Reg. Toxicol. Pharm. 62, 1–6.

Newman, S.P., Wilding, I.R., Hirst, P.H., 2000. Human lung deposition data: the bridge between in vitro and clinical evaluations for inhaled drug products? Int. J. Pharm. 208, 49–60.

Ógáin, O.N., Li, J., Tajber, L., Corrigan, O.I., Healy, A.M., 2011. Particle engineering of materials for oral inhalation by dry powder inhalers. I—Particles of sugar excipients (trehalose and raffinose) for protein delivery. Int. J. Pharm. 405, 23–35.

Ohashi, K., Kabasawa, T., Ozeki, T., Okada, H., 2009. One-step preparation of rifampicin/poly (lactic-*co*-glycolic acid) nanoparticle-containing mannitol microspheres using a four-fluid nozzle spray drier for inhalation therapy of tuberculosis. J. Control. Release 135, 19–24.

Paranjpe, M., Müller-Goymann, C.C., 2014. Nanoparticle-mediated pulmonary drug delivery: a review. Int. J. Mol. Med. 15, 5852–5873.

Paranjpe, M., Finke, J., Richter, C., Gothsch, T., Kwade, A., Büttgenbach, S., Müller-Goymann, C., 2014. Physicochemical characterization of sildenafil-loaded solid lipid nanoparticle dispersions (SLN) for pulmonary application. Int. J. Pharm. 476, 41–49.

Pardeike, J., Weber, S., Haber, T., Wagner, J., Zarfl, H., Plank, H., Zimmer, A., 2011. Development of an itraconazole-loaded nanostructured lipid carrier (NLC) formulation for pulmonary application. Int. J. Pharm. 419, 329–338.

Park, J.H., Lee, S., Kim, J.-H., Park, K., Kim, K., Kwon, I.C., 2008. Polymeric nanomedicine for cancer therapy. Prog. Polym. Sci. 33, 113–137.

Pastor, M., Moreno-Sastre, M., Esquisabel, A., Sans, E., Viñas, M., Bachiller, D., Asensio, V.J., Del Pozo, Á., Gainza, E., Pedraz, J.L., 2014. Sodium colistimethate loaded lipid nanocarriers for the treatment of *Pseudomonas aeruginosa* infections associated with cystic fibrosis. Int. J. Pharm. 477, 485–494.

Patravale, V., Kulkarni, R., 2004. Nanosuspensions: a promising drug delivery strategy. J. Pharm. Pharm. Sci. 56, 827–840.

Patton, J.S., 1996. Mechanisms of macromolecule absorption by the lungs. Adv. Drug Deliv. Rev. 19, 3–36.

Patton, J.S., Byron, P.R., 2007. Inhaling medicines: delivering drugs to the body through the lungs. Nat. Rev. Drug Discov. 6, 67–74.

Patton, J.S., Brain, J.D., Davies, L.A., Fiegel, J., Gumbleton, M., Kim, K.-J., Sakagami, M., Vanbever, R., Ehrhardt, C., 2010. The particle has landed—characterizing the fate of inhaled pharmaceuticals. J. Aerosol. Med. 23, S-71–S-87.

Pilcer, G., Wauthoz, N., Amighi, K., 2012. Lactose characteristics and the generation of the aerosol. Adv. Drug Deliv. Rev. 64, 233–256.

Pilcer, G., De Bueger, V., Traina, K., Traore, H., Sebti, T., Vanderbist, F., Amighi, K., 2013a. Carrier-free combination for dry powder inhalation of antibiotics in the treatment of lung infections in cystic fibrosis. Int. J. Pharm. 451, 112–120.

Pilcer, G., Rosière, R., Traina, K., Sebti, T., Vanderbist, F., Amighi, K., 2013b. New co-spray-dried tobramycin nanoparticles–clarithromycin inhaled powder systems for lung infection therapy in cystic fibrosis patients. J. Pharm. Sci. 102, 1836–1846.

Rosen, J.E., Chan, L., Shieh, D.-B., Gu, F.X., 2012. Iron oxide nanoparticles for targeted cancer imaging and diagnostics. Nanomedicine 8, 275–290.

Rosiere, R., Amighi, K., Vermeersch, M., Wauthoz, N., 2015. New Dry Powders for Inhalation Containing Chitosan Derivative-Coated Solid Lipid Nanoparticles for Targeted Delivery to Lung Cancer Cells. RDD Europe 2015. Virginia Commonwealth University, pp. 447–452.

Rundfeldt, C., Steckel, H., Scherliess, H., 2009. Nanosuspension with antifungal medication to be administered via inhalation with improved impurity profile and safety. Google Patents.

Rundfeldt, C., Steckel, H., Scherliess, H., Wyska, E., Wlaź, P., 2013. Inhalable highly concentrated itraconazole nanosuspension for the treatment of bronchopulmonary aspergillosis. Eur. J. Pharm. Biopharm. 83, 44–53.

Rytting, E., Nguyen, J., Wang, X., Kissel, T., 2008. Biodegradable polymeric nanocarriers for pulmonary drug delivery. Expert Opin. Drug Deliv. 5 (6), 629–639.

Sakagami, M., 2006. In vivo, in vitro and ex vivo models to assess pulmonary absorption and disposition of inhaled therapeutics for systemic delivery. Adv. Drug Deliv. Rev. 58, 1030–1060.

Sayes, C.M., Reed, K.L., Warheit, D.B., 2007. Assessing toxicity of fine and nanoparticles: comparing in vitro measurements to in vivo pulmonary toxicity profiles. Toxicol. Sci. 97, 163–180.

Scheuch, G., Kohlhaeufl, M.J., Brand, P., Siekmeier, R., 2006. Clinical perspectives on pulmonary systemic and macromolecular delivery. Adv. Drug Deliv. Rev. 58, 996–1008.

Schreier, H., Gonzalez-Rothi, R.J., Stecenko, A.A., 1993. Pulmonary delivery of liposomes. J. Control. Release 24, 209–223.

Shaji, J., Bhatia, V., 2013. Proliposomes: a brief overview of novel delivery system. Int. J. Pharm. Bio. Sci. 4, 150–160.

Shoyele, S.A., Cawthorne, S., 2006. Particle engineering techniques for inhaled biopharmaceuticals. Adv. Drug Deliv. Rev. 58, 1009–1029.

Sinha, B., Mukherjee, B., 2012. Development of an inhalation chamber and a dry powder inhaler device for administration of pulmonary medication in animal model. Drug Dev. Ind. Pharm. 38, 171–179.

Stass, H., Nagelschmitz, J., Willmann, S., Delesen, H., Gupta, A., Baumann, S., 2013. Inhalation of a dry powder ciprofloxacin formulation in healthy subjects: a phase I study. Clin. Drug Invest. 33, 419–427.

Steichen, S.D., Caldorera-Moore, M., Peppas, N.A., 2013. A review of current nanoparticle and targeting moieties for the delivery of cancer therapeutics. Eur. J. Pharm. Sci. 48, 416–427.

Steimer, A., Haltner, E., Lehr, C.-M., 2005. Cell culture models of the respiratory tract relevant to pulmonary drug delivery. J. Aerosol Med. 18, 137–182.

Sung, J.C., Pulliam, B.L., Edwards, D.A., 2007. Nanoparticles for drug delivery to the lungs. Trends Biotech. 25 (12), 563–570.

Sung, J.C., Padilla, D.J., Garcia-Contreras, L., Verberkmoes, J.L., Durbin, D., Peloquin, C.A., Elbert, K.J., Hickey, A.J., Edwards, D.A., 2009. Formulation and pharmacokinetics of self-assembled rifampicin nanoparticle systems for pulmonary delivery. Pharm. Res. 26, 1847–1855.

Taratula, O., Kuzmov, A., Shah, M., Garbuzenko, O.B., Minko, T., 2013. Nanostructured lipid carriers as multifunctional nanomedicine platform for pulmonary co-delivery of anticancer drugs and siRNA. J. Control. Release 171, 349–357.

Tikshdeep, C., Sonia, A., Bharat, P., Abhishek, C., 2012. Liposome drug delivery: a review. Int. J. Pharm. Chem. Sci. 1, 1103–1113.

Tripp, J.T., 1987. UNEP Montreal Protocol: industrialized and developing countries sharing the responsibility for protecting the stratospheric ozone layer. N Y UJ Int'l L. Pol. 20, 733.
Ungaro, F., D'angelo, I., Coletta, C., Di Villa Bianca, R.D.E., Sorrentino, R., Perfetto, B., Tufano, M.A., Miro, A., La Rotonda, M.I., Quaglia, F., 2012a. Dry powders based on PLGA nanoparticles for pulmonary delivery of antibiotics: modulation of encapsulation efficiency, release rate and lung deposition pattern by hydrophilic polymers. J. Control. Release 157, 149–159.
Ungaro, F., D'angelo, I., Miro, A., La Rotonda, M.I., Quaglia, F., 2012b. Engineered PLGA nano-and micro-carriers for pulmonary delivery: challenges and promises. J. Pharm. Pharmacol. 64, 1217–1235.
Videira, M.A., Botelho, M., Santos, A.C., Gouveia, L.F., Pedroso de Lima, J., Almeida, A.J., 2002. Lymphatic uptake of pulmonary delivered radiolabelled solid lipid nanoparticles. J. Drug. Target. 10, 607–613.
Walkey, C.D., Olsen, J.B., Guo, H., Emili, A., Chan, W.C., 2012. Nanoparticle size and surface chemistry determine serum protein adsorption and macrophage uptake. J. Am. Chem. Soc. 134, 2139–2147.
Wang, A.Z., Langer, R., Farokhzad, O.C., 2012. Nanoparticle delivery of cancer drugs. Ann. Rev. Med. 63, 185–198.
Wauthoz, N., Amighi, K., 2014. Phospholipids in pulmonary drug delivery. Eur. J. Lipid Sci. Technol. 116, 1114–1128.
Weber, S., Zimmer, A., Pardeike, J., 2014. Solid Lipid Nanoparticles (SLN) and Nanostructured Lipid Carriers (NLC) for pulmonary application: a review of the state of the art. Eur. J. Pharm. Biopharm. 86, 7–22.
Webster, T.J., 2006. Nanomedicine: What's in a definition? Int. J. Nanomed. 1, 115.
Westerman, E., Heijerman, H., Frijlink, H., 2007. Dry powder inhalation versus wet nebulisation delivery of antibiotics in cystic fibrosis patients. Expert Opin. Drug Deliv. 4 (2), 91–94.
Williams, R.O., Taft, D.R., Mcconville, J.T., 2007. Advanced drug formulation design to optimize therapeutic outcomes. CRC Press, Taylor and Francis Group.
Willis, L., Hayes, Jr., D., Mansour, H.M., 2012. Therapeutic liposomal dry powder inhalation aerosols for targeted lung delivery. Lung 190, 251–262.
Xiao, Z., Levy-Nissenbaum, E., Alexis, F., Lupták, A., Teply, B.A., Chan, J.M., Shi, J., Digga, E., Cheng, J., Langer, R., 2012. Engineering of targeted nanoparticles for cancer therapy using internalizing aptamers isolated by cell-uptake selection. ACS Nano 6, 696–704.
Xie, H., Zhu, Y., Jiang, W., Zhou, Q., Yang, H., Gu, N., Zhang, Y., Xu, H., Xu, H., Yang, X., 2011. Lactoferrin-conjugated superparamagnetic iron oxide nanoparticles as a specific MRI contrast agent for detection of brain glioma in vivo. Biomaterials 32, 495–502.
Yang, W., Peters, J.I., Williams, R.O., 2008. Inhaled nanoparticles: a current review. Int. J. Pharm. 356, 239–247.
Yu, J., Chien, Y.W., 1997. Pulmonary drug delivery: physiologic and mechanistic aspects. Crit. Rev. Ther. Drug 14 (4), 395–453.
Zarogouldis, P., Karamanos, N.K., Porpodis, K., Domvri, K., Huang, H., Hohenforst-Schimdt, W., Goldberg, E.P., Zarogoulidis, K., 2012. Vectors for inhaled gene therapy in lung cancer: application for nano oncology and safety of bio nanotechnology. Int. J. Mol. Sci. 13, 10828–10862.
Zatta, P., Favarato, M., Nicolini, M., 1993. Deposition of aluminum in brain tissues of rats exposed to inhalation of aluminum acetylacetonate. Neuroreport 4, 1119–1122.
Zeng, X.M., Martin, G.P., Marriott, C., Pritchard, J., 2000. The influence of carrier morphology on drug delivery by dry powder inhalers. Int. J. Pharm. 200, 93–106.
Zhang, J., Wu, L., Chan, H.-K., Watanabe, W., 2011. Formation, characterization, and fate of inhaled drug nanoparticles. Adv. Drug Deliv. Rev. 63, 441–455.

CHAPTER

21

NANOSTRUCTURES IN TRANSDERMAL DRUG DELIVERY SYSTEMS

Najma Bibi, Naveed Ahmed, Gul Majid Khan
Quaid-i-Azam University, Islamabad, Pakistan

CHAPTER OUTLINE

1 INTRODUCTION

The word *nanoscale* has found an ever-increasing application in different fields of science. In the past few years, nanotechnology and nanoscience have been expanded rapidly by leaps and bounds (Pal et al., 2011). It is a multidisciplinary field that consists of a wide range of devices and systems derived by employing the principles of engineering and science. Nanotechnology represents the design, synthesis, characterization, and application of materials and devices (at an atomic or molecular level), in order to produce new nano-sized materials, preferably 1–100 nm (Sahoo et al., 2007). It also includes all the fundamental processes involved in manufacturing nanoscale devices and their utilization in various research areas (Koo et al., 2005; Walters and Hadgraft, 1993). The greatest breakthrough of nanotechnology in the field of medicines opened up new avenues in targeted drug delivery systems that release therapeutics moieties in a controlled manner at the desired site of action. Nanomedicine has blossomed the pharmaceutical industry because it possesses inherent properties to overcome various issues regarding solubility, stability, permeation, absorption, localized drug delivery, and toxicity (Bawarski et al., 2008). Its application ranges from diagnosis to therapy (Bawarski et al., 2008). Nanomedicine includes utilization of nanochips for better diagnosis, innovative surgical devices, materials for substituting body structures, and various types of novel systems capable of carrying drugs to target specific organs of the body for the treatment of many ailments. All these materials and devices are generally termed nanostructures or nanocarriers (Escobar Chavez, 2012). Nanostructures act as powerful weapons in the treatment of diseases, as they are smaller in size to be detected by the immune system and they deliver a therapeutic agent to the target site using lower drug doses while reducing side effects (Escobar Chavez, 2012). They are composed of different materials in variable sizes and shapes.

The use of nanostructures is rapidly emerging in the topical and transdermal delivery of drugs (Schleich and Pret, 2012). Transdermal drug delivery is the delivery of drugs through the skin in order to achieve systemic effects. Transdermal drug delivery systems (TDDS), also traditionally known as "patches," are the dosage forms designed to deliver a therapeutically effective amount of drug to viable epidermal and dermal tissues of the skin for local and systemic effects (Verma et al., 2012). Although the skin is the most attractive and accessible route of drug administration, it constitutes an excellent barrier due to the presence of the stratum corneum (SC) for the penetration of most of the therapeutic agents, especially hydrophilic drugs (Muzzalupo et al., 2011). The development of a nanotechnological product for percutaneous application is challenging (Guy, 2003; Walters and Hadgraft, 1993). In the past few decades, several strategies have been developed. These include physical (iontophoresis, sonophoresis, electroporation, electroosmosis) (Rastogi and Singh, 2005) and chemical penetration enhancement technologies. Mostly, the conventional methods have some limitations regarding safety and efficacy of drugs as well as patient compliance. Most of the physical techniques are invasive and complicated to use. The use of surfactant and organic solvents in chemical penetration enhancers causes irritation, does damage, and reduces the barrier function of skin (Verma and Pathak, 2010). In the past few years, budding interest in nanopharmaceuticals has emerged to develop novel drug delivery systems through the skin. The current approach pursued in transdermal drug delivery is the application of nanostructures, since this system maintains and restores the barrier function of the skin after the drug application, avoids the side effects, protects the drug from premature physical and chemical degradation in a hostile environment of the body, and achieves the desired goals (Elias, 2005). Hence, these carriers have been devised to transport the quantitative drug delivery specifically to the target site through the transdermal route by enhancing the drug's absorption, penetration, half-life, and bioavailability

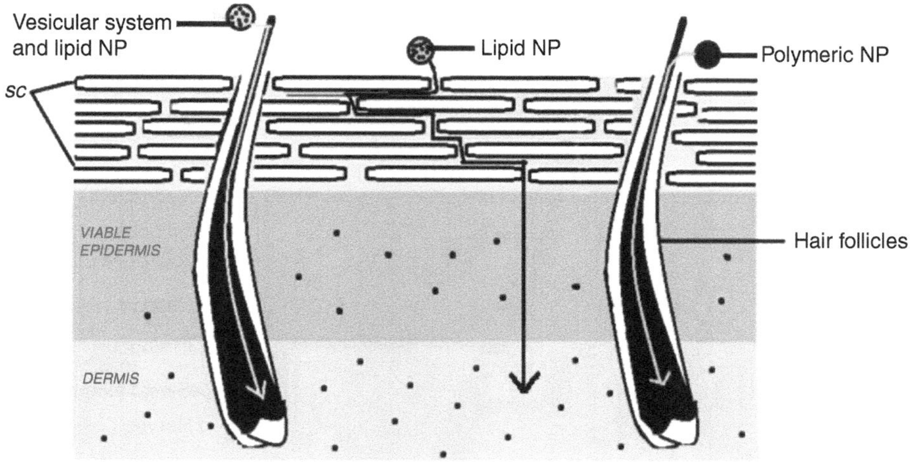

FIGURE 21.1 Nanostructures Penetration in TDDS

(Escobar Chavez et al., 2012). Fig. 21.1 gives an overview of the nanostructures' penetration of the skin for transdermal drug delivery.

In this chapter, primarily skin anatomy is discussed for general understanding; then an overview of absorption through skin, advantages of nanotechnology-based TDDS over conventional delivery systems, physicochemical properties of nanostructures required for transdermal drug delivery, and some of them being used in TDDS are presented.

2 SKIN ANATOMY

In describing skin anatomy, discussion has been restricted to those features of the skin that are pertinent from the drug delivery point of view. The SC, the outermost layer of the epidermis, is a prominent layer in this regard (Hillery et al., 2002). The skin is one of the largest organs of the body and acts as a physical barrier against external environmental threats. It provides protection against physical, chemical, mechanical, and microbial influences (Desai et al., 2010). It is made up of three main layers: the epidermis, dermis, and subcutaneous layer (also termed the hypodermis) (Prausnitz et al., 2004), as shown in Fig. 21.2. Other anatomical features of interest are appendages, including hair follicles and associated sebaceous glands (pilosebaceous units) and sweat glands (eccrine and apocrine glands) (Prow et al., 2011).

2.1 EPIDERMIS

The epidermis is the outermost layer of skin. Histologically, it is a stratified squamous keratinized epithelial layer. It is made of keratinocytes (>90%) that undergo keratinization, resulting in the formation of five layers of strata from the stratum basal to the SC. These layers are the stratum basal, stratum spinosum, stratum granulosum, stratum lucidum, and SC. The undifferentiated columnar-shaped stem cells in the stratum basal are continuously proliferating through mitosis, consequently producing new

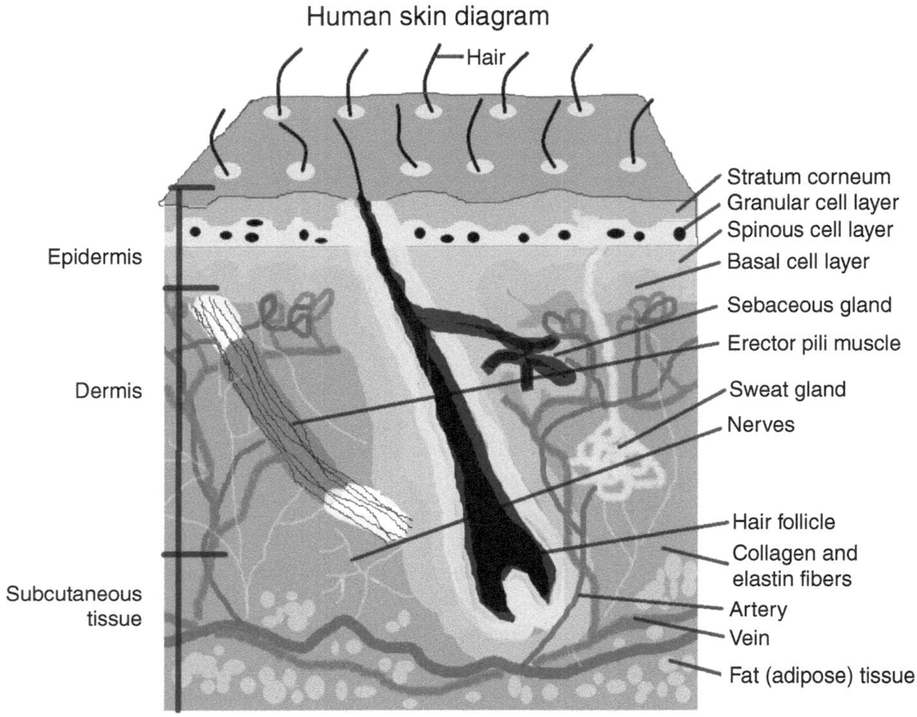

FIGURE 21.2 Skin Structure with Three Layers: Epidermis, Dermis, and Hypodermis

keratinocytes. These new cells push already-existing cells from the stratum spinosum and granulosum, and finally their terminal differentiation results in the formation of the SC (Hillery et al., 2002).

2.1.1 Stratum corneum structure and barrier function

The SC is considered a "brick wall" made up of protein and lipids. These lipids are not dispersed uniformly in the SC. Rather, they are sequestered in the extracellular spaces, where they are arranged into multilamellar bodies that surround the corneocytes (Elias and Menon, 1991). Skin permeability is different at various body sites depending upon variation in the number of lamellar membranes, their structure, and the composition of lipids in the SC (Lampe et al., 1983). The extracellular enriched lipid makes the SC predominately a lipophilic barrier that limits the transdermal delivery of hydrophilic drugs. It also acts as a reservoir within which lipophilic drugs can be accumulated and slowly released over a period of days (Prausnitz et al., 2012). The human SC is typically composed of about 15–20 layers of corneocytes (Prausnitz et al., 2004), which are held together by corneodesmosome. The corneodesmosomes are necessary for maintaining the shape and regular packing of corneocytes (Prow et al., 2011). Each corneocyte is surrounded by a cornified envelope, which in turn is covered by a lipid-enriched extracellular matrix that is arranged into a lamellar membrane. The inner part of the cell is packed with keratin filament (protein), which is embedded in a filigrin matrix and its breakdown product, the natural moisturizing factor. Depending upon the body site, the SC differs in thickness, number of corneodesmosomes, packing of keratin filament, and filagrin content (Prausnitz et al., 2012). Its barrier function is chiefly imparted by lipid bilayers in intercellular spaces that consist of an equimolar mixture of cholesterol, fatty acids, and ceramide (Hillery et al., 2002). It undergoes continuous desquamation once every 2–3 weeks, which further enhances its barrier properties. The major function of the SC is to prevent the loss of internal body components, especially water, to the external environment (Escobar Chavez et al., 2012).

The novel strategies in transdermal drug delivery have mainly focused on manipulation of the extracellular lipid matrix (Prausnitz and Langer, 2008) and aqueous pores present in the extracellular matrix (Prausnitz et al., 2012).

2.2 DERMIS AND SKIN APPENDAGEAL STRUCTURES

The dermis is present below the epidermis, and it covers 90% of the skin surface (Hillery et al., 2002). The main cells present in the dermis are fibroblasts that produce connective tissue, which provides support to the epidermis layer (Escobar Chavez, 2010). It incorporates nerves, blood vessels, and lymphatic vessels. The presence of an extensive vascular network in the dermis is the main area through which transdermally absorbed molecules gain entry into the systemic circulation and reach their target sites (Hillery et al., 2002). Furthermore, the dermis also supports several skin appendageal structures, the pilosebaceous unit and sweat glands. They open into the skin surface and form shunt pathways or transfollicular pathways for ingress of molecules (Kierszenbaum and Tres, 2002). Sweat glands regulate the body temperature and are responsible for the excretion of acids and body wastes. Sweat is excreted at the rate of 2–20 nL/min from each orifice of diameter 60–80 μm. The pilosebaceous units consist of hair follicles associated with the sebaceous glands. The hair follicles provide a route to discharge sebum secreted by the sebaceous glands (Kierszenbaum and Tres, 2002). Sebum is excreted from each orifice of 10–210 μm diameter depending upon body site (Kierszenbaum and Tres, 2002; Otberg et al., 2004). These appendageal structures provide an opportunity for large hydrophilic molecules (Dokka et al., 2005; Toll et al., 2004), as well as small hydrophilic molecules (El Maghraby et al., 2001b; Essa et al., 2002; Hueber et al., 1994; Illel et al., 1991; Kao et al., 1988) to permeate through the skin.

3 PERMEATION PATHWAYS ACROSS SKIN

At present, there is an ongoing debate on the determination of permeation pathways for nanostructures-based drug delivery systems (Baroli, 2010). The transport of the drug across the skin occurs by diffusion that basically follows three pathways: a transfollicular pathway consisting of the hair follicles and associated sweat glands, and a transepidermal pathway, which is subdivided into intercellular and transcellular routes through the SC (Ganesh et al., 2012). The skin appendages form the shunt pathways through the intact epidermis and cover less than 0.1% of the total human skin surface. Previously, they were thought to be an unimportant pathway for the delivery of drugs due to their small surface area. Nowadays, follicular penetration is a major area of interest in the treatment of skin diseases (Escobar Chavez, 2012). The pilosebaceous units are connected with a vast network of blood capillaries below which no mature SC is present. So the drug molecules that follow follicular penetration directly enter the surrounding tissues and, through a dense network of blood capillaries, reach the systemic circulation, thereby avoiding the barrier of the SC. The hair follicles and sweat glands represent an ideal target for drug delivery and may be an alternative penetration pathway to the intercellular route (Wosicka and Cal, 2010).

Several in vivo and in vitro studies have been carried out to show the effective penetration through hair follicles by topical administration of micro- and nanoparticles in the porcine skin (Escobar Chavez, 2012). The follicular pathway, in contrast to the transcellular pathway, is favorable for hydrophilic and high-molecular-weight drugs, as well as for particulate base delivery. Great interest has developed in their utilization as reservoirs for treating local diseases (Wosicka and Cal, 2010). The penetration of drugs through hair follicles is size dependent. The size of hair cuticle is about 300–600 nm, so the optimum penetration of therapeutic agent of such size can occur (Schleich and Pret, 2012).

In the intercellular route, the permeation of drugs occurs through the continuous lipid matrix present between corneocytes. The lipophilic and amphiphilic molecules mostly follow this route by providing the secure free volume needed for permeation through intercellular lipid layers. The free spaces between the outer membrane of corneocytes and lamella are utilized by polar molecules for diffusion (Geinoz et al., 2004). The macromolecular matrix present inside the SC contains hydrated keratin, which provides mechanical stability and intactness. It does not contribute directly to the diffusive barrier, so this pathway is, for practical purposes, unimportant for transdermal drug delivery (Hadgraft and Lane, 2005).

The potential of most of the environmental nano-sized particles such as bacteria, viruses, dust, allergens, or any other particles to penetrate the skin is negligible unless the barriers of the skin are disrupted or diseased (Prow et al., 2011). There is no evidence that the solid nanoparticles (SNPs) of size greater than 100 nm can penetrate through various skin layers (Wosicka and Cal, 2010). Another hypothesis supported by an investigation reveals that nanoparticles of size 10–20 nm are unable to permeate through the SC into viable epidermis and dermis but accumulate into openings of hair follicles after a massage (Wosicka and Cal, 2010). During the past few decades different approaches have been pursued to increase permeation of drugs through the skin, but most of them have several drawbacks regarding safety, inconsistent efficacy, and patient compliance. The use of nanostructures has gained increasing attention of researchers to deep layers of the skin. Attempts to overcome these issues have focused on the delivery of drugs to deeper layers of the skin through the use of nanostructures. These carrier systems have gained increasing attention in transdermal drug delivery owing to their outstanding properties (Desai et al., 2010). They provide opportunities to deliver the therapeutic agent to the skin without compromising its barrier function where significant uptake is achieved for the production of systemic or local effects (Schleich and Pret, 2012).

4 ADVANTAGES OF TRANSDERMAL DRUG DELIVERY COMPARED TO OTHER ROUTES OF ADMINISTRATION

Transdermal drug delivery is an attractive alternative to conventional drug delivery systems such as oral and parenteral routes. It provides many advantages over other routes of administration for five reasons:

1. It is a noninvasive drug delivery system that maintains the steady plasma level of drug and increased drug efficacy (Thomas and Finnin, 2004).
2. It maintains the drug level in the systemic circulation within the therapeutic window (i.e., above the minimum effective concentration but below the level at which side effects become apparent) for prolonged periods of time, so the drug remains within its therapeutic level.
3. The transdermal drug delivery system, compared to an oral and intravenous route of administration, avoids peaks and valleys in plasma level by providing steady blood concentration (Thomas and Finnin, 2004).

4. It also avoids degradation of the drug in the gastrointestinal tract, eliminates the first-pass effect, and improves patient compliance and acceptability of drug therapy.
5. The patch can be removed easily by the patient on an appearance of side effects.

5 ADVANTAGES OF NANOSTRUCTURES IN TDDS

The skin is the most attractive and accessible route of drug administration, particularly for systemic effects. But it constitutes an effective barrier, especially the SC, for the penetration of most of the therapeutic agents like hydrophilic drugs (Muzzalupo et al., 2011). In the past few years, novel drug delivery systems have been widely investigated by providing opportunities to solve issues linked with conventional drug therapy such as instability, high toxicity, unfavorable pharmacokinetics, lower cellular uptake, and drug resistance (Gupta et al., 2012). They also improve patient compliance, safety, and efficacy of drugs. Currently, the development of nanostructure-based drug delivery systems is rapidly growing owing to their spatial and temporal drug delivery in a time-controlled manner. At present a number of such products are available in the market. Some of the significant advantages of nanostructure-based drug delivery systems rendering them ideal compared to conventional systems are:

1. The nanostructures can deliver the therapeutic agent to the target site at the required concentration.
2. Because of their large surface-to-volume ratio, nanostructures could transport a large amount of cargo with less volume.
3. Their application results in the creation of reservoirs in the skin from where they control the release of the therapeutic agent to the affected site of the skin with an enhanced localized effect.
4. Because of their smaller and narrower distribution, they can easily cross smaller networks of blood capillaries of the skin and provide efficient site-specific targeting (Bawarski et al., 2008).
5. The encapsulated drugs in the nanostructures have sustained releasing effect for prolonged periods with least side effects.
6. In addition, the entrapped moieties are protected from premature chemical and enzymatic degradation (Gupta et al., 2012; Zhang et al., 2010).

6 PHYSICOCHEMICAL PROPERTIES OF NANOSTRUCTURES REQUIRED FOR GOOD PERMEATION

In transdermal drug delivery, the nanostructures are used to deliver drugs for both local and systemic effects. Several investigations have shown that a number of factors affect the absorption of drugs through the skin. To produce a therapeutic effect, the nanostructure has to interact with the skin, which depends on its physicochemical properties. When engineering nanostructures, the key characteristics that should be considered are their size, shape, surface charge, lipophilicity, stability, and entrapment efficiency (Desai et al., 2010).

6.1 SIZE, SHAPE, AND MOLECULAR WEIGHT OF NANOSTRUCTURES

The size and shape of nanostructures should be optimum for transdermal drug delivery because those characteristics play an important role in physical stability, cellular uptake, and drug release (Escobar Chavez et al., 2012). Depending upon the anatomical features of the skin, it is concluded that in a healthy individual the structure and porosity of the SC are such that it allows the free diffusion of particles smaller than 5–7 nm in intact form throughout the fluid portion of the lipid bilayer (Bouwstra and Ponec, 2006; Johnson et al., 1997). The penetrating agents of size 36 nm (or smaller) can utilize the aqueous pores for transport across the SC layer (Cevc, 2004; Potts and Francoeur, 1991; Tang et al., 2001), and particles greater than 3–10 μm may penetrate through a transfollicular route. It seems that nano-sized particles concurrently used have the potential to follow different penetration routes (Baroli, 2010). Nanostructures with appropriate lipophilicity, high partition coefficient, and molecular weight <600 Da are the best candidates for transdermal drug delivery, whereas substances with molecular weight >600 Da cannot cross the skin barrier (Baroli, 2010; Desai et al., 2010). These are available in different shapes, such as spherical, ellipsoidal, cubic, triangular, needle-like, and prism-like. The lipid particles are not always rigid, whereas the carbon and metallic particles are not always spherical. The rigid and deformable shapes, as well as the orientation of the nanostructure, will affect their aggregation and movement through narrow penetration pathways across the skin (Baroli, 2010). The shape also influences the selection of penetration route, diffusion coefficient, and potential of the interaction of nanostructure with skin components/structure. Therefore, it is concluded that with new advancements in nanotechnology the nanostructures can be engineered with tunable sizes and shapes with desired therapeutic properties.

6.2 SURFACE CHARGE

As just stated, size, shape, and molecular weight of nanostructures are not the sole determinants of transdermal drug delivery; other factors like surface charge or polarity are also quite important. The surface charge is responsible for attachment of the nanostructure to the cell membrane, selection of route of penetration, and diffusion coefficient of the skin. The cell membrane is known to carry a negative charge due to the presence of negatively charged sulfated proteoglycan molecules, which play an important role in cellular motility, migration, and proliferation (Uchechi et al., 2014). In general, the negatively charged nanoparticles should be repelled by the cell membrane because of the presence of a large negatively charged domain. On the basis of electrostatic interaction, they first adsorb on the cell membrane and then form nanoparticle clusters from which their cellular uptake occurs by endocytosis (Honary and Zahir, 2013). This means that surface properties of the nanostructure influence the cellular uptake and intracellular distribution. However, it is possible to localize the therapeutic action of drugs and other bioactive agents at the desired site through modification of the surface charge of the nanostructures (Patil et al., 2007). Some studies have shown the effect of the surface charge on membrane permeability. For example, more densely charged polymer dendrimers facilitate the transport of dye across membranes (Wolinsky and Grinstaff, 2008). Another study revealed that lipid-coated charged nanoparticles had 3–4 times more permeation across endothelial cell layers of skin than did nonlipid-coated nanoparticles, whereas a coating of neutral charge had no significant effect on permeation (Honary and Zahir, 2013). It is thus expected that the positively charged nanoparticles would predominantly cross the skin layers as compared to nanostructures of neutral charge.

6.3 DIFFUSION COEFFICIENT

The diffusion coefficient of the penetrating agent in its vehicle and the skin is also an important parameter. It may be calculated by the Stokes–Einstein equation, which states diffusion as a function of size, shape, and temperature:

$$D = \frac{kT}{6\pi\eta r}$$

where k is Boltzman's constant, T is absolute temperature, η is the viscosity of vehicle, and r is the radius of the particle, which is assumed to be spherical. However, this equation is not applied for dermal and transdermal delivery where dimension would be an important parameter, though it will influence the entry of agent into the skin, selection of penetration pathway, and coefficient of diffusion (D) in the vehicle and in the skin (Baroli, 2010).

6.4 STABILITY

Stability is one of the most important characteristics that affect polymeric, polysaccharide, and lipid nanoparticles as well as carbon and metallic nanostructures. It represents not only disassembly or rupturing of a particle but also the stability of the nanostructure (core and coating components) against physical reactions (particle disassembly, rupture) and chemical reactions (hydrolysis, oxidation, deamination) (e.g., carbon and metallic nanoparticles). Instability leads to modifying the superficial properties of nanostructures such as interaction with skin layers, self-aggregation, aggregation within a dispersing vehicle, and alteration of the diffusion coefficient of nanomaterial in dispersing vehicle and skin (Baroli, 2010). Stability also refers to particles' ability to form micelles when they come into contact with the skin structure, or decoating of nanoparticles as a consequence of a reaction between the coating component of nanomaterials and the penetration route.

6.5 pH OF VEHICLE AND PK_A OF PENETRATING AGENT

The pH of the vehicle and the PK_a of the penetrating agent are two parameters that should be set accurately, as only the unionized fraction of the penetrating molecule will cross the skin barrier (Barry, 1983; Swarbrick et al., 1984).

6.6 O/W PARTITION COEFFICIENT OF PENETRATING MOLECULE

The nanostructure should have an adequate o/w partition coefficient for optimum penetration through the lipid matrix and aqueous pores of the skin (Baroli, 2010). If it will be more lipophilic. it can easily cross the SC but will have trouble while partitioning through the hydrophilic pores, and the hydrophilic drug will have poor penetration through the lipid portion of the SC (Barry, 1983; Moss et al., 2002). However, it is known that a molecule with high solubility in its vehicle is correlated with the highest thermodynamic activity and thus will have a higher chance to penetrate the skin.

6.7 ENTRAPMENT EFFICIENCY

Entrapment efficiency is an important parameter and is defined as the percentage ratio of the entrapped drug concentration to the total drug concentration (Abdallah, 2013).

It is calculated by this equation:

$$\frac{\text{EE\%} = \text{Total drug concentration} - \text{Free drug concentration} \times 100}{\text{Total drug concentration}}$$

7 TYPES OF NANOSTRUCTURES USED IN TDDS

The nanostructures have been greatly investigated in TDDS owing to their remarkable characteristics, as mentioned earlier (Escobar Chavez et al., 2012; Gupta et al., 2012). These are colloidal drug delivery systems synthesized by employing principles of nanotechnology (Korting and Schafer Korting, 2010). Noteworthy nanostructures used to deliver drugs include vesicular systems (liposomes, transferosomes, niosomes, and ethosomes); lipid nanoparticles (solid lipid nanoparticles and nanostructures lipid carriers); polymeric nanoparticles; and dendrimers. In this section, an updated overview of various types of nanostructures used in TDDS with their applications in cosmetics and pharmaceuticals is provided.

7.1 VESICULAR SYSTEMS

In TDDS, the vesicular systems have attracted great attention of researchers owing to their excellent properties. They have the ability to encapsulate both hydrophilic and lipophilic drug molecules and extend the half-lives of drugs by increasing their systemic bioavailability, targeted drug delivery, biodegradability, and nontoxicity (Modi and Bharadia, 2012). Their structural characteristics like shape, size, and lamellae nature can be tuned by changing their composition. They act as permeation enhancers of dermally active agents by solubilizing the lipid matrix to form a local depot or modulate the systemic absorption of drugs through the skin by acting as a rate-limiting membrane (Muzzalupo et al., 2011). The vesicular systems have the potential to permeate drugs to viable tissues through appendageal units (hair follicles, sebaceous, and sweat glands) (Choi and Maibach, 2005). In this section, four common vesicular systems—liposomes, transferosomes, niosomes, and ethosomes—are briefly discussed.

7.1.1 Liposomes

Among nanopharmaceuticals, liposome is one of the extensively studied nanostructures. These were first introduced by Alice Bengham (Bangham et al., 1965). Liposomes are hollow spherical vesicles made up of cholesterol and phospholipids (Neubert, 2011). The phospholipids form bilayers containing water in their center and disperse it in an aqueous medium. The phospholipid (lecithin) polar head forms an interface to the aqueous medium. The hydrophilic agents are incorporated into the water phase inside the vesicle, whereas lipophilic agents are loaded into the bilayers. However, the hydrophilic agent can also be loaded in a high amount on the outer surface (Korting and Schafer Korting, 2010). Liposomes have great potential to permeate the drugs and other cargo through the epidermal barrier of the skin since the lipid composition in liposome is similar to lipids in the epidermis (Gupta et al., 2012). At present, the liposomes are prepared by two common methods, thin film hydration and solvent injection. Most of the liposomes are prepared by thin film hydration (Li and Zhao, 2007). Depending upon the method of preparation and the technique used, liposomes can be classified into categories: small unilamellar vesicles (SUVs), large unilamellar vesicles (LUVs), and multilamellar vesicles (MLVs).

SUVs and LUVs consist of only one lipid bilayer whereas multilamellar vesicles consist of many concentric bilayers (Neubert, 2011). The sizes of SUVs, LUVs, and MLVs are 10–100 nm, 100–500 nm, and >500 nm in dimension, respectively (Korting and Schafer Korting, 2010). The physicochemical properties of liposomes affect the penetration of the drug across the skin. Among them, size is most important. The skin has small pores of 0.3 nm narrowness, which can be opened without major skin diseases up to 20–40 nm. The size of the liposome affects the transdermal drug delivery through intercellular and transcellular pathways. The penetration of the encapsulated drug is decreased with increasing the size of the liposome. However, if the liposome permeates through appendageal units, then the size will not affect the penetration (Choi and Maibach, 2005). In addition to size, the thermodynamic state of liposomes also plays a vital role in penetration. It depends on the composition and percentage of cholesterol and the nature of the phospholipids used (Schleich and Pret, 2012).

The main purpose of the liposome in transdermal drug delivery is to produce localized or systemic effects (Korting and Schafer Korting, 2010). A lot of research has been going on to study possible interaction between keratinocytes and liposomes using reconstructed human skin (Gupta et al., 2012). The mechanism of liposomal drug delivery into the skin is not yet clear. Several proposed mechanisms have been illustrated in the literature, including: (1) adsorption of the liposomes on the SC and fusion with the lipid domain of the SC where the lipid exchange takes place between the phospholipid of the liposomes and the SC, (2) disintegration of the liposomes on the skin surface by acting as a permeation enhancer via disrupting the cellular packing and fluidity of the lipid bilayer of the SC, (3) liposomes possibly acting as a drug reservoir by penetrating into pilosebaceous units, and (4) liposomes possibly being penetrated into the SC in intact form (El Maghraby and Williams, 2009).

However, despite all their unique properties, the major problems that circumscribe the formulation of liposomes in transdermal drug delivery are stability, low drug loading, difficulty in sterilization, and batch-to-batch reproducibility (Bawarski et al., 2008). They undergo a change in vesicle size due to aggregation, fusion, leakage of vesicles, oxidation, and ester hydrolysis. These problems are generally associated with the liquid state of the liposomes synthesized from nonsaturated phospholipids. These issues can be fixed by using hydrogenated phospholipids in the formulation and at the end incorporate into gel matrix (Gupta et al., 2012). The use of such types of phospholipids although solved problems related to physical stability due to viscous nature but have a low penetration property as compared to liquid state liposomes (Honeywell-Nguyen and Bouwstra, 2005). In addition to instability, another problem related to liposomes is the formation of a reservoir in the SC to produce a localized effect of drugs and cosmetics. To overcome this drawback, new classes of lipid vesicles are investigated that include transferosomes (Benson, 2006), ethosomes, invasosomes, and liposomes containing propylene glycol. They exhibit pronounced elasticity and superior skin penetration through the transdermal route (Palac et al., 2014).

The first liposomal preparation, marketed in 1988, was Econazole (Pevaryl1 Lipogel). It was introduced for topical treatment of dermatomycosis. Currently, liposome loaded with ultraviolet A and ultraviolet B absorbing agent has been introduced with the name of Daylong Actinica, approved by the European Union as a medical device for the treatment of actinic keratosis in high-risk organ transplant recipients on a long-term basis (Ulrich et al., 2008). Another formulation available for topical therapy of osteoarthritis is diclofenac with a trade name of Diclac lipogel (Gupta et al., 2012). Furthermore, the examples of liposomal formulations for transdermal delivery are lignocain (Sharma et al., 1994), clindamycin hydrochloride, estradiol (Essa et al., 2004), ketoprofen (Maestrelli et al., 2005), amphoteracin B (Manosroi et al., 2004), methotrexate (Trotta et al., 2004), indinavir (Sheo et al., 2010), and melatonin (Dubey et al., 2007b).

7.1.2 Transferosomes

Transferosomes are a new class of liposomes also known as deformable, ultraflexible, or elastic liposomes or vesicles (Benson, 2006). They were first introduced in 1990 by Gregor Cevc (Cevc and Blume, 1992) and since have been the subject of numerous patents and literature reports (Benson, 2006). The German company IDEA AG registered the term Transfersome as the trademark of its proprietary drug delivery technology (Modi and Bharadia, 2012). These vesicles have an inner aqueous compartment surrounded by an outer complex of lipid bilayers (Abdallah, 2013). They are elastic in nature due to the presence of an edge activator in the lipid bilayer membrane (Bouwstra et al., 2002; Honeywell Nguyen et al., 2004). The shape and composition of transferosomes assemble them self-optimizing and self-regulating nanostructures (Cevc et al., 1998), which enable them to cross through various transport barriers efficiently (Modi and Bharadia, 2012). These are noninvasive nanostructure for targeted drug delivery and sustained release effect (Walve et al., 2011). Transferosomes are generally composed of two ingredients. The amphipathic agents like phospholipids (phosphatidylcholine) self-assemble into lipid bilayers in an aqueous environment and close to form a vesicle. The second agent is a surfactant (Walve et al., 2011), also known as an edge activator, which may be Span 80, Tween 80, sodium cholate, deoxycholate, or potassium glycyrrhizinate (Abdallah, 2013). The edge activator in transferosomes' membrane destabilizes the lipid bilayers and increases deformability of the vesicle (Bouwstra et al., 2002; Honeywell Nguyen et al., 2004). For the formulation of transferosomes, the main ingredients required are phospholipids (egg phosphatidylcholine, soya phosphatidyl choline), surfactants (sodium cholate, Tween 80), and alcohol (ethanol, methanol). They are developed to prevail over the limitations of conventional liposomes. These vesicles compress themselves to one-tenth of their diameter to squeeze through intercellular lipid bilayers of the SC as compared to liposomes (Cevc, 1996). The mechanism is described in Fig. 21.3.

When these vesicles are applied under nonocclusive conditions, they enter the skin as a result of mechanical stress of surrounding in a self-adapting manner following the natural water gradient in the epidermis (Modi and Bharadia, 2012). Transferosomes act as excellent nanostructures for transdermal delivery of low- and high-molecular-weight drugs (El Maghraby and Williams, 2009) such as

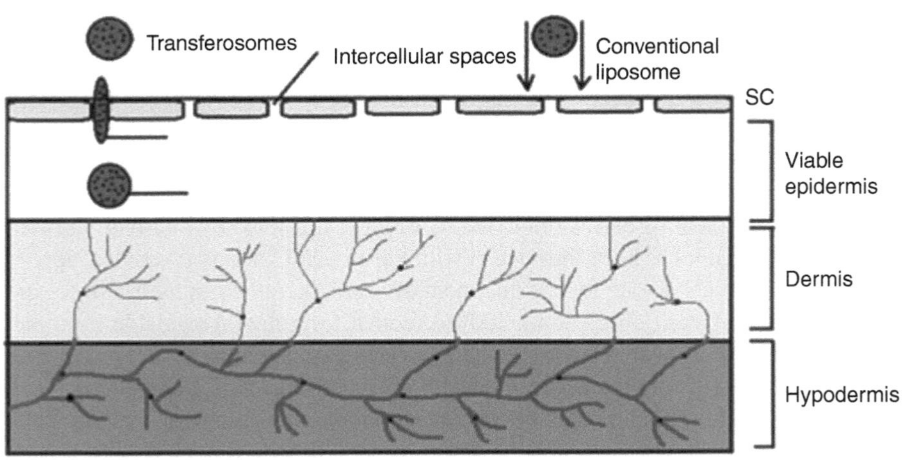

FIGURE 21.3 Mechanism of Penetration of Transferosomes Through Skin Pores

sex hormones, gap junction protein, insulin (Cevc, 2003), albumin, corticosteroids, and anesthetics (Walve et al., 2011), 5-fluorouracil (El Maghraby et al., 2001a), dipotassium glycyrrhizinate (Trotta et al., 2002), ketotifen fumarate (Elsayed et al., 2007), interleukin-2 (Hafer et al., 1999), ketoprofen, triamcinolone acetonide (Cevc and Blume, 2003), DNA, superoxide dimusate, tetnus oxide, and diclofenac (Cevc and Blume, 2001).

7.1.3 Niosomes

Niosomes are nonionic surfactant vesicles widely investigated as an alternative to liposomes. In a number of instances, the formulations based on niosomes have been shown to be superior to liposomes with respect to composition and stability, and they are economical to manufacture (Vora et al., 1998). These vesicles are biocompatible, biodegradable, and nonimmunogenic (Choi and Maibach, 2005; Uchechi et al., 2014). Niosomes are used for targeted and controlled drug delivery (Namdeo and Jain, 1996). Their methods of production include reverse-phase evaporation, ether injection, and hand-shaking technique (Namdeo and Jain, 1999). They enhance permeation of drugs across the skin by disrupting the SC lipids due to the presence of a surfactant and the flexible nature of the vesicle (Fang et al., 2001b; Schreier and Bouwstra, 1994). Another mechanism regarding alteration of the barrier function is the adsorption and fusion with the skin surface followed by modification of lipids of the SC (Kushla et al., 1993). Studies revealed that development of such vesicles reduces the side effects of drugs (Guinedi et al., 2005). The efficiency of transdermal delivery depends on the nature and concentration of the nonionic surfactant and cholesterol content in niosomes. One report on the transdermal delivery of estradiol presented the efficiency of formulation with the following surfactants as Tween 20 > Span 60 > Span 85 > Span 40 (Fang et al., 2001b), while decreased cholesterol content increased its transdermal delivery (Van Hal et al., 1996). Niosomes are the promising drug delivery carriers that find their application in dermal and transdermal drug delivery (Manconi et al., 2011; Mura et al., 2009). They have a significant role in transdermal drug delivery in the treatment of a number of dermatological disorders. Ellagic acid and minoxidil niosomal transdermal formulation are available in the market now. Minoxidil encapsulated in niosomes is a promising approach in hair loss treatment (Ciotti and Weiner, 2002). It is composed of 1:1 molar concentration of cholesterol and Span (sorbitan monoester) (Balakrishnan et al., 2009). Other research on the formulation of niosomes from Span 60 and Tween 60 indicated that these are effective carriers for dermal delivery of ellagic acid (Junyaprasert et al., 2012). Transdermal formulation of enoxacine containing span 80 as a surfactant showed high skin penetration owing to flexible nature of niosomes as compared to its free form (without surfactant) and liposomal formulation (Fang et al., 2001a). Antiacne agent (e.g., tritinone) when entrapped into niosomes showed higher encapsulation efficiency and also protected labile agents from degradation (Naik and Nagarsenker, 2006).

7.1.4 Ethosomes

Ethosomes are novel nanostructures composed of phospholipids, ethanol, and water (Touitou et al., 2000). They were first developed in 1997 by Elka Touitou (Bendas and Tadros, 2007). These vesicles are soft and malleable and can be modulated from tenths of a nanometer to micrometer range (Patel, 2007; Verma and Pathak, 2010). They are the modified form of liposomes with high content of ethanol. The phospholipids present in ethosomes could be phosphatidic acid, phosphatidyl serine, or phosphatidylcholine, and the alcohol may be ethanol or isopropyl alcohol (Ramakrishna et al., 2013).

It has been reported that ethosomes enhance the delivery of drugs through the deep layers of the skin into the systemic circulation, and permeate through the intercellular path of the SC

(Magnusson et al., 1997) by disturbing the organization of lipid bilayers. Owing to the presence of a high concentration of ethanol, the lipid membrane turns out to be less tightly packed, has a more malleable structure, and permits improved distribution of drug through the SC lipids (Verma and Pathak, 2010). It has been studied that ethosomes enhance drug delivery through the skin in vivo and in vitro under both occlusive and nonocclusive conditions (Godin and Touitou, 2004). The mechanism of the penetration of the encapsulated drug from ethosomes is based on synergistic action between components of skin lipids, ethanol, and vesicles (Touitou et al., 2001). The exact mode of interaction between ethosomes and skin lipids is not yet clear. However, it is assumed that this nanostructure delivers the encapsulated drug on the basis of two effects: the ethanol and the ethosome. The ethanol effect encompasses the interaction of ethanol with the polar head region of lipid molecules that causes their increased fluidity and decreased density of multilamellar layer by reducing the transition temperature of lipids in the SC. It is followed by the ethosomes' (vesicle) effect in which the ethosomes fuse with the skin lipids due to their malleable nature at various points, resulting in the opening of new pathways and the release of the encapsulated drugs in the deep layers of the skin (Elsayed et al., 2007).

Many cosmetic and pharmaceutical industries are involved in successfully conducting research in formulating ethosomes-based drug delivery systems for market application. Examples of drugs available in transdermal formulations of ethosomes are tacrolimus, an immunosuppressant used for atopic dermatitis (Li et al., 2012); clotrimazole (Maheshwari et al., 2012); trihexyphenidyl HCl (Dayan and Touitou, 2000); methotrexate (Dubey et al., 2007a); ketoprofen (Chourasia et al., 2011); and testosterone (Ainbinder and Touitou, 2005).

7.2 LIPID NANOPARTICLES

Lipid nanoparticles are innovative drug delivery systems developed for dermatological and cosmetic purposes owing to their remarkable properties such as physical stability, controlled release, and excellent tolerability (Bawarski et al., 2008; Desai et al., 2010). There are two types of lipid nanoparticles: solid lipid nanoparticles (SLNs) and nanostructured lipid carriers (NLCs) (Bhaskar et al., 2009). In 1990, SLNs were developed by two researchers, Rainer H. Muller and Maria Rose, as an attractive alternative nanostructure to emulsion, liposomes, and polymeric nanoparticles (Gasco, 1993; Muller et al., 1995; Uner and Yener, 2007). SLNs are composed of lipids that are solid at body and room temperature (Desai et al., 2010). They are prepared by substituting the liquid lipids of the o/w emulsion by solid lipids or a mixture of solid and liquid lipids (Pardeike et al., 2009). They are composed of three ingredients: (1) solid lipid (0.1%–30% w/w), (2) water, and (3) emulsifier (surfactants 0.5%–5% w/w) (Bleve et al., 2011). The solid lipids are dispersed in the aqueous phase, which is preferably stabilized with surfactants (Bleve et al., 2011; Pardeike et al., 2009). An SLN is a colloidal carrier system for controlled drug delivery that combines the advantages of an emulsion, liposomes, and polymeric nanoparticles in a single system (Bhaskar et al., 2009).

The SLNs have the ability to modulate the release characteristics of active compounds and also increase skin hydration when compared to liposomes and emulsion. However, an SLN has some limitations of low drug loading capacity and uncontrolled drug expulsion from the carrier system during storage. To combat these issues, the second generation of lipid nanoparticles has been developed known as nanostructured lipid carriers (NLCs) (Mehnert and Mader, 2001). They have higher loading capacity and lower leaching properties compared to SLNs (Mehnert and Mader, 2001).

NLCs are composed of solid and liquid lipids in which a solid lipid matrix is embedded in a liquid lipid or the liquid lipid is localized at the surface of the solid lipid matrix with a layer of surfactant (Patlolla et al., 2010). NLCs and SLNs can be loaded with cosmetics and pharmaceuticals. They are frequently used in topical drug delivery because of prevention of the encapsulated agent from enzymatic degradation and transepidermal water loss, and controlled release of the therapeutic agent for a prolonged period of time, therefore enhancing the overall therapeutic action of active compounds (Desai et al., 2010). In TDDS, SLNs and NLCs have several advantages (e.g., they can be administered directly to the skin in sufficient quantity with fewer side effects). They are composed of physiological and biodegradable lipids so are nonirritant and nontoxic, and can be applied to damaged and inflamed skin (Bhaskar et al., 2009). The increased permeation into the skin is described to be a combination of three mechanisms: adhesiveness, occlusiveness, and skin hydration. These lipid nanoparticles with size <200 nm form a monolayer on the skin showing (1) adhesiveness and occlusiveness, (2) prevention of transepidermal water loss, and (3) opening of an intergap between corneocytes, facilitating penetration of therapeutic agent deep into skin layers (Wissing and Muller, 2003). When water content is lost from an SLN, its solid matrix is modified by crystallization that induces expulsion of the drug from the carrier system and results in its penetration into deep layers of the skin (Borgia et al., 2005). A number of factors such as particle size, crystalline nature, lipid concentration, and applied volume affect the occlusive property of lipid nanoparticles (Teeranachaideekul et al., 2008). Nevertheless, it is generally considered that these particles do not penetrate the corneocytes, so permeation through hair follicles is considered one of the routes (Pardeike et al., 2009).

In literature, a number of methods has been extensively reported for the preparation of lipid nanoparticles. These include the microemulsion technique (Gasco, 1993; Priano et al., 2007), emulsification solvent evaporation (Sjostrom and Bergenstahl, 1992), emulsification solvent diffusion method (Hu et al., 2002; Trotta et al., 2003), solvent injection or solvent displacement method (Schubert and Muller Goymann, 2003), multiple emulsion technique (Garcıa-Fuentes et al., 2003), high-pressure homogenization technique (Liedtke et al., 2000; Wissing et al., 2004), membrane contractor technique (Charcosset et al., 2005), and ultrasonication technique (Pietkiewicz and Sznitowska, 2004; Puglia et al., 2008). Lipid nanoparticles, especially SLNs, can also be prepared by the hot and cold homogenization technique (Pardeike et al., 2009). SLNs are widely used in cosmetics for the topical treatment of skin diseases such as psoriasis, acne, and atopic eczema, as well as being used for rheumatoid arthritis (Pardeike et al., 2009). Flurbiprofen (NSAIDs) is a potent inhibitor of platelets aggregation and is used in rheumatoid arthritis, osteoarthritis, sunburn, and gout. The transdermal formulation of flurbiprofen in SLNs and NLCs prepared by the aqueous dispersion method has improved permeation through the SC and overcomes the gastrointestinal tract side effects associated with its oral formulation (Bhaskar et al., 2009). Examples of drugs loaded in lipid nanoparticles for topical and transdermal applications are celecoxib (Joshi and Patravale, 2008), coenzyme Q10 (Teeranachaideekul et al., 2007), podophyllotoxin (Chen et al., 2006), arthemether (Aditya et al., 2010), econazole nitrate (Sanna et al., 2010), and cyclosporine A (Ugazio et al., 2002).

7.3 POLYMERIC NANOPARTICLES

The current progression of nanoparticles proposes promising applications in topical and transdermal drug delivery. They protect the encapsulated drugs from physical/chemical degradation and allow their controlled release. They are particularly significant in dermatology because of their unique characteristics such as good flow properties and soft and transparent nature (Desai et al., 2010; Karande et al., 2005; Wiesenthal et al., 2011). For dermatological applications, nanoparticles have been designed from natural and synthetic polymers.

Natural polymers are composed of polysaccharides or proteins, but are not used frequently in transdermal drug delivery because of their variation in purity and requirement of a linking agent that could denature the embedded drug. Considerable attention has been given to synthetic polymers (Karande et al., 2005). Chitosan, the most widely used natural polymer, is made up of cationic polysaccharide. It is biocompatible and extracted from crustacean shells (Panos et al., 2008). It possesses antimicrobial (Kong et al., 2008), anticoagulant (Huang et al., 2007), and wound healing (Rossi et al., 2007) properties, and is a proficient carrier of drug and gene delivery (Panos et al., 2008).

The synthetic polymers generally used in skin delivery are polyesters such as poly(glycolic acid) (PGA), poly(lactic acid) (PLA), and poly(ε-caprolactone) (PCL), and copolymers such as poly(lacto-*co*-glycolic acid) (PLGA) (Alvarez Roman et al., 2004; Naik et al., 2004; Rancan et al., 2009). All these polymers are biocompatible and biodegradable. PLA nanoparticles loaded with sunscreen agent that were used in cosmetics showed only permeation through the SC (Alvarez Roman et al., 2004). Nanoparticles can be prepared by different methods depending upon their physicochemical properties. These include nanoprecipitation, dispersion polymerization, in situ polymerization, inverse salting out, emulsification polymerization, solvent displacement, and solvent evaporation/extraction and gelation (Escobar Chavez, 2010, 2012).

In these nanostructures, drugs can be distributed in different ways. They can be either adsorbed on the surface of nanoparticles, incorporated into the core of the capsule, or homogeneously dispersed into the polymeric matrix (Schleich and Pret, 2012).

The mechanism of drug release depends on the mode of distribution of active therapeutic agents in nanoparticles. The mode could be (1) diffusion through polymer wall, (2) erosion of nanoparticles matrix, (3) combined erosion and diffusion process (Maestrelli et al., 2005), or (4) desorption of surface bound or surface adsorbed drug (Soppimath et al., 2001). All these mechanisms have different drug release rates, either burst release or controlled release from the polymer for the prolonged time (Korting and Schafer Korting, 2010). The polymeric nanoparticles penetrate the skin through the transappendageal pathway; they generally do not permeate through the SC. In hair follicles, they form a reservoir, and a high local concentration of drug is observed there. This route of drug permeation is of great importance since the storage time of drug incorporated into carriers lasts for only 10 days (Desai et al., 2010; Lademann et al., 2008; Prow et al., 2011; Schneider et al., 2009). Sometimes the *trans*-appendageal pathway is negligible as they constitute only 0.1% of the total skin surface, so the drug can diffuse through the incomplete corneocytes layer below the hair follicles and produce its systemic effect (Schleich and Pret, 2012).

The polymeric nanoparticles for transdermal drug delivery are less studied than are lipid nanoparticles. Research is still ongoing on the development of a stable nanoparticulate system and investigation of permeation pathways through the skin (Korting and Schafer Korting, 2010). Polymeric nanoparticles are frequently used for topical drug delivery. A biodegradable polymer such as PLA is used as a carrier in cosmetics as a sunscreen agent where it stays on the SC by forming a protective film on the skin surface (Alvarez Roman et al., 2004). Polymeric nanoparticles are used to deliver therapeutic agents to various types of tumors, in bone healing therapies, and in vaccinations. Examples of transdermal drugs using nanoparticles are 5-fluorouracil (McCarron and Hall, 2008), DNA (Lee et al., 2008), minoxidil (Shim et al., 2004), Dexamethasone phosphate (Thote and Gupta, 2005), Flufenamic acid (Santander-Ortega et al., 2010), chlorohexidine (Lboutounne et al., 2002), and insulin (Huang et al., 2009).

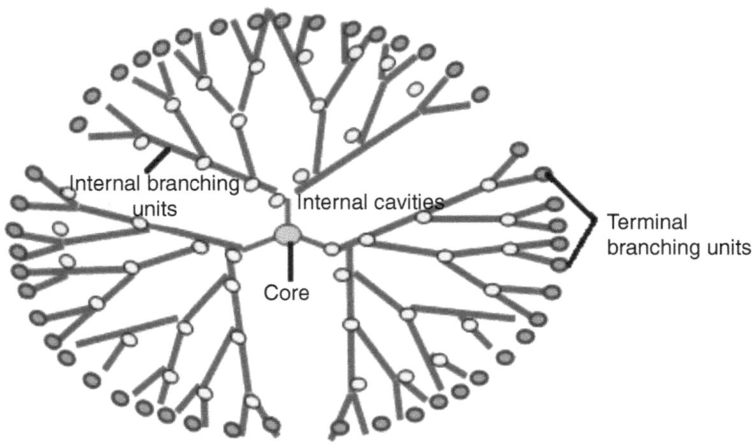

FIGURE 21.4 Three-Dimensional Structure of Dendrimer

7.4 DENDRIMERS

In 1978 Fritz Vögtle and coworkers introduced the chemical nature of dendrimers (Buhleier et al., 1978). Its first family was synthesized by Donald A. Tomalia in 1985 (Tomalia and Dewald, 1986). It is a three-dimensional (3-D) polymeric macromolecule with multiple branches. It is a highly versatile nanostructure, and a number of different functional groups can be attached to its branching units. The structure of dendrimers possesses three main components: the core, the branching units, and the attached functional groups. The core may be a single atom or a group of atoms, from which a number of branches are attached to repeating units, which are radially positioned. To the exterior surface of dendrimers a number of different functional groups can be attached (Escobar Chavez et al., 2012; Verma et al., 2012) as shown in Fig. 21.4.

The structure of a dendrimer is such that it possesses empty internal cavities and open conformation to where a hydrophobic drug can be encapsulated. The possible control on its architecture such as size, shape, charge density, molecules attached makes it an attractive carrier in pharmaceutical and medicinal chemistry (Mane et al., 2014). Dendrimers possess their useful application in topical and transdermal drug delivery, including indomethacin (Chauhan et al., 2003), 5-fluorouracil (Venuganti and Perumal, 2008), ketotifen, diflunisal (Yiyun et al., 2007), peptides (Niederhafner et al., 2005), and tamsulosin (Wang et al., 2003).

8 LIMITATIONS OF NANOSTRUCTURES IN TDDS

The nanostructures used in TDDS may promise endless opportunities regarding stability, penetration, encapsulation efficiency, controlled release, targeted drug delivery, decreased side effects, and improved patient compliance (Bawarski et al., 2008; Gupta et al., 2012; Zhang et al., 2010). In addition to these benefits, they encompass some restrictions. Nanostructures necessitate some significant tests that should be conducted to ensure their analytical evaluations, characterizations, and toxicological and pharmacological profiles. These tests are obligatory in evaluating their efficacy in diagnoses and therapies. The size, shape, surface charge, composition, attached molecules, and surface charge are frequently observed for dendrimers (Esfand and Tomalia, 2001). It was reported that the nanostructure of size 1–10 nm with encapsulated drugs and imaging agents for skin showed better penetration in cell culture and intestinal membrane (Escobar Chavez, 2012). Their safety and long-term use must not be overlooked. Toxicity of nanostructures will arise if transdermal carriers are poorly biodegradable (Parekh, 2007). This is because the surface properties of the material at the nanoscale are different from those at the large scale (Caruthers et al., 2007; Medintz et al., 2005). Therefore, studies should be carried out to investigate the toxicity issues. In addition, the metabolites produced by nanostructures are difficult to predict on account of a lack of necessary techniques. The production of nanocarriers from lab to industrial level is very difficult because materials required for their fabrication are usually very expensive and all these require expertise (Escobar Chavez, 2012).

9 APPLICATIONS OF NANOSTRUCTURES IN TDDS

In the topical and transdermal delivery of drugs, extensive strategies have been devised by employing different types of nanostructures for cosmaceutical applications (Escobar Chavez et al., 2012; Essa et al., 2002; James Smith et al., 2011; Kumar and Rao, 2012; Kumar and Priyanka, 2015). In nanostructures, liposomes have emerged as important carriers for a number of drugs and biomedical products owing to their outstanding pharmacokinetics and biodistribution profile. Primarily they were utilized for the study of membrane-bound proteins and various mechanisms associated with them; currently these vesicles are believed to be excellent carriers for a number of drugs with high entrapment efficiency and fewer side effects. Drug delivery through liposomes is somewhat controversial as it depends on their size and composition. They produce a local effect by forming a local depot into the skin and minimizing systemic side effects, or work to increase transdermal delivery by enhancing drug concentration into the systemic circulation (El Maghraby et al., 2006). They have been successfully employed in cancer therapy and skin melanoma such as doxorubicin in breast carcinoma (Symon et al., 1999), Carboplatin and doxorubicin in ovarian cancer (Ferrero et al., 2007), and liposomal lurtotecan (topoisomerase I inhibitor) in topotecan-resistant ovarian cancer (Seiden et al., 2004). However, liposomes have some drawbacks such as structural instability, expense of preparation, and variable purity of phospholipid composition. To combat these disadvantages, a new class of deformable liposomes has emerged as an invaluable strategy to deliver drugs through the transdermal route (Vora et al., 1998). Ethosomes have the ability to deform and penetrate deep layers of the skin; for example, tacrolimus-loaded ethosomes show excellent effect in atopic dermatitis (Li et al., 2012), 5-aminolevulinic acid–loaded ethosomes are used in hyperproliferative murine skin disease, and they have the potential to produce protoporphyrin IX and also reduce the tumor necrosis factor in disordered skin (Fang et al., 2009). To study skin permeation of ethosomes formulation, confocal microscopy of the skin was carried out, and it was revealed that ethosomal formulation enhanced permeation of the rhodamine 123–loaded formulation as compared to the hydroalcoholic solution. Ketoprofen loaded in ethosomes also showed good permeation through the transdermal route (Chourasia et al., 2011).

Another flexible carrier for transdermal delivery is niosome (Alsarra et al., 2005; Muzzalupo et al., 2011). It has been reported that niosomes increase penetration of the drug through the skin (Ciotti and Weiner, 2002) by increasing its residence time in the SC as well as in the epidermis and reduced its systemic absorption. Studies also showed that development of such vesicles also reduces the side effects of the drugs (Guinedi et al., 2005). Niosomes-encapsulated minoxidil is a promising approach in hair loss treatment (Ciotti and Weiner, 2002). Other research on the preparation of niosomes from Span 60 and Tween 60 revealed that they are good carriers for dermal delivery of ellagic acid (Junyaprasert et al., 2012).

Today nanoparticles have established a new diverse application as a theranostic agent in cancer treatment. The critical bottleneck of the conventional formulation of cancer includes high toxicity following intravenous administration due to indiscriminate distribution in the healthy tissues as well as in the malignant tissues, resulting in loss of useful functions. Therefore, designing of a distinct nanostructure that could encapsulate a large quantity of drugs has found valuable applications in tumor targeting. The delivery of 5-fluorouracil (McCarron and Hall, 2008), DNA (Lee et al., 2008), and minoxidil (Shim et al., 2004) in polymeric nanoparticles has found indispensable application in cancer treatment. The nanoparticles are chemically stable and reproducible and have been used for vaccine development. As a peptide vector, they improved the vaccine delivery through the transdermal route as well as augmented the potency as compared to the conventional adjuvant (Cui et al., 2005). Different types of nanoparticles have been designed such as polymeric nanoparticles, SLNs, NLCs, and recently pH-sensitive nanoparticles (Heffernan and Murthy, 2005). Polymeric nanoparticles have found tremendous applications in the transdermal treatment of tumors, in the healing of bones, and as an excellent route for delivery of various types of vaccinations. Table 21.1 presents drugs loaded inside the nanostructures with possible biomedical indications.

10 CONCLUSIONS AND FUTURE PERSPECTIVES

The greatest breakthrough of nanotechnology in the field of medicine results in the development of novel drug delivery systems for diagnosis and therapy. Transdermal drug delivery is one of the areas that have drawn great attention of researchers. The skin is an easily accessible and attractive route for drug delivery; however, it possesses a formidable barrier in the form of the SC for the penetration of large molecules. By understanding the structure, physiology, and physicochemical properties of the skin, the utilization of a novel delivery system in the form of nanostructures has gained significant importance in the 21st century for transdermal drug delivery. Various types of nanostructures such as liposomes, transferosomes, niosomes, ethosomes, dendrimers, SLNs, and polymeric nanoparticles have been greatly investigated. They possess inherent properties to solve the problems with regard to drug solubility, stability, penetration, absorption, half-life, and toxicity. They have the ability to successfully deliver both hydrophilic and lipophilic drugs as well as macromolecules. The advancement with regard to materials, methods of formulation, and characterization techniques also aid the development of such nanostructures with features to deliver the drug to the targeted organ, especially polymeric nanoparticles for cancer targeting.

It is essential to understand the physicochemical properties of nanostructures used in transdermal delivery. These characteristics include size, shape, molecular weight, surface charge, entrapment efficiency, and physical and chemical stability. These properties are important to evaluate their possible interaction as well as the mechanism of transport across the skin and also to comprehend how to optimize a formulation to achieve a systemic rather than localized effect. The penetration through

Table 21.1 Applications of Nanostructures in Dermal and Transdermal Drug Delivery Systems

Nanostructure	Active Ingredients	Indication	Method of Preparation	Study Findings	References
Liposomes	Ketoprofen	Rheumatoid arthritis and osteoarthritis	Thin layer evaporation technique	High stability; high permeation rate across skin	Maestrelli et al. (2005)
	Estradiol	Osteoporosis	Sonication and homogenized by manual extrusion	High penetration	Essa et al. (2004)
	Lignocaine	Local anesthetic	Thin film hydration (film casting)	High local activity	Sharma et al. (1994)
Transferosomes	Dipotassium glycyrrhizinate (KG)	Acute and chronic dermatitis	Solvent evaporation method	High skin deposition	Trotta et al. (2002)
	Methotrexate (MTX)	Topical treatment of psoriasis	Reverse-phase evaporation technique	three- to fourfold higher penetration as compared to water solution of MTX	Trotta et al. (2004)
	Diclofenac	Rheumatic diseases and actinic keratoses	Thin film hydration	High permeation as compare to conventional cream	Boinpally et al. (2003)
	Halogenated corticosteroid, triamcinolone-acetonide	Murine ear edema	Conventional film method	Prolongs antiinflammatory action, faster onset of action, edema effect, longer action time as compared to conventional creams and ointments	Cevc and Blume (2003)
	Insulin	Antidiabetic	Convention film hydration	Shows bioefficiency of at least 50% of subcutaneous dose action	Cevc et al. (1998)
Ethosomes + transferosomes (ultradeformable)	Clotrimazole (CLT)	Antifungal bioactive against *Candida* spp *albicans*	Mechanical dispersion method	High entrapment efficiency, high penetration, higher zone of inhibition than marketed cream formulation	Maheshwari et al. (2012)
Ethosomes	Indinavir	Anti-HIV protease inhibitor	Reverse-phase evaporation technique	Significantly increased transdermal flux as compared to conventional liposomes, ethanolic drug solution, or plain drug solution	Dubey et al. (2010)

	Melatonin	Delayed sleep phase syndrome; sleep disturbances in blind people; jet lag in travelers, shift workers, and cosmonauts	Reverse-phase evaporation technique	Greater mobility of skin lipids and deep-layer penetration of ethosomes as compared to ethanol or plain liposomes	Dubey et al. (2007b)
Polymeric nanoparticles	DNA delivery	Gene expression	Ionic-gelation method	High penetration	Lee et al. (2008)
	Flufenamic acid	Analgesic	O/w emulsion diffusion technique	High encapsulation efficiency and initial burst effect	Santander-Ortega et al. (2010)
	Triclosan	Preservative used in soaps	Emulsification–solvent displacement technique	High penetration	Rodriguez-Cruz et al. (2013)
NLC	Coenzyme Q_{10} (Q_{10})	Endogenous cellular antioxidant	High-pressure homogenization (HPH) technique	High penetration and efficiency	Teeranachaideekul et al. (2007)
	Celecoxib	Antiinflammatory agent	Microemulsion	Faster onset and prolonged activity until 24 h	Joshi and Patravale (2008)

the skin and systemic absorption of nanostructures in transdermal formulation need to be estimated with the intention to review their in vivo conduit through human skin. This is because size reduction at nanoscale has a number of beneficial effects, but on the other side it also has untoward effects. In vivo toxicity studies are a need of today's nanoscience for inaugurating safe therapeutic profiles of these nanostructures.

REFERENCES

Abdallah, M.H., 2013. Transfersomes as a transdermal drug delivery system for enhancement the antifungal activity of nystatin. Int. J. Pharm. Sci. 5, 560–567.

Aditya, N., Patankar, S., Madhusudhan, B., Murthy, R., Souto, E., 2010. Arthemeter loaded lipid nanoparticles produced by modified thin film hydration: pharmacokinetics, toxicological and in vivo anti malarial activity. Eur. J. Pharmaceut. Sci. 40, 448–455.

Ainbinder, D., Touitou, E., 2005. Testosterone ethosomes for enhanced transdermal delivery. Drug Deliv. 12, 297–303.

Alsarra, I.A., Bosela, A.A., Ahmed, S.M., Mahrous, G., 2005. Proniosomes as a drug carrier for transdermal delivery of ketorolac. Eur. J. Pharm. Biopharm. 59, 485–490.

Alvarez Roman, R., Naik, A., Kalia, Y., Guy, R.H., Fessi, H., 2004. Skin penetration and distribution of polymeric nanoparticles. J. Control. Rel. 99, 53–62.

Balakrishnan, P., Shanmugam, S., Lee, W.S., Lee, W.M., Kim, J.O., Oh, D.H., Kim, D.D., Kim, J.S., Yoo, B.K., Choi, H.G., 2009. Formulation and in vitro assessment of minoxidil niosomes for enhanced skin delivery. Int. J. Pharm. 377, 1–8.

Bangham, A.D., Standish, M.M., Watkins, J.C., 1965. Diffusion of univalent ions across the lamellae of swollen phospholipids. J. Mol. Biol. 13, 238–252.

Baroli, B., 2010. Penetration of nanoparticles and nanomaterials in the skin: fiction or reality? J. Pharm. Sci. 99, 21–50.

Barry, B.W., 1983. Dermatological Formulations: Percutaneous Absorption. Dekker, New York.

Bawarski, W.E., Childlowsky, E., Bharali, D.J., Mousa, S.A., 2008. Emerging nanopharmaceuticals. Nanomed.-Nanotechnol. 4, 273–282.

Bendas, E.R., Tadros, M.I., 2007. Enhanced transdermal delivery of salbutamol sulfate via ethosomes. AAPS PharmSciTech 8, 213–220.

Benson, H.A., 2006. Transfersomes for transdermal drug delivery. Expert Opin. Drug Del. 3, 727–737.

Bhaskar, K., Anbu, J., Ravichandiran, V., Venkateswarlu, V., Rao, Y.M., 2009. Lipid nanoparticles for transdermal delivery of flurbiprofen: formulation, in vitro, ex vivo and in vivo studies. Lipids Health Dis. 8, 1–15.

Bleve, M., Pavanetto, F., Perugini, P., 2011. Solid lipid nanoparticles: technological developments and in vivo techniques to evaluate their interaction with the skin. In: Capri, A. (Ed.), Progress in Molecular and Environmental Engineering—From Analysis and Modeling to Technology Applications. InTech Publisher, Italy, pp. 192–216.

Boinpally, R.R., Zhou, S.L., Poondru, S., Devraj, G., Jasti, B.R., 2003. Lecithin vesicles for topical delivery of diclofenac. Eur. J. Pharm. Biopharm. 56, 389–392.

Borgia, S.L., Regehly, M., Sivaramakrishnan, R., Mehnert, W., Korting, H., Danker, K., Roder, B., Kramer, K., Schafer Korting, M., 2005. Lipid nanoparticles for skin penetration enhancement correlation to drug localization within the particle matrix as determined by fluorescence and parelectric spectroscopy. J. Control. Rel. 110, 151–163.

Bouwstra, J.A., De Graaff, A., Groenink, W., Honeywell, L., 2002. Elastic vesicles: interaction with human skin and drug transport. Drug Deliv. Technol. 7, 222–223.

Bouwstra, J.A., Ponec, M., 2006. The skin barrier in healthy and diseased state. Biochim. Biophys. Acta. 1758, 2080–2095.

Buhleier, E., Wehner, W., Vögtle, F., 1978. "Cascade"- and "nonskid-chain-like" synthesis of molecular cavity topologies. Synthesis 978/2, 155–158.

Caruthers, S.D., Wickline, S.A., Lanza, G.M., 2007. Nanotechnological applications in medicine. Curr. Opin. Biotechnol. 18, 26–30.

Cevc, G., 1996. Transfersomes, liposomes and other lipid suspensions on the skin: permeation enhancement, vesicle penetration, and transdermal drug delivery. Crit. Rev. Ther. Drug 13, 257–388.

Cevc, G., 2003. Transdermal drug delivery of insulin with ultradeformable carriers. Clin. Pharmacokinet. 42, 461–474.

Cevc, G., 2004. Lipid vesicles and other colloids as drug carriers on the skin. Adv. Drug Deliv. Rev. 56, 675–711.

Cevc, G., Blume, G., 1992. Lipid vesicles penetrate into intact skin owing to the transdermal osmotic gradients and hydration force. Biochim. Biophys. Acta 1104, 226–232.

Cevc, G., Blume, G., 2001. New highly efficient formulation of diclofenac for the topical, transdermal administration in ultradeformable drug carriers, transfersomes. Biochim. Biophys. Acta 1514, 191–205.

Cevc, G., Blume, G., 2003. Biological activity and characteristics of triamcinolone-acetonide formulated with the self regulating drug carriers, transfersomes. Biochim. Biophys. Acta 1614, 156–164.

Cevc, G., Gebauer, D., Stieber, J., Schatzlein, A., Blume, G., 1998. Ultraflexible vesicles, transfersomes, have an extremely low pore penetration resistance and transport therapeutic amounts of insulin across the intact mammalian skin. Biochim. Biophys. Acta 1368, 201–215.

Charcosset, C., Elharati, A., Fessi, H., 2005. Preparation of solid lipid nanoparticles using a membrane contactor. J. Control. Release 108, 112–120.

Chauhan, A.S., Sridevi, S., Chalasani, K.B., Jain, A.K., Jain, S.K., Jain, N., Diwan, P.V., 2003. Dendrimer mediated transdermal delivery: enhanced bioavailability of indomethacin. J. Control. Release 90, 335–343.

Chen, H., Chang, X., Du, D., Liu, W., Liu, J., Weng, T., Yang, Y., Xu, H., Yang, X., 2006. Podophyllotoxin loaded solid lipid nanoparticles for epidermal targeting. J. Control. Release 110, 296–306.

Choi, M., Maibach, H., 2005. Liposomes and niosomes as topical drug delivery systems. Skin Pharmacol. Physiol. 18, 209–219.

Chourasia, M.K., Kang, L., Chan, S.Y., 2011. Nanosized ethosomes bearing ketoprofen for improved transdermal delivery. Results Pharma. Sci. 1, 60–67.

Ciotti, S.N., Weiner, N., 2002. Follicular liposomal delivery systems. J. Liposome Res. 12, 143–148.

Cui, Z., Han, S.J., Vangasseri, D.P., Huang, L., 2005. Immunostimulation mechanism of LPD nanoparticle as a vaccine carrier. Mol. Pharm. 2, 22–28.

Dayan, N., Touitou, E., 2000. Carriers for skin delivery of trihexyphenidyl HCL: ethosomes vs liposomes. Biomaterials 21, 1879–1885.

Desai, P., Patlolla, R.R., Singh, M., 2010. Interaction of nanoparticles and cell-penetrating peptides with skin for transdermal drug delivery. Mol. Membr. Biol. 27, 247–259.

Dokka, S., Cooper, S.R., Kelly, S., Hardee, G.E., Karras, J.G., 2005. Dermal delivery of topically applied oligonucleotides via follicular transport in mouse skin. J. Invest. Dermatol. 124, 971–975.

Dubey, V., Mishra, D., Dutta, T., Nahar, M., Saraf, D., Jain, N., 2007a. Dermal and transdermal delivery of an anti-psoriatic agent via ethanolic liposomes. J. Control. Release 123, 148–154.

Dubey, V., Mishra, D., Jain, N., 2007b. Melatonin loaded ethanolic liposomes: physicochemical characterization and enhanced transdermal delivery. Eur. J. Pharm. Biopharm. 67, 398–405.

Dubey, V., Mishra, D., Nahar, M., Jain, V., Jain, N.K., 2010. Enhanced transdermal delivery of an anti-HIV agent via ethanolic liposomes. Nanomedicine 6, 590–596.

El Maghraby, G.M., Williams, A.C., 2009. Vesicular systems for delivering conventional small organic molecules and larger macromolecules to and through human skin. Eur. J. Pharm. Sci. 34, 203–222.

El Maghraby, G.M., Williams, A.C., Barry, B.W., 2001a. Skin delivery of 5-fluorouracil from ultradeformable and standard liposomes in vitro. J. Pharm. Pharmacol. 53, 1069–1077.

El Maghraby, G.M., Williams, A.C., Barry, B.W., 2001b. Skin hydration and possible shunt route penetration in controlled estradiol delivery from ultradeformable and standard liposomes. J. Pharm. Pharmacol. 53, 1311–1322.

El Maghraby, G.M., Williams, A.C., Barry, B.W., 2006. Can drug bearing liposomes penetrate intact skin? J. Pharm. Pharmacol. 58, 415–429.

Elias, P.M., 2005. Stratum corneum defensive functions: an integrated view. J. Gen. Intern. Med. 20, 183–200.

Elias, P.M., Menon, G.K., 1991. Structural and lipid biochemical correlates of the epidermal permeability barrier. Adv. Lipid Res. 24, 1–26.

Elsayed, M.M., Abdallah, O.Y., Naggar, V.F., Khalafallah, N.M., 2007. Lipid vesicles for skin delivery of drugs: reviewing three decades of research. Int. J. Pharm. 332, 1–16.

Escobar Chavez, J.J., 2010. Current Technologies to Increase the Transdermal Delivery of Drugs. Bentham, Mexico.

Escobar Chavez, J.J., 2012. Nanocarriers for transdermal drug delivery. Dove Press J. 1, 1–15.

Escobar Chavez, J.J., Revilla Vazquez, A.L., Dominguez Delgado, C.L., Rodriguez Cruz, I.M., Alencaster, N.C., Diaz Torres, R., 2012. Nanocarrier Systems for Transdermal Drug Delivery. InTech, Mexico.

Esfand, R., Tomalia, D.A., 2001. Poly(amidoamine) (PAMAM) dendrimers: from biomimicry to drug delivery and biomedical applications. Drug Discov. Today 6, 427–436.

Essa, E.A., Bonner, M.C., Barry, B.W., 2002. Human skin sandwich for assessing shunt route penetration during passive and iontophoretic drug and liposome delivery. J. Pharm. Pharmacol. 54, 1481–1490.

Essa, E.A., Bonner, M.C., Barry, B.W., 2004. Electrically assisted skin delivery of liposomal estradiol: phospholipid as damage retardant. J. Control. Rel. 95, 535–546.

Fang, J.Y., Hong, C.T., Chiu, W.T., Wang, Y.Y., 2001a. Effect of liposomes and niosomes on skin permeation of enoxacin. Int. J. Pharm. 219, 61–72.

Fang, J.Y., Yu, S.Y., Wu, P.C., Huang, Y.B., Tsai, Y.H., 2001b. In vitro skin permeation of estradiol from various proniosome formulations. Int. J. Pharm. 215, 91–99.

Fang, Y.P., Huang, Y.B., Wu, P.C., Tsai, Y.H., 2009. Topical delivery of 5-aminolevulinic acid-encapsulated ethosomes in a hyperproliferative skin animal model using the CLSM technique to evaluate the penetration behavior. Eur. J. Pharm. Biopharm. 73, 391–398.

Ferrero, J., Weber, B., Geay, J., Lepille, D., Orfeuvre, H., Combe, M., Mayer, F., Leduc, B., Bourgeois, H., Paraiso, D., 2007. Second-line chemotherapy with pegylated liposomal doxorubicin and carboplatin is highly effective in patients with advanced ovarian cancer in late relapse: a GINECO phase II trial. Ann. Oncol. 18, 263–268.

Ganesh, S., Jyotsana, J., Avinash, K., Abha, D., 2012. Penetration enhancement of medical agent. Int. Res. J. Pharmacy 5, 1–7.

Garcıa-Fuentes, M., Torres, D., Alonso, M., 2003. Design of lipid nanoparticles for the oral delivery of hydrophilic macromolecules. Colloids Surf. B 27, 159–168.

Gasco, M.R., 1993. Method for producing solid lipid microspheres having a narrow size distribution. US 5250236 A.

Geinoz, S., Guy, R.H., Testa, B., Carrupt, P.A., 2004. Quantitative structure–permeation relationships (QSPeRs) to predict skin permeation: a critical evaluation. Pharm. Res. 21, 83–92.

Godin, B., Touitou, E., 2004. Mechanism of bacitracin permeation enhancement through the skin and cellular membranes from an ethosomal carrier. J. Control. Release 94, 365–379.

Guinedi, A.S., Mortada, N.D., Mansour, S., Hathout, R.M., 2005. Preparation and evaluation of reverse-phase evaporation and multilamellar niosomes as ophthalmic carriers of acetazolamide. Int. J. Pharm. 306, 71–82.

Gupta, M., Agrawal, U., Vyas, S.P., 2012. Nanocarrier based topical drug delivery for the treatment of skin diseases. Expert Opin. Drug Deliv. 9, 783–804.

Guy, R.H., 2003. Transdermal Drug Delivery. Dekker, New York.

Hadgraft, J., Lane, M.E., 2005. Skin permeation: the years of enlightenment. Int. J. Pharm. 305, 2–12.

Hafer, C., Goble, R., Deering, P., Lehmer, A., Breut, J., 1999. Formulation of interleukin-2 and interferon-a containing ultradeformable carriers for potential transdermal application. Anticancer Res. 19, 1505–1512.

Heffernan, M.J., Murthy, N., 2005. Polyketal nanoparticles: a new pH-sensitive biodegradable drug delivery vehicle. Bioconjugate Chem. 16, 1340–1342.

Hillery, A.M., Lloyd, A.W., Swarbrick, J., 2002. Drug Delivery and Targeting: For Pharmacists and Pharmaceutical Scientists, second ed. Taylor & Francis, USA and Canada.

Honary, S., Zahir, F., 2013. Effect of zeta potential on the properties of nano-drug delivery systems—a review (part 1). Trop. J. Pharm. Res. 12, 255–264.

Honeywell Nguyen, P.L., Gooris, G.S., Bouwstra, J.A., 2004. Quantitative assessment of the transport of elastic and rigid vesicle components and a model drug from these vesicle formulations into human skin in vivo. J. Invest. Dermatol. 123, 902–910.

Honeywell-Nguyen, P.L., Bouwstra, J.A., 2005. Vesicles as a tool for transdermal and dermal delivery. Drug. Discov. Today Technol. 2, 67–74.

Hu, F., Yuan, H., Zhang, H., Fang, M., 2002. Preparation of solid lipid nanoparticles with clobetasol propionate by a novel solvent diffusion method in aqueous system and physicochemical characterization. Int. J. Pharm. 239, 121–128.

Huang, S.S., Sung, S.H., Chiang, C.E., 2007. Chitosan potentiation of warfarin effect. Ann. Pharmacother. 41, 1912–1914.

Huang, X., Du, Y.Z., Yuan, H., Hu, F.Q., 2009. Preparation and pharmacodynamics of low-molecular-weight chitosan nanoparticles containing insulin. Carbohydr. Polym. 76, 368–373.

Hueber, F., Schaefer, H., Wepierre, J., 1994. Role of transepidermal and transfollicular routes in percutaneous absorption of steroids: in vitro studies on human skin. Skin Pharmacol. Physiol. 7, 237–244.

Illel, B., Schaefer, H., Wepierre, J., Doucet, O., 1991. Follicles play an important role in percutaneous absorption. J. Pharm. Sci. 80, 424–427.

James Smith, M.A., Hellner, B., Annunziato, N., Mitragotri, S., 2011. Effect of surfactant mixtures on skin structure and barrier properties. Ann. Biomed. Eng. 39, 1215–1223.

Johnson, M.E., Blankschtein, D., Langer, R., 1997. Evaluation of solute permeation through the stratum corneum: lateral bilayer diffusion as the primary transport mechanism. J. Pharm. Sci. 86, 1162–1172.

Joshi, M., Patravale, V., 2008. Nanostructured lipid carrier (NLC) based gel of celecoxib. Int. J. Pharm. 346, 124–132.

Junyaprasert, V.B., Singhsa, P., Suksiriworapong, J., Chantasart, D., 2012. Physicochemical properties and skin permeation of Span 60/Tween 60 niosomes of ellagic acid. Int. J. Pharm. 423, 303–311.

Kao, J., Hall, J., Helman, G., 1988. In vitro percutaneous absorption in mouse skin: influence of skin appendages. Toxicol. Appl. Pharm. 94, 93–103.

Karande, P., Jain, A., Ergun, K., Kispersky, V., Mitragotri, S., 2005. Design principles of chemical penetration enhancers for transdermal drug delivery. Proc. Natl. Acad. Sci. USA 102, 4688–4693.

Kierszenbaum, A.L., Tres, L., 2002. Integumentary system. Elsevier Saunders, Philadelphia, PA.

Kong, M., Chen, X.G., Liu, C.S., Liu, C.G., Meng, X.H., Yu, L.J., 2008. Antibacterial mechanism of chitosan microspheres in a solid dispersing system against *E coli*. Colloids Surf. B 65, 197–202.

Koo, O.M., Rubinstein, I., Onyuksel, H., 2005. Role of nanotechnology in targeted drug delivery and imaging: a concise review. Nanomedicine 1, 193–212.

Korting, H.C., Schafer Korting, M., 2010. Drug delivery carriers in the topical treatment of skin disease. In: Korting, M.S. (Ed.), Drug Delivery. Springer, Germany, pp. 435–469.

Kumar, G.P., Priyanka, G., 2015. Cellular permeation pathways: current focus of permeation enhancers for effective drug delivery. BAOJ Pharm. Sci. 1, 1–4.

Kumar, G.P., Rao, P.R., 2012. Ultra deformable niosomes for improved transdermal drug delivery: the future scenario. Asian J. Pharm. Sci. 7, 96–109.

Kushla, G.P., Zatz, J.L., Mills, O.H., Berger, R.S., 1993. Noninvasive assessment of anesthetic activity of topical lidocaine formulations. J. Pharm. Sci. 82, 1118–1122.

Lademann, J., Knorr, F., Richter, H., Blume Peytavi, U., Vogt, A., Antoniou, C., Sterry, W., Patzelt, A., 2008. Hair follicles an efficient storage and penetration pathway for topically applied substances. Skin Pharmacol. Physiol. 21, 150–155.

Lampe, M.A., Burlingame, A., Whitney, J., Williams, M.L., Brown, B.E., Roitman, E., Elias, P.M., 1983. Human stratum corneum lipids: characterization and regional variations. J. Lipid Res. 24, 120–130.

Lboutounne, H., Chaulet, J.F., Ploton, C., Falson, F., Pirot, F., 2002. Sustained ex-vivo skin antiseptic activity of chlorhexidine in poly(ε-caprolactone) nanocapsule encapsulated form and as a digluconate. J. Control. Rel. 82, 319–334.

Lee, P.W., Peng, S.F., Su, C.J., Mi, F.L., Chen, H.L., Wei, M.C., Lin, H.J., Sung, H.W., 2008. The use of biodegradable polymeric nanoparticles in combination with a low-pressure gene gun for transdermal DNA delivery. Biomaterials 29, 742–751.

Li, P., Zhao, L., 2007. Developing early formulations: practice and perspective. Int. J. Pharm. 341, 1–19.

Li, G., Fan, Y., Fan, C., Li, X., Wang, X., Li, M., Liu, Y., 2012. Tacrolimus-loaded ethosomes: physicochemical characterization and in-vivo evaluation. Eur. J. Pharm. Biopharm. 82, 49–57.

Liedtke, S., Wissing, S., Muller, R., Mader, K., 2000. Influence of high pressure homogenisation equipment on nanodispersions characteristics. Int. J. Pharm. 196, 183–185.

Maestrelli, F., Gonzalez Rodriguez, M.L., Rabasco, A.M., Mura, P., 2005. Preparation and characterisation of liposomes encapsulating ketoprofen cyclodextrin complexes for transdermal drug delivery. Int. J. Pharm. 298, 55–67.

Magnusson, B., Runn, P., Karlsson, K., Koskinen, L.O., 1997. Terpenes and ethanol enhance the transdermal permeation of the tripeptide thyrotropin releasing hormone in human epidermis. Int. J. Pharm. 157, 113–121.

Maheshwari, R.G., Tekade, R.K., Sharma, P.A., Darwhekar, G., Tyagi, A., Patel, R.P., Jain, D.K., 2012. Ethosomes and ultradeformable liposomes for transdermal delivery of clotrimazole: a comparative assessment. Saudi Pharmaceut. J. 20, 161–170.

Manconi, M., Caddeo, C., Sinico, C., Valenti, D., Mostallino, M.C., Biggio, G., Fadda, A.M., 2011. Ex vivo skin delivery of diclofenac by Transcutol containing liposomes and suggested mechanism of vesicle skin interaction. Eur. J. Pharm. Biopharm. 78, 27–35.

Mane, P.P., Jadhav, V.S., Humbe, P.V., Hakke, S.C., Kale, B.B., 2014. Dendrimer: a novel polymer and tool for drug delivery. Indo Am. J. Pharmaceut. Res. 5, 1–4.

Manosroi, A., Kongkaneramit, L., Manosroi, J., 2004. Stability and transdermal absorption of topical amphotericin B liposome formulations. Int. J. Pharm. 270, 279–286.

McCarron, P., Hall, M., 2008. Incorporation of novel 1-alkylcarbonyloxymethyl prodrugs of 5-fluorouracil into poly(lactide-*co*-glycolide) nanoparticles. Int. J. Pharm. 348, 115–124.

Medintz, I.L., Uyeda, H.T., Goldman, E.R., Mattoussi, H., 2005. Quantum dot bioconjugates for imaging, labelling and sensing. Nat. Mater. 4, 435–446.

Mehnert, W., Mader, K., 2001. Solid lipid nanoparticles: production, characterization and applications. Adv. Drug Deliv. Rev. 47, 165–196.

Modi, C.D., Bharadia, P.D., 2012. Transfersomes: new dominants for transdermal drug delivery. Am. J. PharmTech Res. 11, 1–21.

Moss, G., Dearden, J., Patel, H., Cronin, M., 2002. Quantitative structure–permeability relationships (QSPRs) for percutaneous absorption. Toxicol. in Vitro 16, 299–317.

Muller, R., Mehnert, W., Lucks, J.S., Schwarz, C., Zur Muhlen, A., Meyhers, H., Freitas, C., Ruhl, D., 1995. Solid lipid nanoparticles (SLN): an alternative colloidal carrier system for controlled drug delivery. Eur. J. Pharm. Biopharm. 41, 62–69.

Mura, S., Manconi, M., Sinico, C., Valenti, D., Fadda, A.M., 2009. Penetration enhancer-containing vesicles (PEVs) as carriers for cutaneous delivery of minoxidil. Int. J. Pharm. 380, 72–79.

Muzzalupo, R., Tavano, L., Cassano, R., Trombino, S., Ferrarelli, T., Picci, N., 2011. A new approach for the evaluation of niosomes as effective transdermal drug delivery systems. Eur. J. Pharm. Biopharm. 79, 28–35.
Naik, B., Nagarsenker, M., 2006. Novel drug delivery systems: potential in improving topical delivery of antiacne agents. Skin Pharmacol. Phys. 19, 2–16.
Naik, A., Kalia, Y.N., Guy, R.H., Fessi, H., 2004. Enhancement of topical delivery from biodegradable nanoparticles. Pharm. Res. 21, 1818–1825.
Namdeo, A., Jain, N., 1996. Niosomes as drug carriers. Indian J. Pharm. Sci. 58, 1–41.
Namdeo, A., Jain, N., 1999. Niosomal delivery of 5-fluorouracil. J. Microencapsul. 16, 731–740.
Neubert, R.H., 2011. Potentials of new nanocarriers for dermal and transdermal drug delivery. Eur. J. Pharm. Biopharm. 77, 1–2.
Niederhafner, P., Sebestík, J., Jezek, J., 2005. Peptide dendrimers. J. Pept. Sci. 11, 757–788.
Otberg, N., Richter, H., Schaefer, H., Blume Peytavi, U., Sterry, W., Lademann, J., 2004. Variations of hair follicle size and distribution in different body sites. J. Invest. Dermatol. 122, 14–19.
Pal, S.L., Jana, U., Manna, P., Mohanta, G., Manavalan, R., 2011. Nanoparticle: an overview of preparation and characterization. J. Appl. Pharmaceut. Sci. 1, 228–234.
Palac, Z., Engesland, A., Flaten, G.E., Skalko Basnet, N., Filipovic Grcic, J., Vanic, Z., 2014. Liposomes for (trans) dermal drug delivery: the skin PVPA as a novel in vitro stratum corneum model in formulation development. J. Liposome Res. 24, 313–322.
Panos, I., Acosta, N., Heras, A., 2008. New drug delivery systems based on chitosan. Curr. Drug Discov Technol. 5, 333–341.
Pardeike, J., Hommoss, A., Muller, R.H., 2009. Lipid nanoparticles (SLN, NLC) in cosmetic and pharmaceutical dermal products. Int. J. Pharm. 366, 170–184.
Parekh, H., 2007. The advance of dendrimers a versatile targeting platform for gene/drug delivery. Curr. Pharm. Design 13, 2837–2850.
Patel, S., 2007. Ethosomes: a promising tool for transdermal delivery of drug. BioPharma 5, 345–350.
Patil, S., Sandberg, A., Heckert, E., Self, W., Seal, S., 2007. Protein adsorption and cellular uptake of cerium oxide nanoparticles as a function of zeta potential. Biomaterials 28, 4600–4607.
Patlolla, R.R., Chougule, M., Patel, A.R., Jackson, T., Tata, P.N., Singh, M., 2010. Formulation, characterization and pulmonary deposition of nebulized celecoxib encapsulated nanostructured lipid carriers. J. Control. Rel. 144, 233–241.
Pietkiewicz, J., Sznitowska, M., 2004. The choice of lipids and surfactants for injectable extravenous microspheres. Pharmazie 59, 325–326.
Potts, R.O., Francoeur, M.L., 1991. The influence of stratum corneum morphology on water permeability. J. Invest. Dermatol. 96, 495–499.
Prausnitz, M.R., Langer, R., 2008. Transdermal drug delivery. Nat. Biotechnol. 26, 1261–1268.
Prausnitz, M.R., Mitragotri, S., Langer, R., 2004. Current status and future potential of transdermal drug delivery. Nat. Rev. Drug Discov. 3, 115–124.
Prausnitz, M.R., Elias, P.M., Franz, T.J., Schmuth, M., Tsai, J.C., Menon, G.K., Holleran, W.M., Feingold, K.R., 2012. Skin barrier and transdermal drug delivery. Dermatology 20, 65–73.
Priano, L., Esposti, D., Esposti, R., Castagna, G., De Medici, C., Fraschini, F., Gasco, M.R., Mauro, A., 2007. Solid lipid nanoparticles incorporating melatonin as new model for sustained oral and transdermal delivery systems. J. Nanosci. Nanotechnol. 7, 3596–3601.
Prow, T.W., Grice, J.E., Lin, L.L., Faye, R., Butler, M., Becker, W., Wurm, E.M., Yoong, C., Robertson, T.A., Soyer, H.P., 2011. Nanoparticles and microparticles for skin drug delivery. Adv. Drug Deliv. Rev. 63, 470–491.
Puglia, C., Blasi, P., Rizza, L., Schoubben, A., Bonina, F., Rossi, C., Ricci, M., 2008. Lipid nanoparticles for prolonged topical delivery: an in vitro and in vivo investigation. Int. J. Pharm. 357, 295–304.
Ramakrishna, G.A., Manohar, S.D., Bhanudas, S.R., 2013. Ethosomes: carrier for enhanced transdermal drug delivery system. Bull. Pharm. Res. 1, 1–8.

Rancan, F., Papakostas, D., Hadam, S., Hackbarth, S., Delair, T., Primard, C., Verrier, B., Sterry, W., Blume Peytavi, U., Vogt, A., 2009. Investigation of polylactic acid (PLA) nanoparticles as drug delivery systems for local dermatotherapy. Pharm. Res. 26, 2027–2036.

Rastogi, S.K., Singh, J., 2005. Effect of chemical penetration enhancer and iontophoresis on the in vitro percutaneous absorption enhancement of insulin through porcine epidermis. Pharm. Dev. Technol. 10, 97–104.

Rodriguez-Cruz, I.M., Merino, V., Merino, M., Diez, O., Nacher, A., Quintanar Guerrero, D., 2013. Polymeric nanospheres as strategy to increase the amount of triclosan retained in the skin: passive diffusion vs. iontophoresis. J. Microencapsul. 30, 72–80.

Rossi, S., Marciello, M., Sandri, G., Ferrari, F., Bonferoni, M.C., Papetti, A., Caramella, C., Dacarro, C., Grisoli, P., 2007. Wound dressings based on chitosans and hyaluronic acid for the release of chlorhexidine diacetate in skin ulcer therapy. Pharm. Dev. Technol. 12, 415–422.

Sahoo, S., Parveen, S., Panda, J., 2007. The present and future of nanotechnology in human health care. Nanomedicine 3, 20–31.

Sanna, V., Caria, G., Mariani, A., 2010. Effect of lipid nanoparticles containing fatty alcohols having different chain length on the ex vivo skin permeability of econazole nitrate. Powder Technol. 201, 32–36.

Santander-Ortega, M., Stauner, T., Loretz, B., Ortega Vinuesa, J., Bastos Gonzalez, D., Wenz, G., Schaefer, U., Lehr, C., 2010. Nanoparticles made from novel starch derivatives for transdermal drug delivery. J. Control. Release 141, 85–92.

Schleich, N., Pret, V., 2012. Nanostructures overcoming the skin barrier: drug delivery strategies. In: Schleich, N., Preat, V., Alonso, M., Csaba, N. (Eds.), Nanostructured Biomaterials for Overcoming Biological Barriers. RSC Publisher, UK, pp. 316–325.

Schneider, M., Stracke, F., Hansen, S., Schaefer, U.F., 2009. Nanoparticles and their interactions with the dermal barrier. Dermato-Endocrinol. 1, 197–206.

Schreier, H., Bouwstra, J., 1994. Liposomes and niosomes as topical drug carriers: dermal and transdermal drug delivery. J. Control. Release 30, 1–15.

Schubert, M., Muller Goymann, C., 2003. Solvent injection as a new approach for manufacturing lipid nanoparticles evaluation of the method and process parameters. Eur. J. Pharm. Biopharm. 55, 125–131.

Seiden, M.V., Muggia, F., Astrow, A., Matulonis, U., Campos, S., Roche, M., Sivret, J., Rusk, J., Barrett, E., 2004. A phase II study of liposomal lurtotecan (OSI-211) in patients with topotecan resistant ovarian cancer. Gynecol. Oncol. 93, 229–232.

Sharma, B., Jain, S., Vyas, S., 1994. Topical liposome system bearing local anaesthetic lignocaine: preparation and evaluation. J. Microencapsul. 11, 279–286.

Sheo, D., Shweta, A., Vijay, K., Ram, C., Aklavya, S., Ghanshyam, M., 2010. Enhanced transdermal delivery of indinavir sulfate via transfersomes. Nanomedicines 1, 1–7.

Shim, J., Kang, H.S., Park, W.S., Han, S.H., Kim, J., Chang, I.S., 2004. Transdermal delivery of minoxidil with block copolymer nanoparticles. J. Control. Rel. 97, 477–484.

Sjostrom, B., Bergenstahl, B., 1992. Preparation of submicron drug particles in lecithin-stabilized o/w emulsions: model studies of the precipitation of cholesteryl acetate. Int. J. Pharm. 88, 53–62.

Soppimath, K.S., Aminabhavi, T.M., Kulkarni, A.R., Rudzinski, W.E., 2001. Biodegradable polymeric nanoparticles as drug delivery devices. J. Control. Rel. 70, 1–20.

Swarbrick, J., Lee, G., Brom, J., Gensmantel, N.P., 1984. Drug permeation through human skin II: permeability of ionizable compounds. J. Pharm. Sci. 73, 1352–1355.

Symon, Z., Peyser, A., Tzemach, D., Lyass, O., Sucher, E., Shezen, E., Gabizon, A., 1999. Selective delivery of doxorubicin to patients with breast carcinoma metastases by stealth liposomes. Cancer 86, 72–78.

Tang, H., Mitragotri, S., Blankschtein, D., Langer, R., 2001. Theoretical description of transdermal transport of hydrophilic permeants: application to low frequency sonophoresis. J. Pharm. Sci. 90, 545–568.

Teeranachaideekul, V., Souto, E.B., Junyaprasert, V.B., Muller, R.H., 2007. Cetyl palmitate-based NLC for topical delivery of coenzyme Q-10 development, physicochemical characterization and in vitro release studies. Eur. J. Pharm. Biopharm. 67, 141–148.

Teeranachaideekul, V., Boonme, P., Souto, E.B., Muller, R.H., Junyaprasert, V.B., 2008. Influence of oil content on physicochemical properties and skin distribution of nile red-loaded NLC. J. Control. Release 128, 134–141.

Thomas, B.J., Finnin, B.C., 2004. The transdermal revolution. Drug Discov. Today 9, 697–703.

Thote, A.J., Gupta, R.B., 2005. Formation of nanoparticles of a hydrophilic drug using supercritical carbon dioxide and microencapsulation for sustained release. Nanomedicine 1, 85–90.

Toll, R., Jacobi, U., Richter, H., Lademann, J., Schaefer, H., Blume Peytavi, U., 2004. Penetration profile of microspheres in follicular targeting of terminal hair follicles. J. Invest. Dermatol. 123, 168–176.

Tomalia, D.A., Dewald, J.R., 1986. Dense star polymers having two dimensional molecular diameter. US 4587329 A.

Touitou, E., Dayan, N., Bergelson, L., Godin, B., Eliaz, M., 2000. Ethosomes—novel vesicular carriers for enhanced delivery: characterization and skin penetration properties. J. Control. Release 65, 403–418.

Touitou, E., Godin, B., Dayan, N., Weiss, C., Piliponsky, A., Levi Schaffer, F., 2001. Intracellular delivery mediated by an ethosomal carrier. Biomaterials 22, 3053–3059.

Trotta, M., Peira, E., Debernardi, F., Gallarate, M., 2002. Elastic liposomes for skin delivery of dipotassium glycyrrhizinate. Int. J. Pharm. 241, 319–327.

Trotta, M., Debernardi, F., Caputo, O., 2003. Preparation of solid lipid nanoparticles by a solvent emulsification diffusion technique. Int. J. Pharm. 257, 153–160.

Trotta, M., Peira, E., Carlotti, M.E., Gallarate, M., 2004. Deformable liposomes for dermal administration of methotrexate. Int. J. Pharm. 270, 119–125.

Uchechi, O., Ogbonna, J.D., Attama, A.A., 2014. Nanoparticles for dermal and transdermal drug delivery. In: Sezer, A.D. (Ed.), Application of Nanotechnology in Drug Delivery. InTech Publisher, Nigeria, pp. 193–235.

Ugazio, E., Cavalli, R., Gasco, M.R., 2002. Incorporation of cyclosporin A in solid lipid nanoparticles (SLN). Int. J. Pharm. 241, 341–344.

Ulrich, C., Degen, A., Patel, M.J., Stockfleth, E., 2008. Sunscreens in organ transplant patients. Nephrol. Dial. Transpl. 23, 1805–1808.

Uner, M., Yener, G., 2007. Importance of solid lipid nanoparticles (SLN) in various administration routes and future perspectives. Int. J. Nanomed. 2, 289–300.

Van Hal, D., Van Rensen, A., De Vringer, T., Junginger, H., Bouwstra, J., 1996. Diffusion of estradiol from non-ionic surfactant vesicles through human stratum corneum in vitro. STP Pharma Sci. 6, 72–78.

Venuganti, V.V.K., Perumal, O.P., 2008. Effect of poly(amidoamine) (PAMAM) dendrimer on skin permeation of 5-fluorouracil. Int. J. Pharm. 361, 230–238.

Verma, P., Pathak, K., 2010. Therapeutic and cosmeceutical potential of ethosomes: an overview. J. Adv. Pharm. Technol. Res. 1, 274.

Verma, P., Prajapati, S.K., Prajapati, R.N., 2012. A review on application of dendrimer in transdermal drug delivery. Int. Res. J. Pharmacy 5, 35–39.

Vora, B., Khopade, A.J., Jain, N., 1998. Proniosome based transdermal delivery of levonorgestrel for effective contraception. J. Control. Rel. 54, 149–165.

Walters, K.A., Hadgraft, J., 1993. Pharmaceutical Skin Penetration Enhancement. Dekker, New York.

Walve, J.R., Balkiwal, S.R., Rane, B.R., Pawar, S.P., 2011. Transfersomes: a surrogated carrier for transdermal drug delivery system. Int. J. Appl. Biol. Pharm. Technol. 2, 1–10.

Wang, Z., Itoh, Y., Hosaka, Y., Kobayashi, I., Nakano, Y., Maeda, I., Umeda, F., Yamakawa, J., Kawase, M., Yag, K., 2003. Novel transdermal drug delivery system with polyhydroxyalkanoate and starburst polyamidoamine dendrimer. J. Biosci. Bioeng. 95, 541–543.

Wiesenthal, A., Hunter, L., Wang, S., Wickliffe, J., Wilkerson, M., 2011. Nanoparticles: small and mighty. Int. J. Dermatol. 50, 247–254.

Wissing, S., Muller, R., 2003. Cosmetic applications for solid lipid nanoparticles (SLN). Int. J. Pharm. 2, 65–68.

Wissing, S., Kayser, O., Muller, R., 2004. Solid lipid nanoparticles for parenteral drug delivery. Adv. Drug Deliv. Rev. 56, 1257–1272.

Wolinsky, J.B., Grinstaff, M.W., 2008. Therapeutic and diagnostic applications of dendrimers for cancer treatment. Adv. Drug Deliv. Rev. 60, 1037–1055.

Wosicka, H., Cal, K., 2010. Targeting to the hair follicles: current status and potential. J. Dermatol. Sci. 57, 83–89.

Yiyun, C., Na, M., Tongwen, X., Rongqiang, F., Xueyuan, W., Xiaomin, W., Longping, W., 2007. Transdermal delivery of nonsteroidal anti-inflammatory drugs mediated by polyamidoamine (PAMAM) dendrimers. J. Pharm. Sci. 96, 595–602.

Zhang, L., Pornpattananangkul, D., Hu, C.M., Huang, C.M., 2010. Development of nanoparticles for antimicrobial drug delivery. Curr. Med. Chem. 17, 585–594.

CHAPTER

ADVANCEMENT IN PULMONARY DRUG DELIVERY SYSTEMS FOR TREATMENT OF TUBERCULOSIS

22

Amit K. Goyal, Tarun Garg, Sourav Bhandari, Goutam Rath
ISF Educational Society, Moga, India

CHAPTER OUTLINE

1 INTRODUCTION

Tuberculosis (TB) is an airborne infectious disease mainly caused by *Mycobacterium tuberculosis*. TB is the second most deadly communicable disease around the world. *M. tuberculosis*, which is an aerobic, nonmotile, rod-shaped, weakly Gram-positive, acid-fast tubercular bacillus is mainly responsible for pulmonary and extrapulmonary TB. *M. tuberculosis* mainly spreads through air from infected persons to healthy persons and utilizes high lipid content of alveolar macrophage walls for their survival (Kaur et al., 2016a). The other modes of TB transmission are transfusion of infected blood from TB patient to healthy person, use of materials and foods contaminated by TB patients, exposure to contaminated saliva, urine, and stool. Manifestations of TB include coughing, blood spots in sputum, chronic coughing, severe chest pain, fatigue, fever, and night sweating. The patient mortality rate is very high in infected people, without treatment about 70% of infected people die within 10 years. The current treatment for TB includes use of four first-line drugs in combination for 6 months; treatment success rate is 85% (Organization, 2013). TB is classified as latent or active. In the latent phase, the bacterium is present inside the human body in inactive form without showing any symptoms and shows a false negative result on a Mantoux tuberculin test. This type of infection is not communicable. In the active phase, the host's immune system becomes weak and the bacterium may become active (i.e., virulent) and multiply. This condition is also called TB disease, which is highly contagious (Kaur et al., 2014a). Additionally, multidrug resistance, MDR (resistant to first line drugs such as isoniazid and rifampicin), single-drug resistance (SDR), and extensive drug resistance, XDR (resistant to flouroquinolones agents and one or more second line injectable ATDs, such as amikacin, kanamycin capreomycin) are the emerging threats in management and treatment of TB. Major factors responsible for drug resistance in TB are ineffective TB control programs, interruption in chemotherapy, use of substandard drugs, massive bacillary load, improper prescription regimen, lack of awareness programs, inadequate drug supply, and ignorance of health-care workers.

The primary treatment of pulmonary infections involves the use of antibiotics by administration via oral/parenteral route. Administration via these routes can lead to manifestation of numerous severe side effects, as higher amount of drug is required in the lungs for curing infection. For example, oral drug administration is recommended for a minimum period of 6–9 months for infected patients with drug-susceptible mycobacterial strains. Therapeutic drug concentration is difficult to maintain in poorly vascularized granulomas or tubercles in the lungs. If patients are infected with drug-resistant mycobacterial strains, then the duration of conventional therapy can be extended for up to 2 or more years and complex combination therapy may be required. However, in comparison to the aforementioned routes of administration, pulmonary route seems to be promising for efficacious delivery of drugs to target site. Pulmonary route is the primary route of entry for mycobacteria that cause TB (Ernst, 1998; Vergne et al., 2004). Therefore, this has been the route of interest for delivery of ATDs for the past few decades. Lungs are becoming an attractive target for local and systemic applications. Lungs possess large surface areas and are covered with a very thin layer of epithelium, which results in an increase in drug absorption and efficacy. Lungs are one of the highly perfused organs, which leads to faster absorption of inhaled drugs and onset of action (Huang and Wang, 2006; Zhang et al., 2011). Pulmonary drug delivery is emerging as an applicable alternative to both oral and parenteral routes of administration. This delivery system bypasses hepatic first-pass metabolism, leading to a reduction in administrated doses, reduction of side effect, and drug bioavailability enhancement (Patton et al., 2004). In addition to this, it is a noninvasive route and can be self-administrated easily without any help in comparison to the parenteral delivery system. Many parameters, such as aerosol particle diameter (mass median aerodynamic diameters), shape, density, hygroscopicity, and the respiratory system complexity, has to be taken into consideration for ensuring deep lung deposition. Particle size is a very critical parameter in pulmonary drug delivery, which plays an important role in the respiratory deposition pattern. This is because particles of diameter more than 5 μm get deposited in oropharynx, smaller particles having diameters below 1 μm get exhaled, while particles in respirable range between 1 and 5 μm reach the lower airways and deposit efficiently in the lung (Kaur et al., 2012). The combination of pulmonary drug delivery systems with nanotechnology provides numerous benefits, which includes uniform distribution of drugs within the lungs and improvement in absorption. Furthermore, the large surface area of nanocarriers enhances the solubility and the dissolution rate of poorly soluble drugs (Keck and Müller, 2006; Kocbek et al., 2006). Due to these merits, different drugs have been administered via pulmonary routes to obtain both local and systemic effects, such as insulin (Al-Qadi et al., 2012), itraconazole (Rundfeldt et al., 2013), tobramycin (Li et al., 2014), azithromycin (Li et al., 2014), celecoxib (Patlolla et al., 2010), thymopentin (Li et al., 2010), phenethyl isothiocyanate (Dharmala et al., 2008), budesonide (El-Gendy et al., 2009), and calcitonin (Makhlof et al., 2010). In recent years, the pulmonary drug delivery is again gaining interest for drug delivery in TB due to numerous reasons. First and foremost, the lung mucosa represents a large surface, which may lead to rapid absorption of drugs into the bloodstream along with protection of hepatic first-pass metabolism. Second, the drugs delivered by this route provide high concentration in the lungs prior to their systemic absorption, which is beneficial in case of pulmonary TB (Kaur et al., 2014a; Pilcer and Amighi, 2010). Fig. 22.1 represents the significance of pulmonary route for administration of ATDs. In addition to this, lung macrophages will phagocytosis the drug due to their evolutionary role of phagocytizing the materials, which enter into the lungs (Misra et al., 2011). It has been established that macromolecular drugs and particulate or vesicular drug delivery systems are more prone to phagocytosis by alveolar macrophages. Therefore, these systems lead to an increase in efficacy of drugs.

Delivery of drugs or bioactive molecules through this route has great potential for achieving maximum therapeutic effect and many effective inhalation devices have been developed. A wide range of

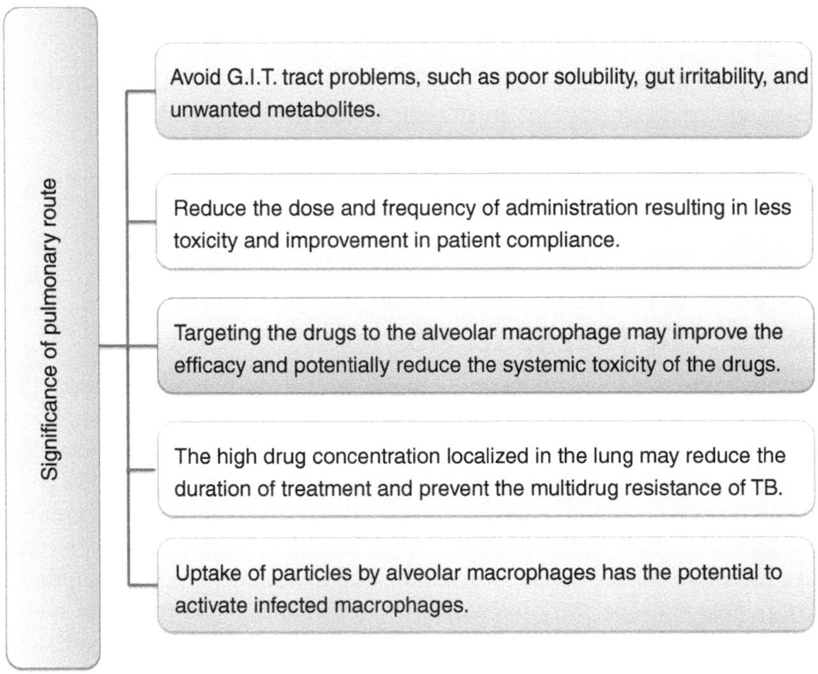

FIGURE 22.1 Significance of Pulmonary Route for Administration of ATDs

aerosol delivery devices are available in the market place for the treatment and management of pulmonary diseases, such as asthma, chronic obstructive pulmonary disease (COPD), and many others currently under development. Various advancements and innovations have improved the performance and efficiency of existing aerosol delivery devices. In this chapter, we review these nanocarrier systems along with inhalation devices (i.e., nebulizers, metered-dose inhalers, and dry powder inhalers) and the recent advances in inhalation technology that have significantly impacted the treatment and potential cure of many acute as well as chronic diseases (Kaur et al., 2014b).

2 BURDEN OF DISEASE AND PROGRESS TOWARD GLOBAL TARGETS

Upgraded figures provide perfect universal images of TB problems; hastening existing amounts of failure is desired to encounter objectives.

- In 2015, 6.1 million TB circumstances stated to the World Health Organization (WHO), from which 5.7 million individuals were newly identified and 0.4 million were previously on treatment.
- In 2015, an estimated 510,000 women died as a result of TB, more than one-third of whom were HIV-positive.
- In 2015, novel (3.5%) and earlier (20.5%) cured TB cases worldwide were predictable to have had MDR-TB.
- In 2013, 58% of lung TB patients reported internationally confirmed by WHO suggested tests, such as Xpert MTB/RIF.
- The Region of the Americas and the Western Pacific Region achieved all three 2015 target occurrences, prevalence, and mortality for declines in TB disease problem.
- There has been persistent improvement in the application of combined TB/HIV actions, however, exaggerated determinations are desirable, particularly to confirm worldwide contact to antiretroviral therapy (ART).
- ART is significant intermediation to diminish death among HIV-positive TB patients. In 2015, 70% of TB patients recognized to be HIV-positive were on ART.

3 PATHOGENESIS OF TUBERCULOSIS

TB is an infectious transmissible ailment and patients who suffer from pulmonary TB are the main cause of contamination. Cell-mediated immunity and hypersensitivity reactions contribute to the disease activity. Infection is commenced through breathing *M. tuberculosis* containing globule foci (0.5–6 μm) by active pulmonary TB patients to healthy person during coughs (Garg et al., 2015f). A lung is the primary route involved in infection. In the lungs, phagocytic antigen presenting cells (APC) is mainly responsible for imitation of TB infection. Specific pathogen recognition receptors (PRRs), such as Toll-like receptors (TLRs), mannose receptor, and complement receptors, initiate and organize host immune responses after the entry of bacilli. *M. tuberculosis* communication with TLRs starts a signaling force and finishes in a proinflammatory response by the expression of both immune and non-immune cells, which is useful or valuable to the individual (Singh et al., 2015). Alveolar macrophages generate inflammatory cytokines and chemokines entrance of *M. tuberculosis.* Infected cells produced chemokines that provide direction of the matured bacilli to the regional lymph node as well as prime T cells (both CD4+ and CD8+). Granuloma development occurs at the infection site due to the gathering of macrophages, fibroblasts, endothelial cells, T cells, dendritic cells, and stromal cells. The CD4+ and CD8+ T cells produce IFN-γ, resulting in switch contagion in granuloma, normal level of nitric oxide synthase (iNOS), apoptosis of infected macrophages, production of IL-2, as well as TNF-α, stimulation of macrophages or dendritic cells, production of IL-10, IL-12, and IL-15, resulting in directly killing *M. tuberculosis*. Activate macrophages comprise production of nitric oxide and Reactive Nitrogen Intermediates (RNI) (formed in tuberculous granuloma) and fusion of the mycobacteria enclosing phagosome with lysosomes use lethal effects on the bacilli (Kaur et al., 2016a). *M. tuberculosis* has changed some approaches to avoid the RNI toxicity. Phagolysosome maturation covers bacilli via succession of merging and division proceedings through numerous endocytic vesicles resulting in employment of vacuolar-proton transporting ATPase (vH + -ATPase) intended for microbicidal effects. Phagosome-enclosed antigens activated CD8+ T cell–involved apoptosis of infected macrophages by producing IFN-γ and RNI. Several *M. tuberculosis*–derived factors expressed of proapoptotic and antiapoptotic genes result in apoptosis of infected macrophages (Ahmad, 2011). *M. tuberculosis* changed numerous tactics to control infected host cells to avoid the resultant immune response to circumvent elimination and thus persevere in the host. In HIV infection, functional abnormalities cause in CD4+ and CD8+ T cells resulting progression to active disease. That's why combination of HIV and tuberculosis is the major reason for death. The other factors responsible for reactivation of a dormant infection involve old age, undernourishment, poor medical circumstances, and therapy with immunosuppressive drugs (Majlessi and Brosch, 2015).

4 CURRENT ANTITUBERCULAR CHEMOTHERAPY

Treatment of active TB is complex due to the emergence of multidrug-resistant TB and HIV infection. Vaccination and chemotherapy seem insufficient to treat TB, so antitubercular drugs (ATDs) suit the simple choice offered. Only 10 drugs are approved by the US Food and Drug Administration for TB treatment. Primarily, oral drug administration is recommended for a minimum period of 6–9 months for infected patients with drug-susceptible mycobacterial strains. In a study, ~99% of the body bacteria were killed after 2 weeks of drug treatment in mice and required more than 3 months for complete eradication of bacteria from the body (Zumla et al., 2015). The extended duration of therapy is moreover restricted by constraints on drug dosage, adverse drug reactions (which are exacerbated in HIV patients), and insufficient drug distribution at the infectious sites (Gerbeaux and Hanoteau, 1970). Table 22.1 represents the classification of anti-TB drugs with their basic mechanism along with their adverse effects.

5 PULMONARY DRUG DELIVERY

The term "pulmonary" or "respiratory" are often used interchangeably. These terms pertain to the lungs or the respiratory system as a whole. The pulmonary system is an integrated system that is involved in the interchange of O_2 and CO_2 between an organism and the environment. At top of the pulmonary system are nostrils, which bring air into the nose where it is warmed and humidified. The respiratory

Table 22.1 Classification of Anti-TB Drugs With Their Basic Mechanism and Their Adverse Effects (Donald and Schaaf, 2007; Grosset et al., 2012; Orme, 2001; Doggrell, 2005)

Antitubercular Drugs	Mechanism of Action	Adverse Effects
Rifampicin	Inhibits DNA dependent RNA polymerase	Stomach upset, bleeding, rashes, hepatitis-1
Isoniazid	Inhibits mycolic acid synthesis	Hepatitis, drug interactions, burning sensation in extremities
Pyrazinamide	Lowers the pH in the immediate surroundings of *M. tuberculosis*; results in inability to grow	Joint aches, hepatitis, rashes
Ethambutol	Inhibits mycobacterial arabinosyl transferases	Visual impairment
Kanamycin, amikacin, capreomycin	Inhibits protein synthesis irreversibly	Nephrotoxicity, ototoxicity, vestibular toxicity
Flouroquinolones: ciprofloxacin, ofloxacin	Inhibits DNA synthesis by inhibition of topoisomerase II (DNA gyrase)	Nausea, bloating, headache, diarrhea, dizziness, insomnia
Linezolid	Disorderly translation of messenger RNA (mRNA) into proteins in the ribosome	Diarrhea, headache, nausea, vomiting and rash
Amoxycillin-clavulanate	Inactivation of active sites of β-lactamases enzymes	Mild diarrhea, nausea, vomiting, and rash
Oxazolidinones	Binds the incoming transfer ribonucleic acid (RNA) with the complementary codon on the messenger RNA and inhibiting translation	Swelling of face, lips, tongue, or throat
Substituted ethylenediamines SQ109	Targeting the cell wall formation but by inhibition of trehalose monophosphate transferase	Swelling of the feet or lower legs, unusual tiredness or weakness
Benzothiazinones and dinitrobenzamides	Targeting the enzymes responsible for the formation of arabinans that are essential parts of the cell wall	Nausea and vomiting, skin rashes, headache

tract is protected by tiny hairs in the nose called cilia, which filter the breathed air. There are two openings of the airways; one is through nostrils as previously described, and the other is through the mouth. Both the openings meet at the pharynx or throat. The pharynx is a funnel-shaped tube that is the common route for air and food. The human pharynx is divided into three parts: nasopharynx (epipharynx), oropharynx (mesopharynx), and laryngopharynx (hypopharynx). The epiglottis, a flap of elastic cartilage tissue, guards the entrance of glottis. It prevents the entry of food into the trachea and directs its entry into the esophagus. The trachea or windpipe is a cartilaginous tube that connects the larynx to the bronchial parts of the lungs. The right lung (average weight 450 g) is divided into three lobes: upper, middle, and lower by two fissures. The left lung (average weight 400 g) is divided into two lobes: upper and lower by one fissure. Within the lungs, bronchi bifurcate into smaller branches called bronchioles. Bronchioles end up into tiny air sacs called alveoli, which are responsible for the exchange of gases. Aerodynamic diameter of the aerosol particles plays an important role in deposition of an inhaled formulation within the respiratory tract. Particle diameters more than 5 micrometers are deposited in upper airways (mouth, trachea, and main bronchi) by inertial impaction, whereas particles less than 1 μm are suspended in the air and are mostly exhaled (Liang et al., 2015). Inhaled particles are immersed into the aqueous surface-lining layer after deposition in conducting airways and alveoli covered by a surfactant film at the air–liquid interface, which promotes the particle displacement from air into the aqueous subphase as according to size, surface chemistry, and topography. Inhaled particles can adhere to mucin via electrostatic interactions, hydrophobic forces, and hydrogen bonding. Large molecules are slowly transported across the alveoli-capillary barrier of macrophage and can remain within the airspaces for several hours. Lung-surface macrophages clear the inhaled microparticles from the lungs by phagocytosis whereas nanoparticles might adhere to epithelial surfaces, might adsorb on the cell membrane of macrophages. Transport of macromolecules occurs across both the airway pseudo stratified columnar epithelium, as well as across the thin alveolar epithelium (Gandhimathi et al., 2015). Fast absorption from the alveoli reduces degradation processes occurring in the airspace and respiratory tissue, resulting in very stable drugs in the lung environment. There are different routes of administration for the delivery of drugs. The choice of route depends on the properties of the drug site of desired action and the condition of the patient (de Boer and Hagedoorn, 2015). Advantages and limitations of various routes of drug administration are represented in Table 22.2. The deposition of drugs in the lungs or pulmonary system, following pulmonary route follow four different mechanisms as shown in Table 22.3.

Ligand-appended pulmonary particles help in the specific drug delivery to the infected site because of favorable interaction of ligand with the specific receptors. Macrophages possess various surface receptors such as Fc, mannosyl, lectin, and galactosyl. These receptors help them in the process of recognition and endocytosis of a particulate carrier. Due to this fact, carriers containing ligands, such as mannosyl, immunoglobulin, fibronectin, and galactosyl are better phagocytozed by macrophages than carriers without such ligands. Table 22.4 enlists the different ligands used in the alveolar targeting.

Table 22.2 Advantages and Limitations of Various Routes of Drug Administration (Cocks et al., 2015; Klinger-Strobel et al., 2015; Mathpal et al., 2015; Zhou et al., 2015)

Parameters	Oral	Parenteral	Topical	Sublingual	Inhalable
Act in	Minutes	Seconds	Minutes	Minutes	Seconds
Drug bioavailability	Low	High	Low	High	High
Safety	High	Low	High	High	High
Convenient	High	Low	High	High	High
Cost	Low	High	Low	Moderate	Moderate
Route	Noninvasive	Invasive	Noninvasive	Noninvasive	Noninvasive
Stability	Low	Moderate	Moderate	Moderate	High
Limitations	Not used in emergencies	Painful	Slow absorption	Irritate buccal lining	Special delivery device needed

Table 22.3 Mechanism of Drug Deposition in the Lungs or Pulmonary System Following Pulmonary Route (Labiris and Dolovich, 2003; Fernandez Tena and Casan Clara, 2012; Kim and Hu, 1998; Vaughan and Vaughan, 1969)

S. No.	Mechanism	Description
1	Gravitational sedimentation	• The process of particles deposition due the effect of gravity. • Particles settle in the smaller airways. • Settling velocity of the particles is more for hygroscopic particles. • Hygroscopic particles may increase the probability of deposition by sedimentation.
2	Inertial impaction	• Large aerosol particles settle on the walls of the airways due to bifurcations in the airway passages. • Particles size greater than 3 μm. • Mechanism mainly depends on the aerodynamic diameter of particles and inspiratory flow rate. • Deposition of particles by impaction is greatest in the bronchial region.
3	Interception	• The process occurs due to the contact of particles with the surface of airways. • Interception is more significant with fibers due to their shape and size. • Deposited particles do not stray from their air streamlines. • Furthermore, fibers have small aerodynamic diameters resulting reach the smallest airways.
4	Diffusion	• The brownian motion of particles from higher concentration to a region of low concentration. • Smaller particles (less than 0.5 μm in diameter) settle into the peripheral region of the lungs. • Particle size must be within 1–5 μm in diameter. • Particles of size less than 2 μm are responsible for deposition in the small peripheral airways whereas particles of size greater than 5 μm result in oropharyngeal delivery and oral absorption.

Table 22.4 Types of Receptors Present on the Alveolar Macrophages With Ligands

Receptors	Ligands
Complement receptors (CR)	Maleylated bovine serum albumin (MBSA)
Mannose receptors (MR)	Lipoarabinomannan (LAM)
CD-14 receptors	Lipoarabinomannan (LAM)
Surfactant protein (SA)-A	Serum mannose binding protein (MBP)
Fc receptors	Immunoglobulin IgG
Insulin growth factor	Lecithin and other protein based ligand

6 ADVANCED DRY POWDER INHALERS (DPI)

DPI devices mainly deliver the medication in a form of dry powder via pulmonary routes for local and systemic effects without using propellants. The inhaled dose depends on the patients inspiratory flow rate. DPIs were first introduced in 1967. The first commercially successful inhaler (Spininhaler) was used for the delivery of sodium cromoglycate (mast cell stabilizer). Turbo inhaler was the first modern multidose DPI. Currently, extensive series of marketed DPI devices are offered; however, insufficient drug deagglomeration, low respirable fraction, low flow dependency rate, and high complexity for power is limited in its applications (Frijlink and De Boer, 2004). To deliver drugs to the lungs effectively, proper techniques must be followed to ensure that maximum doses reach the lungs. Various improvements and some add-on devices have been introduced to improve the efficiency of drug delivery. Recent approaches have mainly focused on improving aerosol dispersion improvement, producing extra fine particles for deep lung targeting, improving delivery efficiency using add-on devices, and improving inhalation techniques. Three important factors are responsible for the development of new inhalation designs (Table 22.5).

The dose received by the patient from the inhaler or DPI depends on several factors: (1) the inspiratory flow rate of the patient, (2) flow properties of formulation, (3) particle size and drug-carrier interaction, and (4) performance of the inhaler device, including aerosol generation, delivery, and inhalation technique. Advanced DPIs overcome the aforementioned limitations and increase the effectiveness of the encapsulated therapeutic molecules. In DPIs, micronized drugs and larger carrier elements are mixed to avoid aggregation and improve flow properties. When the patient inhales, air is led into the powder bed creating shear and turbulence in static powder mixture, resulting in the powder entering the patient's

Table 22.5 Factors Responsible for Development of Advanced Inhalation Designs (Newman, 2004; Heyder, 2004; Mathias et al., 1996; Byron and Patton, 1994)

Factors	Influence for Development of Advanced Inhalation Designs
Pharmaceutical factors	• Include regulatory and pharmacopoeial requirements for the approval of the inhaler device. • Reproducibility of the dose and the particle size distribution are of great concern. • The fraction of the dose delivered must be less than 5 μm. • Include the selection of delivery systems for new chemical entities. • Include optimization of design and efficiency.
Clinical application	• Inhalation devices target a specific area of the lungs. • Provide the desired therapeutic effect. • Easy to use. • Proper training provided to patients. • Minimal inspiratory flow rate required and ensure that the required clinical effect is achieved.
Commercial interests	• Device must be cost-effective. • Easy to use and handle. • Shows better results compared with other marketed devices. • New devices must be able to establish a place in the market and improve patient compliance.

Table 22.6 Different Mechanisms for the Dispersion of Powder Blend in DPI (Canonica et al., 2015; Farkas et al., 2015; Ziffels et al., 2015; Holmes et al., 2015)

Delivery Device	Mechanism for the Dispersion of Powder
Easyhaler (Orion Pharma)	Passage through narrow tubes (venturi tubes)
Clickhaler (Innovata Biomed)	Conduction against impact bodies like baffles, internal inhaler surfaces
Airmax, Taifun	Through circulation or cyclone chambers
Turbuhaler, Twisthaler	Through discharge channels or channels with helical inserts
Inhance, Aspirar	Through pressurized air or vacuum chambers

airways where drug elements get separated from carrier (Capanoglu et al., 2015). The smaller particles deposit deep inside the lungs while larger particles stay in the upper airways and get cleared. DPI mainly depends on inspiratory flow rate, flow properties of formulation, particle size, particle shape, aerodynamic diameter, drug-carrier interaction, device performance, disease state, and age (Hamilton et al., 2015). The finer particle size and better deaggregation can be obtained with forceful inhalation through an inspiratory flow driven inhaler. Optimal size can be obtained by milling or spray drying, which provides more control over the shape, size, and surface properties. However, micronized particles are difficult to handle in the inhaler as they have more tendency to adhere to the surfaces due the van der Waals and electrostatic forces. So, this problem can be solved by the use of large carrier particles, such as lactose or glucose to which smaller particles get adhered at the time of blending. The adhesion must not be too strong as the particles have to detach during inhalation (Ezzati Nazhad Dolatabadi et al., 2015). Several different mechanisms have been adopted for the dispersion of powder blend is discussed in Table 22.6.

Metered dose inhalers (MDI) and nebulizers are also used to deliver the therapeutic molecules to the target site, that is, lungs. Metered-dose inhalers are handheld devices used to deliver a metered drug dose to the lungs in a mist form. The drug is present either in dissolved or suspended form in propellants. On the other hand, a nebulizer is a device used to administer medication in the form of a mist that can be inhaled into the lungs through a mouthpiece or mask. In nebulizers, compressed gas or a compressor (electric/battery operated) is used to convert the drug from a solution to an aerosol form, producing droplets small enough to reach the alveolar region (Grant et al., 2015). Table 22.7 presents a comparative study regarding different aerosol delivery devices.

All aerosol delivery devices have comparative advantages and limitations. Certain carrier systems are not suitable with certain delivery devices. For example, small volume nebulizers are not as portable as pMDIs or DPIs. More than 75% of patients with COPD have difficulty using MDIs. Medication is lost on the nebulizer walls, baffles, on the mask or mouthpiece, and/or the patient's face. The selection of the aerosol device depends on the medication prescribed and the capabilities of the patient. Furthermore, patients need to be well informed and feel comfortable using the selected device. Challenges related to devices are related to device design, and diseases affect the performance of aerosol devices. The selection of the drug delivery device is mainly based on the degree of breathlessness. From a regulatory perspective, container-drug inhalation product interactions (e.g., adhesion, extraction, and/or reaction) must be closely controlled. Alternatively, regulatory authorities may ban the use of inhalers that do not demonstrate good compliance. In addition, regulatory agencies also ensure that inhalers are tested for pharmaceutical stability over the long term (3 months) at 25°C and 60% relative humidity (RH) as

Table 22.7 Comparative Characteristics of Different Aerosol Devices (Zhu et al., 2015; Yang et al., 2015; Young et al., 2015; Weiss et al., 2015)

Characteristics	DPI	MDI	Nebulizer
Particle size	1–5 μm	1–5 μm	1–10 μm
Formulations	Dry powder	Solution or suspension	Solution or suspension
Propellants	Not used	Used	Not used
Principle	Inspiratory flow rate	Propellant pressure	Bernoulli's
Breath hold	Not mandatory	Mandatory	Not mandatory
Patient management	Not mandatory	Mandatory	Not mandatory
Price	Cheap	Cheap	Expensive

well as the short term (1 month) at 40°C and 75% RH. European and American regulatory agencies have now implemented design standards to ensure consistent dosing from DPIs. Specific guidelines have been issued by various regulatory bodies on the subject of DPI performance, manufacturing, and controls for such products. However, some problems, such as moisture protection and dose content uniformity, remain the most challenging issues in Europe and the rest of the world.

7 CARRIERS USED FOR PULMONARY DELIVERY OF ATDs

Carriers are used as vehicles for delivering drugs and other bioactive molecules to the target site (Garg et al., 2012). Various lipid- and polymer-based carriers are used to ensure their effectiveness as well as stability with respect to encapsulated substances (Garg et al., 2013). Lipid-based carriers seem to be a promising ATDs delivery. It encapsulates a wide variety of drugs, provide protection, improved permeability and solubility of hydrophobic materials and exhibit sustained release (Garg and Goyal, 2014). On the other hand, polymer-based carriers are classified into two categories: (1) natural and (2) synthetic. A natural polymer includes chitosan, sodium alginate, cellulose acetate, ethyl cellulose, collagen, gelatin. Natural polymers are biocompatible, biodegradable, commercially available, easy processable, and used as biomaterials for cell/drug/gene delivery purposes for the treatment of TB disease. Natural polymers closely mimic the natural extracellular matrix of tissues, which make it the first choice of selection for fabrication of delivery systems. However, short supply, expensive, batch-to-batch variation, and susceptible to cross-contamination also associated with natural polymers (Garg et al., 2015d). On the other hand, synthetic polymers includes polylactic acid (PLA), polyethylene glycol (PEG), polymethylmethacrylate (PMMA), polyacrylic acid, hydroxypropyl cellulose (HPC), poly(lactic-*co*-glycolic) (PLGA), Poloxamer 407. The major advantage of synthetic polymers is their easily controlled physicochemical properties, no immunogenicity, they are processed with various techniques and consistently supplied in large quantities (Garg et al., 2015a). In some cases, natural and synthetic polymers fail to deliver the ATDs to the target site. In this situation, composite polymers (combination of natural-natural, natural-synthetic, and synthetic-synthetic) are used to prevent and control the various stages of TB disease. These polymers combinations control drug release, target ability, protect the drug from internal as well as external environment, stabilize the structure, enhance the transport of polar drugs across epithelial surfaces, easily converted into derivatives, provide elasticity, mechanical strength and excellent stability (Goyal et al., 2014).

Different lipid- and polymer-based systems can be used for designing different delivery systems and can be efficiently used for encapsulating and controlling release of ATDs via pulmonary routes (Kaur et al., 2016c). Colloidal drug carriers offer numerous advantages over conventional therapy, such as better systemic bioavailability, rapid onset of therapeutic action, avoidance of first-pass metabolism, providing sustained and controlled release, uniform distribution of drug into alveoli, decreased incidence of side-effects, fewer dosing frequencies, and improved patient compliance (Rohilla et al., 2016). An ideal polymer- or lipid-based carrier systems should fulfill the following requirements (Table 22.8). Colloidal

Table 22.8 Ideal Requirements in Polymer or Lipid Based Carrier Systems (Chaudhary et al., 2014; Garg et al., 2014, 2015c; Chaudhary and Garg, 2015)

Ideal Characteristics	Description
Biocompatibility	• Possess acceptable biocompatibility and toxicity profiles. • Without eliciting a harmful immune or inflammatory reaction. • Nontoxic and degradable. • Biologically active.
Biodegradability	• Should be biodegradable. • Degradation products should not be toxic. • Degradation products easily eliminated from body.
Mechanical property	• Match or compatible with tissue. • Survive under physiological conditions.
Structure	• Reproducible microscopic and macroscopic structure. • High surface and volume ratio.
Porosity	• Should have an adequate porosity. • Magnitude of the porosity, the pore size distribution, and its interconnectivity.
Nature	• Mimicking the native extracellular matrix (ECM). • Aid cellular development and morphogenesis.
Processability	• Possess relatively easy processability. • Malleability into the desired shape. • Capable of being produced into a sterile product.
Loading capacity	• Should have a maximum loading capacity. • Should have maximum entrapment efficiency. • Dispersed homogenously throughout the carrier.
Release kinetics	• Released continuously for longer duration after insertion into the body. • Avoid an initial burst effect. • Allow the appropriate dose of drug to reach the target site for prolonged period of time.
Binding affinity	• Binding affinity must be sufficiently low to allow release of the drug.
Stability	• Stable for entire study period. • Should possess dimensional stability, chemical stability, and biological activity over a prolonged period of time.

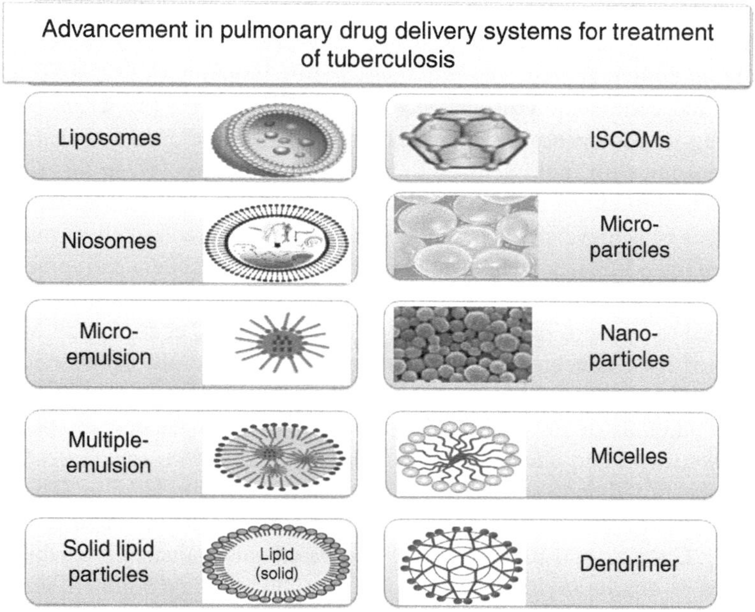

FIGURE 22.2 Advanced Pulmonary Carrier Systems for an Effective Treatment of TB

drug delivery systems, such as liposomes, nanoparticles, microparticles, niosomes, polymeric micelles, solid lipid nanoparticles (SLNs), dendrimers, cyclodextrin inclusion complexes, nanosuspensions, and nanoemulsions have been developed (Fig. 22.2) and are discussed in the next section.

7.1 LIPOSOMES

Liposomes (self-closed spherical structures) composed of one or several concentric curved lipid bilayer membranes (thickness about 4 nm), which enclose solvent into their interior. The liposome size ranges from some 50 nanometers to several micrometers. Liposomes possess unique properties, such as transport ability, permeation, adhesion, fusion as well as amphiphillic character of the lipids, which make them suitable for drug delivery (Swaminathan and Ehrhardt, 2012). It consists of soluble parts in both nonpolar as well as polar solvents, resulting in the promoting the self-assembly. The class of lipids that will be mainly considered is glycerophospholipids, often just called phospholipids. Phospholipids (phosphatidylcholines) have two acyl chains linked to a head group by means of a glycerol-backbone, and are the most widely used lipids in the formation of liposomes. Phosphatidylcholines form lamellar structures all relevant to pH. The presence of cholesterol acts as a fluidity buffer and effects the properties of the lipid bilayer of the liposomes (Iqbal et al., 2012). Different types of liposomes, such as conventional, pH sensitive, cationic, long circulating or stealth, immuno, magnetic, and temperature (or) heat sensitive liposomes, are successfully used to deliver different types of bioactive molecules to the target site (Chen et al., 2013). Liposomes are biocompatible, biodegradable, target a drug to the intended site of action in the body, act as a depot from which the entrapped compound is slowly released over time, and maintain therapeutic (but nontoxic) drug levels in the bloodstream or at the local administration site. Despite the advantages of liposomes, they suffer from some limitations, such as poor stability, short half-life, high production cost, leakage and fusion, quick uptake by cells of reticule endothelial system, allergic reactions, and problems in drug targeting (Vanic et al., 2014). Moreover, lipid concentration and composition also affects the physicochemical properties and aerosol performance of several drug-loaded liposomes. Bhardwaj et al. (2013) developed and characterized inhalable ligand-appended liposomes for multiple drug therapy for pulmonary TB. In this study, ATDs loaded liposomal formulation, formulated via thin film hydration method. Mannan was attached to the surface of prepared liposomes to achieve site-specific targeting, that is, alveolar cell of macrophage. Results showed that prepared ligand-appended liposomal formulation has excellent entrapment efficiency 40%, prolonged drug release 78% up to 24 h, improved recovery of drugs from the target site 70%–80% from 24 h, and proved to be more suitable for use as DPI (Bhardwaj et al., 2013). Manca et al. (2012) studied the effects of lipid concentration and composition on the physicochemical properties and aerosol performance of pulmonary delivery of rifampicin liposomes. Various liposome formulations containing different amounts of soy phosphatidylcholine and hydrogenated soy phosphatidylcholine, with and without cholesterol and oleic acid, were prepared and fully characterized. The result showed that the highest phospholipid concentration led to higher nebulization stability. The drug retention as well as cell toxicity was higher in oleic acid containing vesicles as compared to other formulations (Manca et al., 2012). Chimote and Banerjee (2009) prepared and evaluated inhalable isoniazid-loaded surfactant liposomes of dipalmitoylphosphatidylcholine (DPPC) as an adjunct therapy in pulmonary tuberculosis. The result showed that prepared liposomes were about 750 nm in diameter and excellent entrapment efficiency of 36.7% with sustained drug release up to 24 h. Approximately, 25%–27% isoniazid deposited in the alveolar chamber after one minute nebulization using a jet nebulizer, twin impinger. Overall, drug-loaded liposomes fulfill the dual purpose of pulmonary drug delivery and alveolar stabilization (Chimote and Banerjee, 2009). Similarly, Vyas et al. (2004) designed rifampicin-loaded liposomal aerosols for improved delivery to alveolar macrophages. In this study, liposomes were prepared by using egg phosphatidylcholine and cholesterol followed by coated via maleylated bovine serum albumin and O-steroyl amylopectin. Results showed that both coated liposomal formulations preferentially uptake in the lung macrophages up to 24 h. After 6 h of administration, the drug localization index was 3.5-fold for both ligand-appended liposome-based systems. Results clearly indicated that the ligand-anchored liposomal aerosols are not only effective in rapid attainment of high-drug concentration in lungs with high population of alveolar macrophages but also maintain the same over a prolonged period of time (Vyas et al., 2004).

7.2 NIOSOMES

Nonionic surfactant-based liposomes, known as niosomes, are microscopic lamellar structures obtained on hydration of nonionic surfactant of the alkyl or dialkyl polyglycerol ether class, cholesterol, and other lipids. Various materials, such as polyoxyethylene alkyl ether surfactants, alkyl ester, alkyl amides, fatty acids, and amino acid compound, are widely used for the fabrication of niosomes (Azeem et al., 2009). In aqueous media, nonionic surfactants form a closed bilayer vesicle structure. Hydrophobic parts are concerned with the distance away from the aqueous solvent, whereas the hydrophilic heads remain in contact with the aqueous solvent. Both hydrophilic and hydrophobic drugs are captured within the space enclosed in the vesicle (El-Ridy et al., 2011). The type of surfactant, nature of encapsulated drug, storage temperature, and detergents mainly affected the stability of the niosomes.

Niosomes entrap hydrophilic, lipophilic as well as amphiphillic drug moieties, increase the therapeutic index of drug, osmotically active, stable, and act as a depot to release the drug in a slow and controlled manner. It provides the protection from the biological environment, high patient compliance, stability, and is used as a vehicle for poorly absorbable drugs to design the novel drug delivery system (Abdelkader et al., 2011). Despite the advantages of niosomes, they also exhibit physical instability, aggregation, fusion, inefficient drug loading, and toxicity. Additionally, fabrications of niosomes are a time-consuming process and it may involve very specialized equipment (Cozar-Bernal et al., 2013). El-Ridy et al. (2015) prepared pulmonary ethambutol hydrochloride-loaded niosomal formulation using thin film hydration technique. Entrapment efficiency of ethambutol hydrochloride ranged from 12.20% to 25.81% with biphasic release kinetics. Optimized niosomal formulations targeted mice lungs for a prolonged period of time resulting in decreased bacterial counts in lung homogenates. Finally, we conclude that encapsulation of ethambutol hydrochloride in niosomal formulations for the treatment of TB provides higher efficacy and safety compared with the free drug (El-Ridy et al., 2015). Jain and Vyas (1995) also prepared and characterized rifampicin-loaded nonionic surfactant-based vesicles containing Span-85 and cholesterol in various molar fractions. Results of in vivo distribution studies of the prepared niosomes showed that 65% of the drug localized in the lungs by controlling the niosomes size (Jain and Vyas, 1995).

7.3 MICROEMULSION

Microemulsion is thermodynamically stable translucent dispersion of oil and water stabilized by an interfacial film of surfactant (cationic, anionic, and zwitterionic) and cosurfactant (short chain alcohol) molecules. Surfactants improve bioavailability, increase intestinal epithelial, and inhibition of p-glycoprotein efflux whereas cosurfactants increase the interfacial fluidity and lower the interfacial tension. Microemulsion droplets are in the range of 100–1000 Å with different shapes, that is, spherical, ellipsoid, and cylindrical (Entezari and Ghows, 2011; Mehnert and Mader, 2001). Three important theories, such as thermodynamics theory, mixed film theory, and solubilization theory involve in the formation of microemulsion. Various factors, such as chemical nature of surfactants and oils, concentration of cosurfactants, oil type, alcohol chain length, salt concentration, and HLB values mainly effects the formation of microemulsions. The major advantages of these carrier systems are improved drug solubilization, longer shelf life, entrap both hydrophilic as well as lipophilic compounds, modified drug release characteristics, reduction of intersubject and intrasubject variation, and the reduction of food effects (Arleth et al., 2007; Trotta et al., 2005). Some limitations are also associated with use of microemulsion, such as use of a large concentration of surfactant and cosurfactant necessary for stabilizing the nanodroplets, limited solubilizing capacity for high melting substances, surfactant shoe toxic behavior in some cases and stability is influenced by environmental parameters, such as temperature and pH. Mehta et al. (2010) prepared ATDs, that is, rifampicin (RIF), isoniazid (INH), pyrazinamide (PZA) loaded microemulsion composed of oleic acid, phosphate buffer, ethanol, and Tween (20, 40, 60, and 80). The result showed that formed droplets are found in the 2–12 μm range and percentage recoveries for each drug, that is, RIF (98.7%), PZA (99.0%), and INH (99.2%) are obtained. The release of PZA and INH from Tween 80 microemulsion is non-Fickian whereas RIF is found to follow a Fickian mechanism (Mehta et al., 2010). Mehta et al. (2007) formulated microemulsion composed of oleic acid, phosphate buffer, Tween 80, ethanol, and investigated its potential as drug delivery system for an ATDs, that is, rifampicin. Surfactant/cosurfactant ratio was selected with the help of pseudo-ternary phase diagram. Conductivity and viscosity studies showed the occurrence of structural changes from water-in-oil (w/o) microemulsion to oil-in-water (o/w). Microemulsion remained stable after the incorporation of rifampicin and particle size analysis indicates that the microemulsion changes into o/w emulsion at infinite dilution (Mehta et al., 2007).

7.4 MULTIPLE EMULSION

Multiple emulsions as double or complex emulsions during an inversion of o/w into w/o emulsion. Internal phase itself contains dispersed globules that are miscible with continued phases. Multiple emulsions lead to two types, water-in-oil-in-water (w/o/w) and oil-in-water-in-oil (o/w/o) (Besnard et al., 2013). Nature of oil, surfactants, phase volume, shear/agitation, and temperature play a significant role in multiple emulsion formation. Various advantages of multiple emulsions as drug delivery systems are easy production and scale up, protect bioactive from degradation, easy to administer via oral, nasal, ocular, parenteral, or any other route, that is, pulmonary. They encapsulate both hydrophilic and lipophilic compounds with high encapsulation efficiency (de Luca et al., 2000). The formulation components are economical and drug release pattern prolonged or controlled. However, it suffers from major limitations, such as low thermodynamic stability (coalescence, flocculation, and creaming), bulkier property, and disagreeable taste in some cases (Fattal et al., 1997). Nakhare and Vyas (2008) prepared multiple emulsions containing rifampicin and evaluated for in vitro characterization such as effect of pH of internal and external aqueous phase, in vitro release profile of rifampicin. Results showed that the partition coefficient of rifampicin between internal aqueous phase and liquid paraffin also effect on the release profile (Nakhare and Vyas, 2008). Lack of sufficient knowledge about the multiple-emulsion as a pulmonary drug delivery system limit its application against pulmonary disease.

7.5 SOLID LIPID PARTICLES

Solid lipid particles as lipid suspensions in water and are prepared from lipids with entrapped drugs in a solid matrix. Solid lipids used in solid lipid particles include fatty acids (palmitic acid, behenic acid, etc.), triglycerides (trilaurin, trimyristin, etc.), steroids (cholesterol), glycerides (glycerylmonostearate), waxes (cetylpalmitate), and surfactants (soybean lecithin, phosphatidylcholine, and sodium glycocholate) (Mehnert and Mader, 2001; Muller et al., 2000). Solid lipid particles are solid at room temperature and have diameters ranging from 50 to 1000 nm. Solid lipid particles are expressively taken up by the lymphatic system after inhalation and undergo phagocytosis by macrophages followed by migration to the lymphatic system; hence, inhalation can be an effective route to deliver particles to the lungs. Advantages of solid lipid particles includes controlled drug release, better encapsulation of both hydrophilic and lipophilic drug, longer stability, excellent pulmonary deposition, tolerability in the lungs, systemic bioavailability, nontoxic, and a long shelf-life (Muller et al., 2000). Maretti et al. (2014) prepared and evaluated rifampicin-loaded solid lipid microparticles to target alveolar macrophages. In this study, rifampicin-loaded solid lipid microparticles composed of stearic acid and sodium taurocholate and characterized on the basis of aerodynamic diameter, surface charge, physical state of the components, drug loading, release, cytotoxicity, cell internalization ability on murine macrophages J774 cell lines as well as drug biological activity against *Bacillus subtilis* strain. Results showed that solid lipid particles exhibited proper aerodynamic diameter, negative charged surface, and excellent drug entrapment as well as antimicrobial activity. In vitro studies clearly indicated that solid lipid particles showed noncytotoxicity against J774 cell lines and ability to be taken up by cell cytoplasm. Finally, we can conclude that microparticulate carrier showing features suitable for the inhaled therapy and for inducing endocytosis by alveolar macrophages by means of a dry powder inhaler device (Maretti et al., 2014). Pandey and Khuller (2005) evaluated the chemotherapeutic potential of nebulized solid lipid particles (SLPs) incorporating rifampicin, isoniazid, and pyrazinamide against experimental tuberculosis. In this study, SLPs prepared by the emulsion solvent diffusion technique and characterized on the basis of various in vitro and in vivo study. Results showed that therapeutic drug concentrations were maintained in lungs for 7 days following a single nebulization to guinea pigs. No tubercle bacilli could be detected in the lungs/spleen after 7 doses of treatment. Further, there was no evidence of any biochemical hepatotoxicity. Therefore, nebulization of SLP-based ATDs forms a sound basis for improving drug bioavailability and reducing the dosing frequency for better management of pulmonary TB (Pandey and Khuller, 2005).

7.6 IMMUNOSTIMULATING COMPLEXES (ISCOMs)

One of the best promising tools for vaccine delivery is immunostimulating complexes (ISCOMs). ISCOMs (negatively open cage-like structures) comprised of saponin, cholesterol, and phospholipids molecules mutually held by hydrophobic forces and possess a strong immunostimulant property. Microscopically ISCOMs are void particles of about 40 nm in diameter embedded with the antigen. Hydrophobic regions are required for efficient incorporation of protein antigen into ISCOMs (Kersten and Crommelin, 2003). ISCOM-Matrix induces strong activation of both the cell-mediated and the humoral arms of the immune systems. It's also able to induce an adaptive immune response in the presence of preexisting antibodies. ISCOMs have been widely applied as drug/bioactive molecules delivery systems. Pabreja et al. (2016) prepared pulmonary tubercular vaccine by using an Antigen 85 complex (Ag85)-loaded nanocarrier such ISCOMs. The particle size, zeta potential, and entrapment efficiency of optimized ISCOMs formulation based on Quill A (Ag85-A) were found to be 87.1 nm, −37 mv, and 79.12%, respectively. Approximately 24% of antigen from Quil A-based ISCOMs was released within the first 2 h followed by the release of 57% within 48 h of study. Caco-2 cell line viability was found to be 75% and 30% at 1.5 μg/mL and at 24 μg/mL, respectively. Immunological outcomes clearly indicated significant improvement in humoral as well as cellular immune responses after pulmonary immunization with ISCOMs containing Quil A in mice (Pabreja et al., 2016). Andersen et al. (2007) developed an intranasal vaccine, comprising the antigenic fusion protein Ag85B-ESAT-6 and the mucosal combined adjuvant vector CTA1-DD/ISCOMs. Optimized formulation strongly boosted prior BCG immunity after mucosal administration, leading to a highly increased recruitment of antigen-specific cells to the site of infection. They also observed a significantly ($P < 0.001$) reduced bacterial burden in the lung. From the aforementioned study, one point made clear is that effectiveness of mucosal vaccination with Ag85B-ESAT-6 mixed with CTA1-DD/ISCOMs act as adjuvant for stimulating TB-specific protective immunity in the lung (Andersen et al., 2007).

7.7 MICROPARTICLES

Microparticles (1–1000 μm diameters) are classified into two broad categories on the basis of morphology (1) microsphere and (2) microcapsules (Serrano-Ruiz et al., 2013). Various important natural and synthetic polymers are used in the fabrication of microparticulate systems. Drug release from the microparticles occurs by diffusion, polymer degradation, and hydrolysis/erosion. Pharmaceutical advantages of microparticles include the protection of drugs from the environment, enhancing solubility of poorly soluble drugs, sustained or controlled drug delivery, targeted release of active substances, safe handling of toxic substances, and improvement in flow of powder (Prow et al., 2011). Although the advantages of microparticles are impressive there are certain limitations. These include high cost, less reproducibility, process conditions that affect the stability, and degradation products that are harmful in some cases. Recently, our research group developed and optimized nanoembedded microparticles for pulmonary delivery of ATDs against experimental TB. In this study, rifampicin and isoniazid loaded various formulations (chitosan, guar gum, mannan, and guar gum coated chitosan) were prepared by spray drying technique. Results showed that guar gum batches have uniform size, smooth surface, and excellent flow as well as cell uptake ability. The drug release showed the biphasic pattern of release, that is, an initial burst followed by a sustained release pattern. A biodistribution study showed that guar gum–coated chitosan batches follow prolonged residence at the target site, that is, lungs resulting in almost fivefold reduction of the number of bacilli as compared to the control group. Finally, we concluded that optimized guar gum formulation acts as a promising carrier for selective delivery of ATDs to alveolar macrophages with the interception of minimal side effects, for efficient management of TB (Goyal et al., 2015). Garg et al. (2015b) developed and characterized inhalable spray-dried particles for pulmonary delivery of ATDs, that is, rifampicin and isoniazid. ATDs-loaded alginate particles were prepared by ionotropic gelation technique followed by spray drying. Results showed that the optimized spray-dried particles were found to be spherical shape, excellent flow properties, biphasic release pattern, that is, initial burst (30%–40% up to 4 h) followed by a sustained release pattern (90% up to 60 h), lower cytotoxicity, and excellent lung uptake up to 8 h. On the basis of the aforementioned results, alginate particles could be a promising carrier for targeted delivery of ATDs to alveolar macrophages for efficient management of TB in future (Garg et al., 2015b). Kaur et al. (2016b) developed and characterized of spray-dried porous nanoaggregates for pulmonary delivery of ATDs. In this study, guar gum and HPMC–based porous nanoaggregates were prepared by precipitation technique using two frontline ATDs, that is, isoniazid and rifampicin. Results showed that the powders obtained had excellent aerosolization property and high drug encapsulation efficiency ranging from 50% to 60% and 60% to 70% in case of guar gum and HPMC, respectively. A single pulmonary dose resulted in therapeutic drug concentrations of 30%–50% and 40% to 60% (in the lungs) in case of guar gum and HPMC, respectively, and in other organs (less than 5%) for 24 h. From this study, we can conclude that ATDs delivery through pulmonary route could be beneficial for reducing the drug dose or frequency of dosing, thus helpful in improving patient compliance (Kaur et al., 2016b,d). Mizoe et al. (2008) prepared rifampicin containing mannitol microparticles using four-fluid nozzle spray dryer. The result showed that prepared microparticles had better aerosol performance and approximately 43% deposited in stages 2–7 of cascade impactor. The optimized microparticles were effective for delivery of rifampicin to the lung and improved targeting to alveolar macrophages (Mizoe et al., 2008). Garcia-Contreras et al. (2007) prepared inhaled large porous particles of capreomycin for treatment of TB in a guinea pig model. In this study, capreomycin loaded large porous particles were prepared by spray-drying technique and characterized on the basis of various evaluation parameters. Results showed that capreomycin particles were thin-walled structures with excellent drug content in particles (73.8%). Particles have a volume diameter of 4.19 μm and desirable aerodynamic properties tract, with a mass median aerodynamic diameter of 4.99 μm. The absolute bioavailability of capreomycin after insufflation of 7.2 or 14.5 mg/kg of particles was 59%. Lungs of animals receiving a 14.5-mg/kg dose of capreomycin particles showed significantly lower bacterial burdens, which also supported by histopathological analysis (Garcia-Contreras et al., 2007).

7.8 NANOPARTICLES

Nanoparticles (10–1000 nm) are subnanosized colloidal structures composed of synthetic or semi-synthetic polymers. Nanoparticles could be nanospheres, nanocapsules, nanocrystals, and nanoparticulate (Couvreur, 2013). Pharmaceutical nanoparticles are those where the drug is dissolved, entrapped, encapsulated, or adsorbed on the surface. Pharmaceutical nanoparticles are classified into five major categories based on the type of the nature of polymer, property of final product and its applications. Polymeric nanoparticles placed a very important role in drug delivery due to its unique characteristics, such as controlled delivery of drugs and absolute targeting (Parveen et al., 2012). Solid lipid nanoparticles contain aqueous dispersions of solid lipids or dry powders. They possess excellent biodegradability, low toxicity, and facilitate prolong drug release. Hydrogel nanoparticles are three-dimensional, macromolecular networks containing a large fraction of water and reversibly swell or shrink in response to a variety of external stimuli (Sultana et al., 2013). They can react quickly to an external stimulus, such as temperature, pH. Peptide nanoparticles stability mainly depends on the nature of solvents, polymers, additives, and processing parameters. Nanocrystals and nanosuspension control oral absorption, bioavailability, and solubility of large therapeutic molecules. Nanoparticles suffers from some limitations, such as high manufacturing cost, toxic nature of solvents, chances of allergic reactions, various polymers have toxicity issues like poly vinyl alcohol, and they are difficult to handle in physical form because particle–particle aggregation occurs due to their small size and large surface area (Li et al., 2012). Goyal et al. (2016) developed and evaluated the therapeutic potential of ATDs-loaded natural polysaccharide comprising of galacto mannan subunit in experimental TB. Formulations were prepared by ionotropic gelation technique followed by spray drying. In vitro results showed that optimized nanoparticles were found to be spherical with a particle size distribution range from 230 ± 4.5 nm to 310 ± 6.2 nm. The in vitro drug release behavior indicated the initial burst followed by a sustained release pattern. Guar gum–coated chitosan nanoparticles (CGNPs) exhibited the highest cell uptake potential. The in vivo result showed that significant reduction in CFU count in experimental TB in mice. Histopathology study demonstrated that none of the treated group shows any evidence of lung tissue abnormality. On the basis of results, we can say that CGNPs could be a promising carrier for selective delivery of ATDs to alveolar macrophages for efficient management of TB with the interception of minimal side effects (Goyal et al., 2016). Garg et al. (2015e) prepared and characterized spray-dried inhalable chitosan nanoparticles (CNPs) for sustained delivery of ATDs, that is, isoniazid and rifampicin to the lungs. CNPs were prepared by ionic gelation technique followed spray drying. Results showed that the CNPs have smooth spherical shape with an average size of 230 nm with a poly dispersity index of 0.180. Both drugs were detected in lungs until 24 h post-nebulization. The optimized formulations showed lower cytotoxicity and significant reduction in the number of bacilli in the lungs, as compared to free drug. Finally, it may be concluded that CNPs may be exploited as a prospective tool for drug delivery to direct drugs to the lung tissues for the treatment of TB (Garg et al., 2015e).

7.9 MICELLES

Micelle is an aggregate of surfactant molecules dispersed in a liquid colloid. A typical micelle is often globally, roughly spherical, ellipsoids, cylinders, and bilayer in shape. Surfactant concentration, temperature, pH mainly affected the shape and size of the micelles. In a micelle, the hydrophobic tails of several surfactant molecules assembles into an oil-like core that has less contact with water.

When the concentration of surfactant is greater than critical micelle concentration and the temperature of the systems is greater than critical micelle temperature, micelles are formed (Xu et al., 2013). Polymeric micelles can be formed by amphiphillic block copolymers such as poly(ethylene oxide)-block-poly(L-amino acid), poly(ethylene oxide)-block-poly(ester), and poly(ethylene oxide)-block-poly(propylene oxide)-block-poly(ethylene oxide). Poly(ethylene oxide)-block-poly(L-amino acid) for entrapping water-insoluble drug molecules or enzymes. The main merits are biocompatible, unaltered biological activity, sustained release of encapsulated materials, nontoxic, and increase solubility of hydrophobic drugs. Various drug incorporated into block copolymers via chemical conjugation, physical entrapment, complexation, and cotreatment (Capretto et al., 2013). Vadakkan et al. (2013) prepared dry powder cationic lipopolymer nanomicelle inhalation for targeted delivery of ATDs to alveolar macrophage using spray-drying technique and characterized on the basis of various evaluation parameters. Results showed that spray-dried drug-loaded particles come in the inhalable size range (67.88% ± 3%). Further, the spray-dried drug-loaded nanomicelles also proved to be a better excipient for delivering xenobiotic to lung through dry powder inhalation technology (Vadakkan et al., 2013). Silva et al. (2007) focused on the characterization of polymeric micelle-forming tuberculostatic prodrugs and the antimycobacterial activity of these prodrugs. In this study, hydroxymethylpyrazinamide, isoniazid, and rifampin loaded poly(ethyleneglycol)-poly(aspartic acid) micelle were prepared and characterized on the basis of various evaluation parameters. The results showed that particle size was very low less than 100 nm while the conjugated drug level was very high to 85%. Additionally, optimized formulation exhibited a low level of critical micelle concentration and stronger antimycobacterial effects as comparison to original drugs. Overall, we can say that synthesized micellar formulation incorporated ATDs decrease the toxicity of the tuberculostatic drugs of choice (Silva et al., 2007).

7.10 DENDRIMER

Dendrimers are three-dimensional macromolecular architectural classes (linear, crosslinked, and branched) with an initiator core, several branching units and multiple active terminal groups. Drug molecules can be loaded both in the interior of dendrimers as well as attached to surface groups. Various types of dendrimers, such as radially layered poly (amidoamine-organosilicon), poly (amidoamine), poly (propylene imine), chiral, liquid crystalline, tecto, hybrid, multilingual, micellar, amphiphillic, peptide, and frechet-type were successfully identified for delivery of drug to the target place (Fuchs et al., 2004; Wang et al., 2003). Dendrimers offer various advantages over other polymers, such as less susceptible for reticulum endothelium uptake, easily modified, targeting to particular site in the body, ensure reproductive pharmacokinetic behavior, and provide a large variety of structures with reduced cost of their production. These systems can be administered by various routes including intravenous, oral, pulmonary, and ocular (Gardikis et al., 2011). However, some limitations are associated with using dendrimer as drug delivery systems requires excess monomer loading, difficult to isolate from their side products, low yield, and uncontrollable growth in some case. Bellini et al. (2015) attached the ATDs, that is, rifampicin with a fourth-generation poly (amidoamine) (G4-PAMAM) dendrimer. The complex was more stable at neutral pH as compared to low pH resulting the rapid release of drug molecules under acidic medium provide switch for drug targeting, that is, alveolar macrophage of lungs. On the basis of results, PAMAM-like dendrimers may be considered suitable drug delivery systems for RIF and derivatives (Bellini et al., 2015). Kumar et al. (2006) prepared and characterized mannosylated dendritic architecture for the selective delivery of ATDs, that is, rifampicin to alveolar macrophages. In this study, rifampicin was efficiently loaded into mannosylated dendrimer using dissolution technique and characterized on the basis of various evaluation parameters such as drug loading, pH-dependent in vitro release, hemolytic toxicity, phagocytic alveolar macrophage uptake and cytotoxicity. Results showed that rifampicin loaded mannosylated dendrimer reduced release rate, hemolytic toxicity, and cytotoxicity at pH 7.4 whereas enhanced drug release and alveolar macrophage uptake at pH 5.0. Finally, we can say that the present novel dendritic system displayed biocompatibility and site-specific delivery of ATDs drug (Kumar et al., 2006).

8 CONCLUSIONS AND FUTURE PROSPECTIVES

In this chapter, we have discussed the entire global scenario of TB by introducing promising drug delivery systems or by modifying existing systems such as a chemotherapy regimen for TB. In the treatment of pulmonary diseases, the lung is a very attractive target for drug delivery. It provides direct target the site of disease, provides enormous surface area, and also provides low enzymatic, controlled environment for systemic absorption of medications. In the pharmaceutical field, the pulmonary drug delivery is truly one of the most up-and-coming routes because of the recent launch of various inhaled pharmaceutical products. So at this time, the more information collected related to lung physiology, lung disease, particle deposition behavior in lungs, inhalation devices, drug release behavior in lungs, and so on, is beneficial. From the various studies in this chapter, we can conclude that, at present-day, it is definitely conceivable to deliver larger drug molecules such as ATDs systematically up to therapeutic drug levels. Inhalable nanocarrier systems (colloidal systems) provide many benefits, such as enhanced dissolution rate, uniform distribution of drug to the alveoli, sustained release of the drug to the target site, reduced dosing frequency, reduced side effects associated with a high drug serum concentration after oral or parenteral administration, and improved patient compliance. Various advances in inhalation devices have improved their efficacy, but further improvements are still needed to achieve the desired effect. For successful therapy, proper knowledge of the device is essential, and proper patient education must be provided by the health-care professionals. The final decision regarding the prescription of a device depends on various factors: the disease state, cost and availability of the device, drug interactions, and treatment efficiency. So, the combination of pulmonary delivery via advanced inhalation devices and nanotechnology has the prospective to expand the targeting, release, and therapeutic effects of drugs with substantial impending proficiency of overwhelming the physicochemical and biological obstacles. In the future, advanced or modified technological developments in the pharmaceutical field are trying to take possession of the aforementioned carrier systems and provide easy access to a disease location, or effectiveness, which can play a major role in the treatment of TB. Additionally, development of vaccines using a combined strategic approach, as discussed earlier, can play a major role in the treatment of TB.

ACKNOWLEDGMENT

The authors (AMR/43/2011-ECD-1 dated July 01 2013) are thankful to the Indian Council of Medical Research (ICMR), New Delhi, India, for providing financial support in carrying out similar research work.

REFERENCES

Abdelkader, H., Ismail, S., Kamal, A., Alany, R.G., 2011. Design and evaluation of controlled-release niosomes and discomes for naltrexone hydrochloride ocular delivery. J. Pharm. Sci. 100, 1833–1846.

Ahmad, S., 2011. Pathogenesis, immunology, and diagnosis of latent *Mycobacterium tuberculosis* infection. Clin. Dev. Immunol. 2011, 814943.

Al-Qadi, S., Grenha, A., Carrión-Recio, D., Seijo, B., Remuñán-López, C., 2012. Microencapsulated chitosan nanoparticles for pulmonary protein delivery: in vivo evaluation of insulin-loaded formulations. J. Control. Release 157, 383–390.

Andersen, C.S., Dietrich, J., Agger, E.M., Lycke, N.Y., Lovgren, K., Andersen, P., 2007. The combined CTA1-DD/ISCOMs vector is an effective intranasal adjuvant for boosting prior *Mycobacterium bovis* BCG immunity to *Mycobacterium tuberculosis*. Infect. Immun. 75, 408–416.

Arleth, L., Svensson, B., Mortensen, K., Pedersen, J.S., Olsson, U., 2007. Block-copolymer micro-emulsion with solvent-induced segregation. Langmuir 23, 2117–2125.

Azeem, A., Anwer, M.K., Talegaonkar, S., 2009. Niosomes in sustained and targeted drug delivery: some recent advances. J. Drug Target. 17, 671–689.

Bellini, R.G., Guimaraes, A.P., Pacheco, M.A., Dias, D.M., Furtado, V.R., De Alencastro, R.B., Horta, B.A., 2015. Association of the anti-tuberculosis drug rifampicin with a PAMAM dendrimer. J. Mol. Graph Model 60, 34–42.

Besnard, L., Marchal, F., Paredes, J.F., Daillant, J., Pantoustier, N., Perrin, P., Guenoun, P., 2013. Multiple emulsions controlled by stimuli-responsive polymers. Adv. Mater. 25, 2844–2848.

Bhardwaj, A., Kumar, L., Narang, R.K., Murthy, R.S., 2013. Development and characterization of ligand-appended liposomes for multiple drug therapy for pulmonary tuberculosis. Artif. Cells Nanomed. Biotechnol. 41, 52–59.

Byron, P.R., Patton, J.S., 1994. Drug delivery via the respiratory tract. J. Aerosol. Med. 7, 49–75.

Canonica, G.W., Arp, J., Keegstra, J.R., Chrystyn, H., 2015. Spiromax, a new dry powder inhaler: dose consistency under simulated real-world conditions. J. Aerosol. Med. Pulm. Drug Deliv. 28, 309–319.

Capanoglu, M., Dibek Misirlioglu, E., Toyran, M., Civelek, E., Kocabas, C.N., 2015. Evaluation of inhaler technique, adherence to therapy and their effect on disease control among children with asthma using metered dose or dry powder inhalers. J. Asthma 52, 838–845.

Capretto, L., Mazzitelli, S., Colombo, G., Piva, R., Penolazzi, L., Vecchiatini, R., Zhang, X., Nastruzzi, C., 2013. Production of polymeric micelles by microfluidic technology for combined drug delivery: application to osteogenic differentiation of human periodontal ligament mesenchymal stem cells (hPDLSCs). Int. J. Pharm. 440, 195–206.

Chaudhary, C., Garg, T., 2015. Scaffolds: a novel carrier and potential wound healer. Crit. Rev. Ther. Drug Carrier Syst. 32, 277–321.

Chaudhary, S., Garg, T., Murthy, R.S., Rath, G., Goyal, A.K., 2014. Recent approaches of lipid-based delivery system for lymphatic targeting via oral route. J. Drug Target. 22, 871–882.

Chen, M.H., Soda, Y., Izawa, K., Kobayashi, S., Tani, K., Maruyama, K., Tojo, A., Asano, S., 2013. A versatile drug delivery system using streptavidin-tagged pegylated liposomes and biotinylated biomaterials. Int. J. Pharm. 454, 478–485.

Chimote, G., Banerjee, R., 2009. Evaluation of antitubercular drug-loaded surfactants as inhalable drug-delivery systems for pulmonary tuberculosis. J. Biomed. Mater. Res. A 89, 281–292.

Cocks, E., Alpar, O., Somavarapu, S., Greenleaf, D., 2015. Impact of surfactant selection on the formulation and characterization of microparticles for pulmonary drug delivery. Drug Dev. Ind. Pharm. 41, 522–528.

Couvreur, P., 2013. Nanoparticles in drug delivery: past, present and future. Adv. Drug Deliv. Rev. 65, 21–23.

Cozar-Bernal, M.J., Rabasco, A.M., Gonzalez-Rodriguez, M.L., 2013. Development and validation of a high performance chromatographic method for determining sumatriptan in niosomes. J. Pharm. Biomed. Anal. 72, 251–260.

de Boer, A.H., Hagedoorn, P., 2015. The role of disposable inhalers in pulmonary drug delivery. Exp. Opin. Drug Deliv. 12, 143–157.

de Luca, M., Grossiord, J.L., Seiller, M., 2000. A very stable W/O/W multiple emulsion. Int. J. Cosmet. Sci. 22, 157–161.

Dharmala, K., Yoo, J.W., Lee, C.H., 2008. Development of chitosan–SLN microparticles for chemotherapy: in vitro approach through efflux-transporter modulation. J. Control. Release 131, 190–197.

Doggrell, S.A., 2005. New drugs being developed for the treatment of tuberculosis. Exp. Opin. Investig. Drugs 14, 917–920.

Donald, P.R., Schaaf, H.S., 2007. Old and new drugs for the treatment of tuberculosis in children. Paediatr. Respir. Rev. 8, 134–141.

El-Gendy, N., Gorman, E.M., Munson, E.J., Berkland, C., 2009. Budesonide nanoparticle agglomerates as dry powder aerosols with rapid dissolution. J. Pharm. Sci. 98, 2731–2746.

El-Ridy, M.S., Abdelbary, A., Essam, T., El-Salam, R.M., Kassem, A.A., 2011. Niosomes as a potential drug delivery system for increasing the efficacy and safety of nystatin. Drug Dev. Ind. Pharm. 37, 1491–1508.

El-Ridy, M.S., Yehia, S.A., Kassem, M.A., Mostafa, D.M., Nasr, E.A., Asfour, M.H., 2015. Niosomal encapsulation of ethambutol hydrochloride for increasing its efficacy and safety. Drug Deliv. 22, 21–36.

Entezari, M.H., Ghows, N., 2011. Micro-emulsion under ultrasound facilitates the fast synthesis of quantum dots of CdS at low temperature. Ultrason. Sonochem. 18, 127–134.

Ernst, J.D., 1998. Macrophage receptors for *Mycobacterium tuberculosis*. Infect. Immun. 66, 1277–1281.

Ezzati Nazhad Dolatabadi, J., Hamishehkar, H., Valizadeh, H., 2015. Development of dry powder inhaler formulation loaded with alendronate solid lipid nanoparticles: solid-state characterization and aerosol dispersion performance. Drug Dev. Ind. Pharm. 41, 1431–1437.

Farkas, D.R., Hindle, M., Longest, P.W., 2015. Characterization of a new high-dose dry powder inhaler (DPI) based on a fluidized bed design. Ann. Biomed. Eng. 43 (11), 2804–2815.

Fattal, E., Roques, B., Puisieux, F., Blanco-Prieto, M.J., Couvreur, P., 1997. Multiple emulsion technology for the design of microspheres containing peptides and oligopeptides. Adv. Drug Deliv. Rev. 28, 85–96.

Fernandez Tena, A., Casan Clara, P., 2012. Deposition of inhaled particles in the lungs. Arch. Bronconeumol. 48, 240–246.

Frijlink, H.W., De Boer, A.H., 2004. Dry powder inhalers for pulmonary drug delivery. Exp. Opin. Drug Deliv. 1, 67–86.

Fuchs, S., Kapp, T., Otto, H., Schoneberg, T., Franke, P., Gust, R., Schluter, A.D., 2004. A surface-modified dendrimer set for potential application as drug delivery vehicles: synthesis, in vitro toxicity, and intracellular localization. Chemistry 10, 1167–1192.

Gandhimathi, C., Venugopal, J.R., Sundarrajan, S., Sridhar, R., Tay, S.S., Ramakrishna, S., Kumar, S.D., 2015. Breathable medicine: pulmonary mode of drug delivery. J. Nanosci. Nanotechnol. 15, 2591–2604.

Garcia-Contreras, L., Fiegel, J., Telko, M.J., Elbert, K., Hawi, A., Thomas, M., Verberkmoes, J., Germishuizen, W.A., Fourie, P.B., Hickey, A.J., Edwards, D., 2007. Inhaled large porous particles of capreomycin for treatment of tuberculosis in a guinea pig model. Antimicrob. Agents Chemother. 51, 2830–2836.

Gardikis, K., Fessas, D., Signorelli, M., Dimas, K., Tsimplouli, C., Ionov, M., Demetzos, C., 2011. A new chimeric drug delivery nano system (chi-aDDnS) composed of PAMAM G 3.5 dendrimer and liposomes as doxorubicin's carrier. In vitro pharmacological studies. J. Nanosci. Nanotechnol. 11, 3764–3772.

Garg, T., Goyal, A.K., 2014. Biomaterial-based scaffolds--current status and future directions. Exp. Opin. Drug Deliv. 11, 767–789.

Garg, T., Singh, O., Arora, S., Murthy, R., 2012. Scaffold: a novel carrier for cell and drug delivery. Crit. Rev. Ther. Drug Carrier Syst. 29, 1–63.

Garg, T., Singh, S., Goyal, A.K., 2013. Stimuli-sensitive hydrogels: an excellent carrier for drug and cell delivery. Crit. Rev. Ther. Drug Carrier Syst. 30, 369–409.

Garg, T., Kumar, A., Rath, G., Goyal, A.K., 2014. Gastroretentive drug delivery systems for therapeutic management of peptic ulcer. Crit. Rev. Ther. Drug Carrier Syst. 31, 531–557.

Garg, T., Bhandari, S., Rath, G., Goyal, A.K., 2015a. Current strategies for targeted delivery of bio-active drug molecules in the treatment of brain tumor. J. Drug Target. 23 (10), 865–887.

Garg, T., Goyal, A.K., Rath, G., Murthy, R.S., 2015b. Spray-dried particles as pulmonary delivery system of anti-tubercular drugs: design, optimization, in vitro and in vivo evaluation. Pharm. Dev. Technol. 21 (8), 951–960.

Garg, T., Rath, G., Goyal, A.K., 2015c. Biomaterials-based nanofiber scaffold: targeted and controlled carrier for cell and drug delivery. J. Drug Target. 23, 202–221.

Garg, T., Rath, G., Goyal, A.K., 2015d. Colloidal drug delivery systems: current status and future directions. Crit. Rev. Ther. Drug Carrier Syst. 32, 89–147.

Garg, T., Rath, G., Goyal, A.K., 2015e. Inhalable chitosan nanoparticles as antitubercular drug carriers for an effective treatment of tuberculosis. Artif. Cells Nanomed. Biotechnol. 44 (3), 997–1001.

Garg, T., Rath, G., Murthy, R.R., Gupta, U.D., Vatsala, P.G., Goyal, A.K., 2015f. Current nanotechnological approaches for an effective delivery of bioactive drug molecules to overcome drug resistance tuberculosis. Curr. Pharm. Des. 21, 3076–3089.

Gerbeaux, J., Hanoteau, J., 1970. New drugs in the treatment of primary tuberculosis. Ann. Pediatr. 17, 638–643.

Goyal, G., Garg, T., Rath, G., Goyal, A.K., 2014. Current nanotechnological strategies for an effective delivery of drugs in treatment of periodontal disease. Crit. Rev. Ther. Drug Carrier Syst. 31, 89–119.

Goyal, A.K., Garg, T., Rath, G., Gupta, U.D., Gupta, P., 2015. Development and characterization of nanoembedded microparticles for pulmonary delivery of antitubercular drugs against experimental tuberculosis. Mol. Pharm. 12 (11), 3839–3850.

Goyal, A.K., Garg, T., Rath, G., Gupta, G.D., 2016. Chemotherapeutic evaluation of guar gum coated chitosan nanoparticle against experimental tuberculosis. J. Biomed. Nanotechnol. 12, 1–14.

Grant, A.C., Walker, R., Hamilton, M., Garrill, K., 2015. The ELLIPTA dry powder inhaler: design, functionality, in vitro dosing performance and critical task compliance by patients and caregivers. J. Aerosol. Med. Pulm. Drug Deliv. 28 (6), 474–485.

Grosset, J.H., Singer, T.G., Bishai, W.R., 2012. New drugs for the treatment of tuberculosis: hope and reality. Int. J. Tuberc. Lung Dis. 16, 1005–1014.

Hamilton, M., Leggett, R., Pang, C., Charles, S., Gillett, B., Prime, D., 2015. In vitro dosing performance of the ELLIPTA dry powder inhaler using asthma and copd patient inhalation profiles replicated with the electronic lung (eLung). J. Aerosol. Med. Pulm. Drug Deliv. 28 (6), 498–506.

Heyder, J., 2004. Deposition of inhaled particles in the human respiratory tract and consequences for regional targeting in respiratory drug delivery. Proc. Am. Thorac. Soc. 1, 315–320.

Holmes, M.S., Seheult, J.N., O'connell, P., D'arcy, S., Ehrhardt, C., Healy, A.M., Costello, R.W., Reilly, R.B., 2015. An acoustic-based method to detect and quantify the effect of exhalation into a dry powder inhaler. J. Aerosol. Med. Pulm. Drug Deliv. 28, 247–253.

Huang, Y.-Y., Wang, C.-H., 2006. Pulmonary delivery of insulin by liposomal carriers. J. Control. Release 113, 9–14.

Iqbal, M.A., MD, S., Sahni, J.K., Baboota, S., Dang, S., Ali, J., 2012. Nanostructured lipid carriers system: recent advances in drug delivery. J. Drug Target. 20, 813–830.

Jain, C.P., Vyas, S.P., 1995. Preparation and characterization of niosomes containing rifampicin for lung targeting. J. Microencapsul. 12, 401–407.

Kaur, G., Narang, R., Rath, G., Goyal, A.K., 2012. Advances in pulmonary delivery of nanoparticles. Artif. Cells Blood Sub. Biotechnol. 40, 75–96.

Kaur, M., Garg, T., Rath, G., Goyal, A.K., 2014a. Current nanotechnological strategies for effective delivery of bioactive drug molecules in the treatment of tuberculosis. Crit. Rev. Ther. Drug Carrier Syst. 31, 49–88.

Kaur, R., Garg, T., Rath, G., Goyal, A.K., 2014b. Advanced aerosol delivery devices for potential cure of acute and chronic diseases. Crit. Rev. Ther. Drug Carrier Syst. 31, 495–530.

Kaur, M., Garg, T., Narang, R., 2016a. A review of emerging trends in the treatment of tuberculosis. Artif. Cells Nanomed. Biotechnol. 44 (2), 478–484.

Kaur, R., Garg, T., Das Gupta, U., Gupta, P., Rath, G., Goyal, A.K., 2016b. Preparation and characterization of spray-dried inhalable powders containing nanoaggregates for pulmonary delivery of anti-tubercular drugs. Artif. Cells Nanomed. Biotechnol. 44 (1), 182–187.

Kaur, P., Garg, T., Rath, G., Murthy, R.S., Goyal, A.K., 2016c. Surfactant-based drug delivery systems for treating drug-resistant lung cancer. Drug Deliv. 23 (3), 727–738.

Kaur, R., Garg, T., Malik, B., Gupta, U.D., Gupta, P., Rath, G., Goyal, A.K., 2015d. Development and characterization of spray-dried porous nanoaggregates for pulmonary delivery of anti-tubercular drugs. Drug Deliv. 23 (3), 882–887.

Keck, C.M., Müller, R.H., 2006. Drug nanocrystals of poorly soluble drugs produced by high pressure homogenisation. Eur. J. Pharm. Biopharm. 62, 3–16.

Kersten, G.F., Crommelin, D.J., 2003. Liposomes and ISCOMs. Vaccine 21, 915–920.

Kim, C.S., Hu, S.C., 1998. Regional deposition of inhaled particles in human lungs: comparison between men and women. J. Appl. Physiol. 84, 1834–1844.

Klinger-Strobel, M., Lautenschlager, C., Fischer, D., Mainz, J.G., Bruns, T., Tuchscherr, L., Pletz, M.W., Makarewicz, O., 2015. Aspects of pulmonary drug delivery strategies for infections in cystic fibrosis--where do we stand? Exp. Opin. Drug Deliv. 12, 1351–1374.

Kocbek, P., Baumgartner, S., Kristl, J., 2006. Preparation and evaluation of nanosuspensions for enhancing the dissolution of poorly soluble drugs. Int. J. Pharm. 312, 179–186.

Kumar, P.V., Asthana, A., Dutta, T., Jain, N.K., 2006. Intracellular macrophage uptake of rifampicin loaded mannosylated dendrimers. J. Drug Target. 14, 546–556.

Labiris, N.R., Dolovich, M.B., 2003. Pulmonary drug delivery. Part I: physiological factors affecting therapeutic effectiveness of aerosolized medications. Br. J. Clin. Pharmacol. 56, 588–599.

Li, Y.-Z., Sun, X., Gong, T., Liu, J., Zuo, J., Zhang, Z.-R., 2010. Inhalable microparticles as carriers for pulmonary delivery of thymopentin-loaded solid lipid nanoparticles. Pharm. Res. 27, 1977–1986.

Li, W., Yang, Y., Wang, C., Liu, Z., Zhang, X., An, F., Diao, X., Hao, X., 2012. Carrier-free, functionalized drug nanoparticles for targeted drug delivery. Chem. Commun. 48, 8120–8122.

Li, X., Vogt, F.G., Hayes, D., Mansour, H.M., 2014. Physicochemical characterization and aerosol dispersion performance of organic solution advanced spray-dried microparticulate/nanoparticulate antibiotic dry powders of tobramycin and azithromycin for pulmonary inhalation aerosol delivery. Eur. J. Pharm. Sci. 52, 191–205.

Liang, Z., Ni, R., Zhou, J., Mao, S., 2015. Recent advances in controlled pulmonary drug delivery. Drug Discov. Today 20, 380–389.

Majlessi, L., Brosch, R., 2015. mycobacterium tuberculosis meets the cytosol: the role of cgas in anti-mycobacterial immunity. Cell Host Microbe 17, 733–735.

Makhlof, A., Werle, M., Tozuka, Y., Takeuchi, H., 2010. Nanoparticles of glycol chitosan and its thiolated derivative significantly improved the pulmonary delivery of calcitonin. Int. J. Pharm. 397, 92–95.

Manca, M.L., Sinico, C., Maccioni, A.M., Diez, O., Fadda, A.M., Manconi, M., 2012. Composition influence on pulmonary delivery of rifampicin liposomes. Pharmaceutics 4, 590–606.

Maretti, E., Rossi, T., Bondi, M., Croce, M.A., Hanuskova, M., Leo, E., Sacchetti, F., Iannuccelli, V., 2014. Inhaled solid lipid microparticles to target alveolar macrophages for tuberculosis. Int. J. Pharm. 462, 74–82.

Mathias, N.R., Kim, K.J., Lee, V.H., 1996. Targeted drug delivery to the respiratory tract: solute permeability of air-interface cultured rabbit tracheal epithelial cell monolayers. J. Drug Target 4, 79–86.

Mathpal, D., Garg, T., Rath, G., Goyal, A.K., 2015. Development and characterization of spray dried microparticles for pulmonary delivery of antifungal drug. Curr. Drug Deliv. 12, 464–471.

Mehnert, W., Mader, K., 2001. Solid lipid nanoparticles: production, characterization and applications. Adv. Drug Deliv. Rev. 47, 165–196.

Mehta, S.K., Kaur, G., Bhasin, K.K., 2007. Analysis of Tween based microemulsion in the presence of TB drug rifampicin. Colloids Surf. B 60, 95–104.

Mehta, S.K., Kaur, G., Bhasin, K.K., 2010. Tween-embedded microemulsions--physicochemical and spectroscopic analysis for antitubercular drugs. AAPS PharmSciTech 11, 143–153.

Misra, A., Hickey, A.J., Rossi, C., Borchard, G., Terada, H., Makino, K., Fourie, P.B., Colombo, P., 2011. Inhaled drug therapy for treatment of tuberculosis. Tuberculosis 91, 71–81.

Mizoe, T., Ozeki, T., Okada, H., 2008. Application of a four-fluid nozzle spray drier to prepare inhalable rifampicin-containing mannitol microparticles. AAPS PharmSciTech 9, 755–761.

Muller, R.H., Mader, K., Gohla, S., 2000. Solid lipid nanoparticles (SLN) for controlled drug delivery: a review of the state of the art. Eur. J. Pharm. Biopharm. 50, 161–177.

Nakhare, S., Vyas, S.P., 2008. Prolonged release multiple emulsion based system bearing rifampicin: in vitro characterisation. Drug Devel. Ind. Pharm. 21, 869–878.

Newman, S.P., 2004. Dry powder inhalers for optimal drug delivery. Exp. Opin. Biol. Ther. 4, 23–33.

Organization, W.H., 2013. Global Tuberculosis Report 2013. WHO. HTM/TB, Geneva, Switzerland.

Orme, I., 2001. Search for new drugs for treatment of tuberculosis. Antimicrob. Agents Chemother. 45, 1943–1946.

Pabreja, S., Garg, T., Rath, G., Goyal, A.K., 2016. Mucosal vaccination against tuberculosis using Ag85A-loaded immunostimulating complexes. Artif. Cells Nanomed. Biotechnol. 44 (2), 532–539.

Pandey, R., Khuller, G.K., 2005. Solid lipid particle-based inhalable sustained drug delivery system against experimental tuberculosis. Tuberculosis 85, 227–234.

Parveen, S., Misra, R., Sahoo, S.K., 2012. Nanoparticles: a boon to drug delivery, therapeutics, diagnostics and imaging. Nanomedicine 8, 147–166.

Patlolla, R.R., Chougule, M., Patel, A.R., Jackson, T., Tata, P.N., Singh, M., 2010. Formulation, characterization and pulmonary deposition of nebulized celecoxib encapsulated nanostructured lipid carriers. J. Control. Release 144, 233–241.

Patton, J.S., Fishburn, C.S., Weers, J.G., 2004. The lungs as a portal of entry for systemic drug delivery. Proc. Am. Thor. Soc. 1, 338–344.

Pilcer, G., Amighi, K., 2010. Formulation strategy and use of excipients in pulmonary drug delivery. Int. J. Pharm. 392, 1–19.

Prow, T.W., Grice, J.E., Lin, L.L., Faye, R., Butler, M., Becker, W., Wurm, E.M., Yoong, C., Robertson, T.A., Soyer, H.P., Roberts, M.S., 2011. Nanoparticles and microparticles for skin drug delivery. Adv. Drug Deliv. Rev. 63, 470–491.

Rohilla, R., Garg, T., Goyal, A.K., Rath, G., 2016. Herbal and polymeric approaches for liver-targeting drug delivery: novel strategies and their significance. Drug Deliv. 23 (5), 1645–1661.

Rundfeldt, C., Steckel, H., Scherliess, H., Wyska, E., Wlaź, P., 2013. Inhalable highly concentrated itraconazole nanosuspension for the treatment of bronchopulmonary aspergillosis. Eur. J. Pharm. Biopharm. 83, 44–53.

Serrano-Ruiz, D., Laurenti, M., Ruiz-Cabello, J., Lopez-Cabarcos, E., Rubio-Retama, J., 2013. Hybrid microparticles for drug delivery and magnetic resonance imaging. J. Biomed. Mater. Res. B 101, 498–505.

Silva, M., Ferreria, E., Leite, C., Sato, D., 2007. Preparation of polymeric micelles for use as carriers of tuberculostatic drugs. Trop. J. Pharm. Res. 6, 815–824.

Singh, J., Garg, T., Rath, G., Goyal, A.K., 2015. Advances in nanotechnology-based carrier systems for targeted delivery of bioactive drug molecules with special emphasis on immunotherapy in drug resistant tuberculosis: a critical review. Drug Deliv. 11, 1–23.

Sultana, S., Khan, M.R., Kumar, M., Kumar, S., Ali, M., 2013. Nanoparticles-mediated drug delivery approaches for cancer targeting: a review. J. Drug Target. 21, 107–125.

Swaminathan, J., Ehrhardt, C., 2012. Liposomal delivery of proteins and peptides. Exp. Opin. Drug Deliv. 9, 1489–1503.

Trotta, M., Cavalli, R., Carlotti, M.E., Battaglia, L., Debernardi, F., 2005. Solid lipid micro-particles carrying insulin formed by solvent-in-water emulsion-diffusion technique. Int. J. Pharm. 288, 281–288.

Vadakkan, M.V., Annapoorna, K., Sivakumar, K.C., Mundayoor, S., Kumar, G.S., 2013. Dry powder cationic lipopolymeric nanomicelle inhalation for targeted delivery of antitubercular drug to alveolar macrophage. Int. J. Nanomed. 8, 2871–2885.

Vanic, Z., Hurler, J., Ferderber, K., Golja Gasparovic, P., Skalko-Basnet, N., Filipovic-Grcic, J., 2014. Novel vaginal drug delivery system: deformable propylene glycol liposomes-in-hydrogel. J. Liposome Res. 24 (1), 27–36.

Vaughan, W.J., Vaughan, B.E., 1969. Dep. Inhal. Part. Lungs. Nature 221, 661–662.

Vergne, I., Chua, J., Singh, S.B., Deretic, V., 2004. Cell biology of *Mycobacterium tuberculosis* phagosome. Annu. Rev. Cell Dev. Biol. 20, 367–394.

Vyas, S.P., Kannan, M.E., Jain, S., Mishra, V., Singh, P., 2004. Design of liposomal aerosols for improved delivery of rifampicin to alveolar macrophages. Int. J. Pharm. 269, 37–49.

Wang, Z., Itoh, Y., Hosaka, Y., Kobayashi, I., Nakano, Y., Maeda, I., Umeda, F., Yamakawa, J., Kawase, M., Yag, K., 2003. Novel transdermal drug delivery system with polyhydroxyalkanoate and starburst polyamidoamine dendrimer. J. Biosci. Bioeng. 95, 541–543.

Weiss, C., Mcloughlin, P., Cathcart, H., 2015. Characterisation of dry powder inhaler formulations using atomic force microscopy. Int. J. Pharm. 494, 393–407.

Xu, W., Ling, P., Zhang, T., 2013. Polymeric micelles, a promising drug delivery system to enhance bioavailability of poorly water-soluble drugs. J. Drug Deliv. 2013, 340315.

Yang, S., Lee, L., Mallett, S., Ayer, J., Wolstenholme, A., Pascoe, S., 2015. A randomized, crossover study to investigate the pharmacokinetics and safety of inhaled fluticasone furoate and umeclidinium, administered separately and in combination via dry powder inhaler in healthy adult volunteers. Adv. Ther. 32, 157–171.

Young, P.M., Salama, R.O., Zhu, B., Phillips, G., Crapper, J., Chan, H.K., Traini, D., 2015. Multi-breath dry powder inhaler for delivery of cohesive powders in the treatment of bronchiectasis. Drug Dev. Ind. Pharm. 41, 859–865.

Zhang, J., Wu, L., Chan, H.-K., Watanabe, W., 2011. Formation, characterization, and fate of inhaled drug nanoparticles. Adv. Drug Deliv. Rev. 63, 441–455.

Zhou, Q.T., Leung, S.S., Tang, P., Parumasivam, T., Loh, Z.H., Chan, H.K., 2015. Inhaled formulations and pulmonary drug delivery systems for respiratory infections. Adv. Drug Deliv. Rev. 85, 83–99.

Zhu, B., Young, P.M., Ong, H.X., Crapper, J., Flodin, C., Qiao, E.L., Phillips, G., Traini, D., 2015. Tuning aerosol performance using the multibreath Orbital(R) dry powder inhaler device: controlling delivery parameters and aerosol performance via modification of puck orifice geometry. J. Pharm. Sci. 104, 2169–2176.

Ziffels, S., Bemelmans, N.L., Durham, P.G., Hickey, A.J., 2015. In vitro dry powder inhaler formulation performance considerations. J. Control. Release 199, 45–52.

Zumla, A., Chakaya, J., Centis, R., D'ambrosio, L., Mwaba, P., Bates, M., Kapata, N., Nyirenda, T., Chanda, D., Mfinanga, S., Hoelscher, M., Maeurer, M., Migliori, G.B., 2015. Tuberculosis treatment and management--an update on treatment regimens, trials, new drugs, and adjunct therapies. Lancet Respir. Med. 3, 220–234.

FURTHER READING

Hashim, F., El-Ridy, M., Nasr, M., Abdallah, Y., 2010. Preparation and characterization of niosomes containing ribavirin for liver targeting. Drug Deliv. 17, 282–287.

CHAPTER

23 NANOSIZED DEVICES AS ANTIBIOTICS AND ANTIFUNGALS DELIVERY: PAST, NEWS, AND OUTLOOK

Pio Maria Furneri*, Giulio P. Petronio, Virginia Fuochi*, Sarha Cupri*, Rosario Pignatello***

**University of Catania, Catania, Italy*

***IRCCS San Raffaele Pisana, BioBIM—Multidisciplinary Interinstitutional BioBank, Rome, Italy*

CHAPTER OUTLINE

1 INTRODUCTION

Since the discovery and clinical application of antibiotics, many problems have been revealed in the treatment of bacterial infections (Aminov, 2010; Davies and Davies, 2010). These failures are associated with the low bioavailability of these drugs, their side effects, difficulty, or impossibility to penetrate tissue and cellular barriers (Carryn et al., 2003), biofilm-related infections (Høiby et al., 2010; Smith, 2005), intrinsic resistance (Cox and Wright, 2013), and, the most important problem, emergence of acquired bacterial resistance (Hawkey, 2008; Rossolini et al., 2014).

Antifungal therapy began in the early 20th century with the discovery of griseofulvin (Oxford et al., 1939); however, the therapeutic breakthrough has occurred with the discovery of polyenes at the beginning of 1950s (Oura et al., 1955). Antifungal drugs, just because of their eukaryotic target, more than the antibacterial drugs suffer from problems of systemic toxicity (Dodds Ashley et al., 2006; Lewis, 2011), and, today, also of considerable problems due to the emergence of acquired resistance (Alcazar-Fuoli and Mellado, 2014). At 60 years since its discovery, despite its problems of distribution and toxicity (Laniado-Laborín and Cabrales-Vargas, 2009), amphotericin B remains the drug of choice for treatment of fungal systemic infections due to primary dimorphic fungi pathogens (Russell and Fothergill, 2015). The new triazoles are a safe and good alternative to amphotericin B (Dodds Ashley and Perfect, 2015); they are active against primary and opportunistic pathogens, although because of their teratogenicity should not be used during pregnancy (Pilmis et al., 2015; Russell and Fothergill, 2015). Echinocandines are well tolerated, but their use is limited to treatment of candidiasis and aspergillosis (Dodds Ashley and Perfect, 2015; Lai et al., 2008). Terbinafine and related compounds are, actually, limited to treatment of dermatomycosis (Newland and Abdel-Rahman, 2009).

In order to solve or overflow these limitations, pharmaceutical technology has developed several strategies that have shown that it is possible to operate either on the ineffectiveness or on bioavailability (Fig. 23.1) (Park, 2014), enhancing the one or the other and, in some instances, both, through the use of specific carriers (Briones et al., 2008; Pinto-Alphandary et al., 2000; Xiong et al., 2014). More than 60 years since the first drug delivery systems (DDS) were developed, we have gone from simple systems aimed to control the release of a drug, to technologically complex platforms able to modify the pharmacokinetics and pharmacodynamics of an active compound, including gene materials (Fig. 23.2) (Park, 2014).

DDS are defined as tools able to control the release of the loaded drugs or target their distribution within the body, and are a major opportunity to narrow the biological effect of a molecule into the infected cell directly (Fresta et al., 1995b), reducing the side effects while increasing the therapeutic

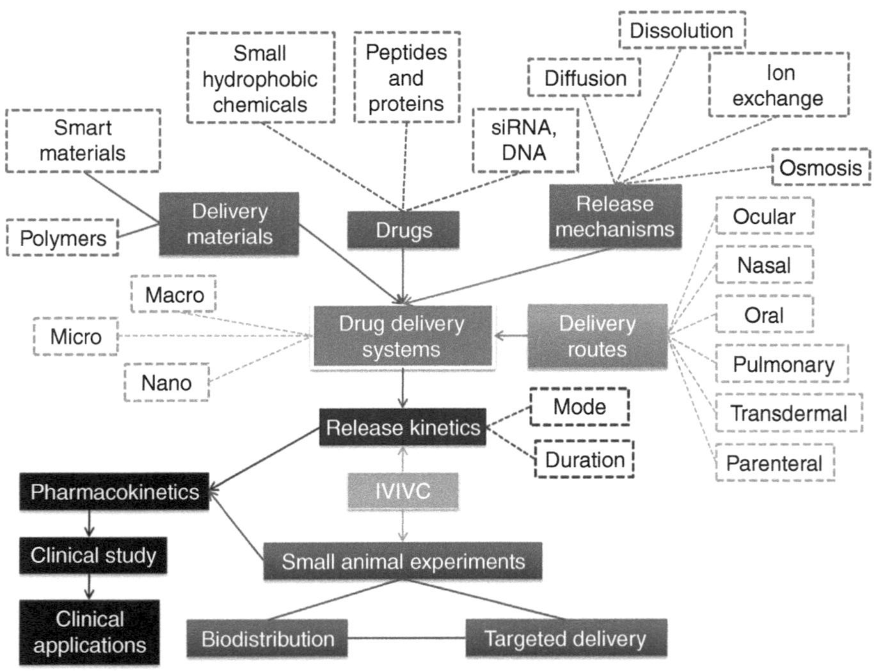

FIGURE 23.1 Overview of Drug Delivery Systems Development From Basic Research to Clinical Applications.

The main components of drug delivery systems and processes are shown in a *full colored box*, and subsections of each component are shown in a *dashed box*. *IVIVC*, *In vitro–in vivo* correlation.

With permission from Park, K., 2014. Controlled drug delivery systems: past forward and future back. J. Control. Release 190, 38.

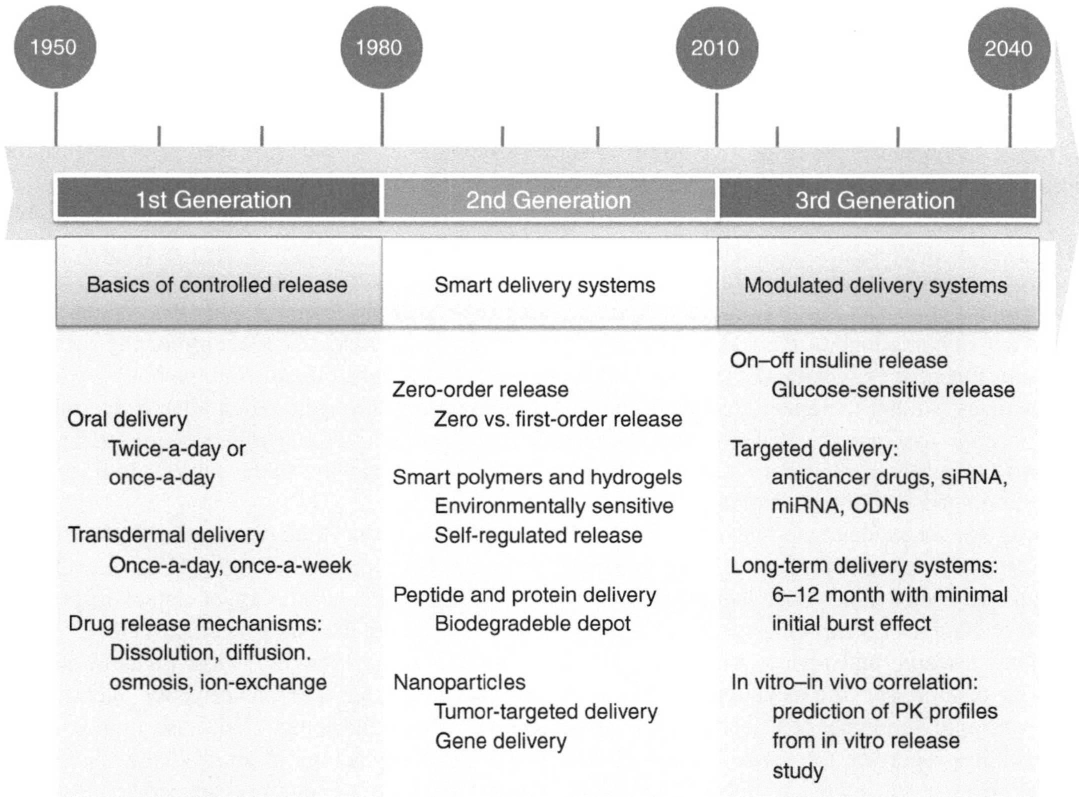

FIGURE 23.2 Evolution of Controlled Drug Delivery Systems Since 1950.

With permission from Park, K., 2014. Controlled drug delivery systems: past forward and future back. J. Control. Release 190, 38.

efficacy, and hence its intrinsic activity (Attama et al., 2012; Fresta et al., 1995a; Furneri et al., 2000; Onoue et al., 2014; Puglisi et al., 1995; Zhang et al., 2010).

In the past few decades, thanks to these studies, novel and efficient DDS have been developed to achieve a more successful clinical application of antibacterial drugs. Some of these products have even reached the pharmaceutical market and clinical use.

While DDS indicate the birth of the "Micro Era," nanotechnological evolution, which rank in the late 1970s had, as its main consequence, the development of different nanocarriers, generally called "nanoscaled drug delivery systems" (NDDS) (Attama et al., 2012; Emeje et al., 2012; Park, 2014). Among the most promising NDDS that meet industrial and clinical requirements, liposomes, polymeric nanospheres and nanocapsules, micelles and lipid-based nanoparticles have reached some therapeutic significance. These systems have become important, thanks to their ability to reduce toxicity and increase the stability of certain drugs, as well as to make soluble compounds that are scarcely soluble in aqueous and biological media (Balmayor et al., 2010; Christophersen et al., 2015; Shrestha et al., 2014).

Therefore, some polymer-based colloidal carriers are able to overcome cell membrane and tissue barriers (Abed et al., 2015; Briones et al., 2008; Drulis-Kawa and Dorotkiewicz-Jach, 2010; Hillaireau and Couvreur, 2009; Xiong et al., 2014), enhancing the antibacterial efficacy against infectious diseases related to bacterial biofilm (Smith, 2005), as well as to overwhelm certain forms of bacterial resistance (Abed et al., 2015; Huh and Kwon, 2011; Pelgrift and Friedman, 2013; Sharma et al., 2012).

Loading antibacterial and antifungal drugs in nanocarriers through physical encapsulation, adsorption or chemical conjugation, it is possible to dramatically improve the pharmacokinetics and therapeutic index, compared to the respective free drug form (Kalhapure et al., 2015), as well as to increase the intrinsic activity of antibiotics (Fresta et al., 1995a).

Nanocarriers can solve the problems of solubility in plasma of sparingly water soluble or highly hydrophobic antimicrobials, usually characterized by a poor oral absorption or drug aggregation after intravenous administration. Administration of antibiotics encapsulated in nanocarriers shows an altered tissue distribution and plasma pharmacokinetics, with the possibility of increasing the concentration of the active compound at the site of infection. Nanoparticles can also provide a protecting mean for antibiotics against enzymatic degradation and deactivation, and protect them from the body's clearing mechanisms, so that therapeutic concentrations can be maintained longer. When a nanoencapsulated drug becomes associated with a slow and sustained release pattern, a suitable drug concentration can be ensured in the bloodstream for relatively long times, reducing dosage and frequency of administration of the antibiotic (and thus its potential side effects and toxicity).

Drug-loaded nanoparticles can enter eukaryotic cells through endocytotic pathways and then release their cargo in the cytosol, to treat microbes-induced intracellular infections (Hillaireau and Couvreur, 2009). It has in fact been demonstrated that to reach a sufficient intracellular concentration of certain drugs, such as fluoroquinolones, a carrier is needed to allow a better and faster uptake (Vallet et al., 2011).

Many research and review articles have been published in the literature, most of them dealing with one or more specific technologies used to encapsulate antibacterial and antimycotic agents (e.g., liposomes, polymeric nanoparticles, lipid nanocarriers, micelles, dendrimers). However, in most cases these studies report in vitro microbiological data, while a general lacking of an in vivo evaluation is evident. Even more, not always the selected technological nanosystem or platform appears to have been rationally dressed on the peculiar physicochemical, pharmacokinetic, or therapeutic limits of the used drug.

Therefore, this chapter, although discussing the more recent development (past 10 years) in the field of pharmaceutical nanotechnologies applied to the release and targeting of antibacterial and antifungal drugs, has been purposely shaped to focus on the pharmaceutical, microbiological, and, whenever possible, clinical outcomes of some specific antimicrobial drugs, once loaded in the various NDDS.

2 LIPOSOMES

The term "liposome" comes from the Greek word "λìποζ" (fat) and σῶμα (body). They were described for the first time early 1960s by Alec D. Bangham (Bangham, 1995). Liposomes are phospholipid-based vesicles, which can have variable dimensions (generally between 50 nm and 5 μm in diameter). They, basically, consist of one or more closed double layers of phospholipids. Unlike micelles, liposomes contain a hydrophilic (aqueous) core. For this reason, a liposomal system can carry hydrophilic, hydrophobic, and amphiphilic compounds. In fact, the polar solutes can be dissolved in the aqueous core, whereas hydrophobic compounds can be allocated within the lipid domains (bilayers). As DDS, they show interesting potentiality (Table 23.1), although their scale-up and industrial production remain difficult and have hampered the translation of the huge lab research into clinical application.

Table 23.1 Benefits of Drug Loading in Liposomes

Improved solubility of lipophilic and amphiphilic drugs
Passive targeting to the cells of the immune system, especially cells of the mononuclear phagocytic system (MPS)
Sustained drug release by systemically or locally administered liposomes
Site-specific targeting
Site-avoidance mechanism
Improved transfer of hydrophilic and charged molecules
Improved penetration into tissues

Modified from Akbarzadeh et al., 2013. Nanoscale Res. Lett. 8, 102.

To transport the encapsulated compounds to their site of action and release their contents inside the cells, the lipid bilayers of liposomes must merge with another double phospholipid layer, either the cell membrane (Briones et al., 2008; Drulis-Kawa and Dorotkiewicz-Jach, 2010; Hillaireau and Couvreur, 2009), or with an outer membrane in the case of Gram-negative bacteria (Seltmann and Holst, 2002), with a polysaccharides structure in the case of fungi (Chavan et al., 2012), with a peptidoglycan structure in the case of Gram-positive bacteria (Seltmann and Holst, 2002), as well as with the capsule for both Gram-positive and Gram-negative bacteria (Sachetelli et al., 2000; Seltmann and Holst, 2002). For this reason, the choice of a (phospho)lipid component is crucial in the liposome preparation.

Liposomes can be composed by natural (e.g., phosphatidylcholine) or synthetic phospholipids, but may include other lipid materials, such as cholesterol, and hydrophilic polymers that can affect the fluidity, surface charge, and permeability to the encapsulated drugs (Allen, 1997; Gregoriadis, 2007; Sahoo and Labhasetwar, 2003).

Liposomal membranes can be characterized by a positive or negative charge (Gregoriadis, 2007). This can be achieved by choosing an appropriate lipid, such as stearylamine to impart a positive charge or phosphatidic acid for a negative one. Charged liposomes have several advantages over neutral vesicles: the charge repulsion increases the space between bilayers, thus raising the aqueous compartment volume, and create a mutual repulsion between the liposomes, avoiding their merging. Negatively charged liposomes do not interact with plasmatic proteins, that are generally negative, too; finally, cationic liposomes can interact with the negatively charged bacterial cell surface (Immordino et al., 2006; Kadry et al., 2004; Trafny et al., 1995).

The basic principle in the preparation of liposomes consists in the amphiphilic proprieties of the phospholipids. In fact, when they are dispersed in an aqueous medium, a reorganization of the water molecules occurs with a free energy enhancement. To minimize this energy cost, phospholipids align in the same direction exposing their polar heads toward the solvent by binding water, and the hydrophobic tails grouped away from the polar solvent, thus minimizing the number of water molecules that surround them (Gregoriadis, 2007).

Various techniques have been proposed to produce liposomes (Akbarzadeh et al., 2013) (Table 23.2). From each technique the type and size of the obtained liposomes will depend. However, some common technical steps are shared among these techniques (Akbarzadeh et al., 2013; Gregoriadis, 2007; Pignatello and Sarpietro, 2013): (1) evaporating the lipids from a solution in an organic solvent (in the presence of the drug to be encapsulated, if liposoluble); (2) hydration of the lipids in an aqueous

Table 23.2 Methods of Liposome Preparation (Akbarzadeh et al., 2013)

General methods of preparation	All the methods of preparing the liposomes involve four basic stages: 1. Drying down lipids from organic solvent 2. Dispersing the lipid in aqueous media 3. Purifying the resultant liposome 4. Analyzing the final product
Method of liposome preparation	1. Passive loading techniques (three different methods): a. mechanical dispersion method: – sonication – french pressure cell: extrusion – freeze-thawed liposomes – lipid film hydration by hand shaking, nonhand shaking or freeze drying – microemulsification. – membrane extrusion – dried reconstituted vesicles b. solvent dispersion method c. detergent removal method (removal of non-encapsulated material) 2. Active loading technique

medium (i.e., buffer solution) (containing any hydrophilic drug); (3) purifying and sizing the resulting vesicles; (4) elimination of the unencapsulated aliquot of the drug.

Liposomes may be classified according to their size and physical structure (lamellarity) in two main classes: unilamellar vesicles (ULV) and multilamellar vesicles (MLV). ULV can be further divided in small (SUV) and large (LUV), with mean size, respectively, below or higher than 100 nm (Drulis-Kawa and Dorotkiewicz-Jach, 2010; Gregoriadis, 2007).

From the early 1970s, liposomes have emerged as an interesting DDS, also for antibacterial drugs (Gregoriadis, 1973), and later for antiparasitic agents (New et al., 1981) and antifungal treatments (Taylor et al., 1982). To improve both the pharmacokinetics (route of administration, drug absorption, release and elimination, intracellular distribution) (Ait-Oudhia et al., 2014; Allen, 1997, 1998; Briones et al., 2008; Fresta et al., 1995a,b; Hillaireau and Couvreur, 2009; Juliano et al., 1978; Song et al., 2012) and pharmacodynamics of the carried drugs (efficacy, toxicity, site-selectivity of release, intrinsic activity) (Ait-Oudhia et al., 2014; Allen, 1997, 1998; Allen and Cullis, 2013; Anwekar et al., 2011; Drulis-Kawa and Dorotkiewicz-Jach, 2010; Fresta et al., 1995a,b; Furneri et al., 2000; Song et al., 2012). In addition, the liposome surface can be functionalized with "stealth" materials (Immordino et al., 2006; Nag and Awasthi, 2013), such as polyethylene glycol (PEG) or its derivatives to enhance their in vivo stability (Basile et al., 2012; Immordino et al., 2006; Nag and Awasthi, 2013), or with targeting moieties, such as antibodies or mAb, aptamers, peptides, and other small molecules, to selectively bind to microorganisms or infected cells and then deliver the loaded drug in specific tissues or cells (Abed and Couvreur, 2014; Allen and Cullis, 2013; Hillaireau and Couvreur, 2009; Immordino et al., 2006; Tan et al., 2010). Unfortunately, liposomal formulations suffer a general physical instability and a short lifetime, which are critical for their industrial scale-up and commercialization. Further production steps are thus required, like freeze-drying or sterilization, techniques that are not always applicable to obtain a product able to cope with the market and regulatory issues and clinical use (Akbarzadeh et al., 2013; Nag and Awasthi, 2013).

2.1 THE LIPOSOMAL DELIVERY OF ANTIBACTERIAL AGENTS

Clinical failure of an antibiotic treatment could be associated with low drug bioavailability (either in plasma or in the intracellular compartments), undesired side effects, production of bacterial biofilm, and development of bacterial resistance. Moreover, antibacterial agents according to their pharmacodynamic characteristics can be divided into two different groups: concentration-dependent and time-dependent; the latter can be further classified into time-dependent ones with minimal or no PAE, and time-dependent ones with PAE (Connors et al., 2013; Levison, 2004; Levison and Levison, 2009; Schentag et al., 2001). However, these parameters are not correlated to other important features, that is, ability to penetrate tissues and cell biomembranes (Abed and Couvreur, 2014; Briones et al., 2008; Carryn et al., 2003; Hillaireau and Couvreur, 2009; Nix et al., 1991): in fact, pharmacokinetics data are commonly based to the description of drug behavior in plasma or in serum (Barger et al., 2003; Nix et al., 1991; Schentag et al., 2001).

Tissue distribution of antibiotic drugs has been extensively described in the past (Carryn et al., 2003; Nix et al., 1991), and recently reviewed (van Bambeke et al., 2006). The study of intracellular pharmacokinetics of antibiotics has shown that their preferred localization inside the cell are cytosol and lysosomes, although for many antibiotics, such as tetracyclines, lincosamides and rifamycins, it is still unknown. Also the intracellular localization of linezolid is unknown, a new drug belonging to the subclass of biaryloxazolidionenes; radezolid is localized to 40% in the lysosomes, and the remaining in the cytosol (Lemaire et al., 2010).

According to such a situation, the gain due to the liposomal delivery can be manifested both in terms of greater bioavailability, either in the blood and in the tissues (Abed and Couvreur, 2014; Allen and Cullis, 2013; Song et al., 2012; Yamashita and Hashida, 2013), along with a greater therapeutic efficacy (Song et al., 2012; Yamashita and Hashida, 2013).

A brief and schematic review of the papers published in the period between 2005 and 2015, and dealing with liposome-based antibacterials is summarized in Table 23.3.

Despite the numerous papers published in the past 50 years and the various liposomal formulations proposed, only one formulation (liposomal amikacin—Arikace) has reached the clinical phase II (Clancy et al., 2013; Ehsan et al., 2014; Okusanya et al., 2014) and probably in the next years the desks of pharmacies.

2.2 LIPOSOMES AND ANTIFUNGALS

The problems in antifungal therapy, in addition to the fact that being eukaryotes they share with animal cells the majority of targets, are the continuous raising of forms of resistance in opportunistic fungi, and in primary pathogen fungi (Alcazar-Fuoli and Mellado, 2014). Today, only, the echinocandins and nikkomicin Z are the only antimycotic drugs that selectively inhibit the formation of the fungal cell wall (Ostrosky-Zeichner et al., 2010).

Indeed, there has been an increase in fungal infections (Chu et al., 2006; Lass-Flörl, 2009; Morrone, 2013; Nucci et al., 2010; Perusquía-Ortiz et al., 2012; Park et al., 2011; Pfaller et al., 2006, 2014; Pfaller and Diekema, 2007, 2010; Rodloff et al., 2011) and especially those systemic or invasive

Table 23.3 Liposomal Antibiotic Formulations

Drug	Formulation	Rationale/ Microorganisms	Study Model	Therapeutic Improvements	References
Amikacin	Liposomes (LAI, Arikace)	Inhalation therapy in cystic fibrosis patients chronically infected by *Pseudomonas aeruginosa.*	In vivo (phase 2 clinical studies)	Efficacy of a once-daily dose for the treatment of cystic fibrosis patients with chronic infections due to *P. aeruginosa.*	Okusanya et al. (2014); Ehsan et al. (2014); Clancy et al. (2013)
	Liposomes	Inhalation therapy in cystic fibrosis patients chronically infected by *P. aeruginosa.*	In vivo pilot study	Estimation of serum exposures for inhalation	Okusanya et al. (2009)
	Liposomes	Inhalation in healthy male volunteers	In vivo pulmonary deposition and clearance	No adverse events and prolonged retention in the lungs	Weers et al. (2009)
	Liposomes	*P. aeruginosa* isolated from sputum of a cystic fibrosis patient	In vitro release and biofilm penetration In vivo antimicrobial activity on rat lung	Sustained and targeted release and increased biofilm penetration	Meers et al. (2008)
	Liposomes	*Mycobacterium avium*	In vivo infection in mice	Successful target drug delivery	de Steenwinkel et al. (2007)
	Liposomes	*Burkholderia cenocepacia*	In vitro antibacterial activity	Enhanced antibacterial activity and penetration for antibiotic resistant strains compared with free drug	Halwani et al. (2007)
	Liposomes	*Bacillus subtilis and S. aureus*	In vitro antibacterial activity and stability	Enhanced stability and encapsulation efficiency	Mugabe et al. (2006a)
	Liposomes	*M. avium* intracellular complex (MAC)	In vivo beige mice and in vitro macrofage cells	Substained and targeted release	Salem et al. (2005)
Azithromycin	Liposomes	*P. aeruginosa* in fibrosis cystis patients	In vitro antimicrobial properties and toxicity studies	Enhanced safety and efficacity: increased antimicrobial activity, lower biofilm forming,and no haemolysis or cell toxicity	Solleti et al. (2015)
	Cationic non-lecithoid niosomes	Therapy of respiratory infections	In vivo bioavailability studies	Improved bioavailabilty and lung targeting by i.e. and oral somministration	Zhong et al. (2014)

		Liposomes modified with cholestin	Therapy of respiratory infections	In vitro release studies	Colistin on liposomes had a permeabilizing effect on bacterial cells	Wallace et al. (2013)
		Positively charged lipidic emulsion	Therapy of ocular infections	Ex vivo pharmacokinetic study on rabbit tear fluid	Formulation suitable for ocular administration	Liu et al. (2009)
		Cationic stearylamine-modified liposomes	Therapy of respiratory infections	Ex vivo on mice tissues	AUC increased in mice lungs	Wang et al. (2005)
		pH-sensitive liposomes	*M. avium* intracellular complex (MAC)	In vivo beige mice and in vitro macrofage cells	Increased antimicrobial activity compared with antibiotic encapsulated conventional liposomes	Salem et al. (2005)
	Clarithromycin	Liposomal codelivery of daptomycin	*S. aureus* (MRSA)	In vitro antibacterial activity and in vivo studies	Enhanced anti-MRSA activity in vitro and reduced MRSA bacterial load, increased host survival in vivo	Li et al. (2014)
		Liposomes	*P. aeruginosa* infections	In vitro antibacterial activity	Reduced bacterial growth, attenuated virulence factor production, reduced bacterial twitching, swarming, and swimming motilities	Alhajlan et al. (2013)
		Liposomes	Improvement patient compliance	In vivo on rats, rabbits and mouse	Less irritation in mouse scratch test, in rat paw lick test and rabbit ear vein	Liu et al. (2013b)
		Liposomes	*M. avium*	In vivo on mice	Enhanced efficacy, reduction of treatment duration	de Steenwinkel et al. (2007)
		Liposomes	*M. avium* intracellular complex (MAC)	In vivo on beige mice and in vitro on macrofage cells	Substained and targeted release	Salem et al. (2005)
	Erythromycin	Oligo-acyl-lysine—lipid mixtures	Bacterial multidrug resistance	In vitro antibacterial activity, in vivo on mice	Incresed efficacy, reduced toxicity	Sarig et al. (2011)
		Liposomes	*Bacillus subtilis and S. aureus*	In vitro antibacterial activity and stability	Enhanced stability and encapsulation efficiency	
	Gatifloxacin	Liposomal hydrogel	Use as ophthalmic formulation	In vitro albino rabbit	Prolonged transcorneal permeation, improved ocular bioavailability, minimized administration frequency, decreased side effect	Hosny (2010)

(*Continued*)

Table 23.3 Liposomal Antibiotic Formulations (*cont.*)

	Drug	Formulation	Rationale/ Microorganisms	Study Model	Therapeutic Improvements	References
	Gentamicin	Nanosized liposomes	*Ralstonia insidiosa*	In vitro biofilm study by ultrasound promoting penetration	Biofilm dissolution and enhanced bacteria killing	Ma et al. (2015)
		Liposomes immobilized in chitosan nanofiber mesh.	*Escherichia coli, P. aeruginosa* and *S. aureus*	In vitro antibacterial activities	Novel formulation for wound dressing application	Monteiro et al. (2015)
		Liposomes	*Rhodococcus equi*	In vivo mouse infection model and ex vivo equine alveolar macrophages	Greater reduction in the liver CFU counting compared to treatment with clarithromycin in combination with rifampin	Burton et al. (2015)
		Piperine nanoliposomes	*S. aureus* MRSA	In vitro antibacterial studies	Enhanced bacterial inhibition and killing compared with vancomycin; inhibition of efflux system	Khameneh et al. (2014)
		Oleic acid nanoliposomes	*S. aureus* MRSA	In vitro antibacterial studies.	Improved antimicrobial activity, reduction of the effective required drug concentration and rapid bacterial killing	Atashbeyk et al. (2014)
		Phospholipon 90H-based PeGylated microscopic lipospheres	*Escherichia coli, P. aeruginosa, Salmonella paratyphii* and *S. aureus*	In vitro antibacterial activities	Enhanced antimicrobial activities especially against *P. aeruginosa*	Momoh and Esimone (2012)
		Liposomes	Pharmacokinetics of free and liposomal forms of gentamicin	In vivo white mice	Gentamicin in liposomes significantly increased most important pharmacokinetic parameters of the antibiotic (AUC, Cmax, MRTpo,T/2, Kel, Vzpo) and reduced Clpo and Kel	Rotov et al. (2012)
		Porous beta-tricalcium phosphate scaffold combined liposomes	*S. aureus*	In vitro biofilm formation	Initial high dose drug release and a further sustained release; in vitro biofilm disruption	Zhu et al. (2010)

		Gallium liposomes	*P. aeruginosa*	In vitro antibacterial activities and biofilm formation	Gallium liposome formulation were more effective in eradicating antibiotic-resistant *P. aeruginosa* growing in planktonic or biofilm community	Halwani et al. (2008a)
		Niosomes	Ophthalmic controlled delivery	In vivo albino rabbits	Enhanced entrapment and release compared to other formulation with no sign of irritation in ocular irritancy test	Abdelbary and El-gendy (2008)
		Liposomes	*Burkholderia cenocepacia* clinical isolated in patients with cystic fibrosis	In vitro antibacterial activity	Enhanced antibacterial activity and penetration for antibiotic resistant strains compared with free drug	Halwani et al. (2007)
		Cationic, neutral and anionic liposomes	Gram- positive and negative bacteria	In vitro antibacterial activity	Cationic formulations resulted in most effective	Drulis-Kawa et al. (2006)
		Labrasol based liposomes	Enhancing mechanism on intestinal membrane permeability	In vitro membrane interaction study	Labrasol formulations increased both membrane lipid fluidity and intestinal membrane permeability	Koga et al. (2006)
		Liposomes	*P. aeruginosa*	In vitro antibacterial activity	enhanced antibacterial activity compared to free drug	Rukholm et al. (2006)
		Liposomes	*Bacillus subtilis and S. aureus*	In vitro antibacterial activity and stability	Enhanced stability and encapsulation efficiency	Mugabe et al. (2006a)
		Liposomes	*P. aeruginosa*	In vitro antibacterial activity	Enhanced drug entry into bacterial cells	Mugabe et al. (2006b)
		Cationic, neutral and anionic liposomes	Organ disposition after intraperitoneal administration	In vivo normal rat	High drug concentration in liver and brain for positive liposomes, while concentration in kidneys and lungs was not affected by liposomes charge	Abraham and Walubo (2005)
		Liposomes	*P. aeruginosa* isolated from cystic fibrosis patients	In vitro antibacterial activity	Enhanced antibacterial activity compared to free drug	Mugabe et al. (2005)
	Levofloxacin	Liposomes	Pulmonary infections treatment	In vitro drug release profiles and in vivo studies on rabbits and mice	Biodistribution improvement, increased efficacy, side effects reduction	Zhang et al. (2009)

(*Continued*)

Table 23.3 Liposomal Antibiotic Formulations (*cont.*)

	Drug	Formulation	Rationale/ Microorganisms	Study Model	Therapeutic Improvements	References
		Liposome-coated lenses	Ocular infections treatment *S. aureus*	In vitro antibacterial activity, in vitro drug release study	Pharmacokinetic improvement, reduced bacterial growth	Danion et al. (2007)
	Ofloxacin	PEGylated niosomes	Comparative study of pharmacokinetics between ofloxacin free and in niosomal forms	In vivo on white mice	Increased efficiency and decreased negative effects compared to free drug	Kulichenko et al. (2014)
		Niosomes	Evaluation as potential carrier system to improve the patient compliance and reduce side effect	In vitro drug release studies	Improved pharmacokinetic parameters, prolonged action and improved bioavailability compared to the free drug	Ramalingam et al. (2013)
		Lipospheres	Melt dispersion technique study	In vitro drug release studies	Controlled drug release for a period of 16 h	Babu Natarajan and Lakshmanan (2013)
		Liposomal hydrogel	Topically effective in situ thermosensitive liposomel hydrogel	In vitro pharmacokinetics studies	Prolonged transcorneal permeation, improved ocular biovailability, minimized administration frequency, decreased side effect	Hosny (2009)
		Liposomes	Pulmonary infection treatment	In vitro drug release profiles and in vivo studies on rabbits and mice	Biodistribution improvement, side effects reduction	Zhang et al. (2009)
Penni-cillins	Amoxicillin	Nanolipobeads	*Helicobacter pylori*	In vitro drug release study, antibacterial activity study and in vivo bacterial clearance study on rats	Pharmacokinetic improvement, reduced bacterial growth, better clearance rate of *H. pylori*	Jain and Jain (2013)
		Double-liposomes	*H. pylori*	In vitro drug release study, antibacterial activity study and *in vivo* bacterial clearance study on rats	Pharmacokinetic improvement, reduced bacterial growth, better clearance rate of *Helicobacter pylori*, enhanced antisecretory and ulcer protective activity	Jain et al. (2012)
	Penicillin G	Liposomes	*Escherichia coli*	In vitro antibacterial activity	Lower MIC values	Colzi et al. (2015)

	Rifampicin	Freeze-dried liposomes	Lung delivery form for the treatment of tuberculosis	In vitro dissolution, and antitubercular activity, in vivo drug disposition study in Wister rat model by intra-tracheal instillation	Drug released in a controlled manner with an enhanced efficiency and a more rapid delivery to macrophages compared to free drug	Patil et al. (2015)
		Anionic, cationic and PEGylated liposomes	*Staphylococcus epidermidis*	In vitro antimicrobial activities and biofilm formation studies	The antibiofilm activity of formulations was concentration- and time-dependent manner	Moghadas-Sharif et al. (2015)
		Polyelectrolyte uncoated and chitosan-coated multilayered vesicles	Inhalable dry powder for lung administration	In vitro cytotoxicity studies on human A549 alveolar cells and cellular uptake studies	Coated formulations had a strong ability to be easily internalized and to greatly prolong the residence time of their components compared to uncoated liposomes that were rapidly internalized and just as quickly removed	Manca et al. (2014)
		Tyloxapol biocompatible niosomes	Drug delivery system for antituberculosis drugs	In vitro dissolution studies	Acceptable release with a non-Fickian release mechanism	Mehta and Jindal (2013)
		Ligand-appended liposomes	Drug therapy for pulmonary tuberculosis	In vitro and in vivo parameters	Prolonged drug release, improved recovery of drugs from the target site	Bhardwaj et al. (2013)
		Liposomes	Effects of lipid concentration and composition on pulmonary delivery	In vitro intracellular drug uptake and toxicity activity	Cell toxicity and drug retention were more evident when oleic acid was used	Manca et al. (2012)
		Triton niosomes	Quantitative investigation, stability and in vitro release studies	In vitro release studies and kinetics	Fickian or diffusional release has been observed	Mehta et al. (2011)
		Liposomes	Pulmonary targeting	Lung deposition studies	Delivery of a consistent lung doses, over modest breath hold periods, even with intrapatient variability in breathing rate	Dandekar et al. (2010)
		Preformed and in situ formed aerosolized liposomes	Drug delivery to the lungs	In vitro drug release profile	In situ formed liposome showed better sustained release profile than the preformed liposome as it released sufficient amount of drug while retaining considerable amount of drug	Gaur et al. (2009)

(*Continued*)

Table 23.3 Liposomal Antibiotic Formulations (*cont.*)

	Drug	Formulation	Rationale/ Microorganisms	Study Model	Therapeutic Improvements	References
		Liposomes	*Mycobacterium bovis*	In vitro toxicity and antimicrobial activity on alveolar macrophages cells	Less negatively charged formulation displayed the greatest activity against intracellular growth of *M. bovis*	Changsan et al. (2009)
		Liposomes	*M. avium* complex (MAC)	In vivo aereosol inhalation in rats and in vitro intracellular activity	Inhibited MAC growth in infected macrophages and reached the lower airways in rats	Zaru et al. (2009b)
		Negatively and noncharged chitosan-coated and uncoated liposomes	Delivery to lungs by nebulization	In vitro mucoadhesive and toxicity studies	All coated formulation showed better mucoadhesive properties compared to uncoated ones and a lower toxicity compared to free drug	Zaru et al. (2009a)
		Liposomes	Evaluation of surfactants as inhalable drug-delivery systems for pulmonary tuberculosis	In vitro alveolar deposition and drug release	Surfactant action had an alveolar stabilization and better alveolar target due to antiatelectatic effect of the surfactant	Chimote and Banerjee (2009)
		Freeze-dried liposomes	Lung delivery	In vitro mucoadhesive, citotoxicity and cellular uptake studies	Lower cytotoxicity compared with free drug and low mucoadhesive proprieties	Zaru et al. (2007)
		Liposomes	*M. avium* intracellular complex (MAC)	In vivo on beige mices and in vitro on macrofage cells	Substained and targeted release	Salem et al. (2005)
	Streptomycin	Liposomes	Activity of liver enzimes	In vivo on albino mice	Decreased inhibitory effect on protease, alkaline phosphatase and phosphorylase	Tikhonov et al. (2010)
		Liposomes	*M. avium* intracellular complex (MAC)	In vivo on beige mices and in vitro on macrofage cells	Substained and target release	Salem et al. (2005)
Tetracy-clines	Doxycycline	Gold nanoparticles-stabilized liposomes	*Helicobacter pylori*	In vitro antibacterial activity	Pharmacokinetic improvement, pH-responsive in the stomach	Thamphiwatana et al. (2013)
	Doxycycline	Nanoliposomal slow-release gel	Biocompatibility	In vivo on rats periodontitis model	Improved cure on rat periodontitis	Jin et al. (2010)

	Minocycline	Liposomal controlled-release gel	Biocompatibility	In vivo on rats periodontitis model	Improved cure on rat periodontitis	Liu et at. (2013a)
	Minocycline	Liposomes	Cytotoxic study	Brain endothelial cells	Lower levels of drug for targeting	Xing et al. (2012)
		Nanoliposomes	Retina pathologies	In vivo on streptozotocin induced diabetic model	Better clearance and distribution	Kaiser et al. (2013)
	Tetracycline	Liposomes	Ocular pathologies	Ex vivo on albino rabbits' cornea and conjunctiva	Improved Schirmer test and tear break up time values, reversed surface ocular pathology	Shafaa (2011)
	Tobramycin	Negatively, positively or neutral liposomes	*Burkholderia cepacia* complex	In vitro biofilm formation studies	Negatively charged formulation showed target delivery to biofilm, while neutral or cationic did not increase biofilm reduction compared to free drug	Messiaen et al. (2013)
		Bismuth-ethanedithiol liposomes	*P. aeruginosa*	In vivo antimicrobial activity by intratracheal administration in infected rats	Reduced production of quorum-sensing molecules and virulence factors	Alhariri and Omri (2013)
		Bismuth-ethanedithiol liposomes	Mucoid *P. aeruginosa* clinical isolated	In vitro antibacterial activity and biofilm formation studies	Enhanced inhibitory and bactericidal activity together with biofilm eradication	Alipour et al. (2011)
		Liposomes	*P. aeruginosa* in cystic fibrosis sputum isolated	In vitro antibacterial activity and antibiotic inhibition studies	Reduced antibiotic inhibition and enhanced antibacterial activity	Alipour et al. (2009)
		Bismuth-thiol liposomes	*P. aeruginosa* isolates in patients with cystic fibrosis	In vitro quorum sensing) and biofilm formation studies	Formulations prevented AHL production and stopped biofilm formation	Halwani et al. (2009)
		Cochleates bridged liposomes	Membrane fusion capability and antimicrobial activity on *Escherichia coli*	Ex vivo human fibroblasts cells and in vitro antimicrobial activity	Cochleates' edges made formulation fuse with the cell surfaces as compared to edge free liposomes with an improved antibacterial activity	Syed et al. (2008)
		Bismuth-ethanedithiol liposomes	*P. aeruginosa* and *Burkholderia cenocepacia* isolates in patients with cystic fibrosis	In vitro antimicrobial activity and ex vivo toxity profile on human lung carcinoma cell line	Enhanced inhibitory and bactericidal activity and reduced toxicity	Halwani et al. (2008b)

(Continued)

Table 23.3 Liposomal Antibiotic Formulations (cont.)

	Drug	Formulation	Rationale/ Microorganisms	Study Model	Therapeutic Improvements	References
		Liposomes	*Burkholderia cenocepacia* clinical isolated in patients with cystic fibrosis	In vitro antibacterial activity	Enhanced antibacterial activity and penetration for antibiotic resistant strains compared with free drug	Halwani et al. (2007)
		Liposomes	*Bacillus subtilis and S. aureus*	In vitro antibacterial activity and stability	Enhanced stability and encapsulation efficiency	Mugabe et al. (2006a)
		Liposomes	*P. aeruginosa*	In vitro antibacterial activity	enhanced drug entry into bacterial cells	Mugabe et al. (2006b)
	Vancomicin	Chitosan-coated liposomes	Treatment of Gram-positive bacterial infections	In vivo intravenous injection to mice, in vitro release profile and hemolysis experiment	Longer retention time and higher AUC, decreased drug distribution in kidney of mice with reduced nephrotoxicity.	Yang et al. (2015)
		Anionic, cationic and PEGylated liposomes	*Staphylococcus epidermidis*	In vitro antimicrobial activities and biofilm formation studies.	The antibiofilm activity of formulations was concentration- and time-dependent manner	Moghadas-Sharif et al. (2015)
		Niosomes	Control of staphylococcal colonization and biofilm formation on abiotic surface	In vitro biofilm forming studies	Inhibition of biofilm formation and growth, eradication of surface-borne biofilms and interference with biofilm induction	Barakat et al. (2014)
		Cationic multivesicular liposomes	Preparation of multivesicular liposomes	In vitro release profiles	Sustained release properties	Yang et al. (2013)
		Nebulized lipid vesicles	Treatment of pulmonary infections	In vivo Wistar rats	Formulations improved drug passage from airway space to the pulmonary tissue and systemic circulation	de Jesús Valle et al. (2013)
		PEGylated liposomes	Treatment in MRSA pneumonia	In vivo pharmacokinetics and biodistribution	Enhanced permeability, retention effect and passive targeting	Pumerantz (2012)
		Liposomes	*S. aureus* (MRSA)	In vivo murine systemic infection model and in vitro antibacterial activity studies	Enhance antibacterial activity and improved kidney clearance	Sande et al. (2012)

		Nano-hydroxyapatite/ chitosan/konjac glucomannan scaffold liposomes	Treatment of bone infected *S. aureus*	In vivo *New Zealend* white rabbits infected model	Enhance therapeutical outcome	Huang et al. (2012)
		PEGylated liposomes	Lung tissue targeting	In vivo mice pharmacokinetic and biodistribution study	prolonged blood circulation time and increased deposition in lung, liver, and spleen and reduced accumulation in kidney tissue	Muppidi et al. (2011)
		Nano-hydroxyapatite/ chitosan/konjac glucomannan scaffolds cationic liposomes	*S. aureus*	In vitro release and biofilm formation studies	Nano-hydroxyapatite/ chitosan/ konjac glucomannan scaffolds provide sustained cationic liposome release and inhibited the formation of *Staphylococcus aureus* biofilms better than scaffolds without cationic liposome antibiotic	Ma et al. (2011)
		PEGylated and non-pegylated liposomes	*S. aureus* (MRSA)	In vitro human macrophage infection studies	Non-pegylated formulations resulted in a sufficient concentration of antibiotic inside the intracellular compartment of macrophages to exert a marked bactericidal effect against MRSA, while surface-pegylated ones gave in no impact on MRSA survival, due to the stealth effect of pegylation	Pumerantz et al. (2011)
		Fusogenic liposomes	*E. coli* and *Acinetobacter baumanii* clinical isolated	In vitro antibacterial activity study	Enhanced antibacterial activity compared to free drug and nonfusogenic formulation	Nicolosi et al. (2010)
		Anionic, cationic and proteoliposomes	*S. aureus*	In vitro biofilm formation	Enhanced drug target to bacteria biofilm	Jones (2005)

(Chu et al., 2006; Laniado-Laborin, 2007; Lass-Flörl, 2009; Nucci et al., 2010; Park et al., 2011; Pfaller et al., 2006, 2014; Pfaller and Diekema, 2007, 2010; Ramos-e-Silva et al., 2012; Richardson, 2005; Rodloff et al., 2011; Skiada et al., 2012). The phenomenon seems to be correlated with:

1. The increase of HIV cases, that resulted in an increase of opportunistic mycoses (Perusquía-Ortiz et al., 2012; Ramos-e-Silva et al., 2012) in developing countries (Park et al., 2011; Richardson, 2005; Warnock, 2007);
2. To migration events (Galimberti et al., 2012; Laniado-Laborin, 2007; Park et al., 2011) and to climatic factors that have contributed to the spread of coccidioidomycosis (Park et al., 2011);
3. To the increased use of broad spectrum antibiotics and cytotoxic chemotherapy (Pfaller and Diekema, 2007);
4. To the use of antifungal prohylaxis that seems to have changed the etiology of invasive fungal infections in Europe (Lass-Flörl, 2009).

Furthermore, according to a recent review, invasive fungal diseases (Pfaller and Diekema, 2010) could be divided in opportunistic and endemic mycoses, and several predisposing factors have been related to an increased incidence in immunocompromised and in hospitalized patients.

This new situation represents a challenge from a therapeutic point of view, because it poses a number of questions on how to approach the problem of resistance to antifungal drugs (Alcazar-Fuoli and Mellado, 2014; Rodloff et al., 2011), as well as in terms of diagnosis (Lai et al., 2008; Leroux and Ullmann, 2013; Miceli and Lee, 2011) and therapy (Chen et al., 2010; Kanafani and Perfect, 2008; Lai et al., 2008; Leroux and Ullmann, 2013; Miceli and Lee, 2011; Pasqualotto and Denning, 2008; Pfaller et al., 2014; Walsh et al., 2004), since clinicians are facing with diseases not adequately known (Kanafani and Perfect, 2008).

The therapeutic arsenal at disposition of therapy is essentially based on drugs in use for many years. The first systemic antifungals griseofulvin (Oxford et al., 1939) and amphotericin B (Laniado-Laborín and Cabrales-Vargas, 2009; Russell and Fothergill, 2015) are still on the market. Interestingly, the clinical use in infectious diseases of liposomes can coincide with the marketing of pharmaceutical amphotericin B lipid formulations (Antoniadou and Dupont, 2005; Hann and Prentice, 2001; Lopez-Berestein et al., 1983; Ostrosky-Zeichner et al., 2003). Nephro-toxicity, hematological toxicity, infusion-related toxicity, and electrolyte abnormalities of amophericin B were markedly reduced by these liposomal formulations (Dodds Ashley et al., 2006), while a better pharmacokinetic profile was given by these carriers (Hamill, 2013; Lewis, 2012).

Triazoles are a safe and good alternative to amphotericin B (Dodds Ashley and Perfect, 2015); they are active against primary and opportunistic pathogens, although, because of their teratogenicity, they should not be used during pregnancy (Pilmis et al., 2015; Russell and Fothergill, 2015). Common side effects consist of gastrointestinal disturbances and hepatotoxicity, with a degree of severity specific for each drug (Dodds Ashley and Perfect, 2015).

Echinocandines are well tolerated, but their use is limited to treatment of candidiasis and aspergillosis (Dodds Ashley and Perfect, 2015; Lai et al., 2008). Terbinafine and related compounds are actually limited to the treatment of dermatomycosis (Newland and Abdel-Rahman, 2009).

Insolubility or limited solubility, rapid development of resistance, and lack of selective toxicity are some of the problems that have characterized and still characterize most antifungal drugs. The use of liposomes also attempted to face these drawbacks; worst adverse symptoms like hepatitis, hypertension, ventricular dysfunction, or photosensitivity rash have been actually reduced. Some liposomal formulations have shown to produce better pharmacokinetics and pharmacodynamics (Table 23.3), as well as to overcome, in some of its issues (site of infection, pharmacokinetics, and pharmacodynamics) the new problematic aspect of antifungal therapy, that is, the "clinical resistance." The latter is a result of several factors: misdiagnosis, immunosuppression, the microbial load at the beginning of the treatment, the infection with a strain with increased virulence, site of infection, pharmacokinetics, and pharmacodynamics, and an underlying disease (Kanafani and Perfect, 2008).

The papers on liposomal antifungal formulations published since 2005 to present are summarized in Table 23.4.

3 SOLID LIPID NANOPARTICLES (SLN)

Solid lipid nanoparticles (SLN) are an important nanotechnological platform for drug delivery (Kumar and Randhawa, 2013; Mehnert and Mäder, 2001; Müller et al., 1995, 2000; Müller and Runge, 1998; Ramya et al., 2015; Üner, 2006; Yadav et al., 2013). Initially described in early 1990s (Müller et al., 1995), SLNs represent an alternative to liposomes and polymeric nanoparticles (Mehnert and Mäder, 2001; Müller et al., 2000). They were born from the need to combine the advantages of different nanocarriers: the sustained release and physical stability of polymeric nanoparticles, with the high biocompatibility and biodegradability profiles of liposomes, due to the use of physiological lipids (Carbone et al., 2013). They are spherical particles with a diameter between 50 and 1000 nm, constituted by a lipid matrix in which the drug can be dissolved or incorporated (Müller et al., 1995, 2000; Zhang et al., 2010); their preparation includes one or more lipids, solid at room temperature, water, and a surfactant (Mehnert and Mäder, 2001; Müller et al., 2000). Among the most commonly used lipids are fatty acids, such as stearic acid or decanoic acid, triglycerides, or steroids (Mehnert and Mäder, 2001). Their properties have been well described elsewhere and are summarized in Table 23.5. They are biocompatible systems that are successfully used in cosmetics and pharmaceutical industry, as well as largely explored in scientific researches (Mehnert and Mäder, 2001; Müller et al., 2002; Pardeike et al., 2009).

SLN properties, such as mean size, surface charge (Zeta potential), drug loading capacity, and release profiles can be accurately adjusted by selecting different lipid structures or blends, the surfactant agent(s), the preparation method, and organic or aqueous media during their production. The surface of these nanoparticles can also be functionalized with different targeting moieties (Rostamia et al., 2014). Lipid-based and polymer-based colloidal carriers have nowadays reached high physicochemical and metabolic stability, along with an even higher biocompatibility.

Their features with encapsulated antimicrobials have been also investigated. The results have shown a sustained drug release pattern, improved pharmacokinetics (Cavalli et al., 2003; Dong et al., 2011; Xie et al., 2011a,b; Xu et al., 2011), a better biodistribution profile (Jain and Banerjee, 2008; Kalam et al., 2013a,b), capacity to overcome cell membranes (Abed and Couvreur, 2014; Vighi et al., 2010; Yu et al., 2010), to pass tissue barriers (Bargoni et al., 2001; Müller et al., 2000; Wissing et al., 2001), to target specific organs (Parveen et al., 2012), and to enhance the antibacterial efficacy against biofilm-related infectious diseases (Singh et al., 2014).

In this review we considered only the nanoparticles constituted by the lipid carrier and an antibiotic or antifungal drug; systems containing metals possessing antimicrobial activity were not discussed. Nevertheless, the latter issue has been recently reviewed and readers can refer to those publications (Allahverdiyev et al., 2011; Huh and Kwon, 2011; Seil and Webster, 2012).

Table 23.4 Liposomes and Antimycotics

Drug	Formulation	Rationale/ Microorganism	Study Model	Therapeutic Improvements	References
	Liposomes (Ambisome)	Toxicity studies	In vitro MTT assay and in vivo rat paraspinal muscles	Ambisome is cytotoxic in vitro but not toxic to tissues in vivo	Roberts et al. (2015)
	Nanoethosomes	Skin targeting for the treatment of cutaneous fungal infections	In vitro delivery studies and antifungal activity, in vivo skin irritation study	Higher steady state transdermal flux and skin deposition of drug, no skin irritation andincrease in the antifungal activity in comparison to marketed formulation	Kaur et al. (2015)
	Angiopep-2 based polymeric micelles	*Cryptococcus neoformans*	In vivo immunosuppressive Male CD-1 mice model with meningoencephalitis	Angiopep-2 modified polymeric micelles can rapidly and efficiently delivery the drug into the brain tissue compared with unmodified polymeric micelles and the free drug	Shao et al. (2012)
	Liposomes	*Aspergillus fumigatus* and *Candida albicans*	In vitro ephae interaction	The presence of β-glucans on the fungal hyphae cell walls of *Aspergillus fumigatus* is one of the factors responsible for mediating the binding of liposome carriers to the hyphae and could explain possible synergy reported between liposomal amphotericin B and echinocanins	Chavan et al. (2012)
	Liposomes from bone cement	*C. albicans*	In vitro antifungal activity and drug release	Enhanced drug release from formulations than amphotericin deoxycholate	Cunningham et al. (2012)
	Chitosan-coated liposomes	*C. albicans* and *Candida tropicalis*	In vitro antifungal activity	Antifungal activities similar to deoxycholate micelles	Albasarah et al. (2010)
	Cationic liposomes in a thermosensitive gel	Vaginal delivery	In vitro MTT assay and in vivo rat paraspinal muscles	Increased in stability and toxicity reduction	Kang et al. (2010)
Amphotericin B	Liposomes (Ambisome)	Tissue drug concentrations following intravenous versus intraperitoneal treatment	In vivo mice stability and toxicity study	Significant differences between formulations tissue distribution depending upon the drug route and these differences that could affect how the drugs perform in fungal infection models	Chang et al. (2010)
	Chitosan-stearic acid conjugate nanomicelles	Delivery by jet nebulization	In vitro antifungal activity	Enhanced antifungal activity	Gilani et al. (2011)

	Aerosolized liposomes	*A. fumigatus*	In vitro antifungal activity and in vivo ICR mice invasive pulmonary aspergillosis	The combination of formulation with intraperitoneal micafungin significantly improved the survival rate, and the fungal burdens and histopathology findings	Takazono et al. (2009)
	Liposomes	*A. fumigatus*	In vitro antifungal activity	Enhanced antifungal activity compared to free drug	Kadimi et al. (2007)
	Tuftsin-bearing liposomes	Systemic cryptococcosis	In vivo leucopenic mice	Formulation increased anticryptococcal activity compared with amphotericin B deoxycholate and conventional liposomal formulations	Khan et al. (2005)
Ciclopiroxolamine	Mucoadhesive nanoliposomes	Vaginal delivery	In vitro release profile and antifungal activity, ex vivo mucoadhesive behavior on sheep vaginal tissue	Formulations showed sustained release, good mucoadhesivity and antifungal activity	Karimunnisa and Atmaram (2013)
Clotrimazole	Chitosan coated liposomes	Vaginal delivery	Ex vivo pregnant sheep vaginal tissue	Lower chitosan concentration increased the system's mucoadhesive potential	Schwarz et al. (2011)
	Surfactant-based nanovesicles	*C. albicans*	In vitro antifungal activity and in vivo dult male albino rabbits	Sustained antifungal effect over 12 h, enhanced AUC with no sign of irritation after testing for ocular tolerance	Basha et al. (2013)
	Nonionic surfactant and cholesterol proniosomal gel	*C. albicans*	In vitro antifungal activity, skin permeation and drug release study	Enhanced permeation, release and antifungal activity compared to marked formulation	Thomas and Viswanad (2012)
	Ethosomes and ultradeformable liposomes	*C. albicans*	In vitro antifungal activity	*C. albicans* growth inhibitions in a 24 h in vitro anticandidal testing	Maheshwari et al. (2012)
	Ethosomal gel	*C. albicans* and *Aspergillus niger*	In vitro antifungal activity and skin permeation	Enhanced drug permeation and antifungal activity	Akhtar and Pathak (2012)
	Liposomes and niosomes	Vaginal delivery	In vitro simulated vaginal fluid, in vivo rat and rabbit vaginal irritation evaluation	Sustained and controlled drug release therapy	Ning et al. (2006)
Clotrimazole/ metronidazole	Liposomes- in-hydrogel	Vaginal delivery	In vitro release studies	Sustained and diffusion-based drug release	Vanic´ et al. (2014)
	Bioadhesive liposome gels	Vaginal delivery	In vitro release studies	Prolonged retention of liposomes in the gel	Pavelic´ et al. (2005)

(*Continued*)

Table 23.4 Liposomes and Antimycotics *(cont.)*

Drug	Formulation	Rationale/ Microorganism	Study Model	Therapeutic Improvements	References
Croconazole	Liposome and micro-emulsion-based gel	Topical delivery	In vitro skin permeations on rat skin and antifungal activity	Microemulsion formulation had superior release/permeation than liposomes and all formulation showed excellent activity against different species of fungi compared with market formulation	El-Badry et al. (2014)
Econazole	Nanosized ethanolic vesicles	Treatment of deep fungal infections through topical gel formulation	Ex vivo rat skin and in vitro antifungal activity	Nanosized ethanolic vesicles exhibited controlled release for 12 h and better antifungal activity compered to liposomal and hydroethanolic gels formulation	Verma and Pathak (2012)
	Polymer-lipid-based mucoadhesive microspheres	*C. albicans*	In vitro antifungal activity and bioadhesion to mucosal tissue	Formulations had similar antifungal activity compared to free drug with a higher mucoadhesive strength	Albertini et al. (2009)
	Lysine derivatives liposomes	Skin permeation	In vitro animal skin permeation study	Enhanced retarding effect on skin permeation	Schwarz et al. (2011)
	Liposomes	*C. albicans*	In vivo: immunosuppressed albino rats	Enhanced therapeutic efficacy	Gupta et al. (2010)
	Liposomes	*C. albicans*	In vivo: keratitis infected adult rabbits	Enhanced therapeutic efficacy	Habib et al. (2010)
	Ethosomes gel	Treatment of topical fungal diseases	Clinical studies and in vitro drug diffusion study	Enhanced reduction in dimension of Candidiasis lesion and drug diffusion	Bhalaria et al. (2009)
Griseofulvin	Deformable membrane vesicles	*Microsporum gypseum Microsporum canis Trichophyton mentagrophytes* and *Trichophyton rubrum*	Ex vivo skin permeation, dermatokinetics, skin sensitivity, in vitro antifungal assay and in vivo antifungal activity using guinea pig model for dermatophytosis	Complete clinical and mycological cure in animals treated with formulation in 10 days, formulation was nonsensitizing and histopathologically safe	Aggarwal and Goindi (2012)
Itraconazole	Nanosized vesicles	*Candida albicans*	In vitro antifungal activity, in vivo ocular toxicity on rabbit eye	Enhanced antifungal activity compared with free drug and signs of acute ocular toxicity	El Meshad and Mohsen (2015)
	Nonionic surfactant vesicles	*Tinea pedis*	In vivo standardized model in Wistar rats	Reduction of fungal burden in comparison to fungal burden of disease control rats after 14 days	Kumar and Goindi (2014)

	Hydroxypropyl-β-cyclodextrin, deformable and conventional liposomes	*C. albicans*	Ex vitro skin penetration studies and antifungal efficacy	Formulations have not negatively impacted on ITZ activity; enhanced skin permeation of hydroxypropyl-β-cyclodextrin formulation compared to deformable and conventional liposomes	Alomrani et al. (2014)
	Niosomes	*C. albicans*	In vitro antifungal activity	Tested formulation were 90 × more effective than marketed formulation	Wagh and Deshmukh (2012)
	Carboxymethyl chitosan-coated liposomes	Antifungal activity, pharmacokinetics and tissue distribution	In vitro antifungal activity, in vivo pharmacokinetics and tissue distribution on mice	Antifungal activity was not inhibited by the entrapment, longer elimination half-life and significant lung targeting efficiency	Wang and Huang (2011)
	Aerosolized nanostructure	*A. fumigatus*	In vivo: Balb/C mice pulmonary inoculation	Enhanced survival and limited invasive disease compared to market formulation	Alvarez et al. (2007)
Ketoconazole	Nanovesicular system	Topical delivery	Ex vivo in mice skin dermal delivery	Formulation targeting deeper mice skin layers which may be translated to human skin with suitable modifications	Kakkar and Kaur (2013)
	Nanovesicular system	Ocular delivery	Ex vivo in corneal permeability studies	Potential carriers for targeting the posterior segment of the eye	Kakkar and Kaur (2011)
	PGylated liposomes	*C. albicans*	In vitro antifungal activity and release study on human skin	Enhanced skin deposition with minimum skin permeation, the presence of polyethylene glycol enhanced the antifungal action	Elmoslemany et al. (2012)
Natamycin	Soybean lecithins liposomal suspensions	*Saccharomyces cerevisiae*	In vitro antifungal activity and release kinetics study	Fine-tuning of sterol concentration allows preparation of liposomal suspensions presenting modulated in vitro release kinetics rates and enhanced antifungal activity	Bouaoud et al. (2016)
Terbinafine	Liposomes	Cutaneous delivery	Ex vivo rat skin permeation study	Enhanced drug retention	Sudhakar et al. (2014)
	Transfersomes	Topical delivery to the nail, nail bed, and surrounding tissue	In vivo clinical trial on patients with onychomycosis in vitro Trichophyton rubrum hyphae exposition	Enhanced drug delivery to its site of action inside the fungus	Ghannoum et al. (2012)
Voriconazole	Liposomes	Several fungi species isolated from clinical cases of corneal keratitis	In vitro mucosal interaction and irritancy levels, ex vivo permeation	Nonirritant formulation with enhanced permeation and drug delivery	de Sá et al. (2015)

Table 23.5 Major Features of SLNs

Aims of SLNs (Mehnert and Mäder, 2001; Rupenagunta et al., 2011)	Advantages of SLNs (Kesharwani et al., 2013; Yadav et al., 2013; Ramya et al., 2015)	Disadvantages of SLNs (Ramya et al., 2015)
• Possibility of controlled drug release and drug targeting (Müller and Runge, 1998) • Increased drug stability • High drug payload • Incorporation of lipophilic and hydrophilic drugs feasible • No biotoxicity of the carrier (Müller et al., 1996; Müller and Runge, 1998) • Avoidance of organic solvents and no problems with respect to large scale production and sterilization	• Small size and relatively narrow size distribution which provide biological opportunities for site specific drug delivery by SLNs (Müller et al., 1995) • Controlled release of active drug over a long period can be achieved (Müller et al., 1995) • Protection of incorporated drug against chemical degradation • Possible sterilization by autoclaving or gamma irradiation and subjected to commercial sterilization procedures (Müller et al., 1995) • SLNs can be lyophilized as well as spray dried • No toxic metabolites are produced • Excellent biocompatibility (Ekambaram et al., 2012; Li et al., 2009) • Much easier to manufacture than biopolymeric nanoparticles (Ekambaram et al., 2012) • No special solvent required (Ekambaram et al., 2012) • Conventional emulsion manufacturing methods applicable (Ekambaram et al., 2012) • Raw materials essential the same as in emulsions (Ekambaram et al., 2012) • Application versatility (Ekambaram et al., 2012)	• Need to remove too much water in tablet/pellet production (Müller and Runge, 1998) • Dosing problems, relatively high water content of the dispersions (70%–99.9%) • Physical stability of aqueous solution, gel formation, particle aggregation (Müller and Runge, 1998) • Poor drug loading capacity, drug expulsion after polymeric transition during storage (Garud et al., 2012) • The low capacity to load hydrophilic drugs due to partitioning effects during the production process

Of the 1616 papers described in PubMed from 2005 on SLNs less than 100 are related to bacteria and fungi. This finding gives us an idea of how little is known and little has been done on the use of SLNs in the treatment of infectious diseases. The undoubted benefits that the SLNs have toward liposomes and polymeric nanoparticles probably did not convince the pharmaceutical industry to invest in this field. Compared to the number of scientific publications, in fact, records of the patents are much lower (Carbone et al., 2013). The only exception to industrial production is represented by topical applications (Müller et al., 2000, 2002; Wissing and Müller, 2003). In fact, it has been demonstrated that the SLNs have the specific characteristics, such as the ability to form a biofilm on the skin, which reduces evaporation and promotes the penetration of the active ingredient (Müller et al., 2000; Wissing et al., 2001), and that are stable in water and in the dermal cream (Wissing and Müller, 2003). Precisely, these features that make SLNs specific for topical applications in mycotic infections (Gupta and Cooper, 2008; Jain et al., 2010; Müller et al., 2000, 2002; Sanna et al., 2007; Souto and Müller, 2006; Wissing et al., 2001; Wissing and Müller, 2003; Yang et al., 2008), as well as, it might be, also useful to create pharmaceutical formulations for some bacterial infections of the skin.

The major finding of the published paper on SLNs and antifungal drugs are summarized in Table 23.6. The formulations with SLNs showed improved therapeutic efficacy, a tissue increases permeation, a sustained release, and an improved half-life.

Antibacterial SLNs formulations have shown a number of improvements regardless of the antibiotic vehicles. The SLNs formulations have been described since 2001. Stearic acid SNLs tobramycin-loaded were the first report on this issue. In vivo studies have shown an increased bioavailability (Cavalli et al., 2002, 2003), blood-brain barrier pass across (Bargoni et al., 2001), and enhanced intracellular concentrations (Cavalli et al., 2003). The most recent results since 2005 are described in Table 23.7. Sustained release was the improvement always shown by SLNs containing quinolones, minocycline, and tetracyclines. Regarding in vitro activity, amikacin, minocycline, and cefuroxime-axetil have shown an improvement of their antibacterial activity. Generally, SLNs formulations carried out an improvement of bioavailability for all investigated drug, as well as a reduced dosage to gain eradication.

4 MUCOADHESIVE DRUG DELIVERY SYSTEMS

A very particular issue is associated with the delivery of antibiotics to the gastrointestinal tract or other mucus tissues for the local treatment of infections, such as that supported by *Helicobacter pylori* or *Candida* spp.

To achieve a complete and effective eradication of *H. pylori*, a local diffusion of the antibiotic into the stomach mucosa is necessary (Gonçalves et al., 2014). One of the most studied approaches is using mucoadhesive micro- or nanocarriers, which can ensure a prolonged retention in the stomach and thus deliver the antibiotic locally in the gastric mucosa region for longer times (Smart, 2005). Mucoadhesive or mucopenetrating properties can be added to nanoparticle surface by using positively charged (cationic) polymers, such as chitosan and its salts and derivatives, PEGs, or synthetic cellulose derivatives, or also natural materials, such as lectins and gliadin (Asane et al., 2008). The mucus membrane possesses a negative charge, due to sialic acid or carboxyl or sulfate groups in the mucus glycoprotein; thereby, these cationic macromolecules can develop adhesiveness through electrostatic interactions, H-bond formation, van der Waals forces and other weak bonds (Lehr et al., 1993). They have the additional feature of protecting acid sensitive drugs against degradation in the gastric lumen and ensure an effective drug diffusion across the mucus layer.

An analogous strategy has been demonstrated to be effective for treating infections in other body districts containing a mucus layers, such as external eye compartment (Ali and Lehmussaari, 2006; Ludwig, 2005) and vaginal mucosa (Acarturk, 2009; Das Neves and Bahia, 2006; Dobaria et al., 2008; Hussain and Ahsan, 2005; Valenta, 2005).

Chitosan is a natural polyaminosaccharide, obtained by alkaline deacetylation of chitin, the major component of the exoskeleton of crustaceans (Xia et al., 2011). The biological properties of chitosans have been recently reviewed (Kong et al., 2010; Xia et al., 2011). In the past decades it has been proposed, together with derivatives and salts, for the production of DDS (Kong et al., 2010; Lv et al., 2013; Pignatello et al., 2008; Zambito, 2013).

Table 23.6 SLN and Antimycotics

Drug	Formulation	Rationale/ Microorganism	Study Model	Therapeutic Improvements	References
Ketoconazole/ clotrimazole	SLNs (PEG-40 stearate)	*Candida albicans*	In vitro studies for drug permeability at the site of action and antifungal activity	Formulations showed good proprieties for treatment of vaginal infections	Cassano et al. (2016)
Itraconazole	Stearic acid (SA) SLNs	*Aspergillus flavus*	Ex vivo drug corneal permeability and in vitro antifungal activity	Formulation showed enhanced drug permeation than one obtained with palmitic acid, maintained antifungal activity	Mohanty et al. (2014)
Clotrimazole, ketoconazole and climbazole	SLNs and NCLs gel formulations	Evaluation of drug-loaded nanoparticle dispersions and their gel formulations	In vitro drug release studies	Formulations showed sustained and prolonged topical drugs delivery	Das et al. (2014)
Griseofulvin	SLNs	*Microsporum canis*	In vivo antifungal activity using guinea pig model	Complete mycological and clinical cure	Aggarwal and Goindi (2013)
Terbinafine	SLNs	*C. albicans*	In vivo cutaneous candidiasis infection on albino rats	Improved contact, adhesion, occlusion, and sustained release compared to market formulation	Vaghasiya et al. (2013)
Fluconazole	SLNs	Topical delivery against cutaneous candidiasis	In vitro release and skin permeation	Sustained drug release	Gupta et al. (2013)
Ketoconazole	SLNs	Develop of a new biocompatible bicephalous dianionic surfactant	In vivo skin irritation test	Development of new and safes excipients for poor soluble drugs	Kalhapure and Akamanchi (2013)
Terbinafine	SLNs	Topical delivery	Ex vivo skin of nude mice	Formulations efficacy were comparable to the market ones	Chen et al. (2012)
Clotrimazole	NLCs and SLNs	Formulation comparison	In vitro drug release study	NLCs showed better stability than tested SLNs and both showed different drug release profile	Das et al. (2012)

Table 23.6 SLN and Antimycotics *(cont.)*

Drug	Formulation	Rationale/ Microorganism	Study Model	Therapeutic Improvements	References
Amphotericin B	SLNs (AmbiOnp)	In vivo pharmacokinetic studies and in vitro stability studies	In vivo pharmacokinetic studies	Formulation showed increased bioavailability compared to the free drug with a greater half-life	Patel and Patravale (2011)
Miconazole	SLN-bearing hydrogel	*Candida* spp.	In vivo infected rats	Enhanced efficiency in candidiasis treatment	Jain et al. (2010)
Miconazole	Nitrate-loaded SLNs	Topical delivery	Ex vivo study on cadaver skins	Enhanced skin targeting and uptake compared to the marketed formulation	Bhalekar et al. (2009)
Econazole	SLN and microparticles (SLMs)	Topical delivery	Ex vivo drug permeation tests on porcine epidermis	Formulation showed good proprieties as carriers for topical administration	Passerini et al. (2009)
Econazole	SLNs	Topical delivery	Ex vivo drug permeation tests by porcine stratum corneum	Enhanced drug diffusion in the deeper skin layers with a rapid penetration through the stratum corneum	Sanna et al. (2007)

NLCs, nanostructured lipid carriers; SLMs, solid lipid microparticles; SLNs, solid lipid nanoparticles.

Using chitosan-based microspheres, an improved mucoadhesive effect can be achieved. Chitosan in fact enhances the absorption through the mucosa of small polar molecules, peptide and protein drugs by means of the combination of bioadhesion and transitory accessibility of the tight junctions of cellular membranes. Cell surface and mucin are negatively charged molecules at physiological pH and therefore good targets for the interaction with the cationic chitosan. Several types of chitosans are commercially available with different molecular weight, degree of deacetylation, presence of free or substituted acid/amine groups (Pignatello et al. 2008). Amoxicillin-loaded chitosan mucoadhesive microspheres have shown a better efficacy than the neat drug for the clearance of *H. pylori* (Patel and Patel, 2007). Mucoadhesive microspheres based on chitosan glutamate (CHG) have been described by some of us to carry the antifungal drug miconazole nitrate (MN). In an in vitro microbiological evaluation against different strains of *Candida*, the freeze-dried dispersions of MN in CHG showed a good inhibitory activity on the growth of yeast, in some cases also improved compared to the free drug (Pignatello et al., 2008).

Various applications of chitosan, not only including mucoadhesive applications, are gathered in Table 23.8.

Table 23.7 SLN and Antibiotics

Antibiotic Class	Drug	Nanosystem Types and Materials	Rationale/ Microorganism	Study Model	Therapeutic Improvements	References
Aminoglycosides	Amikacin	Cholesterol SLNs	Pulmonary delivery/*P. aeruginosa*	In vitro drug delivery studies and antimicrobial activity	Enhanced drug distribution, less side effects, elongation of dosing intervals	Ghaffari et al. (2011)
		Cholesterol SLNs	Pulmonary delivery	In vitro drug delivery studies and in vivo pharmacokinetics on male rates	Better drug distribution, less adverse effects, elongation of dosing intervals	Varshosaz et al. (2013)
Cephalosporin	Cefuroxime	Stearic acid (SA) and Tristearine (TS) SLNs	*Staphylococcus aureus*	In vitro drug release studies and antibacterial activity, biofilm forming studies	Enhanced antibacterial activity, less biofilm formation and prolonged release compared with free drug	Singh et al. (2014)
Tetracyclines	Minocycline	PLGA SLNs	*Aggregatibacter actinomycetocomitans*	In vitro drug release and antibacterial activity	Enhanced antibacterial activity, enhanced drug loading and release	Kashi et al. (2012)
	Oxytetracycline	SLNs	Comparison of three different delivery systems	In vitro drug stability and release studies	Enhanced drug loading and release	Faustino-Vega et al. (2009)
	Tetracycline	SLNs	Bioavailability and biodistribution after subcutaneous injection	In vitro drug delivery studies and in vivo pharmacokinetics studies on mice	Enhanced drug release and distribution with less adverse effects	Xu et al. (2011)
Quinolones		SLNs	Comparison of four different delivery systems	In vitro drug delivery studies and drug release studies	Enhanced drug delivery and release	Jain and Banerjee (2008)
	Ciprofloxacin	Glycerol monostearate SLNs	Delivery system optimization	In vitro drug stability and release studies	Enhanced drug delivery and release	Shah and Agrawal (2012)
		PLGA SLNs	Delivery system optimization	In vitro drug delivery studies and drug release studies	Better preparation process and enhanced drug delivery and release	Mobarak et al. (2014)
	Enrofloxacin	Fatty acids SLNs	Comparison of different delivery systems/*S. aureus*	In vitro drug delivery and drug release studies, in vivo pharmacokinetics studies on mice	Maintained antibacterial activity, enhanced and sustained drug release	Xie et al. (2011a)

	Gatifloxacin	SA SLNs, SA and Compritol SLNs	Ocular delivery, development and delivery system optimization	In vitro drug delivery studies and drug release studies, ex vivo corneal permeation (goats)	Enhanced and sustained drug release, no side effects on cornea	Kalam et al. (2010)
		Cationic SLNs	Ocular delivery, development and delivery system optimization	In vitro drug delivery studies and drug release studies, statistical analysis	Enhanced and sustained drug release	Kalam et al. (2013a)
		Cationic SLNs	Transcorneal delivery, comparison between delivery system and commercial eye drops	In vitro drug delivery studies and drug release studies, in vivo pharmacokinetics studies on albino rabbits	Enhanced and sustained drug release, better bioavailability with less side effects	Kalam et al. (2013b)
	Norfloxacin	SA SLNs	Oral delivery/*S. aureus*	In vitro drug delivery studies, drug release studies, cytotoxicity in BHK-21 cell line and antibacterial activity, in vivo pharmacokinetics studies on rats	Better bioavailability with less side effects, enhanced pharmacological properties	Dong et al. (2011)
	Ofloxacin	Palmitic acid-SLNs	Delivery system optimization/ *S. aureus*	In vitro drug delivery and release studies, antibacterial activity, in vivo pharmacokinetics studies on mice	Enhanced antibacterial activity, enhanced drug release, better bioavailability	Xie et al. (2011b)
		DPPC-SLNs	Pulmonary delivery, development and delivery system optimization	In vitro drug delivery studies, drug release studies	Enhanced and sustained drug release	Duan et al. (2013)
		Cholesterol-SLNs	Ophthalmic delivery/*S. aureus, Escherichia coli*	In vitro drug delivery and release studies, antibacterial activity, in vivo pharmacokinetic studies in rabbits, ex vivo penetration and permeation studies on rabbit cornea	Enhanced antibacterial activity, better bioavailability, enhanced drug delivery and release	Üstündagˇ-Okur et al. (2014)

NLCs, nanostructured lipid carriers; SLMs, solid lipid microparticles; SLNs, solid lipid nanoparticles.

Table 23.8 Applications of Chitosan in Microbiological Studies

Drug/Active Molecule	Formulation	Rationale/ Microorganism	Study Model	Therapeutic Improvements	References
Cationic antimicrobial peptide	Nanoparticles	Local drug delivery for *P. aeruginosa* lung infections	In vitro artificial cystic fibrosis mucus penetration and biofilm eradication studies	Prolonged efficacy in biofilm eradication compared to the free control due to enhanced penetration into bacterial biofilm and sustained release	D'Angelo et al. (2015)
Phenolic compounds	Nanoparticles	Review on bacterial gastrointestinal pathogens	In vitro antimicrobial activity	Loading phenolic compounds into nanoparticles can bring the advantage of combining two compounds with effective antimicrobial activities, acting in a synergetic way	Madureira et al. (2015)
Antimicrobial peptide temporin B	Nanoparticles	*Staphylococcus epidermidis* clinical isolates	In vitro antibacterial activity, toxicity and release studies	Long-lasting antibacterial effect due to chitosan intrinsic antimicrobial activity and temporin B gradual release with reduction of peptide toxicity	Piras et al. (2015)
Berberine	Beads	*Staphylococcus aureus* and *Escherichia coli*	In vitro antibacterial activity and drug release	Sustained drug release to inhibit the growth of the bacteria over 24 h	Yu et al. (2015)
Glycyrrhizin	Nanoparticles	Drug delivery	In vitro drug release and antibacterial activity studies	Sustained drug release and antibacterial activity, with maximum zone of inhibition against *B. ceresus*, which demonstrated higher efficacy for a long time	Rani et al. (2015)
Propolis	Varnish	Evaluation on oral pathogen bacteria	In vitro antibacterial activity and ex vivo adhesion and controlled release	Formulations presented surface adherence and were able to form films with antimicrobial activity similar to or better than chlorhexidine varnish against all oral pathogen bacteria; active components were released for more than 1 week	Franca et al. (2014)
Tobramycin	Nanoparticles	*P. aeruginosa*	In vitro antibacterial activity in vivo Galleria mellonella infection model	The proposed system overcomes the mucus barrier, increases local drug concentrations, avoids systemic adverse effects and improves outcomes for pulmonary infections	Deacon et al. (2015)
Amphotericin B and vancomycin	Sponges	Local antifungal delivery for adjunctive therapy to reduce musculoskeletal wound contamination	In vitro antifungal and antibacterial activity and cytocompatibility studies, in vivo studies on mice	Formulations released amphotericin B in vitro, alone and in combination with vancomycin, well above the amphotericin B MIC against *Candida albicans*	Parker et al. (2015)

Berberine	Nanoparticles	*Helicobacter pylori* eradication therapy	In vitro drug release and in vivo infection model	Formulation effectively controlled the drug release at *H. pylori* site of infection with an in vivo bacterial clearance effect	Lin et al. (2015)
Ofloxacin	Nanostructured lipid carriers (NLC)	Ophthalmic delivery for treatment of bacterial keratitis	In vitro antibacterial activity and ex vivo penetration/ permeation studies on rabbit cornea	The addition of the oligosaccharide lactate improved the preocular residence time, controlled the drug release and enhanced the corneal bioavailability	Üstündag˘-Okur et al. (2014)
Ofloxacin	Mucoadhesive films	Ocular delivery system	In vitro antibacterial activity and drug release studies	Ofloxacin was not inactivated during the process of manufacturing; the formulation showed a highest antibacterial activity compared to reference drugs	Oliveira Fulgêncio et al. (2014)
Daptomycin	Nanoparticles	Delivery in ocular treatment of bacterial endophthalmitis	In vitro antibacterial activity, release and stability studies	Complete drug elution was obtained in 4 h, with confirmed antibacterial activity and mucoadhesive properties	Silva et al. (2013)
Amphotericin B	Self-aggregated nanoparticles	Ocular delivery	In vitro: drug release with mucoadhesive properties and antifungal activity; in vivo: ocular irritation, ocular pharmacokinetics, and corneal penetration studies	Formulation showed a sustained drug release profile, no sign of irritation, similar antifungal activity compared to neat drug, prolonged residence time on the ocular surface and drug penetration into the cornea	Zhou et al. (2013)
Ciprofloxacin and ceftriaxone	Nanocapsule	*Salmonella typhimurium*	In vitro cytotoxicity, proliferation assay and drug release; in vivo antibacterial activity in a murine salmonellosis model	Efficient targeting and killing of the intracellular pathogen at a dosage significantly lower than that of the free antibiotic, due to increased retention time	Gnanadhas et al. (2013)
Ciprofloxacin	Nanoparticles	*E. coli*	In vitro cytotoxicity, antibacterial activity and drug release	Enhanced antibacterial activity compared to the free drug due to an easily uptake by cells	Zhao et al. (2013)
Benzoyl peroxide	Nanoparticles	Targeted therapy for cutaneous pathogens *Propionibacterium acnes*	In vitro antibacterial activity, cytotoxicity and human monocyte stimulation	Formulation showed potent antimicrobial activity due to induction of bacterial cell wall membrane disruption with anti-inflammatory properties	Friedman et al. (2013)
Rifampin	Polyurethane films	*S. epidermidis*, *S. aureus* and *E. coli*	In vitro antibacterial activity, biofilm forming studies, cytotoxicity and drug release	Formulation strongly bounded and then slow released drug for weeks to provide potent antimicrobial and biofilm controlling effects without negatively affecting cell viability, drugs could be recharged to further extend antimicrobial duration	Lv et al. (2013)

(*Continued*)

Table 23.8 Applications of Chitosan in Microbiological Studies (cont.)

Drug/Active Molecule	Formulation	Rationale/ Microorganism	Study Model	Therapeutic Improvements	References
Doxycycline	Nanoparticles	Drug delivery and efficacy	In vitro antibacterial activity, cytotoxicity and drug release	Formulation showed sustained release, enhanced antibacterial activity and lower cytotoxicity compared to free drug	Cover et al. (2012)
Mupirocin	Liposomes-in-hydrogel	Burns therapy	Ex vivo skin penetration studies; in vitro bioadhesion release and antibacterial activity	The size of liposomes affects release, antibacterial activity and bioadhesion of entrapped drug	Hurler et al. (2012)
Tetracycline	Nanoparticles	*S. aureus.*	In vitro drug release, binding, antibacterial activity, cytotoxicity, survival, hemolysis, coagulation, platelet assay studies	Increased drug concentrations at the infection site and enhanced *S. aureus* intracellular killing compared to free drug	Maya et al. (2012)
Ceftriaxone	Nanoparticles	*S. typhimurium*	In vitro hemolysis test, cytotoxicity, cellular uptake and antibacterial activity	Formulations were biocompatible and were internalized by cells with a significant reduction in the count of intracellular bacteria	Zaki and Hafez (2012)
Amoxicillin	Hydrogel nanoparticles	Antibiotic delivery to stomach for treatment of *H. pylori*	In vitro and in vivo drug release studies	Formulation provided therapeutic concentration at a much lower dose that may reduce the adverse effects ofamoxicillin in high doses	Moogooee et al. (2011)
Berberine	Nanoparticles shelled with heparin	*H. pylori*	In vitro release profiles, antibacterial activity, cell viability and adherence study	Formulations significantly increased the suppressive effect of berberine on *H. pylori* growth while efficiently reducing cytotoxic effects in *H. pylori*-infected cells	Chang et al. (2011)
Amikacin and vancomycin	Sponges	Delivery of antibiotics within the wound	In vivo hindlimb of goats contaminated with *P. aeruginosa* or *S. aureus*	Formulations were effective delivering the antibiotic and reducing the bacteria within the wounds	Stinner et al. (2010)
Amoxicillin	Nanoparticles in hydrogels	*H. pylori*	In vitro drug releasing study	Formulation were pH-sensitive, protecting nanoparticles from being destructed by gastric acid and facilitated drug interaction specifically in the site of *H. pylori* infection	Chang et al. (2009)
Amoxicillin, clarithromycin, and omeprazole	Nanoparticles	*H. pylori*	In vivo clearance study and histopathological study on Swiss albino mice; in vitro antibacterial activity and drug release profiles	Formulations showed selectively target and drugs delivery due to bioadhesive, and site-specific, properties with a burst release that may effectively kill and eradicate the causal organism	Ramteke et al. (2009)

Streptomycin, gentamicin and tobramycin	Nanoparticles	*Mycobacterium tuberculosis*	In vivo chronic infection mouse model	Oral formulations induced reduction in growth of the bacilli at the same concentration of subcutaneously injected aqueous solution	Lu et al. (2009)
Rifampin	Films	Gram-positive bacteria including *S. epidermidis* and *S. aureus*	In vitro biofilm-formation study, antimicrobial activity and drug release mechanism	Formulations demonstrate sustained drug releasing with potent, durable and rechargeable antimicrobial and biofilm-controlling functions	Cao and Sun (2009)
Amikacin and daptomycin	Films	Drug delivery system for treating musculoskeletal infections	In vitro elution study and *S. aureus* grow inhibition	Considerable release during the first few hours after testing and drug activity seemed unaffected by the weak acid solvent required to dissolve chitosan	Noel et al. (2008)
Metronidazole	Alginate bead	Local anti-*H. pylori* therapy	In vivo mice study	Groups receiving the formulation were better than the corresponding suspension form, regarding eradication of *H. pylori* infection with a better drug clearance	Ishak et al. (2007)
Ag85B-based fusion protein	Microspheres	Subunit vaccine made from three *M. tuberculosis* genes	In vitro release and in vivo C57BL/6 mice immunization study.	Higher level of IgG (H + L), IgG1 and IgG2a in sera of mice immunized with chitosan microspheres compared to one immunized with free fusion protein	Zhu et al. (2007)
Amoxicillin and clarithromycin	Hydrogels	*H. pylori*	Clinical trial	Site-specific delivery in the stomach: the formulation did not increase the eradication efficacy of a proton pump inhibitor agent, in a regimen with the addition of amoxicillin and clarithromycin	Gisbert et al. (2006)
Ampicillin	Microspheres	Controlled release	In vitro release study and antimicrobial activity	Enhanced stability and slower but sustained drug release, with confirmed antimicrobial activity	Anal et al. (2006)

Noteworthy, chitosan has shown a potent intrinsic antibacterial and antifungal activity (Gadelha de Carvalho et al., 2011; Goy et al., 2009; Pasquantonio et al., 2008; Peña et al., 2013), and an antiviral activity has been also postulated (Rabea et al., 2003). Although its mechanism of action is still debated (Goy et al., 2009; Kong et al., 2010; Rabea et al., 2003; Yadav and Bhise, 2004), a wide number of antimicrobial applications of chitosan and its derivatives have been reported in food, pharmaceutical, and textile industry (Kong et al., 2010).

5 CHEMICAL DRUG DELIVERY SYSTEMS

In the past years, our groups have investigated other systems, such as lipoamino acids (LAA), which are α-amino acids bearing one or more alkyl side chains of various length (Toth, 1994; Ziora et al., 2012). LAA have become increasingly important because of their chemical simplicity and versatility, tenside-like surface activity, aggregation properties, and low toxicity pattern (Pignatello et al., 2014). These promoieties have been used to impart amphiphilic properties to different drugs and peptides (Chen et al., 2015; Goodwin et al., 2014; Pignatello and Pecora, 2007; Pignatello et al., 2006, 2009, 2010) and, more recently, they have been exploited as a strategy of chemical drug delivery, known as hydrophobic ion pairing (HIP), using some antibiotics, such as erythromycin, tobramycin, and kanamycin. The profile of in vitro antibacterial activity of such ion pairs against several bacterial strains, both susceptible and resistant to these antibiotics, was similar in terms of MIC values to one of the parent drugs, or even improved in the case of tobramycin (Pignatello et al., 2011a,b, 2014).

In another study, lyophilized dispersions of miconazole in chitosan glutamate showed a good inhibitory activity of yeast growth, in some instances even improved when compared to the free drug (Pignatello et al., 2008).

6 CONCLUSIONS AND FUTURE OUTLOOKS

This chapter has discussed recent advances in polymer- and lipid-based nanotechnologies for suppression of pathogen microorganisms growth. Along the design and synthesis of new antibacterials and the research for novel naturally available agents, nanotechnology-based antimicrobial and antifungal drug delivery has been widely exploited in the past decades as an alternative way to improve the efficacy and safety of already known molecules. In some instances, these strategies have also reached the end of development chain, up to a clinical use. However, many critical points still remain to be optimized or solved, and this could be obtained only by a close connection between microbiologists and pharmaceutical technologists.

For example, acquired microbial drug resistance remains a major therapeutic challenge to face. The encapsulation of even very potent drugs in the various nanocarriers not always has shown to be able to solve this problem and, in most published cases, this issue has not even been discussed by researchers. A potential approach is combinatorial drug therapy, that is, the coencapsulation of two or more antimicrobial drugs in a single nanovector. This strategy could reach a higher potency than single treatments, because of synergistic effects (Zhang et al., 2010).

To avoid the premature drug release from DDS in areas different than the target cells, and to increase the selectivity of drug release between microbes or infectious cells and healthy cells, nanoparticles must be engineered to that to start drug release by triggering microenvironmental stimuli, such as pH, presence of enzymes or other specific properties of the infected sites. Coating of nanocarriers with bacterial antigen- or infectious cell antigen-specific ligands can be another potential, although hard to realize strategy.

A deeper investigation to the physicochemical and biological aspects of the interaction between microorganisms and engineered nanocarriers is strongly required, to realize new therapeutic platforms that may avoid bacterial persistence and, among others, inhibit the formation of bacterial biofilm (Rizzello et al., 2013), and with a very low affinity and toxicity toward mammalian cells.

An interesting consideration has been made by Hajipour et al. (2012): in most of the published articles, studies have been performed using "classical" nanoparticles and bacteria, rather than developing specific microorganisms—relevant for clinically important infections—or nanocarriers with specific technological features and physicochemical properties. The poor definition and characterization of nanoparticles often makes it difficult to derive general information on their applications in therapy, out from the specific drug carried or microorganisms targeted in that work. For instance, an accurate determination of the biocompatibility and biodegradability of nanoparticles and their associated drugs are important issues before suggesting their clinical application. An agreement on nanocarriers and bacteria/fungi standardization as reference systems would be probably necessary in the near future, in order to make more efficient the basic and applicative studies in this field (Hajipour et al., 2012).

Finally, and as per other applications of nanotechnology in medicine, there is the still open issue of safety of these systems. No ad hoc act from the main authorization bodies regulates at present nanobiotechnology applications in human and veterinary use. However, since many industrial products are ready to enter market, researchers and expert groups are concentrated to realize systematic studies that would make light on these aspects and, above all, remove the banal and populistic conclusion that "nano is dangerous to health and environment" (http://ec.europa.eu/research/industrial_technologies/nanoscience-and-technologies_en.html).

REFERENCES

Abdelbary, G., El-gendy, N., 2008. Niosome-encapsulated gentamicin for ophthalmic controlled delivery. AAPS PharmSciTech 9, 740–747.

Abed, N., Couvreur, P., 2014. Nanocarriers for antibiotics: a promising solution to treat intracellular bacterial infections. Int. J. Antimicrob. Agents 43, 485–496.

Abed, N., Saïd-Hassane, F., Zouhiri, F., Mougin, J., Nicolas, V., Desmaële, D., Gref, R., Couvreur, P., 2015. An efficient system for intracellular delivery of beta-lactam antibiotics to overcome bacterial resistance. Sci. Rep. 5, 13500.

Abraham, A.M., Walubo, A., 2005. The effect of surface charge on the disposition of liposome-encapsulated gentamicin to the rat liver, brain, lungs and kidneys after intraperitoneal administration. Int. J. Antimicrob. Agents 25, 392–397.

Acarturk, F., 2009. Mucoadhesive vaginal drug delivery systems. Recent Pat. Drug Deliv. Formul. 3, 193–205.

Aggarwal, N., Goindi, S., 2012. Preparation and evaluation of antifungal efficacy of griseofulvin loaded deformable membrane vesicles in optimized guinea pig model of *Microsporum canis*—dermatophytosis. Int. J. Pharm. 437, 277–287.

Aggarwal, N., Goindi, S., 2013. Preparation and in vivo evaluation of solid lipid nanoparticles of griseofulvin for dermal use. J. Biomed. Nanotechnol. 9, 564–576.

Ait-Oudhia, S., Mager, D.E., Straubinger, R.M., 2014. Application of pharmacokinetic and pharmacodynamic analysis to the development of liposomal formulations for oncology. Pharmaceutics 6, 137–174.

Akbarzadeh, A., Rezaei-Sadabady, R., Davaran, S., Joo, S.W., Zarghami, N., Hanifehpour, Y., Samiei, M., Kouhi, M., Nejati-Koshki, K., 2013. Liposome: classification, preparation, and applications. Nanoscale Res. Lett. 8, 102.

Akhtar, N., Pathak, K., 2012. Cavamax w7 composite ethosomal gel of clotrimazole for improved topical delivery: development and comparison with ethosomal gel. AAPS PharmSciTech 13, 344–355.

Albasarah, Y.Y., Somavarapu, S., Stapleton, P., Taylor, K.M., 2010. Chitosan-coated antifungal formulations for nebulisation. J. Pharm. Pharmacol. 62, 821–828.

Albertini, B., Passerini, N., Di Sabatino, M., Vitali, B., Brigidi, P., Rodriguez, L., 2009. Polymer–lipid based mucoadhesive microspheres prepared by spray-congealing for the vaginal delivery of econazole nitrate. Eur. J. Pharm. Sci. 36, 591–601.

Alcazar-Fuoli, L., Mellado, E., 2014. Current status of antifungal resistance and its impact on clinical practice. Br. J. Haematol. 166, 471–484.

Alhajlan, M., Alhariri, M., Omri, A., 2013. Efficacy and safety of liposomal clarithromycin and its effect on *Pseudomonas aeruginosa* virulence factors. Antimicrob. Agents Chemother. 57, 2694–2704.

Alhariri, M., Omri, A., 2013. Efficacy of liposomal bismuth-ethanedithiol-loaded tobramycin after intratracheal administration in rats with pulmonary *Pseudomonas aeruginosa* infection. Antimicrob. Agents Chemother. 57, 569–578.

Ali, Y., Lehmussaari, K., 2006. Industrial perspective in ocular drug delivery. Adv. Drug Deliv. Rev. 58, 1258–1268.

Alipour, M., Suntres, Z.E., Halwani, M., Azghani, A.O., Omri, A., 2009. Activity and interactions of liposomal antibiotics in presence of polyanions and sputum of patients with cystic fibrosis. PLoS One 45, e5724.

Alipour, M., Dorval, C., Suntres, Z.E., Omri, A., 2011. Bismuth-ethanedithiol incorporated in a liposome-loaded tobramycin formulation modulates the alginate levels in mucoid *Pseudomonas aeruginosa*. J. Pharm. Pharmacol. 63, 999–1007.

Allahverdiyev, A.M., Kon, V., Abamor, E.S., Bagirova, M., Rafailovich, M., 2011. Coping with antibiotic resistance: combining nanoparticles with antibiotics and other antimicrobial agents. Exp. Rev. Anti Infect. Ther. 9, 1035–1052.

Allen, T.M., 1997. Liposomes: opportunities in drug delivery. Drugs 54 (S4), 8–14.

Allen, T.M., 1998. Liposomal drug formulations. Drugs 56, 747–756.

Allen, T.M., Cullis, P.R., 2013. Liposomal drug delivery systems: from concept to clinical applications. Adv. Drug Deliv. Rev. 65, 36–48.

Alomrani, A.H., Shazly, G.A., Amara, A.A., Badran, M.M., 2014. Itraconazole-hydroxypropyl-β-cyclodextrin loaded deformable liposomes: in vitro skin penetration studies and antifungal efficacy using *Candida albicans* as model. Colloids Surf. B 121, 74–81.

Alvarez, C.A., Wiederhold, N.P., McConville, J.T., Peters, J.I., Najvar, L.K., Graybill, J.R., Coalson, J.J., Talbert, R.L., Burgess, D.S., Bocanegra, R., Johnston, K.P., Williams, 3rd, R.O., 2007. Aerosolized nanostructured itraconazole as prophylaxis against invasive pulmonary aspergillosis. J. Infect. 55, 68–74.

Aminov, R.I., 2010. A brief history of the antibiotic era: lessons learned and challenges for the future. Front. Microbiol. 8 (1), 134.

Anal, A.K., Stevens, W.F., Remunan-Lopez, C., 2006. Ionotropic cross-linked chitosan microspheres for controlled release of ampicillin. Int. J. Pharm. 312, 166–173.

Antoniadou, A., Dupont, B., 2005. Lipid formulations of amphotericin B: where are we today? J. Mycol. Med. 15, 230–238.

Anwekar, H., Patel, S., Singhai, A.K., 2011. Liposome as drug carrier. Int. J. Pharm. Life Sci. 2, 945–951.

Asane, G.S., Nirmal, S.A., Rasal, K.B., Naik, A.A., Mahadik, M.S., Rao, Y.M., 2008. Polymers for mucoadhesive drug delivery system: a current status. Drug Dev. Ind. Pharm. 34, 1246–1266.

Atashbeyk, D.G., Khameneh, B., Tafaghodi, M., Fazly Bazzaz, B.S., 2014. Eradication of methicillin-resistant *Staphylococcus aureus* infection by nanoliposomes loaded with gentamicin and oleic acid. Pharm. Biol. 52, 1423–1428.

Attama, A.A., Momoh, M.A., Builders, P.F., 2012. Lipid nanoparticulate drug delivery systems: a revolution in dosage form design and development. In: Demir Sezer, A. (Ed.), Recent Advances in Novel Drug Carrier Systems. INTECH Publisher, Rijeka, Croatia, pp. 107–139.

Babu Natarajan, S., Lakshmanan, P., 2013. Effect of processing variables on characterization of ofloxacin loaded lipospheres prepared by melt dispersion technique. Curr. Drug Deliv. 10, 517–526.

Balmayor, E.R., Azevedo, H.S., Reis, R.L., 2010. Controlled delivery systems: from pharmaceuticals to cells and genes. Pharm. Res. 28, 1241–1258.

Bangham, A.D., 1995. Surrogate cells or Trojan horses. BioEssays 17, 1081–1088.

Barakat, H.S., Kassem, M.A., El-Khordagui, L.K., Khalafallah, N.M., 2014. Vancomycin-eluting niosomes: a new approach to the Inhibition of staphylococcal biofilm on abiotic surfaces. AAPS PharmSciTech 15, 1263–1274.

Barger, A., Fuhst, C., Wiedemann, B., 2003. Pharmacological indices in antibiotic therapy. J. Antimicrob. Chemother. 52, 893–898.

Bargoni, A., Cavalli, R., Zara, G.P., Fundarò, A., Caputo, O., Gasco, M.R., 2001. Transmucosal transport of tobramycin incorporated in solid lipid nanoparticles (SLN) after duodenal administration to rats. Part II—tissue distribution. Pharmacol. Res. 43, 497–502.

Basha, M., Abd El-Alim, S.H., Shamma, R.N., Awad, G.E., 2013. Design and optimization of surfactant-based nanovesicles for ocular delivery of clotrimazole. J. Lipos. Res. 23, 203–210.

Basile, L., Passirani, C., Huynh, N.-T., Béjaud, J., Benoit, J.-P., Puglisi, G., Pignatello, R., 2012. Serum-stable, long-circulating paclitaxel-loaded colloidal carriers decorated with a new amphiphilic PEG derivative. Int. J. Pharm. 426, 231–238.

Bhalaria, M.K., Naik, S., Misra, A.N., 2009. Ethosomes: a novel delivery system for antifungal drugs in the treatment of topical fungal diseases. Indian J. Exp. Biol. 47, 368–375.

Bhalekar, M.R., Pokharkar, V., Madgulkar, A., Patil, N., Patil, N., 2009. Preparation and evaluation of miconazole nitrate-loaded solid lipid nanoparticles for topical delivery. AAPS PharmSciTech 10, 289–296.

Bhardwaj, A., Kumar, L., Narang, R.K., Murthy, R.S.R., 2013. Development and characterization of ligand-appended liposomes for multiple drug therapy for pulmonary tuberculosis. Artif. Cells Nanomed. Biotechnol. 41, 52–59.

Bouaoud, C., Lebouille, J.G., Mendes, E., De Braal, H.E., Meesters, G.M., 2016. Formulation and antifungal performance of natamycin-loaded liposomal suspensions: the benefits of sterol-enrichment. J. Lipos. Res. 26 (2), 103–112.

Briones, E., Colino, C.I., Lanao, J.M., 2008. Delivery systems to increase the selectivity of antibiotics in phagocytic cells. J. Control. Release 125, 210–227.

Burton, A.J., Giguère, S., Berghaus, L.J., Hondalus, M.K., Arnold, R.D., 2015. Efficacy of liposomal gentamicin against *Rhodococcus equi* in a mouse infection model and colocalization with *R. equi* in equine alveolar macrophages. Vet. Microbiol. 176, 292–300.

Cao, Z., Sun, Y., 2009. Chitosan based rechargeable long term antimicrobial and biofilm controlling systems. J. Biomed. Mater. Res. A 89, 960–967.

Carbone, C., Cupri, S., Leonardi, A., Puglisi, G., Pignatello, R., 2013. Lipid-based nanocarriers for drug delivery and targeting: a patent survey of methods of production and characterization. Pharm. Pat. Anal. 2, 665–677.

Carryn, S., Chanteux, H., Seral, C., Mingeot-Leclercq, M.P., Van Bambeke, F., Tulkens, P.M., 2003. Intracellular pharmacodynamics of antibiotics. Infect. Dis. Clin. N. Am. 17, 615–634.

Cassano, R., Ferrarelli, T., Mauro, M.V., Cavalcanti, P., Picci, N., Trombino, S., 2016. Preparation, characterization and in vitro activities evaluation of solid lipid nanoparticles based on PEG-40 stearate for antifungal drugs vaginal delivery. J. Drug Deliv. 23 (3), 1047–1056.

Cavalli, R., Gasco, M.R., Chetoni, P., Burgalassi, S., Saettone, M.F., 2002. Solid lipid nanoparticles (SLN) as ocular delivery system for tobramycin. Inter. J. Pharm. 238, 241–245.
Cavalli, R., Bargoni, A., Podio, V., Muntoni, E., Zara, G.P., Gasco, M.R., 2003. Duodenal administration of solid lipid nanoparticles loaded with different percentages of tobramycin. J. Pharm. Sci. 92, 1085–1094.
Chang, C.H., Lin, Y.H., Yeh, C.L., Chen, Y.C., Chiou, S.F., Hsu, Y.M., Chen, Y.S., Wang, C.C., 2009. Nanoparticles incorporated in pH-sensitive hydrogels as amoxicillin delivery for eradication of *Helicobacter pylori*. Biomacromolecules 11, 133–142.
Chang, T., Olson, J.A., Proffitt, R.T., Adler-Moore, J.P., 2010. Differences in tissue drug concentrations following intravenous versus intraperitoneal treatment with amphotericin B deoxycholate or liposomal amphotericin B. Med. Mycol. 48, 430–435.
Chang, C.H., Huang, W.Y., Lai, C.H., Hsu, Y.M., Yao, Y.H., Chen, T.Y., Wu, J.Y., Peng, S.F., Lin, Y.H., 2011. Development of novel nanoparticles shelled with heparin for berberine delivery to treat *Helicobacter pylori*. Acta Biomater. 7, 593–603.
Changsan, N., Nilkaeo, A., Pungrassami, P., Srichana, T., 2009. Monitoring safety of liposomes containing rifampicin on respiratory cell lines and in vitro efficacy against *Mycobacterium bovis* in alveolar macrophages. J. Drug Target. 17, 751–762.
Chavan, N.L., Young, J.K., Drezek, R.A., Lewis, R., Bikram, M., 2012. Interactions of liposome carriers with infectious fungal hyphae reveals the role of β-glucans. Mol. Pharm. 9, 2489–2496.
Chen, S.C., Playford, E.G., Sorrell, T.C., 2010. Antifungal therapy in invasive fungal infections. Curr. Opin. Pharmacol. 10, 522–530.
Chen, Y.C., Liu, D.Z., Liu, J.J., Chang, T.W., Ho, H.O., Sheu, M.T., 2012. Development of terbinafine solid lipid nanoparticles as a topical delivery system. Int. J. Nanomed. 7, 4409–4418.
Chen, Y.-S., Green, C.R., Teague, R., Perrett, J., Danesh-Meyer, H.V., Toth, I., Rupenthal, I.D., 2015. Intravitreal injection of lipoamino acid-modified connexin43 mimetic peptide enhances neuroprotection after retinal ischemia. Drug Deliv. Transl. Res. 2015 (5), 480–488.
Chimote, G., Banerjee, R., 2009. Evaluation of antitubercular drug-loaded surfactants as inhalable drug-delivery systems for pulmonary tuberculosis. J. Biomed. Mater. Res. A 89, 281–292.
Christophersen, P.C., Fano, M., Saaby, L., Yang, M., Nielsen, H.M., Mu, H., 2015. Characterization of particulate drug delivery systems for oral delivery of peptide and protein drugs. Curr. Pharm. Des. 21, 2611–2628.
Chu, J.H., Feudtner, C., Heydon, K., Walsh, T.J., Zaoutis, T.E., 2006. Hospitalizations for endemic mycoses: a population-based national study. Clin. Infect. Dis. 42, 822–825.
Clancy, J.P., Dupont, L., Konstan, M.W., Billings, J., Fustik, S., Goss, C.H., Lymp, J., Minic, P., Quittner, P.A.L., Rubenstein, R.C., Young, K.R., Saiman, L., Burns, J.L., Govan, J.R., Ramsey, B., Gupta, R., 2013. Phase II studies of nebulised Arikace in CF patients with *Pseudomonas aeruginosa* infection. Thorax 68, 818–825.
Colzi, I., Troyan, A.N., Perito, B., Casalone, E., Romoli, R., Pieraccini, G., Škalko-Basnet, N., Adessi, A., Rossi, F., Gonnelli, C., Ristori, S., 2015. Antibiotic delivery by liposomes from prokaryotic microorganisms: similia cum similis works better. Eur. J. Pharm. Biopharm. 94, 411–418.
Connors, K.P., Kuti, J.L., Nicolau, D.P., 2013. Optimizing antibiotic pharmacodynamics for clinical practice. Pharm. Anal. Acta 4, 214.
Cover, N.F., Lai-Yuen, S., Parsons, A.K., Kumar, A., 2012. Synergetic effects of doxycycline-loaded chitosan nanoparticles for improving drug delivery and efficacy. Int. J. Nanomed. 7, 2411–2419.
Cox, G., Wright, G.D., 2013. Intrinsic antibiotic resistance: mechanisms, origins, challenges and solutions. Int. J. Med. Microbiol. 303, 287–292.
Cunningham, B., McLaren, A.C., Pauken, C., McLemore, R., 2012. Liposomal formulation increases local delivery of amphotericin from bone cement: a pilot study. Clin. Orthop. Relat. Res. 470, 2671–2676.
D'Angelo, I., Casciaro, B., Miro, A., Quaglia, F., Mangoni, M.L., Ungaro, F., 2015. Overcoming barriers in *Pseudomonas aeruginosa* lung infections: engineered nanoparticles for local delivery of a cationic antimicrobial peptide. Colloids Surf. B 135, 717–725.

Dandekar, P., Venkataraman, C., Mehra, A., 2010. Pulmonary targeting of nanoparticle drug matrices. J. Aerosol. Med. Pulm. Drug Deliv. 23, 343–353.
Danion, A., Arsenault, I., Vermette, P., 2007. Antibacterial activity of contact lenses bearing surface-immobilized layers of intact liposomes loaded with levofloxacin. J. Pharma. Sci. 96, 2350–2363.
Das, S., Ng, W.K., Tan, R.B., 2012. Are nanostructured lipid carriers (NLCs) better than solid lipid nanoparticles (SLNs): development, characterizations and comparative evaluations of clotrimazole-loaded SLNs and NLCs? Eur. J. Pharm. Sci. 47, 139–151.
Das, S., Ng, W.K., Tan, R.B., 2014. Sucrose ester stabilized solid lipid nanoparticles and nanostructured lipid carriers: II. Evaluation of the imidazole antifungal drug-loaded nanoparticle dispersions and their gel formulations. Nanotechnology 25, 105102.
Das Neves, J., Bahia, M.F., 2006. Gels as vaginal drug delivery systems. Inter. J. Pharm. 318, 1–14.
Davies, J., Davies, D., 2010. Origins and evolution of antibiotic resistance. Microbiol. Mol. Biol. Rev. 74, 417–433.
de Jesús Valle, M.J., González, J.G., López, F.G., Navarro, A.S., 2013. Pulmonary disposition of vancomycin nebulized as lipid vesicles in rats. J. Antibiot. 66, 447–451.
de Sá, F.A.P., Taveira, S.F., Gelfuso, G.M., Lima, E.M., Gratieri, T., 2015. Liposomal voriconazole (VOR) formulation for improved ocular delivery. Colloids Surf. B 133, 331–338.
de Steenwinkel, J.E., van Vianen, W., Marian, T., Verbrugh, H.A., van Agtmael, M.A., Schiffelers, R.M., Bakker-Woudenberg, I.A., 2007. Targeted drug delivery to enhance efficacy and shorten treatment duration in disseminated *Mycobacterium avium* infection in mice. J. Antimicrob. Chemother. 60, 1064–1073.
Deacon, J., Abdelghany, S.M., Quinn, D.J., Schmid, D., Megaw, J., Donnelly, R.F., Jones, D.S., Kissenpfennig, A., Elborn, J.S., Gilmore, B.F., Taggart, C.C., Scott, C.J., 2015. Antimicrobial efficacy of tobramycin polymeric nanoparticles for *Pseudomonas aeruginosa* infections in cystic fibrosis: formulation, characterisation and functionalisation with dornase alfa (DNase). J. Control. Rel. 198, 55–61.
Dobaria, N., Mashru, R., Vadia, N.H., 2008. Vaginal drug delivery systems: a review of current status. East Cent. Afr. J. Pharm. Sci. 10, 3–13.
Dodds Ashley, E., Perfect, J.R., 2015. Pharmacology of azoles, In: Kaufmfman, C.A., Thorner, A.R., (Eds.), UpToDate, Waltham, MA. Available from: www.uptodate.com
Dodds Ashley, E.S., Lewis, R., Lewis, J.S., Martin, C., Andes, D., 2006. Pharmacology of systemic antifungal agents. Clin. Infect. Dis. 43 (S1), S28–S39.
Dong, Z., Xie, S., Zhu, L., Wang, Y., Wang, X., Zhou, W., 2011. Preparation and in vitro, in vivo evaluations of norfloxacin-loaded solid lipid nanopartices for oral delivery. Drug Deliv. 18, 441–450.
Drulis-Kawa, Z., Dorotkiewicz-Jach, A., 2010. Liposomes as delivery systems for antibiotics. Int. J. Pharm. 387, 187–198.
Drulis-Kawa, Z., Gubernator, J., Dorotkiewicz-Jach, A., Doroszkiewicz, W., Kozubek, A., 2006. A comparison of the in vitro antimicrobial activity of liposomes containing meropenem and gentamicin. Cell. Mol. Biol. Lett. 11, 360–375.
Duan, J., Vogt, F.G., Li, X., Hayes, Jr., D., Mansour, H.M., 2013. Design, characterization, and aerosolization of organic solution advanced spray-dried moxifloxacin and ofloxacin dipalmitoylphosphatidylcholine (DPPC) microparticulate/nanoparticulate powders for pulmonary inhalation aerosol delivery. Int. J. Nanomed. 8, 3489–3505.
Ehsan, Z., Wetzel, J.D., Clancy, J.P., 2014. Nebulized liposomal amikacin for the treatment of *Pseudomonas aeruginosa* infection in cystic fibrosis patients. Exp. Opin. Investig. Drugs 23, 743–749.
Ekambaram, P., Sathali, A.A.H., Priyanka, K., 2012. Solid lipid nanoparticles: a review. Sci. Rev. Chem. Commun. 2, 80–102.
El Meshad, A.N., Mohsen, A.M., 2015. Enhanced corneal permeation and antimycotic activity of itraconazole against *Candida albicans* via a novel nanosystem vesicle. Drug Deliv. 21, 1–9.
El-Badry, M., Fetih, G., Shakeel, F., 2014. Comparative topical delivery of antifungal drug croconazole using liposome and micro-emulsion-based gel formulations. Drug Deliv. 21, 34–43.

Elmoslemany, R.M., Abdallah, O.Y., El-Khordagui, L.K., Khalafallah, N.M., 2012. Propylene glycol liposomes as a topical delivery system for miconazole nitrate: comparison with conventional liposomes. AAPS PharmSciTech 13, 723–731.

Emeje, M.O., Akpabio, E.I., Obidike, I.C., Ofoefule, S.I., 2012. Nanotechnology in drug delivery. In: Demir Sezer, A. (Ed.), Recent Advances in Novel Drug Carrier Systems. INTECH Publisher, Rijeka, Croatia, pp. 69–106.

Faustino-Vega, A., Alvarez-Polo, M.A., Gasca, B., Bernad-Bernad, M.J., 2009. Influence of three different colloidal systems on the oxytetracycline-lecithin behavior. Pharmazie 64, 505–509.

Franca, J.R., De Luca, M.P., Ribeiro, T.G., Castilho, R.O., Moreira, A.N., Santos, V.R., Faraco, A.A., 2014. Propolis-based chitosan varnish: drug delivery, controlled release and antimicrobial activity against oral pathogen bacteria. BMC Compl. Altern. Med. 14, 478.

Fresta, M., Puglisi, G., Giammona, G., Cavallaro, G., Micali, N., Furneri, P.M., 1995a. Pefloxacine mesilate- and ofloxacin-loaded polyethylcyanoacrylate nanoparticles: characterization of the colloidal drug carrier formulation. J. Pharm. Sci. 84, 895–902.

Fresta, M., Spadaro, A., Cerniglia, G., Ropero, I.M., Puglisi, G., Furneri, P.M., 1995b. Intracellular accumulation of ofloxacin-loaded liposomes in human synovial fibroblasts. Antimicrob. Agents Chemother. 39, 1372–1375.

Friedman, A.J., Phan, J., Schairer, D., Champer, J., Qin, M., Pirouz, A., Blecher-Paz, K., Oren, A., Liu, P.T., Modlin, R.L., Kim, J., 2013. Antimicrobial and anti-inflammatory activity of chitosan-alginate nanoparticles: a targeted therapy for cutaneous pathogens. J. Invest. Dermatol. 133, 1231–1239.

Furneri, P.M., Fresta, M., Puglisi, G., Tempera, G., 2000. Ofloxacin-loaded liposomes: in vitro activity and drug accumulation in bacteria. Antimicrob. Agents Chemother. 44, 2458–2464.

Gadelha de Carvalho, M.M.S., Montenegro Stamford, T.C., Pereira Dos Santos, E., Tenorio, P., Sampaio, F., 2011. Chitosan as an oral antimicrobial agent. In: Mendez-Vilas, A. (Ed.), Science Against Microbial Pathogens: Communicating Current Research and Technological Advances Mendez-Vilas, A. Formatex Research Center, Badajoz, Spain, pp. 542–550.

Galimberti, R., Torre, A.C., Baztán, M.C., Rodriguez-Chiappetta, F., 2012. Emerging systemic fungal infections. Clin. Dermatol. 30, 633–650.

Garud, A., Singh, D., Garud, N., 2012. Solid lipid nanoparticles (SLN): method, characterization and applications. Int. Curr. Pharm. J. 1, 384–393.

Gaur, P.K., Mishra, S., Gupta, V.B., Rathod, M.S., Purohit, S., Savla, B.A., 2009. In-situ formation of liposome of rifampicin: better availability for better treatment. Curr. Drug Deliv. 6, 461–468.

Ghaffari, S., Varshosaz, J., Saadat, A., Atyabi, F., 2011. Stability and antimicrobial effect of amikacin-loaded solid lipid nanoparticles. Int. J. Nanomed. 6, 35–43.

Ghannoum, M., Isham, N., Henry, W., Kroon, H.A., Yurdakul, S., 2012. Evaluation of the morphological effects of TDT 067 (terbinafine in transfersome) and conventional terbinafine on dermatophyte hyphae in vitro and in vivo. Antimicrob. Agents Chemother. 56, 2530–2534.

Gilani, K., Moazeni, E., Ramezanli, T., Amini, M., Fazeli, M.R., Jamalifar, H., 2011. Development of respirable nanomicelle carriers for delivery of amphotericin B by jet nebulization. J. Pharm. Sci. 100, 252–259.

Gisbert, J.P., Torrado, G., Torrado, S., Olivares, D., Pajares, J.M., 2006. Clinical trial evaluating amoxicillin and clarithromycin hydrogels (chitosan-polyacrylic acid polyionic complex) for *H. pylori* eradication. J. Clin. Gastroenterol. 40, 618–622.

Gnanadhas, D.P., Thomas, M.B., Elango, M., Raichur, A.M., Chakravortty, D., 2013. Chitosan–dextran sulphate nanocapsule drug delivery system as an effective therapeutic against intraphagosomal pathogen *Salmonella*. J. Antimicrob. Chemother. 68, 2576–26862.

Gonçalves, I.C., Henriques, P.C., Seabra, C.L., Martins, M.C.L., 2014. The potential utility of chitosan micro/nanoparticles in the treatment of gastric infection. Exp. Rev. Anti Infect. Ther. 12, 981–992.

Goodwin, D., Varamini, P., Simerska, P., Toth, I., 2014. Stability, permeability and growth-inhibitory properties of gonadotropin-releasing hormone liposaccharides. Pharm. Res. 32, 1570–1584.

Goy, R.C., Britto, D.D., Assis, O.B., 2009. A review of the antimicrobial activity of chitosan. Polímeros 19, 241–247.

Gregordiadis, G., 2007. Liposome Technology, vol. 3, Liposome Preparation and Related Techniques. Informa Healthcare USA, New York

Gregoriadis, G., 1973. Drug entrapment in liposomes. FEBS Lett. 36, 292–296.

Gupta, A.K., Cooper, E.A., 2008. Update in antifungal therapy of dermatophytosis. Mycopathologia 166, 353–367.

Gupta, M., Goyal, A.K., Paliwal, S.R., Paliwal, R., Mishra, N., Vaidya, B., Dube, D., Jai, S.K., Vyas, S.P., 2010. Development and characterization of effective topical liposomal system for localized treatment of cutaneous candidiasis. J. Lipos. Res. 20, 341–350.

Gupta, M., Tiwari, S., Vyas, S.P., 2013. Influence of various lipid core on characteristics of SLNs designed for topical delivery of fluconazole against cutaneous candidiasis. Pharm. Dev. Technol. 18, 550–559.

Habib, F.S., Fouad, E.A., Abdel-Rhaman, M.S., Fathalla, D., 2010. Liposomes as an ocular delivery system of fluconazole: in-vitro studies. Acta Ophthalmol. 88, 901–904.

Hajipour, M.J., Fromm, K.M., Ashkarran, A.A., Jimenez de Aberasturi, D., de Larramendi, I.R., Rojo, T., Serpooshan, V., Parak, W.J., Mahmoudi, M., 2012. Antibacterial properties of nanoparticles. Trends Biotechnol. 30 (10), 499–511.

Halwani, M., Mugabe, C., Azghani, A.O., Lafrenie, R.M., Kumar, A., Omri, A., 2007. Bactericidal efficacy of liposomal aminoglycosides against *Burkholderia cenocepacia*. J. Antimicrob. Chemother. 60, 760–769.

Halwani, M., Blomme, S., Suntres, Z.E., Alipour, M., Azghani, A.O., Kumar, A., Omri, A., 2008a. Liposomal bismuth-ethanedithiol formulation enhances antimicrobial activity of tobramycin. Int. J. Pharm. 358, 278–284.

Halwani, M., Yebio, B., Suntres, Z.E., Alipour, M., Azghani, A.O., Omri, A., 2008b. Co-encapsulation of gallium with gentamicin in liposomes enhances antimicrobial activity of gentamicin against *Pseudomonas aeruginosa*. J. Antimicrob. Chemother. 62, 1291–1297.

Halwani, M., Hebert, S., Suntres, Z.E., Lafrenie, R.M., Azghani, A.O., Omri, A., 2009. Bismuth–thiol incorporation enhances biological activities of liposomal tobramycin against bacterial biofilm and quorum sensing molecules production by *Pseudomonas aeruginosa*. Int. J. Pharm. 373, 141–146.

Hamill, R.J., 2013. Amphotericin B formulations: a comparative review of efficacy and toxicity. Drugs 73, 919–934.

Hann, I.M., Prentice, H.G., 2001. Lipid-based amphotericin B: a review of the last 10 years of use. Int. J. Antimicrob. Agents 17, 161–169.

Hawkey, P.M., 2008. The growing burden of antimicrobial resistance. J. Antimicrob. Chemother., i9.

Hillaireau, H., Couvreur, P., 2009. Nanocarriers' entry into the cell: relevance to drug delivery. Cell. Mol. Life Sci. 66, 2873–2896.

Høiby, N., Bjarnsholt, T., Givskov, M., Molin, S., Ciofu, O., 2010. Antibiotic resistance of bacterial biofilms. Int. J. Antimicrob. Agents 35, 322–332.

Hosny, K.M., 2009. Preparation and evaluation of thermosensitive liposomal hydrogel for enhanced transcorneal permeation of ofloxacin. AAPS PharmSciTech 10, 1336–1342.

Hosny, K.M., 2010. Optimization of gatifloxacin liposomal hydrogel for enhanced transcorneal permeation. J. Lipos. Res. 20, 31–37.

Huang, J., Ma, T., Tang, H., Fan, X., Xu, Y., 2012. Vancomycin cationic liposome combined with nano-hydroxyapatite/chitosan/konjac glucomannan scaffold for treatment of infected bone defects in rabbits. Chinese J. Repar. Reconstr. Surg. 26, 190–195.

Huh, A.J., Kwon, Y.J., 2011. "Nanoantibiotics": a new paradigm for treating infectious diseases using nanomaterials in the antibiotics resistant era. J. Control. Release 156, 128–145.

Hurler, J., Berg, O.A., Skar, M., Conradi, A.H., Johnsen, P.J., Škalko-Basnet, N., 2012. Improved burns therapy: liposomes-in-hydrogel delivery system for mupirocin. J. Pharm. Sci. 101, 3906–3915.

Hussain, A., Ahsan, F., 2005. The vagina as a route for systemic drug delivery. J. Control. Release 103, 301–313.

Immordino, M.L., Dosio, F., Cattel, L., 2006. Stealth liposomes: review of the basic science, rationale, and clinical applications, existing and potential. Int. J. Nanomed. 1, 297–315.

Ishak, R.A., Awad, G.A., Mortada, N.D., Nour, S.A., 2007. Preparation, in vitro and in vivo evaluation of stomach-specific metronidazole-loaded alginate beads as local anti-*Helicobacter pylori* therapy. J. Control. Release 119, 207–214.

Jain, D., Banerjee, R., 2008. Comparison of ciprofloxacin hydrochloride-loaded protein, lipid, and chitosan nanoparticles for drug delivery. J. Biomed. Mater. Res. B V 86, 105–112.

Jain, A.K., Jain, S.K., 2013. Development and characterization of nanolipobeads-based dual drug delivery system for *H. pylori* targeting. J. Drug Target. 21, 593–603.

Jain, S., Jain, S., Khare, P., Gulbake, A., Bansal, D., Jain, S.K., 2010. Design and development of solid lipid nanoparticles for topical delivery of an anti-fungal agent. Drug Deliv. 17, 443–451.

Jain, A.K., Agarwal, A., Agrawal, H., Agrawal, G.P., 2012. Double-liposome–based dual-drug delivery system as vectors for effective management of peptic ulcer. J. Lipos. Res. 22, 205–214.

Jin, H.L., Wu, W., Shu, R., 2010. Doxycycline nano-liposome slow-release gel improves rat periodontitis. Shanghai J. Stomatol. 19, 508–511.

Jones, M.N., 2005. Use of liposomes to deliver bactericides to bacterial biofilms. Methods Enzymol. 391, 211–228.

Juliano, R.L., Stamp, D., McCullough, N., 1978. Pharmacokinetics of liposome-encapsulated antitumor drugs and implications for therapy. Ann. N.Y. Acad. Sci. 308, 411–425.

Kadimi, U.S., Balasubramanian, D.R., Ganni, U.R., Balaraman, M., Govindarajulu, V., 2007. In vitro studies on liposomal amphotericin B obtained by supercritical carbon dioxide–mediated process. Nanomedicine 3, 273–280.

Kadry, A.A., Al-Suwayeh, S.A., Abd-Allah, A.R., Bayomi, M.A., 2004. Treatment of experimental osteomyelitis by liposomal antibiotics. J. Antimicrob. Chemother. 54, 1103–1108.

Kaiser, J.M., Imai, H., Haakenson, J.K., Brucklacher, R.M., Fox, T.E., Shanmugavelandy, S.S., Unrath, K.A., Pedersen, M.M., Dai, P., Freeman, W.M., Bronson, S.K., Gardner, T.W., Kester, M., 2013. Nanoliposomal minocycline for ocular drug delivery. Nanomedicine 9, 130–140.

Kakkar, S., Kaur, I.P., 2011. Spanlastics—A novel nanovesicular carrier system for ocular delivery. Int. J. Pharm. 413, 202–210.

Kakkar, S., Kaur, I.P., 2013. A novel nanovesicular carrier system to deliver drug topically. Pharm. Dev. Technol. 18, 673–685.

Kalam, M.A., Sultana, Y., Ali, A., Aqil, M., Mishra, A.K., Chuttani, K., 2010. Preparation, characterization, and evaluation of gatifloxacin loaded solid lipid nanoparticles as colloidal ocular drug delivery system. J. Drug Target. 18, 191–204.

Kalam, M.A., Sultana, Y., Ali, A., Aqil, M., Mishra, A.K., Aljuffali, I.A., Alshamsan, A., 2013a. Part I: Development and optimization of solid-lipid nanoparticles using Box–Behnken statistical design for ocular delivery of gatifloxacin. J. Biomed. Mat. Res. A 101, 1813–1827.

Kalam, M.A., Sultana, Y., Ali, A., Aqil, M., Mishra, A.K., Chuttani, K., Aljuffali, I.A., Alshamsan, A., 2013b. Part II: enhancement of transcorneal delivery of gatifloxacin by solid lipid nanoparticles in comparison to commercial aqueous eye drops. J. Biomed. Mat. Res. A 101, 1828–1836.

Kalhapure, R.S., Akamanchi, K.G., 2013. A novel biocompatible bicephalous dianionic surfactant from oleic acid for solid lipid nanoparticles. Colloids Surf. B 105, 215–222.

Kalhapure, R.S., Suleman, N., Mocktar, C., Seedat, N., Govender, T., 2015. Nanoengineered drug delivery systems for enhancing antibiotic therapy. J. Pharm. Sci. 104, 872–905.

Kanafani, Z.A., Perfect, J.R., 2008. Resistance to antifungal agents: mechanisms and clinical impact. Clin. Infect. Dis. 46, 120–128.

Kang, J.W., Davaa, E., Kim, Y.T., Park, J.S., 2010. A new vaginal delivery system of amphotericin B: a dispersion of cationic liposomes in a thermosensitive gel. J. Drug Target. 18, 637–644.

Karimunnisa, S., Atmaram, P., 2013. Mucoadhesive nanoliposomal formulation for vaginal delivery of an antifungal. Drug Dev. Ind. Pharm. 39, 1328–1337.

Kashi, T.S.J., Eskandarion, S., Esfandyari-Manesh, M., Marashi, S.M.A., Samadi, N., Fatemi, S.M., Atyabi, F., Eshraghi, S., Dinarvand, R., 2012. Improved drug loading and antibacterial activity of minocycline-loaded PLGA nanoparticles prepared by solid/oil/water ion pairing method. Int. J. Nanomed. 7, 221–234.

Kaur, L., Jain, S.K., Manhas, R.K., Sharma, D., 2015. Nanoethosomal formulation for skin targeting of amphotericin B: an in vitro and in vivo assessment. J. Lipos. Res. 25, 294–307.

Kesharwani, R., Nair, S.K., Patel, D., 2013. Solid lipid nanoparticle (SLN): a modern approach for drug delivery. J. Pharm. Res. 2, 13–21.

Khameneh, B., Iranshahy, M., Ghandadi, M., Ghoochi Atashbeyk, D., Fazly Bazzaz, B.S., Iranshahi, M., 2014. Investigation of the antibacterial activity and efflux pump inhibitory effect of co-loaded piperine and gentamicin nanoliposomes in methicillin-resistant *Staphylococcus aureus*. Drug Dev. Ind. Pharm. 41, 989–994.

Khan, M.A., Nasti, T.H., Owais, M., 2005. Incorporation of amphotericin B in tuftsin-bearing liposomes showed enhanced efficacy against systemic cryptococcosis in leucopenic mice. J. Antimicrob. Chemother. 56, 726–731.

Koga, K., Kusawake, Y., Ito, Y., Sugioka, N., Shibata, N., Takada, K., 2006. Enhancing mechanism of Labrasol on intestinal membrane permeability of the hydrophilic drug gentamicin sulfate. Eur. J. Pharm. Biopharm. 64, 82–91.

Kong, M., Chen, X.G., Xing, K., Park, H.J., 2010. Antimicrobial properties of chitosan and mode of action: a state of the art review. Int. J. Food Microbiol. 144, 51–63.

Kulichenko, A.N., Mikhaĭlova, M.E., Kovalev, D.A., Pisarenko, S.V., Siritsa Iu, V., Liapustina, L.V., 2014. Comparative study of pharmacokinetics of ofloxacin in a free and niosomal forms in experiments on white mice when administered per os. Vestn. Ross Akad. Med. Nauk., 80–84.

Kumar, N., Goindi, S., 2014. Statistically designed nonionic surfactant vesicles for dermal delivery of itraconazole: characterization and in vivo evaluation using a standardized *Tinea pedis* infection model. Int. J. Pharm. 472, 224–240.

Kumar, S., Randhawa, J.K., 2013. High melting lipid based approach for drug delivery: solid lipid nanoparticles. Mater. Sci. Eng. C 33, 1842–1852.

Lai, C.C., Tan, C.K., Huang, Y.T., Shao, P.L., Hsueh, P.R., 2008. Current challenges in the management of invasive fungal infections. J. Infect. Chemother. 14, 77–85.

Laniado-Laborin, L., 2007. Expanding understanding of epidemiology of coccidioidomycosis in the Western hemisphere. Ann. N.Y. Acad. Sci. 1111, 19–34.

Laniado-Laborín, R., Cabrales-Vargas, M.N., 2009. Amphotericin B: side effects and toxicity. Rev. Iberoam. Micol. 26, 223–227.

Lass-Flörl, C., 2009. The changing face of epidemiology of invasive fungal disease in Europe. Mycoses 52, 197–205.

Lehr, C.M., Bodde, H.E., Bouwstra, J.A., Junginger, H.E., 1993. A surface energy analysis of mucoadhesion. II. Prediction of mucoadhesive performance by spreading coefficients. Eur. J. Pharm. Sci. 1, 19–30.

Lemaire, S., Tulkens, P.M., Van Bambeke, F., 2010. Cellular pharmacokinetics of the novel biaryloxazolidinone Rradezolid in phagocytic cells: studies with macrophages and polymorphonuclear neutrophils. Antimicrob. Agents Chemother. 54, 2540–2548.

Leroux, S., Ullmann, A.J., 2013. Management and diagnostic guidelines for fungal diseases in infectious diseases and clinical microbiology: critical appraisal. Clin. Microbiol. Infect. 19, 1115–1121.

Levison, M.E., 2004. Pharmacodynamics of antimicrobial drugs. Infect. Dis. Clin. North Am. 18, 451–465.

Levison, M.E., Levison, J.H., 2009. Pharmacokinetics and pharmacodynamics of antibacterial agents. Infect. Dis. Clin. North Am. 23, 791–815.

Lewis, R.E., 2011. Current concepts in antifungal pharmacology. Mayo Clinic Proc. 86, 805–817.

Lewis, R.E., 2012. Importance of pharmacokinetic considerations for selecting therapy in the treatment of invasive fungal infections. Am. J. Ther. 19, 51–63.

Li, H., Zhao, X., Ma, Y., Zhai, G., Li, L., Lou, H., 2009. Enhancement of gastrointestinal absorption of quercetin by solid lipid nanoparticles. J. Control. Release 133, 238–244.

Li, Y., Su, T., Zhang, Y., Huang, X., Li, J., Li, C., 2014. Liposomal co-delivery of daptomycin and clarithromycin at an optimized ratio for treatment of methicillin-resistant *Staphylococcus aureus* infection. Drug Deliv. 22, 627–637.

Lin, Y.H., Lin, J.H., Chou, S.C., Chang, S.J., Chung, C.C., Chen, Y.S., Chang, C.H., 2015. Berberine-loaded targeted nanoparticles as specific *Helicobacter pylori* eradication therapy: in vitro and in vivo study. Nanomedicine 10, 57–71.

Liu, Y., Lin, X., Tang, X., 2009. Lipid emulsions as a potential delivery system for ocular use of azithromycin. Drug Dev. Ind. Pharm. 35, 887–896.

Liu, X., Sun, W., Zhang, B., Tian, B., Tang, X., Qi, N., He, H., Li, H., Jin, X., 2013b. Clarithromycin-loaded liposomes offering high drug loading and less irritation. Int. J. Pharm. 443, 318–327.

Lopez-Berestein, G., Mehta, R., Hopfer, R.L., Mills, K., Kasi, L., Mehta, K., Fainstein, V., Luna, M., Hersh, E.M., Juliano, R., 1983. Treatment and prophylaxis of disseminated infection due to *Candida albicans* in mice with liposome-encapsulated amphotericin B. J. Infect Dis. 147, 939–945.

Lu, E., Franzblau, S., Onyuksel, H., Popescu, C., 2009. Preparation of aminoglycoside-loaded chitosan nanoparticles using dextran sulphate as a counterion. J. Microencapsul. 26, 346–354.

Ludwig, A., 2005. The use of mucoadhesive polymers in ocular drug delivery. Adv. Drug Deliv. Rev. 57, 1595–1639.

Lv, W., Luo, J., Deng, Y., Sun, Y., 2013. Biomaterials immobilized with chitosan for rechargeable antimicrobial drug delivery. J. Biomed. Mater. Res. A 101, 447–455.

Ma, T., Shang, B.C., Tang, H., Zhou, T.H., Xu, G.L., Li, H.L., Chen, Q.H., Xu, Y.Q., 2011. Nano-hydroxyapatite/chitosan/konjac glucomannan scaffolds loaded with cationic liposomal vancomycin: Preparation, in vitro release and activity against *Staphylococcus aureus* biofilms. J. Biomater. Sci. Polym. Ed. 22, 1669–1681.

Ma, D., Green, A.M., Willsey, G.G., Marshall, J.S., Wargo, M.J., Wu, J., 2015. Effects of acoustic streaming from moderate-intensity pulsed ultrasound for enhancing biofilm mitigation effectiveness of drug-loaded liposomes. J. Acoust. Soc. Am. 138, 1043–1051.

Madureira, A.R., Pereira, A., Pintado, M., 2015. Current state on the development of nanoparticles for use against bacterial gastrointestinal pathogens. Focus on chitosan nanoparticles loaded with phenolic compounds. Carbohydr. Polym. 130, 429–439.

Maheshwari, R.G., Tekade, R.K., Sharma, P.A., Darwhekar, G., Tyagi, A., Patel, R.P., Jain, D.K., 2012. Ethosomes and ultradeformable liposomes for transdermal delivery of clotrimazole: a comparative assessment. Saudi Pharm. J. 20, 161–170.

Manca, M.L., Sinico, C., Maccioni, A.M., Diez, O., Fadda, A.M., Manconi, M., 2012. Composition influence on pulmonary delivery of rifampicin liposomes. Pharmaceutics 4, 590–606.

Manca, M.L., Valenti, D., Sales, O.D., Nacher, A., Fadda, A.M., Manconi, M., 2014. Fabrication of polyelectrolyte multilayered vesicles as inhalable dry powder for lung administration of rifampicin. Int. J. Pharm. 472, 102–109.

Maya, S., Indulekha, S., Sukhithasri, V., Smitha, K.T., Nair, S.V., Jayakumar, R., Biswas, R., 2012. Efficacy of tetracycline encapsulated O-carboxymethyl chitosan nanoparticles against intracellular infections of *Staphylococcus aureus*. Int. J. Biol. Macromol. 51, 392–399.

Meers, P., Neville, M., Malinin, V., Scotto, A.W., Sardaryan, G., Kurumunda, R., Mackinson, C., James, G., Fisher, S., Perkins, W.R., 2008. Biofilm penetration, triggered release and in vivo activity of inhaled liposomal amikacin in chronic *Pseudomonas aeruginosa* lung infections. J. Antimicrob. Chemother. 61, 859–868.

Mehnert, W., Mäder, K., 2001. Solid lipid nanoparticles: production, characterization and applications. Adv. Drug Deliv. Rev. 47 (2–3), 165–196.

Mehta, S.K., Jindal, N., 2013. Formulation of tyloxapol niosomes for encapsulation, stabilization and dissolution of anti-tubercular drugs. Colloids Surf. B 101, 434–441.

Mehta, S.K., Jindal, N., Kaur, G., 2011. Quantitative investigation, stability and in vitro release studies of anti-TB drugs in Triton niosomes. Colloids Surf. B 87, 173–179.

Messiaen, A.S., Forier, K., Nelis, H., Braeckmans, K., Coenye, T., 2013. Transport of nanoparticles and tobramycin-loaded liposomes in *Burkholderia cepacia* complex biofilms.

Miceli, M.H., Lee, S.A., 2011. Emerging moulds: epidemiological trends and antifungal resistance. Mycoses 54, e666–e678.

Mobarak, D.H., Salah, S., Elkheshen, S.A., 2014. Formulation of ciprofloxacin hydrochloride loaded biodegradable nanoparticles: optimization of technique and process variables. Pharm. Dev. Technol. 19, 891–900.

Moghadas-Sharif, N., Fazly Bazzaz, B.S., Khameneh, B., Malaekeh-Nikouei, B., 2015. The effect of nanoliposomal formulations on *Staphylococcus epidermidis* biofilm. Drug Dev. Ind. Pharm. 41, 445–450.

Mohanty, B., Majumdar, D.K., Mishra, S.K., Panda, A.K., Patnaik, S., 2014. Development and characterization of itraconazole-loaded solid lipid nanoparticles for ocular delivery. Pharm. Dev. Technol. 20, 458–464.

Momoh, M.A., Esimone, C.O., 2012. Phospholipon 90H (P90H)–based PEGylated microscopic lipospheres delivery system for gentamicin: an antibiotic evaluation. Asian Pac. J. Trop. Biomed. 2, 889–894.

Monteiro, N., Martins, M., Martins, A., Fonseca, N.A., Moreira, J.N., Reis, R.L., Neves, N.M., 2015. Antibacterial activity of chitosan nanofiber meshes with liposomes immobilized releasing gentamicin. Acta Biomater. 18, 196–205.

Moogooee, M., Ramezanzadeh, H., Jasoori, S., Omidi, Y., Davaran, S., 2011. Synthesis and in vitro studies of cross-linked hydrogel nanoparticles containing amoxicillin. J. Pharm. Sci. 100, 1057–1066.

Morrone, A., 2013. Migration and mycoses. G. Ital. Dermatol. Venereol. 148, 573–591.

Mugabe, C., Azghani, A.O., Omri, A., 2005. Liposome-mediated gentamicin delivery: development and activity against resistant strains of *Pseudomonas aeruginosa* isolated from cystic fibrosis patients. J. Antimicrob. Chemother. 55, 269–271.

Mugabe, C., Azghani, A.O., Omri, A., 2006a. Preparation and characterization of dehydration–rehydration vesicles loaded with aminoglycoside and macrolide antibiotics. Int. J. Pharm. 307, 244–250.

Mugabe, C., Halwani, M., Azghani, A.O., Lafrenie, R.M., Omri, A., 2006b. Mechanism of enhanced activity of liposome-entrapped aminoglycosides against resistant strains of *Pseudomonas aeruginosa*. Antimicrob. Agents Chemother. 50, 2016–2022.

Müller, R., Runge, S., 1998. Solid lipid nanoparticles (SLN) for controlled drug delivery. In: Benita, S. (Ed.), Submicron Emulsions in Drug Targeting and Delivery. Harwood Academic Publishers, Amsterdam, pp. 219–234.

Müller, R.H., Mehnert, W., Lucks, J.S., Schwarz, C., zur Mühlen, A., Meyhers, H., Freitas, C., Rühl, D., 1995. Solid lipid nanoparticles (SLN): an alternative colloidal carrier system for controlled drug delivery. Eur. J. Pharm. Biopharm. 41, 62–69.

Müller, R.H., Maaβen, S., Weyhers, H., Specht, F., Lucks, J.S., 1996. Cytotoxicity of magnetite-loaded polylactide, polylactide/glycolide particles and solid lipid nanoparticles. Int. J. Pharm. 138, 85–94.

Müller, R.H., Mäder, K., Gohla, S., 2000. Solid lipid nanoparticles (SLN) for controlled drug delivery: a review of the state of the art. Eur. J. Pharm. Biopharm. 50, 161–177.

Müller, R.H., Radtke, M., Wissing, S.A., 2002. Solid lipid nanoparticles (SLN) and nanostructured lipid carriers (NLC) in cosmetic and dermatological preparations. Adv. Drug Deliv. Rev. 54 (S1), S131–S155.

Muppidi, K., Wang, J., Betageri, G., Pumerantz, A.S., 2011. PEGylated liposome encapsulation increases the lung tissue concentration of vancomycin. Antimicrob. Agents Chemother. 55, 4537–4542.

Nag, O.K., Awasthi, V., 2013. Surface engineering of liposomes for stealth behavior. Pharmaceutics 5, 542–569.

New, R.R.C., Chance, M.L., Heath, S., 1981. Antileishmanial activity of amphotericin and other antifungal agents entrapped in liposomes. J. Antimicrob. Chemother. 8, 371–381.

Newland, J.G., Abdel-Rahman, S.M., 2009. Update on terbinafine with a focus on dermatophytoses. Clin. Cosmet. Investig. Dermatol. 2, 49–63.

Nicolosi, D., Scalia, M., Nicolosi, V.M., Pignatello, R., 2010. Encapsulation in fusogenic liposomes broadens the spectrum of action of vancomycin against Gram-negative bacteria. Int. J. Antimicrob. Agents 35, 553–558.

Ning, M., Guo, Y., Pan, H., Zong, S., Gu, Z., 2006. Preparation and characterization of EP-liposomes and Span 40-niosomes. Pharmazie 61, 208–212.

Nix, D.E., Goodwin, S.D., Peloquin, C.A., Rotella, D.L., Schentag, J.J., 1991. Antibiotic tissue penetration and its relevance: impact of tissue penetration on infection response. Antimicrob. Agents Chemother. 35, 1947–1952.

Noel, S.P., Courtney, H., Bumgardner, J.D., Haggard, W.O., 2008. Chitosan films: a potential local drug delivery system for antibiotics. Clin. Orthop. Relat. Res. 466, 1377–1382.

Nucci, M., Queiroz-Telles, F., Tobón, A.M., Restrepo, A., Colombo, A.L., 2010. Epidemiology of opportunistic fungal infections in Latin America. Clin. Infect. Dis. 51, 561–570.

Okusanya, O.O., Bhavnani, S.M., Hammel, J., Minic, P., Dupont, L.J., Forrest, A., Mulder, G.J., Mackinson, C., Ambrose, P., Gupta, R., 2009. Pharmacokinetic and pharmacodynamic evaluation of liposomal amikacin for inhalation in cystic fibrosis patients with chronic pseudomonal infection. Antimicrob. Agents Chemother. 53, 3847–3854.

Okusanya, O.O., Bhavnani, S.M., Hammel, J.P., Forrest, A., Bulik, C.C., Ambrose, P.G., Gupta, R., 2014. Evaluation of the pharmacokinetics and pharmacodynamics of liposomal amikacin for inhalation in cystic fibrosis patients with chronic pseudomonal infections using data from two phase 2 clinical studies. Antimicrob. Agents Chemother. 58, 5005–5015.

Oliveira Fulgêncio, G., Viana, F.A.B., Silva, R.O.S., Lobato, F.C.F., Ribeiro, R.R., Fanca, J.R., Byrro, R.M., Faraco, A.A., Silva Cunha-Júnior, A., 2014. Mucoadhesive chitosan films as a potential ocular delivery system for ofloxacin: preliminary in vitro studies. Vet. Ophthalmol. 17, 150–155.

Onoue, S., Yamada, S., Chan, H.K., 2014. Nanodrugs: pharmacokinetics and safety. Int. J. Nanomed. 9, 1025–1037.

Ostrosky-Zeichner, L., Marr, K.A., Rex, J.H., Cohen, S.H., 2003. Amphotericin B: time for a new "gold standard". Clin. Infect. Dis. 37, 415–425.

Ostrosky-Zeichner, L., Casadevall, A., Galgiani, J.N., Odds, F.C., Rex, J.H., 2010. An insight into the antifungal pipeline: selected new molecules and beyond. Nat. Rev. Drug Discov. 9, 719–727.

Oura, M., Sternberg, T.H., Wright, E.T., 1955-1956. A new antifungal antibiotic, amphotericin B. Antibiot. Ann. 3, 566–573.

Oxford, A.E., Raistrick, H., Simonart, P., 1939. Studies in the biochemistry of micro-organisms: Griseofulvin, C17H17O6Cl, a metabolic product of *Penicillium griseo-fulvum* Dierckx. Biochem. J. 33, 240–248.

Pardeike, J., Hommoss, A., Müller, R.H., 2009. Lipid nanoparticles (SLN, NLC) in cosmetic and pharmaceutical dermal products. Int. J. Pharm. 366, 170–184.

Park, K., 2014. Controlled drug delivery systems: past forward and future back. J. Control. Release 190, 3–8.

Park, B.J., Chiller, T.M., Brand, M.E., Warnock, D.W., 2011. Epidemiology of systemic fungal diseases: an overview. In: Kauffman, C.A., Pappas, P.G., Sobel, J.D., Dismuskes, W.E. (Eds.), Essentials of Clinical Microbiology. Springer-Verlag, New York, pp. 27–35.

Parker, A.C., Beenken, K.E., Jennings, J.A., Hittle, L., Shirtliff, M.E., Bumgardner, J.D., Smeltzer, M.S., Haggard, W.O., 2015. Characterization of local delivery with amphotericin B and vancomycin from modified chitosan sponges and functional biofilm prevention evaluation. J. Orthop. Res. 33, 439–447.

Parveen, S., Misra, R., Sahoo, S.K., 2012. Nanoparticles: a boon to drug delivery, therapeutics, diagnostics and imaging. Nanomedicine 8, 147–166.

Pasqualotto, A.C., Denning, D.W., 2008. New and emerging treatments for fungal infections. J. Antimicrob. Chemother. 61 (S1), i19–i30.

Pasquantonio, G., Greco, C., Prenna, M., Ripa, C., Vitali, L.A., Petrelli, D., Di Luca, M.C., Ripa, S., 2008. Antibacterial activity and anti-biofilm effect of chitosan against strains of *Streptococcus mutans* isolated in dental plaque. Int. J. Immunopathol. Pharmacol. 21, 993–997.

Passerini, N., Gavini, E., Albertini, B., Rassu, G., Sabatino, M., Sanna, V., Giunchedi, P., Rodriguez, L., 2009. Evaluation of solid lipid microparticles produced by spray congealing for topical application of econazole nitrate. J. Pharm. Pharmacol. 61, 559–567.

Patel, J.K., Patel, M.M., 2007. Stomach specific anti-helicobacter pylori therapy: preparation and evaluation of amoxicillin-loaded chitosan mucoadhesive microspheres. Curr. Drug Deliv. 4, 41–50.

Patel, P.A., Patravale, V.B., 2011. AmbiOnp: solid lipid nanoparticles of amphotericin B for oral administration. J. Biomed. Nanotechnol. 7, 632–639.
Patil, J.S., Devi, V.K., Devi, K., Sarasija, S., 2015. A novel approach for lung delivery of rifampicin-loaded liposomes in dry powder form for the treatment of tuberculosis. Lung India 32, 331–338.
Pavelic´, Ž., Škalko-Basnet, N., Jalšenjak, I., 2005. Characterisation and in vitro evaluation of bioadhesive liposome gels for local therapy of vaginitis. Int. J. Pharm. 301, 140–148.
Pelgrift, R.Y., Friedman, A.J., 2013. Nanotechnology as a therapeutic tool to combat microbial resistance. Adv. Drug Deliv. Rev. 65, 1803–1815.
Peña, A., Sánchez, N.S., Calahorra, M., 2013. Effects of chitosan on *Candida albicans*: conditions for its antifungal activity. Biomed. Res. Int. 2013, 527549.
Perusquía-Ortiz, A.M., Vázquez-González, D., Bonifaz, A., 2012. Opportunistic filamentous mycoses: aspergillosis, mucormycosis, phaeohyphomycosis and hyalohyphomycosis. J. Dtsch. Dermatol. Ges. 10, 611–621.
Pfaller, M.A., Diekema, D.J., 2007. Epidemiology of invasive *Candidiasis*: a persistent public health problem. Clin. Microbiol. Rev. 20, 133–163.
Pfaller, M.A., Diekema, D.J., 2010. Epidemiology of invasive mycoses in North America. Crit. Rev. Microbiol. 36, 1–53.
Pfaller, M.A., Pappas, P.G., Wingard, J.R., 2006. Invasive fungal pathogens: current epidemiological trends. Clin. Infect. Dis. 43, S3–S14.
Pfaller, M.A., Andes, D.R., Diekema, D.J., Horn, D.L., Reboli, A.C., Rotstein, C., Franks, B., Azie, N.E., 2014. Epidemiology and outcomes of invasive candidiasis due to non-albicans species of Candida in 2,496 patients: data from the Prospective Antifungal Therapy (PATH) Registry 2004-2008. PLoS One 9, e101510.
Pignatello, R., Pecora, T.M.G., 2007. Conjugation of thymopentin (TP5) with lipoamino acid residues increases the hydrolytic stability and preserves the biological activity. Pharmazie 62, 663–667.
Pignatello, R., Sarpietro, M.G., 2013. General experimental set-up of liposomal systems for DSC. In: Pignatello, R. (Ed.), Drug-Biomembrane Interaction Studies. The Application of Calorimetric Techniques. Elsevier/ Woodhead Publications Limited, Cambridge (UK), pp. 363–379.
Pignatello, R., Guccione, S., Castelli, F., Sarpietro, M.G., Giurato, L., Lombardo, M., Puglisi, G., Toth, I., 2006. Enhancement of drug affinity for cell membranes by conjugation with lipoamino acids II. Experimental and computational evidence using biomembrane models. Int. J. Pharm. 310, 53–63.
Pignatello, R., Mangiafico, A., Pantò, V., Puglisi, G., Furneri, P.M., 2008. Solid dispersions of chitosan glutamate for the local delivery of miconazole: characterization and in vitro activity. Open Drug Deliv. J. 2, 44–51.
Pignatello, R., Paolino, D., Pantò, V., Pistarà, V., Calvagno, M.G., Russo, D., Puglisi, G., Fresta, M., 2009. Lipoamino acid prodrugs of paclitaxel: synthesis and cytotoxicity evaluation on human anaplastic thyroid carcinoma cells. Curr. Cancer Drug Targets. 9, 202–213.
Pignatello, R., Vicari, L., Pistarà, V., Musumeci, T., Gulisano, M., Puglisi, G., 2010. Synthesis and in vitro cytotoxic activity on human anaplastic thyroid cancer cells of lipoamino acid conjugates of gemcitabine. Drug Dev. Res. 71, 294–302.
Pignatello, R., Mangiafico, A., Basile, L., Ruozi, B., Furneri, P.M., 2011a. Amphiphilic ion pairs of tobramycin with lipoamino acids. Eur. J. Med. Chem. 46, 1665–1671.
Pignatello, R., Mangiafico, A., Ruozi, B., Puglisi, G., Furneri, P.M., 2011b. Amphiphilic erythromycin-lipoamino acid ion pairs: characterization and in vitro microbiological evaluation. AAPS PharmSciTech 12, 468–475.
Pignatello, R., Leonardi, A., Petronio, G.P., Ruozi, B., Puglisi, G., Furneri, P.M., 2014. Preparation and microbiological evaluation of amphiphilic kanamycin-lipoamino acid ion-pairs. Antibiotics 3, 216–232.
Pilmis, B., Jullien, V., Sobel, J., Lecuit, M., Lortholary, O., Charlier, C., 2015. Antifungal drugs during pregnancy: an updated review. J. Antimicrob. Chemother. 70, 14–22.
Pinto-Alphandary, H., Andremont, A., Couvreur, P., 2000. Targeted delivery of antibiotics using liposomes and nanoparticles: research and applications. Int. J. Antimicrob. Agents 13, 155–168.

Piras, A.M., Maisetta, G., Sandreschi, S., Gazzarri, M., Bartoli, C., Grassi, L., Esin, S., Chiellini, F., Batoni, G., 2015. Chitosan nanoparticles loaded with the antimicrobial peptide temporin B exert a long-term antibacterial activity in vitro against clinical isolates of *Staphylococcus epidermidis*. Front. Microbiol. 6, 372.

Puglisi, G., Fresta, M., Mazzone, G., Furneri, P.M., Tempera, G., 1995. Formulation parameters of fluoroquinolone-loaded liposomes and in vitro antimicrobial activity. Int. J. Pharm. 118, 65–76.

Pumerantz, A., 2012. PEGylated liposomal vancomycin: a glimmer of hope for improving treatment outcomes in MRSA pneumonia. Recent Pat. Antiinfect. Drug Discov. 7, 205–212.

Pumerantz, A., Muppidi, K., Agnihotri, S., Guerra, C., Venketaraman, V., Wang, J., Betageri, G., 2011. Preparation of liposomal vancomycin and intracellular killing of meticillin-resistant *Staphylococcus aureus* (MRSA). Int. J. Antimicrob. Agents 37, 140–144.

Rabea, E.I., Badawy, M.E.T., Stevens, C.V., Smagghe, G., Steurbaut, W., 2003. Chitosan as antimicrobial agent: applications and mode of action. Biomacromolecules 4, 1457–1465.

Ramalingam, N., Natesan, G., Dhandayuthapani, B., Perumal, P., Balasundaram, J., Natesan, S., 2013. Design and characterization of ofloxacin niosomes. Pak. J. Pharm. Sci. 26, 1089–1096.

Ramos-e-Silva, M., Lima, C.M.O., Schechtman, R.C., Trope, B.M., Carneiro, S., 2012. Systemic mycoses in immunodepressed patients (AIDS). Clin. Dermatol. 30, 616–627.

Ramteke, S., Ganesh, N., Bhattacharya, S., Jain, N.K., 2009. Amoxicillin, clarithromycin, and omeprazole based targeted nanoparticles for the treatment of H. pylori. J. Drug Target. 17, 225–234.

Ramya, K., Sekhar, C.H.C., Sastry, D.S., 2015. Solid lipid nanoparticles—future technology: a brief review. W.J.P.R. 4, 439–515.

Rani, R., Dilbaghi, N., Dhingra, D., Kumar, S., 2015. Optimization and evaluation of bioactive drug-loaded polymeric nanoparticles for drug delivery. Int. J. Biol. Macromol. 78, 173–179.

Richardson, M.D., 2005. Changing patterns and trends in systemic fungal infections. J. Antimicrob. Chemother. 56 (S 1), i5–i11.

Rizzello, L., Cingolani, R., Pompa, P.P., 2013. Nanotechnology tools for antibacterial materials. Nanomedicine 8 (5), 807–821.

Roberts, J., Bingham, J., McLaren, A.C., McLemore, R., 2015. Liposomal formulation decreases toxicity of amphotericin b in vitro and in vivo. Clin. Orthop. Relat. Res. 473, 2262–2269.

Rodloff, A.C., Kock, D., Schaumann, R., 2011. Epidemiology and antifungal resistance in invasive candidiasis. Eur. J. Med. Res. 16, 187–195.

Rossolini, G.M., Arena, F., Pecile, P., Pollini, S., 2014. Update on the antibiotic resistance crisis. Curr. Opin. Pharmacol. 18, 56–60.

Rostamia, E., Kashanianb, S., Azandaryanic, A.H., Faramarzia, H., Dolatabadie, J.E.N., Omidfarf, K., 2014. Drug targeting using solid lipid nanoparticles. Chem. Phys. Lipids 181, 56–61.

Rotov, K.A., Tikhonov, S.N., Alekseev, V.V., Snatenkov, E.A., 2012. Pharmacokinetics of liposomal gentamicin. Bull. Exp. Biol. Med. 153, 475–477.

Rukholm, G., Mugabe, C., Azghani, A.O., Omri, A., 2006. Antibacterial activity of liposomal gentamicin against *Pseudomonas aeruginosa*: a time–kill study. Int. J. Antimicrob. Agents 27, 247–252.

Rupenagunta, A., Somasundaram, I., Ravichandiram, V., Kausalya, J., Senthilnathan, B., 2011. Solid lipid nanoparticles: a versatile carrier system. J. Pharm. Res. 4, 2069–2075.

Russell, E.L., Fothergill, A., 2015. Antifungal Agents. In: Hospenthal, D.R., Rinaldi, M.G. (Eds.), Diagnosis and Treatment of Fungal Infections. Springer International Publishing, Cham, pp. 79–97.

Sachetelli, S., Khalil, H., Chen, T., Beaulac, C., Sénéchal, S., Lagacé, J., 2000. Demonstration of a fusion mechanism between a fluid bactericidal liposomal formulation and bacterial cells. Biochim. Biophys. Acta 1463, 254–266.

Sahoo, S.K., Labhasetwar, V., 2003. Nanotech approaches to drug delivery and imaging. Drug Discov. Today 8, 1112–1120.

Salem, I.I., Flasher, D.L., Düzgüneş, N., 2005. Liposome-encapsulated antibiotics. Method. Enzymol. 391, 261–291.

Sande, L., Sanchez, M., Montes, J., Wolf, A.J., Morgan, M.A., Omri, A., Liu, G.Y., 2012. Liposomal encapsulation of vancomycin improves killing of methicillin-resistant *Staphylococcus aureus* in a murine infection model. J. Antimicrob. Chemother. 67, 2191–2194.

Sanna, V., Gavini, E., Cossu, M., Rassu, G., Giunchedi, P., 2007. Solid lipid nanoparticles (SLN) as carriers for the topical delivery of econazole nitrate: in-vitro characterization, ex-vivo and in-vivo studies. J. Pharm. Pharmacol. 59, 1057–1064.

Sarig, H., Ohana, D., Epand, R.F., Mor, A., Epand, R.M., 2011. Functional studies of cochleate assemblies of an oligo-acyl-lysyl with lipid mixtures for combating bacterial multidrug resistance. FASEB J. 25, 3336–3343.

Schentag, J.J., Gilliland, K.K., Paladino, J.A., 2001. What have we learned from pharmacokinetic and pharmacodynamic theories? Clin. Infect. Dis. 32 (S1), S39–S46.

Schwarz, J.C., Kählig, H., Matsko, N.B., Kratzel, M., Husa, M., Valenta, C., 2011. Decrease of liposomal size and retarding effect on fluconazole skin permeation by lysine derivatives. J. Pharm. Sci 100, 2911–2919.

Seil, J.T., Webster, T.J., 2012. Antimicrobial applications of nanotechnology: methods and literature. Int. J. Nanomed. 7, 2767–2781.

Seltmann, G., Holst, O., 2002. The Bacterial Cell Wall, first ed. Springer-Verlag, Heidelberg.

Shafaa, M.W., 2011. Efficacy of topically applied liposome-bound tetracycline in the treatment of dry eye model. Vet. Ophthalmol. 14 (1), 18–25.

Shah, M., Agrawal, Y., 2012. Ciprofloxacin hydrochloride-loaded glyceryl monostearate nanoparticle: factorial design of Lutrol F68 and Phospholipon 90G. J. Microencapsul. 29, 331–343.

Shao, K., Wu, J., Chen, Z., Huang, S., Li, J., Ye, L., Lou, J., Zhu, L., Jiang, C., 2012. A brain-vectored angiopep-2 based polymeric micelles for the treatment of intracranial fungal infection. Biomaterials 33, 6898–6907.

Sharma, A., Kumar Arya, D., Dua, M., Chhatwal, G.S., Johri, A.K., 2012. Nano-technology for targeted drug delivery to combat antibiotic resistance. Exp. Opin. Drug Deliv. 9, 1325–1332.

Shrestha, H., Bala, R., Arora, S., 2014. Lipid-based drug delivery systems. J. Pharm. 2014, 801820.

Silva, N.C., Silva, S., Sarmento, B., Pintado, M., 2013. Chitosan nanoparticles for daptomycin delivery in ocular treatment of bacterial endophthalmitis. Drug Deliv. 22, 885–893.

Singh, B., Vuddanda, P.R., Vijayakumar, M.R., Kumar, V., Saxena, P.S., Singh, S., 2014. Cefuroxime axetil loaded solid lipid nanoparticles for enhanced activity against S. aureus biofilm. Colloids Surf. B 121, 92–98.

Skiada, A., Rigopoulos, D., Larios, G., Petrikkos, G., Katsambas, A., 2012. Global epidemiology of cutaneous zygomycosis. Clin. Dermatol. 30, 628–632.

Smart, J.D., 2005. The basics and underlying mechanisms of mucoadhesion. Adv. Drug Deliv. Rev. 57, 1556–1568.

Smith, A.W., 2005. Biofilms and antibiotic therapy: is there a role for combating bacterial resistance by the use of novel drug delivery systems? Adv. Drug Deliv. Rev. 57, 1539–1550.

Solleti, V.S., Alhariri, M., Halwani, M., Omri, A., 2015. Antimicrobial properties of liposomal azithromycin for *Pseudomonas* infections in cystic fibrosis patients. J. Antimicrob. Chemother. 70, 784–796.

Song, G., Wu, H., Yoshino, K., Zamboni, W.C., 2012. Factors affecting the pharmacokinetics and pharmacodynamics of liposomal drugs. J. Lipos. Res. 22, 177–192.

Souto, E.B., Müller, R.H., 2006. The use of SLN and NLC as topical particulate carriers for imidazole antifungal agents. Pharmazie 61, 431–437.

Stinner, D.J., Noel, S.P., Haggard, W.O., Watson, J.T., Wenke, J.C., 2010. Local antibiotic delivery using tailorable chitosan sponges: the future of infection control? Orthop. Trauma 24, 592–597.

Sudhakar, B., Ravi Varma, J.N., Ramana Murthy, K.V., 2014. Formulation, characterization and ex vivo studies of terbinafine HCl liposomes for cutaneous delivery. Curr. Drug Deliv. 11, 521–530.

Syed, U.M., Woo, A.F., Plakogiannis, F., Jin, T., Zhu, H., 2008. Cochleates bridged by drug molecules. Int. J. Pharm. 363 (1), 118–125.

Takazono, T., Izumikawa, K., Mihara, T., Kosai, K., Saijo, T., Imamura, Y., Miyazaki, T., Seki, M., Kakeya, H., Yamamoto, Y., Yanagihara, K., Kohno, S., 2009. Efficacy of combination antifungal therapy with

intraperitoneally administered micafungin and aerosolized liposomal amphotericin B against murine invasive pulmonary aspergillosis. Antimicrob. Agents Chemother. 53, 3508–3510.
Tan, M.L., Choong, P.F., Dass, C.R., 2010. Recent developments in liposomes, microparticles and nanoparticles for protein and peptide drug delivery. Peptides 31, 184–193.
Taylor, R.L., Williams, D.M., Craven, P.C., Graybill, J.R., Drutz, D.J., Magee, W.E., 1982. Amphotericin B in liposomes: a novel therapy for histoplasmosis. Am. Rev. Respir. Dis. 125, 610–611.
Thamphiwatana, S., Fu, V., Zhu, J., Lu, D., Gao, W., Zhang, L., 2013. Nanoparticle-stabilized liposomes for pH-responsive gastric drug delivery. Langmuir 29, 12228–12233.
Thomas, L., Viswanad, V., 2012. Formulation and optimization of clotrimazole-loaded proniosomal gel using 32 factorial design. Sci. Pharm. 80, 731.
Tikhonov, S.N., Rotov, K.A., Alekseev, V.V., Snatenkov, E.A., Khrapova, N.P., 2010. Activity of some enzymes in the liver of experimental animals after treatment with the liposomal formulation of tetracycline and streptomycin. Bull. Exp. Biol. Med. 149, 47–49.
Toth, I., 1994. A novel chemical approach to drug delivery: lipidic amino acid conjugates. J. Drug Target. 2, 217–239.
Trafny, E.A., Stepińska, M., Antos, M., Grzybowski, J., 1995. Effects of free and liposome-encapsulated antibiotics on adherence of *Pseudomonas aeruginosa* to collagen type I. Antimicrob. Agents Chemother. 39, 2645–2649.
Üner, M., 2006. Preparation, characterization and physico-chemical properties of solid lipid nanoparticles (SLN) and nanostructured lipid carriers (NLC): their benefits as colloidal drug carrier systems. Pharmazie 61, 375–386.
Üstündag˘-Okur, N., Gökçe, E.H., Bozbıyık, D.İ., Eğrilmez, S., Özer, Ö., Ertan, G., 2014. Preparation and in vitro–in vivo evaluation of ofloxacin loaded ophthalmic nano structured lipid carriers modified with chitosan oligosaccharide lactate for the treatment of bacterial keratitis. Eur. J. Pharm. Sci. 63, 204–215.
Vaghasiya, H., Kumar, A., Sawant, K., 2013. Development of solid lipid nanoparticles based controlled release system for topical delivery of terbinafine hydrochloride. Eur. J. Pharm. Sci. 49, 311–322.
Valenta, C., 2005. The use of mucoadhesive polymers in vaginal delivery. Adv. Drug Deliv. Rev. 57, 1692–1712.
Vallet, C.M., Marquez, B., Ngabirano, E., Lemaire, S., Mingeot-Leclercq, M.P., Tulkens, P.M., van Bambeke, F., 2011. Cellular accumulation of fluoroquinolones is not predictive of their intracellular activity: studies with gemifloxacin, moxifloxacin and ciprofloxacin in a pharmacokinetic/pharmacodynamic model of uninfected and infected macrophages. Int. J. Antimicrob. Agents 38, 249–256.
van Bambeke, F., Barcia-Macay, M., Lemaire, S., Tulkens, P.M., 2006. Cellular pharmacodynamics and pharmacokinetics of antibiotics: Current views and perspectives. Curr. Opin. Drug Discov. Devel. 9, 218–230.
Vanic´, Ž., Hurler, J., Ferderber, K., Golja Gašparovic´, P., Škalko-Basnet, N., Filipovic´-Grcˇic´, J., 2014. Novel vaginal drug delivery system: deformable propylene glycol liposomes-in-hydrogel. J. Lipos. Res. 24, 27–36.
Varshosaz, J., Ghaffari, S., Mirshojaei, S.F., Jafarian, A., Atyabi, F., Kobarfard, F., Azarmi, S., 2013. Biodistribution of amikacin solid lipid nanoparticles after pulmonary delivery. BioMed Res. Int. 2013, 136859.
Verma, P., Pathak, K., 2012. Nanosized ethanolic vesicles loaded with econazole nitrate for the treatment of deep fungal infections through topical gel formulation. Nanomedicine 8, 489–496.
Vighi, E., Ruozi, B., Montanari, M., Battini, R., Leo, E., 2010. pDNA condensation capacity and in vitro gene delivery properties of cationic solid lipid nanoparticles. Int. J. Pharm. 389, 254–261.
Wagh, V.D., Deshmukh, O.J., 2012. Itraconazole niosomes drug delivery system and its antimycotic activity against *Candida albicans*. ISRN Pharm. 2012, 653465.
Wallace, S.J., Nation, R.L., Li, J., Boyd, B.J., 2013. Physicochemical aspects of the coformulation of colistin and azithromycin using liposomes for combination antibiotic therapies. J. Pharm. Sci. 102, 1578–1587.
Walsh, T.J., Teppler, H., Donowitz, G.R., Maertens, J.A., Baden, L.R., Dmoszynska, A., Cornerly, O.A., Borque, M.R., Lupinacci, R.J., Sable, C.A., dePauw, B.E., 2004. Caspofungin versus liposomal amphotericin B for empirical antifungal therapy in patients with persistent fever and neutropenia. N. Engl. J. Med. 351, 1391–1402.

Wang, J., Huang, G., 2011. Preparation of itraconazole-loaded liposomes coated by carboxymethyl chitosan and its pharmacokinetics and tissue distribution. Drug Deliv. 18, 631–638.

Wang, J.S., Zhu, J.B., Lu, R.Q., Shen, W., 2005. Preparation of lung targeting azithromycin liposomes and its tissue distribution in mice. Acta Pharm. Sinica 40, 274–278.

Warnock, D.W., 2007. Trends in the epidemiology of invasive fungal infections. Japanese J. Med. Mycol. 48, 1–12.

Weers, J., Metzheiser, B., Taylor, G., Warren, S., Meers, P., Perkins, W.R., 2009. A gamma scintigraphy study to investigate lung deposition and clearance of inhaled amikacin-loaded liposomes in healthy male volunteers. J Aerosol. Med. Pulm. Drug Deliv. 22, 131–138.

Wissing, S.A., Müller, R.H., 2003. Cosmetic applications for solid lipid nanoparticles (SLN). Int. J. Pharm. 254, 65–68.

Wissing, S., Lippacher, A., Müller, R., 2001. Investigations on the occlusive properties of solid lipid nanoparticles (SLN). J. Cosmet. Sci. 52, 313–324.

Xia, W., Liu, P., Zhang, J., Chen, J., 2011. Biological activities of chitosan and chitooligosaccharides. Food Hydrocolloid. 25, 170–179.

Xie, S., Zhu, L., Dong, Z., Wang, X., Wang, Y., Li, X., Zhou, W., 2011a. Preparation, characterization and pharmacokinetics of enrofloxacin-loaded solid lipid nanoparticles: influences of fatty acids. Colloids Surf. B 83, 382–387.

Xie, S., Zhu, L., Dong, Z., Wang, Y., Wang, X., Zhou, W., 2011b. Preparation and evaluation of ofloxacin-loaded palmitic acid solid lipid nanoparticles. Int. J. Nanomed. 6, 547–555.

Xing, C., Levchenko, T., Guo, S., Stins, M., Torchilin, V.P., Lo, E.H., 2012. Delivering minocycline into brain endothelial cells with liposome-based technology. J. Cereb. Blood Flow Metab. 32, 983–988.

Xiong, M.H., Bao, Y., Yang, X.Z., Zhu, Y.H., Wang, J., 2014. Delivery of antibiotics with polymeric particles. Adv. Drug Deliv. Rev. 78, 63–76.

Xu, X.M., Wang, Y.S., Chen, R.Y., Feng, C.L., Yao, F., Tong, S.S., Wang, L., Yamashita, F., Yu, J.N., 2011. Formulation and pharmacokinetic evaluation of tetracycline-loaded solid lipid nanoparticles for subcutaneous injection in mice. Chem. Pharm. Bull. 59, 260–265.

Yadav, A.V., Bhise, S.B., 2004. Chitosan: a potential biomaterial effective against typhoid. Curr. Sci. India 87, 1176–1178.

Yadav, N., Khatak, S., Sara, U.V.S., 2013. Solid lipid nanoparticles: a review. I.J.A.P. 5, 8–18.

Yamashita, F., Hashida, M., 2013. Pharmacokinetic considerations for targeted drug delivery. Adv. Drug Deliv. Rev. 65, 139–147.

Yang, W., Wiederhold, N.P., Williams, III, R.O., 2008. Drug delivery strategies for improved azole antifungal action. Exp. Opin. Drug Deliv. 5, 1199–1216.

Yang, D., Xu, Y., Li, F., Liu, H., He, X., 2013. Preparation of cationic vancomycin hydrochloride multivesicular liposomes and its quality. Chinese J. Repar. Reconst. Surg. 27, 443–448.

Yang, Z., Liu, J., Gao, J., Chen, S., Huang, G., 2015. Chitosan coated vancomycin hydrochloride liposomes: characterizations and evaluation. Int. J. Pharm. 495, 508–515.

Yu, W., Liu, C., Liu, Y., Zhang, N., Xu, W., 2010. Mannan-modified solid lipid nanoparticles for targeted gene delivery to alveolar macrophages. Pharm. Res. 27, 1584–1596.

Yu, S.H., Wu, S.J., Wu, J.Y., Wen, D.Y., Mi, F.L., 2015. Preparation of fucoidan-shelled and genipin-crosslinked chitosan beads for antibacterial application. Carbohydr. Polym. 126, 97–107.

Zaki, N.M., Hafez, M.M., 2012. Enhanced antibacterial effect of ceftriaxone sodium-loaded chitosan nanoparticles against intracellular *Salmonella typhimurium*. AAPS PharmSciTech 13, 411–421.

Zambito, Y., 2013. Nanoparticles based on chitosan derivatives. In: Pignatello, R. (Ed.), Advances in Biomaterials Science and Biomedical Applications. InTech Publisher, pp. 244–263.

Zaru, M., Mourtas, S., Klepetsanis, P., Fadda, A.M., Antimisiaris, S.G., 2007. Liposomes for drug delivery to the lungs by nebulization. Eur. J. Pharm. Biopharm. 67, 655–666.

Zaru, M., Manca, M.L., Fadda, A.M., Antimisiaris, S.G., 2009a. Chitosan-coated liposomes for delivery to lungs by nebulisation. Colloids Surf. B 71, 88–95.

Zaru, M., Sinico, C., De Logu, A., Caddeo, C., Lai, F., Manca, M.L., Fadda, A.M., 2009b. Rifampicin-loaded liposomes for the passive targeting to alveolar macrophages: in vitro and in vivo evaluation. J. Lipos. Res. 19, 68–76.

Zhang, X., Sun, P., Bi, R., Wang, J., Zhang, N., Huang, G., 2009. Targeted delivery of levofloxacin-liposomes for the treatment of pulmonary inflammation. J. Drug Target. 17, 399–407.

Zhang, L., Pornpattananangkul, D., Hu, C.M., Huang, C.M., 2010. Development of nanoparticles for antimicrobial drug delivery. Curr. Med. Chem. 17, 585–594.

Zhao, L., Zhu, B., Jia, Y., Hou, W., Su, C., 2013. Preparation of biocompatible carboxymethyl chitosan nanoparticles for delivery of antibiotic drug. BioMed Res. Int. 2013, 236469.

Zhong, M., Feng, Y., Liao, H., Hu, X., Wan, S., Zhu, B., Zhang, M., Xiong, H., Zhou, Y., Zhang, J., 2014. Azithromycin cationic non-lecithoid nano/microparticles improve bioavailability and targeting efficiency. Pharm. Res. 31, 2857–2867.

Zhou, W., Wang, Y., Jian, J., Song, S., 2013. Self-aggregated nanoparticles based on amphiphilic poly (lactic acid)-grafted-chitosan copolymer for ocular delivery of amphotericin B. Int. J. Nanomed. 8, 3715.

Zhu, Bd., Qie, Yq., Wang, Jl., Zhang, Y., Wang, Qz., Xu, Y., Wang, Hh., 2007. Chitosan microspheres enhance the immunogenicity of an Ag85B-based fusion protein containing multiple T-cell epitopes of *Mycobacterium tuberculosis*. Eur. J. Pharm. Biopharm. 66, 318–326.

Zhu, C.T., Xu, Y.Q., Shi, J., Li, J., Ding, J., 2010. Liposome combined porous β-TCP scaffold: Preparation, characterization, and anti-biofilm activity. Drug Deliv. 17, 391–398.

Ziora, Z.M., Blaskovich, M.A., Toth, I., Cooper, M.A., 2012. Lipoamino acids as major components of absorption promoters in drug delivery. Curr. Top. Med. Chem. 14, 1562–1580.

FURTHER READING

Liu, D., Yang, P., Hu, D., Liu, F., 2013a. Minocycline hydrochloride liposome controlled-release gel improves rat experimental periodontitis. W. China J. Stomatol. 31, 592–596.

James, G., Fisher, S., Perkins, W.R., 2008. Biofilm penetration, triggered release and in vivo activity of inhaled liposomal amikacin in chronic *Pseudomonas aeruginosa* lung infections. J. Antimicrob. Chemother. 61, 859–868.

Gangadhar, K.N., Adhikari, K., Srichana, T., 2014. Synthesis and evaluation of sodium deoxycholate sulfate as a lipid drug carrier to enhance the solubility, stability and safety of an amphotericin B inhalation formulation. Int. J. Pharm. 471, 430–438.

Gupta, M., Vyas, S.P., 2012. Development, characterization and in vivo assessment of effective lipidic nanoparticles for dermal delivery of fluconazole against cutaneous candidiasis. Chem. Phys. Lipids 165, 454–461.

Jøraholmen, M.W., Vanic´, Ž., Tho, I., Škalko-Basnet, N., 2014. Chitosan-coated liposomes for topical vaginal therapy: assuring localized drug effect. Int. J. Pharm. 472, 94–101.

Mackinson, C., Ambrose, P.G., Gupta, R., 2009. Pharmacokinetic and pharmacodynamic evaluation of liposomal amikacin for inhalation in cystic fibrosis patients with chronic pseudomonal infection. Antimicrob. Agents Chemother. 53, 3847–3854.

Mathpal, D., Garg, T., Rath, G., Goyal, A.K., 2015. Development and characterization of spray dried microparticles for pulmonary delivery of antifungal drug. Curr. Drug Deliv. 12, 464–471.

Mendes, A.I., Silva, A.C., Catita, J.A.M., Cerqueira, F., Gabriel, C., Lopes, C.M., 2013. Miconazole-loaded nanostructured lipid carriers (NLC) for local delivery to the oral mucosa: improving antifungal activity. Colloids Surf. B 111, 755–763.

CHAPTER

24

DRUG DELIVERY MEDIATED BY CONFINED NANOSYSTEMS: STRUCTURE-ACTIVITY RELATIONS AND FACTORS RESPONSIBLE FOR THE EFFICACY OF FORMULATIONS

Lucia Zakharova, Tatiana Pashirova, Ruslan Kashapov, Dinar Gabdrakhmanov, Oleg Sinyashin

A.E. Arbuzov Institute of Organic and Physical Chemistry, Kazan Scientific Center, Russian Academy of Sciences, Kazan, Russian Federation

CHAPTER OUTLINE

1 INTRODUCTION

Engineering of drug delivery systems has received much attention in recent years. Rational design of drug nanocarriers is of paramount importance due to several reasons, connected with (1) low water solubility of a majority of therapeutics, (2) existence of different biological barriers on the drug molecule way toward a target, (3) protection of drugs against biological environment, (4) targeted delivery of therapeutics to deceased cell, (5) controlled release and modulation of farmacokinetics, and so forth. Generally, two main approaches are used for the development of drug carriers. First way involves binding the drug molecules with amphiphilic or polymer nanocontainers formed through self-association mechanism or coating technology. Second is the so-called prodrug approach, which assumes chemical conjugation of drug molecules to a targeting ligand. This chapter focuses on noncovalent strategy, while a separate research direction devoted to the problem of bioconjugation remains out of this work. Reddy and Bazile (2014) discuss this problem on the example of taxane formulations, with the ADME (absorption, distribution, metabolism, excretion) criteria emphasized. These authors consider three generations of drug formulations, including vehicles based on surfactants, surfactant-free carriers, and formulations with improved pharmacokinetics through noncovalent or chemical drug/carrier binding. Surfactant- or, wider, amphiphile-based formulations are of key significance in the drug delivery engineering. Such nanocarriers as liposomes, niosomes, nano- and microemulsions came into general applications in nanotechnologies. On the other hand, numerous limitations typical of self-assembling confined systems are avoided in the case of polymeric materials. Therefore, separate subsections of the chapter are directed toward amphiphile-mediated drug delivery and polymer-based strategies for biotransport of medicines, especially protein and peptides. Importantly, nanocarrier-mediated drug delivery makes it possible to balance between therapeutic efficacy and possible side effects in the medical treatment. At the same time, nanocarrier-mediated therapy in vivo remains a challenging task due to multifactor mechanism of the drug biotransport. As proposed in Zou et al. (2015) and Blum et al. (2015), one of the promising ways to solve this problem is to develop nanocarriers sensitive to environmental conditions. Stimuli-responsive drug delivery systems are addressed here in a separate subsection.

To give more information on recent trends in formulated drug delivery, design of nanocarriers involving biomolecules and/or naturally derived species should be mentioned. This can be exemplified by review that covered the development of drug delivery systems based on naturally occurring neutroxines to bypass the blood–brain barrier (BBB) (Soddu et al., 2015). In Tashima (2015) transporter-conscious approach is strongly encouraged, which can provide benefits in ADMET criteria. Fuhrmann et al. (2015) review the studies focusing on the extracellular vesicles as targeted drug delivery vectors, classified as exosomes, and their application in diagnostics and therapy.

2 SURFACTANT-BASED FORMULATIONS; STRUCTURE-ACTIVITY RELATIONS AS CRITERIA FOR THE CHOICE OF BUILDING BLOCKS

Surfactant-based technologies are of particular importance in nanomedicine. Proper choice of amphiphilic agents for the fabrication of drug nanocarriers may markedly improve the binding efficacy of drug molecules and their biological fate. Therefore, the elucidation of the structure-activity correlations is of key importance in surfactant-mediated drug delivery.

2.1 NONIONIC AMPHIPHILES

According to green chemistry criteria much attention is paid to surfactants with low toxicity and low aggregation threshold. These trends are taken into account in our recent researches directed toward construction of self-assembling nanosystems capable of incorporating nonpolar bioactive molecules and probes. In this context, particular emphasis is being put on the low toxic formulations based on nonionic amphiphiles, including conventional C_nE_m nonionic surfactants (here n is the number of carbon atom in alkyl tails; E is the amount of oxyethylene fragments in polar group) (Ablakova et al., 2010; Mirgorodskaya et al., 2010; Zakharova et al., 2012a) or widely used Tween (Mirgorodskaya et al., 2014a; Valeeva et al., 2015) and/or Triton (Kudryashova et al., 2011) type surfactants, oligomers (e.g., Tyloxapol) or block-copolymers (e.g., Pluronic, Synperonic) (Ablakova et al., 2010; Mirgorodskaya et al., 2010), as well as oxyethylated calixarenes (Antipin et al., 2009; Kudryashova et al., 2012; Mirgorodskaya et al., 2014b; Zakharova et al., 2010a,b) (see the formulas in Fig. 24.1). Beneficial solubilization properties of nonionic amphiphiles and their mixed systems with cationic analogs in respect of fatty acid esters are demonstrated in Ablakova et al. (2010) and Mirgorodskaya et al. (2010). Maximal solubilizing effect is achieved in microemulsions based on Brij-97, with mutual influence of solute and nanocontainers revealed. In particular, changes in size characteristics of aggregates and pK_a shift of acids are shown. Micelles of oligomeric nonionic amphiphile Tyloxapol demonstrated advanced solubilization efficacy and catalytic activity toward *p*-nitrophenyl esters of carbon acids (Ablakova et al., 2010). Synergetic effect is shown to occur upon addition of even low concentration of cationic surfactant to nonionic amphiphile (Zakharova et al., 2012a). Strategies involving typical nonionic surfactants (Brij-35, Triton-X-100, Tween-20) have recently been developed for the incorporation of synthetic probes and a drug doxorubicin by water insoluble derivatives of biodegradable hyperbranched polyesters polyols Boltorn H(10-30) (Khannanov et al., 2016; Kutyreva et al., 2015).

2.2 GEMINI SURFACTANTS

Such kind of amphiphiles as geminis composed of two head groups with long-chain alkyl tails coupled with a spacer are the focus of our research activity (Mirgorodskaya et al., 2012a, 2014c,d). Gemini surfactants show reduced critical micelle concentration (cmc) values due to enhanced hydrophobicity, while the variation of the structure of spacer can markedly modify morphological behavior, thereby diversifying functional properties of nanoaggregates (Grigoriev et al., 2012). Considerable variation of properties was shown to be achieved by introducing of hydroxyethyl fragment into head group, which can markedly influence aggregation behavior, catalytic activity, and complexation capacity with synthetic and natural polyanions (Mirgorodskaya et al., 2012a,c; Vasilieva et al., 2014; Zakharova et al., 2014b). Effective tool for the optimization of structural characteristics and functionality is noncovalent modification of single gemini solution with cosurfactant or polymer, which resulted in synergetic effects (Mirgorodskaya et al., 2012c, 2014c).

2.3 CATIONIC SURFACTANTS; AMPHIPHILES WITH NATURAL FRAGMENTS

Our researches focus on the design of cationic surfactants with varied structure of head group (Zakharova et al., 2010c; Zakharova and Konovalov, 2012). It was shown that their aggregation capacity is almost unchanged upon transferring from typical alkyl trimethylammonium bromide (TMAB-n) series to cationic surfactants with morpholinium, chinuclidinium, and diazabicyclooctane-based polar fragments

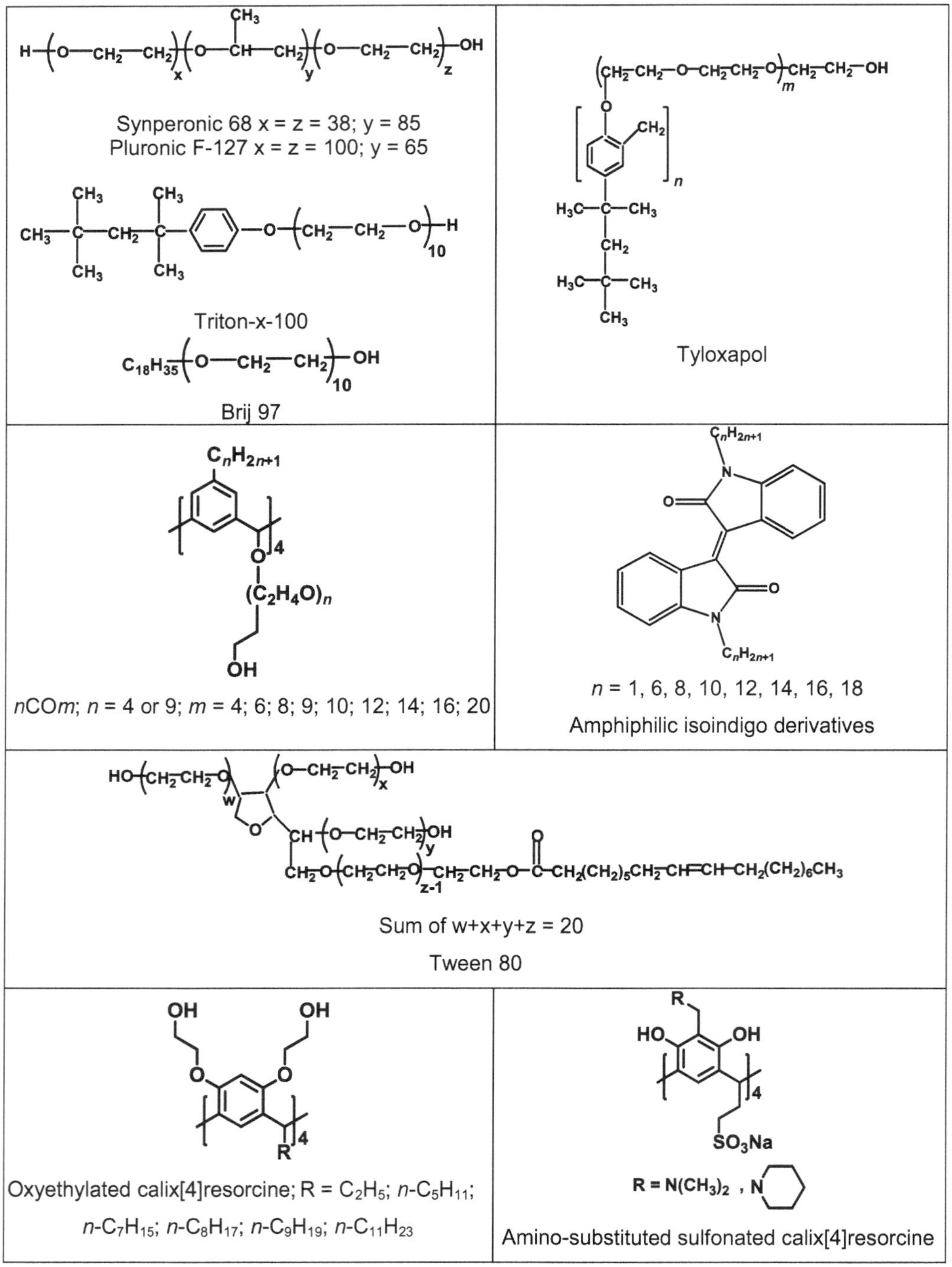

FIGURE 24.1 Structural Formulas of Selected Amphiphiles Used for the Solubilization of Hydrophobic Guests

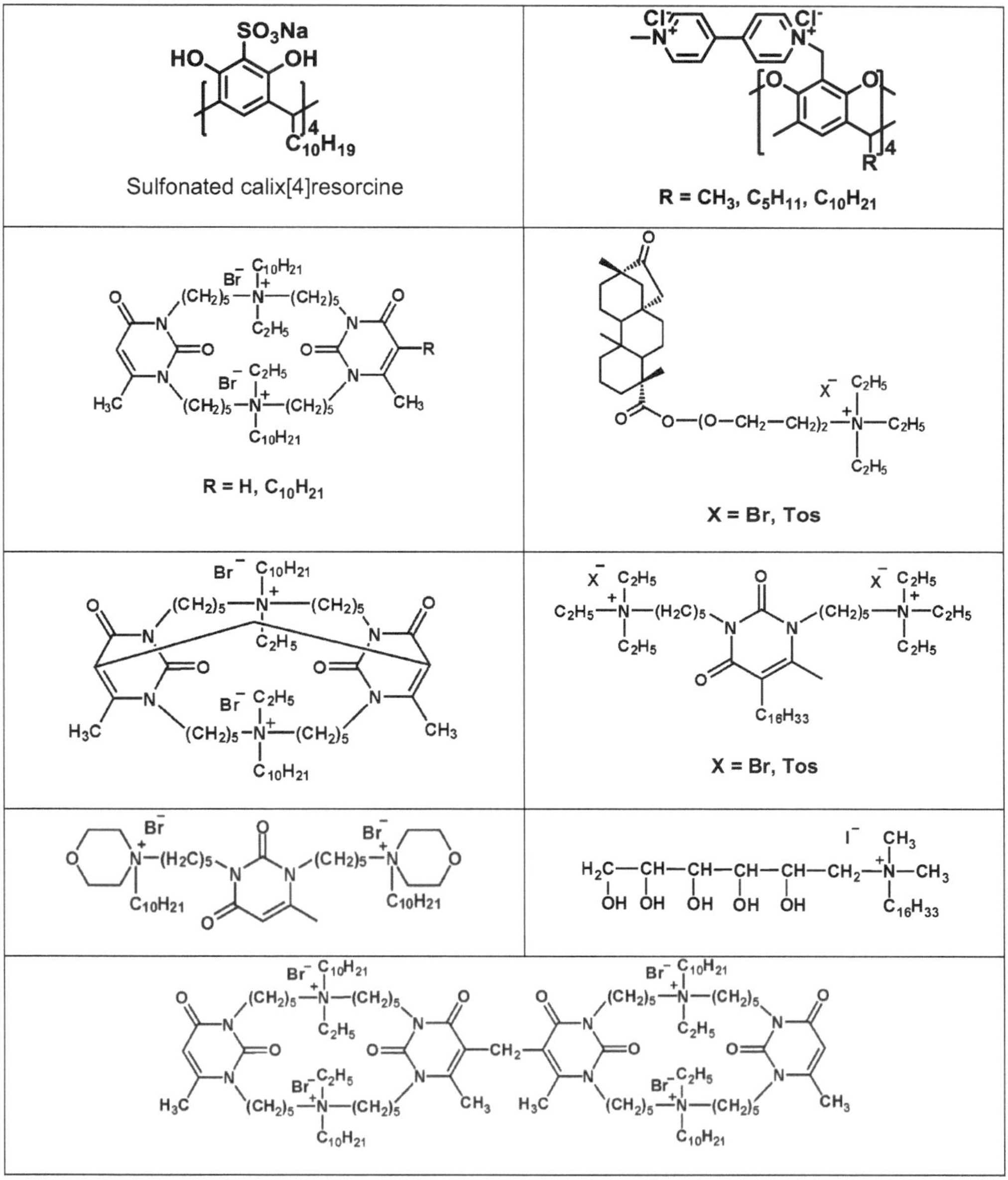

FIGURE 24.1 (*Cont.*)

(Pashirova et al., 2015a; Yackevich et al., 2014; Zakharova et al., 2012c; Zhiltsova et al., 2012). Hydroxyethyl fragments in TMA head group do not markedly affect the aggregation behavior, while dramatically modifies functional activity (Grigoriev et al., 2012; Mirgorodskaya et al., 2012b). Meanwhile for triphenylphosphonium bromide derivatives (TPPB-n), cmc values are ca. 10-fold lower as compared to TMAB-n series (Gaynanova et al., 2012). Beneficial solubilization capacity, catalytic effect, and complexing activity toward oligonucleotide were shown for TPPB-n based nanocontainers (Gaynanova et al., 2012, 2016; Zakharova et al., 2012d).

Promising class of surfactants involves amphiphilic molecules bearing natural fragment, which may increase affinity toward biomolecules and biointerfaces, as well as markedly improve biocompatibility of nanocontainers. Steady research activity is also directed toward cationic amphiphiles due to their effective complexing with negatively charged residues, for example, phosphate backbones of nucleotide unions. Therefore, a separate line of investigations is devoted to cationic surfactants with natural fragments. Specifically, mono- and dicationic amphiphiles bearing uracilic, diterpenoid, aminoacid, isoindigo, and glucamine moieties have been studied in respect of their self-assembly in aqueous solutions and binding the water-insoluble and/or anionic guests. Uracilic amphiphiles including those of macrocycle structure (the so-called uracilophanes) were documented to show specific aggregation characteristics, size and morphological behavior, pH properties, and may interact with oligo- and polynucleotides through multifactor mechanism (Gabdrakhmanov et al., 2015a; Kharlamov et al., 2013a; Voronin et al., 2011; Zakharova et al., 2010b, 2012d,e, 2014a) (Fig. 24.2). In particular, di- and tetracationic uracilophanes were explored by fluorescence spectroscopy, DLS and NMR self-diffusion techniques, which revealed diverse self-assembly behavior. The molecular structure of macrocycles was shown to influence markedly their aggregation ability, with energetic and geometrical factors contributing to the association mechanism (Zakharova et al., 2014a). According to data (Zakharova et al., 2007) association model of uracilophanes is mainly determined by their packing parameters close to 1. Therefore, the rather large aggregates with low surface curvature are formed in aqueous solutions. Such factors as the presence of bulky uracilic fragment and charged nitrogen within the macrocycle result in the noncompact packing mode and low binding degree of counterions.

In work by Bogdanov et al. (2014) amphiphiles with isoindigo moiety (Fig. 24.1) were synthesized with a view of their potential biological activity. Despite the absence of surface active properties these

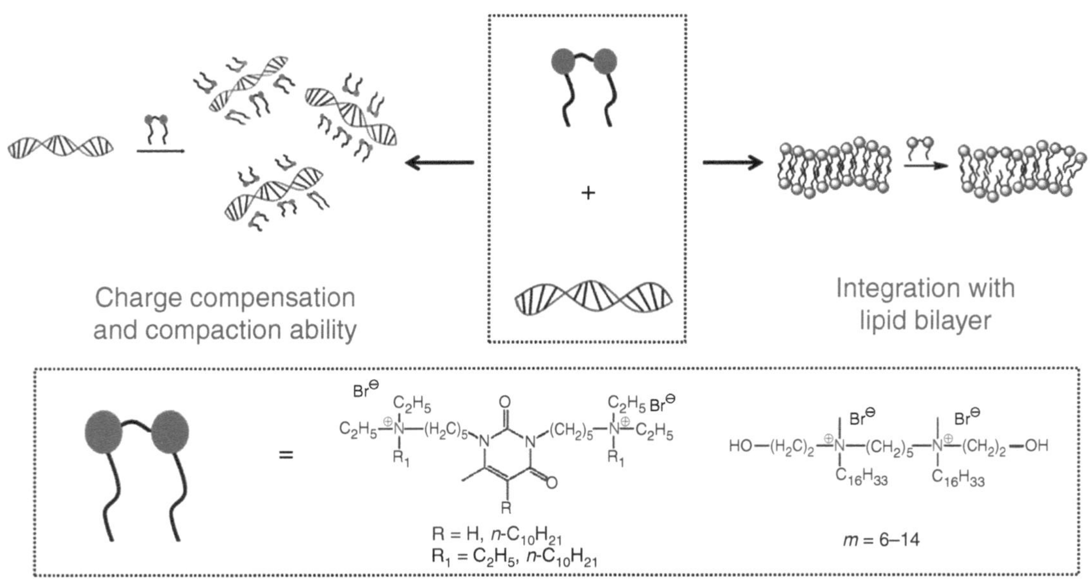

FIGURE 24.2 Schematic Representation of Surfactant-Mediated DNA Complexation

compounds form nanosized aggregates above cmc of ca. 0.1 mM (fluorimetry) (Fig. 24.1; R = $C_{18}H_{37}$) capable of solubilizing hydrophobic probe.

Noteworthy, not only the structure of surface active ion but also the nature of counterion can markedly influence aggregation properties and functional activity of cationic surfactants (Gabdrakhmanov et al., 2013; Voronin et al., 2013). For two diterpenoid surfactants with bromide and tosylate counterions (Fig. 24.1), difference in morphological behavior was revealed by DLS, fluorimetry, and spectrophotometry (Voronin et al., 2013). While similar cmc values occur for both amphiphiles, only tosylate derivative showed typical aggregation behavior with micelle-like particles observed beyond cmc. Surfactant with bromide counterion underwent the micelle-vesicle-micelle transition upon increasing in its concentration (Fig. 24.3). Only bromide analog showed the ability of integrating with lipid bilayer, while both amphiphiles form lipoplexes with DNA decamer (Gabdrakhmanov et al., 2013). As exemplified in Zakharova et al. (2010b) for acyclic uracilic amphiphiles the low value of tosylate counterion binding of 0.51 was measured by NMR against counterion binding of 0.81 for TMA analog with tosylate counterion. Importantly, cationic amphiphiles with sterically hindered charged atoms bring about acidification of bulk solution with an increase in surfactant concentration. This phenomenon observed for TPPB-n series (Gaynanova et al., 2012) and uracilic amphiphiles (Zakharova et al., 2010b) can be of practical importance for the control of microenvironmental conditions.

Specific aggregation and solubilization properties were shown for the glucamine-based amphiphile (Fig. 24.1). Addition of glucamine was shown to decrease the toxicity of surfactant-based formulations (Pashirova et al., 2015b) preserving their biological activity. Therefore, glucamine-containing cationic surfactant is synthesized, that demonstrated fast micellar growth upon increasing in the concentration

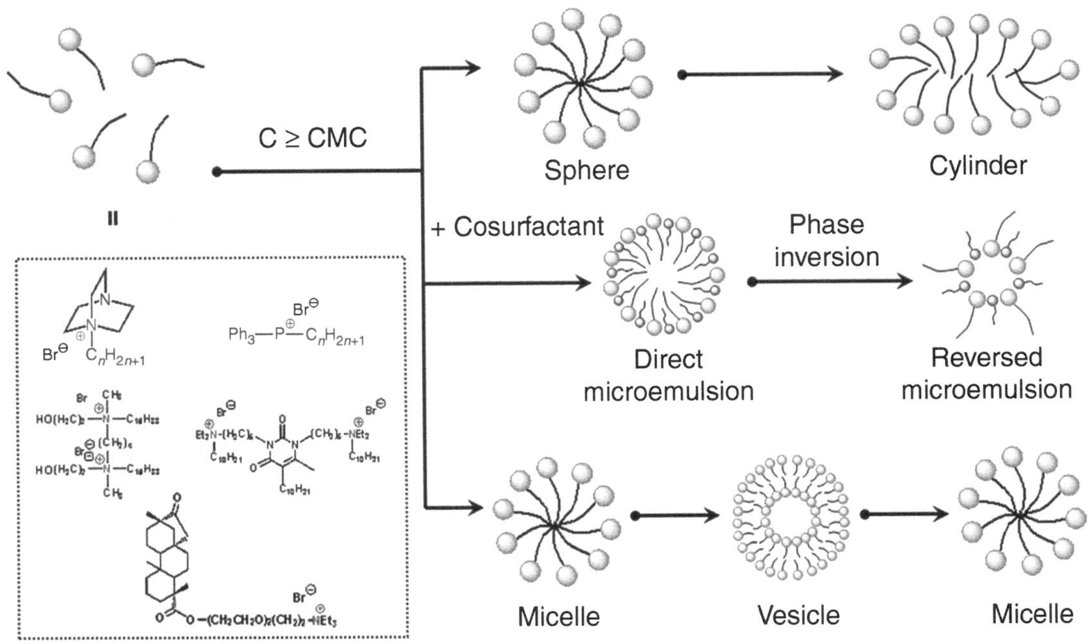

FIGURE 24.3 Schematic Representation of Structural Transitions Occurring in Solution of Surfactants

due to structural rearrangement, as well as potentiality of nanocontainer for hydrophobic substrate (Gabdrakhmanov et al., 2015b,c).

These properties of cationic surfactants (aggregation ability, binding of hydrophobic solutes, and complexation with DNA) may be tuned by the fabrication of mixed systems with helper lipids or polymer–colloid complexes (Kharlamov et al., 2013a; Mirgorodskaya et al., 2012c, 2014b; Vagapova et al., 2013).

2.4 SELF-ASSEMBLING SYSTEMS BASED ON CALIXARENE MATRIX

Even more advantages for the design and engineering of nanocarriers are provided by amphiphiles on calixarene platform. They can be easily functionalized on upper or lower rims, which can markedly modify association model and properties of aggregates formed and guests bound. Importantly, multifactor mechanism of binding the guest molecules can occur in this case. Along with the solubilization binding mode typical for the micelle-like nanocontainers, the guest–host interactions based on the molecular recognition principles can be involved. Additional contribution of H-bonding, π-stacking effect, and van-der-Waals interactions can be engaged through directed functionalization of calixarene cavity. Innovative formulation was developed in Antipin et al. (2009) based on the macrocyclic analogs of nonionic surfactants, amphiphilic oxyethylated calixarenes nCOm (Fig. 24.1). Nanocontainers fabricated demonstrated zero toxicity and enhanced solubilization activity toward hydrophobic drugs, biomolecules, and probes. The structure-activity correlations elucidated that it is the hydrophilic–lipophilic balance that controlled the association model, size, and morphology of aggregates (Zakharova et al., 2010a). The step-by-step design of biomimetic formulations based on nCOm calixarene derivatives made it possible to modify markedly binding capacity and catalytic properties toward ecotoxicants, phosphorus acid esters, and their carbonic acid analogs (Kudryashova et al., 2012; Mirgorodskaya et al., 2014b; Zakharova et al., 2012b). About 200-fold increase in the catalytic efficacy of the formulation was achieved by a stepwise noncovalent modification of 9CO9 with polyethylene imine and lanthanum ions and immobilization of nanoreactors on mesoporous silica (Zakharova et al., 2012b). Amino acid functionalized calix[4]resorcines were synthesized by Kashapov et al. (2015a,b), with their solubility improved by admixing with surfactants. Noteworthy, calixarene-surfactant systems provide unique possibility for the development of tunable nanocarriers and nanoreactors demonstrating versatile structural behavior and synergetic functional activity (Kashapov et al., 2011; Kharlamov et al., 2013b; Pashirova et al., 2014a; Zakharova et al., 2009).

Due to the presence of bulky cavity calixarene molecules demonstrate specific association model, differing from typical surfactants. This makes them promising building blocks for the fabrication of nanocontainers with stimuli responsive properties (Gaynanova et al., 2015; Kashapov et al., 2011, 2014; Kharlamov et al., 2013b; Sultanova et al., 2015). In Pashirova et al. (2014b) amphiphilic oxyethylated calix[4]resorcines were synthesized, with surface activity and their morphological behavior strongly controlled by the hydrophobicity of molecules and the nature of solvent. Strong structure-activity dependence is documented for this homological series of amphiphilic calix[4]arenes differing in alkyl chain length, with pentyl tail holding the intermediate position. The variation of association model occurs upon transition from short-tail (<C5) to long-tail (>C5) calixarene derivatives. Low and high homologies exhibited structural transition between large aggregates of 300–400 nm and small micelle-like nanocontainers with nonpolar interior. Unlike the former, the latter was shown to be capable of incorporating hydrophobic dye and drug nitrofurantoin.

3 CATIONIC SURFACTANTS AS AGENTS FOR DNA TRANSFECTION

Cationic surfactants are widely used as DNA carriers due to their high affinity toward negatively charged phosphate backbones of nucleic acids. Cationic amphiphiles along with positively charged polymers (polyethyleneimine, chitosan, etc.) belong to the so-called nonviral vectors (synthetic compounds for DNA delivery). The latter were designed as alternative agents to viral transfection systems and exhibit such advantages as neutralization of DNA charge, compaction of polyanion, and lack of immune response. However, there are several problems limiting their further application as agents for gene delivery, the main one of which is low transfection efficacy as compared with their viral analogs. Therefore the search for novel cationic surfactants and development of simple protocols for gene delivery based on conventional surfactants remains a challenge.

Development of novel agents for DNA delivery on the basis of amphiphiles has progressed in several directions. Much attention is paid to the enhancement of DNA-surfactant electrostatic interactions by means of using the multicharged cationic amphiphiles. Our recent studies (Fayzullin et al., 2015; Gabdrakhmanov et al., 2013, 2015a; Grigoriev et al., 2012; Zakharova et al., 2012d, 2016) have explored an alternative approach based on the application of uracilic amphiphiles capable of complementary interactions with nucleotide units. Significantly, the presence of natural fragments in nanocarriers provides them biocompatible properties.

Design of novel compositions consisting of monocationic surfactant and helper agent is one of the key trends as well. Recently cationic liposomes based on dioctadecyldimethylammonium bromide (DODAB) and 1-monooleoyl-rac-glycerol (MO) have been reported as a perspective DNA carriers, exhibiting high transfection efficacy of the 293T mammalian cell line with pSV-β-gal plasmid DNA. Researchers reported about structural changes in morphology of complexes DODAB:MO depending on the molar ratio of components and incubation temperature (25 and 50°C). Transfection efficiency of DODAB:MO composition was shown to be almost at the same level as that of commercially available Lipofectamine (Silva et al., 2014). Surfactants with shorter hydrophobic chain length also find their application in the design of agents for DNA delivery. For instance, composition on the basis of didodecyldimethylammonium bromide, oleic acid, and cholesterol was reported. In the case of some cell cultures, this composition demonstrates more pronounced transfection efficacy as compared with Lipofectamine 2000 (Jin et al., 2015). Along with researches focusing on the optimization of formulations based on conventional surfactants, design and synthesis of cationic surfactants of desirable structure occupy a special place in modern literature. In Patra et al. (2014) 3-(dodecylcarbamoyl-2-hydroxypropyl)-trimethylammonium chloride (Fig. 24.4) was reported to self-assemble with vesicle formation with hydrodynamic diameter 30–70 nm. It was shown that the surfactant and cholesterol in 5:1 mole ratio are able to transfer DNA into the COS-1 cells.

In recent studies much attention is devoted to gemini surfactants composed of more than one head group and hydrophobic tail. There are examples of successful application of *m-s-m* type gemini surfactants (Menger and Littau, 1993) and helper lipids (cholesterol). It has been shown that in the case of individual gemini surfactants the transfection efficiency (TE) of HeLa cells is approximately 30%, while addition of cholesterol may increase it up to 50% (Cardoso et al., 2014). For increasing the biocompatibility and biodegradability of surfactant/DNA complexes amino acid-functionalized gemini surfactants are synthesized. For example, in Cardoso et al. (2015) serine-derived gemini surfactants were tested as gene delivery systems. Authors managed to achieve formulations with high transfection efficacy and zero toxicity at low surfactant/DNA charge ratios (1/1–2/1). Principally new type of gene delivery

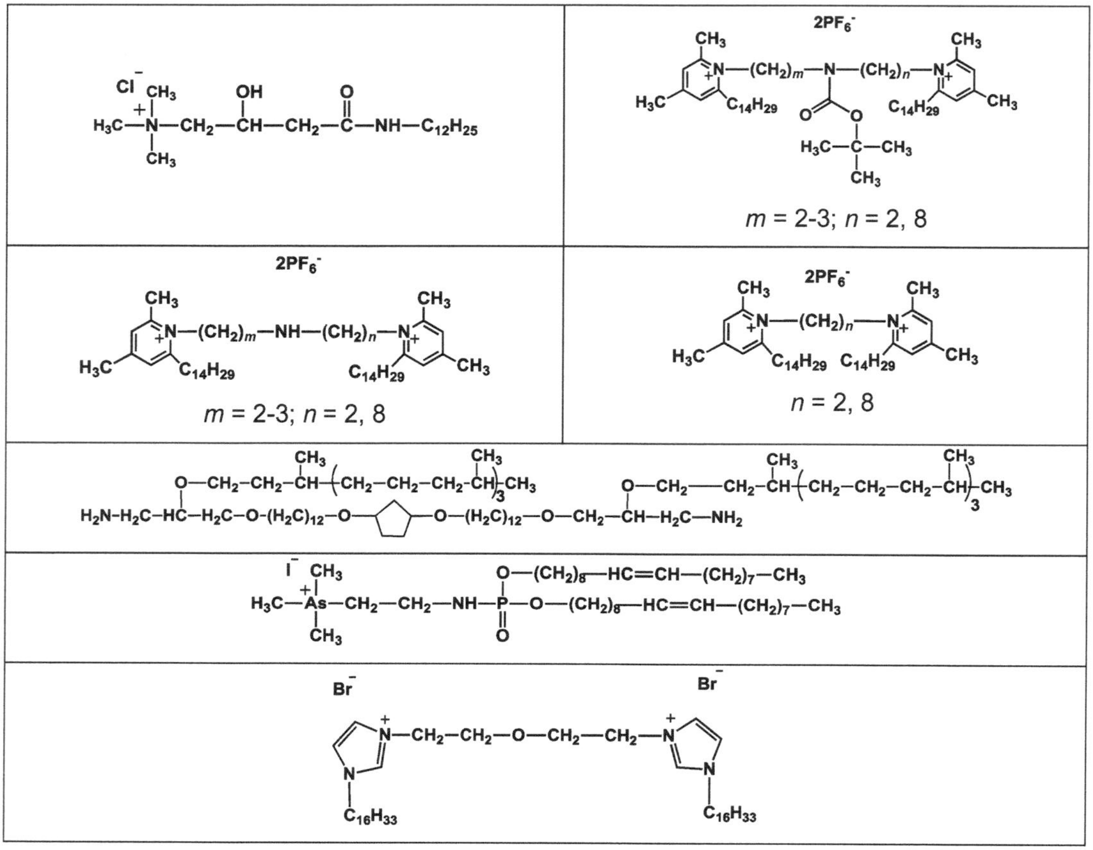

FIGURE 24.4 Structural Formulas of Cationic Amphiphiles Used for DNA Condensation Described in Section 3

agents was designed in Le Gall et al. (2014), in which an archaeal tetraether derivative (Fig. 24.4) was proposed as a colipid in combination with monocationic amphiphiles. The liposomes (the so-called archaeosomes) were examined in gene transfer experiments in mice mediated by lipid/DNA complexes administered through systemic or local routes. Results showed that the tetraether colipid can provide complexes with different in vivo TE controlled by lipid structure, lipid/colipid molar ratio, and route of drug administration.

One of the strategies for the design of effective gene delivery vehicles is an application of surfactants with delocalizable cationic headgroup, which can improve surface charge character and molecular flexibility. According to the study by Sharma et al. (2013), this is a key criterion for efficient gene delivery in vitro and in vivo. This can be exemplified by transfection of NCI-H23 lung cancer cell line mediated by pyridine-containing gemini surfactants (Fig. 24.4) that was 6 times higher as compared with commercially available Lipofectamine.

In Savarala et al. (2013) cationic amphiphiles were synthesized containing cleavable ether fragments served as linkers between dopamine moiety and alkyl tails. Based on the structure-activity correlations revealed in this study, efficient formulation was proposed for gene delivery.

Several formulations are shown to surpass transfection efficacy of well-known vectors, with different types of malignized cells transfected in the presence of serum. Another example of application of pyridine-containing surfactants as gene delivery agent is given in Fisicaro et al. (2014), where bispyridinium dihexadecyl gemini surfactants of varied spacer length have been studied. Authors concluded that the compound with spacer formed by four carbon atoms in combination with L-α-dioleylphosphatidylethanolamine demonstrated highest TE commeasurable with that of the commercial agent. At the same time, the compound with the longest spacer exhibited a zero transfection efficacy due to the formation of incompact structures. Complex study of gemini surfactants with pyridine moiety was carried out in Petrichenko et al. (2015), where amphiphiles containing dodecyloxycarbonyl substituents and cationic moieties have been synthesized. Evaluation of TE was exemplified by delivery of pEGFP-C1 plasmid DNA into the baby hamster kidney-derived (BHK-21) cell line. It was shown that among the tested pyridines the compounds with N-methylpyrrolidinium or pyridinium moieties possessed the highest activity. Helper lipids were shown to increase TE of pyridine-containing gemini surfactants (Sharma et al., 2014). TE of selected ternary lipoplexes derived from cationic gemini surfactants and lipids markedly surpassed commercial transfection agents.

Amphiphiles with delocalizable imidazolium polar group (Fig. 24.4) could be the alternative to pyridine-containing gemini surfactants. The mixed cationic liposomes based on such amphiphile and commonly used lipid dioleoylphosphatidylethanolamine (DOPE) were designed, and their ability to act as gene delivery agents toward plasmid DNA (pDNA) pEGFP-C3 was examined. TE of each lipid mixture was determined in different cell lines (HEK293T, HeLa, Caco-2, and A549) at varied surfactant/DOPE molar ratios, with the serum conditions tested. The formulation optimized in molar and charge ratio in surfactant/colipid pair is shown to excel Lipofectamine 2000 even in the presence of serum (Kumar et al., 2015).

Another effective approach to design gene delivery vehicle is the use of nonionic surfactants. For example, in Puras et al. (2014) cationic lipid 2,3-di(tetradecyloxy)propan-1-amine formulated with squalene and polysorbate 80 was used for the niosome fabrication. Cell lines HEK-293 and ARPE-19 were efficiently transfected with lipoplexes formed at lipid/DNA molar ratio of 15/1. Ojeda et al. (2015) reported on the niosomes fabricated from lipids differing in the structure of polar-head group (with amino and tripeptide moieties), which were used for the transfection of HEK-293, ARPE-19, and MSC-D1 cells. Dependence of TE on the chemical structure of cationic head group was revealed. The highest TE was demonstrated in the case of cationic lipids with a dimethyl amino head group. It was observed that nioplexes based on the cationic lipid with amino head group had a higher deleterious effect compared to its analogs, especially in the case of ARPE-19 cells at cationic lipid/DNA molar ratios of 20/1 and 30/1. In Zhao et al. (2015) several novel tri-peptide cationic lipids with tri-lysine and tri-ornithine head groups, carbamate group as a linker and 12- and 14-carbon atoms alkyl tails were synthesized for delivering DNA and siRNA. It was shown that these lipids can efficiently transfer DNA and siRNA into NCI-H460 and Hep-2 tumor cells. One of the surfactants studied was able to deliver combined siRNAs against c-Myc and VEGF for silencing distinct oncogenic pathways in lung tumors of mice, with low in vitro and in vivo toxicity. Examples of spermine-based lipids using as gene delivery vehicles are also reported. Niyomtham et al. (2015) synthesized cationic lipids with varied polar groups and alkyl tails of different length (C_{12-16}) that were used for the liposome fabrication in combination with DOPE. The influence of the structure of polar fragment on the transfection of HeLa cells was elucidated in the absence and presence of serum.

In Huang et al. (2015) a series of novel cationic lipids based on 1,4,7,10-tetrazacyclododecane (cyclen) with the imidazole group as the pH-sensitive moiety and various aliphatic long chains were designed. Both the hydrophobic chain and the linking bond significantly affected TE, so that the linoleyl-containing lipoplex gave the best TE, which is comparable to Lipofectamine 2000. The imidazole group was demonstrated to play an important role in the transfection, and the imidazole-free analog gave dramatically lower TE. Furthermore, it was also found that Ca^{2+} could largely enhance the TE of these lipids, so that the optimized TE was about 5 times higher than Lipofectamine 2000.

Apart from synthetic surfactants various phospholipids are widely explored. For instance, cationic glycol phospholipids were synthesized and used for the transfection of Chinese hamster ovary (CHO-K1) cells. Authors concluded on unpredictable lipid structure-transfection efficiency correlation for these compounds. However, one of novel lipids exhibited TE and toxicity comparable with control cationic lipid 1,2-dioleoyl-sn-glycero-3-ethylphosphocholine (Øpstad et al., 2014). In Berchel et al. (2015) three new lipophosphoramide-based amphiphilic compounds bearing methoxy, hydroxyl, or dihydroxyl functional group in their cationic part were synthesized. In vitro, all of the cationic lipids evaluated not only efficiently condense plasmids but also transfect two types of human airway epithelial cells. In vivo study showed that the compound with OH-groups was the most efficient in transfecting the lungs compared to other cationic lipids studied.

4 BINDING/RELEASE PROPERTIES OF SELF-ASSEMBLED SYSTEMS

The most important aspects to take into account when choosing the drug delivery vehicle are rate and extent of the drug release. The delay of vehicle degradation and drug release may be achieved by using systems based on surfactants and polymers as drug vehicles. Solubilization by surfactant aggregates may offer an opportunity to control the release rate of certain drugs. Moreover, a sustained release over a prolonged time can be obtained. This is particularly important in cases when the compound is rapidly metabolized (Santos et al., 2013). Polymers are used in drug delivery due to their efficiency as stabilizers, their capacity to form self-assembled stable structures and gels (Chen, 2013). Polymeric nanoparticles allow for versatile modification possibilities and can act as functional platforms for the assembly of well-defined multifunctional structures (Nicolas et al., 2013). Another advantage of polymers is their biodegradability potential.

Polymers and surfactants can be also used as mixtures bringing new and strong advantages into the field of drug delivery. Polymer-surfactant, polymer-polymer and surfactant-surfactant mixtures present several different properties from the individual behavior of polymer or surfactant systems. Of particular interest, it is the fact that these mixtures are capable of forming nanoparticles with controlled release property and improved stability.

There are some basic requirements for design of nanocarriers for targeted drug delivery and controlled release. The nanosized delivery systems should be inherently safe (nontoxic, nonimmunogenic, and biodegradable), and preferably cost-effective and amenable for industrial-scale manufacture. They are also required to deliver bioactive molecules into specific target cells by traversing across various extracellular and intracellular barriers, and release them at a controlled manner at sites of action. Increasingly sophisticated applications in the field of controlled release require that the delivery systems can respond in some way to physical stimuli (i.e., temperature, ultrasound, light, and magnetic and electrical fields) or chemical stimuli (i.e., pH, redox potential, ionic strength, and chemical agents) or both stimuli. These stimuli can be used for the effective release of encapsulating materials from surfactant- and/or polymer-based delivery systems that can be designed and formulated in different forms, such as micelles, vesicles, capsules, emulsions, liposomes, hydrogels, crosslinked core–shell polymers, modified nanoparticles, and aptamer conjugates.

4.1 MICELLES AND VESICLES

Micelles can be used as carriers encapsulating the drugs in order to ensure the transport into cells, to minimize drug degradation and loss, and to allow accumulation in the required pathological area. Micellar systems are considered as model systems for biomembranes, useful to investigate drug-membrane interaction, and the possibility of the drugs passage through cell membranes. For example, the nonionic surfactants could act as interesting drug carriers into cells. Triton X-100 micelles are able to release anticancer drugs to interact with DNA, pointing to a better affinity of the investigated compounds for DNA rather than for micelles (Mazzoli et al., 2015).

Ammonium surfactant micelles embedding tetracycline antibiotics (doxycycline and oxytetracycline) act as smart host-guest supramolecular systems allowing a pH controlled inclusion of antibiotics and a magnesium-induced release of them (Cesaretti et al., 2014). The presence of Mg^{2+} is able to extract part of the antibiotic embedded in the core of the micelles and therefore to induce the release of its biologically active form with magnesium into the aqueous solution.

Gemini surfactants with a degradable or stimuli-sensitive spacer can form micelles for controlled drug release. Gemini polymeric micelles containing disulfide bonds have been prepared for redox-responsive drug delivery (Chen et al., 2013a; Cunningham and Oh, 2013; Khorsand et al., 2013; Lv et al., 2014; Xu et al., 2014a). Kim et al. (2014a) reported on thiol-responsive micelles of nonionic gemini surfactants with cystine disulfide spacers consisting of hydrophilic poly(ethylene glycol) (PEG) and hydrophobic stearoyl groups. The nanosized assemblies of these surfactants showed the adjustable release of encapsulated indomethacinas depending on the concentration of glutathione that acts as a reducing agent. In other work, amphiphilic gemini copolymer consisting of hydrophilic PEG blocks and hydrophobic polylactide blocks joined by a cystine disulfide spacer was synthesized (Kim et al., 2015a). In the presence of cellular glutathione, gemini micelles gradually destabilize into monomeric micelles through cleavage of the cystine disulfide bond. This destabilization of the gemini micelles changed their size distribution, which became bimodal with the appearance of small aggregates, and led to the enhanced release of encapsulated anticancer drug doxorubicin.

The stimuli-responsive micelles self-assembled from amphiphilic polymers have been intensively investigated for controlled drug delivery due to simple preparation and easy incorporation of various stimuli responsive blocks into the structure of polymers. Light as noncontact stimuli is the most suitable for use in delivery systems because it can be manipulated with extremely high spatial and temporal precision. Specifically, spiropyran has been introduced into polymers to develop light-responsive micelles (Haque et al., 2013; Ivashenko et al., 2013; Renkecz et al., 2013). Based on reversible light-responsive behavior of spiropyran, change in the nanocarrier assembly directly or indirectly leads to release of drug from the nanocarrier (Wang et al., 2014a). Han et al. (2013) reported on amphiphilic diselenide-containing block copolymers micelles. In the presence of chromophores, such as porphyrin derivatives, red light can be used to trigger the release of cargo encapsulated in polymeric micelles.

Although drug-loaded polymeric nanoparticles showed many superior features compared with traditional drugs, they often suffer sluggish drug release in the cancer cells and at the tumor sites (Chen et al., 2013b). To achieve a fast release in the cancer cells or at the sites of tumor, polymeric micelles are often designed to respond to the intrinsic environment of tumor with low extracellular pH (Huang et al., 2013a; Jiang et al., 2013; Luo et al., 2013; Sun et al., 2013; Victor and Sharma, 2013; Wu et al., 2013a). Such polymers as poly(2-(diethylamino)ethylmethacrylate) (Cao et al., 2013a) and poly(L-histidine) (Cavus et al., 2013; Wu et al., 2013b) have been demonstrated to possess pH response. Mixed micelles composed by the two amphiphilic copolymers can increase intracellular drug release rates (Cai et al., 2014). In vitro drug release showed that drug-loaded mixed micelles could release the drug faster under low pH and reducing condition.

Lee et al. (2013) developed pH-sensitive amphiphilic poly(ketal adipate)-*co*-poly(ethylene glycol) block copoplymers which can self-assemble into thermodynamically stable micelles and encapsulate anticancer drug camptothecin. At the acidic environment these biocompatible micelles rapidly dissociated to release a payload of anticancer drugs due to the presence of rapidly acid-cleavable ketal linkages. Xu et al. (2013a) also developed pH-sensitive polymeric micelles formed by self-assembled amphiphilic PEG-*co*-poly(*N*-butyl methacrylate-ran-methacrylic acid) for doxorubicin release, simultaneously modifying micellar shell with folate to enhanced the efficiency of cellular uptake by tumor cells. In other work, pH-sensitive biodegradable micelles were fabricated from amphiphilic chondroitin sulfate–histamine conjugate (Yu et al., 2014). A pH-dependent release behavior of these nanoparticles inhibits drug release in neutral surroundings (blood circulation or extracellular matrix) and rapidly disintegrates to release cargos in acidic surroundings (intracellular endosomes). Li et al. (2013) developed a polypeptide-based block ionomer complex formed by anionic methoxy PEG-*b*-poly(L-glutamic acid) and cationic drug doxorubicin hydrochloride that spontaneously self-assembled into spherical nanoparticles in aqueous solutions via electrostatic interaction and hydrophobic stacking, with a high loading efficiency (almost 100%). Doxorubicin hydrochloride release from these nanoparticles was slow at physiological pH, but greatly increased at the acidic pH, mimicking the endosomal/lysosomal environment.

Glucose-responsive systems enable self-regulated drug delivery and regularly monitoring the blood glucose concentration in the patients of diabetes mellitus, thus reducing the discomfort of insulin administration and chemotherapy for the patients (Zhang et al., 2013). A series of studies on the phenylboronic acid-based glucose-responsive materials have been reported for the development of self-regulated insulin delivery systems (Jiang et al., 2014a,b,c).

Stimuli-responsive micelles that can disassemble and release their encapsulated cargo upon external stimuli have gained increasing attention. The substantial advantages could be achieved by utilizing enzymes as stimuli. As many diseases are characterized by imbalances in the expression and activity of specific enzymes in the diseased tissue, this overexpression could potentially be translated into the selective activation of advanced drug delivery platforms (Zelzer et al., 2013). The enzyme-sensitive polymer micelles are based on breaking the amphiphilic block copolymer into a soluble hydrophilic polymer and an insoluble hydrophobic block (Rao and Khan, 2013). Harnoy et al. (2014) reported on amphiphilic block copolymers based on a linear hydrophilic PEG and an enzyme-responsive hydrophobic dendron that can self-assemble in aqueous medium into micellar nanocontainers that can further disassemble and release encapsulated cargo Nile red upon enzymatic activation. Self-assembly was examined by utilizing solubilization experiments with the solvachromic hydrophobic dye Nile red (Leenders et al., 2013). The enzyme-responsive disassembly was further supported by the change in the fluorescence of encapsulated Nile red dyes.

Pluronic can self-assemble into micelles in aqueous medium and has aroused wide concern in drug delivery and drug targeting (Chen et al., 2013c). Song et al. (2014) developed Pluronic F68 derivative, cholesterol-coupled F68; the micelles of this derivative loaded cabazitaxel, a novel antitumor drug had sustained-release properties. The mixed micelles of Pluronics L61 and F127 had the ability to deplete cancer stem cells and decrease the tumorigenic behavior of cancer cells in vivo (Alakhova et al., 2013).

pH-triggered drug release was obtained for paclitaxel containing (1,2-distearoyl-sn-glycero-3-phosphoethanolamine)–PEG immunomicelles by incorporation of pH-sensitive poly(histidine)–PEG. The micelles were stable at neutral and tumor extracellular pH but quickly released paclitaxel after tumor cell-specific internalization by endocytosis at the low endocytic pH (<6.0), resulting in enhanced anticancer efficacy (Wu et al., 2013b).

Polymer vesicles are attractive nanocarriers due to their high stability compared to liposomes and ability to encapsulate both hydrophilic and hydrophobic cargos, contrary to micelles. The robust polymer vesicles with diacid bilayer shell being responsive toward pH in aqueous solution were self-assembled by pH triggering from linear-brush polyoxanorbornene-based diblock copolymers. These cross-linked polymer vesicles are ideal nanocarriers for anticancer drug cis-platinum, because the diacid bilayer shell has the tendency to bind cis-platinum, forming metal complexes that can release drugs by hydrolytically cleaving carboxylic group (Xu et al., 2013b). In other work, the segments of water-soluble methionine sulfoxide residues in the vesicles were found to be substrates for reductase enzymes, resulting in a change in supramolecular morphology that caused vesicle disruption and release of encapsulated cargos (Rodriguez et al., 2013).

Ultrasound and pH dually responsive polymer vesicles were successfully prepared based on copolymers composed of poly(ethylene oxide), 2-(diethylamino)ethyl methacrylate and (2-tetrahydrofuranyloxy)ethyl methacrylate blocks (Chen and Du, 2013). These dually responsive polymer vesicles become smaller upon either ultrasound irradiation or by decreasing pH. The ultrasound induced significant size diminution of vesicles and caused the fast disruption and re-self-assembly of vesicles. The vesicles can also become smaller at acidic conditions, as a result of the shrinkage of vesicles. During both processes, the controlled release of loaded doxorubicin can be reached. In addition, the vesicles show noncytotoxicity at low concentration, which makes their biomedical applications possible.

Soluble aggregates formed in dilute aqueous mixtures of polyacrylate, cationic surfactant decyl betainate and nonionic surfactant octaethylene glycol monododecyl ether are properly designed (Janiak et al., 2013). The polyacrylate is released from the mixed micelles as a consequence of decyl betainate hydrolysis, the rate of which, and thus the rate of polymer release, can be controlled by the pH. Decyl betainate containing an ester bond is susceptible to base-catalyzed hydrolysis at mild alkaline conditions, and degradation of decyl betainate leads to a decreased attraction of polyacrylate to the mixed micelles.

Mixtures of ionic polysaccharides and oppositely charged surfactants show a large diversity in functionality, structure, and stimuli responsiveness (Chiappisi et al., 2014), which can be employed for drug-delivery purposes. It is supported by some recent reviews (Akhlaghi et al., 2013; Chiappisi et al., 2013). The coassembly of surfactant micelles and biopolysaccharide chitosan leads to an effective incorporation of the hydrophobic cargo within the supramolecular complex (Chiappisi et al., 2015). The ability of mixtures of the cationic biopolysaccharide chitosan and an environmentally friendly anionic surfactant to release and/or recover hydrophobic cargos was demonstrated upon small pH changes. This property, arising from the strong structural changes taking place within 1 pH unit, can be exploited for the formulation of selective drug delivery systems.

As mentioned earlier, micelles and vesicles are two important types of discrete spherical assemblies. They also have been fully studied fundamentally and practically, especially their mutual transition. The conversion between them has attracted considerable attention because of their extensive use in drug delivery and release systems in pharmaceutics and gene transport in biological science. Micelle-to-vesicle transitions in surfactant self-assembly systems have been widely studied (Li et al., 2014; Yi et al., 2013; Zou and Yuan, 2015) for stimulus-responsive materials (Chu et al., 2013). There has been recent progress in fabrication for multiple stimuli-responsive systems and the realization of their multifunctions (Li et al., 2015; Wu et al., 2013c; Zou and Yuan, 2015).

4.2 CAPSULES AND EMULSIONS

Liquid-crystalline phases or "mesophases" formed in mixtures of oppositely charged surfactant and polymer solutions offer great versatility in structure manipulation over oppositely charged polymer-based drug delivery systems, such as layer-by-layer engineered capsules. The layer-by-layer technique is the most popular and frequently studied method for the encapsulation of sensitive reagents (Sadovoy et al., 2013). Controlling interfacial structural attributes in complex oppositely charged surfactant and polymer systems may provide an interesting route to tailored release nanomaterials (Tangso et al., 2014).

Study of polyelectrolyte multilayer assembly and lipid bilayer formation on colloidal particles focuses on hollow polyelectrolyte multilayer capsules and release of hydrophilic macromolecules (Trojer et al., 2013a). Layer-by-layer adsorption of the polyelectrolyte pair poly(diallyldimethylammonium chloride) and poly(sodium methacrylate) was successfully made on microcapsule systems with the formation of very thin multilayers for controlled release of hydrophobic actives (Trojer et al., 2013b). Trojer et al. (2013c) also reported on the very significant reduction of the release rate of a hydrophobic probe molecule from microparticles which had been surface modified with up to four polyelectrolyte layers.

Smart hollow multilayer microcapsules made of aliphatic poly(urethane-amine) and sodium poly(styrene sulfonate) are prepared for pH-/thermally responsive drug delivery (Shi et al., 2013). The electrostatic interaction and hydrogen bonding under weak-acid environment between these polymers contribute to the formation of multilayer microcapsules. Drug-release behaviors using doxorubicin hydrochloride demonstrate that the drug release can be increased by decreasing the pH value because of the interaction weakness between polymers in acidic conditions. Moreover, the drug release is higher at 55°C than that at 37°C due to the shrinkage of aliphatic polymer above its lower critical solution temperature.

One of the approaches used for the control of drug release is the fabrication of capsules with the cross-linked multilayers. As reported in Wang et al. (2015), covalent linkage of polyallylamine hydrochloride/polyamidoamine capsules with photoactive agents dramatically differs the release profile of acridine orange, namely, prolongs the diffusion of ca. 98% of loads from 5 to 30 min. In other work, Min et al. (2014) reported on the control over spatial and temporal release of multiple drugs. This approach involves the use of laponite clay barrier separating therapeutics loaded in compartmentalized hybrid film. Original device developed by Song et al. (2013) allows switching from sustained to pulsatile release, with the process electrochemically controlled. In other work, layer-by-layer assembly of tannic acid with temperature-responsive block copolymer micelles demonstrated temperature-triggered release of doxorubicin (Zhu et al., 2013).

Chen et al. (2014a) synthesized well-defined nanocapsules through the interfacial cross-linking of *N*-maleoyl-functionalized chitosan in miniemulsions and investigated their application in the delivery

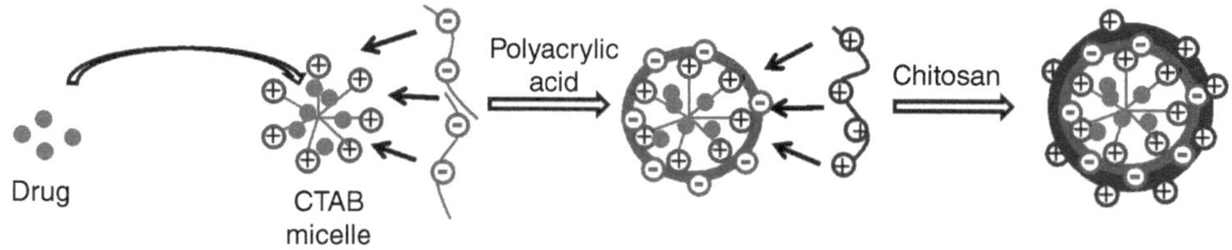

FIGURE 24.5 Algorithm of the Fabrication of Polyelectrolyte Capsules with the Layer-by-Layer Technique, Including the Substrate Treatment with a Cationic Surfactant

of doxorubicin. These chitosan-based capsules possess both significant drug-loading capacity toward doxorubicin and its faster release rate under acidic conditions.

Nanocapsules for protection of bioactive compounds (*P*-nitrophenyl laurate and acetylsalicylic acid, aspirin) and tunable sustained release behavior were fabricated through the layer-by-layer deposition of polyacrylic acid and modified chitosan (or polyethylene imine) with CTAB used to increase the affinity of polyelectrolyte to the substrate (Fig. 24.5). The shell permeability of these capsules and their protective effect may be mediated through the variation of number of layers deposited, the ultrasonication, and the adjustment of solution pH (Zakharova et al., 2012f, 2015). The layer-by-layer technique was applied to emulsions leading to the formation of (oil) core–(polyelectrolyte) shell microcapsules that showed a sustained release of vitamin E within at least 80 h (Sharipova et al., 2016).

Emulsions of oil and water stabilized by solid particles are known as solid-stabilized emulsions (often referred to as Pickering emulsions) (Pan et al., 2013). Compared with normal emulsion, Pickering emulsion is much stable (Yang et al., 2013a) and easy to be recycled. Under sonication, the surfactant (sodium dodecyl benzene sulfonate) stabilized toluene-in-water emulsions were obtained, which were then used as soft templates for the fabrication of microcapsules that loaded with fluorescein (Yang et al., 2015). The fluorescein maximum release percentage 61% was obtained after 36 h at 37°C at pH 7 with one double-layer capsule. In other investigation of Pickering emulsions, rhodamine B release from the microgel-stabilized emulsions of water-in-oil-in-water type depends on the temperature and pH conditions (Zou et al., 2013).

Pickering emulsions were also used as templates to fabricate Janus particles. Amphiphilic Janus particles that have stimuli-responsive properties represent promising surfactant systems. It has been shown that not only highly stable emulsions can be generated with amphiphilic Janus particles, but also the destabilization of emulsions can be achieved by using stimuli-responsive Janus particles as emulsifiers (Ji et al., 2014; Lu et al., 2014; Yamagami et al., 2014; Yang et al., 2013b). More importantly, inversion of emulsion can also be achieved with stimuli-responsive Janus particles (Tu and Lee, 2014a). Tu and Lee (2014b) also demonstrated that highly stable water/oil/water multiple emulsions can be generated with Janus particles via one-step emulsification, enabling the one-step encapsulation of hydrophilic species in multiple emulsions. Noteworthy, increasing the solution pH triggers the destabilization of aqueous compartments and subsequent release of encapsulants from these multiple emulsions.

4.3 HYDROGELS AND CROSSLINKED CORE–SHELL POLYMERS

The interest in molecular hydrogels has exponentially grown in recent years, and their applications have been developed in regenerative medicine and controlled drug release. These materials are obtained by weak noncovalent interactions that may be conveniently reversed by external physicochemical stimuli leading to environment responsive systems (Segarra-Maset et al., 2013).

Microgels can trap appropriate molecules in their swollen state by means of some attraction between the adsorbate and the cross-linked polymer network (Chen and Dai, 2013). Yin et al. (2014) reported on the construction of a polyelectrolyte-responsive system consisting of sterically stabilized protonated poly(2-vinylpyridine) microgels. Negatively charged sodium dodecylbenzenesulfonate surfactants could be readily internalized into these cationic microgels, resulting in microgel collapse and concomitant formation of surfactant micellar domains contained electrostatic complexes. These internal hydrophobic domains conferred the opportunity of fluorescent dyes to be encapsulated. The obtained microgel complexes with an encapsulated fluorescent dye could be further disintegrated upon addition of anionic poly(sodium 4-styrenesulfonate) due to the competitive binding to microgels. The stronger electrostatic attraction between multicharged microgels and polyelectrolyte compared to single-charged surfactant led to triggered release of the encapsulated pyrene dyes from the hydrophobic domains.

Polymer hydrogels have received much attention in controlled drug release due to their biocompatible and biodegradable properties (Cigognini et al., 2014; Contri et al., 2014). Copolymer hydrogels of polymerizable surfactant (surfmer), ω-methoxy poly(ethylene oxide)$_{40}$undecyl-α-methacrylate, and acrylic acid exhibit a linear, strong release of ibuprofen in simulated gastric fluid (pH 1) and phosphate buffer solution (pH 6.8), which can hardly be influenced by the surfmer content of the gel (Kraus and Tieke, 2014). In another work, the monodisperse hydrogel hollow microcapsules were fabricated by the formation of hydrophobic complexes between negatively charged hydrogel chains and oppositely charged surfactants (Lim et al., 2013). The pH-dependent release behavior of methylene blue from these microcapsules is attributed to the degree of electrostatic attraction between the dye and microcapsules.

In addition, increased attention is paid to the study of complex supramolecular systems. One of such complex systems is formed when mixing two different kinds of soft matter building blocks: molecular gelators and surfactants. The presence of SDS micelles provokes dye release due to gel disassembly (Nebot et al., 2013). Interestingly, the gels loaded with the dye did not release their content when a premicellar solution of SDS in distilled water was used as medium.

In (Liu et al., 2014a) an ion-pair amphiphilic molecule composed of anticancer drug (cytarabine-hydrochloride) cation and surfactant (AOT) anion is used for straightforward fabricating vesicles that show sustained drug release properties superior to the pure drug solution. To obtain more prolonged and convenient drug release systems, the obtained vesicles are further incorporated into the thermosensitive poly(lactic-*co*-glycolic acid) (PLGA)–PEG–PLGA hydrogel to prepare a subcutaneous injection.

Core-shell nanostructured polymers also have attracted significant research interest in the drug delivery (Ramli et al., 2013a; Shalviri et al., 2013; Sohrabi et al., 2013). Ramli et al. (2013b) synthesized poly(styrene-*co*-methyl methacrylate) core and poly(acrylamide-coacrylic acid) shell microgels for potential application as drug releasing coatings. Biocompatible and biodegradable polystyrene (core)–biosurfactant (shell) bionanocomposites were synthesized for their feasibility as drug delivery vehicle (Hazra et al., 2013). Similarly, poly(methylmethacrylate) (core)–biosurfactant (shell) particles were designed for drug delivery applications (Hazra et al., 2014) and able to efficiently load the high contents of ibuprofen, anthraquinone, and curcumin with no loss of colloidal stability. The release of these three drugs from the particles was sustained and strongly dependent on the initial drug loading content and medium pH. The release of pyrene from core-shell nanospheres was found to be a highly pH-responsive process over an environment of pH 5, 6, and 7, which covers the pH range of acidic tumor tissues and endosomes-lysosomes (Wang and Rempel, 2013).

The delivery systems with efficient hydrophobic drug loading capacity combined with good stability and controlled release were introduced with the fabrication of PLGA–lecithin–PEG core–shell nanoparticles (Tan et al., 2013; Zheng et al., 2013).

The peptide-based nanomaterials self-assembled from covalent amphiphilic polypeptides have been studied as efficient vehicles for the delivery of drugs and genes (Stephanopoulos et al., 2013), with focus on biological activity, multifunctionality, and tailoring of the materials to clinical applications. The research on smart supramolecular peptide-amphiphile systems may promote the development of stimuli-responsive supramolecular amphiphiles that do not require specific ligands. More importantly, the responsive range of these systems is compatible with biological stimuli. For example, supramolecular peptide-amphiphiles based on dendritic poly (L-lysine) and poly(L-leucine) self-assembled into compact core-shell nanostructures are found to encapsulate guest molecules at pH 7.4, but release them at pH 6.2 (Xu et al., 2014b).

Unique drug release mechanism of hydrogen peroxide-responsive nanocarriers constructed from an amphiphilic hyperbranched polymer with alternative hydrophobic selenide groups and hydrophilic phosphate segments in its backbone framework is based on an amphiphilic-to-hydrophilic transition (Liu et al., 2013a). Because hydrophobic selenide groups are easily oxidized into hydrophilic selenone groups, the amphiphilic hyperbranched polymer becomes totally hydrophilic after oxidation, resulting in the rapid disassembly of the nanocarriers and in fast intracellular drug release.

4.4 LIPID SYSTEMS

Among the various types of drug delivery vehicles developed, lipid-based materials are highly attractive for their excellent biocompatibility (Allen and Cullis, 2013). Liposomal carriers have been reported to show limited immunogenicity and have the ability to respond to surfactant triggers (Kim et al., 2014b). A study by Chana et al. (2015) showed the feasibility of trigger-responsive inhaled delivery of medicines using soft solid shelled nanocapsules. This lipid rich capsule carrier was distended rapidly when mixed with an exogenous nonionic surfactant trigger, Pluronic L62D. Capsule distension was accompanied by solid shell permeabilization, resulting in payload release of rhodamine from the carrier.

A paper from Kauscher et al. (2014) describes phospholipid vesicles that can be equipped with an amphiphilic β-cyclodextrin (CD) functionalized with an azobenzene moiety to obtain an intramolecular host–guest inclusion complex with the CD cavity. Photoisomerization of azobenzene-CD from trans to cis has no influence on size, shape and stability of these vesicles, but regulates the availability of the CD cavities on the vesicles surface for complexation with other guest molecules. UV irradiation results in enhanced permeability compared to vesicles containing CDs without azobenzene and in a faster and more extensive release of entrapped sulforhodamine B.

Biocompatible polymeric cores have been employed to adsorb drugs and prepare supported bilayers. The combination of lipid and polymer takes advantage of the properties of both materials (Hadinoto et al., 2013; Wen et al., 2014; Zou et al., 2014). Polymer/lipid nanoparticles fabricated conjugated to trans-activating transcriptor peptide may function as an effective drug delivery system that crosses the BBB to the brain (Wen et al., 2014). The diameter of nanoparticles can be controlled by adjusting the lipid/polymer ratio (Mandal et al., 2013).

To be delivered to the lungs, liposomal formulations are typically aerosolized by a nebulizer from aqueous dispersion (Cipolla et al., 2013). Joshi et al. (2013) developed paclitaxel-loaded lipid nanovesicles sensitive to both temperature (42°C) and secretory phospholipase A_2 enzyme (a tumor specific enzyme). The use of multiple triggers simultaneously showed a substantial increase in the paclitaxel release leading to the enhanced therapeutic efficacy. Paclitaxel-loaded lipid nanovesicles were also prepared by combining 1,2-dipalmitoyl-*sn*-glycero-3-phosphocholine, the phospholipid and the major constituent of pulmonary surfactant, with 1,2-dioleoyl-*sn*-glycero-3-phosphoethanolamine used as a fusogenic lipid. 100–200 nm–sized nanovesicles showed improved fusogenicity and cytosolic drug release, specifically with cancer cells (Joshi et al., 2014). Significantly, continuous aerosol treatment of these vesicles did not show indications of any adverse pulmonary effects observed with inhaled nanoparticles (Ferreira et al., 2013).

For pulmonary drug delivery, solid lipid particles could be produced by spray-drying. Park et al. (2013) optimized the spray-drying process through statistical experimental design to produce 1,2-dipalmitoyl-*sn*-glycero-3-phosphocholine-based solid lipid particles encapsulating and controlling the release of vancomycin and clarithromycin.

Solid lipid nanoparticles encapsulating resveratrol for brain targeting were successfully prepared using compritol as the lipid core and a combination of Tween 80 and polyvinyl alcohol as the shell material. The in vitro release of resveratrol from these nanoparticles in phosphate buffer pH 7.4 followed a sustained release pattern (Jose et al., 2014a).

The thermoresponsive bubble-generating liposomal systems rely on the creation of permeable defects in the lipid bilayer by means of the generation of carbon dioxide bubbles through decomposition of ammonium bicarbonate at mild hyperthermia (42°C) (Chen et al., 2013d). As a result, bubble-generating liposomal vesicles rapidly release doxorubicin.

4.5 MODIFIED NANOPARTICLES

Surface modification of gold nanoparticles with various capping agents, such as surfactants and polymers has attracted considerable interest because such capped nanoparticles find several applications in drug or nucleic acid delivery (Xu et al., 2013c; Zargar and Hatamie, 2013). Amino acids and their derivatives can be used in the biofunctionalization of gold nanoparticles and considered as protective layers due to the presence of different functional groups with affinity for gold (Liu et al., 2013b). CTAB is a very effective stabilizing agent for the preparation of gold nanoparticles that facilitates electrostatic adsorption of an anionic polypeptide, poly-(γ-glutamic acid), through electrostatic self-assembly (Guan et al., 2014). The change in aggregation behavior with pH is caused within a narrow range of pH from 5 to 6. The designed pH probe is being extensively applied in biological and medical applications.

CTAB has also been utilized in the preparation of template semi-graphitized mesoporous silica nanoparticles, which exhibited pH-dependent and near-infrared-mediated release behavior, which can be attributed to pH-triggered reduction and heat-stimulative dissociation of the strong interaction between drug and semi-graphitized carbon (Wang et al., 2014b). The drug release process in this study has also been exploited as nanocarriers for drug delivery (Feng et al., 2013). Such nanoparticles with large internal cavities were synthesized by using nanoparticles of pharmaceutical polymer Eudragit S100 as both a core template and an assistant in the self-organization of surfactant micelles (She et al., 2015). The modified nanoparticles were found to exhibit a high drug loading and sustained release of anticancer drug 5-fluorouracil, due to their mesoporous structure (Cao et al., 2013b; Kamarudin et al., 2014).

Surfactants are used to stabilize the surface of polymeric or lipid nanoparticles to form stable amphiphilic colloids in physiological fluids (Sadhukha et al., 2013). Polymeric nanoparticles coated with surfactant polysorbate 80 are reported to cross the BBB by mimicking the low density lipoproteins, enabling them to interact with the receptor of low density lipoprotein, resulting in the nanoparticles uptake by brain endothelial cells (Jose et al., 2014b; Nagpal et al., 2013). The in vitro release of Bacoside-A from the PLGA nanoparticles followed a sustained release pattern with a maximum release of up to 83.04 ± 2.55% in 48 h.

Magnetoliposomes, which are formed by encapsulating iron oxide nanoparticles within liposomes, have attracted interest as controlled release drug delivery agents (Spera et al., 2015). The study of intracellular release of drugs using liposomes with membrane-associated magnetic nanoparticles showed that triggered liposome drug release can achieve cancer cell death without hyperthermia effects (Chen et al., 2014b).

4.6 APTAMER CONJUGATES

Aptamers are a class of small nucleic acid ligands composed of RNA or single-stranded DNA oligonucleotides and have high specificity and affinity for their targets in a cell- or tissue-specific manner (Lakhin et al., 2013; Sun et al., 2014). They have recently been investigated for targeted drug delivery systems (Alvarez-Lorenzo and Concheiro, 2014; Edwards et al., 2015). Fig. 24.6 outlines the principal strategies adopted for self-assembled systems conjugated with aptamer for targeted drug delivery. The aptamer-modified mixed micelles possess properties of tumor targeting, pH-sensitive drug release, and efficient drug encapsulation (Zhang et al., 2015). The paclitaxel-loaded aptamer-decorated mixed micelles had a uniform spherical shape and ideal particle size. The resulting paclitaxel-loaded

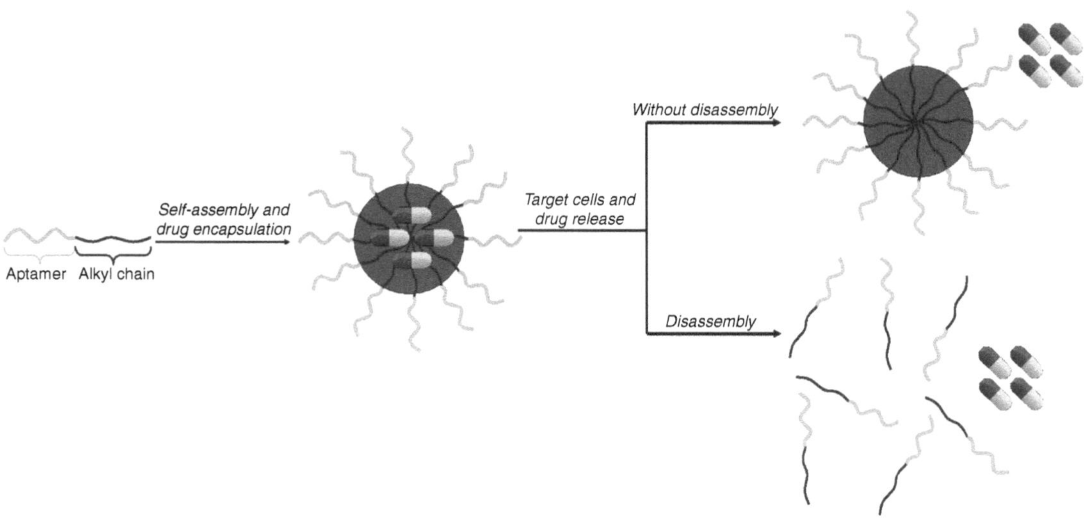

FIGURE 24.6 A Schematic Representation of the Principal Strategies Adopted for Self-Assembled Systems Conjugated With Aptamer for Targeted Drug Delivery

aptamer-mixed micelles were stable at pH 7.4 and disassembled in weak acidic environment (pH 5.5), demonstrating a well-controlled pH-sensitive drug-release behavior.

Aptamer-conjugated polymeric nanoparticles were prepared by conjugation of 1,3-phenylenediamine resin nanoparticles containing rhodamine with aptamers. Unlike the previously mentioned case, the structure of these nanoparticles becomes less compact inside the cells (lower pH environment relative to the phosphate-buffered saline solution), leading to the release of the fluorophores (Ho et al., 2015).

Alibolandi et al. (2015) developed PEG–PLGA nanopolymersomes by covalently coupling the aptamer to the surface of nanopolymersomes loaded with doxorubicin. These polymersomes exhibited a higher release rate at pH 5.5 than at pH 7.4 due to the improved solubility of protonated drug. The surface of PLGA acid nanoparticles can be modified with transferring the major protein in plasma, for specific targeting to reduce side effects (Cui et al., 2013). Due to the biocompatibility of these particles, they serve as a suitable drug delivery system that could be applied for the treatment of brain glioma.

A number of advances have been made to improve the drug encapsulation efficiency and stability of liposomal vehicles. Vinorelbine-loaded lipid-polymer hybrid nanoparticles with covalently linked aptamer were fabricated by a self-assembly method, using PLGA as the core and lipids as the shell (Liu et al., 2015a). Since PLGA is insoluble in the aqueous phase, it self-assembled into nanoparticles, encapsulating vinorelbine, and the surface was also capped by the lipids. In this process the solution temperature was maintained at 65°C, such that the lipid chains were above their phase transition temperature to facilitate the self-assembly process. The accumulative vinorelbine released into phosphate buffer saline (pH 7.4) over 108 h was lower than 50%, indicating a sustained release characteristic.

AS1411 is one of the most investigated DNA aptamer and has been successfully adopted as a targeting ligand in several nanoparticles (Wu et al., 2013d). The human serum albumin nanoparticles were fabricated by a self-assembly method and then modified with this aptamer. These nanoparticles can effectively deliver paclitaxel in a cancer-cell-specific manner. Paclitaxel released continuously during 24 h and the cumulative release percentage of drug reached 70%.

5 BIOMATERIALS FOR HYDROPHILIC PEPTIDES

Numerous peptides and proteins are involved in biomedical strategies due to their therapeutic potential (Adessi and Sotto, 2002, 2004; Al-Tahami and Singh, 2007; Holcenberg and Roberts, 1981; Torchilin, 1991, 2008; Wolf and Ransberger, 1972). Proteins and peptides are the focus of the pharmacy science and technology since 1990, when insulin, thyroid hormones, and coagulation Factor viii became commercially available. Nowadays peptide therapy is even more attractive due to several key points: (1) understanding of the involvement of regulatory proteins and peptides into the molecular mechanism of human diseases, (2) progress in DNA recombinant technology that marks the separation of natural sources and large-scale production of polypeptides; (3) high progress in the physical methods and their applications in biological analysis, which resulted in the widening of the spectrum of the protein and peptide drugs including hormones and growth factors, anticoagulants, bacterial or plant toxins, antibody-based drugs, and so forth. Among therapeutic proteins and peptides, the most investigated and widely used are enzymes, including antitumor enzyme, enzymes for replacement therapy, enzymes for the treatment lysosomal storage diseases, and antibacterial and antiviral enzymes. Peptide hormones, especially insulin and somatostatin derivatives are related to the most popular drugs for the treatment of pituitary and gastrointestinal tumors (Figg et al., 2002; Froidevaux and Eberle, 2002; Kerbel and Folkman, 2002).

Despite the fact that much attention is paid to the development of protocols for hydrophilic macromolecular drugs, there are limitations for their wide application in the practice medicine (Fix, 1996; Saffran et al., 1986, 1997). Unlike low molecular drugs, polymer therapeutic agents have supramolecular structure (secondary, tertiary, and quaternary) that must be preserved upon delivery process, since even a partial destruction of the supramolecular structure may result in the loss of biological function. Two different mechanisms of structural transformation of macromolecule drugs occur, that is, physical and chemical ones. The former is not associated with covalent transformation and assume alteration in the higher order structure, that is, denaturation of proteins and formation of unfolded structures, aggregation, precipitation, and adsorption. The mechanism of chemical modification of protein macromolecules upon supramolecular destruction involves, for example, the formation of disulfide bridged bonds, hydrolytic oxidation, or deamination process, and so forth.

Historically, the first peptide that received research attention was insulin, which plays an important biological function of controlling the blood glucose level. This peptide hormone can be administered by different methods (Matteucci et al., 2015), namely oral, nasal, transdermal, pulmonary, vaginal, and rectal, with the oral administration preferred and uncontrived. Meanwhile, the oral route of insulin administration remains a challenging task due to the low stability of the hormone in the digestive tract, the occurrence of physical and biological barriers, the low solubility, and the insufficient absorption by intestine mucosa. Per-oral administration of peptides meets many difficulties connected with gastric emptying, transit time, presence of food, variation in pH across the gastro-intestinal tract (pH 1~8), digestive enzymes, intestinal flora, and epithelial transport. Consequently, the absolute oral bioavailability levels of most peptides and proteins are less than 1%. To overcome these problems, different strategies are developed, in particular, the use of absorption enhancers and ferment inhibitors, modification of insulin by mucoadhesive polymers, and application of formulated preparation, with different microcapsulation techniques involved. For this purpose lipid- and polymer-based micro- and nanosized delivery systems, such as liposomes, niosomes, polymer capsules and spheres, solid lipid nanoparticles, and so forth, need to be fabricated (Cárdenas-Bailón et al., 2013).

Thus, formulations for the macromolecular drugs should be chemically inert toward loaded macromoleculars and preserve them from physiological factors, such as proteolysis in plasma, low pH in endosomes, enzyme degradation in lysosomes, and so forth. In recent literature, examples are available of application of nanosystems for effective treatment of animals with type I diabetes. Nanoparticles are shown to enhance intestinal absorption of insulin. This theme is the main focus of recent reviews, which summarize novel strategies in both fundamental and practice-oriented researches (Fonte et al., 2013, 2015; Gundogdu and Yurdasiper, 2014; Lopes et al., 2014). Such aspects as optimization of the size and surface properties of nanoparticles are discussed in ref. (Des Rieux et al., 2013), with the strategies for grafting of nonpeptide ligands emphasized. Grafted nanoparticles are shown to be effective for targeted transport of therapeutic proteins through the intestinal epithelium. Biodegradable nanoparticles for the protection of protein structure from physiological environment are described in (Bakhru et al., 2013), with special emphasis placed on systematic delivery of insulin as a model protein.

5.1 LIPOSOMAL FORMULATIONS

A growing number of medications must be administered through parenteral delivery, that is, intravenous, intramuscular, or subcutaneous injection, to ensure effectiveness of the therapeutics. For some therapeutics the use of delivery vehicles in conjunction with this delivery mechanism can improve drug efficacy and patient compliance. Self-assembled materials (synthetic amphiphiles, lipids, and polymers) offer the advantages over conventional crosslinked materials normally used for release, but also provide the ability to tailor specific bulk material properties, such as release profiles, at the molecular level via monomer design. As a result, design of materials according to the "bottom-up" approach has generated a variety of supramolecular devices for biomedical applications (Branco and Schneider, 2009). Numerous systematic reviews of the progress in biotechnological applications of lipid formulations are available (Caracciolo, 2015; Koshkaryev et al., 2013; Noble et al., 2014; Patil and Jadhav, 2014; Torchilin, 2009; Van der Meel et al., 2014).

In many cases, intracellular delivery of drugs is needed in order to produce therapeutic effect inside cell. However, lipophilic microenvironment of cell membranes prevents the transport of hydrophilic macromolecules. Short peptides may be taken into cell by endocytic pathway (Varga et al., 2000). In this case molecules are included in endosomes and ultimately come in lysosomes, where is subjected to the degradation by lysosome enzymes. For this reason only a very small part of active drugs may enter the cell, and many formulations successfully used in vitro fail in vivo. As an alternative way, microinjection and electroporation can be considered. Their main limitation is that they are invasive and may disrupt cell membranes (Arnheiter and Haller, 1988; Chakrabarti et al., 1989).

The simplest and most attractive approach well documented in recent literature is associated with the use of liposomes (Lasic, 1993; Torchilin, 2005). Liposomes are synthetic vesicles based on lipid molecules with hydrodynamic diameter from 50 nm to submicron dimension, while the size $\leq$200 nm should ideally be controlled. They are characterized by many attractive features, such as low toxicity, biological inertness, ideal biocompatibility, proper size, reliable environmental defense of the drug, and so forth, which enables liposomal formulations to be succeeded in the bed-to-bench transition. Loading the therapeutic macromolecules into liposomes restricts their biodistribution and delays their metabolism (Allen et al., 1995).

According to their size characteristics liposomes are divided into small, large, and giant vesicles, while they are classified as uni-, oligo-, and multilamellar vesicles depending on the number of bilayers included. Based on recent progress in liposome fabrication and biotechnological applications, these nanocarriers may be grouped into conventional, stealth, triggered- or sustained-release, and ligand-targeted vehicles. Methods for liposome preparation may be roughly divided into bulk methods and film methods, with such characteristics as size, lamelarity, and polydispersity ideally controlled (Patil and Jadhav, 2014).

5.1.1 Modified Liposomal Carriers

The survey of recent literature reveals a great number of simple liposomal formulations developed, with some of them being of clinical importance. At the same time an in vivo application assumes that a complex of criteria needs to be answered. This idea underlies the concept of multifunctional nanocarriers developed by Torchilin and coworkers (Perche and Torchilin, 2013; Torchilin, 2012). Ideally, the following functions need to be combined: (1) allowing for long circulation in blood; (2) target specificity; (3) ability of cumulating around diseased tissues; (4) stimuli-responsivity; (5) possibility of intracellular drug delivery; (6) contrast/reporter function.

One of the critical points limiting simple liposome formulations is associated with their entrapping by cells of the reticuloendothelial system and fast elimination from the blood. This problem is usually solved through the conjugation of liposomes with hydrophilic biocompatible polymers, for example, PEG. The so-called PEGylation exerts a multifactor effect, in particular, it may serve as a linker.

However, the main role of the PEG surface coating is a shielding or stealth effect, which reduces uptake by macrophages of the mononuclear phagocyte system and leads to relatively long plasma residence times (Cao and Jiang, 2012; Torchilin, 2012). Importantly, PEG provides the possibility for conjugation with other biospecies, for example, nowadays different PEG modifications bearing end functional moieties capable of substituting are commercially available. Moreover, various PEGylated biopolymers, such as interferons (PEGasys and Intron), growth hormone receptor antagonist (Somavert) are used in biomedical practice (Manjappa et al., 2011).

Due to high extent of hydration of oxyethylene groups the PEGylated particles are characterized by increased water solubility, colloid stability, improved blood residence time, and safety against opsonization, thereby showing expressed stealth effect. Noteworthy that apart from high water solubility PEG is soluble in organic media, which is assumed to result from partially amphiphilic character of polyoxyethylene chains (Cao and Jiang, 2012). Along with PEG and other nonionic polymers (dextran, tetraglyme, mannitol), much attention has recently been paid to zwitter-ionic polymers poly(phosphobetaine), poly(sulfobetaine), and poly(carboxybetaine) (Cao and Jiang, 2012). Unlike PEGs these polymers are not soluble in organic solvents, being superhydrophilic due to tight water binding through electrostatic mechanism.

The protection by polymers makes it possible to develop the long-circulating liposomes, which are characterized by such advantages as dose-independent behavior, log-linear kinetics, and increased bioavailability (Manjappa et al., 2011). When the prolonged circulation of encapsulated drug is succeeded, further criteria should be raised, that is, targeting of drugs toward disease sites. This task is usually solved through the conjugation of ligands to the liposomal surface capable of interacting specifically with the diseased cells. Among such ligands, antibodies or antibody fragments, vitamins, glycoproteins, peptides (RGD-sequences), and oligonucleotides are used. Nowadays, one of the promising approaches in targeted therapy is the use of antibody conjugates, which is especially progressed due to the development of the hybridoma technology. This makes it possible to produce a large quantity of a monoclonal antibody (Köhler and Milstein, 1975).

Alternative small peptide molecules, the so-called affibodies, were introduced in biomedical investigations (Alexis et al., 2008; Belousova et al., 2008; Beuttler et al., 2009; Lee et al., 2008; Puri et al., 2008, 2009; Tolmachev et al., 2007; Wikman et al., 2004). They have some advantages over monoclonal antibody, such as enhanced cell penetration ability resulting from the small size of affibody (MW 6 kDa), presence of functional end groups for chemical conjugation, possibility of the large-scale synthesis (Alexis et al., 2008). In Wikman et al. (2004), an affibody (His6-ZHER2/neu:4) is shown to specifically bind to the HER2 extracellular domain with a nanomolar affinity (~50 nM). The His6-ZHER2/neu:4affibody also showed selective binding to native HER2 on breast cancer cells. Due to the short plasma circulation and fast blood clearance, affibodies are optimal for tumor imaging, but not for the affibody drug conjugates and radiotherapeutics (Tolmachev et al., 2007).

Along with the polymer-protection the additional trends should be emphasized, which are typical of the currently developed protocols. In particular, for in vivo used formulations, such functions as (1) therapeutic moiety, (2) recognizing and binding sites, (3) stimuli responsive function need to be combined. The concept of such "smart multifunctional" nanocarriers composed of polymer barrier, targeting ligand, lipid component, and stimuli responsive function is formulated in recent publications (Perche and Torchilin, 2013; Torchilin, 2012) and reflect a key trend in these fields. This is supported by the analysis of recent publication in the field under review (Cao and Jiang, 2012; Koren and Torchilin, 2012; Milosevits et al., 2015; Vieira et al., 2013). Ducat et al. (2011) describe an effective vector with a prolonged lifetime for intracellular therapeutic peptide delivery. The proposed formulations based on long circulating stealth pH-sensitive liposome were compared with classical liposomal carriers. It was shown that both formulations are similar from the viewpoint of size, morphology, and loading capacity, while stealth pH-sensitive liposome demonstrated enhanced efficacy in terms of cellular uptake exemplified by three cell lines: Hs578t human epithelial cells from breast carcinoma, MDA-MB-231 human breast carcinoma cells and WI-26 human diploid lung fibroblast cells.

The pH-sensitive long-circulating PEG-conjugated liposomes modified with cell-penetrating TAT peptide moieties and cancer-specific mAb 2C5 were developed in (Koren et al., 2012). In this work, a shielding effect produced by PEG chains toward TAT-peptide is modeled at different pH. It was found that the multifunctional preparation provides an enhanced cytotoxicity and carrier uptake by cancer cells (B16-F10, HeLa, and MCF-7), as well as demonstrating the potential for intracellular drug delivery after exposure to lowered pH environment, typical of solid tumors.

In Yao et al. (2013) novel pH dependent liposomes are reported for targeted delivery of therapeutic drugs into acidic diseased cell. In particular, pH dependent fusion of liposome into cell membrane mediated by pH-low insertion peptide (pHLIP) is developed. The decoration of PEGylated liposome by pHLIP stimulates a pH dependent membrane fusion, thereby increasing cellular uptake and payload release and retarding the cell proliferation with the help of ceramide-containing liposomes.

Liposomal carriers for anticancer drug delivery were worked out (Jiang et al., 2012) based on pH-responsive cell-penetrating peptide (CPP) and active targeting hyaluronic acid (HA). To fabricate R6H4 (RRRRRRHHHH) modified liposome steraylR6H4 was hydrophobically bound to liposome. To shield the positive charge of R6H4 it was treated with HA for the protection from plasma proteins. At tumor microenvironment enriched by hyaluronidase the HA shielding is switched off due to its fast degradation, which resulted in the initiating of the R6H4cell-penetrating ability.

5.2 OTHER LIPID-BASED FORMULATIONS

5.2.1 Solid lipid nanoparticles

Solid lipid nanoparticles (SLNs) are nanocarriers being solid at ambient and physiological temperature; they represent advanced alternative to other amphiphile- and polymer-based formulations (De Jesus and Zuhorn, 2015; Gastaldi et al., 2014; Liang et al., 2015; Rosiaux et al., 2014; Rostamia et al., 2014; Summerlin et al., 2015). Solid lipid nanoparticles were initially proposed for hydrophobic loads, and this trend is currently preserved as well (Yucel et al., 2013). Meanwhile, analysis of recent publications reveals their binding capacity toward guests of different nature, from small molecules to protein drugs, including both hydrophilic and hydrophobic substances. As reported in Almeida and Souto (2007) SLN technique is successfully used for transport of hydrophilic macromolecules, including therapeutic peptides, which are assumed to be loaded in inner aqueous phase. In the case of hydrophilic biomolecules, factorial design can be used for the fabrication of SLN (Fangueiro et al., 2012) by means of optimization of the concentration of solid lipid, lipophilic, and hydrophilic components, with a number of compositions being screened. Problem of compromise between the limited lipid content and the droplet stability was solved by the use of two emulsifiers with different hydrophilic-hydrophobic balance, capable of stabilizing either oil/water or water/oil emulsions.

Importantly, SLNs preserved benefits of various families of lipid carriers, while being free of some limitations characteristic of them. An attractive feature of these carriers is that SLNs are readily functionalized with specific ligands (e.g., antibody), which allows them to be targeted to deceased cells. In addition, SLNs can be adopted for different routes of administration, and therefore they may be used for a wide range of dangerous diseases (Gastaldi et al., 2014; Liang et al., 2015).

5.2.2 Niosomes

Niosomes are vesicular systems composed of nonionic surfactants admixed with lipid molecules (e.g., cholesterol) and/or ionic additives (charge inducers). According to a large number of characteristics, niosomes are very similar to liposomes, in particular both have bilayer structure and are capable of incorporating the substances of different nature, that is, hydrophilic (in the inner water pool) and lipophilic (in hydrophobic domains formed by alkyl tails). At the same time there are principle advantages of niosomes over liposomal carriers, such as lower cost, easy surface functionalization, reduced toxicity, chemical stability, easy handling, scalability, and so forth (Marianecci et al., 2014; Moghassemi and Hadjizadeh, 2014; Shingitha, 2015). To date, different methods of niosome preparation are developed, which generally can influence the properties of formulations. Among them, two main groups can be considered. First strategy assumes dissolving of all ingredients in organic medium, followed by removing the solvent and thin film deposition. Further hydration of the film results in the niosome formation (Moghassemi and Hadjizadeh, 2014). This approach covers such methods as thin layer hydration, reverse phase evaporation, ether injection, and so forth. According to the second strategy, no organic solvent is needed. Instead, all the components are mixed with hydration medium upon heating, which can be exemplified by bubble technique, promosome strategy, and so forth.

Niosome-mediated drug delivery is reported to be suitable for different routes of administration, including transdermal, oral, ocular, pulmonary, and parenteral, with particularly dermal and transdermal routes being the focus of clinical trials (Marianecci et al., 2014). Recent literature has overviewed numerous successful efforts of academic researchers and technologists directed toward niosomal delivery and targeting of various active therapeutics including proteins (insulin, cyclosporine, trypsin, bacitracin), which characterizes these formulations as prospective nanocarriers for this family of drugs (Shilpa et al., 2011).

5.3 POLYMER-BASED DELIVERY SYSTEMS FOR PEPTIDE DRUGS

Since the mid 1970s polymer nanoparticles (NPs) have received much attention due to their ability to enhance the efficacy of a drug, especially therapeutic macromolecules including polypeptides, nucleic acids, proteins, hormones, and so forth (Yamamoto et al., 2012). Such advantages of polymer nanocarriers as high stability, possibility of simple cellular uptake through endocytosis pathway, targeted delivery to specific tissues by adsorption and/or grafting with ligand have been elucidated. Special attention was paid to biodegradable nanoparticles based on poly(lactic acid) (PLA), poly(glycolic acid), PLGA, alginate, chitosan, and dextran (Kumari et al., 2010; Danhier et al., 2012). Hudson and Margaritis (2014) present detailed review on the influence of polymer structure and medium conditions on physicochemical properties, benefits of natural polymer-based NPs over synthetic polymers, as well as methods of preparations of NPs.

5.3.1 Mucoadhesive material

The attractiveness for patients of the oral transmucosal delivery of drugs stimulated the development of various dosage forms differing in their preparation, morphology, phase state, and composition, that is, solutions, tablets/lozenges, chewing gums, sprays, films, hydrogels, hollow fibers, and microspheres

(Patel et al., 2011; Sudhakar et al., 2006). Mucoadhesive polymers are classified into two types, namely, the first-generation mucoadhesive polymers and second-generation novel mucoadhesive polymers, with the first generation being divided into three basic categories: anionic, cationic, and uncharged polymers. As reported in Ludwig (2005), ionic polymers have improved mucoadhesion.

Due to their high mucoadhesive activity and low toxicity the anionic polymers are widely used in pharmaceutical formulations. Among these polymers polyacrylic acid and sodium carboxymethyl cellulose are characterized by the formation of strong hydrogen bonding interactions with mucin. Among cationic polymers the most popular is chitosan and its functionalized forms, which will be discussed later.

Thiolated polymers (thiomers) are the type of second-generation mucoadhesive material derived from hydrophilic polymers, such as polyacrylates, chitosan, or deacetylated gellan gum (Leitner et al., 2003). The presence of thiol groups allows for the formation of covalent bonds with cysteine rich subdomains of the mucus gel layer, thereby leading to increased residence time and improved bioavailability (Albrecht et al., 2006). While first-generation mucoadhesive platforms are facilitated via noncovalent secondary interactions, the covalent bonding mechanisms involved in second-generation systems lead to interactions that are less susceptible to changes in ionic strength and/or the pH (Mikos and Peppas, 1990; Roldo et al., 2004). As reviewed in Dünnhaupt et al. (2012), S-protected thiomers represent a promising type of mucoadhesive polymers for the development of various mucosal drug delivery systems.

5.3.1.1 Chitosan nanoparticles

Chitosan is nontoxic biocompatible polysaccharide bearing positively charged N-acetylglucosamine and D-glucosamine moieties that provide mucoadhesive properties due to ionic interactions with the negatively charged sialic-acid groups in mucin and the ability to open the tight junctions, thereby improving the cellular uptake of biopharmaceutics (Amidi et al., 2010; Garcia-Fuentes and Alonso 2012; Van der Lubben et al., 2001). Due to pK_a values of 6.4, in neutral and alkali media chitosan becomes insoluble and undergoes precipitation. To avoid this unfavorable behavior, quaternized chitosan is usually synthesized with sufficient solubility within wide range of solution pH (Mahjub et al., 2014). Nanoparticles fabricated through this way demonstrated higher insulin intestinal adsorption compared to unmodified polymer. Application of chitosan nanoparticles for delivery of therapeutic macromolecules across the intestinal canal is reviewed in Chen et al. (2013e), including such aspects as synthesis of water-soluble chitosan derivatives, transport mechanism through intestinal epithelium, as well as different routes of peptide and proteins administration, with oral delivery emphasized.

As mentioned earlier, along with conventional nonfunctionalized chitosan different derivatives are synthesized, for example, trimethylated or thiolated chitosane, as well as hybrid nanoparticles based on chitosan admixed with other saccharides, for example, chitosan/CD systems (Goycoolea et al., 2009; Krauland and Alonso, 2007; Teijeiro-Osorio et al., 2009). Thiolated polymers (thiomers) were introduced in pharmaceutical literature in 1990s (Bernkop-Schnürch et al., 1999; Gum et al., 1992). They provide strong adhesion, which is sufficient to guarantee the localization of dosage form at a given target site for prolonged time. Thus, an enhanced oral bioavailability is expected, resulting in many cases in a reduced dosing frequency and patient-compliance (Bernkop-Schnürch et al., 2004). Thiomers exhibit furthermore enzyme inhibitory, permeation enhancing, and efflux pump inhibitory properties (Borchard et al., 1996). Additionally, thiomers act as protective shield for incorporated drugs, especially for peptides and proteins, toward an enzymatic attack in the intestine (Bernkop-Schnürch, 1999).

The study by Iqbal et al. (2012) summarizes the synthesis and characterization of polymeric excipients with improved mucoadhesive, cohesive, and in situ–gelling properties. In particular, preactivated poly(acrylic acid)-cysteine-2-mercaptonicotinic acid conjugates were designed and obtained, which makes it possible to provide an increased uptake and improved oral bioavailability of therapeutic peptides and proteins. Unlike the earlier known thiomers, the preactivated analog contains disulfide bonds that are unable to further oxidize at high pH.

5.3.1.2 PLGA nanoparticles

PLGA is the FDA-approved copolymer that attracts much research interest due to (1) biocompatibility and ability to biodegrade with the formation of cleavable monomers, lactic and glycolic acids; (2) certification for clinical application via parenteral administration; (3) adaptability toward different types of preparation of both hydrophilic and lipophilic nature; (4) possibility of protecting the preparation against degradation; (5) allowing the prolonged release; and (6) targeted delivery of drug (Danhier et al., 2012). There are numerous examples of encapsulation of Paclitaxel, Xanthones, Cisplatin, diabetes drug insulin, psychotic drugs haloperidol, hormones, and so forth on PLGA nanoparticles (Kumari et al., 2010). Different techniques are developed for the fabrication of PLGA nanoparticles, with the emulsion solvent diffusion (ESD) method the most widely used. ESD technique is based on polymer spherical crystallization (Kawashima et al., 1998) and makes it possible to obtain particles of uniform sizes and to modify their surface. Importantly, this method is simple and can occur under mild conditions without sonication and homogenization. Meanwhile, in accordance with clinical standards (high loading capacity, sustained release, etc.) more innovative methods of preparation can be used. In (Hong et al., 2013) new technique for fabrication of the emulsion type carriers was developed to achieve the effective loading of drug and prolonged release by varying the ratio between hydrophilic oil and PLGA.

Self-healing PLGA-based microspheres were proposed recently as a unique class of nanocarriers capable of effective encapsulation of drugs including proteins. This may be exemplified by a study by Shah and Schwendeman (2014), in which biomimetic approach was developed for the self-microencapsulation of proteins in PLGA/biopolymer microspheres. Double water-oil-water emulsion method was used, with the nature of biopolymers and PLGA/biopolymer ratio optimized. Simple protocol developed by Reinhold and Schwendeman (2013) makes it possible to prepare the drug-free PLGA microspheres of varied porosity ($\in = 0.49$–73) with further heating them in protein solution. The $\in$ value is shown to be a key parameter controlling the efficacy of these formulations.

One of the challenging problems in the process of encapsulation of proteins is the loss of their activity upon the processing (high temperature and/or shear forces) and uncontrolled burst release of loads. Therefore a variety of publications focus on these aspects. Generally, some of disadvantages can be avoided by the PEGylation procedure, which increases the stability of formulated drugs (Lee et al., 2015). Surface decoration of PLGA nanospheres with hydrophobized polyethylene imine was proposed in Andersen et al. (2010) for delivery of siRNA, with increasing stability and lowering toxicity of polycation achieved. Review by Ma (2014) covers the strategy directed toward the achievement of protein/peptide functional activity, uniform size distribution, reproducibility, and controlled drug release upon encapsulating within biodegradable polymer microspheres based on PLA, PLGA, and chitosan. Due to mild conditions, membrane emulsification techniques developed by the author group demonstrated advantages over conventional mechanical stirring methods, thereby guaranteeing high encapsulation efficacy and preserving the bioactivity of therapeutic proteins. Animal tests revealed that high protein concentration in the blood may be maintained within two months in the case of the formulated preparation.

Different ways are developed to modulate the release profile of encapsulated drugs. This may be exemplified by work by Wan et al. (2014), in which injectable BSA-loaded PLGA microcapsules were obtained with the initial burst release of model drug observed. The surface and rheological properties of capsules were modified by hydrophilic polymer HA that probably suppressed the enrichment of drug within the periphery region of PLGA capsules. Improved release profile was achieved in study by Sophocleous et al. (2013) devoted to the engineering of PLGA microparticles bearing terminate COOH groups for the encapsulation of positively charged protein drugs. It was shown that due to electrostatic interaction of leuprolide with low-molecular-weight PLGA acid groups (ca. 70% of COOH moieties were involved), the salt-like structures were formed, thereby promoting the integration of the therapeutic peptides into the capsules' interior and preventing an initial burst release of the drug. Advanced PLGA/Pluronic F-127 formulations were fabricated for sustained delivery of thrombin inhibitors heparin and hirudin (Sellers et al., 2014). The PLGA/Pluronic F-127 gels demonstrated accelerated functional recovery after traumatic injury compared to individual carriers.

5.3.1.3 Alginate nanoparticles

Alginates are anionic polysaccharides, salts of alginic acid that can readily be gelled in the presence of multicharged cations. The most convenient and patient-friendly oral route of drug administration involving alginate particles is overviewed by Sosnik (2014). These strategies were exemplified by a few drugs, especially by insulin, with limitations and perspectives of these formulations emphasized. Unlike PLGA nano- and microparticles, the most alginate-based particles have large sizes of $\geq 100\ \mu m$, which considerably limits their application in biomedicine. Review (Paques et al., 2014) focuses on strategies devoted to the fabrication of nanosized alginate particles. Generally, two ways are used to prepare alginate nanoparticles, that is, (1) complexation method occurring in bulk solution or in the surface, with cross-linking agents (calcium chloride or oppositely charged polyelectrolyte chitosan) involved; and (2) emulsification technique in combination with gelation of alginate emulsion droplet. Advanced complex alginate/quaternized chitosan nanoparticles were fabricated by Mansourpour et al. (2015) for insulin encapsulation, in which beneficial properties of chitosan (mucoadhesiveness) and alginate (Ca-cation induced gelation) were combined for the protection of insulin under acidic conditions. Additional effect was achieved by complexation of insulin with β-CD that prevented denaturation and degradation of the drug.

5.3.1.4 Nanoparticles based on hyaluronic acid

Hyaluronic acid (HA) is a widely used nonbranched polysaccharide with functionalizable, biodegradable, biocompatible, low toxic, and nonimmunogenic characteristics. A variety of reviews (Chen et al., 2014c; Oh et al., 2010) focus on biomedical applications of HA and its conjugates as promising nanocarriers for biopolymers (proteins, peptides, nucleic acids), with different routes of administration and different biotargets involved. Specifically, Zhang et al. (2014) concentrated on the prodrug strategies, with bioconjugates of HA with doxorubicin, curcumin, butyric acid, cisplatin, and other drug molecules considered. Similar to the widely used PEGylation procedure, HA conjugation (HAylation) attracts much research activity resulting in prolonged release and suppression of immunogenicity of protein drugs. HA-insulin conjugates with different loadings were obtained in study by Mero et al. (2013) and compared with blood glucose level controlled by unformulated drug. The six fold higher efficacy of the former was demonstrated, which testified to the high potentiality of the strategy. In other work (Yang et al., 2011), interferon alpha/HA conjugate was synthesized (conjugation efficacy ≥95%) as anti hepatitis C virus preparation, with secondary structure of the protein preserved. An in vitro study demonstrated lower functional activity of the conjugated drug compared to native interferon, however, it was competitive with that of PEGylated preparation. A bioimaging study by Kim et al. (2015b) of HAylated interferon alpha demonstrated the target-specific delivery of the protein to the liver and revealed beneficial release behavior compared to PEGylated drug.

5.3.1.5 Layer-by-layer strategy

One of the advanced approaches to the vehicle fabrication is a layer-by-layer strategy, which makes it possible to control the particle size, loading and release capacity, surface and physicochemical properties. The layer deposition may be contributed by different factors including electrostatic (Halozan et al., 2005; Peng et al., 2010; Yu et al., 2008) or hydrogen (Kozlovskaya and Sukhishvili, 2006; Kozlovskaya et al., 2009) bonding, host–guest interaction (Wang et al., 2008) or chemical modification (Kinnane et al., 2009). The combination of electrostatic mechanism and covalent cross-linking was used for the capsule fabrication based on poly(N-vinyl pyrrolidone) and poly(methacrylic acid) (Chong et al., 2009; Connal et al., 2009; Zelikin et al., 2006). Stimuli-responsive microcapsules based on poly(allylamine hydrochloride) were designed through host-guest interactions of β-CD/ferrocene pair (Wang et al., 2008). This group also fabricated the capsules by alternate deposition of two natural polysaccharides, that is, chitosan and alginate at poly(lactide-coglycolide) nanoparticles (Zhou et al., 2010). The interactions of as-designed nanoparticles with HepG2 cell line were investigated and revealed that the cellular uptake is markedly increased with the additional nanoparticle modification with the folic acid grafted PEG.

5.4 ALTERNATIVE NONINVASIVE ROUTES OF DELIVERY

5.4.1 Buccal delivery of peptides

As mentioned earlier, peptide drugs with rather high molecular weight characterized by biodegradability and low penetration/adsorption ability are difficult to administrate via the oral route. Therefore, alternative transport mechanisms need to be considered. Oral cavity is assumed to be an effective and convenient site for noninvasive delivery of macromolecular drugs due to simplicity of the administration. Importantly, oral mucosa is highly vascularized and characterized by lower enzymatic activity compared to intestinal, rectal, and nasal epithelium. At the same time the buccal route have some limitations, such as low adsorption surface and occurrence of mucosa barrier. Hence, the design of nanocarriers preventing these shortcomings has been the focus of recent research (Rossi et al., 2005). A review by Nair et al. (2013) outlines different techniques of fabrication of buccal films demonstrating such advantages as convenient dosing and patient-friendly drug intake. Authors focus on quantitative criteria for the evaluation of physical characteristics, efficacy, and safety of buccal films, as well as release profile, mucoadhesion, and other functionalities. Nowadays, different strategies are developed for overcoming limitations of the buccal route, including use of chemical agents enhancing the penetration ability (surfactants, fatty acids, and chitosan), physical methods, for example, supersaturation, iontophoresis, electroporation, and so forth (Giannola et al., 2013), as well as application of materials combining mucoadhesive properties, enzyme-inhibition, and penetrating activity (Caon et al., 2015).

5.4.2 Airway delivery of peptides

Respiratory tract provides an easy noninvasive route for delivery of macromolecular drug due to reduced acidity and proteolytic activity, thinner mucus layer, and so forth. Review by Dombu and Betbeder (2013) emphasizes that the majority of advanced examples of airway delivery refer to peptides and proteins with relatively low molecular mass, especially insulin (5.8 kDa), while systemic administration of larger peptide and protein molecules through the respiratory tract is a challenging problem due to the occurrence of biological barriers. These challenges can be overcome by the use of particulate protein drug delivery systems, with nanoparticles offering advantages over larger particles. The small size of nanocarriers allows them to reach different regions of the respiratory tract, avoiding clearance and phagocytosis. Meanwhile, processes of endocytosis and phagocytosis depend on the nature and properties of nanoparticles (Fröhlich, 2015). One of prospective ways of treatment of respiratory deceases is the use of inhaled peptides and proteins. This research direction can be exemplified by the study of peptide/dextran formulations by Kuehl et al. (2014), optimization of their size and composition, and comparison with commercial product. The stability characteristics of the novel dry-powder formulation was shown to be improved compared to control, while their functionalities quite competitive. Pulmonary delivery of protein for both local and systemic applications is a subject of recent publications, since this route may prevent the drug from biodegradation. In particular, Amancha et al. (2014) report on prolonged release of insulin encapsulated by the layer-by-layer technique, which has been achieved via pulmonary administration. Pharmacokinetics and pharmacodynamics of the drug were monitored, with the sustained level of insulin revealed. Other research groups developed lyophilized powders based on amphiphilic polymers for inhaled insulin (Andrade et al., 2015). A formulation based on Pluronic F-127 demonstrated enhanced penetration ability and reduced biodegradability.

5.4.3 Transdermal delivery

Transdermal drug delivery is a promising alternative that provides various advantages over other methods of administration including ease of access, large surface area, reduced side effect, bypassing enzyme action, and so forth. However, this route of drug delivery meets difficulties connected with occurrence of barrier property of the stratum corneum, the outermost layer of the skin. Generally, these limitations are avoided by using different physical or chemical treatments (Nino et al., 2010). Skin permeation was reported to be enhanced by involving electroporation, iontophoresis, sonophoresis, laser technique, and so forth (Wong, 2014). Chemical enhancer, for example, DMSO, terpenes, oleic acid, propylene glycol mediate the drug delivery via different mechanisms, including increase in diffusivity of formulations through stratum corneum, improving of partition coefficient or solubility of drugs (Nino et al., 2010). Transdermal delivery of hydrophilic substances causes especial problems due to their inability of integrating with cell membrane and penetrating through stratum corneum. This problem has been addressed in recent study (Candan et al., 2012) that demonstrated effectiveness of transdermal delivery of antimelanogenesis agent hydroquinone and green fluorescent protein using a cell-penetrating peptide poly-arginine (11R) in combination with 4-(1-pyrenyl)-butyric acid. In (Choi et al., 2012) two hydrophilic proteins fluorescein isothiocyanate-labeled BSA (67 kDa) and insulin (6 kDa) were transdermally delivered with the use of nanogel based on Pluronic F-127–chitosan conjugate. Both permeation essay and bioactivity tests reveal the efficacy of the developed nanocarriers.

Innovative approaches for enhancing the skin permeability involve the use of microneedles that puncture the skin, thereby forming channels for the penetration of macromolecular drugs (Narayan, 2014). Microneedle strategies offer a variety of benefits for patients including painless administration, possibility of unassisted use, convenient fashions of the devices, for example, a microneedle patch, and so forth. Different types of microneedles are used including hollow, solid, and polymer coated. The matter of special importance is the choice of materials for their fabrication, such as silicon, metal, zeolite, and polymers. Marked preferences received dissolving devices based on biocompatible nontoxic polymers (McGrath et al., 2014), for example, polysaccharides, silk protein (Raja et al., 2013), starch and gelatin (Ling and Chen, 2013). A separate line of research activity concentrated on improving the process of microneedle fabrication and administration, in particular on the safety of application, avoiding of organic solvent and high temperature (Liu et al., 2014b), and using of controllable methods to coat microneedles (Chen et al., 2015a).

6 PENETRATION THROUGH BBB

The blood–brain barrier (BBB) is a dynamic barrier that protects the brain from harmful substances and organisms and prevents cerebral drug circulation. Despite a huge breakthrough in the field of nanotechnology and the different approaches in treatment of diseases related to the central nervous system (CNS), the major challenges remain to be overcome, namely, the inability of many therapeutic molecules to cross the BBB. It is known that a large number of newly developed drugs for the treatment of disorders of the CNS were suspended in clinical trials because of the inability to pass through the BBB. Generally, preparations based on macromolecules that are unable to penetrate the BBB are used for the treatment of CNS diseases. It is thus necessary to look for new strategies for delivery of large molecules for the treatment and diagnosis of diseases of the CNS. Currently, there are several reviews on contemporary nanotechnology approaches for macromolecules to cross the BBB (Ajetunmobi et al., 2014; Gendelman et al., 2015). The reviews on the mechanisms of penetration of drugs across the BBB, such as adsorption, receptor-mediated endocytosis followed by transcytosis, and so forth are particularly important (Fig. 24.7) (Kreuter, 2014). Particular attention is paid to cell-mediated approach (Chen and Liu, 2012).

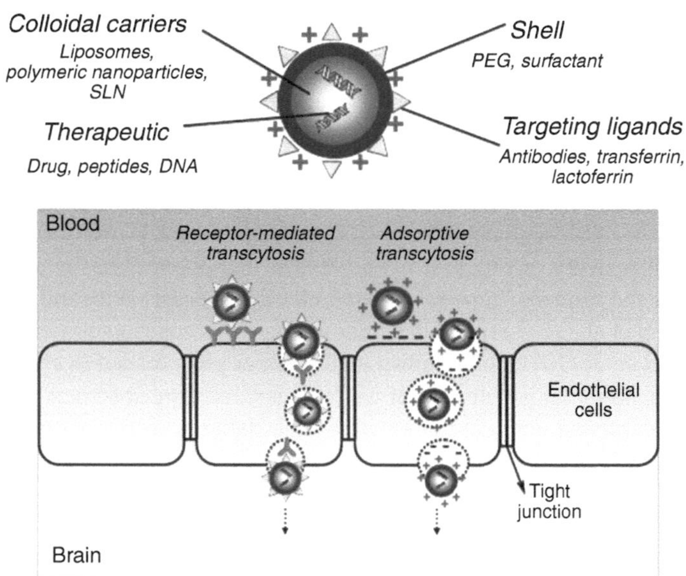

FIGURE 24.7 Schematic Representation of Transport of Colloidal Carriers Across Blood–Brain Barrier

6.1 DIFFERENT PATHWAYS THROUGH THE BBB

BBB is a complex multibarrier, consisting of transport, metabolic or enzymatic and immunological barriers. At present, the various passing mechanisms, which are primarily dependent on the structure of penetrating molecule, are known and described. For example, small water-soluble molecules simply diffuse through the BBB. Substances, such as lipids, hormones, and alcohols penetrate transcellularly, that is, by dissolving and oxidation in the plasma membrane. Special transport system (GLUT-1 transporter) or protein molecules are located on the surface of the cell membrane for directed diffusion of large polar molecules (such as glucose or amino acids). The mass transport takes place as a result of conformational changes in the protein caused by binding of the transporters on one side of the membrane. These transporters can carry out the transfer of substances in one or two directions. However, besides the passive transport, there is the active transport of substances requiring cellular energy costs obtained through the ATP molecule. In this case, the transfer is due to the active substance transporters, such as Efflux transporters.

Several penetration mechanisms for macromolecules are described. One of those mechanisms is the receptor-mediated transcytosis providing selective uptake of macromolecules through receptors that are located in endothelial cells and utilized various types of ligands including growth factors, enzymes, and proteins.

Adsorption-mediated transcytosis or pinocytosys is the electrostatic interaction between the positively charged materials with negatively charged surface of the membrane. In fact, most of the existing nanotechnological drug delivery systems focused on adsorptive-mediated transcytosis. Typically, such carriers contain cationic proteins or cell-penetrating peptides.

Finally, cell-mediated transcytosis is a recently described route for BBB penetration. This transport relies on the absorption of solid eukaryotic cells, bacteria, viruses, and so forth. Cell-mediated transcytosis is unique in that it can be used for virtually any type of molecules or materials, as well as carrier systems.

A review of the literature reveals that currently the various invasive physical methods (intra-cerebroventricular infusion, convection-enhanced delivery, and polymer or microchip systems after implantation) are used for drug delivery. However, they have several disadvantages: poor diffusion, pain, high cost, and so forth. Therefore, such noninvasive ways to deliver drugs that may circumvent the BBB, as the nasal route, are the most commonly used. Recently, the therapeutic effects in the treatment of Alzheimer's disease, when insulin is administered intranasally, were shown (Craft et al., 2012; Freiherr et al., 2013). At the same time, the residence time of drugs in the nasal cavity is limited due to the mucociliary clearance. The use of biodegradable polymers, such as chitosan is one of the possible approaches that enhance intranasal delivery of drugs. It is known that this polymer has mucoadhesive properties, which allow increasing the residence time of nanoparticles in the nasal cavity and improving the absorption of the drug. Trapani et al. (2011) found that chitosan particles contribute to improved penetration of the BBB. Recently, Md et al. (2013) have shown that bromocriptine loaded in chitosan nanoparticles is capable of noninvasive intranasal penetration of the BBB and of being absorbed by brain cells.

It is known that lactoferrin conjugation enhances the delivery to the epithelial cells of both the respiratory tract (Elfinger et al., 2007) and brain (Huang et al., 2013b). The authors synthesized lactoferrin-conjugated PEG-PLGA particles, demonstrating a promising delivery of neuroprotective peptide (NAPVSIPQ) to the brain via the nasal route (Liu et al., 2013c).

All existing noninvasive methods can be subdivided into: (1) pharmaceutical methods, which are based on a modification of the drug or use of drug carriers, such as liposomes or nanoparticles; (2)

barrier methods to change the structure of the BBB, such as temporarily increasing its permeability by ultrasound or electromagnetic modulators; (3) physiological methods in which transporters, receptors, or enzymes occurring in the endothelium of the brain are used to bypass the BBB (Soddu et al., 2015).

6.2 COLLOIDAL SYSTEMS FOR THE DELIVERY OF PEPTIDES THROUGH THE BBB

From a wide range of existing types of nanoparticles the considerable attention for delivery of therapeutic compounds to the brain is paid to the liposomes (Fonseca-Santos et al., 2015), nanoparticles based on biopolymers, such as chitosan (Sarvaiya and Agrawal, 2015), PLA, poly(D,L-lactic acid–glycolic acid) (Kreuter, 2014), and solid lipid nanoparticles (Gastaldi et al., 2014; Patel et al., 2013). Nonbiodegradable particles, such as fullerenes, metal particles, quantum dots, needle delivery system, and carbon nanotubes cannot be used as delivery systems to the brain (Wohlfart et al., 2012).

Gold nanoparticles may have advantages in biocompatibility and nontoxicity compared with other metal particles. Synthetic gold nanoparticles functionalized by transferrin and antibodies have the ability to be transported across the BBB via receptor-mediated endocytosis and can diagnose molecular targets in brain tumors (Tripathi et al., 2015). It is known that the penetration of colloidal particles through the BBB depends on their characteristics (size and charge), methods of preparation, crosslinking agents, and also on the chemical structure of the compounds used (Sarvaiya and Agrawal, 2015). For example, unlike other derivatives of chitosan, the ammonium salt-based delivery system facilitate the oral absorption of the peptide by protecting it from gastrointestinal degradation, increasing the residence time and facilitating its transport into the brain (Lalatsa et al., 2012).

A liposome made of bolaamphiphiles (amphiphilic molecules with two head groups) is another high profile example. Such compounds form monolayer membranes that can overcome some of the disadvantages of liposome with bilayer membrane. Single-layer arrangement of the membrane provides high stability. Furthermore, the encapsulation capacity of the liposomes with a single-layer membrane is higher than those of liposome with bilayer membrane. Another advantage of the liposomes with single-layer membrane is the ability to provide controlled release of the encapsulated agents (Fariya et al., 2014). The apparent advantage of the liposomes obtained from bolaamphiphiles is their ability to penetrate the BBB and deliver a peptide/protein material (Dakwar et al., 2012; Popov et al., 2013) and the therapeutic agent (tenofovir) (Heldman et al., 2014) to endothelial cells of the brain. It is known that surface modification of nanoparticles also plays an important role in the transport of nanoparticles and liposomes through the BBB. In order to improve the penetration of colloidal systems through the BBB the functionalization by PEG or surfactants is carried out (Fig. 24.7), for example, by polysorbate 80 or poloxomer 188. Typically, the PEG-coated nanoparticles increase the circulation time of nanoparticles in the blood. However, PEGylation of liposomes and nanoparticles increases their hydrophilicity, which does not make them ideal for delivery to the brain and the BBB.

The ability of nanoparticulate systems to be transported across the BBB can be improved by using different ligands attached to the nanoparticle surface. The surface modification by antibodies or other ligands is one of the alternative approaches to improve transport of drugs through the BBB (Loureiro et al., 2014; Lu et al., 2014). The pharmacokinetic properties of PEGylated cationic solid lipid nanoparticles, modified by antibodies were investigated. It is found that OX26 conjugation has greatly improved absorption of baicalin through the BBB into the brain (Liu et al., 2015b).

As described earlier, receptor-mediated transcytosis is efficient and site-specific drug transport method to the brain. Endothelial cells of the brain contain receptors for specific recognition of ligands, such as transferrin, lactoferrin, and so forth. Thus, a series of works showed that receptor-mediated transcytosis and rapid passage through the BBB can be due to cross-linking (conjugation) of nanoparticles with transferrin and antitransferrin receptors, monoclonal antibodies (TfRMAb), acting against the transferrin receptor-1 on the endothelium of the brain. Accumulation of antitransferrin receptor Ri7 mate in the brain has been confirmed in a paper from Alata et al. (2014). They used the in situ brain perfusion technique and fluorescently labeled proteins. Other researchers have developed nanoparticles of chitosan modified with PEG and transferrin receptor downloaded by oligonucleotides to achieve the brain parenchyma. It is found that the resultant system may be directed via receptors to the endothelial brain capillary by monoclonal antibody as the result of a streptavidin-biotin binding (Kozlu et al., 2014). High absorption of OX26-conjugated PEG-chitosan nanoparticles in the brain accounts for the circulating properties of PEG, reacting of chitosan with negatively charged sites of the brain, as well as their affinity for the receptor OX26 TfR (Monsalve et al., 2015). Ye et al. (2013) demonstrated that receptor-mediated transcytosis is one of the most effective ways to cellular uptake. Transport across the BBB is greatly improved by lactoferrin conjugated with CD and PEG (Ye et al., 2013).

Efficient transport of peptides (fibroblast growth factor (bFGF) and inhibitor of caspase-3 (z-DEVD-FMK) into the brain and the successful application for neuroprotection against stroke were shown using chitosan nanoparticles conjugated with antibodies (Yemisci et al., 2015).

However, inclusion of large proteins, such as transferrin, in the particle has many disadvantages, including the process of synthesis, stability, and immunological response. In the work (Liu et al., 2013d) it was suggested to use a peptide B6 as a transferrin substitute for its conjugation with PEG-PLA nanoparticles to improve the delivery of neuroprotective peptide. These particles showed a significantly higher accumulation in brain endothelial cells due to lipid-mediated endocytosis.

Short peptides, rabies virus glycoproteins (RVG), were used as the conjugates for binding to the acetylcholine receptor in neural cells and for targeting ligand to brain (Kim et al., 2013). The developed multifunctional system based on Pluronic, chitosan, and RVG-conjugated peptide to deliver the β-galactosidase contributed to a significant increase in the efficiency of targeted delivery and accumulation of β-galactosidase in the brain.

Combined approaches for the transport drugs through the BBB are also described in the literature. A new approach using microbubble-enhanced unfocused ultrasound technology for further delivery to the brain of poly PEG-PLA nanoparticles functionalized by amyloid β-antibody and loaded with fluorescent probe was used. Analysis of transmission electron microscopy showed that the high accumulation of nanoparticles in the brain is largely mediated by ultrasonic cavitation of microbubbles, whereby the mechanical stretching of the vessel wall and, therefore, the induced cellular transcytosis nanoparticles occurs (Yao et al., 2014). Another paper reported that the new noninvasive method enhances delivery of a therapeutic agent (doxorubicin) through BBB by desoctanoyl ghrelin-conjugated microbubbles-carrying TGFβ1 inhibitor (LY364947) in combination with ultrasound. It was shown that the resulting system can overcome two barriers: blood-to-brain and brain-to-glioma, which allows for effective accumulation of doxorubicin in tumor cells (Chen et al., 2015b).

New techniques and new methods of the analytical control are designed to create delivery systems for crossing the BBB. Loperamide-loaded PLGA nanoparticles having incorporated Polysorbate

80, capable of being efficiently transported across the BBB into the CNS, was obtained by the low-energy method, nanoemulsification, and the ability to produce a large scale was shown (Fornaguera et al., 2015). In other work, the use of an analytical method for fixing curcumin loaded in PLGA-particles was demonstrated. Curcumin was determined in liver, heart, spleen, lung, kidney, and brain. It was found that the use of nanocarriers for downloading curcumin can increase the efficiency of accumulation in all organs except the heart (Tsai et al., 2011).

7 CONCLUSIONS

This chapter focuses on recent trends in the field of drug delivery mediated by confined systems based on amphiphiles and polymers. Special emphasis is placed on synthetic surfactants and lipid formulations. The structure-activity correlation is exemplified by cationic surfactant series, with the benefits of amphiphiles with natural fragments emphasized. Survey of recent literature revealed that researchers tend to design amphiphilic compounds corresponding to the criteria of green chemistry, thereby focusing on the dimeric (gemini) or oligomeric surfactants of low toxicity, with low aggregation threshold, capable of multicentered interactions with active molecules. The application of cationic amphiphiles as DNA complexation agents is a special research direction that has progressed most markedly. Along with typical surfactants macrocyclic amphiphilic compounds are widely used as building blocks due to their easy functionalization, specific model of association, low toxicity, versatile morphological behavior, and host–guest selective interaction with therapeutic molecules. Design of drug delivery vehicles assumes the control of their binding/release properties. Therefore, stimuli responsivity is an important criterion that should be taken into account. The binding/release behavior of different formulations is elucidated including micelles, liposomes, polymer capsules, and so forth, and various mechanisms of the control are discussed, such as pH-triggered, heat-stimulative, light-induced release of loads. One of the main themes of the review is the delivery of peptide therapeutics, with the liposomal formulations receiving special attention. Besides simple liposomes, multifunctional nanocarriers are thoroughly discussed based on such criteria as long-time circulation in blood, target specificity, stimuli responsivity, possibility of intracellular delivery, and contrast/reporter functionality. Polymeric delivery systems for peptide drugs are shown to be highly beneficial as well, due to their high stability and targeted delivery to specific tissue through adsorption and grafting with ligands. We focused on biodegradable polysaccharide nanoparticles demonstrating superior mucoadhesive characteristics. Different noninvasive administration routes of peptide delivery are considered in the chapter, including buccal, airway, and transdermal ways. Up to the present, the treatment of diseases of CNS remains a challenging task. Therefore the chapter addressed the problem of the therapeutics penetration across the BBB, with different pathways and various delivery systems considered. Solely biodegradable nanoparticles are needed in this case, which offers numerous benefits for colloidal systems, especially lipid formulations, over other carriers.

ACKNOWLEDGMENT

This work is supported by the Russian Scientific Foundation; Project no. 14-50-00014.

REFERENCES

Ablakova, Yu.R., Mirgorodskaya, A.B., Zakharova, L.Ya., Valeeva, F.G., 2010. Systems based on nonionic amphiphilic compounds: aggregation and catalytic properties. Russ. Chem. Bull. 59 (4), 784–789.

Adessi, C., Sotto, C., 2002. Converting a peptide into a drug: strategies to improve stability and bioavailability. Curr. Med. Chem. 9 (9), 963–978.

Adessi, C., Sotto, C., 2004. Strategies to improve stability and bioavailability of peptide drugs. Front. Med. Chem. 1 (1), 513–527.

Ajetunmobi, A., Prina-Mello, A., Volkov, Y., Corvin, A., Tropea, D., 2014. Nanotechnologies for the study of the central nervous system. Prog. Neurobiol. 123, 18–36.

Akhlaghi, S.P., Peng, B., Yao, Z., Tam, K.C., 2013. Sustainable nanomaterials derived from polysaccharides and amphiphilic compounds. Soft Matt. 9, 7905–7918.

Alakhova, D.Y., Zhao, Y., Li, S., Kabanov, A.V., 2013. Effect of doxorubicin/pluronic SP1049C on tumorigenicity, aggressiveness: DNA methylation and stem cell markers in murine leukemia. PLoS One 8, e72238.

Alata, W., Paris-Robidas, S., Emond, V., Bourasset, F., Calon, F., 2014. Brain uptake of a fluorescent vector targeting the transferrin receptor: a novel application of in situ brain perfusion. Mol. Pharm. 11 (1), 243–253.

Albrecht, K., Greindl, M., Kremser, C., Wolf, C., Debbage, P., Bernkop-Schnürch, A., 2006. Comparative in vivo mucoadhesion studies of thiomer formulations using magnetic resonance imaging and fluorescence detection. J. Control. Rel. 115 (1), 78–84.

Alexis, F., Basto, P., Levy-Nissenbaum, E., Radovic-Moreno, A.F., Zhang, L., Pridgen, E., Wang, A.Z., Marein, S.L., Westerhof, K., Molnar, L.K., Farokhzad, O.C., 2008. HER-2-targeted nanoparticle-affibody bioconjugates for cancer therapy. Chem. Med. Chem. 3 (12), 1839–1843.

Alibolandi, M., Ramezani, M., Sadeghi, F., Abnous, K., Hadizadeh, F., 2015. Epithelial cell adhesion molecule aptamer conjugated PEG–PLGA nanopolymersomes for targeted delivery of doxorubicin to human breast adenocarcinoma cell line in vitro. Int. J. Pharm. 479, 241–251.

Allen, T.M., Cullis, P.R., 2013. Liposomal drug delivery systems: from concept to clinical applications. Adv. Drug Deliv. Rev. 65, 36–48.

Allen, T.M., Hansen, C.B., Lopes de Menezes, D.E., 1995. Pharmacokinetics of long-circulating liposomes. Adv Drug Deliv. Rev. 16, 267–284.

Almeida, A.J., Souto, E., 2007. Solid lipid nanoparticles as a drug delivery system for peptides and proteins. Adv. Drug Deliv. Rev. 59 (6), 478–490.

Al-Tahami, K., Singh, J., 2007. Smart polymer-based delivery systems for peptides and proteins. Recent Pat. Drug Deliv. Formul. 1, 65–71.

Alvarez-Lorenzo, C., Concheiro, A., 2014. Smart drug delivery systems: from fundamentals to the clinic. Chem. Commun. 50, 7743–7765.

Amancha, K.P., Balkundi, S., Lvov, Y., Hussain, A., 2014. Pulmonary sustained release of insulin from microparticles composed of polyelectrolyte layer-by-layer assembly. Int. J. Pharm. 466 (1–2), 96–108.

Amidi, M., Mastrobattista, E., Jiskoot, W., Hennink, W.E., 2010. Chitosan-based delivery systems for protein therapeutics and antigens. Adv. Drug Deliv. Rev. 62 (1), 59–82.

Andersen, M.Ø., Lichawska, A., Arpanaei, A., Rask Jensen, S.M., Kaur, H., Oupicky, D., Besenbacher, F., Kingshott, P., Kjems, J., Howard, K.A., 2010. Surface functionalization of PLGA nanoparticles for gene silencing. Biomaterials 31 (21), 5671–5677.

Andrade, F., Neves, J.D., Gener, P., Schwartz, Jr., S., Ferreira, D., Oliva, M., Sarmento, B., 2015. Biological assessment of self-assembled polymeric micelles for pulmonary administration of insulin. Nanomedicine 11 (7), 1621–1631.

Antipin, I.S., Zakharova, L.Ya, Solovieva, S.E., Valeeva, F.G., Voronin, M.A., Volodina, T.I., Shtirlin, Y.G., Badeev, Y.W., Safina, G.D., Zobov, V.V., Konovalov, A.I., 2009. Aqueous systems containing supramolecular calixarene nanocontainers and methods of its obtaining. Russ. Pat., 2362761.

Arnheiter, H., Haller, O., 1988. Antiviral state against influenza virus neutralized by microinjection of antibodies to interferon-induced Mx proteins. EMBO J. 7, 1315–1320.

Bakhru, S.H., Furtado, S., Morello, A.P., Mathiowitz, E., 2013. Oral delivery of proteins by biodegradable nanoparticles. Adv. Drug Deliv. Rev. 65 (6), 811–821.

Belousova, N., Mikheeva, G., Gelovani, J., Krasnykh, V., 2008. Modification of adenovirus capsid with a designed protein ligand yields a gene vector targeted to a major molecular marker of cancer. J. Virol. 82 (2), 630–637.

Berchel, M., Le Gall, T., Haelters, J.-P., Lehn, P., Montier, T., Jaffrès, P.-A., 2015. Cationic lipophosphoramidates containing a hydroxylated polar headgroup for improving gene delivery. Mol. Pharm. 12 (6), 1902–1910.

Bernkop-Schnürch, A., 1999. Polymer-inhibitor conjugates: a promising strategy to overcome the enzymatic barrier to perorally administered (poly)peptide drugs? STP Pharma. Sci. 9, 78–87.

Bernkop-Schnürch, A., Schwarz, V., Steininger, S., 1999. Polymers with thiol groups: a new generation of mucoadhesive polymers? Pharm. Res. 16 (6), 876–881.

Bernkop-Schnürch, A., Krauland, A.H., Leitner, V.M., Palmberger, T., 2004. Thiomers: potential excipients for noninvasive peptide delivery systems. Eur. J. Pharm. Biopharm. 58 (2), 253–263.

Beuttler, J., Rothdiener, M., Müller, D., Frejd, F.Y., Kontermann, R.E., 2009. Targeting of epidermal growth factor receptor (EGFR)-expressing tumor cells with sterically stabilized affibody liposomes (SAL). Bioconjugate Chem. 20 (6), 1201–1208.

Blum, A.P., Kammeyer, J.K., Rush, A.M., Callmann, C.E., Hahn, M.E., Gianneschi, N.C., 2015. Stimuli-responsive nanomaterials for biomedical applications. J. Am. Chem. Soc. 137 (6), 2140–2154.

Bogdanov, A., Pashirova, T., Musin, L., Krivolapov, D., Zakharova, L., Mironov, V., Konovalov, A., 2014. Novel isoindigo derivatives bearing long-chain N-alkyl substituents: synthesis and self-assemble behavior. Chem. Phys. Lett. 594, 69–73.

Borchard, G., Lueβen, H.L., de Boer, A.G., Verhoef, J.C., Lehr, C.M., Junginger, H.E., 1996. The potential of mucoadhesive polymers in enhancing intestinal peptide drug absorption. III: Effects of chitosan-glutamate and carbomer on epithelial tight junctions in vitro. J. Control. Release 39, 131–138.

Branco, M.C., Schneider, J.P., 2009. Self-assembling materials for therapeutic delivery. Acta Biomat. 5, 817–831.

Cai, M., Zhu, K., Qiu, Y., Liu, X., Chen, Y., Luo, X., 2014. pH and redox-responsive mixed micelles for enhanced intracellular drug release. Colloids Surf. B. 116, 424–431.

Candan, G., Michiue, H., Ishikawa, S., Fujimura, A., Hayashi, K., Uneda, A., Mori, A., Ohmori, I., Nishiki, T., Matsui, H., Tomizawa, K., 2012. Combining poly-arginine with the hydrophobic counter-anion 4-(1-pyrenyl)-butyric acid for protein transduction in transdermal delivery. Biomaterials 33 (27), 6468–6475.

Cao, Z., Jiang, S., 2012. Super-hydrophilic zwitterionic poly(carboxybetaine) and amphiphilic nonionic poly(ethylene glycol) for stealth nanoparticles. Nano Today 7, 404–413.

Cao, J., Zhai, S., Li, C., He, B., Lai, Y., Chen, Y., Luo, X., Gu, Z., 2013a. Novel pH-sensitive micelles generated by star-shape copolymers containing zwitterionic sulfo-betaine for efficient cellular internalization. J. Biomed. Nanotechnol. 9, 1847–1861.

Cao, X., Deng, W., Fu, M., Zhu, Y., Liu, H., Wang, L., Zeng, J., Wei, Y., Xu, X., Yu, J., 2013b. Seventy-two-hour release formulation of the poorly soluble drug silybin based on porous silica nanoparticles: in vitro release kinetics and in vitro/in vivo correlations in beagle dogs. Eur. J. Pharm. Sci. 48, 64–71.

Caon, T., Jin, L., Simões, C.M., Norton, R.S., Nicolazzo, J.A., 2015. Enhancing the buccal mucosal delivery of peptide and protein therapeutics. Pharm. Res. 32 (1), 1–21.

Caracciolo, G., 2015. Liposome-protein corona in a physiological environment: challenges and opportunities for targeted delivery of nanomedicines. Nanomedicine 11 (3), 543–557.

Cárdenas-Bailón, F., Osorio-Revilla, G., Gallardo-Velázquez, T., 2013. Microencapsulation techniques to develop formulations of insulin for oral delivery: a review. J. Microencapsul. 30 (5), 409–424.

Cardoso, A.M., Morais, C.M., Silva, S.G., Marques, E.F., Pedroso de Lima, M.C., Jurado, M.A.S., 2014. Bis-quaternary gemini surfactants as components of nonviral gene delivery systems: a comprehensive study from physicochemical properties to membrane interactions. Int. J. Pharm. 474 (1–2), 57–69.

Cardoso, A.M., Morais, C.M., Cruz, A.R., Silva, S.G., do Vale, M.L., Marques, E.F., Pedroso de Lima, M.C., Jurado, A.S., 2015. New serine-derived gemini surfactants as gene delivery systems. Eur. J. Pharm. Biopharm. 89, 347–356.

Cavus, A., Baysal, Z., Alkan, H., 2013. Preparation of poly(hydroxyethylmethacrylate) cryogels containing l-histidine for insulin recognition. Colloids Surf. B. 107, 84–89.

Cesaretti, A., Carlotti, B., Gentili, P.L., Clementi, C., Germani, R., Elisei, F., 2014. Spectroscopic investigation of the pH controlled inclusion of doxycycline and oxytetracycline antibiotics in cationic micelles and their magnesium driven release. J. Phys. Chem. B. 118, 8601–8613.

Chakrabarti, R., Wylie, D.E., Schuster, S.M., 1989. Transfer of monoclonal antibodies intomammalian cells by electroporation. J. Biol. Chem. 264, 15494–15500.

Chana, J., Forbes, B., Jones, S.A., 2015. Triggered-release nanocapsules for drug delivery to the lungs. Nanomedicine 11, 89–97.

Chen, R., 2013. Polymers in drug delivery: concepts, developments and potential. In: Coelho, J. (Ed.), Drug Delivery Systems: Advanced Technologies Potentially Applicable in Personalised Treatment. Springer, Dordrecht, pp. 35–85.

Chen, H.B., Dai, L.L., 2013. Adsorption and release of active species into and from multifunctional ionic microgel particles. Langmuir 29, 11227–11235.

Chen, W., Du, J., 2013. Ultrasound and pH dually responsive polymer vesicles for anticancer drug delivery. Sci. Rep. 3, 2162–2170.

Chen, Y., Liu, L., 2012. Modern methods for delivery of drugs across the blood-brain barrier. Adv. Drug. Deliv. Rev. 64 (7), 640–665.

Chen, W., Zou, Y., Jia, J., Meng, F., Cheng, R., Deng, C., Feijen, J., Zhong, Z., 2013a. Functional poly(ε-caprolactone)s via copolymerization of ε–caprolactone and pyridyl disulfide-containing cyclic carbonate: controlled synthesis and facile access to reduction-sensitive biodegradable graft copolymer micelles. Macromolecules 46, 699–707.

Chen, W., Zhong, P., Meng, F., Cheng, R., Deng, C., Feijen, J., Zhong, Z., 2013b. Redox and pH-responsive degradable micelles for dually activated intracellular anticancer drug release. J. Control. Release 169, 171–179.

Chen, L., Sha, X., Jiang, X., Chen, Y., Ren, Q., Fang, X., 2013c. Pluronic P105/F127 mixed micelles for the delivery of docetaxel against Taxol-resistant nonsmall cell lung cancer: optimization and in vitro, in vivo evaluation. Int. J. Nanomed. 8, 73–84.

Chen, K.-J., Liang, H.F., Chen, H.L., Wang, Y., Cheng, P.Y., Liu, H.L., Xia, Y., Sung, H.W., 2013d. A thermoresponsive bubble-generating liposomal system for triggering localized extracellular drug delivery. ACS Nano. 7, 438–446.

Chen, M.C., Mi, F.L., Liao, Z.X., Hsiao, C.W., Sonaje, K., Chung, M.F., Hsu, L.W., Sung, H.W., 2013e. Recent advances in chitosan-based nanoparticles for oral delivery of macromolecules. Adv. Drug Deliv. Rev. 65 (6), 865–879.

Chen, C.-K., Wang, Q., Jones, C.H., Yu, Y., Zhang, H., Law, W.-C., Lai, C.K., Zeng, Q., Prasad, P.N., Pfeifer, B.A., Cheng, C., 2014a. Synthesis of pH-responsive chitosan nanocapsules for the controlled delivery of doxorubicin. Langmuir 30, 4111–4119.

Chen, Y., Chen, Y., Xiao, D., Bose, A., Deng, R., Bothun, G.D., 2014b. Low-dose chemotherapy of hepatocellular carcinoma through triggered-release from bilayer-decorated magnetoliposomes. Colloids Surf. B. 116, 452–458.

Chen, B., Miller, R.J., Dhal, P.K., 2014c. Hyaluronic acid-based drug conjugates: state-of-the-art and perspectives. J. Biomed. Nanotechnol. 10, 4–16.

Chen, J., Qiu, Y., Zhang, S., Yang, G., Gao, Y., 2015a. Controllable coating of microneedles for transdermal drug delivery. Drug Dev. Ind. Pharm. 41 (3), 415–422.

Chen, Y.C., Chiang, C.F., Wu, S.K., Chen, L.F., Hsieh, W.Y., Lin, W.L., 2015b. Targeting microbubbles-carrying TGFβ1 inhibitor combined with ultrasound sonication induce BBB/BTB disruption to enhance nanomedicine treatment for brain tumors. J. Control. Release 211, 53–62.

Chiappisi, L., Hoffmann, I., Gradzielski, M., 2013. Complexes of oppositely charged polyelectrolytes and surfactants: recent developments in the field of biologically derived polyelectrolytes. Soft Matt. 9, 3896–3909.

Chiappisi, L., Li, D., Wagner, N.J., Gradzielski, M., 2014. An improved method for analyzing isothermal titration calorimetry data from oppositely charged surfactant–polyelectrolyte mixtures. J. Chem. Thermodyn. 68, 48–52.

Chiappisi, L., Simon, M., Gradzielski, M., 2015. Toward bioderived intelligent nanocarriers for controlled pollutant recovery and pH-sensitive binding. ACS Appl. Mater. Interf. 7, 6139–6145.

Choi, W.I., Lee, J.H., Kim, Ja.-Y., Kim, Jin-Ch., Kim, Y.H., Tae, G., 2012. Efficient skin permeation of soluble proteins via flexible and functional nano-carrier. J. Control. Rel. 157, 272–278.

Chong, S.F., Chandrawati, R., Städler, B., Park, J., Cho, J., Wang, Y., Jia, Z., Bulmus, V., Davis, T.P., Zelikin, A.N., Caruso, F., 2009. Stabilization of polymer-hydrogel capsules via thiol-disulfide exchange. Small 22, 2601–2610.

Chu, Z., Dreiss, C.A., Feng, Y., 2013. Smart wormlike micelles. Chem. Soc. Rev. 42, 7174–7203.

Cigognini, D., Silva, D., Paloppi, S., Gelain, F., 2014. Evaluation of mechanical properties and therapeutic effect of injectable self-assembling hydrogels for spinal cord injury. J. Biomed. Nanotechnol. 10, 309–323.

Cipolla, D., Gonda, I., Chan, H.K., 2013. Liposomal formulations for inhalation. Ther. Deliv. 4, 1047–1072.

Connal, L.A., Kinnane, C.R., Zelikin, A.N., Caruso, F., 2009. Stabilization and functionalization of polymer multilayers and capsules via thiolene click chemistry. Chem. Mater. 21, 576–578.

Contri, R.V., Katzer, T., Ourique, A.F., Silva, A.M., Beck, R.R., Pohlmann, A.R., Guterres, S.S., 2014. Combined effect of polymeric nanocapsules and chitosan hydrogel on the increase of capsaicinoids adhesion to the skin surface. J. Biomed. Nanotechnol. 10, 820–830.

Craft, S., Baker, L.D., Montine, T.J., Minoshima, S., Watson, G.S., Claxton, A., Arbuckle, M., Callaghan, M., Tsai, E., Plymate, S.R., Green, P.S., Leverenz, J., Cross, D., Gerton, B., 2012. Intranasal insulin therapy for Alzheimer disease and amnestic mild cognitive impairment: a pilot clinical trial. Arch. Neurol. 69 (1), 29–38.

Cui, Y.N., Xu, Q.X., Chow, P.K.H., Wang, D.P., Wang, C.H., 2013. Transferrin-conjugated magnetic silica PLGA nanoparticles loaded with doxorubicin and paclitaxel for brain glioma treatment. Biomaterials 34, 8511–8520.

Cunningham, A., Oh, J.K., 2013. New design of thiol-responsive degradable polylactide-based block copolymer micelles. Macromol. Rapid Commun. 34, 163–168.

Dakwar, G.R., Abu Hammad, I., Popov, M., Linder, C., Grinberg, S., Heldman, E., Stepensky, D., 2012. Delivery of proteins to the brain by bolaamphiphilic nano-sized vesicles. J. Control. Release 160 (2), 315–321.

Danhier, F., Ansorena, E., Silva, J.M., Coco, R., Le Breton, A., Préat, V., 2012. PLGA-based nanoparticles: an overview of biomedical applications. J. Control. Release 161, 505–522.

De Jesus, M.B., Zuhorn, I.S., 2015. Solid lipid nanoparticles as nucleic acid delivery system: properties and molecular mechanisms. J. Control. Release 201, 1–13.

Des Rieux, A., Pourcelle, V., Cani, P.D., Marchand-Brynaert, J., Préat, V., 2013. Targeted nanoparticles with novel nonpeptidic ligands for oral delivery. Adv. Drug Deliv. Rev. 65 (6), 833–844.

Dombu, C.Y., Betbeder, D., 2013. Airway delivery of peptides and proteins using nanoparticles. Biomaterials 34, 516–525.

Ducat, E., Deprez, J., Gillet, A., Noël, A., Evrard, B., Peulen, O., Piel, G., 2011. Nuclear delivery of a therapeutic peptide by long circulating pH-sensitive liposomes: benefits over classical vesicles. Int. J. Pharm. 420 (2), 319–332.

Dünnhaupt, S., Barthelmes, J., Thurner, C.C., Waldner, C., Sakloetsakun, D., Bernkop-Schnürch, A., 2012. S-protected thiolated chitosan: synthesis and in vitro characterization. Carbohydr. Polym. 90 (2), 765–772.

Edwards, S.L., Poongavanam, V., Kanwar, J.R., Roy, K., Hillman, K.M., Prasad, N., Leth-Larsen, R., Petersen, M., Marusic, M., Plavec, J., Wengel, J., Veedu, R.N., 2015. Targeting VEGF with LNA-stabilized G-rich oligonucleotide for efficient breast cancer inhibition. Chem. Commun. 51, 9499–9502.

Elfinger, M., Maucksch, C., Rudolph, C., 2007. Characterization of lactoferrin as a targeting ligand for nonviral gene delivery to airway epithelial cells. Biomaterials 28 (23), 3448–3455.

Fangueiro, J.F., Andreani, T., Egea, M.A., Garcia, M.L., Souto, S.B., Souto, E.B., 2012. Experimental factorial design applied to mucoadhesive lipid nanoparticles via multiple emulsion process. Colloids Surf. B. 100, 84–89.

Fariya, M., Jain, A., Dhawan, V., Shah, S., Nagarsenker, M.S., 2014. Bolaamphiphiles: a pharmaceutical review. Adv. Pharm. Bull. 4 (Suppl. 2), 483–491.

Fayzullin, D., Zuev, Y., Zakharova, L., Pokrovsky, A., Korobeynikov, V., Mukhametzyanov, T., Konovalov, A., 2015. Lipoplexes of dicationic gemini surfactants with DNA: structural features of DNA compaction and transfection efficiency. Dokl. Biochem. Biophys. 465, 432–435.

Feng, W., Zhou, X., He, C., Qiu, K., Nie, W., Chen, L., Wang, H., Mo, X., Zhang, Y., 2013. Polyelectrolyte multilayer functionalized mesoporous silica nanoparticles for pH-responsive drug delivery: layer thickness-dependent release profiles and biocompatibility. J. Mater. Chem. B 1, 5886–5898.

Ferreira, A.J., Cemlyn-Jones, J., Robalo Cordeiro, C., 2013. Nanoparticles, nanotechnology and pulmonary nanotoxicology. Rev. Port. Pneumol. 19, 28–37.

Figg, W.D., Kruger, E.A., Price, D.K., Kim, S., Dahut, W.D., 2002. Inhibition of angiogenesis: treatment options for patients with metastatic prostate cancer. Invest. New Drugs 20, 183–194.

Fisicaro, E., Compari, C., Bacciottini, F., Contardi, L., Barbero, N., Viscardi, G., Quagliotto, P., Donofrio, G., Różycka-Roszak, B., Misiak, P., Woźniak, E., Sansone, F., 2014. Nonviral gene delivery: gemini bispyridinium surfactant-based DNA nanoparticles. J. Phys. Chem. B 118 (46), 13183–13191.

Fɪx, J.A., 1996. Oral controlled release technology for peptides: status and future prospects. Pharm. Res. 13 (12), 1760–1764.

Fonseca-Santos, B., Gremião, M.P., Chorilli, M., 2015. Nanotechnology-based drug delivery systems for the treatment of Alzheimer's disease. Int. J. Nanomed. 10, 4981–5003.

Fonte, P., Araújo, F., Reis, S., Sarmento, B., 2013. Oral insulin delivery: how far are we? J. Diabetes Sci. Technol. 7 (2), 520–531.

Fonte, P., Araújo, F., Silva, C., Pereira, C., Reis, S., Santos, H.A., Sarmento, B., 2015. Polymer-based nanoparticles for oral insulin delivery: revisited approaches. Biotechnol. Adv. 33, 1342–1354.

Fornaguera, C., Dols-Perez, A., Calderó, G., García-Celma, M.J., Camarasa, J., Solans, C., 2015. PLGA nanoparticles prepared by nano-emulsion templating using low-energy methods as efficient nanocarriers for drug delivery across the blood-brain barrier. J. Control. Release 211, 134–143.

Freiherr, J., Hallschmid, M., Frey, W.H., Brünner, Y.F., Chapman, C.D., Hölscher, C., Craft, S., De Felice, F.G., Benedict, C., 2013. Intranasal insulin as a treatment for Alzheimer's disease: a review of basic research and clinical evidence. CNS Drugs 27 (7), 505–514.

Fröhlich, E., 2015. Value of phagocyte function screening for immunotoxicity of nanoparticles in vivo. Int. J. Nanomed. 10, 3761–3778.

Froidevaux, S., Eberle, A.N., 2002. Somatostatin analogs andradiopeptides in cancer therapy. Biopolymers 66, 161–183.

Fuhrmann, G., Herrmann, I.K., Stevens, M.M., 2015. Cell-derived vesicles for drug therapy and diagnostics: opportunities and challenges. Nano Today. 10, 397–409.

Gabdrakhmanov, D.R., Voronin, M.A., Zakharova, L.Y., Konovalov, A.I., Khaybullin, R.N., Strobykina, I.Y., Kataev, V.E., Faizullin, D.A., Gogoleva, N.E., Konnova, T.A., Salnikov, V.V., Zuev, Y.F., 2013. Supramolecular design of biocompatible nanocontainers based on amphiphilic derivatives of a natural compound isosteviol. Phys. Chem. Chem. Phys. 15 (39), 16725–16735.

Gabdrakhmanov, D., Samarkina, D., Semenov, V., Syakaev, V., Giniyatullin, R., Gogoleva, N., Reznik, V., Latypov, S., Konovalov, A., Pokrovsky, A., Zuev, Y., Zakharova, L., 2015a. Novel dicationic pyrimidinic surfactant: self-assembly and DNA complexation. Colloids Surf. A. 480, 113–121.

Gabdrakhmanov, D., Valeeva, F., Syakaev, V., Lukashenko, S., Zakharov, S., Kuryashov, D., Bashkirtseva, N., Zakharova, L., Latypov, Sh., Sinyashin, O., 2015b. Novel supramolecular system based on cationic amphiphile bearing glucamine fragment: structural behavior and binding the hydrophobic probe. Mendeleev Commun. 25, 174–176.

Gabdrakhmanov, D.R., Samarkina, D., Valeeva, F.G., Saifina, L.F., Semenov, V.E., Reznik, V.S., Zakharova, L.Ya., Konovalov, A.I., 2015c. Supramolecular systems based on dicationic pyrimidine containing surfactant and polyethyleneimine. Russ. Chem. Bull. 3, 573–578.

Garcia-Fuentes, M., Alonso, M.J., 2012. Chitosan-based drug nanocarriers: where do we stand? J. Control. Release 161 (2), 496–504.

Gastaldi, L., Battaglia, L., Peira, E., Chirio, D., Muntoni, E., Solazzi, I., Gallarate, M., Dosio, F., 2014. Solid lipid nanoparticles as vehicles of drugs to the brain: current state of the art. Eur. J. Pharm. Biopharm. 87, 433–444.

Gaynanova, G.A., Vagapova, G.I., Syakaev, V.V., Ibragimova, A.R., Valeeva, F.G., Tudriy, E.V., Galkina, I.V., Kataeva, O.N., Zakharova, L.Y., Latypov, S.K., Konovalov, A.I., 2012. Self-assembling systems based on amphiphilic alkyltriphenylphosphonium bromides: elucidation of the role of head group. J. Colloid Interface Sci. 367 (1), 327–336.

Gaynanova, G.A., Bekmukhametova, A.M., Mukhitova, R.K., Kharlamov, S.V., Ziganshina, A.Y., Zakharova, L.Ya., Konovalov, A.I., 2015. Pyrene fluorescence quenching in supramolecular systems based on dimethylaminomethylated resorcinarene. J. Mol. Liq. 206, 316–320.

Gaynanova, G.A., Vagapova, G.I., Valeeva, F.G., Vasilieva, E.A., Galkina, I.V., Zakharova, L.Y., Sinyashin, O.G., 2016. A novel supramolecular catalytic system based on amphiphilic triphenylphosphonium bromide for the hydrolysis of phosphorus acid esters. Colloids Surf. A. 489, 95–102.

Gendelman, H.E., Anantharam, V., Bronich, T., Ghaisas, S., Jin, H., Kanthasamy, A.G., Liu, X., McMillan, J., Mosley, R.L., Narasimhan, B., Mallapragada, S.K., 2015. Nanoneuromedicines for degenerative, inflammatory, and infectious nervous system diseases. Nanomedicine 11 (3), 751–767.

Giannola, L.I., Sutera, F.M., De Caro, V., 2013. Physical methods to promote drug delivery on mucosal tissues of the oral cavity. Exp. Opin. Drug Deliv. 10 (10), 1449–1462.

Goycoolea, F.M., Lollo, G., Remuñán-López, C., Quaglia, F., Alonso, M.J., 2009. Chitosan-alginate blended nanoparticles as carriers for the transmucosal delivery of macromolecules. Biomacromolecules 10 (7), 1736–1743.

Grigoriev, I.V., Korobeinikov, V.A., Cheresiz, S.V., Pokrovsky, A.G., Zakharova, L.Y., Voronin, M.A., Lukashenko, S.S., Konovalov, A.I., Zuev, Yu.F., 2012. Cationic gemini surfactants as new agents for plasmid DNA delivery into cells. Dokl. Biochem. Biophys. 445 (1), 197–199.

Guan, H., Wang, W., Liu, X., Liang, J., 2014. Real-time visualization of colorimetric probe for pH-sensitive basedon poly-(γ-glutamic acid)-functionalized gold nanoparticles. Colloids Surf. A. 448, 147–153.

Gum, Jr., J.R., Hicks, J.W., Toribara, N.W., Rothe, E.M., Lagace, R.E., Kim, Y.S., 1992. The human MUC2 intestinal mucin has cysteine-rich subdomains located both upstream and downstream of its central repetitive region. J. Biol. Chem. 267 (30), 21375–21383.

Gundogdu, E., Yurdasiper, A., 2014. Drug transport mechanism of oral antidiabetic nanomedicines. Int. J. Endocrinol. Metab. 12 (1), e8984.

Hadinoto, K., Sundaresan, A., Cheow, W.S., 2013. Lipid-polymer hybrid nanoparticles as a new generation therapeutic delivery platform: a review. Eur. J. Pharm. Biopharm. 85, 427–443.

Halozan, D., Déjugnat, C., Brumen, M., Sukhorukov, G.B., 2005. Entrapment of a weak polyanion and H^+/Na^+ exchange in confined polyelectrolyte microcapsules. J. Chem. Inf. Model. 45 (6), 1589–1592.

Han, P., Li, S., Cao, W., Li, Y., Sun, Z., Wang, Z., Xu, H., 2013. Red light responsive diselenide-containing block copolymer micelles. J. Mater. Chem. B. 1, 740–743.

Haque, H.A., Kakehi, S., Hara, M., Nagano, S., Seki, T., 2013. High density liquid crystalline azobenzene polymer brush attained by surface-initiated ring opening metathesis polymerization. Langmuir 29, 7571–7575.

Harnoy, A.J., Rosenbaum, I., Tirosh, E., Ebenstein, Y., Shaharabani, R., Beck, R., Amir, R.J., 2014. Enzyme-responsive amphiphilic PEG-dendron hybrids and their assembly into smart micellar nanocarriers. J. Am. Chem. Soc. 136, 7531–7534.

Hazra, C., Arunbabu, D., Kundu, D., Chaudhari, A., Jana, T., 2013. Biomimetic fabrication of biocompatible and biodegradable core–shell polystyrene/biosurfactant bionanocomposites for protein drug release. J. Chem. Technol. Biotechnol. 88, 1551–1560.

Hazra, C., Kundu, D., Chatterjee, A., Chaudhari, A., Mishra, S., 2014. Poly(methyl methacrylate) (core)–biosurfactant (shell) nanoparticles: size controlled sub-100 nm synthesis, characterization, antibacterial activity, cytotoxicity and sustained drug release behavior. Colloids Surf. A. 449, 96–113.

Heldman, E., Linder, C., Grinberg, S., Popov, M., Hollander, I., 2014. Delivery of Tenofovir to the brain by novel nano-vesicles for the treatment of neuro-HIV (S10.003). Neurology 82 (10), S10.003.

Ho, L.C., Wu, W.C., Chang, C.Y., Hsieh, H.H., Lee, C.H., Chang, H.T., 2015. Aptamer-conjugated polymeric nanoparticles for the detection of cancer cells through "turn-on" retro-self-quenched fluorescence. Anal. Chem. 87, 4925–4932.

Holcenberg, J.S., Roberts, J., 1981. Enzymes as Drugs. John Wiley & Sons, New York.

Hong, X., Wei, L., Ma, L., Chen, Y., Liu, Z., Yuan, W., 2013. Novel preparation method for sustained-release PLGA microspheres using water-in-oil-in-hydrophilic-oil-in-water emulsion. Int. J. Nanomed. 8, 2433–2441.

Huang, H., Zhang, X., Yu, J., Zeng, J., Chang, P.R., Xu, H., Huang, J., 2013a. Fabrication and reduction-sensitive behavior of polyion complex nano-micelles basedon PEG-conjugated polymer containing disulfide bonds as a potential carrier of antitumor paclitaxel. Colloids Surf. B. 110, 59–65.

Huang, F.Y., Chen, W.J., Lee, W.Y., Lo, S.T., Lee, T.W., Lo, J.M., 2013b. In vitro and in vivo evaluation of lactoferrin-conjugated liposomes as a novel carrier to improve the brain delivery. Int. J. Mol. Sci. 4 (2), 2862–2874.

Huang, Z., Liu, Y.-H., Zhang, Y.-M., Zhang, J., Liu, Q., Yu, X.-Q., 2015. Cyclen-based cationic lipids containing a pH-sensitive moiety as gene delivery vectors. Org. Biomol. Chem. 13, 620–630.

Hudson, D., Margaritis, A., 2014. Biopolymer nanoparticle production for controlled release of biopharmaceuticals. Crit. Rev. Biotechnol. 34 (2), 161–179.

Iqbal, J., Shahnaz, G., Dünnhaupt, S., Müller, C., Hintzen, F., Bernkop-Schnürch, A., 2012. Preactivated thiomers as mucoadhesive polymers for drug delivery. Biomaterials 33 (5), 1528–1535.

Ivashenko, O., van Herpt, J.T., Feringa, B.L., Rudolf, P., Browne, W.R., 2013. Electrochemical write and read functionality through oxidative dimerization of spiropyran self-assembled monolayers on gold. J. Phys. Chem. C. 117, 18567–18577.

Janiak, J., Piculell, L., Schillén, K., Lundberg, D., 2013. Responsive release of polyanions from soluble aggregates formed with a hydrolyzable cationic surfactant and a nonionic surfactant. Soft Matt. 9, 4103–4112.

Ji, X., Zhang, Q., Liang, F., Chen, Q., Qu, X., Zhang, C., Wang, Q., Li, J., Song, X., Yang, Z., 2014. Ionic liquid functionalized Janus nanosheets. Chem. Commun. 50, 5706–5709.

Jiang, T., Zhang, Z., Zhang, Y., Lv, H., Zhou, J., Li, C., Hou, L., Zhang, Q., 2012. Dual-functional liposomes based on pH-responsive cell-penetrating peptide and hyaluronic acid for tumor-targeted anticancer drug delivery. Biomaterials 33 (36), 9246–9258.

Jiang, T., Li, Y.-M., Lv, Y., Cheng, Y.-J., He, F., Zhuo, R.-X., 2013. Amphiphilic polycarbonate conjugates of doxorubicin with pH-sensitive hydrazone linker for controlled release. Colloid Surf. B. 111, 542–548.

Jiang, G., Jiang, T., Li, X., Wei, Z., Du, X., Wang, X., 2014a. Boronic acid functionalized N-doped carbon quantum dots as fluorescent probe for selective and sensitive glucose determination. Mater. Res. Express 1, 025708.

Jiang, G., Jiang, T., Wang, Y., Du, X., Wei, Z., Zhou, H., 2014b. Facile preparation of novel Au-polydopamine nanoparticles modified by 4-mercaptophenylboronic acid for use in a glucose sensor. RSC Adv. 4, 33658–33661.

Jiang, T., Jiang, G., Wang, X., Dong, Y., Wei, Z., Li, X., Tang, B., 2014c. Facile one-pot synthesis of fluorescent hyperbranched polymers for optical detection of glucose. Des Monomers Polym. 17, 576–581.

Jin, Y., Wang, Sh., Tong, L., Du, L., 2015. Rational design of didodecyldimethylammonium bromide-based nanoassemblies for gene delivery. Colloids Surf. B 126, 257–264.

Jose, S., Anju, S.S., Cinu, T.A., Aleykutty, N.A., Thomas, S., Souto, E.B., 2014a. In vivo pharmacokinetics and biodistribution of resveratrol-loaded solid lipid nanoparticles for brain delivery. Int. J. Pharm. 474, 6–13.

Jose, S., Sowmya, S., Cinu, T.A., Aleykutty, N.A., Thomas, S., Souto, E.B., 2014b. Surface modified PLGA nanoparticles for brain targeting of Bacoside-A. Eur. J. Pharm. Sci. 63, 29–35.

Joshi, N., Kaviratna, A., Banerjee, R., 2013. Multi trigger responsive, surface active lipid nanovesicle aerosols for improved efficacy of paclitaxel in lung cancer. Integr. Biol. (Camb). 5, 239–248.

Joshi, N., Shirsath, N., Singh, A., Joshi, K.S., Banerjee, R., 2014. Endogenous lung surfactant inspired pH responsive nanovesicle aerosols: pulmonary compatible and site-specific drug delivery in lung metastases. Sci. Rep. 4, 7085–7095.

Kamarudin, N.H.N., Jalil, A.A., Triwahyono, S., Artika, V., Salleh, N.F.M., Karim, A.H., Jaafar, N.F., Sazegar, M.R., Mukti, R.R., Hameed, B.H., Johari, A., 2014. Variation of the crystal growth of mesoporous silica nanoparticles and the evaluation to ibuprofen loading and release. J. Colloid Interf. Sci. 421, 6–13.

Kashapov, R.R., Pashirova, T.N., Kharlamov, S.V., Ziganshina, A.Y., Ziltsova, E.P., Lukashenko, S.S., Zakharova, L.Y., Habicher, W.D., Latypov, S.K., Konovalov, A.I., 2011. Novel self-assembling system based on resorcinarene and cationic surfactant. Phys. Chem. Chem. Phys. 13 (35), 15891–15898.

Kashapov, R.R., Kharlamov, S.V., Sultanova, E.D., Mukhitova, R.K., Kudryashova, Y.R., Zakharova, L.Y., Ziganshina, A.Y., Konovalov, A.I., 2014. Controlling the size and morphology of supramolecular assemblies of viologen-resorcin[4]arene cavitands. Chemistry 20 (43), 14018–14025.

Kashapov, R.R., Zakharova, L.Ya., Saifutdinova, M.N., Gavrilova, E.L., Sinyashin, O.G., 2015a. Self-assembly strategies for improving the water solubility of new amino acid calix[4]resorcinarenes. Tetrahedron Lett. 56 (19), 2508–2511.

Kashapov, R.R., Zakharova, L.Ya., Saifutdinova, M.N., Kochergin, Y.S., Gavrilova, E.L., Sinyashin, O.G., 2015b. Construction of a water-soluble form of amino acid C-methylcalix[4]resorcinarene. J. Mol. Liq. 208, 58–62.

Kauscher, U., Samanta, A., Ravoo, B.J., 2014. Photoresponsive vesicle permeability based on intramolecular host-guest inclusion. Org. Biomol. Chem. 12, 600–606.

Kawashima, Y., Yamamoto, H., Takeuchi, H., Hino, T., Niwa, T., 1998. Properties of a peptide containing DL-lactide/glycolide copolymer nanospheres prepared by novel emulsion solvent diffusion methods. Eur. J. Pharm. Biopharm. 45 (1), 41–48.

Kerbel, R., Folkman, J., 2002. Clinical translation of angiogenesis inhibitors. Nat. Rev. Cancer 2, 727–739.

Khannanov, A.A., Kutyreva, M.P., Ulakhovich, N.A., Gataulina, A.R., Bondar, O.V., Zakharova, L.Ya., Kutyrev, G.A., 2016. Hyperbranched polyester polyacids and their binary systems with surfactants for doxorubicin encapsulation. Fluid Phase Equilibria. 411, 93–100.

Kharlamov, S.V., Voronin, M.A., Semenov, V.E., Gabdrakhmanov, D.R., Strobykina, A.S., Nikolaev, A.E., Reznik, V.S., Zakharova, L.Y., Konovalov, A.I., 2013a. Tunable biomimetic systems based on a novel amphiphilic pyrimidinophane and a helper nonionic surfactant. Colloids Surf. B. 111, 218–223.

Kharlamov, S., Kashapov, R., Pashirova, T., Zhiltsova, E., Ziganshina, A., Zakharova, L., Latypov, S., Konovalov, A., 2013b. A supramolecular amphiphile based on calix[4]resorcinarene and cationic surfactant for controlled self-assembly. J. Phys. Chem. C. 117 (39), 20280–20288.

Khorsand, B., Lapointe, G., Brett, C., Oh, J.K., 2013. Intracellular drug delivery nanocarriers of glutathione-responsive degradable block copolymers having pendant disulfide linkages. Biomacromolecules 14, 2103–2111.

Kim, J.Y., Choi, W.I., Kim, Y.H., Tae, G., 2013. Brain-targeted delivery of protein using chitosan- and RVG peptide-conjugated, pluronic-based nano-carrier. Biomaterials 34 (4), 1170–1178.

Kim, H.-C., Kim, E., Lee, S.G., Lee, S.J., Kim, H., Jeong, S.W., 2014a. Thiol-responsive micelles based on nonionic gemini surfactants with a cystine disulfide spacer. J. Polym. Sci. A 52, 582–589.

Kim, J.C., Chungt, Y.I., Kim, Y.H., Tae, G., 2014b. The modulation of the permeability and the cellular uptake of liposome by stable anchoring of lipid-conjugated pluronic on liposome. J. Biomed. Nanotechnol. 10, 100–108.

Kim, H.-C., Kim, E., Ha, T.-L., Jeong, S.W., Lee, S.G., Lee, S.J., Lee, B., 2015a. Thiol-responsive gemini poly(ethylene glycol)-poly(lactide) with a cystine disulfide spacer as an intracellular drug delivery nanocarrier. Colloids Surf. B. 127, 206–212.

Kim, K.S., Hyun, H., Yang, J.A., Lee, M.Y., Kim, H., Yun, S.H., Choi, H.S., Hahn, S.K., 2015b. Bioimaging of hyaluronate-interferon α conjugates using a noninterfering zwitterionic fluorophore. Biomacromolecules 16, 3054–3061.

Kinnane, C.R., Wark, K., Such, G.K., Johnston, A.P.R., Caruso, F., 2009. Peptide-functionalized, low-biofouling click multilayers for promoting cell adhesion and growth. Small 5, 444–448.

Köhler, G., Milstein, C., 1975. Continuous cultures of fused cells secreting antibody of predefined specificity. Nature 256, 495–497.

Koren, E., Torchilin, V.P., 2012. Cell-penetrating peptides: breaking through to the other side. Trends Mol. Med. 18 (7), 385–393.

Koren, E., Apte, A., Jani, A., Torchilin, V.P., 2012. Multifunctional PEGylated 2C5-immunoliposomes containing pH-sensitive bonds and TAT peptide for enhanced tumor cell internalization and cytotoxicity. J. Control. Release 160 (2), 264–273.

Koshkaryev, A., Sawant, R., Deshpande, M., Torchilin, V., 2013. Immunoconjugates and long circulating systems: origins, current state of the art and future directions. Adv. Drug Deliv. 65 (1), 24–35.

Kozlovskaya, V., Sukhishvili, S.A., 2006. pH-Controlled permeability of layered hydrogen–bonded polymer capsules. Macromolecules 39, 5569–5572.

Kozlovskaya, V., Kharlampieva, E., Chang, S.H., Muhlbauer, R., Tsukruk, V.V., 2009. pH-Responsive layered hydrogel microcapsules as gold nanoreactors. Chem. Mater. 21, 2158–2167.

Kozlu, S., Caban, S., Yerlikaya, F., Fernandez-Megia, E., Novoa-Carballal, R., Riguera, R., Yemisci, M., Gursoy-Ozdemir, Y., Dalkara, T., Couvreur, P., Capan, Y., 2014. An aquaporin 4 antisense oligonucleotide loaded, brain-targeted nanoparticulate system design. Pharmazie 69 (5), 340–345.

Krauland, A.H., Alonso, M.J., 2007. Chitosan/cyclodextrin nanoparticles as macromolecular drug delivery system. Int. J. Pharm. 340 (1–2), 134–142.

Kraus, K., Tieke, B., 2014. pH- and temperature-responsive hydrogels of acrylic acid, N-isopropylacrylamide and a nonionic surfmer: phase behavior, swelling properties and drug release. Colloid Polym. Sci. 292, 3127–3135.

Kreuter, J., 2014. Drug delivery to the central nervous system by polymeric nanoparticles: what do we know? Adv. Drug Deliv. Rev. 71, 2–14.

Kudryashova, Yu.R., Valeeva, F.G., Ibragimova, A.R., Zakharova, L.Ya., Konovalov, A.I., 2011. Supramolecular catalytic system Triton X-100-polyethyleneimine-La(III) for hydrolysis of phosphonic acid esters. Kinet. Catal. 52 (1), 55–61.

Kudryashova, Yu.R., Valeeva, F.G., Zakharova, L.Ya., Solovieva, S.E., Antipin, I.S., Gubaidullin, A.T., Konovalov, A.I., 2012. New organized systems based on amphiphilic oxyethylated calix[4]arene. Colloid J. 74 (1), 67–77.

Kuehl, P.J., Cherrington, A., Dobry, D.E., Edgerton, D., Friesen, D.T., Hobbs, C., Leach, C.L., Murri, B., Neal, D., Lyon, D.K., Vodak, D.T., Reed, M.D., 2014. Biologic comparison of inhaled insulin formulations: Exubera™ and novel spray-dried engineered particles of dextran-10. AAPS Pharm. Sci. Tech. 15 (6), 1545–1550.

Kumar, K., Barrán-Berdón, A.L., Datta, S., Muñoz-Ubeda, M., Aicart-Ramos, C., Kondaiah, P., Junquera, E., Bhattacharya, S., Aicart, E., 2015. A delocalizable cationic headgroup together with an oligo-oxyethylene spacer in gemini cationic lipids improves their biological activity as vectors of plasmid DNA. J. Mater. Chem. B 3, 1495–1506.

Kumari, A., Kumar, S., Subhash, Y., Yadav, C., 2010. Biodegradable polymeric nanoparticles–based drug delivery systems. Colloids Surf. B 75, 1–18.

Kutyreva, M.P., Khannanov, A.A., Zakharova, L.Ya., Ulakhovich, N.A., Kutyrev, G.A., Gabdrakhmanov, D.R., 2015. Self-organization and solubilization in binary systems based on hyperbranched polyesters polyols. Colloids Surf. A. 468, 40–48.

Lakhin, A.V., Tarantul, V.Z., Gening, L.V., 2013. Aptamers: problems, solutions, and prospects. Acta Naturae 5, 34–43.

Lalatsa, A., Garrett, N.L., Ferrarelli, T., Moger, J., Schätzlein, A.G., Uchegbu, I.F., 2012. Delivery of peptides to the blood and brain after oral uptake of quaternary ammonium palmitoyl glycol chitosan nanoparticles. Mol. Pharm. 9 (6), 1764–1774.

Lasic, D.D., 1993. Liposomes: From Physics to Applications. Elsevier.

Le Gall, T., Barbeau, J., Barrier, S., Berchel, M., Lemiègre, L., Jeftić, J., Meriadec, C., Artzner, F., Gill, D.R., Hyde, S.C., Férec, C., Lehn, P., Jaffrès, P.-A., Benvegnu, T., Montier, T., 2014. Effects of a novel archaeal tetraether-based colipid on the in vivo gene transfer activity of two cationic amphiphiles. Mol. Pharmaceutics 11 (9), 2973–2988.

Lee, S.B., Hassan, M., Fisher, R., Chertov, O., Chernomordik, V., Kramer-Marek, G., Gandjbakhche, A., Capala, J., 2008. Affibody molecules for in vivo characterization of HER2-positive tumors by near-infrared imaging. Clin. Cancer Res. 14 (12), 3840–3849.

Lee, I., Park, M., Kim, Y., Hwang, O., Khang, G., Lee, D., 2013. Ketal containing amphiphilic block copolymer micelles as pH-sensitive drug carriers. Int. J. Pharm. 448, 259–266.

Lee, P., Towslee, J., Maia, J., Pokorski, J., 2015. PEGylation to improve protein stability during melt processing. Macromol. Biosci. 15, 1332–1337.

Leenders, C.M.A., Albertazzi, L., Mes, T., Koenigs, M.M.E., Palmans, R.A., Meijer, E.W., 2013. Supramolecular polymerization in water harnessing both hydrophobic effects and hydrogen bond formation. Chem. Commun. 49, 1963–1965.

Leitner, V.M., Walker, G.F., Bernkop-Schnürch, A., 2003. Thiolated polymers: evidence for the formation of disulphide bonds with mucus glycoproteins. Eur. J. Pharm. Biopharm. 56 (2), 207–214.

Li, M., Song, W., Tang, Z., Lv, S., Lin, L., Sun, H., Li, Q., Yang, Y., Hong, H., Chen, X., 2013. Nanoscaled poly(L-glutamic acid)/doxorubicin-amphiphile complex as pH-responsive drug delivery system for effective treatment of nonsmall cell lung cancer. ACS Appl. Mater. Interf. 5, 1781–1792.

Li, W., Yang, Y., Luo, T., Zhang, J., Han, B., 2014. CO_2-induced micelle to vesicle transition in zwitterionic–anionic surfactant systems. Phys. Chem. Chem. Phys. 16, 3640–3647.

Li, W., Yang, Y., Liu, L., Tan, X., Luo, T., Shen, J., 2015. Dual stimuli-responsive self-assembly transition in zwitterionic/anionic surfactant systems. Soft Matt. 11, 4283–4289.

Liang, Z., Ni, R., Zhou, J., Mao, S., 2015. Recent advances in controlled pulmonary drug delivery. Drug Discov. Today 20 (3), 380–389.

Lim, H.-S., Kwon, E., Lee, M., Moo Lee, Y., Suh, K.-D., 2013. One-pot template-free synthesis of monodisperse hollow hydrogel microspheres and their resulting properties. Macromol. Rapid Commun. 34, 1243–1248.

Ling, M.H., Chen, M.C., 2013. Dissolving polymer microneedle patches for rapid and efficient transdermal delivery of insulin to diabetic rats. Acta Biomater. 9 (11), 8952–8961.

Liu, J., Pang, Y., Zhu, Z., Wang, D., Li, C., Huang, W., Zhu, X., Yan, D., 2013a. Therapeutic nanocarriers with hydrogen peroxide-triggered drug release for cancer treatment. Biomacromolecules 14, 1627–1636.

Liu, Y., Yuan, M., Liu, L.L., Guo, R., 2013b. A facile electrochemical uricase biosensor designed from gold/amino acid nanocomposites. Sens. Actuat. B: Chem. 176, 592–597.

Liu, Z., Jiang, M., Kang, T., Miao, D., Gu, G., Song, Q., Yao, L., Hu, Q., Tu, Y., Pang, Z., Chen, H., Jiang, X., Gao, X., Chen, J., 2013c. Lactoferrin-modified PEG-*co*-PCL nanoparticles for enhanced brain delivery of NAP peptide following intranasal administration. Biomaterials 34 (15), 3870–3881.

Liu, Z., Gao, X., Kang, T., Jiang, M., Miao, D., Gu, G., Hu, Q., Song, Q., Yao, L., Tu, Y., Chen, H., Jiang, X., Chen, J., 2013d. B6 peptide-modified PEG-PLA nanoparticles for enhanced brain delivery of neuroprotective peptide. Bioconj. Chem. 24 (6), 997–1007.

Liu, J., Jiang, Y., Cui, Y., Xu, C., Ji, X., Luan, Y., 2014a. Cytarabine-AOT catanionic vesicle-loaded biodegradable thermosensitive hydrogel as an efficient cytarabine delivery system. Int. J. Pharm. 473, 560–571.

Liu, S., Jin, M.N., Quan, Y.S., Kamiyama, F., Kusamori, K., Katsumi, H., Sakane, T., Yamamoto, A., 2014b. Transdermal delivery of relatively high molecular weight drugs using novel self-dissolving microneedle arrays fabricated from hyaluronic acid and their characteristics and safety after application to the skin. Eur. J. Pharm. Biopharm. 86 (2), 267–276.

Liu, Z., Zhao, H., He, L., Yao, Y., Zhou, Y., Wu, J., Liu, J., Ding, J., 2015a. Aptamer density dependent cellular uptake of lipidcapped polymer nanoparticles for polyvalent targeted delivery of vinorelbine to cancer cells. RSC Adv. 5, 16931–16939.

Liu, Z., Zhao, H., Shu, L., Zhang, Y., Okeke, C., Zhang, L., Li, J., Li, N., 2015b. Preparation and evaluation of Baicalin-loaded cationic solid lipid nanoparticles conjugated with OX26 for improved delivery across the BBB. Drug. Dev. Ind. Pharm. 41 (3), 353–361.

Lopes, M.A., Abrahim, B.A., Cabral, L.M., Rodrigues, C.R., Seiça, R.M., de Baptista Veiga, F.J., Ribeiro, A.J., 2014. Intestinal absorption of insulin nanoparticles: contribution of M cells. Nanomedicine 10 (6), 1139–1151.

Loureiro, J.A., Gomes, B., Coelho, M.A., do Carmo Pereira, M., Rocha, S., 2014. Targeting nanoparticles across the blood-brain barrier with monoclonal antibodies. Nanomedicine (Lond.) 9 (5), 709–722.

Lu, Y.M., Huang, J.Y., Wang, H., Lou, X.F., Liao, M.H., Hong, L.J., Tao, R.R., Ahmed, M.M., Shan, C.L., Wang, X.L., Fukunaga, K., Du, Y.Z., Han, F., 2014. Targeted therapy of brain ischaemia using Fas ligand antibody conjugated PEG-lipid nanoparticles. Biomaterials 35 (1), 530–537.

Ludwig, A., 2005. The use of mucoadhesive polymers in ocular drug delivery. Adv. Drug Deliv. Rev. 57 (11), 1595–1639.

Luo, P., Luo, Y., Huang, J., Lu, W., Luo, D., Yu, J., Liu, S., 2013. Incorporation of camptothecin into reduction-degradable supramolecular micelles for enhancing its stability. Colloids Surf. B 109, 167–175.

Lv, Y., Yang, B., Li, Y.-U., Wu, Y., He, F., Zhuo, R.-X., 2014. Crosslinked triblock copolymeric micelle for redox-responsive drug delivery. Colloids. Surf. B 122, 223–230.

Ma, G., 2014. Microencapsulation of protein drugs for drug delivery: strategy, preparation, and applications. J. Control. Rel. 193, 324–340.

Mahjub, R., Radmehr, M., Dorkoosh, F.A., Ostad, S.N., Rafiee-Tehrani, M., 2014. Lyophilized insulin nanoparticles prepared from quaternized N-aryl derivatives of chitosan as a new strategy for oral delivery of insulin: in vitro, ex vivo, and in vivo characterizations. Drug Dev. Ind. Pharm. 40 (12), 1645–1659.

Mandal, B., Bhattacharjee, H., Mittal, N., Sah, H., Balabathula, P., Thoma, L.A., Wood, G.C., 2013. Core–shell-type lipid–polymer hybrid nanoparticles as a drug delivery platform. Nanomedicine 9, 474–491.

Manjappa, A.S., Chaudhari, K.R., Venkataraju, M.P., Dantuluri, P., Nanda, B., Sidda, C., Sawant, K.K., Murthy, R.S., 2011. Antibody derivatization and conjugation strategies: application in preparation of stealth immunoliposome to target chemotherapeutics to tumor. J. Control. Release 150 (1), 2–22.

Mansourpour, M., Mahjub, R., Amini, M., Ostad, S.N., Shamsa, E.S., Rafiee-Tehrani, M., Dorkoosh, F.A., 2015. Development of acid-resistant alginate/trimethyl chitosan nanoparticles containing cationic β-cyclodextrin polymers for insulin oral delivery. AAPS Pharm. Sci. Tech. 16 (4), 952–962.

Marianecci, C., Di Marzio, L., Rinaldi, F., Celia, C., Paolino, D., Alhaique, F., Esposito, S., Carafa, M., 2014. Niosomes from 80s to present: the state of the art. Adv. Colloid Int. Sci. 205, 187–206.

Matteucci, E., Giampietro, O., Covolan, V., Giustarini, D., Fanti, P., Rossi, R., 2015. Insulin administration: present strategies and future directions for a noninvasive (possibly more physiological) delivery. Drug Des. Devel. Ther. 9, 3109–3118.

Mazzoli, A., Spalletti, A., Carlotti, B., Emiliani, C., Fortuna, C.G., Urbanelli, L., Tarpani, L., Germani, R., 2015. Spectroscopic investigation of interactions of new potential anticancer drugs with DNA and nonionic micelles. J. Phys. Chem. B 119, 1483–1495.

McGrath, M.G., Vucen, S., Vrdoljak, A., Kelly, A., O'Mahony, C., Crean, A.M., Moore, A., 2014. Production of dissolvable microneedles using an atomised spray process: effect of microneedle composition on skin penetration. Eur. J. Pharm. Biopharm. 86 (2), 200–211.

Md, S., Khan, R.A., Mustafa, G., Chuttani, K., Baboota, S., Sahni, J.K., Ali, J., 2013. Bromocriptine loaded chitosan nanoparticles intended for direct nose to brain delivery: pharmacodynamic, pharmacokinetic and scintigraphy study in mice model. Eur. J. Pharm. 48 (3), 393–405.

Menger, F.M., Littau, C.A., 1993. Gemini surfactants: a new class of self-assembling molecules. J. Am. Chem. Soc. 115 (22), 10083–10090.

Mero, A., Pasqualin, M., Campisi, M., Renier, D., Pasut, G., 2013. Conjugation of hyaluronan to proteins. Carbohyd. Polym. 92 (2), 2163–2170.

Mikos, A.G., Peppas, N.A., 1990. Kinetics of mucus–polymer interactions. In: Gurny, R., Junginger, H.E. (Eds.), Bioadhesione: Possibilities and Future Trends. Wissenschaftliche, Verlagsges, Stuttgart, Germany, pp. 65–85.

Milosevits, G., Szebeni, J., Krol, S., 2015. Exosomes: potential model for complement-stealth delivery systems. Eur. J. Nanomed. 7 (3), 207–218.

Min, J., Braatz, R.D., Hammond, P.T., 2014. Tunable staged release of therapeutics from layer-by-layer coatings with clay interlayer barrier. Biomaterials 35, 2507–2517.

Mirgorodskaya, A.B., Yatskevich, E.I., Zakharova, L.Ya., 2010. The solubilization of fatty acids in systems based on block copolymers and nonionic surfactants. Rus. J. Phys. Chem. A. 84 (12), 2066–2070.

Mirgorodskaya, A.B., Yackevich, E.I., Lukashenko, S.S., Zakharova, L.Ya., Konovalov, A.I., 2012a. Solubilization and catalytic behavior of micellar system based on gemini surfactant with hydroxyalkylated head group. J. Mol. Liq. 169, 106–109.

Mirgorodskaya, A., Yackevich, E., Syakaev, V., Latypov, S., Zakharova, L., Konovalov, A., 2012b. Micellization and catalytic properties of cationic surfactants with head groups functionalized with a hydroxyalkyl fragment. J. Chem. Eng. Data. 57 (11), 3153–3163.

Mirgorodskaya, A.B., Yatskevich, E.I., Zakharova, L.Ya., Konovalov, A.I., 2012c. Geminal surfactant–nonionic polymer mixed micellar systems. Colloid J. 74, 91–98.

Mirgorodskaya, A.B., Valeeva, F.G., Yackevich, E.I., Beschastnova, T.N., Zhukova, N.A., Zakharova, L.Ya., Sinyashin, O.G., Mamedov, V.A., 2014a. Effect of surfactant micelles on solubility and spectral characteristics of 2,2'-bibenzimidazole. Russ. Chem. Bull. 63 (12), 2681–2685.

Mirgorodskaya, A.B., Yackevich, E.I., Kudryashova, Y.R., Kashapov, R.R., Solovieva, S.E., Gubaidullin, A.T., Antipin, I.S., Zakharova, L.Y., Konovalov, A.I., 2014b. Design of supramolecular biomimetic catalysts of high substrate specificity by noncovalent self-assembly of calix[4]arenes with amphiphilic and polymeric amines. Colloids Surf. B. 117, 497–504.

Mirgorodskaya, A.B., Karpichev, Y.A., Zakharova, L.Ya., Yackevich, E.I., Kapitanov, I.V., Lukashenko, S.S., Popov, A.F., Konovalov, A.I., 2014c. Aggregation behavior and interface properties of mixed surfactant systems gemini 14-s-14/ CTABr. Colloids Surf. A 457, 425–432.

Mirgorodskaya, A.B., Yackevich, E.I., Valeeva, F.G., Pankratov, V.A., Zakharova, L.Ya., 2014d. Solubilizing and catalytic properties of supramolecular systems based on gemini surfactants. Russ. Chem. Bull. 63, 82–87.

Moghassemi, S., Hadjizadeh, A., 2014. Nano-niosomes as nanoscale drug delivery systems: an illustrated review. J. Control. Release 185, 22–36.

Monsalve, Y., Tosi, G., Ruozi, B., Belletti, D., Vilella, A., Zoli, M., Vandelli, M.A., Forni, F., López, B.L., Sierra, L., 2015. PEG-g-chitosan nanoparticles functionalized with the monoclonal antibody OX26 for brain drug targeting. Nanomedicine (Lond.) 10 (11), 1735–1750.

Nagpal, K., Singh, S.K., Mishra, D.N., 2013. Nanoparticle mediated brain-targeted delivery of gallic acid: in vivo behavioral and biochemical studies for protection against scopolamine-induced amnesia. Drug Deliv. 20, 112–119.

Nair, A.B., Kumria, R., Harsha, S., Attimarad, M., Al-Dhubia, B.E., Alhaider, I.A., 2013. In vitro techniques to evaluate buccal films. J. Control. Release 166 (1), 10–21.

Narayan, R.J., 2014. Transdermal delivery of insulin via microneedles. J. Biomed. Nanotechnol. 10 (9), 2244–2260.

Nebot, V.J., Escuder, B., Miravet, J.F., Smets, J., Fernández-Prieto, S., 2013. Interplay of molecular hydrogelators and SDS affords responsive soft matter systems with tunable properties. Langmuir 29, 9544–9550.

Nicolas, J., Mura, S., Brambilla, D., Mackiewicz, N., Couvreur, P., 2013. Design, functionalization strategies and biomedical applications of targeted biodegradable/biocompatible polymer-based nanocarriers for drug delivery. Chem. Soc. Rev. 42, 1147–1235.

Nino, M., Calabrò, G., Santoianni, P., 2010. Topical delivery of active principles: the field of dermatological research. Dermatol. Online J. 16 (1), 4.

Niyomtham, N., Apiratikul, N., Suksen, K., Opanasopit, P., Yingyongnarongkul, B., 2015. Synthesis and in vitro transfection efficiency of spermine-based cationic lipids with different central core structures and lipophilic tails. Bioorg. Med. Chem. Lett. 25, 496–503.

Noble, G.T., Stefanick, J.F., Ashley, J.D., Kiziltepe, T., Bilgicer, B., 2014. Ligand-targeted liposome design: challenges and fundamental considerations. Trends Biotechnol. 32 (1), 32–45.

Oh, E.J., Park, K., Kim, K.S., Kim, J., Yang, J.A., Kong, J.H., Lee, M.Y., Hoffman, A.S., Hahn, S.K., 2010. Target specific and long-acting delivery of protein, peptide, and nucleotide therapeutics using hyaluronic acid derivatives. J. Control. Rel. 141 (1), 2–12.

Ojeda, E., Puras, G., Agirre, M., Zárate, J., Grijalvo, S., Pons, R., Eritja, R., Martinez-Navarrete, G., Soto-Sanchez, C., Fernández, E., Pedraz, J.L., 2015. Niosomes based on synthetic cationic lipids for gene delivery: the influence of polar headgroups on the transfection efficiency in HEK-293, ARPE-19, and MSC-D1 cells. Org. Biomol. Chem. 13 (4), 1068–1081.

Øpstad, C.L., Zeeshan, M., Zaidi, A., Sliwka, H.-R., Partali, V., Nicholson, D.G., Surve, C., Izower, M.A., Bilchuk, N., Lou, H.H., Leopold, Ph.L., Larsen, H., Liberska, A., Abdul Khaliqued, N., Rajud, L., Flinterman, M., Jubeli, E., Pungente, M.D., 2014. Novel cationic polyene glycol phospholipids as DNA transfer reagents: lack of a structure–activity relationship due to uncontrolled self-assembling processes. Chem. Phys. Lipids. 183, 117–136.

Pan, J.M., Li, L.Z., Hang, H., Wu, R.R., Dai, X.H., Shi, W.D., Yan, Y.S., 2013. Fabrication and evaluation of magnetic/hollow double-shelled imprinted sorbents formed by pickering emulsion polymerization. Langmuir 29, 8170–8178.

Paques, J.P., van der Linden, E., van Rijn, C.J., Sagis, L.M., 2014. Preparation methods of alginate nanoparticles. Adv. Colloid Interf. Sci. 209, 163–171.

Park, C.W., Li, X., Vogt, F.G., Hayes, Jr., D., Zwischenberger, J.B., Park, E.-S., Mansoura, H.M., 2013. Advanced spray-dried design, physicochemical characterization, and aerosol dispersion performance of vancomycin and clarithromycin multifunctional controlled release particles for targeted respiratory delivery as dry powder inhalation aerosols. Int. J. Pharm. 455, 374–392.

Pashirova, T.N., Ziganshina, A.Y.u., Sultanova, E.D., Lukashenko, S.S., Kudryashova, Y.R., Zhiltsova, E.P., Zakharova, L.Ya., Konovalov, A.I., 2014a. Supramolecular systems based on calix[4]resorcine with mono-, di-, and tetracationic surfactants: Synergetic structural and solubilization behavior. Colloids Surf. A. 448, 67–72.

Pashirova, T.N., Gibadullina, E.M., Burilov, A.R., Kashapov, R.R., Zhiltsova, E.P., Syakaev, V.V., Habicher, W.D., Rümmeli, M.H., Latypov, S.K., Zakharova, L.Ya., Konovalov, A.I., 2014b. Amphiphilic O-functionalized calix[4]resocinarenes with tunable structural behavior. RSC Adv. 4 (20), 9912–9919.

Pashirova, T.N., Zhiltsova, E.P., Lukashenko, S.S., Zakharova, L.Ya., Konovalov, A.I., 2015a. The catalytic properties of polymer-colloid complexes based on polyethyleneimines, mono- and diquaternized derivatives of 1,4-diazabicyclo[2.2.2]octane in hydrolysis of phosphorus acid esters. Russ. Chem. Bull. 64, 2879–2884.

Pashirova, T.N., Lukashenko, S.S., Zakharov, S.V., Voloshina, A.D., Zhiltsova, E.P., Zobov, V.V., Souto, E.B., Zakharova, L.Ya., 2015b. Self-assembling systems based on quaternized derivatives of 1,4-diazabicyclo[2.2.2] octane in nutrient broth as antimicrobial agents and carriers for hydrophobic drugs. Colloids Surf. B. 127, 266–273.

Patel, V.F., Liu, F., Brown, M.B., 2011. Advances in oral transmucosal drug delivery. J. Control. Release 153 (2), 106–116.

Patel, M., Souto, E.B., Singh, K.K., 2013. Advances in brain drug targeting and delivery: limitations and challenges of solid lipid nanoparticles. Expert. Opin. Drug. Deliv. 10 (7), 889–905.

Patil, Y.P., Jadhav, S., 2014. Novel methods for liposome preparation. Chem. Phys. Lipids. 177, 8–18.

Patra, T., Ghosh, S., Dey, J., 2014. Cationic vesicles of a carnitine-derived single-tailed surfactant: physicochemical characterization and evaluation of in vitro gene transfection efficiency. J. Colloid Interf. Sci. 436, 138–145.

Peng, C.Y., Zhao, Q.H., Gao, C.Y., 2010. Sustained delivery of doxorubicin by porous $CaCO_3$ and chitosan/alginate multilayers-coated $CaCO_3$ microparticles. Colloid Surf. A. 353, 132–139.

Perche, F., Torchilin, V.P., 2013. Recent trends in multifunctional liposomal nanocarriers for enhanced tumor targeting. J. Drug Deliv. 32 (2013), 705265.

Petrichenko, O., Rucins, M., Vezane, A., Timofejeva, I., Sobolev, A., Cekavicus, B., Pajuste, K., Plotniece, M., Gosteva, M., Kozlovska, T., Plotniece, A., 2015. Studies of the physicochemical and structural properties of self-assembling cationic pyridine derivatives as gene delivery agents. Chem. Phys. Lipids 191, 25–37.

Popov, M., Abu Hammad, I., Bachar, T., Grinberg, S., Linder, C., Stepensky, D., Heldman, E., 2013. Delivery of analgesic peptides to the brain by nano-sized bolaamphiphilic vesicles made of monolayer membranes. Eur. J. Pharm. Biopharm. 85, 381–389.

Puras, G., Mashal, M., Zárate, J., Agirre, M., Ojeda, E., Grijalvo, S., Eritja, R., Diaz-Tahoces, A., Martínez Navarrete, G., Avilés-Trigueros, M., Fernández, E., Pedraz, J.L., 2014. A novel cationic niosome formulation for gene delivery to the retina. J. Control. Rel. 174, 27–36.

Puri, A., Kramer-Marek, G., Campbell-Massa, R., Yavlovich, A., Tele, S.C., Lee, S.B., Clogston, J.D., Patri, A.K., Blumenthal, R., Capala, J., 2008. HER2-specific affibody-conjugated thermosensitive liposomes (Affisomes) for improved delivery of anticancer agents. J. Liposome Res. 18 (4), 293–307.

Puri, A., Loomis, K., Smith, B., Lee, J.H., Yavlovich, A., Heldman, E., Blumenthal, R., 2009. Lipid-based nanoparticles as pharmaceutical drug carriers: from concepts to clinic. Crit. Rev. Ther. Drug Carrier Syst. 26 (6), 523–580.

Raja, W.K., Maccorkle, S., Diwan, I.M., Abdurrob, A., Lu, J., Omenetto, F.G., Kaplan, D.L., 2013. Transdermal delivery devices: fabrication, mechanics and drug release from silk. Small 9 (21), 3704–3713.

Ramli, R.A., Laftah, W.A., Hashim, S., 2013a. Core–shell polymers: a review. RSC Adv. 3, 15543–15565.

Ramli, R.A., Hashim, S., Laftah, W.A., 2013b. Synthesis, characterization, and morphology study of poly(acrylamide-*co*-acrylic acid)-grafted-poly(styrene-*co*-methyl methacrylate) "raspberry"-shape like structure microgels by preemulsified semi-batch emulsion polymerization. J. Colloid Interf. Sci. 391, 86–94.

Rao, J., Khan, A., 2013. Enzyme sensitive synthetic polymer micelles based on the azobenzene motif. J. Am. Chem. Soc. 135, 14056–14059.

Reddy, L.H., Bazile, D., 2014. Drug delivery design for intravenous route with integrated physicochemistry, pharmacokinetics and pharmacodynamics: illustration with the case of taxane therapeutics. Adv. Drug Deliv. Rev. 71, 34–57.

Reinhold, S.E., Schwendeman, S.P., 2013. Effect of polymer porosity on aqueous self-healing encapsulation of proteins in PLGA microspheres. Macromol. Biosci. 13 (12), 1700–1710.

Renkecz, T., Mistlberger, G., Pawlak, M., Horváth, V., Bakker, E., 2013. Molecularly imprinted polymer microspheres containing photoswitchable spiropyran-based binding sites. ACS Appl. Mater. Interf. 5, 8537–8545.

Rodriguez, A.R., Kramer, J.R., Deming, T.J., 2013. Enzyme-triggered cargo release from methionine sulfoxide containing copolypeptide vesicles. Biomacromolecules 14, 3610–3614.

Roldo, M., Hornof, M., Caliceti, P., Bernkop-Schnürch, A., 2004. Mucoadhesive thiolated chitosans as platforms for oral controlled drug delivery: synthesis and in vitro evaluation. Eur. J. Pharm. Biopharm. 57 (1), 115–121.

Rosiaux, Y., Jannin, V., Hughes, S., Marchaud, D., 2014. Solid lipid excipients: matrix agents for sustained drug delivery. J. Control. Release 188, 18–30.

Rossi, S., Sandri, G., Caramella, C.M., 2005. Buccal drug delivery: a challenge already won? Drug Discov. Today Technol. 2 (1), 59–65.

Rostamia, E., Kashanianb, S., Azandaryani, A.H., Faramarzi, H., Dolatabadi, J.E.N., Omidfar, K., 2014. Drug targeting using solid lipid nanoparticles. Chem. Phys. Lipids. 181, 56–61.

Sadhukha, T., Wiedmann, T.S., Panyam, J., 2013. Inhalable magnetic nanoparticles for targeted hyperthermia in lung cancer therapy. Biomaterials 34, 5163–5174.

Sadovoy, A.V., Lomova, M.V., Antipina, M.N., Braun, N.A., Sukhorukov, G.B., Kiryukhin, M.V., 2013. Layer-by-layer assembled multilayer shells for encapsulation and release of fragrance. ACS Appl. Mater. Interf. 5, 8948–8954.

Saffran, M., Kumar, G.S., Savariar, C., Burnham, J.C., Williams, F., Neckers, D.C., 1986. A new approach to the oral administration of insulin and other peptide drugs. Science 233 (4768), 1081–1084.
Saffran, M., Pansky, B., Budd, G.C., Williams, F.E., 1997. Insulin and the gastrointestinal tract. J. Control. Release 46 (1–2), 89–98.
Santos, S., Medronho, B., Santos, T., Antunes, F.E., 2013. Amphiphilic molecules in drug delivery systems. In: Coelho, J. (Ed.), Drug Delivery Systems: Advanced Technologies Potentially Applicable in Personalized Treatment. Springer, Dordrecht, pp. 1–35.
Sarvaiya, J., Agrawal, Y.K., 2015. Chitosan as a suitable nanocarrier material for anti-Alzheimer drug delivery. Int. J. Biol. Macromol. 72, 454–465.
Savarala, S., Brailoiu, E., Wunder, S.L., Ilies, M.A., 2013. Tuning the self-assembling of pyridinium cationic lipids for efficient gene delivery into neuronal cells. Biomacromolecules 14, 2750–2764.
Segarra-Maset, M.D., Nebot, V.J., Miravet, J.F., Escuder, B., 2013. Control of molecular gelation by chemical stimuli. Chem. Soc. Rev. 42, 7086–7098.
Sellers, D.L., Kim, T.H., Mount, C.W., Pun, S.H., Horner, P.J., 2014. Poly(lactic-*co*-glycolic) acid microspheres encapsulated in Pluronic F-127 prolong hirudin delivery and improve functional recovery from a demyelination lesion. Biomaterials 35, 8895–8902.
Shah, R.B., Schwendeman, S.P., 2014. A biomimetic approach to active self-microencapsulation of proteins in PLGA. J. Control. Release 196, 60–70.
Shalviri, A., Chan, H.K., Raval, G., Abdekhodaie, M.J., Liuc, Q., Heerklotz, H., Wu, X.Y., 2013. Design of pH-responsive nanoparticles of terpolymer of poly(methacrylicacid), polysorbate 80 and starch for delivery of doxorubicin. Colloids Surf. B. 101, 405–413.
Sharipova, A.A., Aidarova, S.B., Grigoriev, D., Mutalieva, B., Madibekova, G., Tleuova, A., Miller, R., 2016. Polymer-surfactant complexes for microencapsulation of vitamin E and its release. Colloids Surf. B. 137, 152–157.
Sharma, V.D., Aifuwa, E.O., Heiney, P.A., Ilies, M.A., 2013. Interfacial engineering of pyridinium gemini surfactants for the generation of synthetic transfection systems. Biomaterials 34, 6906–6921.
Sharma, V.D., Lees, J., Hoffman, N.E., Brailoiu, E., Madesh, M., Wunder, S.L., Ilies, M.A., 2014. Modulation of pyridinium cationic lipid−DNA complex properties by pyridinium gemini surfactants and its impact on lipoplex transfection properties. Mol. Pharm. 11 (2), 545–559.
She, X., Chen, L., Velleman, L., Li, C., Zhu, H., He, C., Wang, T., Shigdar, S., Duan, W., Kong, L., 2015. Fabrication of high specificity hollow mesoporous silica nanoparticles assisted by Eudragit for targeted drug delivery. J. Colloid Interf. Sci. 445, 151–160.
Shi, J., Du, C., Shi, J., Wang, Y., Cao, S., 2013. Hollow multilayer microcapsules for pH-/thermally responsive drug delivery using aliphatic poly(urethane-amine) as smart component. Macromol. Biosci. 13, 494–502.
Shilpa, S., Srinivasan, B.P., Chauhan, M., 2011. Niosomes as vesicular carriers for delivery of proteins and biologicals. Int. J. Drug Deliv. 3, 14–24.
Shingitha, K.P., 2015. A review: niosomes a novel tool for drug delivery. Int. J. Pharm. Technol. 6 (3), 3017–3026.
Silva, J.P., Oliveira, A.C., Lúcio, M., Gomes, A.C., Coutinho, P.J., Oliveira, M.E., 2014. Tunable pDNA/DODAB:MO lipoplexes: the effect of incubation temperature on pDNA/DODAB:MO lipoplexes structure and transfection efficiency. Colloids Surf. B. 121, 371–379.
Soddu, E., Rassu, G., Giunchedi, P., Sarmento, B., Gavini, E., 2015. From naturally occurring neurotoxic agents to CNS shuttles for drug delivery. Eur. J. Pharm. Sci. 74, 63–76.
Sohrabi, A., Shaibani, P.M., Etayash, H., Kaur, K., Thundat, T., 2013. Sustained drug release and antibacterial activity of ampicillin incorporated poly(methylmethacrylate)-nylon6 core/shell nanofibers. Polymer 54, 2699–2705.
Song, J., Janczewski, D., Ma, Y., van Ingen, L., Ee Sim, C., Goh, Q., Xu, J., Vancso, G.J., 2013. Electrochemically controlled release of molecular guests from redox responsive polymeric multilayers and devices. Eur. Polym. J. 49, 2477–2484.

Song, Y., Tian, Q., Huang, Z., Fan, D., She, Z., Liu, X., Cheng, X., Yu, B., Deng, Y., 2014. Self-assembled micelles of novel amphiphilic copolymer cholesterol-coupled F68 containing cabazitaxel as a drug delivery system. Int. J. Nanomed. 9, 2307–2317.

Sophocleous, A.M., Desai, K.G., Mazzara, J.M., Tong, L., Cheng, J.X., Olsen, K.F., Schwendeman, S.P., 2013. The nature of peptide interactions with acid end-group PLGAs and facile aqueous-based microencapsulation of therapeutic peptides. J. Control. Rel. 172 (3), 662–670.

Sosnik, A., 2014. Alginate particles as platform for drug delivery by the oral route: state-of-the-art. ISRN Pharm. 9, 926157.

Spera, R., Apollonio, F., Liberti, M., Paffi, A., Merla, C., Pinto, R., Petralito, S., 2015. Controllable release from high-transition temperature magnetoliposomes by low-level magnetic stimulation. Colloids Surf. B. 131, 136–140.

Stephanopoulos, N., Ortony, J.H., Stupp, S.I., 2013. Self-assembly for the synthesis of functional biomaterials. Acta Mater. 61, 912–930.

Sudhakar, Y., Kuotsu, K., Bandyopadhyay, A.K., 2006. Buccal bioadhesive drug delivery-a promising option for orally less efficient drugs. J. Control. Release 114 (1), 15–40.

Sultanova, E.D., Krasnova, E.G., Kharlamov, S.V., Nasybullina, G.R., Yanilkin, V.V., Nizameev, I.R., Kadirov, M.K., Mukhitova, R.K., Zakharova, L.Y., Ziganshina, A.Y., Konovalov, A.I., 2015. Thermoresponsive polymer nanoparticles based on viologen-cavitand. Chem. Plus Chem. 80, 217–222.

Summerlin, N., Soo, E., Thakur, S., Qu, Z., Jambhrunkar, S., Popat, A., 2015. Resveratrol nanoformulations: challenges and opportunities. Int. J. Pharm. 479, 282–290.

Sun, Y., Huang, Y., Bian, S., Liang, J., Fan, Y., Zhang, X., 2013. Reduction-degradable PEG-b–PAA-b–PEG triblockcopolymer micelles incorporated with MTX for cancer chemotherapy. Colloids Surf. B. 112, 197–203.

Sun, H., Zhu, X., Lu, P.Y., Rosato, R.R., Tan, W., Zu, Y., 2014. Oligonucleotide aptamers: new tools for targeted cancer therapy. Mol. Ther. Nucleic Acids 3, e182.

Tan, S.W., Li, X., Guo, Y.J., Zhang, Z.P., 2013. Lipid-enveloped hybrid nanoparticles for drug delivery. Nanoscale 5, 860–872.

Tangso, K.J., Lindberg, S., Hartley, P.G., Knott, R., Spicer, P., Boyd, B.J., 2014. Formation of liquid-crystalline structures in the bile salt–chitosan system and triggered release from lamellar phase bile salt–chitosan capsules. ACS Appl. Mater. Interf. 6, 12363–12371.

Tashima, T., 2015. Intriguing possibilities and beneficial aspects of transporter-conscious drug design. Bioorg. Med. Chem. 23 (15), 4119–4131.

Teijeiro-Osorio, D., Remuñán-López, C., Alonso, M.J., 2009. New generation of hybrid poly/oligosaccharide nanoparticles as carriers for the nasal delivery of macromolecules. Biomacromolecules 10 (2), 243–249.

Tolmachev, V., Orlova, A., Pehrson, R., Galli, J., Baastrup, B., Andersson, K., Sandström, M., Rosik, D., Carlsson, J., Lundqvist, H., Wennborg, A., Nilsson, F.Y., 2007. Radionuclide therapy of HER2-positive microxenografts using a 177Lu-labeled HER2-specific Affibody molecule. Cancer Res. 67 (6), 2773–2782.

Torchilin, V.P. (Ed.), 1991. Immobilized Enzymes in Medicine. Springer-Verlag, Berlin.

Torchilin, V.P., 2005. Recent advances with liposomes as pharmaceutical carriers. Nat. Rev. Drug Discov. 4, 145–160.

Torchilin, V., 2008. Intracellular delivery of protein and peptide therapeutics. Drug Discov. Today 5 (2–3), e95–e103.

Torchilin, V., 2009. Multifunctional and stimuli-sensitive pharmaceutical nanocarriers. Eur. J. Pharm. Biopharm. 71, 431–444.

Torchilin, V.P., 2012. Multifunctional nanocarriers. Adv. Drug Deliv. 64, 302–315.

Trapani, A., De Giglio, E., Cafagna, D., Denora, N., Agrimi, G., Cassano, T., Gaetani, S., Cuomo, V., Trapani, G., 2011. Characterization and evaluation of chitosan nanoparticles for dopamine brain delivery. Int. J. Pharm. 419 (1–2), 296–307.

Tripathi, R.M., Shrivastav, A., Shrivastav, B.R., 2015. Biogenic gold nanoparticles: as a potential candidate for brain tumor directed drug delivery. Artif. Cells Nanomed. Biotechnol. 43 (5), 311–317.

Trojer, M.A, Nordstierna, L., Nordin, M., Nyden, B.M., Holmberg, K., 2013a. Encapsulation of actives for sustained release. Phys. Chem. Chem. Phys. 15, 17727–17741.

Trojer, M.A., Li, Y., Wallin, M., Holmberg, K., Nydén, M., 2013b. Charged microcapsules for controlled release of hydrophobic actives. Part II: Surface modification by LbL adsorption and lipid bilayer formation on properly anchored dispersant layers. J. Colloid Interf. Sci. 409, 8–17.

Trojer, M.A., Andersson, H., Li, Y., Borg, J., Holmberg, K., Nyden, M., Nordstierna, L., 2013c. Charged microcapsules for controlled release of hydrophobic actives. Part III: the effect of polyelectrolyte brush- and multilayers on sustained release. Phys. Chem. Chem. Phys. 15, 6456–6466.

Tsai, Y.M., Chien, C.F., Lin, L.C., Tsai, T.H., 2011. Curcumin and its nano-formulation: the kinetics of tissue distribution and blood-brain barrier penetration. Int. J. Pharm. 416 (1), 331–338.

Tu, F., Lee, D., 2014a. Shape-changing and amphiphilicity-reversing Janus particles with pH-responsive surfactant properties. J. Am. Chem. Soc. 136, 9999–10006.

Tu, F., Lee, D., 2014b. One-step encapsulation and triggered release based on Janus particle-stabilized multiple emulsions. Chem. Commun. 50, 15549–15552.

Vagapova, G.I., Ibragimova, A.R., Zakharov, A.V., Dobrynin, A.B., Galkina, I.V., Zakharova, L.Ya., Konovalov, A.I., 2013. Novel biomimetic systems based on polyethylene glycols and amphiphilic phosphonium salt. Self-organization and solubilization of hydrophobic guest. Eur. Polymer J. 49, 1031–1039.

Valeeva, F.G., Vasilieva, E.A., Gainanova, G.A., Kashapov, R.R., Zakharov, S.V., Kuryashov, D.A., Lukashenko, S.S., Bashkirtseva, N.Y.u., Zakharova, L.Ya., 2015. Structural behavior of aqueous solutions of hydrotropes, related compounds and their mixtures with typical surfactants. J. Mol. Liq. 203, 104–110.

Van der Lubben, I.M., Verhoef, J.C., Borchard, G., Junginger, H.E., 2001. Chitosan and its derivatives in mucosal drug and vaccine delivery. Eur. J. Pharm. Sci. 14 (3), 201–207.

Van der Meel, R., Fens, M.H., Vader, P., van Solinge, W.W., Eniola-Adefeso, O., Schiffelers, R.M., 2014. Extracellular vesicles as drug delivery systems: lessons from the liposome field. J. Control. Release 195, 72–85.

Varga, C.M., Wickham, T.J., Lauffenburger, D.A., 2000. Receptor-mediated targeting of gene delivery vectors: insights from molecular mechanisms for improved vehicle design. Biotechnol. Bioeng. 70 (6), 593–605.

Vasilieva, E., Ibragimova, A., Lukashenko, S., Konovalov, A., Zakharova, L., 2014. Mixed self-assembly of polyacrylic acid and oppositely charged gemini surfactants differing in the structure of head group. Fluid Phase Equilib. 376, 172–180.

Victor, S.P., Sharma, C.P., 2013. Polymethacrylicacid modified CDHA nanocomposites as potential pH responsive drug delivery vehicles. Colloids Surf. B 108, 219–228.

Vieira, F.P., Mesquita, T.L., Lara, P.C.P., Ramaldes, G.A., Beinner, M.A., Silva, J.B.B., Oliveira, M.C., Silveira, J.N., 2013. ET AAS evaluation of the stability and pH-sensitivity of, pH-sensitive stealth liposomes containing cisplatin in mouse plasma. J. Pharm. Biomed. Anal. 84, 135–139.

Voronin, M.A., Gabdrakhmanov, D.R., Semenov, V.E., Valeeva, F.G., Mikhailov, A.S., Nizameev, I.R., Kadirov, M.K., Zakharova, L.Y., Reznik, V.S., Konovalov, A.I., 2011. Novel bolaamphiphilic pyrimidinophane as building block for design of nanosized supramolecular systems with concentration-dependent structural behavior. ACS Appl. Mater. Interf. 3 (2), 402–409.

Voronin, M.A., Gabdrakhmanov, D.R., Khaibullin, R.N., Strobykina, I.Y., Kataev, V.E., Idiyatullin, B.Z., Faizullin, D.A., Zuev, Y.F., Zakharova, L.Y., Konovalov, A.I., 2013. Novel biomimetic systems based on amphiphilic compounds with a diterpenoid fragment: role of counterions in self-assembly. J. Colloid Interf. Sci. 405, 125–133.

Wan, F., Maltesen, M.J., Andersen, S.K., Bjerregaard, S., Baldursdottir, S.G., Foged, C., Rantanen, J., Yang, M., 2014. Modulating protein release profiles by incorporating hyaluronic acid into PLGA microparticles Via a spray dryer equipped with a 3-fluid nozzle. Pharm. Res. 31 (11), 2940–2951.

Wang, H., Rempel, G.L., 2013. pH-responsive polymer core-shell nanospheres for drug delivery. J. Polym. Sci. A 51, 4440–4450.

Wang, Z.P., Feng, Z.Q., Gao, C.Y., 2008. Stepwise assembly of the same polyelectrolytes using host-guest interaction to obtain microcapsules with multiresponsive properties. Chem. Mater. 20, 4194–4199.

Wang, B., Chen, K., Yang, R., Yang, F., Liu, J., 2014a. Stimulus-responsive polymeric micelles for the light-triggered release of drugs. Carbohyd. Polym. 103, 510–519.

Wang, Y., Wang, K., Yan, X., Huang, R.Q., 2014b. A general strategy for dual-triggered combined tumor therapy based on template semi-graphitized mesoporous silica nanoparticles. Adv. Health. Mater. 3, 485–489.

Wang, Y., An, Q., Zhou, Y., Niu, Y., Akram, R., Zhang, Y., Shi, F., 2015. Post-infiltration and subsequent photocrosslinking strategy for layer-by-layer fabrication of stable dendrimers enabling repeated loading and release of hydrophobic molecules. J. Mater. Chem. B, 562–569.

Wen, X.R., Wang, K., Zhao, Z.M., Zhang, Y.F., Sun, T.T., Zhang, F., Wu, J., Fu, Y.Y., Du, Y., Zhang, L., Sun, Y., Liu, Y.H., Ma, K., Liu, H.Z., Song, Y.J., 2014. Brain-targeted delivery of trans-activating transcriptor-conjugated magnetic PLGA/lipid nanoparticles. PLoS One 9, e106652.

Wikman, M., Steffen, A.C., Gunneriusson, E., Tolmachev, V., Adams, G.P., Carlsson, J., Ståhl, S., 2004. Selection and characterization of HER2/neu-binding affibody ligands. Protein Eng. Des. Sel. 17 (5), 455–462.

Wohlfart, S., Gelperina, S., Kreuter, J., 2012. Transport of drugs across the blood-brain barrier by nanoparticles. J. Control. Release 161 (2), 264–273.

Wolf, M., Ransberger, K., 1972. Enzyme Therapy. Vantage Press, New York.

Wong, T.W., 2014. Electrical, magnetic, photomechanical, and cavitational waves to overcome skin barrier for transdermal drug delivery. J. Control. Release 193, 257–269.

Wu, L., Zou, Y., Deng, C., Cheng, R., Meng, F., Zhong, Z., 2013a. Intracellular release of doxorubicin from core-crosslinked polypeptide micelles triggered by both pH and reduction conditions. Biomaterials 34, 5262–5272.

Wu, H., Zhu, L., Torchilin, V.P., 2013b. pH-sensitivepoly(histidine)-PEG/DSPE-PEG copolymer micelles for cytosolic drug delivery. Biomaterials 34, 1213–1222.

Wu, H., Dong, J., Li, C., Liu, Y., Feng, N., Xu, L., Zhan, X., Yang, H., Wang, G., 2013c. Multiresponsive nitrobenzene-based amphiphilic random copolymer assemblies. Chem. Commun. 49, 3516–3518.

Wu, J., Song, C., Jiang, C., Shen, X., Qiao, Q., Hu, Y., 2013d. Nucleolin targeting AS1411 modified protein nanoparticle for antitumor drugs delivery. Mol. Pharm. 10, 3555–3563.

Xu, Z., Guo, M., Yan, H., Liu, K., 2013a. Enhanced loading of doxorubicin into polymeric micelles by a combination of ionic bonding and hydrophobic effect, and the pH-sensitive and ligand-mediated delivery of loaded drug. React. Funct. Polym. 73, 564–572.

Xu, J., Fu, Q., Ren, J.M., Bryant, G., Qiao, G.G., 2013b. Novel drug carriers: from grafted polymers to cross-linked vesicles. Chem. Commun. 49, 33–35.

Xu, C., Yang, D., Mei, L., Li, Q., Zhu, H., Wang, T., 2013c. Targeting chemophotothermal therapy of hepatoma by gold nanorods/graphene oxide core/shell nanocomposites. ACS Appl. Mater. Interf. 5, 12911–12920.

Xu, Z., Wang, D., Xu, S., Liu, X., Zhang, X., Zhang, H., 2014a. Preparation of a camptothecin prodrug with glutathione-responsive disulfide linker for anticancer drug delivery. Chem. Asian J. 9, 199–205.

Xu, X., Li, Y., Li, H., Liu, R., Sheng, M., He, B., Gu, Z., 2014b. Smart nanovehicles based on pH-triggered disassembly of supramolecular peptide-amphiphiles for efficient intracellular drug delivery. Small 10, 1133–1140.

Yackevich, E.I., Mirgorodskaya, A.B., Lukashenko, S.S., Zakharova, L.Ya., 2014. Polyfunctional supramolecular systems based on surfactants containing the hydroxyalkyl moiety in the head group. Russ. Chem. Bull. 63 (8), 1801–1806.

Yamagami, T., Kitayama, Y., Okubo, M., 2014. Preparation of stimuli-responsive "mushroom-like" Janus polymer particles as particulate surfactant by site-selective surface-initiated AGET ATRP in aqueous dispersed systems. Langmuir 30, 7823–7832.

Yamamoto, H., Tahara, K., Kawashima, Y., 2012. Nanomedical system for nucleic acid drugs created with the biodegradable nanoparticle platform. J. Microencapsulation 29 (1), 54–62.

Yang, J.A., Park, K., Jung, H., Kim, H., Hong, S.W., Yoon, S.K., Hahn, S.K., 2011. Target specific hyaluronic acid-interferon alpha conjugate for the treatment of hepatitis C virus infection. Biomaterials 32 (33), 8722–8729.

Yang, Y., Wei, Z.J., Wang, C.Y., Tong, Z., 2013a. Versatile fabrication of nanocomposite microcapsules with controlled shell thickness and low permeability. ACS Appl. Mater. Interf. 5, 2495–2502.

Yang, H., Liang, F., Wang, X., Chen, Y., Zhang, C., Wang, Q., Qu, X., Li, J., Wu, D., Yang, Z., 2013b. Responsive Janus composite nanosheets. Macromolecules 46, 2754–2759.

Yang, F., Ma, S., Zong, W., Luo, N., Lv, M., Hu, Y., Zhou, L., Han, X., 2015. Fabrication of pH sensitive microcapsules using soft templates and their application to drug release. RSC Adv. 5, 51271–51277.

Yao, L., Daniels, J., Wijesinghe, D., Andreev, O.A., Reshetnyak, Y.K., 2013. pHLIP®-mediated delivery of PEGylated liposomes to cancer cells. J. Control. Rel. 167 (3), 228–237.

Yao, L., Song, Q., Bai, W., Zhang, J., Miao, D., Jiang, M., Wang, Y., Shen, Z., Hu, Q., Gu, X., Huang, M., Zheng, G., Gao, X., Hu, B., Chen, J., Chen, H., 2014. Facilitated brain delivery of poly (ethylene glycol)-poly (lactic acid) nanoparticles by microbubble-enhanced unfocused ultrasound. Biomaterials 35 (10), 3384–3395.

Ye, Y., Sun, Y., Zhao, H., Lan, M., Gao, F., Song, C., Lou, K., Li, H., Wang, W., 2013. A novel lactoferrin-modified β-cyclodextrin nanocarrier for brain-targeting drug delivery. Int. J. Pharm. 458 (1), 110–117.

Yemisci, M., Caban, S., Gursoy-Ozdemir, Y., Lule, S., Novoa-Carballal, R., Riguera, R., Fernandez-Megia, E., Andrieux, K., Couvreur, P., Capan, Y., Dalkara, T., 2015. Systemically administered brain-targeted nanoparticles transport peptides across the blood-brain barrier and provide neuroprotection. J. Cereb. Blood Flow Metab. 35 (3), 469–475.

Yi, Y., Jinfeng, D., Bo, C., Zan, J., Li, C., Xuefeng, L., 2013. Environmentally responsive adsorption and assembly behaviors from N-alkyl-1,2-ethylenediamines. Soft Matt. 9, 1458–1467.

Yin, J., Shi, S., Hu, J., Liu, S., 2014. Construction of polyelectrolyte-responsive microgels, and polyelectrolyte concentration and chain length-dependent adsorption kinetics. Langmuir 30, 9551–9559.

Yu, L., Gao, Y., Yue, X., Liu, S., Dai, Z., 2008. Novel hollow microcapsules based on iron-heparin complex multilayers. Langmuir 24 (23), 13723–13729.

Yu, C., Gao, C., Lü, S., Chen, C., Yang, J., Di, X., Liu, M., 2014. Facile preparation of pH-sensitive micelles self-assembled from amphiphilic chondroitin sulfate-histamine conjugate for triggered intracellular drug release. Colloids Surf. B. 115, 331–339.

Yucel, U., Elias, R.J., Coupland, J.N., 2013. Localization and reactivity of a hydrophobic solute in lecithin and caseinate stabilized solid lipid nanoparticles and nanoemulsions. J. Colloid Interf. Sci. 394, 20–25.

Zakharova, L.Y., Semenov, V.E., Voronin, M.A., Valeeva, F.G., Ibragimova, A.R., Giniatullin, R.K.h., Chernova, A.V., Kharlamov, S.V., Kudryavtseva, L.A., Latypov, S.K., Reznik, V.S., Konovalov, A.I., 2007. Nanoreactors based on amphiphilic uracilophanes: self-organization and reactivity study. J. Phys. Chem. B. 111 (51), 14152–14162.

Zakharova, L.Ya., Syakaev, V.V., Voronin, M.A., Valeeva, F.V., Ibragimova, A.R., Ablakova, Y.R., Kazakova, E.K.h., Latypov, Sh.K., Konovalov, A.I., 2009. The NMR and spectrophotometry study of the supramolecular catalytic system based on polyethyleneimine and amphiphilic sulfonatomethylated calix[4]resorcinarene. J. Phys. Chem. C. 113, 6182–6190.

Zakharova, L.Ya., Kudryashova, Yu.R., Selivanova, N.M., Voronin, M.A., Ibragimova, A.I., Solovieva, S.E., Gubaidullin, A.T., Litvinov, A.I., Nizameev, I.R., Kadirov, M.K., Galyametdinov, Yu.G., Antipin, I.S., Konovalov, A.I., 2010a. Novel membrane mimetic systems based on amphiphilic oxyethylated calix[4]arene: aggregative and liquid crystalline behavior. J. Membr. Sci. 364 (1–2), 90–101.

Zakharova, L.Ya., Syakaev, V.V., Voronin, M.A., Semenov, V.E., Valeeva, F.G., Ibragimova, A.R., Bilalov, A.V., Giniyatullin, R.K.h., Latypov, S.h.K., Reznik, V.S., Konovalov, A.I., 2010b. New self-assembling systems based on bola-type pyrimidinic surfactants. J. Colloid Interf. Sci. 342, 119–127.

Zakharova, L.Ya., Mirgorodskaya, A.B., Zhiltsova, T.P., Kudryavtseva, L.A., Konovalov, A.I., 2010c. Reactions in supramolecular systems in molecular encapsulation: organic reactions in constrained systems. John Wiley & Sons, pp. 397-420.

Zakharova, L.Ya., Konovalov, A.I., 2012. Supramolecular systems based on cationic surfactants and amphiphilic macrocycles. Colloid J. 74, 194–206.

Zakharova, L.Ya, Valeeva, F.G., Ibragimova, A.R., Zakharov, A.V., Shtykov, S.N., Bogomolova, I.V., Konovalov, A.I., 2012a. Factors determining the catalytic activity of the mixed micellar system cetyltrimethylammonium bromide-Brij-35 in the hydrolysis of a phosphonic acid ester. Kinet. Catal. 53 (3), 344–352.

Zakharova, L., Kudryashova, Y., Ibragimova, A., Vasilieva, E., Valeeva, F., Popova, E., Solovieva, S., Antipin, I., Ganeeva, Y., Yusupova, T., Konovalov, A., 2012b. Step-by-step design of novel biomimetic nanoreactors based on amphiphilic calixarene immobilized on polymer or mineral platform. Chem Eng. J. 185–186, 285–293.

Zakharova, L.Ya., Gaysin, N.K., Gnezdilov, O.I., Bashirov, F.I., Kashapov, R.R., Zhiltsova, E.P., Pashirova, T.N., Lukashenko, S.S., 2012c. Micellization of alkylated 1.4-diazabicyclo[2.2.2]octane by nuclear magnetic resonance technique using pulsed gradient of static magnetic field. J. Mol. Liq. 167, 89–93.

Zakharova, L.Ya., Kashapov, R.R., Vagapova, G.I., Gabdrakhmanov, D.R., Vasilieva, E.A., 2012d. Comparative study of aqueous solutions of cationic surfactants: structure/activity relation in their aggregation and solubilization behavior and complexation with an oligonucleotide. Chem. Lett. 41, 1226–1228.

Zakharova, L., Voronin, M., Semenov, V., Gabdrakhmanov, D., Syakaev, V., Gogolev, Y., Giniyatullin, R., Lukashenko, S., Reznik, V., Latypov, S., Konovalov, A., Zuev, Y., 2012e. Supramolecular systems based on novel mono- and dicationic pyrimidinic amphiphiles and oligonucleotides: a self-organization and complexation study. Chem. Phys. Chem. 13 (3), 788–796.

Zakharova, L.Ya., Ibragimova, A.R., Vasilieva, E.A., Mirgorodskaya, A.B., Yackevich, E.I., Nizameev, I.R., Kadirov, M.K., Zuev, Y.F., Konovalov, A.I., 2012f. Polyelectrolyte capsules with tunable shell behavior fabricated by the simple layer-by-layer technique for the control of the release and reactivity of small guests. J. Phys. Chem. C 116, 18865–18872.

Zakharova, L.Y., Semenov, V.E., Syakaev, V.V., Voronin, M.A., Gabdrakhmanov, D.R., Valeeva, F.G., Mikhailov, A.S., Voloshina, A.D., Reznik, V.S., Latypov, Sh.K., Konovalov, A.I., 2014a. Amphiphilic macrocycles bearing biofragment: molecular design as factor controlling self-assembly. Mater. Sci. Eng. C 38, 143–150.

Zakharova, L.Ya., Vasilieva, E.A., Gabdrakhmanov, D.R., Konovalov, A.I., Zuev, Y.u.F., 2014b. Complexation of mono- and dicationic surfactants with decanucleotide. Influence of the head group nature. Russ. Chem. Bull. 63 (7), 1615–1618.

Zakharova, L.Ya., Vasilieva, E.A., Gaynanova, G.A., Mirgorodskaya, A.B., Ibragimova, A.R., Salnikov, V.V., Uchegbu, I.F., Konovalov, A.I., Zuev, Y.F., 2015. The polyacrylic acid/modified chitosan capsules with tunable release of small hydrophobic guests. Colloids Surf. A 471, 93–100.

Zakharova, L., Gabdrakhmanov, D., Ibragimova, A., Vasilieva, E., Nizameev, I., Kadirov, M., Pokrovsky, A., Korobeynikov, V., Cheresiz, S., Ermakova, E., Gogoleva, N., Fayzullin, D., Zuev, Y., 2016. Structural, biocomplexation and gene delivery properties of hydroxyethylated gemini surfactants: role of spacer length. Colloids Surf. B 140, 269–277.

Zargar, B., Hatamie, A., 2013. A simple and fast colorimetric method for detection of hydrazine in water samples based on formation of gold nanoparticles as a colorimetric probe. Sens. Actuat. B 182, 706–710.

Zelikin, A.N., Quinn, F., Caruso, F., 2006. Disulfide cross-linked polymer capsules: en route to biodeconstructible systems. Biomacromolecules 7, 27–30.

Zelzer, M., Todd, S.J., Hirst, A.R., McDonald, T.O., Ulijn, R.V., 2013. Enzyme responsive materials: design strategies and future developments. Biomater. Sci. 1, 11–39.

Zhang, M.-J., Wang, W., Xie, R., Ju, X.-J., Liu, L., Gu, Y.-Y., Chu, L.-Y., 2013. Microfluidic fabrication of monodisperse microcapsules for glucoseresponse at physiological temperature. Soft Matt. 9, 4150–4159.

Zhang, H., Huang, S., Yang, X., Zhai, G., 2014. Current research on hyaluronic acid-drug bioconjugates. Eur. J. Med. Chem. 86, 310–317.

Zhang, J., Chen, R., Fang, X., Chen, F., Wang, Y., Chen, M., 2015. Nucleolin targeting AS1411 aptamer modified pH-sensitive micelles for enhanced delivery and antitumor efficacy of paclitaxel. Nano Res. 8, 201–218.

Zhao, Y., Zhang, Sh., Zhang, Y., Cui, Sh., Chen, H., Zhi, D., Zhen, Y., Zhang, Sh., Huang, L., 2015. Tri-peptide cationic lipids for gene delivery. J. Mater. Chem. B 3, 119–126.

Zheng, M.B., Yue, C.X., Ma, Y.F., Gong, P., Zhao, P.F., Zheng, C.F., Sheng, Z.H., Zhang, P.F., Wang, Z.H., Cai, L.T., 2013. Single-step assembly of DOX/ICG loaded lipid-polymer nanoparticles for highly effective chemo-photothermal combination therapy. ACS Nano. 7, 2056–2067.

Zhiltsova, E.P., Pashirova, T.N., Kashapov, R.R., Gaisin, N.K., Gnezdilov, O.I., Lukashenko, S.S., Voloshina, A.D., Kulik, N.V., Zobov, V.V., Zakharova, L.Ya., Konovalov, A.I., 2012. Alkylated 1,4-diazabicyclo[2.2.2]octanes: self-association, catalytic properties, and biological activity. Russ. Chem. Bull., 113–120.

Zhou, J., Romero, G., Rojas, E., Ma, L., Moya, S., Gao, C., 2010. Layer by layer chitosan/alginate coatings on poly(lactide-*co*-glycolide) nanoparticles for antifouling protection and folic acid binding to achieve selective cell targeting. J. Colloid Interf. Sci. 345 (2), 241–247.

Zhu, Z., Gao, N., Wang, H., Sukhishvili, S.A., 2013. Temperature-triggered on-demand drug release enabled by hydrogen-bonded multilayers of block copolymer micelles. J. Control. Rel. 171, 73–80.

Zou, H., Yuan, W., 2015. CO_2- and thermo-responsive vesicles: from expansion–contraction transformation to vesicles–micelles transition. Polym. Chem. 6, 2457–2465.

Zou, S., Wang, C., Gao, Q., Tong, Z., 2013. Surfactant-free multiple pickering emulsions stabilized by combining hydrophobic and hydrophilic nanoparticles. J. Disper. Sci. Technol. 34, 173–181.

Zou, P., Stern, S.T., Sun, D.X., 2014. PLGA/liposome hybrid nanoparticles for short-chain ceramide delivery. Pharm. Res. 31, 684–693.

Zou, H., Wang, Z., Feng, M., 2015. Nanocarriers with tunable surface properties to unblock bottlenecks in systemic drug and gene delivery. J. Control. Rel. 214, 121–133.

CHAPTER

THERAPEUTIC USE OF MONOCLONAL ANTIBODIES: GENERAL ASPECTS AND CHALLENGES FOR DRUG DELIVERY

25

Daniela A. Quinteros*, José M. Bermúdez, Soledad Ravetti[†], Alicia Cid**, Daniel A. Allemandi*, Santiago D. Palma***

**National University of Salta-CONICET, Salta, Argentina*

***National University of Córdoba, Córdoba, Argentina*

[†]National University of Villa Maria, Córdoba, Argentina

CHAPTER OUTLINE

1 GENERAL INTRODUCTION

Monoclonal antibodies are protein molecules made from hybridoma cells by recombinant DNA technology. When applied as a therapeutic, a particular monoclonal antibody is referred to as Mab. The Mabs are defined as "monospecific antibodies that are made by identical immune cells that are all clones of a unique parent cell." Mabs are far superior to polyclonal antibodies with respect to their controlled manufacturing procedures and their reproducible affinity for specific target antigens.

Mabs are indispensable not only to health, but also to prevent food poisoning, and are used to investigate environmental pollution. However, despite their widespread distribution and significance, most people have never heard of Mabs or how they have both transformed healthcare and spawned an entire new industry. Produced in the laboratory, Mabs are derived from the billions of tiny antibodies made every day by our immune systems to combat substances, known as antigens, which are regarded as foreign or potentially dangerous. Millions of different types of antibodies can be found in the blood of humans and other mammals. Made by white blood cells known as B lymphocytes, each antibody is highly specific—that is, it has the ability to bind to only one particular antigen, which may be derived from bacteria, viruses, fungi, parasites, pollen, or nonliving substances, such as toxins, chemicals, drugs, or foreign particles considered alien to the body. Once antibodies have marked their antigen, they and other types of cells produced by the immune system can attack it. The field of genetically engineered therapeutic Mabs has relied on many inventions during decades of research, but two key discoveries in the mid-1970s stand out as seminal events that laid the groundwork for this field to exist as we know it today. Although Mabs were first described in 1975 (Kohler and Milstein, 1975), only when the original rodents forms were replaced by their human equivalents did their potential as therapeutic agents began to be properly appreciated (Lonberg, 2005; Winter et al., 1994). The reasons for this are complex, but related to a combination of perceptions including patentability, immunogenicity, effector function, and a wish to avoid undesirable side effects.

The increasing scientific interest on Mabs can be clearly seen in Fig. 25.1, which shows the rising number of published articles using the keyword *monoclonal antibodies* in different databases (Pubmed and Scopus). Athough fewer published articles, the number of publications is constant over the years to Science Direct and Wiley.

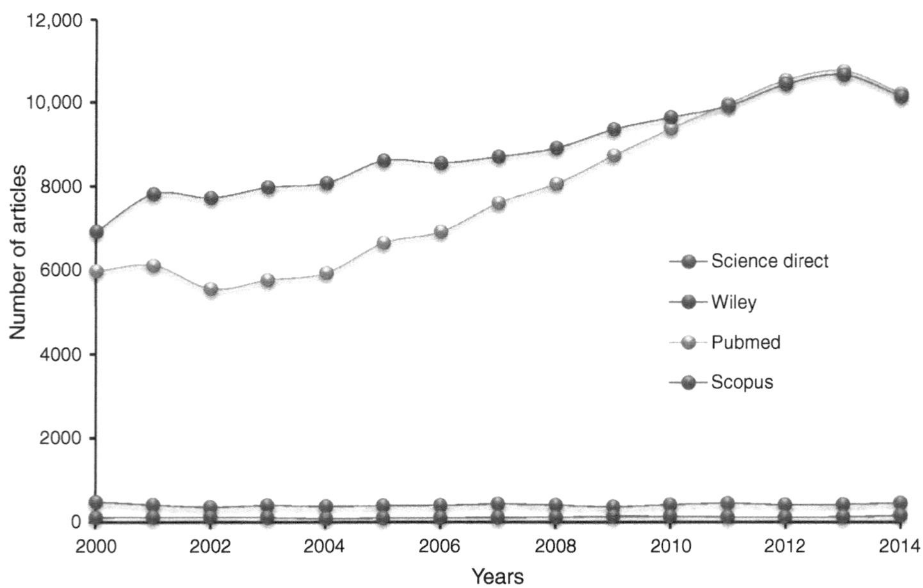

FIGURE 25.1 Number of Publications by Year/Database

Mabs are used not only as drugs for treating various diseases, but are also used as powerful tools for a wide range of medical applications. They are routinely used in hospitals for blood type and tissue, a vital process to ensure safe blood transfusion and organ transplantation. In other cases, they are employed as research probes to determine the pathological pathway and the cause of diseases, such as cancer, autoimmune diseases, and neurological disorders. On the diagnostic front, monoclonal antibodies are intrinsic components of test kits for the detection of ovulation, pregnancy, or menopause. They are also used for analyzing body fluids for medical diagnosis, and to determine whether there has been a heart attack.

Unlike polyclonal antibodies, Mabs are identical antibodies because they are produced by one type of immune cell. Using current hybridoma (mouse/human hybrid cells) technology originally developed by Georges Kohler, Cesar Milstein, and Neils Jerne, Mabs can be produced to bind tightly to virtually any material or antigen, which is defined as a substance that prompts the generation of antibodies that specifically bind to it. Antigens typically consist of proteins or polysaccharides. Epitopes, also known as antigenic determinants, are the part of the antigens through which actual binding to antibodies occurs.

The Mabs technology allowed scientists to produce huge quantities of pure antibodies aimed at specific selected targets, leading to the design of new diagnostic tests and therapeutics. By injecting a payload of Mabs into the bloodstream, the antibodies were headed straight to their disease target.

Today, the growth and profitability of Mabs are outstripping those of earlier types of biotechnology drugs and more traditional pharmaceutical ones. Indeed, their expansion is among the fastest in the global pharmaceutical world. In part, this reflects the sector's embrace of Mabs as an answer to dwindling drugs in the pipeline and reduced revenue streams in the face of the expiration of key patents and the growth of generic medicines.

Drugs based on Mabs technology have names ending in "mab." The first drug that received Food and Drug Administration (FDA) approval was Rituxan (rituximab) (Fig. 25.2) for the treatment of non-Hodgkin's lymphoma, in 1998. It was developed by IDEC, and it has since merged with Biogen. Rituximab is a Mab against the protein CD20, which is primarily found on the surface of immune system B cells. Rituximab destroys B cells and is therefore used to treat diseases that are characterized by excessive numbers of B cells, overactive B cells, or dysfunctional B cells. This includes many lymphomas, leukemias, transplant rejection, and autoimmune disorders. IDEC teamed up with Genentech to get FDA approval and to comarket Rituxan. In parallel, Genentech developed Herceptin (Trastuzumab) for the treatment of breast cancer associated with the HER2/neu receptor. Herceptin was approved by the FDA in 1998 and is the first important example of personalized medicine requiring a diagnostic test to ensure efficacy, since only patients who express the Her2 gene in their breast cancer tissue have a chance to benefit from treatment with this Mab.

Today, six therapeutic Mabs are among the 10 best-selling drugs in the world. In 2013 it was estimated that the global sales of the 10 more successful Mabs were worth more than $58.1 billion. Among these, adalimumab (Humira) is used for the treatment of rheumatoid arthritis and other autoimmune diseases and was at the top of the selling drug list across the world in 2012, reaching an annual income of $9.3 million. Thus, Mab sales are expected to be higher than those for Lipitor, used for cholesterol reduction, which has been the largest prescribed drug historically (King, 2013).

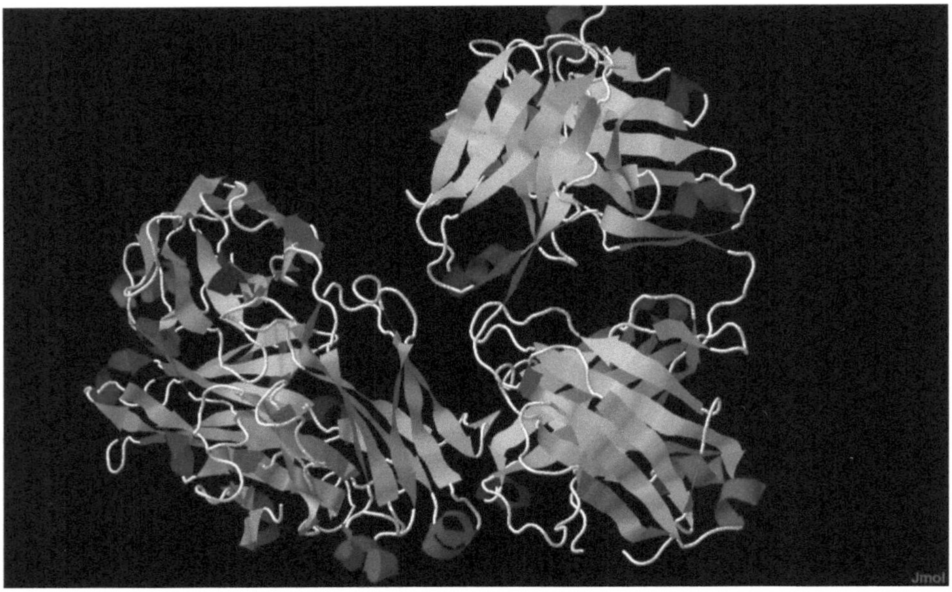

FIGURE 25.2 Monoclonal Antibody Rituximab

2 MABS PRODUCTION

Since the 1970s, the methodologies available for generating Mabs have dramatically evolved from the hybridoma technology used to produce murine Mabs (Kohler and Milstein, 1975) to sophisticated recombinant engineering technologies (Brüggemann et al., 1989a,b; Steinitz et al., 1977) appropriate to design specific Mabs of choice, thus producing humanized, chimeric, and completely (fully) human Mabs (Fig. 25.3) (Brüggemann et al., 1989a,b; Jones et al., 1986; Lonberg, 2005; Riechmann et al., 1998).

These novel methods have had major consequences in the development of clinically applicable Mabs against disease targets. The immunogenicity and safety of Mabs, a substantial limitation of antibodies produced by the hybridoma technology, could be overcome, and Mabs against any antigen specificity and with desirable physicochemical properties could be developed. High-scale production of Mabs at the industrial level could further be achieved. This has sparked a race for the development of antibody-mediated therapeutics in the clinic with the investment of large amounts of money from the biopharmaceutical sector toward that end, which is already paying off. The major success of human Mabs in immune and inflammatory conditions, such as rheumatoid arthritis (RA), Crohn's disease, ulcerative colitis, spondyloarthropathies, juvenile arthritis, psoriasis, psoriatic arthritis, and others (Andreakos et al., 2002), marks merely the beginning of a rapidly evolving field with more than 150 diverse Mabs currently being evaluated in clinical trials or being candidates for approval by the FDA in the USA.

Scaling up Mabs production for therapeutics posed several challenges. The major issue was how to produce high quantities of drugs at a reasonable cost. Most cell lines in the 1980s yielded only half a gram of Mabs per liter, so production was time consuming and expensive. The ideal was to develop

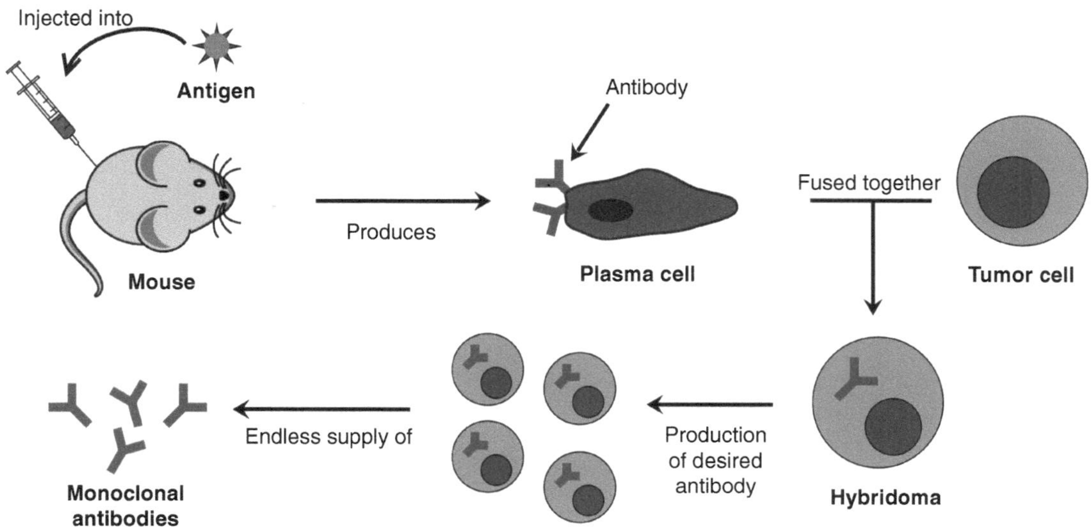

FIGURE 25.3 Schematic Representation of the Production of Monoclonal Antibodies

a hybrid cell line that could produce between 5 and 10 g of Mabs per liter. This demanded several steps, and even more, each requiring skill and patience. First was the creation of a hybrid cell, after which a clone secreting Mabs in high concentrations had to be selected. Then a culture medium had to be developed to encourage the optimal growth of the hybridoma. Scaling up such media was not easy in terms of quality control because they contained fifty or more ingredients, and it was important to determine how many nutrients to add. Hybridomas stop secreting Mabs, for example, if given too much glucose.

3 ANTIBODY–DRUG CONJUGATES

Antibody–drug conjugates (ADCs) are Mabs bearing cytotoxic drugs covalently bound via a chemical linker and can be defined as prodrugs. Prodrugs are inactive or less active derivatives of drug molecules and undergo enzymatic or chemical transformation to regenerate the active forms. ADCs are designed to be superior to either antibody therapeutics or chemotherapy alone by overcoming their limitations while preserving the merits from both. The antibody connected to the cytotoxic warhead (drug) via a linker serves as targeted delivery system to the tumor expressing the antigen/target recognized by the antibody. Ideally, in blood, after systemic administration, this prodrug is nontoxic. Upon binding of the antibody to the targeted tumor antigen and internalization of the complex into the cancer cell, the drug is then released in its active form and in sufficient quantity to kill the cell. On top of the careful choice of a target/antigen expressed in specific tumor indication, it requires finding the best combination between the antibody, the linker, and the drug, which, besides its own characteristics and constraints, are linked and impact each other (Bander, 2013).

The combination of nanoparticles and antibodies can offer versatility together with specificity (Fig. 25.4), allowing a huge potential market.

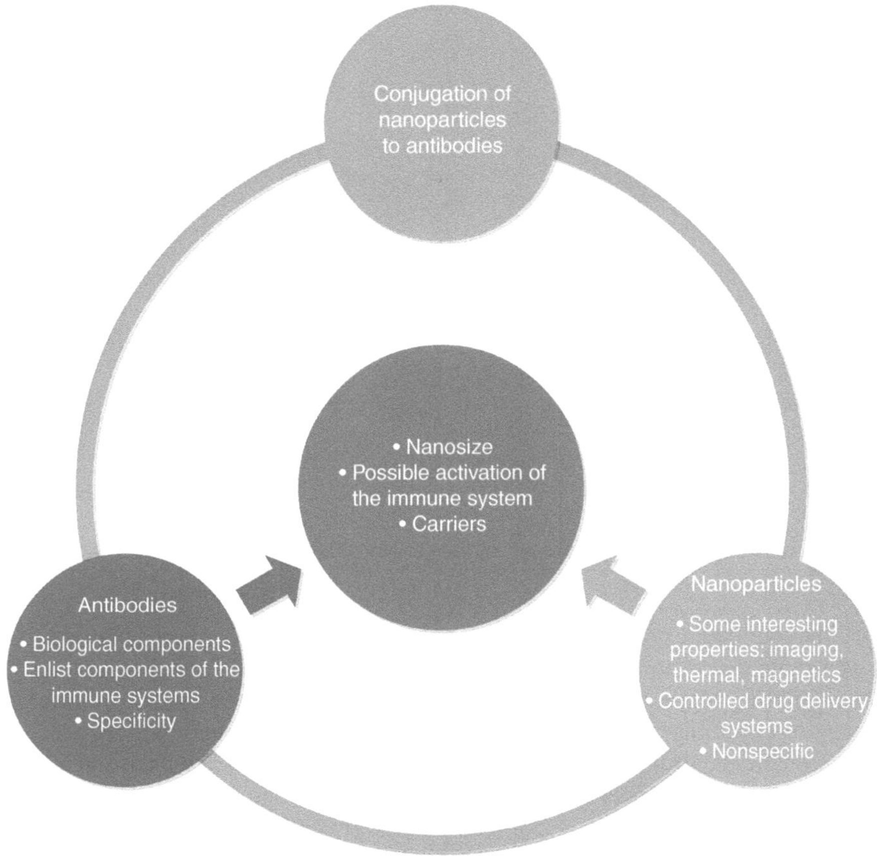

FIGURE 25.4 Conjugation of Nanoparticles to Antibodies and Their Advantages

The target/antigen is the starting point to build an ADC. It first determines which tumor indication will be targeted by the ADC and potentially impacts the choice of the conjugated drug. In addition, the target will also drive the criteria that will be defined for the selection of the targeted patient population within the tumor indication.

3.1 PHYSICAL STABILITY OF ANTIBODY–DRUG CONJUGATES

Mabs are typically susceptible to noncovalent aggregation upon storage. Aggregation is driven by the minimization of free energy achieved when antibodies come into close contact—for instance, at relatively hydrophobic sites on the molecules. Dimerization may be thought of as the first step in aggregation, and sometimes the process proceeds no further. Dimers maybe reversible (Moore et al., 1999) or not, but growth can lead to larger and larger oligomers and even potentially insoluble particulates, as time progresses (Wang et al., 2006). Because ADCs are decorated with one or more small molecule drug moieties, they have a different set of biophysical properties compared with the unconjugated antibody, leading to new or altered intermolecular interactions. An antibody that exhibits acceptable aggregation behavior in its unconjugated form may have different physical behavior in its conjugated form, either because of changes in surface properties, such as hydrophobicity due to drug attachment or because the drugs have altered the higher-order structure of the antibody such that new modes of antibody–antibody interactions are possible. These factors could contribute to a substantially different propensity for aggregation.

This effect is most readily observed in a formulation screen designed to test both conjugated and unconjugated forms in parallel and subjected to the same assays. Such a study does not automatically reveal the exact cause of the effect, only that it is related to the conjugated form or conjugation process. The degree to which the molecule has been physically altered by conjugation may be assessed by techniques such as differential scanning calorimetry (DSC) (Wakankar et al., 2010), with physical perturbations manifested as changes (likely decreases) in the onset of melting or melting temperature, Tm, of the conjugated Mab as compared to the unconjugated antibody. This decrease in thermodynamic stability may translate to a decreased colloidal stability of the conjugate compared to an unconjugated antibody.

3.2 CHEMICAL STABILITY OF ANTIBODY–DRUG CONJUGATES

Part of ADC chemical stability is inherited from the unconjugated antibody. For instance, if the unconjugated antibody is susceptible to deamidation or isomerization with known pH dependence, this degradation mode is also likely to be present in the ADC. These types of degradations may impact the product potency, especially if the affected residues are found in the complementarity-determining region. Much has been learned through years of experience with unconjugated antibodies about the susceptibility of various amino acid residues to chemical degradations, allowing substantial insights to be gained from examination of the protein primary sequence alone, coupled to available information about local solvent exposure and flexibility (Kosky et al., 2009; Robinson and Robinson, 2001). It may be expected that most chemical modifications to the primary sequence of the unconjugated antibody will also occur in the ADC form. Fragmentation is another possible chemical degradation pathway for ADCs, driven by the breakage of covalent bonds between chains or within the peptide backbone. The altered physical state of the molecule following addition of the small molecule drug may result in some differences in the susceptibility of the ADC to fragmentation, but the fundamental susceptibility of the ADC to fragmentation is likely derived mostly from the Mab. A common unconjugated antibody degradation mode involves breaking the heavy chain peptide near the hinge region, leading to free Fab and Fab + Fc products (Cordoba et al., 2005). This is still a feasible degradation pathway for ADCs as well. Comparison of fragmentation rates between conjugated and unconjugated antibodies can reveal whether the ADC form has different stability toward fragmentation.

The covalent bonds linking drugs to antibodies may be broken by a variety of mechanisms over the shelf life of the product and under stressed storage conditions.

3.3 FORMULATION DEVELOPMENT OF ANTIBODY–DRUG CONJUGATES

Formulation development of ADCs aims to ensure that stable, high-quality products are dosed to patients. Though this goal is common to all pharmaceutical formulations, ADCs present a unique set of physicochemical properties that can impact safety, quality, and efficacy as compared to traditional therapeutic proteins. Although there is a body of literature that outlines which product attributes affect the safety, quality, and efficacy for conventional, unconjugated Mabs (Goetze et al., 2010), an understanding of attributes important to ADCs is only now emerging.

The choice of formulation has the potential to affect all categories of ADCs quality attributes. Because conjugation of an ADC requires an unconjugated antibody as an intermediate, the formulation development process for a conjugate necessarily involves formulating a Mab at some stage, likely as an intermediate for manufacturing of the final ADC form. The formulation optimization for the antibody portion of the ADC follows a similar course to that of an unconjugated antibody and, therefore, can leverage the biotechnology industry's considerable experience developing this class of molecule. Many examples of successful commercial antibody formulations, both liquid and lyophilized, have been reported (Wang et al., 2006). However, the complexity of the formulation-development process potentially increases when quality attributes of the small molecule drug and those of the conjugated form are taken into account.

4 NANOTECHNOLOGY'S POTENTIAL IMPACT ON MABS

The antibody-conjugated nanoparticles can be used principally in two biomedical applications: diagnosis and therapy.

In therapy, the development of targeted drug delivery represents, together with tissue repair, the main applications of antibody-conjugated nanoparticles. In diagnosis, the applications can be divided into those using in vivo and those using in vitro experimentation and include contrast agents for magnetic resonance imaging (MRI), sensing, cell sorting, bioseparation, enzyme immobilization, immunoassays, transfection (gene delivery), and purification.

Today, one of the most interesting challenges posed by drug-delivery systems is the development of smart vectors, which are required to be safe, easily administered, and economic, and which allow simultaneous diagnosis and treatment. Moreover, it is intended that the delivery is precisely controlled in term of dose and site of action to reduce adverse side effects. In this regard, there are many important site-selective drugs, such as highly toxic antitumor molecules, that must reach targeted cells or tissues without being released before.

Nanoscience and nanotechnology have found their way into the fields of biotechnology and medicine. Nanoscience is the study of structures and materials on an atomic and molecular scale, in a level that ranges usually below 100 nm.

Nanoparticles by themselves offer specific physicochemical properties that they do not exhibit in bulk form, where materials show constant physical properties regardless of size. Antibodies are nanosize biological products that are part of the specific immune system. In addition to their own properties as pathogens or toxin neutralizers, as well as in the recruitment of immune elements (complement, improving phagocytosis, cytotoxicity antibody dependent by natural killer cells, etc.), they could carry several elements (toxins, drugs, fluorochroms, or even nanoparticles, etc.) and be used in several diagnostic procedures, or even in therapy to destroy a specific target. The conjugation of antibodies to nanoparticles can generate a product that combines the properties of both.

Since the properties of materials differ significantly between atomic or subatomic level and larger scales, nanomaterials have attracted the attention of researchers for pharmaceutical application, especially in the area of drug delivery. This technology overcomes some of the drawbacks of large-size materials, such as poor bioavailability, in vivo stability, solubility, intestinal absorption, sustained and targeted delivery to site of action, and generalized side effects, among others. In this regard, nanostructures have been reported to protect drugs from the degradation in the gastrointestinal tract, and even more, to provide means of bypassing the liver, preventing the first pass metabolism. This technology also has the ability to target delivery of drugs to various areas of the body, and enables the delivery of drugs that are poorly water-soluble. Moreover, nanotechnology improves drugs bioavailability and allows them to remain circulating in the blood for a long time, controlling their release. Nanostructures can also be able to penetrate tissues and be easily taken up by cells, allowing for efficient delivery of drugs to target sites of action. Nanostructures were reported to be taken up by the cells at rates 15–250 times greater than microparticles. Nanotechnology improves performance and acceptability of dosage forms and even may enhance the performance of drugs that fail in clinical trial phases. Nanotechnology will definitely revolutionize the science of drug delivery and help overcome the major challenges of conventional drugs used for the treatment and management of chronic diseases such as, cancer, asthma, hypertension, HIV, and diabetes.

In this context, nanomedicine can be defined as the medical application of nanotechnology and involves research, and diagnostic and therapeutic applications related with nanotechnology. Nanotechnology has been applied in medicine to develop strategies to manage, treat, or even cure virtually every type of human disease.

Mabs are not only successful drugs but also powerful tools for a wide range of medical applications. In this regard, over the last 10 years several nanotechnological platforms for medical applications have converged upon a specialization toward diagnosis, therapy or even both, as in the case of magnetic nanoparticles.

The study of nanoscience and the optimization of technologies, such as ADCs and nanocarriers have driven the design of methodologies that are tailored to address specific diseases and medical anomalies. ADCs, for example, represent excellent nanoscale-based drug delivery vehicles and thus tend to be focused on delivery to certain drugs, as in the treatment of cancer.

The conjugation of different moieties to the nanoparticles widens their application fields and provides them with new or enhanced properties. A range of biomoieties can be conjugated to the nanoparticles including low molecular weight ligands (folic acid, thiamine, dimercaptosuccinic acid), peptides (RGD, LHRD, antigenic peptides, internalization peptides), proteins (BSA, transferrin, antibodies, lectins, cytokines, fibrinogen, thrombin), polysaccharides (hyaluronic acid, chitosan, dextran, oligosaccharides, heparin), polyunsaturated fatty acids (palmitic acid, phospholipids), DNA, and plasmids.

Nanotemplates provide unique access to extremely high-throughput capabilities for rapidly defining the presence of a biomarker and thus lend themselves as excellent platforms for diagnostics applications. Rapid progress is being made to drive these advances in nanomedicine, especially in the area of cancer treatment. Although still in its infancy when compared to other medical technologies, nanomedicine is undoubtedly here to stay and will have an enormous impact on the future health and well-being of man.

4.1 METHODS FOR CONJUGATING ANTIBODIES TO NANOCARRIERS

Nanocarriers are nanometer-sized materials that have the capacity to deliver therapeutic agents at the disease site (Arias, 2011; Schroeder et al., 2012). They are designed to possess unique physicochemical properties, aiming to improve the pharmacokinetic and biodistribution of a drug molecule (Li and Huang, 2008) and to deliver a significant amount of drug molecules. Examples of some therapeutic nanocarriers are lipid-based particles (Al-Jamal and Kostarelos, 2011), micelles (Kedar et al., 2010), nanoparticles (Shi et al., 2010), dendrimers (Svenson, 2009), and polymersomes (Levine et al., 2008). Some of these have been proposed for the treatment of various diseases including cancer (Dhar et al., 2011), coronary artery diseases (Chan et al., 2011), and rheumatoid arthritis (Thomas et al., 2011). In the particular case of cancer, the unique anatomy—that is, the leakiness of the tumoral vasculatures—concedes a passive transport of the nanocarriers by enhanced permeability and retention (EPR) effect (Maeda et al., 2003). However, the porosity of the tumor blood vessels may vary with the tumor type (Bae, 2009). Even with a successful delivery by the EPR effect, the nanocarriers must be able to internalize into the cancer cells (Gradishar et al., 2005). A new paradigm in drug delivery embroils a combination of active and passive targeting. Targeting ligands, such as antibodies (Mamot et al., 2005), peptides (Srinivasan et al., 2009), small molecules (Kukowska-Latallo et al., 2005), or aptamers (Farokhzad et al., 2006) can be attached onto the surface of a nanocarrier. The carriers recognize and bind to the cell-surface receptors and are subsequently taken by the cells via receptor-mediated endocytosis for releasing the therapeutic payloads (Shi et al., 2011). The binding affinity of targeted nanocarriers can also be increased several orders of magnitude by the multivalent effect (Wang et al., 2010).

Among all the targeting ligands, antibodies are well known for their high binding affinity, specificity, and availability for a number of disease biomarkers (Manjappa et al., 2011). An antibody can be simply absorbed on the surface of a nanocarrier via hydrophobic and/or electrostatic interaction (Sokolov et al., 2003). However, using this approach, the absorbed antibody may orient randomly on the surface and result in losing the binding affinity. Furthermore, the antibody may exchange with other endogenous protein in vivo (Nobs et al., 2004). Therefore, antibodies are generally preferred to attach to the nanocarriers covalently (Arruebo et al., 2009). An antibody consists of a number of functional groups that provides many options for bioconjugation (Manjappa et al., 2011). Bioconjugation can take place by means of adsorption (at the isoelectrical point of the antibody via electrostatic interaction), by direct covalent linkage between the surface of the nanoparticle and the antibody, or by using adapter molecules.

The use of adapter molecules generally involves streptavidin and biotin for the formation of the complex. Biotin-labeled polyclonal goat anti-*Escherichia coli* antibodies have been attached to streptavidin-coated magnetic nanoparticles, and used for the separation and selective quantification of *Escherichia coli* O157:H7 in ground beef in the presence of other bacteria (Varshney et al., 2005).

5 THERAPEUTIC APPLICATIONS OF MONOCLONAL ANTIBODIES

Mabs are capable of specifically recognizing and binding to many molecules, mainly those of protein nature. This capability have been widely used in the field of diagnosis for the detection of hormones, vitamins, cytokines, allergens, numerous tumor markers, and a wide range of markers associated with many diseases, including microbial infections. For these properties, the Mabs are the substances most commonly used in the field of clinical diagnosis and in identifying markers and therapeutic targets. Besides applications in disease diagnosis and in drug discovery, therapeutic applications include Mabs, such as vehicles or carriers of other drugs.

Current major therapeutic applications of Mabs include cancer, chronic inflammatory disease, and infection, and they constitute the largest and fastest growing sector of the biological pharmaceutical industry.

In addition to offering a host of new drugs to fight disease, Mabs have provided the means to monitor a patient's response to therapy and helped lead the way in personalized medicine.

Mabs are now marketed not only for cancer and autoimmune disorders, but also for a range of other diseases, including allergic conditions, such as asthma, age-related macular degeneration (an eye disorder), multiple sclerosis (a neurological disorder), and osteoporosis (brittle bones). They are also being investigated for central nervous system disorders, such as Alzheimer's disease (a degenerative brain disease), metabolic diseases such as diabetes, and the prevention of migraines.

Mabs used for therapeutic purposes may bind to a variety of antigens present on the surface of tumor and cancer cells. They may also be utilized to deliver a number of different types of payloads to destroy the targeted cells in question including radioactive ligands, cytokines, toxins, liposomes containing drugs, and specific killer cell types. These immunoconjugates, which are defined as antibodies linked to a second molecule, such as a toxin, radioisotope, or label may act at the cell surface as killing agents or be internalized to release payloads intracellularly.

5.1 DIAGNOSIS AND THERAPEUTIC APPLICATIONS OF ANTIBODY-CONJUGATED NANOPARTICLES

Despite their high sensitivity and reproducibility, conventional diagnostic methods for a microbial infection require cumbersome sample preparation and long readout times (Kaittanis et al., 2009). Unique electrical, magnetic, luminescent, and catalytic properties of nanomaterials enable fast, sensitive, and cost-effective diagnosis, as well as rapid determination of the susceptibility and resistance of antibacterial drugs (Bruchez et al., 1998; Edgar et al., 2006).

Antibody-conjugated nanoparticles amplify the signals for bioanalysis and enumeration of highly pathogenic bacteria, such as *E. coli* O157:H7, resulting in highly selective, convenient, and rapid detection of single bacterium within 20 min (Look et al., 2010).

The rapid and sensitive determination of pathogenic bacteria is extremely important in biotechnology and medical diagnosis. Current methods either lack ultrasensitivity or take a long time for analysis. The use of magnetic nanoparticles also could be a very sensitive and rapid strategy to detect microbial infection. Magnetic nanoparticles, in particular superparamagnetic iron oxide nanoparticles (SPION) with appropriate surface chemistry have been widely used experimentally for numerous in vivo applications, such as magnetic resonance imaging contrast enhancement, tissue repair, immunoassay, detoxification of biological fluids, hyperthermia, drug delivery, in cell separation, and so forth (Neuberger et al., 2005).

Dextrancoated supermagnetic iron oxide nanoparticles were clustered by Con-A treatment, or equipped with Con A-conjugated nanosensors (Zhao et al., 2004).

Supermagnetic iron oxide nanoprobes greatly assisted the identification of *Mycobacterium avium* spp. *paratuberculosis*, as well as the quick quantification of these bacteria in milk and blood with high sensitivity (Basu et al., 2004).

A quick method for detecting infections in the urinary tract has also been developed using gold nano-wire arrays in conjunction with a linker arm attached to specific *E. coli* antibodies (de la Escosura-Muñiz and Merkoçi, 2011).

Li and Huang (2008) have shown that nanoparticles with specific Raman spectroscopic fingerprints can distinguish antibiotic resistant bacteria, by detecting single-nucleotide polymorphisms in microarray-based systems (Li et al., 2009).

A bioassay based on bioconjugated nanoparticle was developed by Zhao et al. (2004) for in situ pathogen quantification. The detection is carried out by a high fluorescent signal provided by the nanoparticle, which is easily attached to an antibody or other biorecognition molecule. These conjugated nanoparticles can be used in the specific identification of a wide range of bacteria, such as *E. coli* O157:H7, through antibody antigen interaction.

These nanoparticles were able to assess the microbial metabolic activity and determine antimicrobial susceptibility in blood, by rapidly quantifying polysaccharides.

The broad absorption spectra of quantum dots (QDs) can be exploited to simultaneously excite QDs emitting different colors using a single wavelength (Cao et al., 2002). These characters suggest that QDs are a promising modality for the analysis of complex samples for histology, pathology and cytology, and can facilitate double or even triple immunostaining of bacterial cells (Kaittanis et al., 2008). Studies have demonstrated that the use of nanotechnology feasibility of achieving fast and reliable pharmaceutical assays for microbial infections in opaque media (e.g., whole blood and milk), without any sample preparations (Alper et al., 2004; Tully et al., 2006).

The ability to track the distribution and differentiation of progenitor and stem cells by high resolution in vivo imaging techniques would have significant clinical and research implications. Lewin et al. (2000) developed a cell labeling approach using short HIV-Tat peptides to derivatize superparamagnetic nanoparticles. These particles were efficiently internalized into hematopoietic and neural progenitor cells in quantities up to 10–30 pg of superparamagnetic iron per cell. Iron incorporation did not affect cell viability, differentiation, or proliferation of $CD34^+$ cells. Following intravenous injection into immunodeficient mice, 4% of magnetically $CD34^+$ cells homed to bone marrow per gram of tissue, and single cells could be detected by magnetic resonance imaging in tissue samples. In addition, magnetically labeled cells that had homed to bone marrow could be recovered by magnetic separation columns. Localization and retrieval of cell populations in vivo enable detailed analysis of specific stem cell and organ interactions critical for advancing the therapeutic use of stem cells (Lewin et al., 2000).

Nanomaterials with fluorescent properties or nanoparticles labeled/encapsulated with fluorescent dyes have also been applied for microbial detection. Using antibody-conjugated silica nanoparticles, these materials can be used as a superior signaling element, to detect cells, proteins and bacteria in an immunoassay (Grifantini et al., 2002; Tan et al., 2004).

Ultrasensitive methods for bioassays have been developed using fluorescent bioconjugated silica nanoparticles (Santra et al., 2001; Zhao et al., 2004). Thousands of fluorescent dye molecules are encapsulated in a protective silica matrix, resulting in a nanoparticle with an amplified and reproducible signal for fluorescence-based bioanalysis. Whereas in conventional immunoassays only one or a few dye molecules are linked to an antibody molecule and then used to signal an antibody-antigen interaction event, the bioconjugated nanoparticles, which are attached to the antibody molecule, are enabled to carry many dye molecules inside, allowing a significant amplification of the analytical signal. Furthermore, bacteria present many surface antigens available for antibody recognition, and, therefore, a greatly amplified signal can be achieved since thousands of nanoparticles can attach to each bacterium. These bioconjugated nanoparticles allow the detection of a bacterium cell per given sample just in 20 min with a spectrofluorometer.

Santra and coworkers used covalent method to attach surface-modified, rubpy-doped silica nanoparticles to mouse antihuman CD10 antibody. This complex was then incubated with mononuclear lymphoid target cells. After washing away the unbound nanoparticles, target leukemia cells could be clearly detected. In comparison to control group, results of this experiment have shown that this technique was very effective to detect leukemia cells selectively (Santra et al., 2001).

Tan and coworkers developed an assay tool for in situ detection of single bacterium cells in less than 20 min and developed multicolored FRET (fluorescence resonance energy transfer) silica nanoparticles by coencapsulating three tandem dyes that emit unique colors upon excitation with a single wavelength (Tan et al., 2004).

Among therapeutic applications of antibody-conjugated nanoparticles, the use of gold nanoparticles was tested in killing *Staphyloccocus aureus* (Zharov et al., 2006). The bacteria were killed by light-adsorbing gold nanoparticles conjugated to antiprotein A antibodies, using laser irradiation at 532 nm. Protein A was chosen because it interacts specifically with the Fc fragment of the antibody. According to the authors, killing efficiency depends on the local overheating effects accompanied by bubble-formation phenomena. Direct irradiation of bacteria with the laser did not damage the bacteria, because of low absorption by natural endogenous cytochromes.

5.1.1 Cancer

The diagnosis of cancer at early stages of growth is a critical factor for obtaining optimal results in therapy and for improving the chances of survival.

Several imaging techniques collaborate with physicians in the diagnosis, including magnetic resonance imaging (MRI), positron emission tomography (PET), computed tomography (CT), ultrasound, radiography, photoacoustic imaging, fluoroscopy, and so forth. In some of these techniques, antibody-conjugated nanoparticles may offer increased selectivity and sensitivity.

Fluorescent semiconductor nanocrystal QDs are a novel class of multifunctional inorganic fluorophores that are promising in biological imaging, including immunofluorescence imaging (Wu, 2003), and are useful for labeling molecules (Lidke et al., 2004; Wu, 2003) and cells in various materials (Gao et al., 2004; Jaiswal et al., 2003; Kim, 2004), even clinical human samples (Tholouli et al., 2006).

The first generation of ADCs presented limitations in relation to the linker instability because the early linkers were either too stable, resulting in low potency and reduced efficacy, or too unstable, resulting in poor targeting and high systemic toxicity. Furthermore, cytotoxics without sufficient potency, inefficient internalization, limited expression of the target antigen and immunogenicity of Mabs (Alley et al., 2009; Chari, 2008).

Second-generation ADCs were designed to deliver potent anticancer agents to tumors in a targeted manner to limit systemic exposure. The success of targeted ADCs depends on antibody selection, potency of the cytotoxin and the method selected to link the antibody to the cytotoxin.

Nanotechnology is a disruptive technology that drives a new generation of cancer preventive diagnosis, and therapeutic products, resulting in dramatically improved cancer outcomes. Nanotechnology in the field of cancer has the potential to improve the monitoring of therapeutic efficacy, provide novel methods for the detection and profiling of cancers at early stages, and allow surgeons to delineate tumor margins and sentinel lymph nodes. Nanomaterials have unique features that are attractive and can be applied to biosensing. The development of various nanomaterials and nanotechnology has enabled detection of cancer biomarkers with great precision and sensitivity that could not be achieved before. Many studies are being conducted on developing sensing mechanisms that will push down the detection limit as far down as possible (Anajwala et al., 2010). Ligand-free nano-formulations generally possess passive targeting property with little tissue specificity. To address these limitations, research has been continued to advance active or specific targeting to enhance the efficacy of anticancer therapeutic agents, as well as to reduce the toxicity to nontargeted healthy tissues. Nanoparticles containing the chemotherapeutic agents are specifically designed to target the cancerous cells either by ligand receptor interaction or antibody-antigen recognition. Highly specific Mabs are used to strengthen the immune response and to intensify the immune system's antitumor capacity. Moreover, Mabs are highly specific when attached to nanoparticles to aid in targeted delivery of various antitumor cytotoxic agents or function themselves as effective therapeutic agents (Sutradhar and Amin, 2014). There are different ways of destroying cancer cells with Mabs: directly by inducing apoptosis, blocking growth factor receptors or inducing the formation of antiidiotype, and indirectly by activating complement-mediated cellular cytotoxicity and antibody dependent cell-mediated cytotoxicity (Praetorius and Mandal, 2007).

The paramagnetic properties of iron-oxide nanoparticles have been harnessed for therapeutic and imaging applications (McCarthy et al., 2007). Chen et al. (2006) have reported that dextran-coated iron-oxide nanoparticles conjugated to radiolabeled (Iodine isotope 131,131I) anti-VEGF Mab significantly increased imaging resolution, as well as destruction of liver cancer in mice (Chen et al., 2006).

Srinivasan, Lakshmikuttyamma, and Shoyele (Srinivasan et al., 2013) studied selective targeting of Mabs to oncoproteins in cancer cells while avoiding their accumulation in normal cells. Results of fluorescence microscopy, TEM, and flow cytometry revealed that bevacizumab nanoparticles were internalized by A549 cells 3 times more than by MRC-5 cells. Macropinocytosis and energy-dependent pathways were elucidated to be involved in their uptake by A549 cells. This study presents the first evidence that uncoated Mab nanoparticles can be selectively delivered to cancer cells while avoiding normal cells.

Mabs have been also employed as an effective active targeted therapy since these antibodies can specifically recognize the over-expressed HER2 positive tumor cells and internalize through receptor mediated endocytosis.

To treat HER2 positive breast cancer, anti-HER2 humanized Mabs are commonly used, although advances can be made in targeted cellular localization via conjugation strategy through a nano-particulate system focusing on surface modified ligand/receptor-mediated nano-therapy to target the tumor cell at the molecular level.

Various nanoparticle systems, such as liposomes, micelles, and dendrimers, have been studied not only in vitro but also in vivo, concluding that the presence of receptor targeting antibody on the nano-carriers demonstrated a higher toxicity than the nano-carriers without an antibody after time dependent incubation (Hamidreza Montazeri et al., 2008; Matsumura, 2008; Yokoyama, 2005).

Antibodies against cell surface biomarkers are the commonly used ligands for the development of targeted nanoparticles. Although mouse Mabs have been used for making targeted nanoparticles, strong cross-species immune responses limit their potential for future clinical translation. Currently only a few types of humanized Mabs, such as HER-2 antibody (Herceptin), are available for the production of targeted nanoparticles (Slamon and Pegram, 2001). Alternatively, high affinity recombinant antibody fragments have been developed as targeting ligands (Yang et al., 2015). For example, a human single chain antibody against the epidermal growth factor receptor (ScFvEGFR) that is highly expressed in the majority of epithelial tumors was conjugated to different types of nanoparticles. Specificity of tumor imaging and targeted therapeutic effects of these nanoparticles have been demonstrated in several animal tumor models (Yang et al., 2015). The major advantages of using natural ligands for tumor targeting are their high binding affinity, specificity, and, most importantly, low immunogenicity.

Owen et al. (2013) reported higher antibody-conjugated lysosomal P (LA-co-TMCC)-g-PEG-furan micellar concentration in HER2 positive cell lines with significant cellular distribution resulting in two- to threefold increase in lysosomal accumulation compared to using the antibody alone (Owen et al., 2013). In addition, Nobs et al. (2004) reported that thiol functionalized anti-HER2 Mab can be conjugated with the maleimide containing liposomes. Internalization of these antibody coupled immunoliposomes by HER2 positive breast cancer cell line (SK-BR-3) was observed with enhanced effects compared to the control liposomes (Nobs et al., 2004). Many authors observed that the use of lisposomes was more effective in vitro than in vivo. To prove targeted drug-loaded liposome as an effective and safe treatment option for HER2 positive breast cancer, more in vivo studies need to be conducted.

When used, anti-HER2 Mab conjugated with dendrimers resulted in the effective delivery to target the overexpressed HER2 receptors of tumor cells in animal models. More recently, Miyano et al. (2010) developed biocompatible sixth generation (G6) anionic lysine (amino acid) dendrimer, on which the surface was modified with glutamate and, thereafter the glutamate modified G6 lysine dendrimer was coupled with trastuzumab Mab and a fluorescent label. Following this, the whole conjugated formulation was evaluated in both HER2 positive and negative cell lines to assess the targeting efficiency and cellular internalization compared to the free antibody application. The results of this study revealed high binding affinity with low cytotoxicity; and cellular internalization or lysosomal trafficking in HER2 positive cells with a dose-dependent profile compared to the HER2 negative cells (Miyano et al., 2010).

In subsequent studies by Chen et al. (2008) and Gao et al. (2009), a model protein toxin (PE38KDEL) was used to develop anti-HER2 antibody functionalized PE38KDEL loaded PLGA nanoparticles utilizing a two phase carbodiimide process. Modified encapsulated nanoparticles showed higher in vitro cytotoxicity and more protective antitumor activity compared to the controls in HER2 overexpressing cells and tumor-bearing mice, respectively. In addition, they reported well tolerability for maximum tolerated dose, as well as reduced systemic toxicity when modified nanoparticles were administered to mice model compared to control (Chen et al., 2008; Gao et al., 2009).

Nanotechnology has promoted greater opportunities for higher specific drug delivery with minimum side effects. Bio-conjugation strategies of therapeutic agents loaded nanoparticles with Mabs have exhibited a targeted drug delivery approach both in vitro and in vivo.

Mabs can act as highly specific probes when they are attached to nanoparticles to aid in targeted delivery of various antitumor cytotoxic agents (Sutradhar and Amin, 2014). The combination therapy of nanopartilces of Mabs and other chemotherapeutic agents, such as doxorubicin, 5-Fluorouracil, paclitaxel, docetaxel, campothecin, and topotecan, could bring about synergistic tumor growth inhibition. Some nanoparticulate systems combining antibody with these chemotherapeutic agents are listed in Table 25.1.

5.1.2 Eye diseases

Neovascularization is a major cause of visual loss in a number of ophthalmic diseases. It consists of new blood vessel growth from preexisting vascular structures. This process occurs in pathologies of the anterior and posterior segments of the eye, such as corneal neovascularization, age-related macular degeneration (AMD), diabetic macular edema, viral retinitis, proliferative vitreoretinopathy, posterior uveitis, retinal vascular occlusions, choroid neovascularization (CNV), and diabetic retinopathy, to name a few.

Anti-VEGF agents have demonstrated efficacy in reducing neovascularization in both animal models and clinical trials. Specifically, anti-VEGF antibodies have shown initial therapeutic success. Bevacizumab is a full-length, humanized murine Mab that recognizes all isoforms of VEGF. Bevacizumab

Table 25.1 Nanoparticulate Systems Combining Antibody With Chemotherapeutic Agents

System	Antibody	Loaded Drug	Pathology/Target	Reference
PEG liposomes	Anti-MT1-MP(Fab′)	Doxorubicin	—	Hatakeyama et al. (2007)
Fe_3O_4/Au magnetic/gold nanoparticles	Bevacizumab anti-VEGF antibody	Doxorubicin	—	Ramos-Tejada et al. (2015)
Liposomes	Anti-HER2 monoclonal antibodies (trastuzumab)	Doxorubicin	Tumor xenograft nude mouse models	Banerjee et al. (2011); Park et al. (2007)
Chitosan nanoparticles	Anti-HER2 monoclonal antibodies	Doxorubicin	Cancer therapy (breast, ovary)	Yousefpour et al. (2011)
Human albumin serum (HAS) protein-based nanoparticle	Trastuzumab	Doxorubicin	Breast cancer therapy	Anhorn et al. (2008)
Chitosan scaffold-PLGA nanoparticles	Bevacizumab anti-VEGF antibody	5-Fluorouracil	Brain cancer	Kutlu et al. (2014)
Poly(propylene imine) dendrimer	Primary antibody (mAbK1)	Paclitaxel	Ovarian cancer therapy	Jain et al. (2015)
PLGA nanoparticles	Cetuximab	Paclitaxel palmitate	Lung cancer therapy	Karra et al. (2013)
Poly(D,L-lactide-*co*-glycolide)/montmorillonite (PLGA/MMT) nanoparticles	Trastuzumab	Paclitaxel	Breast cancer therapy	Sun et al. (2008)
Micellar nanoparticles (vitamin E TPGS and TPGS-siRNA)	AntiHER2 antibody	Docetaxel	Breast cancer therapy	Zhao et al. (2013); Mi et al. (2013)
Polylactide (PLA)-based nanoparticles	AntiHER2 antibody	Docetaxel	Breast cancer therapy	Zhao et al. (2012)
Poly(lactide-*co*-glycolide) (PLGA) nanoparticles	IgG Isotype control antibody anti-Fas mAb	Camptothecin	Colorectal cancer therapy	McCarron et al. (2008)
Liposomes	Antibody fragments scFv antibodies	Topotecan	Prostate cancer therapy	Roth et al. (2007)

was initially approved by the U.S. Food and Drug Administration (FDA) to treat metastatic colon cancer, but it has also shown efficacy in the treatment of various neovascular ocular diseases and is used off-label to treat neovascular age-related macular degeneration.

Topical delivery is a relatively easy and a less risky method of drug administration. However, delivery to the posterior segment via this route is considered inefficient and unsuccessful, since less than 5% of the topically applied dose enters the eye and an even smaller fraction of it (0.001%) is expected to reach the posterior segment. Intravitreal administration involves the direct administration of drug solution/suspension into vitreous humor via pars plana using a 30-G needle. In contrast to the topical and systemic routes, intravitreal injection makes high concentrations of drug locally available to the internal eye tissue, including the choroid and the retina. Similarly, the intravitreal administration of Macugen (pegaptanib sodium; Pfizer) and Lucentis (ranibizumab; Genentech/Novartis), vascular endothelial growth factor (VEGF) inhibitors, is highly successful for the control of AMD. However, agents with molecular weight less than 500 Da when applied intravitreally tend to be drained off from the site of application with a half-life of less than 3 days, indicating a need for repetitive injections. Anyway, the period requiring a repeat dose may extend from a few days to several months for macromolecular antibodies. For example, three is the mean number of injections of bevacizumab (Avastin; Roche) required to be administered per year for the treatment of AMD. On the other hand, the recommended dosing frequency of ranibizumab (Lucentis) is once a month (0.5 mg; 50 μL) for at least 9 months (Rosenfeld et al., 2006), whereas pegaptanib sodium (Macugen) needs to be injected intravitreally at 6-week intervals for 1 year (Kitagawa and Yuzawa, 2013).

Nevertheless, repetitive intravitreal injections, even if spaced widely, are invariably associated with complications, such as vitreous hemorrhage, retinal detachment, cataract, and endophthalmitis. The rate of endophthalmitis and retinal detachment being observed with intravitreal injection is 0.2 and 0.05%, respectively (Edelhauser et al., 2010). Moreover, patient compliance is lower with such regimens because of the painful and invasive procedures requiring hospitalization and specially trained physician for administration, in addition to the high cost of the medicine per se (Kaur and Kakkar, 2014).

Biodegradable polymeric nanoparticles offer properties that make them suitable candidates to overcome these administration issues. In this regard, Abrishami et al. (2009) studied encapsulated bevacizumab into liposomesand and found that the intravitreal injection of liposomes was well tolerated through 42 days in rabbits. The clearance of this drug in vitreous from liposomal formulations was slower than the soluble form. The concentration level of bevacizumab after intravitreal injection suggested that intravitreal injection of drug by these carriers provide sufficient concentration of therapeutic drug for 6 weeks for diabetic neovascularization and probably for other neovascular eye diseases (Abrishami et al., 2009).

Other authors developed nanoparticles of PLGA loaded with bevacizumab and stabilized whit albumin. They found that the vitreous concentration of bevacizumab was sustained for about 8 weeks at values greater than 500 ng mL^{-1} and that the nanoparticles presented a drug vitreous medium retention time (MRT) 3.3 times higher than the control. In addition, it was confirmed the nanoparticles persistence in ocular tissues for 56 days (Varshochian et al., 2013, 2015).

Recently Lu et al. (2014) studied the effects of intravitreal injection of bevacizumab-chitosan nanoparticles on pathological morphology of retina and the expression of vascular endothelial growth factor (VEGF) protein and VEGF mRNA in the retina of diabetic rats. The results have offered a new approach for inhibiting angiogenesis of diabetic retinopathy and indicated that the intravitreal injection of bevacizumab inhibited VEGF expression in retina. In addition, bevacizumab-chitosan nanoparticles presented a longer period of action (Zhou et al., 2015).

5.1.3 Respiratory diseases

Chronic respiratory diseases (CRD) are chronic diseases of the airways and other structures of the lung. Among the most common, asthma, chronic obstructive pulmonary disease (COPD), and respiratory allergies can be mentioned.

The airway inflammation in these diseases cannot always be controlled with conventional therapies. The critical role of the combination therapy of an inhaled corticosteroid (ICS) and a long-acting β-adrenoceptor agonist (LABA) in the treatment of patients also suffering from severe asthma and chronic obstructive pulmonary disease (COPD) explains why there is a strong interest within the pharmaceutical industry in developing new pharmaceutical alternatives.

Biological therapies represented in particular by Mabs against selective targets, could improve the outcome of these diseases.

Many current publications have shown the important role of nanomedicine in the treatment of respiratory diseases (Card et al., 2008; Mansour et al., 2009; Smola et al., 2008). Systems releasing nanoparticles have several possibilities to improve treatment of upper respiratory tract since they can protect drugs from degradation by enzymes present in the epithelial lining fluid, the particle size may permit circumvent pass macrophages, and crossing the endothelium will allow systemic treatment of nanoparticles and pulmonary drug intravenously administration.

Identifying asthma phenotypes has given impetus to the search for biomarkers to help classify patients, pointing to new therapies and predicting different pathological mechanisms of disease progression with strong benefits for affected patients (Moore et al., 2007).

An ideal biomarker is easy to detect and measure, noninvasive, and inexpensive, and it can be used to identify phenotypes either for clinical response or treatment, assessing changes in disease activity, or confirming a diagnosis.

COPD is currently the fourth leading cause of death in the world, and if current trends continue, it is likely to rank as the third leading cause of death by 2030 (World Health Organization, 2008).

Both cytokines and chemokines play a fundamental role in the organization of COPD, so the use of Mabs in the treatment options is very valuable to provide more targeted therapies (Li et al., 2002; Yamagata and Ichinose, 2006).

Numerous antibodies directed to cytokines, chemokines, growth factors, and their receptors are considered for the treatment of COPD and some have already been studied in clinical trials in patients with COPD.

The macrophage imaging using MRI, together with the use of magnetic nanoparticles of iron oxide, has recently emerged as a promising noninvasive technique for preclinical and clinical studies of inflammatory diseases.

However, limited research on inflammation and passage of macrophages in the lung, using imaging technology has been performed due to difficulties in imaging the body (i.e., the signal loss due to the pulsations and heart breathing, low proton density, and susceptibility artifacts). Novel perspectives for imaging, diagnosis, and treatment of respiratory diseases, such as COPD can appear due to technical improvements and the development of techniques of magnetic resonance pulse sequence detection.

Studies reported the possibility of noninvasive monitoring of subsets of macrophages in an inflammatory model using high resolution MRI after intravenous administration. The possibility of noninvasive monitoring of subsets of macrophages was reported in an inflammatory model using high resolution MRI after intravenous administration. Furthermore, other studies have proven that polarization and proliferation was not affected using iron oxide nanoparticles with ex vivo labeled subpopulation of macrophages. The coupling of a specific antibody with the iron oxide nanoparticles directed to a particular subpopulation of macrophages may provide a promising strategy for early diagnosis and improved various inflammatory diseases noninvasively using MRI.

Al Faraj et al. (2014) evaluated the in vivo effect of intrapulmonary administration of SPION on the profiles of alveolar macrophage polarization in a model of COPD and developed a protocol MRI noninvasive to specifically target and monitor a subpopulation of macrophages using specific antibody-conjugated SPION.

An interesting approach was performed to target a subpopulation of macrophages in the lung of a COPD animal model using biocompatible antibody-conjugated iron oxide nanoparticles. This conjugation allowed noninvasive tracking using a free-breathing MRI protocol.

Antibodies to IL-8 Abgenix improved dyspnea but not lung function. This limited clinical benefit is attributed to suboptimal dosing into the airways because of intravenous administration (Mahler et al., 2004). Therefore, the use of nanoparticles may allow administration by inhalation for local concentrations and avoiding side effects. So far, clinical trials with anti-TNFalpha antibody infliximab have been carried out, but no effects in patients with COPD were observed (Dentener et al., 2008; Rennard et al., 2007; Van der Vaart et al., 2005).

Therefore, the release of cetuximab (currently undergoing clinical trials for cancer therapy) could be a valuable option for the treatment of COPD.

Molecularly targeted nanoparticles were chosen as the best strategy for imaging inflammation to make progress in health care (McCarthy and Weissleder, 2008; Weissleder, 2006).

Magnetic nanoparticle's versatility makes them well suited for applications to enable early detection and prevention, and so improving diagnosis, treatment, and follow-up of diseases (Oghabian and Farahbakhsh, 2010).

Al Faraj et al. (2014) developed an assay to quantify the antibodies CD86 or CD206 conjugated to SPION, which involves the reduction of Cu^{2+} to Cu^{+} by proteins and the appearance of a purple-blue copper-protein complex in alkaline medium.

6 FUTURE TRENDS OF MONOCLONAL ANTIBODIES

Some significant trends that appear to be developing include bispecific antibodies, ADCs, and companion diagnostics, which should help to identify the most appropriate patient populations for treatment.

Even though the health benefits of Mab therapeutics are proven, controversies will continue about their economic viability because of their expensive price. High prices, however, are generally associated with early innovative treatments, and Mabs are not the only drugs to have staggering prices. The cost of cancer treatments, for example, has more than doubled in the past two decades, leading to an outcry by many European and American cancer specialists. Another reason why Mabs are so expensive is the fact that many of them are still protected by patents. A drug's patent life is 20 years from the date of filing, in order to help developers recover some of the research and development costs involved in getting a drug to market. Some of the patent life is in fact reduced because some of that time, an average of 8 years, is taken up by clinical trials and regulatory approval (Siddiqui and Rajkumar, 2012).

The slow progress of Mab therapeutics for infectious diseases can be attributed in part to the large arsenal of other antiinfective drugs, such as vaccines and antibiotics. Because they are specific to a single pathogen, Mab drugs are also commercially less attractive than traditional drugs because they cover a narrower spectrum of patients. In addition, Mabs need to be administered by intravenous or subcutaneous injection, unlike other antiinfectives that can be taken orally, so they are unsuitable for patients in developing countries who have limited access to healthcare. Mabs are also more effective at preventing infection rather than treating established ones, and unlike vaccines provide only short-term prophylaxis. And finally, the high development and manufacturing costs associated with Mabs and the poor record in winning approval for treating infectious diseases with Mabs lessen their commercial appeal (Reichert, 2006).

More promising are recent developments to enhance the potency and efficacy of Mabs, so as to make it possible to prescribe lower doses and potentially reduce costs. A number of approaches have been adopted to augment the efficacy of Mabs. One of the most encouraging is the use of genetic engineering to remove glycosylation sites from the variable domain of the antibody. This enhances the effector function of Mabs, such as antibody-dependent cell-mediated cytotoxicity (ADCC), which activates the patient's innate immune cells to kill a target cell like cancer.

Although the new generation of Mabs may greatly improve the treatment of cancer and autoimmune disorders, which are well-established disease targets for Mab therapeutics, it remains to be seen whether Mab therapeutics will be effective in other areas. Nowhere is this question more urgent than in the case of infectious diseases.

The fact that Mab therapeutics is pathogen-specific could hinder its use for treating mixed infections. One solution might lie in the use of a cocktail of Mabs to target the diverse range of antigens that viruses carry. Such a strategy would effectively mimic the natural immune response: once infected, the body tends to develop several antibodies in response to the antigens presented by a virus, each of which attaches to one of the different antigens. It is this diversity of antibodies that helps the immune system in fighting the invader. The use of a cocktail of Mabs is already being investigated for the treatment for rabies

Recent advances in Mab engineering have opened up new opportunities for serum therapy. Importantly, Mabs offer the means to prepare standardized agents, which, when combined in a cocktail, can yield a product that is more precise and more potent than traditional serum therapy. To date, the development of antiinfective Mab products has attracted little commercial investment. In part this is because infectious diseases are short-lived and therefore have a limited market. This is in contrast to chronic conditions like cancer and autoimmune diseases, which require regular treatment and therefore have greater profit potential. Nonetheless, the pharmaceutical climate is changing, fueled by concern over the rise in new pathogens (such as West Nile and corona viruses); the reemergence of old pathogens (like tuberculosis), increasing antibiotic resistance among microorganisms; and the rise of superbugs like MRSA, as well as the growing epidemic of patients who are immunocompromised as a result of HIV infection, organ transplantation, chronic degenerative diseases, and improvements in cancer care.

Likewise, Mabs have been used to investigate and treat cancer, providing powerful tools for identifying and targeting different antigens on tumors. Although these applications did not quickly translate into successful Mab therapeutics for cancer, in more recent years Mabs therapeutics have offered alternatives to drugs with a broad spectrum and high toxicity. This has transformed the care of cancer patients, who no longer face the prospect of losing hair and other serious side effects associated with other cytotoxic drugs. The advantage of Mabs is that they can be given as maintenance therapies. This is reshaping our perceptions of some cancers from what was once seen as inevitably fatal to a chronic condition. Mabs have also enabled the prescription of specific therapeutics for particular tumor antigens in individual patients. This allows a greater degree of personalization in the management of cancer than in the past. Indeed, Mabs are expected to be an increasingly important component in personalized cancer therapies.

In a world of antibiotic-resistant superbugs and an aging population grappling with autoimmune disorders and cancer, Mabs offer the potential for new, targeted treatments and drugs that can offer personalized care—and a window into the complex, overlapping conditions that underlie human disease.

REFERENCES

Abrishami, M., Ganavati, S.Z., Soroush, D., Rouhbakhsh, M., Jaafari, R., Malaekeh Nikouei, P.B., 2009. Preparation, characterization, and in vivo evaluation of nanoliposomes-encapsulated bevacizumab (Avastin) for intravitreal administration. Retina 29 (5), 699–703.

Al Faraj, A., Shaik, A.S., Afzal, S., Al Sayed, B., Halwani, R., 2014. MR imaging and targeting of a specific alveolar macrophage subpopulation in LPS-induced COPD animal model using antibody-conjugated magnetic nanoparticles. Int. J. Nanomed. 24 (9), 1491–1503.

Al-Jamal, W.T., Kostarelos, K., 2011. Liposomes: from a clinically established drug delivery system to a nanoparticle platform for theranostic nanomedicine. Acc. Chem. Res. 44, 1094–1104.

Alley, S.C., Zhang, X., Okeley, N.M., Anderson, M., Law, C.L., Senter, P.D., Benjamin, D.R., 2009. The pharmacologic basis for antibody-auristatin conjugate activity. J. Pharmacol. Exp. Ther. 330 (3), 932–938.

Alper, M.D., Bertozzi, C.R., Clarke, J., 2004. Detection of bacteria in suspension by using a superconducting quantum interference device. Proc. Natl. Acad. Sci. USA 101 (1), 129–134.

Anajwala, C.C., Girish, K.J., Vijayendra Swamy, S.M., 2010. Current trends of nanotechnology for cancer therapy. Int. J. Pharm. Sci. Nanotechnol. 3 (3), 1043–1056.

Andreakos, E., Taylor, P.C., Feldmann, M., 2002. Monoclonal antibodies in immune and inflammatory diseases. Curr. Opin. Biotechnol. 13, 615–620.

Anhorn, M.G., Wagner, S., Kreuter, J., Langer, K., von Briesen, H., 2008. Specific targeting of HER2 overexpressing breast cancer cells with doxorubicin-loaded trastuzumab-modified human serum albuminnanoparticles. Bioconjug. Chem. 19 (12), 2321–2331.

Arias, J.L., 2011. Advanced methodologies to formulate nanotheragnostic agents for combined drug delivery and imaging. Expert Opin. Drug Deliv. 8, 1589–1608.

Arruebo, M., Valladares, M., González-Fernández, A., 2009. Antibody-conjugated nanoparticles for biomedical applications. J. Nanomater., 1–24.

Bae, Y.H., 2009. Drug targeting and tumor heterogeneity. J. Control. Release 133 (1), 2–3.

Bander, N.H., 2013. Antibody–drug conjugate target selection: critical factors. In: Ducry, L. (Ed.), Antibody-Drug Conjugates. Humana Press, Switzerland, pp. 29–40, (Chapter 2).

Banerjee, D., Harfouche, R., Sengupta, S., 2011. Nanotechnology-mediated targeting of tumor angiogenesis. Vasc. Cell. 3 (3), 1–13.

Basu, M., Seggerson, S., Henshaw, J., Jiang, J., Cordna, R., Lefave, C., Boyle, P.J., Miller, A., Pugia, M., Basu, S., 2004. Nano-biosensor development for bacterial detection during human kidney infection: use of glycoconjugate-specific antibody-bound gold nanowire arrays (GNWA). Glycoconj. J. 21, 487–496.

Bruchez, Jr., M., Moronne, M., Gin, P., Weiss, S., Alivisatos, A.P., 1998. Semiconductor nanocrystals as fluorescent biological labels. Science 281, 2013–2016.

Brüggemann, M., Winter, G., Waldmann, H., Neuberger, M.S., 1989a. The immunogenicity of chimeric antibodies. J. Exp. Med. 170, 2153–2157.

Brüggemann, M., Caskey, H.M., Teale, C., Waldmann, H., Williams, G.T., Surani, M.A., Neuberger, M.S., 1989b. A repertoire of monoclonal antibodies with human heavy chains from transgenic mice. Proc. Natl. Acad. Sci. USA 86 (17), 6709–6713.

Cao, Y.C., Jin, R., Mirkin, C.A., 2002. Nanoparticles with Raman spectroscopic fingerprints for DNA and RNA detection. Science 297, 1536–1540.

Card, J.W., Zeldin, D.C., Bonner, J.C., Nestmann, E.R., 2008. Pulmonary applications and toxicity of engineered nanoparticles. Am. J. Physiol. Lung Cell. Mol. Physiol. 295 (3), L400–L411.

Chan, J.M., Rhee, J.W., Drum, C.L., Bronson, R.T., Golomb, G., Langer, R., Farokhzad, O.C., 2011. In vivo prevention of arterial restenosis with paclitaxel-encapsulated targeted lipidpolymeric nanoparticles. Proc. Natl. Acad. Sci. USA 108, 19347–19352.

Chari, R.V., 2008. Targeted cancer therapy: conferring specificity to cytotoxic drugs. Acc. Chem. Res. 41 (1), 98–107.

Chen, H., Gao, J., Lu, Y., Kou, G., Zhang, H., Fan, L., Sun, Z., Guo, Y., Zhong, Y., 2008. Preparation and characterization of PE38KDEL-loaded anti-HER2 nanoparticles for targeted cancer therapy. J. Control. Release 128 (3), 209–216.

Chen, J., Wu, H., Han, D., Xie, C., 2006. Using anti-VEGF McAb and magnetic nanoparticles as double-targeting vector for the radioimmunotherapy of liver cancer. Cancer Lett. 231, 169–175.

Cordoba, A., Shyong, B., Breen, D., Harris, R., 2005. Non-enzymatic hinge region fragmentation of antibodies in solution. J Chromatogr. B. Anal. Technol. Biomed. Life Sci. 818 (2), 115–121.

de la Escosura-Muñiz, A., Merkoçi, A., 2011. A nanochannel/nanoparticle-based filtering and sensing platform for direct detection of a cancer biomarker in blood. Small 7 (5), 675–682.

Dentener, M.A., Creutzberg, E.C., Pennings, H.J., Rijkers, G.T., Mercken, E., Wouters, E.F., 2008. Effect of infliximab on local and systemic inflammation in chronic obstructive pulmonary disease: a pilot study. Respiration 76 (3), 275–282.

Dhar, S., Kolishetti, N., Lippard, S.J., Farokhzad, O.C., 2011. Targeted delivery of a cisplatin prodrug for safer and more effective prostate cancer therapy in vivo. Proc. Natl. Acad. Sci. USA 108, 1850–1855.

Edelhauser, H.F., Rowe-Rendleman, C.L., Robinson, M.R., Dawson, D.G., Chader, G.J., Grossniklaus, H.E., Rittenhouse, K.D., Wilson, C.G., Weber, D.A., Kuppermann, B.D., Csaky, K.G., Olsen, T.W., Kompella, U.B., Holers, M.V., Hageman, G.S., Gilger, B.C., Campochiaro, P.A., Whitcup, S.M., Wong, W.T., 2010. Ophthalmic drug delivery systems for the treatment of retinal diseases: basic research to clinical applications. Invest. Ophthalmol. Vis. Sci. 51, 5403–5420.

Edgar, R., McKinstry, M., Hwang, J., Oppenheim, A.B., Fekete, R.A., Giulian, G., Merril, C., Nagashima, K., Adhya, S., 2006. High-sensitivity bacterial detection using biotin-tagged phage and quantum-dot nanocomplexes. Proc. Natl. Acad. Sci. USA 103, 4841–4845.

Farokhzad, O.C., Cheng, J., Teply, B.A., Sherifi, I., Jon, S., Kantoff, P.W., Richie, J.P., Langer, R., 2006. Targeted nanoparticle-aptamer bioconjugates for cancer chemotherapy in vivo. Proc. Natl. Acad. Sci. USA 103, 6315–6320.

Gao, J., Kou, G., Wang, H., Chen, H., Li, B., Lu, Y., Zhang, D., Wang, S., Hou, S., Qian, W., Dai, J., Zhao, J., Zhong, Y., Guo, Y., 2009. PE38KDEL-loaded anti-HER2 nanoparticles inhibit breast tumor progression with reduced toxicity and immunogenicity. Breast Cancer Res. Treat. 115 (1), 29–41.

Gao, X., Cui, Y., Levenson, R.M., Chung, L.W., Nie, S., 2004. In vivo cancer targeting and imaging with semiconductor quantum dots. Nat. Biotechnol. 22 (8), 969–976.

Goetze, A.M., Schenauer, M.R., Flynn, G.C., 2010. Assessing monoclonal antibody product quality attribute criticality through clinical studies. Mabs 2 (5), 500–507.

Gradishar, W.J., Tjulandin, S., Davidson, N., Shaw, H., Desai, N., Bhar, P., Hawkins, M., O'Shaughnessy, J., 2005. Phase III trial of nanoparticle albumin-bound paclitaxel compared with polyethylated castor oil-based paclitaxel in women with breast cancer. J. Clin. Oncol. 23, 7794–7803.

Grifantini, R., Bartolini, E., Muzzi, A., Draghi, M., Frigimelica, E., Berger, J., Ratti, G., Petracca, R., Galli, G., Agnusdei, M., Giuliani, M.M., Santini, L., Brunelli, B., Tettelin, H., Rappuoli, R., Randazzo, F., Grandi, G., 2002. Previously unrecognized vaccine candidates against group B meningococcus identified by DNA microarrays. Nat. Biotechnol. 20 (9), 914–921.

Hamidreza Montazeri, A., Shahin, M., Brocks, D.R., Lavasanifar, A., 2008. Disposition of drugs in block copolymer micelle delivery systems: from discovery to recovery. Clin. Pharmacokinet. 47, 619–634.

Hatakeyama, H., Akita, H., Ishida, E., Hashimoto, K., Kobayashi, H., Aoki, T., Yasuda, J., Obata, K., Kikuchi, H., Ishida, T., Kiwada, H., Harashima, H., 2007. Tumor targeting of doxorubicin by anti-MT1-MMP antibody-modified PEG liposomes. Int. J. Pharm. 342, 194–200.

Jain, N.K., Tare, M.S., Mishra, V., Tripathi, P.K., 2015. The development, characterization and in vivo anti-ovarian cancer activity of poly(propylene imine) (PPI)-antibody conjugates containing encapsulated paclitaxel. Nanomed. Nanotech. Biol. Med. 11, 207–218.

Jaiswal, J.K., Mattoussi, H., Mauro, J.M., Simon, S.M., 2003. Long-term multiple color imaging of live cells using quantum dot bioconjugates. Nat. Biotechnol. 21 (1), 47–51.

Jones, P.T., Dear, P.H., Foote, J., Neuberger, M.S., Winter, G., 1986. Replacing the complementarity-determining regions in a human antibody with those from a mouse. Nature 321, 522–525.
Kaittanis, C., Nath, S., Perez, J.M., 2008. Rapid nanoparticle-mediated monitoring of bacterial metabolic activity and assessment of antimicrobial susceptibility in blood with magnetic relaxation. PLoS One 3 (9), e3253.
Kaittanis, C., Santra, S., Perez, J.M., 2009. Role of nanoparticle valency in the nondestructive magnetic-relaxation-mediated detection and magnetic isolation of cells in complex media. J. Am. Chem. Soc. 131 (35), 12780–12791.
Karra, N., Nassar, T., Ripin, A.N., Schwob, O., Borlak, J., Benita, S., 2013. Antibody conjugated PLGA nanoparticles for targeted delivery of paclitaxel palmitate: efficacy and biofate in a lung cancer mouse model. Small 9 (24), 4221–4236.
Kaur, I.P., Kakkar, S., 2014. Nanotherapy for posterior eye diseases. J. Control. Release 193, 100–112.
Kedar, U., Phutane, P., Shidhaye, S., Kadam, V., 2010. Advances in polymeric micelles for drug delivery and tumor targeting. Nanomedicine 6, 714–729.
Kim, S., Lim, Y.T., Soltesz, E.G., De Grand, A.M., Lee, J., Nakayama, A., Parker, J.A., Mihaljevic, T., Laurence, R.G., Dor, D.M., Cohn, L.H., Bawendi, M.G., Frangioni, J.V., 2004. Near-infrared fluorescent type II quantum dots for sentinel lymph node mapping. Nat. Biotechnol. 22 (1), 93–97.
King, S., 2013. The best-selling drugs of all time: Humira joins the elite. Forbes. Available from: http://www.forbes.com/sites/simonking/2013/01/28/the-best-selling-drugs-of-all-time-humira-joins-the-elite/#21b9bed86193.
Kitagawa, T., Yuzawa, M., 2013. Intravitreal pegaptanib sodium for myopic choroidal neovascularization: 1 year results of a prospective pilot study. Nihon. Ganka. Gakkai. Zasshi. 117, 344–350.
Kohler, G., Milstein, C., 1975. Continuous cultures of fused cells secreting antibody of predefined specificity. Nature 256, 495–497.
Kosky, A.A., Dharmavaram, V., Ratnaswamy, G., Manning, M.C., 2009. Multivariate analysis of the sequence dependence of asparagine deamidation rates in peptides. Pharm. Res. 26 (11), 2417–2428.
Kukowska-Latallo, J.F., Candido, K.A., Cao, Z., Nigavekar, S.S., Majoros, I.J., Thomas, T.P., Balogh, L.P., Khan, M.K., Baker, J.R., 2005. Nanoparticle targeting of anticancer drug improves therapeutic response in animal model of human epithelial cancer. Cancer Res. 65, 5317–5324.
Kutlu, C., Cakamak, A.S., Gumusdereliglu, M., 2014. Double-effective chitosan scaffold-PLGA nanoparticle system for brain tumour therapy: in vitro study. J. Microencapsul. 31 (7), 700–707.
Levine, D.H., Ghoroghchian, P.P., Freudenberg, J., Zhang, G., Therien, M.J., Greene, M.I., Hammer, D.A., Murali, R., 2008. Polymersomes: a new multi-functional tool for cancer diagnosis and therapy. Methods 46, 25–32.
Lewin, M., Carlesso, N., Tung, C.H., Tang, X.W., Cory, D., Scadden, D.T., Weissleder, R., 2000. Tat peptide-derivatized magnetic nanoparticles allow in vivo tracking and recovery of progenitor cells. Nat. Biotechnol. 18 (4), 410–414.
Li, L., Das, A.M., Torphy, T.J., Griswold, D.E., 2002. What's in the pipeline? Prospects for monoclonal antibodies (mAbs) as therapies for lung diseases. Pulm. Pharmacol. Ther. 15 (5), 409–416.
Li, M., Hu, B., Li, J., Chen, R., Zhang, X., Chen, H., 2009. Extractive electrospray ionization mass spectrometry toward in situ analysis without sample pretreatment. Anal. Chem. 81 (18), 7724–7731.
Li, S.D., Huang, L., 2008. Pharmacokinetics and biodistribution of nanoparticles. Mol. Pharm. 5, 496–504.
Lidke, D.S., Nagy, P., Heintzmann, R., Arndt-Jovin, D.J., Post, J.N., Grecco, H.E., Jares-Erijman, E.A., Jovin, T.M., 2004. Quantum dot ligands provide new insights into erbB/HER receptor–mediated signal transduction. Nat. Biotechnol. 22 (2), 198–203.
Lonberg, N., 2005. Human antibodies from transgenic animals. Nat. Biotechnol. 23, 1117–1125.
Look, M., Bandyopadhyay, A., Blum, J.S., Fahmy, T.M., 2010. Application of nanotechnologies for improved immune response against infectious diseases in the developing world. Adv. Drug Deliv. Rev. 62 (4–5), 378–393.
Lu, Y., Zhou, N., Huang, X., Cheng, J.W., Li, F.Q., Wei, R.L., Cai, J.P., 2014. Effect of intravitreal injection of bevacizumab-chitosan nanoparticles on retina of diabetic rats. Int. J. Ophthalmol. 7 (1), 1–7.
Maeda, H., Fang, J., Inutsuka, T., Kitamoto, Y., 2003. Vascular permeability enhancement in solid tumor: various factors, mechanisms involved and its implications. Int. Immunopharmacol. 3, 319–328.
Mahler, D.A., Huang, S., Tabrizi, M., Bell, G.M., 2004. Efficacy and safety of a monoclonal antibody recognizing interleukin-8 in COPD: a pilot study. Chest 126 (3), 926–934.

Mamot, C., Drummond, D.C., Noble, C.O., Kallab, V., Guo, Z., Hong, K., Kirpotin, D.B., Park, J.W., 2005. Epider mal growth factor receptortargeted immunoliposomes significantly enhance the efficacy of multiple anticancer drugs in vivo. Cancer Res. 65, 11631–11638.

Manjappa, A.S., Chaudhari, K.R., Venkataraju, M.P., Dantuluri, P., Nanda, B., Sidda, C., Sawant, K.K., Murthy, R.S., 2011. Antibody derivatization and conjugation strategies: application in preparation of stealth immunoliposome to target chemotherapeutics to tumor. J. Control. Release.

Mansour, H.M., Rhee, Y.S., Wu, X., 2009. Nanomedicine in pulmonary delivery. Int. J. Nanomedicine 4, 299–319.

Matsumura, Y., 2008. Polymeric micellar delivery systems in oncology. J. Clin. Oncol. 38 (12), 793–802.

McCarron, P.A., Marouf, W.M., Quinn, D.J., Fay, F., Burden, R.E., Olwill, S.A., Scott, C.J., 2008. Antibody targeting of camptothecin-loaded PLGA nanoparticles to tumor cells. Bioconjug. Chem. 19, 1561–1569.

McCarthy, J.R., Weissleder, R., 2008. Multifunctional magnetic nanoparticles for targeted imaging and therapy. Adv. Drug Deliv. Rev. 60 (11), 1241–1251.

McCarthy, J.R., Kelly, K.A., Sun, E.Y., Weissleder, R., 2007. Targeted delivery of multifunctional magnetic nanoparticles. Nanomedicine 2, 153–167.

Mi, Y., Zhao, J., Feng, S.S., 2013. Targeted co-delivery of docetaxel, cisplatin and herceptin by vitamin E TPGS-cisplatin prodrug nanoparticles for multimodality treatment of cancer. J. Control. Release 169 (3), 185–192.

Miyano, T., Wijagkanalan, W., Kawakami, S., Yamashita, F., Hashida, M., 2010. Anionic amino acid dendrimer-trastuzumab conjugates for specific internalization in HER2-positive cancer cells. Mol. Pharm. 7 (4), 1318–1327.

Moore, W.C., Bleecker, E.R., Curran-Everett, D., Erzurum, S.C., Ameredes, B.T., Bacharier, L., et al., 2007. Characterization of the severe asthma phenotype by the National Heart, Lung, and Blood Institute's Severe Asthma Research Program. J. Allergy Clin. Immunol. 119, 405–413.

Moore, J., Patapoff, T., Cromwell, M., 1999. Kinetics and thermodynamics of dimer formation and dissociation for a recombinant humanized monoclonal antibody to vascular endothelial growth factor. Biochemistry 38 (42), 13960–13967.

Neuberger, T., Schöpf, B., Hofmann, H., Hofmann, M., von Rechenberg, B., 2005. Superparamagnetic nanoparticles for biomedical applications: possibilities and limitations of a new drug delivery system. J. Magn. Magn. Mater. 293 (1), 483–496.

Nobs, L., Buchegger, F., Gurny, R., Allemann, E., 2004. Current methods for attaching targeting ligands to liposomes and nanoparticles. J. Pharm. Sci. 93 (8), 1980–1992.

Oghabian, M.A., Farahbakhsh, N.M., 2010. Potential use of nanoparticle based contrast agents in MRI: a molecular imaging perspective. J. Biomed. Nanotechnol. 6 (3), 203–213.

Owen, S.C., Patel, N., Logie, J., Pan, G., Persson, H., Moffat, J., Sidhu, S.S., Shoichet, M.S., 2013. Targeting HER2+ breast cancer cells: lysosomal accumulation of anti-HER2 antibodies is influenced by antibody binding site and conjugation to polymeric nanoparticles. J. Control. Release 172, 395–404.

Park, J.W., Hong, K., Kirpotin, D.B., Colbern, G., Shalaby, R., Baselga, J., Shao, Y., Nielsen, U.B., Marks, J.D., Moore, D., Praetorius, N.P., Mandal, T.K., 2007. Engineered nanoparticles in cancer therapy. Recent Pat. Drug Deliv. Formul. 1 (1), 37–51.

Praetorius, N.P., Mandal, T.K., 2007. Engineered nanoparticles in cancer therapy. Recent Pat. Drug Deliv. Formul. 1 (1), 37–51.

Ramos-Tejada, M.M., Viota, J.L., Rudzka, K., Delgado, A.V., 2015. Preparation of multi-functionalized Fe3O4/Au nanoparticles for medical purposes. Colloids Surf. B Biointerfaces 1 (128), 1–7.

Reichert, J.M., 2006. Anti-infective monoclonal antibodies: perils and promise of development. Nat. Rev. Drug Discov. 5, 191–195.

Rennard, S.I., Fogarty, C., Kelsen, S., Long, W., Ramsdell, J., Allison, J., Mahler, D., Saadeh, C., Siler, T., Snell, P., Korenblat, P., Smith, W., Kaye, M., Mandel, M., Andrews, C., Prabhu, R., Donohue, J.F., Watt, R., Lo, K.H., Schlenker-Herceg, R., Barnathan, E.S., Murray, J., 2007. The safety and efficacy of infliximab in moderate to severe chronic obstructive pulmonary disease. Am. J. Respir. Crit. Care Med. 175 (9), 926–934, 1.

Riechmann, L., Clark, M., Waldmann, H., Winter, G., 1998. Reshaping human antibodies for therapy. Nature 332, 323–327.

Robinson, N.E., Robinson, A.B., 2001. Molecular clocks. Proc. Natl. Acad. Sci. USA 98 (3), 944–949.

Rosenfeld, P.J., Brown, D.M., Heier, J.S., Boyer, D.S., Kaiser, P.K., Chung, C.Y., Kim, R.Y., MARINA Study Group, 2006. Ranibizumab for neovascular age-related macular degeneration. N. Engl. J. Med. 355, 1419–1431.

Roth, A., Drummond, D.C., Conrad, F., Hayes, M.E., Kirpotin, D.B., Benz, C.C., Marks, J.D., Liu, B., 2007. Anti-CD166 single chain antibody-mediated intracellular delivery of liposomal drugs to prostate cancer cells. Mol. Cancer Ther. 6 (10), 2737–2746.

Santra, S., Zhang, P., Wang, K., Tapec, R., Tan, W., 2001. Conjugation of biomolecules with luminophore-doped silica nanoparticles for photostable biomarkers. Anal. Chem. 73 (20), 4988–4993.

Schroeder, A., Heller, D.A., Winslow, M.M., Dahlman, J.E., Pratt, G.W., Langer, R., Jacks, T., Anderson, D.G., 2012. Treating metastatic cancer with nanotechnology. Nat. Rev. Cancer 12, 39–50.

Shi, J., Votruba, A.R., Farokhzad, O.C., Langer, R., 2010. Nanotechnology in drug delivery and tissue engineering: from discovery to applications. Nano Lett. 10, 3223–3230.

Shi, J., Xiao, Z., Kamaly, N., Farokhzad, O.C., 2011. Self-assembled targeted nanoparticles: evolution of technologies and bench to bedside translation. Acc. Chem. Res. 44, 1123–1134.

Siddiqui, M., Rajkumar, S.C., 2012. The high costs of cancer drugs and what we can do about it. Mayo Clin. Proc. 87 (10), 935–943.

Slamon, D., Pegram, M., 2001. Rationale for trastuzumab (Herceptin) in adjuvant breast cancer trials. Semin Oncol. 28, 13–19.

Smola, M., Vandamme, T., Sokolowski, A., 2008. Nanocarriers as pulmonary drug delivery systems to treat and to diagnose respiratory and non respiratory diseases. Int. J. Nanomedicine 3 (1), 1–19.

Sokolov, K., Follen, M., Aaron, J., Pavlova, I., Malpica, A., Lotan, R., Richards-Kortum, R., 2003. Real-time vital optical imaging of precancer using anti-epider mal growth factor receptor antibodies conjugated to gold nanoparticles. Cancer Res. 63, 1999–2004.

Srinivasan, A.R., Lakshmikuttyamma, A., Shoyele, S.A., 2013. Investigation of stability and cellular uptake of self associated monoclonal antibody (MAb) nanoparticles by non-small lung cancer cells. Mol. Pharm. 10 (9), 3275–3284, 3.

Srinivasan, R., Marchant, R.E., Gupta, A.S., 2009. In vitro and in vivo platelet targeting by cyclic RGD-modified liposomes. J. Biomed. Mater. Res. 93, 1004–1015.

Steinitz, M., Klein, G., Koskimies, S., Makel, O., 1977. EB virus-induced B lymphocyte cell lines producing specific antibody. Nature 269, 420–422.

Sun, B., Ranganathan, B., Feng, S.S., 2008. Multifunctional poly (D,L-lactide-co-glycolide)/montmorillonite (PLGA/MMT) nanoparticles decorated by Trastuzumab for targeted chemotherapy of breast cancer. Biomaterials 29 (4), 475–486.

Sutradhar, K.B., Amin, Md.L., 2014. Nanotechnology in cancer drug delivery and selective targeting review. ISRN Nanotechnol., 1–12.

Svenson, S., 2009. Dendrimers as versatile platform in drug delivery applications. Eur. J. Pharm. Biopharm. 71, 445–462.

Tan, M., Wang, G., Hai, X., Ye, Z., Yuan, J., 2004. Development of functionalized fluorescent europium nanoparticles for biolabeling and time-resolved fluorometric applications. J. Mater. Chem. 14 (19), 2896–2901.

Tholouli, E., Hoyland, J.A., Di Vizio, D., O'Connell, F., Macdermott, S.A., Twomey, D., Levenson, R., Yin, J.A., Golub, T.R., Loda, M., Byers, R., 2006. Imaging of multiple mRNA targets using quantum dot based in situ hybridization and spectral deconvolution in clinical biopsies. Biochem. Biophys. Res. Commun. 348 (2), 628–636.

Thomas, T.P., Goonewardena, S.N., Majoros, I.J., Kotlyar, A., Cao, Z., Leroueil, P.R., Baker, J.R., 2011. Folate-targeted nanoparticles show efficacy in the treatment of inflammatory arthritis. Arthritis Rheum. 63, 2671–2680.

Tully, E., Hearty, S., Leonard, P., O'Kennedy, R., 2006. The development of rapid fluorescence-based immunoassays, using quantum dot-labelled antibodies for the detection of Listeria monocytogenes cell surface proteins. Int. J. Biol. Macromol. 39 (1–3), 127–134.

Van der Vaart, H., Koëter, G.H., Postma, D.S., Kauffman, H.F., ten Hacken, N.H., 2005. First study of infliximab treatment in patients with chronic obstructive pulmonary disease. Am. J. Respir. Crit. Care Med. 172 (4), 465–469, 15.

Varshney, M., Yang, L., Su, X.L., Li, Y., 2005. Magnetic nanoparticle-antibody conjugates for the separation of Escherichia coli O157:H7 in ground beef. J. Food Prot. 68 (9), 1804–1811.

Varshochian, R., Jeddi-Tehrani, M., Khoshayand, M.R., Dinarvand, R., 2013. The protective effect of albumin on bevacizumab activity and stability in PLGA nanoparticles intended for retinal and choroidal neovascularization treatments. Eur. J. Pharm. Sci. 50 (3-4), 341–352.

Varshochian, R., Riazi-Esfahani, M., Jeddi-Tehrani, M., Mahmoudi, A.R., Aghazadeh, S., Mahbod, M., Movassat, M., Atyabi, F., Sabzevari, A., Dinarvand, R., 2015. Albuminated PLGA nanoparticles containing bevacizumab intended for ocular neovascularization treatment. J. Biomed. Mater. Res. A 103 (10), 3148–3156.

Wakankar, A.A., Feeney, M.B., Rivera, J., Chen, Y., Kim, M., Sharma, V.K., Wang, Y.J., 2010. Physicochemical stability of the antibody–drug conjugate Trastuzumab-DM1: changes due to modification and conjugation processes. Bioconjug. Chem. 21 (9), 1588–1595.

Wang, W., Singh, S., Zeng, D.L., King, K., Nema, S., 2006. Antibody structure, instability, and formulation. J. Pharm. Sci. 96 (1), 1–26.

Wang, J., Tian, S., Petros, R.A., Napier, M.E., Desimone, J.M., 2010. The complex role of multivalency in nanoparticles targeting the transfer rin receptor for cancer therapies. J. Am. Chem. Soc. 132, 11306–11313.

Weissleder, R., 2006. Molecular imaging in cancer. Science 31 (5777), 1168–1171.

Winter, G., Griffiths, A.D., Hawkins, R.E., Hoogenboom, H.R., 1994. Making antibodies by phage display technology. Annu. Rev. Immunol. 12, 433–455.

World Health Organization, 2008. World Health Statistics.

Wu, X., Liu, H., Liu, J., Haley, K.N., Treadway, J.A., Larson, J.P., Ge, N., Peale, F., Bruchez, M.P., 2003. Immunofluorescent labeling of cancer marker Her2 and other cellular targets with semiconductor quantum dots. Nat. Biotechnol. 21 (1), 41–46.

Yamagata, T., Ichinose, M., 2006. Agents against cytokine synthesis or receptors. Eur. J. Pharmacol. 533 (1–3), 289–301, 8.

Yang, E., Qian, W., Cao, Z., Wang, L., Bozeman, E.N., Ward, C., Yang, B., Selvaraj, P., Lipowska, M., Wang, Y.A., Mao, H., Yang, L., 2015. Theranostic nanoparticles carrying doxorubicin attenuate targeting ligand specific antibody responses following systemic delivery. Theranostics 5 (1), 43–61.

Yokoyama, M., 2005. Drug targeting with nano-sized carrier systems. J. Artif. Organs 8 (2), 77–84.

Yousefpour, P., Atyabi, F., Vasheghani-Farahani, E., Movahedi, A.A., Dinarvand, R., 2011. Targeted delivery of doxorubicin-utilizing chitosan nanoparticles surface-functionalized with anti-Her2 trastuzumab. Int. J. Nanomed. 6, 1977–1990.

Zhao, X., Hilliard, L.R., Mechery, S.J., Wang, Y., Bagwe, R.P., Jin, S., Tan, W., 2004. A rapid bioassay for single bacterial cell quantitation using bioconjugated nanoparticles. Proc. Natl. Acad. Sci. USA 101 (42), 15027–15032.

Zhao, J., Mi, Y., Feng, S.S., 2013. Targeted co-delivery of docetaxel and siPlk1 by herceptin-conjugated vitamin E TPGS based immuno-micelles. Biomaterials 34 (13), 3411–3421.

Zhao, J., Mi, Y., Liu, Y., Feng, S.S., 2012. Quantitative control of targeting effect of anticancer drugs formulated by ligand-conjugated nanoparticles of biodegradable copolymer blend. Biomaterials 33 (6), 1948–1958.

Zharov, V.P., Mercer, K.E., Galitovskaya, E.N., Smeltzer, M.S., 2006. Photothermal nanotherapeutics and nanodiagnostics for selective killing of bacteria targeted with gold nanoparticles. Biophys. J. 90 (2), 619–627.

Zhou, C., Wu, Y.L., Chen, G., Liu, X., Zhu, Y., Lu, S., Feng, J., He, J., Han, B., Wang, J., Jiang, G., Hu, C., Zhang, H., Cheng, G., Song, X., Lu, Y., Pan, H., Zheng, W., Yin, A.Y., 2015. BEYOND: a randomized, double-blind, placebo-controlled, multicenter, phase III study of first-line carboplatin/paclitaxel plus bevacizumab or placebo in chinese patients with advanced or recurrent nonsquamous non-small-cell lung cancer. J. Clin. Oncol. 33 (19), 2197–2204.

FURTHER READING

Daugherty, A.L., Mrsny, R.J., 2006. Formulation and delivery issues for monoclonal antibody therapeutics. Adv. Drug Deliv. Rev. 58 (5-6), 686–706.

Kirpotin, D., Park, J.W., Hong, K., Zalipsky, S., Li, W.L., Carter, P., Benz, C.C., Papahadjopoulos, D., 1997. Sterically stabilized anti-HER2 immunoliposomes: design and targeting to human breast cancer cells in vitro. Biochemistry 36 (1), 66–75.

Papahadjopoulos, D., Benz, C.C., 2002. Anti-HER2 immunoliposomes: enhanced efficacy attributable to targeted delivery. Clin. Cancer Res. 8 (4), 1172–1181.

CHAPTER

TARGETED DRUG DELIVERY VIA CHITOSAN-COATED MAGNETIC NANOPARTICLES

26

Gozde Unsoy, Ufuk Gunduz

Middle East Technical University, Ankara, Turkey

CHAPTER OUTLINE

The devastating effects of chemotherapeutic agents have been observed on healthy cells as well as tumor cells because they are nonspecifically distributed all over the body. This treatment results in hazardous side effects and excessive toxicity. Targeted drug delivery has emerged to overcome the lack of specificity of conventional chemotherapy. Nanoparticles used for drug targeting are promising to circumvent these challenges, by enabling the localization of high drug amounts at the site of disease. When drugs are conjugated on nanoparticles, it is possible to increase their half-life while decreasing the side effects and improving therapeutic efficacy. Several therapeutic nanoparticles that selectively bind and target cancer cells have been approved for clinical use (Davis et al., 2010; Hrkach et al., 2012; Tabernero et al., 2013). Surface modifications of nanoparticles with organic polymers enable the stabilization of nanoparticles, reduce agglomeration, provide functional groups for further alterations, furnish internal cavities for loading of therapeutics, and prevent immediate uptake of drug-loaded nanoparticles by the reticuloendothelial system (RES). Polymer-coated magnetic nanoparticles (MNPs), characterized by high surface to volume ratios, are excellent scaffolds for loading targeting moieties, permeation enhancers, imaging tags, and drugs, simultaneously providing diagnostic and therapeutic capabilities. Multifunctionality of nanoparticle constructs can facilitate combination therapies. Nanoparticles can be routed to the tumor tissue via passive or active targeting. Nanotechnology has the potential to change cancer diagnosis and therapy fundamentally, which may bring new insights to the chemotherapy.

1 NANOTECHNOLOGY-BASED TARGETED DRUG-DELIVERY SYSTEMS

In recent years, innovations in nanotechnology have been translated to medical applications, facilitating diagnosis and targeted drug delivery. Nanomedicine is a multidisciplinary field that gathers knowledge and innovation from several disciplines, ranging from materials science and physics to molecular biology and clinical practice. It allows researchers from these various communities to contribute their diverse knowledge, experience, and creativity to a common goal of designing the best targeted drug-delivery or diagnostic systems.

Nanostructures, used in targeted drug delivery, have many advantages compared with free drugs. Encapsulation of the drug in nanoparticles provides protection against degradation in the liver and prolongs systemic circulation of the drug owing to the small size of nanoparticles (10–100 nm) (Gupta and Gupta, 2005). Thus, fewer fluctuations in plasma drug concentrations have been achieved with an enhancement in drug efficacy. Moreover, through the controlled delivery and release of drugs in the targeted tissues of body, a significant decrease has been achieved in undesirable adverse effects. Nanoparticles can reach remote areas of the body because their size is sufficiently small to pass through narrow capillaries, penetrate tissues, and cross-cell membranes. They can deliver drugs efficiently (hydrophilic or hydrophobic) to specific tissues by passive or active targeting and improve drug concentrations at targeted sites. Nanoparticles are internalized by cells preferentially depending on their surface properties, charge, and relative hydrophobicity. Internalization of nanoparticles by cells has been reported as being 15–250 times higher than that of microparticles (Jain et al., 2008).

Nanostructures possess very high surface-to-volume ratios. Nanomaterials are degraded by cells easily, which helps to sustain the release of drugs. Drug release is increased according to the dissolution rate of nanoparticles and the medium.

2 TARGETING STRATEGIES IN DRUG DELIVERY

Drugs, especially chemotherapeutics, are often highly toxic to healthy cells, so targeting is very important for minimizing adverse effects. Efficient targeting and delivery of drugs significantly reduces toxicity, while decreasing the needed dose of the drug. Utilization of nanoparticles as delivery vehicles ensures that their cargoes perform their effects only inside the targeted cells. Nanoparticles, compatible with biological membranes, can be loaded with high amounts of drugs on their shells. For instance, the drug load can be directed toward the site of the magnetic field by active magnetic targeting. Organ or tissue targeting and accumulation of drugs can be achieved by either passive or active targeting (magnetic field or antibody mediated) (Jaracz et al., 2005; Torchilin, 2004), while intracellular delivery can be mediated by certain ligands (Gabizon et al., 2004; Widera et al., 2003) or by cell-penetrating peptides (Gupta and Gupta, 2005; Lochmann et al., 2004). Once the particles have entered the cell, they can be stimulated by a variety of signals (pH, temperature, light, ultrasound, radiation, thermal conditions, magnetic field, and specific molecules) to release their drug loads (Lochmann et al., 2004).

2.1 PASSIVE TARGETING

Passive targeting occurs via extravasation of drug-loaded nanoparticles, which are able to penetrate tissues due to their small size, owing to the hyperpermeable and leaky nature of tumor microvasculature. Furthermore, poor lymphatic drainage of this microenvironment results in the slow clearance and higher accumulation of nanoparticles at the tumor site as compared to normal tissues. This is called the enhanced permeation and retention (EPR) effect, which means that certain macromolecules accumulate preferentially to tumors. The EPR effect was first reported more than 30 years ago (Matsumura and Maeda, 1986). Angiogenesis, triggered by the insufficiency of oxygen and food supply, promotes the rapid development of new, irregular blood vessels that present a discontinuous epithelium, or lack of normal vascular structures (Jain, 1998; Jain and Stylianopoulos, 2010). The sizes of resulting fenestrations in the capillaries can reach from 200 to 2000 nm, depending on the tumor type, tumor microenvironment, and the site of localization (Hobbs et al., 1998). The lymphatic function is minimal in tumors (Padera et al., 2004); because of that, nanoparticles cannot rely on returning into circulation. By the way, residual lymphatic activity and de novo lymphangiogenesis are believed to be one of the reasons for dissemination of metastases (Stacker et al., 2002). Therefore, nanoparticles that have reached the perivascular space are not cleared efficiently and accumulate in the tumor interstitium. However, passive targeting has some restrictions. Short residence time of nanoparticles in the bloodstream is the main limitation of passive targeting. They are removed from the circulation in a size-dependent manner in healthy capillaries. So, the passive targeting of conventional nanoparticles is restricted to mononuclear phagocyte system organs, including the spleen, liver, and bone marrow. Due to the short circulation time in the bloodstream and insufficient accumulation, active targeting must be applied to increase the drug concentrations at the tumor site.

2.2 ACTIVE TARGETING

Active targeting, depending on the physical characteristics of carriers, is more effective than passive targeting. Nanoparticles sensitive to physical stimuli, including magnetism and temperature, have been developed for active targeting (Moghimi et al., 2005). MNPs are synthesized and incorporated with certain ligands, binding to the cell surface receptors for dual active targeting. These ligands (antibodies, folic acid, ferritin, vitamin B_{12}) conjugated with the nanoparticles are highly expressed in target cells. Active targeting, achieved either by the physical characteristics of nanoparticles or conjugation of ligands, increases the therapeutic efficacy of drugs through the enhanced accumulation and high cellular internalization by receptor-mediated endocytosis.

3 MAGNETIC NANOPARTICLES

In the last decade, synthesis and surface engineering of MNPs have made substantial progress. MNPs offer attractive and versatile applications in the field of biotechnology, such as RNA/DNA and cell separation, targeted drug delivery, magnetic resonance imaging (MRI), and hyperthermia (the selective heating of tumors) (Cheng et al., 2005; Khor and Lim, 2003; Lee et al., 2005; Pankhurst et al., 2003;

Woo et al., 2004). MNPs are being used in an increasing number of biomedical applications and demonstrated great promise for diagnostic and therapeutic purposes. The important properties of mostly preferred iron oxide MNPs are nontoxicity, biocompatibility, injectability, and high levels of accumulation in the targeted tissue or organ. MNPs can be targeted to a specific tissue under a magnetic field, visualized (used as MRI agents), and triggered for drug release or used in hyperthermia (Goya et al., 2008). Moreover, there is an opportunity to carry out both MRI and drug delivery, generating both diagnostics and therapy within a single system called theranostics (Yang et al., 2011). Due to the high accumulation rates of drug-loaded MNPs inside the tumor tissues, MNPs have a great potential to be used as cancer therapeutics. MNPs used in this approach are core-shell structures made of a magnetic core (iron oxide, nickel oxide, or cobalt oxide) coated with either various kinds of polymers (Unsoy et al., 2012a; Yalcin et al., 2014), silica (Zhang et al., 2012), liposomes (Plassat et al., 2011), or dendrimers (Khodadust et al., 2013). These magnetic core-shell nanoparticles are potential candidates for drug targeting systems in cancer therapy because they can deliver appropriate amounts of drugs specifically to the tumor site and allow a selective temperature increase in the cancer mass by an alternating magnetic gradient.

MNPs are used extensively in hyperthermia by generating heat under an alternating magnetic field. Hyperthermia can deliver the highest amounts of thermal (heat) energy as compared to other heating resources (microwaves, radio frequency, ultrasound, and lasers) into the deeply situated tumors without any heat loss to healthy cells (Alvarez-Lorenzo and Concheiro, 2014).

Superparamagnetism (negligible remanence and coercivity) is a desired property for MNPs in targeted drug delivery and MRI because once the external magnetic field is removed, magnetization of nanoparticles disappears (Fig. 26.1). Thus, the possible embolization of capillary vessels due to the agglomeration is avoided. Small MNPs are often superparamagnetic at room temperature, and are very useful for biomedical applications, due to their magnetic properties and comparable size to biological structures (Mürbe et al., 2008). Studies evaluating the adverse effects of static magnetic field exposure up to 8 T revealed no significant or relevant toxicity in MRI applications (Chakeres and de Vocht, 2005; Schenck, 2005).

MNPs including metallic, bimetallic, and iron oxide nanoparticles have reactive surfaces that can be coated with biocompatible polymers to load therapeutic agents (Sun et al., 2008a,b). Metal oxides

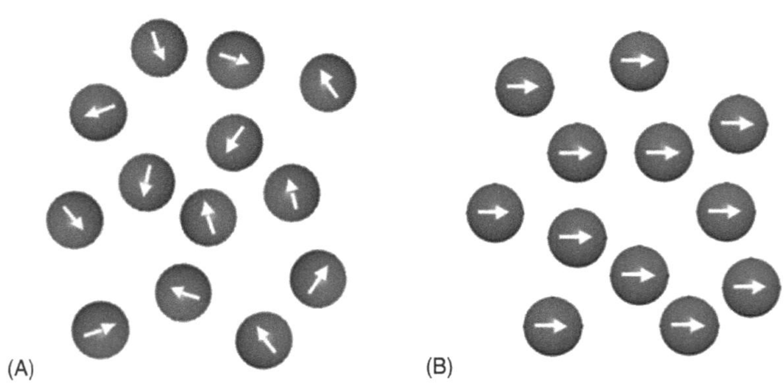

FIGURE 26.1

(A) Schema of the superparamagnetic nanoparticles. (B) There will be a net statistical alignment of magnetic moments in the presence of magnetic field.

of nickel (Ni), cobalt (Co), and iron (Fe) are also used to synthesize superparamagnetic nanoparticles in the size range of 1–100 nm. Composite cobalt- and nickel-iron oxide nanoparticles have also generated great interest, with their superparamagnetic properties and diverse application potentials in many fields (Rutirawut et al., 2015).

Coatings facilitate the stabilization of MNPs in an environment with different pHs or significant salt concentrations. The nature of the coating material is important to increase the stability and reduce the agglomeration of nanoparticles in the body. However, the fate of nanoparticles in the body strongly depends on the application dose and the administration route (oral, intravenous, pulmonary, transdermal, or ocular).

After systemic administration, rapid uptake of MNPs by the mononuclear phagocyte system is likely, depending on the size, surface functionalization, and hydrophilicity of the nanoparticles. The blood serum proteins bind to the surface of foreign bodies to accelerate the phagocytosis of nanoparticles, followed by their clearance in the spleen, liver, and bone marrow. The size (size distribution, diameter of magnetic core, and hydrodynamic diameter of the nanoparticle), magnetic properties (magnetic saturation, coercivity, and remanence), and surface chemistry of nanoparticles are very important in increasing the half-life of nanoparticles in the bloodstream and their chances of escape from the immune system, particularly the RES. Coating of MNPs with hydrophilic compounds increases the blood circulation time from minutes to days. Reducing the particle size is an alternative way to increase the time; however, the complete avoidance of the RES does not seem possible (Gaur et al., 2000). The coating of nanoparticles with biocompatible materials retards their clearance in blood stream by the RES. Various coating materials (organic, inorganic, biodegradable, and nonbiodegradable) can be used to retard the disclosure and capture of nanoparticles by the macrophages. Chitosan is one of the most widely used coating materials, whose attachment to nanoparticle surfaces provides a shielding effect, delaying the action of the RES (Macaroff et al., 2006; Neuberger et al., 2005). Biodistribution of drug-loaded nanoparticles is related to the physicochemical properties of the coating material; properties such as molecular weight, size, surface charge, surface hydrophobicity, stability, hydration behavior, electrophoretic mobility, porosity, and crystallinity of nanoparticles.

Functionalization of nanoparticles with coating materials can also improve their performance via active functional groups on their surfaces. Appropriate functionalization is very important because it can change their size, self-organization, and aggregation tendency. For instance, amine functionalization of chitosan results in enhanced solubility, strength, porosity, absorption efficacy, chemical tolerance, nonimmunogenic, nonantigenic, and biocompatible properties (Bennet and Kim, 2014).

4 CHITOSAN AS A SURFACE COATING MATERIAL FOR MAGNETIC NANOPARTICLES

Different types of molecules are used for covering the surface of MNPs to stabilize and functionalize them for further biomedical applications. The polymeric shell renders MNPs biocompatible, prevents their agglomeration, and functions as drug reservoir (McBain et al., 2008). Superparamagnetic iron oxide nanoparticles (SPIONs) can be functionalized or modified with coatings of hydrophilic, biocompatible, and biodegradable polymers, such as chitosan. Chitosan is preferable as a coating material for surface modification of core structures, to improve either their interaction patterns with adjacent molecules or the biodegradation profile (Andrade et al., 2011; Fonte et al., 2012; Grenha, 2012).

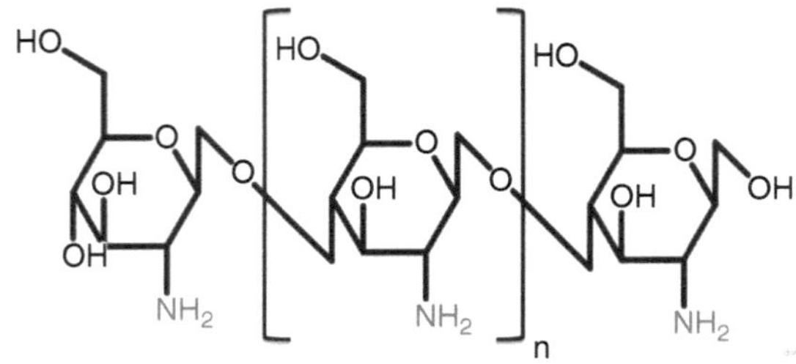

FIGURE 26.2 Molecular Formula of Chitosan

Chitosan is a natural, linear, alkaline, hydrophilic, biodegradable, biocompatible polymer with no toxicity (Andrade et al., 2011; Kumar, 2000) (Fig. 26.2). Chitosan is one of the most abundant polymers in nature. It is obtained by the *N*-deacetylation of chitin, which is the structural element in the shells of crustaceans (crabs, shrimp, lobsters, etc.) and in the cell walls of fungi.

Chitosan has gained attention in pharmaceutical applications due to its nontoxicity, biocompatibility, and biodegradability. It has been used in various applications, including food packaging, antioxidants, hydrogels, membranes, water treatment, adhesives, biosensors, tissue engineering, and drug delivery (Elsabee et al., 2009; Honarkar and Barikani, 2009).

The abundance of reactive hydroxyl and amine functional groups in the unique molecular structure of chitosan is advantageous to provide anchor sites for further conjugation of functional molecules. The protonated amine groups of chitosan also ensure a net cationic character, facilitating the electrostatic complexation with anionic ligands and drugs. All these features of chitosan make it available for further modifications and desirable for a wide range of biomedical applications.

Chitosan has also been utilized to improve nanoparticles for drug and gene delivery (Park et al., 2010). The ease of chitosan conjugation with anionic therapeutics and ligand molecules has enabled the development of many promising carrier formulations, targeting cells and improving the cell permeation of therapeutics (De la Fuente et al., 2008; Park et al., 2010). One of the most relevant properties of chitosan in drug delivery relies on its ability to transiently open epithelial tight junctions, resulting in the permeation enhancement of macromolecules through well-organized epithelia such as nasal (Fernandez-Urrusuno et al., 1999; Vila et al., 2004), intestinal (Artursson et al., 1994; Borchard et al., 1996; Prego et al., 2005), ocular (De Campos et al., 2001), buccal (Portero et al., 2007), and pulmonary (Al-Qadi et al., 2012). Chitosan also has a high retention rate in blood circulation, suggesting that it can be an effective carrier for drugs that are excreted rapidly (Dodane and Vilivalam, 1998).

The anionic character of chitosan facilitates loading of nucleic acids, providing protection against nuclease degradation and easy permeation through cell membranes in gene delivery and therapy (Liu and De Yao, 2002; van der Lubben et al., 2001). These nanocarriers may circumvent all the limitations of conventional viral gene delivery owing to their biocompatibility and nonimmunogenicity (Mao et al., 2010). Chitosan can also be used in delivering genetic vaccines. In tissue engineering, chitosan has been used as scaffolding material for tissue regeneration by promoting cell adhesion and growth due to its bioadhesive properties (Custódio et al., 2010; Isikli et al., 2012; Zan et al., 2008).

Chitosan includes a large number of polymers, which differ in their degree of *N*-deacetylation (40%–98%) and in their molecular weight (50,000–2,000,0000 Daltons) (Hejazi and Amiji, 2003). These two characteristics are very important to chitosan's physicochemical properties and may have a major effect on its biological properties (Sannan et al., 1976). Chitosan salts are soluble in water;

the solubility depends on the degree of deacetylation and the pH of the solution. With a p*Ka* of approximately 6.5, chitosan is soluble in acidic solutions owing to the protonation of amine groups in the polymeric chain (Rinaudo, 2008; Singh and Ray, 2000). In this regard, highly deacetylated chitosan (85%) is readily soluble in solutions with a pH up to 6.5, but as the deacetylation degree decreases, the solubilization becomes more difficult (Cho et al., 2000).

The amine groups are protonated at acidic pH, making chitosan a water-soluble cationic polyelectrolyte. Complete protonation occurs at pH values below 3.4 (Kumar et al., 2004; Tsai et al., 2010). In contrast, at high pH values above 6.5, chitosan's amines become deprotonated and the polymer loses its charge and becomes insoluble. Chitosan can undergo interpolymer associations that can lead to fiber and network (i.e., film and gel) formation at high pH values (Kumar et al., 2004; Yi et al., 2005).

Chitosan nanoparticles have been demonstrated to be pH-responsive systems with a reversible process of swelling and shrinking of particles. In tumor cells pH values are low (pH 5.5) due to high anaerobic glucose metabolism. Moreover, pH values of 3.0–5.5 are seen in acidic intracellular organelles, such as endosomes and lysosomes, within the cancer cells (Lim et al., 2011). After endocytosis, the drug-release rate can be rapidly accelerated with the low pH of the tumor cells, owing to the pH-responsive chitosan nanoparticles. It is expected that they will show an effective drug-release pattern in the tumor microenvironment (Aydin and Pulat, 2012). Additionally, chitosan has an intrinsic antitumor activity. It enhances the production of tumor necrosis factor-α (TNF-α) by monocytes, thus promoting nitric oxide synthesis by macrophages and stimulating neutrophils (Dodane and Vilivalam, 1998).

5 SYNTHESIS METHODS FOR CHITOSAN-COATED MAGNETIC NANOPARTICLES

Chitosan-coated MNPs have gained a prominent role as a drug-delivery nanocarrier due to their various benefits, including high stability, low toxicity, and simple and mild synthesis methods. Chitosan-coated MNPs offer versatile routes of administration and have the ability to control the delivery and release of therapeutics, including anticancer agents. As chitosan polymers are soluble in aqueous acidic solution, the use of hazardous organic solvents can been avoided in the production of chitosan-coated MNPs. Additionally, chitosan is a linear polymer with a number of free amine groups that are readily available for cross-linking, whereas its cationic nature allows for ionic cross-linking with multivalent anions.

In brief, production of chitosan-coated MNPs can be categorized in ex situ or in situ synthesis methods. The in situ synthesis method provides production of chitosan-coated MNPs in one step in the presence of the chitosan polymers, and thus the MNPs are simultaneously produced and coated with the polymer. In ex situ synthesis methods, MNPs are first synthesized then coated and stabilized with the derivatives of chitosan polymer.

5.1 EX SITU SYNTHESIS METHODS FOR CHITOSAN-COATED MAGNETIC NANOPARTICLES

In ex situ synthesis, chitosan is coated onto MNP cores, which are then synthesized either by thermal decomposition or by coprecipitation of iron salts. In the thermal decomposition method, synthesized MNPs are hydrophobic and can be dispersed only in organic solvents, which is not preferred for biological applications (Lee et al., 2010a,b). In contrast, MNPs synthesized by coprecipitation are hydrophilic and can be dispersed in aqueous solvents (Arami et al., 2011; Shi et al., 2009).

Chitosan-coated polyelectrolyte complex nanoparticles, prepared by electrostatic interactions between oppositely charged polyions, have received considerable attention as carrier systems for gene and drug delivery (Il'ina and Varlamov, 2005; Liu et al., 2005; Sun et al., 2008a,b). The complex formation is influenced by many factors, such as degree of ionization of the chitosan and anionic counterparts, chain flexibility, charge distribution over the polymer chain, pH, temperature, time of interaction, ionic strength, and concentration of the polymeric solutions (Drogoz et al., 2007; Park et al., 2010).

The most commonly used method for the stabilization of MNPs is taking advantage of the physical interaction between the MNPs and chitosan or its derivatives. For instance, Zhu et al. (2009) have used various forms of chitosan, O-carboxymethyl chitosan, and *N*-succinyl-O-carboxymethyl chitosan to coat hydrophilic MNPs. The polymer-to-MNP ratio plays an important role in determining the size distribution and uniformity of the coat (Zhu et al., 2008). Lee et al. (2009) mixed an amphiphatic chitosan–linoleic acid copolymer with hydrophobic nanoparticles. While the hydrophobic end of the polymer was self-assembled on the surface of the nanoparticles, the hydrophilic end dispersed the nanoparticles in the aqueous media. In their method, after coating with the copolymer the dispersion of hydrophobic nanoparticles in water was improved.

Cross-linking of the chitosan coat can be used to increase the coating efficiency and the stability of nanoparticles. Yuan et al. (2008) encapsulated iron oxide nanoparticles with chitosan-*g*-poly(*N*-isopropylacrylamide-*co*-*N*,*N*-dimethylacrylamide), a thermosensitive smart copolymer, and stabilized the coating by using a cross-linker agent. On the other hand, when these cross-linkers are used, additional investigations are required to determine their toxicity and tolerable dose (Hou et al., 2008).

To improve the uniformity of chitosan coatings and produce biodegradable chitosan-coated iron oxide/poly(D,L-lactic-*co*-glycolic acid) nanoparticles, the emulsion diffusion evaporation method was used by Lee et al. (2010a,b). In this method, hydrophilic iron oxide nanoparticles and a hydrophobic poly(D,L-lactic-*co*-glycolic acid) polymer were mixed in dichloromethane and the organic solvent was evaporated after being emulsified in a solution mixture of polyvinyl alcohol and positively charged glycol chitosan. The size of synthesized nanoparticles was determined to be about 275 nm, with the formation of a uniform coating around the nanoparticles.

5.2 IN SITU SYNTHESIS METHODS FOR CHITOSAN-COATED MAGNETIC NANOPARTICLES

The in situ synthesis of chitosan-coated MNPs occurs via the coprecipitation of iron salts in the presence of chitosan polymers. Hydrophilic MNPs are precipitated from the solution, forming nuclei of core-shell nanoparticles with high surface energies, which rapidly adsorb the protonated chitosan polymers through physical binding. Under in situ synthesis conditions, uniform layers of chitosan polymers are assembled on MNP cores. These chitosan polymers serve to limit iron oxide core growth during the synthesis, stabilize each discrete nanoparticle via steric repulsions when the nanoparticles disperse in water (and later in the physiological medium), and reduce the opsonization process in vivo (Mornet et al., 2004). The chitosan-coated MNPs, synthesized through the in situ method, are modified with a uniform coat of chitosan and have a monodispersed size distribution (Mizukoshi et al., 2009).

Chitosan is particularly interesting in metal nanoparticle synthesis due to its interaction with metal ions and metal oxide nanoparticles. In this regard, it is worth noting that chelation evenly disperses metal oxides throughout the chitosan polymer. The primary amines of chitosan are involved in metal ion chelation and nanoparticle immobilization (Klepka et al., 2008). Unsoy (2013), Unsoy et al. (2012a), and Kavaz et al. (2010) have demonstrated that the chitosan coat around iron oxide core can be cross-linked with TPP ions because protonated chitosan ionically interacts with the TPP molecules and forms a chitosan–TPP network around the iron oxide core. This mainly occurs due to the simultaneous cross-linking of chitosan with TPP ions and neutralization through deprotonation. The ionic interaction between the chitosan molecules and TPP ions is depicted in Fig. 26.3.

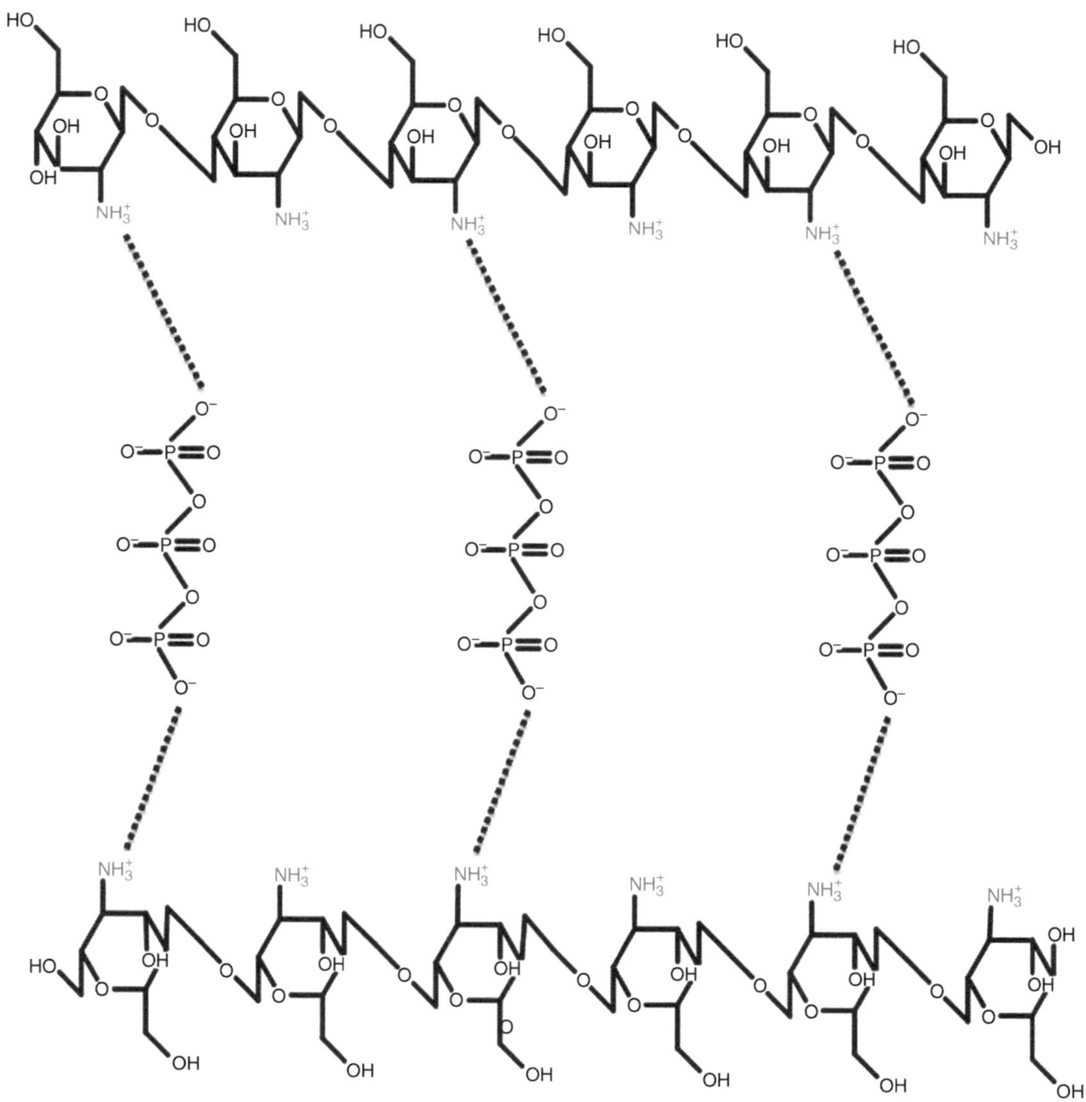

FIGURE 26.3 Schema Showing the Interaction Between the Chitosan and the Cross-Linker, TPP Molecules

TPP and its derivatives can also be used as small molecule–targeting ligands for mitochondria (Marrache and Dhar, 2012; Smith et al., 2003). TPP is a relatively large, cationic, and hydrophobic molecule that can penetrate easily through the cell membrane. An investigation indicated that the positively charged TPP could accumulate several hundred folds within mitochondria (Smith et al., 2003). An in vitro study recently established the mitochondria targeting potential of TPP-conjugated nanoparticles (Marrache and Dhar, 2012), reporting that they could be efficiently internalized and escape the endosomal compartments when their size was around 100 nm and zeta-potential was higher than +22 mV. Although this positive charge, conferred by TPP, seems important for in vitro performance, it might possibly undermine its blood exposure in vivo. This factor will have to be considered in the future developments of this targeting strategy.

The limitation of the in situ synthesis method is that the molecular weight of chitosan is not sufficiently low to solubilize under the basic conditions needed for coprecipitation of MNPs (Tsai et al., 2010). A promising solution for this is the development of chitosan copolymers soluble at high pH values. For instance, to improve the solubility of chitosan under high pH conditions, polyethylene glycol (PEG) can be grafted to chitosan polymers (Fernandez-Megia et al., 2007). In the in situ synthesis method, such copolymers can be used to coat MNPs, as demonstrated by Veiseh et al. (2009, 2010). These MNPs had uniform and small hydrodynamic size (32 nm). Kievit et al. (2009, 2010) and Gunn et al. (2011) have successfully used this approach (Arami et al., 2011). However, the modified versions of chitosan may not be properly assembled on the surface of MNPs, resulting in incomplete coating.

6 MAGNETICALLY TARGETED DRUG DELIVERY VIA CHITOSAN-COATED MAGNETIC NANOPARTICLES

Targeting therapeutic agents to the tumor site by chitosan-coated MNPs can restrict the adverse effects of chemotherapeutics on healthy cells. Firstly, MNPs accumulate in the tumor tissue by passive targeting (Byrne et al., 2008). Passive targeting not only facilitates the accumulation of nanoparticles through the EPR effect in solid tumors but also enables the removal of nanoparticles by Kupffer cells in the liver. The EPR effect, which is the primary route of passive targeting, helps the nanoparticles to extravasate through the leaky vasculature of tumor tissue. Nanoparticles less than 150 nm in size can pass through the vasculature. Poor lymphatic drainage also supports the accumulation of the nanoparticles in the tumor site.

To reach the targeted tumors, the majority of systemically administered drug-loaded nanoparticles depend on the leakiness of the vasculature around the tumor area (Bartlett et al., 2007; Kirpotin et al., 2006; Mamot et al., 2005). Therefore, prolonged circulation and reduced immune clearance are the primary design criteria for tumor-targeted nanoparticles. As the hydrophobic surfaces cause nonspecific interactions with healthy tissues (Moreno et al., 2010) and premature clearance by the RES (Carrstensen et al., 1992), nanoparticles are coated with polymers, forming a hydrophilic shell.

SPIONs can be guided with a magnetic field generated by a permanent Nd-Fe-B magnet up to 10–15 cm depths in the body (Neuberger et al., 2005). The design of the magnetic targeting system must be planned carefully because the form of the magnetic field, applied externally for targeting, is critically important. The influence of the magnetic field decreases as the distance to the target increases.

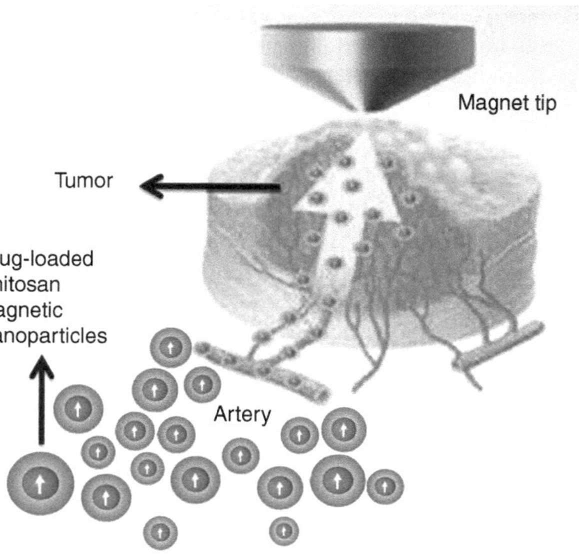

FIGURE 26.4 Illustration of Magnetic Drug Targeting

After the intravenous administration, drug-loaded nanoparticles are magnetically guided to the tumor tissue by external magnetic field. Chitosan-coated MNPs disintegrate and release the drug at the targeted tumor site.

Modified from Barakat, N.S., 2009. Magnetically modulated nanosystems: a unique drug-delivery platform. Nanomedicine 4 (7), 799–812.

Therefore, one of the main challenges in magnetic drug delivery is the strength and duration optimization of externally applied magnetic field to obtain sufficient magnetic gradient for the accumulation of nanoparticles in the tumor tissue/desired area (Fig. 26.4).

Active targeting strategies utilizing the molecular markers of cancer cells, such as overexpressed nutrient receptors and growth factors, provide specific accumulation of drug carrier nanoparticles or imaging agents within the tumors (Gunn et al., 2011). Many ligands, such as antibodies, peptides, and aptamers, are being studied for use in the targeting of molecular markers to improve the diagnosis of cancer and the targeted delivery of therapeutics (Alexis et al., 2008).

Gunn et al. (2011) have developed biotinylated chitosan/PEG-coated MNPs for active targeting of tumor tissue, requiring the pretreatment of a fusion protein composed of streptavidin and an antibody to label cancer cells. Biotinylated MNPs specifically bind to the streptavidin part of the fusion protein on the antibody-labeled cells. Therefore, biotinylated chitosan/PEG-coated MNPs can selectively target desired the tumor tissues.

De Campos et al. (2001) have studied the potential of chitosan nanoparticles for the delivery of immunosuppressive drugs, such as cyclosporin A, to the ocular mucosa. They revealed that drug-loaded chitosan nanoparticles can interact with the corneal and conjunctival surfaces, which increases the external delivery to the eye without requiring drug exposure to the systemic circulation and inner structures of eye. Chitosan-based nanoparticulate systems have high ocular retention and can carry various types of therapeutics, such as cyclosporin A (medium size peptides), indomethacin (lipophilic low molecular weight compounds), and gene molecules (nucleotide sequences). Accordingly, they show great promise for the external treatment of inflammatory or autoimmune ocular diseases, including dry-eye disease and kerato conjunctivitis sicca, because the chitosan nanoparticles increase the penetration of drugs with a good biocompatibility and tolerance.

Park et al. (2010) have embedded MNPs into chitosan polymers to produce magnetic chitosan particles, which are expected to reveal relatively low cytotoxicity owing to the complete coating of MNPs with a biocompatible polymer chitosan. Modifications of MNPs with surface coating materials are required to enable the effective loading and controlled release of therapeutics. Therapeutic agents are loaded to the MNPs via electrostatic interactions, chemical linkage, or entrapped in a biodegradable shell. Functional groups of coating materials such as amine, carboxylic acid, aldehyde, and thiol groups are utilized for the covalent bonding of therapeutic agents with MNPs (McBain et al., 2008). Positively charged amine groups can form amide bonds with negatively charged carboxylic acids. Amine groups can also bind with aldehydes, forming imide bonds, while disulfide bridges can be formed between thiol groups, as in the cysteine residues found on proteins (Dave and Gao, 2009; McBain et al., 2008).

Unsoy et al. (2012b) have synthesized chitosan-coated MNPs, which have a high surface area-to-volume ratio, a magnetic core, and many active amine groups on their shells, to support the efficient loading of much more therapeutics and their targeting to tumor tissue. Chitosan-coated MNPs have been used as nanocarrier systems of various chemotherapeutic agents in vitro (Unsoy, 2013; Unsoy et al., 2012b). Unsoy et al. (2014b), demonstrating the successful loading of the chemotherapeutic drug bortezomib (Velcade) in conjugation with a flourophore (FITC) on chitosan-coated MNPs, which were synthesized using the in situ coprecipitation method (Unsoy, 2013; Unsoy et al., 2012a) (Fig. 26.4). Drug-release ratios were evaluated by UV–vis spectrophotometry. Cellular internalization of bortezomib-loaded chitosan MNPs were visualized by fluorescence microscopy (Unsoy et al., 2014b). According to the results of this study, only 51, 28, and 15% of the bortezomib had been released at pHs of 4.2, 6.0, and 7.4, respectively, during the first 15 h, reflecting the pH-controlled drug-release characteristic of the synthesized chitosan-coated MNPs. Apoptotic gene expression and immunocytochemistry analysis revealed that the efficacy of bortezomib was increased when loaded on chitosan-coated MNPs (Unsoy, 2013; Unsoy et al., 2014b).

In another study by Unsoy et al. (2014a), the use of various sizes of chitosan-coated MNPs for targeted delivery of a chemotherapeutic drug was analyzed using the effect of doxorubicin on breast cancer cells in vitro (Fig. 26.5). The highest loading efficiency (97%–98%) was obtained with chitosan-coated MNPs-S_1 (103 nm). This may be due to the higher chitosan content (obtained as 23% by TGA analysis) of chitosan-coated MNPs-S_1. The FTIR results supported that the doxorubicin loading on chitosan-coated MNPs was successfully achieved. The chitosan-coated MNPs released most of the doxorubicin load at pH 4.2, while the nanoparticles were highly stable at pH 7.4. Chitosan has primary amine groups with p*K*a values of 6.3. At a pH below p*K*a, most of the amine groups are protonated, making chitosan a water-soluble cationic polyelectrolyte. At a pH above the p*K*a, the amine groups of chitosan are deprotonated, and this polymer becomes insoluble (Zemskova et al., 2005). Thus, chitosan-coated MNPs were stable at pH 7.4 and the drug was not released at a neutral pH, which is the pH of blood. The drug-release profiles showed a pH-dependent, slow-release pattern. It is expected that drug is released at targeted cancer cells because the pH of tumor tissue and endosomes are acidic (Unsoy, 2013; Unsoy et al., 2012b, 2014a).

Arias et al. (2012) prepared magnetic core/shell nanoparticles consisting of a magnetite (Fe_3O_4) core and a chitosan shell and used them for the magnetically targeted delivery of the anticancer drug

(A)

(B)

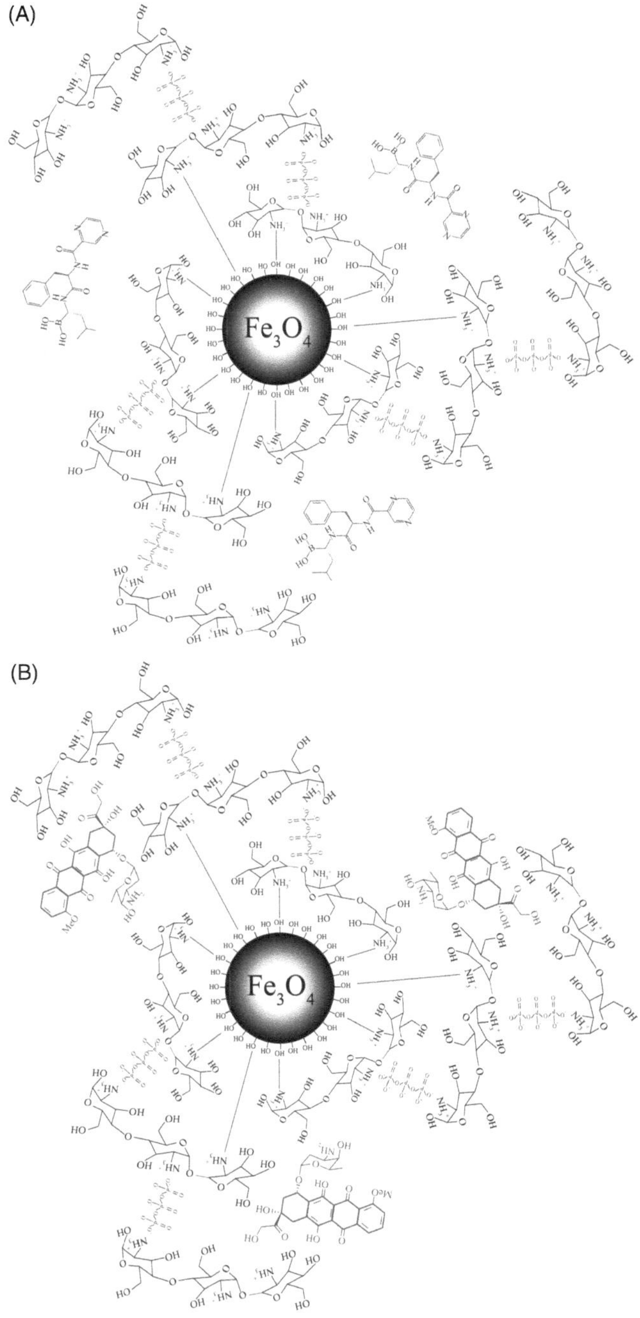

FIGURE 26.5 Schematic Representation of Doxorubicin and Bortezomib-Loaded Chitosan-Coated MNPs Synthesized for Targeted Drug Delivery

From Unsoy, G., 2013. The Synthesis and Characterization of Doxorubicin and Bortezomib Loaded Magnetic Nanoparticles for Targeting Tumor Cells. Doctoral Dissertation, Middle East Technical University, PhD Thesis, 2013. The Graduate School of Natural and Applied Sciences of Middle East Technical University; Unsoy, G., Khodadust, R., Yalcin, S., Mutlu, P., Gunduz, U., 2014a. Synthesis of doxorubicin loaded magnetic chitosan nanoparticles for pH responsive targeted drug delivery. Eur. J. Pharm. Sci. 62, 243–250; Unsoy, G., Yalcin, S., Khodadust, R., Mutlu, P., Onguru, O., Gunduz, U., 2014b. Chitosan magnetic nanoparticles for pH responsive bortezomib release in cancer therapy. Biomed. Pharmacother. 68 (5), 641–648.

gemcitabine for intravenous administration. Gemcitabine was loaded by encapsulation and entrapment procedure into the chitosan shell instead of through surface adsorption with high drug-loading capacity. The release of gemcitabine from the nanoparticles was slow and pH dependent. The pH-dependent release of the drug occurs as a result of the polymer dissolution in acidic environments, such as the tumor microenvironment and the endosomes of cells, which is very advantageous for controlled drug release from chitosan-coated MNPs (Arias et al., 2012)

The loading of another chemotherapeutic agent, camptothecin, was investigated in MNPs stabilized with various forms of chitosan, O-carboxymethyl chitosan, and *N*-succinyl-O-carboxymethyl chitosan by Zhu et al. (2009). MNPs coated with positively charged chitosan were not stable, after the absorbtion of drug, camptothecin, into the chitosan layer. However, MNPs coated with ampholytic O-carboxymethyl chitosan and negatively charged *N*-succinyl-O-carboxymethyl chitosan displayed high stability after drug loading. O-carboxymethyl chitosan-coated MNPs exhibited sustained release of camptothecin over 10 days, while the *N*-succinyl-O-carboxymethyl chitosan-coated MNPs showed a burst release. The drug release from polymer-coated MNPs can occur via simple diffusion or mechanisms requiring enzymatic activity, or proceed through changes in physiological conditions (ionic strength, pH) of solutions (Arruebo et al., 2007).

Another form of chitosan-coated iron oxide nanoparticles was developed by Yuan et al. (2008) for the delivery of doxorubicin. After the conjugation of MNPs with doxorubicin, they were encapsulated by the smart chitosan-*g*-poly(*N*-isopropylacrylamide-*co*-*N*,*N*-dimethylacrylamide)[chitosan-*g*-poly(NIPAAm-*co*-DMAAm)] polymer. The low critical solution temperature (LCST) value of this thermosensitive polymer is 38°C. The temperature responsive chitosan polymer collapses and then releases the entire entrapped doxorubicin payload at or above 38°C (LCST) in physiological pH. Below the LCST, the smart polymer-coated MNPs reveal poor and slow release of doxorubicin, until there is a significant decrease in pH. This form has been designed for the release of drug increases at the low pH of the tumor microenvironment owing to the acid labile bonds of doxorubicin with MNPs.

The amine groups of polycationic chitosan facilitate the conjugation of chitosan-coated MNPs with the negatively charged phosphate backbone of nucleic acids. Thus, chitosan-coated MNPs are capable of carrying therapeutic nucleic acids sequences for targeted gene delivery. Plasmid DNA, oligonucleotides, and small interfering RNA (siRNA) are the most commonly investigated nucleic acids in gene therapy (Lee, 2007; Mansouri et al., 2004; Tan et al., 2009). There is a capacity to treat cellular functions via gene therapy. Specifically, targeted delivery of DNA is employed to introduce required sequences of gene to the cells and treat the diseases with the expression of the introduced gene. The siRNA inhibits the translation of messenger RNA (mRNA) to protein by binding to and cleaving the complementary sequences of mRNA, which results in the silencing of the interested gene. However, conventional viral gene delivery systems have serious drawbacks in terms of safety (Mansouri et al., 2004; Tan et al., 2009); toxicity analysis of these polycationic polymers must be done.

One of the first reported cationic polymers used in the gene therapy is polyethylenimine (PEI). Besides binding with nucleic acids, the positive charge of PEI provides for the release of nucleic acids from lysozome (McBain et al., 2008). Polycationic polymers buffer the intralysozomal pH (pH 3.5–5), lead to an influx of protons, and subsequently cause the rupture of lysozome, delivering the DNA to the cytoplasm. Even though PEI has been a promising polymer for gene transfection, researchers are investigating other polycationic polymers, especially chitosan, in gene therapy applications due to the cytotoxicity of PEI (Park et al., 2006).

Kievit et al. (2009) have reported that MNPs coated with the polymers of chitosan, PEG, and PEI had significantly low cytotoxicity as compared to MNPs coated with only PEI. According to this study, chitosan/PEG/PEI-coated MNPs were not only less toxic to cells than PEI-coated MNPs but also had significantly higher transfection efficiency with high ratios of DNA to MNP.

In a study by Kievit et al. (2010), cellular internalization of nanoparticles was demonstrated by MR imaging in vivo. The results indicated that both targeted and nontargeted nanoparticles were equally internalized by tumor cells. However, the expression of the reporter gene reflecting the transfection efficiency was higher at the tumor sites of mice treated with the targeted nanoparticles.

Cancer cell–specific nanoparticles have been created for the efficient delivery of siRNA by Veiseh et al. (2010). MNPs were synthesized using a coating of PEG–chitosan polymers and grafted with PEI, as with the nanoparticles of Kievit et al. (2010). Then, chlorotoxin and green fluorescent protein (GFP) flourophore-labeled siRNA were loaded on nanoparticles. Chlorotoxin was used to target cancer cells because they bind preferentially to glioma cells compared with healthy cells. It was detected by flow cytometry that the cellular internalization of chlorotoxin targeted siRNA-loaded nanoparticles was higher at GFP positive glioma C6 cells in vitro. Moreover, expression of GFP was reduced in the cells treated with these nanoparticles (Kievit et al., 2010).

MNPs coated with hexanoyl chloride modified chitosan have been developed for the improvement of viral gene therapy by Bhattarai et al. (2008). Then, MNPs were targeted to K562 cells under the magnetic field in vitro. Another group of K562 cells were treated with adenovirus, including reporter gene LacZ. Results of the study revealed that the transduction efficiency of adenovirus was increased when loaded on nanoparticles. This improvement in transduction efficiency was achieved by MNPs coated with a modified chitosan polymer. The increased efficacy was attributed to the interactions of chitosan nanoparticles with the cell membrane or the destabilization of the endosomes and lysosomes. Chitosan-coated MNPs can enormously decrease the required viral doses, prevent their delivery to healthy tissues, and reduce the application frequency and incubation times.

7 CELLULAR INTERNALIZATION OF CHITOSAN-COATED MAGNETIC NANOPARTICLES

Nanoparticles can efficiently intrude into living cells by endocytosis, resulting in permanent cell damage (AshaRani et al., 2008; Lunov et al., 2010). Only specialized cells such as macrophages are capable of phagocytosis, a form of endocytosis in which the cell engulfs large particles. Almost all cells internalize nanoparticles by pinocytosis. Four different basic pinocytic mechanisms are known: macropinocytosis, clathrin-mediated endocytosis, caveolae-mediated endocytosis, and mechanisms independent of clathrin and caveolin (Conner and Schmid, 2003; Jiang et al., 2011). Physicochemical properties of nanoparticles including size (Chithrani et al., 2006; Rejman et al., 2004), shape (Chithrani et al., 2006), and surface charge (Arbab et al., 2003; Labhasetwar et al., 1998; Sun et al., 2005) have been identified as the main modulators of the cellular uptake efficiency (Treuel et al., 2013).

Nanoparticles may also enter cells by passive penetration of the plasma membrane (Wang et al., 2012). It has been shown that the adherence and penetration abilities of nanoparticles depend on physical properties, including size, surface composition, and surface charge (Leroueil et al., 2007; Roiter et al., 2008; Verma et al., 2008). Small and positively charged nanoparticles have been observed as passing through cell membranes, leading to membrane rupture and noticeable cytotoxic effects (Cho et al., 2009; Kostarelos et al., 2007; Leroueil et al., 2008; Yu et al., 2007). Even particles greater than 500 nm in diameter were reported to penetrate cell membranes by inducing strong local membrane deformations (Zhao et al., 2011). Membrane disruption can be reduced or even entirely avoided by a suitable design of the surface structure and charge density (Lin et al., 2010; Verma et al., 2008). Cell adhesion and potential cell uptake of chitosan-coated MNPs should be efficient due to their positive charge, which is attracted by negatively charged cell membranes.

Drugs, whose target molecules are within the cells, have to penetrate the cellular membrane and escape from the endosome before exhibiting their biological effects. *N*-acetyl histidine conjugated glycol chitosan nanoparticles that are pH responsive were developed for the efficient intra cytoplasmic delivery of paclitaxel (Park et al., 2006). Under slightly acidic conditions (similar to endosomes), the amine groups of chitosan get protonated. This may induce the influx of water and ions into endosomes when the nanoparticles are taken up by the cells, causing disruption of endosomal membranes (Fig. 26.6). As a consequence, the disassembled nanoparticles can release the encapsulated drug into the cytosol.

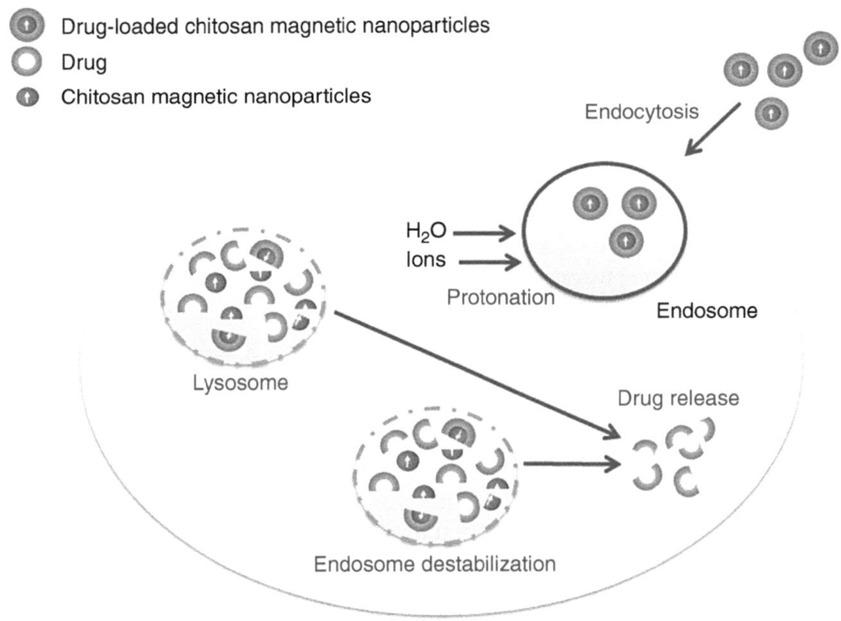

FIGURE 26.6 Schematic Representation of the Cellular Uptake and Drug Release From Citosan Nanoparticles

Internalization of nanoparticles is initiated by interactions between nanoparticles and cell membranes. Drug-loaded nanoparticles are routed to lysosomes, where a high level of lysosomal enzymes is present. Drugs sensitive to these enzymes are degraded and lose their activity. Under slightly acidic environments, the protonated groups cause the disruption of endosomal membranes and simultaneous delivery of drugs into the cytosol.

From Arias, J.L., Reddy, L.H., Couvreur, P., 2012. Fe_3O_4/chitosan nanocomposite for magnetic drug targeting to cancer. J. Mater. Chem. 22 (15), 7622–7632; Aydin, R., Pulat, M., 2012. 5-Fluorouracil encapsulated chitosan nanoparticles for pH-stimulated drug delivery: evaluation of controlled release kinetics. J. Nanomater. 42.

8 DRUG RELEASE FROM CHITOSAN-COATED MAGNETIC NANOPARTICLES

8.1 DRUG RELEASE FROM CHITOSAN-COATED MAGNETIC NANOPARTICLES VIA PH CHANGE

A pH difference occurs in tumor tissue due to the disordered angiogenesis around the rapidly growing tumor, causing lack of nutrients and oxygen, triggering tumor cells shift to the glycolytic metabolism, and resulting in the overproduction of lactic acid in the tumor microenvironment (Aydin and Pulat, 2012). Drug release that is triggered by the pH differences between the blood, healthy tissue (neutral), and the tumor tissue (acidic) is used for controlled drug release. Magnetic drug-delivery systems that are pH responsive can target the tumor site with the help of the magnetic field and release their payloads there owing to the acidic pH of tumor tissue. The pH-responsive polymers have the tendency to protonate or deprotonate at different pH values, which alters the attractive forces between the polymer chains and results in the swelling or deswelling of polymers. For instance, the swelling of chitosan (p*K*a of ~6.3) induced by the protonation of amine groups leads to the release of encapsulated TNF-α in the acidic tumor environment (Yang et al., 2011). Polyacrylate, polymethacrylate, and alginate polymers exhibit the maximum swelling in alkaline pH, while chitosan, polylysine, and polyvinylpyridine polymers display acidic pH-dependent swelling. The charged polymer chains repulse the neighboring chains. To stabilize this repulsion, counter ions in the medium enter the polymer matrix with the water molecules, which results in the swelling of the polymer (Arshady and Kono, 2006). This phenomenon should be utilized for controlled drug release. Cell internalization can also be promoted by means of pH-triggered surface-charge reversal from negative or neutral to positive (Du et al., 2011).

Biological systems have distinct pH values in different regions, tissues, and cellular compartments that can be used to design pH-responsive smart drug-delivery systems for pH-stimulated release in desired locations. The pH values of different tissues are: stomach (pH 1.0–3.0), small intestine (pH 4.8–8.2), and colon (pH 7.0–7.5); those of cellular compartments are: cytosol (pH 7.4), early endosome (pH 6.0–6.5), late endosome (pH 5.0–6.0), and lysosome (pH 4.0–5.0) (Arshady and Kono, 2006)

For oral drug delivery, pH-responsive systems have been designed to protect drugs from the harsh conditions in the stomach and to improve intestinal absorption, owing to the broad range of pH throughout the gastrointestinal tract (Wang et al., 2012). For instance, polymethacrylate has been used as a pH-responsive coating material on the surface of nanocarriers (Qu et al., 2012), as well as copolymer micelles capable of disassembling at the intestinal pH.

At the cellular level, acidification of endosomes (pH 5–6) and their fusion with lysosomes (pH 4–5) is another pH gradient that can be used for effective intracellular drug accumulation. Nanocarriers swelling at a mildly acidic pH that ensure the rapid release of a drug have been designed either by masking the hydroxyl groups with acid labile protecting groups (Griset et al., 2009) or by using protonable monomer units. This provides the tunable swelling of the polymer and controlled release within the endosomal pH range (You and Auguste, 2009). The chemotherapeutic drugs bortezomib and doxorubicin were released in a pH-controlled manner from chitosan MNPs when they were loaded via electrostatic interactions (Unsoy, 2013; Unsoy et al., 2014a,b). The presence of pH-sensitive bonds in the polymer backbone or the use of acid-degradable cross-linkers (Ahmed and Narain, 2012) can lead to the nanocarrier disassembly. The pH-sensitive bonds enable the controlled release of drugs covalently conjugated to polymer backbones (Griset et al., 2009; Zhu et al., 2012) and protein scaffolds (Ren et al., 2011; Siegel and Firestone, 1988).

Bacterial infections are generally characterized by very low pH values because of anaerobic fermentation and subsequent inflammation. In this regard, systemic antibiotic therapy has been achieved by incorporating an ionizable polyhistidine in a triblock copolymer, PLGA-b-polyhistidine-b-PEG nanoparticles. The switch of charge promoted the interactions of nanocarriers with the negatively charged bacterial cell wall, and internalization by both Gram-positive and Gram-negative bacteria (Radovic-Moreno et al., 2012). Lastly, development of pH-responsive systems are based on polymers that bear weak acidic (carboxylic acid) or basic (primary and tertiary amines) groups with a p*Ka* that enables sharp changes in the ionization state at the desired pH. An increase in the degree of ionization can dramatically alter the conformation and the affinity of the polymer chains, resulting in either the disassembly of nanoparticles or the swelling–shrinking of covalent networks (Siegel and Firestone, 1988). Chitosan polymers that are pH sensitive can also be used for this purpose.

8.2 DRUG RELEASE FROM CHITOSAN-COATED MAGNETIC NANOPARTICLES VIA TEMPERATURE CHANGE

Drug delivery via temperature change is one of the most investigated extrinsic stimulus. Thermo-responsive nanocarriers, having the ability to respond to temperature changes, are prepared with at least one thermo-sensitive polymer (Chorny et al., 2010). Thermo-sensitive polymers exhibit a specific lower critical solution temperature (LCST), below which the polymer solution has one phase and above which phases are separated as a result of the collapse and aggregation of polymer chains (Huang et al., 2012).

LCST is affected by the ratio of hydrophilic and hydrophobic monomers in thermo-sensitive polymers. These polymers exhibit major changes in conformation, solubility, and hydrophilic/hydrophobic balance, responding to the environmental temperature. This response to thermal changes triggers the release of the drug from nanocarriers following a change in the environmental temperature. Preferably, thermo-responsive nanocarriers retain their drug load at a physiological temperature of ~37°C, and quickly release at the locally heated tumor area of ~40–42°C. In addition, thermo-responsive smart drug delivery can also take place with a temperature decrease; a thermally reversible swelling or deswelling of the nanocarrier leads to the release of drug load as a consequence of increased porosity (Yuan et al., 2008). The challenge in the design of thermo-responsive nanodevices lies in the use of materials that are both safe and sensitive enough to respond to slight temperature changes in body temperature.

Jaiswal et al. (2014) improved a dual pH and temperature responsive nanohydrogels by incorporating magnetite nanoparticles into a poly(*N*-isopropylacrylamide) chitosan polymer matrix, which has a LCST of 42°C. As a chemotherapeutic drug, doxorubicin was loaded onto the nanohydrogels. The pH- and temperature-controlled release of the drug was achieved at a temperature of 42°C generated by hyperthermia and in an acidic environment of pH 4.6. The cytotoxicity of the doxorubicin-loaded magnetic nanohydrogels was evaluated on MCF-7 (human breast cancer cell line) and HeLa cells (cervical carcinoma cell line). Consequently, this combinational therapy resulted in cell death of the cancer cells up to 85%, which is significantly higher than the cancer cell–killing efficacy of hyperthermia alone (Jaiswal et al., 2014).

9 TOXICITY OF CHITOSAN-COATED MAGNETIC NANOPARTICLES

The principle rule for nanoparticles used in biomedical applications is safety. The nanocarriers should be nontoxic, nonimmunogenic, and at a size that avoids embolization of capillary ducts. Generally, the size, coating material, composition, surface area, and shape of nanoparticles are the most determinant characteristics of cytotoxicity. Surface modifications of nanoparticles are key in minimizing toxicological effects (Lin et al., 2010).

Once a nanoparticle enters the bloodstream, opsonization processes activate the RES response. Circulating mononuclear phagocytes carry the nanoparticles to the liver, spleen, and bone marrow, where they are metabolized by resident cells (e.g., Kupffer cells in the liver). Some of the nanoparticles present in the lysosomal vesicles of Kupffer cells may be incorporated into the bile and excreted in feces, depending on their size and biodegradability. Other ones may be filtered by the kidneys and excreted in urine.

Generally, small nanoparticles with hydrodynamic sizes of up to 5–8 nm (similar to the size of 30–50 kDa serum proteins) and low molecular weight contrast agents can undergo clearance, while larger nanoparticles are taken up by the spleen, liver, and bone marrow (Liu et al., 2013). Due to this characteristic, larger nanoparticles are used in the detection and imaging of spleen tumors (Weissleder et al., 1988), hepatocarcinoma (Weissleder et al., 1987), and bone marrow. Nanoparticles in the size range of 20–150 nm (Elsabahy and Wooley, 2012) have been collected in the bones, kidneys (Lee et al., 2010a,b), and stomach (Banerjee et al., 2002). While smaller nanoparticles are cleared by endocytic cells (i.e., by B and T lymphocytes), larger nanoparticles are cleared by phagocytic cells (i.e., by macrophages or dendritic cells). Generally, nanoparticles up to 100 nm are phagocytosed through the liver cells while ones larger than 200 nm are filtered by the spleen. The decomposition products of biodegradable MNPs are taken up by any cell with pinocytosis (Macaroff et al., 2006; Neuberger et al., 2005).

It has been shown that SPIONs are metabolically degraded in the spleen or liver and thus they are biocompatible (Jain et al., 2008). In addition, surface coating materials have an effect on the biocompatibility, cytotoxicity, cellular internalization, and endocytosis mechanisms of nanoparticles (Gupta and Gupta, 2005). Chitosan-coated MNPs exhibit lower cytotoxicity in comparison with particles without surface coverage (Unsoy, 2013; Unsoy et al., 2012a). The decomposition of MNP follows the lysosomal degradation of nanoparticles with the discharge of iron cations. In the cells, these free cations theoretically promote oxidation reactions. Then, the excess iron is rapidly dissociated and kept in ferritin proteins (Lartigue et al., 2013). The ferritin proteins help to control iron levels in the body (Jain et al., 2008). Therefore, MNPs are considered biodegradable, owing to the fact that iron (Fe) is recycled within the biochemical pathways of Fe metabolism in the cells (Arruebo et al., 2007).

The chemical biodegradation of chitosan occurs in different organs of the body, such as the stomach, where levels of hydrogen (H) ions are high (pH 1). Biological degradation, which is also called enzyme mediated degradation, occurs in the colon of the body. Chitosan polymers can be digested by specific enzymes (chitinases) produced by bacteria found in the colon. Chitinases degrade chitosan to small fragments that can then be readily excreted by the body through renal clearance (Fulzele et al., 2003). The safety of chitosan for humans and other mammals has been confirmed by many studies, and currently chitosan is considered as a GRAS (generally regarded as safe) material for pharmaceutical and food applications by the FDA (Kean and Thanou, 2010). Nevertheless, it has been demonstrated that chitosan may cause hypocholesterolemia in humans at high doses (Lee et al., 2005).

MNPs are considered as biodegradable because the iron (Fe) is reused and recycled within the normal biochemical pathways of Fe metabolism.

Despite all this, before variously modified chitosan-coated MNPs can be used in clinical applications it is necessary to reach a decision about the pharmacokinetics (body distribution, metabolism, bioavailability, organ specific toxicity), biodegradation, hematocompatibility, immunogenicity, and in vitro/in vivo toxicity (teratogenicity, mutagenicity, acute, subacute, and chronic toxicity) of nanoparticles (Macaroff et al., 2006).

10 ADMINISTRATION ROUTES OF CHITOSAN-COATED MAGNETIC NANOPARTICLES

Chitosan-coated MNPs have gained high interest for their wide applications in targeted delivery of loaded proteins, genes, vaccines, and anticancer chemotherapeutic drugs. Chitosan-coated MNPs have also been applied in various drug-delivery systems, such as ocular drug delivery, mucosal drug delivery, brain targeting drug delivery, controlled delivery of drugs, and delivery of insulin (Nagpal et al., 2010). Different administration routes are possible for chitosan nanoparticles, including oral, nasal, intravenous, ocular, and parenteral (Bhattarai et al., 2010). The majority of preclinical studies have involved injection as the route of administration into the tumor. However, low aqueous solubility of drugs limits the possibility of intravenous delivery, which is the most preferred administration route for many anticancer drugs. The route of administration is an important consideration for targeted delivery of drugs and nucleic acids to the cells. This in turn will impact the required dose, frequency, and dosing intervals.

Chitosan-coated MNPs offer versatile routes of administration, especially with respect to noninvasive routes such as oral, intranasal, intratracheal, ocular, and transdermal, which are the most preferable due to the mucoadhesive property of chitosan, which promotes adhesion and absorption of nanoparticles through a mucous surface (Kim et al., 2007). In oral administration, nanoparticles can protect drugs from degradation in the gastrointestinal tract and improve the absorption of drugs. The oral allergen—gene immunization with chitosan—DNA nanoparticles has been found to be effective in modulating murine anaphylactic responses, indicating its prophylactic utility in treating food allergies (Roy et al., 1999). Yin et al. (2009) improved a promising vehicle for oral administration. A trimethyl chitosan–cysteine conjugate was synthesized to combine the advantages of trimethyl chitosan and thiolated polymers, including the mucoadhesion and the permeation enhancing effects in different stages of oral absorption. The trimethyl chitosan–cysteine nanoparticles are obtained via a self-assembly method. They have a spherical morphology, uniform size, positive zeta potential values, and high insulin encapsulation efficiency. The mucoadhesion and permeation-enhancing effects of the trimethyl chitosan–cysteine nanoparticles were significantly higher than those of noncysteine conjugated particles. Biocompatibility assessment of the trimethyl chitosan–cysteine nanoparticles revealed that they have no toxicity and offer a good alternative for insulin delivery (Yin et al., 2009). Additionally, Pan et al. (2002) reported that a hypoglycemic effect was observed in induced diabetic rats after oral administration of chitosan nanoparticles.

In addition to oral administration, chitosan nanoparticles can also be applied to other mucous membrane systems. Nasal and pulmonary administration routes are taken into account as promising for the delivery of peptides and proteins because chitosan nanoparticles have extremely large surface areas and demonstrate less intracellular or extracellular enzymatic degradation. Drug nanoparticles may not need protection against enzymatic degradation when using nasal administration, in contrast to oral administration, because they are not exposed to low pH conditions and degradative enzymes. They remain in the nasal cavity only about 15 min, owing to the small distance they are transported. Therefore, it is not necessary to design the nanoparticles as drugs delivered via systemic circulation are. The administration of drug-loaded nanoparticles can instead be performed by way of a solution or powder. Some of the polymers can act as an absorption enhancer agent to slow down the clearance of nanoparticles through the mucociliary process. Among the polymers, chitosan nanoparticles are one of the most recently explored and extensively studied vaccine nanocarriers. Thus, the contact time between the drug-loaded chitosan nanoparticles and nasal tissue was increased (Fernandez-Urrusuno et al., 1999).

In a recent study, chitosan was confirmed as stimulating the nasal absorption of insulin in rats and sheep. The insulin–chitosan mixture was prepared by grinding the chitosan with insulin with a mortar and pestle. The insulin–chitosan mixture as powder showed higher bioavailability than insulin-loaded chitosan nanoparticles (Fernandez-Urrusuno et al., 1999; Sailaja et al., 2010).

The absorption-enhancing effect of chitosan is also believed to boost the mucosal immune response. The production of chitosan vaccines containing the antigens of influenza, diphtheria, and pertussis was achieved for nasal delivery. The studies revealed that the chitosan vaccines produced a significant level of antibodies in mice. Chitosan has also been considered as an adjuvant for the systemic administration of vaccine delivery as it increases the activation and accumulation of macrophages. The activation of macrophages is initiated after the internalization of chitosan. Moreover, chitosan has been widely used in the mucosal delivery of nucleic acid vaccines. As an example of nasal administration, chitosan-based DNA flu vaccine has been developed. After the intranasal administration of vaccines, antibody levels were significantly increased in mice (van der Lubben et al., 2003). It has been showed that the vaccines loaded on particulate systems such as chitosan nanoparticles boost the systemic and mucosal immune responses against the antigens.

Kumar et al. (2004) reported that the mice, vaccinated with a cocktail of RSV (respiratory syncytial virus) cDNA loaded to the chitosan polymer in the form of nanoparticles by intranasal administration, exhibited a significant reduction in RSV titer and antigen load, as compared to untreated and naked cDNA administered controls.

The pulmonary administration of plasmid DNA loaded on chitosan nanoparticles was also shown to induce increased levels of IFN-γ secretion compared to pulmonary delivery of the plasmid DNA in solution or the more frequently used intramuscular immunization route in HLA-A2 transgenic mice (Bivas-Benita et al., 2004) Another pulmonary administration study in mice of chitosan–DNA complex confirmed that transfection was higher with the powder form of chitosan–DNA complex than solutions containing the same amount of DNA (Mao et al., 2010).

The ocular administration of chitosan-fluorescent nanoparticles was studied in vivo. The results showed that the accumulation of chitosan-fluorescent nanoparticles was significantly higher in the cornea and conjunctiva than for the controls that used the solution of nanoparticles. The amount of the accumulated drug was quite constant up to 24 h (Maeda and Kimura, 2004).

11 CONCLUDING REMARKS

Nanoparticles, which are biodegradable, biocompatible, capable of sufficient retention in the target site, cost effective for commercialization, designed with optimal properties for the highest drug-loading capacity, with a maximum circulation half-life, and sustained drug release, are forecast to be produced for targeted drug delivery.

Chitosan nanoparticles have gained increasing attention as a drug-delivery system owing to their good features, including biocompatibility, degradability, and nontoxicity. Chitosan-coated MNPs show high promise in the diagnosis and treatment of the most devastating diseases, including cancer. Chitosan-coated MNPs have been used in successful demonstrations of targeted drug delivery and imaging applications in cancer cells in vitro and in vivo as well as for packaging and delivery of DNA and siRNA for gene therapy.

A chitosan coat around MNPs reduces agglomeration and provides internal cavities for the loading of therapeutics and surface functional groups for modifications by signaling molecules. MNPs must be superparamagnetic, biocompatible, biodegradable, hydrophilic, and monodisperse for targeted drug delivery, MRI, and hyperthermia. Drugs can be released at the targeted tumor tissue because the pH of the tumor microenvironment and endosomes of the cells is acidic. The swelling of the chitosan around the nanoparticles results from the escape and release of the loaded drug from the cavities of particles.

The chitosan coating also facilitates the stabilization of nanoparticles and retards their clearance by the RES. The coating, size, surface area, composition, and shape of the nanoparticles are the most important characteristics determining cytotoxicity. Surface modifications of nanoparticles are key in minimizing toxicological effects. Chitosan-coated MNPs exhibit lower cytotoxicity in comparison with bare MNPs.

While great progress has been achieved in the application of chitosan-coated MNPs as drug carriers, some problems remain to be resolved. Although chitosan can be modified easily to encapsulate hydrophobic drugs, further investigation is required into the biocompatibility of modified chitosan and its derivatives. In conclusion, chitosan and its derivatives have great potential for wide applications as drug carriers. Eventually nanotechnology has the potential to change the diagnosis and therapy of cancer fundamentally, which may bring new hope.

REFERENCES

Ahmed, M., Narain, R., 2012. Intracellular delivery of DNA and enzyme in active form using degradable carbohydrate-based nanogels. Mol. Pharm. 9 (11), 3160–3170.

Alexis, F., Rhee, J.W., Richie, J.P., Radovic-Moreno, A.F., Langer, R., Farokhzad, O.C., 2008. New frontiers in nanotechnology for cancer treatment. Urol. Oncol. 26 (1), 74–85.

Al-Qadi, S., Grenha, A., Carrión-Recio, D., Seijo, B., Remuñán-López, C., 2012. Microencapsulated chitosan nanoparticles for pulmonary protein delivery: in vivo evaluation of insulin-loaded formulations. J. Control. Release 157 (3), 383–390.

Alvarez-Lorenzo, C., Concheiro, A., 2014. Smart drug delivery systems: from fundamentals to the clinic. Chem. Commun. 50 (58), 7743–7765.

Andrade, F., Antunes, F., Vanessa Nascimento, A., Baptista da Silva, S., das Neves, J., Ferreira, D., Sarmento, B., 2011. Chitosan formulations as carriers for therapeutic proteins. Curr. Drug Discov. Technol. 8 (3), 157–172.

Arami, H., Stephen, Z., Veiseh, O., Zhang, M., 2011. Chitosan-coated iron oxide nanoparticles for molecular imaging and drug delivery. In: Abe, A., Albertsson, A.-C., Coates, G.W., Genzer, J., Kobayashi, S., Lee, K.-S., Leibler, L., Long, T.E., Möller, M., Okay, O., Percec, V., Tang, B.Z., Terentjev, E.M., Theato, P., Vicent, M.J., Voit, B., Wiesner, U., Zhang, X. (Eds.), Chitosan for Biomaterials I. Springer, Berlin, Heidelberg, pp. 163–184.

Arbab, A.S., Bashaw, L.A., Miller, B.R., Jordan, E.K., Lewis, B.K., Kalish, H., Frank, J.A., 2003. Characterization of biophysical and metabolic properties of cells labeled with superparamagnetic iron oxide nanoparticles and transfection agent for cellular MR imaging 1. Radiology 229 (3), 838–846.

Arias, J.L., Reddy, L.H., Couvreur, P., 2012. Fe_3O_4/chitosan nanocomposite for magnetic drug targeting to cancer. J. Mater. Chem. 22 (15), 7622–7632.

Arruebo, M., Fernández-Pacheco, R., Ibarra, M.R., Santamaría, J., 2007. Magnetic nanoparticles for drug delivery. Nano Today 2 (3), 22–32.

Arshady, R., Kono, K. (Eds.), 2006. Smart Nanoparticles in Nanomedicine. Kentus Books, London, UK.

Artursson, P., Lindmark, T., Davis, S.S., Illum, L., 1994. Effect of chitosan on the permeability of monolayers of intestinal epithelial cells (Caco-2). Pharm. Res. 11 (9), 1358–1361.

AshaRani, P.V., Low Kah Mun, G., Hande, M.P., Valiyaveettil, S., 2008. Cytotoxicity and genotoxicity of silver nanoparticles in human cells. ACS Nano 3 (2), 279–290.

Aydin, R., Pulat, M., 2012. 5-Fluorouracil encapsulated chitosan nanoparticles for pH-stimulated drug delivery: evaluation of controlled release kinetics. J. Nanomater. 2010, 10.

Banerjee, T., Mitra, S., Singh, A.K., Sharma, R.K., Maitra, A., 2002. Preparation, characterization and biodistribution of ultrafine chitosan nanoparticles. Int. J. Pharm. 243 (1), 93–105.

Bartlett, D.W., Su, H., Hildebrandt, I.J., Weber, W.A., Davis, M.E., 2007. Impact of tumor-specific targeting on the biodistribution and efficacy of siRNA nanoparticles measured by multimodality in vivo imaging. Proc. Natl. Acad. Sci. USA 104 (39), 15549–15554.

Bennet, D., Kim, S., 2014. Polymer nanoparticles for smart drug delivery. In: Sezer, A.D. (Ed.), Application of nanotechnology in drug delivery. InTech, DOI: 10.5772/58422. Available from: http://www.intechopen.com/books/application-of-nanotechnology-in-drug-delivery/polymer-nanoparticles-for-smart-drug-delivery.

Bhattarai, N., Gunn, J., Zhang, M., 2010. Chitosan-based hydrogels for controlled, localized drug delivery. Adv. Drug Deliv. Rev. 62 (1), 83–99.

Bhattarai, S.R., Kim, S.Y., Jang, K.Y., Lee, K.C., Yi, H.K., Lee, D.Y., et al., 2008. N-hexanoyl chitosan-stabilized magnetic nanoparticles: enhancement of adenoviral-mediated gene expression both in vitro and in vivo. Nanomedicine 4 (2), 146–154.

Bivas-Benita, M., van Meijgaarden, K.E., Franken, K.L., Junginger, H.E., Borchard, G., Ottenhoff, T.H., Geluk, A., 2004. Pulmonary delivery of chitosan-DNA nanoparticles enhances the immunogenicity of a DNA vaccine encoding HLA-A* 0201-restricted T-cell epitopes of Mycobacterium tuberculosis. Vaccine 22 (13), 1609–1615.

Borchard, G., Lueβen, H.L., de Boer, A.G., Verhoef, J.C., Lehr, C.M., Junginger, H.E., 1996. The potential of mucoadhesive polymers in enhancing intestinal peptide drug absorption. III: effects of chitosan-glutamate and carbomer on epithelial tight junctions in vitro. J. Control. Release 39 (2), 131–138.

Byrne, J.D., Betancourt, T., Brannon-Peppas, L., 2008. Active targeting schemes for nanoparticle systems in cancer therapeutics. Adv. Drug Deliv. Rev. 60 (15), 1615–1626.

Carrstensen, H., Mueller, R.H., Müller, B.W., 1992. Particle size, surface hydrophobicity and interaction with serum of parenteral fat emulsions and model drug carriers as parameters related to RES uptake. Clin. Nutr. 11 (5), 289–297.

Chakeres, D.W., de Vocht, F., 2005. Static magnetic field effects on human subjects related to magnetic resonance imaging systems. Prog. Biophys. Mol. Biol. 87 (2), 255–265.

Cheng, F.Y., Su, C.H., Yang, Y.S., Yeh, C.S., Tsai, C.Y., Wu, C.L., et al., 2005. Characterization of aqueous dispersions of Fe_3O_4 nanoparticles and their biomedical applications. Biomaterials 26 (7), 729–738.

Chithrani, B.D., Ghazani, A.A., Chan, W.C., 2006. Determining the size and shape dependence of gold nanoparticle uptake into mammalian cells. Nano Lett. 6 (4), 662–668.

Cho, Y.W., Jang, J., Park, C.R., Ko, S.W., 2000. Preparation and solubility in acid and water of partially deacetylated chitins. Biomacromolecules 1 (4), 609–614.

Cho, E.C., Xie, J., Wurm, P.A., Xia, Y., 2009. Understanding the role of surface charges in cellular adsorption versus internalization by selectively removing gold nanoparticles on the cell surface with a I2/KI etchant. Nano Lett. 9 (3), 1080–1084.

Chorny, M., Hood, E., Levy, R.J., Muzykantov, V.R., 2010. Endothelial delivery of antioxidant enzymes loaded into non-polymeric magnetic nanoparticles. J. Control. Release 146 (1), 144–151.

Conner, S.D., Schmid, S.L., 2003. Regulated portals of entry into the cell. Nature 422 (6927), 37–44.
Custódio, C.A., Alves, C.M., Reis, R.L., Mano, J.F., 2010. Immobilization of fibronectin in chitosan substrates improves cell adhesion and proliferation. J. Tissue Eng. Regen. Med. 4 (4), 316–323.
Dave, S.R., Gao, X., 2009. Monodisperse magnetic nanoparticles for biodetection, imaging, and drug delivery: a versatile and evolving technology. Wiley Interdiscip. Rev. Nanomed. Nanobiotechnol. 1 (6), 583–609.
Davis, M.E., Zuckerman, J.E., Choi, C.H.J., Seligson, D., Tolcher, A., Alabi, C.A., et al., 2010. Evidence of RNAi in humans from systemically administered siRNA via targeted nanoparticles. Nature 464 (7291), 1067–1070.
De Campos, A.M., Sánchez, A., Alonso, M.J., 2001. Chitosan nanoparticles: a new vehicle for the improvement of the delivery of drugs to the ocular surface. Application to cyclosporin A. Int. J. Pharm. 224 (1), 159–168.
De la Fuente, M., Seijo, B., Alonso, M.J., 2008. Novel hyaluronan-based nanocarriers for transmucosal delivery of macromolecules. Macromol. Biosci. 8 (5), 441–450.
Dodane, V., Vilivalam, V.D., 1998. Pharmaceutical applications of chitosan. Pharm. Sci. Technol. Today 1 (6), 246–253.
Drogoz, A., David, L., Rochas, C., Domard, A., Delair, T., 2007. Polyelectrolyte complexes from polysaccharides: formation and stoichiometry monitoring. Langmuir 23 (22), 10950–10958.
Du, J.Z., Du, X.J., Mao, C.Q., Wang, J., 2011. Tailor-made dual pH-sensitive polymer–doxorubicin nanoparticles for efficient anticancer drug delivery. J. Am. Chem. Soc. 133 (44), 17560–17563.
Elsabahy, M., Wooley, K.L., 2012. Design of polymeric nanoparticles for biomedical delivery applications. Chem. Soc. Rev. 41 (7), 2545–2561.
Elsabee, M.Z., Morsi, R.E., Al-Sabagh, A.M., 2009. Surface active properties of chitosan and its derivatives. Colloids Surf. B Biointerfaces 74 (1), 1–16.
Fernandez-Megia, E., Novoa-Carballal, R., Quiñoá, E., Riguera, R., 2007. Conjugation of bioactive ligands to PEG-grafted chitosan at the distal end of PEG. Biomacromolecules 8 (3), 833–842.
Fernandez-Urrusuno, R., Calvo, P., Remuñán-López, C., Vila-Jato, J.L., Alonso, M.J., 1999. Enhancement of nasal absorption of insulin using chitosan nanoparticles. Pharm. Res. 16 (10), 1576–1581.
Fonte, P., Andrade, F., Araujo, F., Andrade, C., Neves, J., Sarmento, B., 2012. Chitosan-coated solid lipid nanoparticles for insulin delivery. Methods Enzymol. 508, 295–314.
Fulzele, S.V., Satturwar, P.M., Dorle, A.K., 2003. Study of the biodegradation and in vivo biocompatibility of novel biomaterials. Eur. J. Pharm. Sci. 20 (1), 53–61.
Gabizon, A., Shmeeda, H., Horowitz, A.T., Zalipsky, S., 2004. Tumor cell targeting of liposome-entrapped drugs with phospholipid-anchored folic acid–PEG conjugates. Adv. Drug Deliv. Rev. 56 (8), 1177–1192.
Gaur, U., Sahoo, S.K., De Tapas, K., Ghosh, P.C., Maitra, A., Ghosh, P.K., 2000. Biodistribution of fluoresceinated dextran using novel nanoparticles evading reticuloendothelial system. Int. J. Pharm. 202 (1), 1–10.
Goya, G.F., Grazu, V., Ibarra, M.R., 2008. Magnetic nanoparticles for cancer therapy. Curr. Nanosci. 4 (1), 1–16.
Grenha, A., 2012. Chitosan nanoparticles: a survey of preparation methods. J. Drug Target. 20 (4), 291–300.
Griset, A.P., Walpole, J., Liu, R., Gaffey, A., Colson, Y.L., Grinstaff, M.W., 2009. Expansile nanoparticles: synthesis, characterization, and in vivo efficacy of an acid-responsive polymeric drug delivery system. J. Am. Chem. Soc. 131 (7), 2469–2471.
Gunn, J., Park, S.I., Veiseh, O., Press, O.W., Zhang, M., 2011. A pretargeted nanoparticle system for tumor cell labeling. Mol. Biosyst. 7 (3), 742–748.
Gupta, A.K., Gupta, M., 2005. Synthesis and surface engineering of iron oxide nanoparticles for biomedical applications. Biomaterials 26 (18), 3995–4021.
Hejazi, R., Amiji, M., 2003. Chitosan-based gastrointestinal delivery systems. J. Control. Release 89 (2), 151–165.
Hobbs, S.K., Monsky, W.L., Yuan, F., Roberts, W.G., Griffith, L., Torchilin, V.P., Jain, R.K., 1998. Regulation of transport pathways in tumor vessels: role of tumor type and microenvironment. Proc. Natl. Acad. Sci. USA 95 (8), 4607–4612.
Honarkar, H., Barikani, M., 2009. Applications of biopolymers I: chitosan. Monatsh. Chem. 140 (12), 1403–1420.

Hou, Y.C., Tsai, S.Y., Lai, P.Y., Chen, Y.S., Chao, P.D.L., 2008. Metabolism and pharmacokinetics of genipin and geniposide in rats. Food Chem. Toxicol. 46 (8), 2764–2769.
Hrkach, J., Von Hoff, D., Ali, M.M., Andrianova, E., Auer, J., Campbell, T., et al., 2012. Preclinical development and clinical translation of a PSMA-targeted docetaxel nanoparticle with a differentiated pharmacological profile. Sci. Transl. Med. 4 (128), 128ra39-128ra39.
Huang, H.Y., Hu, S.H., Chian, C.S., Chen, S.Y., Lai, H.Y., Chen, Y.Y., 2012. Self-assembling PVA-F127 thermosensitive nanocarriers with highly sensitive magnetically-triggered drug release for epilepsy therapy in vivo. J. Mater. Chem. 22 (17), 8566–8573.
Il'ina, A.V., Varlamov, V.P., 2005. Chitosan-based polyelectrolyte complexes: a review. Appl. Biochem. Microbiol. 41 (1), 5–11.
Isikli, C., Hasirci, V., Hasirci, N., 2012. Development of porous chitosan–gelatin/hydroxyapatite composite scaffolds for hard tissue-engineering applications. J. Tissue Eng. Regen. Med. 6 (2), 135–143.
Jain, R.K., 1998. The next frontier of molecular medicine: delivery of therapeutics. Nat. Med. 4 (6), 655–657.
Jain, R.K., Stylianopoulos, T., 2010. Delivering nanomedicine to solid tumors. Nat. Rev. Clin. Oncol. 7 (11), 653–664.
Jain, T.K., Reddy, M.K., Morales, M.A., Leslie-Pelecky, D.L., Labhasetwar, V., 2008. Biodistribution, clearance, and biocompatibility of iron oxide magnetic nanoparticles in rats. Mol. Pharm. 5 (2), 316–327.
Jaiswal, M.K., Pradhan, A., Banerjee, R., Bahadur, D., 2014. Dual pH and temperature stimuli-responsive magnetic nanohydrogels for thermo-chemotherapy. J. Nanosci. Nanotechnol. 14 (6), 4082–4089.
Jaracz, S., Chen, J., Kuznetsova, L.V., Ojima, I., 2005. Recent advances in tumor-targeting anticancer drug conjugates. Bioorg. Med. Chem. 13 (17), 5043–5054.
Jiang, X., Musyanovych, A., Röcker, C., Landfester, K., Mailänder, V., Nienhaus, G.U., 2011. Specific effects of surface carboxyl groups on anionic polystyrene particles in their interactions with mesenchymal stem cells. Nanoscale 3 (5), 2028–2035.
Kavaz, D., Odabaş, S., Güven, E., Demirbilek, M., Denkbaş, E.B., 2010. Bleomycin loaded magnetic chitosan nanoparticles as multifunctional nanocarriers. J. Bioact. Compat. Polym. 25 (3), 305–318.
Kean, T., Thanou, M., 2010. Biodegradation, biodistribution and toxicity of chitosan. Adv. Drug Deliv. Rev. 62 (1), 3–11.
Khodadust, R., Unsoy, G., Yalcın, S., Gunduz, G., Gunduz, U., 2013. PAMAM dendrimer-coated iron oxide nanoparticles: synthesis and characterization of different generations. J. Nanopart. Res. 15 (3), 1–13.
Khor, E., Lim, L.Y., 2003. Implantable applications of chitin and chitosan. Biomaterials 24 (13), 2339–2349.
Kievit, F.M., Veiseh, O., Bhattarai, N., Fang, C., Gunn, J.W., Lee, D., et al., 2009. PEI–PEG–chitosan-copolymer-coated iron oxide nanoparticles for safe gene delivery: synthesis, complexation, and transfection. Adv. Funct. Mater. 19 (14), 2244–2251.
Kievit, F.M., Veiseh, O., Fang, C., Bhattarai, N., Lee, D., Ellenbogen, R.G., Zhang, M., 2010. Chlorotoxin labeled magnetic nanovectors for targeted gene delivery to glioma. ACS Nano 4 (8), 4587–4594.
Kim, T.H., Jiang, H.L., Jere, D., Park, I.K., Cho, M.H., Nah, J.W., et al., 2007. Chemical modification of chitosan as a gene carrier in vitro and in vivo. Prog. Polym. Sci. 32 (7), 726–753.
Kirpotin, D.B., Drummond, D.C., Shao, Y., Shalaby, M.R., Hong, K., Nielsen, U.B., et al., 2006. Antibody targeting of long-circulating lipidic nanoparticles does not increase tumor localization but does increase internalization in animal models. Cancer Res. 66 (13), 6732–6740.
Klepka, M.T., Nedelko, N., Greneche, J.M., Lawniczak-Jablonska, K., Demchenko, I.N., Slawska-Waniewska, A., et al., 2008. Local atomic structure and magnetic ordering of iron in Fe− chitosan complexes. Biomacromolecules 9 (6), 1586–1594.
Kostarelos, K., Lacerda, L., Pastorin, G., Wu, W., Wieckowski, S., Luangsivilay, J., et al., 2007. Cellular uptake of functionalized carbon nanotubes is independent of functional group and cell type. Nat. Nanotechnol. 2 (2), 108–113.
Kumar, M.N.R., 2000. A review of chitin and chitosan applications. React. Funct. Polym. 46 (1), 1–27.

Kumar, M.N., Muzzarelli, R.A., Muzzarelli, C., Sashiwa, H., Domb, A.J., 2004. Chitosan chemistry and pharmaceutical perspectives. Chem. Rev. 104 (12), 6017–6084.

Labhasetwar, V., Song, C., Humphrey, W., Shebuski, R., Levy, R.J., 1998. Arterial uptake of biodegradable nanoparticles: effect of surface modifications. J. Pharm. Sci. 87 (10), 1229–1234.

Lartigue, L., Alloyeau, D., Kolosnjaj-Tabi, J., Javed, Y., Guardia, P., Riedinger, A., et al., 2013. Biodegradation of iron oxide nanocubes: high-resolution in situ monitoring. ACS Nano 7 (5), 3939–3952.

Lee, K.Y., 2007. Chitosan and its derivatives for gene delivery. Macromol. Res. 15 (3), 195–201.

Lee, H.S., Kim, E.H., Shao, H., Kwak, B.K., 2005. Synthesis of SPIO-chitosan microspheres for MRI-detectable embolotherapy. J. Magn. Magn. Mater. 293 (1), 102–105.

Lee, C.M., Jeong, H.J., Kim, S.L., Kim, E.M., Kim, D.W., Lim, S.T., et al., 2009. SPION-loaded chitosan–linoleic acid nanoparticles to target hepatocytes. Int. J. Pharm. 371 (1), 163–169.

Lee, M.J.E., Veiseh, O., Bhattarai, N., Sun, C., Hansen, S.J., Ditzler, S., et al., 2010a. Correction: rapid pharmacokinetic and biodistribution studies using chlorotoxin-conjugated iron oxide nanoparticles: a novel non-radioactive method. PLoS One 5 (3), 10–1371.

Lee, P.W., Hsu, S.H., Wang, J.J., Tsai, J.S., Lin, K.J., Wey, S.P., et al., 2010b. The characteristics, biodistribution, magnetic resonance imaging and biodegradability of superparamagnetic core–shell nanoparticles. Biomaterials 31 (6), 1316–1324.

Leroueil, P.R., Hong, S., Mecke, A., Baker, Jr., J.R., Orr, B.G., Banaszak Holl, M.M., 2007. Nanoparticle interaction with biological membranes: does nanotechnology present a Janus face? Acc. Chem. Res. 40 (5), 335–342.

Leroueil, P.R., Berry, S.A., Duthie, K., Han, G., Rotello, V.M., McNerny, D.Q., et al., 2008. Wide varieties of cationic nanoparticles induce defects in supported lipid bilayers. Nano Lett. 8 (2), 420–424.

Lim, E.K., Huh, Y.M., Yang, J., Lee, K., Suh, J.S., Haam, S., 2011. pH-triggered drug-releasing magnetic nanoparticles for cancer therapy guided by molecular imaging by MRI. Adv. Mater. 23 (21), 2436–2442.

Lin, J., Zhang, H., Chen, Z., Zheng, Y., 2010. Penetration of lipid membranes by gold nanoparticles: insights into cellular uptake, cytotoxicity, and their relationship. ACS Nano 4 (9), 5421–5429.

Liu, W.G., De Yao, K., 2002. Chitosan and its derivatives—a promising non-viral vector for gene transfection. J. Control. Release 83 (1), 1–11.

Liu, W., Sun, S., Cao, Z., Zhang, X., Yao, K., Lu, W.W., Luk, K.D.K., 2005. An investigation on the physicochemical properties of chitosan/DNA polyelectrolyte complexes. Biomaterials 26 (15), 2705–2711.

Liu, J., Yu, M., Zhou, C., Zheng, J., 2013. Renal clearable inorganic nanoparticles: a new frontier of bionanotechnology. Mater. Today 16 (12), 477–486.

Lochmann, D., Jauk, E., Zimmer, A., 2004. Drug delivery of oligonucleotides by peptides. Eur. J. Pharm. Biopharm. 58 (2), 237–251.

Lunov, O., Syrovets, T., Röcker, C., Tron, K., Nienhaus, G.U., Rasche, V., et al., 2010. Lysosomal degradation of the carboxydextran shell of coated superparamagnetic iron oxide nanoparticles and the fate of professional phagocytes. Biomaterials 31 (34), 9015–9022.

Macaroff, P.P., Simioni, A.R., Lacava, Z.G., Lima, E.C., Morais, P.C., Tedesco, A.C., 2006. Studies of cell toxicity and binding of magnetic nanoparticles with blood stream macromolecules. J. Appl. Phys. 99 (8), 08S102.

Maeda, Y., Kimura, Y., 2004. Antitumor effects of various low-molecular-weight chitosans are due to increased natural killer activity of intestinal intraepithelial lymphocytes in sarcoma 180–bearing mice. J. Nutr. 134 (4), 945–950.

Mamot, C., Drummond, D.C., Noble, C.O., Kallab, V., Guo, Z., Hong, K., et al., 2005. Epidermal growth factor receptor–targeted immunoliposomes significantly enhance the efficacy of multiple anticancer drugs in vivo. Cancer Res. 65 (24), 11631–11638.

Mansouri, S., Lavigne, P., Corsi, K., Benderdour, M., Beaumont, E., Fernandes, J.C., 2004. Chitosan-DNA nanoparticles as non-viral vectors in gene therapy: strategies to improve transfection efficacy. Eur. J. Pharm. Biopharm. 57 (1), 1–8.

Mao, S., Sun, W., Kissel, T., 2010. Chitosan-based formulations for delivery of DNA and siRNA. Adv. Drug Deliv. Rev. 62 (1), 12–27.

Marrache, S., Dhar, S., 2012. Engineering of blended nanoparticle platform for delivery of mitochondria-acting therapeutics. Proc. Natl. Acad. Sci. USA 109 (40), 16288–16293.

Matsumura, Y., Maeda, H., 1986. A new concept for macromolecular therapeutics in cancer chemotherapy: mechanism of tumoritropic accumulation of proteins and the antitumor agent smancs. Cancer Res. 46 (12 Part 1), 6387–6392.

McBain, S.C., Yiu, H.H., Dobson, J., 2008. Magnetic nanoparticles for gene and drug delivery. Int. J. Nanomed. 3 (2), 169.

Mizukoshi, Y., Shuto, T., Masahashi, N., Tanabe, S., 2009. Preparation of superparamagnetic magnetite nanoparticles by reverse precipitation method: contribution of sonochemically generated oxidants. Ultrason. Sonochem. 16 (4), 525–531.

Moghimi, S.M., Hunter, A.C., Murray, J.C., 2005. Nanomedicine: current status and future prospects. FASEB J. 19 (3), 311–330.

Moreno, D., Zalba, S., Navarro, I., de Ilarduya, C.T., Garrido, M.J., 2010. Pharmacodynamics of cisplatin-loaded PLGA nanoparticles administered to tumor-bearing mice. Eur. J. Pharm. Biopharm. 74 (2), 265–274.

Mornet, S., Vasseur, S., Grasset, F., Duguet, E., 2004. Magnetic nanoparticle design for medical diagnosis and therapy. J. Mater. Chem. 14 (14), 2161–2175.

Mürbe, J., Rechtenbach, A., Töpfer, J., 2008. Synthesis and physical characterization of magnetite nanoparticles for biomedical applications. Mater. Chem. Phys. 110 (2), 426–433.

Nagpal, K., Singh, S.K., Mishra, D.N., 2010. Chitosan nanoparticles: a promising system in novel drug delivery. Chem. Pharm. Bull. 58 (11), 1423–1430.

Neuberger, T., Schöpf, B., Hofmann, H., Hofmann, M., Von Rechenberg, B., 2005. Superparamagnetic nanoparticles for biomedical applications: possibilities and limitations of a new drug delivery system. J. Magn. Magn. Mater. 293 (1), 483–496.

Padera, T.P., Stoll, B.R., Tooredman, J.B., Capen, D., di Tomaso, E., Jain, R.K., 2004. Pathology: cancer cells compress intratumour vessels. Nature 427 (6976), 695–1695.

Pan, Y., Li, Y.J., Zhao, H.Y., Zheng, J.M., Xu, H., Wei, G., Hao, J.S., 2002. Bioadhesive polysaccharide in protein delivery system: chitosan nanoparticles improve the intestinal absorption of insulin in vivo. Int. J. Pharm. 249 (1), 139–147.

Pankhurst, Q.A., Connolly, J., Jones, S.K., Dobson, J.J., 2003. Applications of magnetic nanoparticles in biomedicine. J. Phys. D 36 (13), R167.

Park, T.G., Jeong, J.H., Kim, S.W., 2006. Current status of polymeric gene delivery systems. Adv. Drug Deliv. Rev. 58 (4), 467–486.

Park, J.H., Saravanakumar, G., Kim, K., Kwon, I.C., 2010. Targeted delivery of low molecular drugs using chitosan and its derivatives. Adv. Drug Deliv. Rev. 62 (1), 28–41.

Plassat, V., Wilhelm, C., Marsaud, V., Ménager, C., Gazeau, F., Renoir, J.M., Lesieur, S., 2011. Anti-estrogen-loaded superparamagnetic liposomes for intracellular magnetic targeting and treatment of breast cancer tumors. Adv. Funct. Mater. 21 (1), 83–92.

Portero, A., Teijeiro-Osorio, D., Alonso, M.J., Remunan-Lopez, C., 2007. Development of chitosan sponges for buccal administration of insulin. Carbohydr. Polym. 68 (4), 617–625.

Prego, C., Garcia, M., Torres, D., Alonso, M.J., 2005. Transmucosal macromolecular drug delivery. J. Control. Release 101 (1), 151–162.

Qu, W., Li, Y., Hovgaard, L., Li, S., Dai, W., Wang, J., et al., 2012. A silica-based pH-sensitive nanomatrix system improves the oral absorption and efficacy of incretin hormone glucagon-like peptide-1. Int. J. Nanomed. 7, 4983.

Radovic-Moreno, A.F., Lu, T.K., Puscasu, V.A., Yoon, C.J., Langer, R., Farokhzad, O.C., 2012. Surface charge-switching polymeric nanoparticles for bacterial cell wall-targeted delivery of antibiotics. ACS Nano 6 (5), 4279–4287.

Rejman, J., Oberle, V., Zuhorn, I., Hoekstra, D., 2004. Size-dependent internalization of particles via the pathways of clathrin-and caveolae-mediated endocytosis. Biochem. J. 377, 159–169.

Ren, D., Kratz, F., Wang, S.W., 2011. Protein nanocapsules containing doxorubicin as a pH-Responsive delivery system. Small 7 (8), 1051–1060.

Rinaudo, M., 2008. Main properties and current applications of some polysaccharides as biomaterials. Polym. Int. 57 (3), 397–430.

Roiter, Y., Ornatska, M., Rammohan, A.R., Balakrishnan, J., Heine, D.R., Minko, S., 2008. Interaction of nanoparticles with lipid membrane. Nano Lett. 8 (3), 941–944.

Roy, K., Mao, H.Q., Huang, S.K., Leong, K.W., 1999. Oral gene delivery with chitosan–DNA nanoparticles generates immunologic protection in a murine model of peanut allergy. Nat. Med. 5 (4), 387–391.

Rutirawut, T., Limphirat, W., Sinsarp, A., Tivakornsasithorn, K., Srikhirin, T., Osotchan, T., 2015. Composition and oxidation state of cobalt and nickel-iron oxide colloidal nanoparticles in liquid phase. Adv. Mater. Res. 1103, 21–27.

Sailaja, A.K., Amareshwar, P., Chakravarty, P., 2010. Chitosan nanoparticles as a drug delivery system. Res. J. Pharm. Biol. Chem. Sci. 1 (3), 474–484.

Sannan, T., Kurita, K., Iwakura, Y., 1976. Studies on chitin, 2. Effect of deacetylation on solubility. Makromol. Chem. 177 (12), 3589–3600.

Schenck, J.F., 2005. Physical interactions of static magnetic fields with living tissues. Prog. Biophys. Mol. Biol. 87 (2), 185–204.

Shi, Z., Neoh, K.G., Kang, E.T., Shuter, B., Wang, S.C., Poh, C., Wang, W., 2009. Chitosan-modified superparamagnetic iron oxide nanoparticles for magnetic resonance imaging of stem cells. ACS Appl. Mater. Interfaces 1 (2), 328–335.

Siegel, R.A., Firestone, B.A., 1988. pH-Dependent equilibrium swelling properties of hydrophobic polyelectrolyte copolymer gels. Macromolecules 21 (11), 3254–3259.

Singh, d.K., Ray, a.R., 2000. Biomedical applications of chitin, chitosan, and their derivatives. J. Macromol. Sci. C Polym. Rev. 40 (1), 69–83.

Smith, R.A., Porteous, C.M., Gane, A.M., Murphy, M.P., 2003. Delivery of bioactive molecules to mitochondria in vivo. Proc. Natl. Acad. Sci. USA 100 (9), 5407–5412.

Stacker, S.A., Achen, M.G., Jussila, L., Baldwin, M.E., Alitalo, K., 2002. Metastasis: lymphangiogenesis and cancer metastasis. Nat. Rev. Cancer 2 (8), 573–583.

Sun, X., Rossin, R., Turner, J.L., Becker, M.L., Joralemon, M.J., Welch, M.J., Wooley, K.L., 2005. An assessment of the effects of shell cross-linked nanoparticle size, core composition, and surface PEGylation on in vivo biodistribution. Biomacromolecules 6 (5), 2541–2554.

Sun, C., Lee, J.S., Zhang, M., 2008a. Magnetic nanoparticles in MR imaging and drug delivery. Adv. Drug Deliv. Rev. 60 (11), 1252–1265.

Sun, W., Mao, S., Mei, D., Kissel, T., 2008b. Self-assembled polyelectrolyte nanocomplexes between chitosan derivatives and enoxaparin. Eur. J. Pharm. Biopharm. 69 (2), 417–425.

Tabernero, J., Shapiro, G.I., LoRusso, P.M., Cervantes, A., Schwartz, G.K., Weiss, G.J., et al., 2013. First-in-humans trial of an RNA interference therapeutic targeting VEGF and KSP in cancer patients with liver involvement. Cancer Discov. 3 (4), 406–417.

Tan, M.L., Choong, P.F., Dass, C.R., 2009. Review: doxorubicin delivery systems based on chitosan for cancer therapy. J. Pharm. Pharmacol. 61 (2), 131–142.

Torchilin, V.P., 2004. Targeted polymeric micelles for delivery of poorly soluble drugs. Cell. Mol. Life Sci. 61 (19–20), 2549–2559.

Treuel, L., Jiang, X., Nienhaus, G.U., 2013. New views on cellular uptake and trafficking of manufactured nanoparticles. J. R. Soc. Interface 10 (82), 20120939.

Tsai, Z.T., Wang, J.F., Kuo, H.Y., Shen, C.R., Wang, J.J., Yen, T.C., 2010. In situ preparation of high relaxivity iron oxide nanoparticles by coating with chitosan: a potential MRI contrast agent useful for cell tracking. J. Magn. Magn. Mater. 322 (2), 208–213.

Unsoy, G., 2013. The Synthesis and Characterization of Doxorubicin and Bortezomib Loaded Magnetic Nanoparticles for Targeting Tumor Cells. Doctoral Dissertation, Middle East Technical University.

Unsoy, G., Yalcin, S., Khodadust, R., Gunduz, G., Gunduz, U., 2012a. Synthesis optimization and characterization of chitosan-coated iron oxide nanoparticles produced for biomedical applications. J. Nanopart. Res. 14 (964), 1–13.

Unsoy, G., Yalcin, S., Khodadust, R., Mutlu, P., Gunduz, U., 2012b. In situ synthesis and characterization of chitosan coated iron oxide nanoparticles and loading of doxorubicin. nanocon-2012, At Brno, Czech Republic, EU, Volume: 23.–25. 10. 2012. Brno, Czech Republic, EU.

Unsoy, G., Khodadust, R., Yalcin, S., Mutlu, P., Gunduz, U., 2014a. Synthesis of doxorubicin loaded magnetic chitosan nanoparticles for pH responsive targeted drug delivery. Eur. J. Pharm. Sci. 62, 243–250.

Unsoy, G., Yalcin, S., Khodadust, R., Mutlu, P., Onguru, O., Gunduz, U., 2014b. Chitosan magnetic nanoparticles for pH responsive bortezomib release in cancer therapy. Biomed. Pharmacother. 68 (5), 641–648.

van der Lubben, I.M., Verhoef, J.C., Borchard, G., Junginger, H.E., 2001. Chitosan and its derivatives in mucosal drug and vaccine delivery. Eur. J. Pharm. Sci. 14 (3), 201–207.

van der Lubben, I.M., Kersten, G., Fretz, M.M., Beuvery, C., Verhoef, J.C., Junginger, H.E., 2003. Chitosan microparticles for mucosal vaccination against diphtheria: oral and nasal efficacy studies in mice. Vaccine 21 (13), 1400–1408.

Veiseh, O., Sun, C., Fang, C., Bhattarai, N., Gunn, J., Kievit, F., et al., 2009. Specific targeting of brain tumors with an optical/magnetic resonance imaging nanoprobe across the blood-brain barrier. Cancer Res. 69 (15), 6200–6207.

Veiseh, O., Kievit, F.M., Fang, C., Mu, N., Jana, S., Leung, M.C., et al., 2010. Chlorotoxin bound magnetic nanovector tailored for cancer cell targeting, imaging, and siRNA delivery. Biomaterials 31 (31), 8032–8042.

Verma, A., Uzun, O., Hu, Y., Hu, Y., Han, H.S., Watson, N., et al., 2008. Surface-structure-regulated cell-membrane penetration by monolayer-protected nanoparticles. Nat. Mater. 7 (7), 588–595.

Vila, A., Sánchez, A., Janes, K., Behrens, I., Kissel, T., Jato, J.L.V., Alonso, M.J., 2004. Low molecular weight chitosan nanoparticles as new carriers for nasal vaccine delivery in mice. Eur. J. Pharm. Biopharm. 57 (1), 123–131.

Wang, T., Bai, J., Jiang, X., Nienhaus, G.U., 2012. Cellular uptake of nanoparticles by membrane penetration: a study combining confocal microscopy with FTIR spectroelectrochemistry. ACS Nano 6 (2), 1251–1259.

Weissleder, R., Stark, D.D., Compton, C.C., Wittenberg, J., Ferrucci, J.T., 1987. Ferrite-enhanced MR imaging of hepatic lymphoma: an experimental study in rats. Am. J. Roentgenol. 149 (6), 1161–1165.

Weissleder, R., Hahn, P.F., Stark, D.D., Elizondo, G., Saini, S., Todd, L.E., et al., 1988. Superparamagnetic iron oxide: enhanced detection of focal splenic tumors with MR imaging. Radiology 169 (2), 399–403.

Widera, A., Norouziyan, F., Shen, W.C., 2003. Mechanisms of TfR-mediated transcytosis and sorting in epithelial cells and applications toward drug delivery. Adv. Drug Deliv. Rev. 55 (11), 1439–1466.

Woo, K., Hong, J., Choi, S., Lee, H.W., Ahn, J.P., Kim, C.S., Lee, S.W., 2004. Easy synthesis and magnetic properties of iron oxide nanoparticles. Chem. Mater. 16 (14), 2814–2818.

Yalcin, S., Unsoy, G., Mutlu, P., Khodadust, R., Gunduz, U., 2014. Polyhydroxybutyrate-Coated magnetic nanoparticles for doxorubicin delivery: cytotoxic effect against doxorubicin-resistant breast cancer cell line. Am. J. Ther. 21 (6), 453–461.

Yang, H.W., Hua, M.Y., Liu, H.L., Huang, C.Y., Tsai, R.Y., Lu, Y.J., et al., 2011. Self-protecting core-shell magnetic nanoparticles for targeted, traceable, long half-life delivery of BCNU to gliomas. Biomaterials 32 (27), 6523–6532.

Yi, H., Wu, L.Q., Bentley, W.E., Ghodssi, R., Rubloff, G.W., Culver, J.N., Payne, G.F., 2005. Biofabrication with chitosan. Biomacromolecules 6 (6), 2881–2894.

Yin, L., Ding, J., He, C., Cui, L., Tang, C., Yin, C., 2009. Drug permeability and mucoadhesion properties of thiolated trimethyl chitosan nanoparticles in oral insulin delivery. Biomaterials 30 (29), 5691–5700.

You, J.O., Auguste, D.T., 2009. Nanocarrier cross-linking density and pH sensitivity regulate intracellular gene transfer. Nano Lett. 9 (12), 4467–4473.

Yu, J., Patel, S.A., Dickson, R.M., 2007. In vitro and intracellular production of peptide-encapsulated fluorescent silver nanoclusters. Angew. Chem. 119 (12), 2074–2076.

Yuan, Q., Venkatasubramanian, R., Hein, S., Misra, R.D.K., 2008. A stimulus-responsive magnetic nanoparticle drug carrier: magnetite encapsulated by chitosan-grafted-copolymer. Acta Biomater. 4 (4), 1024–1037.

Zan, Q., Wang, C., Dong, L., Cheng, P., Tian, J., 2008. Effect of surface roughness of chitosan-based microspheres on cell adhesion. Appl. Surf. Sci. 255 (2), 401–403.

Zemskova, L.A., Voit, A.V., Nikolenko, Y.M., Sergienko, V.I., Troshkina, I.D., Plevaka, A.V., Maiboroda, S.B., Chekmarevb, A.M., 2005. Sorption of rhenium on carbon fibrous materials modified with chitosan. J. Nucl. Radiochem. Sci. 6 (3), 221–222.

Zhang, L., Wang, T., Yang, L., Liu, C., Wang, C., Liu, H., et al., 2012. General route to multifunctional uniform yolk/mesoporous silica shell nanocapsules: a platform for simultaneous cancer-targeted imaging and magnetically guided drug delivery. Chemistry 18 (39), 12512–12521.

Zhao, Y., Sun, X., Zhang, G., Trewyn, B.G., Slowing, I.I., Lin, V.S.Y., 2011. Interaction of mesoporous silica nanoparticles with human red blood cell membranes: size and surface effects. ACS Nano 5 (2), 1366–1375.

Zhu, A., Yuan, L., Dai, S., 2008. Preparation of well-dispersed superparamagnetic iron oxide nanoparticles in aqueous solution with biocompatible N-succinyl-O-carboxymethylchitosan. J. Phys. Chem. C 112 (14), 5432–5438.

Zhu, A., Yuan, L., Jin, W., Dai, S., Wang, Q., Xue, Z., Qin, A., 2009. Polysaccharide surface modified Fe_3O_4 nanoparticles for camptothecin loading and release. Acta Biomater. 5 (5), 1489–1498.

Zhu, S., Lansakara-P, D.S., Li, X., Cui, Z., 2012. Lysosomal delivery of a lipophilic gemcitabine prodrug using novel acid-sensitive micelles improved its antitumor activity. Bioconjug. Chem. 23 (5), 966–980.

CHAPTER

DRUG DELIVERY: ADVANCEMENTS AND CHALLENGES

27

Narinder Singh*, Anjali Joshi*, Amrit Pal Toor*, Gaurav Verma*,**

**Panjab University, Chandigarh, India*

***Massachusetts Institute of Technology, Cambridge, MA, United States*

CHAPTER OUTLINE

1 INTRODUCTION

Nanotechnology paves the way for all new classes of noble engineered nanocarriers and systems that dramatically change the modern ways of drug delivery. Today, researchers attempt to reduce the size of carriers without compromising the drug loading and try to reach the unexplored biological sites for better and quicker treatments, such as the central nervous system (CNS). In pursuit of better drug delivery systems a variety of organic and inorganic material are developed by researchers and have unique physicochemical properties suitable for biodistribution (Farokhzad and Langer, 2009). Manufacturing and engineering of materials at molecular scale enable us to penetrate the very biological sites where even blood is not circulating, such as blood–brain barrier (BBB) (Bidros and Vogelbaum, 2009; Buse and El-Aneed, 2010). Nanotechnology helps us to improve (1) hydrophobic drugs delivery; (2) transport of drugs across biological barriers, such as tight epithelial and endothelial barriers; (3) targeted drug delivery with extreme precision and control over release of drug; (4) combination therapy is possible by delivery of more than one drug together; (5) imaging techniques can be improved by nanomarkers that would also help in visualizing the site of drug delivery by modulating the nanocarriers. The basic features of an ideal nanocarrier for drug delivery are biocompatibility, bioacceptability, supporting cell uptake, and their mass production should be possible at low cost. Cellular uptake of drugs could be increased by attaching targeting ligands or modifying the shape, size, and surface properties of nanocarries (Dobson and Kell, 2008; Zhang et al., 2013). The main advantages of nanocarriers and systems are that they increase efficiency of modern drug delivery by introducing a new class of therapeutics with ultraprecision and overcome the adversities associated with conventional methods by reducing side effects (Bamrungsap et al., 2012; Peer et al., 2007; Shi et al., 2010). In past decades special emphasis was on targeted drug delivery, which increases the precision and reduces the side effects associated with excess drug distribution. Numerous nanomaterials have been developed for modern therapeutics. In this chapter some prominent nanomaterials are discussed in detail. The abrupt growth in the synthesis of nanoparticle for enormous application, including electronic, coatings, and food packaging to biomedical application increase the risk of exposure of nanoparticles to human body and environment. Recent studies suggest that some nanoparticles are toxic to the human body, such as carbon nanotubes, and raise serious concern about their potential threat to human health and environment. A detailed investigation is needed for the toxicological effect of nanoparticles. In this chapter we elaborate the advancements and toxicological challenges in modern therapeutics, which cover almost every aspect of oral to targeted drug delivery systems, having special emphasis on nanocarriers and nanosystems.

2 DRUG DELIVERY INTO THE BRAIN

We adopt two approaches to deliver drugs to the brain; one way is to manipulate either the drug or capillary permeability to increase the drug delivery of intravascular drugs (Cecchelli et al., 2007), and the other way is to localize administration (Wang et al., 2002) of the drugs on the site. Researchers worldwide adopt different approaches (Gabathuler, 2010) to increase the amount of intravascular drugs that reach the tumor cell, such as drug packaging (Dou et al., 2009), barrier disruption (Kullervo et al., 2006), and the intraarterial route (Chertok et al., 2011). The major drawback of intravascular drugs is that the major amount of the drug was wasted in the body and only a small fraction of drug reaches the tumor cell (Guhasarkar and Banerjee, 2010). On the other hand, when the drug is delivered directly into the tumor cell or into the cerebrospinal fluid, 100% drug is delivered but it consists of variable drug concentration and unpredictable spatial distribution, which leads to neurotoxicity (McCann et al., 1998; Muller and Keck, 2004). The movements of most of the drugs are forbidden by BBB (Alyautdin et al., 2014). The reason is that the capillaries of the brain are having interendothelial junctions and few pinocytotic vesicles but no fenestrations (Gaillard et al., 2001). The drugs cross the BBB by simple diffusion (Provenzale et al., 2005). As far as a brain tumor is concerned its permeability is very complex (Groothuis et al., 1982). Schlageter et al. (1999) have suggested that brain tumors may consist of three distinct microvessel populations. The first microvessel consists of nonfenestrated and continuous capillaries similar to a normal brain. The second one consists of continuous and fenestrated capillaries having increased permeability to small molecules, but large molecules are not permitted through these capillaries. The third one consists of capillaries having interendothelial gaps with sizes up to 1 μm. The spatial distribution of capillaries plays an important role in drug diffusion across the barrier. There are many drug delivery methods, each having its own advantages and disadvantages. The following are the few prominent methods of drug delivery to BBB and brain tumor barrier (BTB):

1. Chemical modification of drugs: Diseased cells express differently as compared to their counterparts (healthy cells). Researchers utilize this very property of diseased cells to formulate drug molecules or drug vectors to target the specific cells without harming the surrounding normal cells. Chemically modified drugs are nothing else but the inactive chemical derivatives of the drugs or pharmacological biological species obtained by chemical modification of drug molecules or active biological molecules (Juillerat-Jeanneret and Schmitt, 2007).
2. Intraarterial administration: Intraarterial injection is capable of treating a large volume of tissues by osmotic BBB disruption (Neal et al., 1992). In this technique the drug is directly injected into the arteries of the targeted tissue by disrupting the BBB using hyperosmolar mannitol solution (Foley et al., 2014).
3. Receptor mediated transport: The receptors/proteins expressed on the surface of endothelial cells of brain capillaries interact with molecules during their transport across BBB. The uptake of modern drug molecules by brain cells is significantly increased by specific or nonspecific interaction of drug molecules with the receptors/proteins on the surface of endothelial cells (Xiao and Gan, 2013). This can be achieved by knowing the endogenous transport systems of BBB (Pardridge, 2012).
4. Vector mediated delivery: In general, vector mediated drug delivery approach involves conjugation of targeting vector with the drug molecules that can transport the drug molecules across the BBB by transcytosis (Newton, 2006).
5. Direct injection or infusion of drugs: This method provides us an opportunity to bypass the most concerned BBB of brain and the direct administration of drug within the brain extracellular region. The 100% drug delivery is possible with this method but we need a better understanding of the movement of drug molecules within the brain and their efflux mechanisms (Groothuis, 2000).

2.1 NANOPARTICLES AND CNS: INTERACTION

Nanoparticles have positive, as well as negative effects on contact with human cells such that nanoparticles are capable of killing viruses (Yang et al., 2010) and pathogens (Rai et al., 2009), but above certain limits their toxicological effect prevails in its positive effect (Yang et al., 2010). The small size of nanoparticles makes the transportation of drugs through BBB possible either by passive diffusion or endocytosis (Barbu et al., 2009). Oberdörster (2004) suggests that the uptake of nanoparticles into the brain could be directed by trans-synaptic transport. For instance, we take the example of silver nanoparticles (AgNPs), which easily enters into the brain through BBB and accumulates in different parts of the brain, this seems good for drug transport but may also cause the risk of toxicity (Muthu and Singh, 2008; Trickler et al., 2010).

2.2 TRANSPORT OF DRUG INTO THE BRAIN: NANOCARRIERS

The difficulties faced during the delivery of drug into the brain are due to these two main reasons:

1. BBB penetration for drug molecules is very poor
2. Efflux (back transport) of drugs

To overcome the first problem researchers have worked on various colloidal drug delivery systems, such as nanoparticles (Hughes, 2005), lipid microspheres (Trubetskoy and Torchilin, 1995), solid lipid nanoparticles (SLNs) (Almeida and Souto, 2007), liposomes (Rabinow, 2004), and niosomes (Dufes et al., 2004). The characteristics of a good nanocarrier is that it should be physically and chemically stable, having high drug loading, low toxicity, and the overall cost of development should be less (Hillaireau and Couvreur, 2009; Masserini, 2013). The ability of nanocarriers to overcome the BBB and produce the biological effects on the CNS was shown in various studies (Kreuter, 2014). Researchers have made significant progress in the understanding of mechanisms of the nanocarriers uptake into the brain (Batrakova and Kabanov, 2008; Tiwari and Amiji, 2006). The structural and physicochemical properties, such as surface charge, molecular size, and binding capability plays an important role in diffusion of molecules across the BBB (Pardridge, 2012). The enhancement of the drug concentration into the brain could be achieved by invasive or noninvasive methods (Choi et al., 2007; Kinoshita et al., 2006). In the invasive method we temporarily increase the BBB permeability but complications, such as brain edema, are also associated with this method. Invasive methods are very expensive and a long recovery period is required. The various strategies adopted for noninvasive methods are (1) chemical modification of drug molecules (Juillerat-Jeanneret, 2008), and (2) antibody specific conjugation. Kreuter (2001) review the nanoparticle systems for brain drug delivery and suggest that the uptake of nanoparticles into the brain is most likely by endocytosis by brain capillary endothelial cells (BCEC). Polymeric (Kumari et al., 2010) and SLNs are mostly used for drug transport to the brain. SLNs are easily taken up by the brain as compared to polymeric nanoparticles, as well as they are less toxic than polymeric nanoparticles due to their biodegradability and bioacceptability (Blasi et al., 2007).

2.2.1 Nanocarriers for brain drug delivery: polymers

Polymeric nanoparticles are used for different formulations of drugs by using one these methods, including, adsorption, dissolution, entrapment, encapsulation, or chemical binding of drug molecules on the surface of polymeric nanoparticles (Bhowmik et al., 2015). The drug release kinetics and its characteristics solely depend on the drug trapping method and polymer structure (Juillerat-Jeanneret, 2007). To increase the bioavailability of drugs into the brain by coating polymers with ploysorbates has been reported widely (Patel et al., 2012) and the proposed mechanisms are:

1. The solubilization and fluidization of endothelial cell membranes due to polysorbates surfactant effect.

2. Polymeric nanoparticles interact with BBB endothelial cell membranes and facilitates endocytosis.
3. Restricts efflux phenomenon.

The disadvantages of these polymeric nanoparticles are that toxic degradation, toxic monomers aggregation, residual material associated with them, and toxic degradation process are involved in such systems (Lockman et al., 2002).

2.2.2 Nanocarrier for brain drug delivery: solid lipid nanoparticles

Spherical SLN is a colloidal drug carrier system in nanorange, which can easily disperse in suitable aqueous surfactant or water (Kaur et al., 2008). In this system the drug is dissolved or dispersed in a high melting fat matrix forming solid core and hydrophobic ends of lipid particles attach to it forming spheres enabling them to transport lipophilic and hydrophilic drugs through BBB (Patel et al., 2009). The physicochemical characteristics of SLNs are highly regarded for the formulation of targeted drug delivery to the brains. The SLNs are preferred over polymeric nanoparticles because of their high drug loading with negligible toxicity and high scale production (Blasi et al., 2007). In fact, all the advantages associated with polymeric nanoparticles, emulsions, and liposomes are exhibited by SNLs (Mukherjee et al., 2009). The advantages are:

1. SLNs easily passed by live.
2. Controlled and prolonged release of drugs is possible by SLNs.
3. Targeted delivery of drugs would be possible by attaching ligands to the SLNs.
4. Drug loading could be increased.
5. SLNs can encapsulate both hydrophilic as well as hydrophobic drugs.
6. Stability of SLN systems is very high. They may be stable up to 3 years.
7. Reproducibility of such system is very high.
8. They are safe due to their biodegradability.
9. Large-scale production is feasible.

3 ORAL DRUG DELIVERY SYSTEMS

The oral route for drug delivery is one of the most preferred among other routes because of its high patient compliance, low cost of drug formulation, and easy administration. However, several disadvantages are associated with oral drug delivery systems, such as drugs must first pass through the liver and undergo metabolism and in the gastrointestinal (GI) tract enzymatic degradation takes place, which restricts the use of several classes of drugs that contain proteins and peptides (Rossi et al., 2005). To design effective and efficient drug delivery systems, we have to incorporate the latest technologies, such as drug delivery vectors, enzyme inhibitors, and permeability enhancers.

3.1 ORAL DRUG DELIVERY SYSTEM: ORAL MUCOSA

The oral mucosa provides an attractive site for drug delivery due to its unique physiological properties and high patient compliance. There are different formulations, such as paste, spray, gels, patches, liquid, and tablets used for drug delivery to/across oral mucosal cavity (Hearnden et al., 2012). There are two permeable routes for passive drug delivery across the oral mucosa; one is paracellular (Shaji and Patole, 2008) and the other is transcellular (Taylor, 1996) as shown in Fig. 27.1. These routes could be adopted simultaneously by permeants but most permeants prefer one route over the other depending on the physical, as well

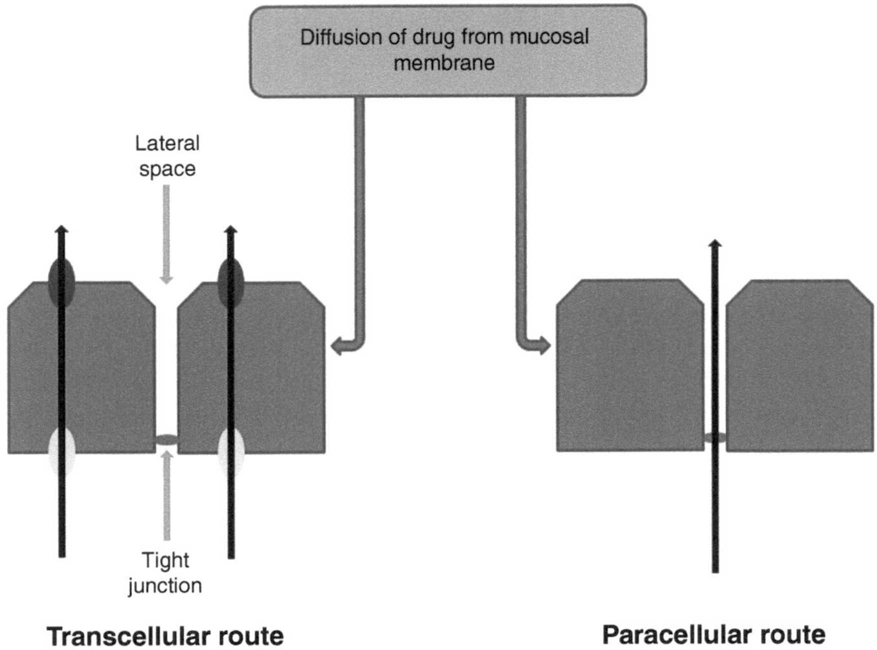

FIGURE 27.1 The Mechanism of Passive Drug Delivery Across Oral Mucosa

as chemical properties of diffusant (Sugano et al., 2010). The intercellular spaces are hydrophilic in nature and they restrict the penetration of lipophilic compounds (Muheem et al., 2016). On the other hand, the lipophilic nature of the cell membrane prevents hydrophilic solute to pass through it (Kalepu et al., 2013). The solutes generally adopt the route that has less hinderance while penetration. The solute penetrates the barrier using the combination of earlier mentioned two routes due to stratified oral epithelium tissues (Holpuch et al., 2010). Drug delivery within the oral mucosal cavity is classified into two broad categories: (a) systematic delivery either via sublingual or buccal mucosa; (b) local delivery (Patel et al., 2011).

3.1.1 Sublingual delivery

In sublingual delivery, the drug is delivered along the floor of the mouth through the mucosal membrane linings (Ahuja et al., 1997). The sublingual route is the most studied one because it is relatively more permeable and leads to rapid absorption of drugs with high acceptability (enel et al., 2012). The formulation of sublingual drugs uses two different designs: one in tablet form and the other in capsule. The tablets are designed in such way that they rapidly disintegrate (Bredenberg et al., 2003) and the capsules generally consist of soft gelatin, which encapsulates liquid drug (Hejazi and Amiji, 2003).

3.1.2 Buccal delivery

In buccal deliver, the drug is administrated through the mucosal membranes lining buccal mucosa (Hao and Heng, 2003). The buccal mucosa is not able to provide rapid absorption and bioacceptability due to less permitivity than sublingual delivery. It is the preferred route for transmucosal drug delivery because the sustainable release of the drug is possible due to its less permitivity and immobile mucosa provides a good retentive system (Martin et al., 2003). Thus, it is more suitable for the sustainable delivery of less permeable drugs, such as peptides.

3.1.3 Local delivery

In local delivery, the drug is directly delivered to oral cavity (Sankar et al., 2011). Local drug delivery is more efficient than systematic drug delivery due to its precise and targeted approach (Sankar et al., 2011). The efficiency of the local drug delivery is increased by the significant reduction in the drug wastage. Targeted delivery reduces the side effects of a drug as only the infected area is exposed to the drug.

3.2 ORAL DRUG DELIVERY: COLON DRUG DELIVERY SYSTEMS (CDDS)

Drug delivery at colon sites could be either local or systematically targeted (Patel, 2015). The treatment of inflammatory disease of the intestine is done by local administration of drugs (Wirtz et al., 1999). However, systematic targeted drug delivery into the colon is preferred to reduce side effects (Hua, 2014). While designing CDDS one has to control the site-specific precise release of the drug and prevent the absorption of the drug in the stomach or intestine (Macleod et al., 1999). Targeted drug delivery to colon would be achieved by multiple coated drug molecules as shown in Fig. 27.2 where each layer dissolves in a specific environment. The peptide and protein drugs are easily absorbed in colon sites

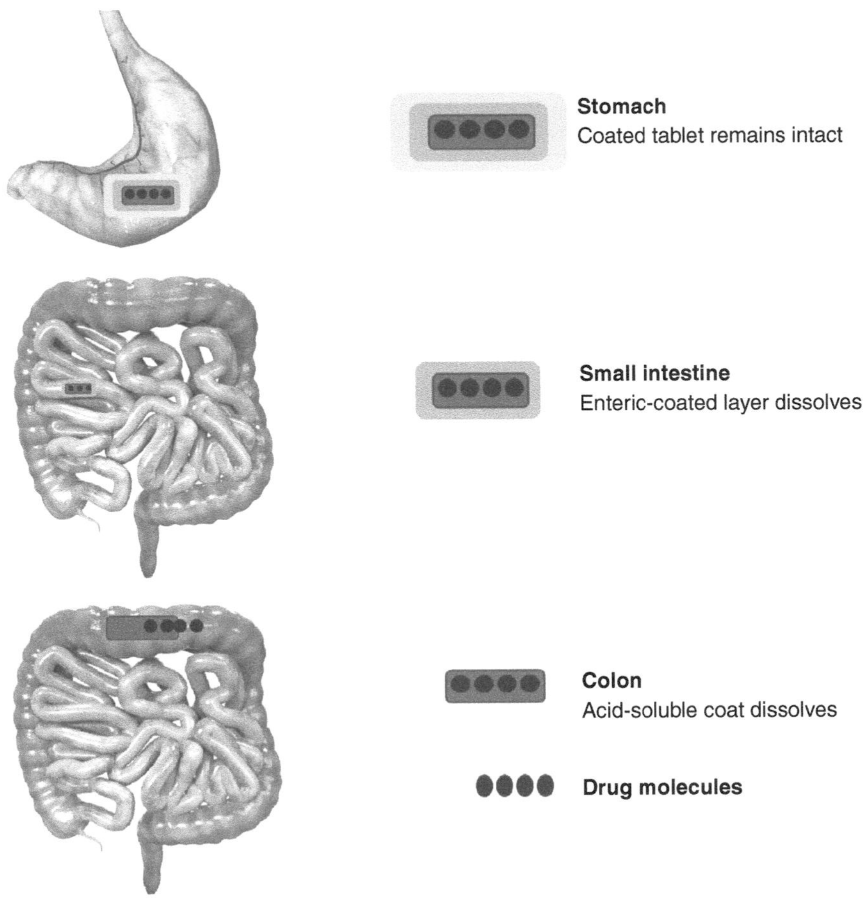

FIGURE 27.2 Colon Targeted Drug Delivery

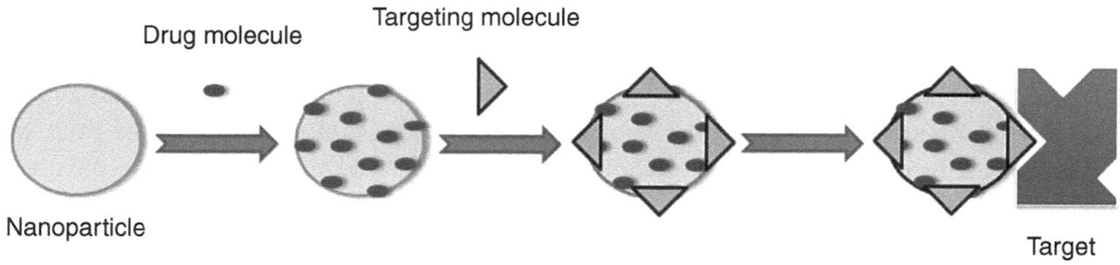

FIGURE 27.3 Basic Steps Involved in Targeted Drug Delivery

due: first, that the intensity and diversity of digestive enzymes is less; and second, colon mucosa's proteolytic activity is significantly less than that observed in the small intestine, which prevents the peptide drug from enzymatic degradation and hydrolysis (Sinha et al., 2007). These three factors named formulation factor, retention time, and retrograde spreading time defines the concentration of the drug that actually reached the colon site (Philip and Philip, 2010). The water absorption capability of colon is very high, which make its content very viscous and leads to reduced mixing efficiency, resulting in low availability of the drug at the absorptive membrane site (Mcrorie et al., 1998). The major advantage of CDDS is that selectivity of this system reduces the quantity of the drug required and also minimizes the systematic side effects of high drug doses (Yang et al., 2002).

4 TARGETED DRUG DELIVERY

Targeted drug delivery systems have improved cancer therapy drastically over the past decade (Kwon et al., 2012). Researchers have successfully developed cost-effective anticancer drugs mostly based on liposomes (Koning et al., 2010) and polymers (Rapoport, 2007). The basic principle of the operation of a targeted drug delivery is explained in Fig. 27.3. In it, nanoparticles act as carriers and drug molecules attached on the surface of nanoparticles by physicochemical interaction, then a targeting molecule is attached to the drug–nanoparticle conjugate, such as antibody or ligands, which direct it to the receptors/antigens on the surface of tumor cells.

4.1 SIMULTANEOUS IMAGING AND DRUG DELIVERY

The basic steps involved in simultaneous imaging and drug delivery are elaborated in Fig. 27.4. First, we attach amphiphilic ligands to hydrophobic drug molecules forming hydrophobic core and hydrophilic surface, which facilitates the diffusion of hydrophobic drug across biological barriers. Then we attach nanomarkers for imaging purposes and antibodies to direct the drug molecules toward tumor cells. At the target tumor cell, the antibody reacts with antigen and releases the drug molecules, which kills the tumor cells.

4.2 TARGETED DRUG DELIVERY: MICELLAR SYSTEMS

Development of the nanosized polymeric micelles opens an all new method of targeted drug delivery system (Nasongkla et al., 2006). The small size, hydrophobic drug loading, long blood circulation time, biodegradable, less toxic, and in vivo stability makes them suitable candidates for targeted drug

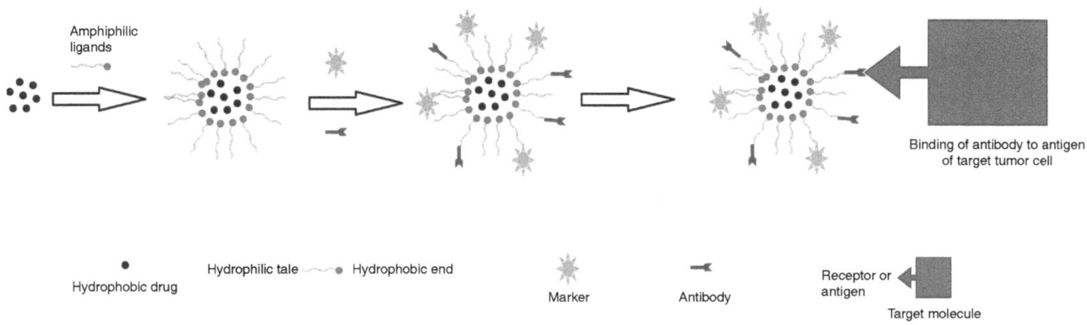

FIGURE 27.4 Simultaneous Imaging and Targeted Drug Delivery

delivery (Uhrich et al., 1999). Recent studies suggest that these systems have less morbidity and toxicity as compared to conventional systems (Torchilin, 2011). Functionalization is one of the ways to make these systems target-specific and have controlled site-specific drug delivery (Sutton et al., 2007).

4.3 TARGETED DRUG DELIVERY: HYBRID SYSTEMS

Hybrid system is a targeted drug delivery that combines more than one system to enhance the overall performance of the systems as a unit. These systems have been engineered to meet the specific requirement of a particular drug delivery system (Singh et al., 2014). In the past decade, researchers have developed many nanotechnological platforms for drug delivery (Huang et al., 2014; Lee et al., 2014; Modak et al., 2013). Particles from gas saturated solution technique is a system that is utilized (Fraile et al., 2013) to study the formulation of S-(+)-ibuprofin with different nanocarriers, such as poloxamers, gelucire, and glyceryl monostearate. Their study suggests that PGSS technique provides a wide possibility of developing new hybrid systems for hydrophobic drugs for the controlled release of drugs. Cancer cells show a characteristic property, named as metastasis, which degrades the extracellular matrix. Kojima et al. (2013) prepared a doxorubicin-conjugated dendrimer hybrid gel for drug delivery for metastatic cells. Beqa et al. (2011) showed gold popcorn-SWCNT hybrid system for targeted imaging of breast cancer cell utilizing the sharp Raman peaks of gold popcorn edges and SWCNT. Pendekal and Tegginamat (2013) formulated a versatile 5-flurorouracil matrix tablets utilizing chitosan–sodium alginate interpolyelectrolyte complex for chemotherapeutic drug delivery system, with less side effects and patient compliance as compared to conventional drug delivery systems.

5 NANOMATERIALS FOR DRUG DELIVERY

Nanomaterials enable us to develop new drug formulations and imaging technique with improved efficiency having fewer side effects as compared to their conventional counterpart (Irvine, 2011; Janib et al., 2010; Wang et al., 2012). Recent studies reported that nanomaterial improve bioavailability and acceptability with very low cytotoxicity by increasing circulation time of drugs with high drug loading, site-specific controlled release, and targeted approach (Kayal and Ramanujan, 2010; Kim et al., 2010; Yao et al., 2014). Some of the prominent nanomaterials for drug delivery are discussed in next sections.

5.1 SINGLE-WALL CARBON NANOTUBES (SWCNT)

The drug loading of SWCNTs is very high due to its high surface-to-volume ratio (Liu et al., 2007). Researchers have developed different strategies to link biological materials, such as proteins, peptides, and DNA to SWCNTs, which makes them soluble in biological fluids to diffuse across cell membranes (Lacerda et al., 2007) by endocytosis with little cytotoxicity (Chen et al., 2008). Carbon nanotubes (CNTs) can be used as vectors for targeted drug delivery system. Bhirde et al. (2009) demonstrated a significant reduction of tumor cells by targeted SWCNTs drug delivery than nontargeted delivery.

5.2 INORGANIC NANOPARTICLES

Nanotechnology revolutionizes the synthesis and modification of inorganic material at nanoscale (Liang et al., 2014). These modified materials are used to make multifunction nanocarriers by incorporating two or more materials, such as drug, marker, and targeting element (ligand or antibodies) (Heneweer et al., 2012; Ma et al., 2015; Sao et al., 2015). The best example is gold nanoparticles, one can easily modify their surface and size to make them compatible with required antibodies for both imaging, as well as targeted drug delivery (Dreaden et al., 2012). Magnetic nanoparticles with suitable fluorescent dye and coated with biodegradable polymer are used for both imaging, as well as target killing of cancer cells (Veiseh et al., 2010). Liong et al. (2008) synthesized inorganic functional nanoparticles for cancer cell–specific targeted drug delivery system having dual imaging capability (optical as well as magnetic resonance) for hydrophobic drugs. They incorporated superparamagnetic iron oxide nanocrystal into mesoporous silica nanoparticles for magnetic resonance imaging and magnetic manipulations. Also they modified the surface of silica nanoparticle with fluorescent dye and targeting ligands and filled the pore with drug molecules. They are better alternatives to conventional chemotherapy due to their capability to simultaneous imaging and drug delivery.

5.3 QUANTUM DOTS (QDs)

QDs are zero-dimensional luminescent semiconductor nanocrystals having versatile surface chemistry enabling them to associate with almost every targeted drug delivery system (Qi and Gao, 2008; Wang and Chen, 2011; Zrazhevskiy et al., 2010). Where they are used for the monitoring of transport (Delehanty et al., 2009) as well as site-specific release of drugs (Gao et al., 2005; Smith et al., 2008). We do not have sufficient knowledge of long-term toxicity of QDs (Shiohara et al., 2004), that is why we do not use them directly for drug delivery systems but we incorporate them into the core of optically or magnetically active nanoparticles, such as gold or iron oxide nanoparticles (Probst et al., 2013). Functionalized QDs in conjugation with targeting ligands could be used to kill specific tumor cells (Al-Jamal et al., 2008; Li et al., 2011). The bioconjugated QDs enable surgeons to map the diseased area and helps them to remove it surgically (Bagalkot et al., 2007; Luo et al., 2012). The cadmium derived QDs with highly tunable surface and fluorescence makes them promising candidates for site specific targeted drug delivery system (Ghaderi et al., 2011). Graphene QDs could be used for pancreatic cancer therapy and imaging applications (Nigam et al., 2014).

5.4 POLYMERIC NANOMATERIALS

The biodegradable polymers and their modification at nanoscale pave the way for controlled targeted drug delivery in modern therapeutics due to their great bioavailability, biocompatibility, controlled

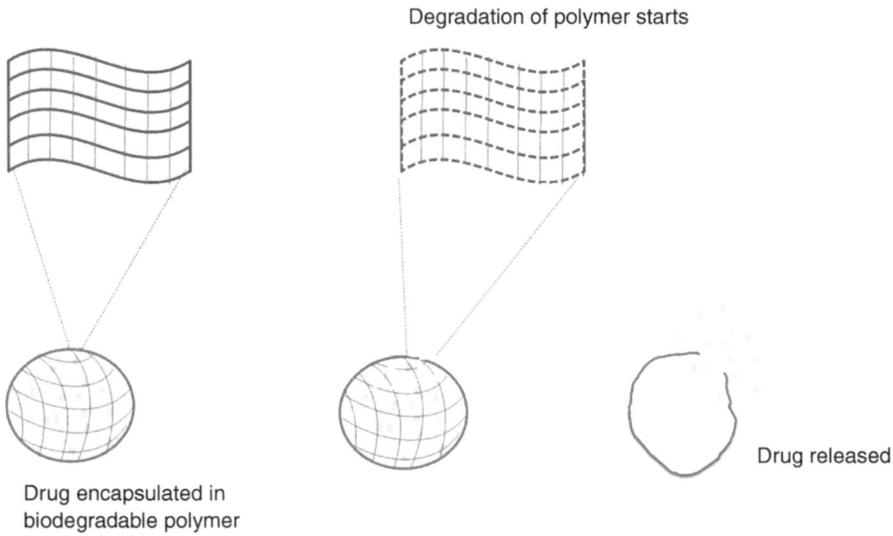

FIGURE 27.5 Polymeric Drug Delivery System

release, and less toxicity (Grobmyer et al., 2010; Pool et al., 2012; Salvia-Trujillo et al., 2013). The basic model of polymeric drug delivery system is shown in Fig. 27.5. Drug molecules are encapsulated in polymeric nanoparticles, which transport them to the tumor cell site where they start degrading due to the surrounding environment of the tumor cell and the drug is being released that kills tumor cells. There is a range of polymeric nanoparticles, such as chitosan, poly(D,L-lactic-*co*-glycolic acid), poly-lactic acid (PLA), poly-alkyl-cyanoacrylates, polycaprolactone, and many more that are used for drug delivery by nanoencapsulation of various drugs (Kumari et al., 2010). The surface functionalization of polymeric nanoparticle improves their drug delivery capability (Cheng et al., 2007). Du et al. (2011) developed a pH-sensitive polymer-drug conjugate nanosystem in which nanoparticles change their surface charge in the vicinity of tumor cell (pH-6.8) from negative to positive to facilitate cell uptake. Subsequently, the acidity in the cellular compartment increases (pH-5), which helps in the release of drug. Thambi et al. (2014) developed a nanocarrrier that releases drugs under hypoxic conditions.

5.5 LIPOSOMES

Liposomes are spherical colloidal structures formed by the self-assembly of lipid bilayers (Samad et al., 2007). For targeted drug delivery various lipid-based systems have been developed (Allen and Cullis, 2013; Malam et al., 2009). Liposomal drug delivery systems modify the kinetic and distribution of cytostatic agents leads to the increased drug release (Al-Jamal and Kostarelos, 2011; Lao et al., 2013). Liposomal drug delivery is of two types: active or triggered (Pattni et al., 2015). In active liposomal drug delivery (Byrne et al., 2008; Danhier et al., 2010) surface modification of liposome is done by ligand and antibodies whereas in triggered liposomal drug delivery (Rijcken et al., 2007; Shum et al., 2001) stimuli sensitive drug release takes place. Liposomal encapsulation of drug molecules reduces the cardio toxicity in breast cancer (Sharma et al., 2006; Zhang et al., 2005). Highly efficient encapsulation was achieved by liposomal anthracyclines and liposomal doxorubicin, but liposomal doxorubicin is better than the former one. Bose et al. (2008) studied the effect of liposomal doxorubicin at very low creatinine clearance levels for the treatment of ovarian carcinoma without any complications.

5.6 GRAPHENE FOR DRUG DELIVERY

Graphene is a single sheet of two-dimensional network of sp^2-bonded carbon atoms (Novoselov et al., 2012) having unique properties, which are promising for biomedical applications (Wang et al., 2011). If we roll graphene it forms structures similar to open-ended single-wall carbon nanotubes (Hersam, 2008) due to this region's research considering it as a promising candidate for modern drug delivery systems (Liu et al., 2013). The delocalized π-electron on the surface of graphene sheets could be used for binding anticancer drugs having aromatic structure as doxorubicin via π-π stacking (Liu et al., 2013). The single-sheet graphene have extremely large surface area-to-volume ratios, which enable high drug loading (Yang et al., 2008). Many studies suggest that functionalized graphene is biocompatible for drug delivery systems (Zhang et al., 2010). The gelatin functionalized graphene nanosheets are prepared via green synthesis (Liu et al., 2011a,b), which show high solubility in water and biological fluids make it an ideal candidate for drug delivery application. Pan et al. (2011) prepared thermo-responsive poly(*N*-isopropylacrylamide)-grafted graphene sheets, which exhibit almost no toxicity and potential to kill cancer cells in vitro. Bao et al. (2011) functionalize the graphene oxide with chitosan, which shows good water solubility, as well as biocompatibility. They show that this nanosystem is capable of loading and delivering a sufficient amount of anticancer drugs.

6 MODERN DRUG DELIVERY: CHALLENGES

Advancements in nanotechnology pave the way for modern therapeutics with its clinical application from markers to drug and gene delivery, but the toxicological aspect associated with the use of nanomaterials pose a serious concern for human health, as well as for environment (Singh and Nalwa, 2007). We discuss the toxicity of some popular drug delivery systems in the next section.

6.1 TOXICITY OF LIPOSOMES AND POLYMERIC NANOPARTICLES

In these systems mostly natural lipids are used, which are highly biocompatible, which is why the body fails to identify them as foreign materials but where we used synthetic polymeric nanoparticles their rate of clearance from body by blood stream varies from material to material leads to their accumulation in body organs or mononuclear phagocytic system (Gong et al., 2012; Murthy, 2007; Zentner et al., 2001). Rapid clearance of polymeric nanoparticles from the body after drug delivery is a matter of concern to reduce toxicity associated with their accumulation (Owens and Peppas, 2006). The removal of synthetic polymeric nanoparticles by the blood steam is very slow, which promotes their accumulation in the human body (Caliceti and Veronese, 2003). To overcome this problem we have to understand the effect of their size, shape, charge, and surface morphology and functionalities (Geng et al., 2007). These factors affect the blood residence time and accumulation of these nanomaterials in body organs. We can control blood clearance and biodistribution of these nanoparticles by enabling them with tissue-specific uptake and accumulations (Alexis et al., 2008).

6.2 TOXICITY OF QUANTUM DOTS

QDs are inorganic material that can be made in different composition such ZnO, CdS, CdSe, CdTe, ZnS. Each type of QD has its different properties depending on the inorganic precursors, synthesis method, shape, size (Moreels et al., 2009; Shiohara et al., 2004). So their respective health risks are different (Lovrić et al., 2005), which makes the toxicological study of QDs very complex. As we mentioned in the targeting drug delivery section, they are used in encapsulated form for imaging, as well as drug delivery (Bruchez et al., 2007; Rosenthal et al., 2011), which makes them biocompatible but in any circumstance if they are exposed to body cells they may cause significant toxic effect due to their inorganic nature (Ryman-Rasmussen et al., 2007). As of now there are not extensive studies done on the toxicity of QDs (Ron, 2006), which is why we are left with one option to ensure its complete encapsulation (Ghaderi et al., 2011).

6.3 TOXICITY OF CARBON MATERIALS

Carbon material at nanoscale has unique physicochemical properties, which motivates researchers to explore it for more applications, leads to increased production of carbon nanomaterials (Nel et al., 2006). This increased production of carbon-based nanomaterials is a matter of concern because it leads to increased exposure of carbon material to the human body, which can cause serious adverse effects on human health (Ge et al., 2012; Kostarelos, 2008). Considerable data on CNTs toxicity was published but there are conflicting results, which makes us unable to develop a systematic understanding of CNT's toxicity (Liu et al, 2013; Monteiro-Riviere and Inman, 2006; Zhao et al., 2008). But there are plenty of studies that show that CNTs are toxic for human health (Hurt et al., 2006; Kayat et al., 2011). Muller et al. (2005) studied the reactivity of MWCNTs in the lungs of rats. Their results show an increase in the production of TNF-α by macrophages, suggesting that MWCNTs are harmful for human health. The cellular toxicity of nanomaterials based on carbon materials was studied by Magrez et al. (2006). They studied the aspect ratio and surface chemistry of MWCNT, carbon fiber, and carbon nanoparticles, which suggest that these materials are toxic and their cytotoxicity increased after functionalization involving acid treatment, but the hazardous effects are size-dependent.

6.4 TOXICITY OF METAL NANOPARTICLES

A variety of metal or metal complexes were successfully synthesized, such as gold nanoparticles, silver nanoparticles, metal oxide nanoparticles, gold nanorods, QDs. To understand their toxicity and bioacceptability we need to know their uptake, clearance, accumulation, and pharmacokinetics (Lasagna-Reeves et al., 2010; Li and Chen, 2011; Meyer et al., 2010). Thorough analysis of published data shows that cellular toxicity is less in chemically stable nanoparticles, whereas nanoparticles on oxidation, reduction, and dissolution become cytotoxic, as well as genotoxic for cellular organisms (Auffan et al., 2009). To understand the adverse effects of metal nanoparticles (Griffitt et al., 2008) studied a variety of metallic nanoparticles of silver, nickel, aluminum, cobalt, and titanium dioxide for their exposure to biological species, such as zebrafish, daphnids, and an algal species. Their results show silver and copper nanoparticles are toxic for all organisms but titanium dioxide is not toxic for any species. They suggested that investigations from other models are needed. The systematic study of TiO_2 nanoparticles suggests that on exposure to lung tissues and cells it induces inflammatory response (Lai et al., 2008; Park et al., 2007).

7 CONCLUSIONS

Nanotechnology provides us a new prospective to exploit the unique properties of material by manipulating them at nanoscale. Fabricating nanoscale materials for drug delivery opens up new methods to target tumors and irregular growths inside the human body. Not only can one have access to remotely reached body parts, but also to portions of the brain where conventional drugs may fail to act. Nanoscale drugs also are being studied for their interfaces with various bacteria, viruses, and infections, such that useful microorganisms are not damaged during the therapy. Pharmacists and chemical engineers in the recent past have joined forces to develop many such unique drug combinations. It enables them to tackle complex and sophisticated health problems with ultraprecision and accuracy, though uncertainty over their toxicity and hazardous effects still exist. Multiple studies on ascertaining toxicity are being undertaken globally to ensure that nanoscale structures are safe for clinical and other trials.

REFERENCES

Ahuja, A., Khar, R.K., Ali, J., 1997. Mucoadhesive drug delivery systems. Drug Dev. Ind. Pharm. 23, 489–515.

Alexis, F., Pridgen, E., Molnar, L.K., Farokhzad, O.C., 2008. Factors affecting the clearance and biodistribution of polymeric nanoparticles. Mol. Pharm. 5, 505–515.

Al-Jamal, W.T., Kostarelos, K., 2011. Liposomes: from a clinically established drug delivery system to a nanoparticle platform for theranostic nanomedicine. Acc. Chem. Res. 44, 1094–1104.

Al-Jamal, W.T., Al-Jamal, K.T., Bomans, P.H., Frederik, P.M., Kostarelos, K., 2008. Functionalized-quantum-dot–liposome hybrids as multimodal nanoparticles for cancer. Small 4, 1406–1415.

Allen, T.M., Cullis, P.R., 2013. Liposomal drug delivery systems: from concept to clinical applications. Adv. Drug Deliv. Rev. 65, 36–48.

Almeida, A.N.J., Souto, E., 2007. Solid lipid nanoparticles as a drug delivery system for peptides and proteins. Adv. Drug Deliv. Rev. 59, 478–490.

Alyautdin, R., Khalin, I., Nafeeza, M.I., Haron, M.H., Kuznetsov, D., 2014. Nanoscale drug delivery systems and the blood brain barrier. Int. J. Nanomed. 9, 795–811.

Auffan, M.L., Rose, J.R.M., Wiesner, M.R., Bottero, J.-Y., 2009. Chemical stability of metallic nanoparticles: a parameter controlling their potential cellular toxicity in vitro. Environ. Pollut. 157, 1127–1133.

Bagalkot, V., Zhang, L., Levy-Nissenbaum, E., Jon, S., Kantoff, P.W., Langer, R., Farokhzad, O.C., 2007. Quantum dotsaptamer conjugates for synchronous cancer imaging, therapy, and sensing of drug delivery based on bi-fluorescence resonance energy transfer. Nano Lett. 7, 3065–3070.

Bamrungsap, S., Zhao, Z., Chen, T., Wang, L., Li, C., Fu, T., Tan, W., 2012. Nanotechnology in therapeutics: a focus on nanoparticles as a drug delivery system. Nanomedicine 7, 1253–1271.

Bao, H., Pan, Y., Ping, Y., Sahoo, N.G., Wu, T., Li, L., Li, J., Gan, L.H., 2011. Chitosan-functionalized graphene oxide as a nanocarrier for drug and gene delivery. Small 7, 1569–1578.

Barbu, E., Molnàr, Ã.V., Tsibouklis, J., GòRecki, D.C., 2009. The potential for nanoparticle-based drug delivery to the brain: overcoming the blood brain barrier. Expert Opin. Drug Deliv. 6, 553–565.

Batrakova, E.V., Kabanov, A.V., 2008. Pluronic block copolymers: evolution of drug delivery concept from inert nanocarriers to biological response modifiers. J. Control. Release 130, 98–106.

Beqa, L., Fan, Z., Singh, A.K., Senapati, D., Ray, P.C., 2011. Gold nano-popcorn attached SWCNT hybrid nanomaterial for targeted diagnosis and photothermal therapy of human breast cancer cells. ACS Appl. Mater. Interfaces 3, 3316–3324.

Bhirde, A.A., Patel, V., Gavard, J., Zhang, G., Sousa, A.A., Masedunskas, A., Leapman, R.D., Weigert, R., Gutkind, J.S., Rusling, J.F., 2009. Targeted killing of cancer cells in vivo and in vitro with EGF-directed carbon nanotube-based drug delivery. ACS Nano 3, 307–316.

Bhowmik, A., Khan, R., Ghosh, M.K., 2015. Blood brain barrier: a challenge for effectual therapy of brain tumors. Biomed. Res. Int. 2015, 20.

Bidros, D.S., Vogelbaum, M.A., 2009. Novel drug delivery strategies in neuro-oncology. Neurotherapeutics 6, 539–546.

Blasi, P., Giovagnoli, S., Schoubben, A.L., Ricci, M., Rossi, C., 2007. Solid lipid nanoparticles for targeted brain drug delivery. Adv. Drug Deliv. Rev. 59, 454–477.

Bose, C.K., Bose, B., Basu, A., Basu, S., 2008. Liposomal doxorubicin in suspected ovarian carcinoma patient with very low creatinine clearance test value. Medscape J. Med. 10, 53.

Bredenberg, S., Duberg, M., Lennernäs, B., Lennernäs, H., Pettersson, A., Westerberg, M., Nyström, C., 2003. In vitro and in vivo evaluation of a new sublingual tablet system for rapid oromucosal absorption using fentanyl citrate as the active substance. Eur. J. Pharm. Sci. 20, 327–334.

Bruchez, M., Hotz, C., Goldman, E., Uyeda, H. T., Hayhurst, A., Mattoussi, H., 2007. Luminescent Biocompatible Quantum Dots. Quantum Dots, vol. 374. Humana Press pp. 207–227.

Buse, J., El-Aneed, A., 2010. Properties, engineering and applications of lipid-based nanoparticle drug-delivery systems: current research and advances. Nanomedicine 5, 1237–1260.

Byrne, J.D., Betancourt, T., Brannon-Peppas, L., 2008. Active targeting schemes for nanoparticle systems in cancer therapeutics. Adv. Drug Deliv. Rev. 60, 1615–1626.

Caliceti, P., Veronese, F.M., 2003. Pharmacokinetic and biodistribution properties of poly(ethylene glycol) protein conjugates. Adv. Drug Deliv. Rev. 55, 1261–1277.

Cecchelli, R., Berezowski, V., Lundquist, S., Culot, M., Renftel, M., Dehouck, M.-P., Fenart, L., 2007. Modelling of the blood-brain barrier in drug discovery and development. Nat. Rev. Drug Discov. 6, 650–661.

Chen, J., Chen, S., Zhao, X., Kuznetsova, L.V., Wong, S.S., Ojima, I., 2008. Functionalized single-walled carbon nanotubes as rationally designed vehicles for tumor-targeted drug delivery. J. Am. Chem. Soc. 130, 16778–16785.

Cheng, J., Teply, B.A., Sherifi, I., Sung, J., Luther, G., Gu, F.X., Levy-Nissenbaum, E., Radovic-Moreno, A.F., Langer, R., Farokhzad, O.C., 2007. Formulation of functionalized PLGAPEG nanoparticles for in vivo targeted drug delivery. Biomaterials 28, 869–876.

Chertok, B., David, A.E., Yang, V.C., 2011. Brain tumor targeting of magnetic nanoparticles for potential drug delivery: effect of administration route and magnetic field topography. J. Control. Release 155, 393–399.

Choi, J.J., Pernot, M., Small, S.A., Konofagou, E.E., 2007. Noninvasive, transcranial and localized opening of the blood-brain barrier using focused ultrasound in mice. Ultrasound Med. Biol. 33, 95–104.

Danhier, F., Feron, O., Préat, V.R., 2010. To exploit the tumor microenvironment: passive and active tumor targeting of nanocarriers for anti-cancer drug delivery. J. Control. Release 148, 135–146.

Delehanty, J., Mattoussi, H., Medintz, I., 2009. Delivering quantum dots into cells: strategies, progress and remaining issues. Anal. Bioanal. Chem. 393, 1091–1105.

Dobson, P.D., Kell, D.B., 2008. Carrier-mediated cellular uptake of pharmaceutical drugs: an exception or the rule? Nat. Rev. Drug Discov. 7, 205–220.

Dou, H., Grotepas, C.B., Mcmillan, J.M., Destache, C.J., Chaubal, M., Werling, J., Kipp, J., Rabinow, B., Gendelman, H.E., 2009. Macrophage delivery of nanoformulated antiretroviral drug to the brain in a murine model of neuroAIDS. J. Immunol. 183, 661–669.

Dreaden, E.C., Austin, L.A., Mackey, M.A., El-Sayed, M.A., 2012. Size matters: gold nanoparticles in targeted cancer drug delivery. Ther. Deliv. 3, 457–478.

Du, J.-Z., Du, X.-J., Mao, C.-Q., Wang, J., 2011. Tailor-made dual pH-sensitive polymer doxorubicin nanoparticles for efficient anticancer drug delivery. J. Am. Chem. Soc. 133, 17560–17563.

Dufes, C., Gaillard, F., Uchegbu, I.F., Schätzlein, A.G., Olivier, J.-C., Muller, J.-M., 2004. Glucose-targeted niosomes deliver vasoactive intestinal peptide (VIP) to the brain. Int. J. Pharm. 285, 77–85.

Farokhzad, O.C., Langer, R., 2009. Impact of nanotechnology on drug delivery. ACS Nano 3, 16–20.

Foley, C.P., Rubin, D.G., Santillan, A., Sondhi, D., Dyke, J.P., Pierre Gobin, Y., Crystal, R.G., Ballon, D.J., 2014. Intra-arterial delivery of AAV vectors to the mouse brain after mannitol mediated blood brain barrier disruption. J. Control. Release 196, 71–78.

Fraile, M., Martín, Á., Deodato, D., Rodriguez-Rojo, S., Nogueira, I.D., Simplã-Cio, A.L., Cocero, M.J., Duarte, C.M.M., 2013. Production of new hybrid systems for drug delivery by PGSS (particles from gas saturated solutions) process. J. Supercrit. Fluids 81, 226–235.

Gabathuler, R., 2010. Approaches to transport therapeutic drugs across the blood brain barrier to treat brain diseases. Neurobiol. Dis. 37, 48–57.

Gaillard, P.J., Voorwinden, L.H., Nielsen, J.L., Ivanov, A., Atsumi, R., Engman, H., Ringbom, C., De Boer, A.G., Breimer, D.D., 2001. Establishment and functional characterization of an in vitro model of the blood brain barrier, comprising a co-culture of brain capillary endothelial cells and astrocytes. Eur. J. Pharm. Sci. 12, 215–222.

Gao, X., Yang, L., Petros, J.A., Marshall, F.F., Simons, J.W., Nie, S., 2005. In vivo molecular and cellular imaging with quantum dots. Curr. Opin. Biotechnol. 16, 63–72.

Ge, C., Li, Y., Yin, J.-J., Liu, Y., Wang, L., Zhao, Y., Chen, C., 2012. The contributions of metal impurities and tube structure to the toxicity of carbon nanotube materials. NPG Asia Mater. 4, e32.

Geng, Y., Dalhaimer, P., Cai, S., Tsai, R., Tewari, M., Minko, T., Discher, D.E., 2007. Shape effects of filaments versus spherical particles in flow and drug delivery. Nat. Nanotechnol. 2, 249–255.

Ghaderi, S., Ramesh, B., Seifalian, A.M., 2011. Fluorescence nanoparticles "quantum dots" as drug delivery system and their toxicity: a review. J. Drug Target. 19, 475–486.

Gong, J., Chen, M., Zheng, Y., Wang, S., Wang, Y., 2012. Polymeric micelles drug delivery system in oncology. J. Control. Release 159, 312–323.

Griffitt, R.J., Luo, J., Gao, J., Bonzongo, J.-C., Barber, D.S., 2008. Effects of particle composition and species on toxicity of metallic nanomaterials in aquatic organisms. Environ. Toxicol. Chem. 27, 1972–1978.

Grobmyer, S.R., Moudgil, B.M., Chan, J., Valencia, P., Zhang, L., Langer, R., Farokhzad, O., 2010. Polymeric nanoparticles for drug delivery. Cancer Nanotechnol 224, 163–175.

Groothuis, D.R., 2000. The blood-brain and blood-tumor barriers: a review of strategies for increasing drug delivery. Neuro Oncol. 2, 45–59.

Groothuis, D.R., Fischer, J.M., Lapin, G., Vick, N.A., Bigner, D.D., 1982. Permeability of different experimental brain tumor models to horseradish peroxidase. J. Neuropathol. Exp. Neurol. 41, 164–185.

Guhasarkar, S., Banerjee, R., 2010. Intravesical drug delivery: challenges, current status, opportunities, and novel strategies. J. Control. Release 148, 147–159.

Hao, J., Heng, P.W.S., 2003. Buccal delivery systems. Drug Dev. Ind. Pharm. 29, 821–832.

Hearnden, V., Sankar, V., Hull, K., Juras, D.V., Greenberg, M., Kerr, A.R., Lockhart, P.B., Patton, L.L., Porter, S., Thornhill, M.H., 2012. New developments and opportunities in oral mucosal drug delivery for local and systemic disease. Adv. Drug Deliv. Rev. 64, 16–28.

Hejazi, R., Amiji, M., 2003. Chitosan-based gastrointestinal delivery systems. J. Control. Release 89, 151–165.

Heneweer, C., Gendy, S.E.M., Peñate-Medina, O., 2012. Liposomes and inorganic nanoparticles for drug delivery and cancer imaging. Ther. Deliv. 3, 645–656.

Hersam, M.C., 2008. Progress towards monodisperse single-walled carbon nanotubes. Nat. Nanotechnol. 3, 387–394.

Hillaireau, H., Couvreur, P., 2009. Nanocarriers entry into the cell: relevance to drug delivery. Cell. Mol. Life Sci. 66, 2873–2896.

Holpuch, A.S., Hummel, G.J., Tong, M., Seghi, G.A., Pei, P., Ma, P., Mumper, R.J., Mallery, S.R., 2010. Nanoparticles for local drug delivery to the oral mucosa: proof of principle studies. Pharm. Res. 27, 1224–1236.

Hua, S., 2014. Orally administered liposomal formulations for colon targeted drug delivery. Front. Pharm. 5, 138.

Huang, F., You, M., Chen, T., Zhu, G., Liang, H., Tan, W., 2014. Self-assembled hybrid nanoparticles for targeted co-delivery of two drugs into cancer cells. Chem. Commun. 50, 3103–3105.

Hughes, G.A., 2005. Nanostructure-mediated drug delivery. Nanomedicine 1, 22–30.

Hurt, R.H., Monthioux, M., Kane, A., 2006. Toxicology of carbon nanomaterials: status, trends, and perspectives on the special issue. Carbon 44, 1028–1033.

Irvine, D.J., 2011. Drug delivery: one nanoparticle, one kill. Nat. Mater. 10, 342–343.

Janib, S.M., Moses, A.S., Mackay, J.A., 2010. Imaging and drug delivery using theranostic nanoparticles. Adv. Drug Deliv. Rev. 62, 1052–1063.

Juillerat-Jeanneret, L., 2007. Critical analysis of cancer therapy using nanomaterials. Nanotechnologies for the Life Sciences. Wiley-VCH Verlag GmbH & Co. KGaA. doi: 10.1002/9783527610419.ntls0066.

Juillerat-Jeanneret, L., 2008. The targeted delivery of cancer drugs across the blood brain barrier: chemical modifications of drugs or drug-nanoparticles? Drug Discov. Today 13, 1099–1106.

Juillerat-Jeanneret, L., Schmitt, F., 2007. Chemical modification of therapeutic drugs or drug vector systems to achieve targeted therapy: looking for the grail. Med. Res. Rev. 27, 574–590.

Kalepu, S., Manthina, M., Padavala, V., 2013. Oral lipid-based drug delivery systems: an overview. Acta Pharm. Sinica B 3, 361–372.

Kaur, I.P., Bhandari, R., Bhandari, S., Kakkar, V., 2008. Potential of solid lipid nanoparticles in brain targeting. J. Control. Release 127, 97–109.

Kayal, S., Ramanujan, R.V., 2010. Doxorubicin loaded PVA coated iron oxide nanoparticles for targeted drug delivery. Mater. Sci. Eng. C 30, 484–490.

Kayat, J., Gajbhiye, V., Tekade, R.K., Jain, N.K., 2011. Pulmonary toxicity of carbon nanotubes: a systematic report. Nanomedicine 7, 40–49.

Kim, K., Kim, J.H., Park, H., Kim, Y.-S., Park, K., Nam, H., Lee, S., Park, J.H., Park, R.-W., Kim, I.-S., Choi, K., Kim, S.Y., Park, K., Kwon, I.C., 2010. Tumor-homing multifunctional nanoparticles for cancer theragnosis: simultaneous diagnosis, drug delivery, and therapeutic monitoring. J. Control. Release 146, 219–227.

Kinoshita, M., Mcdannold, N., Jolesz, F.A., Hynynen, K., 2006. Targeted delivery of antibodies through the blood brain barrier by MRI-guided focused ultrasound. Biochem. Biophys. Res. Commun. 340, 1085–1090.

Kojima, C., Suehiro, T., Watanabe, K., Ogawa, M., Fukuhara, A., Nishisaka, E., Harada, A., Kono, K., Inui, T., Magata, Y., 2013. Doxorubicin-conjugated dendrimer/collagen hybrid gels for metastasis-associated drug delivery systems. Acta Biomater. 9, 5673–5680.

Koning, G., Eggermont, A.M., Lindner, L., Ten Hagen, T.M., 2010. Hyperthermia and thermosensitive liposomes for improved delivery of chemotherapeutic drugs to solid tumors. Pharm. Res. 27, 1750–1754.

Kostarelos, K., 2008. The long and short of carbon nanotube toxicity. Nat. Biotechnol. 26, 774–776.

Kreuter, J.R., 2001. Nanoparticulate systems for brain delivery of drugs. Adv. Drug Deliv. Rev. 47, 65–81.

Kreuter, J.R., 2014. Drug delivery to the central nervous system by polymeric nanoparticles: what do we know? Adv. Drug Deliv. Rev. 71, 2–14.

Kullervo, H., Nathan, M., Natalia, V., Scott, R., Ralph, W., Ferenc, A.J., Nickolai, S., 2006. Focal disruption of the blood brain barrier due to 260-kHz ultrasound bursts: a method for molecular imaging and targeted drug delivery. J. Neurosurg. 105, 445–454.

Kumari, A., Yadav, S.K., Yadav, S.C., 2010. Biodegradable polymeric nanoparticles based drug delivery systems. Colloids Surf. B 75, 1–18.

Kwon, I.K., Lee, S.C., Han, B., Park, K., 2012. Analysis on the current status of targeted drug delivery to tumors. J. Control. Release 164, 108–114.

Lacerda, L., Raffa, S., Prato, M., Bianco, A., Kostarelos, K., 2007. Cell-penetrating CNTs for delivery of therapeutics. Nano Today 2, 38–43.

Lai, J.C.K., Lai, M.B., Jandhyam, S., Dukhande, V.V., Bhushan, A., Daniels, C.K., Leung, S.W., 2008. Exposure to titanium dioxide and other metallic oxide nanoparticles induces cytotoxicity on human neural cells and fibroblasts. Int. J. Nanomed. 3, 533–545.

Lao, J., Madani, J., Puértolas, T., Alvarez, M., Hernández, A., Pazo-Cid, R., Artal, A., Antón Torres, A., 2013. Liposomal doxorubicin in the treatment of breast cancer patients: a review. J. Drug Deliv. 2013, 12.

Lasagna-Reeves, C., Gonzalez-Romero, D., Barria, M.A., Olmedo, I., Clos, A., Sadagopa Ramanujam, V.M., Urayama, A., Vergara, L., Kogan, M.J., Soto, C., 2010. Bioaccumulation and toxicity of gold nanoparticles after repeated administration in mice. Biochem. Biophys. Res. Commun. 393, 649–655.

Lee, E.M., Singh, D., Singh, D., Choi, S.M., Zo, S.M., Park, S.J., Han, S.S., 2014. Novel alginate-gelatin hybrid nanoparticle for drug delivery and tissue engineering applications. J. Nanomater. 2014, 7.

Li, Y.-F., Chen, C., 2011. Fate and toxicity of metallic and metal-containing nanoparticles for biomedical applications. Small 7, 2965–2980.

Li, D., Yu, J., Xu, R., 2011. Mesoporous silica functionalized with an AIE luminogen for drug delivery. Chem. Commun. 47, 11077–11079.

Liang, R., Wei, M., Evans, D.G., Duan, X., 2014. Inorganic nanomaterials for bioimaging, targeted drug delivery and therapeutics. Chem. Commun. 50, 14071–14081.

Liong, M., Lu, J., Kovochich, M., Xia, T., Ruehm, S.G., Nel, A.E., Tamanoi, F., Zink, J.I., 2008. Multifunctional inorganic nanoparticles for imaging, targeting, and drug delivery. ACS Nano 2, 889–896.

Liu, J., Cui, L., Losic, D., 2013. Graphene and graphene oxide as new nanocarriers for drug delivery applications. Acta Biomater. 9, 9243–9257.

Liu, K., Zhang, J.-J., Cheng, F.-F., Zheng, T.-T., Wang, C., Zhu, J.-J., 2011b. Green and facile synthesis of highly biocompatible graphene nanosheets and its application for cellular imaging and drug delivery. J. Mater. Chem. 21, 12034–12040.

Liu, Z., Robinson, J.T., Tabakman, S.M., Yang, K., Dai, H., 2011a. Carbon materials for drug delivery and cancer therapy. Mater. Today 14, 316–323.

Liu, Z., Sun, X., Nakayama-Ratchford, N., Dai, H., 2007. Supramolecular chemistry on water-soluble carbon nanotubes for drug loading and delivery. ACS Nano 1, 50–56.

Lockman, P.R., Mumper, R.J., Khan, M.A., Allen, D.D., 2002. Nanoparticle technology for drug delivery across the blood-brain barrier. Drug Dev. Ind. Pharm. 28, 1–13.

Lovrić, J., Bazzi, H., Cuie, Y., Fortin, G.A., Winnik, F.O., Maysinger, D., 2005. Differences in subcellular distribution and toxicity of green and red emitting CdTe quantum dots. J. Mol. Med. 83, 377–385.

Luo, G., Long, J., Zhang, B., Liu, C., Ji, S., Xu, J., Yu, X., Ni, Q., 2012. Quantum dots in cancer therapy. Expert. Opin. Drug Deliv. 9, 47–58.

Ma, P.A., Xiao, H., Li, C., Dai, Y., Cheng, Z., Hou, Z., Lin, J., 2015. Inorganic nanocarriers for platinum drug delivery. Mater. Today 18, 554–564.

Macleod, G.S., Fell, J.T., Collett, J.H., Sharma, H.L., Smith, A.-M., 1999. Selective drug delivery to the colon using pectin:chitosan:hydroxypropyl methylcellulose film coated tablets. Int. J. Pharm. 187, 251–257.

Magrez, A., Kasas, S., Salicio, V.R., Pasquier, N., Seo, J.W., Celio, M., Catsicas, S., Schwaller, B., Forró, L.S., 2006. Cellular toxicity of carbon-based nanomaterials. Nano Lett. 6, 1121–1125.

Malam, Y., Loizidou, M., Seifalian, A.M., 2009. Liposomes and nanoparticles: nanosized vehicles for drug delivery in cancer. Trends Pharmacol. Sci. 30, 592–599.

Martin, L., Wilson, C.G., Koosha, F., Uchegbu, I.F., 2003. Sustained buccal delivery of the hydrophobic drug denbufylline using physically cross-linked palmitoyl glycol chitosan hydrogels. Eur. J. Pharm. Biopharm. 55, 35–45.

Masserini, M., 2013. Nanoparticles for brain drug delivery. ISRN Biochem. 2013, 18.

McCann, U.D., Szabo, Z., Scheffel, U., Dannals, R.F., Ricaurte, G.A., 1998. Positron emission tomographic evidence of toxic effect of MDMA ("Ecstasy") on brain serotonin neurons in human beings. Lancet 352, 1433–1437.

Mcrorie, J., Pepple, S., Rudolph, C., 1998. Effects of fiber laxatives and calcium docusate on regional water content and viscosity of digesta in the large intestine of the pig. Digest. Dis. Sci. 43, 738–745.
Meyer, J.N., Lord, C.A., Yang, X.Y., Turner, E.A., Badireddy, A.R., Marinakos, S.M., Chilkoti, A., Wiesner, M.R., Auffan, M., 2010. Intracellular uptake and associated toxicity of silver nanoparticles in Caenorhabditis elegans. Aquat. Toxicol. 100, 140–150.
Modak, A., Barui, A.K., Ranjan Patra, C., Bhaumik, A., 2013. A luminescent nanoporous hybrid material based drug delivery system showing excellent theranostics potential for cancer. Chem. Commun. 49, 7644–7646.
Monteiro-Riviere, N.A., Inman, A.O., 2006. Challenges for assessing carbon nanomaterial toxicity to the skin. Carbon 44, 1070–1078.
Moreels, I., Lambert, K., Smeets, D., De Muynck, D., Nollet, T., Martins, J.C., Vanhaecke, F., Vantomme, A., Delerue, C., Allan, G., Hens, Z., 2009. Size-dependent optical properties of colloidal PbS quantum dots. ACS Nano 3, 3023–3030.
Muheem, A., Shakeel, F., Jahangir, M.A., Anwar, M., Mallick, N., Jain, G.K., Warsi, M.H., Ahmad, F.J., 2016. A review on the strategies for oral delivery of proteins and peptides and their clinical perspectives. Saudi Pharm. J. 24 (4), 413–428.
Mukherjee, S., Ray, S., Thakur, R.S., 2009. Solid lipid nanoparticles: a modern formulation approach in drug delivery system. Ind. J. Pharm. Sci. 71, 349–358.
Muller, R.H., Keck, C.M., 2004. Challenges and solutions for the delivery of biotech drugs: a review of drug nanocrystal technology and lipid nanoparticles. J. Biotechnol. 113, 151–170.
Muller, J., Huaux, F.O., Moreau, N., Misson, P., Heilier, J.-F.O., Delos, M., Arras, M., Fonseca, A., Nagy, J.B., Lison, D., 2005. Respiratory toxicity of multi-wall carbon nanotubes. Toxicol. Appl. Pharmacol. 207, 221–231.
Murthy, S.K., 2007. Nanoparticles in modern medicine: state of the art and future challenges. Int. J. Nanomed. 2, 129–141.
Muthu, M.S., Singh, S., 2008. Targeted nanomedicines: effective treatment modalities for cancer. Nanomedicine 4, 105–118.
Nasongkla, N., Bey, E., Ren, J., Ai, H., Khemtong, C., Guthi, J.S., Chin, S.-F., Sherry, A.D., Boothman, D.A., Gao, J., 2006. Multifunctional polymeric micelles as cancer-targeted, MRI-ultrasensitive drug delivery systems. Nano Lett. 6, 2427–2430.
Neal, F.K., Gregory, H., Nathan, S., Phillips, C.D., Wayne, S.C., 1992. Treatment of cerebral vasospasm with intra-arterial papaverine. J. Neurosurg. 77, 848–852.
Nel, A., Xia, T., Mädler, L., Li, N., 2006. Toxic potential of materials at the nanolevel. Science 311, 622–627.
Newton, H.B., 2006. Advances in strategies to improve drug delivery to brain tumors. Exp. Rev. Neurother. 6, 1495–1509.
Nigam, P., Waghmode, S., Louis, M., Wangnoo, S., Chavan, P., Sarkar, D., 2014. Graphene quantum dots conjugated albumin nanoparticles for targeted drug delivery and imaging of pancreatic cancer. J. Mater. Chem. B 2, 3190–3195.
Novoselov, K.S., Falko, V.I., Colombo, L., Gellert, P.R., Schwab, M.G., Kim, K.A., 2012. Roadmap for graphene. Nature 490, 192–200.
Oberdörster, E., 2004. Manufactured nanomaterials (fullerenes, C(60)) induce oxidative stress in the brain of juvenile largemouth bass. Environ. Health Perspect. 112, 1058–1062.
Owens, III, D.E., Peppas, N.A., 2006. Opsonization, biodistribution, and pharmacokinetics of polymeric nanoparticles. Int. J. Pharm. 307, 93–102.
Pan, Y., Bao, H., Sahoo, N.G., Wu, T., Li, L., 2011. Water-soluble poly(*N*-isopropylacrylamide)–graphene sheets synthesized via click chemistry for drug delivery. Adv. Funct. Mater. 21, 2754–2763.
Pardridge, W.M., 2012. Drug transport across the blood-brain barrier. J. Cereb. Blood Flow Metab. 32, 1959–1972.
Park, S., Lee, Y.K., Jung, M., Kim, K.H., Chung, N., Ahn, E.-K., Lim, Y., Lee, K.-H., 2007. Cellular toxicity of various inhalable metal nanoparticles on human alveolar epithelial cells. Inhal. Toxicol. 19, 59–65.
Patel, M.M., 2015. Colon: a gateway for chronotherapeutic drug delivery systems. Expert Opin. Drug Deliv. 12, 1389–1395.

Patel, M., Goyal, B., Bhadada, S., Bhatt, J., Amin, A., 2009. Getting into the Brain. CNS Drugs 23, 35–58.

Patel, T., Zhou, J., Piepmeier, J.M., Saltzman, W.M., 2012. Polymeric nanoparticles for drug delivery to the central nervous system. Adv. Drug Deliv. Rev. 64, 701–705.

Patel, V.F., Liu, F., Brown, M.B., 2011. Advances in oral transmucosal drug delivery. J. Control. Release 153, 106–116.

Pattni, B.S., Chupin, V.V., Torchilin, V.P., 2015. New developments in liposomal drug delivery. Chem. Rev. 115 (19), 10938–10966.

Peer, D., Karp, J.M., Hong, S., Farokhzad, O.C., Margalit, R., Langer, R., 2007. Nanocarriers as an emerging platform for cancer therapy. Nat. Nanotechnol. 2, 751–760.

Pendekal, M.S., Tegginamat, P.K., 2013. Hybrid drug delivery system for oropharyngeal, cervical and colorectal cancer: in vitro and in vivo evaluation. Saudi Pharm. J. 21, 177–186.

Philip, A.K., Philip, B., 2010. Colon targeted drug delivery systems: a review on primary and novel approaches. Oman Med. J. 25, 79–87.

Pool, H., Quintanar, D., Figueroa, J., Bechara, J.E.H., Mcclements, D., Mendoza, S., 2012. Polymeric nanoparticles as oral delivery systems for encapsulation and release of polyphenolic compounds: impact on quercetin antioxidant activity & bioaccessibility. Food Biophys. 7, 276–288.

Probst, C.E., Zrazhevskiy, P., Bagalkot, V., Gao, X., 2013. Quantum dots as a platform for nanoparticle drug delivery vehicle design. Adv. Drug Deliv. Rev. 65, 703–718.

Provenzale, J.M., Mukundan, S., Dewhirst, M., 2005. The role of blood-brain barrier permeability in brain tumor imaging and therapeutics. Am. J. Roentgenol. 185, 763–767.

Qi, L., Gao, X., 2008. Emerging application of quantum dots for drug delivery and therapy. Expert Opin. Drug Deliv. 5, 263–267.

Rabinow, B.E., 2004. Nanosuspensions in drug delivery. Nat. Rev. Drug Discov. 3, 785–796.

Rai, M., Yadav, A., Gade, A., 2009. Silver nanoparticles as a new generation of antimicrobials. Biotechnol. Adv. 27, 76–83.

Rapoport, N., 2007. Physical stimuli-responsive polymeric micelles for anti-cancer drug delivery. Prog. Polym. Sci. 32, 962–990.

Rijcken, C.J.F., Soga, O., Hennink, W.E., Nostrum, C.F.V., 2007. Triggered destabilisation of polymeric micelles and vesicles by changing polymers polarity: an attractive tool for drug delivery. J. Control. Release 120, 131–148.

Ron, H., 2006. A toxicologic review of quantum dots: toxicity depends on physicochemical and environmental factors. Envrion. Health Perspect. 114, 165–172.

Rosenthal, S.J., Chang, J.C., Kovtun, O., Mcbride, J.R., Tomlinson, I.D., 2011. Biocompatible quantum dots for biological applications. Chem. Biol. 18, 10–24.

Rossi, S., Sandri, G., Caramella, C.M., 2005. Buccal drug delivery: a challenge already won? Drug Discov. Today 2, 59–65.

Ryman-Rasmussen, J.P., Riviere, J.E., Monteiro-Riviere, N.A., 2007. Variables influencing interactions of untargeted quantum dot nanoparticles with skin cells and identification of biochemical modulators. Nano Lett. 7, 1344–1348.

Salvia-Trujillo, L., Qian, C., Martín-Belloso, O., McClements, D.J., 2013. Influence of particle size on lipid digestion and β-carotene bioaccessibility in emulsions and nanoemulsions. Food Chem. 141, 1472–1480.

Samad, A., Sultana, Y., Aqil, M., 2007. Liposomal drug delivery systems: an update review. Curr. Drug Deliv. 4, 297–305.

Sankar, V., Hearnden, V., Hull, K., Juras, D.V., Greenberg, M.S., Kerr, A.R., Lockhart, P.B., Patton, L.L., Porter, S., Thornhill, M., 2011. Local drug delivery for oral mucosal diseases: challenges and opportunities. Oral Dis. 17, 73–84.

Sao, R., Vaish, R., Sinha, N., 2015. Multifunctional drug delivery systems using inorganic nanomaterials: a review. J. Nanosci. Nanotechnol. 15, 1960–1972.

Schlageter, K.E., Molnar, P., Lapin, G.D., Groothuis, D.R., 1999. Microvessel organization and structure in experimental brain tumors: microvessel populations with distinctive structural and functional properties. Microvasc. Res. 58, 312–328.

enel, S., Rathbone, M.J., Cansiz, M., Pather, I., 2012. Recent developments in buccal and sublingual delivery systems. Expert Opin. Drug Deliv. 9, 615–628.

Shaji, J., Patole, V., 2008. Protein and peptide drug delivery: oral approaches. Indian J. Pharm. Sci. 70, 269–277.

Sharma, G., Anabousi, S., Ehrhardt, C., Ravi Kumar, M.N.V., 2006. Liposomes as targeted drug delivery systems in the treatment of breast cancer. J. Drug Target. 14, 301–310.

Shi, J., Votruba, A.R., Farokhzad, O.C., Langer, R., 2010. Nanotechnology in drug delivery and tissue engineering: from discovery to applications. Nano Lett. 10, 3223–3230.

Shiohara, A., Hoshino, A., Hanaki, K.-I., Suzuki, K., Yamamoto, K., 2004. On the cyto-toxicity caused by quantum dots. Microbiol. Immunol. 48, 669–675.

Shum, P., Kim, J.-M., Thompson, D.H., 2001. Phototriggering of liposomal drug delivery systems. Adv. Drug Deliv. Rev. 53, 273–284.

Singh, S., Nalwa, H.S., 2007. Nanotechnology and health safety: toxicity and risk assessments of nanostructured materials on human health. J. Nanosci. Nanotechnol. 7, 3048–3070.

Singh, R.K., Patel, K.D., Kim, J.-J., Kim, T.-H., Kim, J.-H., Shin, U.S., Lee, E.-J., Knowles, J.C., Kim, H.-W., 2014. Multifunctional hybrid nanocarrier: magnetic CNTs ensheathed with mesoporous silica for drug delivery and imaging system. ACS Appl. Mater. Interfaces 6, 2201–2208.

Sinha, V.R., Singh, A., Kumar, R.V., Singh, S., Kumria, R., Bhinge, J.R., 2007. Oral colon-specific drug delivery of protein and peptide drugs. Crit. Rev. Ther. Drug Carrier Syst. 24, 63–92.

Smith, A.M., Duan, H., Mohs, A.M., Nie, S., 2008. Bioconjugated quantum dots for in vivo molecular and cellular imaging. Adv. Drug Deliv. Rev. 60, 1226–1240.

Sugano, K., Kansy, M., Artursson, P., Avdeef, A., Bendels, S., Di, L., Ecker, G.F., Faller, B., Fischer, H., Gerebtzoff, G.G., Lennernaes, H., Senner, F., 2010. Coexistence of passive and carrier-mediated processes in drug transport. Nat. Rev. Drug Discov. 9, 597–614.

Sutton, D., Nasongkla, N., Blanco, E., Gao, J., 2007. Functionalized micellar systems for cancer targeted drug delivery. Pharm. Res. 24, 1029–1046.

Taylor, M.D., 1996. Improved passive oral drug delivery via prodrugs. Adv. Drug Deliv. Rev. 19, 131–148.

Thambi, T., Deepagan, V.G., Yoon, H.Y., Han, H.S., Kim, S.-H., Son, S., Jo, D.-G., Ahn, C.-H., Suh, Y.D., Kim, K., Chan Kwon, I., Lee, D.S., Park, J.H., 2014. Hypoxia-responsive polymeric nanoparticles for tumor-targeted drug delivery. Biomaterials 35, 1735–1743.

Tiwari, S.B., Amiji, M.M., 2006. A review of nanocarrier-based CNS delivery systems. Curr. Drug Deliv. 3, 219–232.

Torchilin, V., 2011. Tumor delivery of macromolecular drugs based on the EPR effect. Adv. Drug Deliv. Rev. 63, 131–135.

Trickler, W.J., Lantz, S.M., Murdock, R.C., Schrand, A.M., Robinson, B.L., Newport, G.D., Schlager, J.J., Oldenburg, S.J., Paule, M.G., Slikker, W., Hussain, S.M., Ali, S.F., 2010. Silver nanoparticle induced blood-brain barrier inflammation and increased permeability in primary rat brain microvessel endothelial cells. Toxicol. Sci. 118, 160–170.

Trubetskoy, V.S., Torchilin, V.P., 1995. Use of polyoxyethylene-lipid conjugates as long-circulating carriers for delivery of therapeutic and diagnostic agents. Adv. Drug Deliv. Rev. 16, 311–320.

Uhrich, K.E., Cannizzaro, S.M., Langer, R.S., Shakesheff, K.M., 1999. Polymeric systems for controlled drug release. Chem. Rev. 99, 3181–3198.

Veiseh, O., Gunn, J.W., Zhang, M., 2010. Design and fabrication of magnetic nanoparticles for targeted drug delivery and imaging. Adv. Drug Deliv. Rev. 62, 284–304.

Wang, Y., Chen, L., 2011. Quantum dots, lighting up the research and development of nanomedicine. Nanomedicine 7, 385–402.

Wang, P.P., Frazier, J., Brem, H., 2002. Local drug delivery to the brain. Adv. Drug Deliv. Rev. 54, 987–1013.
Wang, A.Z., Langer, R., Farokhzad, O.C., 2012. Nanoparticle delivery of cancer drugs. Annu. Rev. Med. 63, 185–198.
Wang, Y., Li, Z., Wang, J., Li, J., Lin, Y., 2011. Graphene and graphene oxide: biofunctionalization and applications in biotechnology. Trends Biotechnol. 29, 205–212.
Wirtz, S., Galle, P.R., Neurath, M.F., 1999. Efficient gene delivery to the inflamed colon by local administration of recombinant adenoviruses with normal or modified fibre structure. Gut 44, 800–807.
Xiao, G., Gan, L.-S., 2013. Receptor-mediated endocytosis and brain delivery of therapeutic biologics. Int. J. Cell Biol. 2013, 14.
Yang, L., Chu, J.S., Fix, J.A., 2002. Colon-specific drug delivery: new approaches and in vitro/in vivo evaluation. Int. J. Pharm. 235, 1–15.
Yang, X., Zhang, X., Liu, Z., Ma, Y., Huang, Y., Chen, Y., 2008. High-efficiency loading and controlled release of doxorubicin hydrochloride on graphene oxide. J. Phys. Chem. C 112, 17554–17558.
Yang, Z., Liu, Z.W., Allaker, R.P., Reip, P., Oxford, J., Ahmad, Z., Ren, G., 2010. A review of nanoparticle functionality and toxicity on the central nervous system. J. R. Soc. Interface.
Yao, M., Xiao, H., Mcclements, D.J., 2014. Delivery of lipophilic bioactives: assembly, disassembly, and reassembly of lipid nanoparticles. Annu. Rev. Food Sci. Technol. 5, 53–81.
Zentner, G.M., Rathi, R., Shih, C., Mcrea, J.C., Seo, M.-H., Oh, H., Rhee, B.G., Mestecky, J., Moldoveanu, Z., Morgan, M., Weitman, S., 2001. Biodegradable block copolymers for delivery of proteins and water-insoluble drugs. J. Control. Release 72, 203–215.
Zhang, J.Q., Zhang, Z.R., Yang, H., Tan, Q.Y., Qin, S.R., Qiu, X.L., 2005. Lyophilized paclitaxel magnetoliposomes as a potential drug delivery system for breast carcinoma via parenteral administration: in vitro and in vivo studies. Pharm. Res. 22, 573–583.
Zhang, L., Xia, J., Zhao, Q., Liu, L., Zhang, Z., 2010. Functional graphene oxide as a nanocarrier for controlled loading and targeted delivery of mixed anticancer drugs. Small 6, 537–544.
Zhang, P., Cheetham, A.G., Lock, L.L., Cui, H., 2013. Cellular uptake and cytotoxicity of drug: peptide conjugates regulated by conjugation site. Bioconjug. Chem. 24, 604–613.
Zhao, Y., Xing, G., Chai, Z., 2008. Nanotoxicology: are carbon nanotubes safe? Nat. Nanotechnol. 3, 191–192.
Zrazhevskiy, P., Sena, M., Gao, X., 2010. Designing multifunctional quantum dots for bioimaging, detection, and drug delivery. Chem. Soc. Rev. 39, 4326–4354.

FURTHER READING

Liu, Y., Zhao, Y., Sun, B., Chen, C., 2013. Understanding the toxicity of carbon nanotubes. Acc. Chem. Res. 46, 702–713.

CHAPTER

28

STIMULI-RESPONSIVE LIPOSOME AND CONTROL RELEASE DRUG

Xueqin An*, Rijun Gui*,**

**East China University of Science and Technology, Shanghai, China*

***Qingdao University, Qingdao, China*

CHAPTER OUTLINE

1 INTRODUCTION

Malignant tumor has become a serious threat to human health, which has been considered to be one of the deadliest diseases. Cancer is the second leading cause of death in the USA with projections estimating approximately 1,700,000 new cases occurring and 1600 deaths per day (Siegel et al., 2014). Despite advances in treatment, diagnosis, and prevention options, cancer mortality rates, however, have only dropped 1.6% (DeSantis et al., 2014). These statistics indicate that there is a pressing need to develop novel treatments and approaches that will enhance the survival rates (James, 2015). In most cases, cancer is considered as a disease very different to be cured, and the cure rate is very low. Consistently, the treatment of cancer is always one of the most important research fields for biomedical researchers. Currently, the treatment methods of cancer mainly refer to surgery, chemotherapy, immunotherapy discharge, radiation therapy, and so on (Lubbe et al., 2001).

The therapeutic window for a drug treatment regimen is defined by the drug doses that achieve the highest therapeutic efficacy while keeping the side effects manageable and tolerable. Chemotherapy is the main treatment due to the influence of various factors, such as tumor location, metastasis. After the drugs enter the body, they are distributed to the body's tissues because the drugs are short of affinity to pathological parts. Chemotherapeutics, although successful in reducing deaths directly related to cancer, suffer from poor pharmacokinetic profiles as well as nonspecific mechanisms of action, which in turn leads to a tremendous amount of systemic toxicity for the patient (Szakacs et al., 2006). The nonspecific mechanism of action causes poor biodistribution that reduces the fraction of the drug administered that is actually delivered to the diseased cells of interest, which results in the need for higher and continuous dosages to provide effective treatment. These drugs target and cause damage not only to cancer cells but any highly proliferative cell, including hair follicles, stomach lining, and bone marrow (Sahoo and Labhasetwar, 2003). Recent developments in medical technologies and new insights into cancer biology have advanced cancer care considerably with improved cure rates and extended life expectancies. Yet, the challenge in chemotherapy to achieve sufficient drug levels in the tumor, while reducing off-site drug uptake, remains largely unresolved. Typically, chemotherapeutic drugs are systemically administered and only a small percentage of the drug reaches the cancerous lesion, while the majority of the drug is cleared or taken up in healthy organs causing undesired side effects (Hijnen et al., 2014). Consequently, research has been devoted to drug delivery approaches using larger drug conjugates or drug-loaded nanoparticles that afford higher drug concentrations in the tumor while decreasing off-site toxicity (Allen and Cullis, 2004; Ferrari, 2005; Parveen et al., 2012; Sanhai et al., 2008; Wagner et al., 2006). For instance, chemotherapeutic drugs have been encapsulated in the lumen of liposomes, which prevents the rapid extravasation of the small parental drug (size ca. 1 nm) into healthy tissue (Abraham et al., 2005; Chang and Yeh, 2012; Working et al., 1994). Upon injection, these long-circulating liposomal drug carriers are taken up in the tumor mediated by the enhanced permeability and retention (EPR) effect, which is a characteristic of angiogenic blood vessels in neoplasms (Maeda, 2010, 2012; Matsumura and Maeda, 1986; Wu et al., 1993).

Nanotechnology with its sophistication and advanced techniques has provided us a new tool to understand the scientific developments at a nanoscale (Cheng et al., 2012) and treat dreadful diseases with a greater ease using nanoparticles as drug delivery systems. Nanoscale materials have been developed for a wide range of purposes related to drug delivery, cancer treatment, and diagnosis. These materials present numerous opportunities to improve the current state of cancer diagnosis and treatment available to patients (Cabral et al., 2014). With the rapid development of medical technology and nanometer technology, a newly intelligent nanoparticulate pharmaceutical drug delivery systems (NDDS) to enhance the in vivo effectiveness of drugs is now well established and are presented as modern pharmacy (Lehner et al., 2012; Singh et al., 2012; Ta and Porter, 2013; Torchilin, 2000). It can be entrapped and save drug activity, and also can transport drugs into the site of action. The NDDS is a smart system that could change the way drugs in the body, the distribution of drugs in vivo, and the drug release rate may be controlled by adjusting stimuli response of the delivery.

The biocompatible liposomes in nanoscale have been used widely for drug delivery, and there are several liposome-mediated drug delivery products approved for clinical trials. It is considered to be a suitable dosage of anticancer drugs (Torchilin, 2000). The liposomes NDDS systems have some favorable conditions over conventional drug carriers, such as improving the drug targeting, prolonging time of drug action, increasing drug efficacy, reducing toxicity. After the treatment of surface modification, ideal liposome drug carriers show the capacity of selectively distinguishing targeted cells, the stability in biological media, as well as excellent biocompatibility (Torchilin, 2001). The NDDS function is to control the distribution of the drug in vivo, control release, and adjustment of the interaction between drug molecules and receptors. The NDDS can also respond to local stimuli that are characteristic of the pathological site by releasing an entrapped drug or shedding a protective coating, thus facilitating the interaction between drug-loaded nanocarriers and target cells or tissues. Once the interplay between all of the components has been optimized, the field of theranostics holds the potential to drive the field of nanomedicine into the area of "personalized medicine" (Ryu et al., 2014). The concept of personalized medicine is to attempt to tailor medicinal regiments and treatments to the specific characteristics of each individual patient (Kalia et al., 2013) based on the hypothesis that no single individual undergoes the same response to any therapeutic agent. This therapy may aid in reducing the undesirable side effects of treatment and potentially producing better overall results. The development of multifunctional and stimulus-sensitive NDDSs is an active area of current research. In addition, imaging contrast moieties can be attached to these carriers to track their real-time biodistribution and accumulation in target cells or tissues. Currently, there are several clinically approved formulations that are comprised of nanoparticle drug delivery systems that have been shown to reduce the toxic side effects associated with traditional chemotherapy regimens (Cheng et al., 2012). These therapeutic-loaded nanoparticles have displayed limited long-term success clinically, which has led to the need to attempt to develop new strategies to improve the efficacy of nanoscale treatments through the addition of functional elements.

The goal of this chapter is to focus on some points that play a key role in the development of stimuli-responsive liposomes. We briefly analyze how the efficacy of stimuli-responsive liposomes depends on the physicochemical properties of their membranes, and on their size, surface charge, and lipid organization. In addition, we discuss how the physicochemical properties of liposomes influence drug release, the relationship between the microstructure and the function.

2 LIPOSOMES

In 1965, Bangham first showed that phospholipids dispersed in water spontaneously formed closed bilayer structures (Bangham, 1968; Bangham et al., 1965) and introduced the term "liposomes" to describe these systems, which were initially used as models of biological membranes (Bangham et al., 1965; Deamer, 2010). Liposomes are synthetic spherical vesicles and are concentric closed bilayer membranes of water-insoluble polar lipids (Fig. 28.1), which are an important structural component of the cell membrane (Bozzuto and Molinari, 2015). Liposomes consist of single or multiple concentric lipid bilayers encapsulating an aqueous compartment (Bangham and Horne, 1964; Bangham et al., 1974; Bozzuto and Molinari, 2015).

The size of these nearly spherical lipid vesicles can range from a few nanometers to several micrometers. However, liposomes applied to medical use range between 50 and 450 nm (Etheridge et al., 2013). The vesicle size is a vital parameter in determining the circulation half-life of liposomes, and both size and number of bilayers affect the amount of drug encapsulation in the liposomes. On the basis of their size and number of bilayers, liposomes can be classified into two categories: multilamellar vesicles (MLV) and unilamellar vesicles. Unilamellar vesicles can also be classified into two categories: large unilamellar vesicles (LUV) and small unilamellar vesicles (SUV). The unilamellar liposome has a single phospholipid bilayer sphere enclosing the aqueous solution. The multilamellar liposomes

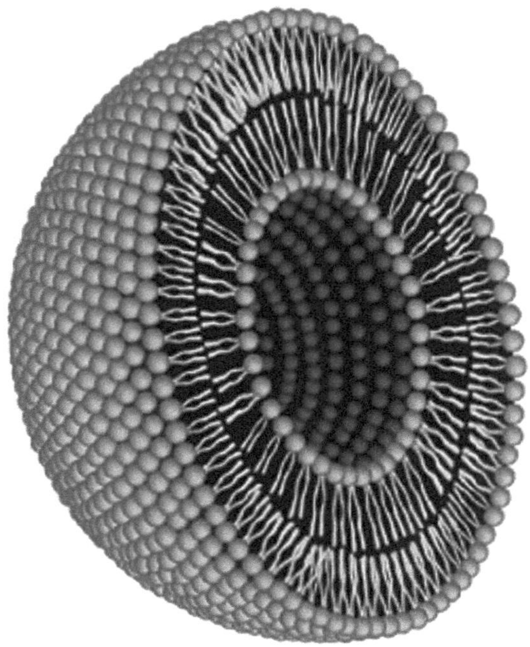

FIGURE 28.1 Representation of the Steric Organization of a Liposome (Bozzuto and Molinari, 2015)

have an onion structure. Several unilamellar vesicles will form on the inside of the other with smaller size, making a multilamellar structure of concentric phospholipid spheres separated by layers of water.

The biomimetic liposomes are a useful tool for drug delivery as a container for the storage, transfer, and controllable release of agents. The liposomes have the capability of entrapping both lipophilic and hydrophilic agents, in the lipid membrane and in the aqueous core, respectively. Liposomes have been recognized as drug nanocarriers for decades, but their clinical applications are often limited by uncontrolled release or poor availability of the encapsulated drug and stability in vivo in humans. A major challenge for drug delivery is controlled release both spatially and temporally. For this reason, many of the current methods to enhance temp oral or spatial control of drug release focus on incorporating components to achieve thermal, pH, photothermal, magnetic, or enzymatic triggered release.

2.1 METHODS FOR THE PREPARATION OF LIPOSOMES

Liposomes can be classified on the basis of the preparation method, size, and lamellarity. The formation of unilamellar liposomes (SUV) or multilamellar liposomes (MLVs) depends on the synthesis methods and postprocessing. There are many different methods for the preparation of liposomes (Bozzuto and Molinari, 2015). To ensure the desired size, lamellarity, and homogeneity properties of liposomes, postformation processing is required. The most common methods for postformation processing are sonication, extrusion, and high-pressure homogenization. Sonication is used to reduce the size of the vesicles and give energy to lipid suspension.

2.1.1 Bangham method

One of the most widely used techniques for liposome preparation is the thin-film hydration or Bangham method (Bangham et al., 1965; New, 1990). This method involves dissolution of the lipid in an organic solvent, evaporation of the solvent, and the dispersion of the obtained lipid film in aqueous media. The hydrophilic and lipophilic drugs can be entrapped in aqueous pool and in lipid film, respectively. The encapsulation efficiency of lipophilic drugs is low (5%~15%), and the encapsulation efficiency of lipophilic drugs is as high as 98%.

2.1.2 Reverse-phase evaporation and solvent-injection methods

The reverse-phase evaporation and solvent-injection methods provide hydration of the lipids directly from an organic solvent, and obtain an aqueous suspension of MLVs and ULVs, respectively (Deamer and Bangham, 1976; Szoka, 1978). These methods are affected by the solubility of lipids in organic solvent. These procedures guarantee higher encapsulation efficiency than thin-film hydration.

2.1.3 Detergent-depletion method

The detergent-depletion method involves the hydration of a lipid film with a detergent solution and leads to the formation of large MLVs (Brunner et al., 1976; Lasch et al., 2003; New, 1990). This method is used rarely because of long preparation time, poor encapsulation efficiency, and complicated process.

2.1.4 Sonication

Sonication is the most extensively used method for the preparation of SUV (Woodbury et al., 2006). MLVs are sonicated either with a bath-type sonicator or a probe sonicator under a passive atmosphere. The main disadvantages of this method are very low internal volume/encapsulation efficacy, degradation of phospholipids, metal pollution from probe tip, and presence of MLV along with SUV. There are two sonication techniques:

1. Probe sonication. The tip of a sonicator is engrossed directly into the liposome dispersion. The energy input into lipid dispersion is very high in this method. The coupling of energy at the tip results in local hotness, and the vessel must be engrossed into a water/ice bath. The disadvantage of this approach is that more than 5% of the lipids can be deesterified with sonication of 1 h, and titanium will slough off and pollute the solution.
2. Bath sonication. The liposome dispersion in a cylinder is placed into a bath sonicator. Usually controlling the temperature of the lipid dispersion is easier in this method, in contrast to sonication by dispersal directly using the tip. The material being sonicated can be protected in a sterile vessel, dissimilar the probe units, or under an inert atmosphere.

2.1.5 French pressure cell: extrusion

French pressure cell involves the extrusion of MLV through a small orifice. An important feature of the method is that the proteins do not seem to be significantly pretentious during the procedure as they are in sonication. The method has several advantages over sonication method and involves gentle handling of unstable materials. The drawbacks of the method are that the high temperature is difficult to attain, the working volumes are comparatively small (about 50 mL as the maximum), and the size of resulting liposomes are larger than that of other preparations.

2.1.6 Solvent vaporization

A solution of lipids dissolved in diethyl ether or ether-methanol mixture is injected gradually to an aqueous solution of the material to be encapsulated under reduced pressure. Then removal of ether leads to the creation of liposomes under vacuum. The main disadvantages of the method are that the population is heterogeneous (70–200 nm) and the exposure of compounds to be encapsulated to organic solvents at high temperature.

2.1.7 Ethanol injection

A lipid ethanol solution is rapidly injected to a huge excess of buffer. The liposomes are at once formed. The following are some of the disadvantages of the method: (1) the population is heterogeneous (30–110 nm); (2) liposomes are very dilute; and (3) the removal all ethanol is difficult because it forms into azeotrope with water.

2.1.8 Reverse phase evaporation method

Reverse-phase evaporation is based on creation of inverted micelles. The inverted micelles are shaped upon sonication of a mixture of a buffered aqueous phase, which contains the water-soluble molecules to be encapsulated into the liposomes and an organic phase in which the amphiphilic molecules are solubilized. The slow elimination of the organic solvent leads to the conversion of these inverted micelles into viscous state and gel form. The gel state collapses, and some of the inverted micelles were disturbed at a critical point in this process. The excess of phospholipids in the environment donates to formation of a complete bilayer around the residual micelles, which results in the creation of liposomes. Liposomes prepared by reverse phase evaporation method can be made from numerous lipid formulations. The ratios of water to lipid in the liposomes are 4 times higher than hand-shaken liposomes or multilamellar liposomes.

Briefly, first, mixture of phospholipids in organic solvent and aqueous buffer is prepared, and water-in-oil emulsion is formed by brief sonication of the mixture. Then, the organic solvents are detached under reduced pressure, resulting in the creation of a viscous gel. The liposomes are shaped when residual solvent is detached during continued rotary evaporation under reduced pressure. High encapsulation efficiency up to 65% can be obtained in a medium of low ionic strength, for example, 0.01 M NaCl by using this technique. The method has been used to encapsulate small-, large-, and macromolecules. The main drawback of the method is the contact of the materials to be encapsulated to organic solvents and to brief periods of sonication. These conditions may possibly result in the breakage of DNA strands or the denaturation of some proteins.

This method provided a progress in liposome preparation technology, since it allowed for the first time the preparation of liposomes with high aqueous space-to-lipid ratio and capability to entrap a large percentage of the aqueous material.

2.1.9 Selection of preparation methods

The choice of appropriate method depends on several factors: (1) the physicochemical characteristics of the liposome; (2) the toxicity and the concentration of the loaded substance; (3) the type of the medium in which the liposomes are dispersed; (4) the additional processes during the application/delivery of the liposomes; (5) the size and the half-life desired for the successful application; and (6) the costs, reproducibility, and applicability regarding large-scale production for clinical purpose and good manufacturing practice-relevant issues (Gómez-Hens and Fernández-Romero, 2006; Mozafari et al., 2008).

2.2 THE MORPHOLOGY, SIZE, AND MICROSTRUCTURE

The liposomes as a carrier system for drugs strictly depends on the physicochemical properties of their membranes, on the nature of their components, on their morphology and size, surface charge, and the microstructure (Bozzuto and Molinari, 2015; Euliss et al., 2006).

2.2.1 Compositions and structures

Liposomes are mainly composed of phospholipids, amphiphilic molecules that have a hydrophilic head and two apolar hydrophobic chains. When phospholipids are dispersed in aqueous solutions, due to their amphipathic nature they have a strong tendency to form membranes (Papahadjopoulos and Kimelberg, 1973). On the one hand, their polar heads prefer to interact with the aqueous environment; on the other, their long apolar aliphatic chains promote interaction with one another. These dual properties favor the formation of two lipid layers in aqueous solution. The hydrophobic chains of each layer face each other and constitute a lipophilic inner compartment that acts as a permeability barrier, both inward and outward. Hydrophobic interactions are behind the formation of these lipid bilayers, and van der Waals forces keep the long hydrocarbon tails together, thus strengthening this architecture.

2.2.2 Size and distribution

The minimal size of liposomes is about 30 nm, which is determined by the difficulty of lipid packing inside a strongly curved geometry. To obtain liposomes with such small sizes, substantial energy transfer to the membranes is required, for example, by extrusion or sonication. The largest liposomes can measure up to several microns, which display opalescence. Especially the light scattering can be an accurate measure of particle size specifically in the 50~200 nm region where scattering is strongly dependent on the liposome size. Considerable efforts have been made to produce small (diameter <100 nm) unilamellar liposomes (Huang, 1969; Schullery et al., 1980). The extrusion technique is now the preferred procedure for making LNP in the 100 nm size range, which uses high pressure to force preformed large multilamellar liposomes through polycarbonate filters of defined pore size (Fenske and Cullis, 2008; Szoka et al., 1980). LNPs can also be formed by diluting organic solutions of lipids with aqueous media under rapid mixing procedures (Batzri and Korn, 1973a,b). Microfluidic mixing techniques also can be used to generate liposomal LNPs using the staggered herringbone mixer (Stroock et al., 2002). "Limit size" LNPs can be formed with diameters of 20–100 nm depending on the lipid composition (Zhigaltsev et al., 2012).

2.2.3 Microstructure and property of bilayer in liposomes

Liposomes are artificially prepared spherically shaped objects consisting of a lipid bilayer surrounding an aqueous core. The cell membrane-like structure provides great capabilities of encapsulation, delivery, and in vitro triggered release of a controlled dose of inclusions at a specific time and location. The functional diversity of liposome renders them one of the most extensively developed drug delivery systems. The microstructure and property of the bilayer is very important for liposomes as drug delivery. Transmission electron microscope (TEM), atomic force microscope (AFM), differential scanning calorimetry (DSC), and fluorescent spectroscopy were applied to probe the microstructure and property of the bilayer in the liposomes.

The samples for TEM measurement were obtained by dispersing a small drop of the suspension onto a copper grid precoated with amorphous carbon. Then the copper grid was dried in vacuum for 1 day before observation.

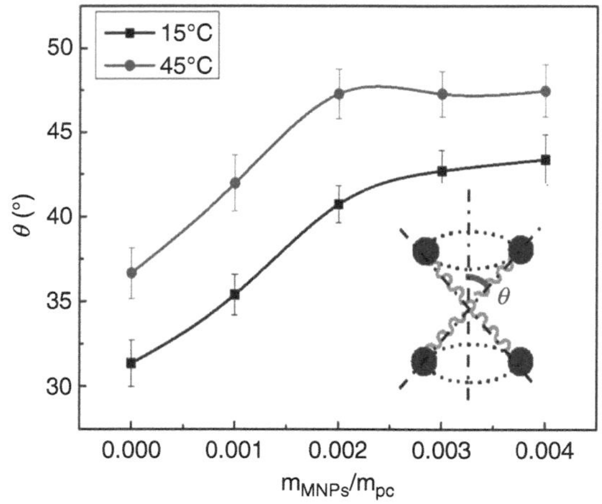

FIGURE 28.2 The extent of angular rotation (*θ*) as a function of the drug concentration in the liposomes at 15°C and 45°C

The inset is the schematic model for θ of DPH in lipid bilayer and the ball and stick molecular models represent DPH (Qiu and An, 2013).

The lipid bilayer for AFM observation was prepared with the vesicle fusion method (Qiu and An, 2013). The AFM topography images were obtained in the tapping mode by an AFM with triangularly shaped silicon cantilever with spring constant 48 N/m. The AFM images were obtained with a maximum scan range 20 × 20 μm and scanning frequencies were usually in the range between 0.6 and 2.5 Hz per line. The measurements were carried out in air at room temperature.

Time-resolved anisotropy measurement (TRAM) was used to study the change of membrane microstructure by using 1,6-diphenyl-1,3,5-hexatriene (DPH) as the hydrophobic fluorescence probe (Qiu and An, 2013). The rotational motion of DPH in the membrane is usually affected by the alkyl chain order, and it can be described by the extent of angular rotation (θ) (see inset in Fig. 28.2). Thus the altering θ can be illustrated for membrane permeability. Fig. 28.2 describes the relationship between θ and hydrophobic drug concentration, which indicates that the value of θ increased with the augment of hydrophobic drug concentration. It means that the hydrophobic drug in the bilayer of the liposome losed the original ordered arrangement of lipid chains and the membrane tended to be with the enhanced permeability.

2.3 LIPOSOMES AS NANOCARRIERS FOR DRUG DELIVERY

Scientific and technological breakthroughs in the late of 20th century have led to the development of synthetic lipid-based carrier systems for the entrapment of therapeutically active molecules to yield nanomedicines (Duncan and Gaspar, 2011; Van der Meel et al., 2014). In 1971, British scientist Rymen prepared closed spherical vesicles with similar biofilm structures. Liposomes are amphiphilic in nature, due to being composed of both hydrophobic and hydrophilic components, which leads to self-assembly into bilayer structures (James, 2015; Malam et al., 2009). Liposomes seem to be an almost ideal drug carrier system, because their morphology is similar to that of cellular membranes and because of their ability to incorporate various biologically active materials (Gregoriadis, 1973, 1976a,b). When compared to the administration of free drugs, advantages, such as improved stability, solubility, and in vivo pharmacokinetics, have resulted in approximately 250 nanomedicines that are currently approved or in various stages of (pre)clinical evaluation (Chauhan and Jain, 2013; Etheridge et al., 2013; Svenson, 2012). These materials possess many advantages for use as drug delivery systems; the first of which is that the surface properties can be easily modified (Zhang et al., 2009) to suit the needs of the system being developed through the addition of targeting functionalization (Noble et al., 2014), and their size can be precisely controlled (Al-Jamal and Kostarelos, 2011). They can also stably encapsulate and protect both water-soluble or insoluble therapeutics in the core (Allen and Cullis, 2013). Therefore, for the past 50 years liposomes have been widely investigated and they continue to be the subject of intense research. They are valued for their biological and technological advantages as optimal delivery systems for biologically active substances, both in vitro and in vivo, and are considered to be the most successful drug-carrier system known to date (Felice et al., 2014). During the two last decades, notable progress has been made, and several biomedical applications of liposomes are either in clinical trials or are about to be put on the market, while others have already been approved for public use (Fanciullino and Ciccolini, 2009). Liposomes have been successfully developed as a carrier for drugs, and several liposomal drugs (e.g., Doxil/Caelyx and Ambisome) have been approved and entered the clinic (Wagner et al., 2006).

Liposomes as drug carriers have the following characteristics: (1) double molecular membrane permeability changes at the phase transition temperature of liposomes; (2) after liposome drug carriers entering into human body, the biocompatibility of drugs increases, while the toxicity or side-effects reduce; (3) the targeting characteristics of liposomes mainly include the physical targeting, natural targeting, specific targeting and compartment targeting; (4) liposomes after entrapping drugs can realize the slow release of drugs, effectively extending the half-life of drugs in human body and increasing the time of maintaining an effective drug concentration at targeting sites time; and (5) after the drugs encapsulated by liposomes, phospholipid bilayer can help to protecting them effectively.

3 STIMULI-RESPONSIVE LIPOSOME

Liposomes have been recognized as drug nanocarriers for decades, the drug delivery systems possess many advantages but their clinical applications are often limited by uncontrolled release or poor availability of the encapsulated drug and stability in vivo in humans. Conventional and long-circulating liposomes may present a slow release of the loaded drug or may be unable to fuse with the endosome after internalization. Moreover, after liposomes into the body circulation, the drug may be released before reaching the target sites, or the drug cannot be timely released in targeting sites. As a result, probably the blood drug concentration in nontargeting sites is too high, losing the targeting meaning. To overcome these problems, stimuli-responsive liposomes have been developed. "Stimuli-responsive" is often called "environment-responsive," and its intelligent behavior is mainly reflected on its various exceptional macroscopic behaviors generated from external stimuli signals (Allen and Cullis, 2004). A major challenge for drug delivery is controlled release both spatially and temporally. For this reason, many of the current methods to enhance temporal or spatial control of drug release focus on incorporating components to achieve thermal, pH, photothermal, magnetic, or enzymatic triggered

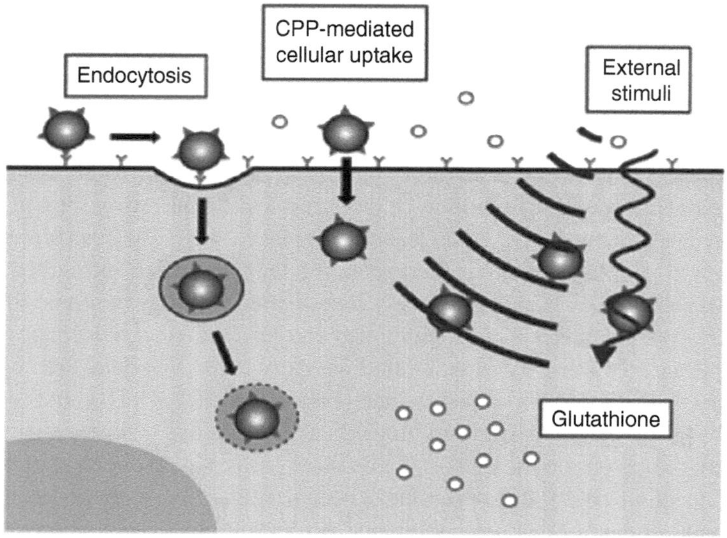

FIGURE 28.3 Principle summary of inside stimuli (e.g., glutathione) in cancer cells and external stimuli (e.g., magnetic, ultrasonic, light, radiation) used for imaging, drug delivery, and treatment through endocytosis and phagocytosis of cell permeable membrane peptide (CPP) cells that can achieve targeted cellular uptake (Lehner et al., 2012)

release. The stimuli-responsive liposomes composed of triggered components and conventional liposome. The triggered components are responsible for gating the stability and/or permeability of the lipid bilayer. According to the difference of stimulus signals, stimuli-responsive nanoparticles can be divided into different types, including temperature (Andresen et al., 2005; Kim and Kim, 2002; Kono et al., 1994a, 2002; Paoli et al., 2010), light (Wardak et al., 2000; Christina et al., 2000), magnetism field (Dandamudi and Campbell, 2007; Nakao et al., 2009; Viroonchatapan et al., 1996), pH (Cho et al., 2009; Kono et al., 1994b), and other types. These signals can change the molecular interaction in bilayer membrane at the molecule level, so as to regulate drug release. Stimuli-responsive liposomes are capable of reacting when triggered by stimuli from target tissues (pH, redox potential) or applied from outside the organism (hyperthermia, ultrasound, and magnetic field) (Torchilin, 2009); pH-sensitive, redox potential-sensitive, temperature-sensitive, magnetic field-sensitive, and ultrasound-sensitive liposomes have been produced. After reaching the targeting sites through blood, stimuli-responsive liposomes are changed by in vivo or in vitro specific physical (e.g., light, sound or heat), chemical (e.g., redox), and biological (e.g., enzymes, nucleic acids, or biomolecules) stimuli. The changes in the compositions and structures of carriers or modified molecules (ligands) occur and lead to the release of loaded drugs or the exposure of ligand molecules (Fig. 28.3), thus realizing drug release at targeting sites. Stimuli-responsive nanodrug delivery system can reduce the premature release of drugs, while perform stimuli-responsive release of drugs at targeting sites. These could improve the targeting efficiency of drug treatment, and reduce the adverse drug reactions. Currently, the investigation of controlled release and release mechanism of liposome drug carriers is still a crucial and challenging problem (Asgharian and Schelly, 1999; Yan and An, 2013).

3.1 TEMPERATURE STIMULI-RESPONSIVE LIPOSOMES

3.1.1 Lower critical solution temperature (LCST) of liposomes

The lower critical solution temperature (LCST) of temperature-sensitive liposomes is phase transition temperature (T_c) of gel to liquid crystalline in the liposomes (An et al., 2010). Temperature-sensitive liposomes release their inner content at the melting phase transition temperature of the lipid bilayer as shown in Fig. 28.4 (Hijnen et al., 2014). The design criteria of the thermosensitive liposomes depend on the heating strategy used (i.e., intra- or extravascular release). When ambient temperature is below the lower critical solution temperature (LCST) of liposomes, the lipid bilayers show a dense arrangement of colloidal crystalline (gel), and their flow ability and permeability are low, which result in the inner contents difficult to diffuse out through penetrating the liposome membrane. When the heating temperature is raised to above LCST, the movement abilities of phospholipid molecules that are used to fabricate liposomal membranes strengthen, such as rotation, movement, flip, and swing. Phospholipid bilayer arrangement changes from the original compact arranged colloidal crystal state into the loose chaos liquid crystalline state (Fig. 28.5A). At this LCST, the lipid membrane changes its permeability because of the transition from the gel to the liquid crystalline phase (Evans and Needham, 1987; Kono, 2001). At this case, the distance between adjacent phospholipid molecules increases and the film thickness decreases. The fluidity and permeability of liposome membranes increase significantly. Thus the inner contents of temperature-sensitive liposomes promptly are discharged upon heating (within seconds to minutes) into the heated targeting sites in a short time at the LCST of the lipid bilayer (Fig. 28.5B), and the drug distribution increase with temperatures in local heated targeting sites (Kong et al., 2000; Li et al., 2010; Needham and Dewhirst, 2001).

In general, the temperature-sensitive liposomes should: (1) stably encapsulate the drug in the aqueous compartment at body temperature; (2) provide a rapid and quantitative release at hyperthermic temperature; and (3) provide high drug plasma levels during the time span of hyperthermia treatment.

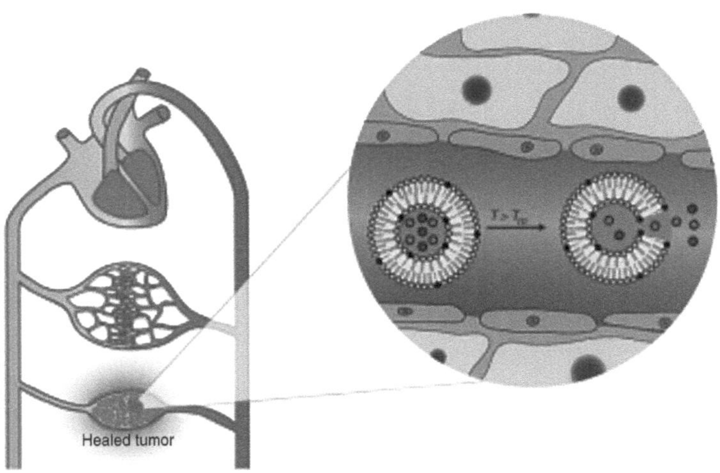

FIGURE 28.4 The Concept of Intravascular Temperature-Induced Drug Delivery Under Image Guidance

Temperature-sensitive liposomes coencapsulating drugs and MRI contrast agents release their inner content at the melting phase transition temperature of the lipid bilayer (Hijnen et al., 2014).

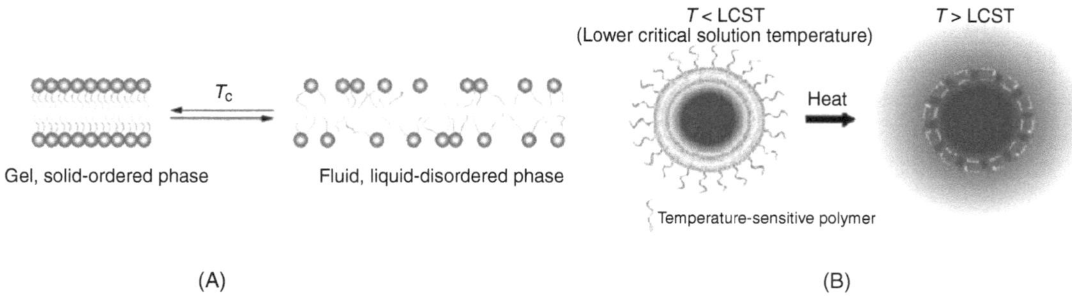

FIGURE 28.5

(A) The bilayer phase transition of thermosensitive liposomes and (B) releasing drugs of temperature stimuli-responsive liposomes at the phase transition temperature (Ta and Porter, 2013).

3.1.2 Temperature-sensitive liposomes

Thermosensitive liposomes have been widely investigated since 1978 (Yatvin et al., 1978). In the past decade, there has been increased interest in delivery mediated by temperature-sensitive liposomes, in part due to advances in image-guided hyperthermia applicators. The temperature-sensitive liposomes in combination with heating the target region can selectively enhance the bioavailability of the drug locally while minimizing systemic exposure. The advantages of temperature-sensitive liposomes for temperature-induced drug delivery are the enhanced drug uptake at the tumor site and decreased drug toxicity profiles in healthy tissue as compared to the systemic administration of low molecular weight parent drugs. Stimuli-responsive liposomes belong to a class of interesting drug carriers because their noninvasive properties are suitable for their biological applications (An et al., 2010). The treatment mechanism of tumors using thermosensitive liposomes is shown in Fig. 28.6. Liposome drug carriers can effuse from tumor blood vessel wall cavity (A). Upon the local heating, the increase of vascular

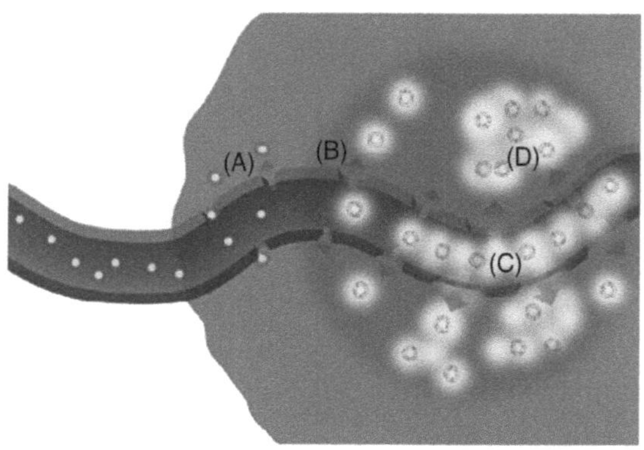

FIGURE 28.6 Mechanism of Thermosensitive Liposomes Used for Cancer Treatment

Liposomal drugs (green stands for the lipid shell, yellow means the drug) can bleed from the tumor vascular wall cavity (A). Upon the local heating, the increase of tumor vascular wall aperture size thereby increases the exudation rate of liposome drugs (B). Temperature-sensitive liposomes release drugs into the tumor vasculature (C) and the tumor cell mass (D) after local heating (Ta and Porter, 2013).

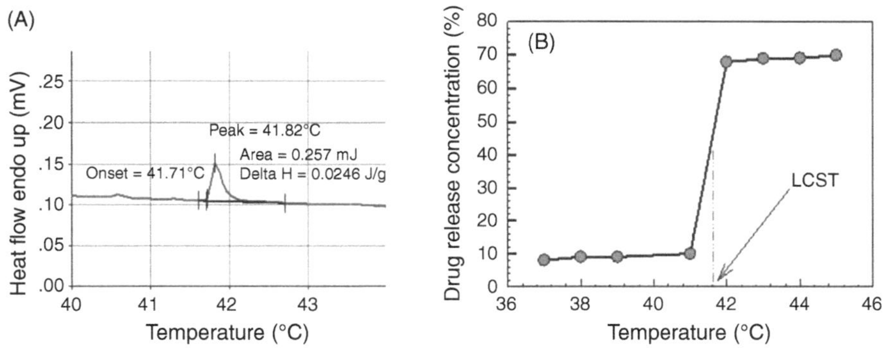

FIGURE 28.7

(A) Liposome LCST was determined by measurement of dynamic laser light scattering (DSC); (B) the drug released from thermosensitive liposome at various temperatures (An et al., 2010).

wall aperture size of tumor, thereby, can increase the exudation rate of liposome drugs (B). After local heating, thermosensitive liposomes release drugs to tumor vasculature (C), and tumor cell mass (D). Mild hyperthermia (HT) has long been used as an auxiliary method of treating cancer (radiation and chemotherapy).

The research findings indicate that thermosensitive liposomes can quickly release drugs in blood vessels under the mild heating conditions. Therefore, based on the principle of liposome carriers releasing drugs above LCST, the concentration of drugs is relatively low in the unheated organs and their action of killing to the normal cells is very low, which probably result in significantly lower side-effects induced chemotherapy drugs, such as sicchasia and vomiting. In other words, these advantages reduce the pain of patients and increase medication compliance. When the local temperature of targeting organs in the body rises to above the LCST, it can cause the rapid release of loading drugs in liposome drug carriers and gives play to the drug efficiency. The LCST for liposomes can be obtained by measurement of differential scanning calorimetry (DSC). A heat exchange flow curve of liposomes obtained from measurements of DSC display the LCST of liposome at about 41.7°C (Fig. 28.7A). The release curve of the liposomes at various temperatures is shown in Fig. 28.7B. The drug released from the thermosensitive liposome at various temperatures demonstrate that drugs encapsulated in the liposome almost cannot be released when the temperature was below 41°C, but it was released quickly when the temperature was raised up to 42°C (Fig. 28.7B). The results reveal that the liposome is thermosensitive and the LCST is about 41.7°C, and the drug encapsulated in the liposome was released above LCST.

In addition, hyperthermia could be used to inhibit or kill the tumor to give assistance for oncotherapy. Hyperthermia means that the tumor is heated to a slightly higher temperature than the body temperature (usually 41~43°C). The sensitivity of the tumor cell to temperature is used for inhibiting or killing tumor cells, and it has a synergistic effect with chemotherapy. It is envisaged that the combination of thermosensitive liposomes loading chemotherapy drugs and tumor hyperthermia will significantly enhance the treatment effect. Thermosensitive liposomes have been prepared by using egg phosphatidylcholine, cholesterol, and ethanol, and their phase transition temperature is 42.7°C (Chelvi and Ralhan, 1997; Zhu et al., 2005). Experiments showed that after combination with tumor localized heating, the prepared thermosensitive liposomes loading mustard can greatly improve the drug

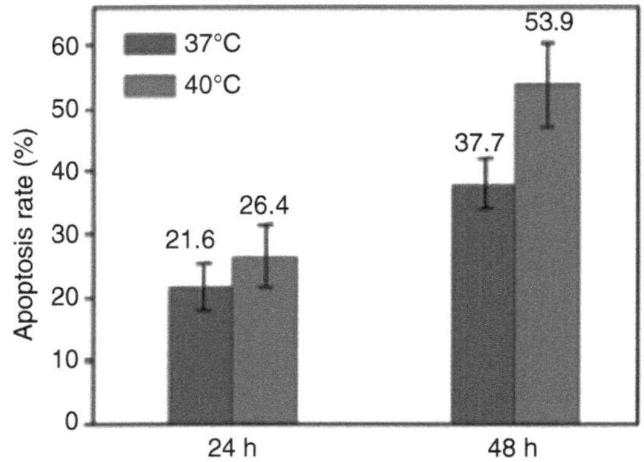

FIGURE 28.8 Temperature-Controlled Drug Release From Thermosensitive Liposomes in Tumor Cells (Bcap-37) (Yan and An, 2014)

dose delivered to the tumor. Thermosensitive liposomes loading cisplatin, combining rat brain tumor localized heating (41°C, 30 min) can significantly increase the drug accumulation in brain tumor sites (Kakinuma et al., 1996). Therefore, the thermal targeting effect of thermosensitive liposomes is obvious. Using thermal excitation control for tumor cells, the function of temperature stimuli-responsive release of drugs from the liposomes can be achieved. Relevant research results are shown in Fig. 28.8. The dark and light gray columns represent the drug release at 37°C (during incubation: 37°C, 24 h) and 40°C (during incubation: 40°C, 30 min and 37°C, 23 h 30 min), respectively (Yan and An, 2014). The research results show that the cell apoptosis rate from thermal stimuli-controlled release of drugs is 1.2 times (24 h) and 1.4 times (48 h) of traditional administration release patterns.

Temperature stimuli-responsive liposomes still have limitations: (1) the release of drugs from thermosensitive liposomes is related to the particle sizes, membrane materials, drugs, and so on. Large unilamellar vesicles (LUV), small unilamellar vesicles body (SUV), and multilamellar liposomes (MLV) also have different phase change temperatures; (2) the temperature can adjust drug release conditions of thermosensitive liposomes, but its targeting behavior is weak, and it is difficult to avoid the phagocytosis by mononuclear macrophage; and (3) although the local temperature can kill tumor cells directly, if the heating time is too long (1 h), it can also cause the damage to normal tissues.

3.2 LIGHT STIMULI-RESPONSIVE LIPOSOMES

As we know, the LCST is a very important parameter for the release of drugs from thermosensitive liposomes. Nevertheless, it is difficult to control the LCST of liposomes using traditional methods because LCST can be affected by the type of drug, additive, medium pH value, and so on. Due to the limitations in temperature stimuli-responsive liposomes for controlled drug release aspect, light stimuli-respective liposomes (photosensitive liposomes) have been proposed by researchers in recent years (Ebrahim et al., 2005). Under various conditions of ambient lighting, photosensitive substances in the interior of liposomes perform different reactions, which result in changes of some properties and the structure of liposomes (Guohui et al., 2008). These further cause the entire morphological changes of nanoparticles, thus resulting in drug release. In the case of azobenzene, the molecule has the *cis*- and *trans*-isomers, and its conformation can be controlled by light. Under the visible light irradiation, azobenzene shows the *cis* structure, but the *cis* structure may transfer into the *trans* structure of azo molecules under UV irradiation. Due to the unique advantages such as cleaning, remote control, the light irradiation is considered one of the best control means. But the use of light as a stimulus signal to regulate and control the behavior of nanoparticles is still in the initial development stages currently.

For common photosensitive liposomes, using the high-power light illumination or UV light irradiation for a long time, the stability of liposomes is undermined, followed by the drug release from the photosensitive liposomes (Volodkin et al., 2009). This typical class of liposomes could be also considered as the light stimuli-response liposomes (Ebrahim et al., 2005). It should be noted that in the process of high-power laser stimulation or UV light prolonging exposure, the release of drugs only resulting from the damage of liposomes cannot realize the really controlled drug release. For ideal photosensitive liposomes, the sensitive substance should have the following requirements: (1) photosensitive materials encapsulated in the interior of liposomes still have a good light-sensitivity, and they should be very sensitive to the external light source; (2) the particles of photosensitive materials are small enough and are relatively stable, and may be encapsulated into liposomes inside aqueous phase or phospholipid bilayer, and can coexist for a long time with liposomes without change; and (3) the photosensitive substances after being added into liposomes cannot cause too much influence on the stability and encapsulation rate of liposomes.

The research has focused on controlled drug release both spatially and temporally. The photosensitization of liposomes has become the research focus (An et al., 2010; Yan and An, 2014). Controlled drug release and the mechanism employed to release the drug are hence of utmost importance and remain a challenging problem. Its principle is that some additives with photosensitive natures are added into liposomes consisting of thermosensitive phospholipid membrane materials. Under outside light irradiation, the light-sensitive materials in the interior of liposomes react with the membrane structure (e.g., phospholipid), and fusion actions occur among liposomes, or between film and vesicle substances, which lead to drug release. In other words, the thermosensitive stimulation response of original thermosensitive liposomes has changed into the light stimuli-responsive characteristic (An et al., 2010, 2013). Generally, the liposomes used for releasing drugs are attributed to the changes of phospholipid bilayer structures and properties, resulting in the loading drugs flowing out from liposomes (Sung et al., 2006; Tapan et al., 2009; Zhu et al., 2005). Nanogold (AuNPs) can strongly absorb light energy and then convert it into heat energy in a very short time (picoseconds) (Au et al., 2008; Chen et al., 2010a; Liu et al., 2006; Mueller et al., 2000; Yavuz et al., 2009). After encapsulation of AuNPs in liposomes, the AuNPs of AuNPs-liposome can absorb a certain wavelength light, convert optical energy into heat, induce phase transition, and release drug, as shown in Fig. 28.9 (Yasuhiro et al., 2007). Meanwhile, AuNPs transmit heat energy to liposomes, which result in some changes in the properties and structures of liposome membranes and the following release of massive drugs (An et al., 2013). The AuNPs encapsulated in the phospholipid bilayers not only directly release energy under the light illumination (affecting phospholipid bilayers), but also the reaction sensitivity under the external lighting conditions is higher than that AuNPs encapsulated in the inner of liposomes. Experiments have shown that the drug release ability for AuNPs in the phospholipid bilayers is higher than that for AuNPs in the internal aqueous phase (An et al., 2010). However, due to the internal volume limitation of the phospholipid bilayer, the size of as-prepared AuNPs is required to be small enough. After the modification with various macromolecules, AuNPs are linked to the outer surface of liposomes. Since the particular external surface of AuNPs, the resulting liposomes can encapsulate not only lipophilic drugs, but also hydrophilic

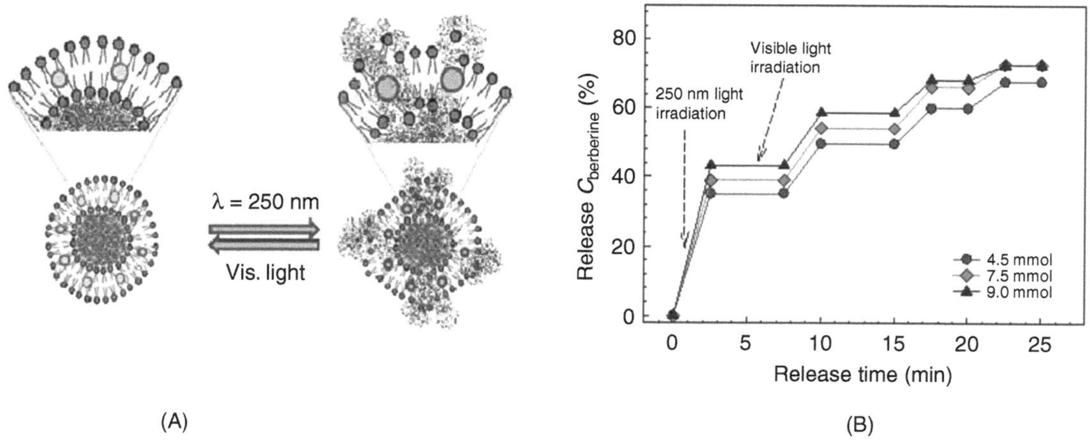

FIGURE 28.9

(A) Drug release from the AuNPs−liposomes by light irradiation at room temperature; (B) drug repetitious release in AuNPs-liposomes with various C_{AuFNPs} by commutative irradiation with UV light and visible light (An et al., 2013).

drugs. Based on AuNPs only linked onto the external surface of liposomes, however, the release ability of drugs inside liposomes is relatively low under the influence from external light irradiation reaction. Because the photothermal conversion of AuNPs occurs in bilayers, which directly results in the phase transition of liposomes and the following release of drugs (An et al., 2010, 2013; Li et al., 2014).

A thermosensitive liposome with embedded AuNPs in a bilayer was prepared using supercritical CO_2 (An et al., 2013). Research results have shown that when the added amount of AuNPs increases, the light stimuli-responsive performance of liposomes becomes more sensitive (Chen et al., 2010a). To prove that the liposomes are thermosensitive and that the drug is released by a thermoinduced phase transition, liposome-encapsulated berberine without AuNPs was prepared by the combination method of film build and $scCO_2$ incubation (An et al., 2013). The LCST was determined by differential scanning calorimetry (DSC) and is shown in Fig. 28.7A. The results reveal that the liposomes were thermosensitive and the LCST was about 41.7°C. The liposome-encapsulated berberine was released at various temperatures, and the results demonstrate that berberine encapsulated in the liposomes almost could not be released when the temperature was below 41°C but was released quickly when the temperature was raised to 42°C as shown in Fig. 28.7B. This means that the liposomes were thermosensitive and that berberine encapsulated in the liposomes was released by thermoinduced phase transition above the LCST.

3.3 MAGNETISM STIMULI-RESPONSIVE LIPOSOMES

As a member of liposome family, magnetic liposomes (MLs) not only have the applied potential in magnetically targeted therapy of tumor, but also can release the loading drugs through magnetism stimulus control. As established by previous studies, when a specific ligand conjugated with the magnetic liposome surface, the liposomes can accurately carry drugs to the target sites, thereby reducing the toxicity to neighboring healthy tissues (Jain et al., 2003). MLs were the first multifunctional hybrid liposome/nanoparticle assembly (Bangham et al., 1958; Hamaguchi et al., 2003). Being an essential constitute of MLs, one of the critical functions of MNPs is to guide MLs to the target site in vivo (Kullberg et al., 2005). Moreover, MNPs make classic liposomes have several novel functions (Babincov et al., 1999; Hodenius et al., 2002; Skouras et al., 2011). Regarding the structure, there have been two types of MLs: hydrophilic nanoparticles encapsulated in the core water of liposomes (MLs-W) (Martina et al., 2005; Skouras et al., 2011; Soenen et al., 2009), and hydrophobic nanoparticles embedded into the lipidbilayer (MLs-M) (Amstad et al., 2011; Chen et al., 2010b; Nappini et al., 2011). High sensitivity and responsiveness of MLs-M for triggered drug release in low or high frequency magnetic fields were presented (Amstad et al., 2011; Nappini et al., 2011). The embedment of hydrophobic MNPs has the ability to stabilize liposomes and suppress spontaneous leakage of cargo in core water (Chen et al., 2010b). Except for these applications, few studies were carried out on the microstructure of MLs-M, especially the influence of hydrophobic MNPs on membrane structure.

Since the internal thermal motion energy of superparamagnetic nanoparticles can be roughly equal to the anisotropic energy, the direction of magnetization is no longer fixed in one of easy magnetization directions under the action of thermal motion energy, and the easy magnetization direction shows irregular changes, thereby leading to the superparamagnetic substances with the nonmagnetic characteristic under normal conditions. Liposomes with superparamagnetic iron oxide nanoparticles embedded in membrane were prepared by the supercritical carbon dioxide (scCO_2) method (Qiu and An, 2013). When an external magnetic field is applied, the original disordered magnetic domain orientation inside superparamagnetic nanoparticles can generate the magnetism for the original nonmagnetic substances. Therefore, the best magnetic liposomes should have the superparamagnetic behavior. When applying a magnetic field, the magnetic thermal effects or disturbances effect of magnetic particles in magnetic liposomes result in the release behavior of liposomes (Hoare et al., 2009; Jackson et al., 1997). Considering that the magnetic particles are superparamagnetic, when the external magnetic field is turned off, the magnetism of nanoparticles will instantly disappear and the disturbance stops. Meanwhile, the phospholipid membrane temperature is cooled to a temperature below the phase transition, and the liposomes come back to the colloidal crystalline state with relatively close structure. And, therefore, the liposomes no longer release the loading substances inside liposomes. The reciprocating release and stop behaviors of liposomes along with the change of the magnetic field indicate that the action of alternating the magnetic field does not destroy the structure of liposomes. The release mechanism was explained from the view of microstructure change investigated by fluorescence probe method. The major reasons may be attributed to the local phase transition caused by magnetic field-induced thermal effect. Under the action of the magnetic field, the disturbance of magnetic particles makes the film localized structure become loose, which leads to the release of substances inside liposomes. Experimental results demonstrate that the magnetism-controlled release of magnetic liposomes is a reversible process, showing potential applications. Fig. 28.10B depicts the controlled release of magnetic liposomes under the action of the magnetic field. Under an alternating electromagnetic field, the release of drugs from liposomes occurs at the "on" state, while it stops at the "off" state (Qiu and An, 2013). The results indicate that the release from the liposomes is due to the magnetocaloric effect resulting in the liposome phase transition and the magnetic-impelled motions leading to the improved bilayer permeability, rather than the destruction of the liposome structure.

With respect to the targeting of drug carriers to tumor tissues, magnetic liposomes exhibit some obvious advantages. After magnetic liposomes with entrapped drugs reach the targeting sites, the drug release can be triggered by using suitable means. For example, when the magnetic liposomes have

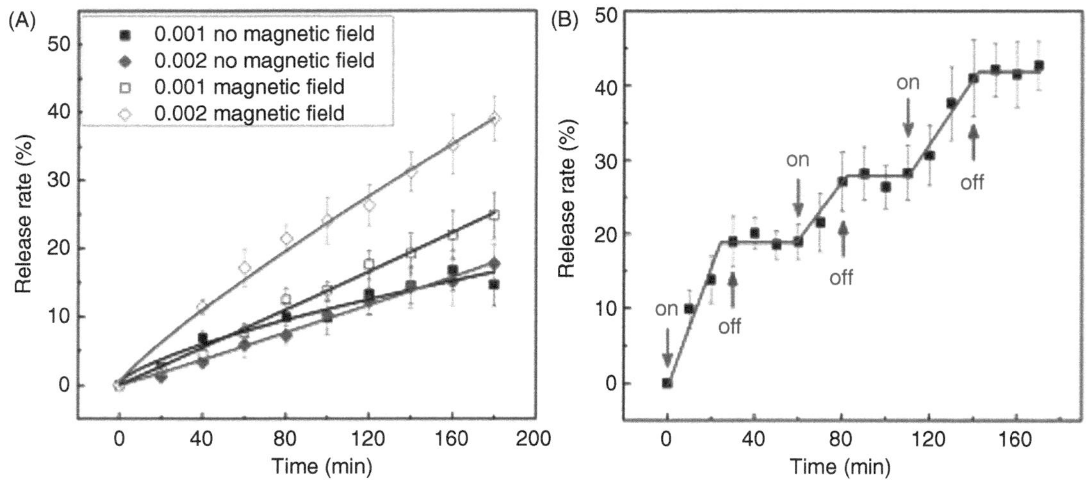

FIGURE 28.10

(A) Release of calcein-loaded magnetoliposomes in the presence of AMF as a function of exposure time and MNPs concentration; (B) release of calcein-loaded magnetoliposomes as a function of time over three successive on/off cycles of the external AMF at room temperature (Qiu and An, 2013).

temperature sensitivity, the phase transition of phospholipid membranes can occur through improving the local temperature, thus triggering drug release (Fortin-Ripoche et al., 2006; Qiu and An, 2013; Wang et al., 2015). The research results show that when magnetic liposomes with loading drugs get into human body by intravenous injection, the magnetic nanoparticles can precisely guide magnetic liposomes toward the in vivo tumor sites under the action of an external magnetic field (Babincova, 1999; Fortin-Ripoche et al., 2006; Qiu and An, 2013; Viroonchatapan et al., 1998; Wang et al., 2015). The magnetic fluid is injected into the site of the mammary gland carcinoma transplanted tumor in mice, and the site is irradiated locally with the alternating magnetic field. The research results reveal that the temperature of magnetic particles after the absorption of energy rises up to 47°C, and consequently the tumor growth is effectively controlled (Chen and Langer, 1997). The ability of tumor cells to absorb nanomagnetic particles is 8–400 times of that of the normal cells (Chen and Langer, 1997). Previous studies have reported the applied effects of magnetic drug carriers (doxorubicin-loaded albumin nanoparticles) in normal liver cells and in rat liver transplanted cancer cells, respectively. The targeting behavior and in vivo distribution of magnetic nanoparticles have been also reported. These reports suggest that the magnetic doxorubicin-loaded albumin nanoparticles possess a very high magnetism targeting, and they can gather together very obviously in cancer cells, and produce a good effect of treatment to transplanted tumor (Wasan, 2005). The fat-soluble anticancer drug, docetaxel, is entrapped in classical magnetic liposomes. Based on this way of offering drugs, the therapeutic effect in vivo of docetaxel to breast cancer is much marked. It can be inferred that magnetic nanoparticles in magnetic liposomes play a role of stabilizing the formulation, avoiding the adverse effects caused by the sudden release of drugs (Zhang et al., 2005). In order to improve the stability of liposomes in organisms, the researchers coat with a layer of polyethylene glycol (PEG) chains on the surface of liposomes to extend the blood circulation time of liposomes, but this modification will prevent the penetration of deep tissues (Cocquyt et al., 2008; Mornet et al., 2004).

3.4 pH STIMULI-RESPONSIVE LIPOSOMES

The pH stimuli-responsive liposome is another type of liposome with the functions of intracellular targeting and controlled drug release. As a consequence, they were found to apply in a variety of drug delivery systems, enzyme and protein immobilization and adsorption, and responsive colloidal arrays. Upon liposomes at a low pH, the protonation of carboxyl groups promotes the transition of liposomes from the "liquid" state into the hexagonal phase, which is the phase transition of pH. Then, the liquidity of film enhances and the drug is released. The normal physiological pH of human body is about 7.2 (Simões et al., 2004). However, in the area of inflammation or infection, the abnormal acidification phenomenon at a certain tumor tissue or when the ischemia of local sites usually occur, which induces a lower pH (generally 7.2–6.5) than the normal physiological pH of the body (Ge and Pei, 1998). Because the pH inside tumor cells is lower than that in surrounding normal tissues, when the pH-sensitive liposomes reach the tumor sites, the fluidity of lipid membranes increases, and loaded drugs are quickly released out from liposomes. In normal tissues, the bilayer membranes of liposomes can remain relatively complete, and the drug does not be released or be released slowly. Therefore, the drug can be much concentrated in the tumor sites, which efficiently improves drug efficacy and reduces toxicity or side effects to surrounding normal tissues. Thus, this type of pH-sensitive liposomes can reduce toxicity and side effects to surrounding normal cells, and meanwhile improve the drug efficacy (Yang et al., 2004).

In early years, the most of relevant studies refer to the inner lipid type of liposomes. The most representative one is the pH-sensitive liposomes composing of dioleoyl phosphatidylethanolamine (DOPE) and single-steroidal alkenyl hemisuccinate (CHEMS). This type of liposome can deliver the drug at neutral pH conditions. However, the circulation time of the liposomes in human body is relatively short because of their instability. The liposomes modified with sensitive copolymers on the surface have been of concern and generated increasing attention (Ringsdorf et al., 1988). The functionalized sensitive copolymers have been designed and favorably synthesized, showing particular structures and physicochemical properties. When they are exposed to the changes or stimuli of the environment, their structures and physicochemical properties will also change. The liposomes are modified with these sensitive copolymers that can induce the sensitivity of liposomes to a particular environment. The stability of sensitive polymeric liposomes is better than that of other liposomes, which could help to prolong the circulation time of drugs in vivo.

In order to solve the instability of DOPE liposomes, PEG (Thomas and Tirrell, 1992) and oleic acid (OAlc) are also used to modify the liposomes (Sudimack et al., 2002). However, earlier studies have found that the interspace stability hinders the aggregates of PEG on the surface of liposomes, or results in the dehydration on the film surface of liposomes. These reduce the integration of liposomes and the release of substances inside liposomes (Holland et al., 1996; Kirpotin et al., 1996).

Polyelectrolytes are an intriguing class of polymers with ionized groups. Stimuli-sensitive polyelectrolytes that contain hydrophobic and hydrophilic segments in the same molecule are attracting researchers' attention. Such a composition tends to lead to an interesting behavior in water. The solution proper ties of polyelectrolytes differ considerably from that of polymers without ionized groups. Researchers have focused on polymer-modified pH-sensitive liposomes, which have the higher stability and longer circulation time in vivo than that of the inner lipid liposomes (Leroux et al., 2001; Maeda et al., 1988). The pH-sensitive block copolymers P(NIPAM-*co*-MAA-*co*-ODA) with different monomer ratios are anchored onto the film layers of liposomes (Fig. 28.11A). Through changing the environmental pH, the controlled release of drugs can be achieved (Zhou et al., 2012). By changing the

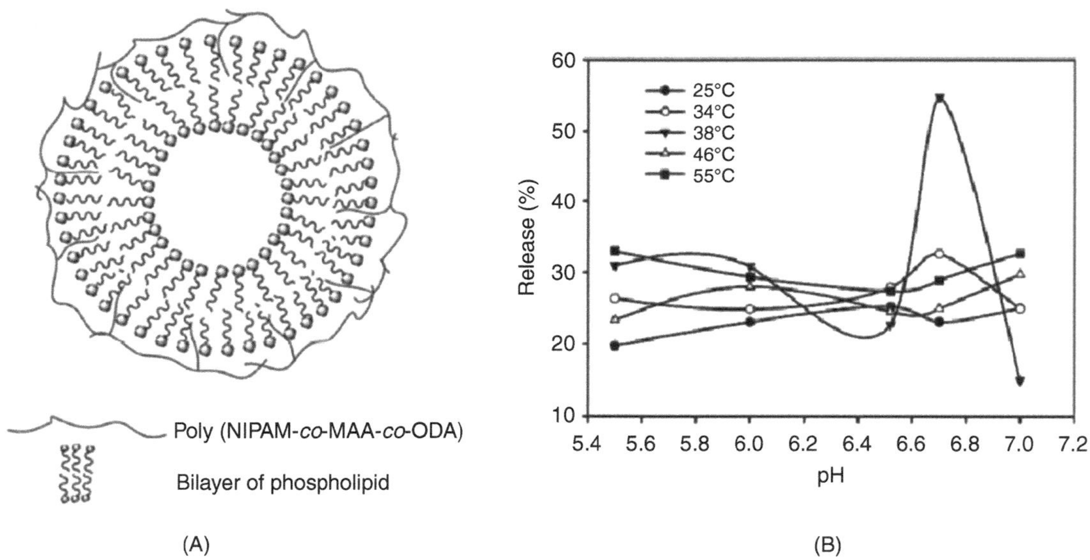

FIGURE 28.11

(A) The structure diagram of pH-sensitive block copolymer-anchored liposomes; (B) drug release of temperature-pH sensitive liposomes at different pH values and temperatures (Zhou et al., 2012).

monomer ratio of block copolymers, this type of pH-sensitive liposomes can achieve controlled drug release in the range of different pH responses.

The studies of hepatocytes-targeted galactose pH-sensitive liposomes have been performed and show that the liposomes have a good targeting, pH-sensitivity, low toxicity, and high transfection rate to liver cells (Santra et al., 2009). In order to improve the stability, bioavailability, and avoiding degradation in the lysosome of antisense oligonucleotides, Hong Wang et al. (Mackiewicz et al., 2011) prepared the antisense oligonucleotide pH-sensitive liposomes, and achieved the optimum preparation method and the law of the release in vitro of drugs. The pH-sensitive liposomes have been used for the delivery of nucleotide molecules within cells, such as DNA, anti oligonucleotides, proteins, and antibiotics (Veiseh et al., 2005).

3.5 TEMPERATURE-pH STIMULI-RESPONSIVE LIPOSOMES

As reported in previous literature, the single sensitive liposomes have shown drawbacks. In order to overcome the drawbacks, the research and development of multisensitive liposomes have become an inevitable trend. The most important stimuli are temperature, pH, light, and ionic strength. Among several stimuli, temperature and pH are commonly used because they are easy to control and have practical advantages in vitro and in vivo. Therefore, temperature-pH sitimuli responsive liposomes have attracted much attention. The liposomes are modified on the surface by copolymers with temperature- and pH-sensitive dual functions. Accompanying with the stability improvement of liposomes, at the same time, the liposomes are given the temperature- and pH-sensitive dual functions. The drug release in the liposomes could be controlled by regulating ambient temperature or pH, which provides more means for the stability of liposomes and the release of drugs. Generally, the polymers that are used to prepare the pH- and temperature-dual sensitive liposomes possess the amphiphilic structure and the hydrophobic portion of polymers can help to their grafting onto the surface of liposomes. The structure of liposomes will rearrange, together with releasing substances inside liposomes at a lower pH. It can increase the carrying capacity of aqueous drugs (Hafez and Cullis, 2001; Kim et al., 2010). The pH-sensitive block copolymers P(NIPAM-*co*-MAA-*co*-ODA) could serve a dual sensitive function for temperature and pH value. Fig. 28.11B presents drug release in the temperature-pH sensitive liposomes at different pH values and temperatures (Zhou et al., 2012). There is optimum temperature and pH value for the copolymers liposomes, which should be 38°C and pH of 6.7. Experimental results reveal that the dual sensitive liposomes achieve controlled drug release through changing the microstructure, which can be further controlled by varying the proportions of as-synthesized temperature- and pH-sensitive blocks (Zhou et al., 2012). To date, there are still few studies that have reported and discussed the phase change and its influence factors for the pH-temperature dual-sensitive liposomes. The phase transition of liposomes not only provides the theoretical basis for improving the stability and targeting of liposomes, efficiency of encapsulated drugs and the release of drugs, but also provides the feasible basis for the clinical applications of polymer liposomes.

3.6 TEMPERATURE STIMULI-RESPONSIVE AND FLUORESCENCE TRACE LIPOSOME

A theranostic is defined as a material that combines the modalities of therapy and diagnostic imaging. So far, it has been greatly developed and facilitated advances in medical research. In contrast to the development and use of separate materials for disease therapy and diagnosis, theranostics combine features of delivering therapeutic drugs and diagnostic imaging into one entity, which has the potential to overcome the undesirable drug release and biological toxicity in current distinct imaging and therapeutic agents. The ultimate goal for the theranosticeld is to gain the ability to image and monitor the diseased tissue, delivery kinetics, and improve the drug efficacy by controlled release. The development of nanomaterials, including liposomes, polymers, micelles, inorganic nanoparticles, and peptide/protein conjugates, offers new opportunities to meet the need of theranostics. The most promising aspects of utilizing nanoparticles as therapeutics, diagnostics, and theranostics are their potential to localize (or be targeted) in a specific manner to the site of disease and reduce or eliminate the possible numerous untoward side effects (Santra et al., 2009). The functional nanoparticles have been investigated in controlled release, (An et al., 2010, 2013) site-specie targeting, (Mackiewicz et al., 2011), or imaging (Kim et al., 2010; Veiseh et al., 2005) for potential applications in cancer therapy and diagnosis. However, there are several inherent disadvantages with current formulations in drug delivery and imaging, including toxicity to living organisms, fluorescence bleaching, severe side effects, uncontrolled drug release, instability in storage, and so on. A biocompatible drug carrier with integrated features of controlled release and imaging is urgently needed for biological and medical research.

In order to effectively utilize the theranostic nanoparticles for diagnostic purposes, molecular imaging is required to characterize biological processes at the cellular and subcellular levels. Utilizing disease state–specific targeting these modalities are aided by contrast agents or molecular probes and can help to detect different disease states at various stages of progression or be utilized to evaluate the efficacy of treatments. The imaging modalities to be utilized in theranostic systems must not interfere with nanoparticle self-assembly or alter the nanoparticle scale of existing particle systems upon their inclusion. The successful utilization of these modalities requires there to be a sufficient accumulation of the contrast agents, which is aided by both the nanoparticle size and the disease state–specific targeting. Optical imaging utilizes photons emitted from bioluminescent or fluorescent probes to detect disease states. It is considered to be an attractive option for imaging due to its high sensitivity, lack of nonionizing radiation, ability to image the spectrum from visible to near-infrared, ability to image in real time, and it is inexpensive to detect the photons. A common probe used in optical imaging is synthetic fluorophores, semiconductor fluorescent crystals, lanthanide-based probes, and fluorescent quantum dots, which have the advantage of being able to have their fluorescent properties tuned in any number of ways (Hafez and Cullis, 2001). Choi et al. utilized PEGylated hyaluronic acid nanoparticles with a near infrared fluorescent dye (Cy 5.5) loaded with irinotecan as a theranostic agent (Wang and Wang, 2000). This system was able to selectively target, treat, and image colon cancer tumors in vivo. Tan et al. used quantum dots as a contrast agent encapsulated in chitosan nanoparticles that utilized HER2/neu siRNA for targeted delivery to HER2 overexpressing SKBR3 cancer cell lines and demonstrated the ability to specifically target these cells as well as gene-silencing behavior that was monitored via the quantum dots (Seki and Tirrell, 1984). Despite the advantages their utility may be hampered by cytotoxicity concerns (Chung et al., 1996). Disadvantages of using optical imaging techniques are poor tissue penetration, susceptibility to noise due to scattering, and potential autofluorescence (Wen et al., 2004).

A multifunctional polydiacetylene-liposome (PDA-liposome) was prepared by self-assembly, which are characterized with controlled drug release and fluorescence tracing (Li et al., 2015; Yan and An, 2014). The temperature-controlled drug release has been realized both in vitro simulation and cancer cells. Significantly, the anticancer drug efficacy was much improved using temperature-controlled drug release in Bcap-37 breast cancer cells. In order to explore the effect of temperature on drug release of PDA-liposome, in vitro release studies were carried out at various temperatures. The cumulative release profiles of PDA-liposomes at different temperatures are shown in Fig. 28.12A. The drug release was found to be greatly dependent on temperature. As shown in Fig. 28.12B, the drug release increased with increasing temperatures at temperature below 40°C, and which decreased with increasing temperatures at temperature above 40°C at release time of 300 min. The drug was released abundantly from PDA-liposomes at an optimal temperature of 40°C. To explain the temperature dependent release mechanism, lower critical solution temperature (T_c) of PDA-liposomes was measured by microdifferential scanning calorimetry (micro-DSC). As shown in Fig. 28.12C, the T_c of PDA-liposomes was 40°C, which was consistent with the optimum drug release temperature of PDA-liposomes within the experimental error. The fluorescence cell imaging was performed using laser scanning confocal microscopy. Transmembrane behavior and distribution of PDA-liposome were monitored by fluorescence cell imaging as shown in Fig. 28.12D–F. It revealed that PDA-liposomes were provided with good membrane permeability. By the transmembrane and endocytic behaviors, the drug was concentrated effectively in cancer cells, and the anticancer efficacy of drug could be improved markedly. The PDA-liposomes were proved with good biocompatibility, membrane permeability, high drug entrapment efficiency, enhanced fluorescence, and thermotriggered drug release. The controlled drug release was realized in both in vitro simulation and cancer cells by temperature control. The internalization and distribution of the PDA-liposomes in cancer cells were presented by fluorescence cell imaging. The intracellular distribution of PDA-liposomes could be traced and monitored by fluorescence imaging in real time. This technology could be realized in oncotherapy of surface tumors. The results showed that there is still great space for developing, and it will be further studied to obtain a novel targeted delivery system in vivo, which is the integration system of both

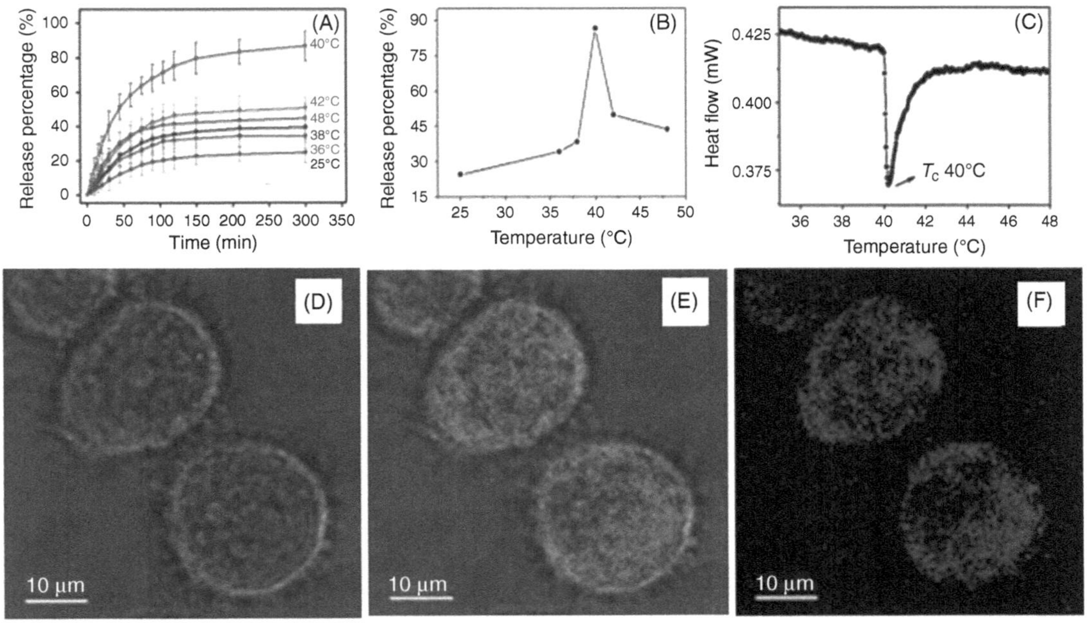

FIGURE 28.12

(A) Cumulative release profiles of PDA-liposomes at different temperatures; (B) the plot of cumulative drug release versus temperature at release time of 300 min; (C) heating curve of PDA-liposomes measured by micro-DSC at scanning rate of 1°C min^{-1}. LCSM images of Bcap-37 cells treated with PDA-liposomes for 2 h; (D) bright-field image; (E) composite bright-field and fluorescence image; (F) fluorescence image. All images were obtained with a 63 X objective (Yan and An, 2014).

treatment and diagnosis. The novel PDA-liposomes could be a promising candidate for drug delivery, disease diagnosis, and therapy.

A novel folate-polydiacetylene-liposome (FP-PL) with both targeted drug delivery and fluorescence tracing was prepared by thin film rehydration method (Li et al., 2015). The simulated drug delivery was performed in Bcap-37 breast cancer cells and Hs578Bst normal cells in vitro. In order to further explore the effect of FP-PLs on targeting and imaging in vitro, Bcap-37 and Hs578Bst cells were incubated with P-PLs and FP-PLs for 6 h at 37°C. The fluorescence microscopy was used to obtain targeting and transmembrane information in vitro, and the images of Bcap-37 and Hs578Bst cells in various solutions were shown in Fig. 28.13A. For Bcap-37 cells, bright fluorescent was observed in FP-PLs solution (Fig. 28.13A,(b)), while only dim light Was presented in P-PLs solution (Fig. 28.13A(a)). This result indicated that the folate receptors overexpressed in Bcap-37 cells could effectively improve the cells' uptaking of FP-PLs via the folate receptor-mediated endocytosis (FRME), while the cells' uptaking of P-PLs could not take advantage of this process. The results indicated that the FP-PLs possess active targeting ability to Bcap-37 cells, but P-PLs do not show it. It means that the FP-PLs are good drug delivery systems with active targeting, which cannot only improve drug efficiency on Bcap-37 cells but also reduce toxicity for Hs578Bst normal cells. Relative fluorescent intensities for various liposomes were compared in Fig. 28.13B, and the intensity in the FP-PLs for Bcap-37 cells was over

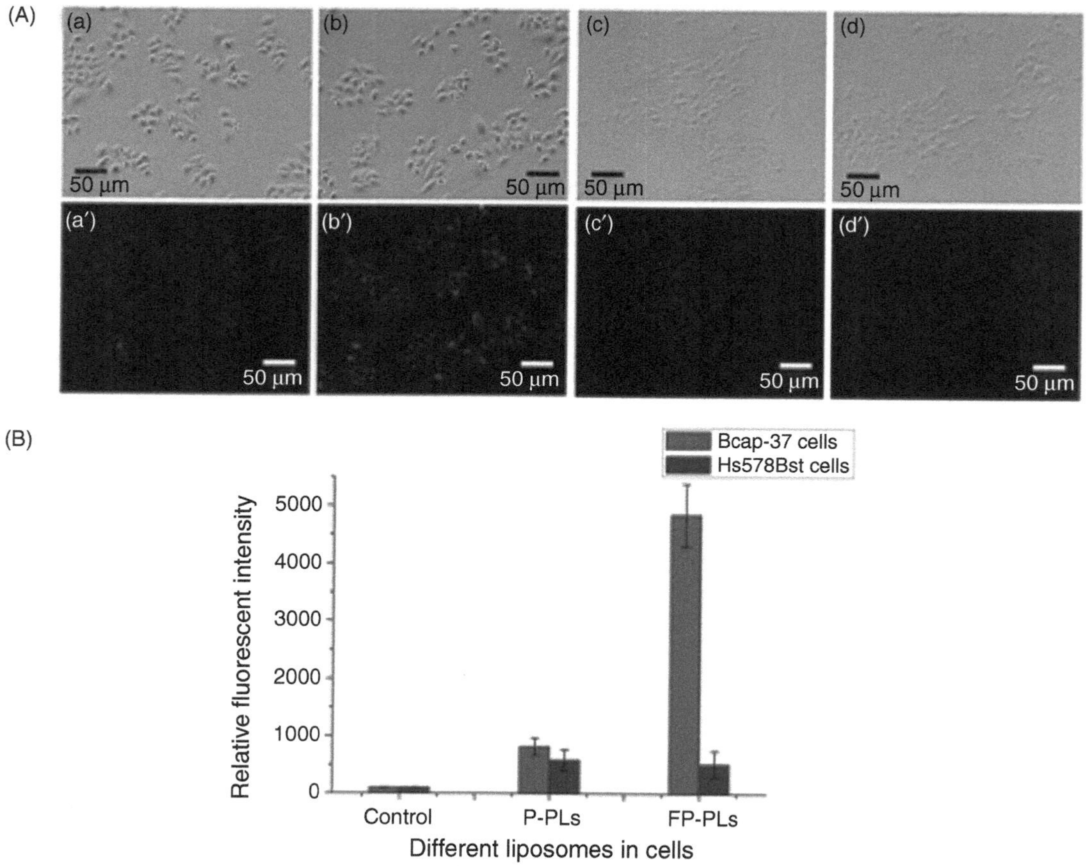

FIGURE 28.13

(A) Fluorescence microscopy images of Bcap-37 (a-b, a′–b′) and Hs578Bst (c-d, c′– d′) cells. Bcap-37 cells incubated with P-PLs (a, a′), FP-PLs (b, b′). Hs578Bst cells incubated with P-PLs (c, c′), FP-PLs (d, d′).
(B) Relative fluorescence intensity of different liposomes in cells, acquired by flow cytometry. Data are shown as the mean ± SD of three separate ($n = 3$). For interpretation of the references to color in the text, the reader is referred to the web version of this chapter (Li et al., 2015).

four times higher than others. In Hs578Bst cells, poor fluorescent intensity was observed for different liposomal formulations (Fig. 28.13B, dark gray columns). These results demonstrated that the coupling of folic acid could significantly improve the cellular uptake of P-PLs by folate receptor positive cells (Bcap-37). It means that FP-PLs were better targeting agents than passively targeted P-PLs, and the results were consistent with that obtained from fluorescent microscope and MTT assay. The internalization and distribution of FP-PLs in the cells were presented by fluorescence cell imaging. The results show that the FP-PLs possess low cytotoxicity, good biocompatibility, and better targeting efficiency in comparison to polydiacetylene-liposomes (P-PLs).

4 CONCLUSIONS

Stimuli-responsive liposomes should be considered as the intelligent drug carriers with good biocompatibility, and provide a means for developing macromolecular drugs and artificial gene fragments to delivery them within cells. Research progresses in liposomes will contribute to the progress of medicine. However, the stimuli-responsive liposomes still have many problems to be solved, such as the sensitivity increase of stimuli responses, the control of drug concentration at the targeting sites, and the clinical applications of control release at the targeting sites. With the development of science and technology, the further research of liposome production engineering, and the development of clinical research will facilitate the resolution of current problems. And eventually, the stimuli-responsive liposomes will become an effective means for the treatment in clinical applications.

ACKNOWLEDGMENTS

This work was financially supported by the National Natural Science Foundation of China (Nos. 21473055, 21273073, 21073063).

REFERENCES

Abraham, S.A., Waterhouse, D.N., Mayer, L.D., Cullis, P.R., Madden, T.D., Bally, M.B., 2005. The liposomal formulation of doxorubicin. Methods Enzymol. 391, 71–97.

Al-Jamal, W.T., Kostarelos, K., 2011. Liposomes: from a clinically established drug delivery system to a nanoparticle platform for theranostic nanomedicine. Acc. Chem. Res. 44 (10), 1094–1104.

Allen, T.M., Cullis, P.R., 2004. Drug delivery systems: entering the mainstream. Science 303, 1818–1822.

Allen, T.M., Cullis, P.R., 2013. Liposomal drug delivery systems: from concept to clinical applications. Adv. Drug Deliv. Rev. 65 (1), 36–48.

Amstad, E., et al., 2011. Triggered release from liposomes through magnetic actuation of iron oxide nanoparticle containing membranes. Nano Lett. 11, 1664–1670.

An, X.Q., Zhang, F., Zhu, Y.Y., Shen, W.G., 2010. Photoinduced drug release from thermosensitive AuNPs-liposome using a AuNPs-switch. Chem. Commun. 46, 7202–7204.

An, X.Q., Zhan, F., Zhu, Y.Y., 2013. Smart photothermal triggered bilayer phase transition in AuNPs liposomes to release drug. Langmuir 29, 1061–1068.

Andresen, T.L., Jensen, S.S., Jørgensen, K., 2005. Advanced strategies in liposomal cancer therapy: problems and prospects of active and tumor specific drug release. Prog. Lipid Res. 44, 68–97.

Asgharian, N., Schelly, Z.A., 1999. Electric field-induced transient birefringence and light scattering of synthetic liposomes. Biochim. Biophys. Acta. 1418, 295–306.

Au, L., Zheng, D.S., Zhou, F., Li, Z.Y., Li, X.D., Xia, Y.N., 2008. A quantitative study on the photothermal effect of immuno gold nanocages targeted to breast cancer cells. ACS Nano 2, 1645–1652.

Babincov, M., et al., 1999. Laser triggered drug release from magnetoliposomes. J. Magn. Magn. Mater. 194, 163–166.

Babincova, M., 1999. Targeted and controlled release of drugs using magneto liposomes. Ceska Slov. Farm. 48, 27–29.

Bangham, A.D., 1968. Membrane models with phospholipids. Prog. Biophys. Mol. Biol. 18, 29–95.

Bangham, A.D., Horne, R.W., 1964. Negative staining of phospholipids and their structural modification by surface-active agents as observed in the electron microscope. J. Mol. Biol. 8, 660–668.

Bangham, A.D., et al., 1958. An apparatus for microelectrophoresis of small particles. Nature 182, 642–644.
Bangham, A.D., Standish, M.M., Watkins, J.C., 1965. Diffusion of univalent ions across the lamellae of swollen phospholipids. J. Mol. Biol. 13 (1), 238–252.
Bangham, A.D., Hill, M.W., Miller, N.G., 1974. Preparation and use of liposomes as models of biological membranes. Korn, E.D. (Ed.), Methods in Membrane Biology, vol 1, Plenum, New York, pp. 1–68.
Batzri, S., Korn, E.D., 1973a. Single bilayer liposomes prepared without sonication. Biochim. Biophys. Acta 298, 1015–1019.
Batzri, S., Korn, E.D., 1973b. Single, bilayer liposomes prepared without sonication. Biochim. Biophys. Acta 298 (4), 1015–1019.
Bozzuto, G., Molinari, A., 2015. Liposomes as nanomedical devices. Int. J. Nanomed. 10, 975–999.
Brunner, J., Skrabal, P., Hausser, H., 1976. Single bilayer vesicles prepared without sonication physicochemical properties. Biochim. Biophys. Acta 455, 322–331.
Cabral, H., Miyata, K., Kishimura, A., 2014. Nanodevices for studying nano-pathophysiology. Adv. Drug Deliv. Rev. 74, 35–52.
Chang, H.I., Yeh, M.K., 2012. Clinical development of liposome-based drugs: formulation, characterization, and therapeutic efficacy. Int. J. Nanomed. 7, 49–60.
Chauhan, V.P., Jain, R.K., 2013. Strategies for advancing cancer nanomedicine. Nat. Mater. 12, 958–962.
Chelvi, T.P., Ralhan, R., 1997. Hyperthermia potentiates antitumor effect of thermosensitive liposome encapsulated mephalan and radiation in murine melanoma. Tumour Biol. 18, 250–260.
Chen, H., Langer, R., 1997. Magnetically-responsive polymerized liposomes as potential oral delivery vehicles. Pharm. Res. 14, 537–540.
Chen, H.J., Shao, L., Ming, T.A., Sun, Z.H., Zhao, C.M., Yang, B.C., Wang, J.F., 2010a. Understanding the photothermal conversion efficiency of gold nanocrystals. Small 6, 2272–2280.
Chen, Y., et al., 2010b. Controlled release from bilayer-decorated magnetoliposomes via electromagnetic heating. ACS Nano 4, 3215–3221.
Cheng, Z., Al Zaki, A., Hui, J.Z., Muzykantov, V.R., Tsourkas, A., 2012. Multifunctional nanoparticles: cost versus benefit of adding targeting and imaging capabilities. Science 338, 903–910.
Cho, E.C., Lim, H.J., Kim, H.J., Son, E.D., Choi, H.J., Park, J.H., Kim, J.W., Kim, J., 2009. Role of pH-sensitive polymer–liposome complex in enhancing cellular uptake of biologically active drugs. Mater. Sci. Eng. C 29, 774–778.
Christina, R.M., Paula, J.C., David, F.O., 2000. Visible light-induced destabilization of endocytosed liposomes. FEBS Lett. 467, 52–56.
Chung, J.C., Gross, D.J., Thomas, J.L., Tirrell, D.A., Opasahl-Ong, L.R., 1996. pH-sensitive, cation-selective channels formed by a simple synthetic polyelectrolyte in artificial bilayer membranes. Macromolecules 29, 4636–4641.
Cocquyt, J., Soenen, S.J.H., Saveyn, P., Meeren, P.V.D., Cuyper, M.D., 2008. Partitioning of propranolol in the phospholipid bilayer coat of anionic magnetoliposomes. J. Phys: Condens. Matter. 20, 204102–204106.
Dandamudi, S., Campbell, R.B., 2007. Development and characterization of magneticcationic liposomes for targeting tumor microvasculature. Biochim. Biophys. Acta 1768, 427–438.
Deamer, D.W., 2010. From "banghasomes" to liposomes: a memoir of Alec Bangham, 1921– 2010. FASEB J. 24 (5), 1308–1310.
Deamer, D., Bangham, A.D., 1976. Large volume liposomes by an ether vaporization method. Biochim. Biophys. Acta 443, 629–634.
DeSantis, C.E., Lin, C.C., Mariotto, A.B., et al., 2014. Cancer treatment and survivorship statistics. Cancer J. Clin. 64 (4), 252–271.
Duncan, R., Gaspar, R., 2011. Nanomedicine(s) under the microscope. Mol. Pharm. 8, 2101–2141.
Ebrahim, S., Peyman, G.A., Lee, P.J., 2005. Applications of liposomes in ophthalmology. Surv. Ophthalmol. 50, 167–182.

Etheridge, M.L., Campbell, S.A., Erdman, A.G., Haynes, C.L., Wolf, S.M., McCullough, J., 2013. The big picture on nanomedicine: the state of investigational and approved nanomedicine products. Nanomedicine 9, 1–14.
Euliss, L.E., DuPont, J.A., Gratton, S., DeSimone, J., 2006. Imparting size, shape, and composition control of materials for nanomedicine. Chem. Soc. Rev. 35, 1095–1104.
Evans, E., Needham, D., 1987. Physical properties of surfactant bilayer membranes: thermal transitions, elasticity, rigidity, cohesion, and colloidal interactions. J. Phys. Chem. 91, 4219–4228.
Fanciullino, R., Ciccolini, J., 2009. Liposome-encapsulated anticancer drugs: still waiting for the magic bullet? Curr. Med. Chem. 16, 4361–4373.
Felice, B., Prabhakaran, M.P., Rodríguez, A.P., Ramakrishna, S., 2014. Drug delivery vehicles on a nano-engineering perspective. Mater. Sci. Eng. C 41, 178–195.
Fenske, D.B., Cullis, P.R., 2008. Liposomal nanomedicines. Exp. Opin. Drug Deliv. 5 (1), 25–44.
Ferrari, M., 2005. Cancer nanotechnology: opportunities and challenges. Nat. Rev. Cancer 5, 161–171.
Fortin-Ripoche, J.P., Martina, M.S., Gazeau, F., Menager, C., Wilhelm, C., Bacri, J.C., Lesieur, S., Clement, O., 2006. Magnetic targeting of magnetoliposomes to solid tumors with MR imaging monitoring in mice: feasibility. Radiology 239, 415–424.
Ge, S., Pei, Y., 1998. Study on drug release mechanism of pH-sensitive liposomes. Chinese J. Pharm. 29, 564–568.
Gómez-Hens, A., Fernández-Romero, J.M., 2006. Analytical methods for the control of liposomal delivery systems. Trends Anal. Chem. 25, 167–178.
Gregoriadis, G., 1973. Drug entrapment in liposomes. FEBS Lett. 36 (3), 292–296.
Gregoriadis, G., 1976a. The carrier potential of liposomes in biology and medicine (second of two parts). N. Engl. J. Med. 295 (14), 765–770.
Gregoriadis, G., 1976b. The carrier potential of liposomes in biology and medicine (first of two parts). N. Engl. J. Med. 295 (13), 704–710.
Guohui, W., Alexander, M., Htet, A., 2008. Remotely triggered liposome release by near-infrared light absorption via hollow gold nanoshells. J. Am. Chem. Soc. 130, 8177–8178.
Hafez, M.I., Cullis, P.R., 2001. Roles of lipid polymorphism in intracellular delivery. Adv. Drug Deliv. Rev. 47, 139–148.
Hamaguchi, S., et al., 2003. Selective hyperthermia using magnetoliposomes to target cervical lymph node metastasis in a rabbit tongue tumor model. Cancer Sci. 94, 834–839.
Hijnen, N., Langereis, S., Grüll, H., 2014. Magnetic resonance guided high-intensity focused ultrasound for image-guided temperature-induced drug delivery. Adv. Drug Deliv. Rev. 72, 65–81.
Hoare, T., Santamaria, J., Goya, G.F., Irusta, S., Lin, D., Lau, S., Padera, R., Langer, R., Kohane, D.S., 2009. A magnetically triggered composite membrane for on-demand drug delivery. Nano Lett. 9, 3651–3657.
Hodenius, M., et al., 2002. Biotinylated stealth magnetoliposomes. Chem. Phys. Lipids 120, 75–85.
Holland, J.W., Hui, C., Cullis, P.K., Madden, T.D., 1996. Poly(ethylene glycol)-lipid conjugates regulate the calcium-induced fusion of liposomes composed of phosphatidylethanolamine and phosphatidylserine. Biochemistry 35, 2618–2624.
Huang, C., 1969. Studies on phosphatidylcholine vesicles: formation and physical characteristics. Biochemistry 8 (1), 344–352.
Jackson, D.K., Leeb, S.B., Mitwalli, A.H., Narvaez, P., Fusco, D., Lupton, Jr., E.C., 1997. Power electronic drives for magnetically triggered gels. Ind. Electron IEEE Trans. 44, 217–225.
Jain, S., Mishra, V., Singh, P., Dubey, P.K., Saraf, D.K., Vyas, S.P., 2003. RGD-anchored magnetic liposomes for monocytes/neutrophils-mediated brain targeting. Int. J. Pharm. 261, 43–55.
James, T., 2015. Multifunctional nanoparticles for use in theranostic applications. Drug Deliv. Transl. Res. 5, 295–309.
Kakinuma, K., Tanaka, R., Takahashi, H., 1996. Drug delivery to the brain using thermosensitive liposome and local hyperthermia. Int. J. Hypertherm. 12, 157–165.

Kalia, M., et al., 2013. Personalized oncology: recent advances and future challenges. Metab. Clin. Exp. 62, S11–S14.

Kim, J.C., Kim, J.D., 2002. Release property of temperature-sensitiveliposome containing poly(*N*-isopropylacrylamide). Colloids Surf. B 24, 45–52.

Kim, D., Jeong, Y.Y., Jon, S., 2010. A drug-loaded aptamer–gold nanoparticle bioconjugate for combined CT imaging and therapy of prostate cancer. ACS Nano 4, 3689–3696.

Kirpotin, D., Hong, K., Mullah, N., Papahadjopoulos, D., 1996. Liposomes with detachable polymer coating: destabilization and fusion of dioleoylphosphatidylethanolamine vesicles triggered by cleavage of surface-grafted poly(ethylene glycol). FEBS Lett. 388, 115–118.

Kong, G., Anyarambhatla, G., Petros, W.P., et al., 2000. Efficacy of liposomes and hyperthermia in a human tumor xenograft model: importance of triggered drug release. Cancer Res. 60, 6950–6957.

Kono, K., 2001. Thermosensitive polymer-modified liposomes. Adv. Drug Deliv. Rev. 53, 307–319.

Kono, K., Hayashi, H., Takagishi, T., 1994a. Temperature-sensitive liposomes: liposomes bearing poly (*N*-isopropylacrylamide). J. Control. Release 30, 69–75.

Kono, K., Zenitani, K.I., Takagishi, T., 1994b. Novel pH-sensitive liposomes: liposomes bearing a poly(ethylene glycol) derivative with carboxyl groups. Biochim. Biophys. Acta 1193, 1–9.

Kono, K., Yoshino, K., Takagishi, T., 2002. Effect of poly(ethylene glycol) grafts on temperature sensitivity of thermosensitive polymer-modified liposomes. J. Control. Release 80, 321–332.

Kullberg, M., et al., 2005. Improved drug delivery to cancer cells: a method using magnetoliposomes that target epidermal growth factor receptors. Med. Hypotheses 64, 468–470.

Lasch, J., Weissig, V., Brandl, M., 2003. Preparation of liposomes. In: Torchilin, V., Weissig, V. (Eds.), Liposomes: A Practical Approach. Oxford University Press, New York.

Lehner, R., Wang, X.Y., Wolf, M., Hunziker, P., 2012. Designing switchable nanosystems for medical application. J. Control. Release 161, 307–316.

Leroux, J.C., Roux, E., Le, G.D.L., Hong, K., Drummond, D.C., 2001. *N*-isopropylacrylamide copolymers for the preparation of pH-sensitive liposomes and polymeric micelles. J. Control. Release 72, 71–84.

Li, L., Hagen, T.L., Schipper, D., et al., 2010. Triggered content release from optimized stealth thermosensitive liposomes using mild hyperthermia. J. Control. Release 143, 274–279.

Li, J.L., An, X.Q., Pan, Z.F., Sun, L.M., 2014. Photoinduced drug release from complexes of liposome and fluorescent silver nanoparticles. RSC Adv. 4, 9476–9479.

Li, L.L., An, X.Q., Yan, X.J., 2015. Folate-polydiacetylene-liposome for tumor targeted drug delivery and fluorescent tracing. Colloids Surf. B 134, 235–239.

Liu, G.L., Kim, J., Lu, Y., Lee, L.P., 2006. Optofluidic control using photothermal nanoparticles. Nat. Mater. 5, 27–32.

Lubbe, A.S., Alexion, C., Bergemann, C., 2001. Clinical application of magnetic drug targeting. J. Surg. Res. 95 (2), 200–206.

Mackiewicz, N., Gravel, E., Garofalakis, A., Ogier, J., John, J., Dupont, D.M., Gombert, K., Tavitian, B., Doris, E., Ducongé, F., 2011. Tumor-targeted polydiacetylene micelles for in vivo imaging and drug delivery. Small 7, 2786–2792.

Maeda, H., 2010. Tumor-selective delivery of macromolecular drugs via the EPR effect: background and future prospects. Bioconjug. Chem. 21, 797–802.

Maeda, H., 2012. Macromolecular therapeutics in cancer treatment: the EPR effect and beyond. J. Control. Release 164, 138–144.

Maeda, M., Kumano, A., Tirrell, D.A., 1988. H^+ induced release of contents of phosphatidylcholine vesicles bearing surface-bound polyelectrolyte chains. J. Am. Chem. Soc. 110, 7455–7459.

Malam, Y., Loizidou, M., Seifalian, A.M., 2009. Liposomes and nanoparticles: nanosized vehicles for drug delivery in cancer. Trends Pharmacol. Sci. 30 (11), 592–599.

Martina, M.S., et al., 2005. Generation of superparamagnetic liposomes revealed as highly efficient MRI contrast agents for in vivo imaging. J. Am. Chem. Soc. 127, 10676–10685.

Matsumura, Y., Maeda, H., 1986. A new concept for macromolecular therapeutics in cancer chemotherapy: mechanism of tumoritropic accumulation of proteins and the antitumoragentsmancs. Cancer Res. 46, 6387–6392.

Mornet, S., Vasseur, S., Grasset, F., Duguet, E., 2004. Magnetic nanoparticle design for medical diagnosis and therapy. J. Mater. Chem. 14, 2161–2175.

Mozafari, M.R., Johnson, C., Hatziantoniou, S., Demetzos, C., 2008. Nanoliposomes and their applications in food nanotechnology. J. Liposome Res. 18, 309–327.

Mueller, A., Bondurant, B., O'Brien, D.F., 2000. Visible-light stimulated destabilization of PEG-liposomes. Macromolecules 33, 4799–4804.

Nakao, R., Matuo, Y., Mishima, F., Taguchi, T., Maenosono, S., Nishijima, S., 2009. Development of magnetic separation system of magnetoliposomes. Phys. C 469, 1840–1844.

Nappini, S., et al., 2011. Structure and permeability of magnetoliposomes loaded with hydrophobic magnetic nanoparticles in the presence of a low frequency magnetic field. Soft Matter 7, 4801–4811.

Needham, D., Dewhirst, M.W., 2001. The development and testing of a new temperature-sensitive drug delivery system for the treatment of solid tumors. Adv. Drug Deliv. Rev. 53, 285–305.

New, R.C., 1990. Preparation of liposomes. In: New, R.C. (Ed.), Liposomes: A Practical Approach. Oxford University Press, New York.

Noble, G.T., Stefanick, J.F., Ashley, J.D., Kiziltepe, T., Bilgicer, B., 2014. Ligand-targeted liposome design: challenges and fundamental considerations. Trends Biotechnol. 32 (1), 32–45.

Paoli, E.E., Kruse, D.E., Seo, J.W., Zhang, H., Kheirolomoom, A., Watson, K.D., Chiu, P., Stahlberg, H., Ferrara, K.W., 2010. An optical and micro PET assessment of thermally-sensitive liposome biodistribution in the Met-1 tumor model: importance of formulation. J. Control. Release 143, 13–22.

Papahadjopoulos, D., Kimelberg, H.K., 1973. Phospholipid vesicles (liposomes) as models for biological membranes: their properties and interactions with cholesterol and proteins. Progress in Surface Science. Oxford, Pergamon, pp. 141–149.

Parveen, S., Misra, R., Sahoo, S.K., 2012. Nanoparticles: a boon to drug delivery, therapeutics, diagnostics and imaging. Nanomedicine 8, 147–166.

Qiu, D., An, X.Q., 2013. Controllable release from magnetoliposomes by magnetic stimulation and thermal stimulation. Colloids Surf. B 104, 326–329.

Ringsdorf, H., Schlarb, J., Venzmer, J., 1988. Molecular architecture and function of polymeric oriented system: models for the study of organization, surface recognition, and dynamics of bilmembrances. Angew. Chem. Int. Ed. 27, 113–158.

Ryu, J.H., Lee, S., Son, S., et al., 2014. Theranostic nanoparticles for future personalized medicine. J. Control. Release 190, 477–484.

Sahoo, S., Labhasetwar, V., 2003. Nanotech approaches to delivery and imaging drug. Drug Discov. Today 8 (24), 1112–1120.

Sanhai, W.R., Sakamoto, J.H., Canady, R., Ferrari, M., 2008. Seven challenges for nanomedicine. Nat. Nanotechnol. 3, 242–244.

Santra, S., Kaittanis, C., Grimm, J., 2009. Drug/dye-loaded, multifunctional iron oxide nanoparticles for combined targeted cancer therapy and dual optical/magnetic resonance imaging. Small 5, 1862–1868.

Schullery, S.E., et al., 1980. Fusion of dipalmitoylphosphatidylcholine vesicles. Biochemistry 19 (17), 3919–3923.

Seki, K., Tirrell, D.A., 1984. pH-dependent complexation of poly(acrylic acid) derivatives with phospholipid vesicle membranes. Macromolecules 17, 1692–1698.

Siegel, R., Ma, J., Zou, Z., Jemal, A., 2014. Cancer statistics. Cancer J. Clin. 64 (1), 9–29.

Simões, S., Moreira, N.J., Fonseca, C., Düzgünes, N., Maria, C., Lima, P.D., 2004. On the formulation of pH-sensitive liposomes with long circulation times. Adv. Drug Deliv. Rev. 56, 947–965.

Singh, M.R., Pradhan, K., Singh, D., 2012. Lipid matrix systems with emphasis on lipid microspheres: potent carriers for transcutaneous delivery of bioactives. Curr. Drug Deliv. 9, 243–254.

Skouras, A., et al., 2011. Magnetoliposomes with high USPIO entrapping efficiency, stability and magnetic properties. NanomedicineV 7, 572–579.

Soenen, S.J.H., et al., 2009. Magnetoliposomes: versatile innovative nanocolloids for use in biotechnology and biomedicine. Nanomedicine 4, 177–191.

Stroock, A.D., et al., 2002. Chaotic mixer for microchannels. Science 295 (5555), 647–651.

Sudimack, J.J., Guo, W.J., Tjarks, W., Lee, R.J., 2002. A novel pH-sensitive liposome formulation containing oleyl alcohol. Biochim. Biophys. Acta 1564, 31–37.

Sung, H.P., Seong, G.O., Ji, Y.M., 2006. Adding of gold nanoparticles inside the DPPC bilayers of liposome and their effects on membrane fluidities. Colloids Surf. B 48, 112–118.

Svenson, S., 2012. Clinical translation of nanomedicines. Curr. Opin. Solid State Mater. Sci. 16, 287–294.

Szakacs, G., Paterson, J.K., Ludwig, J.A., Booth-Genthe, C., Gottesman, M.M., 2006. Targeting multidrug resistance in cancer. Nat. Rev. Drug Discov. 5, 219–234.

Szoka, F., 1978. Procedure for preparation of liposomes with large internal aqueous space and high capture by reverse-phase evaporation. Proc. Natl. Acad. Sci. USA 75, 4194–4198.

Szoka, F., et al., 1980. Preparation of unilamellar liposomes of intermediate size by a combination of reverse phase evaporation and extrusion through polycarbonate membranes. Biochim. Biophys. Acta 601 (3), 559–571.

Ta, T., Porter, T.M., 2013. Thermosensitive liposomes for localized delivery and triggered release of chemotherapy. J. Control. Release 169, 112–125.

Tapan, K., Saua, A.S., Urbana, S.K., 2009. Controlling loading and optical properties of gold nanoparticles on liposome membranes. Colloids Surf. A 342, 92–96.

Thomas, J.L., Tirrell, D.A., 1992. Polyelectrolyte-sensitized phospholipid vesicles. ACC Chem. Res. 25, 336–342.

Torchilin, V.P., 2000. Drug targeting. Eur. J. Pharm. Sci. 11 (2), 81–91.

Torchilin, V.P., 2001. Structure and design of polymeric surfactant-based drug delivery systems. J. Control. Release 73, 137–172.

Torchilin, V., 2009. Multifunctional and stimuli-sensitive pharmaceutical nanocarriers. Eur. J. Pharm. Biopharm. 71, 431–444.

Van der Meel, R., et al., 2014. Extracellular vesicles as drug delivery systems: lessons from the liposome field. J. Control. Release 195, 72–85.

Veiseh, O., Sun, C., Gunn, J., Kohler, N., Gabikian, P., Lee, D., Bhattarai, N., Ellenbogen, R., Sze, R., Hallahan, A., Olson, J., Zhang, M., 2005. Optical and MRI multifunctional nanoprobe for targeting gliomas. Nano Lett. 5, 1003–1008.

Viroonchatapan, E., Sato, H., Ueno, M., Adachi, I., Tazawa, K., Horikoshi, I., 1996. Magnetic targeting of thermo sensittve magneto liposomes to mouse livers in an in situon-line perfusion system. Life Sci. 58, 2251–2261.

Viroonchatapan, E., Sato, H., Ueno, M., Adachi, I., Murata, J., Saiki, I., Tazawa, K., Horikoshi, I., 1998. Micro dialysis assessment of 5-fluorouracil release from thermosensitive magnetoliposomes induced by an electromagnetic field in tumor-bearing mice. J. Drug Target. 5, 379–390.

Volodkin, D.V., Skirtach, A.G., Möhwald, H., 2009. Near-IR remote release from assemblies of liposomes and nanoparticles. Angew. Chem. Int. Ed. 48, 1807–1809.

Wagner, V., Dullaart, A., Bock, A.K., Zweck, A., 2006. The emerging nanomedicine landscape. Nat. Biotechnol. 24, 1211–1217.

Wang, H., Wang, S., 2000. Preparation and properties analysis of antisense oligonucleotides pH-sensitive liposomes. Biochem. Biol. Phys. Prog. 2, 178–181.

Wang, Y.H., Song, S.Y., Liu, J.H., 2015. ZnO-functionalized upconverting nano theranostic agent: multi-modality imaging-guided chemotherapy with on-demand drug release triggered by pH nanotechnology. Angew. Chem. Int. Ed. 54, 536–540.

Wardak, A., Brodowski, R., Krupa, Z., Gruszecki, W.I., 2000. Effect of light-harvesting complex II onion transport across model lipid membranes. J. Photochem. Photobiol. B 56, 12–18.

Wasan, M.B.B., 2005. Encapsulation of doxorubicin into thermosensitive liposomes via complexation with the transition metal manganese. J. Control. Release 104, 271–288.

Wen, S.Y., Wang, X.H., Lin, L., Guan, W., Wang, S.Q., 2004. Preparation and property analysis of a hepatocyte targeting pH-sensitive liposome. World J. Gastroenterol. 2, 244–249.

Woodbury, D.J., Richardson, E.S., Grigg, A.W., Welling, R.D., Knudson, B.H., 2006. Reducing liposome size with ultrasound: bimodal size distributions. J. Liposome Res. 16, 57–80.

Working, P.K., Newma, M.S., Huan, S.K., Mayhew, E., Vaage, J., Lasi, D.D., 1994. Pharmacokinetics, biodistribution and therapeutic efficacy of doxorubicin encapsulated in Stealth liposomes (Doxil). J. Liposome Res. 4, 667–687.

Wu, N.Z., Da, D., Rudoll, T.L., Needham, D., Whorton, A.R., Dewhirst, M.W., 1993. Increased microvascular permeability contributes to preferential accumulation of Stealth liposomes in tumor tissue. Cancer Res. 53, 3765–3770.

Yan, X.J., An, X.Q., 2013. Thermal and photic stimuli-responsive polydiacetylene liposomes with reversible fluorescence. Nanoscale 5 (14), 6280–6283.

Yan, X.J., An, X.Q., 2014. Multifunctional polydiacetylene-liposome with controlled release and fluorescence tracing. RSC Adv. 4, 18604–18607.

Yang, F., Ma, Y., Zhou, H., Qu, G., 2004. Research and application of new type of liposomes. Shandong Med. 44, 59–60.

Yasuhiro, H., Chie, K., Atsushi, H., 2007. Preparation of poly(ethylene glycol)-modified poly(amido amine) dendrimers encapsulating gold nanoparticles and their heat-generating ability. Langmuir 23, 5243–5246.

Yatvin, M.B., Weinstein, J.N., Dennis, W.H., Blumenthal, R., 1978. Design of liposomes for enhanced local release of drugs by hyperthermia. Science 202, 1290–1293.

Yavuz, M.S., Cheng, Y.Y., Chen, J.Y., Cobley, C.M., Zhang, Q., Rycenga, M., Xie, J.W., Kim, C.K., Song, H., Schwartz, A.G., Wang, L.H.V., Xia, Y.N., 2009. Gold nanocages covered by smart polymers for controlled release with near-infrared light. Nat. Mater. 8, 935–939.

Zhang, J.Q., Zhang, Z.R., Yang, H., Tan, Q.Y., Qin, S.R., Qiu, X.L., 2005. Lyophilized paclitaxel magnetoliposomes as a potential drug delivery system for breast carcinoma via parenteral administration: in vitro and in vivo studies. Pharm. Res. 22, 573–583.

Zhang, H., Ma, Y., Sun, X., 2009. Chemically selective surface glyco-functionalization of liposomes through staudinger ligation. Chem. Commun. 21, 3032–3034.

Zhigaltsev, I.V., et al., 2012. Bottom-up design and synthesis of limit size lipid nanoparticle systems with aqueous and triglyceride cores using millisecond microfluidic mixing. Langmuir 28 (7), 3633–3640.

Zhou, W.T., An, X.Q., Wang, J.Z., Shen, W.G., Chen, Z.Y., 2012. Characteristics, phase behaviour and control release for copolymer-liposome with both pH and temperature sensitivities. Colloids Surf. A 395, 225–232.

Zhu, H.F., Tao, C., Zheng, S.P., 2005. One step synthesis and phase transition of phospholipid-modified Au particles into toluene. Colloids Surf. A 257-258, 411–414.

CHAPTER

NANOTECHNOLOGY TO ENHANCE TRANSDERMAL DELIVERY OF HYDROPHILIC HUMECTANTS FOR IMPROVED SKIN CARE: A MODEL FOR THERAPEUTIC APPLICATIONS

29

Steven Dominguez, Gina A. Mackert, Marek K. Dobke
University of California San Diego, San Diego, CA, United States

CHAPTER OUTLINE

1 BASIC CONCEPTS

Any drug, and by analogy any cosmeceutical, is composed of one or more biologically active ingredients classified as active pharmaceutical ingredients (APIs), as well as excipients. "API" also refers to the active core ingredient that affects the diagnosis of the disease or cosmetic condition, including the prevention, the treatment, and/or the cure. Excipients are the inactive or inert substances present in a drug, frequently selected to ensure drug (cosmeceutical) stability, penetration, and controlled release, as well as desired form. The development of submicron particles and other nanostructures in the pharmaceutical and cosmetic fields has advanced in the past 20 years, allowing novel design formulations and delivery systems consisting of both APIs and excipients.

Nanoscience and nanotechnology are the study and application of extremely small elements, which are used across multiple specialty fields, such as medicine, biology, physics, and material science.

Literally, nanoscale refers to elements and molecules sized to a billionth of a meter or 10^{-9} m. Nanoparticles are 1–1000 nm in size. Nanoscience and nanotechnology applications for optimization of drug delivery pertain both to APIs and to excipients. Transcutaneous delivery is a key investigative model for many nanoscience and nanotechnology applications.

Many recent key drug development components are in the class of small molecules, with whole new therapeutic agents having arisen over the past decade driven by novel chemistries, for example, in antivirals and oncology drugs (Shanley, 2015). The Anatomical Therapeutic Chemical Classification System is used for the classification of active ingredients of drugs according to the organ or system on which they act and their therapeutic, pharmacological, and chemical properties. It is controlled by the World Health Organization Collaborating Centre for Drug Statistics Methodology. Both small molecules and (to a lesser degree) biopharmaceutical technologies have represented sources of new revenues for pharmaceutical industries for the past 6 years (Shanley, 2015).

Antiaging and skin-complexion-improving topicals, with their numerous marketing claims, seemingly limitless APIs, and complex formulations, are leading the way in the remarkable growth of the cosmeceutical industry. As society demands new innovative products with scientific endorsements, researchers are producing agents with increasingly sophisticated and effective delivery systems. Research on and practice of transdermal therapeutical drugs and cosmeceuticals for skin care delivery offer opportunities to investigate not only the agents' delivery per se, but also models of how to enhance drug delivery, optimize active ingredient carriers, and improve extended or controlled release technologies (Siew, 2015). This is one of the examples where small molecules and nanoscale technology have already met. Although nanoparticles and nanostructures sound like futuristic technologies, the revolution that triggered these new developments was initiated by the cosmeceutical products consumer, who wants products that not only cover up or camouflage imperfections in personal appearance, but also have some biological benefits (e.g., skin rejuvenation capabilities). Consequently, the latest trend in cosmeceutical products is to combine clinically proven ingredients with patented delivery systems and the aesthetics of fine cosmetics. Cosmeceutical products are considered those poised in the gap between cosmetic products that simply cleanse and beautify, and pharmaceuticals that cure and heal. Researchers in both fields can take lessons from each other and, in fact, in many instances, the two fields do overlap.

Several excipients are able to promote the transport of active substances across the skin barrier by a variety of mechanisms. Water, clinically proven to hydrate, protect, and revitalize all individual skin types, is one of the most important "APIs" in cosmetology and is central for pharmacological effects of humectants (skin care products designed to hydrate the skin). For example, humectants are supposed to retard moisture loss from the tissue or organ and to increase the moisture content in tissue that is in contact with or treated by the product (Loden, 2009). Agents discussed in this review, such as collagen and hyaluronic acid (HA) can be formulated either as nanoparticles, nanocarriers, or moieties subject to transcutaneous delivery by nanotechnological formulations (e.g., by encapsulation). However, the focus of this review is presentation of collagen moieties and HA as percutaneously delivered humectants; they are therefore considered excipients (carriers of water as an API).

2 TRANSDERMAL AGENTS DELIVERY AND NANOTECHNOLOGY

A molecule will use either or both routes to permeate normal, intact human skin: the transepidermal and the appendageal routes. The appendageal route comprises transport via glands (sweat and sebaceous) and hair follicles. Most transdermally delivered drugs are relatively hydrophobic molecules that can penetrate through the lipid pathway (Siew, 2015). The appendageal route seems to be more important for ions, large polar molecules, and hydrophilic (water-soluble) agents, which normally barely penetrate through the stratum corneum (Moghimi et al., 1999).

Following topical application of nanoparticle formulations, absorption of active compounds can also follow the transcellular absorption pathway (Cieslik-Boczula et al., 2013; Moghimi et al., 1999). Nanocarriers can intactly translocate into the skin without being degraded, or they can be degraded near the skin surface and the incorporated therapeutic molecule can penetrate into the skin layers. The interaction of nanoparticles with the skin layers depends on multiple physicochemical properties of nanoparticles, such as size, surface charge, properties of nanomaterial used, drug-loading efficiency, lamellarity, and mode of application. Nonetheless, there is convincing evidence that irrespective of nanomaterial used, most particles do not cross the stratum corneum. The transappendageal route appears to be the dominant pathway of nanoparticle entry into the skin. Although the appendages present only a small percentage of the skin's surface, they are important pathways for transcutaneous API delivery. Skin appendageal structures might even be utilized not only for directed local drug delivery but also for systemic drug delivery because of the depth they reach and the proximity of blood vessel networks available to the API. However, it is important to note that the mere entry of a substance into a skin appendage does not constitute a completed absorption process or ingress into the circulation (Schneider et al., 2009).

Transdermal drug delivery systems are designed to deliver biologically active agents (drugs or cosmeceuticals) through the skin, principally by diffusion, for local internal if not systemic effects. Transdermal API delivery represents an alternative to other forms and routes of drug delivery. However, there is only a limited number of drugs that are suitable for administration by this route. Overall, skin properties allow the transport of a limited number of agents through passive transdermal delivery; for example, small lipophilic APIs, when dosage and consistency of release are not critical, are poised for transcutaneous delivery. Achieving sufficient drug absorption can prove to be a challenge for many molecules, in particular for hydrophilic particles, given that the skin is an efficient barrier against outside environment factors. The stratum corneum is made up of dense, hydrophilic structures surrounded by an intracellular lipid matrix; therefore, transdermal delivery of hydrophilic agents is a challenge (Cieslik-Boczula et al., 2013; Moghimi et al., 1999). Consequently, there is a significant effort to develop physical and chemical enhancers of percutaneous API absorption. Development of chemical and physical enhancers of transdermal delivery systems (e.g., ethanol-enhancing skin flux, iontophoresis, microneedling, ultrasound, etc.) enables delivery of molecules with hydrophilic properties, such as peptides, proteins, and vaccines (Siew, 2015). In skin care practice the most important methods rely on:

- extraction of lipids from the stratum corneum,
- disruption of the lipid bilayer structure,
- loosening of horny cells (e.g., by microneedling), and
- delamination of the stratum corneum (Moghimi et al., 1999; Siew, 2015).

Methods on how to strengthen and prolong the skin's moisturizing process and how to restore and retain the ideal skin hydration levels are tested on countless patients. Empirical observations—for example, findings, such as the improvement of the delivery and efficacy of hydrating agents, or anesthetics by the pretreatment of the skin's surface through superficially ablating the outer epidermal layers with a conventional Erbium YAG laser, which is an easy and practical method to improve cutaneous delivery of APIs—lead to more elaborate research on the previously mentioned topics. However, in light of the fact that there are still limitations and inconsistencies in chemical or mechanical transcutaneous agents delivery enhancements that can be overcome and that the reproducibility can be refined by physical enhancement or nanotechnologies, new areas of research were developed to address these potential opportunities (Kim et al., 2010).

In general, various methods to overcome the impact of humectant solubility on skin permeability have been investigated. These methods include changing the chemical structure of humectants and utilizing proagent approaches whereby a desired polar functional group is introduced into the structure of the API molecule. Particle size reduction to the nanometer range is one of the most widely investigated approaches to enhance dissolution. In the case of therapeutical agents nanonization (i.e., production of drug nanocrystals), it reduces the drug particle size to the submicron range via either bottom-up methods, such as precipitation and self-assembly or top-down technologies, such as milling and high-pressure homogenization.

Nanocrystals dramatically increase the drug particle surface area, thereby enhancing the dissolution rate when the route (e.g., transcutaneous) is incompatible or phobic to an API solubility. In addition to the nanocrystal approach, other types of nanonization strategies have emerged as new nanoplatforms for the delivery of poorly soluble drugs. Typical examples of these nanoplatforms include nanoemulsions and polymeric micelles. A common feature of these nanoplatforms is the ability to solubilize poorly lipid-soluble agents in a hydrophilic/hydrophobic reservoir or core, and encapsulation within the reservoir or core in a dissolved state. Developments in nanotechnology offer two main transepidermal active hydrophilic agent delivery models:

- by reduction of the size of active molecules penetration pathways not available for larger molecules, and
- by the development of capsules with external lipophilic surfaces containing and transporting encapsulated hydrophilic molecules.

Both appear to lead to relatively consistent results and allow improvements in the control of the process (Kim et al., 2010; Schneider et al., 2009). Certain humectants (skin care products designed to hydrate the skin) also appear to be a good model for studying nanoscale transcutaneous agents delivery procedures, as they can be formulated as nanomolecules acting independently (nanoparticles) or be delivered in an encapsulated form. A complex consisting of a biodegradable nanoparticle wrapping around the API can penetrate the epidermis or both epidermis and dermis, and deliver the content (e.g., humectant as a carrier and water as API) to the desired level of dermis. Because of the relative molecular simplicity, reproducibility, nontoxicity, and plurality of forms, as well as the ability to form systems for sustained delivery of other APIs, both collagen and HA are extremely attractive for investigation of delivery models via biologically repellent, relatively nonpermeable barriers, such as the skin (Hanai et al., 2012; Jaswir et al., 2011; Jegasothy et al., 2014; Wallace and Rosenblatt, 2003).

Downsizing molecules to smaller units requires the continued retention of desired pharmacological properties as an API or as an excipient (e.g., when the molecule is used as a carrier of another API). Nanoscale peptides, which can be utilized to form synthetic collagen, can be developed in the laboratory, while small molecules of tropocollagen, which are native to marine life, can bind water molecules and can be used as a humectant with molecules small enough that they easily permeate through alternative pathways that are more friendly to hydrophilic particles (i.e., appendageal route). Natural tropocollagen extracted from the skin of salmon or tilapia is in micro- to nanomolecular fibrillar form; the molecule size is approximately 280 $\times$ 15 nm (Fallas et al., 2012; Pietrzycka and Zaleska-Zylka, 2014).

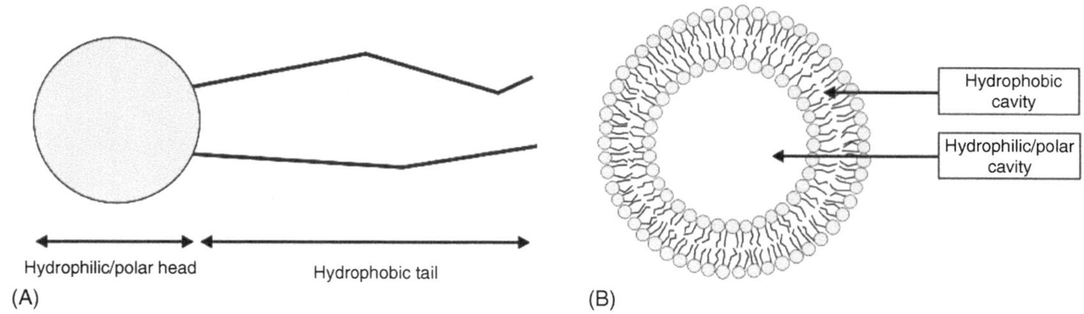

FIGURE 29.1 Liposome: Liposomes Can Encapsulate and Transport Water-Soluble APIs in Their Polar Cavity and Oil-Soluble APIs in Their Hydrophobic Compartment

Collagens from different animals are immunologically distinct. Fish (carp) collagens show no immunologic cross-reactions with phylogenetically more advanced species (Rothbard and Watson, 1965). Overall, marine life species have a higher percentage of simpler collagen units compared to mammals (Kimura et al., 1981; Silva et al., 2014). It has been shown that soluble atelocollagen obtained from tropocollagen, the collagen molecule that makes up collagen fibrils, via the elimination the telopeptide moieties (e.g., by pepsin extraction; technology utilized to produce "Marine Collagen Biologicals, Inc. skin care products" investigated and utilized by the authors) practically abrogates its antigenicity (Hanai et al., 2012; Silva et al., 2014). Recent studies show atelocollagen to be an effective delivery agent for proteins, DNA, and RNA. Many studies, including some involving in vivo tumor models, have evidenced the contribution of atelocollagen to the enhancement of drugs' antitumor activities, and some of them have described the mechanisms (Friess, 1998). Atelocollagen increases endothelial permeability and reduces the level of intercellular sealing (increase of paracellular flux). However, the biological functions of atelocollagen and the mechanism by which it enhances delivery efficiency are still not fully understood. Preliminary data suggest that the three-dimensional structure of the oligonucleotide and atelocollagen complex (with collagen functioning as a carrier) stimulates a signal transduction pathway that acts as a permeability modulator, although the specific pathway that it stimulates remains unknown (Hanai et al., 2012).

Do results on endothelial cell barrier permeability apply to the skin? Neither the path of delivery of atelocollagen into the skin nor the ingredient's ability to stimulate natural collagen production has yet been proven. How about transcutaneous delivery of larger, not necessarily nanoscale molecules? Following the discovery of liposomes in 1960, the cosmetics industry adopted this class of nanotechnology relatively early for the enhancement of transdermal cosmeceuticals or water delivery (Fig. 29.1). Liposomes are particles consisting of a closed bilayer of phospholipids with a hydrophobic external lipid layer, as well as a hydrophilic compartment (inner lining of the liposomal sphere). The wide variety of APIs can be incorporated into the lumen of the liposome and "smuggled" through repellent environments (e.g., epidermis) into deeper skin layers due to the biocompatibility inherently connected with the natural external layer and the lipophilic parts of phospholipids. Liposomes are the largest group of nanoparticulate carriers used for cosmetic applications or for therapeutic drug delivery purposes. For example, liposomes are used as carriers for hydrophilic (water-soluble) anticancer drugs (e.g., doxorubicin). In case of collagen moieties (which can play the role of either carriers or excipients) and water

FIGURE 29.2 "Capture Totale" Antiaging Cream, One of the Earliest Cosmetic Products to Utilize Liposome Technology (Müller-Goymann, 2004)

A special characteristic of liposomes is the ability to adapt to water-soluble and nonwater-soluble ingredients within a product, such as a cosmetic cream or gel without the use of a surfactant (an active ingredient in soap) or an emulsifier (a substance added to oil to help the oil mix with water). The skin care preparations with empty or moisture-loaded liposome or niosome reduce the transdermal water loss and are suitable for the treatment of dry skin. They also enhance the supply of lipids and water to the stratum corneum. Liposome with antiaging complex, such as "Capture Totale" contains thymus extracts, collagen, and elastin (Patravale and Mandawgade, 2008).

as APIs, liposomes may be utilized as carriers of the load consisting of subcarriers, such as collagen or HA and water as the API (Cieslik-Boczula et al., 2013).

The penetration behavior of liposomes is still under discussion in determining if such objects might penetrate (intact) the skin (Schneider et al., 2009). However, ultraflexible liposomal-like structures (e.g., niosomes) are assumed to successfully penetrate whereas conventional liposomes have failed. Since the liposomes are typically larger in size ($\geq$50 nm) than the skin openings, simple penetration is obstructed and a "driving force" or a significant liposome conformability is required. This "driving force" needs to be effective enough to drag the liposomes through the outer skin layer. It has been proposed that the mechanism possibly explaining the observed penetration of dermal liposomes relates to a hydration gradient-driven transport. Thus the different activity of water on the two barrier sides is responsible for aggregate migration (Cieslik-Boczula et al., 2013; Schneider et al., 2009).

Since the introduction of Capture Totale by Dior (Fig. 29.2) and the even more promising ultra-flexible niosomes technology by L'Oréal in 1987, numerous moisturizers, extracts, and creams for skin care have been introduced. Cosmetics manufacturers claim enhanced transepidermal permeability, while others assert that liposomes and niosomes are noninteractive, do not irritate the skin, and provide a water-based matrix for active ingredients released into the deeper layers of the skin (Makeshwar and Wasankar, 2013; Pietrzycka and Zaleska-Zylka, 2014; Vishvakrama and Sharma, 2012). The cosmetics industry typically utilizes liposomes, which are phospholipid vesicles (sized 100–250 nm) that assist in the transepidermal transfer of hydrophilic molecules, such as antibiotics, collagen, elastin, and gene elements (Kowalska-Wojcicka, 2011; Makeshwar and Wasankar, 2013; Pietrzycka and Zaleska-Zylka, 2014; Vishvakrama and Sharma, 2012). Since rules for topical cosmeceuticals application and active agents transcutaneous delivery are less stringent in the cosmetics industry than for drugs, the experience gained is extensive and is accumulated faster than that by the pharmacological industry. Consequently, this research data could be valuable for the design of delivery systems for therapeutic agents.

3 HUMECTANTS FOR IMPROVED SKIN CARE: A MODEL FOR THERAPEUTIC APPLICATIONS

3.1 PEPTIDES, TROPOCOLLAGEN, AND COLLAGEN AS HUMECTANTS AND DERMIS-REGENERATING AGENTS

For millennia, body lotions, creams, and ointments have been placed on the skin for therapeutic and cosmetic effects. Skin hydration, nutrition, and protection from sun, pollution, and stress have been the mainstays of skin care. As cosmetic surgery and dermatology transitioned to become specialties with strong evidence-based practices, skin care products were brought to new efficacy and quality standards. Tradition-driven and empirical potions have been replaced by the most current scientific developments and modern production technologies focused on consistency and safety. New, innovative, and effective skin care products have been developed owing to the integration of additional cosmeceutical characteristics, such as the ability to reinstate the natural skin milieu, alleviate sensitivities, protect from radiation (specifically from the ultraviolet spectrum), lighten the skin's complexion, cleanse and exfoliate, tone and even lift, and change the skin scent. All the aforementioned and nano- and stem cell technologies seem to be the latest technological additions instrumental in enabling continued improvements. Among the cosmeceutical wonders, the best are those developed by coalescing empirical experience with traditional and modern evidence-based principles of hydration, restoration of the skin "water-bank," and retention of the ideal hydration levels of the skin matrix as the core of skin care. The effects of skin-hydrating formulas are heterogeneous, and they are defined as lubrication (increasing skin smoothness, reducing flakiness), boosting skin water content (imparting skin moisture and improving the skin's physical properties such as flexibility), and, last, as enhancers of emolliency (agents filling the spaces between the layers of the epidermis with oils). All are frequently marketed as repair or cell and skin matrix renewal and/or replenishing skin care formulas (Draelos, 1995).

Collagen (and HA) became a top candidate as an ingredient of high-end moisturizers. Considering on one hand the physiology and mechanisms of moisturizing, and on the other hand the physicochemical properties of collagen, it is not surprising that developers of skin care products are focused on collagen. Collagen creams, gels, or sera can not only effectively retard transepidermal water loss but also actively hydrate the skin. The latter effect is due to the fact that collagen molecules (and collagen nanometer range moieties, from basic peptides to tropocollagen) bind molecules of water. The collagen molecule can absorb and adsorb up to 30 times its weight in water (Batieczko and Liedzjewirow, 2010; Draelos, 1995; Fallas et al., 2012; Pietrzycka and Zaleska-Zylka, 2014; Silva et al., 2014; Wallace and Rosenblatt, 2003). Collagen itself is a powerful humectant or it may serve as a hydrophilic matrix or scaffold for other chemicals or cellular products (including stem cells) that have been shown to penetrate and revitalize the skin (Fig. 29.3) (Batieczko and Liedzjewirow, 2010; Draelos, 1995; Gniadecka et al., 1998).

FIGURE 29.3 Commercially Available Marine Tropocollagen (Made Up of Arctic Salmon Skin) for Skin Hydration Is Highly Appreciated by Patients

Regarding cosmeceuticals' penetration across the epidermal barrier in general, transepidermal and appendageal pathways favor hydrophobic compounds, such as lipids; hydrophilic tropocollagen penetration via appendageal mechanisms is similar to water diffusivity and ionic mobility pathways. Simple collagen peptides and tropocollagen permeation is significant enough to enable collagen moieties to deliver drugs or cosmeceuticals, or to function as an effective vehicle for water delivery with striking cosmetic effects with or without liposomal assistance (Batieczko and Liedzjewirow, 2010; Draelos, 1995; Kowalska-Wojcicka, 2011; Pietrzycka and Zaleska-Zylka, 2014). Mechanical enhancement of aqueous compounds, such as collagen or HA penetration across the epidermal barrier by the removal (or thinning) of the protective and lipophilic superficial epidermal layer (stratum corneum) by means of scrubbing or exfoliation is widely practiced in cosmetic medicine. Similarly, other skin-resurfacing procedures, such as chemical peels or laser treatments expose the aqueous-viable epidermal layers and can enhance permeation of hydrophilic agents through the epidermis. In addition, not only will the use of microneedles and microneedle roller skin massage stimulate native dermal collagen formation by creating microdermal puncture wounds, but the procedure may also result in an increase of exogenous collagen infusion into the dermis via man-made temporary skin "pores" (Draelos, 1995; Kowalska-Wojcicka, 2011; Pietrzycka and Zaleska-Zylka, 2014).

Commonly, collagen is an ingredient found in many skin care products. However, very frequently formulations contain negligible amounts of collagen. Therefore, its full hydration-enhancing potential is not maximized because of its low dosage. After injectable bovine collagen utilized as a tissue filler was developed in the 1970s, it also became a popular additive to moisturizers. Typically, complex Type I collagen prepared from bovine tissue by limited proteolysis and purified by differential precipitation has been used for skin care products, as well as in the production of collagen-based, injectable tissue fillers (Draelos, 1995; Wallace and Rosenblatt, 2003). However, purity problems (including prions), antigenicity, and allergic reactions leading to biocompatibility problems have led to a significant setback and decrease in the popularity of bovine-collagen-based skin care products and injectable tissue fillers (Jaswir et al., 2011; Silva et al., 2014). Nowadays, despite technologies allowing for the removal of macromolecular, stable, fibrillar nonhelical collagen ends of molecular chains, which are responsible for the antigenicity and allergic reactions experienced by humans, the safety record of bovine collagen is questioned (Draelos, 1995; Mullins et al., 1996). A more primitive, simpler-structured collagen molecule is free of the immunizing and allergy-causing domains present in collagenous proteins derived from species positioned "higher" in the phylogenetic tree (e.g., bovine) but retains the water-binding capability essential for the subsequent skin hydration effect (Kowalska-Wojcicka, 2011; Pietrzycka and Zaleska-Zylka, 2014; Rothbard and Watson, 1965; Silva et al., 2014; Wallace and Rosenblatt, 2003). As just elucidated, research on phylogenetically less advanced and specialized forms of collagen with a relatively simple molecule structure (e.g., moieties derived from marine life) opened up new opportunities for the cosmetics industry and medicine and sparked a renewed interest in collagen. The scientific fact is that simpler collagen has demonstrated more biocompatibility with more phylogenetically advanced species, such as humans than collagen derivatives from other advanced animals, such as mammals.

Once the technology needed to extract and process marine collagen to obtain simple peptides and to ensure tropocollagen stability was mastered, marine collagen became used in cosmetology (Batieczko and Liedzjewirow, 2010; Draelos, 1995). The discovery that hydrated tropocollagen thermostability may depend on the nature of the source (e.g., products from marine life in cooler water may behave differently than moieties obtained from tropical fish), and that tropocollagen particles from arctic fish break down into smaller components in contact with the 37°C environment of human skin, thus further enhancing its penetrability across the epidermal barrier, and change dermal phospholipid conformation, improving the process of hydration, provided additional impetus for research on the potential of marine collagen in cosmetology and medicine (Batieczko and Liedzjewirow, 2010; Cieslik-Boczula et al., 2013; Gniadecka et al., 1998; Pietrzycka and Zaleska-Zylka, 2014; Silva et al., 2014).

Skin care test applications revealed no adverse inflammatory or allergic reactions even with a pure marine collagen formulation and concentration of almost 100%. Superb skin hydration and volumizing effects of marine-collagen-based products led to their rapidly growing popularity in Europe, Asia, and the Americas (Fig. 29.4). Based on the research findings and clinical practice observations available to date, the striking revitalization effects of marine collagen gel–based skin care is attributed to hydration and thickening of the skin layers by their volumization, clinically resulting in the reduction of small wrinkles and improved skin tone (Draelos, 1995; Gniadecka et al., 1998; Pietrzycka and Zaleska-Zylka, 2014).

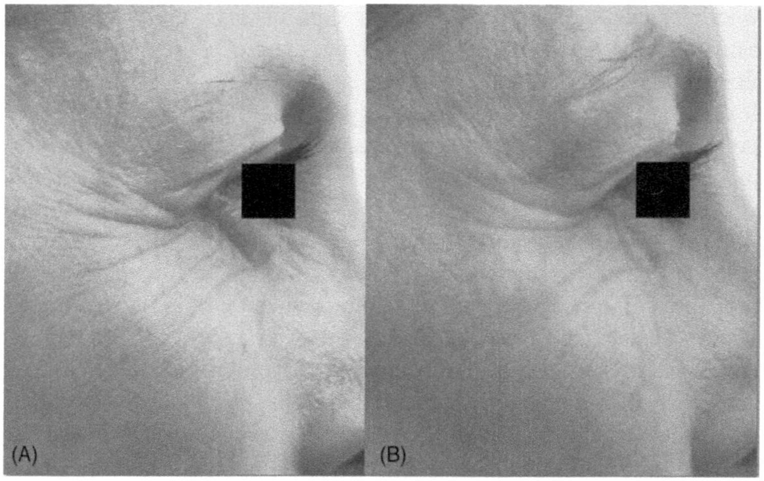

FIGURE 29.4 Notable Volumization Through Hydration of the Lower Eyelid and Lateral Orbital Area Skin, Showing Reduction of Small Wrinkles After Only 2 weeks of Topical Marine Tropocollagen Application

Researchers have no evidence that the treatment of the skin's surface with collagen, with the assumption that some percentage of the applied collagen will reach the dermal layer, would result in an increase of dermal collagen by the incorporation of the externally delivered substrate or by the increased production of native collagen in addition to the increased skin water content effect (Gniadecka et al., 1998; Wallace and Rosenblatt, 2003). However, the fate of the externally supplied collagen is tested through intriguing projects. Such questions as whether exogenous collagen stimulates fibroblasts to produce native dermal collagen, and whether when externally delivered collagen is incorporated into or binds to existing molecules of skin collagen it might reveal antioxidative activity, thus improving aging skin quality as the end point outcome, await answers. Last, investigations on vehicles [microcapsules, liposomes, niosomes, nanostructured lipid carriers (NLCs) technology] that enhance the transepidermal delivery of APIs essential for dermal revitalization and maintenance of youthful characteristics, such as collagen seem to be promising and raise an expectation that in the foreseeable future topical collagen delivery to the dermis and its structural and functional integration with the native skin matrix will become the reality and part of the management in combatting skin aging (Fallas et al., 2012; Pietrzycka and Zaleska-Zylka, 2014).

In addition, progress in supramolecular chemistry, an emerging subdiscipline of chemistry that primarily concerns itself with noncovalent intermolecular interactions, has opened new opportunities for nanocosmetology. These relatively weak linking forces are applied to create, in a massively parallel fashion, out of nanometer-sized objects (e.g., nanoclusters of precollagen peptides) large complex collagen matrices, which form hydrogels suitable for injection, implantation, or bioengineering and regenerative medicine applications (the latter are well beyond nanoscale materials, although they originate in the nanotechnological world) (Wallace and Rosenblatt, 2003).

Depending on the collagen's fibrillar and structural nativity, molecule size, and source, collagen-based products can assure a spectrum of pathways of transdermal delivery and APIs release kinetics. Collagens are excellent, biocompatible, and biodegradable regardless of whether delivered via liposomes/niosomes nanotechnologies or grossly by surgical implantation of the collagen matrix drug carrier materials (Friess, 1998; Ruszczak and Friess, 2003; Wallace and Rosenblatt, 2003).

3.2 HYALURONIC ACID AND NANOTECHNOLOGICAL SKIN CARE

What about HA, which is another popular hydrophilic humectant? Just like collagen, HA can attach and bind large amounts of water (liters can be bound by 1 g of HA) (Jegasothy et al., 2014). HA is said to be the ultimate hydrating agent and the key ingredient responsible for moisture retention in the skin (Fig. 29.5). Youthful skin is hydrated better than aging skin of seniors because it contains large amounts

FIGURE 29.5 One of the Commercially Available Formulations of Hyaluronic Acid Producing Excellent Skin Hydration Outcomes

of endogenous HA. Similar to collagen (and its nano- and nonnanotechnological forms), the concepts of skin hydration and strategy of hydration also apply to HA. The size and molecular weight decide how these molecules interact with the stratum corneum and how effective they are in skin hydration (Pavicic et al., 2011). HA can be low-molecular (0.1–0.3 MDa) or nanomolecular (0.02–0.05 MDa), and both forms are small enough that they can passively penetrate through the epidermis with bound water molecules, whereas macromolecular HA (1.3–1.8 MDa) requires a transepidermal delivery enhancement (e.g., microneedling, thermal ablation epidermal barrier breaching) (Draelos, 1995; Haedersdal et al., 2010; Siew, 2015). Low-molecular HA seems to hydrate skin better than high-molecular-weight formulation (Pavicic et al., 2011). The challenging key characteristic of conventional HA formulations in topically applied preparations historically has been that the molecule is approximately 3,000 nm in diameter whereas the intercellular spaces are only 15–50 nm and just 6–10 nm at the hyaline membrane level (Desai et al., 2010; Moghimi et al., 1999). This makes it impossible for conventional HA products to penetrate into the deep dermis (Desai et al., 2010; Draelos, 1995; Siew, 2015). But history repeats the collagen path: new technologies have managed to reduce the size of an HA molecule without disturbing its subunit structure and water-binding capacity to a nanosize of 5 nm, enabling this type of HA molecule to cross the skin barrier to dermal level even without liposomal and niosomal enhancers (Desai et al., 2010; Jegasothy et al., 2014).

3.3 TESTING OF SKIN WATER INCREASE AND ATTRACTION OF WATER TO EPIDERMIS

Practical methods to test humectants include:

- clinical assessment and evaluation,
- noninvasive biophysical measurements such as the assessment of the electrical properties of the skin and spectroscopic methods, and
- invasive procedures such as skin biopsy, histochemistry, and microdialysis (Darlenski and Fluhr, 2011).

To test the effects of a humectant, spectrophotometric intracutaneous spectroscopy is a skin-imaging technique that allows noninvasive in vivo quantification and assessment of different compounds within the skin, including the construction of mathematical optical models of the skin and assessment of skin hydration (Angelopoulou, 1999).

Different layers of the skin have distinct optical properties. When a light beam incident to the skin penetrates the superficial layers, some of it is absorbed, while some energy is remitted back and can be measured. Depending on the substance subjected to analysis, there are differences in the absorbed and remitted light spectrum. In general, regarding hydration, this technique operates on the principle that direct surface reflection measured by the surface analysis technique is lower in the hydrated stratum corneum (Matts and Cotton, 2009). Regarding humectants and nanotechnology, infrared spectroscopy with attenuated total reflectance (ATR) is particularly useful to assess liposomal dispersion and epidermal lipid layers conformation (*trans* vs. *gauche*) (Cieslik-Boczula et al., 2013). The physics of ATR testing is based on the principle that total internal reflection results in an evanescent wave. During testing, an infrared light beam is passed through the ATR crystal in such a manner that it reflects off the internal surface in contact with the sample at least once, thus creating an evanescent wave extending into the sample. The depth of penetration into the sample (e.g., water) is typically in the range of 0.5–2 μm, with the exact value determined by the wavelength of the light, the angle of incidence, and the refractory indices of the crystal and the tested medium. As the beam exits the crystal, it is recognized and analyzed by a detector. Internal reflection with or without an applied cosmeceutical, pharmaceutical, or ligand will produce a difference in spectra that allows for studying the sample's conformational changes of proteins upon binding (e.g., changes resulting from different levels of hydration). In other words, changing levels of hydration will result in refractional changes. To enable quantifiable measurements, water-soluble molecules may require tags, allowing the molecule to be anchored to bilayered lipids.

As a cosmetic bonus, the mere presence of solid lipid nanoparticles themselves in cream formulations may result in improved skin hydration and viscoelasticity (Patravale and Mandawgade, 2008). On the other hand, methodologically their presence may pose a pitfall as they may influence such electrical properties and measurements as capacitance (measured by a Corneometer) or viscoelasticity (measured by a Cutometer), and cause difficulty in the interpretation of the results (Fig. 29.6) (Wissing and Muller, 2003). Particles sized in the nanometer range (40 nm) are highly crystalline and show a tendency to fuse, whereas particles sized in the micrometer range (2000 μm) form films. Each size can

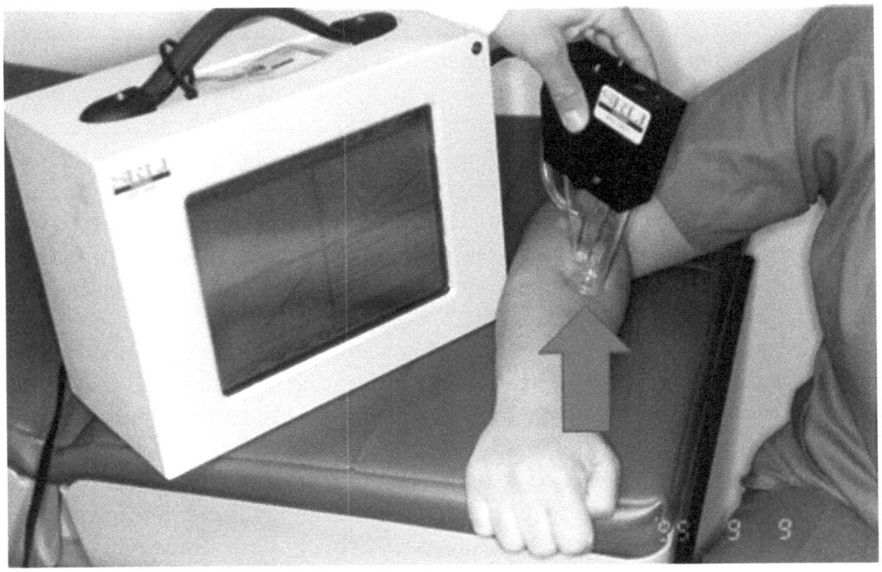

FIGURE 29.6 Noninvasive Testing of the Mechanical and Viscoelastic Skin Properties Incorporates Fixed Boundaries of the Selected Skin Segment (*Arrow*) and Minimal Prestress Skin Alteration

Linear negative pressure (rate 10 mmHg/s) is exerted, and anisotropic skin segment changes are characterized by changes in reflectivity of a calibrated target laser. Beyond water content, skin deformation and its speed reflect rather complex skin characteristics, such as the organization of the dermal collagen and elastic fibers, and the desquamation process. Thus, assessment of skin biomechanical properties can be applicable only as an addition to the established hydration measurements.

From Darlenski, R., Fluhr, J.W., 2011. Mositurizers and emollients. In: Fluhr, J.W. (Ed.), Practical Aspects of Cosmetic Testing. Springer, Heidelberg, pp. 123–141; Dobke, M., DiBernardo, B., Thompson, R.C., Usal, H., 2002. Assessment of biomechanical skin properties: is cellulitic skin different? Aesthet. Surg. J. 22, 260–266.

be incorporated into creams, increasing the degree of skin hydration; however, each is done through different means (Wissing and Muller, 2003). It could be hypothesized that in this particular experiment nanoparticles self-assembled cubosomes (submicron nanostructured particles of bicontinuous cubic liquid crystalline phases). The essence of this process is the formation of liquid crystalline particles of certain surfactants when mixed with water and a microstructure at a certain specific ratio. Cubosomes offer a large surface area and low viscosity, and can exist at almost any dilution level. They have high heat stability and are capable of carrying hydrophilic and hydrophobic molecules, so they are an attractive choice for cosmetic applications, as well as for drug delivery. However, for the simplicity of investigative models and to avoid interfering (although there could be potentially positive problems), humectants should be delivered in vehicles that biodegrade as rapidly as possible, and not by nonbiodegradable capsules (Lohani et al., 2014; Silpa et al., 2012).

While collagen moieties allow chromophore mapping, HA may not scatter the incident light beam and could therefore induce the Tyndall effect, depending on the size of particles, and the effect's severity may be proportional to the amount of hyaluronic molecules within the tissue. Consequently, spectrophotometry-based methods may be more difficult to apply to in vivo investigation of HA transepidermal penetration, attraction of water to the epidermis, and modulation of the overall skin hydration by HA (Hirsch et al., 2006; Matts and Cotton, 2009). Efficacy of skin hydration by HA-based formulations is measured by such devices as a Corneometer (skin capacitance measuring principle) or relies on the assessment of changes in viscoelastic properties of the skin (Dobke et al., 2002; Khazaka, 2011; Pavicic et al., 2011).

Therefore, as transcutaneous APIs delivery is influenced by many biophysical, chemical, and structural characteristics of the epidermis, it is appropriate to apply multiple techniques simultaneously to comprehensively assess their penetration and skin hydration factors. Factors known as causing bias or a predisposition to investigative errors should be taken into account when selecting an investigative method. The variety of different techniques can improve the validity of research studies, especially considering that still little is known about the impact of specific nanomaterials on the physiology of skin (Darlenski and Fluhr, 2011; Gupta et al., 2013; Wissing and Muller, 2003).

Recently, a very practical, small, and light ultrathin wafer-like sensor was introduced. It resembles and is applied like a dressing similar to a Band-Aid. This dressing is applied on a skin segment to measure intrinsic thermal absorption and thermal gradient dispersion properties of the skin, including their relationship to vascularity, thickness of the stratum corneum, and skin hydration level simultaneously, potentially overcoming shortcomings of some other systems (Webb et al., 2015). A cosmetics industry leader (L'Oréal, France; holder of many patents in nanotechnology, the same manufacturer that developed and patented niosomes for transcutaneous APIs delivery years ago) embraced this device to test cosmeceuticals (Kaur and Agrawal, 2007; Sharma and Sharma, 2012). This comprehensive sensor system is superior to cumbersome and traditional approaches, and is made possible by the expanded capabilities of soft, ultrathin, noninvasive, and in vivo measurement systems. In vivo measurement capabilities enable the establishment of a clear relationship between skin hydration and in vivo skin thermoregulatory properties, including heat distribution or heat loss. As expected, the data also reveal that the in vivo thermal energy transport properties of the skin are not only a mere function of hydration, but are also influenced by the local structure and local blood and lymph perfusion characteristics of the skin. To generate similar information with other methods would require either expensive and complex optical thermography and laser heating systems or bulky single-point probes. The latter can have undesired effects potentially leading to measurement errors: skin properties (e.g., resulting from alteration of blood perfusion) may change due the pressure that must be applied to the skin when measured in vivo. The thin wafer technique provides the ability to examine the effectiveness of dermatological APIs. In addition, miniaturization and developments in wireless technology provide a tool for continuous, real-time monitoring of the skin's functional status.

4 CONCLUSIONS AND LESSONS FROM HUMECTANTS FOR SKIN CARE MODELING THERAPEUTIC APPLICATIONS OF NANOTECHNOLOGIES

Developments in skin care and percutaneous delivery of skin care products, skin care APIs, new delivery systems, permeability enhancers, and nanotechnologies all mirror achievements, areas of scientific uncertainty, and practical pitfalls pertaining to therapeutic applications. Therefore, in the authors' opinion, the inclusion of this chapter in the volume on therapeutic nanostructures for drug delivery is very appropriate. The review of nanotechnology and pathways on how to enhance transdermal delivery of hydrophilic humectants for skin care appears to set a stage for a full array of investigative models for therapeutic applications, and almost all currently known applications can be modeled by skin care and transdermal delivery. We advocate that clinical trials and research on cosmetic applications do not involve critically sick patients, so that APIs administration can be interrupted at any time without altering the course of the disease. Also, assuming that most skin care users are in general healthy, research observations have reduced sources of bias and are therefore academically appropriate and ethical. Collagen and its different forms are flagship molecules for research on nanotechnology integrating engineering with biology, chemistry, and physics (Desai et al., 2010; Friess, 1998; Makeshwar and Wasankar, 2013; Vishvakrama and Sharma, 2012). Collagens have their place in multiple nanostructured formulations currently used in cosmetology, such as submicron-sized nanoemulsions and nanosuspensions, nanospheres (with APIs nanoparticles in matrix or gel), nanotubes and nanopowders, liposomes and niosomes, dendrimers (nanoscaled three-dimensional macromolecules) all mirroring therapeutic methodologies (Jansen, 2009).

As a practical note from a clinical practitioner's experience, it is noteworthy to mention that when applying an HA or collagen humectant in very dry climates you have to be careful. The water binding and delivering agents cannot draw enough moisture from the environment; consequently, they may actually pull moisture out from deeper layers of the skin and paradoxically dehydrate the skin. The process of water flux enhanced by humectants is not necessarily only in one direction! Even nanotechnology is subject to guidelines resulting from the experience of medical practice. For instance, shortly before applying humectants one should moisten the skin surface with water to increase the pool of water available for the humectant to be drawn into the skin. Shortly after (e.g., 2 min after) application of a humectant, a sealing emollient should be applied to prevent surface desiccation. Intuitively, this seemingly nonscientific observation (and lesson) has theoretical meaning for the therapeutic applications of nanoproducts delivery: the assumption that an API's passage is always "in" may be naive and not necessarily true; the direction of the transdermal API's passage depends on circumstances and handling of the environment (and perhaps the microenvironment, too). Some humectants (e.g., urea) also have a strong osmotic activity and, if not absorbed, may draw the moisture in an unintended direction, causing transepidermal water loss (Loden, 2009).

The use of collagen and HA themselves as drug delivery systems are very comprehensive and diverse. Collagen can be extracted into an aqueous solution and molded into various forms of delivery systems, including its nanotechnological formulation for gene delivery (Lee et al., 2001; Rossler et al., 1995; Wallace and Rosenblatt, 2003). Although bioengineers can form biomatrices made of collagen, its fragility precludes the building of collagen-made epidermis-piercing microneedles (current designs do not ensure an adequate ratio of microneedle fracture force to skin insertion force) (Lee et al., 2001; Prausnitz, 2004; Wallace and Rosenblatt, 2003). However, designers of a novel transcutaneous drug delivery system took advantage of the observation that hydrating agents can draw and bind interstitial tissue water (Siew, 2015). This feature, which is rather an unfavorable feature from the skin care point of view, was utilized to construct a creative model of therapeutic agents delivery. The device combines concepts of microneedling and the ability of humectant agents to change configuration upon hydration (Haedersdal et al., 2010; Lee et al., 2001; Wallace and Rosenblatt, 2003). Traditional methods of microneedling-mediated drug delivery across the skin use solid microneedles that are applied and removed to create transient micropores, followed by application of a traditional API-containing transdermal patch (Siew, 2015). Similarly, uses of solid microneedles coated with drug molecules for instant delivery, with microneedles removed after the coating material dissolves; uses of soluble polymeric/carbohydrate microneedles containing drugs that dissolve in skin interstitial fluid over time, thereby delivering the drug; and even uses of hollow microneedles for delivery of fluid containing drugs do not ensure high levels of control over the delivery process (Siew, 2015). In brief, a novel hydrogel-made microneedle system for controlled transdermal drug delivery consists of a drug-loaded adhesive patch integrated into a solid, crosslinked hydrogel microneedle array; following application of the integrated hydrogel microneedle patch to the skin's surface, interstitial epidermal/dermal water diffuses into the crosslinked, integrated microneedles, causing their controlled swelling (with or without microlumen), forming an in situ hydrogel conduit. This further results in the liberation and diffusion of drug molecules from the adhesive patch through the hydrogel microneedles or microtubes into the skin. The hydrogel microneedle array remains intact, even after removal from the skin, thereby leaving no polymeric material in the skin following drug delivery, or it may be biodegradable (Donelly et al., 2012; Sharma and Sharma, 2012; Siew, 2015).

Besides differences in water-binding capacity of humectants, their absorption into the dermis (the rate, the speed, effective penetration of repeated doses) is important for the final result: hydration of skin layers. As stated, the permeation of APIs across the skin layers is influenced by a number of physiological and physicochemical factors. Physiological factors that affect API transcutaneous penetration include skin age and general thickness, in particular the thickness of the stratum corneum (which varies between different anatomical sites); the degree of skin hydration; the presence and density of appendages, including hair follicles; and skin temperature. The potential usefulness of increased skin temperature (physical enhancement potentially also applicable to therapy) is another lesson from cosmetic spas and illustrates how simple interventions may facilitate transdermal water-soluble APIs delivery (Figs. 29.7 and 29.8) (Pietrzycka and Zaleska-Zylka, 2014; Siew, 2015). These traditional, empirical, and sometimes casual methods of enhancement of transcutaneous APIs delivery are being validated and augmented by the most current scientific developments (Yosipovitch et al., 1998). Some of these observations (e.g., changes in epidermal permeability, fluctuations in the opening and closing of skin pores depending on the skin temperature) may require further research, as these variations may impact the efficacy of liposomes, which are not thermostable (Ruszczak and Friess, 2003; Wissing and Muller, 2003).

The advent of new active transcutaneous APIs delivery systems enables delivery of larger molecules with hydrophilic properties, such as peptides and proteins. Techniques for controlled breaching of the skin surface for purposes of transcutaneous APIs have been studied to refine and enhance further the

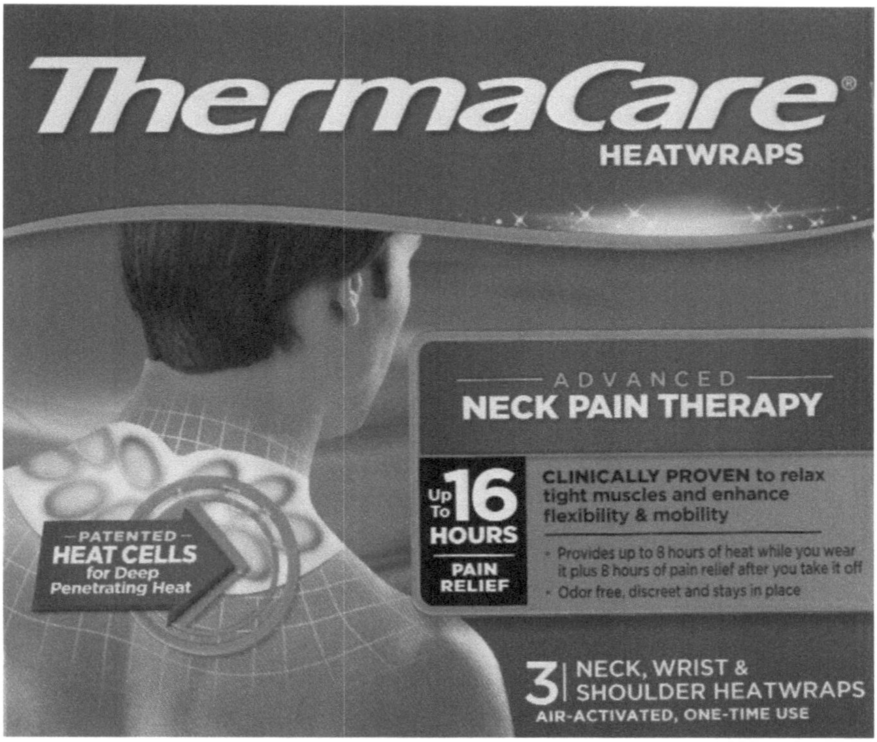

FIGURE 29.7 The Permeation of Drugs Across the Skin Is Influenced by a Number of Factors, Including Temperature

Commercially available warming pads may induce local skin hyperthermia and cause a phase change of the epidermal lipids so that transcutaneous penetration of hydrophilic agents is increased.

effectiveness of micro- and nanotechnology. Many of these techniques were tested in cosmetology for many years. Skin abraders, local energy dischargers (porators), or hard, sharp objects that can punch openings controlled in number, size, and depth into the skin are tested. The family of these tools includes perforators and microneedles, including hollow and/or biodegradable designs. Using these gadgets, one can punch a number ($\sim 10^2$ cm^{-2}) of relatively wide ($\geq 10^3$ nm) openings into the epidermis (reaching the dermis, depending on the needle length and thickness of the epidermis), intentionally violating the protective skin barrier and creating temporary (~ 1 day) penetrating miniwounds that then allow small APIs, drug quantities (≤ 10 mg), and even large molecules to pass. Mechanical skin priming through microneedling allows microscopic ($\geq 10^3$ nm) ballistic droplets or particles to insert small drug amounts (~ 1 mg) into the upper skin through the $\geq 10^6$ cm^2 pores they create. The nanoscaled (5–10 nm) skin-perforating devices for transdermal APIs delivery are still in the experimental phase of development, whereas the larger epidermal barrier-breaching methods are approved for use in humans (Cevc and Vierl, 2010).

Macro- or microneedling and ablative laser devices (e.g., ablative CO_2 laser) allow the creation of vertical channels that may assist the delivery of topical agents applied to the skin's surface into the

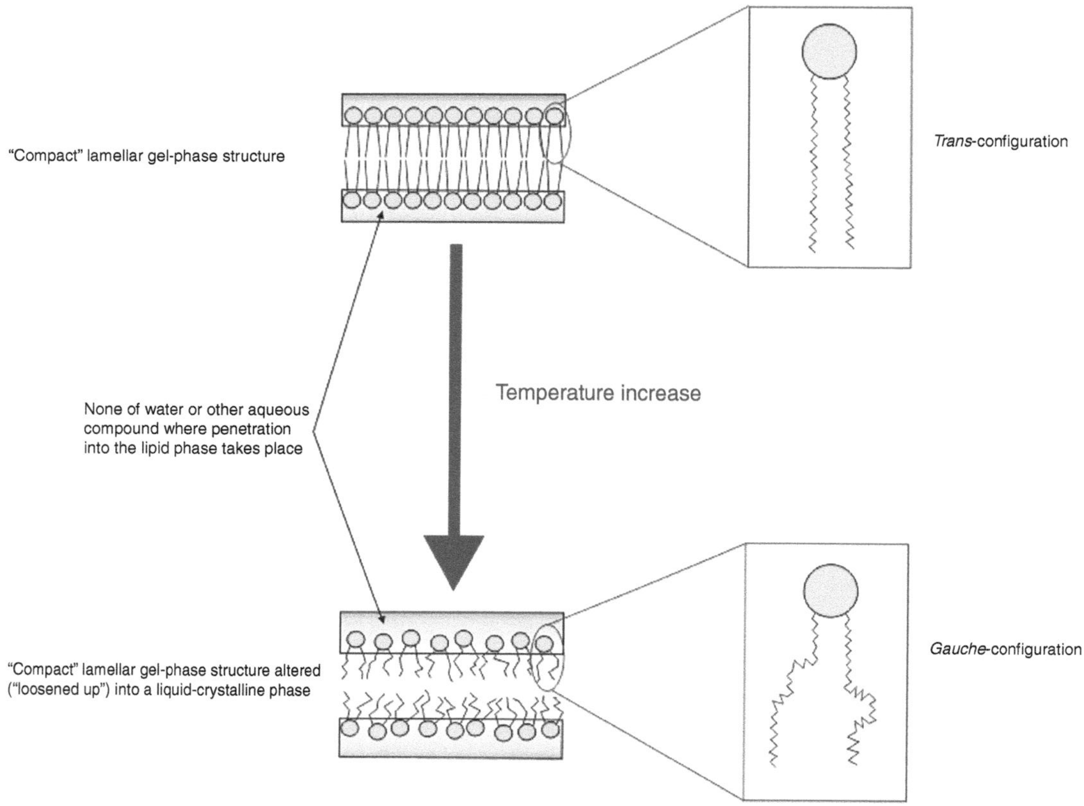

FIGURE 29.8 An Increase in Local Temperature Can Change the Temperature-Dependent Compact Epidermal Lipids Gel Phase Into a Liquid-Crystalline Phase, Which Is More Permeable for Water-Soluble Agents

This happens by changing the conformation of the phospholipids from *trans* to *gauche*.

desired level of the skin, even extending to the dermis if desired and possibly acting as channels for drug uptake, thereby allowing bioavailability of humectants (and water) in the skin to be increased and controlled (Haedersdal et al., 2010; Prausnitz, 2004; Siew, 2015).

Paralleling the research and design of skin barrier breaching devices, which enable transcutaneous APIs delivery, significant effort is directed into the development of micro- and nanosized carriers (e.g., NLC technology). Resilient to challenges posed by physical and chemical barriers, sufficiently deformable and chemically stable nanosized carriers emerge as alternatives, enabling controlled and reliable transcutaneous drug delivery. Deformability of "soft" skin penetrators, which are typically composite colloids, is their key feature enabling epidermal penetration. For medical, especially systemic, applications, they must obtain or retain their ability to act as API carriers on, in, and ideally below the skin barrier(s). Such self-regulating, shape-adaptable, and stable hetero-aggregates can carry drugs through $\leq 10^9$ cm^{-2} cutaneous pores through the primary skin barrier, minimize cutaneous drug clearance, and open spontaneously at the desired level and at the desired time. This permits deep/targeted deposition and prolonged action of the carrier-transported agents. The development of products for clinical or cosmetic applications based on ultraadaptable, self-regulating, nanosized ($\sim 10^2$ nm) carriers is underway (Cevc and Vierl, 2010; Jansen, 2009; Morganti, 2010; Siew, 2015).

Collagen is one of the most promising carrier materials, ensuring formulation stability, nontoxicity, and controlled release rates. Systems consisting of liposomes and collagen hydrogels have been developed. It was found that a coupling of liposomes to the collagen gel matrices additionally enhances the stability of APIs and the whole system due to—so well known from cosmetic research and applications—the antioxidant effect of collagen molecules (Lee et al., 2001; Pietrzycka and Zaleska-Zylka, 2014; Weiner et al., 1985).

In conclusion, little is known about the safety aspects of the nano-engineered materials that are being released into consumer cosmetic products, health care products, and our environment (Gupta et al., 2013; Lohani et al., 2014). Nanotechnology-based cosmeceuticals seem to offer the advantage of the same diversity in products as traditional lines but with different (increased) bioavailability of active ingredients and, most important, an increased aesthetic appeal of cosmeceutical products with prolonged and/or systemic effects. However, the increased use of nanotechnology in cosmeceuticals has raised a concern about the possible penetration of nanoparticles through the skin and potential hazards to human health. Nanoscale drug delivery may lead to the emergence of new categories of irritants, allergens, and so on, and improved API penetration may open the door for unexpected toxicities. Hence, it is critical for professionals intimately involved with the research on nanotechnologies and the health of participants to be aware of this technology, to educate patients about it, and to play an active role in evaluating this technology and setting policies and guidelines for its safe and rewarding use. The Food and Drug Administration agency has established guidelines, "Guidance for Industry: Safety of Nanomaterials in Cosmetic Products," available at (http://www.fda.gov/Cosmetics/GuidanceRegulation/GuidanceDocuments/ucm300886.htm), summarizing data and concepts on the safety assessment of nanomaterials when used in cosmetic products. Key points include the legal requirements for cosmetics manufactured using nanomaterials, which should be the same as those for any other cosmetics. While cosmetics are not subject to premarket approval, companies and individuals who market cosmetics are legally responsible for the safety of their products, which must be properly labeled (Gupta et al., 2013; Morganti, 2010).

REFERENCES

Angelopoulou, E., 1999. The reflectance spectrum of human skin. University of Pennsylvania Department of Computer and Information Science (CIS) Technical Report No. MS-CIS 99-29, Paper 584.

Batieczko, S.A., Liedzjewirow, A.M., 2010. Kolagen (Collagen: new strategy to maintaining good health and youth). Hobbit Plus, Poland (in Polish).

Cevc, G., Vierl, U., 2010. Nanotechnology and the transdermal route: a state of art review and critical appraisal. J. Control. Release 141, 277–299.

Cieslik-Boczula, K., Jaszczyszyn, A., Gasiorowski, K., Swiatek, P., Malinka, W., 2013. The role of lamellar lipid layers in the skin penetration. Aesthet. Cosmetol. 2, 19–22.

Darlenski, R., Fluhr, J.W., 2011. Moisturizers and emollients. In: Fluhr, J.W. (Ed.), Practical Aspects of Cosmetic Testing. Springer, Heidelberg, pp. 123–141.

Desai, P., Patlolla, R., Singh, M., 2010. Interaction of nanoparticles and cell-penetrating peptides with the skin for transdermal drug delivery. Mol. Membr. Biol. 7, 247–259.

Dobke, M., DiBernardo, B., Thompson, R.C., Usal, H., 2002. Assessment of biomechanical skin properties: is cellulitic skin different? Aesthet. Surg. J. 22, 260–266.
Donelly, R.F., Singh, T.R., Garland, M.J., Migalska, K., Majithiya, R., McCrudden, C.M., Kole, P.L., Mahmood, T.M., McCarthy, H.O., Woolfson, A.D., 2012. Hydrogel-forming microneedle arrays for enhanced transdermal drug delivery. Adv. Funct. Mater. 22, 4879–4890.
Draelos, Z., 1995. Cosmetics in Dermatology. Churchill Livingstone, New York.
Fallas, J.A., Dong, J., Tao, Y.J., Hartgerink, J.D., 2012. Structural insights into charge pair interactions in triple helical collagen-like proteins. J. Biol. Chem. 287, 8039–8047.
Friess, W., 1998. Collagen—biomaterial for drug delivery. Eur. J. Pharm. Biopharm. 45, 113–136.
Gniadecka, M., Nielsen, F., Wessel, S., Heidenheim, M., Christensen, D., Wulf, H.C., 1998. Water and protein structure in photoaged and chronically aged skin. J. Invest. Dermatol. 111, 1129–1133.
Gupta, S., Bansal, R., Gupta, S., Jindal, N., Jindal, A., 2013. Nanocarriers and nanoparticles for skin care and dermatological treatments. Indian Dermatol. Online J. 4, 267–272.
Haedersdal, M., Sakamoto, F.H., Farinelli, W.A., Doukas, A.G., Tam, J., Anderson, R.R., 2010. Fractional CO_2 laser-assisted drug delivery. Lasers Surg. Med. 42, 113–122.
Hanai, K., Kojima, T., Ota, M., Oyodene, J., Sawada, N., 2012. Effects of atelocollagen formulation containing oligonucleotide on endothelial permeability. J. Drug Deliv. 2012, 1–9.
Hirsch, R.J., Narukar, V., Carruthers, J., 2006. Management of injected hyaluronic acid induced Tyndall effects. Lasers. Surg. Med. 38, 202–204.
Jansen, J., 2009. Encapsulation to deliver topical actives. In: Barel, A.O., Paye, M., Maibach, H.I. (Eds.), Handbook of Cosmetic Science and Technology. Informa Healthcare, New York, pp. 787–796.
Jaswir, I., Monsur, H.A., Salleh, H.M., 2011. Nano-structural analysis of fish collagen extracts for the new process development. Afr. J. Biotechnol. 10, 18847–18854.
Jegasothy, S.M., Zabolotniaia, V., Bielfeldt, S., 2014. Efficacy of new topical nano-hyaluronic acid in humans. J. Clin. Aesthet. Dermatol. 7, 27–29.
Kaur, I.P., Agrawal, R., 2007. Nanotechnology: a new paradigm in cosmeceuticals. Recent Pat. Drug Deliv. Formul. 1, 171–182.
Khazaka, G., 2011. Compliance check with on-site measurements: advanced ways in product testing. In: Fluhr, J.W. (Ed.), Practical Aspects of Cosmetic Testing. Springer, Berlin, Heidelberg, pp. 261–265.
Kim, B., Rutka, J.T., Chan, W., 2010. Nanomedicine. N. Engl. J. Med. 363, 2434–2443.
Kimura, S., Kamimura, T., Takema, Y., Kubota, M., 1981. Lower vertebrate collagen: evidence for type I-like collagen in the skin of lamprey and shark. Biochim. Biophys. Acta 669, 251–257.
Kowalska-Wojcicka, E., 2011. Trends in modern cosmetology. Cabines 48, 40–46.
Lee, C.H., Singla, A., Lee, Y., 2001. Biomedical applications of collagen. Int. J. Pharm. 221, 1–22.
Loden, M., 2009. Hydrating substances. In: Barel, A.O., Paye, M., Maibach, H.I. (Eds.), Handbook of Cosmetic Science and Technology. Informa Healthcare, New York, pp. 107–119.
Lohani, A., Verma, A., Joshi, H., Yadav, N., Karki, N., 2014. Nanotechnology-based cosmeceuticals. ISRN Dermatol. 2014, 1–14.
Makeshwar, K., Wasankar, S.R., 2013. Niosome: a novel drug delivery system. Asian J. Pharm. Res. 3, 16–20.
Matts, P.J., Cotton, S.D., 2009. Spectrophotometric intracutaneous analysis. In: Barel, A.O., Paye, M., Maibach, H.I. (Eds.), Handbook of Cosmetic Science and Technology. Informa Healthcare, New York, pp. 275–281.
Moghimi, H.R., Barry, B.W., Williams, A.C., 1999. Stratum corneum and barrier performance: a model lamellar structural approach. In: Bronaugh, R.L., Maibach, H.I. (Eds.), Percutaneous Absorption: Drugs–Cosmetics–Mechanisms–Methodology. Marcel Dekker, New York, pp. 515–553.
Morganti, P., 2010. Use and potential of nanotechnology in cosmetic dermatology. Clin. Cosmet. Invest. Dermatol. 3, 5–13.

Müller-Goymann, C.C., 2004. Physicochemical characterization of colloidal drug delivery systems such as reverse micelles, vesicles, liquid crystals and nanoparticles for topical administration. Eur. J. Pharm. Biopharm. 58, 343–356.

Mullins, R.J., Richards, C., Walker, T., 1996. Allergic reaction to oral, surgical and topical bovine collagen: anaphylactic risk for surgeons. Aust. N.Z. J. Ophthalmol. 24, 257–260.

Patravale, V.B., Mandawgade, S.D., 2008. Novel cosmetic delivery systems: an application update. Int. J. Cosmet. Sci. 30, 19–33.

Pavicic, T., Gauglitz, G.G., Lersch, P., Schwach-Abdellaoui, K., Melle, B., Korting, H.C., Farwick, M., 2011. Efficacy of cream-based novel formulation of hyaluronic acid of different molecular weight in anti-wrinkle treatment. J. Drugs Dermatol. 10, 990–1000.

Pietrzycka, A., Zaleska-Zylka, I., 2014. The influence of natural collagen on antioxidant and lipid status of sensitive skin. Aesthet. Cosmetol. 3, 181–184.

Prausnitz, M.R., 2004. Microneedles for transdermal delivery. Adv. Drug Deliv. Rev. 27, 581–587.

Rossler, B., Kreuter, J., Scherer, D., 1995. Collagen microparticles: preparation and properties. J. Microencapsul. 12, 49–57.

Rothbard, S., Watson, R.F., 1965. Immunologic relations among various animal collagens. J. Exp. Med. 122, 441–454.

Ruszczak, Z., Friess, W., 2003. Collagen as a carrier for on-site delivery of antibacterial drugs. Adv. Drug Deliv. Rev. 55, 1679–1698.

Schneider, M., Strache, F., Hansen, S., Schaefer, U., 2009. Nanoparticles and their interaction with dermal barrier. Dermatoendocrinology 1, 197–206.

Shanley, A., 2015. Pharma APIs: it is still a small world. Pharm. Technol. 39, 26–33.

Sharma, B., Sharma, A., 2012. Future prospect of nanotechnology development of anti-ageing formulations. Int. J. Pharm. Pharm. Sci. 4, 57–66.

Siew, A., 2015. Transdermal drug delivery gains traction. Pharm. Technol. 39, 34–36.

Silpa, R., Jose, S., Sumod, U.S., Sabitha, W., 2012. Nanotechnology in cosmetics: opportunities and challenges. J. Pharm. Bioallied Sci. 4, 186–193.

Silva, T.H., Moreira-Silva, J., Marques, A.L., Domingues, A., Bayon, Y., Reis, R.L., 2014. Marine origin collagens and its potential applications. Mar. Drugs 12, 5881–5901.

Vishvakrama, P., Sharma, S., 2012. Liposomes: an overview. Int. J. Res. Pharm. Biomed. Sci. 3, 1074–1084.

Wallace, D.G., Rosenblatt, J., 2003. Collagen gel systems for sustained delivery and tissue engineering. Adv. Drug Deliv. Rev. 55, 1631–1649.

Webb, R.C., Pielak, R.H., Bastien, P., Ayers, J., Niittynen, J., Kurniawan, J., Manco, M., Lin, A., Cho, N., Malyrchuk, V., Balooch, G., Rogers, J.A., Ugaz, V.M., 2015. Thermal transport characteristics of human skin measured in vivo using ultrathin conformal arrays of thermal sensors and actuators. PloS One 10, e0118131.

Weiner, A.L., Carpenter-Green, S.S., Soehngen, E.C., Lenk, R.P., Popescu, M.C., 1985. Liposome-collagen matrix: a novel sustained drug delivery system. J. Pharm. Sci. 74, 922–925.

Wissing, S.A., Muller, R.H., 2003. The influence of solid lipid nanoparticles on skin hydration and viscoelasticity in vivo study. Eur. J. Pharm. Biopharm. 56, 67–72.

Yosipovitch, G., Xiong, G.L., Haus, E., Sachett-Lunden, L., Ashkenazi, I., Maibach, H.I., 1998. Time-dependent variations of the skin barrier function in humans: transepidermal water loss, stratum corneum hydration, skin surface pH, and skin temperature. J. Invest. Dermatol. 110, 20–23.

CHAPTER

NANOSTRUCTURES FOR DRUG DELIVERY: PHARMACOKINETIC AND TOXICOLOGICAL ASPECTS

30

Adrian G. Ciucă*, Cristina I. Grecu*, Petre Rotărescu*, Irina Gheorghe, Alexandra Bolocan[†,‡], Alexandru Mihai Grumezescu*, Alina M. Holban*,**, Ecaterina Andronescu***

**University Politehnica of Bucharest, Bucharest, Romania*

***Research Institute of the University of Bucharest (ICUB), Bucharest, Romania*

[†]Emergency University Hospital, Bucharest, Romania

[‡]"Carol Davila" University of Medicine and Pharmacy, Bucharest, Romania

CHAPTER OUTLINE

1 INTRODUCTION

Nowadays, drug delivery systems represent an interesting field with much support at a global scale. In the classic therapy, the drug is suddenly released, causing a rapid increase of its concentration in the blood, with the possibility of exceeding the toxic levels. Biological systems require controlled drug delivery systems specific for the individualized treatment of certain areas of the human body.

Controlled release systems can be used in most clinical treatments using active substances, such as proteins, opioids, and local anesthetics (Zhang et al., 2013).

The biological activity depends in a large percentage on the administration pathway and the type of metabolism of the substance entered in the body. Active substances that are administered orally have a relatively low bioavailability, requiring their re-administration at different intervals of time, resulting in a decreased efficiency and a risk of overdose (Webster et al., 2013).

The introduction of a new drug is not only expensive, but also time consuming. It involves discovery, clinical testing, development, and approval. Improving safety/efficacy ratio of marketed drugs is more cost-effective. This can be obtained by controlling the time, rate, and place of drug release in the body through a drug-delivery system. Hence, a drug-delivery system could be seen as an interface between the patient and the drug (Butte et al., 2013).

Over the past decades, a growing number of drugs were discovered and were optimized for an enhanced efficiency. However, about 40% of the new drugs, especially those based on biomolecules, such as peptides, nucleotides, or proteins, often present a low bioavailability and are rejected by the pharmaceutical industry (Li et al., 2012).

The ideal materials for controlled release must control certain key issues: easy reach to the target site in the body, ability to transport the necessary volume of active compounds, and a certain level of release with a certain speed, apart from the properties needed to ensure a better and safe interaction with the human body (Argyo et al., 2013).

Compared to generic conventional systems, the controlled release allows the maintenance of optimal concentrations of the active compound, thus overcoming the excess toxicity and passing below the therapeutic action.

2 NANOPARTICLES AND DRUG DELIVERY

Nanotechnology represents the manipulation of matter on an atomic, molecular, and supra-molecular scale involving the characterization, production design, and application of different nanomaterials in several key areas, providing technological advances mainly in the field of medicine. The development and optimization of drug delivery in nanoparticles concerns the early detection of cancer cells or specific tumor biomarkers and the enhancement of the treatments efficacy (Sanna et al., 2014).

Due to their distinctive electrical and physicochemical properties, nanosized materials gained considerable attention in the fields of biotechnology, electronics, and aerospace engineering. In biomedicine, nanoparticles are being employed as a novel delivery system for drugs, peptides, and proteins, macromolecular structures, and antibodies (Shvedova et al., 2012). The nature of nanoparticles can be metallic, nonmetallic, polymeric, and bioceramic. Other nanostructures with medical applications are the liposomes and dendrimers (Pourmand and Abdollahi, 2012).

Nanomaterials are able to provide a high degree of biocompatibility before and after biomolecule conjugation for specific functions and also to translate into nanomedical applications for clinical practice (Lammers et al., 2012). One of the primary goals in the research of nanobiotechnologies for drug delivery systems design is to offer a greater safety and biocompatibility, reduction in toxicity while maintaining therapeutic effects, more specific drug targeting and delivery, faster development of new safe medicines.

Drug delivery systems can improve the properties of conventional drug administration routes by increasing their biodistribution and in vivo stability, solubility, and also by modulating their pharmacokinetics; promoting the transport and releasing in the target site a higher doses of drug in order to be efficient (Drbohlavova et al., 2013).

Nanoparticle composite systems are polymer-coated submicronic-sized colloidal particles with specific therapeutic agents dispersed or encapsulated within their polymeric phase or grafted onto their surface. Biodegradable synthetic polymers are commonly used to reveal nanoparticle drug delivery systems. The nature of the nanoparticles may vary from silica, gold, or other metals, quantum dots and quantum rods, oxides. Nanoparticle systems can have significant enhanced effects when employed in the therapeutic practices due to their site specificity and efficient delivery of a bioactive agent. Nanoparticles can actively target various biological structures when attached with specific ligands on their surface, such as antibodies (Couvreur, 2013).

One of the biggest challenges in drug delivery system design is to achieve targeted delivery in that body place where the drug is needed, avoiding potential side effects to healthy tissue structures. This is challenging especially in cancer treatment where the tumor is placed as distinct metastases in different organs. Local drug delivery or drug targeting results in increased local drug concentrations and provides strategies for more specific therapy. However, the full use of the active potential of chemotherapeutics is limited by their nonrestricted toxicity.

3 MAIN APPLICATIONS

3.1 CANCER THERAPY

Using small molecules as chemotherapeutic agents had many successful outcomes against cancer, but in their unfunctionalized state they are not able to distinguish between cancerous and noncancerous cells. Hence, developing alternative approaches for distinctively targeting cancer cells has been the aim of numerous studies. Besides, many intracellular pathways became deregulated in cancer cells, therefore the use of two or more chemotherapeutic agents aiming at more than one deregulated pathways is a rational therapeutic strategy. A promising cancer treatment involved the use of RNA interference (RNAi) to downregulating multiple targets (Ruiz et al., 2013). Basically, the RNAi strategy uses small RNA molecules, which bind the RNA messenger (mRNA) by complementary base pairing, to induce the degradation of the mRNA and/or block the protein synthesis. The small RNA molecules are incorporated and processed into cellular RNA processing machinery to induce their inhibitory effects. According to a RNAi-based therapy (RNA silencing), a 21–27 base pair double-stranded RNA (siRNAs) is introduced into the cells, where it binds to its specific complementary mRNA sequence and inhibits protein synthesis. SiRNAs are designed to target only one gene, which is generally overexpressed in the cancer cells compared to the normal cells (Rettig and Behlke, 2012). Therefore, it is highly desirable to use siRNAs targeting key genes involved in cancer cell migration, invasion, and metastasis. Another new way concerns the introduction of 22-bp single or double-stranded RNAs into the cells to control the microRNAs (miRNAs). Recent evidence indicates that the human genome may encode more than 1500 miRNAs, which regulate about 40%–60% of the human genes. Many studies involving various types of human cancers have been demonstrated that miRNAs have an essential role in carcinogenesis and drug resistance. A particular miRNA can potentially bind to thousands of mRNAs (Bader, 2012). While some miRNAs are down-regulated and act as tumor suppressor genes, others are up-regulated and may represent novel oncogenes (oncomiRs). Thus, miRNA-targeted therapies are used in two ways, to imitate the function of tumor suppressor genes (microRNA mimics) or to inhibit the action of miRNAs via oncomiRs (antagomirs). One of the most interesting properties of miRNAs as therapeutic agents and probably the most significant advantage in comparison with siRNA is that while siRNA-based approaches target single genes, the miRNA-based approaches target multiple molecules, making them highly efficient in regulating distinct biological processes relevant to normal and malignant cells. However, in some tissues, certain miRNAs could regulate a group of genes, each one with contrary roles. Sequence-specific inhibition of gene expression was achievable using nucleic acid–based antisense technology before RNAi discovery (Marelli et al., 2013). The highest advantage of RNAi over antisense-based approaches is that the RNAi relies on the cellular machinery to target complementary transcripts, resulting in efficient and potent down-regulation of gene expression. Furthermore, the use of the RNAi strategy is highly specific when modulation of a particular target is desired. In this case, lower side effects are expected compared to the conventional chemotherapy. In spite of all the advantages of RNAi as a therapeutic strategy in cancer medicine, the in vivo systemic administration of RNAi has remained a major challenge due to its short half-life, activation of the immune response, lack of ability to penetrate the plasma membrane, and potential toxicity. Therefore, the use of nanoparticles as RNAi carriers has been proposed to address these concerns (Jordheim et al., 2013).

Quantum dots are nanocrystals with less than 10 nm size and with semiconductor behavior that under light stimulation manifest different color and fluorescent emissions. Different biomarkers can be targeted after modulation of the quantum dots and by conjugation with specific biomolecules that can be highly sensitive to probes (Gangrade, 2011). Prior to planning and employing a therapeutic strategy against the cancer, quantum dots can be used to better image the tumor growth; breast cancer was also reported susceptible to quantum dots based upon this treatment. Thus, quantum dots may ensure novel insights into the tumor pathophysiology comprehension as a result of real time imaging and screening.

3.2 TREATMENT OF INFECTIOUS DISEASES

The hydrophilic nature of some antibiotics prevents their capacity to penetrate the cells and, furthermore, the internalized molecules are mostly accumulated in lysosomes, where the bioactivity of the drug is low. This is the reason why limited intracellular activity against sensitive bacteria is often found. Hence, drug delivery systems have been used for passive targeting of infected cells in the mononuclear phagocytic system to enhance the therapeutic index of antimicrobials in the intracellular environment, while minimizing the side effects associated with the systemic administration of the antibiotics.

The delivery of various nanotechnology-based products is preferred for the treatment of pathophysiological and anatomical changes of the affected tissues in a diseased state. There are three main reasons that can determine the gaining of antibiotic resistance in bacteria: (1) the modification of active site of the target resulting in reduction of the efficiency of the bindingdrug; (2) the direct destruction or modification of the antibiotic by enzymes produced by the organism; and (3) the active and massive efflux of the antibiotic from the cell. Recent studies offered significant proof that the resistance rates can be reduced with the aid of nanoparticles, which can target antimicrobial agents at the site of infection, so that higher doses of drug can cluster at the infected site (Leid et al., 2012). As nanoparticles target the site of infection, high concentrations of antimicrobial drugs can be released at the site of infection while keeping low the total dose of administered drug.

Nanoparticles can be actively or passively targeted to the sites of infection. Actively targeted nanoparticles contain ligands (such as antibodies) with receptor binding ability. Passively targeted nanoparticles selectively undergo extravasation at sites of infection, where inflammation has led to enhanced blood vessel permeability (Pelgrift and Friedman, 2013). Passive targeting with nanoparticles, however, faces multiple barriers on the way to their target; these include nonspecific uptake of the particles, nonspecific delivery of the drug (as a result of uncontrolled release), and mucosal barriers. Nanoparticle passive targeting of chemotherapeutics to the cells and organs of the reticulo-endothelial system (RES) has been a significant area of research for the treatment of chronic infectious diseases. The RES comprises monocyte-lineage immune cells, such as macrophages and dendritic cells, as well as the spleen, liver, and kidneys. These components of the RES are implicated as sites of nanoparticle localization and clearance. The few studies that have compared targeted and nontargeted systems have demonstrated that the role of targeting ligands, in localization at the target site, is application-dependent. Targeted delivery to atherosclerotic lesions is greatly enhanced by targeting ligands, which impart an improved ability to accumulate at the target site. Many active targeting strategies use the enhanced permeability and retention effect, so that active and passive targeting mechanisms act synergistically, that lead to higher concentration of nanostructures in the infected region than that in healthy tissues (Heidari et al., 2014). Nanoparticle-based targeted antimicrobial drug delivery to the site of infection, particularly intracellular infections, is a highly desired goal prevision in treating infectious diseases. Antibiotics-loaded nanoparticles can enter host cells through endocytosis, followed by releasing the bioactive compound load to delete intracellular microbes. As a result of therapeutic and economic factors, the need to target drugs to specific sites is rapidly increasing (Rai et al., 2012).

Nanoantibiotics promise significant benefits and advances in addressing the key obstacles in treating infectious diseases, but there are foreseeable challenges in translating this exciting technology for clinical application. Profound knowledge about the potential toxicity of nanoantibiotics is also needed to guarantee successful clinical translation (Emeje et al., 2012). The toxic effects of antimicrobial nanoparticles on the central nervous system are still unknown and the interactions of nanoparticles with the cells and tissues in central nervous system are poorly understood. Furthermore, nanoparticles represent size-specific properties that limit the use of currently available in vitro experiments in a general way and there is no standardized definition for nanoparticles dose in mass, number, surface area, and biological samples (Bonifacio et al., 2014).

Nanoparticles usually have short circulation half-life due to natural defense mechanism; as a result, nanoparticles are eliminated after opsonization by the mononuclear phagocytic system. Therefore, the particles surfaces need to be changed (Szlek et al., 2013). A hydrophilic polymer, such as polyethylene glycol is prevalently utilized for this purpose because it features convenient characteristics, such as low degrees of immunogenicity and antigenicity, chemical inertness of the polymer backbone, and availability of the terminal primary hydroxyl groups for derivatization (Diaz and Vivas-Mejia, 2013).

3.3 TRANSPLANT

Nanopores are incorporated into wafers with high porosity, which allow the free movement and passing through of oxygen and glucose molecules, proteins, drugs, growth factors. Nanopores with a 20 nm diameter could be used to benefit the transplantation procedures, as vessels to protect transplanted organic materials from the immune system of the host (Rabanel et al., 2012). Cell transplants can be facilitated by the use of porous structures (pancreatic β-cells can be embedded within the nanoporous device and then inserted into the body of the recipient).

3.4 GENE DELIVERY

Nanotubes are nanoscaled structures, similar to a tube, which can consist of different types of materials: that is, carbon, inorganic compounds, DNA, membrane. Carbon nanotubes, generally insoluble and hydrophobic, can be made soluble in aqueous solution by carboxylic or ammonium groups functionalization of their surface, thus facilitating the transport of peptides, nucleic acids, and other molecules. Studies in the emerging field of gene therapy often used nanotubes to attached or incorporated DNA through the cellular membrane (Chen et al., 2012a,b).

4 PHARMACOKINETIC CHARACTERISTICS OF NANOPARTICLES IN DRUG DELIVERY SYSTEMS

Nanoparticles can penetrate cells according to endocytotic mechanisms, such as endocytosis, potocytosis, pinocytosis, and patocytosis (Gramoun et al., 2014). On the other hand, larger particles are quickly opsonized and can be removed from the circulatory system due to the macrophages from the RES. For biomedical purposes, means of disabling opsonization and prolonging the time nanoparticles spend in the bloodstream, must be applied. One approach to achieve it implies superficially coating the nanoparticles with hydrophilic polymers or surfactants; formulating the nanoparticles from the start with biodegradable copolymers, which encompass hydrophilic segments, such as polyethylene glycol or polyethylene oxide, is also a possibility (Rosca et al., 2015).

Positively charged nanodrugs are able to interact with the negatively charged mucin and as a result the transportation across mucus is favored and the bioactive compounds can be easier internalized by epithelial cells. The specific functionalization with ligands for the superficial receptors expressed on cells may also increase the transcellular transportation of embedded drugs (Ghaemmaghami et al., 2012). Plus, bioadhesive polymers/chelator equipped nanoparticles could improve the paracellular transportation of the entrapped drugs (Wikswo et al., 2013).

The specific biomolecule–nanodrug modification can be used to design selectively active drug delivery systems, which target particular tissues. Generally, nanotechnology-based drug delivery systems are designed for intravenous, transdermal, or for oral administration; recent studies revealed that pulmonary-administered nanodrug systems can offer better outcomes (Murphy and Atala, 2014). The particles that the patient inhales have to undergo pulmonary clearance: that is, mucociliary clearance and macrophage clearance.

For the last years, the systemic absorption of a wide range of therapeutic agents has been demonstrated after pulmonary application in both animals and humans. Drug administration may take place in two ways, via the pulmonary route: intranasal administration with anatomical limitations, such as narrower airway lumen and administered by oral inhalation. By oral inhalation the administration gives much better results because it allows the administration of very small particles and loss of concentration is much lower compared to the results for the intranasal administration (20 vs. 85%) (Beck-Broichsitter et al., 2013). In preclinical animal studies, intratracheal instillation is a frequently employed assay for assessing pulmonary absorption and also systemic bioavailability, due to the precise dosing and effectiveness this method is associated with (Willis et al., 2012). However, the intratracheal route is not a standard physiological route for application and the date obtained from these studies might not be extrapolated to aerosol applications in humans. On the other hand, this method is an aerosol-based technique that can guarantee uniform distribution and great penetration. However, measuring the exact dose in lungs is more costly and difficult. The deposition of drug by aerosol administration in the pulmonary airway mainly takes place through three types of mechanisms: inertial impaction, gravitational sedimentation, and diffusion.

A relatively high drug particle size implies that the deposition occurs through the first two mechanisms where, either inertial impaction occurs due to hyperventilation or sedimentation occurs due to gravitational force. When the particle size is smaller they deposit mainly by diffusion mechanism, which in turn is based on the Brownian motion. Apart from the ventilator parameters and pulmonary morphological aspects, the size of the particles or droplets and the geometry are quite important (Jaafar-Maalej et al., 2012). The size of particle or droplet in terms of diameter along with the surface electrical charges, shape of the particulate, matter if it is a fiber and hygroscopy also having a huge influence on drug deposition through pulmonary route. The term mass median aerodynamic diameter is used and it depends on size, shape, and density of the particulate system (Patel et al., 2013). The delivery device is crucial in pulmonary administration. Important steps have been made in developing more advanced devices. Nonetheless, these are considerably less employed than powder formulations. A wide range of passive breath driven and active power driven single/-multiple dose dry powder inhalers is available in the market. Ideally, a device should be able to generate and deliver the bioactive particles or droplets of specific aerodynamic size, if the drug is destined to a specific area in the lungs (Tena and Clarà, 2012). Most commonly used for respiratory delivery are devices, such as dry powder inhalers (for protein delivery), metered-dose inhalers, and nebulizers. The nebulizers commercially available are: number jet nebulizers, ultrasonic nebulizers, and vibrating mesh nebulizers. Jet nebulizers function by compressed air, while ultrasonic nebulizers use piezoelectric transducers (Varshosaz et al., 2013). Selecting the adequate device is dependent on parameters like the targeted site, the pathology and physiology of the lung, and drug nature and formulation. Also, aerosols based on mechanically generated vibration mesh technologies are another convenient choice to deliver proteins to lungs; such aerosols are currently used in the clinical trials for protein and peptide-based pharmaceuticals delivery (Trivedi et al., 2012).

5 TOXICOLOGICAL EFFECTS OF NANOPARTICLES

Over the past decades, numerous types of nanoparticles have been developed based on various components, including metal oxides, silica, carbon, nanocrystals, polymers, lipids, dendrimers, and quantum dots, often in mixtures with new developed materials.

The human body is exposed to various unsafe nanoparticles throughout its life and some researchers fear that the field of nanobiotechnology may become another threat to human health. Because of their small size, nanoparticles can easily enter the human body despite its natural barriers, cross the tissues, and reach vital organs, where, according to studies, conglomerate (Chuang et al., 2015). It is a commonly spread concept among the science community that nanoparticles with a diameter of less than 10 nm behave similarly to a gas, hence they can insert themselves with ease into human tissues and modify the cell environment or disrupt the cell membrane (Albanese et al., 2012).

Nanotoxicology is developing as a subdiscipline of nanotechnology. Nanotoxicology investigates the interactions of nanostructures with the biological systems. Nanoparticles can change their physiochemical properties and offer the opportunity to a better uptake and interaction with biological tissues. Nowadays, a comprehensive understanding of how the size, the structure, the shape, the composition and aggregation tendencies of the nanoparticles influence their interactions with biological systems is lacking, thus it is unpredictable whether the contact of living organisms to engineered nanoparticles would cause harmful effects and unwanted biological responses (Smith et al., 2014). Human skin, the gastrointestinal tract, and lungs are the closest in contact with the environment, easing the nanoparticles entrance into the organism. The toxicity of nanoparticles depends on different factors: size, composition, aggregation, crystallinity, and surface functionalization (Cornelia et al., 2013).

Toxicological studies have demonstrated that the small nanoparticles (<100 nm) may cause respiratory health conditions, determining more inflammation than larger particles made from the same material (Alshatwi et al., 2012). The inflammatory effect may be reliant on the surface area of nanoparticles, smaller nanoparticles having a higher surface area and particle number per unit mass compared to larger particles. Larger surface areas lead to enlarged reactivity and to an important source of reactive O_2 species. The higher surface area of nanoparticles causes a dose-dependent increase in oxidation and DNA destruction, much higher than an identical mass dose of larger particles (Sharma et al., 2012a,b).

The dose is defined as the quantity of substance that can influence a biological system, being directly related to exposure or the concentration of substance in the applicable medium, such as water, food, and air increased by the extent of contact. For example, health effects of inhaled TiO_2 nanoparticles with different sizes: the low dose (10 mg/m3) exposure to 20 nm diameter particles give more chances to lung tumor incidence than the high dose (250 mg/m3) exposure of 300 nm diameter particles (Grabowski et al., 2015).

6 DETERMINATION OF TOXICITY

Despite the fact that recently an exponential growth of the production of nanoparticles and their use in different domains has been witnessed, particularly in medicine, research related to the toxicological impact and possible hazard of nanoparticles to human health and the environment, is still in a beginning phase. The concentration and time remain the main parameters in conventional toxicology, as they can be easily measured for individual compounds and the threshold levels can be extrapolated after the evaluation of the nature of the response a dose induces on a bioentity. Thus, it can be decided if a chemical is either safe or dangerous for biomedical use. Linear relationship is expected when investigating the link between dose and the response it determines, according to what Paracelsus stated: "the dose makes the poison" (Venkatraman, 2014).

Soon after the nanotechnology was developed, nanotoxicology has emerged, as more and more suitable materials were put under the spotlight by the science community. The properties manifested by materials under "bulk" delivery have been studied and understood; those of the nanoscale materials are yet to be discovered and confirmed with no shred of questioning. The processing of bulk materials into smaller particles leads to chemistry changes in terms of increasing chemical reactivity (Xu and Sun, 2013). Nanotoxicology focuses on the superior reactive nanomaterial surface and its interaction with the proper environment biochemistry. An urge to keep up with the nanotechnology progress was observed, as most the use of most nanomaterials is yet to be regulated and the chances of human exposure to such compounds is high (Sawdon et al., 2014).

The long experience with asbestos exposure and pollution raised concerns regarding the health conditions individuals could face from nanostructures; more recently, strong alarm signals were voiced by scientists, as the tiny carbon nanotubes could have asbestos-like effects on living cells and possibly more, because their size is smaller compared to the one of the asbestos fibers (Patil et al., 2015). The studies cannot conclude without doubt the sum of their effects and intermediary activity because nanoparticles often intimately interact with the living matter and display changes in superficial features, according to the environment. Thorough studies indicate the cells can incorporate and deposit nanoparticles according to either active or passive mechanisms. After the incorporation, their evaluation is even harder to accomplish as it was reported that identical particles with different coatings, size, or charge behave differently. No straightforward approaches exist so far to determine exactly the intricate behavior of the potentially hazardous nanoparticles. Standard toxicological assays and protocols do not apply to nanoparticles and their classification is more difficult to be done as the dose is not the only parameter to be taken into account. Nanotoxicological standardization is being sketched but it is far from becoming validated internationally as there are many variables, such as the shape, coating, size, charge, aggregation, dispersion, concentration, matrix to be considered (Juillerat-Jeanneret et al., 2015).

The shifting from in vitro to in vivo studies complicates the situation because most in vivo studies of nanomaterials are still to become optimized and standardized. The most important entry routes and agglomeration sites have been assessed but the underlying mechanisms are yet to be determined. An endocytotic pathway, similar to the one activated in viral infections, was observed for insoluble nanoparticles with a diameter of 50 nm or less (Sharma et al., 2012a,b). As the potential spreading throughout the body is a serious risk factor, nanotoxicological safety precautions should be taken.

In order to measure toxicological endpoints, first it is necessary to understand the starting point, the nanomaterial, otherwise possible toxic effects cannot be easily attributed to a certain property of the nanomaterial or even the nanomaterial itself because, for example, impurities and other components could be held responsible. Therefore, it is absolutely critical to know the material and its properties. In the past, this has not always been straightforward for industrially synthesized nanoparticles due to crude production processes and, therefore, variations in material properties existed (size, shape, etc.) (Valdiglesias et al., 2015). As mentioned before, the dose is one of the key parameters in toxicology. In nanotoxicology it is important to evaluate relevant and realistic dose regimes in order to draw meaningful conclusions from in vitro and in vivo experiments for the public health risk assessment. Therefore, nanotoxicologists should test nanoparticles toxicity based on real-world doses rather than unrealistically high doses in order to achieve a biological response. The latter can be useful in elucidating the mechanisms but is unlikely to predict human pathology from environmental exposure, while low dose inhalation experiments are likely to be much more predictive of human harm (Yang et al., 2015a,b). The experiments show that high-dose acute exposures are likely to be detected and corrective/protective measures could be employed. Currently, the most appropriate metric for assessing the dose of nanoparticles, is debated, which is clearly a critical consideration in the field of nanomedicine. In classical toxicology the dose is almost invariably related to mass. However, with nanoparticles, the toxicological effects do not fit with these concepts and have given rise to some level of disbelief among classically trained toxicologists. Because of the nature of nanoparticles, a logical dose metric (Lee et al., 2014) will be related to the number of particles reaching each cell or cellular subcompartment of relevance. The nanoparticles number can be hard to estimate, but there are some indications from the literature that the total surface of nanoparticles, where many of the chemical reactions of interest occur, may be a more discriminatory metric, in some circumstances. While mass alone cannot predict surface area, the number of nanoparticles within a certain size and shape range is predictive of surface area and is, therefore, likely to be a dose metric, which is more predictive of harm than simple mass (Chen et al., 2012a,b).

7 TOXICITY

In biomedicine, the nanoparticles are being employed as a novel delivery system for drugs, peptides, and proteins, macromolecular structures, and antibodies (Shvedova et al., 2012). The nature of nanoparticles has been metallic, nonmetallic, polymeric, and bioceramic (Pourmand and Abdollahi, 2012).

Other nanostructures with medical applications are: liposomes and dendrimers. The human body is exposed to various unsafe nanoparticles throughout its life and some researchers fear that the field of nanobiotechnology may become another threat to human health. Because of their small size, nanoparticles can easily to enter the human body despite its natural barriers, cross the tissues, and reach vital organs, where, according to studies, they agglomerate (Chuang et al., 2015). It is a commonly spread concept among the science community that nanoparticles with a diameter of less than 10 nm behave similarly to a gas, hence they can insert themselves with ease into human tissues and modify the cell environment or disrupt the cell membrane (Albanese et al., 2012).

7.1 METALLIC NANOPARTICLES

7.1.1 *Aluminum oxide*

According to "The Global Market for Aluminum Oxide Nanoparticles," aluminum-based nanoparticles contribute 20% to all nanosized chemicals. Studies about their toxicology provided mixed results. Nanoparticles of aluminum oxide alternate the cell viability, the mitochondrial function and influence the protein expression of the blood–brain barrier (BBB) while also increasing oxidative stress. Radziun et al. (2011) showed that alumina nanoparticles, at concentrations of 10 to 400 μg/mL have no significant toxic effect on the viability of mammalian cells. Another study reported a dose-dependent (25–40 μg/mL) cytotoxicity as the effect of aluminum oxide (160 nm) on human mesenchymal stem cells. Balasubramanyam et al. have reported that alumina nanoparticles (30–40 nm) possess dose-dependent genotoxic properties; they assessed the genotoxicity with comet assay and micronucleus test using rat blood cells. Another study, where mouse lymphoma cells line was investigated, revealed that aluminum oxide nanoparticles (<50 nm) cause genotoxic effects in the form of DNA damage, but without any mutagenic effects (Peng et al., 2015).

7.1.2 *Gold*

Nanoparticles of gold exhibit unique physical and chemical properties. They have the ability of binding to amine and thiol groups with ease. Gold nanoparticles are considered to be quite safe, as their core is inert and display nontoxic effects. Different size gold nanoparticles (4/12/18 nm), with different coatings have been investigated for cytotoxic activity against leukemia cells line and the results of this study argue that spherical nanoparticles could break the cell barriers and are nontoxic to the intracellular media; their cytotoxicity was evaluated by MTT assay (Li et al., 2014). Nonetheless, there are other reports suggesting that the gold nanoparticles cytotoxicity is strongly associated with the dose and the stabilizer employed. Cytotoxicity of gold nanoparticles varies from cell line to cell line and is intimately correlated to their physicochemical properties. The toxicity variation according to different cell lines has been extensively investigated for human lung and liver cancer cell lines (Leung et al., 2012).

7.1.3 *Copper oxide*

Copper oxide nanoparticles are mainly used in heat transfer fluids, antimicrobial reagents, semiconductors, and intrauterine contraceptive devices. The studies have been demonstrated that copper-based nanomaterials have a toxic effect on the kidney and liver. In one in vitro study, 50 nm copper oxide nanoparticles were analyzed and the results indicated that they are genotoxic and cytotoxic, inducing oxidative stress reactions and disturbing the cell membrane integrity.

7.1.4 Silver

Silver nanoparticles are used in the form of prostheses and wound dressings embedded antimicrobial agents and for coatings of surgical instruments. They penetrate the human body via different ways and accumulate in various tissues, crossing the BBB. In in vivo studies, silver nanoparticles have been observed in different organs (liver, kidney, lungs, spleen, and brain) after exposing the subjects to silver nanoparticles via inhalation or by subcutaneous injection (Al Gurabi et al., 2015). Compared to other nanoparticles, silver nanoparticles have shown more toxicity in term of cell viability, lactate dehydrogenase leakage, and generation of reactive oxygen species (ROS). Foldbjerg et al. reported a dose-dependent cytotoxicity and cellular DNA adducts formation after exposing polyvinyl-pyrrolidone-coated silver nanoparticles (6-20 nm) to human lung cancer cell line (Ebabe Elle et al., 2013).

7.1.5 Zinc oxide

Nanoparticles produced from zinc oxide have many applications and are being used as UV detectors, gas sensors, and additives in paints, sunscreens, wave filters, and personal care products. Zinc oxide nanoparticles have been studied for any possible toxic effects they might manifest toward bacteria and mammalian cells. Cytotoxicity, ROS inducement, and the cell membrane disruptions have been registered in experiments on mammalian cell lines as the most significant adverse effects of zinc nanoparticles (Guan et al., 2012). Brunner et al. reported zero cell viability in the cell culture after exposing human mesothelioma cells and rodent fibroblast cells to zinc oxide nanoparticles with high concentration (49 mg/mL). Meyer et al. have reported a decrease in cell viability after exposing the above cell lines to 20 nm zinc oxide nanoparticles using as a model human dermal fibroblast cells and as an assay technique, MTT (Wilhelmi et al., 2013). Both in vitro and in vivo studies have reported various degrees of cytotoxicity and the genotoxicity manifested by the zinc oxide nanoparticles. Osman et al. revealed that zinc oxide nanoparticles manifest its genotoxicity via DNA damage of HEp-2 cell line, according to the comet assay technique. Chronic exposure to zinc oxide nanoparticles (300 mg/kg) result in oxidative DNA damage and liver enzymes altering (Hackenberg et al., 2013).

7.1.6 Iron oxide

Iron oxide nanoparticles are used in the diagnostic fields, biomedical, and drug delivery. In vivo studies have shown that after penetrating the cells, iron oxide nanoparticles remain in cell organelles (endosomes/lysosomes, especially in hepatocytes), contribute to cellular iron poll and release into cytoplasm after decomposing. Magnetic iron oxide nanoparticles have been observed to accumulate in the liver, spleen, lungs, and brain after inhalation, showing its ability to cross the BBB (Formenti et al., 2014). Nanoparticles wield their toxic effect in the form of cell lyses, inflammation, disturbing the blood coagulation system and reducing cell viability. Variable cell viability results were obtained prior to the contact with iron oxide nanoparticles coated with different substances. Murine macrophage cells were susceptible to the Tween-coated 30 nm iron oxide nanoparticles (Guo and Barnard, 2013). On the other hand, dextran-coated iron oxide nanoparticles decreased the cell viability in human macrophages by 20% after 7 days of incubation. In addition, in another study on mouse Neuro-2A cell line, 25 nm iron oxide nanoparticles have been found to exert less toxicity in terms of changing cell morphology, determining cell apoptosis, altering cell permeability and mitochondrial functions (Reynolds et al., 2014). However, rat's mesenchymal stem cells have shown 70% cell viability after 48 h exposure to polymer-coated iron oxide nanoparticles, according to MTS assay. It has been thought that the toxic effects of iron oxide nanoparticles are due to excessive production of ROS. These generated ROS further were able to cause lipid peroxidation and DNA damage (Yang et al., 2014).

7.1.7 Titanium oxide

Titanium oxide is an inert compound, but studies have shown that nanoparticles of titanium dioxide possess some toxic effects in experimental animals, including DNA damage, genotoxicity, and lung inflammation (Peira et al., 2014). Titanium dioxide can induce oxidative stress and determine DNA adducts. Besides genotoxicity, nanometric, and submicronic titanium dioxide particles (5–200 nm) possess toxic effects toward the immune system, kidney, spleen, myocardium, and lipid homeostasis in experimental animals (Jaeger et al., 2012).

7.2 NANOPARTICLES OF NONMETALLIC SUBSTANCES

7.2.1 Carbon-based nanomaterials

One of the most researched nanomaterials are the carbon-based, such as nanotubes, fullerenes, both single- and multiwalled carbon nanotubes. These nanomaterials have been reported in literature as cytotoxic agents. Magrez et al. argues that carbon-based nanomaterials possess size-dependent cytotoxicity; they tested various forms of carbon nanoparticles on lung cancer cells to assess the cell viability. Similarly, results were reported by Herzog et al.; the clonogenic assay technique was used for the cell proliferation. Poland et al. discovered that multiwalled carbon nanotubes produced carcinogenic effects similar to asbestos after in vivo administration by injection. On the other hand, single-walled carbon nanotubes were taken up by macrophages (Cavallo et al., 2012). Nonetheless, single-walled carbon nanotubes have the tendency to accumulate in the liver and unbalance biochemical processes involving glutathione, LDH, aspartate transaminases, alanine transaminase, and malondialdehyde in laboratory animals (Khan et al., 2013a,b). Long-time in vitro studies on human epidermoid-like carcinoma cells, Chinese hamster ovary cells and human embryonic kidney cells (HEK293) conclude that fullerenes can exert a genotoxic behavior with respect to chromosomal damage, DNA strand breakage, and micronucleus formation after 1 ng/mL fullerene incubation. Contrary, other studies reported no significant effect on DNA strand breakage caused by fullerenes, which could be directly related to the different experimental conditions (Khan et al., 2013a,b).

7.2.2 Silica

Employing silica nanoparticles can be an advantageous approach for efficient drug delivery systems. Silica nanoparticles can be functionalized for the purpose of drug carrying. Even though initially nanosilica was thought to manifest superior biocompatible properties, recent studies have been indicated correlations between said nanoparticles and ROS generation and increased oxidative stress (Vivero-Escoto and Vega, 2014). Lin et al. described increased levels of ROS, LDH, and malondialdehyde in human bronchoalveolar carcinoma cells incubated with silica nanoparticles (15–46 nm) at various dosages. Also, increased silica nanoparticles-related inflammatory biomarkers like IL-1, IL-6, IL-8, tumor necrosis factor TNF-α (tumor necrosis factor) and mitochondrial disturbances have been described (Fan et al., 2014; Yang et al., 2015a,b).

7.3 NANOPARTICLES OF POLYMERIC MATERIALS

Biodegradable nanoparticles of polymeric nature are potentially useful for encapsulation bioactive substances for targeted drug delivery, with exceptional prospects for chemotherapeutic agents. Embedding various other molecules (peptides, proteins, nucleic acids) could improve the release sustainability. To enhance their specificity, functionalization or coating approaches are commonly used (Xue et al., 2013). Featuring nontoxic, noninflammatory, and nonimmunologic properties, polymer nanoparticles could be an alternative to less convenient materials for some applications. Poly(D,L-lactide-*co*-glycolide) nanosystems have been successfully employed for aimed release of drugs and other molecules. So far, poly(D,L-lactide-*co*-glycolide)-based nanosystems are described as the least toxic for the environment they are implanted because the polymer undergoes hydrolysis that leads to the production of biocompatible metabolites: lactic and glycolic acids. Despite their remarkable properties, some argue that surface coatings could be the source of toxicity polymer nanoparticles exhibit toward human-like macrophages (Josefsen and Boyle, 2012).

8 CONCLUSIONS

Nanaomaterials, their applications and toxicity represent an area with various branches of approach but when it comes to the biomedical field, this concept takes on a sensitive turn as it involves living tissues, organisms dependent on certain processes that could bring both advantages and disadvantages, some under severe forms, their peak being the death of the organism. Cellular processes have a predominantly biochemical nature and many of them are based on chemical bonds that give a certain look and a certain function to the large chain structure of the proteins. Depending on the chemical conformation, structure, size, and architecture, one could speak of beneficial and harmful effects of therapeutic nanomaterials on the host body, this way appearing the notion of degree of toxicity.

The advantages of these materials are extensively studied; nanoparticles-based systems are considered an effective approach to treat severe diseases, such as cancer, difficult to treat infections, or to eliminate blood clots from the bloodstream. The size of the nanoparticles are always advantageous for these applications because they can diffuse from the blood into the tissue and from there to settle in any type of structure from the interstitial cavity to the cells and organs. Researchers in nanoimmunology often study functionalized particles because they could attach substances with cytotoxic or antimicrobial effect on any surfaces. The particles have the advantage of binding to their surface a large amount of active substance because they have high surface area, thus allowing the attachment and delivery of a pharmacological compound at the target site that can also be released in a controlled manner.

In terms of the toxicity of these significant biomedical new entities in the field of biomaterials, opinions are divided among researchers because, depending on the functionalization agents, on the route of synthesis and properties, the reported data vary. The main disadvantage that can produce toxicity is the inability to control the total amount of material. Often, the question is how to extract from the body the biomaterial, without generating side effects. A degree of toxicity is being manifested toward healthy living tissues due to the concentration of nanoparticles. In addition, they may be unsafe because they can spread through the body generating systemic effects and invading major organs. Nonetheless, nanoparticles will play an important part in the biomedicine of the future, but to be used to their maximum capacity, proper studies are needed to optimize their transit through the body and to avoid their unwanted manifestation of toxicity.

ACKNOWLEDGMENTS

This work has been funded by University Politehnica of Bucharest, through the "Excellence Research Grants" Program, UPB – GEX. Identifier: UPB–EXCELENȚĂ–2016 Research project "Suprafete nanobiostructurate antimicrobiene utilizate pentru stimularea fixarii la interfata os-implant", Contract number (acronym: 554).

REFERENCES

Albanese, A., Tang, P.S., Chan, W.C., 2012. The effect of nanoparticle size, shape, and surface chemistry on biological systems. Annu. Rev. Biomed. Eng. 14, 1–16.

Al Gurabi, M.A., Ali, D., Alkahtani, S., Alarifi, S., 2015. In vivo DNA damaging and apoptotic potential of silver nanoparticles in Swiss albino mice. Onco Targets Ther. 8, 295–302.

Alshatwi, A.A., VaiyapuriSubbarayan, P., Ramesh, E., ALHazzani, A.A., Alsaif, M.A., Alwarthan, A.A., 2012. Al_2O_3 nanoparticles induce mitochondria-mediated cell death and upregulate the expression of signaling genes in human mesenchymal stem cells. J. Biochem. Mol. Toxicol. 26 (11), 460–469.

Argyo, C., Weiss, V., Bräuchle, C., Bein, T., 2013. Multifunctional mesoporous silica nanoparticles as a universal platform for drug delivery. Chem. Mater. 26, 430–434.

Bader, A.G., 2012. miR-34—a microRNA replacement therapy is headed to the clinic. Front. Genet. 3, 120.

Beck-Broichsitter, M., Ruppert, C., Schmehl, T., Günther, A., Seeger, W., 2013. Biophysical inhibition of synthetic vs. naturally-derived pulmonary surfactant preparations by polymeric nanoparticles. Biochim. Biophys. Acta 1838, 474–481.

Bonifacio, B., Silva, P., Aparecido dos, M., Ramos, S., et al., 2014. Nanotechnology-based drug delivery systems and herbal medicines: a review. Int. J. Nanomedicine 9, 1–51.

Butte, K., Momin, M., Kurhade, S., Kar, S., 2013. Intravesical drug delivery system for bladder: an overview. Int. J. Pharm. Chem. Biol. Sci. 3 (3), 680–681.

Cavallo, D., Fanizza, C., Ursini, C.L., et al., 2012. Multiwalled carbon nanotubes induce cytotoxicity and genotoxicity in human lung epithelial cells. J. Appl. Toxicol. 32 (6), 454–464.

Chen, E.Y., Garnica, M., Wang, Y.C., Mintz, A.J., Chen, C.S., Chin, W.C., 2012a. A mixture of anatase and rutile TiO_2 nanoparticles induces histamine secretion in mast cells. Part. Fibre Toxicol. 9, 2–3.

Chen, Z., Yin, J.-J., Zhou, Y.-T., Zhang, Y., Song, L., Song, M., Hu, S., Gu, N., 2012b. Dual enzyme-like activities of iron oxide nanoparticles and their implication for diminishing cytotoxicity. ACS Nano 6, 4001–4012.

Chuang, H.C., Chenc, L.C., Leic, Y.C., Wuc, K.Y., Fengb, P.H., Chengc, T.J., 2015. Surface area as a dose metric for carbon black nanoparticles: a study of oxidative stress. DNA single-strand breakage and inflammation in rats. Atmos. Environ. 106, 10–12.

Cornelia, M., Keck, R., Müller, H., 2013. Nano toxicological classification system (NCS): a guide for the risk-benefit assessment of nanoparticulate drug delivery systems. Eur. J. Pharm. Biopharm. 84 (3), 4–8.

Couvreur, P., 2013. Nanoparticles in drug delivery: past, present and future. Adv. Drug Deliv. Rev. 65 (1), 21–23.

Diaz, M.R., Vivas-Mejia, P., 2013. Nanoparticles as drug delivery systems in cancer medicine: emphasis on RNAi-containing nanoliposomes. Pharmaceuticals 6, 1361–1380.

Drbohlavova, J., Chomoucka, J., Adam, V., Ryvolova, M., Eckschlager, T., Hubalek, J., Kizek, R., 2013. Nanocarriers for anticancer drug—new trends in nanomedicine. Curr. Drug Metab. 14 (5), 547–554.

Ebabe Elle, R., Gaillet, S., Vide, J., Romain, C., Lauret, C., Rugani, N., Cristol, J.P., Rouanet, J.M., 2013. Dietary exposure to silver nanoparticles in Sprague-Dawley rats: effects on oxidative stress and inflammation. Food Chem. Toxicol. 60, 297–301.

Emeje, M.O., Obidike, I.C., Akpabio, E.I., Ofoefule, S.I., 2012. Nanotechnology in Drug Delivery. In Tech, Rijeka, 70–106.

Fan, W., Shen, B., Bu, W., Chen, F., He, Q., Zhao, K., Zhang, S., Zhou, L., Peng, W., Xiao, Q., et al., 2014. A smart upconversion-based mesoporous silica nanotheranostic system for synergetic chemo-/radio-/photodynamic therapy and simultaneous MR/UCL imaging. Biomaterials 35, 8992–9002.

Formenti, P., Caquineau, S., Chevaillier, S., Klaver, A., Desboeufs, K., Rajot, J., Belin, S., Briois, V., 2014. Dominance of goethite over hematite in iron oxides of mineral dust from Western Africa: quantitative partitioning by X-ray absorption spectroscopy. J. Geophys. Res. Atmos. 119, 12740–12754.

Gangrade, S.M., 2011. Nanocrystals: a way for carrier free drug delivery. Pharma Buzz 6, 26–27.

Ghaemmaghami, A.M., Hancock, M.J., Harrington, H., Kaji, H., Khademhosseini, A., 2012. Biomimetic tissues on a chip for drug discovery. Drug Discov. Today 17, 173–181.

Grabowski, N., Hillaireau, H., Vergnaud, J., Tsapis, N., Pallardy, M., Kerdine-Rö me, S., Fattal, E., 2015. Surface coating mediates the toxicity of polymeric nanoparticles towards human-like macrophages. Int. J. Pharm. 482 (1-2), 74–80.

Gramoun, A., Crowe, L.A., Maurizi, L., Wirth, W., Tobalem, F., Grosdemange, K., Coullerez, G., Eckstein, F., Koenders, M.I., van den Berg, W.B., et al., 2014. Monitoring the effects of dexamethasone treatment by MRI using in vivo iron oxide nanoparticle-labeled macrophages. Arthritis Res. Ther. 16, R131.

Guan, R., Kang, T., Lu, F., Zhang, Z., Shen, H., Liu, M., 2012. Cytotoxicity, oxidative stress, and genotoxicity in human hepatocyte and embryonic kidney cells exposed to ZnO nanoparticles. Nanoscale Res. Lett. 7 (1), 602.

Guo, H., Barnard, A.S., 2013. Naturally occurring iron oxide nanoparticles: morphology, surface chemistry and environmental stability. J. Mater. Chem. 1, 27–42.

Hackenberg, S., Scherzed, A., Technau, A., Froelich, K., Hagen, R., Kleinsasser, N., 2013. Functional responses of human adipose tissue-derived mesenchymal stem cells to metal oxide nanoparticles in vitro. J. Biomed. Nanotechnol. 9 (1), 86–95.

Heidari, Z., Sariri, R., Salouti, M., 2014. Gold nanorods-bombesin conjugate as a potential targeted imaging agent for detection of breast cancer. J. Photochem. Photobiol. B 130, 40–46.

Jaafar-Maalej, C., Elaissari, A., Fessi, H., 2012. Lipid-based carriers: manufacturing and applications for pulmonary route. Expert Opin. Drug Deliv. 9, 1111–1127.

Jaeger, A., Weiss, D.G., Jonas, L., Kriehuber, R., 2012. Oxidative stress-induced cytotoxic and genotoxic effects of nano-sized titanium dioxide particles in human HaCaT keratinocytes. Toxicology 296, 27–36.

Jordheim, L.P., Durantel, D., Zoulim, F., Dumontet, C., 2013. Advances in the development of nucleoside and nucleotide analogues for cancer and viral diseases. Nat. Rev. Drug Discov. 12, 447–464.

Josefsen, L.B., Boyle, R.W., 2012. Unique diagnostic and therapeutic roles of porphyrins and phthalocyanines in photodynamic therapy, imaging and theranostics. Theranostics 2, 916–966.

Juillerat-Jeanneret, L., Dusinska, M., Fjellsbo, L.M., Collins, A.R., Handy, R.D., Riediker, M., 2015. Biological impact assessment of nanomaterial used in nanomedicine. Introduction to the NanoTEST project. Nanotoxicology 9 (Suppl. 1), 5–12.

Khan, I.A., Aich, N., Afrooz, A.N., Flora, J.R., Schierz, P.A., Ferguson, P.L., Sabo-Attwood, T., Saleh, N.B., 2013a. Fractal structures of single-walled carbon nanotubes in biologically relevant conditions: role of chirality vs. media conditions. Chemosphere 93, 1997–2003.

Khan, I.A., Afrooz, A.N., Flora, J.R., Schierz, P.A., Ferguson, P.L., Sabo-Attwood, T., Saleh, N.B., 2013b. Chirality affects aggregation kinetics of single-walled carbon nanotubes. Environ. Sci. Technol. 47, 1844–1852.

Lammers, T., Kiessling, F., Hennink, W.E., Storm, G., 2012. Drug targeting to tumors: principles, pitfalls and (pre-) clinical progress. J. Control. Release 161 (2), 175–181.

Lee, J.H., Ju, J.E., Kim, B.I., Pak, P.J., Choi, E.K., Lee, H.S., Chung, N., 2014. Rod-shaped iron oxide nanoparticles are more toxic than sphere-shaped nanoparticles to murine macrophage cells. Environ. Toxicol. Chem. 33, 2759–2766.

Leid, J.G., Ditto, A.J., Knapp, A., Shah, P.N., Wright, B.D., Blust, R., et al., 2012. In vitro antimicrobial studies of silver carbene complexes: activity of free and nanoparticle carbene formulations against clinical isolates of pathogenic bacteria. J. Antimicrob. Chemother. 67 (1), 138–148.

Leung, K.C.F., Xuan, S.H., Zhu, X.M., Wang, D.W., Chak, C.P., Lee, S.F., Ho, W.K.W., Chung, B.C.T., 2012. Gold and iron oxide hybrid nanocomposite materials. Chem. Soc. Rev. 41, 1911–1928.

Li, Z., Barnes, J.C., Bosoy, A., Stoddart, J.F., Zink, J.I., 2012. Mesoporous silica nanoparticles in biomedical applications. Chem. Soc. Rev. 41, 2590–2596.

Li, N., Zhao, P., Astruc, D., 2014. Anisotropic gold nanoparticles: synthesis, properties, applications, and toxicity. Ang. Chem. Int. Ed. Engl. 53, 1756–1789.

Marelli, U.K., Rechenmacher, F., Sobahi, T.R.A., Mas-Moruno, C., Kessler, H., 2013. Tumor targeting via integrin ligands. Front. Oncol. 3, 222.

Murphy, S.V., Atala, A., 2014. 3D bioprinting of tissues and organs. Nat. Biotechnol. 32, 773–785.

Patel, P., Soni, T., Thakkar, V., Gandhi, T., 2013. Nanoparticle as an emerging tool in pulmonary drug delivery system. MNS 5, 288–302.

Patil, U.S., Osorno, L., Ellender, A., Grimm, C., Tarr, M.A., 2015. Cleavable ester-linked magnetic nanoparticles for labeling of solvent-exposed primary amine groups of peptides/proteins. Anal. Biochem. 484, 18–20.

Peira, E., Turci, F., Corazzari, I., Chirio, D., Battaglia, L., Fubini, B., Gallarate, M., 2014. The influence of surface charge and photo-reactivity on skin-permeation enhancer property of nano-TiO_2 in ex vivo pig skin model under indoor light. Int. J. Pharm. 5, 90–99.

Pelgrift, R., Friedman, A., 2013. Nanotechnology as a therapeutic tool to combat microbial resistance. Adv. Drug Deliv. Rev., 1–13.

Peng, M., Li, H., Luo, Z., Kong, J., Wan, Y., Zheng, L., Zhang, Q., Niu, H., Vermorken, A., van de Ven, W., et al., 2015. Dextran-coated superparamagnetic nanoparticles as potential cancer drug carriers in vitro. Nanoscale 7, 11155–11162.

Pourmand, A., Abdollahi, M., 2012. Current opinion on nanotoxicology. Daru 20 (1), 95.

Rabanel, J.M., Aoun, V., Elkin, I., Mokhtar, M., Hildgen, P., 2012. Drug loaded nanocarriers: passive targeting and crossing of biological barriers. Curr. Med. Chem. 19 (19), 3070–3072.

Radziun, E., DudkiewiczWilczyńska, J., Książek, I., Nowak, K., Anuszewska, E.L., Kunicki, A., Olszyna, A., Zabkowski, T., 2011. Assessment of the cytotoxicity of aluminium oxide nanoparticles on selected mammalian cells. Toxicol. In Vitro 25 (8), 1694–1700.

Rai, M., Gade, A., Gaikwad, S., Marcato, P., Duran, N., 2012. Biomedical applications of nanobiosensors: the state-of-the-art. J. Braz. Chem. Soc. 1, 14–24.

Rettig, G.R., Behlke, M.A., 2012. Progress toward in vivo use of siRNAs-II. Mol. Ther. 20, 483–512.

Reynolds, R.L., Cattle, S.R., Moskowitz, B.M., Goldstein, H.L., Yauk, K., Flagg, C.B., Berquó, T.S., Kokaly, R.F., Morman, S., Breit, G.N., 2014. Iron oxide minerals in dust of the Red Dawn event in eastern Australia. Aeolian Res. 15, 1–13.

Rosca, E.V., Wright, M., Gonitel, R., Gedroyc, W., Miller, A.D., Thanou, M., 2015. Thermosensitive, near-infrared-labelled nanoparticles for topotecan delivery to tumours. Mol. Pharm. 12, 1335–1346.

Ruiz, M.E., Gantner, M.E., Talevi, A., 2013. Applications of nanosystems to anticancer drug therapy (Part II. Dendrimers, micelles, lipid-based nanosystems). Recent Pat. Anticancer Drug Discov. 9, 1–30.

Sanna, A., Pala, N., Sechi, M., 2014. Targeted therapy using nanotechnology: focus on cancer. Int. J. Nanomedicine 9, 467–468.

Sawdon, A., Weydemeyer, E., Peng, C.A., 2014. Antitumor therapy using nanomaterial-mediated thermolysis. J. Biomed. Nanotechnol. 10, 1894–1917.

Sharma, V., Anderson, D., Dhawan, A., 2012a. Zinc oxide nanoparticles induce oxidative DNA damage and ROS-triggered mitochondria mediated apoptosis in human liver cells (HepG2). Apoptosis 17 (8), 67–70.

Sharma, P., Jha, A.B., Dubey, R.S., Pessarakli, M., 2012b. Reactive oxygen species, oxidative damage, and antioxidativedefense mechanism in plants under stressful conditions. J. Bot. 2012, 26.

Shvedova, A.A., Pietroiusti, A., Fadeel, B., Kagan, V.E., 2012. Mechanisms of carbon nanotube-induced toxicity: focus on oxidative stress. Toxicol. Appl. Pharmacol. 261 (2), 10–13.

Smith, M.J., Brown, J.M., Zamboni, W.C., Walker, N.J., 2014. From immunotoxicity to nanotherapy: the effects of nanomaterials on the immune system. Toxicol. Sci. 138, 52–55.

Szlek, J., Paclawski, A., Lau, R., Jachowicz, R., Mendyk, A., 2013. Heuristic modeling of macromolecule release from PLGA microspheres. Int. J. Nanomedicine 8, 4601–4611.

Tena, A.F., Clarà, P.C., 2012. Deposition of inhaled particles in the lungs. Arch. Bronconeumol. 48, 240–246.

Trivedi, R., Redente, E.F., Thakur, A., Riches, D.W.H., Kompella, U.B., 2012. Local delivery of biodegradable pirfenidone nanoparticles ameliorates bleomycin-induced pulmonary fibrosis in mice. Nanotechnology 23, 505101.

Valdiglesias, V., Kilic, G., Costa, C., Fernandez-Bertolez, N., Pasaro, E., Teixeira, J.P., Laffon, B., 2015. Effects of iron oxide nanoparticles: cytotoxicity, genotoxicity, developmental toxicity, and neurotoxicity. Environ. Mol. Mutagen. 56, 125–148.

Varshosaz, J., Ghaffari, S., Mirshojaei, S.F., Jafarian, A., Atyabi, F., Kobarfard, F., Azarmi, S., 2013. Biodistribution of amikacin solid lipid nanoparticles after pulmonary delivery. BioMed. Res. Int. 2013, 1–8.

Venkatraman, S., 2014. Has nanomedicine lived up to its promise? Nanotechnology 25, 372501.

Vivero-Escoto, J.L., Vega, D.L., 2014. Stimuli-responsive protoporphyrin IX silica-based nanoparticles for photodynamic therapy in vitro. RSC Adv. 4, 14400–14407.

Webster, D.M., Sundaram, P., Byrne, M.E., 2013. Injectable nanomaterials for drug delivery: carriers, targeting moieties, and therapeutics. Eur. J. Pharm. Biopharm. 84, 1–5.

Wikswo, J.P., Curtis, E.L., Eagleton, Z.E., et al., 2013. Scaling and systems biology for integrating multiple organs-on-a-chip. Lab Chip 13, 3496–3511.

Wilhelmi, V., Fischer, U., Weighardt, H., Schulze-Osthoff, K., Nickel, C., Stahlmecke, B., et al., 2013. Zinc oxide nanoparticles induce necrosis and apoptosis in macrophages in a p47phox- and Nrf2-independent manner. PLoS One 8 (6).

Willis, L., Hayes, D., Mansour, H.M., 2012. Therapeutic liposomal dry powder inhalation aerosols for targeted lung delivery. Lung 190, 251–262.

Xu, C., Sun, S., 2013. New forms of superparamagnetic nanoparticles for biomedical applications. Adv. Drug Deliv. Rev. 65, 732–743.

Xue, X., Hall, M.D., Zhang, Q., Wang, P.C., Gottesman, M.M., Liang, X.-J., 2013. Nanoscale drug delivery platforms overcome platinum-based resistance in cancer cells due to abnormal membrane protein trafficking. ACS Nano 7, 10452–10464.

Yang, G., Gong, H., Qian, X., Tan, P., Li, Z., Liu, T., Liu, J., Li, Y., Liu, Z., 2015a. Mesoporous silica nanorods intrinsically doped with photosensitizers as a multifunctional drug carrier for combination therapy of cancer. Nano Res. 8, 751–764.

Yang, L., Kuang, H., Zhang, W., Aguilar, Z.P., Xiong, Y., Lai, W., Xu, H., Wei, H., 2015b. Size dependent biodistribution and toxicokinetics of iron oxide magnetic nanoparticles in mice. Nanoscale 7, 625–636.

Yang, S., Sheng, Y., Han, Y., Chen, S., 2014. Diffuse reflectance spectroscopic analysis of iron-oxide minerals in dust aerosol from Golmud. J. Lanzhou Univ. Nat. Sci. 50, 710–715.

Zhang, Y., Chan, H.F., Leong, K.W., 2013. Advanced materials and processing for drug delivery: the past and the future. Adv. Drug Deliv. Rev. 65, 104–105.

Index

D

E

F

G

H

M

N

Q

R

U

V

W

X

Z

Kolibri 360 GmbH
Ferdinand-Jühlke-Straße 7
99095 Erfurt, Deutschland
produktsicherheit@kolibri360.de